Life Cycle of a Retrovirus (Figure 6-22) *overview animation*

Preparing Monoclonal Antibodies (Figure 6-10) *technique animation*

Retroviral Gene Expression *focus animation*

Retroviral Genome Integration *focus animation*

T7 DNA Replication Complex *molecular tutorial*

Chapter 7: Recombinant DNA and Genomics

Demonstrating Sequence-Specific Cleavage by a Restriction Enzyme *classic experiment*

Dideoxy Sequencing of DNA (Figure 7-29) *technique animation*

Plasmid Cloning (Figures 7-3, 7-4) *technique animation*

Polymerase Chain Reaction (Figure 7-38) *technique animation*

Screening an Oligonucleotide Array for Patterns of Gene Expression *technique animation*

Synthesizing an Oligonucleotide Array *technique animation*

Unleashing the Power of Exponential Growth: The Polymerase Chain Reaction *classic experiment*

Chapter 8: Genetic Analysis in Cell Biology

Creating a Transgenic Mouse (Figure 8-36) *technique animation*

Expressing Foreign Genes in Mice *classic experiment*

In Vitro Mutagenesis of Cloned Genes (Figure 8-29) *technique animation*

Meiosis *focus animation*

Chapter 9: Molecular Structure of Genes and Chromosomes

Retroviral Reverse Transcription (Figure 9-16) *focus animation*

Three-Dimensional Packing of Nuclear Chromosomes (Figure 9-35) *focus animation*

Two Genes Become One: Somatic Rearrangement of Immunoglobin Genes *classic experiment*

Chapter 10: Regulation of Transcription Initiation

Combinatorial Control of Transcription (Figure 10-61) *focus animation*

I⁻ *lac* Repressor Mutations *focus animation*

Regulation of the *lac* Operon *focus animation*

Chapter 11: RNA Processing, Nuclear Transport, and Post-Transcriptional Control

Catalysis without Proteins: The Discovery of Self-Splicing RNA *classic experiment*

Life Cycle of an mRNA (Figure 11-7) *overview animation*

mRNA Splicing (Figures 11-17, 11-19) *focus animation*

The U1A Spliceosomal Protein *molecular tutorial*

DNA Polymerase Beta (polB) *molecular tutorial*

Chapter 12: DNA Replication, Repair, and Recombination

Subunit of *E. coli* DNA Polymerase III *molecular tutorial*

Bidirectional Replication of DNA (Figures 12-2, 12-9) *focus animation*

Coordination of Leading- and Lagging-Strand Synthesis (Figure 12-11) *focus animation*

Hin Recombinase *molecular tutorial*

Nucleotide Polymerization by DNA Polymerase (Figure 12-10) *focus animation*

Proving That DNA Replication Is Semiconservative *classic experiment*

RecA *(E. coli)* *molecular tutorial*

Telomere Replication (Figure 12-13) *focus animation*

Topoisomerase I *(E. coli)* *molecular tutorial*

Chapter 13: Regulation of the Eukaryotic Cell Cycle

Cell Biology Emerging from the Sea: The Discovery of Cyclins *classic experiment*

Cell Cycle Control (Figure 13-2) *overview animation*

Nuclear Envelope Dynamics during Mitosis *video*

Chapter 14: Gene Control in Development

Gene Control in Embryonic Development (Figures 14-25, 14-32) *overview animation*

Using Lethal Injection to Study Development *classic experiment*

Chapter 15: Transport across Cell Membranes

Biological Energy Interconversions (Figures 15-9, 15-13, 15-19) *overview animation*

Stumbling upon Active Transport *classic experiment*

MOLECULAR
CELL
BIOLOGY

ABOUT THE AUTHORS

 **Harvey Lodish,** Member of the National Academy of Sciences, is a Member of the Whitehead Institute for Biomedical Research and Professor of Biology at Massachusetts Institute of Technology. His laboratory studies signaling by the erythropoietin, TGFβ, and insulin receptors, as well as the functions of a family of mammalian fatty acid transporters.

 **Arnold Berk,** Professor and Director, Molecular Biology Institute, University of California at Los Angeles, does research in the area of transcription control in animal cells and the development of viral vectors for gene transduction.

 **S. Lawrence Zipursky** is a Professor of Biological Chemistry and an Investigator of the Howard Hughes Medical Institute in the School of Medicine at the University of California, Los Angeles. His research is in the area of intercellular signaling and signal transduction in the developing nervous system.

 **Paul Matsudaira,** Member of the Whitehead Institute for Biomedical Research, and Professor of Biology and Bioengineering, Massachusetts Institute of Technology, studies the control of cell shape and motility by the actin cytoskeleton and investigates the applications of microfabrication technology for biological research.

 **David Baltimore,** Winner of the Nobel Prize in Medicine and President of the California Institute of Technology, has an active lab at Caltech that studies molecular events in the immune system and in the growth of viruses.

 **James E. Darnell,** Member of the National Academy of Sciences, Vincent Astor Professor and Head of Molecular Cell Biology Laboratory, The Rockefeller University, directs research on transcriptional responses to cytokines and growth factors and the effects of these responses on cell growth and specialization.

Photographs of Drs. Lodish, Berk, Zipursky, Matsudaira, and Darnell by Margaret Lampert; photograph of Dr. Baltimore by Bob Paz.

FOURTH EDITION

MOLECULAR CELL BIOLOGY

Harvey Lodish

Arnold Berk

S. Lawrence Zipursky

Paul Matsudaira

David Baltimore

James Darnell

MOLECULAR
CELL BIOLOGY
4.0

Paul Matsudaira
Arnold Berk
S. Lawrence Zipursky
David Baltimore
James Darnell
Harvey Lodish

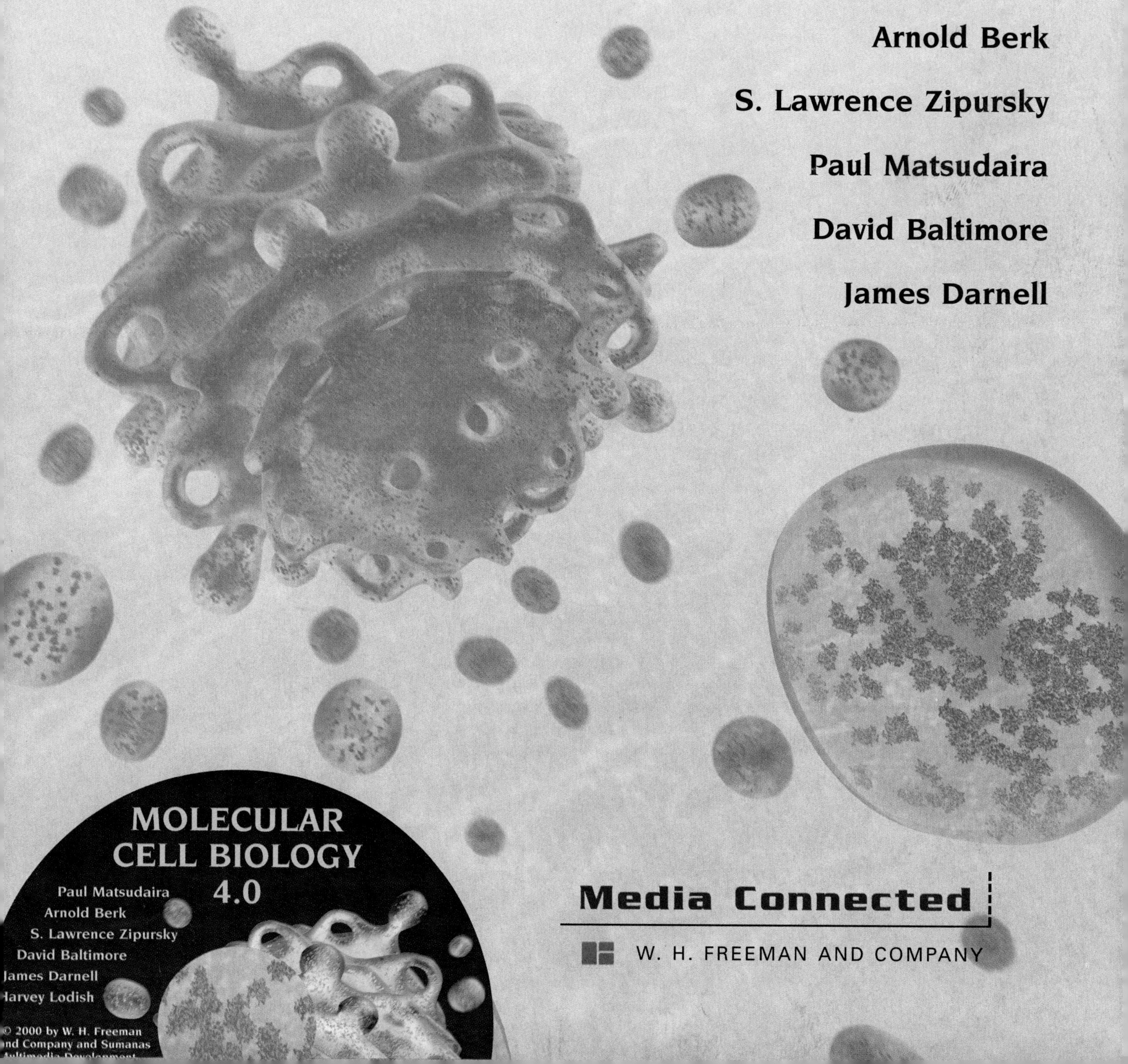

Media Connected

W. H. FREEMAN AND COMPANY

EXECUTIVE EDITOR: Sara Tenney

DEVELOPMENT EDITORS: Katherine Ahr, Ruth Steyn, Kay Ueno

EDITORIAL ASSISTANT: Jessica Olshen

EXECUTIVE MARKETING MANAGER: John A. Britch

PROJECT EDITOR: Katherine Ahr

TEXT AND COVER DESIGNER: Victoria Tomaselli

PAGE MAKEUP: Michael Mendelsohn, Design 2000, Inc.

COVER ILLUSTRATION: Kenneth Eward

ILLUSTRATION COORDINATOR: John Smith, Network Graphics; Tamara Goldman, Bill Page

ILLUSTRATIONS: Network Graphics

PHOTO RESEARCHER: Jennifer MacMillan

PRODUCTION COORDINATOR: Paul W. Rohloff

MEDIA AND SUPPLEMENTS EDITORS: Tanya Awabdy, Adrie Kornasiewicz, Debra Siegel

MEDIA DEVELOPERS: Sumanas, Inc.

COMPOSITION: York Graphics Services, Inc.

MANUFACTURING: Von Hoffman Press

Library of Congress Cataloging-in-Publication Data

Molecular cell biology / Harvey Lodish p [et al.] – 4th ed.
     p.   cm.
     Includes bibliographical references.
     ISBN 0-7167-3136-3
     1. Cytology.   2. Molecular biology.   I. Lodish, Harvey F.
QH581.2.M655   1999
571.6–dc21   99-30831
                                                         CIP

Printed in the United States of America

W. H. Freeman and Company
41 Madison Avenue, New York, New York 10010
Houndsmills, Basingstoke RG21 6XS, England

Second printing, 2000

**To our students and to our teachers,
from whom we continue to learn**

Preface

Molecular Cell Biology for the 21st Century

Modern biology is rooted in an understanding of the molecules within cells and of the interactions between cells that allow construction of multicellular organisms. The more we learn about the structure, function, and development of different organisms, the more we recognize that **all life processes exhibit remarkable similarities.** *Molecular Cell*

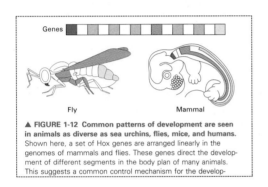

▲ FIGURE 1-12 Common patterns of development are seen in animals as diverse as sea urchins, flies, mice, and humans. Shown here, a set of Hox genes are arranged linearly in the genomes of mammals and flies. These genes direct the development of different segments in the body plan of many animals. This suggests a common control mechanism for the develop-

Biology concentrates on the macromolecules and reactions studied by biochemists, the processes described by cell biologists, and the gene control pathways identified by molecular biologists and geneticists. In this millennium, two gathering forces will reshape molecular cell biology: *genomics*, the complete DNA sequence of many organisms, and *proteomics*, a knowledge of all the possible shapes and functions that proteins employ.

   All the concepts of molecular cell biology continue to be derived from experiments, and powerful experimental tools that allow the study of living cells and organisms at higher and higher levels of resolution are being developed constantly. In this fourth edition, we address the current state of molecular cell biology and look forward to what further exploration will uncover in the twenty-first century.

New Discoveries, New Methodologies

Since the publication of the third edition of this text in 1995, extraordinary developments have taken place:

• The **entire genomes** of yeast, a nematode, and many bacterial species have been sequenced (Chapter 7). Now researchers are racing, with corporate and government sponsorship, to sequence all 3 billion base pairs of the **human genome,** and thus identify the sequence of each of the ~70,000 proteins it encodes, by the year 2001. How do we store and use this new information? A rapidly developing area of computer science, **bioinformatics,** is

devoted to collecting, organizing, and analyzing DNA and protein sequences (Chapter 7).

From Figure 7-39.

• The simultaneous expression of thousands of genes can now be analyzed using newly developed **DNA "chip" microarray technology,** enhancing our understanding of gene control during development and disease states (Chapter 7).

• **Macromolecular synthesis** is now understood to require large multiprotein machines. The detailed structure of the ribosome has advanced our knowledge of the steps in peptide bond formation (Chapter 4), and we now understand many of the proteins that participate in and regulate steps of DNA replication (Chapter 12) and messenger RNA synthesis (Chapter 10).

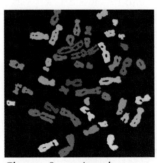

Chapter 9 opening photo; chromosome painting

• **Chromosome "painting,"** using fluorescent in situ hybridization, allows each human chromosome in a metaphase spread to be clearly distinguished on the basis of size and color (Chapter 9). This method greatly increases the sensitivity for detecting chromosomal translocations in cancer cells (Chapter 24).

• We now understand a great deal about how **chromatin structure** is modified by histone acetylation to **regulate gene expression,** and how the RNA polymerase II holoenzyme interacts with activators and coactivators to **regulate transcription** (Chapter 10).

• Significant advances have been made in understanding the processes that regulate **transport into and out of the nucleus** through nuclear pore complexes, and their interactions with importins, exportins, and the Ran-GTPase (Chapter 11).

• The function of many proteins essential for **vesicle budding and fusion,** and for **targeting proteins** to specific subcellular organelles is understood in detail, and our understanding of **vesicular transport through the Golgi** has undergone a major revision (Chapter 17).

• New findings about microfilaments and microtubule dynamics and the role of various motor proteins, including myosin V and kinesin-related proteins, give us a better understanding of **cell motility and mitosis** (Chapters 18 and 19).

- We now understand a great deal more about **apoptosis** (regulated cell death). New discoveries in the molecular mechanisms of cell death inform our understanding of the mechanisms of development; coverage of the pathways relevant to cell death sheds light on cancer and neurological diseases (Chapter 23).

- The essential features of intricate **cell signaling pathways** that start at cell surface receptors and control cell proliferation, growth, and motility have been identified. For example, the structures of G proteins and G-protein complexes have been determined through x-ray crystallography, and are providing important insights into the mechanisms by which G proteins regulate many aspects of cell function (Chapter 20).

Countless other new developments are incorporated in this completely rewritten, fully updated fourth edition.

Streamlined Coverage

The author team working together.

With each edition, and with all that is new, the question arises: **What should we cut out and what should we cover** to provide the most useful text possible to students and professors of this course? We have consulted with professors across the country and worked closely as an author team to streamline our coverage in each chapter and across the book. In streamlining coverage within a chapter we chose, where appropriate, to **illustrate key concepts with one key experiment rather than several.** In streamlining coverage across the book we have **reorganized and combined chapters** to reflect the connections we see between and among topics. And we have **added more pedagogical aids** to help the student succeed in this course.

The four-part structure of the previous edition continues to reflect our broader concept of the organization of this material. Changes in organization and coverage are noted below.

In Part I, "Laying the Groundwork," we present the scope of the book in an all-new Chapter 1. We lay the groundwork for understanding the experimental and conceptual basis of *Molecular Cell Biology* in Chapters 2–8. Coverage of membrane structure and cell organization is now presented early in the book (Chapter 5).

In Part II, "Nuclear Control of Cellular Activity," we teach students how genes work. We now discuss transcription and RNA processing (Chapters 10 and 11) before DNA replication, repair, and recombination (Chapter 12). We have moved the chapter on the eukaryotic cell cycle to this part because of its close relationship to those on DNA replication and control of gene expression. In our chapter on gene control in development (Chapter 14), we have added a full section on plant development in the model organism *Arabidopsis*.

In Part III, "Building and Fueling the Cell," we focus on the ways in which proteins work together to make a living cell. We have combined coverage of cellular energetics in plant and animal cells into a single chapter to focus attention on the commonalities in these processes: generation of a proton motive force and its utilization in ATP synthesis (Chapter 16). We have combined discussion of organelle biogenesis and protein secretion into a single chapter on protein sorting to emphasize the multiple mechanisms by which proteins are targeted to specific subcellular locations (Chapter 17).

In Part IV, "Cell Interactions," we emphasize how cells interact with each other, in normal and abnormal situations. We have consolidated the discussion of cell adhesion molecules and junctions (Chapter 22) and added a new chapter on cell interactions in development (Chapter 23) that includes expanded information on the role of the TGFβ pathway in determining the overall organization of early vertebrate embryos, and also coverage of regulated cell death. We conclude with a completely new chapter on cancer (Chapter 24), which integrates much material presented earlier in the book, on the cell cycle, cell-cell signaling, DNA repair, and interactions of cells with the extracellular matrix. For the first time, we cover immunology in a hot-linked chapter on our Web site to expose students to the original literature, and to gain the flexibility to update coverage of this fast-moving field.

Training the Scientists of Tomorrow

We have always believed that it is critical to present to students the **experimental basis of our understanding**—to show them *how* we know what we know. We hope that this will demonstrate the dynamic nature of science and prepare them not only to engage actively in scientific research and teaching but also to become educated members of a public that increasingly is asked to deal with complex issues such as environmental toxins, genetically modified foods, and human gene technology.

To further this end, we have added a new set of features to this edition: **Perspectives for the Future** and **Perspectives in the Literature**. Perspectives for the Future is a brief essay that gives us an opportunity to discuss potential future developments. Perspectives in the Literature is a related

> **PERSPECTIVES for the Future**
>
> During the past two decades remarkable progress has been made in understanding the mechanisms regulating gene transcription during development. Through a combination of genetic and biochemical studies, the DNA sequences regulating expression of many different genes in simple and higher eukaryotes and the transcription factors that bind to them have been identified. ... grams in various organisms ... only on combinations of tr... specific cells but also on in... extracellular signals acting ... through their direct interac... Continuation of such studie... into the way specific genes... and times in development
>
> p. 573

> **PERSPECTIVES in the Literature**
>
> You are working with a well-characterized in vitro system that allows you to induce myoblasts to differentiate into myotubes synchronously. Your long term goal is to describe the transcriptional regulatory networks that control muscle differentiation. An important step in your studies is to describe patterns of gene expression. You are interested in studying which genes are turned on and off and in what order during the transition from a myoblast to a differentiated myotube. Design a set of experiments that would allow you

critical-thinking question that challenges students to solve a problem working with original research and review articles and resources available on the Web.

Of course, we want to show students not only how we know what we know but why we do what we do. As in previous editions, our coverage of **medical topics, biotechnology, and plant biology** is integrated throughout. We know that these topics may be of particular interest to your students, so we have highlighted them in context with clear icons.

Chemicals, which are thought to be the cause of many human cancers, were originally associated with cancer through experimental studies in animals. The classic experiment is to repeatedly paint a test substance on the back of a mouse and look for development of both local and systemic tumo... the many substances identified as a very broad range of structures features, they can be classified ...

p. 474

Important strides have been made in dissecting the mechanisms controlling the development of plants. These advances have been possible largely due to the choice of *Arabidopsis thaliana* as a model plant. This plant has many of the same advantages as flies and worms for use as a model system. First, *Arabidopsis* is small and ... Second, mutants can be ... has a short generation ... itagenized by treatment ...ion, the small size of the *Arabidopsis* genome facilitates positional cloning methods to isolate the genes defined by mutations (Chapter 8). And

Most proteins used for therapeutic purposes in humans or animals are secretory proteins stabilized by disulfide bonds. When recombinant DNA tech-

p. 708

p. 571

Text, Figures, and Animations Developed Together

In this edition, we have worked simultaneously on the development of the text and of the animations available on our CD-ROM. Through careful planning and collaboration, we have developed a **CD that is an integral part of the text.** We have developed thirty-five new, aesthetically pleasing and pedagogically useful animations that are visually consistent with the figures in the book.

Acknowledgments

In updating, revising, and rewriting this book, we were given invaluable help by many colleagues. We thank the following people who generously gave of their time and expertise by making contributions to specific chapters in their areas of interest, providing us with detailed information about their courses, or by reading and commenting on one or more chapters:

Susan M. Abmayr, *Pennsylvania State University*

Ross S. Anderson, *Lamar University*

Nigel S. Atkinson, *University of Texas at Austin*

Tania A. Baker, *Howard Hughes Medical Institute and Massachusetts Institute of Technology*

Margarida M. Barroso, *University of Virginia at Charlottesville*

John D. Bell, *Brigham Young University*

William Bement, *University of Wisconsin at Madison*

Mark K. Bennett, *University of California at Berkeley*

James Berger, *Massachusetts Institute of Technology*

Lutz Birnbaumer, *University of California at Los Angeles*

Richard Blanton, *Texas Technical University*

Jonathan Bogan, *Whitehead Institute*

David Bourgaize, *Colby College*

William S. Bradshaw, *Brigham Young University*

William J. Brown, *Cornell University*

Robert Bruner, *University of California at Berkeley*

Dennis Buetow, *University of Illinois at Champaign-Urbana*

Stephen K. Burley, *Rockefeller University*

Stanley Caveney, *University of Western Ontario*

Sulie Lin Chang, *Seton Hall University*

Andrew Chess, *Whitehead Institute and Massachusetts Institute of Technology*

Nathan L. Collie, *Texas Technical University*

Duane Compton, *Dartmouth University*

Scott Cooper, *University of Wisconsin at La Crosse*

Dean Danner, *Emory University School of Medicine*

Robin Davies, *Sweet Briar College*

Ann Dell, *Oklahoma School of Science and Mathematics*

Eddy M. De Robertis, *Howard Hughes Medical Institute and University of California at Los Angeles*

Joyce J. Diwan, *Rensselaer Polytechnic Institute*

Paul Evans, *Brigham Young University*

Guy E. Farish, *Adams State University*

Abram Gabriel, *Rutgers University*

Nord Gale, *University of Missouri at Rolla*

Matt Gardner, *Massachusetts Institute of Technology*

David Scott Gilmour, *Pennsylvania State University*

Bruce Greenberg, *University of Waterloo*

Paul Greenwood, *Colby College*

Dennis Goode, *University of Maryland*

Barry M. Gumbiner, *Memorial Sloan-Kettering Cancer Center*

Brian Haarer, *University of Texas at Austin*

David Harris, *University of Arizona*

Greg L. Harris, *San Diego State University*

Michele C. Heath, *University of Toronto*

Merrill Hille, *University of Washington*

Paul Huber, *University of Notre Dame*

Victoria Iwanij, *University of Minnesota*

Tyler E. Jacks, *Howard Hughes Medical Institute and Massachusetts Institute of Technology*

Chris A. Kaiser, *Massachusetts Institute of Technology*

Thomas C. S. Keller III, *Florida State University*

Gregory M. Kelly, *University of Western Ontario*

Michael W. Klymkowsky, *University of Colorado at Boulder*

Joseph Koke, *Southwest Texas State University*

Peter J. Krell, *University of Guelph*

William Langridge, *Loma Linda University*

Jacqueline A. Lees, *Massachusetts Institute of Technology*

Michael Levine, *University of California at Berkeley*

Xuan Liu, *University of California at Riverside*

Ponzy Lu, *University of Pennsylvania*

Paula M. Lutz, *University of Missouri at Rolla*

Sara McCowen, *Virginia Commonwealth University*

Grant McGregor, *Emory University School of Medicine*

Paul Mahoney, *University of Missouri at Rolla*

Christopher Miller, *Brandeis University*

Robert L. Millette, *Portland State University*

Frank Monette, *Boston University*

Deborah K. Morrison, *National Cancer Institute—Frederick Cancer Research and Development Center*

Lawrence Mwasi, *University of Arkansas at Pine Bluff*

Scott A. Ness, *University of New Mexico*

Jeffrey D. Newman, *Lycoming College*

Laura Olsen, *University of Michigan at Ann Arbor*

Charlotte K. Omoto, *Washington State University*

Bruce Parker, *Utah Valley State College*

Paul Richmond, *University of the Pacific*

Austen Riggs, *University of Texas at Austin*

David Sabatini, *Whitehead Institute*

Edmund Samuel, *Southern Connecticut State University*

William S. Saunders, *University of Pittsburgh*

Fred Schreiber, *California State University at Fresno*

Sandra L. Schmid, *Scripps Research Institute*

James R. Sellers, *National Institutes of Health*

Dan Shea, *Eastern Nazarene College*

Nahum Sonenberg, *McGill University*

Hermann Steller, *Howard Hughes Medical Institute and Massachusetts Institute of Technology*

Brian Storrie, *Virginia Polytechnic Institute and State University*

Phyllis R. Strauss, *Northeastern University*

Christine Tachibana, *Pennsylvania State University*

Patrick Thorpe, *Grand Valley State University*

Dean Tolan, *Boston University*

Sharon Tracy, *University of California at San Diego*

Yukiko Ueno, *Massachusetts Institute of Technology*

Angela Wandinger-Ness, *University of New Mexico*

Christopher Watters, *Middlebury College*

Geraldine Weinmaster, *University of California at Los Angeles*

Patrick Weir, *Felician College*

Bruce Wightman, *Muhlenberg College*

David Worcester, *University of Missouri at Columbia*

Michael Wormington, *University of Virginia at Charlottesville*

Michael Yamauchi, *Massachusetts Institute of Technology*

Jonathan P. Zehr, *Rensselaer Polytechnic Institute*

Patty Zwollo, *Occidental College*

Thanks also to Frank Monette and his students at Boston University who offered their feedback on working with the third edition: Miguel Ariza, Alicia Chares, Benjamin Hyatt, and Ira Miner.

This book would not have been possible without the efforts of many, many people. We are grateful to the talented staff we have had the pleasure of working with at W. H. Freeman and Company in this and the three previous editions. We would like to thank Elizabeth Widdicombe, President, for her support of this fourth edition.

We thank Sara Tenney, our acquiring editor, for overseeing the seamless integration of our text and media package and for organizing our "group reads"—opportunities for us to work together on all of the chapters. Thanks to Kate Ahr and Ruth Steyn, our development editors, for their devoted efforts in streamlining the text and clarifying the figures. Thanks also to Kay Ueno for her extensive contributions to the initial planning of this edition. Jennifer MacMillan, our photo researcher, was particularly adroit at tracking down the photographs we needed.

We thank our project editorial team: Kate Ahr, who continued to oversee the implementation of our plan as project editor; Philip McCaffrey, managing editor; and our copyeditors, Bill O'Neal and Ruth Steyn, for clarifying and polishing our chapters. Thanks also to our proofreaders—Jane Elias, Walter Hadler, Michele Kornegay, and Elizabeth Marraffino—for scrutinizing the text and art proofs. Thanks to Victoria Tomaselli for her elegant design and to our layout artist, Michael Mendelsohn, for its successful implementation. Thanks to Paul Rohloff, production coordinator, for overseeing the production of the book. John Smith of Network Graphics coordinated the art program and played a major role in assuring its consistency.

Thanks to media editors Tanya Awabdy, Adrie Kornasiewicz, and Debra Siegel for their indispensable contributions to the CD-ROM, and to Sumanas, Inc. for taking our storyboards and rendering beautiful animations. Thanks also to the editorial assistants who have provided their support on this edition: Michelle Baildon, Julien Devereaux, Robert Jordan, Jennifer King, Jessica Olshen, and Trimmette Roberts.

Our own staff, including Lois Cousseau, Julie Ellis, Carol Eng, Cynthia Petersen, and Karen Ronan provided invaluable assistance and support. Thanks to the librarians at the Dana Library at Dartmouth University, to David Richardson at the Whitehead Institute Library, and to Janet Tannenbaum at W. H. Freeman and Company for their assistance in our research.

Special thanks to our families for inspiring us and for granting us the time it takes to work on such a book.

October 1999

Supplements

In preparing the supplements package for *Molecular Cell Biology*, we have drawn on our collective experience with the instruction of cellular and molecular biology at almost every level taught, both undergraduate and graduate, at Virginia Polytechnic Institute and State University.

The Virginia Tech supplements author team.

Together, we have written the end-of-chapter questions for the textbook, the self-test questions resident on the companion CD-ROM and Web site, the review questions and more challenging data analysis questions for the new study guide/problems book, *Working with Molecular Cell Biology: A Study Companion*, and the test bank questions that parallel the questions found in the study companion. We've worked to produce a package that is not only extremely useful to students and instructors but also closely integrated with the goals of text, and with each individual supplement.

Special thanks to our families for their support and encouragement. Muriel Lederman extends a special thanks to Jill Sible and Bruce Turner.

Brian Storrie
Muriel Lederman
Eric A. Wong
Richard Walker
Glenda Gillaspy

October 1999

For the Instructor

Print and Computer Test Banks NEW

Brian Storrie, Muriel Lederman, Eric A. Wong, Richard Walker, and Glenda Gillaspy, Virginia Polytechnic Institute and State University.

Print Test Bank: 0-7167-3601-2; Computer Test Bank CD-ROM (Windows/Macintosh hybrid) 0-7167-3603-9

Realizing that instructors would appreciate an occasional inspiration when writing tests, we've chosen to add a test bank to our instructor's materials. Questions parallel those posed in both the Study Companion and the end-of-chapter review—the same concept or principle is asked in different ways. Questions are also posed in a number of different formats: short-answer, essay, and multiple choice. The electronic version of the test bank allows instructors to edit and rearrange the questions, or add their own questions.

In addition, instructors can visit the password-protected **Online Instructor Test Bank** on the **Molecular Cell Biology Web Site Companion** to access favorite test questions submitted by their peers, or post their own questions.

Instructor's Resource CD-ROM NEW

© W. H. Freeman and Company, and Sumanas, Inc. 0-7167-3600-4

Contains all art from the text, plus all animations and videos, with a powerful and easy-to-use presentation manager application, Presentation Manager Pro. Source files for the resources are also provided for instructors using other presentation programs.

Overhead Transparency Set

0-7167-3605-5

275 full-color figures from the text, optimized for classroom projection, in one volume. Available free to qualified adopters.

Instructor's Solutions Manual NEW

Brian Storrie, Muriel Lederman, Eric A. Wong, Richard Walker, and Glenda Gillaspy, Virginia Polytechnic Institute and State University. 0-7176-3752-3

Contains answers for all end-of-chapter questions. Also includes a convenient print version of the User's Guide for the Instructor's Resource CD-ROM.

For the Student

Molecular Cell Biology 4.0 CD-ROM Companion *NEW*

(hybrid format for Windows and Macintosh)

Packaged with every copy of the textbook.

Animations authored by Paul Matsudaira, Arnold Berk, S. Lawrence Zipursky, James Darnell, and Harvey Lodish, with Tanya Awabdy.

With contributions from: David Marcey, California Lutheran University; Brian Storrie, Muriel Lederman, Eric A. Wong, Richard Walker, and Glenda Gillaspy, Virginia Polytechnic Institute and State University; Lisa Rezende, Harvard Medical School; Ruth Alscher, Virginia Polytechnic Institute and State University.

© W. H. Freeman and Company, and Sumanas, Inc.

Visualization of cell and molecular processes helps students better understand the relationship between structure, function, and process. In addition to the **book animations,** the CD also features several valuable visual supplements:

- **Videos of cell processes** using cutting-edge techniques that vividly illustrate the dynamic nature of the cell;

- Interactive macromolecular biology **tutorials** by David Marcey, to help students better understand the relationship between structure and function in cell and molecular biology;

- Illustrated, printable **essays on classic experiments** in molecular and cell biology which treat the investigative side of classic groundbreaking experiments by exploring the process of asking questions and devising tests.

To help students put cell and molecular processes in context, the CD also features a visual **"Cell Navigation" interface.** Clicking on structures and organelles brings up animations, videos, and other resources pertaining to that cellular feature.

An interactive review tutorial also resides on the CD; additionally, for students intending graduate study, we provide a **timed MCAT/GRE prep exam** referenced to the text.

Molecular Cell Biology Web Site Companion: *NEW*

www.whfreeman.com/biology

Michael Klymkowsky, University of Colorado at Boulder

© W. H. Freeman and Company, and Sumanas, Inc.

The companion Web site to the text provides students with a **bridge to the world of working cell and molecular biologists.** Updated links to Web sites referenced in the text, plus suggestions for how to explore topics in-depth using Internet resources, are just some of the features of the site.

We've also included valuable enrichment resources to bring students to the real world of molecular cell biology:

- *Working with the Literature*—selected scientific papers on topics covered in the text, with questions based on the data and methods presented, to help students navigate though current scientific literature and better understand the experimental process;

- *Classic Experiments* essays which treat the investigative side of classic groundbreaking experiments by exploring the process of asking questions and devising tests;

- *Analyzing Experiments* questions test students' ability to apply concepts and understand data and experiments;

- *Integrative Biology Topics* presents printable and hypertext optional chapters on subjects related to molecular and cell biology, such as immunology, developmental biology, among others, for instructors who wish to cover these topics in their courses, and for students with an interest in the applications of cell and molecular biology.

- An **online self-test section** with text and study references, to reinforce key terms and concepts;

- **Links** to the Macromolecular Tutorial, plus a molecular viewer and modeler, to help students better understand the relationship between structure and function in Cell and Molecular Biology.

Working with Molecular Cell Biology: A Study Companion

Brian Storrie, Muriel Lederman, Eric A. Wong, Richard Walker, and Glenda Gillaspy, Virginia Polytechnic Institute and State University. 0-7167-3604-7

The study companion has been **reorganized to mirror the text in-chapter organization,** providing greater flexibility for instructors who prefer to teach topics in alternative sequences. Students can easily find the questions in the study companion which correspond to the material they're covering in class.

The study companion has also been **tailored to the needs of students at many levels.** The first part of each chapter, "Reviewing Concepts," serves as a study and review resource, posing questions on key principles and concepts. Students are encouraged to draw together the text, CD-ROM, and their own lecture notes to answer the questions. The second part of the chapter, "Analyzing Experiments," allows students to apply the knowledge they've gained to experimental situations, and to work with data sets. Since Molecular Cell Biology is used by students at many levels, we have coded each "Analyzing Experiments" question by a bullet system that denotes the level of difficulty: one bullet is appropriate for sophomore or junior level students, two bullets for junior or senior level, and three for advanced students. Worked-out answers for all questions are included, as well as answers for every other end-of-chapter question.

To make study more productive, we've also chosen to pose the sorts of questions students are most likely to encounter on tests—**questions in the study companion parallel those in the Test Bank** in terms of content, and the principles and concepts tested.

To the Student

Biology is a living science, in which changing knowledge continually generates fresh perspectives and fresh opportunities for productive impacts on our society. Our goal in this book is to provide you with the **experimental basis** of our current understanding, and to give you the tools to participate in the development of our future knowledge. We have provided 15 new text and visual aids to help you through each chapter.

3

Protein Structure and Function

Proteins, the working molecules of a cell, carry out the program of activities encoded by genes. This program requires the coordinated effort of many different types of proteins, which first evolved as rudimentary molecules that facilitated a limited number of chemical reactions. Gradually, many of these primitive proteins evolved into a wide array of enzymes capable of catalyzing an incredible range of intracellular and extracellular chemical reactions, with a speed and specificity that is nearly impossible to attain in a test tube. Other proteins acquired numerous structural, regulatory, and other functions. For a flavor of the various roles of proteins in today's organisms, we can look to the yeast *Saccharomyces cerevisiae,* a simple unicellular eukaryote. The yeast genome is predicted to encode about 6225 proteins (see Table 7-3). On the basis of their sequences, 17 percent are estimated to be involved in metabolism, the synthesis or degradation of cell building blocks; 30 percent, in cellular organization and biogenesis of cell organelles and membranes; and 10 percent, in transporting molecules across membranes.

A two-dimensional array of α-actinin molecules.

In this chapter, we will study how the structure of a protein gives rise to its function. The first section examines protein architecture: the structure and chemistry of amino acids, the linkage of amino acids to form a linear chain, and the forces that guide folding of the chain into higher orders of structure. In the next section, we learn about special proteins that aid in the folding of proteins, modifications that occur after the protein chain is synthesized, and mechanisms that degrade proteins. In the third section, we illustrate several key concepts in the functional design of proteins, using antibodies and enzymes as examples. A separate section is devoted to the general characteristics of membrane proteins, which reside in the lipid bilayer surrounding cells and organelles.

p. 50

Chapter Outline ▶
The chapter Outline lists the major section headings and the pages on which they can be found.

Numbered Headings ▶
We know that your professor may only assign certain portions of a chapter. We have numbered the major section headings for easy reference.

▲ Media Connections
We list the animations and Classic Experiments available on the CD-ROM that directly pertain to the chapter.

3.3 Functional Design of Proteins

A key concept in biology is that form and function are inseparable. This concept applies equally well to protein design as to other levels of biological organization (e.g., the morphology of cells and the organization of tissues). In fact, we can often guess how a protein works by looking at its structure. Perhaps the best way to illustrate this is by examining a few protein structures. For instance, a barrel-like nuclear pore, a complex of several proteins, sits in the nuclear membrane and acts as a channel through which molecules travel in or out of the nucleus (Figure 3-20a). In the cavity of a different barrel-like structure, the GroEL/ES chaperonin, protein folding takes place (Figure 3-20b). Some proteins have grooves in their surface, which are logical binding sites for a variety of molecules, especially rod-shaped or filamentous ones. An example is reverse transcriptase, which copies RNA into DNA; this enzyme has a groove on one side through which RNA slides along the surface of the protein (Figure 3-20c). Topoisomerase II, a DNA-binding enzyme, is an articulated enzyme that opens and closes at both ends like locks in a canal (Figure 3-20d). A delight in studying protein structure is uncovering the simple but ingenious ways that nature has built each protein to perform a particular function.

p. 68

▲ Overview Paragraphs
Each major section begins with an overview that introduces the topic at hand.

SUMMARY Functional Design of Proteins

• The function of nearly all proteins depends on their ability to bind other molecules (ligands). Ligand-binding sites on proteins and the corresponding ligands are chemically and topologically complementary. The affinity of a protein for a particular ligand refers to the strength of binding; its specificity, to the restriction of binding to one or a few preferred ligands.

• Enzymes are catalytic proteins that accelerate the rate of cellular reactions by lowering the activation energy and stabilizing transition-state intermediates.

• Enzyme active sites comprise two functional parts: a substrate-binding region and a catalytic region. The amino acids composing the active site are not necessarily adjacent in the amino acid sequence, but are brought into proximity in the native conformation.

• The kinetics of many enzymes are described by the Michaelis-Menten equation. From plots of reaction rate versus substrate concentration, two characteristic parameters of an enzyme can be determined: the Michaelis constant K_m, a measure of the enzyme's affinity for substrate, and the maximal velocity V_{max} (see Figure 3-26).

p. 78

◀ Summaries
Each major section concludes with a summary that reviews the important points covered in the section, both in the text and in the figures.

MEDICINE Another example of this approach involves the *BRCA-1* gene. Women who inherit a mutant form of this gene have a high probability of developing breast cancer before age 50. The *BRCA-1* gene was isolated by methods described in Chapter 8, and a cDNA of the *BRCA-1* mRNA was cloned and sequenced, revealing the amino acid sequence of the BRCA-1 protein. Sophisticated methods of sequence comparison revealed that ... tein is distantly, but significantly related to th...

p. 235

▶ Applications in Context
Medicine, biotechnology, and plants play increasingly important roles in our lives. We think it's important to discuss these issues in context, and we've highlighted them with clear icons for ready reference.

PLANTS Unlike animal cells, plant cells are surrounded by a cell wall and lack the extracellular matrix found in animal tissues. As a plant cell matures, new layers of wall are laid down just outside the plasma membrane, which is intimately involved in the assembly of cell walls (Figure 5-41). The walls are built prim...

p. 167

BIOTECH This procedure is commonly used to purify the different types of white blood cells, each of which bears on its surface one or more distinctive proteins

p. 153

▼ Figure Titles
We've started each legend with a clear description of the figure, in blue.

New, Clearer Art ▶
We have stepped out and numbered many experimental processes.

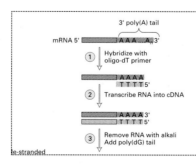

◀ **FIGURE 7-15 Preparation of a bacteriophage λ cDNA library.** A mixture of mRNAs, isolated as shown in Figure 7-14, is used to produce cDNAs corresponding to all the cellular mRNAs (steps ①–③). These single-stranded cDNAs (light green) are then converted into double-stranded cDNAs, which are treated with *Eco*RI methylase to prevent subsequent digestion by *Eco*RI (steps ④–⑥). The protected double-stranded cDNAs are ligated to a synthetic double-stranded *Eco*RI-site linker at both ends and then cleaved with the corresponding restriction enzyme, yielding cDNAs with sticky ends (red letters); these are incorporated into λ phage cloning vectors, and the resulting recombinant λ virions are plated on a lawn of *E. coli* cells (steps ⑦–⑧). See text for further discussion.

A s scientists and educators, we know that some processes, life cycles, and techniques are easier to comprehend if you can see them in motion. We have chosen these subjects to animate.

It's easy to navigate your way from the book to the CD-ROM. In addition to the list of resources provided in the **Media Connections** section at the beginning of each chapter, we have tabbed those figures directly related to animations. We also list all the animations, videos, and macromolecular models on the endpapers of the book.

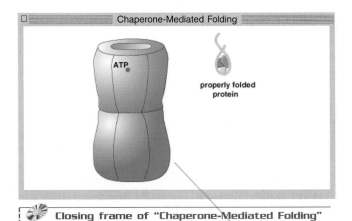

Chaperone-Mediated Folding

ATP

properly folded
protein

◀ Closing frame of "Chaperone-Mediated Folding"

◀ **Figures and Animations Have the Same Look**
The figures in the book and their related animations use a consistent color scheme and set of icons for representing like elements. This means you can use the CD and the book together without confusion.

▼ **Media Connections**
Each figure that has a related animation is tabbed for ready reference. The title of the animation runs alongside the figure.

▶ **FIGURE 3-15 Chaperone-mediated protein folding.**
(a) Many proteins ① fold into their proper three-dimensional structure with the assistance of Hsp70, a molecular chaperone that transiently binds to a nascent polypeptide as it emerges from a ribosome. Proper folding of some proteins ② also depends on the chaperonin TCiP, a large barrel-shaped complex of Hsp60 units. (b) GroEL, the bacterial homolog of TCiP, is a barrel-shaped complex of 14 identical 60,000-MW subunits arranged in two stacked rings. In the absence of ATP or

(a)   Ribosome

①

Hsp 70

②

ATP

Partially folded
protein

Properly folded
protein

ATP

Conformational
change

GroEL/TCiP

Future Possibilities

We have written two new sections at the end of each chapter: **Perspectives for the Future** and **Perspectives in the Literature**. These sections provide an opportunity for us to discuss what we think the future will hold, and give you a chance to explore original literature and Web resources in answering critical-thinking questions.

Testing Yourself on the Concepts

These questions help you review the key concepts covered in the chapter. Questions like this are also available in the Student Companion.

MCAT/GRE-Style Questions

These questions cover key concepts, applications, and experiments discussed in the chapter. Framed in the style of standardized tests, they are another useful method of review.

Key Terms

We have provided a selective list of key terms covered in the chapter, along with the page on which they can be found.

References

References to the literature are organized by major section heading. We also list some pertinent Web sites you may be interested in.

Contents in Brief

Contents

3   Protein Structure and Function

4   Nucleic Acids, the Genetic Code, and the Synthesis of Macromolecules

5 Biomembranes and the Subcellular Organization of Eukaryotic Cells

6   Manipulating Cells and Viruses in Culture

7   Recombinant DNA and Genomics

8 Genetic Analysis in Cell Biology

PART II: Nuclear Control of Cellular Activity

9 Molecular Structure of Genes and Chromosomes

10 Regulation of Transcription Initiation

11 RNA Processing, Nuclear Transport, and Post-Transcriptional Control

12 DNA Replication, Repair, and Recombination

13 Regulation of the Eukaryotic Cell Cycle

14 Gene Control in Development

17 Protein Sorting: Organelle Biogenesis and Protein Secretion

18   Cell Motility and Shape I: Microfilaments

PART IV: Cell Interactions

21 Nerve Cells

22 Integrating Cells into Tissues

23 Cell Interactions in Development

24 Cancer

Chapter Opening Illustrations

Chapter 1 An artist's rendition of the interior of a eukaryotic cell. Depicting a cell's interior is difficult because electron micrographs provide detailed pictures of only a thin slice of a cell, while the cell itself is a three-dimensional object with a very complex interior structure. Thus an artist can create a special sense of the cell's inner workings by using color and shading. Here the artist rendered the organelles inside the cell as he imagined them rather than as a faithful reconstruction from electron micrographs. The blue object is the cell's nucleus with the DNA visible inside as a coil. The red strands emerging from the nucleus are RNA molecules. In the rest of the cell is the cytoplasm, which contains many organelles like the red, kidney-shaped mitochondria and the sectioned orange vesicles. The green stack of flattened vesicles near the nucleus is the Golgi apparatus, and the other flat vesicles represent the cell's endoplasmic reticulum. All of these cellular elements are described in later chapters. [This picture, drawn by Tomo Narashima, originally appeared on the cover of the second edition of this book.]

Chapter 2 Three-dimensional model of an ATP molecule. The atoms are represented by spheres of the appropriate van der Waals radius; carbon atoms are gray, nitrogens are blue, phosphorus are yellow, and oxygen is red. The model was based on the three-dimensional coordinates of the atoms in several nucleotide protein complexes, derived from the crystalline structures of the molecules, in the computerized file of the Protein Data Bank. [Photograph courtesy of Sung Choe.]

Chapter 3 α-Actinin crosslinks actin filaments at the Z-line of muscle cells and at focal adhesions of non-muscle cells. To determine its structure, electron microscopists grow a two-dimensional crystal of the protein and solve its structure from images taken in a cryoelectron microscope. The micrograph shows the protein has the rough outline of a dumb-bell, an elongated molecule with globular domains at either end. [Micrograph courtesy of K. Taylor.]

Chapter 4 Ribosomes plus attached tRNA decode messenger RNA during translation, the process of assembling the amino acids in correct order to make a protein. This figure shows the two-lobed structure (small and large ribosomal subunits) of the *E. coli* ribosome deduced from various physical techniques. Highlighted here are the positions occupied by a tRNA with an attached amino acid (red) and the tRNA to which is attached the growing peptide chain (green). The messenger is not shown. [K. H. Nierhaus et al., 1998, *Proc. Nat'l Acad. Sci. USA* **95**:945–950; photograph courtesy of K. H. Nierhaus.]

Chapter 5 Human squamous carcinoma cells (SqCC/Y1) stained with fluorescent antibodies to the cytoskeletal proteins tubulin (green) and keratin-19 (red), and also with the dye DAPI that causes DNA to fluoresce blue. Note that the keratin fibers appear to concentrate around the nucleus, and terminate at adjacent sites of cell-cell contact. 125X magnification (to the slide). [Photograph courtesy of Nancy Kedersha.]

Chapter 6 Formation of syncytia in NIH 3T3 cells that express truncated Moloney murine leukemia virus envelope proteins. The nuclei are stained with Hoechst dye 33258 (blue) and the location of the envelope protein is indicated by the red rhodamine staining. Cultured NIH 3T3 cells were transfected by electroporation with DNA that encodes a truncated envelope protein that promotes cell-cell membrane fusion and formation of syncytia (multinucleated cells). The truncation is necessary for making the envelope protein competent for membrane fusion. After DNA transfection the cells were grown on glass cover slips and fixed 48 hours later with a 15-minute incubation with 4% paraformaldehyde in phosphate buffered saline. The cells were stained for 30 minutes with a rat monoclonal antibody, 83A25, directed against the gp70 envelope protein, and then with goat anti-rat immunoglobulin G antibodies coupled to rhodamine. The cells were then incubated for 2 minutes with Hoechst dye 33258 and the cover slips were mounted in Fluormount and viewed with a fluorescence microscope. [Photograph courtesy of David Sanders, Whitehead Institute for Biomedical Research.]

Chapter 7 Detection of HIV-1 nucleic acid in human lymphocytes by in situ PCR. Lymphocytes isolated from peripheral blood were fixed, permeabilized, and subjected to PCR with HIV-1 specific primers. Amplified DNA (green) was detected by hybridization to a complementary oligonucleotide probe conjugated with 5-carboxyfluorescein. Nuclei were counterstained (red) with propidium iodide. Green fluorescent cells were isolated with a fluorescence activated cell sorter and visualized by confocal microscopy. [Photograph courtesy of Bruce Patterson, M.D.]

Chapter 8 In mammals, only one X chromosome is active. A specific region of the X chromosome called Xic (X inactivation center) is required in cis for X inactivation. One gene within this region called Xist (in mouse) or XIST (in human) plays a crucial role in silencing expression of genes from the inactive X chromosome. The Xist gene does not encode a protein. Accumulation of Xist RNA is required for the spreading of X inactivation. In this figure the FISH technique was used to analyze the expression pattern of Xist RNA in female E7.0 mouse embryo nuclei. A probe to exonic sequence is shown in red and a probe for intronic sequence is shown in green. The overlap of the two probes appears in yellow. These data demonstrate that at this stage in development Xist RNA is being synthesized from both the inactive and active X chromosomes, but only accumulates over the former. This results, in large part, through selective stabilization of Xist RNA on the inactive X chromosome. [From Panning et al., 1997, *Cell* 90:907–916; photograph courtesy of Rudolf Jaenisch.]

Chapter 9 Male human chromosomes visualized by the method of chromosome "painting." Metaphase chromosomes were hybridized to multiple DNA probes specific for sequences along the length of each chromosome. A different combination of fluorochromes that fluoresce with different spectra was used to label the probes for each chromosome. Following hybridization, digital images of the fluorescently labeled chromosomes were collected using a charge-coupled device (CCD) camera and multiple exposures with separate optical filters specific for each of the fluorochromes. The images were then analyzed by computer and a composite image was generated in which each chromosome can be clearly distinguished from chromosomes of a similar size by a pseudo-color assigned on the basis of its fluorochrome composition. Note that there are two homologs of each chromosome. This same method can be used to recognize abnormal chromosomal translocations with great sensitivity (see Figure 9-38b). [See P. Lichter, 1997, *Trends Genet.* 13:475–479; photograph courtesy of M. Speicher and D. C. Ward.]

Chapter 10 An active region of transcription producing a "puff" in a *Drosophila* polytene chromosome. Chromosomes were stained with fluorescently labeled antibodies against the heat shock transcription factor (red) and RNA polymerase II (green). Regions of overlap appear yellow. The transcription factor is concentrated near the 5′ end of the transcription unit comprising the puff and at additional positions along the polytene chromosomes. [Photograph courtesy of John R. Weeks and Arno L. Greenleaf.]

Chapter 11 The non-snRNP pre-mRNA splicing factor SC35 localizes in a speckled distribution in interphase nuclei (orange regions). HeLa cell SC35 was visualized by immunostaining with a fluorescently labeled antibody. An optical section of the immunostaining pattern is superimposed over a differential interference contrast image of the cells. [Photograph courtesy of David L. Spector, Cold Spring Harbor Laboratory.]

Chapter 12 During S phase of the cell cycle, DNA replication proteins assemble into large complexes or replication foci containing many replication forks (e.g., tens to hundreds) and thousands of replication proteins. During interphase these components are distributed uniformly throughout the nucleoplasm. In this figure antibodies to a DNA methylase were used to visualize replication foci in S-phase cells. Antibodies to other replication factors also show recruitment into these structures during S phase. [From H. Leohardt, A. Page, H.-U Weier, and T. H. Bestor, 1992, *Cell* 71:865–873.]

Chapter 13 Metaphase in a cultured newt lung cell. Microtubules were visualized by indirect immunofluorescence. Chromosomes were stained with Hoechst 33342. [From J. C. Waters, R. W. Cole, and C. L. Reider, 1993, *J. Cell Biol.* 122:361–372. Photograph courtesy of Conly L. Reider.]

Chapter 14 Different proteins are expressed in different cells in the developing spinal cord. The Hedgehog protein (yellow) is an extracellular signal specifically expressed in the ventral-most region of the spinal cord, called the floor plate (see Chapter 23). Hedgehog controls the identity of different neuronal precursor cells in the ventral spinal cord. The more dorsal population expresses the homeobox protein Pax-6 (green), while the more ventral population expresses the homeobox protein Nkx2.2 (red). Hedgehog represses Pax-6 and induces Nkx2.2 in the ventral-most progenitor cells. Motoneurons are derived from both populations of progenitor cells and express IsI1 (blue), another homeobox transcription factor. [See J. Ericson et al., 1997, *Cell* 90: 169–180; photograph courtesy of T. M. Jessell.]

Chapter 15 Three-dimensional structure of a recombinant cardiac gap junction membrane channel determined by electron crystallography. These channels allow the direct exchange of ions and small molecules between adjacent cells. Each channel is formed by association of six connexin subunits, each of which contains four α helices, in one plasma membrane, with a similar structure in the plasma membrane of an adjacent cell. [From V. Unger et al., 1999, *Science* 283:1176; courtesy of Mark Yeager.]

Chapter 16 Ubiquinone (orange) bound to the surface of the photosynthetic reaction center (white) from the bacterium *Rhodobacter spheroides*. Only one of the oxygen atoms in ubiquinone is visible (blue). [After C.-H Cheng et al., 1991, *Biochemistry* 30:5352; courtesy of Dr. Lawren Wu.]

Chapter 17 Firefly luciferase, a peroxisomal matrix protein, is transported to peroxisomes of normal human fibroblasts, but remains cytoplasmic in cells from a Zellweger syndrome patient. The fibroblasts (on coverslips) were microinjected with mRNA encoding the luciferase. After overnight incubation in a humidified CO_2 incubator, the cells were fixed, permeabilized, and labeled with appropriate primary (rabbit anti-luciferase) and secondary (FITC anti-rabbit) antibodies. The punctate immunofluorescence observed in normal human HS68 cells *(left)* is indicative of peroxisomal luciferase. The fibrobrast cell line *(GM6231)* from the human patient *(right)* does not import luciferase into peroxisomes, but shows a cytoplasmic signal instead of the punctate signal. Magnification 165X. [See P. Walton et al., 1993, *Mol. Cell Biol.* 12:531–541; photographs courtesy of Suresh Subramani.]

Chapter 18 A fish scale keratinocyte is one of the fastest moving cells. In this micrograph, the actin (blue) and myosin (red) molecules are labeled with specific fluorescent antibodies. [Courtesy of A. B. Verhovsky.]

Chapter 19 Macrophage cells stained for tubulin (green) and the K14 keratin subunit (red). Immunofluorescence micrograph shows the fibrillar distribution of microtubules and intermediate filaments in the same cell. Regions in which the two proteins are colocalized are colored yellow. [Photograph courtesy of Nancy Kedersha.]

Chapter 20 The compound eye of the fruit fly *Drosophila melanogaster* contains about 750 simple eyes, or ommatidia, each containing eight photoreceptor neurons. The eye develops from an epithelium called the eye imaginal disc. Cells in the disc assemble into clusters in a highly ordered fashion. A wave of morphogenesis sweeps across the disc from posterior (left) to anterior (right). At the leading edge of this wave, called the morphogenetic furrow, cells change shape and stain strongly with phalloidin, giving rise to a continuous band of staining (green). Within the furrow, cells form a reiterated pattern of clusters. This is correlated with activation of the Ras/MAP kinase pathway. Activated MAP kinase (red) was detected using an antibody that specifically recognizes the diphosphorylated (active) but not the unphosphorylated (inactive) form of MAP kinase. The overlap of the green (actin) and red (active MAP kinase) gives rise to the yellow clusters evenly spaced at the morphogenetic furrow. Each one of these clusters will give rise to an ommatidium. [From Kumar et al., 1998, *Development* **125**(19):3875–3885; courtesy of Kevin Moses.]

Chapter 21 Neonatal rat cortical brain cells, cultured for 25 days in vitro, stained with a fluorescent antibody to the cytoskeletal intermediate filament protein GFAP (Glial Fibrillary Acidic Protein, green) and with the dye DAPI that causes DNA to fluoresce blue. Two distinct types of astrocytes (green cells) are present in this culture, along with other types of cells (non-green) that appear as isolated blue nuclei. [Photograph courtesy of Nancy Kedersha.]

Chapter 22 A dense network of elastin and collagen fibers form the extracellular matrix of elastic cartilage. These fibers of the ECM are intimately connected to the plasma membrane of a chondrocyte. The membrane is supported by the network of filaments from the actin cytoskeleton. [Courtesy of R. Mecham and J. Heuser, Washington University School of Medicine.]

Chapter 23 As a single fertilized human egg divides, its progeny give rise to hundreds of different types of cells. During embryonic development, specific interactions between different types of cells, and between cells and different types of extracellular matrices, are essential for formation of each type of differentiated cell and its specific organization into tissues and organs. Shown is Emma Rachel Steinert soon after her birth, and her parents Heidi Lodish Steinert and Eric Steinert. [Photograph courtesy of Stephanie Lodish.]

Chapter 24 Human melanoma cells (cell line Hs695T) stained for a melanoma-specific cell surface glycoprotein in green and counterstained for myosin (antibody 5.15, red) and for DNA (Hoechst, blue). [Photograph courtesy of Nancy Kedersha.]

The Dynamic Cell

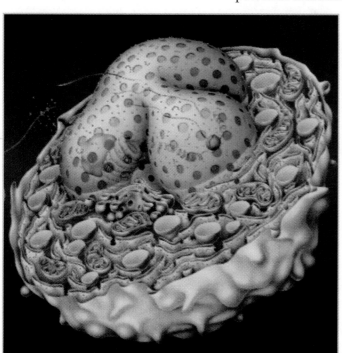

1

An artist's rendering of a eukaryotic cell.

At first glance, the biological universe appears amazingly diverse—from tall palm trees to tiny ferns, from single-cell bacteria and protozoans visible only under a microscope to multicellular animals of all kinds. Yet the bewildering array of outward biological form overlies a powerful uniformity: all biological systems are composed of the same types of chemical molecules and employ similar principles of organization at the cellular level. Although the basic kinds of biological molecules have been conserved during the billion years of evolution, the ways in which they are assembled with one another to form functioning cells and organisms have undergone considerable change.

To study the properties of the molecules of life and the innumerable variations on basic themes that are found in different organisms, modern researchers employ concepts and experimental techniques drawn from biochemistry, molecular biology, genetics, and cell biology. The resulting discipline of *molecular cell biology* investigates how cells develop, operate, communicate, and control their activities, and on occasion go awry. This book contains the authors' attempt to describe systematically the current state of knowledge about cells and to present many of the key experiments that have led to our current understanding of cellular life. It may seem to you, our new reader, a daunting challenge. Our hope is that the overwhelming inventiveness and sheer beauty of construction of biological systems will intrigue you and amply reward your efforts to understand the story we tell.

Living systems, including the human body, consist of such closely interrelated elements that no single element can be fully appreciated in isolation from the others. Organisms contain organs; organs are composed of tissues; tissues consist of cells; and cells are formed from molecules (Figure 1-1). The unity of living systems is coordinated by many levels of interrelationship: molecules carry messages from organ to organ and cell to cell;

MEDIA CONNECTIONS

Overview: Life Cycle of a Cell

tissues are delineated and integrated with other tissues by noncellular membranes secreted by cells; and cells gain identity from contact with other cells. Generally all the levels into which we fragment biological systems interconnect. To learn about biological systems, however, we must take a segment at a time. The biology of cells is a logical starting point because an organism can be viewed as consisting of interacting cells, which are the closest thing to an autonomous biological unit that exists. The integration of cellular activity into tissues, the development of organisms by growth and specialization of cells, and the metabolic events fueling the dynamism of living systems are all topics on which we will touch, but they are all topics that fall within the province of other subdisciplines of biological science.

In this chapter, we provide a framework for understanding the primacy of cells in biological systems and review several fundamental concepts that recur throughout our more detailed discussions in subsequent chapters. We begin with a brief look at the role of evolution and then discuss the general properties of the molecules found in biological systems. Next, we review the main features of cellular architecture, noting the similarities and differences between the three main cell lineages that have emerged over evolutionary time. The remaining topics covered in this chapter focus on the assemblage of cells into organized structures and their dynamic nature, in preparation for later chapters dealing with various processes critical to cellular growth, differentiation, and adaptation to changing circumstances.

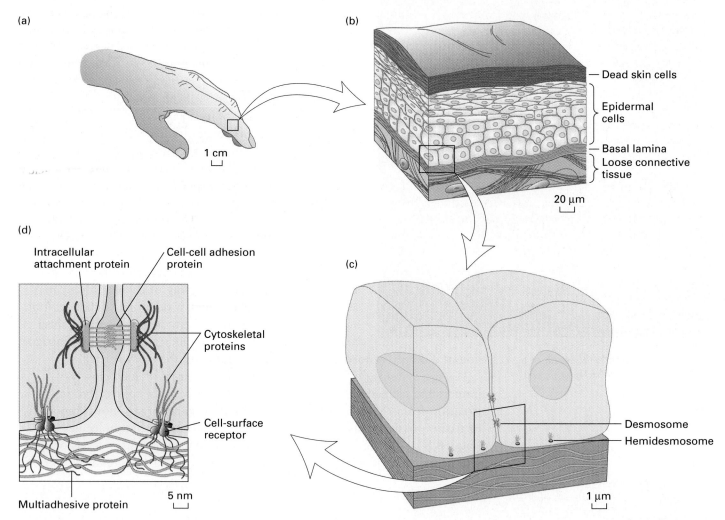

▲ **FIGURE 1-1 Living systems such as the human body consist of closely interrelated elements.** (a) The surface of our hand is covered by a living organ, skin, that is covered by several layers of tissue. (b) An outer covering of hard, dead cells protects the body from injury, infection, and dehydration. This layer is constantly renewed by living epidermal cells, which also give rise to hair and fur. Deeper layers of muscle and connective tissue give skin its tone and firmness. (c) Tissues are formed through subcellular adhesion structures that join cells to each other and to an underlying layer of supporting fibers. (d) At the heart of the adhesion are its structural components: phospholipid molecules that make up the plasma membrane and protein molecules which form strong bonds with molecules on other cells and with internal or external fibers.

1.1 Evolution: At the Core of Molecular Change

The interplay of events played out over billions of years, in the historical process called *evolution*, dictates the form and structure of the living world today. Thus biology, which is the study of the results of these historical events, differs fundamentally from physics and chemistry, which deal with the essential and unchanging properties of matter. The great insight of Charles Darwin was that all organisms are related in a great chain of being extending from the distant past to the present. The Darwinian principle that organisms vary randomly and the fittest are then selected by the forces of their environment guides biological thinking to this day.

We now know that **genes,** which chemically are composed of **deoxyribonucleic acid (DNA),** ultimately define biological structure and maintain the integration of cellular function. The genes encode **proteins,** the primary molecules that make up cell structures and carry out cellular activities. Alterations in the structure and organization of genes thus provide the random variation that nurtures evolutionary change in biological structure and function.

Even scientists brought up in the evolutionary tradition have been surprised to learn in recent years just how closely the genes of different species are related. During evolution, genes have been conserved to such an extent that some human genes will function in a yeast cell and quite a few will function in a fly cell. Clearly, one feature of evolution is the *maintenance* unchanged of many aspects of cellular life even while great changes in external form and capability are occurring. Recent progress in determining the sequences of all the genes in a variety of organisms is revealing the subtle changes that have fueled evolution.

The creative part of the evolutionary process is *adaptation* to rapidly changing environments and the conquest of new environmental niches. During this process, small alterations in cellular structures and functions are selected. Entirely new structures rarely are created; more often, old structures are adapted to new circumstances. More rapid change is possible by rearranging or multiplying previously evolved components rather than by waiting for a wholly new approach to emerge. The cellular organization of organisms plays a fundamental role in this process because it allows change to come about by small alterations in previously evolved cells, giving them new capabilities.

1.2 The Molecules of Life

Among the many events that occur in the life of a cell are a multitude of specific chemical transformations, which provide the cell with usable energy and the molecules needed to form its structure and coordinate its activities. These biochemical reactions and other cellular processes are governed by basic principles of chemistry reviewed in Chapter 2. Here we briefly describe the functions of the main types of chemicals that compose cells. Throughout many later chapters we will focus on the interactions and transformations of these molecules.

Water, inorganic ions, and a large array of relatively small organic molecules (e.g., sugars, vitamins, fatty acids) account for 75–80 percent of living matter by weight. Of these small molecules, water is by far the most abundant. The remainder of living matter consists of **macromolecules,** including proteins, polysaccharides, and DNA (Figure 1-2).

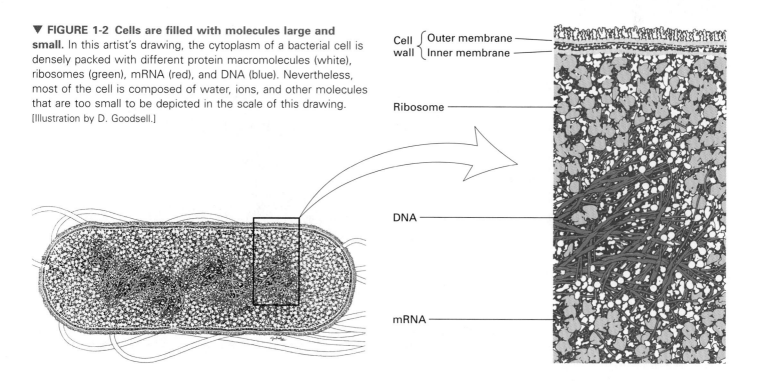

▼ **FIGURE 1-2 Cells are filled with molecules large and small.** In this artist's drawing, the cytoplasm of a bacterial cell is densely packed with different protein macromolecules (white), ribosomes (green), mRNA (red), and DNA (blue). Nevertheless, most of the cell is composed of water, ions, and other molecules that are too small to be depicted in the scale of this drawing. [Illustration by D. Goodsell.]

Cell wall { Outer membrane — Inner membrane —

Ribosome —

DNA —

mRNA —

Cells acquire and use these two size classes of molecules in fundamentally different ways. Ions, water, and many small organic molecules are imported into the cell. Cells also make and alter many small organic molecules by a series of different chemical reactions. In contrast, cells can obtain macromolecules only by making them. Their synthesis entails linking together a specific set of small molecules (**monomers**) to form **polymers** through repetition of a single type of chemical-linkage reaction.

Some small molecules function as precursors for synthesis of macromolecules, and the cell is careful to provide the appropriate mix of small molecules needed. Small molecules also store and distribute the energy for all cellular processes; they are broken down to extract this chemical energy, as when sugar is degraded to carbon dioxide and water with the release of the energy bound up in the molecule (Chapter 16). Other small molecules (e.g., hormones and growth factors) act as signals that direct the activities of cells (Chapter 20), and nerve cells communicate with one another by releasing and sensing certain small signaling molecules (Chapter 21). The powerful effect on our body of a frightening event comes from the instantaneous flooding of the body with a small-molecule hormone that mobilizes the "fight or flight" response.

Macromolecules, though, are the most interesting and characteristic molecules of living systems; in a true sense the evolution of life as we know it is the evolution of macromolecular structures. Proteins, the workhorses of the cell, are the most abundant and functionally versatile of the cellular macromolecules. To appreciate the abundance of protein within a cell, we can estimate the number of protein molecules in a typical eukaryotic cell, such as a hepatocyte in the liver. This cell, roughly a cube 15 μm (0.0015 cm) on

a side, has a volume of 3.4×10^{-9} cm^3 (or milliliters). Assuming a cell density of 1.03 g/ml, the cell would weigh 3.5×10^{-9} g. Since protein accounts for approximately 20 percent of a cell's weight, the total weight of cellular protein is 7×10^{-10} g. The average yeast protein has a molecular weight of 52,700 (g/mol), as noted in Chapter 3. Assuming this value is typical of eukaryotic proteins, we can calculate the total number of protein molecules per liver cell as about 7.9×10^9 from the total protein weight and the number of molecules per mole, which is a constant (Avogadro's number). To carry this calculation one step further, consider that a liver cell contains about 10,000 different proteins; thus, a cell contains close to a million molecules of each protein on average. In actuality, however, the abundance of different proteins varies widely, from the quite rare cell-surface protein that binds the hormone insulin (20,000 molecules) to the abundant structural protein actin (5×10^8 molecules).

Many of the proteins within cells are **enzymes,** which accelerate (catalyze) reactions involving small molecules. Other proteins allow cells to move and do work, maintain internal cell rigidity, and transport molecules across membranes. Proteins even direct their own synthesis and that of other macromolecules. Reflecting their numerous functions, proteins come in many shapes and sizes (Figure 1-3). The elucidation of the structure of proteins and the relation of protein structure to function remain active areas of scientific investigation (Chapter 3). Proteins are formed from only 20 different monomers, the **amino acids.** That such a limited set of building blocks can do so much is a continuous marvel, even to researchers who work with proteins every day. They are the true glory of the biological world.

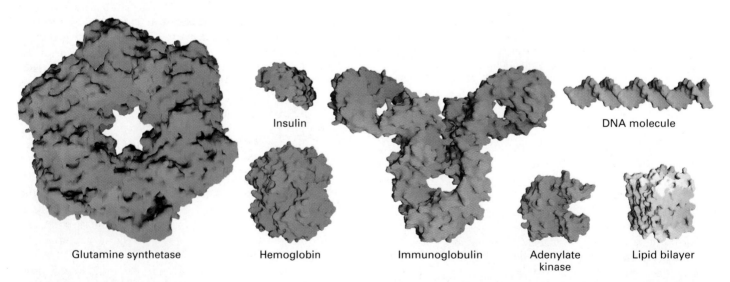

Insulin

DNA molecule

Glutamine synthetase

Hemoglobin

Immunoglobulin

Adenylate kinase

Lipid bilayer

▲ FIGURE 1-3 Models of some representative proteins (pink) drawn to a common scale and compared with a small portion of a lipid bilayer sheet (yellow) and a DNA molecule (blue). Each protein has a defined three-dimensional shape held together by numerous chemical bonds. The illustrated proteins include enzymes (glutamine synthetase and adenylate kinase), an antibody (immunoglobulin), a hormone (insulin), and the blood's oxygen carrier (hemoglobin). [Courtesy of Gareth White.]

▲ **FIGURE 1-4 James D. Watson (left) and Francis H. C. Crick (right) with the double-helical model of DNA they constructed in 1952–1953.** Their model ultimately proved correct in all its essential aspects. [From J. D. Watson, 1968, *The Double Helix*, Atheneum, Copyright 1968, p. 215; Courtesy of A. C. Barrington Brown.]

The macromolecule that garners the most public attention is not protein but deoxyribonucleic acid (DNA), whose functional properties make it the cell's master molecule. The three-dimensional structure of DNA, first proposed by James D. Watson and Francis H. C. Crick about 50 years ago, consists of two long helical strands that are coiled around a common axis forming a **double helix** (Figure 1-4). The double-helical structure of DNA, one of nature's most magnificent constructions, is critical to the phenomenon of heredity, the transfer of genetically determined characteristics from one generation to the next.

Each strand of DNA is composed of just four different types of monomers called **nucleotides.** Genes are simply coded representations of the structures of individual proteins, a code written in four chemical "letters"—the nucleotides—and displayed as a continually varying sequence in DNA. Since cells use proteins (enzymes) to make other molecules like sugars or fats, DNA indirectly directs the synthesis of many small molecules as well as proteins. DNA also contains a coded set of instructions about when various proteins are to be made and in what quantities.

In the common view, DNA is the storage form of genetic information, which protein "machines" read out for use by the cell. But a third macromolecule, **ribonucleic acid (RNA),** is necessary in the process. The *central dogma* of biology states that the coded genetic information hard-wired into DNA is transcribed into individual transportable cassettes, composed of **messenger RNA (mRNA);** each mRNA cassette contains the program for synthesis of a particular protein (or small number of proteins). This critical trio of macromolecules—DNA, RNA, and proteins—is present in all cells. The mechanism whereby the information encoded in DNA is deciphered into proteins is now understood quite well and explained in Chapter 4. How this process of **gene expression** is regulated—that is, how cells "know" to make the right proteins at the right time in the right amounts—is a major focus of current research in molecular cell biology and a recurring theme throughout this book.

1.3 The Architecture of Cells

Although generalizations in biology usually lack the theoretical underpinnings found in physics, there are very clear commonalities among living systems that give biology a unity. One is the style of cellular construction. The biological universe consists of two types of cells—*prokaryotic cells,* which lack a defined nucleus and have a simplified internal organization, and *eukaryotic cells,* which have a more complicated internal structure including a defined, membrane-limited **nucleus.** Detailed analysis of the DNA from a variety of prokaryotic organisms in recent years has revealed two distinct types: bacteria (often called *"true" bacteria* or **eubacteria**) and **archaea** (also called *archaebacteria* or *archaeans*). As we discuss in Chapter 7, the archaea are in some respects more similar to eukaryotic organisms than to the true bacteria.

Based on the assumption that organisms with more similar genes evolved from a common progenitor more recently than those with more dissimilar genes, researchers have developed the lineage tree shown in Figure 1-5. According to this tree, the archaea and eukarya (eukaryotes) are thought to have diverged from the bacteria before they diverged from each other. Despite the differences in the organization of prokaryotic and eukaryotic cells, all cells share certain structural features and carry out many complicated processes in basically the same way.

Cells Are Surrounded by Water-Impermeable Membranes

A cell, because it is a limited space, must have an outer border. The construction of that border represents one of the most fundamental considerations in biological organization. The outer shell of cells, like any shell, is built to keep the interior contents from leaking out into the surrounding environment. The chemical processes of cellular life generally take place in a watery solution, and the intracellular constituents of cells are largely molecules that are easily dissolved in water. Similarly, the environment around cells is a watery one, the blood and other bodily fluids being solutions in water. Cells then, in order to maintain their integrity, need to be surrounded by an environment through which water cannot flow. A membrane composed of fatty molecules serves this purpose.

We all know from common experience that "oil and water don't mix." That maxim is all one needs to appreciate how a cell is constructed. When oil is poured on water, the oil spreads into a thin film; that film is analogous to the film of fat that surrounds cells, called the **plasma membrane** (Figure 1-6). Biological membranes differ from a pure oil film in that the molecules that make the membrane have both oily and watery portions; they have long fatty chains, but they also have a head group that is water-soluble by virtue of being electrically charged. Thus membranes are formed because these bipartite molecules, called **phospholipids,** spontaneously orient themselves to form a double layer, or **bilayer,** having a fatty interior with external surfaces bonded to the surrounding water by the charged head groups. The membrane is given rigidity by interspersion of **cholesterol,** a molecule we have come to hate because of its association with heart disease, but one that is required to build the outer membrane of all our cells. Hence from an understanding of the contrasting properties of watery solutions and oily layers, an understanding of cellular construction emerges.

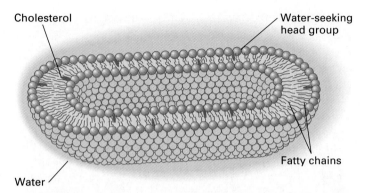

▲ **FIGURE 1-6 The watery interior of cells is surrounded by the plasma membrane, a two-layered shell of phospholipids.** Cholesterol molecules provide some rigidity to the fatty layer. The phospholipid molecules are oriented with their fatty chains facing inward and their water-seeking head groups (red spheres) facing outward. Thus both sides of the membrane are lined by head groups, mainly charged phosphates, adjacent to the watery spaces inside and outside the cell. In actuality, the interior space is much larger relative to the volume of lipid. All biological membranes have the same basic phospholipid bilayer structure.

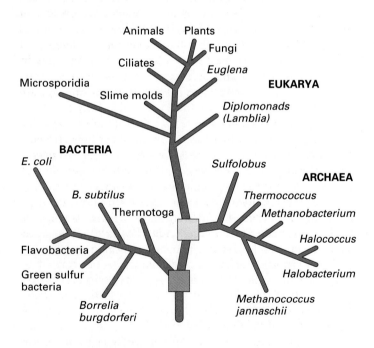

▲ **FIGURE 1-5 The three kingdoms of organisms are related through common sequences of their ribosomal RNAs.** Their lineage depicts a view of how all life on earth, from simple bacteria to complex mammals, evolved from a common, single-celled progenitor.

In spite of the rigidity provided by cholesterol, membranes composed of fat are not very strong, so numerous mechanisms for strengthening the borders of cells have evolved. In plants the plasma membrane is surrounded by a rigid **cell wall.** Although most animal cells lack a cell wall, proteins attached to their exterior surfaces provide some stability; the linking of cells together through these proteins helps maintain the integrity of tissues. Tissues and organs are often covered by strong networks of proteins and other molecules that strengthen and protect them, and also wall off the various compartments of the body. Single-celled organisms, like bacteria, have special outer coats to protect them.

Membranes Serve Functions Other Than Segregation

Although membranes are valuable as a way to segregate the watery interior of the cell from its environment, or to segregate intracellular events from one another, they have other important functions, including energy storage. Because membranes separate watery compartments from one another, if an ion or a molecule dissolved in water is moved through a membrane into a new cellular compartment, it will not be able to diffuse freely out of the compartment into which it was moved. It takes energy to move the molecule, but once moved, the molecule stores that energy by virtue of its entrapment. Formally, this storage of energy is just like the storage of energy in a battery. Therefore, membranes not only delineate compartments, but also serve as active participants in the cell's dynamism.

The functions of many proteins depend on their mode of association with membranes. For instance, the passage of water-soluble molecules through membranes is carried out by protein transporters that are embedded in the membrane. Also, cells send information to one another by releasing signaling molecules. The outer membranes of cells have proteins, known appropriately as *receptors,* that bind the circulating signaling molecules. These signaling molecules allow the individual activities of the many cells in the body to be coordinated. The receipt of a signaling molecule by a receptor causes the transient organization of particular types of intracellular proteins, called *signal-transduction proteins,* into an activated complex at the interior face of the cell's outer membrane, from which it directs alterations of events in the cell's cytoplasm or nucleus (Chapter 20).

Prokaryotes Comprise a Single Membrane-Limited Compartment

All **prokaryotes** are single-celled organisms, or protists. The bacterial lineage includes *Escherichia coli,* found in animal intestines and a favorite experimental organism, and the photosynthetic organisms formerly known as *blue-green algae* but better known today as *cyanobacteria.* (Because most prokaryotes studied in laboratories are bacteria, discussions of prokaryotic structure or metabolism throughout this book refer to these organisms, not archaeans, unless noted otherwise.) Many members of the archaeal lineage grow in unusual, often extreme, environments. For instance, the halophiles require high concentrations of salt to survive, and the thermoacidophiles grow in hot (80°C) sulfur springs, where a pH of less than 2 is common. Other archaeans, called *methanogens,* live in oxygen-free milieus and generate methane (CH_4) by the reduction of carbon dioxide.

Figure 1-7a illustrates the general structure of a typical bacterial cell; archaeal cells have a similar structure. In general, prokaryotes consist of a single closed compartment containing the **cytosol** and bounded by the plasma membrane. Although bacterial cells do not have a defined nucleus, the genetic material, DNA, is condensed into the central region of the cell. In addition, most **ribosomes**—the cell's protein-synthesizing particles—are found in the DNA-free region of the cell. Some bacteria also have an invagination of the cell membrane, called a *mesosome,* which is associated with synthesis of DNA and secretion of proteins. Thus bacterial cells are not completely devoid of internal organization.

Bacterial cells possess a cell wall, which lies adjacent to the external side of the plasma membrane. The cell wall is composed of layers of peptidoglycan, a complex of proteins and oligosaccharides; it helps protect the cell and maintain its shape. Some bacteria (e.g., *E. coli*) have a thin cell wall and an unusual outer membrane separated from the cell wall by the periplasmic space. Such bacteria are not stained by the Gram technique and thus are classified as gram-negative. Other bacteria (e.g., *Bacillus polymyxa*) that have a thicker cell wall and no outer membrane take the Gram stain and thus are classified as gram-positive.

Eukaryotic Cells Contain Many Organelles and a Complex Cytoskeleton

Eukaryotes comprise all members of the plant and animal kingdoms, including the unicellular fungi (e.g., yeasts, mushrooms, molds) and protozoans. Eukaryotic cells, like prokaryotic cells, are surrounded by a plasma membrane. However, unlike prokaryotic cells, most eukaryotic cells also contain extensive internal membranes that enclose specific compartments, the **organelles,** and separate them from the rest of the **cytoplasm,** the region of the cell lying outside the nucleus (Figure 1-7b and chapter opening figure).

Most organelles are surrounded by a single phospholipid membrane, but several, including the nucleus, are enclosed by two membranes. Each type of organelle plays a unique role in the growth and metabolism of the cell, and each contains a collection of specific enzymes that catalyze requisite chemical reactions. The membranes defining these subcellular compartments control their internal ionic composition so that it commonly differs from that of the cytosol (the portion of the cytoplasm outside the organelles) and among the various organelles.

The largest organelle in a eukaryotic cell is generally the nucleus, which houses most of the cellular DNA. In addition to the nucleus, several other organelles are present in nearly all eukaryotic cells: the **mitochondria,** in which much of the cell's energy metabolism is carried out; the rough and smooth **endoplasmic reticula,** a network of membranes in which glycoproteins and lipids are synthesized; **Golgi vesicles,** which direct membrane constituents to appropriate places in the cell; and **peroxisomes,** in which fatty acids and amino acids are degraded. Animal cells, but not plant cells, contain **lysosomes,** which degrade worn-out cell constituents and foreign materials taken in by the cell. **Chloroplasts,** where photosynthesis occurs, are found only in certain leaf cells of plants and some single-celled organisms. Both plant cells and some single-celled eukaryotes contain one or more vacuoles, large, fluid-filled organelles in which nutrients and waste compounds are stored and some degradative reactions occur.

The cytosol of eukaryotic cells contains an array of fibrous proteins collectively called the **cytoskeleton** (Chapters 18 and 19). Three classes of fibers compose the cytoskeleton: **microtubules** (20 nm in diameter), built of polymers of the protein tubulin; **microfilaments** (7 nm in diameter), built of the protein actin; and **intermediate filaments** (10 nm in diameter), built of one or more rod-shaped protein subunits. The cytoskeleton gives the cell strength and rigidity, thereby helping to maintain cell shape. Cytoskeletal fibers also control movement of structures within the cell; for example, some cytoskeletal fibers connect to organelles or provide tracks along which organelles move.

The rigid cell wall, composed of cellulose and other polymers, that surrounds plant cells contributes to their strength and rigidity. Fungi are also surrounded by a cell wall, but its composition differs from that of bacterial or plant cell walls.

(a) Prokaryotic cell

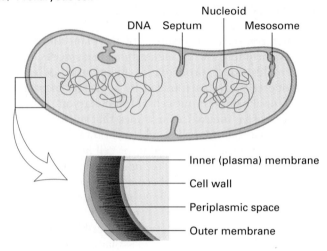

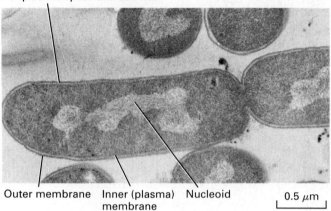

(b) Eukaryotic cell

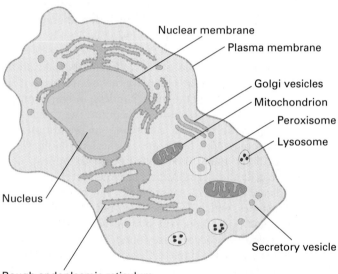

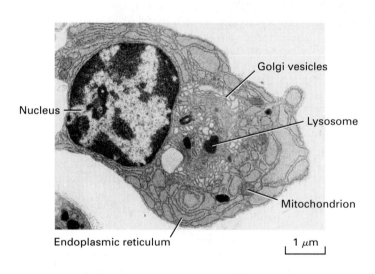

▲ **FIGURE 1-7 Comparison of the structure of prokaryotic and eukaryotic cells.** (a) Drawing of a typical gram-negative prokaryotic (bacterial) cell in the process of dividing and electron micrograph of a thin section of *E. coli,* a common intestinal bacterium. Note the periplasmic space between the inner and outer membranes, and the cell wall adjacent to the inner membrane. (b) Drawing of a eukaryotic cell and electron micrograph of a plasma cell, a type of white blood cell that secretes antibodies. Only a single membrane (the plasma membrane) surrounds the cell, but the interior contains many membrane-limited compartments known as *organelles,* which are described in more detail in Chapter 5. The defining characteristic of eukaryotic cells is segregation of the cellular DNA within a defined nucleus, which is bounded by a double membrane. [Photograph in part (a) courtesy of I. D. J. Burdett and R. G. E. Murray; photograph in part (b) from P. C. Cross and K. L. Mercer, 1993, *Cell and Tissue Ultrastructure: A Functional Perspective,* W. H. Freeman and Company.]

Cellular DNA Is Packaged within Chromosomes

The DNA in the nuclei of eukaryotic cells is distributed among 1 to more than 50 long linear structures called **chromosomes.** The number and size of the chromosomes are the same in all cells of an organism, but vary among different types of organisms. Each chromosome comprises a single DNA molecule associated with numerous proteins, and the total DNA in the chromosomes of an organism is referred to as its **genome.** Chromosomes, which stain intensely with basic dyes, are visible in the light microscope only during cell division when the DNA becomes tightly compacted (Figure 1-8).

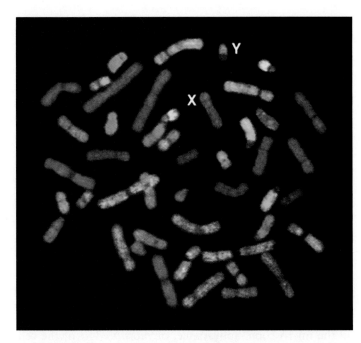

▲ **FIGURE 1-8 Light micrograph of the 46 human chromosomes.** A normal individual has 23 pairs of chromosomes; one member of each pair is inherited from the mother and the other member from the father. In this example, use of a special technique permits each of the chromosome pairs to be shown in a different color. The presence of an X and Y chromosome pair identifies the sex of the individual as male. [Courtesy of K. Heselmeyer-Haddad and H. M. Padilla-Nash.]

In all prokaryotic cells, most of or all the genetic information resides in a single circular DNA molecule, about a millimeter in length; this molecule lies, folded back on itself many times, in the central region of the cell. Although the large genomic DNA molecule in prokaryotes is associated with proteins and often is referred to as a chromosome, the arrangement of DNA within a bacterial chromosome differs greatly from that within the chromosomes of eukaryotic cells.

The concept that genes are like "beads" strung on a long "string," the chromosome, was proposed early in the 1900s based on genetic work with the fruit fly *Drosophila*. The early *Drosophila* workers could position, or map, the genes responsible for various mutant traits on a chromosome, even though they did not yet know that genes were segments of DNA or that the function of a gene was due to a protein whose sequence was encoded by that gene!

1.4 The Life Cycle of Cells

A **cell** in an adult organism can be viewed as a steady-state system. The DNA is constantly read out into a particular set of mRNAs, which specify a particular set of proteins. As these proteins function, they are also being degraded and re-

placed by new ones, and the system is so balanced that the cell neither grows, shrinks, nor changes its function. This static view of the cell, however, misses the all-important dynamic aspects of cellular life.

The dynamics of a cell can best be understood by examining the course of its life. A new cell arises when one cell divides or when two cells, like a sperm and an egg cell, fuse. Either event sets off a cell-replication program that is encoded in the DNA and executed by proteins. This program usually involves a period of cell growth, during which proteins are made and DNA is replicated, followed by **cell division,** when a cell divides into two daughter cells. Whether a given cell will grow and divide is a highly regulated decision of the body, assuring that an adult organism replaces worn out cells or makes more cells in response to a new need. Examples of the latter are the growth of muscle in response to exercise or damage, and the proliferation of red blood cells when a person ascends to a higher altitude and needs more capacity to capture oxygen. However, in one major and devastating disease—cancer—cells multiply even though they are not needed by the body. To understand how cells become cancerous, biologists have intensely studied the mechanisms that control the growth and division of cells.

The Cell Cycle Follows a Regular Timing Mechanism

Most eukaryotic cells live according to an internal clock; that is, they proceed through a sequence of phases, called the **cell cycle,** during which DNA is duplicated during the synthesis (S) phase and the copies are distributed to opposite ends of the cell during mitotic (M) phase (Figure 1-9). Progress along the cycle is controlled at key **checkpoints,** which monitor the status of a cell, for instance, the internal amount of DNA or the presence of extracellular nutrients. When certain conditions are met, the cell proceeds to the next checkpoint. The cycle begins after the cell divides into two daughter cells, each containing an identical copy of the parental cell's genetic material.

The cell cycle of prokaryotes is simple and fast. Replication of the single chromosome begins at a particular DNA sequence, the replication origin, which is anchored to the cell membrane. Once DNA replication is complete, assembly of new membrane and cell wall forms a septum, which eventually divides the cell in two (see Figure 1-7a). Because the origins of the two newly formed chromosomes are anchored to different membrane sites, each daughter cell receives one chromosome. In ideal growth conditions, the bacterial cell cycle is repeated every 30 minutes.

Only a few types of eukaryotic cells can grow and divide as quickly as bacteria. Most growing plant and animal cells take 10–20 hours to double in number, and some duplicate at a much slower rate. Many cells in adult animals, such as nerve cells and striated muscle cells, do not divide at all. They have temporarily exited from the cell cycle

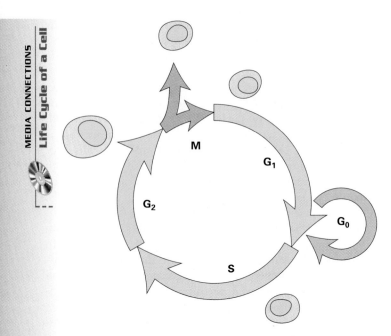

▲ **FIGURE 1-9 The eukaryotic cell cycle.** In most growing cells, the four phases proceed successively, taking from 10–20 hours depending on cell type and developmental state. Interphase comprises the G_1, S, and G_2 phases. DNA is synthesized in S, and other cellular macromolecules are synthesized throughout interphase, so the cell roughly doubles its mass. During G_2 the cell is prepared for the mitotic (M) phase, when the genetic material is evenly partitioned and the cell divides. Nondividing cells exit the normal cycle, entering the quiescent G_0 state.

after mitosis and entered a "paused or quiescent" state called G_0. Because eukaryotic cells are larger and more complex than prokaryotic cells, a specialized mechanism coordinates their replication of genomic DNA, distribution of chromosomes, and cell division. The complex regulatory events that guide eukaryotic cells from phase to phase are described in Chapter 13.

Mitosis Apportions the Duplicated Chromosomes Equally to Daughter Cells

Mitosis is the mechanism in eukaryotes for partitioning the genome equally at cell division. To accomplish this complex task, plant and animal cells build a specialized machine, called the **mitotic apparatus**, which captures the chromosomes and then pushes and pulls them to opposite sides of the dividing cell (Chapter 19). Remarkably, the mitotic apparatus is a temporary structure that exists only during mitosis to distribute the genetic material. Although the events of mitosis unfold continuously, they are conventionally divided into four substages representing phases of chromosome movement. During the first substage, **prophase,** the replicated chromosomes, each comprising two identical **chromatids,** are condensed into compact packets and then

released to the cytoplasm when the nuclear membrane breaks down. During **metaphase** and **anaphase,** the chromosomes are sorted, and each chromatid of a pair moves to opposite sides of the cell (Figure 1-10). The end of mitosis is marked by re-formation of a membrane around each set of chromosomes (**telophase**). Division of the cytoplasm, called **cytokinesis,** then yields two daughter cells, each with a $2n$ complement of genetic material.

Cell division in plant and animal cells differs mainly at cytokinesis. Animal cells divide in two by pinching of the cytoplasm. However, because a plant cell is surrounded by a rigid cell wall, daughter cells are formed by building a new cell membrane and cell wall between the two daughter nuclei, thereby cutting the cytoplasm into two portions.

Cell Differentiation Creates New Types of Cells

The most complicated example of cellular dynamics occurs when a cell changes, or *differentiates,* to carry out a specialized function. This process often is marked by a change in the microscopic appearance, or *morphology,* of the cell. For example, the different structures of a nerve cell and a muscle cell reflect their respective functions in long-distance communication and contraction, highlighting the biological principle that "form follows function."

Cell differentiation creates the diversity of cell types that arise during the development of an organism from a fertilized egg. This is a process of extensive cell multiplication and differentiation. A mammal that starts as one cell becomes an organism with hundreds of diverse cell types such as muscle, nerve, and skin. Here we see at its most dramatic the power of DNA to control cellular behavior: development is a DNA-orchestrated set of cellular changes (easily tens of thousands of them) that occur virtually without fail. The almost perfect resemblance of "identical" twins is a testament to the program encoded by DNA to reproducibly direct the development of a human being.

Nowhere is the variety of cellular activities and responses better illustrated than in the body's immune system. It is there that many cell types come together in organized tissues specifically designed to allow the body to distinguish its own cells from those of foreign invaders. Within the immune system, we see both development of specialized cells that can recognize invading cells and formation of tissues from cells that originate in various parts of the body. The immune-system cells not only actively survey their environment with surface receptor proteins like antibodies, but also change their properties when they encounter a foreign substance, allowing the body to rid itself of invaders.

Cells Die by Suicide

Unchecked cell growth and multiplication produce a mass of cells, a tumor. *Programmed cell death* plays the very important role of population control by balancing cell growth and multiplication. In addition, cell death also eliminates

▶ **FIGURE 1-10 Cell division.** A parental cell in G_1 has two copies of each chromosome ($2n$), one maternal (red) and one paternal (blue). Chromosomes are replicated during the S phase, giving a $4n$ chromosomal complement. At the midpoint of mitosis (metaphase), the replicated chromosomes are aligned and held in position by the mitotic apparatus. The two identical chromatids composing each replicated chromosome then move to opposite ends of the cell, the nuclear membrane re-forms around each set of chromosomes, and finally cytokinesis splits the cell into two genetically identical daughter cells.

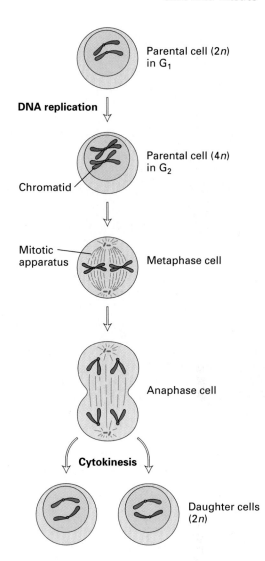

Parental cell ($2n$) in G_1

DNA replication

Chromatid

Parental cell ($4n$) in G_2

Mitotic apparatus

Metaphase cell

Anaphase cell

Cytokinesis

Daughter cells ($2n$)

unnecessary cells. For example, during embryogenesis, the digits of our fingers and toes are sculpted by the death of cells in the intervening spaces. If these cells remained alive, our hands and feet would become webbed. Thus the timing and location of cell death, as well as cell growth and division, must be precisely controlled.

Cell death follows an internal program of events called **apoptosis,** in which all traces of a cell vanish. The first visible sign of apoptosis is condensation of the nucleus and fragmentation of the DNA. The cell soon shrivels and is consumed by macrophages. A cell is directed to commit suicide when an essential factor is removed from the extracellular environment or when an internal signal is activated. Thus, the default state of the cell is to remain alive. The discovery of genes that suppress the growth of tumors by activating cell death stimulated an exciting new line of cancer research that may lead to more effective treatment strategies.

1.5　Cells into Tissues

The evolution of multicellular organisms most likely began when cells remained associated in small colonies after division instead of separating into individual cells. A few prokaryotes and several unicellular eukaryotes exhibit such rudimentary social behavior. The full flowering of multicellularity, however, occurs in eukaryotic organisms whose cells become differentiated and organized into groups, or *tissues*, in which the tissue's cells perform a specialized, common function.

Multicellularity Requires Extracellular Glues

The simplest multicellular organisms are single cells embedded in a jelly of protein and polysaccharide called the **extracellular matrix.** More complicated arrangements of cells into a chain, a ball, or a sheet require other means. The cells of higher plants, for instance, are connected by cytoplasmic bridges, called **plasmodesmata,** and are encased in a network of chambers formed by the interlocking cell walls surrounding the cells. Animal cells, in contrast, are "glued" together by **cell-adhesion molecules** (CAMs) on their surface. Some CAMs bind cells to one another; other types bind cells to the extracellular matrix, forming a cohesive unit. In

animals, the matrix cushions and lubricates cells. A specialized matrix, the **basal lamina,** which is especially tough, forms a supporting layer underlying cell sheets and preventing the cells from ripping apart.

Tissues Are Organized into Organs

The specialized groups of differentiated cells form tissues, which are themselves the major components of organs. For example, the lumen of a blood vessel is lined with a sheet-like layer of endothelial cells, or **endothelium,** which prevents blood cells from leaking out (Figure 1-11). A layer of smooth muscle tissue encircles the lumen and contracts to limit the blood flow. During times of fright, constriction of smaller peripheral vessels forces more blood to the vital organs. The muscle layer of a blood vessel is wrapped in an outer layer of connective tissue, a network of fibers and cells that encase and protect the vessel walls from stretching and rupture. This hierarchy of tissues is copied in other blood vessels, which differ mainly in the thickness of the layers. The wall of a major artery must withstand much stress and

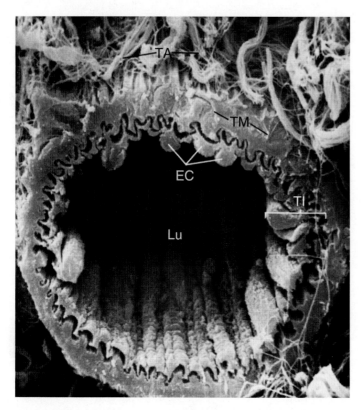

▲ **FIGURE 1-11 All organs are organized arrangements of various tissues, as illustrated in this cross section of a small artery (arteriole).** Blood flows through the vessel lumen (Lu), which is lined by a thin sheet of endothelial cells (EC) forming the endothelium (TI) and by the underlying basal lamina. This tissue adheres to the overlying layer of smooth muscle tissue (TM); contraction of the muscle layer controls blood flow through the vessel. A fibrillar layer of connective tissue (TA) surrounds the vessel and connects it to other tissues. [From R. Kessels and R. Kardon, 1979, *Tissues and Organs: A Text-Atlas of Scanning Electron Microscopy*, W. H. Freeman and Company, p. 42.]

is therefore thicker than a minor vessel. The strategy of grouping and layering of different tissues is used to build other complex organs. In each case the function of the organ is determined by the specific functions of its component tissues.

Body Plan and Rudimentary Tissues Form Early in Embryonic Development

The human body consists of some 100 trillion cells, yet it develops from a single cell, the zygote, resulting from fusion of a sperm and an egg. The early stages in the development of an embryo are characterized by rapid cell division and the differentiation of cells into tissues. The embryonic *body plan*, the spatial pattern of cell types (tissues) and body parts, emerges from two influences: a program of genes that specify the pattern of the body and local cell interactions that induce different parts of the program. Remarkably, the ba-

sic body plan of all animals is very similar (Figure 1-12). This conservation of body plan reflects evolutionary pressure to preserve the commonalities in the molecular and cellular mechanisms controlling development in different organisms. The impressive strides made in understanding these mechanisms are detailed in several later chapters.

With only a few exceptions, most animals display axial symmetry; that is, their left and right sides mirror each other. This most basic of patterns is encoded in the genome. In fact, *patterning genes* specify the general organization of an organism, beginning with the major body axes—anterior-posterior, dorsal-ventral, and left-right—and ending with body segments such as the head, chest, abdomen, and tail. The conservation of axial symmetry from the simplest worms to mammals is explained by the presence of conserved patterning genes in the genomes. Some patterning genes encode proteins that control expression of other genes; other patterning genes encode proteins that are important in cell adhesion or in cell signaling. This broad repertoire of patterning genes permits the integration and coordination of events in different parts of the developing embryo.

The precise timing of developmental events is maintained by the ability of one group of cells to induce or activate differentiation of a second group of cells. Most often **induction** is mediated by direct cell contact or by soluble factors released by the cells. In a typical case, contact between an aggregate of cells, the mesenchyme, with an overlying epithelial cell layer directs the latter cells to differentiate into an embryonic tissue or in later stages of development into a specific type of tissue. For example, the primitive notochord induces the development of embryonic nervous tissue

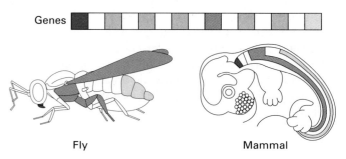

Genes

Fly   Mammal

▲ **FIGURE 1-12 Common patterns of development are seen in animals as diverse as sea urchins, flies, mice, and humans.** Shown here, a set of Hox genes are arranged linearly in the genomes of mammals and flies. These genes direct the development of different segments in the body plan of many animals. This suggests a common control mechanism for the development of body segments. Remarkably, the position of the gene also marks the position of its expression in the embryo. During embryonic development, the first genes (red) in the genome are expressed at the anterior of the embryo while other genes (orange, yellow) are expressed at more distal parts. This pattern of expression in flies gives rise to mouth parts, thorax, wing segments, and the tail.

and brain. Later, an eye forms when contact between a lobe of the developing brain induces the overlying embryonic "skin" to differentiate into a primitive lens.

1.6 Molecular Cell Biology: An Integrated View of Cells at Work

Today's scientific understanding of cellular complexity and dynamism rests on the work of many thousands of scientists over the last century and a half. Modern researchers have fused concepts and experimental techniques drawn from biochemistry, genetics, and molecular biology with those from classical cell biology to produce a dynamic conception of cellular life. In the chapters that follow, we will flesh out this introductory overview of the cell, drawing on insights from all the subdisciplines that contribute to the hybrid science of molecular cell biology.

Our knowledge of cell structure and function at any point in time is only as good as the tools available for investigation; as those tools become more effective, old concepts are sometimes totally reformulated. For this reason, our presentation will be anchored on the experimental foundations supporting various concepts. We hope this approach will encourage the reader to appreciate biology as a living science, one in which changing knowledge continually generates fresh perspectives and fresh opportunities for productive impacts on our society.

In the next ten years, a new view of biology will emerge as the massive endeavor currently under way to sequence the human genome is completed. Knowledge of the sequences of the roughly 100,000 genes in human DNA will add a whole new dimension to biological study, assisting in the even more difficult task of determining the functions of all the genes and bringing further insight about the interplay of genes in the development and differentiation of organisms. Perhaps the major challenge facing cell biologists in the twenty-first century will be to analyze the molecular basis of integrated functions in whole organisms, including learning, behavior, and aging. Astonishing as it may seem, today's young researchers may well achieve the goal stated in 1973 by Francois Jacob in *The Logic of Life:* "to interpret the properties of the organism by the structure of its constituent molecules."

Key Terms

amino acids *4*
anaphase *10*
apoptosis *11*
archaea *5*
bilayer *6*
cell *9*
cell cycle *9*
cell division *9*
cell wall *6*
checkpoints *9*
chloroplasts *7*
chromatids *10*
chromosomes *8*
cytoplasm *7*
cytoskeleton *7*
cytosol *7*
deoxyribonucleic acid (DNA) *3*
double helix *5*
endoplasmic reticula *7*
enzymes *4*
eubacteria *5*
eukaryotes *7*
extracellular matrix *11*
gene expression *5*
genes *3*
genome *8*
Golgi vesicles *7*
induction *12*
lysosomes *7*
macromolecules *3*
messenger RNA (mRNA) *5*
metaphase *10*
mitochondria *7*
mitosis *10*
mitotic apparatus *10*
monomers *4*
nucleotides *5*
nucleus *5*
organelles *7*
peroxisomes *7*
phospholipids *6*
plasma membrane *6*
polymers *4*
prokaryotes *7*
prophase *10*
proteins *3*
ribonucleic acid (RNA) *5*
ribosomes *7*
telophase *10*

2

Chemical Foundations

In this chapter we review many important chemical concepts required to comprehend cellular processes, all of which follow the rules of chemistry. Constituting 70–80 percent by weight of most cells, water is the most abundant of the "chemicals of life." About 7 percent of the weight of living matter is composed of inorganic ions and small molecules such as nucleotides (the building blocks of DNA and RNA), amino acids (the building blocks of proteins), and sugars (Figure 2-1). All these small molecules can be chemically synthesized in the laboratory. The principal cellular macromolecules— DNA, RNA, and protein—compose the remainder of living matter. Like the small molecules found in cells, these very large molecules follow the general rules of chemistry and can be chemically synthesized. Molecular cell biology, however, aims at understanding higher levels of biological organization, that is, explaining the structure and function of organisms and cells in terms of the properties of individual molecules, such as proteins and nucleic acids. In this book, we examine many examples of complex, higher-order processes that are understood, at least partially, at the molecular level.

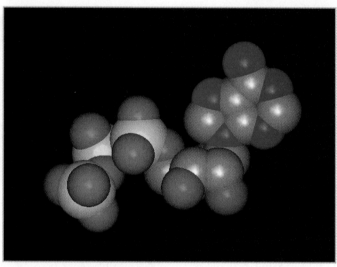

Three-dimensional structure of an ATP molecule.

The realization that complex processes such as evolution, development of an organism from a fertilized egg, motion, perception, and thought follow the rules of chemistry and physics—and that no vital or supernatural force is involved—has profound philosophical and even political implications.

All the topics discussed in this chapter relate to important concepts and experiments presented in later chapters. We begin with a discussion of covalent bonds, which connect individual atoms in a molecule, and the structures of carbohydrates, which illustrate the importance of relatively small differences in the arrangements of covalent bonds. Next we consider noncovalent bonds, important stabilizing forces between

(a) Water, ions, and small molecules (77%)

Water (70%): H—O
                    \
                     H

Inorganic ions (1%): Na^+ Cl^- K^+ $H_2PO_4^-$

Small molecules (6%):

Amino acid
(alanine)

Sugar
(glucose)

Nucleotide
(uradine monophosphate)

$$CH_3-CH_2-CH_2-CH_2-CH_2-CH_2-CH_2-CH_2-CH_2-CH_2-CH_2-CH_2-CH_2-\overset{\overset{O}{\|}}{C}-O^-$$
Fatty acid (myristic acid)

(b) Macromolecules (23%)

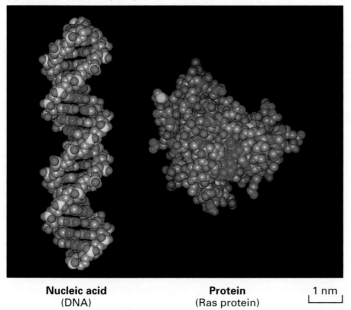

Nucleic acid   **Protein**   1 nm
   (DNA)   (Ras protein)

▲ **FIGURE 2-1 The chemicals of life.** All living matter is
composed of water, inorganic ions, a variety of small molecules,
and macromolecules. The approximate percentage by weight of
each class of chemicals in growing *Escherichia coli* bacteria is
shown in parenthesis. (a) The chemical structures of water and
several ions and small molecules common in biological systems.
(b) Models of the three-dimensional structure of DNA and Ras
protein. In these macromolecules each atom is depicted by a
colored ball (white = H, red = O, gray = C, yellow = P, blue = N,
green = S). Covalent bonds link nucleotides and amino acids into
linear chains, but the folded three-dimensional structures of DNA
and proteins are determined by weaker noncovalent bonds.
Three-dimensional molecular models of these and all other molecules depicted throughout the book are available on the accompanying CD-ROM.

groups of atoms within larger molecules and between
different molecules. The four most important ones are
hydrogen bonds, ionic interactions, van der Waals interactions, and hydrophobic bonds. We discuss the role of
hydrogen bonds in determining the properties of water,
the major constituent of cells and of the spaces between
cells. Phospholipids and their association into a bilayer
structure, stabilized by multiple noncovalent bonds, are
also described. The phospholipid bilayer is the basis of
all biological membranes (biomembranes), which have
numerous functions discussed in later chapters. The
next section reviews the concept of chemical equilibrium,
focusing on the properties of chemicals in aqueous solution; in particular, the pH of a solution, the reactions of
acids and bases, and the buffering capacity of solutions.
We then consider various aspects of biochemical energetics,
including the central concept of free energy, its relationship
to the direction of chemical reactions, and the central role
of ATP (adenosine triphosphate) in capturing and transferring energy in cellular metabolism. In the final section,
we describe the factors that control the rates of chemical
reactions and how enzymes accelerate reaction rates.

2.1 Covalent Bonds

Covalent bonds, which hold the atoms within an individual
molecule together, are formed by the sharing of electrons
in the outer atomic orbitals. The distribution of shared as
well as unshared electrons in outer orbitals is a major
determinant of the three-dimensional shape and chemical
reactivity of molecules. For instance, as we learn in Chapter 3, the shape of proteins is crucial to their function and
their interactions with small molecules. In this section,
we discuss important properties of covalent bonds and describe the structure of carbohydrates to illustrate how the
geometry of bonds determines the shape of small biological
molecules.

Each Atom Can Make a Defined Number of Covalent Bonds

Electrons move around the nucleus of an atom in clouds called *orbitals*, which lie in a series of concentric *shells*, or energy levels; electrons in outer shells have more energy than those in inner shells. Each shell has a maximum number of electrons that it can hold. Electrons fill the innermost shells of an atom first; then the outer shells. The energy level of an atom is lowest when all of its orbitals are filled, and an atom's reactivity depends on how many electrons it needs to complete its outermost orbital. In most cases, in order to fill the outermost orbital, the electrons within it form covalent bonds with other atoms. A covalent bond thus holds two atoms close together because electrons in their outermost orbitals are shared by both atoms.

Most of the molecules in living systems contain only six different atoms: hydrogen, carbon, nitrogen, phosphorus, oxygen, and sulfur. The outermost orbital of each atom has a characteristic number of electrons:

$$H \quad \cdot \overset{\cdot}{C} \cdot \quad \cdot \overset{\cdot \cdot}{\underset{\cdot}{N}} \cdot \quad \cdot \overset{\cdot \cdot}{P} \cdot \quad \cdot \overset{\cdot \cdot}{\underset{\cdot \cdot}{O}} \cdot \quad \cdot \overset{\cdot \cdot}{\underset{\cdot \cdot}{S}} \cdot$$

These atoms readily form covalent bonds with other atoms and rarely exist as isolated entities. As a rule, each type of atom forms a characteristic number of covalent bonds with other atoms.

For example, a hydrogen atom, with one electron in its outer shell, forms only one bond, such that its outermost orbital becomes filled with two electrons. A carbon atom has four electrons in its outermost orbitals; it usually forms four bonds, as in methane (CH_4), in order to fill its outermost orbital with eight electrons. The single bonds in methane that connect the carbon atom with each hydrogen atom contain two shared electrons, one donated from the C and the other from the H, and the outer (s) orbital of each H atom is filled by the two shared electrons:

$$H : \overset{\cdot \cdot}{C} : H \quad \text{or} \quad H - \overset{\overset{\textstyle H}{|}}{\underset{\underset{\textstyle H}{|}}{C}} - H$$

Nitrogen and phosphorus each have five electrons in their outer shells, which can hold up to eight electrons. Nitrogen atoms can form up to four covalent bonds. In ammonia (NH_3), the nitrogen atom forms three covalent bonds; one pair of electrons around the atom (the two dots on the right) are in an orbital not involved in a covalent bond:

$$H : \overset{\cdot \cdot}{\underset{\underset{\textstyle H}{|}}{N}} : \quad \text{or} \quad H - \overset{\overset{\textstyle H}{|}}{\underset{\underset{\textstyle H}{|}}{N}} :$$

In the ammonium ion (NH_4^+), the nitrogen atom forms four covalent bonds, again filling the outermost orbital with eight electrons:

$$H - \overset{\overset{\textstyle H}{|}}{\underset{\underset{\textstyle H}{|}}{N}}^{\scriptstyle +} - H$$

Phosphorus can form up to five covalent bonds, as in phosphoric acid (H_3PO_4). The H_3PO_4 molecule is actually a "resonance hybrid," a structure between the two forms shown below in which nonbonding electrons are shown as pairs of dots:

$$H - \overset{\overset{\textstyle H}{|}}{\underset{\underset{\textstyle \cdot O \cdot}{\|}}{\underset{}{O}}} \overset{\cdot O \cdot}{\underset{}{-}} P - O - H \quad \longleftrightarrow \quad H - \overset{\overset{\textstyle H}{|}}{\underset{\underset{\textstyle O}{\|}}{O}} - P^{+} - O - H^{-}$$

In the resonance hybrid on the right, one of the electrons from the P=O double bond has accumulated around the O atom, giving it a net negative charge and leaving the P atom with a net positive charge. The resonance hybrid on the left, in which the P atom forms the maximum five covalent bonds, has no charged atoms. Esters of phosphoric acid form the backbone of nucleic acids, as discussed in Chapter 4; phosphates also play key roles in cellular energetics (Chapter 16) and in the regulation of cell function (Chapters 13 and 20).

The difference between the bonding patterns of nitrogen and phosphorus is primarily due to the relative sizes of the two atoms: the smaller nitrogen atom has only enough space to accommodate four bonding pairs of electrons around it without creating destructive repulsions between them, whereas the larger sphere of the phosphorus atom allows more electron pairs to be arranged around it without the pairs being too close together.

Both oxygen and sulfur contain six electrons in their outermost orbitals. However, an atom of oxygen usually forms only two covalent bonds, as in molecular oxygen, O_2:

$$\overset{\cdot \cdot}{O} : : \overset{\cdot \cdot}{O} \quad \text{or} \quad \overset{\cdot \cdot}{O} = \overset{\cdot \cdot}{O}$$

Primarily because its outermost orbital is larger than that of oxygen, sulfur can form as few as two covalent bonds, as in hydrogen sulfide (H_2S), or as many as six, as in sulfur trioxide (SO_3) or sulfuric acid (H_2SO_4):

$$H - \overset{\overset{\textstyle H}{|}}{S} \qquad O = S \overset{\displaystyle O}{\underset{\displaystyle O}{<}} \qquad H - \overset{\cdot \cdot}{O} - \overset{\overset{\textstyle \cdot \cdot O \cdot \cdot}{\|}}{\underset{\underset{\textstyle \cdot O \cdot}{\|}}{S}} - \overset{\cdot \cdot}{O} - H$$

Hydrogen sulfide   **Sulfur trioxide**   **Sulfuric acid**

Esters of sulfuric acid are important constituents of the proteoglycans that compose part of the extracellular matrix surrounding most animal cells (Chapter 22).

The Making or Breaking of Covalent Bonds Involves Large Energy Changes

Covalent bonds tend to be very stable because the energies required to break or rearrange them are much greater than the thermal energy available at room temperature (25 °C) or body temperature (37 °C). For example, the thermal energy at 25 °C is less than 1 kilocalorie per mole (kcal/mol), whereas the energy required to break a C—C bond in ethane is about 83 kcal/mol:

$$H_3C : CH_3 \longrightarrow H_3C \cdot + \cdot CH_3 \qquad \Delta H = +83 \text{ kcal/mol}$$

where ΔH represents the difference in the total energy of all of the bonds (the enthalpy) in the reactants and in the products.* The positive value indicates that an input of energy is needed to cause the reaction, and that the products contain more energy than the reactants. The high energy needed for breakage of the ethane bond means that at room temperature (25 °C) well under 1 in 10^{12} ethane molecules exists as a pair of $\cdot CH_3$ radicals. The covalent bonds in biological molecules have ΔH values similar to that of the C—C bond in ethane (Table 2-1).

Covalent Bonds Have Characteristic Geometries

When two or more atoms form covalent bonds with another central atom, these bonds are oriented at precise angles to one another. The angles are determined by the mutual repulsion of the outer electron orbitals of the central atom. These bond angles give each molecule its characteristic shape (Figure 2-2). In methane, for example, the central carbon atom is bonded to four hydrogen atoms, whose positions define the four points of a tetrahedron, so that the angle between any two bonds is 109.5°. Like methane, the ammonium ion also has a tetrahedral shape. In these molecules, each bond is a *single bond,* a single pair of electrons shared between two atoms. When two atoms share two pairs of

---

TABLE 2-1 The Energy Required to Break Some Important Covalent Bonds Found in Biological Molecules*

Type of Bond	Energy (kcal/mol)	Type of Bond	Energy (kcal/mol)
SINGLE BOND		DOUBLE BOND	
O—H	110	C=O	170
H—H	104	C=N	147
P—O	100	C=C	146
C—H	99	P=O	120
C—O	84		
C—C	83	TRIPLE BOND	
S—H	81	C≡O	195
C—N	70		
C—S	62		
N—O	53		
S—S	51		

Note that double and triple bonds are stronger than single bonds.

electrons—for example, when a carbon atom is linked to only three other atoms—the bond is a *double bond:*

$$\diagdown C =$$

In this case, the carbon atom and all three atoms linked to it lie in the same plane (Figure 2-3). Atoms connected by a double bond cannot rotate freely about the bond axis, while those in a single bond generally can. The rigid planarity imposed by double bonds has enormous significance for the shape of large biological molecules such as proteins and

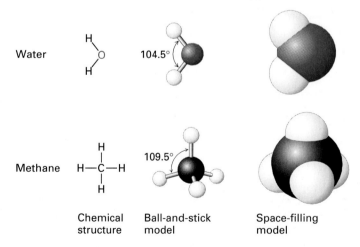

▲ **FIGURE 2-2 Bond angles give these water and methane molecules their distinctive shapes.** Each molecule is represented in three ways. The atoms in the ball-and-stick models are smaller than they actually are in relation to bond length, to show the bond angles clearly. The sizes of the electron clouds in the space-filling models are more accurate.

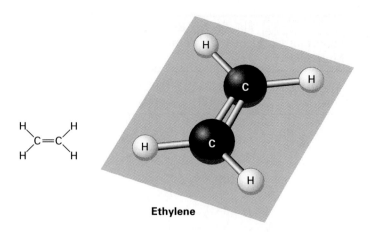

Ethylene

▲ **FIGURE 2-3 In an ethylene molecule, the carbon atoms are connected by a double bond, causing all the atoms to lie in the same plane.** Unlike atoms connected by a single bond, which usually can rotate freely about the bond axis, those connected by a double bond cannot.

nucleic acids. (In *triple bonds,* two atoms share six electrons. These are rare in biological molecules.)

All outer electron orbitals, whether or not they are involved in covalent bond formation, contribute to the properties of a molecule, in particular to its shape. For example, the outer shell of the oxygen atom in a water molecule has two pairs of nonbonding electrons; the two pairs of electrons in the H—O bonds and the two pairs of nonbonding electrons form an almost perfect tetrahedron. However, the orbitals of the nonbonding electrons have a high electron density and thus tend to repel each other, compressing the angle between the covalent H—O—H bonds to 104.5° rather than the 109.5° in a tetrahedron (see Figure 2-2).

Electrons Are Shared Unequally in Polar Covalent Bonds

In a covalent bond, one or more pairs of electrons are shared between two atoms. In certain cases, the bonded atoms exert different attractions for the electrons of the bond, resulting in unequal sharing of the electrons. The power of an atom in a molecule to attract electrons to itself, called *electronegativity,* is measured on a scale from 4.0 (for fluorine, the most electronegative atom) to a hypothetical zero (Figure 2-4). Knowing the electronegativity of two atoms allows us to predict whether a covalent bond can form between them; if the differences in electronegativity are considerable—as in sodium and chloride—an ionic bond, rather than a covalent bond, will form. This type of interaction is discussed in a later section.

In a covalent bond in which the atoms either are identical or have the same electronegativity, the bonding electrons are shared equally. Such a bond is said to be **nonpo-**

lar. This is the case for C—C and C—H bonds. However, if two atoms differ in electronegativity, the bond is said to be **polar.** One end of a polar bond has a partial negative charge (δ^-), and the other end has a partial positive charge (δ^+). In an O—H bond, for example, the oxygen atom, with an electronegativity of 3.4, attracts the bonded electrons more than does the hydrogen atom, which has an electronegativity of 2.2. As a result, the bonding electrons spend more time around the oxygen atom than around the hydrogen. Thus the O—H bond possesses an *electric dipole,* a positive charge separated from an equal but opposite negative charge. We can think of the oxygen atom of the O—H bond as having, on average, a charge of 25 percent of an electron, with the H atom having an equivalent positive charge. The *dipole moment* of the O—H bond is a function of the size of the positive or negative charge and the distance separating the charges.

In a water molecule both hydrogen atoms are on the same side of the oxygen atom. As a result, the side of the molecule with the two H atoms has a slight net positive charge, whereas the other side has a slight net negative charge. Because of this separation of positive and negative charges, the entire

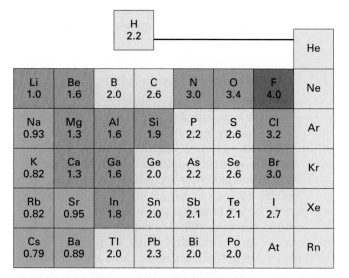

Electronegativity

| ■ 4.0– | ■ 3.0–3.9 | □ 2.0–2.9 | ■ 1.0–1.9 | ■ 0–0.99 |

▲ **FIGURE 2-4 Electronegativity values of main-group elements in the periodic table.** Atoms located to the upper right tend to have high electronegativity, fluorine being the most electronegative. Elements with low electronegativity values, such as the metals lithium, sodium, and potassium, are often called *electropositive.* The electronegativities of several atoms abundant in biological molecules differ enough that they form polar covalent bonds (e.g., O—H, N—H) or ionic bonds (e.g., Na^+Cl^-). Because the inert gases (He, Ne, etc.) have complete outer shells of electrons, they neither attract nor donate electrons, rarely form covalent bonds, and have no electronegativity values.

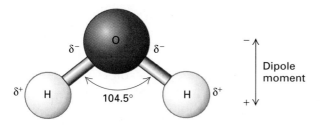

▲ **FIGURE 2-5 The water molecule has two polar O—H bonds and a net dipole moment.** The symbol δ represents a partial charge (a weaker charge than the one on an electron or a proton), and each of the polar H—O bonds has a dipole moment. The net dipole moment of the molecule is determined by the sizes and directions of the dipole moments of each of the bonds.

molecule has a net dipole moment (Figure 2-5). Some molecules, such as the linear molecule CO_2, have two polar bonds:

$$O^{\delta-}\!=\!C^{\delta+}\!=\!O^{\delta-}$$

Because the dipole moments of the two C=O bonds point in opposite directions, they cancel each other out, resulting in a molecule without a net dipole moment.

Asymmetric Carbon Atoms Are Present in Most Biological Molecules

A carbon (or any other) atom bonded to four dissimilar atoms or groups is said to be asymmetric. The bonds formed by an **asymmetric carbon atom** can be arranged in three-dimensional space in two different ways, producing molecules that are mirror images of each other. Such molecules are called *optical isomers,* or **stereoisomers.** One isomer is said to be right-handed and the other left-handed, a property called *chirality.* Most molecules in cells contain at least one asymmetric carbon atom, often called a *chiral carbon* atom. The different stereoisomers of a molecule usually have completely different biological activities.

Amino Acids Except for glycine, all amino acids, the building blocks of the proteins, have one chiral carbon atom, called the α *carbon,* or C_α, which is bonded to four different atoms or groups of atoms. In the amino acid alanine, for instance, this carbon atom is bonded to —NH_2, —COOH, —H, and —CH_3 (Figure 2-6). By convention, the two mirror-image structures are called the D (*dextro*) and the L (*levo*) isomers of the amino acid. The two isomers cannot be interconverted without breaking a chemical bond. With rare exceptions, only the L forms of amino acids are found in proteins. We discuss the properties of amino acids and the covalent peptide bond that links them into long chains in Chapter 3.

Carbohydrates The three-dimensional structures of carbohydrates provide another excellent example of the structural

and biological importance of chiral carbon atoms, even in simple molecules. A carbohydrate is constructed of carbon (*carbo-*) plus hydrogen and oxygen (-*hydrate,* or water). The formula for the simplest carbohydrates—the **monosaccharides,** or simple sugars—is $(CH_2O)_n$, where n equals 3, 4, 5, 6, or 7. All monosaccharides contain hydroxyl (—OH) groups and either an aldehyde or a keto group:

$$\begin{array}{cc} \underset{\textbf{Aldehyde}}{-\overset{|}{\underset{|}{C}}-\overset{O}{\overset{\parallel}{C}}-H} & \underset{\textbf{Keto}}{-\overset{|}{\underset{|}{C}}-\overset{O}{\overset{\parallel}{C}}-\overset{|}{\underset{|}{C}}-} \end{array}$$

In the linear form of D-*glucose* ($C_6H_{12}O_6$), the principal source of energy for most cells in higher organisms, carbon atoms 2, 3, 4, and 5 are asymmetric (Figure 2-7, *top*). If the hydrogen atom and the hydroxyl group attached to carbon atom 2 (C_2) were interchanged, the resulting molecule would be a different sugar, D-mannose, and could not be converted to glucose without breaking and making covalent bonds. Enzymes can distinguish between this single point of difference.

D-Glucose can exist in three different forms: a linear structure and two different hemiacetal ring structures (see Figure 2-7). If the aldehyde group on carbon 1 reacts with the hydroxyl group on carbon 5, the resulting hemiacetal,

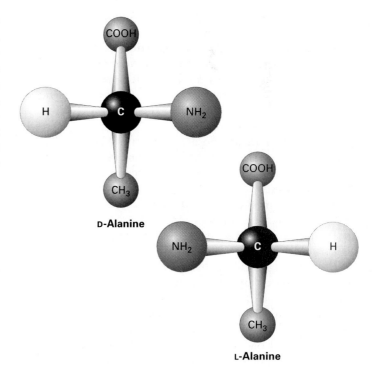

▲ **FIGURE 2-6 Stereoisomers of the amino acid alanine.** The asymmetric α carbon is black. Although the chemical properties of such optical isomers are identical, their biological activities are distinct.

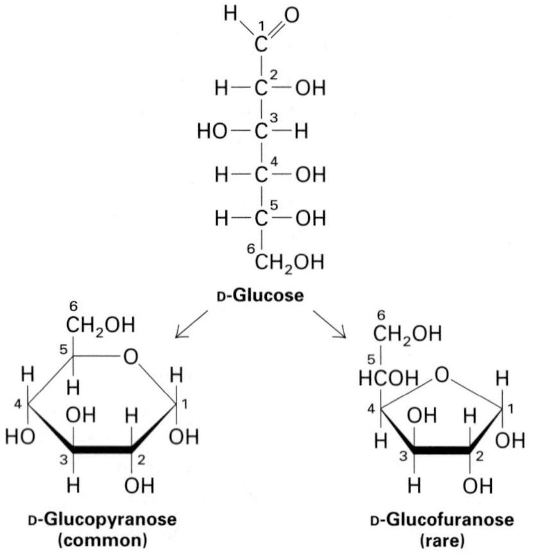

D-Glucose

D-Glucopyranose (common)

D-Glucofuranose (rare)

▲ **FIGURE 2-7 Three alternative configurations of D-glucose.** The ring forms, shown as Haworth projections, are generated from the linear molecule by reaction of the aldehyde at carbon 1 with the hydroxyl on carbon 5 or carbon 4.

α-D-Mannopyranose

α-D-Glucopyranose

α-D-Galactopyranose

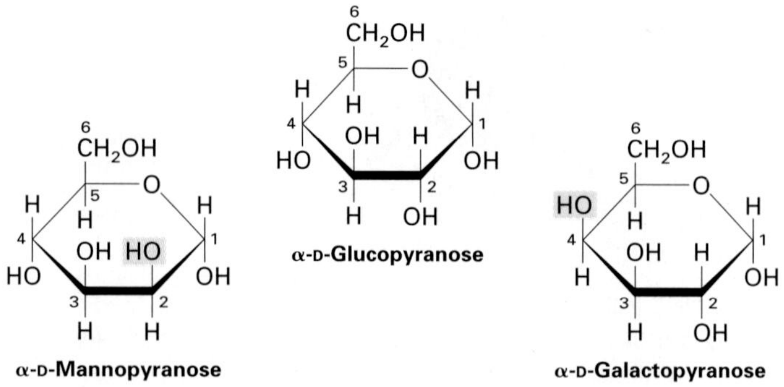

▲ **FIGURE 2-8 Haworth projections of the structures of glucose, mannose, and galactose in their pyranose forms.** The hydroxyl groups with different orientations from those of glucose are highlighted.

D-glucopyranose, contains a six-member ring. Similarly, condensation of the hydroxyl group on carbon 4 with the aldehyde group results in the formation of D-glucofuranose, a hemiacetal containing a five-member ring. Although all three forms of D-glucose exist in biological systems, the pyranose form is by far the most abundant.

The planar depiction of the pyranose ring shown in Figure 2-7 is called a *Haworth projection*. When a linear molecule of D-glucose forms a pyranose ring, carbon 1 becomes asymmetric, so two stereoisomers (called *anomers*) of D-glucopyranose are possible. The hydroxyl group attached to carbon 1 "points" down (below the plane of projection) in α-D-glucopyranose, as shown in Figure 2-7, and points up (above the plane of projection) in the β anomer. In aqueous solution the α and β anomers readily interconvert spontaneously; at equilibrium there is about one-third α anomer and two-thirds β, with very little of the open-chain form. Because enzymes can distinguish between the α and β anomers of D-glucose, these forms have specific biological roles.

Most biologically important sugars are six-carbon sugars, or **hexoses**, that are structurally related to D-glucose. Mannose, as noted, is identical with glucose except for the orientation of the substituents on carbon 2. In Haworth projections of the pyranose forms of glucose and mannose, the hydroxyl group on carbon 2 of glucose points downward, whereas that on mannose points upward (Figure 2-8). Similarly, galactose, another hexose, differs from glucose only in the orientation of the hydroxyl group on carbon 4.

The Haworth projection is an oversimplification because the actual pyranose ring is not planar. Rather, sugar

molecules adopt a conformation in which each of the ring carbons is at the center of a tetrahedron, just like the carbon in methane (see Figure 2-2). The preferred conformation of pyranose structures is the chair (Figure 2-9). In this conformation, the bonds going from a ring carbon to nonring atoms may take two directions: axial (perpendicular to the ring) and equatorial (in the plane of the ring).

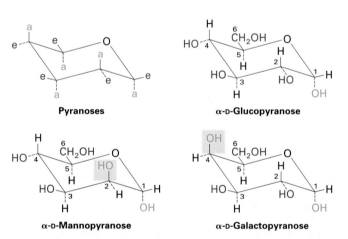

Pyranoses

α-D-Glucopyranose

α-D-Mannopyranose

α-D-Galactopyranose

▲ **FIGURE 2-9 Chair conformations of glucose, mannose, and galactose in their pyranose forms.** The chair is the most stable conformation of a six-membered ring. (In an alternative form, called the *boat*, both carbon 1 and carbon 4 lie above the plane of the ring.) The four bonds at each of the ring carbon atoms are tetrahedral. As shown in the generalized pyranose ring at the top left, bonds that extend nearly perpendicular to the plane of the ring are said to be axial (a); those that extend nearly parallel to the ring are said to be equatorial (e). In α-D-glucopyranose, all the hydroxyl groups except the one bonded to carbon 1 are equatorial. In α-D-mannopyranose, the hydroxyl groups bonded to carbons 1 and 2 are axial. In α-D-galactopyranose, the hydroxyl groups bonded to carbons 1 and 4 are axial. Note that, as in Figure 2-8, the hydroxyl groups with orientations different from those in glucose are highlighted.

The L isomers of sugars are virtually unknown in biological systems except for L-fucose. One of the unsolved mysteries of molecular evolution is why only D isomers of sugars and L isomers of amino acids were utilized, and not the chemically equivalent L sugars and D amino acids.

α and β Glycosidic Bonds Link Monosaccharides

In addition to the monosaccharides discussed above, two common **disaccharides,** lactose and sucrose, occur naturally (Figure 2-10). A disaccharide consists of two monosaccharides linked together by a C—O—C bridge called a **glycosidic bond.** The disaccharide lactose is the major sugar in milk; sucrose is a principal product of plant photosynthesis and is refined into common table sugar.

In the formation of any glycosidic bond, the carbon 1 atom of one sugar molecule reacts with a hydroxyl group of another. As in the formation of most biopolymers, the linkage is accompanied by the loss of water. In principle, a large number of different glycosidic bonds can be formed between two sugar residues. Glucose could be bonded to fructose, for example, by any of the following linkages: $\alpha(1 \rightarrow 1)$, $\alpha(1 \rightarrow 2)$, $\alpha(1 \rightarrow 3)$, $\alpha(1 \rightarrow 4)$, $\alpha(1 \rightarrow 6)$, $\beta(1 \rightarrow 1)$, $\beta(1 \rightarrow 2)$, $\beta(1 \rightarrow 3)$, $\beta(1 \rightarrow 4)$, or $\beta(1 \rightarrow 6)$, where α or β specifies the conformation at carbon 1 in glucose and the number following the arrow indicates the fructose carbon to which the glucose is bound. Only the $\alpha(1 \rightarrow 2)$ linkage occurs in sucrose because of the specificity of the enzyme (the biological catalyst) for the linking reaction.

Glycosidic linkages also join chains of monosaccharides into longer polymers, called **polysaccharides,** some of which function as reservoirs for glucose. The most common storage carbohydrate in animal cells is **glycogen,** a very long, highly branched polymer of glucose units linked together mainly by $\alpha(1 \rightarrow 4)$ glycosidic bonds. As much as 10 percent by weight of the liver can be glycogen. The primary storage carbohydrate in plant cells, **starch,** also is a glucose polymer with $\alpha(1 \rightarrow 4)$ linkages. It occurs in two forms, amylose, which is unbranched, and amylopectin, which has some branches. In contrast to glycogen and starch, some polysaccharides, such as **cellulose,** have structural and other nonstorage functions. An unbranched polymer of glucose linked together by $\beta(1 \rightarrow 4)$ glycosidic bonds, cellulose is the major constituent of plant cell walls and is the most abundant organic chemical on earth. Because of the different linkages between the glucose units, cellulose forms long rods, whereas glycogen and starch form coiled helices. Human digestive enzymes can hydrolyze $\alpha(1 \rightarrow 4)$ glycosidic bonds, but not $\beta(1 \rightarrow 4)$ bonds, between glucose units; for this reason humans can digest starch but not cellulose. The synthesis and utilization of these polysaccharides are described in later chapters.

SUMMARY Covalent Bonds

- Covalent bonds, which bind the atoms composing a molecule in a fixed orientation, consist of pairs of electrons shared by two atoms. Relatively high energies are required to break them (50–200 kcal/mol).

- In covalent bonds between unlike atoms that differ in electronegativity, the bonding electrons are distributed unequally. In such polar bonds, one end has a partial positive charge and the other end has a partial negative charge (see Figure 2-5).

- Most molecules in cells contain at least one chiral (asymmetric) carbon atom, which is bonded to four dissimilar atoms. Such molecules can exist as optical isomers, designated D and L, which have identical chemical properties but completely different biological activities. In biological systems, nearly all amino acids are L isomers and nearly all sugars are D isomers.

- Glucose and other hexoses can exist in three forms: an open-chain linear structure, a six-member (pyranose) ring, and a five-member (furanose) ring (see Figure 2-7). In biological systems, the pyranose form of D-glucose predominates. The two possible stereoisomers of D-glucopyranose (the α and β anomers) differ in the orientation of the hydroxyl group attached to carbon 1.

- Glycosidic bonds link carbon 1 of one monosaccharide to a hydroxyl group on another sugar, leading to formation of disaccharides and polysaccharides. Many different glycosidic bonds are theoretically possible between two sugar residues, but the enzymes that make and break these bonds are specific for the α or β anomer of one sugar and a particular hydroxyl group on the other.

2.2 Noncovalent Bonds

Carbohydrates illustrate the importance of subtle differences in covalent bonds in generating molecules with different biological activities. However, several types of **noncovalent bonds** are critical in maintaining the three-dimensional structures of large molecules such as proteins and nucleic acids (see Figure 2-1b). Noncovalent bonds also enable one large molecule to bind specifically but transiently to another, making them the basis of many dynamic biological processes.

The energy released in the formation of noncovalent bonds is only 1–5 kcal/mol, much less than the bond energies of single covalent bonds (see Table 2-1). Because the average kinetic energy of molecules at room temperature (25 °C) is about 0.6 kcal/mol, many molecules will have enough energy to break noncovalent bonds. Indeed, these weak bonds sometimes are referred to as *interactions* rather than bonds. Although noncovalent bonds are weak and have

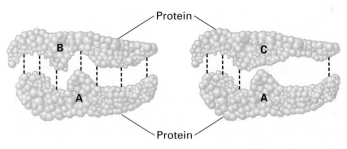

▲ **FIGURE 2-11 Multiple weak bonds stabilize specific associations between large molecules.** *(Left)* In this hypothetical complex, seven noncovalent bonds bind the two protein molecules A and B together, forming a stable complex. *(Right)* Because only four noncovalent bonds can form between proteins A and C, this interaction may be too weak for the A-C complex to exist in cells.

a transient existence at physiological temperatures (25–37 °C), multiple noncovalent bonds often act together to produce highly stable and specific associations between different parts of a large molecule or between different macromolecules (Figure 2-11). In this section we consider the four main types of noncovalent bonds and discuss their role in stabilizing the structure of biomembranes.

The Hydrogen Bond Underlies Water's Chemical and Biological Properties

Hydrogen bonding between water molecules is of crucial importance because all life requires an aqueous environment and water constitutes about 70–80 percent of the weight of most cells. The mutual attraction of its molecules causes water to have melting and boiling points at least 100 °C higher than they would be if water were nonpolar; in the absence of these intermolecular attractions, water on earth would exist primarily as a gas. The exact structure of liquid water is still unknown. It is believed to contain many transient, maximally hydrogen-bonded networks. Most likely, water molecules are in rapid motion, constantly making and breaking hydrogen bonds with adjacent molecules. As the temperature of water increases toward 100 °C, the kinetic energy of its molecules becomes greater than the energy of the hydrogen bonds connecting them, and the gaseous form of water appears.

Properties of Hydrogen Bonds Normally, a hydrogen atom forms a covalent bond with only one other atom. However, a hydrogen atom covalently bonded to a donor atom, D, may form an additional weak association, the **hydrogen bond**, with an acceptor atom, A:

$$D^{\delta-}\!-\!H^{\delta+} + :A^{\delta-} \rightleftharpoons D^{\delta-}\!-\!H^{\delta+}\underbrace{\cdots\cdots}\; :A^{\delta-}$$

Hydrogen bond

In order for a hydrogen bond to form, the donor atom must be electronegative, so that the covalent D—H bond is polar. The acceptor atom also must be electronegative, and its outer shell must have at least one nonbonding pair of electrons that attracts the δ^+ charge of the hydrogen atom. In biological systems, both donors and acceptors are usually nitrogen or oxygen atoms, especially those atoms in amino (—NH$_2$) and hydroxyl (—OH) groups. Because all covalent N—H and O—H bonds are polar, their H atoms can participate in hydrogen bonds. By contrast, C—H bonds are nonpolar, so these H atoms are almost never involved in a hydrogen bond.

Water molecules provide a classic example of hydrogen bonding. The hydrogen atom in one water molecule is attracted to a pair of electrons in the outer shell of an oxygen atom in an adjacent molecule. Not only do water molecules hydrogen-bond with one another, they also form hydrogen bonds with other kinds of molecules, as shown in Figure 2-12. The presence of hydroxyl (—OH) or amino

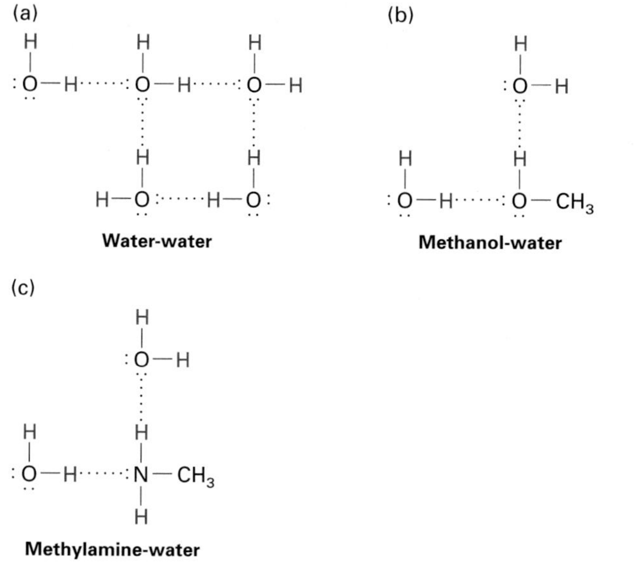

(a)

(b)

Water-water

Methanol-water

(c)

Methylamine-water

▲ **FIGURE 2-12 Water readily forms hydrogen bonds.** In liquid water, each water molecule apparently forms transient hydrogen bonds with several others, creating a fluid network of hydrogen-bonded molecules (a). The precise structure of liquid water is still not known with certainty. Water also can form hydrogen bonds with methanol (b) and methylamine (c). Each of the two pairs of nonbonding electrons in the outer shell of an oxygen atom can accept a hydrogen atom in a hydrogen bond. Similarly, the single pair of unshared electrons in the outer shell of a nitrogen atom is capable of becoming an acceptor in a hydrogen bond. The hydroxyl oxygen and the amino nitrogen can also be the donor in hydrogen bonds to oxygen atoms in water.

($-NH_2$) groups makes many molecules soluble in water. For instance, the hydroxyl group in methanol (CH_3OH) and the amino group in methylamine (CH_3NH_2) can form several hydrogen bonds with water, enabling the molecules to dissolve in water to high concentrations. In general, molecules with polar bonds that easily form hydrogen bonds with water can dissolve in water and are said to be **hydrophilic** (Greek, "water-loving"). Besides the hydroxyl and amino groups, peptide and ester bonds are important chemical groups that interact well with water:

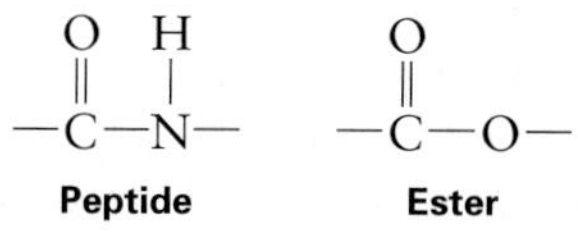

Peptide

Ester

Most hydrogen bonds are 0.26–0.31 nm long, about twice the length of covalent bonds between the same atoms. In particular, the distance between the nuclei of the hydrogen and oxygen atoms of adjacent hydrogen-bonded molecules in water is approximately 0.27 nm, about twice the length of the covalent O—H bonds in water. The hydrogen atom is closer to the donor atom, D, to which it remains covalently bonded, than it is to the acceptor. The length of the covalent D—H bond is a bit longer than it would be if there were no hydrogen bond, because the acceptor "pulls" the hydrogen away from the donor. The strength of a hydrogen bond in water (≈ 5 kcal/mol) is much weaker than a covalent O—H bond (≈ 110 kcal/mol).

Hydrogen Bonds as a Stabilizing Force in Macromolecules An important feature of all hydrogen bonds is directionality. In the strongest hydrogen bonds, the donor atom, the hydrogen atom, and the acceptor atom all lie in a straight line. Nonlinear hydrogen bonds are weaker than linear ones; still, multiple nonlinear hydrogen bonds help to stabilize the three-dimensional structures of many proteins. It is only because of the aggregate strength of multiple hydrogen bonds that they play a central role in the architecture of large biological molecules in aqueous solutions (see Figure 2-11).

The strengths of the hydrogen bonds in proteins and nucleic acids are only 1 to 2 kcal/mol, considerably weaker than the hydrogen bonds between water molecules. The reason for this difference can be seen from Figure 2-13, which depicts the formation of a hydrogen bond between two amino acids in a protein. Initially, both the —OH and —NH₂ groups in the protein are hydrogen-bonded to water, and the formation of a hydrogen bond between these groups involves disruption of their hydrogen bonds with water. Thus the *net* change in energy in forming this —OH⋯N hydrogen bond will be less than the 5 kcal/mol characteristic of hydrogen bonds between water molecules.

Ionic Interactions Are Attractions between Oppositely Charged Ions

In some compounds, the bonded atoms are so different in electronegativity that the bonding electrons are never shared: these electrons are always found around the more

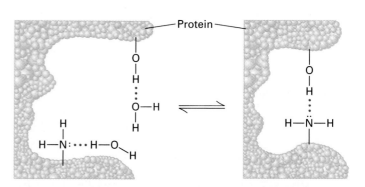

▲ **FIGURE 2-13 In order for a hydrogen bond (red dots) to form between a —OH and an —NH₂ group in a protein** *(right)*, **the hydrogen bonds between these groups and water must be disrupted** *(left)*.

electronegative atom. In sodium chloride (NaCl), for example, the bonding electron contributed by the sodium atom is completely transferred to the chlorine atom. Even in solid crystals of NaCl, the sodium and chlorine atoms are ionized, so it is more accurate to write the formula for the compound as Na^+Cl^-.

Because the electrons are not shared, the bonds in such compounds cannot be considered covalent. They are, rather, **ionic bonds** (or interactions) that result from the attraction of a positively charged ion—a cation—for a negatively charged ion—an anion. Unlike covalent or hydrogen bonds, ionic bonds do not have fixed or specific geometric orientations because the electrostatic field around an ion—its attraction for an opposite charge—is uniform in all directions. However, crystals of salts such as Na^+Cl^- do have very regular structures because that is the energetically most favorable way of packing together positive and negative ions. The force that stabilizes ionic crystals is called the *lattice energy*.

In aqueous solutions, simple ions of biological significance, such as Na^+, K^+, Ca^{2+}, Mg^{2+}, and Cl^-, do not exist as free, isolated entities. Instead, each is surrounded by a stable, tightly held shell of water molecules (Figure 2-14). An ionic interaction occurs between the ion and the oppositely charged end of the water dipole, as shown below for the K^+ ion:

$$K^+ \cdots\cdots \overset{\delta-}{O} \overset{H^{\delta+}}{\underset{H^{\delta+}}{}}$$

Ions play an important biological role when they pass through narrow, protein-lined pores, or channels, in membranes. For example, ionic movements through membranes are essential for the conduction of nerve impulses and for the stimulation of muscle contraction. As we will see in Chapter 21, ions must lose their shell of water molecules in order to pass through ion channel proteins; channel proteins can then selectively admit only Na^+, or K^+, or Ca^{2+} ions, a selectivity essential for nerve function.

Most ionic compounds are quite soluble in water because a large amount of energy is released when ions tightly bind water molecules. This is known as the *energy of hy-dration*. Oppositely charged ions are shielded from one another by the water and tend not to recombine. Salts like Na^+Cl^- dissolve in water because the energy of hydration is greater than the lattice energy that stabilizes the crystal structure. In contrast, certain salts, such as $Ca_3(PO_4)_2$, are virtually insoluble in water; the large charges on the Ca^{2+} and PO_4^{3-} ions generate a formidable lattice energy that is greater than the energy of hydration.

Van der Waals Interactions Are Caused by Transient Dipoles

When any two atoms approach each other closely, they create a weak, nonspecific attractive force that produces a **van der Waals interaction,** named for Dutch physicist Johannes Diderik van der Waals (1837–1923), who first described it. These nonspecific interactions result from the momentary random fluctuations in the distribution of the electrons of any atom, which give rise to a transient unequal distribution of electrons, that is, a transient electric dipole. If two noncovalently bonded atoms are close enough together, the transient dipole in one atom will perturb the electron cloud of the other. This perturbation generates a transient dipole in the second atom, and the two dipoles will attract each other weakly. Similarly, a polar covalent bond in one molecule will attract an oppositely oriented dipole in another.

Van der Waals interactions, involving either transient induced or permanent electric dipoles, occur in all types of molecules, both polar and nonpolar. In particular, van der Waals interactions are responsible for the cohesion between molecules of nonpolar liquids and solids, such as heptane, $CH_3-(CH_2)_5-CH_3$, that cannot form hydrogen bonds or ionic interactions with other molecules. When these stronger interactions are present, they override most of the influence of van der Waals interactions. Heptane, however, would be a gas if van der Waals interactions could not form.

The strength of van der Waals interactions decreases rapidly with increasing distance; thus these noncovalent bonds can form only when atoms are quite close to one another. However, if atoms get too close together, they become repelled by the negative charges in their outer electron shells. When the van der Waals attraction between two atoms exactly balances the repulsion between their two electron clouds, the atoms are said to be in *van der Waals contact* (Figure 2-15). Each type of atom has a van der Waals radius at which it is in van der Waals contact with other atoms. The van der Waals radius of an H atom is 0.1 nm, and the radii of O, N, C, and S atoms are between 0.14 and 0.18 nm. Two covalently bonded atoms are closer together than two atoms that are merely in van der Waals contact. For a van der Waals interaction, the internuclear distance is approximately the sum of the corresponding radii for the two participating atoms. Thus the distance between a C atom and an H atom in van der Waals contact is 0.27 nm, and between two C atoms is 0.34 nm. In general, the van der Waals radius of an atom is about twice as long as its covalent radius.

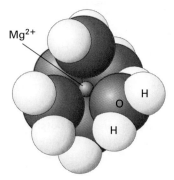

◀ **FIGURE 2-14 In aqueous solutions, a shell of water molecules surrounds ions.** In the case of a magnesium ion (Mg^{2+}), six water molecules are held tightly in place by electrostatic interactions between the two positive charges on the ion and the partial negative charge on the oxygen of each water molecule.

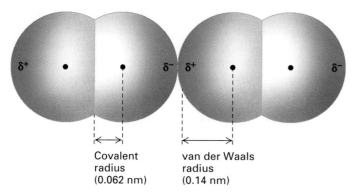

Covalent
radius
(0.062 nm)

van der Waals
radius
(0.14 nm)

▲ **FIGURE 2-15 Two oxygen molecules in van der Waals contact.** Transient dipoles in the electron clouds of all atoms give rise to weak attractive forces, called van der Waals interactions. Each type of atom has a characteristic van der Waals radius at which van der Waals interactions with other atoms are optimal. Because atoms repel one another if they are close enough together for their outer electron shells to overlap, the van der Waals radius is a measure of the size of the electron cloud surrounding an atom. The covalent radius indicated here is for the double bond of O=O; the single-bond covalent radius of oxygen is slightly longer.

For example, a C—H covalent bond is about 0.107 nm long and a C—C covalent bond is about 0.154 nm long.

The energy of the van der Waals interaction is about 1 kcal/mol, only slightly higher than the average thermal energy of molecules at 25 °C. Thus the van der Waals interaction is even weaker than the hydrogen bond, which typically has an energy of 1–2 kcal/mol in aqueous solutions. The attraction between two large molecules can be appreciable, however, if they have precisely complementary shapes, so that they make many van der Waals contacts when they come into proximity. Van der Waals interactions, as well as other noncovalent bonds, mediate the binding of many enzymes with their specific **substrates** (the substances on which an enzyme acts) and of each type of antibody with its specific antigen (Chapter 3).

Hydrophobic Bonds Cause Nonpolar Molecules to Adhere to One Another

Nonpolar molecules do not contain ions, possess a dipole moment, or become hydrated. Because such molecules are insoluble or almost insoluble in water, they are said to be **hydrophobic** (Greek, "water-fearing"). The covalent bonds between two carbon atoms and between carbon and hydrogen atoms are the most common nonpolar bonds in biological systems. Hydrocarbons—molecules made up only of carbon and hydrogen—are virtually insoluble in water. A large triacylglycerol (or triglyceride) such as tristearin, a component of animal fat, is also insoluble in water, even though its six oxygen atoms participate in some slightly polar bonds with adjacent carbon atoms (Figure 2-16). When shaken in water, tristearin forms a separate phase similar to the separation of oil from the water-based vinegar in an oil-and-vinegar salad dressing.

The force that causes hydrophobic molecules or nonpolar portions of molecules to aggregate together rather than to dissolve in water is called the **hydrophobic bond**. This is not a separate bonding force; rather, it is the result of the energy required to insert a nonpolar molecule into water. A nonpolar molecule cannot form hydrogen bonds with water molecules, so it distorts the usual water structure, forcing the water into a rigid cage of hydrogen-bonded molecules around it. Water molecules are normally in constant motion, and the formation of such cages restricts the motion of a number of water molecules; the effect is to increase the structural organization of water. This situation is energetically unfavorable because it decreases the randomness (entropy) of the population of water molecules. The role of entropy in chemical systems is discussed further in a later section.

The opposition of water molecules to having their motion restricted by forming cages around hydrophobic molecules or portions thereof is the major reason molecules such as tristearin and heptane are essentially insoluble in water and interact mainly with other hydrophobic molecules. Nonpolar molecules can also bond together, albeit weakly,

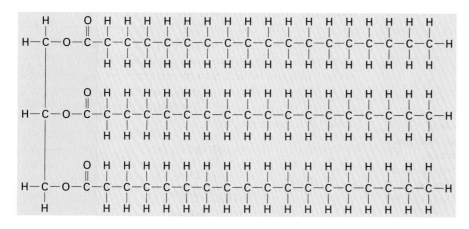

◀ **FIGURE 2-16 The chemical structure of tristearin, or tristearoyl glycerol, a component of natural fats.** It contains three molecules of the fatty acid stearic acid, $CH_3(CH_2)_{16}COOH$, esterified to one molecule of glycerol, $HOCH_2CH(OH)CH_2OH$. One end of the molecule (green) is hydrophilic; the rest of the molecule is highly hydrophobic.

through van der Waals interactions. The net result of the hydrophobic and van der Waals interactions is a very powerful tendency for hydrophobic molecules to interact with one another, and not with water.

Small hydrocarbons like butane (CH_3—CH_2—CH_2—CH_3) are somewhat soluble in water, because they can dissolve without disrupting the water lattice appreciably. However, 1-butanol (CH_3—CH_2—CH_2—CH_2OH) mixes completely with water in all proportions. The replacement of just one hydrogen atom with the polar —OH group allows the molecule to form hydrogen bonds with water and greatly increases its solubility.

Simply put, *like dissolves like*. Polar molecules dissolve in polar solvents such as water, while nonpolar molecules dissolve in nonpolar solvents such as hexane.

Multiple Noncovalent Bonds Can Confer Binding Specificity

Besides contributing to the stability of large biological molecules, multiple noncovalent bonds can also confer *specificity* by determining how large molecules will fold or which regions of different molecules will bind together. All types of these weak interactions are effective only over a short range and require close contact between the reacting groups. For noncovalent bonds to form properly, there must be a complementarity between the sites on the two interacting surfaces. Figure 2-17 illustrates how several different weak bonds can bind two protein chains together. Almost any

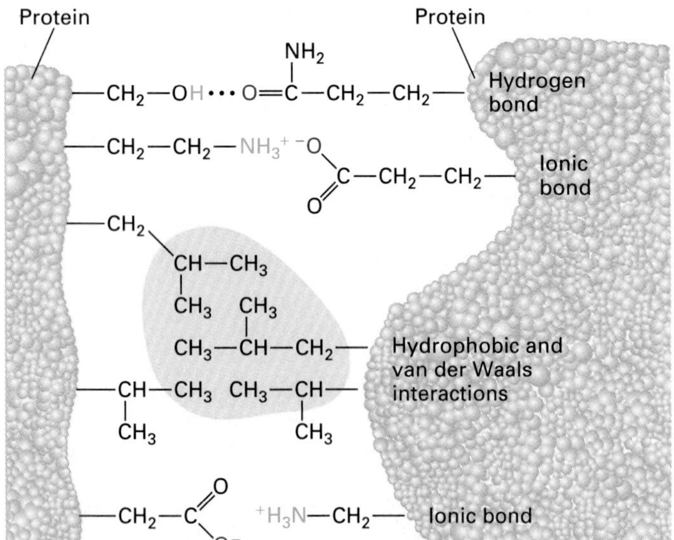

▲ **FIGURE 2-17 The binding of a hypothetical pair of proteins by two ionic bonds, one hydrogen bond, and one large combination of hydrophobic and van der Waals interactions.** The structural complementarity of the surfaces of the two molecules gives rise to this particular combination of weak bonds and hence to the specificity of binding between the molecules.

other arrangement of the same groups on the two surfaces would not allow the molecules to bind so tightly. Such multiple, specific interactions allow protein molecules to fold into a unique three-dimensional shape (Chapter 3) and the two chains of DNA to bind together (Chapter 4).

Phospholipids Are Amphipathic Molecules

Multiple noncovalent bonds also are critical in stabilizing the structure of **biomembranes,** whose primary components are **phospholipids.** Because the essential properties of biomembranes derive from phospholipids, we first examine the chemistry of these compounds and then see how they associate into the sheetlike structures that are the foundation of biomembranes.

All phospholipids contain one or more acyl chains derived from **fatty acids,** which consist of a hydrocarbon chain attached to a carboxyl group (—COOH). Fatty acids are insoluble in water and salt solutions; they differ in length and in the extent and position of their double bonds. Table 2-2 lists the principal fatty acids found in cells. Most fatty acids have an even number of carbon atoms, usually 16, 18, or 20.

Fatty acids with no double bonds are said to be *saturated;* those with at least one double bond are *unsaturated.* Unsaturated fatty acid chains normally have one double bond, but some have two, three, or four. Two stereoisomeric configurations, cis and trans, are possible around each double bond:

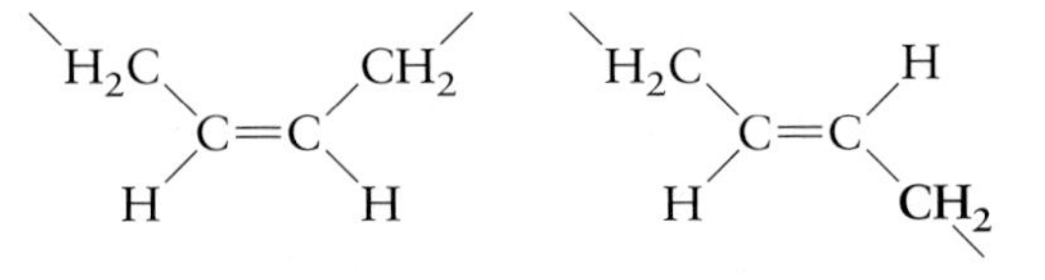

Cis   **Trans**

A cis double bond introduces a rigid kink in the otherwise flexible straight chain of a fatty acid (Figure 2-18). In general, the fatty acids in biological systems contain only cis double bonds.

Phospholipids consist of two long-chain fatty acyl groups linked (usually by an ester bond) to small, highly hydrophilic groups. Consequently, unlike tristearin, phospholipids do not clump together in droplets but orient themselves in sheets, exposing their hydrophilic ends to the aqueous environment. Molecules in which one end (the "head") interacts with water and the other end (the "tail") is hydrophobic are said to be **amphipathic** (Greek, "tolerant of both"). The tendency of amphipathic molecules to form organized structures spontaneously in water is the key to the structure of cell membranes.

In phosphoglycerides, a principal class of phospholipids, fatty acyl side chains are esterified to two of the three hydroxyl groups in glycerol

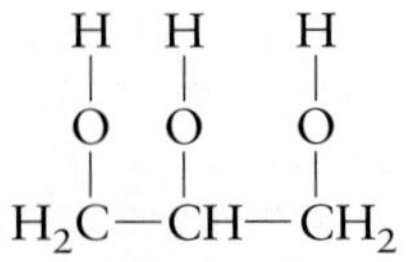

Glycerol

TABLE 2-2 Some Typical Fatty Acids Found in Cells

Chemical Formula	Systematic Name	Common Name
SATURATED FATTY ACIDS		
$CH_3(CH_2)_{10}COOH$	n-Dodecanoic	Lauric
$CH_3(CH_2)_{12}COOH$	n-Tetradecanoic	Myristic
$CH_3(CH_2)_{14}COOH$	n-Hexadecanoic	Palmitic
$CH_3(CH_2)_{16}COOH$	n-Octadecanoic	Stearic
$CH_3(CH_2)_{18}COOH$	n-Eicosanoic	Arachidic
$CH_3(CH_2)_{22}COOH$	n-Tetracosanoic	Lignoceric
UNSATURATED FATTY ACIDS		
$CH_3(CH_2)_5CH{=}CH(CH_2)_7COOH$		Palmitoleic
$CH_3(CH_2)_7CH{=}CH(CH_2)_7COOH$		Oleic
$CH_3(CH_2)_4CH{=}CHCH_2CH{=}CH(CH_2)_7COOH$		Linoleic
$CH_3CH_2CH{=}CHCH_2CH{=}CHCH_2CH{=}CH(CH_2)_7COOH$		Linolenic
$CH_3(CH_2)_4(CH{=}CHCH_2)_3CH{=}CH(CH_2)_3COOH$		Arachidonic

but the third hydroxyl group is esterified to phosphate. The simplest phospholipid, phosphatidic acid, contains only these components:

$$
\begin{array}{c}
\overset{\displaystyle O}{\overset{\|}{R_1{-}C}}{-}O{-}CH_2 \\
\overset{\displaystyle O}{\overset{\|}{R_2{-}C}}{-}O{-}CH \quad\quad O \\
CH_2{-}O{-}\overset{\displaystyle O}{\underset{O^-}{\overset{\|}{P}}}{-}O^-
\end{array}
$$

where R_1 and R_2 are fatty acyl groups.

In most phospholipids, however, the phosphate group is also esterified to a hydroxyl group on another hydrophilic compound. In phosphatidylcholine, for example, choline is attached to the phosphate (Figure 2-19). In other phosphoglycerides, the phosphate group is linked to other molecules, such as ethanolamine, the amino acid serine, or the sugar inositol. The negative charge on the phosphate as well as the charged groups or hydroxyl groups on the alcohol esterified to it interact strongly with water.

The Phospholipid Bilayer Forms the Basic Structure of All Biomembranes

When a suspension of phospholipids is mechanically dispersed in aqueous solution, they can assume three different forms: micelles, bilayer sheets, and liposomes (Figure 2-20). The type of structure formed by a pure phospholipid or a mixture of phospholipids depends on the length of the fatty

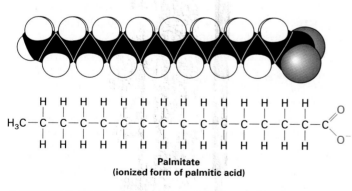

Palmitate
(ionized form of palmitic acid)

▲ **FIGURE 2-18 The effect of a double bond.** Shown are space-filling models and chemical structures of the ionized form of palmitic acid, a saturated fatty acid, and oleic acid, an unsaturated one. In saturated fatty acids, the hydrocarbon chain is linear; the cis double bond in oleate creates a kink in the hydrocarbon chain. [After L. Stryer, 1994, *Biochemistry*, 4th ed., W. H. Freeman and Company, p. 265.]

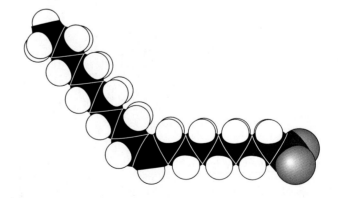

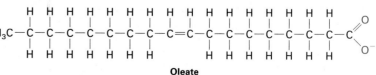

Oleate
(ionized form of oleic acid)

acyl chains and their degree of saturation, on the temperature, on the ionic composition of the aqueous medium, and on the mode of dispersal of the phospholipids in the solution. In all three forms, hydrophobic interactions cause the

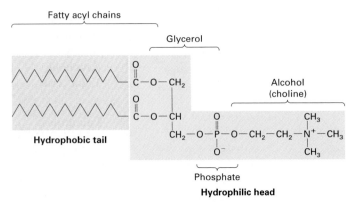

FIGURE 2-19 Phosphatidylcholine, a typical phosphoglyceride, has a hydrophobic tail and a hydrophilic head in which choline is linked to glycerol by phosphate. Either or both of the fatty acyl side chains in a phosphoglyceride may be saturated or unsaturated.

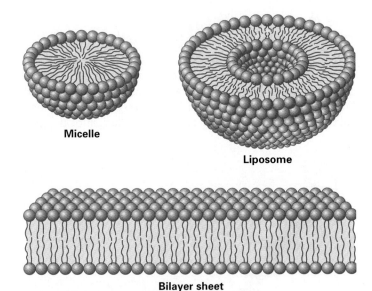

Micelle

Liposome

Bilayer sheet

▲ FIGURE 2-20 Cross-sectional views of the three structures that can be formed by mechanically dispersing a suspension of phospholipids in aqueous solutions. Shown are a spherical micelle with a hydrophobic interior composed entirely of fatty acyl chains; a spherically arranged bilayer structure, called a *liposome*, that is larger than a micelle and has an aqueous center; and a two-molecule-thick sheet of phospholipids in a bilayer, the basic structural unit of biomembranes. The red circles depict the hydrophilic heads of the phospholipids, and the squiggly lines (in the yellow region) the hydrophobic tails.

fatty acyl chains to aggregate and exclude water molecules from the "core." Micelles are rarely formed from natural phosphoglycerides, whose fatty acyl chains generally are too bulky to fit into the interior of a micelle.

Under suitable conditions, phospholipids of the composition present in cells spontaneously form symmetric sheet-like structures, called **phospholipid bilayers,** that are two molecules thick. Each phospholipid layer in this lamellar structure is called a *leaflet.* The hydrocarbon side chains in each leaflet minimize contact with water by aligning themselves tightly together in the center of the bilayer, forming a hydrophobic core that is about 3 nm thick. The close packing of these hydrocarbon side chains is stabilized by van der Waals interactions between them. Ionic and hydrogen bonds stabilize the interaction of the phospholipid polar head groups with each other and with water. At neutral pH, the polar head groups in some phospholipids (e.g., phosphatidylcholine) have no net electric charge, whereas the head groups in others have a net negative charge. Nonetheless, all phospholipids can pack together into the characteristic bilayer structure.

A phospholipid bilayer can be of almost unlimited size—from micrometers (μm) to millimeters (mm) in length or width—and can contain tens of millions of phospholipid molecules. Because of their hydrophobic core, bilayers are impermeable to salts, sugars, and most other small hydrophilic molecules. Like a phospholipid bilayer, all biological membranes have a hydrophobic core, and they all separate two aqueous solutions. The **plasma membrane,** for example, separates the interior of the cell from its surroundings. Similarly, the membranes that surround the organelles of eukaryotic cells separate one aqueous phase—the cell cytosol—from another—the interior of the organelle. Several types of evidence indicate that the phospholipid bilayer is the basic structural unit of nearly all biomembranes (Chapter 5). Associated with membrane phospholipids are various proteins that help confer unique properties on each type of membrane. We describe the general structure of membrane proteins and their association with the phospholipid bilayer in Chapter 3.

SUMMARY Noncovalent Bonds

- Noncovalent bonds determine the shape of many large biological molecules and stabilize complexes composed of two or more different molecules.

- There are four main types of noncovalent bonds in biological systems: hydrogen bonds, ionic bonds, van der Waals interactions, and hydrophobic bonds. The bond energies for these interactions range from about 1 to 5 kcal/mol.

- In a hydrogen bond, a hydrogen atom covalently bonded to an electronegative donor atom associates with an acceptor atom whose nonbonding electrons attract the hydrogen (see Figure 2-12). Hydrogen bonds among water molecules are largely responsible for the properties of both liquid water and the crystalline solid form (ice).

• Ionic bonds result from the electrostatic attraction between the positive and negative charges of ions. In aqueous solutions, all cations and anions are surrounded by a tightly bound shell of water molecules.

• The weak and relatively nonspecific van der Waals interactions are created whenever any two atoms approach each other closely (see Figure 2-15). They result from the attraction between transient dipoles associated with all molecules.

• Hydrophobic bonds occur between nonpolar molecules, such as hydrocarbons, in an aqueous environment. Hydrophobic bonds result mainly because aggregation of the hydrophobic molecules necessitates less organization of water into "cages" (and, hence, less reduction in entropy) than if many cages of water molecules had to surround *individual* hydrophobic molecules.

• Although any single noncovalent bond is quite weak, several such bonds between molecules or between the parts of one molecule can stabilize the three-dimensional structures of proteins and nucleic acids and mediate specific binding interactions.

• Phospholipids, the main components of biomembranes, are amphipathic molecules (see Figure 2-19). Noncovalent bonds are responsible for organizing and stabilizing phospholipids into one of three structures in aqueous solution (see Figure 2-20).

• The basic structure of biomembranes consists of a phospholipid bilayer in which the long hydrocarbon fatty acyl side chains in each leaflet are oriented toward one another, forming a hydrophobic core, and the polar head groups line both surfaces. Natural biomembranes also contain proteins, cholesterol, and other components inserted into the phospholipid bilayer.

2.3 Chemical Equilibrium

We now shift our discussion to chemical reactions in which bonds, primarily covalent bonds, are broken and re-formed. At any one time several hundred different kinds of chemical reactions are occurring simultaneously in every cell, and any chemical can, in principle, undergo multiple chemical reactions. Both the *extent* to which a reaction can proceed and the *rate* at which it actually takes place determine which reactions actually occur in cells.

When reactants first come together—before any products have been formed—their rate of reaction is determined in part by their initial concentrations. As the reaction products accumulate, the concentration of each reactant decreases and so does the reaction rate. Meanwhile, some of the product molecules begin to participate in the reverse

reaction, which re-forms the reactants. This reaction is slow at first but speeds up as the concentration of products increases. Eventually, the rates of the forward and reverse reactions become equal, so that the concentrations of reactants and products stop changing. The mixture is then said to be in **chemical equilibrium.**

At equilibrium the ratio of products to reactants, called the **equilibrium constant,** is a fixed value that is independent of the rate at which the reaction occurs. The rate of a chemical reaction can be increased by a **catalyst,** a substance that brings reactants together and accelerates their interactions but is not permanently changed during a reaction. This function is aptly reflected in the Chinese term for catalyst, *tsoo mei,* which literally means "marriage broker." In this section, we discuss several aspects of chemical equilibria; in the next section, we examine energy changes during reactions and their relationship to equilibria. In the final section, we review the factors that determine reaction rates and how enzymes, the cell's biological catalysts, increase the rates of biochemical reactions.

Equilibrium Constants Reflect the Extent of a Chemical Reaction

The equilibrium constant (K_{eq}) depends on the nature of the reactants and products, the temperature, and the pressure (particularly in reactions involving gases). Under standard physical conditions (25 °C and 1 atm pressure, for biological systems), the K_{eq} is always the same for a given reaction, whether or not a catalyst is present.

For the simple reaction $A + B \rightleftharpoons X + Y$, the equilibrium constant is given by

$$K_{eq} = \frac{[X][Y]}{[A][B]} \qquad (2\text{-}1)$$

where brackets indicate equilibrium concentrations. In general, for a reaction

$$aA + bB + cC + \cdots \rightleftharpoons zZ + yY + xX + \cdots \qquad (2\text{-}2)$$

where capital letters represent particular molecules or atoms and lowercase letters represent the number of each in the reaction formula, the equilibrium constant is given by

$$K_{eq} = \frac{[X]^x[Y]^y[Z]^z}{[A]^a[B]^b[C]^c} \qquad (2\text{-}3)$$

In the generalized reaction (2-2) above, the rate of the forward (left to right) reaction will be

$$\text{Rate}_{\text{forward}} = k_f[A]^a[B]^b[C]^c$$

where k_f is the rate constant for the forward reaction. Similarly, the rate of the reverse (right to left) reaction is

$$\text{Rate}_{\text{reverse}} = k_r[X]^x[Y]^y[Z]^z$$

where k_r is the rate constant for the reverse reaction. Since at equilibrium the rate of the forward reaction equals that of the reverse reaction, we can write

$$k_f[A]^a[B]^b[C]^c = k_r[X]^x[Y]^y[Z]^z$$

which can be rearranged to give

$$\frac{k_f}{k_r} = \frac{[X]^x[Y]^y[Z]^z}{[A]^a[B]^b[C]^c}$$

Comparing this expression with Equation 2-3, we see the important relationship

$$K_{eq} = \frac{k_f}{k_r} \qquad (2\text{-}4)$$

In other words, the equilibrium constant equals the ratio of the forward and reverse rate constants. Because a catalyst accelerates the rates of the forward and reverse reactions by the same factor, it does not change the value of k_f/k_r. Thus, as noted above, catalysts do not alter the equilibrium constant, which depends only on the chemical properties of the molecules involved and on the temperature and pressure.

To illustrate several points concerning equilibrium, we shall use a fairly simple biochemical reaction: the interconversion of the compounds glyceraldehyde 3-phosphate (G3P) and dihydroxyacetone phosphate (DHAP). This reaction, which occurs during the breakdown of glucose, is catalyzed by the enzyme triosephosphate isomerase:

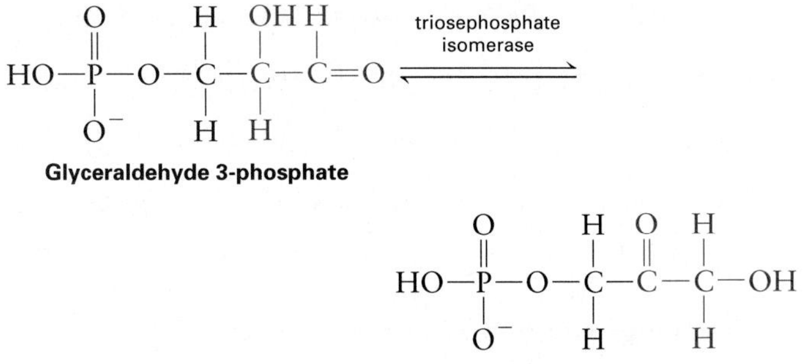

Glyceraldehyde 3-phosphate

Dihydroxyacetone phosphate

The equilibrium constant for this reaction under standard conditions is

$$K_{eq} = \frac{[DHAP]}{[G3P]} = 22.2$$

Thus the ratio of the concentrations of G3P and DHAP is 1:22.2 when the reaction reaches equilibrium. In practice, one measures the concentrations of reactants and products after a reaction has reached equilibrium, and uses these values to calculate the equilibrium constant.

In the presence of an enzyme or other catalyst, the reaction rate may increase, but the final ratio of product to reactant will always be the same. The magnitude of the equilibrium constant has no bearing on the rate of the reaction or on whether the reaction will take place at all under normal conditions. Despite the large equilibrium constant for the conversion of G3P to DHAP, for example, so much energy is required to rearrange the bonds that no detectable reaction actually occurs in an aqueous solution in the absence of an enzyme or other catalyst.

When a reaction involves a *single* reactant and a *single* product, the ratio of the product concentration to reactant concentration at equilibrium is equal to the equilibrium constant K_{eq} and is independent of the initial concentrations. When a reaction involves *multiple* reactants and/or products, the *equilibrium concentration* of any one product or reactant depends on the initial concentrations of all reactants and products as well as on the equilibrium constant. Consider, for example, the hydrolysis (cleavage by addition of water) of the dipeptide glycylalanine (GA) to glycine (G) and alanine (A):

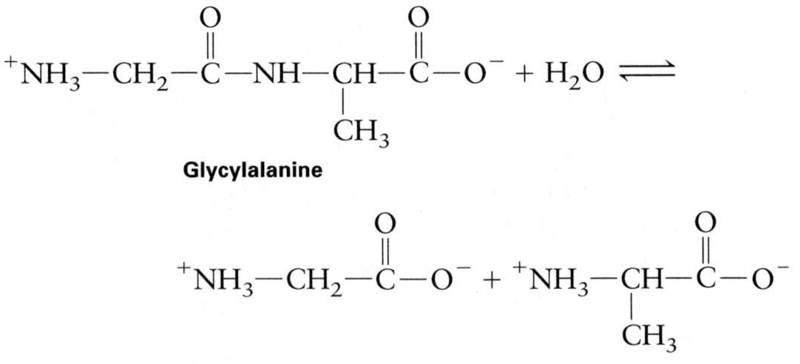

Glycylalanine

Glycine   **Alanine**

Here,

$$K_{eq} = \frac{[G][A]}{[GA]}$$

(The concentration of water does not change significantly during normal aqueous chemical reactions and, by convention, is not included in the calculation of equilibrium ratios.) The equilibrium is strongly in the direction of the formation of glycine and alanine. In other words, most of the glycylalanine is hydrolyzed at equilibrium. However, an excess of one of the products can drive the reaction in the reverse direction. For instance, suppose that the initial reaction mixture contains a small amount of glycylalanine and a large amount of alanine. As the reaction proceeds, the total concentration of alanine [A] will always greatly exceed the concentration of glycine [G] produced by hydrolysis. This must reduce the equilibrium ratio of [G] to [GA] because K_{eq} remains constant. Thus, the reaction GA $\rightleftharpoons$ G + A can be driven to the left by an excess of the product alanine. More generally, in reactions involving more than one reactant or product, changes in the concentration of any one reactant or product will affect the concentrations at equilibrium of all the reactants and products.

Under appropriate conditions, individual biochemical reactions carried out in a test tube eventually will reach equilibrium. Within cells, however, many reactions are linked in pathways in which a product of one reaction is frequently used quickly as a substrate for another. In this more complex situation, reactants and products generally are in *steady state* but not in equilibrium. This feature of cellular biochemistry prevents the buildup of toxic intermediates, which are generated in certain normal pathways. Because these intermediates are immediately consumed in other reactions, their concentrations do not reach equilibrium values but rather are maintained at relatively low steady-state values, which cause no deleterious effects.

The Concentration of Complexes Can Be Estimated from Equilibrium Constants for Binding Reactions

The reactions discussed above exemplify the formation or cleavage of covalent bonds. Many important reactions, however, involve the binding of one molecule to another, mediated by various noncovalent interactions. A common example is the binding of a **ligand** (e.g., the hormone insulin or adrenaline) to its receptor on the surface of a cell, triggering a biological response. Another example is the binding of a protein to a specific sequence of base pairs in a molecule of DNA, which frequently causes the expression of a nearby gene to turn on or off. Such DNA-binding proteins are discussed in detail in Chapter 10. Here we focus on how the equilibrium constant for binding of a protein to DNA can be used to calculate the extent to which the protein is bound to DNA in a cell. In Chapter 20, we will see how similar calculations help us understand the stability of hormone-receptor complexes.

For the binding reaction $P + D \rightleftharpoons PD$, where PD is the specific complex of a protein (P) and DNA (D), the equilibrium constant is customarily defined as the ratio of reactants to products; it is also called the *dissociation constant* and is given by

$$K_{eq} = \frac{[P][D]}{[PD]} \qquad (2\text{-}5)$$

the inverse of the reactions in the previous section. This means that when the concentration of protein, [P], equals K_{eq}, then half the DNA will contain a bound protein and half will not; that is, [D] = [PD]. Typical reactions in which a protein binds to a specific DNA sequence have a K_{eq} of 10^{-10} M, where M symbolizes *molarity*, or moles per liter (mol/L). We can relate the magnitude of this equilibrium constant to events within a bacterial cell, which contains *one molecule* of DNA. Assume, for simplicity, that the cell also contains *ten molecules* of the DNA-binding protein P. Further assume that the cell is a cylinder 2 μm long and 1 μm in diameter. Thus its volume will be 1.5×10^{-15} L, and the total concentration of the one molecule of DNA (D) within the cell, will be

$$\frac{(1 \text{ molecule}/1.5 \times 10^{-15} \text{ L})}{(6.02 \times 10^{23} \text{ molecules/mol})} = 1.1 \times 10^{-9} \text{ M}$$

Similarly, the total concentration of the 10 molecules of P, the protein, will be 1.1×10^{-8} M.

If we define [D] to be the concentration of free DNA, unbound to protein, and [PD] to be the concentration of the DNA-protein complex, then the total concentration of DNA is [D] + [PD]. Similarly, if [P] is defined as the concentration of free protein, unbound to DNA, then the total concentration of protein is [P] + [PD]. By transposing terms and using the total concentrations calculated above, we can write the following equations:

[D] = 1.1×10^{-9} M − [PD]   and
[P] = 1.1×10^{-8} M − [PD]

Substituting these expressions into Equation 2-5 and rearranging gives

$$[PD] = \frac{[P][D]}{K_{eq}} =$$
$$\frac{(1.1 \times 10^{-8} \text{ M} - [PD])(1.1 \times 10^{-9} \text{ M} - [PD])}{10^{-10} \text{ M}}$$

Solving this quadratic equation for [PD] gives 1.089×10^{-9} M, the concentration of the protein-DNA complex at equilibrium. Substituting this value into the above equations for [P], the concentration of free (unbound) protein, and [D], the concentration of unbound DNA, at equilibrium, we get

[P] = 1.1×10^{-8} M − 1.089×10^{-9} M = 9.911×10^{-9} M

[D] = 1.1×10^{-9} M − 1.089×10^{-9} M = 1.1×10^{-11} M

From these values, we can now calculate the ratio of bound to total DNA:

$$\frac{[PD]}{[PD] + [D]} = \frac{1.089 \times 10^{-9} M}{1.1 \times 10^{-9} M} = 0.99$$

Thus 99 percent of the time this specific sequence of DNA will have a molecule of protein bound to it, and 1 percent of the time it will not, even though the cell contains only 10 molecules of the protein! This example illustrates the general approach for predicting the abundance of intracellular complexes if one knows the equilibrium constants for the binding reactions and the numbers of the participating molecules within a cell.

Biological Fluids Have Characteristic pH Values

The solvent inside cells and in all extracellular fluids is water. An important characteristic of any aqueous solution is the concentration of positively charged hydrogen ions (H^+) and negatively charged hydroxyl ions (OH^-). Because these ions are the dissociation products of H_2O, they are constituents of all living systems, and they are liberated by many reactions that take place between organic molecules within cells.

When a water molecule dissociates, one of its polar H—O bonds breaks. The resulting hydrogen ion, often referred to as a *proton*, has a short lifetime as a free particle and quickly combines with a water molecule to form a hydronium ion (H_3O^+). For convenience however, we refer to the concentration of hydrogen ions in a solution, $[H^+]$, even though we really mean the concentration of hydronium ions, $[H_3O^+]$. The dissociation of water is a reversible reaction,

$$H_2O \rightleftharpoons H^+ + OH^-$$

and at 25 °C,

$$[H^+][OH^-] = 10^{-14} \, M^2$$

In pure water, $[H^+] = [OH^-] = 10^{-7}$ M.

The concentration of hydrogen ions in a solution is expressed conventionally as its **pH**:

$$pH = -\log[H^+] = \log \frac{1}{[H^+]}$$

In pure water at 25 °C, $[H^+] = 10^{-7}$ M, so

$$pH = -\log 10^{-7} = 7.0$$

On the pH scale, 7.0 is considered neutral: pH values below 7.0 indicate acidic solutions and values above 7.0 indicate basic (alkaline) solutions (Table 2-3). In a 0.1 M solution of hydrogen chloride (HCl) in water, $[H^+] = 0.1$ M because virtually all the HCl has dissociated into H^+ and Cl^- ions. For this solution $pH = -\log 0.1 = 1.0$. In fact, pH values can be less than zero, since a 10 M solution of HCl will have a pH of -1.

One of the most important properties of a biological fluid is its pH. The cytosol of cells normally has a pH of about 7.2. In certain organelles of eukaryotic cells, such as the lysosomes and vacuoles, the pH is much lower, about 5; this corresponds to a H^+ concentration more than 100 times higher than that in the cytosol. Lysosomes contain many degradative enzymes that function optimally in an acidic environment, whereas their action is inhibited in the near-neutral environment of the cytosol. Maintenance of a specific pH is imperative for some cellular structures to function properly. On the other hand, dramatic shifts in cellular pH may play an important role in controlling cellular activity. For example, the pH of the cytosol of an unfertilized sea urchin egg is 6.6. Within 1 minute of fertilization, however, the pH rises to 7.2; that is, the H^+ concentration decreases to about one-fourth its original value. The change in pH is necessary for subsequent growth and division of the egg.

TABLE 2-3	The pH Scale		
	Concentration of H^+ Ions (mol/L)	pH	Example
⇑ Increasing acidity	10^{-0}	0	
	10^{-1}	1	Gastric fluids
	10^{-2}	2	Lemon juice
	10^{-3}	3	Vinegar
	10^{-4}	4	Acid soil
	10^{-5}	5	Lysosomes
	10^{-6}	6	Cytoplasm of contracting muscle
Neutral	10^{-7}	7	Pure water and cytoplasm
	10^{-8}	8	Sea water
	10^{-9}	9	Very alkaline natural soil
	10^{-10}	10	Alkaline lakes
	10^{-11}	11	Household ammonia
⇓ Increasing alkalinity	10^{-12}	12	Lime (saturated solution)
	10^{-13}	13	
	10^{-14}	14	

Hydrogen Ions Are Released by Acids and Taken Up by Bases

In general, any molecule or ion that tends to release a hydrogen ion is called an **acid**, and any molecule or ion that readily combines with a hydrogen ion is called a **base**. Thus hydrogen chloride is an acid. The hydroxyl ion is a base, as is ammonia (NH_3), which readily picks up a hydrogen ion to become an ammonium ion (NH_4^+). Many organic molecules are acidic because they have a carboxyl group (—COOH), which tends to dissociate to form the negatively charged carboxylate ion (—COO$^-$):

$$X-C{\overset{\displaystyle O}{\underset{\displaystyle OH}{\big\langle}}} \rightleftharpoons X-C{\overset{\displaystyle O}{\underset{\displaystyle O^-}{\big\langle}}} + H^+$$

where X represents the rest of the molecule. The amino group (—NH_2), a part of many important biological molecules, is a base because, like ammonia, it can take up a hydrogen ion:

$$X-NH_2 + H^+ \rightleftharpoons X-NH_3^+$$

When acid is added to a solution, $[H^+]$ increases (the pH goes down). Consequently, $[OH^-]$ decreases because hydroxyl ions readily combine with the hydrogen ions to form water. Conversely, when a base is added to a solution, $[H^+]$ decreases (the pH goes up). Because $[H^+][OH^-] = 10^{-14}$ M^2, any increase in $[H^+]$ is coupled with a decrease in $[OH^-]$, and vice versa. No matter how acidic or alkaline a solution is, it always contains both ions: neither $[OH^-]$ nor $[H^+]$ is ever zero. For example, if $[H^+] = 0.1$ M (pH = 1.0), then $[OH^-] = 10^{-13}$ M.

The degree to which a dissolved acid releases hydrogen ions or a base takes them up depends partly on the pH of the solution. Amino acids provide an example of this phenomenon. These molecules have the general formula

$$\begin{array}{c} NH_2 \\ | \\ H-C-COOH \\ | \\ R \end{array}$$

where R represents the rest of the molecule. In neutral solutions (pH = 7.0), amino acids exist predominantly in the doubly ionized form

$$\begin{array}{c} NH_3^+ \\ | \\ H-C-COO^- \\ | \\ R \end{array}$$

Such a molecule, containing both a positive and a negative ion, is called a *zwitterion*. Zwitterions, having no net charge, are neutral.

In solutions at low pH, carboxylate ions ($-COO^-$) recombine with the abundant hydrogen ions, so that the predominant form of the amino acid molecule is

$$\begin{array}{c} NH_3^+ \\ | \\ H-C-COOH \\ | \\ R \end{array}$$

At high pH, the scarcity of hydrogen ions decreases the chance that an amino group or a carboxylate ion will pick up a hydrogen ion, so that the predominant form of an amino acid molecule is

$$\begin{array}{c} NH_2 \\ | \\ H-C-COO^- \\ | \\ R \end{array}$$

The Henderson-Hasselbalch Equation Relates pH and K_{eq} of an Acid-Base System

Many molecules used by cells have multiple acidic or basic groups, each of which can release or take up a proton. In the laboratory, it is often essential to know the precise state of dissociation of each of these groups at various pH val-

ues. The dissociation of an acid group HA, such as acetic acid (CH_3COOH), is described by

$$HA \rightleftharpoons H^+ + A^-$$

The equilibrium constant K_a for this reaction is

$$K_a = \frac{[H^+][A^-]}{[HA]}$$

By taking the logarithm of both sides and rearranging the result, we can derive a very useful relation between the equilibrium constant and pH as follows:

$$\log K_a = \log \frac{[H^+][A^-]}{[HA]} = \log [H^+] + \log \frac{[A^-]}{[HA]}$$

or

$$-\log [H^+] = -\log K_a + \log \frac{[A^-]}{[HA]}$$

Substituting pH for $-\log [H^+]$ and pK_a for $-\log K_a$, we have

$$pH = pK_a + \log \frac{[A^-]}{[HA]} \tag{2-6}$$

From this expression, commonly known as the *Henderson-Hasselbalch equation*, it can be seen that the pK_a of any acid is equal to the pH at which half the molecules are dissociated and half are neutral (undissociated). This is because when pK_a = pH, then log $([A^-]/[HA])$ = 0, and therefore $[A^-]$ = $[HA]$. The Henderson-Hasselbalch equation allows us to calculate the degree of dissociation of an acid if both the pH of the solution and the pK_a of the acid are known. Experimentally, by measuring the concentration of A^- and of HA as a function of the solution's pH, one can calculate the pK_a of the acid and thus the equilibrium constant for the dissociation reaction.

Buffers Maintain the pH of Intracellular and Extracellular Fluids

A growing cell must maintain a constant pH in the cytoplasm of about 7.2–7.4 despite the production, by metabolism, of many acids, such as lactic acid and CO_2, which reacts with water to form carbonic acid (H_2CO_3). Cells have a reservoir of weak bases and weak acids, called **buffers**, which ensure that the cell's pH remains relatively constant. Buffers do this by "soaking up" H^+ or OH^- when these ions are added to the cell or are produced by metabolism.

If additional acid (or base) is added to a solution of an acid (or a base) at its pK_a value (a 1:1 mixture of HA and A^-), the pH of the solution changes, but it changes less than it would if the original acid (or base) had not been present. This is because protons released by the added acid are taken up by the original A^- form of the acid; likewise, hydroxyl ions generated by the added base are neutralized by protons released by the original HA.

This ability of a buffer to minimize changes in pH, its *buffering capacity*, depends on the relationship between its pK_a value and the pH. To understand this point, we need to recognize the effect of pH on the fraction of molecules in the undissociated form (HA). The *titration curve* for acetic acid shown in Figure 2-21 illustrates these relationships: at one pH unit below the pK_a of an acid, 91 percent of the molecules are in the HA form; at one pH unit above the pK_a, 91 percent are in the A^- form. Thus the buffering capacity of weak acids and bases declines rapidly at more than one pH unit from their pK_a values. In other words, the addition of the same number of moles of acid to a solution containing a mixture of HA and A^- that is at a pH near the pK_a of the acid will cause less of a pH change than it would if the HA and A^- were not present or if the pH were far from the pK_a value.

All biological systems contain one or more buffers. Phosphoric acid (H_3PO_4) is a physiologically important buffer; phosphate ions are present in considerable quantities in cells and are an important factor in maintaining, or buffering, the pH of the cytosol. Phosphoric acid

has three groups that are capable of dissociating, but the three protons do not dissociate simultaneously. Loss of each proton can be described by a discrete dissociation reaction and pK_a as shown in Figure 2-22. The titration curve for phosphoric acid shows that the pK_a for the dissociation of the second proton is pH 7.2, similar to the pH of the cytosol. Because $pK_a = 7.2$ for the reaction $H_2PO_4^- \rightleftharpoons HPO_4^{2-} + H^+$, at pH 7.2 about 50 percent of cellular phosphate is $H_2PO_4^-$ and 50 percent is HPO_4^{2-} according to the Henderson-Hasselbalch equation. (The actual proportions, calculated by exact solution of the Henderson-Hasselbalch equation using all three dissociation constants are 0.499973 as $H_2PO_4^-$, 0.499973 as HPO_4^{2-}, 0.0000039 as H_3PO_4, and 0.0000016 as PO_4^{3-}.) Thus, phosphate is an excellent buffer at pH values around 7.2, the approximate pH of the cytosol of cells, and at pH 7.4, the pH of human blood.

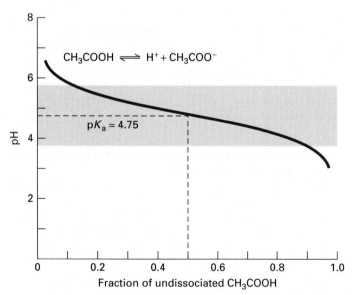

▲ **FIGURE 2-21 The titration curve of acetic acid (CH₃COOH).** The pK_a for the dissociation of acetic acid to hydrogen and acetate ions is 4.75. At this pH, half the acid molecules are dissociated. Because pH is measured on a logarithmic scale, the solution changes from 91 percent CH₃COOH at pH 3.75 to 9 percent CH₃COOH at pH 5.75. The acid has maximum buffering capacity in this pH range.

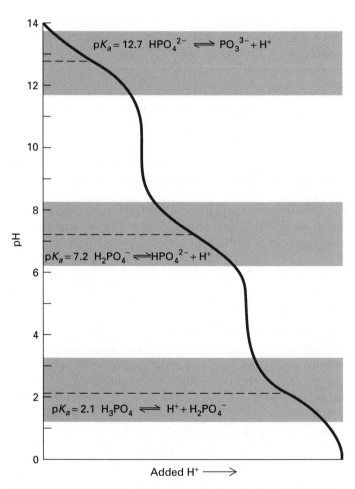

▲ **FIGURE 2-22 The titration curve of phosphoric acid (H₃PO₄).** This biologically ubiquitous molecule has three hydrogen atoms that dissociate at different pH values; thus, phosphoric acid has three pK_a values, as noted on the graph. The shaded areas denote the pH ranges—within one pH unit of the three pK_a values—where the buffering capacity of phosphoric acid is maximum. In these regions the addition of acid (or base) will cause the least change in the pH.

In nucleic acids, phosphate is found as a diester linked to two carbon atoms of adjacent ribose sugars:

$$-\overset{|}{\underset{|}{C}}-O-\overset{O}{\underset{OH}{\overset{\|}{P}}}-O-CH_2-$$

The pK_a for the dissociation of the single —OH proton is about 3, which is similar to the pK_a for the dissociation of the first proton from phosphoric acid. Therefore, each phosphate residue in deoxyribonucleic acid (DNA) or ribonucleic acid (RNA) is dissociated and carries a negative charge at neutral pH, which is why DNA and RNA are called nucleic *acids:*

$$-\overset{|}{\underset{|}{C}}-O-\overset{O}{\underset{O^-}{\overset{\|}{P}}}-O-CH_2-$$

SUMMARY Chemical Equilibrium

• The equilibrium constant K_{eq} of a reaction reflects the ratio of products to reactants at equilibrium and thus is a measure of the extent of the reaction. The K_{eq} depends on the temperature and pressure, but is independent of the reaction rate and of the initial concentrations of reactants and products.

• For a reaction involving a single reactant and single product, the equilibrium concentrations of reactant and product are unaffected by their initial concentrations. For a reaction involving multiple reactants and/or products, however, the equilibrium concentration of a *particular* product or reactant depends on the initial concentrations of all reactants and products as well as on the equilibrium constant.

• For any reaction, the equilibrium constant K_{eq} equals the ratio of the forward rate constant k_f to the reverse rate constant k_r. Because a catalyst accelerates the forward and reverse reactions to the same degree, it does not change the K_{eq} of a reaction.

• Equilibrium constants for reactions involving the noncovalent binding of a ligand to its receptor or of a protein to specific sequences in DNA are a measure of the stability of the ligand-receptor or protein-DNA complexes.

• Biological fluids have characteristic pH values. The pH of the cytosol is normally about 7.2–7.4, whereas the interior of organelles such as the lysosomes and vacuoles has a pH of about 5.

• Acids release hydrogen ions and bases bind them. In biological molecules, the carboxyl group and phosphate group are the most common acidic groups; the amino group is the most common basic group.

• The extent to which weak acids and bases release or take up hydrogen ions depends on the pH. At neutral pH most amino acids are zwitterions, bearing both a positive and a negative charge.

• Biological systems use various buffers to maintain their pH within a very narrow range.

2.4 Biochemical Energetics

The production of energy, its storage, and its use are as central to the economy of the cell as they are to the management of the world's resources. Cells require energy to do all their work, including the synthesis of sugars from carbon dioxide and water in photosynthesis, the contraction of muscles, and the replication of DNA. Energy may be defined as the ability to do work, a concept that is easy to grasp when it is applied to automobile engines and electric power plants. When we consider the energy associated with chemical bonds and chemical reactions within cells, however, the concept of work becomes less intuitive.

Living Systems Use Various Forms of Energy, Which Are Interconvertible

There are two principal forms of energy: kinetic and potential. *Kinetic energy* is the energy of movement—the motion of molecules, for example. The second form of energy, *potential energy,* or stored energy, is more important in the study of biological or chemical systems.

Kinetic Energy Heat, or *thermal energy,* is a form of kinetic energy—the energy of the motion of molecules. For heat to do work, it must flow from a region of higher temperature—where the average speed of molecular motion is greater—to one of lower temperature. Differences in temperature often exist between the internal and external environments of cells; however, cells generally cannot harness these heat differentials to do work. Even in warm-blooded animals that have evolved a mechanism for thermoregulation, the kinetic energy of molecules is used chiefly to maintain constant organismic temperatures.

Radiant energy is the kinetic energy of photons, or waves of light, and is critical to biology. Radiant energy can be converted to thermal energy, for instance when light is absorbed by molecules and the energy is converted to molecular motion. In the process of photosynthesis, light energy is absorbed by chlorophyll and is ultimately converted into other types of energy, such as that stored in covalent chemical bonds.

One of the major forms of *electric energy* is also kinetic—the energy of moving electrons or other charged particles.

Potential Energy Several forms of potential energy are biologically significant. Central to biology is the potential energy *stored in the bonds* connecting atoms in molecules. Indeed, most of the biochemical reactions described in this book involve the making or breaking of at least one covalent chemical bond. We recognize this energy when chemicals undergo energy-releasing reactions. The sugar glucose, for example, is high in potential energy. Cells degrade glucose continuously, and the energy released when glucose is metabolized is harnessed to do many kinds of work.

A second biologically important form of potential energy, to which we shall refer often, is the energy in a *concentration gradient*. When the concentration of a substance on one side of a permeable barrier, such as a membrane, is different from that on the other side, the result is a concentration gradient. All cells form concentration gradients between their interior and the external fluids by selectively exchanging nutrients, waste products, and ions with their surroundings. Also, compartments within cells frequently contain different concentrations of ions and other molecules; the concentration of protons within a lysosome, as we saw in the last section, is about 500 times that of the cytosol.

A third form of potential energy in cells is an *electric potential*—the energy of charge separation. For instance, there is a gradient of electric charge of $\approx 200,000$ volts per cm across the outer, or "plasma," membrane of virtually all cells.

Interconvertibility of All Forms of Energy According to the first law of thermodynamics, energy is neither created nor destroyed, but can be converted from one form to another.* In photosynthesis, for example, as we have just seen, the radiant energy of light is transformed into the chemical potential energy of the covalent bonds between the atoms in a sucrose or starch molecule. In muscles and nerves, chemical potential energy stored in covalent bonds is transformed, respectively, into kinetic and electric energy. In all cells, chemical potential energy, released by breakage of certain chemical bonds, is used to generate potential energy in the form of concentration and electric potential gradients. Similarly, energy stored in chemical concentration gradients or electric potential gradients is used to synthesize chemical bonds, or to transport other molecules "uphill" against a concentration gradient. This latter process occurs during the transport of nutrients such as glucose into certain cells and transport of many waste products out of cells. Because all forms of energy are interconvertible, they can be expressed in the same units of measurement, such as the calorie or kilocalorie.

The Change in Free Energy ΔG Determines the Direction of a Chemical Reaction

Because biological systems are generally held at constant temperature and pressure, it is possible to predict the direction of a chemical reaction by using a measure of potential

energy called **free energy,** or G, after the great American chemist Josiah Willard Gibbs (1839–1903), a founder of the science of thermodynamics. Gibbs showed that under conditions of constant pressure and temperature, as generally found in biological systems, "all systems change in such a way that free energy is minimized." In general, we are interested in what happens to the free energy when one molecule or molecular configuration is changed into another. Thus our concern is with relative, rather than absolute, values of free energy—in particular, with the difference between the values before and after the change. This **free-energy change** ΔG, where Δ stands for difference, is given by

$$\Delta G = G_{\text{products}} - G_{\text{reactants}}$$

In mathematical terms, Gibbs's law—that systems change to minimize free energy—is a set of statements about ΔG:

- If ΔG is negative for a chemical reaction or mechanical process, the forward reaction or process (from left to right as written) will tend to occur spontaneously.
- If ΔG is positive, the reverse reaction (from right to left as written) will tend to occur.
- If ΔG is zero, both forward and reverse reactions occur at equal rates; the reaction is at equilibrium.

The value of ΔG, like the equilibrium constant, is independent of the reaction mechanism and rate. Reactions with negative ΔG values that have very slow rate constants may not occur, for practical purposes, unless a catalyst is present, but the presence of a catalyst does not affect the value of ΔG.

The ΔG of a Reaction Depends on Changes in Enthalpy (Bond Energy) and Entropy

At any constant temperature and pressure, two factors determine the ΔG of a reaction and thus whether the reaction will tend to occur: the change in bond energy between reactants and products and the change in the randomness of the system. Gibbs showed that free energy can be defined as

$$G = H - TS$$

where H is the bond energy, or **enthalpy,** of the system; T is its temperature in degrees Kelvin (K); and S is a measure of randomness, called **entropy.** If temperature remains constant, a reaction proceeds spontaneously only if the free-energy change ΔG in the following equation is negative:

$$\Delta G = \Delta H - T\,\Delta S \qquad (2\text{-}7)$$

The enthalpy H of reactants or of products is equal to their total bond energies; the overall change in enthalpy ΔH is equal to the overall change in bond energies (see Table

---

*The exceptions to this rule are nuclear reactions, in which mass is converted to energy, but these are not relevant to biological systems.

2-1). In an **exothermic** reaction, the products contain less bond energy than the reactants, the liberated energy is usually converted to heat (the energy of molecular motion), and ΔH is negative. In an **endothermic** reaction, the products contain more bond energy than the reactants, heat is absorbed, and ΔH is positive. Reactions tend to proceed if they liberate energy (if $\Delta H < 0$), but this is only one of two important parameters of free energy to consider; the other is entropy.

Entropy S is a measure of the degree of randomness or disorder of a system. Entropy increases as a system becomes more disordered and decreases as it becomes more structured. Consider, for example, the diffusion of solutes from one solution into another one in which their concentration is lower. This important biological reaction is driven only by an increase in entropy; in such a process ΔH is near zero. To see this, suppose that a 0.1 M solution of glucose is separated from a large volume of water by a membrane through which glucose can diffuse. Diffusion of glucose molecules across the membrane will give them more room in which to move, with the result that the randomness, or entropy, of the system is increased. Maximum entropy is achieved when all molecules can diffuse freely over the largest possible volume—that is, when the concentration of glucose molecules is the same on both sides of the membrane. If the degree of hydration of glucose does not change significantly on dilution, ΔH will be approximately zero; the negative free energy of the reaction in which glucose molecules are liberated to diffuse over a larger volume will be due solely to the positive value of ΔS in Equation 2-7.

As mentioned previously, the formation of hydrophobic bonds is driven primarily by a change in entropy. That is, if a long hydrophobic molecule, such as heptane or tristearin, is dissolved in water, the water molecules are forced to form a cage around it, restricting their free motion. This imposes a high degree of order on their arrangement and lowers the entropy of the system ($\Delta S < 0$). Because the entropy change is negative, hydrophobic molecules do not dissolve well in aqueous solutions and tend to stay associated with one another.

We can summarize the relationships between free energy, enthalpy, and entropy as follows:

- An exothermic reaction ($\Delta H < 0$) that increases entropy ($\Delta S > 0$) occurs spontaneously ($\Delta G < 0$).
- An endothermic reaction ($\Delta H > 0$) will occur spontaneously if ΔS increases enough so that the $T \Delta S$ term can overcome the positive ΔH.
- If the conversion of reactants into products results in no change in free energy ($\Delta G = 0$), then the system is at equilibrium; that is, any conversion of reactants to products is balanced by an equal conversion of products to reactants.

Many biological reactions lead to an increase in order, and thus a decrease in entropy ($\Delta S < 0$). An obvious example is the reaction that links amino acids together to form a protein. A solution of protein molecules has a lower entropy than does a solution of the same amino acids unlinked, because the free movement of any amino acid in a protein is restricted when it is bound in a long chain. For the linking reaction to proceed, a compensatory decrease in free energy must occur elsewhere in the system, as is discussed in Chapter 4.

Several Parameters Affect the ΔG of a Reaction

The change in free energy of a reaction (ΔG) is influenced by temperature, pressure, and the initial concentrations of reactants and products. Most biological reactions—like others that take place in aqueous solutions—also are affected by the pH of the solution.

The standard free-energy change of a reaction $\Delta G^{\circ\prime}$ is the value of the change in free energy under the conditions of 298 K (25 °C), 1 atm pressure, pH 7.0 (as in pure water), and initial concentrations of 1 M for all reactants and products except protons, which are kept at pH 7.0. Table 2-4 gives values of $\Delta G^{\circ\prime}$ for some typical biochemical reactions. The sign of $\Delta G^{\circ\prime}$ depends on the direction in which the reaction is written. If the reaction A $\rightarrow$ B has a $\Delta G^{\circ\prime}$ of $-x$ kcal/mol, then the reverse reaction B $\rightarrow$ A will have a $\Delta G^{\circ\prime}$ value of $+x$ kcal/mol.

Most biological reactions differ from standard conditions, particularly in the concentrations of reactants. However, we can estimate free-energy changes for different temperatures and initial concentrations, using the equation

$$\Delta G = \Delta G^{\circ\prime} + RT \ln Q = \Delta G^{\circ\prime} + RT \ln \frac{[\text{products}]}{[\text{reactants}]} \qquad (2\text{-}8)$$

where R is the gas constant of 1.987 cal/(degree · mol), T is the temperature (in degrees Kelvin), and Q is the initial ratio of products to reactants, which is expressed as in Equation 2-1 defining the equilibrium constant. Again using as our example the interconversion of glyceraldehyde 3-phosphate (G3P) and dihydroxyacetone phosphate (DHAP)

$$\text{G3P} \rightleftharpoons \text{DHAP}$$

we have $Q = [\text{DHAP}]/[\text{G3P}]$ and $\Delta G^{\circ\prime} = -1840$ cal/mol (see Table 2-4). Equation 2-8 for ΔG then becomes

$$\Delta G = -1840 + 1.987\, T \ln \frac{[\text{DHAP}]}{[\text{G3P}]}$$

from which we can calculate ΔG for any set of concentrations of DHAP and G3P. If the initial concentrations of both DHAP and G3P are 1 M, then $\Delta G = \Delta G^{\circ\prime} = -1840$ cal/mol, because $RT \ln 1 = 0$. The reaction will tend to proceed from left to right, in the direction of formation of DHAP. If, however, the initial concentration of DHAP is

SOURCE: A. L. Lehninger, 1975, *Biochemistry*, 2d ed., Worth, p. 397.

TABLE 2-4 Values of $\Delta G^{\circ\prime}$, the Standard Free-Energy Change, for Some Important Biochemical Reactions

Reaction	$\Delta G^{\circ\prime}$ (kcal/mol)
HYDROLYSIS	
Acid anhydrides:	
Acetic anhydride + H_2O → 2 acetate	−21.8
PP_i + H_2O → 2 P_i*	−8.0
ATP + H_2O → ADP + P_i	−7.3
Esters:	
Ethylacetate + H_2O → ethanol + acetate	−4.7
Glucose 6-phosphate + H_2O → glucose + P_i	−3.3
Amides:	
Glutamine + H_2O → glutamate + NH_4^+	−3.4
Glycylglycine + H_2O → 2 glycine (hydrolysis of a peptide bond)	−2.2
Glycosides:	
Sucrose + H_2O → glucose + fructose	−7.0
Maltose + H_2O → 2 glucose	−4.0
ESTERIFICATION	
Glucose + P_i → glucose 6-phosphate + H_2O	+3.3
REARRANGEMENT†	
Glucose 1-phosphate → glucose 6-phosphate	−1.7
Fructose 6-phosphate → glucose 6-phosphate	−0.4
Glyceraldehyde 3-phosphate → dihydroxyacetone phosphate	−1.8
ELIMINATION	
Malate → fumarate + H_2O	+0.75
OXIDATION	
Glucose + 6 O_2 → 6 CO_2 + 6 H_2O	−686
Palmitic acid + 23 O_2 → 16 CO_2 + 16 H_2O	−2338
PHOTOSYNTHESIS	
6 CO_2 + 6 H_2O → six-carbon sugars‡ + 6 O_2	+686

*PP_i = pyrophosphate; P_i = phosphate.
†*The three reactions shown occur during glycolysis, the first steps in conversion of glucose to CO_2.*
‡*The two principal products of photosynthesis are sucrose, a disaccharide, and starch, a long polymer of glucose.*

0.1 M and that of G3P is 0.001 M, with other conditions being standard, then $Q = 0.1/0.001 = 100$, and

$$\Delta G = -1840 + (1.987)(298) \ln (100) = +887 \text{ cal/mol}$$

Clearly, the reaction will now proceed in the direction of formation of G3P.

In a reaction A + B ⇌ C, in which two molecules combine to form a third, the equation for ΔG becomes

$$\Delta G = \Delta G^{\circ\prime} + RT \ln Q = \Delta G^{\circ\prime} + RT \ln \frac{[C]}{[A][B]}$$

The direction of the reaction will shift more toward the right (toward formation of C) if either [A] or [B] is increased.

The $\Delta G^{\circ\prime}$ of a Reaction Can Be Calculated from Its K_{eq}

A chemical mixture at equilibrium is already in a state of minimal free energy: no free energy is being generated or released. Thus, for a system at equilibrium, we can write

$$0 = \Delta G = \Delta G^{\circ\prime} + RT \ln Q$$

At equilibrium the value of Q is the equilibrium constant K_{eq}, so that

$$\Delta G^{\circ\prime} = -RT \ln K_{eq}$$

Expressed in terms of base 10 logarithms, this equation becomes

$$\Delta G^{\circ\prime} = -2.3RT \log K_{eq}$$

or

$$\Delta G^{\circ\prime} = -2.3 (1.987) (298) \log K_{eq}$$
$$= -1362 \log K_{eq} \quad (2\text{-}9)$$

under standard conditions. Thus, if the concentrations of reactants and products at equilibrium (i.e., the K_{eq}) are determined, the value of $\Delta G^{\circ\prime}$ can be calculated. For example, we saw earlier that K_{eq} equals 22.2 for the interconversion of glyceraldehyde 3-phosphate to dihydroxyacetone phosphate (G3P ⇌ DHAP) under standard conditions. Substituting this value into Equation 2-9, we can easily calculate the $\Delta G^{\circ\prime}$ for this reaction as −1840 cal/mol.

By rearranging Equation 2-9 and taking the antilogarithm, we obtain

$$K_{eq} = 10^{-(\Delta G^{\circ\prime}/2.3RT)} \quad (2\text{-}10)$$

From this expression, it is clear that if $\Delta G^{\circ\prime}$ is negative, then the exponent will be positive and hence K_{eq} will be greater than 1; that is, the formation of products from reactants is favored (Table 2-5). Conversely, if $\Delta G^{\circ\prime}$ is positive, then the exponent will be negative and K_{eq} will be less than 1.

TABLE 2-5	Values of $\Delta G^{\circ\prime}$ for Some Values of K_{eq}
K_{eq}	$\Delta G^{\circ\prime}$ (cal/mol)*
0.001	+4086
0.01	+2724
0.1	+1362
1.0	0
10	−1362
100	−2724
1000	−4086

*Calculated from the formula $\Delta G^{\circ\prime} = -2.3\ RT \log K_{eq}$.

Although a chemical equilibrium appears to be unchanging and static, it is actually a dynamic state. The forward and the reverse reactions proceed at exactly the same rate, thereby canceling each other out. As noted earlier, when an enzyme or some other catalyst speeds up a reaction, it also speeds up the reverse reaction; thus equilibrium is reached sooner than it is when the reaction is not catalyzed. However, the equilibrium constant and $\Delta G^{\circ\prime}$ of a reaction are the *same* in the presence and absence of a catalyst.

Cells Must Expend Energy to Generate Concentration Gradients

A cell must often accumulate chemicals, such as glucose and K^+ ions, in greater concentrations than exist in its environment. Consequently, the cell must transport these chemicals against a concentration gradient. To find the amount of energy required to transfer 1 mole of a substance from outside the cell to inside the cell, we use Equation 2-8 relating ΔG to the concentration of reactants and products. Because this simple transport reaction does not involve making or breaking covalent bonds and no heat is taken up or released, the $\Delta G^{\circ\prime}$ is 0. Thus Equation 2-8 becomes

$$\Delta G = \Delta G^{\circ\prime} + RT \ln Q = RT \ln \frac{C_2}{C_1}$$

where C_2 is the initial concentration of a substance inside the cell and C_1 is its concentration outside the cell. If the ratio of C_2 to C_1 is 10, then at 25 °C, $\Delta G = RT \ln 10 = +1.36$ kcal per mole of substance transported. Such calculations assume that a molecule of a given substance inside a cell is identical with a molecule of that substance outside and that the substance is not sequestered, bound, or chemically changed by the transport.

Since the "uphill" transport of molecules against a concentration gradient ($C_2 > C_1$) has a positive ΔG, it clearly cannot take place spontaneously. To occur, such transport requires the input of cellular chemical energy, which often is supplied by the hydrolysis of ATP (Chapter 15). Conversely, when a substance moves down its concentration gradient ($C_1 > C_2$) in crossing a membrane, ΔG has a negative value and the transport can be coupled to a reaction that has a positive ΔG, say, the movement of another substance uphill across a membrane.

Many Cellular Processes Involve Oxidation-Reduction Reactions

Many chemical reactions result in the transfer of electrons from one atom or molecule to another; this transfer may or may not accompany the formation of new chemical bonds. The loss of electrons from an atom or a molecule is called **oxidation**, and the gain of electrons by an atom or a molecule is called **reduction**. Because electrons are neither created nor destroyed in a chemical reaction, if one atom or molecule is oxidized, another must be reduced. For example, oxygen draws electrons from Fe^{2+} (ferrous) ions to form Fe^{3+} (ferric) ions, a reaction that occurs as part of the process by which carbohydrates are degraded in mitochondria. Each oxygen atom receives two electrons, one from each of two Fe^{2+} ions:

$$2\ Fe^{2+} + {}^{1}\!/_{2}\ O_2 \longrightarrow 2\ Fe^{3+} + O^{2-}$$

Thus Fe^{2+} is oxidized, and O_2 is reduced. Oxygen similarly accepts electrons in many oxidation reactions in aerobic cells.

The transformation of succinate into fumarate is another oxidation reaction that takes place during carbohydrate breakdown in mitochondria. In this reaction, succinate loses two hydrogen atoms, which is equivalent to a loss of two protons and two electrons (Figure 2-23). Protons are soluble in aqueous solutions (as H_3O^+), but electrons are not and must be transferred directly from one atom or molecule to another. The electrons lost from succinate in its conversion to fumarate are transferred to flavin adenine dinucleotide (FAD), which is reduced to $FADH_2$. Many biologically important oxidation and reduction reactions involve the removal or the addition of hydrogen atoms (protons plus electrons) rather than the transfer of isolated electrons.

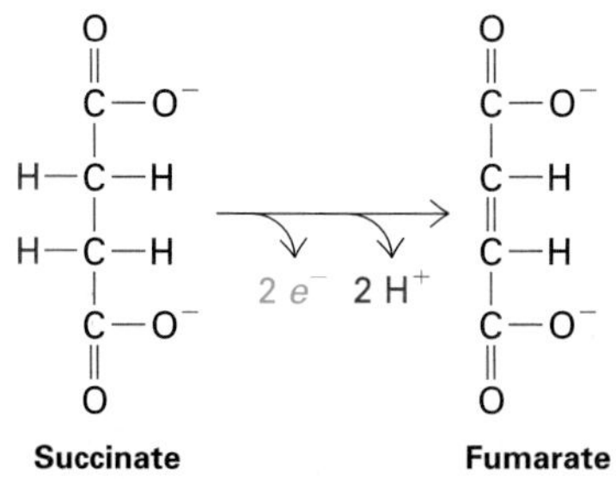

▲ **FIGURE 2-23 Succinate is converted to fumarate by the loss of two electrons and two protons.** This oxidation reaction, which occurs in mitochondria as part of the citric acid cycle, is coupled to reduction of FAD to $FADH_2$.

Standard Reduction Potentials To describe oxidation-reduction reactions, such as the reaction of ferrous ion (Fe^{2+}) and oxygen (O_2), it is easiest to divide them into two half-reactions:

Oxidation of Fe^{2+}: $2\ Fe^{2+} \longrightarrow 2\ Fe^{3+} + 2\ e^-$

Reduction of O_2: $2\ e^- + \frac{1}{2}\ O_2 \longrightarrow O^{2-}$

In this case, the reduced oxygen (O^{2-}) readily reacts with two protons to form one water molecule:

$$2\ H^+ + O^{2-} \rightleftharpoons H_2O$$

Thus if we add two protons to each side of the equation for the half-reaction for reduction of O_2, the half-reaction can be rewritten as

$$2\ e^- + \frac{1}{2}\ O_2 + 2\ H^+ \rightleftharpoons H_2O$$

The readiness with which an atom or a molecule *gains* an electron is its **reduction potential** E. Reduction potentials are measured in volts (V) from an arbitrary zero point set at the reduction potential of the following half-reaction under standard conditions (25 °C, 1 atm, and reactants at 1 M):

$$H^+ + e^- \underset{\text{oxidation}}{\overset{\text{reduction}}{\rightleftharpoons}} \frac{1}{2}\ H_2$$

The value of E for a molecule or an atom under standard conditions is its standard reduction potential, E_0' (Table 2-6). Standard reduction potentials may differ somewhat from those found under the conditions in a cell, because the concentrations of reactants in a cell are not 1 M. A positive reduction potential means that a molecule or ion (say, Fe^{3+}) has a higher affinity for electrons than the H^+ ion does in the standard reaction. A negative reduction potential means that a substance—for example, acetate (CH_3COO^-) in its reduction to acetaldehyde (CH_3CHO)—has a lower affinity for electrons. In an oxidation-reduction reaction, electrons move spontaneously toward atoms or molecules having *more positive* reduction potentials. In other words, a compound having a more negative reduction potential (or more positive oxidation potential) can reduce—or transfer electrons to—one having a more positive reduction potential.

The Relationship between Changes in Free Energy and Reduction Potentials In an oxidation-reduction reaction, the total voltage change (change in electric potential) ΔE is the sum of the voltage changes (reduction potentials) of the individual oxidation or reduction steps. Because all forms of energy are interconvertible, we can express ΔE as a change in chemical free energy (ΔG). The charge in 1 mole (6.02×10^{23}) of electrons is 96,500 coulombs (96,500 joules per volt), a quantity known as the *Faraday constant* ($\mathscr{F}$) after British physicist Michael Faraday (1791–1867). The following formula shows the relationship between free energy and reduction potential:

$$\Delta G\ (\text{cal/mol}) = -n\mathscr{F}\ \Delta E = -n\ (96,500/4.184)\ \Delta E\ (\text{volts})$$

or

$$\Delta G\ (\text{cal/mol}) = -n\ (23,064)\ \Delta E\ (\text{volts}) \qquad (2\text{-}11)$$

TABLE 2-6   Values of the Standard Reduction Potential E_0' and Standard Free Energy $\Delta G^{\circ\prime}$ for Selected Oxidation-Reduction Reactions (pH 7.0, 25 °C)

Oxidant	Reductant	n*	E_0' (volts)[†]	$\Delta G^{\circ\prime}$ (kcal/mole)[‡]
Succinate + CO_2	α-Ketoglutarate	2	−0.67	+30.9
Acetate	Acetaldehyde	2	−0.60	+27.7
Ferredoxin (oxidized)	Ferredoxin (reduced)	1	−0.43	+9.9
2 H^+	H_2	2	−0.42	+19.4
NAD^+	$NADH + H^+$	2	−0.32	+14.8
$NADP^+$	$NADPH + H^+$	2	−0.32	+14.8
Glutathione (oxidized)	Glutathione (reduced)	2	−0.23	+10.6
Acetaldehyde	Ethanol	2	−0.20	+9.2
Pyruvate	Lactate	2	−0.19	+8.7
Fumarate	Succinate	2	+0.03	−1.4
Cytochrome c (+3)	Cytochrome c (+2)	1	+0.22	−5.1
Fe^{3+}	Fe^{2+}	1	+0.77	−17.8
$\frac{1}{2}\ O_2 + 2\ H^+$	H_2O	2	+0.82	−37.8

*n is the number of electrons transferred.
[†]E_0' refers to the partial reaction: oxidant + $e^- \rightarrow$ reductant.
[‡]Calculated from the equation $\Delta G^{\circ\prime} = -n\mathscr{F}\ \Delta E_0'$.

SOURCE: L. Stryer, 1988, *Biochemistry*, 3d ed., W. H. Freeman and Company, p. 400.

where n is the number of electrons transferred and 4.184 is the factor used to convert joules into calories. Note that an oxidation-reduction reaction with a positive ΔE value will have a negative ΔG and thus will tend to proceed from left to right.

The reduction potential is customarily used to describe the electric energy change that occurs when an atom or a molecule gains an electron. In an oxidation-reduction reaction, we also use the **oxidation potential**—the voltage change that takes place when an atom or molecule *loses* an electron—which is simply the negative of the reduction potential:

Reduction: $\quad Cu^{2+} + e^- \longrightarrow Cu^+ \qquad E_0' = +0.35$ V

Oxidation: $\quad Cu^+ \longrightarrow Cu^{2+} + e^- \qquad E_0' = -0.35$ V

The voltage change in a complete oxidation-reduction reaction, in which one molecule is reduced and another is oxidized, is simply the sum of the oxidation potential and the reduction potential of the two partial oxidation and reduction reactions, respectively. Consider, for example, the change in electric potential (and, correspondingly, in standard free energy) when succinate is oxidized by oxygen:

$$\text{Succinate} + \tfrac{1}{2} O_2 \rightleftharpoons \text{fumarate} + H_2O$$

In this case, the partial reactions are

Oxidation:

Succinate $\rightleftharpoons$ $E_0' = -0.03$ V
fumarate $+ 2 H^+ + 2 e^-$ (oxidation potential)
$\Delta G^{\circ\prime} = +1.39$ kcal/mol
($n = 2$ electrons)

Reduction:

$\tfrac{1}{2} O_2 + 2 e^- + 2 H^+ \rightleftharpoons$ $E_0' = +0.82$ V
H_2O (reduction potential)
$\Delta G^{\circ\prime} = -37.88$ kcal/mol
($n = 2$ electrons)

Sum: Succinate $+ \tfrac{1}{2} O_2 \rightleftharpoons$ $\Delta E_0' =$
fumarate $+ H_2O$ -0.03 V $+ 0.82$ V
$= + 0.79$ V
$\Delta G^{\circ\prime} =$
$+1.39$ kcal/mol $- 37.88$ kcal/mol
$= - 36.49$ kcal/mol

The overall reaction has a positive $\Delta E_0'$ or, equivalently, a negative $\Delta G^{\circ\prime}$ and thus, under standard conditions, will tend to occur from left to right.

An Unfavorable Chemical Reaction Can Proceed If It Is Coupled with an Energetically Favorable Reaction

Many chemical reactions in cells are energetically unfavorable ($\Delta G > 0$) and will not proceed spontaneously. One example is the synthesis of small peptides (e.g., glycylalanine) or proteins from amino acids. Cells are able to carry out a reaction that has a positive ΔG by coupling it to a reaction that has a negative ΔG of larger magnitude, so that the sum of the two reactions has a negative ΔG. Suppose that the reaction

$$A \rightleftharpoons B + X$$

has a $\Delta G^{\circ\prime}$ of $+5$ kcal/mol and that the reaction

$$X \rightleftharpoons Y + Z$$

has a $\Delta G^{\circ\prime}$ of -10 kcal/mol. In the absence of the second reaction, there would be much more A than B at equilibrium. The occurrence of the second process, by which X becomes Y + Z, changes that outcome: because it is such a favorable reaction, it will pull the first process toward the formation of B and the consumption of A.

The $\Delta G^{\circ\prime}$ of the overall reaction will be the sum of the $\Delta G^{\circ\prime}$ values of each of the two partial reactions:

A $\rightleftharpoons$ B + X	$\Delta G^{\circ\prime} = +5$ kcal/mol
X $\rightleftharpoons$ Y + Z	$\Delta G^{\circ\prime} = -10$ kcal/mol
Sum: A $\rightleftharpoons$ B + Y + Z	$\Delta G^{\circ\prime} = -5$ kcal/mol

The overall reaction releases energy. In cells, energetically unfavorable reactions of the type A $\rightleftharpoons$ B + X are often coupled to the hydrolysis of the compound adenosine triphosphate (ATP), a reaction with a negative change in free energy ($\Delta G^{\circ\prime} = -7.3$ kcal/mol), so that the overall reaction has a negative $\Delta G^{\circ\prime}$.

Hydrolysis of Phosphoanhydride Bonds in ATP Releases Substantial Free Energy

All cells extract energy from foods through a series of reactions that exhibit negative free-energy changes; plant cells also can extract energy from absorbed light. In both cases, much of the free energy is not allowed to dissipate as heat but is captured in chemical bonds formed by other molecules for use throughout the cell. In almost all organisms, the most important molecule for capturing and transferring free energy is **adenosine triphosphate**, or **ATP** (Figure 2-24).

The useful free energy in an ATP molecule is contained in **phosphoanhydride bonds**, which are formed from the condensation of two molecules of phosphate by the loss of water:

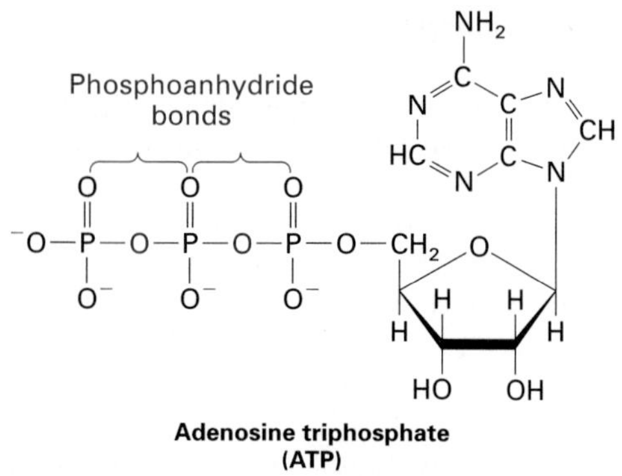

Phosphoanhydride bonds

Adenosine triphosphate (ATP)

▲ FIGURE 2-24 **In adenosine triphosphate (ATP), two high-energy phosphoanhydride bonds (red) link the three phosphate groups.**

An ATP molecule has two phosphoanhydride bonds and is often written as adenosine-p~p~p, or simply Ap~p~p, where p stands for a phosphate group and ~ denotes a high-energy bond.

Hydrolysis of a phosphoanhydride bond in each of the following reactions has a highly negative $\Delta G^{\circ\prime}$ of about -7.3 kcal/mol:

$$Ap{\sim}p{\sim}p + H_2O \longrightarrow Ap{\sim}p + P_i + H^+$$
$$\textbf{(ATP)} \qquad\qquad\qquad \textbf{(ADP)}$$

$$Ap{\sim}p{\sim}p + H_2O \longrightarrow Ap + PP_i + H^+$$
$$\textbf{(ATP)} \qquad\qquad\qquad \textbf{(AMP)}$$

$$Ap{\sim}p + H_2O \longrightarrow Ap + P_i + H^+$$
$$\textbf{(ADP)} \qquad\qquad\quad \textbf{(AMP)}$$

In these reactions, P_i stands for inorganic phosphate and PP_i for inorganic pyrophosphate, two phosphate groups linked by a phosphoanhydride bond. As the top two reactions show, the removal of a phosphate or a pyrophosphate group from ATP leaves adenosine diphosphate (ADP) or adenosine monophosphate (AMP), respectively.

The phosphoanhydride bond is an ordinary covalent bond, but it releases about 7.3 kcal/mol of free energy (under standard biochemical conditions) when it is broken. In contrast, hydrolysis of the phosphoester bond in AMP, forming inorganic phosphate and adenosine, releases only about 2 kcal/mol of free energy. Phosphoanhydride bonds commonly are termed "high-energy" bonds, even though the $\Delta G^{\circ\prime}$ for the reaction of succinate with oxygen is much higher (-37 kcal/mol).

Cells can transfer the free energy released by the hydrolysis of phosphoanhydride bonds to other molecules. This transfer supplies cells with enough free energy to carry out reactions that would otherwise be unfavorable. For example, if the reaction

$$B + C \longrightarrow D$$

is energetically unfavorable ($\Delta G > 0$), it can be made favorable by linking it to the hydrolysis of the terminal phosphoanhydride bond in ATP. Some of the energy in this phosphoanhydride bond is used to transfer a phosphate group to one of the reactants, forming a phosphorylated intermediate, B~p. The intermediate thus has enough free energy to react with C, forming D and free phosphate:

$$B + Ap{\sim}p{\sim}p \longrightarrow B{\sim}p + Ap{\sim}p$$
$$B{\sim}p + C \longrightarrow D + P_i$$

Thus, the overall reaction is

$$B + C + Ap{\sim}p{\sim}p \longrightarrow D + Ap{\sim}p + P_i$$

which is energetically favorable. Chapter 4 illustrates in detail how the hydrolysis of ATP is coupled to protein formation from amino acids; in the above example B and C would represent amino acids and D a dipeptide. Cells keep the ratio of ATP to ADP and AMP high, often as high as 10:1. Thus reactions in which the terminal phosphate group of ATP is transferred to another molecule will be driven even further along.

As shown in Table 2-7, the $\Delta G^{\circ\prime}$ for hydrolysis of a phosphoanhydride bond in ATP (-7.3 kcal/mol) is about twice the $\Delta G^{\circ\prime}$ for hydrolysis of a phosphoester bond, such as that in glucose 6-phosphate (-3.3 kcal/mol). A principal reason for this difference is that ATP and its hydrolysis products ADP and P_i are highly charged at neutral pH. Three of the four ionizable protons in ATP are fully dissociated at pH 7.0, and the fourth, with a pK_a of 6.95, is about 50 percent dissociated. The closely spaced negative charges in ATP repel each other strongly. When the terminal phosphoanhydride bond is hydrolyzed, some of this stress is removed by the separation of the hydrolysis products ADP^{3-} and HPO_4^{2-}; that is, the separated negatively charged ADP^{3-} and HPO_4^{2-} will tend not to recombine to form ATP. In glucose 6-phosphate, by contrast, there is no charge repulsion between the phosphate group and the carbon atom to which it is attached. One of the hydrolysis products, glucose, is uncharged and will not repel the negatively charged HPO_4^{2-} ion; thus there is less resistance to the recombination of glucose and HPO_4^{2-} to form glucose 6-phosphate.

Many other bonds—particularly those between a phosphate group and some other substance—have the same high-energy character as phosphoanhydride bonds. The phosphoanhydride bond of ATP is not the most or the least energetic of these bonds (see Table 2-7). The preeminent role of ATP in capturing and transferring free energy within cells represents a compromise. The free energy of hydrolysis of ATP is sufficiently great that reactions in which the terminal phosphate group is transferred to another molecule have a substantially negative $\Delta G^{\circ\prime}$. However, if hydrolysis of this phosphoanhydride bond liberated considerably more free energy than it does, cells might require too much energy to

TABLE 2-7 Values of $\Delta G°'$ for the Hydrolysis of Various Biologically Important Phosphate Compounds*

Compound	$\Delta G°'$ (kcal/mol)
PHOSPHOENOLPYRUVATE 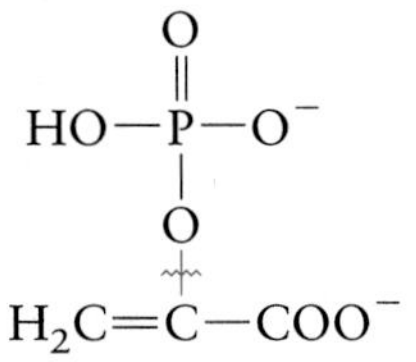	−14.8

$$
\begin{array}{c}
O \\
\parallel \\
HO{-}P{-}O^- \\
\mid \\
O \\
\text{\small(wavy)} \\
H_2C{=}C{-}COO^-
\end{array}
$$

| CREATINE PHOSPHATE 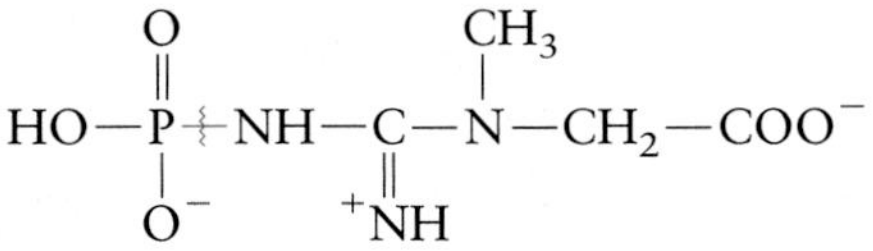 | −10.3 |

$$
\begin{array}{c}
O \qquad\qquad CH_3 \\
\parallel \qquad\qquad \mid \\
HO{-}P{+}NH{-}C{-}N{-}CH_2{-}COO^- \\
\mid \qquad\qquad \parallel \\
O^- \qquad\quad {}^+NH
\end{array}
$$

| PYROPHOSPHATE 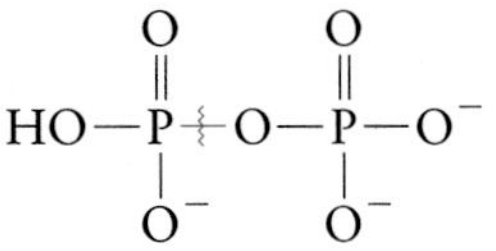 | −8.0 |

$$
\begin{array}{c}
O \qquad\quad O \\
\parallel \qquad\quad \parallel \\
HO{-}P{+}O{-}P{-}O^- \\
\mid \qquad\quad \mid \\
O^- \qquad\quad O^-
\end{array}
$$

ATP (to ADP + P_i)	−7.3
ATP (to AMP + PP_i)	−7.3
GLUCOSE 1-PHOSPHATE 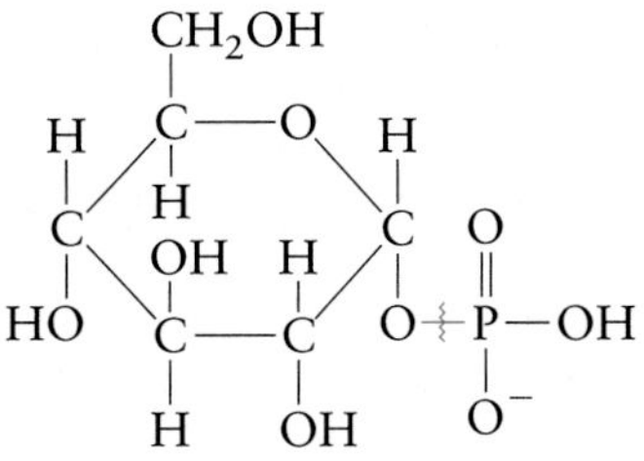	−5.0

$$
\begin{array}{c}
CH_2OH \\
\end{array}
$$

| GLUCOSE 6-PHOSPHATE 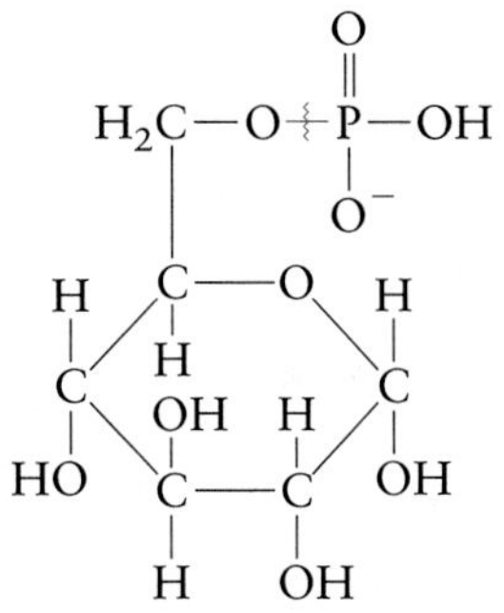 | −3.3 |

| GLYCEROL 3-PHOSPHATE 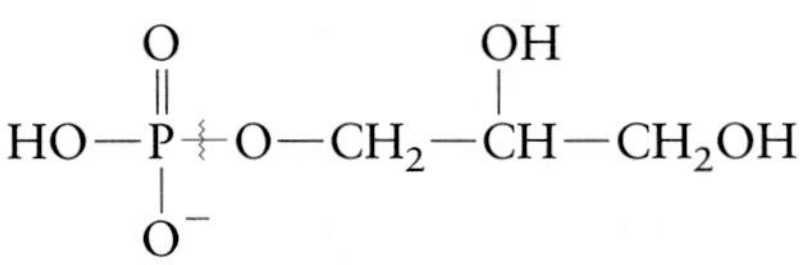 | −2.2 |

$$
\begin{array}{c}
O \qquad\qquad OH \\
\parallel \qquad\qquad \mid \\
HO{-}P{+}O{-}CH_2{-}CH{-}CH_2OH \\
\mid \\
O^-
\end{array}
$$

*The bond that is cleaved is indicated by the wavy line.

form this bond in the first place. In other words, many reactions in cells release enough energy to form ATP, and hydrolysis of ATP releases enough energy to drive many of the cell's energy-requiring reactions and processes.

ATP Is Used to Fuel Many Cellular Processes

If the terminal phosphoanhydride bond of ATP were to rupture by hydrolysis to produce ADP and P_i, energy would be released in the form of heat. However, cells contain various enzymes that can couple ATP hydrolysis to other reactions, so that much of the released energy is converted to more useful forms (Figure 2-25). For instance, cells use energy from ATP to synthesize macromolecules (proteins, nucleic acids, and polysaccharides) and many types of small molecules. The hydrolysis of ATP also supplies the energy needed to move individual cells from one location to another, to contract muscle cells, and to transport molecules into or out of the cell, usually against a concentration gradient. Gradients of ions, such as Na^+ and K^+, across a cellular membrane are produced by the action of membrane-embedded enzymes, called **ion pumps**, that couple the hydrolysis of ATP to the "uphill" movement of ions. The resulting ion concentration gradients are responsible for the generation of an electric potential across the membrane. This potential is the basis for the electric activity of cells and, in particular, for the conduction of impulses by nerves.

Clearly, to continue, functioning cells must constantly replenish their ATP supply. The ultimate energy source for formation of high-energy bonds in ATP and other compounds in nearly all cells is sunlight. Plants and microorganisms trap the energy in light through **photosynthesis.** In this process, chlorophyll pigments absorb the energy of light, which is then used to synthesize ATP from ADP and P_i. Much of the ATP produced in photosynthesis is used to help convert carbon dioxide to six-carbon sugars such as fructose and glucose:

$$6\ CO_2 + 6\ H_2O \xrightarrow[\;]{\text{ATP}\quad\text{ADP} + P_i} C_6H_{12}O_6 + 6\ O_2$$

Additional energy is used to convert hexoses into the disaccharide sucrose and polysaccharides. In animals, the free energy in sugars and other molecules derived from food is released in the process of **respiration.** All synthesis of ATP in animal cells and in nonphotosynthetic microorganisms results from the chemical transformation of energy-rich dietary or storage molecules. We discuss the mechanisms of photosynthesis and cellular respiration in Chapter 16.

As noted earlier, glucose is a major source of energy in most cells. When 1 mole (180 g) of glucose reacts with oxygen under standard conditions according to the following reaction, 686 kcal of energy is released:

$$C_6H_{12}O_6 + 6\ O_2 \longrightarrow 6\ CO_2 + 6\ H_2O$$

$$\Delta G°' = -686\ \text{kcal/mol}$$

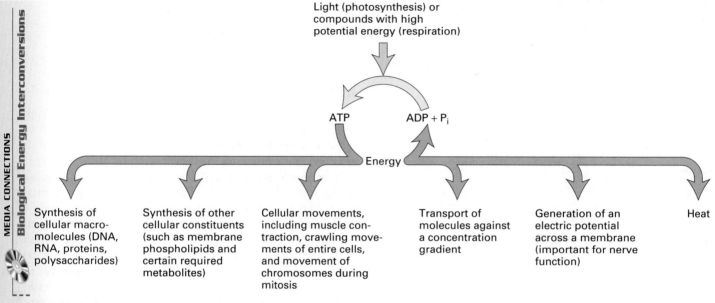

▲ **FIGURE 2-25 The ATP cycle.** ATP is formed from ADP and P_i by photosynthesis in plants and by the metabolism of energy-rich compounds in most cells. The hydrolysis of ATP to ADP and P_i is linked to many key cellular functions; the free energy released by the breaking of the phosphoanhydride bond is trapped as usable energy.

If glucose is simply burned in air, all this energy is released as heat. By an elaborate set of enzyme-catalyzed reactions, cells couple the metabolism of 1 molecule of glucose to the synthesis of as many as 36 molecules of ATP from 36 molecules of ADP:

$$C_6H_{12}O_6 + 6\ O_2 + 36\ P_i + 36\ ADP \longrightarrow$$
$$6\ CO_2 + 6\ H_2O + 36\ ATP$$

Because formation of one high-energy phosphoanhydride bond in ATP, from P_i and ADP, requires an input of 7.3 kcal/mol, about 263 kcal of energy (36×7.3) is conserved in ATP per mole of glucose metabolized (an efficiency of 263/686, or about 38 percent). This type of cellular metabolism is termed **aerobic** because it is dependent on the oxygen in the air. Aerobic **catabolism** (degradation) of glucose is found in all higher plant and animal cells and in many bacterial cells.

The overall reaction of glucose respiration

$$C_6H_{12}O_6 + 6\ O_2 \longrightarrow 6\ CO_2 + 6\ H_2O$$

is the reverse of the photosynthetic reaction in which six-carbon sugars are formed

$$6\ CO_2 + 6\ H_2O \longrightarrow C_6H_{12}O_6 + 6\ O_2$$

The latter reaction requires energy from light, whereas the former releases energy. Respiration and photosynthesis are the two major processes constituting the carbon cycle in nature: sugars and oxygen produced by plants are the raw materials for respiration and the generation of ATP by plant and animal cells alike; the end products of respiration, CO_2 and H_2O, are the raw materials for the photosynthetic production of sugars and oxygen. The only net source of energy in this cycle is sunlight. Thus, directly or indirectly, light energy captured in photosynthesis is the source of chemical energy for almost all cells.

The exceptions to this are certain microorganisms that exist in deep ocean vents where sunlight is completely absent. These unusual bacteria derive the energy for converting ADP and P_i into ATP from the oxidation of reduced inorganic compounds present in the dissolved vent gas that originates in the center of the earth. Unfortunately, little is yet known about the biology of these organisms.

SUMMARY Biochemical Energetics

• The change in free energy ΔG is the most useful measure for predicting the direction of chemical reactions in biological systems. Chemical reactions tend to proceed in the direction for which ΔG is negative.

• The ΔG of a reaction depends on the change in enthalpy ΔH (sum of bond energies), the change in entropy ΔS (the randomness of molecular motion), and the temperature T: $\Delta G = \Delta H - T\,\Delta S$.

• The standard free-energy change $\Delta G^{\circ\prime}$ equals $-2.3\ RT \log K_{eq}$. Thus the value of $\Delta G^{\circ\prime}$ can be calculated from the experimentally determined concentrations of reactants and products at equilibrium.

- The tendency of an atom or molecule to gain electrons is its reduction potential E, which is measured in volts. The tendency to lose electrons is the oxidation potential, which has the same magnitude but opposite sign as the reduction potential for the reverse reaction.

- Oxidation and reduction reactions always occur in pairs. The ΔE for an oxidation-reduction reaction is the sum of the oxidation potential and the reduction potential of the two partial reactions. Oxidation-reduction reactions with a positive ΔE have a negative ΔG and thus tend to proceed spontaneously.

- A chemical reaction having a positive ΔG can proceed if it is coupled with a reaction having a negative ΔG of larger magnitude.

- Many energetically unfavorable cellular reactions are fueled by hydrolysis of one or both of the two phosphoanhydride bonds in ATP.

- Directly or indirectly, light energy captured by photosynthesis in plants and photosynthetic bacteria is the ultimate source of chemical energy for almost all cells.

2.5 Activation Energy and Reaction Rate

Many chemical reactions that exhibit a negative $\Delta G^{\circ\prime}$ do not proceed unaided at a measurable rate. For example, in pure aqueous solutions, glyceraldehyde 3-phosphate (G3P), our recurrent model, is a fairly stable compound that reacts very slowly or not at all, yet potentially it can undergo several different reactions, each of which has a negative $\Delta G^{\circ\prime}$ (Figure 2-26). To understand why only one of several possible reactions generally occurs, we need to consider how reactions proceed and the factors that affect reaction rates.

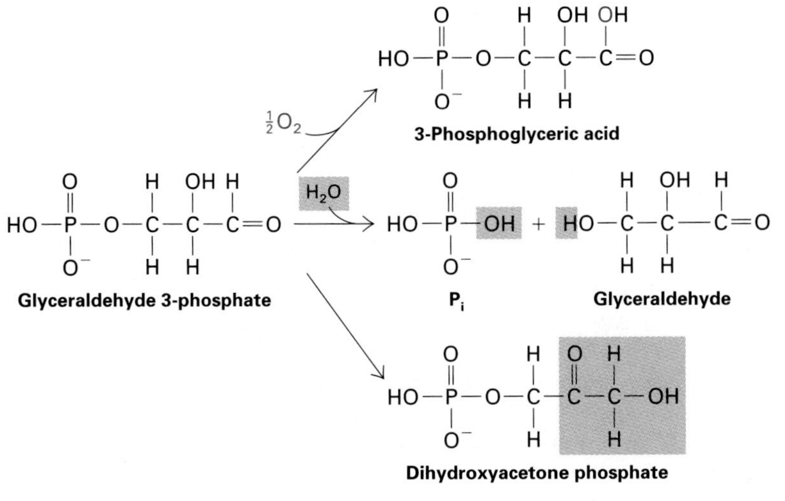

Chemical Reactions Proceed through High-Energy Transition States

All chemical reactions proceed through one or more transition-state intermediates whose content of free energy is greater than that of either the reactants or the products. For the simple reaction R (reactants) $\rightleftharpoons$ P (products), we can write

$$ R \overset{K^{\ddagger}}{\rightleftharpoons} S \overset{v}{\rightarrow} P $$

where S is the reaction intermediate with the highest free *energy*; $K^{\ddagger}$ *is the equilibrium constant for the reaction* $R \rightleftharpoons S$, the conversion of the reactant to the high-energy intermediate S; and v is the rate constant for conversion of S into the product P. The energetic relation between the initial reactants and the products of a reaction can usually be depicted as shown in Figure 2-27. The free energy of activation $\Delta G^{\ddagger}$ is equal to the difference in free energy between the transition-state intermediate S and the reactant R. Because $\Delta G^{\ddagger}$ generally has a very large positive value, only a small fraction of the reactant molecules will at any one time have acquired this free energy, and the overall rate of the reaction will be limited by the rate of formation of S.

The rate V of the overall reaction R $\rightarrow$ S will be proportional to the rate constant v and to the number of molecules in the transition state S, that is, the concentration of the transition-state intermediate, [S]:

$$ V = v\,[S] $$

But since S is in equilibrium with R, the reactant, we can write

$$ K^{\ddagger} = \frac{[S]}{[R]} \qquad \text{or} \qquad [S] = [R]\,K^{\ddagger} $$

As with all equilibrium constants, $K^{\ddagger}$ and $\Delta G^{\ddagger}$ are related as shown in Equation 2-10, so that

$$ V = v\,[R] \times 10^{-(\Delta G^{\ddagger}/2.3RT)} $$

◀ **FIGURE 2-26 Glyceraldehyde 3-phosphate, like most cellular molecules, can undergo any of several reactions with negative $\Delta G^{\circ\prime}$ values: oxidation by O_2 to 3-phosphoglyceric acid, hydrolysis to glyceraldehyde and phosphate (P_i), and rearrangement to dihydroxyacetone phosphate.** In the absence of enzymes or other catalysts, however, these reactions cannot occur in aqueous solution because insufficient energy is available to form the transition states. Different enzymes catalyze each reaction; the presence of a specific enzyme is required for a particular reaction to proceed.

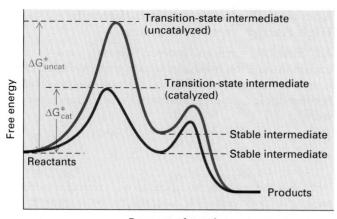

▲ FIGURE 2-27 Hypothetical energy changes in the conversion of a reactant—say, glyceraldehyde 3-phosphate (G3P)— to a product—say, dihydroxyacetone phosphate (DHAP)—in the presence and absence of a catalyst. The trough in each curve represents a stable intermediate in the reaction, as depicted in Figure 2-28. The total activation energy ($\Delta G^{\ddagger}$) is the difference between the free energy of the reactants and that represented by the highest crest along the pathway. A catalyst accelerates the rate of a reaction by reducing the free energy of the transition state, so that the activation energy of the catalyzed reaction, $\Delta G^{\ddagger}_{cat}$, is less than that of the uncatalyzed reaction, $\Delta G^{\ddagger}_{uncat}$. Catalysts do not alter the free energy of reactants or products or affect their equilibrium concentrations.

From this equation, we can see that lowering the activation energy—that is, decreasing the free energy of the transition state $\Delta G^{\ddagger}$—leads to an acceleration of the overall reaction rate V by increasing the concentration of S. A reduction in $\Delta G^{\ddagger}$ of 1.36 kcal/mol leads to a tenfold increase in the concentration of S, and thus a tenfold increase in the rate of the reaction (see Table 2-5). Similarly, a reduction of 2.72 kcal/mol in $\Delta G^{\ddagger}$ leads to a 100-fold increase in the reaction rate. Thus relatively small changes in $\Delta G^{\ddagger}$ can lead to large changes in the overall rate of the reaction.

In some reactions, certain covalent bonds are moved to a strained position in the transition state, and an input of energy—the energy of activation $\Delta G^{\ddagger}$—is essential for this to happen. In other reactions, formation of the transition state involves excitation of electrons, which likewise requires an input of energy; only then can the electrons pair up, forming a covalent bond in the product. In still other reactions, molecules need only enough energy to overcome the mutual repulsion of their electron clouds to get close enough to react.

To illustrate the concept of a transition state we again consider the conversion of glyceraldehyde 3-phosphate (G3P) to dihydroxyacetone phosphate (DHAP), which involves at least one reaction intermediate (Figure 2-28, *top*). As the intermediate forms, the following events take place simultaneously: a proton is removed from carbon 2 of G3P,

another proton is donated to the aldehyde oxygen on carbon 1, and pairs of electrons move from one bond to another. The activation energy required by each of these events contributes to the overall activation energy needed to form this reaction intermediate, which then rearranges through a second transition-state intermediate to generate the final reaction product (Figure 2-28, *bottom*). Each stage in such a multistep reaction has its own activation energy (see Figure 2-27), but for the overall reaction to proceed, the highest activation energy must be achieved.

At room or body temperature, the average kinetic energy, the energy of motion, of a typical molecule is about 1.5 kcal/mol. Although many molecules will have more kinetic energy than this average, the kinetic energy of colliding molecules is generally insufficient to provide the necessary activation energy to convert a reactant to the transition state and thus to allow a particular reaction to proceed. Without some mechanism for accelerating reactions, cells would be able to carry out few, if any, of the biochemical reactions needed to sustain life.

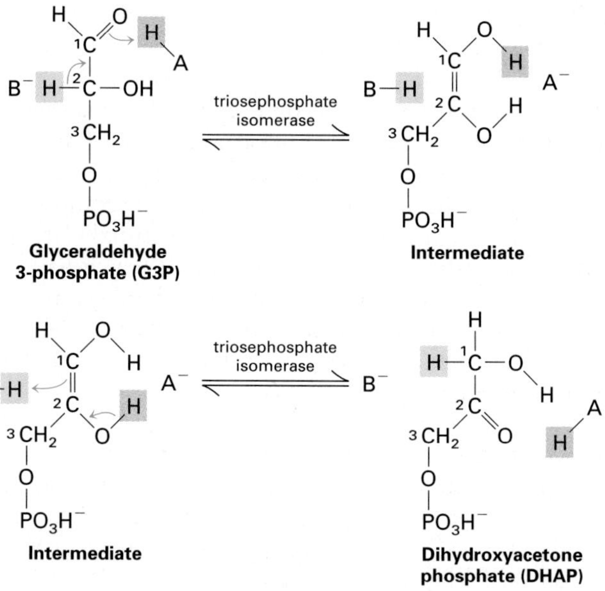

▲ FIGURE 2-28 The conversion of glyceraldehyde 3-phosphate (G3P) to dihydroxyacetone phosphate (DHAP) involves an intermediate. Two groups, a base B⁻ and an acid HA, are parts of triosephosphate isomerase, the enzyme that catalyzes this reaction. To form the intermediate, B⁻ abstracts a proton (blue) from carbon 2 of G3P; HA adds a proton (red) to the aldehyde oxygen on carbon 1. To convert the intermediate to DHAP, BH donates its proton to carbon 1 (regenerating the original B⁻) and A⁻ abstracts a proton from the —OH on carbon 2 (regenerating HA). The curved arrows denote the movements of pairs of electrons that accompany the making and breaking of these bonds. [See D. Straus et al., 1985, *Proc. Nat'l. Acad. Sci.* **82**:2272.]

Enzymes Accelerate Biochemical Reactions by Reducing Transition-State Free Energy

Two significant rate-regulating factors for biological systems are the concentrations of the reactants and the pH. A reaction involving two or more different molecules proceeds faster at high concentrations because the molecules are more likely to encounter one another. The pH determines the dissociation state of the various acidic and basic groups on biological molecules. Since only one of the possible ionic forms (e.g., the zwitterion form of an amino acid) can undergo a particular reaction, the pH can determine the amount of the reactive species present. However, the most important determinants of biochemical reaction rates are **enzymes**, the proteins that act as catalysts. The structure and mechanism of action of enzymes are discussed in detail in the next chapter. Here we examine their general effect on reactions.

As discussed earlier, enzymes, like all catalysts, cause reactions to reach equilibrium faster. A catalyst accelerates the rates of forward and reverse reactions by the same factor; it does not alter the change in free energy or the equilibrium constant. Enzymes and all other catalysts act by reducing the activation energy required to make a reaction proceed (see Figure 2-27). To achieve this, an enzyme binds either a single substrate or a set of similar substrates. Each enzyme catalyzes a single chemical reaction on the bound substrate. Triosephosphate isomerase, the enzyme that catalyzes the conversion of G3P to DHAP, binds G3P as its substrate. Once G3P is bound, the requisite movements of its hydrogen atoms, protons, and electrons are all facilitated by specific chemical groups on the parts of the enzyme adjacent to the bound substrate (see Figure 2-28).

Some enzymes bind a substrate in a way that strains certain of its bonds and makes it easy for these bonds in the substrate to undergo a reaction. That is, the enzyme stabilizes the transition state of the reaction by tightly binding a specific form of the substrate in which certain bonds are strained. Other enzymes form a covalent bond with the substrate that enables a different part of the substrate to undergo a reaction; after this happens, the bond between the enzyme and substrate is broken. Still other enzymes bind multiple substrates in a way that brings them close enough so they can react readily with one another. In each case, the overall effect is to reduce the activation energy needed for formation of the reaction intermediate. Thus the presence or absence of particular enzymes in a cell or in extracellular fluids determines which of many possible chemical reactions will occur.

SUMMARY Activation Energy and Reaction Rate

- All chemical reactions proceed through high-energy transition-state intermediates (see Figure 2-27). The overall reaction rate depends on the activation energy $\Delta G^{\ddagger}$, the difference in free energy between the transition state and the reactant.

- Enzymes, like other catalysts, accelerate specific reactions by decreasing the activation energy $\Delta G^{\ddagger}$. They accomplish this by various mechanisms.

- A catalyst accelerates both the forward and the reverse reactions to the same extent; it does not change ΔG, $\Delta G^{\circ\prime}$, or K_{eq}.

- A particular enzyme generally facilitates only one of the possible transformations that its substrate molecule can undergo.

- The reactions that actually occur in various cells depend on the substrates and enzymes available.

Testing Yourself on the Concepts

1. Under aerobic conditions, glucose is oxidized to CO_2 and H_2O in a series of reactions in which 263 kcal of energy is captured by the conversion of ADP to ATP. Under anaerobic conditions—for example, during intense muscle activity—the energy capture from the metabolism of glucose to lactic acid is only 14.6 kcal. What is the molar yield of ATP in each case? What portion of the total energy inherent in glucose oxidation is captured in the metabolism of glucose to lactic acid? Please see Table 2-4 for values of $\Delta G^{\circ\prime}$, the standard free-energy change.

2. Ammonia (NH_3) is a weak base that under acidic conditions becomes protonated to the ammonium ion by the following reaction:

$$NH_3 + H^+ \longrightarrow NH_4^+$$

Ammonia freely permeates biological membranes, including those of lysosomes. The lysosome is a subcellular organelle with a pH of about 5.0; the pH of the cytosol is about 7.0. What is the effect on lysosomal and cytosolic pH of exposing cells to ammonia?

3. The fatty acyl chains of lipid molecules in biological membranes exist either in an ordered, rigid state or in a relatively disordered, fluid state. Over the temperature range of 27 to 42 °C, the bacteria *E. coli* maintains a fairly constant rigidity to its membranes by varying the ratio of saturated to unsaturated fatty acyl chains. Predict how this ratio changes over this temperature range?

4. The chemical basis of blood group specificity resides in the carbohydrates displayed on the surface of red blood cells. Carbohydrates have the potential for great structural diversity. Indeed, the structural complexity of the oligosaccharides that can be formed from four sugars is greater than that for the oligopeptides from four amino acids. What properties of carbohydrates make this great structural diversity possible?

MCAT/GRE-Style Questions

Key Concept Please read the introductory paragraph to Section 2.2, "Noncovalent Bonds" (p. 22) and study the associated Figure 2-17; then answer the following questions.

1. Noncovalent bonds include all the following *except:*
   a. A carbon-carbon double bond.
   b. An ionic bond.
   c. A hydrogen bond.
   d. A van der Waals interaction.

2. Ligands such as insulin bind to their receptors in a reversible manner. What properties of noncovalent bonds lend themselves to such interactions?
   a. The high energy state of each noncovalent bond.
   b. The ability of noncovalent bonds to form very stable interactions.
   c. The transient dynamic nature of multiple low-energy noncovalent interactions.
   d. The ability of enzymes to catalyze the making and breaking of noncovalent bonds.

3. Biomembranes are stabilized by all the following interactions *except:*
   a. van der Waals interactions between the hydrophobic side chains of phospholipids.
   b. Ionic bonds that stabilize the interaction of the phospholipid polar head groups with water.
   c. Hydrogen bonds that stabilize the interaction of the phospholipid polar head groups with water.
   d. Carbon-carbon bonds between adjacent phospholipids.

4. As shown in Figure 2-17, several types of noncovalent bonds are involved in protein-protein interactions and by extension in interactions between portions of a single protein. Which of the following bonds are apt to be more common in the nonaqueous, interior environment of a protein than in the aqueous, surface environment of a protein?
   a. Ionic bonds.
   b. Hydrophobic bonds.
   c. Hydrogen bonds.
   d. Covalent bonds.

5. Covalent bonds in *contrast* to noncovalent bonds
   a. Have an energy level of about 1 kcal/mol.
   b. Are stable relative to the average kinetic energy of molecules at room temperature.
   c. Are readily broken at room temperature.
   d. Are the main force in maintaining the three-dimensional structure of large molecules such as proteins and nucleic acids.

Key Concept Please read the section titled "The Concentration of Complexes Can Be Estimated from Equilibrium Constants for Binding Reactions" (p. 31) and answer the following questions:

6. Lymphocytes are a cell type circulating in the human blood stream. They are spherical with a diameter of roughly 10 μm. Humans have 22 pairs of somatic chromosomes and 1 pair of sex chromosomes (XX or XY). What is the molar concentration of a DNA sequence mapping to chromosome pair 11?
   a. 3.17×10^{-12} M.
   b. 6.34×10^{-12} M.
   c. 1.27×10^{-11} M.
   d. 1.46×10^{-10} M.

7. What is the molar concentration of a DNA sequence mapping to the Y chromosome?
   a. 3.17×10^{-12} M.
   b. 6.34×10^{-12} M.
   c. 12.7×10^{-12} M.
   d. 146×10^{-12} M.

8. Assuming the same equilibrium constant (dissociation constant) for protein-DNA binding in humans as in bacteria and the same copy number for the protein, what is the ratio of bound to total DNA for a sequence mapping to chromosome pair 11?
   a. 1%.
   b. 5%.
   c. 10%.
   d. 20%.

9. Assuming the same copy number for the protein, how must the equilibrium constant in humans change to give the same 99% ratio of bound to total sequence DNA just calculated?
   a. Decrease.
   b. Stay the same.
   c. Increase.

10. Assuming the same equilibrium constant (dissociation constant) for protein-DNA binding in humans as in bacteria, how must the number of copies of the human protein change to give same 99% ratio of bound to total sequence DNA calculated earlier?
    a. Decrease.
    b. Stay the same.
    c. Increase.

Key Concept Please read the section titled "Enzymes Accelerate Biochemical Reactions by Reducing Transition-State Free Energy" (p. 47) and study the associated Figure 2-28; then answer the following questions:

11. The rearrangement of G3P to DHAP involves the formation of the intermediate illustrated in Figure 2-28. Of all

possible reaction intermediates that G3P can form, why is this the observed intermediate?

a. It is most easily hydrated by water.

b. It is the intermediate with the highest energy of activation.

c. It is the intermediate whose energy of activation most closely corresponds to that released by the hydrolysis of an ATP high-energy phosphoanhydride bond.

d. It is the intermediate favored by the binding relationships between triosephosphate isomerase and G3P.

12. From the answer to the previous question, it can be concluded that all enzymes alter the rate of a chemical reaction through

a. Forming a covalent bond with the substrate.

b. Binding to the substrate.

c. Bringing multiple substrates close to one another.

d. Raising the activation energy for formation of a reaction intermediate.

13. G3P is a natural substrate for triosephosphate isomerase, while glyceraldehyde is not. Why?

a. Glyceraldehyde is too small to fit into the active site of triosephosphate isomerase.

b. Glyceraldehyde is too big to fit into the active site of triosephosphate isomerase.

c. Glyceraldehyde cannot form an ionic interaction with a positively charged group in triosephosphate isomerase.

d. Glyceraldehyde cannot form an ionic interaction with a negatively charged group in triosephosphate isomerase.

14. What is the effect of increasing the concentration of G3P on the final ratio of G3P to DHAP in the reaction mix?

a. The ratio is unchanged.

b. The ratio is increased.

c. The ratio is decreased.

d. The ratio is increased as an exponential power of the concentration of G3P.

Key Terms

acid *32*

base *32*

buffers *33*

catalyst *29*

covalent bonds *15*

endothermic *37*

enthalpy *36*

entropy *36*

enzymes *47*

equilibrium constant *29*

exothermic *37*

free-energy change *36*

glycosidic bond *21*

hydrogen bond *22*

hydrophobic bond *25*

ionic bonds *24*

monosaccharides *19*

oxidation *39*

pH *32*

phospholipid bilayers *28*

polar *18*

polysaccharides *21*

reduction *39*

reduction potential *40*

stereoisomers *19*

van der Waals interaction *24*

References

General References

Alberty, R. A., and R. J. Silbey. 1997. *Physical Chemistry,* 2d ed. Wiley, chaps. 1–5.

Atkins, P. W. 1997. *The Elements of Physical Chemistry,* 2d ed. W. H. Freeman and Company, chaps. 2, 3, 5, 7, 9.

Cantor, P. R., and C. R. Schimmel. 1980. *Biophysical Chemistry.* W. H. Freeman and Company, part. 1, chap. 5.

Davenport, H. W. 1974. *ABC of Acid-Base Chemistry,* 6th ed. University of Chicago Press.

Edsall, J. T., and J. Wyman. 1958. *Biophysical Chemistry,* vol. 1. Academic Press.

Eisenberg, D., and D. Crothers. 1979. *Physical Chemistry with Applications to the Life Sciences.* Benjamin-Cummings.

Gennis, R. B. 1989. *Biomembranes: Molecular Structure and Function.* Springer-Verlag, New York, chaps. 1–3.

Guyton, A. C., and J. E. Hall. 1996. *Textbook of Medical Physiology,* 9th ed. Saunders, chap. 30.

Hill, T. J. 1977. *Free Energy Transduction in Biology.* Academic Press.

Klotz, I. M. 1978. *Energy Changes In Biochemical Reactions.* Academic Press.

Lehninger, A. L., D. L. Nelson, and M. M. Cox. 1993. *Principles of Biochemistry,* 2d ed. Worth, chaps. 1, 3, 4, 13.

Murray, R. K., et al. 1996. *Harper's Biochemistry,* 24th ed. Lange, chaps. 3, 12, 15.

Nicholls, D. G., and S. J. Ferguson. 1992. *Bioenergetics 2.* Academic Press.

Oxtoby, D., H. Gillis, and N. Nachtrieb. 1999. *Principles of Modern Chemistry,* Saunders, chaps. 7–13.

Sharon, N. 1980. Carbohydrates. *Sci. Am.* **243**(5):90–116.

Stryer, L. 1995. *Biochemistry,* 4th ed. W. H. Freeman and Company, chaps. 1, 8, 17, 21.

Tanford, C. 1980. The Hydrophobic Effect: Formation of Micelles and Biological Membranes, 2d ed. Wiley.

Tinoco, I., K. Sauer, and J. Wang. 1995. *Physical Chemistry—Principles and Applications in Biological Sciences.* Prentice-Hall.

Van Holde, K., W. Johnson, and P. Ho. 1998. *Principles of Physical Biochemistry.* Prentice-Hall, chaps. 1–3.

Voet, D., and J. Voet. 1995. *Biochemistry.* Wiley, chaps. 2, 3.

Watson, J. D., et al. 1988. *Molecular Biology of the Gene,* 4th ed. Benjamin-Cummings, chaps. 2, 5.

Wood, W. B., et al. 1981. *Biochemistry: A Problems Approach,* 2d ed. Benjamin-Cummings, chaps. 1, 5, 9.

3

Protein Structure and Function

Proteins, the working molecules of a cell, carry out the program of activities encoded by genes. This program requires the coordinated effort of many different types of proteins, which first evolved as rudimentary molecules that facilitated a limited number of chemical reactions. Gradually, many of these primitive proteins evolved into a wide array of enzymes capable of catalyzing an incredible range of intracellular and extracellular chemical reactions, with a speed and specificity that is nearly impossible to attain in a test tube. Other proteins acquired numerous structural, regulatory, and other functions. For a flavor of the various roles of proteins in today's organisms, we can look to the yeast *Saccharomyces cerevisiae*, a simple unicellular eukaryote. The yeast genome is predicted to encode about 6225 proteins (see Table 7-3). On the basis of their sequences, 17 percent are estimated to be involved in metabolism, the synthesis or degradation of cell building blocks; 30 percent, in cellular organization and biogenesis of cell organelles and membranes; and 10 percent, in transporting molecules across membranes.

In this chapter, we will study how the structure of a protein gives rise to its function. The first section examines protein architecture: the structure and chemistry of amino acids, the linkage of amino acids to form a linear chain, and the forces that guide folding of the chain into higher orders of structure. In the next section, we learn about special proteins that aid in the folding of proteins, modifications that occur after the protein chain is synthesized, and mechanisms that degrade proteins. In the third section, we illustrate several key concepts in the functional design of proteins, using antibodies and enzymes as examples. A separate section is devoted to the general characteristics of membrane proteins, which reside in the lipid bilayer surrounding cells and organelles.

A two-dimensional array of α-actinin molecules.

These functionally diverse proteins play critical roles in transfer of molecules and information across the lipid bilayer and in cell-cell interactions; their structures and functions will be discussed in greater detail in later chapters. We finish the chapter by describing the most commonly used techniques in the biologist's tool kit for isolating proteins and characterizing their properties. Our understanding of biology critically depends on how we can ask a question and test it experimentally.

3.1 Hierarchical Structure of Proteins

Proteins are designed to bind every conceivable molecule—from simple ions to large complex molecules like fats, sugars, nucleic acids, and other proteins. They catalyze an extraordinary range of chemical reactions, provide structural rigidity to the cell, control flow of material through membranes, regulate the concentrations of metabolites, act as sensors and switches, cause motion, and control gene function. The three-dimensional structures of proteins have evolved to carry out these functions efficiently and under precise control. The *spatial* organization of proteins, their shape in three dimensions, is a key to understanding how they work.

One of the major areas of biological research today is how proteins, constructed from only 20 different **amino acids,** carry out the incredible array of diverse tasks that they do. Unlike the intricate branched structure of carbohydrates, proteins are single, unbranched chains of amino acid monomers. The unique shape of proteins arises from noncovalent interactions between regions in the linear sequence of amino acids. Only when a protein is in its correct three-dimensional structure, or **conformation,** is it able to function efficiently. A key concept in understanding how proteins work is that *function is derived from three-dimensional structure, and three-dimensional structure is specified by amino acid sequence.*

The Amino Acids Composing Proteins Differ Only in Their Side Chains

Amino acids are the monomeric building blocks of proteins. The α carbon atom (C_α) of amino acids, which is adjacent to the carboxyl group, is bonded to four different chemical groups: an amino (NH_2) group, a carboxyl (COOH) group, a hydrogen (H) atom, and one variable group, called a *side chain* or *R group* (Figure 3-1). All 20 different amino acids have this same general structure, but their side-chain groups vary in size, shape, charge, hydrophobicity, and reactivity.

The amino acids can be considered the alphabet in which linear proteins are "written." Students of biology must be familiar with the special properties of each letter of this al-

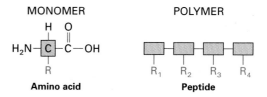

▲ **FIGURE 3-1 Amino acids, the monomeric units that link together to form proteins, have a common structure.** The α carbon atom (green) of each amino acid is bonded to four different chemical groups and thus is asymmetric. The side chain, or R group (red), is unique to each amino acid. The diversity of natural proteins reflects different linear combinations of the 20 naturally occurring amino acids. The short peptide shown here, containing only four amino acids, has 20^4, or 160,000, possible sequences.

phabet, which are determined by the side chain. Amino acids can be classified into a few distinct categories based primarily on their solubility in water, which is influenced by the polarity of their side chains (Figure 3-2). Amino acids with **polar** side groups tend to be on the surface of proteins; by interacting with water, they make proteins soluble in aqueous solutions. In contrast, amino acids with **nonpolar** side groups avoid water and aggregate to form the water-insoluble core of proteins. The polarity of amino acid side chains thus is one of the forces responsible for shaping the final three-dimensional structure of proteins.

Hydrophilic, or water-soluble, amino acids have ionized or polar side chains. At neutral pH, *arginine* and *lysine* are positively charged; *aspartic acid* and *glutamic acid* are negatively charged and exist as aspartate and glutamate. These four amino acids are the prime contributors to the overall charge of a protein. A fifth amino acid, *histidine*, has an imidazole side chain, which has a pK_a of 6.8, the pH of the cytoplasm. As a result, small shifts of cellular pH will change the charge of histidine side chains:

The activities of many proteins are modulated by pH through protonation of histidine side chains. *Asparagine* and *glutamine* are uncharged but have polar amide groups with extensive hydrogen-bonding capacities. Similarly, *serine* and *threonine* are uncharged but have polar hydroxyl groups, which also participate in hydrogen bonds with other polar molecules. Because the charged and polar amino acids are hydrophilic, they are usually found at the surface of a water-soluble protein, where they not only contribute to

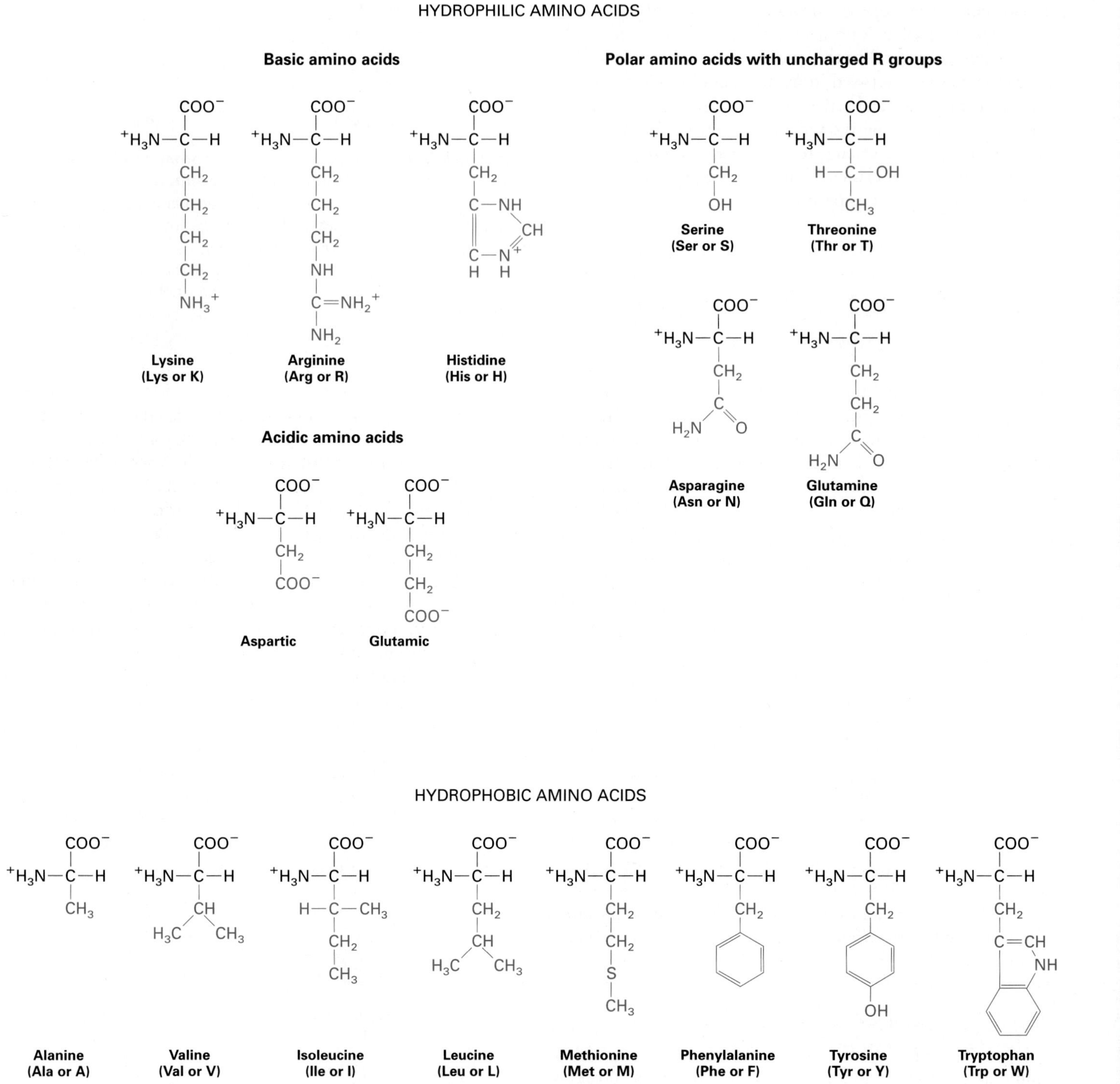

HYDROPHILIC AMINO ACIDS

Basic amino acids

Lysine
(Lys or K)

Arginine
(Arg or R)

Histidine
(His or H)

Acidic amino acids

Aspartic

Glutamic

Polar amino acids with uncharged R groups

Serine
(Ser or S)

Threonine
(Thr or T)

Asparagine
(Asn or N)

Glutamine
(Gln or Q)

HYDROPHOBIC AMINO ACIDS

Alanine
(Ala or A)

Valine
(Val or V)

Isoleucine
(Ile or I)

Leucine
(Leu or L)

Methionine
(Met or M)

Phenylalanine
(Phe or F)

Tyrosine
(Tyr or Y)

Tryptophan
(Trp or W)

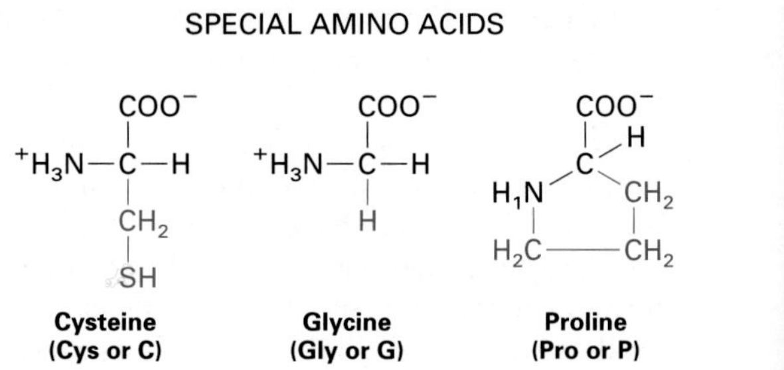

SPECIAL AMINO ACIDS

Cysteine
(Cys or C)

Glycine
(Gly or G)

Proline
(Pro or P)

▲ FIGURE 3-2 The structures of the 20 common amino acids grouped into three categories: hydrophilic, hydrophobic, and special amino acids. The side chain determines the characteristic properties of each amino acid. Shown are the zwitterion forms, which exist at the pH of the cytosol. In parentheses are the three-letter and one-letter abbreviations for each amino acid.

the solubility of the protein in water but also form binding sites for charged molecules.

Hydrophobic amino acids have aliphatic side chains, which are insoluble or only slightly soluble in water. The side chains of *alanine, valine, leucine, isoleucine,* and *methionine* consist entirely of hydrocarbons, except for the sulfur atom in methionine, and all are nonpolar. *Phenylalanine, tyrosine,* and *tryptophan* have large bulky aromatic side groups. As explained in Chapter 2, hydrophobic molecules avoid water by coalescing into an oily or waxy droplet. The same forces cause hydrophobic amino acids to pack in the interior of proteins, away from the aqueous environment. Later in this chapter, we will see in detail how hydrophobic residues line the *surface* of membrane proteins that reside in the hydrophobic environment of the lipid bilayer.

Lastly, *cysteine, glycine,* and *proline* exhibit special roles in proteins because of the unique properties of their side chains. The side chain of cysteine contains a reactive **sulfhydryl group** (—SH), which can oxidize to form a **disulfide bond** (—S—S—) to a second cysteine:

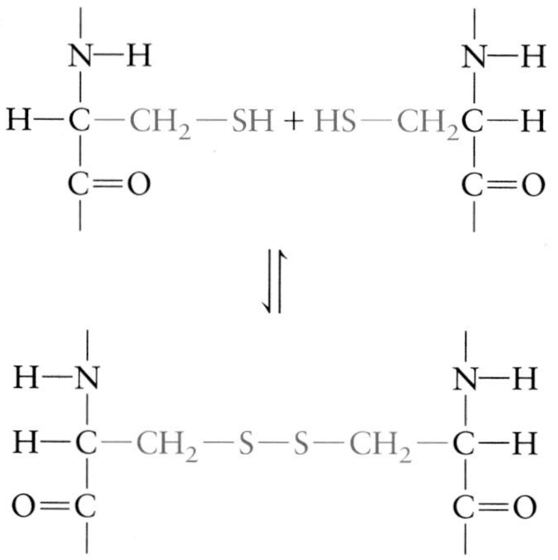

$$\begin{array}{ccc} & \overset{|}{N}-H & & \overset{|}{N}-H \\ H-\overset{|}{\underset{|}{C}}-CH_2-SH & + & HS-CH_2\overset{|}{\underset{|}{C}}-H \\ C=O & & C=O \\ | & & | \end{array}$$

$$\Updownarrow$$

$$\begin{array}{ccc} H-\overset{|}{N} & & \overset{|}{N}-H \\ H-\overset{|}{\underset{|}{C}}-CH_2-S-S-CH_2-\overset{|}{\underset{|}{C}}-H \\ O=C & & C=O \\ | & & | \end{array}$$

Regions within a protein chain or in separate chains sometimes are cross-linked covalently through disulfide bonds. Although disulfide bonds are rare in intracellular proteins, they are commonly found in extracellular proteins, where they help maintain the native, folded structure. The smallest amino acid, glycine, has a single hydrogen atom as its R group. Its small size allows it to fit into tight spaces. Unlike any of the other common amino acids, proline has a cyclic ring that is produced by formation of a covalent bond between its R group and the amino group on C_α. Proline is very rigid, and its presence creates a fixed kink in a protein chain. Proline and glycine are sometimes found at points on a protein's surface where the chain loops back into the protein.

The 6225 known and predicted proteins encoded by the yeast genome have an average molecular weight (MW) of 52,728 and contain, on average, 466 amino acid residues. Assuming that these average values represent a "typical" eukaryotic protein, then the average molecular weight of amino acids is 113, taking their average relative abundance in proteins into account. This is a useful number to re-

member, as we can use it to estimate the number of residues from the molecular weight of a protein or vice versa. Some amino acids are more abundant in proteins than other amino acids. Cysteine, tryptophan, and methionine are rare amino acids; together they constitute approximately 5 percent of the amino acids in a protein. Four amino acids—leucine, serine, lysine, and glutamic acid—are the most abundant amino acids, totaling 32 percent of all the amino acid residues in a typical protein. However, the amino acid composition of proteins can vary widely from these values. For example, as discussed in later sections, proteins that reside in the lipid bilayer are enriched in hydrophobic amino acids.

Peptide Bonds Connect Amino Acids into Linear Chains

Nature has evolved a single chemical linkage, the **peptide bond**, to connect amino acids into a linear, unbranched chain. The peptide bond is formed by a condensation reaction between the amino group of one amino acid and the carboxyl group of another (Figure 3-3a). The repeated amide N, C_α, and carbonyl C atoms of each amino acid residue form the backbone of a protein molecule from which the various side-chain groups project. As a consequence of the peptide linkage, the backbone has polarity, since all the amino groups lie to the same side of the C_α atoms. This leaves at opposite ends of the chain a free (unlinked) amino group (the N-terminus) and a free carboxyl group (the

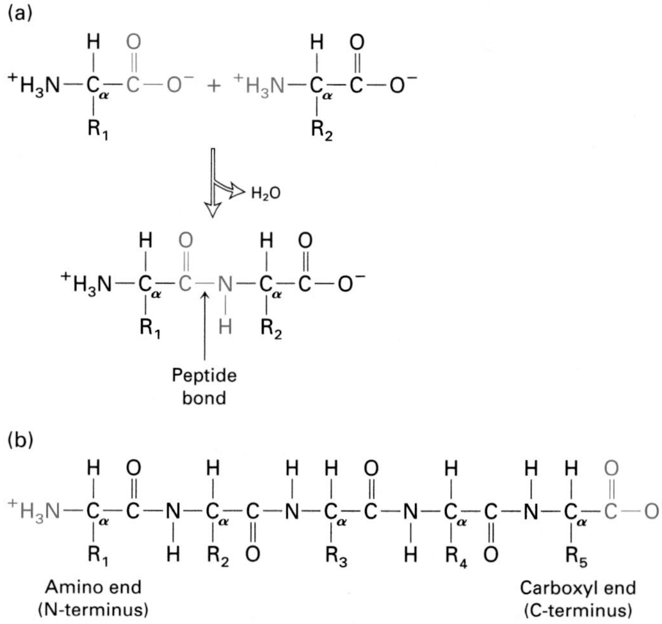

(a)

(b)

▲ **FIGURE 3-3 The peptide bond.** (a) A condensation reaction between two amino acids forms the peptide bond, which links all the adjacent residues in a protein chain. (b) Side-chain groups (R) extend from the backbone of a protein chain, in which the amino N, α carbon, carbonyl carbon sequence is repeated throughout.

C-terminus). A protein chain is conventionally depicted with its N-terminal amino acid on the left and its C-terminal amino acid on the right (Figure 3-3b).

Many terms are used to denote the chains formed by polymerization of amino acids. A short chain of amino acids linked by peptide bonds and having a defined sequence is a **peptide**; longer peptides are referred to as **polypeptides**. Peptides generally contain fewer than 20–30 amino acid residues, whereas polypeptides contain as many as 4000 residues. We reserve the term **protein** for a polypeptide (or a complex of polypeptides) that has a three-dimensional structure. It is implied that proteins and peptides represent natural products of a cell.

The size of a protein or a polypeptide is reported as its mass in **daltons** (a dalton is 1 atomic mass unit) or as its molecular weight (a dimensionless number). For example, a 10,000-MW protein has a mass of 10,000 daltons (Da), or 10 kilodaltons (kDa). In the last section of this chapter, we will discuss different methods for measuring the sizes and other physical characteristics of proteins.

Four Levels of Structure Determine the Shape of Proteins

The structure of proteins commonly is described in terms of four hierarchical levels of organization. These levels are illustrated in Figure 3-4, which depicts the structure of hemagglutinin, a surface protein on the influenza virus. This protein binds to the surface of animal cells, including human cells, and is responsible for the infectivity of the flu virus.

The **primary structure** of a protein is the linear arrangement, or *sequence,* of amino acid residues that constitute the polypeptide chain.

Secondary structure refers to the localized organization of parts of a polypeptide chain, which can assume several different spatial arrangements. A single polypeptide may exhibit all types of secondary structure. Without any stabilizing interactions, a polypeptide assumes a *random-coil* structure. However, when stabilizing hydrogen bonds form between certain residues, the backbone folds periodically into one of two geometric arrangements: an **α helix**, which is a spiral, rodlike structure, or a **β sheet**, a planar structure composed of alignments of two or more *β strands,* which are relatively short, fully extended segments of the backbone. Finally, U-shaped four-residue segments stabilized by hydrogen bonds between their arms are called *turns.* They are located at the surfaces of proteins and redirect the polypeptide chain toward the interior. (These structures will be discussed in greater detail later.)

Tertiary structure, the next-higher level of structure, refers to the overall conformation of a polypeptide chain, that is, the three-dimensional arrangement of all the amino acids residues. In contrast to secondary structure, which is stabilized by hydrogen bonds, tertiary structure is stabilized by hydrophobic interactions between the nonpolar side chains and, in some proteins, by disulfide bonds. These stabilizing forces hold the α helices, β strands, turns, and random coils in a compact internal scaffold. Thus, a protein's size and shape is dependent not only on its sequence but also on the number, size, and arrangement of its secondary structures. For proteins that consist of a single polypeptide chain, **monomeric** proteins, tertiary structure is the highest level of organization.

Multimeric proteins contain two or more polypeptide chains, or *subunits,* held together by noncovalent bonds. **Quaternary structure** describes the number (stoichiometry) and relative positions of the subunits in a multimeric protein. Hemagglutinin is a trimer of three identical subunits; other multimeric proteins can be composed of any number of identical or different subunits.

In a fashion similar to the hierarchy of structures that make up a protein, proteins themselves are part of a hierarchy of cellular structures. Proteins can associate into larger structures termed *macromolecular assemblies.* Examples of such macromolecular assemblies include the protein coat of a virus, a bundle of actin filaments, the nuclear pore complex, and other large submicroscopic objects. Macromolecular assemblies in turn combine with other cell biopolymers like lipids, carbohydrates, and nucleic acids to form complex cell organelles.

Graphic Representations of Proteins Highlight Different Features

Different ways of depicting proteins convey different types of information. The simplest way to represent three-dimensional structure is to trace the course of the backbone atoms with a solid line (Figure 3-5a); the most complex model shows the location of every atom (Figure 3-5b; see also Figure 2-1a). The former shows the overall organization of the polypeptide chain without consideration of the amino acid side chains; the latter details the interactions among atoms that form the backbone and that stabilize the protein's conformation. Even though both views are useful, the elements of secondary structure are not easily discerned in them.

Another type of representation uses common shorthand symbols for depicting secondary structure, cylinders for α helices, arrows for β strands, and a flexible stringlike form for parts of the backbone without any regular structure (Figure 3-5c). This type of representation emphasizes the organization of the secondary structure of a protein, and various combinations of secondary structures are easily seen.

However, none of these three ways of representing protein structure conveys much information about the protein surface, which is of interest because this is where other molecules bind to a protein. Computer analysis in which a water molecule is rolled around the surface of a protein can identify the atoms that are in contact with the watery environment. On this water-accessible surface, regions having a common chemical (hydrophobicity or hydrophilicity) and electrical (basic or acidic) character can be mapped. Such models show the texture of the protein surface and the

(a)

68

DALLGDPHCDVFQNETWDLFVERSKAFSNCYPYDVPDYASLRSLVASSGTLEFITEGFTWTGV

195

TQNGGSNACKRGPGSGFFSRLNWLTKSGSTYPVLNVTMPNNDNFDKLYIWGIHHPSTNQEQTSL

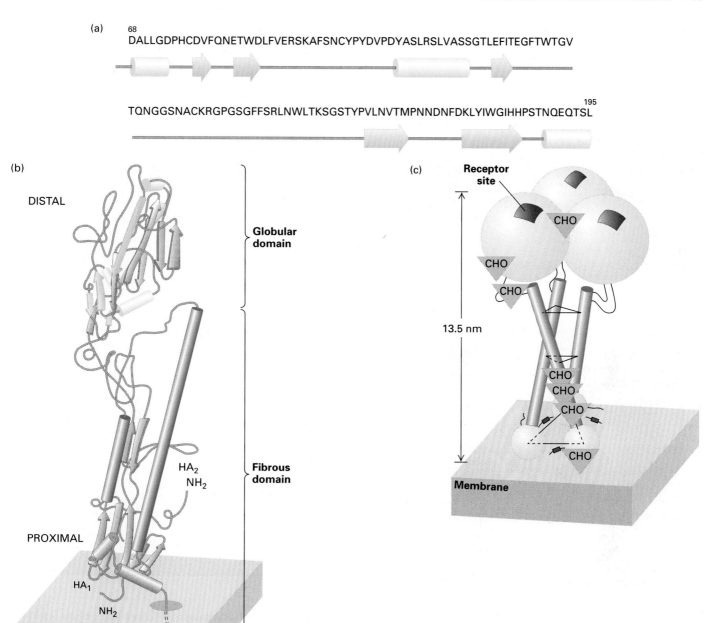

▲ **FIGURE 3-4 Four levels of structure in hemagglutinin, which is a long multimeric molecule whose three identical subunits are each composed of two chains, HA₁ and HA₂.** (a) Primary structure is illustrated by the amino acid sequence of residues 68–195 of HA₁. This region is used by influenza virus to bind to animal cells. The one-letter amino acid code is used. Secondary structure is represented diagrammatically beneath the sequence, showing regions of the polypeptide chain that are folded into α helices (light blue cylinders), β strands (light green arrows), and random coils (white strands). (b) Tertiary structure constitutes the folding of the helices and strands in each HA subunit into a compact structure that is 13.5 nm long and divided into two domains. The membrane-distal domain is folded into a globular conformation. The blue and green segments in this domain correspond to the sequence shown in part (a). The proximal domain, which lies adjacent to the viral membrane, has a stemlike conformation due to alignment of two long helices of HA₂ (dark blue) with β strands in HA₁. Short turns and longer loops, which usually lie at the surface of the molecule, connect the helices and strands in a given chain. (c) The quaternary structure comprises the three subunits of HA; the structure is stabilized by lateral interactions among the long helices (dark blue) in the subunit stems, forming a triple-stranded coiled-coil stalk. Each of the distal globular domains in trimeric hemagglutinin has a site (red) for binding sialic acid molecules on the surface of target cells. Like many membrane proteins, HA has several covalently bound carbohydrate (CHO) chains.

(a)

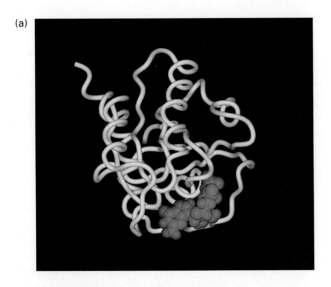

(b)

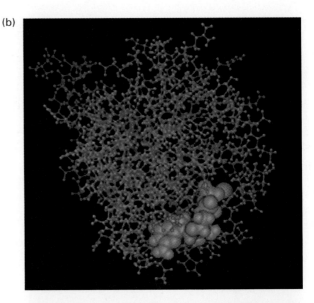

(c)

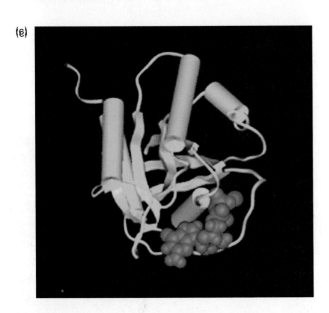

(d)

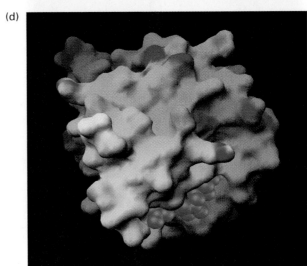

▲ FIGURE 3-5 Various graphic representations of the structure of Ras, a guanine nucleotide–binding protein. Guanosine diphosphate, the substrate that is bound, is shown as a blue space-filling figure in parts (a)–(d). (a) The C_α trace of Ras, which highlights the course of the backbone. Evident from this view is how the polypeptide is packed into the smallest possible volume. (b) Ball-and-stick model of Ras showing the location of all atoms. (c) A schematic diagram of Ras showing how β strands (arrows) and α helices (cylinders) are organized in the protein. Note the turns and loops connecting pairs of helices and strands. (d) The water-accessible surface of Ras. Painted on the surface are regions of positive charge (blue) and negative charge (red). Here we see that the surface of a protein is not smooth but has lumps, bumps, and crevices. The molecular basis for specific binding interactions lies in the uneven distribution of charge over the surface of the protein. [Adapted from L. Tong et al., 1991, *J. Mol. Biol.* **217**:503; courtesy of S. Choe.]

distribution of charge, both of which are important parameters of binding sites (Figure 3-5d). This view represents a protein as seen by another molecule.

Secondary Structures Are Crucial Elements of Protein Architecture

In an average protein, 60 percent of the polypeptide chain exists as two regular secondary structures, α helices and β sheets; the remainder of the molecule is in random coils and turns. Thus, α helices and β sheets are the major internal supportive elements in proteins. In this section, we explore the forces that favor formation of secondary structures. In later sections, we examine how these structures can pack into larger arrays.

The α Helix Polypeptide segments can assume a regular spiral, or helical, conformation, called the α *helix*. In this secondary structure, the carbonyl oxygen of each peptide bond is hydrogen-bonded to the amide hydrogen of the amino acid four residues toward the C-terminus. This uniform arrangement of bonds confers a polarity on a helix because all the hydrogen-bond donors have the same orientation. The peptide backbone twists into a helix having 3.6 amino acids per turn (Figure 3-6). The stable arrangement of amino acids in the α helix holds the backbone as a rodlike cylinder from which the side chains point outward. The hydrophobic or hydrophilic quality of the helix is determined entirely by the side chains, because the polar groups of the peptide backbone are already involved in hydrogen bonding in the helix and thus are unable to affect its hydrophobicity or hydrophilicity.

In many α helices hydrophilic side chains extend from one side of the helix and hydrophobic side chains from the opposite side, making the overall structure **amphipathic.** In such helices the hydrophobic residues, although apparently randomly arranged, occur in a regular pattern (Figure 3-7). One way of visualizing this arrangement is to look down the center of an α helix and then project the amino acid residues onto the plane of the paper. The residues will appear as a wheel, and in the case of an amphipathic helix, the hydrophobic residues all lie on one side of the wheel and the hydrophilic ones on the other side.

Amphipathic α helices are important structural elements in fibrous proteins found in a watery environment. In a coiled-coil region of a protein, the hydrophobic surface of the α helix faces inward to form the hydrophobic core, and the hydrophilic surfaces face outward toward the surrounding fluid. This same orientation matrix of surfaces is also found in most globular proteins. A crucial difference is that the hydrophobic interaction could be with a β strand, random coil, or another α helix. As we discuss later, amphipathic β strands line the walls of an ion channel in the cell membrane.

The β Sheet Another regular secondary structure, the β sheet, consists of laterally packed β strands. Each β strand is a short (5–8-residue), nearly fully extended polypeptide chain. Hydrogen bonding between backbone atoms in adjacent β

3.6 residues/turn

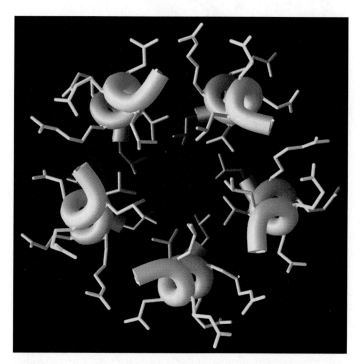

▲ **FIGURE 3-6 Model of the α helix.** The polypeptide backbone is folded into a spiral that is held in place by hydrogen bonds (black dots) between backbone oxygen atoms and hydrogen atoms. Note that all the hydrogen bonds have the same polarity. The outer surface of the helix is covered by the side-chain R groups.

▲ **FIGURE 3-7 Regions of an α helix may be amphipathic.** The five chains of cartilage oligomeric matrix protein associate into a coiled-coil fibrous domain through amphipathic α helices. Seen in cross section through a part of the domain, the hydrophobic residues (gray) face the interior, and the hydrophilic residues (yellow) line the surface. This arrangement of hydrophobic and hydrophilic residues is typical of proteins in an aqueous environment. [Courtesy of V. Malashkevich.]

(a)

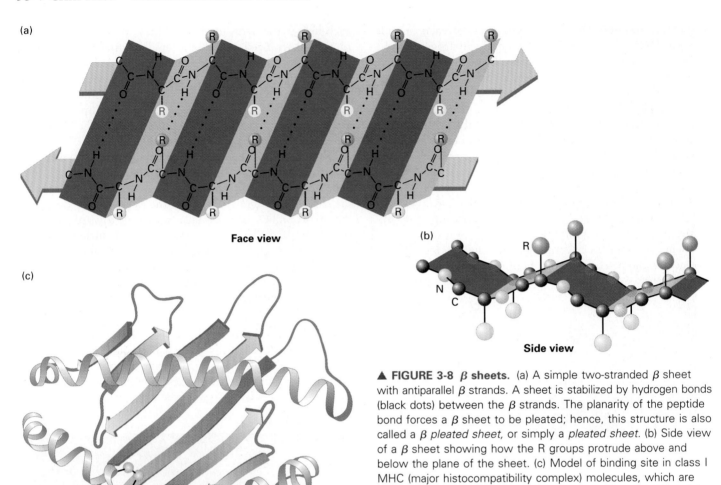

Face view

(b)

Side view

(c)

N

▲ **FIGURE 3-8 β sheets.** (a) A simple two-stranded β sheet with antiparallel β strands. A sheet is stabilized by hydrogen bonds (black dots) between the β strands. The planarity of the peptide bond forces a β sheet to be pleated; hence, this structure is also called a β *pleated sheet,* or simply a *pleated sheet.* (b) Side view of a β sheet showing how the R groups protrude above and below the plane of the sheet. (c) Model of binding site in class I MHC (major histocompatibility complex) molecules, which are involved in graft rejection. A sheet comprising eight antiparallel β strands (green) forms the bottom of the binding cleft, which is lined by a pair of α helices (blue). A disulfide bond is shown as two connected yellow spheres. The MHC binding cleft is large enough to bind a peptide 8–10 residues long. [Part (b) adapted from C. Branden and J. Tooze, 1991, *Introduction to Protein Structure,* Garland.]

strands, within either the same or different polypeptide chains, forms a β sheet (Figure 3-8a). Like α helices, β strands have a polarity defined by the orientation of the peptide bond. Therefore, in a pleated sheet, adjacent β strands can be oriented antiparallel or parallel with respect to each other. In both arrangements of the backbone, the side chains project from both faces of the sheet (Figure 3-8b).

In some proteins, β sheets form the floor of a binding pocket (Figure 3-8c). In many structural proteins, multiple layers of pleated sheets provide toughness. Silk fibers, for example, consist almost entirely of stacks of antiparallel β sheets. The fibers are flexible because the stacks of β sheets can slip over one another. However, they are also resistant to breakage because the peptide backbone is aligned parallel with the fiber axis.

Turns Composed of three or four residues, turns are compact, U-shaped secondary structures stabilized by a hydrogen bond between their end residues. They are located on the surface of a protein, forming a sharp bend that redirects the polypeptide backbone back toward the interior. Glycine and proline are commonly present in turns. The lack of a large side chain in the case of glycine and the presence of a built-in bend in the case of proline allow the polypeptide backbone to fold into a tight U-shaped structure. Without turns, a protein would be large, extended, and loosely packed. A polypeptide backbone also may contain long bends, or *loops.* In contrast to turns, which exhibit a few defined structures, loops can be formed in many different ways.

Motifs Are Regular Combinations of Secondary Structures

Many proteins contain one or more **motifs** built from particular combinations of secondary structures. A motif is defined by a specific combination of secondary structures that has a particular topology and is organized into a charac-

(a)

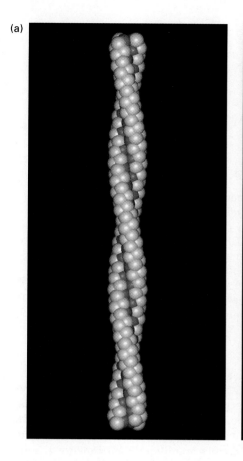 

◀ **FIGURE 3-9 Secondary-structure motifs.** (a) The coiled-coil motif *(left)* is characterized by two or more helices wound around one another. In some DNA-binding proteins, like c-Jun, a two-stranded coiled coil is responsible for dimerization *(right)*. Each helix in a coiled coil has a repeated heptad sequence. **LASTANMLREQVAQL** 1 4 1 4 1 with a leucine or other hydrophobic residue (red) at positions 1 and 4, forming a hydrophobic stripe along the helix surface. The helices pair by binding along their hydrophobic stripes, as seen in both models displayed here, in which the hydrophobic side chains are shown in red. (b) The helix-loop-helix motif occurs in many calcium-binding proteins. Oxygen-containing R groups of residues in the loop form a ring around a Ca^{2+} ion. The 14-aa loop sequence *(right)* is rich in invariant hydrophilic residues. (c) The zinc-finger motif is present in many proteins that bind nucleic acids. A Zn^{2+} ion is held between a pair of β strands (green) and a single α helix (blue) by a pair of cysteine and histidine residues. In the 25-aa sequence of this motif the invariant cysteines usually occur at positions 3 and 6, and the invariant histidines at positions 20 and 24. [Part (a) courtesy of V. Malashkevich and S. Choe.]

(b) Helix–loop–helix motif

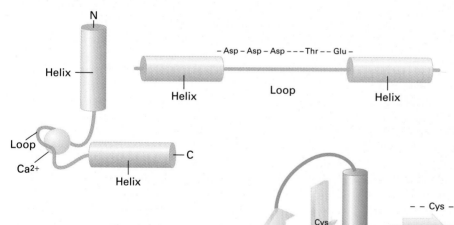

(c) Zinc-finger motif

teristic three-dimensional structure. Three common motifs are depicted in Figure 3-9.

The **coiled-coil** motif comprises two, three, or four amphipathic α helices wrapped around one another. In this motif, hydrophobic side chains project like "knobs" from one helix and interdigitate into the gaps, or "holes," between the hydrophobic side chains of the other helix along the contact surface. The subunits in some multimeric proteins and in rodlike fibers are held together by coiled-coil interactions. The Ca^{2+}-binding **helix-loop-helix** motif is marked by the presence of certain hydrophilic residues at invariant positions in the loop. Oxygen atoms in the invariant residues bind a calcium ion through hydrogen bonds. In another common motif, the **zinc finger,** three secondary structures—an α helix and two β strands with an antiparallel orientation—form a fingerlike bundle held together by a zinc ion. This motif is most commonly found in proteins that bind RNA or DNA.

Additional motifs will be examined in discussions of other proteins. The presence of the same motif in different proteins with similar functions clearly indicates that during evolution these useful combinations of secondary structures have been conserved.

Structural and Functional Domains Are Modules of Tertiary Structure

The tertiary structure of large proteins is often subdivided into distinct globular or fibrous regions called **domains.** Structurally, a domain is a compactly folded region of polypeptide. For large proteins, domains can be recognized in structures determined by **x-ray crystallography** or in images captured by electron microscopy. These discrete regions are well distinguished or physically separated from other parts of the protein, but connected by the polypeptide chain. Hemagglutinin, for example, contains a globular domain and a fibrous domain (see Figure 3-4b).

A structural domain consists of 100–200 residues in various combinations of α helices, β sheets, turns, and random coils. Often a domain is characterized by some interesting structural feature, for example, an unusual abundance of a particular amino acid (a proline-rich domain, an acidic domain, a glycine-rich domain), sequences common to (conserved in) many proteins (SH3, or Src homology region 3), or a particular secondary-structure motif (zinc-finger motif in kringle domain).

Domains sometimes are defined in functional terms based on observations that the activity of a protein is localized to a small region along its length. For instance, a particular region or regions of a protein may be responsible for its catalytic activity (e.g., a kinase domain) or binding ability (e.g., a DNA-binding domain, membrane-binding domain). Functional domains often are identified experimentally by whittling down a protein to its smallest active fragment with the aid of proteases, enzymes that cleave the polypeptide backbone. Alternatively, the DNA encoding a protein can be subjected to mutagenesis, so that segments of the protein's backbone are removed or changed (Chapter 7). The activity of the truncated or altered protein product synthesized from the mutated gene is then monitored.

The functional definition of a domain is less rigorous than a structural definition. However, if the three-dimensional structure of a protein has not been determined, identification of functional domains can provide useful information about the protein. Because the activity of a protein usually depends on a proper three-dimensional structure, a functional domain consists of at least one and often several structural domains.

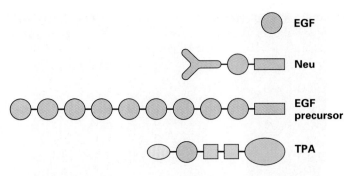

▲ **FIGURE 3-10 Schematic diagrams of various proteins, illustrating their modular nature.** Epidermal growth factor (EGF) is generated by proteolytic cleavage of a precursor protein containing multiple EGF domains (orange). The EGF domain also occurs in Neu protein and in tissue plasminogen activator (TPA). Other domains, or modules, in these proteins include a chymotryptic domain (purple), an immunoglobulin domain (green), a fibronectin domain (yellow), a membrane-spanning domain (pink), and a kringle domain (blue). [Adapted from I. D. Campbell and P. Bork, 1993, *Curr. Opin. Struc. Biol.* **3:**385.]

The organization of tertiary structure into domains further illustrates the principle that complex molecules are built from simpler components. Like secondary-structure motifs, tertiary-structure domains are incorporated as modules into different proteins, thereby modifying their functional activities. The modular approach to protein architecture is particularly easy to recognize in large proteins, which tend to be a mosaic of different domains and thus can perform different functions simultaneously.

The epidermal growth factor (EGF) domain is one example of a module that is present in several proteins (Figure 3-10). EGF is a small soluble peptide hormone that binds to cells in the skin and connective tissue, causing them to divide. It is generated by proteolytic cleavage between repeated EGF domains in the EGF precursor protein, which is anchored in the cell membrane by a membrane-spanning domain. Six conserved cysteine residues form three pairs of disulfide bonds that hold EGF in its native conformation. The EGF domain also occurs in other proteins, including tissue plasminogen activator (TPA), a protease that is used to dissolve blood clots in heart attack victims; Neu protein, which is involved in embryonic differentiation; and Notch protein, a cell-adhesion molecule that glues cells together. Besides the EGF domain, these proteins contain additional domains found in other proteins. For example, TPA possesses a chymotryptic domain, a common feature in proteins that catalyze proteolysis.

Sequence Homology Suggests Functional and Evolutionary Relationships between Proteins

Early evidence supporting the key principle that the amino acid sequence of a protein determines its three-dimensional structure was obtained in the 1960s by Max Perutz. On comparing the structures of myoglobin and hemoglobin determined from x-ray crystallographic analysis, he immediately noted that the subunits of hemoglobin, a tetramer of two α and two β subunits, resembled myoglobin, a monomer (Figure 3-11). Although the sequences of the two proteins were unknown at the time, Perutz proposed that the similar arrangement of α helices in the two proteins is a consequence of their having similar amino acid sequences. Later sequencing of myoglobin and hemoglobin revealed that many identical or chemically similar residues occur in identical positions throughout the sequences of both proteins. The two proteins also exhibit similar functions: myoglobin is the oxygen-carrier protein in muscle, and hemoglobin the oxygen-carrier protein in blood. Most of the conserved residues hold the heme group in place or are responsible for maintaining the hydrophobic interior of the protein.

As data concerning protein sequences and three-dimensional structures accumulated, the concept that similar sequences fold into similar secondary and tertiary structures was confirmed. The propensity of each amino acid to occur in the various types of secondary structures has been calculated from the amino acid sequence of secondary structures extracted from databases of the three-dimensional structures of proteins. This tabulation of the folding information inherent in the sequence is now being used in attempts to predict the three-dimensional structure of various proteins from their amino acid sequences.

In the classical taxonomy of the eighteenth and nineteenth centuries, organisms were classified according to their morphological similarities and differences. In this century, the molecular revolution in biology has given birth to "molecular" taxonomy: the classification of proteins based on similarities and differences in their amino acid sequences. This new taxonomy provides much information about protein function and evolutionary relationships. If the similarity between proteins from different organisms is significant over their entire sequence, then the proteins are homologs of one another, and they probably carry out similar functions. Sequence similarity also suggests an evolutionary relationship between proteins; that is, they evolved from a common ancestor. We can therefore describe homologous proteins as belonging to the same "family" and can trace their lineage from comparisons of sequences. Closely related proteins have the most similar sequences; distantly related proteins have only faintly similar sequences.

The kinship among homologous proteins is most easily visualized from a tree diagram based on sequence analyses. For example, the amino acid sequences of hemoglobins from different species suggest that they evolved from an ancestral monomeric, oxygen-binding protein (Figure 3-12). Over time, this ancestral protein slowly changed, giving rise to myoglobin, which remained a monomeric protein, and to the α and β subunits, which evolved to associate into the tetrameric hemoglobin molecule. As the tree diagram in Figure 3-12 shows, evolution of the globin protein family parallels that of the vertebrates.

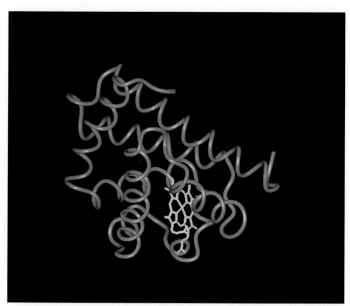

Myoglobin

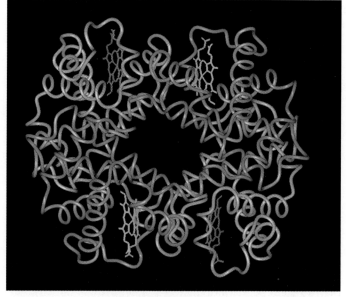

Hemoglobin

▲ **FIGURE 3-11 Models of the tertiary structures of the oxygen-carrier proteins myoglobin and hemoglobin based on x-ray crystallographic analysis.** Note the similarity in the tertiary structures of myoglobin and the two α subunits (blue) and two β subunits (purple) of hemoglobin. The planar white (or gray) structure in the center of each polypeptide chain is the heme prosthetic group. [Myoglobin adapted from S. E. V. Phillips, 1980, *J. Mol. Biol.* **142**:531; hemoglobin adapted from B. Shaanan, 1983, *J. Mol. Biol.* **171**:31; courtesy of S. Choe.

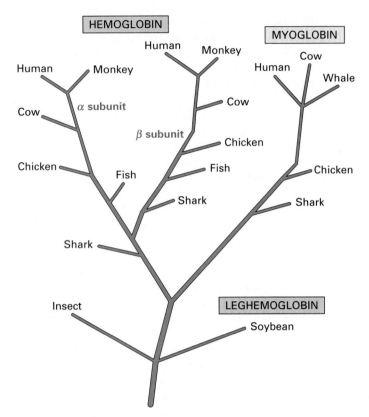

▲ **FIGURE 3-12 Evolutionary tree showing how the globin protein family arose, starting from the most primitive oxygen-binding proteins, leghemoglobins, in plants.** Sequence comparisons have revealed that evolution of the globin proteins parallels the evolution of vertebrates. Major junctions occurred with the divergence of myoglobin from hemoglobin and the later divergence of hemoglobin into the α and β subunits. [Adapted from R. E. Dickerson and I. Geis, 1983, *Hemoglobin: Structure, Function, Evolution, and Pathology*, Benjamin-Cummings.

The power of such comparative analysis and identification of homologous proteins has expanded substantially in recent years by use of the base sequences in an organism's genome to deduce the amino acid sequences of the encoded proteins. As discussed in Chapter 7, this approach permits "sequencing" of proteins that are difficult to purify in significant amounts.

SUMMARY Hierarchical Structure of Proteins

• A protein is a linear polymer of amino acids linked together by peptide bonds. Various, mostly noncovalent, interactions between amino acids in the linear sequence stabilize a specific folded three-dimensional structure (conformation) for each protein.

• The 20 different amino acids found in natural proteins are conveniently grouped into three categories based on the nature of their side (R) groups: hydrophilic

amino acids, with a charged or polar and uncharged R group; hydrophobic amino acids, with an aliphatic or bulky and aromatic R group; and amino acids with a special group, consisting of cysteine, glycine, and proline (see Figure 3-2).

• The α helix, β strand and sheet, and turn are the most prevalent elements of protein secondary structure, which is stabilized by hydrogen bonds between atoms of the peptide backbone. Certain combinations of secondary structures give rise to different motifs, which are found in a variety of proteins and often are associated with specific functions (see Figure 3-9).

• Protein tertiary structure results from hydrophobic interactions and disulfide bonds that stabilize folding of the secondary structure into a compact overall arrangement, or conformation. Large proteins often contain distinct domains, independently folded regions of tertiary structure with characteristic structural and/or functional properties.

• Quaternary structure encompasses the number and organization of subunits in multimeric proteins.

• The sequence of a protein determines its three-dimensional structure, which determines its function. In short, function is derived from structure; structure is derived from sequence.

• Homologous proteins, which have similar sequences, structures, and functions, most likely evolved from a common ancestor.

3.2 Folding, Modification, and Degradation of Proteins

As described in the next chapter, a polypeptide chain is synthesized on large cellular structures, the **ribosomes**, by a complex process in which assembly of amino acids in a particular sequence is dictated by **messenger RNA (mRNA)**. The nascent polypeptide chain undergoes folding and, in many cases, chemical modification to generate the final protein. Any polypeptide chain containing n residues could, in principle, fold into 8^n conformations. This value is based on the fact that only eight bond angles are stereochemically allowed in the polypeptide backbone. In general, however, all molecules of any protein species adopt a single conformation, called the *native state,* which is the most stably folded form of the molecule. Misfolding to non-native conformations is suppressed by two distinct mechanisms. At the molecular level, a protein folds through a pathway that favors only a few intermediate steps. Furthermore, a cellular system prevents misfolded proteins from forming. After a protein has carried out its functions, specific sequences that limit the life span of the protein target it for degradation.

The Information for Protein Folding Is Encoded in the Sequence

The realization that the amino acid sequence of a protein determines its folding came from in vitro studies on protein unfolding and refolding. Thermal energy from heat, extremes of pH that alter the charges on amino acid side chains, and chemicals such as urea or guanidine hydrochloride at concentrations of 6–8 M can disrupt the weak noncovalent bonds that stabilize the native conformation of a protein. The **denaturation** resulting from such treatment causes a protein to lose both its compact conformation and activity. Most denatured proteins precipitate in solution because hydrophobic groups, normally buried inside the molecules, interact with similar regions of other unfolded molecules, causing them to form an insoluble aggregate.

Many proteins that are completely unfolded in 8 M urea and β-mercaptoethanol (which reduces disulfide bonds) can *renature* (refold) into their native state when the denaturing reagents are removed by dialysis. During renaturation, all the disulfide, hydrogen, and hydrophobic bonds that stabilize the native conformation are re-formed. Thus, in this case proteins can be carried through a denaturation-renaturation cycle, which first destroys and then reestablishes their original structure and function (Figure 3-13). Because renaturation requires no cofactors or other proteins, at least in the test tube, protein folding is a self-assembly process.

The observation by Christian Anfinsen of such reversible denaturation and renaturation of ribonuclease, an enzyme that degrades RNA, provided a clue that the information for folding a protein lies in its sequence. A general mechanism by which proteins refold in vitro has been elucidated in experiments in which the renaturing conditions are carefully adjusted and the refolding reaction is interrupted at various time intervals. Such studies have shown that the polypeptide goes through several transient reconfigurations, including a "molten globule" state, before the native tertiary conformation is reached (Figure 3-14). In the case of ribonuclease, which has several internal disulfide bonds, the folding pathway involves rearrangements of disulfide bond pairs to the native conformation.

Folding of Proteins in Vivo Is Promoted by Chaperones

Folding of proteins in vitro is an inefficient process, with only a minority of unfolded molecules undergoing complete folding within a few minutes. Clearly, in vivo most protein molecules must rapidly fold into their correct shape; otherwise, cells would waste much energy in the synthesis of nonfunctional proteins and in the degradation of misfolded or unfolded proteins. More than 95 percent of the proteins present within cells have been shown to be in their native conformation, despite high protein concentrations ($\approx$100 mg/ml), which usually cause proteins to precipitate in vitro.

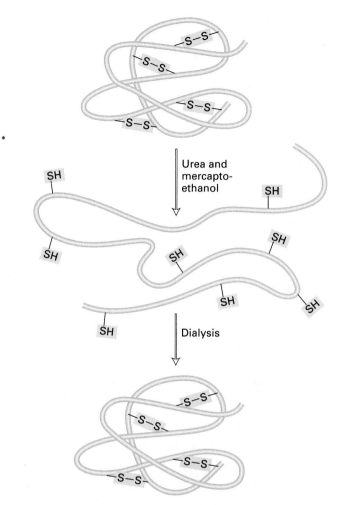

▲ **FIGURE 3-13 In vitro denaturation and renaturation of proteins.** Treatment with an 8 M urea solution containing mercaptoethanol (HSCH$_2$CH$_2$OH) completely denatures most proteins. The urea breaks intramolecular hydrogen and hydrophobic bonds, and the mercaptoethanol reduces each disulfide bridge (–S–S–) to two sulfhydryl (–SH) groups. When these chemicals are removed by dialysis, the –SH groups on the unfolded chain oxidize spontaneously to re-form disulfide bridges, and the polypeptide chain simultaneously refolds into its native conformation.

The explanation for the cell's remarkable efficiency in promoting protein folding probably lies in **chaperones,** a family of proteins found in all organisms from bacteria to humans. Chaperones are located in every cellular compartment, bind a wide range of proteins, and may be part of a general protein-folding mechanism. There are two general families of chaperones: *molecular chaperones,* which bind and stabilize unfolded or partially folded proteins, thereby preventing these proteins from being degraded; and *chaperonins,* which directly facilitate their folding. Chaperones have ATPase activity, and their ability to bind and stabilize their target proteins is specific and dependent on ATP hydrolysis. Binding of chaperones to partially folded proteins suggests that the folding process could be regulated at intermediate steps.

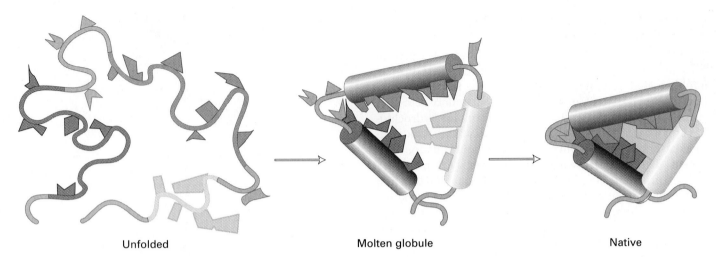

Unfolded
Molten globule
Native

▲ **FIGURE 3-14 Three stages in unassisted protein folding. In its denatured state, the entire polypeptide chain assumes a random conformation.** Under appropriate refolding conditions, the molecule condenses around a hydrophobic core into a compact, but non-native, intermediate, called a *molten globule*. In this folding intermediate, much of the secondary structure is present. Long-range interactions then form the tertiary structure, folding the molecule into its native three-dimensional conformation. [Adapted from L. Wu et al., 1995, *Nature Struc. Biol.* **2**:281; courtesy of J. Harris and P. S. Kim.]

Molecular chaperones consist of the Hsp70 family of proteins, which includes Hsp70 in the cytosol and mitochondrial matrix, Bip in the endoplasmic reticulum, and DnaK, a bacterial chaperone. First identified by its rapid appearance after a cell has been stressed by heat shock, Hsp70 is the major chaperone protein in all organisms. When bound to ATP, Hsp70 assumes an open form in which an exposed hydrophobic pocket transiently binds to exposed hydrophobic regions of the unfolded target protein. Hydrolysis of the bound ATP causes Hsp70 to assume a closed form, releasing the target protein (Figure 3-15a). Molecular chaperones are thought to bind all nascent polypeptide chains as they are being synthesized on ribosomes. In bacteria, 85 percent of the proteins are released from their chaperone and go on to fold normally; an even higher percentage of proteins in eukaryotes follow this pathway.

Proper folding of a small proportion of proteins (e.g., the cytoskeletal proteins actin and tubulin) requires additional assistance, which is provided by chaperonins. Eukaryotic chaperonins, called *TCiP*, are large, barrel-shaped, multimeric complexes composed of eight Hsp60 units. The bacterial homolog, known as *GroEL*, contains 14 identical subunits. The GroEL folding mechanism, which is better understood than TCiP-mediated folding, serves as a reasonable general model. In bacteria, a partially folded or misfolded polypeptide is inserted into the cavity of GroEL, where it binds to the inner wall and folds into its native conformation. In an ATP-dependent step, GroEL expands and the protein exits GroEL, a process assisted by a co-chaperonin, GroES, which caps the ends of GroEL (Figure 3-15b). Because the eukaryotic chaperonin TCiP lacks a GroES-type co-chaperonin, the last step must differ in eukaryotes. More-

over, the size of the cavity in TCiP limits this folding pathway to polypeptides smaller than 55 kDa.

Chemical Modifications and Processing Alter the Biological Activity of Proteins

Nearly every protein in a cell is chemically altered after its synthesis on a ribosome. Such modifications may alter the activity, life span, or cellular location of proteins, depending on the nature of the alteration. Protein alterations fall into two categories: chemical modification and processing. *Chemical modification* involves the linkage of a chemical group to the terminal amino or carboxyl groups or to reactive groups in the side chains of internal residues; in some cases, these modifications are reversible. *Processing* involves the removal of peptide segments and generally is irreversible.

Acetylation, the addition of an acetyl group (CH_3CO) to the amino group of the N-terminal residue is the most common form of chemical modification, involving an estimated 80 percent of all proteins:

$$CH_3-\overset{\overset{\displaystyle O}{\|}}{C}-\overset{\overset{\displaystyle }{\underset{\underset{\displaystyle H}{|}}{N}}}-\overset{\overset{\displaystyle R}{|}}{\underset{\underset{\displaystyle H}{|}}{C}}-\overset{\overset{\displaystyle O}{\|}}{C}-$$

Acetylated N-terminus

This modification may play an important role in controlling the life span of proteins within cells, as nonacetylated proteins are rapidly degraded by intracellular proteases. As discussed later, residues at or near the termini of some membrane proteins are chemically modified by addition of long

▶ **FIGURE 3-15 Chaperone-mediated protein folding.**
(a) Many proteins ① fold into their proper three-dimensional structure with the assistance of Hsp70, a molecular chaperone that transiently binds to a nascent polypeptide as it emerges from a ribosome. Proper folding of some proteins ② also depends on the chaperonin TCiP, a large barrel-shaped complex of Hsp60 units. (b) GroEL, the bacterial homolog of TCiP, is a barrel-shaped complex of 14 identical 60,000-MW subunits arranged in two stacked rings. In the absence of ATP or presence of ADP, GroEL exists in a "tight" conformational state *(left)* that binds partially folded or misfolded proteins. Binding of ATP shifts GroEL to a more open, "relaxed" state *(right),* which releases the folded protein. [Part (b) from A. Roseman et al., 1996, *Cell* **87**:241. Courtesy of Helen Saibil.]

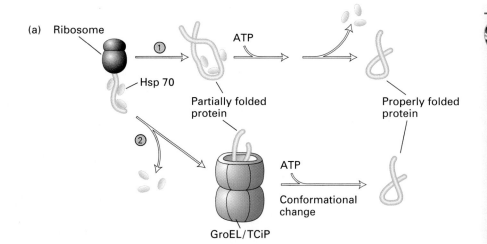

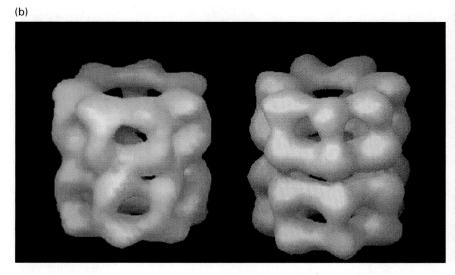

(b)

lipidlike groups. Attachment of these hydrophobic "tails," which function to anchor proteins to the lipid bilayer, constitute one way that cells restrict certain proteins to membranes.

The internal residues in proteins can be modified by attachment of a variety of chemical groups to their side chains. The most important modification is *phosphorylation* of serine, threonine, and tyrosine residues. We will encounter numerous examples of proteins whose activity is regulated by reversible phosphorylation and dephosphorylation. The side chains of asparagine, serine, and threonine are sites for *glycosylation,* the attachment of linear and branched carbohydrate chains. Many secreted proteins and membrane proteins contain glycosylated residues; the synthesis of such proteins is described in Chapter 17. Various less common modifications are found in a limited number of proteins (Figure 3-16).

▶ **FIGURE 3-16 Examples of modified internal residues produced by hydroxylation, methylation, and carboxylation.**
These modifications occur after synthesis of the polypeptide chain.

3-Hydroxyproline
(mainly in collagen)

4-Hydroxyproline
(mainly in collagen)

3-Methylhistidine
(mainly in actin)

5-Hydroxylysine
(mainly in collagen)

γ-Carboxyglutamate
(mainly in prothrombin, an essential blood-clotting factor)

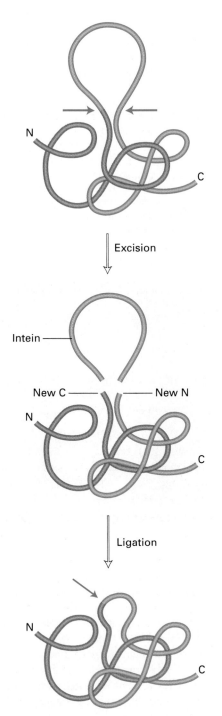

▲ FIGURE 3-17 Self-splicing, a protein-processing mechanism found in bacteria and lower eukaryotes. A segment (red) of a polypeptide, called an *intein,* is removed by cleavage of two peptide bonds (red arrows), leaving two segments (blue and green). One new peptide bond then forms between the two segments (blue arrow), regenerating a continuous polypeptide backbone. The process is autocatalytic and does not depend on enzymes. Consensus splice sites in the polypeptide chain mark the points of breakage and re-formation of the backbone. The excised segment is thought to be exposed at the surface of the folded protein.

Unlike chemical modification of residues, which often is reversible, processing of some proteins causes irreversible changes that alter their activity. In the most common form of processing, residues are removed from the C- or N-terminus of a polypeptide by cleavage of the peptide bond in a reaction catalyzed by proteases. Proteolytic cleavage is a common mechanism of activation or inactivation, especially of enzymes involved in blood coagulation or digestion. As discussed later, the activity of certain digestive enzymes is controlled by this mechanism. Proteolysis also generates active peptide hormones, such as EGF mentioned earlier and insulin, from larger precursor polypeptides. A detailed description of the processing of membrane and secreted proteins is presented in Chapter 17.

An unusual type of processing, termed *protein self-splicing,* occurs in bacteria and primitive eukaryotes. *Splicing* refers to a process analogous to editing film: an internal segment of polypeptide, an **intein,** is removed and the ends of the polypeptide are rejoined (Figure 3-17). Unlike proteolytic processing, protein self-splicing is an autocatalytic process, which proceeds by itself without the involvement of enzymes. The excised peptide appears to eliminate itself from the protein by a mechanism similar to that used in processing of RNA molecules (Chapter 4). In vertebrate cells, processing of some proteins involves self-cleavage, but the subsequent ligation step is absent. One such protein is Hedgehog, which is critical to a number of developmental processes (Chapter 23).

Cells Degrade Proteins via Several Pathways

Cells have both extracellular and intracellular pathways for degrading proteins. The major extracellular pathway is the system of *digestive proteases,* which break down ingested proteins to polypeptides in the intestinal tract. These include *endoproteases* such as trypsin and chymotrypsin, which cleave the protein backbone adjacent to basic and aromatic residues; *exopeptidases,* which sequentially remove residues from the N-terminus (aminopeptidases) or C-terminus (carboxypeptidases) of proteins; and *peptidases,* which split oligopeptides into di- and tripeptides and individual amino acids. These small molecules then are transported across the intestinal lining into the bloodstream.

The life span of intracellular proteins varies from as short as a few minutes for mitotic cyclins, which help regulate passage through mitosis, to as long as the age of an organism for proteins in the lens of the eye. Cells have several intracellular proteolytic pathways for degrading misfolded or denatured proteins, normal proteins whose concentration must be decreased, and foreign proteins taken up by the cell. One major intracellular pathway involves degradation by enzymes within lysosomes, membrane-limited organelles whose interior is acidic (Chapter 5). Distinct from the lysosomal pathway are cytosolic mechanisms for degrading proteins. The best-understood pathway, the ubiquitin-mediated

(a)

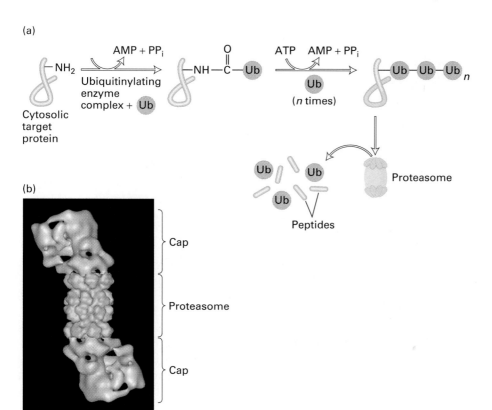

(b)

Cap

Proteasome

Cap

◄ **FIGURE 3-18 Ubiquitin-mediated proteolytic pathway.** (a) A conjugating enzyme catalyzes formation of a peptide bond between ubiquitin (Ub) and the side-chain $-NH_2$ of a lysine residue in a target protein. Additional Ub molecules are added, forming a multiubiquitin chain. This chain is thought to direct the tagged protein to a proteasome, which cleaves the protein into numerous small peptide fragments. (b) Computer-generated image reveals cylindrical structure of a proteasome with a cap at each end of a core. Proteolysis of ubiquitin-tagged proteins occurs along the inner wall of the core. [Part (b) from W. Baumeister et al., 1998, *Cell* **92**:357. Courtesy of W. Baumeister.]

pathway, involves two steps: addition of a chain of **ubiquitin** molecules to an internal lysine side chain of a target protein and proteolysis of the ubiquitinated protein by a **proteasome**, a large, cylindrical multisubunit complex (Figure 3-18). The numerous proteasomes present in the cell cytosol proteolytically cleave ubiquitin-tagged proteins in an ATP-dependent process that yields peptides and intact ubiquitin molecules.

To be targeted for degradation by the ubiquitin-mediated pathway, a protein must contain a structure that is recognized by a ubiquitinating enzyme complex. Different conjugating enzymes recognize different degradation signals in target proteins. For example, the internal sequence Arg-X-X-Leu-Gly-X-Ile-Gly-Asx in mitotic cyclin is recognized by the ubiquitin-conjugating enzyme E1. Internal sequences enriched in proline, glutamic acid, serine, and threonine (PEST sequences) are recognized by other enzymes. The life span of many cytosolic proteins is correlated with the identity of the N-terminal residue, suggesting that certain residues at the N-terminus favor rapid ubiquitination. For example, short-lived proteins that are degraded within 3 minutes in vivo commonly have Arg, Lys, Phe, Leu, or Trp at their N-terminus. In contrast, a stabilizing amino acid such as Cys, Ala, Ser, Thr, Gly, Val, or Met is present at the N-terminus in long-lived proteins that resist proteolytic attack for more than 30 hours. As explained in Chapter 4, all newly synthesized proteins have methionine, a stabilizing amino acid, at the N-terminus. Thus subsequent enzymatic alteration that generates one of the destabilizing amino acids at the N-terminus is necessary to target a protein for degradation.

Aberrantly Folded Proteins Are Implicated in Slowly Developing Diseases

As noted earlier, each protein species normally folds into a single, energetically favorable conformation that is specified by its amino acid sequence. Recent evidence suggests, however, that a protein may fold into an alternative three-dimensional structure for reasons that have not yet been identified. Such "misfolding" not only leads to a loss of the normal function of a protein but also marks it for proteolytic degradation. The subsequent accumulation of proteolytic fragments contributes to certain degenerative diseases characterized by the presence of insoluble protein plaques in various organs including the liver and brain.

Alzheimer's disease, for example, is marked by formation of plaques and tangles in a deteriorating brain. The filaments composing these structures are derived from proteolytic products of abundant natural proteins such as amyloid precursor protein, a transmembrane protein, and Tau, a microtubule-binding protein (Figure 3-19). Plaques in other organs are formed from proteolytic fragments of natural proteins such as gelsolin, an actin-binding protein, and serum albumin, a blood protein. The polypeptide fragments liberated by proteolysis polymerize into very stable filaments. A degeneration of

(a)

(b)

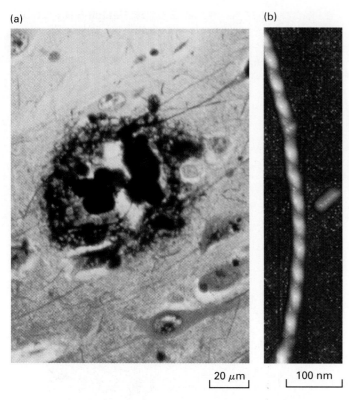

20 μm   100 nm

▲ **FIGURE 3-19 Amyloidosis is characterized by the formation of insoluble protein plaques in various organs of the body.** (a) The amyloid plaque in the brain of an Alzheimer's patient appears as a tangle of filaments. (b) In the atomic force microscope, the filaments are seen to be regular arrangements of a short 47-residue fragment, called *β-amyloid peptide*, produced by proteolysis of amyloid precursor protein. [Courtesy of K. Kosik.]

the brain, similar to that seen in Alzheimer's disease, is thought to be caused by *prions*, an infectious protein agent derived by proteolysis and re-folding of a normal brain protein.

SUMMARY Folding, Modification, and Degradation of Proteins

- The amino acid sequence of a protein dictates its folding into a specific three-dimensional conformation, the native state.

- Folding of denatured proteins in vitro proceeds through intermediates having secondary and non-native tertiary structure.

- Protein folding in vivo occurs with the assistance of two types of special proteins (see Figure 3-15). Molecular chaperones (Hsp70 proteins) bind to nascent polypeptides emerging from ribosomes and prevent their misfolding. Chaperonins, large complexes of Hsp60-like proteins, shelter some partially folded or misfolded proteins in a barrel-like cavity, providing additional time for proper folding.

- Following their synthesis, all proteins are modified in various ways that alter their structure and function.

- The life span of intracellular proteins is largely determined by their susceptibility to proteolytic degradation by various pathways.

- The presence of certain internal sequences or N-terminal residues targets cytosolic proteins for addition of ubiquitin and subsequent proteolysis within a proteasome (see Figure 3-18).

3.3 Functional Design of Proteins

A key concept in biology is that form and function are inseparable. This concept applies equally well to protein design as to other levels of biological organization (e.g., the morphology of cells and the organization of tissues). In fact, we can often guess how a protein works by looking at its structure. Perhaps the best way to illustrate this is by examining a few protein structures. For instance, a barrel-like nuclear pore, a complex of several proteins, sits in the nuclear membrane and acts as a channel through which molecules travel in or out of the nucleus (Figure 3-20a). In the cavity of a different barrel-like structure, the GroEL/ES chaperonin, protein folding takes place (Figure 3-20b). Some proteins have grooves in their surface, which are logical binding sites for a variety of molecules, especially rod-shaped or filamentous ones. An example is reverse transcriptase, which copies RNA into DNA; this enzyme has a groove on one side through which RNA slides along the surface of the protein (Figure 3-20c). Topoisomerase II, a DNA-binding enzyme, is an articulated enzyme that opens and closes at both ends like locks in a canal (Figure 3-20d). A delight in studying protein structure is uncovering the simple but ingenious ways that nature has built each protein to perform a particular function.

In this section, we examine several features of proteins that are critical to their biological activity and the regulation of that activity, focusing on antibodies, enzymes, and membrane proteins as examples. The functioning of many proteins involves some change in their conformation induced by binding of a specific molecule, change in the environment, or chemical modification. As numerous examples in later chapters will illustrate, such induced conformational changes can make proteins into switches and machines. The changes in conformation can be enormous, as seen in proteins like topoisomerase, an enzyme that moves DNA strands across one another, or myosin, a motor protein that moves along actin filaments.

Proteins Are Designed to Bind a Wide Range of Molecules

The function of nearly all proteins depends on their ability to bind other molecules, or **ligands,** with a high degree of specificity. As catalysts of chemical reactions, enzymes must

(a)

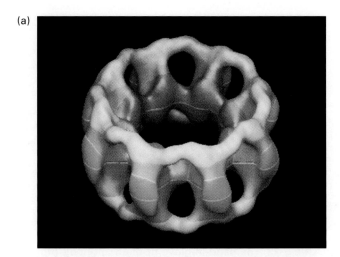

(b)

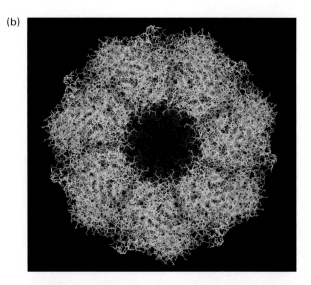

(c)

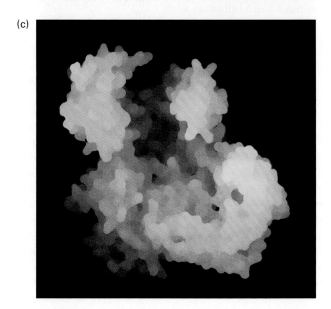

(d)

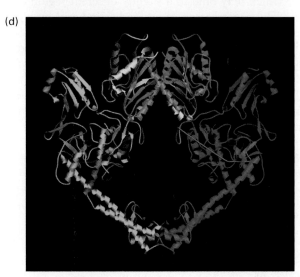

▲ **FIGURE 3-20 Gallery of protein structural models showing the link between structure and function.** (a) The nuclear pore, a complex of proteins with a total molecular weight of 1.2×10^8, an outer diameter of 133 nm, a height of 70 nm, and a central hole 42 nm in diameter. Transport of molecules into and out of the nucleus occurs through such pores located in the nuclear membrane. (b) Protein folding takes place within the cavity of a chaperonin, GroEL/ES. Built of 21 subunits, the chaperonin is 18.4 nm long and has a diameter of 3.3 nm, large enough for small- and medium-sized proteins. (c) Reverse transcriptase, an enzyme present in RNA viruses that copies the viral RNA

genome into DNA. Reverse transcriptase from the AIDS-causing virus HIV measures $11 \times 3.0 \times 4.5$ nm. RNA to be copied lies in a groove on the surface of the enzyme. (d) Topoisomerase II, which prevents DNA from overtwisting during replication, is shaped like a pair of tongs. Opening and closing of Topo II at the top and bottom of the protein permit a nicked strand of DNA to pass through and be repaired. [Part (a) from J. E. Hinshaw, B. O. Carragher, and R. A. Milligan, 1992, *Cell* **69**:1133; part (b) courtesy of S. Choe; part (c) from T. A. Steitz et al., 1992, *Science* **256**:1783; part (d) courtesy of J. Berger.]

first bind tightly and specifically to their target molecules, called **substrates,** which may be a small molecule (e.g., glucose) or a macromolecule. The many different types of hormone receptors on the surface of cells also display a high degree of sensitivity and discrimination for their ligands, which generally are present at low concentrations in blood. These receptors, essential to signaling between cells, are discussed in Chapter 20.

Two properties of a protein characterize its interaction with ligands. *Affinity* refers to the strength of binding between a protein and ligand; the equilibrium constant K_{eq} (Chapter 2) or the dissociation constant K_D for binding is a measure of affinity. *Specificity* refers to the ability of a protein to bind one molecule in preference to other molecules. Both properties depend on the structure of the ligand-binding site on a protein, which is designed to fit its partner like a mold. For high-affinity and highly specific interactions to occur, the shape and chemical surface of the binding site must be complementary to the ligand molecule. To illustrate this critical concept, we consider how an antibody binds an antigen and how an enzyme catalyzes a chemical reaction.

Antibodies Exhibit Precise Ligand-Binding Specificity

The capacity of proteins to distinguish different molecules is highly developed in blood proteins called **antibodies.** Animals produce antibodies in response to the invasion of an infectious agent (e.g., a bacterium or a virus) or after exposure to certain foreign substances (e.g., proteins or polysaccharides in pollens). The antibody-inducing agent is called an **antigen.** The presence of antigen causes an organism to make a large quantity of different antibody proteins, each of which may bind to a slightly different region of the antigen. The constellation of antibodies induced by a given antigen may differ from one member of a species to another.

All antibodies belong to a family of proteins called **immunoglobulins.** These Y-shaped molecules are formed from two types of polypeptides: heavy chains and light chains. The heavy chains run the length of the molecule; their C-terminal regions pair to form a stem. Visually we can dis-

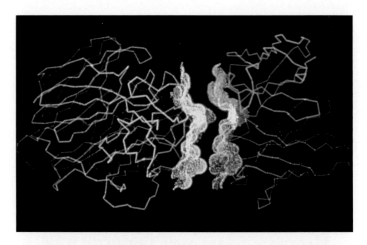

▲ **FIGURE 3-22 The hand-in-glove fit of an antibody *(right)* to an antigen from influenza virus *(left).*** This computer simulation is based on x-ray crystallography of the complex formed between the antigen and the antibody Fab domain. The complementarity of the antigen and antibody surfaces is especially apparent where the "finger" extending from the antigen surface is opposed to the "cleft" in the antibody surface. [From G. J. V. H. Nossal, 1993, *Sci. Am.* **269** (Sept.):54.]

tinguish three globular domains: two identical domains corresponding to each arm and the third composing the stem (Figure 3-21). Each arm of the antibody molecule contains a single light chain linked to a heavy chain by disulfide bonds. The N-terminal regions of both heavy and light chains lie at the tip of each arm and are distinguished by highly variable amino acid sequences. The remaining portions of the sequences in both chains are constant (i.e., nearly identical) among antibodies with different specificities. The arms are the business end of an antibody molecule, since an antigen-binding site lies at the end of each arm. Because of its dimeric structure, each antibody molecule can bind two identical antigen molecules. X-ray crystallographic analysis of antigen-antibody complexes has revealed that the antigenic specificity of an antibody is dependent on three highly variable regions, called *complementarity-determining regions (CDRs),* near the end of each arm. These regions form the antigen-binding site, which physically matches the antigen like a glove.

Most large antigens have multiple different sites, called **epitopes** (or antigenic determinants) that can induce production of specific antibodies; each type of antibody binds to its own inducing epitope. For example, lysozyme, an enzyme that degrades the carbohydrate coat of bacteria, induces several different antibodies, each of which binds to a particular epitope on the lysozyme molecule. Although the different epitopes on lysozyme differ greatly in their chemical properties, the interaction between lysozyme and antibody is complementary in all cases; that is, the surface of the antibody's antigen-binding site fits into that of the corresponding epitope as if they were molded together (Figure 3-22). The intimate contact between these two surfaces,

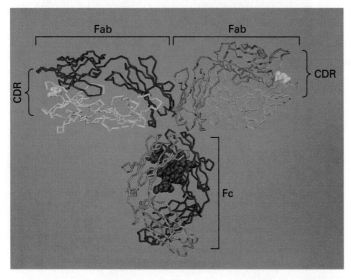

▲ **FIGURE 3-21 Structure of an antibody molecule, which consists of two identical heavy chains (blue and orange) and two identical light chains (yellow and green).** The Y-shaped molecule contains two identical Fab domains, forming the arms, and one Fc domain, forming the stem. In the native molecule, each heavy chain is a continuous polypeptide, with a hinge region connecting the two halves shown in this figure. Antigen molecules (white) bind to the complementarity-determining regions (CDRs), which are highly variable regions located at the ends of each arm. Antibodies contain carbohydrate moieties (red) and thus are glycoproteins. [From A. Levine, 1992, *Viruses,* W. H. Freeman, p. 53.]

stabilized by numerous noncovalent bonds, is responsible for the exquisite binding specificity exhibited by an antibody. Antibodies, for instance, can distinguish between the cells of individual members of a species and in some cases can distinguish between proteins that differ by only a single amino acid. Because of their specificity and the ease with which they can be produced, antibodies are critical reagents in many experiments discussed in the following chapters.

Enzymes Are Highly Efficient and Specific Catalysts

Almost every chemical reaction in a cell is catalyzed by a class of proteins called **enzymes.** As discussed in Chapter 2, catalysts increase the rates of reactions that are already energetically favorable by lowering the activation energy (see Figure 2-27). In the test tube, catalysts such as charcoal and platinum facilitate reactions but often at high temperatures, at extremes of high or low pH, or in organic solvents. In contrast to these harsh conditions, enzymes must catalyze chemical reactions in the mild conditions of a cell: 37 °C, pH 6.5–7.5, and aqueous solvents. As we just discussed, all antibodies belong to the immunoglobulin family of proteins and have a similar structure. Enzymes, however, are a structurally diverse group of proteins that have evolved through unrelated and highly divergent mechanisms.

The ability of enzymes to function as catalysts under conditions where nonbiological catalysts would be ineffectual is exemplified by two striking properties: their enormous reaction rates and their specificity. Quite often, the rate of an enzymatically catalyzed reaction is 10^6–10^{12} times that of an uncatalyzed reaction under otherwise similar conditions. The specificity of an enzyme denotes its ability to act selectively on one substance or a small number of chemically similar substances, the enzyme's substrates. Like antibody specificity, enzyme specificity depends on a close fit between substrate molecules and their binding sites on an enzyme. An example of specificity is provided by the enzymes that act on amino acids. As noted in Chapter 2, amino acids can exist as two stereoisomers, designated L and D, although only L isomers normally are found in biological systems. Not surprisingly, enzyme-catalyzed reactions involving L-amino acids occur much more rapidly than do those involving D-amino acids, even though both stereoisomers of a given amino acid are the same size and possess the same R groups (see Figure 2-6).

The number of different types of chemical reactions that occur in any one cell is very large: an animal cell, for example, normally contains 1000–4000 different types of enzymes, each of which catalyzes a single chemical reaction or set of closely related reactions. Certain enzymes are found in the majority of cells because they catalyze synthesis of common cellular products (e.g., proteins, nucleic acids, and phospholipids) or are involved in the production of energy by the conversion of glucose and oxygen to carbon dioxide and water. Other enzymes are present only in a particular type of cell (e.g., a liver cell or a nerve cell) because they catalyze some chemical reaction unique to that cell type. Although most enzymes are located within cells, some are secreted and function in the blood, lumen of the digestive tract, or other extracellular space. Some microbial enzymes are secreted from and are active outside the organism.

An Enzyme's Active Site Binds Substrates and Carries Out Catalysis

Certain amino acid side chains of an enzyme are important in determining its specificity and its ability to accelerate a chemical reaction. In the native conformation of an enzyme, these side chains are brought into proximity, forming the **active site.** Active sites thus consist of two functionally important regions: one that recognizes and binds the substrate (or substrates), and one that catalyzes the reaction once the substrate has been bound. In some enzymes, the catalytic site is part of the substrate-binding site; in others, the two sites are structurally as well as functionally distinct. The amino acids that make up the active site do not need to be adjacent in the linear polypeptide sequence; rather, folding of the molecule results in juxtaposition of these amino acids, forming a space in which the substrate sits.

To illustrate how the active site binds a specific substrate and then promotes a chemical change in the bound substrate, we examine the action of *cAMP-dependent protein kinase (cAPK).* This enzyme and other protein kinases, which add a phosphate group to serine, threonine, or tyrosine residues in proteins, are critical for regulating the activity of many cellular proteins. Because the structure of the active site and mechanism of phosphorylation are very similar in all kinases, cAPK can serve as a general model for this important class of enzymes.

As discussed later, an unusual nucleotide called *cAMP* induces dissociation of the inactive tetrameric form of cAPK, releasing two catalytic subunits. To aid in understanding the mechanism of binding and catalysis, we focus here on the 260-residue "kinase core" of each catalytic subunit. The kinase core, which is largely conserved in all protein kinases, is responsible for the binding of ATP and a target peptide, followed by transfer of a phosphate group from ATP to a serine, threonine, or tyrosine in the peptide. The kinase core consists of a large and small domain with an intervening deep cleft; the active site comprises residues located in both domains.

Substrate Binding by Protein Kinases The small domain of the kinase core binds ATP, while the large domain binds the target peptide (Figure 3-23). The structure of the ATP-binding site complements the structure of the nucleotide substrate. The adenine ring of ATP sits snugly at the base of the cleft, which is characterized by a highly conserved sequence, Gly-X-Gly-X-X-Gly. This triad of glycine residues, the "glycine lid," is part of a strand-loop-strand motif that closes over the adenine of ATP and holds it in position. The adenine ring sits in a hydrophobic pocket and is positioned by

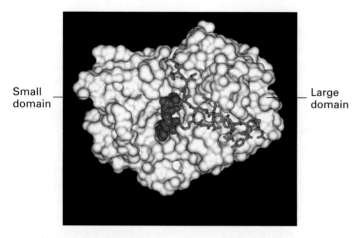

Small domain

Large domain

▲ **FIGURE 3-23 Model of the catalytic kinase core of cAMP-dependent protein kinase (cAPK), which is largely conserved in other kinases.** Residues in the small domain position ATP (red) in a deep cleft between the large and small domains of the core. Residues in the large domain bind the target peptide (green). [Courtesy of L. Wu.]

hydrogen bonds and van der Waals attractions with the glycine residues and backbone amide groups. Two invariant residues, lysine at position 72 and aspartic acid at position 184, stabilize the phosphate groups, which protrude from the nucleotide-binding cleft (step 1 in Figure 3-24). Lys-72 bridges to the α and β phosphates of ATP, while the γ-phosphate group is chelated by a Mg^{2+} ion bound to Asp-184.

ATP is a common substrate for all protein kinases, but the sequence of the target peptide varies among different kinases. The peptide sequence recognized by cAPK is Arg-Arg-X-Ser-Y, where X is any amino acid and Y is a hydrophobic amino acid. The portion of the polypeptide chain containing the target serine, threonine, or tyrosine is bound to a shallow groove in the large domain of the kinase core. The peptide specificity of cAPK is conferred by several glutamic acid residues in the large domain, which bind the two arginine residues in the target peptide. Different residues determine the specificity of other protein kinases.

Phosphoryl Transfer by cAPK Figure 3-24 also summarizes the catalytic mechanism of cAPK. Binding of ATP and then the peptide target positions the γ phosphate of ATP near the target serine residue of the peptide. Catalysis takes place in two stages. First, a bond forms between the serine and phosphate group, yielding a pentavalent phosphate transition state. Second, the phosphodiester bond between the β and γ phosphates is broken, yielding ADP and the phosphorylated peptide. Because the phosphate-serine bond is formed on the opposite side of the phosphodiester bond from the β phosphate, this process is called an *in-line mechanism of phosphoryl transfer*. The products, ADP and phosphorylated peptide, are then released from the active site.

During catalysis by cAPK, two "catalytic residues" appear to participate in formation of the transition state. Asp-

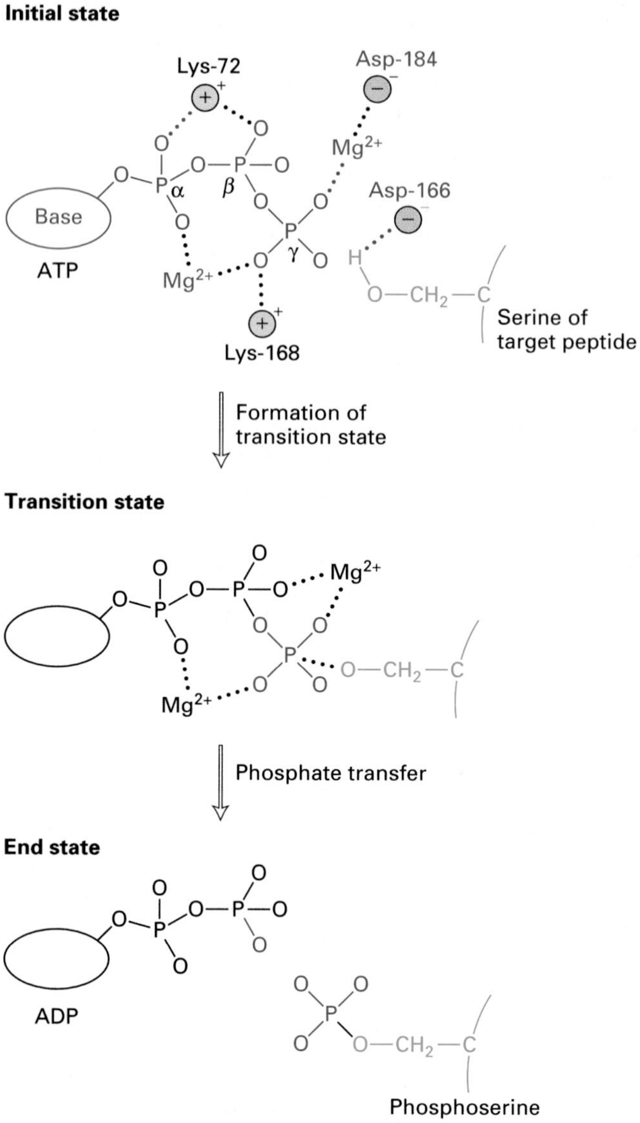

Initial state

Lys-72
Asp-184
Mg²⁺
Asp-166
Base
ATP
Mg²⁺
Lys-168
Serine of target peptide

Formation of transition state

Transition state

O
O
Mg²⁺
O
Mg²⁺
O—CH₂—C

Phosphate transfer

End state

ADP
Phosphoserine
O—CH₂—C

▲ **FIGURE 3-24 The mechanism of phosphorylation by cAMP-dependent protein kinase (cAPK), which catalyzes transfer of a phosphate group from ATP to a serine side chain in a target peptide sequence.** Step 1: Initially, both substrates bind to the active site (see Figure 3-23). Electrons of the phosphate group are delocalized by interactions with lysine residues and Mg^{2+}. Asp166 abstracts a proton from the hydroxyl group of the serine in the bound target peptide. Step 2: A new bond then forms between the serine side-chain oxygen and γ phosphate, yielding a pentavalent transition-state intermediate. Step 3: The phosphoester bond between the β and γ phosphates is broken to form a phosphorylated serine side chain and ADP.

166 is thought to remove a proton from the serine hydroxyl group in the target peptide, while Lys-168 neutralizes the negative charge of the γ phosphate. Then the electrons of the deprotonated serine hydroxyl group are thought to form a bond to the γ phosphorus atom, yielding the pentavalent

transition-state intermediate. The newly created phosphoserine is repelled from the β phosphate of ADP and the catalytic base. The products induce a conformational change in the enzyme, described below, that permits them to diffuse from the active site.

Interactions between residues in the active site of an enzyme and the substrates help stabilize the transition state, thereby allowing more time for the rearrangement of bonds needed to form the products. As explained in Chapter 2, the activation energy is the energy required for formation of the transition state (see Figure 2-27). An enzyme, by virtue of its three-dimensional binding site, reduces the activation energy of a reaction compared with an uncatalyzed reaction involving the same reactants. The ability to bind transition-state intermediates is the one property that distinguishes enzymes from other proteins. If a protein cannot bind a transition-state intermediate, then it cannot catalyze a reaction.

Conformational Changes Induced by Substrate Binding to cAPK

The catalytic subunit of cAPK exists in an "open" and "closed" conformation (Figure 3-25a). In the open position, the large and small domains of the kinase core are separated enough that substrate molecules can bind. Once the active site is occupied by substrate, the domains move together into the closed position. This change in tertiary structure, an example of *induced fit,* brings the bound target peptide close enough to the terminal phosphate group of the bound ATP that phosphoryl transfer can occur. After the phosphorylation reaction is completed, the presence of the products causes the domains to rotate to the open position, from which the products are released.

The rotation from the open to closed position also causes movement of the short glycine-rich sequence over the ATP-binding cleft in the active site. This small finger of the polypeptide chain, the glycine lid, controls the entry of ATP and release of ADP at the active site. In the open position, ATP can enter and bind to the active site cleft. In the closed position, the glycine-rich sequence moves over the nucleotide and acts as a lid that prevents ATP from leaving (Figure 3-25b). Following phosphoryl transfer, the glycine lid must rotate back to the open position before ADP can be released. Kinetic measurements show that the rate of ADP release is 20-fold slower than that of phosphoryl transfer, reflecting the influence of the glycine lid in cAPK. Mutations in the glycine lid that inhibit its flexibility slow catalysis by cAPK even further. Besides trapping ATP in the binding pocket, the glycine lid prevents water from entering the active site. Water would inhibit the reaction by dampening the charge delocalization steps.

Kinetics of an Enzymatic Reaction Are Described by V_{max} and K_m

Enzymatic specificity is usually quantified in relative terms; that is, the reaction with a good substrate may occur, for example, 10,000 times faster than it does with a poor substrate. The catalytic action of an enzyme on a given substrate can be described by two parameters: K_m (the **Michaelis constant**), which measures the affinity of an enzyme for its substrate, and V_{max}, which measures the maximal velocity of the reaction at saturating substrate concentrations. Equations for K_m and V_{max} are most easily derived by considering the simple reaction

$$\text{Substrate} \rightleftharpoons \text{product}$$

in which the rate of product formation v depends on the concentration of substrate, [S], and on the concentration of the enzyme, [E].

For an enzyme with a single catalytic site, Figure 3-26a shows how the rate of product formation depends on [S] when [E] is kept constant. At low concentrations of S, the reaction rate is proportional to [S]. As [S] is increased, the rate does not increase indefinitely in proportion to [S]; rather, it eventually reaches a maximum velocity V_{max}. The value of V_{max} is independent of [S], but is proportional to [E] and to the catalytic constant k_{cat}, which is an intrinsic property of the individual enzyme. Halving [E] reduces the reaction rate at all values of [S] by half. Both V_{max} and K_m for a particular enzyme and substrate can be determined from experimental curves of reaction velocity versus substrate concentration, as illustrated in Figure 3-26.

When interpreting kinetic curves such as those in Figure 3-26, bear in mind that all enzyme-catalyzed reactions include at least three steps: (1) the binding of a substrate (S) to an enzyme (E) to form an enzyme-substrate complex (ES), (2) the conversion of ES to the enzyme-product complex

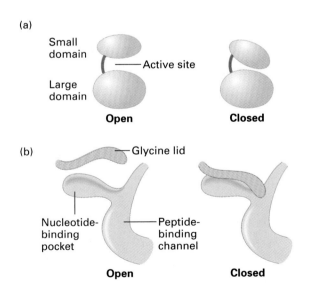

(a)

Small domain — Active site
Large domain

Open   **Closed**

(b) — Glycine lid

Nucleotide-binding pocket — Peptide-binding channel

Open   **Closed**

▲ **FIGURE 3-25 Conformational changes in catalytic subunit of cAPK.** Substrate binding causes a rotation of the large and small domains in the catalytic subunit from the open to the closed state (a). This rotation brings the peptide closer to ATP, and also causes the glycine lid to move over the adenine of ATP, thereby trapping the nucleotide in the cleft (b).

(a)

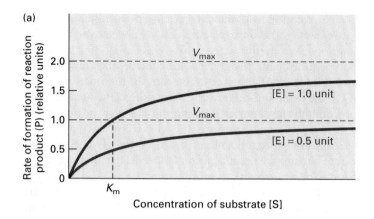

(b)

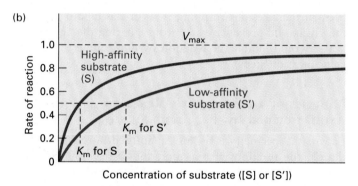

▲ **FIGURE 3-26 Dependence of the velocity of an enzyme-catalyzed reaction on substrate concentration.** (a) The rates of a hypothetical reaction S ⟶ P at two different concentrations of enzyme [E] as a function of substrate concentration [S]. The [S] that yields a half-maximal reaction rate is the Michaelis constant K_m, a measure of the affinity of E for S. Doubling the concentration of enzyme causes a proportional increase in the reaction rate, so that the maximal velocity V_{max} is doubled; the K_m, however, is unaltered. (b) The rates of the reactions catalyzed by an enzyme with substrate S, for which the enzyme has a high affinity, and with substrate S′, for which the enzyme has a low affinity. Note that the V_{max} is the same with both substrates but that K_m is higher for S′, the low-affinity substrate.

(EP), and (3) the release of the product (P) from EP, to yield free P:

$$E + S \underset{}{\overset{\text{binding}}{\rightleftharpoons}} ES \xrightarrow{\text{catalysis}} EP \xrightarrow{\text{release}} P + E$$

In the simplest case, when the release of P is very rapid, we can simplify the reaction equation as follows:

$$E + S \underset{k_2}{\overset{k_1}{\rightleftharpoons}} ES \xrightarrow{k_{cat}} E + P$$

In this case, the rate of product formation v is equal to $k_{cat} \times$ [ES]. Starting from this relationship, we can derive the *Michaelis-Menten equation*

$$v = V_{max} \frac{[S]}{[S] + K_m}$$

where K_m, the Michaelis constant, is defined as $(k_2 + k_{cat})/k_1$. This equation fits the curves shown in Figure 3-26.

The slowest step in most enzymatic reactions is conversion of the enzyme-substrate complex ES to the free enzyme E and product P. In such cases, k_{cat} is much less than k_2, so that

$$K_m \cong \frac{k_2}{k_1} = K_d$$

where K_d is the dissociation constant for binding of S to E. Thus the parameter K_m describes the affinity of an enzyme for its substrate. The smaller the value of K_m, the more avidly the enzyme can bind the substrate from a dilute solution and the smaller the concentration of substrate needed to reach half-maximal velocity (see Figure 3-26b). The concentrations of the various small molecules in a cell vary widely, as do the K_m values for the different enzymes that act on them. Generally, the intracellular concentration of a substrate is approximately the same as or greater than the K_m value of the enzyme to which it binds.

Many Proteins Contain Tightly Bound Prosthetic Groups

The native conformation and activities of some proteins require the presence of a **prosthetic group,** a small nonpeptide molecule or metal that binds tightly to a protein, keeping the protein in a fixed conformation and participating in binding ligands. For example, each of the four subunits of hemoglobin binds and enfolds a prosthetic group called *heme,* which consists of an iron atom held in a cage by protoporphyrin:

The heme groups are the oxygen-binding components of hemoglobin (see Figure 3-11). Heme is also present in the cytochromes of the electron-transport chain; in this case, it functions to bind electrons. Other electron-transport proteins employ sulfur or flavin as prosthetic groups. In addition to acting as carriers of oxygen or electrons, prosthetic groups can act as antennae. For example, proteins involved in vision or photosynthesis contain retinal or chlorophyll, which absorb energy from sunlight. Prosthetic groups can be linked to proteins noncovalently, as in hemoglobin, or covalently, as in cytochrome.

The activity of numerous enzymes also depends on the presence a prosthetic group, commonly referred to as a **coenzyme.** Many coenzymes act to lower the activation energy of biochemical reactions by forming a covalent intermediate with a substrate. For instance, the enzyme that converts the amino acid histidine into histamine (a potent dilator of small blood vessels) requires the coenzyme *pyridoxal phosphate.* In this reaction, a covalent bond first forms between histidine and the enzyme-bound pyridoxal phosphate, forming a Schiff base intermediate. Rearrangement of the bonds in this intermediate yields carbon dioxide, which is released, and a second intermediate. This is then hydrolyzed, producing the product, histamine, and regenerating the coenzyme, pyridoxal phosphate.

A Variety of Regulatory Mechanisms Control Protein Function

Most reactions in cells do not occur independently of one another or at a constant rate. Instead, the catalytic activity of enzymes is so regulated that the amount of reaction product is just sufficient to meet the needs of the cell. As a result, the steady-state concentrations of substrates and products will vary depending on cellular conditions. The flow of material in an enzymatic pathway is controlled by several mechanisms, some of which also regulate the functions of nonenzymatic proteins.

One of the most important mechanisms for regulating protein function entails **allosteric transitions,** changes in the tertiary and/or quaternary structure of a protein induced by binding of a small molecule, which may be an activator, inhibitor, or substrate. Allosteric regulation is particularly prevalent in multimeric (multisubunit) enzymes. Some multimeric enzymes are composed of identical subunits, each containing an active site and, often, a distinct regulatory site. Other enzymes comprise structurally different subunits; in these, active sites and regulatory sites may be located on different subunits.

Allosteric Release of Catalytic Subunits As mentioned previously, cAMP-dependent protein kinase (cAPK) exists as an inactive tetrameric protein composed of two catalytic subunits and two regulatory subunits. Each regulatory subunit contains a *pseudosubstrate* sequence that binds to the active site in a catalytic subunit. By blocking substrate binding, the regulatory subunit inhibits the activity of the catalytic subunit. Binding of the allosteric effector molecule **cyclic AMP (cAMP)** to the regulatory subunit induces a conformational change in the pseudosubstrate sequence so that it no longer can bind the catalytic subunit. Thus the inac-

tive tetramer dissociates into two monomeric active catalytic subunits and a dimeric regulatory subunit (Figure 3-27). As discussed in Chapter 20, binding of various hormones to cell-surface receptors induces a rise in the intracellular concentration of cAMP, leading to activation of cAPK. Once the signaling ceases and the cAMP level decreases, the activity of cAPK is turned off by reassembly of the inactive tetramer.

Allosteric Transition between Active and Inactive States Many multimeric enzymes undergo allosteric transitions that alter the relationship of the subunits to one another but do not cause dissociation as in cAPK. A well-understood enzyme illustrating this mechanism is aspartate transcarbamoylase (ATCase). This bacterial enzyme catalyzes the first step in the pyrimidine biosynthetic pathway:

$$\text{Aspartate} + \text{carbamoyl phosphate} \xrightarrow{\text{ATCase}} N\text{-carbamoylaspartate}$$

ATCase, which is composed of six catalytic subunits and six regulatory subunits, exists in an active R state and inactive T state (Figure 3-28). The equilibrium between these states is shifted toward the inactive T state by binding of cytidine

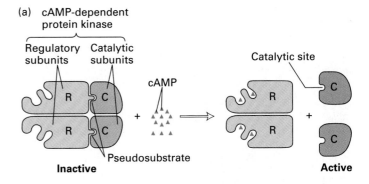

(a) cAMP-dependent protein kinase

Inactive — Regulatory subunits / Catalytic subunits / Pseudosubstrate

+ cAMP →

Active — Catalytic site

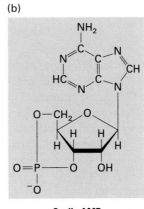

(b)

Cyclic AMP

▲ **FIGURE 3-27 Activation of cAMP-dependent protein kinase (cAPK) by cyclic AMP.**
(a) At low concentrations of cAMP, the enzyme exists as an inactive tetramer composed of two regulatory (R) and two catalytic (C) subunits. The tetrameric protein is inactive because the pseudosubstrate sequences on the R subunits block the active sites on the C subunits. Binding of cAMP to the regulatory subunits causes release of the active monomeric catalytic subunits. (b) Structure of cAMP. This unusual nucleotide, which acts as a "second messenger" in many intracellular signaling pathways, controls the activity of many proteins.

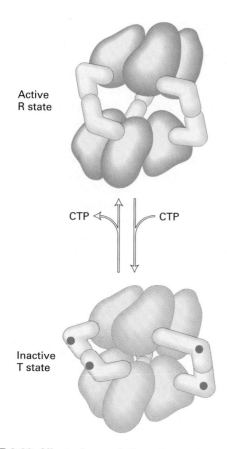

Active
R state

CTP $\rightleftarrows$ CTP

Inactive
T state

▲ **FIGURE 3-28 Allosteric regulation of aspartate transcarbamoylase (ATCase), the initial enzyme in synthesis of pyrimidines.** This enzyme comprises a pair of trimeric catalytic subunits (orange) connected by three pairs of dimeric regulatory subunits (green). Binding of cytidine triphosphate (CTP; the blue dot) to the regulatory subunits causes a conformational transition from the active R state to the inactive T state. The more open conformation of the R state permits substrate binding. Thus an increase in the concentration of CTP, an end product in the pyrimidine pathway, shuts off ATCase, an example of feedback inhibition. [Adapted from B. Mathews and K. E. van Holde, 1996, *Biochemistry*, p. 393.]

triphosphate (CTP), an end product of the pyrimidine pathway, to the regulatory subunits. Thus CTP is an allosteric inhibitor of ATCase. The CTP-induced allosteric transition in ATCase is an example of *feedback inhibition*, whereby an enzyme that catalyzes an early reaction in a multistep pathway is inhibited by an ultimate product of the pathway. Clearly, this type of regulation prevents accumulation of pyrimidines in excess of what the cell needs for DNA synthesis.

The mechanism of feedback inhibition helps regulate most biosynthetic pathways; that is, the final product of the pathway inhibits the enzyme that catalyzes the first step, thus preventing both production of the intermediate products and unnecessary metabolic activity. Feedback inhibition of enzyme function is reversible. If the concentration of free feedback inhibitor (e.g., CTP) falls, the bound inhibitor dis-

sociates from the regulated enzyme, which then reverts to its active conformation. The binding of a feedback inhibitor to an enzyme and its subsequent release can be described by the equilibrium binding constant K_i, which is similar to the Michaelis constant K_m used to describe substrate binding.

Cooperative Binding of Ligands In many cases, especially when a protein binds several molecules of one ligand, the binding is graded; that is, binding of one ligand molecule affects the binding of subsequent ligand molecules. Such *cooperative allostery*, or cooperative binding, permits many multisubunit proteins to respond more efficiently to small changes in ligand concentration than would otherwise be possible. In positive cooperativity, sequential binding is enhanced; in negative cooperativity, sequential binding is inhibited.

Hemoglobin presents a classic example of positive cooperative binding. Each of the four subunits in hemoglobin can bind one oxygen molecule. Binding of oxygen to one subunit induces a local conformational change whose effect spreads to the other subunits, lowering the K_m for binding of additional oxygen molecules.

Many multimeric enzymes, including aspartate transcarbamoylase (ATCase), also exhibit cooperative binding of substrate. For reactions catalyzed by such enzymes, a plot of reaction velocity versus substrate concentration yields a sigmoidal curve rather than the hyperbolic curve characteristic of enzymes with typical Michaelis-Menten kinetics. As a result of cooperative substrate binding, the maximal enzyme activity (V_{max}) is achieved over a narrow range of substrate concentration (Figure 3-29). Other multimeric enzymes exhibit cooperative binding of an allosteric inhibitor. Because of cooperative allostery, a quite small change in ligand concentration can effectively turn an enzyme on or off.

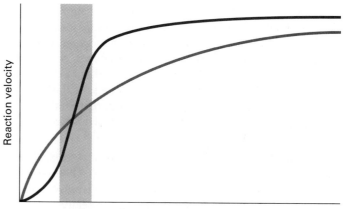

Substrate concentration

▲ **FIGURE 3-29 Enzymes with multiple active sites commonly exhibit sigmoidal kinetics (red curve), indicative of cooperative binding of substrates, rather than typical Michaelis-Menten kinetics (blue curve).** Small changes in substrate concentration (pink-shaded region) can effectively switch such allosteric enzymes on or off.

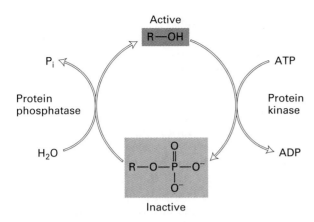

▲ **FIGURE 3-30 Cyclic phosphorylation and dephosphoryla-tion is a common cellular mechanism for regulating protein activity.** In this example, the target protein R (orange) is inactive when phosphorylated and active when dephosphorylated; the opposite pattern occurs in some proteins.

Cyclic Protein Phosphorylation and Dephosphoryla-tion As noted earlier, one of the most common mechanisms for regulating protein activity is the addition and removal of phosphate groups from serine, threonine, or tyrosine residues. Protein **kinases** catalyze phosphorylation, and **phosphatases** catalyze dephosphorylation. Although both re-actions are essentially irreversible, the counteracting activi-ties of kinases and phosphatases provide cells with a "switch" that can turn on or turn off the function of vari-ous proteins (Figure 3-30). Phosphorylation changes a pro-tein's charge and generally leads to a conformational change; these effects can significantly alter ligand binding by a pro-tein and its activity.

Nearly 3 percent of all yeast proteins are protein kinases or phosphatases, reflecting the importance of phosphoryla-tion and dephosphorylation reactions even in simple cells. These enzymes target all classes of proteins including struc-tural proteins, enzymes, membrane channels, and signaling molecules. In later chapters, we will encounter many exam-ples of important cellular functions that are controlled by phosphorylation and dephosphorylation of specific proteins.

Proteolytic Activation The regulatory mechanisms dis-cussed so far can act like switches, turning proteins on and off. Regulation of some enzymes involves the irreversible ac-tivation of an inactive form, commonly at the site where the active enzyme is needed. Good examples of such enzymes are the digestive proteases trypsin and chymotrypsin, which are synthesized in the pancreas and secreted into the small intestine as inactive precursors, or *zymogens*, called *trypsino-gen* and *chymotrypsinogen*, respectively. In the protease-rich environment of the small intestine, the zymogens are con-verted to the active enzymes, which then begin to hydrolyze the peptide bonds of ingested proteins. As shown in Figure 3-31, two irreversible proteolytic cleavages of chymotrypsino-gen yield active chymotrypsin. The delay in activation of

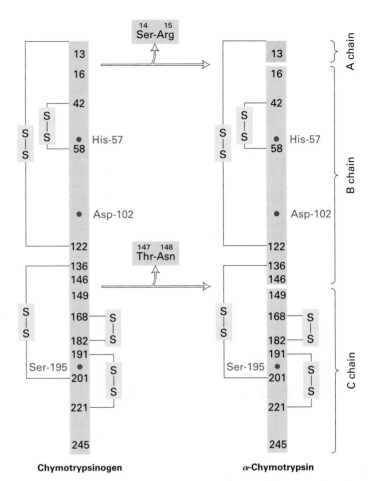

▲ **FIGURE 3-31 A linear representation of the conversion of chymotrypsinogen into chymotrypsin by the excision of two dipeptides.** These reactions yield three separate chains (A, B, and C), which are covalently linked by disulfide bonds (yellow) in the active enzyme. In the folded, native conformation of chymotrypsin, histidine 57, aspartate 102, and serine 195 are located in the active site.

these proteases until they reach the intestine prevents them from digesting the pancreatic tissue in which they are made.

Other Regulatory Mechanisms The activities of enzymes are extensively regulated in order that the numerous enzymes in a cell can work together harmoniously. All metabolic pathways are closely controlled at all times. Synthetic reac-tions occur when the products of these reactions are needed; degradative reactions occur when molecules must be broken down. All the regulatory mechanisms described above affect enzymes locally at their site of action.

Regulation of cellular processes, however, involves more than simply turning enzymes on and off. Some regulation is accomplished by keeping enzymes in compartments where the delivery of substrate or exit of product is controlled. In many cases, the compartments are organelles, such as the mi-tochondria, nuclei, or lysosomes. Compartmentation permits competing reactions to occur simultaneously in different parts of a cell. In addition to compartmentation, cellular processes are regulated by enzyme synthesis and destruction. Often enzymes are synthesized at low rates when the cell has no need for their activities; however, upon increased demands by the cell (for instance, appearance of substrate), new enzyme is synthesized. Later, the pool of enzyme is lowered when levels of substrate decrease or the cell becomes inactive.

SUMMARY   Functional Design of Proteins

- The function of nearly all proteins depends on their ability to bind other molecules (ligands). Ligand-binding sites on proteins and the corresponding ligands are chemically and topologically complementary. The affinity of a protein for a particular ligand refers to the strength of binding; its specificity, to the restriction of binding to one or a few preferred ligands.

- Enzymes are catalytic proteins that accelerate the rate of cellular reactions by lowering the activation energy and stabilizing transition-state intermediates.

- Enzyme active sites comprise two functional parts: a substrate-binding region and a catalytic region. The amino acids composing the active site are not necessarily adjacent in the amino acid sequence, but are brought into proximity in the native conformation.

- The kinetics of many enzymes are described by the Michaelis-Menten equation. From plots of reaction rate versus substrate concentration, two characteristic parameters of an enzyme can be determined: the Michaelis constant K_m, a measure of the enzyme's affinity for substrate, and the maximal velocity V_{max} (see Figure 3-26).

- Many multimeric enzymes and other proteins exhibit allostery. In this phenomenon, binding of one ligand molecule (a substrate, activator, or inhibitor) induces a conformational change, or allosteric transition, that alters the protein's activity or affinity for other ligands.

- In multimeric proteins that bind multiple ligands, binding of one ligand molecule may increase or decrease the binding affinity for subsequent ligand molecules. Enzymes that cooperatively bind substrates exhibit sigmoidal kinetics (see Figure 3-29).

- Allosteric mechanisms can act like switches, turning protein activity on and off. Cyclic phosphorylation and dephosphorylation of amino acid side chains can have the same regulatory effect. Proteolytic cleavage irreversibly converts inactive zymogens into active enzymes.

3.4   Membrane Proteins

As we've seen, all antibodies have a similar structure and function; enzymes are structurally varied, but all have a catalytic function. In contrast, although all membrane proteins are located at the membrane, they otherwise are both structurally and functionally diverse. As we noted in Chapter 2 and discuss in more detail in Chapter 5, every biological membrane has the same basic phospholipid bilayer structure. Associated with each membrane is a set of membrane proteins that enables the membrane to carry out its distinctive activities (Figure 3-32). The complement of proteins attached to a membrane varies depending on cell type and subcellular location.

Some proteins are bound only to the membrane surface, whereas others have one region buried within the membrane and domains on one or both sides of it. Protein domains on the extracellular membrane surface are generally involved in cell-cell signaling or interactions. Domains within the membrane, particularly those that form channels and pores, move molecules across the membrane. Domains lying along the cytosolic face of the membrane have a wide range of functions, from anchoring cytoskeletal proteins to the membrane to triggering intracellular signaling pathways. In many cases, the function of a membrane protein and the topology of its polypeptide chain in the membrane can be predicted based on its homology with another, well-characterized protein. In this section, we examine the characteristic structural features of membrane proteins and some of their basic functions. More complete characterization of the structure and function of various types of membrane proteins is presented in several later chapters. The synthesis and processing of membrane proteins are discussed in Chapter 17.

Proteins Interact with Membranes in Different Ways

Membrane proteins can be classified into two broad categories—integral (intrinsic) and peripheral (extrinsic)—based on the nature of the membrane-protein interactions (see Figure 3-32). Most biomembranes contain both types of membrane proteins.

Integral membrane proteins, also called *intrinsic proteins,* have one or more segments that are embedded in the phospholipid bilayer. Most integral proteins contain residues with hydrophobic side chains that interact with fatty acyl groups of the membrane phospholipids, thus anchoring the protein to the membrane. Most integral proteins span the entire phospholipid bilayer. These *transmembrane* proteins contain one or more membrane-spanning domains as well as domains, from four to several hundred residues long, extending into the aqueous medium on each side of the bilayer. In all the transmembrane proteins examined to date, the membrane-spanning domains are α helices or multiple β strands. In contrast, some integral proteins are anchored to one of the membrane leaflets by covalently bound fatty acids, as discussed later. In these proteins, the bound fatty acid is embedded in the membrane, but the polypeptide chain does not enter the phospholipid bilayer.

Peripheral membrane proteins, or extrinsic proteins, do not interact with the hydrophobic core of the phospholipid

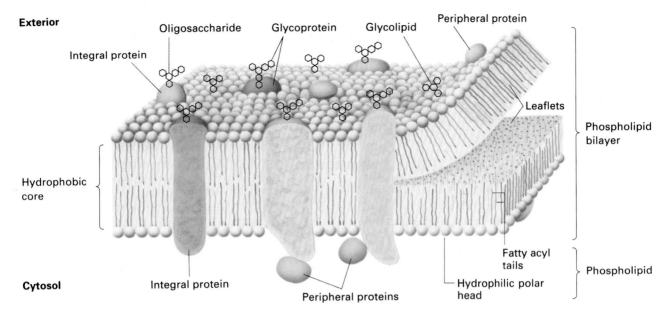

▲ FIGURE 3-32 Schematic diagram of typical membrane proteins in a biological membrane. The phospholipid bilayer, the basic structure of all cellular membranes, consists of two leaflets of phospholipid molecules whose fatty acyl tails form the hydrophobic interior of the bilayer; their polar, hydrophilic head groups line both surfaces. Most integral proteins span the bilayer as shown; a few are tethered to one leaflet by a covalently attached lipid anchor group. Peripheral proteins are primarily associated with the membrane by specific protein-protein interactions. Oligosaccharides bind mainly to membrane proteins; however, some bind to lipids, forming glycolipids.

bilayer. Instead they are usually bound to the membrane indirectly by interactions with integral membrane proteins or directly by interactions with lipid polar head groups. Peripheral proteins localized to the cytosolic face of the plasma membrane include the cytoskeletal proteins spectrin and actin in erythrocytes (Chapter 18) and the enzyme protein kinase C. This enzyme shuttles between the cytosol and the cytosolic face of the plasma membrane and plays a role in signal transduction (Chapter 20). Other peripheral proteins, including certain proteins of the extracellular matrix, are localized to the outer (exoplasmic) surface of the plasma membrane.

Hydrophobic α Helices in Transmembrane Proteins Are Embedded in the Bilayer

Integral proteins containing membrane-spanning α-helical domains are embedded in membranes by hydrophobic interactions with the lipid interior of the bilayer and probably also by ionic interactions with the polar head groups of the phospholipids. *Glycophorin*, a major erythrocyte membrane protein, exhibits both types of interaction. As shown in Figure 3-33, glycophorin contains a membrane-embedded α helix composed entirely of hydrophobic (or uncharged) amino acids. The predicted length of this α helix (3.75 nm) is just sufficient to span the hydrocarbon core of a phospholipid bilayer. The hydrophobic side chains form van der Waals interactions with the fatty acyl chains and shield the polar carbonyl (C=O) and imino (NH) groups of the peptide bond, which are all hydrogen-bonded to one another. This hydrophobic helix is prevented from slipping across the membrane by a flanking set of positively charged amino acids (lysine and arginine) that are thought to interact with negatively charged phospholipid head groups. In glycophorin, most of these charged residues lie adjacent to the cytosolic leaflet.

Many Integral Proteins Contain Multiple Transmembrane α Helices

Although Figure 3-33 depicts glycophorin as a monomer with a single α helix spanning the bilayer, this protein is present in erythrocyte membranes as a dimer of two identical polypeptide chains. The two membrane-spanning α helices of glycophorin are thought to form a coiled-coil structure (see Figure 3-9a) stabilized by specific interactions between the amino acid side chains at the interface of the two helices. It is now known that many other transmembrane proteins contain two or more membrane-spanning α helices. For instance, the *bacterial photosynthetic reaction center (PRC)* comprises four subunits and several prosthetic groups, including four chlorophyll molecules. In this complex protein, three of the four subunits span the membrane; two of these subunits (L and M) each contain five membrane-spanning α helices (see Figure 16-40).

A large and important family of integral proteins is defined by the presence of seven membrane-spanning α helices.

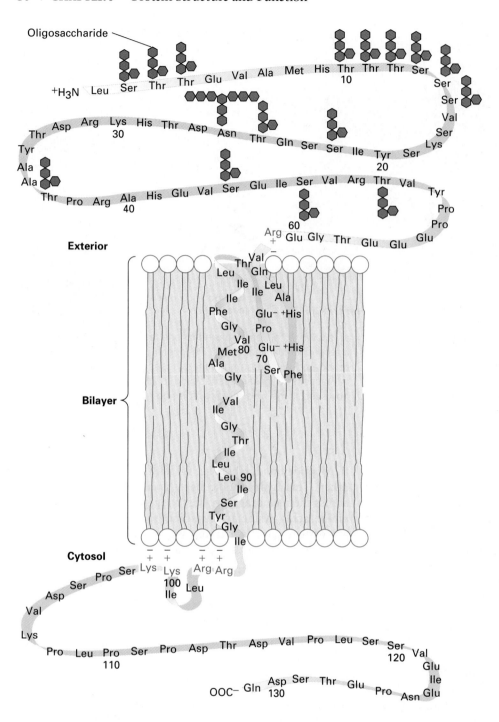

◀ **FIGURE 3-33 Amino acid sequence and transmembrane disposition of glycophorin A from the erythrocyte plasma membrane.** This protein is a homo-dimer, but only one of its polypeptide chains is shown. Residues 62–95 are buried in the membrane, with the sequence from position 73 through 95 forming an α helix. The ionic interactions shown between positively charged arginine and lysine residues and negatively charged phospholipid head groups in the cytosolic and exoplasmic faces of the membrane are hypothetical. Both the amino-terminal segment of the molecule, located outside the cell, and the carboxy-terminal segment, located inside the cell, are rich in charged residues and polar uncharged residues, making these domains water-soluble. Note the numerous carbohydrate residues attached to amino acids in the exoplasmic domain. [See V. T. Marchesi, H. Furthmayr, and M. Tomita, 1976, *Ann. Rev. Biochem.* **45**:667; A. H. Ross et al., 1982, *J. Biol. Chem.* **257**:4152.]

More than 150 such "seven-spanning" membrane proteins have been identified. This class of integral proteins is typified by *bacteriorhodopsin*, a protein found in a photosynthetic bacterium (Figure 3-34). Absorption of light by the retinal group attached to bacteriorhodopsin causes a conformational change in the protein that results in pumping of protons from the cytosol across the bacterial membrane to the extracellular space. The proton concentration gradient thus generated across the membrane is used to synthesize ATP, as discussed in Chapter 16. Both the overall arrangement of the seven α helices in bacteriorhodopsin and the identity of most of the amino acids can be resolved by computer analysis of micrographs of two-dimensional crystals of the membrane-embedded protein taken at various angles to the electron beam.

Other seven-spanning membrane proteins include the opsins (eye proteins that absorb light), cell-surface receptors for many hormones, and receptors for odorous molecules. Amino acid sequence analysis of these proteins has shown that no amino acids are found in the same position in all of

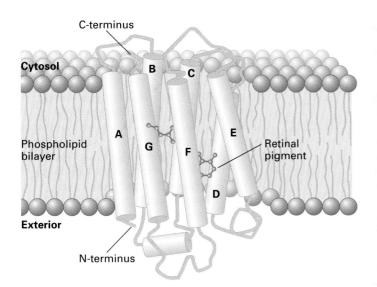

▲ **FIGURE 3-34 Overall structure of bacteriorhodopsin as deduced from electron diffraction analyses of two-dimensional crystals of the protein in the bacterial membrane.** The seven membrane-spanning α helices are labeled A–G. The retinal pigment is covalently attached to lysine 216 in helix G. The approximate position of the protein in the phospholipid bilayer is indicated. [Adapted from R. Henderson et al., 1990, *J. Mol. Biol.* **213**:899.

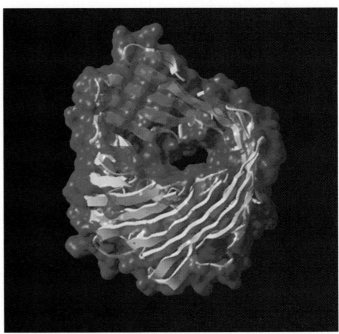

▲ **FIGURE 3-35 Model of the three-dimensional structure of a subunit of OmpF, a porin found in the *E. coli* outer membrane.** All porins are trimeric transmembrane proteins. Each subunit is barrel-shaped with β strands forming the wall and a transmembrane pore in the center. [Courtesy of S. Choe.]

them, and only a few residues are conserved in even a substantial number of them. Nonetheless, each of these proteins contains seven stretches of hydrophobic amino acids long enough (>22 amino acids) to span the phospholipid bilayer. Though direct evidence is lacking, it is thought that all of these proteins adopt a conformation in the membrane similar to that of bacteriorhodopsin. This is one of several examples of how investigators can predict the orientation of proteins in a membrane from the amino acid sequence alone.

Multiple β Strands in Porins Form Membrane-Spanning "Barrels"

The *porins* are a class of transmembrane proteins whose structure differs radically from that of other integral proteins. Several types of porin are found in the outer membrane of gram-negative bacteria such as *E. coli* (see Figure 1-7a). The outer membrane protects an intestinal bacterium from harmful agents (e.g., antibiotics, bile salts, and proteases) but permits the uptake and disposal of small hydrophilic molecules including nutrients and waste products. The porins in the outer membrane of an *E. coli* cell provide channels for passage of disaccharides, phosphate, and similar molecules.

The amino acid sequences of porins are predominantly polar and contain no long hydrophobic segments typical of integral proteins with α-helical membrane-spanning domains. X-ray crystallography has revealed that porins are trimers of identical subunits. In each subunit 16 β strands

form a barrel-shaped structure with a pore in the center (Figure 3-35). As noted earlier, half the amino acid side groups of a β strand point in one direction, and the other half point in the opposite direction (see Figure 3-8). Unlike a typical globular protein, porins have an inside-out arrangement. In a porin monomer, the outward-facing side groups on each of the β strands are hydrophobic and thus can interact with the fatty acyl groups of the membrane lipids or with other porin monomers. The side groups facing the inside of a porin monomer are predominantly hydrophilic; these line the pore through which small water-soluble molecules cross the membrane.

Covalently Attached Hydrocarbon Chains Anchor Some Proteins to the Membrane

In eukaryotic cells, as noted earlier, the polypeptide chain of some integral membrane proteins does not enter the bilayer but rather is anchored in one leaflet by a covalently attached hydrocarbon chain. Several common lipid anchors are shown in Figure 3-36.

Some cell-surface proteins are anchored to the exoplasmic face of the plasma membrane by a complex glycosylated phospholipid that is linked to the C-terminus. A common example of this type of anchor is *glycosylphosphatidylinositol*, which contains two fatty acyl groups, *N*-acetylglucosamine, mannose, and inositol (see Figure 3-36a). Several enzymes,

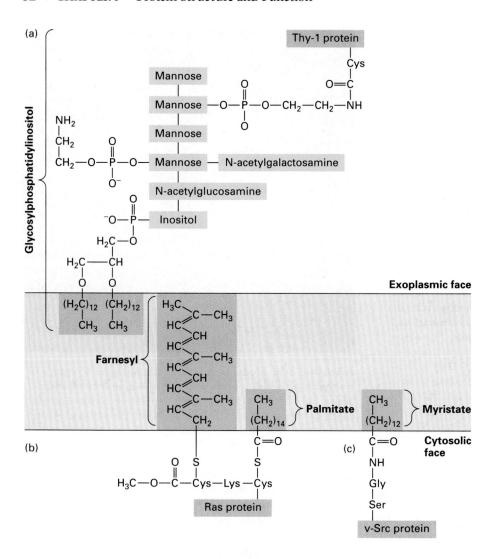

◀ FIGURE 3-36 **Anchoring of integral proteins to the plasma membrane by membrane-embedded hydrocarbon groups (highlighted in red).** (a) Thy-1 protein and several hydrolytic enzymes are anchored by glycosylphosphatidylinositol. This complex anchor is found only on the exoplasmic face. (b) Cytosolic proteins involved in signaling such as Ras are anchored to the cytosolic face of the membrane through farnesyl and palmitoyl groups. (c) Other cytosolic proteins are associated with the membrane through myristate and similar fatty acids attached to an N-terminal glycine residue.

including alkaline phosphatase, fall into this class. Various experiments have shown that the phospholipid anchor is both necessary and sufficient for binding these cell-surface proteins to the membrane. For instance, the enzyme phospholipase C cleaves the phosphate-glycerol bond in phospholipids as well as in glycosylphosphatidylinositol anchors, and treatment of cells with phospholipase C releases glycosylphosphatidylinositol-anchored proteins such as Thy-1 protein and alkaline phosphatase from the cell surface.

Some cytosolic proteins are anchored to the cytosolic face of membranes by a hydrocarbon moiety covalently attached to a cysteine near the C-terminus. The most common anchors are prenyl, farnesyl, and geranylgeranyl groups. These proteins undergo a chemical modification involving several steps. First, the anchor moiety forms a thioether bond with the thiol group of a cysteine that is four residues from the C-terminus of the protein. The modified protein then undergoes proteolysis and methylation; these reactions remove the three terminal residues and add a methyl to the new C-terminus. In some cases, fatty acyl palmitate groups

form thioester bonds to nearby cysteine residues, providing additional anchors that are thought to reinforce the attachment of the protein to the membrane (see Figure 3-36b).

In another group of lipid-anchored cytosolic proteins, a fatty acyl group (e.g., myristate or palmitate) is linked by an amide bond to the N-terminal glycine residue (see Figure 3-36c). In these proteins, the N-terminal anchor is necessary for retention at the membrane and may play an important role in a membrane-associated function. For example, v-Src, a mutant form of a cellular tyrosine kinase, is oncogenic and can transform cells only when it retains a myristylated N-terminus.

Some Peripheral Proteins Are Soluble Enzymes That Act on Membrane Components

An important group of peripheral membrane proteins are water-soluble enzymes that associate with the polar head groups of membrane phospholipids. One well-understood group of such enzymes are the *phospholipases*, which hydrolyze various bonds in the head groups of phospholipids

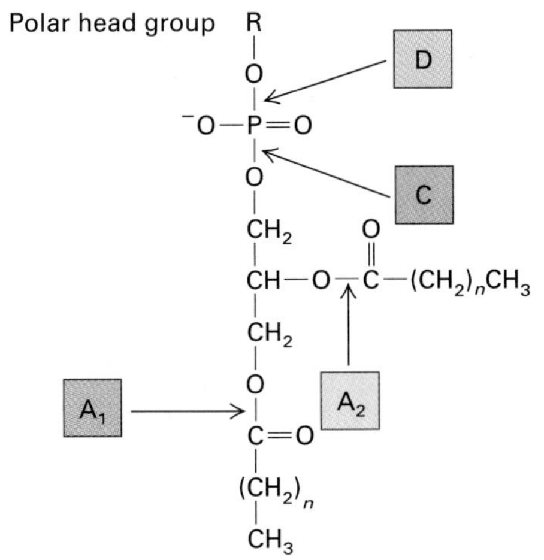

Polar head group   R
|
O
|
$^-$O—P=O
|
O
|
CH_2 O
| ||
CH—O—C—$(CH_2)_n CH_3$
|
CH_2
|
O
|
C=O
|
$(CH_2)_n$
|
CH_3

▲ **FIGURE 3-37 Specificity of cleavage of phospholipids by phospholipases A_1, A_2, C, and D.** Susceptible bonds are shown in red. R denotes the polar group attached to the phosphate, such as choline in phosphatidylcholine (see Figure 5-27a) or inositol in phosphatidylinositol.

(Figure 3-37). These enzymes have an important role in the degradation of damaged or aged cell membranes.

The mechanism of action of phospholipase A_2 illustrates how such water-soluble enzymes can reversibly interact with membranes and catalyze reactions at the interface of an aqueous solution and lipid surface. When this enzyme is in aqueous solution, its Ca^{2+}-containing active site is buried in a channel lined with hydrophobic amino acids. Binding of the enzyme to a phospholipid bilayer induces a small conformational change that fixes the protein to the phospholipid heads and opens the hydrophobic cleft. As a phospholipid molecule moves from the bilayer into the channel, the enzyme-bound Ca^{2+} binds to the phosphate in the head group and positions the ester bond to be cleaved next to the catalytic site.

SUMMARY   Membrane Proteins

- Biological membranes usually contain both integral and peripheral membrane proteins (see Figure 3-32).

- Integral membrane proteins include transmembrane proteins and lipid-anchored proteins.

- Two types of membrane-spanning domains are found in transmembrane proteins: one or more α helices or, less commonly, multiple β strands (as in porins). Proteins containing seven membrane-spanning α helices form a major class that includes bacteriorhodopsin and many cell-surface receptors.

- When the polypeptide chain of a transmembrane protein spans the membrane multiple times, the core of the protein generally is hydrophilic, permitting passage of water-soluble molecules, and the surface is hydrophobic, permitting interaction with the interior of the lipid bilayer.

- Amino acid residues modified with long-chain hydrocarbons anchor some integral proteins to one membrane leaflet (see Figure 3-36).

- Peripheral membrane proteins interact with integral membrane proteins or with the polar head groups of membrane phospholipids. They do not enter the hydrophobic core of the membrane.

3.5   Purifying, Detecting, and Characterizing Proteins

A protein must be purified before its structure and the mechanism of its action can be studied. However, because proteins vary in size, charge, and water solubility, no single method can be used to isolate all proteins. To isolate one particular protein from the estimated 10,000 different proteins in a cell is a daunting task that requires methods both for separating proteins and for detecting the presence of specific proteins.

Any molecule, whether protein, carbohydrate, or nucleic acid, can be separated from other molecules based on large differences in some physical characteristic. Although the sequence of amino acids in a protein uniquely determines its function, the most useful physical characteristic for separation of proteins is *size*, defined as either length or mass. In this section, we briefly outline different techniques for separating proteins based on their size and other properties. These techniques also apply to the separation of nucleic acids and other biomolecules. We then consider general methods for detecting, or *assaying*, specific proteins, including the use of radioactive compounds for tracking biological activity. Finally, we discuss several techniques for characterizing a protein's mass, sequence, and three-dimensional structure.

Proteins Can Be Removed from Membranes by Detergents or High-Salt Solutions

Because water-soluble globular proteins have many exposed hydrophilic groups, they maintain their native conformation and remain individually suspended in an aqueous medium when separated from cells. In contrast, when transmembrane proteins are separated from membranes, their exposed hydrophobic regions interact, causing the protein molecules to aggregate and precipitate from aqueous solutions. Such proteins can be solubilized by detergents, which have affinity both for hydrophobic groups and for water.

Detergents are amphipathic molecules that disrupt membranes by intercalating into phospholipid bilayers and solubilizing lipids and proteins. The hydrophobic part of a detergent molecule is attracted to hydrocarbons and mingles with them readily; the hydrophilic part is strongly attracted to water. Some detergents are natural products, but most are synthetic molecules developed for cleaning and for dispersing mixtures of oil and water (Figure 3-38). Ionic detergents, such as sodium deoxycholate and sodium dodecylsulfate

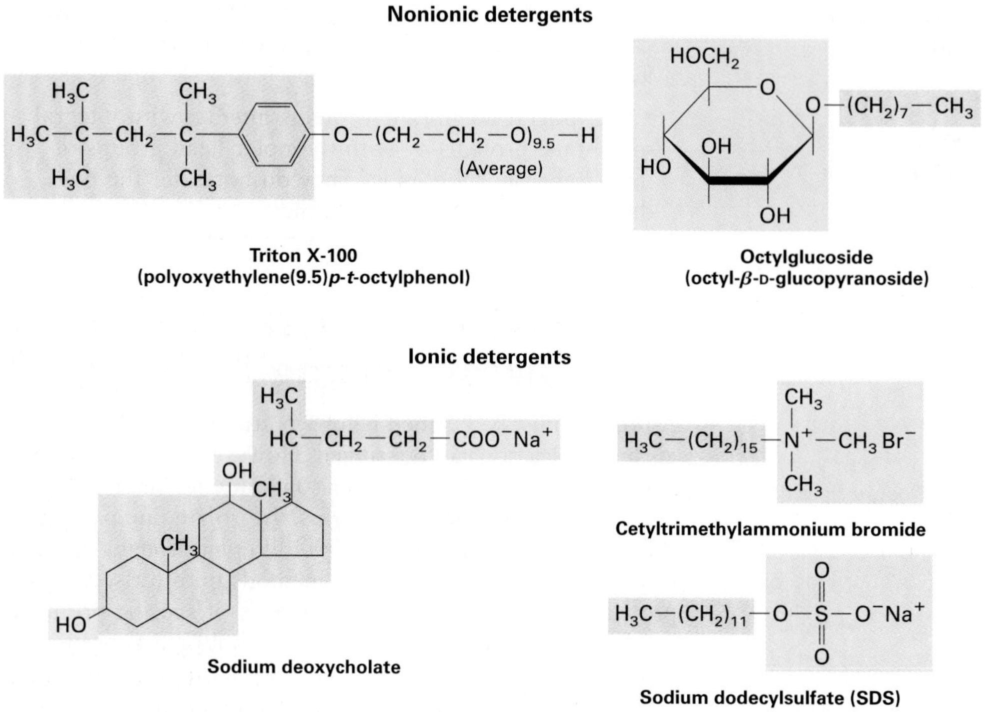

Nonionic detergents

Triton X-100
(polyoxyethylene(9.5)*p-t*-octylphenol)

Octylglucoside
(octyl-β-D-glucopyranoside)

Ionic detergents

Cetyltrimethylammonium bromide

Sodium deoxycholate

Sodium dodecylsulfate (SDS)

◀ **FIGURE 3-38 Structures of five common detergents.** The bile salt sodium deoxycholate is a natural product; the others are synthetic ones. The hydrophobic portion of each molecule is shown in yellow; the hydrophilic portion, in blue.

(SDS), contain a charged group; nonionic detergents, such as Triton X-100 and octylglucoside, lack a charged group.

At very low concentrations, detergents dissolve in pure water as isolated molecules. As the concentration increases, the molecules begin to form *micelles*. These are small, spherical aggregates in which hydrophilic parts of the molecules face outward and the hydrophobic parts cluster in the center (see Figure 2-20). The *critical micelle concentration (CMC)* at which micelles form is characteristic of each detergent and is a function of the structures of its hydrophobic and hydrophilic parts.

Ionic detergents bind to the exposed hydrophobic regions of membrane proteins as well as to the hydrophobic core of water-soluble proteins. Because of their charge, these detergents also disrupt ionic and hydrogen bonds. At high concentrations, for example, sodium dodecylsulfate completely denatures proteins by binding to every side chain. *Nonionic detergents* act in different ways at different concentrations. At high concentrations (above the CMC), they solubilize biological membranes by forming mixed micelles of detergent, phospholipid, and integral membrane proteins (Figure 3-39). At low concentrations (below the CMC), these detergents may

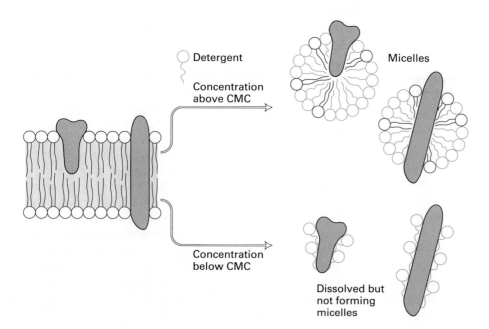

Detergent

Concentration above CMC

Micelles

Concentration below CMC

Dissolved but not forming micelles

◀ **FIGURE 3-39 Solubilization of integral membrane proteins by nonionic detergents.** At a concentration higher than its critical micelle concentration (CMC), a detergent solubilizes lipids and integral membrane proteins, forming mixed micelles containing detergent, protein, and lipid molecules. At concentrations below the CMC, many detergents (e.g., octylglucoside) can dissolve membrane proteins without forming micelles by coating the membrane-spanning regions. Since octylglucoside has a high CMC, it is particularly effective in solubilizing integral membrane proteins without denaturing them or forming mixed micelles.

bind to the hydrophobic regions of most membrane proteins, making them soluble in aqueous solution. In this case, although mixed micelles are not formed, the solubilized protein will not aggregate during subsequent purification steps.

Most peripheral proteins are bound to specific integral membrane proteins by ionic or other weak interactions. Generally they can be removed from the membrane by solutions of high ionic strength (high salt concentrations), which disrupt ionic bonds, or by chemicals that bind divalent cations such as Mg^{2+}. Unlike integral proteins most peripheral proteins are soluble in aqueous solution and are not solubilized by nonionic detergents.

Centrifugation Can Separate Particles and Molecules That Differ in Mass or Density

The first step in a typical protein-purification scheme is centrifugation. The principle behind centrifugation is that two particles in suspension (cells, organelles, or molecules), having different masses or densities will settle to the bottom of a tube at different rates. Remember, *mass* is the weight of a sample (measured in grams), whereas *density* is the ratio of its weight to volume (grams/liter). Proteins vary greatly in mass but not in density. The average density of a protein is 1.37 g/cm^3. Unless a protein has an attached lipid or carbohydrate, its density will not vary by more than 15 percent from this value. Table 3-1 lists the density and other physical characteristics of several blood proteins. Heavier or more dense molecules settle, or sediment, more quickly than lighter or less dense molecules.

A centrifuge speeds sedimentation by subjecting particles in suspension to centrifugal forces as great as 600,000 times the force of gravity g. The centrifugal force is proportional to the rotation rate of the rotor (measured in revolutions per minute, or rpm) and the distance of the tube from the center of the rotor. Modern ultracentrifuges reach speeds of 60,000 rpm or greater and generate forces sufficient to sediment particles with masses greater than 10,000 daltons

(Da). However, small particles with masses of 5 Da or less will not sediment uniformly even at such high rotor speeds.

Centrifugation is used for two basic purposes: (1) as a preparative technique to separate one type of material from others and (2) as an analytical technique to measure physical properties (e.g., molecular weight, density, shape, and equilibrium binding constants) of macromolecules. The *sedimentation constant, s,* of a protein equals its velocity in a centrifugal field divided by the centrifugal force. The value of s depends on a protein's density and shape, as well as the density and viscosity of the medium. Because the centrifugal force and the density and viscosity of the medium are all known, the radius and mass of a molecule can be calculated from measurements of its rate of movement in an analytical ultracentrifuge. The sedimentation constant is commonly expressed in svedbergs (S): $1 \text{ S} = 10^{-13}$ seconds.

Differential Centrifugation The most common initial step in protein purification is separation of soluble proteins from insoluble cellular material by *differential centrifugation* (Figure 3-40a). The centrifugal force and duration of centrifugation are adjusted to ensure that the insoluble materials sediment into a pellet. After a starting mixture of a cell homogenate is poured into a tube and spun in a centrifuge, cell organelles such as nuclei collect into a pellet, but the soluble proteins remain in the supernatant. The supernatant fraction still contains a large mixture of proteins, which can be collected by decanting the supernatant and then subjecting it to further purification methods.

Rate-Zonal Centrifugation Based on differences in their mass, proteins can be separated by centrifugation through a solution, usually containing sucrose (an inert sugar), of increasing density called a *density gradient*. When mixtures of proteins are layered on top of a sucrose gradient in a tube and subjected to centrifugation, they migrate down the tube at a rate controlled by the factors that affect the sedimentation constant. The proteins start from a thin zone at the top of the tube and separate into bands, or zones (actually

TABLE 3-1	Physical Characteristics of Selected Blood Proteins				
Protein	Molecular Weight (kDa)	Sedimentation Constant (S)*	Density (g/cm³)	pI	Concentration in Plasma (mg/mL)
Immunoglobulin M	1000	18–20	1.38	5.1–7.8	0.8–0.9
α₂-Macroglobulin	820	19.6	1.36	5.4	2.65 (men) 3.35 (women)
High-density lipoprotein	435	5.5	0.91	—	0.37–1.17
Immunoglobulin G	153	6.6–7.2	1.35	5.8–7.3	12–18
Transferrin	76	4.9	1.38	5.2	2–4
Serum albumin	69	4.6	1.36	4.9	35–45
Retinol-binding protein	21	2.3	1.39	4.4–4.8	0.04–0.06
Lysozyme	15	2.19	1.39	10.5	0.005

*S = svedberg unit = 10^{-13} seconds

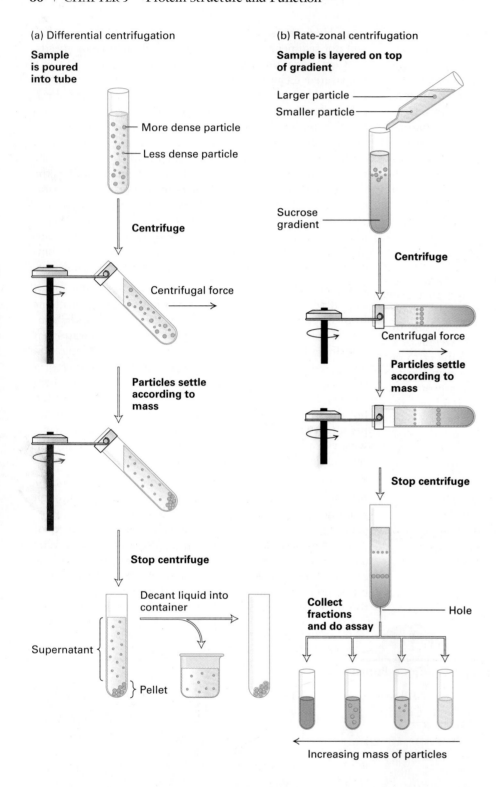

(a) Differential centrifugation

Sample is poured into tube

— More dense particle
— Less dense particle

Centrifuge

Centrifugal force

Particles settle according to mass

Stop centrifuge

Decant liquid into container

Supernatant

Pellet

(b) Rate-zonal centrifugation

Sample is layered on top of gradient

Larger particle
Smaller particle

Sucrose gradient

Centrifuge

Centrifugal force

Particles settle according to mass

Stop centrifuge

Collect fractions and do assay

Hole

Increasing mass of particles

◄ **FIGURE 3-40 Two common centrifugation techniques for separating particles.** (a) Differential centrifugation separates a mixture of particles (macromolecules, cell organelles, and cells) that differ in mass or density. The most dense particles collect at the bottom of the tube as a pellet. The least dense particles remain in the liquid supernatant, which can be transferred to another tube. (b) Rate-zonal centrifugation separates particles or molecules that differ in mass but may be similar in shape and density (e.g., RNA molecules). Here two particles of different mass separate into two zones.

disks), of proteins of different masses. This density-gradient separation technique is called *rate-zonal centrifugation* (Figure 3-40b). Samples are centrifuged just long enough to separate the molecules of interest. If they are centrifuged for too short a time, the molecules will not separate sufficiently. If they are centrifuged much longer than necessary, all the molecules will end up in a pellet at the bottom of the tube.

Although the sedimentation rate is strongly influenced by particle mass, rate-zonal centrifugation is seldom effective in determining *precise* molecular weights because variations in shape also affect sedimentation rate. The exact effects of shape are hard to assess, especially for proteins and single-stranded nucleic acid molecules that can assume many complex shapes. Nevertheless, rate-zonal centrifugation has proved to be the most practical method for separating many different types of polymers and particles. A second density-gradient technique, called *equilibrium density-gradient centrifugation,* is used mainly to separate DNA or organelles (see Figure 5-24).

Electrophoresis Separates Molecules according to Their Charge:Mass Ratio

Electrophoresis is a technique for separating, or *resolving,* molecules in a mixture under the influence of an applied electric field. Dissolved molecules in an electric field move, or migrate, at a speed determined by their charge:mass ratio. For example, if two molecules have the same mass and shape, the one with the greater net charge will move faster toward an electrode. The separation of small molecules, such as amino acids and nucleotides, is one of the many uses of electrophoresis. In this case, a small drop of sample is deposited on a strip of filter paper or other porous substrate, which is then soaked with a conducting solution. When an electric field is applied at the ends of the strip, small molecules dissolved in the conducting solution move along the strip at a rate corresponding to the magnitude of their charge.

SDS-Polyacrylamide Gel Electrophoresis Because many proteins or nucleic acids that differ in size and shape have nearly identical charge:mass ratios, electrophoresis of these macromolecules in solution results in little or no separation of molecules of different lengths. However, successful separation of proteins and nucleic acids can be accomplished by electrophoresis in various *gels* (semisolid suspensions in water) rather than in a liquid solution. Electrophoretic separation of proteins is most commonly performed in *polyacrylamide gels.* These gels are cast between a pair of glass plates by polymerizing a solution of acrylamide monomers into polyacrylamide chains and simultaneously cross-linking the chains into a semisolid matrix. The *pore size* of a gel can be varied by adjusting the concentrations of polyacrylamide and the cross-linking reagent.

When a mixture of proteins is applied to a gel and an electric current applied, smaller proteins migrate faster than larger proteins through the gel. The rate of movement is influenced by the gel's pore size and the strength of the electric field. The pores in a highly cross-linked polyacrylamide gel are quite small. Such a gel could resolve small proteins and peptides, but large proteins would not be able to move through it.

In what is probably the most powerful technique for resolving protein mixtures, proteins are exposed to the ionic detergent SDS (sodium dodecylsulfate) before and during gel electrophoresis (Figure 3-41). SDS denatures proteins, causing multimeric proteins to dissociate into their subunits, and all polypeptide chains are forced into extended conformations with similar charge:mass ratios. SDS treatment thus eliminates the effect of differences in shape, so that chain length, which reflects mass, is the sole determinant of the

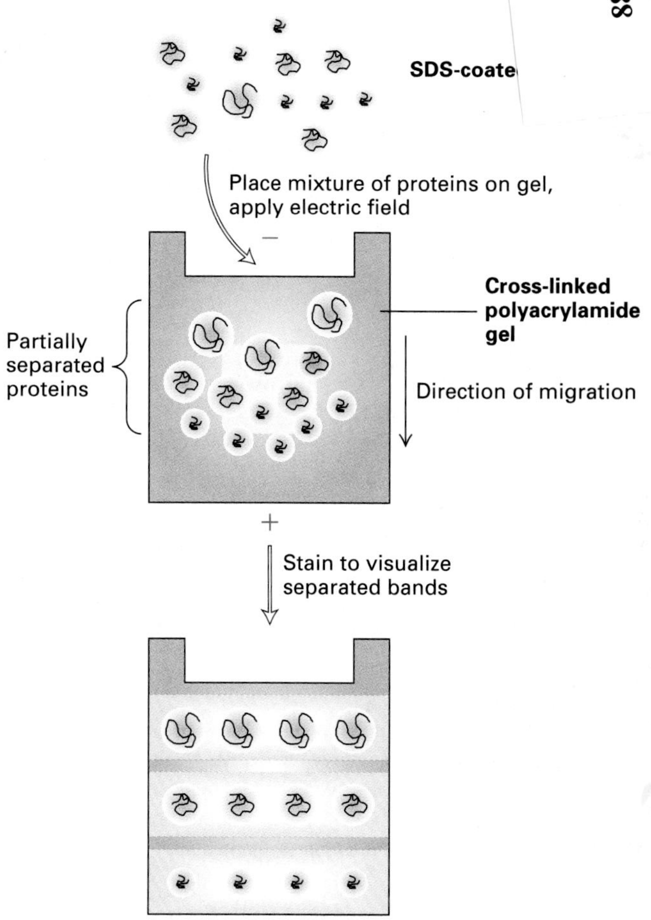

▲ **FIGURE 3-41 SDS-polyacrylamide gel electrophoresis, a common technique for separating proteins at good resolution.** The protein mixture first is treated with SDS, a negatively charged detergent that binds to proteins. This binding dissociates multimeric proteins and forces all polypeptide chains into denatured conformations with nearly identical charge:mass ratios. During electrophoresis, the SDS-protein complexes migrate through the polyacrylamide gel. Small proteins are able to move through the pores more easily, and faster, than larger proteins. Thus the proteins separate into bands according to their size as they migrate through the gel. The separated protein bands are visualized by staining with a dye.

migration rate of proteins in SDS-polyacrylamide electrophoresis. Even chains that differ in molecular weight by less than 10 percent can be separated by this technique. Moreover, the molecular weight of a protein can be estimated by comparing the distance it migrates through a gel with the distances that proteins of known molecular weight migrate.

Two-Dimensional Gel Electrophoresis Electrophoresis of all cellular proteins through an SDS gel can separate proteins having relatively large differences in molecular weight but cannot resolve proteins having similar molecular weights (e.g., a 41-kDa protein from a 42-kDa protein). To separate proteins of similar mass, another physical characteristic must

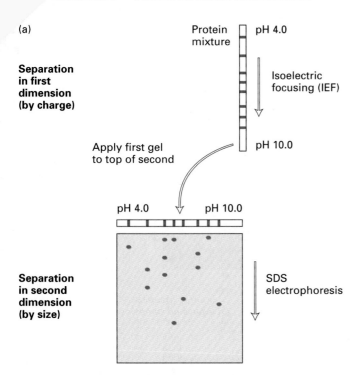

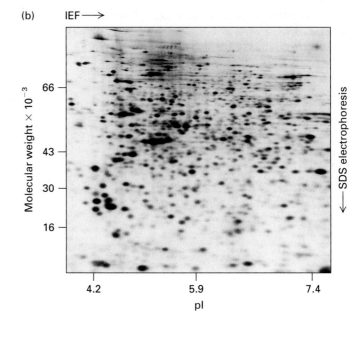

▲ **FIGURE 3-42 Two-dimensional gel electrophoresis, a technique for separating proteins based on their charge and their mass.** (a) Preparation of a two-dimensional protein gel by isoelectric focusing (IEF) followed by SDS electrophoresis. (b) Two-dimensional gel of protein extracts from cells growing in a medium. Each spot represents a single polypeptide; polypeptides can be detected by dyes, as here, or by other techniques such as autoradiography. Each polypeptide is characterized by its pI (isoelectric point) and molecular weight. [Part (b) courtesy of J. Celis.]

be exploited. Most commonly, this is electric charge, which is determined by the number of acidic and basic residues in a protein. Two unrelated proteins having similar masses are unlikely to have identical net charges because their sequences, and thus the number of acid and basic residues, are different.

In two-dimensional electrophoresis, proteins are separated in two sequential steps: first by their charge and then by their mass. In the first step, a cell extract is fully denatured by high concentrations (8 M) of urea and then layered on a glass tube filled with polyacrylamide that is saturated with a solution of *ampholytes,* a mixture of polyanionic and polycationic molecules. When placed in an electric field, the ampholytes will separate and form a continuous gradient based on their net charge (Figure 3-42a). The most highly polyanionic ampholytes will collect at one end of the tube, and the most polycationic ampholytes will collect at the other end. This gradient of ampholytes establishes a pH gradient. Charged proteins will migrate through the gradient until they reach their **pI,** or isoelectric point, the pH at which the net charge of the protein is zero. This technique, called **isoelectric focusing** (IEF), can resolve proteins that differ by only one charge unit.

Proteins that have been separated on an IEF gel can then be separated in a second dimension based on their molecular weights. To accomplish this, the IEF gel is extruded from the tube and placed lengthwise on a second polyacrylamide gel, this time formed as a slab saturated with SDS. When an electric field is imposed, the proteins will migrate from the IEF gel into the SDS slab gel and then separate according to their mass. The sequential resolution of proteins by their charge and mass can achieve excellent separation of cellular proteins (Figure 3-42b). For example, two-dimensional gels have been very useful in studying the expression of various genes in differentiated cells because as many as 1000 proteins can be resolved simultaneously.

Liquid Chromatography Resolves Proteins by Mass, Charge, or Binding Affinity

Liquid chromatography, a third commonly used technique to separate mixtures of proteins, nucleic acids, and other molecules, is based on the principle that molecules dissolved in a solution will interact (bind and dissociate) with a solid surface. If the solution is allowed to flow across the surface, then molecules that interact frequently with the surface will spend more time bound to the surface and thus move more slowly than molecules that interact infrequently with the surface. Liquid chromatography is performed in a column packed tightly with spherical beads. The nature of these beads determines whether separation of proteins depends on differences in mass, charge, or binding affinity.

Gel Filtration Chromatography Proteins that differ in mass can be separated by gel filtration chromatography. In this technique, the column is composed of porous beads made from polyacrylamide, dextran (a bacterial polysaccharide), or agarose (a seaweed derivative). Proteins *flow around* the spherical beads in gel filtration chromatography.

However, the surface of the beads is punctured by large holes, and proteins will spend some time within these holes. Because smaller proteins can penetrate into the beads more easily than larger proteins, they travel through a gel filtration column more slowly than larger proteins (Figure 3-43a). (In contrast, proteins migrate *through* the pores in an electrophoretic

(a) Gel filtration chromatography

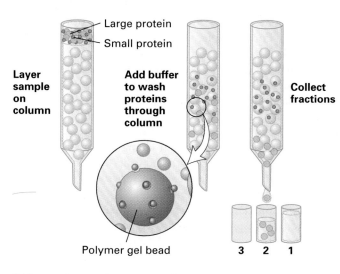

(c) Antibody-affinity chromatography

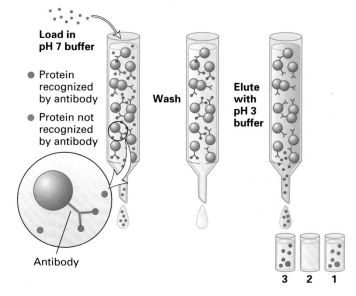

(b) Ion-exchange chromatography

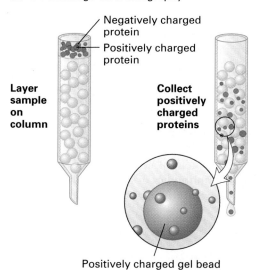

▲ **FIGURE 3-43 Three commonly used liquid chromatographic techniques.** (a) Gel filtration chromatography separates proteins that differ in size. A mixture of proteins is carefully layered on the top of a glass cylinder packed with porous beads. Smaller proteins travel through the column more slowly than larger proteins. Thus different proteins have different elution volumes and can be collected in separate liquid fractions from the bottom. (b) Ion-exchange chromatography separates proteins that differ in net charge in columns packed with special beads that carry either a positive charge (shown here) or a negative charge. Proteins

having the same net charge as the beads are repelled and flow through the column, whereas proteins having the opposite charge bind to the beads. Bound proteins, in this case negatively charged, are eluted by passing a salt gradient (usually of NaCl or KCl) through the column. As the ions bind to the beads, they desorb the protein. (c) In antibody-affinity chromatography, a specific antibody is covalently attached to beads packed in a column. Only protein with high affinity for the antibody is retained by the column; all the nonbinding proteins flow through. The bound protein is eluted with an acidic solution, which disrupts the antigen-antibody complexes.

gel; thus smaller proteins move faster than larger ones.) The total volume of liquid required to elute a protein from the column depends on its mass: the smaller the mass, the greater the elution volume. By use of proteins of known mass, the elution volume can be used to estimate the mass of a protein in a mixture.

Ion-Exchange Chromatography In a second type of liquid chromatography, called *ion-exchange chromatography,* proteins are separated based on differences in their charge. This technique makes use of specially modified beads whose surfaces are covered by amino groups or carboxyl groups and thus carry either a positive charge (NH_3^+) or a negative charge (COO^-) at neutral pH.

The proteins in a mixture carry various net charges at any given pH. When a solution of a protein mixture flows through a column of positively charged beads, only proteins with a net negative charge (acidic proteins) adhere to the beads; neutral and basic proteins flow unimpeded through the column (Figure 3-43b). The acidic proteins are then eluted selectively by passing a gradient of increasing concentrations of salt through the column. At low salt concentrations, protein molecules and beads are attracted by their opposite charges. At higher salt concentrations, negative salt ions bind to the positively charged beads, displacing the negatively charged proteins. In a gradient of increasing salt concentration, weakly charged proteins are eluted first and highly charged proteins are eluted last. Similarly, a negatively charged column can be used to retain and fractionate positively charged (basic) proteins.

Affinity Chromatography A third form of chromatography, called *affinity chromatography,* relies on the ability of a protein to bind specifically to another molecule. Columns are packed with beads to which are covalently attached ligand molecules that bind to the protein of interest. Ligands can be enzyme substrates or other small molecules that bind to specific proteins. In a widely used form of this technique, *antibody-affinity chromatography,* the attached ligand is an antibody specific for the desired protein (Figure 3-43c). An affinity column will retain only the proteins that bind the ligand attached to the beads; the remaining proteins, regardless of their charge or mass, will pass through the column without binding to it. The proteins bound to the affinity column then are eluted by adding an excess of ligand or by changing the salt concentration or pH. Obviously, the ability of this technique to separate particular proteins depends on the selection of appropriate ligands.

Highly Specific Enzyme and Antibody Assays Can Detect Individual Proteins

Purification of a protein, or any other molecule, requires a specific assay that can detect the molecule of interest in column fractions or gel bands. An assay capitalizes on some highly distinctive characteristic of a protein: the ability to bind a particular ligand, to catalyze a particular reaction, or to be recognized by a specific antibody. An assay must also be simple and fast in order to minimize errors and the possibility that the protein of interest is denatured or degraded while the assay is performed. The goal of any purification scheme is to isolate sufficient amounts of a given protein for study; thus a useful assay must also be sensitive enough that only a small proportion of the available material is consumed. Many common protein assays require just 10^{-9} to 10^{-12} g of material.

Chromogenic and Light-Emitting Enzyme Reactions Many assays are tailored to detect some functional aspect of a protein. For example, enzyme assays are based on the ability to detect the loss of substrate or the formation of product. Many enzyme assays utilize *chromogenic* substrates, which change color during the course of the reaction. (Some substrates are naturally chromogenic; if they are not, they can be linked to a chromogenic molecule.) Because of the specificity of an enzyme for its substrate, only samples that contain the enzyme will change color in the presence of a chromogenic substrate and other required reaction components; the rate of the reaction provides a measure of the quantity of enzyme present.

Such chromogenic enzymes also can be fused or chemically linked to an antibody and used to "report" the presence or location of the antigen. Alternatively *luciferase,* an enzyme present in fireflies and some bacteria, can be linked to an antibody. In the presence of ATP and luciferin, this enzyme catalyzes a light-emitting reaction. In either case, after the antibody binds to the protein of interest, substrates of the linked enzyme are added and the appearance of color or emitted light is monitored.

Western Blotting One of the most powerful methods for detecting a particular protein in a complex mixture combines the superior resolving power of gel electrophoresis, the specificity of antibodies, and the sensitivity of enzyme assays. Called **Western blotting,** or immunoblotting, this three-step procedure is commonly used to separate proteins and then identify a specific protein of interest. As shown in Figure 3-44, two different antibodies are used in this method, one specific for the desired protein and the other linked to a reporter enzyme.

Radioisotopes Are Indispensable Tools for Detecting Biological Molecules

Since World War II, when radioactive materials first became widely available as byproducts of work in nuclear physics, chemists and biologists have fashioned an almost limitless variety of radioactive chemicals. Today, radioactively labeled precursors of macromolecules greatly simplify many standard biochemical assays and significantly enhance the ability of researchers to follow biochemical events in whole cells as well as in cell extracts. Almost all experimental biology depends on the use of radioactive compounds.

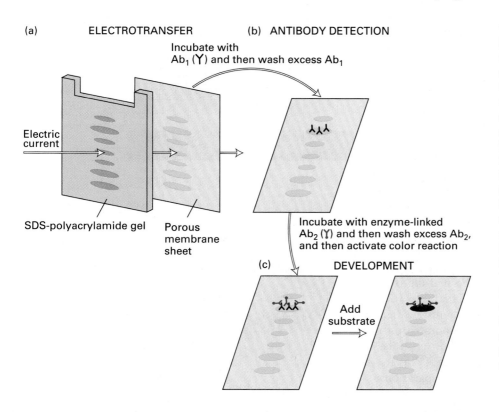

(a) ELECTROTRANSFER

Electric current

SDS-polyacrylamide gel

Porous membrane sheet

(b) ANTIBODY DETECTION

Incubate with Ab₁ (Y) and then wash excess Ab₁

Incubate with enzyme-linked Ab₂ (Y) and then wash excess Ab₂, and then activate color reaction

(c) DEVELOPMENT

Add substrate

◀ **FIGURE 3-44 Western blotting, or immunoblotting.** (a) A protein mixture is electrophoresed through an SDS gel, and then transferred from the gel onto a membrane. (b) The membrane is flooded with a solution of antibody (Ab₁) specific for the desired protein. Only the band containing this protein binds the antibody, forming a layer of antibody molecules (although their position can't be seen at this point). After sufficient time for binding, the membrane is washed to remove unbound Ab₁. (c) In the development step, the membrane first is incubated with a second antibody (Ab₂) that binds to the bound Ab₁. This second antibody is covalently linked to alkaline phosphatase, which catalyzes a chromogenic reaction. Finally, the substrate is added and a deep purple precipitate forms, marking the band containing the desired protein.

MEDIA CONNECTIONS

Immunoblotting

At least one atom in a radiolabeled molecule is present in a radioactive form, called a **radioisotope** (Table 3-2). The presence of a radioisotope does not change the chemical properties of a molecule. For example, enzymes, both in vivo and in vitro, catalyze reactions involving labeled substrates just as readily as those involving nonlabeled substrates. Because radioisotopes emit easily detected particles, the fate of radiolabeled molecules can be traced in cells and cellular extracts.

Characteristics of Different Radiolabels The choice of which labeled compound to use in a particular experiment involves several considerations. Some labeled compounds, for instance, are not suitable for studies with whole cells because they do not enter cells. One prominent example is ATP, as well as most other phosphorylated compounds (e.g., glucose 6-phosphate). Although ^{32}P-labeled ATP can contribute phosphorus-32 during RNA and DNA synthesis in a cell-free system, it cannot do so with whole cells, because

TABLE 3-2 Radioisotopes Commonly Used in Biological Research

Isotope	Half-Life	Energy of Emitted Particle (MeV)[*]	Specific Activity of Labeled Compounds (mCi/mmol)[†]
Tritium (hydrogen-3)	12.35 years	0.0186	10^2–10^5
Carbon-14	5730 years	0.156	1–10^2
Phosphorus-32	14.3 days	1.709	10–10^5
Phosphorus-33	25.5 days	0.248	10–10^4
Sulfur-35	87.5 days	0.167	1–10^6
Iodine-125	60 days	0.035	10^2–10^4
Iodine-131	8.07 days	0.806	10^2–10^4

[*]$MeV = 10^6$ *electronvolts. The maximum energy for each emission is given. The particle emitted is a β particle, except in the case of* ^{125}I *and* ^{131}I, *which emit γ rays.*
[†]*The millicurie (mCi) is a measure of the number of disintegrations per time unit: 1 mCi = 2.2 × 10⁹ disintegrations per minute. These values are for commercially available compounds of interest in biological research that contain the indicated radioisotope.*

SOURCE: New England Nuclear Corporation, Boston.

it never gets into the cells. On the other hand, labeled or-thophosphate ($^{32}PO_4^{3-}$) in the medium does enter both bac-terial cells and animal cells, and then is incorporated into phosphorylated proteins, nucleotides, and eventually into cellular RNA and DNA.

Hundreds of biological compounds (e.g., amino acids, nucleosides, and numerous metabolic intermediates) labeled with various radioisotopes are commercially available. These preparations vary considerably in their *specific activity*, which is the amount of radioactivity per unit of mater-ial, measured in disintegrations per minute (dpm) per mil-limole. The specific activity of a labeled compound depends on the ratio of unstable potentially radioactive atoms to stable nonradioactive atoms. It also depends on the prob-ability of decay of the radioisotope, indicated by its *half-life*, which is the time required for half the atoms to undergo radioactive decay. In general, the shorter the half-life of a radioisotope, the higher its specific activity (see Table 3-2).

The specific activity of a labeled compound must be high enough so that sufficient radioactivity is incorporated into cellular molecules to be accurately detected. For example, methionine and cysteine labeled with sulfur-35 (^{35}S) are widely used to label cellular proteins because preparations of these amino acids with high specific activities ($>10^{15}$ dpm/mmol) are available. Likewise, commercial preparations of ^3H-labeled nucleic acid precursors have much higher specific activities than the corresponding ^{14}C-labeled preparations. In most experiments, the former are preferable because they allow RNA or DNA to be ade-quately labeled after a shorter time of incorporation or re-quire a smaller cell sample.

Various phosphate-containing compounds in which every phosphorus atom is the radioisotope phosphorus-32 are readily available. Because of their high specific activity, ^{32}P-labeled nucleotides are routinely used to label nucleic acids in cell-free systems. The radioisotope iodine-125 (^{125}I), which also is available in almost pure form, can be cova-lently linked to a protein or nucleic acid to yield prepara-tions with a high specific activity. Such attachment of iodine-125 can be achieved enzymatically or chemically and generally does not drastically alter the properties of a macro-molecule.

Labeling Experiments and Detection of Radiolabeled Molecules Depending on the nature of an experiment, la-beled compounds are detected by **autoradiography**, a semi-quantitative visual assay, or their radioactivity is measured in an appropriate "counter," a highly quantitative assay that can determine the concentration of a radiolabeled compound in a sample. In some experiments, both types of detection are used.

In autoradiography, a cell or cell constituent is labeled with a radioactive compound and then overlaid with a pho-tographic emulsion sensitive to radiation. Development of the emulsion yields small silver grains whose distribution cor-responds to that of the radioactive material (Figure 3-45a).

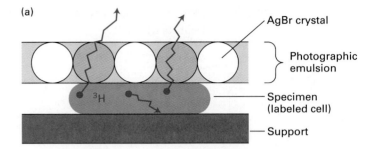

(a)

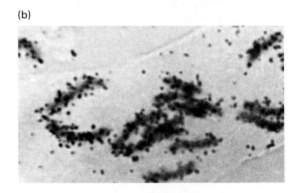

(b)

▲ **FIGURE 3-45 Autoradiography.** (a) A radiation-sensitive photographic emulsion containing silver salts (AgBr) is placed over tritium-labeled cells attached to a glass slide (for the light microscope) or to a carbon-coated grid (for the electron micro-scope). The cell regions containing the labeled molecules emit radioactive particles, along the tracks of which silver is deposited. When the photographic emulsion is developed, the silver deposits appear as dark grains under the light microscope and as curly filaments in the electron microscope. (b) When cells are incubated with [^3H]thymidine for a short period, any DNA synthesized during this labeling period incorporates the labeled precursor and can be localized by autoradiography. The root cells of a lily, shown here, were pulse-labeled; a sample was then taken 8 hours later. During the pulse period, the silver grains lie over both chromatids of the chromosomes. [Part (a) adapted from E. D. P. DeRobertis and E. M. F. DeRobertis, 1979, *Cell and Molecular Biology,* Saunders, p. 62; part (b) courtesy of J. H. Taylor.]

Autoradiographic studies of whole cells have been crucial in determining the intracellular sites where various macromol-ecules are synthesized and their subsequent movements within cells. For example, when cells are incubated for a short time with [^3H]thymidine, a unique DNA precursor, most of the radioactivity is localized to the nucleus, identi-fying the nucleus as the major site of DNA synthesis (Fig-ure 3-45b). Even after cells are incubated for prolonged pe-riods with [^3H]thymidine, virtually all the radioactivity remains in the nucleus, indicating that the DNA remains there. Similarly, the site of synthesis of RNA is revealed by incubating cells for 1 minute with [^3H]uridine, a unique RNA precursor; in this case, all the autoradiographic grains are found over the nucleus. After a longer period of incorporation, however, many autoradiographic grains are located over the cytoplasm, indicating that RNA, in contrast to DNA, is transported from the nucleus.

In autoradiographic studies, the ability of the experimenter to localize the site at which the radioisotope is incorporated is affected by the energy of the particles emitted during radioactive disintegrations. For example, the β particles emitted by phosphorus-32 are so energetic that the streaks they make on a photographic emulsion can be as long as 1 mm, much longer than the diameter of individual cells. In contrast, the β particles emitted by tritium create tracks on a photographic emulsion that are only about 0.47 μm long; thus ^{3}H-labeled structures can be located within cells to an accuracy of about 0.5–1.0 mm, or about one-fifth the diameter of the nucleus of mammalian cells. Because tritium emits the least-energetic particles of all the common radioisotopes, it is highly preferred for locating labeled compounds or structures within cells.

Quantitative measurements of the amount of radioactivity in a labeled material are performed with several different instruments. A *Geiger counter* measures ions produced in a gas by the β particles or γ rays emitted from a radioisotope. In a *scintillation counter,* a radiolabeled sample is mixed with a liquid containing a fluorescent compound that emits a flash of light when it absorbs the energy of the β particles or γ rays released during decay of the radioisotope; a phototube in the instrument detects and counts these light flashes. *Phosphorimagers* are used to detect radiolabeled compounds on a surface, storing digital data on the number of decays in dpm per small pixel of surface area. These instruments commonly are used to quantitate radioactive molecules separated by gel electrophoresis and are replacing photographic film for this purpose.

The usual experimental protocol for determining the cellular location of a particular molecule has three steps:

1. A radioactive precursor is incubated with whole cells or cell-free extracts.

2. The cellular constituents then are isolated and purified in various ways.

3. The radioactivity of the various fractions is measured with a counter.

For example, to identify the site of RNA synthesis, cells can be incubated for a short period with [^{3}H]uridine and then subjected to a fractionation procedure to separate the various organelles (Chapter 5). The specific activity of the nuclear fraction (dpm/mg protein) is found to be much higher than that of any other organelle fraction, thus confirming the nucleus as the site of RNA synthesis.

A combination of labeling and biochemical techniques and of visual and quantitative detection methods is often employed in labeling experiments. For instance, to identify the major proteins synthesized by a particular cell type, a sample of the cells is incubated with a radioactive amino acid (e.g., ^{35}S-labeled methionine) for a few minutes. The mixture of cellular proteins then is resolved by gel electrophoresis, and the gel is subjected to autoradiography or phosphorimager analysis. The radioactive bands correspond to newly synthesized proteins, which have incorporated the radiolabeled amino acid. Alternatively, the proteins can be resolved by liquid chromatography, and the radioactivity in the eluted fractions determined quantitatively with a counter.

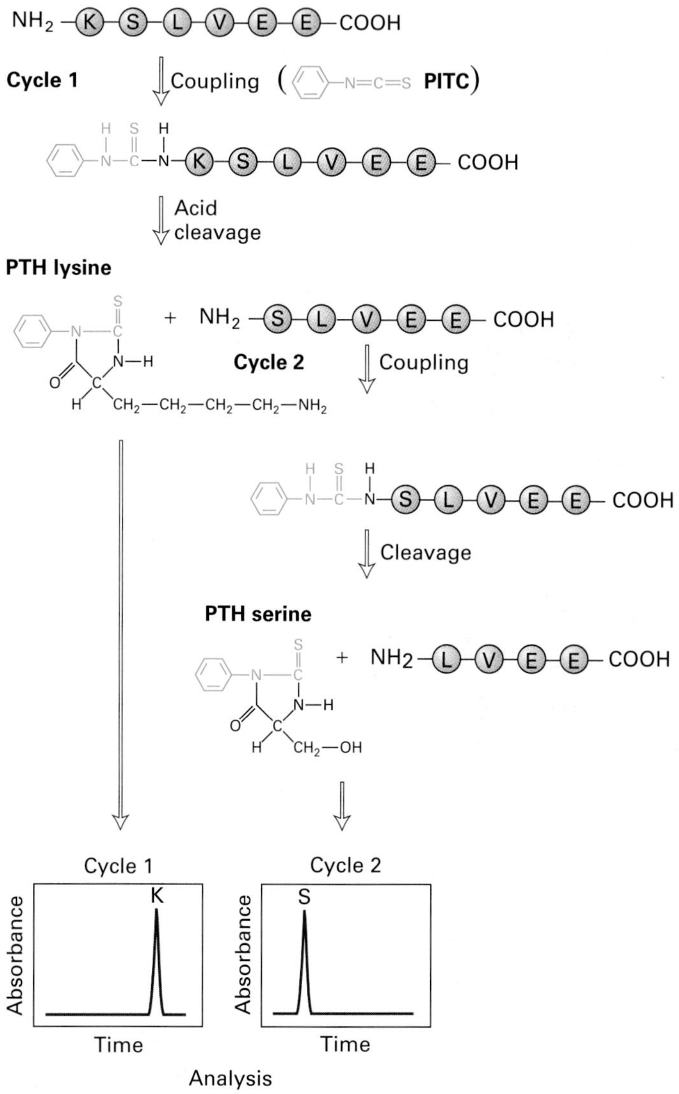

▲ FIGURE 3-46 Chemical determination of the sequence of a protein by Edman degradation, which involves a repetitive three-step procedure. In the first step, the polypeptide N-terminus is reacted with phenylisothiocyanate (PITC). In the second step, the N-terminal amino acid is cleaved from the polypeptide by acid hydrolysis, yielding the cyclic phenylthiohydantoin (PTH) derivative and a polypeptide that is shorter at its N-terminus by one residue. These two steps are then repeated with the shortened polypeptide. The PTH derivative formed in each cycle is identified by liquid chromatography.

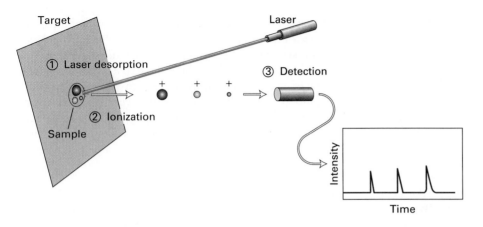

Researchers often use the **pulse-chase** technique in labeling experiments. In this protocol, a cell sample is exposed to a radiolabeled compound for only a brief period of time (the "pulse"), then washed with buffer to remove the label, and finally incubated with a nonlabeled form of the compound (the "chase"). Pulse-chase experiments are particularly useful for tracing changes in the intracellular location of proteins or the transformation of a metabolite into others over time.

Protein Primary Structure Can Be Determined by Chemical Methods and from Gene Sequences

The primary structure of a protein is characterized in two ways: by its overall amino acid composition and by its precise amino acid sequence. The amino acid composition of a protein gives the same information as an elemental analysis of a molecule—the types of amino acids present and their abundance but not their linear order. In contrast, the sequence of a protein is like a fingerprint; it uniquely establishes the identity of a protein—the linear order of amino acids. The composition of a protein is easily calculated from the sequence. Two proteins can differ in their sequence but nonetheless have identical amino acid compositions. Composition and sequence are determined by chemical methods based on the ability to cleave the peptide backbone at the peptide bond.

The classic method for determining the amino acid sequence of a protein involves *Edman degradation* (Figure 3-46). In this procedure the amino group at the N-terminus of a polypeptide is labeled and its amino acid then cleaved from the polypeptide and identified by high-pressure liquid chromatography. The polypeptide is left one residue shorter, with a new amino acid at the N-terminus. The cycle is repeated on the ever shortening polypeptide until all the residues have been identified.

Before about 1985, biologists commonly used the Edman chemical procedure for determining protein sequences. However, recombinant DNA techniques developed in the 1970s and 1980s permit the detection and cloning of the mRNA or gene encoding a specific protein. From the sequence of its mRNA or gene, the protein's amino acid sequence can be deduced. Because sequencing of mRNA and DNA generally is faster than chemical sequencing of proteins, this approach is now the most popular way to determine protein sequences, especially of large proteins. As we describe in Chapter 7, the complete genome sequences of several organisms have already been determined, and the database of genome sequences from humans and numerous model organisms is expanding rapidly.

Time-of-Flight Mass Spectrometry Measures the Mass of Proteins and Peptides

A powerful technique for measuring the mass of molecules like proteins and peptides is *mass spectrometry* (Figure 3-47). Mass spectrometry requires a method for ionizing the sample, usually a mixture of peptides or proteins, accelerating the molecular ions, and then detecting the ions. In a laser desorption mass spectrometer, the proteins are mixed with an organic acid and then dried on a metal target. Light from a laser ionizes the proteins, which "fly" down a tube to a detector. Their time of flight is inversely proportional to their mass and directly proportional to the charge on the protein. As little as 1 femtomole of proteins as large as 200,000 MW can be measured with an error of 0.1 percent.

One powerful use of mass spectrometers is to identify a protein from its *peptide mass fingerprint.* A peptide mass fingerprint is a compilation of the molecular weights of peptides that are generated by a specific protease. The molecular weights of the parent protein and its proteolytic fragments are used to search genome databases for any similarly sized protein with identical or similar peptide mass maps. With the increasing availability of genome sequences, this approach has almost eliminated the need to chemically sequence a protein to determine its primary structure.

Peptides with a Defined Sequence Can Be Synthesized Chemically

Synthetic peptides that are identical with peptides synthesized in vivo are useful experimental tools in studies of proteins and cells. For example, short synthetic peptides of

10–15 residues can function as antigens to trigger production of antibodies in animals. A synthetic peptide can trick the animal into producing antibodies that bind the full-sized, natural protein antigen. As we'll see throughout this book, antibodies are extremely versatile reagents for isolating proteins from mixtures by affinity chromatography (see Figure 3-43c), separating and detecting proteins by Western blotting (see Figure 3-44), and localizing proteins in cells by microscopic techniques described in Chapter 5. Synthetic peptides also have been helpful in elucidating the rules that determine the secondary and tertiary structure of proteins. By systematically varying the sequence of synthetic peptides, researchers have studied the influence of various amino acids on protein conformation.

Peptides are routinely synthesized in a test tube from monomeric amino acids by condensation reactions that form peptide bonds. Peptides are constructed sequentially by coupling the C-terminus of a monomeric amino acid with the N-terminus of the growing peptide, as outlined in Figure 3-48. To prevent unwanted reactions involving the amino groups and carboxyl groups of the side chains during the coupling steps, a protecting (blocking) group is attached to the side chains. Without these protecting groups, branched peptides would be generated. In the last steps of synthesis, the side chain–protecting groups are removed and the peptide is cleaved from the resin on which synthesis occurs.

Protein Conformation Is Determined by Sophisticated Physical Methods

In this chapter we have emphasized that protein function is derived from protein structure. Thus, to figure out how a protein works, its three-dimensional structure must be known. Determining a protein's conformation requires sophisticated physical methods and complex analyses of the experimental data. We briefly describe three methods used to generate three-dimensional models of proteins.

X-Ray Crystallography The use of x-ray crystallography to determine the three-dimensional structures of proteins was pioneered by Max Perutz and John Kendrew in the 1950s. To date, the detailed three-dimensional structures of more than 8000 proteins have been established by this technique in which beams of x-rays are passed through a crystal of protein. The wavelengths of x-rays are about 0.1–0.2 nanometer (nm), short enough to resolve the atoms in the protein crystal. Atoms in the protein crystal scatter the x-rays, which produce a diffraction pattern of discrete spots when they are intercepted by photographic film (Figure 3-49). Such patterns are extremely complex; as many as 25,000 diffraction spots can be obtained from a small protein. Elaborate calculations and modifications of the protein (such as binding of heavy metals) must be made to interpret the diffraction pattern and to solve the structure of the protein. The process is analogous to reconstructing the precise shape of a rock from the ripples it creates in a pond.

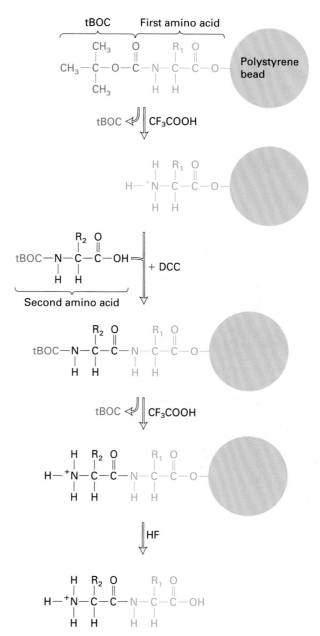

▲ **FIGURE 3-48 Solid-phase peptide synthesis.** The first amino acid (blue) of the desired peptide is attached at its carboxyl end by esterification to a polystyrene bead. The amino group of this amino acid is blocked by the attachment of a tertbutyloxycarbonyl (tBOC) group (red), which is removed by treatment with trifluoroacetic acid (CF₃COOH). The resulting free amino group forms a peptide bond with a second amino acid, which is presented with a reactive carboxyl group and a blocked amino group, together with the coupling agent dicyclohexylcarbodiimide (DCC). The process is repeated until the desired product is obtained; the peptide is then chemically cleaved from the bead with hydrofluoric acid (HF). [See R. B. Merrifield, L. D. Vizioli, and H. G. Boman, 1982, *Biochemistry* **21**:5020.]

(a)

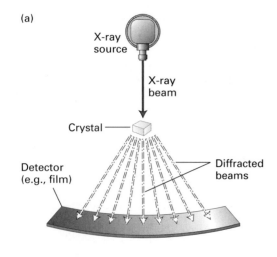

(b)

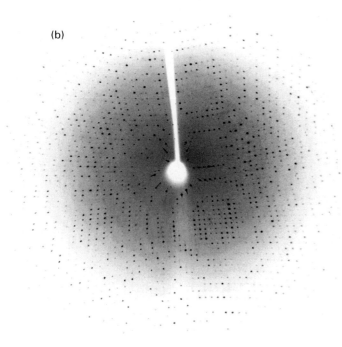

▲ FIGURE 3-49 X-ray diffraction. (a) Basic components of an x-ray crystallographic determination. When a narrow beam of x-rays strikes a crystal, part of it passes straight through and the rest is scattered (diffracted) in various directions. The intensity of the diffracted waves is recorded on an x-ray film or with a solid-state electronic detector. (b) X-ray diffraction pattern for a topoisomerase crystal collected on a solid-state detector. [Part (a) adapted from L. Stryer, 1995, *Biochemistry*, 4th ed., W. H. Freeman, p. 64; part (b) courtesy of J. Berger.]

Cryoelectron Microscopy Although some proteins readily crystallize, obtaining crystals of others—particularly large multisubunit proteins—requires a time-consuming trial-and-error effort to find just the right conditions. Low-resolution views of such difficult-to-crystallize proteins can be obtained by electron microscopy (see Figure 4-34). In this technique, a protein sample is rapidly frozen in liquid helium to preserve its structure and then examined in the frozen, hydrated state in a *cryoelectron microscope*. Pictures are recorded on film using a low dose of electrons to prevent radiation-induced damage to the structure. Sophisticated computer programs analyze the images and reconstruct the protein's structure in three dimensions. With recent advances in cryoelectron microscopy, this technique can generate molecular models that compare with those generated by x-ray crystallography.

NMR Spectroscopy The three-dimensional structures of small proteins containing up to about 200 amino acids can be studied with nuclear magnetic resonance (NMR) spectroscopy. In this technique, a concentrated protein solution is placed in a magnetic field and the effects of different radio frequencies on the resonances of different atoms are measured. However, the behavior of any atom is influenced by neighboring atoms in adjacent residues; the closely spaced residues are more perturbed than distant residues. From the magnitude of the effect, the distances between residues can be calculated; these distances then are used to generate a model of the three-dimensional structure of the protein. Although NMR does not require crystallization of a protein, a definite advantage, this technique is limited to proteins smaller than about 20 kDa. However, NMR analysis also can be applied to protein domains, which tend to be small enough for this technique and often can be obtained as stable structures.

SUMMARY Purifying, Detecting, and Characterizing Proteins

• Proteins can be isolated based on differences in their physical and chemical properties. Centrifugation, electrophoresis, and chromatography are the most common techniques for purifying and analyzing proteins.

• Centrifugation separates proteins based on their rate of sedimentation, which is influenced by their mass and shape.

• Gel electrophoresis separates proteins based on their rate of movement in an applied electric field. SDS-polyacrylamide gel electrophoresis can resolve polypeptide chains differing in molecular weight by 10 percent or less (see Figure 3-41).

• Liquid chromatography separates proteins based on their rate of movement through a column packed with spherical beads. Proteins differing in mass are resolved on gel filtration columns; those differing in charge, on ion-exchange columns; and those differing in ligand-binding properties, on affinity columns (see Figure 3-43).

• Various assays are used to detect, identify, and quantify proteins. The most sensitive assays use a light-producing reaction or radioactivity to generate a signal. Other assays produce an amplified colored signal with enzymes and chromogenic substrates.

- Antibodies are powerful reagents used to detect, quantify, and isolate proteins. They are used in affinity chromatography and combined with gel electrophoresis in Western blotting, a powerful method for separating and detecting a protein in a mixture (see Figure 3-44).

- Autoradiography is a semiquantitative technique for detecting radioactively labeled molecules in cells, tissues, or electrophoretic gels (see Figure 3-45).

- Three-dimensional structures of proteins are obtained by x-ray crystallography, NMR spectroscopy, and cryo-electron microscopy. X-ray crystallography provides the most detailed structures, but requires protein crystallization. Cryoelectron microscopy is particularly useful for large protein complexes, which are difficult to crystallize. Only relatively small proteins are amenable to NMR analysis.

PERSPECTIVES for the Future

Researchers are seeking easier and more widely applicable tools and methods for characterizing the three-dimensional structures of proteins. As more and more sequence data are amassed in large data banks, the sequence and function of new proteins will increasingly be deduced from gene sequences and by comparison with known proteins. Development of an algorithm for predicting the three-dimensional structure of a protein from the sequence of its gene or mRNA would extend this approach even further. Such an algorithm seems attainable in view of our increasing knowledge about the rules that guide protein folding and the recurrent use of domains and motifs as modules of structure and function in different proteins. For novel proteins that exhibit little homology to known proteins, improvements in x-ray crystallography and NMR spectroscopy will permit analysis of ever larger proteins. Advances in cryoelectron microscopy will provide images of uncrystallizable proteins at a resolution comparable with that obtained with x-ray crystallography.

Increased understanding of the detailed relationship between protein structure and function offers the possibility of creating new proteins with specified activities. Industrial engineers use computer-aided design and computer-aided manufacturing (CAD/CAM) systems to create and then manufacture any component by computer control and without the need to build a prototype. This capability requires the principles of design to be codified. In biology, predicting the three-dimensional structure of proteins will provide the codes for protein design. Beginning with genetics and molecular biology, biologists have started to create custom-designed proteins whose new functions are generated by fusion of different protein domain sequences. Perhaps, as our understanding of protein structure becomes more refined, we will produce a virtual reality model of a protein's structure and then simulate its activity. This approach might be used to create new enzymes capable of synthesizing novel compounds, as well as fluorescent proteins that will respond to the activation or inactivation of some cell activity.

Progress also is likely in our ability to study the action of single protein molecules. Most current techniques measure the behavior of a population of protein molecules in solution, providing average data. However, if we could study a single protein molecule, we could more fully understand how it operates. During the last five years, physicists have developed optical traps and single-molecule imaging techniques for detecting and manipulating single molecules. With such techniques, for instance, researchers will be able to measure the V_{max} of a single enzyme molecule and to determine binding energy directly by measuring the force required to pull a protein from its ligand.

PERSPECTIVES in the Literature

Naturally occurring proteins are formed through an organization of modules, or domains, that dictates their structures and functions. We can create new proteins to have desired properties by choosing the combination of modules we use. Among the list of "parts" available to a protein designer are domains that perform structural, enzymatic, anchoring, and transport functions. Perhaps the best-studied type of domain is the multistranded coiled-coil. Using this domain, construct a protein that has a novel function. You might, for example, design the simplest protein that connects the nucleus to the cell membrane. The basic design would call for a long protein that is anchored at either end to the plasma and nuclear membranes. For a "parts" list of domains, including the coiled-coil, please see the following Web site on the structural classification of proteins:

http://scop.mrc-lmb.cam.ac.uk/scop/

Testing Yourself on the Concepts

1. Describe the molecular features of the four hierarchical levels of structure that determine the shape or conformation of a protein.

2. Describe the structural and functional properties of a cellular enzyme.

3. Compare and contrast the properties of integral and peripheral membrane proteins.

4. Describe the methodologies for separating proteins based on their charge or mass.

MCAT/GRE-Style Questions

Key Concept Please read the sections titled "Enzymes Are Highly Efficient and Specific Catalysts" and "An Enzyme's Active Site Binds Substrates and Carries Out Catalysis" (p. 71) and answer the following questions:

1. Which of the following describes a primary property of enzymes?

   a. They contain platinum.

   b. They lower the activation energy of a reaction.

   c. They decrease the rate of a reaction.

   d. They bind to a variety of substrates.

2. A typical cellular enzyme is active

   a. In an aqueous solution at a pH of 4.

   b. In a nonaqueous solution such as chloroform.

   c. In an aqueous solution at 95 °C.

   d. In an aqueous solution at a pH of 7.

3. Which of the following statements describes the active site of an enzyme?

   a. Nonadjacent amino acids can make up the active site.

   b. The active site always consists of a single region that binds the substrate and catalyzes the reaction.

   c. The amino terminus of the protein always contains the active site.

   d. The active site interacts with the substrate through a peptide bond.

4. What feature of a protein can be used to predict its function?

   a. Number of amino acids.

   b. Overall charge of the molecule.

   c. Molecular weight.

   d. Structure.

Key Experiment Please read the section titled "Electrophoresis Separates Molecules according to Their Charge: Mass Ratio" (p. 87) and answer the following questions:

5. The rate of migration of a protein through an SDS-polyacrylamide gel is *not* influenced by

   a. Size of the protein.

   b. Charge of the protein.

   c. Pore size of the gel.

   d. Strength of the electric field.

6. Please refer to Table 3-1. Which of the following proteins would be expected to migrate the fastest through an SDS-polyacrylamide gel:

   a. α_2-Macroglobulin.

   b. Transferrin.

   c. Serum albumin.

   d. Lysozyme.

7. Please refer to Table 3-1. Which of the following proteins would be expected to migrate to a position closest to the pH 4.0 end of an isoelectric focusing gel like that shown in the top of Figure 3-42a:

   a. α_2-Macroglobulin.

   b. Transferrin.

   c. Serum albumin.

   d. Lysozyme.

8. A newly isolated protein has a molecular weight of 95 kDa and a pI of 7.0. Where would this protein be expected to migrate relative to serum albumin (69 kDa, pI of 4.9) on a two-dimensional gel like that shown at the bottom of Figure 3-42a?

   a. To the left of and above serum albumin.

   b. To the left of and below serum albumin.

   c. To the right of and above serum albumin.

   d. To the right of and below serum albumin.

Key Application Please read the section titled "Sequence Homology Suggests Functional and Evolutionary Relationships between Proteins" (p. 60) and answer the following questions:

9. Which statement best describes the basic principle of molecular taxonomy?

   a. All proteins evolved from a single common protein.

   b. Closely related proteins have similar sequences.

   c. Proteins are classified according to the presence or absence of α helices.

   d. Prokaryotic and eukaryotic proteins are all similar in structure.

10. Two proteins with a similar structure are likely to have

   a. A heme group in the interior of the protein.

   b. Little similarity in amino acid sequence.

   c. Evolved from a common ancestral protein.

   d. Different functions.

11. Please refer to Figure 3-12. Indicate which of the following pairs of proteins are most similar in amino acid sequence based on their classification by molecular taxonomy.

   a. α subunit of cow hemoglobin and β subunit of chicken hemoglobin.

   b. Soybean leghemoglobin and chicken myoglobin.

   c. α subunit of chicken hemoglobin and whale myoglobin.

   d. Human myoglobin and β subunit of human hemoglobin.

12. A newly discovered protein has a structure similar to myoglobin. On the basis of this structural similarity, this protein would likely function to

   a. Bind nitrogen.

   b. Regulate the synthesis of muscle proteins.

   c. Carry oxygen.

   d. Provide structural integrity for muscle cells.

Key Terms

α helix *54*

active site *71*

amino acids *51*

autoradiography *92*

β sheet *54*

chaperones *63*

conformation *51*

disulfide bond *53*

domains *60*

electrophoresis *87*

enzymes *71*

ligands *68*

liquid chromatography *88*

Michaelis constant *73*

motifs *58*

peptide bond *53*

polypeptides *54*

primary structure *54*

protein *54*

quaternary structure *54*

radioisotope *91*

secondary structure *54*

substrates *69*

tertiary structure *54*

V_{max} *73*

Western blotting *90*

x-ray crystallography *60*

References

General References

Stryer, L. 1995. *Biochemistry,* 4th ed. W. H. Freeman and Company, chaps. 1–4, 7–9, 11, 12, 14, 16, 31.

Web Sites

Entry site into the proteins, structures, genomes, and taxonomy
http://www.ncbi.nlm.nih.gov/Entrez/

The protein 3D structure database
http://www.rcsb.org/

Structural classifications of proteins
http://scop.mrc-lmb.cam.ac.uk/scop/

Sites containing general information about proteins
http://www.expasy.ch/
http://www.proweb.org/

Hierarchical Structure of Proteins

Branden, C., and J. Tooze. 1999. *Introduction to Protein Structure.* Garland.

Creighton, T. E. 1993. *Proteins: Structures and Molecular Properties,* 2d ed. W. H. Freeman and Company.

Perutz, M. F. 1991. *Protein Structure and Function.* W. H. Freeman and Company.

Sheterline, P., series ed. *Protein Profile.* Academic Press. A series of monographs devoted to compiling facts about different classes of proteins.

Folding, Modification, and Degradation of Proteins

Hochstrasser, M. 1996. Protein degradation or regulation: Ub the judge. *Cell* **84**:813–815.

Levitt, M., et al. 1997. Protein folding: the endgame. *Ann. Rev. Biochem.* **66**:549–579.

Miranker, A. D., and C. M. Dobson. 1996. Collapse and cooperativity in protein folding. *Curr. Opin. Struc. Biol.* **6**:31–42.

Perler, F. B., M. Q. Xu, and H. Paulus. 1997. Protein splicing and autoproteolysis mechanisms. *Curr. Opin. Chem. Biol.* **1**:292–299.

Rechsteiner, M., and S. W. Rogers. 1996. PEST sequences and regulation by proteolysis. *Trends Biochem. Sci.* **21**:267–271.

Functional Design of Proteins

Cox, S., E. Radzio-Andzelm, and S. S. Taylor. 1994. Domain movements in protein kinases. *Curr. Opin. Struc. Biol.* **4**:893–901.

Dressler, D. H., and H. Potter. 1991. *Discovering Enzymes.* Scientific American Library.

Fersht, A. 1999. *Enzyme Structure and Mechanism,* 3d ed. W. H. Freeman and Company.

Kyte, J. 1995. *Mechanism in Protein Chemistry.* Garland.

Kyte, J. 1995. *Structure in Protein Chemistry.* Garland.

Smith, C. M., et al. 1997. The protein kinase resource. *Trends Biochem. Sci.* **22**:444–446.

Taylor, S. S., and E. Radzio-Andzelm. 1994. Three protein kinase structures define a common motif. *Structure* **2**:345–355.

Purifying, Detecting, and Characterizing Proteins

Hames, B. D. *A Practical Approach.* Oxford Press. A methods series that describes protein purification methods and assays.

4

Nucleic Acids, the Genetic Code, and the Synthesis of Macromolecules

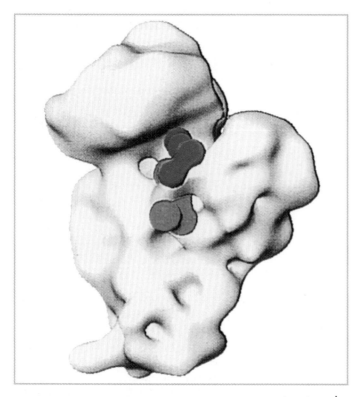

Model of small ribosomal subunit (yellow) showing contacts with two tRNAs (red and green) at one point in the protein chain elongation cycle.

The extraordinary versatility of proteins in catalyzing chemical reactions and building cellular structures was featured in the previous chapter. In this chapter we consider the **nucleic acids,** the molecules that (1) contain the information prescribing amino acid sequence in proteins and (2) serve in the several cellular structures that choose, and then link into the correct order, the amino acids of a protein chain. **Deoxyribonucleic acid (DNA)** is the storehouse, or cellular library, that contains all the information required to build the cells and tissues of an organism. The exact duplication of this information in any species from generation to generation assures the genetic continuity of that species. The information is arranged in units identified by classical geneticists from Gregor Mendel through Thomas Hunt Morgan, and known now as **genes,** hereditary units controlling identifiable traits of an organism. In the process of **transcription,** the information stored in DNA is copied into **ribonucleic acid (RNA),** which has three distinct roles in protein synthesis. **Messenger RNA (mRNA)** carries the instructions from DNA that specify the correct order of amino acids during protein synthesis. The remarkably accurate, stepwise assembly of amino acids into proteins occurs by **translation** of mRNA. In this process, the information in mRNA is interpreted by a second type of RNA called **transfer RNA (tRNA)** with the aid of a third type of RNA, **ribosomal RNA (rRNA),** and its associated proteins. As the correct amino acids are brought into sequence by tRNAs, they are linked by peptide bonds to make proteins.

Discovery of the structure of DNA in 1953 and the subsequent elucidation of the steps in the synthesis of DNA, RNA, and protein are the monumental achievements of the early days of molecular biology. To understand how DNA directs synthesis of RNA, which then directs assembly of proteins—the so-called *central dogma* of molecular biology—we first discuss the building blocks composing DNA and RNA, the cell's two primary nucleic acids, and their general structures. We then introduce the basic mechanisms of DNA and RNA chain synthesis, including a brief discussion of gene structure, which reveals why molecular processing is required to make functional RNA molecules. Next we outline the roles of mRNA, tRNA, and rRNA in protein synthesis. The chapter closes with a detailed description of the components and biochemical steps in the formation of proteins. Since the events of macromolecular synthesis are so central to all biological functions—growth control, differentiation, and the specialized chemical and physical properties of cells—they will arise again and again in later chapters. A firm grasp of the fundamentals of DNA, RNA, and protein synthesis is necessary to follow the subsequent discussions without difficulty.

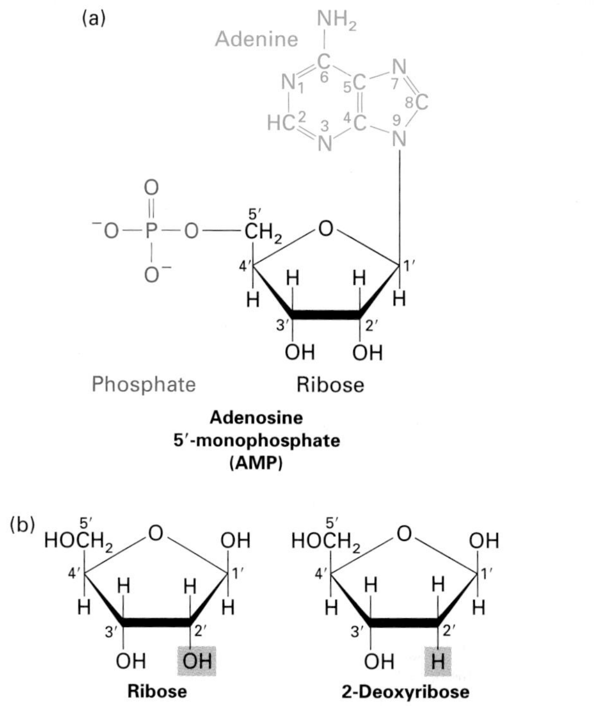

4.1 Structure of Nucleic Acids

DNA and RNA have great chemical similarities. In their *primary structures* both are linear **polymers** (multiple chemical units) composed of **monomers** (single chemical units), called **nucleotides.** Cellular RNAs range in length from less than one hundred to many thousands of nucleotides. Cellular DNA molecules can be as long as several hundred million nucleotides. These large DNA units in association with proteins can be stained with dyes and visualized in the light microscope as chromosomes.

Polymerization of Nucleotides Forms Nucleic Acids

DNA and RNA each consists of only four different nucleotides. All nucleotides have a common structure: a *phosphate* group linked by a phosphoester bond to a *pentose* (a five-carbon sugar molecule) that in turn is linked to an organic *base* (Figure 4-1a). In RNA, the pentose is *ribose;* in DNA, it is *deoxyribose* (Figure 4-1b). The only other difference in the nucleotides of DNA and RNA is that one of the four organic bases differs between the two polymers. The bases adenine, guanine, and cytosine are found in both DNA and RNA; thymine is found only in DNA, and uracil is found only in RNA. The bases are often abbreviated A, G, C, T, and U, respectively. For convenience the single letters are also used when long sequences of nucleotides are written out.

The base components of nucleic acids are heterocyclic compounds with the rings containing nitrogen and carbon. Adenine and guanine are **purines,** which contain a pair of fused rings;

▲ FIGURE 4-1 All nucleotides have a common structure.
(a) Chemical structure of adenosine 5′-monophosphate (AMP), a nucleotide that is present in RNA. All nucleotides are composed of a phosphate moiety, containing up to three phosphate groups, linked to the 5′ hydroxyl of a pentose sugar, whose 1′ carbon is linked to an organic base. By convention, the carbon atoms of the pentoses are numbered with primes. In natural nucleotides, the 1′ carbon is joined by a β linkage to the base, which is in the plane above the furanose ring, as is the phosphate. (b) Haworth projections of ribose and deoxyribose, the pentoses in nucleic acids.

cytosine, thymine, and uracil are **pyrimidines,** which contain a single ring (Figure 4-2). The acidic character of nucleotides is due to the presence of phosphate, which dissociates at the pH found inside cells, freeing hydrogen ions and leaving the phosphate negatively charged (see Figure 2-22). Because these charges attract proteins, most nucleic acids in cells are associated with proteins. In nucleotides, the 1′ carbon atom of the sugar (ribose or deoxyribose) is attached to the nitrogen at position 9 of a purine (N_9) or at position 1 of a pyrimidine (N_1).

Cells and extracellular fluids in organisms contain small concentrations of **nucleosides,** combinations of a base and a sugar without a phosphate. Nucleotides are nucleosides that have one, two, or three phosphate groups esterified at the 5′ hydroxyl. *Nucleoside monophosphates* have a single esterified phosphate (see Figure 4-1a), *diphosphates* contain a prophosphate group

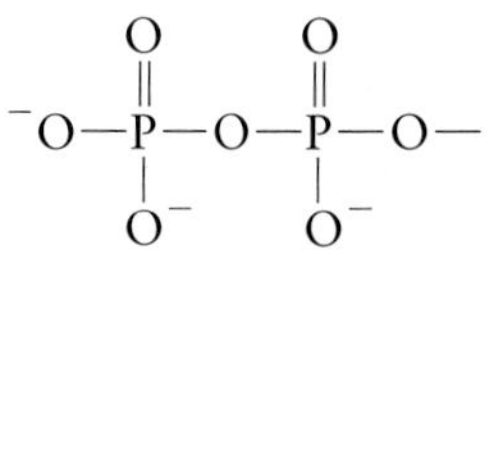

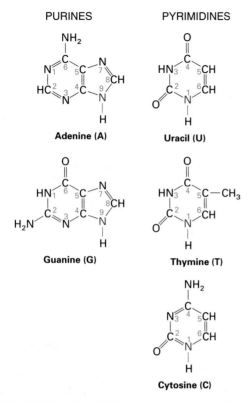

PURINES

PYRIMIDINES

Adenine (A)

Uracil (U)

Guanine (G)

Thymine (T)

Cytosine (C)

▲ **FIGURE 4-2 The chemical structures of the principal bases in nucleic acids.** In nucleic acids and nucleotides, nitrogen 9 of purines and nitrogen 1 of pyrimidines (red) are bonded to the 1' carbon of ribose or deoxyribose.

and *triphosphates* have a third phosphate. Table 4-1 lists the names of the nucleosides and nucleotides in nucleic acids and the various forms of nucleoside phosphates. As we will see later, the nucleoside triphosphates are used in the synthesis of nucleic acids. However, these compounds also serve many other functions in the cell: ATP, for example, is the most widely used energy carrier in the cell (see Figure 2-25), and GTP plays crucial roles in intracellular signaling and acts as an energy reservoir, particularly in protein synthesis.

When nucleotides polymerize to form nucleic acids, the hydroxyl group attached to the 3' carbon of a sugar of one nucleotide forms an ester bond to the phosphate of another nucleotide, eliminating a molecule of water:

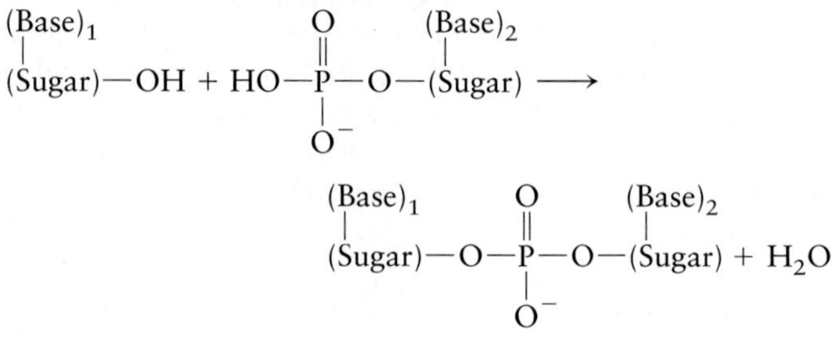

$$\text{(Sugar)}\overset{\text{(Base)}_1}{-}\text{OH} + \text{HO}\overset{\overset{\text{O}}{\|}}{-}\underset{\text{O}^-}{\text{P}}-\text{O}\overset{\text{(Base)}_2}{-}\text{(Sugar)} \longrightarrow$$

$$\text{(Sugar)}\overset{\text{(Base)}_1}{-}\text{O}\overset{\overset{\text{O}}{\|}}{-}\underset{\text{O}^-}{\text{P}}-\text{O}\overset{\text{(Base)}_2}{-}\text{(Sugar)} + \text{H}_2\text{O}$$

This condensation reaction is similar to that in which a peptide bond is formed between two amino acids (Chapter 3).

Thus a single nucleic acid strand is a phosphate-pentose polymer (a polyester) with purine and pyrimidine bases as side groups. The links between the nucleotides are called **phosphodiester bonds.** Like a polypeptide, a nucleic acid strand has an end-to-end chemical orientation: the *5' end* has a free hydroxyl or phosphate group on the 5' carbon of its terminal sugar; the *3' end* has a free hydroxyl group on the 3' carbon of its terminal sugar (Figure 4-3). This directionality, plus the fact that synthesis proceeds 5' to 3', has

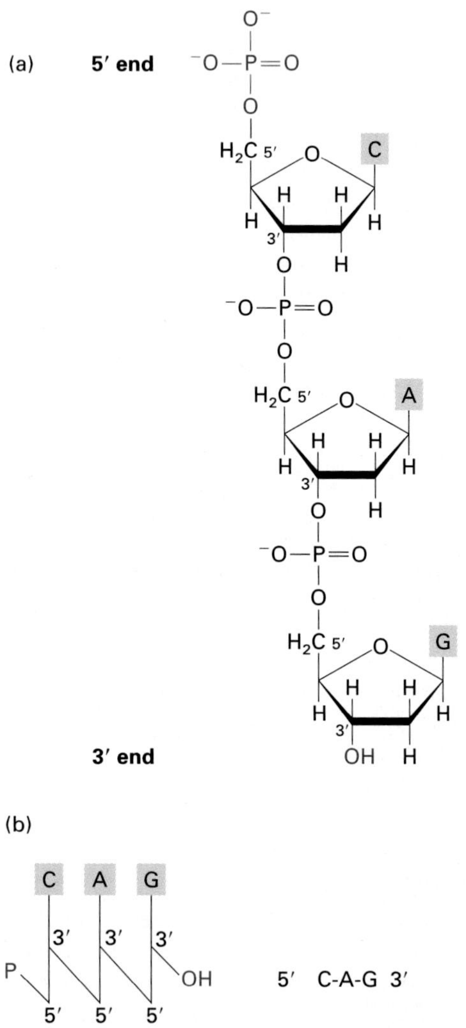

(a)   5' end

3' end

(b)

C A G

P   OH   5' C-A-G 3'

▲ **FIGURE 4-3 Alternative ways of representing nucleic acid chains, in this case a single strand of DNA containing only three bases: cytosine (C), adenine (A), and guanine (G).** (a) Chemical structure of the trinucleotide CAG. Note the free hydroxyl group at the 3' end and free phosphate group at the 5' end. (b) Two common simplified methods of representing polynucleotides. In the "stick" diagram *(left),* the sugars are indicated as vertical lines and the phosphodiester bonds as slanting lines; the bases are denoted by their single-letter abbreviations. In the simplest representation *(right),* the bases are indicated by single letters. By convention, a polynucleotide sequence is always written in the 5' → 3' direction (left to right).

| TABLE 4-1 | Naming Nucleosides and Nucleotides | | | |

		Bases			
		Purines		Pyrimidines	
		Adenine (A)	Guanine (G)	Cytosine (C)	Uracil (U) Thymine [T]
Nucleosides { in RNA		Adenosine	Guanosine	Cytidine	Uridine
in DNA		Deoxyadenosine	Deoxyguanosine	Deoxycytidine	Deoxythymidine
Nucleotides { in RNA		Adenylate	Guanylate	Cytidylate	Uridylate
in DNA		Deoxyadenylate	Deoxyguanylate	Deoxycytidylate	Thymidylate
Nucleoside monophosphates		AMP	GMP	CMP	UMP
Nucleoside diphosphates		ADP	GDP	CDP	UDP
Nucleoside triphosphates		ATP	GTP	CTP	UTP
Deoxynucleoside mono-, di-, and triphosphates		dAMP, etc.			

given rise to the convention that polynucleotide sequences are written and read in the $5' \rightarrow 3'$ direction (from left to right); for example, the sequence AUG is assumed to be $(5')AUG(3')$. (Although, strictly speaking, the letters A, G, C, T, and U stand for bases, they are also often used in diagrams to represent the whole nucleotides containing these bases.) The $5' \rightarrow 3'$ directionality of a nucleic acid strand is an extremely important property of the molecule.

The linear sequence of nucleotides linked by phosphodiester bonds constitutes the primary structure of nucleic acids. As we discuss in the next section, polynucleotides can twist and fold into three-dimensional conformations stabilized by noncovalent bonds; in this respect, they are similar to polypeptides. Although the primary structures of DNA and RNA are generally similar, their conformations are quite different. Unlike RNA, which commonly exists as a single polynucleotide chain, or strand, DNA contains two intertwined polynucleotide strands. This structural difference is critical to the different functions of the two types of nucleic acids.

Native DNA Is a Double Helix of Complementary Antiparallel Chains

The modern era of molecular biology began in 1953 when James D. Watson and Francis H. C. Crick proposed correctly the double-helical structure of DNA, based on the analysis of x-ray diffraction patterns coupled with careful model building. A closer look at the "thread of life," as the DNA molecule is sometimes called, shows why the discovery of its basic structure suggests its function.

DNA consists of two associated polynucleotide strands that wind *together* through space to form a structure often described as a **double helix.** The two sugar-phosphate backbones are on the outside of the double helix, and the bases project into the interior. The adjoining bases in each strand stack on top of one another in parallel planes (Figure 4-4a). The orientation of the two strands is antiparallel; that is, their $5' \rightarrow 3'$ directions are opposite. The strands are held in precise register by a regular base-pairing between the two strands: A is paired with T through two hydrogen bonds; G is paired with C through three hydrogen bonds (Figure 4-4b). This *base-pair complementarity* is a consequence of the size, shape, and chemical composition of the bases. The presence of thousands of such hydrogen bonds in a DNA molecule contributes greatly to the stability of the double helix. Hydrophobic and van der Waals interactions between the stacked adjacent base pairs also contribute to the stability of the DNA structure.

To maintain the geometry of the double-helical structure shown in Figure 4-4a, a larger purine (A or G) must pair with a smaller pyrimidine (C or T). In natural DNA, A almost always hydrogen bonds with T and G with C, forming A·T and G·C base pairs often called *Watson-Crick base pairs.* Two polynucleotide strands, or regions thereof, in which all the nucleotides form such base pairs are said to be **complementary.** However, in theory and in synthetic DNAs other interactions can occur. For example, a guanine (a purine) could theoretically form hydrogen bonds with a thymine (a pyrimidine), causing only a minor distortion in the helix. The space available in the helix also would allow pairing between the two pyrimidines cytosine and thymine. Although the nonstandard G·T and C·T base pairs are normally not found in DNA, G·U base pairs are quite common in double-helical regions that form within otherwise single-stranded RNA.

Two polynucleotide strands can, in principle, form either a right-handed or a left-handed helix (Figure 4-5). Because the geometry of the sugar-phosphate backbone is more compatible with the former, natural DNA is a right-handed

(a)

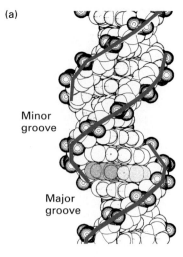

Normal B DNA

▲ **FIGURE 4-4 Two representations of contacts within the DNA double helix.** (a) Space-filling model of B DNA, the most common form of DNA in cells. The sugar and phosphate residues (gray) in each strand form the backbone, which is traced by a red line, showing the helical twist of the overall molecule. The bases project inward, but are accessible through major and minor grooves; a pair of bases from opposite strands in the major groove are highlighted in light and dark blue. The hydrogen bonds between the bases are in the center of the structure. (b) Stick diagram of the chemical structure of double-helical DNA, unraveled to show the sugar-phosphate backbones (sugar rings in green), base-paired bases (light blue and light red), and hydrogen bonds between the bases (dark red dotted lines). The backbones run in opposite directions; the 5' and 3' ends are named for the orientation of the 5' and 3' carbon atoms of the sugar rings. Each base pair has one purine base—adenine (A) or guanine (G)—and one pyrimidine base—thymine (T) or cytosine (C)—connected by hydrogen bonds. In this diagram, carbon atoms occur at the junction of every line with another line and no hydrogen atoms are shown. [Part (a) courtesy of A. Rich; part (b) from R. E. Dickerson, 1983, *Sci. Am.* **249**(6):94.]

helix. The x-ray diffraction pattern of DNA indicates that the stacked bases are regularly spaced 0.34 nm apart along the helix axis. The helix makes a complete turn every 3.4 nm; thus there are about 10 pairs per turn. This is referred to as the *B form* of DNA, the normal form present in most DNA stretches in cells (Figure 4-6a). On the outside of B-form DNA, the spaces between the intertwined strands form two helical grooves of different widths described as the *major* groove and the *minor* groove (see Figure 4-4a). Consequently, part of each base is accessible from outside the helix to both small and large molecules that bind to the DNA by contacting chemical groups within the grooves. These two binding surfaces of the DNA molecule are used by different classes of DNA-binding proteins.

In addition to the major B form of DNA, three additional structures have been described. In very high humidity, the crystallographic structure of B DNA changes to the *A form*; RNA-DNA and RNA-RNA helices also exist in this form. The A form is more compact than the B form,

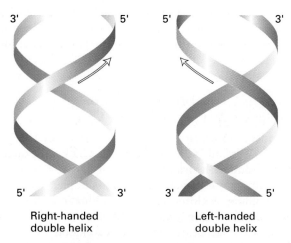

Right-handed double helix   Left-handed double helix

▲ **FIGURE 4-5 Two possible helical forms of DNA are mirror images of each other.** The geometry of the sugar-phosphate backbone of DNA causes natural DNA to be right-handed. (*Right-handed* and *left-handed* are defined by convention.)

(a) B DNA　　(b) A DNA　　(c) Z DNA　　(d) Triple-Helical DNA

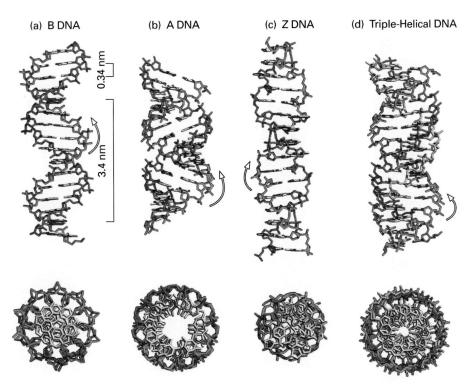

◀ **FIGURE 4-6 Models of various DNA structures that are known to exist.** The sugar-phosphate backbone of each chain is on the outside in all structures (one red and one blue) with the bases (silver) oriented inward. Side views are shown at the top, and views along the helical axis at the bottom. (a) The B form of DNA, the usual form in cells, is characterized by a helical turn every 10 base pairs (3.4 nm); adjacent stacked base pairs are 0.34 nm apart. The major and minor grooves are also visible. (b) The more compact A form of DNA has 11 base pairs per turn and exhibits a large tilt of the base pairs with respect to the helix axis. In addition, the A form has a central hole *(bottom)*. This helical form is adopted by RNA-DNA and RNA-RNA helices. (c) Z DNA is a left-handed helix and has a zig-zag (hence "Z") appearance. (d) A triple-helical structure can occur in stretches of DNA where all purines (A, G) in one strand are matched by all pyrimidines (T, C) in the other strand. Such stretches can accommodate a third polypyrimidine strand (yellow). [Courtesy of C. Kielkopf and P. B. Dervan.]

having 11 bases per turn, and the stacked bases are tilted (Figure 4-6b). Short DNA molecules composed of alternating purine-pyrimidine nucleotides (especially Gs and Cs) adopt an alternative left-handed configuration instead of the normal right-handed helix. This structure is called *Z DNA* because the bases seem to zigzag when viewed from the side (Figure 4-6c). It is entirely possible that both A-form and Z-form stretches of DNA exist in cells.

Finally, a triple-stranded DNA structure can also exist at least in the test tube, and possibly during recombination and DNA repair. For example, when synthetic polymers of poly(A) and polydeoxy(U) are mixed, a three-stranded structure is formed (Figure 4-6d). Further, long homopolymeric stretches of DNA composed of C and T residues in one strand and A and G residues in the other can be targeted by short matching lengths of poly(C+T). The synthetic oligonucleotide can insert as a third strand, binding in a sequence-specific manner by so-called *Hoogsteen base pairs*. Specific cleavage of the DNA at the site where the triple helix ends can be achieved by attaching a chemical cleaving agent (e.g., Fe^{2+}-EDTA) to the short oligodeoxynucleotide that makes up the third strand. Such reactions may be useful in studying site-specific DNA damage in cells.

By far the most important modifications in standard B-form DNA come about as a result of protein binding to specific DNA sequences. Although the multitude of hydrogen and hydrophobic bonds between the polynucleotide strands provide stability to DNA, the double helix is somewhat flexible about its long axis. Unlike the α helix in proteins (see Figure 3-6), there are no hydrogen bonds between successive residues in a DNA strand. This property allows DNA to bend when complexed with a DNA-binding protein. Crystallographic analyses of proteins bound to particular regions of DNA have conclusively demonstrated departures from the standard B-DNA structure in protein-DNA complexes. Two examples of DNA deformed by contact with proteins are shown in Figure 4-7. The specific DNA-protein contacts that occur in these tightly bound complexes have the ability both to untwist the DNA and to bend the axis of the helix. Although DNA in cells likely exists in the B form most of the time, particular regions bound to protein clearly depart from the standard conformation.

DNA Can Undergo Reversible Strand Separation

In DNA replication and in the copying of RNA from DNA, the strands of the helix must separate at least temporarily. As we discuss later, during DNA synthesis two new strands are made (one copied from each of the original strands), resulting in two double helices identical with the original one. In the case of copying the DNA template to make RNA, the RNA is released and the two DNA strands reassociate with each other.

The unwinding and separation of DNA strands, referred to as **denaturation**, or "melting," can be induced experimentally. For example, if a solution of DNA is heated, the thermal energy increases molecular motion, eventually breaking the hydrogen bonds and other forces that stabilize the double helix, and the strands separate (Figure 4-8). This

(a)

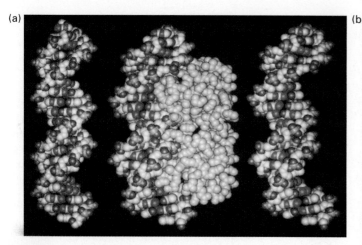

(b)

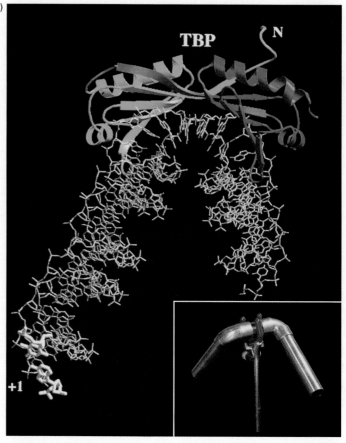

▲ FIGURE 4-7 Bending of DNA resulting from protein binding. (a) A linear DNA *(left)* is shown binding a repressor protein encoded by bacteriophage 434 *(center)*; the resulting bend in the DNA *(right)* is easily seen by comparison with the linear molecule. Binding of this repressor to the viral genome prevents its transcription by bacterial host-cell enzymes. (b) Binding of TATA box–binding protein (TBP) to DNA causes a complete change in the winding and direction of the double helix. Transcription of most eukaryotic genes requires participation of TBP. *(Inset)* Copper pipe bent to mimic the path of the DNA backbone in the DNA-TBP complex. [Part (a) from A. K. Aggarwal et al., 1988, *Science* **242**:899; courtesy of S. C. Harrison. Part (b) from D. B. Nicolov and S. K. Burley, 1997, *Proc. Nat'l. Acad. Sci. USA* **94**:15; courtesy of S. K. Burley.]

melting of DNA changes its absorption of ultraviolet (UV) light (in the 260-nm range), which is routinely used to measure DNA concentration because of the high absorbance of UV light by nucleic acid bases. Native double-stranded DNA absorbs about one-half as much light at 260 nm as does the equivalent amount of single-stranded DNA

(Figure 4-9a). Thus, as DNA denatures, its absorption of UV light increases. Near the denaturation temperature, a small increase in temperature causes an abrupt, near simultaneous, loss of the multiple, weak, cooperative interactions holding the two strands together, so that denaturation rapidly occurs throughout the entire length of the DNA.

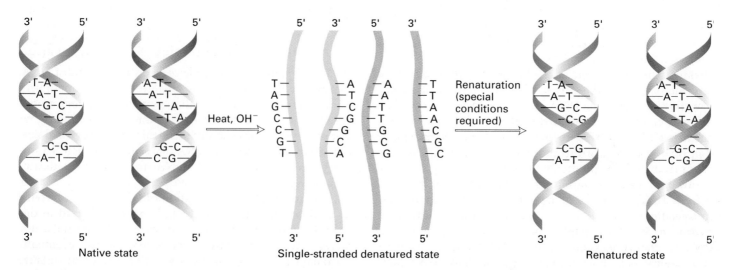

▲ FIGURE 4-8 The denaturation and renaturation of double-stranded DNA molecules.

(a)

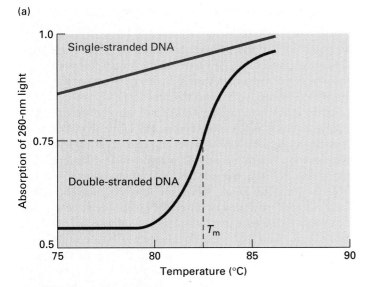

(b)

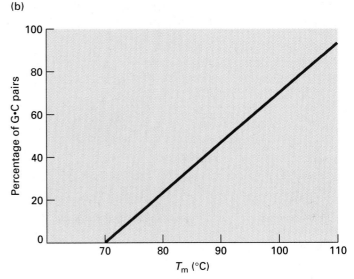

▲ FIGURE 4-9 Light absorption and temperature in DNA denaturation. (a) Melting of doubled-stranded DNA can be monitored by the absorption of ultraviolet light at 260 nm. As regions of double-stranded DNA unpair, the absorption of light by those regions increases almost twofold. The temperature at which half the bases in a double-stranded DNA sample have denatured is denoted T_m (for temperature of melting). Light absorption by single-stranded DNA changes much less as the temperature is increased. (b) The T_m is a function of the G·C content of the DNA; the higher the G·C percentage, the greater the T_m.

The *melting temperature*, T_m, at which the strands of DNA will separate depends on several factors. Molecules that contain a greater proportion of G·C pairs require higher temperatures to denature because the three hydrogen bonds in G·C pairs make them more stable than A·T pairs with two hydrogen bonds (see Figure 4-4b). Indeed, the percentage of G·C base pairs in a DNA sample can be estimated from its T_m (Figure 4-9b). In addition to heat, solutions of low ion concentration destabilize the double helix, causing it to melt at lower temperatures. DNA is also denatured by exposure to other agents that destabilize hydrogen bonds, such as alkaline solutions and concentrated solutions of formamide or urea:

$$
\begin{array}{cc}
\overset{\displaystyle O}{\underset{\displaystyle HC-NH_2}{\parallel}} & \overset{\displaystyle O}{\underset{\displaystyle H_2N-C-NH_2}{\parallel}} \\
\textbf{Formamide} & \textbf{Urea}
\end{array}
$$

The single-stranded DNA molecules that result from denaturation form random coils without a regular structure. Lowering the temperature or increasing the ion concentration causes the two complementary strands to reassociate into a perfect double helix (see Figure 4-8). The extent of such *renaturation* is dependent on time, the DNA concentration, and the ionic content of the solution. Two DNA strands not related in sequence will remain as random coils and will not renature and, most important, will not greatly inhibit complementary DNA partner strands from finding each other. Denaturation and renaturation of DNA are the basis of nucleic acid **hybridization,** a powerful technique used to study the relatedness of two DNA samples and to detect and isolate specific DNA molecules in a mixture containing numerous different DNA sequences (Chapter 7).

Many DNA Molecules Are Circular

All prokaryotic genomic DNAs and many viral DNAs are circular molecules. Circular DNA molecules also occur in mitochondria, which are present in almost all eukaryotic cells, and in chloroplasts, which are present in plants and some unicellular eukaryotes.

Each of the two strands in a circular DNA molecule forms a closed structure without free ends. Just as is the case for linear DNA, elevated temperatures or alkaline pH destroy the hydrogen bonds and other interactions that stabilize double-helical circular DNA molecules. Unlike linear DNA, however, the two strands of circular DNA cannot unwind and separate; attempts to melt such DNA result in an interlocked, tangled mass of single-stranded DNA (Figure 4-10a).

Only if a native circular DNA is *nicked* (i.e., one of the strands is cut), will the two strands unwind and separate when the molecule is denatured. In this case, one of the separated strands is circular, and the other is linear (Figure 4-10b). Nicking of circular DNA occurs naturally during DNA replication and can be induced experimentally with a low concentration of deoxyribonuclease (a DNA-degrading enzyme), so that only a single phosphodiester bond in the molecule is cleaved. The study of circular DNA molecules lacking free ends first uncovered the complicated geometric shape changes that the double-stranded DNA molecule must undergo when the strands are not free to separate.

(a)

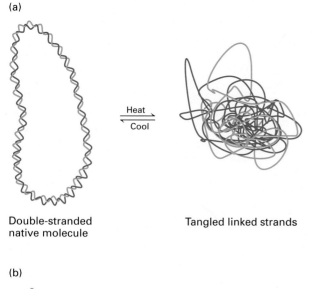

Double-stranded   Tangled linked strands
native molecule

(b)

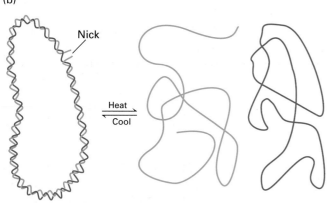

Nick in one strand   Strands competely separated

▲ **FIGURE 4-10 Denaturation of circular DNA.** (a) If both strands are closed circles, denaturation disrupts the double helix, but the two single strands become tangled about each other and cannot separate. (b) If one or both strands are nicked, however, the two strands will separate on thermal denaturation.

Local Unwinding of DNA Induces Supercoiling

So far we have described DNA as a long regular helical structure that can have local perturbations, especially due to protein binding. In addition, when the two ends of a DNA molecule are fixed, the molecule exhibits a superstructure under certain conditions. This occurs when the base pairing is interrupted and a local region unwinds. The stress induced by unwinding is relieved by twisting of the double helix on itself, forming **supercoils** (Figure 4-11). Unwinding and subsequent supercoiling occurs during replication, transcription, and binding of many proteins to circular DNAs or to long DNA loops whose ends are fixed within eukaryotic chromosomes. Supercoiling is recognized and regulated by enzymes called **topoisomerases.** As discussed in later chapters, these enzymes have an important role in both DNA replication and the transcription of DNA into RNA.

RNA Molecules Exhibit Varied Conformations and Functions

As noted earlier, the primary structure of RNA is generally similar to that of DNA; however, the sugar component (ribose) of RNA has an additional hydroxyl group at the 2′ position (see Figure 4-1b), and thymine in DNA is replaced by uracil in RNA (see Figure 4-2). The hydroxyl group on C_2 of ribose makes RNA more chemically labile than DNA and provides a chemically reactive group that takes part in RNA-mediated enzymatic events. As a result of this lability, RNA is cleaved into mononucleotides by alkaline solution, whereas DNA is not. Like DNA, RNA is a long polynucleotide that can be double-stranded or single-stranded, linear or circular. It can also participate in a hybrid helix composed of one RNA strand and one DNA strand; this hybrid has a slightly different conformation than the common B form of DNA.

Unlike DNA, which exists primarily in a single, very long three-dimensional structure, the double helix, the various types of RNA exhibit different conformations. Differences in the sizes and conformations of the various types of RNA permit them to carry out specific functions in a cell. The simplest secondary structures in single-stranded RNAs are formed by pairing of complementary bases. "Hairpins" are formed by pairing of bases within ≈5–10 nucleotides of each other, and "stem-loops" by pairing of bases that are separated by ≈50 to several hundred nucleotides (Figure 4-12a). These simple folds can cooperate to form more complicated tertiary structures, one of which is termed a "pseudoknot" (Figure 4-12b).

As discussed in detail later, tRNA molecules adopt a well-defined three-dimensional architecture in solution that is crucial in protein synthesis. Larger rRNA molecules also have locally well defined three-dimensional structures, with more flexible links in between. Secondary and tertiary structures also have been recognized in mRNA, particularly near the ends of molecules. These recently discovered structures are under active study. Clearly, then, RNA molecules are like proteins in that they have structured domains connected by less structured, flexible stretches.

The folded domains of RNA molecules not only are structurally analogous to the α helices and β strands found in proteins, but in some cases also have catalytic capacities. Such catalytic RNAs, called **ribozymes,** can cut RNA chains. Some RNA domains also can catalyze *RNA splicing,* a remarkable process in which an internal RNA sequence, an **intron,** is cut and removed and the two resulting chains, the **exons,** are sealed together. This process occurs during formation of the majority of functional mRNA molecules in eukaryotic cells, and also occurs in bacteria and archaea. Remarkably, some RNAs carry out *self-splicing,* with the catalytic activity residing in the intron sequence. The mechanisms of splicing and self-splicing are discussed in detail in Chapter 11. As noted later in this chapter, rRNA is thought to play a catalytic role in the formation of peptide bonds during protein synthesis.

Form I   Form II

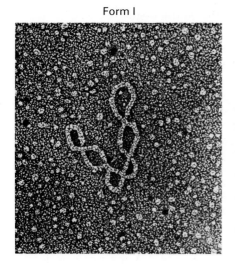

◀ **FIGURE 4-11 Supercoiling in electron micrographs of DNA isolated from the SV40 virus.** When isolated SV40 DNA is separated from its associated protein, the DNA duplex is underwound and assumes the supercoiled configuration (form I). If one strand is nicked, the strands can rewind, producing the relaxed-circle configuration (form II), which lacks super-coils. Only a few of the possible supercoils are visualized in the left photograph.

(a) Secondary structure

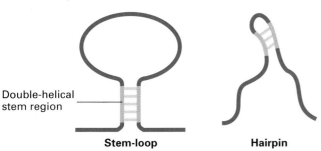

Double-helical
stem region

Stem-loop   **Hairpin**

(b) Tertiary structure

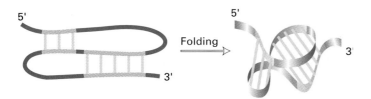

Folding

Pseudoknot

▲ **FIGURE 4-12 RNA secondary and tertiary structures.**
(a) Stem-loops, hairpins, and other secondary structures can form by base pairing between distant complementary segments of an RNA molecule. In stem-loops, the single-stranded loop (dark red) between the base-paired helical stem (light red) may be hundreds or even thousands of nucleotides long, whereas in hairpins, the short turn may contain as few as 6–8 nucleotides. (b) Interactions between the flexible loops may result in further folding to form tertiary structures such as the pseudoknot. This tertiary structure resembles a figure-eight knot, but the free ends do not pass through the loops, so no knot is actually formed. [Part (b) adapted from C. W. A. Pleij et al., 1985, *Nucl. Acids Res.* **13**:1717.]

In this chapter, we focus on the functions of mRNA, tRNA, and rRNA in **gene expression**—the process of getting the information in DNA converted into proteins. In later chapters we will encounter other RNAs, often associated with proteins, that participate in other cell functions.

SUMMARY Structure of Nucleic Acids

- Deoxyribonucleic acid (DNA), the genetic material, carries information to specify the amino acid sequences of proteins. It is transcribed into several types of ribonucleic acid (RNA) including messenger RNA (mRNA), transfer RNA (tRNA), and ribosomal RNA (rRNA), which function in protein synthesis.

- Both DNA and RNA are long, unbranched polymers of nucleotides. Each nucleotide consists of a heterocyclic base linked via a five-carbon sugar (deoxyribose or ribose) to a phosphate group (see Figure 4-1).

- DNA and RNA each contain four different bases (see Figure 4-2). The purines adenine (A) and guanine (G) and the pyrimidine cytosine (C) are present in both DNA and RNA. The pyrimidine thymine (T) present in DNA is replaced by the pyrimidine uracil (U) in RNA.

- The bases in nucleic acids can interact via hydrogen bonds. The standard Watson-Crick base pairs are G·C, A·T (in DNA), and A·U (in RNA). Base pairing stabilizes the native three-dimensional structures of DNA and RNA.

- Adjacent nucleotides in a polynucleotide are linked by phosphodiester bonds. The entire strand has a chemical directionality: the 5′ end with a free hydroxyl or phosphate group on the 5′ carbon of the sugar, and the 3′ end with a free hydroxyl group on the 3′ carbon of the sugar (see Figure 4-3). Polynucleotide sequences are always written in the 5′ → 3′ direction (left to right).

• Natural DNA (B DNA) contains two complementary polynucleotide strands wound together into a regular right-handed double helix with the bases on the inside and the two sugar-phosphate backbones on the outside (see Figure 4-6a). Base pairing (A·T and G·C) and hydrophobic interactions between adjacent bases in the same strand stabilize this native structure.

• Binding of protein to DNA can deform its helical structure, causing local bending or unwinding of the DNA molecule.

• Heat causes the DNA strands to separate (denature). The melting temperature of DNA increases with the percentage of G·C base pairs. Under suitable conditions, separated complementary nucleic acid strands will renature.

• Local unwinding of the DNA helix induces stress, which is relieved by twisting of the molecule on itself, forming supercoils. This process is regulated by topoisomerases, which can add or remove supercoils.

• Natural RNAs are single-stranded polynucleotides that form well-defined secondary and tertiary structures (see Figure 4-12). Some RNAs, called *ribozymes*, have catalytic activity.

4.2 Synthesis of Biopolymers: Rules of Macromolecular Carpentry

According to the central dogma of molecular biology, DNA directs the synthesis of RNA, and RNA then directs the synthesis of proteins. However, the one-way flow of information posited by the central dogma—that is, DNA → RNA → protein—does not reflect the role of proteins in facilitating the information flow. A more accurate way of representing the relationship between the synthesis of DNA, RNA, and proteins in all cells would look like

$$DNA \longrightarrow RNA \longrightarrow protein$$

indicating that special proteins catalyze the synthesis of both RNA and DNA.

Interestingly, although the primary and higher-order structures of nucleic acids and proteins differ radically, several common principles apply to the synthesis of both types of macromolecules. Research over the past 30 years has revealed that giant molecular machines, composed largely of proteins, assemble nucleotides into RNA and DNA and amino acids into proteins. Later in this chapter and in subsequent chapters, we examine in detail the molecular machines that synthesize nucleic acids and proteins. To set the stage for these more detailed discussions, we review four general rules that govern the synthesis of both polypeptide and polynucleotide chains:

• *Proteins and nucleic acids are made up of a limited number of different monomeric building blocks.* Although the number of amino acids is theoretically limitless, and several dozen have been identified as metabolic products in various organisms, only 20 different amino acids are used in making proteins. Likewise, only five nitrogenous bases are used to construct RNA and DNA in cells. Cell-free enzyme preparations and cells in culture can be "fooled" into incorporating chemical relatives of the five bases or 20 amino acids, but this almost never happens in nature.

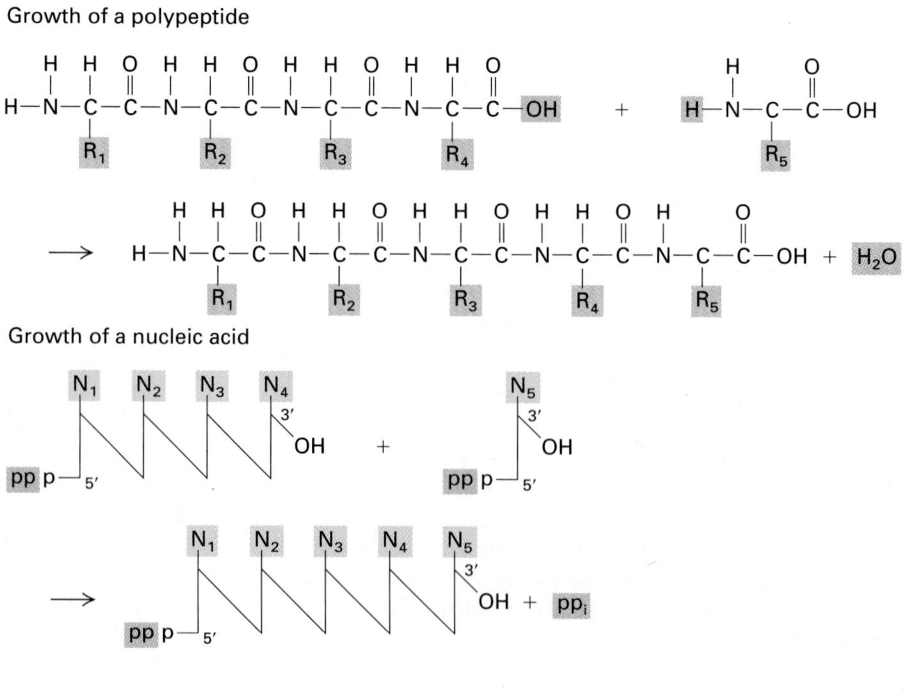

▶ **FIGURE 4-13 Chain elongation in the in vivo synthesis of both proteins and nucleic acids proceeds by sequential addition of monomeric units—amino acids and nucleotides, respectively.** Elongation of a polypeptide chain proceeds from the N-terminus to the C-terminus. (R_1, R_2, etc., denote side chains of amino acids.) Elongation of a polynucleotide chain proceeds from the 5′ end to the 3′ end. As each nucleotide is added, a pyrophosphate group (PP$_i$) is released, but the 5′ end of the first nucleotide in the chain retains its triphosphate group. (N_1, N_2, etc., denote purine and pyrimidine bases.)

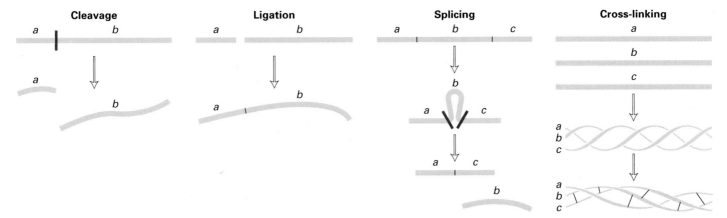

▲ **FIGURE 4-14 Proteins and nucleic acids typically undergo one or more modifications after the basic polymeric chains are formed.** RNA, DNA, and protein can be cleaved; DNA pieces must be ligated during synthesis. Splicing, the removal of an internal sequence and reunion of the end pieces, often occurs in RNA synthesis; a similar reaction (intein removal) occurs in the formation of some proteins. Covalent chemical cross-linking between protein chains is common, but cross-linking is rare in nucleic acids. Besides the modifications to the polymeric chains illustrated here, individual chemical groups (e.g., phosphate, nitrate, sulfate, methyl, and acetyl groups) may be added. Various types of postpolymerization changes are discussed in later chapters.

- *The monomers are added one at a time.* Biological polymers theoretically could be built by aligning their component monomeric units in the correct order on a template, or mold, and *simultaneously* fusing them all. But cells do not use this mechanism. Rather, the assembly of proteins and nucleic acids occurs through the step-by-step addition of monomers.

- *Each polypeptide and polynucleotide chain has a specific starting point, and growth proceeds in one direction to a fixed terminus.* Synthesis of chains begins and ends at well-defined "start" and "stop" signals. Monomer addition proceeds from the amino (NH_2-) terminus to the carboxyl (COOH-) terminus in proteins, and from the 5′ end to the 3′ end in nucleic acids (Figure 4-13).

- *The primary synthetic product is often modified.* The initially synthesized form of a polypeptide or polynucleotide chain is often inactive or incomplete and undergoes modification to yield an active molecule (Figure 4-14). Protein chains are often shortened, and RNA chains almost always are; DNA chains are made in pieces and then linked together. Distant segments of long RNA chains may be spliced together, eliminating the intervening sequence (intron). A similar process occurs in some proteins (see Figure 3-17). Polypeptide chains sometimes are cross-linked by covalent bonds to form a functional protein. Finally, certain chemical groups may be added either during the formation of the chain or after its synthesis is complete: e.g., methyl groups to specific sites in DNA, RNA, and proteins; phosphate groups and a wide variety of oligosaccharides to proteins. The details of this macromolecular carpentry are discussed in later chapters.

4.3   Nucleic Acid Synthesis

The ordered assembly of deoxyribonucleotides into DNA and of ribonucleotides into RNA involves somewhat simpler cellular mechanisms than the correct assembly of the amino acids in a protein chain. Here we consider a few general principles governing the formation of polynucleotide chains in cells and briefly discuss some properties of the enzymes that carry out such synthesis. We also describe the steps in the production of mRNA and examine how and why this process differs in bacteria and eukaryotes. Later chapters cover the mechanism of DNA replication and its control during cell growth and division, and the mechanism and the control of the synthesis of specific mRNAs during differentiation (Chapters 10 and 12).

Both DNA and RNA Chains Are Produced by Copying of Template DNA Strands

The regular pairing of bases in the double-helical DNA structure suggested to Watson and Crick a mechanism of DNA synthesis. Their proposal that new strands of DNA are synthesized by copying of parental strands of DNA has proved to be correct.

The DNA strand that is copied to form a new strand is called a **template**. The information in the template is preserved: although the first copy has a complementary sequence, not an identical one, a copy of the copy produces the original (template) sequence again. In the replication of a double-stranded, or *duplex*, DNA molecule, both original (parental) DNA strands are copied. When copying is finished, the two new duplexes, each consisting of one of the two original strands plus its copy, separate from each other.

In some viruses, single-stranded RNA molecules function as templates for synthesis of complementary RNA or DNA chains (Chapter 7). However, the vast majority of RNA and DNA in cells is synthesized from preexisting duplex DNA.

Nucleic Acid Strands Grow in the 5′ → 3′ Direction

All RNA and DNA synthesis, both cellular and viral, proceeds in the same chemical direction: from the 5′ (phosphate) end to the 3′ (hydroxyl) end (see Figure 4-13). Nucleic acid chains are assembled from 5′ triphosphates of ribonucleosides or deoxyribonucleosides. Strand growth is energetically unfavorable but is driven by the energy available in the triphosphates. The α phosphate of the incoming nucleotide attaches to the 3′ hydroxyl of the ribose (or deoxyribose) of the preceding residue to form a phosphodiester bond, releasing a pyrophosphate (PP_i). The equilibrium of the reaction is driven further toward chain elongation by pyrophosphatase, which catalyzes the cleavage of PP_i into two molecules of inorganic phosphate (see Table 2-7).

RNA Polymerases Can Initiate Strand Growth but DNA Polymerases Cannot

The enzymes that copy (replicate) DNA to make more DNA are **DNA polymerases;** those that copy (transcribe) DNA to form RNA are **RNA polymerases.** Because the two DNA strands are complementary, rather than identical, transcription of a particular DNA segment theoretically could yield two mRNAs with different sequences and hence different protein-coding potentials. Generally, only one strand of the duplex in a particular DNA segment gives rise to usable information when transcribed into mRNA. In unusual cases, though, limited sections of DNA encode proteins on both strands.

An RNA polymerase can find an appropriate initiation site on duplex DNA; bind the DNA; temporarily "melt," or separate, the two strands in that region; and begin generating a new RNA strand (Figure 4-15). As discussed in Chapter 10, the location and regulated use of transcription start sites to produce mRNA requires many dozens of proteins in eukaryotes and several proteins

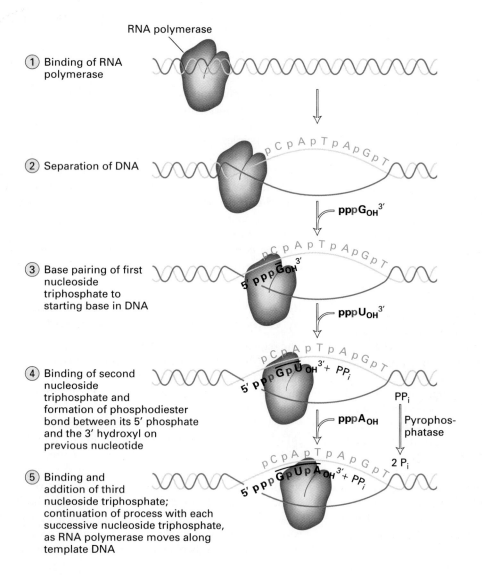

◀ **FIGURE 4-15 Transcription of DNA into RNA is catalyzed by RNA polymerase, which can initiate the synthesis of strands de novo on DNA templates.** The nucleotide at the 5′ end of an RNA strand retains all three of its phosphate groups; all subsequent nucleotides release pyrophosphate (PP_i) when added to the chain and retain only their α phosphate (red). The released PP_i is subsequently hydrolyzed by pyrophosphatase to P_i, driving the equilibrium of the overall reaction toward chain elongation. In most cases, only one DNA strand is transcribed into RNA.

even in bacteria. The nucleotide at the 5' terminus of a growing RNA strand is chemically distinct from the nucleotides within the strand in that it retains all three phosphate groups. When an additional nucleotide is added to the 3' end of the growing strand, only the α-phosphate is retained; the β and γ phosphates are lost as pyrophosphate, which is subsequently hydrolyzed to yield 2 molecules of inorganic phosphate.

Unlike RNA polymerases, DNA polymerases cannot initiate chain synthesis de novo; instead, they require a short, preexisting RNA or DNA strand, called a primer, to begin chain growth. With a primer base-paired to the template strand, a DNA polymerase adds nucleotides to the free hydroxyl group at the 3' end of the primer:

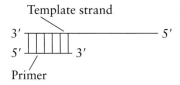

Template strand

If RNA is the primer, the polynucleotide copied from the template is RNA at the 5' end and DNA at the 3' end.

Both prokaryotic and eukaryotic cells have several different types of DNA polymerases. Some polymerases participate in making new DNA to prepare for cell division; other polymerases serve in the repair and recombination of DNA molecules. The structure, mechanism, and physiological role of these enzymes are described in Chapter 12.

Replication of Duplex DNA Requires Assembly of Many Proteins at a Growing Fork

Because duplex DNA consists of two intertwined strands, the base-pair copying of each strand requires unwinding of the original duplex, which is accomplished by specific "unwinding proteins" called **helicases.** As noted earlier, local unwinding of duplex DNA produces torsional stress, leading to formation of supercoils, which are removed by topoisomerases. The action of all these proteins produces a moving, highly specialized region of the DNA called the **growing fork,** at which DNA polymerase carries out nucleotide addition. In order for DNA polymerase to move along and copy a duplex DNA, helicase must sequentially unwind the duplex and topoisomerase must remove the supercoils that form.

DNA replication begins with creation of a growing fork by a protein or proteins that have helicase activity and unwind a short section of parental DNA. A specialized RNA polymerase then forms short RNA primers complementary to the unwound template strands. Each such primer, still bound to its complementary DNA strand, is then elongated by DNA polymerase, thereby forming a new daughter strand. One final major complication in the operation of a DNA growing fork is that although the two

strands of the parental duplex are antiparallel, nucleotides can be added to the growing new strands only in the 5' → 3' direction. As diagrammed in Figure 4-16, synthesis of one daughter strand, called the **leading strand,** proceeds continuously from a single RNA primer in the 5' → 3' direction, *the same direction as movement of the growing fork.* Because growth of the other daughter strand, called the **lagging strand,** also must occur in the 5' → 3' direction, copying of its template strand must somehow occur in the *opposite* direction from the movement of the growing fork. A cell accomplishes this feat by producing additional short RNA primers every 1000 bases or so on the second parental strand, as more of the strand is exposed by unwinding. Each of these primers, base-paired to their template strand, is elongated in the 5' → 3' direction, forming discontinuous segments called **Okazaki fragments** after their discoverer Reiji Okazaki. The RNA primer of each Okazaki fragment is removed and replaced by DNA chain growth from the neighboring Okazaki fragment; finally an enzyme called *DNA ligase* joins the adjacent fragments. At least 30 proteins participate in the formation and operation of a DNA growing fork; this DNA-replication machine is discussed in detail in Chapter 12.

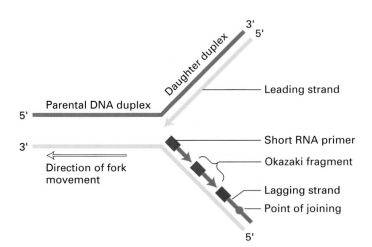

▲ **FIGURE 4-16 Schematic diagram of DNA replication at a growing fork.** Nucleotides are added by DNA polymerase to each daughter strand in the 5' → 3' direction (indicated by arrowheads). Synthesis of the leading strand occurs continuously from a single RNA primer at its 5' end (not shown). Synthesis of the other new strand—the lagging strand—proceeds discontinuously, initially forming Okazaki fragments, from multiple RNA primers that are formed on the parental strand as each new region of DNA is exposed at the growing fork. The RNA primers are elongated by DNA polymerase. As each growing fragment approaches the previous primer, that primer is removed by another enzyme and the fragments are joined by DNA ligase to form a continuous DNA strand. By repetition of this process, the entire lagging strand eventually is completed.

Organization of Genes in DNA Differs in Prokaryotes and Eukaryotes

Having outlined the principles governing the stepwise assembly of polynucleotides, we now focus briefly on the large-scale arrangement of information in DNA and how this arrangement dictates the requirements for RNA manufacture so that information transfer goes smoothly. The simplest definition of a gene is a "unit of DNA that contains the information to specify synthesis of a single polypeptide chain." The number of genes in cells varies widely, with the simpler non-nucleated prokaryotic cells having far fewer genes than eukaryotic cells. The vast majority of genes carry information to build protein molecules, and it is the RNA copies of such *protein-coding genes* that are the mRNA molecules of cells. In recent years, the entire sequence of the DNA genome of several organisms has been determined, providing direct evidence for large differences in their protein-coding capacity (Chapter 7).

The most common arrangement of protein-coding genes in all prokaryotes has a powerful and appealing logic: genes devoted to a single metabolic goal, say, the synthesis of the amino acid tryptophan, are most often found in a contiguous array in the DNA. This gene order makes it possible to produce a continuous strand of mRNA that carries the message for a related series of enzymes devoted to making tryptophan (Figure 4-17a). Each section of the mRNA represents the unit (or gene) that instructs the protein-synthesizing apparatus to make a particular protein. Such an arrangement of genes in a functional group is called an **operon,** because it operates as a unit from a single transcription start site. In prokaryotic DNA the genes are closely packed with very few noncoding gaps, and the DNA is transcribed directly into colinear mRNA, which then is translated into protein, even while stretches of the mRNA closer to the 3′ end are still being produced.

This economic clustering of genes devoted to a single metabolic function does not occur in eukaryotes, even simple ones like yeasts that can be metabolically similar to bacteria. Rather, eukaryotic genes, even those devoted to a single pathway, are most often physically separated in the DNA, sometimes even being located on different chromosomes.

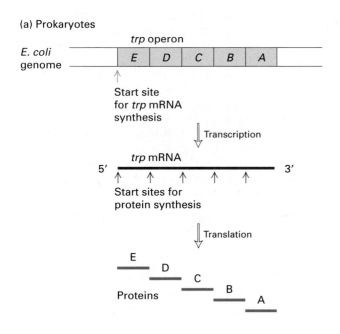

(a) Prokaryotes

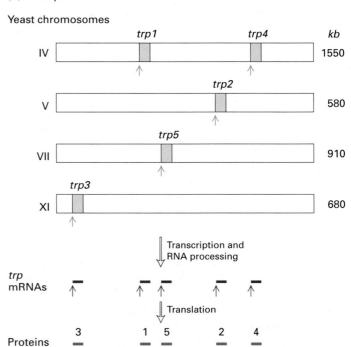

(b) Eukaryotes

▲ **FIGURE 4-17 Comparison of gene organization, transcription, and translation in prokaryotes and eukaryotes.** (a) The tryptophan *(trp)* operon is a continuous segment of the *E. coli* chromosome, containing five genes (blue) that encode the enzymes necessary for the stepwise synthesis of tryptophan. The entire operon is transcribed from one start site (blue arrow) into one long continuous *trp* mRNA (red). Translation of this mRNA begins at five different start sites, yielding five proteins (green). Proteins E and D associate to form the first enzyme in the tryptophan biosynthetic pathway; protein C catalyzes the intermediate step; and proteins A and B form tryptophan synthetase, the final enzyme. Thus the order of the genes in the bacterial genome parallels the sequential function of the encoded proteins in the tryptophan pathway. (b) The five genes encoding the enzymes required for tryptophan synthesis in yeast *(Saccharomyces cerevisiae)* are carried on four different chromosomes. Each gene is transcribed from its own start site to yield a primary transcript that is processed into a functional mRNA encoding a single protein (see Figure 4-19). The length of the yeast chromosomes is given in kilobases (10^3 bases), with all drawn to the same length.

Each gene is transcribed from its own start site, producing one mRNA, which generally is translated to yield a single protein (Figure 4-17b). Moreover, when researchers first compared the nucleotide sequences of eukaryotic mRNAs with the DNAs encoding them, they were astounded to find that the uninterrupted protein-coding sequence of a given mRNA was broken up (discontinuous) in its corresponding section of DNA. They concluded that the eukaryotic gene existed in pieces of coding sequence, the *exons*, separated by non-protein-coding segments, the *introns*. This astonishing finding, first discovered in viruses that infect eukaryotic cells, implied that the long initial RNA copy, called the **primary transcript**, the entire copied DNA sequence, had to be clipped apart to remove the introns and then carefully stitched back together to produce many mRNAs of eukaryotic cells.

Eukaryotic Primary RNA Transcripts Are Processed to Form Functional mRNAs

In prokaryotic cells, which have no nuclei, translation of an mRNA into protein can begin from the 5′ end of the mRNA even while the 3′ end is still being copied from DNA. Thus, transcription and translation can occur concurrently.

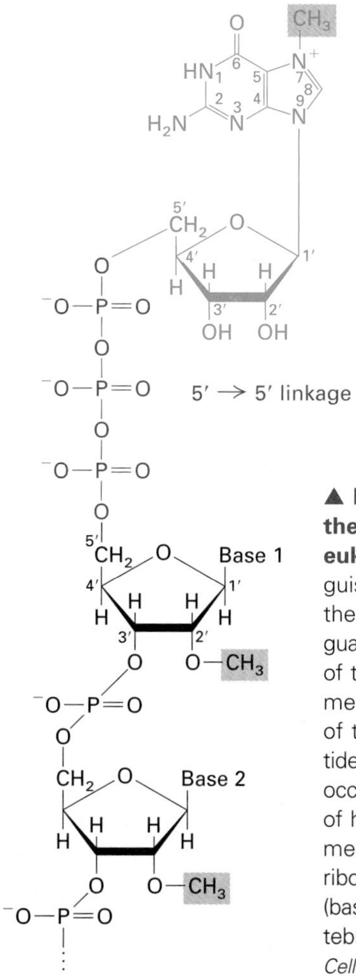

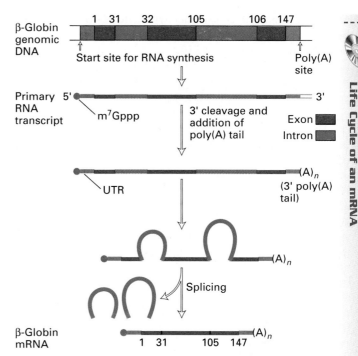

▲ **FIGURE 4-19 Overview of RNA processing in eukaryotes using β-globin gene as an example.** The β-globin gene contains three protein-coding exons (red) and two intervening noncoding introns (blue). The introns interrupt the protein-coding sequence between the codons for amino acids 31 and 32 and 105 and 106. Transcription of this and many other genes starts slightly upstream of the 5′ exon and extends downstream of the 3′ exon, resulting in noncoding regions (gray) at the ends of the primary transcript. These regions, referred to as *untranslated regions (UTRs)*, are retained during processing. The 5′ 7-methylguanylate cap (m⁷Gppp; green dot) is added during formation of the primary RNA transcript, which extends beyond the poly(A) site. After cleavage at the poly(A) site and addition of multiple A residues to the 3′ end, splicing removes the introns and joins the exons. The small numbers refer to positions in the 147-aa sequence of β-globin.

▲ **FIGURE 4-18 Structure of the 5′ methylated cap of eukaryotic mRNA.** The distinguishing chemical features are the 5′ → 5′ linkage of 7-methylguanylate to the initial nucleotide of the mRNA molecule and the methyl group on the 2′ hydroxyl of the ribose of the first nucleotide (base 1). Both these features occur in all animal cells and in cells of higher plants; yeasts lack the methyl group on base 1. The ribose of the second nucleotide (base 2) also is methylated in vertebrates. [See A. J. Shatkin, 1976, *Cell* **9**:645.]

In eukaryotic cells, however, not only is the nucleus separated from the cytoplasm where protein synthesis occurs, but the primary RNA transcript of a protein-coding gene must undergo several modifications, collectively termed **RNA processing,** that yield a functional mRNA. This mRNA then must be transported to the cytoplasm before it can be translated into protein. Thus, transcription and translation cannot occur concurrently in eukaryotic cells.

The initial steps in processing of all eukaryotic primary RNA transcripts occur at the two ends, and these modifications are retained in mRNAs. To the initiating (5′) nucleotide of the primary transcript is added the *5′ cap,* which may serve to protect mRNA from enzymatic degradation (Figure 4-18). This modification occurs before transcription is complete, so the 5′ cap is present in the primary transcript. Processing at the 3′ end of the primary transcript involves cleavage by an endonuclease to yield a free 3′-hydroxyl group to which a string of adenylic acid residues is added by an enzyme called *poly(A) polymerase*. The resulting poly(A) tail contains 100–250 bases, being shorter in yeasts and invertebrates than in vertebrates. Poly(A) polymerase is part of a complex of proteins that adds the poly(A) tail. This complex does not require a template and can determine the correct number of A residues to add in each species.

The final step in the processing of many different eukaryotic mRNA molecules is **splicing**: the internal cleavage of the RNA transcript to excise the introns, followed by ligation of the coding exons. Many eukaryotic mRNAs also contain noncoding regions at each end; these are referred to as the 5' and 3' *untranslated regions* (UTRs). Figure 4-19 summarizes the basic steps in RNA processing. We examine the cellular machinery for carrying out processing of mRNA, as well as tRNA and rRNA, in Chapter 11.

SUMMARY Nucleic Acid Synthesis

- The transfer of information from genes to proteins is assisted by proteins that participate in the synthesis of DNA and RNA.

- Polynucleotide and polypeptide chains are assembled from a limited number of monomeric units that are added one at a time, beginning at the 5' end in nucleic acids and the amino-terminal end in proteins (see Figure 4-13). In both cases, the initial polymeric product generally is modified in some fashion to produce a functional molecule.

- A polynucleotide chain is synthesized by copying of a complementary template strand (usually DNA). In this process, the duplex DNA is locally unwound, revealing the unpaired template strand, and nucleotides are added to the 3'-hydroxyl end of the growing strand by RNA or DNA polymerase.

- RNA polymerase can initiate transcription of DNA into RNA by binding to a specific start site and unwinding the duplex. As the enzyme moves along the DNA, it unwinds sequential segments of the DNA and adds nucleotides to the growing RNA strand (see Figure 4-15). Most commonly, only one DNA strand in any one locus is transcribed into RNA.

- Replication of DNA requires the assistance of helicase to unwind the duplex, topoisomerase to remove supercoils, and a specialized RNA polymerase to form RNA primers because DNA polymerase cannot start chains. Nucleotide addition at the growing fork, a moving region of strand separation produced by sequential unwinding of the duplex, is catalyzed by one type of DNA polymerase.

- During DNA replication, the two new daughter strands are assembled somewhat differently because DNA polymerase can add nucleotides only in the 5' → 3' direction (see Figure 4-16). One new strand, the leading strand, is elongated continuously from a single primer. The other new strand, the lagging strand, is synthesized discontinuously as a series of short segments, called Okazaki fragments, initiated from multiple RNA primers. After removal of the intervening primer, adjacent Okazaki fragments are joined by DNA ligase.

- In prokaryotic DNA, related protein-coding genes are clustered into a functional region, an operon, which is transcribed from a single start site into one mRNA encoding multiple proteins (see Figure 4-17a). Translation of a mRNA can begin before synthesis of the mRNA is complete.

- In eukaryotic DNA, each protein-coding gene is transcribed from its own start site, and very often the coding regions (exons) are separated by noncoding regions (introns). The primary RNA transcript produced from such a gene must undergo processing to yield a functional mRNA. During processing, the ends of all primary transcripts are modified by addition of a 5' cap and 3' poly(A) tail; many transcripts also undergo splicing—removal of the introns and joining of the exons (see Figure 4-19).

4.4 The Three Roles of RNA in Protein Synthesis

Although DNA stores the information for protein synthesis and RNA carries out the instructions encoded in DNA, most biological activities are carried out by proteins. The accurate synthesis of proteins thus is critical to the proper functioning of cells and organisms. We saw in Chapter 3 that the linear order of amino acids in each protein determines its three-dimensional structure and activity. For this reason, assembly of amino acids in their correct order, as encoded in DNA, is the key to production of functional proteins.

Three kinds of RNA molecules perform different but cooperative functions in protein synthesis (Figure 4-20):

1. Messenger RNA (mRNA) carries the genetic information copied from DNA in the form of a series of three-base code "words," each of which specifies a particular amino acid.

2. Transfer RNA (tRNA) is the key to deciphering the code words in mRNA. Each type of amino acid has its own type of tRNA, which binds it and carries it to the growing end of a polypeptide chain if the next code word on mRNA calls for it. The correct tRNA with its attached amino acid is selected at each step because each specific tRNA molecule contains a three-base sequence that can base-pair with its complementary code word in the mRNA.

3. Ribosomal RNA (rRNA) associates with a set of proteins to form **ribosomes**. These complex structures, which physically move along an mRNA molecule, catalyze the assembly

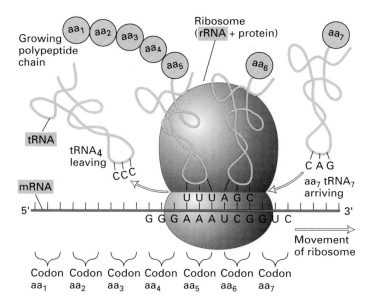

▲ **FIGURE 4-20 The three roles of RNA in protein synthesis.** Messenger RNA (mRNA) is translated into protein by the joint action of transfer RNA (tRNA) and the ribosome, which is composed of numerous proteins and two major ribosomal RNA (rRNA) molecules. [Adapted from A. J. F. Griffiths et al., 1993, *An Introduction to Genetics Analysis*, 5th ed., W. H. Freeman.]

of amino acids into protein chains. They also bind tRNAs and various accessory molecules necessary for protein synthesis. Ribosomes are composed of a large and small subunit, each of which contains its own rRNA molecule or molecules.

Translation is the whole process by which the base sequence of an mRNA is used to order and to join the amino acids in a protein. The three types of RNA participate in this essential protein-synthesizing pathway in all cells; in fact, the development of the three distinct functions of RNA was probably the molecular key to the origin of life. How each RNA carries out its specific task is discussed in this section, while the biochemical events in protein synthesis and the required protein factors are described in the final section of the chapter.

Messenger RNA Carries Information from DNA in a Three-Letter Genetic Code

RNA contains ribonucleotides of adenine, cytidine, guanine, and uracil; DNA contains deoxyribonucleotides of adenine, cytidine, guanine, and thymine. Because 4 nucleotides, taken individually, could represent only 4 of the 20 possible amino acids in coding the linear arrangement in proteins, a *group* of nucleotides is required to represent each amino acid. The code employed must be capable of specifying at least 20 words (i.e., amino acids).

If two nucleotides were used to code for one amino acid, then only 16 (or 4^2) different code words could be formed, which would be an insufficient number. However, if a group

of three nucleotides is used for each code word, then 64 (or 4^3) code words can be formed. Any code using groups of three or more nucleotides will have more than enough units to encode 20 amino acids. Many such coding systems are mathematically possible. However, the actual **genetic code** used by cells is a *triplet* code, with every three nucleotides being "read" from a specified starting point in the mRNA. Each triplet is called a **codon.** Of the 64 possible codons in the genetic code, 61 specify individual amino acids and three are stop codons. Table 4-2 shows that most amino acids are encoded by more than one codon. Only two—methionine and tryptophan—have a single codon; at the other extreme, leucine, serine, and arginine are each specified by six different codons. The different codons for a given amino acid are said to be *synonymous*. The code itself is termed *degenerate*, which means that it contains redundancies.

Synthesis of all protein chains in prokaryotic and eukaryotic cells begins with the amino acid methionine. In most mRNAs, the start (initiator) codon specifying this amino-terminal methionine is AUG. In a few bacterial mRNAs, GUG is used as the initiator codon, and CUG occasionally is used as an initiator codon for methionine in eukaryotes. The three codons UAA, UGA, and UAG do not specify amino acids but constitute stop (*terminator*) signals that mark the carboxyl terminus of protein chains in almost all cells. The sequence of codons that runs from a specific start site to a terminating codon is called a **reading frame.** This precise linear array of ribonucleotides in groups of three in mRNA specifies the precise linear sequence of amino acids in a protein and also signals where synthesis of the protein chain starts and stops.

Because the genetic code is a commaless, overlapping triplet code, a particular mRNA theoretically could be translated in three different reading frames. Indeed some mRNAs have been shown to contain overlapping information that can be translated in different reading frames, yielding different polypeptides (Figure 4-21). The vast majority of

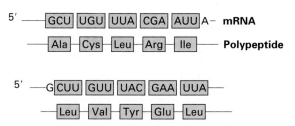

▲ **FIGURE 4-21 Example of how the genetic code—an overlapping, commaless triplet code—can be read in two different frames.** If translation of the mRNA sequence shown begins at two different upstream start sites (not shown), then two overlapping reading frames are possible; in this case, the codons are shifted one base to the right in the lower frame. As a result, different amino acids are encoded by the same nucleotide sequence. Many instances of such overlaps have been discovered in viral and cellular genes of prokaryotes and eukaryotes. It is theoretically possible for the mRNA to have a third reading frame.

TABLE 4-2 The Genetic Code (RNA to Amino Acids)*

First Position (5' end)	Second Position				Third Position (3' end)
	U	C	A	G	
U	Phe	Ser	Tyr	Cys	U
	Phe	Ser	Tyr	Cys	C
	Leu	Ser	Stop (och)	Stop	A
	Leu	Ser	Stop (amb)	Trp	G
C	Leu	Pro	His	Arg	U
	Leu	Pro	His	Arg	C
	Leu	Pro	Gln	Arg	A
	Leu (Met)	Pro	Gln	Arg	G
A	Ile	Thr	Asn	Ser	U
	Ile	Thr	Asn	Ser	C
	Ile	Thr	Lys	Arg	A
	Met (start)	Thr	Lys	Arg	G
G	Val	Ala	Asp	Gly	U
	Val	Ala	Asp	Gly	C
	Val	Ala	Glu	Gly	A
	Val (Met)	Ala	Glu	Gly	G

*"Stop (och)" stands for the ochre termination triplet, and "Stop (amb)" for the amber, named after the bacterial strains in which they were identified. AUG is the most common initiator codon; GUG usually codes for valine, and CUG for leucine, but, rarely, these codons can also code for methionine to initiate an mRNA chain.

mRNAs, however, can be read in only one frame because stop codons encountered in the other two possible reading frames terminate translation before a functional protein is produced. Another unusual coding arrangement occurs because of *frameshifting*. In this case the protein-synthesizing machinery may read four nucleotides as one amino acid and then continue reading triplets, or it may back up one base and read all succeeding triplets in the new frame until termination of the chain occurs. These frameshifts are not common events, but a few dozen such instances are known.

The meaning of each codon is the same in most known organisms—a strong argument that life on earth evolved only once. Recently the genetic code has been found to differ for a few codons in many mitochondria, in ciliated protozoans, and in *Acetabularia*, a single-celled plant. As shown in Table 4-3, most of these changes involve reading of normal stop codons as amino acids, not an exchange of one

TABLE 4-3 Unusual Codon Usage in Nuclear and Mitochondrial Genes

Codon	Universal Code	Unusual Code	Occurrence*
UGA	Stop	Trp	*Mycoplasma, Spiroplasma,* mitochondria of many species
CUG	Leu	Thr	Mitochondria in yeasts
UAA, UAG	Stop	Gln	*Acetabularia, Tetrahymena,* Paramecium, etc.
UGA	Stop	Cys	*Euplotes*

*"Unusual code" is used in nuclear genes of the listed organisms and in mitochondrial genes as indicated.

SOURCE: S. Osawa et al., 1992, *Microbiol. Rev.* 56:229.

amino acid for another. It is now thought that these exceptions to the general code are later evolutionary developments; that is, at no single time was the code immutably fixed, although massive changes were not tolerated once a general code began to function early in evolution.

Experiments with Synthetic mRNAs and Trinucleotides Broke the Genetic Code

Having described the genetic code, we briefly recount how it was deciphered—one of the great triumphs of modern biochemistry. The underlying experimental work was carried out largely with cell-free bacterial extracts containing all the necessary components for protein synthesis except mRNA (i.e., tRNAs, ribosomes, amino acids, and the energy-rich nucleotides ATP and GTP).

Initially, researchers added synthetic mRNAs containing a single type of nucleotide to such extracts and then determined the amino acid incorporated into the polypeptide that was formed. In the first successful experiment, synthetic mRNA composed only of U residues [poly(U)] yielded polypeptides made up only of phenylalanine. Thus it was concluded that a codon for phenylalanine consisted entirely of U's. Likewise, experiments with poly(C) and poly(A) showed that a codon for proline contained only C's and a codon for lysine only A's (Figure 4-22). [Poly(G) did not work in this type of experiment because it assumes an unusable stacked structure that is not translated well.] Next, synthetic mRNAs composed of alternating bases were used. The results of these experiments not only revealed more codons but also demonstrated that codons are three bases long. The example of this approach illustrated in Figure 4-23 led to identification of ACA as the codon for threonine and CAC for histidine. Similar experiments with many such mixed polynucleotides revealed a substantial part of the genetic code.

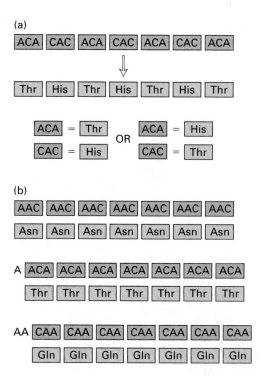

▲ **FIGURE 4-23 Assigning codons using mixed polynucleotides.** (a) When a synthetic mRNA with alternating A and C residues was added to a protein-synthesizing bacterial extract, the resulting polypeptide contained alternating threonine and histidine residues. This finding is compatible with the two alternative codon assignments shown. (Note that alternating residues yield the same sequence of triplets regardless of which reading frame is chosen.) (b) To determine which codon assignment shown in (a) is correct, a second mRNA consisting of AAC repeats was tested. This mRNA, which can be read in three frames, yielded the three types of polypeptides shown. Since only the ACA codon was common to both experiments, it must encode threonine; thus CAC must encode histidine in (a). The assignments AAC = asparagine (Asn) and CAA = glutamine (Gln) were derived from additional experiments. [See H. G. Korana, 1968, reprinted in *Nobel Lectures: Physiology or Medicine (1963–1970),* Elsevier (1973), p. 341.]

The entire genetic code was finally worked out by a second type of experiment conducted by Marshall Nirenberg and his collaborators. In this approach, all the possible trinucleotides were tested for their ability to attract tRNAs attached to the 20 different amino acids found in natural proteins (Figure 4-24). In all, 61 of the 64 possible trinucleotides were found to code for a specific amino acid; the trinucleotides UAA, UGA, and UAG did not encode amino acids.

Although synthetic mRNAs were useful in deciphering the genetic code, in vitro protein synthesis from these mRNAs is very inefficient and yields polypeptides of variable size. Successful in vitro synthesis of a naturally occurring protein was achieved first when mRNA from bacteriophage F2 (a virus) was added to bacterial extracts, leading to formation of the coat, or capsid, protein (the "packaging" protein that covers the virus particle). Studies with such

Bacterial extract

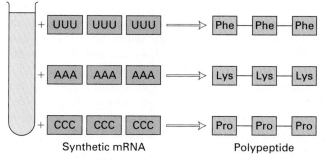

▲ **FIGURE 4-22 Assigning codons using synthetic mRNAs containing a single ribonucleotide.** Addition of such a synthetic mRNA to a bacterial extract that contained all the components necessary for protein synthesis except mRNA resulted in synthesis of polypeptides composed of a single type of amino acid as indicated. [See M. W. Nirenberg and J. H. Matthei, 1961, *Proc. Nat'l. Acad. Sci. USA* **47**:1588.]

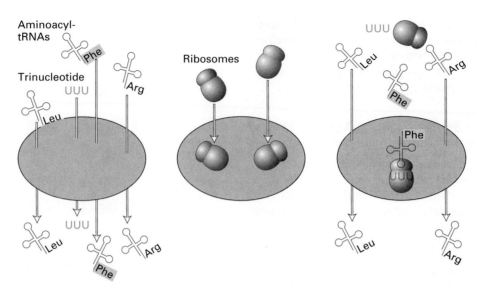

Trinucleotide and all tRNAs pass through filter

Ribosomes stick to filter

Complex of ribosome, UUU, and Phe-tRNA sticks to filter

▲ **FIGURE 4-24 Breaking the entire genetic code by use of chemically synthesized trinucleotides.** Marshall Nirenberg and his collaborators prepared 20 ribosome-free bacterial extracts containing all possible aminoacyl-tRNAs (tRNAs with an amino acid attached). In each sample, a different amino acid was radioactively labeled (green); the other 19 amino acids were bound to tRNAs but were unlabeled. Aminoacyl-tRNAs and trinucleotides passed through a nitrocellulose filter *(left)*, but ribosomes were retained by the filter *(center)* and would bind trinucleotides and their cognate tRNAs *(right)*. Each possible trinucleotide was tested separately for its ability to attract a specific tRNA by adding it with ribosomes to samples from each of the 20 aminoacyl-tRNA mixtures. The sample was then filtered. If the added trinucleotide caused the radiolabeled aminoacyl-tRNA to bind to the ribosome, then radioactivity would be detected on the filter (a positive test); otherwise, the label would pass through the filter (a negative test). By synthesizing and testing all possible trinucleotides, the researchers were able to match all 20 amino acids with one or more codons (e.g., phenylalanine with UUU as shown here). [See M. W. Nirenberg and P. Leder, 1964, *Science* **145**:1399.]

natural mRNAs established that AUG encodes methionine at the start of almost all proteins and is required for efficient initiation of protein synthesis, while the three trinucleotides (UAA, UGA, and UAG) that do not encode any amino acid act as stop codons, necessary for precise termination of synthesis.

The Folded Structure of tRNA Promotes Its Decoding Functions

The next step in understanding the flow of genetic information from DNA to protein was to determine how the nucleotide sequence of mRNA is converted into the amino acid sequence of protein. This decoding process requires two types of adapter molecules: tRNAs and enzymes called *aminoacyl-tRNA synthetases*. First we describe the role of tRNAs in decoding mRNA codons, and then examine how synthetases recognize tRNAs.

All tRNAs have two functions: to be chemically linked to a particular amino acid and to base-pair with a codon in mRNA so that the amino acid can be added to a growing peptide chain. Each tRNA molecule is recognized by one and only one of the 20 aminoacyl-tRNA synthetases. Likewise, each of these enzymes links one and only one of the

20 amino acids to a particular tRNA, forming an **aminoacyl-tRNA.** Once its correct amino acid is attached, a tRNA then recognizes a codon in mRNA, thereby delivering its amino acid to the growing polypeptide (Figure 4-25).

As studies on tRNA proceeded, 30–40 different tRNAs were identified in bacterial cells and as many as 50–100 in animal and plant cells. Thus the number of tRNAs in most cells is more than the number of amino acids found in proteins (20) and also differs from the number of codons in the genetic code (61). Consequently, many amino acids have more than one tRNA to which they can attach (explaining how there can be more tRNAs than amino acids); in addition, many tRNAs can attach to more than one codon (explaining how there can be more codons than tRNAs). As noted previously, most amino acids are encoded by more than one codon, requiring some tRNAs to recognize more than one codon.

The function of tRNA molecules, which are 70–80 nucleotides long, depends on their precise three-dimensional structures. In solution, all tRNA molecules fold into a similar stem-loop arrangement that resembles a cloverleaf when drawn in two dimensions (Figure 4-26a). The four stems are short double helices stabilized by Watson-Crick base pairing; three of the four stems have loops containing seven or

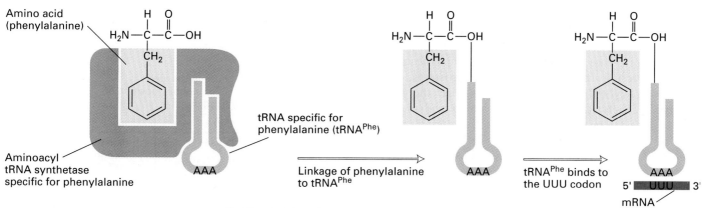

Net Result: Phenylalanine Is Selected by Its Codon

▲ **FIGURE 4-25 Translation of nucleic acid sequences in mRNA into amino acid sequences in proteins requires a two-step decoding process.** First, an aminoacyl-tRNA synthetase couples a specific amino acid to its corresponding tRNA. Second, a three-base sequence in the tRNA (the anticodon) base-pairs with a codon in the mRNA specifying the attached amino acid. If an error occurs in either step, the wrong amino acid may be incorporated into a polypeptide chain.

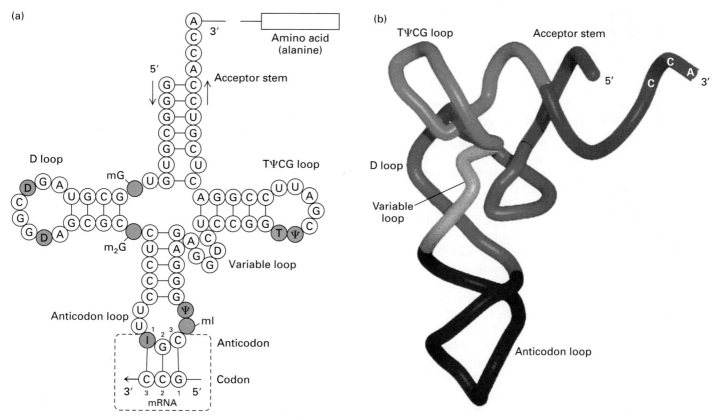

▲ **FIGURE 4-26 Structure of tRNAs.** (a) The primary structure of yeast alanine tRNA (tRNA^{Ala}), the first such sequence determined. This molecule is synthesized from the nucleotides A, C, G, and U, but some of the nucleotides, shown in red, are modified after synthesis: D = dihydrouridine, I = inosine, T = thymine, Ψ = pseudouridine, and m = methyl group. Although the exact sequence varies among tRNAs, they all fold into four base-paired stems and three loops. The partially unfolded molecule is commonly depicted as a cloverleaf. Dihydrouridine is nearly always present in the D loop; likewise, thymidylate, pseudouridylate, cytidylate, and guanylate are almost always present in the TΨCG loop. The triplet at the tip of the anticodon loop base-pairs with the corresponding codon in mRNA. Attachment of an amino acid to the acceptor arm yields an aminoacyl-tRNA. (b) Computer-generated three-dimensional model of the generalized backbone of all tRNAs. Note the L shape of the molecule. [Part (a) see R. W. Holly et al., 1965, *Science* **147**:1462; part (b) from J. G. Arnez and D. Moras, 1997, *Trends Biochem. Sci.* **22**:211.]

eight bases at their ends, while the remaining, unlooped stem contains the free 3′ and 5′ ends of the chain. Three nucleotides termed the **anticodon,** located at the center of one loop, can form base pairs with the three complementary nucleotides forming a codon in mRNA. As discussed later, specific aminoacyl-tRNA synthetases recognize the surface structure of each tRNA for a specific amino acid and covalently attach the proper amino acid to the unlooped *amino acid acceptor stem.* The 3′ end of all tRNAs has the sequence CCA, which in most cases is added after synthesis and processing of the tRNA are complete. Viewed in three dimensions, the folded tRNA molecule has an L shape with the anticodon loop and acceptor stem forming the ends of the two arms (Figure 4-26b).

Besides addition of CCA at the 3′ terminus after a tRNA molecule is synthesized, several of its nucleic acid bases typically are modified. For example, most tRNAs are synthesized with a four-base sequence of UUCG near the middle of the molecule. The first uridylate is methylated to become a thymidylate; the second is rearranged into a pseudouridylate (abbreviated Ψ), in which the ribose is attached to carbon 5 instead of to nitrogen 1 of the uracil. These modifications produce a characteristic TΨCG loop in an unpaired region at approximately the same position in nearly all tRNAs (see Figure 4-26a).

Nonstandard Base Pairing Often Occurs between Codons and Anticodons

If perfect Watson-Crick base pairing were demanded between codons and anticodons, cells would have to contain exactly 61 different tRNA species, one for each codon that specifies an amino acid. As noted above, however, many cells contain fewer than 61 tRNAs. The explanation for the smaller number lies in the capability of a single tRNA anticodon to recognize more than one, but not necessarily every, codon corresponding to a given amino acid. This broader recognition can occur because of nonstandard pairing between bases in the so-called "wobble" position: the third base in a mRNA codon and the corresponding first base in its tRNA anticodon. Although the first and second bases of a codon form standard Watson-Crick base pairs with the third and second bases of the corresponding anticodon, four nonstandard interactions can occur between bases in the wobble position. Particularly important is the G·U base pair, which structurally fits almost as well as the standard G·C pair. Thus, a given anticodon in tRNA with G in the first (wobble) position can base-pair with the two corresponding codons that have either pyrimidine (C or U) in the third position (Figure 4-27). For example, the phenylalanine codons UUU and UUC (5′ → 3′) are both recognized by the tRNA that has GAA (5′ → 3′) as the anticodon. In fact, any two codons of the type NNPyr (N = any base; Pyr = pyrimidine) encode a single amino acid and are decoded by a single tRNA with G in the first (wobble) position of the anticodon.

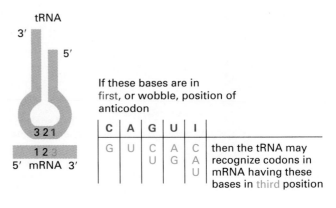

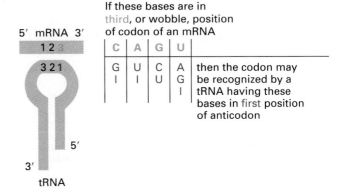

▲ **FIGURE 4-27 The first and second bases in an mRNA codon form Watson-Crick base pairs with the third and second bases, respectively, of a tRNA anticodon.** However, the base in the third (or wobble) position of an mRNA codon often forms a nonstandard base pair with the base in the first (or wobble) position of a tRNA anticodon. Wobble pairing allows a tRNA to recognize more than one mRNA codon *(top);* conversely, it allows a codon to be recognized by more than one kind of tRNA *(bottom),* although each tRNA will bear the same amino acid. Note that a tRNA with I (inosine) in the wobble position can "read" (become paired with) three different codons (see Figure 4-28), and a tRNA with G or U in the wobble position can read two codons. Although A is theoretically possible in the wobble position of the anticodon, it is almost never found in nature.

Although adenine rarely is found in the anticodon wobble position, many tRNAs in plants and animals contain *inosine* (I), a deaminated product of adenine, at this position. Inosine can form nonstandard base pairs with A, C, and U (Figure 4-28). A tRNA with inosine in the wobble position thus can recognize the corresponding mRNA codons with A, C, or U in the third (wobble) position (see Figure 4-27). For this reason, inosine-containing tRNAs are heavily employed in translation of the synonymous codons that specify a single amino acid. For example, four of the six codons for leucine have a 3′ A, C, or U (see Table 4-2); these four codons are all recognized by the same tRNA (3′-GAI-5′), which has inosine in the wobble position of the anticodon (and thus recognizes CUA, CUC, and CUU), and uses a G·U pair in position 1 to recognize the UUA codon.

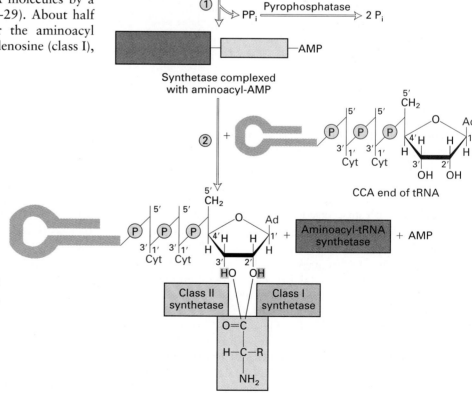

Aminoacyl-tRNA Synthetases Activate Amino Acids by Linking Them to tRNAs

Recognition of the codon or codons specifying a given amino acid by a particular tRNA is actually the second step in decoding the genetic message. The first step, attachment of the appropriate amino acid to a tRNA, is catalyzed by a specific aminoacyl-tRNA synthetase (see Figure 4-25). Each of the 20 different synthetases recognizes *one* amino acid and *all* its compatible, or *cognate*, tRNAs. These coupling enzymes link an amino acid to the free 2′ or 3′ hydroxyl of the adenosine at the 3′ terminus of tRNA molecules by a two-step ATP-requiring reaction (Figure 4-29). About half the aminoacyl-tRNA synthetases transfer the aminoacyl group to the 2′ hydroxyl of the terminal adenosine (class I),

and about half to the 3′ hydroxyl (class II). In this reaction, the amino acid is linked to the tRNA by a high-energy bond and thus is said to be *activated*. The energy of this bond subsequently drives the formation of peptide bonds between adjacent amino acids in a growing polypeptide chain. The

▶ FIGURE 4-29 Aminoacylation of tRNA. Amino acids are covalently linked to tRNAs by aminoacyl-tRNA synthetases. Each of these enzymes recognizes one kind of amino acid and all the cognate tRNAs that recognize codons for that amino acid. The two-step aminoacylation reaction requires energy from the hydrolysis of ATP. The equilibrium of the overall reaction favors the indicated products because the pyrophosphate (PP$_i$) released in step ① is converted to inorganic phosphate (P$_i$) by a pyrophosphatase. The 3′ end of all tRNAs, to which the amino acid attaches, has the sequence CCA. Class I synthetases (purple) attach the amino acid to the 2′ hydroxyl of the terminal adenylate in tRNA; class II synthetases (green) attach the amino acid to the 3′ hydroxyl. (Ad = adenine; Cyt = cytosine.)

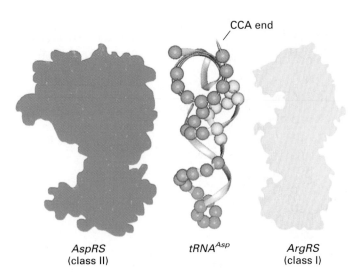

▲ FIGURE 4-30 Recognition of a tRNA by aminoacyl synthetases. Aspartyl-tRNA synthetase (AspRS) is a class II enzyme, and arginyl-tRNA synthetase (ArgRS) is a class I enzyme. Shown here are the outlines of the three-dimensional structures of the two synthetases. The tRNA shown between them as a ribbon diagram will bind to either and is a slightly modified version of tRNA^{Asp}. It is used as an illustration of common surface interactions between tRNA and class I and II enzymes. Sites on the opposite sides of this modified tRNA^{Asp} make contacts with the two enzymes: the blue balls show contacts with the class II AspRS; those that make contact with class I ArgRS are indicated by yellow balls. The synthetases are shown positioned away from the tRNA for clarity, but the fit of the surfaces at close range is obvious. The ability of ArgRS to interact with the noncognate tRNA^{Asp} is lost when residue G37 in tRNA^{Asp} is methylated, a normal modification that occurs in vivo. However, the shape and binding sites of this modified tRNA are characteristic of class I and class II interactions with tRNAs. This molecular graphic picture was produced using the DRAWNA program. [Adapted from M. Sissler et al., 1997, *Nucl. Acids Res.* **25**:4899; courtesy of R. Giegé.]

equilibrium of the aminoacylation reaction is driven further toward activation of the amino acid by hydrolysis of the high-energy phosphoanhydride bond in pyrophosphate. The overall reaction is

$$\text{Amino acid} + \text{ATP} + \text{tRNA} \xrightarrow{\text{enzyme}} \text{aminoacyl-tRNA} + \text{AMP} + 2\,\text{P}_i$$

The amino acid sequences of the aminoacyl-tRNA synthetases (ARSs) from many organisms are now known, and the three-dimensional structures of over a dozen enzymes of both classes have been solved. Each of these enzymes has a rather precise binding site for ATP (GTP is not admitted and CTP and UTP are too small) and binding pockets for its specific amino acid. Class I and class II enzymes bind to opposite faces of the incoming tRNAs. The binding surfaces of class I enzymes tend to be somewhat complementary

to those of class II enzymes. These different binding surfaces and the consequent alignment of bound tRNAs probably account in part for the difference in the hydroxyl group to which the aminoacyl group is transferred (Figure 4-30). Because some amino acids are so similar structurally, aminoacyl-tRNA synthetases sometimes make mistakes. These are corrected, however, by the enzymes themselves, which check the fit in the binding pockets and facilitate deacylation of any misacylated tRNAs. This crucial function helps guarantee that a tRNA delivers the correct amino acid to the protein-synthesizing machinery.

Each tRNA Molecule Is Recognized by a Specific Aminoacyl-tRNA Synthetase

The ability of aminoacyl-tRNA synthetases to recognize their correct cognate tRNAs is just as important to the accurate translation of the genetic code as codon-anticodon pairing. Once a tRNA is loaded with an amino acid, codon-anticodon pairing directs the tRNA into the proper ribosome site; if the wrong amino acid is attached to the tRNA, an error in protein synthesis results.

As noted already, each aminoacyl-tRNA synthetase can aminoacylate all the different tRNAs whose anticodons correspond to the same amino acid. Therefore, all these cognate tRNAs must have a similar binding site, or "identity element," that is recognized by the synthetase. One approach for studying the identity elements in tRNAs that are recognized by aminoacyl-tRNA synthetases is to produce synthetic genes that encode tRNAs with normal and various mutant sequences by techniques discussed in Chapter 7. The normal and mutant tRNAs produced from such synthetic genes then can be tested for their ability to bind purified synthetases.

Very probably no single structure or sequence completely determines a specific tRNA identity. However, some important structural features of several *E. coli* tRNAs that allow their cognate synthetases to recognize them are known. Perhaps the most logical identity element in a tRNA molecule is the anticodon itself. Experiments in which the anticodons of methionine tRNA (tRNA^{Met}) and valine tRNA (tRNA^{Val}) were interchanged showed that the anticodon is of major importance in determining the identity of these two tRNAs. In addition, x-ray crystallographic analysis of the complex between glutamine aminoacyl-tRNA synthetase (GlnRS) and glutamine tRNA (tRNA^{Gln}) showed that each of the anticodon bases neatly fits into a separate, specific "pocket" in the three-dimensional structure of GlnRS. Thus this synthetase specifically recognizes the correct anticodon.

However, the anticodon may not be the principal *identity element* in other tRNAs (see Figure 4-30). Figure 4-31 shows the extent of base sequence conservation in *E. coli* tRNAs that become linked to the same amino acid. Identity elements are found in several regions, particularly the end of the acceptor arm. A simple case is presented by tRNA^{Ala}:

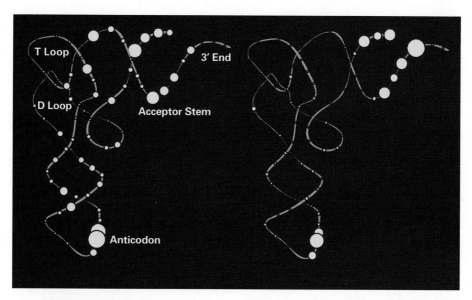

▲ **FIGURE 4-31 Identity elements in tRNA involved in recognition by aminoacyl-tRNA synthetases, as demonstrated by both conservation and experimentation.** The 67 known tRNA sequences in *E. coli* were compared by computer analysis. The conserved nucleotides in different tRNAs that recognize the same amino acid are shown as yellow circles in the left drawing, with the tRNA chain in blue. Increasing size indicates increasing conservation of a base at a given position. It is clear that nucleotides in the anticodon loop and in the acceptor stem are most often similar when a particular amino acid must be recognized. This appreciation is heightened by results shown in the right drawing. Here, nucleotides that have been experimentally demonstrated to have a role in identity (selection of an amino acid by an ARS-tRNA complex) are shown as yellow circles. In this case, the circle size indicates the relative frequency that a given position acts as an identity element. [From W. H. McClain, 1993, *J. Mol. Biol.* **234**:257; also see L. D. H. Schulman and J. Abelson, 1988, *Science* **240**:1590.]

a single G·U base pair (G3·U70) in the acceptor stem is necessary and sufficient for recognition of this tRNA by its cognate aminoacyl-tRNA synthetase. Solution of the three-dimensional structure of additional complexes between aminoacyl-tRNA synthetases and their cognate tRNAs should provide a clear understanding of the rules governing the recognition of tRNAs by specific synthetases.

Ribosomes Are Protein-Synthesizing Machines

If the many components that participate in translating mRNA had to interact in free solution, the likelihood of simultaneous collisions occurring would be so low that the rate of amino acid polymerization would be very slow. The efficiency of translation is greatly increased by the binding of the mRNA and the individual aminoacyl-tRNAs to the most abundant RNA-protein complex in the cell—the ribosome. This two-part machine directs the elongation of a polypeptide at a rate of three to five amino acids added per second. Small proteins of 100–200 amino acids are therefore made in a minute or less. On the other hand, it takes 2 to 3 hours to make the largest known protein, titin, which is found in muscle and contains 30,000 amino acid residues. The machine that accomplishes this task must be precise and persistent.

With the aid of the electron microscope, ribosomes were first discovered as discrete, rounded structures prominent in animal tissues secreting large amounts of protein; initially, however, they were not known to play a role in protein synthesis. Once reasonably pure ribosome preparations were obtained, radiolabeling experiments showed that radioactive amino acids first were incorporated into growing polypeptide chains associated with ribosomes before appearing in finished chains.

A ribosome is composed of several different ribosomal RNA (rRNA) molecules and more than 50 proteins, organized into a large subunit and a small subunit. The proteins in the two subunits differ, as do the molecules of rRNA. The small ribosomal subunit contains a single rRNA molecule, referred to as *small rRNA*; the large subunit contains a molecule of *large rRNA* and one molecule each of two much smaller rRNAs in eukaryotes (Figure 4-32). The ribosomal subunits and the rRNA molecules are commonly designated in svedbergs (S), a measure of the sedimentation rate of suspended particles centrifuged under standard conditions (Chapter 3). The lengths of the rRNA molecules, the quantity of proteins in each subunit, and consequently the sizes of the subunits differ in prokaryotic and eukaryotic cells. (The small and large rRNAs are about 1500 and 3000 nucleotides long in bacteria and about 1800 and 5000 nucleotides long in humans.) Perhaps of more

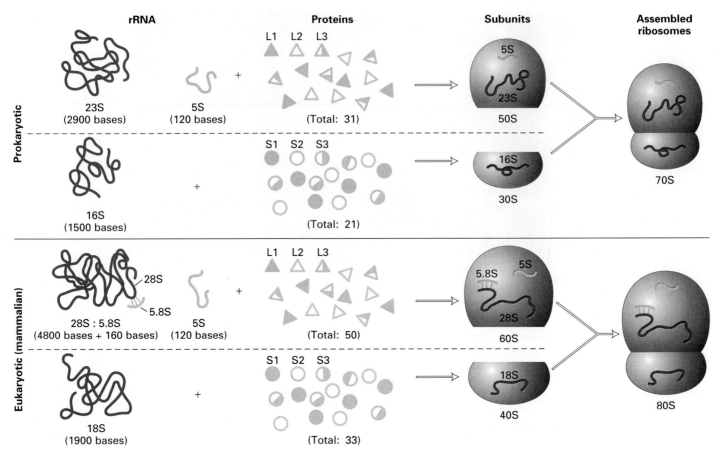

▲ **FIGURE 4-32 The general structure of ribosomes in prokaryotes and eukaryotes.** In all cells, each ribosome consists of a large and a small subunit. The two subunits contain rRNAs of different lengths, as well as a different set of proteins. All ribosomes contain two major rRNA molecules (dark red)—23S and 16S rRNA in bacteria, 28S and 18S rRNA in eukaryotes—and one or two small RNAs (light red). The proteins are named L1, L2, etc., and S1, S2, etc., depending on whether they are found in the large or the small subunit.

interest than these differences are the great structural and functional similarities among ribosomes from all species. This consistency is another reflection of the common evolutionary origin of the most basic constituents of living cells.

The sequences of the small and large rRNAs from several thousand organisms are now known. Although the primary nucleotide sequences of these rRNAs vary considerably, the same parts of each type of rRNA theoretically can form base-paired stem-loops, generating a similar three-dimensional structure for each rRNA in all organisms. Evidence that such stem-loops occur in rRNA was obtained by treating rRNA with chemical agents that cross-link paired bases; the samples then were digested with enzymes that destroy single-stranded rRNA, but not any cross-linked (base-paired) regions. Finally, the intact, cross-linked rRNA that remained was collected and sequenced, thus identifying the stem-loops in the original rRNA. Experiments of this type have located about 45 stem-loops at similar positions in small rRNAs from many different prokaryotes and eukaryotes (Figure 4-33). An even larger number of regularly positioned stem-loops have been demonstrated in large rRNAs. All the ribosomal proteins have been identified and their sequences determined, and many have been shown to bind specific regions of rRNA. It seems clear that the fundamental protein-synthesizing machinery in all present-day cells arose only once and has been modified about a common plan during evolution.

During protein synthesis, a ribosome moves along an mRNA chain, interacting with various protein factors and tRNA and very likely even undergoes shape changes. Despite the complexity of the ribosome, great progress has been made in determining both the overall structure of bacterial ribosomes and in identifying reactive sites that bind specific proteins, mRNA, and tRNA and that participate in important steps in protein synthesis. Quite detailed models of the large and small ribosomal subunits from *E. coli* have been constructed based on cryoelectron microscope and neutron-scattering studies (Figure 4-34). These studies not only have determined the dimensions and

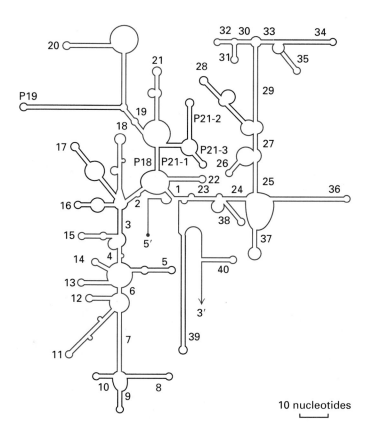

▲ **FIGURE 4-33 Two-dimensional map of the secondary structure of the small (16S) rRNA from bacteria, showing the location of base-paired stems and loops.** In general, the length and position of the stem-loops are very similar in all species, although the exact sequence varies from species to species. The most highly conserved regions are represented as red lines, and the numbered stem-loops unique to prokaryotes are preceded by a P. Eukaryotic small (18S) rRNAs exhibit a generally similar pattern of stem-loops, although, as with prokaryotes, a few are unique. [Adapted from E. Huysmans and R. DeWachter, 1987, *Nucl. Acids Res.* **14**:73].

overall shape of the ribosomal subunits, but also have localized the positions of tRNAs bound to the ribosome during protein chain elongation. Powerful chemical experiments have also helped unravel the complex interactions between proteins and RNAs. In a technique called **footprinting**, for example, ribosomes are treated with chemical reagents that modify single-stranded RNA unprotected by binding either to protein or to another RNA. If the total sequence of the RNA is known, then the location of the modified nucleotides can be located within the molecule. (This technique, which is also useful for locating protein-binding sites in DNA, is described in Chapter 10.) Thus the overall structure and function of ribosomes during protein synthesis is finally, after 40 years, yielding to successful experiments. How these results aid in understanding the specific steps in protein synthesis is described in the next section.

SUMMARY The Three Roles of RNA in Protein Synthesis

• Genetic information is copied into mRNA in the form of a commaless, overlapping, degenerate triplet code. Each amino acid is encoded by one or more three-base sequences, or codons, in mRNA. Each codon specifies one amino acid, but most amino acids are encoded by multiple codons (see Table 4-2).

• The AUG codon for methionine is the most common start codon, specifying the amino acid at the NH_2-terminus of a protein chain. Three codons function as stop codons and specify no amino acids.

• A reading frame, the uninterrupted sequence of codons in mRNA from a specific start codon to a stop codon, is translated into the linear sequence of amino acids in a protein.

• Decoding of the nucleotide sequence in mRNA into the amino acid sequence of proteins depends on transfer RNAs and amino-acyl tRNA synthetases (see Figure 4-25).

• All tRNAs have a similar three-dimensional structure that includes an acceptor arm for attachment of a specific amino acid and a stem-loop with a three-base anticodon sequence at its ends (see Figure 4-26). The anticodon can base-pair with its corresponding codon or codons in mRNA.

• Because of nonstandard interactions, a tRNA may base-pair with more than one mRNA codon, and conversely, a particular codon may base-pair with multiple tRNAs.

• Each of the 20 aminoacyl-tRNA synthetases recognizes a single amino acid and covalently links it to a cognate tRNA, forming an aminoacyl-tRNA (see Figure 4-29). This reaction activates the amino acid, so it can participate in peptide-bond formation.

• The composition of ribosomes—the large ribonucleoprotein complexes on which proteins are synthesized—is quite similar in all organisms (see Figure 4-32). All ribosomes are composed of a small and a large subunit. Each contains numerous different proteins and one rRNA (small or large). The large subunit also contains one accessory RNA (5S).

• Analogous rRNAs from many different species fold into quite similar three-dimensional structures containing numerous stem-loops and binding sites for proteins, mRNA, and tRNAs. As a ribosome moves along an mRNA, a region of the large rRNA molecule in each ribosome sequentially binds the aminoacylated ends of incoming tRNAs and probably catalyzes peptide-bond formation (see Figure 4-34).

(a)

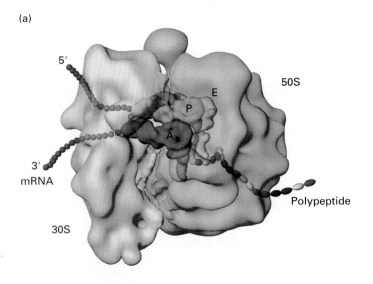

(b)

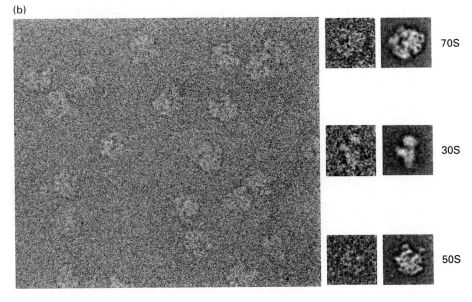

◀ **FIGURE 4-34 Overall structure of the _E. coli_ ribosome at 25-Å resolution inferred from cryoelectron microscopy and three-dimensional reconstruction based on the analysis of 4300 individual projections.** (a) This model shows the shapes of the large (blue) and small (yellow) subunits of the ribosome with three aminoacyl-tRNAs (pink, green, yellow) superimposed at the A, P, and E sites. The roles of these tRNA-binding sites during protein synthesis are discussed later. Chemical cross-linking experiments have demonstrated that the mRNA (orange beads) passes close to the anti-codon loops of the tRNAs and that the nascent polypeptide chain is buried in the tunnel in the large ribosomal subunit that begins within 10–15 Å of the 3′ amino-acylated end of the tRNAs. The tunnel termination site on the ribosome surface has also been accurately mapped. (b) Large panel shows a field of 70S ribosomes. Small panels (left) show cryoelectron microscopy images of a single 70S ribosome, small (30S) subunit, and large (50S) subunit. Small panels (right) show computer-derived averages of many dozens of images in the same orientation. Cryo-electron microscopy is carried out on unstained samples of ribosomes or subunits flash frozen as "vitreous ices" (without ice crystals) in a very thin layer of water (Chapter 5). Individual images are analyzed by computer projections. [See R. K. Agrawal et al., _Cell_, in press; J. Frank, 1995, _Nature_ **356**:441; J. Frank et al., 1995, _Biochem. Cell Biol._ **73**:757. Courtesy of J. Frank.]

4.5 Stepwise Formation of Proteins on Ribosomes

The previous sections have introduced the major participants in protein synthesis—mRNA, aminoacylated tRNAs, and ribosomes containing large and small rRNAs. Here, we take a detailed look at how these components are brought together to carry out the biochemical events leading to formation of proteins on ribosomes. The complex process of translating mRNA into protein can be divided into three stages—initiation, elongation, and termination—which we consider in order.

The AUG Start Codon Is Recognized by Methionyl-tRNA$_i^{Met}$

As noted earlier, the AUG codon for methionine functions as the start codon in the vast majority of mRNAs. A critical aspect of initiation is to begin protein synthesis at the start codon, thereby setting the stage for the correct in-frame translation of the entire mRNA. Both prokaryotes and eukaryotes contain two different methionine tRNAs: tRNA$_i^{Met}$ can initiate protein synthesis, and tRNA^{Met} can incorporate methionine only into a growing protein chain (Figure 4-35). The same aminoacyl-tRNA synthetase (MetRS) charges both tRNAs with methionine. But _only_ Met-tRNA$_i^{Met}$ (i.e., activated methionine attached to tRNA$_i^{Met}$) can bind at the appropriate site on the small ribosomal subunit (the P site; see discussion below) to begin synthesis of a protein chain. (The regular tRNA^{Met} binds only to another ribosomal site, the A site, as described later.) In bacteria, but not in archaeans or eukaryotes, the amino group of the methionine of Met-tRNA$_i^{Met}$ is modified by addition of a formyl group and is sometimes designated fMet-tRNA$_i^{Met}$. However, Met-tRNA$_i^{Met}$ is commonly used to designate the initiator tRNA in all cells.

During the initial stage of protein synthesis in all cells, a ribosome assembles, complexed with a mRNA and an

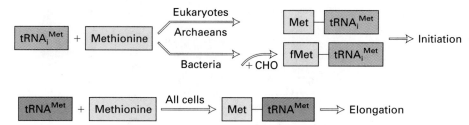

◄ **FIGURE 4-35 Two types of methionine tRNA are found in all cells.** One, designated tRNA$_i^{Met}$, is used exclusively to start protein chains, and the other, designated tRNA^{Met}, delivers methionine to internal sites in a growing protein chain. In bacteria, a formyl group (CHO) is added to methionyl-tRNA$_i^{Met}$, forming fMet-tRNA$_i^{Met}$.

activated initiator tRNA, which is correctly positioned at the start codon. Because the details of initiation and the mechanism for locating the translation start site differ in bacteria and eukaryotes, we discuss the two systems separately.

Bacterial Initiation of Protein Synthesis Begins Near a Shine-Dalgarno Sequence in mRNA

To begin the assembly of a bacterial translation complex, sequential interactions occur between specific proteins referred to as **initiation factors** (IFs) and the small (30S) ribosomal subunit. The resulting preinitiation complex together with fMet-tRNA$_i^{Met}$ then binds to the mRNA at a specific site usually located quite near the AUG initiation codon. This binding, assisted by IF1 and IF3, yields the *30S initiation complex* (Figure 4-36).

In most bacteria, the small ribosomal subunit identifies start codons through interaction of the small (16S) rRNA with an eight-nucleotide sequence in mRNA called the *Shine-Dalgarno sequence* after its discoverers. This sequence, located near the AUG start codon, base-pairs to a sequence at or very near the 3′ end of 16S rRNA, thereby binding the mRNA and small ribosomal subunit to each other. Although the Shine-Dalgarno sequence varies some at different initiation sites, on average six out of eight nucleotides are complementary to the 3′-end sequence of 16S rRNA. Thus bacterial rRNA plays a direct role in recruiting the small ribosomal subunit to a translation start site on the mRNA. The long mRNAs produced from bacterial operons contain multiple *internal* Shine-Dalgarno sequences

located near the start sites for each of the encoded proteins (see Figure 4-17a). Because a small ribosomal subunit can bind to each Shine-Dalgarno sequence, synthesis of the different proteins can occur independently and simultaneously from the multiple start sites on these long mRNAs.

Assembly of a complete *70S initiation complex* is achieved by the recruitment of the large ribosomal subunit,

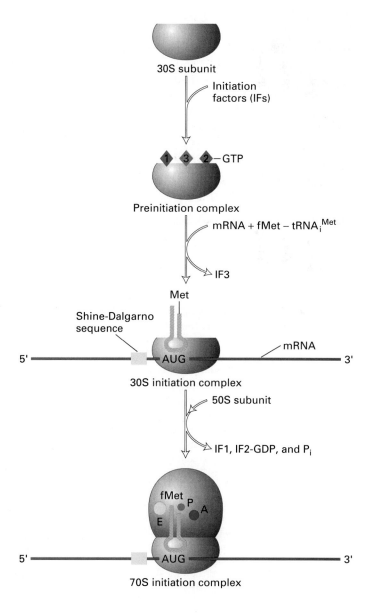

▶ **FIGURE 4-36 Bacterial initiation of protein synthesis.** The three initiation factors (IF1, IF2, and IF3) bind to the small subunit of the ribosome. The 30S preinitiation complex, together with the fMet-tRNA$_i^{Met}$, binds the mRNA, guided by the Shine-Dalgarno sequence, which is complementary to the 3′ terminus of the 16S rRNA. The AUG sequence is then located, from which protein initiation will begin. Addition of the large (50S) ribosomal subunit is accompanied by release of the protein factors and hydrolysis of IF2-GTP to yield the complete 70S initiation complex. The charged initiator tRNA (fMet-tRNA$_i^{Met}$) is positioned in the P site. The ribosome now is ready for entry of the next aminoacyl-tRNA at the A site, which will lead to the formation of the first peptide bond. The initiation process is similar in archaeans, except there is no formylation of the Met-tRNA$_i^{Met}$.

an energy-requiring step that is powered by hydrolysis of GTP bound to IF2; in the process IF1, IF2-GDP, and P_i are discharged. In the completed initiation complex, the charged initiator tRNA is positioned at the *P site* on the ribosome. As discussed later, during elongation the growing peptide chain remains attached to the tRNA located at this site.

Eukaryotic Initiation of Protein Synthesis Occurs at the 5' End and Internal Sites in mRNA

Initiation of protein synthesis in eukaryotic cells, as in bacteria, begins with formation of a preinitiation complex prior to mRNA binding. However, the process differs significantly from initiation in bacteria. Two eukaryotic factors, eIF3 (a large multimeric protein with about eight subunits) and eIF6, serve to keep the ribosomal subunits apart after they have previously finished synthesizing a protein chain so that they can participate again in starting a new chain. To assemble a eukaryotic preinitiation complex, an active "ternary complex" of eIF2 bound to a GTP molecule and Met-tRNA$_i^{Met}$ associates with a small (40S) ribosomal subunit complexed with two other factors, eIF3 and eIF1A, which stabilize binding of the ternary complex (Figure 4-37). Cells can regulate protein synthesis by phosphorylating a serine residue on the eIF2 bound to GDP; this complex is then unable to bind Met-tRNA$_i^{Met}$, thus inhibiting protein synthesis.

Recognition of the translation start site by the small ribosomal subunit in the eukaryotic preinitiation complex requires different types of factors than in prokaryotic cells. Most eukaryotic mRNAs have a single start site near the 5' capped end of the mRNA (see Figure 4-19). Interaction of the small ribosomal subunit with the methylated 5'-cap structure

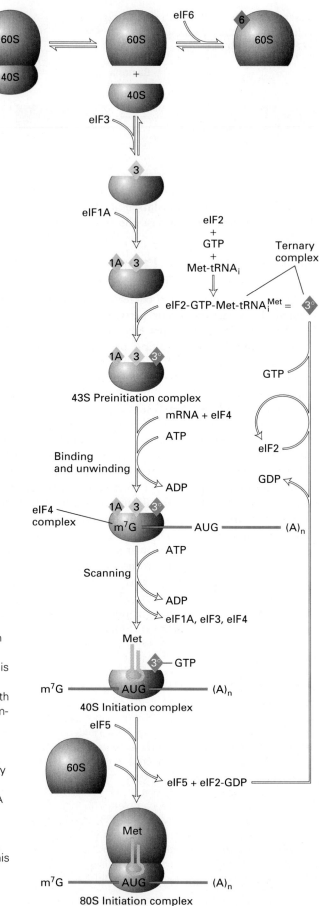

▶ **FIGURE 4-37 Eukaryotic initiation of protein synthesis.** Establishment of the initiation complex must follow a precise sequence of events, and the large and small ribosomal subunits must be kept apart until late in the process. Two eukaryotic factors, eIF3 (a large multimeric protein with about eight subunits) and eIF6 fulfill this role. A 43S preinitiation complex is formed when a ternary complex of eIF2 bound to GTP and Met-tRNA$_i^{Met}$ associates with the small (40S) ribosomal subunit, which is complexed with two other factors, eIF3 and eIF1A, that stabilize binding of the ternary complex. (The inactive eIF2-GDP fails to bind Met-tRNA$_i^{Met}$; it also can be phosphorylated, thereby inhibiting protein initiation.) The 5' cap (m^{7}G) of the mRNA to be translated is guided to the preinitiation complex by a subunit of the multiprotein eIF4 complex, which also unwinds any secondary structure at the 5' end of the mRNA. Subsequent scanning by the small ribosomal subunit, assisted by eIF3 and eIF4G, positions the initiator tRNA at the AUG start codon, yielding the 40S initiation complex and releasing eIF1A, eIF3, and eIF4. With the Met-tRNA$_i^{Met}$ properly positioned at the start codon, another factor, eIF5, assists union of the 40S complex with the 60S subunit. Hydrolysis of GTP in eIF2-GTP provides the energy for this step. Factors eIF5 and eIF2-GDP are then released, yielding the final 80S initiation complex, with Met-tRNA$_i^{Met}$ at the P site. The complex can now accept the second aminoacyl-tRNA (see Figure 4-40).

present on all eukaryotic mRNAs requires a set of proteins collectively called *eIF4*. After the methylated cap structure is recognized by eIF4F, any secondary structure at the 5' end is removed by an associated helicase activity. The bound preinitiation complex, which contains the eIF3 group of proteins, then probably slides along the mRNA, most often stopping at the first AUG. However, selection of the initiating AUG is facilitated by specific surrounding sequences called *Kozak sequences* (for Marilyn Kozak, who defined them):

mRNA 5'–ACCAUGG–

The A preceding the AUG seems to be the most important nucleotide affecting initiation efficiency. During the selection of the starting AUG, there is a proven interaction between the eIF3 complex, the 40S ribosomal subunit, and the eIF4G component of the eIF4 complex. Scanning of the mRNA by the preinitiation complex eventually yields a *40S initiation complex* in which Met-tRNA$_i^{Met}$ is correctly positioned at the translation start site.

Once the small ribosomal subunit with its bound Met-tRNA$_i^{Met}$ is correctly positioned at the start codon, union with the large (60S) ribosomal subunit completes formation of the *80S initiation complex* (see Figure 4-37).

Initiation of translation of most mRNAs by the eukaryotic protein-synthesizing machinery begins near the 5' capped end as just described. However, some viral mRNAs, which are translated by the host-cell machinery in infected eukaryotic cells, lack a 5' cap; translation of these mRNAs is initiated at *internal ribosome entry sites* (IRESs). It is now known that some cellular mRNAs also can be translated beginning at IRESs (Figure 4-38). Many of the same protein factors that assist in ribosome scanning from a 5' cap are required for locating an internal AUG start codon, but exactly how an IRES is recognized is less clear.

During Chain Elongation Each Incoming Aminoacyl-tRNA Moves through Three Ribosomal Sites

The correctly positioned bacterial 70S or eukaryotic 80S ribosome–Met-tRNA$_i^{Met}$ complex is now ready to begin the task of stepwise addition of amino acids by the in-frame translation of the mRNA. As is the case with initiation, a set of special proteins, termed **elongation factors** (EFs), are required to carry out the process. The key steps in elongation are entry of each succeeding aminoacyl-tRNA, formation of a peptide bond, and the movement, or *translocation*, of the ribosome with respect to the mRNA. For example, the second aminoacyl-tRNA is brought into the ribosome as a ternary complex in association with an EF1 α-GTP in eukaryotes (EF-Tu–GTP in bacteria) and becomes bound to the A site on the ribosome (Figure 4-39). Recall that the initiating Met-tRNA$_i^{Met}$ is bound at the P site and base-paired with the AUG start codon. If the anticodon of the incoming (second) aminoacyl-tRNA correctly matches the second codon of the mRNA, a tight binding ensues at the A site. If this second codon does not match the incoming aminoacyl-tRNA, it diffuses away. The choice of the correct aminoacyl-tRNA and its tight binding at the A site requires energy that is supplied by hydrolysis of the EF1 α-GTP (or EF-Tu–GTP) complex.

With the initiating Met-tRNA$_i^{Met}$ at the P site and the second aminoacyl-tRNA tightly bound at the A site, the α amino group of the second amino acid reacts with the "activated" (aminoacylated) methionine on the initiator tRNA, forming a peptide bond. A revolutionary piece of biochemistry has provided insight about this critical *peptidyltransferase reaction*, which actually effects the transfer of the growing peptide chain on the peptidyl-tRNA at the P site to the activated amino acid on the incoming aminoacyl-tRNA. In bacterial protein synthesis, the 23S rRNA in the large ribosomal subunit itself may carry out the peptidyltransferase function. Supporting this catalytic role for 23S rRNA is the finding that when the vast majority of the protein is carefully removed from the large ribosomal subunit and the remaining rRNA, or even a pure segment of 23S rRNA, is mixed with analogs of aminoacylated-tRNA and peptidyl-tRNA, the peptidyltransferase reaction still occurs.

During the process of peptide synthesis and tRNA site changes, the ribosome is moved along the mRNA a distance equal to one codon with the addition of each amino acid. This translocation step is catalyzed by eukaryotic EF2-GTP (bacterial EF-G–GTP), which is hydrolyzed to provide the required energy. Referring again to Figure 4-39, we can see that after peptide linkage tRNA$_i^{Met}$, now without its activated methionine, is moved to an exit (E) site on the ribosome and is soon discharged. Concurrently, another ternary complex, carrying the next amino acid to be added, enters the ribosome, and the cycle continues.

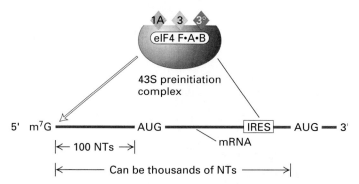

▲ **FIGURE 4-38 Eukaryotic initiation of protein synthesis can occur near the 5' capped end and at internal ribosome entry sites (IRESs) of a mRNA.** In the first case, the 43S preinitiation complex binds to the 5'-cap structure (m^{7}G) and then slides along the mRNA until it reaches an acceptable AUG start codon, usually within about 100 nucleotides (see Figure 4-39). Much less frequently, the preinitiation complex binds to an IRES within the mRNA sequence, far downstream of the 5' end, and then scans downstream for an AUG start codon. (Nts = nucleotides.)

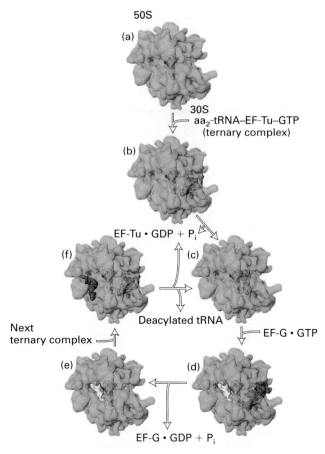

▲ **FIGURE 4-39 The elongation cycle in protein synthesis visualized for *E. coli* ribosomes.** The models of the *E. coli* ribosome associated with tRNAs in various states are based on cryoelectron microscopy studies and cross-linking experiments (see Figure 4-34). (a) Elongation begins on the 70S initiation complex with Met-tRNA$_i^{Met}$ (green) in the P site, assembled as depicted for the eukaryotic system in Figure 4-37. Both ribosomal subunits are shown in blue. (b) A ternary complex consisting of elongation factor EF-Tu bound to GTP (orange and red) plus the correct aminoacyl-tRNA (purple) to decode the second codon then enters the ribosome. (c) Hydrolysis of EF-Tu–GTP to EF-Tu–GDP and P$_i$ brings aa$_2$-tRNA$_2^{aa}$ to the A site (purple). (d, e) After binding of EF-G–GTP (dark blue), the peptidyl synthesis and translocation steps occur: Met-tRNA$_i^{Met}$ (now yellow and moving out of the P site) donates its methionine to the aa$_2$-tRNA$_2^{aa}$ (now green and in the P site), producing aa$_2$-tRNA$_2^{aa}$–Met. Hydrolysis of the bound EF-G–GTP furnishes energy for translocation, and the products are released. (f) tRNA$_i^{Met}$ (now brown) moves completely to the E site and is ejected; a new ternary complex containing aa$_3$-tRNA$_3^{aa}$ arrives and the cycle can continue (c → d → e → f). [Adapted from R. K. Agrawal et al., *Cell*, in press.]

Protein Synthesis Is Terminated by Release Factors When a Stop Codon Is Reached

The final phase of protein synthesis, like initiation and elongation, requires highly specific molecular signals that decide the fate of the mRNA-ribosome-tRNA-peptidyl complex.

Recent research has uncovered two types of specific **termination** (release) **factors** in bacteria and their eukaryotic counterparts. Bacterial RF1 and RF2 (and eukaryotic eRF1), whose shape is thought to be similar to that of tRNAs ("molecular mimicry"), apparently act by recognizing the stop codons themselves. RF1 recognizes UAG, and RF2 recognizes UGA; both these factors recognize UAA. Like some of the initiation factors and the elongation factors discussed previously, the third release factor, RF3 (and its eukaryotic counterpart), is a GTP-binding protein. RF3 acts in concert with the codon-recognizing factors to promote cleavage of the peptidyl-tRNA, thus releasing the completed protein chain (Figure 4-40).

Folding of a newly released protein into its native three-dimensional conformation is facilitated by other proteins

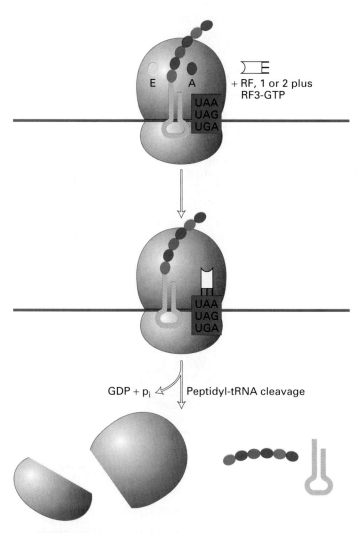

▲ **FIGURE 4-40 Termination of translation.** When a ribosome bearing a nascent protein chain reaches a stop codon (UAA, UGA, UAG), release factors (RFs) enter the ribosomal complex, probably at or near the A site. In bacteria, RF1 or RF2 first recognizes a stop codon. RF3-GTP then catalyzes cleavage of the peptide chain from the tRNA and release of the two ribosomal subunits, reactions that require hydrolysis of GTP.

(a)

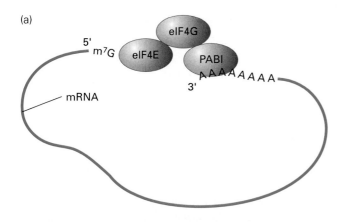

(b)

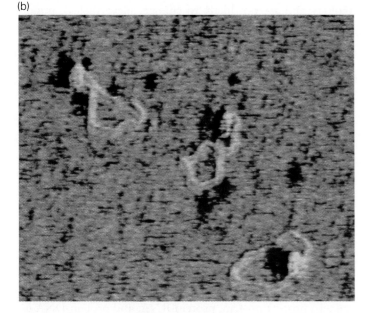

▲ FIGURE 4-41 Formation of circular eukaryotic mRNA by protein-protein interactions bridging the 5′ and 3′ ends.
(a) Poly(A)-binding protein I (PABI) binds both to the 3′ poly(A) tail on a eukaryotic mRNA and to subunits of the initiation factor eIF4, which are bound to the m⁷G cap at the 5′ end of the mRNA. (b) Force-field electron micrographs of mRNA circles held together by interactions between the three purified proteins described in part (a). [Part (a) adapted from A. Sachs et al., 1997, *Cell* **89**:841; part (b) courtesy of A. Sachs.]

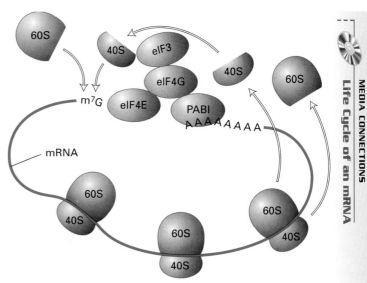

▲ FIGURE 4-42 Model of protein synthesis on circular polysomes and recycling of ribosomal subunits. Multiple individual ribosomes can simultaneously translate a eukaryotic mRNA, shown here in circular form stabilized by interactions between proteins bound at the 3′ and 5′ ends. When a ribosome completes translation and dissociates from the 3′ end, the separated subunits can readily find the nearby 5′ cap (m⁷G) and initiate another round of synthesis. This rapid recycling of subunits on the same mRNA contributes to the formation of polysomes and increases the efficiency of translation.

called *chaperones* (Chapter 3). Additional release factors then promote dissociation of the ribosome, freeing the subunits, mRNA, and terminal tRNA for another round of protein synthesis.

Simultaneous Translation by Multiple Ribosomes and Their Rapid Recycling Increase the Efficiency of Protein Synthesis

As noted earlier, translation of a single eukaryotic mRNA molecule to yield a typical-sized protein takes 30–60 seconds. Two phenomena significantly increase the overall rate at which cells can synthesize a protein: the simultaneous translation of a single mRNA molecule by multiple ribosomes and rapid recycling of ribosomal subunits after they disengage from the 3′ end of a mRNA. Simultaneous translation of an mRNA by multiple ribosomes is readily observable in electron micrographs and by sedimentation analysis, revealing mRNA attached to multiple ribosomes bearing nascent growing peptide chains. These structures are referred to as *polyribosomes* or *polysomes*. However, the second phenomenon—rapid recycling of ribosomal subunits—was not discovered so easily.

Polysomes in some tissues were seen in electron micrographs to be circular. It was not until recently that studies with yeast cells explained the circular shape of polyribosomes and suggested the mode by which ribosomes recycle efficiently.

Researchers initially found that the poly(A) tail at the 3′ end of eukaryotic mRNA stimulates in vitro translation by yeast protein-synthesizing components. This observation at first seems puzzling, since translation is begun upstream (usually near the 5′ end) of the mRNA, at a great distance from the poly(A) tail. A cytosolic protein found in all eukaryotic cells, *poly(A)-binding protein I* (PABI), proved literally to be a missing link in explaining the relationship of the poly(A) tail to translation. In earlier studies, yeast cells that had mutations in the gene encoding PABI were found to translate mRNA poorly. The role of PABI in improving efficiency of protein synthesis was demonstrated by the recent finding that PABI and two components of yeast eIF4

(4G and 4E) can interact on mRNA to circularize the molecule (Figure 4-41). By linking the poly(A) tail to the 5′ end, PABI positions recently disengaged ribosomal subunits near the start site for protein synthesis, thus enhancing their reentry into the process and improving efficiency. Thus in many eukaryotic cells, ribosomal subunits may follow the circular pathway depicted in Figure 4-42, providing another example of the merger of form and function.

SUMMARY Stepwise Formation of Proteins on Ribosomes

- Of the two methionine tRNAs found in all cells, only one (tRNA$_i^{Met}$) functions in initiation of protein synthesis.

- During initiation, the large and small ribosomal subunits assemble at the translation start site in an mRNA molecule, and the tRNA carrying the amino-terminal methionine (Met-tRNA$_i^{Met}$) base-pairs with the start codon, forming a 70S (bacterial) or 80S (eukaryotic) initiation complex (see Figures 4-36 and 4-37).

- The mechanism by which the small ribosomal subunit locates the start codon differs somewhat in bacteria and eukaryotes. In all cells, however, it depends on specific initiation factors and requires energy supplied by hydrolysis of GTP.

- Elongation of a protein chain entails three steps, which occur over and over (see Figure 4-39). First, an aminoacyl-tRNA binds tightly to the A site on the ribosome and base-pairs with its corresponding codon in the mRNA. Second, a peptide bond is formed between the incoming amino acid and the growing chain at the P site, transferring the peptidyl chain to the incoming tRNA. Peptide-bond formation is accompanied by movement of the incoming tRNA to the P site and of the now unloaded tRNA to the E site. Finally, the ribosome translocates to the next codon, and the old tRNA is discharged from the ribosome. As with initiation, elongation requires specific protein factors and GTP hydrolysis.

- The large rRNA in a ribosome most likely catalyzes the peptidyltransferase reaction in which peptide bonds between adjacent amino acids are formed.

- Termination is carried out by two types of termination factors: those that recognize stop codons and those that promote hydrolysis of peptidyl-tRNA (see Figure 4-40).

- In eukaryotic cells, multiple ribosomes commonly are bound to a single mRNA, forming a circular polysome; all the bound ribosomes translate the mRNA concurrently. As each ribosome completes translation and is released from the 3′ end of the mRNA, the subunits probably reassemble quickly at the 5′ end (see Figure 4-42). This circular ribosome pathway greatly increases the efficiency of protein synthesis.

PERSPECTIVES for the Future

In this chapter we first considered the basic chemical makeup of DNA and RNA and some elementary principles about their two- and three-dimensional structures. We then described the elementary properties of RNA and DNA polymerases, which catalyze the synthesis of nucleic acids in cells. The large-scale protein machines that make RNA and DNA according to regulated plans are examined in detail in later chapters. Finally, we discussed the genetic code and the participation of tRNA and the protein-synthesizing machine, the ribosome, in decoding the information in mRNA to allow accurate assembly of protein chains.

These facts form the bedrock principles of molecular cell biology and are not likely to change. Nevertheless, important extensions of these principles are now being made, and these will form a new generation of fundamental principles. For example, sequencing of the entire genomes of many different bacteria and archaea has been completed, and dozens more are under way. The genomic sequence of baker's yeast, a simple eukaryote, also has been determined, as has that of the nematode worm *C. elegans*. The genome of *Drosophila melanogaster* may well be finished by the time this book is released, and early in the next century, the human genome will yield its complete sequence to researchers. As discussed in Chapter 7, once the genomic sequence of an organism is known, all the possible encoded proteins can be deduced with the aid of computer programs.

Paralleling the work on genomic analysis, biochemists and structural biologists are collaborating to determine through structural studies all the possible folds that protein domains assume in order to expand our knowledge of the interactions between the cell's principal macromolecules, namely, protein-protein, protein-RNA, and protein-DNA interactions. The fundamental principles of interaction and a catalogue of typical macromolecular interactions will be in the hands of scientists early in the twenty-first century.

Much of current biological research, and that in the next few years, is focused on discovering how these molecular interactions endow cells with decision-making capacity and their special properties. For this reason several of the following chapters describe current knowledge about how such interactions regulate RNA and DNA synthesis and how such regulation endows cells with the capacity to become specialized and grow into complicated organs. Other chapters deal with how protein-protein interactions underlie the construction of specialized organelles in cells, how cells and cell organelles move, and how these properties of mobility coupled with regulated growth lead to construction of body plans of correct size and shape.

One frontier, neurobiology, will benefit from all these molecular and cellular advances but will require a new level of integrated research for ultimate solutions. Future advances in the molecular biology of the specialized cell types that compose the brain will, no doubt, contribute to understanding brain function. It is clear, however, that neurobiology poses a new challenge in understanding how numerous functions within and between brain cells are integrated. Such integration, which exists at a vastly more complicated level in the brain than elsewhere, is critical to brain functioning.

PERSPECTIVES in the Literature

Macromolecular synthesis is carried out by multiprotein machines. Compare the machines for making RNA, DNA, and protein. What are the similarities and differences among them that seem most cogent to their individual tasks? As we acquire ever deeper knowledge of these machines, this will be a worthwhile project for every student, and indeed for every practicing biologist. Start your exploration of the literature in this broad topic with the following references:

Ban, N., et al. 1998. A 9Å resolution X-ray crystallographic map of the large ribosomal subunit. *Cell* 93:1105–1115.

Holstege, F. C. P., and R. A. Young. 1999. Transcriptional regulation: contending with complexity. *Proc. Nat'l Acad. Sci. USA* 96:2–4.

Holstege, F. C. P., and R. A. Young. 1998. Review Issue, Macromolecular Machines. *Cell* 92:291–423.

Testing Yourself on the Concepts

1. What are the minimal components needed for information flow from DNA to protein?

2. Investigators use the PCR (polymerase chain reaction) to amplify regions of DNA contained between two single-stranded oligonucleotide primers that anneal to opposite strands of a DNA template. In designing primers, it is important to consider sequence information, T_m, and directionality of the primers. Why?

3. Mutations that alter DNA can alter the function of an expressed protein. A mutation is characterized in a gene from a model organism and found to contain a single base change relative to the wild-type gene, yet the encoded protein still functions in this organism. How is this possible?

MCAT/GRE-Style Questions

Key Concept Please read the sections titled "Messenger RNA Carries Information from DNA in a Three-Letter Genetic Code" (p. 117) and "Experiments with Synthetic mRNAs and Trinucleotides Broke the Genetic Code" (p. 119) and answer the following questions. Where relevant, refer to the following DNA sequences, which are coding-strand sequences derived from genes within the nuclear mammalian genome:

I. GTGCTCTTGCAAGACA

II. AGTATTGTTGCAGGATT

III. TAGTTCTTTTACAGGAC

1. Using Table 4-2, determine which of the above sequences has the potential to encode the following protein sequence: Val-Leu-Leu-Gln-Asp.

    a. I only.

    b. I and II only.

    c. I and III only.

    d. I, II, and III.

2. Which of the following occurs when the eukaryotic translational machinery encounters the TAG codon:

    a. The bound preinitiation complex stops scanning and positions the Met-tRNA$_i^{Met}$ at this site.

    b. The termination factors recognize this codon and translation ends.

    c. This codon is recognized by the corresponding anticodon of an empty tRNA molecule that is not linked to an amino acid.

    d. This codon is not recognized by any factors that ultimately cause the translational machinery to stop.

3. In which of the following would translation of sequence II yield a novel polypeptide compared with mammals:

    a. Rice.

    b. Acetabularia.

    c. Yeast.

    d. None of the above.

4. A naturally occurring coding strand composed of alternating C and U residues would result in the formation of

    a. A polypeptide containing alternating Leu and Ser residues.

    b. A polypeptide containing either Leu or Ser residues.

    b. A polypeptide containing only Leu residues.

    c. A polypeptide containing only Ser residues.

Key Experiment Please read the section titled "DNA Can Undergo Reversible Strand Separation" (p. 105) and use Figure 4-9 to answer the following questions:

5. An investigator would be able to distinguish a solution containing RNA from one containing DNA by

    a. Heating the solutions to 82.5 °C and measuring the absorption of light at 260 nm.

    b. Comparing the T_m of each solution.

    c. Monitoring the change in absorption of light at 260 nm while elevating the temperature.

    d. Measuring the absorption of light at 260 nm.

6. The percentage of G·C base pairs in a DNA molecule is related to the T_m because

a. The stability of G·C and A·T base pairs is intrinsically different.

b. A·T base pairs require a higher temperature for denaturation.

c. The triple bonds of G·C base pairs are less stable than the double bonds of A·T base pairs.

d. The G·C content equals the A·T content.

7. A circular DNA molecule will give rise to data as in Figure 4-9 only

a. If one of the strands is cut first.

b. If protein is bound to the DNA.

c. If replication is taking place.

d. If renaturation is allowed to take place.

Key Application Plant ribosome-inactivating proteins (RIPs) are toxins that inhibit eukaryotic translation. RIPs can be used to control viruses in both humans and plants, if they can be targeted specifically to virally infected cells. Catalytically, the RIPs act to remove an adenine residue from a conserved stem loop of the 60S rRNA, thereby preventing the binding of the eEF2-GTP complex during protein elongation. Please refer to the section titled "Eukaryotic Initiation of Protein Synthesis Occurs at the 5′ End and Internal Sites in mRNA" (p. 130) and answer the following questions:

8. Which of the following is not present in cells treated with an RIP:

a. The 43S preinitiation complex.

b. The ternary complex.

c. The 80S initiation complex.

d. The 40S initiation complex.

9. A mutant RIP, when added to eukaryotic cells, can be judged to be inactive by

a. Lack of depurination of rRNA.

b. Lack of union of the 40S and 60S ribosomal subunits.

c. Lack of viral growth.

d. Cytotoxicity.

10. RIPs do not inhibit prokaryotic translation, because

a. They do not recognize the 30S initiation complex.

b. They do not depurinate the 16S rRNA.

c. They do not recognize the Shine-Dalgarno sequence.

d. They do not recognize the large ribosomal subunit of prokaryotes.

11. Recognition of viral mRNAs by the eukaryotic machinery always involves

a. Initiation at the 5′ capped end of the mRNA.

b. Initiation at an internal ribosome entry site.

c. Protein factors that assist in ribosome scanning.

d. Initial binding of the 60S rRNA.

Key Terms

aminoacyl-tRNA *120*
anticodon *122*
codon *117*
complementary *103*
deoxyribonucleic acid (DNA) *100*
DNA polymerases *112*
double helix *103*
elongation factors *131*
exons *108*
genes *100*
genetic code *117*
growing fork *113*
initiation factors *129*
intron *108*
messenger RNA (mRNA) *100*

nucleosides *101*
nucleotides *101*
Okazaki fragments *113*
ribonucleic acid (RNA) *100*
ribosomal RNA (rRNA) *100*
ribosomes *116*
RNA polymerases *112*
RNA processing *115*
splicing *116*
termination factors *132*
transcription *100*
transfer RNA (tRNA) *100*
translation *100*

References

General References

Lehninger, A. L., D. L. Nelson, and M. M. Cox. 1993. *Principles of Biochemistry,* 2d ed. Worth.

The new age of RNA. 1993. *FASEB J.,* vol. 7.

Stryer, L. 1995. *Biochemistry,* 4th ed. W. H. Freeman and Company.

Nucleic Acids: Structure and General Properties

Brenner, S., F. Jacob, and M. Meselson. 1961. An unstable intermediate carrying information from genes to ribosomes for protein synthesis. *Nature* 190:576–581.

Cate, J. H., et al. 1996. Crystal structure of a Group I ribozyme domain: principles of RNA packing. *Science* 273:1678–1685.

Cech, T. R., and B. L. Bass. 1986. Biological catalysis by RNA. *Ann. Rev. Biochem.* 55:599–629.

Crothers, D. M., T. E. Haran, and J. G. Nadeau. 1990. Intrinsically bent DNA. *J. Biol. Chem.* 265:7093–7099.

Dickerson, R. E. 1983. The DNA helix and how it is read. *Sci. Am.* 249(6):94–111.

Felsenfeld, G. 1985. DNA. *Sci. Am.* 253(4):58–66.

Kornberg, A., and T. A. Baker. 1992. *DNA Replication,* 2d ed. W. H. Freeman and Company, chap. 1. A good summary of the principles of DNA structure.

Vasquez, K. M., and J. H. Wilson. 1998. Triplex-directed modification of genes and gene activity. *Trends Biochem. Sci.* 23:4–9.

Wang, J. C. 1980. Superhelical DNA. *Trends Biochem. Sci.* 5:219–221.

Zhang, B., and T. T. Cech. 1997. Peptide bond formation by in vitro selected ribozymes. *Nature* 390:96–100.

Nucleic Acid Synthesis

Doolittle, W. F. 1997. Archaea and the origins of DNA replication proteins. *Cell* 87:995–998.

Hampsey. M. 1998. Molecular genetics of the RNA polymerase II general transcriptional machinery. *Microbiol. Mol. Biol. Rev.* **62**:465–503.

Mattaj, I. W., D. Tollervey, and B. Seraphin. 1993. Small nuclear RNAs in messenger RNA and ribosomal RNA processing. *FASEB J.* **7**:47–53.

McClure, W. R. 1985. Mechanism and control of transcription initiation in prokaryotes. *Ann. Rev. Biochem.* **54**:171–204.

Steitz, J. A. 1988. "Snurps." *Sci. Am.* **258**(6):56–65.

Wahle, E., and W. Keller. 1992. The biochemistry of 3' -end cleavage and polyadenylation of messenger RNA precursors. *Ann. Rev. Biochem.* **61**:419–440.

Young, R. A. 1991. RNA polymerase II. *Ann. Rev. Biochem.* **60**:689–715.

The Genetic Code

Khorana, H. G. 1968. Nucleic acid synthesis in the study of the genetic code. Reprinted in *Nobel Lectures: Physiology or Medicine* (1963–1970). Elsevier (1973).

Nirenberg, M. W., and J. H. Matthei. 1961. The dependence of cell-free protein synthesis in *E. coli* upon naturally occurring or synthetic polyribonucleotides. *Proc. Nat'l. Acad. Sci. USA* **47**:1588–1602.

Nirenberg, M. W., and P. Leder. 1964. RNA codewords and protein synthesis. *Science* **145**:1399–1407.

Initiation

Jackson, R. J. 1991. Initiation without an end. *Nature* **353**:14–15.

Kozak, M. 1991. An analysis of vertebrate mRNA sequences: intimations of translational control. *J. Cell Biol.* **115**:887–903.

Le, S.-Y., and J. B. Maizel. 1997. A common RNA structural motif involved in the internal initiation of translation of cellular RNAs. *Nucl. Acids Res.* **35**:173–369.

Sonenberg, N. 1988. Cap-binding proteins of eukaryotic messenger RNA: functions in initiation and control of translation. *Prog. Nucl. Acid Res. Mol. Biol.* **35**:173–206.

Sonenberg, N. 1991. Picornavirus RNA translation continues to surprise. *Trends Genet.* **7**:105–106.

Transfer RNA and Amino Acids

Arnez, J. G., and D. Moras. 1997. Structural and functional considerations of the aminoacylation reaction. *Trends Biochem. Sci.* **22**:211–216.

Bjork, G. R., et al. 1987. Transfer RNA modification. *Ann. Rev. Biochem.* **56**:263–287.

Cavarelli, J., and D. Moras. 1993. Recognition of tRNAs by aminoacyl-tRNA synthetases. *FASEB J.* **7**:79–86.

Grosjean, H., J. Edqvist, K. B. Stråby, and R. Giegé. 1996. Enzymatic formation of modified nucleosides in tRNA: dependence on tRNA architecture. *J. Mol. Biol.* **255**:67–85.

Hoagland, M. B., et al. 1958. A soluble ribonucleic acid intermediate in protein synthesis. *J. Biol. Chem.* **231**:241–257.

Holley, R. W., et al. 1965. Structure of a ribonucleic acid. *Science* **147**:1462–1465.

McClain, W. H. 1993. Transfer RNA identity. *FASEB J.* **7**:72–78.

Rich, A., and S.-H. Kim. 1978. The three-dimensional structure of transfer RNA. *Sci. Am.* **240**(1):52–62 (offprint 1377).

Rould, M. A., J. J. Perona, and T. A. Steitz. 1991. Structural basis of anticodon loop recognition by glutaminyl-tRNA synthetase. *Nature* **352**:213–218.

Sissler, M., G. Eriani, F. Martin, R. Giegé, and C. Florentz. 1997. Mirror image alternative interaction patterns of the same tRNA with either class I arginyl-tRNA synthetase or class II aspartyl-tRNA synthetase. *Nucl. Acids Res.* **25**:4899–4906.

Ribosomes

De Rijk, P., Y. Van de Peer, and R. De Wachter. 1997. Database on the structure of large ribosomal subunit RNA. *Nucl. Acids Res.* **25**:117–122.

Frank, J. 1998. How the ribosome works. *Am. Scientist* **86**:428–439.

Frank, J., et al. 1995. A model of protein synthesis based on cryo-electron microscopy of the *E. coli* ribosome. *Nature* **376**:441–444.

Moazed, D., and H. F. Noller. 1991. Sites of interaction of the CCA end of peptidyl-tRNA with 23S rRNA. *Proc. Nat'l. Acad. Sci. USA* **88**:3725–3728.

Nitta, I., et al. 1998 Reconstitution of peptide bond formation with *Escherichia coli* 23S ribosomal RNA domains. *Science* **281**:666–669.

Noller, H. F. 1993. tRNA-rRNA interactions and peptidyl transferase. *FASEB J.* **7**:87–89.

Olsen, G. J., and C. R. Woese. 1993. Ribosomal RNA: a key to phylogeny. *FASEB J.* **7**:113–123.

Van de Peer, Y., J. Jansen, P. De Rijk, and R. De Wachter. 1997. Database on the structure of small ribosomal subunit. *Nucl. Acids Res.* **25**:111–116.

Yonath, A., K. R. Leonard, and H. G. Wittman. 1987. A tunnel in the large ribosomal subunit revealed by three-dimensional reconstitution. *Science* **236**:813–816.

The Steps in Protein Synthesis

Agrawal, R. K., et al. 1998. Visualization of elongation factor EF-G on the *E. coli* ribosome: the mechanism of translocation. *Proc. Nat'l Acad. Sci. USA* **95**:6134–6138.

Freistroffer, D. V., et al. 1997. Release factor RF3 in *E. coli* accelerates the dissociation of release factors RF1 and RF2 from the ribosome in a GTP-dependent manner. *EMBO J.* **16**:4126–4133.

Heurgué-Hamard, R., et al. 1998. Ribosome release factor RF4 and termination factor RF3 are involved in dissociation of peptidyl-tRNA from the ribosome. *EMBO J.* **17**:808–816.

Jacob, W. F., M. Santer, and A. E. Dahlberg. 1987. A single base change in the Shine-Dalgarno region of 16S rRNA of *E. coli* affects translation of many proteins. *Proc. Nat'l. Acad. Sci. USA* **84**:4757–4761.

Kozak, M. 1997. Recognition of AUG and alternative initiator codons is augmented by G in position +4 but is not generally affected by the nucleotides in positions +5 or +6. *EMBO J.* **16**:2482–2492.

Merrick, W. C., and J. W. B. Hershey. 1995. The pathway and mechanism of eukaryotic protein synthesis. In *Translational Control*, Cold Spring Harbor Laboratory Press.

Moazed, D., and F. Noller. 1989. Interaction of tRNA with 23S rRNA in the ribosomal A, P, and E sites. *Cell* **57**:585–587.

Nakamura, Y., K. Ito, and L. A. Isaksson. 1996. Emerging understanding of translation termination. *Cell* **87**:147–150.

Noller, H. F., V. Hoffarth, and L. Zimniak. 1992. Unusual resistance of peptidyl transferase to protein extraction procedures. *Science* **256**:1416–1419.

Pavlov, M. Y., et al. 1997. Fast recycling of *Escherichia coli* ribosomes requires both ribosome recycling factor (RRF) and release factor RF3. *EMBO J.* **16**:4134–4141.

Sachs, A. B., P. Sarnow, and M. W. Hentze. 1997. Starting at the beginning, middle, and end: translation initiation in eukaryotes. *Cell* **89**:831–838.

Saks, M. E., J. R. Sampson, and N. N. Abelson. 1994. The transfer RNA identity problem: a search for rules. *Science* **263**:191–197.

Wells, S. E., et al. 1998. Circularization of mRNA by eukaryotic translation initiation factors. *Mol. Cell* **2**:135–140.

5

Biomembranes and the Subcellular Organization of Eukaryotic Cells

Recognition that modern-day organisms fall into three distinct evolutionary lineages came from comparisons of the nucleotide sequences of genes common to all organisms and of ribosomal RNAs, particularly the RNA species found in the small ribosomal subunit (see Figure 1-5). Prokaryotes, which lack a defined nucleus and have a simple subcellular organization, form two of the lineages—the bacteria and archaea. The third lineage is the eukaryotes (eukarya), whose cells have a membrane-limited nucleus containing most of the cellular DNA, numerous specialized organelles, and a complex cytoskeleton. Despite the differences in cellular organization between prokaryotes and eukaryotes, all cells share certain structural features and carry out DNA replication, protein synthesis, production of ATP, and many other complicated metabolic events in basically the same way. At the molecular level, as emphasized in Chapter 3, there is a close relationship between protein structure and function. Likewise, at the cellular level, function and structure are intimately related. Thus this chapter describes the basic structural components of cells and the general functions of each, focusing on eukaryotic cells.

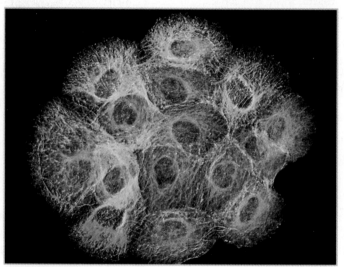

Human skin cells stained for DNA (blue) and the intermediate filament protein GFAP (green).

We begin by describing the capabilities and applications of various techniques of microscopy used to visualize cells and subcellular structures. We then consider the basic methods for purifying biological membranes and subcellular structures. The purified preparations produced by cell fractionation permit detailed studies on the structure and function of biomembranes and other subcellular structures. Following this overview of experimental techniques, we continue the discussion of biomembranes begun in earlier chapters. The surface membrane found on all cells and the membranes that line eukaryotic organelles all have the same basic

MEDIA CONNECTIONS

Overview: Protein Secretion

Technique: Reporter Constructs

Classic Experiment 5.1: Separating Organelles

(a)

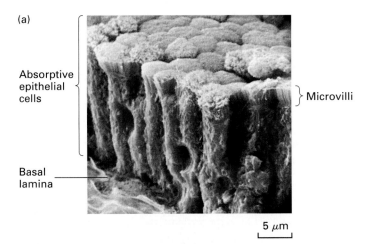

Absorptive epithelial cells

Microvilli

Basal lamina

5 μm

(b)

Plasma membranes of adjacent cells

Brush border

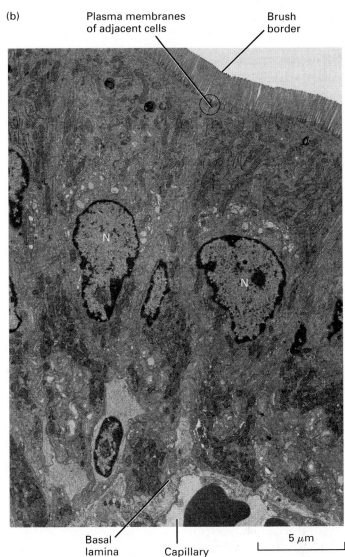

Basal lamina

Capillary

5 μm

(c)

Brush border

Lateral membrane

Lamina propia

20 μm

▲ **FIGURE 5-1 Views of the epithelial cells lining the small intestine, produced by three different microscopic techniques.** (a) Scanning electron micrograph of the intestinal wall. The lumen, or cavity, of the intestine, is lined by a sheet of epithelial cells that rests on a fiber-filled material called the *basal lamina.* Abundant fingerlike microvilli extend from the lumen-facing surface of each cell. The three-dimensional appearance of the cell surface is characteristic of images obtained by this technique. (b) Transmission electron micrograph through two intestinal epithelial cells. Clearly visible are the microvilli, often called the *brush border,* two nuclei (N), and other organelles. Parts of the basal lamina and a capillary (a type of small blood vessel) that courses through the basal lamina are visible at the bottom. Nutrients absorbed by the cells from the lumen find their way into adjacent capillaries, which also provide hormonal signals to the cells. (c) Stained section of the rat intestinal wall viewed in a fluorescence microscope. The tissue section was stained with Evans blue, which generates a nonspecific red fluorescence, and with a yellow-green–fluorescing antibody specific for GLUT2, a glucose transport protein. This technique localizes GLUT2 to the basal and lateral sides of the intestinal cells and shows that it is absent from the brush border. Capillaries run through the lamina propria, a loose connective tissue beneath the epithelial layer. [Part (a) from R. Kessel and R. Kardon, 1979, *Tissues and Organs: A Text-Atlas of Scanning Electron Microscopy,* W. H. Freeman and Company, p. 176; part (b) from P. A. Cross and K. L. Mercer, 1993, *Cell and Tissue Ultrastructure, A Functional Perspective,* W. H. Freeman and Company, p. 293; part (c) see B. Thorens et al., 1990, *Am. J. Physiol.* **259:**C279, courtesy of B. Thorens.]

architecture — a **phospholipid bilayer;** the unique function of each type of membrane is determined primarily by the proteins it contains. The final section briefly describes the structure and function of the main internal organelles and fibers of eukaryotic cells, as well as the extracellular substances that surround cells and give them shape and strength. This review sets the stage for later discussions of cell function at the molecular level.

5.1 Microscopy and Cell Architecture

The modern, detailed understanding of cell architecture is based on several types of microscopy. Because there is no one "correct" view of a cell, it is essential to understand the characteristics of the key cell-viewing techniques, the types of images they produce, and their limitations.

Schleiden and Schwann, using a primitive light microscope, first described individual cells as the fundamental unit of life, and light microscopy has continued to play a major role in biological research. The development of electron microscopes greatly extended the ability to resolve subcellular particles and has yielded much new information on the organization of plant and animal tissues. The nature of the images depends on the type of light or electron microscope employed and on the way in which the cell or tissue has been prepared. Each technique is designed to emphasize particular structural features of the cell. Figure 5-1 shows how a typical cell, the epithelial cell lining the small intestine, appears when viewed by three different microscopic techniques.

In this section, we focus on the most common application of light and electron microscopy—to visualize fixed, killed cells. Although this approach reveals much information, a critical question about such results is how true to life is the image of a biological specimen that has been fixed, stained, and dehydrated before examination? Thus we also consider some of the refinements that allow microscopy of unaltered or less altered specimens.

Light Microscopy Can Distinguish Objects Separated by 0.2 μm or More

The *compound microscope,* the most common microscope in use today, contains several lenses that magnify the image of a specimen under study (Figure 5-2a). The total magnification is a product of the magnification of the individual lenses: if the *objective lens* magnifies 100-fold (a 100X lens, the maximum usually employed) and the *eyepiece* magnifies 10-fold, the final magnification recorded by the human eye or on film will be 1000-fold.

However, the most important property of any microscope is not its magnification but its resolving power, or **resolution**—its ability to distinguish between two very closely

(a)

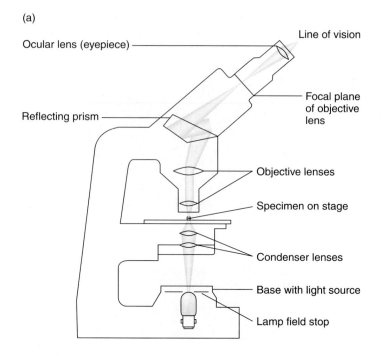

(b)

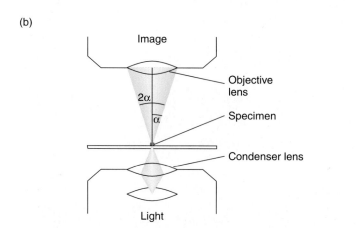

▲ **FIGURE 5-2 The optical pathway in a modern compound optical microscope.** (a) The specimen is usually mounted on a transparent glass slide and positioned on the movable specimen stage of the microscope. Light from a bright source is focused by the condenser lenses onto the specimen. The objective lenses pick up the light transmitted by the specimen and focus it on the focal plane of the objective lens, creating a magnified image of the specimen. Usually the image on the objective focal plane is magnified by the ocular lens, or eyepiece, which is focused on this objective focal plane; it picks up the light emanating from the already magnified image of the specimen and projects it onto the plane of the human eye or onto a piece of photographic film or a video camera. The lamp field stop and other apertures restrict the amount of light entering or leaving a lens. (b) The half-angle, α, of the cone of light entering the objective lens from the specimen is one parameter that determines the resolution of a microscope: the larger the value of α, the finer the resolution the objective lens can provide.

positioned objects. Merely enlarging the image of a specimen accomplishes nothing if the image is blurry. The resolution of a microscope lens is numerically equivalent to D, the minimum distance between two distinguishable objects; the smaller the value of D, the better the resolution. D depends on three parameters, all of which must be considered in order to achieve the best possible resolution: the *angular aperture*, α, or half-angle of the cone of light entering the objective lens from the specimen; the *refractive index*, N, of the air or fluid medium between the specimen and the objective lens; and the *wavelength*, λ, of incident light: $D = (0.61\lambda) \div (N \times \sin \alpha)$. Decreasing the value of λ or increasing either N or α will decrease the value of D and thus improve the resolution. Note that the magnification is not part of this equation.

The angular aperture, α, depends on the width of the objective lens and its distance from the specimen (Figure 5-2b). Moving the objective lens closer to the specimen increases the angle α and thus $\sin \alpha$, and therefore reduces D (i.e., increases the resolution). Intuitively, one can recognize that increasing α allows a greater fraction of the light emanating from the specimen to enter the objective lens. The refractive index N is a measure of the degree to which a medium bends a light ray that passes through it; the refractive index of air is defined as 1.0. Use of immersion oil, which has a refractive index of 1.5, is a simple way to reduce D by 33 percent. An intuitive explanation for this improvement is that a medium with a higher refractive index than air, if placed between the specimen and the objective lens, will "bend" more of the light emanating from the specimen such that it goes into the lens. Finally, the shorter the wavelength of incident light, the lower will be the value of D and the better the resolution.

Due to limitations on the values of α, λ, and N, the *limit of resolution* of a light microscope using visible light is about 0.2 μm (200 nm). No matter how many times the image is magnified, the microscope can never resolve objects that are less than $\approx$0.2 μm apart or reveal details smaller than $\approx$0.2 μm in size. This is true because the maximum angular aperture for the best objective lenses is 70° ($\sin 70° = 0.94$). With the visible light of shortest wavelength (blue, $\lambda = 450$ nm) and with an immersion oil ($N = 1.5$) above the sample, then

$$D = \frac{0.61 \times 450 \text{ nm}}{1.5 \times 0.94} = 194 \text{ nm}$$

or about 0.2 μm.

Despite this limit of resolution, the light microscope can be used to track the location of a small bead of known size to a precision of only a few nanometers! If we know the precise size and shape of an object—say, a 5-nm sphere of gold—and if we use a video camera to record the microscopic image as a digital image, then a computer can calculate the position of the *center* of the object to within a few nanometers. This technique has been used, to nanome-

ter resolution, for tracking the movement of gold particles attached via antibodies to specific proteins on the surface of living cells.

Samples for Light Microscopy Usually Are Fixed, Sectioned, and Stained

Specimens for light microscopy are commonly fixed with a solution containing alcohol or formaldehyde, compounds that denature most proteins and nucleic acids. Formaldehyde also cross-links amino groups on adjacent molecules; these covalent bonds stabilize protein-protein and protein–nucleic acid interactions and render the molecules insoluble and stable for subsequent procedures. Usually the sample is then embedded in paraffin or plastic and cut into thin sections of one or a few micrometers thick (Figure 5-3). Alternatively, the sample can be frozen without prior fixation and then sectioned; this avoids the denaturation of enzymes by fixatives such as formaldehyde.

Since the resolution of the light microscope is $\approx$0.2 μm and mitochondria and chloroplasts are $\approx$1 μm long (about the size of bacteria), theoretically one should be able to see these organelles. However, most cellular constituents are not colored and absorb about the same degree of visible light, so that they are hard to distinguish under a light microscope unless the specimen is stained. Thus the final step in preparing

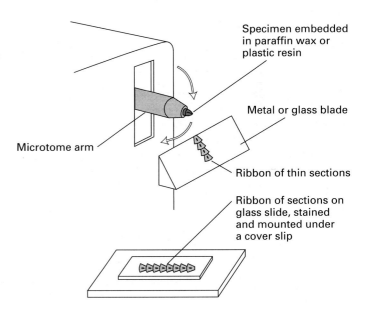

Specimen embedded in paraffin wax or plastic resin

Metal or glass blade

Microtome arm

Ribbon of thin sections

Ribbon of sections on glass slide, stained and mounted under a cover slip

▲ **FIGURE 5-3 Preparation of tissues for light microscopy.** A piece of fixed tissue is dehydrated by soaking it in alcohol-water solutions, then in pure alcohol, and finally in a solvent such as xylene. The specimen is next placed in warm liquid paraffin, which is allowed to harden. A piece of the specimen is mounted on the arm of a microtome. The arm moves up and down over a metal or glass blade, cutting specimen sections a few micrometers (microns) thick.

20 μm

▲ **FIGURE 5-4 Cytochemical staining.** Light micrograph of a cross section of human skeletal muscle stained for succinate dehydrogenase, an enzyme found only in mitochondria. At this low magnification the stained mitochondria appear as purple dots; in skeletal muscle there are several different types of cells differing in the number of mitochondria. [From P. R. Wheater, H. G. Burkitt, and V. C. Daniels, 1987, *Functional Histology; A Text and Colour Atlas,* 2d ed., Churchill Livingstone, Fig. 1.23b, p. 25. Photo Researchers, Inc.]

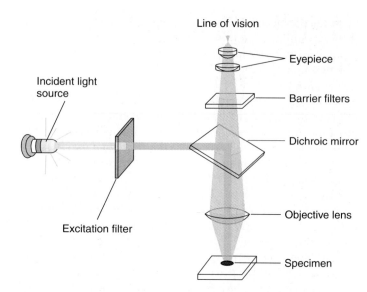

▲ **FIGURE 5-5 The optical pathway in an epi-fluorescence microscope.** Light from a multiwavelength source moves through an excitation filter, which allows only the desired wavelength of exciting radiation to pass. This radiation is reflected downward by the dichroic mirror and focused by the objective lens onto the sample. Fluorescent molecules in the sample are then excited to emit light (fluoresce) at a specific and longer wavelength. This emitted light is focused by the objective lens; most of it passes upward through the dichroic mirror and is not reflected. A final barrier filter blocks any residual light of wavelengths not corresponding to that of the fluorescent substance used to stain the specimen.

a specimen for light microscopy is to stain it, in order to visualize the main structural features of the cell or tissue. Many chemical stains bind to molecules that have specific features. For example, *hematoxylin* binds to basic amino acids (lysine and arginine) on many different kinds of proteins, whereas *eosin* binds to acidic molecules (such as DNA, and aspartate and glutamate side chains). Because of their different binding properties, these dyes stain various cell types sufficiently differently that they are distinguishable visually. Two other common dyes are *benzidine*, which binds to heme-containing proteins and nucleic acids, and *fuchsin*, which binds to DNA and is used in Fuelgen staining.

If an enzyme catalyzes a reaction that produces a colored or otherwise visible precipitate from a colorless precursor, the enzyme may be detected in cell sections by their colored reaction products. This technique is called *cytochemical staining* (Figure 5-4).

Fluorescence Microscopy Can Localize and Quantify Specific Molecules in Cells

Perhaps the most versatile and powerful technique for localizing proteins within a cell by light microscopy is **fluorescent staining** of cells and observation in the *fluorescence microscope.* A chemical is said to be *fluorescent* if it absorbs light

at one wavelength (the *excitation wavelength*) and emits light (fluoresces) at a specific and longer wavelength. Most fluorescent dyes emit visible light, but some (such as Cy5 and Cy7) emit infrared light. In modern fluorescence microscopes, only fluorescent light emitted by the sample is used to form an image; light of the exciting wavelength induces the fluorescence but is then not allowed to pass the filters placed between the objective lens and the eye or camera (Figure 5-5).

Revealing Specific Proteins in Fixed Cells Four very useful dyes for fluorescent staining are rhodamine and Texas red, which emit red light; Cy3, which emits orange light; and fluorescein, which emits green light. These dyes have a low, nonspecific affinity for biological molecules, but they can be chemically coupled to purified antibodies specific for almost any desired macromolecule. When a fluorescent dye–antibody complex is added to a permeabilized cell or tissue section, the complex will bind to the corresponding antigens, which then light up when illuminated by the exciting wavelength, a technique called *immunofluorescence microscopy* (Figure 5-6). By staining a specimen with two or three dyes that fluoresce at different wavelengths, multiple proteins can be localized within a cell, as illustrated in the chapter opening figure and in Figure 5-1c.

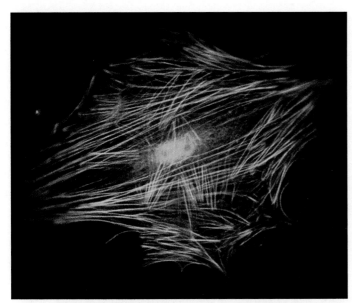

▲ **FIGURE 5-6 Fluorescence micrograph showing the distribution of long actin fibers in a cultured fibroblast cell.** A fixed human skin fibroblast was permeabilized with a detergent and stained with a fluorescent anti-actin antibody before viewing. [Courtesy of E. Lazarides.]

Revealing Specific Proteins in Living Cells Fluorescence microscopy can also be applied to live cells. For example, purified actin may be chemically linked to a fluorescent dye. Careful biochemical studies have established that this "tagged" molecule is indistinguishable in function from its normal

counterpart. If the tagged protein is *microinjected* into a cultured cell, the endogenous cellular and injected tagged actin monomers copolymerize into normal long actin fibers. This technique can also be used to study individual microtubules within a cell.

Another technique for detecting specific proteins within living cells takes advantage of *green fluorescent protein* (GFP), a naturally fluorescent protein found in the jellyfish *Aequorea victoria*. The bioluminescence of this organism, which radiates a green fluorescence, is due to GFP. This 238-aa protein contains serine, tyrosine, and glycine residues whose side chains have spontaneously reacted with one another to form a fluorescent chromophore. By recombinant DNA techniques discussed in Chapter 7, the GFP gene can be introduced into living cultured cells or into specific cells of an entire animal. Because the introduced gene will express GFP, the cells will emit a green fluorescence when irradiated; this GFP fluorescence can be used to localize the cells within a tissue.

Alternatively, the gene for GFP can be fused to the gene for another protein of interest, producing a recombinant DNA encoding one long chimeric protein that contains the entirety of both proteins. Cells in which this recombinant DNA has been introduced will synthesize this chimeric protein, whose green fluorescence will reveal the subcellular localization of the protein. Figure 5-7 illustrates how this technique can demonstrate changes in the localization of a protein within a living cell following treatment with a particular hormone.

(a) (b)

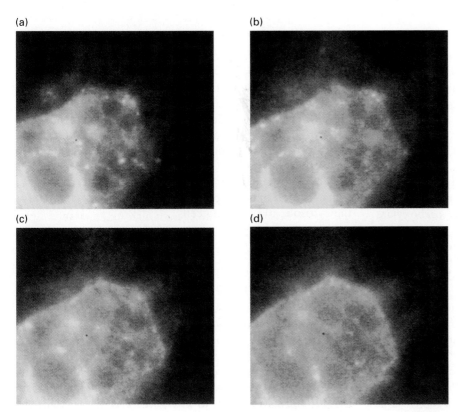

(c) (d)

◀ **FIGURE 5-7 Use of green fluorescent protein (GFP) to localize GLUT4, a glucose transport protein, within living fat cells.** Cells were engineered to express a chimeric protein whose N-terminal end corresponded to the GLUT4 sequence, followed by the entirety of the GFP sequence. When a cell is exposed to light of the exciting wavelength, GFP fluoresces yellow-green, indicating the position of GLUT4 within the cell. In resting cells (a), GLUT4 is in internal membranes that are not connected to the plasma membrane. Successive images of the same cell after treatment with insulin for 2.5, 5, and 10 minutes (panels b, c, and d, respectively) show that, with time, increasing numbers of these GLUT4-containing membranes fuse with the plasma membrane, thereby moving GLUT4 to the cell surface and enabling it to transport glucose from the blood into the cell. As detailed in Chapter 20, this is the principal mechanism by which insulin controls the level of glucose in the blood. [Courtesy of J. Bogan.]

5 μm

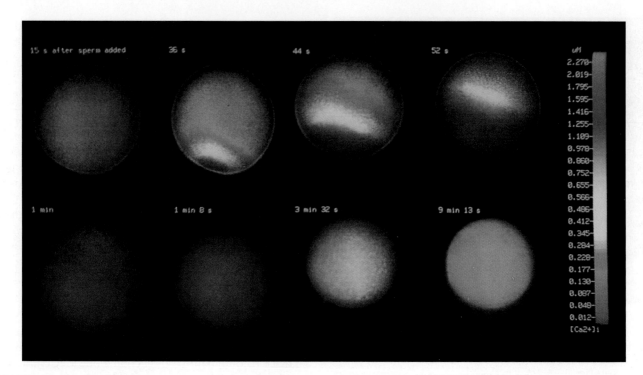

▲ **FIGURE 5-8 Changes in the local concentration of Ca²⁺ in a sea urchin egg following fertilization.** The Ca²⁺ throughout the cell was monitored at different times after fertilization using a fluorescence microscope and fura-2, a Ca²⁺-binding dye whose fluorescence is proportional to the Ca²⁺ concentration. For graphic purposes, the Ca²⁺ concentrations are expressed in a calibrated color scale *(right)* in units of micromolar Ca²⁺. When the sperm penetrates the egg, the level of Ca²⁺ rises initially at the point of sperm entry in the lower left part of the cell and then gradually increases throughout the egg. This spreading increase in cytosolic Ca²⁺ triggers the fusion of small vesicles with the plasma membrane, causing changes in the cell surface that prevent penetration by additional sperm. Eventually, the Ca²⁺ concentration becomes uniformly high and then falls uniformly to the resting state. [See R. Y. Tsien and M. Poenie, 1986, *Trends Biochem. Sci.* **11**:450; courtesy of J. Alderton, M. Poenie, R. A. Steinhardt, and R. Y. Tsien.]

Determining the Intracellular Concentration of Ca²⁺ and H⁺ Ions Changes in the cytosolic concentration of Ca^{2+} ions or pH frequently signal changes in cellular metabolism. The Ca^{2+} concentration in the cytosol of resting cells, for instance, is about 10^{-7} M. Many hormones or other stimuli cause a rise in cytosolic Ca^{2+} to 10^{-6} M; this, in turn, causes changes in cellular metabolism, such as contraction of muscle (Chapter 18).

The fluorescent properties of certain dyes, such as *fura-2*, facilitate measurement of the concentration of free Ca^{2+} in the cytosol. This dye contains five carboxylate groups that form ester linkages with ethanol. The resulting fura-2 ester is lipophilic and can diffuse from the medium across the plasma membrane into cells. Within the cytosol, esterases hydrolyze fura-2 ester yielding fura-2, whose free carboxylate groups render the molecule nonlipophilic, so it cannot cross cellular membranes and remains in the cytosol. Each fura-2 molecule can bind a single Ca^{2+} ion but no other cellular cation, and the amount of fura-2 bound to Ca^{2+} is proportional, over a certain range, to the Ca^{2+} concentration. The fluorescence of fura-2 at one particular wavelength is enhanced when Ca^{2+} is bound, and the fluorescence is proportional to the Ca^{2+} concentration. At another wavelength the fluorescence of fura-2 is the same whether or not Ca^{2+} is bound and provides a measure of the total amount of fura-2 in the segment of the cell. By examining cells continuously in the fluorescence microscope and measuring rapid changes in the ratio of fura-2 fluorescence at these two wavelengths, one can quantify rapid changes in the fraction of fura-2 that has a bound Ca^{2+} ion and thus in the concentration of cytosolic Ca^{2+} (Figure 5-8).

The fluorescence of other dyes is sensitive to the H⁺ concentration and can be used in a similar way to monitor the cytosolic pH of living cells.

Confocal Scanning and Deconvolution Microscopy Provide Sharper Images of Three-Dimensional Objects

Immunofluorescence microscopy has its limitations. The fixatives employed to preserve cell architecture often destroy the *antigenicity* of a protein, that is, its ability to bind to its specific antibody. Also, the method generally gives poor results with thin cell sections, because embedding media often fluoresce themselves, obscuring the specific signal from the antibody. Moreover, in microscopy of whole cells, the

(a)

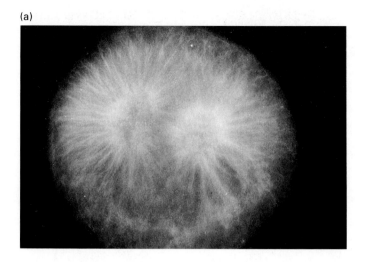

(b)

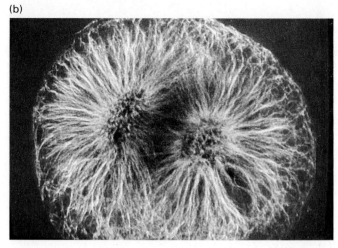

40 μm

▲ **FIGURE 5-9 The advantage of confocal fluorescence microscopy.** A mitotic fertilized egg from a sea urchin *(Psammechinus)* was lysed with a detergent, exposed to an anti-tubulin antibody, and then exposed to a fluorescein-tagged antibody that binds to the first antibody. (a) When viewed by conventional fluorescence microscopy, the mitotic spindle is blurred owing to the background glow of fluorescence from tubulin above and below the plane of focus. (b) The confocal microscopic image is sharp, particularly in the center of the mitotic spindle; fluorescence is detected only from molecules in the focal plane. [From J. G. White, W. Amos, and M. Fordham, 1987, *J. Cell Biol.* **104**:41.

fluorescent light comes from molecules above and below the plane of focus; thus the observer sees a superposition of fluorescent images from molecules at many depths in the cell, making it difficult to determine the actual three-dimensional molecular arrangement (see Figure 5-6).

The *confocal scanning microscope* avoids the last problem by permitting the observer to visualize fluorescent molecules in a single plane of focus, thereby creating a vastly sharper cross-sectional image (Figure 5-9). At any instant during confocal imaging, only a single small part of a sample is illuminated with exciting light from a focused laser beam, which rapidly moves to different spots in the sample focal plane. Images from these spots are recorded by a video camera and stored in a computer, and the composite image is displayed on a computer screen.

Deconvolution microscopy is similar to confocal microscopy in that a cross-sectional image is obtained, but the two techniques differ in the details of how this image is generated. In both cases, the objective lens collects light that originates from above and below the focal plane as well as that which originates from within the focal plane. Confocal microscopes use a pinhole to exclude the out-of-focus light. In contrast, deconvolution microscopes collect all the light from several focal planes, and then mathematically reassign the out-of-focus light to its correct focal plane with the aid of a high-speed computer, a mathematical operation called *deconvolution.*

To understand how a deconvolution microscope works, consider an infinitely small fluorescent source of light, which can be approximated by a fluorescent bead smaller than the resolution of the light microscope (i.e., <0.2 μm in diameter). The emitted light radiates in all directions, and when the source is in the focal plane of the objective, it appearsas a bright point of light. When the point source is outside the focal plane of the objective, some of the light is still collected by the objective lens, and the point source appears as a halo. As the focal plane is moved farther away from the plane containing the point source, the halo becomes larger and more diffuse. Knowing exactly how the light emitted by an infinitely small fluorescent source is collected and distorted by the optics of the sample and microscope, it is possible to reconstruct an individual cross-sectional image (containing only light that originated in the focal plane of interest) from a set of images taken as the objective focal plane is moved through the plane of interest.

Cross-sectional images obtained with a deconvolution microscope may have even greater detail than those obtained with a confocal microscope. Additionally, the fluorescent labeling of the sample does not need to be as intense for deconvolution microscopy as it does for confocal microscopy, since all the light produced by a fluorescent sample is collected and analyzed by the microscope. Three-dimensional images can be obtained by a refinement known as *optical sectioning.* In this method, a computer records individual fluorescent images of planes at different depths of the sample—in effect, serial sections—and combines the stack of images into one three-dimensional image (Figure 5-10).

(a)

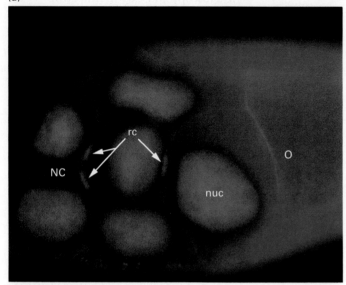

(b)

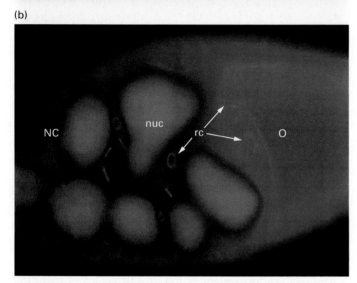

▲ FIGURE 5-10 Optical sectioning of a developing *Drosophila* egg chamber obtained with deconvolution fluorescence microscopy. An egg chamber was labeled with the dye DAPI, which binds to DNA and generates a blue fluorescence from the nuclei. Actin filaments were labeled with the actin-binding chemical phalloidin coupled to red-fluorescing rhodamine. (Nuc = nucleus, RC = ring cell, NC = nurse cell, and O = oocyte.) (a) A single optical plane of an egg chamber. (b) A three-dimensional reconstruction of a portion of the egg chamber shown in (a), consisting of stacked serial optical sections obtained as above. Note the ring canals (surrounded by actin filaments) that connect the nurse cells to one another and to the developing oocyte. [Courtesy of D. Marcey.]

Phase-Contrast and Nomarski Interference Microscopy Visualize Unstained Living Cells

Detailed views of transparent, live, unstained cells and tissues are obtainable with *phase-contrast microscopy* and *Nomarski interference microscopy*. Both techniques take advantage of the phenomena of refraction and diffraction of light waves (Figure 5-11). As a result, small differences

(a)

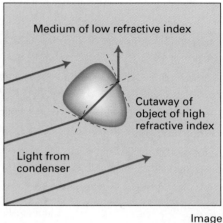

(b)

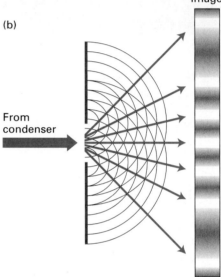

▲ FIGURE 5-11 Light passing through a specimen can be redirected by refraction and diffraction. (a) Refraction: Because light moves at different speeds in different materials (more slowly in a medium of higher refractive index), a beam of light is bent (refracted) as it passes from air into a transparent object and bent again when it departs. Consequently, the part of an incident light wave that passes through a specimen will be refracted and will be out of phase (out of synchrony) with the part of the wave that does not pass through the specimen. The magnitude of the phase difference depends on the difference in refractive index along the two paths and on the thickness of the specimen. If the two parts of the light wave are recombined, the resultant light will be brighter if they are in phase and less bright if they are out of phase. (b) Diffraction: Light waves impinging on a pinhole in an opaque object spread out in all directions. Overlapping waves emanating from different sides of the hole will reinforce one another in the directions (red arrows) where the waves are in the same phase; to an observer in one of those directions, the pinhole will seem bright. In other directions, where the waves are out of phase, peaks of some light waves fall on troughs of others and cancel one another out, producing dark areas. These phenomena are called *constructive* and *destructive interference*, respectively, and explain the resulting diffraction patterns. Similarly, when light impinges on an opaque object, the edges diffract the light waves, producing an image that contains bright areas (white bands) when viewed in some directions and dark areas (gray bands) in other directions.

in refractive index and thickness between parts of the specimen (say, between the nucleus and cytosol) or between the specimen and the surrounding medium can be converted into differences of light and dark in the final image.

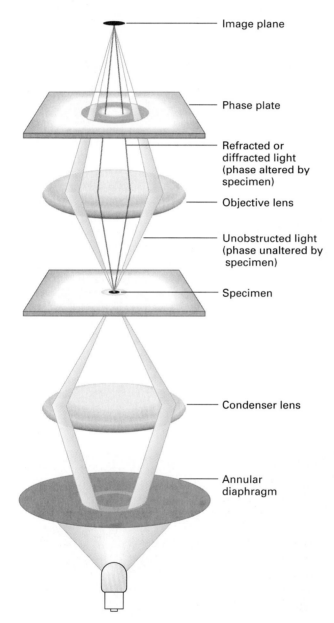

▲ **FIGURE 5-12 The optical pathway of the phase-contrast microscope.** Incident light passes through an annular diaphragm, and the condenser lens focuses a circular annulus (ring) of light on the sample. Light that passes unobstructed through the specimen is focused by the objective lens onto the thick gray ring of the phase plate, which absorbs some of the direct light and alters its phase by one-quarter of a wavelength. If a specimen refracts or diffracts the light, the phase of some light waves is altered and the light waves are redirected through the thin, clear region of the phase plate. The refracted and unrefracted light are recombined at the image plane to form the image.

The phase-contrast microscope generates an image in which the degree of darkness or brightness of a region of the sample depends on the refractive index of that region (Figure 5-12). The improved definition of subcellular structures in live, unstained cells obtained by phase-contrast microscopy compared with standard bright-field microscopy is illustrated in Figure 5-13a.

Nomarski, or differential, interference microscopy generates an image that looks as if the specimen is casting a shadow to one side (Figure 5-13b): the "shadow" primarily represents a difference in refractive index and thickness of a specimen rather than its topography. In this technique, a prism splits an incident beam of plane-polarized light so that one part of the beam passes through one region of a specimen and the other part passes through a closely adjacent region; a second prism then reassembles the two beams. Minute differences in thickness or in the refractive index between adjacent parts of a sample are converted into a bright image (if the two beams are in phase when they recombine) or a dark one (if they are out of phase).

Phase-contrast microscopy is especially useful in examining the structure and movement of larger organelles, such as the nucleus and mitochondria, in live cultured cells. The greatest disadvantage of this technique is that it is suitable for observing only single cells or thin cell layers. Nomarski interference microscopy, in contrast, defines only the outlines of large organelles, such as the nucleus and vacuole. However, thick objects, such as the nuclei in a worm, can be observed by combining this technique with optical sectioning (Figure 5-13c).

Both phase-contrast and Nomarski interference microscopy can be used in *time-lapse microscopy,* in which the same cell is photographed at regular intervals over periods of several hours. This procedure allows the observer to study cell movement, provided the microscope's stage can control the temperature of the specimen and the gas environment (Figure 5-14).

Transmission Electron Microscopy Has a Limit of Resolution of 0.1 nm

The fundamental principles of electron microscopy are similar to those of light microscopy; the major difference is that electromagnetic lenses, not optical lenses, focus a high-velocity electron beam instead of visible light. Because electrons are absorbed by atoms in air, the entire tube between the electron source and the viewing screen is maintained under an ultrahigh vacuum.

The *transmission electron microscope* (TEM) directs a beam of electrons through a specimen. Electrons are emitted by a tungsten cathode when it is electrically heated. The electric potential of the cathode is kept at 50,000–100,000 volts; that of the anode, near the top of the tube, is zero. This drop in voltage causes the electrons to accelerate as they move toward the anode. A condenser lens focuses the electron beam onto the sample; objective and projector

(a)

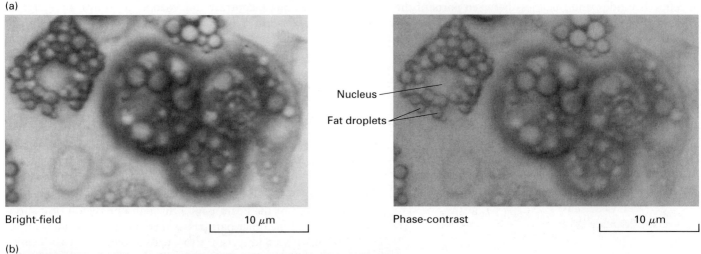

Bright-field   10 μm

Phase-contrast   10 μm

(b)

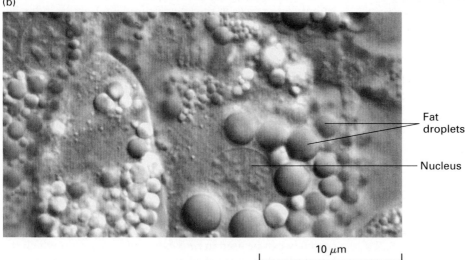

Fat droplets

Nucleus

10 μm

▲ **FIGURE 5-13 Phase contrast and Nomarski optics.** (a) Several live, cultured fat cells (adipocytes) viewed by bright-field microscopy *(left)* and phase-contrast microscopy *(right)*. (b) Another specimen of adipocytes viewed with Nomarski interference (differential interference) microscopy. Thick black lines trace the surface membrane of two cells. (c) A newly hatched larva of the nematode *Caenorhabditis elegans* viewed with Nomarski optics. The individual nuclei of many of the organism's 959 cells are visible. [Part (a) courtesy of J. Bogan; part (b) courtesy of P. Matsudaira and J. Bogan; part (c) from J. E. Sulston and H. R. Horvitz, 1977, *Devel. Biol.* **56**:110.]

(c)

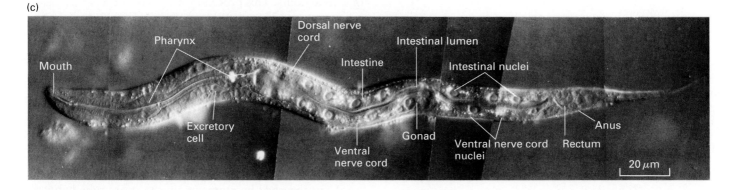

lenses focus the electrons that pass through the specimen and project them onto a viewing screen or a piece of photographic film (Figure 5-15).

In typical electron microscopes, electrons have the properties of a wave with a wavelength of only 0.005 nm. Recall that the minimum distance D at which two objects can be distinguished is proportional to the wavelength λ of the light that illuminates the objects. Thus the limit of resolution for the electron microscope is theoretically 0.005 nm

(less than the diameter of a single atom), or 40,000 times better than the resolution of the light microscope and 2 million times better than that of the unaided human eye. However, the effective resolution of the electron microscope in the study of biological systems is considerably less than this ideal. Under optimal conditions, a resolution of 0.10 nm can be obtained with transmission electron microscopes, about 2000 times better than the best resolution of light microscopes.

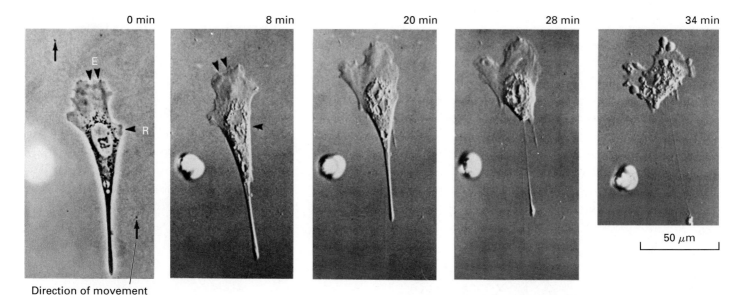

0 min   8 min   20 min   28 min   34 min

50 μm

Direction of movement

▲ **FIGURE 5-14 Time-lapse micrographs show the movement of a cultured fibroblast cell along a glass surface.** A bit of debris on the substratum serves as a reference point. The first image, at 0 min, was obtained by phase-contrast microscopy. Successive images of the same cell, obtained by Nomarski optics, show the lamella at the right of the cell retracting (R) and the lamellipodia at the leading edge of the cell extending (E). In the frame taken at 8 min, the leading edge has moved forward about 9 μm, and the lamellipodia there form a thin flat sheet. By 28 min, the broad leading edge has spread and separated into two lamellae; the thin trailing edge of the cell has begun to retract into the cell body. By 34 min, retraction of the trailing edge is almost complete; only a thin thread of cytoplasm from the trail is left behind, anchored to the substratum. [From W.-T. Chen, 1981, *J. Cell Sci.* **49**:1.]

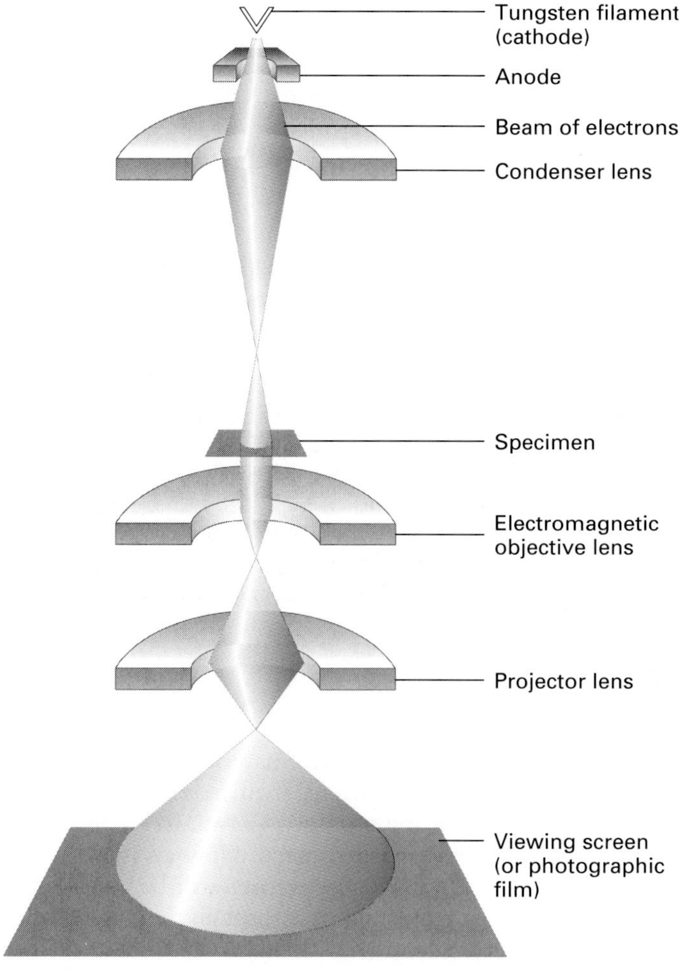

Tungsten filament (cathode)

Anode

Beam of electrons

Condenser lens

Specimen

Electromagnetic objective lens

Projector lens

Viewing screen (or photographic film)

Preparation of Fixed, Stained Samples for TEM Like the light microscope, the transmission electron microscope is used to view thin sections of a specimen, but the fixed sections must be much thinner for electron microscopy (only 50–100 nm, about 0.2 percent of the thickness of a single cell). Clearly, only a small portion of a cell can be observed in any one section. Figure 5-16 depicts the preparation of a sample for transmission electron microscopy. Generation of the image depends on differential scattering of the incident electrons by molecules in the preparation. Without staining, the beam of electrons passes through a cell or tissue sample uniformly, so

◀ **FIGURE 5-15 The optical path in a transmission electron microscope.** A beam of electrons emanating from a heated tungsten filament is focused onto the specimen plane by the magnetic condenser lens. The electrons passing through the specimen are focused by a series of magnetic objective and projector lenses to form a magnified image of the specimen on a fluorescent viewing screen or a piece of photographic film. The entire column, from the electron generator to the screen, is maintained at a very high vacuum.

(a)

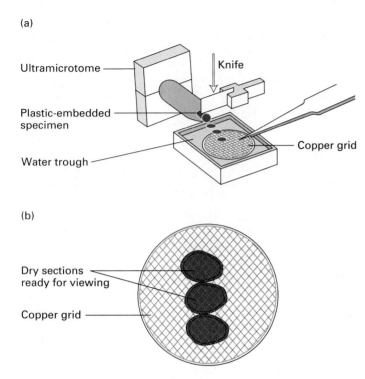

(b)

▲ **FIGURE 5-16 Preparation of a sample of tissue for transmission electron microscopy.** The tissue is dissected, cut into small cubes, and plunged into a fixing solution that cross-links and immobilizes proteins. (Glutaraldehyde is frequently used; osmium tetroxide, another fixing substance, also stains intracellular membranes and certain macromolecules.) The sample is dehydrated by placing it in successively more concentrated solutions of alcohol or acetone; it is then immersed in a solution of plastic embedding medium and put in an oven. Heat causes the solution to polymerize into a hard plastic block, which is trimmed; sections less than 0.1 μm thick are then cut with an ultramicrotome, a fine-slicing instrument with a diamond blade (a). The sections are floated off the blade edge onto the surface of water in a trough. A copper grid coated with carbon or some other material is used to pick up the sections, which are then dried (b).

the entire sample appears uniformly bright with little differentiation of components. Staining techniques are therefore used to reveal the location and distribution of specific materials.

Heavy metals, such as gold or osmium, appear dark on a micrograph because they scatter (diffract) most of the incident electrons; scattered electrons are not focused by the electromagnetic lenses and do not form the image. Osmium tetroxide preferentially stains certain cellular components, such as membranes, which appear black in micrographs. Specific proteins can be detected in thin sections by use of electron-dense gold particles coated with protein A, a bacterial protein that binds antibody molecules nonspecifically (Figure 5-17).

Electron microscopy also is used to obtain information about the shapes of purified viruses, fibers, enzymes, and other subcellular particles. In one technique, called *metal shadowing*, a thin layer of evaporated metal, such as platinum, is laid at an angle on a biological sample (Figure 5-18). An

(a)

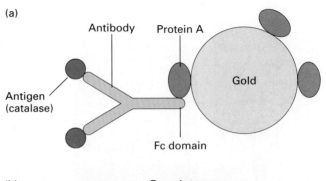

(b)

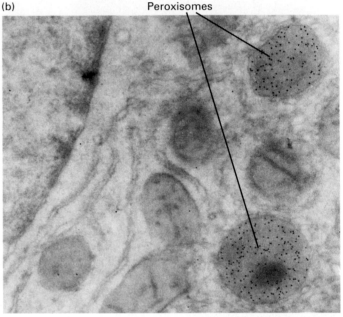

0.5 μm

▲ **FIGURE 5-17 Localization of a specific protein by electron microscopy of antibody-stained samples.** (a) First the antibody is allowed to interact with its specific antigen (e.g., catalase) in a section of fixed tissue. Then the section is treated with a complex of protein A from the bacterium *Staphylococcus aureus* and electron-dense 5–7 nm diameter gold particles. Binding of this complex to the common Fc domain of antibody molecules makes the target protein, catalase in this case, visible in the electron microscope. (b) A slice of liver tissue was fixed with glutaraldehyde, sectioned, and then treated as described in (a) to localize catalase. The gold particles (black dots) indicating the presence of catalase are located exclusively in peroxisomes. [From H. J. Geuze et al., 1981, *J. Cell Biol.* **89**:653. Reproduced from the *Journal of Cell Biology* by copyright permission of The Rockefeller University Press.]

acid bath dissolves the biological material, leaving a metal replica of its surface, which can then be examined in the transmission electron microscope. Variations in the angle and thickness of the deposited metal allow an image to be formed because some incident electrons will be scattered in various directions rather than pass through the preparation. If the metal is deposited mainly on one side of the sample,

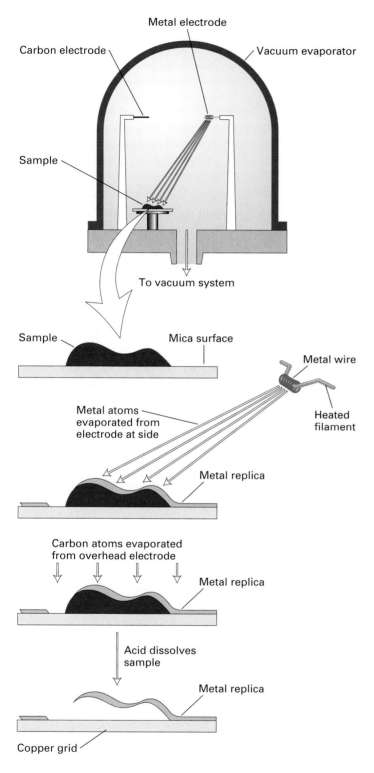

Metal electrode

Carbon electrode

Vacuum evaporator

Sample

To vacuum system

Sample

Mica surface

Metal wire

Metal atoms evaporated from electrode at side

Heated filament

Metal replica

Carbon atoms evaporated from overhead electrode

Metal replica

Acid dissolves sample

Metal replica

Copper grid

◀ FIGURE 5-18 Metal shadowing, a technique that makes surface details on very small particles visible in the electron microscope. The sample is spread on a mica surface and then dried in a vacuum evaporator. A filament of a heavy metal, such as platinum or gold, is heated electrically so that the metal evaporates and some of it falls over the sample grid in a very thin film. In order to stabilize the replica, the specimen is then coated with a carbon film evaporated from an overhead electrode. The biological material is then dissolved by acid, so that the observer views only the metal replica of the sample. In electron micrographs of such preparations, the image is usually reversed: carbon-coated areas appear light and platinum-shadowed areas appear dark.

the technique of *cryoelectron microscopy* allows examination of hydrated, unfixed, and unstained biological specimens directly in the transmission electron microscope. In this technique, an aqueous suspension of a sample is applied in an extremely thin film to a grid. After it has been frozen in liquid nitrogen and maintained in this state by means of a special mount, it is observed in the electron microscope. The very low temperature ($-196\ °C$) keeps the water from evaporating, even in a vacuum, and the sample can be observed in detail in its native, hydrated state without shadowing or fixing it (Figure 5-19). By computer-based averaging of

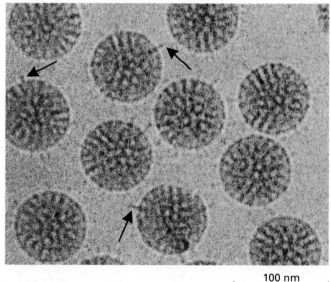

100 nm

▲ FIGURE 5-19 Cryoelectron micrograph of unstained rotavirus particles. A thin suspension of virus particles in water is applied to an electron microscopy grid and frozen. It is then visualized in a transmission electron microscope equipped with a sample stage cooled with liquid nitrogen. The low temperature prevents the ice surrounding the particles from evaporating in the vacuum. Because many biological specimens scatter more electrons than water does, the investigator can observe a very thin specimen without fixing, staining, or dehydrating it. Note the minute spikes (arrows) visible on some particles. [From B. V. Venkataram Prasad et al., 1988, *J. Mol. Biol.* **199**:269.]

for instance, the image seems to have "shadows," where the metal appears dark and the shadows appear light.

Cryoelectron Microscopy Standard electron microscopy cannot be used to study live cells because they are generally too vulnerable to the required conditions and preparatory techniques. In particular, the absence of water causes macromolecules to become denatured and nonfunctional. However,

images of hundreds of particles, a three-dimensional model almost to atomic resolution can be generated.

Scanning Electron Microscopy Visualizes Details on the Surfaces of Cells and Particles

The *scanning electron microscope* allows the investigator to view the surfaces of unsectioned specimens. These cannot be visualized with transmission equipment because the electrons pass through the entire specimen. The sample is fixed, dried, and coated with a thin layer of a heavy metal, such as platinum, by evaporation in a vacuum (see Figure 5-18); in this case, the sample is rotated so that the platinum is deposited uniformly on the surface. An intense electron beam inside the microscope scans rapidly over the sample. Molecules in the specimen are excited and release secondary electrons that are focused onto a scintillation detector; the resulting signal is displayed on a cathode-ray tube. Because the number of secondary electrons produced by any one point on the sample depends on the angle of the electron beam in relation to the surface, the scanning electron micrograph has a three-dimensional appearance (Figure 5-20; see also Figure 5-1a).

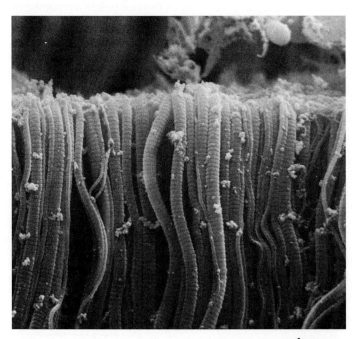

1 μm

▲ **FIGURE 5-20 Scanning electron micrograph of a tendon located within the shoulder of a 29-year-old male.** The sample was rapidly frozen and metal-shadowed before viewing in a scanning electron microscope. These long collagen fibrils are of variable diameter and are oriented parallel to one another; similar fibrils also form the major structural element of bone and similar tissues. The side-by-side arrangement of the linear collagen molecules comprising a fibril gives rise to a characteristic 64-nm repeated pattern visible as parallel striations along the length of each fibril. [Courtesy of D. Keene.]

The resolving power of scanning electron microscopes, which is limited by the thickness of the metal coating, is only about 10 nm, much less than that of transmission instruments.

SUMMARY Microscopy and Cell Architecture

• Various microscopic techniques generate different views of the cell and have different resolutions. The limit of resolution of a light microscope is about 0.2 μm; of a transmission electron microscope, about 0.1 nm; and of a scanning electron microscope, about 10 nm.

• Standard (bright-field) light microscopy is best for stained or colored cells or tissue sections.

• Fluorescence microscopy allows specific proteins and organelles to be detected in fixed cells stained with a fluorescent dye or fluorescent-labeled antibodies (immunofluorescence microscopy). The movements of microinjected or expressed recombinant fluorescent proteins also can be followed in living cells.

• By use of dyes whose fluorescence is proportional to the concentration of Ca^{2+} or H^+ ions, fluorescence microscopy can measure the local concentration of Ca^{2+} ions and intracellular pH in living cells.

• Confocal imaging, which allows the observer to view fluorescent molecules in a single plane of a specimen, permits optical sectioning of the sample and produces very sharp images.

• Phase-contrast and Nomarski optics enable scientists to view the details of live, unstained cells and to monitor cell movement.

• Specimens for electron microscopy generally must be fixed, sectioned, and dehydrated, and then stained with electron-dense heavy metals.

• Surface details of particles such as viruses and collagen fibers can be revealed by electron microscopy of metal-shadowed specimens (see Figure 5-18).

• Unfixed, unstained specimens can be viewed in the electron microscope if they are frozen in hydrated form, a technique called *cryoelectron microscopy.*

• The scanning electron microscope can be used to view unsectioned cells or tissues; it produces images that appear to be three-dimensional.

5.2 Purification of Cells and Their Parts

Most animal and plant tissues contain a mixture of cell types. However, an investigator often wishes to study a pure population of one type of cell. In some cases, cells differ in some physical property that allows different cell types to be separated. White blood cells (leukocytes) and red blood cells (erythrocytes), for instance, have very different densities because erythrocytes have no nucleus; thus these cells can be separated on the basis of density. Since most cell types cannot be differentiated so easily, other cell-separation techniques have had to be developed. Similarly, it is essential to isolate quantities of each of the major subcellular organelles to study their structures and metabolic functions in detail.

Flow Cytometry Separates Different Cell Types

A *flow cytometer* can identify different cells by measuring the light they scatter, or the fluorescence they emit, as they flow through a laser beam; thus it can sort out cells of a particular type from a mixture. Indeed, a *fluorescence-activated cell sorter (FACS)*, an instrument based on flow cytometry, can select one cell from thousands of other cells (Figure 5-21). For example, if an antibody specific to a certain cell-surface molecule is linked to a fluorescent dye, any cell bearing this molecule will bind the antibody and will then be separated from other cells when it fluoresces in the FACS. Once sorted from the other cells, the selected cell can be grown in culture.

 This procedure is commonly used to purify the different types of white blood cells, each of which bears on its surface one or more distinctive proteins

and thus will bind monoclonal antibodies specific for that protein. Such FACS separations are more difficult to conduct on cultured cells or cells from animal tissues, which interact with adjacent cells and are surrounded by an extracellular matrix. Samples must be treated with proteases to degrade the extracellular-matrix proteins and cell-surface proteins that attach cells in tissues to one another; these proteases usually also degrade the distinctive cell-surface "marker" proteins that distinguish one cell type from another.

Other uses of flow cytometry include the measurement of a cell's DNA and RNA content and the determination of its general shape and size. The FACS can make simultaneous measurements of the size of a cell (from the amount of scattered light) and the amount of DNA it contains (from the amount of fluorescence from a DNA-binding dye).

Disruption of Cells Releases Their Organelles and Other Contents

The initial step in purifying subcellular structures is to rupture the plasma membrane and the cell wall, if present. First, the cells are suspended in a solution of appropriate pH and salt content, usually isotonic sucrose (0.25 M) or a combination of salts similar in composition to those in the cell's interior. Many cells can then be broken by stirring the cell suspension in a high-speed blender or by exposing it to high-frequency sound *(sonication)*. Plasma membranes can also be

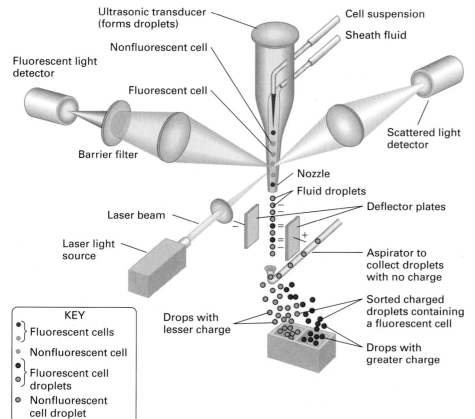

◄ **FIGURE 5-21 Fluorescence-activated cell sorter (FACS).** A concentrated suspension of cells is allowed to react with a fluorescent antibody or a dye that binds to a particle or molecule such as DNA. The suspension is then mixed with a buffer (the sheath fluid), the cells are passed single-file through a laser light beam, and the fluorescent light emitted by each cell is measured. The light scattered by each cell can be measured at the same time as the fluorescence; from measurements of the scattered light, the size and shape of the cell can be determined. The suspension is then forced through a nozzle, which forms tiny droplets containing at most a single cell. At the time of formation, each droplet is given an electric charge proportional to the amount of fluorescence of its cell. Droplets with no charge and with different electric charges (due to different amounts of bound dye) are each separated by an electric field and collected. It takes only milliseconds to sort each droplet, so up to 10 million cells per hour can pass through the machine. In this way, cells that have desired properties can be separated and then grown. [Adapted from D. R. Parks and L. A. Herzenberg, 1982, *Meth. Cell Biol.* **26**:283.]

(a) Isotonic medium

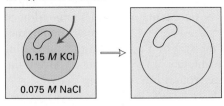

(b) Hypotonic medium

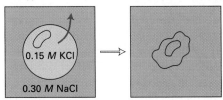

(c) Hypertonic medium

▲ **FIGURE 5-22 Response of animal cells to the osmotic strength of the surrounding medium.** Sodium, potassium, and chloride ions do not move freely across the cell membrane, but water does. (a) When the medium is isotonic, there is no net flux of water into or out of the cell. (b) When the medium is hypotonic, water flows into the cell (red arrow) until the ion concentration inside and outside the cell is the same. Here, the initial cytosolic ion concentration is twice the extracellular ion concentration, so the cell tends to swell to twice its original volume, at which point the internal and external ion concentrations are the same. (c) When the medium is hypertonic, water flows out of the cell until the ion concentration inside and outside the cell is the same. Here, the initial cytosolic ion concentration is half the extracellular ion concentration, so the cell is reduced to about half its original volume.

sheared by special pressurized tissue homogenizers in which the cells are forced through a very narrow space between the plunger and the vessel wall. Generally, the cell solution is kept at 0 °C to best preserve enzymes and other constituents after their release from the stabilizing forces of the cell.

Because the plasma membrane is highly permeable to water but poorly permeable to the salts and other small molecules (solutes) within cells, *osmotic flow* can be enlisted to help rupture cells. Recall that water flows across a semipermeable membrane, such as the plasma membrane, from a solution of high water (low solute) concentration to one of low water (high solute) concentration until the water concentration on both sides is equal. Consequently, when cells are placed in a **hypotonic** solution (i.e., one with a lower salt concentration than

that of the cell interior), water flows into the cells (Figure 5-22). This osmotic flow causes the cells to swell and then more easily rupture. Conversely, in a **hypertonic** solution (i.e., one with a higher salt concentration than that of the cell interior), water flows out of cells, causing them to shrink. When cells are placed in an **isotonic** solution (i.e., one with a salt concentration equal to that of the cell interior), there is no net movement of water in or out of cells. For this reason, an isotonic solution is best for preserving normal cell structure.

Disrupting the cell produces a mix of suspended cellular components, the *homogenate*, from which the desired organelles can be retrieved. Because rat liver contains an abundance of a single cell type, this tissue has been used in many classic studies of cell organelles. However, the same isolation principles apply to virtually all cells and tissues, and modifications of these cell-fractionation techniques can be used to separate and purify any desired components.

Different Organelles Can Be Separated by Centrifugation

In Chapter 3 we discussed the principles of centrifugation and the uses of centrifugation techniques for separating proteins and nucleic acids. Similar approaches are used for separating and purifying the various organelles, which differ in both size and density.

Most fractionation procedures begin with *differential centrifugation* at increasingly higher speeds (Figure 5-23), also called *differential-velocity centrifugation*. The different sedimentation rates of various cellular components make it possible to separate them partially by centrifugation. Nuclei and viral particles can sometimes be purified completely by such a procedure. After centrifugation at each speed for an appropriate time, the supernatant is poured off and centrifuged at higher speed. Each pelleted fraction can be resuspended and further separated by equilibrium density-gradient centrifugation (discussed next).

Differential centrifugation does not yield totally pure organelle fractions. One method for further purifying fractions is *equilibrium density-gradient centrifugation*, which separates cellular components according to their density. The impure organelle fraction is layered on top of a solution that contains a gradient of a dense nonionic substance, such as sucrose or glycerol. The tube is centrifuged at a high speed (about 40,000 rpm) for several hours, allowing each particle to migrate to an equilibrium position where the density of the surrounding liquid is equal to the density of the particle. In typical preparations from animal cells, the rough endoplasmic reticulum (density = 1.20 g/cm³) separates well from the Golgi vesicles (density = 1.14 g/cm³) and from the plasma membrane (density = 1.12 g/cm³). (The higher density of the rough endoplasmic reticulum is due largely to the ribosomes bound to it.) This method also works well for resolving lysosomes, mitochondria, and peroxisomes in the initial mixed fraction obtained by differential centrifugation (Figure 5-24).

Since each organelle has unique morphological features, the purity of organelle preparations can be assessed by

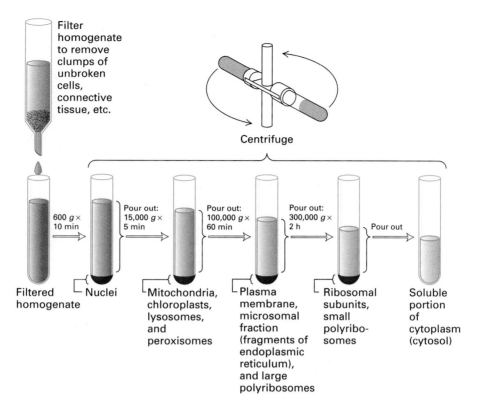

▲ **FIGURE 5-23 Cell fractionation by differential centrifugation.** Generally, the cellular homogenate is first filtered or centrifuged at relatively low speeds to remove unbroken cells. Then centrifugation of the homogenate at a slightly faster speed or for a longer duration will selectively pellet the nucleus—the largest organelle (usually 5–10 μm in diameter). A centrifugal force of 600 g (600 times the force of gravity) is necessary to sediment nuclei; this is generated by a typical centrifuge rotor operating at 500 revolutions per minute (rpm). The undeposited material (the supernatant) is next centrifuged at a higher speed (15,000 $g \times$ 5 min), which deposits the mitochondria, chloroplasts, lysosomes,

and peroxisomes. A subsequent centrifugation in the ultracentrifuge (100,000 $g \times$ 60 min) results in deposition of the plasma membrane, fragments of the endoplasmic reticulum, and large polyribosomes. A force of 100,000 g requires about 50,000 rpm in an ultracentrifuge; at this speed, the rotor chamber is kept in a high vacuum to reduce heating due to friction between air and the spinning rotor. The recovery of ribosomal subunits, small polyribosomes, and particles such as complexes of enzymes requires additional centrifugation at still higher speeds. Only the cytosol—the soluble aqueous portion of the cytoplasm—remains undeposited after centrifugation at 300,000 g for 2 hours.

▶ **FIGURE 5-24 Separation of organelles from rat liver by equilibrium density-gradient centrifugation.** For example, the material deposited as a pellet by centrifugation at 15,000 g (see Figure 5-23) can be resuspended and layered on a density gradient composed of layers of increasingly more dense sucrose solutions in a centrifuge tube. Under centrifugation, each organelle migrates to its appropriate equilibrium density and remains there. To obtain a good separation of lysosomes from mitochondria, the liver is perfused with a solution containing a small amount of detergent before the tissue is disrupted. The detergent is taken into the cells by endocytosis and transferred to the lysosomes, making them less dense than they would normally be, thereby affording a "clean" separation from the mitochondria.

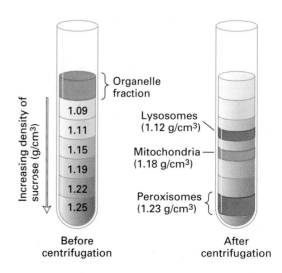

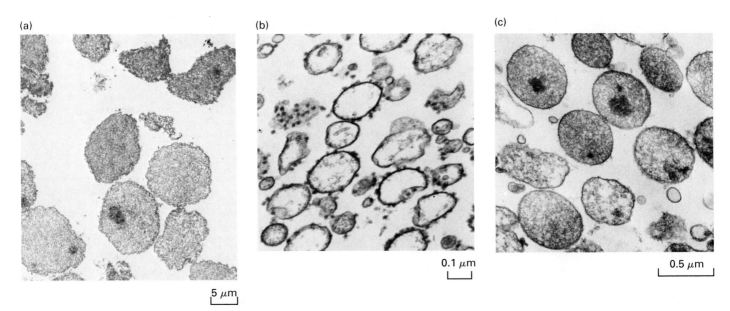

(a)

(b)

(c)

0.1 μm

0.5 μm

5 μm

▲ **FIGURE 5-25 Electron micrographs of purified rat liver organelles:** Seen here are (a) nuclei, (b) rough endoplasmic reticulum, sheared into smaller vesicles termed *microsomes,* and (c) peroxisomes. Note that the rough endoplasmic reticulum is studded with ribosomes on its outer, or cytosolic surface, and that the peroxisomes are filled with an almost crystalline core that consists of catalase and other proteins. [Parts (a) and (b) courtesy of S. Fleischer and B. Fleischer; part (c) courtesy of P. Lazarow.]

(a)

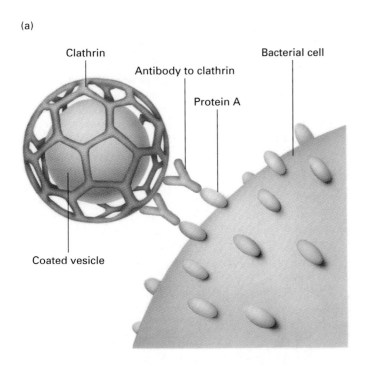

Clathrin

Antibody to clathrin

Protein A

Bacterial cell

Coated vesicle

(b)

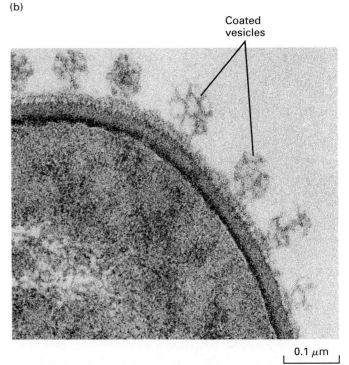

Coated vesicles

0.1 μm

▲ **FIGURE 5-26 Immunological purification of clathrin-coated vesicles.** (a) A suspension of membranes from rat liver is incubated with an antibody specific for clathrin, a protein that coats the outer surface of certain cytoplasmic vesicles. To this mixture is added a suspension of *S. aureus* bacteria whose surface membrane contains protein A. Specific binding of protein A to the constant region of antibodies bound to the clathrin-coated vesicles links the vesicles to the bacterial cells. The vesicle-bacteria complexes then are recovered by low-speed centrifugation. (b) A thin-section electron micrograph of clathrin-coated vesicles bound to these bacteria. [See E. Merisko, M. Farquahr, and G. Palade, 1982, *J. Cell Biol.* **93**:846; part (b) courtesy of G. Palade.]

examination in an electron microscope (Figure 5-25). Alternatively, organelle-specific marker molecules can be quantified. For example, the protein cytochrome *c* is present only in mitochondria, so the presence of this protein in a fraction of lysosomes would indicate its contamination by mitochondria. Similarly, catalase is present only in peroxisomes; acid phosphatase, only in lysosomes; and ribosomes, only in the rough endoplasmic reticulum or the cytosol.

Organelle-Specific Antibodies Are Useful in Preparing Highly Purified Organelles

Cell fractions often contain more than one type of organelle even after differential and equilibrium density-gradient centrifugation. Such fractions can be further purified by immunological techniques, using monoclonal antibodies for various organelle-specific membrane proteins. One example is the purification of a particular class of cellular vesicles whose outer surface is coated with the protein *clathrin* (Figure 5-26). An antibody to clathrin, bound to a bacterial carrier, can selectively bind these vesicles in a crude preparation of membranes, and the whole antibody complex can then be isolated by low-speed centrifugation. A recently developed technique uses tiny metallic beads coated with specific antibodies. Organelles that bind to the antibodies, and thus are linked to the metallic beads, are recovered from the preparation by adhesion to a small magnet on the side of the test tube.

All cells contain a dozen or more different types of small membrane-limited vesicles of about the same size (50–100 nm in diameter) and density. Because of their similar size and density, these vesicles are difficult to separate from one another by centrifugation techniques. Immunological techniques are particularly useful for purifying specific classes of such vesicles. Fat and muscle cells, for instance, contain a particular glucose transporter (GLUT4) that is localized to the membrane of a specific kind of vesicle. When insulin is added to the cells, these vesicles fuse with the cell-surface membrane, a process critical to maintaining the appropriate concentration of sugar in the blood (see Figure 5-7). These vesicles can be purified using an antibody that binds to a segment of the GLUT4 protein that faces the cytosol.

SUMMARY Purification of Cells and Their Parts

- Flow cytometry can identify different cells based on the light they scatter or the fluorescence they emit. The fluorescence-activated cell sorter (FACS) is particularly useful in separating different types of white blood cells (see Figure 5-21). A cell's DNA and RNA content also can be measured with a FACS.

- Disruption of cells by vigorous homogenization, sonication, or other techniques releases their organelles. When placed in a hypotonic solution, cells swell, thereby weakening the plasma membrane and making it easier to rupture.

- Sequential differential-velocity centrifugation of a cell homogenate yields partially purified organelles that differ in mass (see Figure 5-23).

- Equilibrium density-gradient centrifugation, which separates cellular components according to their density, can further purify cell fractions obtained by differential centrifugation.

- Because the membrane surrounding each type of organelle contains organelle-specific proteins, immunological techniques are very useful in purifying organelles and vesicles, particularly those that have a similar size and density.

5.3 Biomembranes: Structural Organization and Basic Functions

Although all biomembranes have the same basic phospholipid bilayer structure and certain common functions, each type of cellular membrane also has certain distinctive activities determined largely by the unique set of proteins associated with that membrane. The two basic categories of membrane proteins were introduced in Chapter 3: integral proteins, all or part of which penetrate or span the phospholipid bilayer, and peripheral proteins, which do not interact with the hydrophobic core of the bilayer (see Figure 3-32). In this section, we first discuss the basic principles that govern the organization of phospholipids and integral proteins in all biological membranes and then outline the functions of the plasma membrane in prokaryotes and eukaryotes.

Phospholipids Are the Main Lipid Constituents of Most Biomembranes

The most abundant lipid components in most membranes are **phospholipids**, which are amphipathic molecules (i.e., they have a hydrophilic and a hydrophobic part). In *phosphoglycerides*, a principal class of phospholipids, fatty acyl side chains are esterified to two of the three hydroxyl groups in glycerol, and the third hydroxyl group is esterified to phosphate. The phosphate group is also esterified to a hydroxyl group on another hydrophilic compound, such as choline in phosphatidylcholine (Figure 5-27a). Instead of choline, alcohols such as ethanolamine, serine, and the sugar derivative inositol are linked to the phosphate in other phosphoglycerides (Figure 5-28). The negative charge on the phosphate as well as the charged groups or hydroxyl groups on the alcohol esterified to it interact strongly with water. Both of the fatty acyl side chains in a phosphoglyceride may be saturated or unsaturated, or one chain may be saturated and the other unsaturated.

Sphingomyelin, a phospholipid that lacks a glycerol backbone, is found mainly in plasma membranes (Figure 5-27b). Instead of a glycerol backbone, it contains *sphingosine*, an amino alcohol with a long unsaturated hydrocarbon chain.

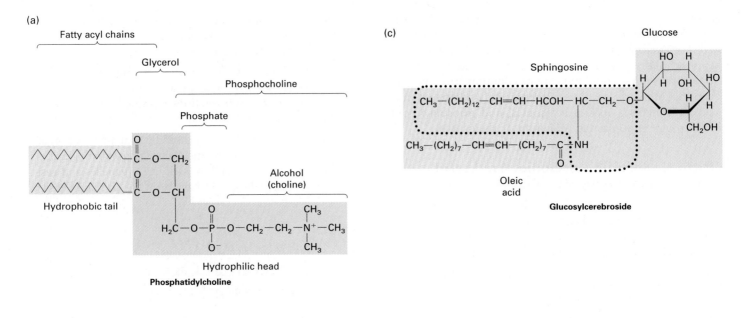

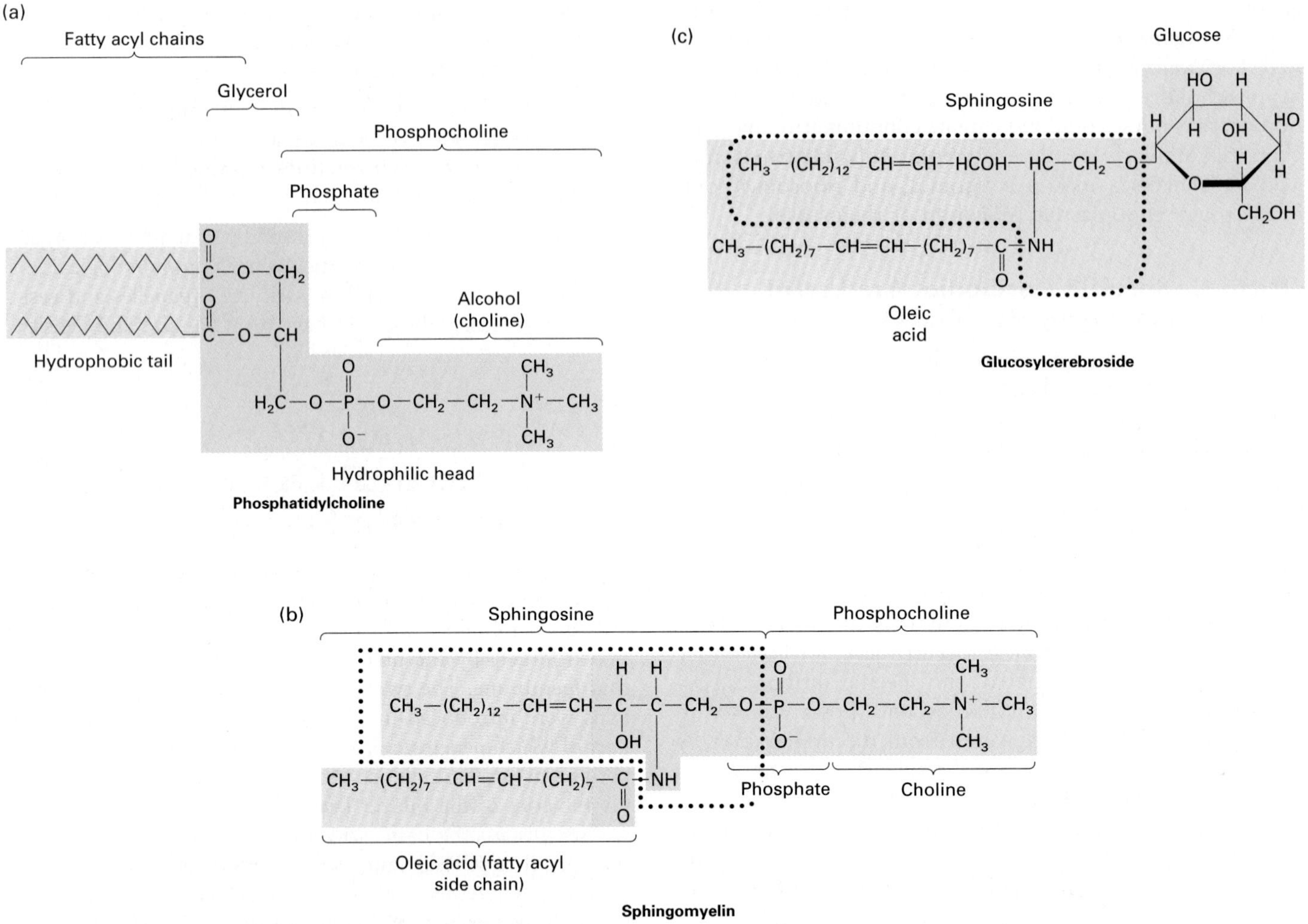

▲ **FIGURE 5-27 Structures of two types of phospholipids and a glycolipid.** The hydrophobic portions of all molecules are shown in yellow; the hydrophilic, in green. (a) Phosphatidylcholine is a typical phosphoglyceride. The fatty acyl side chains can be saturated, or they can contain one or more double bonds. Common alcohols found in these phospholipids are shown in Figure 5-28. (b) Sphingomyelins are a group of phospholipids that lack a glycerol backbone; a sphingomyelin may contain a different fatty acyl side chain than oleic acid (shown here). Linkage of sphingosine (outlined by black dots) to a fatty acid via an amide bond forms a ceramide. (c) Glucosylcerebroside, one of the simplest glycolipids, consists of the ceramide formed from sphingosine and oleic acid linked to a single glucose residue. This glycolipid is abundant in the myelin sheath.

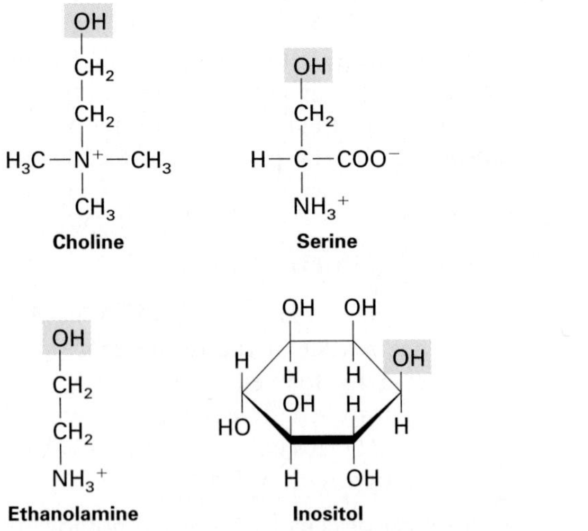

In sphingomyelin, the terminal hydroxyl group of sphingosine is esterified to phosphocholine, so its hydrophilic head is similar to that of phosphatidylcholine.

Cholesterol and its derivatives constitute another important class of membrane lipids, the **steroids.** The basic structure of steroids is the four-ring hydrocarbon shown in Figure 5-29a. Cholesterol, the major steroidal constituent of animal tissues, has a hydroxyl substituent on one ring (Figure 5-29b). Although cholesterol is almost entirely hydrocarbon in composition, it is amphipathic because its hydroxyl

▲ **FIGURE 5-28 Common alcohols found in phosphoglycerides present in cellular membranes.** The indicated —OH groups are linked by a phosphate group to the glycerol backbone, which also is esterified to two fatty acyl chains (see Figure 5-27a).

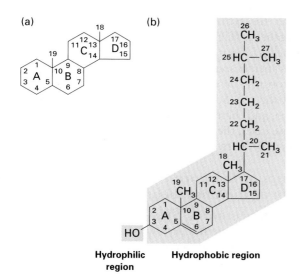

◀ **FIGURE 5-29 (a) The general structure of a steroid.** All steroids contain the same four hydrocarbon rings, conventionally labeled A, B, C, and D, with the carbons numbered as shown. (b) The structure of cholesterol. The major portion of the molecule is hydrophobic (yellow), but the hydroxyl group is hydrophilic (green).

group can interact with water. Cholesterol is especially abundant in the plasma membrane of mammalian cells but is absent from most prokaryotic cells. As much as 30 to 50 percent of the lipids in plant plasma membranes consists of cholesterol and certain steroids unique to plants.

Carbohydrates are found in many membranes, covalently bound either to proteins as constituents of **glycoproteins** or to lipids as constituents of **glycolipids** (see Figure 3-32). Bound carbohydrates increase the hydrophilic character of lipids and proteins and help to stabilize the conformations of many membrane proteins. The simplest glycolipid, *glucosylcerebroside*, contains a single glucose unit attached to a ceramide (Figure 5-27c).

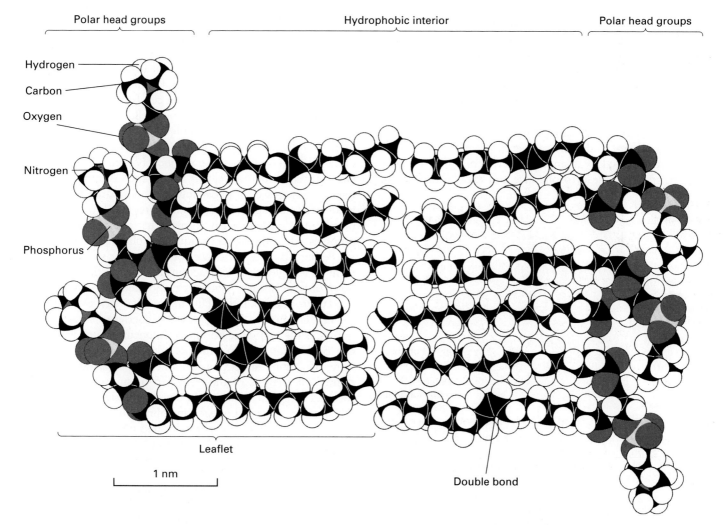

▲ **FIGURE 5-30 A space-filling model of a typical phospholipid bilayer.** The hydrophobic interior is generated by the fatty acyl side chains. Some of these chains have bends caused by double bonds. The different polar head groups all lie on the outer, aqueous surface of the bilayer. [From L. Stryer, 1995, *Biochemistry*, 4th ed., W. H. Freeman and Company, p. 270; courtesy of L. Stryer.]

Every Cellular Membrane Forms a Closed Compartment and Has a Cytosolic and an Exoplasmic Face

Phospholipids of the composition present in cells spontaneously form symmetric sheetlike phospholipid bilayers, which are two molecules thick. The hydrocarbon side chains in each leaflet form a hydrophobic core that is 3–4 nm thick in most biomembranes (Figure 5-30.) The various phospholipids differ in the charge carried by the polar head groups at neutral pH: some phosphoglycerides (e.g., phosphatidylcholine and phosphatidylethanolamine) have no net electric charge; others (e.g., phosphatidylglycerol and phosphatidylserine) have a net negative charge. Nonetheless, the polar head groups in all phospholipids can pack together into the characteristic bilayer structure. Sphingomyelins are similar in shape to phosphoglycerides and can form mixed bilayers with them.

Perhaps the most important lesson gleaned from the study of pure phospholipid bilayer membranes is that they spontaneously seal to form closed structures that separate two aqueous compartments (see Figure 2-20). Were a phospholipid bilayer to form a sheet with ends in which the hydrophobic interior were in contact with water, it would be unstable; thus a spherical structure with no ends is the most stable state of a phospholipid bilayer.

Similarly, all cellular membranes are closed structures, surrounding the cell itself or individual compartments. Cellular membranes thus have an *internal face* (the side oriented toward the interior of the compartment) and an *external face* (the side presented to the environment). Because most organelles are surrounded by a single bilayer membrane, it is also useful to speak of the **cytosolic face** and **exoplasmic face** of

the membrane, the cytosol being the part of the cytoplasm outside of organelles (Figure 5-31). Thus the exoplasmic face of such organelles faces inward. Similarly, the exoplasmic face of the plasma membrane is directed away from the cytosol, in this case toward the extracellular space, and defines the outer limit of the cell. Some organelles, such as the nucleus, mitochondrion, and chloroplast, are surrounded by two membranes; in these cases, the exoplasmic surface faces the lumen, or space, between the two membranes.

Several Types of Evidence Point to the Universality of the Phospholipid Bilayer

A typical cell contains myriad types of membranes, each in turn bearing unique properties bestowed by its particular mix of lipids and proteins. When samples of plasma, nuclear, and mitochondrial membranes are prepared, using the cell-fractionation techniques we have already described, these preparations are often contaminated with the membranes of many other organelles. (The plasma membrane of human erythrocytes, however, can be isolated in near purity because these cells contain no internal membranes.) And all cellular membranes, regardless of their source, possess enormously varied protein-to-lipid ratios. The inner mitochondrial membrane, for example, is 76 percent protein; the myelin membrane, only 18 percent. The high phospholipid content of myelin allows it to electrically insulate the nerve cell from its environment.

Given the variable composition of cellular membranes, how certain are we that the phospholipid bilayer structure is common to all biomembranes? One piece of evidence is that many of the physical properties of pure phospholipid bilayers are similar to those of natural cellular membranes. Another is that either a single species of phospholipid, or a mixture of phospholipids with a composition approximating that found in natural membranes, spontaneously forms either planar bilayers or **liposomes** when dispersed in aqueous solutions (Figure 5-32).

Perhaps the best evidence for the bilayer structure comes from low-angle x-ray diffraction analysis of the multimembrane **myelin sheath,** which is elaborated by Schwann cells and covers and insulates many mammalian nerve cells (Figure 5-33a). The myelin sheath, which is a series of stacked membranes, is the major membrane component of such nerves and can be separated from other cellular membranes in a pure state, permitting direct physical and chemical analyses. X-ray diffraction analysis of these stacked plasma membranes has revealed a regular variation in density that is consistent with a bilayer organization of each membrane unit (Figure 5-33b). In this organization, protein is located mainly on either side of the membrane, which has hydrophilic external faces and a central region of almost pure low-density hydrocarbon.

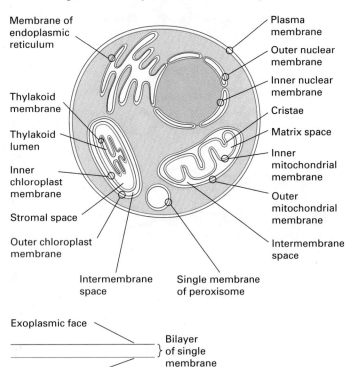

Membrane of endoplasmic reticulum

Thylakoid membrane

Thylakoid lumen

Inner chloroplast membrane

Stromal space

Outer chloroplast membrane

Intermembrane space

Single membrane of peroxisome

Plasma membrane

Outer nuclear membrane

Inner nuclear membrane

Cristae

Matrix space

Inner mitochondrial membrane

Outer mitochondrial membrane

Intermembrane space

Exoplasmic face

Cytosolic face

Bilayer of single membrane

◄ **FIGURE 5-31 Faces of cellular membranes.** For organelles enclosed in two phospholipid membranes (e.g., the nucleus, chloroplast, mitochondrion), the exoplasmic faces (red) border the space between the inner and outer membranes. Chloroplasts also contain a stack of internal thylakoid membranes; the exoplasmic face of these membranes line the thylakoid lumen.

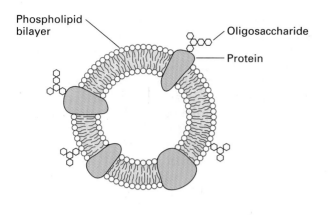

Phospholipid bilayer

Oligosaccharide

Protein

Treat with organic solvent

Proteins and oligosaccharides form insoluble residue that is removed

Phospholipids in solution

Evaporate solvent

Disperse phospholipids in water

Dissolve phospholipids in solvent and apply to small hole in partition

Planar bilayer

Plastic partition

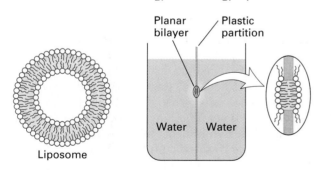

Water   Water

Liposome

▲ **FIGURE 5-32 Experimental formation of pure phospholipid bilayers.** *(Top)* A preparation of biological membranes is treated with an organic solvent, such as a mixture of chloroform and methanol (3:1), which selectively solubilizes the phospholipids and cholesterol. Proteins and carbohydrates remain in an insoluble residue. The solvent is removed by evaporation. *(Bottom left)* If the lipids are mechanically dispersed in water, they spontaneously form a liposome, shown in cross-section, with an internal aqueous compartment. *(Bottom right)* A planar bilayer, also shown in cross-section, can form over a small hole in a partition separating two aqueous phases; such bilayers are often termed "black lipid membranes" because of their appearance.

▼ **FIGURE 5-33 Low-angle x-ray diffraction analysis of myelin membranes.** This technique measures the density of matter and can be used to determine the distribution of lipid and protein in biomembranes. (a) During development of the nervous system, a large Schwann cell envelops the axon of a neuron. The continuous growth of the Schwann cell membrane into its own cytoplasm, together with rotation of the nerve axon, results in a laminated spiral of double plasma membranes around the axon. Mature myelin, a stack of plasma membranes of the Schwann cell, is relatively rich in phospholipids. (b) The profile of electron density — and thus of matter — obtained by x-ray diffraction studies on fresh nerve, and the relation of this profile to the protein and lipid components of the myelin membranes. [Adapted from W. T. Norton, 1981, in G. J. Siegel et al., eds., *Basic Neurochemistry*, 3d ed., Little, Brown, p. 68.]

(a)

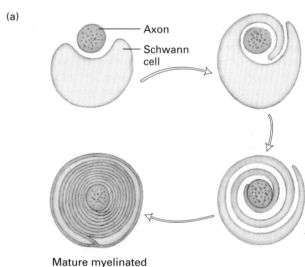

Axon

Schwann cell

Mature myelinated nerve cell

(b)

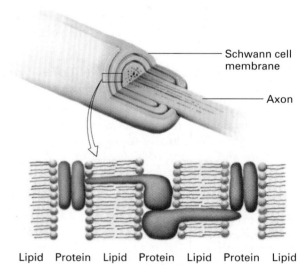

Schwann cell membrane

Axon

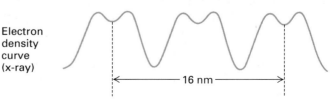

Lipid   Protein   Lipid   Protein   Lipid   Protein   Lipid

Electron density curve (x-ray)

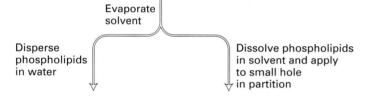

|←——— 16 nm ———→|

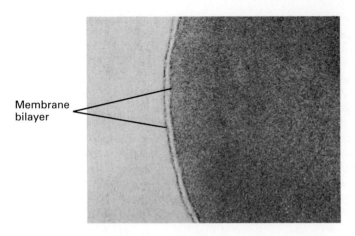

Membrane bilayer

▲ **FIGURE 5-34 Electron micrograph of a thin section of an erythrocyte membrane stained with osmium tetroxide, which binds preferentially to polar groups.** The "railroad track" appearance of the membrane indicates the presence of two polar layers, consistent with the bilayer structure for phospholipid membranes. [Courtesy of J. D. Robertson.]

Although some polypeptide segments pass through the lipid bilayer, these make up less than 10 percent of the inner mass of the membrane and are not detected in this type of analysis.

Electron microscopy of thin membrane sections stained with osmium tetroxide, which binds strongly to the polar head groups of phospholipids, provides the most direct evidence for the universality of the bilayer structure. A cross section of all single membranes stained with osmium tetroxide looks like a railroad track: two thin dark lines (the stain–head group complexes) with a uniform light space of about 2 nm (the hydrophobic tails) between them (Figure 5-34). Although some osmium tetroxide may bind to double bonds in the hydrophobic fatty acyl chains, most of it binds to the polar head groups.

All Integral Proteins and Glycolipids Bind Asymmetrically to the Lipid Bilayer

The spaces inside and outside the closed compartments formed by cellular membranes usually have very different compositions. Such *asymmetry* is an essential aspect of the structure and function of biological membranes, and is reflected in the asymmetric structure of all integral membrane proteins. That is, each type of integral membrane protein has a single, specific orientation with respect to the cytosolic and exoplasmic faces of a cellular membrane, and all molecules of any particular integral membrane protein share this orientation (see Figure 3-33). This absolute asymmetry in protein orientation confers different properties on the two membrane faces. Proteins have never been observed to flip-flop across a membrane; such movement, involving a transient movement of hydrophilic amino acid and sugar residues through the hydrophobic interior of the membrane, would be energetically unfavorable. Accordingly, the asymmetry of a membrane protein, which is established during

its biosynthesis and insertion into a membrane, is maintained throughout the protein's lifetime.

Both glycoproteins and glycolipids are especially abundant in the plasma membrane of eukaryotic cells but are absent from the inner mitochondrial membrane, the chloroplast lamellae, and several other intracellular membranes. Almost invariably, attached carbohydrates are localized to the exoplasmic membrane face. Glycolipids are always found in the exoplasmic leaflet of membranes and are situated mainly, but not exclusively, on the surface membrane of cells. As with glycoproteins, their polar carbohydrate chains face outward toward the environment and away from the cell.

The Phospholipid Composition Differs in Two Membrane Leaflets

Most kinds of phospholipid, as well as cholesterol, are generally present in both membrane leaflets, although they are often more abundant in one or the other. For instance, in plasma membranes from human erythrocytes and certain canine kidney cells grown in culture, almost all the sphingomyelin and phosphatidylcholine, both of which have a positively charged head group (see Figure 5-27a, b), are found in the exoplasmic leaflet. In contrast, lipids with neutral or negative polar head groups (e.g., phosphatidylethanolamine, phosphatidylserine, and phosphatidylinositol) are preferentially located in the cytosolic leaflet. Phosphorylated forms of phosphatidylinositol are cleaved as a result of cell stimulation by certain hormones, generating in the cytosol soluble forms of the "head groups" that affect many aspects of cellular metabolism (Chapter 20).

The relative abundance of a particular phospholipid in the two leaflets of a plasma membrane can be determined based on its susceptibility to hydrolysis by *phospholipases*, enzymes that cleave the phosphoester bonds that connect the phospholipid head groups (see Figure 3-37). Phospholipids in the cytosolic leaflet are resistant to hydrolysis by phospholipases added to the external medium, because the enzymes cannot penetrate to the cytosolic face of the plasma membrane. It is not clear how these differences in lipid composition of the two leaflets arise. One possibility is that certain lipids bind to specific protein domains that occur preferentially in one membrane leaflet.

Most Lipids and Integral Proteins Are Laterally Mobile in Biomembranes

In both pure phospholipid bilayers and natural membranes, thermal motion permits phospholipid and glycolipid molecules to rotate freely around their long axes and to diffuse laterally within the membrane leaflet. Because such movements are lateral or rotational, the fatty acyl chains remain in the hydrophobic interior of the membrane. In both natural and artificial membranes, a typical lipid molecule exchanges places with its neighbors in a leaflet about 10^7 times per second and diffuses several micrometers per second at 37 °C. At this rate, a lipid could diffuse the length of a typical bacterial cell ($\approx$1 μm) in only 1 second and the length of an animal cell in about 20 seconds.

In pure phospholipid bilayers, phospholipids do not migrate, or flip-flop, from one leaflet of the membrane to the other. In some natural membranes, however, they occasionally do so, catalyzed by certain membrane proteins called *flippases* (Chapter 15). Energetically, such movements are extremely unfavorable, because the polar head of a phospholipid must be transported through the hydrophobic interior of the membrane.

Various experiments have shown that many integral membrane proteins, like phospholipids, float quite freely within the plane of a natural membrane. In one such study, outlined in Figure 5-35, two different cells (e.g., mouse and human fibroblasts) are fused and the movement of their distinct surface proteins is then monitored at various times after incubation at 37 °C. Such experiments suggest that many integral proteins are free to diffuse in a sea of lipid in the two-dimensional space of the membrane. According to this concept, known as the *fluid mosaic model,* the membrane is viewed as a two-dimensional mosaic of laterally mobile phospholipid and protein molecules (see Figure 3-32). As discussed in Chapter 3, some integral membrane proteins consist of two or more noncovalently linked subunits; such multimeric membrane proteins float as a unit in the lipid.

The lateral movements of surface proteins and lipids can be quantified by a technique called *fluorescence recovery after photobleaching* (FRAP). With this method, described in Figure 5-36, the rate at which surface protein or lipid molecules move—the diffusion coefficient—can be determined, as well as the proportion of the molecules that are laterally mobile. FRAP studies with fluorescent-labeled phospholipids have shown that in fibroblast plasma membranes, all the phospholipids are freely mobile over distances of about 0.5 μm, but most cannot diffuse over much longer distances. These findings suggest that protein-rich regions of the plasma membrane, about 1 μm in diameter, separate lipid-rich regions containing the bulk of the membrane phospholipid. Phospholipids are free to diffuse within such a region but not from one lipid-rich region to an adjacent one. Furthermore, the rate of lateral diffusion of lipids in the plasma membrane is nearly an order of magnitude slower than in pure phospholipid bilayers: diffusion constants of 10^{-8} cm^2/s and 10^{-7} cm^2/s are characteristic of the plasma membrane and a lipid bilayer, respectively. This difference suggests that lipids may be tightly but not irreversibly bound to certain integral proteins in some membranes.

Numerous experiments similar to those just discussed have shown that depending on the cell type, 30–90 percent of all integral proteins in the plasma membrane are freely mobile. Immobile proteins are permanently attached to the underlying cytoskeleton. The lateral diffusion rate of a mobile protein in an intact membrane is generally 10–30 times lower than that of the same protein embedded in synthetic liposomes. These findings suggest that the mobility of integral proteins in intact membranes is restricted by interactions with the rigid submembrane cytoskeleton. Clearly, such interactions would have to be broken and remade as mobile proteins diffuse laterally in the plasma membrane. Chapter 18 details the interactions of specific plasma membrane proteins with the underlying cytoskeleton, explaining how these affect cell shape and motility.

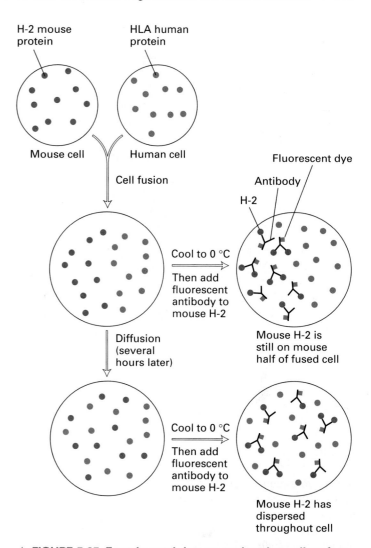

▲ **FIGURE 5-35 Experimental demonstration that cell-surface proteins are laterally mobile.** Human and mouse cells are fused as described in Chapter 6. Immediately after fusion, surface antigens of the two cell types remain localized in their respective halves of the fused cell; they can be detected by fluorescent antibodies (in this case specific for mouse H-2 protein). After several hours of incubation, the mouse and human surface proteins are evenly distributed throughout the membrane of the fused cell, demonstrating that most of the surface H-2 and HLA proteins were not rigidly held in place in the membranes of the original mouse and human cells. Protein movement is stopped by cooling the cells and treating them with a reagent that cross-links lysine residues (e.g., glutaraldehyde). [See L. D. Frye and M. Edidin, 1970, *J. Cell Sci.* **7**:319.]

(a)

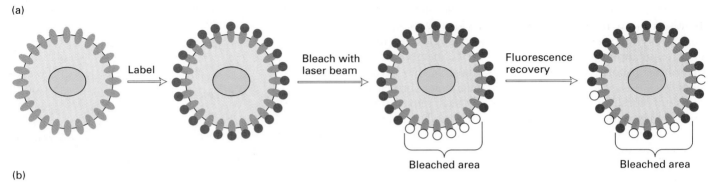

(b)

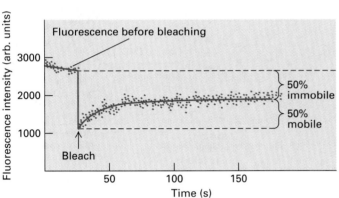

◀ **FIGURE 5-36 Fluorescence recovery after photobleaching (FRAP).** (a) Cells are labeled with a fluorescent reagent that binds to a specific surface protein or lipid, which is uniformly distributed on the surface. A laser light is then focused on a small area of the surface, irreversibly bleaching the bound reagent and thus reducing the fluorescence in the illuminated area. In time, the fluorescence of the bleached area increases as unbleached fluorescent surface molecules diffuse into it and bleached ones diffuse outward. The extent of recovery of fluorescence in the bleached patch is proportional to the fraction of labeled molecules that are mobile in the membrane. (b) Results of FRAP experiment with human hepatoma cells, treated with a fluorescent antibody specific for the asialoglycoprotein receptor, show that 50 percent of the fluorescence returned to the bleached area. Thus 50 percent of the asialoglycoprotein receptor molecules in the illuminated patch of membrane were mobile, and 50 percent were immobile. Because the rate of fluorescence recovery is proportional to the rate at which labeled molecules move into the bleached region, the diffusion coefficient of a protein or lipid in the membrane can be calculated from such data. [See Y. I. Henis et al., 1990, *J. Cell Biol.* **111**:1409.]

Fluidity of Membranes Depends on Temperature and Composition

One consequence of the packing of the fatty acyl chains within the center of a phospholipid bilayer is an abrupt change in its physical properties over a very narrow temperature range. For example, when a suspension of liposomes or a planar bilayer composed of a single type of phospholipid is heated, it passes from a highly ordered, gel-like state to a more mobile fluid state (Figure 5-37). During this *phase transition*, a relatively large amount of heat (thermal energy) is absorbed over a narrow temperature range; the midpoint of this range is the "melting temperature" of the bilayer.

In general, lipids with short or unsaturated fatty acyl chains undergo the phase transition at lower temperatures than do lipids with long or saturated chains. Compared with long chains, short chains have less surface area to form van der Waals interactions with one another. Since the gel state

is stabilized by these interactions, short-chain lipids melt at lower temperatures than long-chain lipids. Likewise, the kinks in unsaturated fatty acyl chains (see Figure 2-18) result in their forming less stable van der Waals interactions with other lipids than do saturated chains. As a result, unsaturated lipids maintain a more random, fluid state at lower temperatures than lipids with saturated fatty acyl chains.

The hydrophobic interior of natural membranes generally has a low viscosity and a fluidlike, rather than gel-like,

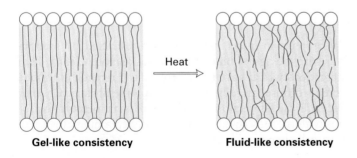

Gel-like consistency **Fluid-like consistency**

◀ **FIGURE 5-37 Alternative forms of the phospholipid bilayer.** Heat induces transition from a gel to a fluid over a temperature range of only a few degrees. The fluid phase is favored by the presence of short fatty acyl chains and by a double bond in the chains; thus these structural features reduce the melting temperature of bilayers.

consistency. Maintenance of this bilayer fluidity appears to be essential for normal cell growth and reproduction. All cell membranes contain a mixture of different fatty acyl chains and are fluid at the temperature at which the cell is grown. Animal and bacterial cells adapt to a decrease in growth temperature by increasing the proportion of unsaturated to saturated fatty acids in the membrane, which tends to maintain a fluid bilayer at the reduced temperature.

Membrane cholesterol is another major determinant of bilayer fluidity. Cholesterol is too hydrophobic to form a sheet structure on its own, but it is intercalated (inserted) among phospholipids. Its polar hydroxyl group is in contact with the aqueous solution near the polar head groups of the phospholipids; the steroid ring interacts with and tends to immobilize their fatty acyl chains. The net effect of cholesterol on membrane fluidity varies, depending on the lipid composition. Cholesterol restricts the random movement of the polar heads of the fatty acyl chains, which are closest to the outer surfaces of the leaflets, but it separates and disperses their tails, causing the inner regions of the bilayer to become slightly more fluid. At the high concentrations found in eukaryotic plasma membranes, cholesterol tends to make the membrane less fluid at growth temperatures near 37 °C. Below the temperature that causes a phase transition, cholesterol keeps the membrane in a fluid state by preventing the hydrocarbon fatty acyl chains of the membrane lipids from binding to one another, thereby offsetting the drastic reduction in fluidity that would otherwise occur at low temperatures.

Membrane Leaflets Can Be Separated and Each Face Viewed Individually

When a frozen tissue specimen is fractured by a sharp blow, the fracture line frequently runs through the hydrophobic interior of cell membranes, separating the two phospholipid leaflets. Integral membrane proteins generally remain associated with one or the other leaflet of membranes subjected to this *freeze-fracturing* technique (Figure 5-38). The fractured specimen then is placed in a vacuum and the surface ice is removed by sublimation, a technique called *deep etching* or *freeze etching*. After metal shadowing with platinum and carbon, the organic material is removed by acid, leaving a carbon-metal replica of the membrane leaflet (see Figure 5-18).

Electron microscopy of membrane samples prepared by these techniques reveals numerous protuberances, most of which are membrane proteins (Figure 5-39). In deep-etching studies, the cytoplasmic face of a membrane is customarily called the *P (protoplasmic) face* and the exoplasmic face is the *E face*. It is not unusual for most or all protuberances to be on one of the two surfaces and their mirror images, in the form of pits or holes, to be on the other. This may occur because the integral proteins are bound more tightly to the lipids in one leaflet than to those in the other.

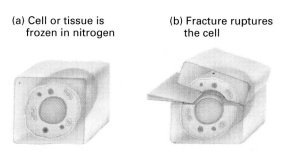

(a) Cell or tissue is frozen in nitrogen

(b) Fracture ruptures the cell

(c) Integral proteins remain embedded in fractured leaflets

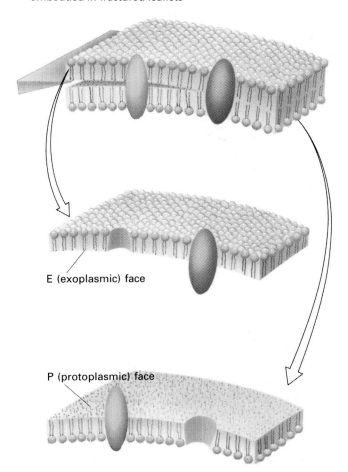

E (exoplasmic) face

P (protoplasmic) face

▶ **FIGURE 5-38 Freeze fracturing can separate the two phospholipid leaflets that form every cellular membrane.**
(a) A preparation of cells or tissues is quickly frozen in liquid nitrogen at −196 °C, which instantly immobilizes cell components. (b) The block of frozen cells is fractured with a sharp blow from a cold knife. The fracture plane is irregular, often between the leaflets of the plasma or an organelle membrane. (c) Membrane proteins and particles remain bound to one leaflet or the other, as illustrated in the expanded view of a fractured membrane.

(a)

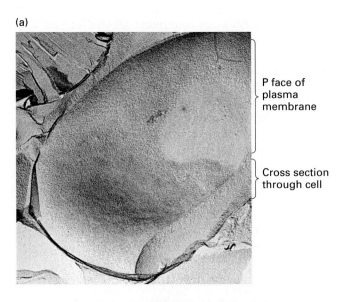

P face of plasma membrane

Cross section through cell

(b) Intramembrane particles (band 3)

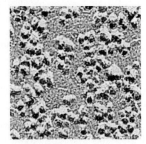

▲ **FIGURE 5-39 Micrographs of freeze-fractured, deep-etched erythrocyte plasma membrane.** (a) The P face of the plasma membrane and a cross section through the cell. The E face, or outer leaflet, has been fractured off, leaving just the P face, or inner leaflet. (b) The intramembrane particles at higher magnification. These particles are composed mainly of AE1, the major intramembrane protein. [Courtesy of D. Branton.]

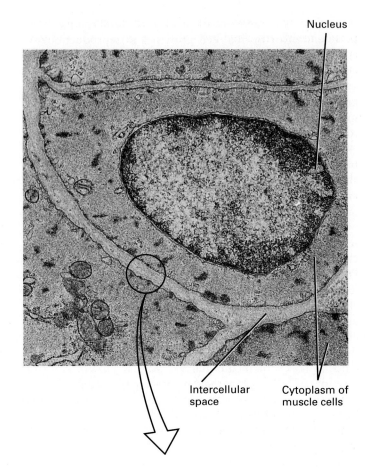

Nucleus

Intercellular space

Cytoplasm of muscle cells

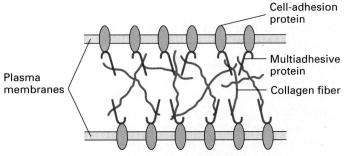

Cell-adhesion protein

Multiadhesive protein

Collagen fiber

Plasma membranes

▲ **FIGURE 5-40 Electron micrograph of smooth muscle in the wall of a small artery.** The muscle cells are separated by relatively wide intercellular spaces that contain a group of glycoproteins, called proteoglycans, and collagen, the most abundant fibrous component of the extracellular matrix. Interactions between cell-adhesive integral membrane proteins and components of the extracellular matrix allow the cells to adhere to one another (as illustrated at the bottom) and give this tissue its strength and resistance to shear forces. [Photograph from D. W. Fawcett, 1981, *The Cell*, 2d ed., Saunders/Photo Researchers, Inc.]

The Plasma Membrane Has Many Common Functions in All Cells

In all cells, the plasma membrane has several essential functions. These include transporting nutrients into and metabolic wastes out of the cell; preventing unwanted materials in the extracellular milieu from entering the cell; preventing loss of needed metabolites and maintaining the proper ionic composition, pH ($\approx$7.2), and osmotic pressure of the cytosol. To carry out these functions, the plasma membrane contains specific *transport proteins* that permit the passage of certain small molecules but not others. Several of these proteins, discussed in detail in Chapter 15, use the energy released by ATP hydrolysis to pump ions and other molecules into or out of the cell against their concentration gradients. Small charged molecules such as ATP and amino acids can diffuse freely within the cytosol but are restricted in their ability to leave or enter it across the plasma membrane.

In addition to these universal functions, the plasma membrane has other crucial roles in multicellular organisms. Few of the cells in multicellular plants and animals exist as isolated entities; rather, groups of cells with related specializations combine to form tissues (Chapter 22). Specialized areas of the plasma membrane contain proteins and glycolipids that form specific contacts and junctions between cells to strengthen tissues and to allow the exchange of metabolites between cells. Other proteins in the plasma membrane act as anchoring points for many of the cytoskeletal fibers that permeate the cytosol, imparting shape and strength to cells. Surrounding most animal cells is a mixture of fibrous proteins and polysaccharides collectively called the **extracellular matrix**. This viscous, water-filled matrix provides a bedding on which most sheets of epithelial cells or small glands lie. Proteins in the plasma membrane anchor cells to many of the matrix components, adding to the strength and rigidity of many tissues (Figure 5-40). In addition, enzymes bound to the plasma membrane catalyze reactions that would occur with difficulty in an aqueous environment. The plasma membrane of many types of eukaryotic cells also contains receptor proteins that bind specific signaling molecules (e.g., hormones, growth factors, neurotransmitters), leading to various cellular responses. These membrane proteins, which are critical for cell development and functioning, are described in later chapters.

Unlike animal cells, plant cells are surrounded by a cell wall and lack the extracellular matrix found in **PLANTS** animal tissues. As a plant cell matures, new layers of wall are laid down just outside the plasma membrane, which is intimately involved in the assembly of cell walls (Figure 5-41). The walls are built primarily of **cellulose**, a rodlike polysaccharide formed from $\beta(1 \rightarrow 4)$-linked glucose monomers. The cellulose molecules aggregate, by hydrogen bonding, into bundles of fibers; other polysaccharides within the wall cross-link the cellulose fibers (Chapter 22). In woody plants, a complex water-insoluble polymer of phenol and other aromatic monomers, called *lignin*, imparts strength and rigidity to the cell walls. Other chemicals also are found in the walls of various plant cells; for example, waxes prevent plant tissues and proteins from drying out.

Like the entire cell, each organelle in eukaryotic cells is bounded by a membrane containing a unique set of proteins essential for its proper functioning. In the next section, we discuss the structure and function of the main organelles found in eukaryotic cells.

SUMMARY Biomembranes: Structural Organization and Basic Functions

- In a phospholipid bilayer, the long fatty acyl side chains in each leaflet are oriented toward one another, forming a hydrophobic core; the polar head groups line both surfaces (see Figure 5-30).

- The phospholipid bilayer forms the basic structure of all biomembranes, which also contain proteins, glycoproteins, cholesterol and other steroids, and glycolipids. The presence of specific sets of membrane proteins permits each type of membrane to carry out distinctive functions.

- All cellular membranes line closed compartments and have a cytosolic and an exoplasmic face (see Figure 5-31).

- The asymmetry of biological membranes is reflected in the specific orientation of each type of integral and peripheral membrane protein with respect to the cytosolic and exoplasmic faces. The presence of glycolipids exclusively in the exoplasmic leaflet also contributes to membrane asymmetry.

- Most integral proteins and lipids are laterally mobile in biomembranes. According to the fluid mosaic model, the membrane is viewed as a two-dimensional mosaic of phospholipid and protein molecules.

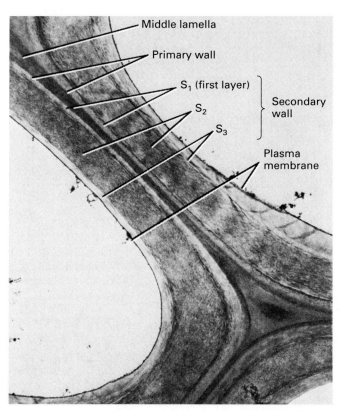

Middle lamella

Primary wall

S₁ (first layer)

S₂

S₃

Secondary wall

Plasma membrane

1 μm

◀ **FIGURE 5-41 Electron micrograph of a thin section showing parts of the cell walls separating three *Taxus canadensis* (plant) cells.** The principal layers of each wall are evident: the middle lamella, the primary wall, and the three layers of secondary wall (S₁, S₂, and S₃). As the cell matures, the layers of cellulose fibers are laid down one by one from the middle lamella inward. The fibers in each layer run in a different direction from those in the preceding layer. The plasma membrane is adjacent to the S₃ layer, the youngest stratum of the cell wall. [Courtesy of Biophoto Associates/Myron C. Ledbetter/Brookhaven National Laboratory.]

• As a phospholipid bilayer is heated, it undergoes a phase transition from a gel-like to a more fluid state over a short temperature range.

• Cholesterol is a major determinant of bilayer fluidity, although its effect depends on the composition of a membrane. Natural biomembranes generally have a fluid-like consistency, and cells adjust their phospholipid composition to maintain bilayer fluidity.

• In all cells, proteins in the plasma membrane selectively absorb nutrients, expel wastes, and maintain the proper intracellular ionic composition. Proteins in the plasma membrane anchor the membrane to intracellular cytoskeletal fibers and the extracellular matrix or cell wall. In multicellular organisms, plasma membrane proteins also act in the interactions and communication between cells, which are critical for proper functioning of multicellular tissues.

• In plants, the cell wall, which is built mainly of cellulose, is the major determinant of cell shape and imparts rigidity to cells.

• Animal cells, which lack a wall, are surrounded by an extracellular matrix consisting of collagen, glycoproteins, and other components that give strength and rigidity to tissues and organs.

5.4 Organelles of the Eukaryotic Cell

The various techniques described earlier have led to an appreciation of the highly organized internal structure of eukaryotic cells, marked by the presence of many different organelles (Figures 5-42 and 5-43). Here we present a brief

(a)

Intercellular space Plasma membrane Endoplasmic reticulum Golgi vesicles

Secretory vesicle Nuclear membrane Nucleus Mitochondrion 2 μm

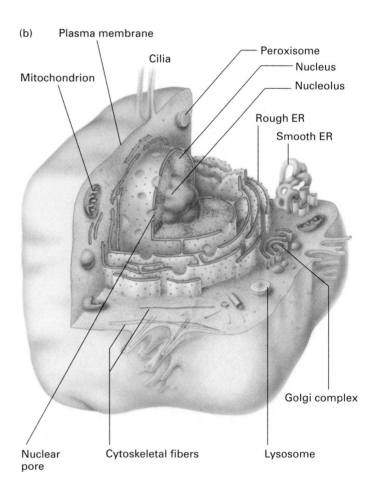

(b) Plasma membrane

Mitochondrion Cilia Peroxisome Nucleus Nucleolus Rough ER Smooth ER

Nuclear pore Cytoskeletal fibers Lysosome Golgi complex

▲ FIGURE 5-42 Structure of animal cells. (a) Electron micrograph of a thin section of a hormone-secreting cell from the rat pituitary stained with osmium tetroxide, which preferentially binds cell membranes. The subcellular features typical of many animal cells are clearly visible. (b) Drawing of a "typical" animal cell. Not every animal cell contains all the organelles, granules, and fibrous structures shown here, and other substructures can be present in some cells. Animal cells also differ considerably in shape and in the prominence of various organelles and substructures. [Part (a) courtesy of Biophoto Associates.]

(a)

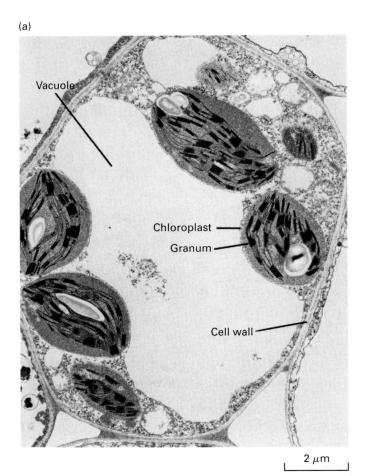

2 μm

(b)

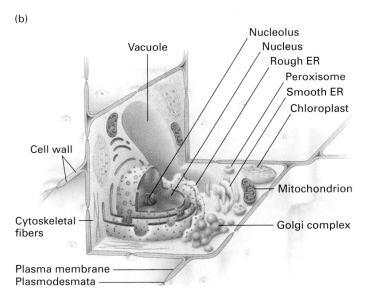

◀ **FIGURE 5-43 Structure of plant cells.** (a) Electron micrograph of a thin section of a leaf cell from *Phleum pratense*, showing a large internal vacuole, parts of five chloroplasts, and the cell wall. Although a nucleus is not evident in this micrograph, plant cells do contain a nucleus and other features of eukaryotic cells. (b) Drawing of a "typical" plant cell. Outside the plasma membrane of plant cells lies a rigid cell wall. Porelike plasmodesmata connect the cytoplasms of adjacent cells. [Part (a) courtesy of Biophoto Associates/Myron C. Ledbetter/Brookhaven National Laboratory.]

Lysosomes Are Acidic Organelles That Contain a Battery of Degradative Enzymes

Lysosomes provide an excellent example of the ability of intracellular membranes to form closed compartments in which the composition of the *lumen* (the aqueous interior of the compartment) differs substantially from that of the surrounding cytosol. Found in animal cells, lysosomes are bounded by a single membrane and are responsible for degrading certain components that have become obsolete for the cell or organism. In some cases, materials taken into a cell by **endocytosis** or **phagocytosis** also are degraded in lysosomes. Endocytosis refers to the process by which extracellular materials are taken up by invagination of a segment of the plasma membrane to form a small membrane-bounded vesicle (endosome). In phagocytosis, relatively large particles are enveloped by the plasma membrane and internalized.

MEDICINE Lysosomes contain a group of enzymes that degrade polymers into their monomeric subunits. For example, nucleases degrade RNA and DNA into their mononucleotide building blocks; proteases degrade a variety of proteins and peptides; phosphatases remove phosphate groups from mononucleotides, phospholipids, and other compounds; still other enzymes degrade complex polysaccharides and lipids into smaller units. *Tay-Sachs disease* is caused by a defect in one enzyme catalyzing a step in the lysosomal breakdown of certain glycolipids called **gangliosides,** which are abundant in nerve cells—with devastating consequences. The symptoms of this inherited disease usually are evident before the age of 1. Affected children commonly become demented and blind by age 2, and die before their third birthday. Nerve cells from such children are greatly enlarged with swollen lipid-filled lysosomes.

All the lysosomal enzymes work most efficiently at acid pH values and collectively are termed *acid hydrolases*. A hydrogen ion pump and a Cl^- channel protein in the lysosomal membrane maintain the pH of the interior at ≈ 4.8. The pump hydrolyzes ATP and uses the released free energy to pump H^+ ions from the cytosol into the lumen of the lysosome; the Cl^- channel allows Cl^- ions to enter. Together they transport HCl (Chapter 15). The acid pH helps to denature proteins, making them accessible to the action of the lysosomal hydrolases, which themselves are resistant to acid

overview of the major organelles. Unique proteins in the interior and membranes of each type of organelle largely determine its specific functional characteristics. Later chapters will examine the key roles that different organelles and the cytosol play in the functioning of eukaryotic cells.

(a)

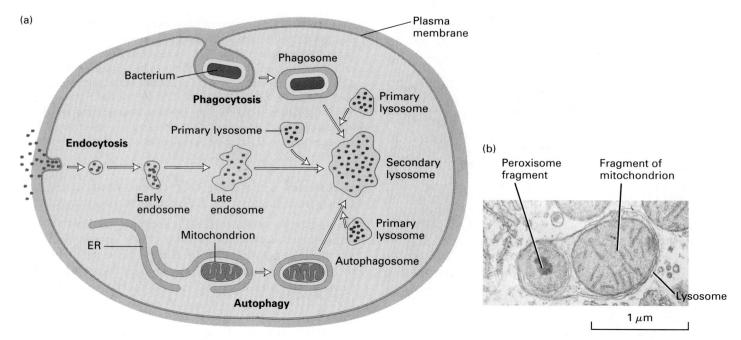

(b)

▲ **FIGURE 5-44 Lysosomal degradation.** (a) Three pathways for delivering materials to lysosomes. This schematic diagram does not depict all the intermediate structures that sometimes can be distinguished morphologically. The autophagic process by which aged or defective organelles (e.g., mitochondria) are transferred to a lysosome is not fully understood. Some evidence suggests that the process begins when a targeted organelle is surrounded by membranes derived from the endoplasmic reticulum, forming an autophagosome. This structure then probably fuses with a primary lysosome, forming a secondary lysosome in which degradation occurs. (b) Section of a rat liver cell showing a secondary lysosome containing fragments of a mitochondrion and a peroxisome. [Part (b) courtesy of D. Friend.]

denaturation. Lysosomal enzymes are poorly active at the neutral pH values of cells and most extracellular fluids. Thus if a lysosome releases its contents into the cytosol, where the pH is between 7.0 and 7.3, little degradation of cytosolic components takes place.

Lysosomes vary in size and shape, and several hundred may be present in a typical animal cell. In effect, they function as sites where various materials to be degraded collect (Figure 5-44a). *Primary lysosomes* are roughly spherical and do not contain obvious particulate or membrane debris. *Secondary lysosomes,* which are larger and irregularly shaped, appear to result from the fusion of primary lysosomes with other membrane organelles; they contain particles or membranes in the process of being digested (Figure 5-44b). The process by which an aged organelle is degraded in a lysosome is called *autophagy* ("eating oneself").

Plant Vacuoles Store Small Molecules and Enable the Cell to Elongate Rapidly

 Most plant cells contain at least one membrane-limited internal vacuole (see Figure 5-43). The number and size of vacuoles depend on both the type of cell and its stage of development; a single vacuole may occupy as much as 80 percent of a mature plant cell. Plant cells store water, ions, and nutrients such as sucrose and amino acids within these vacuoles. We will see in Chapter 15 how such materials are accumulated in vacuoles. Vacuoles also act as receptacles for waste products and excess salts taken up by the plant and may function similarly to lysosomes in animal cells. Like lysosomes, vacuoles have an acidic pH, maintained by a proton pump and a Cl$^-$ channel protein in the vacuole membrane, and contain a battery of degradative enzymes. Similar storage vacuoles are found in green algae and many microorganisms such as yeast.

Like most cellular membranes, the vacuolar membrane is permeable to water but is poorly permeable to the small molecules stored within it. Because the solute concentration is much higher in the vacuole lumen than in the cytosol or extracellular fluids, water tends to move by osmotic flow into vacuoles, just as it moves into cells placed in a hypotonic medium (see Figure 5-22). This influx of water causes both the vacuole to expand and water to move into the cell from the wall, creating hydrostatic pressure, or *turgor,* inside the cell. This pressure is balanced by the mechanical resistance of the cellulose-containing cell wall that surrounds plant cells. Most plant cells have a turgor of 5–20 atmospheres (atm); their cell walls must be strong enough to react to this pressure in a controlled way. Unlike animal cells, plant cells can elongate extremely rapidly—at rates of

20–75 μm/h. This elongation, which usually accompanies plant growth, occurs when a segment of the somewhat elastic cell wall stretches under the pressure created by water taken into the vacuole.

Peroxisomes Degrade Fatty Acids and Toxic Compounds

All animal cells (except erythrocytes) and many plant cells contain **peroxisomes**, a class of small organelles ($\approx$0.2–1 μm in diameter) bounded by a single membrane (see Figure 5-25c). (*Glyoxisomes* are similar organelles found in plant seeds that oxidize stored lipids as a source of carbon and energy for growth. They contain many of the same types of enzymes as peroxisomes as well as additional ones used to convert fatty acids to glucose precursors.) Peroxisomes contain several *oxidases*—enzymes that use molecular oxygen to oxidize organic substances, in the process forming hydrogen peroxide (H_2O_2), a corrosive substance. Peroxisomes also contain copious amounts of the enzyme *catalase*, which degrades hydrogen peroxide to yield water and oxygen:

$$2\ H_2O_2 \xrightarrow{\text{catalase}} 2\ H_2O + O_2$$

In contrast to oxidation of fatty acids in mitochondria, which produces CO_2 and is coupled to generation of ATP, peroxisomal oxidation of fatty acids yields acetyl groups and is not linked to ATP formation. The energy released during peroxisomal oxidation is converted to heat, and the acetyl groups are transported into the cytosol, where they are used in the synthesis of cholesterol and other metabolites. In most eukaryotic cells, the peroxisome is the principal organelle in which fatty acids are oxidized, thereby generating precursors for important biosynthetic pathways. Particularly in liver and kidney cells, various toxic molecules that enter the bloodstream also are degraded in peroxisomes, producing harmless products.

 In the human genetic disease *X-linked adrenoleukodystrophy* (ADL), peroxisomal oxidation of very long chain fatty acids is defective. The *ADL* gene encodes the peroxisomal membrane protein that transports into peroxisomes an enzyme required for oxidation of these fatty acids. Individuals with the severe form of ADL are unaffected until mid-childhood, when severe neurological disorders appear, followed by death within a few years.

Mitochondria Are the Principal Sites of ATP Production in Aerobic Cells

Most eukaryotic cells contain many mitochondria, which occupy up to 25 percent of the volume of the cytoplasm. These complex organelles, the main sites of ATP production during aerobic metabolism, are among the largest organelles, generally exceeded in size only by the nucleus, vacuoles, and chloroplasts.

Mitochondria contain two very different membranes, an outer one and an inner one, separated by the intermembrane space (Figure 5-45; see also 5-31). The outer membrane, composed of about half lipid and half protein, contains proteins that render the membrane permeable to molecules having molecular weights as high as 10,000. In this respect, the outer membrane is similar to the outer membrane of gram-negative bacteria (see Figure 1-7a). The inner membrane, which is much less permeable, is about 20 percent lipid and 80 percent protein—a higher proportion of protein than occurs in other cellular membranes. The surface area of the inner membrane is greatly increased by a large number of infoldings, or *cristae*, that protrude into the *matrix*, or central space.

In nonphotosynthetic cells, the principal fuels for ATP synthesis are fatty acids and glucose. The complete aerobic degradation of glucose to CO_2 and H_2O is coupled to synthesis of as many as 36 molecules of ATP (Chapter 2). In eukaryotic cells, the initial stages of glucose degradation occur in the cytosol, where two ATP molecules per glucose molecule are generated. The terminal stages, including those involving phosphorylation coupled to final oxidation by oxygen, are carried out by enzymes in the mitochondrial matrix and cristae (Chapter 16). As many as 34 ATP molecules per glucose molecule are generated in mitochondria, although this value can vary because much of the energy

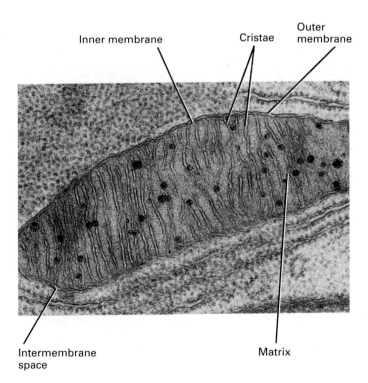

▲ **FIGURE 5-45 Electron micrograph of a mitochondrion in a section from bat pancreas.** This organelle is bounded by a double membrane. The inner membrane, which surrounds the matrix space, has many infoldings, called *cristae*. [From D. W. Fawcett, 1981, *The Cell*, 2d ed., Saunders, p. 421.]

released in mitochondrial oxidation can be used for other purposes (e.g., heat generation and the transport of molecules into or out of the mitochondrion), making less energy available for ATP synthesis. Similarly, virtually all the ATP formed during the oxidation of fatty acids to CO_2 is generated in the mitochondrion. Thus the mitochondrion can be regarded as the "power plant" of the cell.

Chloroplasts, the Sites of Photosynthesis, Contain Three Membrane-Limited Compartments

PLANTS Except for vacuoles, chloroplasts are the largest and most characteristic organelles in the cells of plants and green algae. They can be as long as 10 μm and are typically 0.5–2 μm thick, but they vary in size and shape in different cells, especially among the algae. Like the mitochondrion, the chloroplast is surrounded by an outer and an inner membrane (see Figure 5-31). Chloroplasts also contain an extensive internal system of interconnected membrane-limited sacs called **thylakoids,** which are flattened to form disks; these often are grouped in stacks called *grana* and embedded in a matrix, the *stroma* (Figure 5-46). The thylakoid membranes contain green pigments (**chlorophylls**) and other pigments and enzymes that absorb light and generate ATP during **photosynthesis.** Part of this ATP is used by enzymes located in the stroma to convert CO_2 into three-carbon intermediates; these are then exported to the cytosol and converted to sugars.

Perhaps surprisingly, the molecular mechanisms by which ATP is formed in mitochondria and chloroplasts are very similar. The details of these critical ATP-generating pathways are discussed in Chapter 16. Chloroplasts and mitochondria share other features: Both often migrate from place to place within cells and also contain their own DNA, which encodes some of the key organellar proteins (Chapter 9). The proteins encoded by mitochondrial or chloroplast DNA are synthesized on ribosomes within the organelles. However, most of the proteins in each organelle are encoded in nuclear DNA and are synthesized in the cytosol; these proteins then are incorporated into the organelles by processes described in Chapter 17.

The Endoplasmic Reticulum Is a Network of Interconnected Internal Membranes

Generally, the largest membrane in a eukaryotic cell encloses the **endoplasmic reticulum (ER)**—a compartment comprising a network of interconnected, closed, membrane-bounded vesicles (Figure 5-47). The endoplasmic reticulum has a number of functions in the cell but is particularly important in the synthesis of many membrane lipids and proteins. The *smooth endoplasmic reticulum* is smooth because it lacks ribosomes; regions of the *rough endoplasmic reticulum* are studded with ribosomes.

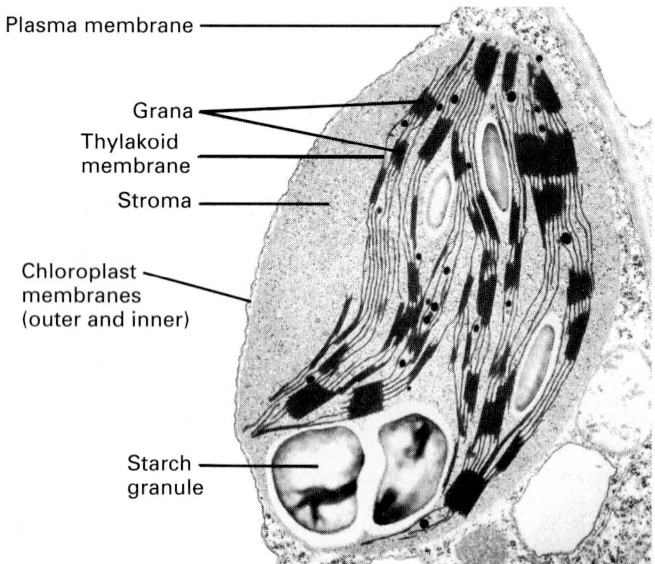

▲ **FIGURE 5-46 Electron micrograph of a chloroplast in a section of a plant cell.** The internal membrane vesicles (thylakoids) are fused into stacks (grana), which reside in a matrix (the stroma). All the chlorophyll in the cell is contained in the thylakoid membranes. [Courtesy of Biophoto Associates/M. C. Ledbetter/Brookhaven National Laboratory.]

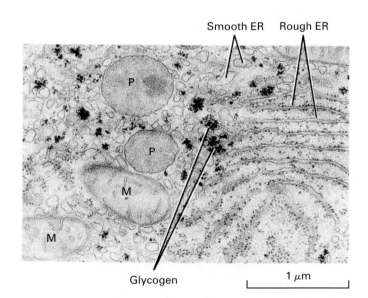

▲ **FIGURE 5-47 Electron micrograph of a section of a rat hepatocyte showing the rough and smooth endoplasmic reticula (ERs).** Note the extensive rough endoplasmic reticulum associated with numerous ribosomes (small black dots). The smooth ER lacks ribosomes. Also visible are two mitochondria (M), two peroxisomes (P), and accumulations of glycogen, a polysaccharide that is the primary glucose-storage molecule in animals. [Courtesy of P. Lazarow.]

The Smooth Endoplasmic Reticulum The synthesis of fatty acids and phospholipids occurs in the smooth ER. Although many cells have very little smooth ER, this organelle is abundant in hepatocytes. Enzymes in the smooth ER of the liver modify or detoxify hydrophobic chemicals such as pesticides and carcinogens by chemically converting them into more water-soluble, conjugated products that can be secreted from the body. High doses of such compounds result in a large proliferation of the smooth ER in liver cells.

The Rough Endoplasmic Reticulum Ribosomes bound to the rough ER synthesize certain membrane and organelle proteins and virtually all proteins to be secreted from the cell, as described in Chapter 17. The ribosomes that fabricate secretory proteins are bound to the rough ER by the nascent polypeptide chain of the protein. As the growing secretory polypeptide emerges from the ribosome, it passes through the rough ER membrane, with the help of specific proteins in the membrane. The newly made secretory proteins accumulate in the lumen (inner cavity) of the rough ER before being transported to their next destination.

All eukaryotic cells contain a discernible amount of rough ER because it is needed for the synthesis of plasma-membrane proteins and proteins of the extracellular matrix. Rough ER is particularly abundant in cells that are specialized to produce secreted proteins. For example, plasma cells produce antibodies, which circulate in the bloodstream, and pancreatic acinar cells synthesize digestive enzymes, which are transported to the intestine via a series of progressively larger ducts (Figure 5-48). In both types of cells, a large part of the cytosol is filled with rough ER.

Golgi Vesicles Process and Sort Secretory and Membrane Proteins

Several minutes after proteins are synthesized in the rough ER, most of them leave the organelle within small membrane-bounded transport vesicles. These vesicles, which bud off from regions of the rough ER not coated with ribosomes, carry the proteins to the luminal cavity of another membrane-limited organelle, the **Golgi complex,** a series of flattened sacs located near the nucleus in many cells (see Figure 5-48a).

Three-dimensional reconstructions from serial sections of a Golgi complex reveal a series of flattened membrane vesicles or sacs, surrounded by a number of more or less spherical membrane vesicles (Figure 5-49). The stack of flattened Golgi sacs has three defined regions—the *cis*, the *medial*, and the *trans*. Transfer vesicles from the rough ER fuse with the cis region of the Golgi complex, where they deposit their proteins. As detailed in Chapter 17, these proteins then progress from the cis to the medial to the trans region. Within each region are different enzymes that modify secretory and membrane proteins differently, depending on their structures and their final destinations.

After secretory proteins are modified in the Golgi sacs, they are transported out of the complex by a second set of

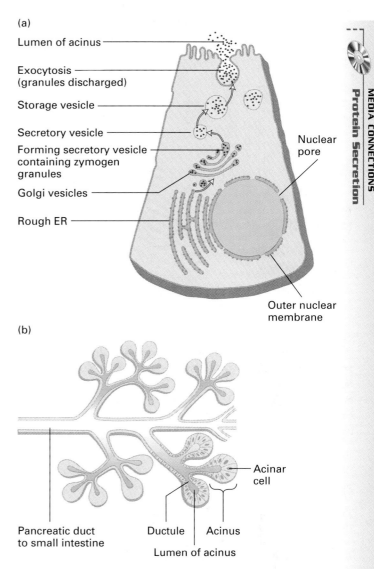

(a)

Lumen of acinus

Exocytosis (granules discharged)

Storage vesicle

Secretory vesicle

Forming secretory vesicle containing zymogen granules

Golgi vesicles

Rough ER

Nuclear pore

Outer nuclear membrane

(b)

Acinar cell

Pancreatic duct to small intestine

Ductule

Acinus

Lumen of acinus

▲ **FIGURE 5-48 The synthesis and release of secretory proteins in acinar cells of the rat pancreas.** (a) Immediately after secretory proteins are made on ribosomes of the rough ER, they are found in the lumen of the rough ER. Transfer vesicles transport them to Golgi vesicles, where they are concentrated and packaged into secretory vesicles containing granules of zymogens (pancreatic enzyme precursors, such as chymotrypsinogen). Several of these coalesce to form a storage vesicle; these accumulate under the apical surface, which faces the lumen of an acinus. Hormone or nerve stimulation triggers fusion of the vesicles with the plasma membrane, releasing the vesicles' contents into the lumen (exocytosis). (b) Once released from acinar cells, the inactive precursors move through ductules and the pancreatic duct to the intestine, where they are proteolytically activated into digestive enzymes.

transport vesicles, which seem to bud off the trans side of the Golgi complex. Some of these transport vesicles, termed *coated vesicles,* are surrounded by an outer protein cage composed primarily of the fibrous protein clathrin (see

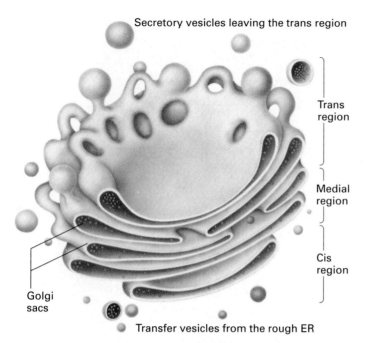

Secretory vesicles leaving the trans region

Trans region

Medial region

Cis region

Golgi sacs

Transfer vesicles from the rough ER

▲ **FIGURE 5-49 Three-dimensional model of the Golgi complex built by analyzing micrographs of serial sections through a secretory cell.** Transfer vesicles that have budded off from the rough ER fuse with the cis membranes of the Golgi complex. In pancreatic acinar cells, the secretory vesicles that form by budding off of sacs on the trans membranes store secretory proteins, such as chymotrypsinogen, in concentrated form. Other vesicles, detailed in Chapter 17, move material from one part of the Golgi to another. [After a model by J. Kephart.]

Figure 5-26). Some vesicles contain membrane proteins destined for the plasma membrane; others, proteins for lysosomes or for other organelles. How intracellular transport vesicles "know" which membranes to fuse with and where to deliver their contents is also discussed in Chapter 17.

The Double-Membraned Nucleus Contains the Nucleolus and a Fibrous Matrix

The nucleus, the largest organelle in eukaryotic cells, is surrounded by two membranes, each one a phospholipid bilayer containing many different types of proteins. The inner nuclear membrane defines the nucleus itself. In many cells, the outer nuclear membrane is continuous with the rough endoplasmic reticulum, and the space between the inner and outer nuclear membranes is continuous with the lumen of the rough endoplasmic reticulum (see Figure 5-31).

The two nuclear membranes appear to fuse at the **nuclear pores** (see Figure 5-42b). The distribution of nuclear pores is particularly vivid when the nucleus is viewed by the freeze-fracture technique described earlier (Figure 5-50). Constructed of a specific set of membrane proteins, these ringlike pores function as channels that regulate the movement of material between the nucleus and the cytosol (see Figure 11-28).

In a growing or differentiating cell, the nucleus is metabolically active, producing DNA and RNA. The latter is exported through nuclear pores to the cytoplasm for use in protein synthesis (Chapter 4). In mature erythrocytes from nonmammalian vertebrates and other types of "resting" cells, the nucleus is inactive or dormant and minimal synthesis of DNA and RNA takes place.

How nuclear DNA is packaged into chromosomes is described in Chapter 9. In a nucleus that is not dividing, the chromosomes are dispersed and not thick enough to be observed in the light microscope. Only during cell division are chromosomes visible by light microscopy (see Figure 1-8). However, a suborganelle of the nucleus, the **nucleolus,** is easily recognized under the light microscope. Most of the cell's ribosomal RNA is synthesized in the nucleolus; some ribosomal proteins are added to ribosomal RNAs within the nucleolus as well (Chapter 11). The finished or partly finished ribosomal subunit passes through a nuclear pore into the cytosol.

In the electron microscope, the nonnucleolar regions of the nucleus, called the *nucleoplasm,* can be seen to have areas of high DNA concentration, often closely associated with the nuclear membrane. Fibrous proteins called **lamins** form a two-dimensional network along the inner surface of the inner membrane, giving it shape and apparently binding DNA to it. The breakdown of this network occurs early in cell division, as we detail in Chapter 13.

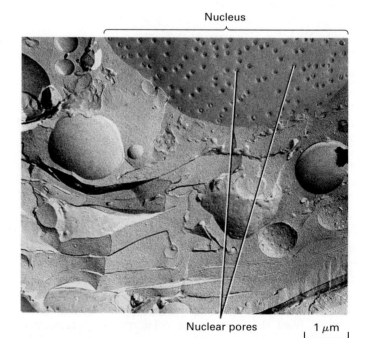

Nucleus

Nuclear pores   1 μm

▲ **FIGURE 5-50 A freeze-fracture preparation of an onion root-tip cell, showing the nucleus and pores in the nuclear membrane, which traverse the inner and outer nuclear membranes.** [Courtesy of D. Branton.]

The Cytosol Contains Many Particles and Cytoskeletal Fibers

Because sections for standard electron microscopy must be thinner than 0.1 μm, fibers in the cytosol, which may be several microns in length, appear as long elements only in sections that by chance happen to be in the plane of the fiber bundles (Figure 5-51). Serial sectioning of a tissue sample can compensate for these shortcomings by tracing a fiber from one image into the next to reconstruct its three-dimensional architecture. Because 200 sections are needed to examine a cell 20 μm thick, serial sectioning is a tedious technique. However, sections up to 1 μm thick can be viewed in high-voltage electron microscopes, considerably reducing the number of sections needed to reconstruct three-dimensional images.

Transmission electron micrographs of cytosolic fibers obtained from unsectioned cells reveal an extensive network of microfilaments, microtubules, and intermediate filaments (Figure 5-52). These cytoskeletal fibers crisscross one another in complex patterns so that different types of fibers contact one another at many points. In cultured cells, actin microfilaments often occur in bundles of long fibers that appear to be connected by small fibrous proteins (see Figure 5-6).

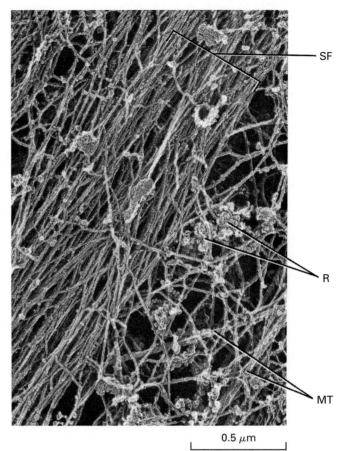

▲ **FIGURE 5-52 Electron micrograph of a platinum replica of a cytoskeleton prepared by quick freezing and deep etching.** A fibroblast cell was immersed in the nonionic detergent Triton X-100, to remove soluble cytoplasmic proteins and dissolve all membranes, and then frozen rapidly to 4° above absolute zero. While it is frozen, water in the cytosol is removed ("etched away") by sublimation in a vacuum, resulting in exposure of the nonvolatile protein fibers. Once coated with a thin layer of platinum, these fibers are visible in the ordinary transmission electron microscope. Prominent are bundles of actin microfilaments termed *stress fibers* (SF), which are thought to connect segments of the plasma membrane and anchor the cell to the substratum. Also visible are two thicker microtubules (MT) and a more diffuse meshwork of filaments studded with grapelike clusters, which are probably polyribosomes (R). [From J. E. Heuser and M. Kirschner, 1980, *J. Cell Biol.* **86**:212. Reproduced from the *Journal of Cell Biology* by copyright permission of The Rockefeller University Press.]

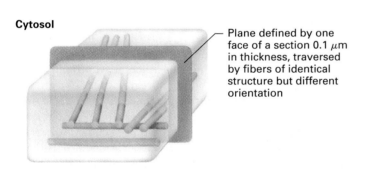

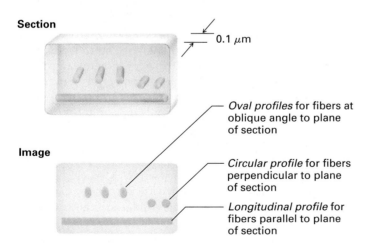

▲ **FIGURE 5-51 A section through a bundle of fibers can generate very different images, depending on the angle of the cut with respect to the plane of the fibers.**

The cytosol of many cells also contains *inclusion bodies,* granules that are not bounded by a membrane. For instance, muscle cells and hepatocytes contain cytosolic granules of glycogen (see Figure 5-47), a glucose polymer that functions as a storage form of usable cellular energy. In well-fed animals, glycogen can account for as much as 10 percent of the wet weight of the liver. The cytosol of the specialized fat cells in adipose tissue contains large droplets of almost pure triacylglycerols, a storage form of fatty acids (see Figure 5-13a and b).

In addition, the cytosol is a major site of cellular metabolism and contains a large number of different enzymes. About 20–30 percent of the cytosol is protein, and from a quarter to half of the total protein within cells is in the cytosol. Because of the high concentration of cytosolic proteins, organized complexes of proteins can form even if the energy that stabilizes them is weak. Many investigators believe that the cytosol is highly organized, with most proteins either bound to fibers or otherwise localized in specific regions.

SUMMARY   Organelles of the Eukaryotic Cell

- The internal architecture of cells and central metabolic pathways are similar in all plants, animals, and unicellular eukaryotic organisms (e.g., yeast). All eukaryotic cells contain a membrane-limited nucleus and numerous other organelles in their cytosol.

- Lysosomes, which are found only in animal cells, have an acidic interior and contain various hydrolases. These degradative enzymes break down some cellular components that are no longer functional or needed by the cell and some ingested materials (see Figure 5-44).

- Plant cells contain one or more large vacuoles, which often fill much of the cell. The vacuole is a storage site for ions and nutrients. Osmotic flow of water into the vacuole generates turgor pressure that pushes the plasma membrane against the cell wall.

- Peroxisomes are small organelles containing enzymes that oxidize various organic compounds, generating hydrogen peroxide. This toxic substance is converted to water and oxygen by catalase, also present in large amounts in these organelles. Oxidation of fatty acids in peroxisomes produces acetyl groups, used in biosynthetic reactions, but no ATP.

- The mitochondrion is bounded by two membranes, with the inner one extensively folded. Enzymes in the inner mitochondrial membrane and central matrix carry out the terminal stages of sugar and lipid oxidation coupled to ATP synthesis.

- Chloroplasts, the sites of photosynthesis, are surrounded by an inner and outer membrane; a complex system of thylakoid membranes in their interior contains the pigments and enzymes that absorb light and produce ATP.

- Secretory proteins and membrane proteins are synthesized on the rough endoplasmic reticulum, a network of interconnected membrane vesicles studded with ribosomes. These proteins then move to the Golgi complex, where they are sorted and processed (see Figure 5-48).

- The nucleus is surrounded by an inner and outer membrane. These contain numerous pores through which materials pass between the nucleus and cytosol. The outer nuclear membrane is continuous with the rough endoplasmic reticulum.

- The cytosol, the protein-rich fraction remaining after removal of all organelles, contains numerous soluble enzymes and three major types of protein filaments: actin microfilaments, microtubules, and intermediate filaments. In all animal and plant cells, these filaments form a complex network, the cytoskeleton, that gives the cell structural stability and contributes to cell movement.

PERSPECTIVES for the Future

Significant improvements in all the techniques discussed in this chapter would enable biologists to carry out experiments to much higher resolution and to conduct entirely new types of studies. The sensitivity of fluorescence microscopy, for instance, is limited by the fluorescence of individual dye molecules; as a consequence, immunofluorescence is useful for detecting proteins that are present in thousands or millions of copies per cell (e.g., abundant cytoskeletal proteins) but not for detecting one molecule or even a small number of molecules per cell. If much more sensitive techniques were available, researchers could image a single molecule of a protein hormone, such as insulin, as it binds to a receptor on the cell surface and generates a signal in the cytosol (e.g., to move GLUT4 to the plasma membrane).

As described in this chapter, current techniques permit two or three proteins to be tagged with different-colored GFP molecules and simultaneous monitoring of the levels of calcium and the pH in living cells. The ability to image dozens of proteins simultaneously and chart their behavior during the cell cycle, during cell differentiation, and in response to external signals will require new types of dyes and new ways of attaching fluorescent molecules to specific proteins. Such studies also require new types of light microscopes, ones merging the high resolution of confocal and deconvolution light microscopes with the speed and sensitivity of CCD cameras and the power of new computers.

Advances in cryoelectron microscopy, which permits viewing of unfixed, frozen samples, and computer-based image analysis should lead to high-resolution images of various objects that are difficult to study by other techniques. Computer averaging of multiple images should lead to construction of structural models of ribosomes, membrane proteins, and important protein complexes to atomic resolution. Indeed, this technique already has been applied to bacteriorhodopsin and a few other membrane proteins; although the images are not yet to atomic resolution, they do allow analyses that define the course of the peptide chain and provide other useful information. Cryoelectron microscopy overlaps in resolution with x-ray crystallography and NMR spectroscopy on one hand and high-resolution light microscopy on the other.

Fairly pure preparations of mitochondria and nuclei can be obtained by current subcellular fractionation procedures, which are based on rather crude parameters, largely size and density. However, no procedure is available for purifying the specific class of vesicles that transport secretory proteins from the ER to the Golgi, or those that move from the cell surface to the lysosome. Immunological techniques would be very useful in such endeavors, but they depend on identifying proteins that are specifically localized to the membranes of these organelles and that might be useful targets for antibody-based organelle purifications.

PERSPECTIVES in the Literature

Cryoelectron microscopy can be used to determine the structure of large proteins and particles at or near the atomic level. The following papers illustrate how this technology has been used to determine the structures of tubulin, the major component of microtubules; of bacteriorhodopsin, a transmembrane protein that uses light energy to pump protons across the membrane; and of the genome of a virus that contains double-stranded-RNA. It has also been used to identify the distinct sites for transfer RNAs in the *Escherichia coli* ribosome. As you read the papers referenced below, consider the following questions:

1. In the Nogales paper, the tubulin structure was determined from 2D crystals, whereas in the other papers the structures were calculated from images of single particles. What two approaches were applied in these studies? How do they compare to x-ray crystallography and nuclear magnetic resonance spectroscopy for determining the molecular structures of proteins and particles?

2. What are the limitations and disadvantages of these techniques?

3. Cryoelectron microscopy requires storing and analyzing images of hundreds of individual proteins or particles. How are these images used to determine the molecular structure?

Agrawal, R. K., et al. 1996. Direct visualization of A-, P-, and E-site transfer RNAs in the *Escherichia coli* ribosome. *Science* 271:1000–1002.

Kimura, Y., et al. 1997. Surface of bacteriorhodopsin revealed by high-resolution electron crystallography. *Nature* 389:206–211.

Nogales, E., et al. 1995. Structure of tubulin at 6.5 Å and location of the taxol-binding site. *Nature* 375:424–427.

Prasad, B. V., et al. 1996. Visualization of ordered genomic RNA and localization of transcriptional complexes in rotavirus. *Nature* 382:471–473.

Testing Yourself on the Concepts

1. Both light and electron microscopy are commonly used to visualize cells, cell structures, and the location of specific molecules. What is the primary advantage of each technique?

2. Much of what we know about cellular function depends on experiments utilizing specific cells and specific parts (e.g., organelles) of cells. What techniques do scientists commonly use to isolate cells and organelles from complex mixtures, and how do these techniques work?

3. Explain the following statement: the structure of all biomembranes depends on the chemical properties of phospholipids, while the function of each specific biomembrane depends on the asymmetric association of different proteins with each face of the membrane.

4. Cell organelles such as mitochondria, chloroplasts, and the Golgi apparatus each have unique structures. How is the structure of each organelle related to its function?

MCAT/GRE-Style Questions

Key Concept Please read the section titled "Fluorescence Microscopy Can Localize and Quantify Specific Molecules in Cells" (p. 142) and answer the following questions:

1. Fluorescence microscopy is based on the ability of certain molecules to
   a. Continuously emit light of a constant wavelength.
   b. Absorb light of many different wavelengths.
   c. Absorb light of a given wavelength and then emit light of a longer wavelength.
   d. Absorb light of a given wavelength and then emit light of a shorter wavelength.

2. Compared with the other types of light microscopy, a distinct advantage of fluorescence microscopy is
   a. Greater magnification of cellular structures.
   b. The ability to determine the location of a certain molecule within a cell.
   c. The ability to determine the location of the nucleus within a cell.
   d. Improved resolution of cellular structures.

3. Unlike the dyes commonly used in immunofluorescence microscopy, the fluorescent nature of green fluorescent protein (GFP) depends on
   a. The antibody to which the GFP is attached.
   b. The correct folding of the polypeptide so that certain amino acid residues are brought together.
   c. The concentration of GFP present in the cell under observation.
   d. The concentration of Ca^{2+} in the cytoplasm.

4. Fluorescence microscopy may be used to observe living cells, provided the fluorescent molecule is
   a. Delivered into the cytoplasm by injection or diffusion.
   b. Able to interact nonspecifically with all types of biological molecules.
   c. Detectable by the human eye.
   d. Used in conjunction with fixatives like glutaraldehyde.

5. Dyes such as fura-2 can report the intracellular concentration of an ion because the fluorescence intensity of the dye at a specific wavelength

   a. Is unaffected by small, rapid changes in ion concentration.

   b. Is the same at all wavelengths.

   c. Increases upon binding the ion.

   d. Decreases upon binding the ion.

Key Experiment Please read the section titled "Most Lipids and Integral Proteins Are Laterally Mobile in Biomembranes" (p. 162) and refer to Figures 5-35 and 5-36; then answer the following questions:

6. The observation that plasma membrane proteins mix after cell fusion provides evidence for

   a. Rotational movement of plasma membrane proteins.

   b. The bilayer structure of biomembranes.

   c. The fluid mosaic model.

   d. Interaction between the plasma membrane proteins from two different cell types.

7. Complete recovery of fluorescence during a FRAP experiment suggests that the fluorescent-tagged molecule

   a. Is not interacting with the cytoskeleton.

   b. Was not completely bleached.

   c. May be interacting with the cytoskeleton.

   d. Has been destroyed by the bleaching process.

8. Based on the results described in Figure 5-35, if a FRAP experiment was conducted at 0 °C instead of 37 °C, the

   a. Rate of fluorescence recovery would be faster at 0 °C than at 37 °C.

   b. Rate of fluorescence recovery would be slower at 0 °C than at 37 °C.

   c. Extent of fluorescence recovery would be less at 0 °C than at 37 °C.

   d. Rate of fluorescence recovery would be identical at 0 °C and 37 °C.

9. Continuous bleaching of a defined area of plasma membrane will eventually

   a. Reduce the fluorescence across the entire surface of the cell.

   b. Reduce the fluorescence of only the area under illumination.

   c. Increase the fluorescence across the entire surface of the cell.

   d. Increase the fluorescence of only the area under illumination.

Key Application Please read the section titled "Lysosomes Are Acidic Organelles That Contain a Battery of Degradative Enzymes" (p. 169) and answer the following questions:

10. Lysosomes function to

   a. Synthesize various cellular macromolecules.

   b. Carry out digestion of all cytosolic components.

   c. Ingest material from the extracellular environment.

   d. Recycle cellular material and digest material taken in from the environment.

11. A defect that inactivates an acid hydrolase required for protein degradation would result in

   a. An increase in the pH of the lysosome lumen.

   b. Degradation of all acid hydrolases.

   c. Accumulation of proteins in the cell's lysosomes.

   d. Tay-Sachs disease.

12. Which of the following would lead to the accumulation of all types of macromolecules in lysosomes:

   a. A defect that inactivates phagocytosis.

   b. A defect that inactivates the H^+ pumps in the lysosomal membrane.

   c. A defect that inactivates a nuclease.

   d. A defect that inactivates autophagy.

13. From the description of Tay-Sachs disease, it can be inferred that cell function is impaired in this disease because

   a. Lysosomes will rupture and release acid hydrolases to the cytosol.

   b. Gangliosides cannot be synthesized.

   c. Acid hydrolases cannot be synthesized.

   d. Lysosomes will eventually occupy the majority of the cell's volume.

Key Terms

cellulose *167*	liposomes *160*
cholesterol *158*	lysosomes *169*
cytosolic face *160*	myelin sheath *160*
endocytosis *169*	nuclear pores *174*
endoplasmic reticulum (ER) *172*	nucleolus *174*
exoplasmic face *160*	peroxisomes *171*
extracellular matrix *167*	phagocytosis *169*
fluorescent staining *142*	phospholipid bilayer *140*
glycolipids *159*	phospholipids *157*
glycoproteins *159*	resolution *140*
Golgi complex *173*	steroids *158*
	thylakoids *172*

References

Light Microscopy

Conn, P. M., ed. 1990. *Quantitative and Qualitative Microscopy.* In *Methods in Neurosciences,* vol. 3. Academic Press.

Gilroy, S. 1997. Fluorescence microscopy of living plant cells. *Ann. Rev. Plant Physiol. and Plant Mol. Biol.* **48**:165–190.

Herman, B., and J. J. Lemasters, eds. 1993. *Light Microscopy: Emerging Methods and Applications.* Academic Press.

Inoué, S., and K. Spring. 1997. *Video Microscopy,* 2d ed. Plenum Press.

Leffel, S. M., S. A. Mabon, and C. N. Stewart, Jr. 1997. Applications of green fluorescent protein in plants. *Biotechniques* **23**:912–918.

Lippincott-Schwartz, J., and C. L. Smith. 1997. Insights into secretory and endocytic membrane traffic using green fluorescent protein chimeras. *Curr. Opin. Neurobiol.* **7**:631–639.

Matsumoto, B., ed. 1993. *Cell Biological Applications of Confocal Microscopy.* In *Methods in Cell Biology,* vol. 38. Academic Press.

Misteli, T., and D. L. Spector. 1997. Applications of the green fluorescent protein in cell biology and biotechnology. *Nature Biotech.* **15**:961–964.

Rizzuto, R., W. Carrington, and R. Tuft. 1998. Digital imaging microscopy of living cells. *Trends Cell Biol.* **8**:288–292.

Slayter, E. M. 1993. *Light and Electron Microscopy.* Cambridge University Press.

Sluder, G., and D. Wolf, eds. 1998. *Video Microscopy.* In *Methods in Cell Biology,* vol. 56.

Smith, R. 1994. *Microscopy and Photomicrography. A Working Manual.* CRC Press, Boca Raton.

Wang, Y.-L., and D. L. Taylor, eds. 1989. *Fluorescence Microscopy of Living Cells in Culture.* Part A: *Fluorescent Analogs, Labeling Cells, and Basic Microscopy.* In *Methods in Cell Biology,* vol. 29. Academic Press.

Electron Microscopy

Chen, I. 1993. *Introduction to Scanning Tunneling Microscopy.* Oxford University Press, New York.

Chiu, W. 1993. What does electron cryomicroscopy provide that x-ray crystallography and NMR spectroscopy cannot? *Ann. Rev. Biophys. Biomol. Struc.* **22**:233–235.

Heuser, J. 1981. Quick-freeze, deep-etch preparation of samples for 3-D electron microscopy. *Trends Biochem. Sci.* **6**:64–68.

Pease, D. C., and K. R. Porter. 1981. Electron microscopy and ultramicrotomy. *J. Cell Biol.* **91**:287s–292s.

Watt, I. M. 1985. *The Principles and Practice of Electron Microscopy.* Cambridge University Press.

Cell Structure: Histology Texts and Atlases

Cormack, D. H. 1992. *Essential Histology.* Lippincott. Another excellent histology text.

Cross, P. A., and K. L. Mercer. 1993. *Cell and Tissue Ultrastructure: A Functional Perspective.* W. H. Freeman and Company.

Fawcett, D. W. 1981. *The Cell,* 2d ed. Saunders.

Fawcett, D. W. 1993. *Bloom and Fawcett: A Textbook of Histology,* 12th ed. Chapman & Hall. An excellent text containing detailed descriptions of the structures of mammalian organs, tissues, and cells.

Ledbetter, M. C., and K. R. Porter. 1970. *Introduction to the Fine Structure of Plant Cells.* Springer-Verlag.

Margulis, L., and K. V. Schwartz. 1988. *Five Kingdoms: An Illustrated Guide to the Phyla of Life on Earth,* 3d ed. W. H. Freeman and Company.

Wheater, P. R., H. G. Burkitt, and V. C. Daniels. 1987. *Functional Histology: A Text and Colour Atlas,* 2d ed. Churchill Livingstone.

Purification of Cells and Their Parts

Battye, F. L., and K. Shortman. 1991. Flow cytometry and cell-separation procedures. *Curr. Opin. Immunol.* **3**:238–241.

de Duve, C. 1975. Exploring cells with a centrifuge. *Science* **189**:186–194. The Nobel prize lecture of a pioneer in the study of cellular organelles.

de Duve, C., and H. Beaufay. 1981. A short history of tissue fractionation. *J. Cell Biol.* **91**:293s–299s.

Howell, K. E., E. Devaney, and J. Gruenberg. 1989. Subcellular fractionation of tissue culture cells. *Trends Biochem. Sci.* **14**:44–48.

Ormerod, M. G., ed. 1990. *Flow Cytometry: A Practical Approach.* IRL Press.

Rickwood, D. 1992. *Preparative Centrifugation. A Practical Approach.* IRL Press, Oxford.

Biomembranes: Structural Organization and Basic Functions

Bretscher, M. S., and S. Munro. 1993. Cholesterol and Golgi apparatus. *Science* **261**:1280–1281.

Dai, J., and M. P. Sheetz. 1998. Cell membrane mechanics. *Meth. Cell Biol.* **55**:157–171.

Edidin, M. 1997. Lipid microdomains in cell surface membranes. *Curr. Opin. Struc. Biol.* **7**:528–532.

Hackenbrock, C. R. 1981. Lateral diffusion and electron transfer in the mitochondrial inner membrane. *Trends Biochem. Sci.* **6**:151–154.

Tanford, C. 1980. *The Hydrophobic Effect,* 2d ed. Wiley. Includes a good discussion of the interactions of proteins and membranes.

Vance, D. E., and J. Vance, eds. 1996. *Biochemistry of Lipids, Lipoproteins, and Membranes.* Elsevier.

Zhang, F., G. M. Lee, and K. Jacobson. 1993. Protein lateral mobility as a reflection of membrane microstructure. *BioEssays* **15**:579–588.

Organelles of the Eukaryotic Cell

Bainton, D. 1981. The discovery of lysosomes. *J. Cell Biol.* **91**:66s–76s.

Carraway, K. L., and C. A. C. Carraway. 1992. *The Cytoskeleton: A Practical Approach.* IRL Press.

Cuervo, A. M., and J. F. Dice. 1998. Lysosomes, a meeting point of proteins, chaperones, and proteases. *J. Mol. Med.* **76**:6–12.

de Duve, C. 1996. The peroxisome in retrospect. *Annals. N. Y. Acad. Sci.* **804**:1–10.

Fowler, V. M., and R. Vale. 1996. Cytoskeleton, a collection of reviews. *Curr. Opin. Cell Biol.,* vol. 8.

Gavin, R. H. 1997. Microtubule-microfilament synergy in the cytoskeleton. *Int'l. Rev. Cytol.* **173**:207–242.

Holtzman, E. 1989. *Lysosomes.* Plenum Press.

Lamond, A., and W. Earnshaw. 1998. Structure and function in the nucleus. *Science* **280**:547–553.

Masters, C., and D. Crane. 1996. Recent developments in peroxisome biology. *Endeavour* **20**:68–73.

Palade, G. 1975. Intracellular aspects of the process of protein synthesis. *Science* **189**:347–358. The Nobel prize lecture of a pioneer in the study of cellular organelles. (See also de Duve, above.)

Sheetz, M. P. 1996. Microtubule motor complexes moving membranous organelles. *Cell Struc. Func.* **21**:369–373.

Subramani, S. 1998. Components involved in peroxisome import, biogenesis, proliferation, turnover, and movement. *Physiol. Rev.* **78**:171–188.

Manipulating Cells and Viruses in Culture

Understanding modern molecular cell biology requires a familiarity with the most commonly used biological materials and knowledge of the latest experimental techniques. In this chapter, we describe the primary techniques used to culture and manipulate certain cells, viruses, and experimental organisms. Biologists choose to study these cells and organisms because they provide experimentally favorable examples of important molecular events or processes such as the control of gene activity and cell replication, formation of organelles, secretion of proteins, and differentiation of cells.

In the first part of the chapter, we discuss the culture of microbial cells, animal cells, and hybrid cells in vitro and give some examples of their experimental uses. Many studies involve cells or organisms with mutations affecting a specific biochemical or developmental pathway. In such studies, an important criterion for using a particular cell type or organism is the ease with which mutants can be isolated and characterized and the mutated genes identified and cloned (Chapter 8). Because viruses contain a small number of genes (4 to ≈100 in different types of viruses), viral mutants commonly are used to study the action of a limited set of genes designed to carry out restricted, specific molecular tasks. To provide a basis for understanding such studies, we consider the basic properties of viruses and their uses in various types of research in the latter part of this chapter.

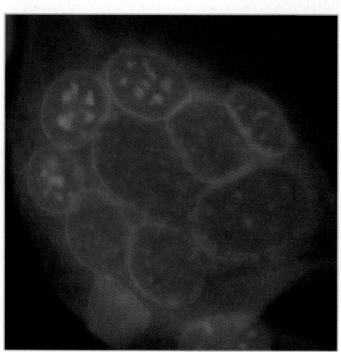

Syncytia of cultured 3T3 fibroblasts induced by expression of a viral envelope glycoprotein.

MEDIA CONNECTIONS

Technique: Preparing Monoclonal Antibodies

Overview: Life Cycle of a Retrovirus

Classic Experiment 6.1: The Discovery of Reverse Transcriptase

6.1 Growth of Microorganisms in Culture

Cultured cells have several advantages over intact organisms for research on fundamental aspects of cell biology. First, most animal and plant tissues comprise a variety of different types of cells, whereas cells of a specific type can be grown in culture; thus cultured cells are more homogeneous in their properties than in vivo cells in tissues. Second, experimental conditions can be controlled much more rigorously in culture than in an organism. By manipulating the growth conditions, for example, one can measure the effects of specific chemicals and growth factors on a particular cell type in culture. A third reason for preferring cultured cells is that in many cases a single cell can be readily grown into a colony, a process called *cell cloning*, or simply *cloning*. The resulting strain of cells, which is genetically homogeneous, is called a **clone**. This simple technique, which is commonly used with many bacteria, yeasts, and mammalian cell types, makes it easy to isolate genetically distinct clones of cells.

Many Microorganisms Can Be Grown in Minimal Medium

Among the advantages of using microorganisms such as the bacterium *Escherichia coli* and the yeast *Saccharomyces cerevisiae* are their rapid growth rate and simple nutritional requirements, which can be met with a minimal medium (Table 6-1, *top*). A minimal medium for such microorganisms can contain glucose as the sole source of carbon; metabolism of glucose to smaller molecules (e.g., CO_2, ethanol, or acetic acid) can generate the ATP necessary for energy-requiring activities of the cells. The sole nitrogen source in a minimal medium can be ammonium (NH_4^+), from which the cells can synthesize all the necessary amino acids and other nitrogen-containing metabolites. Salts and trace elements are the only other components of a minimal medium.

Many prokaryotes (i.e., bacteria) and single-celled eukaryotes such as yeast, both of which grow in nature as single cells, are easily grown in culture dishes—usually on top of *agar*, a semisolid base of plant polysaccharides. The agar is first dissolved in a heated nutrient medium, and the solution is poured into petri dishes; as it cools, the solution solidifies. A dilute suspension of cells then is dispersed on top of the agar; in time each cell grows into a discrete colony (Figure 6-1a). Since the cells in a colony all derive from a single cell, they form a clone and have identical genomes (DNA). All the cells in a clone generally express the same set of genes and contain the same enzymes and other constituents in similar proportions. The division time for *E. coli* and similar microorganisms ranges from 20 minutes to 1 hour. Thus a single *E. coli* cell, which divides approximately every 30 minutes, can grow into a colony containing 10^7–10^8 cells in 12 hours ($2^{24} = 1.7 \times 10^7$).

Because yeasts, unlike most eukaryotic organisms, grow as single cells and can be grown in a simple defined medium

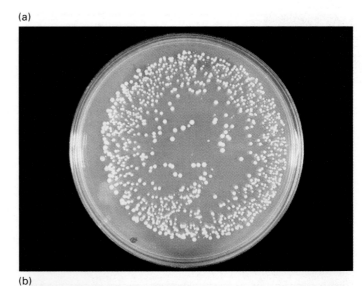

(a)

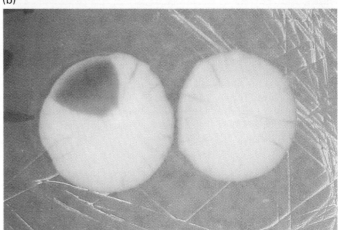

(b)

▲ FIGURE 6-1 Plating of the yeast *Saccharomyces cerevisiae* on agar plates. (a) Colonies of the yeast *S. cerevisiae* growing on a plate of agar containing only glucose, adenine, and salts. All the cells in a colony are descendants of a single cell and thus form a clone. Except for spontaneous mutations, which are rare, the cells in each clone are genetically identical. (b) Close-up view of two yeast colonies. As the cells in the colonies grew, a mutation affecting adenine biosynthesis occurred in one cell in the colony on the left. As a result, all the cells descended from the mutant cell accumulated an orange pigment derived from intermediates in adenine biosynthesis that could not be metabolized further. The orange sector in this colony thus constitutes a "subclone," which differs in one genetic trait from the original clone. [See N. A. Levin et al., 1993, *Genetics* **133**:799; courtesy of Dr. Nikki Levin.]

similarly to bacteria, they are a popular choice for studies of eukaryotic cell function. The entire life cycle of yeasts can be studied in culture, and colonies can be grown from a single vegetative cell (a growing cell) or from a single spore (a dormant cell). Studies with yeasts have provided valuable information on such subjects as the cellular mechanisms for

TABLE 6-1 Growth Media for Common Bacteria and Yeasts

MINIMAL MEDIUM*

Carbon source: glucose or glycerol

Nitrogen source: $NH_4^{(+)}$ (e.g., $NaNH_4HPO_4$) or an organic compound such as histidine

Salts: Na^+, K^+, Mg^+, Ca_2^{2+}, SO_4^+, Cl^-, and PO_4^{3-}

Trace elements

RICH MEDIUM

Partly hydrolyzed animal or plant tissue (rich in amino acids, short peptides, and lipids)

Yeast extract (rich in vitamins and enzyme cofactors, nucleic acid precursors, and amino acids)

Carbon source, nitrogen source, and salts as in minimal medium

NOTE: *For more detailed information, see R. W. Davis, D. Botstein, and J. W. Roth, 1982, A Manual for Genetic Engineering: Advanced Bacterial Genetics, Cold Spring Harbor Laboratory.*
Typical for most bacteria and yeasts. Some photosynthetic bacteria (e.g., Rhodospirillum rubrum, cyanobacteria, and blue-green algae) require CO_2 as the carbon source. Some nitrogen-fixing bacteria (e.g., Azotobacter) require atmospheric N_2. Other organisms have special needs: for example, Hemophilus strains require factors found in whole blood.

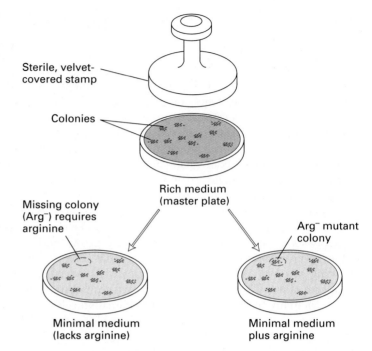

▲ FIGURE 6-2 Replica plating is used to detect random mutations in bacterial and yeast cultures that produce cells differing in a single genetic trait from other cells in the culture. A velvet-covered stamp equal in size to a petri dish is pressed onto the master plate used to culture colonies of bacteria on rich medium. The bacteria picked up on the stamp are then deposited in the same arrangement on new (selective) plates. The media in selective plates can be chosen to test different genetic traits such as nutritional auxotrophy (the inability to grow in the absence of a specific nutrient). In the case shown, the selective plates can detect cells with a mutation in the pathway of arginine biosynthesis.

controlling DNA synthesis, genetic recombination, and the gene products that regulate the cycle of events leading to replication and segregation of chromosomes and ultimately cell division (Chapter 13).

Mutant Strains of Bacteria and Yeast Can Be Isolated by Replica Plating

If a random mutation in a microorganism cultured, or plated, on a minimal agar medium renders cells unable to synthesize a necessary metabolite, and this metabolite is not supplied in the medium (e.g., the amino acid arginine), then the cells will die. But if the cells are grown on a rich medium that supplies the metabolite whose synthesis is prevented by a mutation, then the mutant cell can survive and grow into a colony in which all the cells have the same mutation (Table 6-1, *bottom*).

A mutant colony, say, an Arg⁻ colony, can be detected by *replica plating*, a technique developed by Joshua and Esther Lederberg in the 1940s. The initial plate of multiple colonies growing on rich medium is called the *master plate* (Figure 6-2). A circular, velvet-covered stamp equal in size to the master plate is pressed onto it, so that the stamp picks up cells from each colony. The stamp is then used to deposit cells onto a second plate containing minimal agar medium without arginine (Arg⁻ medium) and also onto a third plate containing minimal agar medium with arginine (Arg⁺ medium). Cells that can synthesize arginine grow into

colonies on both the Arg⁻ and Arg⁺ plates. However, a clone of cells on the master plate with a mutation in a gene required for arginine synthesis cannot grow on the Arg⁻ medium, resulting in a space on the Arg⁻ plate corresponding to the position of this clone on the master plate and Arg⁺ plate. This clone of cells, thus identified as being defective in the synthesis of arginine, then can be isolated from the master plate.

Repeated replica plating of a master plate with paired minimal-medium plates containing or lacking different specific nutrients (e.g., amino acids, nucleic acid bases, vitamins) can identify clones of cells with mutations in the genes required for synthesis of different biological molecules. This approach has been used to isolate thousands of different bacterial and yeast nutritional **auxotrophs**; these are strains that are unable to grow in the absence of a specific nutrient.

Under standard culture conditions, random mutations occur in E. coli genes at a frequency of 1 in 10^6–10^8 base pairs per generation. Thus, by the time a single cell has divided to produce a few hundred or thousand cells, some individual genes will differ among cells in the clone (see Figure 6-1b). To maintain the genetic homogeneity of a clone, it must be recloned frequently under selective conditions that permit only cells with a particular set of desired characteristics to survive and grow. For example, to maintain an Arg⁻ clone, isolated as described above, the cells are replated on minimal medium plus arginine; only cells that can synthesize all the metabolites required for growth except arginine will grow on this selective medium. The use of mutant clones both to map the position of genes on *E. coli* and yeast chromosomes and to unravel various biosynthetic pathways is discussed in Chapter 8.

SUMMARY Growth of Microorganisms in Culture

- Free-living microorganisms like the bacterium *E. coli* and the budding yeast *S. cerevisiae* can be grown in minimal media consisting of glucose as a source of carbon and energy, ammonium as a source of nitrogen, and salts and trace elements.

- Under suitable conditions, a single cell grows into a colony of cells on agar medium. The resulting population of genetically identical cells derived from a single parent cell is called a *clone*.

- Replica plating of colonies of microorganisms on two different types of agar medium allows detection of mutant clones that can form colonies on one type of medium but not another (see Figure 6-2).

- Nutritional auxotrophs, a common type of mutant, cannot synthesize an essential metabolite that wild-type cells produce.

6.2 Growth of Animal Cells in Culture

Animal cells are more difficult to culture than microorganisms because they require many more nutrients and typically grow only when attached to specially coated surfaces. Despite these difficulties, various types of animal cells, including both undifferentiated and differentiated ones, can be cultured successfully.

Rich Media Are Required for Culture of Animal Cells

Nine amino acids, referred to as the *essential amino acids,* cannot be synthesized by adult vertebrate animals and thus must be obtained from their diet. Animal cells grown in culture also must be supplied with these nine amino acids, namely, histidine, isoleucine, leucine, lysine, methionine, phenylalanine, threonine, tryptophan, and valine. In addition, most cultured cells require cysteine, glutamine, and tyrosine. In the intact animal, these three amino acids are synthesized by specialized cells; for example, liver cells make tyrosine from phenylalanine, and both liver and kidney cells can make glutamine. Animal cells both within the organism and in culture can synthesize the 8 remaining amino acids; thus these amino acids need not be present in the diet or culture medium. The other essential components of a medium for culturing animal cells are vitamins, which the cells cannot make at all or in adequate amounts; various salts; glucose; and *serum,* the noncellular part of the blood (Table 6-2, *top*).

Serum, a mixture of hundreds of proteins, contains various factors needed for proliferation of cells in culture. For example, it contains insulin, a hormone required for growth of many cultured vertebrate cells, and transferrin, an iron-transporting protein essential for incorporation of iron by cells in culture. Although many animal cells can grow in a serum-containing medium, such as Eagle's medium, certain cell types require specific protein growth factors that are not present in serum. For instance, precursors of red blood cells require the hormone erythropoietin, and T lymphocytes of the immune system require interleukin 2 (IL-2). These factors bind to receptor proteins that span the plasma membrane, signaling the cells to increase in size and mass and undergo cell division (Chapter 20). A few mammalian cell types can be grown in a completely defined, serum-free medium supplemented with trace minerals, specific protein growth factors, and other components (Table 6-2, *bottom*).

Most Cultured Animal Cells Grow Only on Special Solid Surfaces

Within the tissues of intact animals, most cells tightly contact and interact specifically with other cells via various cellular junctions. The cells also contact the extracellular matrix, a complex network of secreted proteins and carbohydrates that fills the spaces between cells (Chapter 22). The matrix, whose constituents are secreted by cells themselves, helps bind the cells in tissues together; it also provides a lattice through which cells can move, particularly during the early stages of animal differentiation.

The extracellular matrices in various animal tissues consist of several common components: fibrous **collagen** proteins; **hyaluronan** (or hyaluronic acid), a large mucopolysaccharide; and covalently linked polysaccharides and proteins in the form of **proteoglycans** (mostly carbohydrate) and **glycoproteins** (mostly protein). However, the exact composition of the matrix in different tissues varies, reflecting the specialized function of a tissue. In connective tissue, for example, the major protein of the extracellular matrix is a type of collagen that forms insoluble fibers with a very high tensile strength. Fibroblasts, the principal cell type in connective tissue, secrete this type of collagen as well as the other matrix components. Receptor proteins in the plasma membrane of cells bind various matrix elements, imparting strength and rigidity to tissues (see Figure 5-40).

TABLE 6-2   Growth Media for Mammalian Cells

SERUM-CONTAINING MEDIUM (EAGLE'S MEDIUM)

Essential amino acids	The essential amino acids—histidine, isoleucine, leucine, lysine, methionine, phenylalanine, threonine, tryptophan, and valine—plus cysteine, glutamine, and tyrosine (all at 10^{-4} to 10^{-5} M)
Vitamins	Choline, folic acid, nicotinamide, pantothenate, pyridoxal, and thiamine (all at 1 mg/L); inositol (2 mg/L); riboflavin (0.1 mg/L)
Salts	Na^+, K^+, Ca^{2+}, Mg^{2+}, Cl^-, PO_4^{3-}, HCO_3^-
Glucose	0.9 g/L
Dialyzed serum*	5–10% of total volume

DEFINED (SERUM-FREE) MEDIUM

Amino acids	As above plus alanine and asparagine (10^{-4} M)
Vitamins, salts, glucose	As above
Other additions:	
Fatty acids	Linoleic acid, lipoic acid
Nitrogen compounds	Hypoxanthine, thymidine, putrescine
Carbon source	Pyruvate and glucose (0.9 g/L)
Trace elements	Cadmium (Cd), manganese (Mn), molybdenum (Mo), nickel (Ni), tin (Sn), vanadium (V)
Hormones and growth factors	Insulin, transferrin, hydrocortisone, fibroblast growth factor, epidermal growth factor

*Serum contains hundreds of proteins with a total protein concentration of 50–70 mg/ml. Albumin is the most plentiful serum protein (30–50 mg/ml). Various growth factors are present in serum at very low concentrations: for example, growth hormone (34 ng/ml) and insulin (0.2 ng/ml). Serum from fetal calves is used frequently because it contains higher concentrations of certain growth factors than serum from other species.
SOURCE: H. Eagle, 1959, Science 130:432; S. E. Hutchings and G. H. Sato, 1978, Proc. Nat'l., Acad. Sci. USA 75:901.

▼ FIGURE 6-3 Cultured mammalian cells viewed at three magnifications. (a) A single mouse cell attached to a plastic petri dish, viewed through a scanning electron microscope. To separate attached cells so they can be plated individually, a cell culture is treated with a protease such as trypsin. (b) A single colony of human HeLa cells about 1 mm in diameter, produced from a single cell after growth for 2 weeks. (c) After cells have been stained in a 6-cm-diameter petri dish, individual colonies can easily be seen and counted. [See P. I. Marcus et al., 1956, J. Exp. Med. **104**:615. Part (a) courtesy of N. K. Weller; parts (b) and (c) courtesy of T. T. Puck.]

The tendency of animal cells in vivo to interact with one another and with the surrounding extracellular matrix is mimicked in their growth in culture. Unlike bacterial and yeast cells, which can be grown in suspension, most cultured animal cells require a surface to grow on. Many types of cells can adhere to and grow on glass, or on specially treated plastics with negatively charged groups on the surface (e.g., SO_3^{2-}). The cultured cells secrete collagens and other matrix components; these bind to the culture surface and function as a bridge between it and the cells. Cells cultured from

(a)

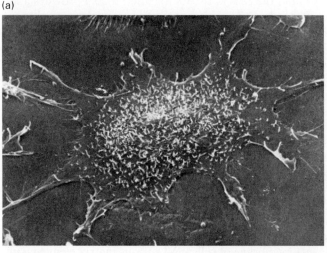

10 μm

(b)

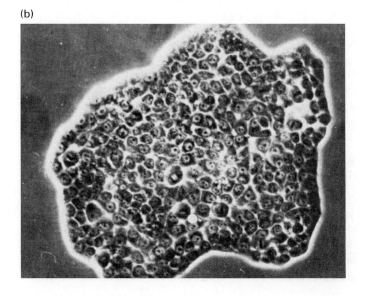

single cells on a glass or a plastic dish form visible colonies in 10–14 days (Figure 6-3). Some tumor cells can be grown in suspension, a considerable experimental advantage because equivalent samples are easier to obtain from suspension cultures than from colonies grown in a dish.

Primary Cell Cultures Are Useful, but Have a Finite Life Span

Normal animal tissues (e.g., skin, kidney, liver) or whole embryos commonly are used to establish *primary cell cultures.* To prepare tissue cells for culture (or to remove adherent cells from a culture dish for biochemical studies), trypsin or another protease is used to destroy the proteins in the junctions that normally interconnect cells. For many years, most cell types were difficult, if not impossible, to culture. But the identification and preparation of various protein growth factors that stimulate the replication of specific cell types, as well as other recent modifications in culture methods, now permit experimenters to grow various types of specialized cells.

Many studies with vertebrate cells, however, still are performed with those few cell types that grow most readily in culture. These are not cells of a defined type; rather, they represent whatever grows when a tissue or an embryo is placed in culture. The cell type that usually predominates in such cultures is called a **fibroblast** because it secretes the types of proteins associated with fibroblasts in fibrous connective tissue of animals. Cultured fibroblasts have the morphology of tissue fibroblasts, but they retain the ability to differentiate into other cell types; thus they are not as differentiated as tissue fibroblasts.

Some studies are conducted with primary cultures of epithelial cells. In general, external and internal surfaces of tissues and organs are covered by a layer of epithelial cells called an **epithelium** (Figure 6-4). These highly differentiated

(c)

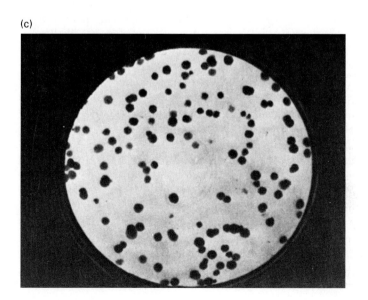

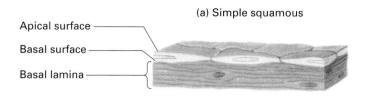

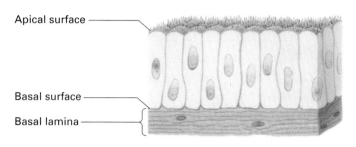

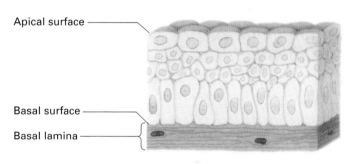

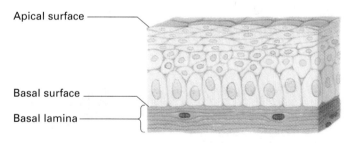

▲ **FIGURE 6-4 Principal types of epithelium.** The apical and basal surfaces of epithelial cells exhibit distinctive characteristics. (a) Simple squamous epithelia, composed of thin cells, line the blood vessels and many body cavities. (b) Simple columnar epithelia consist of elongated cells, including mucus-secreting cells (in the lining of the stomach and cervical tract) and absorptive cells (in the lining of the small intestine). (c) Transitional epithelia, composed of several layers of cells with different shapes, line certain cavities subject to expansion and contraction (e.g., the urinary bladder). (d) Stratified squamous (nonkeratinized) epithelia line surfaces such as the mouth and vagina; these linings resist abrasion and generally do not participate in the absorption or secretion of materials into or out of the cavity.

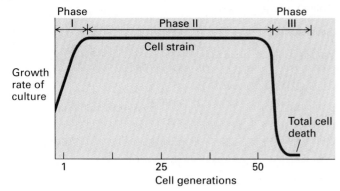

(a) Human cells

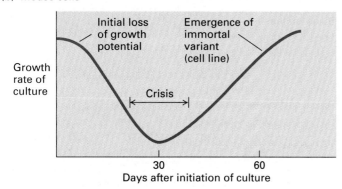

(b) Mouse cells

▲ **FIGURE 6-5 Stages in the establishment of a cell culture.**
(a) When an initial explant is made of human cells, some cells die and others (mainly fibroblasts) start to grow; overall the growth rate increases (phase I). If the remaining cells are continually diluted, the cell strain grows at a constant rate for about 50 cell generations (phase II), after which the growth rate falls rapidly. During the ensuing period of increasing cell death (phase III), all the cells in the culture eventually die. (b) In a culture prepared from mouse or other rodent embryo cells, there is initial cell death coupled with the emergence of healthy growing cells. As these are diluted and allowed to continue growth, they soon begin to lose growth potential and most cells die (the culture goes into crisis). Very rare cells do not die but continue growing until their progeny overgrow the culture. These cells constitute a cell line, which will grow forever if it is appropriately diluted and fed with nutrients: the cells are immortal.

cells are said to be *polarized* because the plasma membrane is organized into at least two discrete regions. For example, the epithelial cells that line the intestine form a simple columnar epithelium (see Figure 6-4b). That portion of the plasma membrane facing the lumen of the intestine, the *apical* surface, is specialized for absorption; the rest of the plasma membrane, the *basolateral* surface, mediates transport of nutrients from the cell to the blood and forms junctions with adjacent cells and the underlying extracellular matrix called the **basal lamina.**

Certain cells cultured from blood, spleen, or bone marrow adhere poorly, if at all, to a culture dish but nonetheless grow well. In the body, such nonadherent cells are held in suspension (in the blood), or they are loosely adherent (in the bone marrow and spleen). Because these cells often come from immature stages in the development of differentiated blood cells, they are very useful for studying normal blood cell differentiation and the abnormal development of leukemias.

When cells are removed from an embryo or an adult animal, most of the adherent ones grow continuously in culture for only a limited time before they spontaneously cease growing. Such a culture eventually dies out after many cell doublings, even if it is provided with fresh supplies of all the known nutrients that cells need to grow, including serum. For instance, when human fetal cells are explanted into cell culture, the majority of cells die within a relatively short time; "fibroblasts," although also destined to die, proliferate for a while and soon become the predominant cell type. They divide about 50 times before they cease growth. Starting with 10^6 cells, 50 doublings can produce $10^6 \times 2^{50}$, or more than 10^{20} cells, which is equivalent to the weight of about 10^5 people. Thus, even though its lifetime is limited, a single culture, if carefully maintained, can be studied through many generations. Such a lineage of cells originating from one initial primary culture is called a **cell strain** (Figure 6-5a).

Transformed Cells Can Grow Indefinitely in Culture

To be able to clone individual cells, modify cell behavior, or select mutants, biologists often want to maintain cell cultures for many more than 100 doublings. This is possible with cells derived from some tumors and with rare cells that arise spontaneously because they have undergone genetic changes that endow them with the ability to grow indefinitely. The genetic changes that allow these cells to grow indefinitely are collectively called *oncogenic* **transformation,** and the cells are said to be *oncogenically transformed,* or simply *transformed.* A culture of cells with an indefinite life span is considered immortal; such a culture is called a **cell line** to distinguish it from an impermanent cell *strain.*

The ability of cultured cells to grow indefinitely or their tendency to be transformed varies depending on the animal species from which the cells originate. Normal chicken cells rarely are transformed and die out after only a few doublings; even tumor cells from chickens almost never exhibit immortality. Among human cells, only tumor cells grow indefinitely. The **HeLa cell,** the first human cell type to be grown in culture, was originally obtained in 1952 from a malignant tumor (carcinoma) of the uterine cervix. This cell line has been invaluable for research on human cells.

In contrast to human and chicken cells, cultures of embryonic adherent cells from rodents routinely give rise to cell lines. When adherent rodent cells are first explanted, they grow well, but after a number of serial replatings they lose growth potential and the culture goes into crisis

Early stage
of myotube

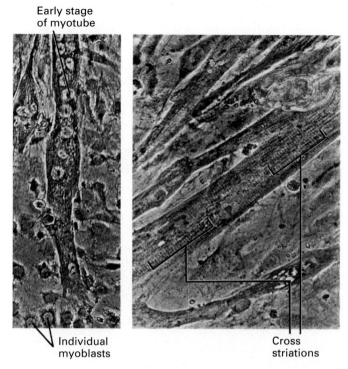

Individual
myoblasts

Cross
striations

▲ **FIGURE 6-6 Cultured transformed line of rat myoblasts.**
(Left) This cell line grows indefinitely as single cells in culture.
(Right) When growth of cultured myoblasts is stopped (e.g., by
removing serum from the medium), the cells fuse to produce
myotubes with the characteristic cross striations of differentiated
muscle cells.

(Figure 6-5b). During this period most of the cells die, but
often a rapidly dividing variant cell arises spontaneously and
takes over the culture. A cell line derived from such a variant will grow forever if it is provided with the necessary nutrients. Cells in spontaneously established rodent cell lines
and in cell lines derived from tumors often have abnormal
chromosomes. In addition, their chromosome number usually is greater than that of the normal cell from which they
arose, and it continually expands and contracts in culture.
Such cells are said to be *aneuploid* (i.e., have an inappropriate number of chromosomes) and are obviously mutants.

Although most cell lines are undifferentiated, some can
carry out many of the functions characteristic of the normal
differentiated cells from which they are derived. One example is certain hepatoma cell lines (e.g., HepG2) that synthesize most of the serum proteins made by normal hepatocytes (the major cell type in the liver) from which they are
derived. These highly differentiated hepatoma cells are often studied as models of normal hepatocytes. Cultured
myoblasts (muscle precursor cells) are another example of
transformed cells that continue to perform many functions
of a specialized, differentiated cell. When grown in culture,
transformed myoblasts can be induced to fuse to form myotubes. These resemble differentiated multinucleated muscle
cells and synthesize many of, if not all, the specialized pro-

teins associated with contraction (Figure 6-6). Certain lines
of epithelial cells also have been cultured successfully. One
such line, Madin-Darby canine kidney (MDCK) cells, forms
a continuous sheet of polarized epithelial cells one cell thick
that exhibits many of the properties of the normal canine
kidney epithelium from which it was derived (Figure 6-7).
This type of preparation has proved valuable as a model for
studying the functions of epithelial cells.

Fusion of Cultured Animal Cells Can Yield Interspecific Hybrids Useful in Somatic-Cell Genetics

Cultured animal cells infrequently undergo **cell fusion** spontaneously. The fusion rate, however, increases greatly in the
presence of certain viruses that have a lipoprotein envelope
similar to the plasma membrane of animal cells. A mutant
viral glycoprotein in the envelope promotes cell fusion (see
the photograph on the first page of this chapter); the mechanism of this effect is discussed at the end of Chapter 17.
Cell fusion also is promoted by polyethylene glycol, which
causes the plasma membranes of adjacent cells to adhere
to each other and to fuse (Figure 6-8). As most fused animal cells undergo cell division, the nuclei eventually fuse,
producing viable cells with a single nucleus that contains
chromosomes from both "parents." The fusion of two
cells that are genetically different yields a hybrid cell
called a **heterokaryon.**

Because some **somatic cells** from animals can be cultured
from single cells in a well-defined medium, it is possible to
select for genetically distinct cultured animal cells, just as is
done with bacterial and yeast cells. Moreover, during mitosis the chromosomes in an animal cell are large and highly
visible after staining, making it easy to distinguish individual chromosomes (Chapter 9). Genetic studies of cultured
animal cells are called *somatic-cell genetics* to distinguish
them from *classical genetics,* which deals with whole organisms derived from **germ cells** (sperm and eggs).

Cultured cells from different mammals can be fused to
produce interspecific hybrids, which have been widely used
in somatic-cell genetics. For instance, hybrids can be prepared from human cells and mutant mouse cells that lack
an enzyme required for synthesis of a particular essential
metabolite. As the human-mouse hybrid cells grow and divide, they gradually lose human chromosomes in random
order, but retain the mouse chromosomes. In a medium that
can support growth of both the human cells and mutant
mouse cells, the hybrids eventually lose all human chromosomes. However, in a medium lacking the essential metabolite that the mouse cells cannot produce, the one human
chromosome that contains the gene encoding the needed enzyme will be retained, because any hybrid cells that lose it
following mitosis will die. All other human chromosomes
eventually are lost.

By using different mutant mouse cells and media in
which they cannot grow, researchers have prepared various

(a)

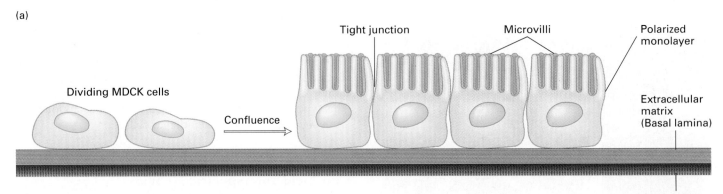

Dividing MDCK cells

Tight junction   Microvilli   Polarized monolayer

Confluence

Extracellular matrix (Basal lamina)

Porous filter

(b)

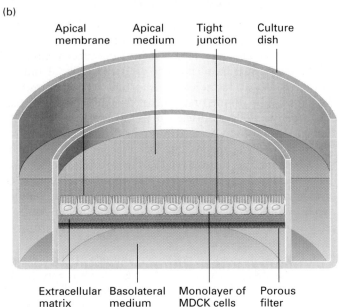

Apical membrane   Apical medium   Tight junction   Culture dish

Extracellular matrix   Basolateral medium   Monolayer of MDCK cells   Porous filter

(c)

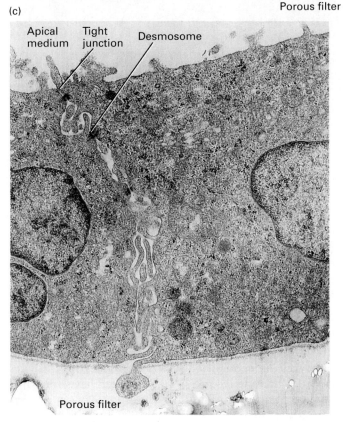

Apical medium   Tight junction   Desmosome

Porous filter

▲ **FIGURE 6-7 Culture of Madin-Darby canine kidney (MDCK) cells, a line of differentiated epithelial cells.** (a) MDCK cells form a polarized epithelium when grown to confluence on a porous membrane filter coated on one side with collagen and other proteins of the basal lamina, the extracellular matrix that supports an epithelial layer. (b) Special culture dishes allow the cells to be bathed with an appropriate medium on each side of the filter; the apical surface faces the medium that bathes the upper side. Note that the tight junctions connecting the epithelial cells form the physical barrier separating the basolateral (blue) from the apical (green) extracellular space. (c) Electron micrograph of parts of two MDCK cells grown in tissue culture on a permeable filter. Like tight junctions, desmosomes are specialized regions of the plasma membrane that connect adjacent cells. [Part (c) courtesy of R. Van Buskirk, J. Cook, J. Gabriels, and H. Eichelberger.]

panels of hybrid cell lines. Each cell line in a panel contains either a single human chromosome or a small number of human chromosomes, and a full set of mouse chromosomes. Because each chromosome can be identified visually under a light microscope, such hybrid cells provide a means for assigning, or "mapping," individual genes to specific chromosomes. For example, suppose a hybrid cell line is shown microscopically to contain a particular human chromosome. That hybrid cell line can then be tested biochemically for the presence of various human enzymes, exposed to specific antibodies to detect human surface antigens, or subjected to DNA hybridization and cloning techniques (Chapter 7) to locate particular human DNA sequences. The genes encoding a human protein or containing a human DNA sequence detected in such tests must be located on the particular human chromosome carried by the cell line being tested. Panels of hybrids between normal mouse and mutant hamster cells also have been established; in these hybrid cells, the majority of mouse chromosomes are lost, allowing mouse genes to be mapped to specific mouse chromosomes.

(a)

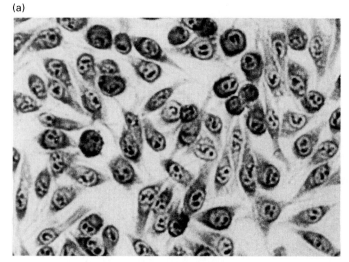

(b)

Heterokaryon

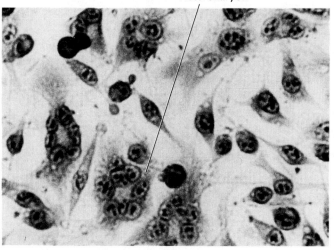

▲ **FIGURE 6-8 Fusion of cultured animal cells.** (a) Unfused growing mouse cells with a single nucleus per cell. (b) Fused mouse cells with 2–5 nuclei per cell. Fusion was induced by treatment with polyethylene glycol (45 percent) for 1 minute. The number of nuclei per fused cell (heterokaryon) is determined by the polyethylene glycol concentration and the time of exposure. By adjustment of these factors, the number of heterokaryons containing only two nuclei can be maximized. [From R. L. Davidson and P. S. Gerald, 1976, *Som. Cell Genet.* **2**:165.]

Hybrid Cells Often Are Selected on HAT Medium

One metabolic pathway has been particularly useful in cell-fusion experiments. Most animal cells can synthesize the purine and pyrimidine nucleotides de novo from simpler carbon and nitrogen compounds, rather than from already formed purines and pyrimidines (Figure 6-9, *top*). The folic acid antagonists amethopterin and aminopterin interfere with the donation of methyl and formyl groups by tetrahydrofolic acid in the early stages of de novo synthesis of glycine, purine nucleoside

monophosphates, and thymidine monophosphate. These drugs are called *antifolates,* since they block reactions involving tetrahydrofolate, an active form of folic acid. Many cells, however, contain enzymes that can synthesize the necessary nucleotides from purine bases and thymidine if they are provided in the medium; these *salvage pathways* bypass the metabolic blocks imposed by antifolates (Figure 6-9, *bottom*).

A number of mutant cell lines lacking the enzyme needed to catalyze one of the steps in a salvage pathway have been isolated. For example, cell lines lacking thymidine kinase (TK) can be selected because such cells are resistant to the otherwise toxic thymidine analog 5-bromodeoxyuridine. Cells containing TK convert 5-bromodeoxyuridine into 5-bromodeoxyuridine monophosphate. This nucleoside monophosphate is then converted into a nucleoside triphosphate by other enzymes and is incorporated by DNA polymerase into DNA, where it exerts its toxic effects. This pathway is blocked in cells with a *TK* mutation that prevents production of functional TK enzyme. Hence, TK$^-$ mutants are resistant to the toxic effects of 5-bromodeoxyuridine. Similarly, cells lacking the HGPRT enzyme have been selected because they are resistant to the otherwise toxic guanine analog 6-thioguanine. As we will see next, HGPRT$^-$ cells and TK$^-$ cells are useful partners in cell fusions with one another or with cells that have salvage-pathway enzymes but that are differentiated and cannot grow in culture by themselves.

The medium most often used to select hybrid cells is called *HAT medium,* because it contains hypoxanthine (a purine), aminopterin, and thymidine. Normal cells can grow in HAT medium because even though aminopterin blocks de novo synthesis of purines and TMP, the thymidine in the media is transported into the cell and converted to TMP by TK and the hypoxanthine is transported and converted into usable purines by HGPRT. On the other hand, neither TK$^-$ nor HGPRT$^-$ cells can grow in HAT medium because each lacks an enzyme of the salvage pathway. However, hybrids formed by fusion of these two mutants will carry a normal *TK* gene from the HGPRT$^-$ parent and a normal *HGPRT* gene from the TK$^-$ parent. The hybrids thus will produce both functional salvage-pathway enzymes and grow on HAT medium. Likewise, hybrids formed by fusion of mutant cells and normal cells can grow in HAT medium.

Hybridomas Are Used to Produce Monoclonal Antibodies

Each normal B **lymphocyte** in an animal is capable of producing a single type of antibody directed against a specific determinant, or **epitope,** on an antigen molecule. If an animal is injected with an antigen, B lymphocytes that make antibody recognizing the antigen are stimulated to grow and proliferate. Each antigen-activated B lymphocyte forms a clone of cells in the spleen or lymph nodes, with each cell of the clone producing identical antibody, termed **monoclonal antibody.** Because most natural antigens contain multiple epitopes, exposure of an animal to an antigen usually stimulates formation

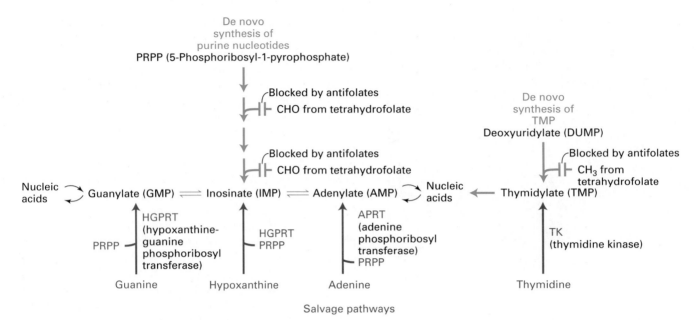

▲ FIGURE 6-9 **De novo and salvage pathways for nucleotide synthesis.** In a normal medium, cultured animal cells synthesize purine nucleotides (AMP, GMP, IMP) and thymidylate (TMP) by de novo pathways (blue). These require the transfer of a methyl or formyl group from an activated form of tetrahydrofolate (e.g., N^5,N^{10}-methylenetetrahydrofolate), as shown in the upper portion of the diagram. Antifolates, such as aminopterin and

amethopterin, block the reactivation of tetrahydrofolate, preventing purine and thymidylate synthesis. Normal cells can also use salvage pathways (red) to incorporate purine bases or nucleosides and thymidine added to the medium. Cultured cells lacking one of the enzymes of the salvage pathways—HGPRT, APRT, or TK—will not survive in media containing antifolates.

of several different B-lymphocyte clones, each producing a different antibody; a mixture of antibodies that recognize different epitopes on the same antigen is said to be *polyclonal.*

For many types of studies involving antibodies, monoclonal antibody is preferable to polyclonal antibody. However, biochemical purification of monoclonal antibody from serum is not feasible, in part because the concentration of any given antibody is quite low. For this reason, researchers looked to culture techniques in order to obtain usable quantities of monoclonal antibody. Because primary cultures of normal B lymphocytes do not grow indefinitely, such cultures have limited usefulness for production of monoclonal antibody. This limitation can be avoided by fusing normal B lymphocytes with oncogenically transformed lymphocytes called *myeloma cells,* which are immortal.

Fusion of a myeloma cell with a normal antibody-producing cell from a rat or mouse spleen yields a hybrid that proliferates into a clone called a **hybridoma.** Like myeloma cells, hybridoma cells are immortal. Each hybridoma produces the monoclonal antibody encoded by its B-lymphocyte partner. Many different myeloma cell lines from mice and rats have been established; from these, HGPRT⁻ lines have been selected based on their resistance to 6-thioguanine as

described above. If such mutant myeloma cells are fused with normal B lymphocytes, any fused cells that result can grow in HAT medium, but the parental cells cannot (Figure 6-10). Each selected hybridoma then is tested for production of the desired antibody; any clone producing that antibody then is grown in large cultures, from which a substantial quantity of pure monoclonal antibody can be obtained.

 Such pure antibodies are very valuable research reagents. For example, a monoclonal antibody that interacts with protein X can be used to label, and thus locate, protein X in specific cells of an organ or in specific cell fractions. Once identified, even very scarce proteins can be isolated by affinity chromatography in columns to which the monoclonal antibody is bound (see Figure 3-43c). Monoclonal antibodies also have become important diagnostic and therapeutic tools in medicine. Monoclonal antibodies that bind to and inactivate toxic proteins (toxins) secreted by bacterial pathogens are used to treat diseases caused by these pathogens. Other monoclonal antibodies are specific for cell-surface proteins expressed by certain types of tumor cells; chemical complexes of such monoclonal antibodies with toxic drugs are being developed for cancer chemotherapy.

▶ **FIGURE 6-10 Procedure for producing a monoclonal antibody to protein X.** Immortal myeloma cells that lack HGPRT, an enzyme of the purine-salvage pathway (see Figure 6-9), are fused with normal antibody-producing spleen cells from an animal that was immunized with protein X. The spleen cells can make HGPRT. When plated in HAT medium, the unfused cells do not grow: the mutant myeloma cells because they cannot make purines via the salvage pathway, and the spleen cells because they have a limited life span in culture. Thus only fused cells, formed from a myeloma cell and a spleen cell, survive on HAT medium, proliferating into clones called *hybridomas*. Each hybridoma produces a single antibody. Once a hybridoma that produces a desired antibody is identified, the clone can be cultured to yield large amounts of that antibody.

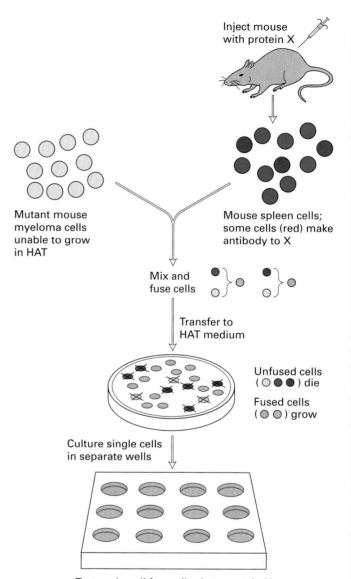

Inject mouse with protein X

Mutant mouse myeloma cells unable to grow in HAT

Mouse spleen cells; some cells (red) make antibody to X

Mix and fuse cells

Transfer to HAT medium

Unfused cells (○ ● ●) die

Fused cells (○ ○) grow

Culture single cells in separate wells

Test each well for antibody to protein X

MEDIA CONNECTIONS — Preparing Monoclonal Antibodies

SUMMARY Growth of Animal Cells in Culture

- Growth of vertebrate cells in culture requires rich media containing essential amino acids, vitamins, and peptide or protein growth factors, frequently provided by serum. Most cultured vertebrate cells will grow only when attached to a negatively charged substratum that mimics the extracellular matrix in animal tissues.

- Primary cells, which are derived directly from animal tissue, have limited growth potential in culture and may give rise to a cell strain.

- Transformed cells, which are derived from animal tumors or arise spontaneously from primary rodent cells, grow indefinitely in culture (see Figure 6-5b). They usually have an unstable, aneuploid complement of chromosomes, including abnormal chromosomes. Transformed cells derived from a single parental cell are called *cell lines*.

- Cultured cells can be induced to fuse into heterokaryons (hybrids) by treatment with certain viruses or polyethylene glycol. Heterokaryons between cells of different species tend to lose the chromosomes of one species as they divide.

- Panels of hybrid lines prepared from mutant mouse cells and normal human cells, each containing different human chromosomes, can be used to map the gene encoding a specific human protein to a specific human chromosome.

- Fusion of an HGPRT⁻ myeloma cell and a single B lymphocyte yields a hybrid cell that can grow on HAT medium and proliferate indefinitely, forming a clone called a *hybridoma* (see Figure 6-10). Since each individual B lymphocyte produces antibodies specific for one antigenic determinant (epitope), a hybridoma produces only the monoclonal antibody synthesized by its original B-lymphocyte parental cell.

6.3 Viruses: Structure, Function, and Uses

A **virus** is a small parasite that cannot reproduce by itself. Once it infects a susceptible cell, however, a virus can direct the cell machinery to produce more viruses. Most viruses have either RNA or DNA as their genetic material. The nucleic acid may be single- or double-stranded. The entire infectious virus particle, called a **virion,** consists of the nucleic acid and an outer shell of protein. The simplest viruses contain only enough RNA or DNA to encode four proteins. The most complex can encode 100–200 proteins.

PLANTS The study of plant viruses inspired some of the first experiments in molecular biology. In 1935, Wendell Stanley purified and partly crystallized tobacco mosaic virus (TMV); other plant viruses were crystallized soon thereafter. Pure proteins had been crystallized only a short time before Stanley's work, and it was considered very surprising at the time that a replicating organism could be crystallized.

A wealth of subsequent research with bacterial viruses and animal viruses has provided detailed understanding of viral structure, and virus-infected cells have proved extremely useful as model systems for the study of basic aspects of cell biology. In many cases, DNA viruses utilize cellular enzymes for synthesis of their DNA genomes and mRNAs; all viruses utilize normal cellular ribosomes, tRNAs, and translation factors for synthesis of their proteins. Most viruses commandeer the cellular machinery for macromolecular synthesis during the late phase of infection, directing it to synthesize large amounts of a small number of viral mRNAs and proteins instead of the thousands of normal cellular macromolecules. For instance, animal cells infected by influenza or vesicular stomatitis virus synthesize only one or two types of glycoproteins, which are encoded by viral genes, whereas uninfected cells produce hundreds of glycoproteins. Such virus-infected cells have been used extensively in studies on synthesis of cell-surface glycoproteins. Similarly, much information about the mechanism of DNA replication has come from studies with bacterial cells and animal cells infected with simple DNA viruses, since these viruses depend almost entirely on cellular proteins to replicate their DNA. Viruses also often express proteins that modify host-cell processes so as to maximize viral replication. For example, the roles of certain cellular factors in initiation of protein synthesis were revealed because viral proteins interrupt their action. Finally, when certain genes carried by cancer-causing viruses integrate into chromosomes of a normal animal cell, the normal cell can be converted to a cancer cell.

Since many viruses can infect a large number of different cell types, genetically modified viruses often are used to carry foreign DNA into a cell. This approach provides the basis for a growing list of experimental gene therapy treatments. Because of the extensive use of viruses in cell biology research and their potential as therapeutic agents, we describe the basic aspects of viral structure and function in this section.

Viral Capsids Are Regular Arrays of One or a Few Types of Protein

The nucleic acid of a virion is enclosed within a protein coat, or **capsid**, composed of multiple copies of one protein or a few different proteins, each of which is encoded by a single viral gene. Because of this structure, a virus is able to encode all the information for making a relatively large capsid in a small number of genes. This efficient use of genetic information is important, since only a limited amount of RNA or DNA, and therefore a limited number of genes, can fit into a virion capsid. A capsid plus the enclosed nucleic acid is called a **nucleocapsid**.

Nature has found two basic ways of arranging the multiple capsid protein subunits and the viral genome into a nucleocapsid. The simpler structure is a protein helix with the RNA or DNA protected within. Tobacco mosaic virus (TMV) is a classic example of the helical nucleocapsid. In TMV the protein subunits form broken disklike structures, like lock washers, which form the helical shell of a long rodlike virus when stacked together (Figure 6-11a).

(a) Section of a helical virus

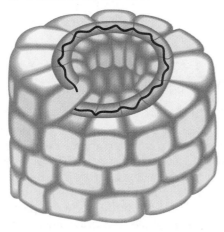

(b) A small icosahedral virus

(c) A large icosahedral virus

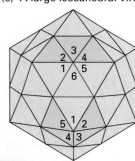

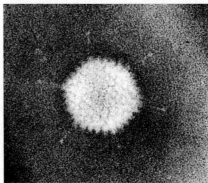

(a)

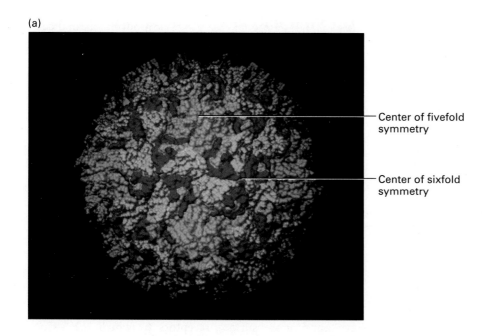

— Center of fivefold symmetry

— Center of sixfold symmetry

(b)

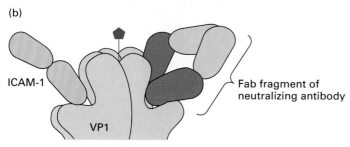

ICAM-1

VP1

Fab fragment of neutralizing antibody

◀ **FIGURE 6-12 Structure of picornaviruses.** These icosahedral viruses include poliovirus and the rhinoviruses, which cause the common cold. (a) The picornavirus capsid is composed of four proteins (VP1, VP2, VP3, and VP4; VP4 is located in the interior). This model of a picornavirus, based on x-ray crystallographic analyses, shows that the vertices with fivefold symmetry contain five VP1 molecules (blue); the surfaces with sixfold symmetry contain three VP2 molecules and three VP3 molecules (red and yellow). (b) Many picornaviruses have an indentation, or "canyon," encircling each vertex of the icosahedron. In rhinoviruses, this canyon interacts with ICAM-1, a cell-adhesion molecule on the surface of respiratory epithelial cells, allowing the virus to bind to these cells in the first step of infection. Neutralizing antibodies also bind to the canyon, thereby preventing ICAM-1 from entering the canyon. Only a portion of the ICAM-1 and antibody molecules are depicted. [Part (a) from J. M. Hogle et al., 1987, *Sci. Am.* **256**(3):42; courtesy of James M. Hogle. Part (b) adapted from T. J. Smith et al., 1996, *Nature* **383**:350.]

◀ **FIGURE 6-11 Two basic geometric shapes of viruses.**
(a) In some viruses, the protein subunits form helical arrays around an RNA or DNA molecule (red), which runs in a helical groove within the enclosing protein tube. The electron micrograph to the right is of tobacco mosaic virus (TMV), illustrating the rodlike shape of this type of virus. (b, c) In other viruses, the capsid proteins associate to form polyhedrons with icosahedral (20-sided) symmetry. In the simplest and smallest of these quasi-spherical viruses (b), three identical capsid protein subunits form each triangular face (red) of the icosahedron. The subunits meet in fivefold symmetry at each vertex. In some larger viruses of this type, each triangular face is composed of four subunits (c). The contact between subunits not at the vertices is quasi-equivalent: the subunits on the vertices maintain fivefold symmetry, but those making up the surfaces in between exhibit sixfold symmetry. Although the actual shape of the protein subunits in these viruses is not a flat triangle as illustrated, the overall effect when the subunits are assembled is of a roughly spherical structure with triangular faces. The electron micrograph beneath (b) and (c) is of an adenovirus. Viral proteins that attach to host-cell receptors project from the vertices. [After S. E. Luria et al., 1978, *General Virology*, 3d ed., Wiley, pp. 39–40. Photograph of TMV courtesy of R. C. Valentine; photograph of adenovirus courtesy of Robley C. Williams, University of California.]

The other major structural class of viruses, called *icosahedral* or *quasi-spherical viruses*, is based on the icosahedron, a solid object built of 20 identical faces, each of which is an equilateral triangle. In the simplest type of icosahedral virion each of the 20 triangular faces is constructed of three identical capsid protein subunits, making a total of 60 subunits per capsid. At each of the 12 vertices, five subunits make contact symmetrically (Figure 6-11b). Thus all protein subunits are in *equivalent* contact with one another. Tobacco satellite necrosis virus has such a simple icosahedral structure. However, most quasi-spherical viruses are larger, requiring the assembly of more than three subunits per face of the icosahedron. These proteins form shells whose subunits are in *quasi-equivalent* contact. Here, the proteins at the icosahedral vertices remain arranged in a fivefold symmetry, but additional subunits cover the surfaces between in a pattern of sixfold symmetry (Figure 6-11c).

The atomic structures of a number of icosahedral viruses have been determined by x-ray crystallography (Figure 6-12a). The first three such viruses to be analyzed—tomato bushy stunt virus, poliovirus, and rhinovirus (the common cold virus)—exhibit a remarkably similar design, in terms of the rules of icosahedral symmetry as well as in the details of their surface proteins. In each virus, at atomic resolution, clefts ("canyons") are observed encircling each of the vertices of the icosahedral structure. Interaction of these

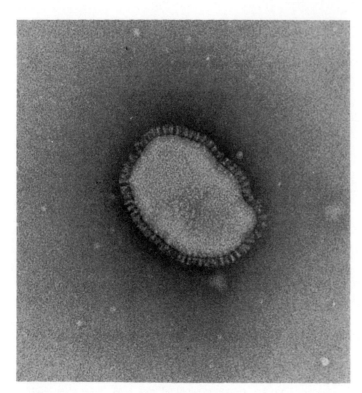

▲ **FIGURE 6-13 Electron micrograph of a negatively stained influenza virus virion.** The virion is surrounded by a phospholipid bilayer; the large spikes protruding outward from the membrane are composed of trimers of hemagglutinin protein and tetramers of neuraminidase protein. Inside is the nucleocapsid. [Courtesy of A. Helenius and J. White.]

clefts with cell-surface receptors attaches the virus to a host cell, the first step in viral infection (Figure 6-12b). Neutralizing antibodies specific for a particular virus also interact with these clefts, thereby inhibiting attachment of the virus to the host cell.

In some viruses, the symmetrically arranged nucleocapsid is covered by an external membrane, or *envelope,* which consists mainly of a phospholipid bilayer but also contains one or two types of virus-encoded glycoproteins (Figure 6-13). The phospholipids in the viral envelope are similar to those in the plasma membrane of an infected host cell. The viral envelope is, in fact, derived by budding from that membrane, but contains mainly viral glycoproteins.

The components of simple viruses such as TMV, which consists of a single RNA molecule and one protein species, undergo self-assembly if they are mixed in solution. More complex viruses containing a dozen or more protein species do not spontaneously assemble in vitro. The multiple components of such viruses assemble within infected cells in stages, first into subviral particles and then into completed virions. The genomes of these complex viruses encode proteins that assist in the assembly of the virion, but the assembly proteins are not themselves components of the completed virion.

Most Viral Host Ranges Are Narrow

The fact that the host range—the group of cell types that a virus can infect—is generally restricted serves as a basis for classifying viruses. A virus that infects only bacteria is called a **bacteriophage,** or simply a **phage.** Viruses that infect animal or plant cells are referred to generally as *animal viruses* or *plant viruses.* A few viruses can grow in both plants and the insects that feed on them. The highly mobile insects serve as vectors for transferring such viruses between susceptible plant hosts. An example is potato yellow dwarf virus, which can grow in leafhoppers (insects that feed on potato plant leaves) as well as in potato plants. Wide host ranges are characteristic of some strictly animal viruses, such as vesicular stomatitis virus, which grows in insects and in many different types of mammalian cells. Most animal viruses, however, do not cross phyla, and some (e.g., poliovirus) infect only closely related species such as primates. The host-cell range of some animal viruses is further restricted to a limited number of cell types because only these cells have appropriate surface receptors to which the virions can attach.

Viruses Can Be Cloned and Counted in Plaque Assays

The number of infectious viral particles in a sample can be quantified by a **plaque assay.** This assay is performed by culturing a dilute sample of viral particles on a plate covered with host cells and then counting the number of local lesions, called *plaques,* that develop (Figure 6-14). A plaque develops on the plate wherever a single virion initially infects a single cell. The virus replicates in this initial host cell and then lyses the cell, releasing many progeny virions that infect the neighboring cells on the plate. After a few such cycles of infection, enough cells are lysed to produce a visible plaque in the layer of remaining uninfected cells.

Since all the progeny virions in a plaque are derived from a single parental virus, they constitute a virus clone. This type of plaque assay is in standard use for bacterial and animal viruses. Plant viruses can be assayed similarly by counting local lesions on plant leaves inoculated with viruses. Analysis of viral mutants, which are commonly isolated by plaque assays, has contributed extensively to current understanding of molecular cellular processes. The plaque assay also is critical in isolating λ bacteriophage clones carrying segments of cellular DNA, as discussed in Chapter 7.

Viral Growth Cycles Are Classified as Lytic or Lysogenic

The surface of viruses includes many copies of one type of protein that binds, or adsorbs, specifically to multiple copies of a receptor protein on a host cell. This interaction determines the host range of a virus and begins the infection process (Figure 6-15). Then, in one of various ways,

(a)

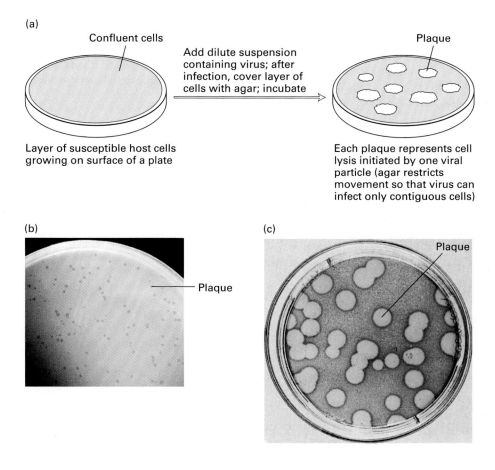

◀ FIGURE 6-14 Plaque assay for **determining number of infectious particles in a viral suspension.** (a) Each lesion, or plaque, which develops where a single virion initially infected a single cell, constitutes a pure viral clone. (b) Plate illuminated from behind shows plaques formed by λ bacteriophage plated on *E. coli.* (c) Plate showing plaques produced by poliovirus plated on HeLa cells. [Part (b) courtesy of Barbara Morris; part (c) from S. E. Luria et al., 1978, *General Virology,* 3d ed., Wiley, p. 26.]

the viral DNA or RNA crosses the plasma membrane into the cytoplasm. The entering genetic material may still be accompanied by inner viral proteins, although in the case of many bacteriophages, all capsid proteins remain outside an infected cell. The genome of most DNA-containing

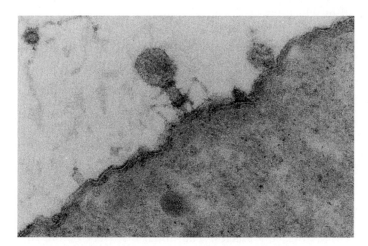

▲ **FIGURE 6-15 Electron micrograph of a T4 bacteriophage adsorbed onto an *E. coli* cell.** Once viral surface proteins interact with receptors on the host cell, the viral DNA is injected into the cell. [From A. Levine, 1991, *Viruses,* Scientific American Library, p. 20.]

viruses that infect eukaryotic cells is transported (with some associated proteins) into the cell nucleus, where the cellular DNA is, of course, also found. Once inside the cell, the viral DNA interacts with the host's machinery for transcribing DNA into mRNA. The viral mRNA that is produced then is translated into viral proteins by host-cell ribosomes, tRNA, and translation factors.

Most viral protein products fall into one of three categories: special enzymes needed for viral replication; inhibitory factors that stop host-cell DNA, RNA, and protein synthesis; and structural proteins used in the construction of new virions. These last proteins generally are made in much larger amounts than the other two types. After the synthesis of hundreds to thousands of new virions has been completed, most infected bacterial cells and some infected plant and animal cells rupture, or lyse, releasing all the virions at once. In many plant and animal viral infections, however, no discrete lytic event occurs; rather, the dead host cell releases the virions as it gradually disintegrates.

These events—adsorption, penetration, replication, and release—describe the **lytic cycle** of viral replication. The outcome is the production of a new round of viral particles and death of the cell. Figure 6-16 illustrates the lytic cycle for T4 bacteriophage. Adsorption and release of enveloped animal viruses are somewhat more complicated processes. In this case, the virions "bud" from the host cell, thereby

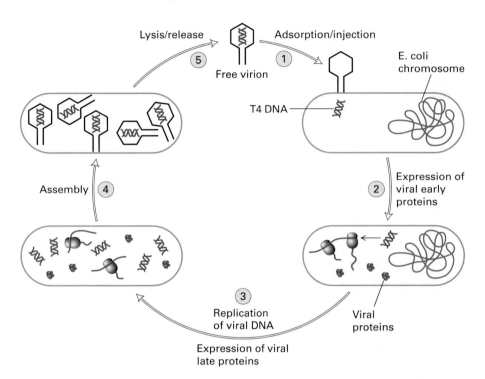

◀ **FIGURE 6-16 The steps in the lytic replication cycle of a nonenveloped virus are illustrated for *E. coli* bacteriophage T4, which has a double-stranded DNA genome.** During adsorption (step ①), viral coat proteins (at the tip of the tail in T4) interact with specific receptor proteins on the exterior of the host cell. The viral genome is then injected into the host. Next, host-cell enzymes transcribe viral "early" genes into mRNAs and subsequently translate these into viral "early" proteins (step ②), which replicate the viral DNA and induce expression of viral "late" proteins by host-cell enzymes (step③). The viral late proteins include capsid and assembly proteins and enzymes that degrade the host-cell DNA, supplying nucleotides for synthesis of viral DNA. Progeny virions are assembled in the cell (step④) and released (step⑤) when the cell is lysed by viral proteins. Newly liberated viruses initiate another cycle of infection in other host cells.

acquiring their outer phospholipid envelope, which contains mostly viral glycoproteins.

We illustrate the lytic cycle of enveloped viruses with the rabies virus, whose nucleocapsid consists of a single-stranded RNA genome surrounded by multiple copies of nucleocapsid protein (Figure 6-17, *upper left*). Within the nucleocapsid of rabies virions are viral enzymes for synthesizing viral mRNA and replicating the viral genome. The envelope around the nucleocapsid is a phospholipid bilayer containing multiple copies of a viral transmembrane glycoprotein. This receptor-binding, or "attachment," protein has a large external folded domain on the outside of the viral envelope, an α-helical transmembrane domain that spans the viral envelope, and a short internal domain. The internal domain interacts with the viral matrix protein, which functions as a bridge between the transmembrane glycoprotein and nucleocapsid protein. Figure 6-17 outlines the events involved in adsorption of a rabies virion, assembly of progeny nucleocapsids, and release of progeny virions by budding from the host-cell plasma membrane. Budding virions are clearly visible in electron micrographs, as illustrated by Figure 6-18.

In some cases, after a bacteriophage DNA molecule enters a bacterial cell, it becomes integrated into the host-cell chromosome, where it remains quiescent and is replicated as part of the cell's DNA from one generation to the next. This association is called **lysogeny**, and the integrated phage DNA is referred to as a *prophage* (Figure 6-19). Under certain conditions, the prophage DNA is activated, leading to its excision from the host-cell chromosome and entrance into the lytic cycle. Bacterial viruses of this type are called *temperate phages*. The genomes of a number of animal viruses also can integrate into the host-cell genome. Probably the

most important are the retroviruses, described briefly later in this chapter.

A few phages and animal viruses can infect a cell and cause new virion production without killing the cell or becoming integrated.

Four Types of Bacterial Viruses Are Widely Used in Biochemical and Genetic Research

Bacterial viruses have played a crucial role in the development of molecular cell biology. Thousands of different bacteriophages have been isolated; many of these are particularly well suited for studies of specific biochemical or genetic events. Here, we briefly describe four types of bacteriophages, all of which infect *E. coli*, that have been especially useful in molecular biology research.

DNA Phages of the T Series The T phages of *E. coli* are large lytic phages that contain a single molecule of double-stranded DNA. This molecule is about 2×10^5 base pairs long in T2, T4, and T6 viruses and about 4×10^4 base pairs long in T1, T3, T5, and T7 viruses. T-phage virions consist of a helical protein "tail" attached to an icosahedral "head" filled with the viral DNA. After the tip of a T-phage tail adsorbs to receptors on the surface of an *E. coli* cell, the DNA in the head enters the cell through the tail (see Figure 6-16). The phage DNA then directs a program of events that produces approximately 100 new phage particles in about 20 minutes, at which time the infected cell lyses and releases the new phages. The initial discovery of the role of messenger RNA in protein synthesis was based on studies of *E. coli* cells infected with bacteriophage T2. By 20 minutes after infection, infected cells synthesize T2 proteins only. The finding that the RNA synthesized at this time

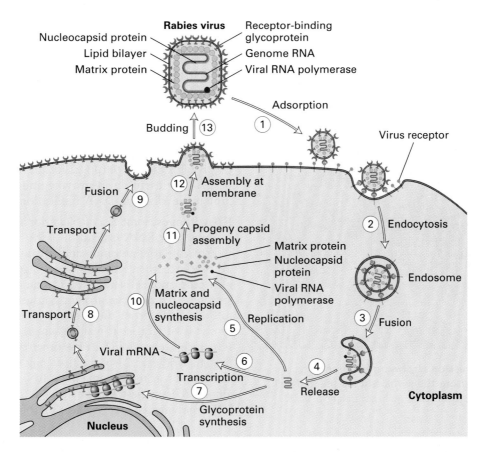

▲ **FIGURE 6-17 The steps in the lytic replication cycle of an enveloped virus are illustrated for rabies virus, which has a single-stranded RNA genome.** The structural components of this virus are depicted at the top. Note that the nucleocapsid of this virus is helical rather than icosahedral. After a virion adsorbs to a specific host membrane protein (step①), the cell engulfs it in an endosome (step②). A protein in the endosome membrane pumps protons from the cytosol into the endosome interior. The resulting decrease in endosomal pH induces a conformational change in the viral glycoprotein, leading to fusion of the viral envelope with the endosomal lipid bilayer membrane and release of the nucleocapsid into the cytosol (steps③ and ④). Viral RNA polymerase uses ribonucleoside triphosphates in the cytosol to replicate the viral RNA genome (step⑤) and synthesize viral mRNAs (step⑥). One of the viral mRNAs encodes the viral transmembrane glycoprotein (blue), which is inserted into the lumen of the endoplasmic reticulum (ER) as it is synthesized on ER-bound ribosomes (step⑦). Carbohydrate is added to the large folded domain inside the ER lumen and is modified as the membrane and the associated glycoprotein pass through the Golgi apparatus (step⑧). Vesicles with mature glycoprotein fuse with the plasma membrane, depositing viral glycoprotein on the cell surface with the large folded domain outside the cell, the transmembrane α helix spanning the plasma membrane, and the small cytoplasmic domain within the cell (step⑨). Meanwhile, other viral mRNAs are translated on host-cell ribosomes into nucleocapsid protein, matrix protein, and viral RNA polymerase (step ⑩). These proteins are assembled with replicated viral genomic RNA (dark red) into progeny nucleocapsids (step ⑪), which then associate with the viral transmembrane glycoprotein in the plasma membrane (step ⑫). As additional copies of the matrix protein on a single nucleocapsid associate with the cytoplasmic domain of additional copies of the viral transmembrane glycoprotein, the plasma membrane is folded around the nucleocapsid, forming a "bud" that eventually is released (step ⑬).

had the same base composition as T2 DNA (not *E. coli* DNA) implied that mRNA copies of T2 DNA were synthesized and used to direct cellular ribosomes to synthesize T2 proteins.

Temperate Phages Bacteriophage λ, which infects *E. coli*, typifies the temperate phages. This phage has one of the most studied genomes and is used extensively in DNA cloning (Chapter 7). On entering an *E. coli* cell, the double-stranded λ DNA assumes a circular form, which can enter either the lytic cycle (as T phages do) or the lysogenic cycle (see Figure 6-19). In the latter case, proteins expressed from the viral DNA bind a specific sequence on the circular viral DNA to a similar specific sequence on the circular bacterial DNA. The viral proteins then break both circular molecules of DNA and rejoin the broken ends, so that the viral DNA becomes inserted into the host DNA. The carefully controlled action of viral genes maintains λ DNA as part of the host chromosome by repressing the lytic functions of the phage. Under appropriate stimulation, the λ prophage is activated and undergoes lytic replication.

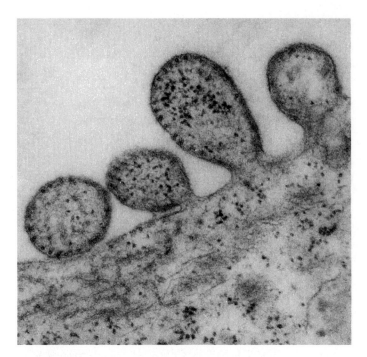

▲ **FIGURE 6-18 Transmission electron micrograph of measles virus budding from the surface of an infected cell.** [From A. Levine, 1991, *Viruses*, Scientific American Library, p. 22.]

Small DNA Phages The genome of some bacteriophages encodes only 10–12 proteins, roughly 5–10 percent of the number encoded by T phages. These small DNA phages are typified by the ΦX174 and the filamentous M13 phages. These were the first organisms in which the entire DNA sequence of a genome was determined, permitting extensive understanding of the viral life cycle. The viruses in this group are so simple that they do not encode most of the proteins required for replication of their DNA but depend on cellular proteins for this purpose. For this reason, they have been particularly useful in identifying and analyzing the cellular proteins involved in DNA replication (Chapter 12).

RNA Phages Some *E. coli* bacteriophages contain a genome composed of RNA instead of DNA. Because they are easy to grow in large amounts and because their RNA genomes also serve as their mRNA, these phages are a ready source of a pure species of mRNA. In one of the earliest demonstrations that cell-free protein synthesis can be mediated by mRNA, RNA from these phages was shown to direct the synthesis of viral coat protein when added to an extract of *E. coli* cells containing all the other components needed for protein synthesis. Also, the first long mRNA molecule to be sequenced was the genome of an RNA phage. These viruses, among the smallest known, encode only four proteins: an RNA polymerase for replication of the viral RNA, two

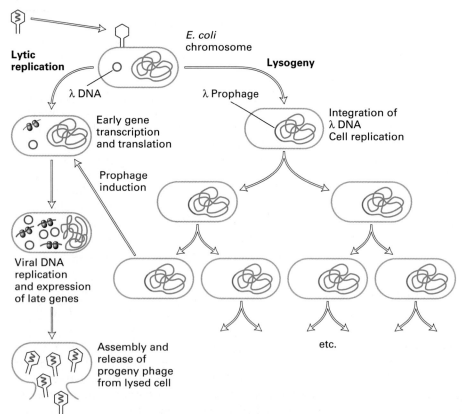

◀ **FIGURE 6-19 λ bacteriophage undergoes either lytic replication or lysogeny following infection of *E. coli*.** The linear double-stranded λ DNA is converted to a circular form immediately after infection. *(Left)* If the nutritional state of the host cell is favorable, most infected cells undergo lytic replication, similar to lytic replication of cells by bacteriophage T4 (see Figure 6-16). *(Right)* If the nutritional state of the host cell cannot support production of large numbers of progeny phages, lysogeny is established. In this case, viral genes required for the lytic cycle are repressed, and host-cell enzymes synthesize viral proteins that integrate the viral DNA into a specific sequence in the host-cell chromosome where no host-cell genes are disrupted. The prophage DNA then is replicated along with the host-cell chromosome as the lysogenized cell (called a *lysogen*) grows and divides. Repression of the viral genes required for lytic replication is maintained in progeny cells. At infrequent intervals, the prophage in a lysogen is induced, or activated, leading to expression of viral proteins that precisely remove the prophage DNA from the host-cell chromosome and to derepression of the genes required for the lytic cycle. As a result, a normal cycle of lytic replication ensues.

capsid proteins, and an enzyme that dissolves the bacterial cell wall and allows release of the intracellular virus particles into the medium.

Animal Viruses Are Classified by Genome Type and mRNA Synthesis Pathway

Animal viruses come in a variety of shapes, sizes, and genetic strategies. In this book, we are concerned with viruses that exhibit at least one of two features: they utilize important cellular pathways to form their molecules, thereby closely mimicking a normal cellular function, or they can integrate their genomes into those of normal cells.

 The names of many viruses are based on the names of the diseases they cause or of the animals or plants they infect. Common examples include poliovirus, which causes poliomyelitis; tobacco mosaic virus, which causes a mottling disease of tobacco leaves; and human immunodeficiency virus (HIV), which causes acquired immunodeficiency syndrome (AIDS). However, many different kinds of viruses often produce the same symptoms or the same apparent disease states; for example, several dozen different viruses can cause the red eyes, runny nose, and sneezing referred to as the common cold. Clearly, any attempt to classify viruses on the basis of the symptoms they produce or their hosts obscures many important differences in their structures and life cycles.

What *are* central to the life cycle of a virus are the types of nucleic acids formed during its replication and the pathway by which mRNA is produced. The relation between the viral mRNA and the nucleic acid of the infectious particle is the basis of a simple means of classifying viruses. In this system, a viral mRNA is designated as a *plus strand* and its complementary sequence, which cannot function as an mRNA, is a *minus strand*. A strand of DNA complementary to a viral mRNA is also a minus strand. Production of a plus strand of mRNA requires that a minus strand of RNA or DNA be used as a template. Using this system, six classes of animal viruses are recognized. Bacteriophages and plant viruses also can be classified in this way, but the system has been used most widely in animal virology because representatives of all six classes have been identified.

The composition of the viral genome and its relationship to the viral mRNA are illustrated in Figure 6-20 for each of the six classes of virus. Table 6-3 summarizes important properties of common animal viruses in each class and the research areas in which they have been widely used. Structural models of several virions are shown in Figure 6-21.

DNA Viruses (Classes I and II) The genomes of both class I and class II viruses consist of DNA. Various types of DNA viruses are commonly used in studies on DNA replication, genome structure, mRNA production, and oncogenic cell transformation.

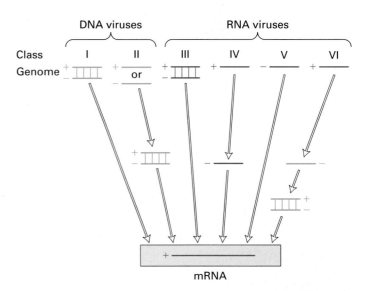

▲ **FIGURE 6-20 Classification of animal viruses based on the composition of their genomes and pathway of mRNA formation.** DNA is shown in blue; RNA, in red. The viral mRNA is designated as a plus strand, which is synthesized from a minus strand of DNA or RNA. Class VI viruses (retroviruses) have two identical plus strands of genomic RNA, although the reason for this is unclear. See Table 6-3 for examples of viruses in each class.

Class I viruses contain a single molecule of double-stranded DNA (dsDNA). In the case of the most common type of class I animal virus, viral DNA enters the cell nucleus, where cellular enzymes transcribe the DNA and process the resulting RNA into viral mRNA. Examples of these viruses include the following:

- *Adenoviruses*, which cause infections in the upper respiratory tract and gastrointestinal tract in many animals
- *SV40* (simian virus 40), a monkey virus that was accidentally discovered in kidney cell cultures from wild monkeys used in the production of poliovirus vaccines
- *Herpesviruses*, which cause various inflammatory skin diseases (e.g., chickenpox) and latent infections that recur after long intervals (e.g., cold sores and shingles)
- *Human papillomaviruses* (HPVs), which cause warts and other insignificant skin lesions and occasionally cause malignant transformation of cervical cells

 Some types of HPV are passed through sexual contact. In some infected women, the HPV genome integrates into the chromosome of a cervical epithelial cell. This rare integration event initiates an intensively studied process that can lead to development of cervical carcinoma, one of the most common types of human cancers. Routine Pap smears performed for early detection of cervical carcinoma are done to identify cells in the early stages of the transformation process initiated by HPV integration.

TABLE 6-3   Animal Viruses Commonly Used in Molecular Biology

Class*/Virus	Known Hosts	Genome Size (kb)[†]	Envelope	Other Properties	Research Areas in Which Virus Is Used
CLASS I (DNA)					
Adenoviruses (class Ia)	Vertebrates	36	No	Replicate in host-cell nucleus; use host enzymes for viral mRNA synthesis	mRNA synthesis and regulation: DNA replication; cell transformation; gene therapy (as vectors)
Herpesviruses (class Ia)	Vertebrates	150	Yes		
SV40 (class Ia)	Primates	5.5	No		
Vaccinia virus (class Ib)	Vertebrates	200	Yes	Replicates in host-cell cytoplasm using viral enzymes	Genome structure; mRNA synthesis by viral enzymes
CLASS II (DNA}					
Parvoviruses	Vertebrates	5	No	Have linear ssDNA genome	DNA replication; gene therapy (as vectors)
CLASS III (RNA)					
Reoviruses	Vertebrates	1.2–4.00[‡]	No	Have a genome of 10 dsRNA segments; use viral enzymes to replicate	mRNA snythesis by viral enzymes; mRNA translation
CLASS IV (RNA)					
Poliovirus (class IVa)	Primates	7	No	Synthesizes a single mRNA, which is translated into a polyprotein that is cleaved to yield functional proteins	Viral RNA replication; interruption of host mRNA translation; polyprotein cleavage
Sindbis virus (class IVb)	Vertebrates, insects	10	Yes	Synthesizes at least two mRNAs, each of which is translated into a polyprotein that is cleaved to yield functional proteins	Membrane formation; glycoprotein biosynthesis and intracellular transport
CLASS V (RNA)					
Vesicular stomatitis virus (class Va)	Vertebrates, insects	12	Yes	Has a virus-specific RNA polymerase that produces several mRNAs from its nonsegmented genome	Membrane formation; glycoprotein biosynthesis and intracellular transport
Influenza virus (class Vb)	Mammals, birds	1.0–3.3[‡]	Yes	Has a genome of 8 ssRNA segments; uses a virus-specific RNA polymerase to produce mRNAs	Membrane formation; glycoprotein biosynthesis and intracellular transport; disease prevention
CLASS VI (RNA)					
Retroviruses	Vertebrates, insects, yeasts	5–8	Yes	Copy RNA genome into DNA with viral reverse transcriptase; integrates viral DNA into host genome	Cell transformation; function of oncogenes; AIDS; gene therapy (as vectors)

*Classification is based on genome composition and the strategy for mRNA synthesis as illustrated in Figure 6-20.
[†]Size is given in kilobases (1 kb = 1000 nucleotides) for single-stranded nucleic acids and kilobase pairs for double-stranded nucleic acids.
[‡]Reoviruses and influenza virus have segmented RNA genomes; the length of each segment is in the range indicated.

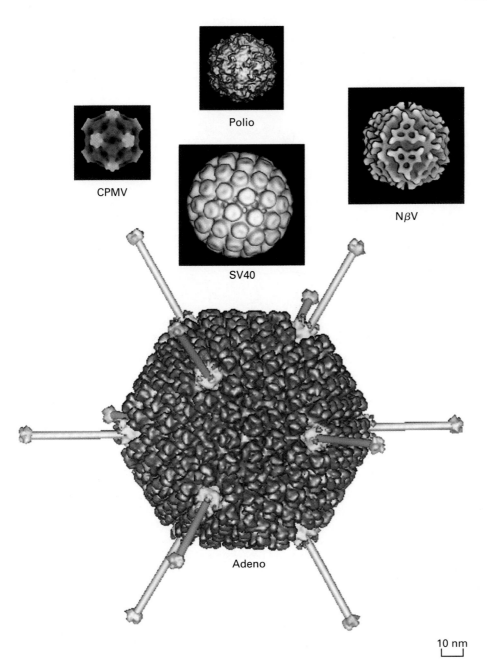

CPMV

Polio

SV40

NβV

Adeno

10 nm

◄ **FIGURE 6-21 Structures of viruses determined by cryo-electron microscopy and image analysis.** Cowpea mosaic virus (CPMV) is a plant RNA virus, poliovirus (polio) a human RNA virus, nudaureila capensis β virus (NβV) an insect RNA virus, simian virus 40 (SV40) a monkey DNA virus, and adenovirus (adeno) a human DNA virus. During infection, adenovirus binds first to cell-surface receptors through the tips of the fibers (green), and then interacts with integrins through mobile portions (red) of the penton base (yellow). All viruses are shown at the same magnification. [See P. L. Stewart et al., 1997, *EMBO J.* **16**:1189; CPMV, poliovirus, NβV, and SV40 courtesy of T. S. Baker; adenovirus courtesy of P. L. Stewart.]

The second type of class I virus, collectively referred to as *poxviruses,* replicates in the host-cell cytoplasm. Typical of class Ib viruses are variola, which causes smallpox, and vaccinia, an attenuated (weakened) poxvirus used in vaccinations to induce immunity to smallpox. These very large, brick-shaped viruses (0.1 × 0.1 × 0.2 μm) carry their own enzymes for synthesizing viral mRNA and DNA in the cytoplasm.

Class II viruses, called *parvoviruses* (from Latin *parvo,* "poor"), are simple viruses that contain one molecule of single-stranded DNA (ssDNA). Some parvoviruses encapsidate (enclose) both plus and minus strands of DNA, but in separate virions; others encapsidate only the minus strand.

In both cases, the ssDNA is copied inside the cell into dsDNA, which is then itself copied into mRNA.

RNA Viruses (Classes III–VI) All the animal viruses belonging to classes III–VI have RNA genomes. A wide range of animals, from insects to human beings, are infected by viruses in each of these classes. These viruses have been particularly useful in studies on mRNA synthesis and translation (class III); glycoprotein synthesis, membrane formation, and intracellular transport (classes IV and V); and cell transformation and oncogenes (class VI).

Class III viruses contain double-stranded genomic RNA (dsRNA). The minus RNA strand acts as a template for the

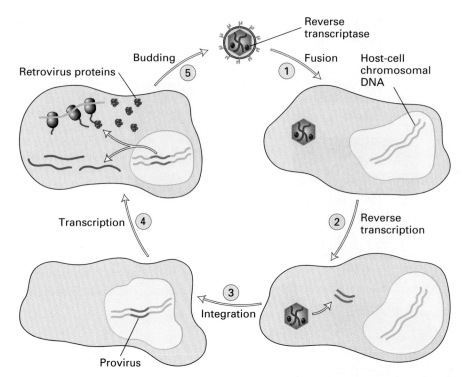

▲ **FIGURE 6-22 Retroviral life cycle.** Retroviruses have two identical copies of a plus single-stranded RNA genome and an outer envelope containing protruding viral glycoproteins. After envelope glycoproteins on a virion interact with a specific host-cell membrane protein or group of proteins, the retroviral envelope fuses directly with the plasma membrane without first undergoing endocytosis (step①). Following fusion, the nucleocapsid enters the cytoplasm of the cell; then deoxynucleoside triphosphates from the cytosol enter the nucleocapsid, where viral reverse transcriptase and other proteins copy the ssRNA genome of the virus into a dsDNA copy (step②). The viral DNA copy is transported into the nucleus (only one host-cell chromosome is depicted) and integrated into one of many possible sites in the host-cell chromosomal DNA (step③). The integrated viral DNA, referred to as a *provirus,* is transcribed by the host-cell RNA polymerase, generating mRNAs (light red) and genomic RNA molecules (dark red). The host-cell machinery translates the viral mRNAs into glycoproteins and nucleocapsid proteins (step④). The latter assemble with genomic RNA to form progeny nucleocapsids, which interact with the membrane-bound viral glycoproteins, as illustrated in Figure 6-17. Eventually the host-cell membrane buds out and progeny virions are pinched off (step⑤). See Figures 9-20 and 9-21 for details of the reverse transcription process and the transcription and processing of viral RNA.

synthesis of plus strands of mRNA. The virions of all class III viruses known to date have genomes containing 10–12 separate double-stranded RNA molecules, each of which encodes one or two polypeptides. Consequently, these viruses are said to have "segmented" genomes. In these viruses, the virion itself contains a complete set of enzymes that can utilize the minus strand of the genomic RNA as a template for synthesis of mRNA in the test tube as well as in the cell cytoplasm after infection. A number of important studies have used class III viruses as a source of pure mRNA.

Class IV viruses contain a single plus strand of genomic RNA, which is identical with the viral mRNA. Since the genomic RNA encodes proteins, it is infectious by itself. During replication of class IV viruses, the genomic RNA is copied into a minus strand, which then acts as a template for synthesis of more plus strands, or mRNA. Two types of class IV viruses are known. In class IVa viruses, typified by poliovirus, viral proteins are first synthesized, from a single

mRNA species, as a long polypeptide chain, or *polyprotein,* which is then cleaved to yield the various functional proteins. Class IVb viruses synthesize at least two species of mRNA in a host cell. One of these mRNAs is the same length as the virion's genomic RNA; the other corresponds to the 3′ third of the genomic RNA. Both mRNAs are translated into polyproteins. Included in class IVb are a large number of rare insect-borne viruses including Sindbis virus and those causing yellow fever and viral encephalitis in human beings. These viruses once were called *arboviruses* (arthropod-borne viruses), but now are called *togaviruses* (from Latin *toga,* cover) because the virions are surrounded by a lipid envelope.

Class V viruses contain a single negative strand of genomic RNA, whose sequence is complementary to that of the viral mRNA. The genomic RNA in the virion acts as a template for synthesis of mRNA but does not itself encode proteins. Two types of class V viruses can be distinguished. The genome in class Va viruses, which include the viruses causing measles and mumps, is a single molecule of RNA. A virus-specific RNA polymerase present in the virion catalyzes synthesis of several mRNAs, each encoding a single protein, from the genomic template strand. Class Vb viruses, typified by influenza virus, have segmented genomes; each segment acts as a template for the synthesis of a different mRNA species. In most cases, each mRNA produced by a class Vb virus encodes a single protein; however, some mRNAs can be read in two different frames to yield two distinct proteins. As with class Va viruses, a class Vb virion contains a virus-specific polymerase that catalyzes synthesis of the viral mRNA. Thus the genomic RNA (a minus strand) in both types of class V viruses is not infectious in the absence of the virus-specific polymerase. The influenza RNA polymerase initiates synthesis of each mRNA by a unique mechanism. In the host-cell nucleus, the polymerase cuts off 12–15 nucleotides from the 5′ end of a cellular mRNA or mRNA precursor; this oligonucleotide acts as a "primer" that is elongated by the polymerase to form viral (+) mRNAs, using the genomic (−) RNA as a template.

Class VI viruses are enveloped viruses whose genome consists of two identical plus strands of RNA. These viruses are also known as **retroviruses** because their RNA genome directs the formation of a DNA molecule. The DNA molecule ultimately acts as the template for synthesis of viral mRNA (Figure 6-22). Initially, a viral enzyme called **reverse transcriptase** copies the viral RNA genome into a single minus strand of DNA; the same enzyme then catalyzes synthesis of a complementary plus strand. (This complex reaction is detailed in Chapter 9.) The resulting dsDNA is integrated into the chromosomal DNA of the infected cell. Finally, the integrated proviral DNA is transcribed by the cell's own machinery into (+) RNA, which either is translated into viral proteins or is packaged within virion coat proteins to form progeny virions, which are released by budding from the host-cell membrane. Because most retroviruses do not kill their host cells, infected cells can replicate, producing daughter cells with integrated proviral DNA. These daughter cells continue to transcribe the proviral DNA and bud progeny virions.

 Some retroviruses contain cancer-causing genes (called **oncogenes**). Cells infected by such retroviruses are oncogenically transformed into tumor cells. Studies of oncogenic retroviruses (mostly viruses of birds and mice) have revealed a great deal about the processes that lead to oncogenic transformation. Among the known human retroviruses are human T-cell lymphotrophic virus (HTLV), which causes a form of leukemia, and human immunodeficiency virus (HIV), which causes acquired immune deficiency syndrome (AIDS). Both of these viruses can infect only specific cell types, primarily certain cells of the immune system and, in the case of HIV, some central nervous system neurons and glial cells. Only these cells have cell-surface receptors that interact with viral proteins, accounting for the host-cell specificity of these viruses.

Viral Vectors Can Be Used to Introduce Specific Genes into Cells

Knowledge about mechanisms of viral replication has allowed virologists to modify viruses for various purposes. For instance, the ability of virions to introduce their contents into the cytoplasm and nuclei of infected cells has been adapted for use in DNA cloning and offers possibilities in the treatment of certain diseases. The introduction of new genes into cells by packaging them into virion particles is called *viral gene transduction,* and the virions used for this purpose are called *viral vectors.*

By use of recombinant DNA techniques described in Chapter 7, it is a relatively straightforward process to construct human adenovirus *recombinants* in which potentially therapeutic genes replace the viral genes required for the lytic cycle of infection. Because adenovirus has a very broad host range for different types of human cells, these vectors can introduce the engineered gene into the cells of tissues where they are applied. If the transduced gene encodes the normal form of a protein that is missing or defective in a particular disease, then such *gene therapy* may successfully treat the disease. One type of adenovirus, for example, efficiently infects cells lining the air passages in the lungs, causing a type of common cold. Researchers have replaced some of the disease-causing genes in this adenovirus with the *CFTR* gene, which is defective in individuals with cystic fibrosis. This recombinant adenovirus currently is being used to introduce a normal *CFTR* gene into the airway-lining cells of cystic fibrosis patients. Unfortunately, with most of the adenovirus vectors currently available, the transduced gene usually is expressed only for a limited period of 2 to 3 weeks. This significantly limits their usefulness in gene therapy.

Viral vectors have also been developed from viruses that integrate their genomes into host-cell chromosomes. Such vectors have the advantage that progeny of the initially infected cell also contain and express the transduced gene because it is replicated and segregated to daughter cells along with the rest of the chromosome into which it is integrated. Retroviral vectors, which can efficiently integrate transduced genes at approximately random positions in host-cell chromosomes are now widely used experimentally to generate cultured cells expressing specific, desired proteins. However, technical limitations in producing the large numbers of retroviral vectors required to infect a significant fraction of cells in the tissues of a human or vertebrate currently limit their use as gene therapy vectors. Another concern with retroviral vectors is that their random integration might disturb the normal expression of cellular genes encoding proteins regulating cellular replication. This type of cellular gene deregulation occurs naturally following infection with certain retroviruses, such as avian leukosis virus and murine leukemia viruses, leading to development of leukemia in birds and mice, respectively.

Adeno-associated virus (AAV) is a "satellite" parvovirus that replicates only in cells that are co-infected with adenovirus or herpes simplex virus. When AAV infects human cells in the absence of these "helper" viruses, its ssDNA genome is copied into dsDNA by host-cell DNA polymerase and then is integrated into a single region on chromosome 19, where it does not have any known deleterious effects. Research is under way to adapt the AAV integration mechanism that operates in the absence of helper virus to the development of a safe and effective integrating viral vector.

SUMMARY Viruses: Structure, Function, and Uses

- Viruses are intracellular parasites that replicate only after infecting specific host cells. Viral infection begins when proteins on the surface of a virion bind to specific receptor proteins on the surface of host cells. The specificity of this interaction determines the host range of a virus.

- Aside from being the causative agents of many diseases, viruses are important tools in cell biology research, particularly in studies on macromolecular synthesis (see Table 6-3).

- Viruses can be counted and cloned by the plaque assay (see Figure 6-14). All the virions in a single plaque compose a clone derived from the single parental virion that infected the first cell at the center of the plaque.

- Individual viral particles (virions) generally contain either an RNA or a DNA genome, surrounded by multiple copies of one or a small number of coat proteins, forming the nucleocapsid. The nucleocapsid of many animal viruses is surrounded by a phospholipid bilayer, or envelope.

- During lytic replication, host-cell ribosomes and enzymes are used to express viral proteins, which then replicate the viral genome and package it into viral coats. The multiple progeny virions produced within a single infected cell eventually are released, following cell lysis or gradual disintegration of the cell (see Figure 6-16). Progeny nucleocapsids of enveloped viruses are released by budding of the host-cell membrane in which viral membrane proteins have been deposited (see Figure 6-17).

- Some bacterial viruses (bacteriophages) may undergo lysogeny following infection of host cells. In this case, the viral genome is integrated into host-cell chromosomes, forming a prophage that is replicated along with the host genome. When suitably activated, a prophage enters the lytic cycle (see Figure 6-19).

- All retroviruses and some other animal viruses can integrate their genomes into host-cell chromosomes (see Figure 6-22). In some cases, this leads to abnormal cell replication and the eventual development of cancers.

- Recombinant viruses can be used as vectors to carry (transduce) selected genes into cells. In this approach, viral genes required for the lytic cycle are replaced by other genes. The use of viral vectors for gene therapy is still in its infancy, but has great potential for treatment of various diseases.

PERSPECTIVES for the Future

Advances in the culture of human cells will allow them to be applied to the treatment of disease and injury. Even today, severe burn patients can be treated by removing some of their undamaged skin, culturing the fibroblasts, and applying them to areas where the skin has been destroyed. As specific growth factors for various cell types are identified and expressed in sufficient quantities by recombinant DNA techniques, other cell types will be cultured and used therapeutically.

In the not too distant future, safe and effective viral vectors will likely be developed to deliver genes encoding therapeutic proteins or RNAs to the specific cells where they are needed. Such gene therapy potentially could successfully treat numerous human diseases caused by the absence of critical cellular molecules.

PERSPECTIVES in the Literature

Viral vectors for gene transduction are a promising approach to gene therapy. Adenovirus vectors have the advantage of replicating to very high numbers. This makes it possible to transduce a large fraction of cells in a target organ—for example, most hepatocytes in the liver of a treated mouse and, potentially, a human patient. However, the expression of therapeutic genes from the adenovirus vectors used in early studies did not persist longer than two weeks. As you read the following articles, consider this question: what modifications of adenovirus vectors might allow long-term expression of a newly introduced therapeutic gene?

Kozarsky, K. F., et al. 1994. In vivo correction of low density lipoprotein receptor deficiency in the Watanabe heritable hyperlipidemic rabbit with recombinant adenoviruses. *J. Biol. Chem.* **269**:13695–13702.

Morsy, M. A., et al. 1998. An adenoviral vector deleted for all viral coding sequences results in enhanced safety and extended expression of a leptin transgene. *Proc. Nat'l Acad. Sci. USA* **95**:7866–7871.

Recchia, A., et al. 1999. Site-specific integration mediated by a hybrid adenovirus/adeno-associated virus vector. *Proc. Nat'l Acad. Sci. USA* **96**:2615–2620.

Testing Yourself on the Concepts

1. What characteristics of hybridomas make them useful cell types for molecular cell biologists?

2. How would you create and isolate a strain of bacteria that could grow on minimal media only in the presence of the amino acid leucine and the purine adenine?

3. It has been claimed that the classification scheme that groups animal viruses into six classes (Table 6-3) was the start of the molecular study of viral replication. What characteristic of this scheme could be the basis of this claim?

MCAT/GRE-Style Questions

Key Concept Please read the first three paragraphs of section 6.3, "Viruses: Structure, Function, and Uses" (p.191) and answer the following questions:

1. Infected cells are useful as model systems for viral and cellular replication because
   a. They produce large amounts of crystalline virus.
   b. They live for only a short time, so normal cellular processes occur more quickly.
   c. Viruses use cellular constituents to replicate.
   d. Virus structure is dependent on the infected cell.

2. Assays of virus-infected cells for factors involved in DNA replication would use as their end point
   a. The ability of the cell to divide.
   b. The replication of chromosomal DNA.
   c. The replication of viral DNA.
   d. The production of chromosome-associated proteins.

3. The minimum component or components of tobacco mosaic virus required to infect a cell and produce progeny are
   a. The genome.
   b. The genome and the proteins of the virion.
   c. Ribosomes and tRNA.
   d. The enzyme to produce mRNA.

4. The glycoproteins produced by cells infected with influenza virus are almost exclusively virus-coded. A possible reason for this might be
   a. mRNAs for cellular glycoproteins do not compete well with viral glycoprotein mRNAs for access to ribosomes.
   b. Portions of cellular mRNAs are incorporated into influenza mRNAs, disabling the cellular mRNAs.
   c. Infection inhibits glycosylation enzymes.
   d. Cells require these glycoproteins for integrity of the cell surface.

5. Viruses with genes that cause cancer in animal cells are useful to scientists for all the reasons given below *except:*
   a. Isolating these genes from viruses is easier than isolating them from chromosomes.
   b. The activity of these genes can be studied in the cells that viruses infect rather than in the whole animal.
   c. The changes caused by these genes in the cell can be readily investigated.
   d. These cells are good models for populations of normal cells.

6. Which of the following would you expect to be encoded by the genome of a virus:
   a. Proteins that regulate the viral life cycle.
   b. Ribosomal RNA.
   c. tRNA.
   d. Ribosomal proteins.

Key Experiment Please read the section titled "Hybrid Cells Often Are Selected on HAT Medium" (p. 189) and refer to Figure 6-9; then answer the following questions:

7. The first goal of a successful cell fusion experiment is
   a. To have only hybrid cells survive.
   b. To have hybrid cells resistant to 5-bromodeoxyuridine.
   c. To have hybrid cells resistant to 6-thioguanine.
   d. To have hybrid cells grow in the presence of aminopterin.

8. Aminopterin is present in HAT medium when fusing TK^- and $HGPRT^-$ cells
   a. To overcome the effect of a TK^- mutation.
   b. To force cells to use the salvage pathways.
   c. As a compound supporting the growth of hybrid cells.
   d. To overcome the effects of the $HGPRT^-$ mutation.

9. Cells containing a mutation in the TK^- gene are selected by
   a. Their ability to grow in the presence of aminopterin.
   b. Their ability to grow in the presence of 5-bromodeoxyuridine.
   c. Their ability to grow in the absence of thymidine.
   d. Their ability to grow in the presence of aminopterin and 5-bromodeoxyuridine.

10. Differentiated cells that do not fuse with TK^- cells in a hybridization procedure will fail to replicate in HAT medium because
   a. Aminopterin will kill them during selection.
   b. They do not replicate in culture.
   c. They are killed by 5-bromodeoxyuridine added to HAT medium in this special circumstance.
   d. They lack the salvage pathway.

11. A hybrid cell produced from TK^- and $HGPRT^-$ parents grows in HAT medium for all the following reasons *except:*
   a. Thymidine is converted to TMP by an enzyme from the $HGPRT^-$ parent.

b. Hypoxanthine is converted to purines by an enzyme from the TK⁻ parent.

c. Aminopterin inhibits the biosynthesis of purines and thymidine from simple carbon and nitrogen compounds.

d. Hybrid cells produce purines and pyrimidines from simple carbon and nitrogen compounds.

Key Application Please read the section titled "Viral Vectors Can Be Used to Introduce Specific Genes into Cells" (p. 203) and answer the following questions:

12. AAV has essentially three genes. What will be the consequences of replacing the gene for the virion proteins with a normal cellular gene in order to create a gene therapy vector?

a. The recombinant virus will be unable to replicate.

b. Site-specific integration will be lost.

c. The recipient of an AAV-based vector will be resistant to adenovirus infection.

d. The transduced gene will now be expressed for only a brief period of time.

13. The advantage of AAV as a vector compared with other viruses mentioned in the section is that

a. It integrates site-specifically.

b. The recipient of an AAV-based vector will be resistant to adenovirus infection.

c. Its helper viruses can be engineered so that they don't cause disease.

d. It can be copied into DNA only in infected cells.

14. Problems associated with gene therapy vectors include all the following *except*:

a. Requirement for special techniques to produce recombinant viruses.

b. Inability to produce enough recombinant virus for widespread use.

c. Deleterious side effects.

d. Ability to infect a wide variety of cells.

Key Terms

bacteriophage *194*	monoclonal antibody *189*
cell fusion *187*	nucleocapsid *192*
cell line *186*	phage *194*
cell strain *186*	plaque assay *194*
clone *181*	retroviruses *203*
fibroblast *185*	reverse transcriptase *203*
heterokaryon *187*	transformation *186*
hybridoma *190*	virion *191*
lysogeny *196*	virus *191*
lytic cycle *195*	

References

Growth of Microorganisms in Culture

Ausubel, F., et al., eds. 1993. *Current Protocols in Molecular Biology.* Part 1: *E. coli,* plasmids, and bacteriophages. Current Protocols.

Guthrie, C., and G. F. Fink, eds. 1991. *Methods in Enzymology.* Vol. 194: *Guide to Yeast Genetics and Molecular Biology.* Academic Press.

Miller, J. 1992. *A Short Course in Bacterial Genetics.* Cold Spring Harbor Laboratory Press.

Neidhardt, F. C., et al. 1987. Escherichia Coli *and* Salmonella Typhimurium: *Cellular and Molecular Biology.* American Society for Microbiology.

Sambrook, J., T. Maniatis, and E. F. Fritsch, eds. 1989. *Molecular Cloning,* 2d ed. Cold Spring Harbor Laboratory Press.

Growth of Animal Cells in Culture

Barnes, D. W., D. A. Sirbasky, and G. H. Sato, eds. 1984. *Cell Culture Methods for Molecular and Cell Biology.* Alan R. Liss.

Birch, J. R., ed. 1995. *Monoclonal Antibodies: Principles and Applications.* Wiley-Liss.

Davis, J. M., ed. 1994. *Basic Cell Culture: A Practical Approach.* IRL Press.

Evans, M. J., and M. H. Kaufman. 1981. Establishment in culture of pluripotential cells from mouse embryos. *Nature* 292:154–156.

Goding, J. W. 1996. *Monoclonal Antibodies: Principles and Practice. Production and Application of Monoclonal Antibodies in Cell Biology, Biochemistry, and Immunology,* 3d ed. Academic Press.

Harlow, E., and D. Lane. 1988. *Antibodies: A Laboratory Manual.* Cold Spring Harbor Laboratory. Chapter 6, *Monoclonal Antibodies,* and Chapter 7, *Growing Hybridomas.*

Harris, H. 1995. *The Cells of the Body: A History of Somatic Cell Genetics.* Cold Spring Harbor Laboratory Press.

Kohler, G., and C. Milstein. 1975. Continuous cultures of fused cells secreting antibody of predefined specificity. *Nature* 256:495–497.

Milstein, C. 1980. Monoclonal antibodies. *Sci. Am.* 243(4):66–74.

Shaw, A. J., ed. 1996. *Epithelial Cell Culture.* IRL Press.

Tyson, C. A., and J. A. Frazier, eds. 1993. *Methods in Toxicology.* Vol. I (Part A): *In Vitro Biological Systems.* Academic Press. Describes methods for growing many types of primary cells in culture.

Watson, J. D., and H. J. McKenna. 1992. Novel factors from stromal cells: bone marrow and thymus microenvironments. *Int'l. J. Cell Cloning* 10:144–252.

Viruses: Structure, Function, and Uses

Brinton, M. A., C. H. Calisher, and R. Rueckert, eds. 1994. *Positive-Strand RNA Viruses.* Springer-Verlag.

Calendar, R., ed. 1988. *The Bacteriophages.* Plenum Press.

Coffin, J. M., S. H. Hughes, and H. E. Varmus, eds. 1997. *Retroviruses.* Cold Spring Harbor Laboratory Press.

Fields, B. N., D. M. Knipe, and P. M. Howley, eds. 1996. *Fundamental Virology.* Lippincott-Raven.

Levine, A. J. 1991. *Viruses.* Scientific American Press Library.

Levy, J. A., H. F. Fraenkel-Conrat, and R. A. Owens. 1994. *Virology,* 3d ed. Prentice Hall.

Recombinant DNA and Genomics

O nce the structure of DNA and the genetic code were unraveled, it became clear that many deep biological secrets were locked up in the sequence of bases in DNA. But identifying the sequences of long regions of DNA, much less altering them at will, seemed a distant dream. An avalanche of technical discoveries in the 1970s drastically changed this perspective and has led to astounding advances in molecular cell biology in the past two decades based on the analysis and manipulation of macromolecules, particularly DNA.

The discovery of two types of enzymes provided the impetus for these developments and permitted the now common technique of **DNA cloning.** One type, called **restriction enzymes,** cuts the DNA from any organism at specific sequences of a few nucleotides, generating a reproducible set of fragments. The other type, called DNA **ligases,** can insert DNA restriction fragments into replicating DNA molecules producing **recombinant DNA.** The recombinant DNA molecules then can be introduced into appropriate cells, most often bacterial cells; all the descendants from a single such cell, called a **clone,** carry the same recombinant DNA molecule. Once a clone of cells bearing a desired segment of DNA is isolated, unlimited quantities of this DNA can be prepared. In addition, DNA sequences up to about 100 bases long can now be chemically synthesized by entirely automated procedures. Recombinant DNAs thus can be produced containing either natural DNA fragments resulting from restriction-enzyme cleavage or any desired chemically synthesized mutant sequences.

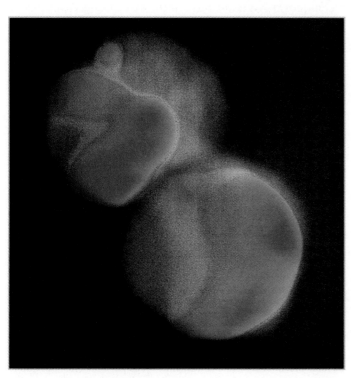

Detection of human immunodeficiency virus (HIV-1) in two human lymphocytes by in situ polymerase chain reaction.

FROM PROTEIN TO GENE

Isolate protein on the basis of its molecular function (e.g., enzymatic or hormonal activity)

Determine partial amino acid sequence of the protein

Synthesize oligonucleotides that correspond to portions of the amino acid sequence

Use oligonucleotides as probes to select cDNA or genomic clone encoding the protein from library

Sequence isolated gene

FROM GENE TO PROTEIN

Isolate genomic clone corresponding to an altered trait in mutants (e.g., nutritional auxotrophy, inherited disease, developmental defect)

Use genomic DNA to isolate a cDNA for the mRNA encoded by the gene

Sequence the cDNA to deduce amino acid sequence of the encoded protein

Compare deduced amino acid sequence with that of known proteins to gain insight into function of the protein

Use expression vector to produce the encoded protein

The availability of restriction enzymes also facilitated development of techniques for rapid DNA sequencing in the late 1970s. A long DNA molecule is first cleaved with restriction enzymes into a reproducible array of fragments, whose order in the original molecule is determined. Procedures also were developed for determining the sequence of bases in fragments up to 500 nucleotides long. Thus there was no longer any obstacle to obtaining the sequence of a DNA containing 10,000 or more nucleotides. Suddenly, any DNA could be isolated and sequenced. With the aid of computer-automated procedures for sequencing DNA and for storing, comparing, and analyzing sequence data, scientists will complete sequencing of the entire human genome in the next few years.

In the past, two basic approaches were available for unraveling the molecular basis of complex biological processes: (a) biochemical purification and analysis of a protein based on its functional characteristics (Chapter 3) and (b) classical genetic analysis for the characterization and mapping of genes defined by mutations (Chapter 8). The group of techniques discussed in this chapter, often collectively referred to as *recombinant DNA technology,* provide a link between these two types of experimental strategies, the analysis of proteins and the analysis of genes. Today's molecular cell biologists can begin with an

isolated protein and clone the gene that encodes it. They also can reinsert cloned DNA, whether natural, modified, or completely synthetic, into cells and test its biological activity. Alternatively, with the techniques described in Chapter 8, researchers can begin with the concept of a gene identified by the characteristics of a mutant organism and isolate a DNA clone containing the gene. Ultimately, the encoded protein can be produced in sufficient quantities for detailed study. The marriage of biochemical and genetic approaches by recombinant DNA technology provides an enormously powerful strategy for studying the role of particular proteins in cellular processes. In this chapter, we describe the various recombinant DNA techniques that permit this fruitful two-pronged approach, which is summarized in the flow diagram on the left.

7.1   DNA Cloning with Plasmid Vectors

The essence of cell chemistry is to isolate a particular cellular component and then analyze its chemical structure and activity. In the case of DNA, this is feasible for relatively short molecules such as the genomes of small viruses. But genomes of even the simplest cells are much too large to directly analyze in detail at the molecular level. The problem is compounded for complex organisms. The human genome, for example, contains about 6×10^9 base pairs (bp) in the 23 pairs of chromosomes. Cleavage of human DNA with restriction enzymes that produce about one cut for every 3000 base pairs yields some 2 million fragments, far too many to separate from each other directly. This obstacle to obtaining pure DNA samples from large genomes has been overcome by recombinant DNA technology. With these methods virtually any gene can be purified, its sequence determined, and the functional regions of the sequence explored by altering it in planned ways and reintroducing the DNA into cells and into whole organisms.

The essence of recombinant DNA technology is the preparation of large numbers of identical DNA molecules. A DNA fragment of interest is linked through standard $3' \rightarrow 5'$ phosphodiester bonds to a **vector** DNA molecule, which can replicate when introduced into a host cell. When a single recombinant DNA molecule, composed of a vector plus an inserted DNA fragment, is introduced into a host cell, the inserted DNA is reproduced along with the vector, producing large numbers of recombinant DNA molecules that include the fragment of DNA originally linked to the vector. Two types of vectors are most commonly used: *E. coli* **plasmid** vectors and bacteriophage λ vectors. Plasmid vectors replicate along with their host cells, while λ vectors replicate as lytic viruses, killing the host cell and packaging the DNA into virions (Chapter 6). In this section, the general procedure for cloning DNA fragments in *E. coli* plasmids is described.

Plasmids Are Extrachromosomal Self-Replicating DNA Molecules

Plasmids are circular, double-stranded DNA (dsDNA) molecules that are separate from a cell's chromosomal DNA. These extrachromosomal DNAs, which occur naturally in bacteria, yeast, and some higher eukaryotic cells, exist in a parasitic or symbiotic relationship with their host cell. Plasmids range in size from a few thousand base pairs to more than 100 kilobases (kb). Like the host-cell chromosomal DNA, plasmid DNA is duplicated before every cell division. During cell division, at least one copy of the plasmid DNA is segregated to each daughter cell, assuring continued propagation of the plasmid through successive generations of the host cell.

 Many naturally occurring plasmids contain genes that provide some benefit to the host cell, fulfilling the plasmid's portion of the symbiotic relationship. For example, some bacterial plasmids encode enzymes that inactivate antibiotics. Such drug-resistance plasmids have become a major problem in the treatment of a number of common bacterial pathogens. As antibiotic use became widespread, plasmids containing several drug-resistance genes evolved, making their host cells resistant to a variety of different antibiotics simultaneously. Many of these plasmids also contain "transfer genes" encoding proteins that can form a macromolecular tube, or *pilus*, through which a copy of the plasmid can be transferred to other host cells of the same or related bacterial species. Such transfer can result in the rapid spread of drug-resistance plasmids, expanding the number of antibiotic-resistant bacteria in an environment such as a hospital. Coping with the spread of drug-resistance plasmids is an important challenge for modern medicine.

E. Coli Plasmids Can Be Engineered for Use as Cloning Vectors

The plasmids most commonly used in recombinant DNA technology replicate in *E. coli*. Generally, these plasmids have been engineered to optimize their use as vectors in DNA cloning. For instance, to simplify working with plasmids, their length is reduced; many plasmid vectors are only ≈3 kb in length, which is much shorter than in naturally occurring *E. coli* plasmids. (The circumference of plasmids usually is referred to as their "length," even though plasmids are almost always circular DNA molecules.) Most plasmid vectors contain little more than the essential nucleotide sequences required for their use in DNA cloning: a replication origin, a drug-resistance gene, and a region in which exogenous DNA fragments can be inserted (Figure 7-1).

Plasmid DNA Replication The replication origin (ORI) is a specific DNA sequence of 50–100 base pairs that must be present in a plasmid for it to replicate. Host-cell enzymes bind to ORI, initiating replication of the circular plasmid. Once DNA replication is initiated at ORI, it continues around the circular plasmid regardless of its nucleotide sequence (Figure 7-2). Thus any DNA sequence inserted into such a plasmid is replicated along with the rest of the plasmid DNA; this property is the basis of molecular DNA cloning.

Selection of Transformed Cells In 1944, O. T. Avery, C. M. Macleod, and M. McCarty first demonstrated gene transfer with isolated DNA obtained from *Streptococcus pneumoniae*. This process involved the genetic alteration of a bacterial cell by the uptake of DNA isolated from a

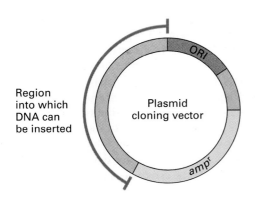

▲ **FIGURE 7-1 Diagram of a simple cloning vector derived from a plasmid, a circular, double-stranded DNA molecule that can replicate within an *E. coli* cell.** Plasmid vectors are ≈1.2–3 kb in length and contain a replication origin (ORI) sequence and a gene that permits selection, usually by conferring resistance to a particular drug. Here the selective gene is *amp^r*; it encodes the enzyme β-lactamase, which inactivates ampicillin. Exogenous DNA can be inserted into the bracketed region without disturbing the ability of the plasmid to replicate or express the *amp^r* gene.

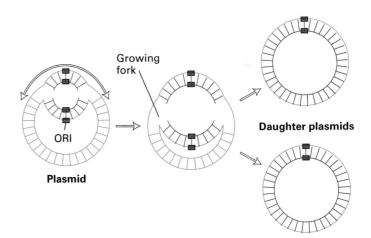

▲ **FIGURE 7-2 Plasmid DNA replication.** The parental strands are shown in blue, and newly synthesized daughter strands are shown in red. The short segments represent the A·T and G·C base pairs connecting the complementary strands. Once DNA replication is initiated at the origin (ORI), it continues in both directions around the circular molecule until the advancing growing forks merge and two daughter molecules are produced. The origin is the only specific nucleotide sequence required for replication of the entire circular DNA molecule.

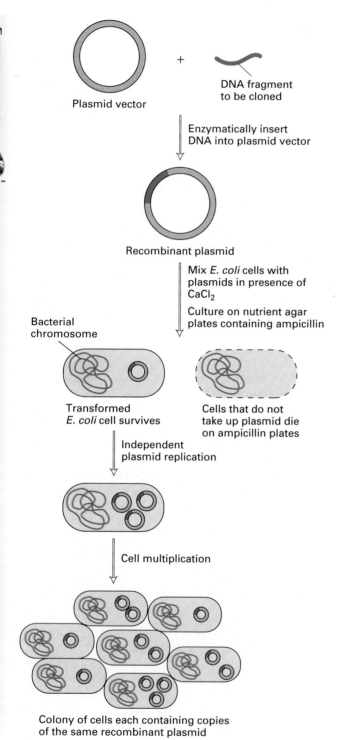

Plasmid vector   +   DNA fragment to be cloned

Enzymatically insert DNA into plasmid vector

Recombinant plasmid

Mix *E. coli* cells with plasmids in presence of CaCl₂

Culture on nutrient agar plates containing ampicillin

Bacterial chromosome

Transformed *E. coli* cell survives

Cells that do not take up plasmid die on ampicillin plates

Independent plasmid replication

Cell multiplication

Colony of cells each containing copies of the same recombinant plasmid

◀ **FIGURE 7-3 General procedure for cloning a DNA fragment in a plasmid vector.** Although not indicated by color, the plasmid contains a replication origin and ampicillin-resistance gene. Uptake of plasmids by *E. coli* cells is stimulated by high concentrations of CaCl₂. Even in the presence of CaCl₂, transformation occurs with a quite low frequency, and only a few cells are transformed by incorporation of a single plasmid molecule. Cells that are not transformed die on ampicillin-containing medium. Once incorporated into a host cell, a plasmid can replicate independently of the host-cell chromosome. As a transformed cell multiplies into a colony, at least one plasmid segregates to each daughter cell.

sion of foreign DNA regardless of the mechanism involved. (Note that *transformation* has a second meaning defined in Chapter 6, namely, the process by which normal cells with a finite life span in culture are converted into continuously growing cells similar to cancer cells.)

The phenomenon of transformation permits plasmid vectors to be introduced into and expressed by *E. coli* cells. In order to be useful in DNA cloning, however, a plasmid vector must contain a *selectable gene*, most commonly a drug-resistance gene encoding an enzyme that inactivates a specific antibiotic. As we've seen, the ampicillin-resistance gene (*amp*[r]) encodes β-lactamase, which inactivates the antibiotic ampicillin. After plasmid vectors are incubated with *E. coli*, those cells that take up the plasmid can be easily selected from the larger number of cells that do not by growing them in an ampicillin-containing medium. The ability to select transformed cells is critical to DNA cloning by plasmid vector technology because the transformation of *E. coli* with isolated plasmid DNA is inefficient.

Normal *E. coli* cells cannot take up plasmid DNA from the medium. Exposure of cells to high concentrations of certain divalent cations, however, makes a small fraction of cells permeable to foreign DNA by a mechanism that is not understood. In a typical procedure, *E. coli* cells are treated with CaCl₂ and mixed with plasmid vectors; commonly, only 1 cell in about 10,000 or more cells becomes competent to take up the foreign DNA. Each competent cell incorporates a *single* plasmid DNA molecule, which carries an antibiotic-resistance gene. When the treated cells are plated on a petri dish of nutrient agar containing the antibiotic, only the rare transformed cells containing the antibiotic-resistance gene on the plasmid vector will survive. All the plasmids in such a colony of selected transformed cells are descended from the single plasmid taken up by the cell that established the colony.

Plasmid Cloning Permits Isolation of DNA Fragments from Complex Mixtures

A DNA fragment of a few base pairs up to ≈20 kb can be inserted into a plasmid vector. When such a recombinant plasmid transforms an *E. coli* cell, all the antibiotic-resistant progeny cells that arise from the initial transformed cell will contain plasmids with the same inserted sequence of DNA (Figure 7-3). The inserted DNA is replicated along with the

genetically different bacterium and its recombination with the host-cell genome. Their experiments provided the first evidence that DNA is the genetic material. Later studies showed that such genetic alteration of a recipient cell can result from the uptake of exogenous extrachromosomal DNA (e.g., plasmids) that does not integrate into the host-cell chromosome. The term **transformation** is used to denote the genetic alteration of a cell caused by the uptake and expres-

▶ **FIGURE 7-4 Isolation of DNA fragments from a mixture by cloning in a plasmid vector.** Four distinct DNA fragments, depicted in different colors, are inserted into plasmid cloning vectors, yielding a mixture of recombinant plasmids each containing a single DNA fragment. *E. coli* cells treated with CaCl₂ are incubated with the mixture of recombinant plasmids and then plated on nutrient agar containing ampicillin. Each colony of transformed, antibiotic-resistant cells that grows (represented by a group of cells) arises from a single cell that took up one or another of the recombinant plasmids; all the cells in a given colony thus carry the same DNA fragment. Overnight incubation of *E. coli* at 37 °C produces visible colonies containing about a million cells. Since the colonies are separated from one another on the culture plate, copies of the DNA fragments in the original mixture are isolated in the individual colonies. Although it's not shown here, the transformed cells contain multiple copies of a given plasmid.

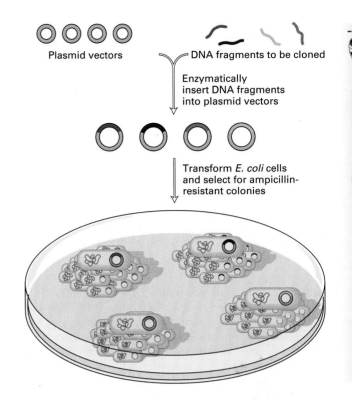

rest of the plasmid DNA and segregates to daughter cells as the colony grows. In this way, the initial fragment of DNA is replicated in the colony of cells into a large number of identical copies. Since all the cells in a colony arise from a single transformed parental cell, they constitute a clone of cells. The initial fragment of DNA inserted into the parental plasmid is referred to as *cloned DNA*, since it can be isolated from the clone of cells.

DNA cloning allows fragments of DNA with a particular nucleotide sequence to be isolated from a complex mixture of fragments with many different sequences. As a simple example, assume you have a solution containing four different types of DNA fragments, each with a unique sequence (Figure 7-4). Each fragment type is individually inserted into a plasmid vector. The resulting mixture of recombinant plasmids is incubated with *E. coli* cells under conditions that facilitate transformation; the cells then are cultured on antibiotic selective plates. Since each colony that develops arose from a single cell that took up a single plasmid, all the cells in a colony harbor the identical type of plasmid characterized by the DNA fragment inserted into it. As a result, copies of the DNA fragments in the initial mixture are isolated from one another in the separate bacterial colonies. DNA cloning thus is a powerful, yet simple method for purifying a particular DNA fragment from a complex mixture of fragments and producing large numbers of the fragment of interest.

Restriction Enzymes Cut DNA Molecules at Specific Sequences

To clone specific DNA fragments in a plasmid vector, as just described, or in other vectors discussed in later sections, the fragments must be produced and then inserted into the vector DNA. As noted in the introduction, restriction enzymes and DNA ligases are utilized to produce such recombinant DNA molecules.

Restriction enzymes are bacterial enzymes that recognize specific 4- to 8-bp sequences, called *restriction sites*, and then cleave both DNA strands at this site. Since these enzymes

cleave DNA within the molecule, they are also called *restriction endonucleases* to distinguish them from exonucleases, which digest nucleic acids from an end. Many restriction sites, like the *Eco*RI site shown in Figure 7-5a, are short inverted repeat sequences; that is, the restriction-site sequence is the same on each DNA strand when read in the 5′ → 3′ direction. Because the DNA isolated from an individual organism has a specific sequence, restriction enzymes cut the DNA into a reproducible set of fragments called **restriction fragments** (Figure 7-6).

The word *restriction* in the name of these enzymes refers to their function in the bacteria from which they are isolated: a restriction endonuclease destroys (restricts) incoming foreign DNA (e.g., bacteriophage DNA or DNA taken up during transformation) by cleaving it at all the restriction sites in the DNA. Another enzyme, called a *modification enzyme*, protects a bacterium's own DNA from cleavage by modifying it at or near each potential cleavage site. The modification enzyme adds a methyl group to one or two bases, usually within the restriction site. When a methyl group is present there, the restriction endonuclease is prevented from cutting the DNA (Figure 7-5b). Together with the restriction endonuclease, the methylating enzyme forms a restriction-modification system that protects the host DNA while it destroys foreign DNA. Restriction enzymes have been purified from several hundred different species of bacteria, allowing DNA molecules to be cut at a large number of different sequences corresponding to the recognition sites of these enzymes (Table 7-1).

(a)

(b)

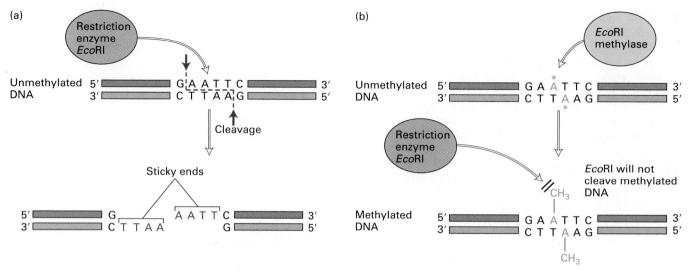

▲ **FIGURE 7-5 Restriction-recognition sites are short DNA sequences recognized and cleaved by various restriction endonucleases.** (a) *Eco*RI, a restriction enzyme from *E. coli,* makes staggered cuts at the specific 6-bp inverted repeat sequence shown. This cleavage yields fragments with single-stranded, complementary "sticky" ends. Many other restriction enzymes also produce fragments with sticky ends. (b) Bacterial cells with restriction endonucleases also contain corresponding

modification enzymes that methylate bases in the restriction-recognition site. For example, *E. coli* cells containing the *Eco*RI restriction enzyme also contain *Eco*RI methylase, a modification enzyme that catalyzes addition of a methyl group to two adenines in the *Eco*RI recognition sequence. The methylated restriction site is not cleaved by *Eco*RI, assuring that a cell making this restriction enzyme does not destroy its own DNA.

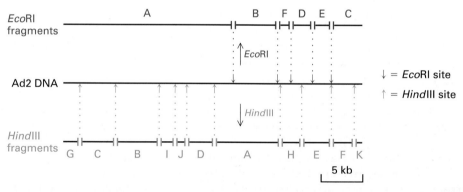

▲ **FIGURE 7-6 Fragments produced by cleavage of the ≈36-kb DNA genome from adenovirus 2 (Ad2) by *Eco*RI and another restriction enzyme, *Hind*III from *Haemophilus influenzae.*** Double-stranded DNA is represented by single black lines in this figure. Digestion of Ad2 DNA *(center)* with *Eco*RI generates 6 *Eco*RI fragments *(top);* these result from cleavage at each *Eco*RI restriction site (GAATTC) in the Ad2 sequence. Digestion with *Hind*III cleaves the Ad2 DNA at each *Hind*III site

(AAGCTT), generating 11 specific fragments *(bottom),* all different from the *Eco*RI fragments. By convention, restriction fragments are labeled A–Z in order of decreasing size. By techniques described later, the order of fragments in the original DNA can be determined, thus mapping the restriction sites on the uncut DNA (indicated by short arrows). Such a "restriction-site map" for various restriction enzymes is a unique characteristic of each DNA.

Restriction Fragments with Complementary "Sticky Ends" Are Ligated Easily

As illustrated in Figure 7-5a, *Eco*RI makes staggered cuts in the two DNA strands. Many other restriction enzymes make similar cuts, generating fragments that have a single-stranded "tail" at both ends. The tails on the fragments generated at a given restriction site are complementary to those

on all other fragments generated by the same restriction enzyme. At room temperature, these single-stranded regions, often called "sticky ends," can transiently base-pair with those on other DNA fragments generated with the same restriction enzyme, regardless of the source of the DNA. This base pairing of sticky ends permits DNA from widely differing species to be ligated, forming chimeric molecules.

TABLE 7-1 Selected Restriction Endonucleases and Their Restriction-Site Sequences

Source Microorganism	Enzyme*	Recognition Site ($\downarrow$)[†]	Ends Produced
Arthrobacter luteus	*Alu*I	AG$\downarrow$CT	Blunt
Bacillus amyloliquefaciens H	*Bam*HI	G$\downarrow$GATCC	Sticky
Escherichia coli	*Eco*RI	G$\downarrow$AATTC	Sticky
Haemophilus gallinarum	*Hga*I	GACGC+5$\downarrow$	[‡]
Haemophilus influenzae	*Hin*dIII	A$\downarrow$AGCTT	Sticky
Haemophilus parahaemolyticus	*Hph*I	GGTGA+8$\downarrow$	[‡]
Nocardia otitiscaviaruns	*Not*I	GC$\downarrow$GGCCGC	Sticky
Staphylococcus aureus 3A	*Sau*3AI	$\downarrow$GATC	Sticky
Serratia marcesens	*Sma*I	CCC$\downarrow$GGG	Blunt
Thermus aquaticus	*Taq*I	T$\downarrow$CGA	Sticky

Enzymes are named with abbreviations of the bacterial strains from which they are isolated; the roman numeral indicates the enzyme's priority of discovery in that strain (for example, AluI was the first restriction enzyme to be isolated from Arthrobacter luteus).

[†]*Recognition sequences are written 5'→3' (only one strand is given), with the cleavage site indicated by an arrow. Enzymes producing blunt ends cut both strands at the indicated site; those producing stick ends make staggered cuts, with cleavage occurring between the same nucleotides in each strand as shown in Figure 7-5a.*

[‡]*The cleavage sites for HphI and HgaI occur several nucleotides away from the recognition sequence. HgaI cuts five nucleotides 3' to the GACGC sequence on the top strand and ten nucleotides 5' to the complementary GTGCG sequence on the bottom strand. HphI cuts eight nucleotides 3' to the GGTGA sequence on the top strand and seven nucleotides 5' to the complementary CCACT sequence on the bottom strand.*

SOURCE: R. J. Roberts, 1988, *Nucl. Acids Res.* **16**(suppl):271.

During in vivo DNA replication, DNA ligase catalyzes formation of $3' \rightarrow 5'$ phosphodiester bonds between the short fragments of the discontinuously synthesized DNA strand at a replication fork (see Figure 4-16). In recombinant DNA technology, purified DNA ligase is used to covalently join the ends of restriction fragments in vitro. This enzyme can catalyze the formation of a $3' \rightarrow 5'$ phosphodiester bond between the 3'-hydroxyl end of one restriction-fragment strand and the 5'-phosphate end of another restriction-fragment strand during the time that the sticky ends are transiently base-paired (Figure 7-7). When DNA ligase and ATP are added to a solution containing restriction fragments with sticky

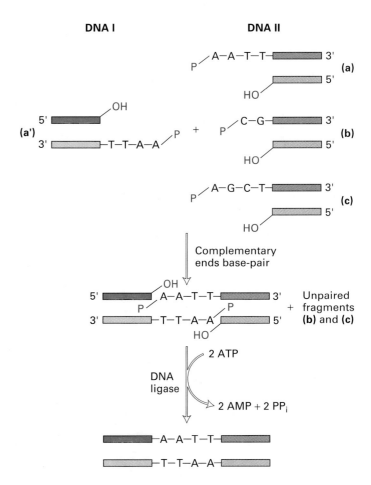

▶ **FIGURE 7-7 Ligation of restriction fragments with complementary sticky ends.** In this example, *Eco*RI fragments from DNA I *(left)* are mixed with several different restriction fragments, including *Eco*RI fragments, produced from DNA II *(right)*. The short DNA sequences composing the sticky ends of each fragment type are shown. The complementary sticky ends on the two types of *Eco*RI fragments, (a') and (a), can transiently base-pair, whereas the *Taq*I fragments (b) and *Hin*dIII fragments (c) with noncomplementary sticky ends do not base-pair to *Eco*RI fragments. The adjacent 3'-hydroxyl and 5'-phosphate groups (red) on the base-paired fragments then are covalently joined (ligated) by T4 DNA ligase. One ATP is consumed for each phosphodiester bond (red) formed.

ends, the restriction fragments are covalently ligated together through the standard $3' \rightarrow 5'$ phosphodiester bonds of DNA.

Some restriction enzymes, such as *Alu*I and *Sma*I, cleave both DNA strands at the same point within the recognition site (see Table 7-1). These restriction enzymes generate DNA restriction fragments with "blunt" (flush) ends in which all the nucleotides at the fragment ends are base-paired to nucleotides in the complementary strand. In addition to ligating complementary sticky ends, the DNA ligase from bacteriophage T4 can ligate any two blunt DNA ends. However, blunt-end ligation requires a higher DNA concentration than ligation of sticky ends.

Polylinkers Facilitate Insertion of Restriction Fragments into Plasmid Vectors

Restriction enzymes to create fragments with sticky ends and DNA ligase to covalently link them allow foreign DNA to be inserted into plasmid vectors in vitro in a straightforward procedure. *E. coli* plasmid vectors can be constructed with a *polylinker*, a synthetic multiple-cloning-site sequence that contains one copy of several different restriction sites (Figure 7-8a). When such a vector is treated with a restriction enzyme that recognizes a recognition sequence in the polylinker, it is cut at that sequence, generating sticky ends. In the presence of DNA ligase, DNA fragments produced with the same

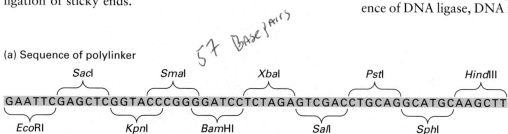

(a) Sequence of polylinker

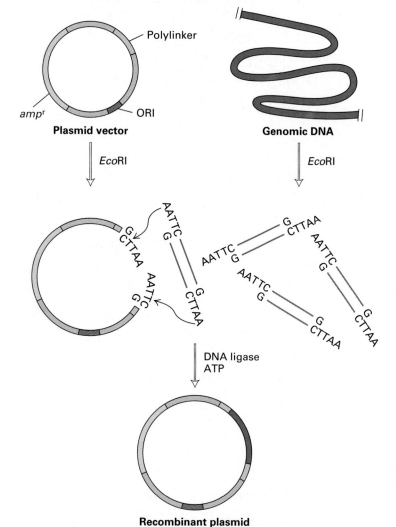

(b) Insertion of *Eco*RI restriction fragments

◀ FIGURE 7-8 Plasmid vectors containing a polylinker, or multiple-cloning-site sequence, commonly are used to produce recombinant plasmids carrying exogenous DNA fragments. (a) Sequence of a polylinker that includes one copy of the recognition site, indicated by brackets, for each of the 10 restriction enzymes indicated. Polylinkers are chemically synthesized and then are inserted into a plasmid vector. Only one strand is shown. (b) Insertion of genomic restriction fragments into the pUC19 plasmid vector, which contains the polylinker shown in (a). (The length of the polylinker in relation to the rest of the plasmid is greatly exaggerated here.) One of the restriction enzymes whose recognition site is in the polylinker is used to cut both the plasmid molecules and genomic DNA, generating singly-cut plasmids and restriction fragments with complementary sticky ends (letters at ends of green fragments). By use of appropriate reaction conditions, insertion of a single restriction fragment per plasmid can be maximized. Note that the restriction sites are reconstituted in the recombinant plasmid. [See C. Yanisch-Perron, J. Vieira, and J. Messing, 1985, *Gene* **33**:103.]

restriction enzyme will be inserted into the plasmid (Figure 7-8b). The ratio of DNA fragments to be inserted to cut vectors and other reaction conditions are chosen to maximize the insertion of one restriction fragment per plasmid vector. The recombinant plasmids produced in in vitro ligation reactions then can be used to transform antibiotic-sensitive *E. coli* cells as shown in Figure 7-4. All the cells in each antibiotic-resistant clone that remains after selection contain plasmids with the same inserted DNA fragment, but different clones carry different fragments.

Small DNA Molecules Can Be Chemically Synthesized

Advances in synthetic chemistry now permit the chemical synthesis of single-stranded DNA (ssDNA) molecules of any sequence up to about 100 nucleotides in length. Synthetic DNA has a number of applications in recombinant DNA technology. Complementary ssDNAs can be synthesized and hybridized to each other to form a dsDNA with sticky ends. Such completely synthetic dsDNAs can be cloned into plasmid vectors just as DNA restriction fragments prepared from living organisms are. For example, the 57-bp polylinker sequence shown in Figure 7-8 was chemically synthesized and then inserted into plasmid vectors to facilitate the cloning of fragments generated by different restriction enzymes. This example illustrates the use of synthetic DNAs to add convenient restriction sites where they otherwise do not occur. As described later in the chapter, synthetic DNAs are used in sequencing DNA and as probes to identify clones of interest. Synthetic DNAs also can be substituted for natural DNA sequences in cloned DNA to study the effects of specific mutations; this topic is examined in Chapter 8.

The technique for chemical synthesis of DNA oligonucleotides is outlined in Figure 7-9. Note that chains grow in the $3' \rightarrow 5'$ direction, opposite to the direction of DNA chain growth catalyzed by DNA polymerases. Once the chemistry for producing synthetic DNA was standardized, automated instruments were developed that allow researchers to program the synthesis of oligonucleotides of specific sequences up to about 100 nucleotides long.

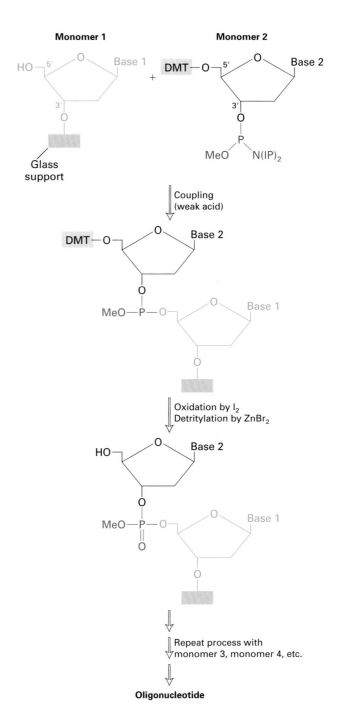

▶ **FIGURE 7-9 Chemical synthesis of oligonucleotides by sequential addition of reactive nucleotide derivatives in the $3' \rightarrow 5'$ direction.** The first nucleotide (monomer 1) is bound to a glass support by its 3′ hydroxyl; its 5′ hydroxyl is available for addition of the second nucleotide. The second nucleotide in the sequence (monomer 2) is derivatized by addition of 4′,4′-dimethoxytrityl (DMT) to its 5′ hydroxyl, thus blocking this hydroxyl from reacting; in addition, a highly reactive methylated diisopropyl phosphoramidite group (red letters) is attached to the 3′ hydroxyl. When the two monomers are mixed in the presence of a weak acid, they form a $5' \rightarrow 3'$ phosphodiester bond with the phosphorus in the trivalent state. Oxidation of this intermediate with iodine (I_2) increases the phosphorus valency to 5, and subsequent removal of the DMT group by detritylation with zinc bromide ($ZnBr_2$) frees the 5′ hydroxyl. Monomer 3 then is added, and the reactions are repeated. Repetition of this process eventually yields the entire oligonucleotide. Finally, all the methyl groups on the phosphates are removed at the same time at alkaline pH, and the bond linking monomer 1 to the glass support is cleaved. [See S. L. Beaucage and M. H. Caruthers, 1981, *Tetrahedron Lett.* **22**:1859.]

SUMMARY DNA Cloning with Plasmid Vectors

• In DNA cloning, recombinant DNA molecules are formed in vitro by inserting DNA fragments of interest into vector DNA molecules. The recombinant DNA molecules are then introduced into host cells, where they replicate, producing large numbers of recombinant DNA molecules that include the fragment of DNA originally linked to the vector.

• The most commonly used cloning vectors are *E. coli* plasmids, small circular DNA molecules that include three functional regions: (1) an origin of replication, (2) a drug-resistance gene, and (3) a region where DNA can be inserted without interfering with plasmid replication or expression of the drug-resistance gene.

• Two enzymes are used to produce recombinant plasmids. Restriction enzymes cut DNA at specific 4- to 8-bp sequences, often leaving self-complementary single-stranded tails (sticky ends). These enzymes are used to cut long DNA molecules into multiple restriction fragments and to cut a plasmid vector at a single site. If a restriction fragment and cut plasmid vector with complementary ends are mixed under the proper conditions, DNA ligase will form phosphodiester bonds between the restriction fragment and vector DNA (see Figure 7-7).

• When recombinant plasmids are incubated with *E. coli* cells first treated with a high concentration of divalent cations, a very small fraction of the cells take up a single recombinant plasmid. These transformed cells, which carry the plasmid drug-resistance gene, can be selected by plating on nutrient agar containing the antibiotic (see Figure 7-3). All the cells in each colony that grows on this medium contain identical plasmids descended from the single plasmid that entered the founder cell of the colony. Isolated colonies thus represent clones of the different restriction fragments originally inserted into the plasmid vector.

• Polylinkers are synthetic oligonucleotides composed of one copy of several different restriction sites. Plasmid vectors that contain a polylinker will be cut only once by multiple restriction enzymes, each acting at its own site. Inclusion of a polylinker in a plasmid vector thus permits cloning of restriction fragments generated by cleavage of DNA with multiple different restriction enzymes.

• Single-stranded DNA containing up to 100 nucleotides of any desired sequence can be chemically synthesized using automated instruments. Synthetic dsDNAs are produced by synthesizing complementary ssDNAs and then hybridizing them.

7.2 Constructing DNA Libraries with λ Phage and Other Cloning Vectors

Most DNA cloning is done with *E. coli* plasmid vectors because of the relative simplicity of the cloning procedure. However, the number of individual clones that can be obtained by plasmid cloning is limited by the relatively low efficiency of *E. coli* transformation and the small number (only a few hundred) of individual transformed colonies that can be grown on a typical culture plate. These limitations make plasmid cloning of all the **genomic DNA** of higher organisms impractical. For example, $\approx 1.5 \times 10^5$ clones carrying 20-kb DNA fragments are required to represent the total human haploid genome, which contains $\approx 3 \times 10^9$ base pairs. Fortunately, cloning vectors derived from various bacteriophages have proved to be a practical means for obtaining the required number of clones to represent large genomes. A collection of clones that includes *all* the DNA sequences of a given species is called a *genomic DNA library,* or simply *genomic library.* Once a genomic library is prepared, it can be screened for clones containing a sequence of interest.

Bacteriophage λ Can Be Modified for Use as a Cloning Vector and Assembled in Vitro

Bacteriophage λ is probably the most extensively studied bacterial virus, and a great deal is known about its molecular biology and genetics. A λ phage virion has a head, which contains the viral DNA genome, and a tail, which functions in infecting *E. coli* host cells (Figure 7-10a). When λ DNA enters the host-cell cytoplasm following infection, it undergoes either lytic or lysogenic growth (see Figure 6-19). In lytic growth, the viral DNA is replicated and assembled into more than 100 progeny virions in each infected cell, killing the cell in the process and releasing the replicated virions. In lysogenic growth, the viral DNA inserts into the bacterial chromosome, where it is passively replicated along with the host-cell chromosome as the cell grows and divides.

The λ genes encoding the head and tail proteins as well as various proteins involved in the lytic and lysogenic growth pathways are clustered in discrete regions of the ≈50-kb viral genome (Figure 7-10b). When bacteriophage λ is used as a cloning vector, it must be capable of lytic growth, but other viral functions are irrelevant. Consequently, the genes involved in the lysogenic pathway and other viral genes not essential for the lytic pathway are removed from the viral DNA and replaced with the DNA to be cloned. Up to ≈25 kb of foreign DNA can be inserted into the λ genome, resulting in a recombinant DNA that can be packaged in vitro to form virions capable of replicating and forming plaques on *E. coli* host cells.

During the in vivo assembly of λ virions within infected host cells, viral heads and tails initially are assembled

(a) λ Phage virion

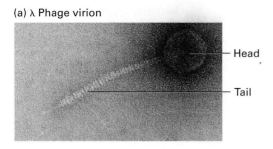

Head

Tail

(b) λ Phage genome

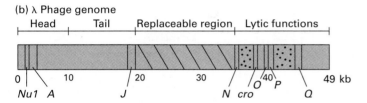

| Head | Tail | Replaceable region | Lytic functions |

0   10   20   30   40   49 kb

Nu1 A   J   N cro O P   Q

▲ **FIGURE 7-10 The bacteriophage genome.** (a) Electron micrograph of bacteriophage λ virion. The genome is contained within the head. (b) Simplified map of the λ phage genome. Genes encoding proteins required for assembly of the head and tail map at the left end; those encoding additional proteins required for the lytic cycle map at the right end. Some regions of the genome can be replaced by exogenous DNA (diagonal lines) or deleted (dotted area) without affecting the ability of λ phage to infect host cells and assemble new virions, permitting insertion of up to ≈25 kb of exogenous DNA between the J and N genes. There are about 60 genes on the λ genome. Only a few individual genes are shown in this diagram. Small numbers indicate positions in kilobases (kb). [Photograph courtesy of R. Duda and R. Hendrix.]

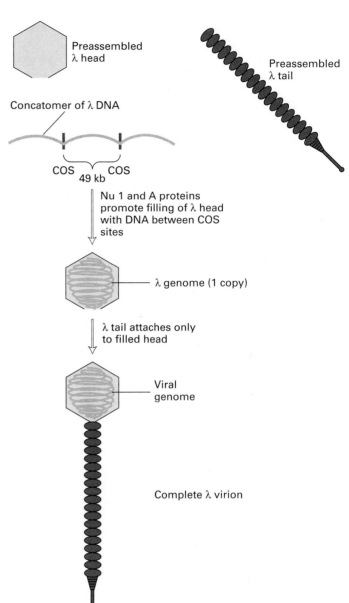

Preassembled λ head

Preassembled λ tail

Concatomer of λ DNA

COS   COS
   49 kb

Nu 1 and A proteins promote filling of λ head with DNA between COS sites

λ genome (1 copy)

λ tail attaches only to filled head

Viral genome

Complete λ virion

▲ **FIGURE 7-11 Assembly of bacteriophage λ virions.** Empty heads and tails are assembled from multiple copies of several different λ proteins. During the late stage of λ infection, long DNA molecules called *concatomers* are formed; these multimeric molecules consist of copies of the λ genome linked end to end and separated by COS sites (red), a protein-binding nucleotide sequence that occurs once in each copy of the λ genome. Binding of the λ proteins Nu1 and A to COS sites promotes insertion of the DNA between two adjacent COS sites into an empty head. After the heads are filled with DNA, preassembled λ tails are attached, producing complete λ virions capable of infecting *E. coli* cells.

separately, from multiple copies of the various proteins that compose these complex structures. Replication of λ DNA in a host cell generates long multimeric DNA molecules, called *concatomers*, that consist of multiple copies of the viral genome linked end to end and separated by specific nucleotide sequences called *COS sites*. Two λ proteins, designated Nu1 and A, bind to COS sites and direct insertion of the DNA lying between two adjacent COS sites into a preassembled head. This process results in the packaging of a single ≈50-kb λ genome from the multimeric concatomer into each preassembled head. Host-cell chromosomal DNA is not inserted into the λ heads because it does not contain any copies of the COS sequence. Once λ DNA is inserted into a preassembled λ head, the preassembled tail is attached, producing a complete virion (Figure 7-11).

To prepare infectious λ virions carrying recombinant DNA, the phage-assembly process is carried out in vitro. In one method, *E. coli* cells are infected with a λ mutant defective in A protein, one of the two proteins required for packaging λ DNA into preassembled phage heads. These

cells accumulate preassembled "empty" heads; since tails attach only to heads "filled" with DNA. Preassembled tails also accumulate in these cells. An extract containing high concentrations of empty heads and tails is prepared by lysing cells infected with the λ A mutant. When this extract is mixed with isolated A protein (obtained from λ-infected cells) and recombinant λ DNA containing a COS site, the DNA is packaged into the empty heads. The tails in the extract then combine with the filled heads, yielding complete virions carrying the recombinant λ DNA.

This procedure produces a high yield of recombinant virions that are fully infectious and can efficiently infect *E. coli* cells. Each virion particle binds to receptors on the surface of a host cell and injects its packaged recombinant DNA into the cell. Infection by λ phage is about a thousand times more efficient than transformation with plasmid vectors, accounting for the high efficiency of λ phage cloning. For instance, $\approx 10^6$ transformed colonies per microgram of recombinant plasmid DNA can be obtained routinely, whereas $\approx 10^9$ plaques representing λ clones can be obtained per microgram of recombinant λ DNA.

Nearly Complete Genomic Libraries of Higher Organisms Can Be Prepared by λ Cloning

With the availability of λ phage cloning vectors, preparation of genomic libraries for higher organisms, including humans, is feasible. A genomic library is a set of λ (or plasmid) clones that collectively contain every DNA sequence in the genome of a particular organism. Figure 7-12 summarizes the general procedure for constructing a λ genomic library. The λ DNA first is treated with a restriction enzyme to produce fragments called λ *vector arms*, which have sticky ends and together contain all the genes necessary for lytic growth. This step frees the nonessential region in the middle of the λ genome; this region is separated from the λ arms and discarded. Genomic DNA then is extracted from a cell type that contains all the genetic information of the organism under study. Sperm cells or cells of an early embryo often are used as sources of mammalian DNA. The extracted DNA then is cleaved by a restriction enzyme to produce ≈ 20-kb fragments with sticky ends complementary to the sticky ends on the λ vector arms being used.

The λ arms and the collection of genomic DNA fragments are mixed in about equal amounts. The complementary sticky ends on the fragments and λ arms hybridize and then are joined covalently by DNA ligase. Each of the resulting recombinant DNA molecules contains a foreign DNA fragment located between the two arms of the λ vector DNA. The ligated recombinant DNAs then are packaged into λ virions in vitro as described above. Only DNA molecules of the correct size can be packaged to produce fully infectious recombinant λ virions.

Finally, the recombinant λ virions are plated on a lawn of *E. coli* cells to generate a large number of recombinant λ plaques. Since each plaque arises from a single recombinant virion, all the progeny λ phages that develop are genetically identical and constitute a clone carrying a particular genomic DNA insert. The different plaques correspond to distinct phage clones, each carrying a different DNA insert, and collectively they constitute a λ genomic library. Multiple λ vectors have been constructed containing different restriction sites, so that restriction fragments generated by a variety of restriction enzymes can be cloned in λ vectors.

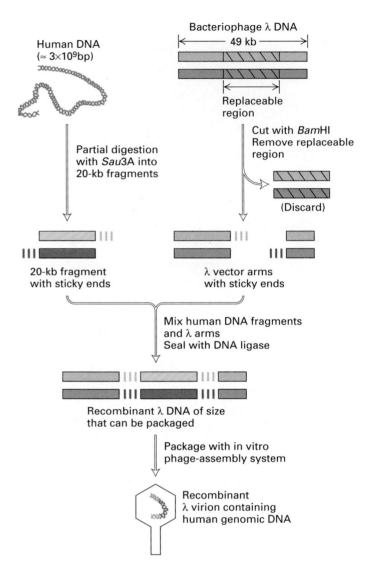

▲ **FIGURE 7-12 Construction of a genomic library of human DNA in a bacteriophage λ vector.** The nonessential regions in the right half of the λ genome (dotted areas in Figure 7-10b) usually are deleted to maximize the size of the exogenous DNA fragment that can be inserted. Then the λ DNA is treated to remove the central replaceable region. In this example, the replaceable region is cut out with *Bam*HI, and the total DNA from human cells is partially digested with *Sau*3A. These two restriction enzymes produce fragments with complementary sticky ends (red lines). The λ vector arms and ≈ 20-kb genomic fragments are mixed, ligated, and packaged in vitro to produce recombinant λ phage virions, which are plated on a lawn of *E. coli* cells. In the diagrams of DNA regions, light and dark shades of the same color indicate complementary strands.

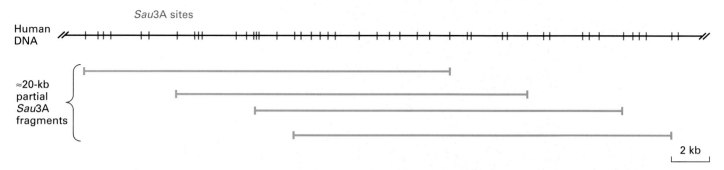

▲ **FIGURE 7-13 Production of overlapping restriction fragments by partial digestion of human genomic DNA with *Sau*3A.** This restriction endonuclease recognizes the 4-bp sequence GATC and produces fragments with single-stranded sticky ends with this sequence on the 5′ end of each strand. A hypothetical region of human genomic DNA showing the *Sau*3A recognition sites (red) is shown at the top. Partial digestion of this region of DNA would yield a variety of overlapping fragments (blue) ≈20 kb long. Use of such overlapping fragments increases the probability that all sequences in the genomic DNA will be represented in a λ library.

In constructing a library of genomic DNA, the DNA commonly is cleaved with a restriction enzyme that recognizes a 4-bp restriction site (e.g., *Sau*3A, as shown in Figure 7-12). A specific 4-bp sequence will occur on average once every $4^4 = 256$ base pairs. Complete digestion of the human haploid genome, which contains $\approx 3 \times 10^9$ base pairs, would yield somewhat more than 10^7 nonoverlapping different fragments. However, to increase the probability that all regions of the genome are successfully cloned and represented in the λ genomic library, the genomic DNA usually is only partially digested to yield overlapping restriction fragments of ≈20 kb (Figure 7-13). In a large λ library constructed from such overlapping restriction fragments, a specific sequence of genomic DNA (e.g., the gene encoding β-globin) often is contained in or extends over several "overlapping" clones. By a technique called chromosome walking, described in the next chapter, overlapping genomic clones can be ordered.

The size of a genomic library for a given organism depends on the amount of DNA in that organism's haploid genome. If the human genome of about 3×10^9 base pairs is cleaved into 20-kb fragments for insertion into a λ vector, then roughly 1.5×10^5 different recombinant λ phage virions would be required to constitute a complete library. Because the restriction fragments of human DNA are incorporated into phages randomly, about 10^6 recombinant phages are necessary to assure that each region of human DNA has a 90–95 percent chance of being included.

Each plaque produced by a recombinant bacteriophage λ contains large numbers of recombinant virions and, consequently, large numbers of identical cloned DNA fragments. Hybridization methods for identifying recombinant λ clones of interest are described in a later section. Because these methods allow specific detection of very small plaques, as many as 5×10^4 plaques can be screened on a single culture plate. Thus only 20–30 petri dishes, each containing about 5×10^4 λ plaques, are sufficient to represent the entire human genome. In contrast, to screen 10^6 recombinant plasmids carrying the entire human genome would require about 5000 petri dishes, because only 200 or so transformed *E. coli* colonies can be detected on a typical petri dish.

cDNA Libraries Are Prepared from Isolated mRNAs

In higher eukaryotes, many genes are transcribed into mRNA only in specialized cell types. For example, mRNAs encoding globin proteins are found only in erythrocyte precursor cells, called *reticulocytes*. Likewise, the mRNA encoding albumin, the major protein in serum, is produced only in liver cells where albumin is synthesized. The specific DNA sequences expressed as mRNAs in a particular cell type can be cloned by synthesizing DNA copies of the mRNAs isolated from that type of cell, and then cloning the DNA copies in plasmid or bacteriophage λ vectors.

DNA copies of mRNAs are called **complementary DNAs (cDNAs)**; clones of such DNA copies of mRNAs are called *cDNA clones*. In addition to representing only the sequences expressed as mRNAs in a particular cell type, cDNA clones lack the noncoding introns present in genomic DNA clones. Thus the amino acid sequence of a protein can be determined directly from the nucleotide sequence of its corresponding cDNA. Many genes in higher eukaryotes are too large to be included in a single λ clone because of their large introns. In contrast, all full-length cDNAs, containing the entire protein-coding sequence, can be included in a single λ clone. However, because of methodological difficulties, not all cDNA clones are full length when initially produced; to obtain a full-length cDNA, it often is necessary to isolate several overlapping cDNA clones and then ligate them at rare restriction sites. Just as a large collection of clones containing fragments of genomic DNA representing the entire genome of a species is called a *genomic library*, a large collection of cDNA copies of all the mRNAs in a cell type is called a *cDNA library*.

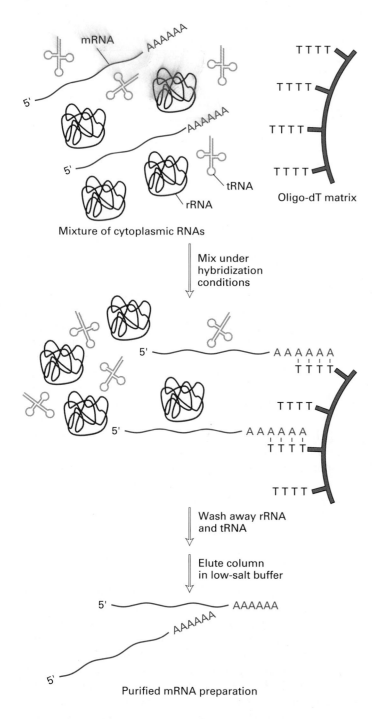

Oligo-dT matrix

Mixture of cytoplasmic RNAs

Mix under
hybridization
conditions

Wash away rRNA
and tRNA

Elute column
in low-salt buffer

Purified mRNA preparation

▲ **FIGURE 7-14 Isolation of eukaryotic mRNA by oligo-dT column affinity chromatography.** Isolated cytoplasmic RNA consists mostly of ribosomal RNAs (rRNAs) and transfer RNAs (tRNAs). The much less abundant mRNAs (red) have 3′ poly(A) tails, which hybridize to oligo-dT covalently coupled to the column matrix. After hybridization, the rRNAs and tRNAs are washed out of the column; then the mRNAs are eluted with a low-salt buffer. The resulting purified mRNA preparation contains many different mRNA molecules encoding different proteins.

Isolation of mRNAs and Synthesis of cDNAs The first step in preparing a cDNA library is to isolate the total mRNA from the cell type or tissue of interest. Nature has greatly simplified the isolation of eukaryotic mRNAs: the 3′ end of nearly all eukaryotic mRNAs consists of a string of 50–250 adenylate residues, the *poly(A) tail.* Because of their poly(A) tail, mRNAs can be easily separated from the much more prevalent rRNAs and tRNAs present in a cell extract by use of a column to which short strings of thymidylate (oligo-dTs) are linked to the matrix (Figure 7-14). When a cell extract is passed through an oligo-dT column, the mRNA poly(A) tails base-pair with the oligo-dTs, binding the mRNAs to the column. Since rRNAs, tRNAs, and other molecules do not bind to the column, they can be washed away. The bound mRNAs are recovered by elution with a low-salt buffer.

The enzyme **reverse transcriptase**, which is found in retroviruses, is then used to synthesize a strand of DNA complementary to each mRNA molecule. This enzyme can polymerize deoxynucleoside triphosphates into a complementary DNA strand using an RNA molecule as template. Like other DNA polymerases, reverse transcriptase can add nucleotides only to the 3′ end of a preexisting primer base-paired to the template. Added free oligo-dT serves this function by hybridizing to the 3′ poly(A) tail of each mRNA template (Figure 7-15, steps 1 and 2).

Conversion of Single-Stranded cDNA to Double-Stranded cDNA After cDNA copies of isolated mRNAs are synthesized, the mRNAs are removed by treatment with alkali, which hydrolyzes RNA but not DNA. The single-stranded cDNAs then are converted to double-stranded DNA molecules. To do this, the 3′ end of each cDNA strand is elongated by adding several residues of a single nucleotide (e.g., dG) through the action of *terminal transferase,* a unique DNA polymerase that does not require a template, but simply adds deoxynucleotides to free 3′ ends. A synthetic oligo-dC primer then is hybridized to this 3′ oligo-dG. DNA polymerase, which uses the oligo-dC as a primer, then is used to synthesize a DNA strand complementary to the original cDNA strand. These reactions produce a complete double-stranded DNA molecule corresponding to each of the mRNA molecules in the original preparation (Figure 7-15, steps 3–5). Each double-stranded DNA, also called cDNA, contains an oligo-dC–oligo-dG double-stranded region at one end and an oligo-dT–oligo-dA double-stranded region at the other end.

Addition of Linkers and Incorporation of cDNA into a Vector To prepare double-stranded cDNAs for cloning, short *restriction-site linkers* first are ligated to both ends. These are double-stranded DNA segments, usually ≈10–12 bp long, that contain the recognition site for a particular restriction enzyme. Restriction-site linkers are prepared by hybridizing chemically synthesized complementary oligonucleotides. The ligation reaction is carried out by DNA ligase from bacteriophage T4, which can join "blunt-ended" double-stranded DNA molecules lacking sticky ends. Al-

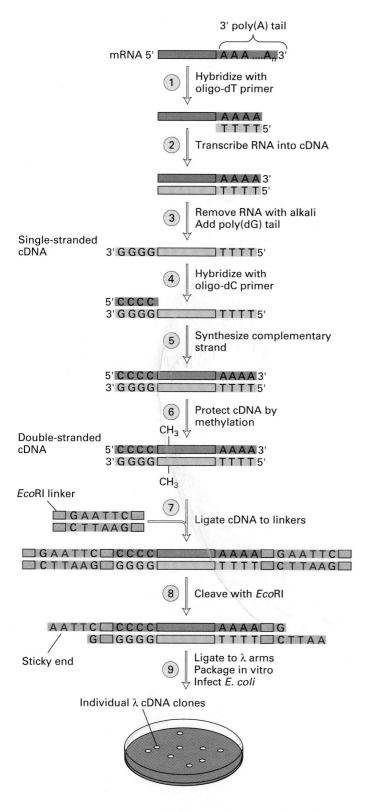

3' poly(A) tail

mRNA 5' ▬▬▬▬ A A A ....A$_n$ 3'

1 | Hybridize with oligo-dT primer

▬▬▬▬ A A A A
T T T T 5'

2 | Transcribe RNA into cDNA

▬▬▬▬ A A A A 3'
▬▬▬▬ T T T T 5'

3 | Remove RNA with alkali / Add poly(dG) tail

Single-stranded cDNA

3' G G G G ▭▭▭▭ T T T T 5'

4 | Hybridize with oligo-dC primer

5' C C C C
3' G G G G ▭▭▭▭ T T T T 5'

5 | Synthesize complementary strand

5' C C C C ▬▬▬▬ A A A A 3'
3' G G G G ▬▬▬▬ T T T T 5'

6 | Protect cDNA by methylation

CH$_3$

Double-stranded cDNA

5' C C C C ▬▬▬▬ A A A A 3'
3' G G G G ▬▬▬▬ T T T T 5'

CH$_3$

EcoRI linker

▭ G A A T T C ▭
▭ C T T A A G ▭

7 | Ligate cDNA to linkers

▭ G A A T T C ▭ C C C C ▬▬▬▬ A A A A ▭ G A A T T C ▭
▭ C T T A A G ▭ G G G G ▬▬▬▬ T T T T ▭ C T T A A G ▭

8 | Cleave with EcoRI

A A T T C ▭ C C C C ▬▬▬▬ A A A A ▭ G
G ▭ G G G G ▬▬▬▬ T T T T ▭ C T T A A

Sticky end

9 | Ligate to λ arms / Package in vitro / Infect E. coli

Individual λ cDNA clones

◀ **FIGURE 7-15 Preparation of a bacteriophage λ cDNA library.** A mixture of mRNAs, isolated as shown in Figure 7-14, is used to produce cDNAs corresponding to all the cellular mRNAs (steps ①–③). These single-stranded cDNAs (light green) are then converted into double-stranded cDNAs, which are treated with *Eco*RI methylase to prevent subsequent digestion by *Eco*RI (steps ④–⑥). The protected double-stranded cDNAs are ligated to a synthetic double-stranded *Eco*RI-site linker at both ends and then cleaved with the corresponding restriction enzyme, yielding cDNAs with sticky ends (red letters); these are incorporated into λ phage cloning vectors, and the resulting recombinant λ virions are plated on a lawn of *E. coli* cells (steps ⑦–⑨). See text for further discussion.

restriction enzyme specific for the linker; this generates cDNA molecules with sticky ends at each end. To prevent digestion of any cDNAs that by chance have a recognition sequence for this restriction enzyme within the cDNA sequence, the mixture of double-stranded cDNAs is treated with the appropriate modification enzyme (see Figure 7-5b) before addition of the linkers. This enzyme methylates specific bases within the restriction-site sequence, preventing the restriction enzyme from digesting the methylated sites.

The final step in construction of a cDNA library is ligation of the restriction-cleaved double-stranded cDNAs, which now have sticky ends, to plasmid or λ phage vectors that have been cut to generate complementary sticky ends. The recombinant vectors then are plated on a lawn of *E. coli* cells, producing a library of plasmid or λ clones (Figure 7-15, steps 6–9). Each clone carries a cDNA derived from a single mRNA.

Larger DNA Fragments Can Be Cloned in Cosmids and Other Vectors

Both λ phage vectors and the more commonly used *E. coli* plasmid vectors are useful for cloning DNA fragments up to ≈20–25 kb. However, cloning of much larger fragments is desirable for sequencing of extremely long DNAs such as the DNA in a eukaryotic chromosome. Also, because of the common occurrence of large introns in genes from higher eukaryotes, it is often necessary to clone DNA fragments greater than 25 kb in order to include an entire gene in one clone. Consequently, additional types of cloning vectors have been developed for cloning larger fragments of DNA (Table 7-2).

One common method for cloning larger fragments makes use of elements of both plasmid and λ phage cloning. In this method, called *cosmid cloning*, recombinant plasmids containing inserted fragments up to 45 kb long can be efficiently introduced into *E. coli* cells. A **cosmid** vector is produced by inserting the COS sequence from λ phage DNA into a small *E. coli* plasmid vector about 5 kb long. Like other plasmid vectors discussed earlier, cosmid vectors contain a replication origin (ORI), an antibiotic-resistance gene (e.g.,

though blunt-end ligation is relatively inefficient, the ligation reaction can be driven to completion by using high concentrations of linkers.

The resulting double-stranded cDNAs, which contain a restriction-site linker at each end, are treated with the

TABLE 7-2 Approximate Maximum Length of DNA That Can Be Cloned in Vectors

Vector Type	Cloned DNA (kb)
Plasmid	20
λ phage	25
Cosmid	45
P1 phage	100
BAC (bacterial artificial chromosome)	300
YAC (yeast artificial chromosome)	1000

amp^r), and a polylinker sequence containing numerous restriction-enzyme recognition sites (Figure 7-16). Next, the cosmid vector is cut with a restriction enzyme and then ligated to 35- to 45-kb restriction fragments of foreign DNA with complementary sticky ends. If the concentration of foreign DNA is high enough, the ligation reaction generates long DNA molecules containing multiple restriction fragments of the foreign DNA separated by the 5-kb cosmid DNA. These ligated DNA molecules, which resemble the concatomers that form during replication of λ phage in a host cell, can be packaged in vitro as described earlier.

In the packaging reaction, the λ Nu1 and A proteins bind to COS sites in the ligated DNA and direct insertion of the DNA between two adjacent COS sites into empty phage heads. Packaging will occur so long as the distance between adjacent COS sites does not exceed about 50 kb (the approximate size of the λ genome). Phage tails then are attached to the filled heads, producing viral particles that contain a recombinant cosmid DNA molecule rather than the λ genome. When these virions are plated on a lawn of *E. coli* cells, they bind to phage receptors on the cell surface and inject the packaged DNA into the cells.

Since the injected DNA does not encode any λ proteins, no viral particles form in infected cells and no plaques develop on the plate. Rather, the injected DNA forms a large circular plasmid, composed of the cosmid vector and an inserted DNA fragment, in each host cell. This plasmid replicates and is segregated to daughter cells like other *E. coli* plasmids (see Figure 7-3), and the colonies that arise from transformed cells can be selected on antibiotic plates. The high efficiency of λ phage infection of *E. coli* cells makes cosmid cloning a practical method of generating plasmid clones carrying DNA fragments up to 45 kb long. Since many genes of higher eukaryotes are on the order of 30–40 kb in length, cosmid cloning increases the chances of obtaining DNA clones containing the entire sequences of genes.

Another strategy similar to cosmid cloning makes use of larger *E. coli* viruses such as bacteriophage P1, whose head can accommodate larger DNA molecules than the λ phage head. Recombinant plasmids containing DNA fragments up to ≈100 kb long can be packaged in vitro with the P1 system.

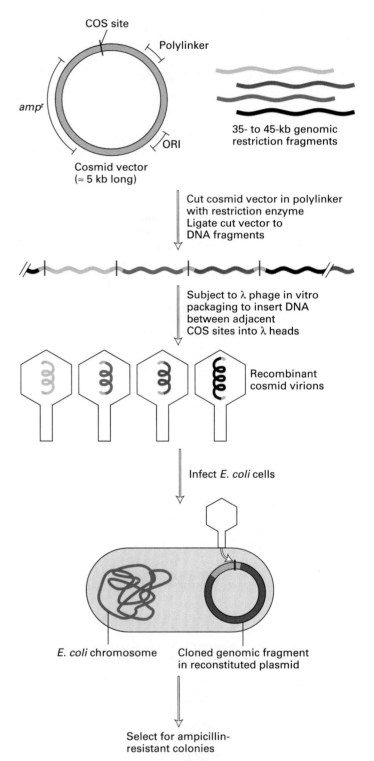

▲ FIGURE 7-16 **General procedure for cloning DNA fragments in cosmid vectors.** This procedure has the high efficiency associated with λ phage cloning and permits cloning of restriction fragments up to ≈45 kb long. In this example, four different types of recombinant cosmid virions could be generated, each carrying one of the genomic fragments indicated by different colors. Plating of the recombinant virions on *E. coli* cells would yield four different types of colonies, but only one is depicted. Note that the lengths of vector DNA and genomic fragments are not to scale. See text for further discussion.

Still larger fragments, up to 300 kb long, can be cloned in *bacterial artificial chromosomes* (BACs). These vectors are similar to standard *E. coli* plasmid vectors except that they contain the origin and genes encoding the ORI-binding proteins required for plasmid replication from a naturally occurring large *E. coli* plasmid called the *F-factor*. Recombinant BACs containing large DNA fragments are produced as in standard plasmid vector cloning. However, an alternative procedure for introducing DNA into cells, called *electroporation*, is used to overcome the very low efficiency of *E. coli* transformation by large plasmids. In this procedure, cells are mixed with DNA and subjected to a brief pulse of extremely high voltage. By an unknown mechanism, DNA molecules as long as 300 kb can be introduced into *E. coli* cells at a sufficiently high frequency to make BAC cloning practical. Large fragments cloned in BACs have been found to be stable through hundreds of generations.

Even larger fragments of DNA, containing up to 1000 kb, can be cloned into *yeast artificial chromosomes* (YACs). To understand how YACs function requires an explanation of the basic elements of eukaryotic chromosomes; this topic is covered in Chapter 9.

SUMMARY **Constructing DNA Libraries with λ Phage and Other Cloning Vectors**

- λ phage cloning is more complex than plasmid cloning, but it has the advantage of producing far more clones per microgram of DNA. Also far more λ plaques than *E. coli* colonies can be plated and detected on a single culture plate, simplifying the storage and screening of large numbers of clones.

- The middle region of λ DNA contains genes that are not required for lytic replication. In λ vector DNA this region is flanked by restriction sites. To produce recombinant virions, λ vector DNA is digested with an appropriate restriction enzyme, removing the nonessential middle region and generating the right and left vector arms with sticky ends. DNA fragments up to ≈25 kb in length then are ligated to the λ arms. Finally, recombinant λ DNA is packaged into λ virions in vitro (see Figure 7-12).

- Recombinant λ virions assembled in vitro are used to infect a lawn of *E. coli* cells, producing one plaque for each recombinant λ phage. Each plaque contains large numbers of recombinant λ virions, all descended from the single recombinant λ phage that initiated the plaque. The high efficiency of λ cloning results from the ability to package recombinant λ DNA in vitro and the high infectivity of the resulting virions.

- A genomic library is a set of λ or other clones prepared from the total genomic DNA of an organism. A genomic library must contain a sufficiently large number of independently derived clones that the probability is high (>95%) that every DNA sequence of the organism is represented in the library. A library of the human genome composed of λ clones contains about 10^6 recombinant phages, which can be concentrated into a single microliter of buffer.

- In cDNA cloning, retroviral reverse transcriptase is used to copy mRNA sequences into complementary DNAs, or cDNAs. By a series of reactions, single-stranded cDNAs are converted into double-stranded DNAs, which then are cloned in λ phage or plasmid vectors (see Figure 7-15).

- Because cDNA clones lack introns, the amino acid sequence of an encoded protein can be determined directly from the DNA sequence of its corresponding cDNA clone. This is a major advantage of cDNA cloning over genomic cloning.

- Nearly all eukaryotic mRNAs contain a 3′ poly(A) tail, which is used twice in cDNA cloning. First, the poly(A) tails of mRNAs are hybridized to an oligo-dT column, thereby separating mRNAs from other RNAs extracted from cells (see Figure 7-14). The poly(A) tails of isolated mRNAs also are hybridized to free short oligo-dT molecules, which serve as primers for copying of mRNAs by reverse transcriptase.

- A cDNA library is a set of cDNA clones prepared from the mRNA isolated from a particular type of tissue. The number of cDNA clones in a cDNA library must be large enough to include the sequence of (almost) every mRNA expressed in the tissue.

- DNA fragments up to ≈25 kb in length can be cloned in plasmid and λ vectors. Other vectors must be used for cloning larger DNA fragments. These include cosmid and P1 vectors, bacterial artificial chromosomes (BACs), and yeast artificial chromosomes (YACs).

7.3 Identifying, Analyzing, and Sequencing Cloned DNA

Suppose you have isolated a particular protein and want to isolate the gene that encodes it. A complete genomic λ library from mammals contains at least a million different clones; a cDNA library must contain as many clones to include the sequences of scarce mRNAs. How are specific clones of interest identified in such large collections? The most common method involves screening a library by **hybridization** with radioactively labeled DNA or RNA **probes.** In an alternative method, a specific clone in a library of cloned DNA is identified based on some property of its encoded protein.

Once a particular genomic DNA or cDNA clone of interest has been identified and isolated from other clones, the cloned DNA can be separated from the vector DNA and analyzed. This separation is achieved by cleaving the recombinant vector with the same restriction enzyme used to insert the DNA fragment initially. During ligation of a cut vector and DNA fragments generated with the same restriction enzyme, the restriction recognition sequence is regenerated between the DNA fragments and vector (see Figure 7-8). Subsequent treatment with the same restriction enzyme will cut the recombinant vector at the same sites, releasing the vector and cloned DNA, which then can be separated by gel electrophoresis.

The most complete characterization of a cloned DNA requires determination of its nucleotide sequence; from this sequence, the amino acid sequence of an encoded protein can be deduced. The sequence of genomic DNA includes introns as well as exons; it also includes regions that control gene expression by determining the type of cell in which the encoded protein is expressed, the stage in development it is expressed, and the amount of protein produced. Genomic DNAs also include replication origins and sequences important in determining how the DNA associates with proteins in chromosomes. In subsequent chapters, we consider how cells use DNA sequences for these functions. In this section, techniques for identifying, characterizing, and finally sequencing cloned DNA are outlined.

Libraries Can Be Screened with Membrane-Hybridization Assay

As discussed in Chapter 4, under the conditions of temperature and ion concentration found in cells, DNA is maintained as a duplex (double-stranded) structure by the hydrogen bonds between A · T and G · C base pairs (see Figure 4-4b). DNA duplexes can be denatured (melted) into single strands by heating them in a dilute salt solution (e.g., 0.01 M NaCl), or by raising the pH above 11. If the temperature is lowered and the ion concentration in the solution is raised, or if the pH is lowered to neutrality, the A · T and G · C base pairs re-form between complementary single strands (see Figure 4-8). This process goes by many names: renaturation, reassociation, hybridization, annealing. In a mixture of nucleic acids, only complementary single strands (or strands containing complementary regions) will reassociate; the extent of their reassociation is virtually unaffected by the presence of noncomplementary strands. Such *molecular hybridization* can take place between two complementary strands of either DNA or RNA, or between an RNA strand and a DNA strand.

To detect specific DNA clones by molecular hybridization, cloned recombinant DNA molecules are denatured and the single strands attached to a solid support, commonly a nitrocellulose filter or treated nylon membrane (Figure 7-17). When a solution containing single-stranded nucleic acids is dried on such a membrane, the single strands become irreversibly bound to the solid support in a manner that leaves most of the bases available for hybridizing to a complementary strand. Although the chemistry of this irreversible binding is not well understood, the procedure is very useful.

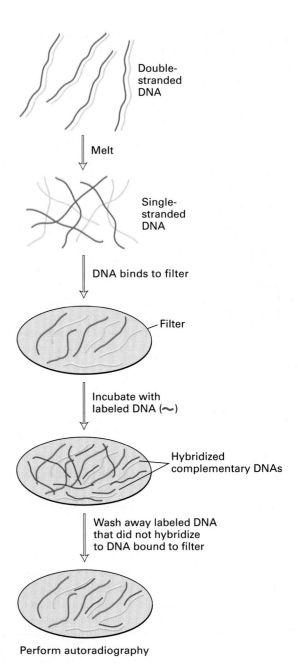

▲ **FIGURE 7-17 Membrane-hybridization assay for detecting nucleic acids.** This assay can be used to detect both DNA and RNA, and the radiolabeled complementary probe can be either DNA or RNA. See text for details.

The membrane is then incubated in a solution containing a radioactively labeled single-stranded DNA (or RNA) *probe* that is complementary to some of the nucleic acid bound to the membrane. Under hybridization conditions (near neutral pH, 40–65 °C, 0.3–0.6 M NaCl), this labeled probe hybridizes to the complementary nucleic acid bound to the membrane. Any excess probe that does not hybridize is washed away, and the labeled hybrids are detected by autoradiography of the filter.

The procedure for screening a λ library with this membrane-hybridization technique is outlined in Figure 7-18. The recombinant λ virions present in plaques on a lawn of *E. coli* are transferred to a nylon membrane by placing the membrane on the surface of the petri dish. Many of the viral particles in each plaque adsorb to the surface of the membrane, but many virions remain in the plaques on the surface of the nutrient agar in the petri dish. In this way a *replica* of the petri dish containing a large number of individual λ clones is reproduced on the surface of the membrane. The original petri dish is refrigerated to store the collection of λ

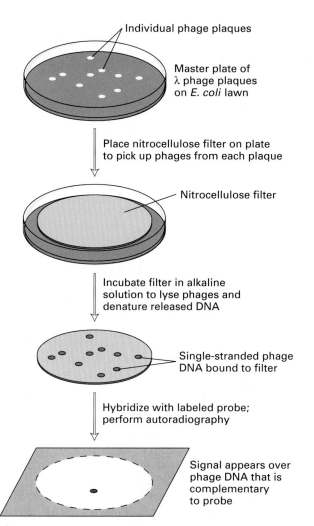

▲ **FIGURE 7-18 Identification of a specific clone from a λ phage library by membrane hybridization to a radiolabeled probe.** The position of the signal on the autoradiogram identifies the desired plaque on the plate. In practice, in the initial plating of a library the plaques are not allowed to develop to a visible size so that up to 50,000 recombinants can be analyzed on a single plate. Phage particles from the identified region of the plate are isolated and replated at low density so that the plaques are well separated. Then pure isolates can be obtained by repeating the plaque hybridization as shown in the figure.

Labels in Figure 7-18:
- Individual phage plaques
- Master plate of λ phage plaques on *E. coli* lawn
- Place nitrocellulose filter on plate to pick up phages from each plaque
- Nitrocellulose filter
- Incubate filter in alkaline solution to lyse phages and denature released DNA
- Single-stranded phage DNA bound to filter
- Hybridize with labeled probe; perform autoradiography
- Signal appears over phage DNA that is complementary to probe

clones. The membrane is then incubated in an alkaline solution, which disrupts the virions, releasing and denaturing the encapsulated DNA. The membrane is then dried, fixing the recombinant λ DNA to the membrane's surface. Next, the membrane is incubated with a radiolabeled probe under hybridization conditions. Unhybridized probe is washed away, and the filter is subjected to autoradiography.

The appearance of a spot on the autoradiogram indicates the presence of a recombinant λ clone containing DNA complementary to the probe. The position of the spot on the autoradiogram corresponds to the position on the original petri dish where that particular clone formed a plaque. Since the original petri dish still contains many infectious virions in each plaque, viral particles from the identified clone can be recovered for replating by aligning the autoradiogram and the petri dish and removing viral particles from the clone corresponding to the spot. A similar technique can be applied for screening a plasmid library in *E. coli* cells.

Oligonucleotide Probes Are Designed Based on Partial Protein Sequences

Identification of specific clones by the membrane-hybridization technique depends on the ability to prepare specific, radiolabeled probes. Specific oligonucleotide probes for the gene encoding a protein of interest can be synthesized chemically if a portion of the amino acid sequence of the protein is determined. For an oligonucleotide to be useful as a probe, it must be long enough for its sequence to occur uniquely in the clone of interest and not in any other clones. For most purposes, this condition is satisfied by oligonucleotides containing about 20 nucleotides. This is because a specific 20-nucleotide sequence occurs once in every 4^{20} ($\approx 10^{12}$) nucleotides. Since all genomes are much smaller ($\approx 3 \times 10^9$ nucleotides for humans), a specific 20-nucleotide sequence in a genome usually occurs only one time. In principle, probes this length can be prepared based on only a 7-aa sequence out of a protein's total sequence (20 nucleotides ÷ 3 nucleotides per codon ≈ 7 amino acids). However, a somewhat longer amino acid sequence usually is determined to allow the preparation of several probes for use in cloning a gene of interest.

Here we describe two different approaches for preparing oligonucleotide probes. Generally, a radiolabeled oligonucleotide probe is used to screen a λ cDNA library using the membrane-hybridization technique. Once a cDNA clone encoding a particular protein is obtained, the full-length radiolabeled cDNA can be used to probe a genomic library for clones that contain fragments of the gene encoding the protein.

Degenerate Probes One method for preparing a specific probe is outlined in Figure 7-19. The purified protein of interest is digested with one or more proteases (e.g., trypsin) into specific peptides, and the N-terminal amino acid sequences of a few of these peptides is determined by sequential Edman degradation or mass spectrometry (see Figures 3-46

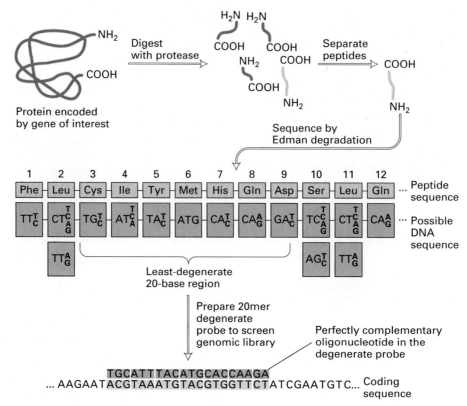

▲ **FIGURE 7-19 Designing oligonucleotide probes based on protein sequence.** An isolated protein is digested with a selective protease such as trypsin, which specifically cleaves peptide bonds on the carboxy-terminal side of lysine and arginine residues. The resulting peptides are separated, and several are partially sequenced from their N-terminus by sequential Edman degradation. The determined sequences then are analyzed to identify the 6- or 7-aa region that can be encoded by the smallest number of possible DNA sequences. Because of the degeneracy of the genetic code, the 12-aa sequence (light green) shown here theoretically could be encoded by any of the DNA triplets below it, with the possible alternative bases at the same position indicated. For example, Phe-1 is encoded by TTT or TTC; Leu-2 is encoded by one of six possible triplets (CTT, CTC, CTA, CTG, TTA, or TTG). The region with the least degeneracy for a sequence of 20 bases (20-mer) is indicated by the red bracket. There are 48 possible DNA sequences in this 20-base region that could encode the peptide sequence 3–9. Since the actual sequence of the gene is unknown, a degenerate 20-mer probe consisting of a mixture of all the possible 20-base oligonucleotides is prepared. If a cDNA or genomic library is screened with this degenerate probe, the one oligonucleotide that is perfectly complementary to the actual coding sequence (blue) will hybridize to it.

and 3-47). Based on the genetic code, the oligonucleotide sequences encoding the determined peptide sequences can be predicted. Recall, however, that the genetic code is degenerate; that is, many amino acids are encoded by multiple codons (see Table 4-2). Since the specific codons used to encode the protein of interest are unknown, oligonucleotides containing all possible combinations of codons must be synthesized to assure that one of them will match the gene perfectly.

Once several peptides have been sequenced, the 6- or 7-aa stretch that can be encoded by the smallest number of possible DNA sequences is determined. For example, as illustrated in Figure 7-19, the amino acids in the sequence extending from position 3 through 8 (Cys-Ile-Tyr-Met-His-Gln) can be encoded by 2, 3, 2, 1, 2, and 2 possible codons, respectively. Consequently, 48 ($= 2 \times 2 \times 1 \times 2 \times 2$) different 18-base DNA sequences could encode this one sequence of amino acids. The GA added at the 3′ end of these 18-base sequences must be complementary to the gene since the next amino acid in this peptide, Asp-9, is encoded by two codons that both start with GA. To be certain of obtaining a probe based on this amino acid sequence that hybridizes perfectly to the unique sequence present in the gene, all 48 of the 20-mer probes must be synthesized. A mixture of 20-mer probes based on any other portion of this peptide sequence would have to contain considerably more than 48 oligonucleotides because of the presence of leucine or serine residues, each encoded by six different codons.

A mixture of all the oligonucleotides that can encode a selected portion of a peptide sequence is called a *degenerate probe*. Such a mixture can be prepared at one time by adding more than one nucleotide precursor to the synthesis reaction at those points in the sequence that can be encoded by alternative bases. The final step in preparing this type of probe is to radiolabel the oligonucleotides, usually by transferring

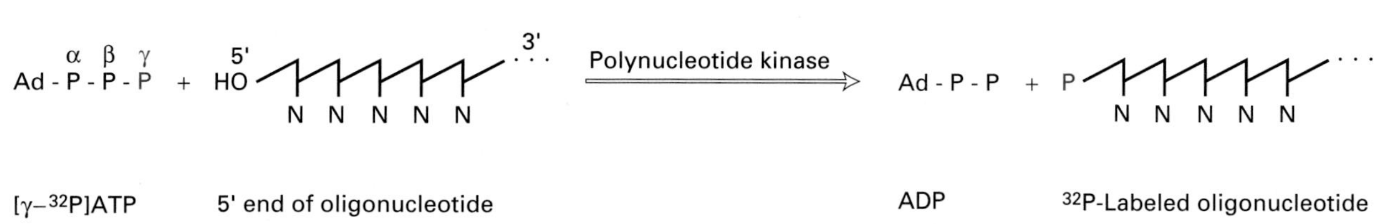

$[\gamma-^{32}P]ATP$ 5' end of oligonucleotide

ADP ^{32}P-Labeled oligonucleotide

▲ **FIGURE 7-20 Radiolabeling of an oligonucleotide at the 5′ end with phosphorus-32.** The three phosphate groups in ATP are designated the α, β, and γ phosphates in order of their position away from the ribose ring of adenosine (Ad). ATP containing the radioactive isotope ^{32}P in the γ-phosphate position is called $[\gamma$-$^{32}P]ATP$. Kinase is the general term for enzymes that transfer the γ-phosphate of ATP to specific substrates. Polynucleotide kinase can transfer the ^{32}P-labeled γ phosphate of $[\gamma$-$^{32}P]ATP$ to the 5′ end of a polynucleotide chain (either DNA or RNA). This reaction is commonly used to radiolabel synthetic oligonucleotides.

a ^{32}P-labeled phosphate group from ATP to the 5′ end of each oligonucleotide using *polynucleotide kinase* (Figure 7-20). Screening of a λ cDNA library with a degenerate probe using the membrane-hybridization technique will identify clones that hybridize to the perfectly complementary oligonucleotide present in the probe mixture. Under the usual experimental conditions, oligonucleotides that differ from the cDNA sequence at one or two bases also will hybridize.

Unique EST-Based Probes In recent years another approach has become available for obtaining a probe based on the partial amino acid sequence determined from an isolated protein. Because this approach utilizes cDNA sequence data, it identifies a single oligonucleotide, rather than a degenerate mixture, that can be used to screen a library for a particular gene. Using methods for DNA sequencing described later, researchers have sequenced portions of vast numbers of cDNAs isolated from human cells and some additional important model organisms such as the mouse, *Drosophila*, and the roundworm *Caenorhabditis elegans*. These partial cDNA sequences, generally 200 to 400 bp in length, have been stored in computers and are available to researchers throughout the world via the Internet, the international computer network. This collection of partial cDNA sequences is called the *expressed sequence tag (EST)* database because it is composed of relatively short portions (tags) of genomic DNA sequence that are expressed in the form of mRNA. The EST database is constantly updated as sequences from increasing numbers of cDNA clones are added to it.

Computer programs that apply the genetic code are used to translate the EST sequences into partial protein amino acid sequences. Using programs that have been developed for the purpose and a personal computer, a researcher can search the current EST database for an EST that encodes a specific partial amino acid sequence in the particular protein under study. If a match is found, then the EST provides the unique DNA sequence of that portion of the full-length cDNA. A single, specific probe up to ≈ 100 bases long that is perfectly complementary to a portion of the EST can then be synthesized and radiolabeled. Alternatively, the polymerase chain reaction, described at the end of the chapter, can be used to synthesize a probe equal to the full length of the EST. By now, the human EST database is so large that ESTs can be identified that encode partial amino acid sequences determined from most isolated human proteins.

Specific Clones Can Be Identified Based on Properties of the Encoded Proteins

Genomic and cDNA libraries can also be screened for the properties of a specific protein encoded in the cloned DNA. This approach uses special cloning vectors, called λ **expression vectors,** in which the cloned DNA is transcribed into mRNA, which in turn is translated into the encoded protein. For example, λ phage vectors have been constructed so that the junction of inserted DNA lies in a region of the vector that is transcribed and translated at a high rate. Cloned DNA inserted at this position is transcribed into mRNA in every cell infected by this type of vector. If the cloned DNA contains a protein-coding sequence inserted in the same reading frame as the vector protein, infected cells will produce a *fusion protein* in which the amino terminus is encoded by the vector DNA and the remainder of the molecule by the cloned DNA (Figure 7-21).

When replica nitrocellulose filters are prepared from a recombinant library constructed in a λ expression vector, fusion proteins expressed from each individual clone are bound to the nitrocellulose filter. The replica filter can be screened by procedures capable of detecting specific fusion proteins. For example, a monoclonal antibody specific for a protein of interest can be incubated with replica filters of a λ cDNA expression library. If one of the λ clones expresses a fusion protein that includes the region of the protein bound by that monoclonal antibody, antibody molecules will bind to the filter at the position of that specific clone. After washing of the filter to remove unbound antibody, the position of the specific clone is detected by incubation with a second radioactively labeled antibody that recognizes the first antibody, followed by autoradiography of the filter.

In this method, termed **expression cloning,** any molecule that binds to a protein of interest with high affinity and specificity can be labeled and used as a probe to identify clones expressing the interacting protein. For instance, expression cloning has been useful in identifying cDNA clones encoding

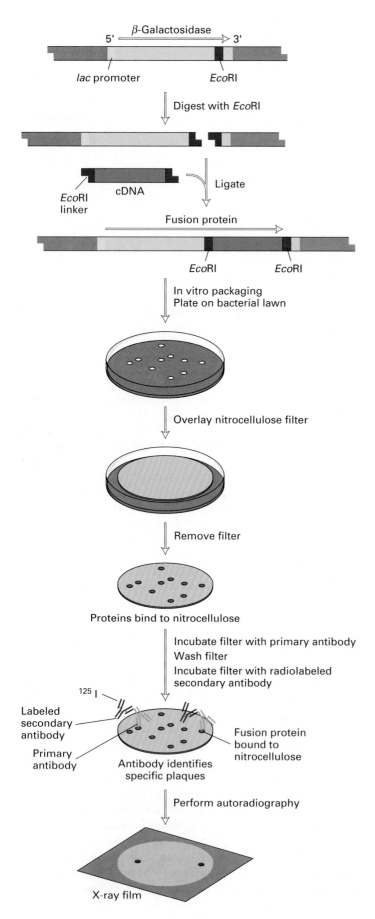

proteins that bind to specific DNA sequences; many such proteins are involved in controlling transcription. In this case, a labeled synthetic double-stranded DNA probe is incubated with replica filters prepared from a cDNA library cloned into a λ expression vector. Binding of the labeled DNA by fusion proteins locates the positions of desired clones on the original filter. As described in a later section, other types of expression vectors can be used to produce large amounts of a protein from a cloned gene.

Gel Electrophoresis Resolves DNA Fragments of Different Size

Once a specific DNA clone has been isolated, the cloned DNA is separated from the vector DNA by cleavage with the restriction enzyme used to form the recombinant plasmid, as described earlier. The cloned DNA and vector DNA then are separated by gel electrophoresis, a powerful method for separating proteins according to size (Chapter 3). Gel electrophoresis also is used to separate DNA and RNA molecules by size and to estimate the size of nucleic acid molecules of unknown length by comparison with the migration of molecules of known length.

DNA and RNA molecules are highly charged near neutral pH because the phosphate group in each nucleotide contributes one negative charge. As a result, DNA and RNA molecules move toward the positive electrode during gel electrophoresis. Smaller molecules move through the gel matrix more readily than larger molecules, so that molecules of different length, such as restriction fragments, separate (Figure 7-22). Because the gel matrix restricts random diffusion of the molecules, molecules of different length separate into "bands"

◄ **FIGURE 7-21 Use of λ expression cloning to identify a cloned DNA based on binding of the encoded protein to a specific antibody.** The λgt11 vector was engineered to express the *E. coli* protein β-galactosidase at high levels. The only *Eco*RI recognition site (red) in this vector lies near the 3′ end of the β-galactosidase gene. If a cDNA (green), or protein-coding fragment of genomic DNA, is inserted into this *Eco*RI site in the correct orientation and proper reading frame, it will be expressed as a fusion protein in which most of the β-galactosidase sequence is at the N-terminal end and the protein sequence encoded by the inserted DNA is at the C-terminal end. Plaques resulting from infection with recombinant λgtll contain high concentrations of such fusion proteins. These proteins can be transferred and bound to a replica filter, which then is incubated with a monoclonal primary antibody (blue) that recognizes the protein of interest. Rinsing the filter washes away antibody molecules that are not bound to the specific fusion protein attached to the filter. Bound antibody usually is detected by incubating the filter with a second radiolabeled antibody (dark red) that binds to the primary antibody. Any signals that appear on the autoradiogram are used to locate plaques on the master plate containing the gene of interest. [Adapted from J. D. Watson et al., 1992, *Recombinant DNA*, 2d ed., Scientific American Books.]

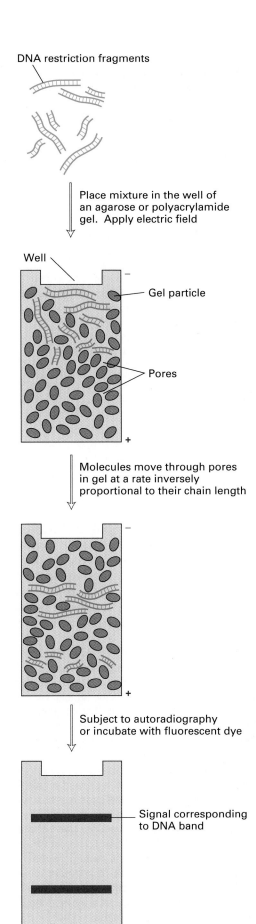

DNA restriction fragments

Place mixture in the well of an agarose or polyacrylamide gel. Apply electric field

Well

Gel particle

Pores

Molecules move through pores in gel at a rate inversely proportional to their chain length

Subject to autoradiography or incubate with fluorescent dye

Signal corresponding to DNA band

◀ **FIGURE 7-22 Separation of DNA fragments of different lengths by gel electrophoresis.** A gel is prepared by pouring a liquid containing either melted agarose or unpolymerized acrylamide between two glass plates a few millimeters apart. As the agarose solidifies or the acrylamide polymerizes into polyacrylamide, a gel matrix (orange ovals) forms consisting of long, tangled chains of polymers. The dimensions of the interconnecting channels, or pores, depend on the concentration of the agarose or acrylamide used to form the gel. Because the pores are larger in agarose gels than in polyacrylamide gels, the former are used to separate large DNA fragments ($\approx$500 bp to $\approx$20 kb) and the latter to separate small DNA fragments (1 nucleotide to $\approx$2 kb). The mixture of DNA fragments to be separated is layered in a well at the top of the gel and an electric current is passed through the gel. DNA fragments move toward the positive pole at a rate inversely proportional to the log of their length, forming bands that can be visualized by autoradiography (if the fragments are radiolabeled) or by addition of a fluorescent dye such as ethidium. Agarose gels can be run in a horizontal orientation; in this case, the melted agarose is allowed to harden on a single horizontal glass or plastic plate. This is not easily done with polyacrylamide gels because oxygen in the atmosphere inhibits polymerization of acrylamide. Gels are generally depicted with the origin at the top and migration downward.

whose width equals that of the well into which the original DNA mixture was placed. The resolving power of gel electrophoresis is so great that single-stranded DNA molecules up to about 500 nucleotides long can be separated if they differ in length by only 1 nucleotide. This high resolution is critical to the DNA-sequencing procedures described later. DNA molecules composed of up to $\approx$2000 nucleotides usually are separated electrophoretically on *polyacrylamide gels*, and molecules from 500 nucleotides to 20 kb on *agarose gels*.

Two methods are common for visualizing separated DNA bands on a gel. If the DNA is not radiolabeled, the gel is incubated in a solution containing the fluorescent dye ethidium:

Ethidium

$$NH_2$$

$$H_2N \qquad N^+ \qquad C_2H_5$$

$$C_6H_5$$

This planar molecule binds to DNA by intercalating between the base pairs. Binding concentrates ethidium in the DNA and also increases its intrinsic fluorescence. As a result, when the gel is illuminated with ultraviolet light, the regions of the gel containing DNA fluoresce much more brightly than the regions of the gel without DNA (Figure 7-23a).

Radioactively labeled DNA can be visualized by autoradiography of the gel. In this case, the gel is laid against a sheet of photographic film in the dark, exposing the film at the positions where labeled DNA is present. When the film is developed, a photographic image of the DNA is observed (Figure 7-23b). Radiolabeled DNA bands also can be detected by laying the gel against a phosphorimager screen, which counts β particles released by labeled molecules in the gel. The resulting data is stored by a computer and can be

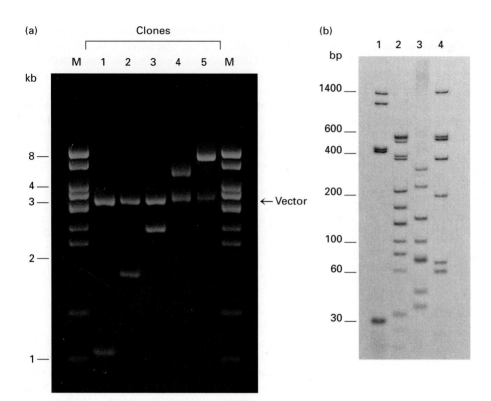

(a) Clones

(b)

▶ **FIGURE 7-23 Visualization of restriction fragments separated by gel electrophoresis.** (a) Several different plasmid clones were digested with *Eco*RI, and the digested DNA was subjected to agarose gel electrophoresis to separate the cloned fragments from the plasmid vector DNA. Each *Eco*RI-cut plasmid was layered on a separate well. *Hind*III fragments produced by digestion of adenovirus 2 were layered in the left and right wells of the gel as size markers (M). The gel was treated with ethidium and then observed under ultraviolet light. The plasmid vector can be seen as a common band in all five clones. (b) Autoradiogram of ^{32}P-labeled fragments separated by polyacrylamide gel electrophoresis. Lengths of fragments are indicated at the left in base pairs. [Courtesy of Carol Eng.]

converted into an image of the gel that looks much like an autoradiogram.

Multiple Restriction Sites Can Be Mapped on a Cloned DNA Fragment

In addition to separating restriction fragments of different lengths, gel electrophoresis provides a means for estimating the length of fragments. As we've seen, the distance that a restriction fragment migrates in a gel is inversely proportional to the logarithm of its length. Thus the length of a

restriction fragment can be determined fairly accurately by comparison with restriction fragments of known length subjected to electrophoresis on the same gel (see Figure 7-23a).

The ability to determine the length of restriction fragments makes it possible to locate the positions of restriction sites relative to one another on a DNA molecule (e.g., a newly cloned DNA fragment). A diagram showing the positions of restriction sites on a DNA molecule is called a *restriction map*, and the process of determining these positions is called *restriction-site mapping*. Figure 7-24 illustrates the procedure for mapping two restriction sites relative to each

(a)

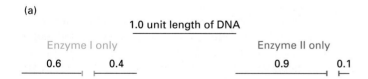

(b)

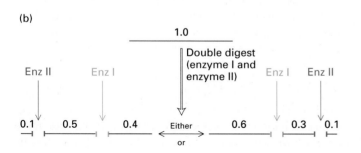

▶ **FIGURE 7-24 Mapping the recognition sites for two restriction enzymes relative to each other in a DNA fragment containing one copy of each site.** For simplicity, double-stranded DNA is represented by a single line. (a) The fragment is exposed separately to two restriction enzymes (I and II). Each enzyme cuts the fragment once, generating two subfragments, whose lengths are determined by gel electrophoresis. (b) The fragment also is digested with both enzymes simultaneously. Since the lengths of the resulting fragments will depend on the relative position of the two restriction sites, the sites can be mapped based on the lengths observed. Here two different possible outcomes are shown, depending on the positions of the two restriction sites relative to each other. If site II lies within the 0.6-kb fragment produced by enzyme I, the pattern on the left would be seen. If site II lies within the 0.4-kb fragment produced by enzyme I, the pattern on the right would be seen. By continuing this process with different pairs of enzymes, the investigator can construct a detailed map of restriction sites.

other when only *one* copy of each site is present in a fragment. In this simple case, three fragment samples are digested: one with enzyme I, one with enzyme II, and one with both enzymes.

When a DNA fragment contains *multiple* copies of the recognition site for one or more restriction enzymes, the mapping procedure is more complicated. In this case, the sites for each enzyme must be mapped before the sites for different enzymes can be mapped relative to one another. The first step in this procedure is radiolabeling the 5' ends of both strands of the fragment with $[\gamma\text{-}^{32}P]ATP$ and polynucleotide kinase (see Figure 7-20). As shown in Figure 7-25,

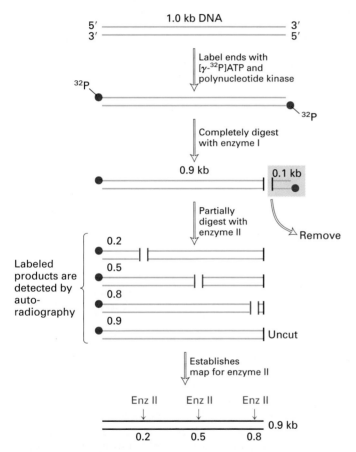

▲ **FIGURE 7-25 Mapping the multiple copies of the recognition site for a given restriction enzyme in a DNA fragment.** The fragment first is doubly end-labeled (red circles) and then is digested completely with an appropriate restriction enzyme (I) to produce fragments labeled at *one* end only. The larger fragment is partially digested with a second enzyme (II), so that no more than one cut is made in any fragment molecule. Complete digestion would generate only one labeled product (here, the 0.2-kb subfragment), whereas partial digestion generates a labeled product for each restriction site. From the lengths of the labeled pieces, the positions of the multiple recognition sites for enzyme II (red arrows) in the original DNA can be inferred. [See H. O. Smith and M. Birnstiel, 1976, *Nucl. Acids Res.* **3**:2387.]

the doubly end-labeled fragment is treated with a restriction enzyme that cuts the fragment just once, and the resulting singly end-labeled fragments are separated. These fragments then are partially digested with the enzyme whose multiple recognition sites are being mapped.

Multiple restriction sites in a cloned DNA fragment can be mapped by use of these two methods with multiple restriction enzymes. Because each distinct DNA sequence has a characteristic restriction-site map, such maps can be used to align partially overlapping cloned DNA fragments (see Figure 7-13). Also, specific small regions within a large cloned DNA fragment can be prepared by digesting the cloned fragment with various combinations of restriction enzymes; the smaller subfragments then can be isolated by gel electrophoresis.

Pulsed-Field Gel Electrophoresis Separates Large DNA Molecules

The gel electrophoretic techniques described so far can resolve DNA fragments up to ≈20 kb in length. Larger DNAs, ranging from 2×10^4 to 10^7 base pairs [20 kb to 10 megabases (Mb)] in length, can be separated by size with *pulsed-field gel electrophoresis*. This technique depends on the unique behavior of large DNAs in an electric field that is turned on and off (pulsed) at short intervals.

When an electric field is applied to large DNA molecules in a gel, the molecules migrate in the direction of the field and also stretch out lengthwise. If the current then is stopped, the molecules begin to "relax" into random coils. The time required for relaxation is directly proportional to the length of a molecule. The electric field then is reapplied at 90° or 180° to the first direction. Longer molecules relax less than shorter ones during the time the current is turned off. Since the molecules must relax into a random coil before moving off in a new direction, longer molecules start moving in the direction imposed by the new field more slowly than shorter ones. Repeated alternation of the field direction gradually forces large DNA molecules of different size farther and farther apart.

Pulsed-field gel electrophoresis is very important for purifying long DNA molecules up to ≈10^7 base pairs in length (Figure 7-26). The technique is required for analyzing cellular chromosomes, which range in size from about 5×10^5 base pairs (smallest yeast chromosomes) to $2–3 \times 10^8$ base pairs (animal and plant chromosomes). Very large chromosomes must be digested into fragments of 10^7 base pairs or less before they can be analyzed. Such large restriction fragments can be generated with restriction enzymes that cut at rarely occurring 8-bp restriction sites.

Purified DNA Molecules Can Be Sequenced Rapidly by Two Methods

Virtually all the information required for the growth and development of an organism is encoded in the DNA of its genome.

The availability of techniques to produce and separate DNA restriction fragments a few hundred nucleotides long led to development of two procedures for determining the exact nucleotide sequence of stretches of DNA up to ≈500 nucleotides long. These DNA sequencing methods, together with the technology for constructing a library representing the entire genome of an organism, make it possible to determine the exact sequence of the entire DNA of that organism.

The total genomes of many viruses, several bacteria and archaeans, the yeast *S. cerevisiae,* and the roundworm *C. elegans* have already been sequenced. Automation of the techniques for sequencing DNA and computerized storage of the sequence data are facilitating the current effort to determine the sequence of the entire human genome. Within the next decade, if not sooner, researchers also are likely to complete sequencing the entire genomes of the fruit fly *Drosophila melanogaster,* the mouse, and other important experimental organisms. Knowledge of these DNA sequences will undoubtedly revolutionize our understanding of how cells and organisms function.

Maxam-Gilbert Method In the late 1970s, A. M. Maxam and W. Gilbert devised the first method for sequencing DNA fragments containing up to ≈500 nucleotides. In this

▶ **FIGURE 7-27 Maxam-Gilbert method for sequencing DNA fragments up to ≈500 nucleotides in length.** The double-stranded fragment to be sequenced is labeled at the 5′ ends with ^{32}P (see Figure 7-20). The label (red circle) is removed from one end, and the fragment then is denatured. Four identical samples of the prepared fragment are subjected to four different sets of chemical reactions that selectively cut the DNA backbone at G, G + A, C + T, or C residues. The reactions are controlled so that each labeled chain is likely to be broken only once. The labeled subfragments created by all four reactions have the label at one end and the chemical cleavage point at the other. Gel electrophoresis and autoradiography of each separate mixture yield one radioactive band for each nucleotide in the original fragment, each separated according to their length. Bands appearing in the G and C lanes can be read directly. Bands in the A + G lane that are not duplicated in the G lane are read as A. Bands in the T + C lane that are not duplicated in the C lane are read as T. The sequence is read from the bottom of the gel up. [See A. Maxam and W. Gilbert, 1977, *Proc. Nat'l. Acad. Sci. USA* **74**:560. Photograph from L. Stryer, 1988, *Biochemistry,* 3d ed., W. H. Freeman and Company, p. 120; courtesy of Dr. David Dressler.]

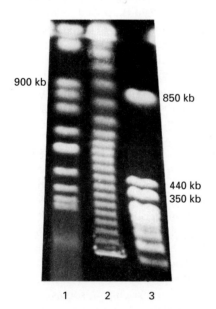

900 kb
850 kb
440 kb
350 kb

1   2   3

▲ **FIGURE 7-26 Pulsed-field gel electrophoretic separation of large DNA molecules.** Lane 1 shows individual DNA molecules; each band represents one chromosome from the yeast *S. cerevisiae.* Lane 3 shows *Not*I restriction fragments of the *E. coli* chromosome, which were used in mapping this genome. Lane 2 shows a "ladder" of concatomers of λ phage DNA in which each unit is ≈48.5 kb long (see Figure 7-11). The band at the bottom is a single unit; the other bands are successively ≈48.5 kb longer. This ladder can be used to estimate the length of long DNA fragments electrophoresed in parallel. [See C. L. Smith et al., 1987, *Nucl. Acids Res.* **15**:4481; photograph courtesy of C. L. Smith.]

method, four samples of an end-labeled DNA restriction fragment are chemically cleaved at different specific nucleotides. The resulting subfragments are separated by gel electrophoresis, and the labeled fragments are detected by autoradiography. As illustrated in Figure 7-27, the sequence of the original end-labeled restriction fragment can be determined directly from parallel **electrophoretograms** of the four samples.

Sanger (Dideoxy) Method A few years later, F. Sanger and his colleagues developed a second method of DNA sequencing, which now is used much more frequently than the Maxam-Gilbert method. The Sanger method is also called *dideoxy sequencing* because it involves use of 2′,3′-dideoxynucleoside triphosphates (ddNTPs), which lack a 3′-hydroxyl group (Figure 7-28). In this method, the single-stranded DNA to be sequenced serves as the template strand for in vitro DNA synthesis; a synthetic 5′-end-labeled oligodeoxynucleotide is used as the primer.

As shown in Figure 7-29, four separate polymerization reactions are performed, each with a low concentration of one of the four ddNTPs in addition to higher concentrations of the normal deoxynucleoside triphosphates (dNTPs). In each reaction, the ddNTP is randomly incorporated at the positions of the corresponding dNTP; such addition of a ddNTP terminates polymerization because the absence of a 3′ hydroxyl prevents addition of the next nucleotide. The mixture of terminated fragments from each of the four reactions is subjected to gel electrophoresis in parallel; the separated fragments then are detected by autoradiography. The sequence of the original DNA template strand can be read directly from the resulting autoradiogram (see Figure 7-29c). Once the sequence for a particular cloned DNA fragment is determined, primers for overlapping fragments can be

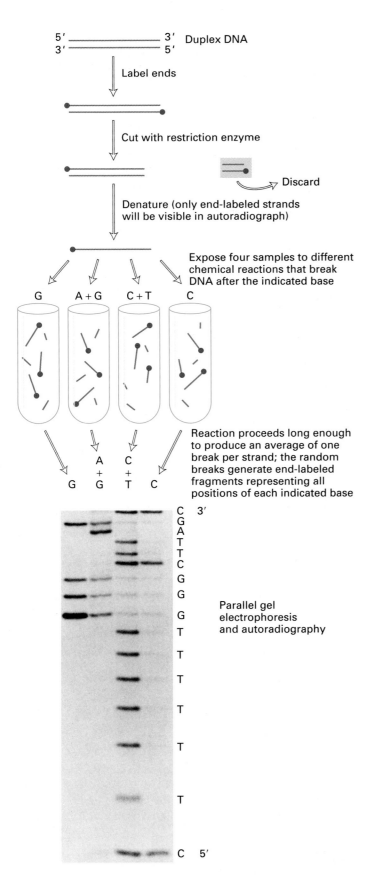

▲ FIGURE 7-28 Structures of ribonucleoside triphosphate (NTP), deoxyribonucleoside triphosphate (dNTP), and dideoxyribonucleoside triphosphate (ddNTP).

chemically synthesized based on that sequence. The sequence of a long continuous stretch of DNA thus can be determined by individually sequencing the overlapping cloned DNA fragments that compose it.

SUMMARY Identifying, Analyzing, and Sequencing Cloned DNA

• Generally, to isolate a DNA clone encoding a protein of interest, it is uniquely identified in a collection of a large number of different clones, such as a total genomic library or a cDNA library. Two common identification methods are (1) hybridization to a radiolabeled DNA probe specific for the clone and detection by autoradiography and (2) expression of the encoded protein and detection of the expressed protein by its biochemical activity or by its binding to a radiolabeled antibody specific for the protein.

• A specific hybridization probe for cloned DNA encoding a protein of interest can be prepared based on a short portion of the amino acid sequence of the protein. Use of a degenerate probe, which is a mixture of all the possible DNA sequences that can encode the determined amino acid sequence, ensures that the radiolabeled probe contains the one sequence exactly complementary to the gene of interest (see Figure 7-19). When this perfectly complementary radiolabeled DNA probe hybridizes to DNA from the specific clone of interest on a replica filter, it can be detected by autoradiography (see Figure 7-18).

(a)

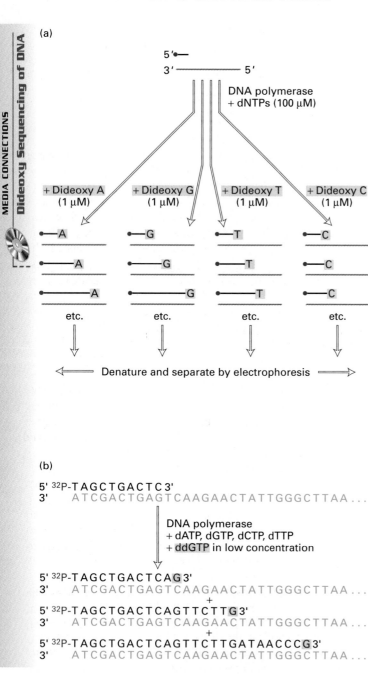

(b)

5' ³²P-TAGCTGACTC 3'
3'   ATCGACTGAGTCAAGAACTATTGGGCTTAA...

DNA polymerase
+ dATP, dGTP, dCTP, dTTP
+ ddGTP in low concentration

5' ³²P-TAGCTGACTCAG 3'
3'   ATCGACTGAGTCAAGAACTATTGGGCTTAA...
+
5' ³²P-TAGCTGACTCAGTTCTTG 3'
3'   ATCGACTGAGTCAAGAACTATTGGGCTTAA...
+
5' ³²P-TAGCTGACTCAGTTCTTGATAACCCG 3'
3'   ATCGACTGAGTCAAGAACTATTGGGCTTAA...

(c)

C A G T C G A T

◄ **FIGURE 7-29 Sanger (dideoxy) method for sequencing DNA fragments.** (a) A single strand of the DNA to be sequenced (blue line) is hybridized to a 5'-end-labeled synthetic deoxyribonucleotide primer. The primer is elongated in four separate reaction mixtures containing the four normal deoxyribonucleoside triphosphates (dNTPs) plus one of the four dideoxyribonucleoside triphosphates (ddNTPs) in a ratio of 100 to 1. A ddNTP molecule can add at the position of the corresponding normal dNTP, but when this occurs, chain elongation stops because the ddNTP lacks a 3' hydroxyl. In time, each reaction mixture will contain a mixture of prematurely terminated chains ending at every occurrence of the ddNTP (yellow). (b) Three of the labeled chains that would be generated in the presence of ddGTP from the specific DNA sequence shown in blue. (c) An actual autoradiogram of a polyacrylamide gel in which more than 300 bases can be read. Each reaction was carried out in duplicate using Sequenase™, a commercial preparation of the DNA polymerase from bacteriophage T7. [Part (c) courtesy of United States Biochemical Corporation.]

• Vast numbers of partial cDNA sequences, or expressed sequence tags (ESTs), from humans and various model organisms are stored in computer databases. If a partial amino acid sequence is determined from an isolated protein, the EST database can be searched for an EST (i.e., a partial cDNA sequence) that encodes it. If found, the EST sequence can be used to prepare a perfectly complementary probe for isolation of the corresponding full-length cDNA from a cDNA library.

• λ expression vectors are designed to express the protein encoded in cloned DNA fragments in *E. coli* cells infected with the vector. Plaques producing the encoded polypeptide can be detected with labeled molecules (e.g., an antibody) that bind to the protein of interest with high affinity (see Figure 7-21).

• Gel electrophoresis separates DNA (and RNA) molecules according to their size. Consequently, a cloned DNA fragment, released from its cloning vector by digestion with the appropriate restriction enzyme, can be separated from the vector DNA by gel electrophoresis.

• The mobility of either a double- or single-stranded DNA molecule during gel electrophoresis is inversely proportional to the logarithm of its length in nucleotides. Thus the size of a DNA molecule of unknown length can be determined by comparison to the electrophoretic migration of molecules of known length.

- DNA molecules from 1 to 2000 nucleotides long are usually separated by electrophoresis in polyacrylamide gels; molecules from 500 nucleotides to 20 kb, by electrophoresis in agarose gels; and molecules from 20 to 10,000 kb, in pulsed-field agarose gels.

- Multiple restriction sites for a particular restriction enzyme can be mapped in cloned DNAs by partially digesting end-labeled DNA molecules and then determining the length of the resulting fragments by gel electrophoresis and autoradiography (see Figure 7-25).

- Once the cleavage sites for two restriction enzymes have been mapped on a DNA, these sites can be mapped relative to each other by comparing the sizes of DNA fragments generated with each enzyme separately and with both enzymes together (see Figure 7-24).

- The sequence of a single-stranded DNA molecule can be determined by either the Maxam-Gilbert method or the Sanger dideoxy method. Sequences of up to ≈500 nucleotides can be determined on a single gel because DNA molecules up to this length can be separated on polyacrylamide gels when they differ in length by a single nucleotide.

7.4 Bioinformatics

Vast amounts of DNA sequence have already been determined, and the pace at which new sequences are characterized is continuously accelerating. Computers are necessary to store and distribute this enormous volume of data. *Bioinformatics* is the rapidly developing area of computer science devoted to collecting, organizing, and analyzing DNA and protein sequences. The principal data banks where such sequences are stored are the GenBank at the National Institutes of Health, Bethesda, Maryland, and the EMBL Sequence Data Base at the European Molecular Biology Laboratory in Heidelberg. These databases continuously exchange newly reported sequences and make them available to molecular cell biologists throughout the world on the Internet. Newly derived sequences can be compared with previously determined sequences to search for similarities, called *homologous sequences*. Protein-coding regions can be translated into amino acid sequences, which also can be compared. Because of degeneracy in the genetic code, related proteins often exhibit more homology than the genes encoding them. As we saw in the previous section, databases of partial cDNA sequences (EST databases) are particularly useful in designing probes for screening libraries.

Stored Sequences Suggest Functions of Newly Identified Genes and Proteins

As discussed in Chapter 3, proteins with similar functions often contain homologous amino acid sequences that correspond to important functional domains in the three-dimensional structure of the proteins. Discovery that a protein encoded by a newly cloned gene exhibits such homologies with proteins of known function can provide revealing insights into the function of the cloned gene.

 To illustrate this approach, we consider *NF1*, a human gene identified and cloned by methods described in Chapter 8. Mutations in *NF1* are associated with the inherited disease neurofibromatosis 1, in which multiple tumors develop in the peripheral nervous system, causing large protuberances in the skin (the "elephant-man" syndrome). Before the *NF1* gene was cloned, there was little understanding of the molecular basis of the disease. After a cDNA clone of *NF1* was isolated and sequenced, the deduced sequence of the NF1 protein was checked against all other protein sequences in GenBank. A region of NF1 protein was discovered to have homology to a yeast protein called Ira (Figure 7-30). Previous studies in yeast had shown that Ira is a GAP-type protein. These proteins regulate the function of a second type of protein called Ras. This finding was significant because the human *RAS* gene was known to be mutated in many human tumors. As we examine in detail in Chapters 20 and 23, GAP and Ras proteins normally function to control cell replication and differentiation in response to signals from neighboring cells. Because of its sequence homology with GAP, researchers hypothesized that NF1 also would regulate Ras. The NF1 protein subsequently was expressed from the cloned gene by methods described later and, indeed, was found to regulate Ras activity. These findings suggest that in individuals with neurofibromatosis, who have a defective *NF1* gene, a mutant NF1 protein is expressed in cells of the peripheral nervous system; the resulting abnormal interaction between the mutant NF1 and Ras leads to inappropriate cell division and tumor formation.

 Another example of this approach involves the *BRCA-1* gene. Women who inherit a mutant form of this gene have a high probability of developing breast cancer before age 50. The *BRCA-1* gene was isolated by methods described in Chapter 8, and a cDNA of the *BRCA-1* mRNA was cloned and sequenced, revealing the amino acid sequence of the BRCA-1 protein. Sophisticated methods of sequence comparison revealed that BRCA-1 protein is distantly, but significantly related to the *S. cerevisiae* Rad9 protein, a cell-cycle checkpoint protein, which functions to control cell division in yeast (Chapter 13). Based on this clue, several laboratories have initiated experiments to determine if the BRCA-1 protein functions similarly to the *S. cerevisiae* Rad9 protein.

These examples illustrate how insight into the molecular basis of inherited human diseases can be gained by identifying and cloning the associated mutant gene and then comparing the sequence of the encoded protein with the sequences of other proteins stored in data banks. This general

```
NF1   841  T R A T F M E V L T K I L Q Q G T E F D T L A E T V L A D R F E R L V E L V T M M G D Q G E L P I A  890
                | |     | |       | |           |         | |       | | | |           | |
Ira  1500  I R I A F L R V F I D I V . . . T N Y P V N P E K H E M D K M L A I D D F L K Y I I K N P I L A F F  1546


      891  M A L A N V V P C S Q W D E L A R V L V T L F D S R H L L Y Q L L W N M F S K E V E L A D S M Q T L  940
                | |             | |     | |   | | | |   | | | |           | |     | |       | | |           |
     1547  G S L A . . C S P A D V D L Y A G G F L N A F D T R N A S H I L V T E L L K Q E I K R A A R S D D I  1594


      941  F R G N S L A S K I M T F C F K V Y G A T Y L Q K L L D P L L R I V I T S S D W Q H V S F E V D P T  990
           | | | | |   | |     | |       | |       | | | |   | | |         | |             | | | | |
     1595  L R R N S C A T R A L S L Y T S R G N K Y L I K T L R P V L Q G I V D N K E . . . . S F E I D . .  1638


      991  R L E P S E S L E E N Q R N L L Q M T E K F . . . . F H A I I S S S S E F P P Q L R S V C H C L Y Q  1036
           | |     | |       | |         | |     | |           | |   | | | |       | |           | |     | |
     1639  K M K P G . . . S E N S E K M L D L F E K Y M T R L I D A I T S S I D D F P I E L V D I C K T I Y N  1685


     1037  V V S Q R F P Q N S I G A V G S A M F L R F I N P A I V S P Y E A G I L D K K P P P R I E R G L K L  1086
           | | |     | |         | | | |   | | | | |   | | | | |   | |                             | |
     1686  A A S V N F P E Y A Y I A V G S F V F L R F I G P A L V S P D S E N I I . I V T H A H D R K P F I T  1734


     1087  M S K I L Q S I A N . . . . . . . . H V L F T K E E H M R P F N D . . . . F V K S N F D A A R R F F  1124
           |   | | | | | | | |                 | |     | | |   | |       | |         | |                   |
     1735  L A K V I Q S L A N G R E N I F K K D I L V S K E E F L K T C S D K I F N F L S E L C K I P T N N F  1784


     1125  L D I A S D C P T S D A V N H S L . . . . . . . . . . . . S F I S D G N V L A L H R L L W N N .  1159
             | |     | |   | |               | |                     | |   | | |   |     | |     | | |
     1785  T V N V R E D P T P I S F D Y S F L H K F F Y L N E F T I R K E I I N E S K L P G E F S F L K N T V  1834


     1160  . . Q E K I G Q Y L S S N R D H K A V G R R P F . . . . . D K M A T L L A Y L G P P E H K P V A  1200
             | | |       | |                   | |                 | |         | |     | | |       | |
     1835  M L N D K I L G V L G Q P S M E I K N E I P P F V V E N R E K Y P S L Y E F M S R Y A F K K V D  1882
```

▲ **FIGURE 7-30 Comparison of the sequences of human NF1 protein and *S. cerevisiae* Ira protein over a region of roughly 180 amino acids near the carboxyl terminus of each protein, where they exhibit significant homology.** The NF1 and the Ira sequences are shown on the top and bottom lines of each row, respectively, in the one-letter amino acid code (see Figure 3-2). Amino acids that are identical in the two proteins are highlighted in yellow. Amino acids with chemically similar but nonidentical side chains are connected by a red line between them. Amino acid numbers in the protein sequences are shown at the left and right ends of each row. Dots indicate "gaps" in the protein sequence inserted in order to maximize the alignment of homologous amino acids. [From Xu et al., 1990, *Cell* **62**:599.]

approach for revealing the molecular function of various proteins simply by sequencing the DNA that encodes them will undoubtedly increase as the sequences of more proteins with known functions are determined and as new computer methods are devised for identifying potentially significant relationships between sequences.

Comparative Analysis of Genomes Reveals Much about an Organism's Biology

The availability of the complete genome sequences of several organisms now permits the comparative analysis of genomes, a branch of bioinformatics known as **genomics**. With this approach, much can be learned about the biology of organisms that are difficult to culture, even when few if any of their proteins have been isolated and studied directly. For example, Table 7-3 lists the approximate number of proteins predicted to function in several basic cellular processes for five microorganisms with fully sequenced genomes. Most proteins from three of these organisms—*Mycoplasma*

genitalium, *Methanococcus jannaschii*, and *Haemophilus influenzae*—have not been studied directly. Rather their functions are inferred from their sequences by comparison with previously studied proteins of model organisms like *E. coli* and *S. cerevisiae*.

M. genitalium has the smallest known genome of any cell. Figure 7-31 shows a diagram of the entire *M. genitalium* genome indicating the positions of all long *open reading frames* (ORFs), sequences that can encode polypeptides containing 100 or more amino acids.* Most of the proteins encoded by these long ORFs have been assigned postulated functions based on their sequence homology with proteins

---

*An ORF is a DNA sequence that can be divided into triplet codons without any intervening stop codons. Although some polypeptides are shorter than 100 amino acids, these are difficult to predict from DNA sequence alone because *short* ORFs occur randomly in a long DNA sequence. *Long* ORFs encoding 100 or more amino acids are unlikely to occur randomly and very likely encode an expressed polypeptide.

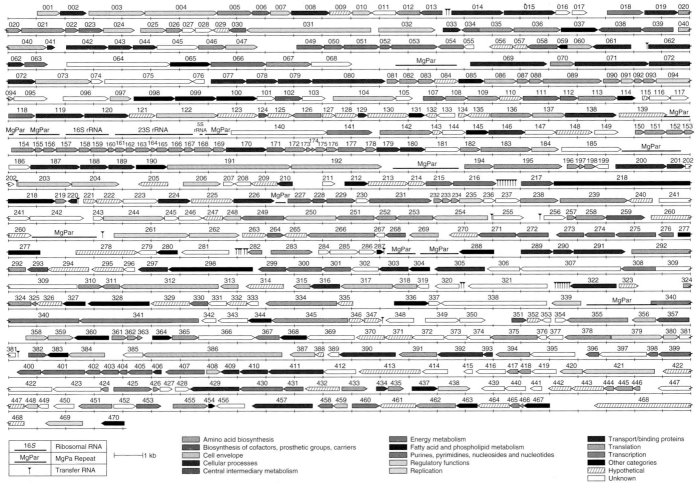

▲ FIGURE 7-31 Diagram of the genome of *Mycoplasma genitalium*. The circular genome is represented as a line; each long open reading frame (ORF), encoding 100 or more amino acids, is depicted as a rectangle with an arrowhead indicating the direction of transcription. Each line of the figure represents 24 kb; the entire genome contains 470 long ORFs. Protein-coding genes are color-coded according to the predicted function of the encoded protein, as indicated in the key at the bottom. Functions have been assigned to about 87 percent of the 470 proteins predicted to be encoded by this genome. [From C. M. Fraser et al., 1995, *Science* **270**:397.]

16S	Ribosomal RNA
MgPar	MgPa Repeat
⊤	Transfer RNA

⊢———⊣ 1 kb

Key:
- Amino acid biosynthesis
- Biosynthesis of cofactors, prosthetic groups, carriers
- Cell envelope
- Cellular processes
- Central intermediary metabolism
- Energy metabolism
- Fatty acid and phospholipid metabolism
- Purines, pyrimidines, nucleosides and nucleotides
- Regulatory functions
- Replication
- Transport/binding proteins
- Translation
- Transcription
- Other categories
- Hypothetical
- Unknown

of known functions from *E. coli* and other better-studied bacteria. *M. genitalium*, which lives as a parasite within epithelial cells lining primate urogenital and respiratory tracts, is extremely difficult to culture separately from its host eukaryotic cells. Nonetheless, a great deal can be inferred about *M. genitalium* from an analysis of its genome sequence.

For example, the *M. genitalium* genome does not encode many of the enzymes of intermediary metabolism found in free-living organisms (e.g., *E. coli*) that can synthesize all their constituents from simple molecules such as acetate and ammonia plus inorganic ions. The lack of these enzymes suggests that *M. genitalium* produces its macromolecules primarily from preformed precursors that are transported into its cytoplasm from the cytoplasm of its eukaryotic host cell. Although the *M. genitalium* genome lacks recognizable genes encoding the enzymes of the tricarboxylic acid (TCA) cycle and the cytochromes required for oxidative phosphorylation, it does encode the enzymes required for glycolysis of glucose to lactate and acetate. Thus this organism most likely produces ATP principally from the comparatively simple and low-yield glycolytic pathway (Chapter 16). Remarkably, the *M. genitalium* genome almost completely lacks the class of proteins that regulate transcription in other bacteria. This implies that the environment within its eukaryotic host cell is so constant that only minimal regulation of gene expression is required for cell growth and division. Alternatively, *M. genitalium* may regulate the expression of its genes by mechanisms that differ substantially from other prokaryotes.

The *M. jannaschii* genome is the first genome from an archaean to be completely sequenced. A strict anaerobe, *M. jannaschii* must live in an oxygen-free environment and derives chemical energy by the reduction of CO_2 with H_2 to produce methane. Genes encoding all the required enzymes for this reductive pathway have been identified in the *M. jannaschii* genome. Comparison of the sequences of *M. jannaschii* proteins with proteins performing equivalent functions from bacteria and eukaryotes has provided insights about the evolution of these three major classes of life on earth, as discussed below.

TABLE 7-3 Estimated Total Number of Proteins and Number Predicted to Function in Certain Cell Processes in Microorganisms with Sequenced Genomes*

	M. genitalium	M. jannaschii	H. influenzae	E. coli	S. cerevisiae
Total genome length (kb)	580	1660	1830	4640	12,050
Total number of proteins[†]	470	1700	1700	4300	6200
Number of proteins with predicted functions in[‡]					
Metabolism	50	230	325	650	650
Energy production and storage	40	130	140	240	175
Transporters	40	60	150	280	250
DNA replication, repair, and recombination	40	90	110	120	175
Transcription	12	20	30	230	400
Translation	100	110	125	180	350
Intracellular protein targeting and secretion	20	25	35	35	430
Cell structure	10	40	110	180	250

*From left to right, the organisms are Mycoplasma genitalium, Methanococcus jannaschii, Haemophilus influenzae, Escherichia coli, and Saccharomyces cerevisiae. M. jannaschii is an archaen; M. genitalium, H. influenzae, and E. coli are bacteria; and S. cerevisiae is a single-celled eukaryote.
[†]Values are the approximate total number of proteins encoded in each genome. They are based on the number of long open reading frames, which can encode proteins containing 100 or more amino acids, plus shorter genes encoding characterized proteins.
[‡]For each organism, only some of the proteins with predicted functions fall into the categories included in the table. The percentages of the estimated total number of proteins that have been assigned predicted functions vary among these organisms as follows: M. genitalium, 87%; M. jannaschii, 38%; H. influenzae, 83%; E. coli, 62%; and S. cerevisiae, 65%.
SOURCES: M. genitalium, C. M. Fraser et al., 1995, Science 270:397; M. jannaschii, C. J. Bult et al., 1996, Science 273:1058; H. influenzae. R. D. Fleischmann et al., 1995, Science 269:496; E. coli, F. R. Blattner et al., 1997, Science 277:1453; S. cerevisiae, A. Goffeau et al., 1996, Science 274:546. See also E. V. Koonin et al., 1997, Mol Microbiol. 25:619.

H. influenzae is one of the first important human pathogens whose genome has been sequenced. It can cause life-threatening bacterial meningitis, an inflammation of the meninges tissue that covers the central nervous system. As a pathogen that normally lives in close association with eukaryotic host cells, H. influenzae lacks many of the enzymes of intermediary metabolism present in E. coli. For this reason, H. influenzae, like M. genitalium, cannot grow on minimal medium. An understanding of the complete set of H. influenzae genes may provide clues for designing more effective therapies against H. influenzae infection. Because of the potential of this approach, scientists are determining the complete genome sequences of a number of pathogenic bacteria.

Homologous Proteins Involved in Genetic Information Processing Are Widely Distributed

As noted in Chapter 1, there are three cell lineages: two prokaryotic lineages, the bacteria and archaea, and one eukaryotic lineage (see Figure 1-5). All cells, regardless of their lineage, process genetic information in three stages: replication, transcription, and translation. Of the three processing systems, translation appears to be the most universal in its overall design in that the majority of proteins associated with ribosomes have clear homologs in bacteria, archaea, and eukaryotes. However, eukaryotic ribosomes, which are larger than prokaryotic ribosomes, contain some proteins not found in either bacteria or archaea (Chapter 4). A small number of bacteria-specific ribosomal proteins also occur. In addition, the archaean M. jannaschii has homologs of a number of eukaryotic ribosomal proteins that bacteria lack. The M. jannaschii translation-initiation and elongation factors and aminoacyl-tRNA synthetases also are more similar to their eukaryotic counterparts. The characteristics of the signal peptides of secreted proteins, discussed in Chapter 17, are also similar in archaea, bacteria, and eukaryotes, suggesting that the basic mechanisms of membrane targeting and translocation are similar among all three cell lineages.

The systems for replicating and transcribing DNA in archaea and eukaryotes are more similar to each other than to the bacterial systems. For example, bacteria lack histones, but the M. jannaschii genome encodes five histones homologous to eukaryotic H1, H2A, H2B, H3, and H4 (Chapter 9). This finding suggests that histones play similar functions in genome packaging and regulation of transcription in both archaea and eukaryotes. Likewise, the genomes of archaea, but not of bacteria, contain genes encoding proteins homologous to the transcription factors that help initiate transcription of nuclear genes in eukaryotic cells (Chapter 10). Using methods described later in this chapter, these archaeal proteins have been produced and found to have functions similar to their eukaryotic counterparts. In contrast, the proteins that transcribe the small DNA genomes found in chloroplasts and mitochondria are homologs of the transcription proteins found in bacteria. Mitochondrial and chloroplast ribosomes also resemble bacterial ribosomes more than the cytoplasmic ribosomes of eukaryotes. These new results from genomics imply that the nuclei of modern eukaryotic cells and modern archaea evolved from a common ancestor, whereas mitochondria and chloroplasts evolved from bacteria that formed a symbiotic relationship with an ancient ancestor of the modern eukaryotic cell.

In contrast to histones and transcription proteins, many *M. jannaschii* proteins involved in cell structure are homologous to bacterial structural proteins and are unrelated to eukaryotic proteins. This pattern reflects the similar structural organization of the two classes of prokaryotic cells and their structural differences from eukaryotic cells.

Many Yeast Genes Function in Intracellular Protein Targeting and Secretion

The genome from the yeast *S. cerevisiae* was the first eukaryotic genome to be sequenced in its entirety. So far, about 65 percent of *S. cerevisiae* genes have known or predicted functions. As shown in Table 7-3, many basic cell processes are carried out by roughly the same number of proteins in *E. coli* and yeast. The greatest difference occurs in the number of proteins involved in intracellular protein targeting and secretion, with yeast having about 10 times as many proteins devoted to these processes as *E. coli*. This difference reflects the complex intracellular membranes and organelles of eukaryotic cells.

To determine the function of the remaining 35 percent of yeast genes, whose function is not yet known or postulated, scientists have organized a worldwide systematic effort to inactivate each of these genes using methods described in Chapter 8. Many of the encoded proteins will probably have equivalents in other eukaryotic cells, including human cells. Indeed, about 45 percent of the human genes thus far sequenced have similarity to yeast genes. The recently completed sequence of the genome from *C. elegans*, a small multicellular organism, suggests that many of the human genes without yeast homologs encode proteins required for multicellular animal life.

The *C. elegans* Genome Encodes Numerous Proteins Specific to Multicellular Organisms

The sequence of the roundworm *C. elegans* genome, which was virtually completed in 1998, represents the first sequenced genome of a multicellular organism. The 97-Mb genome contains ≈19,100 protein-coding genes, about three times the number of yeast genes and one-fifth to one-third the estimated number of human genes. Approximately 40 percent of the predicted *C. elegans* proteins are homologous to genes of known function from other organisms. By now, ≈5000 human genes have been sequenced, mostly as cDNAs. About 75 percent of the sequenced human genes are homologous to *C. elegans* genes and are expected to have similar functions. This extensive homology makes *C. elegans* an excellent model organism for discovering the functions of many human proteins.

Comparison of the genomes of *S. cerevisiae*, a single-celled eukaryote, and *C. elegans*, a simple multicellular eukaryote, reveals that they encode a comparable number of homologous proteins to carry out the basic eukaryotic cellular functions. However, although *C. elegans* is composed of only ≈1000 cells, its genome also encodes a multitude of proteins that participate in several functions critical to the development and functioning of multicellular organisms (Table 7-4). The structures and functions of such proteins, many of which have no homologs in single-celled organisms, are discussed in detail in subsequent chapters.

TABLE 7-4 Proteins and Protein Domains Critical to Multicellularity Encoded in the *C. elegans* Genome

General Function	Proteins and Domains	Approximate Number Encoded
Control of transcription	Zinc-finger and homeobox domains	540
Control of RNA processing	RNA-binding domains	100
Cell-cell signaling	G protein–linked receptors, protein kinases, protein tyrosine phosphatases	1290
Nerve impulse transmission	Neurotransmitter-gated ion channels	80
Tissue formation	Collagens	170
Dynamic cell-cell interactions	EGF-like and immunoglobulin domains, glycosyltransferases, lectins	330

SOURCE: The *C. elegans* Sequencing Consortium, 1998, *Science* **282**:2012.

SUMMARY Bioinformatics

- Computer methods for storing, distributing, and analyzing vast amounts of DNA sequence data are collectively referred to as *bioinformatics*. Scientists around the world now have access via the Internet to computer banks of stored genomic and cDNA sequences.

- The function of a protein that has not been isolated often can be predicted based on the homology of its gene or cDNA with DNA sequences encoding proteins of known function. This approach is proving useful in unraveling the functions of proteins encoded by genes whose mutation causes various hereditary diseases.

- Determination of the complete genome sequences of several organisms has given rise to the discipline of genomics, the analysis and comparison of entire genomes.

- Genomics already has revealed many homologous proteins among the three major cell lineages. In some respects, archaea are more similar to eukaryotes than to bacteria.

- The *S. cerevisiae* genome, the first eukaryotic genome to be entirely sequenced, encodes many more proteins involved in intracellular protein targeting and secretion than do prokaryotic genomes.

- The genome of *C. elegans*, the first to be sequenced from a multicellular organism, encodes many proteins that have no homologs in single-celled organisms. These proteins participate in the complex gene control and diversity of interactions between cells typical of even quite simple multicellular organisms.

7.5 Analyzing Specific Nucleic Acids in Complex Mixtures

Once a specific DNA sequence has been isolated by cloning, the cloned DNA can be used as a probe to detect the presence and the amounts of complementary nucleic acids in complex mixtures such as total cellular DNA or RNA. These procedures depend on the exquisite specificity of nucleic acid hybridization. Related methods are used to locate DNA regions encoding specific mRNAs and transcription start sites.

Southern Blotting Detects Specific DNA Fragments

The technique of **Southern blotting**, named after its originator Edwin Southern, can identify specific restriction fragments in a complex mixture of restriction fragments. The DNA to be analyzed, such as the total DNA of an organism, is digested to completion with a restriction enzyme. For an organism with a complex genome, this digestion may generate millions of specific restriction fragments. The complex mixture of fragments is subjected to gel electrophoresis to separate the fragments according to size. However, many different fragments are of exactly the same length, and these do not separate from one another.

Even though all the fragments are not resolved by gel electrophoresis, an individual fragment that is complementary to a specific DNA clone can be detected. The restriction fragments present in the gel are denatured with alkali and transferred onto a nitrocellulose filter or nylon membrane by blotting (Figure 7-32). This procedure preserves the distribution of the fragments in the gel, creating a replica

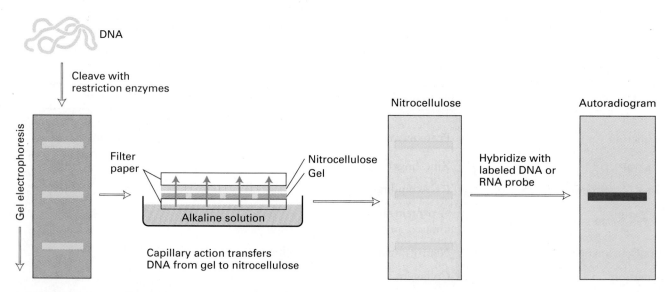

▲ **FIGURE 7-32 The Southern blot technique for detecting the presence of specific DNA sequences following gel electrophoresis of a complex mixture of restriction fragments.** The diagram depicts three restriction fragments in the gel, but the procedure can be applied to a mixture of millions of DNA fragments. A similar procedure, called *Northern blotting*, is used to detect specific RNA sequences. [See E. M. Southern, 1975, *J. Mol. Biol.* **98**:508.]

of the gel on the filter, much like the replica filter produced from plaques of a λ library. (The blot is used because probes do not readily diffuse into the original gel.) The filter then is incubated under hybridization conditions with a specific radiolabeled DNA probe, which usually is generated from a cloned restriction fragment. The DNA restriction fragment that is complementary to the probe hybridizes, and its location on the filter can be revealed by autoradiography.

Southern blotting permits a comparison between the restriction map of DNA isolated directly from an organism and the restriction map of cloned DNA. This comparison is necessary to be certain that no rearrangements have occurred during the cloning procedure such as might happen if two restriction fragments that do not normally lie next to each other were inadvertently ligated together before ligation into a cloning vector. Southern blotting also is used to map restriction sites in genomic DNA next to the sequence of a cloned DNA fragment. This provides a rapid method of comparing the restriction maps of different individual organisms in the region surrounding a cloned fragment. Deletion and insertion mutations are readily detected, as well as sequence differences in specific restriction sites.

Northern Blotting Detects Specific RNAs

Northern blotting, humorously named because it is patterned after Southern blotting, is used to detect a particular RNA in a mixture of RNAs. An RNA sample, often the total cellular RNA, is denatured by treatment with an agent (e.g., formaldehyde) that prevents hydrogen bonding between base pairs, ensuring that all the RNA molecules have an unfolded, linear conformation. The individual RNAs then are separated according to size by gel electrophoresis and transferred to a nitrocellulose filter to which the extended denatured RNAs adhere. The filter then is exposed to a labeled DNA probe and subjected to autoradiography. Because the amount of a specific RNA in a sample can be estimated from a Northern blot, the procedure is widely used to compare the amounts of a particular mRNA in cells under different conditions (Figure 7-33).

Specific RNAs Can Be Quantitated and Mapped on DNA by Nuclease Protection

Another important method for detecting and quantitating specific RNA molecules employs endonucleases that digest single-stranded but not double-stranded nucleic acids. The method was originally designed using endonuclease S1, an enzyme from the mold *Aspergillus oryzae* that digests single-stranded RNA and DNA but not double-stranded molecules. A labeled DNA strand, or probe, complementary to an RNA of interest is prepared from a cloned DNA. A source of RNA, such as the total polyadenylated RNA isolated from a particular tissue or type of cultured cell, is incubated with a high concentration of the labeled DNA probe under conditions in which all the RNA complementary to the probe hybridizes

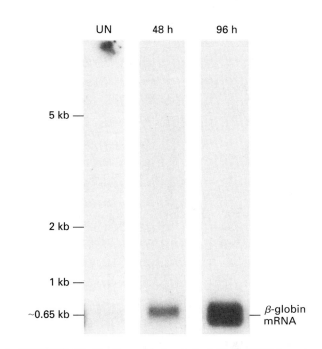

▲ **FIGURE 7-33 Northern blots of β-globin mRNA in extracts of erythroleukemia cells that are growing but uninduced (UN lane) and in cells that are induced to stop growing and allowed to differentiate for 48 hours or 96 hours.** The density of a band is proportional to the amount of mRNA present. The β-globin mRNA is barely detectable in uninduced cells but increases more than 1000-fold by 96 hours after differentiation is induced. [Courtesy of L. Kole.]

to it (Figure 7-34a). The preparation then is treated with endonuclease S1, which digests all the unhybridized RNA and probe molecules, leaving only the double-stranded region in the RNA-DNA hybrids, which is protected from nuclease digestion. Treatment of the digested preparation with alcohol precipitates the probe–target RNA hybrid, which then is subjected to gel electrophoresis followed by autoradiography or phosphorimager analysis of the gel to detect the protected probe. The amount of radioactivity in the resulting band is a measure of the amount of RNA complementary to the probe in the initial sample of RNA. Nuclease protection also can be performed with a complementary labeled RNA and ribonuclease A, a single-strand-specific pancreatic ribonuclease.

The DNA region that encodes a particular RNA can be mapped with the nuclease-protection technique by use of restriction-fragment probes in which one end is complementary to only a portion of the RNA of interest. In this case, the RNA-DNA hybrid protected from S1 digestion is shorter than the RNA being probed; its length corresponds to that of the DNA region extending from one end of the coding region to a restriction site within it (Figure 7-34b). Comparison of the protected doubled-stranded fragments obtained with two or more such "partial" probes can map the RNA sequence relative to restriction sites in the complementary DNA (Figure 7-34c).

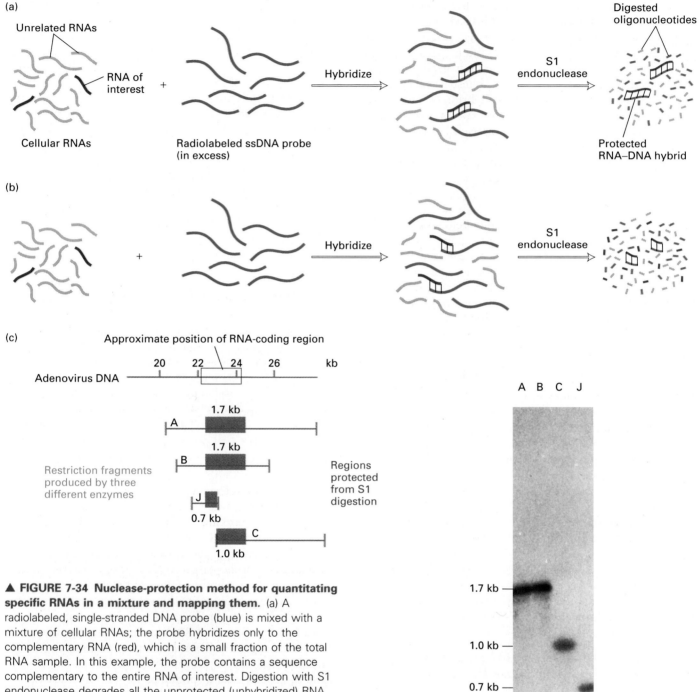

▲ **FIGURE 7-34 Nuclease-protection method for quantitating specific RNAs in a mixture and mapping them.** (a) A radiolabeled, single-stranded DNA probe (blue) is mixed with a mixture of cellular RNAs; the probe hybridizes only to the complementary RNA (red), which is a small fraction of the total RNA sample. In this example, the probe contains a sequence complementary to the entire RNA of interest. Digestion with S1 endonuclease degrades all the unprotected (unhybridized) RNA and DNA sequences, leaving a double-stranded RNA-DNA hybrid equal in length to the RNA. The protected hybrid is detected by gel electrophoresis followed by autoradiography. The density of the resulting band is proportional to the amount of the hybridized RNA in the original mixture. (b) With a "partial" DNA probe, containing only a portion of the DNA sequence complementary to the RNA, the protected S1-digestion product is shorter than the RNA and equal in length to the complementary region of the probe. (c) In this example of mapping an RNA on the genome, a 1.7-kb RNA was approximately mapped to the region between 22.4 and 24.1 from the left end of the 36-kb adenovirus genome. Four radiolabeled restriction-fragment probes (A, B, C, and J)

from this region of the viral DNA were prepared, hybridized with RNA from virus-infected cells, and then treated with S1 endonuclease. An autoradiogram of the S1-digestion products is shown at the right. Probes A and B produced S1-digestion products of 1.7 kb, indicating the RNA sequence maps entirely within these restriction fragments. The results with the partial probes C and J map the RNA sequence relative to the restriction site separating fragments C and J. [Photograph in part (c) from A. J. Berk and P. A. Sharp, 1977, *Cell* **12**:721; copyright M.I.T.]

Transcription Start Sites Can Be Mapped by S1 Protection and Primer Extension

As discussed in Chapter 10, some of the DNA regulatory elements that control transcription of genes into mRNA are located near the transcription *start site*. Mapping the start site for synthesis of a particular mRNA often helps in identifying the DNA regulatory sequences that control its transcription. Two methods are used to map the 5′ end of a particular mRNA on a complementary DNA: *S1 protection* and *primer extension*. The first step in both methods is to identify the general region of a DNA that includes the start site of interest (Figure 7-35a); this can be done by Northern blot analysis or nuclease protection using various cloned restriction fragments as probes.

In the S1-protection method, the identified DNA region is treated with appropriate restriction enzymes to produce a single-stranded DNA fragment that will hybridize with the 5′ portion of the mRNA. This fragment is radiolabeled at the 5′ end, hybridized with the mRNA, and then trimmed with S1 endonuclease (Figure 7-35b). From the length of the labeled probe segment protected from digestion, the position of the start site in the original DNA can be located.

The primer-extension method uses a synthetic oligonucleotide that is complementary to an approximately 20-nucleotide stretch of the mRNA located 50–200 nucleotides from its 5′ end. This synthetic oligonucleotide is end-labeled at the 5′ end and then used to prime DNA synthesis by reverse transcriptase with the mRNA as the template (Figure 7-35c). The position of the start site can be mapped from the length of the resulting extension product.

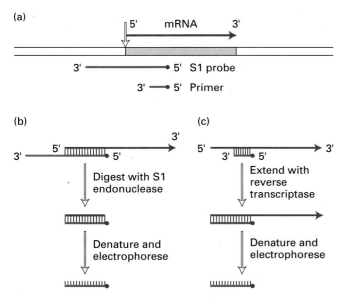

▲ **FIGURE 7-35 Two methods for mapping the start site for transcription of a particular gene in a region of DNA of known sequence.** (a) Diagram of the DNA fragment containing the gene of interest (light blue) and the corresponding mRNA (red). The end-labeled (red dot) single-stranded DNA fragment used as a probe in the S1 mapping technique and the end-labeled oligonucleotide primer used in the primer-extension technique are shown below the position of their sequence in the DNA. (b) In the S1 mapping technique, the probe is hybridized with the mRNA, and unpaired nucleic acid is then digested with S1 endonuclease (see Figure 7-34a). Denaturation leaves a labeled DNA fragment whose length accurately marks the distance of the starting nucleotide of the mRNA from the nucleotide that hybridized with the labeled DNA end. (c) In the primer-extension technique, a short (approximately 20-nucleotide) oligodeoxyribonucleotide is synthesized and end-labeled. After the primer is hybridized to the mRNA, it is extended by reverse transcriptase until it reaches the first nucleotide of the mRNA. The length of the primer-extension product, determined by gel electrophoresis, measures the distance from the 5′ end of the primer to the 5′ end of the mRNA.

SUMMARY Analyzing Specific Nucleic Acids in Complex Mixtures

- Cloned DNA fragments that have been isolated and labeled (usually radioactively) are widely used as probes to identify and quantitate a complementary DNA or RNA in a complex mixture. When added to a mixture of polynucleotide chains, a probe base-pairs only with complementary chains.

- Southern and Northern blots detect specific DNAs or RNAs, respectively, that have been separated by gel electrophoresis. Polynucleotides in the gel are transferred and attached to a filter or membrane replica of the gel (see Figure 7-32). The membrane is then incubated with a probe that hybridizes to the complementary polynucleotide of interest. After excess probe is washed away, the position and amount of labeled probe bound to the filter is determined by autoradiography or phosphorimager analysis.

- The amount of a specific RNA in a mixture can be detected by nuclease-protection analysis. Hybridization of an RNA to a labeled, single-stranded probe protects it from digestion by a nuclease that digests single-stranded but not double-stranded nucleic acids (see Figure 7-34a). The nuclease-resistant probe, hybridized to the target RNA, is recovered, subjected to gel electrophoresis, and detected by autoradiography or phosphorimager analysis. A variant of nuclease-protection analysis can map an RNA relative to restriction sites in complementary DNA (see Figure 7-34b, c).

- A target mRNA can also be quantitated and its transcription start site mapped by nuclease-protection analysis with a 5′-end-labeled DNA probe that hybridizes with the 5′ end of the mRNA (see Figure 7-35a, b). In the alternative primer-extension method, a short 5′-end-labeled DNA probe that hybridizes to the target mRNA acts as a primer for extension by reverse transcriptase to the 5′ end of the mRNA template strand (see Figure 7-35a, c).

7.6 Producing High Levels of Proteins from Cloned cDNAs

Many proteins with interesting or useful functions are normally expressed at very low concentrations. A case in point is *granulocyte colony-stimulating factor* (G-CSF). This human protein hormone stimulates the production of granulocytes, the phagocytic white blood cells critical to defense against bacterial infections. Both granulocytes and the cells that produce G-CSF are very sensitive to chemotherapeutic agents used in the treatment of cancer. As a result, one of the most serious side effects of chemotherapy is a fall in the concentration of granulocytes, making patients prone to life-threatening infections. G-CSF is normally made in very low concentrations in the bone marrow, where granulocytes differentiate and by some cultured human cell lines. Isolation of G-CSF from these sources is a tedious process yielding minuscule amounts of purified protein. By the techniques described in this section, G-CSF can now be produced in large enough amounts for therapeutic use in cancer patients to diminish the impact of chemotherapy on granulocyte production.

The first step in obtaining large amounts of low-abundance proteins such as G-CSF is to obtain a cDNA clone encoding the full length of the protein by methods discussed in previous sections. Once the desired cDNA is cloned, large amounts of the encoded protein often can be synthesized in engineered *E. coli* cells. Previously we discussed a λ phage expression vector that produces fusion proteins consisting of β-galactosidase joined to a protein fragment encoded by a DNA inserted into the vector; this type of expression system can be used to identify DNA clones encoding a specific protein (see Figure 7-21). Here, in contrast, we discuss expression vectors designed to produce full-length proteins at high levels.

E. coli Expression Systems Can Produce Full-Length Proteins

G-CSF and many other low-abundance proteins can be expressed at high levels in *E. coli* through use of specially designed expression vectors. Although many different expression vectors have been constructed, they all take advantage of the molecular mechanisms that control transcription and translation in *E. coli*.

Plasmid Expression Vectors Carrying a Strong, Regulated Promoter The first *E. coli* expression vectors developed were assembled by ligation of a basic plasmid vector, containing a replication origin (ORI) and selectable antibiotic-resistance gene, to a DNA sequence that functions as a strong, regulated **promoter**. A promoter is a DNA sequence where RNA polymerase initiates transcription. At a strong, regulated promoter, transcription is initiated many times per minute under specific environmental conditions. For example,

one expression vector contains a cloned fragment of the *E. coli* chromosome that includes the *lac* promoter and the *lacZ* gene encoding β-galactosidase. Transcription from the *lac* promoter occurs only when lactose, or a lactose analog such as isopropylthiogalactoside (IPTG), is added to the culture medium. IPTG generally is used because it cannot be metabolized, and therefore its concentration does not change as the cells grow. After addition of IPTG, the *lacZ* gene is transcribed into mRNA, which then is translated to yield many copies of the β-galactosidase protein (Figure 7-36a).

To modify this plasmid for production of G-CSF, the *lacZ* gene is replaced with a cDNA encoding G-CSF using restriction enzymes and DNA ligase. In this process, the *lac* promoter, which is required for efficient transcription, must

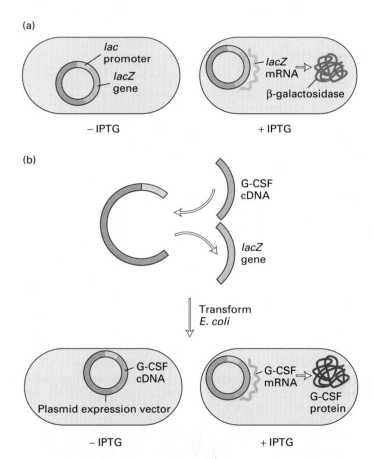

▲ **FIGURE 7-36 A simple *E. coli* expression vector utilizing the *lac* promoter.** (a) The expression vector plasmid contains a fragment of the *E. coli* chromosome containing the *lac* promoter and the neighboring *lacZ* gene. In the presence of the lactose analog IPTG, RNA polymerase normally transcribes the *lacZ* gene, producing *lacZ* mRNA, which is translated into the encoded protein, β-galactosidase. (b) The *lacZ* gene can be cut out of the expression vector with restriction enzymes and replaced by the G-CSF cDNA. When the resulting plasmid is transformed into *E. coli* cells, addition of IPTG and subsequent transcription from the *lac* promoter produces G-CSF mRNA, which is translated into G-CSF protein.

be maintained just before the start site of the inserted cDNA. In *E. coli* cells transformed by the resulting plasmid, transcription of the G-CSF cDNA and expression of G-CSF protein occurs in the presence of IPTG (Figure 7-36b).

Plasmid Expression Vectors Carrying the T7 Late Promoter A more complicated expression system, involving two levels of amplification, can produce larger amounts of a desired protein than the system just described. This second-generation system depends on the regulated expression of *T7 RNA polymerase,* an extremely active enzyme that is encoded in the DNA of bacteriophage T7. The T7 RNA polymerase transcribes DNA beginning within a specific 23-bp promoter sequence called the *T7 late promoter.* Copies of the T7 late promoter are located at several sites on the T7 genome, but none is present in *E. coli* chromosomal DNA. As a result, in T7-infected cells, T7 RNA polymerase catalyzes transcription of viral genes but not of *E. coli* genes.

In this expression system, recombinant *E. coli* cells first are engineered to carry the gene encoding T7 RNA polymerase next to the *lac* promoter. These cells then are transformed with plasmid vectors that carry a copy of the T7 late promoter and, adjacent to it, the cDNA encoding the desired protein (Figure 7-37). When IPTG is added to the culture medium containing these transformed, recombinant *E. coli*

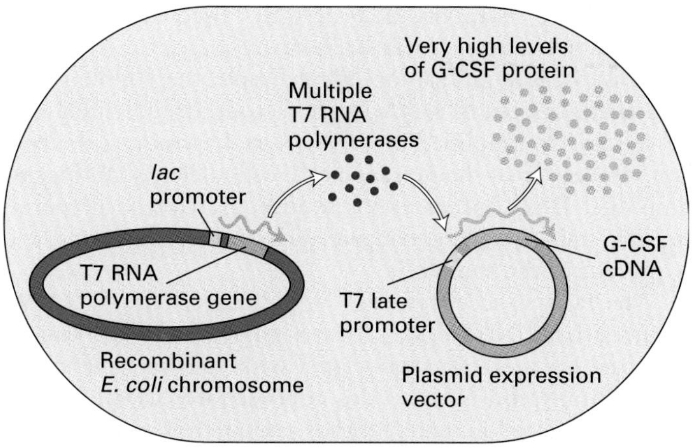

▲ **FIGURE 7-37 Two-step expression vector system based on bacteriophage T7 RNA polymerase and T7 late promoter.** The chromosome of a specially engineered *E. coli* cell contains a copy of the T7 RNA polymerase gene under the transcriptional control of the *lac* promoter. When transcription from the *lac* promoter is induced by addition of IPTG, the T7 RNA polymerase gene is transcribed, and the mRNA is translated into the enzyme. The T7 RNA polymerase molecules produced then initiate transcription at a very high rate from the T7 late promoter on the expression vector. Multiple copies of the expression vector are present in such cells, although only one copy is diagrammed here. The large quantity of mRNA transcribed from the cDNA cloned next to the T7 late promoter is translated into abundant protein product.

cells, T7 RNA polymerase is expressed by transcription from the *lac* promoter. The polymerase then binds to the T7 late promoter on the plasmid expression vectors, catalyzing transcription of the inserted cDNA at a high rate. Since each *E. coli* cell contains many copies of the expression vector, prodigious amounts of mRNA corresponding to the cloned cDNA can be produced in this system. Typically 10–70 percent of the total protein synthesized by these cells after addition of IPTG is the protein of interest. Because of the high yields possible with the T7 two-step expression system, it is often used for producing proteins in *E. coli.*

Eukaryotic Expression Systems Can Produce Proteins with Post-Translational Modifications

Most of the enzymes used in recombinant DNA technology (e.g., restriction enzymes, DNA polymerases, DNA ligases, polynucleotide kinase, and reverse transcriptase) are now produced commercially in *E. coli* expression systems. Large quantities of many eukaryotic proteins of interest, such as G-CSF, also can be produced in these systems. However, some eukaryotic proteins cannot be produced in active form in *E. coli* cells. These include proteins that are extensively modified during or following their synthesis, such as glycoproteins to which carbohydrate groups are added. *E. coli* lacks the enzymes that catalyze many of the post-translational modifications found on eukaryotic proteins.

To overcome this limitation of *E. coli* expression systems, researchers have developed *eukaryotic expression vectors* that permit addition of appropriate post-translational modifications to expressed proteins. Such vectors can be used in various types of eukaryotic cells to direct abundant synthesis of eukaryotic proteins from cloned genes.

Cloned cDNAs Can Be Translated in Vitro to Yield Labeled Proteins

In the *E. coli* and eukaryotic expression systems just described, proteins are synthesized in vivo within living cells. In vitro expression of proteins encoded by cloned cDNA also is possible with the T7 late promoter expression vector diagrammed in Figure 7-37. In this system, the mRNA encoded by the cloned cDNA is synthesized in vitro using the purified expression vector and purified T7 RNA polymerase. The mRNA then is translated in vitro to yield the desired protein. Although an in vitro expression system yields less protein than an in vivo system, it allows the desired protein to be radioactively labeled by addition of labeled amino acids to the in vitro translation reaction. In addition to the T7 system, plasmid vectors containing late promoters and the corresponding RNA polymerases from related bacteriophages (e.g., T3 and SP6) also can be used for in vitro production of proteins from cloned DNA.

Cell extracts are used to translate mRNA synthesized in vitro into protein. These extracts, which usually are prepared from rabbit reticulocytes or wheat germ, are rich in the components required for translation: ribosome subunits; tRNAs; aminoacyl-tRNA synthetases; and initiation, elongation, and termination factors (Chapter 4). Extracts first are treated with low concentrations of micrococcal nuclease to eliminate any endogenous mRNAs. At low levels, this enzyme, which is active only in the presence of Ca^{2+}, digests mRNA but not tRNA or rRNA. After digestion of endogenous mRNAs is complete, the enzyme is inactivated by addition of EGTA, a chelating-agent that binds Ca^{2+} with a much higher affinity than the nuclease does. When an in vitro synthesized mRNA is added to a treated extract, only the corresponding protein is produced, since the endogenous mRNAs have been destroyed. If a radiolabeled amino acid such as [^{35}S]methionine is included in the translation reaction, the protein product will be labeled. The labeled protein can be used in binding experiments to test its interaction with other proteins and in various other experimental systems.

SUMMARY **Producing High Levels of Proteins from Cloned cDNAs**

- *E. coli* expression vectors allow the production of abundant amounts of a protein of interest once a cDNA encoding it has been cloned. One of the most productive *E. coli* expression systems uses engineered *E. coli* cells that express large amounts of the T7 RNA polymerase in the presence of the lactose analog IPTG (see Figure 7-37).

- Eukaryotic expression vectors are used to produce eukaryotic proteins that are post-translationally modified, for example by the addition of carbohydrate or phosphate groups. Because *E. coli* lacks the enzymes catalyzing these modifications, such proteins can be produced only in eukaryotic cells.

- cDNA clones can be transcribed and translated in vitro, but the protein yield generally is less than with in vivo expression systems. The advantage of in vitro expression is that the protein of interest can be radioactively labeled during its synthesis.

7.7 Polymerase Chain Reaction: An Alternative to Cloning

An alternative to cloning, called the **polymerase chain reaction (PCR)**, can be used to directly amplify rare specific DNA sequences in a complex mixture when the ends of the sequence are known. This method of amplifying rare sequences from a mixture has numerous applications in basic research, human genetics testing, and forensics.

A typical PCR is outlined in Figure 7-38. Genomic DNA is digested into large fragments using a restriction enzyme and then is heat-denatured into single strands. Two synthetic oligonucleotides complementary to the 3' ends of the target DNA segment of interest are added in great excess to the denatured DNA, and the temperature is lowered to 50–60 °C. The genomic DNA remains denatured, because the complementary strands are at too low a concentration to encounter each other during the period of incubation, but the specific oligonucleotides, which are at a very high concentration, hybridize with their complementary sequences in the genomic DNA. The hybridized oligonucleotides then serve as primers for DNA chain synthesis, which begins upon addition of a supply of deoxynucleotides and a temperature-resistant DNA polymerase such as that from *Thermus aquaticus* (a bacterium that lives in hot springs). This enzyme, called *Taq polymerase*, can extend the primers at temperatures up to 72 °C. When synthesis is complete, the whole mixture is heated further (to 95 °C) to melt the newly formed DNA duplexes. When the temperature is lowered again, another round of synthesis takes place because excess primer is still present. Repeated cycles of synthesis (cooling) and melting (heating) quickly amplify the sequence of interest. At each round, the number of copies of the sequence between the primer sites is doubled; therefore, the desired sequence increases exponentially.

PCR Amplification of Mutant Alleles Permits Their Detection in Small Samples

 The sensitivity of procedures used in human genetics testing has been vastly increased by use of PCR amplification. For example, the *β*-globin gene in a small sample of DNA isolated from an individual can be specifically amplified by the PCR to determine if the person is a carrier of the mutant sickle-cell allele. Quantities of amplified DNA sufficient for sequencing can be prepared rapidly; subsequent sequencing reveals if the mutant allele is present in the sample.

The PCR is so effective at amplifying specific DNA sequences that DNA isolated from a single human cell can be analyzed for mutations associated with various genetic diseases. In one reported case, this approach was used to screen in vitro fertilized human embryos prepared from sperm and ova from a couple who both were carriers of the genetic disorder cystic fibrosis. This disease results from mutation in the *CFTR* gene, which is located on chromosome 7. The DNA isolated from a single embryonic cell was subjected to PCR amplification and then analyzed for mutations identified in one of the two copies of chromosome 7 in each parent. In this way embryos that had inherited the wild-type chromosome from at least one parent were identified and then transferred to the mother's uterus. (Removal of a single cell from an in vitro fertilized human embryo has no apparent effect on subsequent development of the embryo after it is implanted in a receptive uterus.) By use of this procedure, carrier couples can be assured of having children that will not be at risk for cystic fibrosis.

▶ **FIGURE 7-38 The polymerase chain reaction.** The starting material is a double-stranded DNA. Large numbers of primers are added, each with the sequence found in one strand at the end of the region to be amplified. The thermostable *Taq* polymerase and dNTPs are also added. In the first cycle, heating to 95 °C melts the double-stranded DNA and subsequent cooling to 60 °C then allows the excess primers to hybridize (anneal) to their complementary sequences in the target DNA. The *Taq* polymerase then extends each primer from its 3′ end by polymerization of dNTPs, generating newly synthesized strands (wavy lines) that extend in the 3′ direction to the 5′ end of the template restriction fragment. In the second cycle, the original and newly made DNA strands are separated at 95 °C and primers annealed to their complementary sequences at 60 °C. (For simplicity, subsequent events involving only newly made strands are shown; these soon greatly outnumber the original strands.) Each annealed primer again is extended by *Taq* polymerase to the end of the other primer sequence at the 5′ end of the template strand. Thus the strands (amplimers) synthesized in this cycle exactly equal the length of region to be amplified. In the third cycle, two double-stranded DNA molecules are generated equal to the sequence of the region to be amplified. These two are doubled in the fourth cycle and are doubled again with each successive cycle. [Adapted from J. D. Watson et al., 1992, *Recombinant DNA*, 2d ed., Scientific American Books.]

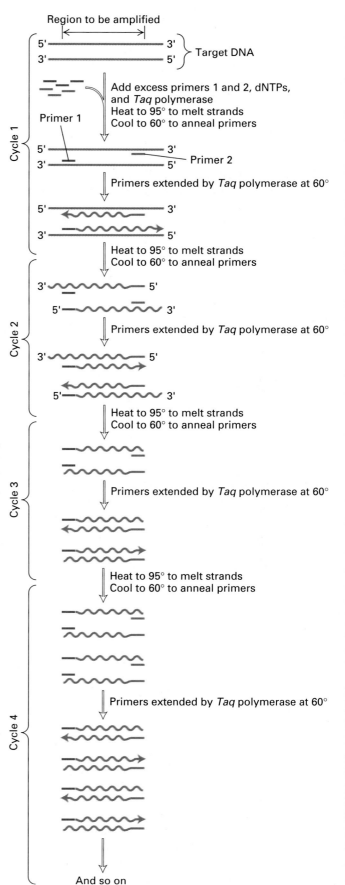

Another medical application of the PCR is early detection of infection with HIV, the virus that causes acquired immunodeficiency syndrome (AIDS). The PCR is so sensitive that it can detect HIV at very early stages in the disease (before symptoms appear) when only a few thousand blood cells in a patient are infected with the virus.

DNA Sequences Can Be Amplified for Use in Cloning and as Probes

In basic research, the PCR also has numerous applications. For example, this procedure allows the recovery and rapid amplification of the entire DNA region between any two ends whose sequences are known; the amplified DNA fragment then can be ligated into standard cloning vectors. Fragments of ≈2 kb or less can be amplified readily, and recent refinements of the technique allow amplification of regions of >30 kb.

The PCR also provides an alternative approach for preparing probes to screen genomic or cDNA libraries for clones encoding a protein of interest. The amino acid sequence of two peptides isolated from the purified protein are used to design two degenerate oligonucleotide mixtures containing all possible DNA sequences encoding the two peptides (see Figure 7-19). Rather than using these oligonucleotides as probes for direct screening of a cDNA library, as described previously, they are used as primers in a PCR. First, cDNA is synthesized from total cellular mRNA using reverse transcriptase. The cDNA is then used as the template for a PCR performed with the two degenerate oligonucleotide primers.

This reaction amplifies the region of the cDNA between the sequences encoding the peptides used to design the degenerate primers. The PCR procedure effectively selects the correct oligonucleotides for priming DNA synthesis from the degenerate oligonucleotide mixtures, because only DNA synthesized from the correct cDNA template will hybridize to oligonucleotides present in both degenerate primer mixtures. For exponential amplification to take place, priming must occur from both ends of a fragment. Even if an oligonucleotide in one of the degenerate mixtures hybridizes to an incorrect cDNA and primes DNA synthesis, the DNA strand that is synthesized will not be amplified, because it will not contain a sequence complementary to one of the oligonucleotides in the second degenerate primer mixture. The cDNA sequence amplified by this procedure contains the unique sequence of the naturally occurring mRNA encoding the region between the two peptides originally sequenced. This unique DNA sequence can then be radioactively labeled and used as a probe for screening a cDNA or genomic library. A probe prepared in this way gives a much stronger and more specific signal than that obtained by direct use of a degenerate probe.

SUMMARY **Polymerase Chain Reaction:**
An Alternative to Cloning

- In the polymerase chain reaction (PCR), the DNA sequence lying between two primers present at high concentration undergoes repeated doublings in an exponential fashion (see Figure 7-38). Large amounts of DNA can be synthesized from just a single initial template DNA molecule by this method. However, the sequence of the ends of the DNA to be amplified must be known.

- PCR amplification permits highly sensitive detection of mutant alleles in human genetics testing and of very low levels of HIV in infected humans.

- In basic research, the PCR is used as both an alternative and an adjunct to standard cloning procedures.

7.8 DNA Microarrays: Analyzing Genome-Wide Expression

Recently invented methods allow researchers to analyze the expression of thousands of genes simultaneously using **DNA microarrays.** Coupling these methods with the results from genome sequencing projects allows researchers to analyze the complete transcriptional program of an organism during specific physiological responses or developmental processes. DNA microarrays consist of thousands of individual gene sequences bound to closely spaced regions on the surface of a glass microscope slide. Two methods have been developed for preparing DNA microarrays.

In one method, an ≈1-kb portion of the coding region of each gene analyzed is individually amplified by PCR. A robotic device is used to apply each amplified DNA sample to closely spaced spots on the surface of a glass microscope slide, which then is processed by chemical and heat treatment to attach the DNA sequences to the glass surface and denature them. Typical arrays are 2 × 2 cm and contain ≈6000 spots of DNA.

In an alternative method, multiple DNA oligonucleotides, usually 20 nucleotides in length, are synthesized from an initial nucleotide that is covalently bound to the surface of a glass slide. Tens of thousands of identical oligonucleotides are synthesized in a small square area on the surface of the slide. Several oligonucleotide sequences from a single gene are synthesized in neighboring regions of the slide to analyze expression of that gene. Thousands of genes can be represented on one glass slide. Because the methods for constructing these arrays of synthetic oligonucleotides were adapted from methods for manufacturing microscopic integrated circuits used in computers, these types of oligonucleotide microarrays are often called *DNA chips.*

In one of the first uses of DNA microarray analysis, gene expression was compared between yeast cells growing on glucose as the source of carbon and energy, and cells growing on ethanol. The total mRNA was separately isolated from cells growing under these two conditions using oligo-dT columns (see Figure 7-14). The total mRNA from the glucose-grown cells was reverse-transcribed into cDNA using reverse transcriptase, an oligo-dT primer, deoxyribonucleoside triphosphates, and a low concentration of a nucleotide analog labeled with a green fluorescent dye. cDNA from the ethanol-grown cells was prepared similarly, but labeled with a red fluorescent dye. Equal amounts of the two cDNA preparations were then mixed and incubated under hybridization conditions with a DNA microarray containing all 6400 yeast genes, so that each gene on the array hybridized to its complementary cDNA in the preparation.

After washing away unhybridized cDNA, each DNA spot on the microarray was analyzed using a scanning laser microscope with a computer-controlled stage. The intensity of green and red fluorescence at each DNA spot was measured with photomultiplier tubes and stored in computer files under the name of each gene according to its known position on the slide. The data is represented in the computer-generated image shown in Figure 7-39. Spots fluorescing bright red are from genes expressed at a high level in ethanol and a low level in glucose media. Spots fluorescing bright green are from genes expressed at a high level in glucose and a low level in ethanol medium. Spots fluorescing bright yellow are from genes expressed at high levels in both types of media. Genes expressed at low to moderate levels do not yield brightly fluorescing spots in the depiction of the data in Figure 7-39 where all genes are compared. However, the relative level of expression of such genes in cells grown in glucose or ethanol can be determined from accurate measurements of the relative intensities of weak red and green

▲ **FIGURE 7-39 Yeast genome microarray.** The array is 18 mm × 18 mm and contains 6400 DNA spots corresponding to each yeast gene. The array was hybridized to cDNA labeled with a green fluorescent dye prepared from cells grown in glucose and with red-labeled cDNA prepared from cells grown in ethanol. cDNA hybridizing to each DNA spot was detected with a scanning confocal laser microscope. [From J. L. DeRisi et al., 1997, *Science* **278**:680.]

This analysis revealed a marked, coordinated change in the global pattern of yeast gene expression. For example, expression from seventeen genes encoding mitochondrial proteins required for respiration increased with a similar time course as glucose was depleted (Figure 7-40b). Analysis of the yeast genome sequence showed that each of these genes contains in their upstream transcription-control region binding sites for the trimeric transcription factor (Chapter 10) encoded by genes *HAP2, 3,* and *4.* Earlier studies showed that this factor regulates the transcription of some of these genes. This genome-wide expression analysis revealed that these seventeen genes are probably chiefly regulated by the trimeric Hap transcription factor. As the rate of cell growth slowed when glucose was depleted from the media, 112 genes encoding ribosomal proteins were repressed with similar kinetics (Figure 7-40c). The upstream control regions (Chapter 10) in all of these genes have binding sites for the Rap1 transcription factor, which earlier had been shown to regulate the expression of a handful of ribosomal protein genes. Seven genes exhibited strong induction only at the last time point analyzed, when glucose was completely depleted from the media. Six of these were found to have an upstream activating sequence in common, the CSRE carbon-source response element. Thus, several distinct temporal patterns of expression were recognized. In most cases, common regulatory mechanisms could be inferred for sets of genes with similar expression profiles.

The results of this study showed that as yeast adapt to decreasing glucose levels, expression of 710 genes increases by a factor of two or more, while expression of 1030 genes decreases by a factor of two or more. About 400 of the differentially expressed genes have no known function. This analysis provides the first clue as to their possible function in yeast biology.

Approximately 15,000 human genes have now been identified. Microarrays of these human genes have been prepared and used to characterize their expression in different tissues and organs, and in cancer cells compared to their normal cell counterparts. When the complete set of genes for humans and model experimental organisms such as the mouse and *Drosophila* become available, microarray technology will be used to analyze the global genome-wide gene expression patterns that occur during cellular differentiation and disease.

fluorescence signals. Genes that are not transcribed under these growth conditions give no detectable signal.

Hybridization of fluorescently labeled cDNA preparations to DNA microarrays such as this provide a means for analyzing gene expression patterns on a genomic scale. One example is the analysis of changes in gene expression as yeast cells convert from growth on a high-sugar medium to growth on ethanol. This adaptation to different nutrients occurs in the natural environment when yeast growing on decaying fruit switch from rapid growth, when they metabolize sugars to ethanol by anaerobic fermentation, to slower growth, when the sugar is depleted and they metabolize the accumulated ethanol to carbon dioxide and water by aerobic respiration (Chapter 16). This situation can be simulated in the laboratory by inoculating yeast into media high in glucose. As shown in Figure 7-40a, growth initially is exponential and rapid until the glucose is exhausted; then the rate of cell doubling is slowed and cells shift from fermentation to respiration. The transcriptional response of each of the 6400 yeast genes during this adaptation to decreasing glucose was analyzed by hybridizing fluorescently labeled cDNAs prepared from cells at several time points after inoculation of the culture to DNA microarrays as shown in Figure 7-39.

SUMMARY   DNA Microarrays: Analyzing Genome-Wide Expression

- DNA microarrays contain microscopic spots of ≈1-kb DNA sequences representing thousands of genes bound to the surface of glass microscope slides.

- Oligonucleotide arrays (DNA chips) contain synthetic oligonucleotides representing thousands of gene sequences synthesized on the surface of small areas of a glass slide.

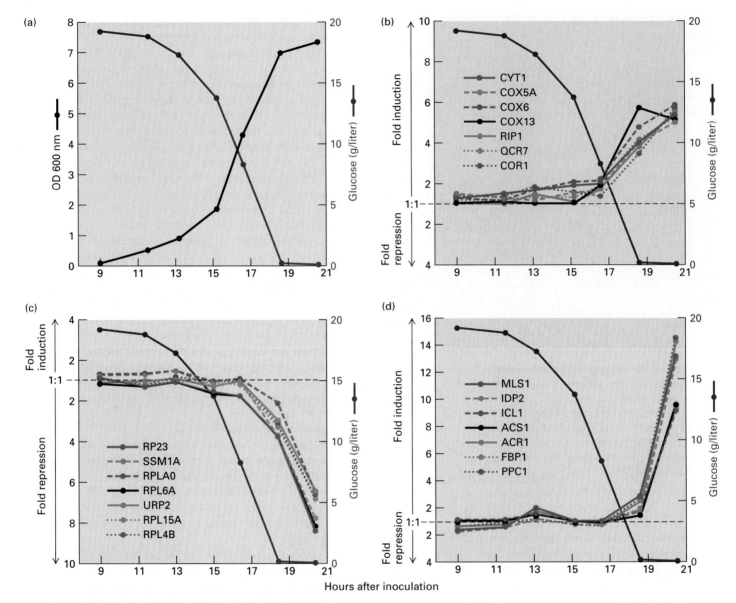

▲ **FIGURE 7-40 Changes in yeast gene expression as growing cells deplete glucose from the growth medium.** (a) Growth of yeast cells and glucose concentration in the medium following inoculation of cells into glucose medium. Cell density was measured at seven time points by the optical density of the medium at a wavelength of 600 nm. Glucose concentration in the medium was measured at each time point. (b) The relative levels of mRNAs for seven mitochondrial proteins required for aerobic respiration. The expression of ten additional mitochondrial protein mRNAs showed similar kinetics. (c) The relative levels of mRNAs encoding seven ribosomal proteins. Expression of 112 total genes encoding ribosomal proteins showed similar kinetics. (d) A total of seven yeast genes showed strong induction only at the last time point. [From J. L. DeRisi et al., 1997, *Science* **278**:680.]

• Expression of the genes represented on a DNA or oligonucleotide array can be assayed by synthesizing fluorescently labeled cDNA from a preparation of mRNA isolated from different cells or from cells under different conditions. The amount of complementary cDNA that hybridizes at each spot on the array is measured using a scanning confocal laser microscope. The data for each gene is stored in computer files and analyzed.

PERSPECTIVES for the Future

The powerful techniques of recombinant DNA technology are applied in virtually every area of biological research. Consequently, a discussion of the future of recombinant DNA technology amounts to a consideration of the future of biological research in general. Perhaps the most dramatic recent impact of this technology is the sequencing of entire genomes of a number of organisms. As DNA sequences are determined, they are being stored in public databases, fully accessible to molecular cell biologists throughout the world.

The genome sequences of additional model multicellular animals and plants are likely to be completed within the next few years. Since the function of a protein is determined by its three-dimensional structure, which is in turn determined by its amino acid sequence, we will in essence have the information to potentially understand the function of every one of the estimated 60,000–100,000 proteins encoded in the human genome and the functions of proteins in organisms important to us such as food crops and pathogens.

The challenge will be to find which of the gene sequences in databases have the greatest interest and utility. As the accumulation of sequence data escalates, the major task facing researchers will shift from discovering new genes and proteins to discovering the functions of genes and proteins whose sequences are already known.

PERSPECTIVES in the Literature

The entire genome sequences of multiple bacteria, archaea, and model eukaryotic organisms are being determined at a rapid rate. The functions of many of the newly sequenced genes can be assigned by identifying homologs in an organism where the function of the encoded protein has been determined from biochemical and genetic studies. However, the functions of a significant fraction of newly sequenced genes remain unknown. How can comparisons between genome sequences beyond straightforward gene-by-gene comparisons be used to develop hypotheses about gene function? To explore this question, consult the following sources:

Genome sequences are listed at the web site of The Institute for Genomic Research: *http://www.tigr.org*

Bock, P., et al. 1998. Predicting function: from genes to genomes and back. *J. Mol. Biol.* **238**:707–725.

Marcotte, E. M., et al. 1999. Genome-wide prediction of protein function for *Saccharomyces cerevisiae. Nature, in press.*

Testing Yourself on the Concepts

1. What particular features of an organism's molecular cell biology are analyzed by the discipline of genomics to gain information about its relationship to other organisms? What has this approach shown when applied to the archaean *M. jannaschii, S. cerevisiae* (as an example of a simple eukaryote), and *C. elegans*?

2. Describe the properties of enzymes that permit cloning of DNA, DNA sequencing, the polymerase chain reaction, and production of cDNA.

3. Recombinant DNA technology is used to produce proteins valued in many different applications. How is high-level expression obtained? What special considerations come into play when producing certain eukaryotic proteins?

4. Section 7.5 is titled "Analyzing Specific Nucleic Acids in Complex Mixtures" (p. 240). How can you ascertain whether a specific DNA or RNA sequence is present in such a mixture? How can you isolate a particular DNA sequence? How can you isolate a nucleic acid containing the sequence of a unique mRNA?

MCAT/GRE-Style Questions

Key Experiment Please read the section titled "DNA Sequences Can Be Amplified for Use in Cloning and as Probes" (p. 247) and answer the following questions:

1. Degenerate oligonucleotides are used in the method because

   a. There are several codons for some amino acids.

   b. They are easier to synthesize.

   c. Their annealing temperature is lower.

   d. They hybridize better to the template.

2. The degeneracy of the primers

   a. Results in multiple PCR products, each with high specificity for the template.

   b. Is overcome because the desired PCR product is further amplified by a subset of the primers.

   c. Is irrelevant since they hybridize as well as perfectly matched primers.

   d. Is irrelevant since the best-matched primer at one end overcomes the mismatch at the other end.

3. A probe prepared in the manner described

   a. Is more specific than a degenerate probe.

   b. Can be used to screen cDNA and genomic libraries.

   c. Is equivalent to an antibody used to screen a cDNA expression library.

   d. All are correct.

Key Application Please read the "Medicine" segments in the section titled "Stored Sequences Suggest Functions of Newly Identified Genes and Proteins" (p. 235) and answer the following questions:

4. The most important step in defining the molecular basis for a human disease is to

   a. Clone and sequence the gene responsible for the condition.

   b. Carry out a karyotype analysis to find abnormal chromosomes in affected individuals.

c. Study relatives of the affected individual to determine the pattern of inheritance of the condition.

d. Isolate the protein mutated in the disease state and prepare monoclonal antibodies.

5. Clues about the function of the gene responsible for the disease can be obtained by

a. Comparing the sequences of the putative gene causing the disease with sequences of other genes stored in a data bank.

b. Carrying out in vitro enzyme activity assays using the isolated protein encoded by the gene.

c. Mutating the gene and looking for expression of the disease phenotype.

d. Expressing the gene and sequencing the mRNA.

6. If a newly discovered human gene has sequence homology to a previously discovered gene in yeast, it can be inferred that

a. The similarity is of no significance.

b. The similarity might reflect a common function for the proteins encoded by the genes.

c. The amino acid code used by these organisms is the same.

d. Humans and yeast are more closely related than previously thought.

7. Correlation of a human gene mutation with a particular disease is supported by all of the following statements *except*:

a. An alteration in the inferred function of the gene is consistent with the disease characteristics.

b. The activity of the protein encoded by the wild-type gene is involved in a pathway whose disruption might cause the disease.

c. Other mutations are also associated with the disease.

d. The mutation can be predicted from other information to cause a change in the function of the protein.

Key Experiment Please read the section titled "Sanger (Dideoxy) Method" (p. 232) and refer to Figure 7-29; then answer the following questions:

8. In the Sanger method for sequencing DNA, the first step is denaturation of the template. Denaturation is required

a. To activate the DNA polymerase.

b. To allow the primer to hybridize to the template.

c. To allow the dNTPs and ddNTPs access to the template.

d. To unwind the DNA so synthesis can proceed.

9. Referring to the four leftmost lanes in Figure 7-29b and reading from the bottom to the top, the sequence of the template for this reaction is

a. 5'TTGCAGGGC3'.

b. 3'TTGCAGGGC5'.

c. 5'AACGTCCCG3'.

d. 3'AACGTCCCG5'.

10. An error has been made in the concentration of ddGTP used in sequencing reactions: the solution is 100 μM rather than 1 μM. Examination of the autoradiogram resulting from using this concentration of compound would show

a. Increased spacing between all bands.

b. Decreased spacing between all bands.

c. A large blank area at the bottom.

d. A decreased number of bands in the lane from the reaction with ddGTP.

Key Terms

clone 207	polymerase chain
complementary DNAs	reaction (PCR) 246
(cDNAs) 219	probes 223
cosmid 221	promoter 244
DNA cloning 207	recombinant DNA 207
expression cloning 227	replication origin 209
expression vectors 227	restriction enzymes 207
genomic DNA 216	restriction fragments 211
genomics 236	reverse transcriptase 220
hybridization 223	Southern blotting 240
ligases 207	transformation 210
Northern blotting 241	vector 208
plasmid 208	

References

DNA Cloning with Plasmid Vectors

Ausubel, F. M., et al. 1987. *Current Protocols in Molecular Biology.* Wiley.

Caruthers, M. H. 1985. Gene synthesis machines: DNA chemistry and its uses. *Science* 230:281–285.

Davies, K. E., ed. 1988. *Genome Analysis: A Practical Approach.* IRL Press (Oxford, Eng.).

Itakura, K., J. J. Rossi, and R. B. Wallace. 1984. Synthesis and use of synthetic oligonucleotides. *Ann. Rev. Biochem.* 53:323–356.

Nathans, D., and H. O. Smith. 1975. Restriction endonucleases in the analysis and restructuring of DNA molecules. *Ann. Rev. Biochem.* 44:273–293.

Nelson, M., and M. McClelland. 1992. The effect of site-specific methylation on restriction-modification enzymes. *Nucl. Acids Res.* 20(suppl.):2145–2157.

Roberts, R. J., and D. Macelis. 1997. REBASE—restriction enzymes and methylases. *Nucl. Acids Res.* 25:248–262. Information on accessing a continuously updated database on restriction and modification enzymes at http://www.neb.com/rebase.

Sambrook, J., E. F. Fritsch, and T. Maniatis. 1989. *Molecular Cloning.* Cold Spring Harbor Laboratory.

Watson, J. D., et al. 1992. *Recombinant DNA*, 2d ed. New York: Scientific American Books.

Constructing DNA Libraries with λ Phage and Other Cloning Vectors

Dunn, I. S., and F. R. Blattner. 1987. Charons 36 to 40: multienzyme, high-capacity, recombination-deficient replacement vectors with polylinkers and polystuffers. *Nucl. Acids Res.* **15**:2677–2701.

Gubler, U., and B. J. Hoffman. 1983. A simple and very efficient method for generating cDNA libraries. *Gene* **25**:263–289.

Han, J. H., C. Stratowa, and W. J. Rutter. 1987. Isolation of full-length putative rat lysophospholipase cDNA using improved methods for mRNA isolation and cDNA cloning. *Biochemistry* **26**:1617–1632.

Hohn, B. 1979. In vitro packaging of λ and cosmid DNA. *Meth. Enzymol.* **68**:299–308.

Jendrisak, J., R. A. Young, and J. D. Engel. 1987. Cloning cDNA into λgt10 and λgt11. *Meth. Enzymol.* **152**:359–385.

Murray, N. E., and K. Murray. 1974. Manipulation of restriction targets in phage λ to form receptor chromosomes for DNA fragments. *Nature* **251**:476–483.

Rambach, A., and P. Tiollais. 1974. Bacteriophage λ having *EcoRI* endonuclease sites only in the nonessential region of the genome. *Proc. Nat'l. Acad. Sci. USA* **71**:3927–3931.

Shizuya, H., et al. 1992. Cloning and stable maintenance of 300-kilobase-pair fragments of human DNA in *Escherichia coli* using an F-factor-based vector. *Proc. Nat'l. Acad. Sci. USA* **89**:8794–8797. Description of bacterial artificial chromosomes.

Sternberg, N. L. 1992. Cloning high molecular weight DNA fragments by the bacteriophage P1 system. *Trends Genet.* **8**:11–16.

Thomas, M., J. R. Cameron, and R. W. Davis. 1974. Viable molecular hybrids of bacteriophage lambda and eukaryotic DNA. *Proc. Nat'l. Acad. Sci. USA* **71**:4579–4583.

Identifying, Analyzing, and Sequencing Cloned DNA

Andrews, A. T. 1986. *Electrophoresis*, 2d ed. Oxford University Press.

Benton, W. D., and R. W. Davis. 1977. Screening λgt recombinant clones by hybridization to single plaques in situ. *Science* **196**:180–183.

Cantor, C. R., C. L. Smith, and M. K. Matthew. 1988. Pulsed-field gel electrophoresis of very large DNA molecules. *Ann. Rev. Biophys. Biophys. Chem.* **17**:41–72.

Carle, G. F., M. Frank, and M. V. Olson. 1986. Electrophoretic separations of large DNA molecules by periodic inversion of the electric field. *Science* **232**:65–70.

Grunstein, M., and D. S. Hogness. 1975. Colony hybridization: a method for the isolation of cloned DNAs that contain a specific gene. *Proc. Nat'l. Acad. Sci. USA* **72**:3961–3965.

Maxam, A. M., and W. Gilbert. 1980. Sequencing end-labeled DNA with base-specific chemical-cleavages. *Methods Enzymol.* **65**:499–560.

Rickwood, D., and B. D. Hames, eds. 1982. *Gel Electrophoresis of Nucleic Acids*. IRL Press (London, Eng.).

Sanger, F. 1981. Determination of nucleotide sequences in DNA. *Science* **214**:1205–1210.

Singh, H., et al. 1988. Molecular cloning of an enhancer binding protein: isolation by screening of an expression library with a recognition site DNA. *Cell* **52**:415–423.

Wallace, R. B., et al. 1981. The use of synthetic oligonucleotides as hybridization probes. II: Hybridization of oligonucleotides of mixed sequence to rabbit β-globin DNA. *Nucl. Acids Res.* **9**:879–887.

Wetmur, J. G., and N. Davidson. 1968. Kinetics of renaturation of DNA. *J. Mol. Biol.* **31**:349–370.

Young, R. A., and R. W. Davis. 1991. Gene isolation with lambda gt11 system. *Meth. Enzymol.* **194**:230–238.

Bioinformatics

Bains, W. 1996. Company strategies for using bioinformatics. *Trends Biotechnol.* **14**:312–317.

Fields, C. 1996. Informatics for ubiquitous sequencing. *Trends Biotechnol.* **14**:286–289.

Koonin, E. V., R. L. Tatusov, and M. Y. Galperin. 1998. Beyond complete genomes: from sequence to structure and function. *Curr. Opin. Struc. Biol.* **8**:355–363.

Analyzing Specific Nucleic Acids in Complex Mixtures

Berk, A. J. 1989. Characterization of RNA molecules by S1 nuclease analysis. *Meth. Enzymol.* **180**:334–347.

Boorstein, W. R., and E. A. Craig. 1989. Primer extension analysis of RNA. *Meth. Enzymol.* **180**:347–369.

Wahl, G. M., J. L. Meinkoth, and A. R. Kimmel. 1987. Northern and Southern blots. *Meth. Enzymol.* **152**:572–581.

Producing High Levels of Proteins from Cloned cDNAs

Leibowitz, M. J., F. P. Barone, and D. E. Georgopoulos. 1991. In vitro protein synthesis. *Meth. Enzymol.* **194**:536–545.

Luckow, V. A., and M. D. Miller. 1989. High level expression of nonfused foreign proteins with *Autographa california* nuclear polyhedrosis virus expression vectors. *Virology* **170**:31–39.

Moss, B. 1993. Poxvirus vectors: cytoplasmic expression of transferred genes. *Curr. Opin. Genet. Devel.* **3**:86–90.

Souza, L. M., et al. 1986. Recombinant human granulocyte-colony stimulating factor: effects on normal and leukemic myeloid cells. *Science* **232**:61–65.

Studier, F. W., and B. A. Moffatt. 1986. Use of bacteriophage T7 RNA polymerase to direct selective high-level expression of cloned genes. *J. Mol. Biol.* **189**:113–134.

Tabor, S., and C. C. Richardson. 1985. A bacteriophage T7 RNA polymerase promoter system for controlled exclusive expression of specific genes. *Proc. Nat'l. Acad. Sci. USA* **82**:1074–1078.

Polymerase Chain Reaction: An Alternative to Cloning

Erlich, H., ed. 1992. *PCR Technology: Principle and Applications for DNA Amplification*. W. H. Freeman and Company.

Erlich, H. E., R. A. Gibbs, and H. H. Kazazian, Jr., eds. 1989. *Polymerase Chain Reaction*. Cold Spring Harbor Laboratory.

Handyside, A. H., et al. 1992. Birth of a normal girl after in vitro fertilization and preimplantation diagnostic testing for cystic fibrosis. *New Eng. J. Med.* **327**:905–909.

Innis, M. A., et al. 1990. *PCR Protocols: A Guide to Methods and Applications*. Academic Press.

Saiki, R. K., et al. 1988. Primer-directed enzymatic amplification of DNA with a thermostable DNA polymerase. *Science* **239**:487–491.

DNA Microarrays: Analyzing Genome-wide Expression

Data from the analysis presented in figure 7-39 can be viewed at http://cmgm.stanford.edu/pbrown/explore/index.html

DeRisi, J. L., V. R. Iyer, and P. O. Brown. 1997. Exploring the metabolic and genetic control of gene expression on a genomic scale. *Science* **278**:680–686.

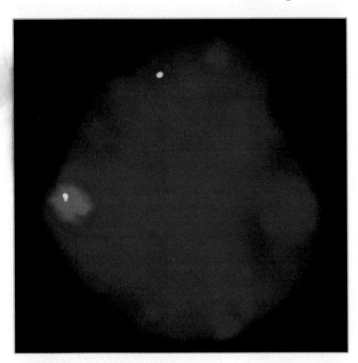

8

Genetic Analysis in Cell Biology

In previous chapters, we learned how proteins—the cell's working molecules—are isolated, how their structures are determined, and how the genes encoding known proteins are cloned. Our primary concern, however, is what a protein does in the organism. In principle, the in vivo function of a protein can be deduced by seeing what effect removal of the protein has on the cell or, in the case of a multicellular organism, on the whole organism. In practice, removing a protein is done indirectly by identifying organisms in which the nucleotide sequence of the gene encoding the protein is altered or deleted. Such changes in the DNA sequence, called **mutations,** can lead to loss of the encoded protein or to a change in its structure. The affected organisms, called *mutants,* are identified by virtue of differences in their appearance, physiology, behavior, or growth properties compared with normal, **wild-type** (nonmutant) organisms. By comparing specific DNA sequences from mutant and normal organisms, researchers can correlate the abnormal features of the mutant organism with differences in the expression or structure of specific proteins. Alternatively, specific mutations can be introduced into cloned genes and the mutated genes then introduced into intact organisms (e.g., yeast and mice); again, comparison of the normal and mutant organisms provides clues about the in vivo functioning of the encoded protein.

In this chapter we consider some basic concepts in genetics and various genetic techniques that are useful in studying how proteins carry out specific cellular processes and in what order specific proteins function. In the first section, we review basic properties of mutations; in the second, we discuss the isolation of mutants and the characterization of mutations using classical

As seen in this female E7.0 mouse embryo nucleus, mature Xist RNA (red probe) accumulates on the inactive X chromosome, while only immature unspliced Xist RNA is associated with the active X chromosome (yellow).

MEDIA CONNECTIONS

Technique: **In Vitro Mutagenesis of Cloned Genes**

Technique: **Creating a Transgenic Mouse**

Classic Experiment 8.1: Expressing Foreign Genes in Mice

genetic analyses. We then describe the various steps in the **mapping** of mutations, that is, the procedures for locating mutations on particular chromosomes, in regions within chromosomes, and in relation to one another. In the fourth section, we discuss the use of recombinant DNA techniques presented in Chapter 7 to isolate and clone mutation-defined genes from relevant DNA libraries. This approach has permitted the identification and cloning of numerous human disease-linked genes since the mid-1980s. Finally, in the last section of the chapter, we consider molecular techniques for introducing cloned genes, including deliberately mutated genes, into the genome of eukaryotes.

As the discussion in this and the previous chapter illustrates, the marriage of two genetic disciplines, classical genetics and recombinant DNA technology, forms a powerful approach for understanding biological function. Chapter 7 focused on the protein-to-gene strategy, that is, using knowledge about the sequence of a normal protein to prepare probes for isolating its corresponding gene from a library of cloned DNA. Most of this chapter focuses on the gene-to-protein strategy, that is, using mutation-defined genes to identify normal proteins:

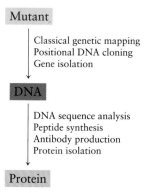

8.1 Mutations: Types and Causes

The development and function of an organism is in large part controlled by genes. Mutations can lead to changes in the structure of an encoded protein or to a decrease or complete loss in its expression. Because a change in the DNA sequence affects all copies of the encoded protein, mutations can be particularly damaging to a cell or organism. In contrast, any alterations in the sequences of RNA or protein molecules that occur during their synthesis are less serious because many copies of each RNA and protein are synthesized.

Geneticists often distinguish between the **genotype** and **phenotype** of an organism. Strictly speaking, the entire set

of genes carried by an individual is its genotype, whereas the function and physical appearance of an individual is referred to as its phenotype. However, the two terms commonly are used in a more restricted sense: genotype usually denotes whether an individual carries mutations in a single gene (or a small number of genes), and phenotype denotes the physical and functional consequences of that genotype.

Mutations Are Recessive or Dominant

A fundamental genetic difference between organisms is whether their cells carry a single set of chromosomes or two copies of each chromosome. The former are referred to as **haploid;** the latter, as **diploid.** Many simple unicellular organisms are haploid, whereas complex multicellular organisms (e.g., fruit flies, mice, humans) are diploid.

Different forms of a gene (e.g., normal and mutant) are referred to as **alleles.** Since diploid organisms carry two copies of each gene, they may carry identical alleles, that is, be **homozygous** for a gene, or carry different alleles, that is, be **heterozygous** for a gene. A **recessive** mutation is one in which both alleles must be mutant in order for the mutant phenotype to be observed; that is, the individual must be homozygous for the mutant allele to show the mutant phenotype. In contrast, the phenotypic consequences of a **dominant** mutation are observed in a heterozygous individual carrying one mutant and one normal allele (Figure 8-1).

Recessive mutations inactivate the affected gene and lead to a *loss of function.* For instance, recessive mutations may remove part of or all the gene from the chromosome, disrupt expression of the gene, or alter the structure of the encoded protein, thereby altering its function. Conversely, dominant mutations often lead to a *gain of function.* For example, dominant mutations may increase the activity of a given gene product, confer a new activity on the gene product, or lead to its inappropriate spatial and temporal expression. Dominant mutations, however, may be associated with a loss of function. In some cases, two copies of a gene are required for normal function, so that removing a single copy leads to mutant phenotype. Such genes are referred to as

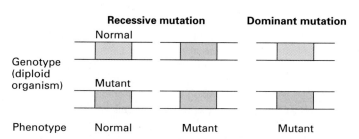

▲ **FIGURE 8-1 For a recessive mutation to give rise to a mutant phenotype in a diploid organism, both alleles must carry the mutation.** However, one copy of a dominant mutant allele leads to a mutant phenotype. Recessive mutations result in a loss of function, whereas dominant mutations often, but not always, result in a gain of function.

as *haplo-insufficient.* In other cases, mutations in one allele may lead to a structural change in the protein that interferes with the function of the wild-type protein encoded by the other allele. These are referred to as *dominant negative mutations.*

Some alleles can be associated with both a recessive and a dominant phenotype. For instance, fruit flies heterozygous for the mutant *Stubble (Sb)* allele have short and stubby body hairs rather than the normal long, slender hairs; the mutant allele is dominant in this case. In contrast, flies homozygous for this allele die during development. Thus the recessive phenotype associated with this allele is lethal, whereas the dominant phenotype is not.

Inheritance Patterns of Recessive and Dominant Mutations Differ

Recessive and dominant mutations can be distinguished because they exhibit different patterns of inheritance. To understand why, we need to review the type of cell division that gives rise to **gametes** (sperm and egg cells in higher plants and animals). The body (somatic) cells of most multicellular organisms divide by mitosis (see Figure 1-10), whereas the **germ cells** that give rise to gametes undergo **meiosis.** Like body cells, premeiotic germ cells are diploid, containing two of each morphologic type of chromosome. Because the two members of each such pair of **homologous chromosomes** are descended from different parents, their genes are similar but not usually identical. Single-celled organisms (e.g., the yeast *S. cerevisiae*) that are diploid at some phase of their life cycle also undergo meiosis (see Figure 10-54).

Figure 8-2 depicts the major events in meiosis. *One* round of DNA replication, which makes the cell 4*n*, is followed by *two* separate cell divisions, yielding four haploid (1*n*) cells that contain only one chromosome of each homologous pair. The apportionment, or **segregation,** of homologous chromosomes to daughter cells during the first meiotic division is random; that is, the maternally and paternally derived mem-

bers of each pair, called homologs, segregate independently, yielding germ cells with different mixes of paternal and maternal chromosomes. Thus parental characteristics are reassorted randomly into each new germ cell during meiosis. The number of possible varieties of meiotic segregants is 2^n,

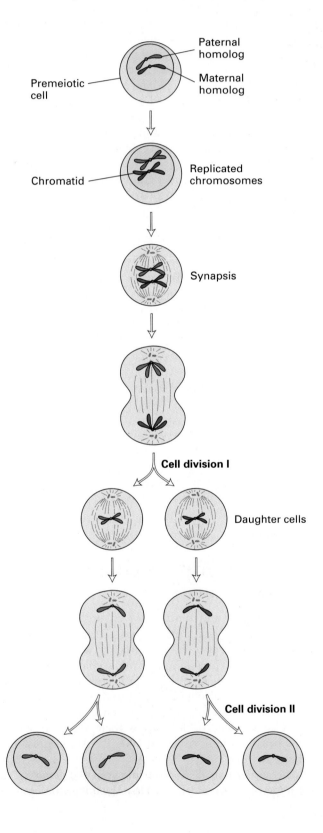

▶ **FIGURE 8-2 Meiosis.** A premeiotic germ cell has two copies of each chromosome (2*n*), one maternal and one paternal. Chromosomes are replicated during the S phase, giving a 4*n* chromosomal complement. During the first meiotic division, each replicated chromosome (actually two sister chromatids) aligns at the cell equator, paired with its homologous partner; this pairing off, referred to as *synapsis,* permits genetic recombination (discussed later). One homolog (both sister chromatids) of each morphologic type goes into one daughter cell, and the other homolog goes into the other cell. The resulting 2*n* cells undergo a second division without intervening DNA replication. During this second meiotic division, the sister chromatids of each morphologic type separate and these now independent chromosomes are randomly apportioned to the daughter cells. Thus, each diploid cell that undergoes meiosis produces four haploid cells, whereas each diploid cell that undergoes mitosis produces two diploid cells (see Figure 1-10).

(a)

Segregation of dominant mutation

A is the dominant mutant allele; *a* is the wild-type allele

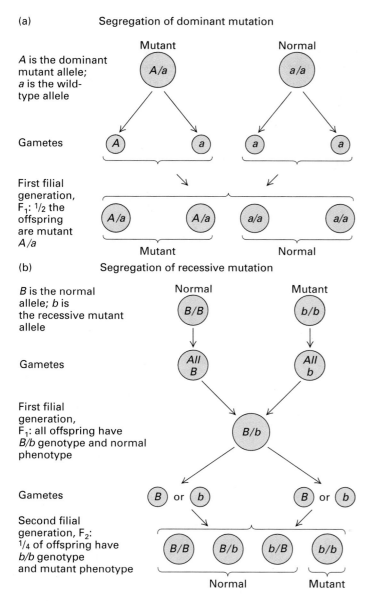

Gametes

First filial generation, F₁: ½ the offspring are mutant *A/a*

(b)

Segregation of recessive mutation

B is the normal allele; *b* is the recessive mutant allele

Gametes

First filial generation, F₁: all offspring have *B/b* genotype and normal phenotype

Gametes

Second filial generation, F₂: ¼ of offspring have *b/b* genotype and mutant phenotype

▲ **FIGURE 8-3 Segregation patterns of dominant and recessive mutations.** Crosses between genotypically normal individuals (blue) and mutants (yellow) that are heterozygous for a dominant mutation (a) or homozygous for a recessive mutation (b) produce different ratios of normal and mutant phenotypes in the F₁ generation. Although all the F₁ progeny from a cross between a normal individual and an individual homozygous for a recessive mutation will have a normal phenotype, one-quarter of the progeny from the intercross between F₁ progeny will have a mutant phenotype. Observation of segregation patterns like these led Gregor Mendel (1822–1884) to conclude that each gamete receives only one of the two parental alleles, a conclusion known as *Mendel's first law.*

where *n* is the haploid number of chromosomes. In the case of a single chromosome, as illustrated in Figure 8-2, meiosis gives rise to two types of gametes; one type carries the maternal homolog and the other carries the paternal homolog.

Now, let's see what phenotypes are generated by mating of wild-type individuals with mutants carrying either a dominant or a recessive mutation. As shown in Figure 8-3a, half the gametes from an individual heterozygous for a dominant mutation in a particular gene will have the wild-type allele, and half will have the mutant allele. Since fertilization of female gametes by male gametes occurs randomly, half the first filial (F₁) progeny resulting from the cross between a normal wild-type individual and a mutant individual carrying a single dominant allele will exhibit the mutant phenotype. In contrast, all the gametes produced by a mutant homozygous for a recessive mutation will carry the mutant allele. Thus, in a cross between a normal individual and one who is homozygous for a recessive mutation, none of the F₁ progeny will exhibit the mutant phenotype (Figure 8-3b). However, one-fourth of the progeny from parents both heterozygous for a recessive mutation will show the mutant phenotype.

Mutations Involve Large or Small DNA Alterations

A mutation involving a change in a single base pair, often called a **point mutation,** or a deletion of a few base pairs generally affects the function of a single gene (Figure 8-4a). Changes in a single base pair may produce one of three types of mutation:

• *Missense mutation,* which results in a protein in which one amino acid is substituted for another
• *Nonsense mutation,* in which a stop codon replaces an amino acid codon, leading to premature termination of translation
• *Frameshift mutation,* which causes a change in the **reading frame,** leading to introduction of unrelated amino acids into the protein, generally followed by a stop codon

Small deletions have effects similar to those of frameshift mutations, although one third of these will be in-frame and result in removal of a small number of contiguous amino acids.

The second major type of mutation involves large-scale changes in chromosome structure and can affect the functioning of numerous genes, resulting in major phenotypic consequences. Such *chromosomal mutations* (or abnormalities) can involve deletion or insertion of several contiguous genes, inversion of genes on a chromosome, or the exchange of large segments of DNA between nonhomologous chromosomes (Figure 8-4b).

Mutations Occur Spontaneously and Can Be Induced

Mutations arise spontaneously at low frequency owing to the chemical instability of purine and pyrimidine bases and to errors during DNA replication. Natural exposure of an organism to certain environmental factors, such as

(a) Point mutations and small deletions

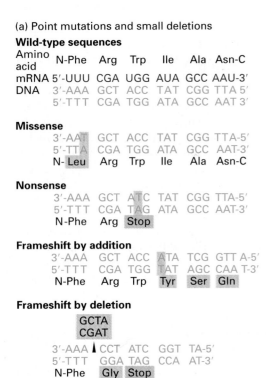

Wild-type sequences

| Amino acid | N-Phe | Arg | Trp | Ile | Ala | Asn-C |

mRNA 5'-UUU CGA UGG AUA GCC AAU-3'

DNA 3'-AAA GCT ACC TAT CGG TTA 5'
    5'-TTT CGA TGG ATA GCC AAT 3'

Missense

3'-AAT GCT ACC TAT CGG TTA-5'
5'-TTA CGA TGG ATA GCC AAT-3'
N- Leu Arg Trp Ile Ala Asn-C

Nonsense

3'-AAA GCT ATC TAT CGG TTA-5'
5'-TTT CGA TAG ATA GCC AAT-3'
N-Phe Arg Stop

Frameshift by addition

3'-AAA GCT ACC ATA TCG GTT A-5'
5'-TTT CGA TGG TAT AGC CAA T-3'
N-Phe Arg Trp Tyr Ser Gln

Frameshift by deletion

GCTA
CGAT

3'-AAA CCT ATC GGT TA-5'
5'-TTT GGA TAG CCA AT-3'
N-Phe Gly Stop

(b) Chromosomal abnormalities

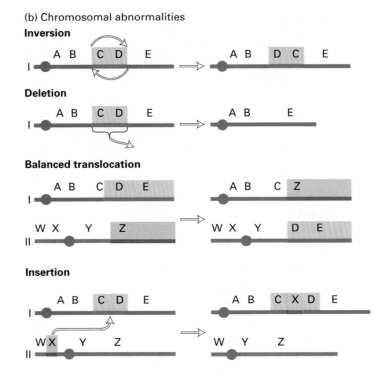

Inversion

Deletion

Balanced translocation

Insertion

▲ **FIGURE 8-4 Different types of mutations.** (a) Point mutations, which involve alteration in a single base pair, and small deletions generally directly affect the function of only one gene. A wild-type peptide sequence and the mRNA and DNA encoding it are shown at the top. Altered nucleotides and amino acid residues are highlighted in green. Missense mutations lead to a change in a single amino acid in the encoded protein. In a nonsense mutation, a nucleotide base change leads to the formation of a stop codon (purple). This results in premature termination of translation, thereby generating a truncated protein. Frameshift mutations involve the addition or deletion of any number of nucleotides that is not a multiple of three, causing a change in the reading frame. Consequently, completely unrelated amino acid residues are incorporated into the protein prior to encountering a stop codon. (b) Chromosomal abnormalities involve alterations in large segments of DNA. Presumably these abnormalities arise owing to errors in the mechanisms for repairing double-strand breaks in DNA. Chromosomes (I or II) are shown as single thick lines with the regions involved in a particular abnormality highlighted in green or purple. Inversions occur when a break is rejoined to the correct chromosome but in an incorrect orientation; deletions, when a segment of DNA is lost; translocations, when breaks are rejoined to the wrong chromosomes; and insertions, when a segment from one chromosome is inserted into another chromosome.

ultraviolet light and chemical carcinogens (e.g., aflatoxin B1), also can cause mutations.

A common cause of spontaneous point mutations is the deamination of cytosine to uracil in the DNA double helix. Subsequent replication leads to a mutant daughter cell in which a T·A base pair replaces the wild-type C·G base pair. Another cause of spontaneous mutations is copying errors during DNA replication. Although replication generally is carried out with high fidelity, errors occasionally occur. Figure 8-5 illustrates how one type of copying error can produce a mutation. In the example shown, the mutant DNA contains nine additional base pairs.

In order to increase the frequency of mutation in experimental organisms, researchers often treat them with high doses of chemical mutagens or expose them to ionizing radiation. Mutations arising in response to such treatments are referred to as *induced* mutations. Generally, chemical mutagens induce point mutations, whereas ionizing radiation gives rise to large chromosomal abnormalities.

Ethylmethane sulfonate (EMS), a commonly used mutagen, alkylates guanine in DNA, forming O^6-ethylguanine (Figure 8-6a). During subsequent DNA replication, O^6-ethylguanine directs incorporation of deoxythymidylate, not deoxycytidylate, resulting in formation of mutant cells in which a G·C base pair is replaced with an A·T base pair (Figure 8-6b). The causes of mutations and the mechanisms cells have for repairing alterations in DNA are discussed further in Chapter 12.

Some Human Diseases Are Caused by Spontaneous Mutations

Many common human diseases, often devastating in their effects, are due to mutations in single genes. Genetic diseases arise by spontaneous mutations in germ cells (egg and sperm), which are transmitted to future generations. For example, *sickle-cell anemia*, which affects 1 in 500 individuals of African descent, is caused by a single

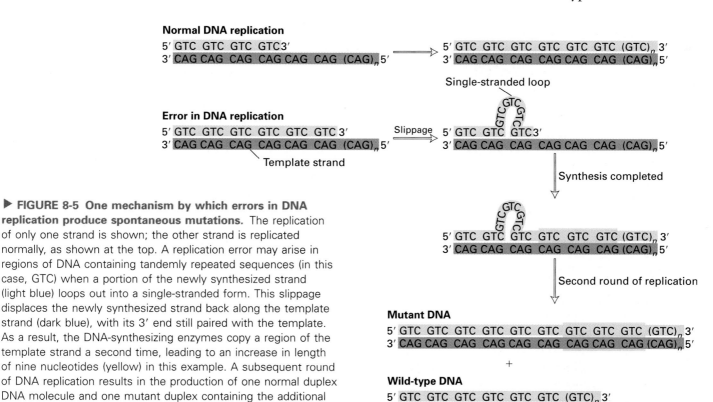

FIGURE 8-5 One mechanism by which errors in DNA replication produce spontaneous mutations. The replication of only one strand is shown; the other strand is replicated normally, as shown at the top. A replication error may arise in regions of DNA containing tandemly repeated sequences (in this case, GTC) when a portion of the newly synthesized strand (light blue) loops out into a single-stranded form. This slippage displaces the newly synthesized strand back along the template strand (dark blue), with its 3′ end still paired with the template. As a result, the DNA-synthesizing enzymes copy a region of the template strand a second time, leading to an increase in length of nine nucleotides (yellow) in this example. A subsequent round of DNA replication results in the production of one normal duplex DNA molecule and one mutant duplex containing the additional nucleotides.

missense mutation at codon 6 of the β-globin gene; as a result of this mutation, the glutamic acid at position 6 in the normal protein is changed to a valine in the mutant protein. This alteration has a profound effect on hemoglobin, the oxygen-carrier protein of erythrocytes, which consists of two α-globin and two β-globin subunits (see Figure 3-11). The deoxygenated form of the mutant protein is insoluble in erythrocytes and forms crystalline arrays. The erythrocytes of affected individuals become rigid and their transit through capillaries is blocked, causing severe pain and tissue damage. Because the erythrocytes of heterozygous individuals are resistant to the parasite causing malaria, which is endemic in Africa, the mutant allele has been maintained. It is not that individuals of African descent are more likely than others to acquire a mutation causing the sickle-cell defect, but rather the mutation has been maintained in this population by interbreeding.

Spontaneous mutation in somatic cells (i.e., non-germ-line body cells) also is an important mechanism in certain human diseases, including *retinoblastoma*, which is associated with retinal tumors in children (see Figure 24-11). The hereditary form of retinoblastoma, for example, results from a germ-line mutation in one *Rb* allele and a second somatically occurring mutation in the other *Rb* allele (Figure 8-7a). When an *Rb* heterozygous retinal cell undergoes somatic mutation, it is left with no normal allele; as a result, the cell proliferates in an uncontrolled manner, giving rise to a retinal tumor. A second form of this disease, called *sporadic retinoblastoma*, results from two independent mutations disrupting both *Rb* alleles (Figure 8-7b). Since only one somatic mutation is required for tumor development in children with hereditary retinoblastoma, it occurs at a much higher frequency than the sporadic form, which requires acquisition of two independently occurring somatic mutations. The Rb protein has been shown to play a critical role in controlling cell division (Chapter 13).

In a later section, we will see how normal copies of disease-related genes can be isolated and cloned.

SUMMARY Mutations: Types and Causes

- Diploid organisms carry two copies (alleles) of each gene, whereas haploid organisms carry only one copy.

- Mutations are alterations in DNA sequences that result in changes in the structure of a gene. Both small and large DNA alterations can occur spontaneously. Treatment with ionizing radiation or various chemical agents increases the frequency of mutations.

- Recessive mutations lead to a loss of function, which is masked if a normal copy of the gene is present. For the mutant phenotype to occur, both alleles must carry the mutation.

- Dominant mutations lead to a mutant phenotype in the presence of a normal copy of the gene. The phenotypes associated with dominant mutations may represent either a loss or a gain of function.

▶ **FIGURE 8-6 Induction of point mutations by ethylmethane sulfonate (EMS), a commonly used mutagen.** (a) EMS alkylates guanine at the oxygen on position 6 of the purine ring, forming O^6-ethylguanine (Et-G), which base-pairs with thymine. (b) Two rounds of DNA replication of a strand containing Et-G yields a mutant DNA in which a G·C base pair is replaced with an A·T pair. Cells also have repair enzymes that can remove the ethyl group from Et-G (Chapter 12).

(a)

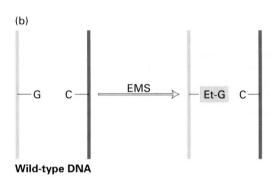

EMS   **Guanine** O^6-**Ethylguanine**   **Thymine**

(b)

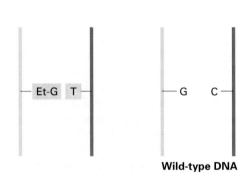

Wild-type DNA

Replication

Wild-type DNA

Replication

Mutant DNA

- In meiosis, a diploid cell undergoes one DNA replication and two cell divisions, yielding four haploid cells (Figure 8-2). The members of each pair of homologous chromosomes segregate independently during meiosis, leading to the random reassortment of maternal and paternal alleles in the gametes.

- Dominant and recessive mutations exhibit characteristic segregation patterns in genetic crosses (see Figure 8-3).

(a) Hereditary retinoblastoma

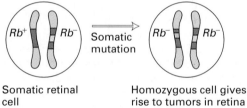

Somatic retinal cell

Homozygous cell gives rise to tumors in retina

(b) Sporadic retinoblastoma

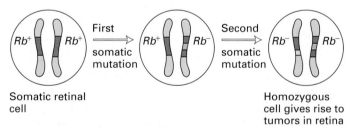

Somatic retinal cell

Homozygous cell gives rise to tumors in retina

▲ **FIGURE 8-7 Role of spontaneous somatic mutation in retinoblastoma, a childhood disease marked by retinal tumors.** Tumors arise from retinal cells that carry two mutant Rb^- alleles. (a) In hereditary retinoblastoma, a child receives a normal Rb^+ allele from one parent and a mutant Rb^- allele from the other parent. A single mutagenic event in a heterozygous somatic retinal cell that inactivates the normal allele will result in a cell homozygous for two mutant Rb^- alleles. (b) In sporadic retinoblastoma, a child receives two normal Rb^+ alleles. Two separate somatic mutations, inactivating both alleles in a particular cell, are required to produce a homozygous Rb^-/Rb^- retinal cell.

8.2 Isolation and Analysis of Mutants

Many different types of mutants have been identified in organisms ranging from bacteria to humans. Mutants can differ from their normal counterparts in a variety of ways. Some mutations cause only subtle changes; for example, certain mutations in the fruit fly, *Drosophila melanogaster*, result in failure of a single type of neuronal cell to develop, but mutant flies otherwise are normal. Other mutations lead to significant changes in development, cellular function, appearance (Figure 8-8), and behavior of an individual. Many mutations are nonlethal, but some result in organismal death.

The procedures used to identify and isolate mutants, referred to as *genetic screens*, depend on whether the experimental organism is haploid or diploid and, if the latter, whether the mutation is recessive or dominant. Usually, mutations are induced by treatment with a mutagen, and the mutagenized population subjected to a genetic screen designed to identify and isolate individuals with mutations affecting a particular process of interest. Genes that encode proteins essential for life are among the most interesting and important ones to study. Since phenotypic expression of mutations in essential genes leads to death of the individual, in-genious screens are needed to isolate and maintain organisms with a lethal mutation.

Characterization of mutants in a variety of experimental organisms has been used to investigate many different fundamental biological processes. Genetic analyses of mutants defective in a particular process can reveal: (a) the number of genes required for the process to occur; (b) the order in which gene products act in the process; and (c) whether the proteins encoded by different genes interact with one another. Genetic studies of this type have helped unravel various metabolic pathways, regulatory mechanisms, and developmental processes.

Temperature-Sensitive Screens Can Isolate Lethal Mutations in Haploids

In haploid organisms (e.g., prokaryotes and yeast), all mutations are in effect dominant so that the mutant phenotype is exhibited immediately in the progeny of the mutagenized population. For instance, mutations that disrupt arginine synthesis lead to cells that require arginine for growth. Such mutations are easily detected by growing mutagenized populations in the presence and absence of arginine (see Figure 6-2).

In prokaryotes and haploid eukaryotes such as yeast, essential genes can be studied through the use of *conditional*

(a)

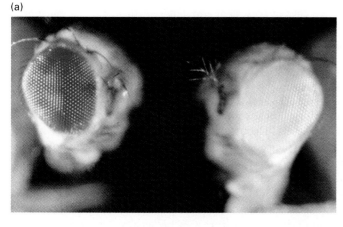

(b)

Normal

Ubx **mutant**

◄ **FIGURE 8-8 Mutants in *Drosophila*.** (a) White-eyed mutants lack the pigment that is present in the normal bright red eyes of fruit flies. The *white* mutant, the first known *Drosophila* mutant, was identified in the early part of the twentieth century. (b) A normal fly has a single pair of wings, which arise from the second thoracic segment, whereas mutation in the *Ultrabithorax (Ubx)* gene leads to flies with a second pair of wings arising from the third thoracic segment. As discussed in Chapter 14, study of *Ubx* mutants has led to remarkable progress in understanding the mechanisms that pattern not only the body of the fruit fly but also the body of mammals. [Part (a) from A. J. F. Griffiths et al., 1996, *An Introduction to Genetic Analysis,* 6th ed., W. H. Freeman and Company, p. 66. Part (b) from E. B. Lewis, 1978, *Nature* **276**:565; photographs courtesy of E. B. Lewis. Reprinted by permission from *Nature;* copyright 1978 Macmillan Journals Limited.]

(a)

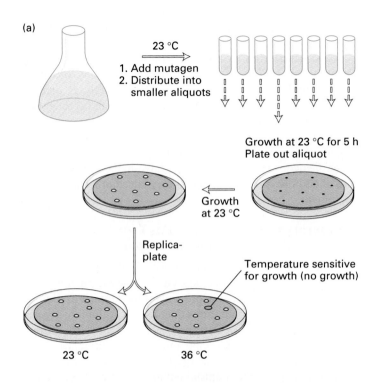

23 °C
1. Add mutagen
2. Distribute into smaller aliquots

Growth at 23 °C for 5 h
Plate out aliquot

Growth at 23 °C

Replica-plate

Temperature sensitive for growth (no growth)

23 °C   36 °C

(b)

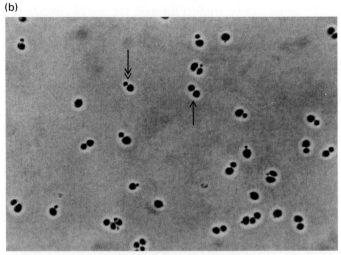

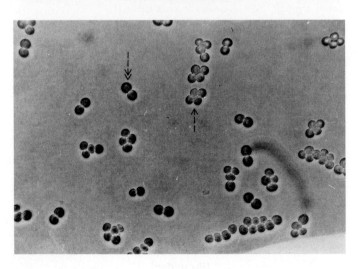

◄ **FIGURE 8-9 Two-step genetic screen used to identify cell-cycle mutants in yeast.** (a) Yeast cells were grown in a large liquid culture, treated with a chemical mutagen, and then subcultured into smaller aliquots. After a 5-h growth period at 23 °C, aliquots from each tube were separately plated onto agar-containing petri dishes and incubated at 23 °C. Colonies that developed on these plates were replica-plated onto two plates: one was incubated at the permissive temperature (23 °C); the other, at the nonpermissive temperature (36 °C). The temperature-sensitive colonies that grew at 23 °C but not at 36 °C were assessed to determine whether they were blocked at specific stages in the cell cycle. (b) Time-lapse photography of temperature-sensitive mutants identified in (a) permitted detection of yeast mutants with cell-cycle defects. The top photograph shows *cdc13* mutant cells growing at the permissive temperature just prior to being shifted to the nonpermissive temperature. The bottom photograph shows exactly the same field after a 6-h incubation at the nonpermissive temperature of 36 °C. By comparison of the size of the buds in the top photograph with the morphology of the corresponding dividing cells in the bottom it is possible to determine the stage in the cell cycle in which the mutated gene is required. For example, when the bud is large (single-headed arrow), the cells will divide at the nonpermissive temperature but then are blocked prior to the next cell division; they thus appear in the bottom photograph as a cluster of four cells of equivalent size. In contrast, cells with very small buds (double-headed arrow) fail to divide at the nonpermissive temperature and appear in the bottom photograph as two cells of equivalent size. Later studies on these mutants indicated that they are blocked in a specific stage of G_2 and cannot progress through mitosis. [Part (a) see L. H. Hartwell, 1967, *J. Bacteriol.* **93**:1662; part (b) from J. Culotti and L. H. Hartwell, 1971, *Exp. Cell Res.* **67**:391.]

mutations. For instance, a mutant protein may be fully functional at 30 °C but completely inactive at 37 °C, whereas the normal protein would be fully functional at both temperatures. A temperature at which the mutant phenotype is observed is called *nonpermissive;* a *permissive* temperature is one at which the phenotype is not observed. Mutant strains can be maintained at a permissive temperature; then, for analysis, a subculture can be set up at a nonpermissive temperature. Such **temperature-sensitive (ts) mutants** can also be generated in *Drosophila* and *C. elegans,* but cannot be isolated in warm-blooded animals.

An example of a particularly important temperature-sensitive screen in the yeast *S. cerevisiae* comes from the studies of L. H. Hartwell and colleagues in the late 1960s and early 1970s. They set out to identify genes important in regulation of the cell cycle. Cell division in this yeast occurs through a budding process, and the size of the bud, which is easily visualized by light microscopy, is an indication of the cell's position in the cell cycle. In these studies, the researchers first identified mutagenized yeast cells that did not grow at 36 °C (Figure 8-9a). Then they used video microscopy to analyze the identified mutants for cell-division defects at the nonpermissive temperature (Figure 8-9b).

These yeast mutants were not simply slow growing as they might be if they carried a mutation affecting general cellular metabolism; rather, they grew normally but showed a stage-specific block in growth at the nonpermissive temperature. The cell-cycle stage at which cell growth was arrested at the nonpermissive temperature indicated when the protein encoded by the mutated gene was required. Cloning and analysis of various genes defined by cell-cycle mutations are described in detail in Chapter 13. This work has provided important insights about the regulation of cell division in organisms ranging from yeast to humans.

Recessive Lethal Mutations in Diploids Can Be Screened by Use of Visible Markers

In diploid organisms, phenotypes resulting from recessive mutations can be observed only in individuals homozygous for the mutant alleles. Figure 8-10 outlines a procedure for inducing, identifying, and maintaining recessive lethal mutations in *Drosophila*, a diploid organism. Male fruit flies are treated with a mutagen and then mated with females, yielding F_1 progeny that are heterozygous for any induced mutations. Because these mutations are recessive, the mutant phenotype is not observed in the F_1 generation, and two additional crosses are needed to reveal the mutant phenotype. By using fly strains carrying known mutations (called *markers*) that give rise to

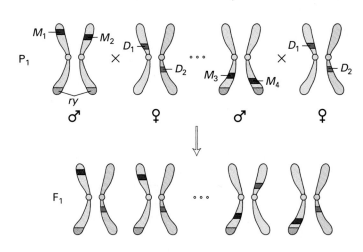

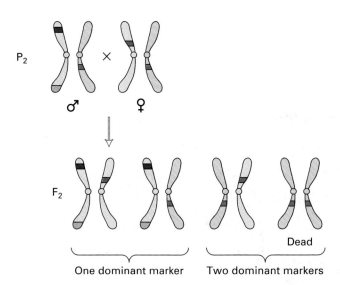

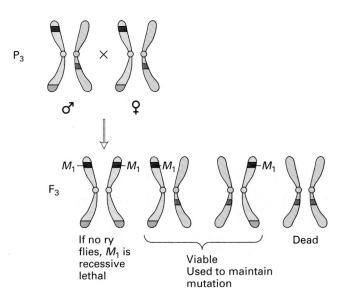

▶ **FIGURE 8-10 Procedure used to identify and maintain recessive lethal mutations on chromosome 3 (an autosome) in *Drosophila*, a diploid organism.** This approach requires three sequential crosses. First, many males are treated with a mutagen (e.g., EMS), producing flies carrying various mutations (M_1, M_2, etc.) in their germ-line cells (sperm). The level of mutagen used is usually sufficient to induce at least one mutation on each chromosome. These males also carry a nonlethal recessive mutation that gives rise to a visible phenotype in homozygotes; the marker in this example is called *rosy (ry)* and confers an altered eye color. In the first cross (P_1), the mutagenized males are mass-mated to a large number of females, traditionally in pint-size milk bottles. The females carry dominant visible mutations (D_1 and D_2) on chromosome 3; these are nonlethal in heterozygotes but are lethal in homozygotes. In the second cross (P_2), individual heterozygous F_1 males carrying mutagenized chromosome 3 are mated individually to nonmutagenized females in small culture vials. The F_2 progeny homozygous for either dominant marker will die; those heterozygous for both markers are easily identified and excluded. The remaining F_2 progeny include males and females that have the identical mutagenized chromosome carrying the *ry* marker and one nonmutagenized chromosome carrying a single dominant visible marker. These heterozygous brothers and sisters are mated individually in the third cross (P_3). The absence of flies with rosy-colored eyes in the F_3 progeny indicates the presence of an induced lethal mutation (M_1 in this example) on chromosome 3. Although flies homozygous for M_1 do not survive, heterozygotes carrying M_1 on one chromosome and one of the dominant markers on the other will survive.

visible phenotypes, researchers can distinguish heterozygous F_2 progeny carrying one mutagenized chromosome and one normal chromosome from siblings with other genotypes. Mating of these F_2 heterozygous siblings produces an F_3 generation in which one-fourth of the flies will be homozygous for any mutation induced on the mutagenized chromosome, and if the mutation is in a gene essential for viability, they will not survive; one-fourth will be homozygous for the normal allele; and half will be heterozygous. The effects of the mutation can then be assessed in the homozygous class that does not survive, and the mutation can be maintained in the flies that are heterozygous for the mutation.

Current understanding of the molecular mechanisms regulating development of multicellular organisms is based, in large part, on this type of genetic screen in *Drosophila*. C. Nüsslein-Volhard, E. Wiechaus, and their colleagues systematically screened for recessive lethal mutations affecting embryogenesis in *Drosophila* using a scheme similar to that shown in Figure 8-10. Dead homozygous embryos carrying lethal recessive mutations identified by this screen were analyzed for specific defects in the cuticular structures on the embryo surface (Chapter 14). A detailed picture of embryonic

development has emerged from the characterization of these defects and the analysis of both the structure of the encoded proteins and their patterns of expression during embryogenesis. We will discuss some of the fundamental discoveries based on these genetic studies in Chapters 14 and 23.

Complementation Analysis Determines If Different Mutations Are in the Same Gene

A common type of genetic analysis can reveal whether different recessive mutations associated with the same phenotype are in the same gene or in different genes. This analysis depends on the phenomenon of genetic **complementation**, that is, the restoration of the wild-type phenotype by mating of two different mutants. If two mutations, *A* and *B*, are in the *same* gene, then a diploid organism heterozygous for both mutations (i.e., carrying one *A* allele and one *B* allele) will exhibit the mutant phenotype. In contrast, if mutation *A* and *B* are in *separate* genes, then heterozygotes carrying a single copy of each mutant allele will exhibit the wild-type (normal) phenotype. In this case, the mutations are said to *complement* each other.

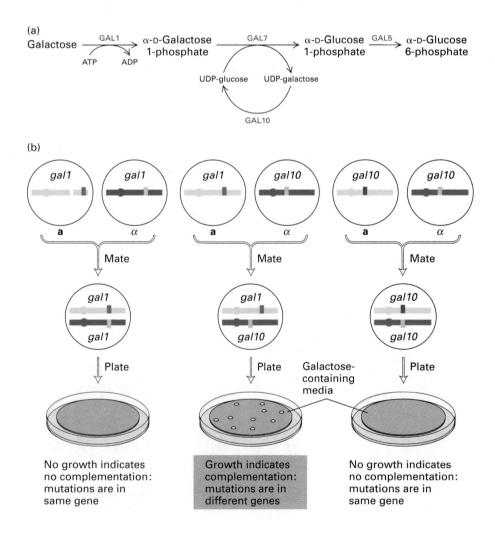

◀ **FIGURE 8-11 Complementation analysis in *S. cerevisiae*.** (a) Pathway used by yeast cells to metabolize galactose to glucose, which then enters the glycolytic pathway. Yeast cells must produce all four enzymes (red) in order to grow on galactose. GAL1=galactokinase; GAL7=galactose 1-phosphate uridyl transferase; GAL10=UDP-galactose 4-epimerase; GAL5=phosphoglucomutase. (b) Complementation tests can be performed with yeast by mating haploid **a** and **α** cells to produce diploid cells (see Figure 10-54). This example shows the results that would be obtained in complementation tests of Gal⁻ strains carrying different mutations (indicated by vertical colored lines) in the *GAL1* and *GAL10* genes, which encode two different enzymes required for galactose metabolism. Both of these genes are located on yeast chromosome II.

Complementation analysis of a set of mutants exhibiting the same phenotype can distinguish the individual genes in a set of functionally related genes, all of which must function to produce a given phenotypic trait. In the yeast *S. cerevisiae*, for example, four enzymes are required for growth on galactose (Figure 8-11a). If any one of these enzymes is absent or defective, yeast cells cannot grow on galactose. Because haploid yeast cells exist in one of two different mating types, **a** or α, which can be mated to yield **a**/α diploids, yeast can be subjected to complementation analysis like other diploid organisms. Figure 8-11b illustrates complementation analysis of Gal$^-$ yeast strains defective for growth on galactose. When Gal$^-$ strains with mutations in different *GAL* genes are mated, the resulting diploid cells will grow on galactose, because the wild-type gene in each strain will compensate for the genetic defect in the other. In contrast, diploids formed from Gal$^-$ strains that are mutated in the same gene will not grow on galactose.

Metabolic and Other Pathways Can Be Genetically Dissected

Various types of analysis can order the genes involved in biochemical pathways and other cellular processes. A fairly straightforward example involves the genetic dissection of the biochemical pathway for synthesis of arginine in the bread mold *Neurospora crassa*. Four different mutant strains that are unable to synthesize arginine and require arginine for growth (called *arginine* auxotrophs) were identified years ago. Each of the steps in biosynthesis of arginine is catalyzed by an enzyme encoded by a separate gene. The order of action of the different genes, hence the order of the biochemical reactions in the pathway, was determined by assessing which mutants could grow on different intermediates (Figure 8-12). Numerous biochemical pathways have been dissected by this type of study.

Other types of cellular processes also are amenable to genetic analysis. For example, the maturation pathway for secretory proteins in yeast has been dissected and ordered by analysis of a set of conditional temperature-sensitive secretion-defective *(sec)* mutants. In these mutant strains, the secretion of all proteins is blocked at the higher (nonpermissive) temperature but is normal at the lower (permissive) temperature. At the higher temperature, *sec* mutants accumulate proteins in the rough endoplasmic reticulum (ER), Golgi complex, or secretory vesicles (see Figure 17-14). At least 60 gene products are required to complete the maturation pathway as defined by the number of genes in which mutations give rise to a secretion defect. The genes can be ordered in a pathway by analyzing double-mutant combinations of *sec* genes. For instance, when ER and Golgi accumulating mutants are combined, proteins accumulate in the ER. These types of studies have shown that the pathway must be ordered in the following sequence: rough ER → Golgi → secretory vesicles. This maturation pathway is believed to apply to all secretory proteins in all eukaryotic organisms, including plants.

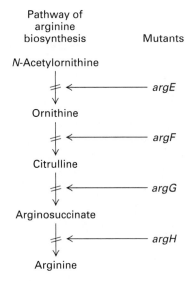

▲ **FIGURE 8-12 Genetic dissection of the arginine biosynthetic pathway.** Mutants in the bread mold *N. crassa* that required arginine for growth were identified many years ago; analysis of these mutants provided the first genetic dissection of a biochemical pathway. By complementation analysis four separate genes were shown to be required for arginine biosynthesis. These were ordered into a pathway by assessing the ability of different mutants to grow on different biosynthetic intermediates. For instance, *argE* mutants can grow on ornithine (or any other intermediates later in the pathway) because their block precedes ornithine in the pathway and is thus bypassed (i.e., they are supplied the intermediates their block would have prevented). On the other hand *argH* mutants, whose block precedes the final product in the pathway, cannot be supplemented by anything but arginine itself.

Suppressor Mutations Can Identify Genes Encoding Interacting Proteins

The phenomenon of genetic suppression can be used to identify proteins that specifically interact with one another in the living cell. The underlying logic is as follows: point mutations may lead to structural changes in protein A that disrupt its ability to associate with another protein (protein B) involved in the same cellular process. Similarly, mutations in protein B might lead to small structural changes that would inhibit its ability to interact with protein A. In rare cases small structural changes in protein A may be suppressed by compensatory changes in protein B. In these rare cases, strains carrying a specific mutant allele of protein A or B would be mutant, but strains carrying both would be normal. This is analogous to changes made in a lock and key.

Identification of such **suppressor mutations** has been elegantly applied in studies of the cytoskeletal protein actin in yeast (Figure 8-13). A strain of yeast that was temperature-sensitive for growth and carried a mutant actin allele called *act1-1* was plated at the nonpermissive temperature. A few cells were capable of growth at this temperature; these *revertants* were shown to have a second mutation in another gene, called *SAC6*, that allowed the *act1-1* mutants to grow.

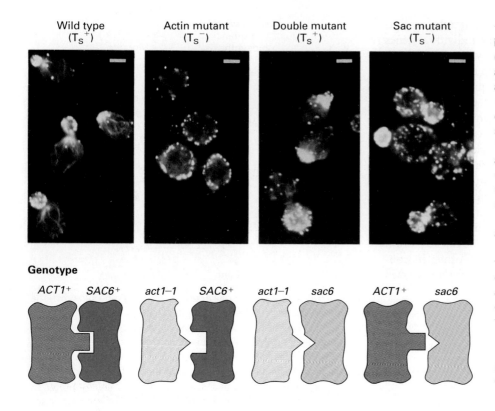

Wild type (Ts⁺)   **Actin mutant (Ts⁻)**   **Double mutant (Ts⁺)**   **Sac mutant (Ts⁻)**

Genotype

ACT1⁺ SAC6⁺   act1–1 SAC6⁺   act1–1 sac6   ACT1⁺ sac6

◄ **FIGURE 8-13 Genetic suppression in yeast involving temperature-sensitive (Ts⁻) mutations in the actin gene (ACT1) and in the SAC6 gene, which encodes an actin-binding protein.** *(Top)* Immunofluorescence micrographs of wild-type yeast, two Ts⁻ single-mutant strains, and a double-mutant revertant. In wild-type yeast, actin is distributed asymmetrically, whereas in the actin mutant (genotype=act1–1 SAC6⁺) and the Sac mutant (ACT1⁺ sac6), it is distributed randomly. The double-mutant revertant (act1–1 sac6) grows at high temperature and shows the wild-type actin distribution. *(Bottom)* Schematic diagrams interpreting these results in terms of compensatory changes in the structure of actin (orange) and the SAC protein (green). The inability of one mutant and one wild-type protein to interact leads to the mutant phenotype. However, the productive mutant-mutant interaction suppresses the mutant phenotype. [Photographs from A. E. M. Adams et al., 1989, *Science* **243**:231.]

This *sac6* mutation acted as a dominant suppressor of the *act1-1* mutation, so that the double mutants (*act1-1 sac6*) exhibited the wild-type phenotype. This suppression was found to be allele-specific; that is, the *sac6* mutation suppressed the *act1-1* mutation but not other *act1* mutations. Single mutants carrying any one of several different *sac6* mutations were, like *act1-1* mutants, temperature-sensitive for growth. Remarkably, some *act1* mutations were found to be dominant suppressors of the recessive temperature-sensitive lethality of various *sac6* mutations.

In summary, then, mutations in either the *SAC6* or *ACT1* gene confer the same recessive lethal phenotype, and specific lethal alleles of each gene can act as dominant suppressors of specific lethal alleles of the other gene, resulting in a viable organism. This reciprocal suppression argues strongly for a direct interaction in vivo between the proteins encoded by the two genes (see Figure 8-13, *bottom*). Indeed, biochemical studies have shown that these two proteins—ACT1 and SAC6—do interact, and immunolocalization studies indicate that the two proteins are present in the same part of the cell.

SUMMARY Isolation and Analysis of Mutants

• Treating experimental organisms with mutagens can produce mutations disrupting organismal function, development, and viability.

• With appropriate genetic screens, lethal mutations can be isolated and maintained in both haploid organisms (see Figure 8-9a) and diploid organisms (see Figure 8-10).

• The number of functionally related genes involved in a process can be defined by complementation analysis (see Figure 8-11). The order in which their encoded proteins act can be deduced from other types of genetic analysis.

• The identification of allele-specific suppressor mutations in two genes suggests that the encoded proteins interact in vivo.

8.3 Genetic Mapping of Mutations

Although genetic analysis can provide important insights into various cellular processes, its most profound impact on molecular cell biology has been to facilitate identification of the proteins that actually carry out these processes. Genetic mapping of a mutation-defined gene is the first step toward isolating and cloning the corresponding normal gene and ultimately identifying its encoded protein. Various techniques are used to produce a *genetic map* of a chromosome, which indicates the positions of genes relative to one another along the length of the chromosome. In a *physical map*, the number of nucleotides between known genes is indicated.* By comparing a genetic

---

*The nucleotide sequence of a chromosome, or portion thereof, constitutes its complete physical map. More common are partial physical maps, such as a set of ordered DNA clones and maps of restriction sites in which the nucleotide distances between sites are known.

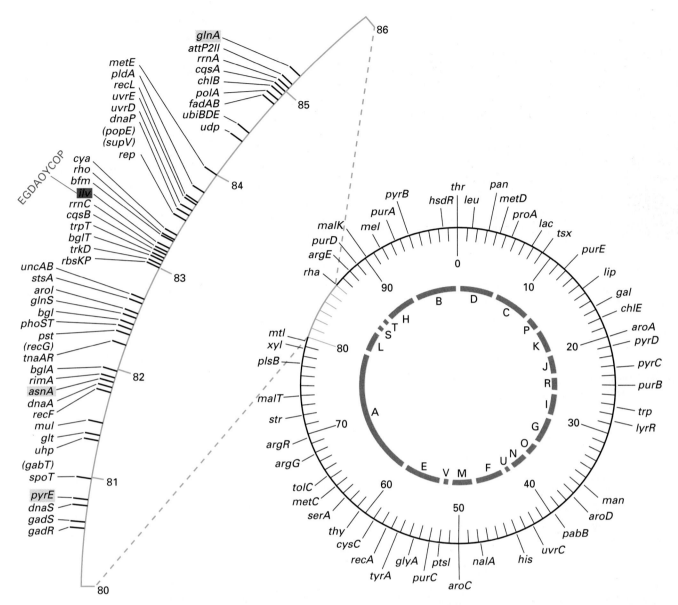

▲ FIGURE 8-14 Genetic and physical maps of the circular chromosome of *E. coli.* The abbreviations on the genetic map (outside circle and enlarged segment) relate to the metabolic properties of the gene products. For example, *pyrE, asnA,* and *glnA* (blue) were each identified in a bacterial clone carrying a mutation that made the addition of pyrimidine, asparagine, or glutamine, respectively, necessary for growth. A mutation-defined genetic trait is termed a *marker* for a strain that exhibits it. A group of such markers can define a strain with great precision. In some regions of the *E. coli* genome, every gene has been mapped. An example is the *ilv* locus (red), which encodes enzymes that catalyze biosynthesis of isoleucine, leucine, and valine; each red capital letter represents a gene encoding a known purified enzyme. The numbers inside the genetic map indicate the approximate percentages of the distances around the circular *E. coli* genome, based on recombination frequencies. The green inner circle represents a physical map of the *E. coli* DNA based on mapping of the restriction sites recognized by *Not*I (Chapter 7). The entire length of the *E. coli* genome is about 4000 kilobases (kb). Thus one genetic map unit, the distance equivalent to a 1 percent recombination frequency, corresponds to about 40 kb. [After C. L. Smith et al., 1986, *Science* **236**:1448; M. Riley, 1993, *Microbiol. Rev.* **57**:862.]

map and corresponding physical map, the actual physical position of any gene can be determined. Figure 8-14 depicts the genetic and physical maps of the single circular chromosome in *E. coli.* We discuss genetic mapping in this section and construction of physical maps in the next section.

Segregation Patterns Indicate Whether Mutations Are on the Same or Different Chromosomes

As discussed earlier, during meiosis each chromosome segregates independently. Therefore, traits controlled by genes on separate chromosomes also segregate independently (Figure 8-15). The observation that two mutant traits segregate independently indicates that the mutations are located on different chromosomes. Conversely, mutant gene loci that segregate together at a higher frequency than predicted by random assortment of chromosomes indicate that these loci are on the same chromosome; such loci are referred to as *linked*. Recombination between loci on the same chromosome provides a basis for mapping them relative to each other along the length of the chromosome, as discussed below.

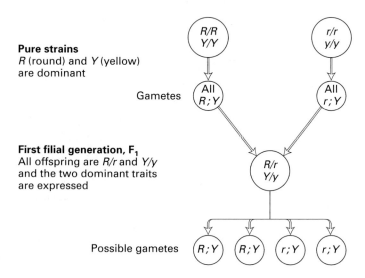

Pure strains
R (round) and *Y* (yellow) are dominant

Gametes

First filial generation, F₁
All offspring are *R/r* and *Y/y* and the two dominant traits are expressed

Possible gametes

Possible combinations in second filial generation, F₂
Each gametic type crossed with each other type
Plants are obtained in the following 9:3:3:1 ratio: 9 round and yellow, 3 round and green, 3 wrinkled and yellow, and 1 wrinkled and green

▲ **FIGURE 8-15 When Mendel crossed strains of peas that were pure (homozygous) for each of two traits—say, round (R/R) and yellow (Y/Y) peas with wrinkled (r/r) and green (y/y) peas—the plants of the F₂ generation exhibited four different combinations of the two traits, which always appeared in the ratio 9:3:3:1.** Mendel accounted for this ratio by proposing that the two traits (form and color) were assorted among the offspring independently of each other. We now know that such independently segregating traits are controlled by genes on separate chromosomes. This conclusion is known as *Mendel's second law*.

Chromosomal Mapping Locates Mutations on Particular Chromosomes

The localization of a mutation to a particular chromosome is the first step in genetic mapping of a mutation-defined gene. A simple example involves recessive mutations on the X chromosome in *Drosophila*. (In fruit flies, as in humans, males are genotypically XY and females are XX.) Such *X-linked recessive* mutations exhibit a distinctive *sex-linked* segregation pattern in various crosses (Figure 8-16). Mutant males mated to normal, homozygous females produce no phenotypically affected progeny. If the female is homozygous for the mutation, all the sons will have the mutant phenotype and all the daughters will be heterozygous and thus unaffected. If the females are heterozygous, they are phenotypically normal, but they act as *carriers*, transmitting the mutant allele to 50 percent of their male progeny. Thus any recessive mutation for which these sex-linked segregation patterns are observed can be mapped to the X chromosome.

 Human genetic diseases exhibit several inheritance patterns depending on the nature of the mutations that cause them. For example, Duchenne muscular dystrophy (DMD), a muscle degenerative disease that specifically affects males, is caused by a recessive mutation on the X chromosome. This clinically important mutation exhibits the typical sex-linked segregation pattern (Figure 8-17a).

In contrast to DMD, cystic fibrosis results from a recessive mutation in an **autosome**. Such *autosomal recessive* mutations exhibit a quite different segregation pattern. First, both males and females can be affected with cystic fibrosis. Second, both parents must be heterozygous carriers of a mutant *CFTR* allele in order for their children to be at risk of being affected with the disease. Each child of heterozygous parents has a 25 percent chance of receiving both mutant *CFTR* alleles and thus being affected; a 50 percent chance of receiving one normal and one mutant allele and thus being a carrier, and a 25 percent chance of receiving two normal alleles (Figure 8-17b).

Autosomal dominant mutations are associated with a third segregation pattern. Huntington's disease, a neural degenerative disease that generally strikes in mid to late life, is caused by this type of mutation. If either parent carries a mutant *HD* allele, each of his or her children (regardless of sex) has a 50 percent chance of inheriting the mutant allele and being affected (Figure 8-17c).

Mapping of a gene to a specific autosome is more complicated than for X-linked genes. Once one mutation has been mapped to a particular chromosome, it can be used as a marker to identify other mutations located on that chromosome. In humans, small differences, or polymorphisms, in DNA sequences between individuals serve as molecular markers.

(a) Mutant male × wild-type homozygous female

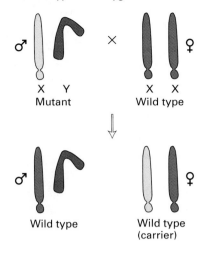

(b) Wild-type male × mutant female

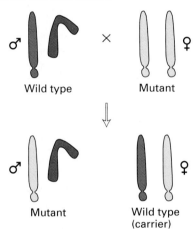

(c) Wild-type male × heterozygous female (carrier)

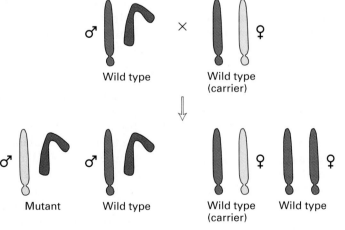

▲ **FIGURE 8-16 Unique segregation pattern of X-linked recessive traits.** In this example, segregation of mutant X chromosomes (yellow) and of wild-type X chromosomes (blue) is shown. Observation of this segregation pattern for a mutant phenotypic trait maps the gene affected by the mutation to the X chromosome.

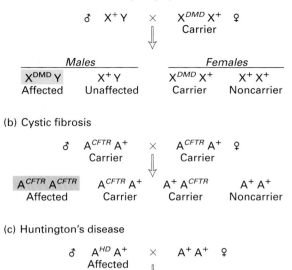

▲ **FIGURE 8-17 Segregation patterns for three human genetic diseases.** Wild-type chromosomes are indicated by superscript plus sign. (a) Duchenne muscular dystrophy is caused by a recessive mutation in the DMD gene on the X chromosome, which exhibits the typical sex-linked segregation pattern. Males born to mothers heterozygous for a DMD mutation have a 50 percent chance of inheriting the mutant allele and being affected. Females born to heterozygous mothers have a 50 percent chance of being carriers. (b) Cystic fibrosis results from a recessive mutation on an autosome (A). Both males and females can be affected but must carry two mutant alleles of the *CFTR* gene to get the disease. Both parents must be heterozygous for their children to be at risk of being affected or being carriers. (c) Huntington's disease is caused by an autosomal dominant mutation. Both males and females can be affected, and only one mutant allele is needed to confer the disease. If either parent is heterozygous for the mutant *HD* allele, his or her children have a 50 percent chance of inheriting the mutant allele and getting the disease.

Recombinational Analysis Can Map Genes Relative to Each Other on a Chromosome

As we've seen, the independent segregation of chromosomes during meiosis provides the basis for determining whether genes are on the same or different chromosomes. Genetic traits that segregate together during meiosis more frequently than expected from random segregation are inferred to be controlled by genes on the same chromosome. (The tendency of genes on the same chromosome to be inherited together is referred to as genetic **linkage**.) Genetic **recombination**, another phenomenon associated with meiosis, provides the basis for mapping a particular gene relative to other genes on the *same* chromosome. Before the first meiotic division, the replicated chromosomes of each homologous pair align with each other, an act called *synapsis* (see Figure 8-2). At this time, the exchange of DNA sequences—that is, recombination—between maternally and paternally derived chromatids can occur (Figure 8-18). This swapping of genetic material between chromosomes is called *crossing over*. The two meiotic phenomena, random segregation and recombination, generate new combinations of genes in interbreeding populations.

The technique of recombinational mapping was devised one night in 1913 by A. Sturtevant while he was an undergraduate working in the laboratory of T. H. Morgan at Columbia University. Originally used in studies on *Drosophila*, this technique is still used today to assess the distance between two genetic loci on the same chromosome in many experimental organisms. The molecular mechanisms underlying meiotic recombination are detailed in Chapter 12. A critical first step in this process is the introduction of breaks in the DNA; the reciprocal exchange of the chromosomal segments occurs at the position of the breaks. These breaks are thought to occur at random along the length of the DNA; thus the farther apart two genes are, the more likely that a DNA break will occur between them at meiosis. In other words, *the more frequently recombination occurs between two genes on the same chromosome, the farther apart they are.* Moreover, the frequency of exchange between two points along the length of a chromosome, called the *recombination frequency*, is proportional to the distance in base pairs separating the two points.

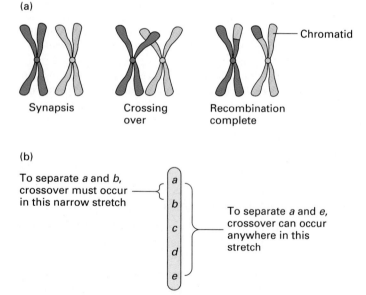

$$\text{Genetic map distance} = \frac{\text{recombinants}}{\text{recombinants} + \text{parentals}} \times 100$$

$$= \frac{104 + 93}{1000} \times 100 = 19.7 \text{ centimorgans}$$

▲ **FIGURE 8-19 Determination of the genetic map distance between two *Drosophila* loci by recombinational analysis.** In *Drosophila*, meiotic recombination occurs only in females. Here, the distance between two recessive mutations causing curled wings *(cu)* and darkly pigmented, or ebony, bodies *(e)* is determined. Wild-type alleles are indicated with a superscript plus sign, mutant alleles with superscript minus sign. Mutant chromosomal regions are diagrammed as light green lines, wild-type regions as dark blue lines. The recombination frequency between *cu* and *e* is 19.7 percent, which by definition is equivalent to a map distance of 19.7 centimorgans.

(a)

Synapsis   Crossing over   Recombination complete

Chromatid

(b)

To separate *a* and *b*, crossover must occur in this narrow stretch

a
b
c
d
e

To separate *a* and *e*, crossover can occur anywhere in this stretch

▲ **FIGURE 8-18 Recombination during meiosis.** (a) Crossing over occurs between chromatids of homologous chromosomes aligned in parallel (synapsis) during metaphase preceding the first meiotic division. (b) The longer the distance between two genes on a chromatid, the more likely they are to be separated by crossing over.

The presence of many different already mapped genetic traits, or markers, distributed along the length of a chromosome permits the position of an unmapped mutation to be determined by assessing its segregation with respect to these marker genes during meiosis. Thus the more markers that are available, the more precisely a mutation can be mapped. As more and more mutations are mapped, the linear order of genes along the length of a chromosome, that is, a genetic map, can be constructed. The more genes that are localized along a chromosome, the more detailed the map. For example, the single *E. coli* chromosome has been mapped in great detail (see Figure 8-14). By convention, one *genetic map unit* is defined as the distance between two positions along a chromosome that results in one recombinant individual in 100 progeny. The distance corresponding to this 1 percent recombination frequency is called a *centimorgan* (cM) in honor of Sturtevant's mentor, Morgan. Determination of the genetic map distance between two loci in *Drosophila* by recombinational analysis is illustrated in Figure 8-19.

Comparison of the physical distances between known genes, determined by molecular analysis, with their recombination frequency indicates that in *Drosophila* a 1 percent recombination frequency on average represents a distance of about 400 kilobases. The relationship between recombination frequency (i.e., genetic map distance) and physical distance varies between different regions of the genome within the same organism and between organisms (e.g., in *E. coli*, 1 cM = ≈40 kb).

DNA Polymorphisms Are Used to Map Human Mutations

In the experimental organisms commonly used in genetic studies, many phenotypic markers are readily available for genetic mapping of mutations. This is not the case for mapping loci associated with genetically transmitted diseases in humans. However, through recombinant DNA technology many molecular DNA markers are now available. One of the first types of commonly used DNA markers to be used is referred to as a *restriction fragment length polymorphism* (RFLP). RFLPs are variations among individuals in the length of restriction fragments produced from identical regions of the genome.

In all organisms natural DNA sequence variations occur throughout the genome. Because most of the human genome does not code for protein, a large amount of sequence variation is acceptable in humans (Chapter 9). Indeed, it has been estimated that in humans nucleotide differences between individuals can be detected every 200 nucleotides or so. These variations in DNA sequence, referred to as *DNA polymorphisms*, may create or destroy restriction-enzyme recognition sites. As a consequence, the pattern of restriction fragment lengths from a region of the genome may differ between homologous chromosomes and between two individuals (Figure 8-20a). Loss of a site will result in the appearance of a larger fragment and the disappearance of two smaller fragments. Formation of a new site will result in the loss of a larger fragment and its replacement with two smaller fragments. These changes in DNA structure are used as *molecular markers* in a fashion analogous to phenotypic markers used in mapping by meiotic recombination. Figure 8-20b illustrates how RFLP analysis of a family can detect allelic forms within a DNA region.

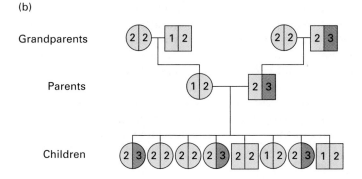

(a) Chromosomal arrangement

Hybridization banding pattern

Enzyme A   Enzyme B

MW

a_1 b_1 b_2 a_2 a_3

a_1 b_1 b_2 a_3

Mutation at site a_2 prevents cleavage

↓ Restriction endonuclease A

〜 Restriction endonuclease B

▨ Probed single-copy region

(b)

Grandparents   Parents   Children

◀ **FIGURE 8-20 Analysis of restriction fragment length polymorphisms (RFLPs).** (a) In the example shown, DNA from an individual is treated with two different restriction enzymes (A and B), which cut DNA at different sequences (*a* and *b*). The resulting fragments are subjected to Southern blot analysis (see Figure 7-32) using a radioactive probe that binds to the indicated DNA region (green) to detect the fragments. Since no differences between the two homologous chromosomes occur in the sequences recognized by the B enzyme, only one fragment is recognized by the probe, as indicated by a single hybridization band. However, treatment with enzyme A produces fragments of two different lengths (two bands are seen), indicating that a mutation has caused the loss of one of the *a* sites in one of the two chromosomes. (b) RFLP analysis of the DNA from eight children, their parents, and grandparents detected the presence of three alleles for a region known to be present on chromosome 5. The DNA samples were cut with the restriction enzyme *Taq*I and analyzed by the Southern blot procedure. In this family, this region exists in three allelic forms characterized by *Taq*I sites spaced 10, 7.7, or 6.5 kb apart. Each individual has two alleles; some contain allele 2 (7.7 kb) on both chromosomes, and others are heterozygous at this site. Circles indicate females; squares indicate males. (The gel at the right corresponds to the 14 subjects shown, beginning and ending with the grandparents and parents, and with the children in the center.) [After H. Donis-Keller et al., 1987, *Cell* **51**:319.]

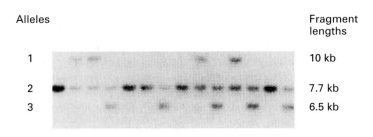

Alleles	Fragment lengths
1	10 kb
2	7.7 kb
3	6.5 kb

How do polymorphisms arise? As we discussed in an earlier section, a finite number of mutations occur spontaneously in the DNA sequence. Some of these will occur in coding sequences, leading to production of defective protein sequences. Others will occur in noncoding sequences; in most cases, these will have no effect on function. For the purposes of genetic markers, the effects of these changes in DNA sequence on organismal function are unimportant. In humans, polymorphisms often appear to be associated with single-base changes; CpG dinucleotide sites are "hot spots" for single-base changes. Consequently, restriction-enzyme cleavage sites that contain this dinucleotide are often polymorphic. The length of DNA between two restriction sites bracketing short sequences of DNA that are repeated in a head-to-tail fashion is often highly variable, providing another form of polymorphism. These variable repeat regions are on the order of 14–70 base pairs long; they are found as single copies or in arrays containing up to 100 repeat units. Recently, even simpler shorter repeat sequences (e.g., C-A repeats) have been used as polymorphic markers. Since these repeats generally contain fewer than 200 base pairs, they can be amplified easily by the PCR, using unique flanking DNA sequences as primers. It is estimated that these repeats arc found on the order of once every 40 kb. Tri- and tetranucleotide repeat sequences are also being increasingly used as polymorphic markers. Presumably, arrays of repeats are formed by recombination or a slippage mechanism of either the template or newly synthesized strands during DNA replication (see Figure 8-5). Polymorphic restriction sites are limited in that there are generally only two alleles at any given locus. In contrast, polymorphic repeat sequences will give rise to many different alleles and hence will be multiallelic and of more general use in mapping disease loci.

In experimental organisms such as *Drosophila*, recombination mapping is facilitated not only by the availability of readily visible markers but by the ability to carry out specifically desired genetic crosses and to obtain large numbers of progeny from any given cross. In human genetic studies, few visible markers are available and planned crosses are not possible, so a different approach is required to map mutations relative to known markers. In this approach, families in which individuals are at risk for a genetic disease (i.e., both parents are heterozygous for an autosomal recessive mutation associated with a particular disease) are identified. DNA samples from various family members are analyzed to determine the frequency with which specific polymorphic markers segregate with the mutant allele causing the disease; this frequency is a measure of the distance between the markers and the mutation-defined locus. The more families afflicted with a particular disease that are available for study and the more DNA polymorphisms identified in proximity of the disease locus, the more precisely the disease-associated gene can be mapped. In most family studies, however, locating more than 100 individuals is highly unlikely, limiting the localization to no less than 1 centimorgan.

An alternative strategy, which can be used in some cases, is called *linkage disequilibrium*. In this approach one assumes that a genetic disease commonly found in a particular community results from a single mutation many generations in the past. This ancestral chromosome will carry closely linked markers that will be conserved through many generations. Markers that are farthest away on the chromosome will tend to become separated from the disease gene by recombination, whereas those closest to the disease gene will remain associated with it. By assessing the distribution of specific markers in all the affected individuals in a population, geneticists can identify DNA markers tightly associated with the disease, thus localizing the disease-associated gene to a relatively small region. The gene causing diastrophic dysplasia in a large Finnish population was localized to a 60-kb region by this approach.

Some Chromosomal Abnormalities Can Be Mapped by Banding Analysis

Many mutations involve large changes in DNA structure that can be mapped by direct light-microscope observation of chromosomes. Normal human metaphase chromosomes exhibit characteristic banding patterns when stained with various dyes (Figure 8-21). The nonuniform staining of chromosomes with these dyes presumably reflects local differences in chromosome structure discussed in Chapter 9.

 Differences in the banding patterns of normal and mutant chromosomes can help map the mutations causing certain human diseases. For instance, the X chromosome from some patients with Duchenne muscular dystrophy has an altered banding pattern reflecting a deletion of part of the chromosome that is large enough to be cytologically visible. Although DMD is X-linked, rare females can also be affected. They carry balanced translocations between the X chromosome and an autosome. Although the autosomal breakpoints are all different, the X-chromosome breaks are at closely linked positions corresponding to cytologically visible deletions in males. Both the deficiencies and balanced translocations aided in the cloning of these genes.

In *Drosophila*, the light-microscope study of chromosomal abnormalities is facilitated by the very large chromosomes of the larval salivary gland (Figure 8-22). These chromosomes have undergone many rounds of DNA replication without cell division, resulting in the colinear arrangement of 1000–2000 copies of each chromosome. The large size of these chromosomes, referred to as *polytene chromosomes*, has permitted compilation of very detailed maps of stained regions (bands) and unstained regions (interbands). Comparison of the banding patterns from normal flies and mutant flies can localize some mutations to quite specific chromosomal regions.

(a)

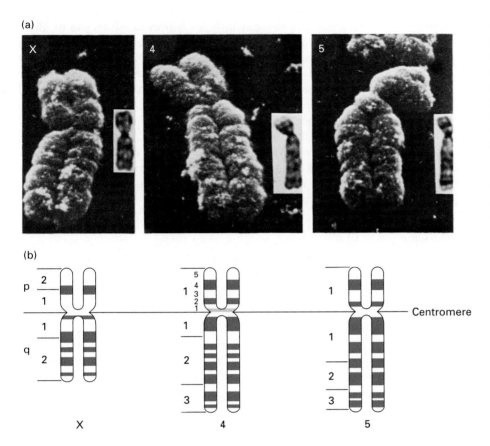

(b)

p   2
     1

     1
q
     2

X

1

2

3

4

1

1

2

3

5

Centromere

◀ FIGURE 8-21 G-banding of human chromosomes. The chromosomes are subjected to brief proteolytic treatment and then stained with Giemsa reagent, producing distinctive bands at characteristic places. (a) Scanning electron micrographs of chromosomes X, 4, and 5 show constrictions at the sites where bands appear in light micrographs (insets). (b) Standard diagrams of G bands (purple). Regions of variable length (green) are most common in the centromere region (e.g., on chromosome 4). The numbering system to indicate positions on the chromosomes is illustrated: p=short arm, q=long arm. Each arm is divided into major sections (1, 2, etc.), and subsections. The short arm of chromosome 4, for example, has five subsections. DNA located in the fourth section would be said to be in p14. [Part (a) from C. J. Harrison et al., 1981, *Exp. Cell Res.* **134**:141; courtesy of C. J. Harrison.]

(a)

Chromocenter

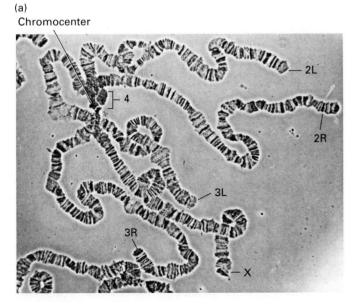

2L

2R

3L

3R

X

(b)

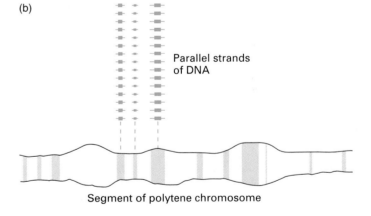

Parallel strands of DNA

Segment of polytene chromosome

▲ FIGURE 8-22 The polytene salivary gland chromosomes from *Drosophila melanogaster* can be stained to reveal their very reproducible banding pattern. (a) In the four chromosomes seen in this light micrograph (X, 2, 3, and 4), a total of approximately 5000 bands can be distinguished. The centromeres of all four chromosomes often appear fused at the chromocenter. The tips of the metacentric 2 and 3 chromosomes are labeled (L=left arm; R=right arm), as is the tip of the acrocentric X chromosome. (b) The DNA in salivary polytene chromosomes is repeated about 1000 times; the duplicated DNA fibers are thought to remain parallel. Therefore, any staining property in a chromosome with one DNA duplex would be amplified 1000 times to produce a transverse band (light blue). [Part (a) courtesy of J. Gall.]

SUMMARY Genetic Mapping of Mutations

• Genes located on different chromosomes segregate independently of one another during meiosis (see Figure 8-15).

• X-linked recessive mutations, autosomal recessive mutations, and autosomal dominant mutations exhibit characteristic segregation patterns.

• Genes and molecular markers previously mapped to a particular autosome can be used to map other genes to that chromosome, as linked genes and markers on the same chromosome segregate together with higher frequency than unlinked genes.

• Genetic recombination between genes on the same chromosomes, which can occur during meiosis, permits the mapping of genes on the same chromosome relative to each other (see Figure 8-18). The farther apart two genes are on a chromosome, the greater their frequency of recombination.

• Numerous phenotypic markers are available in common experimental organisms for use in recombinational analysis. In humans, restriction fragment length polymorphisms (RFLPs) and other types of polymorphisms, particularly in noncoding DNA regions, are useful in genetic mapping of disease genes.

• Abnormalities in chromosome structure, reflected in altered banding patterns, can aid in mapping certain mutations in experimental animals and disease genes in humans.

8.4 Molecular Cloning of Genes Defined by Mutations

In the first three sections of this chapter, we described genetic techniques for isolating, characterizing, and mapping mutations. In this section, we discuss the use of recombinant DNA techniques to isolate mutation-defined genes. This approach, often referred to as **positional cloning,** has revolutionized the study of developmental biology and human genetics by providing a link between phenotypes associated with genetic diseases or developmental abnormalities in experimental model organisms and the specific proteins encoded by the genes inactivated by these mutations.

In this approach, as noted earlier, the chromosomal position of a mutation-defined gene is determined with the aid of molecular markers, the normal gene is isolated and cloned from an appropriate DNA library, and then the function of its encoded protein is investigated. In *E. coli* and yeast a specific gene defined by mutations can be isolated by transforming cells exhibiting the mutant phenotype with various plasmids that collectively represent the total genomic DNA. Those plasmids that carry the gene of interest will restore the mutants to the wild-type phenotype, which can be identified easily. These plasmids can rescue mutants because they carry a normal copy of genes. This *DNA-rescue* approach is feasible in *E. coli* and yeast because collections of plasmids representing the entire genome are available and can be effectively transferred into mutants. Although individual cloned genes can be introduced into genetically tractable higher organisms (e.g., *Drosophila, C. elegans,* and mice), these methods are not efficient enough to be used for screening DNA libraries representing the entire genome.

Four steps are involved in cloning a mutation-defined gene in higher organisms:

1. Identification of a cloned segment of DNA near the gene of interest.

2. Isolation of a contiguous stretch of DNA and construction of a physical map of DNA in the region. This region will contain the gene of interest as well as other neighboring genes. The position of the gene of interest within this cloned segment is not known.

3. Correlation of the physical map with the genetic map to localize the approximate position of the gene of interest along a cloned segment of DNA.

4. Detection of alterations in expression of transcripts or changes in DNA sequence between DNA from mutant and wild-type organisms to determine which region of the cloned DNA corresponds to the normal form of the mutation-defined gene.

The specific experimental techniques used in positional cloning may vary depending on the species and nature of the mutation. In this section, we describe some of the common techniques for carrying out each of the basic steps in this approach.

Cloned DNA Segments Near a Gene of Interest Are Identified by Various Methods

One of the first steps in isolating a particular mutation-defined gene is to identify one or more cloned DNA segments that map in close proximity to that gene. To accomplish this, the gene's segregation patterns either with respect to DNA clones that recognize different polymorphic markers or to specific genetic loci that have been cloned previously are assessed. For example, suppose you have 100 clones scattered throughout the genome to be used as molecular markers. If one of these markers is present in a high percentage of individuals carrying a mutant allele of the gene of interest, but not in normal siblings, then the DNA clone carrying the marker is located near the gene. In most genetically tractable organisms, many DNA clones mapping throughout the genome are already available. As discussed in the previous section, mapping of a large number of polymorphic sites (detected by cloned DNA) relative to a gene of interest increases the likelihood that a clone mapping close to that gene will be identified.

(a)

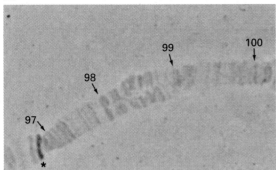

(b)

▲ FIGURE 8-23 Chromosomal mapping of cloned DNA segments by in situ hybridization. In this procedure cloned DNA is labeled by one of several possible methods and then hybridized to chromosomes attached to microscope slides. (a) A *Drosophila* salivary gland chromosome hybridized with a DNA probe labeled with biotin-derivatized nucleotides. Hybridization is detected with a biotin-binding protein called *avidin* that is covalently bound to an enzyme such as alkaline phosphatase. On addition of a soluble substrate, the enzyme catalyzes a reaction that results in formation of an insoluble colored precipitate at the site of hybridization (asterisk). The numbers indicate the band divisions. Note the local separation of the two homologs between bands 98–99. (b) A stretch of ≈400 kb on an interphase (decondensed) human X chromosome hybridized to fluorescent-labeled cosmid probes from the dystrophin gene. The cosmids were alternately labeled red and green. The red probe in the middle contains a deletion and overlaps with a part of the green probe, resulting in yellow. The chromosomes were subjected to the Fiber-FISH method and then visualized by epifluorescence microscopy. [Part (a) courtesy of F. Pignoni; part (b) from A. K. Raap et al., 1998, *Mut. Res.* **400**:287.]

The chromosomal location of a cloned DNA segment can be determined by hybridizing it to metaphase or interphase chromosomes from mice and humans and to interphase polytenized chromosomes from *Drosophila*. In this technique, referred to as *in situ hybridization,* cloned DNA is labeled in vitro with biotin (a naturally occurring prosthetic group that can be used as a molecular tag) or with a fluorescent nucleoside derivative and then denatured into single strands. The labeled DNA is incubated with the chromosomes, which has been treated to separate the DNA strands and attached to a microscope slide. The sample then is prepared to allow visualization by microscopy of regions where the labeled DNA probe hybridizes to the chromosomes.

In situ hybridization with a biotin-labeled probe has been particularly useful in chromosomal mapping of DNA clones in *Drosophila,* because the cytological map of the polytene chromosomes of this organism is known at high resolution (Figure 8-23a). When a fluorescent-labeled probe is used, the technique is referred to as DNA-fluorescence in situ hybridization (FISH). This technique can identify two different DNA segments on the same human chromosome by use of probes specific for each segment. In early studies in which the fluorescent probes were hybridized to metaphase (condensed) chromosomes, the two sites of hybridization had to be about 3 megabases (Mb) apart to be distinguished. Recently this technique has been adapted to localizing DNA segments on decondensed naked DNA fibers immobilized on slides or other supports such as agarose. This variation, called *Fiber-FISH,* has a resolution in the kilobase range and can be used to order clones on relatively short stretches of DNA (Figure 8-23b).

Specific DNA clones also can be mapped to particular regions of human chromosomes by screening for their presence in hybrid cells formed between cultured human and rodent cells. As discussed in Chapter 6, the human chromosomes in these hybrids typically are unstable, but stable hybrids containing one or a few human chromosomes can be produced. Furthermore, if human cells carrying chromosomal translocations are used, stable hybrid lines can be established in which only part of a human chromosome is represented. Detection of specific human DNA clones by Southern blot or PCR screening of such lines localizes the clones to the chromosomes or chromosomal regions present in the hybrid lines.

Chromosome Walking Is Used to Isolate a Limited Region of Contiguous DNA

Although complete physical maps of the genomes of many experimental organisms and humans will soon be completed, many genes are now cloned by constructing a physical map of a limited region of the genome containing the gene of interest, whose precise position is not known at the beginning of the analysis. A molecular technique called *chromosome (DNA) walking* can be used to isolate contiguous regions of genomic DNA beginning with a previously cloned DNA fragment that maps near a gene of interest. In this reiterative process, overlapping DNA clones are isolated; the process is repeated until a clone containing the desired gene is identified.

The first step in a DNA walk is to radiolabel the starting clone and hybridize it to a genomic DNA library (Figure 8-24). After clones that hybridize are identified, their DNA is isolated and mapped by restriction-endonuclease cleavage. Typically, only part of the cloned DNA fragments hybridize with the labeled starting clone; the nonhybridizing parts extend into contiguous DNA regions. Small regions of these overlapping cloned fragments that extend the farthest from the probe in each direction are isolated, radiolabeled, and used to probe the genomic DNA library once again. In this way, DNA that extends farther away in both directions from the starting point can be identified.

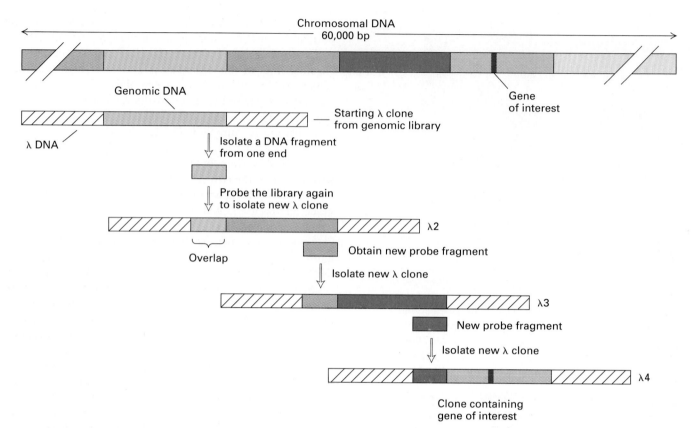

▲ **FIGURE 8-24 Chromosome (DNA) walking.** This technique can be used to isolate overlapping DNA fragments starting with a previously cloned DNA fragment that maps near a gene of interest (dark red). The walk is continued until a clone containing the desired gene is identified. In this example, the chromosomal DNA fragments are cloned in λ phage. The starting clone (green) is used as a probe to screen a genomic DNA library for overlapping sequences; usually a walk is conducted in both directions, since it generally is not known how far the gene of interest is from the starting clone or whether it is to the left or right. For simplicity, a one-way walk is illustrated here. The clones isolated in the walk are used as probes in Southern blot analysis of genomic DNA from mutants. This analysis can detect chromosomal deletions and rearrangements in the mutant DNA, but normally cannot detect point mutations unless they change the restriction-site sequences recognized by the restriction enzymes used in the analysis. In this way, the DNA clone containing the region corresponding to the gene of interest can be identified. Normally, a walk would involve more than the three steps shown here. In some cases, part of the desired gene may already be cloned, and DNA walking may be necessary to isolate the entire gene. By initiating walks from several different starting clones in parallel (these walks will eventually link up), the process of isolating overlapping clones for long contiguous stretches of DNA is accelerated. [From J. D. Watson et al., 1992, *Recombinant DNA*, 2d ed., W. H. Freeman and Company, p. 128.]

The larger the steps taken in a chromosome walk, the faster large segments of contiguous DNA can be isolated. Thus DNA vectors that are used to clone large pieces of DNA (e.g., cosmids and YACs) are particularly valuable in chromosome walking. Although the starting clone may be considered close to a gene of interest by genetic recombination, it may be a considerable distance from the gene at the molecular level.

Physical Maps of Entire Chromosomes Can Be Constructed by Screening YAC Clones for Sequence-Tagged Sites

As mentioned in Chapter 7, segments of human DNA up to 1000 kb long can be cloned in yeast artificial chromosomes (YACs). These clones can be ordered by PCR screening for *sequence-tagged sites* (STSs). These are randomly spaced, unique sequences of DNA known to map to particular chromosomes or parts of chromosomes. Figure 8-25 illustrates this procedure in the simple case of three contiguous YAC clones containing six STSs randomly distributed over the region spanned by the clones.

Detecting and ordering a set of YACs for an entire chromosome is done in a similar way. However, instead of subjecting individual YACs to PCR analysis, which is far too laborious, pools of YACs are screened. Those pools in which PCR products are detected then are divided into smaller pools, and the PCR analysis is repeated. This process is continued until a set of YACs is identified, each of which con-

(a)

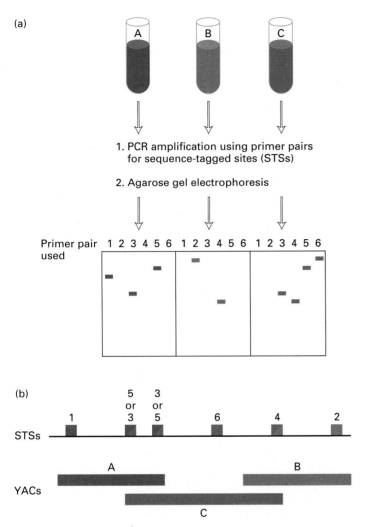

1. PCR amplification using primer pairs for sequence-tagged sites (STSs)

2. Agarose gel electrophoresis

Primer pair used 1 2 3 4 5 6   1 2 3 4 5 6   1 2 3 4 5 6

(b)

STSs

YACs

▲ FIGURE 8-25 Ordering of contiguous overlapping YAC clones (a) Aliquots of DNA prepared from each YAC clone (A, B, and C) are subjected to PCR amplification (see Figure 7-38) using primer pairs (1–6) corresponding to the ends of various sequence-tagged sites (STSs). Only those clones containing STSs with ends complementary to particular primers will be amplified. Electrophoretic analysis then shows which YAC clones contain which STSs. For example, clone A contains STSs 1, 3, and 5. (b) Based on the gel patterns in this example, the three YAC clones can be ordered as shown. However, the positions of STSs 3 and 5 relative to each other cannot be determined from this analysis. This general approach has been used to generate a complete physical map of the human Y chromosome and the long arm of human chromosome 21.

tains at least one STS and that collectively contains all the STSs. The detail of a map constructed with this procedure depends on the number of STSs used in the PCR analyses and the number of DNA clones available for analysis. In the mapping of human chromosome 21, for example, 120,000 clones from two separate YAC libraries were screened in pools; in addition, 14,000 YACs isolated from a library prepared specifically from chromosome 21 were screened individually. By use of 198 STSs, researchers identified 810 positive clones and ordered them into a contiguous map.

Physical and Genetic Maps Can Be Correlated with the Aid of Known Markers

In order for a physical map to be helpful in identifying a segment of DNA that contains a mutation-defined gene, the cloned DNA must contain specific landmarks (markers) that can be identified by genetic linkage. Such landmarks include known genetic loci and DNA polymorphisms. The availability of extensive physical and genetic maps of the genome of the genetically tractable worm *C. elegans* has greatly facilitated the cloning of genes defined by mutation. In this organism, the genetic map position of a mutation immediately defines a physical interval of DNA within which the gene is located.

One strategy for correlating physical and genetic maps is to position polymorphic DNA markers relative to mutation-defined genes. An example is outlined in Figure 8-26 for a defined region of the *Drosophila* genome. In this example, the objective was to localize the *shibire (shi)* locus, which is on the X chromosome, in order to clone it. This locus had already been mapped by recombinational analysis with visible phenotypic markers spread along the entire length of the X chromosome. From a collection of *Drosophila* YAC clones, a 275-kb clone containing the genetic map position of *shi* was identified. To precisely locate the *shi* locus within this clone, recombinational analysis was carried out using 10 RFLPs, *shi,* and a closely linked visible phenotypic marker called *scalloped (sd),* which affects wing appearance. All the 23 male sd^+shi^+ recombinants contained the A–C RFLP alleles from one maternal chromosome and the h–j alleles from the other. Thus these markers lie outside of the region between the *sd* and *shi* genes. The frequency of the remaining four RFLP alleles (D/d, E/e, F/f, and G/g) varied among the recombinants. Since the frequency of recombination between any two DNA sites is directly proportional to the distance in nucleotides between them, *shi* will be more tightly associated with the RFLP closest to it than to RFLPs farther away. For example, the "g" RFLP allele segregates more frequently with shi^+ than does the "f" allele.

The length of DNA to which a specific mutation-defined gene can be localized by this approach depends on the density of DNA polymorphisms in the region of interest. In humans, it also depends on the number of individuals that can be studied: the more individuals, the greater the probability of obtaining a recombination event between polymorphic markers and the locus of interest. However, the desired gene may correspond to only a small segment of the DNA region defined by polymorphic marker analysis and other genes may be located in the same region.

(a) Isolation of recombinants between scalloped (*sd*) and shibire (*shi*) genes

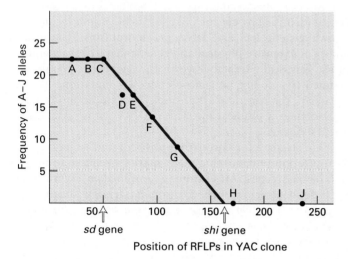

(c) Frequency of A–J alleles from *sd⁺ shi⁻* chromosome in *sd⁺ shi⁺* recombinants

(b) Molecular analysis of recombinants

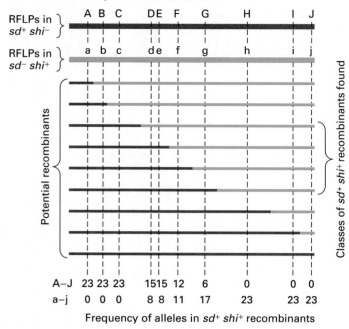

▲ **FIGURE 8-26 Correlation of the physical and genetic maps of the region of the *Drosophila* X chromosome containing the *shibire* (*shi*) locus.** The genetic map position of *shi* was known from recombinational analysis with visible phenotypic markers spread along the length of the X chromosome. In addition, a 275-kb YAC clone containing the *shi* locus had been identified. In order to clone *shi*, its specific location within this YAC clone had to be determined. (a) As a first step, 23 male recombinants carrying wild-type alleles of both the *shibire* locus and a closely linked, visible phenotypic marker called *scalloped* (*sd*) were isolated.

(b) These recombinants then were analyzed for the presence of 10 RFLPs that differed in *shi⁺* chromosomes (a–j) and *shi⁻* chromosomes (A–J). This analysis showed, for example, that none of the *sd⁺ shi⁺* recombinants found contained just A, whereas all contained A–C and 15 of the 23 recombinants contained A–D. (c) The frequency of the A–J alleles in these recombinants was plotted against the known positions of the RFLPs within the 275-kb YAC clone. The intersection of the curve with the x axis predicts the position of the *shi* gene.
[See A. M. van der Bliek and E. M. Meyerowitz, 1991, *Nature* **351**:411.]

Further Analysis Is Needed to Locate a Mutation-Defined Gene in Cloned DNA

Although the analysis described in the previous section may localize a mutation-defined gene of interest to a particular DNA region, it frequently cannot determine which of several neighboring genes corresponds to that specific gene. Here we describe strategies for pinpointing the precise position of the gene of interest within a larger region of cloned DNA.

As a first step, genomic DNA from mutants is often subjected to Southern blot analysis with DNA probes representing different regions of the chromosomal walk of the normal genome (see Figure 8-24). Generally, Southern blotting can successfully identify chromosomal rearrangements and small deletions in the mutant DNA but is not sensitive enough to identify point mutations. Compared with normal DNA, Southern blots of mutant DNA carrying large deletions will

often not contain any sequences homologous to the probe, whereas blots of mutant DNA that have rearrangements or smaller deletions will show fragments of different sizes. Point mutations are not detected by Southern blotting unless they either destroy or create a restriction-enzyme recognition site, creating a RFLP. Creation of a new recognition site will result in the loss of one fragment and its replacement with two smaller ones. Conversely, loss of a site will lead to the loss of two smaller fragments and the appearance of a larger one.

In general, Southern blot analysis will define a region of the genome containing the gene of interest. In many cases the interval defined may not contain the entire gene, or it may contain other genes as well. Additional analyses, including screening for point mutations and the identification of sequences transcribed into mRNAs in the region, are needed to isolate and characterize the gene.

Analysis of mRNA Expression Identification of mRNA encoded by DNA in the region of the desired gene may help to further localize the gene within the genome. Phenotypic analysis of a mutant may suggest tissues in which the mRNA is likely to be expressed. For instance, a mutation that phenotypically affects muscle, but no other tissue, might be in a gene that is expressed only in muscle tissue. The expression of mRNA in both normal and mutant tissues generally is determined by Northern blotting or in situ hybridization of labeled DNA or RNA to tissue sections. Northern blots permit comparison of both the level of expression and the size of mRNAs in mutant and wild-type tissues (see Figure 7-33). Although the sensitivity of in situ hybridization is lower than that of Northern blot analysis, it can be very helpful in identifying an mRNA that is expressed at low levels in a given tissue but at very high levels in a subclass of cells within that tissue (Figure 8-27). An mRNA that is altered or missing in different mutants compared with wild-type individuals is an excellent candidate for encoding the protein whose function is disrupted in the mutants.

Identification of Point Mutations In many cases, point mutations may result in no detectable change in the level of expression or electrophoretic mobility of mRNAs. Thus if comparison of mRNA expression in mutants and normal individuals reveals no detectable differences in the candidate mRNAs, a search for point mutations in the DNA regions encoding the mRNAs is undertaken. The overall strategy is to scan for regions of DNA that carry point mutations and then to sequence that region of the mutant allele. Of several

different methods devised to screen for single-base changes, two are discussed here: one is applicable to DNA fragments as large as 2 kb; the other, to fragments less than about 400 base pairs.

In the first method, a wild-type, single-stranded DNA fragment is radiolabeled at one end and then is hybridized to unlabeled, single-stranded DNA from the mutant. By the procedure outlined in Figure 8-28a, a single-base mismatch between the normal and mutant strands can be detected. The second method for detecting point mutations, applicable to smaller fragments, depends on the property of single-stranded DNA molecules of the same length to assume different conformations depending on their nucleotide sequence. Single-base differences can lead to altered conformations—referred to as *single-stranded conformation polymorphisms* (SSCPs)—that can be detected by electrophoresis on nondenaturing polyacrylamide gels (Figure 8-28b).

One disadvantage of these two methods is that the mutant chromosome is likely to carry naturally occurring polymorphisms unrelated to the gene of interest; these can lead to misidentification of the DNA fragment carrying the gene of interest. For this reason, the more mutant alleles available for analysis, the more likely that a gene will be correctly identified. Now that highly efficient methods for sequencing DNA are available, researchers frequently determine the sequence of candidate regions of DNA isolated from mutants to identify point mutations.

Protein Structure Is Deduced from cDNA Sequence

Once the mRNA encoded by a mutation-defined gene has been identified, the corresponding cDNA can be identified in a cDNA library using specific genomic fragments that hybridize with the mRNA as probes. Sequencing of the cDNA permits the sequence of its encoded protein to be deduced. This is an important step in linking genetics to the cell and molecular biology of development and function, since proteins with similar structures often have similar functions.

As described in Chapter 7, large databases of DNA and protein sequences are maintained and available to biologists throughout the world. Thus one can efficiently search for sequence similarities between a new sequence and all those previously determined. The sequence of a protein deduced from the cDNA of a newly isolated gene is likely to show similarities with the structures of previously determined proteins. In those cases where the biochemical function of a structurally similar class of proteins is known, it is possible to generate biochemically testable hypotheses. For example, if a newly identified protein is similar to proteins known to regulate the transcription of genes, the mutant phenotype may result from lack of expression of a particular gene or set of genes in the affected tissue.

Even if a newly determined sequence does not show similarities with proteins of known function, certain properties

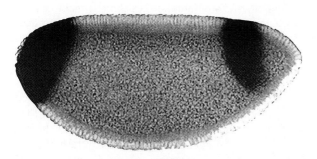

▲ **FIGURE 8-27 In situ hybridization of *Drosophila* embryo with DNA fragments from the tailless gene detect tailless mRNA.** This gene is expressed only in the anterior and posterior termini of fly embryos and is required for development of both termini (Chapter 14). DNA fragments from the tailless gene were labeled with digoxigenin-substituted nucleotides and hybridized to developing embryos to localize *tailless* mRNA. Antibodies to digoxigenin (a steroid derivative isolated from plants) were used to visualize the distribution of hybridized mRNA. [Courtesy of L. Tsuda.]

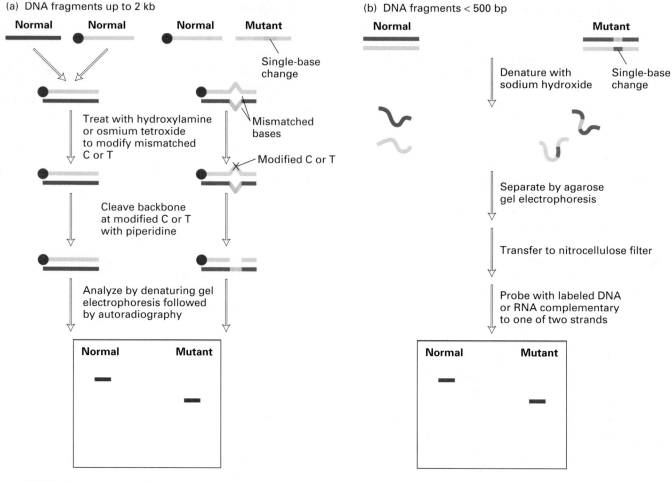

▲ **FIGURE 8-28 Detection of point mutations.** (a) In DNA fragments up to 2 kb in length, single-base changes (green) can be detected by chemical cleavage at the mismatched bases in mutant-normal heteroduplexes. One strand of the normal DNA is radiolabeled at one end (red dot) and then is hybridized with the complementary normal or mutant strand. The resulting heteroduplex DNA is treated with hydroxylamine or osmium tetroxide, which modifies any C or C and T, respectively, in mismatched single-stranded regions; the modified backbone is susceptible to cleavage by piperidine. The shortened labeled fragment is detected by gel electrophoresis and autoradiography in comparison with normal DNA. (b) Single-stranded conformation polymorphisms (SSCPs) can be used to detect single-base changes (green) in DNA restriction fragments less than about 500 bp. Single-stranded DNA molecules that differ by only one base frequently show different electrophoretic mobilities in nondenaturing gels. Differences between normal and mutant DNA mobility are revealed by hybridization with labeled probes.

of the protein (e.g., whether it is a secreted from the cell or spans the cell membrane) may be predicted. In addition, various biochemical and immunological techniques can be used to generate antibodies to proteins predicted from DNA sequence analysis. These antibodies then can be used to determine which cells produce a specific protein and what its subcellular distribution is (e.g., nuclear or cytoplasmic). Again this information can provide a critical link between the phenotype caused by a mutation and the underlying cellular and molecular mechanisms. Many examples of the central importance of this approach to cell and molecular biology are discussed in subsequent chapters.

Before concluding this section, let's return to the *Drosophila shi* gene to illustrate further how the phenotype associated with a mutation-defined gene can be related to a specific protein. We saw earlier that RFLP analysis localized *shi* within a particular YAC clone (see Figure 8-26c). To identify the protein encoded by *shi*, researchers then used a cosmid clone containing the *shi* gene and flanking sequences to probe Northern blots of *Drosophila* mRNA isolated from fly heads. (Compared with the YAC clone, the cosmid clone contained a smaller fragment of genomic DNA and thus is a more precise probe for identifying a specific mRNA.) This analysis identified two mRNAs. However, a DNA fragment isolated from the cosmid hybridized to only one of the mRNAs; when this fragment was transformed into mutant flies, it rescued the mutant phenotype, thus identifying that mRNA as encoding the Shi protein. (We will discuss the method of transforming DNA into flies later in the chapter.) The DNA fragment also was used to screen a *Drosophila* cDNA library. Sequencing of the identified cDNA revealed that it encodes a protein homologous to mammalian dynamin, which plays a key role in endocytosis (Chapter 16).

SUMMARY Molecular Cloning of Genes Defined by Mutations

- The objective of positional cloning is to locate a mutation-defined gene within cloned DNA so that it can be characterized at the molecular level.

- The first step in positional cloning involves identification of cloned DNA fragments near the gene of interest.

- The chromosomal location of cloned DNA fragments can be mapped to particular chromosomes by in situ hybridization or by using them as probes in Southern blot or PCR screening of mouse-human hybrid cells containing one or a few human chromosomes or portions of chromosomes.

- Regions of contiguous DNA can be isolated by chromosome walking starting near a gene of interest (see Figure 8-24) and by ordering of overlapping YACs (see Figure 8-25). Such DNAs can be used to construct physical maps of chromosomal regions of interest.

- Correlating the genetic and physical maps of a specific chromosomal region is a key step to the identification and isolation of a particular mutation-defined gene. Landmarks on the physical map that facilitate correlation with the genetic map include DNA polymorphisms and chromosomal abnormalities such as large deletions and translocations.

- Identification of the gene of interest within a candidate region typically requires comparison of DNA sequences between wild-type and mutant (or disease-affected) individuals.

8.5 Gene Replacement and Transgenic Animals

The goal of modern molecular cell biology is nothing short of understanding the biochemical, cellular, and organismal functions of all the proteins encoded in the genome. In the preceding sections, we have discussed the isolation and analysis of mutants, the genetic mapping of mutations, and finally the isolation and cloning of mutation-defined genes. This approach can provide valuable information about the molecular mechanisms underlying the cellular processes affected by the original mutations and the in vivo functions of the normal proteins encoded by the affected genes. As discussed in Chapter 7, however, many genes have been identified based on the biochemical properties of their encoded protein, the sequence similarity of the encoded protein with proteins of known function, or their interesting patterns of expression in development. In the absence of mutant forms

of such genes, their in vivo functions may be unclear. By mutating a specific gene in vitro and then replacing the normal copy in the genome with a mutant form, scientists can assess its in vivo function. This technique, referred to as **gene-targeted knockout**, or simply "knockout," is in essence the reverse of the approach described in the previous sections. The process of isolating normal genes to be mutated will be greatly simplified as sequencing of the genomes of several model organisms and of the human genome progresses (Chapter 7). Whether starting from a normal protein or sequenced genome, this approach can be summarized as follows:

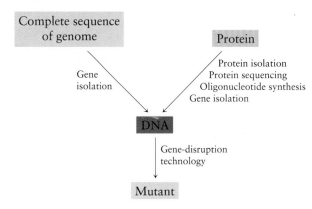

Other techniques permit the introduction of foreign genes or altered forms of an endogenous gene into an organism. For the most part, these techniques do not result in replacement of the endogenous gene, but rather in the integration of additional copies of it. Such introduced genes are called **transgenes**; the organisms carrying them are referred to as **transgenics**. Transgenes can be used to study organismal function and development in a variety of different ways. For instance, genes that are normally expressed at specific times and places during development can be genetically engineered in vitro to be expressed in different tissues at different times and then reintroduced into the animal to assess the cellular and organismal consequences. For example, the *Antennapedia (Antp)* gene in *Drosophila* normally controls leg development, but misexpression of this gene in the developing antenna transforms it into a leg.

The production of both gene-targeted knockout and transgenic animals makes use of techniques for mutagenizing cloned genes in vitro and then transferring them into eukaryotic cells. We briefly describe these procedures first, then discuss the production and uses of knockout and transgenic organisms.

Specific Sites in Cloned Genes Can Be Altered in Vitro

Specific sequences in cloned genes can be altered in vitro and then introduced into experimental organisms. This approach has been exploited primarily to study two questions. First, what is the relationship between the structure of a

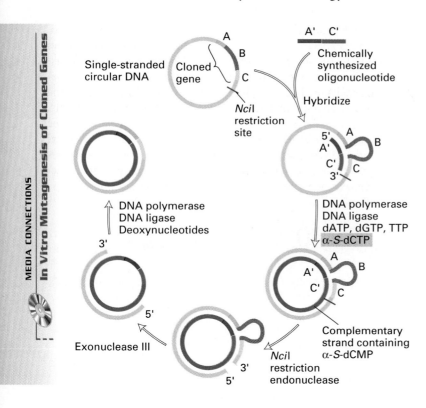

◀ **FIGURE 8-29 In vitro mutagenesis of cloned genes with chemically synthesized oligonucleotides.** In this example, a cloned gene that normally contains three segments (A, B, and C) is mutagenized by deletion of segment B. An oligonucleotide consisting of A' and C' segments, which are complementary to A and C, is chemically synthesized and then is hybridized to the complementary single-stranded DNA containing the gene to be mutagenized. In the hybrid molecule, segment B forms a loop. The A'-C' oligonucleotide serves as a primer for synthesis of a complementary strand; one of the nucleotide triphosphates contains sulfur in place of one of the oxygens at the α position. The heteroduplex DNA is treated with the restriction endonuclease *Nci*I; because this enzyme cannot cleave the sulfur-substituted strand (green), a single-strand nick is created in the original template strand (purple). The nicked molecule then is treated with exonuclease III, which cleaves nucleotides in the 3' to 5' direction, thereby removing the B segment. The resulting gap in the template strand is filled in by DNA polymerase and closed by DNA ligase.

particular protein and its biological function? And second, what are the specific DNA sequences required to determine the expression pattern of a particular gene?

A variety of enzymatic and chemical methods are available for producing site-specific mutations in vitro. In recent years, however, the most common methods use specific oligonucleotides as mutagens. Because oligonucleotides of any desired sequence can be chemically synthesized (see Figure 7-9), oligonucleotide-based mutagenesis can generate precisely designed deletions, insertions, and point mutations in a DNA sequence. Figure 8-29 illustrates the use of this strategy to produce a deletion.

DNA Is Transferred into Eukaryotic Cells in Various Ways

Production of both knockout and transgenic organisms requires the transfer of DNA into eukaryotic cells. Many types of cells can take up DNA from the medium. Yeast cells, for instance, can be treated with enzymes to remove their thick outer walls; the resulting *spheroplasts* will take up DNA added to the medium. Plant cells also can be converted to spheroplasts, which will take up DNA from the medium. Cultured mammalian cells take up DNA directly, particularly if it is first converted to a fine precipitate by treatment with calcium ions. Another popular method for introducing DNA into yeast, plant, and animal cells is called *electroporation*. Cells subjected to a brief electric shock of several thousand volts become transiently permeable to DNA. Presumably the shock briefly opens holes in the cell membrane

allowing the DNA to enter the cells before the holes reseal. DNA also can be injected directly into the nuclei of both cultured cells and developing embryos.

Once the foreign DNA is inside the host cell, enzymes that probably function normally in DNA repair and recombination join the fragments of foreign DNA with the host cell's chromosomes. Since only a relatively small fraction of cells take up DNA, a selective technique must be available to identify the transgenic cells. In most cases the exogenous DNA includes a gene encoding a selectable marker such as drug resistance. The introduced DNA can insert into the host genome in a highly variable fashion showing no site specificity, can replace an endogenous gene by homologous recombination, or can remain as an independent extrachromosomal DNA molecule referred to as an *episome*.

Normal Genes Can Be Replaced with Mutant Alleles in Yeast and Mice

Gene knockout is a technique for selectively inactivating a gene by replacing it with a mutant allele in an otherwise normal organism. This technique of disrupting gene function, which has been widely used in yeast and mice, is a powerful tool for unraveling the mechanisms by which basic cellular processes occur.

Gene Knockout in Yeast After foreign, or exogenous, yeast DNA is taken up by diploid yeast cells, recombination generally occurs between the introduced DNA and the homologous chromosomal site in the recipient cell. Because of this specific, targeted recombination of identical stretches of DNA, called *homologous recombination,* any gene in yeast chromosomes can be replaced with a mutant allele (Figure 8-30). The resulting heterozygous yeast cells, carrying one mutant allele and one wild-type allele, generally grow normally. To determine whether the knocked-out gene controls an obligatory function, recombinants containing the mutant allele on one chromosome are treated to induce meiosis and sporulation; each diploid cell produces four haploid spores, which are tested for viability. One of the first genes tested in this way was the one encoding actin, a prominent cytoskeletal protein in yeast and higher organisms. Haploid yeast spores without a normal actin gene cannot grow (Figure 8-31).

This technique also is useful in assessing the role of proteins identified solely on the basis of DNA sequence. For instance, the entire sequence of the *S. cerevisiae* genome has been determined. As described in Chapter 7, analysis of genomic sequences can identify stretches of DNA that exhibit long open reading frames or homology to genes encoding known proteins; such stretches are likely to be transcribed and translated into as yet unidentified proteins (see Figure 7-31). Gene knockouts can be used to determine whether such regions are important for specific cellular functions that are phenotypically detectable. This technique thus provides a powerful approach to identifying and studying new genes and the proteins that they encode.

This gene-knockout approach already has been used to analyze yeast chromosome III. Analysis of the DNA sequence indicated that this chromosome contains 182 open reading frames of sufficient length to encode proteins

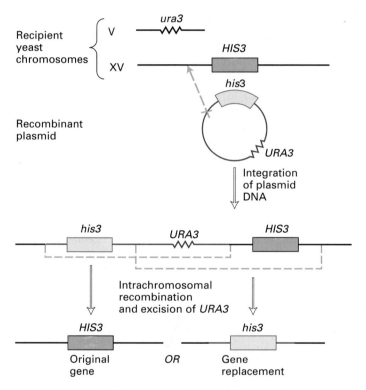

▲ **FIGURE 8-30 Replacement of the normal *HIS3* gene (blue) by homologous recombination with a mutant *his3* gene (yellow) in the yeast *S. cerevisiae.*** Recombinant DNA technology is used to prepare a plasmid containing *his3,* a deletion mutant that encodes only part of the sequence of one enzyme in the histidine biosynthetic pathway, and *URA3,* which encodes an enzyme in the uracil biosynthetic pathway. The *URA3* gene is included simply as a selectable marker. The recipient cell (a uracil-requiring strain) takes up the plasmid, which can integrate into the *ura3* gene (not of interest in this case) or into the *HIS3* gene. Recombinant cells are selected by their ability to grow in the absence of uracil. Subsequent intrachromosomal recombination yields cells that have lost *URA3* and retain either the wild-type *HIS3* gene or the *his3* mutant. Cells carrying the *his3* mutant gene are detected by replica plating in the presence and absence of histidine. [See S. Scherer and R. W. Davis, 1979, *Proc. Nat'l. Acad. Sci. USA* **76**:4951.]

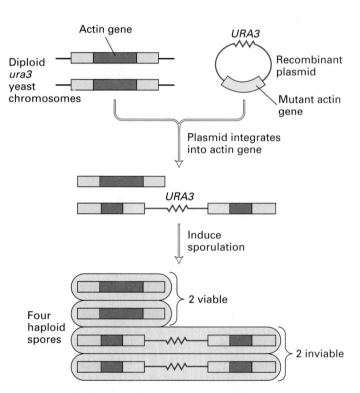

▲ **FIGURE 8-31 Demonstration that actin gene is required for yeast viability by gene-targeted knockout.** A recombinant plasmid containing the *URA3* gene, a selectable marker, and a mutant actin gene (yellow) is introduced into uracil-requiring (Ura3⁻) diploid yeast cells. (The *ura3* gene, which is located on a different chromosome, is not depicted.) Cells that integrate the plasmid are selected by their growth in the absence of uracil. Since the recombinant cells contain one normal and one disrupted actin gene, they can synthesize actin. To determine whether the actin gene is essential, meiosis and sporulation are induced by starving the cells; each diploid cell produces four haploid spores (see Figure 10-54). The wild-type spores, which are Ura3⁻, can grow on uracil, but those with the disrupted actin gene do not. [Adapted from D. Shortle et al., 1982, *Science* **217**:371.]

longer than 100 amino acids, which is assumed in this analysis to be the minimum length of a naturally occurring protein. The sequences of the proteins that could be encoded by 116 of these putative protein-coding regions exhibited no obvious homology to any known proteins. Gene knockout of 55 of these regions showed that 3 were required for viability; further analysis of 42 nonessential genes revealed that 14 showed a mutant phenotype and 28 did not. The large number of putative genes with no detectable mutant phenotype is quite surprising. In some cases the lack of a phenotype could indicate the existence of backup or compensatory pathways in the cell. Alternatively, the mutations may give rise to subtle defects that would require more in-depth phenotypic analysis to uncover.

Gene Knockout in Mice   Gene-targeted knockout mice are a powerful experimental system for studying development, behavior, and physiology; they also may be useful model systems for studying certain human genetic diseases. The procedure for producing gene-targeted knockout mice involves the following steps:

1. Mutant alleles are introduced by homologous recombination into *embryonic stem (ES) cells.*

2. ES cells containing a knockout mutation in one allele of the gene being studied are introduced into early mouse embryos. The resultant mice will be **chimeras** containing tissues derived from both the transplanted ES cells and the host cells. These cells can contribute to both the germ-cell and somatic-cell populations.

3. Chimeric mice are mated to assess whether the mutation is incorporated into the germ line.

4. Mice each heterozygous for the knockout mutation are mated to produce homozygous knockout mice.

The isolation and culture of embryonic stem cells, which are derived from the blastocyst, are illustrated in Figure 8-32. These cells can be grown in culture through many generations. Exogenous DNA containing a mutant allele of the gene being studied is introduced into ES cells by transfection. The introduced DNA recombines with chromosomal sequences in about 1 cell out of 100 (i.e., 1 percent recombination frequency). In some cells, the added DNA recombines with the homologous chromosomal site, but recombination at other chromosomal sites (i.e., nonhomologous recombination) occurs 10^3–10^4 times more frequently. The small fraction of cells in which homologous recombination takes place can be identified by a combination of positive and negative selection: positive selection to identify cells in which any recombination occurs and negative selection to remove cells in which recombination takes place at nonhomologous sites.

For this selection scheme to work, the DNA constructs introduced into ES cells need to include, in addition to sequences used to knock out the gene of interest, two selectable marker genes (Figure 8-33). One of these additional genes *(neo^r)* confers neomycin resistance; it permits positive

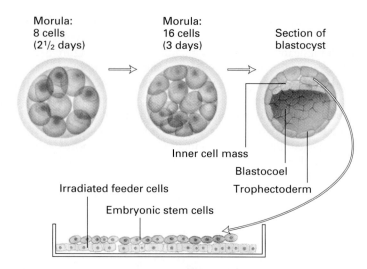

▲ **FIGURE 8-32 Preparation of embryonic stem (ES) cells.** Fertilized mouse eggs divide slowly at first; after $4\frac{1}{2}$ days, they form the blastocyst, a hollow structure composed of about 100 cells surrounding an inner cavity called the *blastocoel*. Only ES cells, which constitute the inner cell mass, actually form the embryo. Other cells form the trophectoderm, which gives rise to the membranes (amnion and placenta) by which the embryo is attached to the uterine wall. Embryonic stem cells can be removed from the blastocyst and grown on lethally irradiated "feeder cells." [See E. Robertson et al., 1986, *Nature* **323**:445.]

selection of cells in which either homologous (specific) or nonhomologous (random) recombination has occurred. The second selective gene, the thymidine kinase gene from herpes simplex virus *(tk^{HSV})* confers sensitivity to ganciclovir, a cytotoxic nucleotide analog; this gene permits negative

▶ **FIGURE 8-33 Isolation of mouse ES cells with a gene-targeted disruption by positive and negative selection.** (a) When exogenous DNA is introduced into ES cells, random insertion via nonhomologous recombination occurs much more frequently than gene-targeted insertion via homologous recombination. Recombinant cells in which one copy of the gene *X* (orange) is disrupted can be obtained by using a recombinant vector that carries gene *X* disrupted with *neo^r* (light red), a neomycin-resistance gene, and, outside the region of homology, *tk^{HSV}* (purple), the thymidine kinase gene from herpes simplex virus. The viral thymidine kinase, unlike the endogenous mouse enzyme, can convert the nucleotide analog ganciclovir into the monophosphate form; this is then modified to the triphosphate form, which inhibits cellular DNA replication in ES cells. Thus ganciclovir is cytotoxic for recombinant ES cells carrying the *tk^{HSV}* gene. Nonhomologous insertion includes the *tk^{HSV}* gene, whereas homologous insertion doesn't; therefore, only cells with nonhomologous insertion are sensitive to ganciclovir. (b) Recombinant cells are selected by treatment with neomycin, since cells that fail to pick up DNA or integrate it into their genome are neomycin-sensitive. The surviving recombinant cells are treated with ganciclovir. Only cells with a targeted disruption in gene *X*, and therefore lacking the *tk^{HSV}* gene, will survive. [See S. L. Mansour et al., 1988, *Nature* **336**:348.]

selection of ES cells in which nonhomologous recombination has occurred. Only ES cells that undergo homologous recombination (i.e., gene-targeted specific insertion of the DNA construct) can survive this selection scheme.

Once ES cells heterozygous for a knockout mutation in the gene of interest are obtained, they are injected into a recipient mouse blastocyst, which subsequently is transferred into a surrogate pseudopregnant mouse (Figure 8-34). If the

(a) Formation of ES cells carrying a knockout mutation

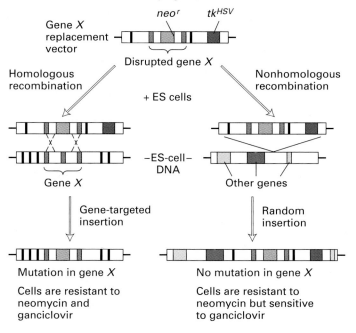

(b) Positive and negative selection of recombinant ES cells

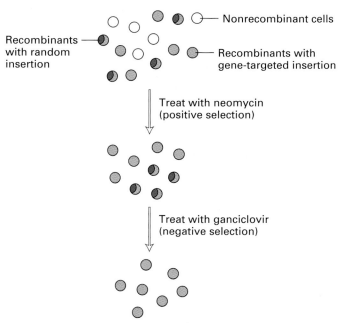

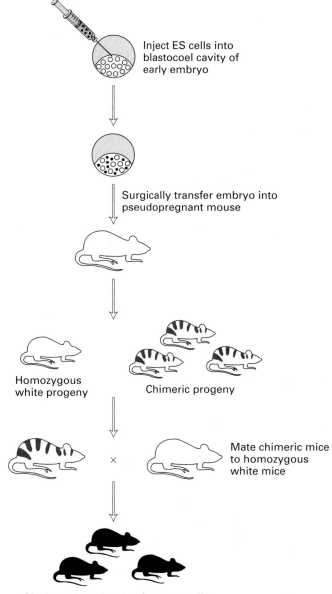

▲ FIGURE 8-34 **General procedure for producing gene-targeted knockout mice.** Embryonic stem (ES) cells heterozygous for a knockout mutation in a gene of interest (X) and homozygous for a marker gene (here, black coat color) are transplanted into the blastocoel cavity of 4.5-day embryos that are homozygous for an alternate marker (here, white coat color). The early embryos then are implanted into a pseudopregnant female. Some of the resulting progeny are chimeras, indicated by their black and white coats. Chimeric mice then are backcrossed to white mice; black progeny from this mating have ES-derived cells in their germ line. By isolating DNA from a small amount of tail tissue, it is possible to identify black mice heterozygous for the knockout allele. Intercrossing of these black mice produces individuals homozygous for the disrupted allele, that is, knockout mice. [Adapted from M. R. Capecchi, 1989, *Trends Genet.* **5**:70.]

ES cells also are homozygous for a visible marker trait (e.g., coat color), then chimeric progeny carrying the knockout mutation can be identified easily. These are then mated with mice homozygous for another allele of the marker trait to determine if the knockout mutation is incorporated into the germ line. Finally, mating mice, each heterozygous for the knockout allele, will produce progeny homozygous for the knockout mutation.

Cell-Type-Specific Gene Knockout in Mice In most cases, investigators are interested in examining the effects of knockout mutations in a particular region of the mouse, at a specific stage in development, or both. Since most genes function in different parts of the organism and at different times, a knockout mouse may die or have defects in various tissues prior to the stage to be analyzed. To address this problem, mouse geneticists have devised a clever technique using site-specific DNA recombination sites (called *loxP sites*) and the enzyme, called *Cre*, that catalyzes recombination between them. The loxP-Cre recombination system is

present in bacteriophage P1, but also promotes recombination when placed in mouse cells.

Homologous recombination strategies discussed in the previous section are used to obtain mice in which loxP sites are inserted so they flank the gene of interest or an essential exon. Since the inserted loxP sites are not within exons, they do not by themselves disrupt gene function. Transgenic mice also are prepared carrying the *cre* gene linked to a cell-type-specific promoter. As depicted in Figure 8-35, mating of these two types of mice will yield progeny that carry the gene of interest modified by insertion of flanking lox P sites and the *cre* gene controlled by a cell-type-specific promoter. In these mice, recombination between the loxP sites, which disrupts the gene of interest, will occur only in those cells in which the promoter is active and therefore producing the Cre protein necessary to induce the recombination.

One important example of this technique comes from studies on learning and memory. Earlier pharmacological

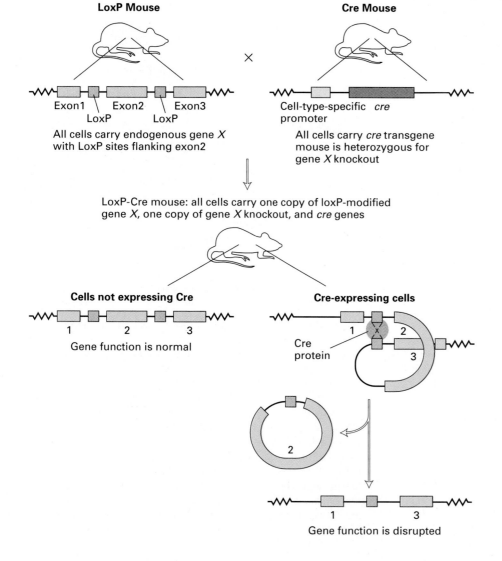

LoxP Mouse

Exon1   Exon2   Exon3
   LoxP   LoxP
All cells carry endogenous gene *X* with LoxP sites flanking exon2

×

Cre Mouse

Cell-type-specific *cre* promoter

All cells carry *cre* transgene mouse is heterozygous for gene *X* knockout

LoxP-Cre mouse: all cells carry one copy of loxP-modified gene *X*, one copy of gene *X* knockout, and *cre* genes

Cells not expressing Cre

1   2   3
Gene function is normal

Cre-expressing cells

1   2
3
Cre protein

2

1   3
Gene function is disrupted

◀ **FIGURE 8-35 Cell-type-specific gene knockouts using the loxP-Cre recombination system.** Two loxP sites are inserted on each side of an essential exon (2) of the gene of interest (i.e., gene X) (blue) by homologous recombination. These sites do not disrupt gene function. The loxP-containing mouse is crossed to a transgenic mouse carrying a cell-type-specific promoter controlling expression of the Cre recombinase, which induces recombination between loxP sites. This mouse is heterozygous for a constitutive gene X knockout. In the resulting loxP-Cre mouse, Cre protein is produced only in those cells in which the promoter is active, and in those cells recombination therefore occurs between the loxP sites, leading to deletion of exon 2. Since the other allele is a constitutive gene X knockout, deletion between the loxP sites results in complete loss of function in all cells expressing Cre.

and physiological studies had indicated that normal learning requires a specific neurotransmitter receptor, the NMDA class of glutamate receptors, in a specific region of the brain called the hippocampus. But mice in which the gene encoding an NMDA receptor subunit was knocked out died neonatally, precluding analysis of the receptor's role in learning. Cell-type-specific inactivation of the receptor was achieved by constructing mice carrying a Cre gene expressed in a subclass of hippocampal neurons and two different alleles of the receptor subunit gene, an allelle containing the loxP sites and a conventional knockout allele. These mice survived to adulthood and showed learning and memory defects, confirming a role for these receptors in normal learning and memory.

Use of Knockout Mice to Study Human Genetic Diseases Gene knockout can produce model systems for studying inherited human diseases. Such model systems are powerful tools for investigating the nature of genetic diseases and the efficacy of different types of treatment, and for developing effective gene therapies to cure these often devastating diseases.

Recent studies on cystic fibrosis illustrate this use of the knockout technique. Cystic fibrosis, which afflicts about 1 in 2000 Caucasians, is caused by an autosomal recessive mutation in the *CFTR* gene (see Figure 8-17b). This gene was cloned by positional cloning strategies, and the biochemical function of its encoded protein studied. Using the human gene, researchers isolated the homologous mouse gene and subsequently introduced mutations in it. The gene-knockout technique was then used to produce homozygous mutant mice, which showed symptoms (i.e., a phenotype) similar to those of humans with cystic fibrosis. These knockout mice are currently being used as a model system for studying this genetic disease and developing effective therapies.

Foreign Genes Can Be Introduced into Plants and Animals

In the previous section we discussed techniques for replacing one form of a gene with another through homologous recombination. In this section we discuss methods for producing transgenic organisms, which carry cloned genes that have integrated randomly into the host genome.

Transgenic technology has numerous experimental applications and potential agricultural and therapeutic value. For instance, dominantly acting alleles of tumor-causing genes can be used to produce transgenic mice, thus providing an animal model for studying cancer. In *Drosophila*, transgenes often are used to determine whether a cloned segment of DNA corresponds to a gene defined by mutation. If the cloned DNA is indeed the gene in question, then introducing it as a transgene into a mutant fly will transform the mutant into a phenotypically normal individual. Transgenic plants may be commercially valuable in agriculture. Plant scientists, for example, have developed transgenic tomatoes that exhibit reduced production of ethylene, which promotes fruit ripening. The ripening process is delayed in these transgenic tomatoes, thus prolonging their shelf life. Finally, transgenic technology is a critical component in the burgeoning field of gene therapy for human genetic diseases.

Transgenic Mice As noted in the discussion of knockout mice, specific integration of exogenous DNA into the genome of mouse cells by homologous recombination occurs at a very low frequency. In contrast, the frequency of random integration of exogenous DNA into the mouse genome at non-homologous sites is very high. Because of this phenomenon, the production of transgenic mice is a highly efficient and straightforward process.

As outlined in Figure 8-36, foreign DNA containing a gene of interest is injected into one of the two pronuclei (the male and female haploid nuclei contributed by the parents) of a fertilized mouse egg before they fuse. The injected DNA has a good likelihood of being randomly integrated into the chromosomes of the diploid zygote. Injected eggs then are transferred to foster mothers in which normal cell growth and differentiation occurs. About 10–30 percent of

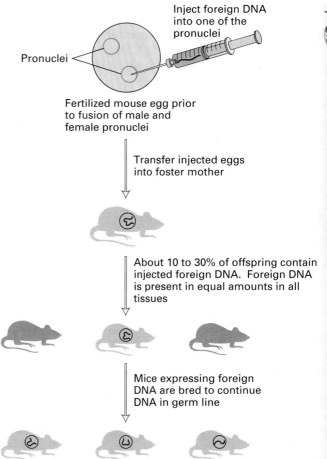

Inject foreign DNA into one of the pronuclei

Pronuclei

Fertilized mouse egg prior to fusion of male and female pronuclei

Transfer injected eggs into foster mother

About 10 to 30% of offspring contain injected foreign DNA. Foreign DNA is present in equal amounts in all tissues

Mice expressing foreign DNA are bred to continue DNA in germ line

▲ **FIGURE 8-36 General procedure for producing transgenic mice.** [See R. L. Brinster et al., 1981, *Cell* **27**:223.]

the progeny will contain the foreign DNA in equal amounts (up to 100 copies per cell) in all tissues, including germ cells. Immediate breeding and backcrossing (parent-offspring mating) of the 10–20 percent of these mice that breed normally can produce pure transgenic strains homozygous for the transgene.

Numerous examples of the use of transgenic mice for studying various aspects of normal mammalian biology are presented in other chapters. They also provide a model system for studying disease processes. For example, many forms of cancer are promoted by normal cellular genes acting in a dominant fashion owing to their misregulated activity. Although transgenic mice carrying one of these genes, called *myc*, develop normally, tumors form at a high frequency. The observation that only a small number of cells expressing the transgene develop tumors supports a model in which additional genetic changes are necessary for tumors to form. These mice may provide an important tool for identifying those changes.

Transgenic Fruit Flies Foreign DNA can be incorporated into the *Drosophila* germ-line genome by the technique of P-element transformation (Figure 8-37). This technique makes use of a segment of the P element, a highly mobile DNA element, which can transpose (jump) from an extrachromosomal element into a chromosome. (Mobile DNA elements are discussed in detail in Chapter 9.) Generally, this procedure results in incorporation of a single copy of the

transgene into the *Drosophila* genome. In contrast, transgenic mice carry multiple copies of the transgene incorporated into their chromosomes. In both organisms, however, the chromosomal insertion site is highly variable.

Flies that develop from injected embryos will carry some germ cells that have incorporated the transgene; some of their progeny will carry the transgene in all somatic and germ-line cells, giving rise to pure transgenic lines. Individuals carrying the transgene are recognized by expression of a marker gene (e.g., one affecting eye color) that is also

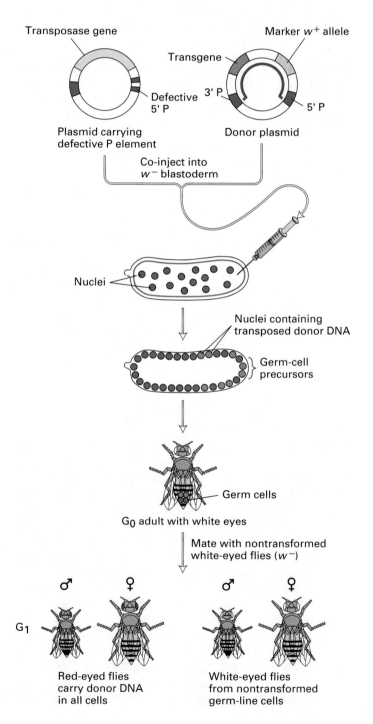

▶ **FIGURE 8-37 Generation of transgenic fruit flies by P-element transformation.** The P element, a mobile genetic element, can move from one place in the genome to another. This movement (transposition) is catalyzed by transposase, which is encoded by the P element; the 3′ and 5′ ends of the P element are recognized by transposase and are required for transposition to occur. To produce transgenic fruit flies by this method, the functionally different regions of the P element are incorporated into two different bacterial plasmids. The donor plasmid contains three necessary elements: the transgene (orange); a marker gene (green) used to indicate flies in which the plasmid DNA is transposed to a recipient chromosome; and both ends of the P element (dark purple)—3′ P and 5′ P— flanking the other two genes. It does not contain transposase. In this example, the marker is the dominant w^+ allele, which confers red eye color. The red bracket indicates the segment of the donor plasmid that can transpose into the fly genome. The other plasmid carries the P element (encoding transposase) with mutations in one end, which prevent it from transposing. The two plasmids are co-injected into blastoderm embryos homozygous for the recessive w^- allele, which confers white eye color. Transposase synthesized from the gene on the P-element plasmid catalyzes transposition of the donor plasmid DNA into the fly genome. Because transposition occurs only in germ-line cells (not in somatic cells), all the G_0 adults that develop from injected embryos have white eyes. Mating of these flies with white-eyed flies will yield some G_1 red-eyed progeny carrying the transgene and the marker allele (w^+) in all cells.

present on the donor DNA. Although the transgenes in *Drosophila* and mice insert in chromosomal sites different from the position of the corresponding endogenous gene, they usually are expressed in the right tissue and at the right time during development. Examples of the importance of this technology for studying development are discussed in Chapter 14.

Transgenic Plants In nature, plant cells often live in close association with certain bacteria, which may provide a convenient vehicle for introducing cloned DNA into plants. *Agrobacterium tumefaciens,* for example, attaches to the cells of dicotyledonous plants and causes the formation of plant tumors known as *galls.* (Plants with two leaflets from each seed are called *dicotyledons,* or *dicots;* plants with one leaflet are called *monocots.*) This bacterium introduces a circular DNA molecule, called the *Ti* (tumor-inducing) *plasmid,* into the plant cell in a manner similar to bacterial conjugation. The plasmid DNA then re-

combines with the plant DNA. Since the Ti plasmid has been isolated, new genes can be inserted into it using recombinant DNA techniques and the Ti genes causing tumors can be disrupted. The resulting recombinant plasmid can then transfer desired genes into plant cells (Figure 8-38).

An especially useful characteristic of plants for transgenic studies is the ability of cultured plant cells to give rise to mature plants. Meristematic (growing) cells from dissected plant tissue or cells within excised parts of a plant will grow in culture to form *callus tissue,* an undifferentiated lump of cells. Under the influence of plant growth hormones, different plant parts (roots, stems, and leaves) develop from the callus and eventually grow into whole, fertile plants. When an agrobacterium containing a recombinant Ti plasmid infects a cultured plant cell, the newly incorporated foreign gene is carried into the plant genome.

As noted above *A. tumefaciens* readily infects dicots (petunia, tobacco, carrot) but not monocots; reliable techniques for introducing genes into monocots are still being developed. Direct introduction of DNA by electroporation has been successful in rice plants, which are monocots, and the future looks bright for the manipulation of other commercially important monocotyledonous crop plants. Also available for gene-transfer experiments are cells of a tiny, rapidly growing member of the mustard family called *Arabidopsis thaliana.* This plant is well-suited to genetic analysis of a variety of developmental and physiological processes. It takes up little space, is easy to grow, and has a small genome, and genes defined by mutations can be cloned by positional cloning strategies.

SUMMARY Gene Replacement and Transgenic Animals

- Genes can be modified in vitro by a variety of enzymatic and chemical methods. Modified genes can be incorporated into the germ line at their original genomic location by homologous recombination, producing knockouts, or at different sites by nonhomologous recombination, producing transgenics.

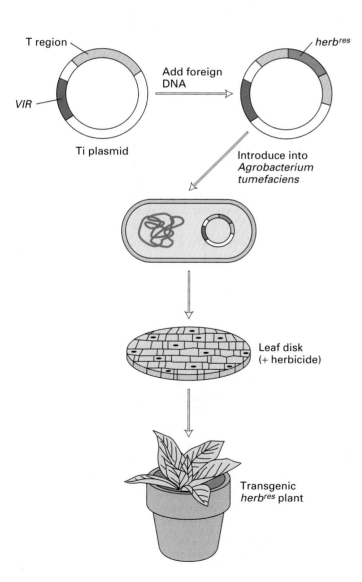

◀ **FIGURE 8-38 Production of transgenic plants with recombinant Ti plasmids.** In nature, the Ti (tumor-inducing) plasmid in *A. tumefaciens* gains entry into a plant and is integrated into the plant DNA owing to the action of the *VIR* (virulent) region of the plasmid. Tumor-like growths (galls) result from action of the T region of the gene. By recombinant DNA techniques, a foreign gene (orange) is introduced within the T region of a Ti plasmid, thus destroying its tumor-inducing ability. An agrobacterium containing such a recombinant Ti plasmid then is used to introduce the foreign gene into plant cells. When a selectable transgene is used—here one conferring resistance to herbicide (*herb*^res)—recombinant plants can be selected. [See R. T. Fraley et al., 1983, *Proc. Nat'l. Acad. Sci. USA* **80**:4803.]

- Any cloned yeast gene or mouse gene can be mutated and used to generate gene-targeted knockouts in these organisms (see Figures 8-27 and 8-31). Use of the loxP-Cre recombination system from bacteriophage P1 permits production of mice in which a gene is knocked out in a specific tissue (see Figure 8-35).

- Clues about the normal function of a gene can be deduced from the observed effects of disrupting it by the knockout technique. Mouse knockouts can provide models for human genetic diseases such as cystic fibrosis.

- Transgenic single-cell organisms, plants, and animals can be produced readily by several different methods. These modified organisms contain one or more copies of a cloned gene integrated into in the genome.

- The transgenic technique can be used to introduce a normal copy of a gene into a mutant organism, thereby identifying a cloned DNA corresponding to a mutation-defined gene. It also is used to study sequences necessary for gene expression, to develop mouse models of dominant forms of human diseases, to modify plants, and to investigate the relationship between the structure of a protein encoded by a gene and its function.

PERSPECTIVES for the Future

During the next 5–10 years, the human genome, as well as the genomes of several model organisms, will be completely sequenced. To understand the function of each gene, it will be essential to isolate mutations in them. The entire genomic sequences from the yeast *S. cerevisiae* and the nematode *C. elegans* are already known, so that each gene can be knocked out. Although mutational analysis of all human genes is not ethically permissible, the vast majority of human genes have mouse homologs, which will be identified and then subjected to knockout analysis. Because of the remarkable conservation of gene function through evolution, homologs of many human genes and their mutations also will be found in lower organisms such as *C. elegans*, *Drosophila*, and yeasts. As you will see throughout this book, genetic studies in these simpler, experimentally assessable organisms have helped to dissect fundamental cellular processes and biochemical pathways in humans. Although genetic studies provide critical insight into gene function, they must be complemented by biochemical and cell biological studies to develop an integrated understanding of the role of specific genes in regulating organismal development, behavior, and function.

The identification and cloning of mutation-defined disease genes by positional cloning will greatly expand our understanding of the biochemical basis of numerous human diseases. In many cases, it will be possible to generate mouse models for human disease through knockout or transgenic technologies. Through improved biochemical knowledge

and model systems, powerful approaches toward developing cures will emerge. The tools and approaches of genetics have not only provided a way to understand disease better, they have also contributed to the exciting prospect of using genes themselves as therapeutic agents to cure disease, an approach called *gene therapy*. For instance, strategies to deliver a normal copy of a gene to patients' homozygous for recessive loss-of-function mutations are being devised to correct the defect. Alternatively, strategies to treat cancer are directed toward introducing specific genes that will induce the cancer cell to kill itself or to be killed effectively by the patient's immune system. This is a new and exciting field with enormous potential. Gene therapy is still in an experimental phase, however, and has not been widely applied to the treatment of any medical condition. Currently, a major challenge is to develop ways of specifically delivering genes to the cells of interest and controlling their expression.

We are just beginning to appreciate the enormous impact of sophisticated genetic knowledge on society. As more and more disease genes are identified and cloned, it will be possible to determine those individuals who are at risk for specific diseases, those who are carriers, and those who, though healthy now, will become afflicted by disease later. In some cases, these developments will benefit society. Even now, DNA from very early human embryos produced by in vitro fertilization of sperm and egg from parents who are carriers for deadly diseases can be genotyped, allowing for selection of only healthy embryos to be implanted into the mother.

The benefits of genetics, however, come with substantial risks. Will employers and medical insurance companies be given access to information about who is at risk for disease? Who will decide how society applies our knowledge about the relationship between specific genes and human traits? Will parents of the future choose their children's eye color, sex, or even disposition? As genetics becomes a more pervasive part of our world, it will become increasingly important for average citizens, not just students of biology, to understand the principles of genetics.

PERSPECTIVES in the Literature

Recessive mutations in a human gene, let us simply call it gene *X*, cause a serious disease for which there is no known treatment. By cloning the gene you have determined its sequence. While the protein encoded by the gene bears no similarity to proteins of known biochemical or cell biological function, it is nearly identical with proteins encoded in other organisms. The sequences for these organisms have been uncovered through efforts to sequence the entire genomes of model organisms. You are particularly excited that a mouse and yeast gene *X* exist. Describe what approaches you would take to devise a model system to study the basic molecular and cell biological function of the protein encoded by gene *X* and how you would generate a model for the human disease to study its progression and possible treatments. For a specific illustration of this general concept, see the following articles on the Werner's syndrome gene:

Gray, M. D., et al. 1997. The Werner syndrome protein is a DNA helicase. *Nature Genet.* 17:100–103.

Sinclair, D. A., K. Mills, and L. Guarente. 1997. Accelerated aging and molecular fragmentation in the yeast sgs1. *Science* 277:1313–1316.

Yu, C. E., et al. 1996. Positional cloning of the Werner's syndrome gene. *Science* 272:258–262.

Testing Yourself on the Concepts

1. What are the similarities and differences between mitosis and meiosis?

2. Diagram the pattern of inheritance of a recessive mutation on the X chromosome of *Drosophila*. Begin with a female heterozygous for the mutation mated with a wild-type male. What are the phenotypes of the female and male F_1 progeny? What are the phenotypes of the progeny of crosses between the F_1 females and males?

3. Describe how knockout mice are used as models for human disease.

4. Describe the principle behind using recombination frequency to order the position of genes on chromosomes.

MCAT/GRE-Style Questions

Key Concept Please read the section titled "Physical and Genetic Maps Can Be Correlated with the Aid of Known Markers" (p. 277) and refer to Figure 8-26; then answer the following questions:

1. The data presented in Figure 8-26 show that
   a. RFLPs A–C are closer to the *shi* gene than to the *sd* gene.
   b. RFLPs A–C are closer to the *sd* gene than to the *shi* gene.
   c. The *sd* gene influences the frequency of the appearance of RFLPs D–G.
   d. The positions of RFLPs D and E are well defined.

2. When analyzing the 23 recombinant sd$^+$ shi$^+$ flies, attention would be focused on
   a. The frequency of the a–j forms of the RFLPs, since these are contributed by the male partner in the cross.
   b. The frequency of the a–j forms of the RFLPs, since these are contributed by the female partner in the cross.
   c. The frequency of the A–J RFLPs, since these forms are the only markers unique to a chromosome.

d. The male and female progeny, to look for differences in recombination frequency on the X and Y chromosomes.

3. In this study, the frequency of the RFLPs in the recombinants is used to
   a. Define the location of the RFLPs on the YAC.
   b. Locate *shi* on the YAC relative to the position of the RFLPs.
   c. Define precisely the location of *shi* on the YAC.
   d. Define the location of the RFLPs on the *Drosophila* chromosome.

4. In the crosses used in this study, females heterozygous for *shi* and *sd* are used because
   a. Meiotic recombination occurs only in male *Drosophila*.
   b. Meiotic recombination occurs only in female *Drosophila*.
   c. *Shi* and *sd* are located on the X chromosome.
   d. Both b and c.

5. The *sd* locus is included in this study to
   a. Provide an additional molecular marker to study recombination frequencies.
   b. Provide a marker that, facilitates identifying recombinants within the region containing the RFLPs.
   c. Provide a marker that is linked to *shi* but not to the RFLPs.
   d. Ensure that recombination is homologous.

Key Experiment Please read the section titled "Recessive Lethal Mutations in Diploids Can Be Screened by Use of Visible Markers" (p. 263) and refer to Figure 8-10; then answer the following questions:

6. The *ry* marker is used to
   a. Identify the F_3 progeny that are homozygous for the induced mutation.
   b. Identify the F_2 progeny that are heterozygous for the induced mutation.
   c. Provide an additional lethal marker when homozygous.
   d. Provide an additional lethal marker when heterozygous.

7. The position of the recessive lethal mutation on chromosome 3 is defined by
   a. Fluorescent in situ hybridization.
   b. Banding analysis.
   c. The location of the dominant D1 and D2 markers.
   d. It is not possible to define the location of the mutation on chromosome 3 based on these data.

8. In order to maintain a stock of the flies with the induced mutation, you must
   a. Mate the viable F_3 progeny with one another.

b. Mate the viable F$_3$ progeny with flies containing the D$_1$ and D$_2$ mutations.

c. Mate the F$_2$ progeny that display the D$_1$ or D$_2$ phenotype with one another.

d. Mate the F$_2$ progeny that display the D$_1$ or D$_2$ phenotype with flies containing both dominant mutations.

9. The P$_2$ cross is carried out with single male flies from the F1 progeny in individual vials to

a. Maintain all the mutations generated by the chemical mutagen.

b. Maintain the stock of flies with both dominant markers.

c. Decrease the number of flies to be further screened.

d. Isolate a strain of flies containing one induced mutation.

Key Application Please read the section titled "DNA Polymorphisms Are Used to Map Human Mutations" (p. 271) and refer to Figure 8-20; then answer the following:

10. Which of the following enzymes will be most useful in defining RFLPs:

a. *Apa*I (GGGCCC).

b. *Hind*III (AAGCTT).

c. *Xba*I (TCTAGA).

d. *Sal*I (GTCGAC).

11. If the top chromosome in Figure 8-20 contained an additional restriction enzyme recognition site for enzyme A, located between sites a$_2$ and a$_3$, the number of bands appearing on the Southern blot after digestion with this enzyme would be

a. 1.

b. 2.

c. 3.

d. 4.

12. RFLPs can be used to study the inheritance of human genetic diseases by

a. Distinguishing the regions of the chromosomes that contain genes from the regions that do not contain genes.

b. Following the co-inheritance of the gene for the disease with random point mutations.

c. Analyzing the co-inheritance of the gene for the disease with other phenotypic markers.

d. Promoting meiotic recombination.

13. Mapping the location of DNA polymorphisms for inherited diseases is *best* done if all the following are true *except*:

a. There are many polymorphisms in the region of the gene.

b. The polymorphic marker is recessive.

c. Many individuals bearing the gene can be tested.

d. PCR can be used to screen for polymorphisms.

Key Terms

alleles *255*
complementation *264*
dominant *255*
gametes *256*
gene-targeted
   knockout *281*
genotype *255*
heterozygous *255*
homologs *256*
homozygous *255*
linkage *269*
mapping *255*
mutations *254*

phenotype *255*
point mutation *257*
positional cloning *274*
reading frame *257*
recessive *255*
recombination *269*
segregation *256*
suppressor mutations *265*
temperature-sensitive (ts)
   mutants *262*
transgenes *281*
wild-type *254*

References

Mutations: Types and Causes

Griffiths, A. G. F., et al. 1997. *An Introduction to Genetic Analysis*, 6th ed. W. H. Freeman and Company.

Strachan, T., and A. P. Read. 1996. *Human Molecular Genetics*. Wiley-Liss.

Isolation and Analysis of Mutants

Adam, A. E. M., D. Botstein, and D. B. Drubin. 1989. A yeast actin-binding protein is encoded by *sac6*, a gene found by suppression of an actin mutation. *Science* **243**:231.

Beadle, G. W., and E. L. Tatum. 1941. Genetic control of biochemical reactions in *Neurospora*. *Proc. Nat'l Acad. Sci USA* **27**:499.

Hartwell, L. H. 1967. Macromolecular synthesis of temperature-sensitive mutants of yeast. *J. Bacteriol.* **93**:1662.

Hartwell, L. H. 1974. Genetic control of the cell division cycle in yeast. *Science* **183**:46.

Jarvik, J., and D Botstein. 1975. Conditional-lethal mutations that suppress genetic defects in morphogenesis by altering structural proteins. *Proc. Nat'l Acad. Sci. USA* **72**:738.

Nüsslein-Volhard, C., and E. Wiechaus. 1980. Mutations affecting segment number and polarity in *Drosophila*. *Nature* **287**:795.

Genetic Mapping of Mutations

The *C. elegans* Consortium. 1998. Genome sequence of the nematode *C elegans*: a platform for investigating biology. *Science* **282**:2012.

Foote, S., et al. 1992. The Y chromosome: overlapping DNA clones spanning the euchromatic region. *Science* **258**:60.

Francke, U., et al. 1985. Minor Xp21 chromosome deletion in a male associated with expression of Duchenne muscular dystrophy, chronic granulomatous disease, retinitis pigmentosa, and McLeod syndrome. *Am. J. Hum. Genet.* **37**:250

Goffeau, A., et al. 1996. Life with 6000 genes. *Science* **274**:546.

Olson, M., et al. 1989. A common language for physical mapping of the human genome. *Science* **245**:1434.

Painter, T. S. 1934. Salivary chromosomes and the attack on the gene. *J. Hered.* **25**:465.

Peters, J. A., ed. 1959. *Classic Papers in Genetics*. Prentice-Hall.

Raap, A. K. 1998. Advances in fluorescence in situ hybridization. *Mut. Res.* **400**:287.

Sturtevant, A. H. 1913. The linear arrangement of six sex-linked factors in *Drosophila,* as shown by their mode of association. *J. Exper. Zool.* **14**:43.

White, R., and J-M. Lalouel. 1988. Sets of linked genetic markers for human chromosomes. *Ann. Rev. Genet.* **22**:259.

Molecular Cloning of Genes Defined by Mutations

Bender, W., P. Spierer, and D. Hogness. 1983. Chromosomal walking and jumping to isolate DNA from the *Ace* and *rosy* loci and the Bithorax complex in *Drosophila melanogaster. J. Mol. Biol.* **168**:17.

Botstein, D., et al. 1980. Construction of a genetic linkage map in man using restriction fragment length polymorphisms. *Am. J. Genet.* **32**:314–331.

Cotton, R. G. H., N. R. Rodrigues, and R. D. Campbell. 1988. Reactivity of cytosine and thymine in single-base-pair mismatches with hydroxyamine and osmium tetroxide and its application to the study of mutations. *Proc. Nat'l Acad. Sci. USA* **85**:4397.

Hastbacka, T., et al. 1994. The diastrophic dysplasia gene encodes a novel sulfate transporter: positional cloning by fine-structure linkage disequilibrium mapping. *Cell* **78**:1073.

Orita, M., et al. 1989. Rapid and sensitive detection of point mutations and DNA polymorphisms using the polymerase chain reaction. *Genomics* **5**:874.

Orita, M., et al. 1989. Detection of polymorphisms of human DNA by gel electrophoresis as single-stranded conformation polymorphisms. *Proc. Nat'l Acad. Sci. USA.* **86**:2766.

Rossiter, B. J. F., and C. T. Caskey. 1990. Molecular scanning methods of mutation detection. *J. Biol. Chem.* **265**:12753.

Gene Replacement and Transgenic Animals

Capecchi, M. R. 1989. Altering the genome by homologous recombination. *Science* **244**:1288.

Davey, M. R., E. L. Rech, and B. J. Mulligan. 1989. Direct DNA transfer to plant cells. *Plant Mol. Biol.* **13**:273.

Evans, M. J., and M. H. Kaufman. 1981. Establishment in culture of pluripotential cells from mouse embryos. *Nature* **292**:154.

Fung-Leung, W-P., and T. W. Mak. 1992. Embryonic stem cells and homologous recombination. *Curr. Opin. Immunol.* **4**:189.

Hasty, P., et al. 1991. Introduction of a subtle mutation into the *Hox-2.6* locus in embryonic stem cells. *Nature* **350**:243.

Hermes, J. D., et al. 1989. A reliable method for random mutagenesis: the generation of mutant libraries using spiked oligonucleotide primers. *Gene* **184**:143.

Horsch, R. B., et al. 1988. Leaf disc transformation. *Plant Mol. Biol. Manual* **A5**:1.

Mansour, S. L., K. R. Thomas, and M. R. Capecchi. 1988. Disruption of the proto-oncogene *int-2* in mouse embryo-derived stem cells: a general strategy for targeting mutations to non-selectable genes. *Nature* **336**:348.

Meyerowitz, E. M. 1989. *Arabidopsis,* a useful weed. *Cell* **56**:263.

Oeller, P. W., et al. 1991. Reversible inhibition of tomato fruit senescence by antisense RNA. *Science* **254**:437.

Rubin, G. M., and A. C. Spradling. 1982. Genetic transformation of *Drosophila* with transposable element vectors. *Science* **218**:341.

Taylor, J., J. Ott, and F. Eckstein. 1985. The rapid generation of oligonucleotide-directed mutations at high frequency using phosphorothioate-modified DNA. *Nucl. Acids Res.* **13**:8765.

Zimmer, A. 1992. Manipulating the genome by homologous recombination in embryonic stem cells. *Ann. Rev. Neurosci.* **15**:115.

9

Molecular Structure of Genes and Chromosomes

Sequencing and other molecular analyses have revealed that a very large fraction of all vertebrate genomes, perhaps well over 90 percent, does not encode precursors to mRNAs or any other RNAs. In multicellular organisms, this noncoding DNA contains many regions that are similar but not identical. Variations within some stretches of this *repetitious DNA* are so great that each single person can be distinguished by a DNA "fingerprint" based on these sequence variations. Moreover, some repetitious DNA sequences are not found in constant positions in the DNA of individuals of the same species. Such "mobile" DNA segments, which are present in both prokaryotic and eukaryotic organisms, can cause mutations when they move to new sites in the genome. These mobile segments probably have played an important role in evolution, even though they generally have no function in the life cycle of an individual organism.

In higher eukaryotes, DNA regions encoding proteins—that is, **genes**—lie amidst this expanse of nonfunctional DNA. In addition to the apparently nonfunctional DNA *between* genes, noncoding **introns** are common *within* genes of multicellular plants and animals. Introns are less common, but sometimes present, in single-celled eukaryotes and very rare in bacteria. Sequencing of the same protein-coding gene in a variety of eukaryotic species has shown that evolutionary pressure selects for maintenance of relatively similar sequences in the coding regions, or **exons.** In contrast, wide sequence variation, even including total loss, occurs among introns, suggesting that most of the sequence of introns is nonfunctional. Cloning and sequencing have also confirmed the

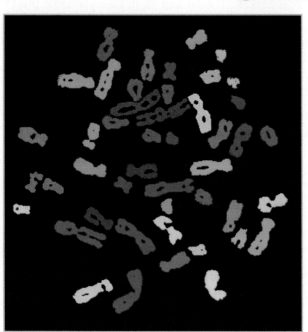

Pseudocolor image of human chromosomes in a metaphase cell from a male visualized by chromosome painting and fluorescent in situ hybridization (FISH).

widespread existence of "families" of similar genes encoding proteins with related, but distinct, specialized functions.

The sheer length of cellular DNA is a significant problem with which cells must contend. The DNA in a typical bacterial cell, which is about 10^3 times longer than the length of the cell, is folded and organized to fit within the cell. The total length of the DNA in eukaryotic cells is even longer compared with the cell diameter. Specialized eukaryotic proteins associated with nuclear DNA organize it into the structures of DNA and protein visualized as individual **chromosomes** during mitosis. Mitochondria and chloroplasts also contain DNA, probably evolutionary remnants of their origins, that encodes essential components of these vital organelles.

In this chapter we first present a molecular definition of genes and then discuss the main classes of eukaryotic DNA and the special properties of mobile DNA. Next we describe several examples of functional rearrangements of chromosomal DNA, including the process for generating functional antibody genes. We also consider the packaging of DNA and proteins into compact complexes, the large-scale structure of chromosomes, and the functional elements required for chromosome duplication and segregation. In the final section, we discuss organelle DNA.

9.1 Molecular Definition of a Gene

As we saw in Chapter 8, genes can be defined by classical genetic analysis (e.g., complementation analysis) of mutant phenotypes. Most such *mutation-defined genes* affect the function of a single protein. However, some mutations affect several proteins simultaneously, and some mutations affect only one of two related proteins encoded in the same region of DNA. Understanding how such phenomena occur requires knowledge of the molecular structure of genes and the mechanisms of mRNA synthesis.

In molecular terms, a gene commonly is defined as *the entire nucleic acid sequence that is necessary for the synthesis of a functional polypeptide*. According to this definition, a gene includes more than the nucleotides encoding the amino acid sequence of a protein, referred to as the *coding region*. A gene also includes all the DNA sequences required for synthesis of a particular RNA transcript. In some prokaryotic genes, DNA sequences controlling the initiation of transcription by RNA polymerase can lie thousands of base pairs from the coding region. In eukaryotic genes, transcription-control regions known as **enhancers** can lie 50 kb or more from the coding region. Other critical noncoding regions in eukaryotic genes are the sequences that specify 3′ cleavage and polyadenylation *[poly(A) sites]* and splicing of primary RNA transcripts. Mutations in these RNA-processing signals prevent expression of a functional mRNA and thus of the encoded polypeptide.

Although most genes are transcribed into mRNAs, which encode proteins, clearly some DNA sequences are transcribed into RNAs that do not encode proteins (e.g., tRNAs and rRNAs). However, because the DNA that encodes tRNAs and rRNAs can cause specific phenotypes when they are mutated, these DNA regions generally are referred to as tRNA and rRNA *genes,* even though the final products of these genes are RNA molecules and not proteins. Many other RNA molecules described in later chapters also are transcribed from non-protein-coding genes.

Bacterial Operons Produce Polycistronic mRNAs

As discussed in Chapter 4, genes encoding enzymes involved in related functions often are located next to each other in bacterial chromosomes. For example, the five genes encoding the enzymes required to synthesize the amino acid tryptophan from simple precursor molecules map in one contiguous stretch of the *E. coli* genome (Figure 9-1a). This cluster of genes comprises a single **transcription unit** referred to as an **operon**. The full set of genes is transcribed to produce a single ≈7-kb mRNA molecule. Ribosomes initiate translation at the beginning of each of the genes in this mRNA producing the five polypeptides required for tryptophan synthesis. Since a **cistron** is defined as a genetic unit that encodes a single polypeptide, *trp* mRNA, which encodes several polypeptides, is said to be *polycistronic.*

One consequence of the arrangement of bacterial genes into operons is that a single mutation can influence the expression of several proteins. For example, if a single point mutation (e.g., a base-pair change) in the transcription-control region of the *trp* operon prevents initiation of transcription by RNA polymerase, then expression of all *five* of the polypeptides required for tryptophan synthesis is eliminated.

Most Eukaryotic mRNAs Are Monocistronic and Contain Introns

In contrast to polycistronic mRNAs, which are common in prokaryotes, most eukaryotic transcription units produce mRNAs that encode only one protein. This distinction correlates with a fundamental difference in mRNA translation in prokaryotes and eukaryotes. As explained in Chapter 4, a bacterial polycistronic mRNA contains multiple ribosome-binding sites located near the start sites for all the protein-coding regions in the mRNA. As a consequence, translation initiation can begin at internal sites in a polycistronic mRNA molecule, producing multiple proteins (see Figure 4-17a). In most eukaryotic mRNAs, however, the 5′-cap structure directs ribosome binding, and translation begins at the closest AUG start codon (see Figure 4-37). As a result, only the sequence following the first AUG in an mRNA is translated. The primary transcripts of eukaryotic protein-coding genes generally are processed into a single type of mRNA, which

(a) Prokaryotic polycistronic transcription unit

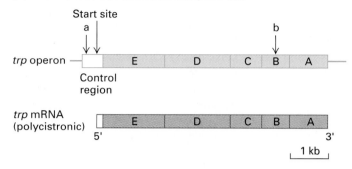

(b) Eukaryotic simple transcription unit

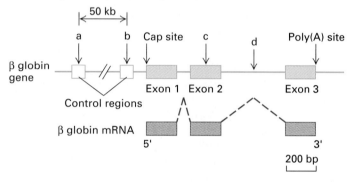

▲ FIGURE 9-1 Comparison of bacterial operons and simple eukaryotic transcription units. (a) The *trp* operon in the *E. coli* genome contains five genes (*A–E*) encoding enzymes required for synthesis of tryptophan. A control region located near the start site regulates transcription of the entire operon, which yields an ≈7-kb polycistronic mRNA. A mutation within the transcription-control region (a) can prevent expression of all the proteins encoded by the *trp* operon. In contrast, a mutation within any one gene of an operon (e.g., *b* in the *trpA* gene) generally affects only the protein encoded by that gene (TrpA protein). (b) A simple eukaryotic transcription unit includes a region that encodes one protein, extending from the 5' cap site to the 3' poly(A) site, and associated control regions. Intron sequences, which lie between exons, are removed during processing of the primary transcripts and thus do not occur in the functional monocistronic mRNA. Dashed lines indicate spliced-out introns. Mutations in a transcription-control region (a, b) may reduce or prevent transcription, thus reducing or eliminating synthesis of the encoded protein. A mutation within an exon (c) may result in an abnormal protein with diminished activity. A mutation within an intron (d) that introduces a new splice site results in an abnormally spliced mRNA encoding a nonfunctional protein.

trons, which are removed during RNA processing. In many cases, the introns in a gene are considerably longer than the exons. Thus of the ≈50,000 base pairs composing many human genes encoding average-size proteins, more than 95 percent are present in introns and noncoding 5' and 3' regions. Many large proteins in higher organisms have repeated domains and are encoded by genes consisting of repeats of similar exons separated by introns of variable length. For instance, fibronectin, a component of the extracellular matrix, is encoded by a gene containing multiple copies of three types of exons. Another example is the largest known human gene, the mutation-defined *DMD* gene, which is associated with Duchenne muscular dystrophy. The *DMD* gene, containing more than two million base pairs, is longer than the entire *Haemophilus influenzae* genome!

Simple and Complex Transcription Units Are Found in Eukaryotic Genomes

Eukaryotic genes that produce a single type of mRNA, encoding a single protein, are called *simple* transcription units. Mutations in exons, introns, and transcription-control regions all may influence the expression of proteins encoded by simple transcription units (Figure 9-1b). Unlike some mutations in bacterial operons, which can affect multiple proteins, mutations in simple eukaryotic transcription units can affect only one protein.

Although many transcription units in eukaryotes are simple, *complex* transcription units are quite common in multicellular organisms. The primary RNA transcript encoded by complex transcription units can be processed in more than one way by use of alternative poly(A) sites or splice sites, leading to formation of mRNAs containing different exons. For example, a transcription unit that contains two or more poly(A) sites can produce different mRNAs, each having the same 5' exons, but distinct 3' exons (Figure 9-2a). Another type of alternative RNA processing, called *exon skipping*, produces mRNAs with the same 5' and 3' exons but different internal exons (Figure 9-2b). Examples of both types of alternative RNA processing occur during sexual differentiation in *Drosophila* (see Figure 11-26). Commonly, one mRNA is produced from a complex transcription unit in some cell types, and an alternative mRNA is made in other cell types. For example, differences in RNA splicing of the primary fibronectin transcript in fibroblasts and hepatocytes determines whether or not the secreted protein includes domains that adhere to cell surfaces (see Figure 11-24).

A mutation in the control region or in an exon shared by alternative mRNAs will affect all the alternative proteins encoded by a given complex transcription unit. On the other hand, mutations in an exon present in only one of the alternative mRNAs will affect only the protein encoded by that mRNA. As a consequence, the relationship between the molecular definition of a gene and a genetic complementation group is not always straightforward for complex

is translated to give a single type of polypeptide (see Figure 4-19). Such eukaryotic mRNAs and their corresponding genes are *monocistronic*. Although the primary transcripts of some protein-coding genes are processed into more than one type of mRNA, each mRNA is translated into a single polypeptide.

Unlike bacterial and yeast genes, which generally lack introns, most genes in higher animals and plants contain in-

(a) Alternative 3' exons

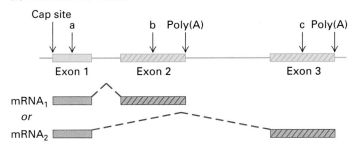

(b) Alternative internal exons

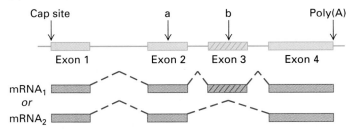

▲ **FIGURE 9-2 Two examples of complex eukaryotic transcription units and the effect of mutations on expression of the encoded proteins.** The RNA transcribed from a complex transcription unit (blue) can be processed in alternative ways to yield two or more functional monocistronic mRNAs. Dashed lines indicate spliced-out introns. (a) A complex transcription unit whose primary transcript has two poly(A) sites produces two mRNAs with alternative 3' exons. (b) A complex transcription unit whose primary transcript undergoes exon skipping during processing produces alternative mRNAs with the same 5' and 3' exons. In this example, some cell types would express the mRNA including exon 3, whereas in other cell types, exon 2 is spliced to exon 4, producing an mRNA lacking exon 3 and the protein sequence it encodes. In (a) and (b), mutations (designated *a*) within exons shared by the alternative mRNAs (solid red) affect the proteins encoded by both alternatively processed mRNAs. In contrast, mutations (designated *b* and *c*) within exons unique to one of the alternatively processed mRNAs (red with diagonal lines) affect only the protein encoded by that mRNA.

transcription units. For example, the complex transcription unit shown in Figure 9-2a encodes two proteins that have the same N-terminal sequence, encoded by their common 5' exons, and different C-terminal sequences, encoded by their unique 3' exons. Mutation *b* affects the protein encoded by mRNA 1, and mutation *c* the protein encoded by mRNA 2. Mutations *b* and *c* complement each other in a genetic complementation test, even though they occur in the same gene, because a chromosome with mutation *b* can express a normal protein encoded by the lower mRNA, and a chromosome with mutation *c* can express a normal protein encoded by the upper mRNA. However, a chromosome with mutation *a* in an exon common to both mRNAs would not complement either mutation *b* or *c*. In other words, mutation *a*

would be in the same complementation groups as mutations *b* and *c*, even though *b* and *c* would not be in the same complementation group!

SUMMARY   Molecular Definition of a Gene

- In molecular terms, a gene is the entire DNA sequence required for synthesis of a functional protein or RNA molecule. In addition to the coding regions (exons), a gene includes transcription-control regions and sometimes introns. Although the majority of genes encode proteins, some encode tRNAs, rRNAs, and other types of RNA.

- Most bacterial genes have no introns, whereas most genes of multicellular organisms do. The introns in human genes encoding average-size proteins are often much longer than the exons.

- Many bacterial proteins with related functions are encoded by contiguous genes regulated by a single transcription-control region. This type of gene cluster, called an operon, is transcribed into a single, polycistronic mRNA (see Figure 9-1a), which is translated to yield several different proteins.

- Most eukaryotic transcription units are transcribed into monocistronic mRNAs, each of which is translated into a single protein.

- The primary transcript produced from a simple eukaryotic transcription unit is processed into a single type of mRNA (see Figure 9-1b).

- The primary transcript produced from a complex eukaryotic transcription unit can be processed into two or more different mRNAs depending on the choice of splice sites and/or polyadenylation sites (see Figure 9-2). In the case of many complex units, one mRNA is produced in one cell type, while an alternative mRNA is produced in a different cell type.

9.2 Chromosomal Organization of Genes and Noncoding DNA

Having reviewed the relationship between transcription units and genes in prokaryotes and eukaryotes, we now consider the organization of genes on chromosomes and the relationship of noncoding DNA sequences to coding sequences.

Genomes of Higher Eukaryotes Contain Much Nonfunctional DNA

The abundance of noncoding sequences in the genomes of higher organisms is illustrated in Figure 9-3, which depicts

(a) *S. cerevisiae* (chromosome III)

tRNA gene   Open reading frame

(b) Human β-globin gene cluster (chromosome 11)

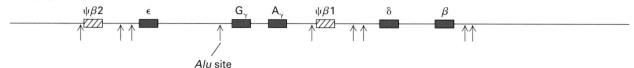

Alu site

▲ **FIGURE 9-3 Diagrams of ≈80-kb region from chromosome III of the yeast *S. cerevisiae* and the β-globin gene cluster on human chromosome 11.** (a) In the yeast DNA, blue boxes indicate open reading frames; it is not clear whether all these potential protein-coding sequences are functional genes. (b) In the human DNA, the blue boxes represent transcribed regions that encode the indicated globin-type proteins. Each globin-type gene has a similar arrangement of exons and introns (not shown). The human β-globin gene cluster contains two pseudogenes (diagonal lines); these regions are related to the functional globin-type genes but are not transcribed. Red arrows indicate the locations of *Alu* sequences, an ≈300-bp noncoding repeated sequence that is abundant in the human genome. Note the much higher proportion of noncoding to coding sequences in the human DNA than in the yeast DNA. [Part (a) see S. G. Oliver et al., 1992, *Nature* **357**:28; part (b) see F. S. Collins and S. M. Weissman, 1984, *Prog. Nucl. Acid Res. Mol. Biol.* **31**:315.]

the protein-coding regions in an 80-kb stretch of DNA from the yeast *S. cerevisiae* and in the β-globin gene cluster of humans, also about 80 kb long. Note that in the single-celled yeast, protein-coding regions are closely spaced along the DNA sequence, whereas only a small fraction of the human DNA encodes protein. DNA sequencing and identification of exons has revealed that in higher organisms there is a considerable amount of DNA that does not encode protein. In fact, the β-globin gene cluster is unusually rich in protein-coding sequences compared with other regions of vertebrate DNA. In the 60-kb region including the chicken lysozyme gene, for example, the coding exons total less than 500 base pairs. Because no function has yet been found for most of the noncoding DNA in higher eukaryotes, it is commonly referred to as nonfunctional.

Different selective pressures during evolution may account, at least in part, for this remarkable difference in the amount of nonfunctional DNA in microorganisms and multicellular organisms. For example, microorganisms must compete for limited amounts of nutrients in their environment, and metabolic economy thus is a critical characteristic. Since synthesis of nonfunctional (i.e., noncoding) DNA requires time and energy, presumably there was selective pressure to lose **nonfunctional DNA** during the evolution of microorganisms. On the other hand, natural selection in vertebrates depends largely on their behavior. The energy invested in DNA synthesis is trivial compared with the metabolic energy required for the movement of muscles; thus there was little selective pressure to eliminate nonfunctional DNA in vertebrates.

Cellular DNA Content Does Not Correlate with Phylogeny

The total amount of chromosomal DNA in different animals and plants does not vary in a consistent manner with the apparent complexity of the organisms. Yeasts, fruit flies, chickens, and humans have successively larger amounts of DNA in their haploid chromosome sets (0.015, 0.15, 1.3, and 3.2 picograms, respectively), in keeping with what we perceive to be the increasing complexity of these organisms. Yet the vertebrates with the greatest amount of DNA per cell are amphibians, which are surely less complex than humans in their structure and behavior. Many plant species also have considerably more DNA per cell than humans have. For example, the DNA content per cell of wheat, broad beans, and garden onions (7.0, 14.6, and 16.8 picograms, respectively) ranges from about two to more than five times that of humans, and tulips have ten times as much DNA per cell as humans.

The DNA content per cell also varies considerably among closely related species. All insects or all amphibians would appear to be similarly complex, but the amount of haploid DNA in species within each of these phyla varies by a factor of 100. The same variation in DNA content per cell is common within groups of plants that have similar structures and life cycles. For example, the broad bean contains about three to four times as much DNA per cell as the kidney bean.

These facts further suggest that much of the DNA in certain organisms is "extra" or expendable—that is, it does not encode RNA or have any regulatory or structural function.

TABLE 9-1	Classification of Eukaryotic DNA

Protein-coding genes
  Solitary genes
  Duplicated and diverged genes (functional gene
    families and nonfunctional pseudogenes)

Tandemly repeated genes encoding rRNA, 5S rRNA,
  tRNA, and histones

Repetitious DNA
  Simple-sequence DNA
  Moderately repeated DNA (mobile DNA elements)
    Transposons
    Viral retrotransposons
    Long interspersed elements (LINES; nonviral
      retrotransposons)
    Short interspersed elements (SINES; nonviral
      retrotransposons)
Unclassified spacer DNA

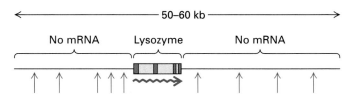

▲ **FIGURE 9-4 The chicken lysozyme gene and its surrounding regions.** This 15-kb simple transcription unit contains four exons (blue) and three introns (tan). The positions indicated by red arrows are repeated *Alu* sequences found at many sites elsewhere in the genome. [See P. Balducci et al., 1981, *Nucleic Acids Res.* **9**:3575.]

The total amount of DNA per haploid cell in an organism is referred to as the *C value;* the failure of C values to correspond to phylogenetic complexity is called the **C-value paradox.** This perplexing variation in genome size occurs mainly because eukaryotic chromosomes contain variable amounts of DNA with no demonstrable function, both between genes and within genes in introns. As discussed later, much of this apparently **nonfunctional DNA** is composed of repetitious DNA sequences, some of which are never transcribed and most all of which are likely dispensable. The different classes of eukaryotic DNA sequences discussed in the following sections are summarized in Table 9-1.

Protein-Coding Genes May Be Solitary or Belong to a Gene Family

In multicellular organisms, roughly 25–50 percent of the protein-coding genes are represented only once in the haploid genome and thus are termed *solitary* genes. The remaining protein-coding genes belong to families comprising two or more similar genes.

A well-studied example of a solitary protein-coding gene is the chicken lysozyme gene mentioned previously. The 15-kb DNA sequence encoding chicken lysozyme constitutes a simple transcription unit (i.e., a single gene) containing four exons and three introns (Figure 9-4). The flanking regions, extending for about 20 kb upstream and downstream from the transcription unit, do not encode any detectable mRNAs. Lysozyme, an enzyme that cleaves the polysaccharides in bacterial cell walls, is an abundant component of chicken egg-white protein and also is found in human tears. Its activity helps to keep the surface of the eye and the chicken egg sterile.

Frequently, the DNA that lies within 5–10 kb of a particular gene contains sequences that are close but inexact

copies of the gene. Such sequences, which are thought to have arisen by duplication of an ancestral gene, are referred to as *duplicated* protein-coding genes; duplicated genes probably constitute half of the protein-coding DNA in vertebrate genomes. A set of duplicated genes that encode proteins with similar but nonidentical amino acid sequences is called a **gene family;** the encoded closely related, homologous proteins constitute a *protein family.* A few protein families, such as protein kinases, transcription factors, and vertebrate immunoglobulins, include hundreds of members. Most families, however, include from just a few to 30 or so members; common examples are cytoskeletal proteins, 70-kDa heat-shock proteins, myosin heavy chain, chicken ovalbumin, and the α- and β-globins in vertebrates.

The genes encoding the β-like globins are a good example of a gene family. As shown in Figure 9-3b, the β-like globin gene family contains five functional genes designated β, δ, A_γ, G_γ, and ϵ; the encoded polypeptides are similarly designated. Two identical β-like globin polypeptides combine with two identical α-globin polypeptides (encoded by another gene family) and with four small heme groups to form a hemoglobin molecule (see Figure 3-10). All the hemoglobins formed from the different β-like globins carry oxygen in the blood, but they exhibit somewhat different properties that are suited to specific roles in human physiology. For example, hemoglobins containing either the A_γ or G_γ polypeptides are expressed only during fetal life. Because these fetal hemoglobins have a higher affinity for oxygen than adult hemoglobins, they can effectively extract oxygen from the maternal circulation in the placenta. The lower oxygen affinity of adult hemoglobins, which are expressed after birth, permits better release of oxygen to the tissues, especially muscles, which have a high demand for oxygen during exercise.

The different β-globin genes probably arose by duplication of an ancestral gene, most likely as the result of an "unequal crossover" during recombination in a germ-cell (egg or sperm) precursor (Figure 9-5). Over evolutionary time the two copies of the gene that resulted accumulated random mutations; beneficial mutations that conferred some refinement in the basic oxygen-carrying function of hemoglobin were retained by natural selection. Repetitions of this process

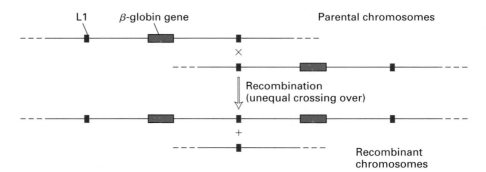

▲ FIGURE 9-5 Gene duplication resulting from unequal crossing over. Each parental chromosome *(top)* contains one ancestral globin gene containing three exons and two introns. Homologous L1 repeated sequences lie 5′ and 3′ of the globin gene. The parental chromosomes are shown displaced relative to each other, so that the L1 sequences are aligned. Homologous recombination between L1 sequences as shown would generate one recombinant chromosome with two copies of the globin gene and one chromosome with a deletion of the globin gene. Subsequent independent mutations in the duplicated genes could lead to slight changes in sequence that might result in slightly different functional properties of the encoded proteins. Unequal crossing over also can result from rare recombinations between unrelated sequences. [See D. H. A. Fitch et al., 1991, *Proc. Nat'l. Acad. Sci. USA* **88**:7396.]

are thought to have resulted in the evolution of the contemporary globin-like genes observed in humans and other complex species today.

Two regions in the human β-like globin gene cluster contain nonfunctional sequences, called **pseudogenes,** similar to those of the functional β-like globin genes (see Figure 9-3b). Sequence analysis shows that these pseudogenes have the same apparent exon-intron structure as the functional β-like globin genes, suggesting that they also arose by duplication of the same ancestral gene. However, *sequence drift* during evolution generated sequences that either terminate translation or block mRNA processing, rendering such regions nonfunctional even if they were transcribed into RNA. Because such pseudogenes are not deleterious, they remain in the genome and mark the location of a gene duplication that occurred in one of our ancestors. As discussed in a later section, other nonfunctional gene copies can arise by reverse transcription of mRNA into cDNA and integration of this intron-less DNA into a chromosome.

Several different gene families encode the various proteins that make up the cytoskeleton. These proteins are present in varying amounts in almost all cells. In vertebrates, the major cytoskeletal proteins are the actins, tubulins, and intermediate filament proteins like the keratins (Chapters 18 and 19). Although the physiologic rational for these protein families is not as obvious as it is for the globins, the different members of a family probably have similar but subtly different functions suited to the particular type of cell in which they are expressed.

Tandemly Repeated Genes Encode rRNAs, tRNAs, and Histones

In invertebrates and some vertebrates, the genes encoding rRNAs, tRNAs, histones (a family of proteins associated with eukaryotic nuclear DNA), and several other proteins occur as *tandemly repeated arrays*. These are distinguished from the duplicated genes of gene families in that the multiple tandemly repeated genes encode identical or nearly identical proteins or functional RNAs. Most often copies of a sequence appear one after the other, in a head-to-tail fashion, over a long stretch of DNA. Within a tandem array of rRNA or tRNA genes, each copy is exactly, or almost exactly, like all the others. Although the transcribed portions of rRNA genes are the same in a given individual, the nontranscribed spacer regions between the transcribed regions can vary. Arrays of tandemly repeated histone DNA are somewhat more complex; however, each histone gene, too, has multiple identical copies.

The tandemly repeated rRNA, tRNA, and histone genes are needed to meet the great cellular demand for their transcripts. Most of the RNA in a cell consists of rRNA and tRNA. Assuming RNA polymerase molecules move at a fixed speed, there must be a limit to the number of RNA copies that transcription of a single gene can provide during one cell generation, even if it is fully loaded with polymerase molecules. If more RNA is required than can be transcribed from one gene, multiple copies of the gene are necessary. For example, during early embryonic development in humans, many embryonic cells have a doubling time of ≈24 hours and contain 5–10 million ribosomes. To produce enough rRNA to form this many ribosomes, an embryonic human cell needs at least 100 copies of the pre-rRNA gene, and most of these must be close to maximally active for the cell to divide every 24 hours (Table 9-2). That is, multiple RNA polymerases must be loaded onto and transcribing each pre-rRNA gene at the same time (see Figure 11-49). The importance of repeated rRNA genes is illustrated by *Drosophila* mutants called *bobbed* (because they have stubby wings), which lack a full complement of the

TABLE 9-2 Effect of Gene Copy Number and Loading with RNA Polymerase on Rate of Pre-rRNA Synthesis in Human Cells

Copies of Pre-RNA Gene	RNA Polymerase Molecules per Gene	Molecules of Pre-rRNA Produced in 24 Hours
1	1	288
1	≈250	≈70,000
100	≈250	≈7,000,000

tandemly repeated rRNA genes. A *bobbed* mutation that reduces the number of rRNA genes to less than ≈50 is a recessive lethal mutation.

All eukaryotes, including yeasts, contain 100 or more copies of the genes encoding 5S rRNA and pre-rRNA. More than 20,000 copies of the 5S rRNA gene are present in frogs. The copy number for individual tRNA genes ranges from 10 to 100.

Reassociation Experiments Reveal Three Major Fractions of Eukaryotic DNA

Besides duplicated protein-coding genes and tandemly repeated genes, eukaryotic cells contain multiple copies of other DNA sequences in the genome, generally referred to as repetitious DNA (see Table 9-1). Some of these sequences are quite short and occur as tandem repeats; others are much longer and are interspersed at many places in the genome. The existence of these repeated sequences was first recognized in reassociation experiments in which denatured eukaryotic DNA was observed to renature nonuniformly; that is, some of it reassociated much more rapidly than the bulk of cellular DNA.

In these studies, the total DNA of an organism was broken into fragments with an average length of about a thousand base pairs. The DNA was then melted into single strands and placed under conditions that allow strand reassociation to occur (e.g., a favorable ion concentration and a favorable temperature). If none of the DNA fragments contained sequences that were repeated in the genome, they all would be expected to re-form duplexes at about the same speed. However, a fragment containing a sequence repeated many times in the genome would find a complementary partner more quickly than a fragment with a sequence that occurred only once per haploid genome, because the repeated sequence would be present at a much higher concentration. Consequently, a fragment containing a repeated sequence would reassociate faster than a fragment with a unique sequence.

About 50–60 percent of mammalian DNA reassociates at a *slow rate* indicating that it consists primarily of *single-copy DNA*. According to Mendelian genetics, only one copy of each gene is contained in the haploid DNA set; thus the single-copy DNA fraction is expected to contain most of the genes encoding mRNA. However, the vast majority of single-copy DNA in the mammalian genome is noncoding DNA between genes and in introns. It appears that only a small fraction of the total DNA in humans, on the order of 5 percent, actually encodes proteins or functional RNA molecules. The remainder of the single-copy DNA, which currently has no known function other than to separate functional DNA sequences, is referred to as *spacer DNA*.

Another 25–40 percent of mammalian DNA reassociates at an *intermediate rate*. Cloning and sequencing of this DNA fraction from many different animals and higher plants have revealed that it is composed primarily of a very large number of copies of a relatively few sequence families in any specific organism. Such repetitious DNA, termed *moderately repeated DNA*, or *intermediate-repeat DNA*, is interspersed throughout mammalian genomes. Because these sequences can be copied and reinserted into new sites in the genome, they are called **mobile DNA elements,** which we describe in the next section. A small portion of this fraction consists of large duplicated gene families and tandemly repeated genes discussed previously.

About 10–15 percent of mammalian DNA reassociates at a *very rapid rate*. This rapidly reassociating type of repetitious DNA, referred to as **simple-sequence DNA,** is composed largely of several different sets of short (5- to 10-bp) sequences repeated in long tandem arrays.

Simple-Sequence DNAs Are Concentrated in Specific Chromosomal Locations

Although much of the simple-sequence DNA of higher organisms is composed of tandemly repeated, 5- to 10-bp sequences, long tandem repeats of simple sequences containing 20–200 nucleotides also occur in some vertebrate and plant genomes. Such tandem repeats generally extend up to 10^5 base pairs in total length. These long stretches of simple-sequence DNA are often referred to as *satellite DNA* because they are separated from the bulk of cellular DNA by equilibrium density-gradient centrifugation. However, not all simple-sequence DNAs separate from the bulk of cellular DNA during centrifugation.

In situ hybridization studies with metaphase chromosomes have localized simple-sequence DNA to specific chromosomal regions. In most mammals, much of the simple-sequence DNA lies near **centromeres,** discrete chromosomal regions that attach to spindle microtubules during mitosis and meiosis (see Figure 19-39). In the chromosomes of *Drosophila melanogaster,* simple-sequence DNA is concentrated in both centromeres and **telomeres,** the ends of chromosomes. Some simple-sequence tandem arrays also are located within chromosome arms in the *Drosophila* genome. In humans, some simple-sequence DNAs are located at a specific location on one chromosome. These sequences are useful for identifying particular chromosomes by fluorescence in situ hybridization (FISH). For example, a particular simple sequence

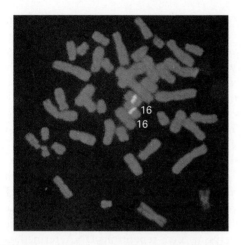

▲ **FIGURE 9-6 Use of simple-sequence DNA as chromosomal marker.** Human metaphase chromosomes stained with a fluorescent dye and hybridized in situ with a particular simple-sequence DNA labeled with a fluorescent biotin derivative. When viewed under the appropriate wavelength of light, the DNA appears red and the hybridized simple-sequence DNA appears as a yellow band on chromosome 16, thus locating this particular simple sequence to one site in the genome. [See R. K. Moyzis et al., 1987, *Chromosoma* **95**:378; courtesy of R. K. Moyzis.]

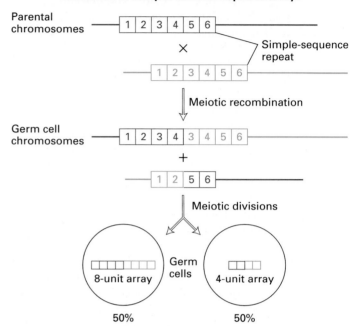

▲ **FIGURE 9-7 Unequal crossing over during meiosis can generate differences in lengths of simple-sequence DNA tandem arrays.** In this example, unequal recombination within a stretch of DNA containing six copies (1–6) of a particular simple-sequence repeat unit yields germ cells containing either an 8-unit or 4-unit tandem array.

in the human genome is present only in the middle of the long arm of chromosome 16 (Figure 9-6).

Simple-sequence DNA located at centromeres is suspected to contribute to the structure and therefore the function of the **kinetochore** of metaphase chromosomes. This large nucleoprotein complex assembles at the centromere and attaches to spindle microtubules during mitosis (Chapter 19). As yet, however, there is little clear-cut experimental evidence demonstrating any function for most simple-sequence DNA.

DNA Fingerprinting Depends on Differences in Length of Simple-Sequence DNAs

Within a species, the nucleotide sequences of the repeat units composing simple-sequence DNA tandem arrays are highly conserved among individuals. In contrast, differences in the *number* of repeats, and thus in the length, of simple-sequence tandem arrays containing the same repeat unit are quite common among individuals. These differences in length result from unequal crossing over within regions of simple-sequence

DNA during development of sperm and oocyte precursors and during meiosis (Figure 9-7). As a consequence of this unequal crossing over, the lengths of some simple-sequence tandem arrays are unique in each individual.

In humans and other mammals, some of the simple-sequence DNA exists in relatively short 1- to 5-kb regions made up of 20–50 repeat units each containing 15 to about 100 base pairs. These regions are called *minisatellites* to distinguish them from the more common regions of tandemly repeated simple-sequence DNA, which are ≈100 kb in length. The sequences of the repeat unit in two human minisatellites are shown in Figure 9-8. Even slight differences in the total lengths of various minisatellites from different individuals can be detected. These differences form the basis of **DNA fingerprinting,** which is superior to conventional fingerprinting for identifying individuals (Figure 9-9).

AAGGGTGGGCAGGAAGTGGAGTGTGTGCCTGCTTCCCTTCCCTGTCTTGTCCTGGAAACTCA

λ 33.1 minisatellite

T
C GGGCAGG•AGGGGGAGG

λ 33.5 minisatellite

▲ **FIGURE 9-8 Consensus sequences of the repeat unit of human minisatellites named λ33.1 and λ33.5 based on analysis of more than ten sets of repeats in each case.** Red letters indicate positions in which base differences have been detected; red solid dot indicates a deletion. The 62-bp repeat unit of λ33.1 is much more highly conserved than the 17-bp unit of λ33.5. [See A. J. Jeffreys et al., 1985, *Nature* **314**:67.]

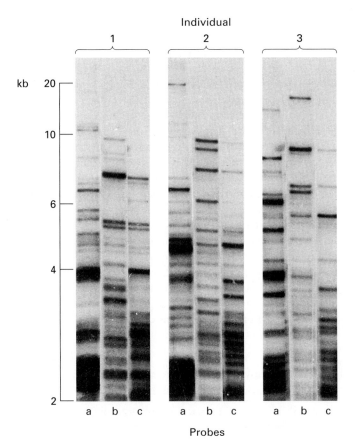

Individual

kb

20
10
6
4
2

a b c   a b c   a b c

Probes

▲ **FIGURE 9-9 Human DNA fingerprints.** DNA samples from three individuals (1, 2, and 3) were subjected to Southern-blot analysis using the restriction enzyme *Hin*f1 and three labeled minisatellites as probes (λ33.6, 33.15, and 33.5; lanes a, b, and c, respectively). DNA from each individual produced a unique band pattern with each probe. Conditions of electrophoresis can be adjusted so that at least 50 bands can be resolved for each person with this restriction enzyme. The nonidentity of these three samples is easily distinguished. [From A. J. Jeffreys et al., 1985, *Nature* **316**:76; courtesy of A. J. Jeffreys.]

SUMMARY Chromosomal Organization of Genes and Noncoding DNA

- The genomes of prokaryotes and lower eukaryotes contain few nonfunctional sequences, whereas vertebrate genomes contain many sequences that do not code for RNAs or have any structural or regulatory function. Only about 5 percent of the genomic DNA in humans encodes proteins or functional RNAs.

- The lack of a consistent relationship between the amount of DNA in the haploid chromosomes of an animal or plant and its phylogenetic complexity is called the C-value paradox.

- About half of the protein-coding genes in vertebrate genomic DNA are solitary genes, whose sequence occurs only once in the haploid genome. The remainder

are duplicated genes, which arose by duplication of an ancestral gene and subsequent independent mutations (see Figure 9-5).

- Duplicated genes, such as those forming the β-like globin gene family, encode closely related proteins and generally appear as a cluster in a particular region of DNA (see Figure 9-3). The proteins encoded by a gene family have homologous but nonidentical amino acid sequences and exhibit similar but slightly different properties.

- In invertebrates and vertebrates, rRNAs, tRNAs, and histone proteins are encoded by multiple copies of genes located in tandem arrays in genomic DNA.

- Single-copy DNA consists of solitary protein-coding genes, small duplicated gene families, and spacer DNA.

- Moderately repeated DNA includes the tandemly repeated genes encoding, rRNA, tRNA genes, and histones; large duplicated gene families; and mobile DNA elements.

- Simple-sequence DNA, which consists largely of very short sequences repeated in long tandem arrays, is preferentially located in centromeres, telomeres, and specific locations within the arms of particular chromosomes.

- The length of a particular simple-sequence tandem array is quite variable among individuals in a species, probably because of unequal crossing over during meiosis (see Figure 9-7). Differences in the lengths of some simple-sequence tandem arrays forms the basis for DNA fingerprinting.

9.3 Mobile DNA

Much of the discussion throughout this book deals with the functions of the gene products expressed from protein-coding genes and RNA genes. However, noncoding repetitious DNA, consisting of simple-sequence DNA and moderately repeated DNA, constitutes a significant fraction of the genomic DNA in higher eukaryotes. In this section, we focus on moderately repeated DNA sequences, or mobile DNA elements, which are interspersed throughout the genomes of higher plants and animals. Although mobile DNA elements, ranging from hundreds to a few thousand base pairs in length, originally were discovered in eukaryotes, they also are found in prokaryotes. The process by which these sequences are copied and inserted into a new site in the genome is called **transposition**. Mobile DNA elements (or simply mobile elements) are essentially molecular parasites, which appear to have no specific function in the biology of their host organisms, but exist only to maintain themselves. For this reason, Francis Crick referred to these sequences as "selfish DNA."

The transposition of mobile DNA elements is believed to have resulted in their slow accumulation in eukaryotic genomes over evolutionary time. These elements also are lost at a very slow rate by deletion of segments of DNA containing them and by accumulation of mutations until they can no longer be recognized to be related to the original mobile DNA element. Since mobile elements are eliminated from eukaryotic genomes so slowly, they have accumulated to the point where they now constitute a significant portion of the genomes of many eukaryotes.

Movement of Mobile Elements Involves a DNA or RNA Intermediate

Barbara McClintock discovered the first mobile elements while doing classical genetic experiments in maize (corn) during the 1940s. She characterized genetic entities that could move into and back out of genes, changing the phenotype of corn kernels. Her theories were very controversial until similar mobile elements were discovered in bacteria, where they were characterized as specific DNA sequences, and the molecular basis of their transposition, was deciphered. When bacterial mobile elements were first discovered, researchers did not initially link them to moderately repeated DNA, which had been previously identified in reassociation experiments with eukaryotic DNA. However, as the wide variations in the amounts and chromosomal positions of eukaryotic intermediate repeats were documented, their similarity with bacterial mobile elements was recognized. Thus the study of moderately repeated DNA in eukaryotes converged with research on mobile DNA elements in bacteria, although at first there was no apparent connection between the two classes of DNA.

As research on mobile elements progressed, they were found to fall into two categories: (1) those that transpose directly as DNA and (2) those that transpose via an RNA intermediate transcribed from the mobile element by an RNA polymerase and then converted back into double-stranded DNA by a **reverse transcriptase** (Figure 9-10). Mobile elements that transpose through a DNA intermediate are generally referred to as **transposons.** (As discussed below, this term has a more specific meaning in reference to bacterial mobile elements.) Mobile elements that transpose to new sites in the genome via an RNA intermediate are called **retrotransposons** because their movement is analogous to the infectious process of retroviruses (see Figure 6-22). Indeed, retroviruses can be thought of as retrotransposons that evolved genes encoding viral coats, thus allowing them to transpose between cells. Both transposons and retrotransposons can be further classified based on their specific mechanism of transposition, as summarized in Table 9-3. We describe the structure and movement of the various types of mobile elements and then consider their likely role in evolution.

Mobile Elements That Move as DNA Are Present in Prokaryotes and Eukaryotes

Most mobile elements in bacteria transpose directly as DNA. In contrast, most mobile elements in eukaryotes are retrotransposons, but some eukaryotic transposons have been identified. Indeed, the original mobile elements discovered by Barbara McClintock are transposons.

Bacterial Insertion Sequences The first molecular understanding of mobile elements came from the study of certain *E. coli* mutations resulting from the spontaneous insertion of a DNA sequence, ≈1–2 kb long, into the middle of a gene. These inserted stretches of DNA—called *insertion sequences*, or *IS elements*—were first visualized by analyzing hybrids (**heteroduplexes**) of wild-type and mutant DNAs in the electron microscope. Because the IS element integrated into the mutant strand has no complement in the wild-type strand, it cannot hybridize and forms a visible single-stranded loop extending from the rest of the double-stranded heteroduplex. So far, more than 20 different IS elements have been found in *E. coli* and other bacteria.

IS elements appear to be molecular parasites of bacterial cells. Transposition of an IS element is a very rare event, occurring in only one in 10^5–10^7 cells per generation,

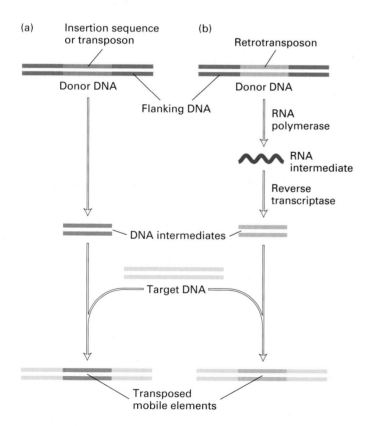

(a) Insertion sequence or transposon

Donor DNA

Flanking DNA

DNA intermediates

Target DNA

Transposed mobile elements

(b) Retrotransposon

Donor DNA

RNA polymerase

RNA intermediate

Reverse transcriptase

DNA intermediates

◀ **FIGURE 9-10 Classification of mobile elements into two major classes.** (a) Insertion sequences and transposons (orange) move via a DNA intermediate. (b) Retrotransposons (green) are first transcribed into an RNA molecule, which then is reverse-transcribed into double-stranded DNA. In both cases, the double-stranded DNA intermediate is integrated into the target-site DNA to complete movement.

TABLE 9-3 Major Types of Mobile DNA Elements

Type	Structural Features	Mechanism of Movement	Examples
DNA-MEDIATED TRANSPOSITION			
Bacterial insertion sequences (IS elements)	≈50-bp inverted repeats flanking region encoding transposase and, in some, resolvase	Excision or copying of DNA and its insertion at target site	IS1, IS10
Bacterial transposons	Central antibiotic-resistance gene flanked by IS elements	Copying of DNA and its insertion at target site	Tn9
Eukaryotic transposons	Inverted repeats flanking coding region with introns	Excision of DNA and its insertion at target site	P element *(Drosophila)* Ac and Ds elements (corn)
RNA-MEDIATED TRANSPOSITION			
Viral retrotransposons	≈250- to 600-bp direct terminal repeats (LTRs) flanking region encoding reverse transcriptase, integrase, and retroviral-like Gag protein	Transcription into RNA from promoter in left LTR by RNA polymerase II followed by reverse transcription and insertion at target site	Ty elements (yeast) *Copia* elements *(Drosophila)*
Nonviral retrotransposons	Of variable length with a 3′ A/T-rich region; full-length copy encodes a reverse transcriptase	Transcription into RNA from internal promoter; folding of transcript to provide primer for reverse transcription followed by insertion at target site	F and G elements *(Drosophila)* LINE and SINE elements (mammals) *Alu* sequences (humans)

depending on the IS element. Higher rates of transposition would probably result in too great a mutation rate for the host cell. At a very low rate of transposition, most host cells survive and therefore propagate the parasitic IS element. Even though many transpositions inactivate essential genes, killing the host cell and the IS elements it carries, other host cells survive. Since IS elements transpose into approximately random sites, some transposed sequences enter nonessential regions of the genome (e.g., regions between genes), thereby expanding the number of IS elements in a cell. IS elements also can insert into plasmids or lysogenic viruses, which can be transferred to other cells. When this happens, IS elements can transpose into the chromosomes of virgin cells.

The general structure of IS elements is diagrammed in Figure 9-11. An *inverted repeat,* usually containing ≈50 base pairs, invariably is present at each end of an insertion sequence. Between the inverted repeats is a protein-coding region, which encodes one or two enzymes required for transposition of an IS element to a new site. In either case, IS-encoded proteins are expressed at a very low rate, accounting for the very low frequency of transposition. An important hallmark of IS elements is the presence of short *direct repeats,* containing

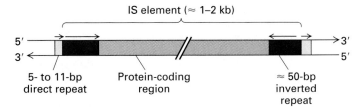

IS element (≈ 1–2 kb)

5′ — 3′ — 5- to 11-bp direct repeat — Protein-coding region — ≈ 50-bp inverted repeat — 3′ — 5′

▲ **FIGURE 9-11 General structure of bacterial IS elements.** The central region, which encodes one or two enzymes required for transposition, is flanked by inverted repeats whose sequence is characteristic of a particular IS element. The 5′ and 3′ short direct repeats are generated from the target-site DNA during insertion of a mobile element. The length of the direct repeats is constant for a given IS element, but their sequence depends on the site of insertion and is not characteristic of the IS element. Arrows indicate sequence orientation. The regions in this diagram are not to scale; the coding region makes up most of the length of an IS element.

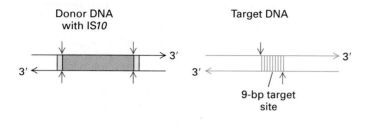

Donor DNA with IS10

Target DNA

9-bp target site

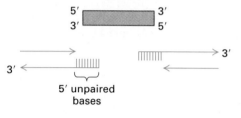

① Transposase makes blunt-ended cuts in donor DNA and staggered cuts in target DNA

5' unpaired bases

② Transposase ligates IS10 to 5' single-stranded ends of target DNA

③ Cellular DNA polymerase extends 3' cut ends and ligase joins extended 3' ends to IS10 5' ends

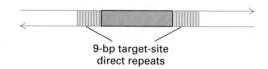

9-bp target-site direct repeats

▲ **FIGURE 9-12 Model for nonreplicative transposition of bacterial insertion sequences.** Step ①: A transposase, which is encoded by the IS element (IS10 in this example), cleaves both strands of the donor DNA between the terminal direct repeats (light blue) and the inverted repeats, excising the IS10 element. At a largely random target site, the transposase makes staggered cuts in the target DNA. In the case of IS10, the two cuts are 9 bp apart. Step ②: Ligation of the 3' ends of the excised IS element to the staggered sites in the target DNA also is catalyzed by transposase. Step ③: The 9-bp gaps of single-stranded DNA left in the resulting intermediate is filled in by cellular DNA polymerase; finally cellular DNA ligase forms the 3' → 5' phosphodiester bonds between the 3' ends of the extended target DNA strands and the 5' ends of the IS10 strands. This process results in duplication of the target-site sequence on each side of the inserted IS element. Note that the length of the target site and IS10 are not to scale. [See H. W. Benjamin and N. Kleckner, 1989, *Cell* **59**:373 and, 1992, *Proc. Nat'l. Acad. Sci. USA* **89**:4648.]

5–11 base pairs, immediately adjacent to both ends of the inserted element. The *length* of the direct repeat is characteristic of each type of IS element, but its *sequence* depends on the target site where a particular copy of the IS element is inserted. When the sequence of a mutated gene containing an IS element is compared to the sequence of the wild-type gene before insertion, only one copy of the short **direct-repeat sequence** is found in the wild-type gene. Duplication of this target-site sequence to create the second direct repeat adjacent to an IS element occurs during the insertion process.

The enzyme that catalyzes transposition of an IS element is called a *transposase*. In the simplest mechanism of transposition, an IS element is excised from one location and inserted at a new position in the bacterial chromosome by a *nonreplicative* process (Figure 9-12). In this mechanism, transposase molecules bind to the inverted-repeat sequences present at each end of the IS element in the donor DNA and cleave the DNA, precisely excising the element. Transposase molecules also bind to and make staggered cuts in a short sequence in the target DNA, generating single-stranded tails. This remarkable enzyme then ligates the 3' termini of the IS element to the 5' ends of the cut donor DNA. A DNA polymerase encoded by the host cell then extends the 3' ends of the target site, filling in the single-stranded gaps and generating a short repeat of the target-site sequence at either end of the newly inserted IS element. This is the origin of the short direct repeats that flank IS elements. Some IS elements transpose by a more complicated *replicative* mechanism; in this case, a copy of the original IS element is generated in the target DNA and the original copy is retained in the donor DNA.

Bacterial Transposons In addition to IS elements, bacteria contain composite mobile genetic elements that are larger than IS elements and contain one or more protein-coding genes in addition to those required for transposition. Referred to as *bacterial transposons*, these elements are composed of an antibiotic-resistance gene flanked by two copies of the same type of IS element (Figure 9-13). Insertion of a transposon into plasmid or chromosomal DNA is readily detectable because of the acquired resistance to an antibiotic. Transposition produces a short direct repeat of the target site on either side of the newly integrated transposon, just as for IS elements.

Transposons are very valuable tools for the bacterial geneticist. They can be introduced into cells on plasmids or viral genomes. Once transferred into a cell, transposons can act as mutagens that affect only a single cellular gene. Although transposition is a rare event, mutagenized cells are readily isolated because of their newly acquired antibiotic resistance gene. The site of the transposon-generated mutation can be determined readily by restriction-enzyme mapping, which reveals the insertion of the large transposon DNA. The precise sequence of bacterial DNA at the site of insertion can then be determined by dideoxy DNA sequencing using a primer complementary to the known sequence

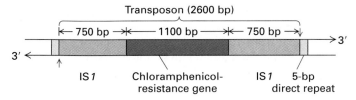

▲ FIGURE 9-13 General structure of bacterial transposons, such as Tn9 of E. coli. This transposon consists of a chloramphenicol-resistance gene (dark blue) flanked by two copies of IS1 (orange), one of the smallest IS elements. Other copies of IS1, without the drug-resistance gene, are located elsewhere in the E. coli chromosome. The internal inverted repeats of IS1 abutting the resistance gene are so mutated that transposase does not recognize them. During transposition, the IS-element transposase makes cuts at the positions indicated by small red arrows, so the entire transposon is moved from the donor DNA (e.g., a plasmid). The target-site sequence at the point of insertion becomes duplicated on either side of the transposon during transposition, which occurs via the replicative mechanism. Note that the 5-bp target-site direct repeat (light blue) is not to scale.

of the inverted repeats at the ends of the transposon (see Figure 7-29).

Eukaryotic Transposons McClintock's original discovery of mobile elements came from observation of certain spontaneous mutations that affect production of any of the several enzymes required to make anthocyanin, a purple pigment. Mutant kernels are white, and wild-type kernels are purple. One class of these mutations is revertible at high frequency, whereas a second class of mutations does not revert unless they occur in the presence of the first class of mutations. McClintock called the agent responsible for the first class of mutations the *activator (Ac) element* and those responsible for the second class *dissociation (Ds) elements* because they also tended to be associated with chromosome breaks.

Many years after McClintock's pioneering discoveries, cloning and sequencing revealed that Ds elements are deleted forms of the Ac element in which a portion of the sequence encoding transposase is missing. Because it does not encode a functional transposase, a Ds element cannot move by itself. However, in plants that carry the Ac element and thus express a functional transposase, Ds elements can move. The structure of these eukaryotic elements are similar to bacterial IS elements, and they appear to move by the non-replicative mechanism shown in Figure 9-12.

Since McClintock's early work on mobile elements in corn, transposons have been identified in other eukaryotes. For instance, approximately half of all the spontaneous mutations observed in *Drosophila* are due to the insertion of mobile elements. Although most of the mobile elements in *Drosophila* function as retrotransposons, at least one—the *P element*—functions as a transposon, moving by a nonreplicative mechanism similar to that used by bacterial

insertion sequences. Current methods for constructing transgenic *Drosophila* depend on engineered, high-level expression of the P-element transposase and use of the terminal repeats as targets for transposition (see Figure 8-37).

Viral Retrotransposons Contain LTRs and Behave Like Retroviruses in the Genome

All eukaryotes studied from yeast to humans contain retrotransposons, mobile DNA elements that transpose through an RNA intermediate utilizing a reverse transcriptase (see Figure 9-10b). These mobile elements are divided into two major categories, *viral* and *nonviral* retrotransposons. Viral retrotransposons, which we discuss in this section, are abundant in yeast (e.g., Ty elements) and in *Drosophila* (e.g., *copia* elements). In mammals, nonviral retrotransposons are the most common type of mobile element; these are described in the next section. Still, viral retrotransposons are estimated to account for ≈4 percent of human DNA.

The general structure of viral retrotransposons found in eukaryotes is depicted in Figure 9-14. In addition to short 5′ and 3′ direct repeats typical of all mobile elements, viral retrotransposons are marked by the presence of ≈250- to 600-bp **long terminal repeats (LTRs)** flanking the central protein-coding region. LTRs are characteristic of integrated retroviral DNA and are critical to the life cycle of retroviruses. Moreover, Ty elements and *copia* encode three of the four proteins encoded by retroviral DNA. These similarities suggest that transposition of mobile elements like Ty and *copia* involves mechanisms similar to those whereby retroviral DNA is integrated into a host-cell genome and the retroviral RNA genome is generated (see Figure 6-22).

We first consider the distinct functions of the two LTRs of integrated retroviral DNA in generating retroviral genomic RNA, which corresponds to the RNA intermediate in transposition of Ty elements and *copia*. As depicted in Figure 9-15, the leftward LTR functions as a promoter that directs host-cell RNA polymerase II to initiate transcription

Viral retrotransposons

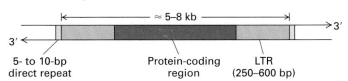

▲ FIGURE 9-14 General structure of eukaryotic viral retrotransposons. The central protein-coding region is flanked by two long terminal repeats (LTRs), which are element-specific direct repeats. LTRs, the hallmark of these mobile elements, also are present in retroviral DNA. Like other mobile elements, integrated retrotransposons have short target-site direct repeats at their 3′ and 5′ ends. Note that the different regions are not drawn to scale; the protein-coding region constitutes 80 percent or more of a retrotransposon.

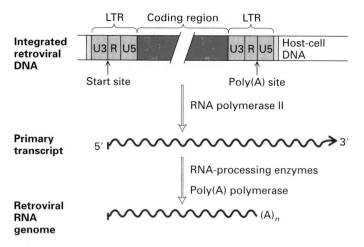

▲ FIGURE 9-15 Generation of retroviral genomic RNA from integrated retroviral DNA. The short direct repeat sequences (light blue) of target-site DNA are generated during integration of the retroviral DNA into the host-cell genome. The left LTR directs cellular RNA polymerase II to initiate transcription at the first nucleotide of the left R region. The resulting primary transcript extends beyond the right LTR. The right LTR, now present in the RNA primary transcript, directs cellular enzymes to cleave the primary transcript at the last nucleotide of the right R region and to add a poly(A) tail, yielding a retroviral RNA genome with the structure shown at the top of Figure 9-16.

at the 5′ nucleotide of the R sequence. After the entire retroviral DNA has been transcribed, the RNA sequence corresponding to the rightward LTR directs host-cell RNA-processing enzymes to cleave the primary transcript and add a poly(A) tail at the 3′ end of the R sequence. The resulting retroviral RNA genome lacks complete LTRs. However, after a virus infects a cell, reverse transcription of the RNA genome by virus-encoded reverse transcriptase yields a double-stranded DNA containing LTRs (Figure 9-16). *Integrase*, another enzyme encoded by retroviruses, then inserts the double-stranded retroviral DNA into the host-cell genome; in this process, short direct repeats of the target-site sequence are generated at either end of the inserted viral DNA sequence. Like retroviral DNA, Ty elements and *copia* encode reverse transcriptase and integrase; these enzymes are thought to function in transposition by converting the RNA intermediate into DNA and inserting the DNA into the target site in a manner similar to retroviruses.

Although these considerations imply that Ty elements transpose through an RNA intermediate, the experiments depicted in Figure 9-17 provided strong functional evidence for this conclusion. Ty elements normally transpose at a very low rate, probably because random insertion of Ty elements into the yeast genome, which contains relatively little spacer and intron DNA, would often inactivate genes. However, when yeast cells were transformed with plasmids containing a Ty element cloned next to a galactose-sensitive promoter, the production of Ty mRNA and Ty-element trans-

position was much higher in the presence of galactose than in its absence (experiment 1). This increased Ty transposition resulted from an increase in the amount of Ty mRNA, which could function as a template for reverse transcription, and in the amount of reverse transcriptase and integrase expressed from the Ty element. An even more revealing result was observed when an unrelated intron was inserted into the Ty DNA sequence (experiment 2). Addition of galactose to the medium stimulated Ty transposition, as in experiment 1, but the resulting newly integrated Ty elements all lacked the inserted intron. Presumably, the intron was spliced out of the Ty mRNA before it was reverse-transcribed into a double-stranded DNA copy, which subsequently inserted into the host-cell genome. The observed removal of the intron from transposed Ty elements strongly implies that transposition occurs by reverse transcription of mRNA produced by transcription of Ty DNA.

In contrast to Ty elements, the coding region of maize Ac elements contains introns. The presence of introns in Ac elements supports the conclusion that they transpose via direct movement of DNA sequences, not by reverse transcription of an RNA intermediate.

Nonviral Retrotransposons Lack LTRs and Move by an Unusual Mechanism

The most abundant mobile elements in mammals are *nonviral retrotransposons*, which lack LTRs. Many of these belong to the two classes of moderately repeated DNA sequences found in mammalian genomes: *long interspersed elements (LINES)* and *short interspersed elements (SINES)*. In humans, full-length LINES are ≈6–7 kb long, and SINES are ≈300 bp long. One major class of SINES and perhaps about ten classes of LINES have been identified in mammals; each class may be present in thousands of copies, which may not be exact repeats. Repeated sequences with characteristics of LINES have been observed in protozoa,

▶ FIGURE 9-16 Generation of LTRs during reverse transcription of retroviral genomic RNA. A complicated series of nine events generates a double-stranded DNA copy of the single-stranded RNA genome of a retrovirus *(top)*. The genomic RNA is packaged in the virion with a retrovirus-specific cellular tRNA hybridized to a complementary sequence near its 5′ end called the primer-binding site (PBS). The retroviral RNA has a short direct-repeat terminal sequence (R) at each end. The overall reaction is catalyzed by reverse transcriptase, which catalyzes polymerization of deoxyribonucleotides and digestion of the RNA strand in a DNA-RNA hybrid. The entire process yields a double-stranded DNA molecule that is longer than the template RNA and has a long terminal repeat (LTR) at each end. The different regions are not shown to scale. The PBS and R regions are actually much shorter than the U5 and U3 regions, and the central coding region is very much longer (≈7500 nucleotides) than the other regions. [See E. Gilboa et al., 1979, *Cell* **18**:93.]

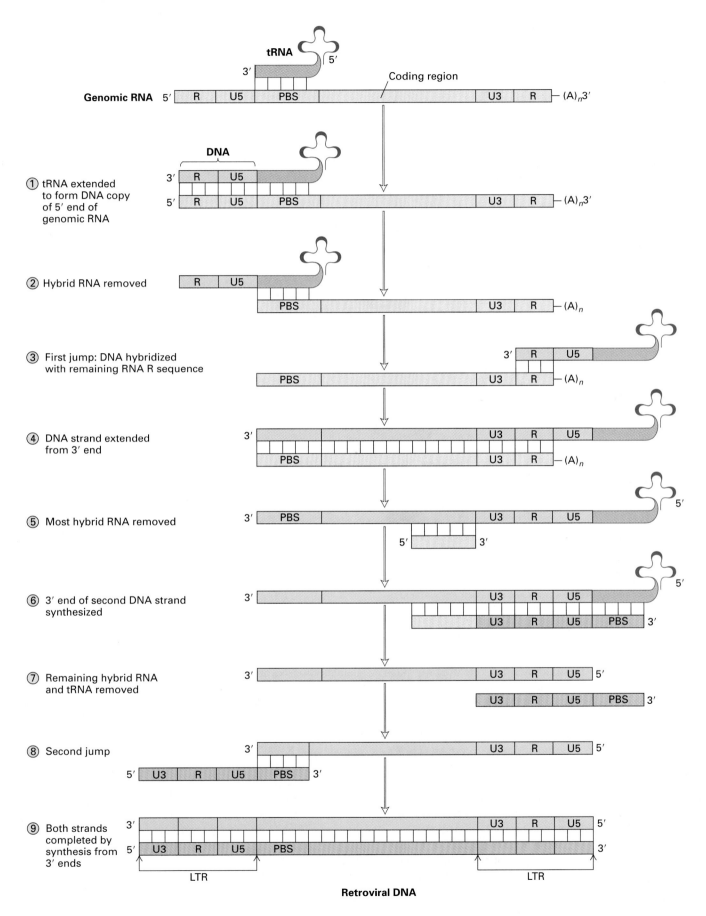

Retroviral DNA

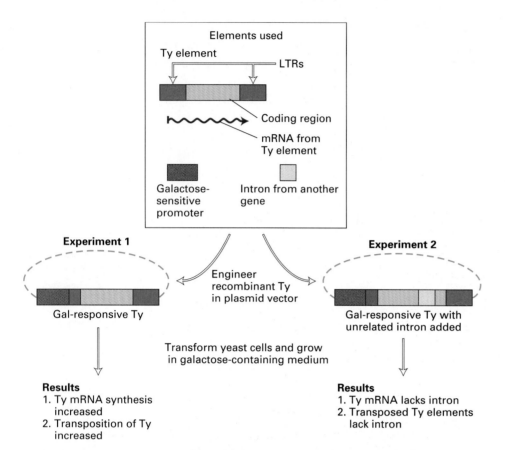

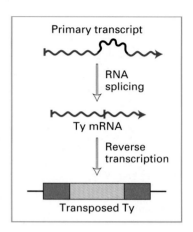

▲ FIGURE 9-17 Experimental demonstration that yeast Ty element moves through an RNA intermediate. When yeast cells are transformed with a Ty-containing plasmid, the Ty element can transpose to new sites, although normally this occurs at a low rate. Using the elements diagrammed at the top, researchers engineered two different plasmid vectors containing recombinant Ty elements adjacent to a galactose-sensitive promoter. These plasmids were transformed into yeast cells, which were grown in a galactose-containing or nongalactose medium. In experiment 1, growth of cells in galactose-containing medium resulted in many more transpositions than in nongalactose medium, indicating that transcription into an mRNA intermediate is required for Ty transposition. In experiment 2, an intron from an unrelated yeast gene was inserted into the putative protein-coding region of the recombinant galactose-responsive Ty element. The observed absence of the intron in transposed Ty elements is strong evidence that transposition involves an mRNA intermediate, as depicted in the box on the right. [See J. Boeke et al., 1985, *Cell* **40**:491.]

insects, and plants, but for unknown reasons they are particularly abundant in the genomes of mammals. SINES also are found primarily in mammalian DNA. The large numbers of LINES and SINES in higher eukaryotes have accumulated over evolutionary time by repeated copying of a sequence at a few positions in the genome and insertion of the copies into new positions. Although these mobile elements do not contain LTRs, the available evidence indicates that they transpose through an RNA intermediate.

L1 LINE Elements The most common LINE elements constitute the L1 LINE family. Some 600,000 copies of L1 elements occur in the human genome, accounting for ≈15 percent of total human DNA. The general structure of L1 elements, based on a "consensus" (average) sequence, is diagrammed in Figure 9-18. L1 elements usually are flanked by short direct repeats, the hallmark of mobile elements. The consensus sequence contains two long open reading frames (ORFs), one ≈1 kb long and the other ≈4 kb long. ORF1 encodes an RNA-binding protein. The protein encoded by ORF2 is similar in sequence to the reverse transcriptases of retroviruses and viral retrotransposons.

Evidence for the mobility of L1 elements first came from analysis of DNA cloned from humans with certain genetic diseases. DNA from these patients was found to carry mutations resulting from insertion of an L1 element into a gene, whereas no such element occurred in the DNA of either parent. Later experiments similar to those just described

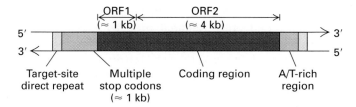

▲ FIGURE 9-18 General structure of an L1 LINE element, a common type of eukaryotic nonviral retrotransposon. The length of the flanking direct repeats varies among copies of the element at different sites in the genome. The sequence of the direct repeats appears to be generated from the target-site sequence during insertion. Although the full-length L1 sequence is ≈6 kb long, variable amounts of the left end are absent at over 90 percent of the sites where this mobile element is found. The shorter open reading frame (ORF1) encodes an RNA-binding protein. The longer ORF2 encodes a protein that is similar to the reverse transcriptases of retroviruses and viral retrotransposons. Although L1 elements do not contain LTRs, the A/T-rich region at the right end is thought to function in retrotransposition.

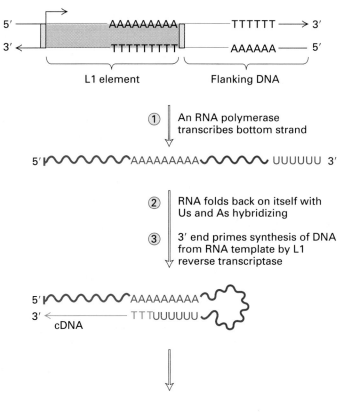

▲ FIGURE 9-19 Proposed mechanism of nonviral retrotransposition of L1 elements. According to this model, transcription of the bottom L1 strand containing a 3′ T-rich region terminates after the first T/A-rich sequence in the flanking DNA. Folding and hybridization of the transcript generates a primer for reverse transcription, which generates a cDNA copy of the L1 sequence starting from the 3′ U in the transcript. After synthesis of a largely double-stranded DNA copy of the L1 sequence, an integrase activity is believed to integrate L1 into a cellular DNA target site. During integration, the RNA loop at the right end presumably is removed and the target-site direct repeats (light green) are generated at both ends. [Adapted from P. Jagadeeswaran et al., 1981, *Cell* **26**:141; S. W. Van Arsdell et al., 1981, Cell **26**:11.]

with yeast Ty elements (see Figure 9-17) confirmed that L1 elements transpose through an RNA intermediate. In these experiments, an intron was introduced into a cloned mouse L1 element, and the recombinant L1 element was stably transformed into cultured hamster cells. After several cell doublings, a PCR-amplified fragment corresponding to the L1 element but lacking the inserted intron was detected in the cells. This finding strongly suggests that over time the recombinant L1 element containing the inserted intron had transposed to new sites in the hamster genome through an RNA intermediate that underwent RNA splicing to remove the intron.

Since L1 elements do not contain LTRs, their mechanism of transposition through an RNA intermediate must differ from viral retrotransposition in which the LTRs play a crucial role. In vitro studies indicate that transcription of L1 is directed by promoter sequences located within the left end of the element. L1 elements also contain an A/T-rich region near their right end, which generates a stretch of A residues in their RNA transcripts. Based on these properties of L1 elements, the model of L1 retrotransposition shown in Figure 9-19 has been proposed.

In the L1 element diagrammed in Figure 9-18, the length of the open reading frames has been maximized. The vast majority of L1 sequences, however, contain stop codons and frameshift mutations in ORF1 and ORF2; these mutations probably have accumulated in most L1 sequences over evolutionary time. Maintenance of L1 transposition requires that only one L1 sequence in the genome maintain intact open reading frames encoding the reverse transcriptase and other proteins required for L1 retrotransposition. Mutations

could accumulate in other L1 elements without interfering with their retrotransposition, which can be directed by the enzymes expressed from only one or a few intact L1 elements in the cell.

SINES and *Alu* Sequences Several hundred examples of SINES, the second major class of moderately repeated DNA in mammals, have been cloned and sequenced. Although no two copies of these intermediate repeats are identical, their general sequence similarity shows that an ancestral relationship exists between these elements both within and among mammalian species. The sequence conservation is about 80 percent within a species, but falls to only about 50–60 percent among different species. Because many of these repetitive sequences in human DNA were found to contain a recognition site for the restriction enzyme *Alu*I, they were collectively called the *Alu* family. However, since these short interspersed elements are not precisely identical, many lack the *Alu* site; nonetheless, the name is widely used to refer to the most abundant type of human SINE.

Alu sequences containing ≈300 base pairs are present at ≈1 million sites in the human genome, accounting for about 10 percent of the total genomic DNA; similar sequences are abundant in other vertebrates. In addition to full-length *Alu* sequences, many partial *Alu*-like sequences, clearly related to the *Alu* family but as short as 10 base pairs, have been found scattered between genes and within introns in human DNA.

Alu sequences are remarkably homologous to 7SL RNA, a small cellular RNA that is part of the signal-recognition particle. This cytoplasmic ribonucleoprotein particle aids in the secretion of newly formed polypeptides through the membranes of the endoplasmic reticulum (Chapter 17). The 7SL sequence is highly conserved even in species as diverse as *Drosophila*, mouse, and man. The discovery of a small (≈100-nucleotide) *E. coli* RNA whose sequence is similar to eukaryotic 7SL RNA indicates that this molecule has existed since early in evolution. However, neither *Drosophila* nor single-celled organisms have any *Alu*-type intermediate repeats (at least in large numbers). These findings suggest that 7SL RNA genes existed before *Alu* sequences and that *Alu* sequences somehow arose fairly late in evolution from the 7SL sequences.

The initial evidence for the mobility of SINES came from analysis of DNA from a patient with neurofibromatosis, a genetic disorder marked by the occurrence of multiple neuronal tumors called neurofibromas due to mutation in the *NF1* gene. Like the retinal tumors that occur in hereditary retinoblastoma (see Figure 8-7), neurofibromas develop only when both *NF1* alleles carry a mutation. In one individual with neurofibromatosis, one *NF1* allele contains an inactivating *Alu* sequence; inactivating somatic mutations in the other *NF1* allele in peripheral neurons lead to the development of neurofibromas. Several other inherited recessive mutations causing disease in humans also have been found to result from insertion of *Alu* sequences in exons, thereby disrupting protein-coding regions.

Like all other mobile elements, *Alu* sequences usually are flanked by direct repeats. Although *Alu* sequences do not encode proteins, they are transcribed by RNA polymerase III and contain an A/T-rich region at one end, similar to L1 elements. Consequently, *Alu* sequences are thought to be retrotransposed by a mechanism similar to that proposed for L1 elements (see Figure 9-19), possibly by the reverse transcriptase and other required proteins expressed from functional L1 elements.

Alu sequences appear to have retrotransposed widely through the human genome and are tolerated, in both possible orientations, at sites where they do not disrupt gene function: flanking solitary genes (see Figure 9-4) and between duplicated genes (see Figure 9-3b), as well as within introns and the regions transcribed into the 5′ and 3′ untranslated regions of mRNAs. Although once postulated to function in controlling gene expression, *Alu* sequences are now thought to have no function, like other mobile elements, despite their widespread occurrence in mammalian genomes.

Retrotransposed Copies of Cellular RNAs Occur in Eukaryotic Chromosomes

In addition to SINES and LINES, which constitute the bulk of the moderately repeated DNA in mammals, other moderately repetitive sequences have been identified. Many of these represent mutated DNA copies of a wide variety of mRNAs that have integrated into chromosomal DNA. These are not duplicates of whole genes that have drifted into nonfunctionality (i.e., the pseudogenes discussed earlier in this chapter) because they lack introns and do not have flanking sequences similar to those of the functional gene copies. Instead, these DNA segments appear to be retrotransposed copies of spliced and polyadenylated (processed) mRNA. Compared with normal genes encoding mRNAs, these inserted segments generally contain multiple mutations, which are thought to have accumulated since their mRNAs were first reverse-transcribed and randomly integrated into the genome of a germ cell in an ancient ancestor. These nonfunctional genomic copies of mRNAs are referred to as *processed pseudogenes*. Most processed pseudogenes are flanked by short direct repeats, supporting the hypothesis that they were generated by rare retrotransposition events involving cellular mRNAs.

Other moderately repetitive sequences representing partial or mutant copies of genes encoding small nuclear RNAs (snRNAs) and tRNAs are found in mammalian genomes. Like processed pseudogenes, these nonfunctional copies of small RNA genes are flanked by short direct repeats and most likely result from rare retrotransposition events that have accumulated through the course of evolution. Enzymes expressed from a LINE or viral retrotransposon are thought to have carried out the retrotransposition of mRNAs, snRNAs, and tRNAs.

Mobile DNA Elements Probably Had a Significant Influence on Evolution

Although mobile DNA elements appear to have no direct function other than to maintain their own existence, their presence probably had a profound impact on the evolution of modern-day organisms. As mentioned earlier, many spontaneous mutations in *Drosophila* result from insertion of a mobile DNA element into or near a transcription unit, and mobile elements also have been found in mutant human genes. In addition, homologous recombination between mobile DNA elements dispersed throughout ancestral genomes may have been important in generating gene duplications and other DNA rearrangements during evolution (see Figure 9-5). Cloning and sequencing of the β-globin gene cluster from various primate species have provided strong evidence that the human G_γ and A_γ genes (see Figure 9-3) arose from an unequal homologous crossover between two L1 sequences. Such duplications and DNA rearrangements contributed greatly to the evolution of new genes. As discussed in an earlier section, gene duplication probably preceded the evolution of a new member of a gene family, which subsequently acquired distinct, beneficial functions.

Mobile DNA most likely also influenced the evolution of genes that contain multiple copies of similar exons encoding similar protein domains (e.g., the fibronectin gene). Homologous recombination between mobile elements inserted into introns probably contributed to the duplication of introns within such genes. Some evidence suggests that during the evolution of higher eukaryotes, recombination between introns of distinct genes occurred, generating new genes made from novel combinations of preexisting exons. For example, tissue plasminogen activator, the Neu receptor, and epidermal growth factor all contain an EGF domain (see Figure 3-10). Evolution of the genes encoding these proteins may have involved recombinations between mobile DNA elements that resulted in the insertion of an EGF-encoding exon into an intron of the ancestral form of each of these genes. The term **exon shuffling** has been coined to refer to this type of evolutionary process.

Recombination between mobile elements also may have played a role in determining which specific genes are expressed in particular cell types and the amount of the encoded protein produced. As noted earlier, eukaryotic genes have transcription-control regions, called enhancers, that can operate over distances of tens of thousands of base pairs. Moreover, as we will learn in the next chapter, the transcription of a gene can be controlled through the combined effects of several enhancers. Recombination between mobile elements inserted randomly near enhancers probably contributed to the evolution of the combinations of enhancers that control gene expression in modern organisms.

So, the early view of mobile DNA elements as completely selfish molecular parasites appears to be premature. Rather, they have probably indirectly made profound contributions to the evolution of higher organisms by serving as sites of recombination, leading to the evolution of novel genes and new controls on gene expression.

SUMMARY Mobile DNA

- Most of the moderately repeated DNA sequences interspersed at multiple sites throughout the genomes of higher eukaryotes arose from mobile DNA elements.

- Mobile DNA elements encode enzymes that can insert their sequence into new sites in genomic DNA.

- Mobile DNA elements that transpose to new sites directly as DNA are called transposons; those that first are transcribed into an RNA copy of the element, which then is reverse-transcribed into DNA, are called retrotransposons (see Figure 9-10 and Table 9-3). Both types generally produce short direct repeats at the site of insertion, which flank the mobile element. The length of the direct repeats depends on the type of mobile element.

- The mobile DNA elements in bacteria—IS elements and bacterial transposons—move via DNA intermediates. Both encode transposase, but the longer transposons also contain at least one other protein-coding gene, generally including a drug-resistance gene.

- Although transposons, similar in structure to bacterial IS elements, occur in eukaryotes (e.g., *Drosophila* P element), retrotransposons generally are much more abundant, especially in higher eukaryotes.

- Viral retrotransposons are flanked by long terminal repeats (LTRs), similar to those in retroviral DNA, and, like retroviruses, encode reverse transcriptase and integrase. They move in the genome by being transcribed into RNA, which then undergoes reverse transcription and integration into the host-cell chromosome (see Figure 9-16).

- Nonviral retrotransposons lack LTRs and have an A/T-rich stretch at one end. These mobile elements are thought to move by an unusual nonviral retrotransposition mechanism (see Figure 9-19).

- The most abundant mobile elements in vertebrates are two types of nonviral retrotransposons called LINES and SINES. Both types appear to have caused mutations associated with human genetic diseases.

- SINES exhibit extensive homology with small cellular RNAs transcribed by RNA polymerase III. The most common SINES in humans frequently contain a site for the restriction enzyme *Alu*I and consequently are called *Alu* sequences. These ≈300-bp sequences are scattered throughout the human genome, constituting ≈5 percent of the total DNA.

- Some moderately repeated DNA sequences are derived from cellular RNAs that were reverse-transcribed and inserted into genomic DNA at some time in evolutionary history. Those derived from mRNAs, called processed pseudogenes, lack introns, a feature that distinguishes them from pseudogenes, which arose by sequence drift of duplicated genes.

- Although mobile DNA elements appear to serve no beneficial function to an individual organism, they most likely influenced evolution significantly.

9.4 Functional Rearrangements in Chromosomal DNA

In contrast to the transposition of mobile elements in genomic DNA, which appears to serve no direct, immediate function for the organism, there are several types of rearrangements of DNA regions that are beneficial to the organism. These functional rearrangements, which have been identified in both prokaryotes and eukaryotes, occur by *inversion* and *deletion* of DNA segments and by *DNA amplification*. Examples of each of these mechanisms, are described in this section. Although functional DNA rearrangements play a role in regulating a few selected genes, the most frequent mechanism of gene control involves the regulation of transcription, discussed in Chapters 10 and 14.

Inversion of a Transcription-Control Region Switches *Salmonella* Flagellar Antigens

When *Salmonella typhimurium*, a type of bacterium closely related to *E. coli*, is ingested, it produces the nausea and diarrhea of common "food poisoning." The protein from which the *Salmonella* flagellum is constructed is one of the major *Salmonella* antigens to which the human immune system responds in eliminating this pathogenic bacterium. However, *Salmonella* cells can express two types of flagellar proteins called H1 and H2, which are encoded at distant sites on the *Salmonella* chromosome. Any one *Salmonella* cell expresses only one type of flagellar protein, but as a clone of cells grows, some progeny spontaneously switch to expression of the other flagellar protein, a process known as *phase variation*. As a result, when individuals respond by making antibody against the major flagellar protein expressed by the *Salmonella* cells that have infected them, a small fraction of the bacterial cells are resistant to the antibody because they express the alternative type of flagellar protein. These cells can then proliferate until the immune system responds a second time to the second flagellar antigen.

The mechanism of *Salmonella* phase variation has been studied by cloning and sequencing the genes involved, by analyzing mutants defective in the switching mechanism, and finally by developing an in vitro switching system using purified proteins that direct the process. The mechanism, outlined in Figure 9-20, involves inversion of a DNA segment that is located adjacent to the *H2* operon and functions as a promoter. When this promoter region is in one 5′ → 3′ orientation (phase I), the two proteins encoded by the *H2* operon are expressed: the H2 flagellar protein and rH1, a specific repressor that inhibits transcription of the *H1* gene, which is located in a different region of the *Salmonella* genome. Approximately once every thousand cell divisions, the segment containing the *H2* promoter is inverted; as a result neither H2 nor rH1 is expressed (phase II). In the absence of the repressor rH1, the *H1* gene is transcribed and the H1 protein is expressed.

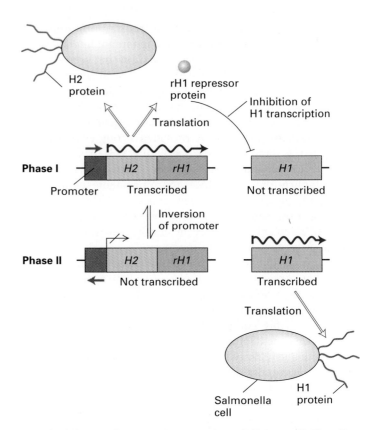

▲ **FIGURE 9-20 Control of expression of *Salmonella* flagellar proteins (H1 and H2) by DNA inversion.** The *H2* operon contains two genes: one encoding H2 and one encoding rH1, which represses transcription of the *H1* gene located in a different region of the *Salmonella* chromosome. When the upstream region containing the *H2* promoter is in one 5′ → 3′ orientation (indicated by blue arrows), the *H2* operon is transcribed. In this phase (I), the cell's flagella contain H2 protein, and rH1 prevents transcription of the *H1* gene. Inversion of the promoter-containing region upstream of *H2* occurs by site-specific recombination catalyzed by an enzyme encoded in the inverting region (see Figure 9-21). As a result, the *H2* operon is not transcribed; in the absence of rH1 protein, the *H1* gene is transcribed. In this phase (II), the cell's flagella contain H1 protein.

The protein that catalyzes inversion of the *H2* promoter is a *site-specific recombinase* encoded by the *Hin* gene, which is completely contained within the inverting segment of DNA. Hin protein binds to a specific recognition sequence located at two sites in the *H2* promoter region and catalyzes recombination, that is, an exchange of DNA strands, at the center of the binding sites. This results in an inversion of the DNA between the two *Hin* recognition sites upstream of the *H2* and *rH1* genes (Figure 9-21). The Hin protein is expressed only very rarely, so that inversion and the resulting phase variation is an infrequent phenomenon. Nonetheless, this process is important to the viability of the *Salmonella* species, because it extends the period of infection and, consequently, increases the number of new hosts that are exposed to *Salmonella*.

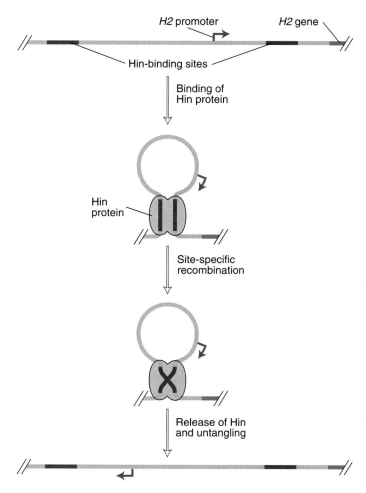

▲ **FIGURE 9-21 Inversion of *H2* promoter mediated by a site-specific recombinase called Hin, a dimeric protein.**
The region upstream of the *H2* gene contains the *H2* promotor, indicated by the blue orientation arrow, and two Hin-binding sites. (Double-stranded DNA is represented by a single line in this figure.) Binding of Hin protein brings the two homologous sites into proximity, allowing the enzyme to catalyze recombination in the middle of the binding sites. Subsequent release of Hin and untangling of the crossed strands results in inversion of the *H2* promoter. The mechanism of recombination is discussed in Chapter 12.

Antibody Genes Are Assembled by Rearrangements of Germ-Line DNA

One of the most remarkable aspects of the immune response is the vast diversity of antibodies that can be elicited against the many possible antigens an individual encounters. It is estimated that an individual can potentially produce many millions of types of antibodies with specificities for as many different antigens. A million different antibody molecules cannot be encoded directly in the human genome, since the genome only contains on the order of one hundred thousand genes. The molecular mechanisms for generating such remarkable diversity from a limited amount of DNA are now

understood. Regulated DNA inversions and deletions figure importantly in the process. To understand how this is accomplished, we must first consider the general structure of antibodies.

Antibody Domain Structure Antibodies belong to a class of proteins called **immunoglobulins,** which constitute about 20 percent of the proteins in the blood. The most abundant type of antibody, immunoglobulin G (IgG) is a symmetrical molecule composed of four polypeptide chains: two identical heavy (H) chains of ≈55 kDa, and two identical light (L) chains of ≈23 kDa. The light chains are composed of two domains, an N-terminal domain called V_L, because it is slightly different or "variable" in sequence and structure in different antibody molecules, and a C-terminal domain called C_L, which has an identical or "constant" sequence in all IgG molecules. Similarly, the heavy chains are composed of four domains, an N-terminal variable V_H domain, and three constant domains designated C_H1, C_H2, and C_H3. The domain structure of IgG is illustrated in Figure 9-22. An antibody molecule binds an antigen molecule through the surfaces created at the interfaces of the V_L and V_H domains at the tips of the Y-shaped molecule (see Figures 3-21 and 3-22). Consequently, the antigen-binding specificity of an antibody is determined by the sequence of its V_L and V_H domains.

Organization of Light-Chain DNA Antibodies are produced by a class of leukocytes (white blood cells) called B **lymphocytes,** or B cells. The genes encoding antibodies with different binding specificities are not directly inherited from the fertilized egg. Rather, they are assembled from a number

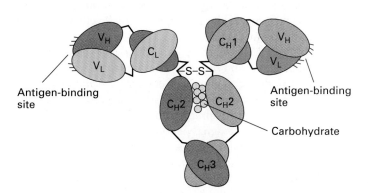

▲ **FIGURE 9-22 Schematic model of an IgG antibody molecule showing its domain structure.** The two identical light (L) chains are shown in shades of green; the two identical heavy (H) chains, in shades of orange. The sequences of the variable domains (V_L and V_H) determine the antigen-binding specificity of an antibody molecule. The sequences of the constant domains (C_L, C_H) are the same in all antibodies belonging to a particular class of immunoglobulins. Differences in the number and sequences of the C_H domains confer unique properties on each Ig class. The heavy chains are linked by disulfide bonds whose number and position vary in different classes. [Adapted from J. Kuby, 1997, *Immunology*, 3d ed., W. H. Freeman and Company, p 114.]

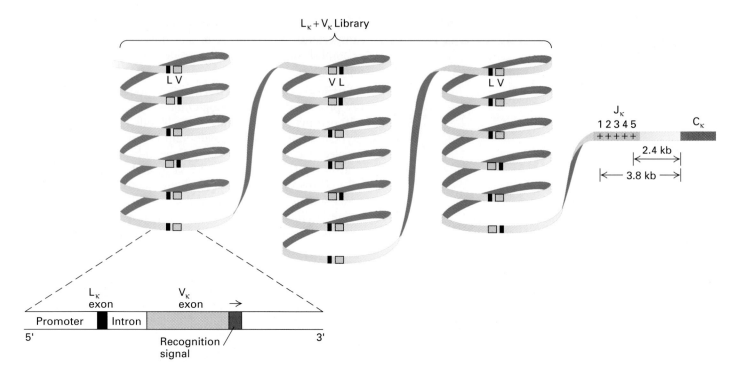

▲ **FIGURE 9-23 Organization of the κ light-chain locus in human germ-line DNA.** In a library spread over hundreds of kilobases of DNA, there are about 100 V_κ segments, each with a unique sequence. The V_κ segments occur in both transcriptional orientations, whereas the five J_κ segments and one C_κ segment are all in the 5′ → 3′ orientation (left to right in the illustration). One L_κ-V_κ unit is shown expanded. It consists of a promoter at the 5′ end (beginning of transcription), an exon encoding a leader peptide (L_κ), an intron, an exon encoding the V_κ region, and a recognition signal. This signal guarantees that the V_κ segment will join to a J_κ segment. The arrow shows the orientation of the recognition site. See text for further discussion. [After Paul D. Gottleib, *Mol. Immunol.* **17**:1423.]

of separated gene segments present in germ-line DNA; this process occurs during the development of B cells from stem cells in the bone marrow. For example, a functional re-arranged gene encoding the κ light chain (the major type of light chain in mice and humans) contains three segments. At the 5′ end is the L_κ segment; it encodes a *leader* or *signal peptide* that directs the newly translated protein into the endoplasmic reticulum in preparation for secretion from the cell (Chapter 17). The signal peptide is removed during post-translational processing of the light chain and is not present in the mature antibody molecule. The second segment encodes the V_L domain of the light chain, and the third segment, at the 3′ end, encodes the C_L domain.

As shown in Figure 9-23, the DNA of germ cells (i.e., sperm and egg cells) and all other cells except mature B lymphocytes contains a κ locus that has a *variable* region at its 5′ end. This region consists of a library of leader (L_κ) and variable (V_κ) segments containing ≈100 L_κ + V_κ units in humans; these units are arrayed in tandem along one long stretch of DNA. (The L_κ segment corresponds to the leader exon in the final gene; the V_κ segment makes up most, but not all, of the final V_L exon, encoding the variable region of the light chain.) Each of the L_κ + V_κ units is about 400 nucleotides long, and they are separated by about 7 kb;

thus 100 L_κ + V_κ units would cover about 740 kb of DNA. The variable region of the κ locus is followed by five *joining* (J_κ) segments in human germ-line DNA and then by the one *constant* (C_κ) segment. The five J_κ segments are tandemly arranged and are separated by about 20 kb from the 3′ end of the variable region. Each of the J_κ segments is about 30 nucleotides long, and they are spread over 1.4 kb of DNA. Between the 3′ J_κ segment and the single C_κ segment lies 2.4 kb of intervening DNA. The number of V_κ and J_κ segments varies with the species of mammal, although there are always many more V_κ than J_κ segments.

Rearrangement of Light-Chain DNA When DNA reorganizes to make a functional κ gene, one V_κ segment joins to one J_κ segment. This joining is performed by a site-specific recombinase that recognizes sequences at the 3′ end of each V_κ segment and the 5′ end of each J_κ segment. Recombination between these sequences results in a deletion or inversion of the intervening sequence, depending on whether the L_κ + V_κ unit has the same or opposite transcriptional orientation as the J_κ segment (Figure 9-24). This recombination forms the completed variable region. So far as is known, any V_κ can join to any J_κ, and the choice is random. Once a V_κ and J_κ are joined, the variable and constant regions are transcribed together into a primary RNA

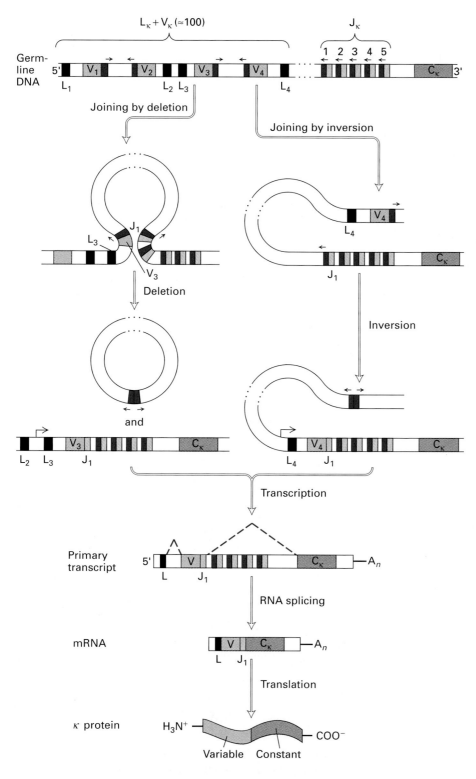

◀ **FIGURE 9-24 Joining of V_κ to J_κ in human germ-line DNA and formation of a κ light chain.** The $L_\kappa + V_\kappa$ segments are oriented in either transcriptional direction relative to C_κ in germ-line DNA. Each V_κ and J_κ segment has a recognition signal (red) to specify the points at which joining should take place. The arrows indicate the orientation of the recognition signals. The joining process either deletes the intervening DNA or inverts it, depending on the relative orientations of the V_κ and J_κ to be joined. In a deletional joining, the deleted DNA is lost from the cell. In an inversional joining, all of the DNA is conserved. Once the V_κ-J_κ joining has occurred, the now complete gene can be transcribed, beginning just 5' of the closest L_κ. The resulting primary RNA transcript is spliced to remove all unwanted segments, as indicated by the inverted V-shaped brackets. The spliced mRNA then is translated into a complete κ light chain. Note that the J segments encode part of the variable region of the protein and that the leader peptide sequence is removed following translation.

transcript. The intervening sequences between L_κ and V_κ and between J_κ and C_κ (including any remaining J_κ regions) then are removed by RNA splicing to produce the mature mRNA for the κ light-chain protein.

Since each V_κ and J_κ segment has a unique nucleotide sequence, V_κ-J_κ joining of 100 V_κ segments and 5 J_κ segments in human germ-line DNA can produce 500 different possible chains. But V_κ-J_κ joining generates even more sequence variability than this calculation would suggest because a small, variable number of nucleotides are lost from the V_κ and J_κ segments when they are joined. The imprecision of the joining process greatly increases the diversity of possible V_L amino acid sequences encoded by the V_κ-J_κ joint region. Significantly this region encodes many of the amino

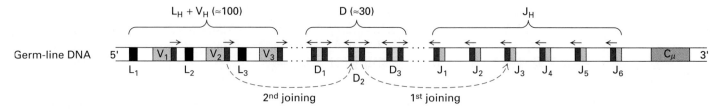

▲ **FIGURE 9-25 Organization of heavy-chain gene segments in human germ-line DNA.** The heavy-chain locus contains a library of D segments in addition to multiple V and J segments analogous to those in the κ light-chain locus. The V_H, D, and J_H segments have recognition signals (red) that assure proper joining. The small arrows indicate the orientation of the recognition signals. By processes similar to those depicted in Figure 9-24, a D segment first joins to a J_H segment with the intervening DNA being deleted; a V_H segment then joins the preformed DJ_H unit, forming the variable region of the heavy-chain gene. Subsequent transcription yields a primary transcript, which is spliced to yield heavy-chain mRNA. The C_μ segment encodes the constant region of the μ heavy chain, which is found in IgM. Other C segments downstream from C_μ (not shown) encode the constant regions of heavy chains in the other Ig classes.

acids in the antigen-binding site at the tip of the V_L domain (see Figure 9-22).

The random loss of nucleotides at the joining site generates significant diversity at that point, but the system pays for its diversity. The cost is evident if we remember the constraints on a coding sequence. Recall that a coding sequence in DNA must be read in three-base (triplet) codons, and that the initiation codon, AUG (methionine), defines a **reading frame** that groups the rest of the coding region into triplets. Reading DNA in one of the two other reading frames would generate a meaningless string of amino acids. Thus, the joining of two pieces of coding DNA, such as a V_κ and J_κ segment, can produce an *in-phase* joint, which maintains a sensible reading frame, or an *out-of-phase* joint, which encodes a nonsense protein. Because the V_κ-J_κ joining process is a random one, two out of every three joinings result in joints that make no sense. Thus the increased diversity permitted by imprecise joining is obtained at the expense of formation of two *nonproductive* joints for each *productive* joint.

We have described three sources of diversity in antibody κ light chains: variability in the sequence of the many V_κ segments in the germ-line κ locus, variability in the sequences of a small number of J_κ segments, and variability in the number of nucleotides deleted at V_κ-J_κ joints. (The actual number of V_κ and J_κ segments varies among species; in the mouse, for instance, there are ≈300 V_κ segments and four functional J_κ segments.) Other processes beyond the scope of our discussion generate even more antibody diversity by randomly altering the DNA sequence in the joint between the V_κ and J_κ segments.

Organization and Rearrangement of Heavy-Chain DNA Functional genes encoding antibody heavy chains are formed by processes similar to those just described for light-chain genes. Analysis of the antigen-binding sites of numerous antibodies suggests that the heavy-chain contribution to contacts with antigen is even greater than the light-chain contribution. Consistent with a need for greater diversity in heavy chains, three libraries of gene segments contribute to the variable region of functional heavy-chain genes, rather than just

the two (V and J) that make up light-chain genes. The third library consists of *diversity* (D) segments, which are located between the other two libraries, whose segments are called V_H and J_H. Thus two joining reactions, V_H to D and D to J_H, are required to assemble the region of a heavy-chain gene encoding the variable region (Figure 9-25). Clearly, having three segments, rather than two, greatly increases the possible combinatorial diversity. Human germ-line DNA contains an estimated 100 V_H segments, 30 D segments, and six functional J_H segments. In addition to the diversity due to random joining of V_H, D, and J_H segments, further diversity is created by loss of nucleotides at the V_H-D and D-J_H joints, as occurs at the V_κ-J_κ joint. The variable domains of heavy chains are diversified even further by the random addition of up to 15 nucleotides when a D segment joins to a J_H, or when a V_H joins to a D. The junctions where this occurs encode most of the amino acids that form the antigen-binding tip of the V_H domain in an antibody molecule (see Figure 9-22). Thus maximum diversity is generated in the portion of the antibody molecule that interacts with antigen.

Heavy-chain germ-line DNA also contains multiple C segments encoding the constant-region domains of the various Ig classes. A combination of alternative RNA processing and additional DNA rearrangements (class switching) determine which Ig class is expressed by a particular B cell.

Generalized DNA Amplification Produces Polytene Chromosomes

All of the DNA rearrangements discussed so far—both functional and nonfunctional—involve changes in the position of sequences within the genome. Another type of rearrangement involves generalized amplification of DNA sequences, or **polytenization.**

The salivary glands of *Drosophila* species contain enlarged interphase chromosomes. When fixed and stained, these chromosomes are characterized by a large number of well-demarcated bands, which can be used to establish the position

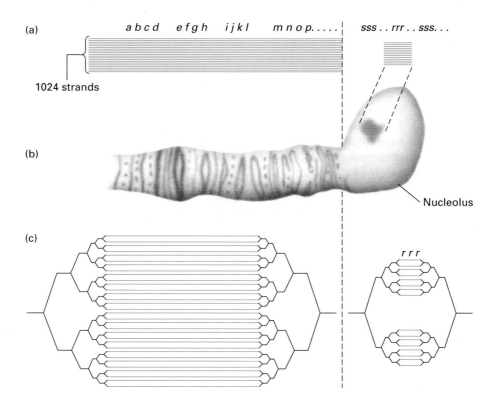

▲ FIGURE 9-26 Model of DNA amplification to produce polytene chromosomes. (a) Sequences along the left arm of the *Drosophila hydei* X chromosome are represented by the letters *a* through *p*. Simple-sequence DNA *(sss)* and ribosomal genes *(rrr)* occur on the short right arm of the chromosome. In the salivary glands, the DNA in the left arm of the chromosome is replicated in parallel many times; ten duplications would give $2^{10} = 1024$ copies. The simple-sequence DNA does not appear to increase at all; the ribosomal genes are replicated, but not as many times as the DNA in the left arm. (b) Morphology of the stained polytene X chromosome as it appears in the light microscope. The banding pattern results from reproducible packing of DNA and protein within each amplified site along the chromosome. (c) The proposed pattern of amplification in the polytene *D. hydei* X chromosome during five replications. Double-stranded DNA is represented by a single line. The simple-sequence DNA at the centromere and telomeres is not amplified, and ribosomal genes are amplified less than sequences in the left arm. [See C. D. Laird et al., 1973, *Cold Spring Harbor Symp. Quant. Biol.* **38**:311.]

and order of genes on *Drosophila* chromosomes (see Figure 8-22). The enlargement of chromosomes in the salivary glands, and in some cells in other *Drosophila* tissues as well, occurs when the DNA repeatedly replicates but the daughter chromosomes do not separate. The result is a *polytene* chromosome composed of many parallel copies of itself. The amplification of chromosomal DNA greatly increases gene copy number, presumably to supply sufficient mRNA for protein synthesis in the massive salivary gland cells.

Although most of a chromosome participates in polytenization, certain sequences, such as the simple-sequence DNAs near the centromere and telomeres, are not amplified. Furthermore, the ribosomal genes tend to be amplified less than other sequences during polytenization (Figure 9-26); as discussed previously, multiple copies of these genes already are present in tandem arrays. The molecular basis for the varying extent of replication along presumably linear chromosomal DNA molecules remains unknown, but the unreplicated simple-sequence DNA probably contributes to alignment of the amplified DNA along the length of polytene chromosomes.

SUMMARY **Functional Rearrangements in Chromosomal DNA**

• Although gene expression is most frequently regulated by DNA-binding proteins that control the initiation of transcription, some cells use specific rearrangements of the DNA sequence to control expression of certain genes.

• *Salmonella typhimurium* controls expression of the H1 or H2 flagellar antigen by rare site-specific recombination between two repeated sequences flanking the promoter of the *H1* gene (see Figure 9-20). The resulting inversion of the promoter causes the cell to shift from expressing one flagellar antigen to expressing the other.

• The remarkable diversity of antibody molecules is achieved by assembly of functional genes, encoding the antibody heavy and light chains, from multiple gene segments with unique sequences present in germ-line

DNA. The segments encoding the variable domains of heavy and light chains are assembled by recombining two or three gene segments from libraries of alternative short coding regions (see Figures 9-24 and 9-25).

• The large number of possible combinations in which light-chain and heavy-chain gene segments can be combined is one mechanism for generating variation in the amino acid sequence, and therefore binding specificity, of antibodies produced by individual B lymphocytes. Additional diversity results from the random loss and addition of nucleotides at the joints between gene segments during the joining process.

• In some organisms, certain specialized cells grow to much larger size than other cells and amplify their chromosomal DNA, producing polytene chromosomes such as those in the larval salivary glands of *Drosophila* species (see Figure 9-26). The chromosomes of these giant cells result from about 10 replications without cell division and without replication of the associated simple-sequence DNA at the centromere and telomeres.

9.5 Organizing Cellular DNA into Chromosomes

We turn now to the question of how DNA molecules are organized within cells into the structures we observe as chromosomes. Because the total length of cellular DNA in cells is up to a hundred thousand times the cell's length, the packing of DNA into chromosomes is crucial to cell architecture.

Most Bacterial Chromosomes Are Circular with One Replication Origin

In most bacterial cells, genes are encoded on large circular chromosomes. The circular nature of bacterial chromosomes was first discovered by analyzing the frequency of genetic recombination between mutant genes that produced easily assayed phenotypes, such as the inability to grow in the absence of a specific amino acid or the inability to grow on a particular sugar. As multiple genes were mapped on the *E. coli* chromosome by recombinational analysis, no ends were found in the single linkage map that developed. Rather, every gene had other identified genes that could be mapped on either side of it. The whole linkage map generated a circle (see Figure 8-14). Eventually, the circular structure of the *E. coli* chromosome was observed directly in autoradiographs of DNA molecules from cells grown in the presence of ^{3}H-labeled thymine, which is incorporated only into DNA. Such studies showed that the *E. coli* chromosome has a total length of 1 mm. The structure of the partially replicated *E. coli* chromosomal DNA molecule shown in Figure 9-27 demonstrates that the chromosome replicates from a single **replication origin**, similar to the small circular plasmid DNA molecule diagrammed in Figure 7-2.

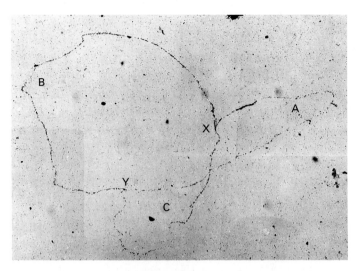

▲ **FIGURE 9-27 Autoradiograph of the *E. coli* chromosome labeled with [^{3}H]-thymine.** Points X and Y indicate the positions of DNA replication forks. Region B is the unreplicated parental chromosome. Regions A and C are the newly replicated nascent daughter chromosomes containing one parental strand and one newly replicated daughter strand. [From J. Cairns, 1964, *Cold Spring Harbor Symp. Quant. Biol.* **28**:43.]

The 1-mm-long DNA molecule of the *E. coli* chromosome is contained within cells that are only about 2 μm long and about 0.5–1 μm wide. A free DNA molecule of this size would form a random coil about 1000 times the volume of an *E. coli* cell. However, several mechanisms operate to compact *E. coli* chromosomal DNA sufficiently to fit inside the bacterial cell. For example, the large volume filled by free DNA is due largely to charge repulsion between the negatively charged phosphate groups. In the cell, this effect is counteracted by association of the DNA with positively charged polyamines, such as spermine and spermidine, which shield the negative charges of the DNA phosphate groups:

$$\overset{+}{H_3N}-(CH_2)_3-\overset{+}{NH}-(CH_2)_4-\overset{+}{NH}-(CH_2)_3-\overset{+}{NH_3}$$
Spermine

$$\overset{+}{H_3N}-(CH_2)_4-\overset{+}{NH}-(CH_2)_3-\overset{+}{NH_3}$$
Spermidine

In addition, numerous small protein molecules associate with the chromosomal DNA, causing it to fold into a more compact structure. The most abundant of these proteins, H-NS, is a dimer of a 15.6-kDa polypeptide. H-NS binds DNA tightly and compacts it considerably, as indicated by the increased sedimentation rate and decreased viscosity of DNA associated with H-NS compared with free DNA. There are about 20,000 H-NS molecules per *E. coli* cell, enough for one H-NS dimer per ≈400 base pairs of DNA.

Finally, *E. coli* chromosomal DNA is tightly supercoiled—that is, twisted upon itself like the circular SV40 DNA shown in Figure 4-11. As discussed in Chapter 12, an *E. coli* enzyme

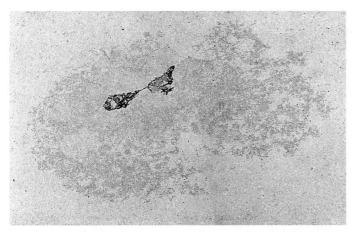

2 μm

▲ **FIGURE 9-28 Electron micrograph of an isolated folded**
E. coli **chromosome.** The highly supercoiled DNA is attached
to a fragment of the cell membrane, which appears as the most
darkly staining material in the micrograph. Although this
chromosome exhibits extensive supercoiling, it actually had
decondensed considerably during isolation. Within the cell, the
chromosome has a diameter of <1 μm. [From H. Delius and
A. Worcel, 1974, *J. Mol. Biol.* **82**:107.]

called DNA gyrase uses energy from ATP hydrolysis to wind
supercoils into DNA. Supercoiling contributes to the compaction necessary to fit chromosomal DNA into the bacterial cell. Figure 9-28 shows an isolated, highly supercoiled
E. coli chromosome attached to a fragment of cell membrane. If all the supercoils were relaxed and the DNA spread
out, it would appear as a single, replicating, circular DNA
molecule similar to that in Figure 9-27.

Eukaryotic Nuclear DNA Associates with Histone Proteins to Form Chromatin

The problem of compacting genomic DNA to fit into the
nucleus of an eukaryotic cell is solved in a different way.
When the DNA from eukaryotic nuclei is isolated in isotonic buffers (i.e., buffers with the same salt concentration
found in cells, ≈0.15 M KCl), it is associated with an equal
mass of protein in a highly compacted complex called chro-

matin. The general structure of chromatin has been found
to be remarkably similar in the cells of all eukaryotes including fungi, plants, and animals.

The most abundant proteins associated with eukaryotic
DNA are **histones,** a family of basic proteins present in all
eukaryotic nuclei. The five major types of histone proteins—
termed H1, H2A, H2B, H3, and H4—are rich in positively
charged basic amino acids, which interact with the negatively charged phosphate groups in DNA. In a fraction of
the histone proteins of most cells, some of the basic amino
acid side chains are modified by post-translational addition
of acetyl (CH$_3$COO−), phosphate, or methyl groups, neutralizing the positive charge of the side chain or converting
it to a negative charge.

The amino acid sequences of four histones (H2A, H2B, H3,
and H4) are remarkably similar among distantly related species.
For example, the sequences of histone H3 from sea urchin
tissue and of H3 from calf thymus are identical except for a
single amino acid, and only four amino acids are different in
H3 from the garden pea and that from calf thymus. Minor
histone variants encoded by genes that differ from the highly
conserved major types also exist, particularly in vertebrates.

The amino acid sequence of H1 varies more from organism to organism than do the sequences of the other major histones. In certain tissues, H1 is replaced by special histones. For example, in the nucleated red blood cells of birds,
a histone termed H5 is present in place of H1. The similarity in sequence among histones from all eukaryotes indicates
that they fold into very similar three-dimensional conformations, which were optimized for histone function early in
evolution in a common ancestor of all modern eukaryotes.

Chromatin Exists in Extended and Condensed Forms

When chromatin is extracted from nuclei and examined in the
electron microscope, its appearance depends on the salt concentration to which it is exposed. At low salt concentration, isolated chromatin resembles "beads on a string" (Figure 9-29a).
In this extended form, the string is a thin filament of
"linker" DNA connecting the beadlike structures termed **nucleosomes.** Composed of DNA and histones, nucleosomes are
about 10 nm in diameter and are the primary structural units
of chromatin. If chromatin is isolated at physiological salt

▶ **FIGURE 9-29 Electron micrographs of**
extended chromatin in extended and
condensed forms. (a) Chromatin isolated in low
ionic strength buffer has an extended "beads-on-
a-string" appearance. The "beads" are
nucleosomes (10-nm diameter) and the "string"
is connecting DNA. (b) Chromatin isolated in buffer
with a physiologic ionic strength (0.15 M KCl)
appears as a condensed fiber 30 nm in diameter.
[Part (a) courtesy of S. McKnight and O. Miller, Jr.; part
(b) courtesy of B. Hamkalo and J. B. Rattner.]

(a)

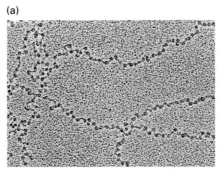

(b)

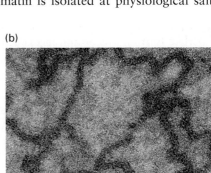

▶ **FIGURE 9-30 Structure of the nucleosome.** (a) Ribbon diagram of the nucleosome shown face-on *(left)* and from the side *(right)*. One DNA strand is shown in green and the other in brown. H2A is yellow; H2B, red; H3, blue; H4, green. (b) Space-filling model shown from the side. DNA is shown in white; histones are colored as in (a). H2A, H2A', H2B, H2B', H3, and H4 indicate the positions of the respective histone N-terminal tails visible in this view. The H2A' N-terminal tail interacts with the upper loop of DNA, while the H2A N-terminal tail (only partially seen in this view) interacts with the bottom loop of DNA. The N-terminal tail of one H4 extends from the bottom of the nucleosome and interacts with the neighboring histone octamer in the crystal lattice (not shown). The N-terminal tails of histones H2B, H2B', H3, and H3' pass between the two loops of DNA. The N-terminal tails of H2A, H4, H3, and H2B include an additional 3, 15, 19, and 23 residues, respectively, that are not visualized in the crystal structure because they are not highly structured. They extend further from the surface of the nucleosome where they may participate in nucleosome-nucleosome interactions in the 30 nm fiber (See Figure 9-31) or interact with other chromatin-associated proteins. [From K. Luger et al., 1997, *Nature* **389**:251; courtesy of T. J. Richmond.]

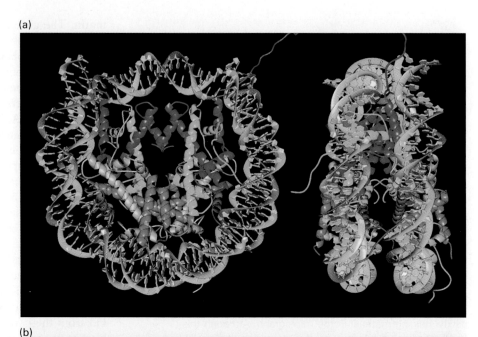

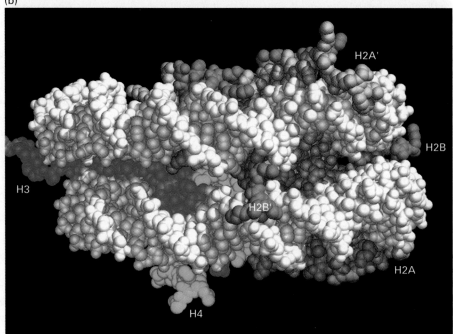

concentration ($\approx$0.15 M KCl), it assumes a more condensed fiberlike form that is 30 nm in diameter (Figure 9-29b).

Structure of Nucleosomes The DNA component of nucleosomes is much less susceptible to digestion than is the linker DNA between them. If the nuclease treatment is carefully controlled, all the linker DNA can be digested releasing individual nucleosomes with their DNA component. A nucleosome comprises a protein core with DNA wound around its surface like thread around a spool. The core is an octamer containing two copies each of histones H2A, H2B, H3, and H4. X-ray crystallography has shown that the octameric histone core is a roughly disk-shaped molecule made of interlocking histone subunits (Figure 9-30). Nucleosomes from all eukaryotes contain about 146 base pairs of DNA wrapped slightly less than two turns around the protein core. The length of the linker DNA is more variable among species, ranging from about 15 to 55 base pairs.

In cells, newly replicated DNA is assembled into nucleosomes shortly after the replication fork passes, but when isolated histones are added to DNA in vitro at physiological salt concentration, nucleosomes do not spontaneously form. However, nuclear proteins that bind histones and assemble

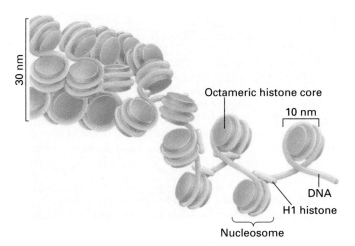

Octameric histone core
(see Figure 9-30)

10 nm

DNA

H1 histone

Nucleosome

▲ **FIGURE 9-31 Solenoid model of the 30-nm condensed chromatin fiber in a side view.** The octameric histone core (see Figure 9-30) is shown as an orange disk. Each nucleosome associates with one H1 molecule, and the fiber coils into a solenoid structure with a diameter of 30 nm. [Adapted from M. Grunstein, 1992, *Sci. Am.* **267**:68]

them with DNA into nucleosomes in vitro have been characterized. Proteins of this type are thought to assemble histones and newly replicated DNA into nucleosomes in vivo as well.

Structure of Condensed Chromatin   When extracted from cells in isotonic buffers, most chromatin appears as fibers ≈30 nm in diameter (see Figure 9-29b). In these condensed fibers, nucleosomes are thought to be packed into a spiral or solenoid arrangement, with six nucleosomes per turn (Figure 9-31). A fifth histone, H1, is bound to the DNA on the inside of the solenoid, with one H1 molecule associated with each nucleosome. Recent electron microscopic studies suggest that the 30-nm fiber is less uniform than the solenoid model predicts. Condensed chromatin may in fact be quite dynamic with regions partially unfolding and then refolding into a solenoid structure occasionally.

The chromatin in chromosomal regions that are not being transcribed exists predominantly in the condensed, 30-nm fiber form. The regions of chromatin actively being transcribed are thought to assume the extended beads-on-a-string form.

Acetylation of Histone N-Termini Reduces Chromatin Condensation

Each of the histone proteins making up the nucleosome core contain flexible amino termini of 20 to 40 residues extending from their globular domains (see Figure 9-30b). The N-termini contain several positively charged lysine groups. Some of these interact with phosphates in DNA of the same nucleosome, and some may interact with linker DNA or with neighboring nucleosomes. These lysines undergo reversible acetylation and deacetylation by enzymes that act on specific lysines in the N-termini of the different histones. In the acetylated form, the positive charge of the lysine ε-amino group is neutralized and its interaction with a DNA phosphate group

is eliminated. Thus the greater the extent of acetylation of histone N-termini, the less likely chromatin is to form condensed 30-nm fibers and possibly higher-order folded structures.

The extent of histone acetylation also is correlated with the relative resistance of chromatin DNA to digestion by nucleases. This phenomenon can be demonstrated by digesting isolated nuclei with DNase I. Following digestion, the DNA is completely separated from chromatin protein, digested to completion with a restriction enzyme, and analyzed by Southern blotting (see Figure 7-32). When a gene is cleaved at random sites by DNase I, the Southern-blot band corresponding to that gene is lost. This method has been used to show that the inactive β-globin gene in nonerythroid cells, where it is associated with relatively unacetylated histones, is much more resistant to DNase I than is the active β-globin gene in erythroid precursor cells, where it is associated with acetylated histones (Figure 9-32). This relative resistance to nuclease indicates that the chromatin structure of nonexpressed DNA is more condensed than that of transcribed DNA. In condensed chromatin, the DNA is largely inaccessible to DNase I because of its close association with histones and possibly other less-abundant chromatin proteins. In contrast, actively transcribed DNA is much more accessible to DNase I digestion because it is present in the extended, beads-on-a-string form of chromatin.

Recent genetic studies in yeast indicate that specific histone acetylases are required for the full activation of transcription of a number of genes. Consequently, as discussed in Chapter 10, the control of acetylation of histone N-termini in specific chromosomal regions is thought to contribute to gene control by regulating the strength of the interaction of histones with DNA and the folding of chromatin into condensed structures. Genes in condensed, folded regions of DNA are inaccessible to RNA polymerase and other proteins required for transcription.

Eukaryotic Chromosomes Contain One Linear DNA Molecule

As discussed in the next section, eukaryotic chromosomes can be visualized when they condense during mitosis. The general belief is that each of the several chromosomes in eukaryotic cells contains a single long DNA molecule. Because the longest DNA molecules in human chromosomes are almost 10 cm long ($2-3 \times 10^8$ base pairs), they are difficult to handle experimentally without breaking. However, in lower eukaryotes, the sizes of the largest DNA molecules that can be extracted are consistent with the hypothesis that each chromosome contains a single DNA molecule. For example, physical analysis of the largest DNA molecules extracted from several genetically different *Drosophila* species and strains shows that they are from 6×10^7 to 1×10^8 base pairs long. These sizes match the DNA content of single stained metaphase chromosomes of *Drosophila melanogaster,* as measured by the amount of DNA-specific stain absorbed. Therefore, each chromosome probably contains a single DNA molecule.

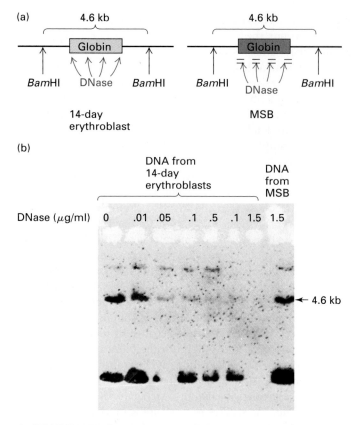

(a)

4.6 kb

Globin

BamHI   DNase   BamHI

14-day
erythroblast

4.6 kb

Globin

BamHI   DNase   BamHI

MSB

(b)

DNA from
14-day
erythroblasts

DNA
from
MSB

DNase (μg/ml)   0   .01 .05   .1   .5   .1 1.5 1.5

← 4.6 kb

▲ **FIGURE 9-32 Demonstration that transcriptionally active genes are more susceptible than inactive genes to DNase I digestion.** Chick embryo erythroblasts at 14 days actively synthesize globin, whereas cultured undifferentiated MSB cells do not. Nuclei from each type of cell were isolated and exposed to increasing concentrations of DNase I. The nuclear DNA was then extracted and treated with the restriction enzyme *Bam*HI, which cleaves the DNA around the globin sequence and normally releases a 4.6-kb globin fragment (a). The DNase I- and *Bam*HI-digested DNA was subjected to Southern-blot analysis with a probe of labeled cloned adult globin DNA, which hybridizes to the 4.6-kb *Bam*HI fragment. As shown in (b) the transcriptionally active DNA from the 14-day globin-synthesizing cells was sensitive to DNase I digestion, indicated by the absence of the 4.6-kb band at higher nuclease concentrations. In contrast, the inactive DNA from MSB cells was resistant to digestion. [See J. Stalder et al., 1980, *Cell* **19**:973; photograph courtesy of H. Weintraub.]

The correspondence between the number of DNA molecules per cell and the number of chromosomes has been conclusively demonstrated in yeast cells. The DNA from each *S. cerevisiae* chromosome can be separated and individually identified by pulsed-field gel electrophoresis (see Figure 7-26). The number of separated DNA molecules equals the number of genetic linkage groups (i.e., chromosomes) in yeast, and the entire sequence of each chromosome has been determined and directly compared with the genetic map. The length of yeast chromosomal DNA ranges from about 1.5×10^5 to 10^6 base pairs.

SUMMARY Organizing Cellular DNA into Chromosomes

• Genomic DNA in both bacteria and eukaryotes must be highly compacted in order to fit within cells.

• Bacterial chromosomes usually are circular DNA molecules that replicate from a single origin. Bacterial DNA is highly supercoiled and associated with polyamines and low-molecular-weight basic proteins, which permit the DNA to fold tightly in the central portion of a bacterial cell.

• In eukaryotic cells, DNA is associated with about an equal mass of histone proteins in a highly condensed structure called chromatin. The building block of chromatin is the nucleosome, consisting of a histone octamer around which is wrapped about 146 bp of DNA (see Figure 9-30).

• When extracted under physiological conditions, chromatin is visualized in the electron microscope as a 30-nm fiber made up of nucleosomes, the linker DNA between them, and histone H1 (see Figure 9-31). Transcriptionally inactive regions of DNA within cells is thought to exist in this condensed form and possibly higher-order structures built from it.

• When extracted at low salt concentrations, chromatin is visualized as an extended beads-on-a-string structure, which lacks histone H1. Transcriptionally active regions of DNA within cells are thought to resemble this extended form of chromatin.

• The reversible acetylation and deacetylation of lysine residues in the N-termini of histones H2A, H2B, H3, and H4 controls how tightly DNA is bound by the histone octamer and affects the assembly of nucleosomes into the condensed forms of chromatin. Hypoacetylated chromatin assumes a more condensed structure than hyperacetylated chromatin.

• The more open chromatin structure of active genes makes them more sensitive to nuclease digestion than inactive genes.

• Each eukaryotic chromosome contains a single linear DNA molecule.

9.6 Morphology and Functional Elements of Eukaryotic Chromosomes

Chromatin is further organized into large units hundreds to thousands of kilobases in length called chromosomes. Microscopic observations on the number and size of chromosomes and their staining patterns led to the discovery of many important general characteristics of chromosome structure.

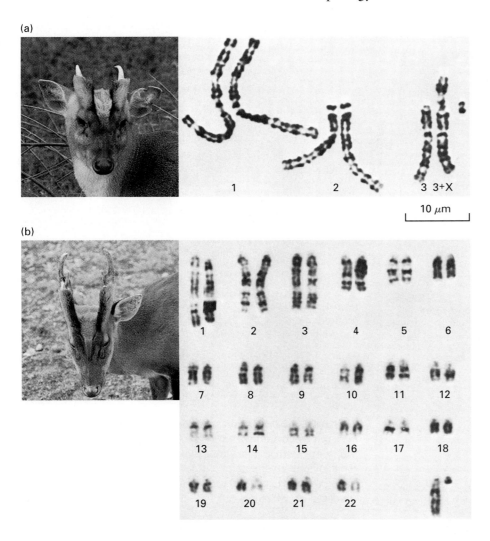

Chromosome Number, Size, and Shape at Metaphase Are Species Specific

In nondividing cells the chromosomes are not visible, even with the aid of histologic stains for DNA (e.g., Feulgen or Giemsa stains) or electron microscopy. During mitosis and meiosis, however, the chromosomes condense and become visible in the light microscope. Therefore, almost all cytogenetic work (i.e., studies of chromosome morphology) has been done with condensed **metaphase** chromosomes obtained from dividing cells—either somatic cells in mitosis or dividing gametes during meiosis.

The condensation of metaphase chromosomes probably results from several orders of folding and coiling of 30-nm chromatin fibers. Recall from Chapter 5 that at the time of mitosis, cells have already progressed through the S phase of the cell cycle and have replicated their DNA, so the chromosomes that become visible during metaphase are duplicated structures (see Figure 5-56). Each metaphase chromosome consists of two sister **chromatids,** which are attached at the centromere. The number, sizes, and shapes of the metaphase chromosomes constitute the **karyotype,** which is distinctive for each species. In most organisms, all cells have the same karyotype. However, species that appear quite similar can have very different karyotypes, indicating that similar genetic potential can be organized on chromosomes in very different ways (Figure 9-33).

Nonhistone Proteins Provide a Structural Scaffold for Long Chromatin Loops

Although histones are the predominant proteins in chromosomes, nonhistone proteins are also involved in organizing the structure of chromosomes. Electron micrographs of histone-depleted metaphase chromosomes from HeLa cells reveal long loops of DNA anchored to a **chromosome scaffold** composed of nonhistone proteins (Figure 9-34). This scaffold has the shape of the metaphase chromosome and persists even when the DNA is digested by nucleases. As depicted schematically in Figure 9-35, megabase long loops of the 30-nm chromatin fiber are thought to associate with the flexible chromosome scaffold, yielding an extended form characteristic of chromosomes during interphase. Coiling of the scaffold into a helix and further packing of this helical

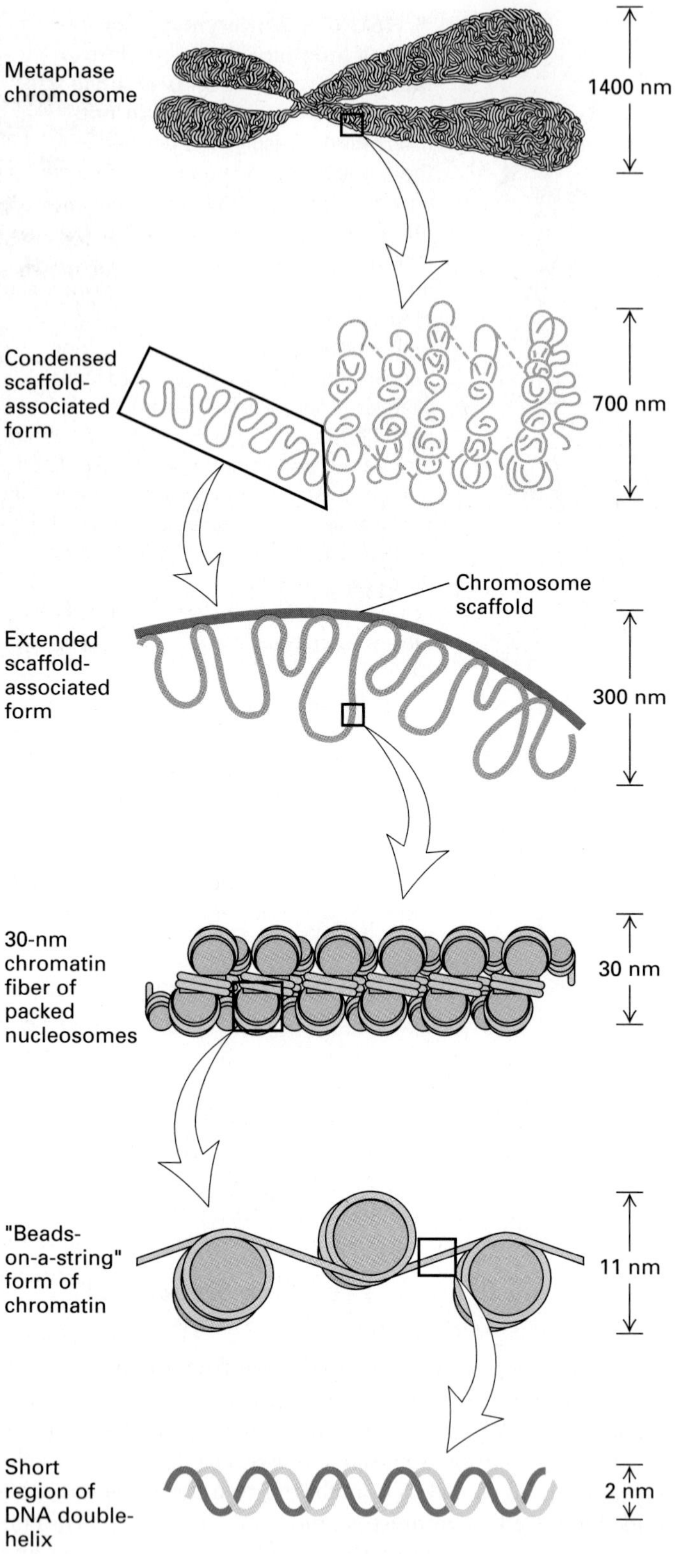

Metaphase
chromosome

1400 nm

Condensed
scaffold-
associated
form

700 nm

Extended
scaffold-
associated
form

Chromosome
scaffold

300 nm

30-nm
chromatin
fiber of
packed
nucleosomes

30 nm

"Beads-
on-a-string"
form of
chromatin

11 nm

Short
region of
DNA double-
helix

2 nm

▲ **FIGURE 9-35 Model for the packing of chromatin and the chromosome scaffold in metaphase chromosomes.** In interphase chromosomes, long stretches of 30-nm chromatin loop out from extended scaffolds. In metaphase chromosomes, the scaffold is folded into a helix and further packed into a highly compacted structure, whose precise geometry has not been determined.

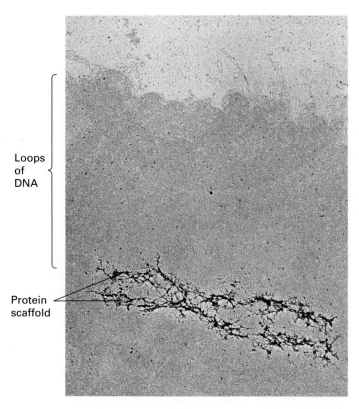

Loops
of
DNA

Protein
scaffold

▲ **FIGURE 9-34 Electron micrograph of a histone-depleted metaphase chromosome prepared from HeLa cells by treatment with a mild detergent.** A nonhistone protein scaffolding (dark structure) is visible from which long loops of DNA extend. [From J. R. Paulson and U. K. Laemmli, 1977, *Cell* **12**:817. Copyright 1977 M.I.T.]

structure produces the highly condensed structure characteristic of metaphase chromosomes.

In situ hybridization experiments with several different fluorescent-labeled probes to DNA in human interphase cells support the loop model shown in Figure 9-35. In these experiments, some probe sequences separated by millions of base pairs in linear DNA appeared reproducibly very close to each other in interphase nuclei from different cells (Figure 9-36). These closely spaced probe sites are postulated to lie close to specific sequences in the DNA, called **scaffold-associated regions (SARs)** or *matrix-attachment regions* (MARs), that are bound to the chromosome scaffold. SARs have been mapped by digesting histone-depleted chromosomes with restriction enzymes and then recovering the fragments that are bound to scaffold proteins.

In general, SARs are found between transcription units. In other words, genes are located primarily within chromatin loops, which are attached at their bases to a chromosome scaffold. Experiments with transgenic mice indicate that in some cases SARs are required for transcription of neighboring genes. In *Drosophila*, some SARs can insulate transcription units from each other, so that proteins regulating

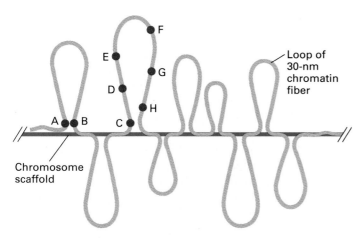

▲ **FIGURE 9-36 Experimental demonstration of chromatin loops in interphase chromosomes.** In situ hybridization of interphase cells was carried out with several different fluorescent-labeled probes specific for sequences separated by known distances in linear, cloned DNA. Lettered circles represent probes. Measurement of the distances between different hybridized probes, which could be distinguished by their color, showed that some sequences (e.g., A, B, and C), separated from each other by millions of base pairs, appear located near each other within nuclei. For some sets of sequences, the measured distances in nuclei between one probe (e.g., C) and sequences successively farther away initially appear to increase (e.g., D, E, and F) and then appear to decrease (e.g., G and H). The measured distances between probes are consistent with loops ranging in size from one to four million base pairs. [Adapted from H. Yokota et al., 1995, *J. Cell Biol.* **130**:1239.]

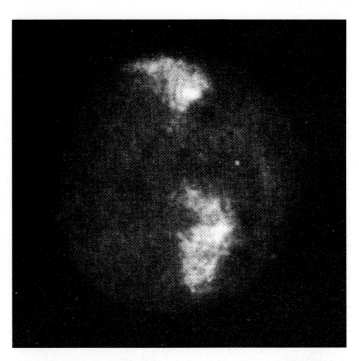

▲ **FIGURE 9-37 Chromosome domains in nuclei of human interphase lymphocytes.** Fixed human lymphocytes were hybridized in situ to biotin-labeled probes specific for sequences along the full length of human chromosome 7, and visualized with fluorescently labeled aviden. In the diploid cell shown here, each of the two chromosome 7s is restricted to a territory or domain within the nucleus, rather than stretching throughout the entire nucleus. [From P. Lichter et al., 1988, *Hum. Genet.* **80**:224.]

transcription of one gene do not influence the transcription of a neighboring gene separated by a SAR.

Individual interphase chromosomes, which are less condensed than metaphase chromosomes, cannot be resolved by standard microscopy or electron microscopy. Nonetheless, the chromatin of interphase cells is associated with extended scaffolds and is further organized into specific domains. This can be demonstrated by the in situ hybridization of interphase nuclei with a large mixture of fluorescent-labeled probes specific for sequences along the length of a particular chromosome. As illustrated in Figure 9-37, the probes are visualized within restricted regions or domains of the nucleus rather than appearing throughout the nucleus. Use of probes specific for different chromosomes shows that there is little overlap between chromosomes in interphase nuclei. However, the precise positions of chromosomes is not reproducible between cells.

Chromatin Contains Small Amounts of Other Proteins in Addition to Histones and Scaffold Proteins

The total mass of the histones associated with DNA in chromatin is about equal to that of the DNA. Interphase chromatin and metaphase chromosomes also contain small

amounts of a complex set of other proteins. For instance, a growing list of DNA-binding **transcription factors** have been identified associated with interphase chromatin. The structure and function of these critical nonhistone proteins, which help regulate transcription, are examined in Chapter 10. Other low-abundance nonhistone proteins associated with chromatin regulate DNA replication (Chapter 12).

A few other nonhistone DNA-binding proteins are present in much larger amounts than the transcription or replication factors. Some of these exhibit high mobility during electrophoretic separation and thus have been designated *HMG* (high-mobility group) *proteins*. When genes encoding the most abundant of these are deleted from yeast cells, normal transcription is disturbed in most genes examined. Some HMG proteins have been found to bind to DNA cooperatively with sequence-specific transcription factors, stabilizing multiprotein complexes that regulate transcription of a neighboring gene (Chapter 10).

Stained Chromosomes Have Characteristic Banding Patterns

Certain dyes selectively stain some regions of metaphase chromosomes more intensely than other regions, producing banding patterns that are specific for individual chromosomes.

Although the molecular basis for the regularity of chromosomal bands remains unknown, they are very useful. When stained chromosomes are viewed microscopically, these bands serve as landmarks along the length of each chromosome and also help to distinguish chromosomes of similar size and shape.

Banding in Metaphase Chromosomes Quinacrine, a fluorescent dye that inserts (intercalates) between base pairs in the DNA helix, produces *Q bands*. However, because Q bands fade with time, other staining techniques are generally preferred in the laboratory. For example, chromosomes can be subjected briefly to mild heat or proteolysis and then stained with Giemsa reagent, a permanent DNA dye, to produce a pattern of *G bands* (see Figure 8-21). Treatment of chromosomes with a hot alkaline solution before staining with Giemsa reagent produces *R bands* in a pattern that is approximately the reverse of the G-band pattern. The distinctiveness of these banding patterns permits cytologists to identify specific parts of a chromosome and to locate the sites of chromosomal breaks and translocations (Figure 9-38a). In addition, cloned

DNA probes that have hybridized to specific sequences in the chromosomes can be located in particular bands.

Banding in Interphase Polytene Chromosomes For nearly a century, cytologists have observed stainable bands in interphase polytene chromosomes in the salivary glands of *Drosophila melanogaster* and other dipteran insects (see Figure 8-22). Although the bands in human chromosomes probably represent very long folded or compacted stretches of DNA containing about 10^7 base pairs, the bands in *Drosophila* chromosomes represent much shorter stretches of only 50,000–100,000 base pairs.

The detailed banding of insect salivary gland chromosomes occurs because of polytenization, which results in thick bundles of parallel interphase chromosomes that all have the same banding pattern across the width of the bundle (see Figure 9-26). Molecular cloning experiments and mRNA mapping studies have suggested that each band contains a limited number of transcription units, perhaps only one unit in some cases. However, transcription units can also be found in interband stretches of DNA; thus the relationship between banding and function is still unclear. Nonetheless, the reproducible pattern of bands seen in *Drosophila* salivary gland chromosomes provides an extremely powerful method for locating specific DNA sequences along the lengths of the chromosomes in this species.

Chromosome Painting Distinguishes Each Homologous Pair by Color

A recently developed method for visualizing each of the human chromosomes in distinct, bright colors, called **chromosome painting**, greatly simplifies the distinction between chromosomes of similar size and shape. This technique makes use of probes specific for sites scattered along the length of each chromosome. The probes are labeled with one of two dyes that fluoresce at different wavelengths. Probes specific for each chromosome are labeled with a predetermined fraction of each of the two dyes. After the probes are hybridized to chromosomes and the excess removed, the sample is placed in a fluorescent microscope in which a detector determines the fraction of each dye present at each fluorescing position in the microscopic field. This information is conveyed to a computer, and a special program assigns a false color image

(a)

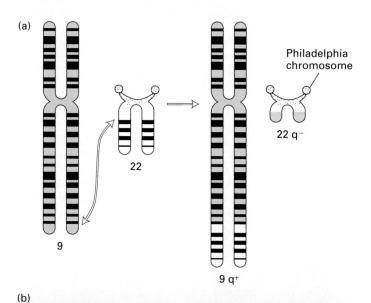

(b)

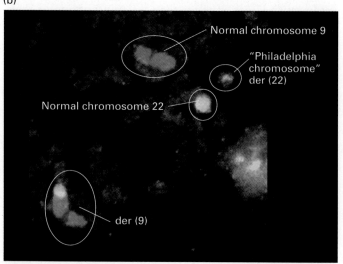

◀ **FIGURE 9-38 Analysis of chromosomal translocations by banding patterns and chromosome painting.** Characteristic chromosomal translocations are associated with certain genetic disorders and specific types of cancers. For example, in nearly all patients with chronic myelogenous leukemia, the leukemic cells contain the Philadelphia chromosome [der(22)] and an abnormal chromosome 9 [der(9)]. These result from a translocation between normal chromosomes 9 and 22. This translocation can be detected by classical banding analysis (a) and by multicolor FISH (b). [Part (a) from J. Kuby, 1997, *Immunology*, 3d ed., W. H. Freeman and Company, p. 578; part (b) courtesy of J. Rowley and R. Espinosa.]

to each type of chromosome (see chapter opening figure). A combination of chromosome painting and fluorescent in situ hybridization, called multicolor FISH, can detect chromosomal translocations (Figure 9-38b). The much more detailed analysis possible with this technique permits detection of chromosomal translocations that banding analysis does not reveal.

Heterochromatin Consists of Chromosome Regions That Do Not Uncoil

As cells exit from mitosis and the condensed chromosomes uncoil, certain sections of the chromosomes remain dark staining. The dark-staining areas, termed **heterochromatin,** are regions of condensed chromatin. The light-staining, less-condensed portions of chromatin are called **euchromatin.** The distinctions between heterochromatin and euchromatin are most clearly visualized in polytene chromosomes such as those of *Drosophila* salivary gland cells. Heterochromatin appears most frequently—but not exclusively—at the centromere and telomeres of chromosomes, and most of the DNA in heterochromatin is simple-sequence DNA.

In mammalian cells, heterochromatin appears as darkly staining regions of the nucleus, often associated with the nuclear envelope (Figure 9-39). Pulse labeling with ^{3}H-uridine

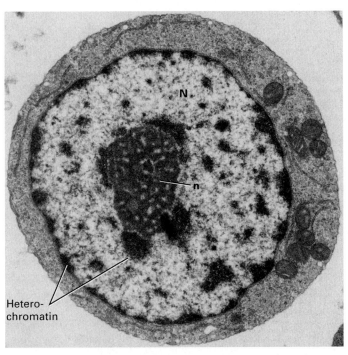

▲ FIGURE 9-39 Electron micrograph of a thin section of a bone-marrow stem cell. The nucleus (N) and nucleolus (n) are indicated. Darkly staining areas in the nucleus outside of the nucleolus are regions of heterochromatin. [From P. C. Cross and K.L. Mercer, 1993, *Cell and Tissue Ultrastructure,* W. H. Freeman and Company, p. 165.]

and autoradiography have shown that most transcription occurs in regions of euchromatin and the nucleolus. Because of this and because heterochromatic regions apparently remain condensed throughout the life cycle of the cell, they have been regarded as sites of inactive genes. However, some genes have been found that are located in regions of heterochromatin. Also, not all inactive genes and nontranscribed regions of DNA are visible as heterochromatin.

Three Functional Elements Are Required for Replication and Stable Inheritance of Chromosomes

So far we have seen that the eukaryotic chromosome is a linear structure composed of an immensely long, single DNA molecule that is wound around histone octamers about every 200 bp, forming strings of closely packed nucleosomes. Nucleosomes fold to form a 30-nm chromatin fiber, which is attached to a flexible protein scaffold at intervals of millions of base pairs, resulting in long loops of chromatin extending from the scaffold (see Figure 9-35). In addition to this general chromosomal structure, a complex set of thousands of low-abundance regulatory proteins are associated with specific sequences in chromosomal DNA.

Although chromosomes differ in length and number among species, cytogenetic studies have shown that they all behave similarly at the time of cell division. Recombinant DNA research with yeast cells has identified all of the chromosomal elements that are necessary for equal segregation of sister chromatids to occur during mitosis. The culmination of this work has been construction of **yeast artificial chromosomes (YACs).** In order to duplicate and segregate correctly, chromosomes must contain three functional elements: (1) origins for initiation of DNA replication; (2) the centromere; and (3) the two ends, or telomeres.

Autonomously Replicating Sequences (Yeast Origins) In eukaryotic DNA, replication origins were first identified by their function in yeast transformation studies. If yeast cells lack a particular gene (e.g., one of the genes that encode an enzyme for synthesis of the amino acid leucine), they can be transformed with cloned plasmids containing the missing DNA (in this case, a *LEU* gene). *LEU*⁺ transformants, which can grow on media lacking leucine, are obtained much more frequently if the plasmid contains one of many sequences from the yeast genome of approximately 100 base pairs called *autonomously replicating sequences* (Figure 9-40a). An autonomously replicating sequence (ARS) acts as an origin of replication, allowing the transformed circular plasmid to replicate in the yeast cell nucleus. The function of ARSs as origins of DNA replication is discussed in Chapter 12.

Centromeres In a culture of leucine-requiring yeast cells transformed with a simple circular *LEU/ARS*-containing plasmid, only about 5–20 percent of progeny cells contain the plasmid because mitotic segregation of the plasmids is

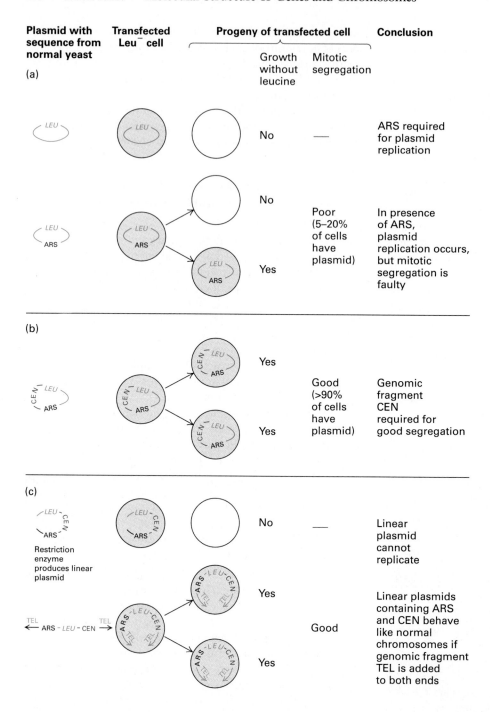

Plasmid with sequence from normal yeast	Transfected Leu⁻ cell	Progeny of transfected cell		Conclusion
		Growth without leucine	Mitotic segregation	
(a)				
LEU	LEU	No	—	ARS required for plasmid replication
LEU ARS	LEU ARS	No / Yes	Poor (5–20% of cells have plasmid)	In presence of ARS, plasmid replication occurs, but mitotic segregation is faulty
(b)				
CEN LEU ARS	CEN LEU ARS	Yes / Yes	Good (>90% of cells have plasmid)	Genomic fragment CEN required for good segregation
(c)				
LEU · CEN ARS Restriction enzyme produces linear plasmid	LEU · CEN ARS	No	—	Linear plasmid cannot replicate
TEL ← ARS – LEU – CEN → TEL	ARS · LEU · CEN	Yes / Yes	Good	Linear plasmids containing ARS and CEN behave like normal chromosomes if genomic fragment TEL is added to both ends

◄ **FIGURE 9-40 Experimental demonstration of functional chromosomal elements in experiments with yeast cells that lack an enzyme necessary for leucine synthesis (leu⁻ cells).** In these experiments, plasmids containing the *LEU* gene from normal yeast cells are constructed and introduced into *leu⁻* cells by transfection. If the plasmid is maintained in the *leu⁻* cells, they are transformed to *LEU⁺* by the *LEU* gene on the plasmid and can form colonies on medium lacking leucine. (a) Sequences that allow autonomous replication (ARS) of a plasmid were identified because their insertion into a plasmid vector containing a cloned *LEU* gene resulted in a high frequency of transformation to *LEU⁺*. However, even plasmids with ARS exhibit poor segregation during mitosis, and therefore do not appear in each of the daughter cells. (b) When randomly broken pieces of genomic yeast DNA are inserted into plasmids containing ARS and *LEU*, some of the transfected cells produce large colonies, indicating that a high rate of mitotic segregation among their plasmids is facilitating the continuous growth of daughter cells. The DNA recovered from plasmids in these large colonies contains yeast centromere (CEN) sequences. (c) When *leu⁻* yeast cells are transfected with linearized plasmids containing *LEU*, ARS, and CEN, no colonies grow. Addition of telomere (TEL) sequences to the ends of the linear DNA gives the linearized plasmids the ability to replicate as new chromosomes that behave very much like a normal chromosome in both mitosis and meiosis. [See A. W. Murray and J. W. Szostak, 1983, *Nature* **305**:89; L. Clarke and J. Carbon, 1985, *Ann. Rev. Genetics* **19**:29.]

faulty; as a result, most of the time plasmid DNA does not enter the bud. However, if restriction fragments of yeast genomic DNA are cloned into such circular plasmids, a small fraction of fragments are found to confer equal segregation of the plasmids to both mother and daughter cells following mitosis (Figure 9-40b). Because the fragments with this effect were found to be derived from the centromeres of yeast chromosomes, they are called *CEN sequences*. Such cloning experiments have led to isolation of sequences that improve mitotic segregation to the extent that >90 percent

of the progeny of transformed Leu⁻ yeast cells contain the *LEU* plasmid. These cells and almost all of their descendants therefore grow well in a medium lacking leucine.

Once the yeast centromere regions that confer mitotic segregation were cloned, their sequences could be determined and compared. Comparison of different yeast centromeres has revealed three regions (I, II, and III) that are necessary for a centromere to function (Figure 9-41). Short, fairly well conserved nucleotide sequences are present in regions I and III. Although region II seems to have a fairly

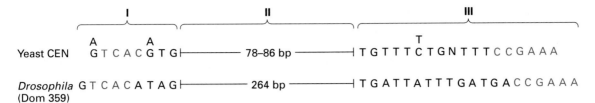

▲ FIGURE 9-41 *(Top)* **Consensus yeast CEN sequence based on analysis of ten yeast centromeres.** All three distinctive regions are required for a centromere to function. Region II, although variable in sequence, is fairly constant in length and is rich in A and T residues. *(Bottom)* One Drosophila simple-sequence DNA that is located near the centromere has a repeat unit with some homology to the yeast consensus CEN, including two identical 4-bp and 6-bp stretches (red). [See L. Clarke and J. Carbon, 1985, *Ann. Rev. Genet.* **19**:29.]

constant length (78–86 base pairs), it contains no definite consensus sequence; however, it is rich in A and T residues. One *Drosophila* simple-sequence DNA, which comes from a centromeric region, has a repeat unit that bears some similarity to yeast CEN regions I and III, suggesting that similar mechanisms may control segregation in yeast and higher eukaryotes. The role of the centromere and proteins that bind to it in the segregation of sister chromatids during mitosis are described in Chapter 19.

Telomeres If circular plasmids containing an ARS and CEN sequence are cut once with a restriction enzyme, they become linear. Such linear plasmids do not replicate in yeast cells unless they contain special telomeric (TEL) sequences ligated to their ends (Figure 9-40c). The first successful experiments involving transfection of yeast cells with linear plasmids were achieved by using the ends of a DNA molecule that was known to replicate as a linear molecule in the ciliated protozoan *Tetrahymena*. During part of the life cycle of *Tetrahymena*, much of the nuclear DNA is repeatedly copied in short pieces to form a so-called macronucleus. One of these repeated fragments was identified as a dimer of ribosomal DNA, the ends of which contained a repeated sequence $(G_4T_2)n$. When a section of this repeated TEL sequence was ligated to the ends of linear yeast plasmids containing ARS and CEN, replication and good segregation of the plasmids occurred.

Considerable research has revealed that telomeric DNA has a characteristic type of sequence and is added to the ends of DNA molecules by a special enzyme termed *telomere terminal transferase,* or *telomerase.* The sequences of telomeres in a dozen or so organisms, including humans, have been determined; most are repetitive oligomers with a high G content in the strand that runs $5' \rightarrow 3'$ toward the telomere. These simple sequences are repeated at the very termini of chromosomes for a total of a few hundred base pairs in yeasts and protozoans and a few thousand base pairs in vertebrates. The 3' end of the G-rich strand extends 12–16 nucleotides beyond the 5' end of the complementary C-rich strand. This region is bound by specific proteins that protect the ends of linear chromosomes from attack by exonucleases.

The enzymes that add telomeric sequences are complexes of both protein and RNA. Because the sequence of the associated RNA serves as the template for addition of deoxyribonucleotides to the ends of telomeres, the source of the enzyme and not the source of the telomeric DNA primer determines the sequence added. This was proven by transforming *Tetrahymena* with a mutated form of the gene encoding the telomerase-associated RNA. The resulting telomerase added a DNA sequence complementary to the mutated RNA sequence to the ends of telomeric primers. Thus telomerase is a specialized form of a reverse transcriptase that carries its own internal RNA template to direct DNA synthesis.

The action of telomerase and its role in preventing chromosome shortening during DNA replication is discussed in more detail in Chapter 12. Knockout mice that cannot produce the RNA associated with telomerase exhibit no telomerase activity and their telomeres shorten successively with each cell generation. Such mice can breed and reproduce normally for three generations before the absence of telomere DNA causes adverse effects including substantial erosion of telomeres, fusion of chromosome termini, and chromosomal loss. By the fourth generation, the reproductive potential of these knockout mice declines and they cannot produce offspring after the sixth generation.

Yeast Artificial Chromosomes Can Be Used to Clone Megabase DNA Fragments

The research on circular and linear plasmids in yeast identified all the basic components of a yeast artificial chromosome (YAC). To construct YACs, TEL sequences from yeast cells or from the protozoan *Tetrahymena* are combined with yeast CEN and ARS sequences; to these are added DNA with selectable yeast genes and enough DNA from any source to make a total of more than 50 kb. (Smaller DNA segments do not work as well.) Such artificial chromosomes replicate in yeast cells and segregate almost perfectly, with only 1 daughter cell in 1000 to 10,000 failing to receive an artificial chromosome. During meiosis, the two sister chromatids of the artificial chromosome separate correctly to produce haploid spores.

Studies such as those depicted in Figure 9-40 strongly support the conclusion that yeast chromosomes, and probably all eukaryotic chromosomes, are linear, double-stranded DNA molecules with special regions—including the centromere (CEN), telomeres (TEL), and autonomously replicating sequences (ARSs)—that ensure replication and proper segregation. A technical point of considerable importance is that YACs can be used to clone very long chromosomal pieces from other species. For instance, YACs have been used extensively for cloning fragments of human DNA up to 1000 kb (1 megabase) in length, permitting the isolation of overlapping YAC clones that collectively encompass nearly the entire length of individual human chromosomes.

SUMMARY **Morphology and Functional Elements of Eukaryotic Chromosomes**

• Eukaryotic chromosomes can be visualized during mitosis when they condense into highly folded metaphase chromosomes. The set of metaphase chromosomes from a cell is its karyotype. Closely related species can have dramatically different karyotypes, indicating that similar genetic information can be organized on chromosomes in different ways.

• Each chromosome is composed of a single DNA molecule packaged into nucleosomes and folded into a 30-nm fiber, which is attached to a protein scaffold at specific sites (see Figure 9-35). Additional folding of the scaffold further compacts the structure into the highly condensed form of metaphase chromosomes.

• When metaphase chromosomes decondense during interphase, certain regions, termed heterochromatin, remain much more condensed than the bulk of chromatin, called euchromatin.

• Three types of DNA sequences are required for a long linear DNA molecule to function as a chromosome in yeast cells: a replication origin, called ARS in yeast; a centromere (CEN) sequence; and two telomere (TEL) sequences at the ends of the DNA (see Figure 9-40).

• All eukaryotic chromosomes are thought to include one or more replication origins and TEL sequences, which are required for DNA replication to occur, and a CEN sequence, which is needed for efficient segregation of chromosomes to both daughter cells.

• Discovery of the functional elements in yeast chromosomes has allowed construction of yeast artificial chromosome (YAC) vectors that make it possible to clone DNAs up to one million base pairs in length.

9.7 Organelle DNAs

Although the vast majority of DNA in most eukaryotes is found in the nucleus, some DNA is present within the mitochondria of animals, plants, and fungi and within the chloroplasts of plants. These organelles are the main cellular sites for ATP formation, during oxidative phosphorylation in mitochondria and photosynthesis in chloroplasts (Chapter 16). Many lines of evidence indicate that mitochondria and chloroplasts evolved from bacteria that were endocytosed into ancestral cells containing a eukaryotic nucleus, forming **endosymbionts.** Over evolutionary time, most of the bacterial genes encoding components of the present-day organelles were transferred to the nucleus. However, mitochondria and chloroplasts in today's eukaryotes retain circular DNAs encoding proteins essential for organellar function as well as the ribosomal and transfer RNAs required for their translation. Thus eukaryotic cells have multiple genetic systems: a predominant nuclear system and secondary systems with their own DNA in the mitochondria and chloroplasts.

Mitochondria Contain Multiple mtDNA Molecules

Individual mitochondria are large enough to be seen under the light microscope and even the mitochondrial DNA (mtDNA) can be detected by fluorescence microscopy. The mtDNA is located in the interior of the mitochondrion, the region known as the matrix. As judged by the number of yellow fluorescent "dots" of mtDNA, a *Euglena gracilis* cell contains at least thirty mtDNA molecules (Figure 9-42).

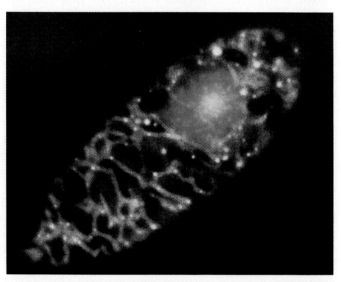

|————| 10 μm

▲ **FIGURE 9-42 Visualization of mitochondrial DNA in a growing *Euglena gracilis* cell.** Cells were treated with a mixture of two dyes: ethidium bromide, which binds to DNA and emits a vermilion fluorescence, and DiOC6, which is incorporated specifically into mitochondria and emits a green fluorescence. Thus the nucleus emits a vermilion fluorescence, and areas rich in mitochondrial DNA fluoresce yellow—a combination of vermilion DNA and green mitochondrial fluorescence. [From Y. Huyashi and K. Veda, 1989, *J. Cell Sci.* **93:**565.]

Since the dyes used to visualize nuclear and mitochondrial DNA do not affect cell growth or division, replication of mtDNA and division of the mitochondrial network can be followed in living cells using time-lapse microscopy. Such studies show that mtDNA replicates throughout interphase. At mitosis each daughter cell receives approximately the same number of mitochondria, but since there is no mechanism for apportioning exactly equal numbers of mitochondria to the daughter cells, some cells contain more mtDNA than others. All the mitochondria in eukaryotic cells contain multiple mtDNA molecules. Thus the total amount of mtDNA in a cell depends on the number of mitochondria, the size of the mtDNA, and the number of mtDNA molecules per mitochondrion.

Genes in mtDNA Exhibit Cytoplasmic Inheritance and Encode rRNAs, tRNAs, and Some Mitochondrial Proteins

Studies of mutants in yeasts and other single-celled organisms first indicated that mitochondria exhibit **cytoplasmic inheritance** and thus must contain their own genetic system (Figure 9-43). For instance, *petite* yeast mutants exhibit structurally abnormal mitochondria and are incapable of oxidative phosphorylation. As a result, petite cells grow more slowly than wild-type yeasts and form smaller colonies (hence the name "petite"). Genetic crosses between different (haploid) yeast strains showed that the *petite* mutation does not segregate with any known nuclear gene or chromosome. In later studies, most petite mutants were found to contain deletions of mtDNA.

Mitochondrial inheritance in yeasts is biparental: during the fusion of haploid cells, both parents contribute equally to the cytoplasm of the diploid. In mammals and most other animals, however, the sperm contributes little (if any) cytoplasm to the zygote, and virtually all of the mitochondria in the embryo are derived from those in the egg, not the sperm. Studies in mice have shown that 99.99 percent of mtDNA is maternally inherited, but a small part (0.01 percent) is inherited from the male parent. In higher plants,

▶ **FIGURE 9-43 Cytoplasmic inheritance of the *petite* mutation in yeast.** Petite-strain mitochondria are defective in oxidative phosphorylation due to a deletion in mtDNA. (a) Haploid cells fuse to produce a diploid cell that undergoes meiosis, during which random segregation of parental chromosomes and mitochondria containing mtDNA occurs. Since yeast normally contain ≈50 mtDNA molecules per cell, all products of meiosis usually contain both normal and petite mtDNAs and are capable of respiration. (b) As these cells grow and divide mitotically, the cytoplasm (including the mitochondria) is randomly distributed to the daughter cells. Occasionally, a cell is generated that contains only defective petite mtDNA and yields a petite colony. Thus formation of such petite cells is independent of any nuclear genetic marker.

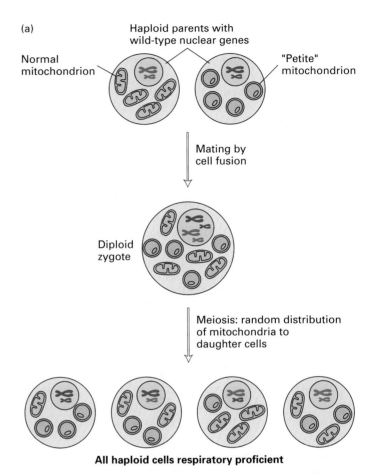

(a) Haploid parents with wild-type nuclear genes

Normal mitochondrion

"Petite" mitochondrion

Mating by cell fusion

Diploid zygote

Meiosis: random distribution of mitochondria to daughter cells

All haploid cells respiratory proficient

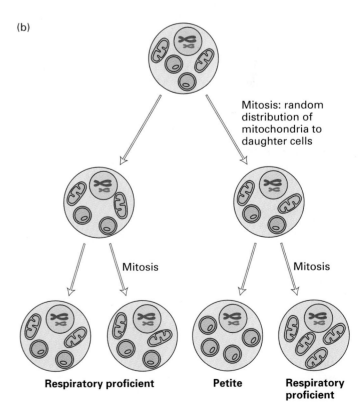

(b)

Mitosis: random distribution of mitochondria to daughter cells

Mitosis

Mitosis

Respiratory proficient

Petite

Respiratory proficient

mtDNA is inherited exclusively in a uniparental fashion through the female parent (egg), not the male (pollen).

The entire mitochondrial genome from a number of different organisms has now been cloned and sequenced, and mtDNAs from all these sources have been found to encode rRNAs, tRNAs, and essential mitochondrial proteins. All proteins encoded by mtDNA are synthesized on mitochondrial ribosomes. All mitochondrially synthesized polypeptides identified thus far (with one possible exception) are not complete enzymes but subunits of multimeric complexes used in electron transport or ATP synthesis. Most proteins localized in mitochondria, such as the mitochondrial RNA and DNA polymerases, are synthesized on cytoplasmic ribosomes and are imported into the organelle by processes discussed in Chapter 17.

The Size and Coding Capacity of mtDNA Vary Considerably in Different Organisms

Surprisingly, the size of the mtDNA, the number and nature of the proteins it encodes, and even the mitochondrial genetic code itself, vary greatly among different organisms. Human mtDNA, a circular molecule that has been completely sequenced, is among the smallest known mtDNAs, containing 16,569 base pairs (Figure 9-44). It encodes the two rRNAs

found in mitochondrial ribosomes and the 22 tRNAs used to translate mitochondrial mRNAs. Human mtDNA has 13 sequences that begin with an ATG (methionine) codon, end with a stop codon, and are long enough to encode a polypeptide of more than 50 amino acids; all of the possible proteins encoded by these open reading frames have been identified. Mammalian mtDNA, in contrast to nuclear DNA, lacks introns and contains no long noncoding sequences.

Invertebrate mtDNA is about the same size as human mtDNA, but yeast mtDNA is almost five times as large ($\approx$78,000 bp). The mtDNAs from yeast and other lower eukaryotes encode many of the same gene products as mammalian mtDNA, as well as others whose genes are found in the nuclei of mammalian cells.

In contrast to other eukaryotes, which contain a single type of mtDNA, plants contain several types of mtDNA that appear to recombine with each other. Plant mtDNAs are much larger and more variable in size than the mtDNAs of other organisms. Even in a single family of plants, mtDNAs can vary as much as eightfold in size (watermelon = 330,000 bp; muskmelon = 2,500,000 bp). Unlike animal, yeast, and fungal mtDNAs, plant mtDNAs contain genes encoding a 5S mitochondrial rRNA, which is present only in the mitochondrial ribosomes of plants, and the α subunit of the F_1 ATPase. The mitochondrial rRNAs of plants are also considerably larger than those of other eukaryotes. The recent

▶ **FIGURE 9-44 Human mitochondrial DNA (mtDNA), which has been sequenced in its entirety.** Proteins and RNAs encoded by each of the two strands are shown separately. Transcription of the outer (H) strand occurs in the clockwise direction and of the inner (L) strand in the counterclockwise direction. The abbreviations for amino acids denote the corresponding tRNA genes. ND1, ND2, etc., denote genes encoding subunits of the NADH-CoQ reductase complex. The 207-bp gene encoding F_0 ATPase subunit 8 overlaps, out of frame, with the N-terminal portion of the segment encoding F_0 ATPase subunit 6. No mammalian mtDNA genes contain introns, although intervening DNA lies between some genes. [See D. A. Clayton, 1991, *Ann. Rev. Cell Biol.* **7**:453.]

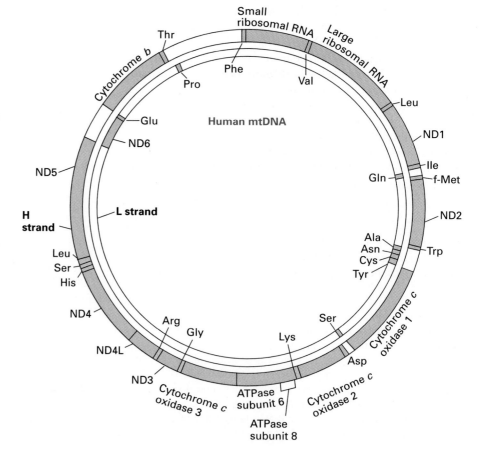

sequencing of one of the smallest plant mtDNAs has revealed that long, noncoding regions and duplicated sequences are largely responsible for the greater length of plant mtDNAs.

Differences in the size and coding capacity of mtDNA from various organisms most likely reflect the movement of DNA between mitochondria and the nucleus during evolution. Direct evidence for this movement comes from the observation that several proteins encoded by mtDNA in some species are encoded by nuclear DNA in others. It thus appears that entire genes moved from the mitochondrion to the nucleus, or vice versa, during evolution.

The most striking example of this phenomenon involves the gene *cox II,* which encodes subunit 2 of cytochrome *c* oxidase. This gene is found in mtDNA in all organisms studied except for one species of legume, the mung bean: in this organism only, the *cox II* gene is nuclear. Many RNA transcripts of plant mitochondrial genes are edited, mainly by the enzyme-catalyzed conversion of selected C residues to U, and occasionally U to C. (RNA editing is discussed in Chapter 11.) The nuclear *cox II* gene of mung bean corresponds more closely to the edited cox II RNA transcripts than to the mitochondrial *cox II* genes found in other legumes. These facts are strong evidence that the *cox II* gene moved from mitochondrion to the nucleus during mung bean evolution by a process that involved an RNA intermediate. Presumably this movement involved a reverse-transcription mechanism similar to that by which processed pseudogenes are generated in the nuclear genome from nuclear-encoded mRNAs.

Products of Mitochondrial Genes Are Not Exported

As far as is known, all RNA transcripts of mtDNA and their translation products remain in the mitochondrion, and all mtDNA-encoded proteins are synthesized on mitochondrial ribosomes. Mitochondria encode the rRNAs that form mitochondrial ribosomes, although all but one or two of the ribosomal proteins (depending on the species) are imported from the cytosol. In most eukaryotes, all of the tRNAs used for protein synthesis in mitochondria are encoded by mtDNAs. However, in wheat, in the parasitic protozoan *Trypanosoma brucei* (the cause of African sleeping sickness), and in ciliated protozoa, all mitochondrial tRNAs are encoded by the nuclear DNA and imported into the mitochondrion.

Reflecting the bacterial ancestry of mitochondria, mitochondrial ribosomes resemble bacterial ribosomes and differ from cytoplasmic ribosomes in their RNA and protein compositions, their size, and their sensitivity to certain antibiotics (see Figure 4-32). For instance, chloramphenicol blocks protein synthesis by bacterial and most mitochondrial ribosomes, but not by cytoplasmic ribosomes. Conversely, cycloheximide inhibits protein synthesis by eukaryotic cytoplasmic ribosomes but does not affect protein synthesis by mitochondrial ribosomes or bacterial ribosomes. In cultured mammalian cells the only proteins synthesized in the presence of cycloheximide are encoded by mtDNA and produced by mitochondrial ribosomes.

Mitochondrial Genetic Codes Differ from the Standard Nuclear Code

The genetic code used in animal and fungal mitochondria is different from the standard code used in all prokaryotic and eukaryotic nuclear genes; remarkably, the code even differs in mitochondria from different species (Table 9-4). Why and how this phenomenon happened during evolution is mysterious. UGA, for example, is normally a stop codon, but is read as tryptophan by human and fungal mitochondrial translation systems; however, in plant mitochondria, UGA is still a stop codon. AGA and AGG, the standard nuclear codons for arginine also code for arginine in fungal and plant mtDNA, but they are stop codons in mammalian mtDNA and serine codons in *Drosophila* mtDNA.

As shown in Table 9-4, plant mitochondria appear to utilize the standard genetic code. However, comparisons of the amino acid sequences of plant mitochondrial proteins with the nucleotide sequences of plant mtDNAs suggested

	Standard Code: Nuclear-Encoded Proteins	Mitochondria				
Codon		Mammals	*Drosophila*	*Neurospora*	Yeasts	Plants
UGA	Stop	Trp	Trp	Trp	Trp	Stop
AGA, AGG	Arg	Stop	Ser	Arg	Arg	Arg
AUA	Ile	Met	Met	Ile	Met	Ile
AUU	Ile	Met	Met	Met	Met	Ile
CUU, CUC, CUA, CUG	Leu	Leu	Leu	Leu	Thr	Leu

TABLE 9-4 Alterations in the Standard Genetic Code in Mitochondria

SOURCE: S. Anderson et al., 1981, *Nature* 290:457; P. Borst, in *International Cell Biology 1980–1981,* H. G. Schweiger, ed., Springer-Verlag, p. 239; C. Breitenberger and U. L. Raj Bhandary, 1985, *Trends Biochem. Sci.* 10:478–483; V. K. Eckenrode and C. S. Levings, 1986, *In Vitro Cell Dev. Biol.* 22:169–176; J. M. Gualber et al., 1989, *Nature* 341:660–662; and P. S. Covello and M. W. Gray, 1989, *Nature* 341:662–666.

that CGG could code for *either* arginine (the "standard" amino acid) or tryptophan. This apparent nonspecificity of the plant mitochondrial code is explained by editing of mitochondrial RNA transcripts, which can convert cytosine residues to uracil residues. If a CGG sequence is edited to UGG, the codon specifies tryptophan, the standard amino acid for UGG, whereas unedited CGG codons encode the standard arginine. Thus the translation system in plant mitochondria does utilize the standard genetic code.

Mutations in Mitochondrial DNA Cause Several Genetic Diseases in Man

 The severity of disease caused by a mutation in mtDNA depends on the nature of the mutation and on the proportion of mutant and wild-type DNAs present in a particular cell type. Generally, when mutations in mtDNA are found, cells contain mixtures of wild-type and mutant mtDNAs—a condition known as *heteroplasmy*. Each time a mammalian somatic or germ-line cell divides, the mutant and wild-type mtDNAs will segregate randomly into the daughter cells, as occurs in yeast cells (see Figure 9-43). Thus, the mtDNA genotype fluctuates from one generation and from one cell division to the next, and can drift toward predominantly wild-type or predominantly mutant mtDNAs. Since all enzymes for the replication and growth of mitochondria, such as DNA and RNA polymerases, are imported from the cytosol, a mutant mtDNA should not be at a "replication disadvantage"; mutants that involve large deletions of mtDNA might even be at a selective advantage in replication.

All cells have mitochondria, yet mutations in mtDNA only affect some tissues. Those most usually affected are tissues that have a high requirement for ATP produced by oxidative phosphorylation, and tissues that require most or all of the mtDNA in the cell to synthesize functional mitochondrial proteins. Leber's hereditary optic neuropathy (degeneration of the optic nerve, accompanied by increasing blindness), for instance, is caused by a missense mutation in the mtDNA gene encoding subunit 4 of the NADH-CoQ reductase. Any of several large deletions in mtDNA cause another set of diseases including chronic progressive external ophthalmoplegia and Kearns-Sayre syndrome, which are characterized by eye defects and, in Kearns-Sayre syndrome, also by abnormal heartbeat and central nervous system degeneration. A third condition, causing "ragged" muscle fibers (with improperly assembled mitochondria) and associated uncontrolled jerky movements, is due to a single mutation in the TYCG loop of the mitochondrial lysine tRNA. As a result of this mutation, the translation of several mitochondrial proteins apparently is blocked.

Chloroplasts Contain Large Circular DNAs Encoding More Than a Hundred Proteins

As we discuss in Chapter 16, the structure of chloroplasts is similar in many respects to that of mitochondria. Like mitochondria, chloroplasts contain multiple copies of the organellar DNA and ribosomes, which synthesize some chloroplast-encoded proteins using the "standard" genetic code. Other chloroplast proteins are fabricated on cytosolic ribosomes and are incorporated into the organelle after translation.

Chloroplast DNAs are circular molecules of 120,000–160,000 bp, depending on the species. The complete sequences of several chloroplast DNAs have been determined, including those from liverwort (121,024 bp) and tobacco (155,844 bp). The liverwort chloroplast genome has two inverted repeats, each consisting of 10,058 bp, that contain the rRNA genes and a few other duplicated genes. Despite the difference in size, the overall organization and gene composition of the liverwort and tobacco DNAs are very similar; the size differential is due primarily to the length of the inverted repeat in which some genes are duplicated.

Of the ≈120 genes in chloroplast DNA, about 60 are involved in RNA transcription and translation, including genes for rRNAs, tRNAs, RNA polymerase subunits, and ribosomal proteins. About 20 genes encode subunits of the chloroplast photosynthetic electron transport complexes and the F_0F_1 ATPase complex. Also encoded in the chloroplast genome is the larger of the two subunits of ribulose 1,5-bisphosphate carboxylase, which is involved in the fixation of carbon dioxide during photosynthesis.

Reflecting the endosymbiotic origin of chloroplasts, some regions of chloroplast DNA are strikingly similar to those of the DNA of present-day bacteria. For instance, chloroplast DNA encodes four subunits of RNA polymerase that are highly homologous to the subunits of *E. coli* RNA polymerase. One segment of chloroplast DNA encodes eight proteins that are homologous to eight *E. coli* ribosomal proteins; the order of these genes is the same in the two DNAs.

Liverwort chloroplast DNA has some genes that are not detected in the larger tobacco chloroplast DNA, and vice versa. Since the two types of chloroplasts contain virtually the same set of proteins, these data suggest that some genes are present in the chloroplast DNA of one species and in the nuclear DNA of the other, indicating that some exchange of genes between chloroplast and nucleus has occurred during evolution.

SUMMARY   Organelle DNAs

• Mitochondria and chloroplasts are believed to have evolved from bacteria that formed a symbiotic relationship with ancestral cells containing a eukaryotic nucleus. Most of the genes originally within these organelles have been transferred to the nuclear genome over evolutionary time, leaving different genes in the organelle DNAs of different organisms.

• Mammalian mtDNAs are only ≈16 kb in length; they contain no introns and very little noncoding DNA. Yeast and plant mtDNAs are much longer. All mtDNAs encode rRNAs, tRNAs, and some of the proteins involved in mitochondrial electron transport and ATP synthesis.

- Most mtDNA is inherited from egg cells rather than sperm, and mutations in mtDNA result in a maternal cytoplasmic pattern of inheritance.

- Mitochondrial ribosomes resemble bacterial ribosomes in their structure, sensitivity to chloramphenicol, and resistance to cycloheximide.

- The genetic code of animal and fungal mtDNAs differs from that of bacteria and the nuclear genome in that several codons encode alternative amino acids or stop signals. The mitochondrial code differs between different animals and fungi. Plant mitochondria appear to use the standard nuclear and bacterial genetic code.

- Mutations in mtDNA can cause human neuromuscular disorders, probably because of the high demand for ATP in these tissues. Patients generally have a mixture of wild-type and mutant mtDNA in their cells (heteroplasmy). The severity of the phenotype is greater, the higher the fraction of mutant mtDNA.

- Chloroplast DNA is circular and contains ≈120–160 kb, depending on the plant species. It encodes ≈120 proteins and uses the standard genetic code.

PERSPECTIVES for the Future

Functionally related genes in bacteria often are organized into operons, which are transcribed into polycistronic mRNAs. In contrast, the genomes of higher eukaryotes rarely contain clusters of functionally related genes, and most genes are transcribed and processed into monocistronic mRNAs. It is interesting that lower eukaryotes such as *Trypanosomes* have an operon-like organization of genes into polycistronic transcription units, which generate monocistronic mRNAs by trans-splicing (discussed in Chapter 11) and RNA processing. Even the nematode *C. elegans* has a significant number of bicistronic transcription units processed into two mRNAs by trans-splicing. Future studies may reveal other variations in gene organization and perhaps lead to additional insights about early eukaryotic evolution.

The function of most simple-sequence DNA remains an intriguing question. It seems likely that simple-sequence DNA in centromeres contributes in some way to their function, but so far no direct evidence for this has been obtained. Indeed, elucidating the structure of centromeres in higher eukaryotes is a significant challenge for future researchers. The chromosomes of higher eukaryotes are immense compared with yeast chromosomes, and their centromeres span hundreds of kilobases, whereas *S. cerevisiae* centromeres contain only about 100 base pairs. Much remains to be learned about the structure of the complex and massive kinetochore, which assembles at the centromere during mitosis and associates with spindle microtubules (Chapter 19).

Many questions about the structure of metaphase and interphase chromosomes are still unresolved. For instance, a clear picture of how chromatin is folded in the compact metaphase chromosomes is not yet available and few of the components of the chromosome scaffold have been identified. Few details are known about the attachment of chromatin loops to the chromosome scaffold, or about the mechanism of chromosome condensation during mitosis and decondensation following mitosis (Chapter 13). Many other questions remain: How does the structure of heterochromatin differ from euchromatin? What is the molecular basis of the banding pattern in polytene chromosomes and how does this relate to the structure of normal diploid interphase chromosomes? How does the structure of interphase chromosomes influence gene function? What influence does the higher-order structure of chromatin have on DNA transcription, replication, and recombination, and on the repair of damaged DNA?

Although a good deal is now known about the molecular anatomy of genes and chromosomes, clearly many significant challenges face future researchers. Important details about how the information stored in genes is expressed in proteins and RNAs also remain to be answered. Our current understanding of the fundamental processes of DNA transcription and RNA processing is considered in the next two chapters.

PERSPECTIVES in the Literature

Transposons and retrotransposons are thought to have little influence on the function of individual cells and organisms. In contrast, they are thought to have had an immense impact on the evolution of organisms. The related sequences dispersed throughout genomes by transposons and retrotransposons have probably contributed to gene duplications by serving as recombination sites for unequal crossing over (see Figure 9-5). A duplicated gene can then evolve a new function. In addition, retrotranspons may have lead to the evolution of new genes by causing the insertion of exons into pre-existing genes, and the evolution of new transcription control regions by causing the insertion of transcription factor binding sites into pre-exisiting control regions. As you read the following articles, consider this question: what mechanisms can account for the mobilization of exons and other sequences by retrotransposons?

Boeke, J. D., and O. K. Pickeral. 1999. Retroshuffling the genomic deck. *Nature* **398**:108–111.

Moran, J. V., et al. 1996. High frequency retrotransposition in cultured mammalian cells. *Cell* **87**:917–927.

Moran, J. V., R. J. DeBerardinis, and H. H. Kazazian, Jr. 1999. Exon shuffling by L1 retrotransposition. *Science* **283**:1530–1534.

Testing Yourself on the Concepts

1. Describe the differences between simple and complex transcription units.

2. Describe the structural organization and mechanism for movement of transposons and retrotransposons.

3. Describe the genomic organization of antibody genes and the different mechanisms for generating antibody diversity.

4. Describe the different levels of DNA organization in chromosomes and the functional elements required for replication and stable inheritance of chromosomes.

MCAT/GRE-Style Questions

Key Experiment Please read the section titled "Viral Retrotransposons Contain LTRs and Behave Like Retroviruses in the Genome" (p. 307) and refer to Figure 9-17; then answer the following questions:

1. The yeast Ty element and retroviruses share the following features *except*:
    a. Long-terminal repeats.
    b. A primer binding site.
    c. An encoded integrase.
    d. An encoded reverse transcriptase.

2. A mutation in the integrase protein of the Ty element would result in
    a. A decrease in the rate of synthesis of Ty mRNA.
    b. An increase in the rate of synthesis of Ty mRNA.
    c. An increase in the number of integrated Ty elements.
    d. A decrease in the number of integrated Ty elements.

3. In Figure 9-17, the addition of the Gal-responsive promoter to the Ty element increased the frequency of transposition for all of the following reasons *except*:
    a. An increase in the utilization of galactose.
    b. An increase in the amount of Ty mRNA.
    c. An increase in the level of integrase.
    d. An increase in the level of reverse transcriptase.

4. In Figure 9-17, the inclusion of the intron in the Ty element demonstrated that
    a. Transposition is a process that required splicing.
    b. Transposition only occurs in the presence of galactose.
    c. Transposition is mediated through a DNA intermediate.
    d. Transposition is mediated through an RNA intermediate.

Key Application Please read the section titled "Inversion of a Transcription-Control Region Switches *Salmonella* Flagellar Antigens" (p. 314) and refer to Figures 9-20 and 9-21; then answer the following questions:

5. This section describes the molecular mechanism for
    a. The replication of the *Salmonella typhimurium* genome.

b. Synthesis of different *Salmonella typhimurium* flagellar proteins.
    c. Food poisoning by *E. coli*.
    d. Motility in both *Salmonella typhimurium* and *E. coli*.

6. The process of phase variation requires all of the following elements *except*:
    a. A protease.
    b. A site-specific recombinase.
    c. A promoter for RNA polymerase.
    d. A repressor protein.

7. A *Salmonella typhimurium* mutant that contains a mutation that inactivates the recombinase function of the Hin protein has been isolated. This mutation would result in a bacterium that
    a. Lacks a flagellum.
    b. Synthesizes only one kind of flagellar protein.
    c. Is no longer pathogenic.
    d. Cannot cause food poisoning.

8. A *Salmonella typhimurium* mutant that contains a mutation that inactivates the function of the rH1 repressor protein has been isolated. In phase I, this mutation would result in a bacterium that
    a. Expresses only the H1 flagellar protein.
    b. Expresses only the H2 flagellar protein.
    c. Expresses both the H1 and H2 flagellar proteins.
    d. Expresses neither the H1 nor H2 flagellar proteins.

Key Concept Please read the section titled "Nonhistone Proteins Provide a Structural Scaffold for Long Chromatin Loops" (p. 325) and answer the following questions:

9. This section describes
    a. The molecular organization of chromosomes.
    b. The mechanism of chromosome segregation.
    c. Replication of genes in chromosomes.
    d. Transcription of genes in chromosomes.

10. All of the following statements are true about metaphase chromosomes *except*:
    a. Metaphase chromosomes contain histone proteins.
    b. Metaphase chromosomes contain nonhistone proteins.
    c. Metaphase chromosomes are less condensed than interphase chromosomes.
    d. Metaphase chromosomes contain a protein scaffold.

11. All of the following statements about scaffold-associated regions (SARs) are true *except*:
    a. SARs often separate transcription units.
    b. SARs can insulate a transcription unit from the influence of an adjacent transcription unit.
    c. SARs are sequences in the DNA that bind to the chromosome scaffold.

d. SARs are proteins that make up the chromosome scaffold.

12. Which of the following statements is true for the loop model for interphase chromosomes:

a. The DNA loops are all the same size.

b. The DNA loops can vary in size.

c. Large loops represent more actively transcribed genes.

d. Loops mainly consist of nongene regions of DNA.

Key Terms

chromatin *321*

chromosome painting *328*

chromosome scaffold *325*

C-value paradox *299*

cytoplasmic inheritance *333*

direct-repeat sequence *306*

DNA fingerprinting *302*

exon shuffling *313*

gene family *299*

histones *321*

karyotype *325*

long terminal repeats (LTRs) *307*

mobile DNA elements *301*

nonfunctional DNA *298*

nucleosome *321*

operon *295*

polytenization *318*

retrotransposons *304*

Salmonella phase variation *314*

scaffold-associated regions (SARs) *326*

simple-sequence DNA *301*

transcription unit *295*

transposition *303*

transposons *304*

yeast artificial chromosome (YAC) *329*

References

Chromosomal Organization of Genes and Noncoding DNA

Efstratiadis, A., et al. 1980. The structure and evolution of the human β-globin gene family. *Cell* 21:653–668.

Fitch, D. H. A., et al. 1991. Duplication of the gamma-globin gene mediated by L1 long interspersed repetitive elements in an early ancestor of simian primates. *Proc. Nat'l Acad. Sci. USA* 88:7396–7400.

Jeffreys, A. J., et al. 1988. Spontaneous mutation rates to new length alleles at tandem-repetitive hypervariable loci in human DNA. *Nature* 332:278–280.

Moyzis, R. K., et al. 1987. Human chromosome-specific repetitive DNA sequences: novel markers for genetic analysis. *Chromosoma* 95:375–386.

Ohta, T. 1983. On the evolution of multigene families. *Theor. Popul. Biol.* 23:216–240.

Weiss, A., and L. A. Leinwand. 1996. The mammalian myosin heavy chain gene family. *Ann. Rev. Cell Devel. Biol.* 12:417–439.

Mobile DNA

Amariglio, N., and G. Rechavi. 1993. Insertional mutagenesis by transposable elements in the mammalian genome. *Environ. Mol. Mutagenesis* 21:212–218.

Berg, D. E., and M. M. Howe, eds. 1989. *Mobile DNA*. American Society for Microbiology.

Britten, R. J., and D. E. Kohne. 1968. Repeated sequences in DNA. *Science* 161:529–540.

Kurose, K., K. Hata, M. Hattori, and Y. Sakaki. 1995. RNA polymerase III dependence of the human L1 promoter and possible participation of the RNA polymerase II factor YY1 in the RNA polymerase III transcription system. *Nucl. Acids Res.* 23:3704–3709.

McClintock, B. 1956. Controlling elements and the gene. *Cold Spring Harbor Symp. Quant. Biol.* 21:197–216.

Moran, J. V., R. J. DeBererdinis, H. H. Kazazian Jr. 1999. Exon shuffling by L1 retrotransposition. *Science* 283:1530–1534.

Narita, N., et al. 1993. Insertion of a 5' truncated L1 element into the 3' end of exon 44 of the dystrophin gene resulted in skipping of the exon during splicing in a case of Duchenne muscular dystrophy. *J. Clin. Invest.* 91:1862–1867.

Singer, M. F., et al. 1993. LINE1: a human transposable element. *Gene* 135:183–188.

Smit, A. F. A. 1996. The origin of interspersed repeats in the human genome. *Curr. Opin. Genet. Dev.* 6:743–748.

Van Arsdell, S. W., et al. 1981. Direct repeats flank three small nuclear RNA pseudogenes in the human genome. *Cell* 26:11–17.

Wallace, M. R., et al. 1991. A de novo *Alu* insertion results in neurofibromatosis type 1. *Nature* 353:864–868.

Functional Rearrangements in Chromosomal DNA

Borst, P., and D. R. Greaves. 1987. Programmed gene rearrangements altering gene expression. *Science* 235:658–667.

Heichman, K. A., I. P. Moskowitz, and R. C. Johnson. 1991. Configuration of DNA strands and mechanism of strand exchange in the Hin invertasome as revealed by analysis of recombinant knots. *Genes and Dev.* 5:1622–1634.

Kurosawa, Y., and S. Tonegawa. 1982. Organization, structure, and assembly of immunoglobulin heavy-chain diversity DNA segments. *J. Exp. Med.* 155:201.

Leder, P. 1982. The genetics of antibody diversity. *Sci. Am.* 246(5):102.

Max, E. 1993. Immunoglobulins: molecular genetics. In W. Paul, ed., *Fundamental Immunology*, 3d ed., Raven, p. 315.

Silverman, M., and M. Simon. 1980. Phase variation: genetic analysis of switching mutants. *Cell* 19:845–854.

Organizing Cellular DNA into Chromosomes

Felsenfeld, G., et al. 1996. Chromatin structure and gene expression. *Proc. Nat'l. Acad. Sci. USA* 93:9384–9388.

Gasser, S. M. 1995. Chromosome structure. Coiling up chromosomes. *Curr. Biol.* 5:357–360.

Grunstein, M. 1997. Molecular model for telomeric heterochromatin in yeast. *Curr. Opin. Cell Biol.* 9:383–387.

Grunstein, M. 1997. Histone acetylation in chromatic structure and transription. *Nature* 389:349–352.

Marcand, S., S. M. Gasser, and E. Gilson. 1996. Chromatin: a sticky silence. *Curr. Biol* 6:1222–1225.

Schmid, M. B. 1990. More than just "histone-like" proteins. *Cell* 63:451–453.

Van Holde, K., and J. Zlatanova. 1996. What determines the folding of the chromatin fiber? *Proc. Nat'l Acad. Sci. USA* 93:10548–10555.

Wade, P. A., D. Pruss, and A. P. Wolffe. 1997. Histone acetylation: chromatin in action. *Trends Biochem. Sci.* 22:128–132.

Widom, J., and A. Klug. 1985. Structure of the 300 Å chromatin filament: x-ray diffraction from oriented samples. *Cell* 43:207–213.

Morphology and Functional Elements of Eukaryotic Chromosomes

Autexier. C., and C. W. Greider. 1996. Telomerase and cancer: revisiting the telomere hypothesis. *Trends Biochem. Sci.* **21**:387–391.

Blackburn, E. H. 1994. Telomeres: no end in sight. *Cell* **77**:621–623.

Burke, D. T., G. F. Carle, and M. V. Olson. 1987. Cloning of large exogenous DNA into yeast by means of artificial chromosome vectors. *Science* **236**:806–812.

Carbon, J., and L. Clark. 1990. Centromere structure and function in budding and fission yeasts. *New Biologist* **2**:10–19.

Clarke, L., and J. Carbon. 1985. The structure and function of yeast centromeres. *Ann. Rev. Genet.* **19**:29–56.

Diller, J. D., and M. K. Raghuraman. 1994. Eukaryotic replication origins: control in space and time. *Trends Biochem. Sci.* **19**:320–325.

Gall, J. G. 1981. Chromosome structure and the C-value paradox. *J. Cell Biol.* **91**:3s–14s.

Greider, C. W., and E. H. Blackburn. 1996. Telomeres, telomerase and cancer. *Sci. Am.* **274**(2):92–97.

Hyman, A. A., and P. K. Sorger. 1995. Structure and function of kinetochores in budding yeast. *Ann. Rev. Cell Devel. Biol.* **11**:471–495.

Kavenoff, R., L. C. Klotz, and B. H. Zimm. 1974. On the nature of chromosome-sized DNA molecules. *Cold Spring Harbor Symp. Quant. Biol.* **38**:1–8.

Larin, A., A. P. Monaco, and H. Lehrach. 1991. Yeast artificial chromosome libraries containing large inserts from mouse and human DNA. *Proc. Nat'l. Acad. Sci. USA* **88**:4123–4127.

Lee, H.-W., et al. 1998. Essential role of mouse telomerase in highly proliferative organs. *Nature* **392**:569–574.

Pluta, A. F., et al. 1995. The centromere: hub of chromosomal activities. *Science* **270**:1591–1594.

Vazquez, J., and P. Schedl. 1994. Sequences required for enhancer-blocking activity of scs are located within two nuclease-hypersensitive regions. *Embo J.* **13**:5984–5993.

Zakian, V. A. 1996. Structure, function, and replication of *Saccharomyces cerevisiae* telomeres. *Ann. Rev. Genet.* **30**:141–172.

Organelle DNAs

Clayton, D. A. 1992. Transcription and replication of animal mitochondrial DNAs. *Intern. Rev. Cytology* **141**:217–232.

Leblanc, C., et al., 1997. Origin and evolution of mitochondria: what have we learnt from red algae? *Curr. Genet.* **31**:193–207.

Ohyama, K., et al. 1986. Chloroplast gene organization deduced from complete sequence of liverwort Marchantia polymorpha chloroplast DNA. *Nature* **322**:572–574.

Saccone, C. 1994. The evolution of mitochondrial DNA. *Curr. Opin. Genet. Devel.* **4**:875–881.

Regulation of Transcription Initiation

One of the underlying principles of molecular cell biology is that *the actions and properties of each cell type are determined by the proteins it contains.* But what determines the types and amounts of the various proteins that characterize a particular cell type? Or that allow a single-celled organism to respond to changes in its environment? The determining factors are the concentration of each protein's corresponding mRNA, the frequency at which the mRNA is translated, and the stability of the protein itself. The concentration of various mRNAs is, in turn, determined largely by which genes are transcribed and their rate of transcription in a particular cell type. Thus the differential transcription of different genes largely determines the actions and properties of cells.

The term **gene expression** commonly refers to the entire process whereby the information encoded in a particular gene is decoded into a particular protein. Theoretically, regulation at any one of the various steps in this process could lead to *differential* gene expression in different cell types or developmental stages or in response to external conditions. Synthesis of mRNA requires that an **RNA polymerase** initiate transcription, polymerize ribonucleoside triphosphates complementary to the DNA coding strand, and then terminate transcription (see Figure 4-15). In prokaryotes, ribosomes and translation-initiation factors have immediate access to newly formed RNA transcripts, which function as mRNA without further modification. In eukaryotes, the initial RNA transcript is processed by addition of a poly(A)

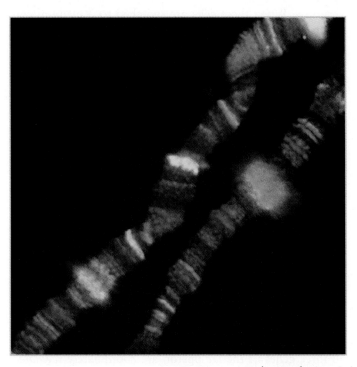

An active region of transcription producing a "puff" in a *Drosophila* polytene chromosome.

MEDIA CONNECTIONS

Focus: Combinatorial Control of Transcription

tail and splicing, which removes noncoding introns, yielding a functional mRNA (see Figure 4-19). The mRNA then is transported from its site of synthesis in the nucleus to the cytoplasm where translation occurs. Finally, the stability of an mRNA affects its concentration in both prokaryotic and eukaryotic cells.

Although examples of regulation at each of the steps in gene expression have been found, control of transcription initiation—the first step—is the most important mechanism for determining whether or not most genes are expressed and how much of the encoded mRNAs, and consequently proteins, are produced. In this chapter, we review current understanding of the molecular events that determine when transcription is initiated. Chapter 11 considers other regulatory mechanisms that control the subsequent steps in gene expression. Chapter 14 presents an overview of how these multiple levels of **gene control** contribute to the development of specific types of cells in multicellular organisms.

10.1 Bacterial Gene Control: The Jacob-Monod Model

A combination of genetic and biochemical experiments in bacteria led to the initial recognition of (1) protein-binding regulatory sequences associated with genes and (2) proteins whose binding to a gene's regulatory sequences either activate or repress its transcription. These key components underlie the ability of both prokaryotic and eukaryotic cells to turn genes on and off, although innumerable variations on the basic process have been discovered. In this section, we first describe some of the early experimental findings leading to a general model of bacterial transcription control. In the next section, we take a closer look at how bacterial RNA polymerase initiates transcription and the mechanisms controlling its ability to do so.

In bacteria, gene control serves mainly to allow a single cell to adjust to changes in its nutritional environment so that its growth and division can be optimized. Thus, the prime focus of research has been on genes that encode *inducible* proteins whose production varies depending on the nutritional status of the cells. Although gene control in multicellular organisms often involves response to environmental changes, its most characteristic and biologically far-reaching purpose is the regulation of a genetic program that underlies embryological development and tissue differentiation. Nonetheless, many of the principles of transcription control first discovered in bacteria also apply to eukaryotic cells.

Enzymes Encoded at the *lac* Operon Can Be Induced and Repressed

E. coli can use either glucose or other sugars such as the disaccharide lactose as the sole source of carbon and energy. When *E. coli* cells are grown in a glucose-containing medium,

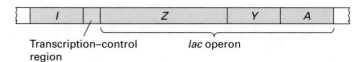

Transcription–control region · · · *lac* operon

▲ **FIGURE 10-1 The *lac* operon includes three genes: *lacZ*, which encodes β-galactosidase; *lacY*, which encodes lactose permease; and *lacA*, which encodes thiogalactoside transacetylase.** Binding of regulatory proteins to sites in the control region immediately upstream (to the left) of the *lacZ* gene regulate its transcription. The *lacI* gene, which encodes the *lac* repressor, maps adjacent to the *lac* control region.

the activity of the enzymes needed to metabolize lactose is very low. When these cells are switched to a medium containing lactose but no glucose, the activities of the lactose-metabolizing enzymes increase. Early studies showed that the increase in the activity of these enzymes resulted from the synthesis of new enzyme molecules, a phenomenon termed **induction.** The enzymes induced in the presence of lactose are encoded by the *lac* operon, which includes two genes, Z and Y, that are required for metabolism of lactose and a third gene, A (Figure 10-1). The *lacY* gene encodes *lactose permease*, which spans the *E. coli* cell membrane and uses the energy available from the electrochemical gradient across the membrane to pump lactose into the cell (Section 15.5). The *lacZ* gene encodes β-*galactosidase*, which splits the disaccharide lactose into the monosaccharides glucose and galactose (see Figure 2-10); these sugars are further metabolized through the action of enzymes encoded in other operons. The *lacA* gene encodes *thiogalactoside transacetylase*, an enzyme whose physiological function is not well understood.

Synthesis of all three enzymes encoded in the *lac* operon is rapidly induced when *E. coli* cells are placed in a medium containing lactose as the only carbon source and repressed when the cells are switched to a medium without lactose. Thus all three genes of the *lac* operon are *coordinately regulated.* The *lac* operon in *E. coli* provides one of the earliest and still best-understood examples of gene control. Much of the pioneering research on the *lac* operon was conducted by Francois Jacob, Jacques Monod, and their colleagues in the 1960s.

Some molecules similar in structure to lactose can induce expression of the *lac*-operon genes even though they cannot be hydrolyzed by β-galactosidase. Such small molecules (i.e., smaller than proteins) are called **inducers.** One of these, isopropyl-β-D-thiogalactoside, abbreviated *IPTG*,

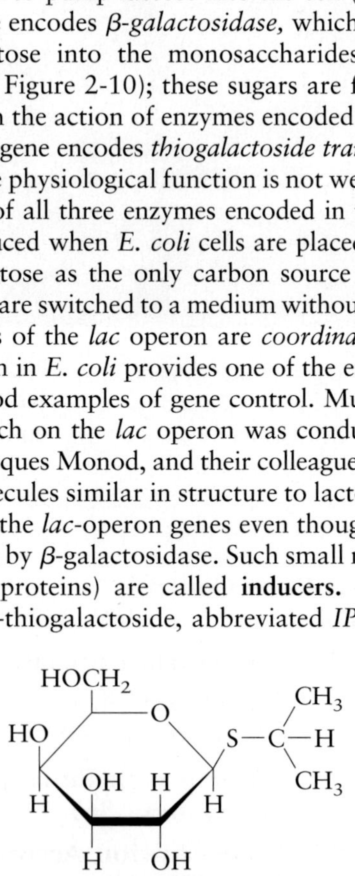

Isopropyl-β-D-thiogalactoside (IPTG)

is particularly useful in genetic studies of the *lac* operon, because it can diffuse into cells and, since it is not metabolized, its concentration remains constant throughout an experiment.

Mutations in *lacI* Cause Constitutive Expression of *lac* Operon

Insight into the mechanisms controlling synthesis of β-galactosidase and lactose permease first came from the study of mutants in which control of β-galactosidase expression was abnormal. A sensitive colorimetric assay for β-galactosidase uses *X-gal* (5-bromo-4-chloro-3-indolyl-β-D-galactoside) as substrate:

X-gal

Hydrolysis of this colorless analog of lactose by β-galactosidase yields an intensely blue product. When wild-type *E. coli* cells are plated on media containing X-gal plus lactose as the major carbon source, all the colonies that grow appear blue. When the cells are plated on media containing X-gal plus glucose as the carbon source, the resulting colonies appear white; in this case, β-galactosidase synthesis is repressed and there is not sufficient β-galactosidase in the cells to hydrolyze the X-gal to its colored product. However, when the cells are exposed to chemical mutagens before plating on X-gal/glucose plates, rare blue colonies appear. In most cases, when cells from these blue colonies are recovered and grown in media containing glucose, they are found to express all the genes of the *lac* operon at much higher levels than wild-type cells in the same medium. Such cells are called **constitutive mutants** because they fail to repress the *lac* operon in media lacking lactose and instead continuously, or constitutively, express the enzymes. By recombinational analysis (Section 8.3), these mutations were mapped to a region on the *E. coli* chromosome to the left of the *lacZ* gene, a region called the *lacI* gene (see Figure 10-1).

Jacob and Monod reasoned that such constitutive mutants probably had a defect in a protein that normally repressed expression of the *lac* operon in the absence of lactose. Hence they called the protein encoded by the *lacI* gene the *lac repressor* and proposed that it binds to a site on the *E. coli* genome where transcription of the *lac* operon is initiated, thereby blocking transcription. They further hypothesized that when lactose is present in the cell, it binds to the *lac* repressor, decreasing its affinity for the repressor-binding site on the DNA. As a result, the repressor falls off the DNA and transcription of the *lac* operon is initiated, leading to

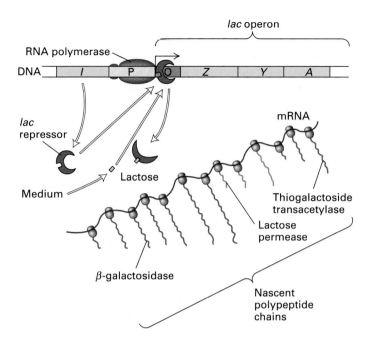

▲ **FIGURE 10-2 Jacob and Monod model of transcriptional regulation of the *lac* operon by *lac* repressor.** When *lac* repressor binds to a DNA sequence called the operator (O), which lies just upstream of the *lacZ* gene, transcription of the operon by RNA polymerase is blocked. Binding of lactose to the repressor causes a conformational change in the repressor, so that it no longer binds to the operator. RNA polymerase then is free to bind to the promoter (P) and initiate transcription of the *lac* genes; the resulting polycistronic mRNA is translated into the encoded proteins. [Adapted from A. J. F. Griffiths et al., 1993, *An Introduction to Genetic Analysis,* 5th ed., W. H. Freeman and Co.]

synthesis of β-galactosidase, lactose permease, and thiogalactoside transacetylase (Figure 10-2).

Isolation of Operator Constitutive and Promoter Mutants Support Jacob-Monod Model

The model proposed by Jacob and Monod predicted that a specific DNA sequence near the transcription start site of the *lac* operon is a binding site for *lac* repressor. They reasoned that mutations in this sequence, which they termed the **operator** (O), would prevent the repressor from binding, thus yielding constitutive mutants that could be identified on X-gal/glucose indicator plates. To distinguish between mutations in the *lacI* gene, which inactivate the repressor, and mutations in the operator, which prevent repressor binding, Jacob and Monod mutagenized cells carrying two copies of the wild-type *lacI* gene, one on the bacterial chromosome and one on a plasmid. In this system, separate mutations in both copies of *lacI* in a given cell are required to generate a *lacI⁻* constitutive mutant, a low-probability event. In contrast, only a single mutation in the operator of one copy of the *lac* operon is required to yield a constitutive mutant. Using this approach, Jacob and Monod isolated mutants that expressed the *lac* operon constitutively even when two copies of the wild-type *lacI* gene encoding the *lac* repressor were present in the same cell. These *operator constitutive (O^c) mutations* mapped to one end of the *lac* operon, as the model predicted (see Figure 10-2).

Most mutations that prevent expression of β-galactosidase in cells exposed to an inducer such as IPTG map in the *lacZ* gene itself. But a rare class of mutations map to a region between *lacI* and the operator, in a region termed the **promoter** (P). Cells carrying these mutations also cannot induce expression of the *lacY* and *lacA* genes; that is, these mutations prevent expression of the entire *lac* operon. According to the Jacob and Monod model, such promoter mutations block initiation of transcription by RNA polymerase (see Figure 10-2). Consequently, no *lac* mRNA and therefore no *lac* proteins are synthesized, even when *lac* repressor binds IPTG and comes off the *lac* operator.

Regulation of *lac* Operon Depends on Cis-Acting DNA Sequences and Trans-Acting Proteins

Subsequent analyses of the effects of various mutations in *E. coli* cells containing one or two copies of *lac* DNA provided further insight into regulation of *lac*-operon expression. In these experiments, assays for β-galactosidase and lactose permease activity were conducted in the presence and absence of inducer (IPTG). These analyses showed that the O^c mutation is dominant over O^+ (the wild-type *lac* O sequence). In addition, the O^c mutation only affects expression of *lac* genes on the same DNA molecule (i.e., genes in cis to the mutation). Experimental demonstration of the **cis-acting** nature of the O^c mutation is illustrated in Figure 10-3.

As noted earlier, mutations in *lacI* (in cells with a single *lac* operon) cause constitutive expression of β-galactosidase and lactose permease because no functional repressor is made. Unlike the O^c mutation, which is dominant, the *lacI^-* mutation is recessive to the wild-type *lacI^+* gene. Furthermore, the wild-type *lacI^+* gene can exert control over the *lacZ* and *lacY* genes on a different DNA molecule (i.e., genes in trans to *lacI^+*). The **trans-acting** ability of *lacI^+* is easy to understand since this gene encodes a protein, which is free to diffuse through the cell and bind to any *lac* operator in the cell (Figure 10-4).

In general, cis-acting mutations are in DNA sequences that function as binding sites for proteins that control the expression of nearby genes. For example, the cis-acting O^c mutations prevent binding of the *lac* repressor to the operator. Similarly, mutations in the *lac* promoter are cis-acting, since they alter the binding site for RNA polymerase. When RNA polymerase cannot initiate transcription of the *lac* operon, none of the genes in the operon can be expressed irrespective of the function of the repressor. In general, trans-acting genes that regulate expression of genes on other DNA

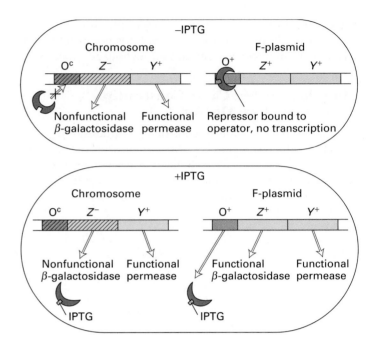

▲ FIGURE 10-3 Experimental demonstration that O^c mutations are cis-acting. *E. coli* cells containing two copies of the *lac* operon are diagrammed. Diagonal lines indicate genes and control regions carrying mutations. In these cells, the *lac* operon on the bacterial chromosome has an O^c mutation in the operator and a mutation in the *lacZ* gene (*lacZ^-*), which inactivates the β-galactosidase enzyme. The *lac* operon on the F-plasmid is wild-type. *(Top)* In the absence of the inducer IPTG, these cells constitutively express a functional lactose permease but no functional β-galactosidase. These results indicate that the O^c mutation only affects the genes on the same DNA (i.e., the chromosomal *lac* operon). If the O^c mutation were trans-acting, then transcription of the *lacZ^+* gene on the F-plasmid would have yielded observable β-galactosidase activity. *(Bottom)* In the presence of IPTG, both permease and β-galactosidase activity are observed. As diagrammed, transcription of the wild-type *lacZ* gene on the plasmid yields functional β-galactosidase.

molecules encode diffusible products. In most cases these are proteins, but in some cases RNA molecules can act in trans to regulate gene expression.

Biochemical Experiments Confirm That Induction of the *lac* Operon Leads to Increased Synthesis of *lac* mRNA

The Jacob and Monod model of repressor control of *lac* operon transcription, which was based on genetic experiments with *E. coli* mutants, proposes that addition of inducer causes an increase in transcription of the *lac* operon. This prediction was tested directly through pulse-labeling experiments that measured the rate of *lac* mRNA synthesis in *E. coli* cells grown initially in glucose media and then after addition of IPTG. The results of such experiments showed that little *lac*

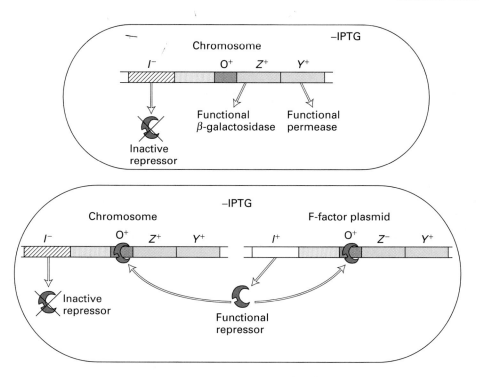

◄ FIGURE 10-4 Experimental demonstration that the *lacI*⁺ gene is trans-acting. **◄ FIGURE 10-4 Experimental demonstration that the *lacI*⁺ gene is trans-acting.** *(Top)* Cells carrying a single *lacI*⁻ gene produce an inactive repressor; as a result, they express β-galactosidase and lactose permease constitutively. *(Bottom)* When a wild-type *lacI*⁺ gene is introduced into *lacI*⁻ cells on a F-factor plasmid, the transformed cells produce functional repressor, which can bind to both *lac* operators. As a result, these cells do not express β-galactosidase or lactose permease in the absence of inducer.

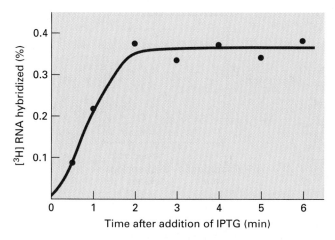

▲ **FIGURE 10-5 Biochemical demonstration that inducer leads to an increase in *lac* operon transcription.** Small samples of an *E. coli* culture growing in glucose medium were removed just before and at short intervals after addition of IPTG. [³H]uridine was added to each sample immediately after it was removed; after 20 seconds the cells were lysed and RNA was isolated. Each such pulse-labeled sample was hybridized to an excess of membrane-bound cloned *lac* DNA (see Figure 7-17). Radioactivity that is retained on the membrane is a measure of the rate of mRNA synthesis during the short pulse-labeling period. A plot of the percentage of total [³H]RNA that hybridized to *lac* DNA versus time after addition of IPTG indicates that the rate of mRNA synthesis increased rapidly after addition of IPTG, reaching a maximal rate in about 2 minutes. [Adapted from G. Contesse, M. Crepin, and F. Gros, 1970, in J. R. Beckwith and D. Zipser, eds., *The Lactose Operon,* Cold Spring Harbor Laboratory Press.]

mRNA is synthesized before the addition of IPTG, but *lac* mRNA synthesis is detectable within 1 minute after the addition of IPTG and reaches a maximal rate by 2 minutes (Figure 10-5). At later times, *lac* mRNA synthesis is maintained at this maximal rate as long as inducer is present. These findings demonstrated directly that inducer does indeed cause an increase in transcription of the *lac* operon.

SUMMARY Bacterial Gene Control: The Jacob-Monod Model

- Many proteins in bacteria are inducible, that is, their synthesis is regulated depending on the cell's nutritional status. Differential expression of genes encoding such proteins most commonly occurs at the level of transcription initiation.

- According to the Jacob and Monod model of transcriptional control, transcription of the *lac* operon, which encodes three inducible proteins, is repressed by binding of *lac* repressor protein to the operator sequence (see Figure 10-2). In the presence of lactose or other inducer, this repression is relieved and the *lac* operon is transcribed.

- Mutations in the promoter, which binds RNA polymerase, or the operator are cis-acting; that is, they only affect expression of genes on the same DNA molecule in which the mutation occurs.

- Mutations in an operator sequence that decrease repressor binding result in constitutive transcription. Mutations in a promoter sequence, which affect the affinity of RNA polymerase binding, can either decrease (down-mutation) or increase (up-mutation) transcription.

- Repressors and activators are trans-acting; that is, they affect expression of their regulated genes no matter on which DNA molecule in the cell these are located.

10.2 Bacterial Transcription Initiation

As noted in previous chapters, RNA polymerase initiates transcription of most genes at a unique position (a single base) in the template DNA lying upstream of the coding sequence. (At some promoters, the polymerase initiates at two or three alternative neighboring bases.) The ribonucleoside triphosphate used by RNA polymerase to initiate transcription in an in vitro reaction can be recognized because it retains its 5′ triphosphate, whereas all other nucleotides in the RNA chain contain a single phosphate in ester linkage to the preceding nucleotide (see Figure 4-15). The base pair where transcription initiates is called the transcription-initiation site, or *start site*. By convention, the transcription-initiation site in the DNA sequence of a transcription unit is usually numbered +1. Base pairs extending in the direction of transcription (**downstream**) are assigned positive numbers and those extending in the opposite direction (**upstream**) are assigned negative numbers.

According to the Jacob and Monod model, regulatory proteins (repressors and activators) and *E. coli* RNA polymerase work together to regulate transcription initiation at bacterial promoters. In this section, we discuss studies of the structure of these proteins and of their interactions with regulatory sequences in DNA. We begin by describing two assays that are commonly used in studying DNA-protein complexes and then present various experimental results that have led to the current detailed understanding of transcription initiation and its control in bacteria.

Footprinting and Gel-Shift Assays Identify Protein-DNA Interactions

When a protein is bound to a region of DNA, it protects the DNA sequence from digestion by DNase. This phenomenon is the basis of a powerful technique, called **DNase I footprinting**, for detecting protein-binding sites in DNA. As illustrated in Figure 10-6, when samples of a DNA fragment that is labeled at one end are digested in the presence and absence of a DNA-binding protein and then denatured and electrophoresed, the region protected by the bound protein appears as a gap, or "footprint," in the array of bands resulting from digestion in the absence of protein. The products of DNA sequencing reactions of the DNA sample being

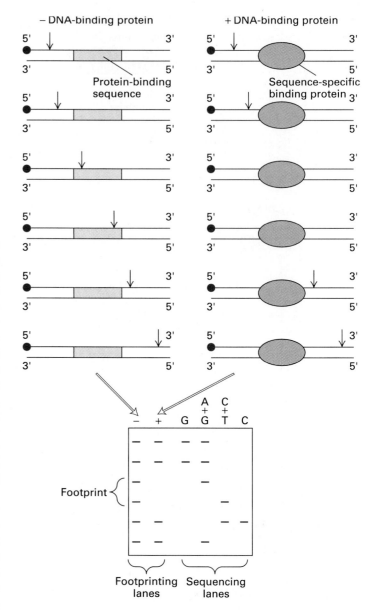

footprinted generally are analyzed on the same gel in order to determine the precise sequence protected from DNase I digestion by the bound protein. (As mentioned in Chapter 4, footprinting experiments using chemical digestion agents can detect protein-binding sites in RNA.)

The *electrophoretic mobility shift assay* (EMSA) is more useful than the footprinting assay for quantitative analysis of DNA-protein binding reactions. In this assay, the electrophoretic mobility of a radiolabeled DNA fragment is determined in the presence and absence of a sequence-specific DNA-binding protein. Protein binding generally reduces the mobility of a DNA fragment, causing a shift in the location of the fragment band detected by autoradiography. This technique, also referred to as the gel-shift or band-shift assay, is widely used to assay sequence-specific DNA-binding proteins during their purification (Figure 10-7). One advantage

◄ **FIGURE 10-6 DNase I footprinting, a common technique for identifying protein-binding sites in DNA.** *(Top)* A DNA fragment is labeled at one end with ^{32}P (red dot) as in the Maxam-Gilbert sequencing method (see Figure 7-25). Portions of the sample then are digested with DNase I in the presence and absence of a protein that binds to a specific sequence in the fragment. DNase I randomly hydrolyzes the phosphodiester bonds of DNA between the 3′ oxygen on the deoxyribose of one nucleotide and the 5′ phosphate of the next nucleotide. A low concentration of DNase I is used so that on average each DNA molecule is cleaved just once (vertical arrows). In the absence of a DNA-binding protein, the sample is cleaved at all possible positions between the labeled and unlabeled ends of the original fragment. The two samples of DNA then are separated from protein, denatured to separate the strands, and electrophoresed. The resulting gel is analyzed by autoradiography, which detects only labeled strands and reveals fragments extending from the labeled end to the site of cleavage by DNase I. *(Bottom)* Diagram of hypothetical autoradiogram of the gel for the minus protein sample above reveals bands corresponding to all possible fragments produced by DNase I cleavage (−lane). In the sample digested in the presence of a DNA-binding protein, two bands are missing (+lane); these correspond to the DNA region protected from digestion by bound protein and are referred to as the footprint of that protein. This protected region can be precisely aligned with the DNA sequence if sequencing reactions are performed on the original end-labeled DNA and the products electrophoresed on the same gel. In this example, the products of four Maxam-Gilbert sequencing reactions are shown.

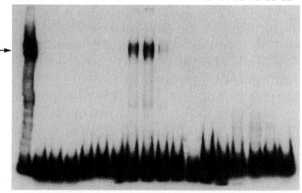

Fraction ON 1 2 3 4 5 6 7 8 9 10 11 12 14 16 18 20 22

▲ **FIGURE 10-7 Results of an electrophoretic mobility shift assay (EMSA) for DNA-binding proteins.** In this example, fractions from an ion-exchange column were assayed for a protein that binds to the promoter region of a eukaryotic tRNA gene. An aliquot of the protein fraction loaded onto the column (ON) and of fractions eluted off the column with increasing salt concentration (numbers) were incubated with a radiolabeled restriction-fragment probe that included the promoter region; each sample was then subjected to electrophoresis through a polyacrylamide gel. Free probe not bound by protein migrated to the bottom of the gel. A protein in column fractions 7 and 8 bound to the probe, forming a DNA-protein complex that migrated more slowly than the free probe (arrow). [From S. Yoshinaga et al., 1989, *J. Biol. Chem.* **264**:10529.]

of the EMSA over footprinting assays is that binding can be detected when only a small fraction of the labeled DNA fragment is bound.

The *lac* Control Region Contains Three Critical Cis-Acting Sites

The original experimental data of the footprint of *E. coli* RNA polymerase and *lac* repressor on the *lac* control region is shown in Figure 10-8. RNA polymerase produced a clear footprint over a region of ≈70 base pairs. Some sites on each strand within this 70-bp region were not protected (note spaces between the brackets in Figure 10-8). This finding indicates that a polymerase molecule lies along one surface of the DNA double helix in the region between approximately −20 and −50 leaving the phosphodiester bonds on the other side of the helix exposed to DNase I. The repressor footprint overlaps the downstream end of the RNA polymerase footprint but is not as large. Similar footprinting experiments have shown that a third protein, termed CAP, complexed to cAMP (cAMP-CAP) also binds to the *lac* control region. As discussed later, cAMP-CAP activates transcription from the *lac* promoter. The position of these three footprints, which define protein-binding sites in the *lac* control region, are diagrammed in Figure 10-9.

These footprinting results extended and are consistent with earlier studies in which the nucleotide sequence of the *lac* control region from wild-type cells was compared with the sequences from operator constitutive (O^c) and promoter mutants. The base changes producing promoter and O^c mutants fall within the footprints for RNA polymerase and *lac* repressor.

RNA Polymerase Binds to Specific Promoter Sequences to Initiate Transcription

We now shift our attention to the proteins that bind to the *lac* control region, starting with RNA polymerase. Unlike eukaryotes, bacteria have a single type of RNA polymerase that catalyzes synthesis of mRNAs, rRNAs, and tRNAs encoded in the bacterial chromosome. The major form of this enzyme is composed of five subunits (Figure 10-10): two large subunits called β′ (156 kDa) and β (151 kDa), two copies of a smaller subunit called α (37 kDa), and one copy of a subunit called σ^{70} (70 kDa). The different subunits have distinct functions:

- Interaction of the σ subunit with a promoter signals the polymerase to initiate transcription at a specific sequence in template DNA.
- The β and β′ subunits polymerize ribonucleoside triphosphates (NTPs) as directed by the template strand.
- The α subunits interact with regulatory proteins and, in some cases, with DNA to control how frequently RNA polymerase initiates transcription from a specific promoter.

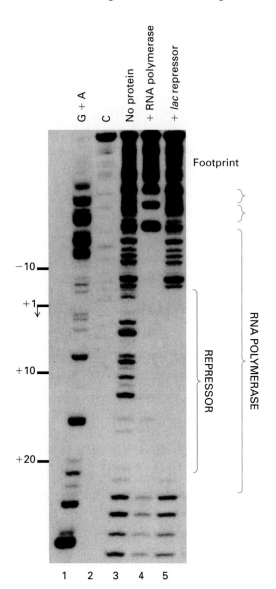

1 2 3 4 5

◀ **FIGURE 10-8 Footprints of RNA polymerase and *lac* repressor on *lac* control-region DNA.** Because DNase I cleaves some phosphodiester bonds more frequently than others, the density of bands in the absence of protein (lane 3) is not as uniform as diagrammed in Figure 10-6. The brackets on the right indicate the DNA regions protected from DNase I digestion by bound RNA polymerase (lane 4) or *lac* repressor (lane 5). Lanes 1 and 2 show the products of two Maxam-Gilbert sequencing reactions: from these lanes, the gel bands can be correlated with the nucleotide sequence of the *lac* control region. The positions in the sequence are noted on the left; the arrowhead indicates the direction of transcription. This footprinting experiment was performed with the 5′ label on the right end of the bottom strand as shown in Figure 10-9. [From A. Schmitz and D. J. Galas, 1979, *Nucl. Acids Res.* **6**:111.]

In vitro binding studies with purified RNA polymerase and *lac* promoter sequences from promoter mutants that exhibit decreased *lac*-operon expression showed that the enzyme has a greatly reduced affinity for the mutant DNA. This finding indicates that the polymerase molecule contacts bases in wild-type *lac* DNA when it binds to the promoter. Of the ≈3000 RNA polymerase molecules in an *E. coli* cell, about half are actively engaged in transcribing DNA into RNA at any one time in rapidly growing cells. Most of the remainder are loosely associated with DNA, moving randomly along the DNA until they encounter a promoter sequence.

As discussed in more detail below, two regions in most *E. coli* promoters, referred to as the −10 and −35 regions, are critical for binding of RNA polymerase via the σ^{70} subunit. The initial complex formed when RNA polymerase binds to a promoter region is called a *closed complex* because the DNA strands near the transcription start site still are base-paired. Before polymerization (chain elongation) can proceed, the polymerase must separate the hydrogen-bonded base pairs in the region of the start site, forming an *open complex*. Treatment of such open complexes with chemical reagents that react with single-stranded but not double-stranded DNA have indicated that the separated region

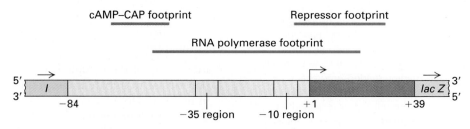

▲ **FIGURE 10-9 Diagram of the *lac*-operon transcription-control region.** The region coding the N-terminus of β-galactosidase is to the right, and the region coding the C-terminus of *lac* repressor is to the left. Regions protected from DNase I digestion (footprints) by cAMP-CAP, RNA polymerase, and *lac* repressor are indicated by colored bars at the top.

Mutational analyses identified three critical cis-acting regions: two in the promoter (yellow) near −35 and −10 and one between +1 and +20 in the operator (brown). [See W. S. Reznikoff and J. N. Abelson, 1978, in J. H. Miller and W. S. Reznikoff, eds., *The Operon*, Cold Spring Harbor Laboratory Press.]

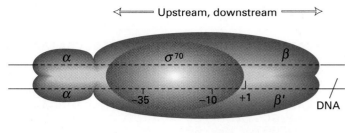

▲ **FIGURE 10-10 Schematic representation of the major form of *E. coli* RNA polymerase bound to DNA.** By convention, the transcription-initiation site is generally numbered +1. Base pairs extending in the direction of transcription are said to be downstream of the start site; those extending in the opposite direction are upstream. The σ^{70} subunit binds to specific sequences near the −10 and −35 positions in the promoter. The α subunits lie close to the DNA in the upstream direction. The β and β' subunits associate with the start site.

extends for about 12 base pairs. Once this separation is achieved, the polymerase can mediate base pairing between the template strand and the incoming ribonucleoside triphosphates to be added to the 3' end of the growing RNA strand.

After the polymerase transcribes approximately 10 base pairs, the σ^{70} subunit is released. Consequently the σ^{70} subunit acts as an *initiation factor* required for transcription initiation, but not for RNA-strand elongation once initiation has taken place. The form of the polymerase that continues transcribing, consisting of the β', β, and two α subunits, is called the *core polymerase*. The form of the polymerase that also contains the σ^{70} subunit and can initiate transcription is called the *holoenzyme*.

Differences in *E. coli* Promoter Sequences Affect Frequency of Transcription Initiation

The relative frequency of transcription initiation (times per minute) at different promoters by *E. coli* RNA polymerase depends largely on the affinity of the enzyme for the promoter sequence. Promoters at which RNA polymerase initiates transcription at a high frequency are classified as "strong" promoters; those with a low frequency of initiation, as "weak" promoters.

Sequencing of the promoters adjacent to hundreds of *E. coli* operons have revealed that the −10 and −35 regions are critical in determining their strength. The sequences of several particularly strong *E. coli* promoters, shown in Figure 10-11a, are similar in these two regions. (The −10 region sometimes is called the *Pribnow box* after an early discoverer.) The consensus sequences of the −10 and −35 regions based on comparison of numerous strong σ^{70} promoters are shown in Figure 10-11b, along with single base-pair mutations that cause a significant decrease in the

frequency of transcription initiation from several different promoters. All of these damaging mutations cause deviations from the −10 and −35 consensus sequences. Footprinting experiments with these mutant promoters show that RNA polymerase binds to them with much lower affinity than to the wild-type promoters.

The relatively weak *lac* promoter deviates at several positions from the −35 and −10 consensus sequences (Figure 10-11c). Two mutations in the −10 region of the *lac* promoter increase expression of the lac operon; as can be seen these up-mutations increase the match to the consensus sequence. Similar up-mutations have been identified in other weak promoters.

The ability of RNA polymerase to initiate transcription at *E. coli* promoters with different sequences in the −10 and −35 regions indicates that this enzyme has a rather broad specificity. This feature contrasts with restriction enzymes, each of which cuts DNA at a single unique nucleotide sequence.

Binding of *lac* Repressor to the *lac* Operator Blocks Transcription Initiation

The Jacob and Monod model predicts that *lac* repressor will block transcription initiation at the *lac* promoter by binding to the *lac* operator. As noted earlier, DNase I footprinting experiments with purified *lac* repressor showed that it binds to a specific sequence in the *lac* control region (see Figure 10-8, lane 5). Moreover, the region of DNA protected from DNase I digestion by *lac* repressor corresponds with the location of O^c mutations, and purified *lac* repressor has a greatly decreased affinity for *lac* control-region DNA containing O^c mutations. These results indicate that interactions between the *lac* repressor and bases in the wild-type operator sequence are responsible for the tight binding of *lac* repressor to *lac*-operator DNA.

The binding sites for RNA polymerase and *lac* repressor in the *lac*-operon control region overlap between about −5 and +20 (see Figures 10-8 and 10-9). In vitro studies have shown that when *lac* repressor is bound to the *lac* operator, it blocks RNA polymerase from interacting with DNA at the start site, thereby preventing transcription initiation, as predicted by the Jacob and Monod model. Finally, DNA-footprinting experiments also demonstrate that addition of lactose or IPTG to the binding reaction vastly decreases the affinity of the repressor for operator DNA, directly confirming the Jacob and Monod model of *lac* repressor transcription control.

Most Bacterial Repressors Are Dimers Containing α Helices That Insert into Adjacent Major Grooves of Operator DNA

The three-dimensional structures of several bacterial repressors that control different operons bound to their specific operator DNA sequences have been solved by x-ray crystallography. Most of these proteins are homodimers, and each

(a) Strong *E. coli* promoters

```
  tyr tRNA    TCTCAACGTAACACTTTACAGCGGCG·•CGTCATTTGATATGATGC·GCCCCGCTTCCCGATAAGGG
  rrn D1      GATCAAAAAAATACTTGTGCAAAAAA··TTGGGATCCCTATAATGCGCCTCCGTTGAGACGACAACG
  rrn X1      ATGCATTTTTCCGCTTGTCTTCCTGA··GCCGACTCCCTATAATGCGCCTCCATCGACACGGCGGAT
  rrn (DXE)₂  CCTGAAATTCAGGGTTGACTCTGAAA··GAGGAAAGCGTAATATAC·GCCACCTCGCGACAGTGAGC
  rrn E1      CTGCAATTTTTCTATTGCGGCCTGCG·•GAGAACTCCCTATAATGCGCCTCCATCGACACGGCGGAT
  rrn A1      TTTTAAATTTCCTCTTGTCAGGCCGG·•AATAACTCCCTATAATGCGCCACCACTGACACGGAACAA
  rrn A2      GCAAAAATAAATGCTTGACTCTGTAG··CGGGAAGGCGTATTATGC·ACACCCCGCGCCGCTGAGAA
  λ Pᴿ        TAACACCGTGCGTGTTGACTATTTTA·CCTCTGGCGGTGATAATGG··TTGCATGTACTAAGGAGGT
  λ Pᴸ        TATCTCTGGCGGTGTTGACATAAATA·CCACTGGCGGTGATACTGA··GCACATCAGCAGGACGCAC
  T7 A3       GTGAAACAAAACGGTTGACAACATGA·AGTAAACACGGTACGATGT·ACCACATGAAACGACAGTGA
  T7 A1       TATCAAAAAGAGTATTGACTTAAAGT·CTAACCTATAGGATACTTA·CAGCCATCGAGAGGGACACG
  T7 A2       ACGAAAAACAGGTATTGACAACATGAAGTAACATCGCAGTAAGATAC·AAATCGCTAGGTAACACTAG
  fd VIII     GATACAAATCTCCGTTGTACTTTGTT··TCGCGCTTGGTATAATCG·CTGGGGGTCAAAGATGAGTG
                            -35                              -10              +1 ⟶
```

(b) Consensus sequences of σ⁷⁰ promoters

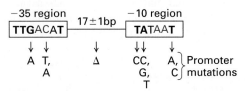

(c) *Lac* promoter sequence

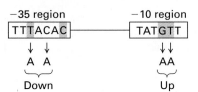

▲ **FIGURE 10-11 Promoters recognized by *E. coli* RNA polymerase containing σ⁷⁰.** (a) Sequences of some strong promoters with spaces (dots) introduced to maximize homology in the −35 region and −10 region. These sequences correspond to the top strand of the promoter with transcription proceeding to the right (see Figure 10-9). Bases that match the −35 and −10 consensus sequences are highlighted in yellow. The six rrn sequences control genes encoding rRNA. The λ, T7, and fd sequences, which are on viral genomes, direct transcription by the host-cell RNA polymerase. (b) Consensus sequences of −35 and −10 regions, which are separated by 15–17 base pairs. Mutations known to significantly decrease the frequency of transcription from a number of different promoters are indicated.

In the consensus sequences, the frequency with which the indicated base occurs at each position in different σ⁷⁰ promoters is indicated as follows: red letters, >75 percent; boldface black letters, 50–75 percent; black letters, 40–50 percent. (c) Sequences of the −35 and −10 regions of *lac* promoter, which deviates from the consensus sequences in four positions (blue highlight). Down mutations cause a decrease in *lac*-operon expression. The two up mutations, which increase the match to the −10 consensus sequence, increase expression. [Part (a) see W. Siebenlist, R. B. Simpson, and W. Gilbert, 1980, *Cell* **20**:269. Part (b) see W. R. McClure, 1985, *Ann. Rev. Biochem.* **54**:171. Part (c) see R. C. Dickson et al., 1975, *Science* **187**:27.]

monomer of the dimeric molecule makes similar contacts with bases in half of the operator sequence. Given the symmetry of most dimeric repressors, it is not surprising that most operator sequences are short **inverted repeats** (Figure 10-12). Each half of the inverted-repeat sequence in an operator is called a *half-site*. Consequently, an operator can be thought of as two half-sites, each of which binds one monomer of a dimeric repressor protein.

In many dimeric repressors, an α helix in each monomer inserts into a major groove in the DNA helix. This α helix is referred to as the *recognition helix* or *sequence-reading helix* because most of the protein side chains that contact DNA extend from this helix. Figure 10-13a shows a space-filling model of bacteriophage 434 repressor bound to its specific operator sequence. A ribbon diagram of one of the repressor monomers in this complex illustrates how an α helix inserts into the major groove of DNA (Figure 10-13b), allowing atoms in the protein to make multiple, specific interactions

with atoms in the DNA (Figure 10-13c). Most monomeric repressor molecules dimerize in such a way that the two recognition helices are positioned precisely to enter the major grooves in the two symmetrical halves of the operator DNA.

In general, a single repressor monomer makes too few interactions to bind to DNA stably. However, a repressor dimer makes twice as many interactions as the monomer, a sufficient number for repressors to bind to their operator sites with high affinity ($K_d = \approx 10^{-8}–10^{-11}$ M). The recognition helix that protrudes from the surface of bacterial repressors to enter the DNA major groove and "read" the DNA sequence is usually supported in the protein structure in part by hydrophobic interactions with a second α helix just N-terminal to it. This recurring structural element in many bacterial repressors is called a **helix-turn-helix** motif.

Although most bacterial repressors have this dimeric structure, some repressors exhibit alternative protein structures. For example, the Arc repressor of bacteriophage P22

▶ **FIGURE 10-12 The *lac* operator sequence is a nearly perfect inverted repeat centered around the GC base pair at position +11.** The 17-bp sequence of the top strand beginning at −7 is identical to the 17-bp sequence of the bottom strand beginning at +28, reading in the 5′ → 3′ direction in both cases, except for the nucleotides indicated by italic letters. Each half of the inverted-repeat sequence is called an operator half-site (yellow highlight).

```
       −10          +1         +10         +20         +30
        •            •           •           •           •
      ──────────────────────────────▶
5′ ATGTTGTGTGGAATTGTGAGCGGATAACAATTTCACACAGGAA 3′
3′ TACAACACACCTTAACACTCGCCTATTGTTAAAGTGTGTCCTT 5′
                          ◀──────────────────
                            Half-site
```

(a)

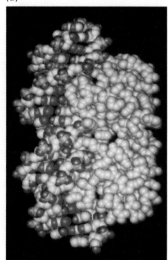

◀ **FIGURE 10-13 Models of bacteriophage 434 repressor and its interaction with DNA.** This and many other bacterial repressors are homodimers that insert an α helix (the recognition helix) of each monomer into a major groove of the operator DNA. Bacteriophage repressors bind to operator sequences in the viral genomes, thereby preventing transcription by the host-cell RNA polymerase. (a) Space-filling model of the dimeric 434 repressor *(right)* bound to its specific operator DNA. (b) Ribbon diagram of the upper monomer of the dimeric 434 repressor *(right)* shown interacting with DNA. Three glutamine (Q) residues, one serine (S), and one threonine (T) in the recognition helix (purple) form hydrogen bonds with specific bases in the major groove of operator DNA. (c) Wire diagram showing interactions between DNA and the lower monomer and dimer interface in the 434 repressor *(right)*.
Red dotted lines indicate hydrogen and ionic bonds between the repressor and DNA. Small circles represent bound water molecules. [Parts (a) and (c) adapted from A. K. Aggarwal et al., 1988, *Science* **242**:899. Part (b) adapted from S. C. Harrison, 1991, *Nature* **353**:715.]

(c)

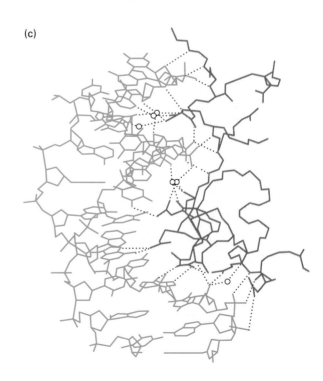

(b) Bacteriophage 434 repressor

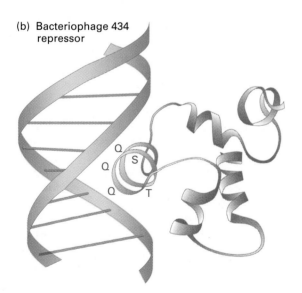

is a tetramer that makes contacts with the bases of its operator half-site through the side chains of two antiparallel β strands that fit into the DNA major groove.

Ligand-Induced Conformational Changes Alter Affinity of Many Repressors for DNA

The DNA-binding affinity of many bacterial repressors, including the *lac* repressor, is altered when the protein binds small molecules. The *allosteric* conformational change responsible for this alteration in affinity has been observed for the *trp* repressor by x-ray crystallography of the protein in the absence and presence of tryptophan.

The *trp* repressor regulates expression of genes required for the synthesis of the amino acid tryptophan (see Figure 9-1a). Whereas *lac* repressor binds operator DNA in the absence of its ligand (lactose), *trp* repressor binds DNA only when it has bound its ligand (tryptophan). In this way *trp*

repressor blocks production of the several enzymes required for tryptophan synthesis when sources of tryptophan in the environment are high. In the absence of tryptophan, the protein is referred to as the *trp aporepressor*. The recognition helices of the *trp* aporepressor are too close together to fit into the major grooves of the *trp*-operator half-sites (Figure 10-14). Binding of tryptophan to the aporepressor causes it to undergo a conformational change that moves these α helices apart, so that they are separated by the precise distance required to fit into the adjacent major grooves of the *trp* operator half-sites.

Positive Control of the *lac* Operon Is Exerted by cAMP-CAP

When glucose enters an *E. coli* cell, it is utilized directly without the induction of any new enzymes. Thus the cell is primed to use this sugar, a central molecule in carbohydrate metabolism, above all others. As *E. coli* becomes starved for glucose, it begins synthesizing an unusual nucleotide named cyclic adenosine-3′,5′-monophosphate, generally referred to as **cyclic AMP**, or cAMP (Figure 10-15). In bacteria, an increase in the cAMP level seems to be an "alert" signal indicating a low glucose level. For instance, when bacteria are grown in a medium containing both lactose and glucose, the intracellular cAMP level initially is low, and the enzymes encoded by the *lac* operon are expressed only at low levels. Once most of the glucose has been metabolized, cAMP levels in the cells rise, and β-galactosidase is induced to high levels. When dibutyryl cyclic AMP, an analog of cAMP that can pass through the cell membrane, is added to media containing lactose and glucose, β-galactosidase synthesis occurs at a high level before all the glucose has been metabolized. Similar results for enzymes required to metabolize other sugars (e.g., galactose, maltose, and arabinose) suggested that cAMP activates transcription of the various operons encoding these enzymes.

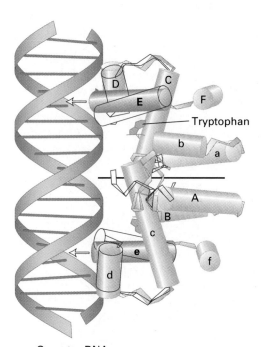

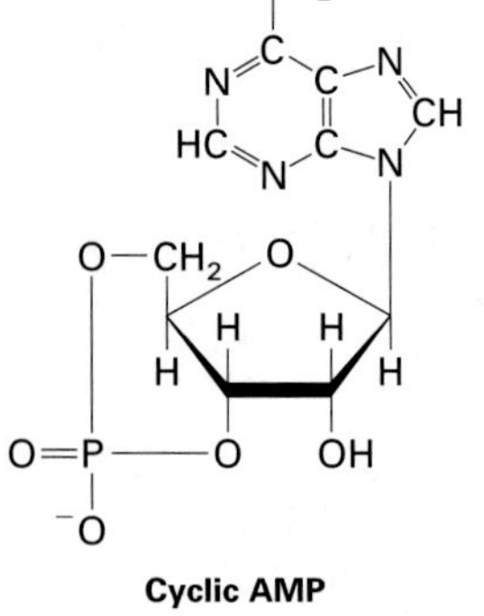

Operator DNA

▲ **FIGURE 10-14 Conformational change in the *trp* aporepressor caused by the binding of tryptophan.** The α helices in the dimeric protein, represented as cylinders, are identified by upper-case letters in one monomer and lower-case letters in the other. The recognition helices (E and e) in the aporepressor (transparent with red holding line) are too close together to fit into adjacent major grooves in the operator DNA. When the aporepressor binds tryptophan, the N-terminal ends of helices E and e move apart by 8 Å. As a result, helices E and e in the repressor with bound tryptophan (orange) fit neatly into the operator DNA. [Adapted from R.-G. Zhang et al., 1987, *Nature* **327**:591.]

Cyclic AMP

▲ **FIGURE 10-15 Structure of cyclic AMP (cAMP).**

(a) Glucose present (cAMP low); no lactose

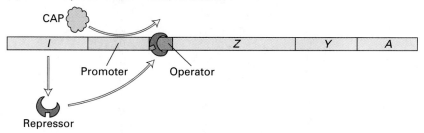

(b) Glucose present (cAMP low); lactose present

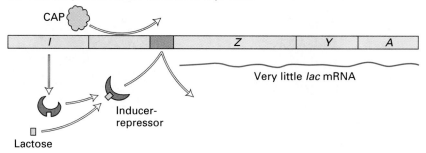

(c) No glucose present (cAMP high); lactose present

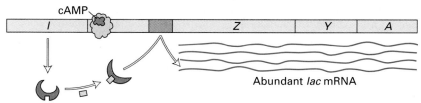

◀ FIGURE 10-16 Negative and positive transcriptional control of the *lac* operon by the *lac* repressor and cAMP-CAP, respectively. (a) In the absence of lactose, no *lac* mRNA is formed because repressor bound to the *lac* operator prevents transcription. (b) In the presence of glucose and lactose, the *lac* repressor binds lactose and undergoes a conformational change, so that it does not bind to the *lac* operator. However, cAMP is low, because glucose is present, and thus cAMP-CAP does not bind to the CAP site in the operator. As a result, RNA polymerase does not bind efficiently to the *lac* promoter and only a little *lac* mRNA is synthesized. (c) In the presence of lactose and the absence of glucose, maximal transcription of the *lac* operon occurs. In this situation, the *lac* repressor does not bind to the *lac* operator, the concentration of cAMP increases, and the cAMP-CAP complex that forms binds at the CAP site, stimulating binding and initiation by RNA polymerase.

Initial understanding of these effects of cAMP on transcription came from study of mutants that had simultaneously lost the ability to metabolize the alternative sugars lactose, galactose, maltose, and arabinose. These mutants fell into two classes. One class did not make cAMP in the absence of glucose; these were defective in the enzyme **adenylate cyclase,** which converts ATP into cAMP. Although mutants in the other class could make cAMP, their synthesis of the enzymes for metabolizing any of the alternative sugars did not increase greatly in the presence of those sugars. This finding suggested that the second group of mutants contained a defective protein that could not respond to cAMP. Subsequently, a protein that could bind to cAMP was isolated from normal cells and was found to be absent in the second group of mutants. This protein now is generally called *catabolite activator protein* (CAP), but sometimes is referred to as cAMP-receptor protein (CRP). In in vitro transcription experiments with cloned DNA carrying the *lac* operon and protein extracts prepared from *E. coli* cells, the rate of synthesis of *lac* mRNA in the absence of either cAMP or CAP was only about 5 percent of the rate in the presence of both. Thus, maximum tran-

scription from the *lac* operon requires the presence of the cAMP-CAP complex.

The DNase I footprinting experiments described earlier revealed that purified cAMP-CAP binds to a specific sequence in the *lac* control region called the CAP site, which lies just upstream from the RNA polymerase–binding site (see Figure 10-9). Mutations in the CAP site that prevent in vitro binding of cAMP-CAP also prevent high-level expression of the *lac* operon in vivo. These findings indicate that the cAMP-CAP complex must bind to this site to stimulate transcription. Thus, bound cAMP-CAP *activates* transcription (positive control), whereas bound *lac* repressor *inhibits* transcription (negative control), as summarized in Figure 10-16.

Cooperative Binding of cAMP-CAP and RNA Polymerase to *lac* Control Region Activates Transcription

We saw earlier that *lac* repressor inhibits transcription of the *lac* operon by blocking access of RNA polymerase to the *lac* promoter. In contrast, cAMP-CAP activates transcription by, in effect, increasing the affinity of the polymerase for the *lac* promoter. By itself RNA polymerase binds fairly weakly to the *lac* promoter, which deviates at several positions from the −10 and −35 consensus sequences (see Figure 10-12). Similarly, cAMP-CAP does not have high affinity for the CAP site in the *lac* control region compared with its affinity for an optimal DNA sequence. However, when cAMP-CAP and RNA polymerase bind to the *lac* control region simultaneously, they stimulate each other's binding (Figure 10-17). This phenomenon, referred to as **cooperativity**, occurs for the same reason that a repressor dimer has higher DNA-binding affinity than a repressor monomer. When cAMP-CAP and RNA polymerase are bound to their adjacent binding sites, they interact with each other and form a cAMP-CAP/RNA polymerase complex, which makes more contacts with the DNA than either cAMP-CAP or RNA polymerase alone.

Strong evidence that direct contact between cAMP-CAP and RNA polymerase is required for activation came from *E. coli* strains producing a mutant CAP. Like most bacterial repressors, CAP is a dimer containing two helices that insert into adjacent major grooves of an approximate inverted repeat sequence. Binding of cAMP-CAP bends the flexible DNA molecule around itself, so that the DNA is curved through an angle of approximately 90 degrees (Figure 10-18). CAP proteins from certain mutant strains form a cAMP-CAP complex that binds and bends DNA to the same extent as wild-type cAMP-CAP but fails to activate *lac* transcription. The amino acid residues mutated in these defective CAP proteins are thought to be the ones that contact the surface of an RNA polymerase. When these residues are changed, the protein-protein interaction between cAMP-CAP and RNA

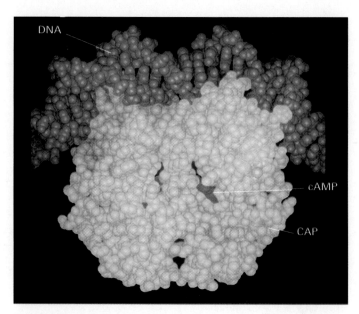

▲ **FIGURE 10-18 Space-filling model of cAMP-CAP bound to *lac* promoter DNA.** CAP bends the DNA double helix around its surface. When the amino acid residues shown in yellow are mutated, cAMP-CAP no longer activates transcription from the *lac* promoter. The DNA sequence near the right edge of the figure corresponds to the region (near −50) where the C-terminus of an α subunit of RNA polymerase is thought to interact with the yellow residues in CAP (see Figure 10-17). [See S. Schultz et al., 1991, *Science* **253**:1001; Y. Zhou et al., 1993, *Proc. Nat'l. Acad. Sci. USA* **90**:6081. Photograph courtesy of R. H. Ebright.]

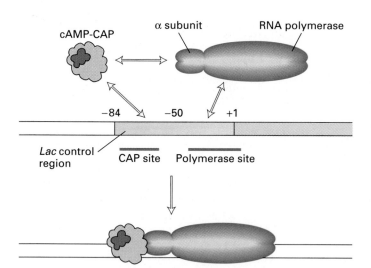

▲ **FIGURE 10-17 Cooperative binding of *E. coli* RNA polymerase and cAMP-CAP to the *lac* promoter.** By themselves, both RNA polymerase and cAMP-CAP have relatively low affinity for their respective binding sites in *lac* promoter DNA. However, interaction between residues in CAP and an α subunit of RNA polymerase forms a protein-protein complex that binds much more stably to promoter DNA than either protein alone.

polymerase no longer occurs, and neither does activation.

Experiments with a mutant form of RNA polymerase that has a deletion of the α-subunit C-termini showed that it does not respond to cAMP-CAP activation of the *lac* promoter. This finding suggests that CAP binds to the C-terminus of one of the polymerase α subunits (see Figure 10-17). The resulting stabilization of polymerase binding to the *lac* promoter is thought to activate *lac* transcription. The interaction between cAMP-CAP and RNA polymerase also may induce conformational changes in the polymerase that stimulate formation of the open complex and other steps in transcription initiation subsequent to polymerase binding.

Transcription Control at All Bacterial Promoters Involves Similar but Distinct Mechanisms

Although the mechanisms of transcriptional control revealed in studies with the *lac* operon and its promoter generally apply to most *E. coli* operons and promoters, the regulation of transcription from each promoter has its unique aspects. For example, cAMP-CAP activates transcription from both the *lac* operon and *gal* operon, which encodes enzymes required for metabolism of galactose. However, to activate *gal*-operon transcription, cAMP-CAP must interact with a

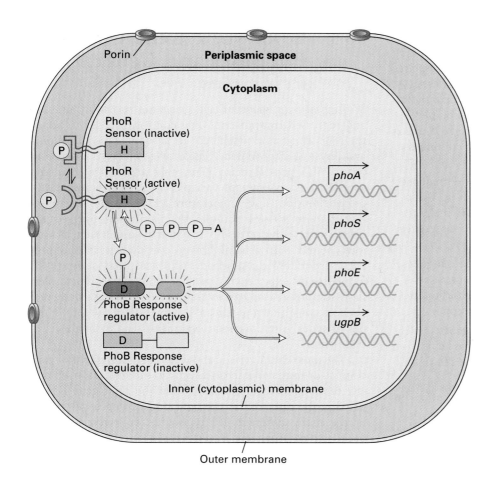

◀ **FIGURE 10-21 The PhoR/PhoB two-component regulatory system in *E. coli*.** In response to low phosphate concentrations in the environment and periplasmic space, a phosphate ion dissociates from the periplasmic domain of the sensor protein PhoR (orange). This causes a conformational change that activates a protein kinase transmitter domain in the cytosolic region of PhoR. The activated transmitter domain transfers an ATP γ-phosphate to a conserved histidine in the transmitter domain. This phosphate is then transferred to an aspartic acid in the receiver domain of the response regulator PhoB (purple). Several PhoB proteins can be phosphorylated by one activated PhoR. Phosphorylated PhoB proteins then activate transcription from genes encoding proteins that help the cell to respond to low phosphate, including *phoA*, *phoS*, *phoE*, and *ugpB*.

transmembrane protein, located in the inner (plasma) membrane, whose periplasmic domain binds phosphate with moderate affinity and whose cytosolic domain has protein kinase activity; PhoB is a cytosolic protein (Figure 10-21). Large protein pores in the *E. coli* outer membrane allow ions to diffuse freely between the external environment and the periplasmic space. Consequently, when the phosphate concentration in the environment falls, it also falls in the periplasmic space, causing phosphate to dissociate from the PhoR periplasmic domain. This causes a conformational change in the PhoR cytoplasmic domain, that activates its protein kinase activity. The activated PhoR initially transfers a γ-phosphate from ATP to a histidine side chain in the PhoR kinase domain itself. The same phosphate is then transferred to a specific aspartic acid side chain in PhoB, converting PhoB from an inactive to an active transcriptional activator. Phosphorylated, active PhoB then induces transcription from several genes that help the cell cope with low phosphate conditions.

Many other bacterial responses are regulated by two proteins with homology to PhoR and PhoB. In each of these regulatory systems, one protein, called a *sensor*, contains a *transmitter* domain homologous to the PhoR protein kinase domain. The transmitter domain of the sensor protein is regulated by a second unique protein domain (e.g., the peri-

plasmic domain of PhoR) that senses environmental changes. The second protein, called a *response regulator*, contains a *receiver* domain homologous to the region of PhoB that is phosphorylated by activated PhoR. The receiver domain of the response regulator is associated with a second domain that determines the protein's function. The activity of this second functional domain is regulated by phosphorylation of the receiver domain. Although all transmitter domains are homologous (as are receiver domains), the transmitter domain of a specific sensor protein will phosphorylate only specific receiver domains of specific response regulators, allowing specific responses to different environmental changes. Note that NtrB and NtrC, discussed above, function as sensor and response regulator proteins, respectively, in the two-component regulatory system that controls transcription of *glnA* (see Figure 10-19).

SUMMARY Bacterial Transcription Initiation

• Bacterial RNA polymerases are large proteins composed of β′, β, and two α subunits, forming the core polymerase, and one of a few alternative σ subunits, which function as initiation factors.

- Initiation begins when the σ subunit of a polymerase molecule binds to a promoter DNA sequence, forming a closed complex in which the DNA strands are base paired (see Figure 10-8). The polymerase then separates the strands over a distance of 12–13 base pairs surrounding the transcription-start site, forming an open complex. After ≈ 10 ribonucleotides have been polymerized, the σ subunit is released and the core polymerase continues transcription of the template.

- The "strength" of a promoter refers to how frequently RNA polymerase initiates transcription from it. The σ^{70} subunit, the major initiation factor in E. coli, interacts with promoter sequences near -10 and -35 base pairs from the start site (see Figure 10-11).

- Repressors bind to DNA sequences called operators, which overlap the promoter region that contacts RNA polymerase. A bound repressor interferes with binding of RNA polymerase and transcription initiation.

- Activators of σ^{70}-containing RNA polymerase generally bind to DNA on the opposite side of the helix from the polymerase, in the region from -20 to -50, or just upstream of the polymerase near -60.

- The activator cAMP-CAP stimulates transcription by forming a complex with RNA polymerase that has a greater affinity for specific DNA sites than the individual proteins (see Figure 10-17). In addition to enhancing polymerase binding, activators may also stimulate open-complex formation and transcription away from the promoter region.

- Many bacterial repressors and activators are dimers. Each monomer contains a surface α helix that inserts into a major groove of operator DNA, so that the dimer binds to two successive major grooves (see Figure 10-13). High-affinity binding results from multiple hydrogen and ionic bonds and van der Waals interactions between the protein and specific DNA sequences.

- DNA sequences that bind dimeric regulatory proteins are approximate inverted repeats, with each half-site binding one monomer.

- Operons transcribed by σ^{54}-containing RNA polymerase are regulated by activators that bind to enhancer sequences ≈ 100 base pairs upstream from the initiation site. Enhancer-bound activators interact transiently with the poised polymerase bound to the promoter, stimulating formation of an open complex and initiation by the bound polymerase (see Figure 10-19).

- In two-component regulatory systems, one protein acts as a sensor, monitoring the level of nutrients in the environment. Under appropriate conditions, the sensor protein activates a second protein, the response regulator, which then binds to DNA regulatory sequences, thereby stimulating or repressing transcription of specific genes (see Figure 10-21).

10.3 Eukaryotic Gene Control: Purposes and General Principles

Within a *single* bacterial cell, genes are reversibly induced and repressed by transcriptional control in order to adjust the cell's enzymatic machinery to its immediate nutritional and physical environment. Single-celled eukaryotes, such as yeasts, also possess many genes that are controlled in response to environmental variables (e.g., nutritional status, oxygen tension, and temperature). Even in the organs of higher animals —for example, the mammalian liver—some genes can respond reversibly to external stimuli such as noxious chemicals. In general, however, metazoan cells are protected from immediate outside influences; that is, most cells in metazoans experience a fairly constant environment. Perhaps for this reason, genes that respond to environmental changes constitute a much smaller fraction of the total number of genes in multicellular organisms than in single-celled organisms.

The most characteristic and exacting requirement of gene control in multicellular organisms is the execution of precise developmental decisions so that the right gene is activated in the right cell at the right time during development of the many different cell types that collectively form a multicellular organism. In most cases, once a developmental step has been taken by a cell, it is not reversed. Thus these decisions are fundamentally different from bacterial induction and repression. In executing their genetic programs, many differentiated cells (e.g., skin cells, red blood cells, lens cells of the eye, and antibody-producing cells) march down a pathway to final cell death, leaving no progeny behind. The fixed patterns of gene control leading to differentiation serve the needs of the whole organism and not the survival of an individual cell.

Most Genes in Higher Eukaryotes Are Regulated by Controlling Their Transcription

Direct measurements of the transcription rates of multiple genes in different cell types have shown that regulation of transcription initiation is the most widespread form of gene control in eukaryotes, as it is in bacteria. The simplest and most direct method of measuring transcription rates is to expose cells for a brief time (e.g., 5 minutes or less) to a labeled RNA precursor and then determine the amount of labeled nuclear RNA formed by its hybridization to a cloned DNA. This method can be used to determine transcription rates in cultured cells but is not practical in whole animals because the amount of labeled RNA obtained is insufficient for accurate measurements. Even with cultured cells, a second method—called *nascent-chain analysis* (or "run-on" analysis)—often is preferred. In this method, nuclei are isolated from cells and allowed to incorporate ^{32}P from

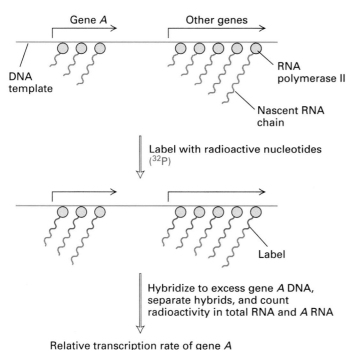

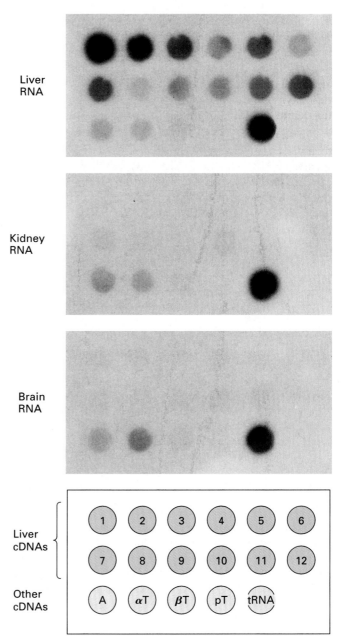

labeled ribonucleoside triphosphates directly into nascent (growing) RNA chains to produce highly labeled RNA preparations (Figure 10-22). Relative transcription rates determined either by labeling of cultured cells or of isolated nuclei are nearly the same, indicating that the polymerase molecules that are active at the time nuclei are taken from intact cells continue to function during incubation of isolated nuclei.

The results of run-on analyses illustrated in Figure 10-23 show that transcription of many genes encoding proteins expressed specifically in hepatocytes (the major cell type in mammalian liver) is readily detected in nuclei prepared from liver, but not in nuclei from brain or kidney. Since the run-on assay measures RNA synthesis, these results indicate that

▲ FIGURE 10-22 Nascent-chain (run-on) assay for transcription rate of a gene. Isolated nuclei are incubated with ^{32}P-labeled ribonucleoside triphosphates for a brief period. During this period RNA polymerase molecules that were transcribing a gene when the nuclei were isolated add 300–500 nucleotides to nascent RNA chains. Very little new initiation occurs. By hybridizing the labeled RNA to the cloned DNA for a specific gene (A in this case), the fraction of total RNA produced from that gene (i.e., its relative transcription rate) can be measured. [See J. Weber et al., 1977, *Cell* **10**:611.]

▶ FIGURE 10-23 Experimental demonstration of differential synthesis of 12 mRNAs encoding liver-specific proteins. Nuclei from mouse liver, kidney, and brain cells were exposed to ^{32}P-UTP, and the resulting labeled RNA was hybridized to various cDNAs fixed to nitrocellulose. After removal of unhybridized RNAs, the hybrids were revealed by autoradiography. The cDNAs labeled 1–12 encode proteins synthesized actively in liver (e.g., 4 = albumin; 3 = α_1-antitrypsin; 6 = transferrin) but not in most other tissues. The other cDNAs tested were actin (A) and α- and β-tubulin (αT, βT), which are proteins found in almost all cell types. Methionine tRNA and the plasmid DNA (pB) in which the cDNAs were cloned were included as controls. The intensity of spots generated by hybridization of RNA synthesized during in vitro run-on transcription in nuclei isolated from the three tissues indicates that genes expressed specifically in hepatocytes are transcribed in liver and are not transcribed in the cells of other tissues. [See E. Derman et al., 1981, *Cell* **23**:731; D. J. Powell et al., 1984, *J. Mol. Biol.* **197**:21.]

differential synthesis of liver-specific proteins is regulated by controlling transcription of the corresponding genes in different tissues. Similar results have been obtained in run-on experiments with other cell types and a wide variety of tissue-specific proteins, indicating that transcriptional control is common in complex organisms.

Regulatory Elements in Eukaryotic DNA Often Are Many Kilobases from Start Sites

The basic principles shown to control transcription in bacteria apply to eukaryotic organisms as well. In most cases transcription is initiated at a specific base pair in the template DNA or at alternative sites within a few base pairs. Transcription is controlled by trans-acting proteins, **transcription factors,** binding at cis-acting regulatory DNA sequences. These proteins are equivalent to the repressors and activators that control transcription of bacterial operons. However, transcriptional control in eukaryotes involves considerably more complex processes than in prokaryotes. In particular, cis-acting DNA control elements in eukaryotic genomes often are located much farther from the promoter they regulate than is the case in prokaryotes. In some cases, transcription factors that regulate expression of protein-coding genes in higher eukaryotes bind at regulatory sites tens of thousands of base pairs either upstream or downstream from the promoter. As a result of this arrangement, transcription from a single promoter may be regulated by binding of multiple transcription factors to alternative control elements, permitting complex control of gene expression.

For example, alternative transcription-control elements regulate expression of the mammalian gene that encodes transthyretin (TTR), which transports thyroid hormone in blood, the cerebrospinal fluid, and extracellular fluid. Transthyretin is expressed in hepatocytes, which synthesize and

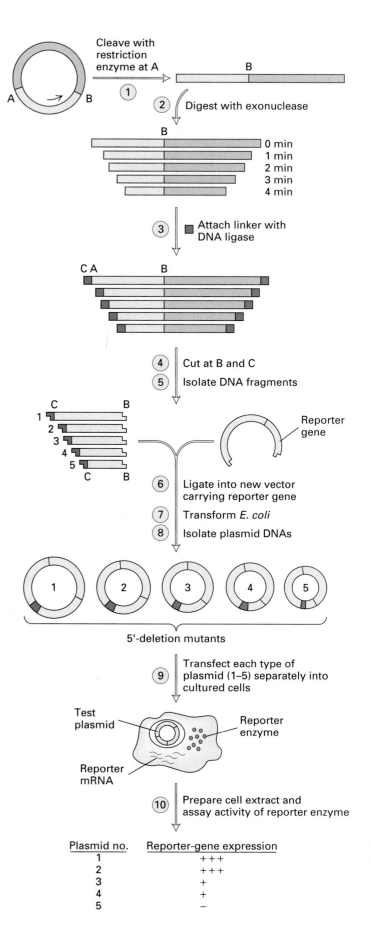

▶ FIGURE 10-24 Construction and analysis of a 5′-deletion series to locate transcription-control sequences in DNA upstream of a eukaryotic gene. A DNA fragment (yellow) containing a transcription-start site is cloned into a plasmid vector (upper left). The plasmid is linearized by digestion with a restriction enzyme (A) that cleaves at the upstream end of the fragment being analyzed. The linearized DNA is then digested with an exonuclease for different periods of time so that increasing lengths of DNA are removed from each end. After addition of a synthetic oligonucleotide linker and digestion with restriction enzymes B and C, the deleted fragments are cloned into a plasmid vector with an easily assayed reporter gene (light blue). Plasmids with deletions of various lengths 5′ to the transcription-start site are then transfected into cultured cells (or used to prepare transgenic organisms) and expression of the reporter gene is assayed. The results of this hypothetical example (bottom) indicate that the test fragment contains two control elements. The 5′ end of one lies between deletions 2 and 3; the 5′ end of the other lies between deletions 4 and 5.

secrete most of the blood serum proteins, and in choroid plexus cells in the brain, which secrete cerebrospinal fluid and its constituent proteins. The control elements required for transcription of the *TTR* gene were identified by the procedure outlined in Figure 10-24. In this experimental approach, DNA fragments with varying extents of sequence upstream of a start site are cloned in front of a *reporter gene* in a bacterial plasmid using recombinant DNA techniques. Reporter genes express enzymes that are easily assayed in cell extracts. Commonly used reporter genes include the *E. coli lacZ* gene encoding β-galactosidase, an *E. coli* gene encoding the enzyme chloramphenicol acetyl transferase (CAT), and the firefly gene encoding luciferase, which converts energy from ATP hydrolysis into light.

By constructing and analyzing a *5'-deletion series* upstream of the *TTR* gene, researchers identified two control elements that stimulate reporter gene expression in hepatocytes, but not in other cell types. One region mapped between ≈1.85 and 2.0 kb upstream of the *TTR* gene start site; the other mapped between the start site and ≈200 base pairs upstream. Further studies demonstrated that alternative DNA sequences control *TTR* transcription in choroid plexus cells. Thus, alternative control elements regulate transcription of the *TTR* gene in two different cell types. We examine the basic molecular events underlying this type of eukaryotic transcriptional control in later sections.

Three Eukaryotic Polymerases Catalyze Formation of Different RNAs

The nuclei of all eukaryotic cells examined so far (e.g., vertebrate, *Drosophila*, yeast, and plant cells) contain three different RNA polymerases, designated I, II, and III. These enzymes initially were recognized as distinct proteins when chromatographic purification of RNA polymerase resulted in three fractions eluting at different salt concentrations during ion-exchange chromatography (Figure 10-25). The eukaryotic RNA polymerases also differ in their sensitivity to α-amanitin, a poisonous cyclic octapeptide produced by some mushrooms. Polymerase I is very insensitive to α-amanitin; polymerase II is very sensitive; and polymerase III has intermediate sensitivity.

Each eukaryotic RNA polymerase catalyzes transcription of genes encoding different classes of RNA. Polymerase I is located in the nucleolus and is responsible for synthesis of precursor rRNA (**pre-rRNA**), which is processed into 28S, 5.8S, and 18S rRNAs (Chapter 11). Polymerase III functions outside the nucleolus and transcribes the genes encoding tRNAs, 5S rRNA, and a whole array of small, stable RNAs. These latter RNAs include one involved in RNA splicing (U6) and the 7S RNA of the signal-recognition particle involved in the transport of proteins into the endoplasmic reticulum (Chapter 17); the functions of many other small RNAs produced by polymerase III are unknown at present. Polymerase II catalyzes transcription of all protein-coding genes; that is, it functions in production of mRNAs. RNA

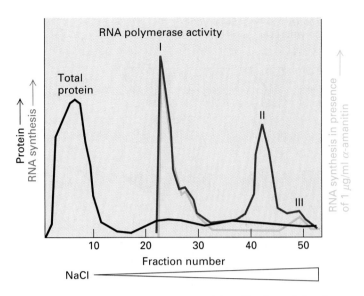

▲ FIGURE 10-25 The separation and identification of the three eukaryotic RNA polymerases by column chromatography. A protein extract from the nuclei of cultured frog cells was passed through a DEAE Sephadex column to which charged proteins absorb differentially. Adsorbed proteins were eluted (black curve) with a solution of constantly increasing NaCl concentration. Fractions containing the eluted proteins were assayed for the ability to transcribe DNA (red curve) in the presence of the four ribonucleoside triphosphates. The synthesis of RNA by each fraction in the presence of 1 μg/ml of α-amanitin also was measured (blue curve). At this concentration, α-amanitin inhibits polymerase II activity but has no effect on polymerases I and III. (Polymerase III is sensitive to 10 μg/ml of α-amanitin, whereas polymerase I is unaffected even at this higher concentration). [See R. G. Roeder, 1974, *J. Biol. Chem.* **249**:241.]

polymerase II also produces four small nuclear RNAs that take part in RNA splicing.

Each of the three eukaryotic RNA polymerases is considerably more complex than *E. coli* RNA polymerase. All three contain two large subunits and 12–15 smaller subunits, some of which are present in two or all three of the polymerases. The best-characterized eukaryotic RNA polymerases are from the yeast *S. cerevisiae*, and the yeast genes encoding the polymerase subunits have been cloned and sequenced. The three nuclear RNA polymerases from all eukaryotes so far examined are very similar to those of yeast; polymerase genes that have been cloned and sequenced from other eukaryotes exhibit considerable homology with the yeast genes.

All three eukaryotic RNA polymerases contain core subunits with some sequence homology to the *E. coli* core polymerase ($\alpha_2\beta\beta'$). The largest subunit (160–220 kDa) and second largest subunit (128–150 kDa) of each eukaryotic polymerase are related though distinct from each other and are similar to the *E. coli* β' and β subunits, respectively (Figure 10-26). Both yeast RNA polymerase I and III contain two subunits (19 and 40 kDa) that have regions of

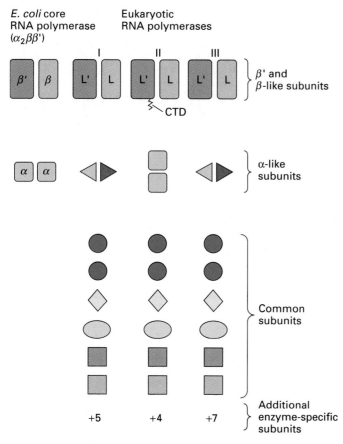

▲ FIGURE 10-26 Schematic representation of the subunit structure of yeast nuclear RNA polymerases and comparison with *E. coli* RNA core polymerase. All three yeast polymerases have four core subunits that exhibit some homology with the β, β', and α subunits in *E. coli* RNA polymerase. The largest subunit (L') of RNA polymerase II also contains an essential C-terminal domain (CTD). RNA polymerases I and III contain the same two nonidentical α-like subunits, whereas polymerase II has two copies of a different α-like subunit. All three polymerases share five other common subunits (two copies of the largest of these). In addition, each yeast polymerase contains four to seven unique smaller subunits.

homology with the *E. coli* α subunit. Yeast RNA polymerase II contains two copies of a different subunit (44 kDa) that exhibits a somewhat more distant sequence similarity to the *E. coli* α subunit. This extensive homology in the amino acid sequences of the core subunits in RNA polymerases from various sources indicates that this enzyme arose early in evolution and was largely conserved. This seems logical for an enzyme catalyzing a process so basic as copying RNA from DNA.

In addition to their core subunits related to the *E. coli* polymerase subunits, all three yeast RNA polymerases contain five small common subunits (10–27 kDa); two copies of one of these and a single copy of each of the others is present in each enzyme molecule. Finally, each eukaryotic

polymerase has four to seven enzyme-specific subunits that are not present in the other two polymerases. The functions of these multiple polymerase subunits are not understood, but gene-knockout experiments in yeast indicate that most of them are essential for cell viability. Disruption of the few polymerase subunit genes that are not absolutely essential for viability results in very poorly growing cells. Thus, it seems likely that every one of the multiple subunits must be present for eukaryotic RNA polymerases to function normally.

The Largest Subunit in RNA Polymerase II Has an Essential Carboxyl-Terminal Repeat

The carboxyl end of the largest subunit of RNA polymerase II (L') contains a stretch of seven amino acids that is nearly precisely repeated multiple times. Neither polymerase I nor III contains these repeating units. This heptapeptide repeat, with a consensus sequence of Tyr-Ser-Pro-Thr-Ser-Pro-Ser, is known as the *carboxyl-terminal domain* (CTD). Yeast RNA polymerase II contains 26 or more repeats; the mammalian enzyme has 52 repeats; and an intermediate number of repeats occurs in RNA polymerase II from nearly all other eukaryotes. The CTD is critical for viability, and at least 10 copies of the repeat must be present for yeast to survive.

In cells, many of the serine and some tyrosine residues in the CTD are phosphorylated. In vitro experiments with model promoters show that RNA polymerase II with an unphosphorylated CTD is used to initiate transcription, but that the CTD becomes phosphorylated as the polymerase transcribes away from the promoter. Analysis of polytene chromosomes from *Drosophila* salivary glands prepared just before molting of the larva indicate that the CTD also is phosphorylated during in vivo transcription. The large chromosomal "puffs" induced at this time in development are regions where the genome is very actively transcribed. Staining with antibodies specific for the phosphorylated or unphosphorylated CTD demonstrated that the highly transcribed puffed regions contain RNA polymerase II with a phosphorylated CTD (Figure 10-27).

RNA Polymerase II Initiates Transcription at DNA Sequences Corresponding to the 5' Cap of mRNAs

As discussed earlier, bacterial RNA polymerases generally initiate transcription at a specific base within promoters lying just upstream of coding regions in the genome. Moreover, binding of repressors and activators to nearby sites are largely responsible for the transcriptional control of bacterial operons. Consequently, once eukaryotic genes began to be cloned and sequenced, determining the site of transcription initiation was a prerequisite for studying the mechanisms of transcriptional control in eukaryotes.

As we saw earlier, if cultured cells are exposed to a radioactive label for very brief times, only growing (nascent) RNA chains are labeled (see Figure 10-22). Analysis of these

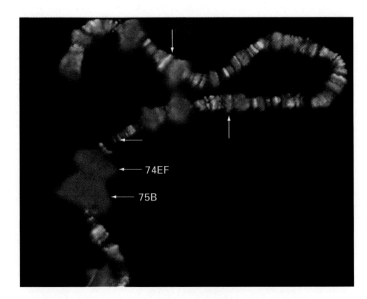

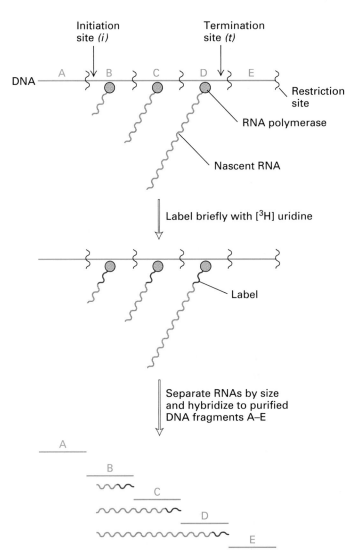

▲ **FIGURE 10-27 Experimental demonstration that carboxyl-terminal domain (CTD) of RNA polymerase II is phosphorylated during in vivo transcription.** Salivary gland polytene chromosomes were prepared from *Drosophila* larvae just before molting. The preparation was treated with a rabbit antibody specific for phosphorylated CTD and with a goat antibody specific for unphosphorylated CTD. The preparation then was stained with fluorescein-labeled anti-goat antibody (green) and rhodamine-labeled anti-rabbit antibody (red). Thus polymerase molecules with an unphosphorylated CTD stain green and those with a phosphorylated CTD stain red. The molting hormone ecdysone induces very high rates of transcription in the puffed regions labeled 74EF and 75B; note that only phosphorylated CTD is present in these regions. Smaller puffed regions transcribed at high rates also are visible. Nonpuffed sites that stain red (up arrow) or green (horizontal arrow) also are indicated, as is a site staining both red and green, producing a yellow color (down arrow). [From J. R. Weeks, 1993, *Genes & Dev.* **7**:2329; courtesy of J. R. Weeks and A. L. Greenleaf.]

▲ **FIGURE 10-28 Approximate mapping of transcription-initiation site by analysis of nascent transcripts synthesized in vivo.** A section of DNA in the act of being transcribed is depicted at the top. After cells are labeled for a brief time, producing labeled nascent RNAs, the cells are disrupted and the isolated RNA is separated on the basis of chain length by velocity sedimentation centrifugation. Each size fraction is then hybridized to restriction fragments A–E shown in the top diagram. The shortest labeled RNA transcripts hybridize to the restriction fragment (B) that contains the initiation site; the exact nucleotide at which transcription starts cannot be determined. Successively longer RNAs hybridize to fragments downstream from the initiation site. The longest labeled RNAs hybridize to the fragment (D) containing the termination site.

labeled nascent transcripts permits approximate mapping of the start site, as outlined in Figure 10-28. One of the first transcription units to be analyzed in this way was the major late transcription unit of adenovirus. This analysis indicated that transcription was initiated at a single start site ≈6 kb from the left end of the viral genome.

The precise base pair where transcription initiates in the adenovirus late transcription unit was determined by analyzing the RNAs synthesized during in vitro transcription of adenovirus restriction fragments that extended somewhat upstream and downstream of the approximate initiation region determined by nascent-transcript analysis. The rationale of this experiment and typical results are illustrated in Figure 10-29. All of the RNA transcripts synthesized in vitro contained a RNA cap structure identical to that present at the 5' end of nearly all eukaryotic mRNAs. This 5' cap was added by enzymes in the nuclear extract, which can add a cap

only to an RNA that has a 5' tri- or diphosphate (see Figure 11-8). Because a 5' end generated by cleavage of a longer RNA would have a 5' monophosphate, it could not be capped. Thus the nucleotide capped in the in vitro RNA transcripts corresponds to the base pair at which transcription is

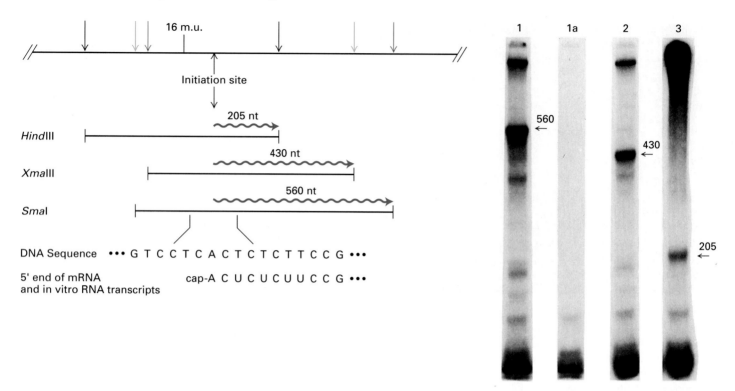

▲ **FIGURE 10-29 Precise mapping of initiation site of adenovirus late transcription unit by in vitro transcription.** *(Left)* The top line shows restriction sites for *Hind*III (black), *Xma*III (blue), and *Sma*I (red) in the region of the adenovirus genome where the transcription-initiation site was located by nascent-transcript analysis (near 16 map units). The *Hind*III, *Xma*III, and *Sma*I restriction fragments that encompass the initiation site were individually incubated with a nuclear extract prepared from HeLa cells and α-[^{32}P]-labeled ribonucleoside triphosphates. Transcription of each fragment begins at the start site; when an RNA polymerase II molecule transcribing a fragment reaches a cut end, it "runs off" the template. *(Right)* The resulting run-off transcripts were then subjected to gel electrophoresis and autoradiography. Since the positions of the restriction sites in the DNA template were known, the lengths of the run-off transcripts

(in nucleotides, nt) produced from the *Sma*I (lane 1), *Xma*III (lane 2), and *Hind*III (lane 3) fragments precisely map the start site at 16.4 map units on the adenovirus genome. The sequences of this region of adenovirus DNA and of the capped 5′ end of the corresponding mRNAs and RNA transcripts produced in vitro are shown in the lower left. Thus the starting point for in vitro transcription by RNA polymerase II corresponds to the cap site in mRNA. The sample in lane 1a is the same as that in lane 1, except that α-amanitin, an inhibitor of RNA polymerase II, was included in the transcription mixture. The bands at the top and bottom of the gel represent high- and low-molecular-weight RNA transcripts that are formed under the conditions of this experiment. [See R. M. Evans and E. Ziff, 1978, *Cell* **15**:1463; P. A. Weil et al., 1979, *Cell* **18**:469. Photograph courtesy of R. G. Roeder.]

initiated by RNA polymerase II. The sequence at the 5′ end of the RNA transcripts produced in vitro was found to be the same as that at the 5′ end of late adenovirus mRNAs isolated from cells (see Figure 10-29, *lower left*). This finding confirmed that the capped nucleotide of adenovirus late mRNAs coincides with the transcription-initiation site.

Similar in vitro transcription assays with other cloned eukaryotic genes have produced similar results. In each case, the start site was found to be equivalent to the 5′ sequence of the corresponding mRNA. Thus, synthesis of eukaryotic mRNAs by RNA polymerase II begins at the DNA sequence encoding the 5′ end of the mRNA. Today, the start site for synthesis of a newly characterized mRNA generally is determined simply by identifying the DNA sequence corresponding to the 5′ end of the mRNA. This is most often done with the primer-extension and nuclease-protection assays (see Figure 7-35).

SUMMARY **Eukaryotic Gene Control: Purposes and General Principles**

• The primary purpose of gene control in multicellular organisms is the execution of precise developmental decisions so that the proper genes are expressed in the proper cells during development and cellular differentiation.

• Transcriptional control is the primary means of regulating gene expression in eukaryotes, as it is in bacteria.

• In eukaryotic genomes, cis-acting control elements that regulate transcription from a promoter often are located many kilobases away from the start site. In contrast, bacterial control elements generally lie within 60 base pairs of the promoters they regulate.

- Eukaryotes contain three types of nuclear RNA polymerases. All three contain two large and two smaller core subunits with homology to the β', β, and α subunits of *E. coli* RNA polymerase, as well several additional small subunits. Some of these small subunits are shared; others are unique to each polymerase (Figure 10-26).

- RNA polymerase I synthesizes only pre-rRNA. RNA polymerase II synthesizes mRNAs and some of the small nuclear RNAs that participate in mRNA splicing. RNA polymerase III synthesizes tRNAs, 5S rRNA, and several relatively short, stable RNAs.

- A repeated heptapeptide sequence, the carboxy-terminal domain (CTD), in the largest subunit of RNA polymerase II becomes phosphorylated during transcription initiation and remains phosphorylated as the enzyme transcribes the template.

- Similar to bacterial RNA polymerase, RNA polymerase II usually initiates transcription of genes at a specific base pair or alternative neighboring base pairs in a DNA template. The 5′ nucleotide that is capped in an mRNA corresponds to the nucleotide in the template strand at which transcription is initiated.

10.4 Regulatory Sequences in Eukaryotic Protein-Coding Genes

Once transcription start sites in eukaryotic DNA had been identified, analysis of the DNA sequences controlling initiation of transcription could begin. In this section, we take a closer look at various elements in the **transcription-control regions** that regulate transcription of eukaryotic protein-coding genes.

TATA Box, Initiators, and CpG Islands Function as Promoters in Eukaryotic DNA

The first eukaryotic genes to be sequenced and studied in in vitro transcription systems were viral genes and cellular protein-coding genes that are very actively transcribed either at particular times of the cell cycle or in specific differentiated cell types. In all of these rapidly transcribed genes, a highly conserved sequence called the **TATA box** was found ≈ 25–35 base pairs upstream of the start site (Figure 10-30). Mutagenesis studies have shown that a single-base change in this nucleotide sequence drastically decreases in vitro transcription by RNA polymerase II of genes adjacent to a TATA box. In most cases, sequence changes between the TATA box and start site do not significantly affect the transcription rate. If the base pairs between the TATA box and the normal start site are deleted, transcription of the altered, shortened template begins at a new site ≈ 25 base pairs downstream from the TATA box. Consequently, the TATA box acts similarly to an *E. coli* promoter to position RNA polymerase II for transcription initiation.

Instead of a TATA box, some eukaryotic genes contain an alternative promoter element called an **initiator**. Most naturally occurring initiator elements have a cytosine (C) at the -1 position and an adenine (A) residue at the transcription-start site ($+1$). Directed mutagenesis of mammalian genes with an initiator-containing promoter has revealed that the nucleotide sequence immediately surrounding the start site determines the strength of such promoters. Unlike the highly conserved TATA box sequence, however, only an extremely degenerate initiator consensus sequence has been defined:

$$(5')\text{Y-Y-A}^{+1}\text{-N-T/A-Y-Y-Y} (3')$$

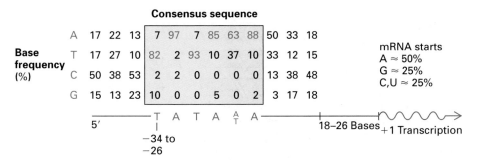

Base frequency (%)		Consensus sequence										
A	17	22	13	7	97	7	85	63	88	50	33	18
T	17	27	10	82	2	93	10	37	10	33	12	15
C	50	38	53	2	2	0	0	0	0	13	38	48
G	15	13	23	10	0	0	5	0	2	3	17	18

mRNA starts
A ≈ 50%
G ≈ 25%
C,U ≈ 25%

5′ —— T A T A A(A/T) A —— 18–26 Bases +1 Transcription
−34 to −26

▲ **FIGURE 10-30 Comparison of nucleotide sequences upstream of the start site in 60 different vertebrate protein-coding genes.** Each sequence was aligned to maximize homology in the region from −35 to −20. The tabulated numbers are the percentage frequency of each base at each position. Maximum homology occurs over a six-base region, referred to as the TATA box, whose consensus sequence is shown at the bottom. The initial base in mRNAs encoded by genes containing a TATA box most frequently is an A. [See R. Breathnach and P. Chambon, 1981, *Ann. Rev. Biochem.* **50**:349; P. Bucher, 1990, *J. Mol. Biol.* **212**:563.]

where A^{+1} is the base at which transcription starts, Y is either pyrimidine (C or T), N is any of the four bases, and T/A is T or A at position +3.

Transcription of genes with promoters containing a TATA box or initiator element begins at a well-defined initiation site. However, transcription of many protein-coding genes has been shown to begin at any one of multiple possible sites over an extended region, often 20–200 base pairs in length. As a result, such genes give rise to mRNAs with multiple alternative 5′ ends. These genes, which generally are transcribed at low rates (e.g., genes encoding the enzymes of intermediary metabolism, often called "housekeeping genes"), do not contain a TATA box or an initiator. Most genes of this type contain a CG-rich stretch of 20–50 nucleotides within ≈100 base pairs upstream of the start-site region. As we discuss later, a transcription factor called SP1 recognizes these CG-rich sequences. The dinucleotide CG is statistically underrepresented in vertebrate DNAs, and the presence of CG-rich regions just upstream from start sites is a distinctly nonrandom distribution. Such *CpG islands,* as they often are called, can be identified by their susceptibility to restriction enzymes (e.g., *Hpa*II) that have CG in their recognition sequences. The presence of a CpG island in a newly cloned DNA fragment suggests that it may contain a transcription-initiation region.

Promoter-Proximal Elements Help Regulate Eukaryotic Genes

Recombinant DNA techniques have been used to systematically mutate the nucleotide sequences upstream of the start sites of various eukaryotic genes in order to identify transcription-control regions. By now, hundreds of eukaryotic genes have been analyzed and scores of transcription-control regions have been identified. These control elements, together with the TATA-box and/or initiator often are referred to as the promoter of the gene they regulate. However, we prefer to reserve the term *promoter* for the TATA-box or initiator sequences that determine the initiation site in the template. We use the term **promoter-proximal elements** for control regions lying within 100–200 base pairs upstream of the start site. In some cases, promoter-proximal elements are cell-type specific; that is, they function only in specific differentiated cell types such as B-lymphocytes.

One approach frequently taken to determine the upstream border of a transcription-control region for a mammalian gene involves constructing a set of 5′ deletions as discussed earlier (see Figure 10-24). Once the 5′ border of a transcription-control region is determined, analysis of *linker scanning mutations* can pinpoint the sequences with regulatory functions that lie between the border and the transcription-start site. In this approach, a systemic set of constructs with contiguous overlapping mutations are assayed for their effect on expression of a reporter gene or production of a specific mRNA (Figure 10-31). One of the first uses of this type of analysis identified promoter-proximal elements of the thymidine kinase (*tk*) gene from herpes simplex virus (HSV), as illustrated in Figure 10-32. The results demonstrated that the DNA region upstream of the HSV *tk* gene contains three separate transcription-control sequences: a TATA box in the interval from −32 to −16, and two other control elements further upstream (from −105 to −80 and from −61 to −47).

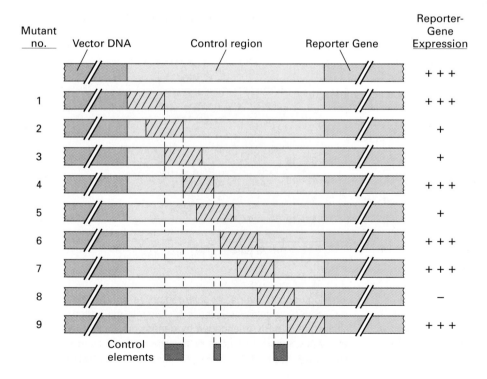

◀ **FIGURE 10-31 Analysis of linker scanning mutations to identify transcription-control elements.** A region of eukaryotic DNA (yellow) that supports high level expression of a reporter gene (light blue) is cloned in a plasmid vector as diagrammed at the top. Overlapping linker scanning (LS) mutations (crosshatch) are introduced from one end of the region being analyzed to the other. These mutations result from scrambling the nucleotide sequence in a short stretch of the DNA. The mutant plasmids are transfected separately into cultured cells, and the activity of the reporter-gene product is assayed. In the hypothetical example shown here, LS mutations 1, 4, 6, 7, and 9 have little or no effect on expression of the reporter gene, indicating that the regions altered in these mutants contain no control elements. Reporter-gene expression is significantly reduced in mutants 2, 3, 5, and 8, indicating that control elements (brown) lie in the intervals shown at the bottom.

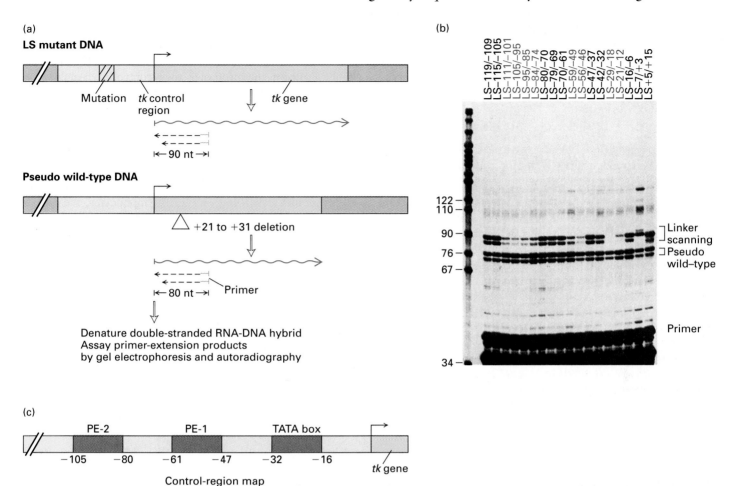

▲ FIGURE 10-32 Identification of promoter-proximal elements controlling the thymidine kinase *(tk)* gene of herpes simplex virus (HSV) by analysis of linker scanning (LS) mutations.
(a) A series of DNA constructs were prepared in which ≈10-bp regions upstream of the *tk* start site were mutated by replacement with synthetic linker DNA containing a restriction site. Each construct was coinjected with a pseudo wild-type *tk* gene into *Xenopus laevis* oocytes in which the HSV *tk* gene is transcribed at a high rate. After 24 hours, RNA was isolated and assayed by the primer-extension method using a labeled RNA primer (light red) complementary to a short section of *tk* mRNA (see Figure 7-36c). The pseudo wild-type gene, which has a 10-bp deletion mutation, served as an internal control on the transcriptional activity of injected oocytes and the recovery of RNA. (b) The primer-extension products were analyzed by gel electrophoresis and autoradiography. Assay of the RNA produced from the LS mutants yielded two main labeled products: one contained 90 nucleotides (nt) corresponding to extension all the way to the 5' end of the RNA; the other was slightly shorter due to incomplete extension. Assay of the RNA transcribed from the pseudo wild-type gene likewise yielded two products: one 80 nucleotides long, and the other slightly shorter. The labeled bands from each linker scanning mutant are compared with those from the pseudo wild-type gene in the same injected oocytes. (Weak background bands seen in multiple lanes are ignored.) The LS mutants are named by the upstream interval in which the wild-type *tk* sequence is mutated. Note the decreased density of bands, indicating decreased *tk* transcription, for the LS mutant DNAs labeled in red type. (c) Based on these results, promoter-proximal elements (brown) in the control region of the *tk* gene were mapped. [Photograph from S. L. McKnight and R. Kingsbury, 1982, *Science* **217**:316.]

As discussed earlier, repressor-binding and activator-binding sites are located within about 50 base pairs of the −10 and −35 promoter sequences in *E. coli* operons transcribed by σ^{70}-polymerase. In the *lac* control region, insertion of five base pairs between the CAP site and the promoter eliminates activation by the cAMP-CAP complex, because the bound complex is not aligned properly to interact with a polymerase molecule bound to the promoter (see Figure 10-17).

To test the spacing constraints on control elements in the HSV *tk* promoter region identified by analysis of linker scanning mutations, researchers prepared and assayed constructs containing small deletions and insertions between the elements. In contrast to the situation in bacteria, changes in spacing between the promoter and promoter-proximal control elements of 20 nucleotides or less had little effect. However, insertions of 30 to 50 base pairs between a promoter-proximal element and the TATA box was equivalent to deleting the element. Similar analyses of other eukaryotic promoters have also indicated that considerable flexibility in the spacing between promoter-proximal elements is generally tolerated, but separations of several tens of base pairs may decrease transcription.

Transcription by RNA Polymerase II Often Is Stimulated by Distant Enhancer Sites

As noted earlier, transcription from many eukaryotic promoters can be stimulated by control elements called enhancers, which are located thousands of base pairs away from the start site. Such long-distance transcriptional control is relatively rare in *E. coli*, occurring principally in the case of operons transcribed by σ^{54}-polymerase (see Figure 10-19).

The first enhancer to be discovered that stimulated transcription of eukaryotic genes was in a 366-bp fragment of the simian virus 40 (SV40) genome (Figure 10-33). Further analysis of this region of SV40 DNA revealed that an ≈100-bp sequence lying ≈100 base pairs upstream of the SV40 early transcription-start site was responsible for its ability to enhance transcription. In SV40, this enhancer sequence functions to stimulate transcription from viral promoters. The SV40 enhancer, however, stimulates transcription from all mammalian promoters that have been tested when it is inserted in either orientation anywhere on a plasmid carrying the test promoter, even when it is thousands of base pairs from the start site. An extensive linker scanning mutational analysis of the SV40 enhancer indicated that it is composed of multiple individual elements each of which contributes to the total activity of the enhancer. As discussed later, each of these regulatory elements is a protein-binding site.

Soon after discovery of the SV40 enhancer, enhancers were identified in other viral genomes and in eukaryotic cellular DNA. Some of these control elements were located 50 or more kilobases from the promoter they controlled. Analyses of many different eukaryotic cellular enhancers have shown that they can occur upstream from a promoter, downstream from a promoter within an intron, or even downstream from the final exon of a gene. Like promoter-proximal elements, many enhancers are cell-type specific. For example, the genes encoding antibodies (immunoglobulins) contain an enhancer within the second intron that can stimulate transcription from all promoters tested, but only in B lymphocytes, the type of cells that normally express antibodies. Analyses of the effects of deletions and linker scanning mutations in cellular enhancers have shown that, like the SV40 enhancer, they generally are composed of multiple elements that contribute to the overall activity of the enhancer.

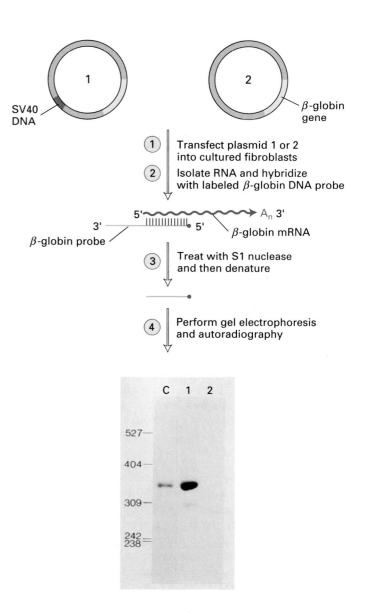

▶ **FIGURE 10-33 Identification of SV40 enhancer region.** Plasmids containing the β-globin gene with or without a 366-bp fragment of SV40 DNA were constructed. Each plasmid was transfected separately into cultured fibroblasts, which do not normally express β-globin (step ①). The amount of β-globin mRNA synthesized by transfected cells was assayed by the S1 nuclease-protection method (see Figure 7-36b). The probe used in this assay was a restriction fragment, generated from a β-globin cDNA clone, that was complementary to the 5′ end of β-globin mRNA (step ②). The 5′ end of the probe was labeled with ^{32}P (red dot). When β-globin mRNA hybridized to the probe, an ≈340-nucleotide fragment of the probe was protected from digestion by S1 nuclease (step ③). Autoradiography of electrophoresed S1-protected fragments (step ④) revealed that cells transfected with plasmid 1 (lane 1) produced much more β-globin mRNA than those transfected with plasmid 2 (lane 2). Lane C is a control assay of β-globin mRNA isolated from reticulocytes, which actively synthesize β-globin. These results show that the SV40 DNA fragment contains an element, called an enhancer, that greatly stimulates synthesis of β-globin mRNA. [Adapted from J. Banerji et al., 1981, *Cell* **27**:299.]

Most Eukaryotic Genes Are Regulated by Multiple Transcription-Control Elements

Initially, enhancers and promoter-proximal elements were thought to be distinct types of transcription-control elements. However, as more enhancers and promoter-proximal elements were analyzed, the distinctions between them became less clear. For example, both types of element generally can stimulate transcription even when inverted, and both types often are cell-type specific. The general consensus now is that a spectrum of control elements regulate transcription by RNA polymerase II. At one extreme are enhancers, which can stimulate transcription from a promoter tens of thousands of base pairs away (e.g., the SV40 enhancer). At the other extreme are promoter-proximal elements, such as the upstream elements controlling the HSV *tk* gene, which lose their influence when moved an additional 30–50 base pairs farther from the promoter. Researchers have identified a large number of transcription-control elements that can stimulate transcription from distances between these two extremes.

Figure 10-34a summarizes the locations of transcription-control sequences for a hypothetical mammalian gene. Transcription initiates at the cap site encoding the first nucleotide of the first exon of an mRNA. For many genes, especially those encoding abundantly expressed proteins, a TATA box located 25–30 base pairs upstream from the cap site directs RNA polymerase II to the start site. Promoter-proximal elements, which are relatively short (≈15–30 base pairs), are located within the first ≈200 base pairs upstream of the cap site. Enhancers, in contrast, usually are 100–200 base pairs long and may be located up to 50 kilobases upstream or downstream from the cap site or within an intron. Many

mammalian genes are controlled by more than one enhancer region.

The *S. cerevisiae* genome contains regulatory elements called **upstream activating sequences (UASs),** which function similarly to enhancers and promoter-proximal elements in higher eukaryotes. Most yeast genes contain only one UAS, which generally lies within a few hundred base pairs of the cap site. In addition, yeast genes contain a TATA box ≈90 base pairs upstream from the transcription start site (Figure 10-34b).

SUMMARY Regulatory Sequences in Eukaryotic Protein-Coding Genes

- Expression of eukaryotic protein-coding genes generally is regulated through multiple cis-acting transcription-control regions (see Figure 10-34). Some control elements are located close to the start site (promoter-proximal elements), whereas others lie more distant (enhancers).

- Promoters determine the site of transcription initiation and direct binding of RNA polymerase II. Three types of promoter sequences have been identified in eukaryotic DNA. The TATA box, the most common, is prevalent in rapidly transcribed genes. Initiator promoters infrequently are found in some genes, and CpG islands are characteristic of transcribed genes.

- Promoter-proximal elements occur within ≈200 base pairs of the start site. Several such elements, containing up to ≈20 base pairs, may help regulate a particular gene.

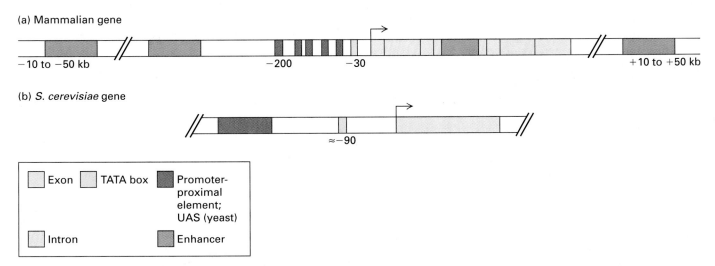

(a) Mammalian gene

−10 to −50 kb   −200   −30   +10 to +50 kb

(b) S. cerevisiae gene

≈−90

| | Exon | | TATA box | | Promoter-proximal element; UAS (yeast) |
| | Intron | | Enhancer | | |

▲ **FIGURE 10-34 General pattern of cis-acting control elements that regulate gene expression in yeast and multicellular organisms (invertebrates, vertebrates, and plants).** (a) Genes of multicellular organisms contain both promoter-proximal elements and enhancers as well as a TATA box or other promoter element. The latter positions RNA polymerase II to initiate transcription at the start site and influences the rate of transcription. Enhancers may be either upstream or downstream and as far away as 50 kb from the transcription start site. In some cases, promoter-proximal elements occur downstream from the start site as well. (b) Most yeast genes contain only one regulatory region, called an upstream activating sequence (UAS), and a TATA box, which is ≈90 base pairs upstream from the start site.

- Enhancers, which are usually ≈100–200 base pairs in length, contain multiple 8- to 20-bp control elements. They may be located from 200 base pairs to tens of kilobases upstream or downstream from a promoter, within an intron, or downstream from the final exon of a gene.

- Promoter-proximal elements and enhancers often are cell-type specific, functioning only in specific differentiated cell types.

10.5   Eukaryotic Transcription Activators and Repressors

As in *E. coli,* the various transcription-control elements found in eukaryotic DNA are binding sites for trans-acting regulatory proteins. In this section, we discuss the identification, purification, and structures of these transcription factors, which function as **activators** and **repressors** of eukaryotic protein-coding genes. Our discussion focuses on activators, as these have been studied most extensively.

Biochemical and Genetic Techniques Have Been Used to Identify Transcription Factors

In yeast, *Drosophila,* and other genetically tractable eukaryotes, classical genetic studies have identified genes encoding transcription factors. However, in mammals and vertebrates, which are less amenable to such genetic analysis, most transcription factors have been identified by biochemical purification.

Biochemical Isolation of Transcription Factors Once a DNA regulatory element has been identified by the kinds of mutational analyses described in the previous section, it can be used to identify *cognate* proteins that bind specifically to it. In this approach, an extract of cell nuclei is subjected to several chromatographic steps (Figure 10-35a); fractions are assayed by DNase I footprinting or an electrophoretic mobility shift assay using DNA fragments containing the identified regulatory element (see Figures 10-6 and 10-7). Fractions containing protein that binds to the regulatory element in these assays probably contain a putative transcription factor (Figure 10-35b). A powerful technique commonly used for the final step in purifying transcription factors is *sequence-specific DNA affinity chromatography,* a particular type of affinity chromatography (Section 3.5). As a final test, the ability of the isolated protein to stimulate transcription of a template containing the corresponding protein-binding sites is assayed in an in vitro transcription reaction (Figure 10-36).

Once a transcription factor is isolated, its partial amino acid sequence can be determined and used to clone the gene or cDNA encoding it, as outlined in Chapter 7. The isolated gene can then be used to test the ability of the encoded protein to stimulate transcription in an in vivo transfection assay (Figure 10-37).

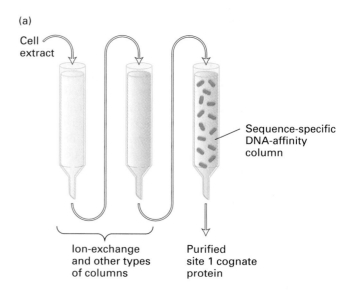

(a)

Cell extract

Sequence-specific DNA-affinity column

Ion-exchange and other types of columns

Purified site 1 cognate protein

(b)

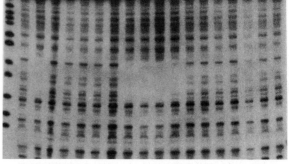

M NE O FT 1 6 7 8 9 10 11 12 13 14 15 16 18 20 22   Fraction

▲ **FIGURE 10-35 Transcription-factor purification.** (a) Several chromatographic steps are used to purify a transcription factor (cognate protein) that binds to a specific regulatory element in DNA. Final purification is obtained by affinity chromatography on a column that is coupled to long DNA strands containing multiple copies of the transcription factor–binding site. A partially purified protein preparation containing the transcription factor is applied to the column in a low-salt buffer (100 mM KCl). Proteins that do not bind to the specific binding site are washed off the column with additional low-salt buffer. Proteins with low affinity for the binding site are eluted with an intermediate salt concentration buffer (e.g., 300 mM KCl). Finally, highly purified transcription factor is eluted with a high salt concentration buffer (e.g., 1 M KCl). (b) Proteins separated by column chromatography are assayed for their ability to bind to an identified regulatory element. In this example, DNase I footprinting assays of protein fractions eluted from an ion-exchange column indicate that the transcription factor of interest is in fractions 9–12. Specific DNA-binding proteins can also be conveniently assayed by EMSA (see Figure 10-7). [For sequence-specific DNA-affinity chromatography see J. T. Kadonaga and R. Tjian, 1986, *Proc. Nat'l. Acad. Sci. USA* **83:**5889. Part (b) from S. Yoshinaga et al., 1989, *J. Biol. Chem.* **264:**10529.]

(a) SP1-binding sites in SV40 genome

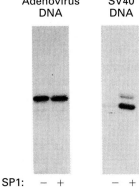

GGGGCGGGGC

TATA box

(b) SP1 transcription-activating assay

Adenovirus DNA   SV40 DNA

SP1:   −   +   −   +

▲ **FIGURE 10-36 In vitro transcription activation by SP1, which binds to 10-bp GC-rich sequences.** (a) The SV40 genome contains six copies of a GC-rich promoter-proximal element up-stream of the early promoter. The optimal SP1-binding sequence is shown for one of these sites, but SP1 also binds with high affinity to related GC-rich sequences. SP1 was isolated based on its ability to bind to this region of the SV40 genome and purified as illustrated in Figure 10-35. (b) To test the activating ability of purified SP1 in vitro, template DNA and a partially purified nuclear extract from cultured HeLa cells containing RNA polymerase II and associated general transcription factors are incubated with labeled ribonucleoside triphosphates. The labeled RNA products are subjected to electrophoresis and autoradiography. Shown here are autoradiograms from assays with adenovirus and SV40 DNA in the absence (−) and presence (+) of SP1. SP1 had no significant effect on transcription from the adenovirus promoter, which contains no SP1-binding sites. In contrast, SP1 stimulated transcription from the SV40 promoter about tenfold. [Adapted from M. R. Briggs et al., 1986, *Science* **234**:47.]

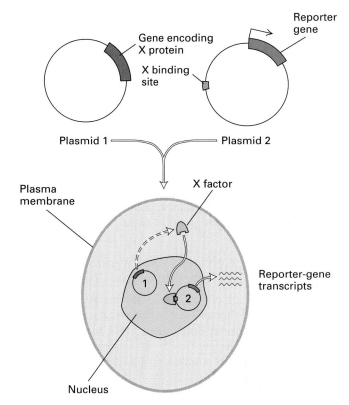

▲ **FIGURE 10-37 In vivo assay for transcription factor activity.** The assay system requires two plasmids. One plasmid contains the gene encoding the putative transcription factor (X protein). The second plasmid contains a reporter gene and one or more binding sites for X protein. Both plasmids are simultaneously introduced into host cells that lack the gene encoding X protein and the reporter gene. The production of reporter-gene RNA transcripts is measured; alternatively, the activity of the encoded protein can be assayed. If reporter-gene transcription is greater in the presence of the X-encoding plasmid, then the protein is an activator; if transcription is less, then it is a repressor. By use of plasmids encoding a mutated or rearranged transcription factor, important domains of the protein can be identified.

Genetic Identification of Genes Encoding Transcription Factors In yeast, genes encoding transcription factors were first identified through classical genetic analysis. For example, one of the yeast genes required for growth on galactose is called *GAL4*. Incubation of wild-type yeast cells in galactose media results in more than a thousand-fold increase in the concentration of mRNAs encoding the enzymes catalyzing galactose metabolism. This activation of mRNA expression is not observed in *gal4* mutants. (In *S. cerevisiae*, wild-type genes are designated with capital letters in italics, and recessive mutant alleles of the gene are indicated with lowercase letters in italics. The encoded protein is designated by the name of the gene in Roman type, with the first letter capitalized, e.g., Gal4.) Directed mutagenesis studies like those described previously identified UASs for the induced genes. Each of these UASs was found to contain one or more copies of a related 17-bp sequence called UAS_{GAL}. When a copy of UAS_{GAL} was cloned upstream of a TATA box followed by a *lacZ* reporter gene, expression of *lacZ* was activated in galactose media in wild-type cells, but not in *gal4* mutants. This indicated that UAS_{GAL} is a transcription-control element activated by the Gal4 protein in galactose media.

The *GAL4* gene was isolated by complementation of a *gal4* mutant with a library of wild-type yeast DNA (Section 8.2). By use of recombinant DNA techniques, the Gal4 protein was expressed in *E. coli* and found to bind to UAS_{GAL}. Thus, the Gal4 protein binds to UAS_{GAL} sequences and activates transcription from a nearby promoter when cells are placed in galactose media.

Classical genetic studies in a number of other organisms including *Drosophila*, the nematode *C. elegans*, and higher plants have uncovered several genes encoding transcription factors. For example, many mutations that interfere with normal *Drosophila* development have been identified. One of these inactivates the *Ultrabithorax* (*Ubx*) gene, causing an extra pair of wings to develop from the third thoracic segment (see Figure 8-8b). The protein encoded by wild-type *Ubx* has been shown to function as a transcription factor. The remarkable change in phenotype observed in *Ubx* mutants indicates that Ubx protein influences transcription of a large number of *Drosophila* genes.

Transcription Activators Are Modular Proteins Composed of Distinct Functional Domains

A remarkable set of experiments with the yeast Gal4 protein demonstrated that this transcription factor is composed of separable functional domains: a **DNA-binding domain**, which interacts with specific DNA sequences, and an **activation domain**, which interacts with other proteins to stim-ulate transcription from a nearby promoter. In these exper-iments, a series of *gal4* deletion mutants were tested for their ability to activate transcription of a reporter gene (*lacZ*) linked to UAS$_{GAL}$ (Figure 10-38a) in an in vivo assay like that depicted in Figure 10-37. Transcription activation was measured in cells lacking the *GAL4* gene, so that wild-type Gal4 protein would not interfere with the analysis. Binding of the mutant Gal4 proteins to the UAS$_{GAL}$ sequence also was assayed. The results of these experiments, outlined in Figure 10-38b, demonstrate that Gal4 contains a N-terminal 74-amino acid DNA-binding domain and a C-terminal acti-vation domain. When the N-terminal DNA-binding domain of Gal4 was fused directly to various C-terminal fragments, the resulting truncated proteins retained the ability to stim-ulate expression of the reporter gene. Thus the internal por-tion of the protein is not required for functioning of Gal4 as a transcription factor.

Similar experiments with the yeast transcription factor Gcn4, which regulates genes required for synthesis of many amino acids, indicated that it contains an ≈60-aa DNA-binding domain at its C-terminus and an ≈20-aa activation domain near the middle of its sequence. Further evidence for the existence of distinct activation domains in Gal4 and Gcn4

(a) Reporter-gene construct

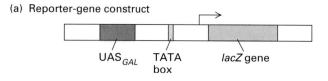

(b) Wild-type and mutant Gal4 proteins

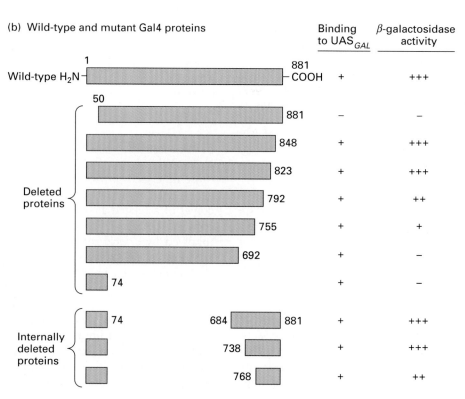

came from experiments in which their activation domains were fused to a DNA-binding domain from an entirely unrelated *E. coli* repressor. Introduction of a reporter gene construct containing the cognate site for the *E. coli* repressor upstream from a TATA-box and *lacZ*, and an expression vector for the repressor DNA-binding domain fused to the coding sequence for either the Gal4 or Gcn4 activation domain, led to expression of the reporter gene in yeast cells. In this case, a fusion protein consisting of the DNA-binding domain from one transcription factor and the activation domain from a different factor was expressed in vivo and activated transcription. Thus, entirely novel transcription factors composed of prokaryotic and eukaryotic elements can be constructed.

Studies such as these have now been carried out with many eukaryotic transcription factors. Activation domains in mammalian transcription factors are frequently assayed by fusing them to the Gal4 DNA-binding domain since mammalian cells do not contain an endogenous transcription factor that binds to the UAS$_{GAL}$ sequence. The structural model of eukaryotic activators that has emerged from these studies is a modular one in which one or more activation domains is connected to a sequence-specific DNA-binding domain through relatively flexible protein domains (Figure 10-39). In some cases, amino acids included in the DNA-binding domain also contribute to transcriptional activation. As discussed in a later section, activation domains are thought to function through protein-protein interactions with transcription factors bound at the promoter. The flexible protein domains in activators, which connect the DNA-binding domains to activation domains, may explain why alterations in the spacing between control elements is so well tolerated in eukaryotic control regions. When the DNA-binding domains of neighboring transcription factors are shifted in their relative positions on the DNA, their activation domains may still be able to interact because they are attached to their DNA-binding domains through flexible protein regions.

DNA-Binding Domains Can Be Classified into Numerous Structural Types

Eukaryotic transcription factors contain a variety of structural **motifs** that interact with specific DNA sequences. As with most bacterial activators and repressors, α helices in the DNA-binding domain of eukaryotic transcription factors are oriented so that they lie in the major groove of DNA where protein atoms make specific hydrogen bonds and van der Waals interactions with atoms in the DNA. Interactions with sugar-phosphate backbone atoms and, in some cases, with atoms in the DNA minor groove also contribute to binding. X-ray crystallographic analyses of complexes between specific protein-binding sites in DNA and isolated transcription-factor DNA-binding domains have revealed a

◀ **FIGURE 10-38 Experimental demonstration of separate functional domains in yeast Gal4 protein.** (a) Diagram of DNA construct containing a *lacZ* reporter gene with an added TATA box ligated to UAS$_{GAL}$, a regulatory element that contains several Gal4-binding sites. The reporter-gene construct and DNA encoding wild-type or mutant Gal4 were simultaneously introduced into mutant *(gal4)* yeast cells and the β-galactosidase activity was assayed. Activity will be high if the introduced *GAL4* DNA encodes a functional protein. (b) Schematic diagrams of wild-type Gal4 and various mutant forms. Small numbers refer to positions in the wild-type sequence. Deletion of 50 amino acids from the N-terminal end destroyed the ability of Gal4 to bind to UAS$_{GAL}$ and to stimulate expression of β-galactosidase from the reporter gene. Proteins with extensive deletions from the C-terminal end still bound to UAS$_{GAL}$. These results localize the DNA-binding domain to the N-terminal end of Gal4. Proteins with a deletion of 58 amino acids from the C-terminal end had no decrease in the ability to stimulate expression of β-galactosidase; those with a deletion of 89 amino acids exhibited a small decrease. The ability to activate β-galactosidase expression was not entirely eliminated unless 126–189 or more amino acids were deleted. Thus, the activation domain lies in the C-terminal region of Gal4. Deleted proteins containing the N-terminal and C-terminal segments also were able to stimulate expression of β-galactosidase, indicating that the central region of Gal4 is not crucial for its function in this assay. [See J. Ma and M. Ptashne, 1987, *Cell* **48**:847; I. A. Hope and K. Struhl, 1986, *Cell* **46**:885; and R. Brent and M. Ptashne, 1985, *Cell* **43**:729.]

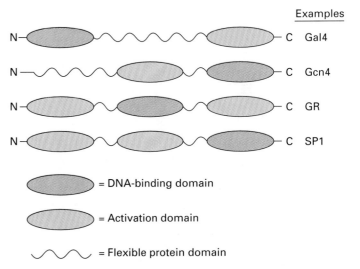

Examples

= DNA-binding domain

= Activation domain

= Flexible protein domain

▲ **FIGURE 10-39 Schematic diagrams illustrating the modular structure of eukaryotic transcription activators.** These transcription factors may contain more than one activation domain but rarely contain more than one DNA-binding domain. Gal4 and Gcn4 are yeast transcription activators. The glucocorticoid receptor (GR), which also contains a hormone-binding domain (not shown), activates transcription of target genes in the presence of certain hormones. SP1 binds to GC-rich promoter elements in a large number of mammalian genes. The relatively unstructured, highly flexible protein domains in activators are extremely sensitive to digestion by proteases.

number of structural motifs that can present an α helix to the major groove.

Transcription factors often are classified according to the type of DNA-binding domain they contain. Most of the structural classes of DNA-binding domains have characteristic consensus amino acid sequences. Consequently, newly characterized transcription factors frequently can be classified once the corresponding genes or cDNAs are cloned and sequenced. Several common classes of DNA-binding domains whose three-dimensional structures have been determined are described and illustrated here. Many additional classes are recognized, and new classes are still being characterized. The genomes of higher eukaryotes may encode dozens of classes of DNA-binding domains and literally hundreds of transcription factors.

Homeodomain Proteins The structure of the DNA-binding domain from the *Drosophila* Engrailed protein is depicted in Figure 10-40. Transcription factors with this type of DNA-binding domain are called **homeodomain** proteins, a name derived from a group of *Drosophila* genes in which the conserved sequence encoding this structural motif was first noted. Mutations in these genes, called **homeotic genes,** result in the transformation of one body part into another during development (Section 14.3). Two of the most-studied of these genes are designated *Antennapedia (Antp)* and *Ultrabithorax*

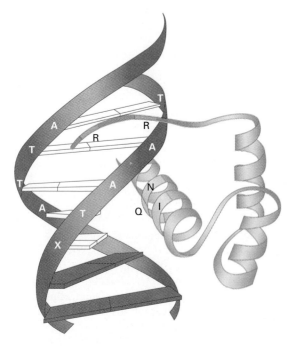

▲ **FIGURE 10-40 Homeodomain from Engrailed protein interacting with its specific DNA recognition site.** The Engrailed transcription factor is expressed during *Drosophila* embryogenesis. Base pairs in the recognition site that directly contact the protein are shown in white type. Lighter regions in the protein contain residues that contact the major groove. [Adapted from S. C. Harrison, 1991, *Nature* **353**:715.]

(Ubx). The proteins encoded by these genes share a highly conserved 60-aa region; this same conserved region was subsequently identified in the proteins encoded by other homeotic genes. Because the conserved DNA sequence encoding this region was often diagrammed in a box when sequences from different genes were compared, it came to be known as the **homeobox.** The conserved sequence has also been found in vertebrate genes, including human genes, that have similar master control functions in development.

Zinc-Finger Proteins A number of different proteins have regions that fold around a central Zn^{2+} ion, producing a compact domain from a relatively short length of the polypeptide chain. Termed a **zinc finger,** this structural motif was first recognized in DNA-binding domains but now is known to occur in proteins that do not bind to DNA. We describe three of the several classes of zinc-finger motifs that have been identified.

The interaction between DNA and a transcription factor containing five C_2H_2 *zinc-finger* domains is shown in Figure 10-41a. Each C_2H_2 finger has the consensus sequence Tyr/Phe-X-Cys-X_{2-4}-Cys-X_3-Phe/Tyr-X_5-Leu-X_2-His-X_{3-4}-His, where X is any amino acid. This sequence binds one Zn^{2+} ion through the two cysteine (C) and two histidine (H) side chains. The name "zinc finger" was coined because a two-dimensional diagram of the structure resembles a finger (see Figure 3-9c). When the three-dimensional structure was solved, it became clear that the binding of the Zn^{2+} ion by the two cysteine and two histidine residues folds the relatively short polypeptide sequence into a compact domain, which can insert its α helix into the major groove of DNA. The C_2H_2 zinc finger is one of the most common DNA-binding motifs in eukaryotic transcription factors. More than a thousand of these consensus sequences are in the current protein sequence data base. The repeating units in these proteins can interact with successive groups of base pairs, primarily within the major groove, as the protein wraps around the DNA double helix.

A second type of zinc-finger structure, designated the C_4 *zinc finger,* is found in more than 100 transcription factors. The first members of this class were identified as specific intracellular high-affinity binding proteins, or "receptors," for steroid **hormones,** leading to the name *steroid receptor superfamily.* Because similar intracellular receptors for nonsteroid hormones subsequently were found, these transcription factors are now commonly called **nuclear receptors.** The DNA-binding domain of these proteins has the consensus sequence Cys-X_2-Cys-X_{13}-Cys-X_2-Cys-X_{14-15}-Cys-X_5-Cys-X_9-Cys-X_2-Cys. The two groups of four critical cysteines in this region each binds a Zn^{2+} ion. Although the C_4 zinc-finger motif initially was named by analogy with the C_2H_2 zinc-finger motif, the three-dimensional structures of these DNA-binding domains later were found to be quite distinct. A particularly important difference between the two is that C_2H_2 zinc-finger proteins generally contain three or more repeating finger units and bind as monomers, whereas C_4 zinc-finger proteins generally contain only two finger units

(a)

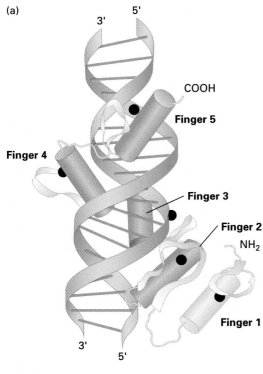

(b)

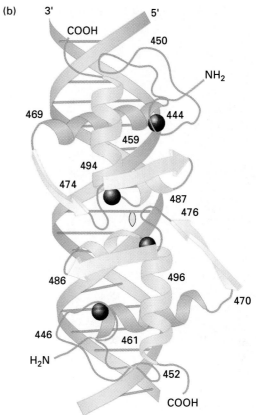

▲ **FIGURE 10-41 Interaction of C₂H₂ and nuclear receptor (C₄) zinc-finger domains with DNA (blue).** (a) A five-finger C₂H₂ protein called GL1. This monomeric protein is encoded by a gene that is amplified in a number of human tumors. Helical regions are represented by cylinders, and β strands by broad ribbons. Finger 1 does not interact with DNA, whereas the other four fingers do. Small black circles indicate Zn²⁺ ions. (b) The glucocorticoid receptor, a homodimeric C₄ protein. Helical regions are shown as spirals, and β strands as broad arrows. Two α helices (darker shade), one in each monomer, interact with the DNA. Like all C₄ zinc-finger homodimers, this transcription factor has twofold rotational symmetry; the center of symmetry is shown by the yellow ellipse. Black circles indicate Zn²⁺ ions. [Part (a) adapted from N. P. Pavletich and C. O. Pabo, 1993, *Science* **261**:1701; part (b) adapted from B. F. Luisi et al., 1991, *Nature* **352**:497.]

and bind to DNA as homodimers or heterodimers. Like bacterial homodimeric helix-turn-helix DNA-binding domains, homodimers of C₄ DNA-binding domains have twofold rotational symmetry (Figure 10-41b). Consequently, homodimeric nuclear receptors bind to consensus DNA sequences that are inverted repeats, another similarity with bacterial systems. Heterodimeric nuclear receptors do not exhibit rotational symmetry; in these proteins one C₄ monomer is inverted relative to the lower monomer in Figure 10-41b.

The DNA-binding domain in the yeast Gal4 protein exhibits a third type of zinc-finger motif, known as the *C₆ zinc finger*. Proteins of this class have the consensus sequence Cys-X₂-Cys-X₆-Cys-X₅₋₆-Cys-X₂-Cys-X₆-Cys. The six cysteines bind two Zn²⁺ ions, folding the region into a compact globular domain (Figure 10-42). The Gal4 protein binds DNA as a homodimer in which the monomers associate through hydrophobic interactions along one face of their α-helical regions. This type of interaction between α helices, to form a coiled coil, also occurs in dimeric leucine-zipper proteins and is discussed in more detail below.

Winged-Helix (Forkhead) Proteins The DNA-binding domains in histone H5 and several transcription factors that function during early development of *Drosophila* and mammals have the *winged-helix* motif, also called the *forkhead* motif. Like C₂H₂ zinc-finger proteins, winged-helix proteins generally bind to DNA as monomers.

Leucine-Zipper Proteins Another structural motif present in a large class of transcription factors is exemplified by the DNA-binding domain of yeast Gcn4. The first transcription factors recognized in this class contained the hydrophobic amino acid leucine at every seventh position in the C-terminal portion of their DNA-binding domains. These proteins bind to DNA as dimers, and mutagenesis of the leucines showed that they were required for dimerization. Consequently, the name **leucine zipper** was coined to denote this structural motif.

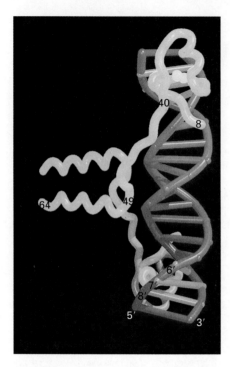

▲ FIGURE 10-42 Trace diagram of interaction between Gal4, a C₆ zinc-finger protein, and DNA. This protein binds DNA as a homodimer with the monomers interacting to form a coiled coil that lies perpendicular to the DNA helix. Binding of cysteine residues to two Zn^{2+} ions (yellow) in each monomer forms two globular domains that interact with DNA. [From R. Marmorstein et al., 1992, *Nature* **356**:408.]

X-ray crystallographic analysis of complexes between DNA and the Gcn4 DNA-binding domain has shown that the dimeric protein contains two extended α helices that "grip" the DNA molecule, much like a pair of scissors, at two adjacent major grooves separated by about half a turn of the double helix (Figure 10-43). The portions of the α helices contacting the DNA include basic residues that interact with phosphates in the DNA backbone and additional residues that interact with specific bases in the major groove.

Gcn4 forms dimers via hydrophobic interactions between the C-terminal regions of the α helices, forming a coiled-coil structure. This structure is common in proteins containing amphipathic α helices in which hydrophobic amino acid residues are regularly spaced alternately three or four positions apart in the sequence. As a result of this characteristic spacing, the hydrophobic side chains form a stripe down one side of the α helix. These hydrophobic stripes make up the interacting surfaces between the α-helical monomers in a coiled-coil dimer (see Figure 3-9a).

As noted above, the first transcription factors in this class to be analyzed contained leucine residues at every seventh position in the dimerization region and thus were named leucine-zipper proteins. However, additional DNA-binding proteins containing other hydrophobic amino acids in these positions subsequently were identified. Like leucine-zipper proteins, they form dimers containing a C-terminal coiled-coil dimerization region and N-terminal DNA-binding domain. The term *basic zipper* (bZip) now is frequently used to refer to all proteins with these common structural features. Many basic-zipper transcription factors are heterodimers of two different polypeptide chains, each containing one basic-zipper domain.

Helix-Loop-Helix Proteins The DNA-binding domain of another class of dimeric transcription factors contains a structural motif very similar to the basic-zipper motif except that a nonhelical loop of the polypeptide chain separates two α-helical regions in each monomer (Figure 10-44). Termed a **helix-loop-helix** (HLH), this motif was predicted from the amino acid sequences of these proteins, which contain an N-terminal α helix with basic residues that interact with DNA, a middle loop region, and a C-terminal region with hydrophobic amino acids spaced at intervals characteristic of an amphipathic α helix. Because of the basic amino acids characteristic of this motif, transcription factors containing it sometimes are referred to as basic helix-loop-helix (bHLH) proteins. As with basic-zipper proteins, different helix-loop-helix proteins can form heterodimers.

Heterodimeric Transcription Factors Increase Gene-Control Options

Three types of DNA-binding proteins discussed in the previous section can form heterodimers: C₄ zinc-finger proteins, basic-zipper proteins, and helix-loop-helix proteins. Other

▶ FIGURE 10-43 Two views of the interaction of yeast Gcn4, a homodimeric leucine-zipper protein, with DNA. The extended α-helical regions of the monomers grip the DNA at adjacent major grooves. The coiled-coil dimerization domain in Gcn4 contains precisely spaced leucine residues. Some DNA-binding proteins with this same general motif contain other hydrophobic amino acids in these positions; hence, this structural motif is generally called a basic zipper. [From T. E. Ellenberger et al., 1992, *Cell* **71**:1223.]

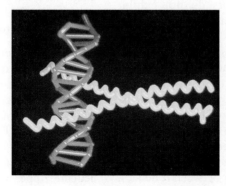
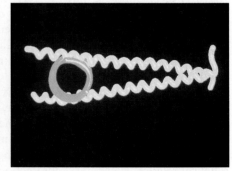

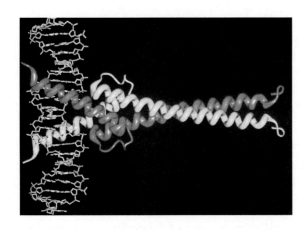

◀ **FIGURE 10-44 Interaction of the helix-loop-helix domain in the homodimeric Max protein with DNA.** The helix-loop-helix motif extends from the DNA-binding helices on the left (N-termini of the monomers) to approximately where the chains first cross; this is followed immediately by a leucine-zipper coiled-coil region in the dimeric protein. [See A. R. Ferre-D'Amare et al., 1993, *Nature* **363**:38; courtesy of A. R. Ferre-D'Amare and S. K. Burley.]

classes of transcription factors whose structures are not specifically considered here also form heterodimeric proteins. In some heterodimeric transcription factors, each monomer has a DNA-binding domain with equivalent sequence specificity. In these proteins, the formation of heterodimers does not influence DNA-binding specificity, but rather allows the activation domains associated with each monomer to be brought together in a single transcription factor. However, if the monomers have different DNA-binding specificity, the formation of heterodimers increases the number of potential DNA sequences that a family of factors can bind, as illustrated in Figure 10-45a. In addition, there are examples of inhibitory basic-zipper and helix-loop-helix proteins that block DNA binding when they dimerize with a partner polypeptide normally capable of binding DNA. When these inhibitory factors are expressed, they repress transcriptional activation by the factors with which they interact (Figure 10-45b).

The rules governing the interactions of members of a transcription-factor class are complex. This combinatorial complexity expands both the number of DNA sites from which these factors can activate transcription and the ways in which they can be regulated. This is not possible for transcription factors that bind only as monomers or homodimers.

Activation Domains Exhibit Considerable Structural Diversity

An activation domain is a polypeptide sequence that activates transcription when it is fused to a DNA-binding domain. For example, a large number of diverse peptide sequences can activate transcription in eukaryotic cells from a promoter with upstream UAS$_{GAL}$ binding sites for the Gal4 DNA-binding domain. In one experiment, random fragments of *E. coli* DNA were ligated to the portion of the *GAL4* gene encoding the DNA-binding domain of Gal4. Remarkably, ≈1 percent of all the resulting fusion proteins, composed of the Gal4 DNA-binding domain and random segments of *E. coli* proteins activated transcription from promoters with an upstream UAS$_{GAL}$ in yeast and mammalian cells. This finding demonstrated that a diverse group of amino acid sequences can function as activation domains, even though they evolved to perform other functions. The actual mechanism of transcription activation is considered in a later section.

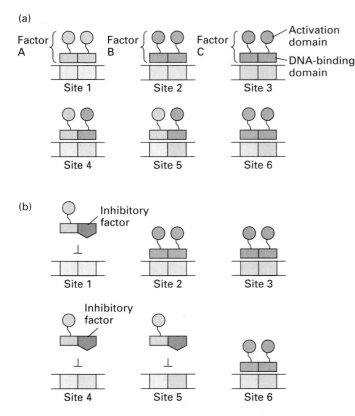

▲ **FIGURE 10-45 Combinatorial possibilities due to formation of heterodimeric transcription factors.** (a) In the hypothetical example shown, transcription factors A, B, and C can each interact with each other, permitting the three factors to bind to six different DNA sequences (sites 1–6) and creating six combinations of activation domains. (Note that each binding site is divided into two half-sites, and that a single heterodimeric factor contains the activation domains of each of its constituent monomers.) Four different factor monomers could combine to make 10 homo- and heterodimeric factors; five monomers could make 15 dimeric factors; and so forth. (b) When an inhibitory factor (green) is expressed that interacts only with factor A, binding to sites 1, 4, and 5 is inhibited, but binding to sites 2, 3, and 6 is unaffected.

Although numerous diverse amino acid sequences can function as activation domains, many activation domains have an unusually high percentage of particular amino acids. Gal4, Gcn4, and most other yeast transcription factors have activation domains that are rich in acidic amino acids (aspartic and glutamic acids). These so-called *acidic activation domains* generally are capable of stimulating transcription in nearly all types of eukaryotic cells—fungal, animal, and plant cells. Activation domains from some *Drosophila* and mammalian transcription factors are glutamine rich, and some are proline rich; still others are rich in the closely related amino acids serine and threonine, both of which have hydroxyl groups. However, some strong activation domains are not particularly rich in any specific amino acid.

Recent biophysical studies on model acidic activation domains show that they exist as unstructured, random-coil regions of polypeptide until they interact with a *co-activator* protein. This interaction induces the activation domain to fold into an amphipathic α helix that contacts a complementary surface of the co-activator protein. For instance, NMR spectra of the isolated activation domain in the mammalian CREB (*c*AMP *r*esponse *e*lement–*b*inding) protein show that it is an unstructured random coil. In response to elevated levels of cAMP, protein kinase A phosphorylates a specific serine residue in the CREB activation domain, allowing it to interact with a specific region in its co-activator CBP (CREB-*b*inding *p*rotein). In the three-dimensional structure of the complex between these two proteins, two α helices at right angles to each other in the CREB activation domain wrap around the interacting domain of CBP (Figure 10-46).

In contrast to the relatively short, random-coil acidic activation domains, some activation domains are larger and more structured. For example, the ligand-binding domains of some nuclear receptors function as activation domains when they bind their specific ligand. Binding of ligand is thought to induce a large conformational change that allows the ligand-binding domain with bound hormone to interact with other proteins (Figure 10-47).

Multiprotein Complexes Form on Enhancers

Enhancers generally range in length from about 50 to 200 base pairs and include binding sites for several transcription factors. The multiple transcription factors that bind to a single enhancer are thought to interact. Analysis of the enhancer that regulates expression of β-**interferon,** an important protein in defense against viral infections in humans, provides a good example of such transcription-factor interactions. Four control elements have been identified in this ≈70-bp enhancer by analysis of linker scanning mutations. Cognate proteins that bind to each of these four sites were identified by techniques described earlier. Once the cDNAs encoding these proteins were isolated, they were shown to activate transcription from the β-interferon enhancer in transfection experiments (see Figure 10-37).

Subsequent studies showed that these transcription factors bind to the β-interferon enhancer simultaneously. In the presence of a small, abundant protein associated with chromatin called HMGI, binding of the transcription factors is highly cooperative, similar to the binding of *E. coli* CAP protein and RNA polymerase to neighboring sites in the *lac* operon (see Figure 10-17). This cooperative binding produces a multiprotein complex on the enhancer DNA (Figure 10-48). The term **enhancesome** has been coined to describe such large nucleoprotein complexes that assemble from transcription factors as they bind cooperatively to their multiple binding sites in an enhancer.

HMGI binds to the minor groove of DNA regardless of the sequence and, as a result, bends the DNA molecule sharply. This bending of the enhancer DNA permits the transcription factors to interact properly. The relatively weak interactions between the bound proteins are enhanced because the transcription factors are bound to neighboring sites, keeping the proteins at very high relative concentration. In studies

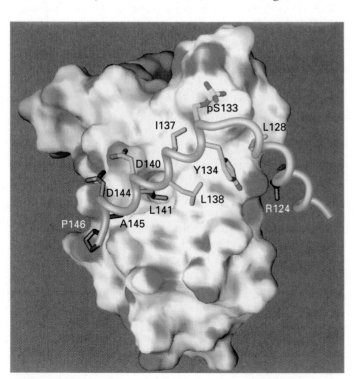

◀ **FIGURE 10-46 Structure of the phosphorylated CREB acidic activation domain complexed to the interacting domain of its co-activator, CBP.** The polypeptide backbone of the phosphorylated CREB acidic activation domain is represented as a pink spiral. Amino acid residues in CREB that interact with the surface of CBP are indicated by the single-letter abbreviations. The water-accessible surface of the interacting domain of CBP is represented at the back of the figure with regions of positive and negative electrical potential shaded blue and red, respectively (see Figure 3-4 for a comparison of this type of surface representation to other representations of protein structure). [From I. Radhakrishnan et al., 1997, *Cell* **91**:741; courtesy of Peter Wright.]

(a) RARα

(b) RARγ

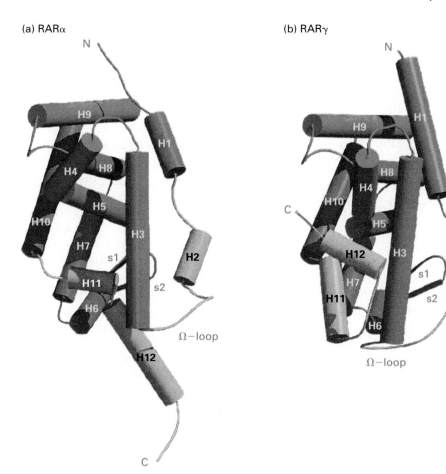

◀ **FIGURE 10-47 Effect of ligand binding on conformation of the ligand-binding domains in two related human nuclear receptors determined by x-ray crystallography.** (a) In the absence of bound ligand (9-*cis* retinoic acid), the ligand-binding domain of human RXRα has a fairly open conformation. In this form, it functions as a repression domain. (b) When bound to all-*trans* retinoic acid, the ligand-binding domain of human RARγ has a fairly compact conformation. In this form, it can stimulate transcription. Cylinders represent α helices. Regions of the ligand-binding domain that do not change significantly in conformation due to ligand binding are shown in green; regions that do, in yellow. As discussed later, RXR associates with several different nuclear-receptor monomers (e.g., RAR), forming heterodimeric transcription factors. [From J. M. Wurtz et al., 1996, *Nature Structure* **3**:87; courtesy of Hinrich Gronemeyer.]

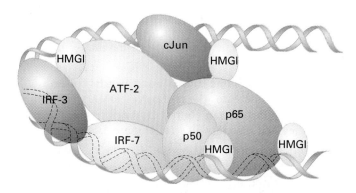

▲ **FIGURE 10-48 Model of the enhancesome that forms on the β-interferon enhancer.** Heterodimeric cJun/ATF-2, IRF-3, IRF-7, and NF-κB (a heterodimer of p50 and p65) bind to the four control elements in the ≈70-bp enhancer. Cooperative binding of these transcription factors is facilitated by HMGI, which binds to the minor groove of DNA. The cJun, ATF-2, p50, and p65 proteins all appear to interact directly with an HMGI bound adjacent to them. Bending of the enhancer sequence resulting from HMGI binding is critical to formation of an enhancesome. Different DNA-bending proteins act similarly at other enhancers. [Adapted from D. Thanos and T. Maniatis, 1995, *Cell* **83**:1091 and M. A. Wathel et al., 1998, *Mol. Cell* **1**:507.]

with the enhancer from the T-cell receptor α gene, also important to the immune response, enhancesome assembly likewise was found to depend on a protein (different from HMGI) that interacts with the minor groove of DNA and introduces a bend. Such DNA-bending proteins have been referred to as **architectural proteins**, because they are required to build these nucleoprotein complexes.

Many Repressors Are the Functional Converse of Activators

Genetic and biochemical studies have shown that eukaryotic transcription is regulated by repressor proteins as well as the more-common activator proteins. For example, geneticists have identified mutations in yeast that result in constitutive expression of certain genes, indicating that these genes normally are regulated by a repressor. In another approach, repressor-binding sites were identified in systematic mutational analyses of eukaryotic transcription-control regions similar to the experiments depicted in Figure 10-32. While mutation of an activator-binding site leads to decreased expression of the linked reporter gene, mutation of a repressor-binding site leads to increased expression of a reporter gene. Repressor proteins that bind such sites have been purified and characterized using sequence-specific DNA affinity chromatography, as for activator proteins (see Figure 10-35).

The absence of appropriate repressor activity can have devastating consequences. For instance, the protein encoded by the *Wilms' tumor (WT1)* gene is a repressor that is expressed preferentially in the developing kidney. Children who inherit mutations in both the maternal and paternal *WT1* genes, so that they produce no functional WT1 protein, invariably develop kidney tumors early in life. The WT1 protein, which has a C_2H_2 zinc-finger DNA-binding domain, binds to the control region of the gene encoding a transcription activator called EGR-1 (Figure 10-49). This gene, like many other eukaryotic genes, is subject to both repression and activation. Binding of WT1 represses transcription of the *EGR-1* gene without inhibiting binding of the two activators that normally stimulate expression of this gene. Eukaryotic transcription repressors like WT1 appear to be the functional converse of activators. They can inhibit transcription from a gene they do not normally regulate when their cognate binding sites are placed within a few hundred base pairs of the gene's start site. The various mechanisms whereby repressor proteins exert their effects are described later.

Like activators, many eukaryotic repressors have two functional domains: a DNA-binding domain and a **repression domain**. As is true for activation domains, a variety of amino acid sequences can function as repression domains. Many of these are relatively short ($\approx$20 amino acids) and contain high proportions of hydrophobic residues. Other repression domains contain a high proportion of basic residues. In some cases, repression domains are larger, well-structured protein domains. For example, in the absence of ligand, the RXRα ligand-binding domain functions as a repression domain (see Figure 10-47a). When the same domain binds its cognate ligand, 9-*cis* retinoic acid, it is converted into an activation domain. As for activation domains, the diverse structures of repression domains is probably a reflection of several possible molecular mechanisms for regulating eukaryotic transcription. To begin to understand how activation and repression domains of eukaryotic transcription factors regulate

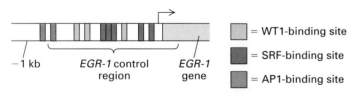

▲ FIGURE 10-49 Diagram of the control region of the gene encoding EGR-1, a transcription activator. The binding sites for WT1, an eukaryotic repressor protein, do not overlap the binding sites for SRF and AP1, two ubiquitous activators, or the start site. Thus repression by WT1 does not involve interference with binding of other proteins.

gene expression, we must first discuss the complexities of transcription initiation by RNA polymerase II.

SUMMARY Eukaryotic Transcription Activators and Repressors

- Transcription factors, which stimulate or repress transcription, bind to promoter-proximal elements and enhancers in eukaryotic DNA.

- Activators are generally modular proteins containing a single DNA-binding domain and one or a few activation domains; the different domains frequently are linked through flexible polypeptide regions (see Figure 10-39). This may allow activation domains in different activators to interact even when their DNA-binding domains are bound to sites separated by tens of base pairs.

- Enhancers generally contain multiple clustered binding sites for transcription factors. Cooperative binding of multiple activators to nearby sites in an enhancer forms a multiprotein complex called an enhancesome (see Figure 10-48). Assembly of enhancesomes often requires small proteins that bind to the DNA minor groove and bend the DNA sharply, allowing proteins on either side of the bend to interact more readily.

- Most eukaryotic repressors also are modular proteins. Similar to activators, they usually contain a single DNA-binding domain, one or a few repression domains, and can control transcription when they are bound at sites hundreds to thousands of base pairs from a start site.

- DNA-binding domains in eukaryotic transcription factors exhibit a variety of structures. Among the most common structural motifs are the homeodomain, basic zipper (leucine zipper), helix-loop-helix, and several types of zinc finger. In general, one or more α helices in a DNA-binding domain interacts with the major groove in its cognate site.

- The ability of some transcription factors to form heterodimers increases the number of DNA sites from which these factors can control transcription and the ways they can be controlled (see Figure 10-45).

- Although some activation and repression domains are rich in particular amino acids, these functional domains exhibit a variety of amino acid sequences and protein structures in different transcription factors.

10.6 RNA Polymerase II Transcription-Initiation Complex

In previous sections many of the eukaryotic proteins and DNA sequences that participate in transcription and its control have been introduced. In this section, we focus on assembly of *transcription-initiation complexes* involving RNA polymerase II (Pol II). Recall that this type of eukaryotic RNA polymerase catalyzes synthesis of mRNAs and a few small nuclear RNAs (snRNAs). Various mechanisms for controlling formation of Pol II transcription-initiation complexes, and hence the rate of transcription, are considered in the next section.

Initiation by Pol II Requires General Transcription Factors

As discussed earlier, the purified *E. coli* core RNA polymerase, which lacks the σ^{70} subunit, cannot initiate transcription. However, when the core polymerase is associated with σ^{70}, the resulting holoenzyme can initiate transcription in vitro from strong promoters. Thus, the σ^{70} subunit of *E. coli* RNA polymerase functions as an initiation factor; that is, it is required for transcription to begin but is released from the template after polymerization of the initial 10 or so ribonucleotide triphosphates. In contrast, in vitro transcription by purified eukaryotic RNA polymerase II requires the addition of several initiation factors that are separated from the polymerase during purification. These initiation factors, which position Pol II at transcription-initiation sites, are called **general transcription factors,** because they are thought to be required for transcription of most genes that are transcribed by this type of polymerase. In contrast, the transcription factors discussed in the previous section bind to specific sites in a limited number of genes. A transcription-initiation complex comprises an RNA polymerase and various general transcription factors bound to the promoter region.

Many general transcription factors required for Pol II to initiate transcription from most TATA-box promoters in vitro have been isolated and characterized. These proteins are designated *TFIIA, TFIIB,* etc., and most are multimeric proteins. TFIID is the largest with a mass of ≈750 kDa. It consists of a single 38-kDa *TATA box–binding protein (TBP)* and eleven TBP-associated factors (TAFs), which have not been extensively characterized. General transcription factors with similar activities have been isolated from cultured human cells, rat liver, *Drosophila* embryos, and yeast. The genes encoding these proteins in yeast have been sequenced as part of the complete yeast genome sequence, and many of the cDNAs encoding human and *Drosophila* general transcription factors have been cloned and sequenced. In all cases, equivalent general transcription factors from different eukaryotes are highly conserved. Although general transcription factors allow Pol II to initiate transcription in vitro at the same start sites used in vivo, additional proteins are required for transcription initiation in vivo.

Proteins Comprising the Pol II Transcription-Initiation Complex Assemble in a Specific Order in Vitro

In most biochemical studies on assembly of the Pol II initiation complex, researchers have used isolated TBP rather than the complete, multisubunit TFIID, which is difficult to purify. Binding of TBP and the other general transcription factors on promoters with consensus TATA boxes has been analyzed by DNase I footprinting and electrophoretic mobility shift assays. These studies demonstrate that the Pol II initiation complex is assembled in vitro in the stepwise sequence depicted in Figure 10-50.

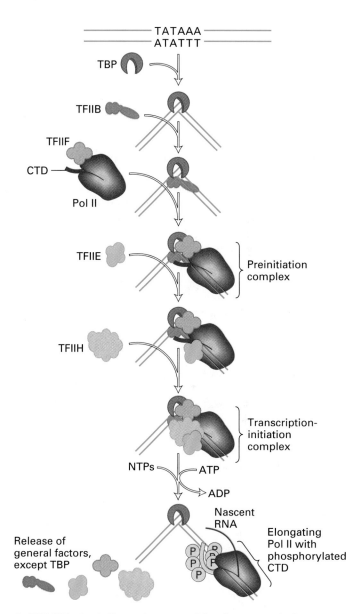

▲ **FIGURE 10-50 Stepwise assembly of a transcription-initiation complex from isolated RNA polymerase II (Pol II) and general transcription factors.** Once the complete transcription-initiation complex has assembled, separation of the DNA strands at the start site to form an open-complex requires ATP hydrolysis. As transcription initiates and the polymerase transcribes away from the promoter, the CTD becomes phosphorylated and the general transcription factors dissociate from the TBP-promoter complex. Numerous other proteins participate in transcription initiation in vivo.

In in vitro experiments with isolated general transcription factors, TBP is the first protein to bind to a TATA-box promoter. All eukaryotic TBPs analyzed to date have very similar C-terminal domains of 180 residues. The sequence of this region is 80 percent identical in the yeast and human proteins, and most differences are conservative substitutions. This conserved C-terminal domain functions as well as the full-length protein in in vitro transcription. (The N-terminal domain of TBP, which varies greatly in sequence and length among different eukaryotes, functions in the transcription of genes encoding snRNAs, which are discussed Chapter 11.) Figure 10-51 shows a model of the conserved C-terminal domain of TBP complexed to TATA-box DNA, based on x-ray crystallographic analysis. TBP is a monomer that folds into a saddle-shape structure; the two halves of the molecule exhibit an overall dyad symmetry but are not identical, unlike dimeric transcription factors. Like the HMGI and other DNA-bending proteins that participate in formation of enhancesomes, TBP interacts with the minor groove in DNA, bending the helix considerably.

Once TBP has bound to the TATA box, TFIIB can bind (see Figure 10-50). TFIIB is a monomeric protein, slightly smaller than TBP. Its C-terminal domain makes contact with both DNA and the bound TBP (Figure 10-52). The N-terminal domain of TFIIB extends toward the start site, but its three-dimensional structure is not yet known. Following TFIIB

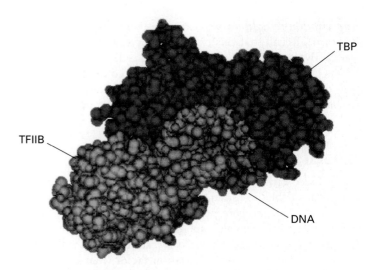

▲ **FIGURE 10-52 Structure of the complex formed between TBP, promoter DNA, and TFIIB.** In in vitro transcription systems, TFIIB binds to the assembled TBP–promoter DNA complex. Shown here are the C-terminal domain of *Arabidopsis* TBP and the C-terminal domain of human TFIIB. Transcription initiation in vivo also requires TFIIA, which binds to the TBP–promoter DNA complex on the side opposite to where TFIIB binds. TFIIA is thought to bind before TFIIB does. [Adapted from D. B. Nikolov et al., 1995, *Nature* **377**:119.]

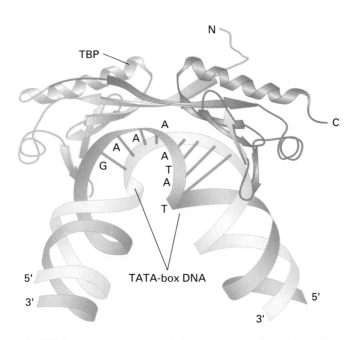

▲ **FIGURE 10-51 Structure of the conserved C-terminal domain of TBP bound to TATA-box DNA.** Although TBP is a monomer, the polypeptide backbone of this domain is folded into a conformation that has an approximate twofold symmetry. However, the surface residues of the protein clearly distinguish the two halves. TBP binds to the minor groove of TATA-box DNA, distorting the normal duplex structure and bending the DNA dramatically. [Adapted from J. L. Kim et al., 1993, *Nature* **365**:520.]

binding, a pre-formed complex of TFIIF (an $\alpha_2\beta_2$ tetramer) and Pol II binds, positioning the polymerase over the start site. In the resulting preinitiation complex, which is stable, the two largest subunits of Pol II (L and L′) interact with the promoter DNA along an ≈240 Å channel extending upstream and downstream of the start site (Figure 10-53). At most promoters, two more general transcription factors must bind before the DNA duplex can be separated to expose the template strand. First to bind is TFIIE (an $\alpha_2\beta_2$ tetramer), creating a docking site for TFIIH, another multimeric factor containing nine subunits. Binding of TFIIH completes assembly of the transcription-initiation complex in vitro (see Figure 10-50).

In the presence of ATP, the **helicase** activities of two TFIIH subunits unwind the DNA duplex at the start site, allowing Pol II to form an open complex. If the remaining ribonucleoside triphosphates are added, Pol II begins transcribing the template strand. As the polymerase transcribes away from the promoter region, another subunit of TFIIH phosphorylates the Pol II CTD at multiple sites (see Figure 10-50). In the minimal in vitro transcription assay containing only these general transcription factors and purified RNA polymerase II, TBP remains bound to the TATA-box as the polymerase transcribes away from the promoter region, but the other general transcription factors dissociate. As discussed below, transcription initiation in vivo requires additional factors; it is not yet clear which factors remain associated with promoter regions following each round of transcription initiation in the cell.

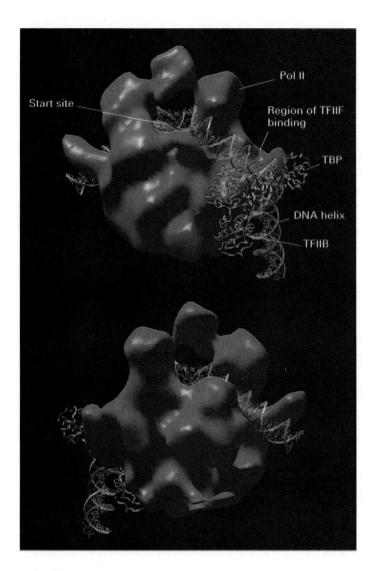

◀ FIGURE 10-53 Structural model of the complex composed of promoter DNA, TBP, TFIIB, and RNA polymerase II indicating the relative sizes of the components. Two views of the complex, rotated ≈180° front to back, are shown. The model shows a hypothetical path of the bound DNA along the surface of Pol II and its interaction with the core domains of TBP and TFIIB. The DNA region where TFIIF subunits can be chemically cross-linked in an initiation complex are indicated, but TFIIF is not depicted. In addition to the sharp bend in the DNA duplex introduced by binding of TBP, it is thought the DNA bends sharply again as it associates with deep clefts on the surface of Pol II. The active site of Pol II, and thus the transcription-start site, is between two finger-like projections at the top of the model. This model was developed from electron-microscope data, x-ray crystallographic analysis, and protein-DNA cross-linking studies. [Adapted from T.-K. Kim et al., 1997, *Proc. Nat'l. Acad. Sci. USA* **94**:12268; courtesy of Richard H. Ebright. See also R. Kornberg, 1996, *Trends Biochem. Sci.* **21**:325.]

function, these individuals may suffer from diseases such as *xeroderma pigmentosum, trichothiodystrophy*, or Cockayne syndrome.

A Pol II Holoenzyme Multiprotein Complex Functions in Vivo

The model shown in Figure 10-50 of eukaryotic transcription initiation in vitro seems far more complex than bacterial transcription initiation. Moreover, genetic and biochemical studies in yeast have revealed additional proteins called Srbs and Meds, which form a complex of approximately 20 polypeptides. This complex, termed Mediator, associates with the CTD region of Pol II. Gene knockout experiments subsequently demonstrated that most of the Srb/Med proteins in this complex are required for cell viability. Further experiments with temperature-sensitive mutants showed that some Srb proteins are essential for transcription initiation by Pol II in vivo. For example, shifting yeast cells carrying a temperature-sensitive mutation in an *Srb* gene from the permissive to the nonpermissive temperature resulted in an immediate cessation of initiation by Pol II.

Researchers also have isolated an ≈2-MDa multiprotein complex from yeast cells that includes Pol II, the Mediator complex, TFIIB, TFIIF, and TFIIH. A total of ≈50 polypeptides are present in this very large protein complex, which is stable in the absence of promoter DNA. The Pol II in this complex has an unphosphorylated CTD tail; this is the form of the enzyme that initiates transcription, as opposed to the chain-elongating form, which has a phosphorylated tail. Similar high-molecular-weight complexes containing a portion of the total cellular Pol II have been observed in extracts from nuclei of cultured human cells. These observations have led to the hypothesis that most of the proteins required for transcription initiation in cells are preassembled in an ≈2-MDa **holoenzyme** complex that then binds to promoter DNA in a single binding step.

MEDICINE Remarkably, many of the subunits of the complex TFIIH protein are also required for two other distinct processes in eukaryotic cells: the activation of protein kinases required for entry into the S phase of the cell cycle (Chapter 13) and transcription-linked repair of DNA damage by the excision-repair pathway (Section 12.4). Indeed the first subunits of TFIIH to be cloned from humans were identified because mutations in the genes encoding them cause defects in the repair of damaged DNA. In normal individuals, when a transcribing RNA polymerase becomes stalled at a region of damaged template DNA, a subcomplex of TFIIH is thought to recognize the stalled polymerase and then recruit other proteins that function with TFIIH in the excision-repair process. In the absence of functional TFIIH, the excision-repair of damaged DNA in transcriptionally active genes is impaired. As a result, affected individuals have extreme skin sensitivity to sunlight (a common cause of DNA damage) and exhibit a high incidence of cancer. Depending on the severity of the defect in TFIIH

Another general transcription factor, TFIIA, which is not required for initiation in vitro, is required for initiation by Pol II in vivo. Purified TFIIA forms a complex with TBP and TATA-box DNA. X-ray crystallography of the complex shows that TFIIA interacts with TBP and DNA on the opposite side of TBP from where the other general transcription factors and Pol II bind (see Figure 10-52). It is likely that in assembly of transcription-initiation complexes in higher eukaryotic cells TFIIA and TFIID, with its multiple TAF subunits, bind first to the promoter DNA and then all the components of the holoenzyme mentioned above subsequently bind. This amounts to a multiprotein complex of some 60–70 polypeptides with a mass of ≈ 3 MDa, nearly as large as a eukaryotic ribosome.

SUMMARY RNA Polymerase II Transcription-Initiation Complex

- RNA polymerase II requires several general transcription factors to locate the proper start site in a DNA template and initiate transcription. These include TFIID, which binds to a TATA-box through its TATA-box binding subunit, TBP.

- Transcription of protein-coding genes by RNA polymerase II can be initiated in vitro by sequential binding of the following in the indicated order: TBP, which binds to TATA-box DNA; TFIIB; a complex of Pol II and TFIIF; TFIIE; and finally TFIIH (see Figure 10-50).

- The helicase activities of two TFIIH subunits separate the template strands at the start site in most promoters, a process that requires hydrolysis of ATP. As Pol II begins transcribing away from the start site, its CTD is phosphorylated by another TFIIH subunit.

- Initiation by Pol II in vivo requires the multiprotein Mediator complex, which associates with the unphosphorylated CTD of Pol II, forming a very large holoenzyme complex that also includes most of the general transcription factors. This pre-assembled holoenzyme is thought to bind to promoter DNA in a single step in vivo.

- The Pol II transcription-initiation complex that assembles on promoters in vivo may comprise as many as 60–70 polypeptides with a total mass similar to that of a ribosome.

10.7 Molecular Mechanisms of Eukaryotic Transcriptional Control

Transcriptional control in eukaryotic cells can be visualized as involving several levels of regulation. The concentrations and activities of activators and repressors that control transcription of many protein-coding genes are regulated during cellular differentiation and in response to hormones and signals from neighboring cells. These activators and repressors in turn regulate changes in chromatin structure and **histone acetylation** and **deacetylation,** thereby influencing the ability of general transcriptions factors to bind to promoters. In addition, activators and repressors directly regulate assembly of transcription-initiation complexes and the rate at which they initiate transcription. In this section, we review current understanding of how activators and repressors control chromatin structure and initiation-complex assembly and how these molecular events work together to regulate gene expression according to the needs of the cell and organism.

N-Termini of Histones in Chromatin Can Be Modified

As discussed in Chapter 9, the DNA in eukaryotic cells is not free, but is associated with a roughly equal mass of protein in the form of **chromatin.** In some cases, the ability of transcription factors to interact with long stretches of DNA sequence is regulated by controlling chromatin structure. The basic structural unit of chromatin is the **nucleosome,** which is composed of ≈ 146 base pairs of DNA wrapped tightly around a disk-shaped core of histone proteins. In condensed chromatin, the nucleosomes associate with each other into a 30-nm fiber (see Figure 9-31). The amino acid residues at the N-terminus of each histone (≈ 20–60 residues depending on the histone) extend from the surface of the nucleosome (see Figure 9-30). These histone N-termini are rich in lysine residues, which can be reversibly modified by acetylation, phosphorylation, and methylation, as well as by the addition of a single ubiquitin molecule, a highly conserved 76-residue protein.

Phosphorylation has an important, but as yet poorly understood, function in the condensation of chromosomes during mitosis. The functions of the other modifications are also poorly understood, except in the case of acetylation. Acetylation of the histone N-termini is associated with gene control during interphase. Other proteins less abundant than histones, but nonetheless present in large numbers, also are associated with chromatin. These include HMG proteins, which participate in formation of enhancesomes (see Figure 10-48).

Formation of Heterochromatin Silences Gene Expression at Telomeres and Other Regions

For many years it has been clear that inactive genes in eukaryotic cells are often associated with heterochromatin, regions of chromatin that stain more darkly with DNA dyes than euchromatin where most transcribed genes are located (see Figure 9-39). Heterochromatin stains more darkly than euchromatin because it is more highly condensed. The DNA in heterochromatin is less accessible to externally added proteins than DNA in euchromatin. For instance, in an experiment described in the last chapter, the DNA of inactive genes was found to be far more resistant to digestion by DNase I than the DNA of transcribed genes (see Figure 9-32).

Silencing of Yeast Silent Mating-Type Loci and Telomeric Regions Study of DNA regions in *S. cerevisiae* that behave like the heterochromatin of higher eukaryotes have provided insight about chromatin-mediated repression of transcription. This yeast can grow either as haploid or diploid cells. Haploid cells exhibit one of two possible mating types, called **a** and α. Cells of different mating type can "mate," or fuse, to generate a diploid cell (Figure 10-54). When diploid cells are starved, they sporulate, forming four haploid spores, each of which can germinate when supplied with nutrients, generating haploid cells. A normal haploid cell switches its mating type each generation. Genetic and molecular analyses have traced this remarkable phenomenon to regulated changes in the DNA sequence of chromosome III.

Three genetic loci directly involved in mating-type switching have been located (Figure 10-55). The central locus is termed *MAT*, the mating-type locus. Genes at the *MAT* locus are actively transcribed into mRNA. The mRNAs expressed from the *MAT* locus encode transcription factors that regulate other genes that give the cell its **a** or α phenotype (Section 14.1). Two additional "silent" (nontranscribed) copies of the genes for these transcription factors are "stored" at loci termed *HML* and *HMR*, near the left and right telomere, respectively, of chromosome III. These sequences are transferred alternately from *HML*α or *HMR*a into the *MAT*

locus, once during each cell generation, by a type of recombination called **gene conversion** (Section 12.5). When the *MAT* locus contains the DNA sequence from *HML*α, the cells behave as α cells. When the *MAT* locus contains the DNA sequence from *HMR*a, the cells behave like **a** cells.

Repression of the silent mating-type loci is critical to haploid cells. If the silent loci are expressed, as they are in yeast mutants with defects in the repressing mechanism, both **a** and α transcription factors are expressed, causing the cells to behave like diploid cells, which cannot mate. The promoters and UASs controlling transcription of the **a** and α genes lie near the center of the DNA sequence that is transferred and are identical whether the sequences are at the *MAT* locus or one of the silent loci. Consequently, the function of the transcription factors that interact with these sequences is somehow blocked at *HML* and *HMR*. This repression of the silent loci depends on **silencer sequences** located next to the region of transferred DNA at *HML* and *HMR* (see Figure 10-55). If the silencer is deleted, the adjacent silent locus is transcribed. Remarkably, any gene placed near the yeast mating-type silencer by recombinant DNA techniques is repressed, or "silenced," even a tRNA gene transcribed by RNA polymerase III, which uses a different set of general transcription factors from RNA polymerase II.

Several lines of evidence indicate that repression of the *HML* and *HMR* loci results from a condensed chromatin structure that sterically blocks transcription factors from interacting with the DNA. In one telling experiment, the gene encoding a DNA methylase of *E. coli* was introduced into yeast cells under the control of a yeast promoter so that the enzyme was expressed. This enzyme methylates adenine residues in the sequence GATC. Methylation at this sequence can be assayed easily with restriction enzymes that digest either the methylated or unmethylated sequence. Using these methods, researchers demonstrated that the *E. coli* methylase expressed in yeast cells was able to methylate GATC sequences within the *MAT* locus and most other regions of the yeast genome, but not within the *HML* and *HMR* loci. These results indicate that the DNA of the silent loci is inaccessible to proteins in general, including transcription factors and RNA polymerase. In similar experiments conducted with various yeast histone mutants, mutations in the N-terminal region of histones H3 and H4 were found to derepress the silent loci, allowing the *E. coli* DNA methylase to gain access to GATC sequences in *HML* and *HMR*. This result suggested that specific interactions involving the H3 and H4 N-termini are required for formation of a fully repressing chromatin structure.

Other studies indicate that the telomeres of every yeast chromosome also behave like silencers. For instance, when a gene is placed within a few kilobases of any of the yeast telomeres, its expression is repressed. In addition, this repression is relieved by the same mutations in the H3 and H4 N-termini that interfere with repression at the silent mating-type loci.

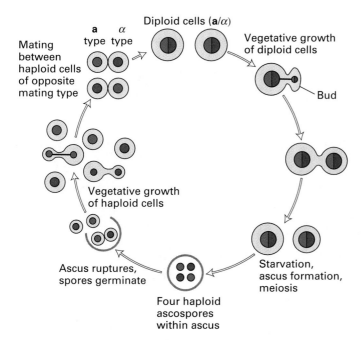

▲ **FIGURE 10-54 Life cycle of *S. cerevisiae*.** Two haploid cells that differ in mating type, called **a** and α, can mate to form a diploid **a**/α cell, which multiplies by budding. Under starvation conditions, diploid cells undergo meiosis, forming haploid ascospores. Rupture of an ascus releases four haploid spores, which can germinate into haploid cells. Once each generation a haploid cell is converted to the opposite mating type.

Labels in figure: Mating between haploid cells of opposite mating type; **a** type; α type; Diploid cells (**a**/α); Vegetative growth of diploid cells; Bud; Starvation, ascus formation, meiosis; Four haploid ascospores within ascus; Ascus ruptures, spores germinate; Vegetative growth of haploid cells

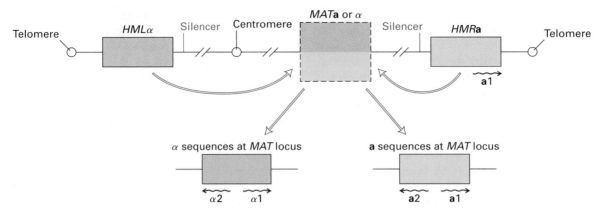

▲ **FIGURE 10-55 Genes on chromosome III involved in mating-type control in the yeast *S. cerevisiae*.** Silent (unexpressed) mating-type genes (either **a** or α, depending on the strain) are located at *HML*. The opposite mating-type genes are present at the silent *HMR* locus. Once every other cell division, the DNA sequence at *HML* is transferred to the *MAT* locus; in the alternate cell divisions, the DNA sequence from *HMR* is transferred to the *MAT* locus. When the α or **a** sequences are present at the *MAT* locus, they can be transcribed into mRNAs whose encoded proteins are transcription factors that regulate the expression of mating-type specific genes (see Figure 14-1). The silencer sequences near *HML* and *HMR* bind proteins that are critical for repression of these silent loci.

Additional genetic studies have revealed several genes— *RAP1* and three *SIR* (*silent information regulator*) genes— that are required for repression of the silent mating-type loci and the telomeres in yeast. *RAP1* encodes a protein that binds within the silencer DNA sequences associated with *HML* and *HMR* and to a sequence that is repeated multiple times at each yeast chromosome telomere. Further biochemical studies of these proteins have shown that they bind to each other and that two bind to the N-termini of H3 and H4. Immunofluorescence confocal microscopy of yeast cells stained with antibody to the Sir and Rap proteins and hybridized to a labeled telomeric DNA probe revealed that these proteins form large telomeric nucleoprotein structures resembling the heterochromatin found in higher eukaryotes (Figure 10-56). These results have led to the model for silencing at yeast telomeres depicted in Figure 10-57. An important feature of this model, which has been experimentally demonstrated, is that the histone N-termini are hypoacetylated.

In this model, formation of heterochromatin is nucleated by the multiple Rap1 proteins bound to repeated sequences in the nucleosome-free region at the telomere. Rap1 binds Sir3 and Sir4, which then form a network of protein-protein interactions with Sir2, hypoacetylated histones H3 and H4,

(a)

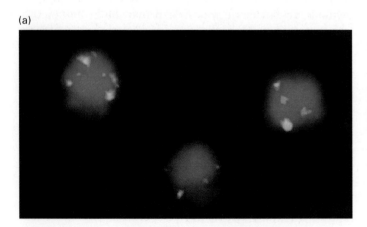

(b)   (c)

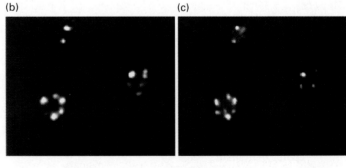

▲ **FIGURE 10-56 Co-localization of Sir3 protein with telomeric heterochromatin in yeast nuclei.** (a) Confocal micrograph 0.3 μm thick through three diploid yeast cells, each containing 34 telomeres. Telomeres were labeled by hybridization to a fluorescent telomere-specific probe (yellow). DNA was stained red to reveal the nuclei. The 34 telomeres coalesce into a much smaller number of regions near the nuclear periphery.

(b,c) Confocal micrographs of yeast cells labeled with a telomere-specific hybridization probe (b) and a fluorescent-labeled antibody specific for Sir3 (c). Note that Sir3 is localized in the repressed telomeric heterochromatin. Similar experiments with Rap1, Sir2, and Sir4 have shown that these proteins also co-localize with the repressed telomeric heterochromatin. [From M. Gotta et al., 1996, *J. Cell Biol.* **134**:1349; courtesy of M. Gotta, T. Laroche, and S. M. Gasser.]

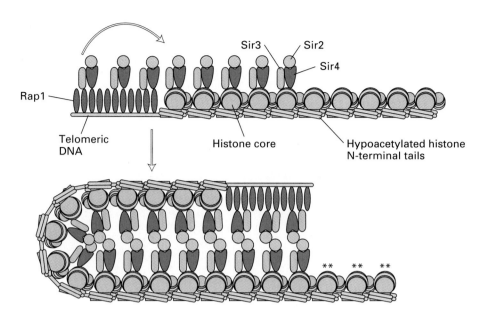

◀ **FIGURE 10-57 Schematic model of silencing mechanism at yeast telomeres.** Multiple copies of Rap1 bind to a simple repeated sequence at each telomere region, which lacks nucleosomes *(top)*. This nucleates the assembly of a multiprotein complex *(bottom)* through protein-protein interactions between Rap1, Sir2, Sir3, Sir4, and the hypoacetylated amino-terminal tails of histones H3 and H4 of nearby nucleosomes. Asterisks represent hyperacetylated histone amino-terminal tails. The heterochromatin structure encompasses ≈4 kb of DNA neighboring the Rap1-binding sites, irrespective of its sequence. The actual structure of the higher-order heterochromatin is not yet understood. See text. [Adapted from M. Grunstein, 1997, *Curr. Opin. Cell Biol.* **9**:383.]

and additional Sir3 and Sir4 proteins, creating a stable, higher-order nucleoprotein complex in which the DNA is largely inaccessible to external proteins. One additional protein, Sir1, is also required for silencing of the silent mating-type loci. Although the function of Sir1 is not yet understood, it is thought to allow the telomeric silencing mechanism to encompass *HML* and *HMR*. The dependence of the silencing mechanism on histone hypoacetylation was shown in experiments in which arginines and glutamines were substituted for lysines in histone N-termini of constructed yeast mutants. Arginine is positively charged like lysine, but cannot be acetylated, and glutamine stimulates lysine acetylation. Substitution to arginine was compatible with silencing, whereas substitution with glutamine was not.

Silencing in Higher Eukaryotes Regulation of transcription through **heterochromatin-mediated repression** is also important in multicellular eukaryotes. For example, expression of Hox transcription factors, which regulate development of the "body plan" (i. e., normal anatomy) in nearly all animals (Section 14.3), is subject to heterochromatin repression. The mechanism of this repression is still being worked out, but genetic analysis in *Drosophila* has revealed that multiple proteins nucleate formation of heterochromatin regions at specific sites within the Hox genes. In *Drosophila*, some of these proteins can be visualized binding to multiple, specific locations in the chromosome by in situ binding of specific-labeled antibodies to salivary gland polytene chromosomes.

Repressors Can Direct Histone Deacetylation at Specific Genes

The role of **histone deacetylation** in chromatin-mediated gene repression has been further supported by the discovery of yeast proteins that repress transcription of multiple genes at internal chromosomal positions. These proteins are now known to act in part by causing deacetylation of histone N-termini in nucleosomes that bind to the TATA box of the genes they repress. In vitro studies have shown that when promoter DNA is assembled onto a nucleosome with unacetylated histones, the general transcription factors cannot bind to the TATA box and initiation region. In unacetylated histones, the N-terminal lysines are positively charged and interact strongly with DNA phosphates, increasing the affinity of DNA for the nucleosome surface. This strong interaction may prevent access of general transcription factors to the promoter region. In contrast, binding of general transcription factors is repressed much less by histones with hyperacetylated N-termini in which the positively charged lysines are neutralized and electrostatic interactions with DNA phosphates are eliminated. Moreover, the binding of transcription factors to promoter DNA in this case is greatly stimulated by transcriptional activators.

The connection between histone deacetylation and repression of transcription at nearby yeast promoters became clearer when the first *histone deacetylase* was purified (from human cells), and the cDNA encoding it was cloned based on amino acid microsequencing. The cDNA sequence showed high homology to the yeast *RPD3* gene, known to be required for the normal repression of a number of yeast genes. Further work showed that the function of the Rpd3 protein at a number of promoters depends on two other proteins: Ume6, a repressor that binds to a specific upstream regulatory sequence (URS1), and Sin3, which is part of a large, multiprotein complex that also contains Rpd3. Sin3 also interacts with the repressor domain of Ume6, thus positioning the Rpd3 histone deacetylase in the complex so it can interact with nearby nucleosomes and remove acetyl groups from specific N-terminal lysines (Figure 10-58a). Additional experiments, using the technique outlined in Figure 10-59,

(a) Repressor-directed histone deacetylation

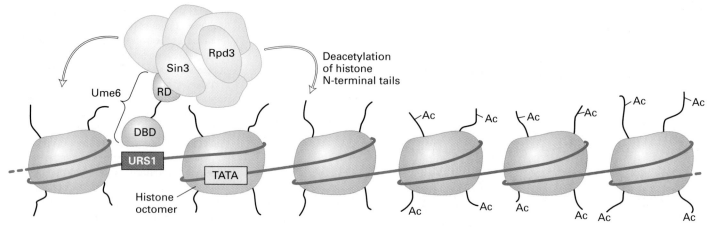

(b) Activator-directed histone hyperacetylation

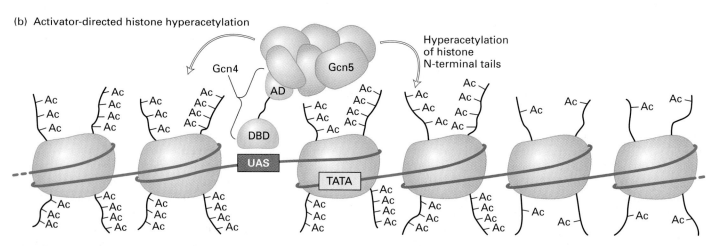

▲ **FIGURE 10-58 Role of deacetylation and hyperacetylation of histone N-terminal tails in yeast transcription control.**
(a) Repressor-directed deacetylation of histone N-terminal tails. The DNA-binding domain (DBD) of the repressor Ume6 interacts with a specific upstream control element (URS1) of the genes it regulates. The Ume6 repression domain (RD) binds Sin3, a subunit of a multiprotein complex that includes Rpd3, a histone deacetylase. Deacetylation of histone N-terminal tails on nucleosomes in the region of the Ume6-binding site inhibits binding of general transcription factors at the TATA box, thereby repressing gene expression. (b) Activator-directed hyperacetylation of histone N-terminal tails. The DNA-binding domain of Gcn4 interacts with specific upstream-activating sequences (UAS) of the genes it regulates. The Gcn4 activation domain (AD) then interacts with a multiprotein histone acetylase complex that includes the Gcn5 catalytic subunit. Subsequent hyperacetylation of histone N-terminal tails on nucleosomes in the vicinity of the Gcn4-binding site facilitates access of the general transcription factors required for initiation. Repression and activation of some genes in higher eukaryotes occurs by similar mechanisms.

demonstrated that in wild-type yeast, one or two nucleosomes in the immediate vicinity of Ume6-binding sites are hypoacetylated. These DNA regions include the promoters of genes repressed by Ume6. In *sin3* and *rpd3* deletion mutants, not only were these promoters derepressed, but the nucleosomes neighboring the Ume6-binding sites were hyperacetylated. This finding provides considerable support for the model shown in Figure 10-58a.

The *SIN3* and *RPD3* genes are required for complete repression by a number of other yeast repressors, which bind to DNA at different sites than does Ume6. These repressors are thought to function by the same mechanism as the

Ume6-Sin3-Rpd3 system. Also, three other histone deacetylase complexes have been identified in yeast extracts. Some of these may also be targeted to specific promoters to repress transcription through deacetylation of histones in specific nucleosomes.

Histone deacetylases also have been found associated with repressors from higher eukaryotes. These include two heterodimeric repressors that participate in regulation of the cell cycle in mammals and a group of nuclear receptors that are regulated by lipid-soluble hormones. These sequence-specific DNA-binding repressors interact with mammalian homologs of Sin3 (mSin3), which are found in large, multiprotein

Isolated, cross-linked chromatin

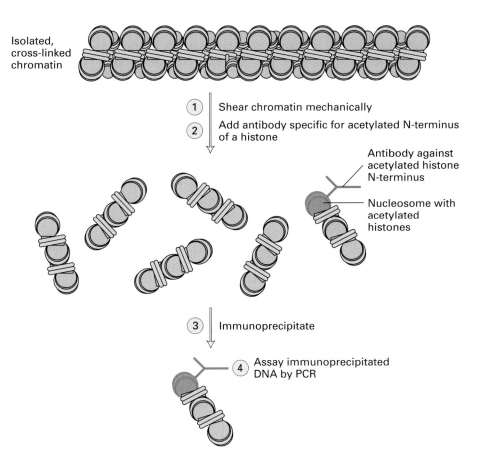

1. Shear chromatin mechanically
2. Add antibody specific for acetylated N-terminus of a histone

Antibody against acetylated histone N-terminus

Nucleosome with acetylated histones

3. Immunoprecipitate

4. Assay immunoprecipitated DNA by PCR

◀ **FIGURE 10-59 Experimental method for analyzing the acetylation state of histones in chromatin associated with a specific region of the genome.** Nucleosomes are lightly cross-linked to DNA in vivo using a cell-permeable, reversible, chemical cross-linking agent. Chromatin is then isolated, sheared to an average length of three nucleosomes, and subjected to immunoprecipitation with an antibody specific for a particular acetylated N-terminal histone sequence. The DNA in the immunoprecipitated chromatin fragment is released by reversing the cross-link and then is quantitated using a sensitive PCR method. [See S. E. Rundlett et al., 1998, *Nature* **392**:831.]

histone deacetylases complexes, and are thought to direct the specific hypoacetylation of nucleosomes in the vicinity of their binding sites, similar to the proposed yeast mechanism (see Figure 10-58a).

These recent findings provide an explanation for earlier observations that in vertebrates transcriptionally inactive DNA regions often contain the modified cytidine residue **5-methylcytidine (mC)** followed immediately by a G, whereas transcriptionally active DNA regions lack mC residues. DNA containing 5-methylcytidine has been found to bind a specific protein that in turn interacts specifically with mSin3. This finding suggests that association of mSin3-containing histone deacetylase complexes with methylated sites in DNA leads to deacetylation of histones in neighboring nucleosomes, making these regions inaccessible to general transcription factors and Pol II, and hence transcriptionally inactive.

Activators Can Direct Histone Acetylation at Specific Genes

The yeast gene *GCN5* is known from genetic studies to be required for maximal activation by the yeast activator Gcn4 and several other activators with acidic activation domains. As in the case of histone deacetylases, purification, microsequencing, and cloning of the gene encoding a histone

acetylase from another source (*Tetrahymena*, a rich source), which has a strong homology to yeast *GCN5*, suggested how the Gcn5 protein functions. Subsequent studies showed that Gcn5 is present in two large multiprotein complexes that have histone acetylase activity. Another subunit of these histone acetylase complexes binds to acidic activation domains. The model shown in Figure 10-58b is consistent with the observation that nucleosomes near the promoter region of a gene regulated by *GCN5* are specifically hyperacetylated, as determined by the nucleosome immunoprecipitation method (see Figure 10-59). The activator-directed hyperacetylation of nucleosomes near a promoter region changes the chromatin structure so as to facilitate the binding of other proteins required for transcription initiation.

A similar activation mechanism operates in higher eukaryotes. In mammals, for instance, there is a small family of ≈400-kDa, multidomain CBP proteins. As noted earlier, one domain of CBP binds the phosphorylated CREB transcription factor (see Figure 10-46). Other domains of CRB interact with distinct classes of activation domains in other transcription factors. Interaction between CBP and various activators is required for their maximal activity, reflecting the function of CBP as a co-activator. Yet another domain of CBP has histone acetylase activity, and large multiprotein CBP–histone acetylase complexes, functionally analogous to the yeast Gcn5-containing complexes, have been identified in nuclear extracts from mammalian cells. Activators are thought to function in part by directing a CBP–histone acetylase complex to specific nucleosomes, where it acetylates histone N-terminal tails, facilitating the interaction of general transcription factors with promoter DNA. In addition, the largest TFIID subunit (TAF$_{II}$145 in yeast and TAF$_{II}$250 in higher eukaryotes) has been shown to interact with a number of activation domains. This TFIID subunit also has histone acetylase activity and may function by acetylating histone N-terminal tails in the vicinity of the TATA box.

As noted previously, chromosomal DNA in the region of a transcribed gene is more sensitive to digestion by DNase I than DNA in a transcriptionally silent region (see Figure 9-32). Nonetheless, transcriptionally active DNA is more resistant to DNase I digestion than "naked" DNA because most of the DNA is bound to the surface of histone octamers in nucleosomes. However, within transcriptionally active regions of chromatin, some sites are nearly as sensitive to DNase I digestion as naked DNA. These **DNase I–hypersensitive sites** occur in regions where transcription factors are bound, and probably result from digestion at sites immediately adjacent to the bound factors. DNase I–hypersensitive sites can be mapped by Southern blotting and may be useful in identifying transcription factor–binding sites for a gene of interest.

Chromatin-Remodeling Factors Participate in Activation at Some Promoters

Genetic analyses in yeast first revealed another type of multiprotein complex, called the *Swi/Snf chromatin-remodeling complex,* required for activation at some promoters. Several of the largest Swi/Snf subunits have homology to DNA and RNA helicases, enzymes that use energy released by ATP hydrolysis to disrupt interactions between base-paired nucleic acids. Other enzymes with the helicase homology disrupt nucleic acid–protein interactions. Even in transcriptionally active regions of chromatin, which are more susceptible to DNase than inactive regions, bound histone octamers partially protect the DNA from digestion. However, in the presence of the purified Swi/Snf complex, nucleosomal DNA becomes more susceptible to DNase I digestion. This finding suggests that the ability of the Swi/Snf complex to facilitate the in vitro binding of some transcription factors to sites in nucleosomal DNA results from transient dissociation of the DNA from the surface of nucleosomes. Yeast also contains other homologous multiprotein complexes with similar "chromatin-remodeling" activities, raising the possibility that different chromatin-remodeling complexes may be required by distinct families of activators.

Higher eukaryotes also contain multiprotein complexes with homology to the yeast Swi/Snf chromatin-remodeling factors. *Drosophila* genetic studies have revealed that some of these are required for the normal regulation of Hox genes through chromatin structure. Also, protein complexes isolated from nuclear extracts of mammalian and *Drosophila*

cells have been found to assist binding of transcription factors to their cognate sites in nucleosomal DNA in an ATP-requiring process. Consequently, it seems clear that chromatin-remodeling factors participate in the regulation of genes in higher eukaryotes as well as in yeast.

Activators Stimulate the Highly Cooperative Assembly of Initiation Complexes

After participating in the hyperacetylation of chromatin in the vicinity of a promoter region (see Figure 10-58b), transcriptional activators are thought to stimulate the assembly of an initiation complex and regulate the frequency at which new Pol II molecules reinitiate transcription. This function of activators, which provides a second level of transcriptional control, often can be demonstrated in in vitro reactions lacking histones. For example, some activators stimulate the binding of TFIID or the simultaneous binding of TFIID plus TFIIA to the TATA box in vitro. Other activators interact with other general transcription factors and with subunits in the multiprotein Mediator complex associated with the CTD of the largest Pol II subunit. As discussed earlier, many of these general transcription factors and the Mediator complex may occur in a preassembled holoenzyme complex that can bind to a TFIID–promoter DNA complex in a single step. CBP and other co-activators also participate in the network of protein-protein and protein-DNA interactions in the large nucleoprotein complexes that assemble at eukaryotic promoters.

Assembly of the multiprotein initiation complex on promoter DNA is thought to result from multiple cooperative interactions such as those illustrated for the binding of *E. coli* CAP-cAMP and RNA polymerase to the *lac* promoter region (see Figure 10-17). In higher organisms, the strong cooperativity of initiation-complex assembly is in part responsible for cell type–specific gene expression. The *TTR* gene, which encodes transthyretin in mammals, is a good example of this. As noted earlier, transthyretin is expressed in hepatocytes and in choroid plexus cells. Transcription of the *TTR* gene in hepatocytes is controlled by at least five different transcriptional activators (Figure 10-60):

- HNF1, a hepatocyte-specific homeobox protein
- HNF3, a hepatocyte-specific winged-helix protein
- HNF4, a nuclear receptor that also is expressed in intestinal epithelial cells and kidney tubule cells
- C/EBP, a basic-zipper heterodimer that also is expressed in intestinal epithelial cells, fat cells, and some neurons
- AP1, a small family of basic-zipper heterodimeric proteins that are expressed in virtually all cell types

Even though three of these activators are expressed in other cell types, the *TTR* gene is not transcribed in these cells. Thus hepatocyte-specific transcription of *TTR* occurs because the complete set of activators is expressed only in hepatocytes. All of the activators must be present to contribute to the highly cooperative assembly of an initiation complex at the *TTR* promoter (Figure 10-61).

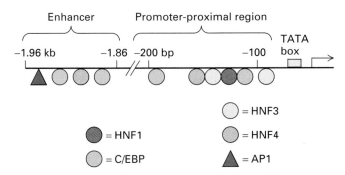

▲ **FIGURE 10-60 Binding sites for activators that control transcription of the mouse transthyretin (TTR) gene in hepatocytes.** HNF = hepatocyte nuclear factor. [See R. Costa et al., 1989, *Mol. Cell Biol.* **9:**1415; K. Xanthopoulus et al., 1989, *Proc. Nat'l. Acad. Sci. USA* **86:**4117.]

Different genes that encode prominent hepatocyte-specific proteins, such as serum albumin or α_1-antitrypsin, have different arrangements of protein-binding sites and use overlapping but not identical sets of factors. Thus there is no single arrangement of sites that dictates hepatocyte-specific gene expression. Serum albumin is expressed at far higher levels than transthyretin because the serum albumin gene is transcribed much more frequently in hepatocytes than the transthyretin gene. This difference reveals another level of control by transcription factors, regulation of the

frequency of transcription initiation for those genes that are transcribed in a specific cell type. Much remains to be learned about the mechanisms that result in differential transcription-initiation frequency within a given cell type.

Repressors Interfere Directly with Transcription Initiation in Several Ways

A repressor is any protein that interferes with transcription initiation when it is bound to a specific site on DNA. As discussed above, some eukaryotic repressors can direct deacetylation of histones in nucleosomes near their cognate binding sites (see Figure 10-58a). Histone deacetylation, in turn, inhibits the interaction of general transcription factors with their binding sites in nucleosomal DNA, thereby repressing transcription. However, the finding that a number of eukaryotic repressor proteins repress in vitro transcription in the absence of histones indicates that more direct repression mechanisms also operate.

Although repression mechanisms are not well understood, different repressor proteins probably exert their effects in different ways (Figure 10-62). Two mechanisms involve competitive binding between a repressor and activator or general transcription factor. In both cases, binding of a repressor molecule to a specific DNA site blocks binding of proteins required to initiate transcription. In many cases, however, eukaryotic repressors inhibit transcription without interfering with the binding of an activator or general transcription factors. In such cases, the bound repressor may interact with a nearby

▶ **FIGURE 10-61 Model for cooperative assembly of an activated transcription-initiation complex at the TTR promoter in hepatocytes.** Four activators enriched in hepatocytes plus the ubiquitous AP1 factor bind to sites in the hepatocyte-specific enhancer and promoter-proximal region of the TTR gene. The activation domains of the bound activators interact extensively with co-activators, TAF subunits of TFIID, Srb/Mediator proteins, and general transcription factors, resulting in looping of the DNA and formation of a stable activated initiation complex. Because of the highly cooperative nature of complex assembly, an initiation complex does not form on the TTR promoter in intestinal epithelial cells, which contain only two of the four hepatocyte-enriched transcription factors. Many of the general transcription factors, Srb/Mediator proteins, and RNA polymerase II (Pol II) may be pre-assembled into a holoenzyme complex.

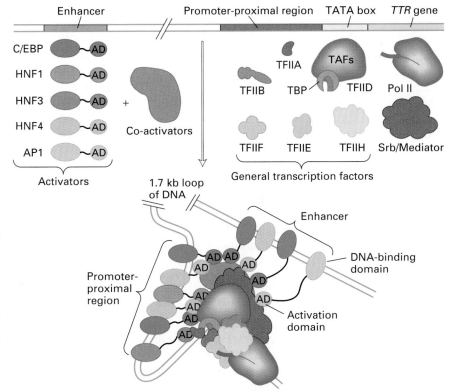

(a) Competitive binding with activator

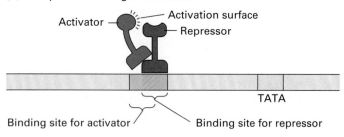

(b) Interaction with activation domain of bound activator

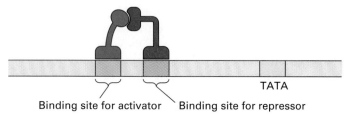

(c) Interaction with general transcription factors

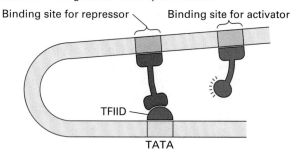

▲ **FIGURE 10-62 Various eukaryotic repressors can inhibit transcription by mechanisms that do not involve histone deacetylation.** In the three mechanisms shown, the repressor either inhibits activation or directly interferes with formation of the initiation complex. In addition, some repressors interact with "co-repressor" proteins, that are thought to interact in turn with general transcription factors to inhibit initiation.

activator, preventing its function, or with general transcription factors bound at the promoter, preventing their assembly into an initiation complex. Presumably, repression of the *EGR-1* gene by WT1 protein, discussed earlier, operates by one of the latter two mechanisms, since WT1 binding does not interfere with activator binding (see Figure 10-49).

Regulation of Transcription-Factor Expression Contributes to Gene Control

We have seen in the preceding discussion that transcription of eukaryotic genes is regulated by combinations of activators and repressors that bind to specific DNA regulatory sequences. Whether or not a specific gene in a multicellular organism is expressed in a particular cell at a particular time

is largely a consequence of the binding and activity of the transcription factors that interact with the regulatory sequences of that gene. Clearly, since different proteins are expressed in different cells at different times in development, the activity of transcription factors must be controlled.

An obvious critical control point for cells is transcription of the genes encoding transcription factors themselves. Hepatocyte-specific expression of transthyretin provides an example: The complete set of activators required for transcription of the *TTR* gene are expressed only in hepatocytes. The transcription factors expressed in a particular cell type, and the amounts produced, are a consequence of multiple regulatory interactions between transcription-factor genes that occur during the development and differentiation of a particular cell type. In Chapters 14, 20, and 23, we present examples of such regulatory interactions during development and discuss the principles of development and differentiation that have emerged from these examples.

Expression of a particular gene is further controlled by regulating the activities of the factors required for its transcription. In the remainder of this section, we discuss two important mechanisms for regulating transcription-factor activity: interaction of transcription factors with small effector molecules (e.g., lipid-soluble hormones) and post-translational modifications (e.g., phosphorylation).

Lipid-Soluble Hormones Control the Activities of Nuclear Receptors

The activities of many transcription factors are regulated by hormones, which function as extracellular signals in multicellular organisms (Chapter 20). Hormones are secreted from one cell type and travel through extracellular fluids to affect the function of cells at a different location in the organism. One class of hormones comprises small, lipid-soluble molecules, which can diffuse through plasma and nuclear membranes (Figure 10-63). As discussed earlier, these lipid-soluble hormones, including many different steroid hormones, retinoids, and thyroid hormones, bind to and regulate specific transcription factors belonging to the nuclear-receptor superfamily.

Domain Structure of Nuclear Receptors Cloning and sequencing of the genes encoding several nuclear receptors permitted comparison of their amino acid sequences. Such studies revealed a remarkable conservation in both the amino acid sequences and different functional regions of various nuclear receptors (Figure 10-64). All the nuclear receptors have a unique N-terminal region of variable length (100–500 amino acids) containing regions that function as transcription-activation domains. The DNA-binding domain maps near the center of the primary sequence and has the C_4 zinc-finger motif. The hormone-binding domain lies near the C-terminal end of these receptors and contains a hormone-dependent activation domain. In some cases the hormone-binding domain functions as a repression domain in the absence of ligand.

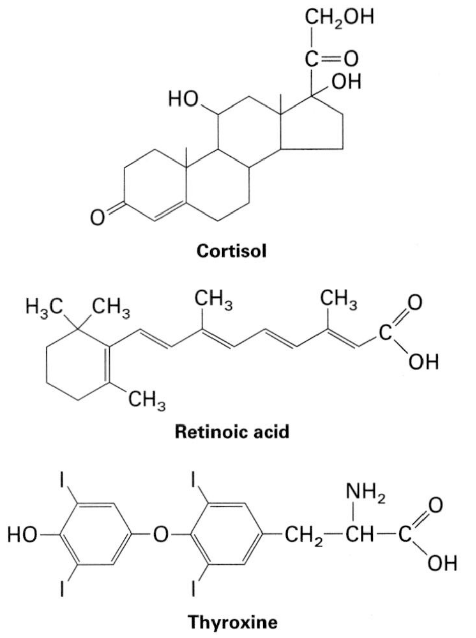

◀ FIGURE 10-63 Examples of lipid-soluble hormones that bind to members of the nuclear-receptor superfamily of transcription factors. Cortisol is a steroid hormone that binds to the glucocorticoid receptor (GR). Like other steroid hormones, it is synthesized from cholesterol. Retinoic acid is a metabolic derivative of vitamin A that has powerful effects on limb bud development in embryos and skin renewal in adult mammals. It is the ligand for the retinoic acid A receptor (RAR). Thyroxine is synthesized from tyrosine residues in the protein thyroglobulin in the thyroid gland. It is a ligand for the thyroid hormone receptor (TR).

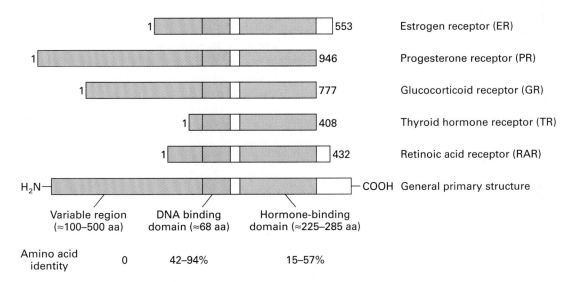

▲ FIGURE 10-64 General design of transcription factors in nuclear-receptor superfamily. The centrally located DNA-binding domain exhibits considerable sequence homology among different receptors and has the C_4 zinc-finger motif. The C-terminal hormone-binding domain exhibits somewhat less homology. The N-terminal regions in various receptors vary in length, have unique sequences, and may contain one or more activation domains. This general pattern has been found in the estrogen receptor (553 amino acids [aa]), progesterone receptor (946 aa), glucocorticoid receptor (777 aa), thyroid hormone receptor (408 aa), and retinoic acid receptor (432 aa). [See R. M. Evans, 1988, *Science* **240**:889.]

Nuclear-Receptor Response Elements The characteristic nucleotide sequences of the DNA sites, called *response elements*, that bind several major nuclear receptors have been determined. The sequences of the consensus response elements for the glucocorticoid and estrogen receptors are 6-bp inverted repeats separated by any three base pairs (Figure 10-65a,b). This finding suggested that these steroid hormone receptors would bind to DNA as symmetrical dimers, as was later shown from the x-ray crystallographic analysis of the homodimeric glucocorticoid receptor's C_4 zinc-finger DNA-binding domain (see Figure 10-41b).

Some nuclear-receptor response elements, such as those for the vitamin D_3, thyroid hormone, and retinoic acid receptors, are direct repeats of the same sequence recognized by the estrogen receptor, separated by three to five base pairs (Figure 10-65c–e). The receptors that bind to such direct-repeat

(a) GRE
5′ $\overrightarrow{\text{AGAACA}}(N)_3$ TGTTCT 3′
3′ TCTTGT(N)$_3$ $\overleftarrow{\text{ACAAGA}}$ 5′

(b) ERE
5′ $\overrightarrow{\text{AGGTCA}}(N)_3$ TGACCT 3′
3′ TCCAGT(N)$_3$ $\overleftarrow{\text{ACTGGA}}$ 5′

(c) VDRE
5′ $\overrightarrow{\text{AGGTCA}}(N)_3$ $\overrightarrow{\text{AGGTCA}}$ 3′
3′ TCCAGT(N)$_3$ TCCAGT 5′

(d) TRE
5′ $\overrightarrow{\text{AGGTCA}}(N)_4$ $\overrightarrow{\text{AGGTCA}}$ 3′
3′ TCCAGT(N)$_4$ TCCAGT 5′

(e) RARE
5′ $\overrightarrow{\text{AGGTCA}}(N)_5$ $\overrightarrow{\text{AGGTCA}}$ 3′
3′ TCCAGT(N)$_5$ TCCAGT 5′

▲ FIGURE 10-65 Consensus sequences of DNA sites, called response elements, that bind the glucocorticoid receptor (GRE), estrogen receptor (ERE), vitamin D$_3$ receptor (VDRE), thyroid hormone receptor (TRE), and retinoic acid receptor (RARE). The inverted repeats in GRE and ERE and direct repeats in VDRE, TRE, and RARE are indicated by red arrows. [See K. Umesono et al., 1991, *Cell* **65**:1255; A. M. Naar et al., 1991, *Cell* **65**:1267.]

response elements do so as heterodimers with a common nuclear-receptor monomer called RXR. The vitamin D$_3$ response element, for example, is bound by the RXR-VDR heterodimer, and the retinoic acid response element is bound by RXR-RAR. The monomers composing these heterodimers interact with each other in such a way that the two DNA-binding domains lie in the same rather than inverted orientation, allowing the RXR heterodimers to bind to direct repeats of the binding site for each monomer. In contrast, the monomers in homodimeric nuclear receptors (e.g., GRE and ERE) have an inverted orientation.

Mechanisms of Hormonal Control of Nuclear-Receptor Activity Hormone binding to a nuclear receptor regulates its activity as a transcription factor. This regulation differs in some respects for heterodimeric and homodimeric nuclear receptors.

When heterodimeric nuclear receptors (e.g., RXR-VDR, RXR-TR, and RXR-RAR) are bound to their cognate sites in DNA, they act as repressors or activators of transcription depending on whether hormone occupies the ligand-binding site. In the absence of hormone, these nuclear receptors direct histone deacetylation at nearby nucleosomes by the mechanism described earlier (see Figure 10-58a). As we saw earlier, in the presence of hormone, the ligand-binding domain undergoes a dramatic conformational change (see Figure 10-47). In the ligand-bound conformation, these nuclear receptors can direct hyperacetylation of histones in nearby nucleosomes, thereby reversing the repressing effects of the free ligand-binding domain. The N-terminal activation domain in these

nuclear receptors then probably interacts with additional factors, stimulating the cooperative assembly of an initiation complex, as described earlier.

In contrast to heterodimeric nuclear receptors, which are located exclusively in the nucleus, homodimeric receptors are found both in the cytoplasm and nucleus, and their activity is regulated by controlling their transport from the cytoplasm to the nucleus. The hormone-dependent translocation of the homodimeric glucocorticoid receptor (GR) was demonstrated in the transfection experiments shown in Figure 10-66. The GR hormone-binding domain alone mediates this transport. Subsequent studies showed that, in the absence of hormone, the glucocorticoid receptor is anchored in the cytoplasm as a large protein aggregate complexed with inhibitor proteins, including Hsp90, a protein related to Hsp70, the major heat-shock chaperone. In this situation, the receptor cannot interact with target genes; hence, no transcriptional activation occurs. Binding of hormone releases the glucocorticoid receptor from its cytoplasmic anchor, allowing it to enter the nucleus where it can bind to response elements associated with target genes (Figure 10-67). Once the receptor with bound hormone interacts with a response element, it activates transcription by directing histone hyperacetylation and facilitating cooperative assembly of an initiation complex.

Orphan Receptors The ligands for the hormone-binding domains in many members of the nuclear-receptor super-family are as-yet unknown. An example is HNF4, which participates in hepatocyte-specific expression of the transthyretin gene (see Figure 10-60). Most of these DNA-binding proteins, referred to as *orphan receptors*, were discovered by screening cDNA libraries with probes specific for the nucleotide sequence encoding the highly conserved DNA-binding domain characteristic of the nuclear receptors. The precise role of orphan receptors and identification of the unknown hormones that presumably regulate their activity are important subjects of current research.

Polypeptide Hormones Signal Phosphorylation of Some Transcription Factors

Although lipid-soluble hormones can diffuse through the plasma membrane and interact directly with transcription factors in the cytoplasm or nucleus, the second major class of hormones, peptide and protein hormones, cannot. Instead, these hormones function by binding to specific cell-surface receptors, which then pass the signal that they have bound hormone to proteins within the cell, a process called **signal transduction** (Chapter 20).

In many cases, the mechanism by which a hormonal signal is transduced into an activating signal for transcription factors involves phosphorylation. A simple example is provided by γ-interferon (IFNγ), a hormone released by antigen-stimulated T-helper lymphocytes, which are critical in the immune response. When IFNγ binds to a specific receptor protein that is present on the surface of most cells, it induces expression of a number of genes, producing an

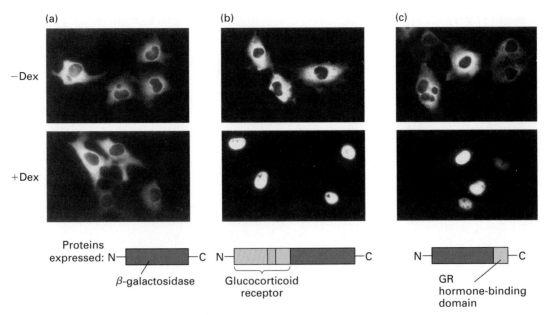

▲ **FIGURE 10-66 Experimental demonstration that hormone-binding domain of the glucocorticoid receptor (GR) mediates translocation to the nucleus in the presence of hormone.** Cultured animal cells were transfected with expression vectors encoding the proteins diagrammed at the bottom. Immunofluorescence with a labeled antibody specific for β-galactosidase was used to detect the expressed proteins in transfected cells. (a) When cells were transfected with β-galactosidase alone, the expressed enzyme was localized to the cytoplasm in the presence and absence of the glucocorti-coid hormone dexamethasone (Dex). (b) When a fusion protein consisting of β-galactosidase and the entire 794-aa rat glucocorticoid receptor (GR) was expressed in the cultured cells, it was present in the cytoplasm in the absence of hormone but was transported to the nucleus in the presence of hormone. (c) A fusion protein composed of a 382-aa region of GR including the ligand-binding domain (light purple) and β-galactosidase also exhibited hormone-dependent transport to the nucleus. [From D. Picard and K. R. Yamamoto, 1987, *EMBO J.* **6**:3333; courtesy of the authors.]

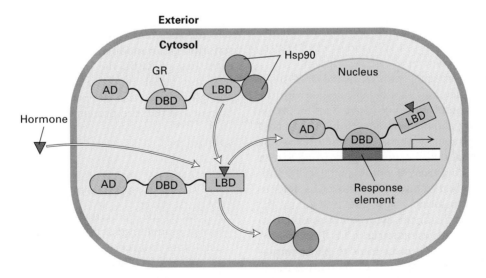

▲ **FIGURE 10-67 Model of hormone-dependent gene activation by the glucocorticoid receptor (GR).** In the absence of hormone, GR is bound in a complex with Hsp90 in the cytoplasm via its ligand-binding domain (light purple). When hormone is present, it diffuses through the plasma membrane and binds to the GR ligand-binding domain, causing a conformational change in the ligand-binding domain that releases the receptor from Hsp90. The receptor with bound ligand is then translocated into the nucleus where its DNA-binding domain (orange) binds to response elements, allowing the activation domain (green) to stimulate transcription of target genes.

▶ **FIGURE 10-68 Model of IFNγ-mediated gene activation by phosphorylation and dimerization of Stat1α.** JAK kinase is activated when the IFNγ receptor dimerizes by binding to IFNγ. Activated JAK kinase phosphorylates a specific tyrosine residue in inactive Stat1α monomers in the cytoplasm. Phosphorylated Stat1α dimerizes, and the phosphorylated dimer then translocates to the nucleus where it binds to corresponding response elements, promoting transcription of IFNγ-regulated genes. [See K. Shuai et al., 1992, *Science* **258**:1808.]

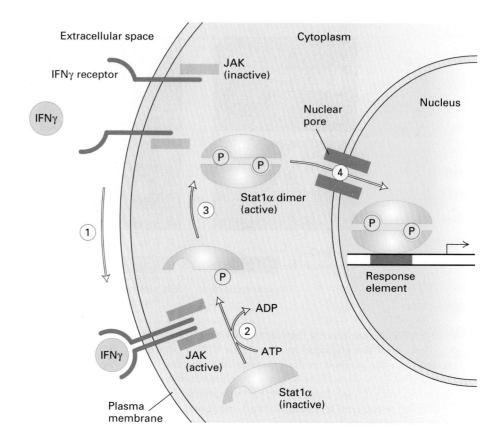

antiviral state that decreases the susceptibility of the cells to infection by a broad variety of viruses. IFNγ also stimulates the function of other cells that participate in the immune response. To analyze how this hormone causes induction of a specific set of genes, researchers first identified the IFNγ response element and then purified a protein from the nuclei of IFNγ-treated cells that binds to that sequence (see Figure 10-35). The isolated protein is ≈91 kDa and is called *Stat1α,* for signal transducer and activator of transcription.

After cultured cells are treated with IFNγ, the DNA-binding activity of Stat1α increases rapidly, in parallel with the rapid rise of transcription of inducible genes. This induction of Stat1α DNA-binding activity occurs even in cells treated with an inhibitor of protein synthesis, indicating that some type of post-translational modification of preexisting Stat1α activates its DNA-binding activity. By staining cells with fluorescein-labeled anti-Stat1α antibody, researchers demonstrated that Stat1α translocates from the cytosol to the nucleus following IFNγ treatment, with kinetics similar to that of gene induction. Analysis of Stat1α from IFNγ-treated cells showed that hormone treatment leads to phosphorylation of a specific tyrosine residue in the protein. Furthermore, phosphorylated Stat1α was found to form a homodimer, whereas the unphosphorylated protein is a monomer. When the critical tyrosine was changed to a phenylalanine by site-specific mutagenesis, the mutant Stat1α failed to activate target genes in a transfection experiment, and failed to translocate to the nucleus.

The model for IFNγ-mediated activation of Stat1α suggested by these results is illustrated in Figure 10-68. Phosphorylation of specific serine or threonine residues also regulates the activity of a number of other transcription factors. Various signal-transduction pathways that regulate transcription factors in this way are considered in Chapter 20. In some cases (e.g., Stat1α), phosphorylation of the free transcription factor modulates its DNA-binding activity. In other cases (e.g., CREB), the inactive, nonphosphorylated transcription factor binds to its DNA recognition sequence; phosphorylation then alters the functioning of the activation domain, so that the protein can stimulate transcription.

SUMMARY Molecular Mechanisms of Eukaryotic Transcriptional Control

• Eukaryotic transcriptional control operates at three levels: modulation of the levels and/or activities of activators and repressors; changes in chromatin structure directed by activators and repressors; and direct influence of activators and repressors on assembly of initiation complexes.

• Heterochromatin refers to condensed regions of chromatin in which the DNA is relatively inaccessible to transcription factors and other proteins, so that gene expression is repressed.

• Heterochromatin-mediated repression occurs in the telomeres and the silent mating-type loci in *S. cerevisiae*. The interactions of several proteins with each other and the hypoacetylated N-termini of histones H3 and H4 are responsible for the repressing chromatin structure in these regions (see Figure 10-57).

• Some repressors function in part by interacting with histone deacetylase complexes, resulting in the deacetylation of histones in nucleosomes near the repressor-binding site (see Figure 10-58a). This inhibits interaction between the promoter DNA and general transcription factors, thereby repressing transcription initiation.

• Some activators function in part by interacting with histone acetylase complexes, resulting in the hyperacetylation of histones in nucleosomes near the activator-binding site (see Figure 10-58b). This facilitates interaction between the promoter DNA and general transcription factors, thereby stimulating transcription initiation.

• Chromatin-remodeling factors cause transient dissociation of DNA from histone cores in an ATP-dependent reaction. These factors thereby promote binding of other DNA-binding proteins needed for initiation to occur at some promoters.

• In vitro, combinations of activators can stimulate the assembly of initiation complexes on a nearby promoter. This direct effect of activators is thought to occur in vivo subsequent to histone acetylation.

• The highly cooperative assembly of initiation complexes in vivo generally requires several activators (see Figure 10-61). A cell must produce the specific set of activators required for transcription of a particular gene in order to express that gene.

• Some repressors competitively inhibit binding of activators or general transcription factors. Others interact directly with general transcription factors or with activators.

• The activities of the nuclear-receptor superfamily of transcription factors are regulated by lipid-soluble hormones (see Figure 10-67). Hormone binding to these transcription factors induces conformational changes that modify their interactions with other proteins.

• The activities of some transcription factors are regulated by phosphorylation induced by binding of polypeptide hormones to their cell-surface receptors (see Figure 10-68).

10.8 Other Transcription Systems

In the previous sections, we have focused on transcription initiation by *E. coli* RNA polymerase and eukaryotic RNA polymerase II, as well as various mechanisms for controlling initiation. In this last section, we briefly discuss transcription initiation by several other RNA polymerases. Although these systems, particularly their regulation, are less thoroughly understood than transcription by *E. coli* RNA polymerase and eukaryotic RNA polymerase II, they have interesting parallels and relationships to the more extensively studied systems described earlier in the chapter.

Transcription Initiation by Pol I and Pol III Is Analogous to That by Pol II

Recall that RNA polymerase I (Pol I) is dedicated to synthesis of pre-rRNA, and RNA polymerase III (Pol III) transcribes tRNA genes, 5S-rRNA genes, and genes encoding several other small RNAs. The formation of transcription-initiation complexes involving Pol I and Pol III is similar in some respects to the assembly process for Pol II (see Figure 10-50). However, each of the three eukaryotic nuclear RNA polymerases requires polymerase-specific general transcription (or initiation) factors and recognizes different DNA control elements. Moreover, neither Pol I nor Pol III requires ATP hydrolysis to initiate transcription, whereas Pol II does.

Initiation by Pol I The human pre-rRNA gene has two transcription-control regions: a *core element,* which includes the start site and is essential for transcription, and an ≈50-bp long *upstream control element* (UCE), beginning at about −100 from the start site, which stimulates in vitro transcription tenfold to a hundredfold (Figure 10-69a). Two DNA-binding proteins assist Pol I to correctly initiate and transcribe pre-rRNA genes. The first one to assemble is upstream binding factor (UBF), which binds to both the UCE and the upstream portion of the core element, even though these sites have little apparent sequence similarity. Selectivity factor 1 (SL1), a multimeric protein, then binds and stabilizes the complex. After SL1 binding, Pol I binds and initiates transcription. One of the SL1 subunits is TBP, the same protein that is central to transcription initiation by Pol II.

Initiation by Pol III Unlike protein-coding genes and pre-rRNA genes, the promoter regions of tRNA and 5S-rRNA genes lie entirely within the transcribed sequence. Two such *internal* promoter elements, termed the *A box* and *B box,* are present in all tRNA genes (Figure 10-69b). These highly conserved sequences not only function as promoters but also encode two portions of eukaryotic tRNAs that are required for protein synthesis. In 5S-rRNA genes, a single internal control region, the *C box,* acts as a promoter (Figure 10-69c).

The general transcription factors required for Pol III to initiate transcription of tRNA and 5S-rRNA genes have been best characterized in *S. cerevisiae.* Two multimeric factors, TFIIIC and TFIIIB, participate in initiation at both tRNA and 5S-rRNA promoters in yeast. Assembly of the initiation complex on tRNA genes begins by binding of a single TFIIIC to the A box *and* B box. Interaction of the bound TFIIIC with TFIIIB then directs the latter to bind to DNA sequences ≈30 base pairs upstream of the transcription-start site. Assembly

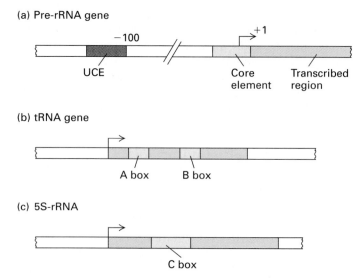

(a) Pre-rRNA gene

UCE

Core element

Transcribed region

(b) tRNA gene

A box   B box

(c) 5S-rRNA

C box

▲ FIGURE 10-69 Transcription-control elements in genes transcribed by RNA polymerase I (a) and III (b,c). Assembly of transcription-initiation complexes on these genes begins with the binding of specific general transcription factors to the control elements.

of the initiation complex on 5S-rRNA genes begins with binding of a third factor, TFIIIA, to the C box. TFIIIC then binds to TFIIIA and is held in place via protein-protein interactions at a similar position relative to the start site as when TFIIIC binds to a tRNA gene. TFIIIB then binds, interacting analogously with TFIIIC and binding to the DNA at about −30 base pairs, as it does in a tRNA gene.

The N-terminal half of one TFIIIB subunit, called BRF (for TFIIB-*related* *factor*), is similar in sequence to TFIIB (a Pol II factor). This similarity suggests that BRF and TFIIB perform a similar function in initiation, namely to direct the polymerase to the correct start site. Once TFIIIB has bound to either a tRNA or 5S-rRNA gene, Pol III can bind and initiate transcription in the presence of ribonucleoside triphosphates. The BRF subunit of TFIIIB interacts specifically with one of the polymerase subunits unique to Pol III, accounting for initiation by this specific nuclear RNA polymerase. Following binding of TFIIIB, TFIIIC can be removed without affecting initiation by Pol III. Thus, TFIIIC can be thought of as an assembly factor for the critical initiation factor, TFIIIB.

Another of the three subunits composing TFIIIB is TBP, which we can now see is a component of a general transcription factor for all three eukaryotic nuclear RNA polymerases. The finding that TBP participates in transcription initiation by Pol I and Pol III was surprising, since the promoters recognized by these enzymes do not contain TATA boxes. Nonetheless, recent studies indicate that the TBP subunit of TFIIIB interacts with DNA similarly to the way it interacts with TATA boxes.

T7 and Related Bacteriophages Express Monomeric, Largely Unregulated RNA Polymerases

Bacteriophage T7 completely dominates macromolecular synthesis in its host *E. coli* cell within a few minutes after infection. All ribosomes in the cell rapidly become committed to the synthesis of virus-encoded proteins exclusively. The virus achieves this by initially directing the synthesis of a virus-encoded RNA polymerase, which does not recognize *E. coli* promoters, and a set of proteins that inactivate the *E. coli* RNA polymerase. Consequently, when the host RNA polymerase is inactivated, transcription of the host genome stops; because *E. coli* mRNAs are very short-lived, the synthesis of most *E. coli* proteins soon ceases. The T7 RNA polymerase initiates transcription from promoters that overlap the start sites in the viral DNA. The sequence of the 23-bp T7 promoter is completely different from *E. coli* promoters.

The T7 RNA polymerase—a single polypeptide chain of 98 kDa—is the simplest RNA polymerase known. The enzyme is probably close to the simplest possible protein molecule capable of carrying out the functions of an RNA polymerase: specific promoter binding, initiation, elongation, and termination. Because the virus is committed to expressing the proteins that make up the virion coat at the maximal level, T7 RNA polymerase is not regulated significantly. An enormously active, unregulated enzyme, T7 RNA polymerase, and viral polymerases from the related bacteriophages such as T3 and SP6, is extremely useful for in vitro RNA synthesis and in bacterial expression systems (Chapter 7).

Mitochondrial DNA Is Transcribed by RNA Polymerases with Similarities to Bacteriophage and Bacterial Enzymes

Mitochondria contain a distinct DNA genome and protein-synthesis system within the inner mitochondrial membrane, in the compartment known as the mitochondrial matrix (see Figure 5-45). This DNA encodes a subset of the proteins essential to the mitochondrion's principle function: ATP synthesis by oxidative phosphorylation (see Figure 9-44). The mitochondrial DNA (mtDNA) is transcribed by RNA polymerases that are encoded in the nuclear DNA. These mitochondrial RNA polymerases are far simpler than the three eukaryotic RNA polymerases that transcribe nuclear DNA or even the RNA polymerases of modern-day bacteria.

The mitochondrial RNA polymerase from *S. cerevisiae* consists of a 145-kDa subunit with ribonucleotide-polymerizing activity and a 43-kDa specificity factor essential for initiating transcription at the start sites in mtDNA used in the cell. Purified mitochondrial RNA polymerase from the frog *Xenopus laevis* has a similar structure. The large subunit of yeast mitochondrial RNA polymerase clearly is related to the monomeric RNA polymerases of bacteriophage T7 and similar bacteriophages. However, the mitochondrial enzyme is functionally distinct from the bacteriophage enzyme in its dependence on the small subunit for transcription from the proper start sites. This small subunit is related to σ factors in bacterial RNA polymerases, which interact with promoter DNA. Thus mitochondrial RNA polymerase appears to be a hybrid of the simple bacteriophage RNA polymerases and the bacterial RNA polymerases of intermediate complexity.

Similar to bacteriophage RNA polymerases, the promoter sequences recognized by mitochondrial RNA polymerases include the transcription start site. A-rich promoter sequences have been characterized in the mtDNA from yeast, plants, and animals. The circular, human mitochondrial genome contains two related 15-bp promoter sequences, one for the transcription of each strand. Each strand is transcribed in its entirety; the long primary transcripts are then processed to yield mitochondrial mRNAs, rRNAs, and tRNAs. A 22-kDa basic protein called mtTF1, which binds immediately upstream from the two mitochondrial promoters, greatly stimulates transcription. A homologous protein found in yeast mitochondria is required for maintenance of mitochondrial DNA and probably performs a similar function.

Transcription of Chloroplast DNA Resembles Bacterial Transcription

The circular DNA found in chloroplasts is considerably larger than mitochondrial DNA, ranging from 120 to 160 kb in different plants (Section 9.7). The RNA polymerase that transcribes chloroplast DNA is encoded in the chloroplast genome itself. The enzyme has subunits with considerable homology to *E. coli* RNA polymerase α, β, and β' subunits, but apparently lacks a subunit equivalent to the *E. coli* σ factor.

Some chloroplast promoters are quite reminiscent of the *E. coli* σ^{70}-promoter, with similar sequences in the -10 and -35 regions. Transcription from one chloroplast promoter, however, depends on sequences from about -20 to $+60$, quite different from most *E. coli* promoters. This promoter may be recognized by a second RNA polymerase, which most likely is encoded in the nuclear genome and imported into the organelle. Analysis of chloroplast transcription is still in its infancy, but at this point it is clear that at least one transcription system is highly homologous to transcription in *E. coli* and other bacteria.

Transcription by Archaeans Is Closer to Eukaryotic Than to Bacterial Transcription

As discussed in previous chapters, the archaea (or archaebacteria) constitute a third lineage of organisms along with bacteria, such as *E. coli,* and eukaryotes (see Figure 1-5). Recent genomic analyses strongly suggest that the bacterial evolutionary line split off from a common ancestor before the split between the archaea and eukaryotes, making modern eukaryotes more closely related to archaea than to bacteria (Section 7.4).

Like bacteria, archaeans have a single RNA polymerase, but it contains 13 distinct subunits (14 in some), a complexity equivalent to that of eukaryotic nuclear RNA polymerases. The deduced amino acid sequences of seven subunits show significant homology to eukaryotic subunits, including the two largest, which are common to all three eukaryotic nuclear RNA polymerases. However, some of the archaeal subunits show little detectable similarity in sequence to either bacterial or eukaryotic RNA polymerase subunits.

Although archaeans, like bacteria, transcribe operons into polycistronic mRNAs, archaeal promoters are similar to eukaryotic promoters. An A/T-rich promoter consensus sequence occurs 26 base pairs upstream from the transcription-start site in archaeal DNA. As in eukaryotic TATA-box promoters, insertions or deletions following this sequence lead to a shift in the start site to maintain the spacing downstream from the promoter element. In addition, archaeans contain initiation factors highly homologous to TBP and TFIIB. Consequently, the mechanism of transcription initiation in archaeans is probably similar to that in eukaryotes. So far, little is known concerning how transcription is regulated in this fascinating group of organisms.

SUMMARY Other Transcription Systems

- Transcription initiation by Pol I is directed by a core promoter element, which overlaps the start site, and an upstream control element (UCE). It requires two general transcription factors, SL1 and UBF.

- Transcription initiation by RNA polymerase III is most often directed by internal promoter elements. Two general transcription factors are required to initiate transcription of tRNA genes (TFIIIC and TFIIIB); an additional factor (TFIIIA) also is required to initiate transcription of 5S-rRNA genes.

- Transcription initiation by all three eukaryotic nuclear RNA polymerases requires a polymerase-specific general transcription factor that contains TBP as a subunit.

- Initiation by Pol I and Pol III does not require ATP hydrolysis, whereas initiation by Pol II depends on ATP hydrolysis.

- Bacteriophage T7 and related bacteriophage express a simple monomeric RNA polymerase. These polymerases recognize a 23-bp promoter region that includes the start site.

- Mitochondria contain circular DNAs transcribed by a nuclear-encoded RNA polymerase composed of two subunits. One subunit is homologous to the simple bacteriophage T7 RNA Polymerase; the other resembles bacterial σ factors.

- Chloroplasts contain DNA that is transcribed by a chloroplast-encoded RNA polymerase homologous to bacterial RNA polymerases, except that it lacks a σ factor.

• Archaeans utilize a multisubunit RNA polymerase that resembles the eukaryotic nuclear RNA polymerases. Transcription initiation in archaeal cells requires a protein, homologous to eukaryotic TBP, that binds to an A/T-rich promoter element just upstream from the start site and a protein homologous to TFIIB. These findings support the hypothesis that the eukaryotic nucleus and modern-day archaea evolved from a common ancestor.

PERSPECTIVES for the Future

Nearly all biological processes—including normal cellular development and differentiation, the abnormal growth and behavior of cancer cells, and memory and learning—involve the regulated transcription of specific sets of genes. We have seen that the mechanisms regulating transcription are intricate, particularly in eukaryotes. The high cooperativity of activation mechanisms in the context of chromatin allows combinatorial control of transcription. Because of this, a given number of transcription factors can regulate the differential expression of a much larger set of genes. Transcription of a particular gene occurs only in those cells that have the complete set of factors required to assemble an initiation complex at the gene's promoter.

The intricacy of transcriptional regulation in eukaryotic cells allows multiple signals to be integrated in the decision to initiate transcription of a specific gene. However, much remains to be learned about how combinations of activators and repressors, the location and affinities of their binding sites, and their interactions with co-activators and co-repressors regulate the final rate of gene transcription. Many of the co-activator and co-repressor proteins in higher organisms remain to be identified. The higher-order structure of heterochromatin and the processes controlling the assembly of a particular gene into heterochromatin remain to be elucidated.

The multiple inputs that regulate expression of each gene associated with complex biological processes are ultimately integrated by the multiple transcription-control mechanisms operating in eukaryotic cells. Detailed understanding of these fundamental biological processes depends on understanding how this final level of signal integration occurs. Extracting general principles from the study of individual examples of transcriptional control and applying these to better understand complex biological processes pose exciting challenges for researchers.

PERSPECTIVES in the Literature

The activation domains of regulatory transcription factors have been found to interact in vitro with chromatin remodeling factors such as the Swi/Snf complex, histone acetylase complexes such as the yeast SAGA complex, the Srb/Mediator complex associated with the Pol II CTD, TAF subunits of TFIID, and other general transcription factors. As yet, there is little understanding of the temporal order in vivo of these multiple potential interactions or how this order might influence transcription initiation. As you read the following articles, consider these questions: what methods are available to determine (1) whether an interaction observed in vitro does in fact contribute to transcriptional regulation in vivo, and (2) whether there is a specific order of different interactions in vivo:

Boyer, T. G., et al. 1999. Mammalian SRB/Mediator complex is targeted by adenovirus E1A protein. *Nature* 399:276–279.

Cosma, M. P., T. Tanaka, and K. Nasmyth. 1999. Ordered recruitment of transcription and chromatin remodeling factors to a cell cycle- and developmentally regulated promoter. *Cell* 97:299–311.

Testing Yourself on the Concepts

1. Describe the Jacob-Monod model for regulation of the *E. coli lac* operon.

2. Describe the methods used to identify DNA-binding proteins and the location of DNA control elements in regulatory regions of genes.

3. Compare the events involved in transcription initiation for prokaryotic and eukaryotic mRNA synthesis.

4. Describe the structural features of activator and repressor proteins.

MCAT/GRE-Style Questions

Key Concept Please read the section titled "Positive Control of the *lac* Operon Is Exerted by cAMP-CAP" (p. 352) and refer to Figure 10-16; then answer the following questions:

1. This section describes the molecular mechanism in *E. coli* for

   a. Responding to starvation conditions.

   b. The utilization of glucose.

   c. Induction of the enzymes for converting lactose to glucose.

   d. The synthesis of cyclic AMP.

2. Regulation of β-galactosidase synthesis is

   a. Under neither positive nor negative control.

   b. Under only positive control.

   c. Under only negative control.

   d. Under both positive and negative control.

3. Which of the following statements describes a step in the regulation of the *lac* operon:

   a. A cyclic AMP-CAP complex binds to the CAP site.

b. The repressor binds to CAP to form an inactive repressor.

c. Cyclic AMP competes with the repressor for binding to the operator.

d. The repressor undergoes a conformational change upon binding cyclic AMP.

4. All the following mutations can result in a reduction of β-galactosidase except:

a. A mutation in adenylate cyclase.

b. A mutation in catabolite activator protein (CAP).

c. A mutation in the CAP site in the *lac* control region.

d. A mutation in the repressor binding site in the operator.

Key Experiment Please read the section titled "Transcription Activators Are Modular Proteins Composed of Distinct Functional Domains" (p. 372) and refer to Figure 10-38; then answer the following questions:

5. Which of the following is true about the DNA-binding and activation domains of transcription activators:

a. The DNA-binding domain is always at the N-terminus, and the activation domain is at the C-terminus.

b. The activation domain of one protein can be fused to the DNA-binding domain of another protein to generate a functional activator protein.

c. The DNA-binding domain is always at the C-terminus, and the activation domain is at the N-terminus.

d. The distance between a DNA-binding domain and activation domain cannot be altered.

6. In Figure 10-38, which of the following Gal4 mutants provides the best evidence that the DNA-binding and activation domains can act independently:

a. The Gal4 mutant containing amino acids 50–881.

b. The Gal4 mutant containing amino acids 1–692.

c. The Gal4 mutant containing amino acids 1–74.

d. The Gal4 mutant containing amino acids 1–74 and 738–881.

7. Which of the following new Gal4 mutants could be used to further define the location of the DNA binding site:

a. A Gal4 mutant containing amino acids 1–66.

b. A Gal4 mutant containing amino acids 700–881.

c. A Gal4 mutant containing amino acids 1–74 and 93–123.

d. A Gal4 mutant containing amino acids 1–325.

8. A fusion protein, which consists of the DNA-binding domain from an activator protein fused to the suppression domain of a repressor protein, would be expected to

a. Be nonfunctional because it contains amino acid sequences from two different proteins.

b. Act as a repressor protein.

c. Act as an activator protein.

d. Act as both an activator and a repressor protein.

Key Application

Please read the section titled "Many Repressors Are the Functional Converse of Activators" (p. 379) and refer to Figure 10-49; then answer the following questions:

9. All repressor and activator proteins share which of the following structural features:

a. A DNA-binding domain.

b. A repression domain.

c. An activation domain.

d. A ligand-binding domain.

10. Which of the following statements is true about repressor and activator proteins:

a. Only repressor proteins contain hydrophobic residues.

b. The binding of a repressor protein to a gene control region always blocks the subsequent binding of an activator protein.

c. A repression domain can be changed to an activation domain by the binding of a ligand.

d. The repression domain of a repressor protein binds to DNA.

11. Development of Wilms's tumors is due to

a. Overexpression of the activators SRF-1 and AP-1.

b. Loss of suppression of the transcription activator gene EGR-1.

c. Inhibition of SRF-1 and AP-1 binding to the EGR-1 gene.

d. Mutation of the SRF-1 and AP-1 genes.

12. Which of the following mutations could also lead to Wilms's tumors:

a. A mutation in the SRF-binding site in the EGR-1 control region.

b. A mutation in the AP-1 protein.

c. A mutation in the WT1-binding site in the EGR-1 control region.

d. A mutation in the EGR-1 protein.

Key Terms

5-methylcytidine (mC) *389*	DNase I footprinting *346*
activation domain *372*	DNase I–hypersensitive sites *390*
activators *370*	
architectural proteins *379*	enhancer *355*
cis-acting *344*	enhancesome *378*
cooperativity *354*	general transcription factors *381*
DNA-binding domain *372*	

References

Bacterial Gene Control: The Jacob-Monod Model

Gralla, J.D. 1992. *lac* Repressor. In S. L. McKnight and K. R. Yamamoto, eds., *Transcriptional Regulation.* Cold Spring Harbor Laboratory Press, pp. 629–642.

Jacob, F., and J. Monod. 1961. Genetic regulatory mechanisms in the synthesis of proteins. *J. Mol. Biol.* 3:318–356.

Muller-Hill, B. 1996. *The* lac *Operon: A Short History of Genetic Paradigm.* Walter de Gruyter (Berlin; New York).

Pace, H. C., et al. 1997. Lac repressor genetic map in real space. *Trends Biochem. Sci.* 22:334–339.

Bacterial Transcription Initiation

Brown, J. M., and R. A. Firtel. 1998. Phosphorelay signalling: new tricks for an ancient pathway. *Curr. Biol.* 8:R662–665.

Busby, S., and R. H. Ebright. 1997. Transcription activation at class II CAP-dependent promoters. *Mol. Microbiol.* 23:853–859.

Chan, C. L., M. A. Lonetto, and C. A. Gross. 1996. Sigma domain structure: one down, one to go. *Structure* 4:1235–1238.

Gralla, J. D. 1991. Transcription control–lessons from an *E. coli* promoter data base. *Cell* 66:415–418.

Gross, C. A., M. Lonetto, and R. Losick. 1992. Bacterial sigma factors. In S. L. McKnight and K. R. Yamamoto, eds., *Transcriptional Regulation.* Cold Spring Harbor Laboratory Press, pp. 129–178.

Kustu, S., A. K. North, and D. S. Weiss. 1991. Prokaryotic transcriptional enhancers and enhancer-binding proteins. *Trends Biochem.* 16:397–402.

Pabo, C. O, and R. T. Sauer. 1992. Transcription factors: structural families and principles of DNA recognition. *Ann. Rev. Biochem.* 61:1053–1095.

Sigler, P. B. 1992. The molecular mechanism of *trp* repression. In S. L. McKnight and K. R. Yamamoto, eds., *Transcriptional Regulation.* Cold Spring Harbor Laboratory Press, pp. 475–500.

Stragier, P., and R. Losick. 1996. Molecular genetics of sporulation in *Bacillus subtilis. Ann. Rev. Genet.* 30:297–341.

Eukaryotic Gene Control: Purposes and General Principles

Darst, S. A., A. M. Edwards, E. W. Kubalek, and R. D. Kornberg. 1991. Three-dimensional structure of yeast RNA polymerase II at 16 Å resolution. *Cell* 66:121–128.

Sentenac, A., et al. 1992. Yeast RNA polymerase subunits and genes. In S. L. McKnight and K. R. Yamamoto, eds., *Transcriptional Regulation.* Cold Spring Harbor Laboratory Press, pp. 27–54.

Shuman, S. 1997. Origins of mRNA identity: capping enzymes bind to the phosphorylated C-terminal domain of RNA polymerase II. *Proc. Nat'l. Acad. Sci. USA* 94:12758–12760.

Regulatory Sequences in Eukaryotic Protein-Coding Genes

Blackwood, E. M., and J. T. Kadonaga. 1998. Going the distance: a current view of enhancer action. *Science* 281:60–63.

Macleod, D., R. R. Bird, and A. Bird. 1998. An alternative promoter in the mouse major histocompatibility complex class II I-Abeta gene: implications for the origin of CpG islands. *Mol. Cell. Biol.* 18:4433–4443.

Périer, R. C., T. Junier, C. Bonnard, and P. Bucher. 1999. The eukaryotic promoter database (EPD): recent developments. *Nucl. Acids Res.* 27:307–309. Access the database at http://www.epd.isb-sib.ch/

Eukaryotic Transcription Activators and Repressor

Hanna-Rose, W., and U. Hansen. 1996. Active repression mechanisms of eukaryotic transcription repressors. *Trends Genet.* 12:229–234.

Harrison, S. C. 1991. A structural taxonomy of DNA-binding domains. *Nature* 353:715–719.

Latchman, D. S. 1996. Transcription-factor mutations and disease. *N. Engl. J. Med.* 334:28–33.

Morimoto, R. I. 1992. Transcription factors: positive and negative regulators of cell growth and disease. *Curr. Opin. Cell Biol.* 4:480–487.

RNA Polymerase II Transcription-Initiation Complex

Asturias, F. J., and R. D. Kornberg. 1999. Protein crystallization on lipid layers and structure determination of the RNA polymerase II transcription initiation complex. *J. Biol. Chem.* 274:6813–6816.

Burley, S. K., and R. G. Roeder. 1996. Biochemistry and structural biology of transcription factor IID (TFIID). *Ann. Rev. Biochem.* 65:769–799.

Myer, V. E., and R. A. Young. 1998. RNA polymerase II holoenzymes and subcomplexes. *J. Biol. Chem.* 273:27757–27760.

Orphanides, G., T. Lagrange, and D. Reinberg. 1996. The general transcription factors of RNA polymerase II. *Genes & Devel.* 10:2657–2683.

Patikoglou, G., and S. K. Burley. 1997. Eukaryotic transcription factor-DNA complexes. *Ann. Rev. Biophy. Biomol. Struct.* 26:289–325.

Roeder, R. G. 1996. The role of general initiation factors in transcription by RNA polymerase II. *Trends Biochem. Sci.* 21:327–335.

Tan, S., and T. J. Richmond. Eukaryotic transcription factors. *Curr. Opin. Struct. Biol.* 8:41–48.

Molecular Mechanisms of Eukaryotic Transcription Control

Berger, S. L. 1999. Gene activation by histone and factor acetyltransferases. *Curr. Opin. Cell Biol.* 11:336–341.

Berk, A. J. 1999. Activation of Pol II transcription. *Curr. Opin. Cell Biol.* 11:330–335.

Cavalli, G., and R. Paro. 1998. Chromo-domain proteins: linking chromatin structure to epigenetic regulation. *Curr. Opin. Cell Biol.* 10:354–360.

Darnell, J. E., Jr. 1997. STATs and gene regulation. *Science* 277:1630–1635.

Grunstein, M. 1998. Yeast heterochromatin: regulation of its assembly and inheritance by histones. *Cell* **93**:325–328.

Kouzarides, T. 1999. Histone acetylases and deacetylases in cell proliferation. *Curr. Opin. Genet. Devel.* **9**:40–48.

Kuo, M. H., and C. D. Allis. 1998. Roles of histone acetyltransferases and deacetylases in gene regulation. *Bioessays* **20**:615–626.

Mangelsdorf, D. J., et al. 1995. The nuclear receptor superfamily: the second decade. *Cell* **83**:835–839.

Moras, D., and H. Gronemeyer. 1998. The nuclear receptor ligand-binding domain: structure and function. *Curr. Opin. Cell Biol.* **10**:384–391.

Pirrotta, V. 1998. Polycombing the genome: PcG, trxG, and chromatin silencing. *Cell* **93**:333–336.

Pollard, K. J., and C. L. Peterson. 1998. Chromatin remodeling: a marriage between two families? *Bioessays* **20**:771–780.

Torchia, J., C. Glass, and M. G. Rosenfeld. 1998. Co-activators and co-repressors in the integration of transcriptional responses. *Curr. Opin. Cell Biol.* **10**:373–383.

Other Transcription Systems

Geiduschek, E. P., and G. A. Kassavetis. 1995. Comparing transcriptional initiation by RNA polymerases I and III. *Curr. Opin. Cell Biol.* **7**:344–351.

Grummt, I. 1999. Regulation of mammalian ribosomal gene transcription by RNA polymerase I. *Prog. Nucl. Acid Res. Mol. Biol.* **62**:109–154.

McAllister, W. T. 1993. Structure and function of the bacteriophage T7 RNA polymerase (or, the virtues of simplicity). *Cell. Mol. Biol. Res.* **39**:385–391.

Mullet, J. E. 1993. Dynamic regulation of chloroplast transcription. *Plant Physiol.* **103**:309–313.

Reeder, R. H. 1999. Regulation of RNA polymerase I transcription in yeast and vertebrates. *Prog. Nucl. Acid Res. Mol. Biol.* **62**:293–327.

Shadel, G. S., and D. A. Clayton. 1993. Mitochondrial transcription initiation. Variation and conservation. *J. Biol. Chem.* **268**:16083–16086.

11

RNA Processing, Nuclear Transport, and Post-Transcriptional Control

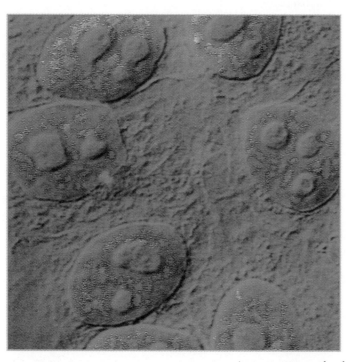

Localization of an RNA splicing factor (orange) to discrete regions of interphase nuclei.

In the previous chapter, we saw that regulation of most genes occurs at the first step in gene expression, namely, initiation of transcription. However, once transcription has been initiated, synthesis of the encoded RNA requires that RNA polymerase transcribe the entire gene and not terminate prematurely. The initial **primary transcripts** produced from eukaryotic genes are not functional and undergo various processing reactions to yield the corresponding functional RNAs. The mature, functional RNAs in the nucleus, then are actively transported to the cytoplasm, as components of ribonucleoproteins.

The multiple steps in the production of RNAs provide opportunities for additional levels of gene control beyond the regulation of transcription initiation. In the case of a protein-coding gene, the amount of protein expressed also can be regulated by controlling the stability of the corresponding mRNA in the cytoplasm and the rate of its translation. In addition, the cellular locations of some mRNAs are regulated, so that newly synthesized protein is concentrated where it is needed. All of the regulatory mechanisms that control gene expression following transcription initiation are referred to as *post-transcriptional control*. In this chapter, we consider the various steps in

MEDIA CONNECTIONS

Overview: **Life Cycle of an mRNA**

Focus: **mRNA Splicing**

Classic Experiment 11.1: **The Discovery of Self-Splicing RNA**

the synthesis of mRNA, tRNA, and rRNA following transcription initiation and the mechanisms by which RNAs and proteins are transported in and out of the nucleus in eukaryotic cells. We also present relevant examples of how regulation of these steps contributes to the control of gene expression.

11.1 Transcription Termination

Several mechanisms of regulating transcription termination have been discovered in bacteria and eukaryotes. RNA polymerase itself plays a role in the two principal mechanisms of transcription termination that occur in *E. coli*. An additional protein, a transcription-termination factor called *Rho*, is required in one mechanism but not the other. These mechanisms are commonly referred to as *Rho-independent* and *Rho-dependent termination*. In eukaryotes, the mechanisms for terminating transcription appear to differ for each of the three RNA polymerases. In this section, we first focus on termination mechanisms in bacteria, which are better understood than eukaryotic mechanisms, and then describe two well-studied examples of transcription termination in eukaryotes.

Rho-Independent Termination Occurs at Characteristic Sequences in *E. coli* DNA

Sequencing of the entire *E. coli* genome has shown that most operons have Rho-independent termination sites. These termination sequences have two characteristic features: a series of U residues in the transcribed RNA and, preceding these, a GC-rich self-complementary region with several intervening nucleotides. When such a self-complementary region of a growing RNA chain is synthesized, the complementary sequences base-pair with one another, forming a **stem-loop** structure (Figure 11-1). This stem-loop structure is thought to interact with the transcribing *E. coli* RNA polymerase, causing it to pause. In addition, the base pairs between the U residues at the 3′ end of the nascent RNA chain and the A residues in the template DNA strand are extremely un-

stable compared to other types of Watson-Crick base pairs. These two features probably permit release of the RNA chain from the transcription complex at Rho-independent termination sites. Mutations that weaken the dyad symmetry or that result in fewer consecutive U residues decrease termination at these sites.

Premature Termination by Attenuation Helps Regulate Expression of Some Bacterial Operons

Another mechanism for regulating bacterial gene expression, called *attenuation*, was initially recognized by analysis of certain *E. coli* strains with mutations in the *trp* operon. Recall that in the presence of tryptophan, the *trp* repressor inhibits transcription initiation at the *trp* operon (see Figure 10-14). This inhibition, however, is not complete, and is supplemented by attenuation, which provides an additional level for regulating tryptophan synthesis according to the requirements for protein synthesis.

In essence, attenuation results from the premature termination of transcription at an **attenuator site** located just downstream from the promoter-operator region and before the coding sequences for the *trp* enzymes. *E. coli* mutants with small deletions in this region of the *trp* operon manufacture greater-than-normal amounts of tryptophan-synthesizing enzymes in the presence and absence of tryptophan. To understand how such deletions could cause increased synthesis of *trp* mRNA, researchers compared the sequences at the beginning of the *trp* mRNA molecule and the DNA region near the *trp* promoter in both normal and mutant *E. coli* strains. A molecule of normal *trp* mRNA contains 162 nucleotides upstream (i.e., toward the 5′ end of the molecule) from the AUG start codon for the first of the five *trp*-encoded enzymes (Figure 11-2). This stretch of mRNA is called the *leader sequence* (L). Due to the action of the *trp* repressor, normal *E. coli* cells produce only a few *trp* mRNA transcripts when tryptophan is abundant, and these transcripts consist of leader RNA only; the remainder of the ≈7-kb *trp* operon is not transcribed. When tryptophan is scarce, however, the entire *trp* operon is transcribed, including the leader sequence and all the coding sequences for

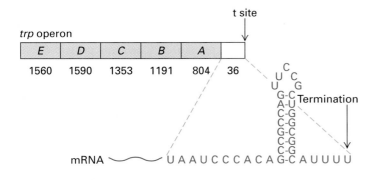

◀ **FIGURE 11-1 Sequence of *trp* termination (t) site, a Rho-independent site.** The *trp* operon, composed of five genes (blue), is followed by a 36-bp sequence at the end of which termination occurs. The 3′ end of the corresponding mRNA has a GC-rich stem-loop structure preceding the final four U residues, a characteristic feature of mRNAs produced from genes with Rho-independent termination sites. In general, bacterial termination sites occur between operons but not between genes within an operon. For this reason, genes within an operon can be coordinately regulated, whereas those in different operons are independently regulated. [See T. Platt, 1981, *Cell* **24**:10.]

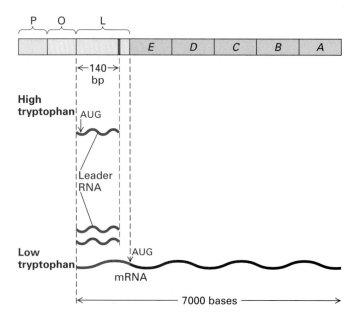

▲ **FIGURE 11-2 Attenuation provides a secondary mechanism for controlling expression of the *trp* operon.** The leader sequence (L), which lies between the operator (O) and the first structural gene *(E)*, contains an attenuator site (red band) at which transcription is terminated depending on the cytosolic concentration of tryptophan. The promoter-operator-leader region, which is ≈200 bp long, is not drawn to scale with the structural genes. AUG indicates the start codons for translation of the leader sequence (red) and *trpE* mRNA (black). [See T. Platt, 1978, in J. H. Miller and W. S. Reznikoff, eds., *The Operon*, Cold Spring Harbor Laboratory.]

the *trp*-encoded enzymes. Mutant cells from which the attenuator region has been deleted produce more *trp* mRNA under all conditions than do normal cells.

An attenuator site, in effect, is a DNA sequence where a choice is made by RNA polymerase between continued transcription and termination. When tryptophan is abundant, little initiation of transcription takes place, and virtually all of the transcripts that are initiated terminate at the attenuator. In contrast, when tryptophan is scarce, initiation occurs at a high rate and many RNA polymerase molecules continue transcribing past the attenuator. Thus regulation of the *trp* operon is not an all-or-none affair involving only the repressor-operator mechanism discussed in Chapter 10; rather, secondary control by attenuation permits a bacterium to finely balance the amount of the tryptophan-synthesizing enzymes produced with the cell's need for tryptophan.

Attenuation requires a particular stem-loop structure in the mRNA leader sequence. Formation of this structure depends on the rate of ribosomal translation of the leader sequence, which is engaged by a ribosome soon after it is synthesized. The rate of translation of the leader sequence depends, in turn, on the supply of **aminoacyl-tRNAs** charged with the amino acids encoded by the sequence. In the case of *trp* mRNA, the supply of charged tryptophan tRNA

(Trp-tRNA^{Trp}), which depends largely on the cytosolic concentration of tryptophan, is critical. To understand how attenuation of the *trp* operon occurs, note that four regions in the *trp* mRNA leader sequence can base-pair (Figure 11-3a). The sequence of region 2 can base-pair with either region 1 or with region 3. Similarly, region 3 can base-pair with either region 2 or region 4.

Region 1 of the leader contains two successive codons for tryptophan. When tryptophan is present in sufficient quantity, the ribosome translates the leader transcript rapidly, melting the base pairs between regions 1 and 2. Then when regions 3 and 4 of the RNA are synthesized, they pair forming a stem-loop followed by a series of U residues, so that transcription is terminated by the Rho-independent mechanism described earlier (Figure 11-3b, *left*). On the other hand, when the supply of tryptophan is low, the ribosome pauses at each tryptophan codon in region 1 of the leader. Since the ribosome is paused over region 1, region 1 cannot base-pair with region 2, which then is free to base-pair with region 3 as soon as it is synthesized (Figure 11-3b, *right*). The resulting 2-3 stem-loop does not induce termination because it is not followed closely by a string of U residues. In this situation, region 3 is sequestered in the 2-3 hybrid region and thus cannot base-pair with region 4 when it is synthesized. Consequently, the terminating RNA structure does not form and RNA polymerase continues transcribing the remainder of the *trp* operon. This mechanism maximizes attenuation when Trp-tRNA^{Trp} is at a high enough concentration to support rapid translation of tryptophan codons and minimizes attenuation when tryptophan is scarce and the concentration of Trp-tRNA^{Trp} falls. The small leader peptide translated from the leader sequence is rapidly degraded.

A similar attenuation mechanism occurs in several operons encoding enzymes that synthesize other amino acids including phenylalanine, histidine, isoleucine, leucine, and valine. In each case, the leader sequence of the mRNA contains multiple codons for the amino acid product of a biosynthetic pathway, and an abundance of the amino acid in the cytosol promotes Rho-independent termination within the leader sequence.

Attenuation of other operons that do not encode amino acid biosynthetic enzymes depends on specific RNA-binding proteins that stabilize one alternative RNA secondary structure over another. The *E. coli bgl* operon, which encodes enzymes that metabolize glucose-containing polysaccharides, provides an example of this mechanism. Binding of a specific RNA-binding protein to a sequence near the 5' end of nascent *bgl* transcripts stabilizes a stem-loop structure that does not support attenuation. Since this protein is activated by glucose-induced phosphorylation, attenuation is reduced and expression of the *bgl* operon is increased in the presence of glucose. In this way, a sequence-specific RNA-binding protein can control gene expression in *E. coli* by controlling attenuation, just as sequence-specific DNA-binding proteins control transcription initiation by acting as repressors or activators.

(a) *trp* leader RNA

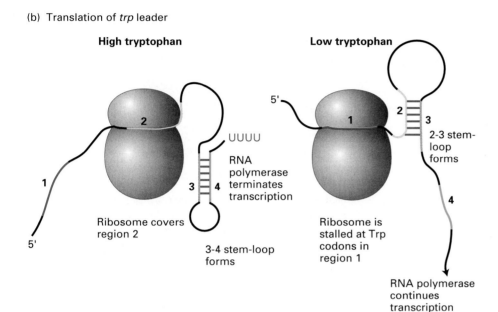

(b) Translation of *trp* leader

High tryptophan

Low tryptophan

◀ **FIGURE 11-3 Mechanism of attenuation of *trp*-operon transcription.** (a) Diagram of the 140-nucleotide *trp* leader RNA. Four regions shown in color can form alternative stem-loop structures, only one of which leads to attenuation. (b) Translation of the *trp* leader sequence begins from the 5′ end soon after it is synthesized, while synthesis of the rest of the *trp* mRNA molecule continues. At high tryptophan concentrations, formation of the 3–4 stem-loop followed by a series of 3′ U residues causes termination by the Rho-independent mechanism. At low tryptophan concentrations, region 3 is sequestered in the 2–3 stem-loop and cannot base-pair with region 4. In the absence of the stem-loop structure required for termination, transcription of the *trp* coding sequences continues. See text for discussion. [Part (b) adapted from T. Platt, 1981, *Cell* **24**:10.]

Rho-Dependent Termination Sites Are Present in Some λ-Phage and *E. coli* Genes

Rho-dependent termination was first recognized during in vitro studies of transcription of λ-phage DNA. Immediately after infection with bacteriophage λ, host-cell RNA polymerases initiate transcription in opposite directions from two promoters in λ DNA, P_L and P_R, yielding two fairly short RNA transcripts resulting from termination at sites designated t_L and t_R, respectively. In contrast, when λ DNA is transcribed in vitro with purified *E. coli* RNA polymerase, the transcripts made from P_L and P_R are many thousands of nucleotides long. Addition of an extract of uninfected cells to the transcription reaction with pure RNA polymerase leads to formation of two discrete products identical to the RNA transcripts found inside infected cells immediately after infection. In other words, in the absence of some factor in the uninfected cell extract, transcription of λ DNA in vitro does not terminate at the same sites as it does in vivo. The protein in the uninfected cell extract responsible for termination at these sites in λ DNA was purified and is now called Rho.

The Rho factor is a hexameric protein around which a 70- to 80-base segment of the growing RNA transcript wraps. This interaction activates an ATPase activity of Rho that is associated with its movement along the RNA in the 3′ direction until it eventually unwinds the RNA-DNA hybrid at the active site of RNA polymerase. Whether transcription

is terminated or not seems to depend on whether Rho moves sufficiently fast to "catch up" with the polymerase. Consequently, pausing of polymerase during elongation is thought to be an important component of Rho-dependent termination, as it is in Rho-independent termination.

Unlike Rho-independent termination sites, the Rho-dependent sites identified in λ and *E. coli* DNA exhibit no obvious sequence similarities. Recent analysis of the *E. coli* genome indicates that Rho-dependent termination operates at relatively few operons.

Sequence-Specific RNA-Binding Proteins Can Regulate Termination by *E. coli* RNA Polymerase

More detailed studies of λ-phage gene expression immediately after infection revealed that the transcript produced from P_L encodes a viral protein called N, which *prevents* termination of transcription at t_L and t_R (Figure 11-4a). Thus once the concentration of N rises sufficiently, synthesis of long RNA transcripts corresponding to other portions of the λ genome can proceed (Figure 11-4b). Studies with λ mutants demonstrated that for N to prevent termination, a specific sequence called *nut* (for N-utilization) must occur between the promoter and the termination site.

The isolation of *E. coli* mutants that do not support antitermination by N, and consequently are resistant to infection

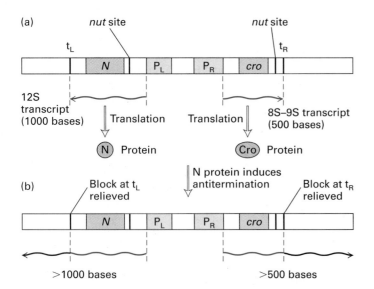

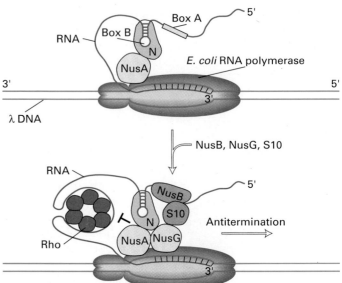

▲ FIGURE 11-4 Transcription from P$_L$ and P$_R$ of the λ-phage genome. (a) Immediately after λ-phage infection, two relatively short transcripts are synthesized from P$_L$ and P$_R$; transcription terminates at the Rho-dependent terminators t$_L$ and t$_R$. These transcripts encode Cro, a DNA-binding protein that induces the lytic pathway, and N protein, which acts with other proteins to prevent termination at t$_L$ and t$_R$. (b) Later in infection, when N accumulates, antitermination complexes assemble at the *nut* sites in the RNA, and much longer transcripts, encoding other proteins, are produced.

▲ FIGURE 11-5 Antitermination by λ-phage N protein and *E. coli* cellular proteins. After N protein recognizes and binds to the B box in the *nut* site, it interacts with the NusA-polymerase complex. Subsequent rapid binding of NusB, NusG, and S10 produces an antitermination complex stabilized by multiple protein-protein contacts. As the polymerase moves along the template DNA away from the *nut* site, the antitermination complex remains associated with the enzyme and the *nut* RNA sequence, so a RNA loop of increasing size forms. The complex prevents termination, and transcription proceeds. Inhibition of the terminating action of hexameric Rho factor is diagrammed; antitermination also occurs at Rho-independent sites. [Adapted from J. Greenblatt et al., 1993, *Nature* **364**:401.]

by λ, led to the identification of several cellular proteins, or N utilization substances (Nus), required for N function. These proteins—designated NusA, NusB, NusE, and NusG —have other functions in uninfected cells. NusA is an *elongation factor*, and NusE (also called S10) is one of the proteins of the *E. coli* small ribosomal subunit. These cellular proteins function together to prevent termination during transcription of the ribosomal RNA genes.

Protein-binding studies have shown that N protein functions by binding to the *nut* sequence in newly transcribed RNA. The *nut* sequence contains two protein-binding regions: a stem-loop called box B, which interacts with N protein, and a 12-base linear sequence called box A, which interacts with the cellular proteins NusB and S10. Based on binding studies and mutant analyses, the model of **N-mediated antitermination** shown in Figure 11-5 has been proposed. N protein binds to the *nut* box B in the nascent transcript and then interacts with NusA complexed with RNA polymerase. Once N and NusA interact, NusB, NusG, and S10 rapidly bind, yielding a stable antitermination complex. This complex can move along the DNA for many kilobases, blocking termination at both Rho-dependent and Rho-independent termination sites by an unknown mechanism, so transcription can proceed.

The elaborate antitermination mechanism, employing a multiprotein complex, permits λ phage to regulate the temporal expression of genes transcribed from P$_L$ and P$_R$.

This regulatory mechanism along with other mechanisms permits λ to undergo either a lytic or lysogenic mode of infection (see Figure 6-19). This example illustrates yet another way that gene expression can be regulated by controlling termination early in a transcription unit.

Three Eukaryotic RNA Polymerases Employ Different Termination Mechanisms

Although considerably less is known about transcription termination in eukaryotes than in bacteria, some basic features and their similarities to and differences from bacterial termination are understood. Transcription of pre-rRNA genes by RNA polymerase I is terminated by a mechanism that requires a polymerase-specific termination factor. This DNA-binding protein binds downstream of the transcription unit, unlike the *E. coli* Rho factor, which is a RNA-binding termination factor. Purified RNA polymerase III terminates after polymerizing a series of U residues; unlike Rho-independent termination in bacteria, however, termination by RNA polymerase III does not require an upstream stem-loop secondary structure in the RNA transcript.

In most mammalian protein-coding genes, RNA polymerase II can terminate at multiple sites located over a distance of 0.5–2 kb beyond the poly(A) addition site. Experiments with mutant genes show that termination is coupled to the process that cleaves and polyadenylates the 3' end of a transcript, which is discussed in the next section. Recent biochemical experiments suggest that the protein complex that cleaves and polyadenylates the nascent mRNA transcript at specific sequences associates with the phosphorylated carboxyl-terminal domain (CTD) of RNA polymerase II following initiation (see Figure 10-50). This cleavage/polyadenylation complex may suppress termination similarly to λ-phage N protein until the sequence signaling cleavage and polyadenylation is transcribed by the polymerase.

As we've seen in previous sections, bacterial transcription termination can be regulated by attenuation and antitermination mechanisms. Analogous control mechanisms, involving a choice between chain elongation or termination, occur in some genes transcribed by eukaryotic RNA polymerase II. We discuss two examples of such regulation next.

Transcription of HIV Genome Is Regulated by an Antitermination Mechanism

MEDICINE Currently, transcription of the human immunodeficiency virus (HIV) genome by RNA polymerase II provides the best-understood example of regulated transcription termination in eukaryotes. Efficient expression of HIV genes requires a small viral protein encoded at the *tat* locus. Cells infected with *tat⁻* mutants produce short viral transcripts that hybridize to restriction fragments containing promoter-proximal regions of the HIV DNA but not to restriction fragments farther downstream from the promoter. In contrast, cells infected with wild-type HIV synthesize long viral transcripts that hybridize to restriction fragments throughout the single HIV transcription unit. Thus Tat protein functions as an antitermination factor, permitting RNA polymerase II to read through a transcriptional block, much the same as the λ-phage N protein. Since antitermination by Tat protein is required for HIV replication, further understanding of this gene-control mechanism may offer possibilities for designing effective therapies for acquired immunodeficiency syndrome (AIDS).

Like N, Tat is a sequence-specific RNA-binding protein. It binds to the RNA copy of a sequence called TAR, which is located near the 5' end of the HIV transcript. Like the *nut* sequence in λ transcripts, TAR contains two binding sites: one that interacts with Tat and one that interacts with a cellular protein called cyclin T. The secondary structures of *nut* and TAR, both of which include a stem-loop, are analogous and must be intact for protein binding to occur. Moreover, the RNA-binding domain of Tat, like that of N protein, contains an arginine-rich region.

As depicted in Figure 11-6, the HIV Tat protein and cellular cyclin T bind to TAR RNA cooperatively, much like the

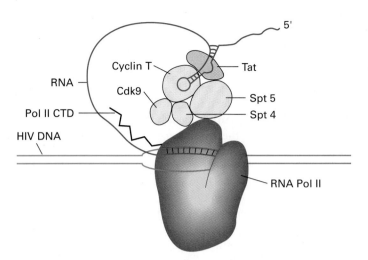

▲ **FIGURE 11-6 Antitermination complex composed of HIV Tat protein and several eukaryotic cellular proteins.** The TAR element in the HIV transcript contains sequences recognized by Tat and the cellular protein cyclin T. Cyclin T activates and helps position Cdk9, a protein kinase, near its substrate, the CTD of RNA polymerase II. See text for discussion. [Adapted from K. A. Jones, 1997, *Genes & Devel.* **11**:2593. See also P. Wei et al., 1998, *Cell* **92**:451; T. Wada et al., 1998, *Genes & Devel.* **12**:357.]

interaction of λ-phage N protein and *E. coli* NusB and S10 with *nut* RNA (see Figure 11-5). Interaction of cyclin T with a protein kinase called Cdk9 activates the kinase, whose substrate is the CTD of RNA polymerase II. (The cyclins are a family of homologous proteins that bind and regulate cyclin-dependent kinases, or Cdks. The important function of cyclins and Cdks in regulating cell division is discussed in Chapter 13.) In vitro transcription studies using a specific inhibitor of Cdk9 suggest that RNA polymerase II molecules that initiate transcription on the HIV promoter terminate after transcribing ≈50 bases, unless the CTD is hyperphosphorylated by Cdk9. Cooperative binding of cyclin T and Tat to the TAR sequence at the 5' end of the HIV transcript positions Cdk9 so that it can phosphorylate the CTD, thereby preventing termination and permitting the polymerase to continue chain elongation.

Remarkably, Spt5, one of the proteins in the cyclinT-Cdk9 complex, has a region of amino acid sequence homology to NusG, one of the proteins involved in bacteriophage λ N protein antitermination. This homology suggests that there are similarities in the mechanisms regulating RNA-chain elongation from bacteria to mammals.

Promoter-Proximal Pausing of RNA Polymerase II Occurs in Some Rapidly Induced Genes

The *Drosophila* heat-shock genes (e.g., *hsp70*) illustrate another mechanism for regulating RNA-chain elongation in eukaryotes. In this case, RNA polymerase II pauses after transcribing ≈25 nucleotides, rather than terminating transcription (as it does when transcribing the HIV genome in the absence of Tat protein). The paused polymerase remains associated with the nascent RNA and template DNA, until conditions occur that lead to activation of *HSTF (heat-shock transcription factor)*. Subsequent binding of activated HSTF to specific sites in the promoter-proximal region of heat-shock genes stimulates the paused polymerase to continue chain elongation and promotes rapid re-initiation by additional RNA polymerase II molecules.

The heat-shock genes are induced by intracellular conditions that denature proteins (such as elevated temperature, "heat shock"). Some encode proteins that are relatively resistant to denaturing conditions and act to protect other proteins from denaturation; others are chaperonins that re-fold denatured proteins (see Chapter 3). The mechanism of transcriptional control that evolved to regulate expression of these genes permits a rapid response because no time is required to assemble transcription-initiation complexes.

SUMMARY Transcription Termination

- Purified bacterial RNA polymerases terminate transcription by a Rho-independent mechanism at sites that encode a string of several U residues, preceded by a stable stem-loop secondary structure in the RNA transcript.

- Expression of the *trp* operon and other biosynthetic operons is regulated in part by attenuation. In this process, rapid translation of the leader sequence in an mRNA favors a RNA secondary structure that terminates transcription prematurely by a Rho-independent mechanism. Slow translation, in contrast, favors an alternative RNA secondary structure that does not cause termination (see Figure 11-3). The rate of leader translation depends on the supply of tryptophan or other product synthesized by the enzymes encoded by the operon.

- Some termination sites in *E. coli* require the action of Rho protein, an ATPase that dislodges the 3' end of a growing RNA chain from the active site of RNA polymerase.

- The temporal expression of λ-phage genes following infection is regulated by a process called antitermination. A multiprotein complex comprising λ N protein and several *E. coli* proteins binds to a specific RNA sequence as soon as it is transcribed, and then interacts with the transcribing RNA polymerase, permitting it to continue chain elongation beyond termination sites (see Figure 11-5).

- Different mechanisms of transcription termination are employed by each of the eukaryotic nuclear RNA polymerases. Transcription of most protein-coding genes is not terminated until an RNA sequence is synthesized that specifies a site of cleavage and polyadenylation.

- Transcription of the HIV genome by RNA polymerase II is regulated by an antitermination mechanism, analogous to that found in *E. coli*, in which a regulated choice is made between termination or further elongation after transcription of ≈20–50 bases. During transcription of *Drosophila* heat-shock genes, RNA polymerase II pauses within the promoter-proximal region; this interruption in transcription is released under certain conditions, leading to rapid expression of the heat-shock genes. Both mechanisms probably operate to control other genes transcribed by eukaryotic RNA polymerases.

11.2 Processing of Eukaryotic mRNA

As discussed in Chapter 4, the initial primary transcript synthesized by RNA polymerase II undergoes several processing steps before a functional mRNA is produced. In this section, we take a closer look at how eukaryotic cells carry out mRNA processing, which includes three major processes: **5' capping, 3' cleavage/polyadenylation,** and **RNA splicing** (Figure 11-7). Processing occurs in the nucleus, and the functional mRNA produced is transported to the cytoplasm by mechanisms discussed later.

The 5'-Cap Is Added to Nascent RNAs Shortly after Initiation by RNA Polymerase II

After nascent RNA molecules produced by RNA polymerase II reach a length of 25–30 nucleotides, *7-methylguanosine* is added to their 5' end. This initial step in RNA processing is catalyzed by a dimeric capping enzyme, which associates with the phosphorylated carboxyl-terminal tail domain (CTD) of RNA polymerase II. Recall that the CTD becomes phosphorylated during transcription initiation (see Figure 10-50). Because the capping enzyme does not associate with polymerase I or III, capping is specific for transcripts produced by RNA polymerase II.

One subunit of the capping enzyme removes the γ-phosphate from the 5' end of the nascent RNA emerging from the surface of a RNA polymerase II (Figure 11-8). The other subunit transfers the GMP moiety from GTP to the 5'-diphosphate of the nascent transcript, creating the guanosine 5'-5'-triphosphate structure. In the final steps, separate enzymes transfer methyl groups from *S*-adenosylmethionine to the N_7 position of the guanine and the 2' oxygens of riboses at the 5' end of the nascent RNA.

Pre-mRNAs Are Associated with hnRNP Proteins Containing Conserved RNA-Binding Domains

Nascent RNA transcripts from protein-coding genes and mRNA processing intermediates, collectively referred to as **pre-mRNA,** do not exist as free RNA molecules in the nuclei

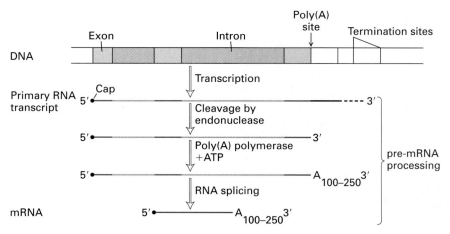

▲ FIGURE 11-7 Overview of mRNA processing in eukaryotes.
Shortly after RNA polymerase II initiates transcription at the first nucleotide of the first exon of a gene, the 5′ end of the nascent RNA is capped with 7-methylguanylate. Transcription by RNA polymerase II terminates at any one of multiple termination sites downstream from the poly(A) site, which is located at the 3′ end of the final exon. After the primary transcript is cleaved at the poly(A) site, a string of adenine (A) residues is added. The poly(A) tail contains ≈250 A residues in mammals, ≈150 in insects, and ≈100 in yeasts. For short primary transcripts with few introns, polyadenylation, cleavage, and splicing usually follows termination, as shown. For large genes with multiple introns, introns often are spliced out of the nascent RNA before transcription of the gene is complete. Note that the 5′ cap is retained in mature mRNAs.

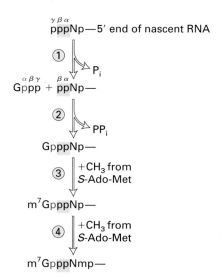

▲ FIGURE 11-8 Capping of the 5′ end of nascent RNA transcripts with 7′-methylguanylate (m⁷G). The first two reactions are catalyzed by a capping enzyme that associates with the phosphorylated CTD of RNA polymerase II shortly after transcription initiation. Two different methyltransferases catalyze reactions 3 and 4. S-adenosylmethionine (S-Ado-Met) is the source of the methyl (CH_3) group for the two methylation steps; the guanylate (G) is methylated first, then the 2′ hydroxyl of the first one or two nucleotides (N) in the transcript. See Figure 4-18 for structure of the resulting 5′ cap. [See S. Venkatesan and B. Moss, 1982, *Proc. Nat'l. Acad. Sci. USA* **79**:304.]

of eukaryotic cells. From the time nascent transcripts first emerge from RNA polymerase II until mature mRNAs are transported into the cytoplasm, the RNA molecules are associated with an abundant set of nuclear proteins, as numerous in growing eukaryotic cells as histones. These proteins are the major protein components of heterogeneous ribonucleoprotein particles (hnRNPs), which contain *heterogeneous nuclear RNA* (hnRNA), a collective term referring to pre-mRNA and other nuclear RNAs of various sizes. The proteins in these ribonucleoprotein particles can be dramatically visualized with fluorescent-labeled monoclonal antibodies (Figure 11-9).

To identify **hnRNP proteins,** researchers exposed cells to high-dose UV irradiation, which causes covalent cross-links to form between RNA bases and closely associated proteins. Chromatography of nuclear extracts from treated cells on an oligo-dT cellulose column, which binds RNAs with a poly(A) tail, was used to recover proteins that had become cross-linked to nuclear mRNA in living cells (i.e., hnRNP proteins). Subsequent treatment of cell extracts from unirradiated human cells with monoclonal antibodies specific for the major hnRNP proteins identified by this cross-linking technique revealed a complex set of abundant hnRNP proteins ranging in size from 34 to 120 kDa. Characterization of the mRNAs encoding these proteins has shown that some of them (e.g., A2 and B1) are related proteins derived by alternative splicing of exons from the same transcription unit.

Binding studies with purified hnRNP proteins suggest that different hnRNP proteins associate with different regions of a newly made pre-mRNA molecule as determined by the sequence of the RNA. For example, the hnRNP proteins A1, C, and D bind preferentially to the pyrimidine-rich sequences at the 3′ ends of introns, discussed in a later section. Like

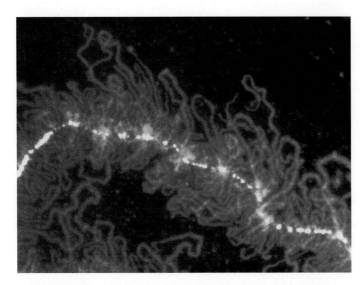

▲ **FIGURE 11-9 Visualization of hnRNP protein associated with nascent transcripts in an oocyte of the newt *Nophthalmus viridescens*.** A portion of a "lampbrush" chromosome is shown. DNA at the chromosome axis stains white with the DNA-specific dye DAPI. The long red filaments are nascent transcripts bound by hnRNP proteins, which fluoresce red after staining with a monoclonal antibody against a specific hnRNP protein. [Courtesy of M. Roth and J. Gall.]

transcription factors, most hnRNP proteins have a modular structure. They contain one or more RNA-binding domains and at least one other domain that is thought to interact with other proteins. Several different RNA-binding motifs have been identified by constructing deletions of hnRNP proteins and testing their ability to bind RNA. Although some RNA-binding proteins contain domains with the zinc-finger motif

common in DNA-binding proteins (see Figure 10-41), this motif has not yet been described in any hnRNP proteins.

The *RNP motif*, also called the RNA-binding domain (RBD), is the most common RNA-binding domain in hnRNP proteins. This ≈80-residue motif, which occurs in many other RNA-binding proteins, contains two highly conserved regions (RNP1 and RNP2) that allow the motif to be recognized in newly sequenced proteins. X-ray crystallographic analysis has shown that the RNP motif consists of a four-stranded β sheet flanked on one side by two α helices. The conserved RNP1 and RNP2 sequences lie side by side on the two central β strands, and their side chains make multiple contacts with a single-stranded region of RNA. The single-stranded RNA loop lies across the surface of the β sheet and fits into a groove between the protein loop connecting strands β_2 and β_3 and the C-terminal region (Figure 11-10).

The *RGG box*, another RNA-binding motif found in hnRNP proteins, contains five Arg-Gly-Gly (RGG) repeats with several interspersed aromatic amino acids. Although the structure of this motif has not yet been determined, its arginine-rich nature is similar to the RNA-binding domains of the λ-phage N and HIV Tat proteins.

The 45-residue *KH motif* is found in the hnRNP K protein and several other RNA-binding proteins; commonly two or more copies of the KH motif are interspersed with RGG repeats. The three-dimensional structure of a representative KH motif, determined by NMR methods (Section 3.5), is similar to that of the RNP motif but smaller, consisting of a three-stranded β sheet supported from one side by a single α helix. It is not yet clear how this motif binds RNA. Mutations in the fragile-X gene *(FMR1)*, which encodes a protein containing the KH motif, are associated with the most common form of heritable mental retardation. Although the molecular function of the Fmr1 protein is unknown, it presumably involves RNA binding.

(a)

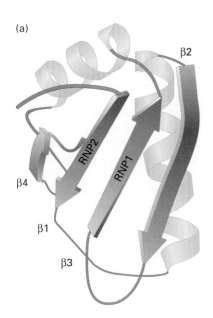

(b)

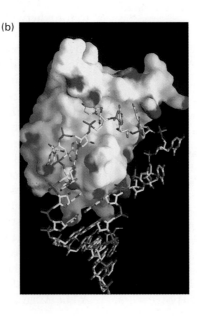

◀ **FIGURE 11-10 Structure of complex between an RNP motif from U1A protein and RNA.** (a) Diagram of the RNP motif domain. The conserved RNP1 and RNP2 regions are located in the two middle β–strands. (b) Surface representation of the U1A protein-RNA complex determined by X-ray crystallography. The RNA forms a stem-loop with the single-stranded portion of the loop bound to the surface of the protein. The N- and C-termini are at the upper left and right, respectively. Acidic and basic amino acids are colored red and blue, respectively. [From K. Nagai et al., 1995, *Trends Biochem. Sci.* **20:** 235.]

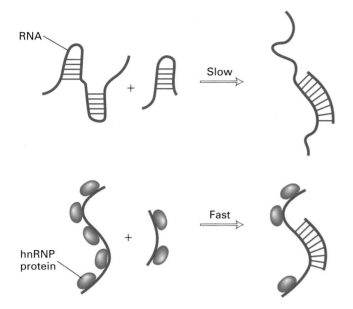

RNA

Slow

hnRNP
protein

Fast

▲ **FIGURE 11-11 Hybridization of RNA molecules in vitro is accelerated by hnRNP proteins.** The presence of complex secondary structures within RNA molecules inhibits hybridization between long complementary sequences in separate molecules. Association of hnRNP proteins with RNA is thought to prevent formation of RNA secondary structures, thereby facilitating base-pairing between different complementary molecules. These proteins may have a similar function in vivo. [Adapted from D. S. Portman and G. Dreyfuss, 1994, *EMBO J.* **13**:213.]

hnRNP Proteins May Assist in Processing and Transport of mRNAs

The association of pre-mRNAs with hnRNP proteins may prevent formation of short secondary structures dependent on base-pairing of complementary regions, thereby making the pre-mRNAs accessible for interaction with other macromolecules (Figure 11-11). Moreover, pre-mRNAs associated with hnRNP proteins present a more uniform substrate for further processing steps than would free, unbound pre-mRNAs each type of which forms a unique secondary structure dependent on its specific sequence.

The diversity of hnRNP proteins suggests that they probably have other functions as well. For example, various hnRNP proteins may interact with the RNA sequences that specify RNA splicing or cleavage/polyadenylation and contribute to the structure recognized by RNA-processing factors. Finally, cell-fusion experiments have shown that some hnRNP proteins remain localized in the nucleus, whereas others cycle in and out of the cytoplasm, suggesting that they function in the transport of mRNA (see later section).

Pre-mRNAs Are Cleaved at Specific 3′ Sites and Rapidly Polyadenylated

In animal cells, all mRNAs, except histone mRNAs, have a 3′ poly(A) tail. Early studies of pulse-labeled adenovirus and SV40 RNA demonstrated that the viral primary transcripts extend beyond the poly(A) site in the viral mRNAs. These results suggested that A residues are added to a 3′ hydroxyl generated by endonucleolytic cleavage, but the predicted downstream RNA fragments are degraded so rapidly in vivo that they cannot be detected. However, this cleavage mechanism was firmly established by detection of both predicted cleavage products in in vitro processing reactions performed with extracts of HeLa-cell nuclei.

Early sequencing of cDNA clones from animal cells showed that nearly all mRNAs contain the sequence AAUAAA 10–35 nucleotides upstream from the poly(A) tail. Polyadenylation of RNA transcripts from transfected genes is virtually eliminated when template DNA encoding the AAUAAA sequence is mutated to any other sequence except one encoding AUUAAA. The unprocessed RNA transcripts produced from such mutant templates do not accumulate in nuclei, but are rapidly degraded. Further mutagenesis of sequences within a few hundred bases of poly(A) sites revealed that a second signal downstream from the cleavage site is required for efficient **cleavage and polyadenylation** of most pre-mRNAs in animal cells. This downstream poly(A) signal is not a specific sequence but rather a GU-rich or simply a U-rich region within ≈50 nucleotides of the cleavage site.

Identification and purification of the proteins required for cleavage and polyadenylation of pre-mRNA has led to the model shown in Figure 11-12. According to this model, a 360-kDa *cleavage and polyadenylation specificity factor* (CPSF), composed of four different polypeptides, first forms an unstable complex with the upstream AU-rich poly(A) signal. Then at least three additional proteins—a 200-kDa heterotrimer called *cleavage stimulatory factor* (CStF), a 150-kDa heterotrimer called *cleavage factor I* (CFI), and a second cleavage factor (CFII), as-yet poorly characterized—bind to the CPSF-RNA complex. Interaction between CStF and the GU- or U-rich downstream poly(A) signal stabilizes the multiprotein complex. Finally, a *poly(A) polymerase* (PAP) binds to the complex before cleavage can occur. This requirement for PAP binding links cleavage and polyadenylation, so that the free 3′ ends generated are rapidly polyadenylated. Assembly of this large, multiprotein cleavage-polyadenylation complex around the AU-rich poly(A) signal in a pre-mRNA is analogous in many ways to formation of the transcription-initiation complex at the AT-rich TATA box of a template DNA molecule (see Figure 10-50). In both cases, multiprotein complexes assemble cooperatively through a network of specific protein–nucleic acid and protein-protein interactions.

Following cleavage at the poly(A) site, polyadenylation proceeds in two phases. Addition of the first 12 or so A residues occurs slowly, followed by rapid addition of up to 200–250 more A residues. The rapid phase requires the binding of multiple copies of a poly(A)-binding protein containing the RNP motif. This protein is designated *PABII* to distinguish it from the poly(A)-binding protein that binds to the poly(A) tail of cytoplasmic mRNAs. PABII binds to the short A tail initially added by PAP, stimulating polymerization of

▶ **FIGURE 11-12 Model for cleavage and polyadenylation of pre-mRNAs in mammalian cells.** Cleavage-and-polyadenylation specificity factor (CPSF) binds to an upstream AAUAAA polyadenylation signal. CStF interacts with a downstream GU- or U-rich sequence and with bound CPSF, forming a loop in the RNA; binding of CFI and CFII help stabilize the complex. Binding of poly(A) polymerase (PAP) then stimulates cleavage at a poly(A) site, which usually is 10–35 nucleotides 3′ of the upstream polyadenylation signal. The cleavage factors are released, as is the downstream RNA cleavage product, which is rapidly degraded. Bound PAP then adds ≈12 A residues at a slow rate to the 3′-hydroxyl group generated by the cleavage reaction. Binding of poly(A)-binding protein II (PABII) to the initial short poly(A) tail accelerates the rate of addition by PAP. After 200–250 A residues have been added, PABII signals PAP to stop polymerization.

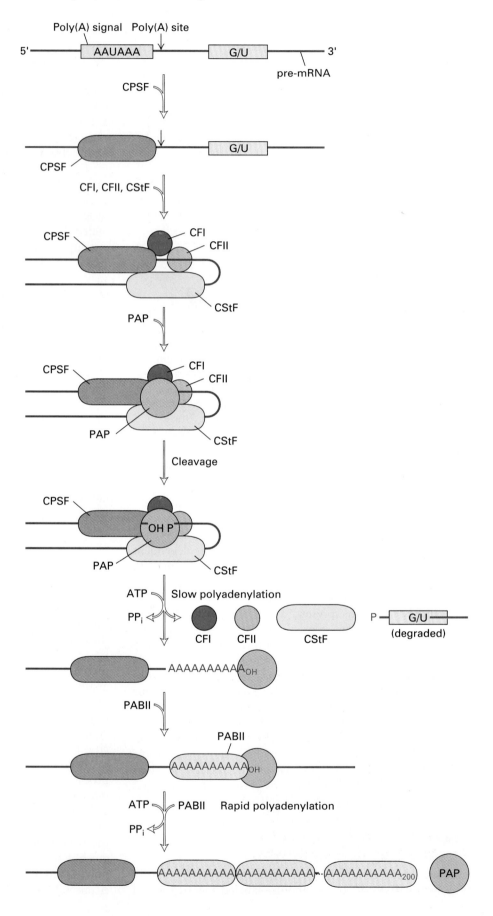

additional A residues by PAP (see Figure 11-12). PABII is also responsible for signaling poly(A) polymerase to terminate polymerization when the poly(A) tail reaches a length of 200–250 residues, although the mechanism for measuring this length is not yet understood.

Splicing Occurs at Short, Conserved Sequences in Pre-mRNAs via Two Transesterification Reactions

During the final step in formation of a mature, functional mRNA, the introns are removed and exons are spliced together (see Figure 11-7). The discovery that introns are removed during splicing came from electron microscopy of RNA-DNA hybrids between adenovirus DNA and the mRNA encoding hexon, a major virion capsid protein (Figure 11-13).* Similar analyses of hybrids between RNA isolated from the nuclei of infected cells and viral DNA revealed RNAs that were colinear with the viral DNA (primary transcripts) and RNAs with one or two of the introns removed (processing intermediates). These results, together with the findings that the 59 cap and 39 poly(A) tail of mRNA precursors are retained in mature cytoplasmic mRNAs, led to the realization that introns are removed from primary transcripts as exons are spliced together. For short transcription units, RNA splicing

---

*Few genes in bacterial or bacteriophage DNAs contain introns, whereas most protein-coding genes in animals and plants, as well as in DNA viruses infecting them, contain introns. Thus DNA viruses generally have a similar intron content as their host cells. Most yeast genes, like bacterial genes, lack introns.

usually follows cleavage and polyadenylation of the 3' end of the primary transcript. But for long transcription units containing multiple exons, splicing of exons in the nascent RNA usually begins before transcription of the gene is complete.

The location of exon-intron junctions (i.e., **splice sites**) in a pre-mRNA can be determined by comparing the sequence of genomic DNA with that of the cDNA prepared from the corresponding mRNA. Sequences that are present in the genomic DNA but absent from the cDNA represent introns and indicate the positions of splice sites. Such analysis of a large number of different mRNAs revealed moderately conserved, short consensus sequences at intron-exon boundaries in eukaryotic pre-mRNAs; in higher organisms, a pyrimidine-rich region just upstream of the 3' splice site also is common (Figure 11-14). The most conserved nucleotides are the (5')GU and (3')AG found at the ends of most introns. Deletion analyses of the center portion of introns in various pre-mRNAs have shown that generally only 30–40 nucleotides at each end of an intron are necessary for splicing to occur at normal rates.

Recombinant DNAs containing the 5' splice site of one transcription unit (e.g., SV40 late region) and the 3' splice site of another (e.g., mouse β-globin gene) have been prepared and introduced into cultured cells. Spliced mRNA molecules are formed in which the two exon sequences are joined and the chimeric intron is deleted precisely. The formation of correctly spliced mRNAs in such experiments indicates that the cell's splicing machinery can recognize 5' and 3' splice sites and correctly splice them together, with little influence from the intervening sequence in most cases.

Analysis of the intermediates formed during splicing of pre-mRNAs in vitro led to the conclusion that introns are

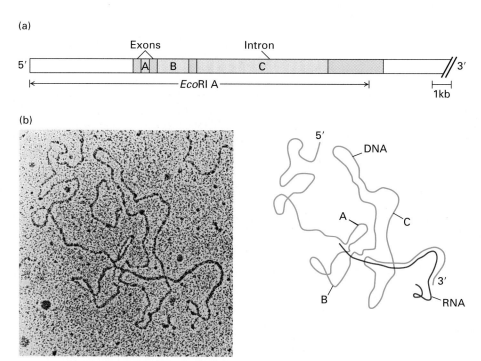

(a)

Exons   Intron

5' [[A] B | C] 3'

|←————— EcoRI A —————→| 1kb

(b)

5'   DNA

A   C

B   3'   RNA

◀ **FIGURE 11-13 Demonstration that introns are spliced out by electron microscopy of RNA-DNA hybrid between adenovirus DNA and the mRNA encoding hexon, a major viral protein.** (a) Diagram of the *Eco*RI A fragment of adenovirus DNA, which extends from the left end of the genome to just before the end of the final exon of the hexon gene. The gene consists of three short exons and one long (≈3.5-kb) exon separated by three introns of ≈1, 2.5, and 9 kb. (b) Electron micrograph *(left)* and schematic drawing *(right)* of hybrid between an *Eco*RI A fragment and hexon mRNA. The loops marked A, B, and C correspond to the introns indicated in (a). Since these intron sequences in the viral genomic DNA are not present in mature hexon mRNA, they loop out between the exon sequences that hybridize to their complementary sequences in the mRNA. [Micrograph from S. M. Berget et al., 1977, *Proc. Nat'l. Acad. Sci. USA* **74**:3171; courtesy of P. A. Sharp.]

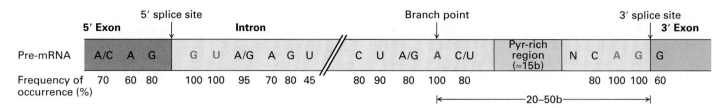

	5' splice site								Branch point						3' splice site		
5' Exon			**Intron**													**3' Exon**		
Pre-mRNA	A/C	A	G	G	U	A/G	A	G	U	C	U	A/G	A	C/U	Pyr-rich region (≈15b)	N	C	A	G	G

Frequency of occurrence (%): 70 60 80   100 100 95 70 80 45   80 90 80 100 80   80 100 100 60

|← 20–50b →|

▲ **FIGURE 11-14 Consensus sequences around 5' and 3' splice sites in vertebrate pre-mRNAs.** The only nearly invariant bases are the (5')GU and (3')AG of the intron, although the flanking bases indicated are found at frequencies higher than expected based on a random distribution. A pyrimidine-rich region (light blue) near the 3' end of the intron is found in most

cases. The branch-point adenosine, also invariant, usually is 20–50 bases from the 3' splice site. The central region of the intron, which may range from 40 bases to 50 kilobases in length, generally is unnecessary for splicing to occur. [See R. A. Padgett et al., 1986, *Ann. Rev. Biochem.* **55**:1119; E. B. Keller and W. A. Noon, 1984, *Proc. Nat'l. Acad. Sci. USA* **81**:7417.]

removed as a *lariat* structure in which the 5' G of the intron is joined in an unusual 2',5'-phosphodiester bond to an adenosine near the 3' end of the intron (Figure 11-15). This A residue is called the **branch point** because it forms an RNA branch in the lariat structure.

The finding that excised introns have a branched lariat structure led to the discovery that splicing of exons proceeds via two sequential *transesterification reactions* (Figure 11-16). In each reaction, one phosphate-ester bond is exchanged for another. Since the number of phosphate-ester bonds in the molecule is not changed in either reaction, no energy is

consumed. The net result of these two transesterification reactions is that two exons are ligated and the intervening intron is released as a branched lariat structure.

Spliceosomes, Assembled from snRNPs and a Pre-mRNA, Carry Out Splicing

Even before splicing was accomplished in vitro, several observations led to the suggestion that **small nuclear RNAs (snRNAs)** assist in the splicing reaction. First, the short consensus sequence at the 5' end of introns was found to be

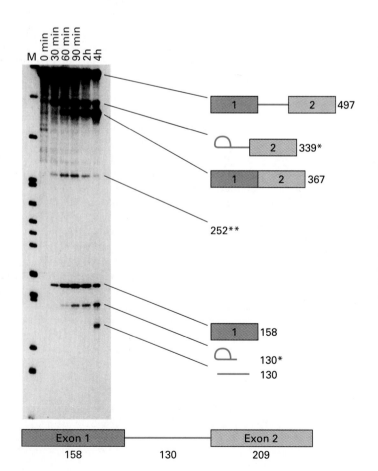

◄ **FIGURE 11-15 Analysis of RNA products formed in an in vitro splicing reaction.** A nuclear extract from HeLa cells was incubated with a 497-nucleotide radiolabeled RNA *(bottom)* that contained portions of two exons (orange and tan) from human β-globin mRNA separated by a 130-nucleotide intron (blue). After incubation for various times, the RNA was purified and subjected to electrophoresis and autoradiography, along with RNA markers (lane M). The number of nucleotides in the various species is indicated. Much of the slower-migrating starting RNA (497) was correctly spliced, yielding a 367-nucleotide product. The excised intron (130*) migrated slower than expected based on its molecular weight, indicating that it is not a linear molecule. Likewise, one of the reaction intermediates (339*) exhibited an anomalously slow electrophoretic mobility. Additional analysis indicated that in both cases the intron had a lariat structure resulting in the slow mobility. The 252** band, an aberrant product of the in vitro reaction, is greatly reduced in reactions in which the RNA is capped. [From B. Ruskin et al., 1984, *Cell* **38**:317; photograph courtesy of Michael R. Green. See also R. A. Padgett et al., 1984, *Science* **225**:898.]

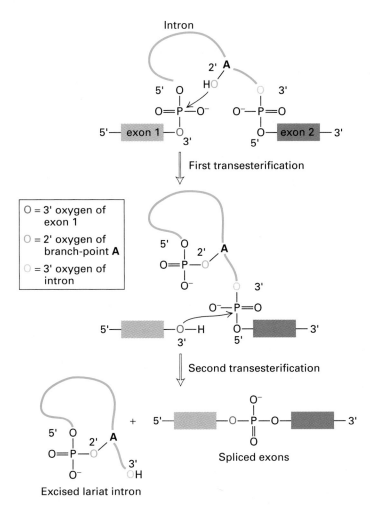

▲ FIGURE 11-16 Splicing of exons in pre-mRNA occurs via two transesterification reactions. In the first reaction, the ester bond between the 5′ phosphorus of the intron and the 3′ oxygen (red) of exon 1 is exchanged for an ester bond with the 2′ oxygen (dark blue) of the branch-site **A** residue. In the second reaction, the ester bond between the 5′ phosphorus of exon 2 and the 3′ oxygen (light blue) of the intron is exchanged for an ester bond with the 3′ oxygen of exon 1, releasing the intron as a lariat structure and joining the two exons. Arrows show where the activated hydroxyl oxygens react with phosphorus atoms.

complementary to a sequence near the 5′ end of the snRNA called U1. Second, snRNAs were found associated with hnRNPs in nuclear extracts. Five U-rich snRNAs (U1, U2, U4, U5, and U6), ranging in length from 107 to 210 nucleotides, participate in RNA splicing.

In the nucleus of eukaryotic cells, snRNAs are associated with six to ten proteins in *small nuclear ribonucleoprotein particles (snRNPs).* Some of these proteins are common to all snRNPs, and some are specific for individual snRNPs. Experiments with a synthetic oligonucleotide that hybridizes with the 5′-end region of U1 snRNA and later studies with pre-mRNAs that were mutated in the 5′ splice-site consensus sequence provided strong evidence that base

pairing between the 5′ splice site of a pre-mRNA and the 5′ region of U1 snRNA is required for RNA splicing.

Involvement of U2 snRNA in splicing initially was suspected when it was found to have an internal sequence that is largely complementary to the consensus sequence flanking the branch point in pre-mRNAs (see Figure 11-14). Mutation experiments, similar to those conducted with U1 snRNA and 5′ splice sites, demonstrated that base pairing between U2 snRNA and the branch-point sequence in pre-mRNA is critical to splicing. These studies with U1 and U2 snRNAs indicate that during splicing they base-pair with pre-mRNA as shown in Figure 11-17. Significantly, the branch-point A itself, which is not base-paired to U2 snRNA, "bulges out," allowing its 2′ hydroxyl to participate in the first transesterification reaction of RNA splicing (see Figure 11-16).

Similar studies with other snRNAs demonstrated that RNA-RNA interactions involving them also occur during splicing. For example, an internal region of U6 snRNA initially base-pairs with the 5′ end of U4 snRNA. Rearrangements later in the splicing process result in U6 snRNA base pairing with the 5′ end of U2 snRNA, which remains base-paired to the branch-point sequence in the intron. Later in the splicing process, base pairing of U5 snRNA with four exon nucleotides adjacent to the splice sites displaces U1 snRNA from the pre-mRNA.

Based on the results of these experiments, identification of reaction intermediates, and other biochemical analyses, the five splicing snRNPs are thought to sequentially assemble on the pre-mRNA forming a large ribonucleoprotein complex called a **spliceosome,** which is roughly the size of a ribosome (Figure 11-18). According to the model depicted in Figure 11-19, assembly of a spliceosome begins with the base pairing of U1 and U2 snRNAs, as part of the U1 and U2 snRNPs, to the pre-mRNA (see Figure 11-17). Extensive base pairing between the snRNAs in the U4 and U6 snRNPs forms a complex that associates with U5 snRNP. The U4/U6/U5 complex then associates, presumably via protein-protein interactions, with the previously formed complex consisting of a pre-mRNA base-paired to U1 and U2 snRNPs to yield a spliceosome.

After formation of the spliceosome, extensive rearrangements occur in the pairing of snRNAs and the pre-mRNA, as noted previously. The rearranged spliceosome then catalyzes the two transesterification reactions that result in RNA splicing. After the second transesterification reaction, the ligated exons are released from the spliceosome while the lariat intron remains associated with the snRNPs. This final intron-snRNP complex is unstable and dissociates. The individual snRNPs released participate in a new cycle of splicing. The excised intron is rapidly degraded by a "debranching enzyme," which hydrolyzes the 5′,2′-phosphodiester bond at the branch point, and other nuclear RNases.

It is estimated that at least one hundred proteins are involved in RNA splicing, making this process comparable in complexity to protein synthesis and initiation of transcription. Some of these splicing factors are associated with

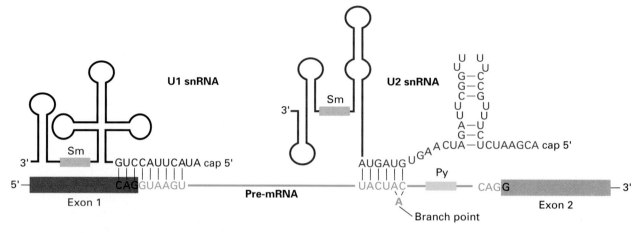

▲ **FIGURE 11-17 Diagram of interactions between pre-mRNA, U1 snRNA, and U2 snRNA early in the splicing process.** The 5′ region of U1 snRNA initially base-pairs with nucleotides at the 5′ end of the intron (blue) and 3′ end of the 5′ exon (dark red) of the pre-mRNA; U2 snRNA base-pairs with a sequence that includes the branch-point A, although this residue is not base-paired. The yeast branch-point sequence is shown here. Secondary structures in the snRNAs that are not altered during splicing are shown in diagrammatic line form. The purple rectangles represent sequences that bind snRNP proteins recognized by anti-Sm antibodies. For unknown reasons, antisera from patients with the autoimmune disease systemic lupus erythematosus (SLE) contain these antibodies. Such antisera have been useful in characterizing components of the splicing reaction. [See E. J. Sontheimer and J. A. Steitz, 1993, *Science* **262**:1989; adapted from M. J. Moore et al., 1993, in R. Gesteland and J. Atkins, eds., *The RNA World*, Cold Spring Harbor Press, pp. 303–357.]

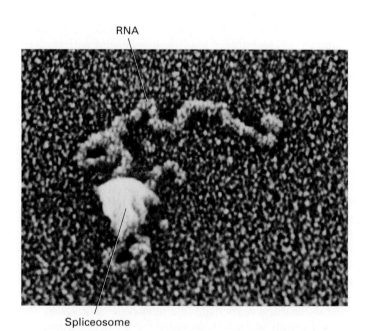

RNA

Spliceosome

▲ **FIGURE 11-18 Electron micrograph of a spliceosome.** Extracts of HeLa cells were mixed with a β-globin pre-mRNA; the reaction was interrupted before splicing was completed, so that the spliceosomes, containing snRNPs and the pre-mRNA substrate, could be purified. [From R. Reid et al., 1988, *Cell* **53**:949; courtesy of J. Griffith.]

snRNPs, but others are not. Sequencing of yeast genes encoding splicing factors has revealed that they contain domains with the RNP motif, which interacts with RNA, and the SR motif, which interacts with other proteins and may contribute to RNA binding. Some splicing factors also exhibit sequence homologies to known RNA helicases; these may be necessary for the base-pairing rearrangements that occur in snRNAs during the spliceosomal splicing cycle.

Introns whose splice sites do not conform to the standard consensus sequence recently were identified in some pre-mRNAs. This class of introns begins with AU and ends with AC rather than following the usual "GU–AG rule" (see Figure 11-14). Research on the biochemistry of splicing for this special class of introns soon identified four novel snRNPs. Together with the standard U5 snRNP, these snRNPs appear to participate in a splicing cycle analogous to that discussed above.

Portions of Two Different RNAs Are Trans-Spliced in Some Organisms

Virtually all functional mRNAs in vertebrate and insect cells are derived from a single molecule of the corresponding pre-mRNA by removal of internal introns and splicing of exons. However, in two types of protozoa—trypanosomes and euglenoids—mRNAs are constructed by splicing together separate RNA molecules. This process, referred to as **trans-splicing,** is also used in the synthesis of 10–15 percent of the mRNAs in the round worm *Caenorhabditis elegans*, an important model organism for studying embryonic development.

▶ **FIGURE 11-19 The spliceosomal splicing cycle.** The splicing snRNPs (U1, U2, U4, U5, and U6) associate with the pre-mRNA and with each other in an ordered sequence to form the spliceosome. This large ribonucleoprotein complex then catalyzes the two transesterification reactions that result in splicing of the exons (light and dark red) and excision of the intron (blue) as a lariat structure (see Figure 11-16). Although ATP hydrolysis is not required for the transesterification reactions, it is thought to provide the energy necessary for rearrangements of the spliceosome structure that occur during the cycle. Note that the snRNP proteins in the spliceosome are distinct from the hnRNP proteins discussed earlier. In higher eukaryotes, the association of U2 snRNP with pre-mRNA is assisted by an hnRNP protein called U2AF, which binds to the pyrimidine-rich region near the 3′ splice site. U2AF also probably interacts with other proteins required for splicing through a domain containing repeats of the dipeptide serine-arginine (the SR motif). The branch-point A in pre-mRNA is indicated in boldface. [See S. W. Ruby and J. Abelson, 1991, *Trends Genet.* **7**:79; adapted from M. J. Moore et al., 1993, in R. Gesteland and J. Atkins, eds., *The RNA World,* Cold Spring Harbor Press, pp. 303–357.]

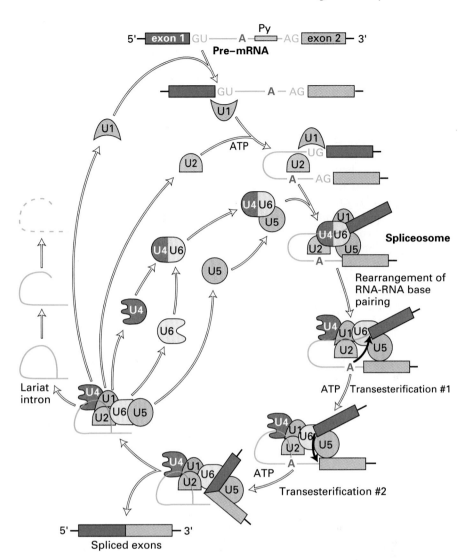

The parasitic trypanosomes produce abundant amounts of a single 140-nucleotide leader RNA from tandemly repeated transcription units. In a two-step reaction analogous to spliceosomal pre-mRNA splicing, a 39-nucleotide portion of the leader RNA, termed a *mini-exon*, is spliced to the 5′ end of protein-coding exons in primary transcripts, which lack internal introns. The 5′ mini-exon, present in all trypanosome mRNAs, is thought to assist in initiation of translation. Because of trans-splicing, polycistronic protein-coding transcription units in trypanosomes, which are common, yield monocistronic mRNAs from their polycistronic primary transcripts. Splicing of a 5′ mini-exon to a coding region in a primary transcript triggers cleavage and polyadenylation at the 3′ end of the exon. Consequently, trypanosomes use trans-splicing and linked cleavage and polyadenylation to combine the operon organization of polycistronic transcription units characteristic of bacteria with the monocistronic organization of mRNAs characteristic of eukaryotes.

Self-Splicing Group II Introns Provide Clues to the Evolution of snRNAs

Under certain nonphysiological in vitro conditions, pure preparations of some RNA transcripts slowly splice out introns in the absence of any protein. This observation led to recognition that some introns are *self-splicing.* Two types of self-splicing introns have been discovered: **group I introns,** present in nuclear rRNA genes of protozoans, and **group II introns,** present in protein-coding genes and some rRNA and tRNA genes of mitochondria and chloroplasts in plants and fungi. Discovery of the catalytic activity of self-splicing introns revolutionized concepts about the functions of RNA.

As discussed in Chapter 4, RNA is now thought to catalyze peptide-bond formation during protein synthesis in ribosomes. Here we discuss the probable role of group II introns, now found only in mitochondrial and chloroplast DNA, in the evolution of snRNAs; the functioning of group I introns is considered in the later section on rRNA processing.

Even though their precise sequences are not highly conserved, all group II introns fold into a conserved, complex secondary structure containing numerous stem-loops (Figure 11-20a). Self-splicing by a group II intron occurs via two transesterification reactions, involving intermediates and products analogous to those found in nuclear pre-mRNA splicing. The mechanistic similarities between group II intron self-splicing and spliceosomal splicing led to the hypothesis that snRNAs function analogously to the stem-loops in the secondary structure of group II introns. According to this hypothesis, snRNAs interact with 5′ and 3′ splice sites of pre-mRNAs and with each other to produce an RNA structure functionally analogous to that of group II self-splicing introns (Figure 11-20b).

An extension of this hypothesis is that introns in present-day nuclear pre-mRNAs evolved from ancient group II self-splicing introns through the progressive loss of internal RNA structures, which concurrently evolved into transacting snRNAs that perform the same functions. In support of this kind of evolutionary model, group II intron mutants have been constructed in which domain V and part of domain I are deleted. Such mutants are defective in self-splicing, but when RNA molecules equivalent to the deleted regions are added to the in vitro reaction, self-splicing occurs. This finding demonstrates that these domains in group II introns can be trans-acting, like snRNAs.

The similarity in the mechanisms of group II intron self-splicing and spliceosomal splicing of pre-mRNAs also suggests that the splicing reaction is catalyzed by the snRNA, not the protein, components of spliceosomes. Although group II introns can self-splice in vitro at elevated temperatures and Mg^{2+} concentrations, under in vivo conditions proteins called *maturases*, which bind to group II intron RNA, are required for rapid splicing. Maturases, encoded by group II introns themselves, are thought to stabilize the precise three-dimensional interactions of the intron RNA required to catalyze the two splicing transesterification reactions. By analogy, snRNP proteins in spliceosomes are thought to stabilize the precise geometry of snRNAs and intron nucleotides required to catalyze pre-mRNA splicing.

The evolution of snRNAs may have been an important step in the rapid evolution of higher eukaryotes. As internal intron sequences were lost and their functions in RNA splicing supplanted by trans-acting snRNAs, the remaining intron sequences would be free to diverge. This in turn likely facilitated the evolution of new genes through exon shuffling (Section 9.3). It also permitted the increase in protein diversity that results from alternative RNA splicing and an additional level of gene control resulting from regulated RNA splicing.

One more remarkable property of group II introns deserves mention, namely, their ability to behave as mobile DNA elements in the genome. The maturases that increase the rate of self-splicing of these introns also contain a domain that is homologous to reverse transcriptase. Thus group II introns can move in the genome like other nonviral retrotransposons discussed in Chapter 9. As is generally true for mobile DNA elements, transposition of group II introns is rare. However, when a group II intron does transpose, it does not inactivate the gene into which it inserts, because the inserted intron is spliced out of the transcript produced from the target gene by self-splicing!

Most Transcription and RNA Processing Occur in a Limited Number of Domains in Mammalian Cell Nuclei

The digital imaging micrographs in Figure 11-21 demonstrate that most of the nuclear polyadenylated RNA (including unspliced and partially spliced pre-mRNA and nuclear mRNA) occurs in discrete foci lying between dense regions of chromatin and that a required protein splicing factor (SC-35) is localized to the center of these same loci. The results of these and other studies suggest that transcription and RNA processing do not occur randomly throughout the eukaryotic nucleus; rather, the nucleus is organized into discreet domains ($\approx$20–100 in human fibroblasts) where the bulk of transcription and RNA processing occurs.

This highly organized view of the nucleus implies that there is an underlying nuclear substructure. It has been known for many years that when mammalian cells are treated with a mild nonionic detergent, DNase I, and high concentrations

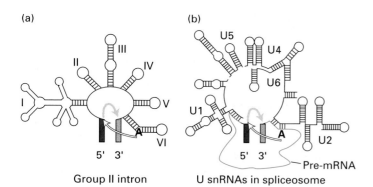

(a)   (b)

Group II intron   U snRNAs in spliceosome

▲ **FIGURE 11-20 Schematic diagrams comparing the secondary structures of group II self-splicing introns (a) and U snRNAs present in the spliceosome (b).** The first transesterification reaction is indicated by black arrows; the second reaction, by blue arrows. The branch-point A is boldfaced. The similarity in these structures suggests that the spliceosomal snRNAs evolved from group II introns, with the trans-acting snRNAs being functionally analogous to the corresponding domains in group II introns. [Adapted from P. A. Sharp, 1991, *Science* **254**:663.]

(a)

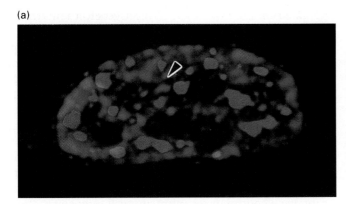

(b)

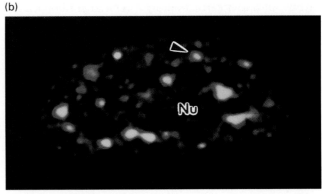

▲ **FIGURE 11-21 Localization of polyadenylated RNA and RNA splicing factors in the nucleus of a mammalian fibroblast.** Digital imaging microscopy was used to reconstruct a 1-μm-thick section of a stained human fibroblast nucleus. (a) Section stained with red rhodamine-labeled poly-dT to detect polyadenylated RNA (red) and with DAPI to detect DNA (blue). Polyadenylated RNA is localized to a limited number of discrete foci (speckles) between regions of chromatin, although not all regions containing low levels of DNA contained detectable polyadenylated RNA (arrow). (b) The same section shown in (a) stained to detect polyadenylated RNA (red) and the essential RNA-splicing protein SC-35, which was visualized with a green fluorescein-labeled monoclonal antibody. Regions where the stains overlap appear yellow. SC-35 is present in the center of many foci (arrow). Nu = nucleolus. [From K. C. Carter et al., 1993, *Science* **259**:1330.]

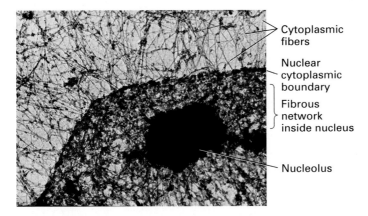

> Cytoplasmic fibers

> Nuclear cytoplasmic boundary

} Fibrous network inside nucleus

~ Nucleolus

▲ **FIGURE 11-22 Transmission electron micrograph showing the nuclear matrix (skeleton) of a HeLa cell.** Cells were treated with a nonionic detergent to remove membranes; digested with DNase to remove most of the DNA; and then extracted with 0.25 M ammonium sulfate to remove histones and chromatin-associated protein. A whole mount of the remaining material was prepared. [From S. Penman et al., 1982, *Cold Spring Harbor Symp. Quant. Biol.* **46**:1013.]

of salt, a fibrillar network of protein and RNA remains in the region of the nucleus (Figure 11-22). This protein network has been called the *nuclear matrix,* or *nuclear skeleton.* It is composed of actin and numerous other protein components that have not been fully characterized, including components of the chromosomal scaffold that rearranges and condenses to form metaphase chromosomes during mitosis (see Figure 9-34). However, snRNPs remain associated with the nuclear matrix prepared from detergent-extracted, DNase I–treated cells. Moreover, when the nuclear matrix is prepared with a low concentration of salt, pre-mRNAs associated with the matrix undergo splicing when ATP is added. These results suggest that the RNA-processing foci observed microscopically may be associated with specific regions of the nuclear matrix.

SUMMARY Processing of Eukaryotic mRNA

- Eukaryotic mRNA precursors are processed by 5′ capping, 3′ cleavage and polyadenylation, and RNA splicing to remove introns before being transported to the cytoplasm where they are translated by ribosomes.

- The cap is added to the 5′ end of a pre-mRNA nascent transcript by a capping enzyme that associates with the phosphorylated CTD of RNA polymerase II shortly after transcription initiation.

- Nascent pre-mRNA transcripts are associated with a class of abundant RNA-binding proteins called hnRNP proteins.

- In most protein-coding genes, a conserved polyadenylation signal (AAUAAA) lies 10–30 nucleotides upstream from a poly(A) site where cleavage and polyadenylation occur. A GU- or U-rich sequence downstream from the poly(A) site contributes to the efficiency of cleavage/polyadenylation.

- A multiprotein complex that includes poly(A) polymerase (PAP) carries out the cleavage and polyadenylation of a pre-mRNA. A nuclear poly(A)-binding protein, PABII, stimulates addition of A residues by PAP and stops addition once the poly(A) tail reaches 200–250 residues (see Figure 11-12).

• RNA splicing is carried out by a very large ribonucleo-protein complex, the spliceosome, that is assembled by interactions of five different snRNP particles with each other and with pre-mRNA (see Figure 11-19). The spliceosome catalyzes two transesterification reactions that join the exons and remove the intron as a lariat structure, which is subsequently degraded (see Figure 11-16).

• Group II self-splicing introns, which are found in chloroplast genes and mitochondrial genes of plants and fungi, exhibit a largely conserved secondary structure, which is necessary for self-splicing. The snRNAs in the spliceosome are thought to have an overall secondary structure similar to that of group II introns.

• Most transcription and RNA processing in a mammalian cell nucleus takes place in a limited number of domains. A nuclear matrix or scaffold is formed by a fibrous protein network throughout the nucleus. This nuclear matrix may help to organize the foci of RNA transcription and processing.

11.3 Regulation of mRNA Processing

As explained in the previous sections, conversion of a 5′ capped RNA transcript into a functional mRNA involves two primary steps: (1) cleavage and polyadenylation at the 3′ end and (2) ligation of exons with the concomitant excision of introns, or RNA splicing. To understand how regulation of RNA processing can control gene expression, we need to recall that higher eukaryotes contain both simple and complex transcription units. The primary transcripts produced from the former contain one poly(A) site and exhibit only one pattern of RNA splicing, even if multiple introns are present; thus simple transcription units encode a single mRNA (see Figure 9-1b). In contrast, the primary transcripts produced from complex transcription units can be processed in alternative ways to yield different mRNAs (see Figure 9-2).

In theory, the expression of both simple and complex transcription units could be controlled by on-off regulation of the cleavage and polyadenylation of their pre-mRNAs. Although this step probably occurs constitutively (i.e., is not regulated) in the expression of most genes, a few examples of this type of post-transcriptional regulation have been discovered. In the case of pre-mRNAs produced from complex transcription units, cleavage and polyadenylation at different poly(A) sites and/or splicing of different exons yields mRNAs encoding distinct proteins. Such alternative processing pathways of complex pre-mRNAs usually are regulated, often in a cell type–specific manner. That is, one possible mRNA is expressed in one type of cell or tissue, while another is expressed in different cells or tissues. Since about 5 percent of all transcription units in higher eukaryotes are complex units, this type of post-transcriptional regulation is a significant gene-control mechanism in such organisms. In this section, we describe several examples of regulated RNA processing.

U1A Protein Inhibits Polyadenylation of Its Pre-mRNA

The best-understood example of on-off regulation of the cleavage and polyadenylation step in RNA processing involves U1A protein, one of the several proteins in U1 snRNP, which plays a critical role in RNA splicing. U1A protein binds to a 7-base sequence in the snRNA component of U1 snRNP. The pre-mRNA encoding U1A itself also contains two copies of this 7-base sequence (one with a single-base change) just upstream of its polyadenylation signal (Figure 11-23). When a U1A protein molecule is bound to each copy of this 7-base sequence in a U1A pre-mRNA in an in vitro system containing poly(A) polymerase and all the factors necessary for cleavage and polyadenylation, cleavage occurs but addition of the poly(A) tail does not.

The finding that U1A protein blocks polyadenylation of its own pre-mRNA in vitro can account for the autoregulation of this protein observed in vivo. When the level of U1A exceeds that of U1 snRNA, the excess U1A protein binds to its two binding sites in U1A pre-mRNAs. Since U1A binding does not prevent cleavage of the pre-mRNAs, free 3′ ends are generated in the normal fashion (see Figure 11-12). In the absence of polyadenylation, however, both cleavage products are rapidly degraded by an exonuclease, so no functional U1A mRNA is produced. As a result, synthesis of U1A protein is decreased until all the excess is used in formation of new U1 snRNPs. Once this occurs, no free U1A protein is available to bind to newly made U1A pre-mRNAs, which

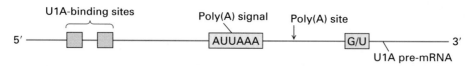

▲ **FIGURE 11-23 Diagram of 3′ end of the pre-mRNA encoding U1A protein, a sequence-specific RNA-binding protein that is part of the U1 snRNP.** Binding of U1A protein to both binding sites upstream of the poly(A) site blocks polyadenylation, but not cleavage, of the pre-mRNA. Note that the sequence of the upstream polyadenylation signal differs slightly from the usual one (AAUAAA). This unusual sequence is found in rare pre-mRNAs.

then can be processed normally. This on-off mechanism of post-transcriptional regulation thus coordinates expression of U1A protein with U1 snRNP assembly.

Tissue-Specific RNA Splicing Controls Expression of Alternative Fibronectins

The fibronectin gene provides a good example of gene control by tissue-specific RNA splicing. The multiple exons in the fibronectin primary transcript are spliced in different ways to produce two **isoforms** of this protein, one secreted by hepatocytes and the other by fibroblasts as well as other types of cells (Figure 11-24). Splicing of fibronectin pre-mRNA in fibroblasts yields an mRNA containing two exons, designated EIIIA and EIIIB, which encode protein domains that interact with cell-surface receptors in many cell types, thus making fibroblast fibronectin adherent to cell surfaces. In contrast, splicing of fibronectin pre-mRNA in hepatocytes "skips" these two exons, yielding an mRNA that does not encode the corresponding domains. As a result, the fibronectin secreted by hepatocytes does not adhere strongly to cell surfaces, allowing it to circulate in the serum.

A Cascade of Regulated RNA Splicing Controls *Drosophila* Sexual Differentiation

Another well-studied example of regulated splicing of alternative exons involves sexual differentiation in *Drosophila*. In this case, the differential expression of a cascade of three genes in male and female embryos depends on the regulated splicing of specific primary transcripts. The first of these genes to be expressed, called *sex-lethal (sxl)*, is transcribed from an early promoter (P_E) that is active only in very early female

embryos (Figure 11-25a). Thus Sex-lethal (Sxl) protein, a sequence-specific RNA-binding protein that interacts with RNA through an RNP motif, is produced in early female embryos but not in early male embryos.

As development of the *Drosophila* embryo proceeds, transcription of the *sxl* gene from P_E is repressed, while transcription from an upstream late promoter (P_L) is induced, in both males and females, and continues from this time in development onward. As depicted in Figure 11-25b, the presence of the early Sxl protein in female embryos directs splicing of the late *sxl* transcript so that a functional late Sxl protein is produced. In contrast, male embryos produce a nonfunctional Sxl protein that lacks the RNP motif. The late female Sxl protein, which differs from the early female Sxl protein at the N-terminus, but has the identical RNP motif at its C-terminus, also can bind to and direct splicing of the late *sxl* primary transcript in females. In this way, the late female Sxl protein autoregulates its own production, ensuring continued expression of a functional Sxl protein after repression of the early promoter and activation of the late promoter.

The second gene to be expressed in the regulatory cascade leading to sexual differentiation in *Drosophila* is called *transformer (tra)*. Again, Sxl protein controls splicing of *tra* pre-mRNA in female embryos so that they express functional Transformer (Tra) protein (Figure 11-26). Because they lack Sxl protein, male embryos cannot express functional Tra protein. Although Tra protein does not contain an RNA-binding domain, it forms a complex with another protein, Transformer 2 (Tra2), that does. In female embryos, splicing of the pre-mRNA synthesized from the third gene in the cascade—*double-sex (dsx)*—is controlled by the Tra-Tra2 complex. As a result, females express a female-specific *dsx* mRNA

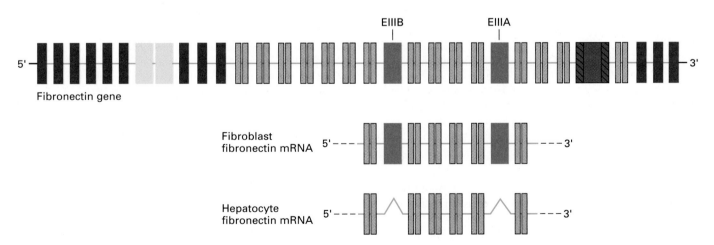

▲ **FIGURE 11-24 Cell type-specific splicing of fibronectin pre-mRNA in fibroblasts and hepatocytes.** The ≈75-kb fibronectin gene *(top)* contains multiple exons that likely evolved from an ancestral gene through a number of duplications of exons. Exons with homologous sequences are shown in the same color. In this diagram, introns (thin blue lines) are not drawn to scale; most of them are much longer than any of the exons. The fibronectin mRNA produced in fibroblasts includes the EIIIA and EIIIB exons, whereas these exons are spliced out of fibronectin mRNA in hepatocytes.

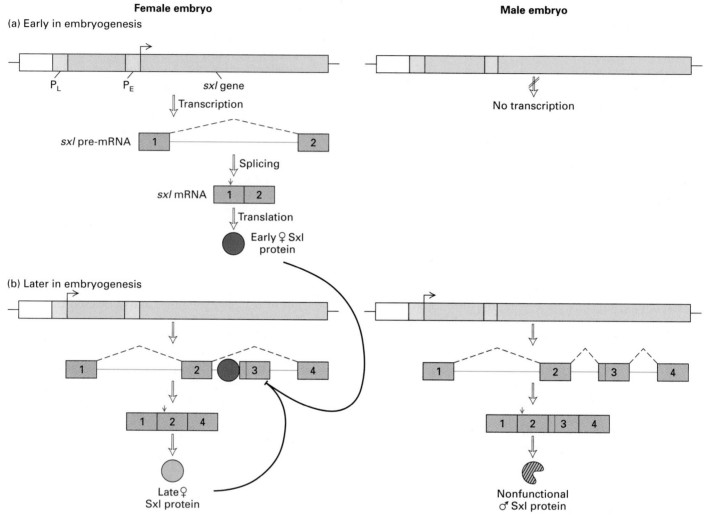

▲ **FIGURE 11-25 Expression of Sex-lethal (Sxl) protein during** ***Drosophila*** **embryogenesis.** In pre-mRNAs, exons are shown as red boxes and introns by a blue line; splicing is indicated by dashed lines. A red vertical arrow indicates the AUG start codon at which translation of an mRNA begins. (a) Early in development, *sxl* pre-mRNA is synthesized from P$_E$, which is active only in female embryos. This pre-mRNA is always spliced as shown. (b) Later in development, a different *sxl* pre-mRNA is synthesized from P$_L$ in both males and females. In males, this transcript is spliced to yield an mRNA with four exons. Because exon 3 contains an in-frame stop codon (dark red band), no functional Sxl protein is produced in males. In females, binding of the early Sxl protein to the late *sxl* pre-mRNA prevents splicing of exons 2 and 3. The resulting mRNA, containing three exons, is translated into functional late Sxl protein, which also binds to the late *sxl* pre-mRNA, ensuring its continued production in females. Note that exon 2 of the early transcript shown in (a) is identical to exon 4 of the late transcript (b). This exon encodes the RNP domain that mediates binding of Sxl protein to RNA.

and Dsx protein, which represses transcription of genes required for male sexual development. Lacking functional Tra protein, male embryos produce an alternative male-specific *dsx* mRNA and Dsx protein, which represses female development genes.

As indicated in Figure 11-26, Sxl protein and the Tra-Tra2 complex have opposite effects on splicing. Binding of Sxl to pre-mRNA *inhibits* splicing between two exons that in the absence of Sxl (as in male embryos) would be joined together. Most likely, bound Sxl prevents the assembly of required splicing factors at a regulated splice site. Conversely,

the Tra-Tra2 complex *activates* splicing between two exons that otherwise would not be joined together. In addition to binding to pre-mRNA, the Tra-Tra2 complex interacts with one of the general splicing factors, thereby promoting assembly of a spliceosome at the regulated splice site.

Numerous other cases of regulated splicing of alternative exons have been discovered. For example, Tra and Tra2 homologs recently have been isolated from vertebrate cells and shown to regulate splicing of some pre-mRNAs. This type of regulation is especially common in the vertebrate nervous system, as we discuss in the next section.

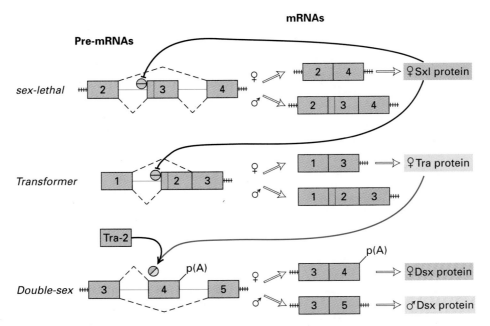

▲ FIGURE 11-26 Cascade of regulated splicing that controls expression of *sex-lethal (sxl), transformer (tra),* and *double-sex (dsx)* genes in *Drosophila* embryos. For clarity, only the exons (light red boxes) and introns (blue lines) where regulated splicing occurs are shown. Splicing is indicated by dashed lines above (female) and below (male) the pre-mRNAs. Dark red bands contain in-frame stop codons, which prevent synthesis of functional protein. Only female embryos produce functional Sxl protein, which represses splicing between exons 2 and 3 in *sxl*

pre-mRNA and between exons 1 and 2 in *tra* pre-mRNA. In contrast, binding of the Tra-Tra2 complex to *dsx* pre-mRNA activates splicing between exons 3 and 4. As a result of this cascade of regulated splicing, distinct Dsx proteins are produced in female and male embryos. These repress transcription of genes required for sexual differentiation of the opposite sex. See text for further discussion. [Adapted from M. J. Moore et al., 1993, in R. Gesteland and J. Atkins, eds., *The RNA World,* Cold Spring Harbor Press, pp. 303–357.]

Multiple Protein Isoforms Are Common in the Vertebrate Nervous System

Multiple isoforms of many proteins required for neuronal development and function are produced as the result of alternative patterns of RNA splicing. These include proteins involved in neurotransmitter storage and release, neurotransmitter receptors, and ion channels. The primary transcripts from these genes often show quite complex splicing patterns that can generate several different mRNAs, with different spliced forms expressed in different anatomical locations within the central nervous system.

One of the most remarkable examples of regulated RNA splicing occurs in the sound-sensing inner ear cells of vertebrates. Individual "hair cells" (ciliated neurons) respond most strongly to a specific frequency of sound. Cells tuned to low frequency ($\approx$50 Hz) are found at one end of the tubular cochlea that makes up the inner ear, and cells responding to high frequency ($\approx$5000 Hz) are found at the other end (Figure 11-27a). Cells in between respond to a gradient of frequencies between these extremes. One component in the tuning of hair cells in reptiles and birds is the opening of K^+ ion channels in response to increased intracellular Ca^{+2} concentrations. The Ca^{+2} concentration at which the channel opens deter-

mines the frequency with which the membrane potential oscillates, and hence the frequency to which the cell is tuned.

The gene encoding this channel (called *slo,* after the homologous *Drosophila* gene) is expressed as multiple alternatively spliced mRNAs. The various Slo proteins encoded by these alternative mRNAs open at different Ca^{2+} concentrations. Hair cells with different response frequencies express different versions of the channel (i.e., different Slo isoforms) depending upon their position along the length of the cochlea. The sequence variation in the protein is very complex: there are at least eight positions in the mRNA where alternative exons are utilized, permitting the expression of 576 possible isoforms (Figure 11-27b). It is not known, however, if all possible isoforms actually are expressed.

PCR analysis of *slo* mRNAs from individual hair cells has shown that each hair cell expresses a mixture of different alternative *slo* mRNAs, with different forms predominating in different cells according to their position along the cochlea. This remarkable arrangement suggests that splicing of the *slo* pre-mRNA is regulated in response to extracellular signals that inform the cell of its position along the cochlea. What these signals are and how they literally "fine tune" the *slo* pre-mRNA splicing pattern are not yet known.

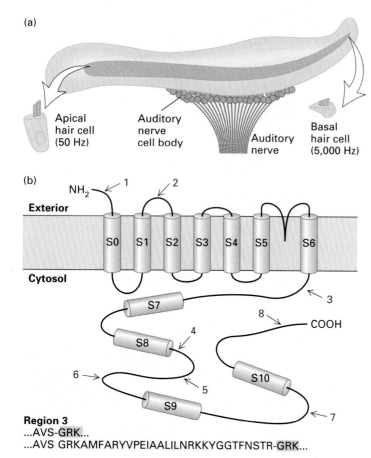

(a)

Apical hair cell (50 Hz)

Auditory nerve cell body

Auditory nerve

Basal hair cell (5,000 Hz)

(b)

NH₂

Exterior

S0 S1 S2 S3 S4 S5 S6

Cytosol

S7

S8

S9

S10

COOH

Region 3
...AVS-GRK...
...AVS GRKAMFARYVPEIAALILNRKKYGGTFNSTR-GRK...

▲ **FIGURE 11-27 Alternative splicing of *slo* mRNA, which encodes a Ca²⁺-gated K⁺ channel, in auditory hair cells contributes to the perception of sounds of different frequency.** (a) The chicken cochlea, a 5-mm-long tube, contains an epithelium of auditory hair cells (stippled area) that are tuned to a gradient of vibrational frequencies from 50 Hz at the apical end *(left)* to 5000 Hz at the basal end *(right)*. (b) The Slo protein contains seven transmembrane α helices (S0–S6), which associate to form the K⁺ channel. The cytosolic domain, which includes four hydrophobic regions (S7-S10), regulates opening of the channel in response to Ca²⁺. Isoforms of the Slo channel, encoded by alternatively spliced mRNAs produced from the same primary transcript, open at different Ca²⁺ concentrations. Red numbers refer to regions where alternative splicing produces different amino acid sequences in the various Slo isoforms. For example, two amino acid sequences (in one-letter code) resulting from alternative splicing in region 3 are shown at the bottom. Dashes indicate exon junctions. Splicing at one splice site joins . . . AVS encoded in one exon to GRK . . . encoded in the next exon. An alternative splice site at this position includes additional bases in the upstream exon so that the longer sequence shown is spliced to the exon encoding GRK. . . . Hair cells at the apical end of the cochlea make only the splice encoding the shorter sequence, whereas hair cells at the basal end make both alternative splices. Other alternatively spliced forms are enriched in hair cells at different specific locations along the length of the cochlea. [Adapted from K.P. Rosenblatt et al., 1997, *Neuron* **19:**1061; region 3 sequences from D.S. Navaratnam et al., 1997, *Neuron* **19:**1077.]

The molecular basis of neuron-specific **alternative RNA splicing** appears to be similar in principle to the regulation of RNA splicing in *Drosophila* sex determination discussed above. That is, binding of sequence-specific RNA-binding proteins near particular splice sites in one type of cell either activate or inhibit splicing. The absence of these RNA-binding proteins, or their inactivation, in other cell types results in alternative splicing patterns. Analysis of in vitro splicing of pre-mRNA from complex transcription units has shown that differences in the relative concentrations of hnRNP proteins and essential splicing factors can influence the selection of alternative splice sites. This observation suggests that differences in the concentrations of general splicing factors and hnRNP proteins among different cell types also may contribute to cell type–specific splicing.

SUMMARY Regulation of mRNA Processing

- The expression of some proteins is regulated by controlling the processing of the primary transcript from the gene encoding them. This type of gene regulation is especially common for genes encoding proteins important for the function of the nervous systems of vertebrates.

- Polyadenylation of the pre-mRNA encoding the U1A protein is inhibited by binding of U1A protein itself to two identical sites slightly upstream of the poly(A) site in U1A pre-mRNA. As a result, no functional mRNA is produced, and expression of U1A protein is repressed. Regulation of this type is uncommon.

- Alternative splicing of the primary transcripts produced from complex transcription units often is regulated. As a result, different mRNAs may be expressed from the same gene in different cell types or at different developmental stages.

- Alternative splicing can be regulated by RNA-binding proteins that bind to specific sequences near regulated splice sites. Splicing inhibitors are thought to sterically block access of splicing factors. Splicing activators may enhance splicing by interacting with splicing factors, thus promoting their association with the regulated splice site.

11.4 Signal-Mediated Transport through Nuclear Pore Complexes

The nucleus is separated from the cytoplasm by two membranes, which form the **nuclear envelope** (see Figure 5-42). Like the plasma membrane surrounding cells, each nuclear membrane consists of a water-impermeable phospholipid bilayer and various associated proteins. In all eukaryotic cells, the nuclear envelope is perforated by many pores (see Figure 5-50) through which water-soluble molecules enter and leave the nucleus. Each pore is formed from an elaborate structure termed the **nuclear pore complex (NPC)**, which can selectively transport macromolecules across the nuclear envelope.

Nuclear Pore Complexes Actively Transport Macromolecules between the Nucleus and Cytoplasm

The nuclear pore complex is immense by molecular standards, ≈125 million daltons in vertebrates, or about 30 times larger than a ribosome. It is made up of multiple copies of some 50–100 different proteins. Nuclear pore complexes are well visualized in electron micrographs of the nuclear envelope microdissected from the large nuclei of amphibian oocytes

(Figure 11-28a). Electron micrographs such as these have led to the model for the nuclear pore complex shown in Figure 11-28b. As illustrated in this model, the NPC nuclear ring supports eight ≈100-nm-long filaments whose distal ends are joined by the *terminal ring*, forming a structure called the *nuclear basket.* The nuclear ring is also attached directly to the **nuclear lamina,** a network of **intermediate filaments** that extends over the inner surface of the nuclear envelope (Section 19.6).

Ions, small metabolites, and globular proteins up to ≈60 kDa can diffuse through water-filled channels in the nuclear pore complex; these channels behave as if they are ≈9 Å in diameter. However, large proteins and ribonucleoprotein complexes, up to ≈25 nm in diameter, cannot diffuse in and out of the nucleus; rather, they are actively transported through the central plug of the nuclear pore complex (see Figure 11-28b). All RNAs synthesized in the nucleus must be

(a)

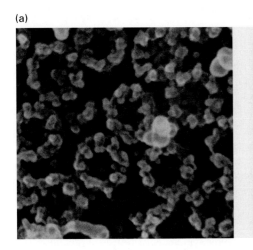
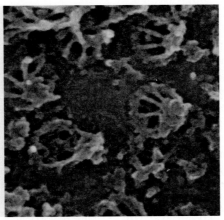
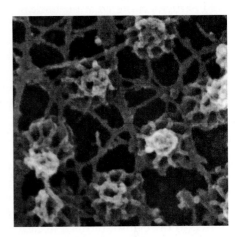

(b)

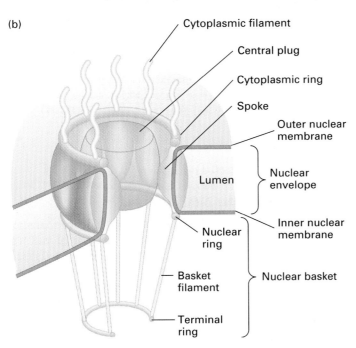

Cytoplasmic filament

Central plug

Cytoplasmic ring

Spoke

Outer nuclear membrane

Lumen

Nuclear envelope

Nuclear ring

Inner nuclear membrane

Basket filament

Nuclear basket

Terminal ring

▲ **FIGURE 11-28 Nuclear pore complex.** (a) Nuclear envelopes of *Xenopus* oocytes visualized by field emission in-lens scanning electron microscopy. *Left:* Cytoplasmic face of nuclear pore complexes (NPCs). *Middle:* Nucleoplasmic face of NPCs, showing the "basket" structure. *Right:* Nucleoplasmic face of the nuclear envelope after removal of the nuclear membrane by mild detergent treatment. The nuclear lamin network, which inserts into the nuclear ring of the NPC, is exposed by this treatment. (b) Cutaway model of the NPC. [Part (a) from V. Doye and E. Hurt, 1997, *Curr. Opin. Cell Biol.* **9**:401; courtesy of M. W. Goldberg and T. D. Allen. Part (b) adapted from M. Ohno et al., 1998, *Cell* **92**:327.]

exported to the cytosol before they can function in protein synthesis. Conversely, all proteins found in the nucleus must be imported from the cytoplasm where they are synthesized on ribosomes. The nuclear pore complex acts as a gated channel through which these macromolecules are selectively transported in and out of the nucleus.

When nuclear pores were first visualized by electron microscopy, researchers assumed that they served as portals of entry into and exit from the nucleus. The first studies of transport through nuclear pore complexes analyzed the import of proteins into the nucleus. These early experiments made use of nucleoplasmin, an abundant nuclear protein in *Xenopus* oocytes that assists in the assembly of chromatin during the rapid cell replication that follows fertilization. In one key study, small gold particles coated with nucleoplasmin or with a non-nuclear protein were microinjected into the cytosol of frog oocytes. The location of the gold particles was determined by electron microscopy after sectioning the oocytes. Shortly after injection, the nucleoplasmin-coated gold particles clustered at the nuclear pore complexes; later, they accumulated in the nucleus after passing through the pores. Gold particles coated with non-nuclear proteins, by contrast, remained in the cytoplasm and did not bind to nuclear pores complexes. These experiments not only demonstrated definitively that nuclear pores are routes for protein import but also suggested that some signal present in proteins like nucleoplasmin is required for transport to occur.

Some of the earliest studies on nuclear export analyzed the transport of mRNA from the nucleus to the cytosol. In the nucleus, as discussed previously, nascent RNA transcripts associate with various proteins forming heterogeneous ribonucleoproteins (hnRNPs). RNPs that contain fully processed mRNAs, referred to as *messenger RNPs* (**mRNPs**), are exported to the cytosol through nuclear pore complexes. The salivary glands of larvae of the insect *Chironomous tentans* are a good model system for studying both the formation of hnRNPs and export of mRNPs. In these larvae, genes in large chromosomal puffs called Balbiani rings are abundantly transcribed into nascent pre-mRNAs that associate with hnRNP proteins to form 50-nm coiled hnRNPs (Figure 11-29).

After processing of Balbiani ring pre-mRNA, the resulting mRNPs move through nuclear pores to the cytosol. Electron micrographs of sections of these cells show mRNPs uncoiling during their passage through nuclear pores and binding to ribosomes as they enter the cytosol (Figure 11-30).The observation that mRNPs become associated with ribosomes during transport indicates that the 5′ end leads the way through the nuclear pore complex. Electron microscopic studies such as these have lead to the model depicted in Figure 11-31 for passage of mRNPs through nuclear pore complexes.

Subsequent experiments have revealed many details about how macromolecules are transported through nuclear pore complexes. We first discuss nuclear export and then consider nuclear import; as we will see the mechanisms of the two processes are similar in many respects.

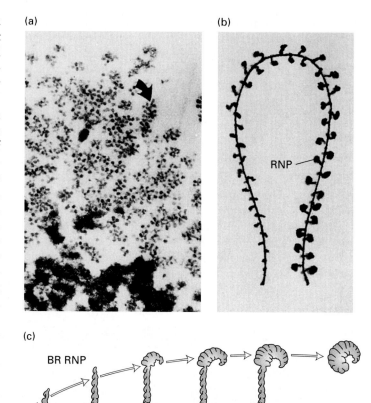

▲ **FIGURE 11-29 Formation of coiled heterogeneous ribonucleoprotein (hnRNP) during synthesis of the *Chironomous tentans* Balbiani ring (BR) mRNA.** (a) Electron micrograph of active transcription loops of chromatin with characteristic granules of ribonucleoprotein (arrow). The transcribed loops of chromatin extend above the dark staining, dense, non-transcribed chromatin. (b) Reconstruction of a single chromatin transcription loop from serial thin sections reveals a gradual increase in size of the associated ribonucleoprotein (RNP) particles that reflects the increasing length of the RNA transcripts being synthesized. (c) A model for the structure and biogenesis of BR hnRNP. The BR gene, which contains four exons and four introns, encodes a secreted protein that glues the insect larvae to a solid support, such as a twig, in preparation for metamorphosis. [Parts (a) and (b) from C. Erricson et al., 1989, *Cell* **56**:631; courtesy of B. Daneholt. Part (c) adapted from B. Daneholt, 1997, *Cell* **88**:585.]

Receptors for Nuclear-Export Signals Transport Proteins and mRNPs out of the Nucleus

Cell-fusion experiments provided the first evidence that specific hnRNP proteins participate in the export of mRNA from the nucleus. The heterokaryon experiments, described in Figure 11-32, demonstrated that some hnRNP proteins cycle in and out of the cytoplasm, whereas others remain localized in the nucleus.

More recent studies have revealed that certain hnRNP proteins contain a *nuclear-export signal (NES)* that stimu-

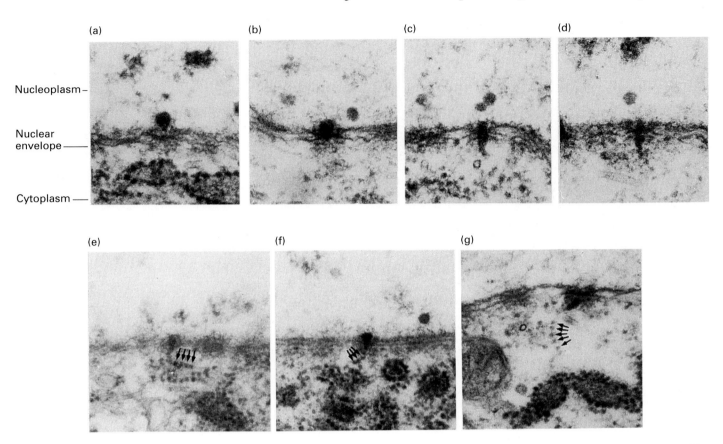

▲ FIGURE 11-30 Electron micrographs of Balbiani ring mRNPs passing through nuclear pore complexes in salivary glands of a *Chironomous tentans* larva. The mRNPs appear to uncoil as they pass through a nuclear pore (a–d). As they enter the cytoplasm (e–g), the mRNPs appear to associate with ribosomes (arrows). [From H. Mehlin et al., 1992, *Cell* **69**:605; courtesy of B. Daneholt.]

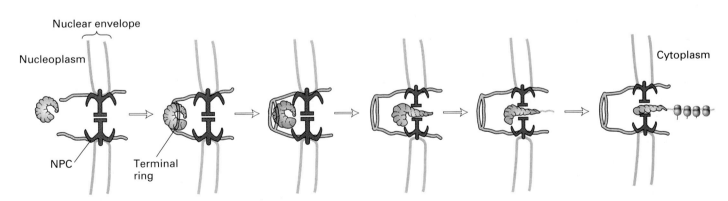

▲ FIGURE 11-31 Model for passage of mRNPs through the nuclear pore complex (NPC) based on electron microscopic studies of Balbiani ring (BR) mRNP transport in *Chironomous tentans* (see Figure 11-30). After the coiled mRNP moves through the terminal ring, it uncoils as it passes through the central plug of the NPC, with the 5' end leading the way and becoming associated with ribosomes in the cytoplasm. The NPC is thought to undergo a conformational change during transport. [Adapted from B. Daneholt, 1997, *Cell* **88**:585.]

(a)

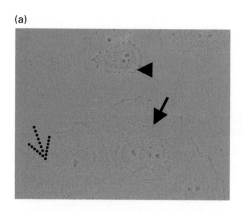

(b)

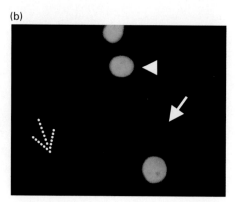

(c)

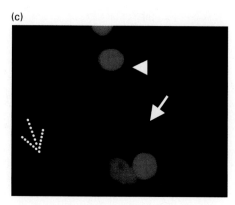

▲ **FIGURE 11-32 Heterokaryon assay demonstrating that human hnRNP A1 protein can cycle in and out of the cytoplasm, but human hnRNP C protein cannot.** Heterokaryons were prepared by treating HeLa cells and cultured *Xenopus* cells with polyethylene glycol. The cells were treated with cycloheximide immediately after fusion to prevent protein synthesis. After 2 hours, the cells were fixed and stained with fluorescent-labeled antibodies specific for hnRNP C and hnRNP A1. These antibodies do not bind to the homologous *Xenopus* proteins. (a) A fixed preparation viewed by phase-contrast microscopy includes unfused HeLa cells (arrowhead) and *Xenopus* cells (dotted arrow), as well as fused heterokaryons (solid arrow). In the heterokaryon in this micrograph, the round HeLa-cell nucleus is to the right of the oval-shaped *Xenopus* nucleus. (b,c) When the same preparation was viewed by fluorescence microscopy, the stained hnRNP C protein appeared green and the stained hnRNP A1 protein appeared red. Note that the unfused *Xenopus* cell on the left is unstained, confirming that the antibodies are specific for the human proteins. In the heterokaryon, hnRNP C only appears in HeLa-cell nuclei (b), whereas hnRNP A1 appears in both nuclei (c). Since protein synthesis was blocked after cell fusion, some of the human hnRNP A1 must have left the HeLa-cell nucleus, moved through the cytoplasm, and entered the *Xenopus* nucleus in the heterokaryon. [See S. Pinol-Roma and G. Dreyfuss, 1992, *Nature* **355**:730; courtesy of G. Dreyfuss.]

lates their active transport through nuclear pores. In these studies, a gene encoding a nucleus-restricted protein such as nucleoplasmin is fused to various segments of a gene encoding a protein (e.g., human hnRNP A1) that shuttles in and out of the nucleus. The engineered gene then is transfected into HeLa cells before fusion to *Xenopus* cells in the heterokaryon assay. Observation of the expressed fusion protein (e.g., nucleoplasmin-hnRNP A1) in the *Xenopus* nucleus indicates that the short fused segment functions as a NES directing transport of the fusion protein (similar to shuttling of hnRNP A1 in Figure 11-32). Experiments of this type have identified at least three different classes of NESs: a 38-residue sequence in hnRNP A1, one in hnRNP K, and a leucine-rich sequence found in PKI (an inhibitor of protein kinase A) and in the Rev protein of human immunodeficiency virus (HIV) discussed in a later section.

The mechanism of export of shuttling proteins is best understood for those containing a leucine-rich NES. According to the current model shown in Figure 11-33, this NES promotes export of a "cargo" protein from the nucleus by binding to a specific *nuclear-export receptor* in the nucleus, a protein called *exportin 1*. Nuclear export also requires a third protein, **Ran,** a small GTPase that exists in two conformations, one when complexed with GTP and an alternative one when complexed with GDP. The simultaneous interaction of exportin 1 with Ran · GTP and the NES of a cargo protein in the nucleus forms a trimolecular **cargo complex.** Interaction of a cargo complex with proteins in the nuclear pore

complex leads to movement of the cargo complex through the pore by mechanisms that are not yet understood.

Once the Ran · GTP/exportin 1/cargo protein complex reaches the cytosol, *Ran GTPase-activating protein* (RanGAP) stimulates Ran to hydrolyze its bound GTP to GDP. The resulting conformational change in Ran causes dissociation of the cargo complex, releasing the free NES-containing cargo protein (see Figure 11-33). The free exportin 1 and Ran · GDP are transported back into the nucleus where a *Ran nucleotide-exchange factor,* called RCC1, causes Ran to release its GDP and rebind GTP, which is present in much higher concentration. The regenerated Ran · GTP and exportin 1 then can transport another NES-containing cargo protein to the cytosol. (As discussed in Chapter 20, other GTPases that cycle between GTP-bound and GDP-bound forms, such as $G_{s\alpha}$ and Ras, function in many signal-transduction pathways.) Localization of RanGAP and RCC1 to the cytosol and nucleus, respectively, is the basis for the unidirectional transport of cargo proteins containing a leucine-rich NES.

A similar nuclear-cytosolic "shuttle" is thought to export fully processed mRNA–hnRNP protein complexes (mRNPs) from the nucleus, as depicted in Figure 11-34. According to this model, a nuclear cap–binding complex (CBC), which associates with the 5' cap, leads the way through the nuclear pore complex. Receptors like exportin 1 are postulated to make successive interactions with proteins of the nuclear pore complex as they are transported through a pore. Through mechanisms that are not fully understood, the

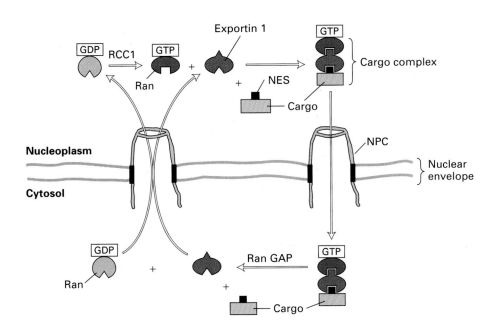

◀ **FIGURE 11-33 Proposed mechanism for the transport of "cargo" proteins containing a leucine-rich nuclear-export signal (NES) from the nucleus to the cytosol.** In the nucleoplasm, the protein exportin 1 binds cooperatively to the NES of the cargo protein to be transported and to Ran · GTP. After the resulting cargo complex passes through a nuclear pore complex (NPC), RanGAP, localized to the cytoplasm, stimulates conversion of Ran · GTP to Ran · GDP. The accompanying conformational change in Ran leads to dissociation of the complex. The NES-containing cargo protein is left free in the cytosol, while exportin 1 and Ran · GDP are transported back into the nucleus through NPCs. RCC1, localized to the nucleus, stimulates conversion of Ran · GDP to Ran · GTP. Repetition of this cycle leads to export of multiple molecules of the cargo protein. [Adapted from M. Ohno et al., 1998, *Cell* **92**:327.]

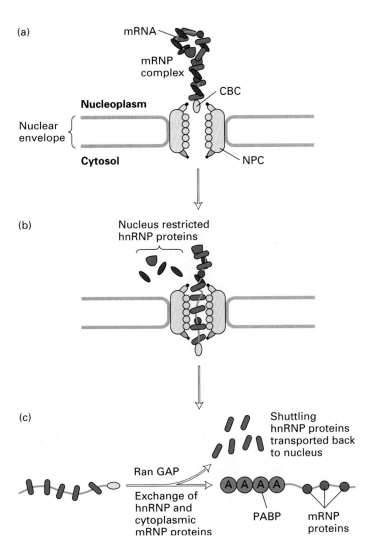

◀ **FIGURE 11-34 Proposed mechanism for hnRNP protein–mediated export of mRNA from the nucleus.** (a) The 5′ end of the fully processed mRNA–hnRNP protein complex (mRNP) associates with cap-binding complex (CBC), which passes through the nuclear pore complex (NPC) first. (b) Nucleus-restricted hnRNPs (orange and dark blue) are removed as a mRNP is transported through the NPC; these proteins, which lack a NES, would hold the mRNA (red) in the nucleus. NES-bearing hnRNPs, such as hnRNP A1, are transported through the NPC by the mechanism diagrammed in Figure 11-33, carrying the associated mRNA into the cytoplasm. (c) Cytoplasmic RanGAP stimulates hydrolysis of GTP by Ran. The shuttling hnRNP proteins then dissociate from the receptor proteins (exportin 1 for leucine-rich NESs) and are transported back into the nucleus. The mRNA is then available to interact with cytosolic mRNP proteins, including poly(A)-binding protein (PABP), which binds to the 3′ poly(A) tail of mRNA. [Adapted from S. Nakielny and G. Dreyfuss, 1997, *Curr. Opin. Cell Biol.* **9**:420.]

shuttling hnRNP proteins are displaced from the mRNA and then replaced with cytosolic mRNA-binding proteins. The released hnRNP proteins, nuclear-export receptor, and Ran · GDP are then transported back into the nucleus by a mechanism that is analogous to their export.

Pre-mRNAs in Spliceosomes Are Not Exported from the Nucleus

It is critical that only fully processed mature mRNAs be exported from the nucleus because translation of incompletely processed pre-mRNAs containing introns would produce defective proteins, which might interfere with the functioning of the cell. By mechanisms that are not fully understood, pre-mRNAs associated with snRNPs in spliceosomes are prevented from being transported to the cytosol. In one type of experiment, for instance, a gene encoding a pre-mRNA with a single intron that is efficiently spliced out was mutated to introduce deviations from the consensus splice-site sequences. Mutation of either the 5′ or 3′ invariant splice site at the ends of the intron resulted in pre-mRNAs that were bound by snRNPs to form spliceosomes; however, RNA splicing was blocked and the pre-mRNA was retained in the nucleus. In contrast, mutation of *both* the 5′ and 3′ splice sites in the same pre-mRNA resulted in efficient export of the unspliced pre-mRNA. In this case, the pre-mRNAs were not efficiently bound by snRNPs.

Many cases of *thalassemia*, an inherited disease that results in abnormally low levels of globin proteins, are due to mutations in globin-gene splice sites that decrease the efficiency of splicing but do not prevent association of the pre-mRNA with snRNPs. The resulting unspliced globin pre-mRNAs are retained in reticulocyte nuclei and are rapidly degraded.

Receptors for Nuclear-Localization Signals Transport Proteins into the Nucleus

All proteins found in the nucleus are synthesized in the cytoplasm and actively imported into the nucleus. These include nucleus-restricted proteins (e.g., histones, lamins, DNA and RNA polymerases, replication and transcription factors, splicing proteins, and some hnRNPs), as well as proteins that shuttle between the nucleus and cytoplasm (e.g., hnRNP A1

protein, exportin 1). All proteins actively imported through nuclear pore complexes contain a *nuclear-localization signal (NLS)*. Shuttling proteins contain both a nuclear-export signal (NES) and NLS.

NLSs were first discovered during the analysis of mutants of simian virus 40 (SV40) that produced an abnormal form of the early viral protein called large T-antigen. The wild-type form of this protein is localized to the nucleus in virus-infected cells, whereas mutated forms of large T-antigen accumulate in the cytosol. The mutations responsible for this altered cellular localization all occur within five consecutive basic amino acids in the sequence Pro-Lys-Lys-Lys-Arg-Lys-Val. Remarkably, when this region of SV40 large T-antigen was fused to pyruvate kinase, a very large cytosolic protein involved in carbohydrate metabolism, the fusion protein was transported into nuclei (Figure 11-35). The minimal amino acid sequence that directs pyruvate kinase to the nucleus is the seven-residue sequence shown above. Moreover, 5-nm gold particles coated with this synthetic peptide are transported through nuclear pores after microinjection into the cytoplasm of cultured cells. These experiments demonstrated that this short sequence from SV40 large T-antigen acts as a signal that causes the transport of associated macromolecules into the nucleus, analogous to the nuclear-export signals discussed above.

Similar methods have been used to identify NLS sequences in numerous other proteins imported into the nucleus. Many are similar to the SV40 large T-antigen NLS, containing several consecutive basic amino acids. Other NLSs are chemically quite different. For instance, the NLS in the hnRNP A1 protein, which shuttles between the nucleus and cytosol, is relatively hydrophobic and overlaps with the NES in this protein.

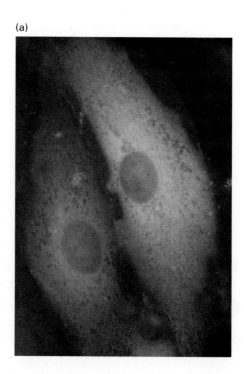 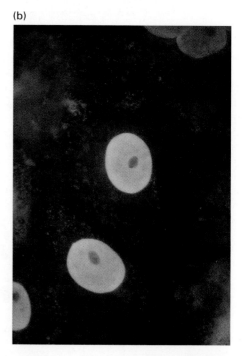

(a)   (b)

▶ FIGURE 11-35 Demonstration that the nuclear-localization signal (NLS) of the SV40 large T-antigen can direct a cytoplasmic protein to the cell nucleus. (a) Normal pyruvate kinase, visualized by immunofluorescence after treating cultured cells with a specific antibody, is localized to the cytoplasm. (b) A chimeric pyruvate kinase protein, containing the SV40 NLS at its N-terminus, is directed to the nucleus. The chimeric protein was expressed from a transfected engineered gene produced by fusing a viral gene fragment encoding the SV40 NLS to the pyruvate kinase gene. [From D. Kalderon et al., 1984, *Cell* **39**:499; courtesy Dr. Alan Smith.]

(a) Effect of digitonin

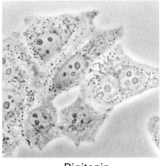 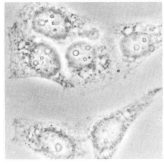

    – Digitonin   + Digitonin

(b) Nuclear import by permeabilized cells

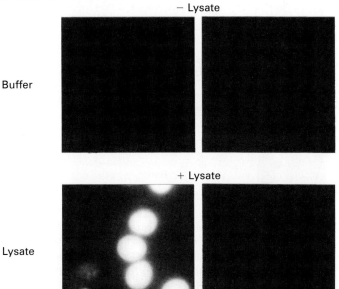

▲ **FIGURE 11-36 Experimental demonstration that nuclear transport in permeabilized cultured cells requires soluble cytosolic components and ATP.** (a) Phase-contrast micrographs of untreated and digitonin-permeabilized HeLa cells. Treatment of a monolayer of cultured cells with the mild, nonionic detergent digitonin permeabilizes the plasma membrane so that cytosolic constituents leak out, but leaves the nuclear envelope and NPCs intact. (b) Fluorescence micrographs of digitonin-permeabilized HeLa cells incubated with a fluorescent protein chemically coupled to a synthetic SV40 T-antigen NLS peptide in the presence and absence of ATP and cytosol (lysate). Accumulation of this transport substrate in the nucleus occurred only when both cytosol and ATP were included in the incubation *(lower left)*. [From S. Adam et al., 1990, *J. Cell. Biol.* **111**:807; courtesy of Dr. Larry Gerace.]

Development of a digitonin-permeabilized cell system provided a convenient in vitro assay for demonstrating that soluble cytosolic components and ATP are required for nuclear import (Figure 11-36). Using this assay, researchers subsequently purified and characterized four cytosolic proteins required for nuclear import of a protein containing a basic NLS: importin α; importin β; nuclear transport factor 2 (NTF2); and Ran, the same GTPase involved in nuclear export (see Figure 11-33). Further biochemical studies showed that importin α and β form a heterodimeric *nuclear-import receptor* that binds to a basic NLS through the α subunit. The β subunit of dimeric importin interacts with NPC cytosolic filaments in the absence of ATP; in the presence of ATP, the β subunit is thought to interact with other components of the NPC during transport through the nuclear pore complex.

A model for the import of cytosolic "cargo" proteins bearing a basic NLS is shown in Figure 11-37. By comparing this model for nuclear import with that in Figure 11-33 for nuclear export, we can see the similarities in the two transport processes. Although the mechanisms differ slightly, the unidirectional nature of both import and export depends on the asymmetric distribution of Ran GTP-activating protein (RanGAP), which is restricted to the cytoplasm, and Ran nucleotide-exchange factor (RCC1), which is restricted to the nucleoplasm. The models of nuclear export and import each require that Ran · GDP be transported from the cytoplasm to the nucleus, that Ran · GTP be transported from the nucleus to the cytoplasm, and that other components (e.g., exportin 1, importin α, and importin β) be selectively transported into or out of the nucleus depending on their associations with other proteins.

There are two obvious differences in the export of leucine-rich NES-bearing proteins and import of basic-NLS–bearing proteins: (1) Ran · GTP is part of the cargo complex during export but not during import, and (2) the receptor that directs transport of a cargo protein through a nuclear pore is a monomer in the case of export (e.g., exportin 1) but a dimer in the case of import (e.g., importin $\alpha\beta$). Another difference between the two transport processes is that import of cargo proteins with a basic NLS also requires NTF2. This soluble cytosolic factor interacts in vitro with Ran · GDP, importin β, and NPCs, but its precise function is not clear. Deletion of the yeast homolog of NTF2 is lethal, but the effects of this mutation can be suppressed by overexpression of yeast Ran. This finding suggests that NTF2 normally functions to enhance the activities of Ran during nuclear import.

The distance between the tip of the NPC cytosolic filaments and the nuclear basket in the nucleoplasm is ≈200 nm (see Figure 11-28b). It is likely that as **nuclear-transport receptors** like exportin 1 and importin β traverse this distance, they make multiple contacts with distinct NPC proteins (called *nucleoporins*). Many nucleoporins contain multiple short repeats of the sequences Phe-X-Phe-Gly and Gly-Leu-Phe-Gly, which have affinity for importin β and may serve

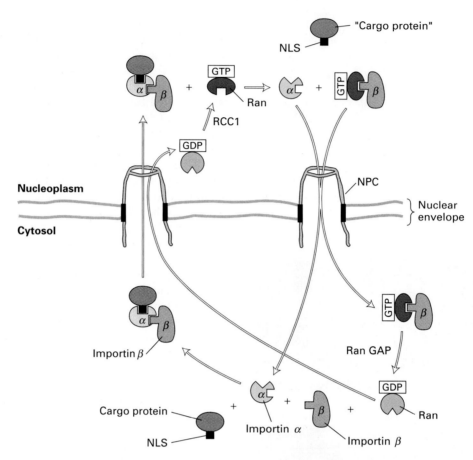

▲ **FIGURE 11-37 Proposed mechanism for the transport of "cargo" proteins containing a basic nuclear-localization signal (NLS) from the cytoplasm to the nucleus.** In the cytoplasm *(bottom)*, importin α and β interact cooperatively with the cargo protein to be transported, with the NLS binding to importin α. The importin β subunit of the resulting trimeric cargo complex interacts with components of the NPC, translocating the complex into the nucleoplasm by a poorly understood mechanism that requires ATP hydrolysis. In the nucleoplasm, Ran · GTP interacts with importin β, causing dissociation of the cargo complex, thereby delivering free cargo protein to the nucleoplasm. To support another cycle of import, monomeric importin α and the importin β – Ran · GTP complex are transported back to the cytoplasm. Ran GTP-activating protein (RanGAP) in the cytoplasm stimulates conversion of Ran · GTP to Ran · GDP resulting in a conformational change in Ran that causes it to dissociate from importin β. The free importin β can now interact with importin α and a new cargo protein bearing a basic NLS, initiating another round of nuclear import. Presumably, Ran · GDP is also transported through nuclear pores from the cytoplasm to the nucleoplasm, where the Ran nucleotide-exchange factor (RCC1) causes it to release GDP and rebind GTP.

as docking sites for interactions during transport. However, it is not yet known how many contacts occur between receptor proteins and nucleoporins or which nucleoporins participate in these interactions.

Various Nuclear-Transport Systems Utilize Similar Proteins

As noted earlier, hnRNP A1 protein, which cycles between the nucleus and cytosol, has an NLS that is hydrophobic rather than basic. In the cytoplasm, hnRNP A1 interacts with a monomeric nuclear-import receptor, called *transportin*, that both binds the NLS and mediates interactions with the NPC resulting in transport into the nucleoplasm. The NES on hnRNP A1 has a sequence distinct from that of the leucine-rich NESs and interacts with a nuclear-export receptor different from exportin 1. The export and import of shuttling proteins like hnRNP A1 is thought to occur by mechanisms similar to those depicted in Figures 11-33 and 11-37, and presumably requires some type of regulation of the NLS and NES signals in the same protein.

Another nuclear-transport system that has been identified and partially characterized imports snRNPs, which are critical to splicing of pre-mRNAs (see Figure 11-19). U1, U2, U4, and U5 snRNAs are transported from their site of synthesis in the nucleus to the cytosol where their 5′ methylguanylate cap is methylated twice more at specific positions and where snRNP proteins bind to form mature snRNP particles.

After the mature snRNPs interact with a nuclear-import receptor that is distinct from importin β and transportin, they are transported into the nucleus through NPCs.

The three nuclear-transport receptors characterized thus far—importin β, exportin 1, and transportin—share sequence homology, which probably reflects their homologous interactions with Ran and nucleoporins. Sequencing of the yeast genome has revealed fourteen genes encoding proteins with significant homology to these three nuclear-transport receptors, including their yeast homologs. This finding suggests that several distinct receptor proteins participate in both nuclear import and export. Additional nuclear-transport systems will likely be characterized in the future.

HIV Rev Protein Regulates the Transport of Unspliced Viral mRNAs

As discussed earlier, transport of mRNPs containing mature, functional mRNAs from the nucleus to the cytoplasm entails a complex mechanism that is crucial to gene expression (see Figure 11-34). Regulation of this transport theoretically could provide another means of gene control, although it appears to be relatively rare. Indeed, the only examples of regulated mRNA export discovered to date occur during the cellular response to conditions (e.g., heat shock) that cause protein denaturation and during viral infection when virus-induced alterations in nuclear transport maximize viral replication. Here we describe the regulation of mRNP export mediated by Rev, a protein encoded by human immunodeficiency virus (HIV).

A retrovirus, HIV integrates a DNA copy of its RNA genome into the host-cell DNA (see Figure 6-22). The integrated viral DNA, or *provirus,* contains a single transcription unit. The single primary transcript produced from the HIV provirus by cellular enzymes can be spliced in alternative ways to yield three classes of mRNAs: a 9-kb unspliced mRNA; ≈4-kb mRNAs formed by removal of one intron; and ≈2-kb mRNAs formed by removal of two or more introns (Figure 11-38). After their synthesis in the host-cell nucleus, all three classes of HIV mRNAs are transported to the cytoplasm and translated into viral proteins; some of the 9-kb unspliced RNA is used as the viral genome in progeny virions that bud from the cell surface. Since the 9-kb and 4-kb HIV mRNAs contain splice sites, they can be viewed as incompletely spliced mRNAs. However, as discussed earlier, association of such incompletely spliced mRNAs with snRNPs in spliceosomes normally blocks their export from the nucleus. Thus HIV must have evolved a mechanism for overcoming this block, permitting export of the longer HIV mRNAs.

Studies with HIV mutants showed that transport of unspliced 9-kb and singly spliced 4-kb viral mRNAs from the nucleus to the cytoplasm does not occur in infected cells

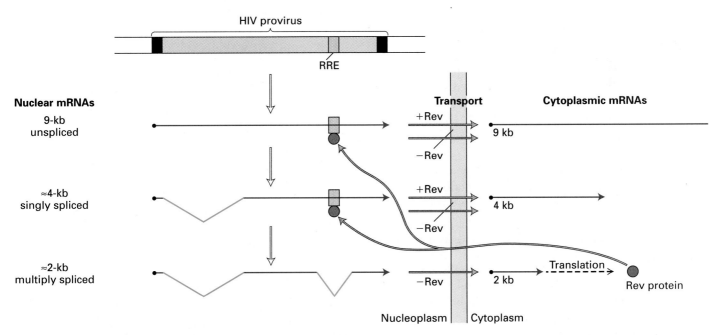

▲ FIGURE 11-38 Role of Rev protein in transport of HIV mRNAs from the nucleus to the cytoplasm. The HIV genome, which contains overlapping coding regions, is transcribed into a single 9-kb primary transcript. Several ≈4-kb mRNAs result from splicing out of one intron (blue angled line), and several ≈2-kb mRNAs from splicing out of two or more introns. After transport to the cytoplasm, each RNA species is translated into different viral proteins. Rev protein, encoded by a 2-kb mRNA, interacts with the Rev-response element (RRE) in the unspliced and singly spliced mRNAs, stimulating their transport to the cytoplasm. [Adapted from B. R. Cullen and M. H. Malim, 1991, *Trends Biochem. Sci.* **16:**346.]

unless Rev protein is expressed. Subsequent biochemical experiments demonstrated that Rev binds to a specific *Rev-response element* (RRE) present in HIV RNA. In cells infected with HIV mutants lacking the RRE, unspliced and singly spliced viral mRNAs remain in the nucleus, demonstrating that the RRE is required for Rev-mediated stimulation of transport.

Subsequent experiments, based on those described earlier with genes mutated at the 5′ and/or 3′ splice sites, have given insight into how binding of Rev protein to the RRE stimulates transport of certain HIV mRNAs. Recall that pre-mRNAs with a mutation in the 5′ or 3′ splice site are assembled into spliceosomes but cannot complete the splicing reaction; because the mutant pre-mRNAs are not released from spliceosomes, they are not transported into the cytoplasm. When a RRE is engineered into a pre-mRNA with a mutation in the 5′ or 3′ splice site, the unspliced pre-mRNA is transported into the cytoplasm in cells expressing Rev. This finding indicates that binding of Rev to a pre-mRNA somehow permits the spliceosome-associated RNA to be transported to the cytoplasm. Recently Rev has been shown to contain a leucine-rich NES that interacts with exportin 1 complexed with Ran · GTP. Consequently, Rev is thought to promote the export of unspliced and singly spliced HIV mRNAs through interactions with exportin 1 and the nuclear pore complex (see Figure 11-33). By an unknown mechanism these interactions overcome the block to RNA export imposed by association with spliceosomes.

SUMMARY Signal-Mediated Transport through Nuclear Pore Complexes

- The nuclear envelope contains numerous nuclear pore complexes (NPCs), large, complicated structures composed of multiple copies of ≈50–100 proteins called nucleoporins (see Figure 11-30b).

- Ions, metabolites, and small proteins diffuse freely through nuclear pores, but macromolecules larger than ≈60 kDa must be actively transported by a process that requires ATP hydrolysis and probably entails substantial conformational changes in the nuclear pore complex.

- In both nuclear export and import, the protein to be transported contains a specific amino acid sequence that functions as a nuclear-export signal (NES) or a nuclear-localization signal (NLS). Nucleus-restricted proteins contain a NLS but not a NES, whereas proteins that shuttle between the nucleus and cytosol contain both signals.

- According to current models, the NES or NLS on a "cargo" protein interacts with a specific nuclear-transport receptor protein located in the nucleus in the case of export or in the cytosol in the case of import. Both transport processes also require participation of Ran, a GTPase that exists in different conformations when bound to GTP or GDP.

- Once a cargo complex is assembled, the receptor protein in the complex is thought to make multiple contacts with nucleoporins, thereby transporting the complex through a nuclear pore. After a cargo complex reaches its destination (the cytoplasm during export and the nucleus during import), it dissociates, freeing the cargo protein and other components. The latter then are transported through nuclear pores in the reverse direction to participate in transporting additional molecules of cargo protein (see Figures 11-33 and 11-37).

- The unidirectional nature of both nuclear export and import is thought to result from the localization of the Ran nucleotide-exchange factor (RCC1) in the nucleus and of Ran GTPase-activating protein (RanGAP) in the cytoplasm.

- During export of a nuclear mRNP, composed of a mature, functional mRNA and hnRNP proteins, nucleus-restricted hnRNPs are removed, while multiple NES-bearing hnRNPs bound to the mRNA are thought to carry it through the nuclear pore complex (see Figure 11-34). Once in the cytosol, these shuttling hnRNPs, which also contain a NLS, are removed from the mRNA and transported back into the nucleus to participate in another round of nuclear export.

- Several different types of NES and NLS have been identified. Each class of nuclear-transport signal is thought to interact with a specific receptor protein. The nuclear-transport receptors characterized so far have homologous regions that interact with Ran and certain nucleoporins.

- Pre-mRNAs within a spliceosome normally are not exported from the nucleus. Although the mechanism of this transport inhibition is not understood, it ensures that only properly processed, functional mRNAs are transported into the cytoplasm for translation.

- The HIV Rev protein, which contains a NES, can override the restriction against transporting pre-mRNAs with unspliced splice sites (see Figure 11-38).

11.5 Other Mechanisms of Post-Transcriptional Control

For most genes, regulation of transcription initiation, discussed in the previous chapter, is the principle mechanism for controlling gene expression. As we've seen in the preceding sections, however, expression of many highly regulated genes is also controlled at one or more additional steps: elongation and termination, alternative RNA splicing, 3′ cleavage and polyadenylation, and even nuclear export. Other forms of post-transcriptional control have been discovered to help regulate expression of some genes. These include alterations in the protein-coding sequence of a pre-mRNA after its synthesis, regulation of the stability and translation of mRNAs, and regulation of the subcellular location of specific mRNAs.

RNA Editing Alters the Sequences of Pre-mRNAs

Sequencing of numerous cDNA clones and of the corresponding genomic DNAs from multiple organisms led in the mid-1980s to the unexpected discovery of a previously unrecognized type of pre-mRNA processing. In this type of processing, called **RNA editing**, the sequence of a pre-mRNA is altered; as a result, the sequence of the corresponding mature mRNA differs from the exons encoding it in genomic DNA.

RNA editing is widespread in the mitochondria of protozoans and plants and in chloroplasts; in these organelles, more than half the sequence of some mRNAs is altered from the sequence of the corresponding primary transcripts! In higher eukaryotes, RNA editing is much rarer, and thus far only single-base changes have been observed. Such minor editing, however, turns out to have important functional consequences in some cases.

RNA Editing in Mammalian Cells In mammals, the *apo-B* gene encodes two alternative forms of the serum protein apolipoprotein B (Apo-B): Apo-B100 expressed in hepatocytes and Apo-B48 expressed in intestinal epithelial cells. The ≈240-kDa Apo-B48 corresponds to the N-terminal region of the ≈500-kDa Apo-B100. Both Apo-B proteins transport lipids in the serum in the form of large lipoprotein complexes. However, only the large protein made in the liver delivers cholesterol to body tissues by binding to the low-density lipoprotein (LDL) receptor present on all cells.

The cell-type specific expression of the two forms of Apo-B results from editing of *apo-B* pre-mRNA so as to change the nucleotide at position 6666 in the sequence from a C to a U. This alteration, which occurs only in intestinal cells, converts a CAA codon for glutamine to a UAA stop codon, leading to synthesis of the shorter Apo-B48 (Figure 11-39). Studies with the partially purified enzyme that performs the post-transcriptional deamination of C_{6666} to U shows that it can recognize and edit an RNA as short as 26 nucleotides with the sequence surrounding C_{6666} in the *apo-B* primary transcript.

A second example of mRNA editing in mammals involves the pre-mRNAs encoding the subunits of receptors for the neurotransmitter glutamate. Glutamate receptors, which span the cell membrane of neurons, have important functions in memory and learning. In response to release of the neurotransmitter glutamate, these transmembrane proteins open ion channels initiating nerve impulses (Section 21.5). Several different glutamate receptor subtypes are assembled from a family of related subunits. Some glutamate receptor channels conduct both Na^+ and Ca^{2+} ions, whereas others only conduct Na^+ ions. This functional difference, which has enormously important consequences for nerve function, is due to editing of the pre-mRNAs encoding one type of subunit. In this process, a specific A residue is converted to inosine (I), changing a CAG codon, translated as glutamine, to a CIG codon, translated as arginine. The edited codon encodes an amino acid in the wall of the ion channel. When arginine is present at this position in the receptor, Ca^{2+} ions cannot pass through the channel. In mRNA preparations from brain, some of the glutamate receptor subunit mRNAs are

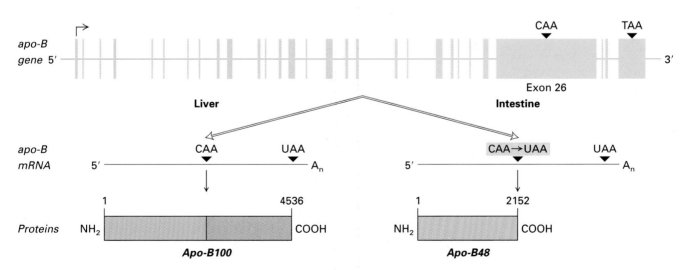

▲ **FIGURE 11-39 RNA editing of *apo-B* pre-mRNA.** The *apo-B* mRNA produced in the liver has the same sequence as the exons in the primary transcript. This mRNA is translated into Apo-B100, which has two functional domains: a N-terminal domain (green) that associates with lipids and a C-terminal domain (orange) that binds to LDL receptors on cell membranes. In the *apo-B* mRNA produced in the intestine, the CAA codon in exon 26 is edited to a UAA stop codon. As a result, intestinal cells produce Apo-B48, which corresponds to the N-terminal domain of Apo-B100.
[Adapted from P. Hodges and J. Scott, 1992, *Trends Biochem. Sci.* **17**:77.]

only partially edited, suggesting that the RNA-editing activity may be higher in some neurons than others. If so, regulation of the editing enzyme may contribute to memory and learning.

Both of these RNA-editing enzymes are deaminases. Discovery of other editing enzymes, which recognize sequences in specific pre-mRNAs, may explain the expression of functionally different forms of certain proteins in some cell types but not others.

RNA Editing in Protozoans The most extreme form of RNA editing uncovered so far occurs in the mitochondria of trypanosomes. These parasitic protozoa have a single large mitochondrion called a *kinetoplast*. Many of the mRNA precursors produced from kinetoplast DNA are extensively edited by the insertion and deletion of U residues, causing substantial alterations in the polypeptide sequence encoded. Insertions predominate and in some cases are so extensive that the length of the transcript is almost doubled.

The extensive RNA editing that occurs in trypanosomes is directed by small *guide RNAs* (gRNAs) encoded at distant regions of the kinetoplast genome. The 5′ end of a gRNA hybridizes to a short region of an unedited pre-mRNA, called an *anchor sequence*, while its 3′ end functions as a template for the editing process (Figure 11-40). Many gRNAs do not hybridize to anchor sequences in the primary transcript, but rather to sequences in partially edited intermediates. Thus editing of a trypanosome pre-mRNA generally starts near the 3′ end and progresses towards the 5′ end in a repetitive process that requires several different gRNAs, which bind sequentially to anchor sequences in previously edited sections.

It is not clear why trypanosome kinetoplasts utilize such an elaborate mechanism to produce mRNAs. The finding that RNA editing is most extensive in the earliest trypanosomes to have evolved suggests that this process may be a "molecular fossil" of the mechanism of RNA synthesis during an early stage in the evolution of modern cells.

Some mRNAs Are Associated with Cytoplasmic Structures or Localized to Specific Regions

As nuclear mRNPs enter the cytoplasm, their hnRNP proteins are replaced by cytosolic mRNA-binding proteins. However, the ratio of protein to RNA is lower in cytoplasmic mRNPs than in nuclear mRNPs (see Figure 11-34). The most abundant cytoplasmic mRNP protein, *poly(A)–binding protein (PABP)*, is tightly associated with the poly(A) tails of mRNAs. In mammalian cells, PABP is distinct from the nuclear poly(A)–binding protein, PABII, involved in the synthesis of the poly(A) tail (see Figure 11-12).

The translation of many mRNAs in the cytoplasm does not take place free in solution. For example, polyribosomes often are intimately associated with membranes of the rough endoplasmic reticulum (see Figure 5-47). The proteins synthesized at these sites are exported from the cell or become components of new membranes (Chapter 17). Other polyribosomes appear to be associated with the cytoskeleton. Experiments with detergent-permeabilized cells suggest that the poly(A)-binding protein associated with mRNA poly(A) tails interacts with the portion of the cytoskeleton made of actin microfilaments.

The association of some mRNAs with cytoskeletal elements leads to localization of the mRNA to specific regions of the cell cytoplasm. In all cases examined thus far, this localization is specified by sequences in the 3′ untranslated region of the mRNA. Such mRNA localization can be visualized in mammalian myoblasts (muscle precursor cells) as they differentiate into myotubes, the fused, multinucleated cells that make up muscle fibers. Myoblasts are motile cells that extend cytoplasmic regions, called lamellipodia, from the leading edge in the direction of movement. Extension of lamellipodia during cell movement requires polymerization of β-actin (Section 18.6). Sensibly, β-actin mRNA is concentrated in

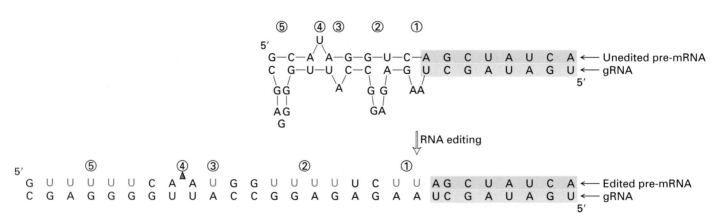

▲ **FIGURE 11-40 Mechanism of RNA editing in kinetoplast pre-mRNAs of trypanosomes.** An anchor sequence (blue) in the unedited pre-mRNA hybridizes to the 5′ anchor sequence (yellow) of a guide RNA (gRNA), which then directs the addition or deletion of U residues. Only a small region of the pre-mRNA is diagrammed. Added U residues are shown in red; Δ = deleted U. Circled numbers identify sites where editing occurs. [Adapted from B. Bass, 1993, in R. Gesteland and J. Atkins, eds., *The RNA World*, Cold Spring Harbor Press, pp. 383–418.]

the leading edges of myoblasts, the region of the cell cytoplasm where the encoded protein is needed for motility. When myoblasts fuse into syncytial myotubes, β-actin expression is repressed and the muscle-specific α-actin is induced. In contrast to β-actin mRNA, α-actin mRNA is restricted to the perinuclear regions of myotubes. Staining of cultured myoblasts in the process of differentiating with fluorescent probes specific for α- or β-actin mRNA reveals both mRNAs localized to their respective cellular regions.

To test the ability of actin mRNA sequences to direct the cytoplasmic localization of an mRNA, fragments of α- and β-actin cDNAs were inserted into separate plasmid vectors that express β-galactosidase from a strong viral promoter. The results of transfecting these plasmids into cultured cells, described in Figure 11-41, showed that inclusion of the 3' untranslated end of α- or β-actin cDNAs directs localization of the expressed β-galactosidase, whereas the 5' untranslated and coding regions do not.

Treatment of cultured myoblasts with cytochalasin D, which disrupts actin microfilaments, leads to a rapid delocalization of actin mRNAs, indicating that cytoskeletal actin microfilaments participate in the localization process. Disruption of other cytoskeletal components, however, does not alter the localization of actin mRNAs. Presumably certain RNA-binding proteins bind to specific sequences in the 3' untranslated regions of actin mRNAs and to specific components of microfilaments, possibly including motor proteins that move cargo (from individual proteins to organelles) along the length of microfilaments (Section 18.5).

This hypothesis is supported by recent studies with yeast mutants that fail to localize the mRNA encoding Ash1, a transcriptional regulator. In wild-type *S. cerevisiae*, Ash1 is found preferentially in the daughter-cell nucleus following mitosis (see Figure 14-6). Ash1 localization depends on actin, profilin and tropomyosin, which are required for the normal function of actin filaments, a myosin motor protein, and several other proteins. Recent studies show that Ash1 mRNA associates with the myosin motor, which then moves the bound mRNA along actin filaments of the cytoskeleton, correctly aimed at the appropriate location in the daughter cell (Figure 11-42). One or more sequence-specific RNA-binding proteins probably link Ash1 mRNA to the myosin motor, but as yet, this key component has not been identified. Other studies suggest that microtubules (rather than

TRANSFECTED ACTIN SEQUENCES

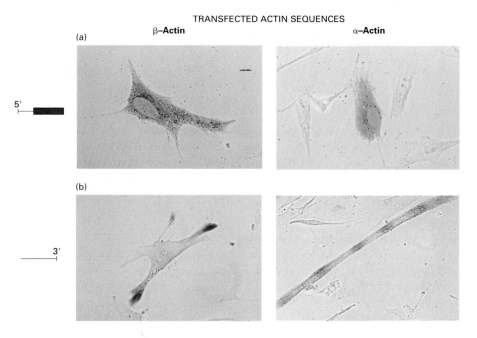

▲ **FIGURE 11-41 Experimental demonstration that the 3' untranslated region of α- and β-actin mRNAs direct localization of a β-galactosidase reporter mRNA.** Expression plasmid vectors were constructed encoding β-galactosidase mRNA containing one of four different 3' untranslated sequences. The recombinant plasmids were transfected separately into differentiating myoblasts. After a period of expression, the cells were fixed and then assayed for β-galactosidase activity by treating them with X-gal. This compound is hydrolyzed by β-galactosidase to yield a blue product. (a) Transfected cells that expressed engineered β-galactosidase mRNAs whose 3' untranslated region corresponded to the 5' untranslated sequence and coding region of α- or β-actin mRNA. These actin sequences did not cause localization of the β-galactosidase mRNA, as evidenced by the diffuse blue staining. (b) Transfected cells that expressed engineered β-galactosidase mRNAs whose 3' untranslated region corresponded to the 3' untranslated sequences of α- or β-actin mRNA. These sequences led to localization of β-galactosidase mRNA to lamellipodia (β-actin) or perinuclear regions (α-actin), as evidenced by the deep staining in these regions. [Micrographs from E. H. Kislaukis et al., 1993, *J. Cell. Biol.* **123**:165.]

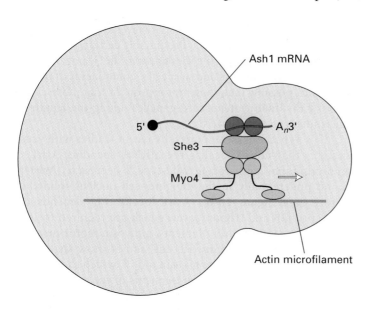

▲ **FIGURE 11-42 Model for localization of Ash1 mRNA to the bud tip in yeast.** Ash1 mRNA is associated with She3 protein and Myo4, a myosin motor protein that tracks along actin microfilaments (Chapter 18). Movement of fluorescently labeled Ash1 mRNA into the bud has been observed by video microscopy. She2 and She4 proteins (not shown) also are required for association of Myo4 with Ash1 mRNA, but a postulated RNA-binding protein (green) that binds directly to the 3′ untranslated region has not yet been identified. [See E. Bertrand et al., 1998 *Mol. Cell* **2**:437.]

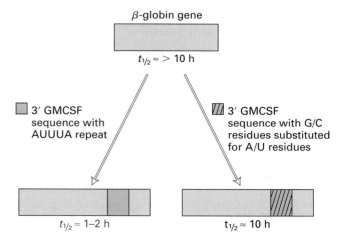

▲ **FIGURE 11-43 Experimental demonstration of the destabilizing effect of AUUUA sequences on mRNA half-life ($t_{\frac{1}{2}}$).** Cultured cells were transfected separately with expression vectors containing the diagrammed β-globin sequences and the half-lives of the expressed mRNAs were determined. The AUUUA sequences (red) were from the gene encoding a cytokine called granulocyte-macrophage colony-stimulating factor (GMCSF), whose mRNA has a $t_{\frac{1}{2}}$ of about 1 hour. Their insertion into the β-globin gene, which normally expresses a stable mRNA, resulted in a short-lived recombinant β-globin mRNA. [See G. Shaw and R. Kamen, 1986, *Cell* **46**:659.]

actin filaments) and a kinesin-like motor protein, which tracks along microtubules, are key components in the cytoplasmic localization of some mRNAs.

Stability of Cytoplasmic mRNAs Varies Widely

The concentration of an mRNA is a function of both its rate of synthesis and its rate of degradation. For this reason, if two genes are transcribed at the same rate, the steady-state concentration of the corresponding mRNA that is more stable will be higher than the concentration of the other. The stability of an mRNA also determines how rapidly the synthesis of the encoded protein can be shut down. For a stable mRNA, synthesis of the encoded protein persists long after transcription of the gene is repressed. Most bacterial mRNAs are unstable, decaying exponentially with a typical half-life of a few minutes (Table 11-1). This allows the cell to rapidly readjust the synthesis of proteins to accommodate changes in the cellular environment. Most cells in multicellular organisms, on the other hand, exist in a fairly constant environment and carry out a specific set of functions over periods of days to months or even the lifetime of the organism (nerve cells, for example). Accordingly, most mRNAs of higher eukaryotes have half-lives of many hours.

However, some proteins in eukaryotic cells are required only for short periods of time and must be expressed in bursts. For example, certain hormones called **cytokines,** which are involved in the immune response of mammals, are synthesized and secreted in short bursts. Similarly, many of the transcription factors that regulate the onset of the S phase of the cell cycle, such as c-Fos and c-Jun, are synthesized for brief periods only. Expression of such proteins occurs in short bursts because transcription of their genes can be rapidly turned on and off and their mRNAs have unusually short half-lives.

Many short-lived mRNAs in eukaryotic cells contain multiple, sometimes overlapping, copies of the sequence AUUUA in their 3′ untranslated region. When such AU-rich sequences are inserted into the 3′ untranslated region of genes encoding stable mRNAs, the resulting recombinant mRNAs are unstable (Figure 11-43). The mechanism by which these sequences destabilize mRNAs is not yet understood.

Degradation Rate of Some Eukaryotic mRNAs Is Regulated

In several cases, the degradation rate of specific eukaryotic mRNAs changes in response to extracellular signals. For example, when rat mammary tissue is cultured in the presence of the hormone prolactin, the concentration of the mRNA encoding the milk protein casein is about 30,000 molecules per cell. When the medium lacks prolactin, the level of casein

TABLE 11-1   Half-Lives of Messenger RNAs

Cell	Cell Generation Time	mRNA Half-Lives*	
		Average	Range Known for Individual Cases
Escherichia coli	20–60 min	3–5 min	2–10 min
Saccharomyces cerevisiae (yeast)	3 h	22 min	4–40 min
Cultured human or rodent cells	16–24 h	10 h	30 min or less (histone and *c-myc* mRNAs) 0.3–24 h (specific mRNAs of cultured cells)

*For information on specific mRNA half-lives for E. coli, see A. Hirashima, G. Childs, and M. Inouye, 1973, J. Mol. Biol. **119**:373; for yeast, see L.-L. Chia and C. McLaughlin, 1979, Mol. Gen. Genet. **170**:137; and for mammalian cells, see M. M. Harpold, M. Wilson, and J. E. Darnell, 1981, Mol. Cell Biol. **1**:188.

mRNA falls a hundred fold to 300 molecules per cell. In vitro run-on analysis has shown that prolactin treatment of cultured breast tissue causes only a threefold increase in the rate of synthesis of casein mRNA, suggesting that most of the prolactin-induced increase in the concentration of casein mRNA results from an enormous increase in its stability. Determination of the half-life of casein mRNA by pulse-chase labeling experiments in the presence and absence of prolactin have directly demonstrated this stabilization.

In contrast to prolactin, which increases the stability of casein mRNA, high levels of iron decrease the stability of the mRNA encoding the receptor that brings iron into cells. In vertebrates, ingested iron is carried through the circulation bound to a protein called *transferrin*. The transferrin-iron complex is brought into cells by interacting with the transferrin receptor (TfR) in the plasma membrane. Intracellular concentrations of iron are regulated with great precision because both too much and too little iron is detrimental to cells. When iron stores in the cell are sufficient, the import

of the transferrin-iron complex is reduced by increasing the degradation rate of TfR mRNA, which quickly leads to a decrease in the level of transferrin receptor. When iron stores in the cell fall, TfR mRNA is stabilized, leading to increased synthesis of the receptor protein.

This regulation of TfR mRNA stability depends on *iron-response elements* (IREs) in the 3' untranslated region of the mRNA and on a cytosolic protein called *IRE-binding protein* (IRE-BP), whose conformation differs at high and low iron levels (Figure 11-44). Each IRE forms a stem-loop structure in which the loop contains five specific bases and the stem contains AU-rich sequences similar to those that destabilize cytokine mRNAs. When iron concentrations are adequate, these AU-rich sequences are thought to promote degradation of TfR mRNA by the same mechanism that leads to rapid degradation of cytokine mRNAs. When the iron concentration falls slightly, the conformation of IRE-BP changes so that it can bind to the IREs. Binding of IRE-BP to the IREs is thought to block recognition of the destabilizing AU-rich

▶ **FIGURE 11-44 Iron-dependent regulation of the stability of transferrin-receptor (TfR) mRNA.** The 3' iron-response elements (IREs) in this mRNA have a stem-loop structure containing AU-rich sequences (yellow) that promote mRNA degradation. At low intracellular iron concentrations, the conformation of IRE-BP (dark green) is such that it binds to the IREs, thereby inhibiting degradation. As a result, the level of transferrin receptor increases, so that more iron can be brought into the cell.

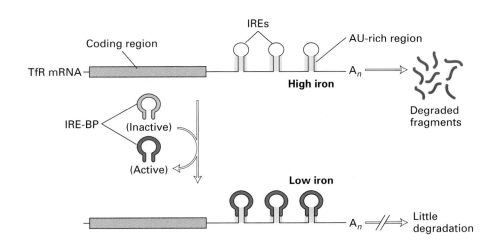

sequences by the proteins that would otherwise degrade TfR mRNA. Other mRNAs whose stability is known to be regulated also may contain response elements that interact with specific proteins, causing a decrease in their degradation rate.

Translation of Some mRNAs Is Regulated by Specific RNA-Binding Proteins

Regulation of the final step in gene expression, mRNA translation, has been demonstrated in a number of cases. One mechanism involves IRE-BP, the same protein that regulates the stability of TfR mRNA. This RNA-binding protein also regulates the translation of mRNAs encoding the two chains of *ferritin*, an intracellular protein that binds iron ions, thereby preventing accumulation of toxic levels of free ions. When intracellular stores of iron are low, translation of ferritin mRNAs is repressed, so that the iron transported into the cell by the transferrin receptor is available for iron-requiring enzymes; when iron is in excess, ferritin synthesis is derepressed so that free iron ions are bound by newly synthesized ferritin.

Translation of the ferritin mRNAs is regulated by binding of IRE-BP to IREs (iron-response elements) at the 5′ ends of the mRNAs (Figure 11-45). Unlike the 3′ IREs in TfR mRNA, the 5′ IREs in ferritin do not have AU-rich stems and do not promote mRNA degradation. When iron concentrations are low, IRE-BP is active and binds to the IREs, inhibiting translation initiation by blocking the ability of the 40S ribosomal subunit to scan for the AUG start codon where it initially binds (see Figure 4-37). When iron concentrations are high, IRE-BP is inactive and does not bind to the 5′ IREs, so that translation initiation can proceed.

Antisense RNA Regulates Translation of Transposase mRNA in Bacteria

Another type of translational control is mediated by **antisense RNA**, which contains sequences complementary to the region of an mRNA containing an initiation codon. In this mechanism, which occurs naturally in bacterial cells, hybridization of the complementary antisense RNA to an mRNA blocks recognition of the initiation codon and binding of the 30S ribosomal subunit to the Shine-Delgarno sequence, thereby preventing initiation of translation (see Figure 4-36).

One example of a protein whose expression is regulated by antisense control is the transposase encoded by the bacterial insertion sequence IS*10*. As discussed in Chapter 9, transposase catalyzes transposition of this mobile DNA element (see Figure 9-12). If too much transposase were expressed, so many mutations would result from IS*10* transposition that the host cell would not survive. Normally, this does not occur because IS*10* also encodes an antisense RNA that controls the rate of translation of transposase mRNA (Figure 11-46).

The strategy of antisense control has been adapted by researchers to inhibit gene expression experimentally in eukaryotic cells. In this approach, vectors are constructed that express high levels of a RNA complementary to the RNA transcript of a target gene. In some cases, introduction of such an antisense vector into cells inactivates the target gene. This repression of gene expression, however, is not due to direct inhibition of translation as in bacterial systems. Rather, hybridization of the antisense RNA to the target-gene transcript interferes with RNA processing in the nucleus. In other cases, antisense expression vectors fail to completely eliminate the activity of proteins encoded by target genes, perhaps because even low levels of some proteins, particularly enzymes, may be sufficient to perform their required functions in the cell. Also, detailed analysis of antisense control in bacteria has revealed that the secondary structures of the complementary RNAs greatly influence the rate at which they hybridize, and thus the efficiency of antisense control. The RNAs expressed from P_{OUT} and P_{IN} in IS*10* have been optimized by natural selection to hybridize at extremely high rates at physiological temperature and salt concentrations, so transposase expression is inhibited very effectively.

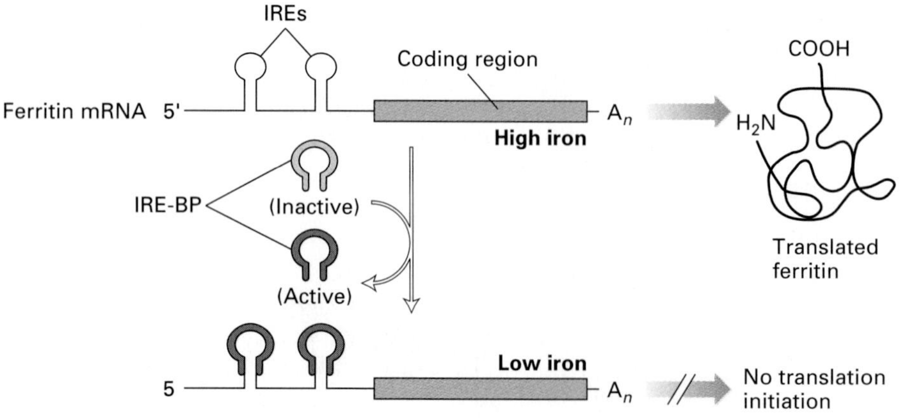

◀ **FIGURE 11-45 Iron-dependent regulation of translation of ferritin mRNA.** At low iron concentrations, binding of active IRE-BP blocks translation initiation. The same mechanism controls translation of the mRNA encoding ALA synthase. The IREs in these two mRNAs are located at the 5′ end and do not contain the degradation-promoting AU-rich sequences present in the 3′ IREs of transferrin receptor mRNA (see Figure 11-44).

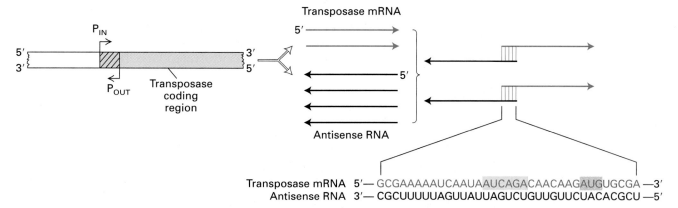

Transposase mRNA 5′— GCGAAAAAUCAAUAAUCAGACAACAAGAUGUGCGA —3′
Antisense RNA 3′— CGCUUUUUAGUUAUUAGUCUGUUGUUCUACACGCU —5′

▲ **FIGURE 11-46 Antisense control of translation of transposase mRNA encoded by IS10, a bacterial mobile element.** The 5′ end of transposase mRNA (red), produced from the lower strand beginning at P_{IN}, is complementary to the 5′ end of antisense RNA (black), produced from the top strand beginning at P_{OUT}. Because P_{OUT} is a much stronger promoter than P_{IN}, more antisense RNA than transposase mRNA is produced, so that nearly all of the transposase mRNA hybridizes to the more abundant antisense RNA. Since the AUG start codon (green highlight) and the ribosome-binding Shine-Delgarno sequence (yellow highlight) are in the hybridized region, initiation of translation of transposase mRNA is blocked. Very rarely, translation of a transposase mRNA is initiated before hybridization to an antisense RNA, leading to a low rate of IS10 transposition.

SUMMARY Other Mechanisms of Post-Transcriptional Control

- In RNA editing the nucleotide sequence of a pre-mRNA is altered in the nucleus. The few examples of editing discovered in vertebrates involve deamination of a single base in the mRNA sequence, resulting in a change in the amino acid specified by the corresponding codon and production of a functionally different protein (see Figure 11-39). Much more extensive editing occurs in the mitochondrial mRNAs of some protozoa.

- Some mRNAs are directed to specific subcellular locations by sequences usually found in the 3′ untranslated region. These sequences are thought to be targets for sequence-specific mRNA-binding proteins that associate with molecular motor proteins. The motor proteins then transport the mRNAs to specific locations by moving along cytoskeletal actin filaments or microtubules.

- The stability of different mRNAs in the cytoplasm varies widely. Although many eukaryotic mRNAs are quite stable, some have unusually short half-lives. The latter mRNAs, which often encode proteins that are expressed in short bursts, generally have repeated copies of the sequence AUUUA in their 3′ untranslated region. By an unknown mechanism these AU-rich sequences stimulate degradation of an mRNA.

- Eukaryotic cells can regulate the rate of degradation or translation of some mRNAs. For example, interaction of a specific RNA-binding protein with sites in the 3′ untranslated region of transferrin-receptor mRNA protects the mRNA from degradation mediated by AU-rich sequences. The same RNA-binding protein, whose activity is regulated by the intracellular iron concentration, blocks translation of several mRNAs by binding to specific sequences in the 5′ untranslated region.

- In bacteria, the translation of some mRNAs is repressed by antisense RNA, which hybridizes to the 5′ end of the mRNA, thereby preventing translation by blocking access of the small ribosomal subunit to the initiation codon.

11.6 Processing of rRNA and tRNA

Approximately 80 percent of the total RNA in rapidly growing mammalian cells (e.g., cultured HeLa cells) is rRNA, and 15 percent is tRNA; protein-coding mRNA thus constitutes only a small portion of the total RNA. The primary transcripts produced from most rRNA genes and from tRNA genes, like pre-mRNAs, are extensively processed to yield the mature, functional forms of these RNAs.

Pre-rRNA Genes Are Similar in All Eukaryotes and Function as Nucleolar Organizers

The 28S and 5.8S rRNAs associated with the large (60S) ribosomal subunit and the 18S rRNA associated with the small (40S) ribosomal subunit in higher eukaryotes are encoded by a single type of **pre-rRNA** transcription unit. Transcription by RNA polymerase I yields a 45S (13.7-kb) primary transcript (pre-rRNA), which is processed into the mature

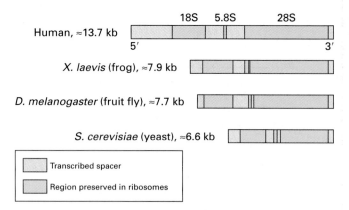

▲ FIGURE 11-47 **General structure of eukaryotic pre-rRNA transcription units.** The three coding regions (blue) encode the 18S, 5.8S, and 28S rRNAs found in ribosomes of higher eukaryotes or their equivalents in other species. The order of these coding regions in the genome is always $5' \rightarrow 3'$. Variations in the lengths of the transcribed spacer regions (tan) account for the major difference in the lengths of pre-rRNA transcription units in different organisms.

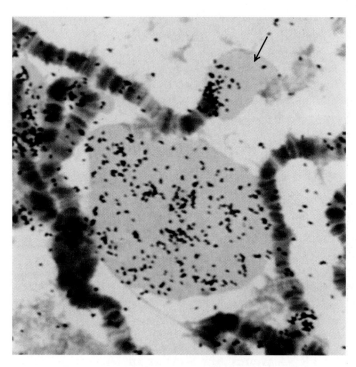

▲ FIGURE 11-48 **Micrograph of a polytene chromosome preparation from a fruit fly carrying a single pre-rRNA transgene.** The cells were pulse-labeled with [³H]uridine before subjecting them to Giemsa staining and autoradiography; the silver grains (black dots) represent newly synthesized RNA. The large pale structure in the center is a normal nucleolus. The smaller one (arrow) is a new nucleolus formed at the end of the transgene. [See G. H. Karpen et al., 1988, *Genes & Devel.* **2**:1745; courtesy of G. H. Karpen.]

28S, 18S, and 5.8S rRNAs found in cytoplasmic ribosomes. Cloning and sequencing of the DNA encoding pre-rRNA from many species showed that this DNA shares several properties in all eukaryotes. First, the pre-rRNA genes are arranged in long tandem arrays separated by nontranscribed spacer regions ranging in length from ≈ 2 kb in frogs to ≈ 30 kb in humans. Second, the genomic regions corresponding to the three finished rRNAs are always arranged in the same $5' \rightarrow 3'$ order: 18S, 5.8S, and 28S (Figure 11-47). Third, in all eukaryotic cells (and even in bacteria), the pre-rRNA gene, as well as the corresponding primary transcript, is considerably longer than the sum of the three finished rRNA molecules. For example, in human cells only about half of the 45S pre-rRNA primary transcript appears in the final rRNA products, whose combined length is about 7.2 kb. The other half, called *transcribed spacer RNA*, is removed during processing and is rapidly degraded. Discovery of pre-rRNA processing was the first indication that mature cytoplasmic RNAs are derived from larger precursor RNAs synthesized in the nucleus. Both the synthesis and processing of pre-mRNA occurs in the nucleolus.

When pre-rRNA genes initially were identified in the nucleolus by in situ hybridization, it was not known whether any other DNA was required to form the nucleolus. Subsequent experiments with transgenic *Drosophila* strains demonstrated that a single complete pre-rRNA transcription unit induces formation of a small nucleolus (Figure 11-48). Thus a single pre-rRNA gene is sufficient to be a **nucleolar organizer,** and all the other components of the ribosome diffuse to the newly formed pre-rRNA. The structure of the induced nucleolus appears, at least by light microscopy, to be the same as, except smaller than, a normal *Drosophila* nucleolus containing 200 or so pre-rRNA genes.

Small Nucleolar RNAs (snoRNAs) Assist in Processing rRNAs and Assembling Ribosome Subunits

Following their synthesis in the nucleolus, nascent pre-rRNA transcripts are immediately bound by proteins, forming *pre-ribonucleoprotein particles*, or pre-rRNPs (Figure 11-49). Several ribonucleoprotein particles of different sizes have been extracted from mammalian nucleoli. The largest of these (80S) contains an intact 45S pre-rRNA molecule, which is cut in a series of cleavage and exonucleolytic steps that ultimately yield the mature rRNAs found in ribosomes (Figure 11-50). During processing, pre-rRNA also is extensively modified, mostly by methylation of the 2′-hydroxyl group of specific riboses and conversion of specific uridine residues to pseudouridine. Some of the proteins in the pre-rRNPs found in nucleoli remain associated with the mature ribosomal subunits, whereas others are restricted to the nucleolus and assist in assembly of the subunits.

The positions of cleavage sites in pre-rRNA and the specific sites of 2′-O-methylation and pseudouridine formation are determined by approximately 150 different small

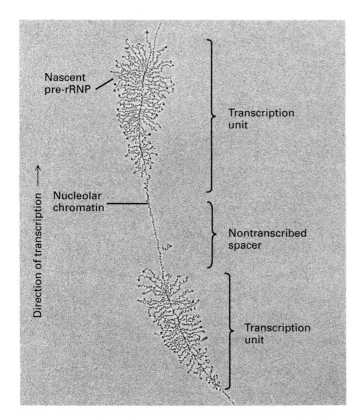

▲ FIGURE 11-49 Electron micrograph of pre-rRNA transcription units from nucleolus of a frog oocyte. Each "feather" represents a pre-rRNA molecule associated with protein in a pre-rRNP emerging from a transcription unit. Pre-rRNA transcription units are arranged in tandem, separated by nontranscribed spacer regions of nucleolar chromatin. [Courtesy of Y. Osheim and O. J. Miller, Jr.]

nucleolus-restricted RNA species, called **small nucleolar RNAs (snoRNAs)**, which hybridize transiently to pre-rRNA molecules. Like snRNAs, snoRNAs associate with proteins, forming *snoRNPs*. One large class of snoRNAs, involved in 2'-O-methylation, contain common sequences bound by the nucleolus-restricted protein *fibrillarin*. A conserved sequence in these snoRNAs, which is invariably positioned close to methylation sites in the pre-rRNA, is thought to bind a methyltransferase enzyme that modifies the ribose moiety. Another snoRNP, called RNase MRP, catalyzes one of the cleavages by which transcribed spacer sequences are removed from pre-rRNA (see Figure 11-50b). The associated snoRNA is homologous to the RNA of RNase P involved in tRNA processing (see below). Based on this homology, the cleavage reaction is thought to be catalyzed by the MRP snoRNA. There is strong evidence that RNase P performs one of the pre-rRNA cleavages as well.

Some snoRNAs are expressed from their own promoters by RNA polymerase II or III. Remarkably, however, the large majority of snoRNAs are spliced-out introns of genes encoding functional mRNAs. These genes invariably encode proteins involved in ribosome synthesis or translation. Some snoRNAs are introns spliced from apparently nonfunctional mRNAs. The genes encoding these mRNAs seem to exist only to express snoRNAs from excised introns!

Unlike pre-rRNA genes, 5S-rRNA genes are transcribed by RNA polymerase III in the nucleoplasm outside of the nucleolus. Without further processing, 5S RNA diffuses to the nucleolus, where it assembles with the 28S and 5.8S rRNAs and proteins into large ribosomal subunits (see Figure 11-50a). When assembly of ribosomal subunits in the nucleolus is complete, they are transported through nuclear pore complexes to the cytoplasm, where they appear first as free subunits.

Self-Splicing Group I Introns Were the First Examples of Catalytic RNA

The DNA in the protozoan *Tetrahymena thermophila* contains an intervening intron in the region that encodes the large pre-rRNA molecule. Careful searches failed to uncover even one pre-rRNA gene without the extra sequence, indicating that splicing is required to produce mature rRNA in these organisms. Subsequent studies showing that the pre-rRNA was spliced at the correct sites when incubated by itself, without assistance from any protein, provided the first indication that RNA can function as a catalyst, like enzymes.

Following the discovery of self-splicing in *Tetrahymena* pre-rRNA, a whole raft of self-splicing sequences were found in pre-rRNAs from other single-celled organisms, in mitochondrial and chloroplast pre-rRNAs, in several pre-mRNAs from certain *E. coli* bacteriophages, and in some bacterial tRNA primary transcripts. The self-splicing sequences in all these precursors, referred to as group I introns, use guanosine as a cofactor and can fold by internal base pairing to juxtapose closely the two exons that must be joined. As discussed earlier, certain mitochondrial and chloroplast pre-mRNAs and tRNAs contain a second type of self-splicing intron, designated group II. As illustrated in Figure 11-51, the splicing mechanisms used by group I introns, group II introns, and spliceosomes are generally similar, involving two **transesterification** reactions, which require no input of energy. Clearly, in the self-splicing introns, RNA functions as a **ribozyme**, an RNA sequence with catalytic ability.

The group I intron within the pre-rRNA of *Tetrahymena* and certain other organisms is unrelated to the transcribed spacer sequences that separate the 18S, 5.8S, and 28S regions in the majority of organisms (see Figure 11-50b). In particular, the self-splicing mechanism that removes group I introns differs from the cleavage mechanism by which spacer sequences are removed during processing of pre-rRNA discussed in the previous section. The three-dimensional structure of the group I intron from *Tetrahymena* pre-rRNA has been solved. Mutational and biochemical experiments are under way to determine which residues are critical in catalyzing the two transesterification reactions leading to splicing.

(a)

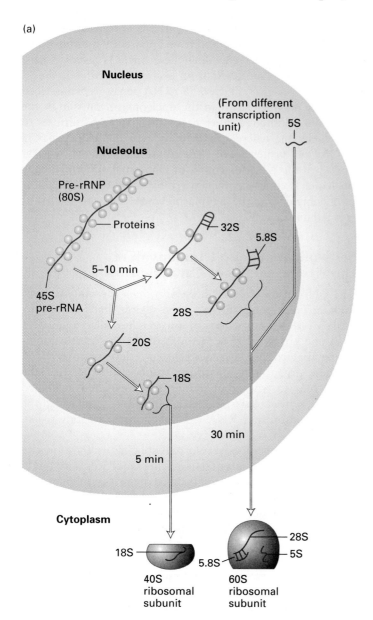

(b)

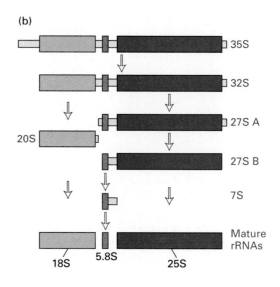

◀ FIGURE 11-50 **Processing of pre-rRNA and assembly of ribosomes in eukaryotes.** (a) Major intermediates and times required for various steps in pre-rRNA processing in higher eukaryotes. Ribosomal and nucleolar proteins associate with 45S pre-rRNA soon after its synthesis, forming an 80S pre-rRNP. Synthesis of 5S rRNA occurs outside of the nucleolus. The extensive secondary structure of rRNAs (see Figure 4-33) is not represented here. Note that RNA constitutes about two-thirds of the mass of the ribosomal subunits, and protein about one-third. (b) Pathway for processing of 6.6-kb (35S) pre-rRNA primary transcript in *S. cerevisiae*. The transcribed spacer regions (tan), which are discarded during processing, separate the regions corresponding to the mature 18S, 5.8S, and 25S rRNAs. All of the intermediates diagrammed have been identified; their sizes are indicated in red type. [Part (b) adapted from S. Chu et al., 1994, *Proc. Nat'l. Acad. Sci. USA* **91**:659.]

All Pre-tRNAs Undergo Cleavage and Base Modification

Mature cytosolic tRNAs, which average 75–80 nucleotides in length, are produced from larger precursors (pre-tRNAs) synthesized by RNA polymerase III in the nucleoplasm. Mature tRNAs also contain numerous modified bases that are not present in tRNA primary transcripts. Cleavage and base modification occur during processing of all pre-tRNAs. As discussed in the next section, some pre-tRNAs contain one or more introns that are spliced out during processing.

A 5′ sequence of variable length that is absent from mature tRNAs is present in all pre-tRNAs (Figure 11-52). These extra 5′ nucleotides are removed by the ribonuclease P (RNase P), a ribonucleoprotein endonuclease. *E. coli* RNase P consists of a 14-kDa polypeptide and a 377-nucleotide

RNA called M1 RNA, which was one of the earliest catalytic RNA molecules to be recognized. At high Mg^{2+} concentrations, isolated M1 RNA recognizes and cleaves *E. coli* pre-tRNAs. The RNase P polypeptide increases the rate of cleavage by M1 RNA, allowing it to proceed at physiological Mg^{2+} concentrations.

About 10 percent of the bases in pre-tRNAs are modified enzymatically during processing. Three types of base modifications occur (see Figure 11-52): replacement of U residues at the 3′ end of pre-tRNA with a CCA sequence, which is found at the 3′ end of all tRNAs; addition of methyl and isopentenyl groups to the heterocyclic ring of purine bases and methylation of the 2′-OH group in the ribose of any residue; and conversion of specific uridines to dihydrouridine, pseudouridine, or ribothymidine residues.

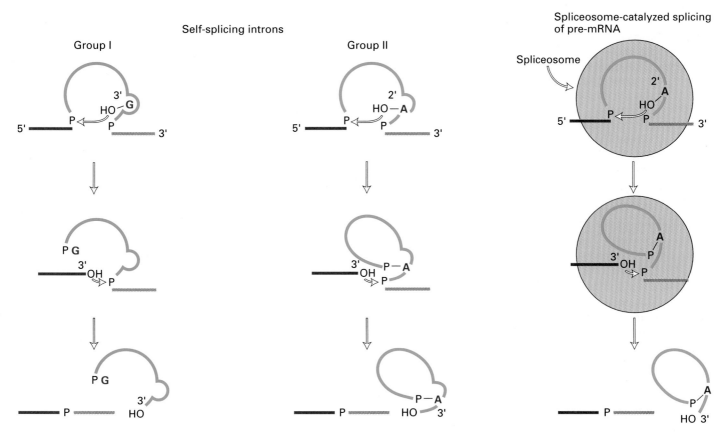

Self-splicing introns

Group I

Group II

Spliceosome-catalyzed splicing of pre-mRNA

Spliceosome

▲ **FIGURE 11-51 Splicing mechanisms in group I and group II self-splicing introns and spliceosome-catalyzed splicing of pre-mRNA.** The intron is shown in blue; the exons to be joined in red. In group I introns, a guanosine cofactor (G) that is not part of the RNA chain associates with the active site. The 3′-hydroxyl group of this guanosine participates in a transesterification reaction with the phosphate at the 5′ end of the intron; this reaction is analogous to that involving the 2′-hydroxyl groups of the branch-site A in group II introns and pre-mRNA introns spliced in splice-osomes (see Figure 11-16). The subsequent transesterification that links the 5′ and 3′ exons is similar in all three splicing mechanisms. Note that spliced-out group I introns are linear structures, unlike the branched intron products in the other two cases. [Adapted from P. A. Sharp, 1987, *Science* **235**:769.]

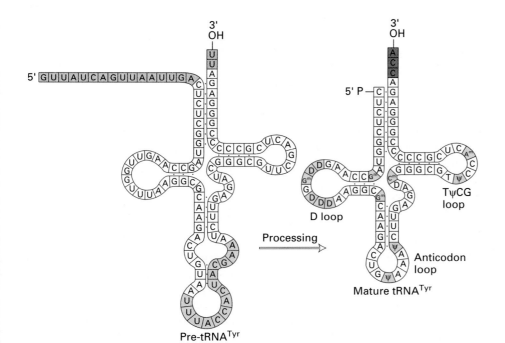

Pre-tRNA^Tyr

Mature tRNA^Tyr

◄ **FIGURE 11-52 Processing of tyrosine pre-tRNA involves four types of changes.** A 14-nucleotide intron (blue) in the anticodon loop is removed by splicing. A 16-nucleotide sequence (green) at the 5′ end is cleaved by RNase P. U residues at the 3′ end are replaced by the CCA sequence (red) found in all mature tRNAs. Numerous bases in the stem-loops are converted to characteristic modified bases (yellow). Not all pre-tRNAs contain introns that are spliced out during processing, but they all undergo the other types of changes shown here. D = dihydrouridine; Ψ = pseudouridine.

▶ **FIGURE 11-53 Mechanism of splicing in pre-tRNA.** First, the pre-tRNA is cleaved at two places, on each side of the intron, thereby excising the intron. The cleavage mechanism generates a 2′,3′-cyclic phosphomonoester at the 3′ end of the 5′ exon. The multistep reaction joining the two exons requires two nucleoside triphosphates: a GTP, which contributes the phosphate group (yellow) for the 3′ → 5′ linkage in the finished tRNA molecule; and an ATP, which forms an activated ligase-AMP intermediate. The 2′-phosphate on the 5′ exon is removed in the final step. [See E. M. Phizicky and C. L. Greer, 1993, *Trends Biochem. Sci.* **18**:31.]

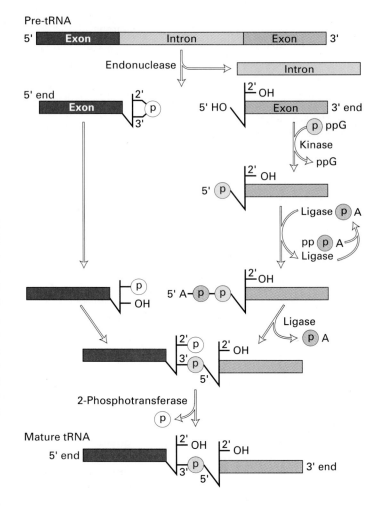

Processing of pre-tRNA, like mRNA processing, occurs in the nucleoplasm. The mature tRNAs then are transported to the cytoplasm through nuclear pore complexes. Interestingly, U6 snRNA, another RNA synthesized by RNA polymerase III, is a component of the spliceosome and remains in the nucleus, whereas tRNAs are efficiently transported to the cytoplasm. Most likely, mature tRNAs in the nucleus, like mature mRNAs and rRNAs, are bound by specific proteins that facilitate their transport through nuclear pores. Once in the cytoplasm, tRNAs are passed between aminoacyl-tRNA synthetases, elongation factors, and ribosomes during protein synthesis (Chapter 4). Thus tRNAs generally are associated with proteins and spend little time free in the cell, as is also the case for mRNAs and rRNAs.

Splicing of Pre-tRNAs Differs from Other Splicing Mechanisms

Comparison of the sequences of tRNA genes with the sequences of the corresponding cytosolic tRNAs has shown that some eukaryotic nuclear tRNA genes contain introns. For example, the pre-tRNA expressed from the yeast tyrosine tRNA (tRNA^{Tyr}) gene contains a 14-base intron that is not present in mature tRNA^{Tyr} (see Figure 11-52). Some archaeal tRNA genes also contain introns. The introns in nuclear pre-tRNAs are shorter than those in pre-mRNAs, and they do not contain the splice-site consensus sequences found in pre-mRNAs (see Figure 11-14). Pre-tRNA introns also are clearly distinct from the much longer self-splicing group I and group II introns found in chloroplast and mitochondrial pre-rRNAs.

The mechanism of pre-tRNA splicing, outlined in Figure 11-53, differs in several ways from the mechanisms utilized by self-splicing introns and spliceosomes (see Figure 11-51). For instance, during pre-tRNA splicing the intron is excised in one step rather than two; GTP and ATP are required; a 2′,3′-cyclic monophosphate forms on the cleaved end of the 5′ exon; and the process is catalyzed by proteins (enzymes) rather than RNA. Certain mutations in pre-tRNA that change its secondary structure prevent the splicing reaction, indicating that pre-tRNA molecules must be folded into a particular secondary structure for intron excision to occur. Since introns always are found in the anticodon loop of pre-tRNAs, pre-tRNAs most likely are folded similarly to mature tRNAs,

thereby bringing the two intron-exon junctions into proximity (see Figure 11-52 and Figure 4-26b).

When intron-containing yeast tRNA genes are microinjected into *Xenopus* oocyte nuclei, correctly processed tRNAs are produced. This finding indicates that enzymatic systems for cleaving, modifying, and splicing pre-tRNAs have been conserved over a wide evolutionary range.

SUMMARY Processing of rRNA and tRNA

- Like pre-mRNAs, the primary transcripts produced from pre-rRNA and tRNA genes undergo extensive processing.

- Synthesis of a large precursor pre-rRNA (45S in higher eukaryotes) by RNA polymerase I and its subsequent processing occur in the nucleolus.

- Cleavage, exonucleolytic digestion, and base modifications of 45S pre-rRNA yields mature 28S, 18S, and 5.8S rRNAs, which associate with ribosomal proteins into ribosomal subunits (see Figure 11-50). Numerous small nucleolar RNAs (snoRNAs) participate in pre-rRNA processing.

- 5S rRNA, which is synthesized in the nucleoplasm by RNA polymerase III, is not extensively processed before assembling with other rRNAs and proteins to form the large ribosomal subunit.

- The first catalytic RNAs (ribozymes) discovered were the group I introns in *Tetrahymena* rRNA. Self-splicing of group I and group II introns, and pre-mRNA spliceosomal splicing all proceed via two analogous transesterification reactions.

- Transcription of tRNA genes by RNA polymerase III and processing of the primary transcripts occur in the nucleoplasm. In all pre-tRNAs, the 5′-end sequence is removed by RNase P, a ribonucleoprotein containing a catalytic RNA; CCA is added to the 3′ end; and multiple internal bases are modified (see Figure 11-52).

- Some pre-tRNAs contain a short intron within the anticodon loop. This is removed by protein enzymes via a mechanism distinct from the splicing of pre-mRNA and self-splicing introns.

- Ribosomal subunits containing mature rRNAs and fully processed tRNAs, probably also complexed with proteins, are transported to the cytoplasm through nuclear pore complexes.

PERSPECTIVES for the Future

In this chapter, we have seen that in eukaryotic cells, mRNAs are synthesized and processed in the nucleus, transported through nuclear pore complexes to the cytoplasm, and then, in some cases, transported to specific areas of the cytoplasm before being translated by ribosomes. Each of these fundamental processes are carried out by complex macromolecular machines composed of scores of proteins and in many cases RNAs as well. The complexity of these macromolecular machines ensures accuracy in finding promoters and splice sites in the long length of DNA and RNA sequences, and provides various avenues for regulating synthesis of a polypeptide chain. Much remains to be learned about the structure, operation, and regulation of such complex machines as spliceosomes and the cleavage/polyadenylation apparatus.

Recent examples of the regulation of pre-RNA splicing raise the question of how extracellular signals might control such events, especially in the nervous system of vertebrates. A case in point is the remarkable situation in the chick inner ear where multiple isoforms of the Ca^{2+}-activated K^+ channel called Slo are produced by alternative RNA splicing. Cell-cell interactions appear to inform cells of their position in the cochlea, leading to alternative splicing of Slo pre-mRNA. The challenging task facing researchers is to discover how such cell-cell interactions regulate the activity of RNA-processing factors.

Future research will likely reveal additional activities of hnRNP proteins and clarify their mechanisms of action. For instance, one of the mRNAs transiently induced in response to growth factors encodes an arginine methyltransferase that modifies arginines at specific sites in a subset of hnRNP proteins. What could be the function of such a transiently induced modification? Some hnRNP proteins contain nuclear-retention signals that prevent nuclear export when fused to hnRNP proteins with nuclear-export signals (NESs). How are these hnRNP proteins selectively removed from processed mRNAs in the nucleus, allowing the mRNAs to be transported to the cytoplasm?

The mechanism of transport through nuclear pore complexes poses many intriguing questions. How are cargo complexes moved in one direction through the nuclear pore complex? Are there multiple distinct binding sites along the way with a mechanism that passes the complex from one site to the next in a unidirectional manner? Or, are there large, transient conformational changes in the nuclear pore complex or central transporter that push or pull the complex long distances in a small number of steps? How are traffic jams avoided between proteins moving into and those moving out of the nucleus? Answering these questions will require ingenious experimental designs.

The localization of certain mRNAs to specific subcellular locations is fundamental to the development of multicellular organisms. During development, an individual cell frequently divides into daughter cells that function differently from each other. In the language of developmental biology, the two daughter cells are said to have different developmental fates. In many cases, this difference in developmental fate results from the localization of an mRNA to one region of the cell before mitosis so that after cell division, it is present in one daughter cell and not the other. Much exciting work remains to be done to fully understand the molecular mechanisms controlling mRNA localization that are critical for the normal development of multicellular organisms.

These are just a few of the fascinating questions concerning RNA processing, post-transcriptional control, and nuclear transport that will challenge molecular cell biologists in the coming decades.

PERSPECTIVES in the Literature

Following nuclear import through nuclear pore complexes, importin proteins release their cargo protein when they interact with Ran-GTP (see Figure 11-37). In contrast, in the process of nuclear export, exportins bind their cargo proteins only when they are bound to Ran-GTP (see Figure 11-33). How does the binding of Ran-GTP affect the conformations of the receptors and, consequently, their ability to bind cargo proteins? Why does Ran-GDP in the cytoplasm not cause the same conformational changes in the receptors? To explore these questions, consult the following articles:

Chook, Y. M., and G. Blobel. 1999. Structure of the nuclear transport complex karyopherin-β2-Ran·GppNHp. *Nature* **399**:230–237.

Cingoli, G., et al. 1999. Structure of importin-β bound to the IBB domain of importin-α. *Nature* 399:221–229.

Mattaj, I. W., and E. Conti. 1999. Snail mail to the nucleus. *Nature* 399:208–210.

Testing Yourself on the Concepts

1. Describe three types of post-transcriptional regulation.

2. Investigators often use in vitro translation systems utilizing wheat germ or rabbit reticulocyte extracts to study proteins that result from the translation of an mRNA molecule. The mRNA molecules used in such experiments are often synthesized in vitro using bacteriophage T7 RNA polymerase to transcribe a plasmid clone isolated from bacteria, as described in Section 7.6. What type of clone and what additional factors are necessary to obtain efficient translation of a human gene in a wheat germ or reticulocyte extract following in vitro transcription by T7 RNA polymerase?

3. Compare and contrast rRNA, mRNA, and tRNA processing in eukaryotes.

4. What is the evidence that HIV alters the normal process of nuclear export of RNA?

MCAT/GRE-Style Questions

Key Application It has been suggested that manipulation of HIV antitermination might provide for effective therapies in combating AIDS. Please read Section 11.1 and answer the following questions:

1. Which of the following would not decrease the amount of full-length HIV transcript produced after HIV infection:

   a. A mutation in the TAR sequence that abolishes Tat binding.

   b. Inhibition of Cdk9 activity.

   c. A mutation in Tat that abolishes interaction with cyclin T.

   d. Activation of Cdk9 activity.

2. When Tat protein fails to bind the TAR region of the HIV transcript, which of the following is true:

   a. RNA polymerase II terminates transcription.

   b. Cyclin T does not interact with Cdk9.

   c. Cdk9 acts on the RNA polymerase II molecule transcribing the HIV genome.

   d. The CTD of RNA polymerase II is phosphorylated.

3. The role of Tat in HIV transcription is

   a. To allow transport of HIV mRNAs to the cytoplasm.

   b. To prevent early termination by RNA polymerase II.

   c. To bind a complex containing cyclin T and Cdk9 to the TAR sequence so that it can phosphorylate the CTD of the RNA polymerase II transcribing the HIV provirus.

   d. All of the above.

   e. a and b.

   f. b and c.

   g. a and c.

Key Experiment Please read Section 11.2 and refer to Figures 11-13 and 11-15; then answer the following questions:

4. If a bacteriophage T7 RNA molecule was hybridized to bacteriophage T7 DNA, which of the following would be evident from the electron microscopic analysis of this hybrid:

   a. Looping of introns.

   b. Looping of 5′ and 3′ UTRs.

   c. No looping of introns.

   d. No looping of 5′ and 3′ UTRs.

   e. a and b.

   f. c and d.

5. Which of the following is evidence that the cell's splicing machinery recognizes similar features of the 5′ and 3′ splice sites common to all pre-mRNAs?

   a. 5′ and 3′ splice sites in chimeric pre-mRNAs are joined by RNA splicing.

   b. The 5′ and 3′ splice sites are identical in all mRNAs.

   c. Deletion of all but 30 nucleotides from introns allows splicing to occur.

   d. All of the above.

6. In Figure 11-15, the reaction intermediate (339*) consists of

   a. Exons 1 and 2 and the intervening intron.

   b. Exon 1 and part of exon 2.

   c. Exon 1 spliced to exon 2.

   d. Exon 1 and a lariat form of the intron.

   e. Exon 2 and a lariat form of the intron.

7. Polyacrylamide gel electrophoresis as used in Figure 11-15 allows investigators to determine

   a. The sizes of various RNA molecules.

   b. The structure of various RNA molecules.

   c. Whether splicing has occurred.

   d. All of the above.

8. If the branch point A residue of the 497-nucleotide radiolabeled RNA molecule used in Figure 11-15 was mutated to a C residue, polyacrylamide gel electrophoresis of the products of an in vitro splicing reaction would exhibit

   a. 497, 339, 130, and 252 bp bands.

b. 497, 339, and 130 bp bands.

c. 497 and 130 bp bands.

d. 497 and 252 bp bands.

e. 497 bp band.

Key Concept Please read Section 11.3 and answer the following questions:

9. Which of the following is true about fibronectin mRNA synthesis in hepatocytes as compared to fibroblasts:

   a. An alternative start site of transcription is used.

   b. Two introns are not spliced out.

   c. Two exons are not spliced out.

   d. Two exons are spliced out.

10. Differential splicing of the *slo* gene in ciliated neurons results in

   a. Hair cells with different response frequencies.

   b. Different *slo* isoforms.

   c. *slo* channels that open at different calcium concentrations.

   d. All of the above.

   e. None of the above.

11. Which of the following does not contribute to differential splicing in eukaryotes:

   a. RNA binding proteins.

   b. hnRNPs.

   c. Self-splicing group II introns.

   d. The spliceosome.

Key Terms

5′ capping *410*

alternative RNA splicing *426*

antisense RNA *442*

attenuator site *405*

branch-point A *416*

cargo complex *430*

group I and II introns *419*

hnRNP proteins *411*

mRNPs *428*

N-mediated antitermination *408*

nuclear pore complex *427*

nuclear-transport receptors *433*

nucleolar organizer *444*

polyadenylation *413*

Ran GTPase *430*

Rho-dependent termination *407*

ribozyme *445*

RNA editing *437*

small nuclear RNAs (snRNAs) *416*

small nucleolar RNAs (snoRNAs) *445*

spliceosome *417*

splice sites *415*

stem-loops *405*

transesterification *445*

trans-splicing *418*

References

Transcription Termination

Geiduschek, E. P., and G. P. Tocchini-Valentini. 1988. Transcription by RNA polymerase III. *Ann. Rev. Biochem.* 57:873–914.

Greenblatt, J., J. R. Nodwell, and S. W. Mason. 1993. Transcription antitermination. *Nature* 364:401–406.

Henkin, T. M. 1996. Control of transcription termination in prokaryotes. *Ann. Rev. Genet.* 30:35–57.

Landick, R. 1997. RNA polymerase slides home: pause and termination site recognition. *Cell* 88:741–744.

Nudler, E., et al. 1998. Spatial organization of transcription elongation complex in *Escherichia coli.*, *Science* 281:424–428.

Reeder, R. H., and W. H. Lang. 1997. Terminating transcription in eukaryotes: lessons learned from RNA polymerase I. *Trends Biochem. Sci.* 22:473–477.

Shilatifard, A. 1998. Factors regulating the transcriptional elongation activity of RNA polymerase II. *FASEB J.* 12:1437–1446.

von Hippel, P. H. 1998. An integrated model of the transcription complex in elongation, termination, and editing. *Science* 281:660–665.

Yamaguchi, Y., T. Wada, and H. Handa. 1998. Interplay between positive and negative elongation factors: drawing a new view of DRB. *Genes To Cells* 1:9–15.

Yankulov, K., and D. Bentley. 1998. Transcriptional control: Tat cofactors and transcriptional elongation. *Curr. Biol.* 8:R447–449.

Yanofsky, C., K. V. Konan, and J. P. Sarsero. 1996. Some novel transcription attenuation mechanisms used by bacteria. *Biochimie* 78:1017–1024.

Processing of Eukaryotic mRNA

Birse, C. E., et al. 1998. Coupling termination of transcription to messenger RNA maturation in yeast. *Science* 280:298–301.

Black, D. L. 1995. Finding splice sites within a wilderness of RNA. *RNA* 8:763–771.

Blumenthal, T. 1998. Gene clusters and polycistronic transcription in eukaryotes. *Bioessays* 20:480–487.

Cho, E. J., C. R. Rodriguez, T. Takagi, and S. Buratowski. 1998. Allosteric interactions between capping enzyme subunits and the RNA polymerase II carboxy-terminal domain. *Genes & Devel.* 12:3482–3487.

Dreyfuss, G., M. J. Matunis, S. Pinol-Roma, and C. G. Burd. 1993. hnRNP proteins and the biogenesis of mRNA. *Ann. Rev. Biochem.* 62:289–321.

Hamm, J., and A. I. Lamond. 1998. Spliceosome assembly: the unwinding role of DEAD-box proteins. *Curr. Biol.* 8:R532–534.

Keller, W., and L. Minvielle-Sebastia. 1997. A comparison of mammalian and yeast pre-mRNA 3′-end processing. *Curr. Opin. Cell Biol.* 9:329–336.

Misteli, T., and D. L. Spector. 1998. The cellular organization of gene expression. *Curr. Opin. Cell Biol.* 10:323–331.

Moore, M. J., C. C. Query, and P. A. Sharp. 1993. Splicing of precursors to messenger RNAs by the spliceosome. In R. Gesteland and J. Atkins, eds., *The RNA World*, Cold Spring Harbor Press, pp. 303–357.

Neugebauer, K. M., and M. B. Roth. 1997. Transcription units as RNA processing units. *Genes & Devel.* 11:3279–3285.

Newman, A. 1998. RNA splicing. *Curr. Biol.* 8:R903–905.

Staley, J. P., and C. Guthrie. 1998. Mechanical devices of the spliceosome: motors, clocks, springs, and things. *Cell* 92:315–326.

Tarn, W. Y., and J. A. Steitz. 1997. Pre-mRNA splicing: the discovery of a new spliceosome doubles the challenge. *Trends Biochem. Sci.* 22:132–137.

Varani, G., and K. Nagai. 1998. RNA recognition by RNP proteins during RNA processing. *Ann. Rev. Biophys. Biomol. Struc.* 27:407–445.

Wahle, E., and U. Keuhn. 1997. The mechanism of 3′ cleavage and polyadenylation of eukaryotic pre-mRNA. *Prog. Nucl. Acid Res. Mol. Biol.* 57:41–71.

Will, C. L., and R. Leuhrmann. 1997. Protein functions in pre-mRNA splicing. *Curr. Opin. Cell Biol.* 9:320–328.

Regulation of mRNA Processing

Black, D. L. 1998. Splicing in the inner ear: a familiar tune, but what are the instruments? *Neuron* 20:165–168.

Grabowski, P. J. 1998. Splicing regulation in neurons: tinkering with cell-specific control. *Cell* 92:709–712.

Gunderson, S. I., M. Polycarpou-Schwarz, and I. W. Mattaj. 1998. U1 snRNP inhibits pre-mRNA polyadenylation through a direct interaction between U1 70K and poly A polymerase. *Mol. Cell* 1:255–264.

Hertel, K. J., K. W. Lynch, and T. Maniatis. 1997. Common themes in the function of transcription and splicing enhancers. *Curr. Opin. Cell Biol.* 9:350–357.

Lopez, A. J. 1998. Alternative splicing of pre-mRNA: developmental consequences and mechanisms of regulation. *Ann. Rev. Genet.* 32:279–305.

Tacke, R., and J. L. Manley. 1999. Functions of SR and Tra2 proteins in pre-mRNA splicing regulation. *Proc. Soc. Exp. Biol. Med.* 220:59–63.

Signal-Mediated Transport through Nuclear Pore Complexes

Arts, G. J., et al. 1998. The role of exportin-t in selective nuclear export of mature tRNAs. *Embo J.* 17:7430–7441.

Cole, C. N., and C. M. Hammell. 1998. Nucleocytoplasmic transport: driving and directing transport. *Curr. Biol.* 8:R368–372.

Corbett, A. H., and P. A. Silver. 1997. Nucleocytoplasmic transport of macromolecules. *Microbiol. Mol. Biol. Rev.* 61:193–211.

Cullen, B. R. 1998. Retroviruses as model systems for the study of nuclear RNA export pathways. *Virology* 249:203–210.

Daneholt, B. 1997. A look at messenger RNP moving through the nuclear pore. *Cell* 88:585–588.

Doye, V. and E. Hurt. 1997. From nucleoporins to nuclear pore complexes. *Curr. Opin. Cell Biol.* 9:401–411.

Georlich, D. 1998. Transport into and out of the cell nucleus. *Embo J.* 17:2721–2727.

Mattaj, I. W., and L. Englmeier. 1998. Nucleocytoplasmic transport: the soluble phase. *Ann. Rev. Biochem.* 67:265–306.

Melchior, F., and L. Gerace. 1998. Two-way trafficking with Ran. *Trends Cell Biol.* 8:175–179.

Ohno, M., M. Fornerod, and I. W. Mattaj. 1998. Nucleocytoplasmic transport: the last 200 nanometers. *Cell* 92:327–336.

Pemberton, L. F., G. Blobel, and J. S. Rosenblum. 1998. Transport routes through the nuclear pore complex. *Curr. Opin. Cell Biol.* 10:392–399.

Stutz, F., and M. Rosbash. 1998. Nuclear RNA export. *Genes & Devel.* 12:3303–3319.

Other Mechanisms of Post-Transcriptional Control

Alfonzo, J. D., O. Thiemann, and L. Simpson. 1997. The mechanism of U insertion/deletion RNA editing in kinetoplastid mitochondria. *Nucl. Acids Res.* 25:3751–3759.

Arn, E. A., and P. M. Macdonald. 1998. Motors driving mRNA localization: new insights from in vivo imaging. *Cell* 95:151–154.

Bashirullah, A., R. L. Cooperstock, and H. D. Lipshitz. 1998. RNA localization in development. *Ann. Rev. Biochem.* 67:335–394.

Bass, B. L. 1997. RNA editing and hypermutation by adenosine deamination. *Trends Biochem. Sci.* 22:157–162.

Hentze, M. W., and L. C. Keuhn. 1996. Molecular control of vertebrate iron metabolism: mRNA-based regulatory circuits operated by iron, nitric oxide, and oxidative stress. *Proc. Nat'l. Acad. Sci. USA* 93:8175–8182.

Kable, M. L., S. Heidmann, and K. D. Stuart. 1997. RNA editing: getting U into RNA. *Trends Biochem. Sci.* 22:162–166.

Maas, S., T. Melcher, and P. H. Seeburg. 1997. Mammalian RNA-dependent deaminases and edited mRNAs. *Curr. Opin. Cell Biol.* 9:343–349.

Oleynikov, Y., and R. H. Singer. 1998. RNA localization: different zipcodes, same postman? *Trends Cell Biol.* 8:381–383.

Ross, J. 1996. Control of messenger RNA stability in higher eukaryotes. *Trends Genet.* 5:171–175.

Rouault, T., and R. Klausner. 1997. Regulation of iron metabolism in eukaryotes. *Curr. Topics Cell. Regulation* 35:1–19.

Wickens, M., P. Anderson, and R. J. Jackson. 1997. Life and death in the cytoplasm: messages from the 3′ end. *Curr. Opin. Genet. Devel.* 7:220–232.

Processing of rRNA and tRNA

Abelson, J., C. R. Trotta, and H. Li. 1998. tRNA splicing. *J. Biol. Chem.* 273:12685–12688.

Dahlberg, J. E., and E. Lund. 1998. Proofreading and aminoacylation of tRNAs before export from the nucleus. *Science* 282:2082–2085.

Frank, D. N., and N. R. Pace. 1998. Ribonuclease P: unity and diversity in a tRNA processing ribozyme. *Ann. Rev. Biochem.* 67:153–180.

Golden, B. L., A. R. Gooding, E. R. Podell, and T. R. Cech. 1998. A preorganized active site in the crystal structure of the *Tetrahymena* ribozyme. *Science* 282:259–264.

Morrissey, J. P., and D. Tollervey. 1995. Birth of the snoRNPs: the evolution of RNase MRP and the eukaryotic pre-rRNA-processing system. *Trends Biochem. Sci.* 20:78–82.

Venema, J., and D. Tollervey. 1995. Processing of pre-ribosomal RNA in *Saccharomyces cerevisiae*. *Yeast* 16:1629–1650.

Wolin, S. L., and A. G. Matera. 1999. The trials and travels of tRNA. *Genes & Devel.* 13:1–10.

DNA Replication, Repair, and Recombination

12

The Watson-Crick duplex structure of DNA immediately suggested how genetic material was duplicated from one generation to the next. The realization that bacterial genomes and eukaryotic chromosomes consist of single DNA molecules millimeters to centimeters in length raised a host of structural and biochemical questions about *DNA replication*. How does replication begin, and how does it progress along the chromosome? What mechanisms ensure that only one round of replication occurs before cell division? Which enzymes take part in DNA synthesis, and what are their functions? How does duplication of the long helical duplex occur without the strands becoming tangled?

To unravel the process of DNA replication, researchers study not only DNA synthesis (i.e., the assembly of new DNA strands by sequential addition of deoxyribonucleotides) but also the initiation and termination of chromosome replication. In addition, studies of replication are concerned with how DNA synthesis starts and stops in such a way that each chromosome is duplicated exactly. Another crucial issue in replication is how the two new chromosomes separate. We discuss these processes in the first half of this chapter. The events of mitosis (or meiosis), during which the chromosomes are distributed to daughter cells, are introduced in Chapters 1 and 8 and discussed in detail in Chapters 13 and 19.

For DNA to serve as the genetic link between generations, the base sequence must be not only copied correctly during replication, but also maintained throughout the life span of cells. To keep DNA sequences

DNA methyltransferase localizes to sites of DNA replication in 3T3 fibroblasts.

OUTLINE

MEDIA CONNECTIONS

accurate, both prokaryotic and eukaryotic cells possess enzymes that catalyze *DNA repair.* Inactivation of the cellular DNA-repair machinery frequently occurs during early stages of tumor formation, allowing additional mutations to accumulate, so that eventually cells escape normal growth regulatory mechanisms. *DNA recombination* — the exchange of DNA segments between homologous chromosomes — provides a mechanism for generating genetic diversity beyond that achieved by the independent assortment of chromosomes. The mechanisms of DNA repair and recombination are discussed in the second half of this chapter.

As we will see, to carry out DNA replication, all cells use the same kinds of enzymes, including **DNA polymerases,** which assemble deoxyribonucleotides into a new strand; **helicases,** which unwind duplex DNA; single-stranded DNA-binding proteins; and exonucleases. Bacteria and yeast contain three different DNA polymerases, and mammalian cells contain five. Each type of polymerase has unique functions during DNA replication and repair. Additionally, some of the enzymes that cut and repair damaged DNA also participate in genetic recombination. Another common feature of the three processes discussed in this chapter is involvement of large multisubunit complexes, each containing many enzymes and structural proteins. These molecular machines have evolved to ensure that DNA replication, repair, and recombination occur very rapidly and with exquisite precision.

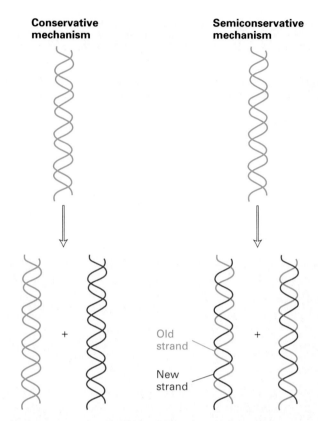

▲ **FIGURE 12-1 Conservative and semiconservative mechanisms of DNA replication differ in whether the newly synthesized strands pair with each other (conservative) or with an old strand (semiconservative).** All available evidence supports semiconservative replication in both prokaryotic and eukaryotic cells.

12.1 General Features of Chromosomal Replication

We first consider three general features of DNA replication: the *semiconservative* and *bidirectional* growth of new strands from a *common site.* Understanding these properties of the replication process provides the foundation for our discussion in the next section of the complex protein machinery that carries out replication.

DNA Replication Is Semiconservative

The base-pairing principle inherent in the Watson-Crick model suggested that the two new DNA strands are copied from the two old strands. Although this mechanism provides for exact copying of genetic information, it raised a new question: Is replication a *conservative* or *semiconservative* process? In the first mechanism, the two new strands form a new duplex and the old duplex remains intact, whereas in the second mechanism, each old strand becomes paired with a new strand copied from it (Figure 12-1).

The first definitive evidence supporting a semiconservative mechanism came from a classic experiment by M. Meselson and W. F. Stahl. *E. coli* cells initially were grown in a medium containing ammonium salts prepared with "heavy" nitrogen (^{15}N) until all the cellular DNA contained the isotope. The cells were then transferred to a medium containing the normal "light" isotope (^{14}N), and samples were removed periodically from the cultures. The DNA in each sample was analyzed by density-gradient equilibrium centrifugation, which can separate heavy-heavy (H-H), light-light (L-L), and heavy-light (H-L) duplexes into distinct bands. The actual banding patterns observed were consistent with semiconservative replication and inconsistent with conservative replication. Subsequent experiments of a different design with cultured plant cells demonstrated for the first time semiconservative DNA replication in eukaryotic chromosomes. Apparently all cellular DNA in both prokaryotic and eukaryotic cells is replicated by a semiconservative mechanism. For further information on these early experiments on semiconservative replication, see Classic Experiment 12.1 on the accompanying CD-ROM.

Most DNA Replication Is Bidirectional

Several possible molecular mechanisms of DNA strand growth would result in semiconservative DNA replication. In one of the simplest possibilities, one new strand derives from one *origin* and the other new strand derives from another origin (Figure 12-2a). Only one strand of the duplex grows at each *growing point*. In this mechanism, which operates in linear DNA viruses such as adenovirus, the ends of the DNA molecules serve as fixed sites for the initiation and termination of replication. A second possibility entails one origin and one **growing fork** (the point where DNA replication occurs), which moves along the DNA in one direction with both strands of DNA being copied (Figure 12-2b). Certain bacterial plasmids replicate in this manner. A third possibility is that synthesis might start at a single origin and proceed in both directions, so that both strands are copied at each of *two* growing forks (Figure 12-2c). The available evidence suggests that the third alternative is generally employed by prokaryotic and eukaryotic cells: that is, DNA replication proceeds *bidirectionally* from a given starting site, with both strands being copied at each fork. Thus two growing forks emerge from a single origin site.

In the circular DNA molecules present in bacteria, plasmids, and some viruses, one origin often suffices, and the two resulting growing forks merge on the opposite side of the circle to complete replication (see Figure 7-2). However, the long linear chromosomes of eukaryotes contain multiple origins; the two growing forks from a particular origin continue to advance until they meet the advancing growing forks from neighboring origins. Each region served by one DNA origin is called a **replicon.**

Evidence for Bidirectional Replication The first experimental support for bidirectional replication in eukaryotic cells was obtained by fiber autoradiography of labeled DNA molecules from cultured mammalian cells (Figure 12-3). Such studies have revealed clusters of active replicons, each of which contains two growing forks moving away from a central origin, thus providing unambiguous evidence of

(a) Unidirectional growth of single strands from two origins

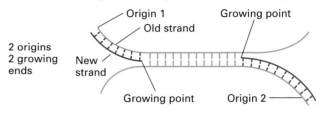

(b) Unidirectional growth of both strands from one origin

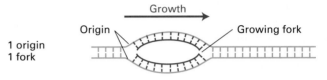

(c) Bidirectional growth of both strands from one origin

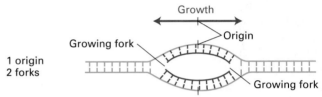

▲ **FIGURE 12-2 Three mechanisms of DNA strand growth that are consistent with semiconservative replication.** The third mechanism—bidirectional growth of both strands from a single origin—appears to be the most common in both eukaryotes and prokaryotes.

(a) Predicted fiber autoradiographic pattern

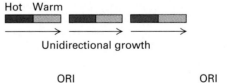

Unidirectional growth

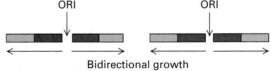

Bidirectional growth

(b) Actual fiber autoradiographic pattern

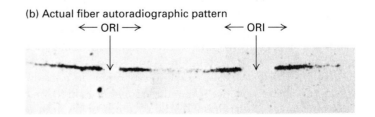

▶ **FIGURE 12-3 Demonstration of bidirectional growth of cellular DNA chains by fiber autoradiography.** If cultured replicating mammalian cells are exposed first to high and then to low concentrations of [³H]thymidine, the resulting DNA will be heavily labeled ("hot") near replication origins (ORI) and lightly labeled ("warm") farther away. When such labeled DNA is dried on a microscope slide as long linear molecules (fibers) and then exposed to a radiation-sensitive emulsion, autoradiographic signals should be produced corresponding to the hot-warm DNA regions. (a) Predicted patterns of autoradiographic bands for uni- and bidirectional DNA synthesis. (b) Actual fiber autoradiograph of DNA from cultured mammalian cells shows autoradiographic signals consistent with bidirectional synthesis. [See J. A. Huberman and A. D. Riggs, 1968, *J. Mol. Biol.* **32**:327; J. A. Huberman and A. Tsai, 1973, *J. Mol. Biol.* **75**:5; photographs courtesy of J. A. Huberman.]

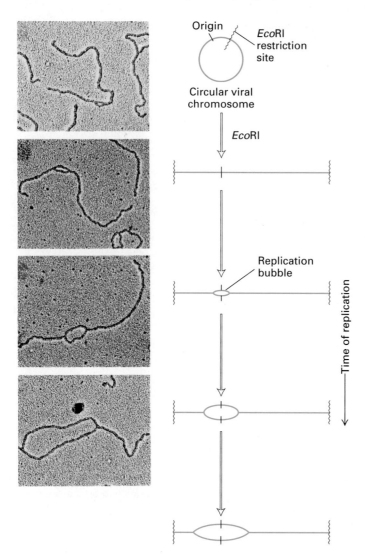

▲ **FIGURE 12-4 Demonstration of bidirectional chain growth from a single origin in viral DNA.** The replicating viral DNA from SV40-infected cells was cut by the restriction enzyme *Eco*RI, which recognizes a single site, and examined by electron microscopy. The electron micrographs and corresponding diagrams show a collection of increasingly longer replication bubbles, the centers of which are a constant distance from each end of the cut molecules, thus indicating that chain growth occurs in two directions from a common origin. [See G. C. Fareed, C. F. Garon, and N. P. Salzman, 1972, *J. Virol.* **10**:484; photographs courtesy of N. P. Salzman.]

bi-directional growth. Most cellular DNA and many viral DNA molecules replicate bidirectionally. Such viruses serve as excellent models for the study of cellular DNA replication.

If DNA from replicating eukaryotic cells is extracted and examined by electron microscopy, so-called replication "bubbles," or "eyes," extending from multiple replication origins are visible. Although such micrographs do not constitute conclusive evidence for unidirectional or bidirectional fork movement, electron-microscope studies of bubbles in

viral DNA have provided evidence for bidirectional replication. If circular viral DNA molecules at different stages of replication are cut with a restriction endonuclease that recognizes a single site, the positions at the center of the replication bubble with respect to the restriction site can be determined (Figure 12-4). The most common result from such analyses is a series of ever larger bubbles whose centers map to the same site, indicating bidirectional replication of both DNA strands from that site. Thus both fiber autoradiography and electron microscopy have indicated that bidirectional DNA replication is the general rule.

Number of Growing Forks and Their Rate of Movement In *E. coli* cells, it takes about 42 minutes to replicate the single circular chromosome, which contains exactly 4,639,221 base pairs and is about 1.4 mm in length. Since the chromosome is duplicated from one origin by two growing forks, we can calculate that the rate of fork movement is about 1000 base pairs per second per fork. The rate of growing fork movement determined from fiber autoradiography of *E. coli* cells labeled for various times agrees with this calculated value, indicating that the fiber-labeling technique can provide a reasonable estimate of the rate of growing fork movement in vivo.

The rate of fork movement in human cells, based on fiber-labeling experiments, is only about 100 base pairs per second per fork. The entire human genome of 3×10^9 base pairs replicates in 8 hours; in this time, one fork theoretically could replicate $\approx 3 \times 10^6$ base pairs, suggesting that the human genome must contain a minimum of 1000 growing forks. However, fiber autoradiography and electron microscopy indicate that growing forks are spaced closer than 3×10^6 base pairs apart. A more likely estimate is that the human genome contains 10,000–100,000 replicons, each of which is actively replicating for only part of the 8 hours required for replication of the entire genome.

DNA Replication Begins at Specific Chromosomal Sites

Perhaps the most important decision every cell has to make is whether, and when, to replicate its DNA. DNA replication, like RNA synthesis and many other biological processes, is controlled at the *initiation* step. Such control would be most efficient if there are specific sites on chromosomes at which DNA replication *always* begins in vivo. As noted already, electron-microscope studies have shown that animal viruses have replication bubbles whose centers are always in the same approximate site (see Figure 12-4). Similar results have been obtained with circular bacterial and plant viruses and for bacterial, yeast, and mammalian plasmids. More detailed molecular studies indicate that replication of these DNAs actually begins at a defined sequence of base pairs near the center of these bubbles, called the **replication origin.**

Genetic and recombinant DNA experiments provide another way to define a replication origin experimentally as a stretch of DNA that is necessary and sufficient for replication of a circular DNA molecule, usually a plasmid or virus, in an appropriate host cell. In yeast, this definition has been refined to include sequences that direct replication once per S phase, the period of the cell cycle in which chromosomal duplication takes place. This important characteristic of DNA replication in eukaryotic cells will be considered in Chapter 13. We discuss three types of replication origins to illustrate some general conclusions about their nature: the *E. coli oriC*, yeast autonomously replicating sequences, and the simian virus 40 (SV40) origin. The detailed knowledge now available about the proteins required to start replication at the *E. coli* origin and the accumulating information about other origins and their use in vitro all suggest that most cellular DNA replication begins at specific sequences, possibly using similar mechanisms.

***E. coli* Replication Origin** The *E. coli* replication origin *oriC* is an ≈240-bp DNA segment present at the start site for replication of *E. coli* chromosomal DNA. Plasmids or any other circular DNA containing *oriC* are capable of independent and controlled replication in *E. coli* cells. Comparison of *oriC* with the origins of five other bacterial species including the distant species *Vibrio harveyi*, a marine bacterium, revealed that all contain repetitive 9-bp and AT-rich 13-bp sequences, referred to as *9-mers* (dnaA boxes) and *13-mers*, respectively (Figure 12-5). As we will see later, these are binding sites for the DnaA protein that initiates replication. In addition, the *E. coli* genome contains a segment of DNA with a relatively high A+T content adjacent to *oriC*; this sequence appears to be important in facilitating local melting of the duplex to reveal the two single-stranded DNA segments onto which the DNA replication machinery is loaded.

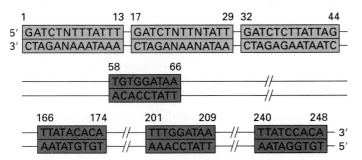

▲ **FIGURE 12-5 Consensus sequence of the minimal bacterial replication origin based on analyses of genomes from six bacterial species.** Similar sequences constitute each origin; the 13-bp repetitive sequences (orange) are rich in adenine and thymine residues. The 9-bp repetitive sequences (brown) exist in both orientations; that is, the lower-right sequence, read right to left, is the same as that of the upper-left sequence, read left to right. These sequences are referred to as 13-mers and 9-mers, respectively. Indicated nucleotide position numbers are arbitrary. [See J. Zyskind et al., 1983, *Proc. Nat'l. Acad. Sci. USA* **80**:1164.]

After *E. coli* DNA replication has initiated, the replication origins in the two daughter DNA duplexes become linked to specific proteins at different points on the plasma membrane. As the cell wall that divides the cell in two forms, this linkage ensures that one of the new DNA duplexes is delivered to each daughter cell (Figure 12-6). There is no visible condensation and decondensation of the DNA in bacterial cells, as there is in eukaryotic cells during mitosis.

Yeast Autonomously Replicating Sequences Each yeast chromosome, like all eukaryotic chromosomes, has multiple origins of replication. Cloning experiments indicate that about 400 origins exist in the 17 chromosomes of *S. cerevisiae*; more than a dozen of these have been characterized in detail. Each yeast origin sequence, called an **autonomously replicating sequence (ARS)**, confers on a plasmid the ability to replicate in yeast and is a required element in yeast artificial chromosomes (see Figure 9-40). Detailed mutational analysis of one ≈180-bp ARS called *ARS1* revealed only one essential element, a 15-bp segment, designated *element A*, stretching from position 114 through 128. Three other short segments—the B$_1$, B$_2$, and B$_3$ elements—increase the efficiency of ARS1 functioning. Comparison of the sequences required for functioning of many different DNA segments that act as ARSs led to recognition of an 11-bp consensus sequence:

(5') A/T-T-T-T-A-T-A/G-T-T-T-A/T (3')
Consensus ARS sequence

Element A in ARS1 is identical at 10 of 11 positions of this consensus sequence, and element B$_2$ at 9 of 11.

DNase footprinting (see Figure 10-6) has shown that a complex of six different proteins called the *ORC (origin-recognition complex)* binds specifically to element A in ARS1 in an ATP-dependent manner. This complex also binds specifically to other ARSs tested. The ORC remains bound to an ARS throughout the cell cycle and during replication becomes associated with other proteins, an event that apparently triggers initiation of DNA synthesis. Yeast cells with mutations in any of these six ORC proteins are defective in DNA replication. All eukaryotic cells express homologs of these proteins, attesting to their importance in initiation of DNA replication. Chapter 13 details how initiation of DNA replication is coupled to specific steps in the cell cycle.

SV40 Replication Origin A 65-bp region in the SV40 chromosome is sufficient to promote DNA replication both in animal cells and in vitro. Three segments of the SV40 origins are required for activity, as demonstrated by testing of origins containing specific mutations. As discussed below, researchers have used mammalian proteins and plasmids carrying the SV40 replication origin to study the molecular mechanisms of mammalian DNA replication.

Three Common Features of Replication Origins Although the specific nucleotide sequences of replication origins from *E. coli*, yeast, and SV40 are very different, they

(a)

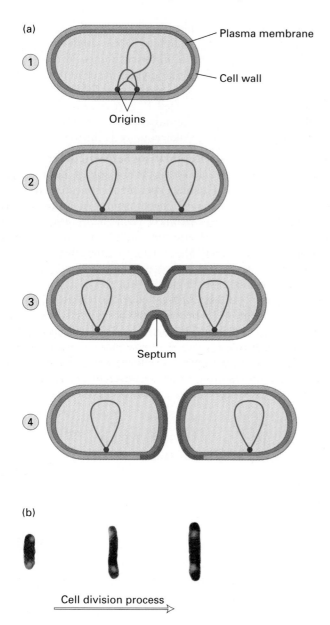

(b)

Cell division process →

◄ **FIGURE 12-6 DNA replication and cell division in a prokaryote.** (a) In a bacterial cell, the partially replicated circular chromosome (blue) is attached to the plasma membrane at the origins of the two daughter DNAs (step ①). The origins of the replicated chromosomes have independent points of attachment to the membrane and thus move farther apart as new membrane and cell wall forms midway along the length of the cell (step ②). Continued formation of more sections of membrane and cell wall gives rise to a septum dividing the cell (step ③), leading to division of the cytoplasm into two daughter cells, each with a chromosome attached to the plasma membrane (step ④). (b) The origin of each DNA molecule can be localized to a particular region of each daughter cell at sequentially later stages in the cell-division process. *B. subtilis* was engineered to express a chimeric protein consisting of a bacterial protein that binds to a segment near the origin of DNA replication fused to GFP, a green-fluorescing protein from the jellyfish *A. victoria* (see Figure 5-7). Shown here are images of growing engineered cells viewed in the fluorescence microscope. The two well-separated green fluorescence spots indicate the location of the daughter origins; these separate from each other as division proceeds. [Part (b) from C. D. Webb et al., 1997, *Cell* **88**:667; courtesy Dr. Richard Losick, Harvard University.]

SUMMARY   General Features of Chromosomal Replication

• The general principles of DNA replication seem to apply, with little modification, to all cells.

• Viewed at the level of whole chromosomes, DNA synthesis is initiated at special regions called *replication origins*. A bacterial chromosome has one origin, whereas each eukaryotic chromosome has many.

• Copying of the DNA duplex at the growing fork is semiconservative: that is, each daughter duplex contains one old strand and its newly made complementary partner (see Figure 12-1).

• Most chromosomes have bidirectional origins. DNA synthesis usually proceeds bidirectionally away from an origin via two growing forks moving in opposite directions; this movement produces a replication bubble (see Figure 12-2c).

• Replication origins typically contain multiple short repeated sequences. These unique DNA segments are recognized by multimeric origin-binding proteins, which in turn assemble other replication enzymes at the origin.

share several properties. First, replication origins are unique DNA segments that contain multiple short repeated sequences. Second, these short repeat units are recognized by multimeric origin-binding proteins. These proteins play a key role in assembling DNA polymerases and other replication enzymes at the sites where replication begins. And third, origin regions usually contain an AT-rich stretch. This property facilitates unwinding of duplex DNA because less energy is required to melt A·T base pairs than G·C base pairs (see Figure 4-9b). Origin-binding proteins control the initiation of DNA replication by directing assembly of the replication machinery to specific sites on the DNA chromosome.

12.2 The DNA Replication Machinery

The cellular mechanisms responsible for DNA replication were uncovered first in bacterial systems; more recently, they have been studied with proteins isolated from yeasts and cultured eukaryotic cells. Because the proteins and reactions involved in *E. coli* DNA replication are now understood in considerable detail, our discussion focuses mainly on these; however, replication in eukaryotes is carried out by analogous proteins and proceeds via similar reactions.

Before plunging into a description of the various enzymes and numerous other factors that participate in DNA replication, we need to recall certain elementary problems in the copying of DNA by DNA polymerases, mentioned in Chapter 4:

- DNA polymerases are unable to melt duplex DNA (i.e., break the interchain hydrogen bonds) in order to separate the two strands that are to be copied.
- All DNA polymerases so far discovered can only elongate a preexisting DNA or RNA strand, the **primer;** they cannot initiate chains.
- The two strands in the DNA duplex are opposite ($5' \rightarrow 3'$ and $3' \rightarrow 5'$) in chemical polarity, but all DNA polymerases catalyze nucleotide addition at the 3'-hydroxyl end of a growing chain, so strands can grow only in the $5' \rightarrow 3'$ direction.

In this section, we describe the cell's solutions to the unwinding, priming, and directionality problems resulting from the structure of DNA and the properties of DNA polymerases.

DnaA Protein Initiates Replication in *E. coli*

Genetic studies first suggested that initiation of replication at *oriC* most likely depended on the protein encoded by a gene designated *dnaA*. Initially, mutant strains carrying temperature-sensitive mutations in *dnaA* were isolated; these cells grew at permissive temperatures (e.g., 30 °C) but not at nonpermissive temperatures (39–42 °C). When *E. coli* cells carrying such conditional lethal mutations had begun DNA replication at the permissive temperature and then were shifted to the higher temperature, they completed the round of DNA synthesis already under way; however, they did not start another round of replication at the nonpermissive temperature. Subsequent genetic studies with recombinant *E. coli* further pinpointed the DnaA protein as a prime candidate for interaction with *oriC*. In vitro studies showed that pure DnaA protein binds to the four 9-mers in *oriC*, forming an *initial complex* that contains 10–20 protein subunits (Figure 12-7). Furthermore, although DnaA can bind to duplex *E. coli* origin DNA in the relaxed-circle form, it can initiate replication only if the DNA is negatively supercoiled. The reason for this specificity is that DNA molecules with negative supercoils are tightly wound and are easier to melt locally (thus providing a single-stranded template region) than

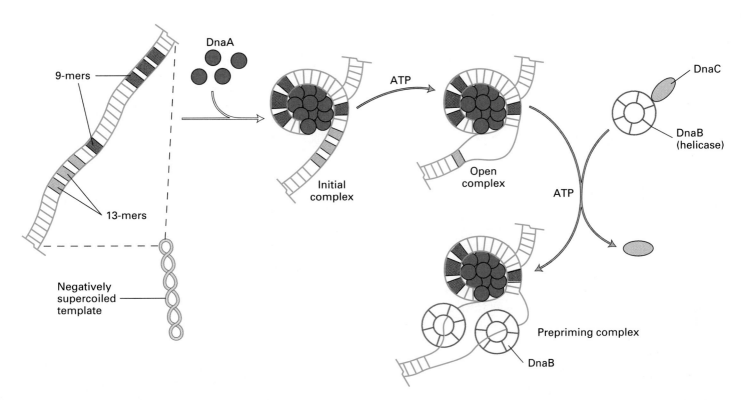

▲ **FIGURE 12-7 Model of initiation of replication at *E. coli* oriC.** The 9-mers and 13-mers are the repetitive sequences shown in Figure 12-5. Multiple copies of DnaA protein bind to the 9-mers at the origin and then "melt" (separate the strands of) the 13-mer segments. The sole function of DnaC is to deliver DnaB, which is composed of six identical subunits, to the template. One DnaB hexamer clamps around each single strand of DNA at *oriC*, forming the prepriming complex. DnaB is a helicase, and the two molecules then proceed to unwind the DNA in opposite directions away from the origin. [Adapted from C. Bramhill and A. Kornberg, 1988, *Cell* **52**:743, and S. West, 1996, *Cell* **86**:177.]

are DNA molecules without supercoils. Supercoiling of DNA and the enzymes that control the degree of DNA supercoiling, called **topoisomerases,** are discussed in detail later.

Binding of DnaA to the *oriC* 9-mers facilitates the initial strand separation, or "melting," of *E. coli* duplex DNA, which occurs at the *oriC* 13-mers. This process requires ATP and yields a so-called *open complex*. When a mixture of *E. coli* DNA and DnaA protein is treated with an endonuclease that specifically recognizes single-stranded DNA, the DNA is cut in the origin region, demonstrating that it is melted.

DnaB Is an *E. coli* Helicase That Melts Duplex DNA

Further melting of the two strands of the *E. coli* chromosome to generate unpaired template strands is mediated by the protein product of the *dnaB* locus, a helicase that is essential for DNA replication. One molecule of DnaB, a hexamer of identical subunits, clamps around each of the two single strands in the open complex formed between DnaA and *oriC*. This binding requires ATP and the protein product encoded by the *dnaC* locus, which "escorts" DnaB to the DnaA proteins, yielding the *prepriming complex* (see Figure 12-7).

Helicases constitute a class of enzymes that can move along a DNA duplex utilizing the energy of ATP hydrolysis to separate the strands. In *E. coli*, the separated strands are inhibited from subsequently reannealing by a *single-strand-binding protein* (SSB protein), which binds to both separated strands. When temperature-sensitive *dnaB* mutants are shifted to nonpermissive temperatures, unwinding ceases; as a result, DNA synthesis stops immediately for lack of single-stranded templates. Helicases like DnaB bind to a single-stranded segment of DNA, then move along that strand melting the hydrogen bonds that link it to its complementary strand. Like many proteins that bind to DNA, helicases exhibit a directionality with respect to the unwinding reaction. DnaB moves along the single strand of DNA to which it binds in the direction of its free 3′ end, and in this sense it is said to unwind DNA in the 5′ → 3′ direction (Figure 12-8). DnaB, like many other proteins that act on DNA, is *processive*. Because it forms a clamp around a single strand of DNA, DnaB does not "fall off" until it reaches the end of that strand or is "unloaded" from DNA by another protein. Other kinds of DNA helicases unwind in the opposite direction, moving along the strand to which they are bound toward the free 5′ end.

E. coli Primase Catalyzes Formation of RNA Primers for DNA Synthesis

As noted earlier, DNA polymerases can only elongate existing primer strands of DNA or RNA. The primers used during DNA replication in both prokaryotes and eukaryotes are short RNA molecules whose synthesis is catalyzed by the RNA polymerase **primase.** *E. coli* strains with temperature-sensitive mutations in *dnaG*, which encodes primase, cannot replicate their DNA at the nonpermissive temperature, thereby establishing the essential role of primase. Primase is usually recruited to a segment of single-stranded DNA by first binding to a DnaB hexamer already attached at that site. The term *primosome* is now generally used to denote a complex between primase and helicase, sometimes with other accessory proteins. In initiation of *E. coli* DNA replication, a primosome is formed by binding of primases to DnaB in the prepriming complex (see Figure 12-7). After the bound primases synthesize short primer RNAs complementary to both strands of duplex DNA, they dissociate from the single-stranded template.

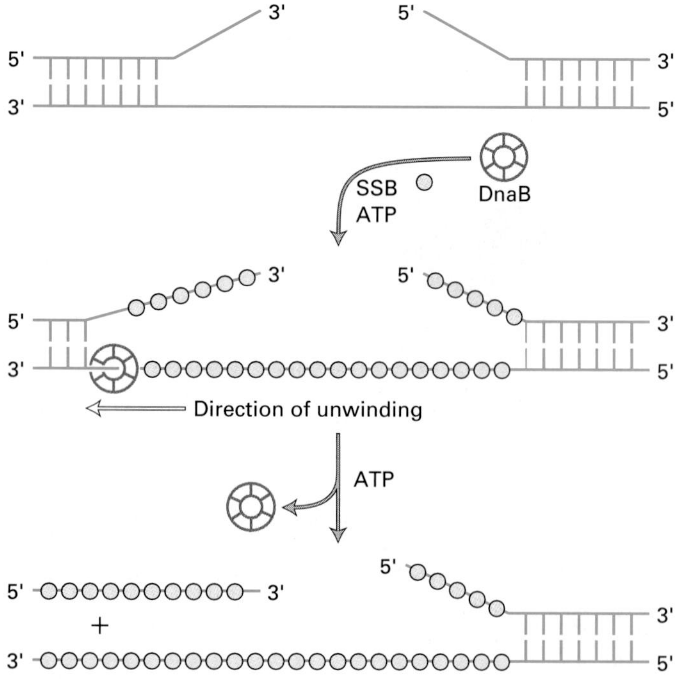

◀ **FIGURE 12-8 Helicase activity of *E. coli* DnaB protein.**
In the presence of ATP and single-strand-binding (SSB) protein, purified DnaB can unwind a gapped DNA duplex in vitro. Unwinding occurs predominantly in the direction of the 3′ end of the strand to which a DnaB molecule is attached. As discussed later, this strand acts as the template for synthesis of the lagging DNA strand. The SSB protein binds to unpaired DNA strands and prevents them from reannealing. For simplicity, binding of DnaB to the complementary strand to the right of the gap and its movement rightward are not depicted. [After A. Kornberg and T. Baker, 1992, *DNA Replication*, 2d ed., W. H. Freeman and Company, and S. Matson, D. Bean, and J. George, 1994, *BioEssays* **16**:13.]

At a Growing Fork One Strand Is Synthesized Discontinuously from Multiple Primers

We have seen how the activities of helicase and primase solve two of the problems inherent to DNA replication—unwinding of the duplex template and the requirement of DNA polymerases for a primer. Remember, though, that both strands of the DNA template are copied as the replication bubble enlarges. Each end of the bubble represents a growing fork where both new strands are synthesized (see Figure 12-2c).

At each growing fork, one strand, called the **leading strand,** is synthesized *continuously* from a single primer on the leading-strand template and grows in the $5' \rightarrow 3'$ direction. Growth of the leading strand proceeds in the same direction as movement of the growing fork (Figure 12-9a). Synthesis of the **lagging strand** is more complicated, because DNA polymerases can add nucleotides only to the 3' end of a primer or growing DNA strand. Movement of the growing fork unveils the template strand for lagging-strand synthesis in the $5' \rightarrow 3'$ direction; thus the *overall* direction of growth of the lagging strand must be from its 3' end toward its 5' end, complementary to the polarity of its template but opposite to the direction of nucleotide addition by DNA polymerases. In both prokaryotes and eukaryotes these apparently incompatible requirements are met by the *discontinuous* copying of the lagging strand from multiple primers, a process involving several steps.

As synthesis of the leading strand progresses, sites uncovered on the single-stranded template of the lagging strand are copied into short RNA primers (<15 nucleotides) by primase (Figure 12-9b). Each of these primers is then elongated by addition of deoxyribonucleotides to its 3' end. In *E. coli*, this reaction is catalyzed by DNA polymerase III (Pol III), one of three DNA polymerases produced by *E. coli*. Thus each lagging strand grows in a direction opposite to that in which the growing fork is moving. The resulting short fragments, containing RNA covalently linked to DNA, are called **Okazaki fragments,** after their discoverer Reiji Okazaki. In bacteria and bacteriophages, Okazaki fragments contain 1000–2000 nucleotides, and a cycle of Okazaki-strand synthesis takes about 2 seconds to complete. In eukaryotic cells, Okazaki fragments are much shorter (100–200 nucleotides).

▶ **FIGURE 12-9 At a growing fork, one strand is synthesized from multiple primers.** (a) The overall structure of a growing fork. Synthesis of the leading strand, catalyzed by DNA polymerase III, occurs by sequential addition of deoxyribonucleotides in the same direction as movement of the growing fork. (b) Steps in the discontinuous synthesis of the lagging strand. This process requires multiple primers, two DNA polymerases, and ligase, which joins the 3'-hydroxyl end of one (Okazaki) fragment to the 5'-phosphate end of the adjacent fragment. (c) DNA ligation. During this reaction, ligase transiently attaches covalently to the 5' phosphate of one DNA strand, thus activating the phosphate group. *E. coli* DNA ligase uses NAD$^+$ as cofactor, generating NMN and AMP, whereas bacteriophage T4 ligase, commonly used in DNA cloning, uses ATP, generating PP$_i$ and AMP.

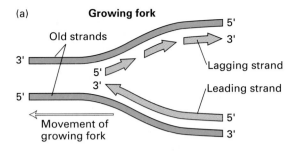

(a) **Growing fork**

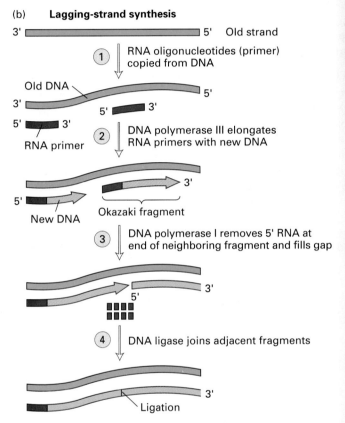

(b) **Lagging-strand synthesis**

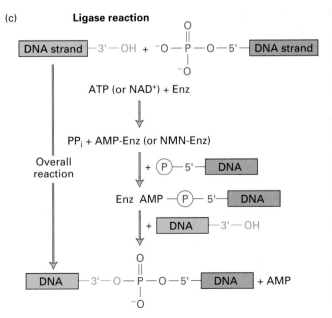

(c) **Ligase reaction**

As each newly formed segment of the lagging strand approaches the 5' end of the adjacent Okazaki fragment (the one just completed), E. coli DNA polymerase I takes over. Unlike polymerase III, polymerase I has 5' → 3' *exonuclease activity*, which removes the RNA primer of the adjacent fragment; the polymerization activity of polymerase I simultaneously fills in the gap between the fragments by addition of deoxyribonucleotides. Finally, another critical enzyme, DNA ligase, joins adjacent completed fragments (Figure 12-9c).

E. coli DNA Polymerase III Catalyzes Nucleotide Addition at the Growing Fork

Three DNA polymerases (I, II, and III) have been purified from E. coli (Table 12-1). In addition to its role in filling the gaps between Okazaki fragments, DNA polymerase I probably is the most important enzyme for gap filling during DNA repair. DNA polymerase II functions in the inducible SOS response discussed later; this polymerase also fills gaps and appears to facilitate DNA synthesis directed by damaged templates. Our discussion here focuses on DNA polymerase III, which catalyzes chain elongation at the growing fork in E. coli.

The DNA polymerase III holoenzyme is a very large (>600 kDa), highly complex protein composed of 10 different polypeptides. The so-called *core polymerase* is composed of three subunits. The α subunit contains the active site for nucleotide addition, and the ϵ subunit is a 3' → 5' exonuclease that removes incorrectly added (mispaired) nucleotides from the end of the growing chain. (This "proofreading" activity of DNA polymerase III is described later.) The function of the Θ subunit is not known.

The central role of the remaining subunits is to convert the core polymerase from a *distributive* enzyme, which falls off the template strand after forming relatively short stretches of DNA containing 10–50 nucleotides, to a *processive* enzyme, which can form stretches of DNA containing up to 5×10^5 nucleotides without being released from the template. This latter activity is necessary for efficient synthesis of both leading and lagging strands. The key to the processive nature of DNA polymerase III is the ability of the β subunit to form a donut-shaped dimer around duplex DNA and then associate with and hold the catalytic core poly-

---

TABLE 12-1 Properties of DNA Polymerases

E. coli	I	II	III		
Polymerization: 5' → 3'	+	+	+		
Exonuclease activity:					
3' → 5'	+	+	+		
5' → 3'	+	−	−		
Synthesis from:					
Intact DNA	−	−	−		
Primed single strands	+	−	−		
Primed single strands plus single-strand-binding protein	+	−	+		
In vitro chain elongation rate (nucleotides per minute)	600	?	30,000		
Molecules present per cell	400	?	10–20		
Mutation lethal?	+	−	+		

Mammalian Cells*	α	$\beta^\dagger$	γ	δ	ϵ
Polymerization: 5' → 3'	+	+	+	+	+
Exonuclease proofreading activity:‡ 3 → 5'	−	−	+	+	+
Synthesis from:					
RNA primer	+	−	−	+	?
DNA primer	+	+	+	+	+
Associated DNA primase	+	−	−	−	−
Sensitive to aphidicolin (inhibitor of cell DNA synthesis)	+	−	−	+	+
Cell location:					
Nuclei	+	+	−	+	+
Mitochondria	−	−	+	−	−

*Yeast DNA polymerase I, II, and III are equivalent to polymerase α, β, and δ, respectively. I and III are essential for cell viability.

†Polymerase β is most active on DNA molecules with gaps of about 20 nucleotides and is thought to play a role in DNA repair.

‡FEN1 is the eukaryotic 5' → 3' exonuclease that removes RNA primers; it is similar in structure and function to the domain of E. coli polymerase I that contains the 5' → 3' exonuclease activity.

(a)

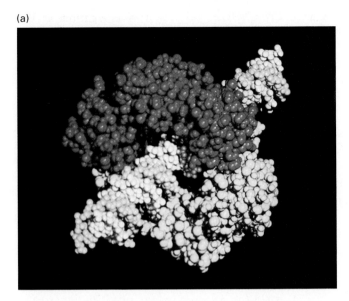

(b)

Nucleotide being added to 3' end

Direction of synthesis

β-subunit clamp

5'

3'

Core of DNA polymerase III

Newly formed DNA strand

5'

3'

Template DNA strand

▲ FIGURE 12-10 A β-subunit dimer tethers the core of *E. coli* DNA polymerase III to DNA, thereby increasing its processivity. (a) Space-filling model based on x-ray crystallographic studies of the dimeric β subunit binding to a DNA duplex. Two β subunits (red and yellow) form a donut-like clamp that remains tightly bound to a closed circular DNA molecule, but readily slides off the ends of a linear DNA molecule. (b) Schematic diagram of proposed association of the core polymerase (green) with the β subunit clamp at the primer-template terminus. This interaction keeps the core from "falling off" the template and positions it near the point of nucleotide addition, increasing the processivity of the core polymerase by more than a thousandfold. [Part (a) from X-P. Kong et al., 1992, *Cell* **69**:425; part (b) adapted from S. Kim et al., 1996, *Cell* **84**:643.]

merase near the 3' terminus of the growing strand (Figure 12-10). Once tightly associated with the DNA, the β-subunit dimer functions like a "clamp," which can slide freely along the DNA, like a ring on a string, as the associated core polymerase moves. In this way, the active sites of the core polymerase remain near the growing fork and the processivity of the enzyme is maximized. Remarkably, of the six remaining subunits, five (γ, δ, δ′, χ, and ψ) form the so-called γ complex, which mediates two essential tasks: (1) loading of the β-subunit clamp onto the duplex DNA–primer substrate in a reaction that requires hydrolysis of ATP and (2) unloading of the β-subunit clamp after a strand of DNA has been completed. Loading and unloading of the β-subunit clamp requires opening the clamp ring, but exactly how the γ complex accomplishes this feat is not known. The final subunit (τ) acts to dimerize two core polymerases and, as summarized in the next section, is essential for coordinating the synthesis of the leading and lagging strands at each growing fork.

The Leading and Lagging Strands Are Synthesized Concurrently

Once the prepriming complex and an RNA primer are formed at the *E. coli* replication origin, chain elongation to yield the leading strand proceeds with little difficulty. As we've seen, however, lagging-strand synthesis proceeds discontinuously from multiple primers. Two molecules of core DNA polymerase III are bound at each growing fork; one adds nucleotides to the leading strand, and the other adds nucleotides to the lagging strand. Coordination between elongation of the leading and lagging strand is essential; otherwise one template strand would be incorporated into a duplex with a newly synthesized complementary strand while large parts of the other template strand would remain single-stranded.

Figure 12-11 shows how this coordination is achieved. The two core-polymerase molecules at the fork are linked together by a τ-subunit dimer. The core polymerase synthesizing the leading strand moves, together with its β-subunit clamp, along its template in the direction of the movement of the fork, elongating the leading strand. It follows closely the movement of the DnaB helicase bound to the lagging-strand template as the helicase melts the duplex DNA at the fork. Since this core-polymerase molecule remains attached to the DNA template, leading-strand synthesis occurs continuously.

The other core-polymerase molecule, which elongates the lagging strand, moves with its β-subunit clamp in the direction opposite to that of the fork movement. As elongation of the lagging strand proceeds, the size of the DNA "loop" between this core polymerase and the fork increases. One way to see this is to imagine the core 2 polymerase fixed in space, linked to core 1; double-stranded DNA newly synthesized by core 2 would be "pushed" into the loop. Eventually the core polymerase synthesizing the lagging strand will complete an Okazaki fragment; it then dissociates from the DNA template, but the τ-subunit dimer continues to tether it to the fork-protein complex. Simultaneously, a primase binds to a site adjacent to the DnaB helicase on the single-stranded segment of the lagging-strand template and initiates synthesis of

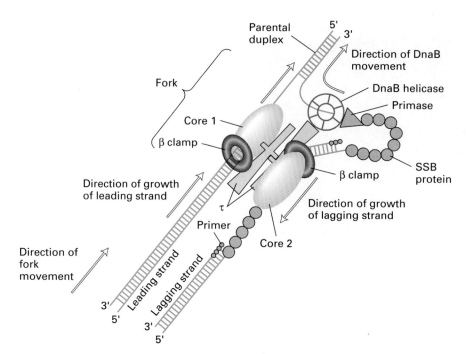

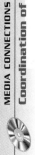

▲ **FIGURE 12-11 Schematic model of the relationship between *E. coli* replication proteins at a growing fork.** (1) A single DnaB helicase moves along the lagging-strand template toward its 3′ end, thereby melting the duplex DNA at the fork. (2) One core polymerase (core 1) quickly adds nucleotides to the 3′ end of the leading strand as its single-stranded template is uncovered by the helicase action of DnaB. This leading-strand polymerase, together with its β-subunit clamp, remains bound to the DNA, synthesizing the leading strand continuously. (3) A second core polymerase (core 2) synthesizes the lagging strand discontinuously as an Okazaki fragment (see Figure 12-9b). The two core polymerase molecules are linked via a dimeric τ protein. (4) As each segment of the single-stranded template for the lagging strand is uncovered, it becomes coated with SSB protein and forms a loop. Once synthesis of an Okazaki fragment is completed, the lagging-strand polymerase dissociates from the DNA but the core remains bound to the τ-subunit dimer. The released core polymerase subsequently rebinds with the assistance of another β clamp in the region of the primer for the next Okazaki fragment. See the text for additional details. [Adapted from A. Kornberg, 1988, *J. Biol. Chem.* **263**:1; S. Kim et al., 1996, *Cell* **84**:643.]

another RNA primer. The resulting DNA-primer complex attracts another β clamp to this segment of the lagging-strand template, followed by re-binding of the core polymerase, which is still attached to the fork complex. This polymerase molecule then proceeds to elongate the RNA primer to form another Okazaki fragment. As mentioned earlier, as each Okazaki fragment nears completion, the RNA primer of the previous fragment is removed by the $5′ \rightarrow 3′$ exonuclease activity of DNA polymerase I. This enzyme also fills in the gaps between the lagging-strand fragments, which then are ligated together by DNA ligase (see Figure 12-9b).

Although the two core polymerase molecules are linked by the τ-subunit dimer, they are oriented in opposite directions (see Figure 12-11). Thus, the 3′ growing ends of both the leading and lagging strands are close together but offset from each other. For this reason, the point in the template at which the lagging strand is being copied is displaced from the point in the template at which leading-strand copying is occurring. Nonetheless, the two core polymerases can add deoxyribonucleotides to the growing strands at the same time and rate, so that leading- and lagging-strand synthesis occurs concurrently.

One τ subunit also contacts the DnaB helicase at the fork. Experiments with purified replication proteins have shown that this interaction increases the normally slow unwinding rate of the helicase (≈ 35 bp/s) over tenfold, thereby enabling the fork to move at rates up to 1000 bp/s. Thus, there is a physical and functional link between the two major replication machines at the fork—the two core polymerases and the primosome complex of primase and DnaB. By closely coordinating all the events depicted in Figures 12-9 and 12-11, the growing fork moves 500–1000 bp/s while both strands are being replicated.

Eukaryotic Replication Machinery Is Generally Similar to That of *E. coli*

As in *E. coli*, researchers investigating DNA replication in eukaryotes initially concentrated on characterizing the different DNA polymerases present in eukaryotic cells (see Table 12-1). This work was followed by development of in vitro systems for copying small chromosomes from animal viruses (e.g., SV40) whose replication is dependent almost entirely on host-cell proteins. As a result of these studies, the SV40 chromosome now can be replicated in vitro using only eight purified components from mammalian cells. The specific functions of these proteins are highly reminiscent of the *E. coli* proteins required for replication of plasmids carrying *oriC*. Thus, the mechanistic problems involved in DNA replication, which are similar in all organisms, have been solved in most cases by use of similar types of proteins. Like DNA replication in *E. coli*, eukaryotic DNA replication occurs bidirectionally from RNA primers made by a primase; synthesis of the leading strand is continuous, while synthesis of the lagging strand is discontinuous. In contrast to the situation in *E. coli*, however, two distinct DNA polymerases, α and either δ or ε, function at the eukaryotic growing fork.

As depicted in Figure 12-12, replication of SV40 DNA is initiated at a unique site, the replication origin, by binding of a virus-encoded protein called *T antigen*, or T_{ag} (step 1). This multifunctional, site-specific DNA-binding protein locally melts duplex DNA through its helicase activity. Opening of the duplex at the SV40 origin also requires ATP and replication protein A (RPA), a host-cell single-strand-binding protein with a function similar to that of SSB protein in *E. coli* cells (step 2). One molecule of polymerase α (Pol α), tightly associated with a primase, then binds to each unwound template strand. The primases form RNA primers, which are elongated for a short stretch by Pol α, forming the first part of the leading strands, which grow from the origins on the two template strands in opposite directions (step 3). The activity of Pol α is stimulated by replication factor C (RFC).

PCNA (proliferating cell nuclear antigen) then binds at the primer-template 3′ termini, displacing Pol α from both leading-strand templates and thus interrupting leading-strand synthesis (Figure 12-12, step 4). Next, Pol δ binds to

▶ **FIGURE 12-12 Model of in vitro replication of SV40 DNA by eukaryotic enzymes.** This figure depicts both growing forks originating from the SV40 origin: the top strand is the leading-strand template for the leftward-moving fork, and the bottom strand is the leading-strand template for the rightward-moving fork. Initiation begins by binding of T antigen (T_{ag}) and other proteins to the origin①, which induces local melting of the duplex DNA. This is followed by binding of a hexameric form of T_{ag} to the leading-strand template at each growing fork②. Functioning as a helicase, each T_{ag} moves along the strand to which it is bound toward the 5′ end, melting the double-stranded DNA as it goes; thus T_{ag} helicase and *E. coli* DnaB move in opposite directions along DNA. Polymerase α (Pol α), which is tightly associated with a primase, forms the 5′ ends of the leading strands ③ and then is displaced from the template ④. Association of polymerase δ (Pol δ) with PCNA increases the processivity of this polymerase, so that it can synthesize the remainder of the leading strands ⑤. Lagging-strand synthesis downstream from the leading-strand primers is thought to be carried out by the combined action of primase and Pol α. RFC stimulates the activity of Pol α. See the text for more details. RPA=replication protein A; RFC=replication factor C; PCNA=proliferating cell nuclear antigen. [See T. Tsurimoto et al., 1990, *Nature* **346**:534.]

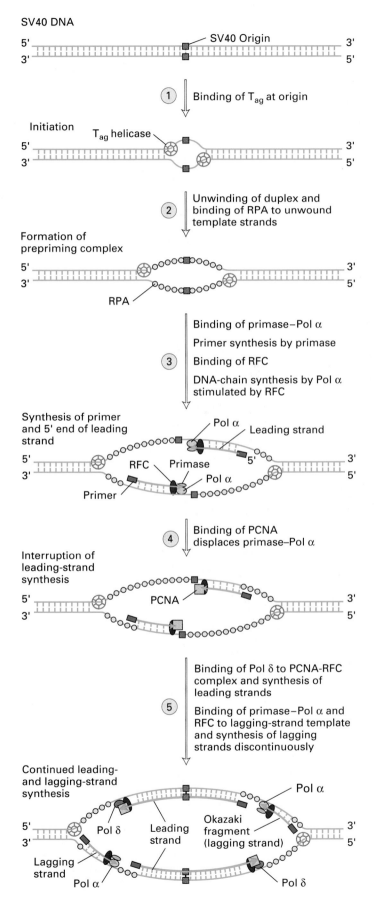

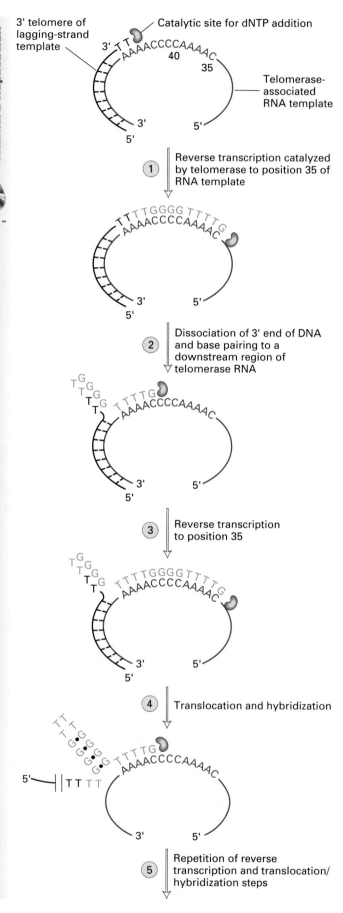

3' telomere of lagging-strand template

Catalytic site for dNTP addition

Telomerase-associated RNA template

1 Reverse transcription catalyzed by telomerase to position 35 of RNA template

2 Dissociation of 3' end of DNA and base pairing to a downstream region of telomerase RNA

3 Reverse transcription to position 35

4 Translocation and hybridization

5 Repetition of reverse transcription and translocation/hybridization steps

PCNA at the 3' ends of the growing strands. The association of Pol δ with PCNA increases the processivity of the polymerase, so that it can continue synthesis of the leading strands without further interruption (step 5). The function of PCNA thus is highly analogous to that of the β-subunit clamp of *E. coli* polymerase III, as both proteins form similar "rings" through which the DNA slides. However, their amino acid sequences are dissimilar, and the β clamp is a dimer, whereas PCNA is a trimer.

As melting of the duplex DNA, catalyzed by a hexameric form of T$_{ag}$, progresses farther away from the origin, the primase–Pol α complex associates with the melted template strands downstream from the leading-strand primers. Synthesis of the lagging strand then is carried out by combined action of primase and Pol α, along with RFC, while leading-strand synthesis on the other side of the origin also proceeds (see Figure 12-12, step 5). Finally, in eukaryotes, as in *E. coli*, topoisomerases play an important role in relieving torsional stress induced by growing-fork movement and in separating the two daughter chromosomes.

Much has been learned about the eukaryotic proteins that can carry out replication of SV40 viral DNA in vitro. As noted above, initiation of SV40 replication in vitro requires the viral protein T antigen. Studies on replication of *eukaryotic cellular* DNA in vitro have been hampered by the lack of eukaryotic experimental systems that can sustain in vitro replication initiated at cellular origins and by the lack of in vitro replication systems prepared from extracts of genetically tractable organisms such as yeast. As discussed in Chapter 13, the recent identification of a protein complex that binds to yeast chromosomal origins may be an important first step toward detailed research on cellular DNA

◀ **FIGURE 12-13 Mechanism of action of telomerase.** This ribonucleoprotein complex elongates the 3' telomeric end of the lagging-strand DNA template by a reiterative reverse transcription mechanism. The action of the telomerase from *Oxytricha*, which adds a T$_4$G$_4$ repeat unit, is depicted; other telomerases add slightly different sequences. The telomerase contains an RNA template (red) that base-pairs to the 3' end of the lagging-strand template. The telomerase catalytic site (green) then adds deoxyribonucleotides (blue) using the RNA molecule as a template; this reverse transcription proceeds to position 35 of the RNA template (step ①). The strands of the resulting DNA-RNA duplex are then thought to slip relative to one another, leading to displacement of a single-stranded region of the telomeric DNA strand and to uncovering of part of the RNA template sequence (step ②). The lagging-strand telomeric sequence is again extended to position 35 by telomerase, and the DNA-RNA duplex undergoes translocation and hybridization as before (steps ③ and ④). The slippage mechanism is thought to be facilitated by the unusual base pairing (black dots) between the displaced G residues, which is less stable than Watson-Crick base pairing. Telomerases can add very long stretches of repeats by repetition of steps ④ and ⑤. [Adapted from D. Shippen-Lentz and E. H. Blackburn, 1990, *Nature* **247**:550.]

replication using a combined genetic and biochemical strategy, which has proved so profitable in *E. coli.*

Telomerase Prevents Progressive Shortening of Lagging Strands during Eukaryotic DNA Replication

Unlike bacterial chromosomes, which are circular, eukaryotic chromosomes are linear and carry specialized ends called **telomeres.** As discussed in Chapter 9, telomeres consist of repetitive oligomeric sequences; for example, the yeast telomeric repeat sequence is $5'$-G_{1-3}T-$3'$. The need for a specialized region at the ends of eukaryotic chromosomes is apparent when we consider that all known DNA polymerases elongate DNA chains from the $3'$ end, and all require an RNA or DNA primer. As the growing fork approaches the end of a linear chromosome, synthesis of the leading strand continues to the end of the DNA template strand; the resulting completely replicated daughter DNA double helix then is released. However, because the lagging-strand template is copied in a discontinuous fashion, it cannot be replicated in its entirety (see Figure 12-9a). When the final RNA primer is removed, there is no upstream strand onto which DNA polymerase can build to fill the resulting gap. Without some special mechanism, the daughter DNA strand resulting from lagging-strand synthesis would be shortened at each cell division.

The enzyme that prevents this progressive shortening of the lagging strand is a modified reverse transcriptase called *telomerase,* which can elongate the lagging-strand *template* from its $3'$-hydroxyl end. This unusual enzyme contains a catalytic site that polymerizes deoxyribonucleotides directed by an RNA template, and the RNA template itself, which is brought to the site of catalysis as part of the enzyme (Figure 12-13). The repetitive sequence added by telomerase is determined by the RNA associated with the enzyme, which varies among telomerases from different sources. Once the $3'$ end of the lagging-strand template is sufficiently elongated, synthesis of the lagging strand can take place, presumably from additional primers.

For cells in many organisms, telomere length is increased many times over early in development. Most human somatic cells replicate in the absence of telomerase activity and thus gradually consume the telomeric repeats added earlier in development. The progressive shortening of the chromosome ends and eventual loss of genetic information that results has been linked to cell death, and it has even been suggested that life span is determined by the number of telomeres with which an individual starts. Indeed, an inverse relationship between age and telomeric length has been observed.

SUMMARY The DNA Replication Machinery

- The enzymes and other protein factors that carry out DNA replication in *E. coli* and in eukaryotic cells are analogous, suggesting that the biochemical mechanism of DNA replication is similar in all cells.

- The enzymatic events at the growing fork are a consequence of two properties of the DNA double helix and two of DNA polymerases. The DNA helix contains antiparallel strands (i.e., the $5' \rightarrow 3'$ direction of one strand is opposite to the $5' \rightarrow 3'$ direction of the other), and the two strands are interwound, so they cannot simply be melted along their entire lengths all at once. DNA polymerases require a nucleic acid primer—either a DNA or an RNA molecule—to begin synthesis, and all DNA chain growth occurs by nucleotide addition at the $3'$ end.

- In all cells, one new DNA strand, the leading strand, is synthesized continuously in the direction of movement of the growing fork by elongation from the $3'$ end of an RNA primer base-paired to a template strand. Synthesis of the other strand, the lagging strand, occurs in the direction opposite to the overall direction of growing fork movement from a series of short RNA primers formed by primase at multiple sites on the second template strand. The resulting segments of RNA plus DNA are called *Okazaki fragments.* After the primers are removed and the gaps between fragments are filled, they are joined.

- Initiation of DNA replication in *E. coli* occurs by binding of DnaA to *oriC,* followed by attachment of DnaB, a helicase that melts DNA at the fork (see Figure 12-7). Association of primase with this complex forms a primosome. After primer synthesis, primase dissociates.

- *E. coli* DNA polymerase III catalyzes nucleotide addition to both the leading and the lagging strands. DNA polymerase I removes the RNA primers from Okazaki fragments and fills in the gaps on the lagging strand. Finally, DNA ligase joins adjacent completed Okazaki fragments (see Figure 12-9).

- Eukaryotic proteins that replicate SV40 DNA in vitro exhibit similarities with *E. coli* replication proteins. A viral protein called *T antigen* functions similarly to the *E. coli* DnaB helicase, and host-cell PCNA (proliferating cell nuclear antigen) is similar to the β-subunit clamp associated with *E. coli* DNA polymerase III. However, two distinct mammalian polymerases, α and δ or ϵ, function at the eukaryotic growing fork (see Figure 12-12).

- The processivity of DNA polymerases is essential for efficient polymerization and is facilitated by their association with the β-subunit clamp in *E. coli* and PCNA in eukaryotes.

- Telomerase, a reverse transcriptase that contains an RNA template, adds nucleotides to the $3'$ end of the lagging-strand template and thus prevents shortening of lagging strands during replication of linear DNA molecules such as those of eukaryotic chromosomes.

12.3 The Role of Topoisomerases in DNA Replication

DNA molecules can coil and bend in space, leading to changes in topology, including formation of negative or positive supercoils. For example, as discussed in Chapter 4, local unwinding of a DNA duplex whose ends are fixed causes stress that is relieved by supercoiling. The enzymes that control the topology of DNA function at several different steps in replication in both prokaryotic and eukaryotic cells. In this section we describe the two different classes of topoisomerases and their role in DNA replication.

Type I Topoisomerases Relax DNA by Nicking and Then Closing One Strand of Duplex DNA

The first topoisomerase to be discovered, *E. coli* topoisomerase I, can remove negative supercoils without leaving nicks in the DNA molecule (Figure 12-14). After the enzyme binds to a DNA molecule, it cuts one strand, simultaneously generating a covalent phosphoester bond between the released 5′ phosphate on the DNA and a tyrosine residue in the enzyme. Formation of this phosphotyrosine bond does not

require ATP or another source of energy. The free 3′-hydroxyl end of the DNA is held noncovalently by the enzyme. The DNA strand that has not been cleaved is then passed through the single-stranded break. The cleaved strand is then resealed, forming a structure with the same *chemical* bonds as the starting DNA, but with one less negative supercoil. By this mechanism, the enzyme removes one negative supercoil at a time.

Any enzyme that cleaves only one strand of a DNA duplex and then reseals it is classified as a *type I topoisomerase* (Topo I). The Topo I from *E. coli* acts on negative, but not positive, supercoils. In contrast, Topo I enzymes from eukaryotic cells can remove both positive and negative supercoils. Because the relaxation (removal) of DNA supercoils by Topo I is energetically favorable, the reaction proceeds without an energy requirement. The sequential action of *E. coli* Topo I can remove essentially all the supercoils in a DNA molecule (Figure 12-15).

Studies with gene-targeted knockout strains of *E. coli* have established that Topo I is essential for viability. The enzyme is thought to help maintain the proper superhelical density of the *E. coli* chromosome by removing negative supercoils formed by action of type II topoisomerase (DNA

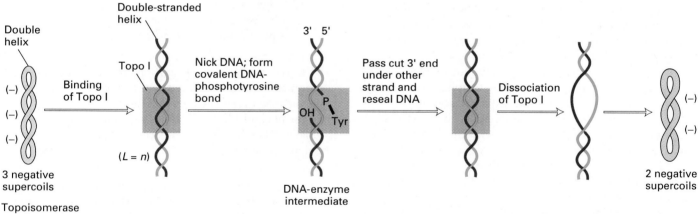

▲ **FIGURE 12-14 Action of *E. coli* type I topoisomerase (Topo I).** The DNA-enzyme intermediate contains a covalent bond between the 5′-phosphoryl end of the nicked DNA and a tyrosine residue in the protein (inset). After the free 3′-hydroxyl end of the red cut strand passes under the uncut strand, it attacks the DNA-enzyme phosphoester bond, rejoining the DNA strand. During each round of nicking and resealing catalyzed by *E. coli* Topo I, one negative supercoil is removed. (The assignment of sign to supercoils is by convention with the helix stood on its end; in a negative supercoil the "front" strand falls from right to left as it passes over the back strand (as here); in a positive supercoil, the front strand falls from left to right.)

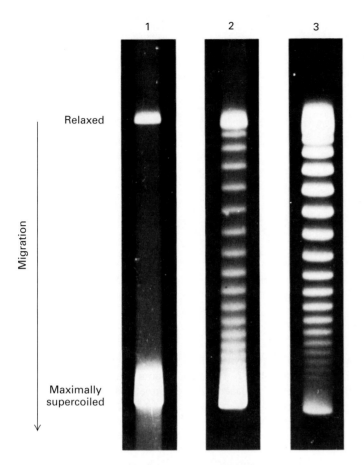

▲ FIGURE 12-15 Separation of SV40 DNA topoisomers containing different numbers of supercoils by gel electrophoresis. DNA was extracted from SV40 virions under conditions that ensure the maximal number of supercoils (lane 1). Aliquots were treated with *E. coli* type I topoisomerase for 3 min (lane 2) or 30 min (lane 3). About 25 bands, equal to the number of possible topoisomers, are visible in the electrophoretograms after Topo I treatment, including the fully relaxed form, which has no supercoils. Because DNAs with few supercoils tend to assume an extended, rodlike conformation, they migrate more slowly in gels than do the more compact molecules that have extensive supercoils. [From W. Keller and I. Wendel, 1974, *Cold Spring Harbor Symp. Quant. Biol.* **39**:199; courtesy of W. Keller.]

gyrase), which is discussed below. Since *E. coli* Topo I cannot remove positive supercoils, it is unlikely to play a role in growing fork progression, which generates positive supercoils. In yeast, either Topo I or Topo II, both of which can relax positive and negative supercoils, functions in the movement of growing forks. Mutations in the gene encoding Topo I affect the growth rate of yeast cells but are not lethal; thus, this enzyme is not essential for viability. However, yeast Topo II, which separates the two DNA duplexes following replication, is essential for viability. In both yeast and *E. coli*, Topo I enzymes may be important in regulating aspects of transcription, as well as in repairing damaged DNA.

Type II Topoisomerases Change DNA Topology by Breaking and Rejoining Double-Stranded DNA

The first *type II topoisomerase* (Topo II) to be described was isolated from *E. coli.* and named *DNA gyrase*. Topo II enzymes have the ability to cut both strands of a double-stranded DNA molecule, pass another portion of the duplex through the cut, and reseal the cut in a process that utilizes ATP (Figure 12-16a). Depending on the DNA substrate, these maneuvers will have the effect of changing a positive supercoil into a negative supercoil or of increasing the number of negative supercoils by 2. The Topo II enzymes from mammalian cells cannot, like *E. coli* DNA gyrase, increase the superhelical density at the expense of ATP; presumably no such activity is required in eukaryotes, since binding of histones increases the potential superhelicity. All type II topoisomerases catalyze *catenation* and *decatenation*, that is, the linking and unlinking, of two different DNA duplexes (Figure 12-16b).

DNA gyrase is composed of two identical subunits. Hydrolysis of ATP by gyrase's inherent ATPase activity powers the conformational changes that are critical to the enzyme's operation (Figure 12-17). The enzyme functions to introduce negative supercoils at or near the *oriC* site in the DNA template; as noted earlier, DnaA can initiate replication only

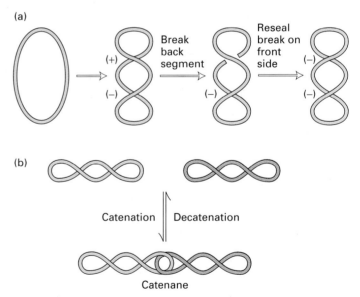

▲ FIGURE 12-16 Action of *E. coli* DNA gyrase, a type II topoisomerase. (a) Introduction of negative supercoils. The initial folding introduces no stable change, but the subsequent activity of gyrase produces a stable structure with two negative supercoils. Eukaryotic Topo II enzymes cannot introduce supercoils but can remove negative supercoils from DNA. (b) Catenation and decatenation of two different DNA duplexes. Both prokaryotic and eukaryotic Topo II enzymes can catalyze this reaction. [See N. R. Cozzarelli, 1980, *Science* **207**:953.]

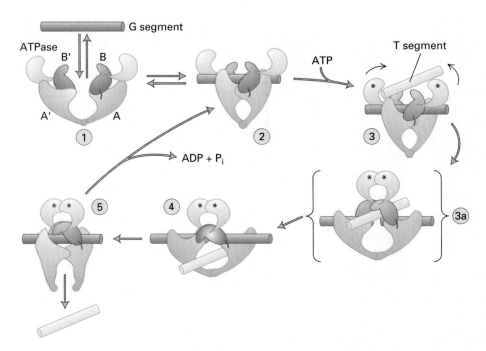

▲ **FIGURE 12-17 Molecular model for the catalytic activity of *E. coli* topoisomerase II (DNA gyrase).** The enzyme is a dimer of two identical subunits. Initially, the enzyme binds one part of a DNA strand, the G segment (dark blue), inducing a conformational change in the B, B', A, and A' domains of the enzyme ②. After binding of ATP (indicated by the asterisks) and another part of the DNA strand, the T segment (light blue), a series of reactions occur in which the G segment is cut by the A and A' domains (light orange) of the enzyme and the ends of the G DNA become covalently linked to tyrosine residues in these domains ③ and ③ₐ. Simultaneously, the ATP-binding domains (green) move toward each other, transporting the T segment through the break and into the central hole ④. The cut G segment is then resealed, and the T segment is released by a conformational change that separates the A and A' domains at the bottom of the enzyme ⑤. The interface between the A and A' domains then re-forms, a reaction that requires ATP hydrolysis and regenerates the starting state ②. At this point, the G segment can dissociate from the enzyme by conversion of ② into ①. Alternatively, the enzyme can proceed through another cycle, again passing the T segment through the G segment and thus removing two more supercoils. [From J. Berger et al., 1996, *Nature* **379**:225.]

on a negatively supercoiled template. Measurements of the degree of DNA supercoiling in *E. coli* suggest that there is one negative supercoil for each 15–20 turns of the DNA helix. A second crucial function of gyrase is to remove the positive supercoils that form ahead of the growing fork during elongation of the growing strands (Figure 12-18).

Replicated Circular DNA Molecules Are Separated by Type II Topoisomerases

During DNA replication the parental strands remain intact and retain their superhelicity. This poses steric and topological constraints to the completion of replication of a circular DNA molecule as the two growing forks approach

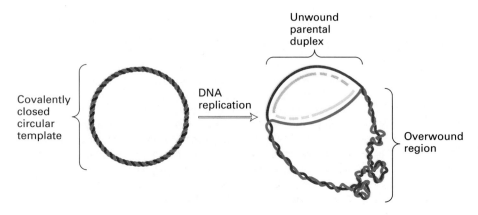

◀ **FIGURE 12-18 Movement of the growing fork during DNA replication induces formation of positive supercoils in the duplex DNA ahead of the fork.** In order for extensive DNA synthesis to proceed, the positive supercoils must be removed (relaxed). This can be accomplished by *E. coli* DNA gyrase and by eukaryotic type I and type II topoisomerases. [Adapted from A. Kornberg and T. Baker, 1992, *DNA Replication*, 2d ed., W. H. Freeman and Company, p. 380.]

each other. (A similar situation arises wherever two growing forks in a linear eukaryotic chromosome approach each other.) Figure 12-19 illustrates how the last few helical turns in the parental DNA could be removed by changing the topology of the already replicated regions, leaving the two nearly complete daughter helices linked together as *catenanes,* covalently linked but not yet completely finished circles. Replication then could be completed before or after decatenation to yield two separated complete daughter helices.

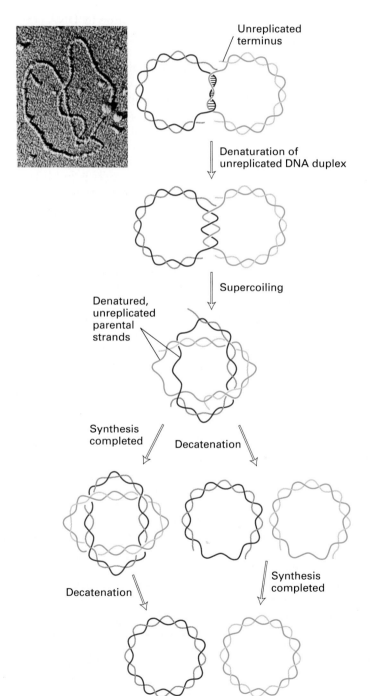

Unreplicated terminus

Denaturation of unreplicated DNA duplex

Supercoiling

Denatured, unreplicated parental strands

Synthesis completed

Decatenation

Decatenation

Synthesis completed

In *E. coli,* decatenation is catalyzed by DNA gyrase and a second type II enzyme, called *topoisomerase IV,* which genetic studies suggest is responsible for separating newly replicated molecules in vivo. Furthermore, temperature-sensitive Topo IV mutants carrying plasmids accumulate interlocked plasmids—that is, catenanes of plasmid DNA, which appear similar to the SV40 catenane shown in Figure 12-19 *(inset).* To separate DNA catenanes, Topo IV presumably binds to the interlocked duplexes and makes a double-stranded break in one molecule; while remaining attached to the substrate, it then passes the other molecule through the break and finally reseals the break in the cut molecule (see Figure12-17). Interestingly, although DNA gyrase can carry out decatenation in vitro, it cannot fully substitute for Topo IV in vivo, as demonstrated by the lethal effects of Topo IV mutations.

Linear Daughter Chromatids Also Are Separated by Type II Topoisomerases

The use of temperature-sensitive mutations in Topo II has shown the importance of this protein in yeast-cell viability; furthermore, light-microscopic studies have revealed the participation of Topo II in separating the linear chromosomes of yeast cells. Although individual yeast chromosomes are too small to be visualized by light microscopy, a structure called the *nuclear body,* which contains all the chromosomes clumped together, can be seen. When temperature-sensitive Topo II mutants are shifted to a nonpermissive temperature, the nuclear body, which usually divides at the junction of the mother and daughter cell, appears to get stuck in the passageway between the two cells.

Fluorescent antibody staining of metaphase chromosomes reveals that topoisomerase II is a principal component of the nonhistone protein scaffolding to which long DNA loops are attached. This finding and the genetic studies in yeast strongly suggest that eukaryotic Topo II resolves tangles that exist in newly replicated linear chromosomes. And finally, the similarity between the phenotypes of Topo II mutants in yeast and Topo IV mutants in bacteria argues that control of topological domains may be analogous in eukaryotic chromosomes and in small circular DNA molecules.

◀ **FIGURE 12-19 Completion of replication of circular DNA molecules.** Denaturation of the unreplicated terminus followed by supercoiling overcomes the steric and topological constraints of copying the terminus. At least with SV40 DNA, the final two steps (synthesis and decatenation) can occur in either order depending on experimental conditions. Parental strands are in dark colors; daughter strands in light colors. *(Inset)* Electron micrograph of two fully replicated SV40 DNA molecules interlocked twice. This structure would result if synthesis was completed before decatenation. Topo II can catalyze decatenation of such interlocked circles in vitro. [Drawing adapted from S. Wasserman and N. Cozzarelli, 1986, *Science* **232**:951. Micrograph from O. Sundin and A. Varshavsky, 1981, *Cell* **25**:659; courtesy of A. Varshavsky.]

SUMMARY The Role of Topoisomerases in DNA Replication

• Among the proteins involved in DNA replication are several that change the topology of DNA: helicases, which can unwind the DNA duplex, thereby inducing formation of supercoils, and topoisomerases, which catalyze addition or removal of supercoils.

• Type I topoisomerases relax DNA (i.e., remove supercoils) by nicking and closing one strand of duplex DNA (see Figure 12-14).

• Type II topoisomerases change DNA topology by breaking and rejoining double-stranded DNA. These enzymes can introduce or remove supercoils and can separate two DNA duplexes that are intertwined (see Figure 12-16).

• Topoisomerases are important both in growing fork movement and in resolving (untangling) finished chromosomes after DNA duplication. Both replicated circular and linear DNA chromosomes are separated by type II topoisomerases.

12.4 DNA Damage and Repair and Their Role in Carcinogenesis

The DNA sequence can be changed as the result of copying errors introduced by DNA polymerases during replication and by environmental agents such as mutagenic chemicals and certain types of radiation. If DNA sequence changes, whatever their cause, are left uncorrected, both growing and nongrowing somatic cells might accumulate so many mutations that they could no longer function. In addition, the DNA in germ cells might incur too many mutations for viable offspring to be formed. Thus the correction of DNA sequence errors in all types of cells is important for survival.

The relevance of DNA damage and repair to the generation of cancer (carcinogenesis) became evident when it was recognized that all agents that cause cancer (**carcinogens**) also cause a change in the DNA sequence and thus are **mutagens.** All the effects of carcinogenic chemicals on tumor production can be accounted for by the DNA damage that they cause and by the errors introduced into DNA during the cells' efforts to repair this damage. Likewise, ultraviolet (UV) radiation and ionizing radiation (x-rays and atomic particles) not only modify DNA, but also can cause cancer in animals and can transform normal cells in culture into rapidly proliferating, cancer-type cells. The ability of ionizing radiation to cause human cancer, especially leukemia, was dramatically shown by the increased rates of leukemia among survivors of the atomic bombs dropped in World War II, and more recently by the increase in melanoma (skin cancer) in individuals exposed to too much sunlight.

Table 12-2 lists the general types of DNA damage and their causes. In this section, we first describe the very efficient mechanism whereby cells correct copying errors and then examine how various chemical carcinogens cause damage to DNA. Finally, we consider several cellular mechanisms for repairing chemical- or radiation-damaged DNA. The role of these DNA-repair systems in carcinogenesis is introduced here and discussed in more detail in Chapter 24.

Proofreading by DNA Polymerase Corrects Copying Errors

Because the specificity of nucleotide addition by DNA polymerases is determined by Watson-Crick base pairing, a wrong base (e.g., A instead of G) occasionally is inserted during DNA synthesis. Indeed, the α subunit of E. coli DNA polymerase III introduces about 1 incorrect base in 10^4 internucleotide linkages during replication in vitro. Since an average E. coli gene is about 10^3 bases long, an error frequency of 1 in 10^4 base pairs would cause a potentially harmful mutation in every tenth gene during each replication, or 10^{-1} mutations per gene per generation. However, the measured mutation rate in bacterial cells is much less, about 1 mistake in 10^9 nucleotide polymerization events or, equivalently, 10^{-5} to 10^{-6} mutations per gene per generation (assuming ≈ 1000 base pairs per gene).

This increased accuracy in vivo is largely due to the *proofreading function* of E. coli DNA polymerases. Figure 12-20 depicts an experiment demonstrating that the $3' \rightarrow 5'$ exonuclease activity of E. coli DNA polymerase I can remove a mismatched base at the $3'$ growing end of a synthetic primer-template complex. In DNA polymerase III, this function resides in the ϵ subunit of the core polymerase. When an incorrect base is incorporated during DNA synthesis, the polymerase pauses, then transfers the $3'$ end of the growing chain to the exonuclease site where the mispaired base is removed. Then the $3'$ end is transferred back to the polymerase site, where this region is copied correctly (Figure 12-21). Proofreading is a property of almost all bacterial DNA polymerases. Both the δ and ϵ DNA polymerases of animal cells, but not the α polymerase, also have proofreading activity. It seems likely that this function is indispensable for all cells to avoid excessive genetic damage.

Genetic studies in E. coli have shown that proofreading does, indeed, play a critical role in maintaining sequence fidelity during replication. Mutations in the gene encoding the ϵ subunit of DNA polymerase III inactivate the proofreading function and lead to a thousandfold increase in the rate of spontaneous mutations. E. coli possesses an additional mechanism for checking the fidelity of DNA replication by identifying mispaired bases in newly replicated DNA. This mismatch-repair machinery, discussed later, determines which strand is to be repaired by distinguishing the newly replicated strand (the one in which an error occurred during replication) from the template strand.

TABLE 12-2   **DNA Lesions That Require Repair**

DNA Lesion	Example/Cause
Missing base	Removal of purines by acid and heat (under physiological conditions $\approx 10^4$ purines/day/cell in a mammalian genome); removal of altered bases (e.g., uracil) by DNA glycosylases
Altered base	Ionizing radiation; alkylating agents (e.g., ethylmethane sulfonate)
Incorrect base	Mutations affecting $3' \rightarrow 5'$ exonuclease proofreading of incorrectly incorporated bases
Bulge due to deletion or insertion of a nucleotide	Intercalating agents (e.g., acridines) that cause addition or loss of a nucleotide during recombination or replication
Linked pyrimidines	Cyclotubyl dimers (usually thymine dimers) resulting from UV irradiation
Single- or double-strand breaks	Breakage of phosphodiester bonds by ionizing radiation or chemical agents (e.g., bleomycin)
Cross-linked strands	Covalent linkage of two strands by bifunctional alkylating agents (e.g., mitomycin C)
3'-deoxyribose fragments	Disruption of deoxyribose structure by free radicals leading to strand breaks

SOURCE: Adapted from A. Kornberg and T. Baker, 1992, *DNA Replication*, 2d ed., W. H. Freeman and Company, pp. 771–773.

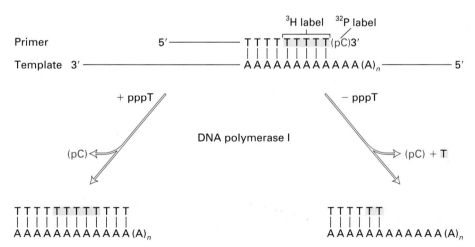

▲ **FIGURE 12-20 Experimental demonstration of the proofreading function of *E. coli* DNA polymerase I.** An artificial template [poly(dA)] and a corresponding primer end-labeled with [³H]thymidine residues were constructed. An "incorrect" cytidine labeled with ³²P was then added to the 3' end of the primer. The template-primer complex was incubated with purified DNA polymerase I. In the presence of thymidine triphosphate (pppT), there was a rapid loss of the [³²P]cytidine and retention of all the [³H]thymidine radioactivity. This indicated that the enzyme removed only the terminal incorrect C and then proceeded to add more T residues complementary to the template. In the absence of pppT, however, both [³H]thymidine and [³²P]cytidine were lost, indicating that if the enzyme lacks pppT to polymerize, its $3' \rightarrow 5'$ exonuclease activity will proceed to remove "correct" bases. [See A. Kornberg and T. A. Baker, 1992, *DNA Replication*, 2d ed., W. H. Freeman and Company.]

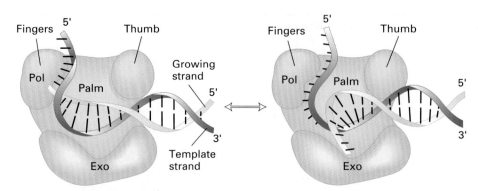

▲ FIGURE 12-21 Schematic model of the proofreading function of DNA polymerases. All DNA polymerases have a similar three-dimensional structure, which resembles a half-opened right hand. The "fingers" bind the single-stranded segment of the template, and the polymerase catalytic activity (Pol) lies in the junction between the fingers and palm. So long as the correct nucleotides are added to the 3' end of the growing strand, the 3' end remains in the polymerase site.

Incorporation of an incorrect base at the 3' end causes a melting of the end of the duplex. As a result, the polymerase pauses and the 3' end of the growing strand is transferred to the 3' → 5' exonuclease site (Exo) about 3 nm away, where the mispaired base and probably other bases are removed. Subsequently, the 3' end flips back into the polymerase site and elongation resumes. [Adapted from C. M. Joyce and T. T. Steitz 1995, *J. Bacteriol.* **177**:6321; S. Bell and T. Baker, 1998, *Cell* **92**:295.]

Chemical Carcinogens React with DNA Directly or after Activation

Chemicals, which are thought to be the cause of many human cancers, were originally associated with cancer through experimental studies in animals. The classic experiment is to repeatedly paint a test substance on the back of a mouse and look for development of both local and systemic tumors in the animal. Although the many substances identified as chemical carcinogens have a very broad range of structures with no obvious unifying features, they can be classified into two broad categories: direct-acting and indirect-acting (Figure 12-22).

Direct-acting carcinogens, of which there are only a few, are reactive electrophiles (compounds that seek out and react with negatively charged centers in other compounds). By chemically reacting with nitrogen and oxygen atoms in DNA, these compounds modify certain nucleotides so as to distort the normal pattern of base pairing. If these modified nucleotides were not repaired, they would allow an incorrect nucleotide to be incorporated during replication. Figure 8-6 shows how one chemical carcinogen, ethyl methanesulfonate (EMS), causes mutations.

Indirect-acting carcinogens generally are unreactive, water-insoluble compounds. They can act as potent cancer inducers only after conversion to *ultimate carcinogens* by introduction of electrophilic centers. Such metabolic activation of carcinogens is carried out by enzymes that are normal body constituents. In animals, activation of indirect-acting carcinogens often is carried out by liver enzymes that normally function to detoxify noxious chemicals (e.g., therapeutic drugs, insecticides, polycyclic hydrocarbons, and some natural products). Many of these compounds are so fat-soluble that they would accumulate continually in fat cells and lipid membranes and not be excreted from the body. The detox-

ification system works by converting such compounds to water-soluble derivatives, which the body can excrete.

Detoxification begins with a powerful series of oxidation reactions catalyzed by a set of proteins called *cytochrome P-450.* These enzymes, which are bound to endoplasmic reticulum membranes, can oxidize even highly unreactive compounds such as polycyclic aromatic hydrocarbons. Oxidation of polycyclic aromatics produces an epoxide, a very reactive electrophilic group:

$$\underset{O}{CH - CH}$$

Usually these epoxides are rapidly hydrolyzed into hydroxyl groups, which are then coupled to glucuronic acid or other groups, producing compounds soluble enough in water to be excreted. Some intermediate epoxides, however, are only slowly hydrolyzed to hydroxyl groups, probably because the relevant enzyme (epoxide hydratase) cannot get to the epoxide to act on it. For example, the indirect-acting carcinogen benzo(a)pyrene, shown in Figure 12-22, undergoes two epoxidation reactions to yield a highly reactive electrophilic ultimate carcinogen:

7,8-Diol-9,10-oxide
(ultimate carcinogen)

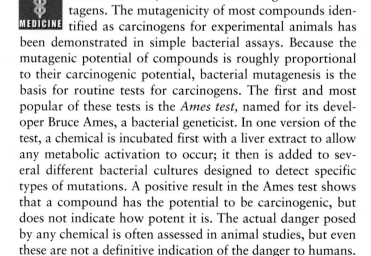

DIRECT-ACTING CARCINOGENS

β-Propiolactone

Ethylmethane sulfonate (EMS)

Dimethyl sulfate (DMS)

Nitrogen mustard

Methyl nitrosourea (MNU)

INDIRECT-ACTING CARCINOGENS

Benzo(*a*)pyrene
(3,4-benzpyrene)

Dibenz(*a,h*)anthracene

2-Naphthylamine

Dimethylnitrosamine

Vinyl chloride

2-Acetylaminofluorene

Safrole
(sassafras)

Aflatoxin B₁
(*Aspergillus flavus*)

◄ **FIGURE 12-22 Structures of some chemical carcinogens.** Direct-acting carcinogens are highly electrophilic compounds that can react with DNA. Indirect-acting carcinogens must be metabolized before they can react with DNA. All these chemicals act as mutagens.

Other types of indirect-acting carcinogens are activated by different oxidative pathways, which also involve P-450 enzymes.

The Carcinogenic Effect of Chemicals Correlates with Their Mutagenicity

MEDICINE As noted earlier, all chemical carcinogens act as mutagens. The mutagenicity of most compounds identified as carcinogens for experimental animals has been demonstrated in simple bacterial assays. Because the mutagenic potential of compounds is roughly proportional to their carcinogenic potential, bacterial mutagenesis is the basis for routine tests for carcinogens. The first and most popular of these tests is the *Ames test,* named for its developer Bruce Ames, a bacterial geneticist. In one version of the test, a chemical is incubated first with a liver extract to allow any metabolic activation to occur; it then is added to several different bacterial cultures designed to detect specific types of mutations. A positive result in the Ames test shows that a compound has the potential to be carcinogenic, but does not indicate how potent it is. The actual danger posed by any chemical is often assessed in animal studies, but even these are not a definitive indication of the danger to humans.

The strongest evidence that carcinogens act as mutagens comes from the observation that cellular DNA altered by exposure of cells to carcinogens can change cultured cells into fast-growing cancer-type cells. This very important result

was first obtained by extracting DNA from human cells exposed to a carcinogen or from human colon tumors. The extracted DNA then was applied to normal mouse 3T3 fibroblast cells, which grow as a monolayer on plastic culture dishes and stop dividing when they contact other cells. A small fraction of the treated 3T3 cells took up the DNA and began rapidly proliferating and growing on top of one another to form a pile of cells on the culture dish (see Figure 24-4). Analysis of the DNA from such *oncogenically transformed* 3T3 cells showed that it had incorporated a segment of human DNA that had undergone a mutation in a normal cellular gene, called a **proto-oncogene,** involved in the control of cell growth or division. Expression of the mutated human gene caused abnormal proliferation of the 3T3 cells. Mutated forms of proto-oncogenes that cause abnormal cell proliferation are called **oncogenes;** these can be passed on to other cells and also are carried by certain cancer-causing retroviruses. As discussed in Chapter 24, this type of study has identified a mutation in the *ras* proto-oncogene as a major contributor to human colon carcinoma. The role of this and other proto-oncogenes in controlling cell growth is described in later chapters.

DNA Damage Can Be Repaired by Several Mechanisms

In addition to the proofreading activity of DNA polymerases that can correct miscopied bases during replication, cells have evolved mechanisms for repairing DNA damaged by chemicals or radiation. Complex organisms with large genomes and relatively long generation times contain many cells that divide very slowly or not at all (e.g., liver and brain cells). Such cells must use the information in their DNA for weeks, months, or even years, greatly increasing their chances for sustaining damage to their DNA. If repair processes were 100 percent effective, chemicals and radiation would pose no threat to cellular DNA. Unfortunately, repair of lesions caused by some environmental agents is relatively inefficient, and such lesions can lead to mutations that ultimately cause cancer. In theory, a carcinogen could act by binding to DNA and causing a change in the sequence that is perpetuated during DNA replication. Current evidence suggests, however, that many permanent DNA sequence changes are induced by the very repair processes cells use to rid themselves of DNA damage.

DNA-repair mechanisms have been studied most extensively in *E. coli*, using a combination of genetic and biochemical approaches. The remarkably diverse collection of enzymatic repair mechanisms revealed by these studies can be divided into three broad categories:

- Mismatch repair, which occurs immediately after DNA synthesis, uses the parental strand as a template to correct an incorrect nucleotide incorporated into the newly synthesized strand.
- Excision repair entails removal of a damaged region by specialized nuclease systems and then DNA synthesis to fill the gap.
- Repair of double-strand DNA breaks in multicellular organisms occurs primarily by an end-joining process.

We discuss each of these DNA-repair mechanisms in order and then see how defective or error-prone DNA-repair systems can preserve mutations and contribute to tumor formation.

Mismatch Repair of Single-Base Mispairs Many spontaneous mutations are point mutations, which involve a change in a single base pair in the DNA sequence (see Figure 8-4). These can arise from errors in replication, during genetic recombination, and, particularly, by *base deamination* whereby a C residue is converted into a U residue (Figure 12-23).

Bacterial and eukaryotic cells have a mismatch-repair system that recognizes and repairs all single-base mispairs

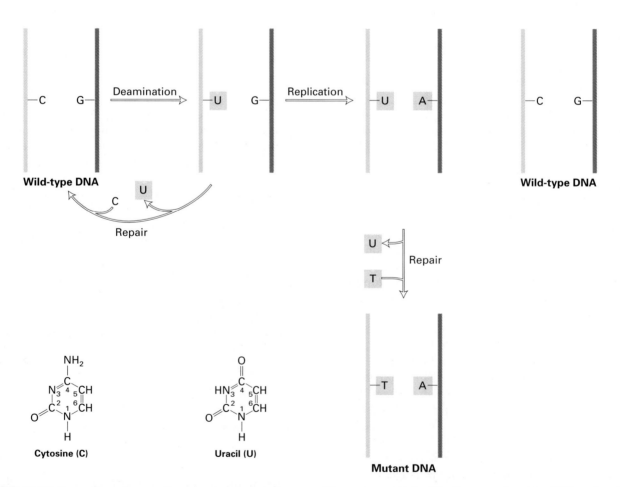

▲ FIGURE 12-23 Formation of a spontaneous point mutation by deamination of cytosine (C) to form uracil (U). If the resulting U·G base pair is not restored to the normal C·G base pair by repair mechanisms, it will be fixed in the DNA during replication.

After one round of replication, one daughter DNA molecule will have the mutant U·A base pair and the other will have the wild-type C·G base pair. The uracil is removed and replaced by thymine, generating a mutant DNA in which a T·A pair replaces a C·G pair.

except C·C, as well as small insertions and deletions. The conceptual problem with mismatch repair is determining which is the normal and which is the mutant DNA strand, and repairing the latter so that it is properly base-paired with the normal strand. How this is accomplished has been elucidated in considerable detail for the *E. coli* methyl-directed mismatch-repair system, often referred to as the *MutHLS system.*

In *E. coli* DNA, adenine residues in a GATC sequence are methylated at the 6 position. Since DNA polymerases incorporate adenine, not methyl-adenine, into DNA, adenine residues in newly replicated DNA are methylated only on the parental strand. The adenines in GATC sequences on the daughter strands are methylated by a specific enzyme, called *Dam methyltransferase,* only after a lag of several minutes. During this lag period, the newly replicated DNA contains hemimethylated GATC sequences:

$$
\begin{array}{c}
\quad\ \ \overset{\displaystyle CH_3}{\underset{\displaystyle |}{}} \\
5'-G-A-T-C-3' \quad \textbf{Parental strand} \\
3'-C-T-A-G-5' \quad \textbf{Daughter strand}
\end{array}
$$

An *E. coli* protein designated *MutH,* which binds specifically to hemimethylated sequences, is able to distinguish the methylated parental strand from the unmethylated daughter strand. If an error occurs during DNA replication, resulting in a mismatched base pair near a GATC sequence, another protein, MutS, binds to this abnormally paired segment (Figure 12-24). Binding of MutS triggers binding of MutL, a linking protein that connects MutS with a nearby MutH. This cross-linking activates a latent endonuclease activity of MutH, which then cleaves specifically the unmethylated daughter strand. Following this initial incision, the segment of the daughter strand containing the misincorporated base is excised and replaced with the correct DNA sequence.

E. coli strains that lack the MutS, MutH, or MutL protein have a higher rate of spontaneous mutations than wild-type cells. Strains that cannot synthesize the Dam methyltransferase also have a high rate of spontaneous mutations. Because

Dam⁻ strains cannot methylate adenines within GATC sequences, the MutHLS mismatch-repair system cannot distinguish between the template and newly synthesized strand and therefore cannot efficiently repair mismatched bases.

A similar mechanism repairs lesions resulting from *depurination,* the loss of a guanine or adenine base from DNA resulting from cleavage of the glycosidic bond between deoxyribose and the base. Depurination occurs spontaneously and is fairly common in mammals. The resulting *apurinic sites,* if left unrepaired, generate mutations during DNA replication because they cannot specify the appropriate paired base. All cells possess apurinic (AP) endonucleases that cut a DNA strand near an apurinic site. As with mismatch repair, the cut is extended by exonucleases, and the resulting gap then is repaired by DNA polymerase and ligase.

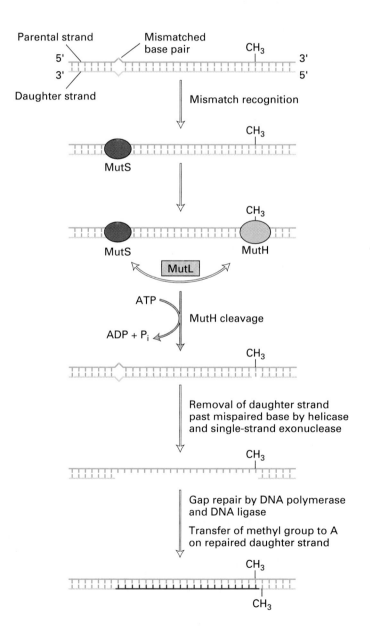

► **FIGURE 12-24 Model of mismatch repair by the *E. coli* MutHLS system.** This repair system operates soon after incorporation of a wrong base, before the newly synthesized daughter strand becomes methylated. MutH binds specifically to a hemimethylated GATC sequence, and MutS binds to the site of a mismatch. Binding of MutL protein simultaneously to MutS and to a nearby MutH activates the endonuclease activity of MutH, which then cuts the unmethylated (daughter) strand in the GATC sequence. A stretch of the daughter strand containing the mispaired base is excised, followed by gap repair and ligation and then methylation of the daughter strand. [Adapted from R. Kolodner, 1996, *Genes and Develop.* **10**:1433; see also A. Sancar and J. Hearst, 1993, *Science* **259**:1415.]

Excision Repair Cells use excision repair to fix DNA regions containing chemically modified bases, often called *chemical adducts,* that distort the normal shape of DNA locally. A key to this type of repair is the ability of certain proteins to slide along the surface of a double-stranded DNA molecule looking for bulges or other irregularities in the shape of the double helix. For example, this mechanism repairs *thymine-thymine dimers,* the most common type of damage caused by UV light (Figure 12-25); these dimers interfere with both replication and transcription of DNA. Excision repair also can correct DNA regions containing bases altered by the attachment of large chemical groups [e.g., carcinogens such as benzo*(a)*pyrene; see Figure 12-22].

Perhaps the best-understood example of excision repair is the *UvrABC* system from *E. coli.* Cells carrying mutations in the *uvrA, B,* or *C* locus are very sensitive to UV light and chemicals that add large groups to DNA. Figure 12-26 illustrates how the UvrABC system repairs damaged DNA. Initially, a complex comprising two molecules of UvrA and one molecule of UvrB forms and then binds to DNA. Both the formation and binding of this complex to DNA requires

ATP. It seems likely that the UvrA-UvrB complex initially binds to an undamaged segment and translocates along the DNA helix until a distortion caused by an adduct is recognized; this translocation along the helix also requires ATP. An ATP-dependent conformational change in the damaged DNA region bound to the UvrA-UvrB complex then produces a bend, or kink, in the DNA backbone. After the UvrA dimer has dissociated, the UvrC protein, which has endonuclease activity, binds to the damaged site. The interaction of UvrC and the bent DNA is thought to open up space within the DNA, allowing the catalytic residues of the enzyme to access their target (Figure 12-27). The precise position of the cleavage sites is determined by the nature of the DNA damage. In the case of thymine dimers, UvrC cleaves two phosphodiester bonds: one is located eight nucleotides 5′ to the lesion, and one is located four or five nucleotides 3′ to the lesion. After UvrC has cleaved the damaged strand at two points, the fragment with the adduct is removed by a helicase and degraded; the gap left in the strand then is repaired by the combined actions of DNA polymerase I and DNA ligase (see Figure 12-26, steps 7 and 8).

End-Joining Repair of Nonhomologous DNA A cell that has suffered a particular double-strand break usually contains other breaks; such breaks can be repaired by joining the free DNA ends. The joining of broken ends from different chromosomes, however, leads to translocation of pieces of DNA from one chromosome to another (see Figure 8-4b). Such translocations may trigger abnormal cell growth by placing a proto-oncogene next to, and thus under the inappropriate control of, a promoter from another gene. Double-strand breaks are caused by ionizing radiation and by anticancer drugs, such as bleomycin, which is why these drugs are used to kill rapidly growing cells. The devastating effects of double-strand breaks make these the "most unkindest cuts of all," to paraphrase Shakespeare's *Julius Caesar.*

Double-strand breaks can be correctly repaired only if the free ends of the DNA rejoin exactly. Such repair is complicated by the absence of single-stranded regions that can direct base-pairing during the joining process. One of the two mechanisms that have evolved to repair double-strand breaks is homologous recombination. In this process, as described later, the double-strand break on one chromosome is repaired using the information on the homologous, intact chromosome.

In multicellular organisms, however, the predominant mechanism for repairing double-strand breaks involves rejoining the ends of the two DNA molecules. Although this process, outlined in Figure 12-28, yields a continuous double-stranded molecule, it results in loss of several base pairs at the joining point. Formation of such a possibly mutagenic deletion is one example of how repair of DNA damage can introduce mutations. A similar process can link together any two DNA molecules, even those cut from different chromosomes.

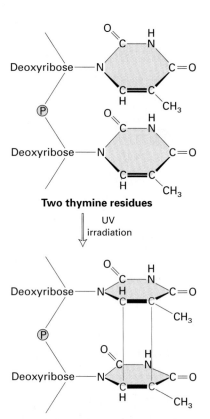

Two thymine residues

UV irradiation

Thymine-thymine dimer residue

▲ **FIGURE 12-25 UV irradiation can cause adjacent thymine residues in the same DNA strand to become covalently attached.** The resulting thymine-thymine dimer (cyclobutylthymine) may be repaired by an excision-repair mechanism.

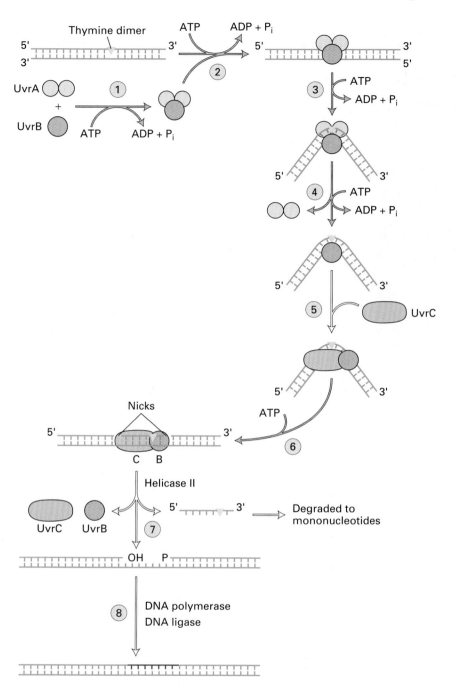

◀ **FIGURE 12-26 Excision repair of DNA by *E. coli* UvrABC mechanism.** Two molecules of UvrA and one of UvrB form a complex that moves randomly along DNA (steps ① and ②). Once the complex encounters a lesion, conformational changes in DNA, powered by ATP hydrolysis, cause the helix to become locally denatured and kinked by 130° (step ③). After the UvrA dimer dissociates (step ④), the UvrC endonuclease binds and cuts the damaged strand at two sites separated by 12 or 13 bases (steps ⑤ and ⑥). UvrB and UvrC then dissociate, and helicase II unwinds the damaged region (step ⑦), releasing the single-stranded fragment with the lesion, which is degraded to mononucleotides. The gap is filled by DNA polymerase I, and the remaining nick is sealed by DNA ligase (step ⑧). [Adapted from A. Sancar and J. Hearst, 1993, *Science* **259**:1415.]

Eukaryotes Have DNA-Repair Systems Analogous to Those of *E. coli*

Evidence is accumulating that the basic mechanisms of DNA repair have been conserved during evolution. For example, recent biochemical studies have shown that human cells carry out mismatch repair by a process similar to that in *E. coli* (see Figure 12-24). The repair process can be initiated by the human MutSα protein, a homolog of bacterial MutS, which binds both to mismatched base pairs and to small insertions or deletions, or by MutSβ, which binds mainly to insertions and deletions. The human MutL protein is recruited to the DNA by MutSα or MutSβ, but the identity of the human nuclease (equivalent to MutH in *E. coli*) that actually cuts the DNA is unknown. Following cleavage, which can occur either 3' or 5' to the mismatch, an exonuclease removes 100–200 nucleotides from the cut strand, spanning the mismatch. DNA polymerase δ is principally responsible for filling in the gap, following which the strands are sealed by the action of a DNA ligase.

Genetic studies in eukaryotes ranging from yeast to humans suggest that quite similar excision-repair mechanisms are employed by different organisms. The basic strategy is

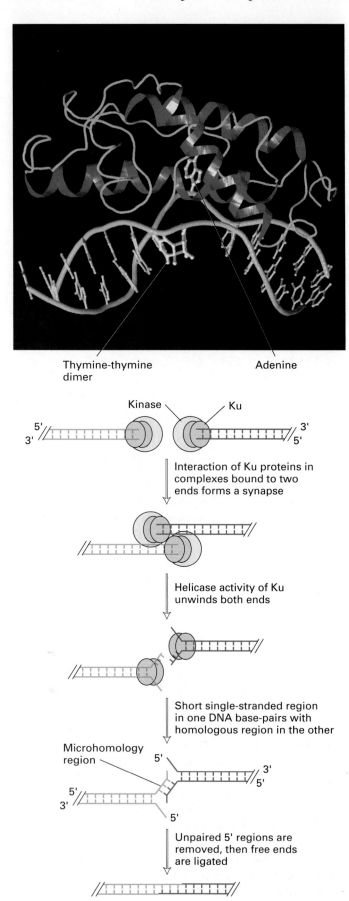

◀ FIGURE 12-27 Model of complex formed between bacteriophage T4 endonuclease V and a 13-bp DNA fragment containing a thymine-thymine dimer based on x-ray crystallographic analysis. Interaction of the enzyme *(top)* with the kinked DNA *(bottom)* causes one of the adenines normally paired to one of the thymines to move so that it can bind to a cavity of the enzyme. The resulting opening of the DNA structure permits the enzyme access to its target sites. UvrC is thought to interact with damaged DNA in a similar fashion. [From D. Vassylyev and K. Morikawa, 1997, *Curr. Opin. Struc. Biol.* **7**:103; courtesy of Dr. Morikawa.]

to search for mutants that exhibit increased sensitivity to UV light or other agents that produce shape-distorting lesions in DNA. Such mutants presumably are deficient in the wild-type excision-repair mechanisms that normally repair damage caused by such agents. In the yeast *S. cerevisiae*, for instance, numerous UV-sensitive mutants and 10 radiation-sensitive (*RAD*) genes have been identified. Techniques of somatic-cell genetics have been used to generate mutant Chinese hamster ovary (CHO) cell lines that are abnormally sensitive to UV light or to DNA damage caused by mitomycin C, which forms bulky adducts with DNA. These mutations fall into eight complementation groups, suggesting that at least eight different genes are involved in excision repair in hamsters. DNA transfection experiments have revealed that certain human genes can rescue some UV-sensitive CHO cell mutants. Two of the human genes identified in this way have been shown to be related to the yeast *RAD3* and *RAD10* genes. The human protein with partial homology to the RAD10 protein also contains a region that is similar to part of *E. coli* UvrC.

Remarkably, five polypeptides required for excision repair in eukaryotic cells, including two with homology to helicases, are also subunits of TFIIH, a general transcription factor required by RNA polymerase II (see Figure 10-50). It appears that Nature has used a similar protein assembly in

◀ FIGURE 12-28 Repair of double-strand breaks by end-joining of nonhomologous DNAs (dark and light blue), that is, DNAs with dissimilar sequences at their ends. These DNAs could be cut fragments from a single gene, or DNAs cut from different chromosomes. A complex of two proteins, Ku and DNA-dependent protein kinase, binds to the ends of a double-strand break. After formation of a synapse in which the broken ends overlap, Ku unwinds the ends, by chance revealing short homologous sequences in the two DNAs, which base-pair to yield a region of microhomology. The unpaired single-stranded 5′ ends are removed by mechanisms that are not well understood, and the two double-stranded molecules ligated together. As a result, the double-strand break is repaired, but several base pairs at the site of the break are removed. [Adapted from G. Chu, 1997, *J. Biol. Chem.* **272**:24097; M. Lieber et al., 1997, *Curr. Opin. Genet. Devel.* **7**:99.]

two different cellular processes that require helicase activity. The use of shared subunits in transcription and DNA repair may help explain *transcription-coupled repair*. This phenomenon is the observed removal of DNA damage in higher eukaryotes at a much faster rate from regions of the genome being actively transcribed than from nontranscribed regions of the genome. (Recall that in higher eukaryotes, only a small fraction of the genome is transcribed in any one cell.)

Inducible DNA-Repair Systems Are Error-Prone

When a cell suffers so much DNA damage over a short time that its repair systems are saturated, it runs the danger of extensively replicating unrepaired lesions, thereby perpetuating mutations. In such situations, both bacterial and animal cells use inducible repair systems in an attempt to catch up. Such systems are not expressed in undamaged cells, but some aspect of the accumulated damage causes their derepression (induction) and expression.

One such inducible system is the *SOS repair system* of bacteria. Because this system generates many errors in the DNA as it repairs lesions, it is referred to as *error-prone*. The SOS system, which repairs UV-induced damage, differs from the constitutive UvrABC system already discussed in that its activity is dependent on RecA protein; as discussed later, RecA also participates in homologous recombination. The errors induced by the SOS system are at the site of lesions, suggesting that the mechanism of repair is insertion of random nucleotides in place of the damaged ones in the DNA. This inducible system is used only as a last resort when error-free mechanisms of repair cannot cope with damage. Bacteria lacking an SOS system will retain mutations caused by UV damage or chemicals, but still will have greatly reduced rates of mutation induced by these agents. This supports the idea that most of the mutations produced by treating bacteria with radiation or chemicals are caused by the error-prone SOS repair system, not by the original lesions themselves.

Animal cells also have inducible repair systems, although it is not known whether these are error-prone. As noted earlier, however, the main mechanism for repairing double-strand breaks in eukaryotes clearly is error-prone (see Figure 12-28). Thus, double-strand repair and, perhaps, error-prone inducible repair likely play a role in mutagenesis and therefore in carcinogenesis in animals. In any case, many investigators believe that in animal cells, as in bacteria, most mutations are an indirect, not direct, consequence of DNA damage.

Because radiation- or carcinogen-induced DNA damage must be repaired before the DNA is replicated, cells have sensing mechanisms that react to DNA damage and stop DNA replication. These mechanisms, which are discussed in detail in the next chapter, involve checkpoint control proteins such as the p53 protein, which acts to stop the cell cycle if DNA is damaged, and thus to suppress production of tumors. Cells that do not express functional p53 protein exhibit high rates of mutation in response to DNA damage, accelerating the formation of tumors.

SUMMARY DNA Damage and Repair and Their Role in Carcinogenesis

- Changes in the DNA sequence result from copying errors and the effects of various physical and chemical agents, or carcinogens. All carcinogens are mutagens; that is, they alter one or more nucleotides in DNA.

- The varied structures of chemical carcinogens have one unifying characteristic: electrophilic reactivity (either they are electrophiles or they are metabolized in the body to become electrophiles). Metabolic activation occurs via the cytochrome P-450 system, a pathway generally used by cells to rid themselves of noxious chemicals.

- Many copying errors that occur during DNA replication are corrected by the proofreading function of DNA polymerases that can recognize incorrect (mispaired) bases at the 3′ end of the growing strand and then remove them by an inherent 3′ → 5′ exonuclease activity (see Figure 12-21).

- In mismatch repair, a short section of a newly synthesized DNA strand containing an incorrect base is identified, removed, and replaced by DNA synthesis directed by the correct template (see Figure 12-24).

- In excision repair, bulky lesions in DNA resulting from exposure to UV light and various chemicals are removed (excised) by specialized nuclease systems; DNA synthesis by a polymerase fills the gap and ligase joins the free ends (see Figure 12-26).

- Double-strand breaks can be repaired by homologous recombination and by an end-joining of nonhomologous DNA duplexes. In the latter process, several bases at the site of the break are removed (see Figure 12-28).

- Both bacterial and eukaryotic cells have inducible DNA-repair systems, which are expressed when DNA damage is so extensive that replication may occur before constitutive mechanisms can repair all the damage. The inducible SOS repair system in bacteria is error-prone and thus generates and perpetuates mutations.

- DNA-repair mechanisms that are ineffective or error-prone may perpetuate mutations. This is a major way by which DNA damage, caused by radiation or chemical carcinogens, induces tumor formation. Thus, cellular DNA-repair processes have been implicated both in protecting against and contributing to the development of cancer.

12.5 Recombination between Homologous DNA Sites

In the previous sections of this chapter, we have discussed how the genome is faithfully reproduced from one generation to another during DNA replication and how the correct sequence is maintained by DNA-repair processes throughout the life of a cell and organism. In this section, we examine the mechanisms by which the genome can change to generate new combinations of genes.

Soon after Mendel's rules of independent gene segregation were rediscovered and the segregation of linked groups of genes on individual chromosomes was widely recognized, another great genetic discovery was made in *D. melanogaster*: blocks of genes from homologous chromosomes could be exchanged by the process of crossing over, or homologous recombination (often referred to simply as **recombination**). Homologous recombination takes place during meiosis in sexually reproducing organisms. Recall that each homologous paternal and maternal chromosome contains a different combination of alleles. By generating new chromosomes that contain part of each homologous paternal and maternal chromosome, recombination results in new combinations of alleles on a given chromosome. Thus recombination provides a mechanism for generating genetic diversity beyond that achieved by the independent segregation of chromosomes (see Figure 8-3). Genetic exchange by recombination occurs not only in animals and plants but also in prokaryotes, viruses, plasmids, and even in the DNA of cell organelles such as mitochondria.

The events in a *reciprocal recombination* are equivalent to the breakage of two homologous duplex DNA molecules, an exchange of *both* strands at the break, and a resolution of the two duplexes so that no tangles remain. To a good approximation, recombination occurs randomly between two homologous DNA segments, and thus the frequency of recombination between two sites is proportional to the distance between the sites. (As discussed in Chapter 8, this phenomenon is the basis of genetic mapping of genes defined by mutations.) In the remainder of this chapter, we describe how several types of proteins catalyze steps in recombination.

The Crossed-Strand Holliday Structure Is an Intermediate in Recombination

In 1964, Robin Holliday proposed the recombination model depicted in Figure 12-29. Except for the initial steps, this model appears to accurately describe the molecular events that lead to genetic recombination. According to the original Holliday model, after two homologous double-stranded DNA molecules (i.e., cellular or viral chromosomes) become aligned, a nick is made in one strand of each of the recombining DNAs (step 1). The two nicked strands then invade each other, a process called *strand exchange*, at the site of the nicks, and the cut 3' ends are joined to the 5' ends

of the homologous strand, producing a crossed-strand **Holliday structure** (step 2). The branch point then migrates, creating a **heteroduplex** region containing one strand from each parental DNA molecule (step 3).

Two mechanisms have been proposed for separation, or *resolution*, of the connected duplexes. According to Holliday's original proposal, all four strands are cut at the crossover site (Figure 12-29, step 4). If the left side of chromosome I joins the right side of chromosome II, and vice versa (step 5a), then both strands in each of the resulting duplexes are *recombinant*; that is, all markers to the left and right of the crossover site have undergone reciprocal recombination. However, over a short region—the B and b genes in the figure—the chromosomes are termed *heteroduplex*; they have genetic material from one chromosome on one strand and material from the other chromosome on the opposite strand. As each of these chromosomes duplicates, half the progeny will have the genetic marker B from one initial chromosome, and half will have the b allele. In contrast, if the left sides of chromosomes I and II rejoin with their own respective right sides (step 5 b), then both strands in each of the resulting duplexes are termed *nonrecombinant*; that is, all markers to the left and right of the crossover site are derived from the same initial chromosome. However, these molecules are also heteroduplex for the B or b segment.

A later proposal simplifies the enzymatic cutting that is necessary to resolve the crossed-strand intermediate. Rotation of the Holliday structure at the crossover site forms a rotational isomer, or isomeric Holliday structure (step 6). (This can be visualized by imagining that the dotted line passing through the crossover point in step 3 serves as an axis around which the bottom duplex spins 180°, untwisting the crossed strands in the process.) Note that no strands are cut or ligated in going from 3 to 6. The two connected duplexes of this structure can be resolved (i.e., disconnected) by cutting and rejoining of only two strands. If this involves the two strands that were not cut to generate the original Holliday intermediate, the resulting "spliced" products are recombinant duplex chromosomes containing a heteroduplex region (steps 7a and 8a). However, if resolution involves cutting of the two strands that were originally cut, the resulting "patched" products are duplex chromosomes that contain a heteroduplex B region but are not termed recombinants (step 7b and 8b), since all the markers to the left of A and to the right of C are derived from the same original chromosome.

Evidence that Holliday intermediates actually exist has come from electron microscopy of viral and plasmid DNA molecules extracted from both bacterial and animal cells. Electron micrographs of such molecules in the act of recombining have revealed structures similar to the crossed-strand and isomeric Holliday structures (Figure 12-30). Thus, regardless of the mechanism initiating recombination, the process seems to involve the kinds of intermediates predicted by the Holliday model.

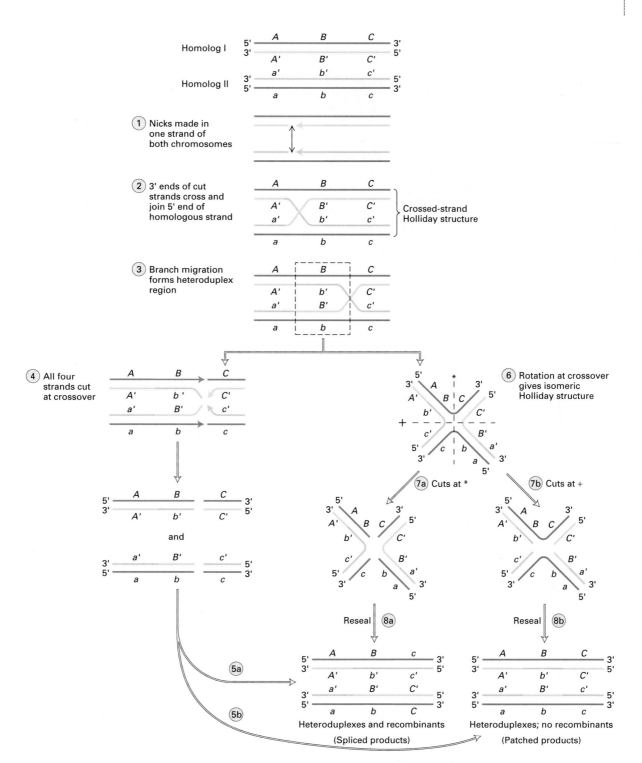

▲ **FIGURE 12-29 Holliday model of genetic recombination.**
Genetically distinct homologous chromosomes (i.e., double-stranded DNA molecules) are indicated by red and blue; alleles are indicated by capital and lowercase letters *(A, a)*. Complementary DNA strands are distinguished by darker and lighter shades and by the presence or absence of prime signs *(A, A'; a, a')*. Resolution of the crossed-strand Holliday structure could occur by two different pathways. Steps ④ and ⑤a or ⑥, ⑦a, and ⑧a would yield spliced products that exhibit recombination from *AC/ac* to *Ac/aC*, with heteroduplex DNA containing the *B* locus in between. Although step ⑤b or steps ⑦b and ⑧b also resolve the connected strands of the Holliday structure, the resulting patched products are not recombinants, since all markers surrounding the crossover site, that is, to the left of *A* and to the right of *C*, are derived from the same initial chromosome. However, these molecules are heteroduplex for the *B* segment, and as each of these DNA molecules is duplicated, half the progeny will have the *B* genetic marker and half will have *b*. [Steps 1–5, see R. Holliday, 1964, *Genet. Res.* **5**:282; steps 6, 7a, 7b, 8a, and 8b see D. Dressler and H. Potter, 1982, *Ann. Rev. Biochem.* **51**:727; also see M. Meselson and C. M. Radding, 1975, *Proc. Nat'l Acad. Sci. USA* **72**:358; and N. Sigal and B. Alberts, 1972, *J. Mol. Biol.* **71**:769.]

(a) (b)

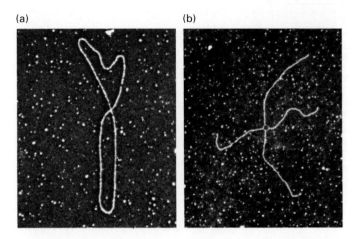

▲ **FIGURE 12-30 Electron micrographs of plasmid DNA in the process of recombination.** (a) Circular plasmid DNA in crossed-strand Holliday structure. (b) More highly magnified view reveals single-stranded ring in center of isomeric Holliday structure that results from rotation about the crossover point. [See H. Potter and D. Dressler, 1978, *Cold Spring Harbor Symp. Quant. Biol.* **43**:969; courtesy of D. Dressler.]

Double-Strand Breaks in DNA Initiate Recombination

Many variations of the Holliday model have been proposed; these differ in how the DNA strands are cut to initiate strand exchange and whether or not DNA synthesis is involved in the recombination process. Most evidence now favors a model in which homologous recombination is initiated by a double-strand break in one of the DNA duplexes and the homologous, intact chromosome is used as a template to repair the break by DNA synthesis.

If left unrepaired, double-strand breaks lead to broken chromosomes and cell death. The repair mechanism that joins the cut ends of nonhomologous chromosomes reconstitutes double-stranded DNA but deletes several nucleotides (see Figure 12-28). The mechanism outlined in Figure 12-31, however, completely repairs a double-strand break, yielding a molecule of the same length as the original. This model is thought to apply to homologous recombination in prokaryotes and during meiosis in yeast and probably other eukaryotes, as well as to the error-free repair of double-strand breaks in DNA induced by radiation and other agents.

As depicted in Figure 12-31, a double-strand break occurs in one of a pair of aligned homologous double-stranded DNA molecules (the **a** chromatid); the break is enlarged to gaps by action of 5′ → 3′ exonucleases, resulting in single-stranded 3′ ends (step 1). The 3′ end of one **a** strand then invades the homologous α chromatid (step 2) and is elongated at its 3′ end by DNA polymerase, using the D complementary strand of the α chromatid as template (step 3). Then the 3′ end of the other cut **a** strand is elongated using the other strand (D′) of the α chromatid as template, producing an intermediate structure that contains two regions

of heteroduplex DNA and two Holliday junctions (step 4). Following rotation at the crossover points, each Holliday junction is resolved by cleavage of two single strands and ligation, yielding two intact double-stranded DNA molecules (step 5a or 5b). Note that the original double-strand cut in the **a** chromatid has been fully repaired.

This double-strand break model can account for the nonmendelian segregation of certain markers that has been observed during meiotic recombination. This phenomenon is most easily studied in yeast in which all four meiotic products can be scored in the haploid progeny spores. In a cross of multiply-marked yeast strains that undergo recombination, most allelic markers segregate according to the Mendelian 2:2 ratio, but those located near the crossover point exhibit 3:1 or 1:3 segregation. Such a nonreciprocal event is called **gene conversion** because one allele is apparently "converted" into another. Gene conversion is thought to result from the process for resolving the crossed-strand Holliday intermediate produced in the double-break model of recombination.

To see how gene conversion occurs, consider the example depicted in Figure 12-31 in which the D (or complementary

▶ **FIGURE 12-31 Double-strand break model of meiotic recombination developed from studies in the yeast *S. cerevisiae*.** A pair of homologous chromatids (double-stranded DNA molecules) are shown, one in blue and the other red. The darker and lighter shades indicate complementary DNA strands. Alleles are indicated by capital and lowercase letters *(D, d)*. Complementary DNA strands are also indicated by the presence or absence of prime signs (for instance D and D′, and c and c′). In this example, the initial double-strand break and resection of 5′ ends occurs on the **a** chromatid, removing the *d′* marker ①. This is followed by strand invasion ② and DNA synthesis with the α chromatid D strand as the template ③. Repair synthesis of the other **a** strand (using its complementary section on the α strand) and ligation result in formation of a Holliday structure with two crossovers ④ⓐ and ④ⓑ. (Repaired regions are marked by black dashed lines.) Resolution of this crossed-strand intermediate can occur in two ways. Cleavage at sites 2 and 4 (step ⑤ⓑ), or at sites 1 and 3 (not shown here), yields nonrecombinant chromosomes, since all markers surrounding the crossover site (i.e., to the left of c and to the right of *E*) are derived from the same initial chromosome. One duplex contains a complementary *D/D′* region, but the other contains a heteroduplex mismatched *d/D′* region (yellow). In contrast, cleavage (step ⑤ⓐ) at sites 2 and 3, or at sites 1 and 4 (not shown here), yields recombinant double-stranded DNA molecules, since all markers to the left and right of the crossover site have undergone reciprocal recombination. Note that one duplex contains a complementary *D/D′* region, but the other contains a heteroduplex, mismatched *d/D′* region (yellow). Cells can repair such mismatched heteroduplex regions by excising a single-strand segment containing the mismatch and using the other strand as a template for synthesis of a matching strand ⑥ (see Figure 12-24). In this example, *d* is removed and *D* is synthesized (jagged red segment), thus "converting" *d* to *D*. The opposite D → *d* conversion occurs with equal frequency.

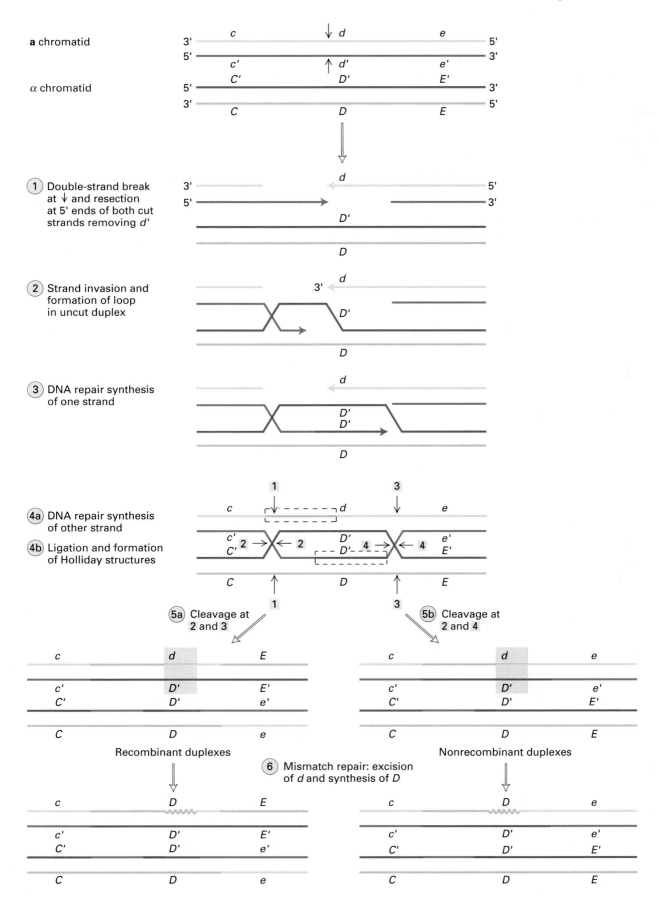

a chromatid

α chromatid

1. Double-strand break at ↓ and resection at 5' ends of both cut strands removing *d'*

2. Strand invasion and formation of loop in uncut duplex

3. DNA repair synthesis of one strand

4a. DNA repair synthesis of other strand

4b. Ligation and formation of Holliday structures

5a. Cleavage at **2** and **3**

5b. Cleavage at **2** and **4**

Recombinant duplexes

Nonrecombinant duplexes

6. Mismatch repair: excision of *d* and synthesis of *D*

D') and d markers lie between the two crossovers in the intermediate. Of the two recombinant DNA duplexes that result from cleavage and ligation of the intermediate (i.e., the two products of step 5a), one contains the D marker in one strand and the complementary D' in the other, whereas the other is a heteroduplex with D' in one strand and the allelic d in the other. Similarly, of the two nonrecombinant DNAs that would result from cleavage of the intermediate at different sites (step 5b), one is also heteroduplex in the d/D segment. Mispairing of the D' and d alleles in the heteroduplexes is recognized by the mismatch-repair system. Half of the time, mismatch repair will convert the D'/d heteroduplex into D'/D (step 6); in this case the d'/d marker in the original **a** chromatid is "converted" into D'/D. The other half of the time, the D'/d heteroduplex will be converted into d'/d, in which case there is no gene conversion. Thus three out of four haploid yeast progeny spores, each carrying one meiotic product, will show the D phenotype, that is, the 3:1 segregation ratio that typifies gene conversion.

The Activities of *E. coli* Recombination Proteins Have Been Determined

Three different but related enzymatic pathways carry out homologous recombination in *E. coli*. All three pathways utilize the basic double-strand break mechanism depicted in Figure 12-31 to generate a Holliday-type structure, which undergoes branch migration followed by endonuclease cleavage and then ligation to yield recombinants. In this section, we describe the enzymes that catalyze the primary *E. coli* recombination pathway and briefly consider homologous eukaryotic enzymes. These enzymes were uncovered through genetic analysis of recombination in *E. coli* and subsequently purified.

Initiation of Recombination (RecBCD Enzyme) The most common way that *E. coli* cells generate a recombinogenic single-stranded region of DNA, equivalent to step 1 in Figure 12-31, probably is by action of the RecBCD enzyme, which has helicase and exonuclease activities. This enzyme complex, composed of proteins encoded by the *recB*, *C*, and *D* genes, specifically recognizes double-strand breaks. Such breaks occur naturally during bacterial conjugation, a process in which chromosomal DNA is transferred from one bacterium to another through direct cell contact, and during bacteriophage-mediated transduction. Double-strand breaks also can be generated by exposure to x-rays and certain chemicals.

The mechanism of action of RecBCD was worked out in studies with bacteriophage λ. Certain regions of λ phage DNA, termed *CHI sites,* undergo recombination at higher frequencies than other regions in normal *E. coli* host cells but not in *recBCD* mutant host cells. Experiments with purified RecBCD enzyme and λ DNA indicate that the protein complex recognizes and binds to a free blunt end of the λ phage chromosome (equivalent to a double-strand break).

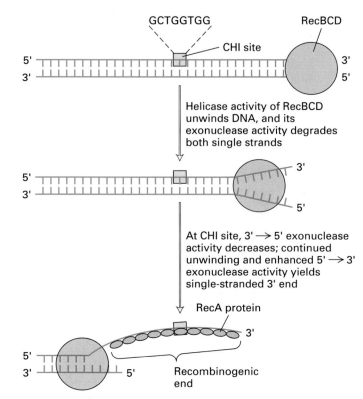

▲ **FIGURE 12-32 Initiation of recombination by *E. coli* RecBCD enzyme.** This multifunction enzyme, which binds to free blunt ends of DNA, acts as a helicase, 3′→ 5′ exonuclease, and 5′→ 3′ exonuclease. When the enzyme encounters a CHI site, its 3′→ 5′ activity is inhibited and its 5′→ 3′ activity enhanced, yielding a single-stranded 3′-hydroxyl end. This recombinogenic 3′ end becomes coated with multiple RecA proteins, which then catalyze strand invasion and formation of a Holliday structure. [See D. G. Anderson and S. C. Kowalczykowski, 1997, *Cell* **90**:77.]

The enzyme then moves along the DNA, its helicase activity unwinding the duplex as it goes (Figure 12-32). Initially, RecBCD degrades both single strands one nucleotide at a time using its dual 5′→ 3′ and 3′→ 5′ exonuclease activities. However, when RecBCD encounters a CHI site, its 3′→ 5′ exonuclease activity is inhibited and its 5′→ 3′ exonuclease activity is enhanced. Thus, after passing a CHI site, RecBCD begins to generate a single-stranded 3′-hydroxyl end. After the resulting *recombinogenic end* becomes coated with multiple RecA proteins, it can participate in the process of strand invasion.

Strand Invasion, Homologous Pairing, and Formation of Holliday-Type Structure (RecA Protein) Biochemical experiments showed that the protein encoded by the *recA* gene can bind to any single-stranded DNA (ssDNA). In the presence of a homologous target duplex DNA, the RecA-ssDNA complex can carry out two remarkable functions. First, RecA aligns the ssDNA with its homologous target double-stranded DNA region and forms a complex with

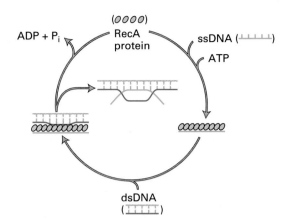

▲ **FIGURE 12-33 Formation of Holliday-type structure by** *E. coli* **RecA.** In the presence of ATP, RecA binds single-stranded DNA (ssDNA) and promotes insertion of the bound strand at a homologous region of double-stranded DNA (dsDNA), yielding a crossed-strand Holliday-type structure. The insertion reaction requires the ATPase activity of RecA. [See S. S. Flory et al., 1984, *Cold Spring Harbor Symp. Quant. Biol.* **49**:513.]

it. Second, RecA inserts the ssDNA into the target DNA, displacing one of the preexisting strands and forming a heteroduplex Holliday-type structure (Figure 12-33).

In subsequent studies, RecA was shown to bind to the single-stranded recombinogenic ends generated by action of the RecBCD enzyme. In fact, the ability of RecA to load onto single-stranded DNA is stimulated by RecBCD once it is activated by traversing a CHI site in the DNA (see Figure 12-32). *E. coli* SSB protein stimulates this reaction by binding to the single-stranded region and preventing intrastrand base pairing, which would inhibit binding of RecA. In the presence of ATP, RecA coats the single-stranded region and polymerizes, forming a filament that wraps around the entire length of the recombinogenic end. Because the polymerization of RecA occurs in the $5' \rightarrow 3'$ direction along the DNA, coating takes place in a discontinuous fashion as a region of the duplex is unwound. X-ray crystallographic analysis of RecA suggests that each molecule has two DNA-binding sites, both of which may lie within the core of the filament.

Branch Migration and Resolution of Holliday Structures (Ruv Proteins) Although formation of Holliday structures depends on RecA, their maintenance does not. Removal of RecA leaves stable Holliday structures, which have been used as substrates in studies on branch migration and resolution. Migration of the crossover point, which can be detected in synthetic Holliday structures, is efficiently catalyzed by *E. coli* RuvA and RuvB proteins (Figure 12-34). Further analysis showed that RuvA specifically recognizes the Holliday junction, whereas RuvB has the helicase activity necessary for promoting the observed branch migration.

Recent studies have clarified the action of these two proteins. The active tetrameric form of RuvA binds to the

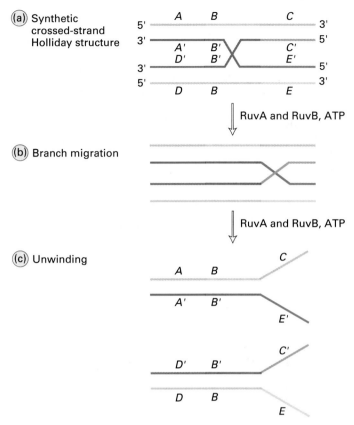

▲ **FIGURE 12-34 Experimental demonstration of branch migration catalyzed by** *E. coli* **RuvA and RuvB proteins.** Complementary strands are indicated by darker and lighter shades and the presence or absence of prime signs (*A, A′; B, B′*); segments with different sequences are indicated by color. (a) A synthetic Holliday structure was produced by annealing four synthetic single-stranded oligonucleotides in which only the center crossover region (green) was homologous and the opposite ends of any given strand are, respectively, complementary to two different strands. (b, c) Treatment of the Holliday structure with RuvA and RuvB in the presence of ATP leads to branch migration followed by unwinding to yield cruciform structures with nonhomologous single-stranded ends. Branch migration in the other direction (towards *A* and *D*) yields similar cruciform structures. [Adapted from H. Iwasaki et al., 1992, *Genes & Dev.* **6**:2214.]

center of a Holliday junction, unfolding the junction into a square planar configuration and keeping the four single-stranded segments apart. This induces binding of two ring-like hexameric RuvB proteins, which surround the double-stranded DNA exiting from opposite sides of the RuvA complex (Figure 12-35a). Powered by ATP hydrolysis, the RuvB rings act as molecular pumps, pulling two double-stranded DNAs into the RuvA complex, separating the strands, and then extruding two double-stranded heteroduplexes out of the RuvA complex. Following branch migration, two RuvC endonuclease proteins bind to the RuvA/

(a)

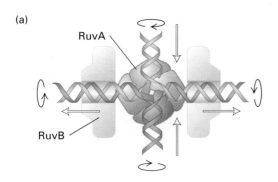

(b)

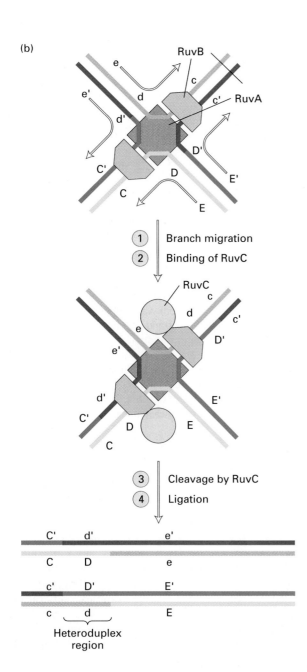

① Branch migration
② Binding of RuvC

③ Cleavage by RuvC
④ Ligation

Heteroduplex
region

◀ **FIGURE 12-35 Action of *E. coli* proteins in branch migration and resolution of Holliday junctions.** (a) Model of the association of RuvA and RuvB with a Holliday junction as determined by electron microscopy. For clarity the proteins are drawn behind the DNA; in reality the DNA passes through the central hole of each RuvB hexamer. Powered by ATP hydrolysis, the ringlike RuvB molecules rotate the DNA double helices inside them, much in the way a screw is rotated inside a nut. The two RuvB proteins impart equal and opposite rotational forces to the DNA, as indicated by the black circular arrows. The straight arrows indicate the direction of movement of DNA through the RuvA/RuvB complex. (b) Diagram illustrating migration and resolution of a Holliday junction. (Alleles and complementary strands are indicated as in Figure 12-31.) Movement of a junction through the RuvA/RuvB complex generates heteroduplex regions, exactly as observed in the experiment shown in Figure 12-34. Following branch migration ①, two molecules of the RuvC endonuclease bind to RuvA ② and cut the intermediate at two points that are 180° apart ③. Ligation of the cut ends completes resolution ④. In this example, two recombinant molecules are formed, but if RuvC cleaved at the two other points, nonrecombinant DNAs would be produced (see Figure 12-31). [Adapted from S. C. West, 1996, *Cell* **86**:177; J. Raftery et al., 1996, *Science* **274**:415; and A. Kuzminov, 1996, *BioEssays* **18**:757.]

RuvB complex and then cut the DNA intermediate at two sites 180° apart; subsequent ligation generates recombinant (or nonrecombinant) molecules containing a segment of heteroduplex DNA (Figure 12-35b).

Homologous Eukaryotic Recombination Enzymes Although RecA, RecBCD, and RuvA, B, and C were initially identified in *E. coli,* all eukaryotic cells, including human cells, produce proteins of similar structure and function. For instance, the human and yeast RAD51 proteins, which are homologous in sequence, catalyze pairing of homologous DNA segments and DNA strand insertion similarly to RecA. A Topo II–like protein encoded by the yeast *Spo11* gene generates the double-strand breaks that occur during meiotic recombination, and homologous proteins are found in bacterial and other eukaryotic cells. Thus the molecular mechanism of homologous recombination most likely is similar in all types of cells.

Cre Protein and Other Recombinases Catalyze Site-Specific Recombination

Homologous recombination occurs randomly between two homologous DNA segments, and there is relatively little specificity as to the site at which the actual crossover occurs. In *site-specific* recombination, a different type of process, relatively short, unique nucleotide sequences in two DNA molecules are recognized by enzymes called *recombinases*, which then catalyze the joining of the two molecules. Several examples of site-specific recombination have been discovered in both prokaryotic and eukaryotic cells.

One well-studied example, the integration of bacterio-phage λ into a particular site in the *E. coli* chromosome (see Figure 6-19), is catalyzed by a viral enzyme called *integrase*. The genome of λ phage contains a 15-bp attachment site whose 7-bp core sequence is identical with the integration (or attachment) site in the host-cell DNA. Integration can be carried out in a cell-free reaction system with only two purified proteins, λ integrase and integration host factor, a cellular protein. Integrase also catalyzes the reverse reaction, excision of the circular λ phage DNA from a bacterial chromosome.

The site-specific recombination reaction that is best un-derstood at the molecular level is catalyzed by Cre, a pro-tein encoded in the genome of bacteriophage P1. During phage P1 DNA replication, long multimeric DNAs are pro-duced; these are resolved into monomeric P1 DNAs by re-combination at loxP sites, which separate the P1 DNA monomers composing a multimeric DNA (Figure 12-36).

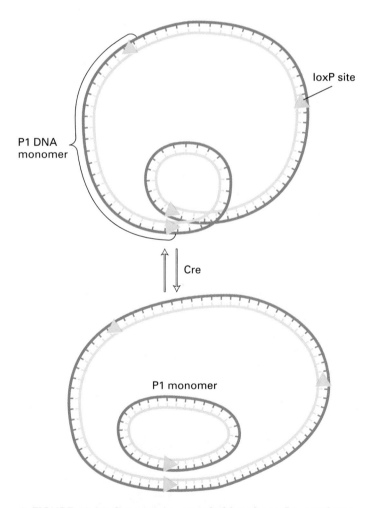

▲ FIGURE 12-36 Cre protein, encoded by phage P1, catalyzes site-specific recombination between loxP sites on multimeric circles of P1 DNA, generating a circular monomeric DNA. Repetition of this process eventually converts the entire multimeric DNA into circular P1 monomers.

The Cre protein that catalyzes this reaction is similar in structure and function to λ integrase. Moreover, the exci-sion of monomeric P1 DNAs from a multimeric circle by Cre is mechanistically similar to the excision of λ DNA from the bacterial chromosome by integrase.

The mechanism of Cre-catalyzed recombination is depicted in Figure 12-37. This mechanism involves the sequential for-mation of two transient intermediates in which Cre and DNA are covalently linked by phosphotyrosine bonds, similar to those between DNA and topoisomerase II (see Figure 12-14, *inset*). Reaction of the first Cre-DNA intermediate generates a Holliday-type structure; reaction of the second one yields the recombinant double-stranded DNA products. By using DNA molecules with mutations or single-strand breaks in the short loxP homologous recognition sites, researchers have been able to stop the Cre-catalyzed reaction at several stages and collect intermediates (Figure 12-38).

The P1 recombinase system has proved useful to mouse geneticists. Because many genes are required at multiple stages of development, "classical" gene-knockout mice, pro-duced by the procedure depicted in Figure 8-34, frequently die as early embryos. However, the P1 loxP-Cre recombi-nase system has enabled geneticists to generate animals in which a particular gene of interest is deleted only in one spe-cific tissue (see Figure 8-35). Such tissue-specific knockout mice enable researchers to study the function of any gene in just one tissue of the adult animal. Thus, an understanding of how one site-specific recombination system excises a DNA segment out of a specific chromosomal site led to the development of an important tool for mouse geneticists.

SUMMARY Recombination between Homologous DNA Sites

• During homologous recombination, two duplex DNA molecules are broken and strands are exchanged. This process, which occurs randomly along the genomes of all organisms, plays an important role in generating genetic diversity.

• Double-strand breaks in DNA initiate most cases of homologous recombination. The break becomes enlarged to gaps, forming single-stranded 3′ recombinogenic ends that invade the other duplex. Repair synthesis of the missing regions forms an intermediate containing two crossed-strand Holliday junctions. Resolution of this intermediate occurs by rotation followed by cleavage and ligation of two strands at each Holliday junction (see Figure 12-31).

• In *E. coli*, a recombinogenic end created by the RecBCD enzyme complex is stabilized by binding of RecA protein (see Figure 12-32). Catalyzed by RecA, the single-stranded 3′ end then pairs with and invades a homologous duplex DNA segment, forming an inter-mediate containing two regions of heteroduplex DNA

(a)

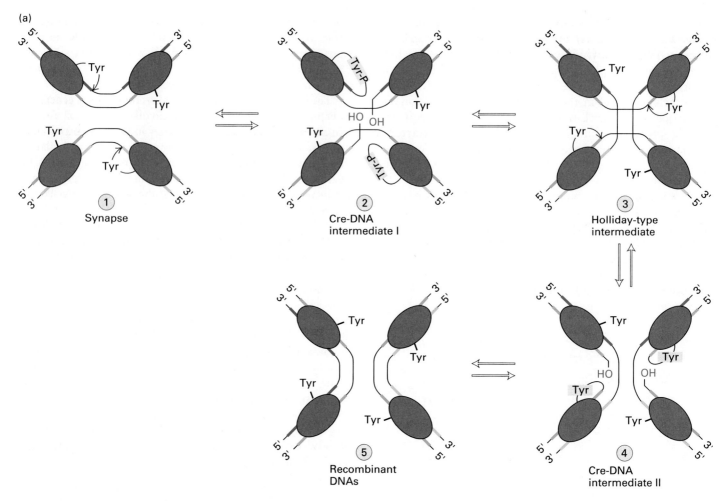

① Synapse

② Cre-DNA intermediate I

③ Holliday-type intermediate

④ Cre-DNA intermediate II

⑤ Recombinant DNAs

▲ FIGURE 12-37 **Mechanism of Cre-loxP site-specific recombination.** Two Cre proteins bind on either side of each of two loxP sites, and then associate together, forming a synapse composed of two DNA molecules and four Cre proteins ①. A tyrosine (Tyr) residue on opposite Cre proteins cleaves the DNA strand to which the proteins are bound, forming two covalent 3′ phosphotyrosine bonds and two free 5′-hydroxyl ends ②. The free 5′-hydroxyl ends in this Cre-DNA intermediate then react with the opposite phosphotyrosine bonds, yielding a Holliday-type intermediate and regenerating native Cre proteins ③. Next, a tyrosine residue on the other two Cre proteins at the synapse cleave the DNA strands to which they are bound, again forming covalent 3′ phosphotyrosine bonds and free 5′-hydroxyl ends to give a second Cre-DNA intermediate ④. Finally, the free 5′-hydroxyl ends of this intermediate attack the opposite phosphotyrosine bonds, yielding the recombinant DNA products still bound to regenerated native Cre proteins ⑤. [Adapted from F. Guo, D. Gopaul, and G. Van Duyne, 1997, *Nature* **389**:40.]

▶ FIGURE 12-38 **Ribbon model based on x-ray crystallography of the Cre-DNA intermediate II in site-specific recombination (see Figure 12-37).** Each of the cleaving subunits is covalently linked to a DNA strand by a phosphotyrosine bond. [Courtesy of Dr. G. Van Duyne.]

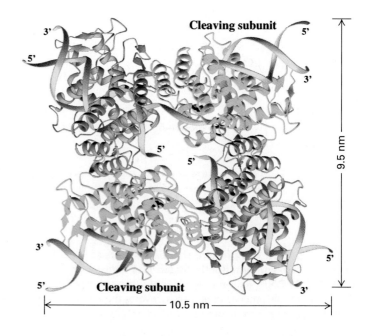

Cleaving subunit

Cleaving subunit

9.5 nm

10.5 nm

and two Holliday junctions. After branch migration, catalyzed by RuvA and RuvB, two strands at each junction are cut by RuvC, an endonuclease, and then ligated to yield two duplex DNA molecules (see Figure 12-35).

- Because eukaryotic cells express proteins homologous to the *E. coli* recombination proteins, all cells are thought to carry out homologous recombination by a similar molecular mechanism.

- Site-specific recombinases recognize, cleave, and recombine short homologous DNA sequences in two different DNA molecules.

- During site-specific recombination catalyzed by phage P1 Cre, transient phosphotyrosine bonds are formed between bound Cre molecules and the cut 3'-hydroxyl ends of the DNA duplexes (see Figure 12-37). Integration of phage λ, catalyzed by integrase, is thought to occur by a similar mechanism.

PERSPECTIVES for the Future

Large multisubunit complexes, often called *molecular machines,* catalyze each of the steps of DNA replication, recombination, and repair discussed in this chapter. Because the three-dimensional structure of none of these complexes has been determined, we do not know how the various proteins in each complex actually interact and catalyze very complex reactions rapidly and with exquisite precision. For example, proteins that bind to replication origins melt the DNA and then recruit helicases, which in turn recruit primases. Multisubunit DNA polymerases that extend RNA primers interact both with "clamp" proteins that ensure processivity and with DNA helicases. Yet we have little idea whether our schematic diagrams depicting DNA replication (see Figures 12-7, 12-11, and 12-12) actually represent the way in which all these proteins physically interact to generate functional complexes. Nor do we know what causes these complexes to form and then disassemble continually in accord with the needs of the cell to replicate its DNA.

For many years DNA mutation and repair was considered somewhat of a backwater science, of interest mainly to radiologists and to those focused on the biological effects of radiation. This science has moved to center stage in recent years with the realization that many chemicals in our environment have the potential to modify DNA and cause tumors. It is obviously essential to understand precisely how these chemicals are metabolized by the body, how they modify DNA, and how the damage can be repaired. Equally important is the quantitative assessment of possible genetic effects caused by trace amounts of certain chemicals, that is, whether they actually cause harm to humans. Such assessments, for example, have shown that supposed hazards such as radiation from power lines and insecticides such as Alar

have no demonstrable genetic effects, despite initial claims to the contrary in the mass media. Other possible genetic hazards deserve to be studied with the same rigor.

Eukaryotic cells have developed elaborate systems for continually monitoring the integrity of genomic DNA. Single-stranded gaps in otherwise duplex DNA generally signify some sort of trouble. After DNA replication in the S phase, mammalian cells pause in order to repair such gaps, and generally do not undergo mitosis unless all lesions have been fixed. If the extent of DNA damage is too great, the cells commit suicide, a process called *apoptosis,* rather than allow multiple mutations to persist in the organism. Many of these "checkpoints" in the cell cycle are described in the following chapter, but we have much to learn about how DNA replication and repair are integrated into the process of cell duplication, and how controls over these process are lost in many tumors.

PERSPECTIVES in the Literature

In this chapter we have seen how large, multi-protein complexes catalyze specific steps in DNA replication, recombination, and repair. These complexes need to be assembled and disassembled in a highly regulated manner, and current research indicates that chaperones are essential for both the assembly and disassembly processes. The four papers below are examples of recent research in this area. As you read them, consider the following questions:

1. How is this function of chaperones similar to and different from their function in catalyzing protein folding (Chapters 3 and 17)?

2. What features of the substrate protein might you expect to be recognized by the chaperones that disassemble protein complexes? Compare these features with those recognized by chaperones involved in protein folding.

3. How does the involvement of chaperones in these reactions influence how assembly and disassembly of the complexes are regulated?

Levchenko, I., L. Luo, and T. A. Baker. 1995. Disassembly of the Mu transposase tetramer by the ClpX chaperone. *Genes Dev.* **9:**2399–2408.

Levchenko, I., M. Yamauchi, and T. A. Baker. 1997. ClpX and MuB interact with overlapping regions of Mu transposase: implications for control of the transposition pathway. *Genes Dev.* **11:**1561–1572.

Schirmer, E. C., J. R. Glover, M. A. Singer, and S. Lindquist. 1996. HSP100/Clp proteins: a common mechanism explains diverse functions. *Trends Biochem.Sci.* **21:**289–296.

Wickner, S., S. Gottesman, D. Skowyra, J. Hoskins, K. McKenney, and M. R. Maurizi. 1994. A molecular chaperone, ClpA, functions like DnaK and DnaJ. *Proc. Nat'l. Acad. Sci. USA* **91:**12218–12222.

Testing Yourself on the Concepts

1. Compare the enzymatic machinery and mechanism of DNA replication in prokaryotes to that in eukaryotes.

2. Describe the mechanism in bacteria for repair of DNA containing mismatches, chemically modified bases, and double strand breaks.

3. Describe the events in the formation and resolution of Holliday structures.

MCAT/GRE-Style Questions

Key Application Please read the section titled "Telomerase Prevents Progressive Shortening of Lagging Strands during Eukaryotic DNA Replication" (p. 467) and refer to Figure 12-13; then answer the following questions:

1. Special structures called telomeres are needed in eukaryotic cells but not bacteria because

   a. Eukaryotic cells contain linear chromosomes.

   b. Eukaryotic cells have more than one chromosome.

   c. Eukaryotic cells contain a nucleus.

   d. Eukaryotic cells contain more forms of DNA polymerase.

2. Telomerase is a ribonucleoprotein complex that

   a. Synthesizes DNA in the absence of a DNA or RNA template.

   b. Synthesizes DNA using a DNA template that is part of the ribonucleoprotein complex.

   c. Synthesizes DNA using an RNA template that is part of the ribonucleoprotein complex.

   d. Synthesizes DNA using ribosomal RNA as template.

3. In the absence of telomerase activity

   a. The ends of chromosomes would lengthen.

   b. The ends of chromosomes would shorten.

   c. Chromosomes would not segregate to daughter cells during mitosis.

   d. Replication of chromosomes would be inhibited.

4. Compared to a wild-type cell, a cell lacking telomerase activity would likely

   a. Die sooner.

   b. Grow longer.

   c. Show lower rates of recombination.

   d. Contain more somatic mutations.

5. Telomerase activity in a cell would be unnecessary if

   a. DNA could be synthesized with shorter RNA primers.

   b. DNA could be synthesized in the absence of ligase.

   c. DNA could be synthesized using deoxyribonucleotide monophosphates.

   d. DNA could be synthesized in the absence of RNA primers.

Key Concept Please read the section titled "Mismatch Repair of Single-Base Mispairs" (p. 476) and answer the following questions:

6. Base deamination can cause single base pair mutations because

   a. Deaminated G structurally resembles U.

   b. Deaminated G can base pair with U.

   c. Deaminated C structurally resembles A.

   d. Deaminated C can base pair with A.

7. In *E. coli*, parental DNA strands are distinguishable from newly synthesized daughter DNA strands because

   a. Parental strands are methylated and daughter strands are unmethylated.

   b. Parental strands are unmethylated and daughter strands are methylated.

   c. Parental strands are depurinated and daughter strands are not.

   d. Parental strands contain point mutations and daughter strands contain deletions.

8. A bacterial strain containing a mutant MutS protein that binds equally well to methylated and unmethylated DNA would be expected to

   a. Repair a single-base mismatch predominantly to the wild-type nucleotide.

   b. Repair a single-base mismatch predominantly to the mutant nucleotide.

   c. Repair a single-base mismatch half of the time to the wild type and half of the time to the mutant nucleotide.

   d. Be unable to correct a base pair mismatch.

9. Which of the following enzymatic activities does not play a role in mismatch repair:

   a. Helicase.

   b. Single-stranded exonuclease.

   c. DNA ligase.

   d. Primase.

Key Experiment Please read the section titled "Double-Strand Breaks in DNA Initiate Recombination" (p. 484) and refer to Figure 12-31; then answer the following questions:

10. Which of the following enzymatic activities is not required for the double-strand break repair model:

   a. DNA endonuclease.

   b. $5' \rightarrow 3'$ exonuclease.

   c. $3' \rightarrow 5'$ exonuclease.

   d. DNA synthesis.

11. The evidence for a gene conversion event during double-strand break repair is

a. The segregation of allelic markers in a typical Mendelian ratio.

b. The segregation of allelic markers in a 3:1 ratio.

c. The presence of Holliday junctions.

d. The presence of both recombinant and nonrecombinant genotypes.

12. Yeast are a convenient system to examine gene conversion events because

a. Yeast are eukaryotes with a small genome.

b. Holliday structures can be visualized using electron microscopy.

c. All four meiotic products can be scored in progeny yeast spores.

d. The enzymatic activities of yeast DNA polymerases have been defined.

13. Cleavage at sites 1 and 3 of the molecule shown in step 4 of Figure 12-31 would result in a DNA strand containing which of the following alleles:

a. c-D-E.

b. c-d-e.

c. c-D-e.

d. c-d-E.

Key Terms

autonomously replicating sequence (ARS) 457

carcinogens 472

DNA polymerases 454

gene conversion 484

growing fork 455

helicases 454

heteroduplex 482

Holliday structure 482

lagging strand 461

leading strand 461

mutagens 472

Okazaki fragments 461

oncogenes 475

primase 460

primer 459

proto-oncogene 475

recombination 482

replication origin 456

telomeres 467

topoisomerases 460

References

General Features of Chromosomal Replication

DePamphilis, M., ed. 1996. *DNA Replication in Eukaryotic Cells.* Cold Spring Harbor Laboratory Press.

Kornberg, A., and T. A. Baker. 1992. *DNA Replication,* 2d ed. W. H. Freeman and Company.

Marians, K. J. 1992. Prokaryotic DNA replication. *Ann. Rev. Biochem.* **61**:673–719.

The DNA Replication Machinery

Bambara, R., R. Murante, and L. Henrickson. 1997. Enzymes and reactions at the eukaryotic DNA replication fork. *J. Biol. Chem.* **272**:4647–4650.

Bell, S., and T. A. Baker. 1998. Polymerases and the replisome: machines within machines. *Cell* **92**:295–305.

Blackburn, E., and C. Greider. 1995. *Telomeres.* Cold Spring Harbor Laboratory Press.

Borowiec, J. A., et al. 1990. Binding and unwinding—how T antigen engages the SV40 origin of DNA replication. *Cell* **60**:181–184.

Bramhill, D., and A. Kornberg. 1988. A model for initiation at origins of DNA replication. *Cell* **54**:915–918.

Burgers, P. M. 1998. Eukaryotic DNA polymerases in DNA replication and DNA repair. *Chromosoma* **107**:218–227.

Cozzarelli, N. R. 1980. DNA gyrase and the supercoiling of DNA. *Science* **207**:953–960.

De Pamphilis, M. L. 1993. Eukaryotic DNA replication: anatomy of an origin. *Ann. Rev. Biochem.* **62**:29–63.

Dutta, A., and S. Bell. 1997. Initiation of DNA replication in mammalian cells. *Ann. Rev. Cell Devel. Biol.* **13**:293–332.

Edwards, A. M., A. Bochkarev, and L. Frappier. 1998. Origin DNA-binding proteins. *Curr. Opin. Struc. Biol.* **8**:49–53.

Egelman, E. 1996. Homomorphous hexameric helicases: tales from the ring cycle. *Structure* **4**:759–762.

Greider, C. 1996. Telomere length regulation. *Ann. Rev. Biochem.* **66**:337–366.

Hübscher, U., and P. Thömmes. 1992. DNA polymerase ε: in search of a function. *Trends Biochem. Sci.* **17**:55–58.

Joyce, C., and T. Steitz. 1995. Polymerase structures and functions: variations on a theme? *J. Bacteriol.* **177**:6321–6329.

Lingner, J., and T. R. Cech. 1998. Telomerase and chromosome end maintenance. *Curr. Opin. Genet. Devel.* **8**:226–232.

Lohman, T., and K. Bjornson. 1996. Mechanisms of helicase-catalyzed DNA unwinding. *Ann. Rev. Biochem.* **66**:61–92.

Mossi, R., and U. Hübscher. 1998. Clamping down on clamps and clamp loaders—the eukaryotic replication factor C. *Eur. J. Biochem.* **254**:209–216.

Waga, S., and B. Stillman. 1998. The DNA replication fork in eukaryotic cells. *Ann. Rev. Biochem.* **67**:721–751.

West, S. 1997. DNA helicases: new breeds of translocating motors and molecular pumps. *Cell* **86**:177–180.

Wold, M. 1997. Replication protein A: a heterotrimeric, single-stranded DNA-binding protein required for eukaryotic DNA metabolism. *Ann. Rev. Biochem.* **66**:61–92.

Zakian, V. 1996. Structure, function, and replication of *Saccharomyces cerevisiae* telomeres. *Ann. Rev. Genetics* **30**:141–172.

Role of Topoisomerases in DNA Replication

Berger, J., S. Gamblin, S. Harrison, and J. Wang. 1996. Structure and mechanism of DNA topoisomerase II. *Nature* **379**:225–232.

Cozzarelli, N. R. 1980. DNA gyrase and the supercoiling of DNA. *Science* **207**:953–960.

Pulleybank, D. 1997. Of Topo and Maxwell's Dream. *Science* **277**:648–649.

Wang, J. 1996. DNA topoisomerases. *Ann. Rev. Biochem.* **66**:635–692.

DNA Damage and Repair and Their Role in Carcinogenesis

Chu, G. 1997. Double strand break repair. *J. Biol. Chem.* **272**:24097–24100.

Clark, A. J. 1991. *rec* genes and homologous recombination proteins in *Escherichia coli*. *Biochimie* **73**:523–532.

Fishel, R., and T. Wilson. 1997. MutS homologs in mammalian cells. *Curr. Opin. Genet. Devel.* **7**:105–113.

Friedberg, E. 1997. *Correcting the Blueprint of Life: An Historical Account of the Discovery of DNA Repair Mechanisms*. Cold Spring Harbor Laboratory Press.

Friedberg, E., G. Walker, and W. Siede. 1995. *DNA Repair and Mutagenesis*. ASM Press.

Jeggo, P. 1998. DNA breakage and repair. *Adv. Genetics* 38:185–218.

Kolodner, R. 1996. Biochemistry and genetics of eukaryotic mismatch repair. *Genes Devel.* 10:1433–1442.

Lindahl, T., P. Karran, and R. D. Wood. 1997. DNA excision repair pathways. *Curr. Opin. Genet. Devel.* 7:158–169.

Loeb, L. 1998. Cancer cells exhibit a mutator phenotype. *Adv. Cancer Res.* 72:26–55.

Modrich, P. 1997. Strand-specific mismatch repair in mammalian cells. *J. Biol. Chem.* 272:24727–24730.

Modrich, P., and R. Lahue. 1996. Mismatch repair in replication, fidelity, genetic recombination, and cancer biology. *Ann. Rev. Biochem.* 66:101–134.

Prolla, T. A. 1998. DNA mismatch repair and cancer. *Curr. Opin. Cell Biol.* 10:311–316.

Vassylyev, D., and K. Morikawa. 1997. DNA-repair enzymes. *Curr. Opin. Struc. Biol.* 7:103–109.

Wood, R. D. 1996. DNA repair in eukaryotes. *Ann. Rev. Biochem.* 65:135–168.

Wood, R. D. 1997. Nucleotide excision repair in mammalian cells. *J. Biol. Chem.* 272:23465–23468.

Recombination between Homologous DNA Sites

Cox, M. M., and I. R. Lehman. 1987. Enzymes of general recombination. *Ann. Rev. Biochem.* 56:229–262.

Haber, J. E. 1997. A super twist on the initiation of meiotic recombination. *Cell* 89:163–166.

Jayaram, M. 1997. The cis-trans paradox of integrase. *Science* 276:49–50.

Kogoma, T. 1996. Recombination by recombination. *Cell* 85:625–627.

Kuzimov, A. 1996. Unraveling the late stages of recombinational repair: metabolism of DNA junctions in *E. coli*. *BioEssays* 18:757–765.

Landy, A. 1989. Dynamic, structural, and regulatory aspects of lambda site-specific recombinations. *Ann. Rev. Biochem.* 58:913–949.

Sancar, A. 1993. Molecular matchmakers. *Science* 259:1415–1420.

Smith, G. R. 1989. Homologous recombination in *E. coli*: multiple pathways for multiple reasons. *Cell* 58:807–809.

Stahl, F. 1996. Meiotic recombination in yeast: coronation of the double-strand-break repair model. *Cell* 87:965–968.

West, S. F. 1992. Enzymes and molecular mechanisms of genetic recombinations. *Ann. Rev. Biochem.* 61:603–640.

West, S. 1997. Processing of recombination intermediates by the Ruv ABC proteins. *Ann. Rev. Genet.* 31:213–244.

Regulation of the Eukaryotic Cell Cycle

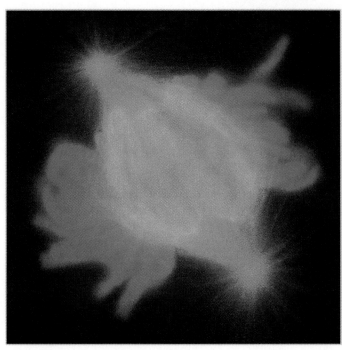

A newt lung cell in metaphase. Microtubules were detected by indirect immunofluorescence (green) and chromosomes were visualized by staining DNA with an intercalating dye (blue).

Most eukaryotic cells proceed through an ordered series of events, constituting the **cell cycle,** during which their chromosomes are duplicated and one copy of each duplicated chromosome segregates to each of two daughter cells (see Figure 1-9). Regulation of the cell cycle is critical for the normal **development** of multicellular organisms. Loss of control ultimately leads to cancer, an all-too-familiar disease that kills one in every six people in the developed world. In the late 1980s, it became clear that the molecular processes regulating the main events in the cell cycle— chromosome replication and **cell division**—are fundamentally similar in all eukaryotic cells. Because of this similarity, research with diverse organisms, each with its own particular experimental advantages, has contributed to a growing understanding of how these events are coordinated and controlled.

Biochemical and genetic techniques, as well as recombinant DNA technology, have been employed in studying various aspects of the eukaryotic cell cycle. These studies have revealed that cell replication is primarily controlled by regulating the timing of two critical events in the cell cycle: nuclear DNA replication and **mitosis.** The master controllers of these events are a small number of *heterodimeric protein kinases* that contain a regulatory subunit and catalytic subunit. These kinases regulate the activities of multiple proteins involved in DNA replication and mitosis by phosphorylating them at specific regulatory sites, activating some and inhibiting others to coordinate their activities. The molecular machinery that replicates DNA was described in Chapter 12, and the microtubule-dependent events that segregate each daughter chromatid during mitosis are considered in

MEDIA CONNECTIONS

Overview: Cell Cycle Control

Classic Experiment 13.1: Cell Biology Emerging from the Sea: The Discovery of Cyclins

Chapter 19. In this chapter we focus on how the cell cycle is regulated and the experimental systems that have led to our current understanding of these crucial regulatory mechanisms.

13.1 Overview of the Cell Cycle and Its Control

We begin our discussion by reviewing the stages of the eukaryotic cell cycle, presenting a summary of the current model of how the cycle is regulated, and briefly describing key experimental systems that have provided revealing information about cell-cycle regulation.

The Cell Cycle Is an Ordered Series of Events Leading to Replication of Cells

As illustrated in Figure 13-1, the cell cycle is divided into four major phases. In cycling (replicating) somatic cells, chromosomes are replicated during the **S (synthesis) phase.** After progressing through the **G_2 phase,** cells begin the complicated process of mitosis, also called the **M phase,** which is divided into several stages (see Figure 19-34). Chromosomes condense during the **prophase** period of mitosis, by tightly folding loops of the 30-nm chromatin fiber attached to the chromosome scaffold (see Figure 9-35). Sister **chromatids,** produced by DNA replication during the S phase, remain attached at the centromere and multiple points along their length and become aligned in the center of the cell during **metaphase.** During the **anaphase** portion of mitosis, sister chromatids separate and move to opposite poles of the **mitotic apparatus,** or spindle (see Figure 19-36), segregating one of the two sister chromatids to each daughter cell.

In most cells from higher eukaryotes, the **nuclear envelope** breaks down into multiple small vesicles early in mitosis and re-forms around the segregated chromosomes as they decondense during **telophase,** the last mitotic stage. The physical division of the cytoplasm, called **cytokinesis,** then yields two daughter cells. The Golgi complex and endoplasmic reticulum also vesiculate during mitosis and re-form in the two daughter cells after cell division. In yeasts and other fungi, the nuclear envelope does not break down. In these organisms, the mitotic spindle forms within the nuclear envelope, which then pinches off, forming two nuclei at the time of cytokinesis. Following mitosis, cycling cells enter the **G_1 phase,** the period before DNA synthesis is reinitiated in the S phase.

In vertebrates and diploid yeasts, cells in G_1 have a diploid number of chromosomes ($2n$), one inherited from each parent. In haploid yeasts, cells in G_1 have one of each chromosome ($1n$). Rapidly replicating human cells progress through the full cell cycle in about 24 hours: mitosis takes $\approx$30 minutes; G_1, 9 hours; the S phase, 10 hours; and G_2, 4.5 hours. In contrast, the full cycle takes only $\approx$90 minutes in rapidly growing yeast cells.

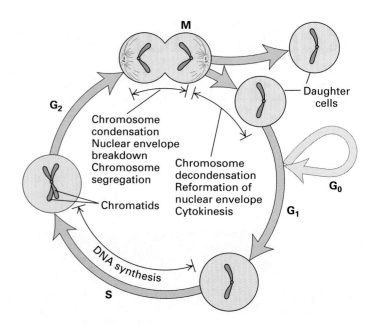

▲ **FIGURE 13-1 The fate of a single parental chromosome throughout the eukaryotic cell cycle.** Although chromosomes condense only during mitosis, they are shown in condensed form to emphasize the number of chromosomes at different cell-cycle stages. The nuclear envelope is not depicted. Following mitosis (M), daughter cells contain $2n$ chromosomes in diploid organisms and $1n$ chromosomes in haploid organisms including yeasts maintained in the haploid state. In proliferating cells, G_1 is the period between "birth" of a cell following mitosis and the initiation of DNA synthesis, which marks the beginning of the S phase. At the end of the S phase, cells enter G_2 containing twice the number of chromosomes as G_1 cells ($4n$ in diploid organisms). The end of G_2 is marked by the onset of mitosis, during which numerous events leading to cell division occur. The G_1, S, and G_2 phases are collectively referred to as interphase, the period between one mitosis and the next. Most nonproliferating cells in vertebrates leave the cell cycle in G_1, entering the G_0 state. See also Figure 1-10.

Postmitotic cells in multicellular organisms can "exit" the cell cycle and remain for days, weeks, or in some cases (e.g., nerve cells and cells of the eye lens) even the lifetime of the organism without proliferating further. Most postmitotic cells in vertebrates exit the cell cycle in G_1, entering a phase called G_0 (see Figure 13-1). G_0 cells returning to the cell cycle enter into the S phase; this reentry is regulated, thereby providing control of cell proliferation.

Regulated Protein Phosphorylation and Degradation Control Passage through the Cell Cycle

As mentioned in the chapter introduction, the complex macromolecular events of the eukaryotic cell cycle are regulated by a small number of heterodimeric protein kinases. The concentrations of the regulatory subunits of these kinases, called **cyclins,** increase and decrease in phase with the cell cycle. Their

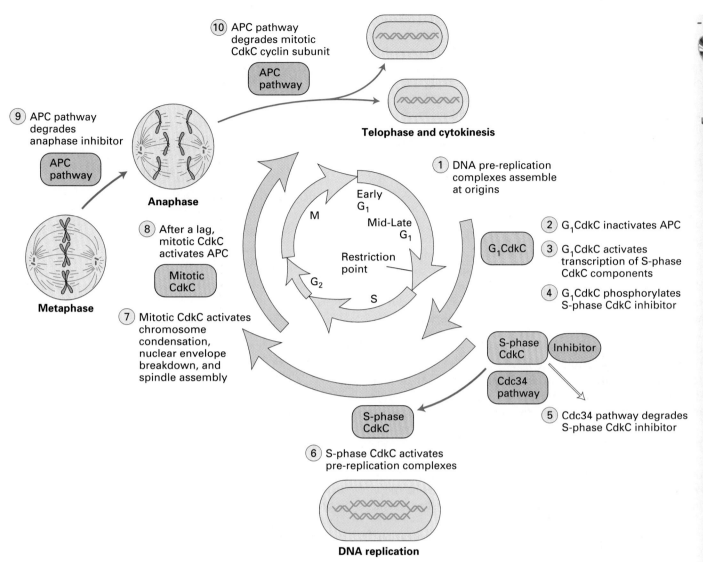

▲ FIGURE 13-2 Current model for regulation of the eukaryotic cell cycle. Passage through the cycle is controlled by G_1, S-phase, and mitotic cyclin-dependent kinase complexes (CdkCs) highlighted in green. These are composed of a regulatory cyclin subunit and a catalytic cyclin-dependent kinase subunit. Protein complexes (orange) in the Cdc34 pathway and APC pathway polyubiquitinate specific substrates including the S-phase inhibitor, anaphase inhibitor, and mitotic cyclins, marking these substrates for degradation by proteasomes (see Figure 3-18). These pathways thus drive the cycle in one direction because of the irreversibility of protein degradation. Proteolysis of anaphase inhibitors inactivates the protein complexes that connect sister chromatids at metaphase (not shown), thereby initiating anaphase.

catalytic subunits are called **cyclin-dependent kinases (Cdks)** because they have no kinase activity unless they are associated with a cyclin. Each Cdk catalytic subunit can associate with different cyclins, and the associated cyclin determines which proteins are phosphorylated by the Cdk-cyclin complex.

Figure 13-2 outlines the role of the three classes of cyclin-Cdk complexes that control passage through the cell cycle: the G_1, S-phase, and mitotic Cdk complexes. When cells are stimulated to replicate, G_1 Cdk complexes are expressed first. These prepare the cell for the S phase by activating transcription factors that cause expression of enzymes re-

quired for DNA synthesis and the genes encoding S-phase Cdk complexes. The activity of S-phase Cdk complexes is initially held in check by a specific inhibitor. Then, in late G_1, G_1 Cdk complexes induce the degradation of the S-phase inhibitor, releasing the activity of the S-phase Cdk complexes, which stimulate entry into the S phase.

Once activated by degradation of the S-phase inhibitor, the S-phase Cdk complexes phosphorylate regulatory sites in the proteins that form DNA **pre-replication complexes**, which are assembled on replication origins during G_1. Phosphorylation of these proteins by S-phase Cdk complexes not only activates initiation of DNA replication but also prevents re-assembly of new pre-replication complexes. Because of this inhibition, each chromosome is replicated just once during passage through the cell cycle, ensuring that the proper chromosome number is maintained in the daughter cells.

Mitotic Cdk complexes are synthesized during the S phase and G_2, but their activities are held in check until DNA synthesis is completed. Once activated, mitotic Cdk complexes induce chromosome condensation, breakdown of the nuclear envelope, assembly of the mitotic spindle apparatus, and alignment of condensed chromosomes at the metaphase plate (see Figure 19-34). After the proper association of all chromosomes with spindle microtubules has occurred, the mitotic Cdk complexes activate the **anaphase-promoting complex (APC)**. This multiprotein complex directs the ubiquitin-mediated proteolysis of anaphase inhibitors, leading to inactivation of the protein complexes that connect sister chromatids at metaphase. Degradation of these inhibitors thus permits the onset of anaphase, during which sister chromatids segregate to opposite spindle poles. Later in anaphase, the APC also directs proteolytic degradation of the mitotic cyclins. The resulting decrease in mitotic Cdk activity permits the now separated chromosomes to decondense, the nuclear envelope to re-form around daughter-cell nuclei during telophase, and the cytoplasm to divide at cytokinesis, yielding the two daughter cells.

During early G_1 of the next cell cycle, phosphatases dephosphorylate the proteins that form pre-replication complexes. As a result, these complexes can assemble at replication origins in preparation for the next S phase. Phosphorylation of APC by G_1 Cdk complexes in late G_1 inactivates it, allowing the subsequent accumulation of mitotic cyclins during the S phase and G_2 of the ensuing cycle.

Passage through three critical cell-cycle transitions, $G_1 \rightarrow$ S phase, metaphase $\rightarrow$ anaphase, and anaphase $\rightarrow$ telophase and cytokinesis, is irreversible because these transitions are triggered by the regulated degradation of proteins, an irreversible process. As a consequence, cells are forced to traverse the cell cycle in one direction only.

In higher organisms, control of the cell cycle is achieved primarily by regulating the synthesis and activity of G_1 Cdk complexes. Extracellular **growth factors**, called **mitogens**, induce the synthesis of G_1 Cdk complexes. The activity of these and other Cdk complexes is regulated by phosphorylation at specific inhibitory and activating sites in the catalytic subunit. Once mitogens have acted for a sufficient period, the cell cycle continues through mitosis even when they are removed. The point in late G_1 where passage through the cell cycle becomes independent of mitogens is called the **restriction point** (see Figure 13-2).

Diverse Experimental Systems Have Been Used to Identify and Isolate Cell-Cycle Control Proteins

The first evidence that diffusible factors regulate the cell cycle came from cell-fusion experiments with cultured

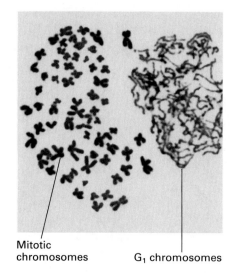

Mitotic chromosomes

G_1 chromosomes

▲ **FIGURE 13-3 Fusion of mitotic cells with interphase cells in G_1.** In unfused interphase cells, the nuclear envelope is intact and the chromosomes are not condensed, so individual chromosomes cannot be distinguished (see Figure 5-42). In mitotic cells, the nuclear envelope is absent and the individual replicated chromosomes are highly condensed. In the fused cell shown here, the nuclear envelope of the G_1 cell has broken down and is not visible. The chromosomes are partially condensed, although homologous chromosomes do not associate. The two sister chromatids of the mitotic chromosomes are joined at the centromere and are distinct. [From R. T. Johnson and P. N. Rao, 1970, *Biol. Rev.* **46**:97.]

mammalian cells. When **interphase** cells in the G_1, S, or G_2 stage of the cell cycle were fused to cells in mitosis, their nuclear envelopes vesiculated and their chromosomes condensed (Figure 13-3). This finding indicates that some diffusible component or components in the cytoplasm of the mitotic cells forced interphase nuclei to undergo many of the processes associated with early mitosis. We now know that these factors are the mitotic Cdk complexes. Similarly, when cells in G_1 were fused to cells in the S phase and the fused cells exposed to radiolabeled thymidine, the label was incorporated into the DNA of the G_1 nucleus, indicating that DNA synthesis began in the G_1 nucleus shortly after fusion. However, when cells in G_2 were fused to S-phase cells, no incorporation of labeled thymidine occurred in the G_2 nuclei. Thus diffusible factors in an S-phase cell can enter the nucleus of a G_1 cell and stimulate DNA synthesis, but these factors cannot induce DNA synthesis in a G_2 nucleus. We now know that these factors are S-phase Cdk complexes, which can activate the pre-replication complexes assembled on DNA replication origins in early G_1 nuclei. Although these cell-fusion experiments demonstrated that diffusible factors control entry into the S and M phases of the cell cycle, genetic and biochemical experiments were needed to identify these factors.

The budding yeast *Saccharomyces cerevisiae* and the distantly related fission yeast *Schizosaccharomyces pombe* have been especially useful for isolation of mutants that are blocked at specific steps in the cell cycle or that exhibit altered regulation of the cycle. In both of these yeasts, temperature-sensitive mutants with defects in specific proteins required to progress through the cell cycle are readily recognized microscopically and therefore easily isolated. *S. cerevisiae* daughter cells form from a growing bud, whose size relative to the parental cell increases during the cell cycle. Mutant *S. cerevisiae* cells with a cell-cycle defect are easily identified because at the nonpermissive temperature they are arrested in the budding process (see Figure 8-9). Such cells are called *cdc* (cell-division cycle) mutants. *S. pombe* cells, in contrast, increase in length and then divide in the middle to form daughter cells. In this yeast, *cdc* mutants grow without dividing and form enormously elongated cells at the nonpermissive temperature. Other *S. pombe* mutants, called *wee* (from the Scottish word for small), divide before the parental cell has grown to the normal length, forming cells that are shorter than normal.

Temperature-sensitive mutations that block progression through the cell cycle at the nonpermissive temperature obviously prevent colony formation from a single haploid yeast cell. The wild-type alleles of recessive temperature-sensitive mutant alleles can be isolated readily by transforming haploid mutant cells with a plasmid library prepared from wild-type cells and then plating the transformed cells at the nonpermissive temperature (Figure 13-4). Complementation of

the recessive mutation by the wild-type allele on one of the plasmids in the library allows a transformed mutant cell to grow into a colony; the plasmid bearing the wild-type allele can then be recovered from those cells. Because many of the proteins that regulate the cell cycle are highly conserved, human cDNAs cloned into yeast expression vectors often can complement yeast cell-cycle mutants, leading to the rapid isolation of human genes encoding cell-cycle control proteins.

Biochemical studies require the preparation of cell extracts from many cells. For biochemical studies of the cell cycle, the eggs and early embryos of amphibians and marine invertebrates are particularly suitable. In these organisms, multiple synchronous cell cycles follow fertilization of a large egg. By isolating large numbers of eggs from females and fertilizing them simultaneously by addition of sperm (or treating them in ways that mimic fertilization), researchers can obtain extracts for analysis of proteins and enzymatic activities that occur at specific points in the cell cycle.

In the following sections we describe critical experiments that led to the current model of eukaryotic cell-cycle regulation summarized in Figure 13-2 and present further details of the various regulatory events. As we will see, results obtained with different experimental systems and approaches have provided insights about each of the key transition points in the cell cycle.

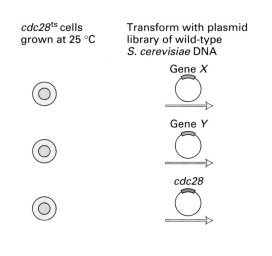
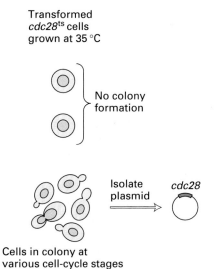

▲ **FIGURE 13-4 Isolation of wild-type cell-division cycle (CDC) genes from S. cerevisiae cells carrying temperature-sensitive mutations in these genes.** After mutant cells are transformed with a genomic library prepared from wild-type cells, they are cultured at the permissive temperature; the transformed cells are then plated at the nonpermissive temperature (35° C). Each transformed cell takes up a single plasmid containing one genomic DNA fragment. Most such fragments include genes

(e.g., *X* and *Y*) that do not encode the defective Cdc protein; transformed cells that take up such fragments do not form colonies at the nonpermissive temperature. The rare cell that takes up a plasmid containing the wild-type version of the mutant gene (in this case *cdc28*^ts^) is complemented, allowing the cell to replicate and form a colony at the nonpermissive temperature. Plasmid DNA isolated from this colony carries the wild-type *CDC* gene. The same procedure is used to isolate wild-type cdc genes in *S. pombe*.

SUMMARY Overview of the Cell Cycle and Its Control

- The eukaryotic cell cycle is divided into four phases: M (mitosis), G_1 (the period between mitosis and the initiation of nuclear DNA replication), S (the period of nuclear DNA replication), and G_2 (the period between the completion of nuclear DNA replication and mitosis) (see Figure 13-1).

- Cdk complexes, composed of a regulatory cyclin subunit and a catalytic cyclin-dependent kinase subunit, regulate progress of a cell through the cell cycle (see Figure 13-2). Large protein complexes also mark specific inhibitors of cell-cycle events for proteolytic degradation by **proteasomes**.

- Diffusible factors in mitotic cells, now known to be mitotic Cdk complexes, cause chromosome condensation and vesiculation of the nuclear envelope in G_1 and G_2 cells when they are fused to mitotic cells. Similarly, S-phase Cdk complexes stimulate DNA replication in the nuclei of G_1 cells when they are fused to S-phase cells.

- Amphibian and invertebrate eggs and early embryos from synchronously fertilized eggs provide sources of extracts for biochemical studies of cell-cycle events.

- The isolation of yeast cell-division cycle (*cdc*) mutants led to the identification of genes that regulate the cell cycle (see Figure 13-4).

13.2 Biochemical Studies with Oocytes, Eggs, and Early Embryos

A breakthrough in identification of the factor that induces mitosis came from studies of oocyte development in the frog *Xenopus laevis*. To understand these experiments, we must first lay out the events of oocyte maturation. As **oocytes** develop in the frog ovary, they replicate their DNA and become arrested in G_2 for 8 months as they grow in size to a diameter of 1 mm, stockpiling all the materials needed for the multiple cell divisions required to generate a swimming, feeding tadpole. When stimulated by a male, an adult female's ovarian cells secrete the steroid hormone progesterone, which induces the G_2-arrested oocytes to enter meiosis I, the first cell division of meiosis (see Figure 8-2). Following this exposure to progesterone, frog oocytes continue through meiosis I, the succeeding interphase, and then arrest during the second meiotic metaphase. At this stage the cells are called *eggs*. When fertilized by sperm, the egg nucleus is released from its metaphase arrest and completes meiosis. The resulting haploid egg nucleus then fuses with the haploid sperm nucleus, producing a diploid **zygote**, and the mitotic divisions of early embryogenesis begin.

MPF Promotes Maturation of Oocytes and Mitosis in Somatic Cells

The process of oocyte maturation, from G_2-arrested oocyte to the egg arrested in metaphase of meiosis II, can be studied in vitro by surgically removing G_2-arrested oocytes from the ovary of an adult female frog and treating them with progesterone (Figure 13-5a). When cytoplasm from eggs arrested in metaphase of meiosis II is microinjected into G_2-arrested oocytes, the oocytes mature into eggs in the absence of progesterone (Figure 13-5b). This system not only led to the initial identification of a factor in egg cytoplasm that stimulates maturation of oocytes in vitro in the absence of progesterone but also provided an assay for this factor, called **maturation-promoting factor (MPF)**. As we will see shortly, MPF turned out to be the key factor that regulates the initiation of mitosis in all eukaryotic cells.

Using the microinjection system to assay MPF activity at different times during oocyte maturation in vitro, researchers found that untreated G_2-arrested oocytes have low levels of MPF activity; treatment with progesterone induces MPF activity as the cells enter meiosis I (Figure 13-6). As the cells enter the interphase between meiosis I and II, MPF activity falls; it then rises as the cells enter meiosis II and are arrested. Following fertilization, MPF activity falls again until the zygote (fertilized egg) enters the first mitosis of embryonic development. All the cells in early frog embryos undergo 12 synchronous cycles of mitosis. Throughout these cycles MPF activity is low in the interphase periods between mitoses and then rises as the cells enter mitosis.

Although initially discovered in frogs, MPF activity has been found in mitotic cells from all species assayed. For example, cultured mammalian cells can be arrested in mitosis by treatment with compounds (e.g., *colchicine*) that inhibit assembly of microtubules. When cytoplasm from such mitotically arrested mammalian cells was injected into G_2-arrested *Xenopus* oocytes, the oocytes matured into eggs; that is, the mammalian somatic mitotic cells contained a cytosolic factor that exhibited frog MPF activity. This finding suggested that MPF controls the entry of mammalian somatic cells into mitosis as well as the entry of frog oocytes into meiosis. When cytoplasm from mitotically arrested mammalian somatic cells was injected into interphase cells, the interphase cells entered mitosis; that is, their nuclear membranes broke down into small vesicles and their chromosomes condensed. Thus MPF is the diffusible factor, first revealed in cell-fusion experiments (see Figure 13-3), that promotes entry of cells into mitosis. Conveniently, the acronym MPF also can stand for **mitosis-promoting factor**, a name that denotes the more general activity of this factor.

Because the assay for MPF is cumbersome, several years passed before MPF was purified by column chromatography and the MPF proteins were characterized. MPF is in fact one of the heterodimeric complexes composed of a cyclin and cyclin-dependent protein kinase (Cdk) now known to

(a) Oocyte maturation in vitro

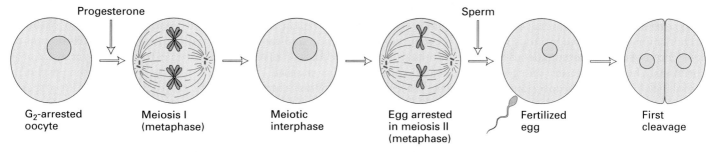

(b) Assay for MPF

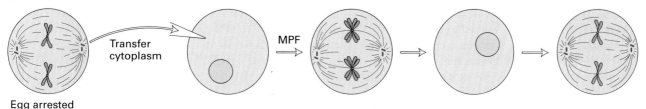

▲ **FIGURE 13-5 In vitro maturation of *Xenopus* oocytes and assay of maturation-promoting factor (MPF).** (a) Treatment of G$_2$-arrested *Xenopus* oocytes with progesterone stimulates them to proceed through meiosis I, interphase, and the first half of meiosis II before arresting in the metaphase of meiosis II. Three pairs of duplicated homologous chromosomes (blue) connected to mitotic spindle microtubules (red) are shown schematically to represent metaphase cells. After addition of sperm and fertilization, fertilized eggs complete meiosis II. The resulting

haploid egg nucleus fuses with the haploid sperm nucleus to produce a diploid zygote, which undergoes the first of 12 synchronous early embryonic cleavages. (b) When cytoplasm from unfertilized eggs arrested in metaphase of meiosis II is injected into G$_2$-arrested oocytes, the oocytes mature into eggs in the absence of progesterone. This process can be repeated multiple times without further addition of progesterone. [See Y. Masui and C. L. Markert, 1971, *J. Exp. Zool.* **177**:129.]

▶ **FIGURE 13-6 Oscillation of MPF activity during meiotic and mitotic cell cycles of *Xenopus* oocytes and early frog embryos.** Diagrams of the cell structures corresponding to each stage are shown in Figure 13-5a. See text for discussion. [See J. Gerhart et al., 1984, *J. Cell Biol.* **98**:1247; adapted from A. Murray and M. W. Kirschner, 1989, *Nature* **339**:275.]

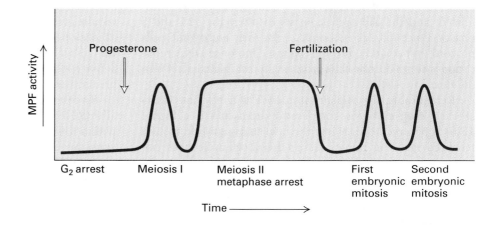

regulate the cell cycle (see Figure 13-2). Each MPF subunit was recognized through different experimental approaches. First we discuss how the regulatory cyclin subunit was identified and then describe how yeast genetic experiments led to discovery of the Cdk catalytic subunit.

Mitotic Cyclin Was First Identified in Early Sea Urchin Embryos

Experiments with inhibitors showed that new protein synthesis is required for the increase in MPF during the mitotic phase of each cell cycle in early frog embryos (see Figure

13-6). Biochemical studies with sea urchin eggs and embryos led to identification of the cyclin component of MPF. As in early frog embryos, the initial cell cycles in the early sea urchin embryo occur synchronously, with all the embryonic cells entering mitosis simultaneously. In these studies, synchronously fertilized sea urchin eggs were incubated with a radiolabeled amino acid and samples were removed every 10 minutes. Protein was isolated from each sample and analyzed by gel electrophoresis followed by autoradiography. The amount of radiolabel in the vast majority of proteins increased steadily through several cell cycles. However, one protein peaked in intensity early in mitosis, fell abruptly during anaphase, and then slowly accumulated during the following interphase to peak early in the next mitosis. Careful analysis showed that this protein, named *cyclin B,* is synthesized continuously during the embryonic cell cycles and is abruptly destroyed at the onset of anaphase.

In subsequent experiments, a cDNA clone encoding sea urchin cyclin B was used as a probe to isolate a homologous cyclin B cDNA from *Xenopus laevis.* Western blotting of MPF purified from *Xenopus* eggs (see Figure 3-44), using antibody prepared against the protein encoded by cyclin B cDNA, showed that one subunit of MPF is indeed cyclin B. The other subunit is the catalytic Cdk subunit, first identified in genetic experiments with yeasts discussed later.

Cyclin B Levels and MPF Activity Change Together in Cycling *Xenopus* Egg Extracts

Some unusual aspects of the rapid cell cycles in early animal embryos provided a way to study the role of mitotic cyclin in controlling MPF activity. Of particular importance, in the 12 rapid, synchronous cell cycles that occur following fertilization of *Xenopus* eggs, the G_1 and G_2 periods are minimized, and the cell cycle consists of alternating M and S phases. Once mitosis is complete, the early embryonic cells proceed immediately into the S phase, and once DNA replication is complete, the cells progress almost immediately into the next mitosis.

Remarkably, the oscillation in MPF activity that occurs as early frog embryos enter and exit mitosis (see Figure 13-6) is observed even when the nucleus is removed from a fertilized egg. This finding shows that a cell-cycle clock operates in the cytoplasm of early frog embryos completely independently of nuclear events. This phenomenon occurs only in synchronously dividing cells of early animal embryos. No transcription occurs during these rapid cell cycles, indicating that all the cellular components required for progress through the truncated cell cycles are stored in the unfertilized egg. In somatic cells generated later in development and in yeasts considered in later sections, specific mRNAs must be produced at particular points in the cell cycle for progress through the cycle to proceed. But in early animal embryos, all the mRNAs necessary for the early cell divisions are present in the unfertilized egg. Extracts prepared from unfertilized frog eggs thus contain all the materials required for multiple

cell cycles, including the enzymes and precursors needed for DNA replication, the histones and other chromatin proteins involved in assembling the replicated DNA into chromosomes, and the proteins and lipids required in formation of the nuclear envelope. These egg extracts also synthesize proteins encoded by mRNAs in the extract, including cyclin B.

When chromatin prepared from interphase frog sperm is added to a *Xenopus* egg extract, a nuclear envelope develops around the chromatin, forming a haploid nucleus. Following formation of a nuclear envelope, the sperm DNA replicates one time. Following DNA replication, the sperm chromosomes condense and the nuclear envelope breaks down into vesicles, just as it does in intact cells entering mitosis. About 10 minutes after the nuclear envelope breaks down, all the cyclin B in the extract suddenly is degraded, as it is in intact cells during anaphase. Following cyclin B degradation, the sperm chromosomes decondense and a nuclear envelope re-forms around them, as in an intact cell at the end of mitosis. After about 20 minutes, the cycle begins again. DNA within the nuclei formed after the first mitotic period (now 2n) replicates, forming 4n nuclei. Cyclin B, synthesized from the cyclin B mRNA present in the extract, accumulates. As cyclin B approaches peak levels, the chromosomes condense once again, the nuclear envelopes break down, and about 10 minutes later cyclin B is once again suddenly destroyed. These remarkable *Xenopus* egg extracts can mediate several of these cycles, which mimic the rapid synchronous cycles of an early frog embryo.

Using this experimental system, researchers found that MPF activity, assayed by its ability to phosphorylate histone H1, rises and falls in synchrony with the concentration of cyclin B (Figure 13-7a). The early events of mitosis—chromosome condensation and nuclear envelope breakdown—occurred when MPF activity reached its highest levels in parallel with the rise in cyclin B concentration. Addition of cycloheximide, an inhibitor of protein synthesis, prevented cyclin B synthesis and also prevented the rise in MPF activity, chromosome condensation, and nuclear envelope breakdown.

To test the functions of cyclin B in these cell-cycle events, all mRNAs in the egg extract were degraded by digestion with a low concentration of RNase, which then was inactivated by addition of a specific inhibitor. This treatment destroys mRNAs without affecting the tRNAs and rRNAs required for protein synthesis, since their degradation requires much higher concentrations of RNase. When sperm chromatin was added to the RNase-treated extracts, nuclear envelopes assembled around the sperm chromatin and the resulting 1n nuclei replicated their DNA, but the increase in MPF activity and the early mitotic events (chromosome condensation and nuclear envelope breakdown), which the untreated extract supports, did not occur (Figure 13-7b). Addition of cyclin B mRNA, produced in vitro from cloned cyclin B cDNA, to the RNase-treated egg extract and sperm chromatin restored the parallel oscillations in MPF activity and cyclin B level and the characteristic early and late

(a) Untreated extract

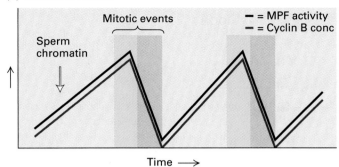

(b) RNase-treated extract

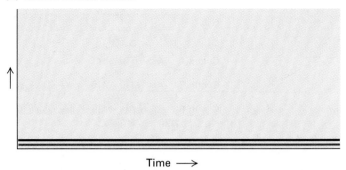

(c) RNase-treated extract + wild-type cyclin B mRNA

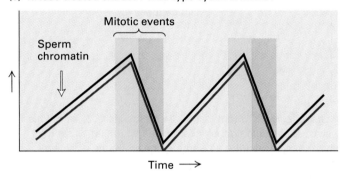

(d) RNase-treated extract + nondegradable cyclin B mRNA

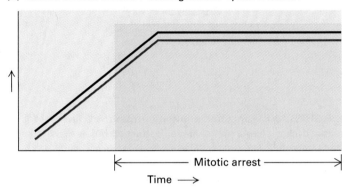

▲ FIGURE 13-7 **Experimental demonstration that the synthesis and degradation of cyclin B are required for the cycling of MPF activity and mitotic events in *Xenopus* egg extracts.** In all cases, MPF activity and cyclin B concentration were determined at various times after addition of sperm chromatin to an extract treated as indicated. Microscopic observations determined the occurrence of early mitotic events (blue shading), including chromosome condensation and nuclear envelope breakdown, and of late events (orange shading), including chromosome decondensation and nuclear envelope reformation. See text for discussion. [See A. W. Murray et al., 1989, *Nature* **339**:275; adapted from A. Murray and T. Hunt, 1993, *The Cell Cycle: An Introduction,* W. H. Freeman and Company.]

mitotic events as observed with the untreated egg extract (Figure 13-7c). Since cyclin B is the only protein synthesized under these conditions, these results demonstrate that it is the crucial protein whose synthesis is required to regulate MPF activity and the cycles of chromosome condensation and nuclear envelope breakdown mediated by cycling egg extracts.

In these experiments, chromosome decondensation and nuclear envelope formation (late mitotic events) coincided with decreases in MPF activity and the cyclin B level. As mentioned earlier and described in detail below, mitotic cyclins can be polyubiquitinated and subsequently degraded. To determine whether degradation of cyclin B is required for exit from mitosis, researchers added a mutant mRNA encoding a nondegradable cyclin B to a mixture of RNase-treated *Xenopus* egg extract and sperm chromatin. As shown in Figure 13-7d, MPF activity increased in parallel with the level of the mutant cyclin B, triggering condensation of the sperm chromatin and nuclear envelope breakdown (early mitotic events). However, the mutant cyclin B synthesized

in this reaction never was degraded as in the reaction with wild-type cyclin B mRNA (see Figure 13-7c). As a consequence, MPF activity continued to increase and the late mitotic events of chromosome decondensation and nuclear envelope formation were both blocked. This experiment demonstrates that the fall in MPF activity and exit from mitosis depends on degradation of cyclin B.

Ubiquitin-Mediated Degradation of Mitotic Cyclins Promotes Exit from Mitosis

Animal cells actually contain three cyclins that can function like cyclin B to stimulate *Xenopus* oocyte maturation: cyclin A (which was the first cyclin shown to have this function) and two closely related cyclin Bs. Sequencing of cDNAs encoding several mitotic cyclins from various eukaryotes has shown that all the encoded proteins contain a homologous sequence near the N-terminus called the *destruction box* (Figure 13-8a). In intact cells, cyclin degradation begins shortly after the onset of anaphase (late anaphase), the

(a) Mitotic cyclin destruction box

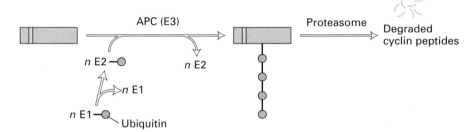

Cyclin A	Arg—Thr—Val—Leu—Gly—Val—Ile—Gly—Asp
Cyclin B1	Arg—Thr—Ala—Leu—Gly—Asp—Ile—Gly—Asn
Cyclin B2	Arg—Ala—Ala—Leu—Gly—Glu—Ile—Gly—Asn

(b) Polyubiquitination of mitotic cyclin

▲ **FIGURE 13-8 Polyubiquitination of mitotic cyclins.** (a) All mitotic cyclins have a homologous destruction box (yellow) near the N terminus. The destruction-box sequences of *Xenopus* mitotic cyclins are shown with amino acid residues conserved in all three proteins in red; residues conserved in two of the three proteins are in black boldface. *Xenopus* and other vertebrates contain two closely related, functionally equivalent B-type cyclins called B1 and B2. Cyclin A, a third mitotic cyclin discussed later, also is present at low levels during early mitosis. The mitotic cyclins are 60–63 kDa and contain ≈550–580 residues. (b) Late in anaphase, multiple ubiquitin molecules are added one at a time to lysine residues C-terminal to the destruction box in mitotic cyclins. A ubiquitin first is linked to the activating enzyme E1, which then passes the activated ubiquitin to one of several ubiquitin-conjugating enzymes (E2). After E2, together with a ubiquitin ligase (E3), called the anaphase-promoting complex (APC), recognizes a destruction-box sequence, the attached ubiquitin is transferred to the substrate protein. Repetition of this process polyubiquitinates mitotic cyclins, which then are rapidly degraded by a multiprotein proteasome complex. [See M. Glotzer et al., 1991, *Nature* **349**:132; adapted from A. Murray and T. Hunt, 1993, *The Cell Cycle: An Introduction,* W. H. Freeman and Company.]

period of mitosis when sister chromatids are separated and pulled toward opposite spindle poles.

Biochemical studies with *Xenopus* egg extracts showed that after their synthesis, wild-type mitotic cyclins are modified by addition of **ubiquitin,** a highly conserved, 76-residue protein. As discussed in Chapter 3, covalent attachment of chains of ubiquitin, a process called **polyubiquitination,** marks proteins for rapid degradation in eukaryotic cells by proteasomes, multiprotein cylindrical structures containing numerous proteases (see Figure 3-18).

Addition of ubiquitin to a mitotic cyclin or other target protein requires three types of enzymes (Figure 13-8b). Ubiquitin is first activated at its carboxyl-terminus by formation of a thioester bond with the cystine residue of *ubiquitin-activating enzyme,* E1. Ubiquitin is subsequently transferred from E1 to the cystine of one of a class of related enzymes called *ubiquitin-conjugating enzymes,* E2. The specific E2 determines, along with a third protein, *ubiquitin ligase* (E3), the substrate protein to which multiple ubiquitins will be covalently linked via a lysine residue, marking the substrate protein for rapid degradation by a proteasome. E3 proteins are frequently complex, multisubunit proteins; for instance, the

E3 for cyclin B purified from *Xenopus* eggs contains at least eight different subunits. This E3 that targets mitotic cyclins for polyubiquitination is the anaphase-promoting complex (APC) mentioned earlier (see Figure 13-2). The APC targets E2-ubiquitin complexes to the destruction box in mitotic cyclins, and then stimulates transfer of the ubiquitin to a lysine residue on the C-terminal side of the destruction box. Further cycles of ubiquitination result in chains of polyubiquitin, which are recognized by proteasomes (see Figure 13-8b). Mutant cyclins that lack a destruction box have been constructed using recombinant DNA techniques; because they lack a destruction box, these mutant proteins are not rapidly degraded.

Regulation of APC Activity Controls Degradation of Cyclin B

The degradation of cyclin B in late anaphase is regulated by controlling APC activity. The APC that is isolated from *Xenopus* eggs arrested in metaphase has low activity for stimulating polyubiquitination of cyclin B. In contrast, APC isolated from eggs stimulated to complete mitosis has high ubiquitination-stimulating activity. Several of the subunits in

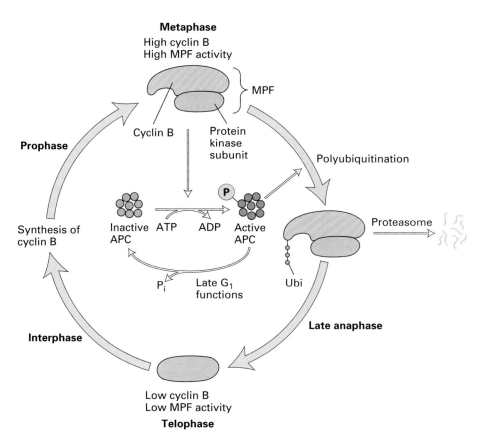

Metaphase
High cyclin B
High MPF activity

Prophase

Cyclin B

Protein
kinase
subunit

MPF

Polyubiquitination

Synthesis of
cyclin B

Inactive
APC

ATP

ADP

Active
APC

Proteasome

Ubi

P_i

Late G_1
functions

Interphase

Late anaphase

Low cyclin B
Low MPF activity

Telophase

◄ **FIGURE 13-9 Regulation of mitotic cyclin levels in cycling cells.** The anaphase-promoting complex (APC) is activated only when MPF activity is high. Binding of the active APC and E2 covalently linked to a ubiquitin (not shown) to the cyclin B destruction box leads to the addition of multiple ubiquitin (Ubi) molecules. As the poly-ubiquitinated cyclin B is degraded, MPF activity declines, triggering the onset of telophase. Following cytokinesis, synthesis of cyclin B occurs in the interphase daughter cells. APC activity remains high until late in the G_1 of the next cell cycle when it is inactivated by a G_1 Cdk complex. When the MPF activity rises enough, another mitoses ensues.

APC with high activity are phosphorylated; removal of these phosphates with a protein phosphatase decreases APC activity. These findings led to the model for regulating APC activity depicted in Figure 13-9.

When MPF activity reaches its peak at metaphase, it phosphorylates and thereby activates APC. Polyubiquitination of cyclin B then occurs, leading to the degradation of cyclin B. Since cyclin B is an essential subunit of MPF, its degradation causes inactivation of MPF activity. APC is deactivated late in G_1, permitting a rise in the cyclin B level and the concomitant increase in MPF activity needed to enter another mitotic cycle. Since cyclin B is synthesized continuously during the cell cycle, this mechanism accounts for the rise in the cyclin B levels following mitosis (during interphase) and the sudden fall in cyclin B levels late in mitosis.

SUMMARY Biochemical Studies with Oocytes, Eggs, and Early Embryos

• MPF is a heterodimer composed of a mitotic cyclin and a cyclin-dependent protein kinase (Cdk). The protein kinase activity of MPF stimulates the onset of mitosis by phosphorylating multiple specific protein substrates, most of which remain to be identified.

• In the synchronously dividing cells of early *Xenopus* embryos, the concentration of mitotic cyclins (e.g., cyclin B) and MPF activity increase as cells enter mitosis and then fall precipitously during late anaphase (see Figure 13-7).

• Proteolysis of mitotic cyclins, which leads to a decrease in MPF activity, is required for the completion of mitosis.

• Mitotic cyclins contain a nine-residue sequence, the destruction box, that is recognized by ubiquitinating enzymes. The multisubunit anaphase-promoting complex (APC) directs specific ubiquitin-conjugating enzymes to polyubiquitinate mitotic cyclins, marking the proteins for rapid degradation by proteasomes.

• The concentration of mitotic cyclins, which are synthesized continuously in early *Xenopus* embryos, is regulated by controlling APC activity. APC activity rises in response to elevated MPF activity, possibly due to direct phosphorylation of APC subunits by MPF. Activated APC then promotes the ubiquitin-dependent degradation of mitotic cyclins in late anaphase (see Figure 13-9). Deactivation of APC in late G_1 permits accumulation of mitotic cyclins.

• The cyclical increases and decreases in MPF activity, resulting in entry into and exit from mitosis in early *Xenopus* embryos, depends on cyclical decreases and increases in the rate of mitotic cyclin degradation.

13.3   Genetic Studies with *S. pombe*

The studies with *Xenopus* egg extracts described in the previous section clearly show that continuous synthesis of cyclin B followed by its periodic degradation at late anaphase is required for the rapid cycles of mitosis observed in early animal embryos. Identification of the catalytic subunit of MPF and further insight into its regulation came from genetic analysis of the cell cycle in the fission yeast *S. pombe*. This yeast grows as a rod-shaped cell that increases in length as it grows and then divides in the middle during mitosis to produce two daughter cells of equal size (Figure 13-10).

Two Classes of Mutations in *S. pombe* Produce Either Elongated or Very Small Cells

In wild-type *S. pombe*, entry into mitosis is carefully regulated in order to properly coordinate cell division with cell growth. Temperature-sensitive mutants of *S. pombe* with conditional defects in the ability to progress through the cell cycle are easily recognized because they cause characteristic changes in cell length at the nonpermissive temperature. Many such mutants with defects in mechanisms regulating the cell cycle have been isolated and found to fall into two groups. In the first group are *cdc* mutants, which fail to progress through one of the phases of the cell cycle at the nonpermissive temperature; they form extremely long cells because they continue to grow in length, but fail to divide. In contrast, *wee* mutants form smaller-than-normal cells because they are defective in the proteins that normally prevent cells from dividing when they are too small. Studies with these and other *S. pombe* mutants have revealed additional mechanisms for controlling entry into mitosis.

Complementation and recombination analyses of *S. pombe* mutants have identified a number of different *cdc* and *wee* genes, designated with individual numbers. Wild-type genes are indicated in italics with a superscript plus sign (e.g., $cdc2^+$); genes with a recessive mutation, in italics with a superscript minus sign (e.g., $cdc2^-$). The protein encoded by a particular gene is designated by the gene symbol in Roman type with an initial capital letter (e.g., Cdc2).

S. pombe Cdc2-Cdc13 Heterodimer Is Equivalent to *Xenopus* MPF

Temperature-sensitive recessive mutations in several different *cdc* genes in *S. pombe* prevent cells from entering mitosis and thus they grow much longer than normal (Figure 13-11). Dominant mutations in one of these genes, designated *cdc2*, gives rise to the wee phenotype. Generally, recessive phenotypes result from the absence of wild-type protein function, whereas dominant phenotypes are due to increased protein function, either because of overproduction or lack of regulation.

(a)

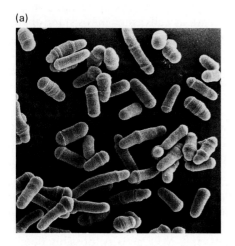

(b)

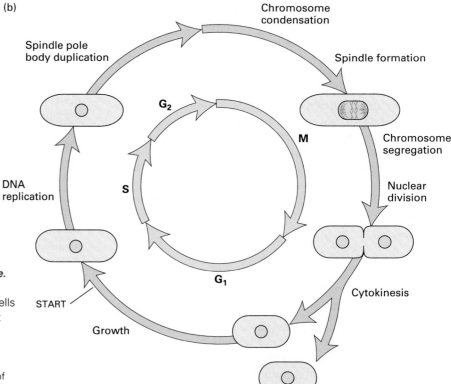

▲ **FIGURE 13-10 The fission yeast *S. pombe*.** (a) Scanning electron micrograph of *S. pombe* cells at various stages of the cell cycle. Long cells are about to enter mitosis; short cells have just passed through cytokinesis. (b) Main events in the *S. pombe* cell cycle. Note that the nuclear envelope does not break down during mitosis in *S. pombe* and other yeasts. [Part (a) courtesy of N. Hajibagheri.]

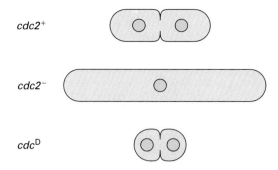

$cdc2^+$

$cdc2^-$

cdc^D

▲ **FIGURE 13-11 Schematic diagrams of phenotypes of *S. pombe cdc2* mutants.** Wild-type cell *(cdc2⁺)* is depicted just before cytokinesis with two normal-size daughter cells. A recessive *cdc2⁻* mutant cannot enter mitosis at the nonpermissive temperature and appears as an elongated cell with a single nucleus, which contains duplicated chromosomes. A dominant $cdc2^D$ mutant enters mitosis prematurely before reaching normal size in G_2; thus, the two daughter cells resulting from cytokinesis are smaller than normal, that is, they have the wee phenotype.

Isolation of these mutants indicates that an absence of Cdc2 activity prevents entry into mitosis, while an excess of Cdc2 activity brings on mitosis earlier than normal. These findings identified Cdc2 as a key regulator of entry into mitosis in *S. pombe*. The wild-type *cdc2⁺* gene contained in a *S. pombe* plasmid library was identified and isolated by its ability to complement *cdc2⁻* mutants (see Figure 13-3). Sequencing showed that it encodes a 34-kDa protein with homology to eukaryotic protein kinases.

To search for genes related to *S. pombe cdc2⁺* in other eukaryotes, researchers tested cDNA clones from other organisms for their ability to complement *S. pombe cdc2⁻* mutants. Remarkably, they isolated a human cDNA encoding a protein identical to *S. pombe* Cdc2 in 63 percent of its residues. This was one of the first demonstrations that complicated functions, such as induction of mitosis, often are carried out by proteins that are highly conserved during evolution. The ability of the human protein to perform all the functions of *S. pombe* Cdc2 also indicates that the mechanism for controlling entry into mitosis is fundamentally similar among all eukaryotes.

The discovery that Cdc2 from *S. pombe* is a protein kinase was one of the important clues that led researchers to test MPF purified from *Xenopus* eggs for protein kinase activity. As discussed earlier, one of the subunits of MPF is cyclin B. The other *Xenopus* MPF subunit not only is the same size as Cdc2 but also reacts with antibody prepared against the region of Cdc2 that is most highly conserved between the human and yeast Cdc2 proteins. These findings, which demonstrated that *Xenopus* MPF is a heterodimer composed of cyclin B and a protein kinase similar to *S. pombe* Cdc2, linked the genetic studies of the cell cycle in yeasts with the biochemical analysis of early embryonic cell cycles in frogs.

Isolation and sequencing of a second *S. pombe cdc* gene (*cdc13⁺*), which also is required for entry into mitosis, revealed that it encodes a protein with homology to sea urchin and *Xenopus* cyclin B. Further studies showed that a heterodimer of Cdc13 and Cdc2 form the *S. pombe* MPF; like *Xenopus* MPF, this heterodimer has protein kinase activity. Moreover, Cdc2 protein kinase activity rises as *S. pombe* cells enter mitosis and falls as they exit mitosis in parallel with the rise and fall in the Cdc13 level. These findings are completely analogous to the results obtained with early *Xenopus* embryos (see Figure 13-7a).

Phosphorylation of the Catalytic Subunit Regulates MPF Kinase Activity

Analysis of other *cdc* and *wee* mutants indicated that proteins encoded by other genes influence the protein kinase activity of *S. pombe* MPF (the Cdc2-Cdc13 heterodimer). For example, temperature-sensitive *cdc25⁻* mutants do not enter mitosis at the nonpermissive temperature. On the other hand, overexpression of Cdc25 from a plasmid present in multiple copies per cell decreases the length of G_2 causing premature entry into mitosis and small cells (Figure 13-12a). Conversely, *wee1⁻* mutants exhibit premature entry into mitosis indicated by their small cell size, whereas overproduction of Wee1 protein increases the length of G_2 resulting in elongated cells. A logical interpretation of these findings is that **Cdc25 protein** stimulates the activity of *S. pombe* MPF, whereas **Wee1 protein** inhibits MPF activity (Figure 13-12b).

In subsequent studies, the wild-type *cdc25⁺* and *wee1⁺* genes were isolated, sequenced, and used to produce the encoded proteins with suitable expression vectors. The deduced sequences of Cdc25 and Wee1 and biochemical studies of the proteins demonstrated that they regulate the activity of *S. pombe* MPF by phosphorylating and dephosphorylating specific regulatory sites in Cdc2, the catalytic subunit.

Cdc2 is active as a protein kinase only when it is associated with a cyclin such as Cdc13. Phosphorylation at one residue (threonine-161) in the Cdc2 subunit activates MPF and phosphorylation at another residue (tyrosine-15) in Cdc2 inactivates MPF, even when the activating site (threonine-161) is phosphorylated. These regulatory phosphorylations of Cdc2 only occur after it is bound by the mitotic cyclin Cdc13. As illustrated in Figure 13-13, Wee1 is the protein kinase that phosphorylates the inhibitory Tyr-15 residue; another kinase, designated *Cdc2-activating kinase (CAK)*, phosphorylates the activating Thr-161 residue. The resulting diphosphorylated MPF is still inactive. Finally, Cdc25, which has protein phosphatase activity, removes the phosphate from Tyr-15, yielding an active MPF. Site-specific mutagenesis that changed the Tyr-15 in Cdc2 to a phenylalanine, which cannot be phosphorylated, produced mutants with the wee phenotype, similar to that of *wee1⁻* mutants. Both mutations prevent the inhibitory phosphorylation at Tyr-15, leading to increased MPF activity, resulting in premature entry into mitosis.

(a)

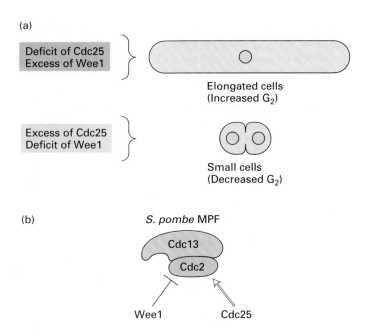

(b)

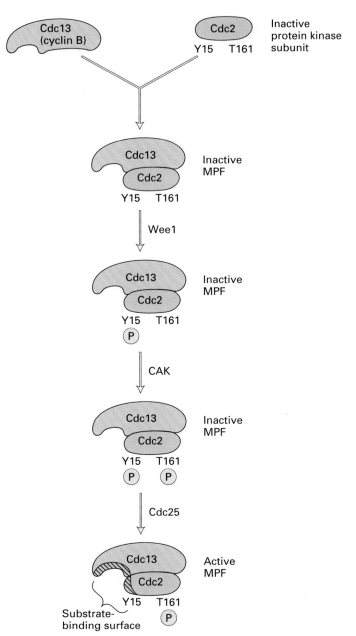

▲ FIGURE 13-12 Effects of mutations in *cdc25* and *wee1* genes on *S. pombe* phenotypes and MPF activity. (a) Cells that lack Cdc25 or Wee1 activity, as a result of recessive temperature-sensitive mutations in the corresponding genes, have the opposite phenotype. Likewise, cells with multiple copies of plasmids containing *cdc25*⁺ or *wee1*⁺, and which thus produce an excess of the encoded proteins, have opposite phenotypes. (b) These phenotypes imply that the Cdc2-Cdc13 complex is activated (→) by Cdc25 and inhibited (⊣) by Wee1. In *cdc25⁻* cells, the inhibitory activity is unopposed and MPF activity is inhibited, blocking entry into mitosis and resulting in elongated cells. When Cdc25 is produced at higher-than-normal levels, it offsets the inhibitory effect of Wee1, so MPF activity rises faster than in wild-type cells, causing premature entry into mitosis, which results in small (wee) cells. In *wee1⁻* mutants, the stimulatory effect of Cdc25 is unopposed, so MPF activity rises faster than normal, leading to premature entry into mitosis and small (wee) cells. Conversely, overproduction of Wee1 inhibits MPF activity more than normal, delaying entry into mitosis and producing elongated cells.

▲ FIGURE 13-13 Regulation of MPF protein kinase activity in *S. pombe* by Cdc13 (cyclin B), Wee1, CAK (Cdc2-activating kinase), and Cdc25. Wee1 and CAK are protein kinases, and Cdc25 is a protein phosphatase. Once bound by Cdc13, the catalytic Cdc2 subunit can be phosphorylated at two regulatory sites, tyrosine-15 (Y15) and threonine-161(T161). Only when Cdc2 is monophosphorylated at T161 is MPF active. Cdc13 contributes to the specificity of substrate binding, probably by forming part of the substrate-binding surface (cross-hatch), which also includes the inhibitory Y15 residue.

Conformational Changes Induced by Cyclin Binding and Phosphorylation Increase MPF Activity

The three-dimensional structure of human cyclin-dependent kinase 2 (Cdk2), which is discussed in a later section, provides insight into how phosphorylation of the Cdc2 subunit of *S. pombe* MPF regulates its protein kinase activity. Although the three-dimensional structures of Cdc2 and most other cyclin-dependent kinases have not been determined, their extensive sequence homology with human Cdk2 suggests that all these cyclin-dependent kinases have a similar structure and are regulated by a similar mechanism.

The three-dimensional structure of unphosphorylated, inactive Cdk2 complexed with ATP, as determined by x-ray crystallography, is shown in Figure 13-14a. A flexible region

of inactive Cdk2, called the *T-loop*, blocks access of protein substrates to the active site where ATP is bound, largely explaining why free Cdk2, unbound to cyclin, has no protein kinase activity. A threonine residue located at the top of this

(a)

(b)

(c)

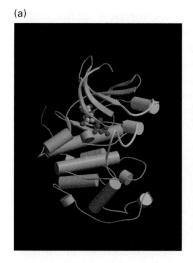
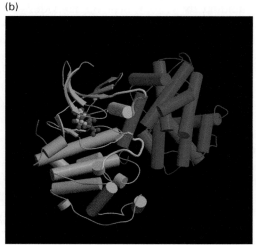
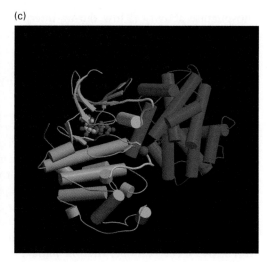

▲ **FIGURE 13-14 Structures of human Cdk2, which is homologous to the Cdc2 catalytic subunit of MPF.** (a) Free, inactive Cdk2 unbound to cyclin A. α-helices are shown as cylinders and β-strands as arrows. Regions of the molecule whose conformation change in the complex with cyclin A are shown in yellow. The T-loop contains the activating threonine whose phosphorylation fully activates the kinase activity of the Cdk2–cyclin A complex. The α1 helix contains a seven-residue sequence that is found at this position in all Cdks. ATP is shown in a ball and stick model, C=gray, O=red, N=blue, P=purple. The T-loop region of Cdk2 is positioned in front of the active site where it blocks access of protein substrates to the γ-phosphate of the bound ATP. (b) Unphosphorylated, low-activity Cdk2–cyclin A complex. Interactions between cyclin A (purple) and the T-loop (yellow) cause the T-loop to pull away from the active site of Cdk2, so that substrate proteins can bind. Helix α1, which interacts extensively with cyclin A, moves several angstroms into the catalytic cleft where it probably contributes to substrate specificity. Conformational changes induced by cyclin A binding also reposition key catalytic side chains leading to the correct alignment of the ATP phosphates for the phosphotransfer reaction. This complex has low kinase activity. (c) Phosphorylated, high-activity Cdk2–cyclin A complex. The conformational changes induced by phosphorylation of the activating threonine alter the shape of the substrate-binding surface, greatly increasing the affinity for protein substrates. [Courtesy of P. D. Jeffrey. See A. A. Russo et al., 1996, *Nature Stuct. Biol.* **3**:696.]

loop corresponds to the activating site in Cdc2 (Thr-161) that is phosphorylated by CAK (see Figure 13-13). Unphosphorylated Cdk2 bound to one of its cyclin partners, cyclin A, has minimal, but detectable protein kinase activity in vitro. The structure of the complex shows that extensive interactions between cyclin A and the T-loop cause a dramatic shift in the position of the T-loop, thereby exposing the Cdk2 active site (Figure 13-14b). Binding of cyclin A also shifts the position of the α1 helix in Cdk2, modifying the substrate-binding surface of Cdk2. Phosphorylation of the activating threonine in the T-loop in Cdk2–cyclin A complexes causes additional conformational changes that further modify the substrate-binding surface of the complex, greatly increasing its affinity for protein substrates (Figure 13-14c). As a result, the kinase activity of the phosphorylated complex is a hundredfold greater than that of the unphosphorylated complex.

The inhibitory Tyr-15 of Cdc2 is in the region of the protein that binds the ATP phosphates. In vertebrate Cdk2 proteins, a second inhibitory site (Thr-14) is located in the same region of the protein. Phosphorylation of Tyr-15 and Thr-14 in these proteins prevents binding of ATP because of electrostatic repulsion between the phosphates linked to the protein and the phosphates of ATP. Thus, these phosphorylations inhibit protein kinase activity even when the

Cdk protein is bound by a cyclin and the activating site is phosphorylated.

Other Mechanisms Also Control Entry into Mitosis by Regulating MPF Activity

So far we have discussed two mechanisms for controlling entry into mitosis: (a) regulation of the concentration of mitotic cyclins as outlined in Figure 13-9 and (b) regulation of the activity of MPF as outlined in Figure 13-13. Further studies of *S. pombe* mutants with altered cell cycles revealed additional complexities in the regulation of MPF activity. The Wee1 protein kinase that inhibits MPF is in turn inhibited by another protein kinase encoded by the *nim1*⁺ gene. Another gene, *mik1*⁺, encodes a protein kinase, very similar to Wee1, that can also phosphorylate the inhibitory Tyr-15 of Cdc2 (see Figure 13-13). Several other genes currently being studied also are thought to influence MPF activity. At present it is clear that multiple mechanisms regulate MPF activity in *S. pombe* in order to control the timing of mitosis and therefore the size of daughter cells.

Enzymes with activities equivalent to *S. pombe* Wee1 and Cdc25 have been found in cycling *Xenopus* egg extracts. The *Xenopus* Wee1 tyrosine kinase activity is high and Cdc25 phosphatase activity is low during interphase, holding the MPF assembled from *Xenopus* Cdc2 and newly synthesized cyclin B in the inactive state with the Cdc2 Tyr-15 phosphorylated. As the extract initiates the events of mitosis, Wee1 activity diminishes and Cdc25 activity increases so that MPF is converted into its active form. As a result, although cyclin B is the only protein whose *synthesis* is required for the cycling of early *Xenopus* embryos, the *activities* of other proteins, including *Xenopus* Wee1 and Cdc25, must be properly regulated for cycling to occur. In its active form, Cdc25 is phosphorylated, suggesting that its activity is also controlled by one or more additional protein kinases and phosphatases.

MPF activity also can be regulated by controlling transcription of the genes encoding the proteins that regulate MPF activity. For example, after the initial rapid synchronous cell divisions of the early *Drosophila* embryo, all the mRNAs are degraded, and the cells become arrested in G_2. This arrest occurs because the *Drosophila* homolog of Cdc25, called String, is unstable. Since *string* mRNA is degraded along with other mRNAs, synthesis of String ceases. Because of the resulting decrease in String phosphatase activity, MPF is maintained in its inhibited state. The subsequent regulated entry into mitosis by specific groups of cells is then triggered by the regulated transcription of the *string* gene.

SUMMARY Genetic Studies with *S. pombe*

- In the fission yeast, *S. pombe*, *cdc* mutations delay progress through one or more phases of the cell cycle, resulting in elongated cells that continue to grow in length but fail to divide. In contrast, *wee* mutations accelerate progress through a phase of the cell cycle, resulting in cells that are shorter than normal.

- *S. pombe* mutants that lack Cdc2 function fail to enter mitosis, while mutants that have greater-than-normal Cdc2 function enter mitosis sooner than normal (see Figure 13-11). Thus Cdc2 is a key regulator of entry into mitosis.

- The *S. pombe cdc2* gene encodes a cyclin-dependent protein kinase. The Cdc2 protein associates with a B-type mitotic cyclin encoded by the *cdc13* gene to form a heterodimer that is equivalent to MPF.

- Two residues in the Cdc2 subunit of the Cdc2-Cdc13 complex can be phosphorylated. The protein kinase activity of the complex is greatest when threonine 161 is phosphorylated. However, phosphorylation at tyrosine 15 by Wee1 inhibits the protein kinase activity of Cdc2-Cdc13 by interfering with binding of ATP. This inhibitory phosphate is removed by Cdc25, a protein phosphatase. As cells enter mitosis, Wee1 activity falls while Cdc25 activity increases, resulting in activation of Cdc2-Cdc13 activity (see Figure 13-13).

- The human Cdk2–cyclin A complex is similar to *Xenopus* MPF and the *S. pombe* Cdc2-Cdc13 complex. Structural studies with the human proteins reveal that cyclin binding to Cdk2 and phosphorylation of the activating threonine (equivalent to Thr-161 in Cdc2) cause conformational changes that expose the active site and modify the substrate-binding surface so that it has high affinity for protein substrates.

13.4 Molecular Mechanisms for Regulating Mitotic Events

So far, we have seen that a regulated increase in MPF activity induces entry into mitosis. Presumably, the entry into mitosis is a consequence of the phosphorylation of specific proteins by the protein kinase activity of MPF. For the most part, however, the proteins phosphorylated by MPF in vivo have not been identified. The active phosphorylated forms of these proteins are thought to mediate the many remarkable events of mitosis including chromosome condensation, formation of the mitotic spindle, and breakdown of the nuclear envelope (see Figure 19-34).

As demonstrated in the studies with *Xenopus* egg cycling extracts described earlier (see Figure 13-7), a decrease in mitotic cyclins and the associated inactivation of MPF coincides with the later stages of mitosis (late anaphase and telophase). Just before this, in early anaphase, sister chromatids separate and move to opposite spindle poles. During telophase, microtubule dynamics return to interphase conditions, the chromosomes decondense, the nuclear envelope reforms, the endoplasmic reticulum and Golgi complex are remodeled, and cytokinesis occurs. Some of these processes are triggered by dephosphorylation; others, by protein degradation.

In this section, we discuss the molecular mechanisms and specific proteins associated with some of the events that characterize early and late mitosis.

Phosphorylation of Nuclear Lamins by MPF Leads to Nuclear-Envelope Breakdown

The nuclear envelope is a double-membrane extension of the rough endoplasmic reticulum containing many nuclear pore complexes (see Figure 5-42). The lipid bilayer of the inner nuclear membrane is supported by the **nuclear lamina**, a meshwork of lamin filaments located adjacent to the inside face of the nuclear envelope (Figure 13-15a). The three nuclear **lamins** (A, B, and C) present in vertebrate cells belong to the class of cytoskeletal proteins, the intermediate filaments, that are critical in supporting cellular membranes (Chapter 19). Lamins A and C, which are encoded by the same transcription unit and produced by alternative splicing of a single pre-mRNA, are identical except for a 133-residue region at the C-terminus of lamin A, which is absent in lamin

(a)

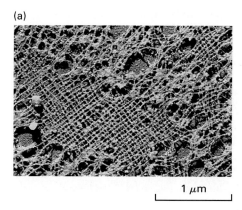

1 μm

▲ **FIGURE 13-15 The nuclear lamina and its depolymerization.**
(a) Electron micrograph of the nuclear lamina. A nuclear membrane
from a hand-dissected *Xenopus* oocyte was fixed to an electron
microscope grid and then extracted with a nonionic detergent to
remove the lipid membranes and nonpolymerized proteins. Note
the regular meshlike network of fibers. (b) Schematic diagram of
the nuclear lamina associated with the inner membrane of the
double-membrane nuclear envelope of an interphase cell. The
nuclear lamina (red) consists of two orthogonal sets of 10-nm-
diameter filaments built of lamins A, B, and C. Individual lamin
filaments are formed by end-to-end polymerization of lamin
tetramers, which consist of two lamin dimers. The red circles
represent the globular N-terminal domains. Phosphorylation of
specific serine residues near the ends of the coiled-coil rodlike
central section of lamin dimers causes the filaments and tetramers
to depolymerize, leading to breakdown of the nuclear envelope.
[Part (a) from U. Aebi et al., 1986, *Nature* **323**:560; courtesy of U. Aebi.
Part (b) adapted from A. Murray and T. Hunt, 1993, *The Cell Cycle: An
Introduction*, W. H. Freeman and Company.]

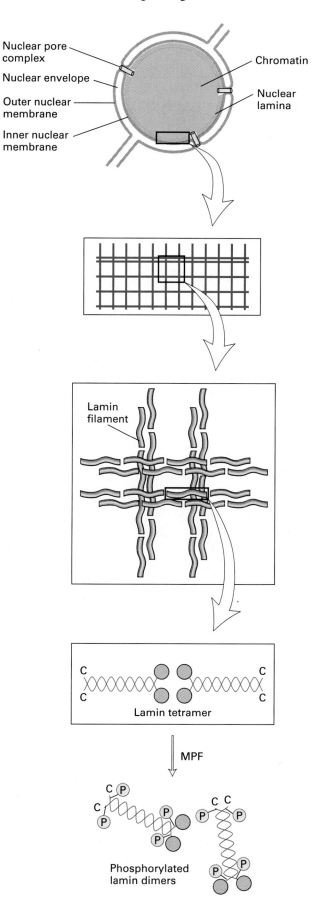

C. Lamin B, encoded by a different transcription unit, is
post-transcriptionally modified by the addition of a hydrophobic isoprenyl group near its carboxyl-terminus. This
fatty acid is incorporated into the inner leaflet of the lipid
bilayer that forms the inner nuclear membrane, thereby anchoring the nuclear lamina to the membrane. All three nuclear lamins form dimers containing a rodlike α-helical
coiled-coil central section and globular head and tail domains; polymerization of these dimers through head-to-head
and tail-to-tail associations generates the intermediate filaments that compose the nuclear lamina (see Figure 19-51).

Early in mitosis, MPF phosphorylates specific serine
residues in all three nuclear lamins, causing depolymerization of the lamin intermediate filaments (Figure 13-15b).
The phosphorylated lamin A and C dimers are released into
solution, whereas the phosphorylated lamin B dimers remain
associated with the nuclear membrane via their isoprenyl
anchor. Depolymerization of the nuclear lamins leads to disintegration of the nuclear lamina meshwork and contributes
to the breakdown of the nuclear envelope into small vesicles.
The experiment summarized in Figure 13-16 shows that the
breakdown of the nuclear envelope, which normally occurs
early in mitosis, depends on phosphorylation of lamin A.

Wild-type lamin A   **Mutant lamin A**

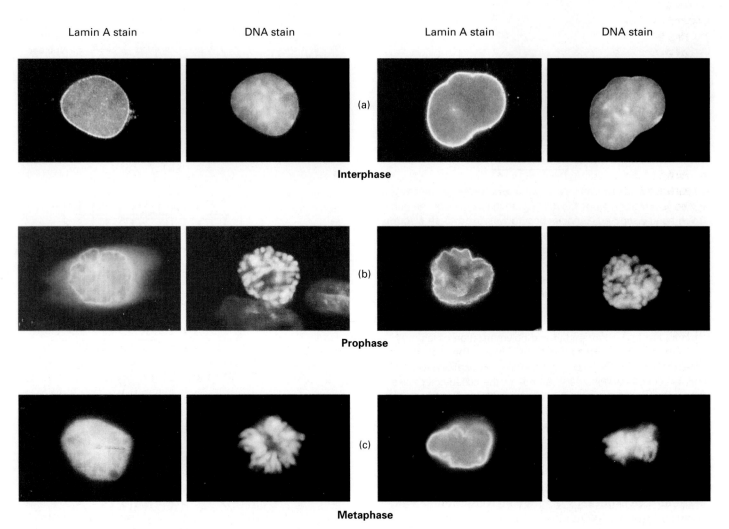

Interphase

Prophase

Metaphase

▲ **FIGURE 13-16 Experimental demonstration that phosphorylation of human nuclear lamin A is required for lamin depolymerization, which contributes to nuclear-envelope breakdown during mitosis.** Site-directed mutagenesis was used to prepare a mutant human lamin A gene encoding a protein in which alanines replace the serines that normally are phosphorylated in wild-type lamin A (see Figure 13-15b). As a result, the mutant lamin A cannot be phosphorylated. Expression vectors carrying the wild-type or mutant gene were separately transfected into cultured hamster cells. Transfected cells at various stages in the cell cycle then were stained with a fluorescent-labeled monoclonal antibody specific for human lamin A and with a fluorescent dye that binds to DNA. In these photomicrographs of cells during interphase, prophase, and metaphase, lamin A staining visualized polymerized (unphosphorylated) lamin A as a bright band of fluorescence around the perimeter of the nucleus, whereas diffuse staining and the absence of the bright peripheral band indicates depolymerized lamin A. In cells with both wild-type and mutant lamin A, the chromosomes were fully condensed by metaphase. Depolymerization of lamin A, however, occurred only in the wild-type cells. The presence of intact lamin A in the mutant cells prevents disintegration of the nuclear lamina and envelope. [From R. Heald and F. McKeon, 1990, *Cell* **61**:579.]

Other Early Mitotic Events May Be Controlled Directly or Indirectly by MPF

The demonstration that nuclear-envelope breakdown depends on phosphorylation of nuclear lamins suggests that MPF-catalyzed phosphorylation of other proteins may play a role in other early mitotic events, such as chromosome condensation. For instance, genetic experiments in the budding yeast *S. cerevisiae* identified a family of *SMC* (*s*tructural *m*aintenance of *c*hromosomes) *proteins* that are required for normal chromosome segregation. Biochemical studies of the homologous *Xenopus* proteins showed that these large proteins (≈1200 amino acids) contain long regions predicted to participate in coiled-coil structures (see Figure 3-9) and characteristic ATPase domains at their C-terminus. Homologs of yeast SMC proteins were cloned from a *Xenopus* cDNA library, and antibodies were raised against the encoded proteins. Immunoprecipitation studies with these antibodies revealed that in a *Xenopus* egg extract the SMC proteins are part of a protein complex, called **condensin,** that includes three additional proteins, which become phosphorylated as cells enter mitosis. When the anti-SMC antibodies were used to deplete condensin from an egg extract, the extract lost its ability to condense added sperm chromatin.

In experiments with purified condensin and DNA, phosphorylated condensin binds to DNA and winds it into **supercoils** in a reaction requiring the hydrolysis of ATP. These results have lead to the model that individual condensin complexes, activated by MPF or another protein kinase regulated by MPF, bind to DNA at intervals along the chromosome scaffold. Self-association of the bound complexes via their coiled-coil domains and supercoiling of the DNA segments between them is proposed to cause chromosome condensation.

Phosphorylation of microtubule-associated proteins by MPF probably is required for the dramatic changes in microtubule dynamics that result in the formation of the mitotic spindle and asters (Section 19.5). In addition, all vesicular traffic in the cell ceases during mitosis, and the endoplasmic reticulum and Golgi complex break down into small vesicles as the nuclear membrane does. Phosphorylation of proteins associated with these membranous organelles, by MPF or other protein kinases activated by MPF-catalyzed phosphorylation, likely is responsible for these mitotic events as well.

APC-Dependent Unlinking of Sister Chromatids Initiates Anaphase

We saw earlier that in the late anaphase and telophase stages of mitosis, APC-mediated polyubiquitination of cyclin B targets it for destruction (see Figure 13-9). Additional experiments with RNase-treated *Xenopus* egg extracts provided evidence that polyubiquitination and subsequent degradation of noncyclin proteins also is required to initiate anaphase. In these studies, the mitotic spindle, which is formed from tubulin-containing microtubules, was visualized by including fluorescent-labeled tubulin in the reaction mixtures. When RNase-treated egg extracts and sperm chromatin were incubated in the presence of mRNA encoding wild-type cyclin B, the mitotic spindle apparatus and condensed sperm chromosomes aligned between the spindle poles were visible, similar to their appearance during metaphase in intact cells. After 15 minutes of incubation, the chromosomes were seen to move toward the spindle poles, just as they do during anaphase in intact cells. Cyclin B degradation and the resulting precipitous decrease in MPF activity began after this point, and over the next half hour the spindle depolymerized and the chromosomes decondensed (Figure 13-17a).

When mRNA encoding a nondegradable cyclin B was substituted for wild-type mRNA, MPF activity remained high as in the experiments described earlier (see Figure 13-7d). As before, chromosome decondensation did not occur, but

chromosome segregation was observed to occur normally, indicating that chromosome segregation during anaphase does not require MPF inactivation (Figure 13-17b).

Researchers then prepared a peptide corresponding to residues 13–110 of cyclin B, which contains the destruction-box sequence and the site of polyubiquitination. When this peptide was added to a reaction mixture containing untreated egg extract and sperm chromatin, movement of chromosomes toward the spindle poles was greatly delayed at peptide concentrations of 20–40 μg/ml and blocked altogether at higher concentrations (Figure 13-17c). The added excess destruction-box peptide is thought to act as a substrate for the APC-directed polyubiquitination system, competing with the normal endogenous target proteins and thereby delaying or preventing their degradation by proteasomes.

As discussed in Chapter 19, each sister chromatid of a metaphase chromosome is attached to microtubules via its **kinetochore,** a complex of proteins assembled at the centromere (see Figure 19-39); the opposite ends of these kinetochore microtubules associate with one of the spindle poles. At metaphase, the spindle is in a state of tension with forces pulling the two kinetochores towards the opposite spindle poles balanced by forces pushing the spindle poles apart. Sister chromatids do not separate because they are held together at their centromeres and multiple positions along the chromosome arms by recently discovered multiprotein complexes called **cohesins.** Among the proteins composing the cohesin complex are members of the SMC protein family discussed earlier. Thus SMC proteins function in multiprotein complexes with other proteins in both chromosome condensation and the cohesion of sister chromatids. The function of the cohesin complex was demonstrated in experiments in which antibodies specific for the cohesin SMC proteins were used to deplete the cohesin complex from *Xenopus* egg extracts. The depleted extract was able to replicate the DNA in added sperm chromatin, but the resulting sister chromatids were defective in their association with each other.

Cohesin function is regulated by the **anaphase inhibitor,** a protein that is a target of APC-directed polyubiquitination. In yeast, anaphase inhibitor functions together with another protein to stimulate the proper association of cohesin with daughter chromosomes. As cells enter anaphase, anaphase inhibitor is polyubiquitinated by the APC and degraded by proteasomes (see Figure 13-2). As a consequence, cohesin function is inactivated, allowing the poleward force exerted on kinetochores to move sister chromatids toward opposite spindle poles (Figure 13-18).

After APC is activated, it initially acts on certain target proteins such as the anaphase inhibitor, but it does not act on cyclin B until late in anaphase (Figure 13-19). This is necessary in order to maintain MPF activity until late anaphase, keeping chromosomes in their condensed state until they have segregated to opposite spindle poles. Recent genetic studies in budding yeast indicate that this stage-dependent

(a) RNase treated extract + mRNA encoding wild-type cyclin B

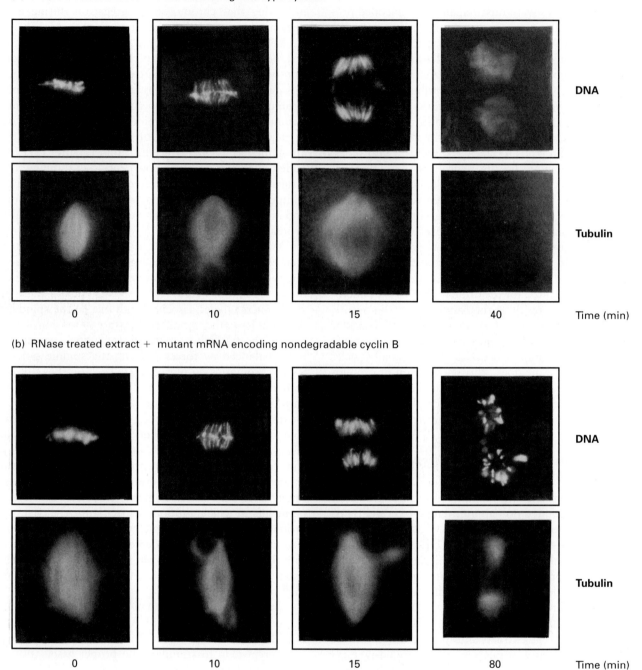

(b) RNase treated extract + mutant mRNA encoding nondegradable cyclin B

targeting of APC activity is due to the regulated activity of two APC-associated proteins.

Phosphatase Activity Is Required for Reassembly of the Nuclear Envelope and Cytokinesis

Earlier we saw that phosphorylation of nuclear lamins results in their depolymerization, leading to breakdown of the nuclear envelope, a crucial event of early mitosis. Removal of these phosphates coincides with lamin repolymerization and re-formation of the nuclear lamina associated with the daughter-cell nuclei during telophase. Studies with *Xenopus* egg extracts and analyses of various organisms with temperature-sensitive mutations in protein phosphatases indicate that specific protein phosphatases indeed are required for reassembly of the nuclear lamina and the nuclear envelope. When MPF is inactivated by the

(c) Untreated extract + cylin B destruction-box peptide

15-min reaction time

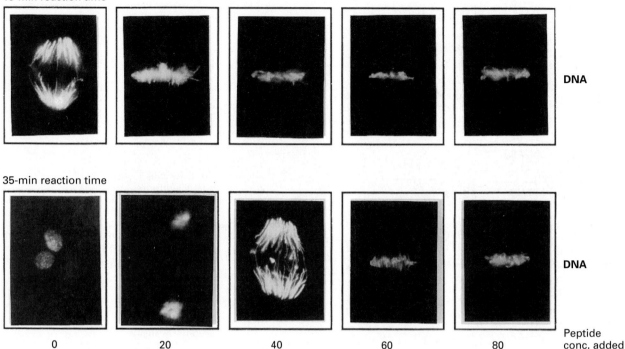

35-min reaction time

| 0 | 20 | 40 | 60 | 80 | Peptide conc. added (μg/ml) |

◀ **FIGURE 13-17 Experimental evidence that onset of anaphase depends on polyubiquitination of proteins other than cyclin B.** The reaction mixtures contained an untreated or RNase-treated *Xenopus* egg extract and isolated *Xenopus* sperm nuclei, plus other components indicated below. Chromosomes were visualized with a fluorescent DNA-binding dye. Addition of fluorescent rhodamine-labeled tubulin to the reactions permitted observation of the microtubules forming the spindle apparatus. (a,b) The egg extract was treated with RNase to destroy endogenous mRNAs, and an RNase inhibitor was added before addition of mRNA encoding wild-type cyclin B or nondegradable cyclin B. In both samples, the condensed chromosomes and assembled spindle apparatus are visible at 0 minutes. In the presence of wild-type cyclin B (a), condensed chromosomes attached to the spindle microtubules and segregated toward the poles of the spindle as cyclin B was degraded. By 40 minutes, the spindle had depolymerized (thus is not visible), and the chromosomes had decondensed. In the presence of nondegradable cyclin B

(b), chromosomes segregated to the spindle poles (15 minutes), as in (a), but the spindle microtubules did not depolymerize and the chromosomes did not decondense even after 80 minutes. These observations indicate that degradation of cyclin B is not required for chromosome segregation during anaphase, although it is required for depolymerization of spindle microtubules and chromosome decondensation during telophase. (c) Various concentrations of a cyclin B peptide containing the destruction box were added to extracts that had not been treated with RNase; the samples were stained for DNA at 15 and 35 minutes after formation of the spindle apparatus. The two lowest peptide concentrations delayed chromosome segregation, and the higher concentrations completely inhibited chromosome segregation. In this experiment, the added cyclin B peptide is thought to function as a competitive inhibitor for polyubiquitination of target proteins whose degradation is required for onset of anaphase. [From S. L. Holloway et al., 1993. *Cell* **73**:1393; courtesy of A. W. Murray.]

degradation of cyclin B late in anaphase, the action of these phosphatases, which remove the lamin regulatory phosphates, is unopposed; consequently, the lamins are rapidly dephosphorylated.

Figure 13-20 schematically depicts reassembly of the nuclear envelope, which occurs late in mitosis. Vesicles derived from the breakdown of the nuclear envelope during prophase associate with the surface of the decondensing chromosomes during telophase. These vesicles fuse to form continuous double membranes around each chromosome.

Nuclear pore complexes, which disassemble into subpore complexes during prophase, reassemble into the nuclear membrane around each chromosome, forming individual mininuclei called *karyomeres*. Subsequent fusion of the karyomeres associated with each spindle pole generates the two daughter-cell nuclei, each one containing a full set of chromosomes. Lamins A and C appear to be imported through the reassembled nuclear pore complexes during this period and reassemble into a new nuclear lamina. Reassembly of the nuclear lamina in the daughter nuclei probably

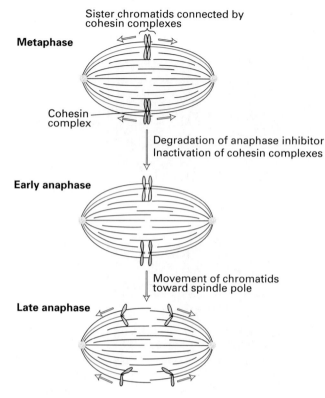

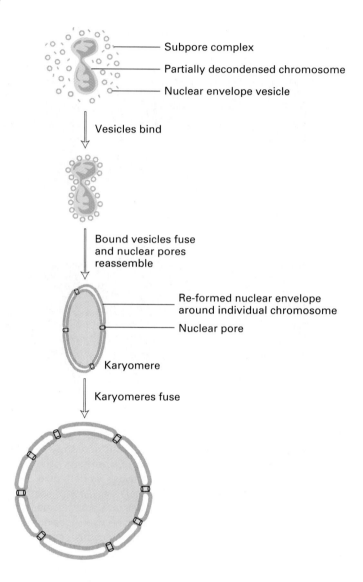

▲ **FIGURE 13-18 Model for induction of anaphase by regulation of cohesin complexes.** Arrows indicate direction of the forces acting on the kinetochores. Cohesin complexes are shown connecting centromeres, but they also occur along the arms of sister chromatids. Cohesin function is regulated by anaphase inhibitor, a protein that is polyubiquitinated by the APC during early anaphase. The subsequent degradation of anaphase inhibitor by proteasomes results in the inactivation of cohesins, permitting the polewise movement of the chromatids. See text for discussion.

▶ **FIGURE 13-19 Control of entry into anaphase and exit from mitosis by the anaphase-promoting complex (APC), which directs the degradation of at least two classes of proteins.** Inactive APC (light orange) is activated (dark purple) directly or indirectly by MPF (Cdc2–cyclin B) after MPF has induced the early events of mitosis through metaphase (see Figure 13-9). Activated APC first polyubiquitinates anaphase inhibitor (brown), which prevents separation of sister chromatids, thereby initiating anaphase. Subsequently, APC polyubiquitinates cyclin B, resulting in the inactivation of MPF and exit from mitosis. [Adapted from R. W. King et al.,1996, *Science* **274**:1652.]

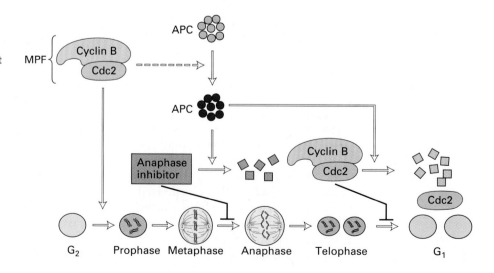

◀ **FIGURE 13-20 Assembly of the nuclear envelope during telophase.** Nuclear envelope vesicles, generated by the breakdown of the envelope during prophase, associate with decondensing chromosomes and then fuse. Subpore complexes reassemble into nuclear pores, forming individual mininuclei called karyomeres. The enclosed chromosome further decondenses, and subsequent fusion of the nuclear envelopes of all the karyomeres at each spindle pole forms a single nucleus containing a full set of chromosomes. Reassembly of the nuclear lamina is not shown. [See G. P. Vigers and M. J. Lohka, 1991, *J. Cell Biol.* **112**:545; adapted from A. Murray and T. Hunt, 1993, *The Cell Cycle: An Introduction,* W. H. Freeman and Company.]

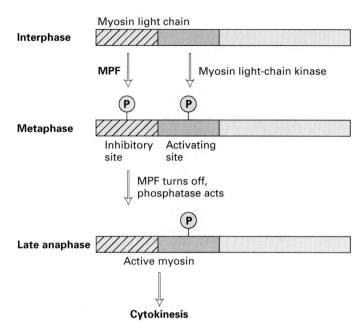

▲ **FIGURE 13-21 Regulation of myosin light chain by MPF.** Phosphorylation of inhibitory sites on the myosin light chain by MPF early in mitosis prevents active myosin heavy chains from interacting with and sliding along actin filaments, a process required for cytokinesis. When MPF activity falls during anaphase, a constitutive phosphatase dephosphorylates the inhibitory sites, permitting cytokinesis to proceed. A different enzyme, myosin light-chain kinase, phosphorylates the activating site. [See L. L. Satterwhite et al., 1992, *J. Cell Biol.* **118**:595; adapted from A. Murray and T. Hunt, 1993, *The Cell Cycle: An Introduction,* W. H. Freeman and Company.]

is initiated on lamin B molecules, which remain associated with the nuclear-envelope vesicles throughout mitosis via the isoprenyl anchors covalently linked to the C-terminal region of lamin B.

During cytokinesis, the final step in cell division, the actin and myosin filaments composing the contractile ring slide past each other to form a cleavage furrow of steadily decreasing diameter (see Figure 18-37). As MPF activity rises early in mitosis, it phosphorylates the myosin light chain, thereby inhibiting its ability to associate with actin filaments (Figure 13-21). Inactivation of MPF toward the end of anaphase, due to the degradation of cyclin B, permits the unopposed action of protein phosphatases to dephosphorylate myosin light chain. As a result, the contractile machinery is activated, the cleavage furrow can form, and cytokinesis proceeds. This regulatory mechanism assures that cytokinesis does not occur before the completion of anaphase when mitotic cyclins are degraded and MPF activity falls.

SUMMARY Molecular Mechanisms for Regulating Mitotic Events

- Although most substrates of MPF remain to be identified, nuclear lamins, subunits of condensin, and myosin light chain are three identified substrates.

- MPF-catalyzed phosphorylation of specific lamin serines early in mitosis causes depolymerization of lamin filaments, leading to breakdown of the nuclear envelope (see Figure 13-15). In addition, phosphorylation of condensin complexes by MPF or a kinase regulated by MPF is thought to promote chromosome condensation.

- When MPF activity falls in late anaphase and telophase, protein phosphatases remove the regulatory phosphates from lamins A, B, and C, permitting reassembly of the nuclear lamina in the two daughter cell nuclei.

- MPF-catalyzed phosphorylation of the myosin light chain prevents cytokinesis. Since MPF activity does not

fall until the completion of anaphase, cytokinesis is delayed until sister chromatids have been segregated to opposite poles of the spindle apparatus.

- The APC-directed degradation of the anaphase inhibitor causes inactivation of the cohesin complexes that connect sister chromatids. This unlinking of sister chromatids heralds the onset of anaphase and allows sister chromatids to move apart (see Figure 13-19). Later, the same APC targets cyclin B for destruction, causing the decrease in MPF activity that marks the onset of telophase.

13.5 Genetic Studies with *S. cerevisiae*

As mentioned earlier, in most vertebrate cells the key decision determining whether or not a cell will divide is the decision to enter the S phase. In most cases, once a vertebrate cell has become committed to entering the S phase, it does

(a)

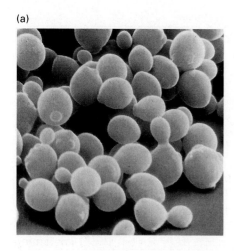

(b)

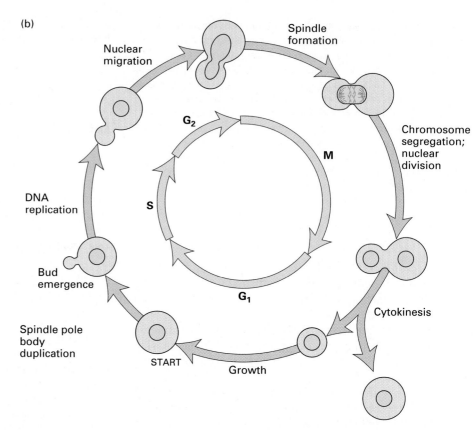

▲ FIGURE 13-22 The budding yeast *S. cerevisiae.* (a) Scanning electron micrograph of *S. cerevisiae* cells at various stages of the cell cycle. The larger the bud, which emerges at the end of the G_1 period, the further along in the cycle the cell is. (b) Main events in *S. cerevisiae* cell cycle. Daughter cells are born smaller than mother cells and must grow longer in G_1 before they are large enough to enter the S phase. As in *S. pombe,* the nuclear envelope does not break down during mitosis. Unlike *S. pombe* chromosomes, *S. cerevisiae* chromosomes do not condense sufficiently to be visible by light microscopy. [Part (a) courtesy of E. Schachtbach and I. Herskowitz.]

so several hours later and progresses through the remainder of the cell cycle until it completes mitosis. *S. cerevisiae* cells regulate their proliferation similarly, and much of our current understanding of the molecular mechanisms controlling entry into the S phase and the control of DNA replication originates with genetic studies of *S. cerevisiae.*

S. cerevisiae cells replicate by budding (Figure 13-22). The daughter cell initially is smaller than the mother cell and must grow in size considerably before it attempts to divide. Both mother and daughter cells remain in the G_1 period of the cell cycle while growing, although it takes mother cells a shorter time to reach a size compatible with cell division. When *S. cerevisiae* cells in G_1 have grown sufficiently, they begin a program of gene expression that leads to entry into the S phase. If G_1 cells are shifted from a rich medium to a medium low in nutrients before they reach a critical size, they remain in G_1 and grow slowly until they are large enough to enter the S phase. However, once G_1 cells reach the critical size, they become committed to completing the cell cycle, entering the S phase and proceeding through G_2 and mitosis, even if they are shifted to a medium low in nu-

trients. This point in late G_1 of growing *S. cerevisiae* cells when they become irrevocably committed to entering the S phase and traversing the cell cycle is called START.

S. cerevisiae Cdc28 Is Functionally Equivalent to *S. pombe* Cdc2

All *S. cerevisiae* cells carrying a mutation in a particular *cdc* gene arrest with the same size bud at the nonpermissive temperature (see Figure 8-9). Each type of mutant has a terminal phenotype with a particular bud size: no bud (*cdc28*), intermediate-sized buds, or large buds (*cdc13*). Note that in *S. cerevisiae* wild-type genes are indicated in italic capital letters (e.g., *CDC28*) and recessive mutant genes in italic lowercase letters (e.g., *cdc28*); the corresponding wild-type protein is written in Roman letters with an initial capital (e.g., Cdc28), similar to *S. pombe* proteins.

The phenotypic behavior of temperature-sensitive *cdc28* mutants indicates that Cdc28 function is critical for entry into the S phase. When these mutants are shifted to the nonpermissive temperature, they behave like wild-type cells suddenly deprived of nutrients. That is, *cdc28* cells that have grown large enough to pass START at the time of the temperature shift continue through the cell cycle normally and undergo mitosis, whereas those that are too small to have passed START when shifted to the nonpermissive temperature do not enter the S phase even though nutrients are plentiful. Although *cdc28* cells blocked in G_1 continue to grow in size at the nonpermissive temperature, they do not initiate formation of a bud, synthesize DNA, or duplicate their spindle pole body. In other words, at a temperature that inactivates their Cdc28, these cells cannot pass START and enter the S phase.

The wild-type *CDC28* gene was isolated by its ability to complement mutant *cdc28* cells at the nonpermissive temperature (see Figure 13-4). Sequencing of *CDC28* showed that the encoded protein is homologous to known protein kinases, and when Cdc28 protein was expressed in *E. coli*, it exhibited protein kinase activity. Actually, Cdc28 from *S. cerevisiae* was the first cell-cycle protein shown to be a protein kinase. When the *S. pombe cdc2*$^+$ gene was cloned shortly afterwards, it was found to be highly homologous to the *S. cerevisiae CDC28* gene. In fact, the Cdc2 and Cdc28 proteins are functionally analogous, and *S. cerevisiae CDC28* can complement a *S. pombe cdc2*$^-$ mutant. Each type of yeast contains a single cyclin-dependent protein kinase, which can substitute for each other: Cdc2 in *S. pombe* and Cdc28 in *S. cerevisiae*.

Even though *cdc2*$^+$ and *CDC28* encode cyclin-dependent protein kinases, the mutant phenotypes at the nonpermissive temperature differ: most *cdc2*$^-$ *S. pombe* cells are arrested in G_2, whereas most *cdc28 S. cerevisiae* cells are arrested in G_1. This difference can be explained in terms of the physiology of the two yeasts. In *S. pombe* cells growing in rich media, cell-cycle control is exerted primarily at the $G_2 \rightarrow M$ transition (i.e., entry to mitosis). In many *cdc2*$^-$ mutants, including those isolated first, enough Cdc2 activity is maintained at the nonpermissive temperature to permit cells to enter the S phase, but not enough to permit entry into mitosis. Such mutant cells are observed to be arrested in G_2. At the nonpermissive temperature, cultures of completely defective *cdc2*$^-$ mutants include some cells arrested in G_1 and some arrested in G_2, depending on their location in the cell cycle at the time of the temperature shift. Conversely, as noted earlier, cell-cycle regulation in *S. cerevisiae* is exerted primarily at the $G_1 \rightarrow S$ transition (i.e., entry to the S phase). Therefore, partially defective *cdc28* cells are arrested in G_1, but completely defective *cdc28* cells are arrested in either G_1 or G_2. These observations demonstrate that Cdc2 and Cdc28 are required for entry into both the S phase and mitosis.

Three G_1 Cyclins Associate with Cdc28 to form S Phase—Promoting Factors

By the late 1980s, it was clear that mitosis-promoting factor (MPF) is composed of two subunits: a cyclin-dependent protein kinase (Cdk) and a mitotic cyclin required to activate the catalytic subunit. By analogy, it seemed likely that *S. cerevisiae* contains an **S phase–promoting factor (SPF)** that phosphorylates and regulates proteins required for DNA synthesis. Similar to MPF, SPF was proposed to be a heterodimer composed of Cdc28 and a cyclin, in this case one that acts in G_1.

To identify this putative G_1 cyclin, researchers looked for a gene that, when expressed at high levels, could suppress certain temperature-sensitive *cdc28* mutations. The rationale of this approach was that some *cdc28* mutants might be temperature sensitive because of decreased affinity of the mutant Cdc28 for a G_1 cyclin at the nonpermissive temperature. In this case, if the G_1 cyclin was present at high enough levels, it might drive formation of enough SPF, containing G_1 cyclin and the mutant Cdc28, to promote entry into the S phase at the nonpermissive temperature. By using a library of *S. cerevisiae* genomic DNA cloned in a high-copy plasmid vector, researchers isolated two genes, *CLN1* and *CLN2*, that suppressed some *cdc28* mutations in this way (Figure 13-23). Using a different approach, researchers identified a dominant mutation in a third G_1 cyclin gene and subsequently isolated the corresponding wild-type *CLN3* gene.

Sequencing of the three *CLN* genes showed that they encoded related proteins. Each **Cln protein** contains an ≈ 100-residue domain exhibiting significant homology with mitotic cyclins from sea urchin, clam, human, and *S. pombe*. This domain is included in the region of human cyclin A shown in Figure 13-14b,c.

Gene knockout experiments showed that *S. cerevisiae* cells can grow in rich medium if they carry any one of the three *CLN* genes. Knockout of all three genes, however, is lethal, and deletion of *CLN3* extends G_1 for several hours. As the data presented in Figure 13-24 indicate, overproduction of one Cln protein decreases the fraction of cells in G_1, demonstrating that high levels of the Cdc28–G_1 cyclin complex drives cells through start prematurely. Moreover, in the absence of any Cln proteins, cells become arrested in G_1, indicating that a Cdc28–G_1 cyclin heterodimer is required for *S. cerevisiae* cells to enter the S phase.

The complexes formed between Cdc28 and the three G_1 cyclins (Cln1, Cln2, and Cln3) in *S. cerevisiae* have protein kinase activity and constitute the hypothesized S phase–promoting factors (SPFs). In wild-type yeast cells, Cln3 is expressed at a nearly constant level throughout the cell cycle. Cln1 and Cln2 are expressed during the second half of G_1 when they increase in concentration rapidly, peaking at the onset of the S phase and diminishing thereafter until they are eliminated by the time of mitosis.

Kinase Activity of Cdc28—G_1 Cyclin Complexes Prepares Cells for the S Phase

A variety of experiments indicate that the activity of Cdc28-Cln3 is regulated in response to cell size, although the

(a) Wild-type cells

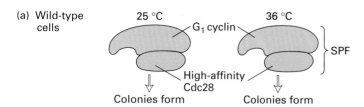

(b) cdc28^{ts} cells

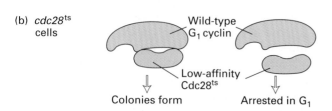

(c) cdc28^{ts} cells transformed with high-copy G₁ cyclin plasmid

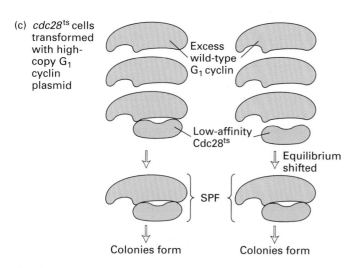

▲ FIGURE 13-23 Genetic screen for G₁ cyclin genes based on proposed interactions between Cdc28 and G₁ cyclins in wild-type and temperature-sensitive (ts) *S. cerevisiae* cells. (a) Wild-type cells produce a high-affinity Cdc28 that associates with G₁ cyclins, forming SPF, at 25 and 36°C. (b) Some *cdc28*^{ts} mutants are thought to express a Cdc28ts protein with low affinity for G₁ cyclin. At the permissive temperature (25° C), enough Cdc28ts–G₁ cyclin (SPF) forms to support growth and colony development, but at the nonpermissive temperature (36° C), SPF does not form, and consequently no colonies develop. (c) When *cdc28*^{ts} cells were transformed with a *S. cerevisiae* genomic library cloned in high-copy plasmids, three types of colonies formed at 36°C: one contained a plasmid carrying the wild-type *CDC28* gene; the other two contained plasmids carrying two genes, *CLN1* and *CLN2*, which encode proteins with homology to mitotic cyclins. In transformed cells carrying the *CLN1* or *CLN2* gene, the concentration of G₁ cyclin is thought to be high enough to offset the low affinity of the altered Cdc28^{ts} at 36°C, so enough SPF forms to support entry into the S phase and subsequent mitosis. Untransformed *cdc28* mutant cells and cells transformed with plasmids carrying other genes are arrested in G₁ and do not form colonies. [See J. A. Hadwiger et al., 1989, *Proc. Nat'l. Acad. Sci. USA* **86**:6255.]

mechanism of this regulation is not currently understood. Once activated, Cdc28-Cln3 phosphorylates and activates two related transcription factors, SBF and MBF. These induce transcription of the *CLN1* and *CLN2* genes as well as several other genes required for DNA replication, including genes encoding DNA polymerase subunits, RPA (the eukaryotic ssDNA-binding protein), DNA ligase, and enzymes required for deoxyribonucleoside triphosphate synthesis.

One of the important substrates of the Cdc28-Cln1 and Cdc28-Cln2 complexes is the yeast anaphase-promoting-complex, APC. Recall that the APC is activated during anaphase of the previous mitosis, probably by MPF phosphorylation. Activated APC then directs polyubiquitination and hence proteasome degradation of the anaphase inhibitor, yeast B-type cyclins, and components of the spindle apparatus (see Figure 13-2). Phosphorylation of APC by Cdc28-Cln1 or Cdc28-Cln2 inactivates the complex in late G₁. Two B-type cyclin genes, called *CLB5* and *CLB6*, also are regulated by MBF and transcribed beginning in late G₁. The corresponding proteins, Clb5 and Clb6, accumulate because of the inactivation of the APC, which would otherwise cause the degradation of these B-type cyclins.

Degradation of the S-Phase Inhibitor Sic1 Triggers DNA Replication

As Cdc28-Clb5 and Cdc28-Clb6 heterodimers accumulate in late G₁, they are immediately inactivated by binding of Sic1, which is expressed late in mitosis and early G₁. Sic1 functions as an *S-phase inhibitor*, specifically inhibiting Cdc28–B-type cyclin complexes but having no effect on the Cdc28–G₁ cyclin (Cln) complexes.

Cells enter the S phase (i.e., DNA replication initiates at yeast origins) when the Sic1 inhibitor is precipitously degraded following its polyubiquitination by a distinct E2 ubiquitin-conjugating enzyme (Cdc34) associated with an E3 ubiquitin ligase called SCF (Figure 13-25). Once Sic1 is degraded, the Cdc28-Clb5 and Cdc28-Clb6 kinases induce DNA replication from yeast origins by phosphorylating as-yet unidentified substrates. This mechanism for activating these Cdk-cyclin complexes—that is, inhibiting them as they are synthesized and then precipitously degrading the inhibitor—permits the sudden activation of large numbers of complexes, as opposed to the gradual increase in kinase activity that would result if no inhibitor were present during synthesis of the S-phase cyclins.

We can now see that regulated proteolysis directed by two ubiquitinating complexes, Cdc34-SCF and APC, controls three major transitions in the cell cycle: onset of the S phase, beginning of anaphase, and exit from mitosis. As discussed earlier, the APC must be activated before anaphase can proceed (see Figure 13-19). In contrast, the activity of Cdc34-SCF is not directly regulated. Rather, control is exerted by marking its substrate, Sic1, for polyubiquitination by phosphorylation by a Cdc28-G₁ cyclin (see Figure

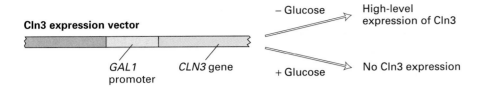

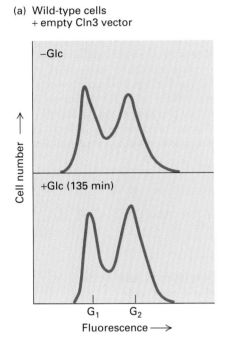

(a) Wild-type cells
+ empty Cln3 vector

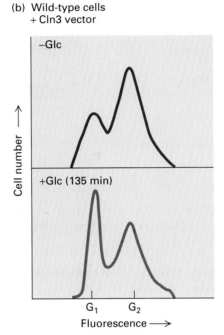

(b) Wild-type cells
+ Cln3 vector

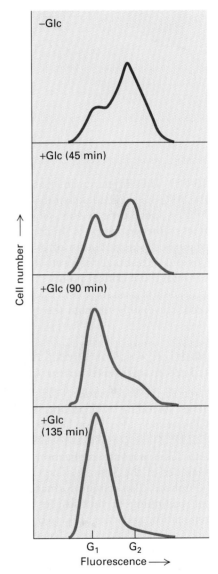

(c) *cln1⁻/cln2⁻/cln3⁻* cells
+ Cln3 vector

▲ **FIGURE 13-24 Experimental demonstration that G₁ cyclins (Cln1, Cln2, and Cln3) regulate entry of *S. cerevisiae* cells into the S phase.** The yeast expression vector used in these experiments *(top)* carried the *CLN3* gene linked to the strong *GAL1* promoter, which is turned off when glucose is present in the medium. To determine the proportion of cells in G₁ and G₂, cells were exposed to a fluorescent dye that binds to DNA and then were passed through a fluorescence-activated cell sorter (see Figure 5-21). Since the DNA content of G₂ cells is twice that of G₁ cells, this procedure can distinguish cells in the two cell-cycle phases. (a) Wild-type cells transformed with an empty expression vector displayed the normal distribution of cells in G₁ and G₂ in the absence of glucose (Glc) and after addition of glucose (bottom curve). (b) Wild-type cells transformed with the Cln3 expression vector displayed a higher-than-normal percentage of cells in the S phase and G₂ because overexpression of Cln3 decreased the G₁ period (red curve). When expression of Cln3 from the vector was shut off by addition of glucose, the cell distribution returned to normal (bottom curve). (c) Cells with mutations in all three *CLN* genes and transformed with the Cln3 expression vector also showed a high percentage of cells in S and G₂ in the absence of glucose (red curve). When expression of Cln3 from the vector was shut off by addition of glucose, the cells completed the cell cycle and arrested in G₁ (bottom curve) because cells contained no functional Cln proteins. [Adapted from H. E. Richardson et al., 1989, *Cell* **59**:1127.]

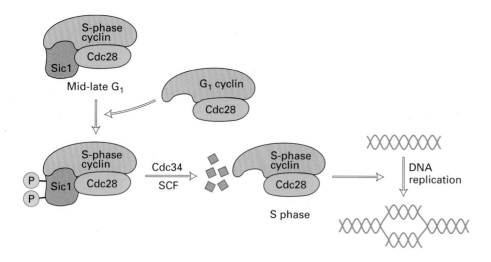

◀ **FIGURE 13-25 Control of the G₁ → S phase transition in *S. cerevisiae* by regulated proteolysis of Sic1.** Although the S-phase Cdk-cyclin complexes (Cdc28-Clb5 and Cdc28-Clb6) begin to accumulate in G₁, they are inhibited by Sic1, preventing initiation of DNA replication until the cell is fully prepared. Cdk-cyclin complexes assembled in late G₁ (Cdc28-Cln1 and Cdc28-Cln2) phosphorylate Sic1, enabling its recognition and polyubiquitination by the Cdc34 ubiquitin-conjugating enzyme and a trimeric ubiquitin ligase called SCF. The active S-phase Cdk complexes then induce initiation of DNA synthesis at origins by phosphorylating substrates that remain to be identified. [Adapted from R. W. King et al., 1996, *Science* **274**:1652.]

13-25). This difference in strategy probably occurs because the APC has several substrates, including the anaphase inhibitor and B-type cyclins, which must be degraded at different times in the cycle. In contrast, entry into the S phase requires the degradation of only a single protein, Sic1. An obvious advantage of proteolysis for controlling passage through these critical points in the cell cycle is that protein degradation is an irreversible process, ensuring that cells proceed irreversibly through the cycle.

Multiple Cyclins Direct Kinase Activity of Cdc28 during Different Cell-Cycle Phases

Later in the *S. cerevisiae* S phase, transcription begins of the genes encoding two additional B-type cyclins, Clb3 and Clb4, which also form heterodimeric protein kinases with Cdc28. These complexes together with those including Clb5 and Clb6 activate replication origins throughout the remainder of the S phase. The Cdc28-Clb3 and Cdc28-Clb4 complexes also initiate formation of the mitotic spindle at the beginning of mitosis. As cells complete chromosome replication and enter G₂, two more B-type cyclins are expressed, Clb1 and Clb2. These function as mitotic cyclins, associating with Cdc28 to form complexes that are required for chromosome segregation and nuclear division.

Each group of cyclins thus directs the Cdc28 kinase activity to specific functions associated with various cell-cycle phases, as outlined in Figure 13-26. Cdc28-Cln3 activates transcription in late G₁ by phosphorylating and activating transcription factors SBF and MBF. Cdc28-Cln1 and -Cln2 inhibit the APC, allowing B-type cyclins to accumulate, and activate proteolysis of the S-phase inhibitor Sic1. Cdc28-Clb5, -Clb6, -Clb3, and -Clb4 trigger DNA synthesis. Cdc28-Clb3 and -Clb4 also trigger

formation of mitotic spindles. And Cdc28-Clb1 and -Clb2 trigger nuclear division.

Replication at Each Origin Is Initiated Only Once during the Cell Cycle

As discussed in Chapter 12, eukaryotic chromosomes are replicated from multiple origins, some of which initiate DNA replication early in the S phase and some of which initiate late. However, none of the multiple replication origins initiate more than once per S phase, ensuring proper maintenance of gene copy number. Recent experiments indicate that this restriction of origin "firing" to once and only once per cell cycle in *S. cerevisiae* is enforced by the alternating cycle of Cdk-Clb activities throughout the cell cycle: low in teleophase through G₁ and high in S, G₂, and M through anaphase (see Figure 13-26).

Yeast replication origins contain an 11-bp conserved core sequence to which is bound a hexameric protein, the *origin-recognition complex* (ORC), required for initiation of DNA synthesis. DNase I footprinting analysis and immunoprecipitation of chromatin proteins cross-linked to specific DNA sequences during the various phases of the cell cycle indicate that the ORC remains associated with origins during all phases of the cycle. Several proteins required for the initiation of DNA synthesis in *S. cerevisiae*, including Cdc6, Cdc45, and a complex of six Mcm proteins, also associate with origins during G₁, but not during G₂ or M. This complex of ORC plus initiation proteins that assembles in G₁ is called a *pre-replication complex*.

As discussed in the previous section, Cdc28-Clb5 and -Clb6 complexes become active at the beginning of the S phase when their specific inhibitor, Sic1, is degraded (see Figure 13-25). In the current model for *S. cerevisiae*

▶ **FIGURE 13-26 Activity of *S. cerevisiae* Cdc28-cyclin complexes through the course of the cell cycle.** The width of the colored bands is approximately proportional to the demonstrated or proposed protein kinase activity of the indicated Cdc28-cyclin complexes.

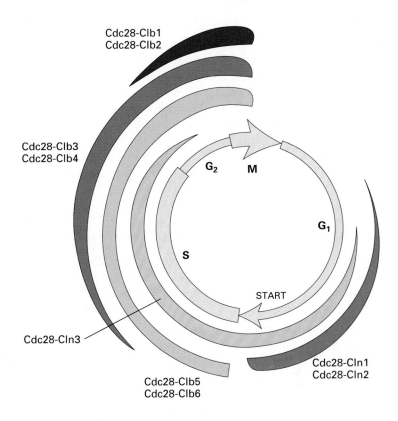

replication, active Cdc28-Clb complexes phosphorylate proteins that simultaneously activate initiation of DNA replication by pre-replication complexes and prevent assembly of new pre-replication complexes needed for another round of initiation (Figure 13-27). Initiation of DNA replication requires both the assembly of a pre-replication complex and an active Cdc28-Clb complex. A second heterodimeric protein kinase, Cdc7-Dbf4, which is expressed in G_1, is also required

to trigger initiation. Each origin fires only once during the S phase because Cdc28-Clb complexes remain active throughout the S phase and G_2, preventing the assembly of new pre-replication complexes.

Activation of the anaphase-promoting complex (APC) at the anaphase $\rightarrow$ telophase transition leads to proteolysis of B-type cyclins (Clbs), eliminating the block to assembly of pre-replication complexes. APC activity remains high during

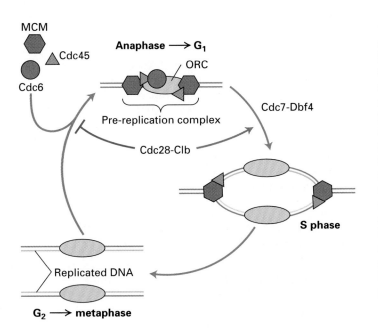

◀ **FIGURE 13-27 Assembly of the pre-replication complex and its regulation by Cdc28–Clb (Cdk–cyclin B) complexes in *S. cerevisiae*.** Pre-replication complexes assemble at origins when Cdc28-Clb activity is absent, that is, following anaphase and in early G_1. The increase in Cdc28-Clb activity at the beginning of the S phase triggers initiation of DNA replication from pre-replication complexes. The heterodimeric protein kinase Cdc7-Dbf4 also is required to trigger DNA replication. The presence of active Cdk-cyclin B complexes during S, G_2, and up to metaphase in mitosis prevents assembly of new pre-replication complexes. This dual function of Cdk-cyclin B complexes ensures that replication does not re-initiate at an origin until the cell undergoes anaphase (when Cdc28-Clb activity falls) and segregates sister chromosomes to daughter cells. Once replication has initiated, Mcm proteins and Cdc45 move away from the origin along with DNA polymerases (not shown). Mcm proteins are homologous to helicases and may form the hexameric helicase associated with replication forks (see Figure 12-11). [Adapted from C. S. Newlon 1997 *Cell* **91**:717.]

early G_1 when the Cdc6 protein, required for pre-replication complex assembly, is synthesized. Pre-replication complexes assemble on origins during early G_1 because Cdc28-Clb activity is low while Cdc6 activity is high (see Figure 13-27). As discussed previously, APC activity is inhibited during late G_1, setting the stage for the synthesis and subsequent activation of Clb5 and Clb6, which depend on Cdc28–G_1 cyclin activity. The release of Cdc28–Clb activity triggers origin firing at the onset of the S phase and inhibition of new pre-initiation complex assembly until the next G_1 phase when active APC eliminates Cdc28–Clb activity.

SUMMARY Genetic Studies with *S. cerevisiae*

- *S. cerevisiae* expresses a single cyclin-dependent protein kinase (Cdk), encoded by *CDC28*, which interacts with several different cyclins during different phases of the cell cycle (see Figure 13-26).

- Three cyclins are active in G_1: Cln1, Cln2 and Cln3. Cln3 concentration does not vary significantly through the cell cycle, but Cdc28-Cln3 activity is regulated by cell size. Active Cdc28–Cln3 phosphorylates two transcription factors that induce expression in late G_1 of Cln1 and Cln2, enzymes and other proteins required for DNA replication, and the S-phase cyclins Clb5 and Clb6.

- Cln1 and Cln2 form complexes with Cdc28 that inhibit the anaphase promoting complex (APC), thus permitting accumulation of Clb5 and Clb6 in late G_1.

- Cdc28–Clb5 and Cdc28–Clb6 formed in late G_1 initially are inhibited by Sic1, expressed in early G_1. Ubiquitin-mediated degradation of Sic1 activates Cdc28–Clb5 and Cdc28–Clb6, triggering initiation of DNA replication, that is, onset of the S phase (see Figure 13-25).

- Cyclins Clb3 and Clb4, expressed later in S, form heterodimers with Cdc28 that also trigger initiation at origins and initiate spindle formation early in mitosis.

- Cyclins Clb1 and Clb2, expressed in G_2, form heterodimers with Cdc28 that stimulate mitotic processes. All of the B-type cyclins (Clbs) are degraded in late anaphase by activated APC.

- DNA replication is initiated from pre-replication complexes assembled at origins during early G_1. S-phase Cdc28-Clb complexes simultaneously trigger initiation from pre-replication complexes and inhibit assembly of new pre-replication complexes (see Figure 13-27). This results in initiation at each origin only once until cells proceed through anaphase when activation of APC leads to a decrease Cdc28-Clb activity.

- The block on reinitiation of DNA replication until cells have completed anaphase and segregated their replicated chromosomes to daughter cells maintains the proper number of chromosomes per cell.

13.6 Cell-Cycle Control in Mammalian Cells

In multicellular organisms, precise control of the cell cycle during development and growth is critical for determining the size and shape of each tissue. Cell replication is controlled by a complex network of signaling pathways that integrate extracellular signals about the identity and numbers of neighboring cells and intracellular cues about cell size and developmental program. Most cells withdraw from the cell cycle during G_1, entering the G_0 state, to differentiate (see Figure 13-1). Some differentiated cells (e.g., fibroblasts and lymphocytes) can be stimulated to reenter the cycle and replicate. Many differentiated cells, however, never reenter the cell cycle to replicate again; they are referred to as *postmitotic* cells. As we discuss in this section, the cell-cycle regulatory mechanisms uncovered in yeasts and *Xenopus* eggs and early embryos also operate in the somatic cells of higher eukaryotes including humans and other mammals.

Mammalian Restriction Point is Analogous to START in Yeast Cells

Most studies of mammalian cell-cycle control have been done with cultured cells that require certain polypeptide growth factors (mitogens) to stimulate cell proliferation. Binding of these growth factors to specific receptor proteins that span the plasma membrane initiates a cascade of intracellular molecular events, referred to as **signal transduction**, that ultimately influence transcription and cell-cycle control (Chapter 20).

Mammalian cells cultured in the absence of growth factors are arrested with a diploid complement of chromosomes in the G_0 period of the cell cycle. If growth factors are added to the culture medium, these **quiescent cells** pass through the restriction point 14–16 hours later, enter the S phase 6–8 hours after that, and traverse the remainder of the cell cycle. Like START in yeast cells, the restriction point is the point in the cell cycle at which mammalian cells become committed to entering the S phase and completing the cell cycle. If mammalian cells are moved from a medium containing growth factors to one lacking growth factors before they have passed the restriction point, the cells do not enter the S phase. But once cells have passed the restriction point, they are committed to entering the S phase and progressing through the entire cell cycle, which takes about 24 hours for most cultured mammalian cells.

Multiple Cdks and Cyclins Regulate Passage of Mammalian Cells through the Cell Cycle

Unlike *S. pombe* and *S. cerevisiae*, which each produce a single cyclin-dependent kinase (Cdk) to regulate the cell cycle, mammalian cells use a small family of related Cdks to regulate progression through the cell cycle. The principle Cdks active in most mammalian cells have been named

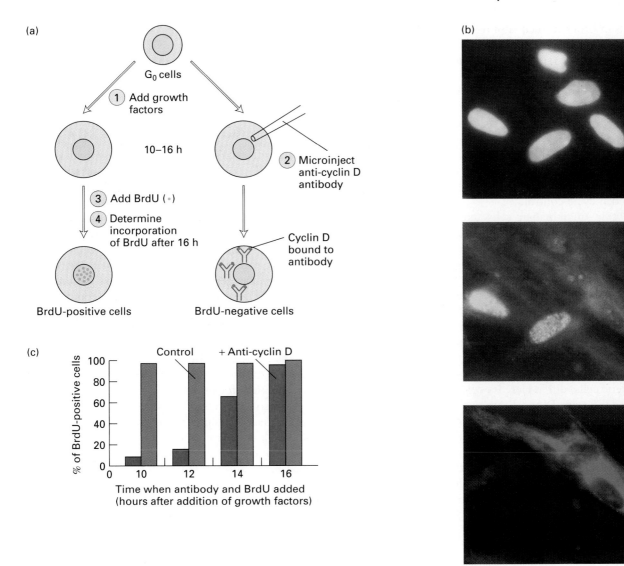

▲ **FIGURE 13-28 Experimental demonstration that cyclin D is required for passage through the restriction point in the mammalian cell cycle.** The G_0-arrested cells used in these experiments pass the restriction point 14–16 hours after addition of growth factors, and enter the S phase 6–8 hours later. (a) Outline of experimental protocol. At various times 10–16 hours after addition of growth factors, some cells were microinjected with rabbit antibodies against cyclin D. Bromodeoxyuridine (BrdU), a thymidine analog, was then added to the medium, and the control and microinjected cells were incubated separately for an additional 16 hours. Each sample then was analyzed to determine the percentage of cells that had incorporated BrdU, indicating that they had entered the S phase. (b) Analysis of cells microinjected with rabbit anti-cyclin D antibody 8 hours after addition of growth factors. The three panels show the same field of cells stained 16 hours after microinjection and addition of BrdU to the medium. Cells were stained with a DNA stain that fluoresces blue *(top)*, a fluorescein-conjugated mouse monoclonal antibody specific for BrdU *(middle)*, and Texas red-conjugated goat antibodies against rabbit antibodies to identify the cells that had been microinjected with the rabbit anti-cyclin D antibody *(bottom)*. Note that the two cells injected with anti-cyclin D antibody (the red cells in the right bottom) did not incorporate BrdU into nuclear DNA. (c) Percentage of control cells (blue) and cells injected with anti-cyclin D antibodies (red) that had incorporated BrdU. Most cells injected with anti-cyclin D antibodies 10 or 12 hours after addition of growth factors failed to enter the S phase. In contrast, anti-cyclin D antibodies had little effect on entry into the S phase and DNA synthesis when injected at 14 or 16 hours, after cells had passed the restriction point. These results indicate that cyclin D is required to pass the restriction point, but once cells have passed the restriction point, they do not require cyclin D to enter the S phase 6–8 hours later. [Parts (b) and (c) adapted from V. Baldin et al., 1993 *Genes & Devel.* **7**:812.]

Cdk1, 2, 4, and 6 in order of their discovery. The first human Cdk to be discovered was identified by the ability of a cDNA clone encoding it to complement *S. pombe cdc2⁻* mutants; now designated Cdk1, this protein initially was called human Cdc2, a term that continues to be commonly used. A cDNA encoding mammalian Cdk2 was isolated by its ability to complement *S. cerevisiae cdc28* mutants. Cdk4 and Cdk6 were isolated based on their homology to other Cdks. cDNAs for Cdk3 and Cdk5 also have been isolated, but the encoded proteins are not expressed at significant levels in most mammalian cells, and they function in processes other than cell cycle control.

Like *S. cerevisiae*, mammalian cells express multiple cyclins. Cyclins A and B, which function in the S phase, G_2, and early mitosis, initially were detected as proteins whose concentration oscillates in experiments with synchronously cycling early sea urchin and clam embryos. Homologous cyclin A and B proteins have been found in all higher eukaryotes examined. cDNAs encoding three related human **D-type cyclins** and cyclin E were isolated based on their ability to complement *S. cerevisiae* cells mutant in all three *CLN* genes encoding G_1 cyclins. The relative amounts of the three D-type cyclins expressed in various cell types (e.g., fibroblasts, hematopoietic cells) differ. Here we refer to them collectively as cyclin D. Cyclin D and E are expressed during G_1. Experiments such as those shown in Figure 13-28 demonstrate that these G_1 cyclins are essential for passage through the restriction point.

Figure 13-29 presents a current model for the periods of the cell cycle in which different Cdk-cyclin complexes act in G_0-arrested mammalian cells stimulated to divide by the addition of growth factors. Neither cyclins nor Cdks are ex-

pressed in G_0 cells cultured in the absence of growth factors. The absence of these proteins critical for stimulating the many molecular processes required for cell-cycle progression explains why G_0 cells do not replicate.

Regulated Expression of Two Classes of Genes Returns G_0 Mammalian Cells to the Cell Cycle

Addition of growth factors to G_0-arrested mammalian cells induces transcription of multiple genes, most of which fall into one of two classes—**early-response** or **delayed-response genes**—depending on how soon their encoded mRNAs appear (Figure 13-30a). Transcription of early-response genes is induced within a few minutes after addition of growth factors by a signal-transduction cascade that leads to activation of specific transcription factors (Chapter 20). Induction of early-response genes is not blocked by inhibitors of protein synthesis (Figure 13-30b), because the required transcription factors are present in G_0 cells and are activated by post-translational modifications such as phosphorylation. Many of the early-response genes encode transcription factors, such as c-Fos and c-Jun, that stimulate transcription of the delayed-response genes. As discussed in Chapter 24, mutant, unregulated forms of both c-Fos and c-Jun are expressed by oncogenic retroviruses; the discovery that the viral forms of these proteins (v-Fos and v-Jun) can transform normal cells into cancer cells led to identification of the regulated cellular forms of these transcription factors.

After peaking at about 30 minutes following addition of growth factors, the concentrations of the early-response mRNAs fall to a lower level that is maintained as long as growth factors are present in the medium. Most of the imme-

▶ **FIGURE 13-29 Activity of mammalian Cdk-cyclin complexes through the course of the cell cycle in G_0 cells induced to divide by treatment with growth factors.** The width of the colored bands is approximately proportional to the protein kinase activity of the indicated complexes. Cyclin D refers to all three D-type cyclins.

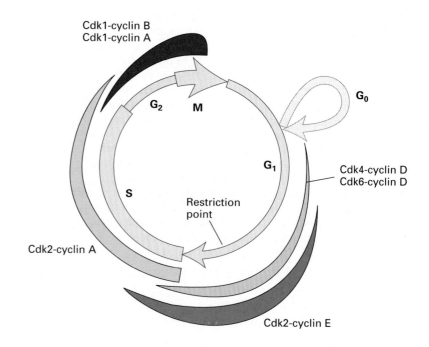

(a)

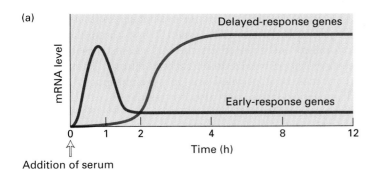

(b)

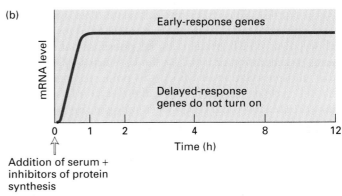

▲ FIGURE 13-30 Time course of expression of early- and delayed-response genes in G_0-arrested mammalian cells after addition of serum containing growth factors in the absence

(a) and presence (b) of inhibitors of protein synthesis. See text for discussion. [Adapted from A. Murray and T. Hunt, 1993, *The Cell Cycle: An Introduction*, W. H. Freeman and Company.]

diate early proteins are unstable and, consequently, fall in concentration as the level of their mRNAs, and hence their rate of synthesis, falls. This drop in transcription is blocked by inhibitors of protein synthesis (see Figure 13-30b), indicating that it depends on production of one or more of the early-response proteins.

Proteins encoded by early response genes induce the transcription and expression of delayed response genes. Delayed response genes are not expressed when mitogens are added to G_0-arrested cells in the presence of an inhibitor of protein synthesis. This is because their expression requires the synthesis of early-response proteins.

Some delayed-response genes encode additional transcription factors such as E2Fs, discussed in the next section. Other delayed-response genes encode the D-type cyclins, cyclin E, Cdk2, Cdk4, and Cdk6. Cdk4, Cdk6, and the D-type cyclins are expressed first, followed by cyclin E and Cdk2 (see Figure 13-29). If growth factors are withdrawn before passage through the restriction point, transcription of these G_1-phase Cdks and cyclins ceases. Since these proteins and the mRNAs encoding them are unstable, their concentrations fall precipitously. As a consequence, the cells do not pass the restriction point and do not replicate.

Passage through the Restriction Point Depends on Activation of E2F Transcription Factors

As in *S. cerevisiae*, mammalian genes encoding many of the proteins involved in DNA and deoxyribonucleotide synthesis are induced as cells pass through the $G_1 \rightarrow S$ transition. A small family of E2F transcription factors is required for transcription of several of these genes and those encoding Cdk2 and cyclins A and E. In addition, E2Fs autostimulate transcription of the genes encoding themselves.

The transcription-activating ability of E2Fs is inhibited by their binding of **Rb protein** and two related proteins, p107 and p130. Indeed, binding of Rb to E2Fs converts them from transcriptional activators to repressors because Rb interacts with histone deacetylase complexes (see Figure

10-58). Rb protein was initially identified as the product of the prototype tumor-suppressor gene *RB* (Chapter 24). As discussed in Chapter 8, a child with hereditary retinoblastoma receives one normal RB^+ allele and one mutant RB^- allele. If the RB^+ allele in a retinal cell is somatically mutated to a RB^- allele, then no functional protein is expressed and the cell or one of its descendants is likely to become cancerous, leading to the retinal tumors that characterize this disease (see Figure 8-7).

Phosphorylation of Rb protein inhibits its repressing function, permitting activation of the genes required for entry into the S phase by E2Fs. As shown in Figure 13-31, phosphorylation of Rb protein is initiated by Cdk4–cyclin D and Cdk6–cyclin D in mid G_1; once expression of Cdk2 and cyclin E is stimulated, Cdk2–cyclin E further phosphorylates

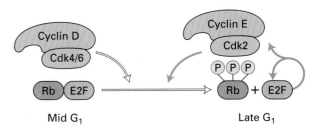

Mid G_1 Late G_1

▲ FIGURE 13-31 Regulation of Rb and E2F activities in late G_1. Stimulation of G_0 cells with mitogens induces expression of Cdk4, Cdk6, D-type cyclins and E2F transcription factors (E2Fs), all encoded by delayed-response genes (see Figure 13-30). Interaction of E2Fs with hypophosphorylated Rb protein initially inhibits E2F activity. When signaling from mitogens is sustained, the resulting Cdk4–cyclin D and Cdk6–cyclin D complexes (Cdk4/6–cyclin D) initiate the phosphorylation of Rb, converting some E2F to the active form. Active E2F then stimulates its own synthesis and the synthesis of Cdk2 and cyclin E. Cdk2–cyclin E further stimulates Rb phosphorylation releasing more E2F activity. These processes result in positive feedback loops (blue arrows) leading to a rapid rise in both E2F and Cdk2–cyclin E activity as the cell approaches the $G_1 \rightarrow S$ transition.

Rb in late G_1. Since E2Fs also stimulate their own expression, these processes form positive feedback loops for phosphorylation of Rb protein. Initial phosphorylation of Rb leads to generation of Cdk2–cyclin E, which accelerates further phosphorylation of Rb. At this point, passage through the cell cycle is independent of Cdk4/6–cyclin D activity, so that progression occurs even when mitogens are withdrawn and cyclin D levels fall—that is, the restriction point is passed. The positive cross-regulation of E2F and Cdk2–cyclin E produces a rapid rise of both activities as cells approach the $G_1 \rightarrow S$ transition point. Events in addition to Rb phosphorylation also contribute to control of passage through the restriction point as evidenced by the finding that $RB^{-/-}$ cells retain some dependence on mitogens, although greatly reduced mitogen concentrations suffice.

Rb protein is maintained in the phosphorylated state throughout the S, G_2, and M phases by Ckd2 -and Cdk1-cyclin complexes. After cells complete mitosis and enter early G_1 or G_0, the fall in Cdk-cyclin levels leads to dephosphorylation of Rb protein by the action of unopposed phosphatases. As a consequence, hypophosphorylated Rb protein is available to inhibit E2F activity during early G_1 of the next cycle.

Cyclin A Is Required for DNA Synthesis and Cdk1 for Entry into Mitosis

Synthesis of cyclin A begins as cells approach the $G_1 \rightarrow S$ transition, and the protein is immediately transported into the nucleus. Disruption of cyclin A function inhibits DNA synthesis in mammalian cells. Consequently, like *S. cerevisiae* Cdc28-Clb5/6 complexes, Cdk2–cyclin A may trigger initiation of DNA synthesis by pre-replication complexes assembled in early G_1 and inhibit the assembly of new pre-replication complexes (see Figure 13-27). At present, it is not clear which mammalian protein performs the function of *S. cerevisiae* Sic1 (see Figure 13-25); by analogy this protein would inhibit the activity of Cdk2–cyclin A until it is degraded precipitously at the onset of S phase.

The principle mammalian Cdk in G_2 is Cdk1 (see Figure 13-29). As mentioned previously, Cdk1 is highly homologous with *S. pombe* Cdc2. Cdk1 associates with cyclins A and B, and mRNAs encoding either of these mammalian cyclins can promote oocyte maturation when injected into *Xenopus* oocytes. In mammalian cells, the MPF activity of Cdk1–cyclin A and Cdk1–cyclin B, which induces entry into mitosis, appears to be regulated by proteins analogous to *S. pombe* Wee1, CAK, and Cdc25 (see Figure 13-13). For instance, activation of each of the mammalian Cdks requires phosphorylation of a T-loop threonine (see Figure 13-14), and there is evidence that each Cdk is inhibited by protein kinases analogous to *S. pombe* Wee1 and that phosphatases analogous to *S. pombe* Cdc25 remove the inhibitory phosphate.

In cycling mammalian cells, cyclin B is first synthesized during the S phase and increases in concentration as cells proceed through G_2, peaking in early mitosis and dropping after anaphase, as in *Xenopus* cycling egg extracts (see Figure 13-7). In human cells, cyclin B first accumulates in the cytoplasm and then enters the nucleus just before the nuclear envelope breaks down early in mitosis. Thus MPF activity may be controlled not only by phosphorylation and dephosphorylation but also by regulation of the nuclear transport of cyclin B.

The APC-directed polyubiquitination of cyclins A and B and their subsequent degradation in proteasomes causes a rapid fall in MPF activity, permitting completion of mitosis and setting the stage for the next cell cycle (see Figure 13-2).

Mammalian Cyclin-Kinase Inhibitors Contribute to Cell-Cycle Control

As noted above, mammalian cells are thought to express a *cyclin-kinase inhibitor (CKI)* that functions like *S. cerevisiae* Sic1. Although this inhibitor has not yet been identified, mammalian cells are known to express several CKIs that contribute to cell-cycle control. These are grouped into two classes: those in the *CIP* (Cdk inhibitory protein) family bind and inhibit all Cdk1-, Cdk2-, Cdk4-, and Cdk6-cyclin complexes; those in the *INK4* (inhibitors of kinase 4) family bind and inhibit only Cdk4–cyclin D and Cdk6–cyclin D complexes.

Experimental overexpression of INK4 proteins inhibits progression through G_1, the expected result for inhibition of Cdk4/6–cyclin D phosphorylation of Rb protein. As discussed in Chapter 24, one INK4 protein, designated p16 after its molecular weight, appears to function primarily as a *tumor suppressor,* but the functions of the other INK4 proteins still are unclear. All INK4 proteins contain four tandem repeats of an ≈ 32-residue consensus sequence called the ankyrin repeat, after the first protein in which this repeated sequence was observed.

The three known mammalian **CIP proteins** (p21, p27, and p57) also are closely related in sequence. As discussed later, p21^{CIP} functions in the response of most mammalian cells to DNA damage. CIP function also has recently been shown to be required for normal embryonic development in *Drosophila*. The single CIP homolog thus far identified in *Drosophila*, called Dacapo, specifically binds to and inhibits Cdk2–cyclin E. During normal *Drosophila* embryogenesis, the fertilized egg and its daughter cells undergo 16 rapid cell cycles. Cells that form the epidermis of the embryo then express Dacapo transiently beginning in G_2 of cycle 16 and persisting into G_1 of cycle 17 when they exhibit an extended G_1 phase. In *dacapo* mutants, these cells fail to arrest in G_1 of cycle 17 and complete one extra cycle. These mutant cells may be limited to only one extra cycle because expression of cyclin E is normally repressed at this time in development, limiting further cell cycles. Thus during normal embryonic development, both Dacapo induction and cyclin E repression contribute to the proper control of cell cycling, thereby limiting the number of cell cycles in the developing epidermis.

The functions of mammalian p27^{CIP} and p57^{CIP} have been assessed in mouse knockout experiments. Homozygous p27$^{-/-}$ mouse embryos develop normally through birth. However, by several weeks of age the mutant mice are 30 percent larger than wild-type animals because of a general overproliferation of cells in most organs. Although p27$^{-/-}$ cells undergo more cell cycles than normal, most eventually arrest in G$_1$ and develop normally. Thus, although p27^{CIP} clearly contributes to the normal control of cell proliferation, other mechanisms also restrain cell cycling, especially during embryonic and fetal development.

p57^{CIP} is normally expressed in newly differentiated cells and in many adult tissues. Knockout mice that do not produce p57^{CIP} die shortly after birth with a number of developmental defects in a subset of the tissues that normally express p57^{CIP}. In these tissues, failure to become postmitotic leads to defects in cell differentiation. In many cases, instead of normal differentiation, the abnormal cells undergo a process of programmed cell death, called **apoptosis** (Chapter 23).

SUMMARY Cell-Cycle Control in Mammalian Cells

- In multicellular organisms, cell replication is controlled by a complex network of signaling pathways that integrate signals from the extracellular environment with intracellular cues about cell size and developmental program.

- Polypeptide growth factors called mitogens stimulate cultured mammalian cells to cycle. Once cycling cells pass the restriction point, they can enter the S phase and complete S, G$_2$, and mitosis in the absence of growth factors.

- Mammalian cells use several Cdks and cyclins to regulate passage through the cell cycle. Cdk4–cyclin D and Cdk6–cyclin D function in mid to late G$_1$; Cdk2–cyclin E in late G$_1$; Cdk2–cyclin A in S; and Cdk1–cyclin A and Cdk1–cyclin B in G$_2$ and M through anaphase (see Figure 13-29).

- Growth factors induce expression of early-response genes, many of which encode transcription factors that stimulate expression of delayed-response genes. Cdks2, 4, and 6, cyclins D and E, and E2F transcription factors are encoded by delayed-response genes.

- Passage through the restriction point requires activation of E2F, which stimulates transcription of genes encoding proteins required for DNA replication and deoxyribonucleotide synthesis as well as Cdk2, cyclin E, and cyclin A.

- E2F activity is inhibited by binding of hypophosphorylated Rb protein generated during mitosis. Polyphosphorylation of Rb protein, first by Cdk4/6–cyclin D and then by Cdk2–cyclin E, in mid and late G$_1$ frees E2F so it can activate transcription (see Figure 13-31).

- Cdk2–cyclin A is required for DNA replication and may be the kinase that activates pre-replication complexes to initiate DNA synthesis.

- Cdk1–cyclin A and Cdk1–cyclin B induce the events of mitosis through metaphase. Cyclins A and B are polyubiquitinated by the anaphase-promoting complex (APC) during late anaphase and then are degraded by proteasomes.

- The activity of all mammalian Cdk-cyclin complexes appears to be regulated by phosphorylation and dephosphorylation similar to the mechanism in *S. pombe* (see Figure 13-13).

- The activity of mammalian Cdk-cyclin complexes is further regulated by two classes of cyclin-kinase inhibitors, CIP proteins and INK4 proteins.

13.7 Checkpoints in Cell-Cycle Regulation

Catastrophic genetic damage can occur if cells progress to the next phase of the cell cycle before the previous phase is properly completed. For example, when S-phase cells are induced to enter mitosis by fusion to a cell in mitosis, the MPF present in the mitotic cell forces the chromosomes of the S-phase cell to condense. However, since the replicating chromosomes are fragmented by the process (Figure 13-32), such premature entry into mitosis is disastrous for a cell. Another example concerns attachment of kinetochores to microtubules

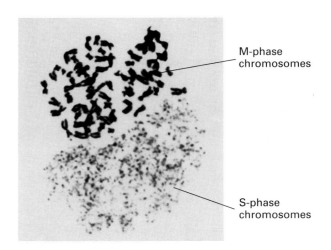

▲ **FIGURE 13-32 Micrograph of chromosomes from a hybrid cell resulting from fusion of an S-phase cell and a M-phase cell.** Chromosomes from the mitotic cell are highly condensed with distinct sister chromatids visible, as expected. The chromosomes from the S-phase cell are fragmented into pieces of condensed chromosomes. [From R. T. Johnson and P. N. Rao, 1971, *Biol. Rev.* **46**:97.]

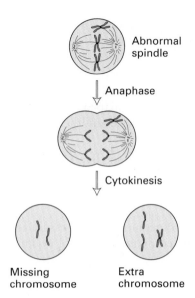

▲ FIGURE 13-33 Nondisjunction occurs when chromosomes segregate in anaphase before the kinetochore of each sister chromatid has attached to microtubules (red lines) from the opposite spindle poles. As a result, one daughter cell contains two copies of one chromosome, while the other daughter cell lacks that chromosome. [Adapted from A. Murray and T. Hunt, 1993, *The Cell Cycle: An Introduction*, W. H. Freeman and Company.]

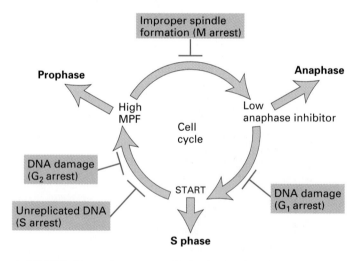

▲ FIGURE 13-34 Stages at which checkpoint controls can arrest passage through the cell cycle. DNA damage due to irradiation or chemical modification prevents G_1 cells from entering the S phase and G_2 cells from entering mitosis. Unreplicated DNA prevents entry into mitosis. Defects in assembly of the mitotic spindle or the attachment of kinetochores to spindle microtubules prevent activation of the APC polyubiquitination system that leads to degradation of the anaphase inhibitor. Consequently, cells do not enter anaphase until all kinetochores are bound to spindle microtubules. [Adapted from A. Murray and T. Hunt, 1993, *The Cell Cycle: An Introduction*, W. H. Freeman and Company.]

of the mitotic spindle during metaphase. If anaphase is initiated before both kinetochores of a replicated chromosome become attached to microtubules from opposite spindle poles, daughter cells are produced that have a missing or extra chromosome (Figure 13-33). When this process, called **nondisjunction**, occurs during the meiosis that generates a human egg, Down syndrome can occur from trisomy of chromosome 23, resulting in developmental abnormalities and mental retardation.

To minimize the occurrence of such mistakes in cell-cycle events, a cell's progress through the cycle is monitored at four key **checkpoints** (Figure 13-34). Control mechanisms that operate at these checkpoints ensure that chromosomes are intact and that each stage of the cell cycle is completed before the following stage is initiated.

The Presence of Unreplicated DNA Prevents Entry into Mitosis

Cells that fail to replicate all their chromosomes do not enter mitosis. Operation of this checkpoint control involves the recognition of unreplicated DNA and inhibition of MPF activation. Although the ability of unreplicated DNA to inhibit entry into mitosis is well documented, little is yet known about the proteins that mediate this checkpoint control. However, yeast mutants defective in this checkpoint control have been isolated; analysis of such mutants should help reveal how cells recognize the presence of unreplicated DNA and how it prevents activation of MPF.

Improper Assembly of the Mitotic Spindle Leads to Arrest in Anaphase

The effect of colchicine, which inhibits microtubule polymerization and thus assembly of the mitotic spindle, demonstrates the presence of another checkpoint in the cell cycle. When colchicine is added to cultured cells, the cells enter mitosis and arrest with condensed chromosomes. With increasing time, a large fraction of the cells in a culture become arrested, thus permitting determination of the size, shape, and number of mitotic chromosomes—that is, the **karyotype**—in multiple cells. A checkpoint control somehow senses when the mitotic spindle has not assembled properly and prevents activation of the APC polyubiquitination system that normally leads to degradation of the anaphase inhibitor, required for onset of anaphase, and later to the degradation of mitotic cyclins, required for the exit from mitosis (see Figure 13-9). As a result, MPF activity remains high, chromosomes remain condensed, and the nuclear envelope does not re-form.

A microtubule-depolymerizing drug called benomyl has been used to isolate yeast mutants defective in this mitotic checkpoint. Low concentrations of benomyl increase the time required for yeast cells to assemble the mitotic spindle and attach kinetochores to microtubules. Wild-type cells exposed to benomyl delay in anaphase until these processes are completed and then proceed on through mitosis, producing normal daughter cells. In contrast, mutants defective in this cell-cycle checkpoint proceed through anaphase before assembly of the spindle and attachment of kinetochores is complete; consequently, they mis-segregate their chromosomes, producing abnormal daughter cells that die. Analysis of these mutants, called *bub* (*b*udding *u*ninhibited by *b*enomyl) and *mad* (*m*itotic *a*rrest *d*eficient), should shed light on the mechanism by which the mitotic checkpoint operates. The sequence of one of the *BUB* genes indicates that it encodes a protein kinase, which may influence the activities of multiple proteins.

G₁ and G₂ Arrest in Cells with Damaged DNA Depends on a Tumor Suppressor and Cyclin-Kinase Inhibitor

Cells whose DNA is damaged by irradiation with UV light or γ-rays or by chemical modification become arrested in G_1 and G_2 until the damage is repaired. Arrest in G_1 prevents copying of damaged bases, which would fix mutations in the genome. Replication of damaged DNA also causes chromosomal rearrangements at high frequency by as-yet unknown mechanisms. Arrest in G_2 allows DNA double-stranded breaks to be repaired before mitosis by mechanisms discussed in Section 12.4. If a double-stranded break is not repaired, the broken distal portion of the damaged chromosome is not properly segregated because it is not physically linked to a centromere, which is pulled toward a spindle pole during anaphase.

As discussed in Chapter 24, genes whose inactivation contributes to the development of a tumor are called **tumor-suppressor genes.** The most commonly mutated tumor-suppressor gene associated with human cancers is *p53*, so named because it encodes a phosphorylated protein with an apparent molecular weight of 53 kDa as estimated from SDS-polyacrylamide gel electrophoresis. The **p53 protein** functions in the checkpoint control that arrests human cells with damaged DNA in G_1, and it contributes to arrest in G_2. Cells with functional p53 arrest in G_1 or G_2 when exposed to γ-irradiation, whereas cells lacking functional p53 do not arrest in G_1 (Figure 13-35).

Although the p53 protein is a transcription factor, under normal conditions, it is extremely unstable; thus it generally does not accumulate to high enough levels to bind to p53-control elements and activate transcription. Damaged DNA somehow stabilizes p53, leading to an increase in its concentration. One of the genes whose transcription is stimulated by p53 encodes p21[CIP], a cyclin-kinase inhibitor (CKI) that binds and inhibits all mammalian Cdk-cyclin complexes. As a result, cells are arrested in G_1 (and G_2) until the DNA damage is repaired and p53 and subsequently p21[CIP] levels fall (Figure 13-36).

(a) Wild-type cells

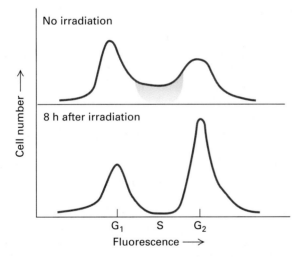

(b) p53⁻ mutant cells

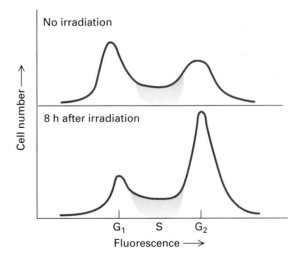

▲ **FIGURE 13-35 Effect of mutation of the *p53* tumor-suppressor gene on G₁ DNA-damage checkpoint control.** The distribution of cultured human cells in G_1, S, and G_2 was determined by analysis of their DNA content with a fluorescence-activated cell sorter as described in Figure 13-24. (a) By 8 hours following exposure of wild-type cells to γ-radiation, cells that were in the S phase (red shading) had completed DNA synthesis, entered G_2, and then arrested, accounting for the rise in the G_2 peak. The absence of S-phase cells indicates that irradiation prevented new cells from entering the S phase, causing them to arrest in G_1. (b) The presence of an S peak 8 hours after irradiation of *p53⁻* mutant cells indicates that the G_1 checkpoint does not operate in these cells. The increase in the G_2 peak indicates that the checkpoint blocking entry of irradiated cells into mitosis still operates in these mutant cells. [See S. J. Kuerbitz et al., 1992, *Proc. Nat'l. Acad. Sci. USA* **89**:7491; adapted from A. Murray and T. Hunt, 1993, *The Cell Cycle: An Introduction*, W. H. Freeman and Company.]

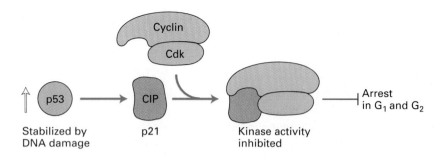

◄ **FIGURE 13-36 p53-induced cell-cycle arrest in response to DNA damage.** The normally unstable p53 protein is stabilized by damaged DNA, so its concentration increases. Acting as a transcription factor, p53 induces expression of p21^{CIP}, a cyclin-kinase inhibitor that inhibits all Cdk1-, Cdk2-, Cdk4-, and Cdk6-cyclin complexes (see Figure 13-29). Binding of p21^{CIP} to these Cdk-cyclin complexes leads to cell cycle arrest in G_1 and G_2.

If DNA damage is extensive, p53 also activates the expression of genes that lead to apoptosis. This process of programmed cell death normally occurs in specific cells during the development of multicellular animals (Chapter 23). In vertebrates, the p53 response evolved to induce apoptosis in the face of extensive DNA damage, presumably to prevent the accumulation of multiple mutations that might result in the development of a cancer cell.

Cells with mutations in both *p53* alleles do not exhibit delayed entry into the S phase following low levels of DNA damage (see Figure 13-35) and do not undergo apoptosis following more extensive DNA damage. If such cells suffer DNA damage, the damaged DNA can replicate, producing mutations and DNA rearrangements that contribute to the development of a highly transformed, metastatic cell (Section 24.5). The consequences of mutations in *p53* provide a dramatic example of the significance of cell-cycle checkpoints to the health of a multicellular organism.

SUMMARY Checkpoints in Cell-Cycle Regulation

- Checkpoint controls function to ensure that chromosomes are intact and that critical stages of the cell cycle are completed before the following stage is initiated.

- One checkpoint operates during S and G_2 to prevent the activation of MPF before DNA synthesis is complete.

- Another checkpoint operates during early mitosis to prevent activation of APC and the initiation of anaphase until the mitotic spindle apparatus is completely assembled and all chromosome kinetochores are properly attached to spindle fibers.

- Checkpoints that function in response to DNA damage prevent entry into S or M until the damage is repaired.

- Moderate DNA damage activates p53, a transcription factor that stimulates expression of p21^{CIP}. This cyclin-kinase inhibitor then binds to and inhibits all Cdk-cyclin complexes, causing arrest in G_1 and G_2.

- In response to extensive DNA damage, p53 activates genes that induce apoptosis.

PERSPECTIVES for the Future

The remarkable pace of cell-cycle research over the last two decades has led to the model of eukaryotic cell-cycle control outlined in Figure 13-2. A beautiful logic underlies these molecular controls. Each regulatory event has two important functions: to activate a step of the cell cycle and to prepare the cell for the next event of the cycle. This strategy ensures that the phases of the cycle occur in the proper order. DNA replication is activated by G_1 Cdks, which stimulate the transcription of S-phase Cdk components and induce degradation of the S-phase inhibitor. G_1 Cdks also prepare cells for mitosis by inactivating the APC and allowing mitotic cyclins to accumulate. Mitosis is triggered by mitotic Cdk activity, which induces chromosome condensation, assembly of the mitotic spindle, and attachment of metaphase chromosomes to spindle fibers resulting in metaphase. Mitotic Cdks also carry out activation of the APC, which is required for initiating anaphase and the degradation of mitotic cyclins. The resulting decrease in mitotic Cdk activity not only allows exit from mitosis but also permits assembly of DNA pre-replication complexes and expression of G_1 cyclins, thus initiating another cycle.

Although the general logic of cell-cycle regulation now seems well established, many critical details remain to be discovered. The components of the pre-replication complex that must be phosphorylated by S-phase Cdk-cyclin complexes to initiate DNA replication remain to be determined, as does the mechanism of initiation. The targets of mitotic Cdk-cyclin activity that cause mitotic spindle assembly also remain to be characterized. Current understanding of the structure of condensed, mitotic chromosomes remains vague. Much remains to be learned about how the APC is activated during anaphase and how its activity is first directed toward anaphase inhibitors and only subsequently toward mitotic cyclins. Considerable progress has been made in understanding how DNA damage leads to G_1 and G_2 arrest, but the mechanisms by which the other two checkpoints operate are poorly understood.

Understanding these detailed aspects of cell-cycle control will have significant consequences, particularly for the treatment of cancers. Radiation therapy works because it causes DNA damage in the target cells that induces their apoptosis. But this induced apoptosis depends on p53 function. For this reason, human cancers associated with mutations of both *p53* alleles, which is fairly common, are particularly resistant to radiotherapy. If more were understood about cell-cycle controls and checkpoints, new strategies for treating p53-minus cancers might be possible. For instance, some chemotherapeutic agents inhibit microtubule function, interfering with mitosis. Resistant cells selected during the course of treatment may be defective in the mitotic checkpoint as the result of mutations in the genes encoding the proteins involved. Can the loss of this checkpoint be turned to advantage? Only better understanding of the molecular processes involved can answer the question.

A great deal remains to be learned about the control of cell replication during the development of multicellular organisms. Regulation of cell replication is of obvious importance for the proper structure and function of the organ systems of vertebrates. Not only is the timing of cell division crucial, but also the position of the plane of cell cleavage, which recapitulates the position of the metaphase plate. In many cases during development, cell division is highly asymmetric. How is this achieved? Clearly, there are many fundamental and significant questions yet to be answered concerning eukaryotic cell-cycle control.

PERSPECTIVES in the Literature

Exit from mitosis requires the regulated activity of the anaphase-promoting-complex (APC). The APC must be inactive during S-phase and early mitosis, and then must be activated to polyubiquitinate the anaphase-inhibitor first and only subsequently B-type cyclins. What processes regulate the activity of the APC? Consult the following articles as you develop an answer to this question:

Fang, G., H. Yu, and M. W. Kirschner. 1998. Direct binding of CDC20 protein family members activates the anaphase-promoting complex in mitosis and G1. *Mol. Cell* **2**:163–171.

Shirayama, M., et al. 1998. The Polo-like kinase Cdc5p and the WD-repeat protein Cdc20p/fizzy are regulators and substrates of the anaphase promoting complex in *Saccharomyces cerevisiae*. *Embo J.* **17**:1336–1349.

Shou, W., et al. 1999. Exit from mitosis is triggered by Tem1-dependent release of the protein phosphatase Cdc14 from nucleolar RENT complex. *Cell* **97**:233–244.

Visintin, R., E. S. Hwang, and A. Amon. 1999. Cfi1 prevents premature exit from mitosis by anchoring Cdc14 phosphatase in the nucleolus. *Nature* **398**:818–823.

Testing Yourself on the Concepts

1. What strategy ensures that passage through the cell cycle is unidirectional and irreversible? What mechanism underlies this strategy?

2. What is the wee phenotype? What mistiming of the cell cycle does it reflect? What protein activities can cause this phenotype?

3. What are the mechanisms that regulate the distribution of the correct number of "normal" chromosomes to daughter cells?

4. Describe how the phosphorylation state of proteins regulates passage through critical points in the cell cycle.

5. What is the functional definition of START/restriction point?

MCAT/GRE-Style Questions

Key Concept Please read the section titled "*S. pombe* Cdc2-Cdc13 Heterodimer Is Equivalent to *Xenopus* MPF" (p. 506) and refer to Figures 13-4 and 13-11; then answer the following questions:

1. The best description for the content of this section is
   a. An analysis of the functions of Cdc2.
   b. The evolutionary conservation of the proteins regulating M phase of the cell cycle.
   c. The demonstration of multiple alleles of *cdc2*.
   d. The use of complementation analysis to analyze *cdc*s.

2. The complementation analysis described in this section demonstrated
   a. The immunological similarity between human and yeast Cdc2.
   b. The homology between *cdc2* and *cdc13*.
   c. The homology between yeast and human cyclin B.
   d. The homology between yeast and human cyclin-dependent kinases.

3. The experiments described here linked biochemical and genetic evidence for regulation of M phase by showing that
   a. A yeast *cdc2* gene, defined by mutation, encoded a protein equivalent to a *Xenopus* protein with kinase activity.
   b. MPF activity parallels the concentration of Cdc13.
   c. Mutation in *cdc2* causes the wee phenotype.
   d. *cdc2* and *cdc13* encode similar proteins.

4. The conclusion that the MPFs of yeast and *Xenopus* are homologous is supported by all of the following *except*:
   a. The demonstration that mutations in both organisms result in early entry into M phase.
   b. A positive reaction with one subunit of Xenopus MPF using an antibody against an amino acid sequence conserved in human and yeast Cdc2.
   c. Amino acid homology between Cdc13 and *Xenopus* cyclin B.

d. The temporal coordination between MPF activity and either Cdc13 concentration in yeast or cyclin B concentration in *Xenopus*.

Key Experiment Please read the section titled "Cyclin B Levels and MPF Activity Change Together in Cyclin G *Xenopus* Egg Extracts" (p. 502) and refer to Figure 13-7; then answer the following questions:

5. All of the following are true about the experimental systems described in the section *except*:

   a. *Xenopus* eggs carry out all phases of the cell cycle.

   b. Extracts from *Xenopus* eggs replicate haploid sperm DNA and carry out the events of mitosis.

   c. Extracts from *Xenopus* eggs contain all the components necessary for passage through M and S phases.

   d. Cyclin B concentration and MPF activity rise and fall in concert with mitotic events in *Xenopus* extracts.

6. The *best* evidence for the involvement of cyclin B in MPF activity is

   a. Absence of cyclin B and MPF activity in activated *Xenopus* extracts treated with RNase.

   b. The decrease in MPF activity at the time of mitotic events.

   c. Data showing that cyclin B concentration and MPF activity rise and fall in parallel.

   d. Observation of MPF cycling in *Xenopus* egg extracts when the only mRNA present encodes cyclin B.

7. The *best* evidence for the requirement of degradation of cyclin B for exit from mitosis and loss of MPF activity in *Xenopus* cytoplamic extracts supplemented with chromatin is

   a. The decrease in cyclin B concentration at the time of mitotic events.

   b. The decrease in cyclin B concentration at the time of mitotic events in extracts expressing only cyclin B mRNA.

   c. The maintenance of high levels of MPF in an extract expressing a mutant cyclin B that cannot be poly-ubiquitinated.

   d. The data showing that cyclin B and MPF activity increase and decrease coordinately.

8. Assays of a cytoplasmic extract such as that shown in Figure 13-7a to which cycloheximide was added

   a. Would give data like those shown in Figure 13-7a.

   b. Would give data like those shown in Figure 13-7b.

   c. Would give data like those shown in Figure 13-7d.

   d. Would show lower amounts of cyclin mRNA than extracts without added cycloheximide.

Key Application Please read the section titled "G1 and G2 Arrest in Cells with Damaged DNA Depends on a Tumor-Suppressor and Cyclin-Kinase Inhibitor" (p. 531) and refer to Figures 13-35 and 13-36; then answer the following questions:

9. Wild-type p53 acts as a tumor-suppressor gene in response to radiation by all of the following mechanisms *except*:

   a. Arresting cells at the G_1 checkpoint to repair DNA damage.

   b. Stimulating transcription of an inhibitor of Cdks.

   c. Interacting with p21^{CIP}.

   d. Activating apoptosis.

10. Mutations in p53 can cause cancer by

   a. Increasing the activity of SPF and MPF, leading to unregulated cell growth.

   b. Abrogating the activity of this protein to repair DNA damage.

   c. Stabilizing the protein and fostering the transcription of genes that can cause cancer.

   d. Preventing checkpoint activity.

11. In Figure 13-35, the absence of a checkpoint is *best* demonstrated by

   a. The increased number of cells in G_2 in a population of irradiated, wild-type cells.

   b. The increased number of G_2 cells in a population of irradiated cells that are mutant for *p53*.

   c. The absence of cells in S phase in a population of irradiated wild-type cells.

   d. The presence of cells in S phase in irradiated *p53⁻* cells.

Key Terms

anaphase inhibitor *513*

anaphase-promoting complex (APC) *498*

Cdc25 protein *507*

checkpoints *530*

CIP proteins *528*

Cln proteins *519*

cohesins *513*

condensin *513*

cyclin-dependent kinases *497*

cyclins *497*

cytokinesis *496*

D-type cyclins (mammalian cells) *526*

delayed-response genes *526*

early-response genes *526*

mitogens *498*

mitosis *495*

mitosis-promoting factor (MPF) *500*

nondisjunction *530*

nuclear lamina *510*

p53 protein *531*

polyubiquitination *504*

pre-replication complex *497*

proteasome *500*

quiescent cells *524*

Rb protein *527*

restriction point *498*

S phase–promoting factors (SPFs) *519*

Wee1 protein *507*

References

Overview of the Cell Cycle and Its Control

King, R. W., et al. 1996. How proteolysis drives the cell cycle. *Science* 274:1652–1659.

Murray, A., and T. Hunt. 1993. *The Cell Cycle: An Introduction.* W. H. Freeman and Company.

Nasmyth, K. 1996. Viewpoint: putting the cell cycle in order. *Science* 274:1643–1645.

Biochemical Studies with Oocytes, Eggs, and Early Embryos

Dunphy, W. G., et al. 1988. The *Xenopus* Cdc2 protein is a component of MPF, a cytoplasmic regulator of mitosis. *Cell* 54:423–431.

Evans, T., et al. 1983. Cyclin: a protein specified by maternal mRNA in sea urchin eggs that is destroyed at each cleavage division. *Cell* 33:389–396.

Gautier, J., et al. 1988. Purification of maturation-promoting factor from *Xenopus* eggs: the factor contains the product of a *Xenopus* homolog of the fission yeast cell cycle control gene *cdc2*⁺. *Cell* 54:433–439.

Gautier, J., et al. 1990. Cyclin is a component of maturation-promoting-factor from *Xenopus*. *Cell* 60:487–494.

Gerhart, J., M. Wu, and M. J. Kirschner. 1984. Cell cycle dynamics of an M-phase specific cytoplasmic factor in *Xenopus laevis* oocytes and eggs. *J. Cell. Biol.* 98:1247–1255.

Masui, Y., and C. L. Markert. 1971. Cytoplasmic control of nuclear behavior during meiotic maturation of frog oocytes. *J. Exp. Zool.* 177:129–145.

Murray, A. W., and M. W. Kirschner. 1989. Cyclin synthesis drives the early embryonic cell cycle. *Nature* 339:275–280.

Swenson, K. I., K. M. Farrell, and J. V. Ruderman. 1986. The clam embryo cyclin A induces entry into M phase and the resumption of meiosis in *Xenopus* oocytes. *Cell* 47:861–870.

Genetic Studies with *S. pombe*

Featherstone, C., and P. Russell. 1991. Fission yeast p107^{wee1} mitotic inhibitor is a tyrosine/serine kinase. *Nature* 349:808–811.

Gautier, J., et al. 1991. Cdc25 is a specific tyrosine phosphatase that directly activates p34cdc2. *Cell* 67:197–211.

Gould, K.L., et al. 1991. Phosphorylation at Thr167 is required for *Schizosaccharomyces pombe* p34cdc2 function. *EMBO J.* 10:3297–3309.

Gould, K. L., and P. Nurse. 1989. Tyrosine phosphorylation of the fission yeast *cdc2*⁺ protein kinase regulates entry into mitosis. *Nature* 342:39–45.

Harper, J. W., and S. J. Elledge. 1998. The role of Cdk7 in CAK function, a retro-retrospective. *Genes & Devel.* 12:285–289.

Lee, M. G., and P. Nurse. 1987. Complementation used to clone a human homologue of the fission yeast cell cycle control gene *cdc2*. *Nature* 327:31–35.

Nurse, P., and Y. Bissett. 1981. Gene required in G1 for commitment to cell cycle and in G2 for control of mitosis in fission yeast. *Nature* 292:558–560.

Nurse, P., and P. Thuriaux. 1980. Regulatory genes controlling mitosis in the fission yeast *Schizosaccharomyces pombe*. *Genetics* 96:627–637.

Stern, B., and P. Nurse. 1996. A quantitative model for the Cdc2 control of S phase and mitosis in fission yeast. *Trends Genet.* 12:345–350.

Molecular Mechanisms for Regulating Mitotic Events

Biggins, S., and A. W. Murray. 1998. Sister chromatid cohesion in mitosis. *Curr. Opin. Cell Biol.* 10:769–775.

Field, C., R. Li, and K. Oegema. 1999. Cytokinesis in eukaryotes: a mechanistic comparison. *Curr. Opin. Cell Biol.* 11:68–80.

Hershko, A. 1997. Roles of ubiquitin-mediated proteolysis in cell cycle control. *Curr. Opin. Cell Biol.* 9:788–799.

Hoyt, M. A. 1997. Eliminating all obstacles: regulated proteolysis in the eukaryotic cell cycle. *Cell* 91:149–151.

Nigg, E. A. 1998. Polo-like kinases: positive regulators of cell division from start to finish. *Curr. Opin. Cell Biol.* 10:776–783.

Nigg, E. A., A. Blangy, and H. A. Lane. 1996. Dynamic changes in nuclear architecture during mitosis: on the role of protein phosphorylation in spindle assembly and chromosome segregation. *Exp. Cell Res.* 229:174–180.

Peters, J. M. 1998. SCF and APC: the Yin and Yang of cell cycle regulated proteolysis. *Curr. Opin. Cell Biol.* 10:759–768.

Strunnikov, A. V. 1998. SMC proteins and chromosome structure. *Trends Cell Biol.* 8:454–459.

Yanagida, M. 1998. Fission yeast *cut* mutations revisited: control of anaphase. *Trends Cell Biol.* 8:144–149.

Genetic Studies with *S. cerevisiae*

Andrews, B., and V. Measday. 1998. The cyclin family of budding yeast: abundant use of a good idea. *Trends Genet.* 14:66–72.

Cross, F. R. 1995. Starting the cell cycle: what's the point? *Curr. Opin. Cell Biol.* 7:790–797.

Hadwiger, J. A., et al. 1989. A family of cyclin homologs that control G₁ phase in yeast. *Proc. Nat'l. Acad. Sci. USA* 86:6255–6259.

Hartwell, L. H., et al. 1974. Genetic control of the cell division cycle in yeast. *Science* 183:46–51.

Koch, C., and K. Nasmyth. 1994. Cell cycle regulated transcription in yeast. *Curr. Opin. Cell Biol.* 6:451–459.

Leatherwood, J. 1998. Emerging mechanisms of eukaryotic DNA replication initiation. *Curr. Opin. Cell Biol.* 10:742–748.

Mendenhall, M. D., and A. E. Hodge. 1998. Regulation of Cdc28 cyclin-dependent protein kinase activity during the cell cycle of the yeast *Saccharomyces cerevisiae*. *Microbiol. Molec. Biol. Rev.* 62:1191–1243.

Nasmyth, K. 1996. At the heart of the budding yeast cell cycle. *Trends Genet.* 12:405–412.

Newlon, C. S. 1997. Putting it all together: building a prereplicative complex. *Cell* 91:717–720.

Reed, S. I. 1980. Selection of *S. cerevisiae* mutants defective in the start event of cell division. *Genetics* 95:561–577.

Reed, S. I., J. A. Hadwiger, and A. T. Lorincz. 1985. Protein kinase activity associated with the product of the yeast cell division cycle gene CDC28. *Proc. Nat'l. Acad. Sci. USA* 82:4055–4059.

Cell-Cycle Control in Mammalian Cells

Donaldson, A. D., and J. J. Blow. 1999. The regulation of replication origin activation. *Curr. Opin. Genet. Devel.* 9:62–68.

Dynlacht, B. D. 1997. Regulation of transcription by proteins that control the cell cycle. *Nature* 389:149–152.

Ford, H. L., and A. B. Pardee. 1998. The S phase: beginning, middle, and end: a perspective. *J. Cell. Biochem. Suppl.* 30-31:1–7.

Hengst, L., and S. I. Reed. 1998. Inhibitors of the Cip/Kip family. *Curr. Topics Microbiol. Immunol.* 227:25–41.

Morgan, D. O. 1997. Cyclin-dependent kinases: engines, clocks, and microprocessors. *Annu. Rev. Cell Devel. Biol.* 13:261–291.

Mulligan, G., and T. Jacks. 1998. The retinoblastoma gene family:

cousins with overlapping interests. *Trends Genet.* **14**:223–229.

Nevins, J. R. 1998. Toward an understanding of the functional complexity of the E2F and retinoblastoma families. *Cell Growth Differentiation* **9**:585–593.

Planas-Silva, M. D., and R. A. Weinberg. 1997. The restriction point and control of cell proliferation. *Curr. Opin. Cell Biol.* **9**:768–772.

Reed, S. I. 1997. Control of the G1/S transition. *Cancer Surveys* **29**:7–23.

Sherr, C. J. 1996. Cancer cell cycles. *Science* **274**:1672–1677.

Zavitz, K. H., and S. L. Zipursky. 1997. Controlling cell proliferation in differentiating tissues: genetic analysis of negative regulators of G1→ S-phase progression. *Curr. Opin. Cell Biol.* **9**:773–781.

Checkpoints in Cell-Cycle Regulation

Adams, P. D., and W. G. Kaelin, Jr. 1998. Negative control elements of the cell cycle in human tumors. *Curr. Opin. Cell Biol.* **10**:791–797.

Amon, A. 1999. The spindle checkpoint. *Curr. Opin. Genet. Devel.* **9**:69–75.

Chin, L., J. Pomerantz, and R. A. DePinho. 1998. The INK4a/ARF tumor suppressor: one gene—two products—two pathways. *Trends Biochem. Sci.* **23**:291–296.

Elledge, S. J. 1996. Cell cycle checkpoints: preventing an identity crisis. *Science* **274**:1664–1672.

Giaccia, A. J., and M. B. Kastan. 1998. The complexity of p53 modulation: emerging patterns from divergent signals. *Genes & Devel.* **12**:2973–2983.

Huberman, J. A. 1999. DNA damage and replication checkpoints in the fission yeast, *Schizosaccharomyces pombe*. *Progress Nucl. Acid Res. Molec. Biol.* **62**:369–395.

Levine, A. J. 1997. p53, the cellular gatekeeper for growth and division. *Cell* **88**:323–331.

Nurse, P. 1997. Checkpoint pathways come of age. *Cell* **91**:865–867.

Paulovich, A. G., D. P. Toczyski, and L. H. Hartwell. 1997. When checkpoints fail. *Cell* **88**:315–321.

Rhind, N., and P. Russell. 1998. Mitotic DNA damage and replication checkpoints in yeast. *Curr. Opin. Cell Biol.* **10**:749–758.

Sherr, C. J. 1998. Tumor surveillance via the ARF-p53 pathway. *Genes & Devel.* **12**:2984–2991.

Skibbens, R. V., and P. Hieter. 1998. Kinetochores and the checkpoint mechanism that monitors for defects in the chromosome segregation machinery. *Annu. Rev. Genet.* **32**:307–337.

Weinert, T. 1998. DNA damage checkpoints update: getting molecular. *Curr. Opin. Genet. Devel.* **8**:185–193.

Gene Control in Development

14

The ultimate goal of developmental biologists is to unravel the mystery of how a fertilized egg is transformed into a complex multicellular organism. This process requires execution of a complex developmental program whereby specific genes are activated in a precise time sequence and in the correct location, generating different types of tissues and the specific cell types composing them. Classical and molecular geneticists have discovered numerous genes that participate in the highly regulated programs that result in the development of plants and animals. Understanding the molecular basis for the action of such genes is one of the most actively studied areas in all of biology.

In Chapters 10 and 11, we examined various mechanisms for regulating gene expression. As a result of these controls, cells can respond to changes in their environment and different cell types can produce characteristic proteins. By far the most prominent mechanism for regulating the production of different proteins in different cells entails a vast array of DNA-binding proteins that act, often in various combinations, to either activate or repress gene transcription.

In this chapter, we take a closer look at the spatial and temporal control of gene expression during development. Genetic and molecular studies show that different cells express different sets of genes based on their developmental history, their patterns of cell division, their position in the developing organism, and their interactions with other cells. We focus on several well-studied cases of differential gene transcription to specify different cell types in yeast, animals, and plants to illustrate the general transcription-control strategies that regulate development. The interrelationships between transcriptional programs in development and signaling between cells are considered in Chapter 23.

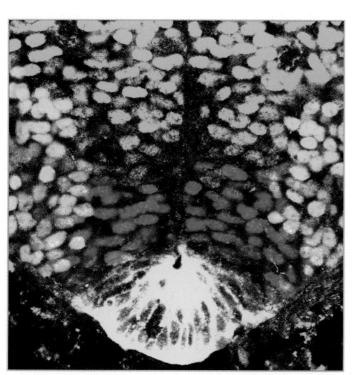

Different proteins are expressed in different cells in the developing spinal cord.

MEDIA CONNECTIONS

Overview: Gene Control in Embryonic Development

Classic Experiment 14.1: Using Lethal Injection to Study Development

14.1 Cell-Type Specification and Mating-Type Conversion in Yeast

We begin our discussion of **cell-type specification** with the yeast *S. cerevisiae*. There are three different cell types in this unicellular eukaryote: haploid **a** and α cells, and diploid **a**/α cells (see Figure 10-54). Because of the simplicity of yeast and the ease of studying it, our understanding of the transcription-control mechanisms specifying its three cell types is much more complete than our understanding of similar processes in animals and plants. It is likely that the mechanisms generating different cell types in the various organs and tissues in higher organisms evolved from mechanisms leading to diversification of cell types in simple unicellular organisms such as yeast.

Combinations of DNA-Binding Proteins Regulate Cell-Type Specification in Yeast

Each of the three *S. cerevisiae* cell types expresses a unique set of genes. All haploid cells express certain haploid-specific genes; in addition, **a** cells express **a**-specific genes, and α cells express α-specific genes. In **a**/α diploid cells, diploid-specific genes are expressed, whereas haploid-specific, **a**-specific, and α-specific genes are not. As illustrated in Figure 14-1, regulation of this cell type–specific transcription is mediated by three cell type–specific transcription factors (α1, α2, and **a**1) encoded at the *MAT* locus in combination with a general transcription factor called *MCM1*, which is expressed in all three cell types.

MCM1 was the first member of the MADS family of transcription factors to be discovered. The DNA-binding proteins in this family dimerize and contain a common homologous N-terminal MADS domain. (MADS is an acronym for the initial four factors identified in this family.) In later sections we will encounter other MADS transcription factors that participate in development of skeletal muscle and floral organs. As outlined in Figure 14-2, MCM1 exhibits different activity in haploid **a** and α cells due to its association with α1 or α2 protein in α cells.

Gene Activation by MCM1 and α1-MCM1 Complex In **a** cells, homodimeric MCM1 binds to the P box in **a**-specific upstream regulatory sequences (URSs), thereby stimulating transcription of the associated **a**-specific genes. The URSs associated with α-specific genes contain two adjacent DNA sequences, the so-called P box and Q box. Although MCM1 alone binds to the P box in **a**-specific URSs, it does not bind to the P box in α-specific URSs. Similarly, α1 does not bind alone to the Q box in an α-specific URS. The simultaneous binding of these proteins, however, occurs with high affinity and turns on transcription from the PQ site.

Gene Repression by α2-MCM1 and α2-a1 Complexes Flanking the P box in each **a**-specific URS are two α2-binding sites. Both MCM1 and α2 can bind independently

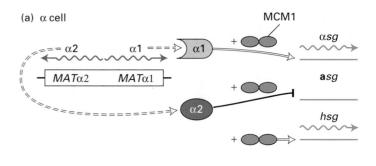

(a) α cell

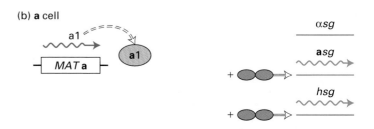

(b) **a** cell

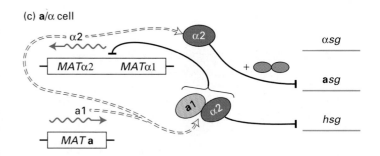

(c) **a**/α cell

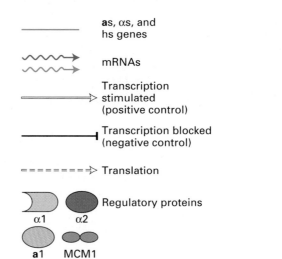

— **a**s, αs, and hs genes

mRNAs

⟹ Transcription stimulated (positive control)

⊣ Transcription blocked (negative control)

⇒ Translation

Regulatory proteins
α1 α2
a1 MCM1

▲ **FIGURE 14-1 Regulation of cell type–specific genes in *S. cerevisiae* by regulatory proteins encoded at the *MAT* locus together with MCM1, a constitutive transcription factor produced by all three cell types.** As a result of this regulation, each cell type exhibits a distinctive pattern of gene expression: **a**s = **a**-specific genes/mRNAs; αs = α-specific genes/mRNAs; hs = haploid-specific genes/mRNAs.

► **FIGURE 14-2 Activity of MCM1 in a and α yeast cells.** MCM1 binds as a dimer to the P site in α-specific and **a**-specific upstream regulatory sequences (URSs). These sequences lie upstream of and control transcription of α-specific genes and **a**-specific genes, respectively. The consensus sequences of the P box in α- and **a**-specific URSs differ slightly. (a) In **a** cells, MCM1 stimulates transcription of **a**-specific genes. MCM1 does not bind efficiently to the P site in α-specific URSs in the absence of α1 protein. (b) In α cells, the activity of MCM1 is modified by its association with α1 or α2. The α1-MCM1 complex stimulates transcription of α-specific genes, whereas the α2-MCM1 complex blocks transcription of **a**-specific genes. The α2-MCM1 complex also is produced in diploid cells, where it has the same blocking effect on transcription of **a**-specific genes (see Figure 14-1c).

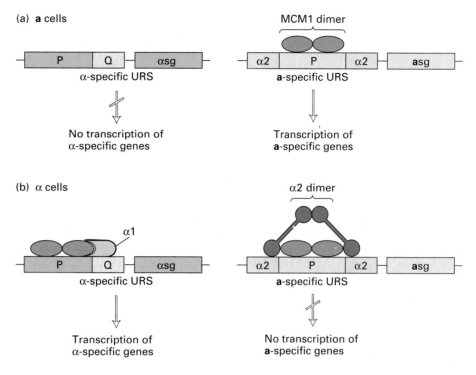

to an **a**-specific URS with relatively low affinity. However, the highly cooperative, simultaneous binding of both proteins occurs with high affinity. This high-affinity binding represses transcription of **a**-specific genes, ensuring that they are not expressed in α cells and diploid cells (see Figure 14-2b).

MCM1 promotes binding of α2 to an **a**-specific URS by orienting the two DNA-binding domains of the α2 dimer to the α2-binding sequences in this regulatory sequence. Since a dimeric α2 molecule binds to both sites in an α-specific URS, each DNA site is referred to as a *half-site*. The relative position of both half-sites and their orientation are highly conserved among different **a**-specific URSs. Experiments with mutant URSs have shown that changing the orientation or spacing of the half-sites has little effect on the binding affinity of isolated α2 dimers in the absence of MCM1, suggesting that the two monomeric α2 subunits have considerable flexibility. However, the highly cooperative, high-affinity binding of α2 to an **a**-specific URS in the presence of MCM1 requires a precise spacing and orientation of the α2 half-sites (Figure 14-3). Presumably, the interaction between MCM1 and α2 constrains the flexibility of the α2 dimer, so that it binds with high affinity only to uniquely oriented and spaced α2 half-sites. Thus the affinity of α2 for sites in an **a**-specific URS is influenced by its association with MCM1.

The presence of numerous α2-binding sites in the genome and the "relaxed" specificity of α2 protein may expand the range of genes that it can regulate. For instance, in **a**/α diploid cells, α2 forms a heterodimer with **a**1 that acts to repress both haploid-specific genes and the gene encoding α1 (see Figure 14-1c). The example of α2 suggests that

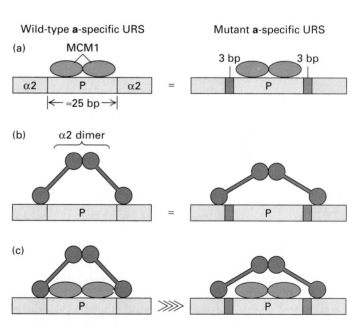

▲ **FIGURE 14-3 Relative binding affinities of MCM1, α2, and MCM1-α2 complex to wild-type and mutant a-specific upstream regulatory sequences (URSs).** (a,b) Insertion of three base pairs (blue) on either side of the P site does not affect the affinity of the independent binding of MCM1 to the P box or α2 to the flanking α2-binding sites. (c) The high-affinity, cooperative binding of these two dimeric proteins requires correct spacing of the P site and α2-binding site, as demonstrated by the much lower affinity of the MCM1-α2 complex for the mutant URS. [See D. L. Smith and A. D. Johnson, 1992, *Cell* **68**:133.]

relaxed specificity may be a general strategy for increasing the regulatory range of a single transcription factor. Highly specific binding, then, occurs as a consequence of the interaction of α2 with other transcription factors at different sites in DNA.

As discussed in Chapter 10, eukaryotic repressor proteins exert their effects via several different mechanisms. The MCM1-α2 or a1-α2 complex in yeast interacts with two proteins designated Tup1 and Ssn6, which do not themselves bind to DNA, to form a large complex that then represses transcription of many genes. Although this repressor complex clearly inhibits formation of a transcription-initiation complex at the promoter, the precise mechanism of transcriptional repression is not known.

Mating of α and a Cells Is Induced by Pheromone-Stimulated Gene Expression

An important feature of the yeast life cycle is the ability of haploid **a** and α cells to mate, that is, attach and fuse giving rise to a diploid **a**/α cell. Each haploid cell type secretes a different mating factor, a small polypeptide called a *pheromone*, and expresses a cell-surface receptor that recognizes the pheromone secreted by cells of the other type. Thus **a** and α cells both secrete and respond to pheromones. Binding of the mating factors to their receptors leads to expression of a set of genes encoding proteins that direct arrest of the cell cycle in G_1 and promote attachment/fusion of haploid cells to form diploid cells. In the presence of sufficient nutrients, the diploid cells will continue to grow. Starvation, however, induces diploid cells to progress through meiosis, each yielding four haploid spores (Figure 14-4). If the environmental conditions become conducive to vegetative growth, the spores will germinate and undergo mitotic division.

Studies with yeast mutants have provided insights into how the **a** and α pheromones induce mating. For instance, haploid yeast cells carrying mutations in the sterile 12 *(STE12)* locus cannot respond to pheromones and do not mate. The *STE12* gene encodes a transcription factor that binds to a DNA sequence referred to as the *pheromone-responsive element* (PRE), which is present in many different **a**- and α-specific URSs. Binding of mating factors to cell-surface receptors induces a cascade of signaling events, resulting in phosphorylation of various proteins including the Ste12 protein (see Figure 20-31). This rapid phosphorylation is correlated with an increase in the ability of Ste12 to stimulate transcription. It is not yet known, however, whether Ste12 must be phosphorylated to stimulate transcription in response to pheromone.

Interaction of Ste12 protein with DNA has been studied most extensively at the URS controlling transcription of *STE2*, an **a**-specific gene encoding the receptor for the α pheromone. Adjacent to this **a**-specific URS is a PRE that binds Ste12. When **a** cells are treated with α pheromone, transcription of *STE2* increases in a process that requires Ste12 protein.

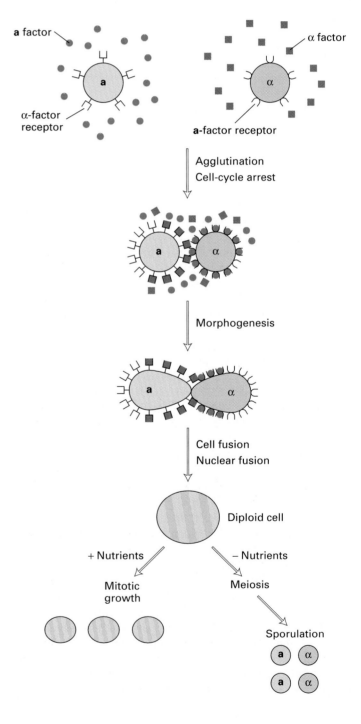

▶ **FIGURE 14-4 Haploid yeast cells produce pheromones, or mating factors, and pheromone receptors.** The α cells produce α factor and **a** receptor; the **a** cells produce **a** factor and α receptor. Binding of the mating factors to their cognate receptors on cells of the opposite type leads to gene activation, resulting in mating and production of diploid cells. In the presence of sufficient nutrients, these cells will grow as diploids. Without sufficient nutrients, cells will undergo meiosis and form four haploid spores.

Presumably, pheromone-induced up-regulation of the α receptor encoded by *STE2* increases the efficiency of the mating process. Ste12 protein has been found to bind most efficiently to PRE in the *STE2* URS when MCM1 is simultaneously bound to the adjacent P site. We saw previously that MCM1 can act as an activator or a repressor at different URSs depending on whether it complexes with α1 or α2. In this case, the function of MCM1 as an activator is stimulated by binding to yet another transcription factor, Ste12, whose activity is modified by extracellular signals.

As discussed in Chapter 20, surprising similarities have been uncovered between the mechanisms by which yeast respond to mating factors and higher eukaryotes respond to various extracellular factors that promote growth and differentiation.

Multiple Regulation of *HO* Transcription Controls Mating-Type Conversion

Recall from Chapter 10 that two silent (nontranscribed) copies of the *MAT* locus—designated *HML* and *HMR*—are located on yeast chromosome 3 in addition to the active (transcribed) **MAT locus**. The phenotype of haploid yeast cells is determined by the mating-type sequence (a or α) that they carry in the central *MAT* locus. Once each generation, the sequences at *HML* or *HMR* are transferred to the central *MAT* locus, thereby converting an a cell to an α cell or vice versa (see Figure 10-55). This process of **mating-type conversion** begins with a site-specific cleavage at *MAT* by the HO endonuclease.

Mating-type conversion in yeast exhibits three types of specificity: it occurs only in haploid cells, during the late G₁ phase of the cell cycle, and in one of the two mitotic products, the so-called mother cell (Figure 14-5). This three-fold specificity results from the complex transcriptional regulation of the *HO* locus, which is controlled through two adjacent regulatory sequences—referred to as *URS1* and *URS2*—that lie ≈110 base pairs upstream of the *HO* locus. Switching occurs only in cells that express the HO endonuclease.

Transcription of the *HO* locus is repressed when a heterodimeric complex comprising α2 and a1 binds to multiple sites within both URS1 and URS2. These proteins are encoded by *MATα* and *MATa*, respectively (see Figure 14-1). Since both loci are present only in diploid cells, repression of HO transcription by the α2-a1 complex does not occur in haploid cells.

URS2 contains 10 repeats of a sequence, termed the *cell-cycle box*, that binds CCBF (cell-cycle box factor), a transcriptional activator. CCBF is composed of two proteins, designated Swi4 and Swi6, both of which are required for mating-type switching; they also help regulate the cell cycle–specific expression of other genes. Swi4 binds specifically to the cell-cycle box in the absence of Swi6, suggesting that Swi6 is necessary for the activating function of CCBF but not for its site-specific DNA-binding ability. The activity

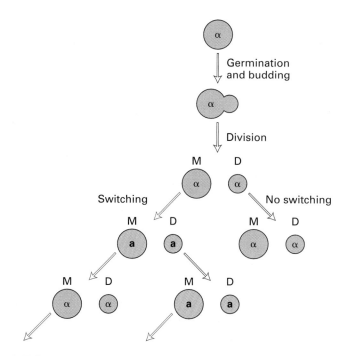

▲ **FIGURE 14-5 Specificity of mating-type conversion.** Under appropriate conditions, haploid yeast spores germinate and undergo mitotic division by budding. The mother cell (M), which is larger than the daughter cell (D), can undergo a switch in the DNA sequences at the *MAT* locus before the next DNA duplication (i.e., during the G₁ phase). The first step in this gene conversion is catalyzed by HO endonuclease. Switching can occur in both directions (a → α and α → a) but only in haploid mother cells in the G₁ phase of the cell cycle.

and/or expression of CCBF is responsive to Cdc28–G₁ cyclin complexes, whose protein kinase activity peaks in the late G₁ phase of the cell cycle (see Figure 13-26). Thus CCBF-mediated activation of *HO* transcription is limited to this phase of the cell cycle.

A key protein in restricting mating-type conversion to mother cells is Swi5, which binds to two short sequences within URS1 and is required for transcription of the *HO* locus. Although *HO* transcription occurs only in late G₁, Swi5 protein is synthesized in all stages of the cell cycle except G₁. Presumably, Swi5 synthesized in the previous cell cycle is selectively functional in mother cells following division. The finding that both mother and daughter cells stain with antibodies to Swi5 suggests that Swi5 is inactivated in daughter cells or, alternatively, selectively activated in mother cells. Recent studies have identified an inhibitor of Swi5, called Ash1, that selectively segregates to daughter cells at mitosis (Figure 14-6). Whether Ash1 specifically interacts with Swi5, the URS1, or both, is not known. A set of genes have been identified that are necessary to promote Ash1 accumulation in daughter cells. In mutants lacking this machinery, neither mother nor daughter cells undergo mating-type switching because Ash1 accumulates in both cells.

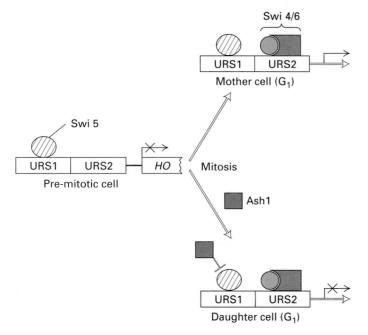

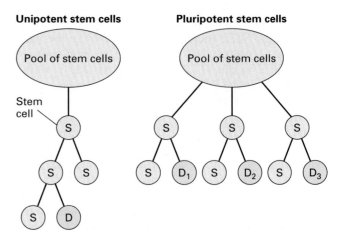

▲ FIGURE 14-7 **The production of differentiated cells (D) from stem cells (S).** Unipotent stem cells produce a single type of differentiated cell, whereas pluripotent stem cells may produce two or more types of differentiated cells.

▲ FIGURE 14-6 **Model for restriction of mating-type conversion to mother cells and the G₁ phase of the cell cycle.** Binding of both Swi5 and the Swi4/6 complex (or CCBF) to the upstream regulatory regions (URS1 and URS2) is required for transcription of the *HO* locus, which encodes the endonuclease that initiates conversion. Swi4/6 is expressed or activated only in late G₁. Although Swi5 in pre-mitotic cells segregates to both mother and daughter cells, Ash1 (or perhaps its mRNA) segregates only to daughter cells. The Ash1-mediated inhibition of Swi5 activity prevents transcription of *HO* in daughter cells during the subsequent G₁ phase. The precise mechanisms by which this inhibition occurs are not known.

The asymmetric fate of two mitotic products, illustrated by the mother-cell specificity of mating-type conversion, is a common occurrence in developmental pathways of multicellular organisms. For example, many differentiated cells are generated from **stem cells**, which can divide asymmetrically to yield a stem cell plus a sibling cell that is more specialized and has lost some of its developmental potential (Figure 14-7). The mechanisms that have been shown to control mating-type conversion in yeast may provide insight into stem-cell development more generally.

Control of *HO* transcription, and hence of mating-type conversion, is even more complex than described above. Several other proteins, including the Swi1, 2, and 3 proteins, also are required for mating-type conversion. In addition, components of chromatin encoded by several *SIN* genes repress *HO* transcription, perhaps by maintaining the *HO* regulatory region in a configuration that prevents binding of the positively acting Swi proteins.

The molecular mechanisms by which these different levels of regulation converge on the *HO* locus to precisely

control its transcription are not known. One model proposes that binding of Swi5 to URS1 promotes the activity of Swi1, 2, and 3. These proteins in turn somehow counteract the effect of the proteins encoded by the *SIN* genes, thereby permitting binding of CCBF (Swi4/6) to URS2. Once Swi5 and CCBF are bound to the *HO* regulatory region in G₁, expression of the HO endonuclease and mating-type switching proceed (see Figure 14-6).

Silencer Elements Repress Expression at *HML* and *HMR*

As noted earlier, the *HML* and *HMR* loci on yeast chromosome 3 contain "extra" silent (nontranscribed) copies of the α and a sequences (see Figure 10-55). The location of the extra a and α sequence in *HML* and *HMR*, respectively, varies in different yeast strains. If these extra copies were transcribed during haploid growth, then haploid-specific genes would be repressed by the α2-a1 complex and the haploid cells could not mate (i.e., the haploid cells would be phenotypically like diploid cells). Hence, it is not only necessary to promote expression of genes required to specify one cell type, it is also necessary to repress other pathways leading to specification of different cell types.

As discussed in Chapter 10, *silencer elements* are responsible for specific repression of the a and α sequences associated with *HML* and *HMR*. Recent studies indicate that these elements, in conjunction with specific proteins, are required to assemble silencer-associated regions into higher-order chromatin structures inaccessible to the transcriptional machinery (see Figure 10-57). Similar mechanisms exist in higher eukaryotic cells, though the precise molecular mechanisms controlling this process are not as well understood as in yeast.

SUMMARY   *Cell-Type Specification and Mating-Type Conversion in Yeast*

• Specification of each of the three yeast cell types—the a and α haploid cells and the diploid a/α cells—is determined by a unique set of transcription factors acting in different combinations at specific cis-acting regulatory sites (see Figure 14-1).

• Some transcription factors can act as repressors or activators depending on the specific cis-acting regulatory sites they bind and the presence or absence of other transcription factors bound to neighboring sites.

• Chromatin structure can play an important role in regulating gene expression in development. In haploid cells, the opposite mating-type locus is silenced by packaging it into a higher-order chromatin structure inaccessible to transcriptional activators.

• Gene expression can be modified by specific extracellular signals through covalent modification (e.g., phosphorylation) of specific transcription factors. Binding of mating-type pheromones by haploid yeast cells activates expression of genes encoding proteins that mediate mating (see Figure 14-4).

• The asymmetric distribution of certain proteins during cell division can lead to changes in gene expression. The Ash1 protein, which is preferentially localized to daughter cells, prevents the Swi5 transcription factor from activating expression of the HO endonuclease and, hence, restricts the potential to switch mating types to mother cells (see Figure 14-6).

14.2 Cell-Type Specification in Animals

Each of the hundreds of different cell types found in multicellular organisms must be generated in the right number and in the appropriate region of the developing embryo and must be integrated into the framework of other cells to form discrete tissues. Specialized cells often have a distinctive morphology and express proteins devoted to the specific biochemical functions carried out by a particular cell type. The extensive cell specification that occurs during development of animals and plants depends on both quantitative and qualitative differences in gene expression, controlled largely at the level of transcription. An impressive array of molecular strategies, some analogous to those found in yeast cell-type specification, have evolved to carry out the complex developmental pathways that characterize multicellular organisms.

Cell biologists do not yet understand the complete set of regulatory molecules for any unique cell type in a multicellular organism that makes it different from other cells. In recent years, however, a family of related regulatory proteins have been shown to play analogous roles in the development of skeletal muscle cells (myogenesis), neural cells (neurogenesis), and perhaps other cell types. Mammalian skeletal muscle is a favorable system for investigating the role of transcription factors in controlling cell-type specification because its development can be studied both in the intact organism and in in vitro systems. In this section, we first describe the network of proteins that control specification of muscle cells and then consider how similar mechanisms operate during neurogenesis.

Embryonic Somites Give Rise to Myoblasts, the Precursors of Skeletal Muscle Cells

Mammalian skeletal myogenesis proceeds through three stages (Figure 14-8): **determination** of precursor muscle cells, called **myoblasts;** proliferation and in some cases migration of myoblasts; and their subsequent **differentiation** into mature muscle. In the first stage, myoblasts arise from blocks of mesodermal cells, called *somites,* found lateral to the neural tube in the embryo (Figure 14-9). Somites also give rise to tissues other than muscle including skeletal tissue and connective tissue in the skin. Specific signals from surrounding tissue including the neural tube and the lateral

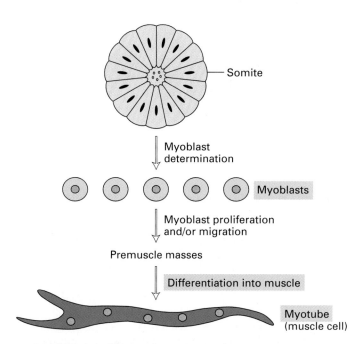

▲ **FIGURE 14-8 Schematic diagram of three stages in development of skeletal muscle in mammals.** Somites are collections of embryonic mesodermal cells, some of which become determined as myoblasts. Myoblasts, which are precursor skeletal muscle cells, are distinct from other somite-derived precursor cells. A subclass of myoblasts migrates to form premuscle masses in the limbs and elsewhere, where they differentiate into multinucleate skeletal muscle cells, called myotubes. Other myoblasts will proliferate and differentiate to form axial muscles.

(a)

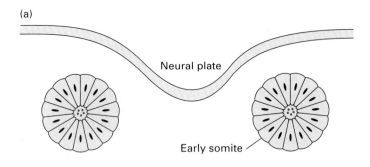

(b)

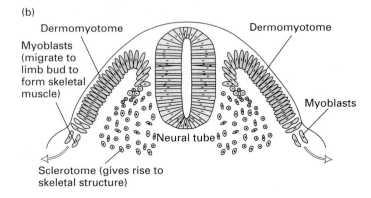

(c)

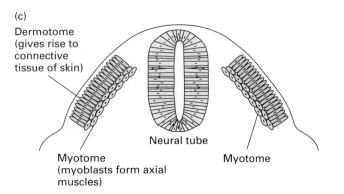

▲ FIGURE 14-9 Embryonic determination and migration of myoblasts in mammals. (a) Skeletal muscle is derived from embryonic structures called somites, which are blocks of mesodermal cells. (b) After formation of the neural tube, each somite forms a dermomyotome, which gives rise to skin and muscle, and a sclerotome, which develops into skeletal structures. Myoblasts form at each edge of a dermomyotome. Lateral myoblasts migrate to the limb bud. Axial myoblasts form the myotome. (c) The dermotome gives rise to skin elements (dermis), and the myotome to axial muscle. [See B. A. Williams and C. P. Ordahl, 1994, *Development* **120**:785. Adapted from M. Buckingham, 1992, *Trends Genet.* **8**:144.]

ectoderm play an important role in determining where myoblasts will form in the developing somite.

Myoblasts are committed to become muscle but have not yet differentiated; hence, they are referred to as *determined*. For instance, myoblasts that will form muscles in the limb migrate from the lateral side of the *myotome* to the

developing limb bud. Here the cells align, stop dividing, fuse to form a **syncytium** (a cell containing many nuclei but sharing a common cytoplasm), and differentiate into muscle. We refer to this multinucleate skeletal muscle cell as a **myotube.** Concomitant with cellular fusion there is a dramatic rise in the expression of genes necessary for muscle development and function. Other myoblasts from the more dorsal and medial regions of the myotome do not migrate and instead form cells of trunk muscles.

The specific extracellular signals that induce determination of each group of myoblasts are expressed only transiently. These signals trigger expression of numerous intracellular factors that can maintain the myogenic program in the absence of the inducing signals. We discuss the identification and functions of these myogenic proteins, and their interactions, in the next several sections.

Myogenic Genes Were First Identified in Studies with Cultured Fibroblasts

In vitro studies with the fibroblast cell line designated C3H 10T$\frac{1}{2}$ have played a central role in dissecting the transcription-control mechanisms regulating skeletal myogenesis. When these cells are incubated in the presence of 5-azacytidine, a cytidine derivative that cannot be methylated, they differentiate into myotubes. Upon entry into cells, 5-azacytidine is converted to 5-azadeoxycytidine triphosphate and then is incorporated into DNA in place of deoxycytidine. As noted in Chapter 10, methylated deoxycytidine residues commonly are present in transcriptionally inactive DNA regions. Thus replacement of cytidine residues with a derivative that cannot be methylated may permit activation of genes previously repressed by methylation. The high frequency at which treated C3H 10T$\frac{1}{2}$ cells are converted into myotubes suggested to early workers that reactivation of one or a small number of closely linked genes is sufficient to drive a program of muscle development.

To test this hypothesis, researchers isolated DNA from C3H 10T$\frac{1}{2}$ cells grown in the presence of 5-azacytidine, so-called **azamyoblasts,** and transfected it into untreated cells (Figure 14-10). The observation that 1 in 10^4 cells transfected with DNA isolated from azamyoblasts was converted into myotubes was consistent with the hypothesis that one or a small set of closely linked genes is responsible for converting fibroblasts into myotubes.

Subsequent studies led to the isolation and characterization of four genes that can convert C3H 10T$\frac{1}{2}$ cells into muscle. Figure 14-11 outlines the experimental protocol for identifying and assaying one of these genes, called *myoD* for *myogenic determination gene D*. Colonies of *myoD*-transfected cells were indistinguishable from C3H 10T$\frac{1}{2}$ cells treated with 5-azacytidine, and both types of cells exhibited myotube-like properties. The *myoD* cDNA also was found to convert a number of other cell lines into muscle. Based on these findings, the *myoD* gene was proposed to play a key role in muscle development. A similar approach has identified three other genes— the *myogenin, myf5,* and *mrf4* genes—that also function in

► **FIGURE 14-10 Experimental system for studying mammalian myogenesis.** A fibroblast cell line called C3H 10T$\frac{1}{2}$ can be converted into muscle cells by incubating them with 5-azacytidine. Under appropriate conditions, intermediate precursor cells, termed azamyoblasts, accumulate. DNA isolated from azamyoblasts can drive conversion of untreated C3H 10T$\frac{1}{2}$ cells into muscle cells. [See S. M. Taylor and P. A. Jones, 1979, *Cell* **17**:771; A. B. Lasser et al., 1986, *Cell* **47**:649.]

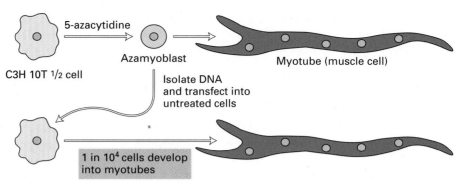

(a) Screen for azamyoblast-specific genes

Azamyoblast mRNAs

① Incubate with reverse transcriptase and [^{32}P]dNTPs

② Remove mRNAs

^{32}P-labeled cDNAs

③ Hybridize with excess of mRNAs from untreated C3H 10T $^{1/2}$ cells

^{32}P-labeled azamyoblast-specific cDNAs

Discard

Clone enriched in azamyoblast-specific cDNA

④ Screen myoblast cDNA library

◄ **FIGURE 14-11 Identification and assay of genes that drive myogenesis.** (a) Azamyoblast mRNAs were isolated from cell extracts on an oligo-dT column (see Figure 7-14). Incubation with reverse transcriptase and [^{32}P]dNTPs yielded radiolabeled cDNAs. The cDNAs were mixed with mRNAs from *untreated* C3H 10T$\frac{1}{2}$ cells; only cDNAs derived from mRNAs (light red) produced by both azamyoblasts and untreated cells hybridized. This technique of subtractive hybridization yielded labeled azamyoblast-specific cDNAs (dark blue), at least some of which correspond to genes required for myogenesis. These cDNAs then were used as probes to screen an azamyoblast cDNA library. (b) Each of the cDNA clones identified as shown in part (a) was incorporated into a plasmid carrying a strong promoter. C3H 10T$\frac{1}{2}$ cells were cotransfected with each recombinant plasmid plus a second plasmid carrying a gene conferring resistance to an antibiotic called G418 and then selected on a medium containing G418. One of the clones, designated *myoD*, was shown to drive conversion of C3H 10T$\frac{1}{2}$ cells into muscle cells, identified by their binding of labeled antibodies against myosin, a muscle-specific protein. [See R. L. Davis et al., 1987, *Cell* **51**:987.]

(b) Assay for myogenic activity of *myoD* cDNA

C3H 10T $^{1/2}$ cell

① Cotransfect with a plasmid carrying *myoD* cDNA and a G418-resistance plasmid

② Select on G418-containing medium

③ Stain with labeled antimyosin antibody (Ⱶ)

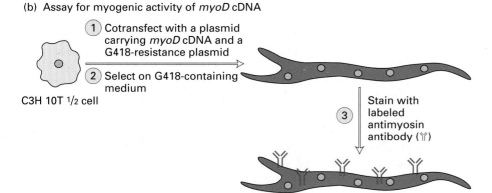

muscle development. As discussed in a later section, knockout experiments have demonstrated the importance of three of these genes in muscle development in the intact mouse.

Myogenic Proteins Are Transcription Factors Containing a Common bHLH Domain

The four myogenic proteins—MyoD, Myf5, myogenin, and Mrf4—are all members of the basic **helix-loop-helix** (bHLH) family of DNA-binding transcription factors (see Figure 10-44). Near the center of these proteins is a DNA-binding basic region adjacent to the HLH domain, which mediates dimer formation. Flanking this central DNA-binding/dimerization region are two activation domains. We refer to the four myogenic bHLH proteins collectively as **muscle-regulatory factors**, or **MRFs** (Figure 14-12a).

bHLH proteins form homo- and heterodimers that bind to a 6-bp DNA site with the consensus sequence C-A-N-N-T-G. Referred to as the *E box*, this sequence is present in many different locations within the genome (i.e., on a purely random basis the E box will be found every 256 nucleotides). Thus some mechanism(s) must ensure that MRFs specifically regulate muscle-specific genes and not other genes containing E boxes in their transcription-control regions. We will examine how this myogenic specificity may be achieved using MyoD as an example.

The affinity of MyoD for DNA is tenfold greater when it binds as a heterodimer complexed with E2A, another bHLH protein, than when it binds as a homodimer. Indeed, in developing azamyoblasts, MyoD is found as a heterodimer complexed with E2A, and E2A, as well as MyoD,

has been shown to be required for myogenesis in C3H $10T\frac{1}{2}$ cells. The DNA-binding domains of E2A and MyoD have similar but not identical amino acid sequences, and both proteins recognize the E-box sequence in DNA. However, since E2A is expressed in other tissues, the requirement for E2A is not sufficient to confer myogenic specificity.

MyoD stimulates transcription only when two or more molecules of MyoD are bound to multiple E boxes; this multiple binding occurs cooperatively. Although multiple copies of E-box sequences are found in most muscle-specific enhancers, they also are present in enhancers that promote expression of genes specifically expressed in other tissues. For instance, E2A is required for normal development of B cells (the white blood cells that produce antibodies), and it regulates B-cell specific genes through enhancers containing multiple E boxes. Thus neither the requirement for multiple E boxes nor the requirement for E2A is sufficient to confer myogenic specificity. Rather, as in yeast cell-type specification, the key to myogenic specificity lies in the combinatorial association of different transcription factors at different transcription-control sites.

MEFs Function in Concert with MRFs to Confer Myogenic Specificity

Some insight into how skeletal muscle cells are specified has come from in vitro mutagenesis studies in which variant E2A proteins were produced. Wild-type E2A protein cannot by itself drive C3H $10T\frac{1}{2}$ cells to a myogenic fate, although it can bind to E boxes controlling muscle-specific genes. To identify which features of MyoD confer myogenic specificity, researchers produced altered E2A molecules in which specific amino acids present in MyoD were substituted into the equivalent positions in the E2A molecule. An E2A variant with three amino acid substitutions corresponding to residues present in MyoD was found to convert C3H $10T\frac{1}{2}$ cells to myotubes. Two of these substitutions are in the central basic DNA-binding region of E2A, and one is just adjacent to this region. Although these substitutions allow E2A to drive myogenic conversion, they do not affect the DNA-binding specificity of E2A. This finding suggests that myogenic specificity is likely to reside in specific interactions between MyoD and other proteins. Recent studies indicate that specific amino acids in the bHLH domain of all the MRFs may confer myogenic specificity by allowing MRF-E2A complexes to bind specifically to another family of DNA-binding proteins called **muscle enhancer–binding factors**, or **MEFs**.

First identified in biochemical experiments as homodimeric proteins that bind to certain DNA sequences in a muscle-specific enhancer, MEFs later were shown to belong to the MADS family of transcription factors. In addition to containing an N-terminal MADS domain, these proteins contain a short stretch of amino acids just C-terminal to the MADS domain, called the MEF domain, and a C-terminal transcription-activation domain (Figure 14-12b). MEFs were considered excellent candidates for interaction with MRFs for two reasons: First, many muscle-specific genes contain

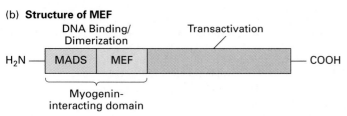

(a) Structure of MRF

(b) Structure of MEF

▲ **FIGURE 14-12 Schematic diagrams of the general structures of two classes of transcription factors that participate in myogenesis.** MRFs are produced only in muscle, whereas MEFs are expressed in several tissues in addition to developing muscle. The myogenic activity of MRFs is enhanced by their interaction with MEFs.

both MEF– and MRF–recognition sequences; second, although MEFs cannot induce myogenic conversion of C3H $10T\frac{1}{2}$ cells, they enhance the ability of MRFs to do so. This enhancement requires physical interaction between a MEF and MRF-E2A heterodimer.

The interaction between these different transcription factors requires both the MADS and MEF domains in the MEF homodimer and the myogenic-specific amino acids in the bHLH domain of the MRF-E2A heterodimer. Crystallographic analysis of MRF-E2A bound to DNA indicates that these amino acids are buried within the major groove of DNA and are unable to directly contact MEFs. Hence, it seems likely that these amino acids confer a particular conformation to other regions of the MRF-E2A heterodimer that, in turn, specifically interact with MEFs. Surprisingly, although both classes of proteins can bind individually to specific DNA sequences, a single DNA site that recognizes one or the other protein is sufficient to act as a platform for assembly of a MRF-E2A-MEF complex. This finding suggests that different configurations of the DNA sites recognized by these factors may drive high levels of muscle-specific gene expression. Indeed, some muscle-specific genes contain bHLH-binding sites (i.e., E boxes); others contain MEF-binding sites; and yet others contain both types of protein-binding sites (Figure 14-13). In contrast to MRFs, which are expressed only in developing muscle, MEFs are expressed in other tissues including the developing central nervous system.

Myogenic Stages at Which MRFs and MEFs Function in Vivo Have Been Identified

Expression of any one of the four MRFs in C3H $10T\frac{1}{2}$ cells can induce the cells to differentiate into muscle in vitro. The functions of these proteins in the intact animal during normal myogenesis have been studied in gene-knockout experiments. In these studies, mice were prepared with gene-targeted knockout mutations in the genes encoding MyoD, Myf5, or myogenin (see Figure 8-34). By analyzing the effects of knocking out these genes, developmental biologists could determine which genes are required for myogenesis and the stage at which they act.

Mice with either the *myoD* or *myf5* gene knocked out have normal muscle, whereas those with the *myogenin* gene knocked out are missing the vast majority of skeletal muscle (Table 14-1). In mice that lack myogenin, myoblasts accumulate at sites normally occupied by skeletal muscle, indicating that myogenin is not required for formation of myoblasts but is required for their differentiation into myotubes. The simple, but erroneous, conclusion from these findings is that

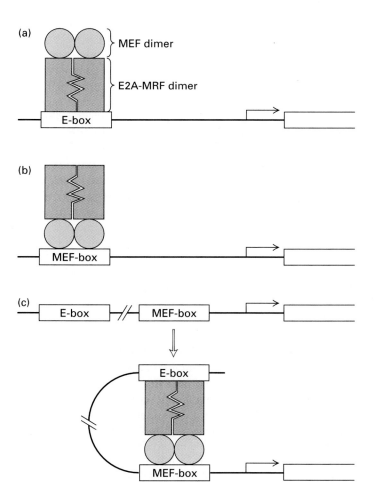

(a) MEF dimer / E2A-MRF dimer / E-box

(b) MEF-box

(c) E-box // MEF-box → E-box / MEF-box

Gene Knocked Out	Phenotype*			Role of Myogenic Protein
	Viable	Myoblasts	Muscle	
myoD	Yes	+	+	?
myf5	Yes	+	+	?
myoD; myf5	No	−	−	Required for myoblast formation or survival
myogenin	No	+	−	Required for myoblast differentiation into muscle

TABLE 14-1 Effect of Knockout of Myogenic Genes in Mice

*+ *sign indicates that myoblasts or mature muscle cells are found at normal sites; − sign indicates that they are not.*
SOURCE: T. G. Braun et al., 1992, *Cell* **71**:369; P. Hasty et al., 1993, *Nature* **364**:501; M. A. Rudnicki et al., 1992, *Cell* **71**:383; M. A. Rudnicki et al., 1993, *Cell* **75**:1351.

◄ **FIGURE 14-13 An MRF-E2A-MEF complex can assemble in the transcription-control region of muscle-specific genes containing an E box, MEF box, or both.** The synergistic action of the MEF homodimer and MRF-E2A heterodimer, which directly interact, drives high-level expression of muscle-specific proteins. [Adapted from K. Yun and B. Wold, 1996, *Current Opinion in Cell Biology* **8**:877.]

Myf5 and MyoD are not required for muscle development. However, since either protein can drive a myogenic program in cell culture, the loss of one gene may be compensated by the function of the other. Indeed, mice homozygous for mutations in both *myf5* and *myoD* die shortly after birth and lack skeletal muscle. In contrast to the *myogenin* mutants, myoblasts do not accumulate in the *myf5; myoD* double mutants, suggesting that the Myf5 and MyoD proteins are required for the formation or survival of myoblasts. Overlapping functions such as these of MyoD and Myf5 are often referred to as redundant. Redundancy provides a more robust developmental program and may allow for more flexibility in the response of cells in different regions of the developing organism to extracellular signals regulating myogenesis.

The results of these gene-knockout experiments are consistent with the observation that azacytidine-treated C3H 10T$\frac{1}{2}$ cells express Myf5 and MyoD prior to fusion but express myogenin only as they fuse to form a syncytium, which then differentiates to form a myotube. As the model in Figure 14-14 illustrates, MyoD and Myf5 are thought to have similar but overlapping functions in selecting cells from developing somites to become myoblasts; that is, they are required for myoblast determination during normal myogenesis. Myogenin, then, is required for the differentiation of myoblasts into myotubes. The fourth MRF protein, Mrf4, is expressed later in development and may play a role in the maintenance of muscle cells.

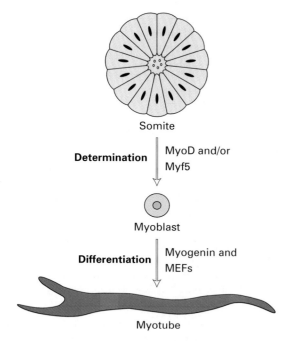

Somite

Determination | MyoD and/or Myf5

Myoblast

Differentiation | Myogenin and MEFs

Myotube

▲ **FIGURE 14-14 Model of genetic control of mammalian skeletal muscle in vivo based on knockout experiments in mice and loss-of-function mutations in *Drosophila*.** According to this model, MyoD and Myf5 serve a redundant function in myoblast determination, while myogenin and MEFs have distinct functions in the differentiation of myoblasts into myotubes (see text).

The role of E2A and MEFs in myogenesis have been assessed in more recent studies. Muscle development is normal in mice with a knockout mutation in the gene encoding E2A, although B-cell development is disrupted. Presumably, during muscle development, E2A-related genes may compensate for loss of E2A, much as *myoD* and *myf5* can compensate for each other. To assess this redundancy, researchers will have to knock out the E2A-related genes and generate mice lacking both E2A and its related genes.

Because mice express multiple MEF proteins, scientists turned to *Drosophila*, which expresses a single MEF, to determine the function of MEFs in muscle development. In flies carrying loss-of-function mutations in the *MEF* gene, no differentiated muscle forms, although myoblasts appear to form normally. Hence, MEFs are required for differentiation, but not for determination (see Figure 14-14).

Multiple MRFs Exhibit Functional Diversity and Permit Flexibility in Regulating Development

The expression of four myogenic bHLH proteins (MRFs) in mice raises intriguing questions. Do these proteins have intrinsically different biochemical properties that correlate with distinctive roles in muscle development? That is, did functionally different MRFs evolve independently? Or have multiple MRFs evolved to facilitate the demands of gene expression in more complex organisms? That is, was duplication of an ancestral MRF gene and the subsequent evolution of divergent transcription-control elements more efficient than incorporation of different control elements into a single gene? Since many mouse genes that regulate development are found in multiple copies, understanding the role of the apparent duplication of the MRFs may provide generally applicable insights about developmental processes.

Scientists have begun to assess these possibilities using a variation of gene-knockout technology called **knockin**. In this technique, the coding sequences of one gene (e.g., *myf5*) are replaced by those of another (e.g., the *myogenin* gene). The experiments combining knockout and knockin technology depicted in Figure 14-15 demonstrate that myogenin and Myf5 are not functionally equivalent in vivo. Indeed, recent biochemical studies have shown that the chromatin-remodeling ability of Myf5 (and MyoD) is much greater than that of myogenin. As we discuss in more detail later, remodeling of chromatin is critical for normal development of most tissues.

As noted earlier, mice with a homozygous knockout of the *myogenin* gene accumulate myoblasts and are not viable (see Table 14-1). By creating mice homozygous for the *myogenin* knockout and carrying one copy of the *myogenin* knockin at the *myf5* locus (i.e., under control of the *myf5* regulatory sequences), scientists could assess the importance of the myogenin-specific transcriptional regulation. The failure of this knockin to rescue the myogenin defect indicated that the unique expression pattern conferred by the

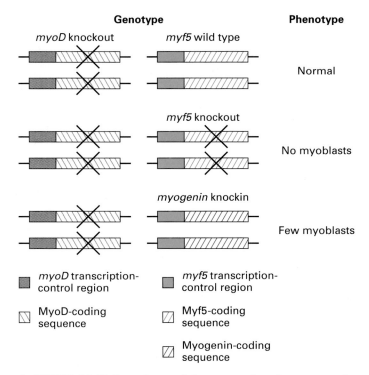

Genotype   **Phenotype**

myoD knockout   *myf5* wild type

                                                    Normal

                         myf5 knockout

                                                    No myoblasts

                         myogenin knockin

                                                    Few myoblasts

▪ *myoD* transcription-   ▪ *myf5* transcription-
  control region   control region

▨ MyoD-coding   ▨ Myf5-coding
  sequence   sequence

                         ▨ Myogenin-coding
                           sequence

▲ **FIGURE 14-15 Experimental demonstration that myogenin cannot substitute for Myf5 in vivo.** Expression of either Myf5 or myogenin in C3H 10T$\frac{1}{2}$ cells can drive myogenesis. To test whether these proteins are functionally equivalent in vivo, researchers inserted the myogenin-coding sequences in place of the sequences encoding Myf5 within the *myf5* gene, forming a *myogenin* knockin. Myogenin will be expressed from the knockin gene in the same spatiotemporal fashion as Myf5 in a wild-type mouse. In knockout mice lacking MyoD and carrying the *myogenin* knockin in place of wild-type *myf5*, some muscle fibers formed (about 10 percent of normal), but the mice died. The inability of the *myogenin* knockin to correct the mutant phenotype resulting from the lack of both Myf5 and MyoD establishes that the functional properties of myogenin and Myf5 differ in vivo.

myogenin regulatory sequences is also critical for normal development. In summary, these studies suggest that gene duplication led to evolution of genes encoding functionally diverse MRFs whose expression is regulated by different transcription-control elements.

Terminal Differentiation of Myoblasts Is under Positive and Negative Control

The determined, yet undifferentiated, myoblast can respond to extracellular signals in the developing embryo that control proliferation (hence, the number of cells that form) and cell migration (hence, the precise location of muscle). In contrast, the differentiated muscle cell, or myotube, cannot respond to such signals. Regulation of the transition from the determined to the differentiated state thus permits the precise spatial and temporal control of cellular differentiation that is necessary to ensure normal morphogenesis in

complex multicellular organisms. The factors that regulate this critical step in various developmental pathways are still poorly understood. However, in vitro experiments have revealed several specific factors that promote or inhibit differentiation during myogenesis.

Inhibitory Proteins Screens for genes related to *myoD* led to identification of a related protein that retains the dimerization helices but lacks the DNA-binding basic region and hence is unable to bind to E-box sequences in DNA. However, this protein interacts with MyoD and E2A, thereby inhibiting formation of MyoD-E2A heterodimers and hence their high-affinity binding to DNA. Accordingly, this protein is referred to as *Id* for *inhibitor of DNA binding*. Analysis of DNA from proliferating azamyoblasts, which express MyoD, E2A, and Id, has shown that the MyoD-binding (or E2A-binding) site in the promoter of the muscle-specific gene encoding creatine kinase is not occupied. This finding presumably reflects the formation of inactive MyoD-Id or E2A-Id complexes and indicates that Id can maintain cells in a determined state during proliferative growth. When these cells are induced to differentiate into muscle (for instance by the removal of serum-containing growth factors required for proliferative growth), the Id concentration falls. As a result, MyoD-E2A dimers can form and bind to the promoters of target genes driving differentiation of azamyoblasts into myotubes. We can see from these results that dimerization of transcription factors with different partners not only can modulate the specificity or affinity of their binding to specific DNA sites, but also may prevent their binding entirely.

Cell-Cycle Proteins The onset of differentiation in many cell types is associated with arrest of the cell cycle, most commonly in G_1, suggesting that cell-cycle proteins (e.g., cyclins and Cdks) may influence the transition from the determined to differentiated state. Researchers recently have found that certain inhibitors of cyclin-Cdk protein kinase activity can induce muscle differentiation in cell culture and that these inhibitors are markedly up-regulated in differentiating muscles in vivo. Conversely, differentiation of cultured myoblasts, under conditions in which they would normally differentiate, can be inhibited by transfecting the cells with DNA encoding cyclin D1 under the control of a constitutively active promoter. Expression of cyclin D1, which normally occurs only during G_1, is up-regulated by mitogenic factors in many cell types (see Figure 13-29). The ability of cyclin D1 to prevent myoblast differentiation in vitro may mimic aspects of the in vivo signals that antagonize the differentiation pathway. The antagonism between negative and positive regulators of G_1 progression is likely to play an important role in controlling myogenesis in vivo.

A Network of Cross-Regulatory Interactions Maintains the Myogenic Program

Precursor cells in different regions of the myotome give rise to different muscles: dorsal medial precursors to epaxial

▶ **FIGURE 14-16 Maintenance of the myogenic program.** Transient signals from the developing spinal cord and ectoderm induce a subset of cells in the developing somite to become myoblasts. Induction is marked by the expression of MRFs and MEFs. These proteins cross-regulate each other's expression and also directly interact to control transcription of other myogenic genes. This network of interactions maintains the myogenic program after the transient inductive signals disappear. Various factors control the final decision to become postmitotic and to differentiate into muscle. See text for further discussion.

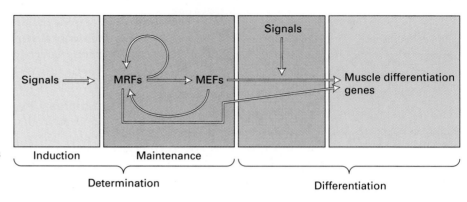

muscles, lateral precursors to hypaxial muscles, and ventro-lateral precursors (after migrating) to limb muscles (see Figure 14-9). Each group of precursor cells shows a distinct pathway of myogenic gene activation induced by different signals from surrounding tissues. Once the myogenic program is activated in a region of the somite, an extensive array of cross-regulatory interactions acts to maintain it (Figure 14-16). These cross-regulatory interactions occur at two levels. First, myogenic factors, both MRFs and MEFs, positively regulate each other's expression by binding to cis-acting regulatory sites. Second, MEFs and MRFs physically interact, thereby acting synergistically to promote expression of myogenic factors that drive differentiation. Thus, although the myogenic program is induced by extracellular signals transiently expressed in tissues surrounding the somite, a network of intracellular interactions maintains the myogenic program in the absence of these signals.

Neurogenesis Requires Regulatory Proteins Analogous to bHLH Myogenic Proteins

Four bHLH proteins that are remarkably similar to the myogenic bHLH proteins control neurogenesis in *Drosophila*. These *Drosophila* proteins are encoded by an ≈100-kb stretch of genomic DNA, termed the *achaete-scute complex* (AS-C), containing four genes designated *achaete (ac), scute (sc), lethal of scute (l'sc),* and *asense (a)*. Analysis of the effects of loss-of-function mutations indicate that the Achaete (Ac) and Scute (Sc) proteins participate in determination of neural precursor cells, while the Asense (As) protein is required for neural differentiation. These functions are analogous to the roles of MyoD and Myf5 in muscle determination and of myogenin in differentiation (Figure 14-17). Two other *Drosophila* proteins, designated Da and Emc, are analogous in structure and function to mammalian E2A and Id, respectively. For example, heterodimeric complexes of

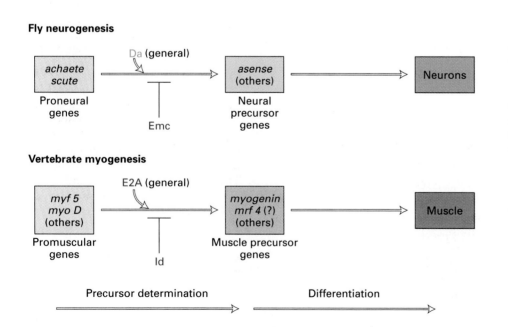

◀ **FIGURE 14-17 Comparison of genes that regulate *Drosophila* neurogenesis and mammalian myogenesis.** bHLH transcription factors have analogous functions in precursor determination and subsequent differentiation into mature muscle cells and neurons. In both cases, the proteins encoded by the earliest-acting genes *(left)* are under both positive and negative control by other related proteins (red type). [Adapted from Y. N. Jan and L. Y. Jan, 1993, *Cell* **75**:827.]

Da with Ac or Sc bind to DNA better than the homodimeric forms of Ac and Sc. Emc, like Id, lacks a DNA-binding basic domain; it binds to Ac and Sc proteins, thus inhibiting their association with Da and binding to DNA.

A family of bHLH proteins related to the *Drosophila* Achaete and Scute proteins have been identified in vertebrates. One of these, called *neurogenin*, which has been isolated from the rat, mouse, and frog, appears to function in determination of neuronal precursor cells. In situ hybridization experiments have shown that neurogenin is expressed at an early stage in the developing nervous system and may induce expression of NeuroD, another bHLH protein that acts later (Figure 14-18a). Injection of large amounts of *neurogenin* mRNA into *Xenopus* embryos further demonstrated the ability of neurogenin to induce neurogenesis (Figure 14-18b). These studies suggest that the function of neurogenin is analogous to that of the Achaete and Scute in *Drosophila*; likewise, NeuroD and Asense may have analogous functions in vertebrates and *Drosophila*.

Gene knockout studies in mice, which have two *neurogenin* genes, have confirmed the essential role of neurogenin in vertebrate neurogenesis. In mice embryos that cannot express neurogenin-1, the trigeminal ganglion in the head region does not develop. However, other regions of the nervous system develop normally in *neurogenin-1* knockouts, suggesting that neurogenin-2 or other bHLH proteins regulate neurogenesis in these regions. In the region of the nervous system affected by the loss of neurogenin-1, development is arrested before expression of NeuroD begins.

Progressive Restriction of Neural Potential Requires Inhibitory HLH Proteins and Local Cell-Cell Interactions

The regulatory mechanisms responsible for restriction of particular developmental pathways to specific cells within an embryo are very complex and not thoroughly understood for any system. For instance, specific cells within somites are selected to become myoblasts, while other cells are destined to become nonmuscle tissues (see Figure 14-9). The best-understood example of such developmental restriction occurs during formation of sensory bristles in *Drosophila*. In this case, the proteins encoded by the proneural genes *achaete* and *scute* must be expressed and active in cells selected to become neural precursor cells, but not in surrounding cells.

The sensory bristles located on the second thoracic segment of the adult fly arise from a monolayer of cells (a columnar epithelium) called the *wing imaginal disc*, from which the epidermis of this segment and the associated wing also are derived. Each bristle is part of a sensory organ that contains four cells: a bristle that protrudes from the epidermis; the socket into which it is inserted; a neuron that transmits the sensory information; and, finally, a cell associated

(a)

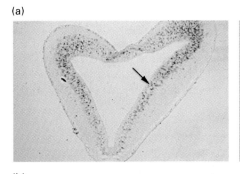

(b)

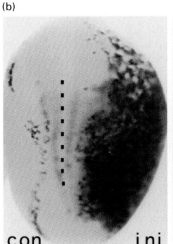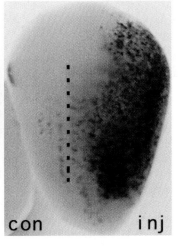

con   inj   con   inj

◀ **FIGURE 14-18 Experimental demonstration that neurogenin acts before NeuroD in vertebrate neurogenesis.** (a) Neurogenin mRNA and neuroD mRNA were detected in the rat neural tube by in situ hybridization. Neurogenin mRNA is expressed in proliferating neuronal precursor cells in the ventricular layer, whereas neuroD mRNA is expressed in migrating neuroblasts that have left the ventricular zone. (b) One of the two cells in early *Xenopus* embryos was injected with neurogenin mRNA (inj) and then stained with a probe specific for neuron-specific β-tubulin mRNA *(left)* or neuroD mRNA *(right)*. The region of the embryo derived from the uninjected cell served as a control (con). The neurogenin mRNA induced a massive increase in the number of neuroblasts expressing neuroD mRNA and neurons expressing β-tubulin mRNA in the region of the neural tube derived from the injected cell. [From Q. Ma et al., 1996, *Cell* **87**:43; Courtesy of D. J. Anderson.]

with the neuron referred to as a support cell. These cells are derived from a single cell by two sequential divisions. This "grandmother cell" is referred to as a **sensory organ precursor (SOP),** or more generally as a neural precursor cell. The pattern of SOPs in the developing imaginal disc presages the pattern of bristles in the adult, which is highly reproducible. These cells do not migrate from another location, but rather arise in distinct positions within the columnar epithelium of the imaginal disc. Each SOP emerges from a cluster of cells, the **proneural cluster,** that express the Achaete and Scute proteins. The pattern of proneural clusters (i.e., of Ac and Sc expression) is determined by earlier-acting patterning genes whose encoded proteins function to divide the epithelium into developmental domains. During normal development, only one cell in each proneural cluster is selected to become an SOP; the remaining cells develop into epidermal structures. Restriction of the neural program to one cell results from down-regulation of the activity and expression of the proneural proteins Achaete and Scute.

As noted earlier, Emc inhibits the binding of Achaete and Scute to DNA and hence their ability to determine neural precursor cells (see Figure 14-17). Loss-of-function mutations in *emc* lead to formation of multiple sensory bristles from a single proneural cluster, whereas gain-of-function mutations suppress SOP formation. In wild-type embryos, expression of Emc is lower in the region of each proneural cluster from which a SOP will arise than in the regions giving rise to epidermal structures. Like *achaete* and *scute*, *emc* is under complex regulation so that Emc is expressed in a specific pattern within the developing epithelium. The resulting variation in the relative level of proteins promoting neurogenesis (e.g., Achaete and/or Scute) and the level of proteins inhibiting it (e.g., Emc) in the cells of the wing imaginal disc limits SOP-forming potential to a small group of neighboring cells within each proneural cluster (Figure 14-19).

Short-range cell-cell interactions further restrict SOP formation by inhibiting the expression of Achaete and/or Scute in all but one cell of a proneural cluster. These local interactions are mediated by two cell-surface proteins: Notch, a cell-surface receptor, and Delta, its specific ligand (Chapter 23). Loss-of-function mutations in either the *Notch* or *Delta* locus result in the formation of multiple SOPs from a proneural cluster and the appearance of multiple bristles arising from a single proneural cluster, indicating that these genes act to inhibit SOP formation (Figure 14-20). Local asymmetry in the expression of Notch and Delta, reinforced by variations in the levels of Achaete, Scute, and Emc, permits one cell, and only one cell, in a proneural cluster to retain its neural potential and become an SOP (Chapter 23; see Figure 23-28). The selected SOP then begins to express Asense and other differentiation proteins, which determine the type of neuron that develops, a process termed *neuronal specification*. As production of Asense protein increases in an SOP, synthesis of Achaete and Scute proteins decreases.

bHLH Regulatory Circuitry May Operate to Specify Other Cell Types

The important role of bHLH proteins in myogenesis and neurogenesis is supported by discovery of similar highly conserved regulatory proteins in *C. elegans*. Moreover, considerable evidence indicates that a bHLH protein, called SCL, participates in determination of hematopoietic stem cells, which differentiate to generate the many different types of blood cells. SCL is expressed in the ventral mesoderm of the developing embryo, in a region of the mesoderm giving rise to hematopoietic stem cells; like MyoD and Myf5, SCL forms a complex with the more generally expressed E2A protein.

The specification of cell type may be an ancient function of bHLH proteins. In the cnidarian *Hydra vulgaris*, which

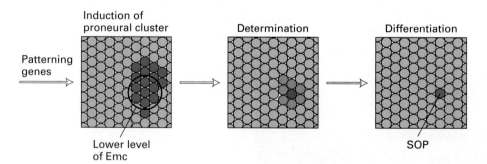

▲ **FIGURE 14-19 Formation of sensory organ precursors (SOPs) in the wing imaginal disc of *Drosophila*.** A set of extracellular signaling molecules and transcription factors, encoded by so-called patterning genes, control the precise spatiotemporal pattern of expression of proneural bHLH proteins such as Achaete and Scute (green) and related proteins that antagonize their function (e.g., Emc). Most cells within the disc express Emc (light red), but only small groups of cells, the proneural clusters, express proneural bHLH proteins. The region of the proneural cluster from which an SOP will form expresses lower levels of Emc, giving these cells a bias towards SOP formation. Interactions between these cells, leading to accumulation of E(spl) repressor proteins in neighboring cells (orange), then restrict SOP formation to a single cell (see Figure 14-20). The activity of Achaete in the SOP promotes expression of Asense (blue), which is required for further differentiation.

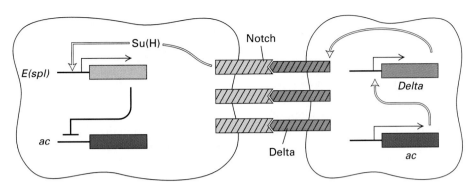

▲ **FIGURE 14-20 Cell-cell interactions that down-regulate proneural genes are critical in determination of a single SOP within a proneural cluster.** Notch, a cell-surface receptor, and Delta, another cell-surface protein that binds to Notch, are initially expressed in all cells within the developing epithelium. Because expression of Delta is promoted by Achaete and other proneural bHLH proteins, those cells within a proneural cluster with the highest Achaete activity *(right)* will express more Delta and thus will provide the strongest signal to neighboring cells *(left)* through the Notch receptor. Interaction of Delta with Notch triggers an intracellular signaling pathway that activates a transcription factor designated Su(H), which in turn promotes expression of *E(spl)* genes. These genes encode a family of bHLH proteins that specifically bind to and repress transcription of proneural genes, such as *achaete (ac)*. The resulting decrease in Achaete leads to a decrease in Delta expression, thus reinforcing the direction of cell-cell signaling. As a consequence of these interactions and others, neural potential is gradually decreased in all but one cell of a proneural cluster. Some workers have proposed that at a critical level of proneural gene expression, an autoregulatory circuit becomes self-staining in one cell, leading to SOP determination. The role of Notch and Delta in developmental programs is discussed further in Chapter 23.

diverged from arthropods some 600 million years ago, a bHLH protein is specifically expressed in the nematocyte, one of some 20 different cell types found in this organism. Nevertheless, it is unlikely that all cell types will be controlled by this regulatory circuitry. Future research of diverse cell types will most likely uncover additional strategies of cell specification using different networks of transcription factors.

SUMMARY   Cell-Type Specification in Animals

- In some systems, transient extracellular signals induce a cell-specification program, and an intracellular network of regulatory proteins maintains it.

- Skeletal muscle cells arise from a subclass of cells in the developing somite that are induced to express myogenic bHLH proteins, or MRFs. MyoD and Myf5 are required for commitment of mesodermal cells to myoblasts, and myogenin is required for myoblasts to differentiate into myotubes (see Figure 14-14).

- MEFs and MRFs bind to each other and act synergistically to control transcription of muscle-specific genes.

- Since myoblasts continue to proliferate and, in some cases, migrate to different regions of the developing embryo, specific mechanisms must maintain the determined state and prevent differentiation until an appropriate time. These mechanisms include inhibitory proteins (e.g., Id) that prevent the formation of bHLH dimers and proteins that promote cell-cycle progression.

- Neurogenesis in flies is controlled by a network of bHLH proteins analogous to those controlling skeletal myogenesis (see Figure 14-17). A similar network probably controls vertebrate neurogenesis.

- Neuronal precursor cells (SOPs) in *Drosophila* arise from equipotent groups of cells, called proneural clusters, which all express Achaete and/or Scute. The ability of these cells to give rise to neuronal precursors is progressively restricted (see Figure 14-19).

- Emc inhibits expression of Achaete/Scute in many cells in a proneural cluster, leaving a small number of cells competent to form SOPs. Interactions between these cells mediated by two transmembrane proteins restricts neuronal specification to a single cell (see Figure 14-20).

14.3 Anteroposterior Specification during Embryogenesis

In previous sections of this chapter, we discussed the transcription-control mechanisms that specify different cell types in yeast and animals. Each cell type expresses specific subsets of genes encoding proteins that determine biochemical and morphologic properties characteristic of that cell type. In addition to different cell types, multicellular organisms exhibit striking regional differences in their cellular organization. For instance, the tissue in hands and feet are composed of the same cells organized in very different ways.

What mechanisms determine how cells are organized in different parts of an organism? Or specify that one end of the developing embryo will become a head and the other the tail? What controls the size and position of different organs? These features of an organism are often collectively referred to as the *body plan.*

Although each phylum has a different body plan, molecular studies have revealed a striking relationship between the body plans of many phyla including the millions of different species of arthropods (e.g., insects) and the 50,000 or so chordate species (e.g., mammals). The body plan along the anteroposterior axis of these and other phyla is specified by a set of highly related transcription factors encoded by discrete clusters of *Hox genes.* The first Hox genes were discovered through classical and molecular genetic studies in *Drosophila* and have since been identified in many other organisms in different phyla through their homology to the fly genes. Mutations in the Hox genes often cause **homeosis**, that is, the formation of a body part having the characteristics normally found in another part at a different site. The proteins encoded by Hox genes are expressed in broad regions along the anteroposterior axis and act to specify the patterning of cells characteristic of tissues within that region in a process sometimes referred to as *regionalization.*

In this section, we examine the mechanisms that determine where Hox genes are expressed in the developing body and how these genes specify regional patterns of development. Because these processes are understood in most detail in *Drosophila,* our discussion focuses on anteroposterior specification in this organism, beginning with how the fly embryo is progressively divided up into smaller and smaller domains characterized by unique patterns of expression of different combinations of transcription factors, which in turn control expression of the Hox genes. Studies of these processes in *Drosophila* not only have provided important insights into regionalization but also have uncovered widely used transcription-control mechanisms that direct various developmental pathways. After we discuss current understanding of early patterning in the fly embryo and how Hox genes function in flies, we consider the roles of Hox genes in mammalian development.

A set of conserved genes also controls regionalization along the dorsoventral axis. Because studies on dorsoventral specification have focused on the role of intercellular communication, we consider this topic in Chapter 23.

Drosophila Has Two Life Forms

The entire life cycle of *Drosophila* occurs within only 9–10 days (Figure 14-21a). The organism has two forms, a worm-like *larval* form and the adult fly form, separated by a period of metamorphosis called *pupation.* Within 1 day the fertilized egg develops into a larva; three subsequent larval stages, or *instars* as they are called, require about 4 more days. The larva is actually a separate animal from the adult. During early **embryogenesis** about a dozen groups of cells, termed

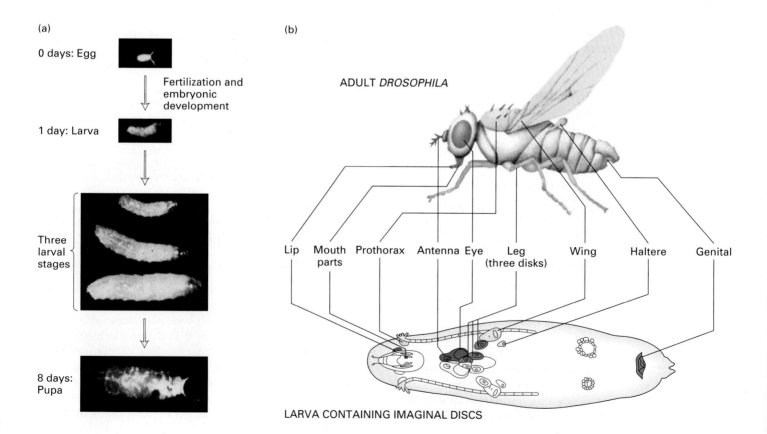

(a)

0 days: Egg

Fertilization and embryonic development

1 day: Larva

Three larval stages

8 days: Pupa

(b)

ADULT *DROSOPHILA*

Lip Mouth parts Prothorax Antenna Eye Leg (three disks) Wing Haltere Genital

LARVA CONTAINING IMAGINAL DISCS

imaginal discs, are set aside and are carried inside the larva (Figure 14-21b). These groups of cells give rise to the adult epidermal structures (wings, legs, etc.). Other groups of precursor cells give rise to adult internal organs such as portions of the gut, the vast majority of the musculature, and the central nervous system. Some larval cells also are conserved in the adult. After the last larval stage, an outer shell is formed. The larval cells are broken down and nutrients derived therefrom are used in the growth and development of the cells that give rise to the different body parts of the adult fly. Pupation takes another 4 days or so. At the end of pupation the shell splits and an adult fly emerges.

Patterning Information Is Generated during Oogenesis and Early Embryogenesis

The blueprint for constructing a fruit fly, including critical spatial information, is in some respects laid down in the egg before **fertilization**. Production of an egg (*oogenesis*) occurs in *ovarioles*, which collectively form the fly ovary. At the distal end of an ovariole, a stem cell divides asymmetrically generating a single germ cell, which divides four times to generate 16 cells. One of these cells completes meiosis, becoming an *oocyte*; the other 15 cells become *nurse cells*, which synthesize proteins and mRNAs that are transported by a series of cytoplasmic bridges into the oocyte (Figure 14-22). These molecules are necessary for maturation of the oocyte and early stages of embryogenesis. Each group of 16 cells is surrounded by a single layer of *follicle cells*, which form the egg shell. As an oocyte matures within an ovariole, new germ cells are produced from a stem cell, displacing the previously generated oocyte, nurse cells, and surrounding follicle cells. The egg is released into the oviduct, where it is fertilized by sperm from a previous mating, which are stored in the seminal vesicle. The fertilized egg, or **zygote**, then is laid through the vulva.

Embryogenesis is activated by fertilization. Interestingly, the polarity of the early embryo is presaged in the mature oocyte, which has distinct anterior and posterior ends. In

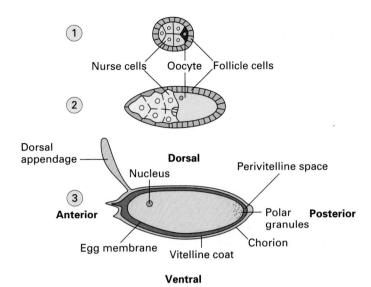

▲ **FIGURE 14-22 Structure of the developing *Drosophila* oocyte at three stages in its maturation.** Each developing unit, or follicle, consists of a developing oocyte, nurse cells, and a layer of somatic cells called follicle cells. Stage ①: Early in oogenesis, the oocyte is about the same size as the neighboring nurse cells. Stage ②: The nurse cells begin to synthesize mRNAs and proteins necessary for oocyte maturation, and the follicle cells begin to form the egg shell. Midway through oogenesis, the oocyte has increased in size considerably. Stage ③: The mature egg is surrounded by the vitelline coat and chorion, which compose the egg shell. The nurse cells and follicle cells have been discarded, but some of the mRNAs synthesized by nurse cells, which become localized in discrete spatial domains of the oocyte, function in early patterning of the embryo. Polar granules are distinct cytoplasmic structures located in the posterior region of the egg. This is the region in which germ cells arise. [Adapted from A. J. F. Griffiths et al., 1993, *An Introduction to Genetic Analysis*, 5th ed., W. H. Freeman and Company, p. 643.]

◀ **FIGURE 14-21 The development of *D. melanogaster*.** (a) The fertilized egg develops into a blastoderm and undergoes cellularization in a few hours. The larva, a segmented form, appears in about 1 day and passes through three stages (instars) over a 4-day period, developing into a prepupa. Pupation takes ≈4–5 days ending with the emergence of the adult fly from the pupal case. (b) Groups of ectodermal cells called imaginal discs are set aside at specific sites in the larval body cavity. From these the various body parts indicated develop during pupation. Other precursor cells give rise to adult muscle, the nervous system, and other internal structures. [Part (a) from M. W. Strickberger, 1985, *Genetics*, 3d ed., Macmillan, p. 38; reprinted with permission of Macmillan Publishing Company. Part (b) adapted from same source and J. W. Fristrom et al., 1969, in E. W. Hanly, ed., *Park City Symposium on Problems in Biology*, University of Utah Press, p. 381.]

addition, some of the mRNAs produced by nurse cells become localized in very discrete spatial domains of the oocyte. The first 13 nuclear divisions of the fertilized ovum are synchronous and rapid, each division occurring about every 10 minutes. Because nuclear division is not accompanied by cell division, a syncytium forms. As the nuclei divide, they begin to migrate outward toward the plasma membrane of the embryo. By 3 hours after fertilization, the nuclei have reached the surface of the embryo, which at this stage is referred to as the *syncytial blastoderm* (Figure 14-23). Cell membranes then form around the nuclei, generating the **blastula**.

Early-patterning events in *Drosophila* occur before formation of the blastula, that is, before cellularization of the embryo. The mechanisms regulating spatial organization in the fly embryo rely in large part on the diffusion of developmentally important proteins, largely transcription factors, within the developing syncytium. This regulatory strategy is very different from that of other organisms, including

(a)

(b)

(c)

(d)

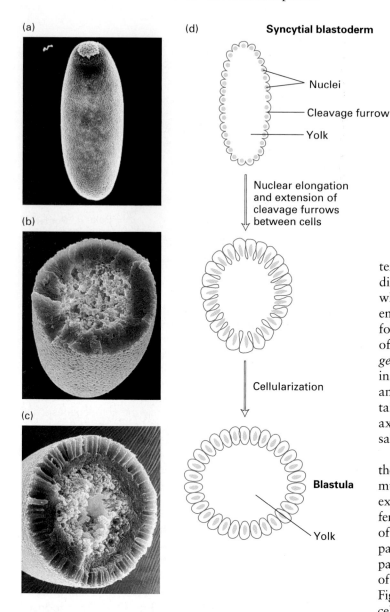

Syncytial blastoderm

Nuclei

Cleavage furrow

Yolk

Nuclear elongation
and extension of
cleavage furrows
between cells

Cellularization

Blastula

Yolk

◄ **FIGURE 14-23 Formation of the blastula during early embryogenesis in *Drosophila*.** Nuclear division is not accompanied by cell division until about 2000–4000 nuclei have formed. Electron micrographs of embryos before cellularization show surface bulges overlying individual nuclei (a) and absence of cell membranes (b), which are evident after cellularization (c). Note separation of the nuclei of so-called pole cells, which give rise to germ cells, at the posterior end *(top)* of the embryo in (a). Change of syncytial blastoderm into a blastula is illustrated in corresponding diagrams (d), in which pole cells are not shown. [See R. R. Turner and A. P. Mahowald, 1976, *Devel. Biol.* **50**:95; photographs courtesy of A. P. Mahowald.]

Mutations in maternal genes necessary for early patterning were identified and classified on the basis of their disruption of the outer (cuticular) structures of the embryo, which occur in a highly reproducible pattern in wild-type embryos (Figure 14-24). These studies led to recognition of four groups of maternal genes, each controlling development of different regions of the embryo as follows: *anterior-group genes*, the head and thorax; *posterior-group genes*, abdominal segments; *terminal-group genes*, the extreme anterior and posterior regions that give rise to the extreme head and tail regions; and *dorsoventral-group genes*, the dorsoventral axis. Some mutations in the latter group lead to loss of dorsal structures, and others, to loss of ventral structures.

In order to decipher the molecular and cellular basis of these patterning mechanisms, investigators had to (1) clone the mutation-defined genes; (2) determine the pattern of mRNA expression and the distribution of the encoded proteins in different spatial domains of the embryo; and (3) assess the effects of mutations on patterning and on the expression of other patterning genes. As just described, the effect of mutations on patterning can be assessed visually by examining the patterns of cuticular structures on the surface of dead embryos (see Figure 14-24). These cuticular structures are formed by the cells directly beneath them. Early gene products can be visualized microscopically by in situ hybridization to detect specific mRNAs and staining with antibody to detect specific proteins. Figure 14-25 illustrates the use of these techniques to localize several early gene products formed before infolding of the blastula to form the **gastrula**. Gene expression in fly embryos also can be detected by use of a reporter-gene construct in transgenic flies. In this method, the *E. coli lacZ* gene, encoding β-galactosidase, is fused to a promoter element that normally controls transcription of a *Drosophila* gene of interest. Expression of β-galactosidase, which is easily assayed, thus serves as a "stand-in" or "reporter" for expression of the fly gene product (see Figure 10-24).

Morphogens Regulate Development as a Function of Their Concentration

A central concept in developmental biology is that of a **morphogen**, a substance that specifies cell identity as a function

mammals, in which early-patterning events are regulated by interactions between cells mediated by signaling molecules (Chapter 23). Thus, although Hox genes play a central role in specifying regional identity in both insects and mammals, the early-patterning mechanisms that control where Hox genes are expressed along the body axis are different.

Four Maternal Gene Systems Control Early Patterning in Fly Embryos

Systematic analysis of mutations that affect early development in *Drosophila* has led to the characterization of four different systems of *maternal* genes that regulate axis determination in the embryo. These genes are transcribed in nurse cells during oogenesis, and the corresponding mRNAs are transported into the oocyte. Translation of these mRNAs yields proteins that regulate the later transcription of *zygotic* genes within the early embryo.

▶ **FIGURE 14-24 Abnormal patterns in the outer (cuticular) structures of the *Drosophila* embryo resulting from mutations in two of the four maternal gene systems that regulate axis determination during early embryogenesis.** Because the pattern of external cuticular structures is highly reproducible, these structures serve as indicators, or markers, of regional identity along the axes in the fly embryo. Embryos derived from mothers homozygous for mutations in these early-patterning genes exhibit various types of abnormal cuticular patterns; these embryos do not survive. In these preparations, anterior is toward the top and ventral is toward the left. (a) The wild-type pattern. (b) Mutations in the *bicoid* locus (anterior system) disrupt development of the anterior abdominal segments, the thorax, and regions of the head. (c) Mutations in the posterior system lead to loss of abdominal segments. An *oskar* mutant is shown. [From D. St. Johnston and C. Nüsslein-Volhard, 1992, *Cell* **68**:201.]

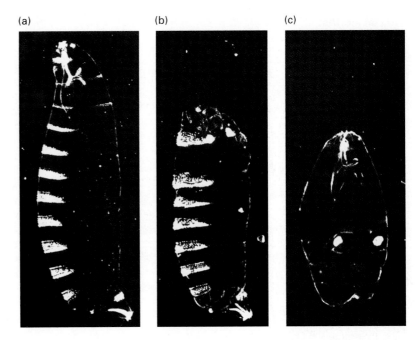
(a)   (b)   (c)

of its concentration. A continuous gradient of morphogen concentration can elicit a set of unique cellular responses at a finite number of *threshold* concentrations: above the threshold, one response is elicited; below it, cells respond differently. As an example, suppose that at the site in the embryo where a particular morphogen is synthesized, its concentration is high enough to establish fate A for cells in the immediate vicinity. As the distance from this site increases, the concentration of the morphogen decreases. At some distance (e.g., 10 cell diameters), a threshold concentration is reached; cells that experience morphogen levels below the threshold are consigned to cell fate B. Still farther away, the morphogen concentration may reach another threshold below which cell fate C is established, and so on.

Several proteins in the developing *Drosophila* embryo have been shown to function as morphogens. Molecular analysis of mutants, for instance, has revealed that gradients of transcription factors produced from maternal mRNAs deposited in the developing oocyte control the transcription of zygotic genes in the early embryo in spatially restricted domains. The products of these zygotic genes, in turn, control the transcription of other zygotic genes, further refining specific spatial domains of gene activity in the embryo. Morphogens controlling development along the anteroposterior axis diffuse within the shared cytoplasm of the syncytial blastoderm. In contrast, patterning along the dorsoventral axis, as well as at the extreme anterior and posterior termini, occurs in response to graded extracellular signals located in the perivitelline space between the plasma membrane of the blastoderm and the vitelline membrane surrounding it. In embryos in which the cells are separated by membranes, early patterning along the anteroposterior axis also depends on

extracellular morphogens, which are considered in Chapter 23. Our discussion here focuses on how transcription activators and repressors function as morphogens within the *Drosophila* syncytial blastoderm.

Maternal *bicoid* Gene Specifies Anterior Region in *Drosophila*

The first morphogen to be described at the molecular level was the protein encoded by the *bicoid* locus in *Drosophila*. The *bicoid* mRNA, which is synthesized in nurse cells during oogenesis and transported to the maturing oocyte, is localized to the most anterior region, or anterior pole, of the early fly embryo (see Figure 14-25a). Embryos produced by female flies that are homozygous for *bicoid* mutations lack head and thoracic tissue (see Figure 14-24b). Other maternal mRNAs either are unlocalized or are localized to different regions of the embryo (e.g., *gurken* mRNA at the antero-dorsal edge and *nanos* mRNA posteriorly). The anterior localization of *bicoid* mRNA depends on its 3′-untranslated end and the products of three other maternal genes. Mutations that result in failure to localize *bicoid* mRNA produce a phenotype similar to, though less severe than, the phenotype associated with mutations in the *bicoid* gene itself.

The *bicoid* gene encodes a transcription factor whose DNA-binding region is a **homeodomain** (see Figure 10-40). In the early fly embryo before cellularization, Bicoid diffuses away from the anterior end where it is produced through the common cytoplasm, forming a protein gradient along the anteroposterior axis. Evidence that the Bicoid protein gradient determines anterior structures was obtained through injection of synthetic *bicoid* mRNA at different locations in

(a) mRNAs

Time
(min)

150

160

180

210

bicoid   *Krüppel*

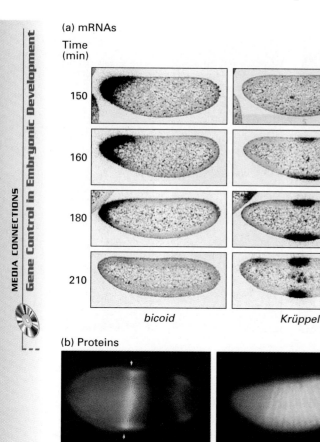

(b) Proteins

Hunchback and Krüppel   Even-skipped and fushi tarazu

◀ **FIGURE 14-25 Localization of developmentally important gene products in early *Drosophila* embryos.** All embryos are positioned with anterior to the left and dorsal at the top. (a) In situ hybridization with labeled RNA probes of whole embryo sections 2.5–3.5 hours after fertilization, which covers the period from the syncytial blastoderm to the beginning of gastrulation. The dark silver grains show the positions of the mRNAs encoded by *bicoid*, a maternal gene and *Krüppel*, a gap gene. These and other early-gene products are expressed in characteristic reproducible temporal and spatial patterns. (b) Fixed embryos stained with antibodies that are coupled to different fluorescent dyes. *(Left)* Hunchback protein (red) and Krüppel protein (green) in a syncytial blastoderm. Both are gap-gene products. The yellow band is a region of overlap of the two proteins. *(Right)* Even-skipped protein (yellow) and Fushi tarazu protein (orange) in alternating bands at beginning of gastrulation. Both are pair-rule gene products. [Part (a) from P. W. Ingham, 1988, *Nature* **335**:25; photographs courtesy of P. W. Ingham. Part (b) courtesy of M. Levine.]

the embryo. This treatment led to formation of anterior structures at the site of injection with progressively more posterior structures forming at increasing distances from the injection site.

The concentration gradient of Bicoid protein, which promotes transcription of the zygotic *hunchback* gene, determines the region in which the Hunchback protein is expressed. Mutations in *hunchback* and several other zygotic genes lead to large gaps in the anteroposterior pattern of the early embryo; hence these genes are collectively called *gap genes*. Several types of evidence indicate that Bicoid protein directly regulates transcription of *hunchback*. For example, the spatial distribution of expressed Hunchback protein parallels that of the Bicoid protein gradient (Figure 14-26a–c). Moreover, analysis of the *hunchback* promoter just upstream of the transcription-start site has shown that it contains three low-affinity and three high-affinity binding sites for Bicoid protein. Studies with transgenic flies carrying reporter genes driven by synthetic promoters containing either all high-affinity or low-affinity Bicoid-binding sites have demonstrated that the affinity of the site determines the threshold concentration of Bicoid at which gene transcription is activated. That is, in response to the same Bicoid protein gradient in the embryo, expression of a reporter gene controlled by a promoter carrying high-affinity Bicoid-binding sites extends

more posteriorly than does transcription of a reporter gene carrying low-affinity sites (Figure 14-26d,e). In addition, the number of Bicoid-binding sites occupied at a given concentration has been shown to determine the amplitude, or level, of the response.

These studies on the ability of Bicoid to regulate transcription of the *hunchback* gene show that variations in the levels of transcription factors, as well as in the number and/or affinity of specific regulatory sequences controlling different target genes, contribute to generating diverse patterns of gene expression during development. Similar mechanisms are found in other developing organisms.

Maternally Derived Inhibitors of Translation Contribute to Early *Drosophila* Patterning

There are two sources of *hunchback (hb)* mRNA in the early fly embryo: that derived from zygotic transcription of *hunchback* under the control of Bicoid protein, which is localized anteriorly, and that derived from the mother, which is uniformly distributed. However, even though *hunchback* mRNA is present throughout the embryo, Hunchback protein is not observed in the posterior region. This exclusion of Hunchback protein from the posterior region depends on the posterior-group maternal gene called *nanos*. Maternal *nanos* mRNA, which is localized to the posterior pole, encodes a morphogen that functions in repressing translation of maternal *hunchback* mRNA in the posterior region (Figure 14-27). Other posterior-group maternal genes are necessary for the specific synthesis and localization of *nanos* mRNA to the posterior pole of the embryo. Mutations in any of these genes (e.g., *oskar*) lead to severe defects in development of the embryo (see Figure 14-24c).

Nanos protein acts in conjunction with Pumilio protein, also encoded by a posterior-group maternal gene, to repress

▶ **FIGURE 14-26 Experimental demonstration that transcription of target genes, such as *hunchback*, directly regulated by Bicoid protein reflects the concentration of Bicoid and its affinity for protein-binding sites in the promoter of the target gene.** (a–c) Increasing the number of *bicoid* genes in mother flies changed the Bicoid gradient in the early embryo, leading to a corresponding change in the gradient of Hunchback protein (red) expressed from the zygotic *hunchback* gene. (d,e) The *hunchback* promoter has been shown to contain three high-affinity and three low-affinity Bicoid-binding sites. Transgenic flies carrying a reporter gene linked to a synthetic promoter containing either four high-affinity sites (d) or four low-affinity sites (e) were prepared. Expression of the reporter-gene product was dependent on the affinity of the Bicoid-binding sites in the promoter. [Adapted from D. St. Johnston and C. Nüsslein-Volhard, 1992, *Cell* **68**:201.]

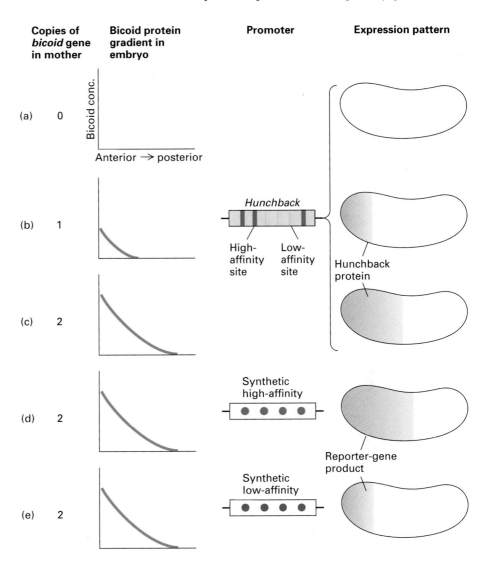

translation of maternal *hunchback* mRNA. Repression also depends on specific sequences in the 3′-untranslated region of *hunchback* mRNA, called Nanos-response elements (NREs). Although the precise mechanism by which repression is achieved is not known, it appears to be correlated with the length of the poly(A) tail in *hunchback* mRNA. In wild-type embryos, the length of the poly(A) tail increases immediately prior to translation of *hunchback* mRNA. The length of the poly(A) tail reflects the activities of antagonistic processes of polyadenylation and deadenylation. Recent genetic and molecular studies suggest that Nanos promotes deadenylation of *hunchback* mRNA and thereby decreases its translation. Figure 14-28 illustrates how this translational regulation helps to establish the Hunchback gradient needed for normal development.

Recent findings raise the prospect that translational control may be a widely used strategy for regulating development. For instance, similar mechanisms have been shown to occur during development of *C. elegans*, and a Nanos-related protein has been identified in *Xenopus* embryos. Even more intriguing is the discovery that Bicoid, which promotes zygotic transcription of the *hunchback* gene, also functions to regulate translation of another *Drosophila* early-patterning gene called *caudal*. This gene plays a crucial role in patterning the posterior region of the embryo; like *bicoid*, it encodes a homeodomain-containing transcription factor. Maternal *caudal* mRNA, like maternal *hunchback* mRNA, is uniformly distributed in the early embryo. Biochemical and genetic studies have shown that Bicoid binds through its homeodomain to a specific sequence in the 3′-untranslated region of *caudal* mRNA. This binding specifically inhibits translation and hence expression of Caudal protein. Since the Bicoid concentration gradient is highest at the anterior end, its repression of *caudal* mRNA translation generates a Caudal concentration gradient that is highest at the posterior end. Future research may well identify other homeodomain proteins that influence gene expression by controlling both transcription and translation.

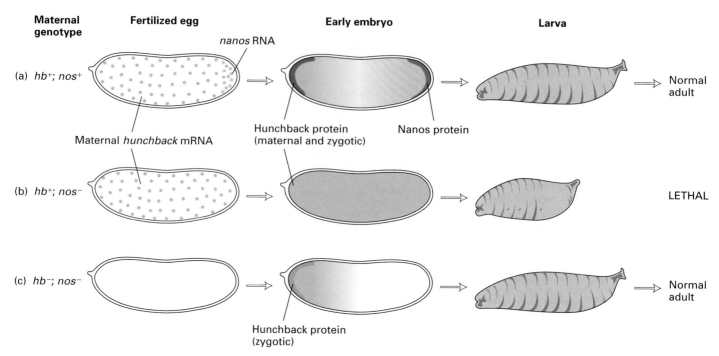

▲ FIGURE 14-27 Role of maternally derived Nanos protein in excluding Hunchback protein from the posterior region of *Drosophila* embryos. mRNAs are indicated by colored dots and relative protein concentrations by color shading. (a) In embryos produced by wild-type female flies, maternal *nanos* mRNA is localized posteriorly. Once this mRNA is translated, the resulting Nanos protein inhibits translation of maternal *hunchback* mRNA in the posterior region of the embryo. (b) In the absence of Nanos, translation of maternal *hunchback* mRNA in the posterior region leads to a failure of the posterior structures to form normally and the embryo dies. (c) Because zygotic transcription of *hunchback* in the anterior region is required for normal development, viable

flies homozygous for mutations in both *hunchback* (*hb⁻*) and *nanos* (*nos⁻*) cannot be produced. However, pole cells (germ-line precursors) that are homozygous for both mutations can be transplanted from an early embryo into a surrogate embryo. Female germ-line cells that carry mutations in both the *hunchback* and *nanos* genes produce embryos that develop normally; zygotic expression of Hunchback in the anterior is controlled by Bicoid. This finding shows that Nanos functions solely to prevent translation of maternal *hunchback* mRNA in the posterior region. [Adapted from P. Lawrence, 1992, *The Making of a Fly: The Genetics of Animal Design*, Blackwell Scientific Publications.]

Graded Expression of Several Gap Genes Further Subdivides the Fly Embryo into Unique Spatial Domains

As noted previously, gap genes participate in early patterning along the anteroposterior axis of *Drosophila* embryos. All of these zygotic genes—including *hunchback*, *Krüppel*, *knirps*, and *giant*—are expressed in specific spatial domains within 2 hours following fertilization and just before cellularization of the embryo (Figure 14-29; see also Figure 14-25). Maternally derived Bicoid plays a key role in activating expression of gap genes (e.g., *hunchback*) in the anterior region, whereas both Bicoid and Caudal control gap-gene expression in more posterior regions. Although a loss-of-function mutation in either *bicoid* or *caudal* has little effect on posterior segmentation, posterior segmentation does not occur in embryos with mutations in both genes.

Genetic and molecular experiments suggest that the boundaries of expression of Krüppel, Knirps, and Giant reflect a balance of transcriptional activation and repression involving maternally derived Bicoid and Caudal as well as Hunchback, the first gap-gene protein to be expressed. For instance, high concentrations of Hunchback protein repress

transcription of *Krüppel*, but below a critical threshold concentration, Hunchback acts as a transcriptional activator of *Krüppel*. This threshold establishes the anterior boundary of the Krüppel protein domain. More posteriorly, the Hunchback concentration falls below the threshold at which it activates transcription of *Krüppel*, thereby setting the posterior Krüppel boundary (see Figure 14-29). The Knirps and Giant proteins are each located in two domains. Expression in their anterior domains is activated by Bicoid, while expression in their posterior domains is activated by the combined action of Bicoid and Caudal. The anterior boundaries of the posterior Knirps and Giant domains are determined by Hunchback-mediated transcriptional repression, and the posterior boundaries are determined by the product of another gap gene, *tailless*.

Expression of Three Groups of Zygotic Genes Completes Early Patterning in *Drosophila*

Through the combined action of the anterior- and posterior-group maternal genes, as well as the zygotic gap genes, the early *Drosophila* embryo becomes divided into broad

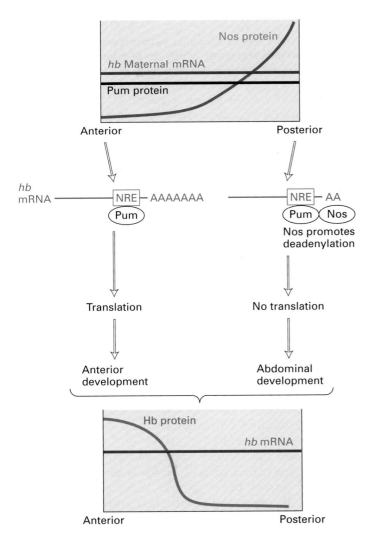

▲ **FIGURE 14-28 Contribution of translational repression to formation of the anterior → posterior Hunchback gradient in the early *Drosophila* embryo.** Maternal *hunchback (hb)* and *pumilio (pum)* mRNAs are uniformly distributed in the early fly embryo, whereas *nanos (nos)* mRNA is localized to the posterior. As *nos* mRNA is translated, Nanos (Nos) protein diffuses through the embryo, establishing a posterior → anterior concentration gradient. Simultaneous binding of Nos and Pum proteins to the NRE sequence of *hb* mRNA promotes deadenylation, which inhibits its translation. As a consequence, maternally derived Hunchback (Hb) protein is expressed in a graded fashion that parallels and reinforces the Hb protein gradient resulting from zygotic transcription of *hb* controlled by Bicoid (see Figure 14-26). In mutants lacking either Nos or Pum, this translational repression is decreased; as a result Hb accumulates posteriorly. [See C. Wreden et al., 1997, *Development* **124**:3015.]

and *selector* genes. The pair-rule and segment-polarity genes, like the maternal and gap genes discussed previously, are expressed transiently and act to establish spatial domains where selector genes are expressed. Selector genes, which are expressed continuously from the embryo into the adult, are required to specify and maintain regional identity along the anteroposterior axis throughout the remainder of the developmental process.

Before describing the final stages in patterning along the anteroposterior axis of *Drosophila*, we need to define the terms *segment* and **parasegment** (Figure 14-30). The segments of the adult fly originally were named based on visual examination of the adult fly; thus each segment corresponds to a visually distinct unit, not a developmental unit. In contrast, parasegments correspond to the actual spatial domain along the anteroposterior axis over which a specific set of selector genes exert their patterning function. Imagine trying to divide any repetitive structure aligned head to tail; regardless of the specific borders chosen, the pattern will remain repetitive (with the exception of the two ends). During early development, the embryo is divided into 14 parasegments with the anterior border of each parasegment demarcated by a sharp band of cells expressing a particular

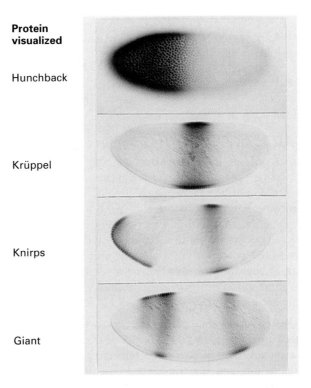

▲ **FIGURE 14-29 Localization of gap-gene products in early *Drosophila* embryos visualized by staining with specific antibodies against Hunchback, Krüppel, Knirps, and Giant proteins.** Anterior is to the left. Hunchback protein functions as both a transcriptional activator and repressor. Transcription of the *Krüppel, knirps,* and *giant* genes is regulated by Hunchback, Bicoid, and Caudal. See text for discussion. [Adapted from G. Struhl et al., 1992, *Cell* **69**:237.]

expression domains characterized by different combinations of transcription factors. These factors are expressed at different levels and act in various combinations to further subdivide the embryo into specific developmental domains. They do so by activating transcription of three additional groups of zygotic genes, termed *pair-rule, segment-polarity,*

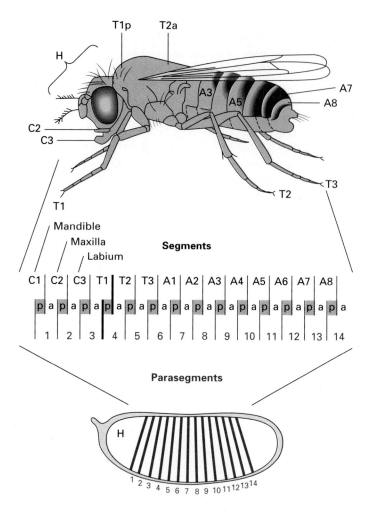

◀ **FIGURE 14-30 The relationship between segments in the adult fly and parasegments, which are developmental units corresponding to the domains of activity of selector genes.** The head segments are designated C1–C3; thoracic segments, T1–T3; and abdominal segments, A1–A8. The anterior border of each parasegment is marked by a sharp band of cells expressing Engrailed protein (red). Each segment is divided into an anterior (a) and posterior (p) compartment. Engrailed expression marks cells in the posterior compartment. There is no mixing of cells across the anterior/posterior compartment boundary of each segment. Each parasegment roughly corresponds to the posterior compartment of one segment and the anterior compartment of the adjacent segment to the rear. [From P. A. Lawrence, 1992, *The Making of a Fly: The Genetics of Animal Design,* Blackwell Scientific Publications.]

stripes results from transcriptional activation in a broad domain followed by delineation of sharp borders by spatially restricted repressors.

To see how this strategy works, we consider expression of Even-skipped (Eve) stripe 2, which is controlled by the maternally derived Bicoid protein and the gap proteins Hunchback, Krüppel, and Giant. These proteins exert their effect by binding to a clustered set of regulatory sites located upstream of the *eve* promoter (Figure 14-31a). Hunchback and Bicoid activate transcription of *eve* in a broad spatial domain, whereas Krüppel and Giant repress *eve* transcription at sharp posterior and anterior boundaries. The segment of DNA containing these regulatory sites will drive expression of a *lacZ* reporter gene specifically in stripe 2. The coordinated effect of these proteins, each of which has a unique concentration gradient along the anteroposterior axis, precisely regulates the boundaries of stripe 2 expression (Figure 14-31b).

Segment-Polarity Genes Next, each parasegment is further subdivided through the action of segment-polarity genes such as *wingless* and *engrailed*. By this stage in embryogenesis, cellularization is complete, and all cells are surrounded by a plasma membrane. Further patterning depends on intercellular signaling, and several segment-polarity genes encode secreted proteins that signal developmental events in neighboring cells. Segment-polarity genes are responsible for generating the patterns of cells within each segment. Patterning of the embryonic cuticle is defined by cuticular protrusions (denticles) and regions devoid of them, which are smooth. Each cuticular pattern reflects a different combination of

segment-polarity gene called *engrailed*. Each parasegment roughly contains the posterior portion of one segment and the anterior portion of the segment located just posterior to it. A segment, then, is roughly divided in half by expression of the Engrailed protein. This boundary defines a developmental unit within each segment called a *compartment*: cells posterior to this boundary will become part of the so-called posterior compartment and those anterior to it part of the anterior compartment. Once the border between these compartments is established, there is no mixing of cells across it.

Pair-Rule Genes The pair-rule genes, which include *fushi tarazu, hairy,* and *even-skipped,* encode transcription factors that are expressed in stripes of cells, corresponding to the parasegments, covering the central part of the embryo (see Figure 14-25b). Each pair-rule gene product is expressed in seven parasegments, either the even or odd ones. Gene expression in each stripe appears to be controlled independently by the action of different transcription factors encoded by gap and maternal genes. The results of various studies suggest that expression of pair-rule genes in discrete

pair-rule and segment-polarity genes. These genes also are frequently responsible for determining the polarity of the pattern within a parasegment; the orientation of the denticles also can be altered by the segment-polarity genes. The *engrailed* gene is expressed in the most anterior band of cells of each of the well-defined parasegments (see Figure 14-30).

Selector Genes The next genes to be transcribed in the regulatory hierarchy that controls regionalization of the *Drosophila* embryo are the selector genes. These genes correspond to the Hox genes mentioned earlier, and their encoded proteins regulate development within parasegmental domains. By this stage in embryogenesis, each band of cells along the anteroposterior axis expresses a unique combination of transcription factors, which control subsequent cell development (Figure 14-32).

The patterns of selector-gene expression are determined in the early embryo and maintained into the adult fly. Continuous expression of selector genes is required to determine the structures of the various body parts along the anteroposterior axis. Two of the best-studied selector genes in *Drosophila* are *Antennapedia (Antp)*, which dominates development of the fourth parasegment, and *Ultrabithorax (Ubx)*, which largely controls the sixth parasegment. Transcription of the *Antp* and *Ubx* genes begins during the third hour of embryogenesis in domains determined in part by various gap-gene products.

Mutations in selector genes often (but not always) cause transformation of one body part into another (homeosis), providing particularly vivid evidence of the role of genes in regional specification. For instance, misexpression of *Antp* in the primordium of the antenna results in its development into a leg rather than an antenna (Figure 14-33). Both struc-

tures are appendages covered with sensory structures although the type and distribution of these structures are quite different. Conversely, loss of *Ubx* leads to development of a pair of wings in place of balancing organs, called halteres, on the third thoracic segment (see Figure 8-8b). Genes in which mutations have such effects are referred to as **homeotic genes**.

Selector genes act in specification of epidermal structures (the external surface), the musculature, neural tissue, and gut tissue along the anteroposterior axis. Because the role of selector genes in regulating development of the epidermis is understood best, we discuss these genes in some detail. These genes were first identified in *Drosophila*, but similar genes have analogous functions in mammalian development.

Selector (Hox) Genes Occur in Clusters in the Genome

Two clusters of selector genes — the bithorax complex (BX-C) and the antennapedia complex (ANT-C) — play a central role in controlling regionalization of external structures along the anteroposterior axis in Drosophila. The BX-C contains three structural genes, called *Ultrabithorax (Ubx)*, *abdominal A (abdA)*, and *Abdominal B (AbdB)*, each of which encodes a homeodomain-containing transcription factor. All three genes have extensive noncoding sequences (e.g., introns) that are critical in regulating their individual expression patterns within different parasegments (Figure 14-34). The linear order of the genes in the BX-C parallels their pattern of expression along the body axis. The same phenomenon is true of the ANT-C, which contains five genes called *labial (lab)*, *proboscipedia (pb)*, *Deformed (Dfd)*, *Sex combs reduced (Scr)*, and *Antennapedia (Antp)*. Both gene clusters are located on chromosome III, separated by some 10 cM. The proteins

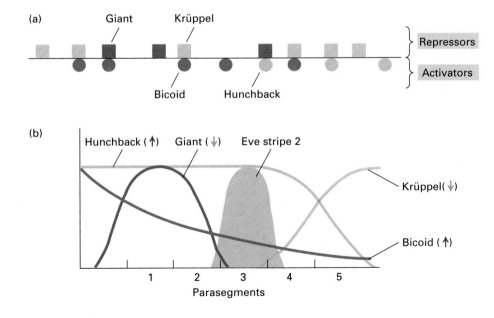

◀ **FIGURE 14-31 Expression of the Even-skipped (Eve) stripe 2 in the *Drosophila* embryo.** (a) Diagram of the 815-bp regulatory region controlling transcription of the pair-rule gene *eve*. This region contains binding sites for Bicoid and Hunchback proteins, which activate transcription of *eve*, and for Giant and Krüppel proteins, which repress transcription. (b) Concentration gradients of Eve stripe 2 and of the four proteins that regulate its expression. The coordinated effect of the two repressors ($\downarrow$) and two activators ($\uparrow$) determine the precise boundaries of the second anterior Eve stripe. Expression of other stripes is regulated independently by other combinations of transcription factors encoded by maternal and gap genes. [See S. Small et al., 1991, *Genes & Devel.* **8**:827.]

MEDIA CONNECTIONS
Gene Control in Embryonic Development

(a)

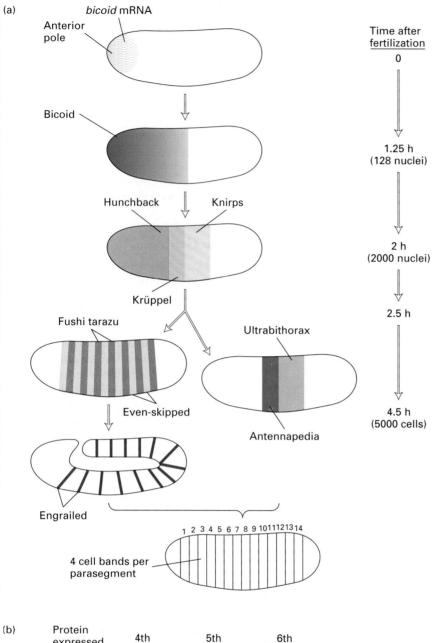

(b)

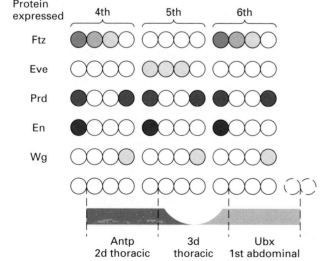

◀ **FIGURE 14-32 Summary of sequential expression of various genes during early development of the *Drosophila* embryo and localization of their gene products within the embryo.** (a) Maternal *bicoid* mRNA is localized at the anterior pole of the egg, but Bicoid protein, which is synthesized soon after fertilization, diffuses to form a gradient. In most cases, an mRNA and its corresponding protein are present in the same regions of the embryo. The expression domains of three gap-gene products— Hunchback, Krüppel, and Knirps—are shown. Not shown is Nanos protein, which represses translation of *hunchback* mRNA in the posterior region (see Figure 14-27a). Expression of the proteins encoded by the pair-rule genes *fushi tarazu (ftz), even-skipped (eve),* and *paired (prd)* is determined by specific combinations of Bicoid and various gap-gene products (see Figure 14-31). These proteins demarcate 14 stripes corresponding to the parasegments. The segment-polarity gene *engrailed (en)* is expressed at the anterior end of each parasegment; it plays an important role along with other segment-polarity genes such as *wingless (wg)* in patterning of each parasegment. Cellularization occurs after 2.5 hours, and gastrulation (including folding of the embryo) occurs at about 4.5 hours. By this time, each parasegment consists of four bands of cells. (b) Within a parasegment, each band of cells (represented by a circle) is characterized by expression of a unique set of proteins encoded by pair-rule and segment-polarity genes. Shown here are several expression patterns in three parasegments (4–6). These expression patterns act as positional values distinguishing each cell band in a parasegment. The gap-gene products largely determine the expression domains of selector genes such as *Antennapedia (Antp)* and *Ultrabithorax (Ubx).* The proteins encoded by selector genes specify the organization of larval and adult structures within the context of the positional identity of cells within each parasegment.

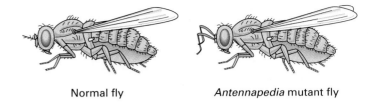

► **FIGURE 14-33 Misexpression of Antp protein in the developing antenna leads to its transformation into a leg, a structure whose development normally is controlled by Antp in thoracic segments.** This is an example of a homeotic transformation. Such misexpression can result from a regulatory mutation or induced expression of an *Antp* transgene. See also Figure 8-8b. [From W. McGinnis and M. Kuziora, 1994, *Sci. Am.* **270** (2):58.]

encoded by the ANT-C and BX-C control development of parasegments 0 – 5 and parasegments 5 – 14, respectively.

The organization of genes within both ANT-C and BX-C is not likely to be serendipitous, as it is observed in homologous gene clusters in organisms in different phyla. As we discuss later, mammalian homologs of the ANT-C and BX-C genes occur in four gene clusters, located on different chromosomes, that are collectively referred to as the **Hox complex** (Hox-C). The interrelated functions of the genes in the ANT-C and BX-C and their expression patterns suggest that these gene complexes were initially adjacent to each other in the genome and became separated during evolution. For this reason, these two *Drosophila* gene clusters are often collectively referred to as the *Hom complex* (Hom-C). For simplicity, we will refer to these complexes of selector genes in both flies and mammals as Hox complexes. The development of the brain in both flies and mammals is controlled by another set of homologous selector genes. The function of these genes is not as well understood as the Hox genes and will not be discussed here.

Combinations of Different Hox Proteins Contribute to Specifying Parasegment Identity in *Drosophila*

To illustrate the function of Hox genes in *Drosophila*, we focus our discussion on the genes in the bithorax complex

(BX-C). Specification of the identity of parasegments 5–14 in the fly embryo requires the BX-C, and removal of the entire complex leads to embryonic death. However, by analyzing the cuticle of dead mutant larvae, investigators have assessed the role of the three BX-C gene products in specifying various parasegments along the anteroposterior axis. Deletion of the entire BX-C causes transformation of parasegments 5–13 into parasegment 4 (Figure 14-35a). Since other genes also contribute to the specificity of parasegment 14 identity, it develops abnormally but does not assume a parasegment 4 identity. In a sense, then, the BX-C represses parasegment 4 identity and allows the more posterior parasegments to be specified.

Analysis of various double and single mutants permits the contribution of the individual BX-C genes to be assessed. For instance, if both *abdA* and *AbdB* functions are removed by mutation leaving only *Ubx*, parasegments 4–6 develop normally, whereas parasegments 7–13 are transformed into parasegment 6 (Figure 14-35b). In contrast, loss-of-function mutations in *AbdB* permit normal development of parasegments 4–9 but cause parasegments 10–13 to assume parasegment 9 identity (Figure 14-35c).

As noted already, the *Ubx*, *abdA*, and *AbdB* genes are transcribed in the same direction and their order of expression along the body axis corresponds to their order within the complex itself (see Figure 14-34b). Although the significance

► **FIGURE 14-34 (a) Organization of genes within the bithorax complex (BX-C) on *Drosophila* chromosome III.** Transcription of all three genes is from right to left. The exons (blue) make up a relatively small part of the BX-C. The large introns (tan) play an important role in regulating the specific spatial and temporal patterns of transcription of the BX-C genes along the anteroposterior axis. The intron/exon structure of AbdB is uncertain. (b) Expression patterns of the products (mRNAs and/or proteins) of the *Ultrabithorax (Ubx)*, *abdominal A (abdA)*, and *Abdominal B (AbdB)* genes in parasegments 4–14. Darker levels of shading indicate higher concentrations of gene products. [Part (a) adapted from M. Peifer et al., 1987, *Genes Devel.* **1**:891; part (b) adapted from T. A. Kaufman et al., 1990, *Adv. Genet.* **27**:309.]

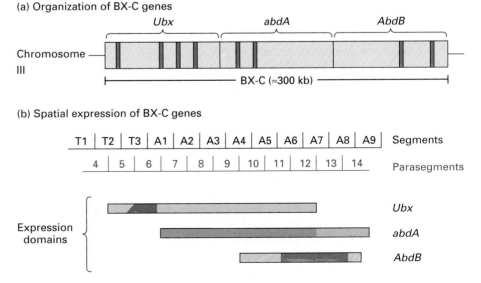

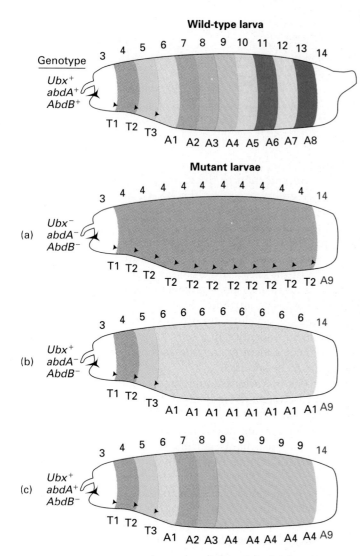

▲ FIGURE 14-35 Contribution of BX-C genes—*Ubx, abdA,*
and *AdbB*—to determination of parasegment identity. The
numbers above each larva indicate the parasegments; those
below, the corresponding segments. The cuticular pattern of
larvae is used to assign an identity to each parasegment (PS),
which is indicated by color, as depicted in the wild type at the
top. Red PS and segment labels indicate abnormal patterns that
do not correspond exactly to any found in wild-type larvae. See
text for discussion. [Adapted from P. A. Lawrence, 1992, *The Making
of a Fly: Genetics of Animal Design*, Blackwell Scientific Publications.]

of this correlation is unclear, the conservation of this order
in vertebrate Hox complexes argues that it plays an important role in controlling patterning. Indeed, the "out-of-
order" expression that occurs in single mutants lacking a
functional *abdA* leads to marked defects in parasegments
10–14; that is, they do not correspond morphologically to
any wild-type parasegment. This finding suggests that the
products of *Ubx* and *AdbB*, in the absence of the *abdA* gene
product, do not provide recognizable patterning informa-

tion in parasegments 10–14. Presumably, during normal
development of the epidermis, *Ubx* and *AdbB* are never
expressed together without *abdA*. Indeed, analysis of other
"out-of-order" mutants indicates that the BC-X genes must
be expressed along the body axis in the order *Ubx, abdA,*
and *AdbB*—their order in the genome—for normal patterning information to be generated.

The cuticular patterns illustrated in Figure 14-35 result
from loss-of-function mutations in BX-C genes. Researchers
also have assessed the effect of gain-of-function mutations
in transgenic flies. For instance, transgenic embryos carrying
Ubx under control of a heat-shock promoter express the Ubx
protein uniformly along the anteroposterior axis, whereas
in wild-type embryos Ubx expression is concentrated in parasegments 5 and 6 (see Figure 14-34b). In these transgenic
embryos, parasegments 6–14 form normally, but parasegments 1–5 are transformed into parasegment 6. During normal development, the identity of these anterior parasegments
is controlled by the ANT-C.

These gain- and loss-of-function studies reveal a consistent relation between the *Drosophila* selector genes: Genes
that are expressed more posteriorly suppress the action of
genes that are expressed more anteriorly. Thus ectopic expression of a Hox gene in a region more anterior to its normal expression domain results in an anterior → posterior
transformation in morphology. The conservation of this phenomenon in the mouse indicates that it must have evolutionary significance.

Although it is clear that BX-C genes and other Hox genes
control specification of tissue along the anteroposterior
axis, the mechanisms by which this occurs are poorly understood. Recent studies indicate that the different combinations of Hox proteins in various cells of the embryo contribute to this regionalization. Since Hox genes are
transcribed after cell membranes have formed, these genes
contribute to patterning in the context of signaling between
cells. Patterning of tissues, then, is a dynamic process in
which gene expression is tightly linked to the context of cells
within tissues (Chapter 23).

Specificity of *Drosophila* Hox-Protein Function Is Mediated by Exd Protein

Similar to transcription factors expressed earlier in embryogenesis, the Hox proteins encoded by BX-C and ANT-C
genes most likely control different developmental pathways
by regulating the expression of different sets of target genes.
However, the discovery that different Hox proteins bind
with high affinity to the same short DNA sequences, which
are found on average once every kilobase, seemed incompatible with this mechanism. Recent genetic and molecular
experiments show that the ability of Hox proteins to control expression of different genes depends on the product of
the *extradenticle (exd)* gene.

In *Drosophila* embryos with loss-of-function mutations
in *exd*, the Hox genes are expressed as in wild-type embryos,

but the structures controlled by them do not develop normally. This finding suggested that the homeodomain protein encoded by *exd* may act in combination with Hox proteins to control transcription of specific target genes. Exd protein has been shown to dimerize with different Hox proteins, forming heterodimers that exhibit different DNA-binding specificities. For example, the Hox proteins Labial (Lab) and Deformed (Dfd), which are encoded in the ANT-C, both bind to the same DNA sites, but Lab-Exd and Dfd-Exd heterodimers each bind selectively to a specific unique sequence (Figure 14-36). Thus, specific interactions between different Hox proteins and Exd may lead to conformational changes unique to each heterodimer, leading to different DNA-binding specificities. Recent studies have shown that Exd may also contribute to the specificity of Hox function by converting bound Hox proteins from repressors to activators. In addition, Exd may act in a Hox-independent fashion to repress yet other genes.

Hox-Gene Expression Is Maintained by Autoregulation and Changes in Chromatin Structure

As mentioned earlier, many of the early-patterning proteins necessary for establishment of specific patterns of Hox-gene expression are produced only transiently during embryogenesis. However, once Hox genes are turned on, they must continue to be transcribed in specific regions, as Hox proteins are required throughout development and into adult stages.

The transcription-control regions of some Hox genes contain binding sites for their encoded proteins. Thus these Hox proteins help maintain their own expression through an **autoregulatory loop**. The *lab* and *Dfd* genes discussed in the previous section provide examples of this phenomenon. The DNA site specific for Lab-Exd heterodimer (GG site) is present in the transcription-control region of the *lab* gene; likewise, the Dfd-Exd–binding site (TA site) is present in the *Dfd* control region. To demonstrate autoregulation of *lab* and *Dfd* in vivo, transgenes carrying the *lacZ* gene linked to tandem arrays of either the GG site or TA site were constructed. When these transgenes were introduced into flies, β-galactosidase was expressed in the same regions of fly embryos as the Lab and Dfd proteins (Figure 14-37).

Another mechanism for maintaining normal patterns of Hox-gene expression requires proteins that modulate chromatin structure. These proteins are encoded by two classes of genes referred to as the *trithorax* group and *polycomb*

(a) Hox-binding consensus sequence

$$TGATNNAT\begin{array}{c}GG\\TA\end{array}$$

↑
Hox-specificity
determinant

(b) Binding of Hox-Exd heterodimers

TGATTAATGG
ATTAATTACC   TA site

Exd   Dfd

TGATGGATGG
ACTACCTACC   GG site

Exd   Lab

▲ **FIGURE 14-36 Role of Exd protein in conferring DNA-binding specificity on *Drosophila* Hox proteins.** (a) Various Hox proteins, including Dfd and Lab, bind to a 10-bp consensus sequence that differs in the nucleotides (N) at the two central positions. (b) Exd-Dfd and Exd-Lab heterodimers specifically recognize Hox-binding sites in which the central dinucleotide is TA or GG, respectively.

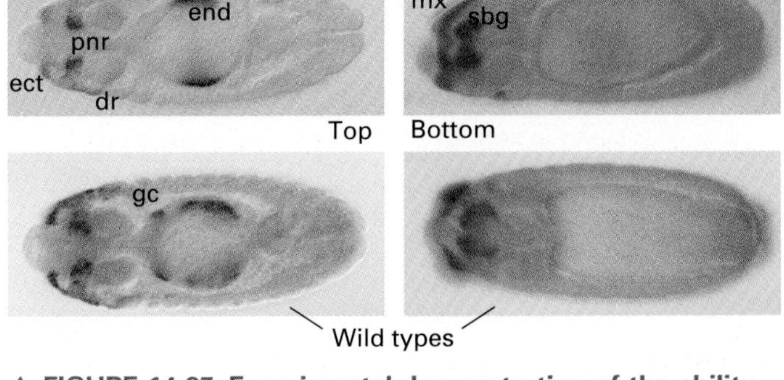

Analysis of lac Z transgene expression in embryos

(GC-Repeat)₃—Lac Z   (TA-Repeat)₃—Lac Z

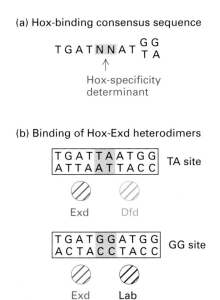

▲ **FIGURE 14-37 Experimental demonstration of the ability of Hox-binding sites to direct gene expression in specific regions of *Drosophila* embryos.** Transgenes containing the *lacZ* gene linked to three copies of the Lab-specific GG site or Dfd-specific TA site from *Drosophila* were introduced into flies. The pattern of expression of β-galactosidase in the transgenic fly embryos mimicked the Lab and Dfd expression patterns, respectively, in wild-type embryos. [From S.-K Chan et al., 1997, *Development* **124**(1):2007; courtesy of R. S. Mann.]

group. Early patterning, as we have noted previously, depends on gene repression as well as gene activation (see Figure 14-31). The pattern of Hox-gene expression is initially normal in *polycomb*-group mutants, but eventually they undergo multiple homeotic transformations, suggesting that Polycomb proteins have a repressive effect on expression of Hox genes. Recent immunohistological and biochemical studies have shown that Polycomb proteins bind to multiple chromosomal locations and form large macromolecular complexes containing different proteins of the Polycomb group. The current view is that the transient repression of loci mediated by patterning proteins expressed earlier in development is "locked in" by Polycomb proteins. This stable Polycomb-dependent repression may result from the ability of these proteins to assemble inactive chromatin structures (Chapter 10). It is not yet known just how changes in chromatin structure resulting from binding of Polycomb proteins leads to repression, nor whether these proteins directly interact with histones.

Whereas Polycomb proteins act to repress expression of certain Hox genes, Trithorax proteins are necessary for maintaining expression of many homeotic genes. Like Polycomb-group proteins, Trithorax-group proteins bind to multiple chromosomal sites and form large multiprotein complexes. Some of the Trithorax proteins are homologous to the yeast Swi proteins, which play an important role in activating expression of many yeast genes. Trithorax proteins are thought to stimulate gene expression by modulating the chromatin structure at certain loci to a transcriptionally active form.

Mammalian Homologs of *Drosophila* ANT-C and BX-C Genes Occur in Four Hox Complexes

A milestone in developmental biology was the discovery that all the structural genes in ANT-C and BX-C contain a region of homology, the **homeobox** sequence, which encodes the DNA-binding homeodomain motif. Shortly after this discovery, mammalian homologs of these selector genes were isolated using the homeobox-containing regions of the *Drosophila* genes as DNA probes. Gene-knockout studies have shown that these mammalian genes, like the homologous *Drosophila* genes, play a critical role in regulating the development of specific regions along the anteroposterior axis.

Molecular cloning studies in the mouse and human have identified four Hox gene clusters (Figure 14-38). These mammalian gene clusters are designated HoxA–HoxD in humans and Hoxa–Hoxd in mice. Each cluster is located on a different chromosome. Homologous genes located in different Hox clusters are referred to as *paralogs*. While there are 13 different sets of Hox genes, no complex contains a complete set of paralogs.

As discussed earlier, the genomic order of the BX-C and ANT-C genes is colinear with the order of expression of these genes along the anteroposterior axis of *Drosophila*. This same feature of genomic organization has been demonstrated for the mouse Hox genes (Figure 14-39). The specific mechanisms by which Hox expression patterns are established in vertebrates are largely unknown. These mechanisms most likely are quite different from those in *Drosophila* because early vertebrate embryos are cellularized and do not exist as a syncytium.

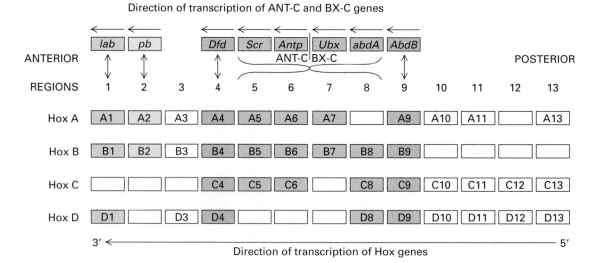

▲ **FIGURE 14-38 Comparison of *Drosophila* ANT-C and BX-C genes and the four human Hox complexes.** For purposes of alignment, the ANT-C and BX-C *(top)* are shown adjacent to each other, although they are separated on chromosome III in the fly genome. Each Hox complex (HoxA–HoxD) can be divided into 13 regions numbered 1–13. Genes in different Hox complexes with the same number are homologs called paralogs (e.g., *HoxA1* and *HoxB1*). Empty boxes indicate that the corresponding genes have not been identified. Similarities in expression patterns and sequences strongly suggest that the fly *labial (lab)*, *proboscipedia (pb)*, *Deformed (Dfd)*, and *Abdominal B (AbdB)* genes are analogous to regions 1, 2, 4, and 9, respectively. The set of fly genes including *Sex combs reduced (Scr)*, *Antennapedia (Antp)*, *Ultrabithorax (Ubx)*, and *abdominal A (abdA)* (purple) is similar to regions 5–8, but precise correlations between the fly genes and Hox regions are not possible at present. Horizontal arrows indicate direction of transcription. Anterior and posterior refer to the order of expression of these genes along the body axis. [Adapted from M. P. Scott, 1992, *Cell* **71**:551.]

► **FIGURE 14-39 Schematic diagram depicting expression domains of the indicated Hox genes in *Drosophila* and mouse embryos.** Note that in the domain labeled *Antennapedia*, four genes are expressed; these are analogous to *Hox5–Hox8*. Dashed lines connect corresponding regions in *Drosophila* and mouse. [Adapted from M. McGinnis and M. Kuziora, 1994, *Sci. Am.* **270**(2):58; drawing by Tomo Narashima, © 1994, Scientific American, Inc. All rights reserved.]

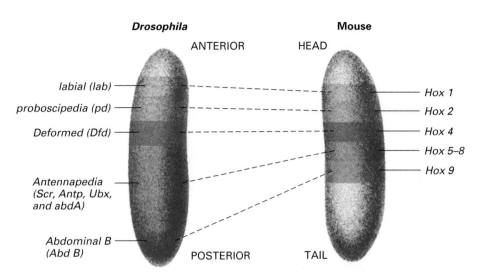

However, the discovery of mammalian homologs of *exd* (called *Pbx* in mammals) and of *polycomb*-group and *trithorax*-group genes suggests that the mechanisms for controlling the DNA-binding specificity of Hox proteins and for maintaining their expression may be similar in *Drosophila* and higher organisms.

Mutations in Hox Genes Result in Homeotic Transformations in the Developing Mouse

The genes composing the *Drosophila* Hox complexes were isolated based on developmental defects observed in flies carrying mutations within these genes. In contrast, the mammalian Hox genes were isolated based on their homology with the fly genes not on their function. If mammalian Hox genes are functionally equivalent to the fly genes, then mutations in the mammalian genes would be expected to produce homeotic transformations along the body axis. Gene-knockout and transgenic technology have been used to assess the functional role of several Hox genes in controlling regional identity in the mouse. The results of these studies support the view that the Hox genes in mammals and flies play qualitatively similar roles in controlling regional identity along the anteroposterior axis.

Gain-of-Function Mutations in *Hoxd-4* Gene The Hoxd-4 protein is a homolog of the *Drosophila* Dfd protein, which is encoded within the ANT-C (see Figure 14-38). The anterior border of Hoxd-4 expression in the mouse embryo includes

precursor cells that normally give rise to cervical vertebrae. Ectopic expression of Hoxd-4 in a domain anterior to this region (in a region giving rise to occipital vertebrae) was achieved by fusing *Hoxd-4* cDNA to the regulatory sequences that control transcription of *Hoxa-1*, which normally is expressed more anteriorly. In newborn transgenic mice carrying this construct, which has the effect of a gain-of-function mutation, anterior (occipital) vertebrae are morphologically similar to more posterior (cervical) vertebrae (Figure 14-40).

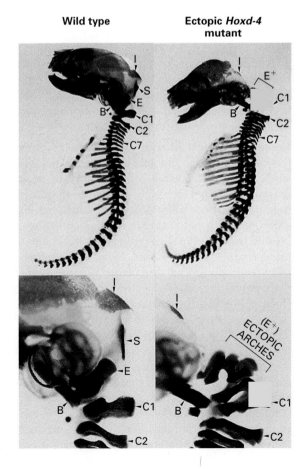

► **FIGURE 14-40 Effect of ectopic expression of Hoxd-4 protein anterior to its usual expression domain on development of anterior structures in mouse embryos.** The photographs show skeletal preparations of a normal newborn mouse and a transgenic newborn that misexpresses Hoxd-4; the lower photographs are higher magnifications of the upper ones. The transgenic mouse exhibits a reduction or loss of the supraoccipital (S) and extraoccipital (E) bones of the skull and the appearance of bony structures called ectopic arches (E⁺), which are fused to the basioccipital (B) bone; these structures appear similar to neural arches characteristic of more posterior vertebrae. The cervical vertebrae are indicated by C followed by a number. [From T. Lufkin et al., 1992, *Nature* **359**:835.]

▶ **FIGURE 14-41 Effect of loss-of-function mutations in *Hox-4* paralogs on development of cervical vertebrae in mice.** Double and triple mutants were obtained by interbreeding mice carrying a knockout mutation in a single paralog. (a) Expression domains of the Hox-4 paralogous proteins in prevertebrae with respect to the cervical vertebrae (C1–C7). Gray indicates lower expression than black. (b) Schematic diagrams showing that sequential removal of *Hox-4* paralogs leads to progressive posterior → anterior transformations. In embryos with homozygous mutations in *Hoxb-4* or *Hoxd-4*, C2 is transformed into C1; in mutants lacking both these paralogs, C2 and C3 are transformed into C1. In triple mutants lacking functional *Hoxb-4, Hoxd-4,* and *Hoxa-4,* C2–C5 are transformed into C1. In double and triple mutants abnormalities are seen in the structure of C1. These are indicated by the shorter bar. (c) Skeletal structures of a wild-type mouse embryo and a triple mutant homozygous for loss-of-function mutations in the *Hox-4a, Hox-4b,* and *Hox-4d* paralogs. The mutant has multiple anterior arch-like structure (arrows) that extend more posterior than normal. aaa = anterior arch of the atlas; ex = exoccipital bone. [From G. S. B. Horan et al., 1995, Genes & Devel. **9**:1667.]

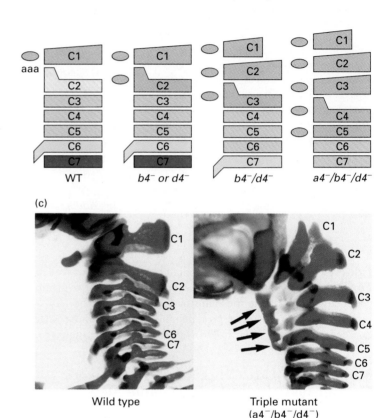

(a) Wild-type expression of *Hox-4* paralogs

(b) Transformations due to loss of *Hox-4* paralogs

(c) Wild type   Triple mutant (a4⁻/b4⁻/d4⁻)

This homeotic transformation is similar to those resulting from ectopic expression of *Drosophila* Hox genes discussed previously; that is, ectopic expression of a posteriorly localized Hox gene product in more anterior regions leads to an anterior → posterior transformation.

Loss-of-Function Mutations in *Hox-4* Genes Assessing the function of mammalian Hox genes from loss-of-function studies is complicated by the presence of multiple copies (paralogs) of these genes. For instance, to adequately assess the individual contributions of the four *Hox-4* paralogs to anteroposterior patterning in mice, scientists generated double and triple mutants carrying various combinations of mutated *Hoxa-4, Hoxb-4,* and *Hoxd-4.* The *Hox-4* paralogs have overlapping but not identical expression domains in the cervical region of the prevertebrae of mouse embryos (Figure 14-41a). By analyzing the morphologic characteristics of the cervical vertebrae in various mutants, researchers have found that vertebrae assume a progressively more anterior morphology as the dosage of *hox-4* paralogs decreases in a particular prevertebral region (Figure 14-41b,c). Therefore, as in flies, Hox genes repress the activity of other, more anteriorly expressed Hox genes. In mouse, this phenomenon is referred to as *posterior prevalence.* Careful analyses of the transformations seen in different single and double mutants demonstrate that different paralogs control overlapping, but

distinct, patterning functions; that is, development of some morphologic features requires two paralogs, whereas development of others requires only a single paralog.

Genetic studies also suggest that, in addition to interactions between paralogous genes, there are similarities in the phenotypes of knockouts of some nonparalogous genes. Hence, Hox genes from different paralogous groups act in a combinatorial fashion to specify different axial patterns. As in flies, the mechanism by which these genes coordinate cellular patterning by controlling the expression of specific target genes remains largely unknown.

SUMMARY Anteroposterior Specification during Embryogenesis

- Gradients of transcription factors are generated in the early *Drosophila* embryo through translation of spatially restricted maternal mRNAs and subsequent diffusion of the encoded proteins through the common cytoplasm of the syncytial blastoderm. These proteins, in turn, control the patterned expression of specific target genes along the anteroposterior axis.

- Different target genes contain different combinations of transcription-control sequences that bind specific transcriptional activators and repressors. The expression patterns characteristic of different target genes can be controlled by particular combinations of protein-binding sites in cis-acting control regions, as well as by differences in the number and affinity of DNA-binding sites for the same transcription factor (see Figure 14-31).

• Early-patterning events, utilizing maternal, gap, pair-rule, and segment-polarity genes, generate a unique pattern of transcription factors expressed in different regions along the anteroposterior axis of *Drosophila* embryos (see Figure 14-32). These transcription factors are expressed transiently and play an essential role in establishing the domains in which different selector genes are expressed.

• Selector genes, which are expressed and required continuously throughout development, direct further development of tissues to form the structures and organs characteristic of each part of the body. The Hox selector genes control the unique morphologic characteristics of different regions along the anteroposterior axis of the major phyla, including arthropods (e.g., insects) and chordates (e.g., mammals).

• Misexpression of Hox genes causes homeotic transformation—the development of body parts in abnormal positions.

• The specificity of Hox proteins is controlled in part by the formation of heterodimers between individual Hox proteins and Exd protein in *Drosophila* or Pbx in mammals (see Figure 14-36). Since different Hox-Exd dimers have different DNA-binding specificities, they control expression of different sets of genes.

• Maintenance of the Hox expression patterns occurs through positive autoregulatory loops involving the Hox genes themselves, and through modulation of chromatin by proteins encoded by both *polycomb*-group and *trithorax*-group genes.

14.4 Specification of Floral-Organ Identity in *Arabidopsis*

Important strides have been made in dissecting the mechanisms controlling the development of plants. **PLANTS** These advances have been possible largely due to the choice of *Arabidopsis thaliana* as a model plant. This plant has many of the same advantages as flies and worms for use as a model system. First, *Arabidopsis* is small and easy to maintain in the laboratory. Second, mutants can be easily produced because *Arabidopsis* has a short generation time (8 weeks) and seeds can be mutagenized by treatment with chemicals or radiation. In addition, the small size of the *Arabidopsis* genome facilitates positional cloning methods to isolate the genes defined by mutations (Chapter 8). And finally, transgenic *Arabidopsis* plants are readily made (see Figure 8-38). Although many different aspects of development are being investigated in this plant, our discussion will focus on the transcription-control mechanisms regulating the formation of flowers. As we will see, these mechanisms are strikingly similar to those controlling cell-type and anteroposterior regional specification in yeast and animals.

Flowers Contain Four Different Organs

Flowers comprise four different organs called sepals, petals, stamens and carpels, arranged in concentric circles called whorls. The number and type of floral organs and the number of whorls vary among plant species. *Arabidopsis* has a complete set of floral organs, including four sepals in whorl 1, four petals in whorl 2, six stamens in whorl 3, and two carpels containing ovaries in whorl 4 (Figure 14-42a). These organs form from a collection of undifferentiated, morphologically indistinguishable cells called the *floral meristem*. As cells within the center of the floral meristem divide, four concentric rings of primordia form sequentially. The outer ring primordia, which give rise to the sepals of whorl 1,

(a)

(b)

◀ **FIGURE 14-42 (a) Flowers of wild-type *Arabidopsis thaliana* have four sepals in whorl 1, four petals in whorl 2, six stamens in whorl 3, and two carpels in whorl 4.** (b) In *Arabidopsis* with mutations in all three classes of floral organ–identity genes, the four floral organs are transformed into leaf-like structures. [From D. Weigel and E. M. Meyerowitz, 1994, *Cell* **78**:203. Courtesy of E. M. Meyerowitz.]

form first, followed by the primordia giving rise to the petals, then the stamens and carpels primordia. Genetic studies have shown that normal flower development requires three classes of *floral organ–identity genes*. In plants with mutations in all three classes, concentric whorls of leaf-like structures replace the floral organs, indicating that these structures are modified leaves (Figure 14-42b).

Three Classes of Genes Control Floral-Organ Identity

Analyses of mutations causing transformation of one floral organ into another led to identification of floral organ–identity genes. These homeotic mutations are equivalent to the homeotic mutations in flies and mammals in which one part of the body is replaced by another. In these studies, researchers mutagenized *Arabidopsis* seeds. Each seed contains two cells, each of which contributes to all the tissues in the adult plant. In a mutagenized seed, mutations are randomly dispersed in the genome of each seed. Thus, after germinating, a seed gives rise to a mosaic plant. Self-fertilization generates plants in which one-quarter are homozygous for the induced mutations. Mutant flowers in which transformation in floral-organ identity has taken place are easily recognized.

As illustrated in Figure 14-43, three classes of loss-of-function mutations (A, B, and C) cause floral-organ transformations in *Arabidopsis*. In plants lacking all A, B, and C function, the floral organs develop as leaves. Based on the phenotypes of these loss-of-function mutants, scientists have proposed a model to explain how three classes of genes control floral-organ identity. According to this ABC model, class A genes specify sepal identity in whorl 1 and do not require either class B or C genes to do so. Similarly, class C genes

specify carpel identity in whorl 4 and, again, do so independently of class B and C genes. In contrast to these structures, which are specified by only a single class of genes, the petals in whorl 2 are specified by class A and B genes, and the stamens in whorl 3 are specified by class B and C genes. To account for the observed effects of removing A genes or C genes, the model also postulates that A genes repress C in whorls 1 and 2 and, conversely, C genes repress A genes in whorls 3 and 4.

To determine if the actual expression patterns of class A, B, and C genes are consistent with this model, researchers cloned these genes and assessed the expression patterns of their mRNAs in the four whorls in wild-type *Arabidopsis* plants and in loss-of-function mutants (Figure 14-44a,b). Consistent with the ABC model, A genes are expressed in whorls 1 and 2, B genes in whorls 2 and 3, and C genes in whorls 3 and 4. Furthermore, in class A mutants, C class genes are also expressed in organ primordia of whorls 1 and 2; similarly, in class C mutants, class A genes are also expressed in whorls 3 and 4. These findings are consistent with the homeotic transformations observed in these mutants.

To test whether these patterns of expression are functionally important, scientists produced transgenic *Arabidopsis* plants in which floral organ–identity genes were expressed in inappropriate whorls. For instance, introduction of a transgene carrying class B genes linked to an A-class promoter leads to ubiquitous expression of class B genes in all whorls (Figure 14-44c). In such transgenics, whorl 1, now under the control of class A and B genes, develops into petals instead of sepals; likewise, whorl 4, under the control of both class B and C genes, gives rise to stamens instead of carpels. These results support the functional importance of the ABC model for specifying floral identity.

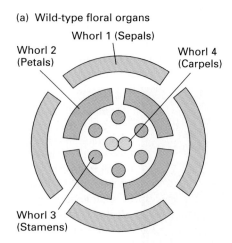

(a) Wild-type floral organs

Whorl 1 (Sepals)
Whorl 2 (Petals)
Whorl 4 (Carpels)
Whorl 3 (Stamens)

(b) Loss-of-function homeotic mutations

	Whorl			
	1	2	3	4
Wild type	▢	▢	▢	▢
Class A mutants	▢	▢	▢	▢
Class B mutants	▢	▢	▢	▢
Class C mutants	▢	▢	▢	▢

▲ **FIGURE 14-43 Identification of three classes of genes that control specification of floral organs in *Arabidopsis*.** (a) Schematic diagram of the arrangement of wild-type floral organs, which are found in concentric circles called whorls. (b) Effect of loss-of-function mutations leading to transformations of one organ into another. Class A mutations affect organ identity in whorls 1 and 2: sepals (green) become carpels (blue) and petals (orange) become stamens (pink). Class B mutations cause transformation of whorls 2 and 3: petals become sepals and stamens become carpels. In class C mutations, whorls 3 and 4 are transformed: stamens become petals and carpels become sepals. [See D. Wiegel and E. M. Meyerowitz, 1994, *Cell* **78**:203.]

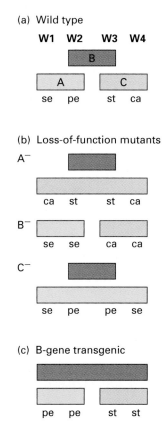

(a) Wild type

W1   W2   W3   W4

B

A   C

se   pe   st   ca

(b) Loss-of-function mutants

A⁻

ca   st   st   ca

B⁻

se   se   ca   ca

C⁻

se   pe   pe   se

(c) B-gene transgenic

pe   pe   st   st

▲ FIGURE 14-44 Expression patterns for three classes of floral organ–identity genes in wild-type *Arabidopsis* (a), loss-of-function mutants (b), and a transgenic that misexpresses class B genes (c). Colored bars represent the A, B, and C mRNAs in each whorl (W1, W2, W3, W4). The observed floral organ in each whorl is indicated below: sepal = se; petals = pe; stamens = st; and carpels = ca. See text for discussion. [See D. Wiegel and E. M. Meyerowitz, 1994, *Cell* **78**:203; B. A. Krizek and E. M. Meyerowitz, 1996, *Development* **122**:11.]

Many Floral Organ—Identity Genes Encode MADS Family Transcription Factors

Sequencing of floral organ–identity genes has revealed that many encode proteins belonging to the MADS family of transcription factors. As discussed in earlier sections, members of this class of proteins, including MCM1 in yeast and MEFs in muscle, form homo- and heterodimers. Thus floral-organ identity may be specified by a combinatorial mechanism in which differences in the activities of different homo-and heterodimeric forms of various A, B, and C proteins regulate the expression of subordinate downstream genes necessary for the formation of the different cell types in each organ. Alternatively, these proteins may act solely as homodimers with organ identity reflecting the activities of different combinations of genes regulated by these homodimers in different organ primordia.

Biochemical experiments have not yet resolved the mechanisms by which organ diversity is regulated by floral organ–

identity genes. The proteins encoded by two class B genes—*pistillata* and *apetala3*—have been shown to form heterodimers that bind to specific DNA sequences. However, neither the Pistillata nor Apetala3 protein forms DNA-binding heterodimers with class A gene products (e.g., Apetela1) or class C gene products (e.g., Agamous). Hence, the differences between the identity of whorls 2 and 3 (giving rise to petals and stamens, respectively) does not appear to represent the formation of different heterodimeric combinations of the proteins encoded by the three classes of mutation-defined floral organ–identity genes.

SUMMARY   **Specification of Floral-Organ Identity in *Arabidopsis***

- Flowers have four different organs—petals, sepals, stamens and carpels—which develop in concentric whorls. These organs are modified leaves.

- Genetic studies have identified three classes of genes that participate in specifying floral-organ identity. The ABC model proposes how these classes of genes work together to specify organ identity in each whorl.

- Class A, B, and C genes encode transcription factors of the MADS family. The patterns of expression of these gene products are consistent with the ABC genetic model (see Figure 14-44). However, the mechanisms by which these proteins regulate development of the cell types composing each organ remain to be elucidated.

PERSPECTIVES   for the Future

During the past two decades remarkable progress has been made in understanding the mechanisms regulating gene transcription during development. Through a combination of genetic and biochemical studies, the DNA sequences regulating expression of many different genes in simple and higher eukaryotes and the transcription factors that bind to them have been identified. Dissection of developmental programs in various organisms has shown that they depend not only on combinations of transcription factors expressed by specific cells but also on interaction between these cells and extracellular signals acting through membrane receptors or through their direct interaction with transcription factors. Continuation of such studies will provide additional insight into the way specific genes are regulated in different places and times in development and the strategies by which the organism utilizes the genome to specify the large number of different cells and tissues in an organism.

Two important developments are likely to change radically the nature and extent of our knowledge of transcriptional regulatory programs controlling development. First, sequencing of the entire genomes of several higher eukaryotes, including *Drosophila*, the mouse, and *Arabidopsis*, most likely will be completed within a few years. Indeed, the sequence of *C. elegans* genome has been completed already. The entire sequence of the human genome should be determined within the next 5 years. Second, new technology will provide remarkably efficient ways to monitor expression of many genes at one time, indeed in principle, the entire transcriptional program of an individual cell.

The prototype for this level of analysis comes from studies in the yeast *S. cerevisiae,* whose genome has been sequenced in its entirety (Chapter 7). Monitoring of the transcriptional program of a vast number of genes as a function of cell type and physiologic state has already been accomplished using DNA microarray technology (Chapter 7). In this approach, robotics technology is used to attach an array of thousands of different DNA sequences to a microscope slide. cDNAs from small numbers of cells synthesized in the presence of fluorescent nucleotides then are hybridized to the DNA sequences attached to the microscope slide. The amount of fluorescent-labeled DNA bound to each sequence attached to the slide is read using a confocal scanning microscope. Gene expression in yeasts of different genotypes, of different mating types, or under different physiologic conditions can be compared by labeling the cDNAs from different populations of cells with different fluorescent tags. The ratio of these probes bound to specific sequences in the microarray will reveal differences in expression at the same loci under different conditions. The microarray technology is rapidly advancing and is already being used to assess gene expression in higher eukaryotes.

Where will these studies lead? In the not-too-distant future, researchers will be able to document the expression of perhaps all the genes in a given cell type, the putative regulatory sequences controlling their expression, and the complete set of DNA-binding transcription factors that the cell expresses. This new knowledge will provide starting points for the biochemical and genetic dissection of many hitherto unstudied transcriptional regulatory regions. It will also provide an enormous data base for eventually deciphering the genetic strategy used by an organism as a whole to generate the many different cell types characteristic of higher eukaryotes.

PERSPECTIVES in the Literature

You are working with a well-characterized in vitro system that allows you to induce myoblasts to differentiate into myotubes synchronously. Your long term goal is to describe the transcriptional regulatory networks that control muscle differentiation. An important step in your studies is to describe patterns of gene expression. You are interested in studying which genes are turned on and off and in what order during the transition from a myoblast to a differentiated myotube. Design a set of experiments that would allow you to carry out these studies. To get started, see the following references:

Brown, P. O., and D. Botstein. 1999. Exploring the new world of the genome with DNA microarrays. *Nature Genetics* (1 Suppl):33–37.

Chu, S., et al. 1998. The transcriptional program of sporulation in budding yeast. *Science* 282:699–705.

DeRisi, J., V. R. Iyer, and P. O. Brown. 1997. Exploring the metabolic and genetic control of gene expression on a genomic scale. *Science* 278:680–686.

Testing Yourself on the Concepts

1. To determine the molecular and cellular basis of patterning mechanisms, what strategies do investigators take in model organisms and in humans?

2. Describe two specific examples of syncytial cells and how transcription and translation during development are affected by these structures.

3. Describe three examples of MADS box transcription factors and their roles in development.

4. Predict the phenotypes of double mutants in *Arabidopsis* that cause a loss of (a) A and B function; (b) B and C function; and (c) A and C function.

5. Describe two specific examples of the effects of chromatin on the transcriptional control of development.

6. Discuss how inhibitory proteins such as Id and Emc function and affect development.

MCAT/GRE-Style Questions

Key Concept Please read the section titled "Multiple Regulation of *HO* Transcription Controls Mating-Type Conversion" (p. 541) and refer to Figure 14-6; then answer the following questions:

1. The fact that Cdc28–G_1 cyclin complexes are required for CCBF activity implies that
   a. CCBF phosphorylation by the Cdc28–G_1 cyclin complex activates *HO* transcription.
   b. CCBF dephosphorylation by the Cdc28–G_1 cyclin complex activates *HO* transcription.
   c. CCBF phosphorylation by the Cdc28–G_1 cyclin complex induces binding of swi6 binding to the URS2.
   d. CCBF dephosphorylation by the Cdc28–G_1 cyclin complex induces binding of swi6 binding to the URS2.

2. Ash1 accumulation in daughter cells
   a. Requires inhibition of swi5.
   b. Requires expression of other yeast genes.
   c. Inhibits swi5.
   d. Degrades swi5 protein.

3. What is the phenotype of a *swi1* mutant?

   a. It is defective in mating.

   b. It is defective in the mating-type switch.

   c. The mating-type switch occurs in both progeny.

   d. Lethality.

4. What is the phenotype of α2 or a1 mutants?

   a. They are defective in mating.

   b. They are defective in the mating-type switch.

   c. The mating-type switch occurs in both progeny.

   d. Lethality.

5. A promoter-reporter gene construct carrying the URS1 and URS2 sequences would be expressed in

   a. Mother cells only.

   b. Daughter cells only.

   c. Both mother and daughter cells.

   d. Yeast not undergoing mating-type switching.

Key Experiment Please read the section titled "Maternal *bicoid* Gene Specifies Anterior Region in *Drosophila*" (p. 557) and refer to Figure 14-26; then answer the following questions:

6. One maternal copy of a mutated *bicoid* gene containing a deletion of the 3′ untranslated end would result in an embryo with *hunchback* transcribed

   a. At the anterior end only.

   b. Throughout the embryo.

   c. At the posterior end only.

   d. Not at all.

7. What is the resulting phenotype of the embryo in Figure 14-26d?

   a. The boundary of anterior structures is increased.

   b. The boundary of posterior structures is increased.

   c. Loss of anterior structures.

   d. None.

8. The Bicoid protein gradient as compared to the *bicoid* RNA gradient

   a. Is confined to a smaller area of the embryo.

   b. Extends more posteriorly.

   c. Is less dense.

   d. Is less stable.

9. The boundary of *hunchback* expression shifts posteriorly in an embryo from a mother with two copies of *bicoid* compared to an embryo produced by a mother with one copy of *bicoid* because

   a. More *bicoid* RNA diffuses throughout the embryo.

   b. More Bicoid protein diffuses throughout the embryo.

   c. More Hunchback protein diffuses throughout the embryo.

   d. More high-affinity sites are present in the *hunchback* promoter.

10. If multiple copies of the synthetic high-affinity promoter-reporter gene construct were placed in an embryo produced by a *bicoid* mutant mother, what would the expression of the reporter gene look like?

    a. A small amount localized to the anterior end.

    b. A large amount localized to the anterior end.

    c. A large amount extended more posteriorly.

    d. No expression.

Key Application A prenatal diagnosis strategy for detecting muscular diseases that involves testing the capability of embryonic cells to form muscle after inducing MyoD expression in vitro has been suggested. Please read the sections titled "Myogenic Genes Were First Identified in Studies with Cultured Fibroblasts" (p. 544) and "Myogenic Proteins Are Transcription Factors Containing a Common bHLH Domain" (p. 546) and answer the following questions:

11. Normal embryonic cells from humans would likely form myoblasts in culture after induction of MyoD expression because

    a. MyoD is known to convert fibroblasts to muscle in vitro.

    b. Embryonic cells are similar in nature to azamyoblasts.

    c. The B-cell development pathway cannot be stimulated in these cells.

    d. E2A is not expressed in these cells.

12. Promoters containing E boxes

    a. Are only found in muscle-specific genes.

    b. Are found upstream of the promoters of many genes not expressed in muscle.

    c. Are always occupied by E2A/MyoD heterodimers.

    d. Always confer expression in muscle cells.

13. Myogenin, Myf5, Mrf4, and MyoD

    a. Are functionally interchangeable in vivo.

    b. Dimerize with E2A and bind E boxes.

    c. Are expressed at the same time during development.

    d. Bind to DNA as heterodimers with Id.

14. For a fibroblast to form myotubes in culture, which of the following must be present:

    a. MyoD.

    b. E2A.

    c. MyoD and E2A.

    d. An MRF and E2A.

Key Terms

References

Cell-Type Specification and Mating-Type Conversion in Yeast

Bobola, N., et al. 1996. Asymmetric accumulation of Ash1p in postanaphase nuclei depends on a myosin and restricts yeast mating-type switching to mother cells. *Cell* **84**:699–709.

Herskowitz, I., J. Rine, and J. Strathern. 1992. Mating-type determination and mating-type interconversion in *Saccharomyces cerevisiae*. In *The Molecular and Cellular Biology of the Yeast Saccharomyces: Gene Expression*, Vol. II. Cold Spring Harbor Laboratory Press.

Keleher, C. A., et al. 1992. Ssn6-Tup1 is a general repressor of transcription in yeast. *Cell* **68**:709–719.

Li, T., et al. 1995. Crystal structure of the MATa1/MATα2 homeodomain heterodimer bound to DNA. *Science* **270**:262–269.

MacKay, V., and T. R. Manney. 1974. Mutations affecting sexual conjugation and related processes in *Saccharomyces cerevisiae*. I. Isolation and phenotypic characterization of nonmating mutants. *Genetics* **76**:255–271.

Sil, A. and I. Herskowitz. 1996. Identification of an asymmetrically localized determinant, Ash1p, required for lineage-specific transcription of the yeast *HO* gene. *Cell* **84**:711–722.

Smith, D. L., and A. D. Johnson. 1992. A molecular mechanism for combinatorial control in yeast: MCM1 protein sets the spacing and orientation of the homeodomains of an alpha 2 dimer. *Cell* **68**:133–142.

Song, D., et al. 1991. Pheromone-dependent phosphorylation of the yeast STE12 protein correlates with transcriptional activation. *Genes & Devel.* **5**:741–750.

Takizawa, P. A., et al. 1997. Actin-dependent localization of an RNA encoding a cell-fate determinant in yeast. *Nature* **389**:90–93.

Vershon, A. K. and A. D. Johnson. 1993. A short, disordered protein region mediates interactions between the homeodomain of the yeast alpha 2 protein and the MCM1 protein. *Cell* **72**:105–112.

Wolberger, C., et al. 1991. Crystal structure of a MAT alpha 2 homeodomain-operator complex suggests a general model for homeodomain-DNA interactions. *Cell* **67**:517–528.

Cell-Type Specification in Animals

Benezra, R., et al. 1990. The protein Id: a negative regulator of helix-loop-helix DNA-binding proteins. *Cell* **61**:49–59.

Bour, B. A., et al. 1995. Drosophila MEF2, a transcription factor that is essential for myogenesis. *Genes & Devel.* **9**:730–741.

Davis, R. L., H. Weintraub, and A. B. Lassar. 1987. Expression of a single transfected cDNA converts fibroblasts to myoblasts. *Cell* **51**:987–1000.

Ellis, H. M., D. R. Spann, and J. W. Posakony. 1990. Extramacrochaetae, a negative regulator of sensory organ development in *Drosophila*, defines a new class of helix-loop-helix proteins. *Cell* **61**:27–38.

Lee, J. L., et al. 1995. Conversion of *Xenopus* ectoderm into neurons by NeuroD, a basic helix-loop-helix protein. *Science* **268**:836–844.

Ma, Q., C. Kintner, and D. J. Anderson. 1996. Identification of neurogenin, a vertebrate neuronal determination gene. *Cell* **87**:43–52.

Mead, P. E., et al. 1998. SCL specifies hematopoietic mesoderm in *Xenopus* embryos. *Development* **125**:2611–2620.

Molkentin, J. D., et al. 1995. Cooperative activation of muscle gene expression by MEF2 and myogenic bHLH proteins. *Cell* **83**:1125–1136.

Nabeshima, Y., et al. 1993. Myogenin gene disruption results in perinatal lethality because of severe muscle defect. *Nature* **364**:532–535.

Rawls, A., and E. N. Olson. 1997. MyoD meets its maker. *Cell* **89**:5–8.

Rudnicki, M. A., et al. 1993. MyoD or Myf-5 is required for the formation of skeletal muscle. *Cell* **75**:1351–1359.

Skeath, J. B., and S. B. Carroll. 1991. Regulation of *achaete-scute* gene expression and sensory organ pattern formation in the *Drosophila* wing. *Genes & Devel.* **5**:984–995.

Wang, Y., et al. 1996. Functional redundancy of the muscle-specific transcription factors Myf5 and myogenin. *Nature* **379**:823–825.

Yun, K., and B. Wold. 1996. Skeletal muscle determination and differentiation: story of a core regulatory network and its context. *Curr. Opin. Cell Biol.* **8**:877–889.

Anteroposterior Specification during Embryogenesis

Capecchi, M. R. 1997. The role of Hox genes in hindbrain development. In W. M. Cowan, T. Jessel, and S. L. Zipursky, eds., *Molecular and Cellular Approaches to Neural Development*, Oxford University Press.

Castelli-Gair, J., and M. Akam. 1995. How the Hox gene *Ultrabithorax* specifies two different segments: the significance of spatial and temporal regulation within metameres. *Development* **121**:2973–2982.

Chan, S-K., et al. 1994. The DNA-binding specificity of Ultrabithorax is modulated by cooperative interactions with Extradenticle, another homeoprotein. *Cell* **78**:603–615.

Driever, W., and C. Nüsslein-Volhard. 1988. A gradient of bicoid protein in *Drosophila* embryos. *Cell* **54**:83–93.

Driever, W., and C. Nüsslein-Volhard. 1988. The bicoid protein determines position in the *Drosophila* embryo in a concentration-dependent manner. *Cell* **54**:95–104.

Dubnau, J., and G. Struhl. 1996. RNA recognition and translational regulation by a homeodomain protein. *Nature* **379**:694–699.

Gavis, E. R., and R. Lehmann. 1992. Localization of nanos RNA controls embryonic polarity. *Cell* **71**:301–313.

Lawrence, P. A. 1992. *The Making of the Fly: The Genetics of Animal Design*. Blackwell Scientific Publications.

Lufkin, T. 1996. Transcriptional control of Hox genes in the vertebrate nervous system. *Curr. Opin. Genet. Devel.* **6**:575–580.

Murata, Y., and R. P. Wharton. 1995. Binding of Pumilio to maternal hunchback mRNA is required for posterior patterning in *Drosophila* embryos. *Cell* **80**:747–756.

Nüsslein-Volhard, C., and E. Wieschaus. 1980. Mutations affecting segment number and polarity in *Drosophila*. *Nature* **287**:795–801.

Orlando, V., and R. Paro. 1995. Chromatin multiprotein complexes involved in the maintenance of transcription patterns. *Curr. Opin. Genet. Devel.* **5**:174–179.

Peifer, M., F. Karch, and W. Bender. 1987. The bithorax complex: control of segmental identity. *Genes & Devel.* **1**:891–898.

Rivera-Pomar, R., et al. 1995. Activation of posterior gap gene expression in the *Drosophila* blastoderm. *Nature* **376**:253–256.

Small, S., et al. 1991. Transcriptional regulation of a pair-rule stripe in *Drosophila*. *Genes & Devel.* 5:827–839.

St. Johnston, D., and C. Nüsslein-Volhard. 1992. The origin of pattern and polarity in the *Drosophila* embryo. *Cell* 68:201–219.

Stanojevic, D., S. Small, and M. Levine. 1991. Regulation of a segmentation stripe by overlapping activators and repressors in the *Drosophila* embryo. *Science* 254:1385–1387.

Struhl, G., P. Johnston, and P. A. Lawrence. 1992. Control of *Drosophila* body pattern by the hunchback morphogen gradient. *Cell* 69:237–249.

Wreden, C., et al. 1997. Nanos and Pumilio establish embryonic polarity in *Drosophila* by promoting posterior deadenylation of hunchback mRNA. *Development* 124:3015–3023.

Specification of Floral-Organ Identity in *Arabidopsis*

Fosket, D. E. 1994. *Plant Growth and Development: A Molecular Approach.* Academic Press.

Krizek, B. A., and E. M. Meyerowitz. 1996. Mapping the protein regions responsible for the functional specificities of the *Arabidopsis* MADS domain organ–identity proteins. *Proc. Nat'l. Acad. Sci. USA* 93:4063–4070.

Krizek, B. A., and E. M. Meyerowitz. 1996. The *Arabidopsis* homeotic genes APETALA3 and PISTELLATA are sufficient to provide the B class organ identity function. *Development* 122:11–22.

Reichman, J. L., and E. M. Meyerowitz. 1997. MADs domain proteins in plant development. *Biol. Chem.* 378:1079–1101.

Sablowski, R. W., and E. M. Meyerowitz. 1998. A homolog of NO APICAL MERISTEM is an immediate target of the floral homeotic genes APETALA3/PISTILLATA. *Cell* 92:93–103.

Weigel, D. 1995. The genetics of flower development: from floral induction to ovule morphogenesis. Ann. Rev. Genet. 29:19–39.

Weigel, D., and E. M. Meyerowitz. 1994. The ABCs of floral homeotic genes. *Cell* 78:203–209.

Wolpert, W. 1998. *Principles of Development.* Oxford University Press (Oxford, Eng.).

15

Transport across Cell Membranes

The plasma membrane is a selectively permeable barrier between the cell and the extracellular environment. Its permeability properties ensure that essential molecules such as glucose, amino acids, and lipids readily enter the cell, metabolic intermediates remain in the cell, and waste compounds leave the cell. In short, the selective permeability of the plasma membrane allows the cell to maintain a constant internal environment. In several earlier chapters, we examined the components and structural organization of cell membranes (see Figures 3-32 and 5-30). The phospholipid bilayer—the basic structural unit of biomembranes—is essentially impermeable to most water-soluble molecules, such as glucose and amino acids, and to ions. Transport of such molecules and ions across all cellular membranes is mediated by *transport proteins* associated with the underlying bilayer. Because different cell types require different mixtures of low-molecular-weight compounds, the plasma membrane of each cell type contains a specific set of transport proteins that allow only certain ions or molecules to cross. Similarly, organelles within the cell often have a different internal environment from that of the surrounding cytosol, and organelle membranes contain specific transport proteins that maintain this difference.

In animals, sheets of epithelial cells line all the body cavities (e.g., the stomach, intestines, urinary bladder) and the skin (see Figure 6-4). Epithelial cells frequently transport ions or small molecules from one side to the other. Those lining the small intestine, for instance, transport products of digestion (e.g., glucose and amino acids)

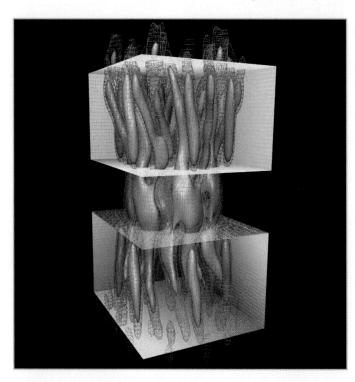

Three-dimensional structure of a gap junction membrane channel connecting two adjacent cells.

MEDIA CONNECTIONS

Overview: Biological Energy Interconversions

Classic Experiment 15.1: Stumbling upon Active Transport

into the blood, and those lining the stomach secrete hydrochloric acid into the stomach lumen. In order for epithelial cells to carry out these transport functions, their plasma membrane must be organized into at least two discrete regions, each with different sets of transport proteins. In addition, specialized regions of the plasma membrane interconnect epithelial cells, imparting strength and rigidity to the sheet and preventing material on one side from moving between the cells to the other.

In the first two sections of this chapter, we discuss the protein-independent movement of small hydrophobic molecules across phospholipid bilayers and present an overview of the various types of transport proteins present in cell membranes. We then describe each of the main types of transport proteins. We also explain how specific combinations of transport proteins in different subcellular membranes enable cells to carry out essential physiological processes, including the maintenance of cytosolic pH, the transport of glucose across the absorptive intestinal epithelium, the accumulation of sucrose and salts in plant-cell vacuoles, and the directed flow of water in both plants and animals. Often the same type of transport protein is involved in quite different physiological processes.

15.1 Diffusion of Small Molecules across Phospholipid Bilayers

An artificial membrane composed of pure phospholipid or of phospholipid and cholesterol is permeable to gases, such as O_2 and CO_2, and small, uncharged polar molecules, such as urea and ethanol (Figure 15-1). Such molecules also can cross cellular membranes by **passive diffusion** unaided by transport proteins. No metabolic energy is expended because movement is from a high to a low concentration of the molecule, down its chemical concentration gradient. As noted in Chapter 2, such transport reactions have a positive ΔS value (increase in entropy) and a negative ΔG (decrease in free energy). The relative diffusion rate of a substance across the bilayer is proportional to its concentration gradient across the layer and to its hydrophobicity. There is little specificity to the process, in that any small hydrophobic molecule will be transported.

The first step in transport by passive diffusion is movement of a molecule from the aqueous solution into the hydrophobic interior of the phospholipid bilayer. The hydrophobicity of a substance is measured by its *partition coefficient, K,* the equilibrium constant for its partition between oil and water. Since the composition of the interior of the phospholipid bilayer resembles that of oil, the partition coefficient of a substance moving across a bilayer equals the ratio of its concentration just inside the hydrophobic core of the bilayer C^m to its concentration in the aqueous solution C^{aq}:

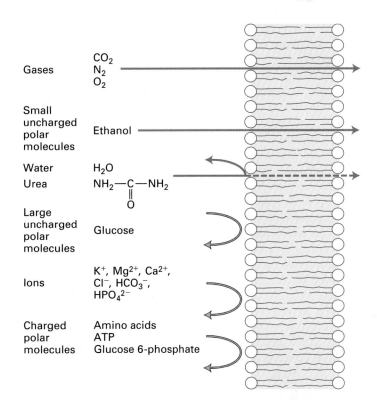

▲ **FIGURE 15-1 A pure artificial phospholipid bilayer is permeable to small hydrophobic molecules and small uncharged polar molecules.** It is slightly permeable to water and urea and impermeable to ions and to large uncharged polar molecules. When a small phospholipid bilayer separates two aqueous compartments, membrane permeability can be easily determined by adding a small amount of radioactive material to one compartment and measuring its rate of appearance in the other compartment.

$$K = \frac{C^m}{C^{aq}} \qquad (15\text{-}1)$$

The partition coefficient is a measure of the relative affinity of a substance for lipid versus water: the higher a substance's partition coefficient, the more lipid-soluble it is. For example, urea

$$\underset{\displaystyle \text{O}}{\overset{\displaystyle \text{O}}{NH_2 - \overset{\|}{C} - NH_2}}$$

has a K of 0.0002, whereas diethylurea (with two ethyl groups)

$$CH_3 - CH_2 - NH - \overset{\overset{\displaystyle O}{\|}}{C} - NH - CH_2 - CH_3$$

has a K of 0.01. Diethylurea, which is 50 times (0.01 ÷ 0.0002) more hydrophobic than urea, will diffuse through phospholipid bilayer membranes about 50 times faster than urea. Diethylurea also enters cells about 50 times faster than urea.

Once a molecule moves into the hydrophobic interior of a bilayer, it diffuses across it; finally, the molecule moves from the bilayer into the aqueous medium on the other side of the membrane. Because the hydrophobic core of a typical cell membrane is 100–1000 times more viscous than water, the diffusion rate of all substances across a phospholipid membrane is very much slower than the diffusion rate of the same molecule in water. Thus, movement across the hydrophobic portion of a membrane is the rate-limiting step in the passive diffusion of molecules across cell membranes.

Now let's consider the passive diffusion of small molecules through a membrane more quantitatively. Suppose a membrane of surface area A and thickness x separates two solutions of concentrations C_1^{aq} and C_2^{aq}, where $C_1^{aq} > C_2^{aq}$ (Figure 15-2). In this case, the diffusion rate dn/dt (in mol/s) is given by a modification of *Fick's law*, which states that the diffusion rate across the membrane is directly proportional to the *permeability coefficient P*, to the difference in solution concentrations $C_1^{aq} - C_2^{aq}$, and to the area A, or

$$\frac{dn}{dt} = PA(C_1^{aq} - C_2^{aq}) \tag{15-2}$$

For any molecule, the value of P, and thus its rate of passive diffusion, is proportional to its partition coefficient K:

$$P = \frac{KD}{x} \tag{15-3}$$

where D is the diffusion coefficient of the substance within the membrane and x is the membrane thickness. By substituting Equation 15-3 into 15-2, we obtain

$$\frac{dn}{dt} = A\frac{KD}{x}(C_1^{aq} - C_2^{aq})$$

Thus we can see that the rate of diffusion is proportional to both the partition coefficient and the diffusion constant and is inversely proportional to the membrane thickness. However, the thickness of the hydrophobic interior of all phospholipid bilayer membranes is approximately the same, about 2.5 to 3 nm, and the diffusion coefficient D is the same for most substances. Thus differences in the rate at which molecules passively diffuse across membranes depends largely on differences in their partition coefficients. The greater the hydrophobicity of a water-soluble molecule, the faster it diffuses across a phospholipid bilayer.

Gases and some small, uncharged molecules, such as ethanol and urea, enter and leave cells by passive diffusion across the plasma membrane. This transport is described by Fick's law. In the following sections, we will see how movement of other molecules and ions across cell membranes differs from simple diffusion.

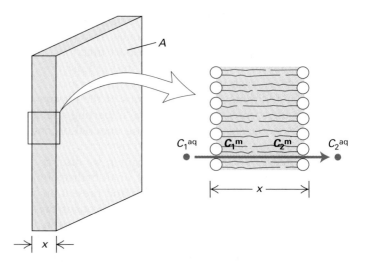

▲ **FIGURE 15-2 A simple model for passive diffusion of small hydrophobic molecules directly across the hydrocarbon core of a pure phospholipid bilayer of thickness x in centimeters and area A in square centimeters.** C_1^{aq} and C_2^{aq} are the concentrations of two solutions on sides 1 and 2 of the membrane; C_1^{m} and C_2^{m} are the corresponding concentrations just within the hydrocarbon core of the bilayer. Movement of a solute molecule is indicated by the blue arrow.

15.2 Overview of Membrane Transport Proteins

Very few molecules enter or leave cells, or cross organellar membranes, unaided by proteins. Even transport of molecules, such as water and urea, that can diffuse across pure phospholipid bilayers is frequently accelerated by transport proteins. The three major classes of membrane transport proteins are depicted in Figure 15-3a. All are integral transmembrane proteins and exhibit a high degree of specificity for the substance transported. The rate of transport by the three types differs considerably owing to differences in their mechanism of action.

ATP-powered pumps (or simply *pumps*) are ATPases that use the energy of ATP hydrolysis to move ions or small molecules across a membrane *against* a chemical concentration gradient or electric potential. This process, referred to as **active transport**, is an example of a coupled chemical reaction (Chapter 2). In this case, transport of ions or small molecules "uphill" against a concentration gradient or electric potential across a membrane, which requires energy, is coupled to the hydrolysis of ATP to ADP and P_i, which releases energy. The overall reaction—ATP hydrolysis and the "uphill" movement of ions or small molecules—is energetically favorable. Such pumps maintain the low calcium (Ca^{2+}) and sodium (Na^+) ion concentrations inside virtually all animal cells relative to that in the medium, and generate the low pH inside animal-cell lysosomes, plant-cell vacuoles, and the lumen of the stomach.

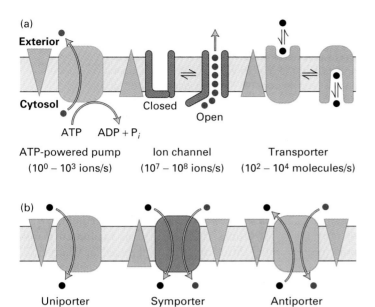

(a)

Exterior

Cytosol

Closed   Open

ATP → ADP + P$_i$

ATP-powered pump
($10^0 - 10^3$ ions/s)

Ion channel
($10^7 - 10^8$ ions/s)

Transporter
($10^2 - 10^4$ molecules/s)

(b)

Uniporter   Symporter   Antiporter

▲ **FIGURE 15-3 Schematic diagrams illustrating action of membrane transport proteins.** Gradients are indicated by triangles with the tip pointing toward lower concentration, electrical potential, or both. (a) The three major types of transport proteins. Pumps utilize the energy released by ATP hydrolysis to power movement of specific ions (red circles) or small molecules against their electrochemical gradient. Channels catalyze movement of specific ions (or water) down their electrochemical gradient. Transporters, which fall into three groups, facilitate movement of specific small molecules or ions (black circles). (b) The three groups of transporters. Uniporters, also shown in part (a), transport a single type of molecule down its concentration gradient. Cotransport proteins (symporters and antiporters) catalyze the movement of one molecule *against* its concentration gradient (black circles), driven by movement of one or more ions down an electrochemical gradient (red circles). The two types of cotransporters differ in the relative direction of movement of the transported molecule and cotransported ion.

Channel proteins transport water or specific types of ions down their concentration or electric potential gradients, an energetically favorable reaction. They form a protein-lined passageway across the membrane through which multiple water molecules or ions move simultaneously, single file at a very rapid rate—up to 10^8 per second. As discussed in a later section, the plasma membrane of all animal cells contains potassium-specific channel proteins that are generally open and are critical to generating the normal, resting electric potential across the plasma membrane. Many other types of channel proteins are usually closed, and open only in response to specific signals. Because these types of ion channels play a fundamental role in the functioning of nerve cells, they will be discussed in detail in Chapter 21.

Transporters, a third class of membrane transport proteins, move a wide variety of ions and molecules across cell membranes. In contrast to channel proteins, transporters bind only one (or a few) substrate molecules at a time; after binding substrate molecules, the transporter undergoes a conformational change such that the bound substrate molecules, and only these molecules, are transported across the membrane. Because movement of each substrate molecule (or small number of molecules) requires a conformational change in the transporter, transporters move only about 10^2-10^4 molecules per second, a lower rate than that associated with channel proteins.

Three types of transporters have been identified (Figure 15-3b). *Uniporters* transport one molecule at a time down a concentration gradient. This type of transporter, for example, moves glucose or amino acids across the plasma membrane into mammalian cells. In contrast, *antiporters* and *symporters* couple the movement of one type of ion or molecule *against* its concentration gradient to the movement of a different ion or molecule *down* its concentration gradient. Like ATP pumps, antiporters and symporters mediate coupled reactions in which an energetically unfavorable reaction is coupled to an energetically favorable reaction. Because symporters and antiporters catalyze "uphill" movement of certain molecules, they are often referred to as "active transporters," but unlike pumps, they do not hydrolyze ATP (or any other molecule) during transport. A better term for these proteins is *cotransporters*, referring to their ability to transport two different solutes simultaneously.

To study the functional properties of the different kinds of membrane-transport proteins, researchers need experimental systems in which a particular transport protein predominates. In one common approach, a specific transport protein is extracted and purified; the purified protein then is reincorporated into pure phospholipid bilayer membranes, such as **liposomes** (Figure 15-4). Alternatively, the gene encoding a transport protein can be expressed at high levels in a cell normally not expressing it; the difference in transport of a substance by the transfected and nontransfected cells will be due to the expressed transport protein. In these systems, the functional properties of the various membrane proteins can be examined without ambiguity.

SUMMARY *Overview of Membrane Transport Proteins*

- The plasma membrane regulates the traffic of molecules into and out of the cell.

- Gases and small hydrophobic molecules diffuse directly across the phospholipid bilayer at a rate proportional to their ability to dissolve in a liquid hydrocarbon.

- Ions, sugars, amino acids, and sometimes water cannot diffuse across the phospholipid bilayer at sufficient rates to meet the cell's needs and must be transported by a group of integral membrane proteins including channels, transporters, and ATP-powered ion pumps (see Figure 15-3).

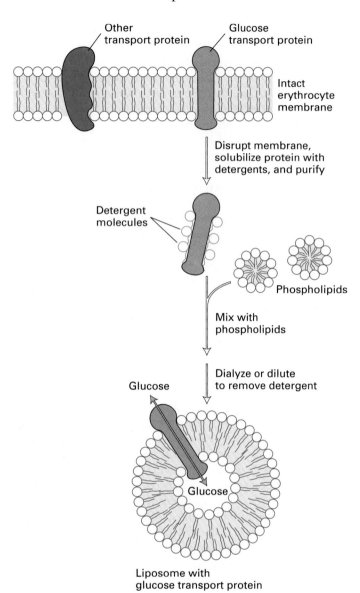

Other transport protein
Glucose transport protein
Intact erythrocyte membrane

Disrupt membrane, solubilize protein with detergents, and purify

Detergent molecules

Phospholipids

Mix with phospholipids

Glucose

Dialyze or dilute to remove detergent

Glucose

Liposome with glucose transport protein

◀ FIGURE 15-4 Liposomes containing a single type of transport protein can be used to investigate properties of the transport process. Here, all the integral proteins of the erythrocyte membrane are solubilized by a nonionic detergent, such as octylglucoside. The glucose transport protein, a uniporter, can be purified by chromatography on a column containing a specific monoclonal antibody and then incorporated into liposomes made of pure phospholipids.

movement of a substance across a membrane down its concentration gradient will have the same negative ΔG value whether or not a protein transporter is involved. This type of movement sometimes is referred to as **facilitated transport** (or facilitated diffusion). As we stressed in Chapter 2, many chemical reactions that are thermodynamically favored will not occur unless an appropriate enzyme is present; such is also the case with movement of hydrophilic molecules across biological membranes. Unlike the substrates of enzymatic reactions, however, transported substances undergo no chemical change during movement across a membrane.

Three Main Features Distinguish Uniport Transport from Passive Diffusion

Three properties of uniporter-catalyzed movement of glucose and other small hydrophilic molecules across a membrane distinguish this type of transport from passive diffusion:

1. The rate of facilitated transport by uniporters is far higher than predicted by Fick's equation describing passive diffusion (Figure 15-5). Because the transported molecules never enter the hydrophobic core of the phospholipid bilayer, the partition coefficient K is irrelevant.

2. Transport is specific. Each uniporter transports only a single species of molecule or a single group of closely related molecules.

3. Transport occurs via a limited number of uniporter molecules, rather *than throughout the phospholipid bilayer.* Consequently, there is a maximum transport rate V_{max} that is achieved when the concentration gradient across the membrane is very large and each uniporter is working at its maximal rate.

Figure 15-5 shows the initial rate of glucose uptake by erythrocytes at different external glucose concentrations. Since the concentration of glucose is usually higher in the extracellular medium than in the cell, the plasma-membrane glucose transporters usually catalyze net movement of glucose in one direction: from the medium into the cell. Under this condition, V_{max} is achieved at high external glucose concentrations. However, if the concentration gradient is reversed, the glucose transporter, like all uniporters, is equally able to catalyze net movement in the reverse direction: from the cell into the medium. Such a situation occurs in liver cells during periods of starvation, when these cells synthe-

● Two common experimental systems for studying the functions of transport proteins are liposomes containing a purified transport protein (see Figure 15-4) and cells transfected with the gene encoding a particular transport protein.

15.3 Uniporter-Catalyzed Transport

We begin our discussion of membrane transport proteins with the simplest type, which catalyze **uniport** transport. The plasma membrane of most cells contains several uniporters that enable amino acids, nucleosides, sugars, and other small molecules to enter and leave cells down their concentration gradients. Similar to enzymes, uniporters accelerate a reaction that is already thermodynamically favored, and the

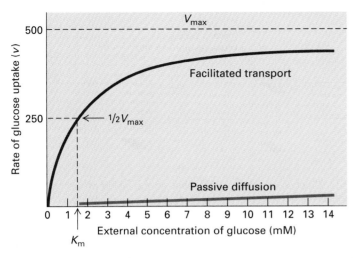

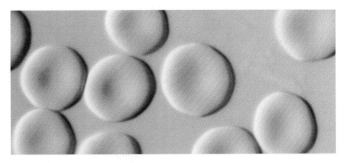

▲ **FIGURE 15-6 Normal human erythrocytes, viewed by differential interference light microscopy, are disk shaped and contain no internal membranes.** The opposite surface also is concave. [Courtesy of M. Murayama, Biological Photo Service.]

▲ **FIGURE 15-5 Comparison of the observed uptake rate of glucose by erythrocytes (red curve) with the calculated rate if glucose were to enter solely by passive diffusion through the phospholipid bilayer (blue curve).** The rate of glucose uptake (measured as micromoles per milliliter of cells per hour) in the first few seconds is plotted against the glucose concentration in the extracellular medium. In this experiment the initial concentration of glucose in the erythrocyte is zero, so that the concentration gradient of glucose across the membrane is the same as the external concentration. The glucose transporter in the erythrocyte membrane clearly increases the rate of glucose transport, compared with that associated with passive diffusion, at all glucose concentrations. Like enzymes, the transporter-catalyzed uptake of glucose exhibits a maximum transport rate V_{max} and is said to be saturable. The K_m is the concentration at which the rate of glucose uptake is half-maximal.

size glucose (from fatty acids, amino acids, and other small molecules) and release it into the blood, and in intestinal epithelial cells during transport of glucose from the intestine to the blood.

GLUT1 Transports Glucose into Most Mammalian Cells

Virtually all mammalian cells use blood glucose as the major source of cellular energy, and most express *GLUT1*, a plasma-membrane uniporter that catalyzes movement of glucose down its concentration gradient. The properties of GLUT1, as well as of many other transport proteins, have been extensively studied in the mammalian erythrocyte, since this cell has no nucleus and no internal membranes; it is essentially a "bag" of hemoglobin containing relatively few other intracellular proteins (Figure 15-6). We discuss GLUT1 in some detail as an example of the uniport type of transport protein.

The glucose transporter GLUT1 alternates between two conformational states: in one, a glucose-binding site faces the outside of the membrane; in the other, a glucose-bind-

ing site faces the inside. Figure 15-7 depicts the sequence of events occurring during the unidirectional transport of glucose from the cell exterior inward to the cytosol. GLUT1 also can catalyze the net movement of glucose from the cytosol outward by reversal of steps 1–4 shown in Figure 15-7. Experimental support for this model, which is thought to apply to other uniport proteins as well, has come from kinetic experiments discussed below.

Kinetics of GLUT1-Catalyzed Movement of Glucose As noted previously, a plot of the entry rate of glucose into erythrocytes versus external glucose concentration is not linear; rather, it is a curve that levels off at V_{max} at high external glucose concentrations (see Figure 15-5). The kinetics of the unidirectional transport of glucose (and other small molecules) from the outside of a cell inward via a uniporter can be described by the same type of equation used to describe a simple enzyme-catalyzed chemical reaction.

For simplicity, let's assume that a substance S (say, glucose) is present initially only on the outside of the membrane. In this case, we can write

$$S_{out} + \text{transporter} \underset{}{\overset{K_m}{\rightleftharpoons}} \text{S-transporter complex} \overset{V_{max}}{\rightleftharpoons} S_{in} + \text{transporter}$$

where K_m is the substance-transporter binding constant and V_{max} is the maximum transport rate of S into the cell. By a similar derivation used to arrive at the Michaelis-Menten equation in Chapter 3, we can derive the following expression for v, the transport rate for S into the cell:

$$v = \frac{V_{max}}{1 + \dfrac{K_m}{C}} \qquad (15\text{-}4)$$

where C is the concentration of S_{out} (initially, the concentration of $S_{in} = 0$); V_{max} is the rate of transport if all molecules of the transporter contain a bound S, which occurs at high

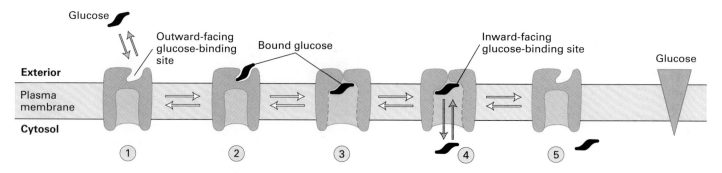

▲ FIGURE 15-7 Model of the mechanism of uniport transport by GLUT1, which is believed to shuttle between two conformational states. In one conformation (①, ②, and ⑤), the glucose-binding site faces outward; in the other (③, ④), the binding site faces inward. Binding of glucose to the outward-facing binding site (① → ②) triggers a conformational change in the transporter (② → ③), moving the bound glucose through the protein such that it is now bound to the inward-facing binding site. Glucose can then be released to the inside of the cell (③ → ④). Finally, the transporter undergoes the reverse conformational change (④ → ⑤), inactivating the inward-facing glucose binding site and regenerating the outward-facing one. If the concentration of glucose is higher inside the cell than outside, the cycle will work in reverse (④ → ①), catalyzing net movement of glucose from inside to out.

S_{out} concentrations; and K_m is the substrate concentration at which half-maximal transport occurs across the membrane. The lower the value of K_m, the more tightly the substrate binds to the transporter, and the greater the transport rate. Equation 15-4 describes the curve for glucose uptake shown in Figure 15-5.

For GLUT1 in the erythrocyte membrane, the K_m for glucose transport is 1.5 millimolar (mM); at this concentration roughly half the transporters with outward-facing binding sites would have a bound glucose. Blood glucose is normally 5 mM, or 0.9 g/L. At this concentration, the erythrocyte glucose transporter is functioning at 77 percent of the maximal rate V_{max}, as can be seen from Figure 15-5.

The kinetics of glucose transport are more complex and more revealing than this simple analysis suggests. For instance, if [^{14}C]glucose is added to a suspension of erythrocytes whose intracellular glucose concentration is zero, the labeled glucose is transported inward at a particular initial rate proportional to the concentration of labeled glucose, as described by Equation 15-4. This *initial* rate of [^{14}C]glucose transport is *accelerated* severalfold if unlabeled glucose is present inside the cells before addition of the labeled glucose. This unexpected experimental observation indicates that the slow (rate-determining) step in the inward transport of glucose is the change in GLUT1 from a conformation with an unoccupied inward-facing glucose-binding site to a conformation with an unoccupied outward-facing binding site (step 4 → step 5 in Figure 15-7). This conformational change is accelerated severalfold when an unlabeled glucose molecule binds to the inward-facing site and is transported outward. This result adds strong support to the conformational-change model of GLUT1 depicted in Figure 15-7.

Specificity and Structure of GLUT1 As noted above, the K_m for glucose transport by GLUT1 is 1.5 mM. The K_m for the nonbiological L-isomer of glucose is >3000 mM; thus at concentrations at which D-glucose is readily transported into the erythrocyte, L-glucose does not enter at a measurable rate. The isomeric sugars D-mannose and D-galactose, which differ from D-glucose in the configuration at only one carbon atom (see Figure 2-8), also are transported by GLUT1 at measurable rates. However, the K_m for D-mannose is 20 mM and for D-galactose is 30 mM, so that considerably higher concentrations of these substrates than of D-glucose are needed to half-saturate the transport reaction. Thus GLUT1 is quite specific, having a much higher affinity (indicated by a lower K_m) for the normal substrate D-glucose than for other substrates.

After glucose is transported into the erythrocyte, it is rapidly phosphorylated, forming glucose 6-phosphate, which cannot leave the cell (see Figure 16-3). Because this reaction is the first step in the metabolism of glucose, the intracellular concentration of free glucose does not increase as glucose is taken up by the cell. Consequently, the glucose concentration gradient across the membrane is maintained, as is the rate of glucose entry into the cell.

GLUT1 is an integral, transmembrane protein with a molecular weight of 45,000. It accounts for 2 percent of the protein in the plasma membrane of erythrocytes. Insertion of purified GLUT1 into artificial liposomes dramatically increases their permeability to D-glucose (see Figure 15-4). This artificial system exhibits all the properties of glucose entry into erythrocytes: in particular, D-glucose, D-mannose, and D-galactose are taken up, but L-glucose is not.

Amino acid sequence and biophysical studies on the glucose transporter indicate that it contains 12 α helices that span the phospholipid bilayer. Although the amino acid residues in the transmembrane α helices are predominantly hydrophobic, several helices bear amino acid residues (e.g., serine, threonine, asparagine, and glutamine) whose side chains can form hydrogen bonds with the hydroxyl groups on glucose. These residues are thought to form the inward-facing and outward-facing glucose-binding sites in the interior of the protein.

SUMMARY Uniporter-Catalyzed Transport

- Uniport-type membrane transport proteins operate to import many types of molecules into the cell driven only by a concentration gradient, a process termed *facilitated transport* or *facilitated diffusion*.

- Three main features distinguish uniport transport from passive diffusion: the rate of transport is far higher than predicted by Fick's equation, transport is specific, and transport occurs via a limited number of transporter proteins rather than throughout the phospholipid bilayer.

- The kinetics of uniporter-catalyzed transport reactions, similar to those of simple enzyme-catalyzed reactions, are characterized by a K_m and a V_{max} (see Figure 15-5).

- The glucose transporter GLUT1, a uniport protein in the plasma membrane of most mammalian cells, allows only glucose and closely related sugars to cross the bilayer down their concentration gradients.

- GLUT1 shuttles between two conformational states, one in which the glucose-binding site faces outward and one in which the binding site faces inward (see Figure 15-7). Transport by other uniporters is thought to involve a similar conformational-change mechanism.

15.4 Intracellular Ion Environment and Membrane Electric Potential

The movement of ions across the plasma membrane and organelle membranes is mediated by several types of transport proteins: all symporters and certain antiporters cotransport ions simultaneously along with specific small molecules, whereas ion channels, ion pumps, and some antiporters transport only ions. In all cases, the rate and extent of ion transport across membranes is influenced not only by the ion concentrations on the two sides of the membrane but also by the voltage (i.e., the electric potential) that exists across the membrane. Here we discuss the origin of the electric potential across the plasma membrane and its relationship to ion channels within the membrane.

Ionic Gradients and an Electric Potential Are Maintained across the Plasma Membrane

The specific ionic composition of the cytosol usually differs greatly from that of the surrounding fluid. In virtually all cells—including microbial, plant, and animal cells—the cytosolic pH is kept near 7.2 and the cytosolic concentration of K^+ is much higher than that of Na^+. In addition, in both invertebrates and vertebrates, the concentration of K^+ is

20–40 times higher in cells than in the blood, while the concentration of Na^+ is 8–12 times lower in cells than in the blood (Table 15-1). The concentration of Ca^{2+} free in the cytosol is generally less than 0.2 micromolar (2×10^{-7} M), a thousand or more times lower than that in the blood. Plant cells and many microorganisms maintain similarly high cytosolic concentrations of K^+ and low concentrations of Ca^{2+} and Na^+ even if the cells are cultured in very dilute salt solutions. The ATP-driven ion pumps that generate and maintain these ionic gradients are discussed later.

In addition to ion pumps, which transport ions against their concentration gradients, the plasma membrane contains channel proteins that allow the principal cellular ions (Na^+, K^+, Ca^{2+}, and Cl^-) to move through it at different rates down their concentration gradients. Ion concentration gradients and selective movements of ions through channels create a difference in voltage across the plasma membrane. The magnitude of this electric potential is ≈ 70 millivolts (mV) with the inside of the cell always negative with respect to the outside. This value does not seem like much until we realize that the plasma membrane is only about 3.5 nm thick. Thus the voltage gradient across the plasma membrane is 0.07 V per 3.5×10^{-7} cm, or 200,000 volts per centimeter! (To appreciate what this means, consider that high-voltage transmission lines for electricity utilize gradients of about 200,000 volts per kilometer!) As explained below, the plasma membrane, like all biological membranes, acts like a *capacitor*—a device consisting of a thin sheet of nonconducting material (the hydrophobic interior) surrounded on both sides by electrically conducting material (the polar head

	TABLE 15-1	Typical Ion Concentrations in Invertebrates and Vertebrates	
Ion		Cell (mM)	Blood (mM)
SQUID AXON*			
K^+		400	20
Na^+		50	440
Cl^-		40–150	560
Ca^{2+}		0.0003	10
$X^{-\dagger}$		300–400	5–10
MAMMALIAN CELL			
K^+		139	4
Na^+		12	145
Cl^-		4	116
HCO_3^-		12	29
X^-		138	9
Mg^{2+}		0.8	1.5
Ca^{2+}		<0.0002	1.8

*The large nerve axon of the squid, an invertebrate cell, has been widely used in studies of the mechanism of conduction of electric impulses.
†X^- represents proteins, which have a net negative charge at the neutral pH of blood and cells.

groups and the ions in the surrounding aqueous medium)—that can store positive charges on one side and negative charges on the other.

The ionic gradients and electric potential across the plasma membrane drive many biological processes. Opening and closing of Na^+, K^+, and Ca^{2+} channels are essential to the conduction of an electric impulse down the axon of a nerve cell (Chapter 21). In many animal cells, the Na^+ concentration gradient and the membrane electric potential power the uptake of amino acids and other molecules against their concentration gradient; this transport is catalyzed by ion-linked symport and antiport proteins. In most cells, a rise in the cytosolic Ca^{2+} concentration is an important regulatory signal, initiating contraction in muscle cells and triggering secretion of digestive enzymes in the exocrine pancreatic cells.

Here we discuss the role of ion channels in generating the membrane electric potential. Later we examine the ATP-powered ion pumps that generate ion concentration gradients, and ion-linked cotransport proteins.

The Membrane Potential in Animal Cells Depends Largely on Resting K^+ Channels

In the experimental system outlined in Figure 15-8a, the distribution of K^+, Na^+, and Cl^- ions is similar to that between an animal cell and its aqueous environment. A membrane separates a 15 mM KCl/150 mM NaCl solution on the right side (representing the "outside" of the cell) from a 150 mM KCl/15 mM NaCl solution on the left side (the "inside"). A potentiometer (voltmeter) is connected to the solution on each side to measure any difference in electric potential across the membrane. If the membrane is impermeable to all ions, no ions will flow across it; there will be no electric potential across it.

▶ **FIGURE 15-8 Experimental system for generating a transmembrane voltage potential across a membrane separating a 150 mM KCl/15 mM NaCl solution (a similar composition to that of the cell cytosol) from a 15 mM KCl/150 mM NaCl solution (concentrations similar to those in blood).** (a) An impermeable membrane prevents ion movement across the membrane, and thus no difference in electric potential is registered on the potentiometer connecting the two solutions. (b) If the membrane is selectively permeable only to Na^+, then Na^+ ions diffuse from right to left, through Na^+ channels. As a consequence, a net positive charge builds up on the left side and a net negative charge builds up on the right side of the membrane. At equilibrium, the membrane potential caused by the charge separation becomes equal to the Nernst potential E_{Na} registered on the potentiometer, and the movement of Na^+ ions in the two directions becomes equal. (c) If the membrane is selectively permeable only to K^+, diffusion of K^+ ions from left to right through K^+ channels causes accumulation of a net negative charge on the left side and a net positive charge on the right side. At equilibrium, the membrane electric potential is equal to E_K.

(a) Membrane impermeable to Na^+, K^+, and Cl^-

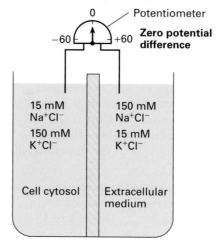

(b) Membrane permeable to Na^+ only

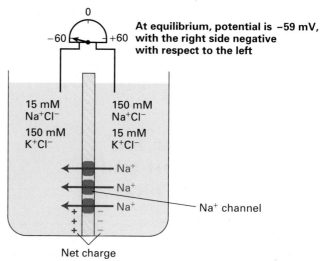

(c) Membrane permeable to K^+ only

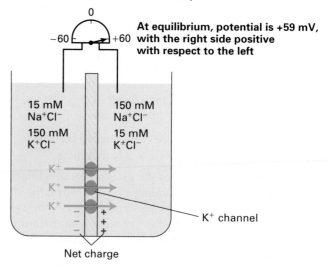

Now suppose that the membrane contains Na^+-channel proteins that accommodate Na^+ ions but exclude K^+ and Cl^- ions. Na^+ ions then tend to move down their concentration gradient from the right side to the left, leaving an excess of negative Cl^- ions compared with Na^+ ions on the right side and generating an excess of positive Na^+ ions compared with Cl^- ions on the left side. The excess Na^+ on the left and Cl^- on the right remain near the respective surfaces of the membrane, since, as in a capacitor, the excess positive charges on one side of the membrane are attracted to the excess negative charges on the other side. The resulting separation of charge across the membrane can be measured by a potentiometer as an electric potential, or voltage, with the right side of the membrane negative (having excess negative charge) with respect to the left (Figure 15-8b).

As more and more Na^+ ions move through channels across the membrane, the magnitude of this charge difference (i.e., voltage) increases. However, continued right-to-left movement of the Na^+ ions eventually is inhibited by the mutual repulsion between the excess positive (Na^+) charges accumulated on the left side of the membrane and by the attraction of Na^+ ions to the excess negative charges built up on the right side. The system soon reaches an equilibrium point at which the two opposing factors that determine the movement of Na^+ ions—the membrane electric potential and the ion concentration gradient—balance each other out. At equilibrium, no net movement of Na^+ ions occurs across the membrane. Thus the excess negative (Cl^-) charges bound to the right surface of the membrane are separated from and attracted to the excess positive (Na^+) ones on the left. In this way, the phospholipid membrane, with its nonconducting hydrophobic interior bounded by the conducting polar head groups and adjacent aqueous medium, stores the charge across it exactly as does a capacitor in an electric circuit.

If a membrane is permeable only to Na^+ ions, then the measured electric potential across the membrane equals the sodium equilibrium potential in volts, E_{Na}. The magnitude of E_{Na} is given by the **Nernst equation**, which is derived from basic principles of physical chemistry:

$$E_{Na} = \frac{RT}{ZF} \ln \frac{[Na_l]}{[Na_r]} \qquad (15\text{-}5)$$

where R (the gas constant) = 1.987 cal/(degree · mol), or 8.28 joules/(degree · mol); T (the absolute temperature) = 293 K at 20 °C, Z (the valency) = +1, F (the Faraday constant) = 23,062 cal/(mol · V), or 96,000 coulombs/(mol · V), and $[Na_l]$ and $[Na_r]$ are the Na^+ concentrations on the left and right sides, respectively, at equilibrium. The Nernst equation is similar to the equations used to calculate the voltage change associated with oxidation or reduction reactions (Chapter 2), which also involve movement of electric charges. At 20 °C, Equation 15-5 reduces to

$$E_{Na} = 0.059 \log_{10} \frac{[Na_l]}{[Na_r]} \qquad (15\text{-}6)$$

If $[Na_l]/[Na_r] = 0.1$, as in Figure 15-8b, then $E_{Na} = -0.059$ V (-59 mV), with the right side negative with respect to the left.

If the membrane is permeable only to K^+ ions and not to Na^+ or Cl^- ions, then a similar equation describes the potassium equilibrium potential E_K:

$$E_K = \frac{RT}{ZF} \ln \frac{[K_l]}{[K_r]} \qquad (15\text{-}7)$$

The *magnitude* of the membrane electric potential is the same (59 mV), except that the right side is now positive with respect to the left (Figure 15-8c), opposite to the polarity obtained with selective Na^+ permeability.

As noted earlier, the membrane potential across the plasma membrane of animal cells is about -70 mV; that is, the cytosolic face is negative with respect to the exoplasmic (outside) face. These membranes contain many open K^+ channels but few open Na^+ or Ca^{2+} channels. As a result, the major ionic movement across the plasma membrane is that of K^+ from the inside outward, leaving an excess of negative charge on the inside and creating an excess of positive charge on the outside. Thus the flow of K^+ ions through these open channels, called *K^+ leak channels* or *resting K^+ channels,* is the major determinant of the inside-negative membrane potential. Quantitatively, the usual resting membrane potential of -70 mV is close to but less than that of the potassium equilibrium potential calculated from the Nernst equation. The K^+ concentration gradient that drives the flow of ions through resting K^+ channels is generated by an ion pump that transports K^+ ions into the cytosol from the extracellular medium and Na^+ ions out. In the absence of this pump, which is discussed later, the K^+ concentration gradient could not be maintained and eventually the membrane potential would fall.

Recent cloning and molecular characterization of resting K^+ channels show that the channel protein is built of four identical subunits. Each subunit contains two membrane-spanning α helices, which partially line the ion-conducting pore in the middle of the protein, and a shorter looped P segment, which acts as a filter to allow K^+ but not other ions to enter the pore and cross the membrane. As we discuss in Chapter 21, the structure of resting K^+ channels is generally similar to the structures of other ion channels that are critical to the function of nerve cells.

Although resting K^+ channels play the dominant role in generating the electric potential across the plasma membrane of animal cells, this is not the case in plant and fungal cells. The inside-negative membrane potential in these cells is generated by transport of H^+ ions out of the cell by an ATP-powered proton pump.

Na$^+$ Entry into Mammalian Cells Has a Negative ΔG

As we've seen, two forces govern the movement of such ions as K^+, Cl^-, and Na^+ across selectively permeable membranes:

the voltage and the ion concentration gradient across the membrane. These forces may act in the same direction or in opposite directions. To calculate the free-energy change ΔG corresponding to the transport of any ion across a membrane, we need to consider the contribution from each of these forces independent of the other.

For example, in a reaction where Na^+ moves from outside to inside the cell, the free-energy change generated from the Na^+ concentration gradient is given by

$$\Delta G_c = RT \ln \frac{[Na_{in}]}{[Na_{out}]} \qquad (15\text{-}8)$$

At the concentrations of Na_{in} and Na_{out} shown in Figure 15-9, which are typical for many mammalian cells, ΔG_c would be -1.45 kcal/mol, the change in free energy for the thermodynamically favored transport of 1 mol of Na^+ ions from outside to inside the cell if there were no membrane electric potential. The free-energy change generated from the membrane electric potential is given by

$$\Delta G_m = FE \qquad (15\text{-}9)$$

where F is the Faraday constant and E is the membrane electric potential. If $E = -70$ mV, then ΔG_m would be -1.6 kcal/mol, the change in free energy for the thermody-namically favored transport of 1 mol of Na^+ ions from outside to inside the cell if there were no Na^+ concentration gradient. Given both forces acting on Na^+ ions, the total ΔG will be the sum of the two partial values:

$$\Delta G = \Delta G_c + \Delta G_m = (-1.45) + (-1.61) = -3.06 \text{ kcal/mol}$$

In this typical example, the Na^+ concentration gradient and the membrane electric potential contribute almost equally to the total ΔG for transport of Na^+ ions. Since ΔG is <0, the inward movement of Na^+ ions is thermodynamically favored. As discussed later, certain cotransport proteins use the inward movement of Na^+ to power the uphill movement of several ions and small molecules into or out of animal cells.

SUMMARY Intracellular Ion Environment and Membrane Electric Potential

- ATP-driven ion pumps generate and maintain ionic gradients across the plasma membrane. As a result, the ionic composition of the cytosol usually differs greatly from that of the surrounding fluid (see Table 15-1).

- In both invertebrates and vertebrates, the K^+ concentration is higher and the Na^+ concentration is lower in cells than in the blood. The cytosolic Ca^{2+} concentration is maintained at less than 0.2 μM.

- An inside-negative electric potential (voltage) of 50–70 mV exists across the plasma membrane of all cells; this is equivalent to a voltage gradient of 200,000 volts per centimeter.

- In animal cells, the electric potential across the plasma membrane is generated primarily by movement of cytosolic K^+ ions through resting K^+ channels to the external medium. Unlike most other ion channels, which open only in response to various signals, these K^+ channels are usually open.

- In plants and fungi, the membrane potential is maintained by the ATP-driven pumping of protons from the cytosol across the membrane.

- Two forces govern the movement of ions across selectively permeable membranes: the membrane electric potential and the ion concentration gradient, which may act in the same or opposite directions. For the thermodynamically favored inward movement of Na^+ into animal cells, these forces act in the same direction (see Figure 15-9).

MEDIA CONNECTIONS
Biological Energy Interconversions

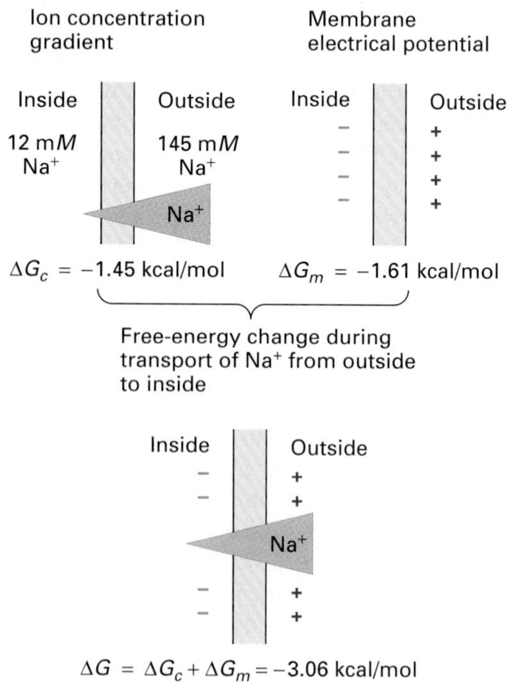

▲ FIGURE 15-9 Transmembrane forces acting on Na^+ ions. As with all ions, the movement of Na^+ ions across the plasma membrane is governed by the sum of two separate forces—the membrane electric potential and the ion concentration gradient. In the case of Na^+ ions, these forces usually act in the same direction.

15.5 Active Transport by ATP-Powered Pumps

We turn now to the ATP-powered pumps that transport ions and various small molecules against their concentration gradients. The general structures of the four principal classes

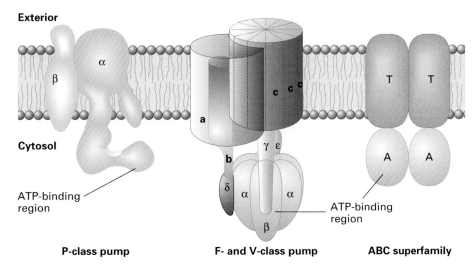

▲ FIGURE 15-10 The four classes of ATP-powered transport proteins. P-class pumps are composed of two different polypeptides, α and β, and become phosphorylated as part of the transport cycle. The sequence around the phosphorylated residue, located in the larger α subunits, is homologous among different pumps. F-class and V-class pumps do not form phosphoprotein intermediates. Their structures are similar and contain similar proteins, but none of their subunits are related to those of P-class pumps. All members of the large ABC superfamily of proteins contain four core domains: two transmembrane (T) domains and two cytosolic ATP-binding (A) domains that couple ATP hydrolysis to solute movement. These core domains are present as separate subunits in some ABC proteins (depicted here), but are fused into a single polypeptide in other ABC proteins. [Adapted from C. H. Higgins, 1995, *Cell* **82**:693; P. Zhang et al., 1998, *Nature* **392**:835; Y. Zhou, T. Duncan, and R. Cross, 1997, *Proc. Nat'l. Acad. Sci. USA* **94**:10583; and T. Elston, H. Wang, and G. Oster, 1998, *Nature* **391**:510.]

of these transport proteins are depicted in Figure 15-10, and their properties are summarized in Table 15-2. Note that the P, F, and V classes transport ions only, whereas the ABC superfamily class transports small molecules as well as ions.

P-class ion pumps contain a transmembrane catalytic α subunit, which contains an ATP-binding site, and usually a smaller β subunit, which may have regulatory functions. Many of these pumps are tetramers composed of two α and two β subunits. During the transport process, at least one of the α subunits is phosphorylated (hence the label "P"), and the transported ions are thought to move through the phosphorylated subunit. This class includes the Na^+/K^+ ATPase in the plasma membrane, which maintains the Na^+ and K^+ gradients typical of animal cells, and several Ca^{2+} ATPases, which pump Ca^{2+} ions out of the cytosol into the external medium or into the lumen of the sarcoplasmic reticulum (SR) of muscle cells. Another member of the P class, found in acid-secreting cells of the mammalian stomach, transports protons (H^+ ions) out of and K^+ ions into the cell. The H^+ pump that maintains the membrane electric potential in plant, fungal, and bacterial cells also belongs to this class.

The structures of *F-class* and *V-class ion pumps* are similar to each other but unrelated to and more complicated than P-class pumps. F- and V-class pumps contain at least three kinds of transmembrane proteins and five kinds of extrinsic polypeptides that form the cytosolic domain. Several of the transmembrane and extrinsic subunits in F-class and V-class pumps exhibit sequence homology, and each pair of homologous subunits is thought to have evolved from a common polypeptide.

All known V and F pumps transport only protons in a process that does not involve a phosphoprotein intermediate. V-class pumps generally function to maintain the low pH of plant vacuoles and of lysosomes and other acidic vesicles in animal cells by using the energy released by ATP hydrolysis to pump protons from the cytosolic to the exoplasmic face of the membrane against the proton electrochemical gradient. F-class pumps are found in bacterial plasma membranes and in mitochondria and chloroplasts. In contrast to V pumps, they generally function to power the synthesis of ATP from ADP and P_i by movement of protons from the exoplasmic to the cytosolic face of the membrane down the proton electrochemical gradient. Because of their importance in ATP synthesis in chloroplasts and mitochondria, F-class proton pumps are treated separately in the next chapter.

The final class of ATP-powered transport proteins is larger and more diverse than the other classes. Referred to as the *ABC (ATP-binding cassette) superfamily*, this class includes more than 100 different transport proteins found in organisms ranging from bacteria to humans. Each ABC protein is specific for a single substrate or group of related substrates including ions, sugars, peptides, polysaccharides, and even proteins. All ABC transport proteins share a common organization consisting of four "core" domains: two transmembrane (T) domains, forming the passageway through

TABLE 15-2 Comparison of Major Classes of ATP-Powered Ion and Small-Molecule Pumps

P Class	F Class	V Class	ABC Class
Substances Transported			
H^+, Na^+, K^+, Ca^{2+}	H^+ only	H^+ only	Ions and various small molecules
Structural and Functional Features			
Large catalytic α subunits (often two) become phosphorylated during solute transport; smaller β subunits may regulate transport.	Multiple transmembrane and cytosolic subunits generally function to synthesize ATP on β cytosolic subunits powered by movement of H^+ down an electrochemical gradient.	Multiple transmembrane and cytosolic subunits generally use energy released by ATP hydrolysis to pump H^+ ions from cytosol to organelle lumens, acidifying them.	Two transmembrane domains form the pathway for solute; two cytosolic ATP-binding domains couple ATP hydrolysis to solute movement. Domains may be in one or separate subunits.
Location of Specific Pumps			
Plasma membrane of plants, fungi, bacteria (H^+ pump)	Bacterial plasma membranes	Vacuolar membranes in plants, yeast, other fungi	Bacterial plasma membranes (amino acid, sugar, and peptide transporters)
Plasma membrane of higher eukaryotes (Na^+/K^+ pump)	Inner mitochondrial membrane	Endosomal and lysosomal membrane in animal cells	Mammalian endoplasmic reticulum (transporters of peptides associated with antigen presentation by MHC proteins)
Apical plasma membrane of mammalian stomach cells (H^+/K^+ pump)	Thylakoid membrane of chloroplast	Plasma membrane of certain acid-secreting animal cells (e.g., osteoclasts and some kidney tubule cells)	
Plasma membrane of all eukaryotic cells (Ca^{2+} pump)			Mammalian plasma membranes (transporters of small molecules, phospholipids, small lipidlike drugs)
Sarcoplasmic reticulum membrane in muscle cells (Ca^{2+} pump)			

which transported molecules cross the membrane, and two cytosolic ATP-binding (A) domains. In some ABC proteins, the core domains are present in four separate polypeptides; in others, the core domains are fused into one or two multidomain polypeptides.

All classes of ATP-powered pumps have one or more binding sites for ATP, and these are always on the cytosolic face of the membrane (see Figure 15-10). Although these proteins are often called ATPases, they normally do not hydrolyze ATP into ADP and P_i unless ions or other molecules are simultaneously transported. Because of the tight coupling between ATP hydrolysis and transport, the energy stored in the phosphoanhydride bond is not dissipated. Thus ATP-powered transport proteins are able to collect the free energy released during ATP hydrolysis and use it to move ions or other molecules uphill against a potential or concentration gradient.

The energy expended by cells to maintain the concentration gradients of Na^+, K^+, H^+, and Ca^{2+} across the plasma and intracellular membranes is considerable. In nerve and kidney cells, for example, up to 25 percent of the ATP produced by the cell is used for ion transport; in human erythrocytes, up to 50 percent of the available ATP is used for this purpose. In cells treated with poisons that inhibit the aerobic production of ATP (e.g., 2,4-dinitrophenol), the ion concentration inside the cell gradually approaches that of the exterior environment as the ions move through plasma membrane channels down their electric and concentration gradients. Eventually treated cells die: partly because protein synthesis requires a high concentration of K^+ ions and partly because in the absence of a Na^+ gradient across the cell membrane, a cell cannot import certain nutrients such as amino acids. Studies on the effects of such poisons provided early evidence for the existence of ion pumps. In this section, we discuss in some detail examples of the P, V, and ABC classes of ATP-powered pumps.

Plasma-Membrane Ca^{2+} ATPase Exports Ca^{2+} Ions from Cells

As discussed in Chapter 20, small increases in the concentration of free Ca^{2+} ions in the cytosol trigger a variety of cellular responses. In order for Ca^{2+} to function in intracellular signaling, its cytosolic concentration usually must be kept below 0.1–0.2 μM. (Although some cytosolic Ca^{2+} is bound to negatively charged groups, it is the concentration of *free*, unbound Ca^{2+} that is critical to its signaling function.) The plasma membranes of animal, yeast, and probably plant cells contain Ca^{2+} ATPases that transport Ca^{2+} out of the cell against its electrochemical gradient. These P-class ion pumps help maintain the concentration of free Ca^{2+} ions in the cytosol at a low level.

In addition to a catalytic α subunit containing an ATP-binding site, as found in other P-class pumps, plasma-membrane Ca^{2+} ATPases also contain the Ca^{2+}-binding regulatory protein **calmodulin**. A rise in cytosolic Ca^{2+} induces the binding of Ca^{2+} ions to calmodulin, which triggers an allosteric activation of the Ca^{2+} ATPase; as a result, the export of Ca^{2+} ions from the cell accelerates, and the original low cytosolic concentration of free Ca^{2+} is restored rapidly.

Muscle Ca^{2+} ATPase Pumps Ca^{2+} Ions from the Cytosol into the Sarcoplasmic Reticulum

Besides the plasma-membrane Ca^{2+} ATPase, muscle cells contain a second, different Ca^{2+} ATPase that transports Ca^{2+} from the cytosol into the lumen of the sarcoplasmic reticulum (SR), an internal organelle that concentrates and stores Ca^{2+} ions. As discussed in Chapter 18, the SR and its calcium pump (referred to as the *muscle calcium pump*) are critical in muscle contraction and relaxation: release of Ca^{2+} ions from the SR into the muscle cytosol causes contraction, and the rapid removal of Ca^{2+} ions from the cytosol by the muscle calcium pump induces relaxation.

Because the muscle calcium pump constitutes more than 80 percent of the integral protein in SR membranes, it is easily purified and characterized. Each transmembrane catalytic α subunit has a molecular weight of 100,000 and transports two Ca^{2+} ions per ATP hydrolyzed. In the cytosol of muscle cells, the free Ca^{2+} concentration ranges from 10^{-7} M (resting cells) to more than 10^{-6} M (contracting cells), whereas the *total* Ca^{2+} concentration in the SR lumen can be as high as 10^{-2} M. Sites on the cytosolic surface of the muscle calcium pump have a very high affinity for Ca^{2+} ($K_m = 10^{-7}$ M), allowing the pump to transport Ca^{2+} efficiently from the cytosol into the SR against the steep concentration gradient.

The concentration of free Ca^{2+} within the sarcoplasmic reticulum is actually much less than the total concentration of 10^{-2} M. Two soluble proteins in the lumen of SR vesicles bind Ca^{2+} and serve as a reservoir for intracellular Ca^{2+}, thereby reducing the concentration of free Ca^{2+} ions in the SR vesicles, and consequently decreasing the energy needed to pump Ca^{2+} ions into them from the cytosol. The activity of the muscle Ca^{2+} ATPase is so regulated that if the free Ca^{2+} concentration in the cytosol becomes too high, the rate of calcium pumping increases until the cytosolic Ca^{2+} concentration is reduced to less than 1 μM. Thus in muscle cells, the calcium pump in the SR membrane can supplement the activity of the plasma-membrane pump, assuring that the cytosolic concentration of free Ca^{2+} remains below 1 μM.

The current model of the mechanism of action of the Ca^{2+} ATPase in the SR membrane is outlined in Figure 15-11. Coupling of ATP hydrolysis with ion pumping involves several steps that must occur in a defined order. When the protein is in one conformation, termed *E1*, two Ca^{2+} ions bind in sequence to high-affinity sites on the cytosolic surface (step 1). Then an ATP binds to its site on the cytosolic surface; in a reaction requiring that a Mg^{2+} ion be tightly complexed to the ATP, the bound ATP is hydrolyzed to ADP and the liberated phosphate is transferred to a specific aspartate residue in the protein, forming a high-energy acyl phosphate bond, denoted by E1~P (step 2). The protein then changes its conformation to E2–P, generating two low-affinity Ca^{2+}-binding sites on the exoplasmic surface, which faces the SR lumen; this conformational change simultaneously propels the two Ca^{2+} ions through the protein to these sites (step 3) and inactivates the high-affinity Ca^{2+}-binding sites on the cytosolic face. The Ca^{2+} ions then dissociate from the exoplasmic surface of the protein (step 4). Following this, the aspartyl-phosphate bond in E2–P is hydrolyzed, causing E2 to revert to E1, a change that inactivates the exoplasmic-facing Ca^{2+}-binding sites and regenerates the cytosolic-facing Ca^{2+}-binding sites (step 5).

Thus phosphorylation of the muscle calcium pump by ATP favors conversion of E1 to E2, and dephosphorylation favors the conversion of E2 to E1. While only E2–P, not E1~P, is actually hydrolyzed, the free energy of hydrolysis of the aspartyl-phosphate bond in E1~P is greater than that for E2–P. The reduction in free energy of the aspartyl-phosphate bond in E2–P, relative to E1~P, can be said to power the E1 → E2 conformational change. The affinity of Ca^{2+} for the cytosolic-facing binding sites in E1 is a thousandfold greater than the affinity of Ca^{2+} for the exoplasmic-facing

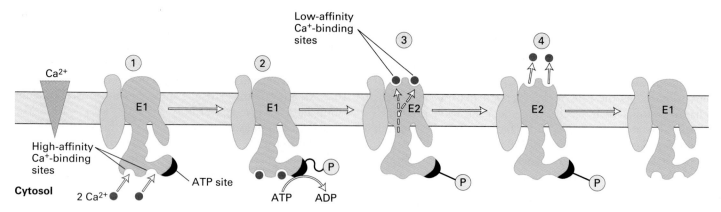

▲ **FIGURE 15-11 Model of the mechanism of action of muscle Ca²⁺ ATPase, which is located in the sarcoplasmic reticulum (SR) membrane.** Only one of the two α subunits of this P-class pump is depicted. E1 and E2 are alternate conformational forms of the protein in which the Ca²⁺-binding sites are on the cytosolic and exoplasmic faces, respectively. An ordered sequence of steps, as diagrammed here, is essential for coupling ATP hydrolysis and the transport of Ca²⁺ ions (red circles) across the membrane. ~P indicates a high-energy acyl phosphate bond; —P indicates a low-energy phosphoester bond. See the text for more details. [Adapted from W. P. Jencks, 1980, *Adv. Enzymol.* **51**:75; W. P. Jencks, 1989, *J. Biol. Chem.* **264**:18855; and P. Zhang et al., 1998, *Nature* **392**:835.]

sites in E2; this difference enables the protein to transport Ca²⁺ unidirectionally from the cytosol, where it binds tightly to the pump, to the exoplasm, where it is released.

Much evidence supports the model depicted in Figure 15-11. For instance, the muscle calcium pump has been isolated with phosphate linked to an aspartate residue, and spectroscopic studies have detected slight alterations in protein conformation during the E1 → E2 conversion. On the basis of the protein's amino acid sequence and various biochemical studies, investigators proposed the structural model

for the catalytic α subunit shown in Figure 15-12. The membrane-spanning α helices are thought to form the passageway through which Ca²⁺ ions move. The bulk of the subunit consists of cytosolic globular domains that are involved in ATP binding, phosphorylation of aspartate, and energy transduction. These domains are connected by "stalks" to the membrane-embedded domain.

As noted previously, all P-class ion pumps, regardless of which ion they transport, are phosphorylated during the transport process. The amino acid sequences around the

▶ **FIGURE 15-12 Schematic structural model for the catalytic (α) subunit of muscle Ca²⁺ ATPase.** The 10 transmembrane α helices are thought to form a channel through which Ca²⁺ ions move. Site-specific mutagenesis studies have identified four residues (red dots), located in four of the transmembrane helices, that participate in Ca²⁺ binding. Trypsin digestion releases three cytosolic globular domains, which constitute the bulk of the protein. One cytosolic domain functions in ATP binding; a second bears the aspartate that is phosphorylated/ dephosphorylated; and the third is involved in energy transduction. [After D. H. MacLennan et al., 1985, *Nature* **316**:696; T. Toyofuku et al., 1992, *J. Biol. Chem.* **267**:14490.]

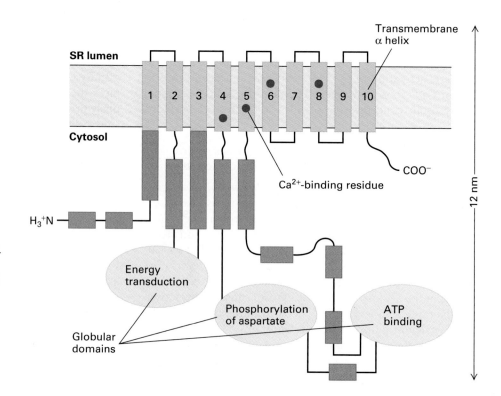

phosphorylated aspartate in the catalytic α subunit are highly conserved in all proteins of this type. Thus the mechanistic model in Figure 15-11 probably is generally applicable to all these ATP-powered ion pumps. In addition, the α subunits of all the P pumps examined to date have a similar molecular weight and, as deduced from their amino acid sequences derived from cDNA clones, have a similar arrangement of transmembrane α helices (see Figure 15-12). These findings strongly suggest that all these proteins evolved from a common precursor, although they now transport different ions.

Na⁺/K⁺ ATPase Maintains the Intracellular Na⁺ and K⁺ Concentrations in Animal Cells

A second P-class ion pump that has been studied in considerable detail is the Na⁺/K⁺ ATPase present in the plasma membrane of all animal cells. This ion pump is a tetramer of subunit composition $\alpha_2\beta_2$. (Classic Experiment 15.1 describes the discovery of this enzyme.) The β polypeptide is required for newly synthesized α subunits to fold properly in the endoplasmic reticulum but apparently is not involved

directly in ion pumping. The α subunit is a 120,000-MW nonglycosylated polypeptide whose amino acid sequence and predicted membrane structure are very similar to those of the muscle Ca²⁺ ATPase. In particular, the Na⁺/K⁺ ATPase has a stalk on the cytosolic face that links domains containing the ATP-binding site and the phosphorylated aspartate to the membrane-embedded domain. The overall process of transport moves three Na⁺ ions out of and two K⁺ ions into the cell per ATP molecule split (Figure 15-13a).

Several lines of evidence indicate that the Na⁺/K⁺ ATPase is responsible for the coupled movement of K⁺ and Na⁺ into and out of the cell, respectively. For example, the drug *ouabain*, which binds to a specific region on the exoplasmic surface of the protein and specifically inhibits its ATPase activity, also prevents cells from maintaining their Na⁺/K⁺ balance. Any doubt that the Na⁺/K⁺ ATPase is responsible for ion movement was dispelled by the demonstration that the enzyme, when purified from the membrane and inserted into liposomes, propels K⁺ and Na⁺ transport in the presence of ATP.

The mechanism of action of the Na⁺/K⁺ ATPase, outlined in Figure 15-13b, is similar to that of the muscle calcium

▶ **FIGURE 15-13 Models for the structure and function of the Na⁺/K⁺ ATPase in the plasma membrane.**
(a) This P-class pump comprises two copies each of a small glycosylated β subunit and a large α subunit, which performs ion transport. Hydrolysis of one molecule of ATP to ADP and Pᵢ is coupled to export of three Na⁺ ions (blue circles) and import of two K⁺ ions (dark red triangles) against their concentration gradients (large triangles). It is not known whether only one α subunit, or both, in a single ATPase molecule transports ions.
(b) Ion pumping by the Na⁺/K⁺ ATPase involves a high-energy acyl phosphate intermediate (E1~P) and conformational changes, similar to transport by the muscle Ca²⁺ ATPase. In this case, hydrolysis of the E2–P intermediate powers transport of a second ion (K⁺) inward. Na⁺ ions are indicated by blue circles; K⁺ ions, by red triangles. See text for details. [Adapted from P. Läuger, 1991, *Electrogenic Ion Pumps*, Sinauer Associates, p. 178.]

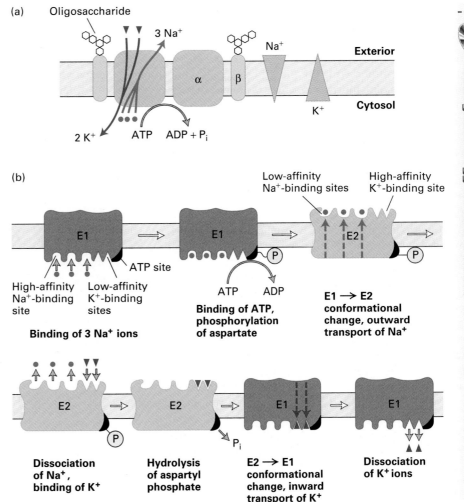

pump, except that ions are pumped in both directions across the membrane. In its E1 conformation, the Na^+/K^+ ATPase has three high-affinity Na^+-binding sites and two low-affinity K^+-binding sites on the cytosolic-facing surface of the protein. The K_m for binding of Na^+ to these cytosolic sites is 0.6 mM, a value considerably lower than the intracellular Na^+ concentration of ≈ 12 mM; as a result, Na^+ ions normally will fill these sites. Conversely, the affinity of the cytosolic K^+-binding sites is low enough that K^+ ions, transported inward through the protein, dissociate from E1 into the cytosol despite the high intracellular K^+ concentration. During the E1 → E2 transition, the three bound Na^+ ions move outward through the protein. Transition to the E2 conformation also generates two high-affinity K^+ sites and three low-affinity Na^+ sites on the exoplasmic face. Because the K_m for K^+ binding to these sites (0.2 mM) is considerably lower than the extracellular K^+ concentration (4 mM), these sites will fill quickly with K^+ ions. In contrast, the three Na^+ ions, transported outward through the protein, will dissociate into the extracellular medium from the low-affinity Na^+ sites on the exoplasmic surface despite the high extracellular Na^+ concentration. Similarly, during the E2 → E1 transition, the two bound K^+ ions are transported inward.

V-Class H^+ ATPases Pump Protons across Lysosomal and Vacuolar Membranes

All V-class ATPases transport H^+ ions only. These proton pumps, present in the membranes of lysosomes, endosomes, and plant vacuoles, function to acidify the lumen of these organelles. The acidity of the lysosomal lumen, usually $\approx 4.5-5.0$, can be measured precisely in living cells by use of particles labeled with a pH-sensitive fluorescent dye. Cells phagocytose these particles (see Figure 5-44a) and transfer them to the lysosomes. The ability of different wavelengths of visible light to excite fluorescence is highly dependent on pH, and the lysosomal pH can be calculated from the spectrum of the fluorescence emitted. Maintenance of the 100-fold or more proton gradient between the lysosomal lumen (pH $\approx 4.5-5.0$) and the cytosol (pH ≈ 7.0) depends on ATP production by the cell.

The ATP-powered proton pumps in lysosomal and vacuolar membranes have been isolated, purified, and incorporated into liposomes. As illustrated in Figure 15-10, these V-class proton pumps contain two discrete domains: a cytosolic-facing hydrophilic domain (V_1) composed of five different polypeptides and a transmembrane domain (V_0) containing 9–12 copies of proteolipid c, one copy of protein b, and one copy of protein a. The subunit composition of the cytosolic domain is $\alpha_3\beta_3\gamma\delta\epsilon$; the α and β subunits contain the sites where ATP binding and hydrolysis occur. Each transmembrane c subunit is thought to span the membrane two times; the c and a subunits together form the proton-conducting channel. Unlike P-class ion pumps, the V-class H^+ ATPases are not phosphorylated and dephosphorylated during proton transport.

Similar V-class ATPases are found in the plasma membrane of certain acid-secreting cells. These include osteoclasts, bone-resorbing macrophagelike cells, which bind to a bone and seal off a small segment of extracellular space between the plasma membrane and the surface of the bone. HCl secreted into this space by osteoclasts dissolves the calcium phosphate crystals that give bone its rigidity and strength.

Another example is the mitochondria-rich epithelial cells lining the toad bladder; the apical plasma membrane of these cells contain many V-class H^+ ATPases, which function to acidify the urine (Figure 15-14). As we discuss later, the membrane of plant vacuoles contains two proton pumps: a typical V-class H^+ ATPase and another one that utilizes the energy released by hydrolysis of inorganic pyrophosphate (PP_i) to pump protons into the vacuole. This PP_i-hydrolyzing proton pump, believed to be unique to plants, has an amino acid sequence different from any other ion-transporting proteins.

ATP-powered proton pumps cannot acidify the lumen of an organelle (or the extracellular space) by themselves. The reason for this is that pumping of protons would rapidly cause a buildup of positive charge on the exoplasmic face of the membrane on the inside of the vesicle membrane and a corresponding buildup of negative charges on the cytosolic face. In other words, the pump would generate a voltage across the membrane, exoplasmic face positive, which would prevent movement of protons into the vesicle before a significant H^+ concentration gradient had been established. In fact, this is the way that H^+ pumps generate an inside-negative potential across plant and yeast plasma membranes. In order for an organelle lumen or an extracellular space

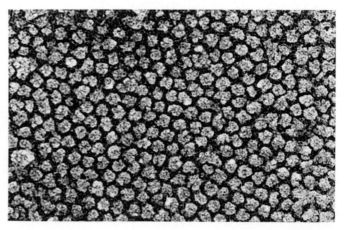

0.02 μm

▲ **FIGURE 15-14 The plasma membrane of certain acid-secreting cells contains an almost crystalline array of V-class H^+ ATPases.** This electron micrograph is of a platinum replica of the cytosolic surface of the apical plasma membrane of a toad bladder epithelial cell. Each stud is a single V-class H^+ ATPase ($\approx 600,000$ MW) composed of several polypeptide subunits surrounding a central channel. [From D. Brown, S. Gluck, and J. Hartwig, 1987, *J. Cell Biol.* **105**:1637.]

(e.g., the outside of an osteoclast) to become acidic, movement of H^+ up its concentration gradient must be accompanied by (1) movement of an equal number of anions in the same direction or (2) movement of equal numbers of a different cation in the opposite direction. The first process occurs in lysosomes and plant vacuoles whose membranes contain V-class H^+ ATPases and ion channels through which accompanying anions (e.g., Cl^-) move. The second occurs in the lining of the stomach, which contains a P-class H^+/K^+ ATPase that pumps one H^+ outward and one K^+ inward.

The ABC Superfamily Transports a Wide Variety of Substrates

As noted earlier, all members of the very large and diverse ABC superfamily of transport proteins contain two transmembrane (T) domains and two cytosolic ATP-binding (A) domains (see Figure 15-10). The T domains, each built of six membrane-spanning α helices, form the pathway through which the transported substance (substrate) crosses the membrane and determine the substrate specificity of each ABC protein. The sequence of the A domains is ≈30 to 40 percent homologous in all members of this superfamily, indicating a common evolutionary origin. Some ABC proteins also contain a substrate-binding subunit or regulatory subunit.

Bacterial Plasma-Membrane Permeases The plasma membrane of many bacteria contain numerous *permeases* that belong to the ABC superfamily. These proteins use the energy released by hydrolysis of ATP to transport specific amino acids, sugars, vitamins, or even peptides into the cell. Since bacteria frequently grow in soil or pond water where the concentration of nutrients is low, these ABC transport proteins allow the cells to concentrate amino acids and other nutrients in the cell against a substantial concentration gradient. Bacterial permeases generally are *inducible;* that is, the quantity of a transport protein in the cell membrane is regulated by both the concentration of the nutrient in the medium and the metabolic needs of the cell.

In *E. coli* histidine permease, a typical bacterial ABC protein, the two transmembrane domains and two cytosolic ATP-binding domains are formed by four separate subunits. In gram-negative bacteria such as *E. coli*, which have an outer membrane, a soluble histidine-binding protein in the periplasmic space assists in transport (Figure 15-15). This soluble protein binds histidine tightly and directs it to the T subunits, through which histidine crosses the membrane powered by ATP hydrolysis. Mutant *E. coli* cells that are defective in any of the histidine-permease subunits or the soluble binding protein are unable to transport histidine into the cell, but are able to transport other amino acids whose uptake is facilitated by other transport proteins. Such genetic analyses provide strong evidence that histidine permease and similar ABC proteins function to transport solutes into the cell.

Mammalian MDR Transport Proteins A series of rather unexpected observations led to discovery of the first eukaryotic ABC protein. Oncologists noted that tumor cells often became simultaneously resistant to several chemotherapeutic drugs with unrelated chemical structures; similarly, cell biologists observed that cultured cells selected for resistance to one toxic substance (e.g., colchicine, a microtubule inhibitor) frequently became resistant to several other drugs, including the anticancer drug adriamycin. Subsequent studies showed that this resistance is due to enhanced expression of a *multidrug-resistance (MDR) transport protein* known as *MDR1*. In this member of the ABC superfamily, all four domains are "fused" into a single 170,000-MW protein (Figure 15-16). This protein

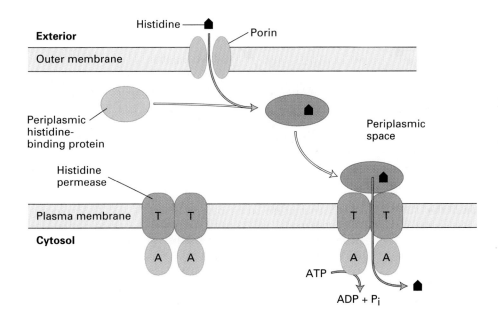

◀ **FIGURE 15-15 Gram-negative bacteria import many solutes by means of ABC proteins (permeases) that utilize a soluble substrate-binding protein present in the periplasmic space.** Depicted here is the import of the amino acid histidine. After diffusing through porins in the outer membrane, histidine is bound by a specific periplasmic histidine-binding protein, which undergoes a conformational change. The histidine-protein complex binds to the exoplasmic surface of a T subunit in histidine permease located in the plasma membrane. Hydrolysis of ATP bound to the A subunit then powers movement of histidine through the protein into the cytosol. The transport process does not appear to involve a phosphoprotein intermediate.

▶ **FIGURE 15-16 Schematic structural model for mammalian MDR1 protein. In this member of the ABC superfamily, the two transmembrane domains and two cytosolic ATP-binding domains are part of a single polypeptide.** Each transmembrane domain contains six α helices. The two halves of this 1280-aa protein have similar amino acid sequences. A variety of lipid-soluble molecules that diffuse across the plasma membrane into the cell are transported outward by MDR1.

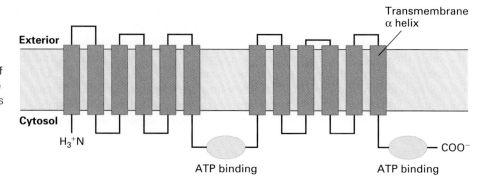

uses the energy derived from ATP hydrolysis to *export* a large variety of drugs from the cytosol to the extracellular medium. The *Mdr1* gene is frequently amplified in multidrug-resistant cells, resulting in a large overproduction of the MDR1 protein.

Most drugs transported by MDR1 are small hydrophobic molecules, which diffuse from the culture medium across the plasma membrane into the cell. The ATP-powered export of such drugs from the cytosol by MDR1 means a much higher extracellular drug concentration is required to kill cells. That MDR1 is an ATP-powered small-molecule pump has been demonstrated with liposomes containing the purified protein (see Figure 15-4). The ATPase activity of these liposomes is enhanced by different drugs in a dose-dependent manner corresponding to their ability to be transported by MDR1.

Not only does MDR1 transport a varied group of molecules, but all these substrates compete with one another for transport by MDR1. Although the mechanism of action of MDR1-assisted transport has not been definitively demonstrated, the *flippase model*, depicted in Figure 15-17a, is a

likely candidate. Substrates of MDR1 are primarily planar, lipid-soluble molecules with one or more positive charges, and they move spontaneously from the cytosol into the cytosolic-facing leaflet of the plasma membrane. The hydrophobic portion of a substrate molecule is oriented toward the hydrophobic core of the membrane, and the charged portion toward the polar cytosolic face of the membrane and is still in the cytosol. The substrate diffuses laterally until encountering and binding to a site on the MDR1 protein that is within the bilayer. The protein then "flips" the charged substrate molecule into the exoplasmic leaflet, an energetically unfavorable reaction powered by the coupled ATPase activity of MDR1. Once in the exoplasmic face, the substrate diffuses into the aqueous phase on the outside of the cell. Support for the flippase model of transport by MDR1 comes from MDR2, a homologous protein present in the region of the liver cell plasma membrane that faces the bile duct. MDR2 has been shown to flip phospholipids from the cytosolic-facing leaflet of the plasma membrane to the exoplasmic leaflet, thereby generating an excess of phospholipids

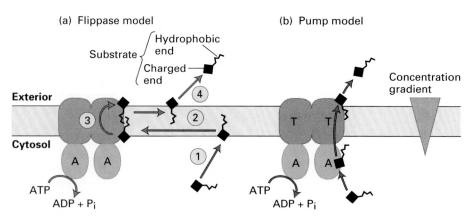

▲ **FIGURE 15-17 Possible mechanisms of action of the MDR1 protein.** (a) The flippase model proposes that a lipid-soluble molecule first dissolves in the cytosolic-facing leaflet of the plasma membrane (①) and then diffuses in the membrane until binding to a site on the MDR1 protein that is within the bilayer (②). Powered by ATP hydrolysis, the substrate molecule flips into the exoplasmic leaflet (③), from which it can move directly into

the aqueous phase on the outside of the cell (④). (b) According to the pump model, MDR1 has a single multisubstrate binding site and transports molecules by a mechanism similar to that of other ATP-powered pumps. [Adapted from G. Ferro-Luzzi Ames and H. Legar, 1992, *FASEB J.* **6**:2660; N. Nelson, 1992, *Curr. Opin. Cell Biol.* **4**:654; C. F. Higgins and M. M. Gottesman, 1992, *Trends Biochem. Sci.* **17**:18; and C. F. Higgins, 1995, *Cell* **82**:693.]

in the exoplasmic leaflet; these phospholipids peel off into the bile duct and form an essential part of the bile. An alternative *pump model* also has been proposed for MDR1 (Figure 15-17b). According to this model, drug molecules in the cytosol bind directly to a single small-molecule binding site on the cytosolic face of the MDR1 protein; subsequent ATP hydrolysis powers movement of the bound drug through the protein to the aqueous phase on the outside of the cell by a mechanism similar to that of other ATP-powered pumps.

MDR1 protein is expressed in abundance in the liver, intestines, and kidney—sites from which natural toxic products are removed from the body. Thus the natural function of MDR1 may be to transport a variety of natural and metabolic toxins into the bile, intestinal lumen, or forming urine. During the course of its evolution, MDR1 appears to have coincidentally acquired the ability to transport drugs whose structures are similar to those of these toxins. Tumors derived from these cell types, such as hepatomas (liver cancers), frequently are resistant to virtually all chemotherapeutic agents and thus difficult to treat, presumably because the tumors exhibit increased expression of the MDR1 or MDR2 proteins.

Cystic Fibrosis Transmembrane Regulator (CFTR) Protein Discovery of another ABC transport protein came from studies of cystic fibrosis (CF), the most common lethal autosomal recessive genetic disease of Caucasians. This disease is caused by a mutation in the *CFTR* gene, which encodes a chloride-channel protein that is regulated by cyclic AMP (cAMP), an intracellular second messenger. These Cl^- channels are present in the apical plasma membranes of epithelial cells in the lung, sweat glands, pancreas, and other tissues. An increase in cAMP stimulates Cl^- transport by such cells from normal individuals, but not from CF individuals who have a defective CFTR protein.

The sequence and predicted structure of the encoded CFTR protein, based on analysis of the cloned gene, are very similar to those of MDR1 protein except for the presence of an additional domain, the regulatory (R) domain, on the cytosolic face. The Cl^--channel activity of CFTR protein clearly is enhanced by binding of ATP. Moreover, as detailed in Chapter 20, cAMP activates a protein kinase that phosphorylates, and thereby activates, CFTR. When purified CFTR protein is incorporated into liposomes, it forms Cl^- channels with properties similar to those in normal epithelial cells. And when the wild-type CFTR protein is expressed by recombinant techniques in cultured epithelial cells from CF patients, the cells recover normal Cl^--channel activity. This latter result raises the possibility that gene therapy might reverse the course of cystic fibrosis.

Since CFTR protein is similar to MDR1 in structure, it may also function as an ATP-powered pump of some as-yet unidentified molecule. In any case, much remains to be learned about this fascinating class of ABC transport proteins.

SUMMARY Active Transport by ATP-Powered Pumps

- Four types of membrane transport proteins couple the energy-releasing hydrolysis of ATP with the energy-requiring transport of substances against their concentration gradient (see Figure 15-10 and Table 15-2).

- In P-class pumps, phosphorylation of the α subunit and a change in conformational states are essential for coupled transport of H^+, Na^+, K^+, or Ca^{2+} ions (see Figures 15-11 and 15-13).

- The P-class Na^+/K^+ ATPase pumps three Na^+ ions out of and two K^+ ions into the cell per ATP hydrolyzed. A homolog, the Ca^{2+} ATPase, pumps two Ca^{2+} ions out of the cell or, in muscle, into the sarcoplasmic reticulum per ATP hydrolyzed. The combined action of these pumps in animal cells creates an intracellular ion milieu of high K^+, low Ca^{2+}, and low Na^+ very different from the extracellular fluid milieu of high Na^+, high Ca^{2+}, and low K^+.

- In the multisubunit V- and F-class ATPases, which pump protons exclusively, a phosphorylated protein is not an intermediate in transport.

- A V-class H^+ pump in animal lysosomal and endosomal membranes and plant vacuole membranes is responsible for maintaining a lower pH inside the organelles than in the surrounding cytosol.

- All members of the large and diverse ABC superfamily of transport proteins contain four core domains: two transmembrane domains, which form a pathway for solute movement and determine substrate specificity, and two cytosolic ATP-binding domains.

- The ABC superfamily includes bacterial amino acid and sugar permeases (see Figure 15-15); the mammalian MDR1 protein, which exports a wide array of drugs from cells; and CFTR protein, a Cl^- channel that is defective in cystic fibrosis.

- According to the flippase model of MDR1 activity, a substrate molecule diffuses into the cytosolic leaflet of the plasma membrane, then is flipped to the exoplasmic leaflet in an ATP-powered process, and finally diffuses from the membrane into the extracellular space (see Figure 15-17a).

15.6 Cotransport by Symporters and Antiporters

Besides ATP-powered pumps, cells have a second, discrete class of proteins that import or export ions and small molecules, such as glucose and amino acids, against a concentration gradient. These proteins use the energy stored in the electrochemical gradient of Na^+ or H^+ ions to power the uphill movement of another substance, a process called **co-transport**. For instance, the energetically favored movement of a Na^+ ion (the "cotransported" ion) into a cell across the plasma membrane, driven both by its concentration gradient and by the transmembrane voltage gradient (see Figure 15-9), can be coupled obligatorily to movement of the "transported" molecule (e.g., glucose) against its concentration gradient. When the transported molecule and cotransported ion move in the same direction, the process is called **symport**; when they move in opposite directions, the process is called **antiport** (see Figure 15-2b).

Na⁺-Linked Symporters Import Amino Acids and Glucose into Many Animal Cells

Many cells, such as those lining the small intestine and the kidney tubules, need to concentrate glucose against a very large concentration gradient. Such cells utilize a *two Na⁺/one-glucose symporter*; a protein that couples transmembrane movement of one glucose molecule to the transport of two Na^+ ions:

$$2\ Na^+_{out} + glucose_{out} \rightleftharpoons 2\ Na^+_{in} + glucose_{in}$$

As discussed earlier, movement of Na^+ from the external medium into the cell is driven by two forces: by the Na^+ concentration gradient (the Na^+ concentration is lower inside the cell than in the medium) and by the inside-negative membrane electric potential (see Figure 15-9). Quantitatively, the free-energy change for the symport transport of two Na^+ ions and one glucose molecule can be written

$$\Delta G = RT \ln \frac{[glucose_{in}]}{[glucose_{out}]} + 2RT \ln \frac{[Na^+_{in}]}{[Na^+_{out}]} + 2FE \quad (15\text{-}10)$$

Glucose concentration gradient (1 molecule transported)   **Na⁺ concentration gradient (2 Na⁺ ions transported)**   **Membrane potential (2 Na⁺ ions transported)**

where F is the Faraday constant, E is the electric potential across the plasma membrane, and the other parameters are

as defined previously. According to our previous calculations, the membrane electric potential and the Na^+ concentration gradient each contribute about -1.5 kcal per mole of Na^+ transported inward, or a total of about 3 kcal/mol (see Equations 15-8 and 15-9). Thus the change in free energy ΔG for transport of two moles of Na^+ inward is about -6 kcal. By substituting this value into the partial equation for glucose transport

$$\Delta G = RT \ln \frac{[glucose_{in}]}{[glucose_{out}]}$$

we can calculate that a ΔG of -6 kcal/mol can generate an equilibrium concentration of glucose inside the cell that is $\approx 30,000$ times greater than the exterior concentration. By transporting two Na^+ ions per glucose, this Na^+/glucose symport protein can accumulate glucose against a much steeper concentration gradient than if only one Na^+ ion were transported per glucose.

The Na^+/glucose symporter contains 14 transmembrane α helices (Figure 15-18). A recombinant protein consisting of only the five C-terminal transmembrane α helices has been shown to transport glucose across the plasma membrane, *down* its concentration gradient, independently of Na^+. This portion of the molecule thus functions as a glucose-permeation pathway. The N-terminal portion of the protein, including helices 1–9, is required to couple Na^+ binding and glucose transport against a concentration gradient. Figure 15-19 depicts the current model of transport by Na^+/glucose symporters. This model, which has not yet been experimentally supported, entails conformational changes analogous to those that occur in uniport transporters such as GLUT1, which do not require a cotransported ion (see Figure 15-7).

Na⁺-Linked Antiporter Exports Ca²⁺ from Cardiac Muscle Cells

The plasma membrane of most cells contains one or more types of antiporters, which couple movement of a cotrans-

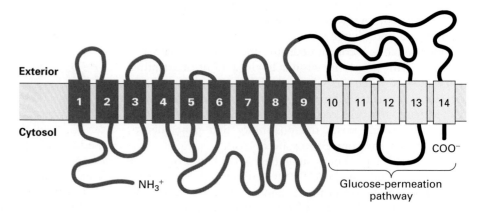

▶ **FIGURE 15-18 Structural model for the two-Na⁺/one-glucose symporter.** This 662-aa protein forms 14 transmembrane α helices with the N-and C-termini facing the cytosol. The five C-terminal helices form the sugar-permeation pathway; the rest of the protein may be required to couple Na^+ binding and glucose transport. The exoplasmic surface of the protein has binding sites for two Na^+ ions and one glucose ≈ 3.5 nm apart, but the location of these sites has not yet been determined. [Adapted from M. Panayotova-Heiermann et al., 1997, *J. Biol. Chem.* **272**:20324.]

Exterior

Cytosol

1 2 3 4 5 6 7 8 9 10 11 12 13 14

NH₃⁺

COO⁻

Glucose-permeation pathway

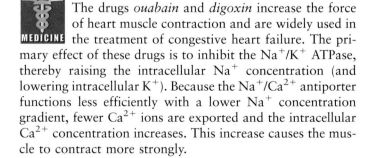

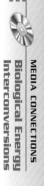

▲ FIGURE 15-19 Proposed model for operation of the two-Na⁺/one-glucose symporter. The simultaneous binding of Na⁺ and glucose to sites on the exoplasmic surface induces a conformational change, generating a transmembrane pore or tunnel that allows both bound Na⁺ and glucose to move through the protein to binding sites on the cytosolic domain and then to pass into the cytosol. After this passage, the protein reverts to its original conformation. [See E. Wright, K. Hager, and E. Turk, 1992, *Curr. Opin. Cell Biol.* **4**:696 for details on the structure and function of this and related transporters.]

ported ion (often Na⁺) down its electrochemical gradient to movement of a different molecule in the opposite direction against a concentration gradient (see Figure 15-3b). In cardiac muscle cells, for example, a *Na⁺/Ca²⁺ antiporter*, rather than a plasma membrane Ca²⁺ ATPase, plays the principal role in maintaining a low concentration of Ca²⁺ in the cytosol. The reaction of this *cation antiporter* can be written

$$3\ Na^+_{out} + Ca^{2+}_{in} \rightleftharpoons 3\ Na^+_{in} + Ca^{2+}_{out}$$

Note that the movement of three Na⁺ ions is required to power the export of one Ca²⁺ ion against the greater than 10,000-fold concentration gradient between the cell interior (2×10^{-7} M) and cell exterior (2×10^{-3} M). As in other muscle cells, a rise in the intracellular Ca²⁺ concentration in cardiac muscle triggers contraction. Thus the operation of the Na⁺/Ca²⁺ antiporter lowers the cytosolic concentration of Ca²⁺ and reduces the strength of heart muscle contraction. The Na⁺/K⁺ ATPase in the plasma membrane of cardiac cells, as in other body cells, creates the Na⁺ concentration gradient used to power export of Ca²⁺ ions.

The drugs *ouabain* and *digoxin* increase the force of heart muscle contraction and are widely used in the treatment of congestive heart failure. The primary effect of these drugs is to inhibit the Na⁺/K⁺ ATPase, thereby raising the intracellular Na⁺ concentration (and lowering intracellular K⁺). Because the Na⁺/Ca²⁺ antiporter functions less efficiently with a lower Na⁺ concentration gradient, fewer Ca²⁺ ions are exported and the intracellular Ca²⁺ concentration increases. This increase causes the muscle to contract more strongly.

AE1 Protein, a Cl⁻/HCO₃⁻ Antiporter, Is Crucial to CO₂ Transport by Erythrocytes

In addition to cation antiporters, which transport only positive ions, many cells also contain *anion transporters*, which transport only negative ions. An important example is *AE1 protein*, the predominant integral protein of the mammalian erythrocyte. This anion antiporter catalyzes the one-for-one exchange of Cl⁻ and HCO₃⁻ across the plasma membrane. Since one singly charged negative ion is exchanged for another, there is no net movement of electric charge and the reaction is not affected by the membrane potential. Thus, the direction of the reaction is dependent only on the concentration gradients of the transported ions.

Transmembrane anion exchange is essential to an important function of the erythrocyte—the transport of waste carbon dioxide (CO_2), which is generated in peripheral tissues, to the lungs for excretion by respiratory exhalation (Figure 15-20). Waste CO_2 released from cells into the capillary blood diffuses across the erythrocyte membrane. In its gaseous form, CO_2 dissolves poorly in aqueous solutions, such as the cytosol or blood plasma, but the potent enzyme *carbonic anhydrase* inside the erythrocyte converts CO_2 to the water-soluble bicarbonate (HCO_3^-) anion:

$$OH^- + CO_2 \rightleftharpoons HCO_3^-$$

Since

$$H^+ + OH^- \rightleftharpoons H_2O$$

we can write the overall reaction for carbonic anhydrase as

$$H_2O + CO_2 \rightleftharpoons HCO_3^- + H^+$$

The release of oxygen from hemoglobin into the peripheral capillaries induces a conformational change in the globin polypeptide that enables a histidine side chain to bind the proton produced by the carbonic anhydrase reaction. Meanwhile, the HCO_3^- formed by carbonic anhydrase is transported out of the erythrocyte in exchange for an entering Cl⁻ via AE1 protein (see Figure 15-20, *top*).

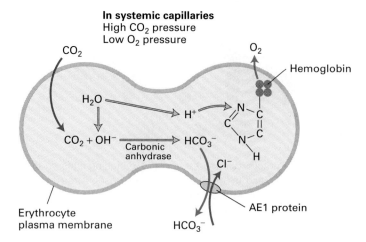

In systemic capillaries
High CO_2 pressure
Low O_2 pressure

Erythrocyte
plasma membrane

Carbonic
anhydrase

H_2O

$CO_2 + OH^-$

H^+

HCO_3^-

Cl^-

AE1 protein

HCO_3^-

O_2

Hemoglobin

CO_2

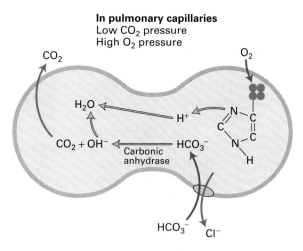

In pulmonary capillaries
Low CO_2 pressure
High O_2 pressure

CO_2

H_2O

$CO_2 + OH^-$

Carbonic
anhydrase

H^+

HCO_3^-

O_2

HCO_3^- Cl^-

▲ **FIGURE 15-20 Schematic drawings showing anion transport across the erythrocyte membrane in systemic and pulmonary capillaries.** AE1 protein (purple)—an anion antiporter—catalyzes the reversible exchange of Cl^- and HCO_3^- ions across the membrane and works in conjunction with carbonic anhydrase. In systemic capillaries, the overall reaction causes HCO_3^- to be released from the cell, which is essential for CO_2 transport from the tissues to the lungs. In the lungs, the overall reaction is reversed. See the text for further discussion.

The entire anion-exchange process in peripheral-blood erythrocytes is completed within 50 milliseconds (ms), during which time 5×10^9 HCO_3^- ions are exported from the cell down its concentration gradient. If anion exchange did not occur, HCO_3^- would accumulate inside the erythrocyte to toxic levels during periods of exercise, when much CO_2 is generated. About 80 percent of the CO_2 in blood is transported as HCO_3^- generated inside erythrocytes; anion exchange allows about two-thirds of this HCO_3^- to be transported by blood plasma external to the cells, increasing the amount of CO_2 that can be transported from tissues to the lungs. Also, without anion exchange, the increased HCO_3^- concentration in the erythrocyte would cause the cytosol to become alkaline. The exchange of HCO_3^- (which we can

think of as equal to $OH^- + CO_2$) for Cl^- causes the cytosolic pH to remain near neutrality.

The overall direction of this anion-exchange process is reversed in the lungs. CO_2 diffuses out of the erythrocyte and is eventually expelled in breathing. The lowered concentration of CO_2 within the cytosol drives the carbonic anhydrase reaction, as written above, from right to left: HCO_3^- reacts to yield CO_2 and OH^-. At the same time, oxygen binding to hemoglobin causes a proton to be released from hemoglobin; the proton combines with the OH^- to form H_2O. The lowered intracellular HCO_3^- concentration causes HCO_3^- to enter the erythrocyte in exchange for Cl^- (see Figure 15-20, *bottom*).

AE1, which has been studied extensively, carries out the precise one-for-one sequential exchange of anions on opposite sides of the membrane required to preserve electroneutrality in the cell; only once every 10,000 or so transport cycles does an anion move unidirectionally from one side of the membrane to the other. AE1 has a large membrane-embedded domain, folded into at least 12 transmembrane α helices, which carry out anion transport, and a cytosolic-facing domain, which anchors certain cytoskeletal proteins to the membrane. Although the precise transport mechanism is not known, conformational changes most likely have a key role as in other membrane transport proteins.

Several Cotransporters Regulate Cytosolic pH

The anaerobic metabolism of glucose yields lactic acid, and aerobic metabolism yields CO_2, which is hydrated by carbonic anhydrase to carbonic acid (H_2CO_3). These weak acids dissociate, yielding H^+ ions (protons); if these protons were not exported from cells, the cytosolic pH would drop precipitously, endangering cellular functioning. Two types of cotransport proteins are employed to remove some of the "excess" protons generated during metabolism of animal cells. One is a *$Na^+ HCO_3^-/Cl^-$ cotransporter*, which imports one Na^+ ion down its concentration gradient, together with one HCO_3^-, in exchange for export of one Cl^- ion against its concentration gradient. The imported HCO_3^- ions combine with protons generated by metabolism to produce CO_2, which diffuses out of the cell. Thus the overall action of this transporter raises the cytosolic pH (reduces the H^+ concentration). Also important in removing excess protons is a *Na^+/H^+ antiporter*, which couples entry of one Na^+ ion into the cell down its concentration gradient to export one H^+ ion.

The plasma membranes of most animal cells also contain a *Na^+-independent Cl^-/HCO_3^- antiporter* similar to the erythrocyte AE1 protein discussed previously. This anion-exchange protein functions to lower the cytosolic pH, in effect removing "excess" OH^- ions. Recall that a HCO_3^- ion can be viewed as a complex of OH^- and CO_2, so export of HCO_3^- lowers the cytosolic pH. Exchange of cytosolic HCO_3^- for extracellular Cl^- is powered by the import of Cl^- down its concentration gradient ($Cl^-_{out} > Cl^-_{in}$).

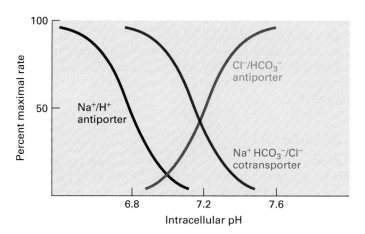

▲ FIGURE 15-21 Effect of intracellular pH on activity of membrane transport proteins that regulate the cytosolic pH of mammalian cells. See the text for discussion. [After S. L. Alper, 1991, *Ann. Rev. Physiol.* **53**:549.]

The activity of all three of these antiport proteins depends on pH, providing cells with a fine-tuned mechanism for controlling the cytosolic pH (Figure 15-21). The proton-exporting transporters, which are activated when the pH of the cytosol falls, act to raise the cytosolic pH. Similarly, a rise in pH above 7 stimulates the Cl^-/HCO_3^- antiporter, leading to a more rapid export of HCO_3^- and decrease in the cytosolic pH. In this manner the cytosolic pH of growing cells is maintained very close to pH 7.4.

Small changes in the cytosolic pH may have profound effects on the overall cellular metabolic rate. For instance, primary fibroblast cells grown to maximal density (confluence) in tissue culture generally become quiescent: DNA synthesis stops; the rates of RNA synthesis, glucose catabolism, and protein synthesis are reduced; and the cytosolic pH drops from the characteristic 7.4 of growing cells to ≈7.2. Treatment of quiescent cells with a mixture of serum growth factors restimulates cell growth and DNA synthesis. An early effect of these growth factors is a marked increase in the cytosolic pH to 7.4; this dramatic change is caused in part by stimulation of the Na^+/H^+ antiport, which expels protons into the medium. The rise in cytosolic pH is believed to help activate certain metabolic pathways required for cell growth and division.

Numerous Transport Proteins Enable Plant Vacuoles to Accumulate Metabolites and Ions

The lumen of plant vacuoles is much more acidic (pH 3 to 6) than is the cytosol (pH 7.5). As noted earlier, the vacuolar membrane contains a V-class ATP-powered pump and a unique PP_i-powered pump, both of which function to pump H^+ ions into the vacuolar lumen against a concentration gradient. As illustrated in Figure 15-22, the vacuolar membrane also contains Cl^- and NO_3^-

channels that transport these anions from the cytosol into the vacuole. Entry of these anions against their concentration gradients is driven by the inside-positive potential generated by the H^+ pumps. Operation of both types of proton pumps in conjunction with these anion channels produces an inside-positive electric potential of about 20 mV across the vacuolar membrane and also a substantial pH gradient.

The proton gradient and electric potential across the plant vacuole membrane are used in much the same way as the Na^+ gradient and electric potential across the animal-cell plasma membrane: to power the selective uptake or extrusion of ions and small molecules. In the leaf, for example, excess sucrose generated during photosynthesis in the day is stored in the vacuole; during the night the stored sucrose moves into the cytoplasm and is metabolized to CO_2 and H_2O with concomitant generation of ATP from ADP and P_i. A *proton-sucrose antiporter* in the vacuolar membrane operates to accumulate sucrose in plant vacuoles. The inward movement of sucrose is powered by the outward movement of H^+, which is favored by its concentration gradient (lumen > cytosol) and by the outward-negative potential across the vacuolar membrane (see Figure 15-22). Uptake of Ca^{2+} and Na^+ into the vacuole from the cytosol against their concentration gradients is similarly mediated by proton antiporters.

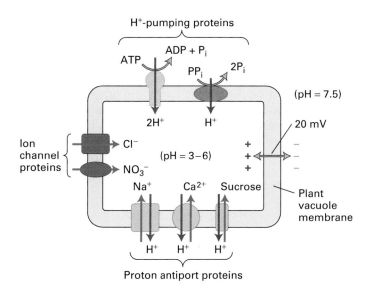

▲ FIGURE 15-22 Concentration of ions and sucrose by the plant vacuole. The vacuolar membrane contains two types of proton pumps: a V-class H^+ ATPase (light green) and a unique pyrophosphate-hydrolyzing proton pump (dark green). These pumps generate a lowered luminal pH as well as an inside-positive electric potential across the vacuolar membrane owing to the inward pumping of H^+ ions. The inside-positive potential powers the movement of Cl^- and NO_3^- from the cytosol through separate channel proteins (dark purple). Proton antiporters (light purple), powered by the H^+ gradient, accumulate Na^+, Ca^{2+}, and sucrose inside the vacuole. [After P. Rea and D. Sanders, 1987, *Physiol. Plant* **71**:131; J. M. Maathuis and D. Sanders, 1992, *Curr. Opin. Cell Biol.* **4**:661; P. A. Rea et al., 1992, *Trends Biochem. Sci.* **17**:348.]

SUMMARY Cotransport by Symporters and Antiporters

- A small molecule or ion may be imported or exported against its concentration gradient by coupling its movement to that of another molecule or ion, usually H^+ or Na^+, down its electrochemical gradient.

- Two forces power the movement of H^+ or Na^+ across a membrane: the electric potential and the ion concentration gradient.

- Entry of glucose and amino acids into certain cells against their concentration gradient is coupled by symport proteins to the energetically favorable entry of Na^+ (see Figure 15-19).

- In cardiac muscle cells, the export of Ca^{2+} is coupled to the import of Na^+ by a cation antiporter, which transports 3 Na^+ ions inward for each Ca^{2+} ion exported.

- The erythrocyte membrane contains a Cl^-/HCO_3^- anion antiporter (AE1 protein) that facilitates transport of CO_2 by the blood (see Figure 15-20).

- Two proton-exporting transporters—a Na^+Cl^-/HCO_3^- cotransporter and a Na^+/H^+ antiporter—maintain the cytosolic pH in animal cells very close to 7.4 despite metabolic production of carbonic and lactic acids. A Na^+-independent Cl^-/HCO_3^- antiporter, similar to AE1 protein, functions to export HCO_3^- when the cytosolic pH rises above normal, causing a decrease in pH.

- Uptake of sucrose, Na^+, Ca^{2+}, and other substances into plant vacuoles is carried out by proton antiporters in the vacuolar membrane. Ion channels in the membrane are critical in generating a proton concentration gradient large enough to power accumulation of ions and metabolites in vacuoles by these proton antiporters (see Figure 15-22).

15.7 Transport across Epithelia

With few exceptions, all the internal and external body surfaces of animals, such as the skin, stomach, and intestines, are covered with a layer of epithelial cells called an **epithelium** (see Figure 6-4). Many epithelial cells transport ions or small molecules from one side to the other of the epithelium. Those lining the stomach, for instance, secrete hydrochloric acid into the stomach lumen, which after a meal becomes pH 1, while those lining the small intestine transport products of digestion (e.g., glucose and amino acids) into the blood. All epithelial cells in a sheet are interconnected by several types of specialized regions of the plasma membrane called **cell junctions**. These impart strength and rigidity to the tissue and prevent water-soluble material on one side of the sheet (as in the intestinal lumen) from moving across to the other side. In this section we first describe the polarized nature of epithelia and how different combinations of membrane proteins enable epithelial cells to carry out their transport or secretory functions. Then we discuss the structure and function of the junctions that interconnect epithelial cells.

The Intestinal Epithelium Is Highly Polarized

An epithelial cell is said to be *polarized* because one side differs in structure and function from the other. In particular, its plasma membrane is organized into at least two discrete regions, each with different sets of transport proteins. In the epithelial cells that line the intestine, for example, that portion of the plasma membrane facing the intestine, the *apical* surface, is specialized for absorption; the rest of the plasma membrane, the lateral and basal surfaces, often referred to as the *basolateral surface*, mediates transport of nutrients from the cell to the surrounding fluids which lead to the blood and forms junctions with adjacent cells and the underlying extracellular matrix called the **basal lamina** (Figure 15-23).

Extending from the lumenal (apical) surface of intestinal epithelial cells are numerous fingerlike projections (100 nm in diameter) called **microvilli** (singular, **microvillus**). Often referred to collectively as the *brush border* because of their appearance, microvilli greatly increase the area of the apical surface and thus the number of transport proteins it can contain, enhancing the absorptive capacity of the intestinal epithelium. A bundle of actin filaments that runs down the center of each microvillus gives rigidity to the projection. Overlying the brush border is the **glycocalyx**, a loose network composed of the oligosaccharide side chains of integral membrane glycoproteins, glycolipids, and enzymes that catalyze the final stages in the digestion of ingested carbohydrates and proteins (Figure 15-24). The action of these enzymes produces monosaccharides and amino acids, which are transported across the intestinal epithelium and eventually into the bloodstream.

Transepithelial Movement of Glucose and Amino Acids Requires Multiple Transport Proteins

Movement of monosaccharides and amino acids from the intestinal lumen into the blood is a two-stage transcellular process. The first stage, *import* of substances from the lumen into intestinal epithelial cells, is carried out by membrane transport proteins in the microvilli on the apical surface of intestinal cells. The second stage, *export* of substances from the cells into the fluid surrounding the basolateral surface, is carried out by other transport proteins on the basolateral plasma membrane. In order for such transepithelial transport to occur, the epithelial cell must be polarized, with different sets of transport proteins localized in the basolateral and apical surfaces. To illustrate this process, we examine the membrane transport proteins required to move glucose

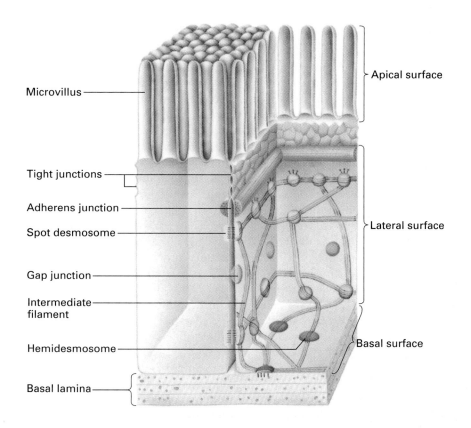

Microvillus

Tight junctions

Adherens junction

Spot desmosome

Gap junction

Intermediate filament

Hemidesmosome

Basal lamina

Apical surface

Lateral surface

Basal surface

◀ **FIGURE 15-23 Schematic diagram of epithelial cells lining the small intestine and the principal types of cell junctions that connect them.** As in all epithelia, the basal surface of the cells rests on the basal lamina, a fibrous network of collagen and proteoglycans that supports the epithelial cell layer. The apical surface faces the intestinal lumen. Tight junctions, lying just under the microvilli, prevent diffusion of substances between the intestinal lumen and the blood via the extracellular space between cells. Gap junctions allow movement of small molecules and ions between the cytosol of adjacent cells. The remaining three types of junctions, adherens junctions, spot desmosomes, and hemidesmosomes are critical to cell-cell and cell-matrix adhesion.

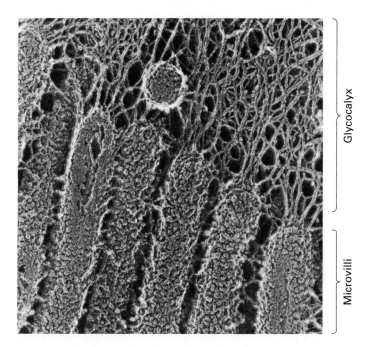

Glycocalyx

Microvilli

▲ **FIGURE 15-24 Micrograph of the microvilli that form the lumenal surface of intestinal epithelial cells, obtained by the deep-etching technique.** The surface of each microvillus is covered with a series of bumps believed to be integral membrane proteins. The glycocalyx, which covers the apices (tips) of the microvilli, is composed of a network of glycoproteins and digestive enzymes. [From N. Hirokawa and J. E. Heuser, 1981, *J. Cell Biol.* **91**:399; courtesy of N. Hirokawa and J. E. Heuser.]

across the epithelial cells lining the intestine and kidney. Similar proteins are used to transport amino acids across these epithelia.

Figure 15-25 depicts the transport of glucose from the intestinal lumen to the blood. Glucose is imported against its concentration gradient from the intestinal lumen across the apical surface of the epithelial cells by a two-Na$^+$/one-glucose symporter located in the microvillar membranes. As noted above, this symporter couples the energetically unfavorable inward movement of one glucose molecule to the energetically favorable inward transport of two Na$^+$ ions (see Figure 15-19). In the steady state, all the Na$^+$ ions transported from the intestinal lumen into the cell during Na$^+$/glucose symport, or the similar process of Na$^+$/amino acid symport, are pumped out across the basolateral membrane, often called the *serosal (blood-facing) membrane.* Thus the low intracellular Na$^+$ concentration is maintained. The Na$^+$/K$^+$ ATPase that accomplishes this is found in these cells exclusively on the basolateral surface of the plasma membrane. The coordinated operation of these transporters allows uphill movement of glucose and amino acids from the intestine into the cell, and ultimately is powered by ATP hydrolysis by the Na$^+$/K$^+$ ATPase.

Glucose and amino acids concentrated inside intestinal cells by symporters are exported down their concentration gradients into the blood via uniport proteins in the basolateral membrane. In the case of glucose, this movement is mediated by *GLUT2*, a glucose transporter that is localized in the basal and lateral membranes of intestinal cells (see

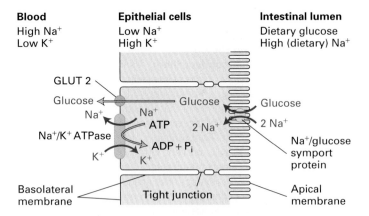

Blood
High Na$^+$
Low K$^+$

Epithelial cells
Low Na$^+$
High K$^+$

Intestinal lumen
Dietary glucose
High (dietary) Na$^+$

▲ **FIGURE 15-25 Transport of glucose from the intestinal lumen into the blood.** Activity of the Na$^+$/K$^+$ ATPase (green) in the basolateral surface membrane generates Na$^+$ and K$^+$ concentration gradients, and the K$^+$ gradient generates an inside-negative membrane potential. Both the Na$^+$ concentration gradient and the membrane potential are used to drive the uptake of glucose from the intestinal lumen by the two-Na$^+$/one-glucose symporter (blue) located in the apical surface membrane. Glucose leaves the cell via facilitated diffusion catalyzed by GLUT2 (orange), a glucose uniporter located in the basolateral membrane.

Figure 5-1c). (GLUT2 is a homolog of GLUT1; as discussed earlier; however, GLUT1 generally functions to import glucose into many body cells.) The net result of the operation of these various transport proteins is movement of Na$^+$ ions, amino acids, and glucose from the intestinal lumen across the intestinal epithelium into the interstitial spaces surrounding the cells, and eventually into the blood. Tight junctions between the epithelial cells prevent these molecules from diffusing back into the intestinal lumen.

The epithelial cells lining kidney tubules, which have an architecture similar to that of intestinal epithelial cells, reabsorb glucose from the blood filtrate that is the forming urine and return it to the blood. In the first part of a kidney tubule, the epithelial cells transport glucose against a relatively small glucose concentration gradient. These cells utilize a second type of Na$^+$/glucose symport protein—a *one-Na$^+$/one-glucose symporter*, which has a high transport rate but cannot transport glucose against a steep concentration gradient. At the intracellular Na$^+$ concentration and membrane potential depicted in Figure 15-9, this symporter can generate an intracellular glucose concentration ≈ 100 times that of the extracellular medium (here the forming urine). In the latter part of a kidney tubule, however, the epithelial cells take up the remaining glucose against a more than 100-fold glucose concentration gradient. To accomplish this, these cells contain in their apical membrane the same two-Na$^+$/one-glucose symporter found in intestinal epithelial cells. The two types of Na$^+$/glucose symport proteins are similar in amino acid sequence, predicted structure, and mechanism but have evolved to transport glucose under different conditions.

Parietal Cells Acidify the Stomach Contents While Maintaining a Neutral Cytosolic pH

The mammalian stomach contains a 0.1 M solution of hydrochloric acid (H$^+$Cl$^-$). This strongly acidic medium denatures many ingested proteins before they are degraded by proteolytic enzymes in the stomach (e.g., pepsin) that function at acidic pH. Hydrochloric acid is secreted into the stomach by *parietal cells* (also known as *oxyntic cells*) in the gastric lining. These cells contain a *H$^+$/K$^+$ ATPase* in their apical membrane, which faces the stomach lumen and generates a concentration of H$^+$ ions 10^6 times greater in the stomach lumen than in the cell cytosol (pH = 1.0 versus pH = 7.0). This enzyme is a P-class ATPase, similar in structure and function to the Na$^+$/K$^+$ ATPase discussed earlier. Operation of the Na$^+$/K$^+$ ATPase results in a *net* outward movement of one charged ion per ATP (see Figure 15-13). In contrast, the action of the H$^+$/K$^+$ ATPase, which exports one H$^+$ ion and imports one K$^+$ ion for each ATP hydrolyzed, produces no *net* movement of electric charge. The numerous mitochondria in parietal cells produce abundant ATP for use by the H$^+$/K$^+$ ATPase.

If parietal cells simply exported H$^+$ ions in exchange for K$^+$ ions, a rise in the concentration of OH$^-$ ions and thus a marked rise in cytosolic pH would occur, since in the cytosol, as in all aqueous solutions, the product of the H$^+$ and OH$^-$ concentrations is a constant (10^{-14} M^2). However, during acidification of the stomach lumen, the pH of the parietal-cell cytosol remains neutral. Parietal cells accomplish this feat by means of a Cl$^-$/HCO$_3^-$ antiporter in the basolateral membrane (Figure 15-26). The "excess" cytosolic OH$^-$, generated by exporting protons, combines with CO$_2$ that diffuses into the cell from the blood, forming HCO$_3^-$ in a reaction catalyzed by cytosolic carbonic anhydrase. The HCO$_3^-$ ion is transported across the basolateral membrane of the cell into the blood in exchange for an incoming Cl$^-$ ion by means of an anion antiporter that is similar in structure and function to the erythrocyte AE1. The Cl$^-$ ions thus imported into the cell exit through Cl$^-$ channels in the apical membrane, entering the stomach lumen. To preserve electroneutrality, each Cl$^-$ ion that moves into the stomach lumen across the apical membrane is accompanied by a K$^+$ ion that moves outward through a separate K$^+$ channel. In this way, the excess K$^+$ ions pumped inward by the H$^+$/K$^+$ ATPase are returned to the stomach lumen, thus maintaining the intracellular K$^+$ concentration. The net result is accumulation of both H$^+$ and Cl$^-$ ions (i.e., HCl) in the stomach lumen, while the pH of the cytosol remains neutral and the excess OH$^-$ ions, as HCO$_3^-$, are transported into the blood.

Tight Junctions Seal Off Body Cavities and Restrict Diffusion of Membrane Components

For polarized epithelial cells to carry out their transport functions, extracellular fluids surrounding their apical and basolateral membranes must be kept separate. This is

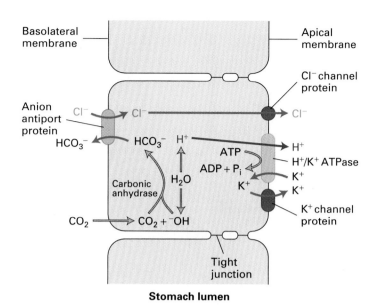

▲ FIGURE 15-26 Acidification of the stomach lumen by parietal cells in the gastric lining. The apical membrane of parietal cells contains a H^+/K^+ ATPase (a P-class pump) as well as Cl^- and K^+ channel proteins. Note the cyclic K^+ transport across the apical membrane: K^+ ions are pumped inward by the H^+/K^+ ATPase and exit via a K^+ channel. The basolateral membrane contains an anion antiporter that exchanges HCO_3^- and Cl^- ions. The combined operation of these four different transport proteins acidifies the stomach lumen while maintaining the neutral pH and electroneutrality of the cytosol. See the text for more details.

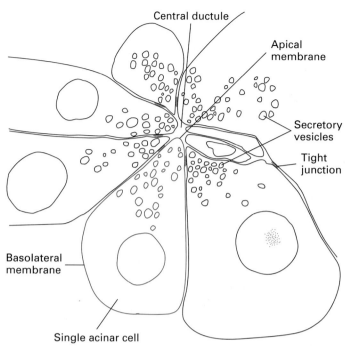

▲ FIGURE 15-27 Diagram of pancreatic acinar cells. An acinus is a spherical aggregate of about a dozen cells; the lumen of an acinus is connected to a ductule that merges with other ductules and eventually leads into a main pancreatic duct, which empties into the lumen of the small intestine (Figure 5-48). Acinar cells synthesize degradative enzymes and store them as inactive precursors (zymogens) in secretory vesicles, which cluster under the apical region of the plasma membrane adjacent to the ductule. The basolateral membrane covers the sides of an acinar cell below the apical (lumen-facing) surface and extends along the base of the cell; nutrients in the blood in the surrounding vessels are transported through this region of the plasma membrane into the cell. Note the tight junctions (orange) just below the apical region between adjacent cells; they prevent movement of substances between the central ductule and the blood.

accomplished by **tight junctions**, which connect adjacent epithelial cells and usually are located just below the apical surface (see Figure 15-23). These specialized regions of the plasma membrane form a barrier that seals off body cavities such as the intestine, the stomach lumen, ductules in pancreatic acini, and the bile duct in the liver. For example, tight junctions prevent diffusion of small molecules directly from the intestinal lumen into the interstitial spaces that surround the basolateral plasma membrane and that lead to the blood. Thus intestinal epithelial cells must transport nutrients through the cells as previously described. In the pancreas, tight junctions between acinar cells likewise prevent leakage of secreted proteins, including digestive enzymes, from the central ductules into the blood (Figure 15-27). Tight junctions also prevent diffusion of membrane proteins and glycolipids between the apical and basolateral regions of the plasma membrane, ensuring that these regions contain different membrane components.

Structure of Tight Junctions Tight junctions are composed of thin bands of plasma-membrane proteins that completely encircle a polarized cell and are in contact with similar thin bands on adjacent cells. When thin sections of cells are viewed in an electron microscope, the plasma membranes of adjacent cells appear to touch each other at intervals and

even to fuse (Figure 15-28a). Freeze-fracture electron microscopy affords a striking view of the tight junction. The microvillar tight junction shown in Figure 15-28b appears to comprise an interlocking network of ridges in the plasma membrane. More specifically, there appear to be ridges on the cytosolic face of the plasma membrane of each of the two contacting cells. (Corresponding grooves not shown here are found on the exoplasmic face.) High magnification reveals that these ridges are made up of protein particles 3–4 nm in diameter. In the model shown in Figure 15-28c, the tight junction is formed by a double row of these particles, one row donated by each cell.

The two principal integral membrane proteins found in tight junctions are *occludin* and *claudin*. Each of these proteins has four membrane-spanning α helices. Although the molecular structure of the junction is not known, the extracellular domains of rows of occludin and claudin proteins

(a)

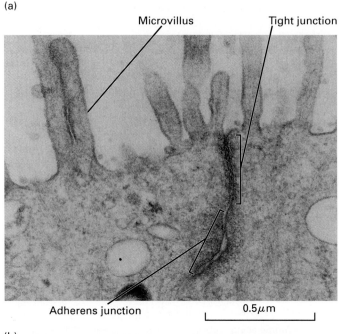

Microvillus   Tight junction

Adherens junction   0.5μm

(b)

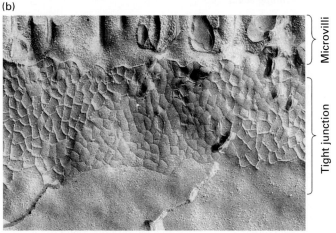

Microvilli

Tight junction

(c)

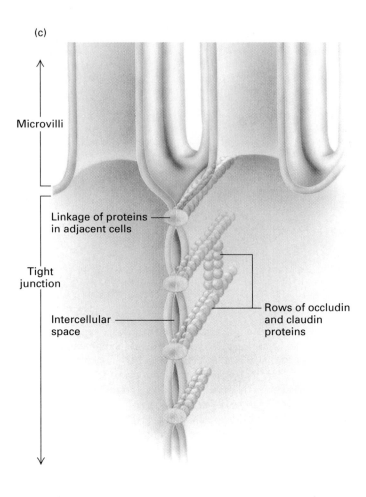

Microvilli

Linkage of proteins in adjacent cells

Tight junction

Intercellular space

Rows of occludin and claudin proteins

▲ **FIGURE 15-28 Tight junctions.** (a) Thin-section electron micrograph of the apical region of two liver epithelial cells, illustrating the tight junction just below the microvilli and the adherens junction. From the apical region of these liver cells, which faces the lumen of the bile duct, phospholipids and other components of bile are secreted into the duct. (b) Freeze-fracture electron micrograph of a tight junction between two intestinal epithelial cells. The fracture plane passes through the plasma membrane of one of the two adjacent cells. The honeycomblike network of ridges of particles below the microvilli forms the tight junction. (c) A model showing how a tight junction might be formed by linkage of rows of protein particles in adjacent cells (see also Figure 15-23). [Part (a) from P. A. Cross and K. L. Mercer, 1993, *Cell and Tissue Ultrastructure, A Functional Perspective*, W. H. Freeman and Company, p. 50; part (b) courtesy of L. A. Staehelin; part (c) adapted from L. A. Staehelin and B. E. Hull, 1978, *Sci. Am.* **238**(5):140, and D. Goodenough, 1999, *Proc. Natl. Acad. Sci. USA* **96**:319.]

in the plasma membrane of one cell probably form extremely tight links with similar rows of claudin and occludin in the adjacent cell, essentially fusing two adjacent cells and creating an impenetrable seal. Treatment of an epithelium with the protease trypsin destroys the tight junctions, supporting the proposal that proteins are essential structural components of these junctions.

The long C-terminal cytosolic-facing domain of occludin is bound to one of a group of large cytosolic proteins (ZO-1,

ZO-2, and ZO-3) that, in turn, are bound to other cytoskeletal proteins and to actin fibers. These interactions appear to stabilize the linkage between occludin molecules that is essential for integrity of the tight junction (Chapter 22).

Impermeability of Tight Junctions to Aqueous Solutions That tight junctions are impermeable to most water-soluble substances can be demonstrated in an experiment in which lanthanum hydroxide (an electron-dense colloid of high molecular weight) is injected into the pancreatic blood

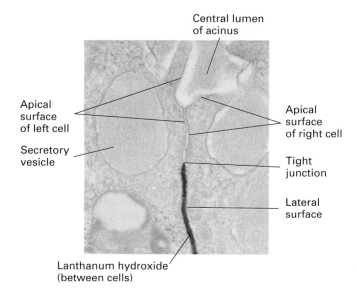

Central lumen
of acinus

Apical
surface
of left cell

Secretory
vesicle

Apical
surface
of right cell

Tight
junction

Lateral
surface

Lanthanum hydroxide
(between cells)

▲ **FIGURE 15-29 Experimental demonstration that tight junctions prevent passage of water-soluble substances.** Pancreatic acinar tissue is fixed and prepared for microscopy a few minutes after electron-opaque lanthanum hydroxide is injected into the blood of an experimental animal. As shown in this electron micrograph of adjacent acinar cells, the lanthanum hydroxide can penetrate between the cells but is arrested at the level of the tight junction. [Courtesy of D. Friend.]

vessel of an experimental animal; a few minutes later the pancreatic acinar cells are fixed and prepared for microscopy. As shown in Figure 15-29, the lanthanum hydroxide diffuses from the blood into the space that separates the lateral surfaces of adjacent acinar cells, but cannot penetrate past the outermost tight junction.

Other studies have shown that tight junctions also are impermeable to salts. For instance, when MDCK cells are grown in a medium containing very low concentrations of Ca^{2+}, they form a monolayer in which the cells are not connected by tight junctions; as a result, fluids and salts flow freely across the cell layer. When Ca^{2+} is added to such a monolayer, tight junctions form within an hour, and the cell layer becomes impermeable to fluids and salts (see Figure 6-7).

Ability of Tight Junctions to Block Diffusion of Proteins and Lipids in the Plane of the Plasma Membrane When liposomes containing a fluorescent-tagged glycoprotein are added to the medium in contact with the apical surface of a monolayer of MDCK cells, some spontaneously fuse with the plasma membrane. Fluorescent glycoprotein is detectable in the apical but not in the basolateral surface of the cells so long as the tight junctions between adjacent cells are intact. However, if the tight junctions are destroyed by removing Ca^{2+} from the medium, the fluorescent protein is soon detectable in the basolateral surface, indicating that it can diffuse from the apical to the basolateral regions of the plasma membrane. These results indicate that plasma membrane proteins cannot diffuse through tight junctions.

Lipids in the cytosolic leaflets of the apical and basolateral membranes of epithelial cells have the same composition and apparently can diffuse from one region of the membrane to the other. In contrast, the lipid compositions of the exoplasmic leaflets of the apical and basolateral membrane regions are very different, and membrane lipids in the exoplasmic leaflets cannot diffuse through tight junctions. All the glycolipid in MDCK cells, for instance, is present in the exoplasmic face of the apical membrane, as are all proteins anchored to the membrane by fatty acids linked to a glycosylphosphatidylinositol group (see Figure 3-36a). In fact, the only lipids in the exoplasmic leaflet of the apical plasma membrane are glycolipids, fatty acid components of glycosylphosphatidylinositol anchors, and cholesterol. Phosphatidylcholine, conversely, is present almost exclusively in the exoplasmic face of the basolateral plasma membrane.

Other Junctions Interconnect Epithelial Cells and Control Passage of Molecules between Them

In order to function in an integrated manner, the individual cells composing epithelia and other organized tissues must adhere to one another and to the surrounding extracellular matrix and also control the movement of ions and small molecules between them. Several specialized cell junctions are critical to adhesion and passage of molecules between cells in tissues (see Figure 15-23).

Three types of cell junctions, called **desmosomes**, function in cell-cell and cell-matrix adhesion. Epithelial and some other types of cells, such as smooth muscle, are bound tightly together by *spot desmosomes*. These are buttonlike points of contact between cells, often thought of as a "spot-weld" between adjacent plasma membranes, that confer mechanical strength on these tissues. *Hemidesmosomes*, similar in structure to spot desmosomes, anchor the plasma membrane to regions of the extracellular matrix. Bundles of intermediate filaments course through the cell, interconnecting spot desmosomes and hemidesmosomes. Finally, *adherens junctions* (also known as *belt desmosomes*), which are found primarily in epithelial cells, form a belt of cell-cell adhesion just under the tight junctions.

The lateral surfaces of adjacent cells contain numerous **gap junctions**. These junctions help to integrate the metabolic activities of all cells in a tissue by allowing the direct passage of ions and small molecules from the cytosol of one cell to that of another (see the chapter opening figure). Among these are intracellular signaling molecules (e.g., cyclic AMP and Ca^{2+}) and precursors of DNA and RNA.

Electron micrographs of animal tissue sections have shown that a space of about 20 nm ordinarily is present between the nonjunctional regions of plasma membranes of adjacent cells. This space contains integral membrane and extracellular surface glycoproteins that assist junctions in intercellular adhesion.

An understanding of the structure and function of desmosomes requires knowledge about actin microfilaments and intermediate filaments. Likewise, an understanding of gap junctions and their equivalent in plant cells (plasmodesmata) depends on knowledge of cellular metabolism and signaling. Therefore, we defer detailed discussion of these junctions until later chapters when these related topics are examined.

SUMMARY Transport across Epithelia

- The apical and basolateral plasma membrane domains of epithelial cells contain different transport proteins and carry out quite different transport processes.

- In the intestinal epithelial cell, Na^+/glucose and Na^+/amino acid symporters are in the apical membrane region facing the intestinal lumen, while Na^+/K^+ ATPases and glucose and amino acid uniporters are in the basolateral membrane region facing the blood capillaries. The coordinated operation of these membrane transport proteins allows the uphill transepithelial movement of amino acids and glucose from the lumen to the blood, powered by ATP hydrolysis by the Na^+/K^+ ATPase (see Figure 15-25).

- Parietal cells in the stomach lining, which secrete HCl into the lumen, have ATP-powered H^+/K^+ pumps, K^+ channels, and Cl^- channels on the apical membrane and pH-sensitive Cl^-/HCO_3^- antiporters on the basolateral membrane. The combined action of these proteins allows the cytosolic pH to be maintained near neutrality, despite the active export of protons from the cells into the stomach lumen, causing its acidification (see Figure 15-26).

- The plasma membrane contains specialized regions that form various types of cell junctions between adjacent cells (see Figure 15-23).

- Tight junctions interconnecting epithelial and other polarized cells seal off body cavities and restrict diffusion of plasma-membrane proteins from the apical to the basolateral surfaces. Tight junctions also prevent diffusion of lipids in the exoplasmic (but not the cytosolic leaflet) from the apical to the basolateral domains of the plasma membrane.

- Adherens junctions and spot desmosomes bind the plasma membranes of adjacent cells in a way that gives strength and rigidity to the entire tissue. Hemidesmosomes help connect cells to the extracellular matrix.

- Gap junctions in animal cells and plasmodesmata in plant cells interconnect the cytosol of two adjacent cells, allowing small molecules and ions to pass between them.

15.8 Osmosis, Water Channels, and the Regulation of Cell Volume

In this section, we examine two types of transport phenomena that, at first glance, may seem unrelated: the regulation of cell volume in both plant and animal cells, and the *bulk flow* of water (the movement of water containing dissolved solutes) across one or more layers of cells. In humans, for example, water moves from the blood filtrate that will form urine across a layer of epithelial cells lining the kidney tubules and into the blood, thus concentrating the urine. (If this did not happen, one would excrete several liters of urine a day!) In higher plants, water and minerals are absorbed by the roots and move up the plant through conducting tubes (the *xylem*); water is lost from the plant mainly by evaporation from the leaves. What these processes have in common is **osmosis**—the movement of water from a region of lower solute concentration to a region of higher solute concentration. We begin with a consideration of some basic facts about osmosis, and then show how they explain several physiological properties of animals and plants.

Osmotic Pressure Causes Water to Move across Membranes

As noted early in this chapter, most biological membranes are relatively impermeable to ions and other solutes, but like all phospholipid bilayers, they are somewhat permeable to water (see Figure 15-1). Permeability to water is increased by water-channel proteins discussed below. Water tends to move across a membrane from a solution of low solute concentration to one of high. Or, in other words, since solutions with a high amount of dissolved solute have a lower concentration of water, water will move from a solution of high water concentration to one of lower. This process is known as *osmotic flow*.

Osmotic pressure is defined as the hydrostatic pressure required to stop the net flow of water across a membrane separating solutions of different compositions (Figure 15-30). In this context, the "membrane" may be a layer of cells or a plasma membrane. If the membrane is permeable to water but not to solutes, the osmotic pressure across the membrane is given by

$$\pi = RT(C_B - C_A) = RT\,\Delta C \qquad (15\text{-}11)$$

where π is the osmotic pressure in atmospheres (atm) or millimeters of mercury (mmHg); R is the gas constant; T is the absolute temperature; and ΔC is the difference in total solute concentrations, C_A and C_B, on each side of the membrane. It is the *total* number of solute molecules that is important. For example, a 0.5 M NaCl solution is actually 0.5 M Na^+ ions and 0.5 M Cl^- ions and has approximately the same osmotic pressure as a 1 M solution of glucose or lactose.

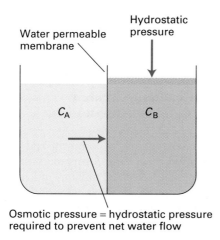

▲ FIGURE 15-30 Experimental system for demonstrating osmotic pressure. Solutions A and B are separated by a membrane that is permeable to water but impermeable to all solutes. If C_B (the total concentration of solutes in solution B) is greater than C_A, water will tend to flow across the membrane from solution A to solution B. The osmotic pressure π between the solutions is the hydrostatic pressure that would have to be applied to solution B to prevent this water flow. From the van't Hoff equation, $\pi = RT(C_B - C_A)$.

From Equation 15-11 we can calculate that a hydrostatic pressure of 0.22 atm (167 mmHg) would just balance the water flow across a semipermeable membrane produced by a concentration gradient of 10 mM sucrose or 5 mM NaCl.

Different Cells Have Various Mechanisms for Controlling Cell Volume

Animal cells will swell when they are placed in a **hypotonic** solution (i.e., one in which the concentration of solutes is *lower* than it is in the cytosol). Some cells, such as erythrocytes, will actually burst as water enters them by osmotic flow. Rupture of the plasma membrane by a flow of water into the cytosol is termed *osmotic lysis*. Immersion of all animal cells in a **hypertonic** solution (i.e., one in which the concentration of solutes is *higher* than it is in the cytosol) causes them to shrink as water leaves them by osmotic flow.

Consequently, it is essential that animal cells be maintained in an **isotonic** medium, which has a solute concentration close to that of the cell cytosol (see Figure 5-22).

Even in an isotonic environment, all animal cells face a problem in maintaining their cell volume. Cells contain a large number of charged macromolecules and small metabolites that attract ions of opposite charge (e.g., K^+, Ca^{2+}, PO_4^{3-}). Also recall that there is a slow leakage of extracellular ions, particularly Na^+ and Cl^-, into cells down their concentration gradient. As a result of these factors, in the absence of some countervailing mechanism, the cytosolic solute concentration would increase, causing an osmotic influx of water and eventually cell lysis. To prevent this, animal cells actively export inorganic ions as rapidly as they leak in. The export of Na^+ by the ATP-powered Na^+/K^+ pump plays the major role in this mechanism for preventing cell swelling. If cultured cells are treated with an inhibitor that prevents production of ATP, they swell and eventually burst, demonstrating the importance of active transport in maintaining cell volume.

PLANTS Unlike animal cells, plant, algal, fungal, and bacterial cells are surrounded by a rigid cell wall. Because of the cell wall, the osmotic influx of water that occurs when such cells are placed in a hypotonic solution (even pure water) leads to an increase in intracellular pressure but not in cell volume. In plant cells, the concentration of solutes (e.g., sugars and salts) usually is higher in the vacuole than in the cytosol, which in turn has a higher solute concentration than the extracellular space. The osmotic pressure, called *turgor pressure*, generated from the entry of water into the cytosol and then into the vacuole pushes the cytosol and the plasma membrane against the resistant cell wall. Cell elongation during growth occurs by a hormone-induced localized loosening of a region of the cell wall, followed by influx of water into the vacuole, increasing its size (see Figure 22-33).

Although most protozoans (like animal cells) do not have a rigid cell wall, many contain a contractile vacuole that permits them to avoid osmotic lysis. A contractile vacuole takes up water from the cytosol and, unlike a plant vacuole, periodically discharges its contents through fusion with the plasma membrane (Figure 15-31). Thus,

(a)

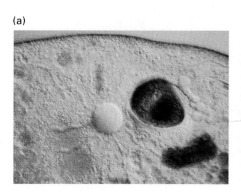

(b)

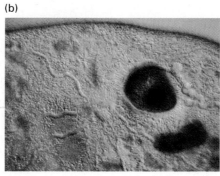

◀ FIGURE 15-31 The contractile vacuole in *Paramecium caudatum,* a typical ciliated protozoan, as revealed by Nomarski microscopy of a live organism. The vacuole is filled by radiating canals that collect fluid from the cytosol. When the vacuole is full, it fuses for a brief period with the plasma membrane and expels its contents. (a) A full vacuole and system of radiating canals. (b) A nearly empty vacuole; the radiating canals are collecting more fluid from the cytosol to refill it.

even though water continuously enters the protozoan cell by osmotic flow, the contractile vacuole prevents too much water from accumulating in the cell and swelling it to the bursting point.

Water Channels Are Necessary for Bulk Flow of Water across Cell Membranes

Even though a pure phospholipid bilayer is only slightly permeable to water, small changes in extracellular osmotic strength cause most animal cells to swell or shrink rapidly. In contrast, frog oocytes and eggs, which have an internal salt concentration comparable to other cells (≈ 150 mM), do not swell when placed in pond water of very low osmotic strength. These observations led investigators to suspect that the plasma membranes of erythrocytes and other cell types contain water-channel proteins that accelerate the osmotic flow of water. The absence of these water channels in frog oocytes and eggs protects them from osmotic lysis.

Microinjection experiments with mRNA encoding *aquaporin*, an erythrocyte membrane protein, provided convincing evidence that this protein increases the permeability of cells to water (Figure 15-32). In its functional form, aquaporin is a tetramer of identical 28-kDa subunits, each of which contains six transmembrane α helices that form three pairs of homologs in an unusual orientation (Figure 15-33a). The channel through which water moves is thought to be lined by eight transmembrane α helices, two from each subunit (Figure 15-33b). Aquaporin or homologous proteins are expressed in abundance in erythrocytes and in other cells

(e.g., the kidney cells that resorb water from the urine) that exhibit high permeability for water.

Simple Rehydration Therapy Depends on Osmotic Gradient Created by Absorption of Glucose and Na$^+$

An understanding of osmosis and the intestinal absorption of glucose forms the basis for a simple therapy that has saved millions of lives, particularly in less-developed countries. In these countries, diarrhea caused by cholera and other intestinal pathogens is a major cause of death of young children. A cure demands not only killing the bacteria with antibiotics, but also *rehydration*—replacement of the water that is lost from the blood and other tissues.

Simply drinking water does not help, because it is excreted from the gastrointestinal tract almost as soon as it enters. To understand the simple therapy that is used, recall that absorption of glucose by the small intestine involves the coordinated movement of Na$^+$; one cannot be transported without the other (see Figure 15-25). The movement of NaCl and glucose from the intestinal lumen, across the epithelial cells, and into the blood creates a transepithelial osmotic gradient, forcing movement of water from the intestinal lumen into the blood. Thus, giving affected children a solution of sugar and salt to drink (but not sugar or salt alone) causes the bulk flow of water into the blood from the intestinal lumen and leads to rehydration.

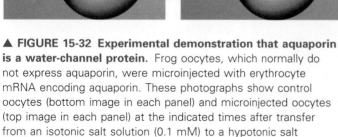

   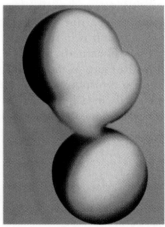

▲ FIGURE 15-32 Experimental demonstration that aquaporin is a water-channel protein. Frog oocytes, which normally do not express aquaporin, were microinjected with erythrocyte mRNA encoding aquaporin. These photographs show control oocytes (bottom image in each panel) and microinjected oocytes (top image in each panel) at the indicated times after transfer from an isotonic salt solution (0.1 mM) to a hypotonic salt solution (0.035 M). The volume of the control oocytes remained unchanged, because they are poorly permeable to water. In contrast, the microinjected oocytes expressing aquaporin swelled because of an osmotic influx of water, indicating that aquaporin increases their permeability to water. [Courtesy of Gregory M. Preston and Peter Agre, Johns Hopkins University School of Medicine.]

(a)

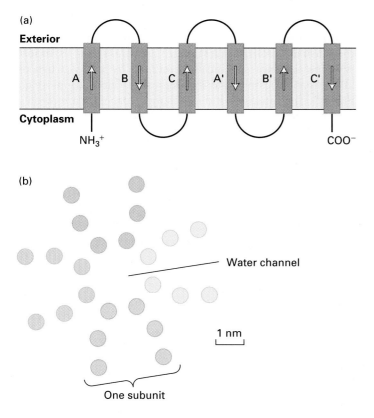

(b)

Water channel

1 nm

One subunit

◄ FIGURE 15-33 The structure of aquaporin, a water-channel protein in the erythrocyte plasma membrane. This tetrameric protein has four identical subunits. (a) Schematic model of an aquaporin subunit showing the three pairs of homologous transmembrane α helices, A and A′, B and B′, and C and C′. As indicated by the arrows showing the N-terminal → C-terminal directionality of the helices, the homologous segments are oriented in the opposite direction. (b) Head-on view of tetrameric aquaporin showing the packing of the transmembrane α helices in the plane of the membrane, as determined by x-ray crystallography. The helices (represented as circles) in each of the four subunits are shown in different colors. Although the identity of the two helices from each subunit that line the central channel is not known, they probably are a pair of homologous segments. The opposite orientation of the two helices in a pair within the membrane would account for the ability of the channel to transport water equally in both directions across the membrane. [Adapted from A. Chang et al., 1997, *Nature* **387**:627.]

chlorophyll-laden mesophyll cells in the leaf interior. As CO_2 enters a leaf, water vapor is simultaneously lost—a process that can be injurious to the plant. Thus it is essential that the stomata open only during periods of light, when photosynthesis occurs; even then, they must close if too much water vapor is lost.

Two guard cells surround each stomate (Figure 15-34a). Changes in turgor pressure lead to changes in the shape of these guard cells, thereby opening or closing the pores. Stomatal opening is caused by an increase in the concentration of ions or other solutes within the guard cells because of (1) opening of K^+ and Cl^- channels and the subsequent influx of K^+ and Cl^- ions from the environment, (2) the metabolism of stored sucrose to smaller compounds, or (3) a combination of these two processes. The resulting increase in the intracellular solute concentration causes water to enter the guard cells osmotically, increasing their

Changes in Intracellular Osmotic Pressure Cause Leaf Stomata to Open

Although most plants cells do not change their volume or shape because of the osmotic movement of water, the opening and closing of *stomata*—the pores through which CO_2 enters a leaf—provides an important exception. The external epidermal cells of a leaf are covered by a waxy cuticle that is largely impenetrable to water and to CO_2, a gas required for photosynthesis by the

(a)

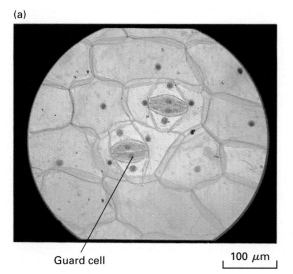

Guard cell

100 μm

(b)

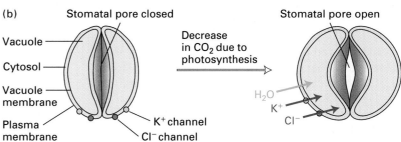

◄ FIGURE 15-34 The opening and closing of stomata. (a) Light micrograph of a leaf of a wandering Jew (*Tradescantia* sp) plant shows two stomata, each surrounded by a pair of guard cells. (b) Opening of K^+ and Cl^- channels in the plasma membrane of the guard cells is followed by an influx of K^+ and Cl^- into the cytosol and then into the vacuole. This triggers the osmotic influx of water, causing the cells to bulge and opening the stomatal pore. [See D. J. Cosgrove and R. Hedrich, 1991, *Planta* **186**:143. Part (a) courtesy Runk/Schoenberger, from Grant Heilman.]

turgor pressure (Figure 15-34b). Since the guard cells are connected to each other only at their ends, the turgor pressure causes the cells to bulge outward, opening the stomatal pore between them. Stomatal closing is caused by the reverse process—a decrease in solute concentration and turgor pressure within the guard cells.

Stomatal opening is under tight physiological control by at least two mechanisms. A drop in CO_2 within the leaf, resulting from active photosynthesis, causes the stomata to open, permitting additional CO_2 to enter the leaf interior so that photosynthesis can continue. When more water exits the leaf than enters it from the roots, the mesophyll cells produce the hormone abscissic acid, which causes K^+ efflux from the guard cells; water then exits the cells osmotically, and the stomata close, protecting the leaf from further dehydration.

SUMMARY Osmosis, Water Channels, and the Regulation of Cell Volume

- Most biological membranes are more permeable to water than to ions or other solutes, and water moves across them by osmosis from a solution of lower solute concentration to one of higher solute concentration.

- Animal cells swell or shrink when placed in hypotonic or hypertonic solutions, respectively. To maintain their normal cytosolic osmolarity and hence cell volume, animal cells must export Na^+ and other ions that leak or are transported from the extracellular space into the cytosol.

- The rigid cell wall surrounding plant cells prevents their swelling and leads to generation of turgor pressure in response to the osmotic influx of water.

- In response to the entry of water, protozoans maintain their normal cell volume by extruding water from contractile vacuoles.

- Aquaporin in the erythrocyte plasma membrane and other water-channel proteins increase the water permeability of biomembranes (see Figure 15-33).

- Opening and closing of K^+ and Cl^- channels and the resulting changes in cytosolic solute concentrations of guard cells cause stomata in leaves to open and close (see Figure 15-34).

PERSPECTIVES for the Future

In this chapter, we have explained certain aspects of human physiology in terms of the action of specific membrane transport proteins. Such a molecular physiology approach has many medical applications. Even today, specific inhibitors or activators of channels, pumps, and transporters constitute the largest single class of drugs. For instance, an inhibitor of the gastric H^+/K^+

ATPase that acidifies the stomach is the most widely used drug for treating stomach ulcers. As we discuss in Chapter 21, the plasma membrane of nerve cells contains Na^+-symport proteins that are specifically inhibited by many drugs of abuse (e.g., cocaine) and antidepression medications (e.g., Prozac). Inhibitors of kidney channel proteins are widely used to control hypertension (high blood pressure); by blocking resorption of water from the forming urine into the blood, these drugs reduce blood volume and thus blood pressure. Calcium-channel blockers are widely employed to control the intensity of contraction of the heart.

Within the next years, the human genome project will generate the sequences of all human membrane transport proteins. Soon after, researchers will discover in which types of cells and tissues these proteins are expressed. Using recombinant DNA techniques, scientists will be able to generate lines of cultured cells that express these in abundance, so that their molecular properties can be studied. Gene-knockout studies in mice will provide clues to their role in human physiology and disease.

This basic knowledge will enable drug company researchers to identify new types of compounds that inhibit or activate just one of these transport proteins and not its homologs expressed in other types of cells. In this way, new and highly specific drugs will be developed to treat a variety of diseases. Physicians will also be able to identify individuals who may be at risk for certain types of diseases (e.g., hypertension or diabetes) because they have mutations in certain membrane transport proteins. And at the level of basic biology, we will all learn precisely how the human body digests and metabolizes all kinds of food and controls the levels of sugars, salts, fats, and other essential molecules in the blood and tissues.

PERSPECTIVES in the Literature

The *Escherichia coli* lactose permease is one of the most extensively studied membrane transport proteins. Encoded by the y gene of the lac operon (see Figure 10-1), this proton–lactose symporter is essential for transport of lactose into the bacterial cell against its concentration gradient. Despite extensive efforts, the lactose permease has not been crystallized, and thus no three-dimensional structure is available. Nonetheless, Cysteine-scanning mutagenesis, in which every amino acid, in turn, is changed to cysteine, has been particularly revealing. The "new" cysteine residue can be subjected to a number of chemical modifications to determine its location relative to other amino acids and its involvement in lactose binding and transport. As you read the recent papers listed below, and earlier ones referenced in Frillingos et al., focus on the following questions:

1. What types of chemical, biochemical, and biophysical techniques were applied to the mutant proteins? What types of data do these different techniques produce?

2. How did such studies lead to a detailed three-dimensional model of the arrangements of the twelve membrane-spanning α-helices?

3. How did these studies identify amino acid residues crucial for binding of lactose at the exoplasmic surface and residues thought to conduct protons through the protein?

4. How did the authors determine that binding of the substrate induced tilting of several membrane-spanning α-helices that might accompany transport of the proton and sugar?

5. What other experimental techniques might be applied to shed additional light on the structure and function of this protein?

Frillingos, S., et al. 1998. Cys-scanning mutagenesis: a novel approach to structure function relationships in polytopic membrane proteins. *FASEB J.* **12**:1281–1299. (A review.)

Venkatesan, P., and H. R. Kaback. 1998. The substrate-binding site in the lactose permease of *Escherichia coli. Proc. Nat'l Acad. Sci. USA* **95**:9802–9807.

Wang, Q., et al. 1998. Proximity of helices VIII (Ala 273) and IX (Met 299) in the lactose permease of *Escherichia coli. Biochemistry* **37**:4910–4915.

Wu, J., D. Hardy, and H. R. Kaback. 1998. Transmembrane helix tilting and ligand-induced conformational changes in the lactose permease determined by site-directed chemical crosslinking in situ. *J. Mol. Biol.* **282**:959–967.

Testing Yourself on the Concepts

1. The concentration of glucose in the mammalian bloodstream is about 5 mM and varies within the range of 3 mM after a few days of fasting to 7 mM after a feast. Considering these to be typical glucose levels, do you expect the K_m for a glucose transporter to be 10^{-6} M or 10^{-3} M. Why?

2. Chemically, steroid hormones are lipids. On the basis of their expected permeability properties in biological membranes, predict whether receptor proteins for steroid hormones would be expected to be cell-surface or intracellular proteins.

3. Membrane transport proteins, be they uniports, symports, antiports, or ATPases, all have several transmembrane helices. How does this contribute to the function of the transport protein?

4. A Na^+/glucose symport can transport glucose against a concentration gradient. How is this energetically unfavorable process linked to ATP consumption, be it direct or indirect?

5. How might water channels be important in the opening of leaf stomata?

MCAT/GRE-Style Questions

Key Concept Please read the section titled "Na^+ Entry into Mammalian Cells Has a Negative ΔG" (p. 587) and answer the following questions (assume a membrane potential of -70 mV and the Na^+ and K^+ concentrations shown in Figure 15-8):

1. The Na^+/K^+ ATPase pumps 2 moles of Na^+ out of the cell for every 3 moles of K^+ pumped into the cell. What is the ΔG for pumping 1 mole of Na^+ out of the cell?

    a. -3.03 kcal/mol.

    b. 1.41 kcal/mol.

    c. 1.61 kcal/mol.

    d. 3.03 kcal/mol.

2. What is the ΔG for pumping 1 mole of K^+ into the cell?

    a. -0.19 kcal/mol.

    b. 1.41 kcal/mol.

    c. -1.61 kcal/mol.

    d. 3.03 kcal/mol.

3. What is the overall energetics of one complete round of transport of 2 moles of Na^+ and 3 moles of K^+?

    a. -6.63 kcal.

    b. 5.49 kcal.

    c. 6.03 kcal.

    d. 6.63 kcal.

4. How many ATPs are minimally consumed during one complete round of transport?

    a. 1.

    b. 2.

    c. 3.

    d. 5.

5. You treat the cell with a drug, a K^+ ionophore, that selectively equilibrates K^+ concentrations across the membrane. What now is the ΔG for K^+ transport by the ATPase? Assume that the membrane potential stays the same.

    a. -1.61 kcal/mol.

    b. -0.19 kcal/mol.

    c. 1.42 kcal/mol.

    d. 3.03 kcal/mol.

Key Experiment Please read the section titled "Tight Junctions Seal Off Body Cavities and Restrict Diffusion of Membrane Components" (p. 604) and answer the following questions:

6. Functions of tight junctions include all the following *except*:

    a. Separation of extracellular fluids.

    b. Sealing of body cavities.

c. Prevention of diffusion of membrane proteins and lipids between apical and basolateral regions.

d. Tight communication and exchange of small molecules between neighboring cells.

7. Molecules present in or associated with tight junctions include all the following *except*:

a. Connexin.

b. Occludin and claudin.

c. ZO-1, ZO-2, and ZO-3.

d. Cytoskeletal linking proteins and actin.

8. Tight junctions may be reversibly dissociated by

a. Mg^{2+} removal and addition.

b. Ca^{2+} removal and addition.

c. Glycosidase treatment.

d. Trypsin treatment.

9. What is the expected effect on the distribution of plasma-membrane proteins between the apical and basolateral regions if tight junctions are dissociated?

a. The distribution stays the same.

b. Apical and basolateral proteins intermix.

c. The distribution becomes even more distinct.

d. The proteins are degraded.

10. Fluorescent lipids may be selectively introduced into the cytosolic leaflet of the apical membrane of epithelial cells in a two-step procedure. After completing the procedure, what is the expected distribution of the fluorescent lipids?

a. Restricted to the apical surface.

b. Restricted to the basolateral surface.

c. Distributed in equal concentrations in the cytosolic and exoplasmic leaflets of the membrane.

d. Distributed in equal concentrations in the apical and basolateral regions of the cell.

Key Application Please read the section titled "Cystic Fibrosis Transmembrane Regulator (CFTR) Protein" (p. 597) and answer the following questions:

11. Given the genetic and phenotypic traits of CF patients, the likely molecular defect in CF is

a. A multigene trait.

b. A defect in cAMP regulation of CFTR.

c. Due to a mutation in the Q domain of CFTR.

d. Due to a mutation in the protein with which CFTR interacts and which it regulates.

12. The name CFTR implies that as originally described the protein was thought not to be a Cl^- channel itself but rather a regulator of the Cl^- channel. What experimental observation shows that the CFTR protein itself is a Cl^- channel?

a. The ability of recombinant CFTR expressed in the epithelial cells of cystic fibrosis patients to restore the Cl^- transport properties of these cells.

b. The ability of recombinant CFTR expressed in test COS cells to alter the Cl^- transport properties of these cells.

c. The ability of recombinant CFTR expressed to confer cAMP sensitivity to C^- transport.

d. The ability of recombinant CFTR inserted into liposomes to form Cl^- transport channels.

13. Introduction of CFTR into the lung is an attractive route for genetic correction of at least some of the symptoms of cystic fibrosis. This is true for all the following reasons *except*:

a. DNA introduced into the lung rather than the bloodstream easily crosses the blood-brain barrier.

b. The lung is easily accessible to DNA-containing aerosols.

c. The lung is one of the major organs affected in CF.

d. The introduction of DNA into lung cells does not alter the DNA of germ cells.

Key Terms

active transport *580*
antiport *598*
basal lamina *602*
cell junctions *602*
cotransport *598*
epithelium *602*
hypertonic *609*
hypotonic *609*
isotonic *609*
Nernst equation *587*
osmosis *608*
passive diffusion *579*
symport *598*
tight junctions *605*
uniport *582*

References

Uniporter-Catalyzed Transport of Specific Molecules

Bell, G., et al. 1993. Structure and function of mammalian facilitative sugar transporters. *J. Biol. Chem.* **268**:19161–19164.

Henderson, P. J. 1993. The 12-transmembrane helix transporters. *Curr. Opin. Cell Biol.* 5:708–721.

Mueckler, M. 1994. Facilitative glucose transporters. *Eur. J. Biochem.* **219**:713–725.

Malandro, M., and M. Kilberg. 1996. Molecular biology of mammalian amino acid transporters. *Ann. Rev. Biochem.* 66:305–336.

Ion Channels, Intracellular Ion Environment, and Membrane Electric Potential

Clapham, D., and B. Ehrlich, eds. 1996. Organellar Ion Channels and Transporters. Society of General Physiologists. Marine Biological Laboratory and Rockefeller University Press, New York.

Choe, S., and R. Robinson. 1998. An ingenious filter: the structural basis for ion selectivity. *Neuron* 20:821–823.

Dawson, D., and R. Frizzell, eds. 1995. Ion Channels and Genetic Diseases. Society of General Physiologists. Marine Biological Laboratory and Rockefeller University Press, New York.

Doyle, D. A., et al. 1998. The structure of the potassium channel: molecular basis of K$^+$ conduction and selectivity. *Science* 280:69–77.

Hille, B. 1991. Ion Channels of Excitable Membranes. 2d ed. Sinauer Associates (Sunderland, Mass.).

Racker, E. 1985. Reconstitutions of Transporters, Receptors, and Pathological States. Academic Press.

Stein, W. D., and W. R. Leib. 1986. Transport and Diffusion across Cell Membranes. Academic Press.

Active Transport and ATP Hydrolysis

Boyer, P. D. 1997. The ATP synthase—a splendid molecular machine. *Ann. Rev. Biochem.* **66**:717–749.

Carafoli, E. 1992. The Ca^{2+} pump of the plasma membrane. *J. Biol. Chem.* **267**:2115–2118.

Doige, C. A., and G. F. Ames. 1993. ATP-dependent transport systems in bacteria and humans: relevance to cystic fibrosis and multidrug resistance. *Ann. Rev. Microbiol.* **47**:291–319.

Gottesman, M. M., I. Pastan, and S. V. Ambudkar. 1996. P-glycoprotein and multidrug resistance. Curr. Opin. Genet. Devel. **6**:610–617.

Glynn, I. M. 1993. All hands to the sodium pump. *J. Physiol. (London)* **462**:1–30.

Higgins, C. F. 1995. The ABC of channel regulation. *Cell* **82**:693–696.

Jencks, W. P. 1995. The mechanism of coupling chemical and physical reactions by the calcium ATPase of sarcoplasmic reticulum and other coupled vectorial systems. *Biosci. Rep.* **15**:283–287.

Lingrel, J. B., and T. Kuntzweiler. 1994. Na^+,K^+-ATPase. *J. Biol. Chem.* **269**:19659–19662.

Linton, K. J., and C. F. Higgins. 1998. The *Escherichia coli* ATP-binding cassette (ABC) proteins. *Mol. Microbiol.* **28**:5–13.

Lutsenko, S., and J. Kaplan. 1995. Organization of P-type ATPases: significance of structural diversity. *Biochemistry* **34**:15607–15613.

MacLennan, D. H., W. J. Rice, and N. M. Green. 1997. The mechanism of Ca^{2+} transport by sarco(endo)plasmic reticulum Ca^{2+}-ATPases. *J. Biol. Chem.* **272**:28815–28818.

Nelson, N., and D. J. Klionsky. 1996. Vacuolar H^+-ATPase: from mammals to yeast and back. *Experientia* **52**:1101–1110.

Rea, P. A., et al. 1992. Vacuolar H^+-translocating pyrophosphatases: a new category of ion translocase. *Trends Biochem. Sci.* **17**:348–353.

Sachs, G. 1997. Proton pump inhibitors and acid-related diseases. *Pharmacotherapy* **17**:22–37.

Shapiro, A. B., and V. Ling. 1995. Using purified P-glycoprotein to understand multidrug resistance. *J. Bioenerg. Biomemb.* **27**:7–13.

Stevens, T., and M. Forgac. 1997. Structure, function, and regulation of the vacuolar (H^+) ATPase. *Ann. Rev. Cell Devel. Biol.* **13**:779–808.

Sze, H., J. M. Ward, and S. Lai. 1992. Vacuolar H^+-translocating ATPases from plants: structure, function, and isoforms. *J. Bioenerg. Biomemb.* **24**:371–381.

Welsh, M. J., and A. E. Smith. 1993. Molecular mechanisms of CFTR chloride channel dysfunction in cystic fibrosis. *Cell* **73**:1251–1254.

Zhang, P., et al. 1998. Structure of the calcium pump from sarcoplasmic reticulum at 8 Å resolution. *Nature* **392**:835.

Cotransport Catalyzed by Symporters and Antiporters

Alper, S. L. 1991. The band 3-related anion exchanger (AE) gene family. *Ann. Rev. Physiol.* **53**:549–564.

Baldwin, S. A. 1993. Mammalian passive glucose transporters: members of an ubiquitous family of active and passive transport proteins. *Biochim. Biophys. Acta* **1154**:17–49.

Barkla, B., and O. Pantoja. 1996. Physiology of ion transport across the tonoplast of higher plants. *Ann. Rev. Plant Physiol. and Plant Mol. Biol.* **47**:159–184.

Jay, D. G. 1996. Role of band 3 in homeostasis and cell shape. *Cell* **86**:853–854.

Maathuis, F. J., and D. Sanders. 1992. Plant membrane transport. *Curr. Opin. Cell Biol.* **4**:661–669.

Orlowski, J., and S. Grinstein. 1997. Na^+/H^+ exchangers of mammalian cells. *J. Biol. Chem.* **272**:22373–22376.

Ruetz, S., A. E. Lindsey, and R. R. Kopito. 1993. Function and biosynthesis of erythroid and nonerythroid anion exchangers. *Soc. Gen. Physiol. Ser.* **48**:193–200.

Shrode, L. D., H. Tapper, and S. Grinstein. 1997. Role of intracellular pH in proliferation, transformation, and apoptosis. *J. Bioenerg. Biomemb.* **29**:393–399.

Turk, E., and E. Wright. 1996. Membrane topology motifs in the SGLT cotransporter family. *J. Memb. Biol.* **159**:1–20.

Wakabayashi, S., M. Shigekawa, and J. Pouyssegur. 1997. Molecular physiology of vertebrate Na^+/H^+ exchangers. *Physiol. Rev.* **77**:51–74.

Wright, E. M., D. D. Loo, E. Turk, and B. A. Hirayama. 1996. Sodium cotransporters. *Curr. Opin. Cell Biol.* **8**:468–473.

Transport across Epithelia

Anderson, J. M., and C. M. Van Itallie. 1995. Tight junctions and the molecular basis for regulation of paracellular permeability. *Am. J. Physiol.* **269**:G467–G475.

Cereijido, M., J. Valdés, L. Shoshani, and R. Conteras. 1998. Role of tight junctions in establishing and maintaining cell polarity. *Ann. Rev. Physiol.* **60**:161–177.

Goodenough, D. A. 1999. Plugging the leaks. *Proc. Natl. Acad. Sci. USA* **96**:319.

Gumbiner, B. M. 1996. Cell adhesion: the molecular basis of tissue architecture and morphogenesis. *Cell* **84**:345–357.

Le Gall, A. H., C. Yeaman, A. Muesch, and E. Rodriquez-Boulan. 1995. Epithelial cell polarity: new perspectives. *Semin. Nephrol.* **15**:272–284.

Mitic, L., and J. Anderson. 1998. Molecular architecture of tight junctions. *Ann. Rev. Physiol.* **60**:121–142.

Nelson, W. J. 1992. Regulation of cell surface polarity from bacteria to mammals. *Science* **258**:948–955.

Rubin, L. L. 1992. Endothelial cells: adhesion and tight junctions. *Curr. Opin. Cell Biol.* **4**:830–833.

Schultz, S., et al., eds. 1997. Molecular Biology of Membrane Transport Disorders. Plenum Press.

Osmosis, Water Channels, and the Regulation of Cell Volume

Agre, P., M. Bonhivers, and M. Borgina. 1998. The aquaporins, blueprints for cellular plumbing systems. *J. Biol. Chem.* **273**:14659.

Chrispeels, M. J., and P. Agre. 1994. Aquaporins: water channel proteins of plant and animal cells. *Trends Biochem. Sci.* **19**:421–425.

Maathuis, F. J., A. M. Ichida, D. Sanders, and J. I. Schroeder. 1997. Roles of higher plant K^+ channels. *Plant Physiol.* **114**:1141–1149.

Maurel, C. 1997. Aquaporins and water permeability of plant membranes. *Ann. Rev. Plant Physiol. and Plant Mol. Biol.* **48**:399–430.

Sarkadi, B., and J. C. Parker. 1991. Activation of ion transport pathways by changes in cell volume. *Biochim. Biophys. Acta* **1071**:407–427.

Verkman, A. S., et al. 1996. Water transport across mammalian cell membranes. *Am. J. Physiol.* **270**:C12–C30.

16

Cellular Energetics: Glycolysis, Aerobic Oxidation, and Photosynthesis

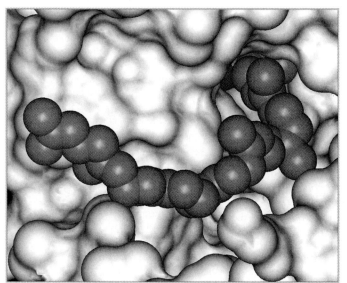

Ubiquinone (orange) bound to the surface of the photosynthetic reaction center (white) from the bacterium *Rhodobacter spheroides*. Only one of the oxygen atoms (blue) in ubiquinone is visible.

The most important molecule for capturing and transferring free energy in biological systems is **adenosine triphosphate,** or **ATP** (see Figure 2-24). Under standard conditions, hydrolysis of the terminal high-energy phosphoanhydride bond in ATP to yield adenosine diphosphate (ADP) and inorganic phosphate (P_i) releases 7.3 kcal/mol of free energy. Cells can use the energy released during this reaction to power many otherwise energetically unfavorable processes, such as the transport of molecules against a concentration gradient by ATP-powered pumps (Chapter 15), the movement (beating) of cilia (Chapter 19), the contraction of muscle (Chapter 18), and the synthesis of proteins from amino acids (Chapter 4) and of nucleic acids from nucleotides (Chapter 12). Although other high-energy molecules occur in cells, ATP is the universal "currency" of chemical energy; it is found in all types of organisms and must have occurred in the earliest life-forms.

This chapter focuses on how cells generate the high-energy phosphoanhydride bond of ATP from ADP and P_i. This endergonic reaction, which is the reverse of ATP hydrolysis and requires an input of 7.3 kcal/mol to proceed, can be written as

$$P_i^{2-} + H^+ + ADP^{3-} \longrightarrow ATP^{4-} + H_2O$$

where P_i^{2-} represents inorganic phosphate (HPO_4^{2-}). The energy to drive this reaction is produced primarily by two main processes—**aerobic**

MEDIA CONNECTIONS

Focus: Electron Transport

Focus: ATP Synthesis

Focus: Photosynthesis

oxidation, which occurs in nearly all cells, and **photosynthesis**, which occurs only in leaf cells of plants and certain single-celled organisms.

In aerobic oxidation, fatty acids and sugars, principally glucose, are metabolized to CO_2 and H_2O, and the released energy is converted to the chemical energy of phosphoanhydride bonds in ATP. In animal cells and most other nonphotosynthetic cells, ATP is generated mainly by this process. The initial steps in the oxidation of glucose, called **glycolysis**, occur in the cytosol in both eukaryotes and prokaryotes and do not require O_2. The final steps, which require O_2, generate most of the ATP. In eukaryotes, the later stages of aerobic oxidation occur in **mitochondria**, whereas in prokaryotes, which lack mitochondria, many of the final steps occur on the plasma membrane. The final stages of fatty acid metabolism sometimes occur in mitochondria and generate ATP; in most eukaryotic cells, however, fatty acids are metabolized in **peroxisomes** without production of ATP.

In photosynthesis, light energy is converted to the chemical energy of phosphoanhydride bonds in ATP and stored in the chemical bonds of carbohydrates (primarily sucrose and starch). Oxygen also is formed during photosynthesis. In plants and eukaryotic single-celled algae, photosynthesis occurs in **chloroplasts**. Although they lack chloroplasts, several prokaryotes also carry out photosynthesis by a mechanism similar to that in chloroplasts. The oxygen generated during photosynthesis is the source of virtually all the oxygen in the air, and the carbohydrates produced are the ultimate source of energy for virtually all nonphotosynthetic organisms.*

At first glance, photosynthesis and aerobic oxidation appear to have little in common. However, a revolutionary discovery in cell biology is that bacteria, mitochondria, and chloroplasts all use the same (or very nearly the same) process, called **chemiosmosis** (or chemiosmotic coupling), to generate ATP from ADP and P_i (Figure 16-1). The immediate energy sources that power ATP synthesis are the transmembrane proton concentration gradient and electric potential (voltage gradient), collectively termed the **proton-motive force**. The proton-motive force is generated by the stepwise movement of electrons from higher to lower energy states via membrane-bound **electron carriers**. In mitochondria and nonphotosynthetic bacterial cells, electrons from NADH (produced during the metabolism of sugars, fatty acids, and other substances) are transferred to O_2, the ultimate

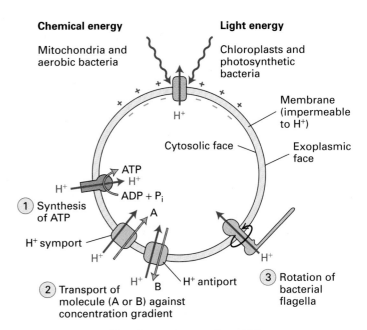

Chemical energy

Mitochondria and aerobic bacteria

Light energy

Chloroplasts and photosynthetic bacteria

Membrane (impermeable to H^+)

Exoplasmic face

Cytosolic face

H^+

ATP
H^+
ADP + P_i

① Synthesis of ATP

H^+

H^+ symport

A

H^+

H^+ antiport

③ Rotation of bacterial flagella

② Transport of molecule (A or B) against concentration gradient

H^+

B

▲ **FIGURE 16-1 Chemiosmotic coupling.** This process can occur only in sealed, closed, membrane-limited compartments that are impermeable to H^+. In photosynthesis, energy absorbed from light is used to move protons across the membrane, generating a transmembrane proton concentration gradient and a voltage gradient, collectively called the *proton-motive force*. In mitochondria and aerobic bacteria, energy liberated by the oxidation of carbon compounds is used to move protons across the membrane, again generating a proton-motive force. However generated, a proton-motive force can be used to power ATP synthesis ①, transport of metabolites across the membrane against their concentration gradient ②, and rotation of bacterial flagella ③.

electron acceptor. In the thylakoid membrane of chloroplasts, energy absorbed from light strips electrons from water (forming O_2) and powers their movement to other electron carriers, particularly $NADP^+$; eventually these electrons are donated to CO_2 to synthesize carbohydrates. All these systems, however, contain some similar carriers that couple **electron transport** to the pumping of protons (always from the cytosolic face to the exoplasmic face of the membrane), thereby generating the proton-motive force (Figure 16-2).

Moreover, all cells utilize essentially the same kind of membrane protein, the F_0F_1 **complex**, to synthesize ATP. The F_0F_1 complex, also called **ATP synthase** and F_0F_1 ATPase, is a member of the F class of ATP-powered proton pumps (see Table 15-2). In all cases, the F_0F_1 complex is positioned with the globular F_1 segment, which catalyzes ATP synthesis, on the cytosolic face of the membrane, so ATP is always formed on the cytosolic face of the membrane (see Figure 16-2). Protons always flow through the F_0F_1 complex from the exoplasmic to the cytosolic face of the membrane, driven by a combination of the

*Contrary to the common popular view, sunlight is not the ultimate source of energy for all organisms on earth. As noted in Chapter 2, bacteria in deep ocean vents, where there is no sunlight, obtain energy for converting CO_2 into carbohydrates and other cellular constituents by oxidation of reduced inorganic compounds in dissolved vent gas.

proton concentration gradient (exoplasmic face > cytosolic face) and the membrane electric potential (exoplasmic face positive with respect to the cytosolic face).

In addition to powering ATP synthesis, the proton-motive force can supply energy for the transport of small molecules across a membrane against a concentration gradient (see Figure 16-1). For example, the uptake of lactose by certain bacteria is catalyzed by a H^+/sugar symport protein, and the accumulation of ions and sucrose by plant vacuoles is catalyzed by proton-driven antiporters (see Figure 15-22). The rotation of bacterial flagella is also powered by the proton-motive force; in contrast, the beating of eukaryotic cilia is powered by ATP hydrolysis. Conversely, hydrolysis of ATP by V-class ATP-powered

proton pumps, which are similar in structure to P-class pumps (see Figure 15-10), provides the energy for transporting protons against a concentration gradient. Chemiosmotic coupling thus illustrates an important principle introduced in our discussion of active transport in Chapter 15: *the membrane potential, the concentration gradients of protons (and other ions) across a membrane, and the phosphoanhydride bonds in ATP are equivalent and interconvertible forms of chemical potential energy.*

In this brief overview, we've seen that oxygen and carbohydrates are produced during photosynthesis, whereas they are consumed during aerobic oxidation. In both processes, the flow of electrons creates a H^+ electrochemical gradient, or proton-motive force, that powers ATP synthesis. As we examine these two processes at the molecular level, focusing first on aerobic oxidation and then on photosynthesis, the striking parallels between them will become evident.

16.1 Oxidation of Glucose and Fatty Acids to CO_2

The complete aerobic oxidation of glucose is coupled to the synthesis of as many as 36 molecules of ATP:

$$C_6H_{12}O_6 + 6\ O_2 + 36\ P_i^{2-} + 36\ ADP^{3-} + 36\ H^+ \longrightarrow$$
$$6\ CO_2 + 36\ ATP^{4-} + 42\ H_2O$$

Glycolysis, the initial stage of glucose metabolism, takes place in the cytosol and does not involve molecular O_2. It produces a small amount of ATP and the three-carbon compound *pyruvate*. In aerobic cells, pyruvate formed in glycolysis is transported into the mitochondria, where it is oxidized by O_2 to CO_2. Via chemiosmotic coupling, the

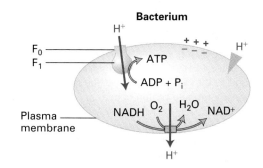

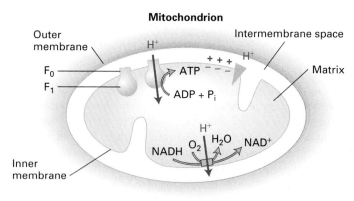

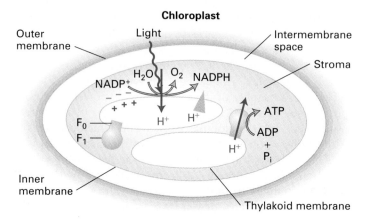

◀ **FIGURE 16-2 Membrane orientation and the direction of proton movement in bacteria, mitochondria, and chloroplasts.** The membrane surface facing a shaded area is a cytosolic face; the surface facing an unshaded area is an exoplasmic face. Note that the cytosolic face of the bacterial plasma membrane, the matrix face of the inner mitochondrial membrane, and the stromal face of the thylakoid membrane are all equivalent. In bacteria, mitochondria, and chloroplasts, the F_0F_1 complexes that synthesize ATP always protrude from the cytosolic face of the membrane. During electron transport, protons are always pumped from the cytosolic face to the exoplasmic face, creating a proton concentration gradient (exoplasmic face > cytosolic face) and an electric potential (negative cytosolic face) across the membrane. During the generation of ATP, protons flow in the reverse direction (down their electrochemical gradient from the exoplasmic to the cytosolic face) through the F_0F_1 complexes.

oxidation of pyruvate in the mitochondria generates the bulk of the ATP produced during the conversion of glucose to CO_2. In this section, we discuss the biochemical pathways that oxidize glucose and fatty acids to CO_2 and H_2O; the fate of the released electrons is described in the next section.

Cytosolic Enzymes Convert Glucose to Pyruvate

A set of 10 enzymes catalyze the reactions, constituting the *glycolytic pathway,* that degrade one molecule of glucose to two molecules of pyruvate (Figure 16-3). All the metabolic intermediates between glucose and pyruvate are water-soluble phosphorylated compounds. Four molecules of ATP are formed from ADP in glycolysis (reactions 6 and 9). However, two ATP molecules are consumed during earlier steps of this pathway: the first by the addition of a phosphate residue to glucose in the reaction catalyzed by *hexokinase* (reaction 1), and the second by the addition of a second phosphate to fructose 6-phosphate in the reaction catalyzed by *phosphofructokinase-1* (reaction 3). Thus there is a net gain of two ATP molecules.

The balanced chemical equation for the conversion of glucose to pyruvate shows that four hydrogen atoms (four protons and four electrons) are also formed:

$$C_6H_{12}O_6 \longrightarrow 2\ CH_3-\overset{\overset{O}{\|}}{C}-\overset{\overset{O}{\|}}{C}-OH + 4\ H^+ + 4\ e^-$$
$$\text{Glucose} \qquad\qquad \text{Pyruvate}$$

(For convenience, we show pyruvate in its un-ionized form, pyruvic acid, although at physiological pH it would be largely dissociated.) All four electrons and two of the four protons are transferred to two molecules of the oxidized form of the electron carrier **nicotinamide adenine dinucleotide (NAD^+)** to produce the reduced form, NADH (Figure 16-4):

$$2\ H^+ + 4\ e^- + 2\ NAD^+ \rightleftharpoons 2\ NADH$$

The reaction that generates these hydrogen atoms and transfers them to NAD^+ is catalyzed by *glyceraldehyde 3-phosphate dehydrogenase* (see Figure 16-3, reaction 5).

Thus the overall reaction for this first stage of glucose metabolism is

$$C_6H_{12}O_6 + 2\ NAD^+ + 2\ ADP^{3-} + 2\ P_i^{2-} \longrightarrow$$
$$2\ C_3H_4O_3 + 2\ NADH + 2\ ATP^{4-}$$

Substrate-Level Phosphorylation Generates ATP during Glycolysis

As noted earlier, the immediate energy source for ATP synthesis in chloroplasts and mitochondria is provided by the proton-motive force across a membrane. Cells also can produce ATP by a process called **substrate-level phosphorylation,** which is catalyzed by water-soluble enzymes in the cytosol; membranes and ion gradients are not involved.

Substrate-level phosphorylation occurs twice in the glycolytic pathway. The first results from the pair of reactions that convert glyceraldehyde 3-phosphate to 3-phosphoglycerate (see Figure 16-3, steps 5 and 6). In the first of these reactions, oxidation of the aldehyde (CHO) group on glyceraldehyde 3-phosphate by NAD^+ is coupled to addition of a phosphate group, forming 1,3-bisphosphoglycerate with a single high-energy phosphoanhydride bond to carbon 1. In this reaction, catalyzed by *glyceraldehyde 3-phosphate dehydrogenase,* a high-energy thioester enzyme intermediate is formed (Figure 16-5). The high-energy phosphate group of 1,3-bisphosphoglycerate then is transferred to ADP, forming ATP and 3-phosphoglycerate, a reaction catalyzed by *phosphoglycerate kinase.* Although the first reaction requires free energy ($\Delta G^{\circ\prime} = +1.5$ kcal/mol), the second releases even more free energy ($\Delta G^{\circ\prime} = -4.5$ kcal/mol). Thus the net change in standard free energy of these two reactions is -3.0 kcal/mol, so the two reactions overall are strongly exergonic.

The three final reactions of glycolysis, in which 3-phosphoglycerate is converted to pyruvate, also result in substrate-level phosphorylation (see Figure 16-3, steps 7–9). The first two of these reactions, which each have a small positive $\Delta G^{\circ\prime}$, lead to formation of phosphoenolpyruvate. In the third reaction, catalyzed by *pyruvate kinase,* the high-energy phosphate group in phosphoenolpyruvate is transferred to ADP, yielding pyruvate and ATP. This reaction is strongly exergonic ($\Delta G^{\circ\prime} = -7.5$ kcal/mol), as is the net free-energy change for the three reactions overall ($\Delta G^{\circ\prime} = -6.0$ kcal/mol).

The glycolytic conversion of one molecule of glucose to two molecules of glyceraldehyde 3-phosphate consumes two ATPs (see Figure 16-3). The two subsequent substrate-level phosphorylations yield two ATPs for each molecule of glyceraldehyde, or a total of four ATPs per glucose molecule. Thus the net yield is two ATPs per glucose molecule. The remaining 34 molecules of ATP generated during the complete aerobic metabolism of glucose are synthesized during oxidation of pyruvate to CO_2 and H_2O in mitochondria. Before discussing what occurs in mitochondria, however, we digress briefly to describe the **anaerobic** metabolism of glucose.

Anaerobic Metabolism of Each Glucose Molecule Yields Only Two ATP Molecules

Most eukaryotes are *obligate aerobes:* they grow only in the presence of oxygen and metabolize glucose (or related sugars) completely to CO_2, with the concomitant production of a large amount of ATP. Most eukaryotes, however, can generate some ATP by anaerobic metabolism. A few eukaryotes are *facultative anaerobes:* they grow in either the

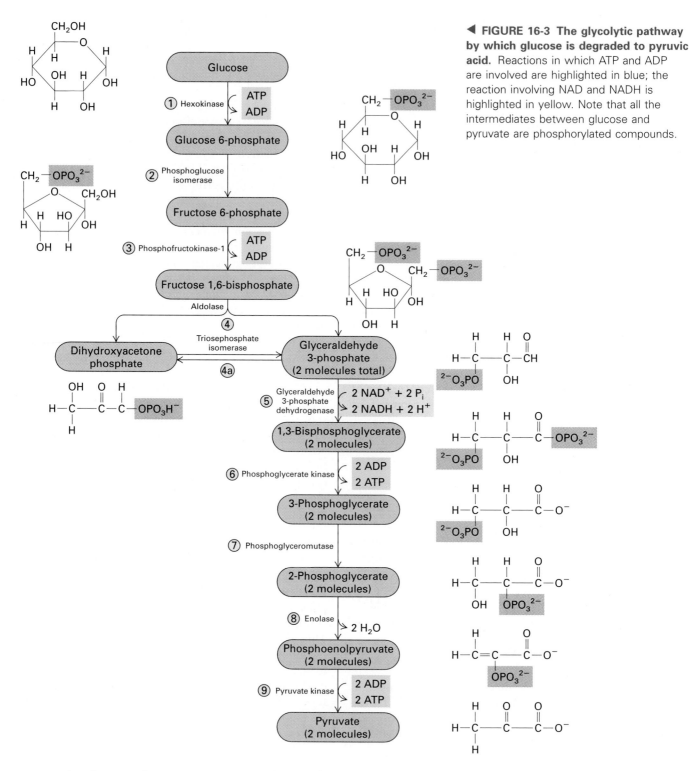

◀ FIGURE 16-3 **The glycolytic pathway by which glucose is degraded to pyruvic acid.** Reactions in which ATP and ADP are involved are highlighted in blue; the reaction involving NAD and NADH is highlighted in yellow. Note that all the intermediates between glucose and pyruvate are phosphorylated compounds.

presence or the absence of oxygen. For example, annelids, mollusks, and some yeasts can live and grow for days without oxygen. Certain prokaryotes are *obligate anaerobes*: they cannot grow in the presence of oxygen, and they metabolize glucose only anaerobically.

In the absence of oxygen, glucose is not converted entirely to CO_2 (as it is in obligate aerobes) but to one or more two- or three-carbon compounds, and only in some cases to

CO_2. For instance, yeasts degrade glucose to two pyruvate molecules via glycolysis, generating a net of two ATP and two NADH molecules per glucose molecule. If necessary, yeasts can anaerobically convert pyruvate to ethanol and CO_2; two NADH molecules are oxidized to NAD^+ for each two pyruvates converted to ethanol, thereby regenerating the supply of NAD^+ (Figure 16-6, *right*). This *anaerobic fermentation* is the basis of beer and wine production.

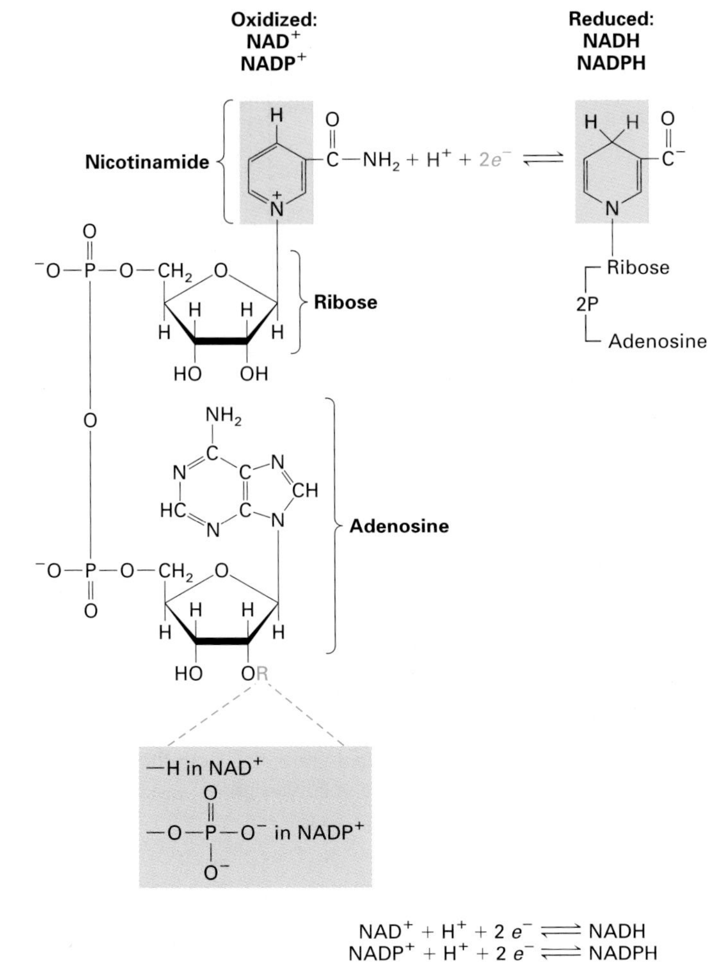

◀ FIGURE 16-4 Structures of the electron-carrying coenzymes NAD⁺ and NADH. Nicotinamide adenine dinucleotide (NAD⁺) and the related nicotinamide adenine dinucleotide phosphate (NADP⁺) accept only pairs of electrons; reduction to NADH or NADPH involves the transfer of two electrons simultaneously. In most oxidation-reduction reactions in biological systems, a pair of hydrogen atoms (two protons and two electrons) are removed from a molecule. One of the protons and both electrons are transferred to NAD⁺; the other proton is released into solution. Thus the overall reaction is sometimes written $NAD^+ + 2 H^+ + 2 e^- \longleftrightarrow NADH + H^+$. NADP is identical in structure with NAD except for the presence of an additional phosphate group. However, NAD and NADP participate in different sets of enzymatically catalyzed reactions.

During the prolonged contraction of mammalian skeletal muscle cells, oxygen becomes limited and glucose cannot be oxidized completely to CO_2 and H_2O. In this situation, muscle cells ferment glucose to two molecules of lactic acid—again, with the net production of only two molecules of ATP per glucose molecule (Figure 16-6, left). The lactic acid causes muscle and joint aches. It is largely secreted into the blood; some passes into the liver, where it is reoxidized to pyruvate and either further metabolized to CO_2 or converted to glucose. Much lactate is metabolized to CO_2 by the heart, which is highly perfused by blood and can continue aerobic metabolism at times when exercising skeletal muscles secrete lactate.

Lactic acid bacteria (the organisms that "spoil" milk) and other prokaryotes also generate ATP by the fermentation of glucose to lactate.

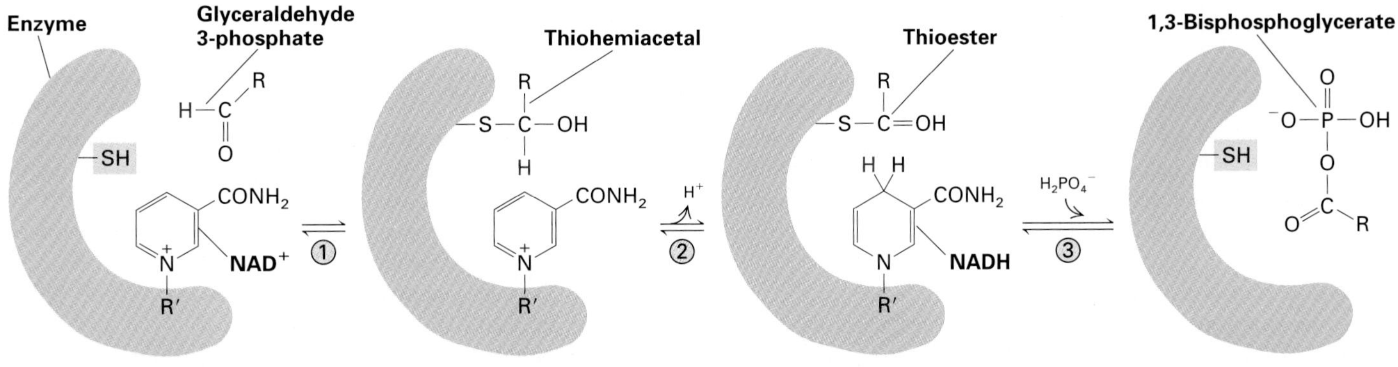

▲ FIGURE 16-5 The mechanism of action of glyceraldehyde 3-phosphate dehydrogenase. The sulfhydryl (SH) group is the side chain of cysteine at the active site of the enzyme; R symbolizes the rest of the glyceraldehyde 3-phosphate molecule. R′ is the rest of the NAD molecule. Step ①: After the enzyme has bound NAD⁺, the —SH group on the enzyme reacts with glyceraldehyde 3-phosphate to form a thiohemiacetal. Step ②: A hydrogen atom (red) and two electrons are transferred to NAD⁺, forming the reduced form NADH, and a proton from the O atom of the thiohemiacetal is simultaneously lost to the medium. The other product is a high-energy enzyme-bound thioester. Step ③: The thioester reacts with phosphate to produce 1,3-bisphosphoglycerate; NADH is freed from the enzyme surface, and the free enzyme with its —SH group is regenerated.

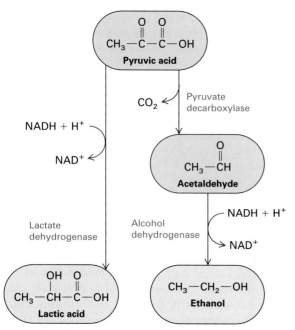

Overall reactions of anaerobic metabolism:

Glucose + 2 ADP + 2 P$_i$ $\longrightarrow$ 2 lactate + 2 ATP

Glucose + 2 ADP + 2 P$_i$ $\longrightarrow$ 2 ethanol + 2 CO$_2$ + 2 ATP

▲ **FIGURE 16-6 The anaerobic metabolism of glucose.** In the formation of pyruvate from glucose, one molecule of NAD$^+$ is reduced to NADH for each molecule of pyruvate formed (see Figure 16-3). To regenerate NAD$^+$, two electrons are transferred from each NADH molecule to an acceptor molecule. When oxygen supplies are low in muscle cells, the acceptor is pyruvic acid, and lactic acid is formed. In yeasts, acetaldehyde is the acceptor, and ethanol is formed.

Mitochondria Possess Two Structurally and Functionally Distinct Membranes

In the second stage of aerobic oxidation, pyruvate formed in glycolysis is transported into mitochondria, where it is oxidized by O$_2$ to CO$_2$. These mitochondrial oxidation reactions generate 34 of the 36 ATP molecules produced from the conversion of glucose to CO$_2$. Mitochondria thus are the "power plants" of eukaryotic cells. To understand how mitochondria operate, we first need to be familiar with their structure.

Mitochondria are among the larger organelles in the cell, each one being about the size of an *E. coli* bacterium. Most eukaryotic cells contain many mitochondria, which collectively can occupy as much as 25 percent of the volume of the cytoplasm. They are large enough to be seen under a light microscope, but the details of their structure can be viewed only with the electron microscope (see Figure 5-45). The *outer* and the *inner* mitochondrial *membranes* define two submitochondrial compartments: the *intermembrane space* between the two membranes, and the *matrix*, or central compartment, (Figure 16-7).

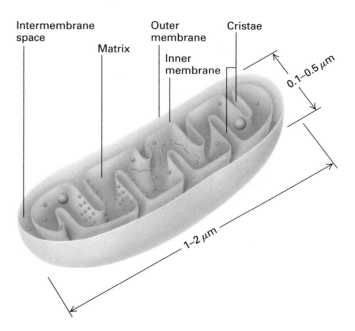

▲ **FIGURE 16-7 A three-dimensional diagram of a mitochondrion cut longitudinally.** The F$_0$F$_1$ complexes (small red spheres), which synthesize ATP, are intramembrane particles that protrude from the inner membrane into the matrix. The matrix contains the mitochondrial DNA (blue strand), ribosomes (small blue spheres), and granules (large yellow spheres).

The fractionation and purification of these membranes and compartments has made it possible to determine their protein and phospholipid compositions and to localize each enzyme-catalyzed reaction to a specific membrane or space.

The outer membrane defines the smooth outer perimeter of the mitochondrion and contains mitochondrial *porin*, a transmembrane channel protein similar in structure to bacterial porins (see Figure 3-35). Ions and most small molecules (up to about 5000 MW) can readily pass through these channel proteins. Although the flow of metabolites across the outer membrane may limit their rate of mitochondrial oxidation, the inner membrane is the major permeability barrier between the cytosol and the mitochondrial matrix.

Freeze-fracture studies indicate that the inner membrane contains many protein-rich intramembrane particles. Some are the F$_0$F$_1$ complexes that synthesize ATP; others function in transporting electrons to O$_2$ from NADH or reduced **flavin adenine dinucleotide** (FADH$_2$) (Figure 16-8). Various transport proteins located in the inner membrane allow otherwise impermeable molecules, such as ADP and P$_i$, to pass from the cytosol to the matrix, and other molecules, such as ATP, to move from the matrix into the cytosol. Protein constitutes 76 percent of the total inner membrane weight—a higher fraction than in any other cellular membrane. Cardiolipin (diphosphatidylglycerol), a lipid concentrated in the inner membrane, sufficiently reduces the membrane's permeability to protons that a proton-motive force can be established across it.

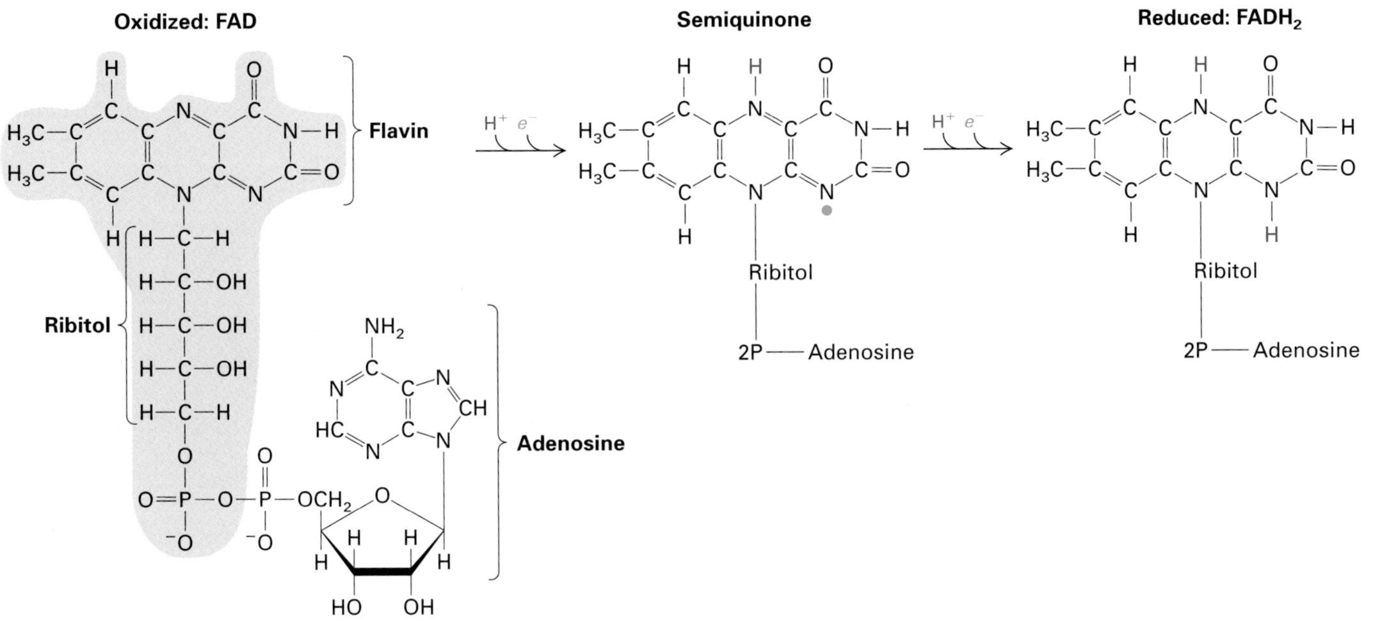

Oxidized: FAD   **Semiquinone**   **Reduced: FADH₂**

▲ **FIGURE 16-8 Structure of FAD and its reduction to FADH₂.**
The coenzyme flavin adenine dinucleotide (FAD) can accept one
or two hydrogen atoms. The addition of one electron together
with a proton (i.e., a hydrogen atom) generates a semiquinone
intermediate. The semiquinone is a free radical because it
contains an unpaired electron (denoted by a blue dot), which is
delocalized by resonance to all the flavin ring atoms. The addition
of a second electron and proton (i.e., a second hydrogen atom)
generates the reduced form, FADH₂. Flavin mononucleotide
(FMN) is a related coenzyme that contains only the flavin–ribitol
phosphate part of FAD (highlighted in blue).

The mitochondrial inner membrane and matrix are the
sites of most reactions involving the oxidation of pyruvate
and fatty acids to CO_2 and H_2O and the coupled synthesis
of ATP from ADP and P_i. These complex processes involve
many steps but can be subdivided into three groups of reactions, each of which occurs in a discrete membrane or
space in the mitochondrion (Figure 16-9):

1. Oxidation of pyruvate and fatty acids to CO_2, coupled
to the reduction of the coenzymes NAD^+ and FAD to
NADH and FADH₂, respectively. These reactions occur in
the matrix or on inner-membrane proteins facing it.

2. Electron transfer from NADH and FADH₂ to O_2. These
reactions occur in the inner membrane and are coupled to
the generation of a proton-motive force across it.

3. Harnessing of the energy stored in the electrochemical
proton gradient for ATP synthesis by the F_0F_1 complex in
the inner membrane.

The inner mitochondrial membrane has numerous infoldings, or *cristae*, that greatly expand its surface area, enhancing its ability to generate ATP (see Figure 16-7). In typical
liver mitochondria, for example, the area of the inner membrane is about five times that of the outer membrane. In fact,
the total area of all inner mitochondrial membranes in liver
cells is about 17 times that of the plasma membrane. The
mitochondria in heart and skeletal muscles contain three times
as many cristae as are found in typical liver mitochondria—

presumably reflecting the greater demand for ATP by muscle cells.

 In plants, stored carbohydrates, mostly in the form
of starch, are hydrolyzed to glucose. Glycolysis
then produces pyruvate, which is transported into
mitochondria, as in animal cells. Mitochondrial oxidation
of pyruvate and concomitant formation of ATP occurs in
photosynthetic cells during dark periods when photosynthesis is not possible, and in roots and other nonphotosynthetic tissues all the time.

Mitochondrial Oxidation of Pyruvate Begins with the Formation of Acetyl CoA

Immediately after pyruvate is transported from the cytosol
across the mitochondrial membranes to the matrix, it reacts
with coenzyme A, forming CO_2 and the intermediate acetyl
CoA (Figure 16-10). This reaction, catalyzed by *pyruvate
dehydrogenase*, is highly exergonic ($\Delta G^{\circ\prime} = -8.0$ kcal/mol)
and essentially irreversible. Pyruvate dehydrogenase, which is
located in the mitochondrial matrix, is a giant multienzyme
complex 30 nm in diameter (4.6×10^6 MW), even larger than
a ribosome. Composed of three different enzymes, pyruvate
dehydrogenase contains 60 subunits, several regulatory
polypeptides, and five different coenzymes (Figure 16-11).

As discussed later, acetyl CoA plays a central role in the
oxidation of fatty acids and many amino acids. In addition,

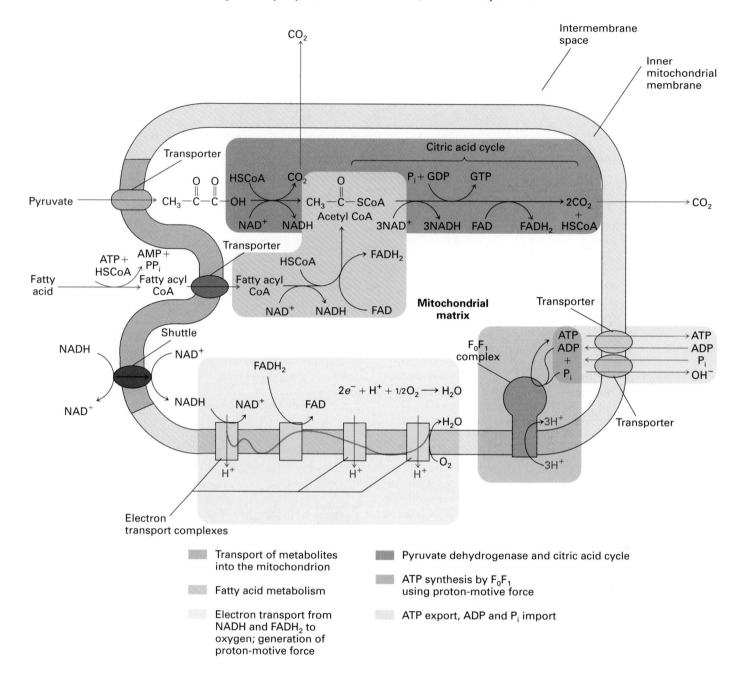

▲ **FIGURE 16-9 Summary of the aerobic oxidation of pyruvate in mitochondria.** The outer membrane is not shown because it is freely permeable to all metabolites. Specific transport proteins (ovals) in the inner membrane import pyruvate (tan), ADP (green), and P_i (purple) into the matrix and export ATP. NADH generated in the cytosol is not transported directly to the matrix because the inner membrane is impermeable to NAD^+ and NADH; instead, a shuttle system (red oval) transports electrons from cytosolic NADH to NAD^+ in the matrix (see Figure 16-13). O_2 diffuses into the matrix and CO_2 diffuses out. HSCoA denotes free coenzyme A (CoA), and SCoA denotes CoA when it is esterified. Fatty acids are linked to CoA on the outer

mitochondrial membrane. Subsequently, the fatty acyl group is removed from the CoA, linked to a carnitine carrier that transports it across the inner membrane, and then the fatty acid is reattached to a CoA on the matrix side of the inner membrane (blue oval). Oxidation of pyruvate in the citric acid cycle generates NADH and $FADH_2$. Electrons from these reduced coenzymes are transferred via four electron transport complexes (blue rectangles) to O_2 concomitant with transport of H^+ ions from the matrix to the intermembrane space, generating the proton-motive force. The F_0F_1 complex (orange) then harnesses the proton-motive force to synthesize ATP. Blue arrows indicate electron flow; red arrows indicate transmembrane movement of metabolites.

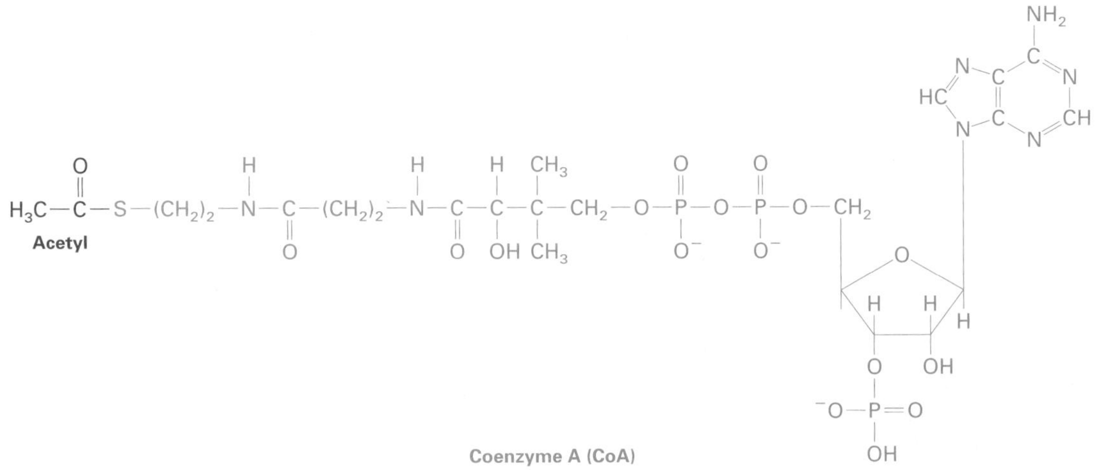

▶ **FIGURE 16-10 The structure of acetyl CoA.** This compound is an important intermediate in the aerobic oxidation of pyruvate, fatty acids, and many amino acids. It also contributes acetyl groups in many biosynthetic pathways.

Coenzyme A (CoA)

(a)

E_1
E_2
E_3

(b)

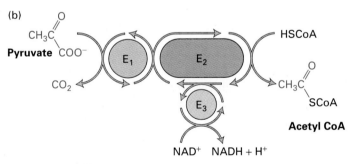

Pyruvate

E_1 E_2 E_3

HSCoA

CO_2

Acetyl CoA

NAD^+ $NADH + H^+$

$$CH_3\overset{O}{\underset{\|}{C}}-COO^- + NAD^+ + HSCoA \longrightarrow CO_2 + NADH + CH_3\overset{O}{\underset{\|}{C}}-SCoA$$

▲ **FIGURE 16-11 Structure of pyruvate dehydrogenase and its catalytic activities.** (a) This very large complex contains three distinct enzymes: E_1 is pyruvate decarboxylase (24 subunits); E_2 is lipoamide transacetylase (24 subunits); and E_3 is dihydrolipoyl dehydrogenase (12 subunits). The E_1 and E_3 subunits are bound to the outside of a transacetylase (E_2) core. Not shown are several subunits involved in regulating the enzyme by reversible phosphorylation and dephosphorylation. (b) The reaction catalyzed by pyruvate dehydrogenase proceeds in three stages and involves several enzyme-bound intermediates. The tight structural integration of E_1, E_2, and E_3 in the complex increases the rate of the overall reaction and minimizes possible side reactions.

it is an intermediate in numerous biosynthetic reactions, such as the transfer of an acetyl group to lysine residues in histone proteins and to the N-termini of many mammalian proteins. Acetyl CoA also is a biosynthetic precursor of cholesterol and other steroids, and of the farnesyl and related groups that anchor proteins such as Ras to membranes (see Figure 3-36b). In respiring mitochondria, however, the acetyl group of acetyl CoA is almost always oxidized to CO_2.

Oxidation of the Acetyl Group of Acetyl CoA in the Citric Acid Cycle Yields CO_2 and Reduced Coenzymes

The final stage in the oxidation of glucose entails a set of nine reactions in which the acetyl group of acetyl CoA is oxidized to CO_2. These reactions operate in a cycle that is referred to by several names: the **citric acid cycle**, the **tricarboxylic acid cycle**, and the **Krebs cycle**. The net result is that for each acetyl group entering the cycle as acetyl CoA, two molecules of CO_2 are produced:

$$CH_3\overset{O}{\overset{\|}{C}}CoA + 3\,NAD^+ + FAD + GDP^{3-} + P_i^{2-} + 2\,H_2O \longrightarrow$$

$$2\,CO_2 + 3\,NADH + FADH_2 + GTP^{4-} + 2\,H^+ + HSCoA$$

As shown in Figure 16-12, the cycle begins with the condensation of the two-carbon acetyl group from acetyl CoA with the four-carbon molecule *oxaloacetate* to yield the six-carbon *citric acid*. In both reactions 4 and 5, a CO_2 molecule is released; in reactions 4, 5, 7, and 9, oxidation of cycle intermediates generates reduced electron carriers (three NADH molecules and one $FADH_2$ molecule). In reaction 6, hydrolysis of the high-energy thioester bond in succinyl CoA is coupled to synthesis of one GTP by substrate-level phosphorylation. (GTP and ATP are interconvertible). The final reaction (9) also regenerates oxaloacetate, so the cycle can

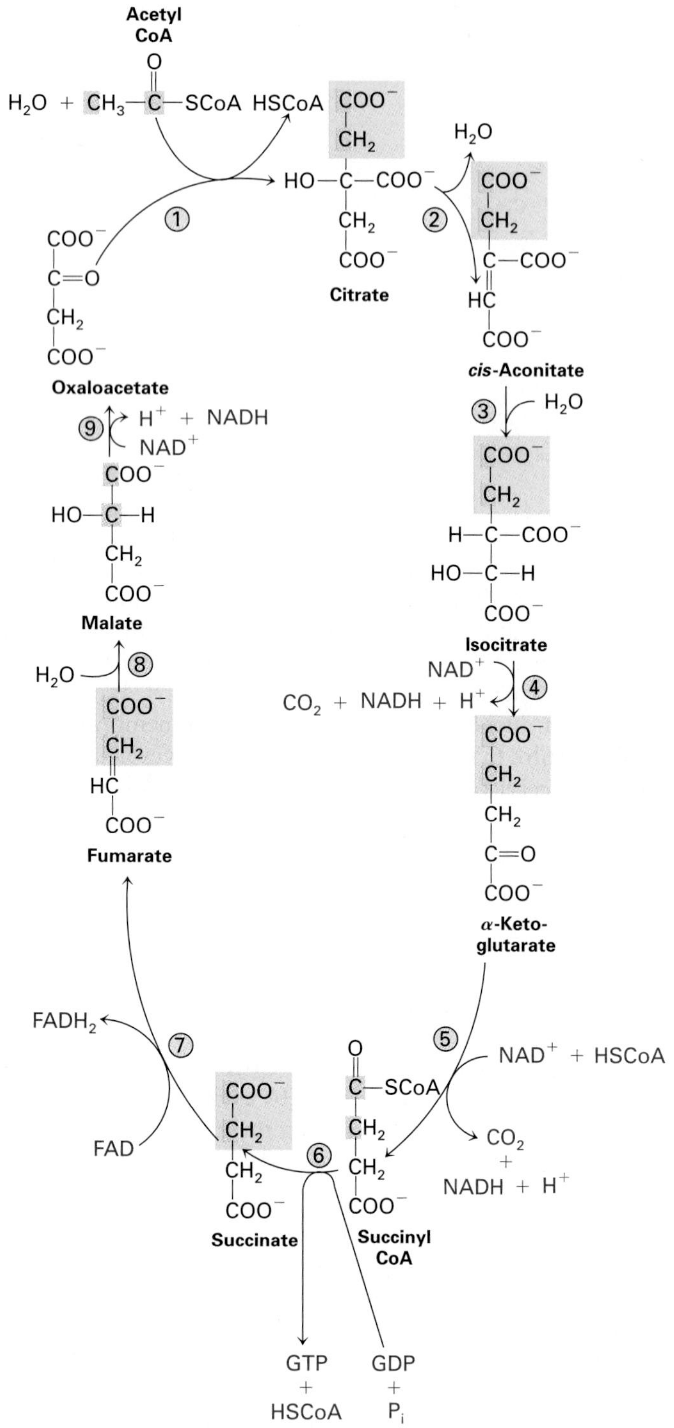

◀ **FIGURE 16-12 The citric acid cycle, in which acetyl groups transferred from acetyl CoA are oxidized to CO_2.** In reaction ①, a two-carbon acetyl residue from acetyl CoA condenses with the four-carbon molecule oxaloacetate to form the six-carbon molecule citrate. In the remaining reactions (②–⑨), each molecule of citrate is eventually converted back to oxaloacetate, losing two CO_2 molecules in the process. In four of the reactions, four pairs of electrons are removed from the carbon atoms: three pairs are transferred to three molecules of NAD^+ to form three NADH and three H^+; one pair is transferred to the acceptor FAD to form $FADH_2$. The two carbon atoms added from acetyl CoA are highlighted in blue. Note that they are *not* lost in the turn of the cycle in which they enter. Because fumarate is a symmetric molecule, these two carbon atoms will be equally distributed among the four in oxaloacetate; one will be lost as CO_2 during the next turn of the cycle and the other in subsequent turns.

dehydrogenase (reaction 5) are localized to the inner membrane with active sites facing the matrix.

The protein concentration of the mitochondrial matrix is 500 mg/ml (a 50 percent protein solution), and the matrix must have a viscous, gel-like consistency. When mitochondria are disrupted by gentle ultrasonic vibration or osmotic lysis, the six non-membrane-bound enzymes in the citric acid cycle are released as a very large multiprotein complex. The reaction product of one enzyme, it is believed, passes directly to the next enzyme without diffusing through the solution. However, much work is needed to determine the structure of the enzyme complex: biochemists generally study the properties of enzymes in dilute aqueous solutions of less than 1 mg/ml, and weak interactions between enzymes are often difficult to detect.

Since glycolysis of one glucose molecule generates two acetyl CoA molecules, the reactions in the glycolytic pathway and citric acid cycle produce six CO_2 molecules, 10 NADH molecules, and two $FADH_2$ molecules per glucose molecule (Table 16-1). Although these reactions also generate four high-energy phosphoanhydride bonds in the form of two ATP and two GTP molecules, this represents only a small fraction of the available energy released in the complete aerobic oxidation of glucose. The remaining energy is stored in the reduced coenzymes, NADH and $FADH_2$. Synthesis of most of the ATP generated in aerobic oxidation is coupled to the reoxidation of these compounds by O_2 in a stepwise process involving the electron transport chain. Moreover, even though molecular O_2 is not involved in any reaction of the citric acid cycle, in the absence of O_2 the cycle soon stops operating as the supply of NAD^+ and FAD dwindles. Before considering electron transport and the coupled formation of ATP in detail, we first discuss how the supply of NAD^+ in the cytosol is regenerated and then the oxidation of fatty acids to CO_2.

begin again. Note that molecular O_2 does not participate in the citric acid cycle.

Most enzymes and small molecules involved in the citric acid cycle are soluble in aqueous solution and are localized to the matrix of the mitochondrion. This includes the water-soluble molecules CoA, acetyl CoA, and succinyl CoA, as well as NAD^+ and NADH. Succinate dehydrogenase together with $FAD/FADH_2$ (reaction 7) and α-ketoglutarate

Inner-Membrane Proteins Allow the Uptake of Electrons from Cytosolic NADH

For aerobic oxidation to continue, the NADH produced during glycolysis in the cytosol must be regenerated. As with

TABLE 16-1 Net Result of the Glycolytic Pathway and the Citric Acid Cycle

Reaction	CO_2 Molecules Produced	NAD Molecules Reduced to NADH	FAD Molecules Reduced to $FADH_2$
1 glucose molecule to 2 pyruvate molecules	0	2	0
2 pyruvates to 2 acetyl CoA molecules	2	2	0
2 acetyl CoA to 4 CO_2 molecules	4	6	2
Total	6	10	2

NADH generated in the mitochondrial matrix, electrons from cytosolic NADH are ultimately transferred to O_2 via the electron transport chain, concomitant with the generation of a proton-motive force. Although the inner mitochondrial membrane is impermeable to NADH itself, several *electron shuttles* can transfer electrons from cytosolic NADH to the matrix.

In the most widespread shuttle—the *malate shuttle*—cytosolic NADH reduces oxaloacetate to malate (Figure 16-13, reaction 1). An antiport protein in the mitochondrial inner membrane then transports malate into the matrix in exchange for α-ketoglutarate (reaction 2). A soluble dehydrogenase in the matrix then converts malate to oxaloacetate, reducing NAD^+ to NADH in the process (reaction 3). Since oxaloacetate, an intermediate in the Krebs cycle cannot cross the inner membrane, it is first converted to the amino acid aspartate, which crosses the inner membrane in exchange for glutamate (reactions 4 and 5). Once in the cytosol, the aspartate is reconverted to oxaloacetate in a reaction in which α-ketoglutarate is converted to glutamate (reaction 6). The net effect of this complex cycle is the oxidation of cytosolic NADH to NAD^+, together with the reduction of matrix NAD^+ to NADH.

Mitochondrial Oxidation of Fatty Acids Is Coupled to ATP Formation

Fatty acids are stored as *triacylglycerols*, primarily as droplets in adipose (fat-storing) cells. In response to hormones such as adrenaline, triacylglycerols are hydrolyzed in the cytosol to free fatty acids and glycerol:

$$CH_3-(CH_2)_n-\overset{\displaystyle O}{\overset{\|}{C}}-O-CH_2$$
$$CH_3-(CH_2)_n-\overset{\displaystyle O}{\overset{\|}{C}}-O-CH + 3\ H_2O \longrightarrow$$
$$CH_3-(CH_2)_n-\overset{\displaystyle O}{\overset{\|}{C}}-O-CH_2$$
Triacylglycerol

$$3\ CH_3-(CH_2)_n-\overset{\displaystyle O}{\overset{\|}{C}}-OH + \begin{matrix} HO-CH_2 \\ HO-CH \\ HO-CH_2 \end{matrix}$$
Fatty acid   **Glycerol**

Fatty acids are released into the blood, from which they are taken up and oxidized by most cells. They are the major energy source for many tissues, particularly heart muscle. In humans, the oxidation of fats is quantitatively more important than the oxidation of glucose as a source of ATP. In part, this is because the oxidation of 1 g of triacylglycerol to CO_2 generates about six times as much ATP as does the oxidation of 1 g of hydrated glycogen, the polymeric storage form of glucose in animals.

In the cytosol, free fatty acids are linked to coenzyme A to form a fatty acyl CoA in an exergonic reaction coupled to the hydrolysis of ATP to AMP and PP_i (inorganic pyrophosphate):

$$R-\overset{\displaystyle O}{\overset{\|}{C}}-O^- + HSCoA + ATP \longrightarrow$$
Fatty acid

$$R-\overset{\displaystyle O}{\overset{\|}{C}}-SCoA + AMP + PP_i$$
Fatty acyl CoA

Subsequent hydrolysis of PP_i to two molecules of phosphate (P_i) drives this reaction to completion. Then the fatty acyl group is transferred to carnitine, moved across the inner mitochondrial membrane by a transporter protein, and is released from carnitine and reattached to another CoA molecule on the matrix side. Each molecule of a fatty acyl CoA in the mitochondrion is oxidized to form one molecule of acetyl CoA and an acyl CoA shortened by two carbon atoms (Figure 16-14). Concomitantly, one molecule apiece of NAD^+ and FAD are reduced, respectively, to NADH and $FADH_2$. This set of reactions is repeated on the shortened acyl CoA until all the carbon atoms are converted to acetyl CoA.

For example, mitochondrial oxidation of each molecule of the 18-carbon stearic acid, $CH_3(CH_2)_{16}COOH$, yields nine molecules of acetyl CoA and eight molecules each of NADH and $FADH_2$. As with acetyl CoA generated from pyruvate, these acetyl groups enter the citric acid cycle and are oxidized to CO_2. Electrons from the reduced coenzymes produced in the oxidation of fatty acyl CoA to acetyl CoA

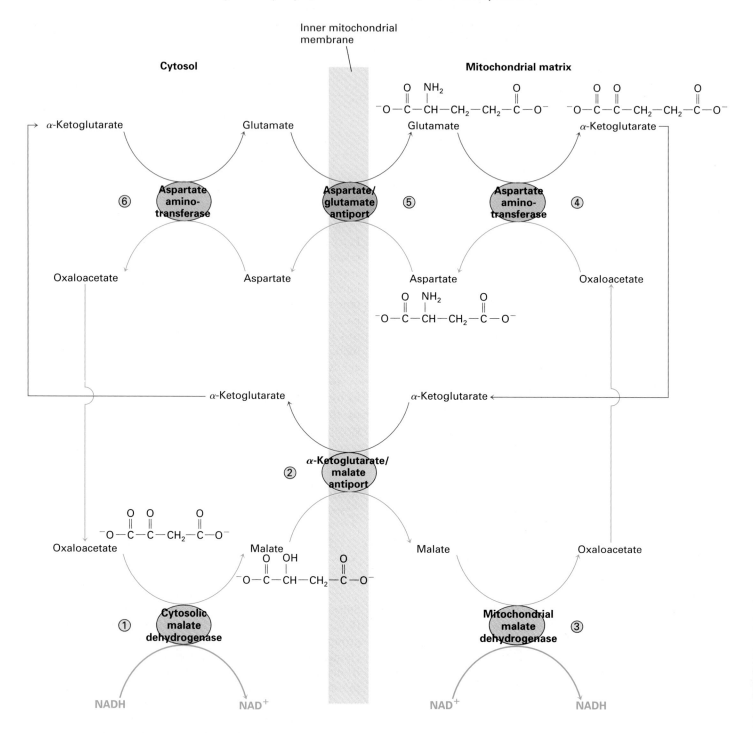

▲ **FIGURE 16-13 The malate shuttle.** Because the inner mitochondrial membrane is impermeable to NADH, the cell uses an indirect mechanism to transfer electrons from cytosolic NADH to NAD^+ in the matrix. Two antiport proteins in the membrane and two soluble enzymes present in both the cytosol and the matrix carry out the cycle reactions. One counterclockwise turn of the entire cycle can be summarized as:

$$NADH_{cytosol} + NAD^+_{matrix} \longleftrightarrow NAD^+_{cytosol} + NADH_{matrix}.$$

[Courtesy of B. Trumpower.]

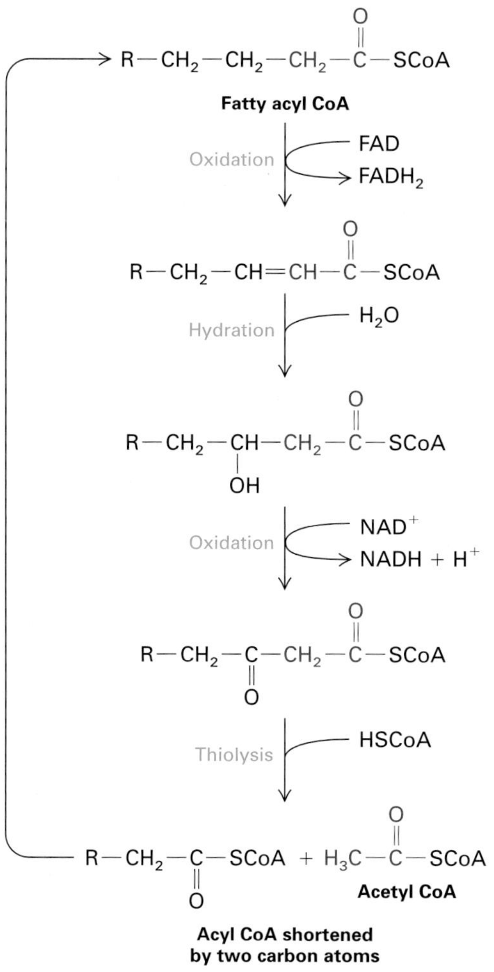

Fatty acyl CoA

$R-CH_2-CH_2-CH_2-\overset{\overset{\displaystyle O}{\|}}{C}-SCoA$

Fatty acyl CoA

Oxidation — FAD → FADH$_2$

$R-CH_2-CH=CH-\overset{\overset{\displaystyle O}{\|}}{C}-SCoA$

Hydration — H$_2$O

$R-CH_2-\underset{\underset{\displaystyle OH}{|}}{CH}-CH_2-\overset{\overset{\displaystyle O}{\|}}{C}-SCoA$

Oxidation — NAD$^+$ → NADH + H$^+$

$R-CH_2-\overset{\overset{\displaystyle O}{\|}}{C}-CH_2-\overset{\overset{\displaystyle O}{\|}}{C}-SCoA$

Thiolysis — HSCoA

$R-CH_2-\overset{\overset{\displaystyle O}{\|}}{C}-SCoA \;+\; H_3C-\overset{\overset{\displaystyle O}{\|}}{C}-SCoA$

Acetyl CoA

**Acyl CoA shortened
by two carbon atoms**

▲ **FIGURE 16-14 Oxidation of fatty acids in mitochondria.**
Four enzyme-catalyzed reactions convert a fatty acyl CoA
molecule to acetyl CoA and a fatty acyl CoA shortened by two
carbon atoms. Concomitantly, one NAD$^+$ molecule is reduced to
NADH and one FAD molecule is reduced to FADH$_2$. The cycle is
repeated on the shortened acyl CoA until fatty acids with an
even number of carbon atoms are completely converted to acetyl
CoA. Fatty acids with an odd number of C atoms are rare; they
are metabolized to one molecule of propionyl CoA and multiple
acetyl CoA molecules.

Oxidation of Fatty Acids in Peroxisomes Generates No ATP

Mitochondrial oxidation of fatty acids is the major source
of ATP in mammalian liver cells, and biochemists at one
time believed this was true in all cell types. However, rats
treated with clofibrate, a drug used to reduce the level of
blood lipoproteins, were found to exhibit an increased rate
of fatty acid oxidation and a large increase in the number
of peroxisomes in their liver cells. This finding suggested
that peroxisomes, as well as mitochondria, can oxidize fatty
acids. These small organelles, $\approx 0.2-1$ μm in diameter, are
lined by a single membrane (see Figure 5-47). Also called
microbodies, peroxisomes are present in all mammalian cells
except erythrocytes and are also found in plant cells, yeasts,
and probably most eukaryotic cells. The peroxisome is now
recognized as the principal organelle in which fatty acids are
oxidized in most cell types. Indeed, very long chain fatty
acids containing more than about 20 CH$_2$ groups are degraded only in peroxisomes; in mammalian cells, mid-length
fatty acids containing 10–20 CH$_2$ groups can be degraded
in both peroxisomes and mitochondria.

Peroxisomes contain several *oxidases*—enzymes that use
oxygen as an electron acceptor to oxidize organic substances, in the process forming hydrogen peroxide (H$_2$O$_2$),
which is then degraded by *catalase*:

$$O_2 \xrightarrow{\text{oxidases}} H_2O_2$$
$$RH_2 \qquad R$$

$$2\,H_2O_2 \xrightarrow{\text{catalase}} 2\,H_2O + O_2$$

The pathway of peroxisomal degradation of fatty acids is
similar to that used in liver mitochondria. However, peroxisomes lack an electron transport chain, and electrons from
the FADH$_2$ and NADH produced during the oxidation of
fatty acids are immediately transferred to O$_2$, regenerating
FAD and NAD$^+$, and forming H$_2$O$_2$ (Figure 16-15). Catalase quickly decomposes the H$_2$O$_2$, which is highly toxic to
the cell.

In contrast to mitochondrial fatty acid oxidation, which
is coupled to generation of ATP, peroxisomal oxidation of
fatty acids is not linked to ATP formation, and the released
energy is converted to heat. The acetyl group of acetyl CoA
generated during peroxisomal oxidation of fatty acids (see
Figure 16-15) is transported into the cytosol, where it is used
in the synthesis of cholesterol and other metabolites.

MEDICINE Before fatty acids can be degraded in the peroxisome, they must first be transported into the organelle from the cytosol. Mid-length fatty acids are esterified to coenzyme A in the cytosol; the resulting fatty acyl
CoA is then transported into the peroxisome by a specific
transporter. However, very long chain fatty acids enter the
peroxisome by another transporter, and then are esterified
to CoA once inside. In the human genetic disease *X-linked
adrenoleukodystrophy (ALD)*, peroxisomal oxidation of very
long chain fatty acids is specifically defective, while the oxidation of mid-length fatty acids is normal. In ALD, very long
chain fatty acids are transported normally into peroxisomes,

and in the subsequent oxidation of acetyl CoA in the citric
acid cycle move via the electron transport chain to O$_2$, coupled to regeneration of a proton-motive force that is used
to power ATP synthesis (see Figure 16-9).

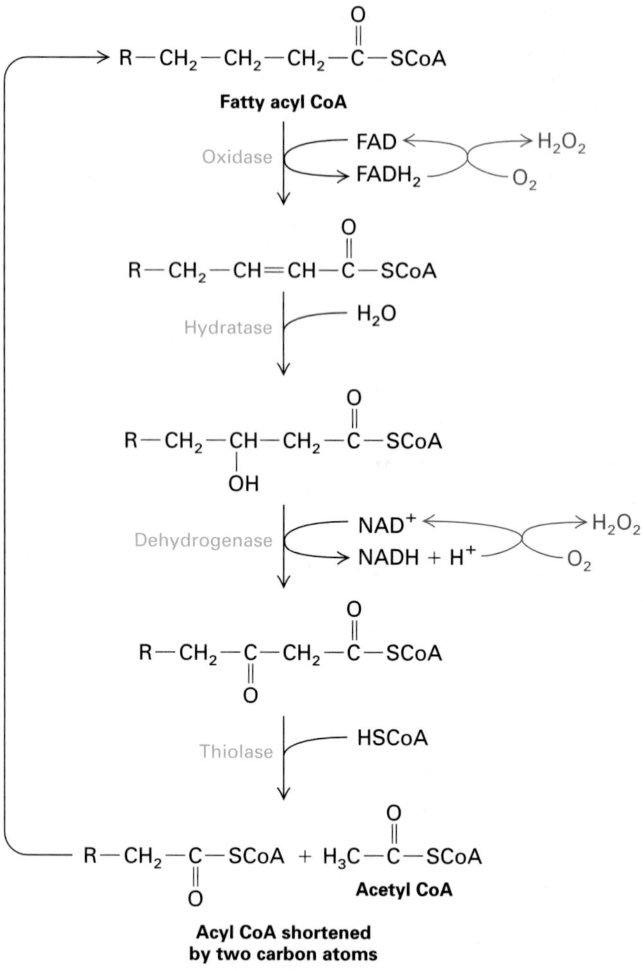

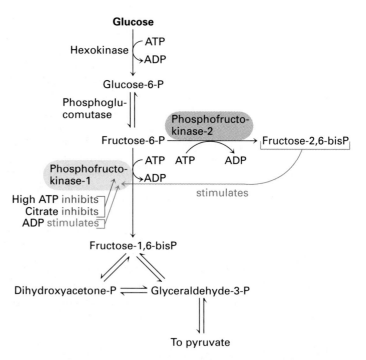

▲ FIGURE 16-15 Oxidation of fatty acids by peroxisomes.
Peroxisomes degrade fatty acids with more than 12 carbon atoms by a series of reactions similar to those used by liver mitochondria (see Figure 16-14). In peroxisomes, however, the electrons and protons transferred to FAD and NAD^+ during the oxidation reactions are subsequently transferred to oxygen, forming H_2O_2.

but are not esterified to CoA and so cannot be oxidized. The enzyme that catalyzes this esterification is synthesized in the cytosol; as we discuss in Chapter 17, the *ADL* gene encodes the peroxisomal membrane protein required for uptake of this enzyme into peroxisomes. Patients with the severe form of ADL are unaffected until mid-childhood, when severe neurological disorders appear, followed by death within a few years.

The Rate of Glucose Oxidation Is Adjusted to Meet the Cell's Need for ATP

All enzyme-catalyzed reactions and metabolic pathways are regulated by cells so as to produce the needed amounts of

metabolites but not an excess. The primary function of the oxidation of glucose to CO_2 in the glycolytic pathway and the citric acid cycle is to produce NADH and $FADH_2$, whose oxidation in the mitochondria generates ATP. The operation of both pathways is continuously regulated, primarily by allosteric mechanisms, to meet the cell's need for ATP (Chapter 3).

Three glycolytic enzymes that are allosterically controlled play a key role in regulating the entire glycolytic pathway (see Figure 16-3). *Hexokinase,* which catalyzes the first step, is inhibited by its reaction product, glucose 6-phosphate. *Pyruvate kinase,* which catalyzes the last step, is inhibited by ATP, so glycolysis slows down if too much ATP is present. The third enzyme, *phosphofructokinase-1,* catalyzes the third reaction in the conversion of glucose to pyruvate and is the principal rate-limiting enzyme of the glycolytic pathway. Emblematic of its critical role in regulating the rate of glycolysis, this enzyme is controlled by four allosteric molecules (Figure 16-16).

If citrate—the product of the first step of the citric acid cycle—accumulates, it allosterically inhibits the activity of phosphofructokinase-1, thereby reducing the generation of pyruvate and acetyl CoA, so that less citrate is formed. This feedback inhibition of phosphofructokinase-1 by citrate allows the activities of the glycolytic pathway to be coordinated with those of the citric acid cycle. Intermediates in the citric acid cycle are also used in biosynthesis of amino acids;

▲ FIGURE 16-16 Enzymatic control of glucose metabolism in the cytosol. Phosphofructokinase-1 is the main control point in the regulation of the glycolytic pathway. It is allosterically inhibited by ATP and citrate, and stimulated by ADP and fructose 2,6-bisphosphate. See the text for discussion.

a buildup of citrate indicates that these intermediates are plentiful and that glucose need not be degraded for this purpose.

Phosphofructokinase-1 also is allosterically *activated* by ADP and allosterically *inhibited* by ATP. This arrangement makes the rate of glycolysis very sensitive to intracellular levels of ATP and ADP. The allosteric inhibition of phosphofructokinase-1 by ATP may seem unusual, since ATP is also a substrate of this enzyme. But the affinity of the substrate-binding site for ATP is much higher (has a lower K_m) than that of the allosteric site. Thus at low concentrations, ATP binds to the catalytic but not to the inhibitory allosteric site and enzymatic catalysis proceeds at near maximal rates. At high concentrations, ATP binds to the allosteric site, inducing a conformational change that reduces the affinity of the enzyme for the other substrate, fructose 6-phosphate, and thus inhibits the rate of this reaction and the overall rate of glycolysis.

The metabolite *fructose 2,6-bisphosphate* is another important allosteric activator of phosphofructokinase-1 (see Figure 16-16). Fructose 2,6-bisphosphate is formed from the glycolytic intermediate fructose 6-phosphate; the catalyst is phosphofructokinase-2, an enzyme different from phosphofructokinase-1. Fructose 6-phosphate accelerates the formation of fructose 2,6-bisphosphate, which, in turn, activates phosphofructokinase-1. This type of control, by analogy with feedback control, is known as *feed-forward activation,* in which the abundance of a metabolite (here, fructose 6-phosphate) induces an acceleration in its metabolism. Fructose 2,6-bisphosphate allosterically activates phosphofructokinase-1 in liver cells by decreasing the inhibitory effect of ATP and by increasing the affinity of phosphofructokinase-1 for one of its substrates, fructose 6-phosphate.

The three glycolytic enzymes that are regulated by allosteric molecules catalyze reactions with large negative $\Delta G°'$ values—reactions that are essentially irreversible under ordinary conditions. These enzymes thus are particularly suitable for regulating the entire glycolytic pathway. Additional control is exerted by glyceraldehyde 3-phosphate dehydrogenase, which catalyzes the reduction of NAD⁺ to NADH (see Figure 16-3). If cytosolic NADH builds up owing to a slowdown in mitochondrial oxidation, this step in glycolysis will be slowed by mass action. As we discuss later, mitochondrial oxidation of NADH and FADH₂, produced in the glycolytic pathway and citric acid cycle, also is tightly controlled to produce the appropriate amount of ATP required by the cell.

Glucose metabolism is controlled differently in various mammalian tissues to meet the metabolic needs of the organism as a whole. During periods of carbohydrate starvation, for instance, glycogen in the liver is converted directly to glucose 6-phosphate (without involvement of lexokinase). Under these conditions, however, phosphofructokinase-1 is inhibited and thus glucose 6-phosphate is not metabolized to pyruvate; rather, it is converted to glucose by a phosphatase and released into the blood to nourish the brain and muscles, which then oxidize the bulk of the available glucose.

(Chapter 20 contains a more detailed discussion of the control of glucose metabolism in the liver and muscles.) In all cases, the activity of these enzymes is regulated by the level of small-molecule metabolites, generally by allosteric interactions or by phosphorylation.

Additional control of glucose oxidation occurs in the mitochondria. Pyruvate dehydrogenase is deactivated by phosphorylation, which is stimulated by high levels of ATP, NADH, and acetyl CoA. Three enzymes of the citric acid cycle also are regulated. As a consequence of these multiple sites of regulation, the entry of two-carbon units into the citric acid cycle and the rate of the cycle are decreased when the cell has sufficient ATP. In this case, extra acetyl CoA is used to synthesize fatty acids, which are stored as triacylglycerols in adipose tissue.

SUMMARY Oxidation of Glucose and Fatty Acids to CO₂

- In the cytosol of eukaryotic cells, glucose is converted to pyruvate via the glycolytic pathway, with the net formation of two ATPs and the net reduction of two NAD⁺ molecules to NADH (see Figure 16-3). ATP is formed by two substrate-level phosphorylation reactions in the conversion of glyceraldehyde 3-phosphate to pyruvate.

- In anaerobic cells, pyruvate can be metabolized further to lactate or to ethanol plus CO₂, with the reoxidation of NADH.

- Mitochondria have a permeable outer membrane and an inner membrane, which is the site of electron transport and ATP synthesis (see Figure 16-9).

- Pyruvate dehydrogenase, a very large, multienzyme complex in the mitochondrial matrix converts pyruvate into acetyl CoA and CO₂ (see Figure 16-11).

- In each turn of the citric acid cycle, acetyl CoA condenses with the four-carbon molecule oxaloacetate to form the six-carbon citrate, which is converted back to oxaloacetate by a series of reactions that release two molecules of CO₂ and generate three NADH molecules and one FADH₂ molecule (see Figure 16-12).

- The NADH generated in the cytosol during glycolysis is oxidized to NAD⁺, with the concomitant reduction of NAD⁺ to NADH in the mitochondrial matrix, by a set of enzymes and transport proteins that form an electron shuttle.

- Electrons from NADH and FADH₂ move via a series of membrane-bound electron carriers in the inner mitochondrial membrane to O₂, regenerating NAD⁺ and FAD. This stepwise movement of electrons is coupled to pumping of protons across the inner membrane. The resulting proton-motive force powers ATP synthesis and generates most of the ATP resulting from aerobic oxidation of glucose.

- Oxidation of fatty acids in mitochondria yields acetyl CoA, which enters the citric acid cycle, and the reduced coenzymes NADH and $FADH_2$. Subsequent oxidation of acetyl CoA and the reduced coenzymes is coupled to the formation of a proton-motive force that powers ATP formation.

- In most eukaryotic cells, oxidation of fatty acids, especially very long chain fatty acids, occurs primarily in peroxisomes and is not linked to ATP production; the released energy is converted to heat. The electrons released during peroxisomal oxidation of fatty acids are used to form H_2O_2, which is decomposed to H_2O and O_2 by catalase.

- The rate of glycolysis and the citric acid cycle, which depends on the cell's need for ATP, is controlled by the inhibition and stimulation of several enzymes (see Figure 16-16). This complex regulation coordinates the activities of the glycolytic pathway and the citric acid cycle and results in the storage of glucose (as glycogen) or fat when ATP is abundant.

16.2 Electron Transport and Oxidative Phosphorylation

Most of the free energy released during the oxidation of glucose to CO_2 is retained in the reduced coenzymes NADH and $FADH_2$ generated during glycolysis and the citric acid cycle. During **respiration,** electrons are released from NADH and $FADH_2$ and eventually are transferred to O_2, forming H_2O according to the following overall reactions:

$$NADH + H^+ + \tfrac{1}{2} O_2 \longrightarrow NAD^+ + H_2O$$

$$FADH_2 + \tfrac{1}{2} O_2 \longrightarrow FAD + H_2O$$

The $\Delta G°'$ values for these strongly exergonic reactions are -52.6 kcal/mol (NADH) and -43.4 kcal/mol ($FADH_2$). Recall that the conversion of one glucose molecule to CO_2 via the glycolytic pathway and citric acid cycle yields 10 NADH and 2 $FADH_2$ molecules (see Table 16-1). Oxidation of these reduced coenzymes has a total $\Delta G°'$ of -613 kcal/mol [$10(-52.6) + 2(-43.4)$]. Thus, of the potential free energy present in the chemical bonds of glucose (-680 kcal/mol), about 90 percent is conserved in the reduced coenzymes.

The free energy released during oxidation of a single NADH or $FADH_2$ molecule by O_2 is sufficient to drive the synthesis of several molecules of ATP from ADP and P_i, a reaction with a $\Delta G°'$ of $+7.3$ kcal/mol. The mitochondrion maximizes the production of ATP by transferring electrons from NADH and $FADH_2$ through a series of electron carriers all but one of which are integral components of the inner membrane. This step-by-step transfer of electrons via the *electron transport chain* (also known as the *respiratory*

▶ **FIGURE 16-17 Stepwise flow of electrons through the electron transport chain from NADH, succinate, and $FADH_2$ to O_2 (blue arrows).** Each of the four large multiprotein complexes in the chain is located in the inner mitochondrial membrane and contains several specific electron carriers. Coenzyme Q (CoQ) and cytochrome *c* transport electrons between the complexes. As shown by the redox scale, electrons pass in sequence from carriers with a lower reduction potential to those with a higher (more positive) potential. The free-energy scale shows the corresponding reduction in free energy as a pair of electrons moves through the chain. The energy released as electrons flow through three of the complexes is sufficient to power the pumping of H^+ ions across the membrane, establishing a proton-motive force.

chain) allows the free energy in NADH and $FADH_2$ to be released in small increments. At several sites during electron transport from NADH to O_2, protons from the mitochondrial matrix are transported uphill across the inner mitochondrial membrane and a proton concentration gradient forms across it (Figure 16-17). Because the outer membrane is freely permeable to protons, the pH of the mitochondrial matrix is higher (i.e., the proton concentration is lower) than that of the cytosol and intermembrane space. An electric potential across the inner membrane also results from the uphill pumping of positively charged protons outward from the matrix, which becomes negative with respect to the intermembrane space. Thus free energy released during the oxidation of NADH or $FADH_2$ is stored both as an electric potential and a proton concentration gradient—collectively, the proton-motive force—across the inner membrane. The movement of protons back across the inner membrane, driven by this force, is coupled to the synthesis of ATP from ADP and P_i by the F_0F_1 complex (see Figure 16-9).

The synthesis of ATP from ADP and P_i, driven by the transfer of electrons from NADH or $FADH_2$ to O_2, is the major source of ATP in aerobic nonphotosynthetic cells. Much evidence shows that in mitochondria and bacteria this process, called **oxidative phosphorylation,** depends on generation of an electrochemical proton gradient (i.e., proton-motive force) across the inner membrane, with electron transport, proton pumping, and ATP formation occurring simultaneously. In the laboratory, for instance, addition of O_2 and an oxidizable substrate such as pyruvate or succinate to isolated intact mitochondria results in a net synthesis of ATP if the inner mitochondrial membrane is intact. In the presence of minute amounts of detergents that make the membrane leaky, the oxidation of these metabolites by O_2 still occurs, but no ATP is made. Under these conditions, no transmembrane proton concentration gradient or membrane electric potential can be maintained.

In this section we first discuss the magnitude of the proton-motive force, then the components of the electron transport chain and the translocation of protons across the inner membrane. Next we describe the structure of the F_0F_1

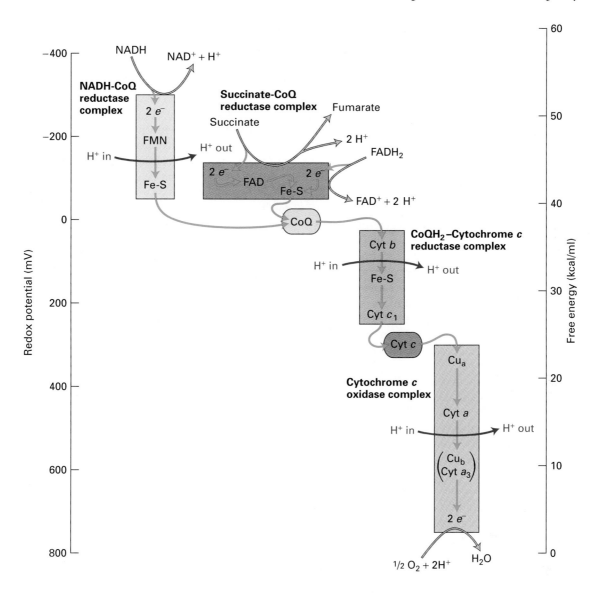

complex and how it uses the proton-motive force to synthe-size ATP. In the last section, we consider how mitochondrial oxidation of NADH and FADH$_2$ is controlled to meet the cell's need for ATP.

The Proton-Motive Force in Mitochondria Is Due Largely to a Voltage Gradient across the Inner Membrane

As we've seen, the proton-motive force (pmf) is the sum of a transmembrane proton concentration (pH) gradient and elec-tric potential, or voltage gradient.* The relative contribution of the two components to the total pmf depends on the per-meability of the membrane to ions other than H$^+$. A signi-ficant voltage gradient can develop only if the membrane is poorly permeable to other cations and to anions, as is the in-ner mitochondrial membrane. In this case, the developing volt-age gradient (i.e., excess H$^+$ ions on the intermembrane face and excess anions on the cytosolic face) soon prevents further

proton movement, so only a small pH gradient is generated. In contrast, a significant pH gradient can develop only if the membrane is also permeable to a major anion (e.g., Cl$^-$), or if the protons are exchanged for another cation (e.g., K$^+$). In either case, proton movement does not lead to a voltage gra-dient across the membrane because there is always an equal concentration of positive and negative ions on each side of the membrane. This is the situation in the chloroplast thyla-koid membrane during photosynthesis, as we discuss later.

---

*Note that the transmembrane electric potential that contributes to the proton-motive force and the resting electric potential across the plasma membrane, discussed in Chapter 15, are generated by fun-damentally different mechanisms. The first results from the transport of H$^+$ ions *against* their concentration gradient powered by electron transport; the second results primarily from the movement of K$^+$ ions from the cytosol to the cell exterior, *down* their concentration gra-dient, through open potassium channels.

Compared with chloroplasts, then, a greater portion of the pmf in mitochondria is due to the membrane electric potential, and the actual pH gradient is smaller.

Since a difference of one pH unit represents a tenfold difference in H^+ concentration, a pH gradient of one unit across a membrane is equivalent to an electric potential of 59 mV (at 20 °C). Thus we can define the proton-motive force, pmf, as

$$\text{pmf} = \Psi - \left(\frac{RT}{\mathcal{F}}\right) \times \Delta\text{pH} = \Psi - 59\Delta\text{pH}$$

where R is the gas constant of 1.987 cal/(degree·mol), T is the temperature (in degrees Kelvin), $\mathcal{F}$ is the Faraday constant [23,062 cal/ (V·mol)], and Ψ is the transmembrane electric potential; Ψ and pmf are measured in millivolts. Measurements on respiring mitochondria have shown that the electric potential (Ψ) across the inner membrane is $\approx$160 mV (negative inside matrix) and that ΔpH is $\approx$1.0 (equivalent to $\approx$60 mV). Thus the total pmf is $\approx$220 mV, with the transmembrane electric potential responsible for about 73 percent.

Because mitochondria are much too small to be impaled with electrodes, the electric potential and pH gradient across the inner mitochondrial membrane cannot be determined by direct measurement. By trapping fluorescent pH-sensitive dyes inside vesicles formed from the inner mitochondrial membrane, researchers can measure the inside pH during oxidative phosphorylation. The electric potential can be determined by adding radioactive K^+ ($^{42}K^+$) ions and a trace amount of valinomycin (K^+ **ionophore**) to a suspension of respiring mitochondria. Although the inner membrane is normally impermeable to K^+, valinomycin is a lipid-soluble peptide that can selectively bind K^+ in its hydrophilic interior and carry it across otherwise impermeable membranes. $^{42}K^+$ will equilibrate across the membrane in accordance with the electric potential; the more negative the matrix side of the inner membrane, the more $^{42}K^+$ will accumulate. Thus, the concentration of radioactive K^+ ions in the matrix and cytosol is measured after equilibrium is reached. From the measured $^{42}K^+_{\text{matrix}}$:$^{42}K^+_{\text{cytosol}}$ ratio, the electric potential E (in mV) across the inner membrane can be calculated using the Nernst equation (Equation 15-5). When trace amounts of valinomycin and radioactive potassium ($^{42}K^+$) ions are added to a suspension of respiring mitochondria, oxidative phosphorylation proceeds and is largely unaffected. The $^{42}K^+$ ions accumulate inside the mitochondria in a K_{matrix}:K_{cytosol} ratio of about 500. By substituting this value in the Nernst equation,

$$E = -59 \log \frac{[K_{\text{in}}]}{[K_{\text{out}}]} = -59 \log 500 = -160 \text{ mV}$$

we can see that the electric potential across the inner membrane is $\approx$160 mV, with the inside negative.

Electron Transport in Mitochondria Is Coupled to Proton Translocation

As noted earlier, the stepwise movement of electrons from NADH and $FADH_2$ to O_2 via a series of electron carriers is coupled to translocation of protons from the mitochondrial matrix to the intermembrane space. This proton movement generates the proton-motive force that directly powers ATP synthesis. That electron transport from NADH (or $FADH_2$) to O_2 is coupled to proton transport across the membrane is demonstrated by the experiment depicted in Figure 16-18. As soon as O_2 is added to a suspension of mitochondria, the medium outside the mitochondria becomes acidic. During electron transport from NADH to O_2, protons translocate from the matrix to the intermembrane space; since the outer membrane is freely permeable to protons, the pH of the outside medium is lowered briefly. The measured change in pH indicates that 10 protons are transported out of the matrix for every electron pair transferred from NADH to O_2.

When this experiment is repeated with succinate rather than NADH as the reduced substrate, the medium outside the mitochondria again becomes acidic, but less so. Recall that oxidation of succinate to fumarate in the citric acid cycle generates $FADH_2$(see Figure 16-12). Because $FADH_2$ transfers electrons to the electron transport chain at a later point than NADH does, electron transport from $FADH_2$ (or succinate) results in translocation of fewer protons from the matrix, and thus a smaller change in pH (see Figure 16-17).

Electrons Flow from FADH$_2$ and NADH to O$_2$ via a Series of Multiprotein Complexes

We now examine more closely the energetically favored movement of electrons from the coenzymes NADH and $FADH_2$ to the final electron acceptor, O_2. In respiring mitochondria, each NADH molecule releases two electrons to the electron transport chain; these electrons ultimately reduce one oxygen atom (half of an O_2 molecule), forming one molecule of water:

$$\text{NADH} \longrightarrow \text{NAD}^+ + \text{H}^+ + 2e^-$$
$$2e^- + 2H^+ + \tfrac{1}{2} O_2 \longrightarrow H_2O$$

As electrons move from NADH to O_2, their potential declines by 1.14 V, which corresponds to 26.2 kcal/mol of electrons transferred, or $\approx$53 kcal/mol for a pair of electrons. Much of this energy is conserved at three stages of electron transport by the movement of protons across the inner mitochondrial membrane from the matrix to the intermembrane space (see Figure 16-17).

Most mitochondrial electron carriers comprise prosthetic groups, such as flavins, heme, iron-sulfur clusters, and copper, bound to four multiprotein complexes, each of which spans the inner mitochondrial membrane (Figure 16-19). Table 16-2 lists the prosthetic groups in each multiprotein complex. Before considering the function of each complex,

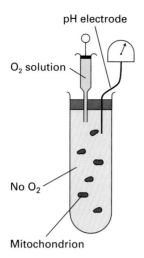

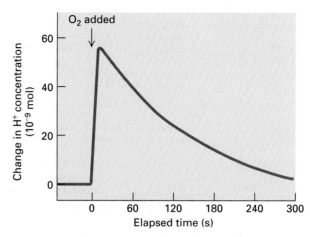

▲ **FIGURE 16-18 Experimental demonstration that electron transport from NADH or FADH₂ to O₂ is coupled to proton transport across the membrane.** If a source of electrons for respiration, such as NADH, is added to a suspension of mitochondria depleted of O_2, no NADH is oxidized. When a small amount of O_2 is added to the system (arrow), the pH of the surrounding medium drops sharply—a change that corresponds to an *increase* in protons outside the mitochondria.

(The presence of a large amount of valinomycin and K^+ in the reaction dissipates the voltage gradient generated by H^+ translocation, so that all pumped H^+ ions contribute to the pH change.) Thus the oxidation of NADH by O_2 is coupled to the movement of protons out of the matrix. Once the O_2 is depleted, the excess protons slowly move back into the mitochondria (powering the synthesis of ATP) and the pH of the extracellular medium returns to its initial value.

we take a detailed look at some of the individual carriers in the electron transport chain.

Iron-sulfur clusters, Fe_2S_2 and Fe_4S_4, are nonheme prosthetic groups consisting of Fe atoms bonded both to inorganic S atoms and to four S atoms on cysteine residues on the protein (Figure 16-20). Some Fe atoms in the cluster bear a +2 charge while others have a +3 charge. However, the net charge of each Fe atom is actually between +2 and +3, because electrons in the outermost orbits are dispersed among the Fe atoms and move rapidly from one atom to another. Iron-sulfur clusters accept and release electrons one at a time; the additional electron is also dispersed over all the Fe atoms in the cluster.

The **cytochromes** are proteins covalently linked to a heme molecule, an iron-containing prosthetic group similar to that in hemoglobin or myoglobin. Electron transport occurs by oxidation and reduction of the Fe atom in the center of the heme:

$$Fe_{ox}^{3+} + e^- \rightleftharpoons Fe_{red}^{2+}$$

In the electron transport chain, electrons move through the cytochromes in the following order: b_{566}, b_{562}, c_1, c, a, and a_3 (see Figure 16-19). The various cytochromes have slightly different heme groups and axial ligands, which generate different environments for the Fe ion (Figure 16-21). Therefore, each cytochrome has a different reduction potential, or tendency to accept an electron—an important property dictating the unidirectional electron flow along the chain. Because the heme ring in cytochromes consists of alternating double-

and single-bonded atoms, a large number of resonance forms exist, and the extra electron is delocalized to the heme carbon and nitrogen atoms as well as to the Fe ion.

Coenzyme Q (CoQ), also called *ubiquinone*, is the only electron carrier that is not a protein-bound prosthetic group. It is a carrier of hydrogen atoms (protons plus electrons).

TABLE 16-2	**Electron Transport Chain**
Enzyme Complex	Prosthetic Groups*
NADH-CoQ reductase	FMN
	FeS
Succinate-CoQ reductase	FAD
	FeS
CoQH₂–cytochrome c reductase	Heme b_{566}
	Heme b_{562}
	Heme c_1
	FeS
Cytochrome c	Heme c
Cytochrome c oxidase	Heme a
	Heme a_3
	Cu_a^+, Cu_b^+

*Not included is coenzyme Q, an electron carrier that is not permanently bound to a protein complex.

SOURCE: J. W. De Pierre and L. Ernster, 1977, *Ann. Rev. Biochem.* 46:201.

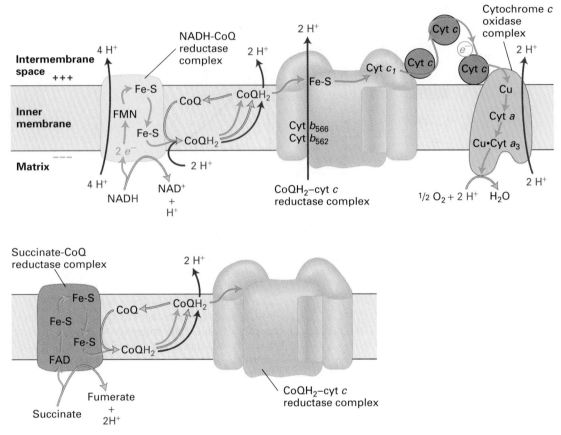

▲ **FIGURE 16-19 The pathway of electron transport (blue arrows) and proton transport (red arrows) in the inner mitochondrial membrane.** Bound to each of the electron transport complexes are several prosthetic groups, which carry electrons through the complex. *(Top)* A pair of electrons from one NADH first move through the NADH-CoQ reductase complex to one coenzyme Q (CoQ) molecule, which binds reversibly to the complex. The reduced CoH_2 diffuses in the membrane and donates electrons to the $CoQH_2$–Cyt *c* reductase complex. The peripheral protein cytochrome *c* diffuses in the intermembrane space, transporting electrons (one at a time) from the $CoQH_2$–Cyt *c* reductase complex to the cytochrome *c* oxidase complex and ultimately to O_2. Protons are pumped from the matrix into the intermembrane space at the indicated sites. The protons released into the matrix space by the oxidation of NADH are consumed in the formation of water from O_2, resulting in no net proton translocation from these reactions. Thus a total of 10 protons are translocated per pair of electrons moved from NADH to O_2. *(Bottom)* The succinate-CoQ reductase complex oxidizes succinate to fumarate, reducing one CoQ to $CoQH_2$. No protons are translocated by the succinate-CoQ reductase complex. The remainder of electron transport from succinate proceeds by the same path as in the top diagram. Thus for every pair of electrons transported from succinate to O_2, six protons are translocated from the matrix into the intermembrane space.

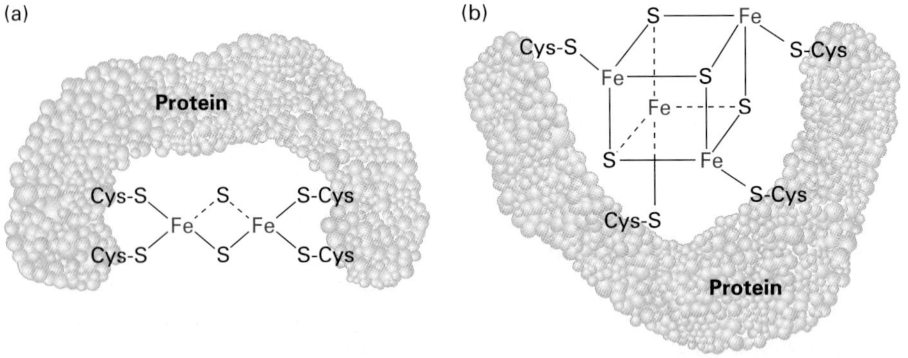

▶ **FIGURE 16-20 Three-dimensional structures of some iron-sulfur clusters in electron-transporting proteins: (a) a dimeric (Fe_2S_2) cluster; (b) a tetrameric (Fe_4S_4) cluster.** In both types of clusters, each Fe atom is bonded to four S atoms: some S atoms are molecular sulfur; others occur in the cysteine side chains of a protein. Such Fe-S clusters accept and release one electron at a time. [See W. H. Orme-Johnson, 1973, *Ann. Rev. Biochem.* **42**:159.]

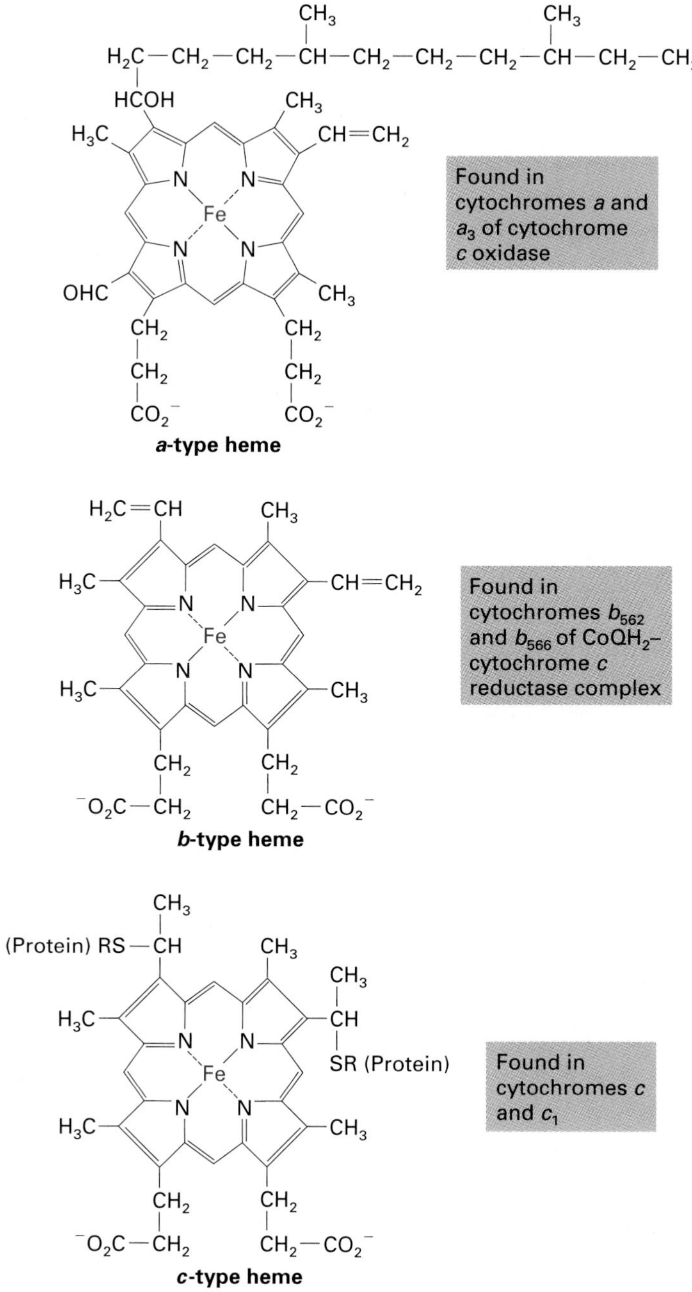

a-type heme

Found in cytochromes *a* and *a₃* of cytochrome *c* oxidase

b-type heme

Found in cytochromes b_{562} and b_{566} of $CoQH_2$–cytochrome *c* reductase complex

c-type heme

Found in cytochromes *c* and c_1

Ubiquinone (CoQ) (oxidized form)

e^-

Semiquinone (CoQ·⁻) (free radical)

$2H^+ + e^-$

Dihydroquinone (CoQH₂) (fully reduced form)

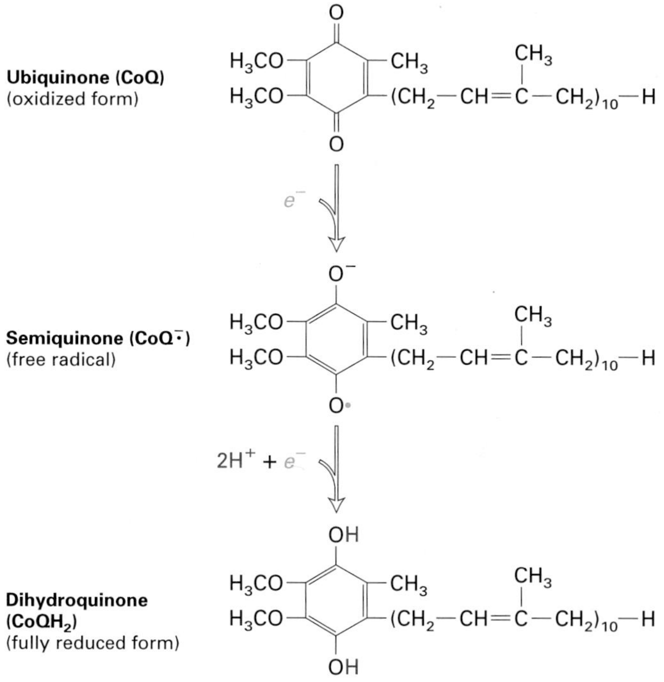

▲ FIGURE 16-22 The structure of coenzyme Q (CoQ), also called *ubiquinone,* illustrating its ability to carry two protons and two electrons. Found in bacterial and mitochondrial membranes, CoQ is the only carrier in the electron transport chain that is not tightly bound or covalently linked to a protein. Because of its long hydrocarbon "tail" of isoprene units, CoQ is soluble in the hydrophobic core of phospholipid bilayers and is very mobile. The addition of one electron to oxidized CoQ results in a half-reduced (semiquinone) form. The semiquinone is a free radical; the unpaired electron (blue dot) is delocalized by resonance over the benzene ring and attached oxygen atoms.

The oxidized quinone form of CoQ can accept a single electron to form a semiquinone, and then a second electron and two protons to form the fully reduced form, dihydroubiquinone (Figure 16-22). Both CoQ and the reduced form $CoQH_2$ are soluble in phospholipids and diffuse freely in the inner mitochondrial membrane. CoQ accepts electrons released from the NADH-CoQ reductase complex and the succinate-CoQ reductase complex (see Figure 16-17).

NADH-CoQ Reductase Complex Electrons are carried from NADH to CoQ by the NADH-CoQ reductase complex. NAD^+ is exclusively a two-electron carrier: it accepts or releases a pair of electrons at a time. In the NADH-CoQ

reductase complex, electrons first flow from NADH to FMN (flavin mononucleotide), a cofactor related to FAD, and then to an iron-sulfur protein (see Figure 16-19). FMN, like FAD, can accept two electrons, but does so one electron at a time (see Figure 16-8).

The overall reaction catalyzed by this complex is

$$NADH + CoQ + 2H^+ \longrightarrow NAD^+ + H^+ + CoQH_2$$

$\quad$ **(Reduced)** $\quad$ **(Oxidized)** $\qquad\qquad$ **(Oxidized)** $\quad$ **(Reduced)**

Each transported electron undergoes a drop in potential of ≈ 360 mV, equivalent to a $\Delta G^{\circ\prime}$ of -16.6 kcal/mol for the two electrons transported (see Figure 16-17). Much of this released energy is used to transport four protons across the inner membrane per molecule of NADH oxidized by the NADH-CoQ reductase complex, but how this protein transport occurs is not know.

Succinate-CoQ Reductase Complex As mentioned earlier, succinate dehydrogenase, the enzyme that oxidizes a molecule of succinate to fumarate in the citric acid cycle, is localized to the inner mitochondrial membrane (see Figure 16-12, step 7). Actually, this enzyme is an integral component of the succinate-CoQ reductase complex. The two electrons released in conversion of succinate to fumarate are transferred first to FAD, then to an iron-sulfur carrier, and finally to CoQ, forming the reduced $CoQH_2$ (see Figure 16-19). The overall reaction catalyzed by this complex is

$$\text{Succinate} + \text{CoQ} \longrightarrow \text{fumarate} + CoQH_2$$
(Reduced) **(Oxidized)** **(Oxidized)** **(Reduced)**

Although the $\Delta G^{\circ\prime}$ for this reaction is negative, the released energy is insufficient for proton pumping. Thus no protons are translocated across the membrane by the succinate-CoQ reductase complex, and no proton-motive force is generated in this part of the electron transport chain.

$CoQH_2$—Cytochrome c Reductase Complex Reduced $CoQH_2$, generated by either oxidation of NADH or succinate, donates two electrons to the $CoQH_2$—cytochrome c reductase complex, regenerating oxidized CoQ. Within this complex the released electrons are transferred to an iron-sulfur protein and to two b-type cytochromes, then to cytochrome c_1. Finally, the two electrons are transferred to two molecules of the oxidized form of cytochrome c (a water-soluble intermembrane-space protein), forming reduced cytochrome c (see Figure 16-19). For each pair of electrons transferred, the overall reaction catalyzed by the $CoQH_2$–cytochrome c reductase complex is

$$CoQH_2 + 2 \text{ Cyt } c^{3+} \longrightarrow \text{CoQ} + 2 \text{ H}^+ + 2 \text{ Cyt } c^{2+}$$
(Reduced) **(Oxidized)** **(Oxidized)** **(Reduced)**

The $\Delta G^{\circ\prime}$ for this reaction is sufficiently negative that four protons are translocated from the mitochondrial matrix across the inner membrane for each pair of electrons transferred; this involves the proton-motive Q cycle discussed later.

Cytochrome c Oxidase Complex Cytochrome c, after being reduced by the $CoQH_2$–cytochrome c reductase complex, transports electrons, one at a time, to the cytochrome c oxidase complex (Figure 16-23). Within this complex, electrons are transferred, again one at a time, first to a pair of copper ions (Cu_a^{2+}), then to cytochrome a, then to a complex of a second copper ion (Cu_b^{2+}), and cytochrome a_3, and finally to O_2, the ultimate electron acceptor, yielding H_2O. For each pair of electrons transferred, the overall reaction catalyzed by the cytochrome c oxidase complex is

$$2 \text{ Cyt } c^{2+} + 2 \text{ H}^+ + \tfrac{1}{2} O_2 \longrightarrow 2 \text{ Cyt } c^{3+} + H_2O$$
(Reduced) **(Oxidized)**

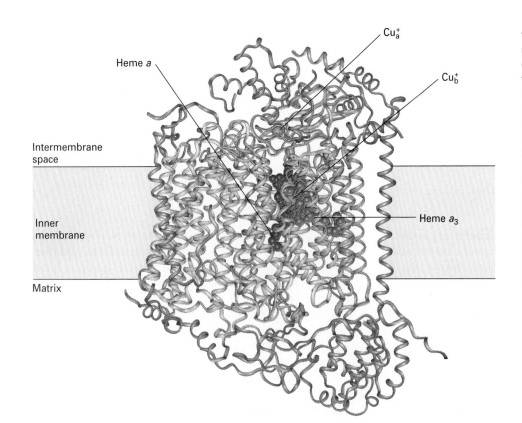

◀ **FIGURE 16-23 Molecular structure of the core of the cytochrome c oxidase complex in the inner mitochondrial membrane.** Mitochondrial cytochrome c oxidases contain 13 different subunits, but the catalytic core of the enzyme consists of only 3 subunits: I (yellow) II (blue), and III (pink). The function of the remaining subunits is not known. Bacterial cytochrome c oxidases contain only the three catalytic subunits. Hemes a and a_3 are shown as blue and red space-filling models, respectively, and the copper atoms are green. [Adapted from T. Tsukihara et al., 1996, *Science* **272**:1136; rendered by Dr. Lawren Wu.]

During transport of each pair of electrons through the cytochrome c oxidase complex, two protons are translocated across the membrane.

CoQ and Cytochrome c Shuttle Electrons from One Electron Transport Complex to Another

Each of the four electron transport complexes just described are laterally mobile in the inner mitochondrial membrane. The complexes are present in nonequal amounts: for each NADH-CoQ reductase complex, there are about three $CoQH_2$–cytochrome c reductase complexes and seven cytochrome c oxidase complexes. Furthermore, there do not appear to be stable contacts between any two complexes; rather, electrons are transported from one complex to another only by diffusion of CoQ and cytochrome c, which act as electron shuttles (see Figure 16-19). Because CoQ is lipid-soluble, it can diffuse in the membrane, shuttling electrons picked up from the NADH-CoQ reductase and succinate-CoQ reductase complexes to the $CoQH_2$–cytochrome c reductase complex. After oxidized cytochrome c picks up an electron from the $CoQH_2$–cytochrome c reductase complex, the reduced carrier diffuses in the intermembrane space until it encounters a cytochrome c oxidase complex, to which it donates an electron. Mitochondrial electron flow, in summary, does not resemble an electric current through a wire, with each electron following the previous one. Rather, electrons are picked up by a carrier, one or two at a time, and then passed along to the next carrier in the pathway.

Reduction Potentials of Electron Carriers Favor Electron Flow from NADH to O_2

As we saw in Chapter 2, the **reduction potential E** for a partial reduction reaction

$$\text{oxidized molecule} + e^- \rightleftharpoons \text{reduced molecule}$$

is a measure of the equilibrium constant of that partial reaction. With the exception of the b cytochromes in the $CoQH_2$–cytochrome c reductase complex, the standard reduction potential E'_0 of the carriers in the mitochondrial electron transport chain increases steadily from NADH to O_2.

For instance, for the partial reaction

$$NAD^+ + H^+ + 2\,e^- \rightleftharpoons NADH$$

the value of the standard reduction potential is -320 mV, which is equivalent to a $\Delta G°'$ of $+14.8$ kcal/mol for transfer of two electrons (see Table 2-7). Thus this partial reaction tends to proceed toward the left; that is, toward the oxidation of NADH to NAD^+.

By contrast, the standard reduction potential for the partial reaction

$$\text{Cyt } c_{ox}(Fe^{3+}) + e^- \rightleftharpoons \text{Cyt } c_{red}(Fe^{2+})$$

is $+220$ mV ($\Delta G°' = -5.1$ kcal/mol) for transfer of one electron. Thus this partial reaction tends to proceed toward the right; that is, toward the reduction of cytochrome c (Fe^{3+}) to c (Fe^{2+}).

The final reaction in the electron transport chain, the reduction of O_2 to H_2O

$$2\,H^+ + \tfrac{1}{2}\,O_2 + 2\,e^- \longrightarrow H_2O$$

has a standard reduction potential of $+816$ mV ($\Delta G°' = -37.8$ kcal/mol for transfer of two electrons), the most positive in the whole series, and thus also tends to proceed toward the right.

As illustrated in Figure 16-17, the steady increase in E'_0 values, and the corresponding decrease in ΔG values, of the carriers in the electron transport chain favors the flow of electrons from NADH and succinate to oxygen.

CoQ and Three Electron Transport Complexes Pump Protons out of the Mitochondrial Matrix

The experiment depicted in Figure 16-18 demonstrated that electron transport in intact mitochondria is coupled to proton export, leading to an increase in the H^+ concentration of the surrounding medium. The multiprotein electron transport complexes responsible for proton pumping have been identified by selectively extracting mitochondrial membranes with detergents, isolating each of the complexes in near purity, and then preparing artificial phospholipid vesicles (liposomes) containing each complex (see Figure 15-4). When an appropriate electron donor and electron acceptor are added to such liposomes, a change in pH of the medium will occur if the embedded complex transports protons.

For example, the cytochrome c oxidase complex can be incorporated into liposomes so that the binding site for cytochrome c is on the outside (Figure 16-24). Direct measurements of the pH change that occurs when reduced cytochrome c and O_2 are added indicate that two protons are transported out of the vesicles for every electron pair transported (or, equivalently, for every two molecules of cytochrome c oxidized). Similar studies indicate that the NADH-CoQ reductase complex and $CoQH_2$–cytochrome c reductase complex each translocates four protons per pair of electrons transported.

Current evidence thus suggests that a total of 10 protons are transported from the matrix space across the inner mitochondrial membrane for every electron pair that is transferred from NADH to O_2. Since the succinate-CoQ reductase complex does not transport protons, only six protons are transported across the membrane for every electron pair that is transferred from succinate (or $FADH_2$) to O_2. (Note that the protons generated in the matrix by the oxidation of NADH to NAD^+ and H^+ are consumed by the cytochrome c oxidase complex during the formation of H_2O, resulting in no net movement of protons across the

(a)

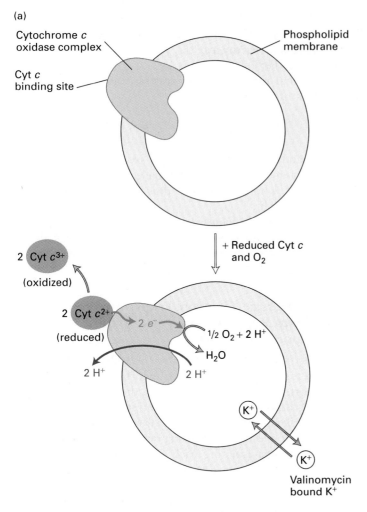

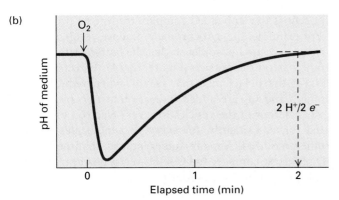

(b)

◀ **FIGURE 16-24 Experimental demonstration that oxidation of reduced cytochrome c (Cyt c²⁺) by the cytochrome c oxidase complex is coupled to proton transport.** (a) The oxidase complex is incorporated into liposomes with the binding site for cytochrome c positioned on the outer surface. When O_2 and reduced cytochrome c are added, electrons are transferred to O_2 to form H_2O, and protons are transported from the inside to the outside of the vesicles. Valinomycin and K^+ are added to the system to dissipate the voltage gradient generated by the translocation of H^+, which would reduce the number of protons moved across the membrane. (b) The transport of protons in this system is indicated by the sharp drop in medium pH following addition of O_2. As the reduced cytochrome c becomes fully oxidized, protons leak back into the vesicles, and the pH of the medium returns to its initial value. Measurements show that two protons are transported per O atom reduced. Two electrons are needed to reduce one O atom, but cytochrome c transfers only one electron; thus two molecules of Cyt c^{2+} are oxidized for each O reduced. [Adapted from B. Reynafarje et al., 1986, *J. Biol. Chem.* **261**:8254.]

this process, CoQ cycles between its reduced and oxidized states by accepting and releasing two protons and two electrons together:

$$CoQ + 2\,H^+ + 2\,e^- \rightleftharpoons CoQH_2$$

Figure 16-25 depicts the cycle, which begins when a molecule of CoQ binds to a site, located near the matrix surface of the NADH-CoQ reductase complex (or succinate-CoQ reductase complex), and picks up two protons from the *matrix space* and two electrons. The reduced $CoQH_2$ diffuses randomly in the membrane, but eventually binds to a site on the *intermembrane side* of the $CoQH_2$–cytochrome c reductase complex, where it releases two protons into the intermembrane space (step 1); these represent two of the four protons translocated from the matrix to the intermembrane space per pair of electrons transported. Simultaneously, one of the two electrons from $CoQH_2$ is transported, via an iron-sulfur protein and cytochrome c_1, directly to cytochrome c (steps 2a, 3, and 4). The other electron released from the $CoQH_2$, called a *cycling electron*, moves through cytochromes b_{566} and b_{562} to another CoQ binding site on the matrix surface, where it reduces a bound oxidized CoQ molecule (steps 2b, 5, and 6a). This reaction forms the partially reduced CoQ semiquinone anion, denoted by CoQ^- (see Figure 16-22). When a second cycling electron, released from a second $CoQH_2$, is similarly transported through cytochromes b_{566} and b_{562}, it reduces the bound CoQ^- in a reaction that uses two protons picked up from the matrix space to form $CoQH_2$ (step 6b). This $CoQH_2$ molecule then is released from the $CoQH_2$–cytochrome c reductase complex and diffuses through the membrane to the CoQ-binding site on the intermembrane surface of the same complex, where it releases its two protons into the intermembrane

membrane.) Relatively little is known about the mechanism of proton translocation by the NADH-CoQ reductase complex, but the mechanism of proton translocation by the two cytochrome complexes is well understood.

Proton-Motive Q Cycle Four protons are translocated across the membrane per electron pair transported through the $CoQH_2$–cytochrome c reductase complex. Coenzyme Q (CoQ) plays a key role in this translocation process, which is known as the *proton-motive Q cycle*, or *Q cycle*. During

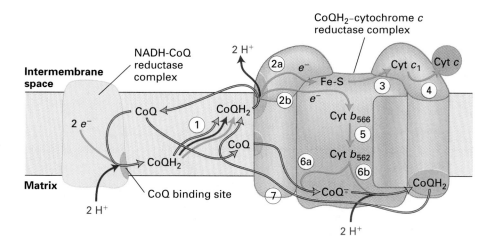

◀ **FIGURE 16-25 The proton-motive Q cycle.** Binding sites for CoQ on the protein complexes are indicated by gray shading. One electron from $CoQH_2$ travels to cytochrome c via steps 2a, 3, and 4; the other cycles through the b cytochromes. The net result is that four protons are translocated from the matrix to the intermembrane space for each pair of electrons transported through the $CoQH_2$–cytochrome c reductase complex. See text for details. [Adapted from B. Trumpower, 1990, *J. Biol. Chem.* **265**:11409–11412.]

space (step 7), as well as one electron directly to cytochromes c_1 and c, and one electron to recycle through cytochromes b_{566} and b_{562}.

To calculate the number of protons transported from the matrix space to the intermembrane space per pair of electrons transported through the $CoQH_2$–cytochrome c reductase complex, imagine 100 molecules of $CoQH_2$ interacting with $CoQH_2$–cytochrome c reductase complexes. In the first passage through the cycle, 200 protons will be transported across the membrane into the intermembrane space and 200 electrons will be released. Of the released electrons, 100 will be transported directly to cytochrome c_1 and 100 will cycle through the b cytochromes, generating 50 new molecules of reduced $CoQH_2$. In the second cycle, these 50 molecules of $CoQH_2$ will transport 100 protons into the intermembrane space and generate 50 cycling electrons, which will in turn generate 25 new molecules of reduced $CoQH_2$. Continuing this process ad infinitum, the number of protons transported as a result of oxidation of the original 100 molecules of $CoQH_2$ by 200 electrons will be

$$200 + 100 + 50 + 25 + 12.5 + 6.25 + 3.125 + \cdots = 400$$

Since we started with 100 molecules of $CoQH_2$, this exercise demonstrates that for every *pair of electrons* transported from $CoQH_2$ through the $CoQH_2$–cytochrome c reductase complex to cytochrome c, *four protons* are translocated across the membrane—two released by the initial $CoQH_2$, and two by electrons cycling through the b cytochromes. The ability of electrons to cycle through the b cytochromes thus doubles the number of protons translocated.

Coupling of H^+ Pumping and O_2 Reduction by Cytochrome c Oxidase After cytochrome c is reduced by the $CoQH_2$–cytochrome c reductase complex, it is reoxidized by the cytochrome c oxidase complex. The oxidation of four cytochrome c molecules is coupled to the reduction of one molecule of O_2, forming two molecules of water. As we saw earlier, cytochrome c oxidase contains three copper ions and

two heme groups; the oxygen-reduction center consists of one molecule of heme a_3 and one copper ion bound to subunit I of the cytochrome c oxidase complex (see Figure 16-23).

The flow of electrons through cytochrome c oxidase is shown in Figure 16-26a. Four molecules of reduced cytochrome c bind, one at a time, to a site on subunit II of the oxidase. An electron is transferred from the heme of each cytochrome c, first to Cu_a^{2+} bound to subunit II, then to the heme a bound to subunit I, and finally to the Cu_b^{2+} and heme a_3 in the oxygen-reduction center. The cyclic oxidation and reduction of the iron and copper in the reduction center, together with the uptake of four protons from the matrix space, is coupled to the transfer of the four electrons to oxygen and the formation of water (Figure 16-26b). Proposed intermediates in oxygen reduction include the peroxide anion (O_2^{2-}) and probably the hydroxyl radical (OH•). These intermediates would be harmful if they escaped from the reaction center, but they do so only rarely.

For every four electrons transferred from reduced cytochrome c through the cytochrome c oxidase complex (i.e., for every molecule of O_2 reduced to two H_2O molecules), four additional protons are translocated from the matrix space to the intermembrane space (two protons per electron pair). The four protons move during steps 2, 5, and 6 of the cycle depicted in Figure 16-26b, but the mechanism by which these protons are translocated is not known.

Experiments with Membrane Vesicles Support the Chemiosmotic Mechanism of ATP Formation

The hypothesis that a proton-motive force is the immediate source of energy for ATP synthesis, introduced in 1961 by Peter Mitchell, was initially opposed by virtually all researchers working in oxidative phosphorylation and photosynthesis. They favored a mechanism similar to the well-elucidated substrate-level phosphorylation in glycolysis, in which oxidation of a substrate molecule is directly coupled to ATP synthesis (see Figure 16-3, steps 6 and 9). By analogy, electron

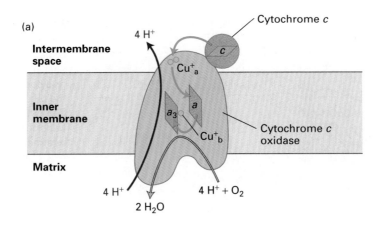

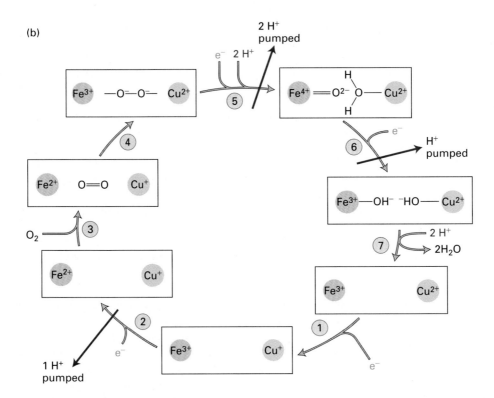

▲ **FIGURE 16-26 Electron transport through the cytochrome c oxidase complex and coupled proton transport.**
(a) Arrangement of electron carriers within the oxidase complex and flow of electrons from reduced cytochrome c to O_2. All the heme groups are shown in red. From the heme of reduced cytochrome c, one electron is transferred, via the Cu_a copper ions (green dots) and heme a, to the oxygen-reduction center, which consists of heme a_3 and the Cu_b copper ion. Four electrons, released from four molecules of reduced cytochrome c, together with four protons from the matrix, combine with one O_2 molecule to form two water molecules. Additionally, for each electron transferred from cytochrome c to oxygen, one proton is transported from the matrix to the intermembrane space, or a total of four for each molecule of O_2 reduced to two of H_2O.
(b) Proposed intermediates in the reduction of oxygen to water. Four electrons, released sequentially from four molecules of reduced cytochrome c, are transferred sequentially to the bimetallic $Fe \cdot Cu_b$ oxygen-reduction center. In reactions ① and ②, two electrons are added, reducing the Fe^{3+} in the heme to Fe^{2+}, and the Cu^{2+} to Cu^{+}. Then an O_2 molecule binds to the oxygen-reduction center (reaction ③), immediately followed by the transfer of two electrons, one from Fe^{2+} and one from Cu^{+}, forming the peroxide anion O_2^{2-} and regenerating Fe^{3+} and Cu^{2+} (reaction ④). In reaction ⑤, one electron and two protons are added, forming an unusual $Fe^{4+}=O^{2-}$ intermediate. The addition of the fourth electron (reaction ⑥) and two more protons (reaction ⑦) forms two molecules of water and regenerates the initial $Cu^{2+} \cdot Fe^{3+}$ oxygen-reduction center. The translocation of four protons from the matrix space to the intermembrane space occurs during reactions ②, ⑤, to ⑥. [Part (a) adapted from R. Williams, 1996, *Nature* **376**:643. Part (b) adapted from H. Michel, 1998, *Proc. Nat'l. Acad. Sci. USA* **95**:12819.

transport through the membranes of chloroplasts or mitochondria was believed to generate an intermediate containing a high-energy chemical bond (e.g., a phosphate linked to an enzyme by an ester bond), which was then used to convert P_i and ADP to ATP. Despite intense efforts by a large number of investigators, however, no such intermediate could ever be identified.

Definitive evidence supporting the role of the proton-motive force in ATP synthesis awaited the development of techniques to purify and reconstitute organelle membranes and membrane proteins. One experiment that led to general acceptance of Mitchell's chemiosmotic mechanism is outlined in Figure 16-27. Chloroplast thylakoid vesicles containing F_0F_1 particles were equilibrated in the dark with a buffered solution at pH 4.0. When the pH in the thylakoid lumen became 4.0, the vesicles were rapidly mixed with a solution at pH 8.0 containing ADP and P_i. A burst of ATP synthesis accompanied the transmembrane movement of protons driven by the 10,000-fold concentration gradient (10^{-4} M versus 10^{-8} M). In reciprocal experiments using

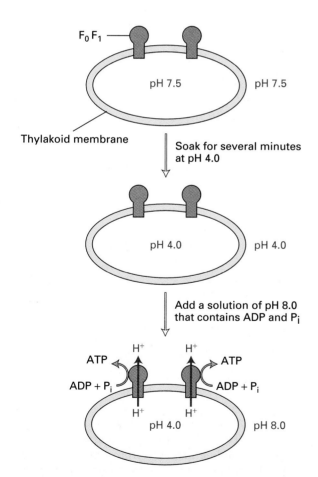

▲ **FIGURE 16-27 Experimental demonstration that ATP synthesis from ADP and P_i in chloroplast thylakoid membranes results from an artificially imposed pH gradient.**

"inside-out" preparations of submitochondrial vesicles, an artificially generated membrane electric potential also resulted in ATP synthesis. These findings left no doubt that the F_0F_1 complex is the ATP-generating enzyme and that ATP generation is dependent on proton movement down its electrochemical gradient.

Bacterial Plasma-Membrane Proteins Catalyze Electron Transport and Coupled ATP Synthesis

Although bacteria lack mitochondria, aerobic bacteria nonetheless carry out oxidative phosphorylation by the same processes that occur in eukaryotic mitochondria. Enzymes that catalyze the reactions of both the glycolytic pathway and the citric acid cycle are present in the cytosol of bacteria; enzymes that oxidize NADH to NAD^+ and transfer the electrons to the ultimate acceptor O_2 are localized to the bacterial plasma membrane.

The movement of electrons through these membrane carriers is coupled to the pumping of protons out of the cell (see Figure 16-2). The movement of protons back into the cell, down their concentration gradient, is coupled to the synthesis of ATP. Bacterial F_0F_1 complexes are essentially identical in structure and function with the mitochondrial F_0F_1 complex, which we examine in some detail below. The proton-motive force across the bacterial plasma membrane is also used to power the uptake of nutrients such as sugars and the rotation of bacterial flagella (see Figure 16-1).

ATP Synthase Comprises a Proton Channel (F_0) and ATPase (F_1)

The F_0F_1 complex, or ATP synthase, has two principal components, F_0 and F_1, both of which are oligomeric proteins (Figure 16-28). F_0 is located within the membrane and contains a transmembrane channel through which protons flow toward F_1, most of which extends into the mitochondrial matrix (or cytosol in bacteria).

The F_0 component contains three types of subunits, **a**, **b**, and **c**; in bacteria, the subunit composition is $a_1b_2c_{9-12}$. The **c** subunits form a donut-shaped ring in the plane of the membrane. Each **a** subunit is thought to span the membrane eight times, each **b** once, and each **c** twice. In mitochondria, each F_0 complex also contains, depending on the species, two to five additional peptides of unknown function. When F_0 is experimentally incorporated into liposomes, the permeability of the vesicles to H^+ is greatly stimulated, indicating that it indeed forms a proton channel. Each copy of subunit **c** contains two membrane-spanning α helices; an aspartate residue in one of these helices is thought to participate in proton movement, since chemical modification of this aspartate by dicyclohexylcarbodiimide or its mutation specifically blocks H^+ translocation. The proton channel lies at the interface of the **a** and **c** subunits (see Figure 16-28).

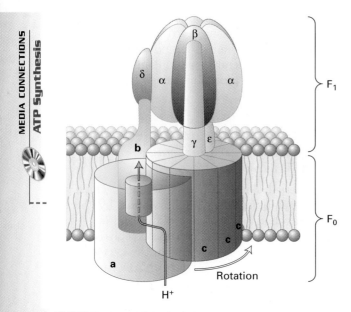

▲ **FIGURE 16-28 Model of the structure of ATP synthase (the F₀F₁ ATPase complex) in the bacterial plasma membrane.** The F₀ portion is built of three integral membrane proteins: **a**, **b**, and 9–12 copies of **c** arranged in a ring in the plane of the membrane. The proton channel lies at the interface between the **a** subunit and *c* ring. The F₁ portion contains three copies each of subunits α and β that form a hexamer resting atop the single rod-shaped γ subunit, which is inserted into the c ring of F₀. The ε subunit is attached to the γ and probably also contacts the **c** subunits, and the δ subunit links the F₁ complex to the **b** subunit of F₀. ATP synthases in the inner mitochondrial membrane and chloroplast thylakoid membranes have a similar structure, although the number of **c** units may vary and some additional subunits may be present, depending on species. [Adapted from Y. Zhou et al., 1997, *Proc. Nat'l. Acad. Sci. USA* **94**:10583; T. Elston et al., 1998, *Nature* **391**:510; and J. Abrahams et al., 1994, *Nature* **370**:621.]

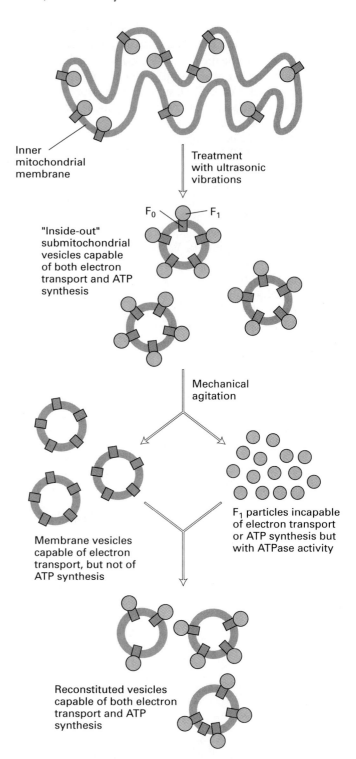

The F₁ portion is a water-soluble complex of five distinct polypeptides with the composition $\alpha_3\beta_3\gamma\delta\epsilon$. The α and β subunits associate in alternating order to form a hexamer, $\alpha\beta\alpha\beta\alpha\beta$ or $(\alpha\beta)_3$. This hexamer rests atop the single long γ subunit, whose lower part is a coiled coil that fits into the c-subunit ring of F₀. The ε subunit is attached to γ and probably also contacts the **c** subunits of F₀. The δ subunit of the F₁ complex contacts the **b** subunit of the F₀ complex; together these subunits form a rigid "stator" that prevents the $(\alpha\beta)_3$ hexamer from rotating while it rests on the γ subunit (see Figure 16-28).

F₁ forms the knobs that protrude from the matrix side of the inner membrane. When physically separated from the membrane by mechanical agitation, F₁ is capable only of catalyzing ATP hydrolysis. Hence, it has been called the F₁ ATPase, but its natural function is synthesis of ATP. Submitochondrial vesicles from which F₁ is removed cannot catalyze ATP synthesis; when F₁ particles reassociate with these vesicles, they once again become fully active in ATP synthesis (Figure 16-29).

Each of the three β subunits in the complete F₀F₁ complex can bind ATP, ADP, and Pᵢ, and catalyze ATP synthesis. However, the coupling between proton flow and ATP synthesis must be indirect, since the nucleotide-binding sites on the β subunits of F₁, where ATP synthesis occurs, are 9–10 nm from the surface of the mitochondrial membrane. In the next section, we examine this coupling mechanism.

◀ **FIGURE 16-29 Experimental demonstration that mitochondrial F₁ particles are required for ATP synthesis, but not for electron transport.** Exposure of the inner mitochondrial membrane to ultrasonic vibration disrupts the membrane; the fragments reseal with the F₁ particles facing outside. These "Inside-out" vesicles can transfer electrons from added NADH to O₂ with the concomitant synthesis of ATP from ADP and Pᵢ. Mechanical agitation of inside-out vesicles causes the F₁ particles to dissociate from the inner membrane. The membrane vesicles, which retain the electron transport complexes, can still transfer electrons from NADH to O₂ but cannot synthesize ATP. The subsequent addition of F₁ particles reconstitutes the native membrane structure, restoring the capacity for ATP synthesis.

The F₀F₁ Complex Harnesses the Proton-Motive Force to Power ATP Synthesis

The most widely accepted model for ATP synthesis by the F₀F₁ complex—the *binding-change mechanism*—posits that the energy released by the downhill movement of protons through F₀ directly powers rotation of the γ subunit and attached ε subunit. Most likely the γ and ε subunits rotate together with the ring of c subunits, relative to the fixed **a** subunit, but it is possible that the γ subunit rotates in the center of a fixed ring of c subunits. In either case, the γ subunit acts as cam, a rotating shaft within F₁ whose movement causes cyclical changes in the conformations of the β subunits. As schematically depicted in Figure 16-30, rotation of the γ subunit relative to the fixed (αβ)₃ hexamer causes the nucleotide-binding site of each β subunit to cycle through three conformational states in the following order:

1. An O state that binds ATP very poorly and ADP and Pᵢ weakly

2. An L state that binds ADP and Pᵢ more strongly

3. A T state that binds ADP and Pᵢ so tightly that they spontaneously form ATP and that binds ATP very strongly

A final rotation of γ returns the β subunit to the O state, thereby releasing ATP and beginning the cycle again. (ATP or ADP also bind to regulatory or allosteric sites on the three α subunits; this binding modifies the rate of ATP synthesis according to the level of ATP and ADP in the matrix, but is not directly involved in synthesis of ATP from ADP and Pᵢ.)

When first proposed, the binding-change mechanism was not generally accepted, but much evidence has accumulated to support it. First, biochemical studies showed that one of the three β subunits on isolated F₁ particles can tightly bind ADP and Pᵢ and then form ATP, which remains tightly bound. The measured $\Delta G°'$ for this reaction is near zero, indicating that once ADP and Pᵢ are bound to what is now called the T state of a β subunit, they spontaneously form ATP. Importantly, dissociation of the bound ATP from the β subunit on isolated F₁ particles occurs extremely slowly. This finding suggested that dissociation of ATP would have to be powered by a conformational change in the β subunit, which, in turn, would be caused by proton movement.

Later x-ray crystallographic analysis of the (αβ)₃ hexamer yielded a striking conclusion: although the three β subunits are identical in sequence and overall structure, the ADP/ATP-binding sites have different conformations in each subunit. The most reasonable conclusion was that the three β subunits cycle between three conformational states, with different nucleotide-binding sites, in an energy-dependent reaction. Recent experiments, such as that depicted in Figure 16-31, have directly demonstrated that the γ subunit

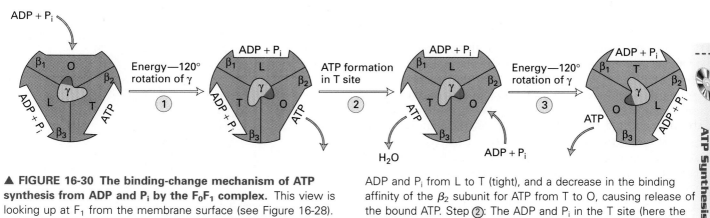

▲ **FIGURE 16-30 The binding-change mechanism of ATP synthesis from ADP and Pᵢ by the F₀F₁ complex.** This view is looking up at F₁ from the membrane surface (see Figure 16-28). The three β subunits alternate between three conformational states that differ in their binding affinities for ATP, ADP, and Pᵢ. Step ①: After ADP and Pᵢ bind to one of the three β subunits (here, arbitrarily designated β₁) whose nucleotide-binding site is in the O (open) conformation, proton flux powers a 120° rotation of the γ subunit (relative to the fixed β subunits). The causes an increase in the binding affinity of the β₁ subunit for ADP and Pᵢ to L (low), an increase in the binding affinity of the β₃ subunit for ADP and Pᵢ from L to T (tight), and a decrease in the binding affinity of the β₂ subunit for ATP from T to O, causing release of the bound ATP. Step ②: The ADP and Pᵢ in the T site (here the β₃ subunit) form ATP, a reaction that does not require an input of energy, and ADP and Pᵢ bind to the β₂ subunit, which is in the O state. This generates an F₁ complex identical with that which started the process (upper left) except that it is rotated 120°. Step ① now occurs again, and the cycling of the O → L → T → O conformations of each β subunit continues. [Adapted from P. Boyer, 1989, *FASEB J.* **3**:2164, and Y. Zhou et al., 1997, *Proc. Nat'l. Acad. Sci. USA* **94**:10583.]

MEDIA CONNECTIONS
ATP Synthesis

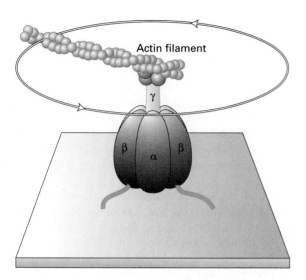

▲ **FIGURE 16-31 Demonstration that the γ subunit of the F_0 complex rotates relative to the $(αβ)_3$ hexamer in an energy-requiring step.** F_1 complexes were engineered that contained β subunits with an additional His_6 sequence, which caused them to adhere to a glass plate coated with a metal reagent that binds histidine. The γ subunit was engineered to attach to a fluorescently labeled actin filament. Using a fluorescence microscope, the actin filaments were seen to rotate counterclockwise in discrete 120° steps in the presence of ATP, powered by ATP hydrolysis by the β subunits. [Adapted from H. Noji et al., 1997, *Nature* **386**:299; and R. Yasuda et al., 1998, *Cell* **93**:1117.]

and attached ϵ subunit rotates relative to the fixed $(αβ)_3$ hexamer in discrete 120° steps, a reaction requiring hydrolysis of ATP. This experiment established that rotation of the γ and ϵ subunits, normally powered by proton movement through the F_0 complex, drives the conformational changes that are required for binding of ADP and P_i, followed by synthesis and subsequent release of ATP.

A simple calculation indicates that the passage of more than one proton is required to synthesize one molecule of ATP from ADP and P_i. Although the ΔG for this reaction under standard conditions is +7.3 kcal/mol, at the concentrations of reactants in the mitochondrion, ΔG is probably higher (+10 to +12 kcal/mol). We can calculate the amount of free energy released by the passage of 1 mol of protons down an electrochemical gradient of 220 mV (0.22 V) from the Nernst equation, setting $n = 1$ and measuring ΔE in volts:

$$\Delta G \text{ (cal/mol)} = -n\mathscr{F}\Delta E = -(23,062 \text{ cal} \cdot \text{V}^{-1} \cdot \text{mol}^{-1}) \Delta E$$
$$= (23,062 \text{ cal} \cdot \text{V}^{-1} \cdot \text{mol}^{-1})(0.22 \text{ V})$$
$$= -5,080 \text{ cal/mol, or } -5.1 \text{ kcal/mol}$$

Thus, since just over 5 kcal/mol of free energy is made available, the passage of at least two, and more likely three or four, protons is essential for the synthesis of each molecule of ATP from ADP and P_i. This calculation has been confirmed

by experimental data, which generally indicate that the passage of four protons through the F_0F_1 complex is coupled to a 120° rotation of the γ subunit and thus to the synthesis of one high-energy phosphate bond in ATP.

Transporters in the Inner Mitochondrial Membrane Are Powered by the Proton-Motive Force

The inner mitochondrial membrane contains a number of proteins that transport various metabolites into and out of the organelle, including pyruvate, malate, and the amino acids aspartate and glutamate. Two such proteins transport ADP and P_i from the cytosol to the mitochondrial matrix in exchange for ATP formed by oxidative phosphorylation inside the mitochondrion. The proton-motive force generated during electron transport is used to power this uphill exchange of ATP for ADP and P_i, which is carried out by an HPO_4^{2-}/OH^- antiporter (*phosphate transporter*) and an *ATP/ADP antiporter* (Figure 16-32).

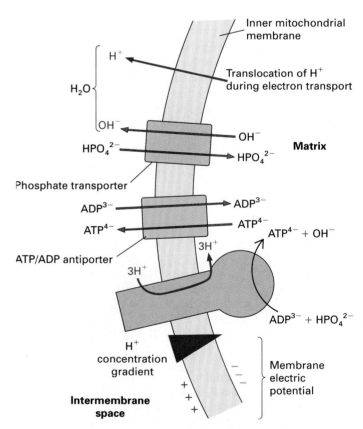

▲ **FIGURE 16-32 The phosphate and ATP/ADP transport system in the inner mitochondrial membrane.** The coordinated action of two antiporters (purple and green) results in the uptake of one ADP^{3-} and one HPO_4^{2-} in exchange for one ATP^{4-}, powered by the outward translocation of one proton during electron transport. The outer membrane is not shown here because it is permeable to molecules smaller than 5000 MW.

The phosphate transporter catalyzes the import of one HPO_4^{2-} coupled to the export of one OH^-. Likewise, the ATP/ADP antiporter allows one molecule of ADP to enter only if one molecule of ATP exits simultaneously. The ATP/ADP antiporter, a dimer of two 30,000-MW subunits, makes up 10–15 percent of the protein in the inner membrane, so it is one of the more abundant mitochondrial proteins. Functioning of the two antiporters together produces an influx of ADP and P_i and efflux of ATP and OH^-. Each OH^- transported outward combines with a proton, translocated during electron transport to the intermembrane space, to form H_2O. This drives the overall reaction in the direction of ATP export and ADP and P_i import.

Because some of the protons translocated out of the mitochondrion during electron transport provide the power (by combining with the exported OH^-) for the ATP-ADP exchange, fewer protons are available for ATP synthesis. It is estimated that for every five protons translocated out, four are used to synthesize one ATP molecule and one is used to power the export of ATP from the mitochondrion in exchange for ADP and P_i. This expenditure of energy from the proton concentration gradient to export ATP from the mitochondrion in exchange for ADP and P_i ensures a high ratio of ATP to ADP in the cytosol, where the phosphoanhydride-bond energy of ATP is utilized to power many energy-requiring reactions.

Rate of Mitochondrial Oxidation Normally Depends on ADP Levels

If intact isolated mitochondria are provided with NADH (or $FADH_2$), O_2, and P_i, but not with ADP, the oxidation of NADH and the reduction of O_2 rapidly cease as the amount of endogenous ADP is depleted by ATP formation. If ADP then is added, the oxidation of NADH is rapidly restored. Thus mitochondria can oxidize $FADH_2$ and NADH only as long as there is a source of ADP and P_i to generate ATP. This phenomenon, termed *respiratory control,* illustrates how one key reactant can limit the rate of a complex set of interrelated reactions. Intact cells and tissues also employ respiratory control. Stimulation of a metabolic activity that utilizes ATP, such as muscle contraction, results in an increased level of cellular ADP; this, in turn, increases the rate of glucose breakdown in the glycolytic pathway and citric acid cycle, as discussed earlier, and the subsequent oxidation of the metabolic products in the mitochondrion.

The molecular nature of respiratory control is now well understood. Recall that the oxidation of NADH, succinate, or $FADH_2$ is *obligatorily* coupled to proton transport across the inner mitochondrial membrane. If the resulting proton-motive force is not dissipated by harnessing it to synthesize ATP from ADP and P_i (or for some other purpose), both the transmembrane proton concentration gradient and the membrane electric potential will increase to very high levels. NADH oxidation will eventually cease, because it will require too much energy to move additional protons across the

inner membrane against the existing proton-motive force. Although the availability of ADP for ATP synthesis—that is, respiratory control—is the primary way that mitochondrial oxidation is regulated in intact cells, it is not the only way. For example, a rise in cytosolic Ca^{2+} ions, as occurs in muscle cells during contraction, also triggers an increase in mitochondrial oxidation and ATP production in many cells.

Certain poisons, called **uncouplers,** render the inner mitochondrial membrane permeable to protons. Uncouplers allow the oxidation of NADH and the reduction of O_2 to continue at high levels but do not permit ATP synthesis. In the uncoupler *2,4-dinitrophenol (DNP),* two electron-withdrawing nitro (NO_2) groups stabilize the negatively charged phenolate form:

Both the neutral and negatively charged forms of DNP are soluble in phospholipid membranes and in aqueous solution, so DNP can act as a proton shuttle. By transporting protons across the inner membrane into the matrix, DNP short-circuits both the transmembrane proton concentration gradient and the membrane electric potential, thereby dissipating the proton-motive force. Uncouplers such as DNP thus abolish ATP synthesis and overcome respiratory control, allowing NADH oxidation to occur regardless of the ADP level. The energy released by the oxidation of NADH in the presence of DNP is converted to heat.

Brown-Fat Mitochondria Contain an Uncoupler of Oxidative Phosphorylation

Brown-fat tissue, whose color is due to the presence of abundant mitochondria, is specialized for the generation of heat. In contrast, *white-fat tissue* is specialized for the storage of fat and contains relatively few mitochondria.

In the inner membrane of brown-fat mitochondria, a 33,000-MW inner-membrane protein called *thermogenin* functions as a natural uncoupler of oxidative phosphorylation. Thermogenin does not form a proton channel like the F_0 complex and typical channel proteins. Rather, thermogenin is a proton transporter whose amino acid sequence is similar to that of the mitochondrial ATP/ADP antiporter (see Figure 16-32); it functions at a rate that is characteristic of transporters, but is 1-million-fold slower than typical channel proteins (see Figure 15-3). Like other uncouplers, thermogenin dissipates the proton-motive force across the inner mitochondrial membrane, converting energy released by NADH oxidation to heat.

The amount of thermogenin is regulated depending on environmental conditions. For instance, during the adaptation of rats to cold, the ability of their tissues to generate heat (*thermogenesis*) is increased by the induction of thermogenin synthesis. In cold-adapted animals, thermogenin may constitute up 15 percent of the total protein in the inner mitochondrial membrane.

Adult humans have little brown fat, but human infants have a great deal. In the newborn, thermogenesis by brown-fat mitochondria is vital to survival, as it also is in hibernating mammals. In fur seals and other animals naturally acclimated to the cold, muscle-cell mitochondria contain thermogenin; as a result, much of the proton-motive force is used for generating heat, thereby maintaining body temperature.

SUMMARY Electron Transport and Oxidative Phosphorylation

- The proton-motive force, a combination of a proton concentration (pH) gradient (exoplasmic face > cytosolic face) and an electric potential (negative cytosolic face), can be generated across the inner mitochondrial membrane, chloroplast thylakoid membrane, and bacterial plasma membrane. It is the energy source for ATP synthesis by the F_0F_1 complexes located in these membranes (see Figure 16-2).

- In the mitochondrion, the flow of electrons from NADH and $FADH_2$ to O_2 is coupled to the uphill transport of protons from the matrix across the inner membrane to the intermembrane space, generating the proton-motive force (pmf).

- The major components of the electron transport chain are four inner membrane multiprotein complexes: succinate-CoQ reductase, NADH-CoQ reductase, $CoQH_2$–cytochrome c reductase, and cytochrome c oxidase. The last complex transfers electrons to O_2 to form H_2O.

- Electrons are transferred along the electron transport chain by the reversible reduction and oxidation of iron-sulfur clusters, ubiquinone (CoQ), cytochromes, and copper ions. Each carrier accepts an electron or electron pair from a carrier with a less positive reduction potential and transfers the electron to a carrier with a more positive reduction potential. Thus the reduction potentials of electron carriers favor unidirectional electron flow from NADH and $FADH_2$ to O_2 (see Figure 16-17).

- A total of 10 H^+ ions are translocated from the matrix across the inner membrane per electron pair flowing from NADH to O_2 (see Figure 16-19). Proton movement occurs at three points in the electron transport chain: the NADH-CoQ reductase (four H^+), $CoQH_2$–cytochrome c reductase (four H^+), and cytochrome c oxidase (two H^+).

- CoQ functions as a lipid-soluble transporter of electrons and protons across the inner membrane. The Q cycle allows additional protons to be translocated per pair of electrons moving to cytochrome c.

- The multiprotein F_0F_1 complex catalyzes ATP synthesis as protons flow back through the inner membrane down their electrochemical proton gradient. F_0 is a transmembrane complex that forms a regulated H^+ channel. F_1 is tightly bound to F_0 and protrudes into the matrix; it contains three β subunits that are the sites of ATP synthesis (see Figure 16-28).

- Proton translocation through F_0 powers the rotation of the γ subunit of F_1, leading to changes in the conformation of the nucleotide-binding sites in the F_1 β subunits (see Figure 16-30). By means of this binding-change mechanism, the F_0F_1 complex harnesses the proton-motive force to power ATP synthesis.

- The F_0F_1 complexes in bacterial plasma membranes and chloroplast thylakoid membranes are very similar in structure to the mitochondrial F_0F_1 complex.

- The proton-motive force also powers the uptake of P_i and ADP from the cytosol in exchange for mitochondrial ATP and OH^- (see Figure 16-32). Import of ADP and P_i to the mitochondrion and the export of ATP from it coordinate and limit the rate of ATP synthesis to meet the cell's needs.

- Mitochondrial oxidation of NADH and the reduction of O_2 continue to proceed only if sufficient ADP is present. This phenomenon of respiratory control is an important, but not the only, mechanism for controlling oxidation and ATP synthesis in mitochondria.

- In brown fat, the inner mitochondrial membrane contains thermogenin, a proton transport protein that converts the proton-motive force into heat. Certain chemicals (e.g., DNP) have the same effect, uncoupling oxidative phosphorylation from electron transport.

16.3 Photosynthetic Stages and Light-Absorbing Pigments

PLANTS We now shift our attention to photosynthesis, the second main process for synthesizing ATP. In plants, photosynthesis occurs in chloroplasts, large organelles found mainly in leaf cells. The principal end products are two carbohydrates that are polymers of hexose (six-carbon) sugars: the disaccharide sucrose (see Figure 2-10) and leaf starch, a large, insoluble glucose polymer (Figure 16-33). Leaf starch is synthesized and stored in the chloroplast. Sucrose is synthesized in the cytosol from three-carbon precursors generated in the chloroplast and is transported from the leaf to other parts of the plant. Nonphotosynthetic (nongreen) plant tissues like roots and seeds metabolize sucrose for energy by the pathways described in the previous sections. Photosynthesis in plants, as well as in eukaryotic single-celled algae and in several photosynthetic prokaryotes

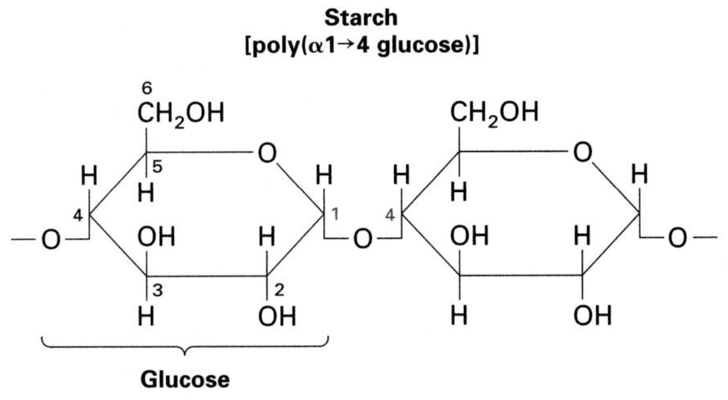

Starch
[poly(α1→4 glucose)]

▲ **FIGURE 16-33 Structure of starch.** This large glucose polymer and the disaccharide sucrose (see Figure 2-10) are the principal end products of photosynthesis. Both are built of six-carbon sugars.

(the *cyanobacteria* and *prochlorophytes*), also generates oxygen. The overall reaction of oxygen-generating photosynthesis,

$$6 CO_2 + 6 H_2O \longrightarrow 6 O_2 + C_6H_{12}O_6$$

is the reverse of the overall reaction by which carbohydrates are oxidized to CO_2 and H_2O.

Our emphasis is on photosynthesis in plant chloroplasts, but we also discuss a simpler photosynthetic process that occurs in green and purple bacteria. Although photosynthesis in these bacteria does not generate oxygen, detailed analysis of their photosynthetic systems has provided insights about the first stages in oxygen-generating photosynthesis—how light energy is converted to a separation of negative and positive charges across the thylakoid membrane, with the simultaneous generation of a strong oxidant and a strong reductant. In this section, we provide an overview of the stages in photosynthesis and introduce the main components, including the **chlorophylls**, the principal light-absorbing pigments.

Photosynthesis Occurs on Thylakoid Membranes

Chloroplasts are bounded by two membranes, which do not contain chlorophyll and do not participate directly in photosynthesis (Figure 16-34). Of these two membranes, the outer one, like the outer mitochondrial membrane, is permeable to metabolites of small molecular weight; it contains proteins that form very large aqueous channels. The inner membrane, conversely, is the permeability barrier of the chloroplast; it contains transporters that regulate the movement of metabolites into and out of the organelle.

Unlike mitochondria, chloroplasts contain a third membrane—the thylakoid membrane—that is the site of photosynthesis. In each chloroplast, the thylakoid membrane is believed to constitute a single, interconnected sheet that forms numerous small flattened vesicles, the **thylakoids,** which

commonly are arranged in stacks termed *grana* (see Figure 16-34). The spaces within all the thylakoids constitute a single continuous compartment, the *thylakoid lumen*. The thylakoid membrane contains a number of integral membrane proteins to which are bound several important prosthetic groups and light-absorbing pigments, most notably chlorophyll. Carbohydrate synthesis occurs in the *stroma*, the soluble phase between the thylakoid membrane and the inner membrane. In photosynthetic bacteria extensive invaginations of the plasma membrane form a set of internal membranes, also termed *thylakoid membranes*, or simply *thylakoids*, where photosynthesis occurs.

Three of the Four Stages in Photosynthesis Occur Only during Illumination

It is convenient to divide the photosynthetic process in plants into four stages, each occurring in a defined area of the chloroplast: (1) absorption of light, (2) electron transport leading to the reduction of $NADP^+$ to NADPH, (3) generation of ATP, and (4) conversion of CO_2 into carbohydrates (**carbon fixation**). All four stages of photosynthesis are tightly coupled and controlled so as to produce the amount of carbohydrate required by the plant. All the reactions in stages 1–3 are catalyzed by proteins in the thylakoid membrane. The enzymes that incorporate CO_2 into chemical intermediates and then convert it to starch are soluble constituents of the chloroplast stroma (see Figure 16-34). The enzymes that form sucrose from three-carbon intermediates are in the cytosol.

Absorption of Light The initial step in photosynthesis is the absorption of light by chlorophylls attached to proteins in the thylakoid membranes. Like cytochromes, chlorophylls consist of a porphyrin ring attached to a long hydrocarbon side chain (Figure 16-35). They differ from cytochromes (and heme) in containing a central Mg^{2+} ion (rather than Fe atom) and having an additional five-membered ring. The energy of the absorbed light is used to remove electrons from an unwilling donor (water, in green plants), forming oxygen,

$$2 H_2O \xrightarrow{\text{light}} O_2 + 4 H^+ + 4 e^-$$

and then to transfer the electrons to a *primary electron acceptor*, a quinone designated Q, which is similar to CoQ.

Electron Transport Electrons move from the quinone primary electron acceptor through a chain of electron transport molecules in the thylakoid membrane until they reach the ultimate electron acceptor, usually **$NADP^+$**, reducing it to NADPH (see Figure 16-4). The transport of electrons is coupled to the movement of protons from the stroma to the thylakoid lumen, forming a pH gradient across the thylakoid membrane ($pH_{lumen} < pH_{stroma}$), in much the same way that a proton-motive force is established across the mitochondrial inner membrane during electron transport (see Figure 16-2).

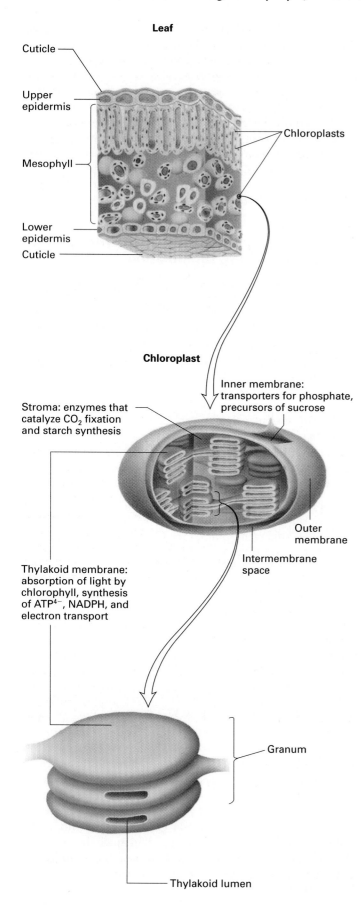

◄ FIGURE 16-34 The structure of a leaf and chloroplast. The chloroplast is bounded by a double membrane: the outer membrane contains proteins that render it permeable to small molecules (MW < 6000); the inner membrane forms the permeability barrier of the organelle. Photosynthesis occurs on the thylakoid membrane, which forms a series of flattened vesicles (thylakoids) enclosing a single interconnected luminal space. The green color of plants is due to the green color of chlorophyll, all of which is localized to the thylakoid membrane. A granum is a stack of adjacent thylakoids. The stroma is the space within the inner membrane surrounding the thylakoids. See also Figure 5-46.

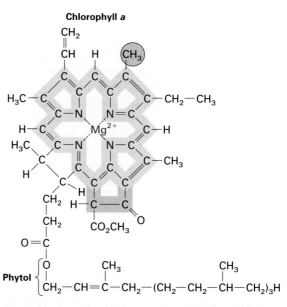

▲ FIGURE 16-35 The structure of chlorophyll a, the principal pigment that traps light energy. Chlorophyll b differs from chlorophyll a by having a CHO group in place of the CH_3 group (green). In the porphyrin ring, a highly conjugated system, electrons are delocalized among three of the four central rings and the atoms that interconnect them in the molecule (yellow). In chlorophyll, a Mg^{2+} ion, rather than an Fe^{3+} ion, is in the center of the porphyrin ring and an additional five-membered ring (blue) is present; otherwise, its structure is similar to that of heme found in molecules such as hemoglobin and cytochromes (see Figure 16-21). The hydrocarbon phytol "tail" facilitates the binding of chlorophyll to hydrophobic regions of chlorophyll-binding proteins.

Thus the overall reaction of stages 1 and 2 can be summarized as

$$2 \, H_2O + 2 \, NADP^+ \xrightarrow{\text{light}} 2 \, H^+ + 2 \, NADPH + O_2$$

Many photosynthetic bacteria do not use water as the donor of electrons. Rather, they use molecules such as hydrogen gas (H_2) or hydrogen sulfide (H_2S) as the ultimate source of electrons to reduce the ultimate electron acceptor (NAD^+ rather than $NADP^+$).

Generation of ATP Protons move down their concentration gradient from the thylakoid lumen to the stroma through the F_0F_1 complex which couples proton movement to the synthesis of ATP from ADP and P_i. This use of the proton-motive force to synthesize ATP is identical with the analogous process occurring during oxidative phosphorylation in the mitochondrion (see Figures 16-28 and 16-30).

Carbon Fixation The ATP^{4-} and NADPH generated by the second and third stages of photosynthesis provide the energy and the electrons to drive the synthesis of polymers of six-carbon sugars from CO_2 and H_2O. The overall balanced equation is written as

$$6 \, CO_2 + 18 \, ATP^{4-} + 12 \, NADPH + 12 \, H_2O \longrightarrow$$

$$C_6H_{12}O_6 + 18 \, ADP^{3-} + 18 \, P_i^{2-} + 12 \, NADP^+ + 6 \, H^+$$

The reactions that generate the ATP and NADPH used in carbon fixation are *directly* dependent on light energy; thus stages 1–3 are called the *light reactions* of photosynthesis. The reactions in stage 4 are *indirectly* dependent on light energy; they are sometimes called the *dark reactions* of photosynthesis because they can occur in the dark, utilizing the supplies of ATP and NADPH generated by light energy. However, the reactions in stage 4 are not confined to the dark; in fact, they primarily occur during illumination.

Each Photon of Light Has a Defined Amount of Energy

Quantum mechanics established that light, a form of electromagnetic radiation, has properties of both waves and particles. When light interacts with matter, it behaves as discrete packets of energy *(quanta)* called *photons*. The energy of a photon, ϵ, is proportional to the frequency of the light wave:

$\epsilon = h\gamma$, where h is Planck's constant (1.58×10^{-34} cal·s, or 6.63×10^{-34} J·s), and γ is the frequency of the light wave. It is customary in biology to refer to the wavelength of the light wave, λ, rather than to its frequency, γ. The two are related by the simple equation $\gamma = c \div \lambda$, where c is the velocity of light (3×10^{10} cm/s in a vacuum). Note that photons of *shorter* wavelength have *higher* energies.

Also, the energy in 1 mol of photons can be denoted by $E = N\epsilon$, where N is Avogadro's number (6.02×10^{23} molecules or photons/mol). Thus

$$E = Nh\gamma = \frac{Nhc}{\lambda}$$

The energy of light is considerable, as we can calculate for light with a wavelength of 550 nm (550×10^{-7} cm), typical of sunlight:

$$E = \frac{(6.02 \times 10^{23} \, \text{photons/mol})(1.58 \times 10^{-34} \, \text{cal·s})(3 \times 10^{10} \, \text{cm/s})}{(550 \times 10^{-7} \, \text{cm})}$$

$$= 51{,}881 \, \text{cal/mol}$$

or about 52 kcal/mol, enough energy to synthesize several moles of ATP from ADP and P_i if all the energy were used for this purpose.

Chlorophyll *a* Is Present in Both Components of a Photosystem

The absorption of light energy and its conversion into chemical energy occurs in multiprotein complexes, called *photosystems*, located in the thylakoid membrane. A photosystem has two closely linked components, an *antenna* containing light-absorbing pigments and a *reaction center* comprising a complex of proteins and two *chlorophyll a* molecules. Each antenna (named by analogy with radio antennas) contains one or more *light-harvesting complexes* (LHCs). The energy of the light captured by LHCs is funneled to the two chlorophylls in the reaction center, where the primary events of photosynthesis occur.

Found in all photosynthetic organisms, both eukaryotic and prokaryotic, chlorophyll *a* is the principal pigment involved in photosynthesis, being present in both antennas and reaction centers. In addition to chlorophyll *a*, antennas contain other light-absorbing pigments: *chlorophyll b* in vascular plants, and *carotenoids* in both plants and photosynthetic bacteria (Figure 16-36). The presence of various antenna

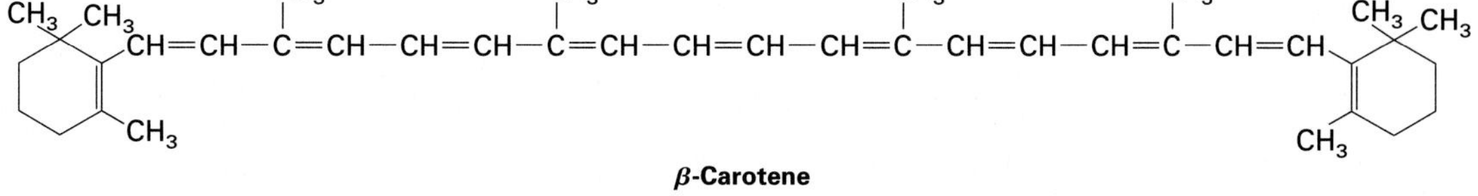

β-Carotene

▲ **FIGURE 16-36 The structure of β-carotene, a pigment that assists in light absorption by chloroplasts.** β-Carotene, which is related to the visual pigment retinal (see Figure 21-47), is one of a family of carotenoids containing long hydrocarbon chains with alternating single and double bonds.

pigments, which absorb light at different wavelengths, greatly extends the range of light that can be absorbed and used for photosynthesis.

One of the strongest pieces of evidence for the involvement of chlorophylls and β-carotene in photosynthesis is that the *absorption spectrum* of these pigments is similar to the *action spectrum* of photosynthesis (Figure 16-37). The latter is a measure of the relative ability of light of different wavelengths to support photosynthesis.

When chlorophyll *a* (or any other molecule) absorbs visible light, the absorbed light energy raises the chlorophyll *a* to a higher energy state, termed an *excited state*. This differs from the ground (unexcited) state largely in the distribution of electrons around the C and N atoms of the porphyrin ring (see Figure 16-35). Excited states are unstable, and will return to the ground state by one of several competing processes. For chlorophyll *a* molecules dissolved in organic solvents, such as ethanol, the principal reactions that dissipate the excited-state energy are the emission of light (fluorescence and phosphorescence) and thermal emission (heat). The situation is quite different when the same chlorophyll *a* is bound to the unique protein environment of the reaction center.

Light Absorption by Reaction-Center Chlorophylls Causes a Charge Separation across the Thylakoid Membrane

The absorption of a quantum of light of wavelength ≈680 nm causes a chlorophyll *a* molecule to enter the *first excited state*. The energy of such photons increases the energy of chlorophyll *a* by 42 kcal/mol. In the reaction center, this excited-state energy is used to promote a charge separation across the thylakoid membrane: an electron is transported from a chlorophyll molecule to the primary electron acceptor, the *quinone Q*, on the stromal surface of the membrane, leaving a positive charge on the chlorophyll close to the luminal surface (Figure 16-38). The reduced primary electron acceptor becomes a powerful reducing agent, with a strong tendency to transfer the electron to another molecule. The positively charged chlorophyll, a strong oxidizing agent, will attract an electron from an electron donor on the luminal surface. These potent biological reductants and oxidants provide all the energy needed to drive all subsequent reactions of photosynthesis: electron transport, ATP synthesis, and CO_2 fixation.

The significant features of the primary reactions of photosynthesis are summarized in the following model, in which P represents the chlorophyll *a* in the reaction center, and Q represents the primary electron acceptor:

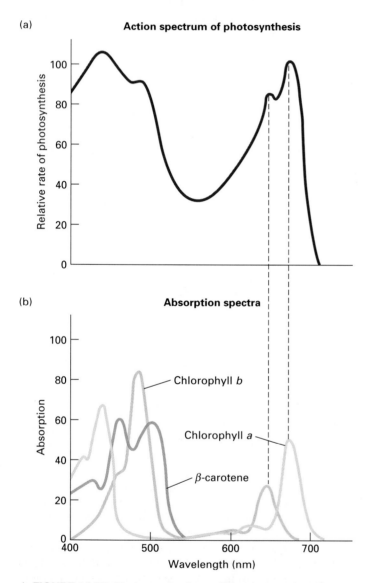

(a)

Action spectrum of photosynthesis

Relative rate of photosynthesis

(b)

Absorption spectra

Absorption

Chlorophyll *b*

Chlorophyll *a*

β-carotene

Wavelength (nm)

▲ **FIGURE 16-37 Photosynthesis at different wavelengths.**
(a) The action spectrum of photosynthesis in plants; that is, the ability of light of different wavelengths to support photosynthesis. (b) The absorption spectra for three photosynthetic pigments: chlorophyll *a*, chlorophyll *b*, and β-carotene. Each spectrum shows how well light of different wavelengths is absorbed by one of the pigments. A comparison of the action spectrum with the individual absorption spectra suggests that photosynthesis at 680 nm is primarily due to light absorbed in the antenna complex by chlorophyll *a*; at 650 nm, to light absorbed by chlorophyll *b*; and at shorter wavelengths, to light absorbed by chlorophyll *b* and by carotenoid pigments, including β-carotene.

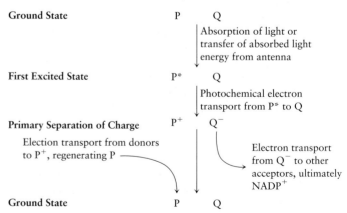

According to this model, the ground state of the reaction-center chlorophyll, P, is not a strong enough reductant to reduce Q; that is, an electron will not move spontaneously from P to Q. However, the excited state of the reaction-center chlorophyll, P*, is an excellent reductant and rapidly (in about 10^{-10} seconds) donates an electron to Q, generating P^+ and Q^-. This *photochemical electron movement*, which depends on the unique environment of both the chlorophylls and the acceptor within the reaction center, occurs nearly every time a photon is absorbed. The acceptor, Q^-, is a powerful reducing agent capable of transferring the electron to still other molecules, ultimately to $NADP^+$. The powerful oxidant P^+ can remove electrons from other molecules to regenerate the original P. In plants, the oxidizing power of four molecules of P^+ is used, by way of intermediates, to remove four electrons from H_2O to form O_2:

$$2\ H_2O + 4\ P^+ \longrightarrow 4\ H^+ + O_2 + 4\ P$$

Chlorophyll *a* also absorbs light at discrete wavelengths shorter than 680 nm (see Figure 16-37b). Such absorption raises the molecule into one of several higher excited states, which decay within 10^{-12} seconds (1 picosecond, ps) to the first excited state P*, with loss of the extra energy as heat. Photochemical charge separation occurs only from the first excited state of the reaction-center chlorophyll *a*, P*. This means that the quantum yield—the amount of photosynthesis per absorbed photon—is the same for all wavelengths of visible light shorter than 680 nm.

The chlorophyll *a* molecules within reaction centers are capable of directly absorbing light and initiating photosynthesis. However, even at the maximum light intensity encountered by photosynthetic organisms (tropical noontime sun, $\approx 1.2 \times 10^{20}$ photons/m²/s), each reaction-center chlorophyll *a* absorbs about one photon per second, which is not enough to support photosynthesis sufficient for the needs of the plant. To increase the efficiency of photosynthesis, especially at more typical light intensities, organisms utilize additional light-absorbing pigments.

Light-Harvesting Complexes Increase the Efficiency of Photosynthesis

As noted earlier, each reaction center is associated with an antenna, which contains several light-harvesting complexes (LHCs), packed with chlorophyll *a* and, depending on the species, chlorophyll *b* and other pigments. LHCs promote photosynthesis by increasing absorption of 680-nm light and by extending the range of wavelengths of light that can be absorbed (see Figure 16-37).

Photons can be absorbed by any of the pigment molecules in each LHC. The absorbed energy is then rapidly transferred (in $<10^{-9}$ seconds) to one of the two chlorophyll *a* molecules in the associated reaction center, where it promotes the primary photosynthetic charge separation (see

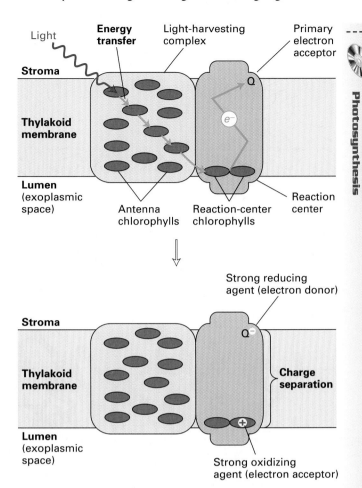

▲ **FIGURE 16-38 The primary event in photosynthesis.** After a photon of light of wavelength ≈680 nm is absorbed by one of the many chlorophyll molecules in one of the light-harvesting complexes (LHCs) of an antenna (only one is shown), some of the absorbed energy is transferred to the pair of chlorophyll molecules in the reaction center *(top)*. The resultant energized reaction-center chlorophylls donate an electron to a loosely bound acceptor molecule, the quinone Q, on the stromal surface of the thylakoid membrane. The result is a negative charge on the stromal side of the thylakoid membrane and a positive charge on the luminal side *(bottom)*. This charge separation is essentially irreversible; the electron cannot easily return through the reaction center to neutralize the positive charge. The electron acceptor with its extra electron (Q^-) is a strong reducing agent, tending to transfer the electron to another molecule. The positively charged chlorophyll is a strong oxidizing agent, attracting an electron from an electron donor on the luminal surface.

Figure 16-38). Within an LHC are several transmembrane proteins whose role is to maintain the pigment molecules in the precise orientation and position that are optimal for light absorption and energy transfer, thereby maximizing the very rapid and efficient process known as *resonance transfer* of energy from antenna pigments to reaction-center chlorophylls. As depicted in Figure 16-39a, some photosynthetic bacteria contain two types of LHCs: the larger type (LH1)

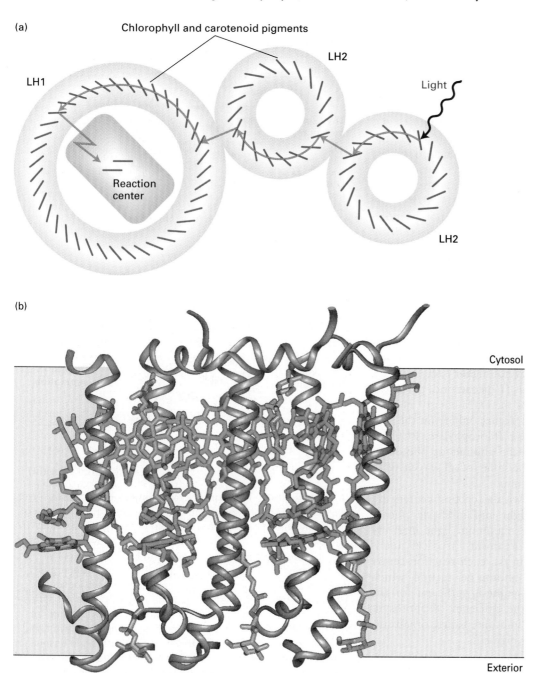

(a)

Chlorophyll and carotenoid pigments

LH2

LH1

Light

Reaction center

LH2

(b)

Cytosol

Exterior

◀ **FIGURE 16-39 Light-harvesting complexes from the photosynthetic bacterium *Rhodopseudomonas acidophila*.** (a) Schematic depiction of the cylindrical LHCs and the reaction center as viewed from above the plane of the membrane. Each LH2 complex consists of nine subunits and a total of 27 chlorophyll and 9 carotenoid molecules. The arrows trace the probable path by which light energy absorbed by an LH2 complex is transferred to the similar but larger LH1 complex and then to its final destination, the special pair of chlorophyll *a* molecules in the reaction center. (b) Structure of two of the nine subunits of the LH2 complex viewed perpendicular to the plane of the membrane. Each subunit contains two small α-helical polypeptides (gray), one carotenoid molecule (orange), and three chlorophylls (green), all of which are within and nearly perpendicular to the membrane plane. [Part (a) adapted from W. Kühlbrandt, 1995, *Nature* **374**:497. Part (b) after G. McDermott et al., 1995, *Nature* **364**:517; courtesy Dr. Lawren Wu.]

is intimately associated with a reaction center; the smaller type (LH2) can transfer absorbed light energy to an LH1. Figure 16-39b shows the structure of the subunits that make up the LH2 complex in *Rhodopseudomonas acidophila*. Surprisingly, the molecular structures of plant light-harvesting complexes are completely different from those in bacteria, even though both types contain carotenoids and chlorophylls in a clustered geometric arrangement within the membrane.

Although antenna chlorophylls can transfer absorbed light energy, they cannot release an electron. As we've seen already, reaction-center chlorophylls are able to release an electron after absorbing a quantum of light. To understand

their electron-releasing ability, we examine the structure and function of the reaction center in bacterial and plant photosystems in the next section.

SUMMARY Photosynthetic Stages and Light-Absorbing Pigments

• The principal end products of photosynthesis are polymers of six-carbon sugars: starch and sucrose. The overall reaction of oxygen-generating photosynthesis is $6 \ CO_2 + 6 \ H_2O \rightarrow 6 \ O_2 + C_6H_{12}O_6$.

- Chloroplasts are surrounded by a permeable outer membrane and an inner membrane that forms the permeability barrier; neither of these membranes participates in photosynthesis. In the chloroplast interior, the thylakoid membrane is folded into numerous flattened vesicles; the light-capturing and ATP-generating reactions of photosynthesis occur on this membrane (see Figure 16-34).

- Photosynthesis in plants can be described in four stages, which occur in specific parts of the chloroplast.

- In stage 1, light is absorbed by chlorophyll a molecules bound to reaction-center proteins in the thylakoid membrane. The energized chlorophylls donate an electron to a quinone on the opposite side of the membrane, creating a charge separation (see Figure 16-38). In green plants, the positively charged chlorophylls then remove electrons from water, forming oxygen.

- In stage 2, electrons move from the quinone through a chain of electron transport molecules in the thylakoid membrane until they reach the ultimate electron acceptor, usually $NADP^+$, reducing it to NADPH. Transport of electrons is coupled to the movement of protons across the membrane from the stroma to the thylakoid lumen, forming a pH gradient across the thylakoid membrane.

- In stage 3, movement of protons down their electrochemical gradient through F_0F_1 complexes powers the synthesis of ATP from ADP and P_i.

- In stage 4, the ATP and NADPH generated in stages 2 and 3 provide the energy and the electrons to drive the fixation of CO_2 and synthesis of carbohydrates. These reactions occur in the thylakoid stroma and cytosol.

- Chlorophyll a is the only light-absorbing pigment in reaction centers. Associated with each reaction center are multiple light-harvesting complexes (LHCs), which contain chlorophylls a and b, carotenoids, and other pigments that absorb light at multiple wavelengths.

- Energy is transferred from the LHC chlorophyll molecules to reaction-center chlorophylls by resonance energy transfer (see Figure 16-39).

16.4 Molecular Analysis of Photosystems

As noted in the previous section, photosynthesis in the green and purple bacteria* does not generate oxygen, whereas photosynthesis in cyanobacteria, algae, and higher plants does. This difference is attributable to the presence of two types of photosystem in the latter organisms: one splits H_2O into O_2 and the other reduces NADP to NADPH. In contrast, photosynthetic bacteria have only one type of photosystem. The three-dimensional structures of the reaction centers in the photosystems from two purple bacteria (*Rhodopseudomonas viridis* and *Rhodobacter spheroides*) have been determined, permitting scientists to trace the detailed paths of electrons during and after the absorption of light. We first discuss the simpler bacterial photosystem and then consider the more complicated photosynthetic mechanism in chloroplasts, which requires two photosystems.

Photoelectron Transport in Purple Bacteria Produces a Charge Separation

The photosynthetic reaction center of purple bacteria has three integral proteins, L, M, and H (Figure 16-40). The reaction center protein contains a total of 11 transmembrane α helices, to which are bound the prosthetic groups that absorb light and transport electrons during photosynthesis. The prosthetic groups include a "special pair" of light-absorbing bacterio-chlorophyll a molecules that are equivalent to P in our earlier model, the site of the primary photochemical reaction. There are, in addition, two "voyeur" bacteriochlorophyll molecules, two bacteriopheophytins (bacteriochlorophyll molecules without a Mg^{2+}), one nonheme Fe atom, and two quinones, termed Q_A and Q_B, that are structurally similar to mitochondrial ubiquinone.

In the photosystem of purple bacteria, as in other photosystems, energy from absorbed light is used to strip an electron from a reaction-center bacteriochlorophyll a molecule and transfer it, via several different pigments, to an acceptor quinone, in this case Q_B, which is located on the cytosolic membrane face. The chlorophyll thereby acquires a positive charge (and thus is converted from P to P^+). Each pigment in the reaction center absorbs light of only certain wavelengths, and its absorption spectrum changes when it possesses an extra electron. The pathway traversed by an electron can be determined by monitoring the changes in the absorption spectra of the various pigments as a function of time after the absorption of a light photon. Because these electron movements are completed in less than 1 millisecond(ms), a special technique called *picosecond absorption spectroscopy* (1 ps = 10^{-12} s) is required to see the early changes.

When a preparation of bacterial membrane vesicles is exposed to an intense pulse of laser light lasting less than 1 ps, each reaction center absorbs one photon. Light is directly absorbed by the special-pair chlorophylls in each reaction center, converting them to the excited state, P^*, and the subsequent electron transfer processes are synchronized in all reaction centers. The pathway of electron flow, as determined by picosecond absorption spectroscopy during the millisecond following the light pulse, is traced in the left portion of Figure 16-41. Within 4×10^{-12} seconds, an

---

*A very different type of bacterial photosynthesis, which occurs only in certain archaebacteria, is not discussed here because it is unrelated to photosynthesis in higher plants. In this type of photosynthesis, the plasma-membrane protein bacteriorhodopsin pumps one proton from the cytosol to the extracellular space for every quantum of light absorbed. This small protein has seven membrane-spanning segments and a covalently attached retinal pigment (see Figure 3–34).

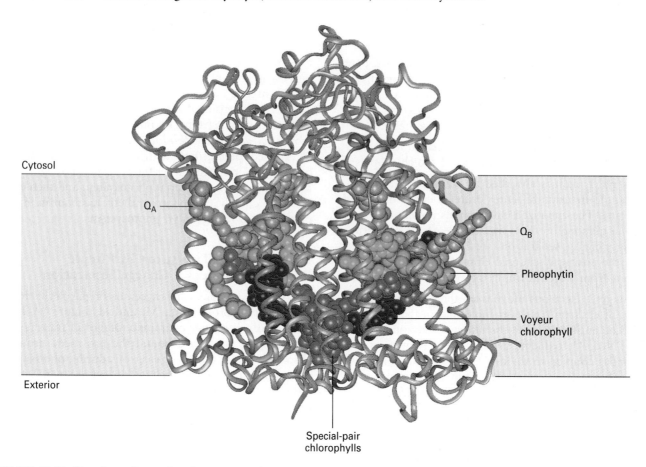

▲ FIGURE 16-40 The three-dimensional structure of the photosynthetic reaction center from *Rhodobacter spheroides*, showing all the pigments. The L (gray) and the M subunits (pink) each form five transmembrane α helices and have a very similar structure overall; the H subunit (light blue) is anchored to the membrane only by a single transmembrane α helix. A fourth subunit (not shown) is a peripheral protein that binds to the exoplasmic sequences of the other subunits. Within each reaction center are a special pair of bacteriochlorophyll *a* molecules (dark blue), two voyeur chlorophylls (purple), two pheophytins (green), and two quinones, Q_A and Q_B (orange). The opening figure of this chapter depicts the binding of the exchangeable quinone, Q_B, to the reaction-center protein. [After C.-H Cheng et al., 1991, *Biochemistry* **30**:5352; courtesy of Dr. Lawren Wu.]

electron moves to one of the pheophytin molecules (Ph), leaving a positive charge on the special-pair chlorophylls. After ≈2×10^{-10} seconds, the electron moves to the permanently bound quinone A (Q_A) molecule near the cytosolic face of the reaction center, and after 2×10^{-4} seconds, the electron is passed from Q_A to the second acceptor, Q_B, forming the semiquinone Q_B^-. The end result is a *charge separation*: a positively charged chlorophyll on the exoplasmic membrane face and a negatively charged quinone Q_B^- on the cytosolic face.

Both Cyclic and Noncyclic Electron Transport Occur in Bacterial Photosynthesis

After the primary electron acceptor, Q_B, in the bacterial reaction center accepts one electron, forming Q_B^-, it accepts a second electron from the same reaction-center chlorophyll following its absorption of a second photon. The quinone then binds two protons from the cytosol, forming the reduced quinone QH_2, which is then released from the reaction center. The flow of electrons from QH_2 to other acceptors can follow either a cyclic or noncyclic path.

Cyclic Electron Transport The pathway of electron transport depicted in Figure 16-41 is cyclic. In this case, QH_2 diffuses within the bacterial membrane to a *cytochrome bc_1 complex* where it releases its two electrons to a site on the exoplasmic (external) surface. Simultaneously, QH_2 releases two protons into the periplasmic space (the space between the plasma membrane and the bacterial cell wall), thereby generating a proton-motive force. (This process is very similar to the proton-motive Q cycle that occurs in the inner mitochondrial membrane; see Figure 16-25.) Electrons are then transferred from the cytochrome bc_1 complex to a soluble cytochrome, a one-electron carrier, in the periplasmic space; this reduces the cytochrome from an Fe^{3+} to an Fe^{2+} state. The cytochrome then diffuses to a reaction center,

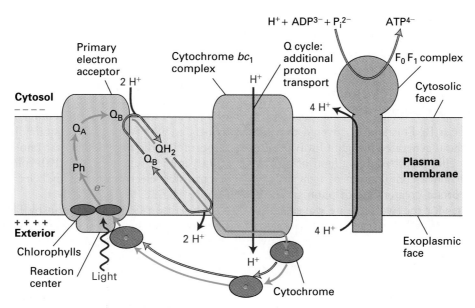

▲ FIGURE 16-41 Summary of the cyclic pathway of photosynthesis in purple bacteria. The blue arrows trace the movement of electrons. LHCs (not illustrated here) absorb light and funnel the energy to one of the special-pair chlorophylls in the reaction center. The energized chlorophyll rapidly loses an electron, which is transferred via pheophytin (Ph) and quinone A (Q_A) to Q_B, forming the semiquinone $Q_B^{\bar{\cdot}}$ and leaving a positive charge on the chlorophyll. Absorption of a second photon and transfer of a second electron generates the reduced quinone QH_2, which dissociates from the reaction center and diffuses randomly through the membrane until it binds to a site on the exoplasmic face of the cytochrome bc_1 complex. Here, QH_2 donates its two electrons and simultaneously gives up its two

protons to the external medium, restoring the oxidized quinone, Q_B, and generating a proton-motive force. The oxidized Q_B diffuses back through the membrane to the Q_B-binding site on the cytosolic face of the reaction center. Electrons are transported back to the reaction-center chlorophyll via the heme (blue dot) of a soluble cytochrome. Electrons also may be transported through the bc_1 oxidoreductase complex during a Q cycle that transports additional protons across the membrane to the external medium. The proton-motive force is used by the $F_0 F_1$ complex (light orange) to synthesize ATP and, as in other bacteria, to transport molecules in and out of the cell. [Adapted from J. Deisenhofer and H. Michael, 1991, *Ann. Rev. Cell Biol.* **7**:1.]

where it releases its electron to the positively charged (P^+) chlorophyll, returning the chlorophyll to the ground state P.

As in other systems, the proton-motive force generated in the bacterial photosystem is used by the F_0F_1 ATPase to synthesize ATP and also to transport molecules across the membrane (see Figure 16-41, *right*). However, this photosynthetic process does not evolve O_2 or split H_2O. Electron flow is cyclic, and there is no reduction of NAD^+ to NADH.

Noncyclic Electron Transport A noncyclic pattern of electron transport can also occur during photosynthesis by purple bacteria. In this case, electrons removed from reaction-center chlorophylls ultimately are transferred to NAD^+ (rather than $NADP^+$ as in plants), forming NADH. To reduce the oxidized reaction-center chlorophyll (P^+) back to its ground state P, an electron is transferred from a reduced cytochrome *c*, and to reduce the resultant oxidized cytochrome *c*, electrons are removed from hydrogen sulfide (H_2S) or hydrogen gas (H_2). Since H_2O is not the electron donor, no O_2 is formed. In the noncyclic pathway, as in the cyclic pathway, a proton-motive force is generated and used to synthesize ATP.

The partial reactions for light-powered noncyclic electron transport in purple bacteria are

$$H_2S + NAD^+ \xrightarrow{\text{light}} H^+ + NADH + S$$

and

$$H_2 + NAD^+ \xrightarrow{\text{light}} H^+ + NADH$$

The NADH produced is used in the fixation of CO_2 and synthesis of carbohydrates and other molecules. The overall reactions for the light-driven reduction of CO_2 via the noncyclic pathway in purple bacteria are

$$12\ H_2S + 6\ CO_2 \xrightarrow{\text{light}} C_6H_{12}O_6 + 12\ S + 6\ H_2O$$

and

$$12\ H_2 + 6\ CO_2 \xrightarrow{\text{light}} C_6H_{12}O_6 + 6\ H_2O$$

Other photosynthetic bacteria use a variety of organic compounds as electron sources for the light-dependent reduction of CO_2.

Chloroplasts Contain Two Functionally and Spatially Distinct Photosystems

In the 1940s, biophysicist R. Emerson discovered that the rate of plant photosynthesis generated by light of wavelength 700 nm can be greatly enhanced by adding light of shorter wavelength. He found that a combination of light at, say, 600 and 700 nm supports a greater rate of photosynthesis than the sum of the rates for the two separate wavelengths. This so-called *Emerson effect* led researchers to conclude that photosynthesis in plants involves the interaction of two separate photosystems referred to as *PSI* and *PSII*. PSI is driven by light of wavelength 700 nm or less; PSII, only by light of shorter wavelength (<680 nm).

As in the bacterial reaction center, at the center of each chloroplast photosystem is a pair of specialized reaction-center chlorophyll *a* molecules, which are capable of under-

going light-driven electron transfer. The chlorophylls in the two reaction centers differ in their light-absorption maxima because of differences in their protein environment. For this reason, the reaction-center chlorophylls are often denoted P_{680} (PSII) and P_{700} (PSI). As in photosynthetic bacteria, each reaction center has an associated antenna that consists of a group of light-harvesting complexes (LHCs); the LHCs associated with PSII and PSI contain different proteins. More importantly, the two chloroplast photosystems differ significantly in their functions: only PSII splits water, whereas only PSI transfers electrons to the final electron acceptor, $NADP^+$ (Figure 16-42).

The absorption of a photon by PSII causes an electron to move from P_{680} to an acceptor plastoquinone (Q_B) on the stromal surface; the resultant positive charge on the P_{680} strips electrons from the highly unwilling donor H_2O, forming O_2 and protons, which remain in the thylakoid lumen and contribute to the proton-motive force. After P_{680} absorbs a second photon, Q_B^- accepts a second electron and picks up two protons from the stromal space, generating reduced quinone (QH_2). The reduced QH_2 dissociates from

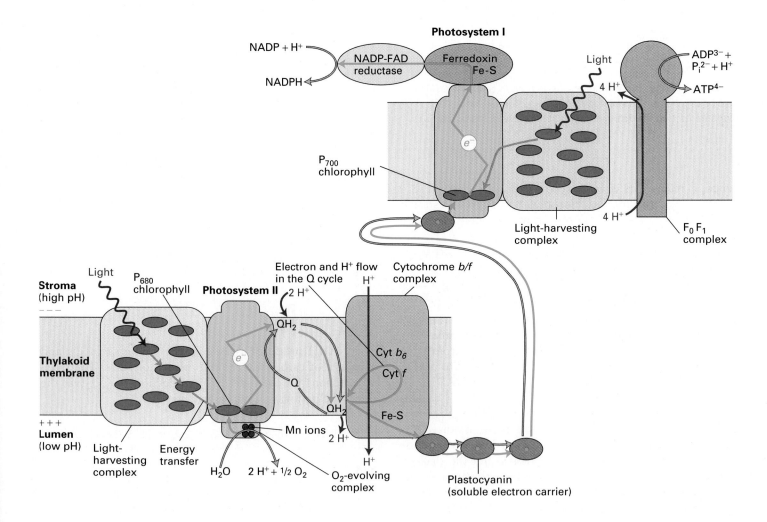

its binding site on the PSII reaction center and is replaced with an oxidized Q. QH_2 diffuses randomly in the thylakoid membrane until it encounters its binding site on the luminal side of the *cytochrome b/f complex*, where it releases its two electrons. (The cytochrome *b/f* complex is similar in structure and function to the cytochrome bc_1 complex in photosynthetic bacteria and to the $CoQH_2$–cytochrome *c* reductase in mitochondria.) Simultaneously, the two protons that QH_2 picked up from the stroma are released into the thylakoid lumen, generating a proton-motive force (see Figure 16-42).

As in PSII, the absorption of a photon by P_{700} in PSI leads to removal of an electron. The resulting oxidized chlorophyll P_{700}^+ is reduced by an electron passed from PSII via *plastocyanin*, a soluble electron carrier. The electron given up at the luminal surface by P_{700} moves to the stromal surface of the thylakoid membrane, where it is accepted by ferredoxin, an iron-sulfur (Fe-S) protein. (Several electron carriers intervene between P_{700} and ferredoxin, including two other iron-sulfur complexes and vitamin K). Electrons excited by PSI face one of two fates. They can be transferred from ferredoxin, via the electron carrier FAD, to $NADP^+$, forming, together with one proton picked up from the stroma, the reduced molecule NADPH (see Figure 16-42). This process, called *linear electron flow*, involves PSII and PSI in an obligate series in which electrons are transferred from H_2O to $NADP^+$. As we discuss below, PSI can also engage in a cyclical type of electron flow.

The transfer of electrons from PSII to PSI requires plastocyanin, a small soluble protein with a single copper (Cu) atom, because the two photosystems are spatially separated in the thylakoid membrane (Figure 16-43). PSII is located preferentially in the stacked regions of the thylakoid membrane (the grana), and PSI in the nonstacked regions; the cytochrome *b/f* complex is present in both regions of the thylakoid membrane. After this cytochrome complex accepts electrons from QH_2, it transfers them, one at a time, to the Cu^{2+} form of plastocyanin, reducing it to the Cu^+ form. Reduced plastocyanin then diffuses in the thylakoid lumen, carrying the electron to P_{700} in PSI (see Figure 16-42).

An Oxygen-Evolving Complex in PSII Regenerates P_{680}

As noted above, PSII removes electrons from H_2O, forming O_2. Somewhat surprisingly, the structure of PSII resembles that of the reaction center of a photosynthetic bacterium, which, as we've seen previously, does not form O_2. Like bacterial reaction centers, PSII contains two reaction-center chlorophylls, two other chlorophylls, two pheophytins, two quinones, called Q_A and Q_B, and one nonheme iron atom. These molecules are bound to two 32,000-MW PSII proteins, called D1 and D2, whose sequences are remarkably similar to the sequences of the L and M peptides of the bacterial reaction center (see Figure 16-40). When PSII absorbs a photon with a wavelength of <680 nm, it triggers the loss of an electron from a P_{680} chlorophyll *a* molecule, generating P_{680}^+. As in photosynthetic bacteria, the electron is transported via a pheophytin and a quinone (Q_A) to the primary electron acceptor, Q_B, on the outer (stromal) surface of the thylakoid membrane (Figure 16-44).

P_{680}^+, the photochemically oxidized reaction-center chlorophyll of PSII, is the strongest biological oxidant known. The reduction potential of P_{680}^+ is more positive than that of water, and thus it can oxidize water to give O_2 and H^+ ions. Photosynthetic bacteria cannot oxidize water because the P^+ of the bacterial reaction center is not a sufficiently strong oxidant. (As noted earlier, bacteria use H_2S and H_2 as electron donors to reduce P^+ in noncyclic electron transport.) The splitting of H_2O, which provides the electrons for reduction of P_{680}^+, is catalyzed by a three-protein complex, the *oxygen-evolving complex*, located on the luminal surface

◄ **FIGURE 16-42 Summary of photosynthesis in plants, which utilize two photosystems, PSI and PSII, during linear electron flow.** Both photosystems use the energy of absorbed light to move an electron from a reaction-center chlorophyll across the thylakoid membrane (blue lines indicate electron movement). *Photosystem II:* Absorbed light energy releases an electron from P_{680} in PSII, forming P_{680}^+; the electron moves to an acceptor quinone (Q) on the stromal surface. Q accepts a second electron released from P_{680}, following absorption of a second photon of light, and then adds two protons from the stroma. The oxygen-evolving complex in PSII removes electrons, one at a time, from H_2O in the thylakoid lumen and transfers them to P_{680}^+, restoring the reaction-center chlorophylls to the ground state and generating O_2. The protons resulting from the splitting of H_2O remain in the lumen, contributing to the proton-motive force. QH_2 diffuses through the membrane to the cytochrome *b/f* complex, where it simultaneously releases its two electrons at a site on the luminal face and its two protons

into the lumen; the released protons add to the proton-motive force. Electrons may also be transported through the cytochrome *b/f* complex during a Q cycle (circular blue arrow), transporting additional protons across the membrane to the thylakoid lumen. *Photosystem I:* Each electron released from P_{700} after absorption of light in PSI is transported, by a series of carriers in the reaction center, to the stromal surface, where soluble ferredoxin (an Fe-S protein) transfers the electron to FAD and finally to $NADP^+$. Two electrons, together with one proton removed from the stromal space, convert each $NADP^+$ to NADPH. P_{700}^+ is restored to its ground state by addition of an electron carried from PSII via plastocyanin, a soluble electron carrier that diffuses through the thylakoid lumen. As in mitochondria, the movement of three or four protons down their concentration gradient through an F_0F_1 complex is coupled to the synthesis of one ATP from ADP and P_i. Thus both NADPH and ATP are generated in the stroma of the chloroplast, where they are utilized for CO_2 fixation.

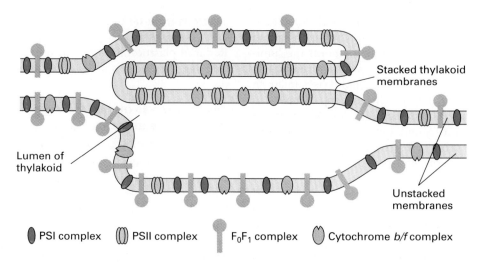

▲ FIGURE 16-43 Distribution of multiprotein complexes in the thylakoid membrane. PSI is located primarily in nonstacked regions; PSII, primarily in stacked regions. The cytochrome *b/f* complex, which aids in the transport of electrons from PSII to PSI, is found in both stacked and nonstacked regions. The stacking of the thylakoid membranes may be due to the binding properties of the proteins in PSII. Evidence for this distribution came from studies in which thylakoid membranes were gently fragmented into vesicles by ultrasound. Stacked and unstacked thylakoid vesicles were then fractionated by density-gradient centrifugation to determine their protein and chlorophyll compositions. [Adapted from J. M. Anderson and B. Andersson, 1982, *Trends Biochem. Sci.* **7**:288.]

of the thylakoid membrane. The oxygen-evolving complex contains four manganese (Mn) ions as well as bound Cl^- and Ca^{2+} ions (see Figure 16-44); this is one of the very few cases in which Mn plays a role in a biological system. The Mn ions together with the three extrinsic proteins can be removed from the reaction center by treatment with solutions of concentrated salts; this abolishes O_2 formation but does not affect light absorption or the initial stages of electron transport.

The oxidation of two molecules of H_2O to form O_2 requires the removal of four electrons, but absorption of each photon by PSII results in the transfer of just one electron. A simple experiment resolved whether the formation of O_2 depends on a single PSII or multiple ones acting in concert. When chloroplasts were illuminated with short (5-μs) saturating pulses of light, O_2 was evolved after every fourth pulse. This finding means that a single PSII must lose an electron and then oxidize the oxygen-evolving complex four times in a row for an O_2 molecule to be formed. Subsequent spectroscopic studies showed that the bound Mn ions in the oxygen-evolving complex cycle through five different oxidation states, S_0–S_4 (Figure 16-45). In this S cycle, a total of two H_2O molecules are split into four protons, four electrons, and one O_2 molecule. The electrons from H_2O are transferred, one at a time, via the Mn ions and a nearby tyrosine side chain (Z in Figure 16-44) on the D1 polypeptide, to the P_{680} reaction center, where they regenerate the reduced chlorophyll. The protons released from H_2O remain in the thylakoid lumen; they represent two of the four protons that are released into the lumen by the transport of each pair of electrons.

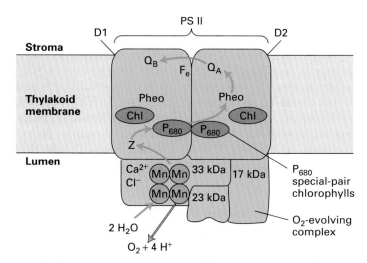

▲ FIGURE 16-44 Electron flow through reaction center of chloroplast PSII. In a reaction center, two integral proteins, D1 and D2, bind the special-pair chlorophylls (P_{680}), two other chlorophylls (Chl), two pheophytins (Pheo), one Fe atom, and two quinones (Q_A and Q_B). All of these are used for electron transport following light absorption by an associated light-harvesting complex (see Figure 16-42). Three extrinsic proteins (33, 23, and 17 kDa) comprise the oxygen-evolving complex; they bind the four Mn^{2+} ions and the Ca^{2+} and Cl^- ions that function in the splitting of H_2O, and they maintain the environment essential for high rates of O_2 evolution. Z is tyrosine residue 161 of the D1 polypeptide; it conducts electrons from the Mn atoms to the oxidized reaction-center chlorophyll (P_{680}^+), reducing it to the ground state P_{680}. [Adapted from G. Babcock, 1993, *Proc. Nat'l. Acad. Sci. USA* **90**:10893.]

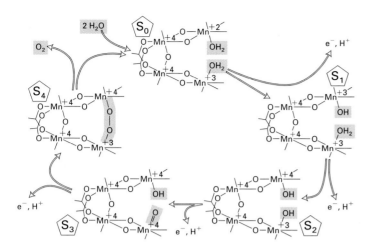

▲ **FIGURE 16-45 Model for the S-state cycle of the oxygen-evolving complex in the reaction center of chloroplast PSII.** Each of the four manganese (Mn) atoms in the complex has six ligands. Two molecules of water bind to the complex when it is in the most reduced state, S_0. At each step of the cycle an electron and proton are removed, causing one of the Mn ions to achieve a higher oxidation state. These electrons are transferred, one at a time, via a tyrosine residue in D1, to the oxidized reaction-center chlorophylls (P_{680}^+) of PSII (see Figure 16-44). After the most oxidized S_4 stage is generated, O_2 is released, lowering the oxidation state of the Mn complex by two positive charges and regenerating the starting S_0. Not shown here are the one Cl^- and one Ca^{2+} ion that are essential components of the oxygen-evolving complex.
[Adapted from C. Hoganson and G. Babcock, 1997, *Science* **277**:1953.]

Cyclic Electron Flow in PSI Generates ATP but No NADPH

In linear electron flow, electrons from ferredoxin in PSI are transferred to NADP, forming NADPH (see Figure 16-42). Alternatively, ferredoxin can reduce a plastoquinone Q, forming, after binding two protons from the stroma, QH_2 on the stromal surface. This QH_2 then diffuses through the thylakoid membrane to its binding site on the luminal side of the cytochrome b/f complex, where it releases two electrons to the b cytochromes and two protons to the thylakoid lumen, generating a proton-motive force. As in linear electron flow, these electrons return to PSI via plastocyanin. This process, called *cyclic electron flow,* is similar to the cyclic process that occurs during photosynthesis in purple bacteria (see Figure 16-41).

In cyclic electron flow, a Q cycle involving plastoquinone and the cytochrome b/f complex is thought to transport two additional protons into the lumen for each pair of electrons transported by a mechanism similar to the analogous process in oxidative phosphorylation (see Figure 16-25). Several lines of evidence indicate that a Q cycle generally does not operate during linear electron flow. Thus, the greater pH gradient generated by cyclic electron flow permits the synthesis of additional ATP, but no NADPH is produced. PSII

is not involved, and no O_2 evolves. During cyclic electron flow, PSI is used solely to generate a pH gradient and ATP; its function in this case is similar to that of the photosystem in purple bacteria.

PSI and PSII Are Functionally Coupled

The oxidized and reduced states of the various electron carriers in chloroplasts have different absorption maxima. Spectroscopic studies have helped to determine the pathway of electron flow and also have provided evidence supporting the obligatory transport of electrons from PSII to PSI in linear electron flow. As discussed earlier, red light of wavelength 700 nm excites only PSI; the wavelength of this light is too long to be absorbed by PSII. Shining only red light on chloroplasts causes cytochromes b_6 and f to become more oxidized. Under these conditions, electrons are drawn into PSI; none are provided from PSII. The addition of light of shorter wavelength, however, activates PSII; the b_6 and f cytochromes immediately become partly reduced as electrons flow to them.

Studies of the effects of herbicides that inhibit photosynthesis also have proved useful in dissecting the pathway of photoelectron movement. One such class of herbicides—the *s*-triazines, such as atrazine—bind specifically to D1, one of the two proteins that compose the PSII reaction center. Recall that the plastoquinone Q_B in PSII dissociates after it has been reduced to QH_2. *s*-Triazines block electron transfer by inhibiting the binding of an oxidized plastoquinone to its binding site on PSII. When added to illuminated chloroplasts, these inhibitors cause all downstream electron carriers to accumulate in the oxidized form, since no electrons can be provided to the electron transport system from PSII. In atrazine-resistant mutants, a single amino acid change in D1 renders it unable to bind the herbicide, so photosynthesis proceeds at normal rates. Such resistant weeds are prevalent and present a major agricultural problem.

Because PSII and PSI act in sequence during linear electron flow, the amount of light energy delivered to the two reaction centers must be controlled so that each center activates the same number of electrons. One control mechanism regulates the rates of phosphorylation and dephosphorylation of the proteins associated with the PSII light-harvesting complex (LHC). A membrane-bound protein kinase senses the relative activities of the two photosystems by recognizing the oxidized-reduced state of the plastoquinone pool that transfers electrons from PSII to the cytochrome b/f complex en route to PSI. If too much plastoquinone is reduced (indicating a higher activity of PSII relative to PSI), the kinase is activated and the PSII LHCs are phosphorylated. Phosphorylation causes the LHCs to dissociate from PSII and migrate to the unstacked thylakoid membranes; this decreases the antenna size of PSII and thus the rate of generation of QH_2 by PSII. In this manner the relative rates of PSII and PSI are balanced; the reverse occurs if the plastoquinone pool becomes too oxidized.

The oxidants and reductants formed in PSII can lead to dangerous side reactions. Oxidative damage to the PSII photosystem could occur if PSII had no available electron acceptor, as would occur if all of the quinones were reduced, or if the light intensity were too high for the electron transport systems to handle. Under these conditions protective structural changes occur in the PSII LHCs, induced in part by a large pH gradient across the thylakoid membrane, so that the energy of absorbed light is dissipated as heat and not transferred to P_{680}.

Both Plant Photosystems Are Essential for Formation of NADPH and O_2

Part of the light energy absorbed by P_{680} in PSII (≈ 42 kcal/mol of photons) is used to stabilize early reactive intermediates, so that not enough is left over to permit the direct movement of electrons from H_2O to NADPH. Light absorption by P_{700} in PSI causes an additional increase in electron energy sufficient to support reduction of NADP to NADPH (Figure 16-46a). Thus each of the two photosystems boosts the electrons part of the way and are essential for use of H_2O as an electron donor in linear electron flow. Although cyclic electron flow between PSI and the cytochrome b/f complex is also energetically favorable, this pathway does not generate NADPH (Figure 16-46b).

Cyanobacteria also use H_2O as an electron donor and also have two photosystems. Purple (and green) photosynthetic bacteria, in contrast, have only one photosystem, which can carry out cyclic or noncyclic electron flow. During noncyclic electron flow, photosynthetic bacteria extract electrons from molecules that have a lower (more negative) reduction potential than that of H_2O, so a single photon can boost these electrons to the reduction potential of the NAD^+-NADH reaction. For instance, for the partial reaction

$$H_2S \longrightarrow S + 2\,H^+ + 2\,e^-$$

that occurs in bacteria with one photosystem, E_0 is -0.25 V; in contrast, E_0 is $+0.8$ V for the analogous oxidation of H_2O:

$$H_2O \longrightarrow \tfrac{1}{2}\,O_2 + 2\,H^+ + 2\,e^-$$

Thus less energy is required to boost electrons removed from H_2S to a level sufficient to reduce $NADP^+$ (or NAD^+) to NADPH (or NADH).

Another reason often cited for the presence of two photosystems in some organisms is that PSII cannot directly transfer electrons to $NADP^+$ and form NADPH. However, mutants of the algae *Chlamydomonas reinhardti* have been isolated that lack PSI yet still carry out normal photosynthetic oxygen and sugar formation. It is possible that, at least in these mutants, electrons can pass from PSII to the cytochrome b/f complex and then directly to ferredoxin, which then reduces $NADP^+$. Whether this process also occurs in normal algae that contain PSI is not known.

Because thylakoid membranes are permeable to anions and cations, a pH gradient (inside pH ≈ 5.0 versus stromal pH $= 7.8$), rather than a membrane electric potential, is the principal component of the proton-motive force. The thylakoid-membrane pH gradient is primarily used in ATP synthesis. Plants use photosynthesis-generated ATP and NADPH in a series of enzymatic reactions to convert CO_2 to sucrose or starch—the principal products of photosynthesis. This fourth stage in photosynthesis is covered in the last section of this chapter.

SUMMARY Molecular Analysis of Photosystems

• The mechanism of photosynthetic energy transduction is best understood in purple bacteria, which have a single type of photosystem whose structure is known in molecular detail (see Figure 16-40).

• In the bacterial photosystem, two electrons are transferred, one at a time, from light-excited chlorophyll a molecules on the exterior surface of the plasma membrane to a quinone (Q_B) located on the cytosolic face; Q_B then picks up two protons from the cytosol, forming QH_2. After QH_2 diffuses to the cytochrome bc_1 complex, it releases its protons to the extracellular space, generating a proton-motive force, and its electrons move through the complex to a soluble cytochrome. This cytochrome transports the electrons back to the reaction center, where they reduce the oxidized chlorophylls, completing the cyclic flow of electrons (see Figure 16-41).

• As in other systems, a Q cycle through the bc_1 oxidoreductase complex transports additional protons across the membrane to the external medium, and the proton-motive force is used mainly to power ATP synthesis through a F_0F_1 complex.

• Photosynthetic bacteria also carry out noncyclic electron transport. In this case, electrons removed from the reaction-center chlorophyll are ultimately transferred to NAD^+ and replaced by electrons derived from H_2 or H_2S, rather than from H_2O as in plants.

• Plants and cyanobacteria contain two photosystems, PSI and PSII. They have different functions and are physically separated in the thylakoid membrane. PSII, which is similar in structure and function to the bacterial photosystem, splits H_2O into O_2. PSI reduces $NADP^+$ to NADPH.

• In chloroplasts, light energy absorbed by light-harvesting complexes (LHCs) is transferred to a special pair of chlorophyll a molecules in the reaction centers (P_{680} in PSII and P_{700} in PSI).

• In PSII (as in the bacterial photosystem), electrons are transferred, one at a time, from light-excited P_{680} to an acceptor quinone, Q_B, on the stromal side of the

(a)

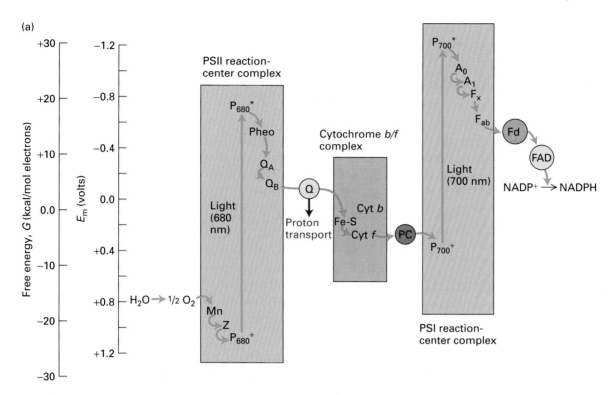

(b)

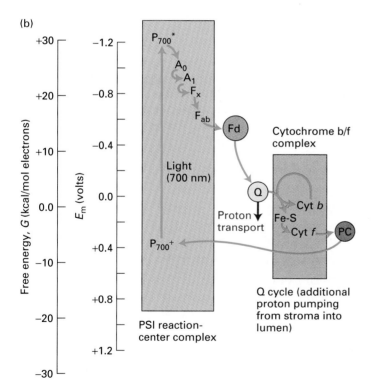

▲ **FIGURE 16-46 Energetics of electron flow through the photosynthetic system in plants.** Electrons will tend to flow from a state of higher to lower free energy, G, or, equivalently, from a more negative to a more positive reduction potential, E. (a) Linear electron flow. Absorption of a photon by P_{680} and P_{700} converts both chlorophylls to strong reductants (P_{680}^* and P_{700}^*). In both PSII and PSI, electrons in the light-excited chlorophylls are transferred through a series of electron carriers with successively more positive E values. Thus all these electron-transfer reactions have a negative ΔG and are energetically favored. The transfer of electrons from H_2O to P_{680}^+, to regenerate P_{680}, also is energetically favored. As electrons move from QH_2 through the cytochrome b/f complex, some of the acquired electron energy is used to transport protons into the thylakoid lumen (see Figure 16-42). Soluble plastocyanin (PC) transports electrons from the cytochrome b/f complex to P_{700}^+, thereby regenerating P_{700}. The process of linear electron flow, in which PSII and PSI are coupled, generates O_2, NADPH, and a proton gradient across the thylakoid membrane. Q=plastoquinone; Pheo=pheophytin; A_0=a monomeric chlorophyll a; A_1 = a derivative of vitamin K; F_x and F_{ab}=nonheme $4Fe \cdot 4S$ complexes; Fd=ferredoxin. (b) Cyclic electron flow, which involves PSI and the cytochrome b/f complex. Electrons from light-excited P_{700}^* are transferred, via a series of carriers in PSI, to ferredoxin. Rather than transferring its electrons to FAD, as in linear electron flow, ferredoxin donates electrons to a quinone (Q). The reduced quinone diffuses in the membrane and donates its electrons to the cytochrome b/f complex. Plastocyanin (PC) then transports the electrons from this complex back to PSI. During this process, protons are transported from the stroma into the thylakoid lumen, both directly by a reduced quinone, and indirectly via a Q cycle involving the cytochrome b/f complex. [Courtesy of Thomas Owens.]

thylakoid membrane. After Q_B accepts two electrons and has picked up two protons from the stroma, the resultant QH_2 diffuses to the cytochrome b/f complex, where it releases its protons to the thylakoid lumen, contributing to the proton-motive force (see Figure 16-42, *bottom*).

- Photochemically oxidized P_{680}^+ in PSII is reduced, regenerating P_{680}, by electrons derived from the splitting of H_2O. In this reaction, catalyzed by the oxygen-evolving complex on the luminal surface of the thylakoid, O_2 is formed and the resulting protons remain in the lumen, adding to the proton-motive force.

- Photochemically oxidized P_{700}^+ in PSI is reduced, regenerating P_{700}, in one of two ways. In linear electron flow, electrons are transferred from the cytochrome b/f complex in PSII, via soluble plastocyanin, to the PSI reaction center. Electrons lost from P_{700} following excitation of PSI are transported via several carriers ultimately to $NADP^+$, generating NADPH (see Figure 16-42, *top*).

- In cyclic electron flow, electrons released from P_{700} are transported through Q, the cytochrome b/f complex, and plastocyanin back to PSI. A Q cycle concomitantly transports additional protons from the stroma across the thylakoid membrane to the lumen. In this pathway, neither NADPH nor O_2 is formed, and PSII does not participate.

- PSI and PSII are functionally coupled and act in sequence during linear electron flow. Several mechanisms ensure that the amount of light energy delivered to the two reaction centers is controlled so that each center activates the same number of electrons.

16.5 CO_2 Metabolism during Photosynthesis

 Chloroplasts perform many metabolic reactions in green leaves. In addition to CO_2 fixation, the synthesis of almost all amino acids, all fatty acids and carotenes, all pyrimidines, and probably all purines occurs in chloroplasts. However, the synthesis of sugars from CO_2 is the most extensively studied biosynthetic reaction in plant cells—and it is certainly a unique one. We now turn to the series of energy-requiring and enzymatically catalyzed reactions known as the **Calvin cycle** (after discoverer Melvin Calvin). These reactions, which fix CO_2 and convert it to hexose sugars, are powered by energy released by ATP hydrolysis and by the reducing agent NADPH. The enzymes that catalyze the Calvin cycle reactions are rapidly inactivated in the dark, so that carbohydrate formation generally ceases when light is absent.

CO_2 Fixation Occurs in the Chloroplast Stroma

The reaction that actually fixes CO_2 into carbohydrates is catalyzed by *ribulose 1,5-bisphosphate carboxylase* (often called *rubisco*), which is located in the stromal space of the chloroplast. This enzyme adds CO_2 to the five-carbon sugar ribulose 1,5-bisphosphate to form two molecules of 3-phosphoglycerate (Figure 16-47). Rubisco is a large enzyme (≈ 500 kDa) composed of eight large subunits and eight small subunits. One subunit is encoded in chloroplast DNA; the other, in nuclear DNA. Because the catalytic rate of rubisco is quite slow, many copies of the enzyme are needed to fix sufficient CO_2. Indeed, this enzyme makes up almost

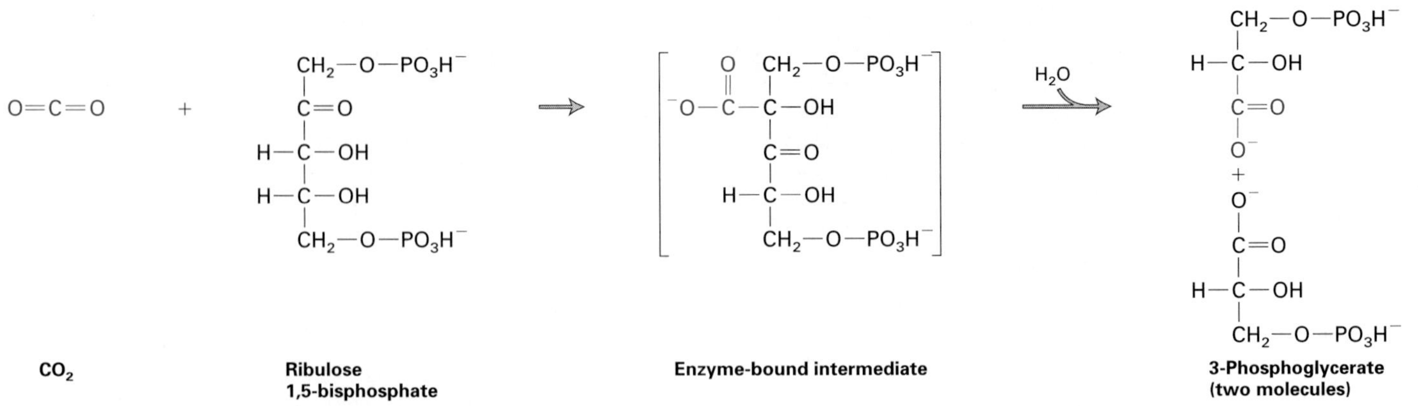

▲ FIGURE 16-47 The initial reaction that fixes CO_2 into organic compounds. In this reaction, catalyzed by ribulose 1,5-bisphosphate carboxylase, CO_2 condenses with the five-carbon sugar ribulose 1,5-bisphosphate. The products are two molecules of 3-phosphoglycerate. When photosynthetic algae are exposed to a brief pulse of ^{14}C-labeled CO_2 and the cells are then quickly disrupted, 3-phosphoglycerate is radiolabeled most rapidly, and all the radioactivity is found in the carboxyl group. This finding establishes that ribulose 1,5-bisphosphate carboxylase fixes the CO_2.

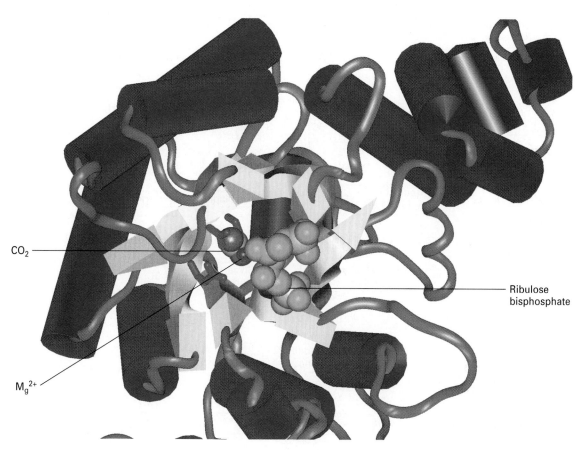

CO₂

Ribulose
bisphosphate

Mg²⁺

▲ FIGURE 16-48 Structure of the catalytic domain of the active form of ribulose 1,5-bisphosphate carboxylase. Dark blue cylinders represent α helices and yellow arrows represent β sheets in the polypeptide. The key residues in the active site are carbamylated lysine 191, aspartate 193, and glutamate 194 (all in red); a Mg²⁺ ion is bound to carbamylated lysine 191. The substrates CO₂ (gray) and ribulose 1,5-bisphosphate (green) are shown bound to the active site. [After T. Lundqvist and G. Schneider, 1991, *J. Biol. Chem.* **266**:12604; courtesy of Dr. Lawren Wu.]

50 percent of the chloroplast protein and is believed to be the most abundant protein on earth. Rubisco is activated by covalent addition of CO₂ to the side-chain amino group of lysine 191, forming a carbamate group, which then binds a Mg²⁺ ion. The structure of the catalytic site of the active enzyme is shown in Figure 16-48.

The fate of the 3-phosphoglycerate formed by this reaction is complex: some is converted to starch or sucrose, but some is used to regenerate ribulose 1,5-bisphosphate. At least nine enzymes are required to regenerate ribulose 1,5-bisphosphate from 3-phosphoglycerate. Quantitatively, for every 12 molecules of 3-phosphoglycerate generated by rubisco (a total of 36 C atoms), 2 molecules (6 C atoms) are converted to 2 molecules of glyceraldehyde 3-phosphate (and later to one hexose), while 10 molecules (30 C atoms) are converted to 6 molecules of ribulose 1,5-bisphosphate (Figure 16-49, *top*). The fixation of six CO₂ molecules and the net formation of two glyceraldehyde 3-phosphate molecules require the consumption of 18 ATPs and 12 NADPHs, generated by the light-requiring processes of photosynthesis.

Synthesis of Sucrose Incorporating Fixed CO₂ Is Completed in the Cytosol

After its formation in the chloroplast stroma, glyceraldehyde 3-phosphate is transported to the cytosol in exchange for phosphate. The final steps of sucrose synthesis occur in the cytosol (Figure 16-49, *bottom*). In these reactions, one molecule of glyceraldehyde 3-phosphate is isomerized to dihydroxyacetone phosphate. This compound condenses with a second molecule of glyceraldehyde 3-phosphate to form fructose 1,6-bisphosphate, an intermediate in both glycolysis (see Figure 16-3) and glucose biosynthesis. In leaf cells, however, most of the fructose 1,6-bisphosphate is converted to sucrose. Half is converted to fructose 6-phosphate; half is converted to glucose 1-phosphate, which then forms uridine diphosphate glucose (UDP-glucose). These two compounds condense to form sucrose 6-phosphate; a final, irreversible removal of phosphate then generates the exportable sucrose.

The transport protein in the chloroplast membrane that brings fixed CO₂ (as glyceraldehyde 3-phosphate) into the

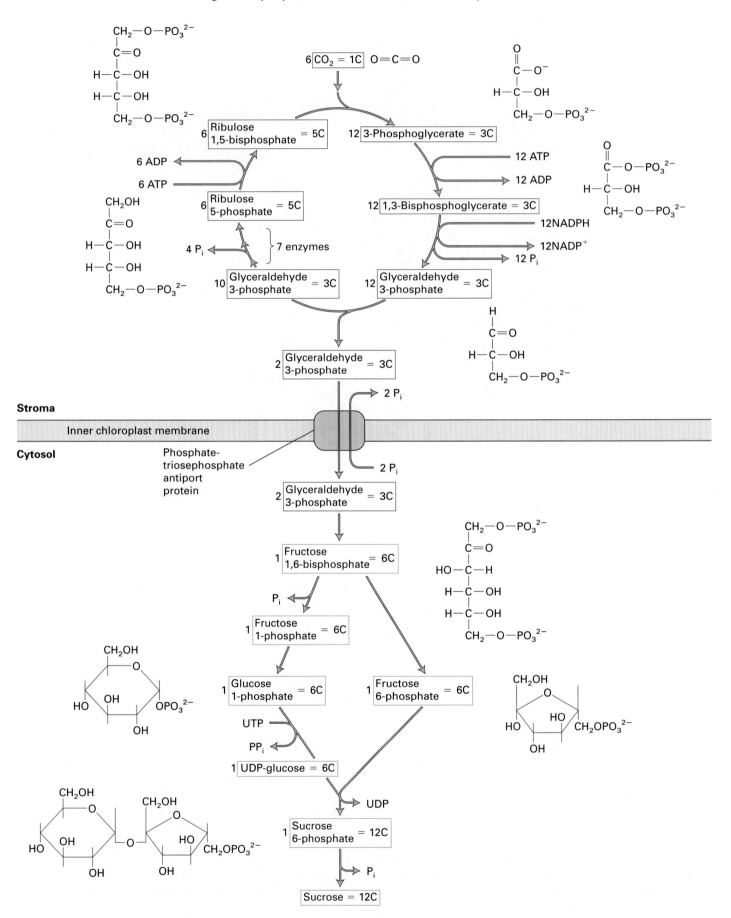

◀ **FIGURE 16-49 The pathway of carbon during photosynthesis.** *(Top)* Six molecules of CO$_2$ are converted into two molecules of glyceraldehyde 3-phosphate. These reactions, which constitute the Calvin cycle, occur in the stroma of the chloroplast. Via the phosphate-triosephosphate antiporter, some glyceraldehyde 3-phosphate is transported to the cytosol in exchange for phosphate. *(Bottom)* In the cytosol, an exergonic series of reactions converts glyceraldehyde 3-phosphate to fructose 1,6-bisphosphate and, ultimately, to the disaccharide sucrose. Some glyceraldehyde 3-phosphate (not shown here) is also converted to amino acids and fats, compounds essential to plant growth.

As noted earlier, the active form of rubisco contains a carbamylated lysine residue, which binds Mg^{2+}, an essential cofactor (see Figure 16-48). The activating reaction

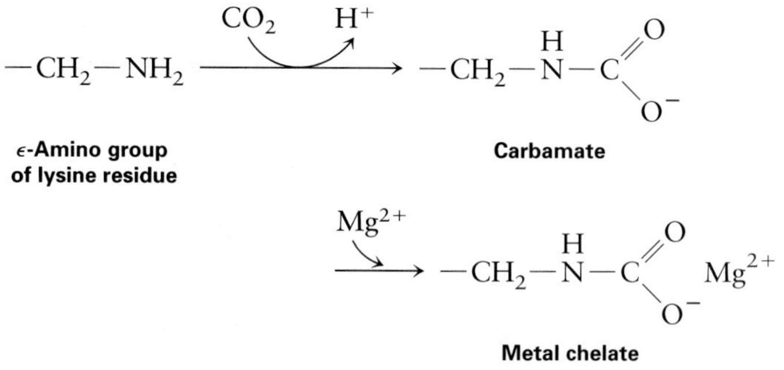

occurs spontaneously in the presence of high CO$_2$ and Mg^{2+} concentrations. Under normal conditions, however, with ambient levels of CO$_2$, the reaction requires catalysis by *rubisco activase*, an enzyme that simultaneously hydrolyzes ATP and uses the energy to attach a CO$_2$ to the lysine. This enzyme was discovered during the study of a mutant strain of *Arabidopsis thaliana* that required high CO$_2$ levels to grow and that did not exhibit light activation of ribulose 1,5-bisphosphate carboxylase; the mutant had a defective rubisco activase.

cytosol when the cell is exporting sucrose vigorously is a strict antiporter: No fixed CO$_2$ leaves the chloroplast unless phosphate is fed into it. The phosphate is generated in the cytosol, primarily during the formation of sucrose, from phosphorylated three-carbon intermediates (see Figure 16-49, *bottom*). Thus the synthesis of sucrose and its export from the cytosol to other cells encourages the export of additional glyceraldehyde 3-phosphate from the chloroplast.

Light Stimulates CO$_2$ Fixation by Several Mechanisms

The Calvin cycle enzymes that catalyze CO$_2$ fixation turn off rapidly in the dark, thereby conserving ATP that is generated in the dark for other synthetic reactions, such as lipid and amino acid biosynthesis. Several mechanisms contribute to this control. The pH of the stroma is ≈ 7 in the dark and ≈ 8 in the light (due to the light-driven transport of protons from the stroma into the thylakoid lumen), and several of the Calvin cycle enzymes function better at the higher pH. During illumination, the stromal concentration of Mg^{2+}, needed for functioning of the ATP-requiring enzymes, increases due to the light-enhanced activity of a thylakoid membrane Mg^{2+} transporter.

A stromal protein called *thioredoxin* (Tx) also plays a role in controlling some Calvin cycle enzymes. In the dark, thioredoxin contains a disulfide bond; in the light, electrons are transferred from PSI, via ferredoxin, to thioredoxin, reducing its disulfide bond:

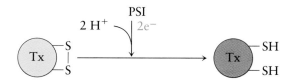

Reduced thioredoxin then activates several Calvin cycle enzymes by reducing disulfide bonds in them. In the dark, when thioredoxin becomes reoxidized, these enzymes are reoxidized and thus inactivated.

Photorespiration, Which Consumes O$_2$ and Liberates CO$_2$, Competes with Photosynthesis

Photosynthesis is always accompanied by *photorespiration*—a process that takes place in light, consumes O$_2$, and converts ribulose 1,5-bisphosphate in part to CO$_2$. As Figure 16-50 shows, rubisco catalyzes two competing reactions: the addition of CO$_2$ to ribulose 1,5-bisphosphate to form two molecules of 3-phosphoglycerate, and the addition of O$_2$ to form one molecule of 3-phosphoglycerate and one molecule of the two-carbon compound phosphoglycolate.

Phosphoglycolate is recycled via a complex pathway that involves reactions in peroxisomes and mitochondria as well as chloroplasts. The net result of these reactions is that for every two molecules of phosphoglycolate formed by photorespiration (four C atoms), one molecule of 3-phosphoglycerate is ultimately formed and recycled, and one molecule of CO$_2$ is lost.

Photorespiration is wasteful to the energy economy of the plant: it consumes ATP and O$_2$, and it generates CO$_2$. It is surprising, therefore, that all known rubiscos catalyze photorespiration. Probably the necessary structure of the active site of rubisco precluded evolution of an enzyme that does not catalyze photorespiration.

The C$_4$ Pathway for CO$_2$ Fixation Is Used by Many Tropical Plants

In a hot, dry environment, plants must keep the gas-exchange pores (stomata) in their leaves closed much of

2. How does one determine the likely amino acid residues that conduct the transported protons through cytochrome *c* oxidase?

3. What roles do conformational changes in cytochrome c oxidase play in coupling electron transport with proton pumping?

Gennis, R. B. 1998. How does cytochrome oxidase pump protons? *Proc. Nat'l. Acad. Sci. USA* 95:12747–12749. (A commentary on the Michel paper.)

Michel, H. 1998. The mechanism of proton pumping by cytochrome *c* oxidase x127e. *Proc. Nat'l. Acad. Sci. USA* 95: 12819–12824.

Tsukihara, T., et al. 1996. The whole structure of the 13-subunit oxidized cytochrome *c* oxidase at 2.8 Å. *Science* 272:1136–1144.

Yoshikawa, S., et al. 1998. Redox-coupled crystal structural changes in bovine heart cytochrome *c* oxidase. Science 280:1723–1729.

Testing Yourself on the Concepts

1. The proton motive force (pmf) is a necessary component for both mitochondrial and chloroplast function. What produces the pmf, and what is its relationship to ATP?

2. An important function of the mitochondrial inner membrane is to provide a selectively permeable barrier to the movement of water-soluble molecules and thus generate different chemical environments on either side of the membrane. However, many of the substrates and products involved in oxidative phosphorylation are water soluble and must cross the membrane. How does this transport occur?

3. Explain the following statement: The O_2 generated by photosynthesis is simply a by-product of the reaction's generation of carbohydrates and ATP.

4. The Calvin cycle reactions that fix CO_2 do not function in the dark. What are the likely reasons for this, and how is this regulation implemented?

MCAT/GRE-Style Questions

Key Concept Please read the section titled "Electrons Flow from $FADH_2$ and NADH to O_2 via a Series of Multiprotein Complexes" (p. 634) and answer the following questions:

1. Prosthetic groups such as heme and iron-sulfur clusters function to
   a. Donate electrons to NADH.
   b. Allow proteins to diffuse within the mitochondrial inner membrane.
   c. Both accept and donate electrons during electron transport.

   d. Transport protons across the mitochondrial inner membrane.

2. The succinate-CoQ reductase complex
   a. Readily accepts electrons from NADH or $FADH_2$.
   b. Contains one of the key enzymes of the citric acid cycle.
   c. Pumps only two protons into the mitochondrial matrix.
   d. Contains multiple cytochrome complexes.

3. During electron flow from NADH to O_2, each of the major enzyme complexes in the chain
   a. Uses all of each electron's potential energy to transport protons into the matrix.
   b. Uses all of each electron's potential energy to transport protons out of the matrix.
   c. Uses a portion of each electron's potential energy to transport protons into the matrix.
   d. Uses a portion of each electron's potential energy to transport protons out of the matrix.

4. Which of the following does not function as an electron carrier?
   a. Coenzyme Q.
   b. Cytochrome c.
   c. Cytochrome a.
   d. H_2O.

5. Which of the following is necessary for electron flow from both $FADH_2$ and NADH to O_2?
   a. Succinate-CoQ reductase complex.
   b. Coenzyme Q.
   c. Flavin mononucleotide.
   d. Fumarate.

Key Experiment Please refer to Figure 16-29 to answer the following questions:

6. The observation that isolated F_1 particles have ATPase activity suggests that
   a. Electron transport requires energy derived from ATP hydrolysis.
   b. The normal enzymatic activity of the particle is reversible.
   c. These particles have been irreversibly inactivated.
   d. Electron transport is still functioning.

7. The results presented in the figure support the hypothesis that the
   a. F_0 complex is a peripheral membrane protein.
   b. F_1 complex is capable of electron transport.
   c. F_0 complex is capable of ATP synthesis.
   d. F_0 complex is an integral membrane protein.

8. Treatment of the inside-out vesicles with an uncoupler that allows protons to freely move across the vesicle membrane would be expected to

a. Increase ATP synthesis.

b. Inhibit ATP synthesis.

c. Increase O_2 consumption.

d. Inhibit electron transport.

9. The ability of the inside-out vesicles to carry out electron transport and ATP synthesis requires

a. That the solution inside the vesicles has a higher pH than the solution surrounding the vesicles.

b. The presence of O_2 in the enclosed space of the inside-out vesicles.

c. The presence of ADP and P_i in the solution surrounding the inside-out vesicles.

d. The presence of ADP and P_i in the enclosed space of the inside-out vesicles.

Key Concept Please read the section titled "Chloroplasts Contain Two Functionally and Spatially Distinct Photosystems" (p. 658) and answer the following questions:

10. Linear electron flow describes the movement of electrons from

a. PSI to PSII to $NADP^+$.

b. PSII to PSI to $NADP^+$.

c. PSII to $NADP^+$ to PSI.

d. PSI to PSII to H_2O.

11. Although PSI and PSII differ in many ways, one common feature of both photosystems is that both

a. Donate electrons to $NADP^+$.

b. Remove electrons from H_2O to generate protons and O_2.

c. Are present in the same region of the thylakoid membrane.

d. Contain chlorophyll a.

12. PSI and PSII absorb light of different wavelengths due to

a. The presence of different soluble electron carriers.

b. Different locations in the chloroplast.

c. The protein environment around each reaction center chlorophyll.

d. Different types of reaction center chlorophyll in each photosystem.

13. The protons created by the splitting of H_2O

a. Are transferred to PSII.

b. Are transferred to PSI.

c. Are freely soluble in the thylakoid membrane.

d. Contribute to a proton gradient across the thylakoid membrane.

14. A mutation that inactivates the cytochrome *b/f* complex would

a. Inhibit movement of electrons from PSII to PSI.

b. Inhibit movement of electrons from PSI to PSII.

c. Inhibit reduction of quinone.

d. Promote formation of NADPH.

Key Terms

adenosine triphosphate 616

aerobic oxidation 616

anaerobic 619

ATP 616

ATP synthase 617

Calvin cycle 664

carbon fixation 649

chemiosmosis 617

chlorophylls 649

chloroplasts 617

citric acid cycle 625

cytochromes 635

electron carriers 617

electron transport 617

F_0F_1 complex 617

flavin adenine dinucleotide 622

glycolysis 617

Krebs cycle 625

mitochondria 617

NAD^+ 619

$NADP^+$ 649

oxidative phosphorylation 632

photosynthesis 617

proton-motive force 617

respiration 632

substrate-level phosphorylation 619

thylakoids 649

tricarboxylic acid cycle 625

References

Oxidation of Glucose and Fatty Acids to CO_2

Fell, D. 1997. *Understanding the Control of Metabolism.* Portland Press.

Fersht, A. 1985. *Enzyme Structure and Mechanism,* 2d ed. W. H. Freeman and Company. Contains an excellent discussion of the reaction mechanisms of key enzymes.

Fothergill-Gilmore, L. A., and P. A. Michels. 1993. Evolution of glycolysis. *Prog. Biophys. Mol. Biol.* **59:**105–135.

Guest, J. R., and G. C. Russell. 1992. Complexes and complexities of the citric acid cycle in *Escherichia coli. Curr. Top. Cell Regul.* **33:**231–247.

Krebs, H. A. 1970. The history of the tricarboxylic acid cycle. *Perspect. Biol. Med.* **14:**154–170.

Lehninger, A. L., D. L. Nelson, and M. M. Cox. 1993. *Principles of Biochemistry.* Worth, chaps. 13–16, 18.

Mannaerts, G. P., and P. P. Van Veldhoven. 1993. Metabolic pathways in mammalian peroxisomes. *Biochimie* **75:**147–158.

Ochs, R. S., R. W. Hanson, and J. Hall, eds. 1985. *Metabolic Regulation.* Elsevier. A. collection of articles originally published in *Trends Biochem. Sci.*

Pilkis, S. J., T. H. Claus, I. J. Kurland, and A. J. Lange. 1995. 6-phosphofructo-2-kinase/fructose-2,6-bisphosphatase: a metabolic signaling enzyme. *Ann. Rev. Biochem.* **64:**799–835.

Stryer, L. 1995. *Biochemistry,* 4th ed. W. H. Freeman and Company, chaps. 19 and 20.

Valle, D., and J. Gartner. 1993. Human genetics. Penetrating the peroxisome. *Nature* **361:**682–683.

Electron Transport and Oxidative Phosphorylation

Beinert, H., R. Holm, and E. Münck. 1997. Iron-sulfur clusters: nature's modular, multipurpose structures. *Science* **277:**653–659.

Bianchet, M. A., J. Hullihen, P. Pedersen, and M. Amzel. The 2.8-Å structure of rat liver F1-ATPase: configuration of a critical intermediate in ATP synthesis/hydrolysis. *Proc. Nat'l. Acad. Sci. USA* **95**:11065–11070.

Boyer, P. D. 1989. A perspective of the binding change mechanism for ATP synthesis. *FASEB J.* **3**:2164–2178.

Boyer, P. D. 1997. The ATP synthase—a splendid molecular machine. *Ann. Rev. Biochem.* **66**:717–749.

Block, S. 1997. Real engines of creation. *Nature* **386**:217–218.

Brandt, U., and B. Trumpower. 1994. The protonmotive q cycle in mitochondria and bacteria. *Crit. Rev. Biochem. Mol. Biol.* **29**:165–197.

Elston, T., H. Wang, and G. Oster. 1998. Energy transduction in ATP synthase. *Nature* 391:510–512.

Gennis, R. B. 1998. How does cytochrome oxidase pump protons? *Proc. Nat'l. Acad. Sci. USA* **99**:12747–12749

Harold, F. M. 1986. *The Vital Force: A Study of Bioenergetics.* W. H. Freeman and Company.

Kinosita, K., R. Yasuda, H., Noji, S. Ishiwata, and M. Yoshida. 1998. F_1-ATPase: a rotary motor made of a single molecule. *Cell* **93**:21–24.

Klingenberg, M. 1993. Mitochondrial carrier family: ADP-ATP carrier as a carrier paradigm. *Soc. Gen. Physiol. Ser.* **48**:201–212.

Läuger, P. 1991. *Electrogenic Ion Pumps.* Sinauer Associates.

Michel, H., J. Behr, A. Harrenga, and A. Kannt. 1998. Cytochrome *c* oxidase. *Ann Rev. Biophys. Biomol. Struct.* **27**:329–356.

Mitchell, P. 1979. Keilin's respiratory chain concept and its chemiosmotic consequences. *Science* **206**:1148–1159. (Nobel Prize Lecture).

Ostermeier, C., S. Iwata, and H. Michel. 1996. Cytochrome *c* oxidase. *Curr. Opin. Struc. Biol.* **6**:460–466.

Ramirez, B. E., B. Malmström, J. R. Winkler, and H. B. Gray. 1995. The currents of life: the terminal electron-transfer complex of respiration. *Proc. Nat'l. Acad. Sci. USA* **92**:11949–11951.

Saraste, M. 1999. Oxidative phosphorylation at the *fin de siecle*. *Science* **283**:1488–1492.

Scandalios, J., ed. 1997. *Oxidative Stress and the Molecular Biology of Antioxidant Defenses.* Cold Spring Harbor Laboratory Press.

Skulachev, V. P. 1992. The laws of cell energetics. *Eur. J. Biochem.* **208**:203–209.

Smith, J. 1998. Secret life of cytochrome bc_1. *Science* **281**:58–59.

Tsukihara, T., et al. 1996. The whole structure of the 13-subunit oxidized cytochrome *c* oxidase at 2.8 Å. *Science* **272**:1136–1144.

Walker, J. E. 1995. Determination of the structures of respiratory enzyme complexes from mammalian mitochondria. *Biochim. Biophys. Acta.* **1271**:221–227.

Williams, R. 1996. Bioenergetics. Purpose of proton pathways. *Nature* 376:643.

Xia, D., et al. 1997. Crystal structure of the cytochrome bc_1 complex from bovine heart mitochondria. *Science* **277**:60–66.

Zhang, Z., et al. 1998. Electron transfer by domain movement in cytochrome bc_1. *Nature* **392**:677–684.

Zhou, Y., T. Duncan, and R. Cross. 1997. Subunit rotation in Escherichia coli F_0F_1-ATP synthase during oxidative phosphorylation. *Proc. Nat'l. Acad. Sci.* **94**:10583–10587.

Photosynthetic Stages and Light Absorbing Pigments.

Deisenhofer, J., and J. R. Norris, eds. 1993. *The Photosynthetic Reaction Center.* Academic Press. Vols. 1 and 2.

Govindjee, and W. J. Coleman. 1990. How plants make oxygen. *Sci. Am.* 262(2):50–58.

Harold, F. M. 1986. *The Vital Force: A Study of Bioenergetics.* W. H. Freeman and Company, chap. 8.

Prince, R. 1996. Photosynthesis: the Z-scheme revisited. *Trends. Biochem. Sci.* **21**:121–122.

McDermott, G., et al. 1995. Crystal structure of an integral membrane light-harvesting complex from photosynthetic bacteria. *Nature* **364**:517.

Molecular Analysis of Photosystems

Aro, E. M., I. Virgin, and B. Andersson. 1993. Photoinhibition of photosystem II. Inactivation, protein damage, and turnover. *Biochim. Biophys. Acta.* **1143**:113–134.

Barber, J., and B. Andersson. 1992. Too much of a good thing: light can be bad for photosynthesis. *Trends Biochem. Sci.* **17**:61–66.

Deisenhofer, J., and H. Michel. 1989. The photosynthetic reaction center from the purple bacterium rhodopseudomonas viridis. *Science* **245**:1463–1473. (The Nobel Prize Lecture.)

Deisenhofer, J., and H. Michel. 1991. Structures of bacterial photosynthetic reaction centers. *Ann. Rev. Cell Biol.* **7**:1–23.

Golbeck, J. H. 1993. Shared thematic elements in photochemical reaction centers. *Proc. Nat'l. Acad. Sci. USA* **90**:1642–1646.

Hankamer, B., J. Barber, and E. Boekema. 1997. Structure and membrane organization of photosystem II from green plants. *Ann. Rev. Plant Physiol. and Plant Mol. Biol.* **48**:641–672.

Hoganson, C. W., and G. T. Babcock. 1997. A metalloradical mechanism for the generation of oxygen from water in photosynthesis. *Science* **277**:1953–1956.

Horton, P., A. Ruban, and R. Walters. 1996. Regulation of light harvesting in green plants. *Ann. Rev. Plant Physiol. and Plant Mol. Biol.* **47**:655–684.

Huber, R. 1989. A structure basis of light energy and electron transfer in biology. *Embo J.* **8**:2125–2147. (The Nobel Prize Lecture.)

Knaff, D. 1991. Regulatory phosphorylation of chloroplast antenna proteins. *Trends Biochem. Sci.* **16**:82–83.

Kühlbrandt, W. 1995. Many wheels make light work. *Nature* **374**:497–498.

Martin, J. L., and M. H. Vos. 1992. Femtosecond biology. *Ann. Rev. Biophys. Biomol. Struct.* **21**:199–222.

Penner-Hahn, J. 1998. Structural characterization of the Mn site in the photosynthetic oxygen-evolving complex. *Structure and bonding* **90**:1–36.

Tommos, C., and G. Babcock. 1998. Oxygen production in nature: a light-driven metalloradical enzyme process. *Accounts. Chem. Res.* **31**:18–25.

Tsiotis, G., G. McDermott, and D. Ghanotakis. 1996. Progress toward structural elucidation of photosystem II. *Photosyn. Res.* **50**:93–101.

CO_2 Metabolism During Photosynthesis

Bassham, J. A. 1962. The path of carbon in photosynthesis. *Sci. Am.* **206**(6):88–100.

Buchanan, B. B. 1991. Regulation of CO_2 assimilation in oxygenic photosynthesis: the ferredoxin/thioredoxin system. Perspective on its discovery, present status, and future development. *Arch. Biochem. Biophys.* **288**:1–9.

Portis, A. 1992. Regulation of ribulose 1,5-bisphosphate carboxylase/oxygenase activity. *Ann. Rev. Plant Physiol. Plant Mol. Biol.* **43**:415–437.

Rawsthorne, S. 1992. Towards an understanding of C_3-C_4 photosynthesis. *Essays Biochem.* **27**:135–146.

Schneider, G., Y. Lindqvist, and C. I. Branden. 1992. Rubisco: structure and mechanism. *Annu. Rev. Biophys. Biomol. Struct.* **21**:119–153.

Wolosiuk, R. A., M. A. Ballicora, and K. Hagelin. 1993. The reductive pentose phosphate cycle for photosynthetic CO_2 assimilation: enzyme modulation. *FASEB J.* **7**:622–637.

17

Protein Sorting: Organelle Biogenesis and Protein Secretion

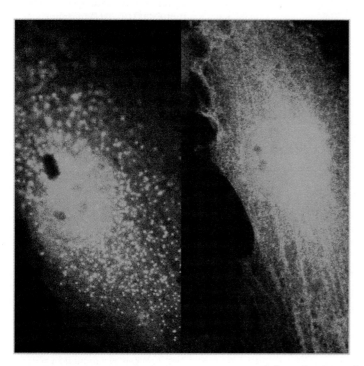

Normal *(left)* but not mutant *(right)* human cells incorporate Firefly luciferase into peroxisomes.

A typical mammalian cell contains up to 10,000 different kinds of proteins; a yeast cell, about 5000. For a cell to function properly, each of its numerous proteins must be localized to the correct cellular membrane or aqueous compartment (e.g., the mitochondrial matrix, chloroplast stroma, lysosomal lumen, or cytosol). Hormone receptor proteins, for example, must be delivered to the plasma membrane if the cell is to recognize hormones, and specific ion-channel and transporter proteins are needed if the cell is to import or export the corresponding ions and small molecules. Water-soluble enzymes such as RNA and DNA polymerases must be targeted to the nucleus; still others, such as proteolytic enzymes or catalase, must go to lysosomes or peroxisomes, respectively. Many proteins, such as hormones and components of the extracellular matrix, must be directed to the cell surface and secreted.

The process of directing each newly made polypeptide to a particular destination—referred to as protein targeting, or protein sorting—is critical to the organization and functioning of eukaryotic cells. This process occurs at several levels. As discussed in Section 9.7, a few proteins, encoded by the DNA present in

MEDIA CONNECTIONS

Overview: Protein Sorting

Overview: Protein Secretion

Focus: Synthesis of Secreted and Membrane-Bound Proteins

Classic Experiment 17.1: Following a Protein out of the Cell

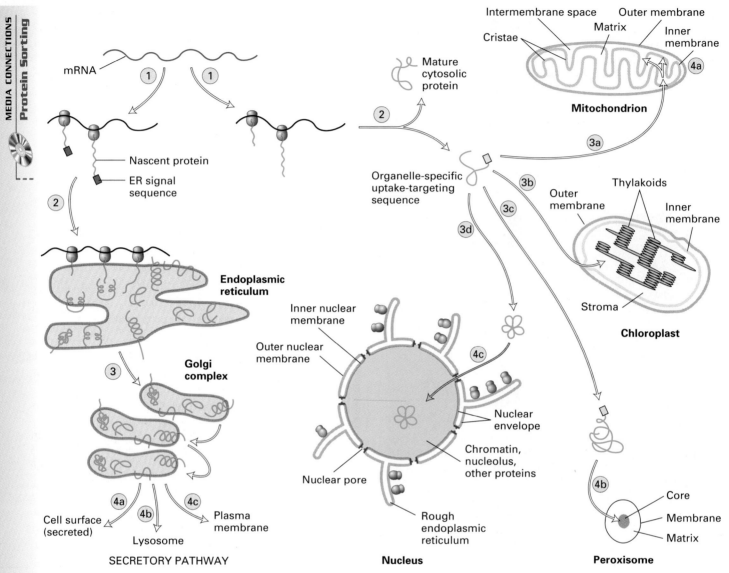

▲ **FIGURE 17-1 Overview of sorting of nuclear-encoded proteins in eukaryotic cells.** All nuclear-encoded mRNAs are translated on cytosolic ribosomes. Ribosomes synthesizing nascent proteins in the secretory pathway ① are directed to the rough endoplasmic reticulum (ER) by an ER signal sequence ②. After translation is completed in the ER, these proteins move via transport vesicles to the Golgi complex ③ from whence they are further sorted to several destinations ④a, ④b, ④c. After synthesis of proteins lacking an ER signal sequence is completed on free ribosomes ①, the proteins are released into the cytosol ②. Those with an organelle- specific uptake-targeting sequence are imported into the mitochondrion ③a, chloroplast ③b, peroxisome ③c, or nucleus ③d. Mitochondrial and chloroplast proteins typically pass through the outer and inner membranes to enter the matrix or stromal space, respectively. Some remain there, and some ④a are sorted to other organellar compartments. Unlike mitochondrial and chloroplast proteins, which are imported in a partially unfolded form, most peroxisomal proteins cross the peroxisome membrane as fully folded proteins ④b. Folded nuclear proteins, often in the form of ribonucleoprotein particles, enter through visible nuclear pores by processes discussed in Chapter 11 ④c.

mitochondria and chloroplasts, are synthesized on ribosomes in these organelles and are incorporated directly into compartments within these organelles. However, most mitochondrial and chloroplast proteins and all of the proteins in the other organelles, particles, and membranes of a eukaryotic cell are encoded by nuclear DNA, are synthesized on ribosomes in the cytosol, and are distributed to their correct destinations via the sequential action of several sorting signals and multiple sorting events. How nuclear-encoded organelle, membrane, and secretory proteins are sorted to their correct destinations are the major subjects of this chapter.

The first sorting event occurs during initial growth of nascent polypeptide chains on cytosolic ribosomes (Figure 17-1). Some nascent proteins contain, generally at the amino terminus, a specific signal, or targeting, sequence that directs the ribosomes synthesizing them to the endoplasmic reticulum (ER). Protein synthesis is completed on ribosomes attached to the rough ER membrane (the presence of these bound ribosomes distinguishes the rough ER from the smooth ER). The completed polypeptide chains then move to the Golgi complex and subsequently are sorted to various destinations. Proteins synthesized and sorted in this pathway, referred to as the **secretory pathway**, include not only those that are secreted from the cell but also enzymes and other resident proteins in the lumen of the ER, Golgi, and lysosomes as well as integral proteins in the membranes of these organelles and the plasma membrane.

Synthesis of all other nuclear-encoded proteins is completed on "free" cytosolic ribosomes, and the completed proteins are released into the cytosol. These proteins remain in the cytosol unless they contain a specific signal sequence that directs them to the mitochondrion, chloroplast, peroxisome, or nucleus (see Figure 17-1). Many of these proteins are subsequently sorted further to reach their correct destinations within these organelles; such sorting events depend on yet other signal sequences within the protein. Each sorting event involves binding of a signal sequence to one or more receptor proteins on the surface or interior of the organelle.

In this chapter, we detail the mechanisms whereby proteins are sorted to the major organelles and compartments of the cell. (The transport of proteins in and out of the nucleus through nuclear pores, which occurs by somewhat different mechanisms, is discussed in Section 11.4.) The first two sections cover targeting of proteins to mitochondria, chloroplasts, and peroxisomes. The next several sections describe the various components and events in the secretory pathway, including the post-translational modifications that occur to proteins as they move through this pathway. We then discuss how proteins are internalized into cells following binding to specific cell-surface receptors and the fate of such internalized proteins. In the final section, we describe how the various small membrane-bounded vesicles that carry proteins within cells are formed and deliver their contents to specific destinations.

17.1 Synthesis and Targeting of Mitochondrial and Chloroplast Proteins

Mitochondria and chloroplasts are surprisingly similar. Both are bounded by two membranes; chloroplasts contain, in addition, an internal membrane compartment—the **thylakoids**—on which photosynthesis takes place (see Figure 16-34). Both organelles use a proton-motive force and the same type of protein—an F-class ATPase—to synthesize

ATP (see Figure 16-2); they also contain similar types of electron-transport proteins. Growth and division of mitochondria and chloroplasts is not coupled to nuclear division. These organelles grow by the incorporation of proteins and lipids, a process that occurs continuously during the interphase period of the cell cycle. As the organelles increase in size, one or more daughters pinch off in a manner similar to the way in which bacterial cells grow and divide. Although the biogenesis of both organelles is similar in many respects, our discussion focuses on mitochondrial biogenesis, about which more is known.

Mitochondria and chloroplasts probably arose by the incorporation of photosynthetic or nonphotosynthetic bacteria into ancestral eukaryotic cells, about 1,500 million years ago, and their subsequent replication in the cytoplasm. Over eons of evolution much of the bacterial DNA in these endosymbionts moved to the nucleus, so that in present-day cells many mitochondrial and chloroplast proteins are imported into the organelles after their synthesis in the cytosol. The mitochondrial and chloroplast DNA found in extant organisms encodes organelle rRNAs and tRNAs but relatively few proteins, mainly subunits of integral membrane proteins essential to organelle function (see Figure 9-44). These proteins are synthesized on ribosomes within the organelles and directed to the correct compartment immediately after synthesis, and current work is elucidating how this happens.

Most chloroplast and mitochondrial proteins, however, are synthesized outside the organelle on cytosolic ribosomes that are not bound to the rough endoplasmic reticulum. The newly made proteins are released into the cytosol and are then taken up specifically into the proper organelle by binding to receptor proteins on the organelle surface that recognize specific *uptake-targeting sequences* in the new proteins (Table 17-1). The mitochondrion and chloroplast contain multiple membranes and membrane-limited spaces. Thus targeting of some proteins requires the sequential action of two targeting sequences and two membrane-bound receptor systems: one to direct the protein into the organelle, and the other to direct it into the correct organellar compartment or membrane (see Figure 17-1). In general, protein uptake into mitochondria or chloroplasts is an energy-requiring process that depends on integral proteins in the organellar membranes. The biogenesis of both organelles is similar in many respects.

Most Mitochondrial Proteins Are Synthesized as Cytosolic Precursors Containing Uptake-Targeting Sequences

Some of the proteins that are synthesized in the cytosol and incorporated into the mitochondrion are listed in Table 17-2. The largest number are transported to the matrix, but some are transported to the intermembrane space or inserted into the outer or inner membrane. Pulse-chase experiments with yeast and *Neurospora* cells clearly demonstrate that

TABLE 17-1 Properties of Uptake-Targeting Signal Sequences That Direct Proteins from the Cytosol to Organelles

Target Organelle	Usual Signal Location within Protein	Signal Removal*	Nature of Signal
Endoplasmic reticulum	N-terminal	(+)	"Core" of 6–12 mostly hydrophobic amino acids, often preceded by one or more basic amino acids
Mitochondrion[†]	N-terminal	(+)	3–5 nonconsecutive Arg or Lys residues, often with Ser and Thr; no Glu or Asp residues
Chloroplast[†]	N-terminal	(+)	No common sequence motifs; generally rich in Ser, Thr, and small hydrophobic amino acid residues and poor in Glu and Asp residues
Peroxisome	C-terminal	(−)	Usually Ser-Lys-Leu at extreme C-terminus
Nucleus	Internal	(−)	One cluster of 5 basic amino acids, or two smaller clusters of basic residues separated by ≈10 amino acids

*Indicates whether signal sequence usually is (+) or is not (−) removed after a protein enters its target organelle.
[†]These signals direct the protein from the cytosol into the matrix space of the mitochondrion or the corresponding stroma of the chloroplast; other signals discussed in the text redirect proteins into other subcompartments of these organelles.

TABLE 17-2 Selected Mitochondrial Proteins That Are Synthesized in the Cytosol

Mitochondrial Location	Protein*
Matrix	Alcohol dehydrogenase (yeast) Carbamoyl phosphate synthase (mammals) Citrate synthase and other citric acid enzymes DNA polymerase F_1 ATPase subunits α (except in plants), β, γ, and δ (in certain fungi) Mn^{2+}-superoxide dismutase Ornithine aminotransferase (mammals) Ornithine transcarbamoylase (mammals) Ribosomal proteins RNA polymerase
Inner membrane	ADP/ATP antiporter $CoQH_2$–cytochrome c reductase complex: subunits 1, 2, 5 (Fe-S protein), 6, 7, and 8 Cytochrome c oxidase subunits 4, 5, 6, and 7 Phosphate/OH^- antiporter Proteolipid of F_0 ATPase Thermogenin (brown fat)
Intermembrane space	Cytochrome c Cytochrome c peroxidase Cytochrome b_2 and c_1 (subunits of $CoQH_2$–cytochrome c reductase complex)
Outer membrane	Mitochondrial porin (P70)

*Most proteins (except the ADP-ATP antiporter, cytochrome c, and porin) are fabricated as longer precursors.

▶ **FIGURE 17-2 Uptake-targeting sequences of imported mitochondrial proteins.** All such proteins have an N-terminal matrix-targeting sequence (red) that is similar but not identical in different proteins. Proteins destined for the inter-membrane space and outer membrane have an additional targeting sequence that functions to direct proteins to these locations by several different mechanisms described later. Except in the case of outer-membrane proteins, the targeting sequences are removed from the remainder of the polypeptide chain (blue wavy line), leaving the mature protein sequence. [After E. C. Hurt and A. P. G. M. van Loon, 1986, *Trends Biochem. Sci.* **11**:204, and F. U. Hartl et al., 1989, *Biochem. Biophys. Acta* **988**:1.]

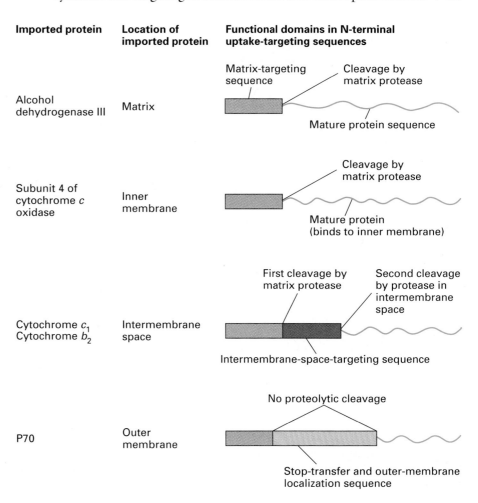

Imported protein	Location of imported protein	Functional domains in N-terminal uptake-targeting sequences
Alcohol dehydrogenase III	Matrix	
Subunit 4 of cytochrome *c* oxidase	Inner membrane	
Cytochrome *c*₁ Cytochrome *b*₂	Intermembrane space	
P70	Outer membrane	

proteins made in the cytosol can be incorporated into mitochondria. When these cells are treated with a radiolabeled amino acid (e.g., [^{35}S]methionine) for a few seconds (the "pulse" label), all newly made mitochondrial proteins are initially located in the cytosol outside the mitochondria. The incorporation of radioactivity is then blocked by incubating the cells with a drug that inhibits protein synthesis or with abundant unlabeled methionine (the "chase" period); subsequently, the radiolabeled proteins in the cytosol gradually accumulate at their proper destinations in the mitochondrion.

Furthermore, most proteins imported into the mitochondrion begin as precursors containing amino acids at the N-terminus that are not present in the mature protein. These N-terminal residues comprise one or more targeting sequences that direct the protein to its proper destination within the mitochondrion (Figure 17-2). Mitochondrial precursor proteins can be synthesized in cell-free systems in the absence of mitochondria. When mitochondria are subsequently added, the precursors are incorporated into the organelle and in most cases, the N-terminal uptake-targeting sequences are removed (Figure 17-3).

All proteins that travel from the cytosol to the same mitochondrial destination have targeting signals that share common motifs, although the signal sequences are generally not identical (see Table 17-1 and Figure 17-2). Thus, the receptors that recognize such sequences, including the mitochondrial outer-membrane receptors that bind matrix-targeting sequences, are able to bind to a number of different but related sequences. The matrix-targeting sequences of the various mitochondrial proteins are rich in positively charged amino acids—arginine and lysine—and hydroxylated ones—serine and threonine; they also are devoid of the acidic residues aspartate and glutamate. These sequences contain all the information required to target precursor proteins from the cytosol to the mitochondrial matrix.

Cytosolic Chaperones Deliver Proteins to Channel-Linked Receptors in the Mitochondrial Membrane

Figure 17-4 presents an overview of protein import from the cytosol into the mitochondrial matrix, the route into the mitochondrion followed by most imported proteins. We will discuss in detail each step in protein transport into the matrix and then consider how proteins are targeted to other compartments of the mitochondrion.

In the cytosol, the soluble precursors of mitochondrial proteins (including hydrophobic integral membrane proteins)

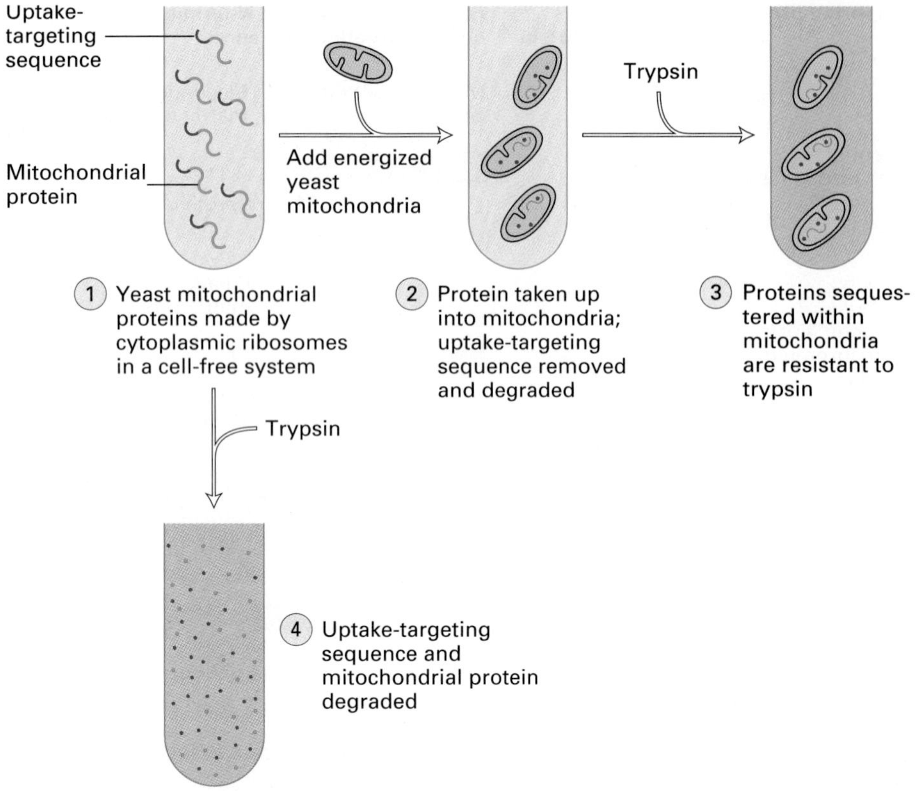

◀ **FIGURE 17-3 Cell-free system for studying post-translational uptake of mitochondrial proteins.** Precursor proteins with an N-terminal uptake-targeting sequence can be produced in an in vitro protein-synthesizing system from the corresponding mRNAs (step ①). Protein uptake occurs in the presence of respiring (energized) mitochondria, which have a proton-motive force across the inner membrane (step ②). Proteins sequestered in the mitochondrion are resistant to the protease (step ③) because it cannot penetrate the mitochondrial membranes. In contrast, precursor proteins in the cytosol are totally destroyed by the protease (step ④).

bind to one or more **chaperones**. These proteins use the energy released by ATP hydrolysis to keep nascent and newly made proteins in an unfolded state (see Figure 3-15). Two chaperones, *cytosolic Hsc70* and *mitochondrial-import stimulation factor (MSF)*, have been shown to prevent the misfolding or aggregation of mitochondrial precursor proteins so that they can be taken up by mitochondria. MSF is also able to disperse aggregates of proteins.

Some precursor proteins, such as the inner-membrane ATP/ADP antiporter, bind to MSF and the resulting complex then binds to a set of receptors called Tom37 and Tom70 on the outer membrane; Tom37 and Tom70 then transfer the precursor to a second set of receptors (Tom20 and Tom22) with release of MSF (Figure 17-4, steps 1a, 2, and 3a). Most precursor proteins bind to cytosolic Hsc70, which delivers the protein directly to the Tom20 and Tom22 receptors (steps 1b and 3b). These receptors are linked to Tom40, the actual channel in the outer membrane. When purified and incorporated into liposomes, Tom40 forms a transmembrane channel with a pore wide enough—about 1.5 to 2.5 nm in diameter—to accommodate an unfolded polypeptide chain.

A precursor protein destined for the mitochondrial matrix passes through the Tom40 channel in the outer membrane and another channel in the inner membrane (Figure 17-4, steps 4 and 5). Translocation into the matrix occurs only at "contact sites" where the outer and inner membranes are in close proximity and requires a proton-motive force across the inner membrane.

Matrix Chaperones and Chaperonins Are Essential for the Import and Folding of Mitochondrial Proteins

As proteins enter the mitochondrial matrix, they are bound by *matrix Hsc70,* a chaperone similar in structure and function to cytosolic Hsc70 (see Figure 3-15). Matrix Hsc70 is bound to the matrix surface of the inner mitochondrial membrane close to the channels traversed by imported proteins. Like cytosolic Hsc70, matrix Hsc70 binds the unfolded form of imported proteins as they emerge into the matrix, preventing protein aggregation or precipitation and premature folding (Figure 17-4, step 5). This is particularly important for subunits of multiprotein complexes, since the proper folding of any one subunit requires the presence of all of the protein's subunits. The energy released by Hsc70-catalyzed ATP hydrolysis also may help power translocation of proteins into the matrix.

Soon after a protein arrives in the mitochondrial matrix, a protease removes its N-terminal matrix-targeting sequence (Figure 17-4, step 6). During in vitro reactions, this matrix enzyme, a dimeric metal-containing protease, specifically cleaves the N-terminal matrix-targeting sequence from several different precursor proteins.

Some imported proteins can fold into their final, active conformation without further assistance (Figure 17-4, step 7a). Final folding of many matrix proteins, however, requires *Hsc60*, a matrix *chaperonin* (step 7b). Chaperonin proteins such as bacterial GroEL, to which Hsc60 is related, actively

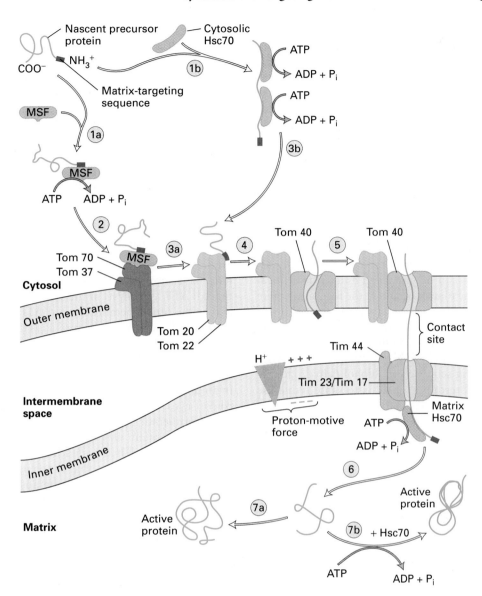

▲ **FIGURE 17-4 Protein import into the mitochondrial matrix.**
As a precursor protein, with its N-terminal matrix-targeting
sequence (red), emerges from cytosolic ribosomes, it binds to
chaperone proteins, such as mitochondrial-import stimulating
factor (MSF) and cytosolic Hsc70. These chaperones use energy
released by ATP hydrolysis to keep the bound precursor in an
unfolded or partially folded state. Steps ①a, ②, and ③a: MSF
binds to the matrix-targeting sequence in some precursors and
uses energy released by ATP hydrolysis to keep the precursors
unfolded. MSF then binds to a Tom37/Tom70 receptor on the
outer membrane. The bound precursor, in turn, is delivered to a
second receptor, Tom20/Tom22, which recognizes the matrix-
targeting sequence. Steps ①b and ③b: Other precursor proteins
bind to cytosolic Hsc70, which also uses energy released by ATP
hydrolysis to keep the precursors unfolded. These are delivered
directly to Tom20/Tom22 receptors. Step ④: Once a precursor
protein is bound to a Tom20/Tom22 receptor near a site of
contact with the inner membrane, it is transported across the
outer membrane through a transport channel composed of
Tom40 and three smaller but essential subunits (not depicted

here). Step ⑤: The precursor protein is then translocated across
the inner membrane through another transport channel composed
of several different Tim proteins. This process requires a
proton-motive force (pmf), a combination of a membrane electric
potential and a pH gradient, across the inner membrane. Note
that translocation occurs at rare "contact sites" at which the
inner and outer membranes appear to touch. The newly imported
protein binds to the matrix chaperone Hsc70, itself bound to
the Tim44 subunit of the inner-membrane transport channel.
Hsc70 uses the energy of ATP hydrolysis to assist import into
the matrix and to prevent aggregation or premature folding.
Step ⑥: After Hsc70 is released, the uptake-targeting sequence
is removed by a matrix protease. Step ⑦a: Within the matrix,
some proteins fold into their mature, active conformation
without the aid of a chaperone. Step ⑦b: Other proteins bind to
the chaperonin Hsc60, which assists in the final folding in a
process that requires energy derived from ATP hydrolysis.
[See K. R. Ryan and R. E. Jensen, 1995, *Cell* **83**:517; G. Schatz, 1996,
J. Biol. Chem. **271**:31763; and N. Pfanner et al., 1997, *Annu. Rev. Cell.
Devel. Biol.* **13**:25.]

facilitate protein folding (see Figure 3-15). Evidence for the importance of Hsc60 in folding of imported mitochondrial proteins comes from yeast mutants defective in this chaperonin. In these mutants, import of matrix proteins (e.g., the β subunit of the F_1 ATPase) and cleavage of the uptake-targeting sequence occur normally, but the imported polypeptide fails to assemble into a normal multiprotein complex.

Studies with Chimeric Proteins Confirm Major Features of Mitochondrial Import

Three properties characterize protein import into mitochondria: (1) all the information required to target a precursor protein from the cytosol to the mitochondrial matrix is contained within its N-terminal matrix-targeting sequence; (2) only unfolded proteins can be imported; and (3) translocation of precursors to the matrix occurs at the rare sites where the outer and inner membranes are close together.

Dramatic evidence for the ability of matrix-targeting sequences to direct import was obtained with chimeric proteins produced by recombinant DNA techniques. For example, the matrix-targeting sequence of alcohol dehydrogenase can be fused to the N-terminus of dihydrofolate reductase (DHFR), which normally resides in the cytosol (Figure 17-5, top). In the presence of chaperones, which prevent the C-terminal DHFR segment from folding in the cytosol, the chimeric protein is transported into the matrix of energized mitochondria, and the targeting sequence is cleaved normally (Figure 17-5a). Conversely, deleting or mutating its matrix-targeting sequence causes a protein normally transported to the matrix to remain in the cytosol.

Binding of the DHFR inhibitor methotrexate to the active site on the DHFR segment of the chimera causes it to become locked in a folded state. The finding that the chimeric protein does not cross into the matrix in the presence of methotrexate indicates that only unfolded proteins can enter the matrix. However, the N-terminal matrix-targeting sequence of the chimera still enters the matrix space, where it is cleaved, leaving the rest of the chimera stuck in the membrane as a stable *translocation intermediate* (Figure 17-5b). The chimeric protein in this experiment contains a spacer sequence of 50 amino acids between the targeting sequence and DHFR segment. In the translocation intermediate, the spacer sequence spans both membranes; it is long enough to do so only if it is in an extended conformation, stretched to its maximum possible length. If the chimera contains a shorter spacer—say 35 amino acids—no stable translocation intermediate is obtained, because the spacer can not traverse both membranes. These observations provide further evidence that translocated proteins traverse the membrane in an unfolded state.

Microscopic studies of stable translocation intermediates show that they accumulate at sites where the inner and outer mitochondrial membranes are close together, evidence that precursor proteins enter only at such sites (Figure 17-5c).

Outer-membrane receptors and other components of the mitochondrial protein-importing machinery are localized at or near these *contact sites*. Moreover, stable translocation intermediates can be chemically cross-linked to integral proteins of both the outer and inner membranes, strongly suggesting that imported proteins traverse both the outer and inner mitochondrial membranes in protein-lined channels as depicted in Figure 17-4. Since ≈ 1000 stuck translocation intermediates can be observed in a typical yeast mitochondrion, it is thought that mitochondria have ≈ 1000 transport channels for the uptake of mitochondrial proteins.

The Uptake of Mitochondrial Proteins Requires Energy

Studies on import of mitochondrial proteins in cell-free systems have shown that three separate inputs of energy are required: ATP hydrolysis in the cytosol, a proton-motive force across the inner membrane, and ATP hydrolysis in the mitochondrial matrix (see Figure 17-4).

As noted earlier, cytosolic chaperone proteins bound to precursor mitochondrial proteins couple the energy released by ATP hydrolysis to the maintenance of the bound proteins in an unfolded state. One study showing that this is the only role of cytosolic ATP in mitochondrial protein import used an experimental protocol similar to that in Figure 17-3. A precursor protein was purified and then denatured (unfolded) by urea. When added to a mixture of yeast cytosol and energized mitochondria, the denatured protein was incorporated into the matrix in the absence of cytosolic ATP. In contrast, import of the native, undenatured precursor required ATP for the normal unfolding function of cytosolic chaperones. In a second study, recombinant DNA techniques were used to construct mutant versions of certain matrix or inner-membrane proteins that were unable to fold normally into a functional conformation. These mutant proteins could be incorporated into energized mitochondria in the absence of cytosolic ATP, presumably because a chaperone was unnecessary to prevent the mutant proteins from folding.

Import of precursor proteins from the cytosol into the mitochondrial matrix always requires a proton-motive force across the inner membrane, but this electrochemical gradient is not required for binding of precursor proteins to receptors in the outer membrane. This can be demonstrated by "poisoning" mitochondria with inhibitors or uncouplers of oxidative phosphorylation such as cyanide or dinitrophenol, which dissipate the proton-motive force across the inner membrane. Although precursor proteins still can bind tightly to receptors on the poisoned mitochondria, the proteins can not be imported, either in intact cells or in cell-free systems, even in the presence of ATP and chaperone proteins.

Scientists still do not understand exactly how the membrane electric potential is used to "pull" a receptor-bound precursor protein into the matrix (see Figure 17-4, step 5). Once a protein is partially inserted into the inner membrane,

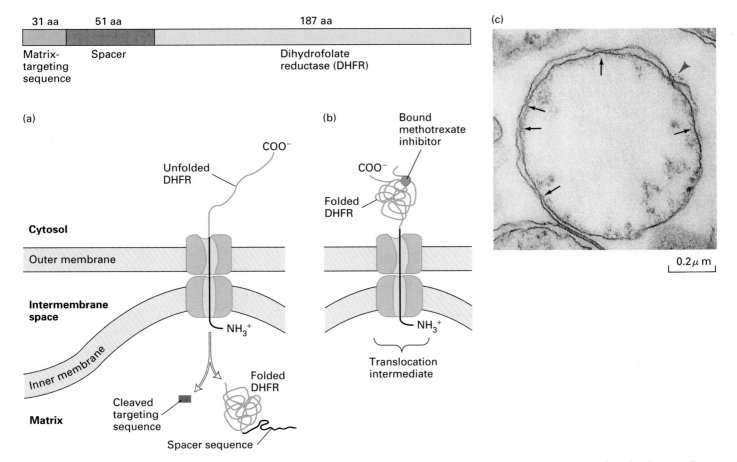

▲ FIGURE 17-5 Experiments with chimeric proteins elucidate mitochondrial protein import. *(Top)* The chimeric protein used contained at its N-terminus a matrix-targeting signal, followed by a spacer sequence of no particular function, and then by dihydrofolate reductase (DHFR), an enzyme normally present only in the cytosol. (a) After synthesis in a cell-free system, the chimeric protein translocates through the transport channels in the outer and inner membranes to the matrix of energized mitochondria. The matrix-targeting signal then is removed by a protease and the protein folds (see Figure 17-4). Thus the matrix-targeting sequence is sufficient to direct a normally cytosolic protein, DHFR, from the cytosol to the matrix. (b) Methotrexate prevents translocation of the DHFR segment at the C-terminus of the chimera by binding to it and locking it in the folded state. This finding indicates that a protein must be unfolded to be translocated. If the spacer sequence is long enough to extend across the transport channels, a stable translocation intermediate, with the targeting sequence cleaved off, is generated in the presence of methotrexate. When methotrexate is removed, the DHFR on the cytosolic surface unfolds sufficiently so that the entire chimera is translocated into the matrix space as shown in part (a). (c) The C-terminus of the translocation intermediate in (b) can be detected by incubating the mitochondria with antibodies that bind to the DHFR segment, followed by gold particles coated with bacterial protein A, which binds nonspecifically to antibody molecules (see Figure 5-17). An electron micrograph of a sectioned sample reveals gold particles (red arrowhead) bound to the translocation intermediate at a contact site between the inner and outer membranes. Other contacts sites (black arrows) also are evident. [Parts (a) and (b) adapted from J. Rassow et al., 1990, *FEBS Letters* **275**:190. Part (c) from M. Schweiger et al., 1987, *J. Cell Biol.* **105**:235; courtesy of W. Neupert.]

it is subjected to a transmembrane potential of 200 mV (matrix space negative), which is equivalent to an electric gradient of about 400,000 V/cm. One hypothesis is that this electric potential alters the conformation of the translocation intermediate, in much the same way that a membrane electric potential affects the conformation of voltage-dependent ion channels in nerve cells. The change in protein folding could pull the precursor protein across the energized inner membrane. A related possibility is that the N-terminal matrix-targeting sequence, with its many positively charged side chains, could be "electrophoresed," or pulled, into the matrix space by the inside-negative membrane electric potential.

ATP hydrolysis by chaperone proteins in the matrix also is required for mitochondrial protein import. These chaperones are essential for proper folding of imported proteins once they reach the matrix. In addition, the sequential binding and ATP-driven release of multiple matrix Hsc70 molecules to a precursor protein, as it emerges from the inner-membrane channel, may provide a force for pulling the un-folded protein into the matrix.

Proteins Are Targeted to Submitochondrial Compartments by Multiple Signals and Several Pathways

We turn now to targeting of proteins to submitochondrial compartments other than the matrix, namely the inter-membrane space, the inner membrane, and the outer mem-brane (see Table 17-2). Such targeting occurs via several pathways, and for many proteins requires a second signal sequence in addition to the matrix-targeting sequence.

Intermembrane-Space Proteins Precursors to such inter-membrane-space proteins as cytochrome c_1 and cytochrome b_2 (subunits of the $CoQH_2$–cytochrome c reductase complex) carry two different N-terminal uptake-targeting sequences that get them to the intermembrane space (see Figure 17-2). The most N-terminal of the two sequences directs the N-terminus of the precursor to the matrix, where this ma-trix-targeting sequence is removed by the matrix protease. What happens next depends on the protein, as depicted in Figure 17-6. Some, such as cytochrome c_1, follow a *conser-vative* sorting mechanism. In this case, the entire protein enters the matrix, and then the second targeting sequence directs the protein, presumably bound to matrix Hsc70,

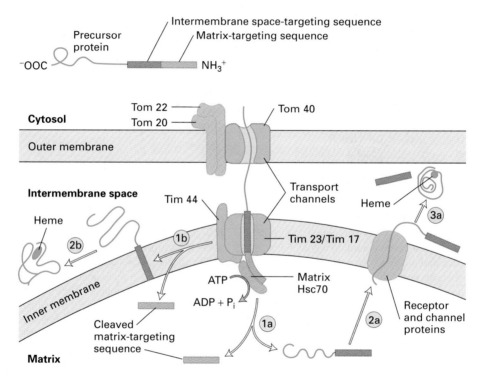

▲ **FIGURE 17-6 Two pathways by which different proteins are transported from the cytosol to the mitochondrial inter-membrane space.** In both pathways, the precursor protein contains two N-terminal targeting sequences *(top)*. It is delivered to Tom20/Tom22 receptors in the outer membrane and begins translocation across the transport channels as detailed in Figure 17-4. *Conservative pathway:* The entire precursor enters the matrix exactly as if it were a typical mitochondrial matrix protein, and the matrix-targeting sequence (red) is cleaved by the matrix protease (step ①a). The protein (e.g., cytochrome c_1) remains unfolded, presumably bound to matrix Hsc70 (not shown here). The intermembrane space-targeting sequence (brown) then targets the protein to the inner membrane by binding to a receptor (green) on the matrix side of the inner membrane, after which the protein is translocated across the inner membrane through an associated protein-lined transport channel into the intermembrane space (step ②a). (Neither the receptor nor the channel proteins have been characterized.) In the intermembrane

space, the targeting sequence is cleaved by a specific protease that is related to the ER signal peptidase, and heme is added, enabling the cytochrome to fold into its mature configuration (step ③a). *Nonconservative pathway:* The matrix-targeting sequence moves across both the outer and inner membranes, but the hydrophobic intermembrane space-targeting sequence becomes anchored in the inner membrane. This prevents translocation of the C-terminus of the protein (e.g., cytochrome b_2) through the inner membrane and apparently causes disassembly of the transport channel. The targeting sequence anchored in the inner membrane diffuses away from the translocation site (step ①b), as the rest of the protein traverses the outer membrane into the intermembrane space, and the matrix-targeting sequence is cleaved. Cleavage of the intermembrane space-targeting sequence by a specific protease releases the protein to which heme is added, followed by folding of the cytochrome into its mature conformation (step ②b). [See B. S. Glick et al., 1992, *Trends Biochem. Sci.* **17**:453, and H. Koll et al., 1992, *Cell* **68**:1163.]

across the inner membrane to the intermembrane space. Other proteins, such as cytochrome b_2, follow a *nonconservative* mechanism, in which the second uptake-targeting sequence functions as a **stop-transfer sequence** blocking translocation of the C-terminus of the precursor protein through the inner membrane. The membrane-anchored intermediate then diffuses, within the inner membrane, away from the translocation site. Finally, cleavage of the second uptake-targeting sequence by a protease in the intermembrane space releases the protein into that space. The importance of the intermembrane space–targeting sequence is evidenced by experiments with a precursor cytochrome b_2 protein that lacks only this sequence. This precursor, produced by recombinant DNA techniques, accumulates in the matrix space, not in the intermembrane space.

Unlike other mitochondrial proteins localized to the intermembrane space, cytochrome c is imported directly from the cytosol to the intermembrane space without the assistance of any targeting signals. The cytosolic form of cytochrome c, called *apocytochrome c*, has the same amino acid sequence as the mature protein, but lacks the covalently bound heme group found in the mature protein. Apocytochrome c, but not the mature *holo* form, can diffuse freely through the outer mitochondrial membrane, probably by traveling through P70. This porin-like protein (see Figure 3-35) forms channels through the phospholipid bilayer and accounts for the unusual permeability of the outer membrane to small proteins. Once apocytochrome c is in the intermembrane space of the mitochondrion, the heme group is added by the enzyme cytochrome c heme lyase. The addition of heme causes a conformational change in the protein, so that it cannot diffuse back through the outer-membrane channels, thus "locking" it into the intermembrane space.

Outer- and Inner-Membrane Proteins Experiments with P70, a well-studied outer-membrane protein, provide clues about how proteins are targeted to the outer mitochondrial membrane. At the N-terminus of P70 there is a short matrix-targeting sequence followed by a long stretch of hydrophobic amino acids (see Figure 17-2). If the hydrophobic sequence is experimentally deleted from P70, the protein accumulates in the matrix space with its matrix-targeting sequence still attached. This finding suggests that the long hydrophobic sequence functions as a stop-transfer sequence that both prevents transfer of the protein into the matrix and anchors it as an integral protein in the outer membrane. Normally, neither the matrix-targeting nor stop-transfer sequence is cleaved from the anchored protein.

Inner-membrane proteins, in contrast, are first imported into the matrix space where the matrix-targeting sequence is removed. How these proteins then are incorporated into the inner membrane as part of multiprotein complexes is not yet known, but it probably follows the same pathway by which mtDNA-encoded inner membrane proteins are incorporated into this membrane.

The Synthesis of Mitochondrial Proteins Is Coordinated

Several proteins critical to mitochondrial function (e.g., cytochrome c oxidase and the F_0F_1 ATPase) are multienzyme complexes composed of some subunits synthesized in the cytosol and others synthesized in the mitochondrion. Since all of the subunits must be fabricated in appropriate ratios, the close coordination of nuclear and mitochondrial genome expression is required for the assembly of a mitochondrion. As yet, little is known about how the expression of these two genomes is coordinated in animals or plants. Studies with petite yeast strains, which lack some or all functional mtDNA (see Figure 9-43), have shown that they contain normal amounts of all nuclear-encoded mitochondrial proteins, such as cytochrome c and the F_1 ATPase. Thus in yeasts, mitochondrial gene products are not essential for the expression of nuclear genes encoding mitochondrial proteins. Since mtDNA cannot be readily deleted in other organisms, it is impossible to say whether this result is true in all cells.

Yeasts and other eukaryotic microorganisms that can grow anaerobically or aerobically—in the absence or presence of O_2, respectively—provide a striking example of the coordination of nuclear and mitochondrial gene expression. When grown anaerobically with glucose as a carbon source, these organisms generate ATP solely by the Embden-Meyerhof pathway (see Figures 16-3 and 16-6). When viewed under an electron microscope, anaerobically grown cells lack typical mitochondria, although they do contain some small mitochondria that have inner and outer membranes but no cristae. These mitochondria contain sufficient amounts of imported mitochondrial DNA polymerase to replicate the mtDNA normally. However, these organelles lack heme as well as cytochromes a, a_3, b, c, and c_1 and the F_0F_1 ATPase complex.

Addition of oxygen to anaerobic yeast directly activates heme synthesis. In turn, heme (not O_2 itself) binds to and activates transcriptional regulatory proteins that enhance transcription of the genes for cytochrome c and many other nuclear-encoded mitochondrial proteins. Since the synthesis of heme requires oxygen, this process may be representative of a general mechanism for down-regulating expression of multiple nuclear genes that encode mitochondrial proteins during anaerobic fermentation.

Several Uptake-Targeting Sequences Direct Proteins Synthesized in the Cytosol to the Appropriate Chloroplast Compartment

The import of proteins from the cytosol to chloroplasts shares several characteristics with mitochondrial import. In both processes, imported proteins are synthesized as cytosolic precursors containing N-terminal uptake-targeting sequences that direct each protein to its correct subcompartment and are subsequently cleaved (see Table 17-1 and Figure 17-1). Protein import from the cytosol into the chloroplast stroma (equivalent to the mitochondrial matrix) occurs, as in mitochondria, at points where the outer and inner organelle membranes are in close contact. Finally, protein import into both organelles requires energy. Here, we briefly discuss targeting of proteins to the chloroplast stroma and thylakoids. We will see that despite the similarities just noted, the mechanisms of chloroplast and mitochondrial protein import differ in various ways.

Targeting to the Chloroplast Stromal Space Among the proteins found in the chloroplast stroma are the enzymes of the Calvin cycle (see Figure 16-49). The large (L) subunit of ribulose 1,5-bisphosphate carboxylase (rubisco) is encoded by chloroplast DNA and synthesized on chloroplast ribosomes in the stromal space. The small (S) subunit of rubisco and all the other Calvin-cycle enzymes are encoded by nuclear genes and transported to chloroplasts after their synthesis in the cytosol. Rubisco, the most abundant protein in chloroplasts, is composed of eight identical L subunits, which contain the catalytic sites, and eight identical S subunits, which are necessary for full enzyme activity. This key enzyme provides an excellent example of how proteins are

imported into the chloroplast stroma and assembled there into a multisubunit protein.

The S subunit of rubisco is synthesized on free cytosolic polyribosomes in a precursor form that has an N-terminal stromal-import sequence of about 44 amino acids. It is maintained in an unfolded state by binding to cytosolic chaperones. Experiments with isolated chloroplasts, similar to those with mitochondria illustrated in Figure 17-3, have shown that they can import the S-subunit precursor after its synthesis. After the unfolded precursor enters the stromal space, it binds transiently to a stromal Hsc70 chaperone, and the N-terminal sequence is cleaved. In reactions facilitated by Hsc60 chaperonins, eight S subunits combine with the eight L subunits to yield the active rubisco enzyme.

At least three chloroplast outer-membrane proteins, including a receptor that binds the stromal-targeting sequence and a transport channel protein, and five inner-membrane proteins are known to be essential for the import process (Figure 17-7). Perhaps surprisingly, these proteins are not homologous to those of the receptor or channel proteins in

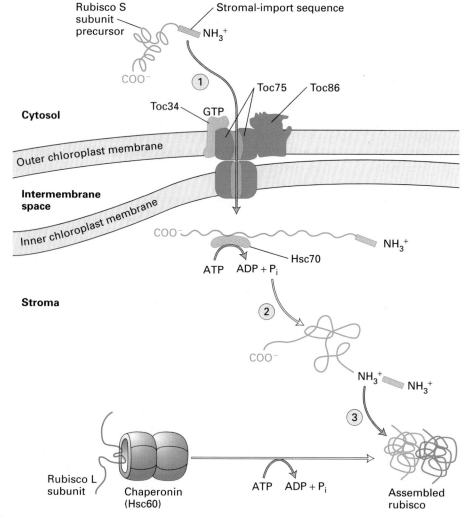

◀ **FIGURE 17-7 Import of rubisco small (S) subunits into the chloroplast stroma and the assembly of small and large (L) subunits into the active rubisco enzyme.** Step ①: After an S-subunit precursor is fabricated in the cytosol, it presumably binds to cytosolic chaperones (not depicted here) that keep it in an unfolded state. A set of receptor and channel proteins located at the point of contact between the outer and inner chloroplast membranes mediates translocation from the cytosol to the stroma. The Toc86 receptor binds the stromal-import sequence, and the precursor then passes through an associated transport channel, composed mainly of the Toc75 protein, into the intermembrane space. Binding of GTP by Toc34 probably causes a conformational change that affects the gating properties of the Toc75 channel. A number of inner-membrane proteins then facilitate entry into the stroma. As the precursor enters the stroma, it transiently binds to an Hsc70 chaperone. Several other proteins that participate in translocation are not shown. Step ②: Release of Hsc70 from the precursor yields the partially folded S subunit from which the stromal-import sequence is cleaved. Step ③: L subunits are synthesized in the stroma by chloroplast ribosomes and are stored complexed to Hsc60 chaperonins. Eight L subunits, released by the chaperonins, bind to eight imported S subunits, forming the mature rubisco enzyme. [See L. Heins et al., 1998, *Trends Plant Sci.* **3**:56.]

the mitochondrial membrane, although their functions are analogous. The nonhomology of these chloroplast and mitochondrial proteins suggests that they may have evolved independently during evolution.

The available evidence suggests that proteins are imported in the unfolded state and that import into the stroma depends on ATP hydrolysis catalyzed by stromal chaperones, whose function is similar to Hsc70 in the mitochondrial matrix. Unlike mitochondria, however, chloroplasts cannot generate an electrochemical gradient (proton-motive force) across their inner membrane. Thus protein import into the chloroplast stroma appears to be powered solely by ATP hydrolysis.

Targeting to the Thylakoids Proteins targeted to the thylakoid membrane or lumen not only must traverse both the outer and inner chloroplast membranes to enter the stroma, but then must travel through the stroma and either be inserted into the thylakoid membrane or cross that membrane and enter the thylakoid lumen. Proteins that are destined for the thylakoid lumen, such as plastocyanin, require the successive action of two uptake-targeting sequences. The first, like that in the rubisco S subunit, targets the protein to the stroma; the second targets the protein from the stroma to the thylakoid lumen. The role of these targeting sequences has been shown in cell-free experiments measuring the uptake into chloroplasts of mutant proteins generated by recombinant DNA techniques. For instance, when the thylakoid uptake-targeting sequence is deleted from plastocyanin, the mutant protein accumulates in the stroma and is not transported into the thylakoid lumen.

Four separate thylakoid-import systems, each transporting a different set of proteins from the stroma into the thylakoid lumen, have been identified. Two of these are illustrated in Figure 17-8. The system that transports proteins such as plastocyanin employs a translocation mechanism similar to the one that translocates unfolded proteins into the endoplasmic reticulum (see later discussion). This system functions even in the absence of a pH gradient across the thylakoid membrane. Other thylakoid proteins, including a 23-kDa protein of photosystem II, are folded in the stroma and bind metal-containing redox cofactors there. In this system, the thylakoid-membrane protein Hef106 assists in translocating such folded proteins and their bound cofactors into the thylakoid lumen, and uptake is powered by the pH gradient normally maintained across the thylakoid membrane. The molecular mechanism whereby these large folded globular proteins are transported across the thylakoid membrane is currently under intense study.

In many respects, movement of proteins from the stroma to the thylakoid lumen resembles secretion of bacterial proteins across the bacterial plasma membrane, a similarity consistent with the evolution of chloroplasts from ancestral photosynthetic bacteria. For instance, thylakoid vesicles form by inward budding from the inner membrane of protoplastids, small organelles composed only of an inner and outer membrane and a small stromal space that contains chloroplast DNA (Figure 17-9). Thus the inner proplastid membrane corresponds to the plasma membrane of the ancestral bacterium, and the chloroplast stroma corresponds to the bacterial cytoplasm. Moreover, the plasma membrane of many bacterial cells contains three protein-secreting systems that are homologs of chloroplast systems for transporting proteins from the stroma to thylakoid membranes. Secretion of most bacterial proteins occurs post-translationally and requires binding of the precursor proteins to cytosolic chaperones, much as in the post-translational uptake of unfolded proteins such as plastocyanin into the thylakoid from the stroma. Additionally, the thylakoid-targeting sequences of plastocyanin and certain other thylakoid proteins resemble the signal sequences that target bacterial proteins to cross the bacterial plasma membrane. Another bacterial system uses the pH gradient normally found across bacterial membranes to transport certain folded globular proteins from the cytosol to the cell exterior, similar to the pH-dependent uptake of folded globular proteins from the stroma into the thylakoid lumen. A final similarity occurs in the peptidases that remove certain targeting sequences. In both bacteria and chloroplasts, the sites at which these peptidases bind and cleave a target protein have a small amino acid such as glycine or alanine just N-terminal to the cleavage site and also at the position two residues toward the N-terminus.

SUMMARY Synthesis and Targeting of Mitochondrial and Chloroplast Proteins

- Both the mitochondrion and chloroplast contain organelle DNA, which encodes organelle rRNAs and tRNAs but only a few organelle proteins.

- The vast majority of mitochondrial and chloroplast proteins are encoded by nuclear genes, synthesized on cytosolic ribosomes, and imported post-translationally into the organelles.

- Proteins destined for the mitochondrial matrix or chloroplast stroma have organelle-specific N-terminal uptake-targeting sequences that direct their entry into the organelle. After protein import, the targeting sequence is removed by proteases within the matrix or stroma.

- Protein import into both mitochondria and chloroplasts occurs only at sites where the inner and outer organellar membranes are in close contact.

- Cytosolic Hsc70 and mitochondrial import stimulation factor (MSF) are chaperones that bind the cytosolic precursors of mitochondrial proteins, keeping them in a partially unfolded form that can be translocated into mitochondria. After the unfolded proteins bind to receptors on the outer mitochondrial membrane, they are translocated through a protein-lined channel into the organelle by a process dependent on the proton-motive

▶ **FIGURE 17-8 Two of the four known pathways for transporting proteins from the cytosol to the thylakoid lumen.** All such proteins are synthesized as precursors in the cytosol with two uptake-targeting sequences at their N-terminus. Shown here are the pathways for targeting proteins such as plastocyanin (a) and metal-binding proteins such as a 23-kDa protein from photosystem II (b). Step ①: The ≈30-residue N-terminal stromal-import sequence (red) directs the precursor across the outer and inner membranes into the chloroplast stroma via the set of receptor and channel proteins depicted in Figure 17-7. A stromal signal protease removes the stromal-import sequence, leaving the thylakoid-targeting sequence at the N-terminus of the precursor. At this point the two pathways diverge. Steps ②ⓐ and ③ⓐ: Plastocyanin and similar proteins are kept unfolded in the stromal space by a set of chaperones that are not shown. The ≈25-residue thylakoid signal peptide (brown) has a core of hydrophobic amino acids. It directs the unfolded stromal precursor into the thylakoid lumen by binding to specific receptor and channel-transport proteins on the thylakoid membrane. Translocation does not require a pH gradient across the thylakoid membrane. After the thylakoid signal peptide is removed in the thylakoid lumen by a separate endoprotease, the protein folds into its mature conformation. Steps ②ⓑ–④ⓑ: Metal-binding proteins fold in the stroma, and complex redox cofactors such as FeS and molybdenum-containing pterins are added. These proteins contain an ≈25-residue thylakoid transfer peptide (yellow) that has two arginine residues (RR) at the N-terminus followed by a core of hydrophobic amino acids. Translocation of these large globular proteins into the thylakoid lumen is powered by the pH gradient maintained across the thylakoid membrane. Removal of the thylakoid-targeting sequence in the lumen leaves the mature, folded protein. [See R. Dalbey and C. Robinson, 1999, *Trends Biochem. Sci.* **24**:17.]

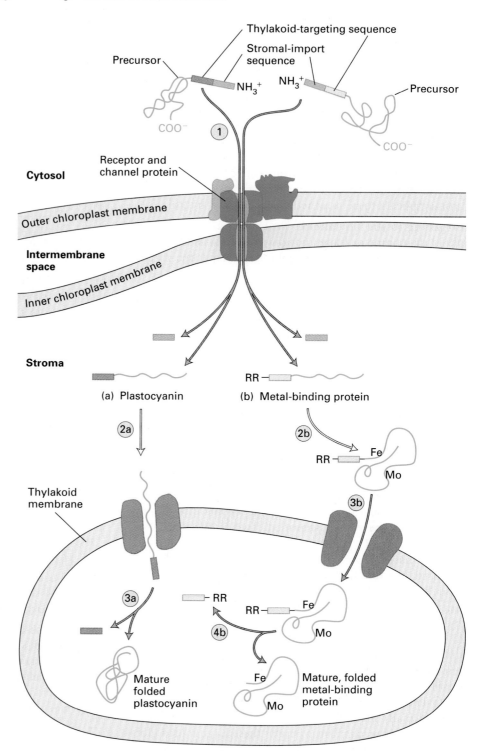

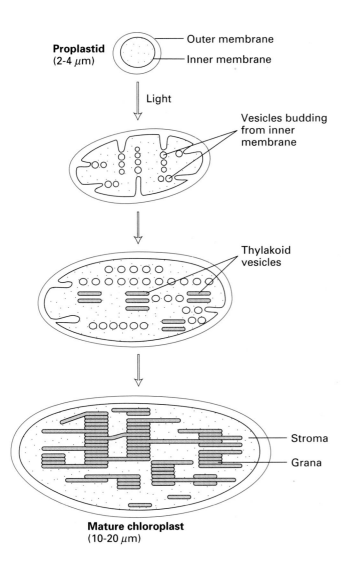

Proplastid (2-4 μm)
— Outer membrane
— Inner membrane

Light

Vesicles budding from inner membrane

Thylakoid vesicles

Stroma

Grana

Mature chloroplast (10-20 μm)

◄ **FIGURE 17-9 Formation of chloroplasts from proplastids begins by the light-induced budding of the inner membrane.** This membrane is equivalent to the plasma membrane of the ancestral photosynthetic bacterium thought to have been incorporated into an early eukaryotic cell. The proplastid in dark-adapted cells contains only the outer and inner chloroplast membranes. Light triggers the synthesis of chlorophyll, phospholipids, and chloroplast stroma and thylakoid proteins, and the budding of small vesicles from the inner chloroplast membrane. As the proplastid enlarges, some of the spherical vesicles fuse, eventually forming one continuous set of flattened thylakoid vesicles, some of which stack into grana. Thus, in the mature chloroplast, the stroma is equivalent to the bacterial cytoplasm; the thylakoid membrane, to the bacterial plasma membrane; and the thylakoid lumen, to the exterior of the bacterial cell. [See D. von Wettstein, 1959, *J. Ultrastruct. Res.* **3**:235.]

17.2 Synthesis and Targeting of Peroxisomal Proteins

Peroxisomes are small organelles, approximately 0.2–1 μm in diameter (see Figure 5-25c). Unlike mitochondria and chloroplasts, peroxisomes lack DNA and ribosomes and are lined by a single membrane. Thus all peroxisomal proteins are encoded by nuclear genes, synthesized on ribosomes free in the cytosol, and then incorporated into pre-existing peroxisomes. As peroxisomes are enlarged by addition of protein (and lipid), they eventually divide, forming new ones, similar to mitochondria and chloroplasts.

The size and enzyme composition of peroxisomes vary considerably, but all contain enzymes that use molecular oxygen to oxidize various substrates, forming hydrogen peroxide (H_2O_2). Catalase, a peroxisome-localized enzyme, efficiently decomposes H_2O_2 into H_2O. Peroxisomes are most abundant in liver cells, where they constitute about one to two percent of the cell volume.

C- and N-Terminal Targeting Sequences Direct Entry of Folded Proteins into the Peroxisomal Matrix

After being released from cytosolic ribosomes, newly synthesized peroxisomal proteins, unlike mitochondrial and chloroplast proteins, generally fold into their mature conformation in the cytosol before import into the organelle. The import of catalase and other proteins into rat liver peroxisomes can be studied in a cell-free system similar to that used for studying mitochondrial protein import (see Figure 17-3). Protein import into peroxisomes requires ATP hydrolysis, but there is no electrochemical gradient across the peroxisomal membrane.

The uptake-targeting signal in catalase is a Ser-Lys-Leu sequence (SKL in one-letter code) or a related sequence at the very C-terminus. After catalase monomers associate with heme and assemble into the mature tetrameric protein, the

force across the inner membrane (see Figure 17-4). Final folding of many imported proteins is facilitated by an ATP-hydrolyzing Hsc60 chaperonin.

- The N-terminus of proteins imported to mitochondrial destinations other than the matrix usually contains a second targeting sequence (see Figure 17-2). Some proteins destined for the intermembrane space or inner membrane are first imported into the matrix and then redirected; others never enter the matrix but go directly to their final location (see Figure 17-6).

- Import of precursor proteins into the chloroplast stroma involves receptors in the outer membrane that recognize stromal-targeting sequences, channel proteins in the outer and inner membranes, and stromal chaperones (see Figure 17-7). Import requires ATP hydrolysis but not a proton-motive force. After their entry into the stroma, precursors with a second targeting sequence are redirected to the proper subcompartment, such as the thylakoid membrane or thylakoid lumen.

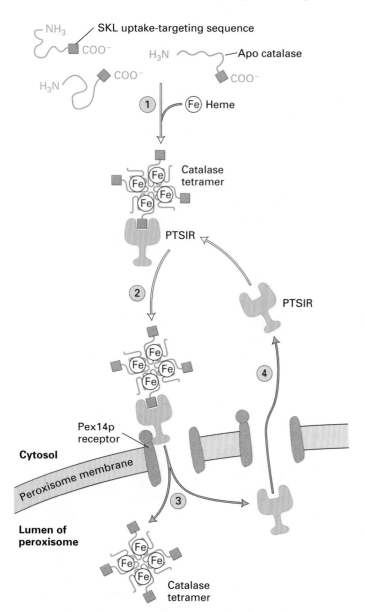

▲ FIGURE 17-10 Synthesis of catalase and its incorporation into peroxisomes. Newly made catalase subunits are released from polyribosomes into the cytosol as *apocatalase*, a monomer that contains a C-terminal SKL uptake-targeting sequence (red) but lacks an iron-containing heme group. Step ①: Four monomers are assembled and heme is added, forming the mature tetrameric catalase molecule. Steps ② and ③: The cytosolic receptor protein PTS1R binds an SKL signal and escorts the catalase tetramer to the Pex14p receptor on the peroxisome membrane. In the model depicted here, the catalase-PTS1R complex is transported across the membrane and then dissociates in the lumen. As-yet uncharacterized proteins on the peroxisomal membrane probably form part of the receptor-transport channel. Step ④: PTS1R returns to the cytosol to pick up another peroxisome-destined protein. In an alternative model, PTS1R delivers catalase to the Pex14p receptor but does not enter the lumen itself. [See R. A. Ruchubinski and S. Subramani, 1995, *Cell* **83**:525; J. A. McNew and J. M. Goodman, 1996, *Trends Biochem. Sci.* **21**:54; and M. Albertini et al., 1997, *Cell* **89**:83.]

SKL signal binds to a soluble receptor protein in the cytosol named PTS1R (Figure 17-10). The resulting PTS1R-catalase complex then binds to a receptor protein, Pex14p, in the peroxisome membrane, following which catalase is transported inwards. The SKL sequence is not cleaved from catalase after its entry into a peroxisome.

Besides catalase, other peroxisomal proteins with a SKL signal include fatty acyl CoA oxidase in rat liver, urate oxidase in cucumber, and firefly luciferase (the enzyme that generates the flashes of light). The model in Figure 17-10 is thought to apply to all these proteins. When luciferase is expressed (by recombinant DNA techniques) in cultured mammalian or plant cells, it is efficiently targeted to peroxisomes, attesting to the universality of peroxisome-targeting mechanisms. Mutant luciferase lacking the SKL sequence remains in the cytosol, and addition of the SKL sequence to the C-terminus of a protein that normally remains in the cytosol leads to uptake of the altered protein by peroxisomes in cultured monkey cells. These findings show that the SKL sequence is both necessary and sufficient for protein uptake into the peroxisomal matrix.

Other peroxisomal matrix proteins are synthesized as precursors with an N-terminal uptake-targeting sequence; in thiolase, this sequence contains 26 amino acids. Proteins with this type of uptake-targeting signal bind to a cytosolic receptor protein named PTS2R that, like PTS1R, escorts the precursor protein to the Pex14p receptor on the peroxisomal membrane. Following import of such proteins, the N-terminal targeting sequence is cleaved. Peroxisomal membrane proteins are also synthesized on free polyribosomes and incorporated into peroxisomes after their synthesis. The signals that target proteins to the peroxisomal membrane do not contain an SKL sequence, but little else is known about this uptake process.

Peroxisomal Protein Import Is Defective in Some Genetic Diseases

Autosomal recessive mutations that cause defective peroxisome assembly occur naturally in the human population. Such defects can lead to severe impairment of many organs and death. In *Zellweger syndrome* and related disorders, for example, the transport of many or all proteins into the peroxisomal matrix is impaired; newly synthesized peroxisomal enzymes remain in the cytosol and are eventually degraded. A remarkable feature of Zellweger syndrome is that cells contain empty peroxisomes that have a normal complement of peroxisomal membrane proteins. These findings demonstrate that peroxisomes from patients with Zellweger syndrome are defective in the uptake of matrix proteins, but not peroxisomal membrane proteins, from the cytosol. Thus, Zellweger mutations cause defects in a peroxisomal receptor or transport protein for peroxisomal matrix proteins but not membrane proteins.

A genetic study of cultured cells from different Zellweger patients has shown that mutations in any of eight different genes can cause this phenotype, so at least eight proteins are involved in the uptake of proteins into the peroxisomal matrix. In this study, cultured fibroblasts from different patients were fused together. In many cases normal peroxisome function was restored—that is, complementation occurred (Section 8.2). This finding indicates that the cells from different patients were defective in different proteins required for the uptake of peroxisomal proteins. Similar kinds of peroxisome-assembly mutants have also been generated in yeast.

Cloning of the wild-type allele of the mutant gene from one Zellweger patient revealed that it encodes PTS1R, the receptor essential for the post-translational peroxisomal uptake of proteins with an SKL sequence. Other Zellweger cells are deficient in uptake of proteins normally escorted by both PTS1R and PTS2R; these cells may be defective in receptor or transporter proteins in the peroxisomal membrane. Eventually, cloning of the wild-type alleles of the various Zellweger mutations and of the similar yeast mutations should elucidate the identity of all proteins involved in the transport of proteins into the peroxisome matrix.

As noted in Section 16.1, the peroxisomal oxidation of very-long-chain fatty acids is defective in *X-linked adrenoleukodystrophy (ALD)*, another genetic disease that affects peroxisome functioning. Peroxisomes from ALD patients lack long-chain fatty acyl CoA synthase, the matrix enzyme that normally links coenzyme A to very-long-chain fatty acids within the peroxisome. The protein encoded by the *ALD* gene has a structure similar to that of the CFTR (cystic fibrosis transmembrane regulator) membrane transport protein and of multidrug resistance proteins (see Figure 15-16). Apparently the ALD protein is the peroxisomal membrane transporter specific for uptake of long-chain fatty acyl CoA synthase from the cytosol.

SUMMARY Synthesis and Targeting of Peroxisomal Proteins

- All peroxisomal membrane and matrix proteins are incorporated into the organelle post-translationally; most, such as catalase, fold in the cytosol and are incorporated as a folded protein.

- Many peroxisomal matrix proteins utilize a C-terminal SKL targeting sequence that is not cleaved after import. Such proteins bind to the cytosolic receptor PTS1R, which escorts an imported protein to receptor and transporter proteins on the peroxisomal membrane, following which the protein is transported inwards (see Figure 17-14).

- Patients with Zellweger syndrome are defective in the import of most or all proteins into the peroxisomal matrix. Genetic analysis of patients' cells indicated that at least eight proteins are involved in the import process, including PTS1R.

17.3 Overview of the Secretory Pathway

We turn our attention now to the very large class of proteins that are synthesized and sorted in the secretory pathway (see Figure 17-1). Once the ribosomes synthesizing these proteins become bound to the rough ER, the proteins enter or cross the ER membrane *cotranslationally*—that is, during their synthesis. Soluble proteins in this class first are localized in the ER lumen and subsequently are sorted to the lumen of other organelles or are secreted from the cell. Likewise, the integral membrane proteins in this class initially are inserted into the rough ER membrane during their synthesis; some remain there, but many eventually become localized to the plasma membrane or membranes of the smooth ER, Golgi complex, lysosomes, or endosomes.

The rough ER is an extensive interconnected series of flattened sacs, generally lying in layers (Figure 17-11). When cells are homogenized, the rough ER breaks up into small closed vesicles, termed *rough microsomes*, with the same orientation (ribosomes on the outside) as that found in the intact cell. The simple experiment outlined in Figure 17-12 shows that immediately after their synthesis secretory proteins are localized in the lumen of ER vesicles, although they have been synthesized on ribosomes bound to the cytosolic face of the ER membrane.

As already noted, all of the proteins that enter the secretory pathway contain an ER signal sequence, generally at the N-terminus (see Table 17-1). This sequence directs the ribosomes that are synthesizing these proteins to the rough ER. Membrane-bound ribosomes and ribosomes free in the cytosol can be separated from other cellular constituents and from each other by a combination of differential and sucrose

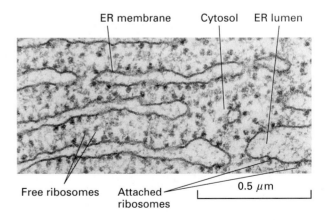

▲ **FIGURE 17-11 Electron micrograph of ribosomes attached to the rough ER in a pancreatic exocrine cell.** Most of the proteins synthesized by this cell are to be secreted and are formed on membrane-attached ribosomes. A few membrane-unattached (free) ribosomes are evident; presumably, these are synthesizing cytosolic or other nonsecretory proteins. [Courtesy of G. Palade.]

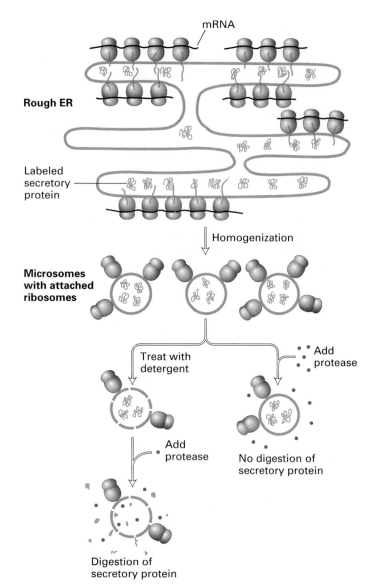

▲ **FIGURE 17-12 Experimental demonstration of location of secretory proteins just after synthesis.** Cells are incubated with radiolabeled amino acids and then are homogenized, which fractures the plasma membrane and shears the rough ER into small vesicles (microsomes). The microsomes are purified and treated with a protease in the presence and absence of a detergent. The newly synthesized labeled secretory proteins associated with the microsomes are digested by added proteases only if the permeability barrier of the microsomal membrane is destroyed by detergent. Thus, the newly made proteins are inside the microsomes, equivalent to the lumen of the rough ER.

density-gradient centrifugation (see Figures 5-23 and 5-24). Due to the low buoyant density of phospholipids, membrane-bound ribosomes "band" at a lighter density than do free ribosomes. Biochemical analyses of purified membrane-bound and free ribosomes show that they contain exactly the same proteins and ribosomal RNAs and are functionally indistinguishable. These findings are consistent with the notion that all information for intracellular protein distribution is located in the amino acid sequence of the newly synthesized protein itself.

Many important experiments on the secretory pathway take advantage of cells that are specialized for the secretion of specific proteins (Table 17-3). These cells contain organelles such as the rough ER and Golgi **cisternae** in abundance. For example, of the total protein made by hepatocytes (the principal cells of the liver), about 70 percent consists of proteins, such as albumin and transferrin, that are secreted into the blood. Likewise, pancreatic acinar cells synthesize several digestive enzymes that are packaged into zymogen vesicles and secreted into ductules that lead to the intestine (see Figure 5-48). All cells, however, secrete *some* proteins. Extracellular matrix proteins such as collagens, proteoglycans, and fibronectin, for example, constitute about 5 percent of the protein made by most cultured cells. All eukaryotic cells use essentially the same pathway for synthesis and sorting of secretory proteins.

Secretory Proteins Move from the Rough ER Lumen through the Golgi Complex and Then to the Cell Surface

Figure 17-13 outlines the movement of proteins within the secretory pathway. Most newly made proteins in the ER lumen or membrane are incorporated into small, ≈50-nm-diameter **transport vesicles.** These either fuse with the *cis*-Golgi or with each other to form the membrane stacks known as the *cis*-Golgi reticulum (network). From the *cis*-Golgi certain proteins, mainly ER-localized proteins, are retrieved to the ER via a different set of *retrograde* transport vesicles. In the process called cisternal migration, or **cisternal progression,** a new *cis*-Golgi stack with its cargo of luminal protein physically moves from the *cis* position (nearest the ER) to the *trans* position (farthest from the ER), successively becoming first a *medial*-Golgi cisterna and then a *trans*-Golgi cisterna. As this happens, membrane and luminal proteins are constantly being retrieved from later to earlier Golgi cisternae by small retrograde transport vesicles. By this process enzymes and other Golgi resident proteins come to be localized either in the *cis*- or *medial*- or *trans*-Golgi cisternae.

Proteins destined to be secreted move by cisternal migration to the *trans* face of the Golgi and then into a complex network of vesicles termed the *trans*-Golgi reticulum. From there a secretory protein is sorted into one of two types of vesicles. In all cell types, at least some of the secretory proteins are secreted continuously. Examples of such **constitutive** (or continuous) **secretion** include collagen secretion by fibroblasts and secretion of serum proteins by hepatocytes (see Table 17-3). These proteins are sorted in the *trans*-Golgi network into transport vesicles that immediately move to and fuse with the plasma membrane, releasing their contents by **exocytosis.**

In certain cells, the secretion of a specific set of proteins is not continuous; these proteins are sorted in the *trans*-Golgi

MEDIA CONNECTIONS
Protein Secretion

▶ **FIGURE 17-13 The secretory pathway of protein synthesis and sorting.** Ribosomes synthesizing proteins bearing an ER signal sequence become bound to the rough ER. As translation is completed on the ER, the polypeptide chains are inserted into the ER membrane or cross it into the lumen. Some proteins (e.g., rough ER enzymes or structural proteins) remain resident in the ER. The remainder move into transport vesicles that fuse together to form new *cis*-Golgi vesicles. Each *cis*-Golgi cisterna, with its protein content, physically moves from the *cis* to the *trans* face of the Golgi stack (red arrows). As this cisternal progression occurs, many luminal and membrane proteins undergo modifications, primarily to attached oligosaccharide chains. Some proteins remain in the *trans*-Golgi cisternae, while others move via small vesicles to the cell surface or to lysosomes. In certain cell types (e.g., nerve cells and pancreatic acinar cells), some soluble proteins are stored in secretory vesicles and are released only after the cell receives an appropriate neural or hormonal signal (regulated secretion). In all cells, certain proteins move to the cell surface in transport vesicles and are secreted continuously (constitutive secretion). Like soluble proteins, integral membrane proteins move via transport vesicles from the rough ER to the *cis*-Golgi and then on to their final destinations. The orientation of a membrane protein, established when it is inserted into the ER membrane, is retained during all the sorting steps: Some segments always face the cytosol; others always face the exoplasmic space (i.e., the lumen of the ER, Golgi cisternae, and vesicles or the cell exterior). Retrograde movement via small transport vesicles retrieves ER proteins that migrate to the *cis*-Golgi and returns them to the ER. Similarly, *cis*- or *medial*-Golgi proteins that migrate to a later compartment are retrieved by small retrograde transport vesicles. [See B. Glick and V. Malhotra, 1988, *Cell* **95**:883.]

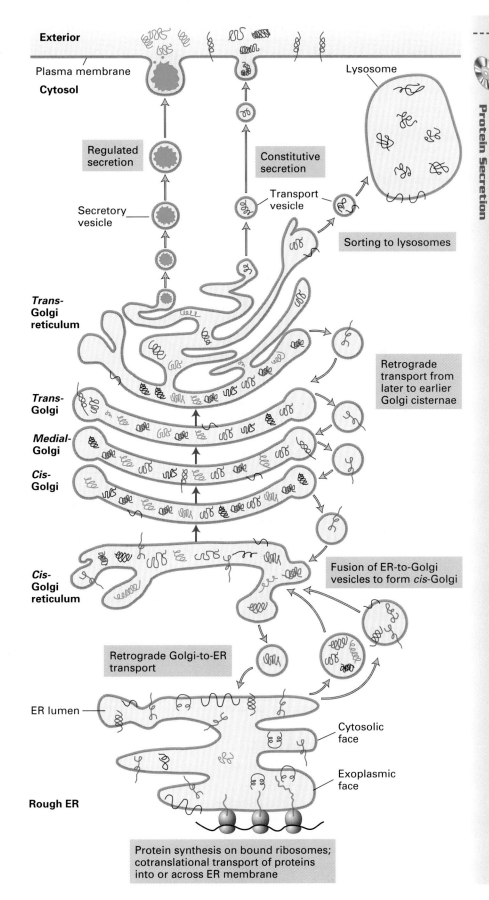

TABLE 17-3	Classes of Secretory Proteins in Vertebrates	
Protein Type	Example	Site of Synthesis
Constitutive Secretory Proteins		
Serum proteins	Albumin	Liver (hepatocyte)
	Transferrin (Fe transporter)	Liver
	Lipoproteins	Liver, intestine
	Immunoglobulins	Lymphocytes
Extracellular matrix proteins	Collagen	Fibroblasts, others
	Fibronectin	Fibroblasts, liver
	Proteoglycans	Fibroblasts, others
Regulated Secretory Proteins		
Peptide hormones	Insulin	Pancreatic β-islet cells
	Glucagon	Pancreatic α-islet cells
	Endorphins	Neurosecretory cells
	Enkephalins	Neurosecretory cells
	ACTH	Anterior pituitary lobe
Digestive enzymes	Trypsin	Pancreatic acini
	Chymotrypsin	Pancreatic acini
	Amylase	Pancreatic acini, salivary glands
	Ribonuclease	Pancreatic acini
	Deoxyribonuclease	Pancreatic acini
Milk proteins	Casein	Mammary gland
	Lactalbumin	Mammary gland

network into **secretory vesicles** that are stored inside the cell awaiting a stimulus for exocytosis. Such **regulated secretion** occurs in pancreatic acinar cells, which secrete precursors of digestive enzymes, and hormone-secreting endocrine cells (see Table 17-3). The release of each of these stored proteins is initiated by different neural and hormonal stimuli. In most cases of regulated secretion studied so far, a rise in the cytosolic Ca^{2+} concentration, induced by binding of the hormone to its receptor, triggers fusion of the secretory-vesicle membrane with the plasma membrane and release of the vesicle contents by exocytosis. As we discuss in Chapter 21, nerve cells also store neurotransmitters in similar types of vesicles, which also fuse with the membrane in response to an elevation in cytosolic Ca^{2+}, releasing their contents.

Analysis of Yeast Mutants Defined Major Steps in the Secretory Pathway

The sequential movement of secretory proteins from the cytosol → the rough ER lumen → Golgi cisternae → secretory vesicles was first elucidated by classical pulse-chase autoradiography studies with pancreatic acinar cells (see Classic Experiment 17.1 on the accompanying CD-ROM). Subsequent experiments with yeast mutants further defined the pathway by which secretory proteins mature. Although

yeasts secrete few proteins into the growth medium, they continuously secrete a number of enzymes that remain localized in the narrow space between the plasma membrane and the cell wall. The best-studied of these, invertase, hydrolyzes the disaccharide sucrose to glucose and fructose. A large number of temperature-sensitive mutant yeast strains were identified in which the secretion of all proteins is blocked at the higher, *nonpermissive* temperature (at which the cells cannot grow) but is normal at the lower, *permissive* temperature (at which the cells grow normally). When transferred from the lower to the higher temperature, these so-called *sec* mutants accumulate secretory proteins at the point in the pathway that is blocked. Analysis of such mutants identified five classes (A–E), corresponding to five steps in the secretory pathway, in which secretory proteins accumulate in the cytosol, rough ER, small vesicles taking proteins from the ER to the Golgi complex, Golgi cisternae, or secretory vesicles (Figure 17-14).

To determine the order of the steps in the pathway, researchers analyzed double *sec* mutants. For instance, when yeast cells contain mutations in both class B and class D functions, proteins accumulate in the rough ER, not in the Golgi cisternae. Since proteins accumulate at the earliest blocked step, this finding shows that class B mutations must act at an earlier point in the maturation pathway than class

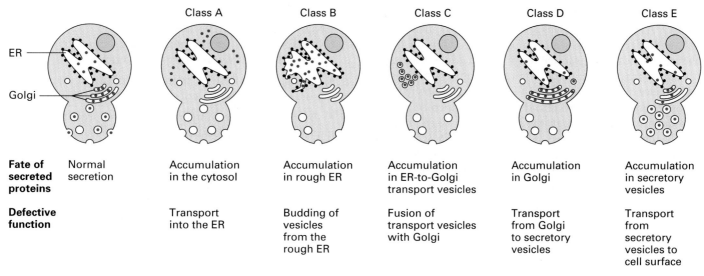

		Class A	Class B	Class C	Class D	Class E
Fate of secreted proteins	Normal secretion	Accumulation in the cytosol	Accumulation in rough ER	Accumulation in ER-to-Golgi transport vesicles	Accumulation in Golgi	Accumulation in secretory vesicles
Defective function		Transport into the ER	Budding of vesicles from the rough ER	Fusion of transport vesicles with Golgi	Transport from Golgi to secretory vesicles	Transport from secretory vesicles to cell surface

▲ **FIGURE 17-14 Five steps in the maturation of secretory proteins as defined by yeast *sec* mutants that are temperature-sensitive for protein secretion.** These mutants can be grouped into five classes, corresponding to the site where newly made secretory proteins (red dots) accumulate when cells are shifted from the growing (permissive) temperature to the higher nonpermissive one. Analysis of double mutants permitted the sequential order of the steps to be determined. [See P. Novick et al., 1981, *Cell* **25**:461; C. A. Kaiser and R. Schekman, 1990, *Cell* **61**:723; and N. Green et al., 1992. *J. Cell Biol.* **116**:597.]

D mutations do. These studies confirmed that as a secretory protein matures it moves sequentially from the cytosol → rough ER → ER-to-Golgi transport vesicles → Golgi cisternae → secretory vesicles and finally is exocytosed.

Anterograde Transport through the Golgi Occurs by Cisternal Progression

As noted above, a newly formed *cis*-Golgi vesicle, with its luminal protein cargo, progresses from the *cis* face to the *trans* face of the Golgi complex and then into the *trans*-Golgi reticulum. At one time it was thought that secreted proteins move from the *cis*- to the *medial*-Golgi, and from the *medial*- to the *trans*-Golgi, via small transport vesicles. Indeed there are many small vesicles that move proteins from one Golgi compartment to another, but they appear to do so in the reverse, or retrograde, direction; these vesicles retrieve ER or Golgi enzymes to an earlier compartment in the secretory pathway. In this way enzymes that modify secretory proteins come to be localized in the correct organelle.

The first evidence for the cisternal progression model of Golgi function came from careful microscopic analysis of the synthesis of algal scales. These are cell-wall glyco-proteins that are assembled in the *cis*-Golgi into large complexes visible in the electron microscope. Like other secretory proteins, newly-made scales move from the *cis*- to the *trans*-Golgi, but they can be 20 times the size of the ≈50-nm-diameter transport vesicles that bud from Golgi cisternae. Thus it was thought unlikely that these and other secretory proteins move from one Golgi compartment to another via small vesicles.

Similarly, in the synthesis of collagen by fibroblasts, large aggregates of the procollagen precursor often form in the lumen of the *cis*-Golgi. These aggregates are too large to be incorporated into small transport vesicles, and investigators could never find such aggregates in transport vesicles. In one test of the cisternal progression model, collagen folding was blocked by an inhibitor of proline hydroxylation, and soon all pre-made, folded, procollagen aggregates were secreted from the cell. When the inhibitor was removed, newly made procollagen peptides folded and then formed aggregates in the *cis*-Golgi that subsequently could be seen to move as a "wave" from the *cis*- through the *medial*-Golgi cisternae to the *trans*-Golgi, followed by secretion and incorporation into the extracellular matrix. Procollagen aggregates were never seen in small transport vesicles. Together with other evidence, decribed later, that the small transport vesicles near the Golgi are moving proteins in the retrograde direction, most researchers in the field have come to favor the cisternal progression model.

The pathway for the maturation of secretory proteins elucidated by autoradiographic, genetic, and electron microscope studies in yeasts, algae, fibroblasts, and pancreatic acinar cells is thought to function in all eukaryotic cells. As we detail in later sections, each step in the pathway requires the action of multiple proteins.

Plasma-Membrane Glycoproteins Mature via the Same Pathway as Continuously Secreted Proteins

The maturation pathway taken by continuously secreted proteins is also followed by plasma-membrane glyco-proteins. Well-studied examples include viral glycoproteins destined for the plasma membranes of infected cells, glycophorin in the erythrocyte plasma membrane, the plasma-membrane Na^+/K^+ ATPase, and enzymes in plant plasma membranes that synthesize such cell-wall components as cellulose. Pulse-labeling studies using radioactive amino acids, followed by subcellular fractionation and immunoprecipitation to detect radiolabeled proteins, have established that the newly made glycoproteins are inserted into the rough ER membrane and subsequently move through the Golgi cisternae en route to the plasma membrane (see Figure 17-13). These plasma-membrane glycoproteins also have been shown to undergo the same types of modifications in the same ER and Golgi compartments that secretory proteins do.

SUMMARY Overview of the Secretory Pathway

- Although all cytosolic ribosomes are functionally equivalent, membrane-attached and membrane-unattached ribosomes synthesize different classes of proteins, depending on a signal sequence in the protein itself.

- Newly made secretory proteins are localized to the lumen of the rough ER.

- All mammalian cells continuously secrete certain proteins, such as those in the extracellular matrix.

- Certain cell types store proteins such as hormones and digestive enzymes in secretory vesicles, awaiting a neural or hormonal signal that triggers an elevation in cytosolic Ca^{2+} and then protein secretion.

- Proteins destined to be secreted move through the secretory pathway in the following order: rough ER → ER-to-Golgi transport vesicles → Golgi cisternae → secretory or transport vesicles → cell surface (exocytosis) (see Figure 17-13).

- Small transport vesicles bud off from the ER and fuse to form the *cis*-Golgi reticulum. By cisternal migration, *cis*-Golgi vesicles with their luminal protein cargo move through the Golgi complex to the *trans*-Golgi reticulum. Proteins are retrieved from the *cis*-Golgi to the ER and also from later Golgi cisternae to earlier ones by small retrograde transport vesicles.

- Plasma-membrane glycoproteins follow the same maturation pathway as continuously secreted proteins.

- Both secreted and integral membrane proteins undergo various modifications as they mature in the secretory pathway.

17.4 Translocation of Secretory Proteins across the ER Membrane

In the next several sections, we examine in detail the events occurring in the secretory pathway (see Figure 17-13). We begin in this section by focusing on three key questions: How are certain proteins targeted to the secretory pathway? How do secretory proteins actually cross the membrane? And what is the source of energy for the transport of proteins across the ER membrane?

A Signal Sequence on Nascent Secretory Proteins Targets Them to the ER and Is Then Cleaved Off

As noted earlier, synthesis of most secretory proteins begins on free ribosomes in the cytosol. The presence of a 16- to 30-residue ER signal sequence directs the ribosome to the ER membrane and initiates transport of the growing polypeptide across the ER membrane. An ER signal sequence typically is located at the N-terminus of the protein and contains one or more positively charged amino acids followed by a continuous stretch of 6–12 hydrophobic residues (Table 17-4); except for these features, the signal sequences of

TABLE 17-4	Amino Acid Sequences of ER Signal Peptides in Three Eukaryotic Proteins
Protein	Amino Acid Sequence*
Preproalbumin	Met-Lys-Trp-Val-Thr-**Phe-Leu-Leu-Leu-Leu-Phe-Ile-Ser-Gly-Ser-Ala-Phe-Ser** ↓ Arg . . .
Pre-IgG light chain	Met-Asp-Met-Arg-Ala-Pro-Ala-Gln-**Ile-Phe-Gly-Phe-Leu-Leu-Leu-Leu-Phe**-Pro-Gly-Thr-Arg-Cys ↓ Asp . . .
Prelysozyme	Met-Arg-Ser-**Leu-Leu-Ile-Leu-Val-Leu-Cys-Phe-Leu**-Pro-Leu-Ala-Ala-Leu-Gly ↓ Lys . . .

Hydrophobic residues are in boldface; arrows (↓) indicate the site of cleavage by signal peptidase.

SOURCE: D. P. Leader, 1979, *Trends Biochem. Sci.* **4**:205 and T. A. Rapoport, 1985, *Curr. Topics Membrane Transport* **24**:1.

different secretory proteins have little in common. Signal sequences are not normally found on complete polypeptides made in cells, implying that the signal sequence is cleaved from the protein while it is still growing on the ribosome.

The hydrophobic "core" of ER signal sequences is essential for their function. For instance, the specific deletion of several of the hydrophobic amino acids from a signal sequence, or the mutation of one of them to a charged amino acid, abolishes the ability of the protein to cross the ER membrane into the lumen. Other experiments indicate that addition (by recombinant DNA techniques) of any random N-terminal amino acid sequence, provided it is sufficiently long and hydrophobic, will cause a normally cytosolic protein to be translocated to the ER lumen. These hydrophobic residues form a binding site that is critical for the interaction of signal sequences with receptor proteins on the ER membrane.

Experiments with microsomes have clarified the function and fate of ER signal sequences. When the mRNA encoding a secretory protein is translated in a cell-free system containing ribosomes, tRNAs, ATP and GTP, and cytosolic enzymes but with no ER membranes, the protein, with its signal sequence attached, is released into the cytosol. If microsomes that have been stripped of their own ribosomes are then added to the reaction mixture, the protein generally is not incorporated into the ER lumen and its signal sequence remains attached (Figure 17-15a). However, if these membranes are present during the cell-free synthesis of a secretory protein, the protein is found in the ER lumen with the signal sequence removed (Figure 17-15b). Thus, a newly made secretory protein can cross the ER membrane and have its signal sequence removed only if the membrane is present during protein synthesis. The enzyme *signal peptidase,* which normally cleaves off the signal sequences, is localized to the ER lumen.

In subsequent experiments, microsomes were added at different times after the synchronized cell-free synthesis of a secretory protein began. In order for the subsequently completed secretory protein to be localized in the ER lumen, microsomes must be added before the first 70 or so amino acids are polymerized. At this point, about 40 amino acids protrude from the ribosomes, including the signal sequence that later will be cleaved off, and about 30 amino acids are buried in a channel or tunnel in the ribosome. Thus the transport of most secretory proteins into the ER lumen—particularly those with more than 100 amino acids—must occur during translation, a process referred to as **cotranslational transport.**

Some small secretory proteins, such as the yeast α mating factor (≈70 amino acids), exhibit post-translational transport into the ER lumen. Such proteins are synthesized in their entirety on free cytosolic ribosomes and then released into the cytosol. Here they are bound by chaperones, which keep them in an unfolded state, and subsequently are translocated across the ER membrane. Thus the targeting of these proteins to the ER lumen is similar to targeting of proteins to the mitochondrial matrix or chloroplast stroma.

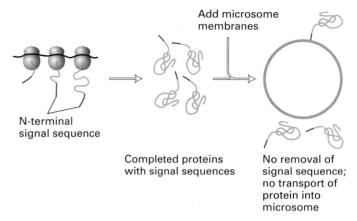

(a) Cell-free protein synthesis; no microsomes present

Add microsome membranes

N-terminal signal sequence

Completed proteins with signal sequences

No removal of signal sequence; no transport of protein into microsome

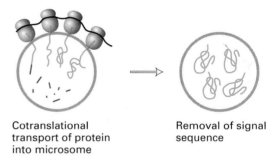

(b) Cell-free protein synthesis; microsomes present

Cotranslational transport of protein into microsome

Removal of signal sequence

▲ **FIGURE 17-15 The cotranslational insertion of secretory proteins into microsomes, which are formed by shearing of the ER into small vesicles.** Treatment of microsomes with EDTA, which chelates Mg^{2+} ions, strips them of associated ribosomes, leaving "clean" microsome membranes. Synthesis is carried out in a cell-free system, such as an extract of wheat germ, which contains functional ribosomes but no functional mRNA. Although synthesis of a secretory protein occurs in the absence of microsomes (a), the protein is transported across the vesicle membrane and the signal sequence is cleaved only if microsomes are present during protein synthesis (b).

Two Proteins Initiate the Interaction of Signal Sequences with the ER Membrane

Since secretory proteins are synthesized in association with the ER membrane but not with any other cellular membrane, some signal-sequence recognition mechanism must target them there. As depicted in Figure 17-16, the two key components in this targeting are the **signal-recognition particle (SRP)** and the **SRP receptor.** SRP is a cytosolic particle that transiently binds to the ER signal sequence in a nascent protein, to the large ribosomal unit, and to the SRP receptor in the ER membrane.

A simple experiment provided the first clue to the existence of the SRP. When microsomes, stripped of their ribosomes, were exposed to a solution of 0.5 M NaCl and then

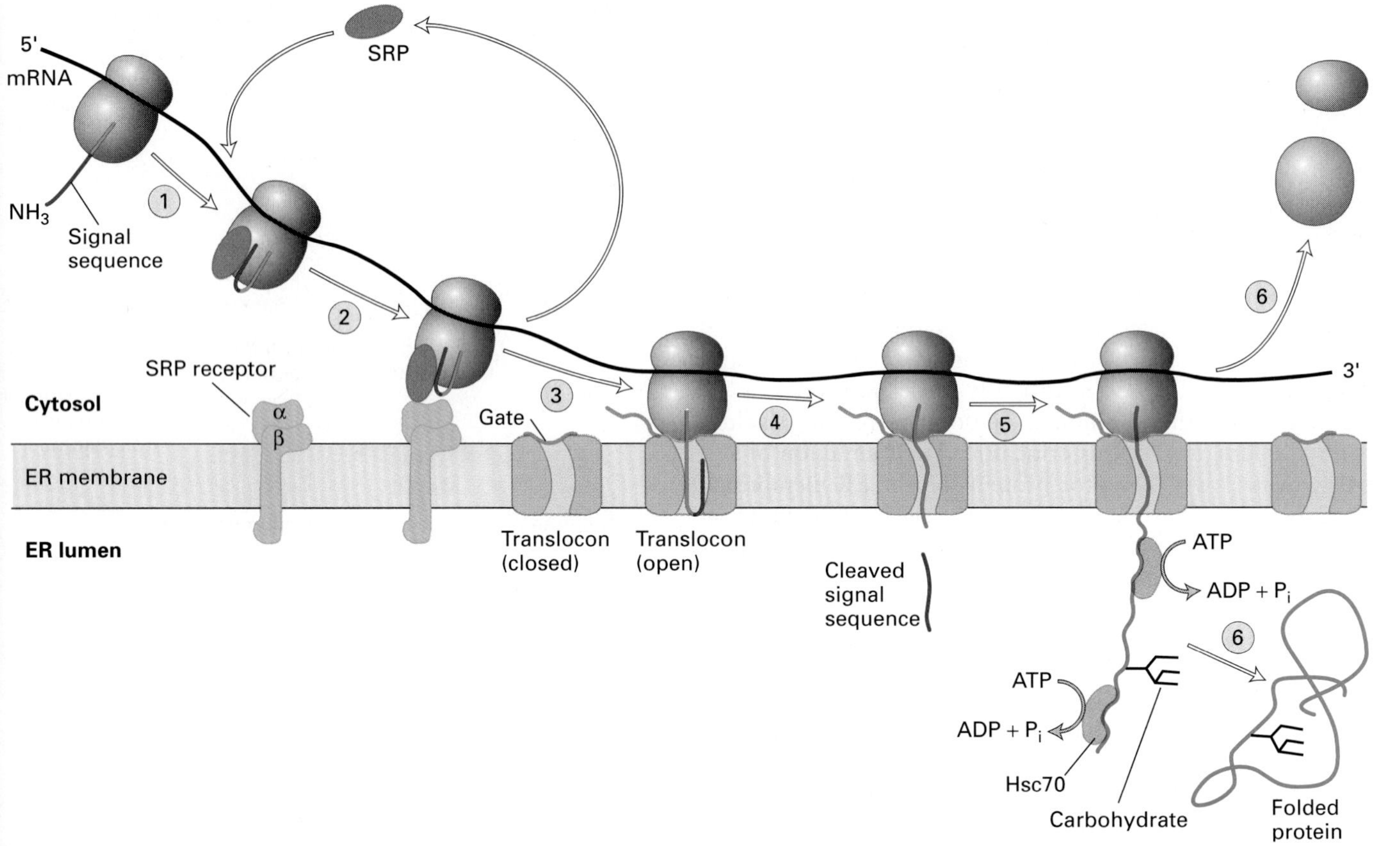

▲ **FIGURE 17-16 Synthesis of secretory proteins on the rough ER.** Synthesis begins on an unattached ribosome in the cytosol. The complete N-terminal signal sequence emerges from the ribosome only when the polypeptide is about 70 amino acids long, because about 30 amino acids remain buried in the ribosome. Steps ① and ②: A signal-recognition particle (SRP) binds to the signal sequence of a nascent protein chain, and the complex of SRP, nascent polypeptide, and ribosome then binds to the α subunit of the SRP receptor in the ER membrane. Step ③: As SRP and its receptor dissociate from the nascent chain, accompanied by hydrolysis of GTP, the signal sequence binds to the translocon, and the translocon gate that blocks the internal channel opens. The signal sequence and the adjacent segment of the growing polypeptide then insert as a loop into the central cavity of the translocon, with the N-terminus facing the cytosol. Both SRP and the SRP receptor are freed to initiate the insertion of another secretory protein. Step ④: The polypeptide chain elongates; then the signal sequence is cleaved by a signal peptidase in the ER lumen and is rapidly degraded. Step ⑤: The peptide chain continues to elongate and is extruded into the ER lumen through the translocon. In yeasts but not in mammalian cells, the ER chaperone protein Hsc70 binds to the growing chain on the luminal surface and then, powered by the hydrolysis of ATP, releases the nascent chain. This cycle of binding and release of the nascent chain by yeast Hsc70 is essential for translocation to continue; it also facilitates the eventual folding of many secreted polypeptides. The ATP needed for functioning of luminal chaperones is imported from the cytosol by an ATP/ADP antiporter in the rough ER membrane, similar to the one in the inner mitochondrial membrane (see Figure 16-32). As discussed later, enzymes on the luminal surface add carbohydrates to the chain at asparagine, serine, and threonine residues. Step ⑥: The peptide chain continues to elongate until translation is completed. Then, the ribosomes are released, the remaining C-terminus of the secreted protein is drawn into the ER lumen, the translocon gate shuts, and the secreted protein assumes its final conformation. [See T. A. Rapoport, 1991, *FASEB J.* **5**:2792; S. L. Sanders et al., 1992, *Cell* **69**:353; T. Powers and P. Walter, 1996, *Nature* **381**:191; B. Martoglio and B. Dobberstein, 1996, *Trends Cell Biol.* **6**:142; and G. Bacher et al., 1996, *Nature* **381**:248.]

recovered by centrifugation, the treated microsomes were unable to support the transmembrane movement of newly synthesized secretory proteins in a cell-free system. However, when the proteins removed by the NaCl treatment were returned to the preparation, secretory proteins were incorporated into the lumen of the vesicles as in Figure 17-15b. A single active component—now called the signal-recognition particle (SRP)—subsequently was purified from the mixture of proteins extracted by NaCl. The SRP is a ribonucleoprotein composed of six discrete polypeptides and a 300-nucleotide RNA (Figure 17-17).

One of the SRP proteins (P54) can be chemically cross-linked to ER signal sequences, evidence that this 54,000-MW protein is the one that binds to the signal sequence in a

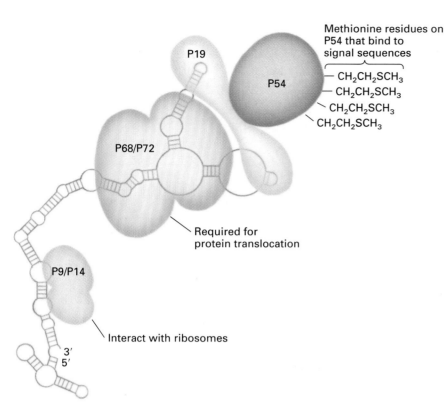

Methionine residues on P54 that bind to signal sequences

— CH₂CH₂SCH₃
— CH₂CH₂SCH₃
— CH₂CH₂SCH₃
— CH₂CH₂SCH₃

P19

P54

P68/P72

Required for protein translocation

P9/P14

Interact with ribosomes

3′
5′

◄ **FIGURE 17-17 Structure of the signal-recognition particle (SRP).** SRP comprises one 300-nucleotide RNA and six proteins designated P9, P14, P19, P54, P68, and P72. (The numeral indicates the molecular weight $\times\ 10^{-3}$). All proteins except P54 bind to the RNA; the precise binding site of P9 and P14 is not known. Different functions have been assigned to the different polypeptides as indicated. In addition to binding the hydrophobic side chains of signal peptide sequences, P54 functions with the α subunit of the SRP receptor to hydrolyze GTP. [See K. Strub et al., 1991, *Mol. Cell. Biol.* **11**:3949; S. High and B. Dobberstein, 1991, *J. Cell Biol.* **113**:229; and D. Zopf et al., 1993, *J. Cell Biol.* **120**:1113.]

nascent secretory protein. One region of P54 contains a large number of clustered methionine residues whose hydrophobic side chains are thought to protrude outward and bind to the hydrophobic side chains that form the "core" of an ER signal sequence. Other SRP proteins, P9 and P14, interact with the ribosome, while P68 and P72 are required for protein translocation. Experiments in the wheat-germ cell-free system (see Figure 17-15) have shown that SRP arrests secretory-protein synthesis after polymerization of about 70 amino acids when microsomes are absent. Thus SRP not only helps mediate interaction of a nascent protein with the ER membrane but also prevents synthesis of a complete protein in the absence of rough ER membranes.

The complex of a SRP, nascent chain, and ribosome binds to the SRP receptor in the ER membrane (see Figure 17-16). This receptor contains two polypeptide subunits: a transmembrane β subunit of about 300 amino acids, and a peripheral α subunit of about 640 amino acids. The treatment of ER membranes with minute amounts of protease cleaves the β subunit very near its site of attachment to the membrane, releasing a soluble form of the SRP receptor. This soluble fragment can relieve the block in chain elongation that is imposed by SRP in cell-free systems, evidence that the SRP receptor binds to the SRP and possibly also to the ribosome.

Polypeptides Move through the Translocon into the ER Lumen

SRP and its receptor only initiate the transfer of the nascent chain across the ER membrane; then they dissociate from the chain, which is transferred to a set of transmembrane proteins, collectively called the **translocon** (see Figure 17-16). Once the nascent chain–translocon complex is assembled, the elongating chain passes directly from the large ribosomal subunit into the centers of the translocon, a protein-lined channel within the membrane. The growing chain is never exposed to the cytosol and does not fold until it reaches the ER lumen. Thus the protein crosses the membrane in an unfolded state; if newly made secretory proteins were allowed to fold in the cytosol, they probably would be unable to cross the ER membrane.

The experiment described in Figure 17-18 identified two abundant ER membrane proteins that can be chemically cross-linked to a nascent chain, indicating that they form part of the mammalian translocon. One of these proteins, named *TRAM* (*t*ranslocating chain–*a*ssociated *m*embrane) protein, can be chemically cross-linked to the ER signal sequence; this protein may bind the signal sequence as it is "handed over" by the SRP to the translocon. TRAM protein, which spans the ER membrane at least eight times, is essential for protein translocation. The second mammalian protein identified in this cross-linking experiment, Sec61p, is similar in amino acid sequence to the protein encoded by the yeast *sec61* gene. Analysis of yeast *sec61* mutants show they are class A mutants, defective in the translocation of secretory proteins into the ER lumen (see Figure 17-14), further evidence that the homologous mammalian protein, Sec61p, has a similar function. Mammalian Sec61p, which has 10 membrane-spanning α helices, is tightly bound to two other proteins, termed Sec61β and Sec61γ; the three

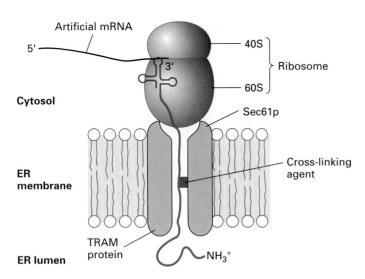

▲ **FIGURE 17-18 Identification of the proteins that form the translocon, the transmembrane channel through which a nascent secretory protein passes into the ER lumen.** An mRNA was synthesized that encoded the first 70 amino acids of the secretory protein prolactin; the mRNA lacked a chain-termination codon and contained only one lysine codon, near the middle of the sequence. This mRNA was translated in a cell-free system containing microsomes (see Figure 17-15b). The reactions contained a chemically modified lysyl tRNA in which a light-activated cross-linking reagent was attached to the lysine $CH_2-CH_2-NH_2$ group. Although the entire mRNA was translated, the completed polypeptide could not be released from the ribosome and thus became "stuck" crossing the ER membrane. An intense light source then was administered to activate the cross-linking agent, whereupon the nascent chain became chemically linked to whatever proteins were near it in the channel. When the experiment was performed using microsomes from mammalian cells, the nascent chain became covalently linked primarily to two proteins, designated TRAM and Sec61p. [Adapted from T. A. Rapoport, 1992, *Science* **258**:931, and D. Görlich and T. A. Rapoport, 1993, *Cell* **75**:615.]

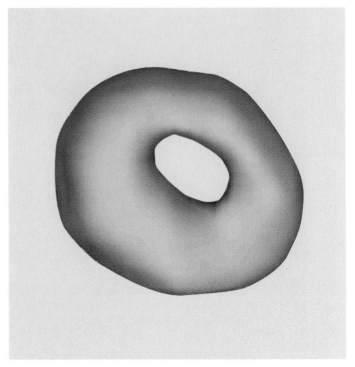

▲ **FIGURE 17-19 Electron microscopic view of a translocon channel.** Purified Sec61p translocon subunits were solubilized in detergents; ribosomes were added and the preparation was reconstituted into phospholipid bilayers (see Figure 15-4). The preparation was subjected to freeze fracture (see Figure 5-38) and viewed in the plane of the membrane by electron microscopy. To generate this reconstructed image, micrographs of a number of particles were generated, stored in a computer, and then averaged to produce a single image. [Courtesy of Dr. Christopher Akey, Boston University School of Medicine.]

polypeptides form the *Sec61 complex*. This protein complex, which also binds tightly to ribosomes, may function to attach the 60S subunit of ribosomes that are synthesizing secretory proteins to the ER membrane.

Subsequent biochemical experiments showed that phospholipid vesicles reconstituted in vitro and containing only the SRP receptor, TRAM, and the Sec61 complex are functional in translocating nascent secretory proteins. Thus these are the only mammalian ER-membrane proteins required for translocation. The actual channel is lined by three or four Sec61 complexes. Images of the channel have been generated by computer averaging of serial reconstructions of electron micrographs of freeze-fractured purified Sec61p channels reconstituted in liposomes. These show the channel as a roughly pentagonal cylinder, 5–6 nm high and 8.5 nm in diameter, with a central pore, ≈2 nm in diameter, that

extends throughout the protein complex perpendicular to the plane of the membrane (Figure 17-19).

In the absence of attached ribosomes and protein translocation, the translocon channel is closed at the cytosolic end by a segment of the Sec61p protein. Binding of a ribosome-nascent chain complex causes this "gate" to open and then form a tight seal between the ribosome and translocon such that small molecules cannot pass though the translocon pore. Opening of the gate allows a loop of the nascent chain, containing the signal sequence and ≈30 adjacent amino acids to insert into the translocon pore (see Figure 17-16). After cleavage of the signal sequence, the growing polypeptide moves through the pore into the ER lumen.

GTP Hydrolysis Powers Protein Transport into the ER in Mammalian Cells

As we learned earlier in this chapter, the uptake of proteins into mitochondria and chloroplasts requires energy. Likewise, energy is required for insertion of the signal sequence

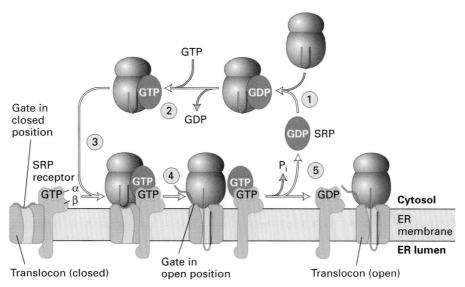

▲ **FIGURE 17-20 Cycles of GDP-GTP exchange and GTP hydrolysis that drive insertion of nascent secretory proteins into the translocon.** Both the P54 subunit of SRP and the α subunit of the SRP receptor bind and hydrolyze GTP. In step ①, SRP containing a bound GDP binds to the complex of a signal sequence and ribosome, triggering release of GDP by SRP and binding of GTP (step ②). The complex of SRP, ribosome, and signal sequence binds to the form of the SRP receptor in which the α subunit has a bound GTP (step ③). Hydrolysis of GTP by SRP and the SRP receptor then powers transfer of the nascent polypeptide, with its signal sequence, to the translocon and opening of the translocon "gate," as well as release of SRP and dissociation of the SRP receptor from the translocon. An unknown protein then promotes release of GDP from the α subunit of the SRP receptor and binding of GTP (not depicted). [Adapted from T. Powers and P. Walter, 1996, *Nature* **381**:191, and G. Bacher et al., 1996, *Nature* **381**:248. For another model of how GTP hydrolysis powers assembly of the nascent chain–translocon complex see J. S. Millman and D. Andrews, 1997, *Cell* **89**:673, and P. Rapiejko and R. Gilmore, 1997, *Cell* **89**:703.]

into the translocon and for the subsequent translocation of the growing polypeptide chain across the ER membrane.

As Figure 17-20 shows, release of GDP and binding of GTP, by the P54 subunit of SRP and by the α subunit of the SRP receptor, promote formation of the complex comprising a ribosome, nascent chain, SRP, and SRP receptor. Subsequent hydrolysis of the bound GTP promotes dissociation of this complex and formation of the nascent chain–translocon complex. After dissociating, SRP and its receptor recycle to initiate the interaction of additional nascent chains with the ER membrane.

In yeast, a homolog of the chaperone Hsc70 is localized to the ER lumen and binds nascent chains as they pass through the translocon (see Figure 17-16). Studies with yeasts expressing mutant forms of this Hsc70 protein have shown that ATP hydrolysis by Hsc70 is essential for co-translational transport of proteins across the ER membrane. ATP hydrolysis by this luminal Hsc70 chaperone also powers the post-translational uptake of the small yeast α factor into the ER lumen. (Recall that small secretory proteins, such as α factor, are synthesized in the cytosol and then imported.) However, in mammalian cells cotranslational translocation does not require ATP hydrolysis. Instead, hydrolysis of GTP during protein synthesis itself is thought to drive the nascent polypeptide from the membrane-attached ribosome across the ER membrane.

SUMMARY **Translocation of Secretory Proteins across the ER Membrane**

• Secretory proteins, enzymes destined for the ER, Golgi complex, and lysosomes, and integral plasma-membrane proteins are synthesized on ribosomes bound to the rough ER. Synthesis of this class of proteins is initiated on membrane-unattached ribosomes.

• A sequence of hydrophobic amino acids, the ER signal sequence, is recognized and bound by a signal-recognition particle (SRP), which in turn is bound by an SRP receptor on the rough ER membrane.

• Generally, ER signal sequences are located at the N-terminus and are cleaved from the protein in the rough ER lumen.

• The SRP directs both the binding of the ribosome to the ER membrane and the insertion of the nascent protein into the transmembrane channel (see Figure 17-16). Hydrolysis of GTP by the P54 subunit of SRP and the α subunit of the SRP receptor is essential for this process.

• Proteins cross the ER membrane in an unfolded state through a protein-lined channel, the translocon, a multiprotein complex composed of TRAM protein and the Sec61 complex.

• In mammalian cells, the only integral ER proteins required for translocation of nascent secretory proteins are the SRP receptor, TRAM, and the Sec61 complex.

17.5 Insertion of Membrane Proteins into the ER Membrane

In Chapters 3 and 15 we were introduced to several of the vast array of integral proteins that are present in the plasma membrane and other cellular membranes. Each such protein has a unique orientation with respect to the membrane's phospholipid bilayer (see Figure 3-32). As noted earlier, these membrane glycoproteins are synthesized on the rough ER membrane and remain membrane associated as they move to their final destinations along the same pathway followed by continuously secreted proteins (see Figure 17-13). During this transport, the topology of a membrane protein is preserved; the same segments of the protein, for instance, always face the cytosol. Thus the orientation of these membrane proteins in their final sites is established during biosynthesis on the ER membrane.

As illustrated in Figure 17-21, integral proteins interact with membranes in many different ways. Many such proteins contain a single membrane-spanning segment: a sequence of 20–25 hydrophobic amino acids that forms a transmembrane α helix, anchoring the protein in the phospholipid bilayer. Most of these *single-pass* transmembrane proteins have their hydrophilic N-terminal segment on the exoplasmic face and their hydrophilic C-terminal segment on the cytosolic face; other single-pass proteins have the reverse orientation. Many plasma-membrane proteins have multiple membrane-spanning α-helical segments. Such *multipass* transmembrane proteins include the glucose transporter GLUT1 and numerous G protein–linked cell-surface receptors (Chapter 20). Still other membrane proteins, which lack a hydrophobic membrane-spanning segment, are linked to a glycosylphosphatidylinositol (GPI) anchor that is embedded in the phospholipid bilayer (see Figure 3-36a).

Despite the variation in how integral proteins associate with membranes, insertion of all of them into the ER membrane depends on specific **topogenic sequences**. These sequences of up to 25 amino acids ensure that a protein acquires the proper orientation during its insertion into the ER membrane. In this section, we describe insertion of several types of proteins into the ER membrane to illustrate the function of topogenic sequences.

Most Nominal Cytosolic Transmembrane Proteins Have an N-Terminal Signal Sequence and an Internal Topogenic Sequence

We can begin our discussion with the large class of single-pass transmembrane proteins that have a cytosolic C-terminus

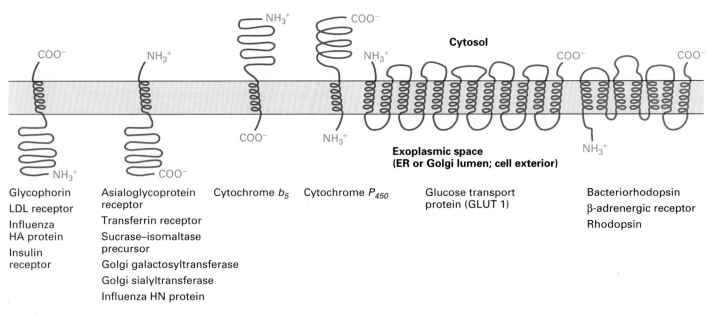

Glycophorin
LDL receptor
Influenza HA protein
Insulin receptor

Asialoglycoprotein receptor
Transferrin receptor
Sucrase–isomaltase precursor
Golgi galactosyltransferase
Golgi sialyltransferase
Influenza HN protein

Cytochrome b_5

Cytochrome P_{450}

Glucose transport protein (GLUT 1)

Bacteriorhodopsin
β-adrenergic receptor
Rhodopsin

▲ FIGURE 17-21 Topologies of some integral membrane proteins synthesized on the rough ER. Segments of the protein chain in the membrane bilayer are shown as transmembrane α helices; the portions outside the membrane are folded. Topogenic sequences in the protein act during biosynthesis to ensure its proper orientation in the ER membrane, which is retained during transport to the cell surface. [See W. Wickner and H. F. Lodish, 1985, *Science* **230**:400; E. Hartmann et al., 1989, *Proc. Nat'l. Acad. Sci.* **86**:5786; and C. A. Brown and S. D. Black, 1989, *J. Biol. Chem.* **264**:4442.]

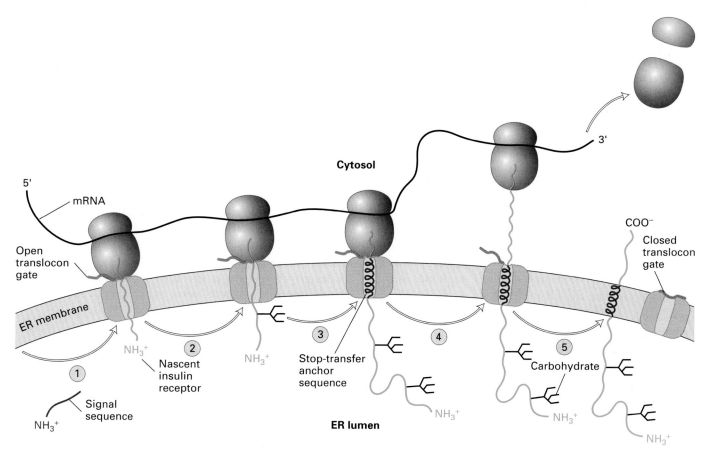

5'

mRNA

Cytosol

3'

Open translocon gate

ER membrane

COO⁻

Closed translocon gate

NH_3^+

① ②

NH_3^+

③ Stop-transfer anchor sequence

④

⑤ Carbohydrate

Nascent insulin receptor

NH_3^+

NH_3^+

NH_3^+

NH_3^+

Signal sequence

NH_3^+

ER lumen

▲ **FIGURE 17-22 Synthesis and insertion into the ER membrane of the insulin receptor and similar proteins that employ a cleaved ER signal sequence and an internal stop-transfer membrane-anchor sequence.** Steps ① and ②: After the nascent chain and ribosome become associated with a translocon in the ER membrane, the N-terminal signal sequence is cleaved and the chain is elongated. This process occurs by the same mechanism as the one for secretory proteins (see Figure 17-16, steps ① – ④). Step ③: Once the hydrophobic stop-transfer membrane-anchor sequence of about 22 amino acids is synthesized, it remains in the translocon, preventing the nascent chain from extruding farther into the ER lumen. Step ④: The tight association between the ribosome and translocon is

disrupted, and synthesis of the protein is completed by the ribosome in the cytosol. All or part of the cytosolic domain may enter the translocon pore (a step not shown here). Step ⑤: During or just after synthesis is complete, the stop-transfer sequence moves in several steps through the proteins that form the translocon into the phospholipid bilayer where it becomes anchored in the membrane. Probably at this time the translocon gate moves into the closed position. Other single-pass transmembrane proteins with an N-terminal ER signal sequence and internal stop-transfer membrane-anchor sequence are inserted in a similar manner. [See H. Do et al., 1996, *Cell* **85**:369, and W. Mothes et al., 1997, *Cell* **89**:523.]

and an exoplasmic N-terminus. Examples include the erythrocyte protein glycophorin, the insulin receptor, and proteins that form the spikes on the outside of viruses whose surfaces are derived by budding from the plasma membrane (see Figure 6-17), including the G protein from vesicular stomatitis virus (VSV) and hemagglutinin (HA) from influenza virus. Immediately after synthesis, each of these proteins spans the ER membrane with the same orientation that it will have when it appears on the plasma membrane (see Figure 17-21). This class of proteins possess an N-terminal signal sequence that targets them to the ER and an internal **stop-transfer membrane-anchor sequence** that becomes the membrane-spanning α helix. The insulin receptor illustrates

how these two distinct sequences function to insert and orient proteins of this class into the ER membrane.

The ER signal sequence on a nascent insulin receptor, like that of a secretory protein, is cleaved while the chain is still growing, and the new N-terminus of the growing polypeptide is extruded across the ER membrane into the lumen. However, unlike the case with secretory proteins, a sequence of about 22 hydrophobic amino acids in the middle of the insulin receptor stops the transfer of the protein through the translocon (Figure 17-22). This internal sequence prevents further extrusion of the nascent chain into the ER lumen, and it remains in the translocon until synthesis of the nascent chain is finished. The sequence then moves laterally through the proteins that line the side of the translocon and becomes anchored in the phospholipid bilayer of the membrane, where it remains. Hence this hydrophobic sequence functions as both a stop-transfer and membrane-anchor sequence. The C-terminus of the nascent chain remains in the cytosol and is not transferred across the ER membrane.

Support for this model, depicted in Figure 17-22, has come from studies in which cDNAs encoding various mutant insulin receptor proteins are expressed in cultured mammalian cells. The wild-type receptor is transported normally to the plasma membrane. However, a mutant receptor with charged residues inserted into the membrane-spanning segment, or missing most of this segment, is transported into the ER lumen and is eventually secreted from the cell. This establishes that the membrane-spanning α helix of the insulin receptor (and of other proteins in its class) functions as a stop-transfer sequence that prevents the C-terminus of the protein from crossing the ER membrane.

A Single Internal Topogenic Sequence Directs Insertion of Some Single-Pass Transmembrane Proteins

The orientation of the asialoglycoprotein and transferrin receptors in the membrane is opposite to that of the insulin receptor and similar proteins (see Figure 17-21). In this class of single-pass transmembrane proteins, the N-terminus is on the cytosolic face and the C-terminus is on the exoplasmic (luminal) face. Proteins with this $N_{cytosol}$-$C_{luminal}$ orientation have a single internal hydrophobic *signal-anchor sequence* that functions as both an ER signal sequence and membrane-anchor sequence (Figure 17-23). This internal signal-anchor sequence directs insertion of the nascent chain into the ER membrane so that the N-terminus of the signal sequence faces the cytosol. The internal signal-anchor sequence is *not* cleaved and remains in the translocon while the C-terminus of the growing chain is extruded into the ER lumen via cotranslational transport. After synthesis is complete, the

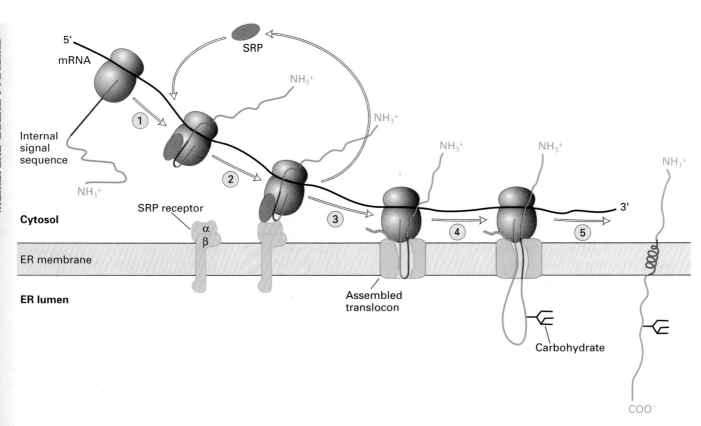

▲ **FIGURE 17-23 Synthesis and membrane insertion of the asialoglycoprotein receptor and proteins with a similar orientation, which employ an internal signal-anchor sequence.** Steps ①–③: The hydrophobic internal signal-anchor sequence (which becomes a transmembrane α helix), like the N-terminal ER signal sequence of a secretory protein, directs insertion of the growing chain into the ER membrane. Step ④: With the uncleaved internal signal remaining in the open translocon, the

nascent chain is elongated and the portion on the C-terminal side of the signal sequence is extruded through the translocon into the ER lumen. Step ⑤: After protein synthesis is completed, the internal signal-anchor sequence moves from the translocon into the phospholipid bilayer. Because it is sufficiently long and hydrophobic, this sequence anchors the protein in the membrane. [See M. Spiess and H. F. Lodish, 1986, *Cell* **44**:177; A. S. Shaw et al., 1988, *Proc. Nat'l. Acad. Sci. USA* **85**:7592; and H. Do et al., 1996, *Cell* **85**:369.]

signal-anchor sequence moves through the proteins that line the walls of the translocon into the phospholipid bilayer, where it functions as a membrane anchor.

Membrane insertion of other single-pass proteins, such as cytochrome P450, also is directed by an internal uncleaved signal-anchor sequence located near the N-terminus. These proteins, however, assume an $N_{luminal}$-$C_{cytosol}$ orientation in the ER membrane, opposite to that of the asialoglycoprotein receptor class (see Figure 17-21). In this case, the signal-anchor sequence inserts into the ER membrane with its N-terminus facing the lumen. It also prevents further extrusion of the nascent chain into the ER lumen, functioning as a stop-transfer sequence. As with the insulin receptor, which has the same $N_{luminal}$-$C_{cytosol}$ orientation as cytochrome P450, the rest of the protein is synthesized without the ribosome becoming attached to the ER membrane, and the C-terminus remains in the cytosol (see Figure 17-22). After synthesis is complete, the internal uncleaved hydrophobic sequence in cytochrome P450 moves out of the translocon into the phospholipid bilayer where it anchors the protein. Thus this topogenic sequence functions as an ER signal sequence, a stop-transfer sequence, and a membrane-anchor sequence.

The opposite membrane orientation of the asialoglycoprotein receptor and cytochrome P450 results from differences in the nature of the hydrophilic amino acids flanking the internal signal-anchor sequences in these proteins. As a general rule, the flanking segment that carries the greatest net positive charge remains on the cytosolic face of the membrane. A striking demonstration of the importance of the flanking charge in determining membrane orientation is provided by neuraminidase, an enzyme in the surface coat of influenza virus that has an $N_{cytosol}$-$C_{luminal}$ orientation similar to the asialoglycoprotein receptor. Three arginine residues are located just N-terminal to the internal signal-anchor sequence in neuraminidase. Mutation of these three positively charged residues to glutamate residues causes neuraminidase to acquire the reverse orientation: $N_{luminal}$-$C_{cytosol}$. Other proteins, too, can be made to "flip" their orientation in the ER membrane by mutating charged residues that flank the internal signal-anchor segment.

The membrane orientation of proteins with internal signal-anchor sequences also is influenced by the length and amino acid composition of the internal hydrophobic segment. Proteins with long hydrophobic segments (>20 residues) tend to adopt a $N_{luminal}$-$C_{cytosol}$ orientation, while the opposite orientation is preferred by proteins with short hydrophobic segments. Thus, the topology of these proteins is determined by both the internal hydrophobic signal-anchor sequences and the flanking hydrophilic residues.

Multipass Transmembrane Proteins Have Multiple Topogenic Sequences

Many important proteins, such as ion pumps, ion channels, and transporters, span the membrane multiple times (see Figure 17-21). Each membrane-spanning α helix in these multipass transmembrane proteins is thought to act as a topogenic sequence. The first α-helical segment initiates insertion of the growing chain into the ER membrane (Figure 17-24). This sequence functions like the internal signal-anchor sequence in the asialoglycoprotein receptor (see Figure 17-23). In both cases, the SRP and the SRP receptor are involved in assembly of the initial nascent chain–ribosome–translocon complex and insertion of the nascent chain as a hairpin loop with the N-terminus extending into the cytosol.

As the nascent chain following the first α helix in a multipass protein elongates, it moves through the translocon until the second hydrophobic α helix is formed. This helix acts as a stop-transfer membrane-anchor sequence, preventing further extrusion of the nascent chain through the translocon; its function is similar to the stop-transfer membrane-anchor sequence in the insulin receptor (see Figure 17-22). Thus after synthesis of the first two transmembrane α helices, both ends of the nascent chain are in the cytosol and a loop faces the lumen (see Figure 17-24). The C-terminus of the nascent chain then continues to grow into the cytosol. According to this model, the third α helix is another internal uncleaved signal-anchor sequence, and the fourth is another stop-transfer membrane-anchor sequence. After their synthesis on the ribosome in the cytosol, helices 3 and 4 insert into the membrane as a hairpin, just as helices 1 and 2 do. However, neither SRP nor the SRP receptor participate in their insertion. All subsequent transmembrane α helices could insert into the ER bilayer in this way; thus the number of topogenic sequences equals the number of transmembrane α helices.

Studies with proteins prepared by recombinant DNA techniques have clarified the function of the hydrophobic topogenic sequences in multipass proteins such as the GLUT1 glucose transporter. For example, if the first membrane-spanning α helix of GLUT1 is placed in the middle of an otherwise cytosolic protein, the chimera will insert into the ER membrane with an $N_{cytosol}$-$C_{luminal}$ orientation similar to that of the asialoglycoprotein receptor (see Figure 17-23). Thus this GLUT1 helix acts as an internal signal-anchor sequence. Other experiments with recombinant multipass proteins suggest that the order of the hydrophobic α helices relative to each other in the growing chain determines whether a given helix functions as a signal-anchor sequence or stop-transfer membrane-anchor sequence. The results of these experiments indicate that the first N-terminal α helix and the subsequent odd numbered ones function as signal-anchor sequences, whereas the intervening even-numbered helices function as stop-transfer membrane-anchor sequences (see Figure 17-24).

After Insertion in the ER Membrane, Some Proteins Are Transferred to a GPI Anchor

Many cell-surface proteins are anchored to the phospholipid bilayer not by a sequence of hydrophobic amino acids, but

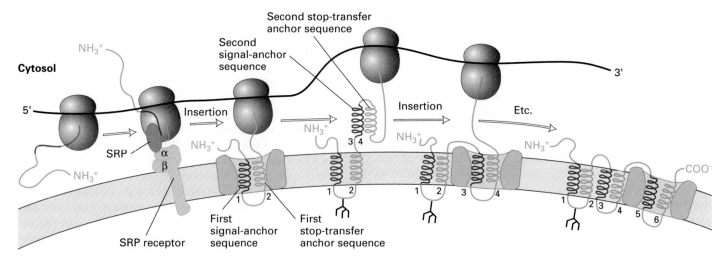

ER lumen

▲ **FIGURE 17-24 Synthesis and insertion into the ER membrane of the GLUT1 glucose transporter and other proteins with multiple transmembrane α-helical segments.** The N-terminal α helix functions as an internal, uncleaved signal-anchor sequence (red), directing binding of the nascent polypeptide chain to the rough ER membrane and initiating cotranslational insertion. Both SRP and the SRP receptor are involved in this step. Following synthesis of helix 2, which functions as a stop-transfer membrane-anchor sequence, extrusion of the chain through the translocon into the ER lumen ceases. The first two α helices then move out of the translocon into the ER bilayer, anchoring the nascent chain as an α-helical hairpin. The C-terminus of the nascent chain continues to grow in the cytosol. Subsequent α-helical hairpins could insert similarly, although SRP and the SRP receptor are required only for insertion of the first signal-anchor sequence. Although only six transmembrane α helices are depicted here, GLUT1 and proteins of similar structure have twelve or more. [See H. P. Wessels and M. Spiess, 1988, *Cell* **55**:61.]

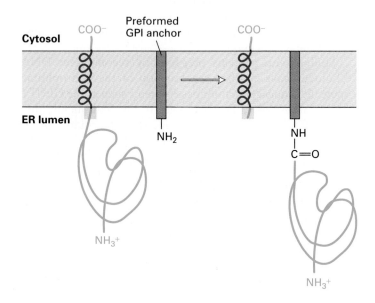

◀ **FIGURE 17-25 Formation of glycosylphosphatidylinositol (GPI)-anchored proteins in the ER membrane.** The protein chain is synthesized and initially inserted into the ER membrane as shown in Figure 17-22. A sequence (yellow highlight) in the exoplasmic-facing domain of the protein, near the hydrophobic stop-transfer membrane-anchor sequence (red), signals cleavage by an endoprotease that simultaneously transfers the carboxyl group of the new C-terminus to the terminal amino group of a preformed GPI anchor. See Figure 3-36a for structure of GPI. [See C. Abeijon and C. B. Hirschberg, 1992, *Trends Biochem. Sci.* **17**:32; I. W. Caras, 1991, *J. Cell Biol.* **113**:77; and K. Kodukula et al., 1992, *Proc. Nat'l. Acad. Sci. USA* **89**:4982.]

by a covalently attached glycosylphosphatidylinositol (GPI) membrane anchor (see Figure 3-36a). These proteins are synthesized and initially anchored to the ER membrane exactly like the insulin receptor, with a cleaved N-terminal signal sequence and internal stop-transfer membrane-anchor sequence directing the process (see Figure 17-22). However, a short sequence of amino acids in the exoplasmic (luminal) domain, adjacent to the membrane-spanning domain, is recognized by an endoprotease that simultaneously cleaves off the original stop-transfer membrane-anchor sequence and transfers the remainder of the protein to a preformed GPI anchor in the membrane (Figure 17-25).

Why change one type of membrane anchor for another? For one thing, attachment of the GPI anchor results in removal of the cytosol-facing hydrophilic domain from the protein. Proteins with GPI anchors can diffuse in the plane of the phospholipid bilayer membrane, whereas proteins anchored by membrane-spanning α helices are frequently attached to the cytoskeleton by their cytosol-facing segments. Also, in certain polarized epithelial cells, the GPI anchor targets the attached protein to the apical domain of the plasma membrane, as we discuss later in this chapter.

SUMMARY *Insertion of Membrane Proteins into the ER Membrane*

- Topogenic sequences direct membrane proteins synthesized on the rough ER to assume their appropriate orientation in the ER membrane. This orientation is retained during transport of a membrane protein to its final destination.

- Topogenic sequences include N-terminal cleaved signal sequences; stop-transfer membrane-anchor sequences; internal uncleaved signal-anchor sequences; and GPI-attachment sequences.

- Many proteins have several membrane-spanning α helices. Each α-helical segment in such multipass proteins functions as an internal uncleaved signal-anchor sequence or a stop-transfer membrane-anchor sequence depending on its location in the polypeptide chain (see Figure 17-24).

17.6 Post-Translational Modifications and Quality Control in the Rough ER

Newly synthesized polypeptides in the membrane and lumen of the ER undergo five principal modifications before they reach their final destinations:

1. Formation of disulfide bonds

2. Proper folding

3. Addition and processing of carbohydrates

4. Specific proteolytic cleavages

5. Assembly into multimeric proteins

The first two and the last of these modifications, which take place exclusively in the rough ER, are discussed in this section. Although addition of some carbohydrates and some proteolytic cleavages also occur in this organelle, many such modifications take place in the Golgi complex or forming secretory vesicles; we discuss these in later sections.

Only properly folded and assembled proteins are transported from the rough ER to the Golgi complex and ultimately to the cell surface or other final destination. Unfolded, misfolded, or partly folded and assembled proteins are selectively retained in the rough ER, or are retrieved from the *cis*-Golgi network and returned to the ER. Misfolded proteins and unassembled subunits of multimeric proteins often move from the ER lumen back through the translocon into the cytosol where they are degraded in proteasomes. We consider several examples of such "quality control" in the second half of this section.

Disulfide Bonds Are Formed and Rearranged in the ER Lumen

In Chapter 3 we learned that both intramolecular and intermolecular **disulfide bonds** (–S–S–) help stabilize the tertiary and quaternary structure of many proteins. These covalent bonds form by the oxidative linkage of **sulfhydryl groups** (–SH), also known as *thiol* groups, on two cysteine residues in the same or different polypeptide chains (see page 53). This reaction can proceed spontaneously only when sufficient oxidant is present. In eukaryotic cells, disulfide bonds are formed in the lumen of the rough ER but not in the cytosol. Thus disulfide bonds are found only in secretory proteins and in the exoplasmic domains of membrane proteins synthesized on the rough ER; soluble cytosolic proteins, synthesized on free ribosomes, lack disulfide bonds and depend on other interactions to stabilize their structures.

Localization of disulfide-bond formation in cells to the ER lumen indicates that this organelle has a favorable redox environment for oxidation of –SH groups, whereas the cytosol does not. The intracellular oxidant required for this reaction has not been identified. However, a mutation in one ER membrane protein renders yeast cells unable to generate disulfide bonds, suggesting that this protein may participate in oxidation of –SH groups in the ER lumen.

The tripeptide *glutathione*, often abbreviated G,

is the major thiol-containing molecule in eukaryotic cells, and serves to prevent the formation of disulfide bonds in the cytosol. Glutathione shuttles between the reduced form (GSH) and the oxidized form, a disulfide-linked dimer (GSSG). The GSH:GSSG ratio is over 50:1 in the cytosol; oxidized GSSG in the cytosol is reduced by the enzyme glutathione reductase, using electrons from the potent reducing agent NADPH (see Figure 16-4):

$$\text{NADPH} + \text{H}^+ + \text{GSSG} \longrightarrow \text{NADP}^+ + 2\,\text{GSH}$$

Thus cytosolic proteins in bacterial and eukaryotic cells do not utilize the disulfide bond as a stabilizing force because the high GSH:GSSG ratio would drive the system in the direction of Cys–SH and away from Cys–S–S–Cys.

In proteins that contain more than one disulfide bond, the proper pairing of cysteine residues is essential for normal structure and activity. Disulfide bonds sometimes are formed sequentially while a polypeptide is still growing on the ribosome. For instance, during synthesis of the immunoglobulin (Ig) light chain, which contains two disulfide bonds, the first and second cysteines closest to the N-terminus form a disulfide bond before the third cysteine has even been added to the nascent chain, automatically ensuring the correct pairing of cysteines. Similarly, the third cysteine pairs with the fourth to create the second disulfide bond.

The disulfide bonds in some proteins, however, do not link cysteines that occur sequentially in the amino acid sequence. For example, proinsulin has three disulfide bonds that link cysteines 1 and 4, 2 and 6, and 3 and 5. In this case, the first disulfide bonds that form spontaneously by oxidation of –SH groups may have to undergo rearrangement for the protein to achieve its proper folded conformation. In cells, the rearrangement of disulfide bonds is accelerated by the enzyme *protein disulfide isomerase* (PDI), which is found in abundance in the ER of secretory tissues in such organs as the liver and pancreas. In catalyzing rearrangement, PDI forms a disulfide-bonded substrate-enzyme intermediate (Figure 17-26a). This enzyme acts on a broad range of protein substrates, allowing them to reach their thermodynamically most stable

conformations. Disulfide bonds generally form in a specific order, first stabilizing small domains of a polypeptide, then stabilizing the interactions of more distant segments; this phenomenon is illustrated by the folding of influenza HA protein discussed below. Although PDI occasionally binds to a protein that is irreversibly misfolded, it can escape from this useless intermediate as shown in Figure 17-26b.

Most proteins used for therapeutic purposes in humans or animals are secretory proteins stabilized by disulfide bonds. When recombinant DNA techniques are used to synthesize mammalian secretory proteins in bacterial cells, the proteins generally are not secreted (even when a bacterial signal sequence replaces the normal one). Rather, they accumulate in the cytosol where they often denature and precipitate, in part because disulfide bonds do not form. Sophisticated chemical methods are required to refold such bacterially produced proteins, an expensive process. Once it became clear that disulfide-bond formation occurs spontaneously only in the ER lumen, biotechnologists realized that bacterial cells are not suitable hosts for the synthesis of mammalian proteins that are normally stabilized by disulfide bonds. Nowadays, cultured animal cells generally are preferred for large-scale production of therapeutic proteins such as monoclonal antibodies, tissue plasminogen activator (an anticlotting agent), and erythropoietin, a hormone that stimulates production of red blood cells.

Correct Folding of Newly Made Proteins Is Facilitated by Several ER Proteins

As discussed in Chapter 3, many reduced, denatured proteins can spontaneously refold into their native state in vitro (see Figure 3-13). In most cases such refolding requires hours to reach completion, yet new secretory proteins generally fold

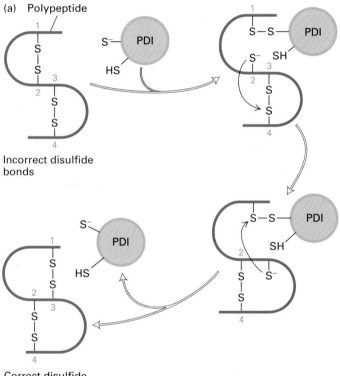

(a) Polypeptide

Incorrect disulfide bonds

Correct disulfide bonds

(b)

Irreversibly misfolded protein

◄ **FIGURE 17-26 (a) Rearrangement of disulfide bonds by protein disulfide isomerase (PDI).** PDI contains an active-site with two reduced cysteine sulfhydryl (–SH) groups. The ionized (–S⁻) form of one of these groups reacts with disulfide (S–S) bonds on nascent or newly completed proteins to form a disulfide-bonded PDI-substrate protein intermediate. This generates a free –S⁻ group on the protein, which, in turn, can react with another disulfide bond in the protein to form a new disulfide bond and another free –S⁻ group. In this way, the disulfide bonds on a protein can rearrange themselves until the most stable conformation for the protein is achieved, and free PDI is released. (b) An escape hatch for PDI. Occasionally PDI forms a disulfide bond with a protein that is irreversibly misfolded. PDI eventually escapes when the second –S⁻ group in the active site attacks the disulfide bond between PDI and the protein, releasing PDI with an internal (intramolecular) disulfide bond. [Part (a) after R. B. Freedman, 1984, *Trends Biochem. Sci.* **9**:438, and R. Noiva and W. Lennarz, 1992, *J. Biol. Chem.* **267**:3553. Part (b) after K. Walker et al., 1997, *J. Biol. Chem.* **272**:8845.]

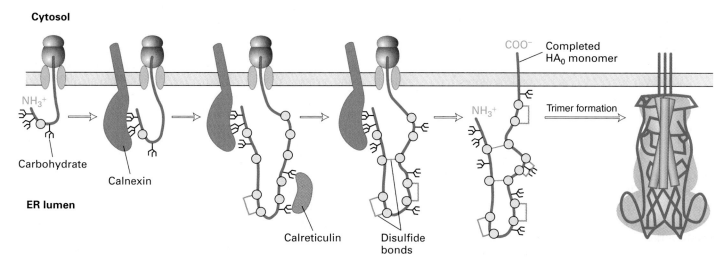

Cytosol

NH₃⁺

Carbohydrate

Calnexin

ER lumen

Calreticulin

COO⁻ Completed HA₀ monomer

NH₃⁺

Trimer formation

Disulfide bonds

▲ **FIGURE 17-27 Folding of the hemagglutinin (HA) precursor polypeptide HA₀ and formation of an HA₀ trimer within the ER.** While the nascent chain is still growing, two protein-folding catalysts, calnexin and calreticulin, associate with it, and three disulfide bonds form in the globular head domain. Following completion of translation, three additional disulfide bonds form and possibly rearrange in the monomer. Three HA₀ chains then interact with each other, initially via their transmembrane α helices; this association apparently triggers the formation of a long stem containing one α helix (dark rod) from the luminal part of each HA₀ polypeptide. Finally, interactions between the three globular heads occur, generating the mature trimeric spike. Calnexin and calreticulin bind to N-linked oligosaccharides with a single glucose residue on unfolded protein segments, thereby promoting the proper folding and assembly of newly synthesized glycoproteins such as HA (see Figure 17-36). [Adapted from sketch by Dan Hebert and Ari Helenius. See M-J. Gething et al., 1986, *Cell* **46**:939; U. Tatu et al., 1995, *EMBO J.* **14**:1340; and D. Hebert et al., 1997, *J. Cell. Biol.* **139**:613.]

into their proper conformation in the ER lumen within minutes after their synthesis. The ER contains several proteins that accelerate the folding of newly synthesized proteins within the ER lumen. Protein disulfide isomerase (PDI) is one such folding catalyst; the chaperone Hsc70 is another (see Figure 17-16). Like cytosolic Hsc70, this ER chaperone transiently binds to proteins and prevents them from misfolding or forming aggregates, thereby enhancing their ability to fold into the proper conformation. Two other ER proteins, the homologous **lectins** *calnexin* and *calreticulin*, bind to certain carbohydrates (discussed later) attached to newly made proteins and aid in protein folding (Figure 17-27).

Other important protein-folding catalysts are *peptidyl-prolyl isomerases*, a family of enzymes that accelerate the rotation about peptidyl-prolyl bonds in unfolded segments of a polypeptide:

Rotation about peptide bond **Prolyl**

cis

trans

Prolyl

Such isomerizations frequently are the rate-limiting step in the folding of protein domains. Many peptidyl-prolyl isomerases can catalyze the rotation of any exposed peptidyl-prolyl bond, but some have very specific substrates. For example, in *Drosophila* an ER peptidyl-prolyl isomerase called NinaA is required for the folding of opsin, the membrane protein that absorbs light and triggers the visual response (see Figure 21-47). Flies with mutations in NinaA lack proper visual responses but otherwise are normal. This finding illustrates the remarkable specificity of some folding catalysts.

Assembly of Subunits into Multimeric Proteins Occurs in the ER

Many important secretory and membrane proteins are built of two or more polypeptides (or subunits). In all cases, these *multimeric* proteins are assembled in the ER. The number and relative positions of the subunits in a multimeric protein constitutes its quaternary structure. One important example is provided by the immunoglobulins, which contain two heavy (H) and two light (L) chains, all linked by S–S bonds (see Figure 3-21). Hemagglutinin (HA), the trimeric protein that forms the spikes protruding from the surface of the influenza virus particle (see Figure 6-13), provides another good illustration of protein folding and subunit assembly in the ER. Each spike is formed within the ER of an infected host cell from three copies of a precursor protein termed HA₀, which has a single membrane-spanning α helix. In the Golgi complex, each of the three HA₀ proteins is cleaved to form two polypeptides, HA₁ and HA₂; thus each spike in the virus particle contains three copies of HA₁ and three of HA$_2$ (see Figure 3-4). The trimer is stabilized by interactions between the exoplasmic domains of the constituent polypeptides as well as by interactions between the three cytosolic and membrane-spanning domains.

Each newly made HA$_0$ polypeptide requires approximately 7 minutes to fold and be incorporated into a trimer. Several approaches have been used to detect different intermediates during the folding of HA$_0$ monomers and their subsequent assembly into trimers in virus-infected cells (see Figure 17-27). One approach employs two monoclonal antibodies: one specific for HA$_0$ monomers and one specific for the correctly folded trimer. In a typical experiment, virus-infected cells are pulse-labeled with a radioactive amino acid; at various times during the subsequent chase period, membranes are solubilized by detergent and exposed to the monomer- or trimer-specific antibody to precipitate HA$_0$ protein. Immediately after the pulse, all of the radioactive HA$_0$ protein is immunoprecipitated by the monomer-specific antibody. During the chase period, increasing proportions of the total radioactive HA$_0$ protein react with the trimer-specific monoclonal antibody. Thus the time course of the monomer-to-trimer conversion within living cells can be followed, and the pattern of disulfide bonds in the immunoprecipitated monomers can be determined to identify various intermediates.

Only Properly Folded Proteins Are Transported from the Rough ER to the Golgi Complex

Almost any mutation in a secretory or membrane protein that prevents it from folding properly also blocks movement of the polypeptide from the lumen or membrane of the rough ER to the Golgi complex. In most cases, improperly folded proteins are permanently bound to the ER chaperones Hsc70 or calnexin. Thus these luminal folding catalysts perform two related functions: assisting in the folding of normal proteins by preventing their aggregation and binding to irreversibly misfolded proteins.

 One medically important example is a mutation in the secretory protein α$_1$-antiprotease, which is secreted by hepatocytes and macrophages; this protein binds to and inhibits trypsin and also the blood protease elastase. In the absence of α$_1$-antiprotease, elastase degrades the fine tissue in the lung that participates in the absorption of oxygen. A genetic inability to produce functional α$_1$-antiprotease, widespread in Caucasians, is the major genetic cause of emphysema (destruction of lung tissue by unchecked elastase). The defect is due to a single mutation in α$_1$-antiprotease that causes lysine to replace glutamate 342. Although the mutant α$_1$-antiprotease is synthesized in the rough ER, it does not fold properly, forming an almost crystalline aggregate that is not exported from the ER. In hepatocytes, the secretion of other proteins also becomes impaired as the rough ER is filled with aggregated α$_1$-antiprotease.

Both mammalian cells and yeasts respond to the presence of unfolded proteins in the rough ER by increasing transcription of several genes encoding ER chaperones and other folding catalysts, including Hsc70, peptidyl-prolyl isomerase, and protein disulfide isomerase. A key element in this *unfolded-protein response* is IRE1, a transmembrane protein in the inner nuclear membrane, which is continuous with the ER membrane. As shown in Figure 17-28, binding of unfolded proteins in the ER lumen to the luminal surface of IRE1 promotes formation of the functional mRNA encoding transcription factor HAC1, which activates transcription of the genes induced in the unfolded-protein response. Thus, the increase in the level of HAC1 in response to an accumulation of unfolded proteins in the ER lumen leads to increased synthesis of precisely those proteins that assist in protein folding.

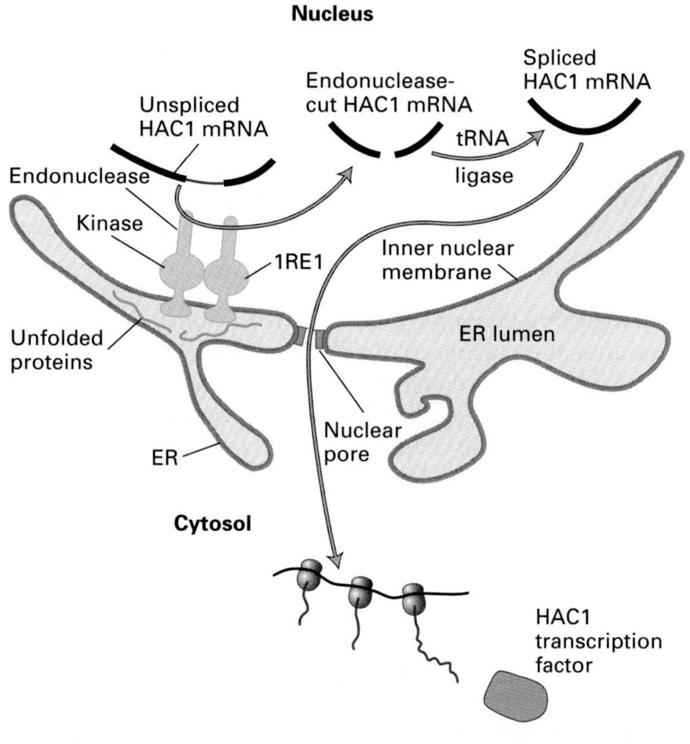

▲ **FIGURE 17-28 The unfolded-protein response.** IRE1 is a transmembrane protein in the inner nuclear membrane, a membrane that is continuous with the ER membrane. This multifunctional protein (green) has a binding site for unfolded proteins (blue) on its luminal surface; its nuclear-facing domain contains a protein kinase of unknown function and a specific RNA endonuclease. Binding of unfolded proteins in the ER lumen dimerizes the receptor and somehow activates the endonuclease, which cleaves the unspliced mRNA precursor encoding the transcription factor HAC1. The two exons of HAC1 mRNA then are linked together by tRNA ligase, which usually splices tRNA precursors, forming a functional HAC1 mRNA. Following its synthesis in the cytosol, HAC1 protein moves back into the nucleus and activates transcription of genes encoding several chaperones and other proteins that assist in folding unfolded proteins in the ER lumen. [See C. Sidrauski and P. Walter, 1997, *Cell* **90**:1031.]

Many Unassembled or Misfolded Proteins in the ER Are Transported to the Cytosol and Degraded

Mutant misfolded secretory and membrane proteins, as well as the unassembled subunits of multimeric proteins, often are degraded within an hour or two after their synthesis in the rough ER. For many years, researchers thought that proteolytic enzymes within the ER catalyzed degradation of misfolded or unassembled polypeptides, but such proteases were never found. Recent studies have shown that misfolded membrane and secretory proteins are transported from the ER lumen, "backwards" through the translocon, into the cytosol where they are degraded by the ubiquitin-mediated proteolytic pathway (see Figure 3-18). Two ubiquitin-conjugating enzymes form a complex localized to the cytosolic face of the ER and recruit a proteasome to the ER membrane where it can degrade misfolded proteins as they are extruded back through the translocon. The addition of ubiquitin to misfolded ER proteins, which is coupled to hydrolysis of ATP, may provide some of the energy required to drag these proteins back to the cytosol.

How misfolded membrane proteins are recognized and targeted to the translocon for export to the cytosol is not known. However, studies with the T-cell receptor suggest that the membrane-spanning region is critical to recognition. The α subunit of the T-cell receptor has a single membrane-spanning segment with two charged amino acids that are normally bound by ionic linkage (within the phospholipid bilayer) to residues of opposite charge in two other subunits of the T-cell receptor. If these other subunits are absent, the charged residues in the α subunit cause the entire polypeptide to assume an abnormal conformation, triggering its export via the translocon into the cytosol. Mutation of the charged residues to neutral ones prevents degradation of the α subunit when the other T-cell receptor subunits are absent—evidence that the membrane-spanning segment is recognized by the machinery that targets a protein back to the cytosol for degradation.

ER-Resident Proteins Often Are Retrieved from the *Cis*-Golgi

Another aspect of ER quality control is the retention in the ER lumen of soluble "resident" proteins such as Hsc70 and PDI, which catalyze the folding of newly made proteins. Most soluble proteins synthesized on the rough ER eventually are secreted from the cell surface or transported to the lumen of lysosomes (see Figure 17-13). How, then, are resident proteins retained in the ER lumen to carry out their work?

The answer lies in a specific C-terminal sequence present in resident ER proteins and in a receptor that recognizes this sequence. PDI, luminal Hsc70, and many other ER-resident proteins have a Lys-Asp-Glu-Leu (KDEL in the one-letter code) sequence at their C-terminus. Several experiments demonstrated that this *KDEL sequence* is both necessary and sufficient for retention in the ER. For instance, when a mutant

PDI protein lacking these four residues is synthesized in a fibroblast, the protein is secreted. Moreover, if a protein that normally is secreted is altered so that it contains these four amino acids at its C-terminus, the protein is retained in the ER.

The *KDEL receptor* acts mainly to retrieve proteins with the KDEL recognition sequence that have escaped to the *cis*-Golgi network and return them to the ER. In support of this concept is the finding that most KDEL receptors are localized to the membranes of small transport vesicles shuttling between the ER and the *cis*-Golgi and to the membranes of the *cis*-Golgi reticulum (Figure 17-29). In addition,

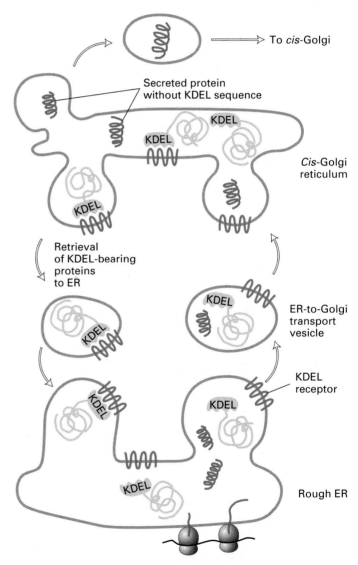

▲ **FIGURE 17-29 Role of the KDEL receptor in the retrieval of ER-resident proteins.** Many resident proteins in the ER lumen bear a C-terminal KDEL (Lys-Asp-Glu-Leu) sequence that localizes them to the ER. The KDEL receptor is located mainly in the *cis*-Golgi network and in ER-to-Golgi transport vesicles; its chief function is to bind proteins with the KDEL recognition sequence and return them to the ER. [After J. Semenza et al., 1990, *Cell* **61**:1349.]

ER-localized proteins that carry the KDEL recognition sequence have oligosaccharide chains with modifications that are made by enzymes found only in the *cis*-Golgi or *cis*-Golgi reticulum; thus at some time these proteins must have been transported at least to the *cis*-Golgi network. Unfortunately, we do not yet know how the movements of the KDEL receptor are controlled.

Clearly, the transport of newly made proteins from the rough ER to the Golgi cisternae is a highly selective and regulated process; the selective entry of proteins into membrane-bounded transport vesicles is an important feature of protein targeting—one we will encounter several times in our study of the subsequent stages in the maturation of secretory and membrane proteins.

SUMMARY Post-Translational Modifications and Quality Control in the Rough ER

- The redox environment in the lumen of the rough ER is favorable for oxidation of cysteine sulfhydryl groups (–SH) to disulfide bonds (–S–S–), whereas this reaction does not occur in the cytosol. Thus disulfide bonds are common in secretory proteins and exoplasmic domains of membrane proteins, but are absent from soluble cytosolic proteins.

- Protein disulfide isomerase (PDI), an enzyme localized to the ER lumen, catalyzes the rearrangement of disulfide bonds, accelerating the folding of newly synthesized secretory and membrane proteins in the ER.

- The folding of many newly made proteins within the ER is facilitated by other folding catalysts such as peptidyl-prolyl isomerases and the lectins calnexin and calreticulin (see Figure 17-27).

- Assembly of subunits to form multimeric secretory and membrane proteins occurs in the ER.

- Only properly folded proteins are transported from the rough ER to the Golgi complex of vesicles.

- Abnormally folded proteins and unassembled subunits are selectively retained in the ER, either because they form aggregates or because they are permanently bound to Hsc70 or other ER chaperones.

- Unassembled or misfolded proteins in the ER often are transported back through the translocon to the cytosol where they are degraded in the ubiquitin/ proteasome pathway.

- Accumulation of misfolded proteins in the ER lumen induces increased production of ER protein-folding catalysts via the unfolded-protein response (see Figure 17-28).

- Certain resident ER proteins are retained in the ER by a C-terminal KDEL sequence or are retrieved there from the *cis*-Golgi network by the KDEL receptor (see Figure 17-29).

17.7 Protein Glycosylation in the ER and Golgi Complex

As noted previously, most plasma-membrane and secretory proteins contain one or more carbohydrate chains; indeed, the addition and subsequent processing of carbohydrates (*glycosylation*) is the principal chemical modification to most such proteins. Some glycosylation reactions occur in the lumen of the ER; others, in the lumina of the *cis*-, *medial*-, or *trans*-Golgi cisternae. Thus the presence of certain carbohydrate residues on proteins provide useful markers for following their movement from the ER and through the Golgi cisternae. In this section we first review the structures of the oligosaccharide chains commonly found in glycoproteins and then discuss their synthesis and function.

Different Structures Characterize *N*- and *O*-Linked Oligosaccharides

The structures of **N- and O-linked oligosaccharides** are very different, and different sugar residues are usually found in each type (Figure 17-30). For instance, *O*-linked oligosaccharides are linked to the hydroxyl group of serine or threonine via *N*-acetylgalactosamine (GalNac) or (in collagens) to the hydroxyl group of hydroxylysine via galactose. In *all* *N*-linked oligosaccharides, *N*-acetylglucosamine (GlcNAc) is linked to the amide nitrogen of asparagine. *O*-linked oligosaccharides are generally short, often containing only one to four sugar residues. Typical *N*-linked oligosaccharides, in contrast, always contain mannose as well as *N*-acetylglucosamine and usually have several branches each terminating with a negatively charged sialic acid residue. Most cytosolic and nuclear proteins are not glycosylated; the exceptions are several transcription factors and a protein localized to the nuclear-pore complex, which have a single *N*-acetylglucosamine residue linked to a serine or threonine hydroxyl group.

The different structures of *N*- and *O*-linked oligosaccharides reflect differences in their biosynthesis. *O*-linked sugars are added one at a time, and each sugar transfer is catalyzed by a different **glycosyltransferase** enzyme. In contrast, biosynthesis of *N*-linked oligosaccharides begins with the addition of a large preformed oligosaccharide, containing 14 sugar residues; subsequently certain sugar residues are removed and others are added, one at a time, in a defined order with each reaction catalyzed by a different enzyme. As described below, the various steps in the formation of both *O*- and *N*-linked oligosaccharides occur in specific organelles.

O-Linked Oligosaccharides Are Formed by the Sequential Transfer of Sugars from Nucleotide Precursors

The immediate precursors used in the biosynthesis of oligosaccharides are nucleoside diphosphate or monophosphate

(a) *O*-linked oligosaccharides

(b) *N*-linked complex oligosaccharides

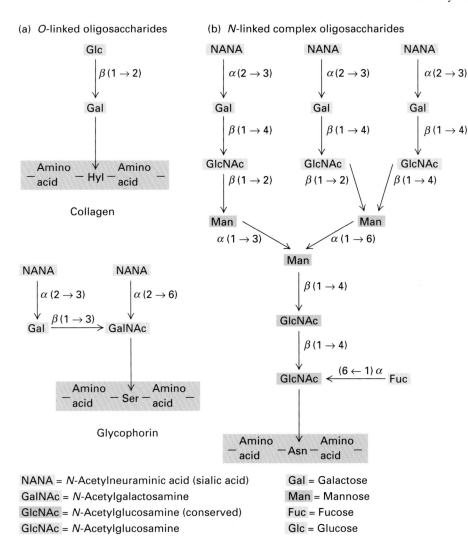

FIGURE 17-30 Structures of typical *O*-linked and *N*-linked oligosaccharides. (a) The *O*-linked oligosaccharides in glycophorin and many other glycoproteins are linked to the hydroxyl group in serine (Ser) and threonine residues by *N*-acetyl-galactosamine. Collagens contain a characteristic glucose → galactose disaccharide attached to hydroxylysine (Hyl) residues. (b) The *N*-linked oligosaccharides found in mammalian serum glycoproteins exhibit various structures, but all contain the five sugars highlighted in purple, are branched, and are linked to the amide nitrogen of asparagine (Asn). [After R. Kornfeld and S. Kornfeld, 1985, *Annu. Rev. Biochem.* **45**:631.]

NANA = *N*-Acetylneuraminic acid (sialic acid) Gal = Galactose
GalNAc = *N*-Acetylgalactosamine   Man = Mannose
GlcNAc = *N*-Acetylglucosamine (conserved)   Fuc = Fucose
GlcNAc = *N*-Acetylglucosamine   Glc = Glucose

sugars (Figure 17-31). The ester bond between the phosphate residue and the carbon atom in the sugar is a high-energy bond with a $\Delta G°'$ of hydrolysis of about −5 kcal/mol. Thus the transfer of the sugar residue to an acceptor hydroxyl group, on a serine or threonine residue or on another sugar residue, is energetically favored.

All known glycosyltransferases that act on secretory proteins are integral membrane proteins with active sites facing the lumen of the organelle. Each glycosyltransferase is specific for both the donor sugar nucleotide and the acceptor molecule. The galactosyltransferase depicted in Figure 17-32, for instance, only transfers a galactose residue (from UDP-galactose), and only to the 3 carbon atom of an acceptor *N*-acetylgalactosamine residue. A different enzyme transfers galactose to the 4 carbon of *N*-acetylglucosamine, and yet another transfers galactose to the 3 carbon of galactose.

Biosynthesis of the *O*-linked oligosaccharide in glycophorin and similar glycoproteins (see Figure 17-30a) begins with transfer of *N*-acetylgalactosamine (GalNAc) from UDP −*N*-acetylgalactosamine to the hydroxyl group of a serine or threonine residue in the protein. This reaction is catalyzed

by a GalNAc transferase that is localized to the rough ER or the *cis*-Golgi network. After the protein has moved to the *trans*-Golgi vesicles, a galactose residue is added to the N-acetylgalactosamine by a specific *trans*-Golgi galactosyltransferase. In vertebrate cells biosynthesis of typical *O*-linked oligosaccharides is completed by the addition of two negatively charged N-acetylneuraminic acid (also called sialic acid) residues from a CMP precursor (see Figure 17-31); these reactions also occur in the *trans*-Golgi or the *trans*-Golgi network.

All the sugar nucleotides used in the synthesis of glycoproteins and glycolipids are made in the cytosol from nucleoside triphosphates and sugar phosphates. Specific antiport proteins in the membranes of the rough ER and Golgi cisternae catalyze the import of the sugar nucleotides into the lumina of these organelles and the export of free nucleotides (UMP, CMP, and GMP) generated within the organelles (Figure 17-33). The one-for-one exchanges catalyzed by these antiporters maintain the concentration of sugar nucleotides in the rough ER and Golgi lumina at a constant level, a requirement for oligosaccharide synthesis.

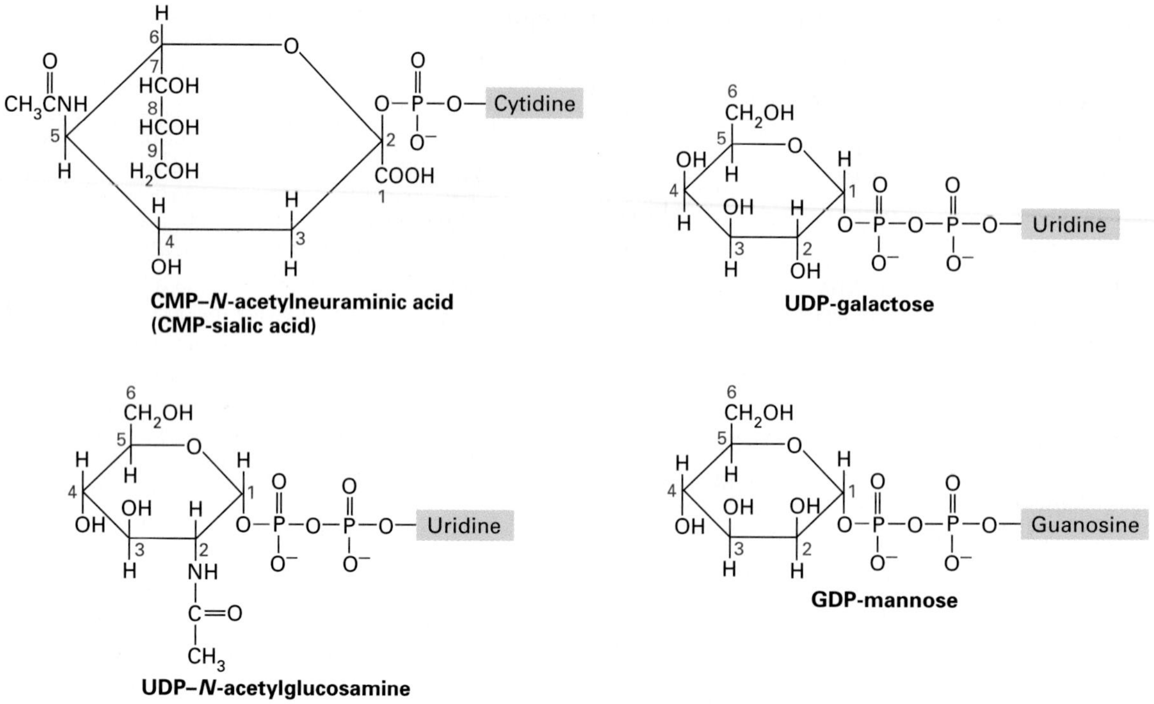

▲ FIGURE 17-31 **Structures of four sugar nucleotides used in the biosynthesis of oligosaccharides found in glycoproteins.** Cleavage of the high-energy phosphoester bond indicated in red provides the energy for transfer of the sugar residue to an acceptor group.

▶ FIGURE 17-32 **A specific glycosyltransferase catalyzes addition of a galactose residue from UDP-galactose to carbon atom 3 of *N*-acetylgalactosamine attached to a protein forming a β1→3 linkage.** This reaction is the second step in the formation of typical *O*-linked oligosaccharides in proteins such as glycophorin (see Figure 17-30a). Glycosyltransferases are specific both for the nucleoside sugar donor and for the carbon atom of the acceptor sugar to which it is transferred.

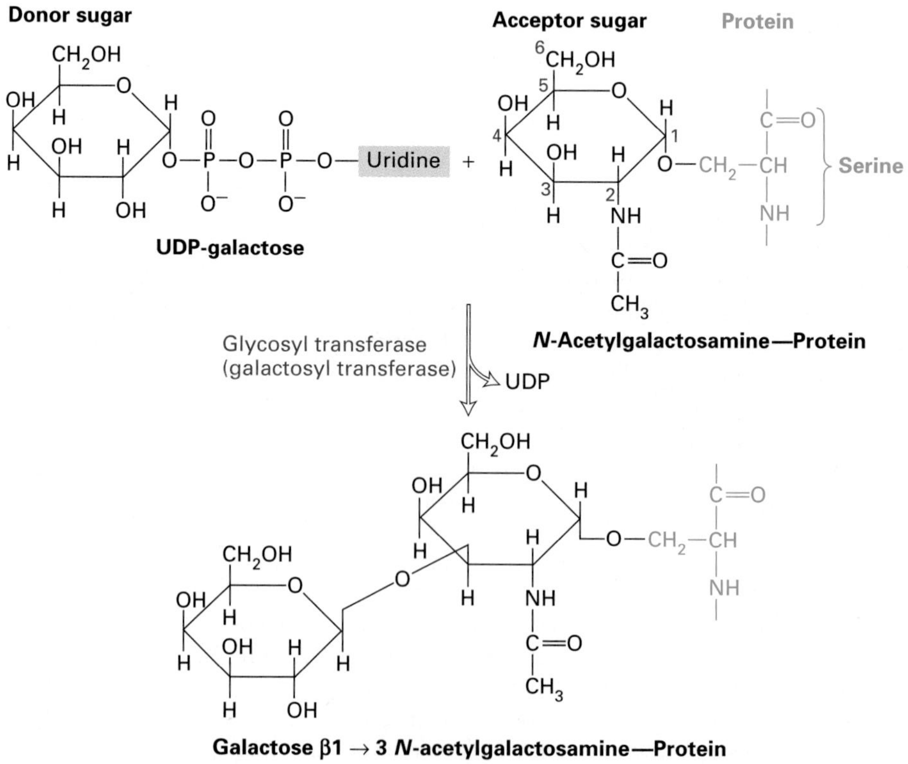

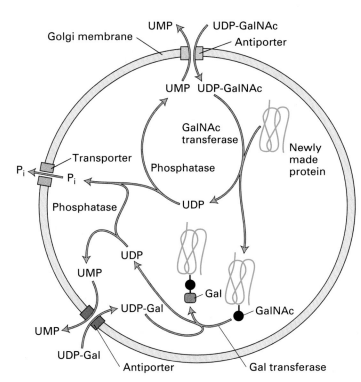

▲ FIGURE 17-33 The antiport uptake of nucleotide sugars into Golgi cisternae. Both UDP-galactose (UDP-Gal) and UDP–N-acetylgalactosamine (UDP-GalNAc) enter from the cytosol in exchange for UMP; these transport processes are mediated by two different antiporters located in the Golgi membrane. UMP is produced by phosphatase action on UDP, a product of the galactosyltransferase and N-acetylgalactosaminyltransferase reactions. A different transporter allows the inorganic phosphate (Pᵢ) formed from UDP to exit the Golgi cisterna. Other known antiporters allow CMP–N-acetylneuraminic acid to enter in exchange for CMP. [See C. Abeijon et al., 1997, *Trends Biochem. Sci.* **22**:203.]

ABO Blood Type Is Determined by Two Glycosyltransferases

The human **A, B, and O blood-group antigens** illustrate the importance of specific glycosyltransferases. These antigens, which can trigger harmful immune reactions, are oligosaccharides present on both glycoproteins and glycolipids on the surface of erythrocytes and many other types of cells. Each antigenic determinant consists of one of three structurally related oligosaccharides attached to a ceramide lipid or a serine or threonine residue on a protein (Figure 17-34). The A antigen is similar to O, except that the A antigen contains an N-acetylgalactosamine attached to the outer galactose residue; the B antigen is also similar to O, except for an extra galactose residue attached to the outer galactose.

All people have the enzymes needed to synthesize the O antigen. People with type A blood also have the GalNAc transferase that adds the extra N-acetylgalactosamine; those with type B blood have the Gal transferase that adds the extra galactose. People with type AB blood have both transferases and synthesize both the A and B antigens; those with type O make only the O antigen. Interestingly, the sequences of GalNAc (A antigen) and Gal (B antigen) transferases differ in just three amino acids, which determine whether the enzyme binds to UDP-Gal or UDP-GalNAc as substrate. Clearly the genes encoding the two enzymes evolved from a common ancestor.

Table 17-5 summarizes the relevance of the A, B, and O antigens to blood transfusions. For example, people who lack the Gal transferase and thus cannot synthesize the B antigen (blood types A and O) normally have antibodies against the B antigen in their serum. Thus, when type B or AB blood is transfused into a person with blood type A or O, the anti-B antibodies of the recipient bind to the transfused erythrocytes and trigger an immune reaction leading

▶ FIGURE 17-34 The human ABO blood-group antigens. The structure of the terminal sugars in the oligosaccharide component of these glycolipids and glycoproteins distinguish the three antigens. The presence or absence of particular glycosyltransferases determine an individual's blood type. [See T. Feizi, 1990, *Trends Biochem. Sci.* **15**:330 for a review.]

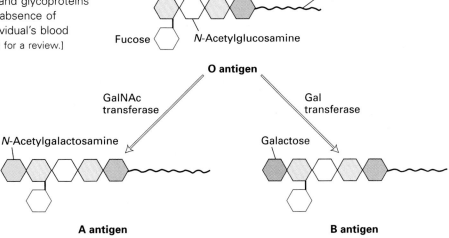

to their destruction. To avoid such harmful reactions, blood-group typing and appropriate matching of blood donors and recipients is required in all transfusions.

A Common Preformed N-Linked Oligosaccharide Is Added to Many Proteins in the Rough ER

As noted earlier, biosynthesis of all N-linked oligosaccharides begins in the rough ER with addition of a large preformed oligosaccharide precursor. This precursor oligosaccharide is linked by a pyrophosphoryl residue to *dolichol*, a long-chain (75–95 carbon atoms) polyisoprenoid lipid that is firmly embedded in the ER membrane and acts as a carrier for the oligosaccharide. The dolichol pyrophosphoryl oligosaccharide is formed on the ER membrane in a complex set of reactions catalyzed by enzymes attached to the cytosolic and luminal faces of the rough ER membrane (Figure 17-35). The final dolichol pyrophosphoryl oligosaccharide is oriented so that the oligosaccharide portion faces the ER lumen.

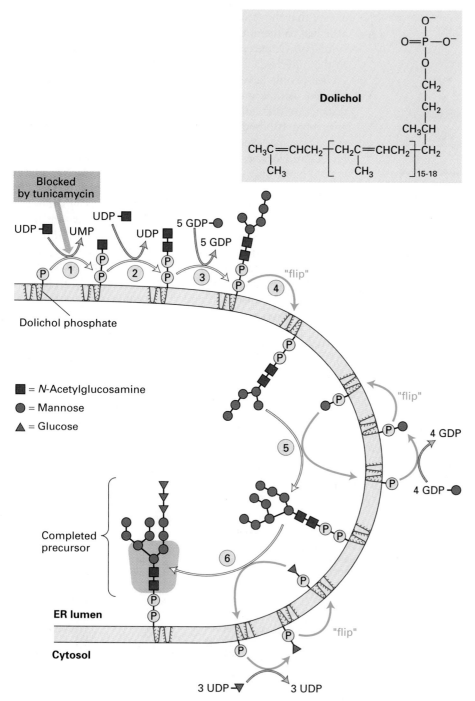

◀ **FIGURE 17-35 Biosynthesis of the dolichol pyrophosphoryl oligosaccharide precursor of N-linked oligosaccharides.** Dolichol phosphate *(inset)* is strongly hydrophobic and long enough to span a phospholipid bilayer membrane four or five times. Two N-acetylglucosamine, one phosphate, and five mannose residues from UDP sugars are added one at a time to a dolichol phosphate on the cytosolic face of the ER membrane ①–③. Tunicamycin, which blocks the first enzyme in this pathway, inhibits the synthesis of all N-linked oligosaccharides in cells. Then the dolichol pyrophosphoryl oligosaccharide is flipped to the luminal face (step ④) where the remaining four mannose and all three glucose residues are added (steps ⑤, ⑥), yielding the completed $Glc_3Man_9(GlcNAc)_2$ precursor, which is transferred to proteins. In the latter reactions the mannose or glucose is first transferred from a nucleotide sugar to a carrier dolichol phosphate on the cytosolic face of the ER; the carrier is then flipped to the luminal face where the glucose or mannose is transferred to the growing oligosaccharide, and the carrier is flipped back again to the cytosolic face. Five residues in the final precursor (highlighted in purple) are conserved in the structures of all N-linked oligosaccharides on secreted proteins (see Figure 17-30b). [After C. Abeijon and C. B. Hirschberg, 1992, *Trends Biochem. Sci.* **17**:32.]

TABLE 17-5 ABO Blood Groups

Blood-group Type	Antigens on RBCs	Serum Antibodies	Can Receive Blood Types
A	A	Anti-B	A and O
B	B	Anti-A	B and O
AB	A and B	None	All
O	None	Anti-A and anti-B	O

The structure of this precursor is the same in plants, animals, and single-celled eukaryotes—a branched oligosaccharide, containing three glucose (Glc), nine mannose (Man), and two *N*-acetylglucosamine (GlcNAc) molecules, which can be written as $Glc_3Man_9(GlcNAc)_2$. Five of its 14 residues are conserved in the structures of all *N*-linked oligosaccharides on secretory and membrane proteins, as can be seen by comparing Figures 17-35 and 17-30b.

The entire $Glc_3Man_9(GlcNAc)_2$ oligosaccharide is transferred en bloc from the dolichol carrier to an asparagine residue on a nascent polypeptide, a reaction catalyzed by *oligosaccharide-protein transferase* (Figure 17-36, step 1). Only asparagine residues in the tripeptide sequences Asn-X-Ser and Asn-X-Thr (where X is any amino acid except proline) are substrates for this transferase. Two of the three subunits of the transferase are *ribophorins*, abundant integral ER membrane proteins whose cytosol-facing domains bind tightly to the larger subunit of the ribosome. This binding localizes the third subunit, which is located within the ER lumen and carries out the transfer reaction, near the growing polypeptide chain. Not all Asn-X-Ser/Thr sequences become glycosylated; for instance, the rapid folding of a segment of a protein containing an Asn-X-Ser/Thr sequence may prevent the transfer of $Glc_3Man_9(GlcNAc)_2$ to it.

Immediately after the oligosaccharide is transferred to a nascent polypeptide, all three glucose residues and one particular mannose residue are removed by three different enzymes (Figure 17-36, steps 2–4). The three glucose residues, which are the last residues added in synthesis of $(Glc)_3(Man)_9(GlcNAc)_2$ on the dolichol carrier, appear to act as a signal that the oligosaccharide is complete and ready to be transferred to a protein.

The ER lumen also contains a glucosyltransferase that adds back one glucose residue to a protein-linked $Man_{7-9}(GlcNAc)_2$ oligosaccharide (Figure 17-36, step 3a). This enzyme glucosylates unfolded and misfolded, but not native, folded glycoproteins. The ER also contains two related lectins (carbohydrate-binding proteins)—membrane-attached calnexin and luminal calreticulin—that selectively bind reglucosylated $Glc_1Man_{7-9}(GlcNAc)_2$ oligosaccharides and prevent folding of the adjacent amino acid segments (see Figure 17-27). Occasionally proteins spontaneously dissociate from calnexin or calreticulin and immediately are deglucosylated; if they then fold properly, they will not be reglucosylated nor rebind to a lectin, and will pass to the Golgi. Thus, like Hsc70, calnexin and calreticulin help prevent premature folding of segments of a newly made protein and also retain unfolded or misfolded proteins within the ER.

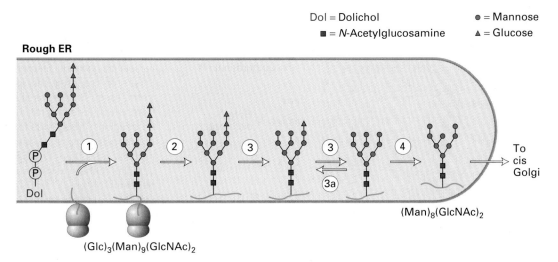

Dol = Dolichol   ● = Mannose
■ = *N*-Acetylglucosamine   ▲ = Glucose

Rough ER

To cis Golgi

$(Man)_8(GlcNAc)_2$

$(Glc)_3(Man)_9(GlcNAc)_2$

▲ **FIGURE 17-36 Addition and initial processing of *N*-linked oligosaccharides in the rough ER of vertebrate cells.** The $Glc_3Man_9(GlcNAc)_2$ precursor is transferred from the dolichol carrier to a susceptible asparagine residue on a nascent protein as soon as the asparagine crosses to the luminal side of the ER (step ①). In three separate reactions, first one glucose residue (step ②), then two glucose residues (step ③), and finally one mannose residue ④ are removed. Re-addition of one glucose residue (step ③a) is part of the ER quality-control process (see text). Following these reactions, the newly made protein is transported in a vesicle to the Golgi for further processing of the oligosaccharide (see Figure 17-38). [See R. Kornfeld and S. Kornfeld, 1985, *Annu. Rev. Biochem.* **45**:631, and M. Sousa and A. J. Parodi, 1995, *EMBO J.* **14**:4196.]

Modifications to *N*-Linked Oligosaccharides Are Completed in the Golgi Complex

Newly made proteins that undergo *N*-linked glycosylation in the ER enter the Golgi complex bearing one or more $Man_8(GlcNAc)_2$ oligosaccharide chains. Biologists traditionally have considered the series of flattened and spherical sacs (cisternae) composing the Golgi complex as a single organelle (Figure 17-37). However, the *cis, medial,* and *trans* cisternae of the Golgi contain different sets of enzymes that introduce different modifications to secretory and membrane proteins; thus each region in effect functions as a distinct organelle.

Among the enzymes localized to specific regions of the Golgi are those that catalyze additional modifications to the $Man_8(GlcNAc)_2$ oligosaccharide chains in glycoproteins produced in the rough ER. Figure 17-38 depicts the sequential reactions that add and remove specific sugar residues to yield a typical *N*-linked complex oligosaccharide in vertebrate cells. Different enzymes localized to the *cis-, medial-,* and *trans-*Golgi cisternae catalyze these reactions as a protein moves through the Golgi complex en route to the cell's exterior. For instance, galactosyltransferase (reaction 6) is localized to the *trans*-most Golgi cisternae, and sialyltransferase (reaction 7) is found in the *trans*-Golgi and the *trans*-Golgi network. In fact, galactosyltransferase is frequently used as a marker enzyme for *trans*-Golgi vesicles during subcellular fractionation procedures (see Figures 5-23 and 5-24).

Variations in the structures of *N*-linked oligosaccharides occur as a result of differences in oligosaccharide processing within the ER and Golgi. In some cases, for instance, reactions 1 or 2 in Figure 17-38 may not occur, perhaps because the relevant part of the *N*-linked oligosaccharide is not accessible to the enzymes catalyzing these reactions. As a consequence, no further carbohydrate modifications will occur since the substrate for the next enzyme in the pathway is not generated. The resulting glycoproteins, which can be secreted, contain a *high-mannose* oligosaccharide, either $Man_8(GlcNAc)_2$ or $Man_5(GlcNAc)_2$, rather than the complex oligosaccharide yielded by the complete pathway shown in Figure 17-38. Other differences in processing yield still other *N*-linked oligosaccharides in some vertebrate glycoproteins.

Processing of the common *N*-linked oligosaccharide precursor, $Glc_3Man_9(GlcNAc)_2$, often differs among species, posing difficulties for the biotechnology and xenotransplantation industries. For example, most mammals, with the exception of humans and other Old World primates, occasionally add a galactose residue, instead of a *N*-acetylneuraminic acid residue, to galactose, forming a terminal $Gal(\alpha1 \rightarrow 3)Gal$ disaccharide on some branches of an *N*-linked oligosaccharide. Because of persistent infection by microorganisms that contain $Gal(\alpha1 \rightarrow 3)Gal$ disaccharides, human blood always contains antibodies to this disaccharide epitope. Thus recombinant proteins produced in nonprimate cells frequently are unsuitable for therapeutic use in humans. Moreover, transplantation of organs—such as a heart, pancreas, or liver—from pigs or other animals into humans is thwarted by reaction of human $Gal(\alpha1 \rightarrow 3)Gal$ antibodies to these epitopes on membrane proteins of the organ, resulting in immediate immune destruction of the transplanted tissue.

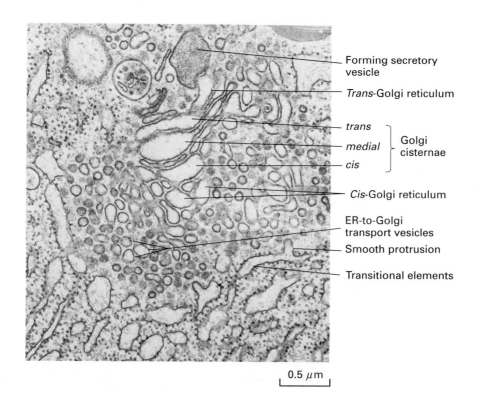

0.5 μm

◀ **FIGURE 17-37 Electron micrograph of the Golgi complex in an exocrine pancreatic cell.** The stacked cisternae of the Golgi complex and a forming secretory vesicle are evident. Elements of the rough ER are on the left in this image. Adjacent to the rough ER are transitional elements from which smooth protrusions appear to be budding. These buds form the vesicles that transport secretory and membrane proteins from the rough ER to the Golgi complex. Other vesicles seen at the periphery of the *cis* side of the Golgi complex fuse to form the *cis*-Golgi reticulum, which lies between the ER and the *cis*-Golgi. See also Figure 5-49. [Courtesy G. Palade.]

Labels on figure:
- Forming secretory vesicle
- *Trans*-Golgi reticulum
- trans
- medial
- cis
- Golgi cisternae
- *Cis*-Golgi reticulum
- ER-to-Golgi transport vesicles
- Smooth protrusion
- Transitional elements

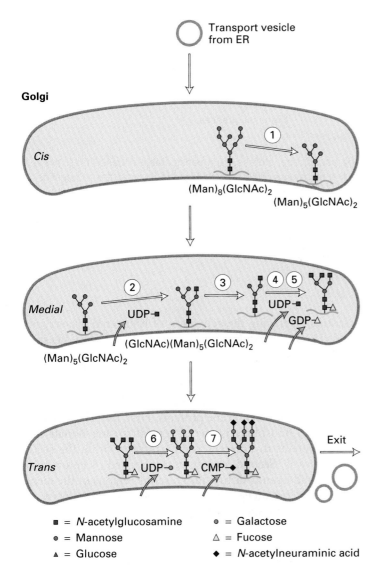

◀ **FIGURE 17-38 Processing of glycoproteins within *cis*-, *medial*-, and *trans*-Golgi cisternae to yield *N*-linked complex oligosaccharides in vertebrate cells.** The enzymes catalyzing each step are localized to the indicated compartments. After removal of three mannose residues in the *cis*-Golgi (step ①), the protein moves by cisternal progression to the *medial*-Golgi. Here, two more mannose residues are removed (step ③), three GlcNAc residues are added (steps ②, ④), and a single fucose is added (step ⑤). Processing is completed in the *trans*-Golgi by addition of three galactose residues (step ⑥) and finally linkage of an N-acetylneuraminic acid residue to each of the galactose residues (step ⑦). Note that added sugars are transferred to the oligosaccharide, one at a time, from sugar nucleotide precursors by specific transferase enzymes, as in the synthesis of *O*-linked oligosaccharides (see Figure 17-32). The sugar nucleotides are imported from the cytosol. [See R. Kornfeld and S. Kornfeld, 1985, *Annu. Rev. Biochem.* **45**:631.]

Oligosaccharides May Promote Folding and Stability of Glycoproteins

As indicated in Figure 17-35, the antibiotic tunicamycin blocks the first step in formation of the dolichol-linked precursor of N-linked oligosaccharides. Studies with this antibiotic indicate that some proteins require N-linked oligosaccharides in order to fold properly in the ER. In the presence of tunicamycin, for instance, the hemagglutinin precursor polypeptide (HA_0) is synthesized, but it cannot fold properly and form a normal trimer (see Figure 17-28); in this case, the protein remains, misfolded, in the rough ER. Moreover, mutation of just one asparagine that normally is glycosylated to a glutamine residue in the HA sequence, thereby preventing addition of an N-linked oligosaccharide to that site, causes the protein to accumulate in the ER in an unfolded state.

Many secretory proteins, however, fold properly and are transported to their final destination even if the addition of

all N-linked oligosaccharides is blocked. For example, both glycosylated and nonglycosylated fibronectin (a constituent of the extracellular matrix) are secreted at the same rate and to the same extent by fibroblasts. But nonglycosylated fibronectin, produced in the presence of tunicamycin, is degraded more rapidly by tissue proteases than is normal glycosylated fibronectin. Similarly, recombinant erythropoietin that lacks its normal N-linked oligosaccharides is as potent as the normal hormone in stimulating the growth of erythrocyte precursors in culture. However, when injected into humans, the nonglycosylated hormone is much less potent than the normal protein because it is degraded much faster than normal. These results establish that oligosaccharide chains confer stability on many extracellular glycoproteins.

Oligosaccharides on cell-surface glycoproteins also play a role in cell-cell adhesion (Chapter 22.3). For example, the plasma membrane of white blood cells (leukocytes) contains cell-adhesion molecules (CAMs) that are extensively glycosylated. The oligosaccharides in these molecules interact with a lectin-type domain in certain CAMs found on endothelial cells lining blood vessels. This interaction tethers the leukocytes to the endothelium and assists in their movement into tissues during an inflammatory response to infection.

Mannose 6-Phosphate Residues Target Proteins to Lysosomes

Another function of some N-linked oligosaccharides is to target lysosomal enzymes to lysosomes and prevent their secretion. The addition and initial processing of the preformed N-linked oligosaccharide precursor in the rough ER is the same for lysosomal enzymes as for membrane and secretory proteins (see Figure 17-36). In the *cis*-Golgi, one or more mannose residues in the resulting $Man_8(GlcNAc)_2$ oligosaccharides become phosphorylated via two sequential reactions. In the first reaction, an N-acetylglucosamine phosphate residue is added to the 6-carbon atom of a mannose

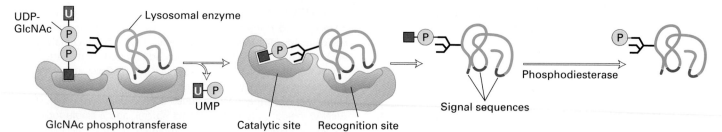

▲ **FIGURE 17-39 Phosphorylation of mannose residues on lysosomal enzymes.** In the first reaction, an *N*-acetylglucosamine (GlcNAc) phosphotransferase in the *cis*-Golgi transfers an *N*-acetylglucosamine phosphate group to carbon atom 6 of one or more mannose residues. This enzyme has a recognition site that binds to signal segments (red) present only in cathepsin D and other lysosomal enzymes. The sugar residues modified by the phosphotransferase are a considerable distance from these signal segments, indicating that in this enzyme the recognition site and the catalytic site are distinct. In the second reaction, a phosphodiesterase removes the GlcNAc group, leaving a phosphorylated mannose residue on the lysosomal enzyme. [See A. B. Cantor et al., 1992, *J. Biol. Chem.* **267**:23349, and S. Kornfeld, 1987, *FASEB J.* **1**:462.]

in the *N*-linked oligosaccharide by *N*-acetylglucosamine phosphotransferase, an enzyme specific for lysosomal enzymes (Figure 17-39). The *N*-acetylglucosamine residue then is removed by a phosphodiesterase in the second reaction, leaving a mannose 6-phosphate (M6P) residue.

Multiple *N*-linked oligosaccharides are added to most lysosomal enzymes in the rough ER and become phosphorylated in the *cis*-Golgi. The many M6P residues that are formed then bind to *mannose 6-phosphate receptors*, which are found primarily in the *trans*-Golgi network. The luminal domain of this transmembrane protein contains a region that binds M6P very tightly and specifically. As shown in Figure 17-40, vesicles containing the M6P receptor and bound lysosomal enzyme bud from the *trans*-Golgi network and then fuse with a sorting vesicle, an organelle often termed the **late endosome**, which has an internal pH of ≈5.5. Because M6P receptors can bind M6P at the slightly acidic pH (≈6.5–7) of the *trans*-Golgi network but not at a pH less than ≈6, the bound lysosomal enzymes are released within late endosomes. Furthermore, a phosphatase within late endosomes generally removes the phosphate from lysosomal enzymes, preventing their rebinding to the M6P receptor.

Two types of vesicles bud from late endosomes. One type contains lysosomal enzymes but not the M6P receptor; after these vesicles bud from late endosomes, they fuse with lysosomes, delivering the lysosomal enzyme to their final destination. The other type of vesicle recycles the M6P receptor back to the *trans*-Golgi network or, on occasion, to the cell surface.

The lysosomal enzymes we have talked about so far are actually precursors, or *proenzymes*. These catalytically inactive proenzymes are the ones sorted by the M6P receptor. Late in its maturation a proenzyme undergoes a proteolytic cleavage that causes a conformational change in the protein, forming a smaller but enzymatically active polypeptide. This cleavage occurs in either the acidic late endosome or the

lysosome. Delaying the activation of lysosomal proenzymes until they reach the lysosome prevents them from digesting macromolecules in earlier compartments of the secretory pathway.

The **M6P sorting pathway** for lysosomal enzymes illustrates several important principles that also apply to sorting of secretory and membrane proteins. First, as discussed later, mannose 6-phosphate is one of several sorting signals that target proteins to different compartments within the secretory pathway. Second, membrane-embedded receptors with their bound ligands diffuse into discrete regions of the membrane of an organelle—in this case the *trans*-Golgi network—where they are specifically incorporated into budding transport vesicles. Third, these transport vesicles fuse only with one specific organelle, here the late endosome. And finally, cellular transport receptors are recycled after dissociating from their bound ligand.

Lysosomal Storage Diseases Provided Clues to Sorting of Lysosomal Enzymes

A group of genetic disorders, termed *lysosomal storage diseases*, are due to the absence of one or more lysosomal enzymes. As a result, undigested glycolipids and extracellular components that would normally be degraded by lysosomal enzymes accumulate in lysosomes as large inclusions. *I-cell disease* is a particularly severe type of lysosomal storage disease in which multiple enzymes are missing from the lysosomes. Cells from affected individuals lack the GlcNAc phosphotransferase that is required for formation of M6P residues on lysosomal enzymes in the *cis*-Golgi (see Figure 17-39). Biochemical comparison of lysosomal enzymes from normal individuals with those from patients with I-cell disease led to the initial discovery of mannose 6-phosphate as the lysosomal sorting signal. Lacking the M6P sorting signal, the lysosomal enzymes in I-cell patients are secreted rather than sequestered in lysosomes.

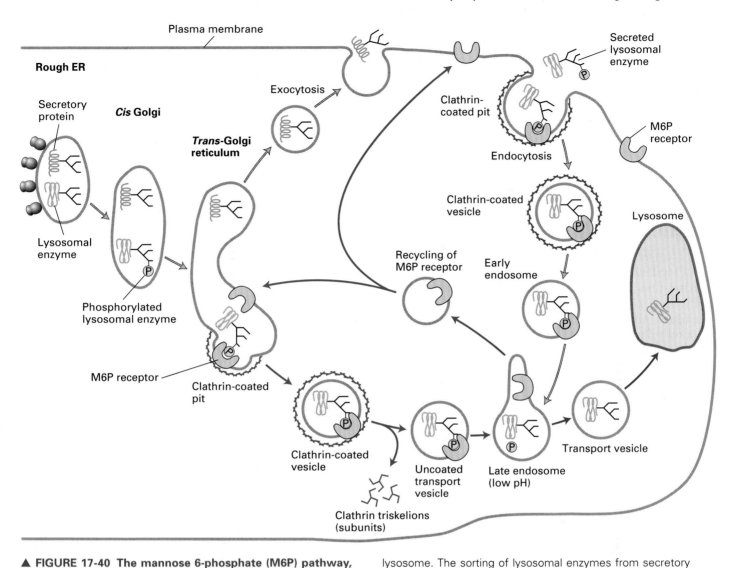

▲ FIGURE 17-40 The mannose 6-phosphate (M6P) pathway, the major route for targeting lysosomal enzymes to lysosomes. Precursors of lysosomal enzymes migrate from the rough ER *(left)* to the *cis*-Golgi where mannose residues are phosphorylated. In the *trans*-Golgi network, the phosphorylated enzymes bind to M6P receptors, which direct the enzymes into vesicles coated with the fibrous protein clathrin. The clathrin lattice surrounding these vesicles is rapidly depolymerized to its subunits, and the uncoated transport vesicles fuse with late endosomes. Within this low-pH compartment, the phosphorylated enzymes dissociate from the M6P receptors and then are dephosphorylated. The receptors recycle back to the Golgi, and the enzymes are incorporated into a different transport vesicle that buds from the late endosome and soon fuses with a lysosome. The sorting of lysosomal enzymes from secretory proteins thus occurs in the *trans*-Golgi network, and these two classes of proteins are incorporated into different vesicles, which take different routes after they bud from the Golgi. For simplicity, only one oligosaccharide side chain is shown on the lysosomal enzyme and secretory protein, although most glycoproteins contain multiple side chains. The M6P receptor is also found on the plasma membrane *(upper right),* where it binds extracellular phosphorylated lysosomal enzymes that are occasionally secreted. These receptor-ligand complexes are internalized from the cell surface in clathrin-coated vesicles, which also lose their clathrin coats and fuse with the late endosome organelle. [See G. Griffiths et al., 1988, *Cell* **52**:329; S. Kornfeld, 1992, *Annu. Rev. Biochem.* **61**:307; and G. Griffiths and J. Gruenberg, 1991, *Trends Cell Biol.* **1**:5.]

When fibroblasts from patients with I-cell disease are grown in a medium containing lysosomal enzymes bearing M6P residues, the diseased cells acquire a nearly normal intracellular content of lysosomal enzymes. This finding indicates that the plasma membrane contains M6P receptors, which can internalize phosphorylated lysosomal enzymes by *receptor-mediated endocytosis* (see Figure 17-40). This process, used by many cell-surface receptors to bring bound proteins or particles into the cell, is discussed in detail in a later section. It is now known that even in normal cells, some M6P receptors are recycled to the plasma membrane and some phosphorylated lysosomal enzymes are secreted. The secreted enzymes can be retrieved by receptor-mediated endocytosis and directed to lysosomes. This pathway thus scavenges any lysosomal enzymes that escape the usual M6P sorting pathway.

Hepatocytes, the predominant type of liver cells, from patients with I-cell disease contain a normal complement of lysosomal enzymes and no inclusions, even though these cells are defective in mannose phosphorylation. This finding implies that hepatocytes, and perhaps other cell types, employ a M6P-independent pathway for sorting lysosomal enzymes. The nature of this pathway is unknown.

SUMMARY   Protein Glycosylation in
                the ER and Golgi Complex

• *O*-linked oligosaccharides, which are bound to serine, threonine, or hydroxylysine residues, are generally short, often containing only one to four sugar residues. All *N*-linked oligosaccharides, which are bound to asparagine residues, contain a core of three mannose and two *N*-acetylglucosamine residues and usually have several branches (see Figure 17-30).

• Sugar nucleotides synthesized in the cytosol are imported into the rough ER and Golgi cisternae by specific transporters in the membranes of these organelles.

• *O*-linked oligosaccharides are formed by the sequential addition of sugars in the ER and Golgi. Addition is catalyzed by various glycosyltransferases that are specific for the donor sugar nucleotide and acceptor molecule (see Figure 17-32).

• ABO blood type is determined by whether an individual produces either one or both of two specific glycosyltransferases.

• Formation of *N*-linked oligosaccharides begins with assembly of a ubiquitous 14-residue high-mannose precursor on dolichol, a lipid in the membrane of the rough ER. This preformed oligosaccharide then is transferred to specific asparagine residues of nascent polypeptide chains in the ER lumen.

• Enzymes localized in the rough ER and *cis*-, *medial*-, and *trans*-Golgi cisternae remove or add sugar residues to the high-mannose precursor yielding a finished *N*-linked oligosaccharide (see Figures 17-36 and 17-38). Differences in processing in different proteins, as well as in different cell types and species, produce *N*-linked oligosaccharides with a variety of structures.

• Oligosaccharides attached to glycoproteins may assist in their proper folding, help protect the mature proteins from proteolysis, and in some cases participate in cell-cell adhesion.

• Enzymes destined for lysosomes are phosphorylated in the *cis*-Golgi, yielding multiple mannose 6-phosphate (M6P) residues. M6P receptors in the *trans*-Golgi

network bind the phosphorylated proteins and direct their transfer to late endosomes, where receptors and proteins dissociate. The receptors then are recycled to the Golgi or plasma membrane, and the lysosomal enzymes are delivered to lysosomes (see Figure 17-40). M6P receptors on the cell surface bind extracellular, phosphorylated lysosomal enzymes and, by receptor-mediated endocytosis, deliver them to lysosomes.

17.8   Golgi and Post-Golgi Protein Sorting and Proteolytic Processing

The previous section dealt with the glycosylation of proteins in the rough ER and Golgi and the use of a carbohydrate signal, mannose 6-phosphate, to target lysosomal enzymes to lysosomes. In this section, we consider the sorting of enzymes to the correct Golgi membrane; the sorting of secretory proteins to regulated and continuously exocytosed vesicles, as well as the specific proteolytic cleavages that occur in these vesicles; and the sorting of proteins to the apical or basolateral domain of polarized epithelial cells. Our focus will be on determining the sequences in a particular protein that direct it to its destination and keep it there. In addition to mannose 6-phosphate, which acts as a sorting sequence for lysosomal enzymes (see Figure 17-40), relatively short segments of a polypeptide, containing ≈10–20 amino acids, function as sorting signals in the secretory pathway.

Sequences in the Membrane-Spanning Domain Cause the Retention of Proteins in the Golgi

All known Golgi enzymes are inserted into the rough ER membrane and move by transport vesicles to the Golgi, where they remain. The various Golgi-localized carbohydrate-modifying enzymes, such as galactosyltransferases and sialyltransferases, have a similar structure: a short N-terminal domain that faces the cytosol, a single transmembrane α helix, and a large C-terminal domain that faces the Golgi lumen and that contains the catalytic site (see Figure 17-21). Although the sequences of the transmembrane segments of these Golgi enzymes are not homologous, several experiments have shown that the single transmembrane α helix in each of these proteins is both necessary and sufficient to cause them to remain in the Golgi. For example, if the membrane-spanning α helix of the transferrin receptor, a plasma-membrane protein with a similar overall structure, is replaced by that of the Golgi enzyme galactosyltransferase, the resultant chimeric protein, when expressed in cultured cells, is localized to the Golgi. Conversely, if the membrane-spanning α helix of galactosyltransferase is replaced by that of the transferrin receptor, the resultant chimeric protein is found on the plasma membrane, not in the Golgi.

How the membrane-spanning α helix in a Golgi enzyme causes its localization and prevents its movement to the plasma membrane is not known. The membrane-spanning α helices in Golgi enzymes tend to be two or three amino acids shorter than those in plasma-membrane proteins. One popular theory is that the membrane-spanning α helices of several Golgi enzymes bind specifically together within the phospholipid bilayer, in much the same way that the 11 membrane-spanning α helices of the photosynthetic reaction center form very specific and stable interactions with each other (see Figure 16-40). The resulting complex of many Golgi enzymes might bind to cytoskeletal proteins, such as actin or tubulin, or in some other way be prevented from diffusing into the transport vesicles that would otherwise take them to the cell surface.

Different Vesicles Are Used for Continuous and Regulated Protein Secretion

As mentioned earlier, in all eukaryotic cells certain secretory proteins are moved via transport vesicles from the *trans*-Golgi to the plasma membrane, where they are continuously secreted. In contrast, certain specialized secretory cells store some secretory proteins in vesicles and secrete them only when triggered by a specific stimulus. One example of such regulated secretion occurs in pancreatic β-islet cells, which store newly made insulin in special secretory vesicles and secrete insulin only in response to an elevation in blood glucose (see Table 17-3 for other examples). These and many other cells simultaneously utilize two different classes of vesicles to move proteins from the *trans*-Golgi to the cell surface: secretory vesicles for regulated secretion and transport vesicles for constitutive (continuous) secretion (see Figure 17-13). How are the two different types of secretory proteins, all of which are soluble in the lumen of the *trans*-Golgi network, sorted to the correct type of vesicle?

A common mechanism appears to sort regulated secretory proteins as diverse as ACTH (adrenocorticotropic hormone), which is normally made in pituitary cells; insulin, made in pancreatic β-islet cells; and trypsinogen, made in pancreatic acinar cells. For example, when recombinant DNA techniques are used to induce the synthesis of insulin and trypsinogen in pituitary tumor cells already synthesizing ACTH, all three proteins segregate in the same regulated secretory vesicles and are secreted together when a hormone binds to a receptor on the pituitary cells and causes a rise in cytosolic Ca^{2+}. Although these three proteins share no identical amino acid sequences that might serve as a sorting sequence, they obviously have some common feature that signals their incorporation into the regulated secretory vesicles.

Morphologic evidence suggests that sorting into the regulated pathway is controlled by selective protein aggregation. Immature vesicles of this pathway—those that have just budded from the *trans*-Golgi network—are coated with **clathrin**, a fibrous protein, and contain a core, consisting of aggregated secretory protein, that is visible in the electron microscope. These aggregates are found in vesicles that are in the process of budding from the *trans*-Golgi network, indicating that proteins destined for regulated secretory vesicles selectively aggregate together before their incorporation into the vesicles (Figure 17-41a). Indeed, regulated secretory vesicles from many mammalian cells contain two proteins, *chromogranin B* and *secretogranin II*, that together form aggregates when incubated at the ionic conditions (pH ≈ 6.5 and 1 mM Ca^{2+}) thought to occur in the *trans*-Golgi network; such aggregates do not form at the neutral pH of the ER. The selective aggregation of regulated secretory proteins together with chromogranin B or secretogranin II could be the basis for sorting of these proteins into regulated secretory vesicles. Secretory proteins that do not associate with these proteins in such ways would be secreted continuously by default.

Proproteins Undergo Proteolytic Processing Late in Maturation

For some secretory proteins (e.g., growth hormone) and certain viral membrane proteins (e.g., the VSV glycoprotein), removal of the N-terminal ER signal sequence from the nascent chain is the only known proteolytic cleavage required to convert the polypeptide to the mature, active species (see Figure 17-16). However, some plasma-membrane and most secretory proteins initially are synthesized as relatively long-lived, inactive precursors, termed *proproteins*, that require further proteolytic processing to generate the mature, active proteins. Examples of proteins that undergo such processing are serum albumin, insulin, glucagon, and the yeast α mating factor, all of which are secretory proteins, and membrane proteins such as influenza HA.

In general, the proteolytic conversion of a proprotein to the corresponding mature protein occurs in secretory vesicles as they move away from the *trans*-Golgi. Normally, mature secretory vesicles are formed by fusion of several immature ones containing proprotein, which is cleaved during this maturation process. That conversion of proinsulin to insulin occurs in newly formed secretory vesicles is demonstrated by comparison of the electron micrographs in Figure 17-41.

Some proproteins, including proalbumin, are cut once at a site C-terminal to a dibasic recognition sequence such as Arg-Arg or Lys-Arg (Figure 17-42a). In other proproteins, additional amino acids are cleaved at the N-terminus or at both ends of the proprotein. In proinsulin, the extra amino acids, collectively termed the *C peptide*, are located internally in the polypeptide; the N-terminal B chain and the C-terminal A chain of mature insulin are linked by disulfide bonds and remain attached when the C peptide is removed (Figure 17-42b).

The breakthrough in identifying the proteases responsible for such maturational processing came from analysis of a yeast mutant, termed *kex2*, that synthesized the precursor of the α mating factor but could not process it to the

(a)

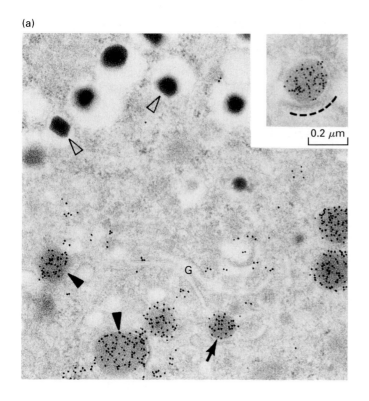

(b)

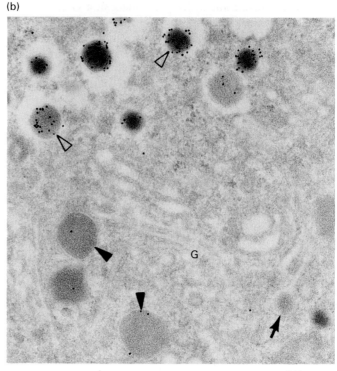

▲ **FIGURE 17-41 Electron micrographs revealing aggregation and cleavage of proinsulin.** This proprotein is packaged into clathrin-coated secretory vesicles before it is cleaved to insulin. Serial sections of the Golgi region of an insulin-secreting cell are stained with (a) a monoclonal antibody that recognizes proinsulin (and not insulin) and (b) a different antibody that recognizes insulin (and not proinsulin). The antibodies, which are bound to electron-opaque gold particles, appear as dark dots in electron micrographs (see Figure 5-17). Immature secretory vesicles (closed arrowheads) and vesicles budding from the *trans*-Golgi (arrows) stain with the proinsulin antibody (a) but not with anti-insulin (b). These vesicles contain diffuse protein aggregates that include proinsulin and other regulated secretory proteins. Mature vesicles (open arrowheads) stain intensely with the anti-insulin antibody (b) but not with anti-proinsulin (a) and have a dense core of almost crystalline insulin. The inset in (a) shows the clathrin coat (dashed line) on a proinsulin-rich secretory vesicle. Since immature secretory vesicles contain proinsulin (not insulin), the proteolytic conversion of proinsulin to insulin must take place after proinsulin is transported from the *trans*-Golgi network to these vesicles. [From L. Orci et al., 1987, *Cell* **49**:865; courtesy of L. Orci.]

functional hormone; the mutant cells thus were defective in mating (see Figure 14-4). The product of the wild-type *KEX2* gene is the endoprotease that cleaves the α factor's precursor at a site C-terminal to Arg-Arg and Lys-Arg residues. Using the *KEX2* gene as a DNA probe, workers were able to clone a family of mammalian endoproteases, all of which cut after an Arg-Arg or Lys-Arg sequence (see Figure 17-42). One, called *furin*, is found in all mammalian cells; it processes proteins such as albumin that are secreted by the continuous pathway. Two, termed *PC2* and *PC3*, which are found only in cells that exhibit regulated secretion, are localized to regulated secretory vesicles and proteolytically cleave the precursors of many hormones at specific sites. PC2 and PC3 catalyze different cleavages in the conversion of proinsulin to insulin. A carboxypeptidase catalyzes the final step in processing of proinsulin and many other proproteins.

Some Proteins Are Sorted from the Golgi Complex to the Apical or Basolateral Plasma Membrane

The plasma membrane of polarized epithelial cells is divided into two domains, *apical* and *basolateral*, which are separated by tight junctions and contain different species of proteins (see Figure 15-23). Membrane proteins are sorted to either the apical or basolateral domains by several mechanisms, ensuring that all plasma-membrane proteins are localized to the correct domain; any one protein may be targeted by more than one mechanism. Although the sorting of plasma-membrane proteins in polarized cells is understood in general terms, the molecular mechanisms underlying the selective movements of these membrane proteins are not yet known.

One mechanism for targeting proteins to the appropriate domain of the plasma membrane involves sorting in the *trans*-Golgi network. A variety of microscopic and cell-fractionation

(a) Constitutive secretory proteins
Proalbumin

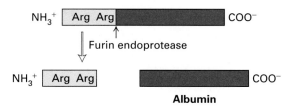

(b) Regulated secretory proteins
Proinsulin

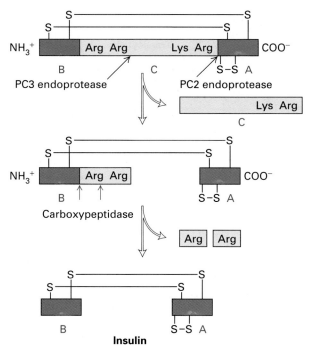

▲ **FIGURE 17-42 Proteolytic processing of proalbumin and proinsulin, typical of processing in the continuous and regulated secretion pathways, respectively.** The endoproteases that function in such processing cleave C-terminal to sequences of two consecutive basic amino acids. (a) The endoprotease furin acts on the precursors of constitutive secretory proteins. (b) Two endoproteases, PC2 and PC3, act on the precursors of regulated secretory proteins. The final processing of many such proteins is catalyzed by a carboxypeptidase that sequentially removes two basic amino acid residues at the C-terminus of a polypeptide. [See D. Steiner et al., 1992, *J. Biol. Chem.* **267**:23435.]

propriate plasma-membrane region and induce fusion, as discussed later.

Cultured MDCK epithelial cells have been useful in investigating this aspect of protein sorting (see Figure 6-7). When MDCK cells are infected with the influenza virus, a membrane-enveloped virus, progeny viruses bud only from the apical membrane; on the other hand, in cells infected with VSV, another enveloped virus, the virus buds only from the basolateral membrane. This difference occurs because the HA glycoprotein of influenza virus is transported from the Golgi complex exclusively to the apical membrane, and the VSV glycoprotein (G protein) is transported only to the basolateral membrane (Figure 17-43). Furthermore, when the gene encoding HA protein is introduced into uninfected cells by recombinant DNA techniques, all the expressed HA accumulates in the apical membrane, indicating that the targeting signal resides in the HA glycoprotein itself and not in other viral proteins produced during viral infection.

Among the cellular proteins that undergo apical-basolateral sorting in the Golgi are those with a **glycosylphosphatidylinositol (GPI) membrane anchor.** In MDCK cells and most other types of epithelial cells, GPI-anchored proteins are targeted to the apical membrane, whereas in thyroid cells and some others, they are targeted to the basolateral membrane. However, aminopeptidase, which is anchored by a single transmembrane α helix, is always sorted to the apical membrane and the Na$^+$/K$^+$ ATPase to the basolateral membrane. Except for the GPI anchor, which acts as an apical or basolateral targeting signal, no unique sequences have been identified that target proteins to the apical or basolateral domain. Each protein may contain multiple sorting sequences, any one of which can target it to the appropriate plasma-membrane domain. It is also not known how the sorting of integral plasma-membrane proteins to the appropriate transport vesicles occurs in the *trans*-Golgi network.

Another mechanism for sorting apical and basolateral proteins, also illustrated in Figure 17-43, operates in hepatocytes, where the basolateral membranes face the blood as in intestinal epithelial cells, and the apical membranes form the bile canaliculus. In hepatocytes, all newly made apical and basolateral proteins are first delivered together from the *trans*-Golgi network to the basolateral membrane. From there, both basolateral and apical proteins are endocytosed in the same vesicles, but then their paths diverge. The endocytosed basolateral proteins are sorted into transport vesicles that recycle them to the basolateral membrane. In contrast, the apically destined endocytosed proteins are sorted into a class of transport vesicles that move across the cell and fuse with the apical membrane, a process called **transcytosis.** As discussed later, transcytosis also is a means of moving extracellular materials from one side of a sheet of epithelial cells to another. Even in epithelial cells, such as MDCK cells, in which apical-basolateral protein sorting occurs in the Golgi, transcytosis may provide a "fail-safe" sorting mechanism. That is, an apical protein sorted incorrectly to the basolateral membrane would be subjected to endocytosis and delivered to the apical membrane.

studies all indicate that proteins destined for the apical and basolateral membranes are found together in the *same* membranes of the *trans*-Golgi network. However, two types of transport vesicles, containing different membrane proteins, bud from the *trans*-Golgi network. One type moves to and fuses with the apical plasma membrane; the other goes to the basolateral region. These vesicles contain distinct Rab and V-SNARE proteins, which may target them to the ap-

► **FIGURE 17-43 The sorting of proteins destined for the apical and basolateral plasma membranes of epithelial cells.** When cultured MDCK cells are infected simultaneously with VSV and influenza virus, the VSV glycoprotein (G protein) is found only on the basolateral membrane, whereas the HA glycoprotein of the influenza virus is found only on the apical membrane. Like these viral proteins, some cellular proteins are sorted directly to the apical membrane and others to the basolateral membrane via specific transport vesicles that bud from the *trans*-Golgi network. In certain other polarized cells, some apical and basolateral proteins are transported together to the basolateral surface; the apical proteins then move selectively, by endocytosis and transcytosis, to the apical membrane. [After K. Simons and A. Wandinger-Ness, 1990, *Cell* **62**:207, and K. Mostov et al., 1992, *J. Cell Biol.* **116**:577.]

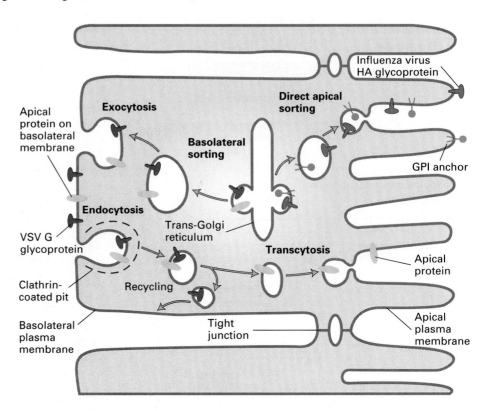

The attachment of integral membrane proteins to the cytoskeleton serves as a *retention signal* and may assist in the apical-basolateral sorting of some proteins. In polarized epithelial cells, for instance, the basolateral membrane, but not the apical membrane, is underlain by a fibrous cytoskeleton containing ankyrin and a spectrin-like protein called fodrin (see Figure 18-7). Several integral proteins that are selectively localized to the basolateral membrane can bind tightly to ankyrin. These include the Na^+/K^+ ATPase in intestinal cells (see Figure 15-25) and the Cl^-/HCO_3^- antiporter (AE2 protein) in parietal cells of the stomach (see Figure 15-26). Once such integral proteins are transported from the Golgi cisternae to the basolateral membrane, they become locked into the ankyrin-fodrin cytoskeleton and cannot undergo endocytosis and subsequent transcytosis to the apical membrane. However, basolateral proteins that are not locked to the cytoskeleton can be endocytosed and recycled back to the basolateral surface (see Figure 17-43).

SUMMARY Golgi and Post-Golgi Protein Sorting and Proteolytic Processing

• The Golgi complex plays a key role in sorting newly made secretory and membrane proteins; proper sorting is thought to be directed by specific amino acid sequences in the proteins themselves.

• Enzymes that remain in the Golgi are membrane proteins, which are retained by sequences within their membrane-spanning domain.

• Some secretory proteins and most membrane proteins are targeted to transport vesicles that continuously fuse with the plasma membrane and exocytose their contents. Other secretory proteins are directed to regulated secretory vesicles, apparently by forming aggregates with secretogranins in the *trans*-Golgi network.

• Regulated secretory proteins are concentrated and stored in secretory vesicles to await a neural or hormonal signal for exocytosis.

• Most secretory proteins and some plasma-membrane proteins undergo post-Golgi proteolytic cleavages that yield the mature, active proteins. Proteolytic processing is catalyzed by various proteases, some of which are unique to the regulated secretory pathway (see Figure 17-42). Generally, proteolytic maturation occurs in secretory or transport vesicles carrying proteins from the *trans*-Golgi network to the cell surface.

• In polarized epithelial cells, membrane proteins destined for the apical or basolateral domains of the plasma membrane are sorted in the *trans*-Golgi or *trans*-Golgi network into different transport vesicles (see Figure 17-43). The GPI anchor is the only apical-basolateral targeting signal identified so far.

• In hepatocytes and some other polarized cells, all plasma-membrane proteins are directed first to the basolateral membrane. Apically destined ones then are endocytosed and moved across the cell to the apical membrane (transcytosis).

17.9 Receptor-Mediated Endocytosis and the Sorting of Internalized Proteins

In previous sections, we've followed the main pathways whereby secretory and membrane proteins are synthesized within cells and then targeted to the cell surface or other destination. However, cells also can internalize materials from their surroundings by **phagocytosis,** an actin-mediated process in which cells envelop bacteria and other large particles and then internalize them, and **endocytosis,** a process in which a small region of the plasma membrane invaginates to form a new intracellular membrane-limited vesicle about 0.05 to 0.1 μm in diameter (see Figure 5-44).

Relatively few cell types carry out phagocytosis, whereas most eukaryotic cells continually engage in endocytosis. In **pinocytosis,** endocytic vesicles nonspecifically take up small droplets of extracellular fluid and any material dissolved in it. In *receptor-mediated endocytosis,* a specific receptor on the cell surface binds tightly to the extracellular macromolecule (the **ligand**) that it recognizes; the plasma-membrane region containing the receptor-ligand complex then undergoes endocytosis, becoming a transport vesicle. Receptor-

ligand complexes are selectively incorporated into the intracellular transport vesicles; most other plasma-membrane proteins are excluded. Moreover, the rate at which a ligand is internalized is limited by the amount of its corresponding receptor on the cell surface. Among the common macromolecular ligands that vertebrate cells internalize by receptor-mediated endocytosis are cholesterol-containing particles called *low-density lipoprotein* (LDL); *transferrin,* an iron-binding protein; insulin and most other protein hormones; and glycoproteins whose oligosaccharide side chains contain terminal glucose, mannose, or galactose residues rather than the normal sialic acid (see Figure 17-30).

Small invaginations of the plasma membrane termed *caveolae,* lined with the membrane protein *caveolin,* contain some receptor proteins and are used for certain types of receptor-mediated endocytosis. However, receptor-mediated endocytosis generally occurs via clathrin-coated pits and vesicles (Figure 17-44). In this respect, the process is similar to the packaging of lysosomal enzymes by mannose 6-phosphate (M6P) in the *trans-*Golgi (see Figure 17-40). As noted earlier, although most M6P receptors are localized to the *trans-*Golgi, some are found on the cell surface. Lysosomal enzymes that are secreted bind to these receptors and are returned to cells via receptor-mediated endocytosis. In general, transmembrane receptor proteins that are internalized from the cell

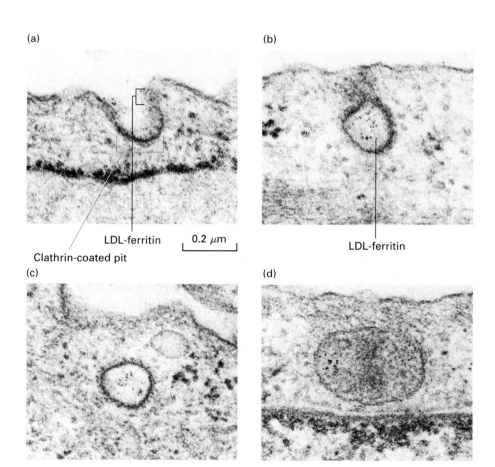

(a)

LDL-ferritin   0.2 μm

Clathrin-coated pit

(c)

(b)

LDL-ferritin

(d)

◄ **FIGURE 17-44 The initial stages of receptor-mediated endocytosis of low-density lipoprotein (LDL) particles by cultured human fibroblasts, revealed by electron microscopy.** The LDL particles were visualized by covalently linking them to the iron-containing protein ferritin; each small iron particle in ferritin is visible as a small dot under the electron microscope. (a) A coated pit, showing the clathrin coat on the inner (cytosolic) surface of the pit, soon after ferritin-tagged LDL particles were added to cells. (b) A pit containing LDL apparently closing on itself to form a coated vesicle. (c) A coated vesicle containing ferritin-tagged LDL particles. (d) Ferritin-tagged LDL particles in a smooth-surfaced early endosome 6 minutes after being added to cells. [Photographs courtesy of R. Anderson. Reprinted by permission from J. Goldstein et al., *Nature* **279**:679. Copyright 1979, Macmillan Journals Limited. See also M. S. Brown and J. Goldstein, 1986, *Science* **232**:34.]

surface during endocytosis are sorted and recycled back to the cell surface, much like the recycling of M6P receptors to the plasma membrane and *trans*-Golgi.

Clathrin-coated pits make up about 2 percent of the surface of cells such as hepatocytes and fibroblasts. Many internalized ligands have been observed in clathrin-coated pits and vesicles, and researchers believe that these structures function as intermediates in the endocytosis of most (though not all) ligands bound to cell-surface receptors. Some receptors are clustered over clathrin-coated pits even in the absence of ligand. Other receptors diffuse freely in the plane of the plasma membrane but undergo a conformational change when binding to ligand, so that when the receptor-ligand complex diffuses into a clathrin-coated pit, it is retained there. Two or more types of receptor-bound ligands, such as LDL and transferrin, can be seen in the same coated pit or vesicle. As discussed later, the regulated polymerization of clathrin is thought to cause the pits to expand and eventually to form clathrin-coated vesicles. Here we consider the endocytosis and subsequent sorting of cell-surface receptors and their ligands. We first describe the most common pathway exemplified by the LDL system and then briefly discuss several variations.

The LDL Receptor Binds and Internalizes Cholesterol-Containing Particles

Whether ingested in foodstuffs or synthesized in the liver, cholesterol is insoluble in body fluids and must be transported by a water-soluble carrier. Low-density lipoprotein (LDL) is one of several complexes that carry cholesterol through the bloodstream. An LDL particle is a sphere 20–25 nm in diameter (Figure 17-45). Its outer surface is a *monolayer* membrane of phospholipids and cholesterol, in which one molecule of a very large protein, called *apo-B*, is embedded. Inside is an extremely nonpolar core of cholesterol, all of which is esterified through the single hydroxyl group of cholesterol to a long-chain fatty acid, mainly linoleic acid. Most mammalian cells produce cell-surface receptors that specifically bind and internalize LDL by receptor-mediated endocytosis (see Figure 17-44). After endocytosis, the LDL particles are transported to lysosomes where lysosomal hydrolases degrade the apo-B protein to amino acids and cleave the cholesterol esters to cholesterol and fatty acids. The cholesterol is incorporated directly into cell membranes or is re-esterified and stored as lipid droplets in the cell for later use; the fatty acids are used to make new phospholipids or triglycerides. Cholesterol also is converted to steroid hormones in adrenal cortical cells and to bile acids in hepatocytes.

The LDL receptor is a single-chain glycoprotein of 839 amino acids with a long N-terminal exoplasmic domain and short C-terminal cytosolic domain (see Figure 17-21). A sequence of 22 hydrophobic amino acids spans the plasma membrane once, presumably as an α helix. The large exoplasmic domain has an N-terminal segment of about 320 residues that is extremely rich in disulfide-bonded cysteine

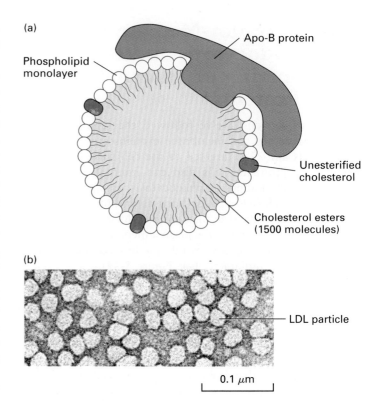

▲ FIGURE 17-45 (a) Schematic diagram of an LDL particle. A monolayer of phospholipid and unesterified cholesterol forms the surface membrane, and fatty acid esters of cholesterol make up the hydrophobic core. One copy of the hydrophobic apo-B protein is embedded in the membrane. This protein mediates binding of LDL particles to specific cell-surface receptors. (b) Electron micrograph of a negatively stained preparation of LDL particles. [See R. Anderson, 1979, *Nature* **279**:679. Part (b) courtesy of R. Anderson. Reprinted by permission from *Nature*. Copyright 1979, Macmillan Journals Limited.]

residues. This segment includes a sevenfold repeat of a sequence of 40 amino acids that contains the LDL-binding site. As explained below, residues in the C-terminal cytosolic domain are involved in trapping the LDL receptor in clathrin-coated pits.

Cytosolic Sequences in Some Cell-Surface Receptors Target Them for Endocytosis

 As we learn in Chapter 20, cells possess many different cell-surface receptors with binding sites for specific ligands. Only some of these receptors, however, form receptor-ligand complexes that are internalized into clathrin-coated vesicles. Insight into what distinguishes receptors that are endocytosed from those that are not has come from studies with mutant receptors. For example, mutant forms of the LDL receptor protein are available from persons who have the inherited disorder *familial hypercholesterolemia*. This disease is characterized by high

TABLE 17-6 Sorting Signals That Direct Secreted and Membrane Proteins to Specific Transport Vesicles

Signal Sequence*	Type of Protein†	Transport Step	Vesicle Type	Signal Receptor
Lys-Asp-Glu-Leu (KDEL)	Secreted	Golgi to ER	COP I	KDEL receptor (ERD2 protein) in Golgi membrane
Lys-Lys-X-X (KKXX)	Membrane	Golgi to ER	COP I	COP α and β subunits
Di-acidic (e.g., Asp-X-Glu)	Membrane	ER to Golgi	COP II	Not known
Mannose 6-phosphate (M6P)	Secreted	*Trans*-Golgi and plasma membrane to late endosome	Clathrin	M6P receptor in Golgi and plasma membrane; AP1 and AP2 adapter proteins
Tyr-X-X-ϕ (YXXϕ)	Membrane	Plasma membrane to endosome	Clathrin	AP2 adapter proteins
Leu-Leu (LL)	Membrane	Plasma membrane to endosome	Clathrin	AP2 adapter proteins

*X = any amino acid; ϕ = bulky hydrophobic residues. Single-letter abbreviations are shown in parentheses.
†Signal sequences are located in the cytosolic domains of membrane proteins.

levels of cholesterol in the blood, and persons homozygous for the mutant alleles often die at an early age from heart attacks caused by atherosclerosis, a buildup of cholesterol deposits that ultimately block the arteries.

In some persons with this disorder, the LDL receptor is simply not produced; in others, it binds LDL poorly or not at all. In one especially instructive case, the mutant receptor binds LDL normally but the LDL-receptor complex cannot be internalized by the cell and is distributed evenly over the cell surface rather than confined to clathrin-coated pits. In individuals with this particular defect, plasma-membrane receptors for other ligands are internalized normally in clathrin-coated pits, but the mutant LDL receptor apparently cannot bind properly to coated pits. The mutant receptor has a single tyrosine-to-cysteine change in its cytosolic domain.

This and other mutant forms of the LDL receptor have been experimentally generated and expressed in fibroblasts. Analysis of these proteins has shown that a four-residue sequence in the cytosolic domain is crucial for internalization: Tyr-X-X-ϕ, where X can be any amino acid and ϕ is a bulky hydrophobic amino acid, such as Phe, Leu, or Met. As we discuss later, this sequence binds to one of the subunits (μ2) of the protein complex that links the clathrin coat to the cytosolic domain of a membrane protein in forming coated pits. Because the tyrosine and ϕ residues mediate this binding, a mutation in either one reduces or abolishes the ability of the receptor to be incorporated into clathrin-coated pits.

Many other plasma-membrane receptors that are internalized into clathrin-coated pits, such as the transferrin receptor, contain a similar amino acid sequence in their cytosolic domains. Mutagenesis studies on these receptors have confirmed that these four amino acids form a general recognition signal for binding to clathrin-coated pits. As further evidence for the importance of this sequence, a cell-surface protein that is not normally internalized into clathrin-coated pits can be made to internalize if these four amino acids are added to its cytosolic domain. For example, a mutant influenza HA protein, genetically engineered to contain such a four-amino-acid recognition sequence in its cytosolic domain, is internalized into clathrin-coated pits.

However, in other proteins different amino acids sequences, such as Leu-Leu, signal endocytosis (Table 17-6). Yet other membrane proteins, such as the yeast α factor receptor, uracil permease, and the human growth hormone receptor require covalent addition of ubiquitin to their cytosolic domain for endocytosis to occur. At present we do not know how endocytosis of such proteins is controlled, nor the identity of the proteins that might bind to these signals.

The Acidic pH of Late Endosomes Causes Most Receptors and Ligands to Dissociate

The overall rate of endocytic internalization of the plasma membrane is quite high; cultured fibroblasts regularly internalize 50 percent of their cell-surface proteins and phospholipids each hour. Most cell-surface receptors that undergo endocytosis will repeatedly deposit their ligands within the cell and then recycle to the plasma membrane, once again to mediate the internalization of ligand molecules. For instance, the LDL receptor makes one round trip into and out of the cell every 10–20 minutes, for a total of several hundred trips in its 20-hour life span. In contrast, after binding its protein ligand the receptors for insulin and other growth factors generally cycle only two or three times before the complex of receptor and ligand is degraded in the lysosome—reducing the number of cell-surface receptors and thus the sensitivity of the cells to hormone signaling.

Regardless of how many times a particular receptor is recycled, internalized receptor-ligand complexes commonly follow the pathway depicted in Figure 17-46. Endocytosed cell-surface receptors dissociate from their ligands within late endosomes. These acidic spherical vesicles with tubular branching membranes are found a few micrometers from the cell surface. (Similar acidic sorting vesicles recycle M6P receptors back to the Golgi complex; see Figure 17-40.)

The original experiments that defined the late endosome sorting vesicle utilized the *asialoglycoprotein receptor*. This liver-specific protein mediates the binding and internalization of abnormal glycoproteins whose oligosaccharides terminate in galactose rather than the normal sialic acid, hence the name *asialo*glycoprotein. Electron microscopy of liver cells perfused with asialoglycoprotein reveal that between 5 and 10 minutes after internalization both the receptors and their ligands accumulate in late endosome vesicles (Figure 17-47). Ligand molecules are found in the lumen of the spherical part of these vesicles, while the tubular membrane extensions are rich in receptor and rarely contain ligand. Thus these membranes contain receptors that have dissociated from their ligands, indicating that the late endosome is the organelle in which receptors and ligands are uncoupled.

Beginning 15 minutes after internalization, ligands are transferred to lysosomes, but the intact receptors themselves usually are not found in these organelles. Instead, the receptor-rich elongated membrane vesicles that bud from the late endosomes mediate the recycling of receptors back to the cell surface (see Figure 17-46). The spherical part of the

▶ **FIGURE 17-46 Fate of an LDL particle and its receptor after endocytosis.** The same pathway is followed by other ligands, such as insulin and other protein hormones, that are internalized by receptor-mediated endocytosis and degraded in the lysosome. After an LDL particle binds to an LDL receptor on the plasma membrane, the receptor-ligand complex is internalized in a clathrin-coated pit that pinches off to become a coated vesicle. The clathrin coat then depolymerizes to triskelions, resulting in an early endosome. This endosome fuses with a sorting vesicle, known as a late endosome, where the low pH (≈5) causes the LDL particles to dissociate from the LDL receptors. A receptor-rich region buds off to form a separate vesicle that recycles the LDL receptors back to the plasma membrane. A vesicle containing an LDL particle may fuse with another late endosome but ultimately fuses with a lysosome to form a larger lysosome. There, the apo-B protein of the LDL particle is degraded to amino acids and the cholesterol esters are hydrolyzed to fatty acids and cholesterol. Abundant imported cholesterol inhibits synthesis by the cell of both cholesterol and LDL receptor protein.

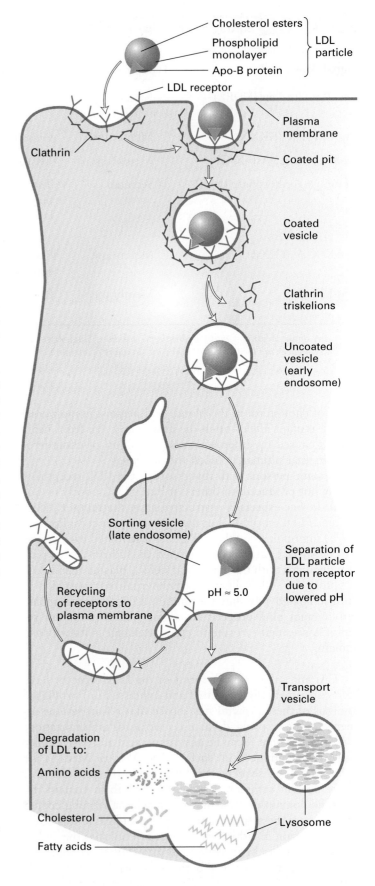

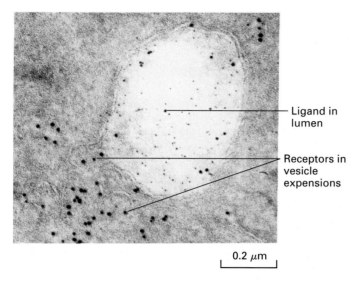

Ligand in lumen

Receptors in vesicle expensions

0.2 μm

▲ **FIGURE 17-47 Experimental demonstration that internalized receptor-ligand complexes dissociate in late endosomes.** Liver cells were perfused with an asialoglycoprotein ligand and then were fixed and sectioned for electron microscopy. This electron micrograph of a late endosome from a perfused hepatocyte reveals that the ligand (smaller dark grains) is localized in the vesicle lumen and the asialoglycoprotein receptor (larger dark grains) is localized in the tubular extensions budding off from the vesicle. The sections were stained with receptor-specific antibodies, tagged with gold particles 8 nm in diameter, to localize the receptor and with asialoglycoprotein-specific antibody, linked to gold particles 5 nm in diameter, to localize the ligand. [Courtesy of H. J. Geuze. Copyright 1983, M.I.T. See H. J. Geuze et al., 1983, *Cell* **32**:277.]

late endosome eventually buds off transport vesicles that, with their cargo of ligand, soon fuse with lysosomes. The LDL receptor, for example, is never directed to a lysosome—to be degraded by potent lysosome proteases—until it becomes damaged in some way.

The key to why receptors release their ligands in the late endosome lies in the progressively decreasing pH encountered by internalized receptor-ligand complexes as they move through clathrin-coated vesicles and various early and late endosomes. Like the very acidic lysosomes, with an internal pH of ≈4.5–5.0, clathrin-coated vesicles and endosomes contain a V-class ATP-dependent proton pump (see Figure 15-10). These vesicles also contain a Cl^- channel, allowing the proton pump to generate a significant H^+ concentration gradient, rather than the transmembrane electric potential that would form if only protons were transferred from the cytosol to the vesicle lumen. Most receptors, including the asialoglycoprotein, insulin, and LDL receptors, bind their ligands tightly at neutral pH but release their ligands if the pH is lowered to 5.0 or below. The late endosome is the first vesicle encountered by receptor-ligand complexes with a pH this low and hence is the organelle in which these and most other receptors dissociate from their tightly bound ligands.

The Endocytic Pathway Delivers Transferrin-Bound Iron to Cells

The endocytic pathway involving the transferrin receptor and its ligand differs from the LDL pathway in that the receptor-ligand complex does not dissociate in late endosomes. Nonetheless, changes in pH also mediate the sorting of receptors and ligands in the transferrin pathway, which functions to deliver iron to cells.

Transferrin, a major glycoprotein in the blood, transports iron to all tissue cells from the liver (the main site of iron storage in the body) and from the intestine (the site of iron absorption). The iron-free form, *apotransferrin*, binds two Fe^{3+} ions very tightly to form *ferrotransferrin*. All growing cells contain surface transferrin receptors that avidly bind ferrotransferrin at neutral pH, after which the receptor-bound ferrotransferrin is subjected to endocytosis. Like the components of LDL, the two bound Fe^{3+} atoms remain in the cell, but there the similarity with the fate of other endocytosed ligands, including LDL, ends: the apotransferrin part of the ligand is secreted from the cell within minutes, carried in the bloodstream to the liver or intestine, and reloaded with iron.

As depicted in Figure 17-48, the explanation for the behavior of the transferrin receptor–ligand complex lies in the unique ability of apotransferrin to remain bound to the transferrin receptor at the low pH (5.0–5.5) of late endosomes. At a pH of less than 6.0, the two bound Fe^{3+} atoms dissociate from ferrotransferrin and are transported from the late endosome vesicle into the cytosol (in an unknown manner). The apotransferrin formed by the dissociation of the iron atoms remains bound to the transferrin receptor and is recycled back to the surface along with the receptor. Remarkably, although apotransferrin binds tightly to its receptor at a pH of 5.0 or 6.0, it does not bind at neutral pH. Hence the bound apotransferrin dissociates from its receptor when the recycling vesicles fuse with the plasma membrane and the receptor-ligand complex encounters the neutral pH of the extracellular interstitial fluid or growth medium. The surface receptor is then free to bind another molecule of ferrotransferrin.

Some Endocytosed Proteins Remain within the Cell

In several receptor-ligand systems, found mainly in oocytes (egg cells), endocytosed material simply remains in the cells and is minimally processed. Developing insect and avian oocytes, for example, internalize yolk proteins and other proteins from the blood or surrounding cells. (Coated pits were first discovered in insect eggs, where they occupy a large portion of the plasma membrane.) A hen's egg is a single cell containing several grams of protein, virtually all of which is imported from the bloodstream by receptor-mediated endocytosis. Vitellogenin, a precursor of several yolk proteins, is synthesized by the liver and secreted into the bloodstream, from which it is endocytosed into the

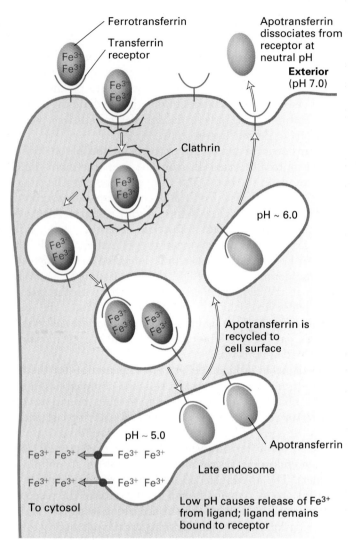

▲ FIGURE 17-48 The transferrin cycle, which operates in all growing mammalian cells. After endocytosis, iron is released from the receptor-ferrotransferrin complex in the acidic late endosome compartment. The apotransferrin protein remains bound to its receptor at this pH, and they recycle to the cell surface together where the neutral pH of the exterior medium causes release of the iron-free apotransferrin. [See A. Ciechanover et al., 1983, *J. Biol. Chem.* **258**:9681.]

developing egg. Yolk proteins remain in storage granules within the egg and are used after fertilization as a source of amino acids and energy by the developing embryo. Egg-white proteins (e.g., ovalbumin, lysozyme, and conalbumin) that are secreted by cells lining the hen oviduct also are endocytosed by the egg cell.

Transcytosis Moves Some Ligands across Cells

As noted previously, transcytosis is used by some cells in the apical-basolateral sorting of certain membrane proteins (see

Figure 17-43). This process of transcellular transport, which combines endocytosis and exocytosis, also can be employed to import an extracellular ligand from one side of a cell, transport it across the cytoplasm, and secrete it from the plasma membrane at the opposite side. Transcytosis occurs mainly in sheets of polarized epithelial cells. An example of transcytosis is the movement of maternal immunoglobulins (antibodies) across the intestinal epithelial cells of the newborn mouse and human. The F_c receptor that mediates the transcytosis of immunoglobulins has the property of binding to its ligand at an acidic pH of 6 but not at neutral pH. Figure 17-49 shows how a difference in the pH of the extracellular media on the two sides of intestinal epithelial cells in newborn mice allows immunoglobulins to move in one direction—from the lumen to the blood. The same process also moves maternal immunoglobulins across mammalian yolk-sac cells into the fetus.

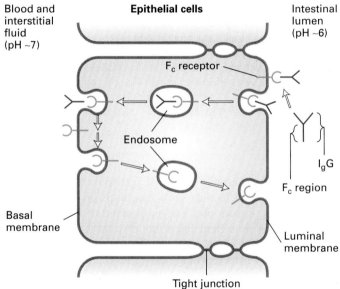

▲ FIGURE 17-49 Transcytosis of maternal IgG immunoglobulins across the intestinal epithelial cells of newborn mice. This transcellular movement of a ligand involves both endocytosis and exocytosis. In newborn mice, the intestinal lumen has a pH of ≈6, whereas the opposite (blood-facing) side of the epithelium has a pH of ≈7. The particular F_c receptors on these epithelial cells bind to the F_c region of IgG molecules only at pH values of 6 or lower, not at a pH of 7.0. Vesicles (endosomes) containing the F_c receptor–IgG complex form just under the luminal surface, move across the cell, and fuse with the basal membrane, where they release the IgG. Unloaded receptors are recycled by transcytosis in the opposite direction: endosomes form from the basal membrane, move across the cell, and fuse with the luminal membrane. Some polarized cells use transcytosis to sort membrane proteins from the basal to the apical surface (see Figure 17-43).

SUMMARY Receptor-Mediated Endocytosis and
the Sorting of Internalized Proteins

- Some extracellular ligands that bind to specific cell-surface receptors are internalized, along with their receptors, in clathrin-coated vesicles. Internalized receptor-ligand complexes undergo various fates.

- Many receptors, such as the LDL receptor, release their ligand in the acidic milieu of the late endosome; the receptors are sorted into vesicles that recycle them to the plasma membrane, while the ligands are sorted into vesicles that fuse with lysosomes (see Figure 17-46). In this endocytic pathway, the released ligands are degraded by lysosomal enzymes.

- Studies with mutant LDL receptors in humans with familial hypercholesterolemia revealed a Tyr-X-X-ϕ signal for internalizing receptors into clathrin-coated pits. Other membrane proteins contain different endocytosis signals.

- Unlike the LDL receptor, the transferrin receptor does not release its ligand following endocytosis. Rather, the two iron atoms bound to ferrotransferrin are released in the acidic late endosome, but the resulting apotransferrin remains tightly bound to its receptor and is recycled back to the plasma membrane (see Figure 17-48). At the neutral pH on the cell surface, aprotransferrin is released from the receptor.

- In several receptor-ligand systems, found mainly in oocytes, endocytosed yolk proteins and other proteins remain in the cells and are minimally processed.

- In transcytosis, endocytosed material passes all the way through the cells and is exocytosed from the plasma membrane at the opposite side. An example is the movement of maternal immunoglobulins across mammalian yolk-sac cells into the fetus and across the intestinal epithelial cells of the newborn mouse.

17.10 Molecular Mechanisms of Vesicular Traffic

We return now to a common element in both the secretory and endocytic pathways: the various small vesicles that transport proteins from one organelle to another. These vesicles bud from the membrane of a particular "parent" organelle and fuse with the membrane of a particular "target" (destination) organelle. They are critical to the sorting of proteins newly made in the rough endoplasmic reticulum and of proteins internalized from the cell surface (see Figures 17-13, 17-40, and 17-46). In this section, we examine three crucial issues concerning vesicular traffic within cells:

- What is the mechanism by which transport vesicles are formed? Why are certain soluble luminal proteins and integral membrane proteins incorporated selectively into these vesicles and others not?

- What is the molecular signal on a particular transport vesicle that causes it to bind only to a particular type of organellar membrane? How, for instance, does a vesicle containing secretory proteins from the rough ER "know" to move to and fuse with cis-Golgi and not with trans-Golgi membranes?

- What is the mechanism by which the membranes of a transport vesicle and the destination organelle fuse with each other?

At Least Three Types of Coated Vesicles Transport Proteins from Organelle to Organelle

All eukaryotic cells contain a plethora of small membrane-limited vesicles, many with a protein coat on their cytosolic surface. Three types of coated vesicles are known, each with a different type of protein coat and each formed by reversible polymerization of a distinct set of protein subunits under carefully regulated conditions. Each type of vesicle transports proteins from particular parent organelles to particular destination organelles (Figure 17-50). For instance, vesicles with a clathrin coat form from the plasma membrane and the trans-Golgi network and move to late endosomes. Vesicles with a **COP II** *coat* transport proteins from the rough ER to the Golgi. Finally, vesicles with a **COP I** *coat* mainly transport proteins in the retrograde direction between Golgi cisternae and from the cis-Golgi back to the rough ER. Researchers have not yet identified the coat proteins surrounding many types of transport vesicles, such as those moving from late endosomes to lysosomes or the constitutive secretory vesicles that move proteins from the trans-Golgi to the plasma membrane.

The general scheme of vesicle budding shown in Figure 17-51 applies to all three known types of coated vesicles. During formation of these vesicles, the coat subunit proteins polymerize around the cytosolic face of a budding vesicle, thereby helping the vesicle to pinch off from the parent organelle. Some coat-protein subunits or associated *adapter proteins* select which membrane and soluble proteins will enter the transport vesicles as *cargo proteins*. These cargo proteins contain short signal sequences that direct them to a specific type of transport vesicle (see Table 17-6). A third feature common to formation of clathrin, COP I, and COP II vesicles is involvement of a GTP-binding protein that regulates the rate of vesicle formation. We begin our study of vesicular transport with a detailed look at the similarities and differences in the formation of the three types of coated vesicles, followed by a discussion of the mechanism of membrane fusion.

Clathrin Vesicles Mediate Several Types of Intracellular Transport

As noted earlier, cells that engage in extensive endocytosis (e.g., hepatocytes and fibroblasts) have numerous clathrin-

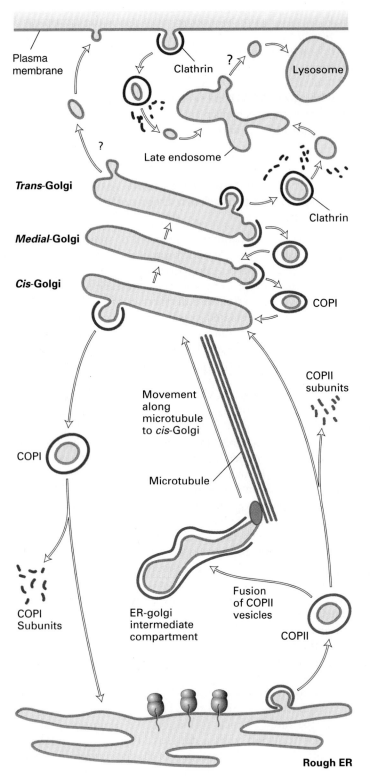

◀ **FIGURE 17-50 Involvement of the three known types of coat proteins—COP I, COP II, and clathrin—in vesicular traffic in the secretory and endocytic pathways.** After formation of vesicles by membrane budding, the coats depolymerize into their subunits (small colored dots), which are re-used to form additional vesicles. COP II mediates transfer of vesicles from the rough ER to the *cis*-Golgi/*cis*-Golgi network; in some cases COP II-coated vesicles fuse with each other to form larger "ER-to-Golgi intermediate compartments" that are transported along a microtubule and eventually fuse to form the *cis*-Golgi. COP I mediates retrograde transport from the *trans*- to the *medial*- to the *cis*-Golgi, as well as from the *cis*-Golgi/*cis*-Golgi network to the rough ER. It may also mediate forward transfer of vesicles from the rough ER to the *cis*-Golgi network (not shown). Clathrin mediates transfer of vesicles that bud from the *trans*-Golgi network and the plasma membrane and that then fuse with late endosomes. Question marks indicate that the nature of the coat is unknown. [See R. Schekman and L. Orci, 1996, *Science* **271**:1526; H. Pelham, 1994, *Cell* **79**:1125; H. Pelham, 1997, *Nature* **389**:17; and J. F. Presley et al., 1997, *Nature* **389**:81.]

coated pits on the cytosolic face of their plasma membrane (Figure 17-52). The formation of these pits requires various adapter proteins as well as clathrin, and their final pinching off requires a GTP-binding protein, called *dynamin*. Here we describe the role of each of these components in vesicle formation from the plasma membrane and *trans*-Golgi.

Clathrin Typical **clathrin-coated vesicles** are 50–100 nm in diameter, with a membrane-bounded vesicle inside a coat composed primarily of the fibrous protein clathrin (Figure 17-53a). Purified clathrin molecules, which have a three-limbed shape, are called *triskelions* from the Greek for three-legged (Figure 17-53b). Each limb contains one clathrin heavy chain (180,000 MW) and one clathrin light chain ($\approx$35,000–40,000 MW). There are two types of light chains, α and β, whose amino acid sequences are 60 percent identical; their functional differences are not known. Even in the absence of membrane vesicles, clathrin triskelions can polymerize to form the cage-like structure that is found around a coated vesicle. When clathrin polymerizes, it forms a polygonal lattice with an intrinsic curvature (Figure 17-53c).

Adapter Proteins Between the fibrous clathrin and the membrane of a clathrin-coated pit lies a 20-nm space containing **assembly particles**. Each particle (340,000 MW) contains one copy each of four different adapter proteins. Assembly particles bind to the globular domain at the end of each clathrin heavy chain in a triskelion (see Figure 17-53b) and promote the polymerization of clathrin triskelions into cages. By also binding to the cytosolic face of membrane proteins, assembly particles determine which proteins are specifically included in (or excluded from) the budding transport vesicle (Figure 17-54). Three types of assembly particles—AP1, AP2, and AP3—composed of different,

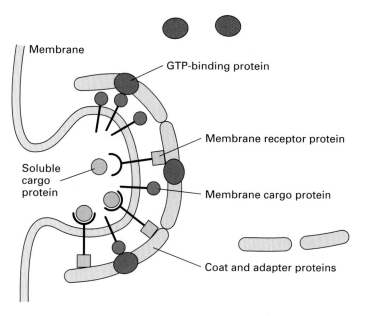

Membrane

GTP-binding protein

Soluble cargo protein

Membrane receptor protein

Membrane cargo protein

Coat and adapter proteins

◀ **FIGURE 17-51 Components that participate in budding of coated vesicles.** Budding is initiated by recruitment of a small GTP-binding protein to a patch of membrane. Then complexes of coat and adapter proteins bind to the cytosolic domains of membrane cargo and receptor proteins; the latter bring soluble luminal cargo proteins into the budding vesicle. [Adapted from a sketch by C. Kaiser.]

Vesicle	Coat and Adapter Proteins	Small GTP-Binding Protein	Transport Step
Clathrin	Clathrin heavy and light chains; AP2	ARF	Plasma membrane → endosome (endocytosis)
	Clathrin heavy and light chains; AP1	ARF	Golgi → endosome
	Clathrin heavy and light chains; AP3	ARF	Golgi → lysosome, vacuole, melanosome, or platelet vesicles
COPI	COP α, β, β', γ, δ, ε, ς	ARF	Golgi → ER Retrograde transport between Golgi cisternae
COPII	Sec23/Sec24 complex; Sec13/Sec31 complex; Sec16	Sar1	ER → Golgi

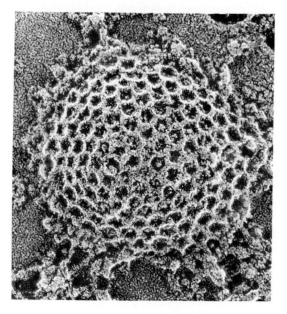

◀ **FIGURE 17-52 Electron micrograph of a clathrin-coated pit on the cytosolic face of the plasma membrane of a fibroblast.** Note the polygonal network of the fibrous clathrin lattice. The cell was rapidly frozen in liquid helium, freeze-fractured, and then treated by the deep-etching technique. [Courtesy of J. Heuser.]

(a) Coated vesicle

(b) Triskelion structure

(c) Assembly intermediate

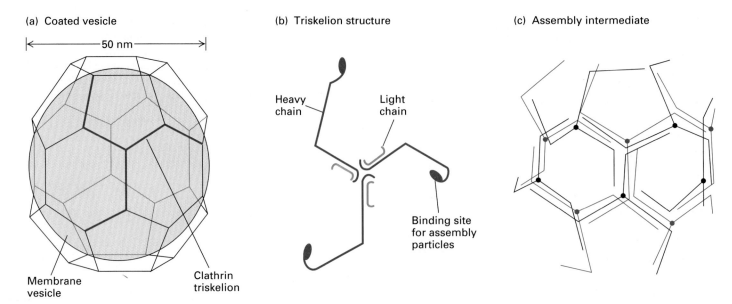

▲ **FIGURE 17-53 Structure of a clathrin-coated vesicle.** (a) A typical clathrin-coated vesicle comprises a membrane-bounded vesicle (tan) about 40 nm in diameter surrounded by a fibrous network of 12 pentagons and 8 hexagons. The fibrous coat is constructed of 36 clathrin triskelions, one of which is shown here in red. One clathrin triskelion is centered on each of the 36 vertices of the coat. Coated vesicles having other sizes and shapes are believed to be constructed similarly: each vesicle contains 12 pentagons but a variable number of hexagons. (b) Detail of a clathrin triskelion. Each of the three clathrin heavy chains has a specific bent structure. A clathrin light chain is attached to each heavy chain near the center; a globular domain is at each distal (outer) tip. Although it is not obvious in (a) or (b), each triskelion has an intrinsic curvature; when triskelions polymerize, they form a curved (not flat) structure. (c) An intermediate in the assembly of a clathrin coat, containing 10 of the final 36 triskelions, illustrates the intrinsic curvature and the packing of the clathrin triskelions. [Part (a) see B. M. F. Pearse, 1987, *EMBO J.* **6**:2507; part (b) see B. Pishvaee and G. Payne, 1998, *Cell* **95**:443.]

though related, adapter proteins, have been identified and found to mediate specific transport steps (see Figure 17-51).

As noted earlier, many cell-surface receptors that are internalized into clathrin-coated pits contain a Tyr-X-X-ϕ sequence (ϕ = bulky hydrophobic amino acid) or a Leu-Leu sequence in their cytosolic domains (see Table 17-6). These signal sequences have been shown to interact with the μ2 subunit of AP2 assembly particles. Plasma-membrane proteins lacking such signals do not interact with AP2 and thus are not internalized in these pits. The LDL and transferrin receptors are examples of plasma-membrane proteins that interact with the AP2 assembly particle (see Figures 17-46 and 17-48). AP1 assembly particles, in contrast, bind to the cytosolic domains of proteins that bud from the *trans*-Golgi network. As discussed previously, the mannose 6-phosphate (M6P) receptor is found in clathrin-coated pits that bud from the plasma membrane or the *trans*-Golgi network; not surprisingly, the cytosolic tail of this receptor binds to both AP1 and AP2 assembly particles.

The most recently identified assembly particle, AP3, has been studied mainly in yeast and mice. Yeasts with mutant AP3 proteins are defective in transport of certain luminal proteins from the *trans*-Golgi to the vacuole, a lysosome-like organelle, but exhibit normal transport of other proteins to the vacuole. Thus yeasts use at least two types of vesicles—one involving AP3 and the other not—for sorting proteins from the *trans*-Golgi to the vacuole. In vertebrates, AP3 is used in budding from the *trans*-Golgi of vesicles destined to fuse with lysosomes. AP3 also participates in the budding of specific types of storage vesicles from the *trans*-Golgi in certain specialized cells. Such vesicles include melanosomes, which contain the black pigment melanin in skin cells, and platelet storage vesicles in megakaryocytes, a large cell that fragments into dozens of platelets. Evidence supporting the involvement of AP3 in formation of both these vesicles comes from mice with mutations in either the β or δ subunit of AP3. Such mutants not only have abnormal skin pigmentation but also exhibit bleeding disorders; the later occur because tears in blood vessels cannot be repaired without platelets that contain normal storage vesicles.

Dynamin Essential to formation of a completed clathrin-coated pit, dynamin is an ≈900-amino-acid cytosolic protein that binds and then hydrolyzes GTP. After dynamin subunits polymerize around the neck of a pit, hydrolysis of GTP is thought to regulate contraction of the polymeric dynamin until the vesicle pinches off (see Figure 17-54).

Incubation of cell extracts with a derivative of GTP that cannot be hydrolyzed provides dramatic evidence for the importance of dynamin in endocytosis. Such treatment leads to accumulation of clathrin-coated pits with excessively long

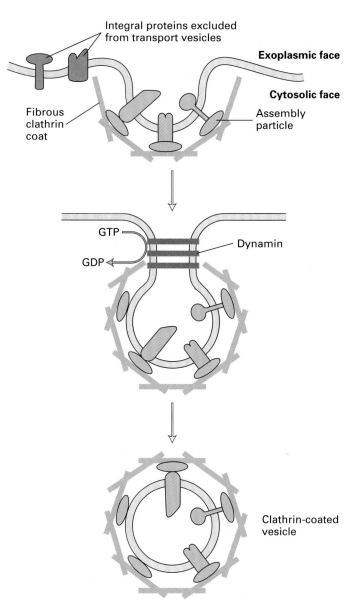

Integral proteins excluded from transport vesicles

Exoplasmic face

Cytosolic face

Fibrous clathrin coat

Assembly particle

GTP

GDP

Dynamin

Clathrin-coated vesicle

◄ **FIGURE 17-54 Model for the formation of a clathrin-coated pit and the selective incorporation of integral membrane proteins into clathrin-coated vesicles.** The cytosolic domains of certain membrane proteins bind specifically to assembly particles that, in turn, bind to clathrin as it polymerizes spontaneously over a region of membrane. Proteins that do not bind to assembly particles are excluded from these vesicles. Dynamin then polymerizes over the neck of the pit; regulated by dynamin-catalyzed hydrolysis of GTP, the neck pinches off, forming a clathrin-coated vesicle. Not depicted here is binding of ARF-GTP to the membrane, which is thought to initiate the process of vesicle budding as in COP I vesicles (see Figure 17-58). [Adapted from K. Takel et al., 1995, *Nature* **374**:186.]

necks that are surrounded by polymeric dynamin but do not pinch off (Figure 17-55). Likewise, the cellular expression of mutant dynamins that cannot bind GTP blocks the formation of clathrin-coated vesicles, resulting in accumulation of similar long-necked pits encased with polymerized dynamin.

Dynamin was discovered by analysis of a temperature-sensitive mutant of *Drosophila*, called *shibere*^ts (*shi*^ts). At low temperature (20°C), the flies were normal, but at 30°C they were paralyzed (shibere means paralyzed in Japanese) because endocytosis of clathrin-coated pits in nerve (and other) cells was blocked. When viewed in the electron microscope, the *shi*^ts neurons at 30°C had abundant clathrin-coated pits with long necks but few clathrin-coated vesicles. Positional cloning of the mutation-defined *shi* gene revealed that it encodes dynamin.

▶ **FIGURE 17-55 Evidence that GTP hydrolysis by dynamin is required for pinching off of clathrin-coated vesicles.** A preparation of nerve terminals, which undergo extensive endocytosis, was lysed by treatment with distilled water and incubated with GTP-γ-S, a derivative of GTP that cannot be hydrolyzed. After sectioning, the preparation was treated with gold-tagged anti-dynamin antibody and viewed in the electron microscope. This image, which shows a long-necked clathrin-coated pit with polymerized dynamin lining the neck, reveals that although pits can form in the absence of GTP hydrolysis, vesicles cannot pinch off. [From K. Takel et al., 1995, *Nature* **374**:186; courtesy of Dr. Pietro De Camilli.]

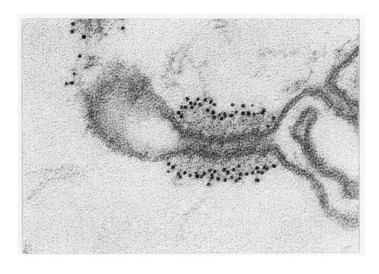

Two other proteins have been identified as essential to formation of clathrin-coated vesicles: *amphyphysin*, which binds to dynamin and assembly particles, and *synaptojanin*, which binds to amphyphysin and dynamin. Amphyphysin is thought to recruit dynamin to the neck of budding vesicles, but precisely how these two proteins function with dynamin to finally pinch off a vesicle is not known.

Depolymerization of Clathrin Coats Clathrin-coated vesicles are stable at the pH and ionic composition of the cell cytosol. As we've seen already, however, coated vesicles normally lose their clathrin coat and the assembly particles just after their formation (see Figure 17-46). Cytosolic Hsc70, a chaperone protein found in all eukaryotic cells, is thought to catalyze depolymerization of the clathrin coat into triskelions; these then can be re-used in the formation of additional pits and vesicles. Both the formation of clathrin-coated vesicles and the depolymerization of the coat must be highly regulated in the cell, as both processes occur simultaneously. Much remains to be learned about the regulation of these processes at the molecular level.

COP I Vesicles Mediate Retrograde Transport within the Golgi and from the Golgi Back to the ER

Budding vesicles surrounded by a nonclathrin coat were first observed in micrographs of isolated Golgi fractions incubated in a solution containing ATP and cytosol. Although these fractions did not contain clathrin, they formed a large number of buds and vesicles with a distinct outer coat on their cytosolic face (Figure 17-56). Subsequent analysis of the coat protein, now called COP I, from these vesicles

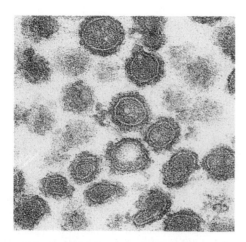

▲ **FIGURE 17-56 Electron micrograph of COP I vesicles purified from the Golgi fraction of rat hepatocytes after incubation with cytosol and ATP.** The distinct coat associated with the cytosolic face of the membrane is easily visible. Such coated vesicles transport proteins in the retrograde direction from one Golgi cisterna to another and from the *cis*-Golgi network back to the rough ER. [Courtesy of L. Orci.]

showed that it is formed from *coatomers*, cytosolic complexes each containing seven polypeptide subunits, α, β, β', γ, δ, ϵ, and ζ. Like clathrin triskelions, COP I coatomers polymerize on the cytosolic surface of a budding vesicle and then dissociate from the vesicles soon after they form. Some of the COP I subunits serve the same functions as the assembly particles in clathrin-coated vesicles—acting as a bridge between the cytosolic tails of membrane proteins and the fibrous cage that surrounds the vesicles, and mediating the specific incorporation of proteins into these coated vesicles (see Figure 17-51). For instance, β COP has amino acid sequences that are distantly related to one of the adapter proteins composing the assembly particles found in clathrin-coated vesicles.

The cell-free system depicted in Figure 17-57 has been useful in documenting the movement of proteins between Golgi cisternae. In this system, cultured mutant cells missing one of the enzymes that modify N-linked oligosaccharides in the Golgi are infected with vesicular stomatitis virus (VSV). This virus is chosen because it synthesizes only *one* predominant glycoprotein, the VSV G protein, which accumulates in the plasma membrane. (Like the influenza HA protein, G protein forms the surface spikes of the virus.) In addition to demonstrating transport of proteins from various donor organelles to acceptor organelles, this general type of assay has been used to discover the identity and function of the various proteins required for formation of different transport vesicles, their targeting to appropriate organelles, and their fusion with acceptor membranes. Here we discuss formation of COP I vesicles, which are thought to mediate retrograde (backward) transport from the *cis*-Golgi to the rough ER and retrograde transport between the various Golgi cisternae (see Figure 17-50).

Formation of COP I Vesicles Fractionation of the cytosol used in the cell-free system just described has shown that several proteins are required for the budding of COP I transport vesicles, as depicted in Figure 17-58. First, in the cytosol a small GTP-binding protein known as ARF releases its bound GDP and binds GTP. The Golgi-attached enzyme that catalyzes this GDP-GTP exchange apparently receives and integrates multiple, as yet unknown, signals from the cytosol, such that the amount of ARF-GTP, and thus the rate of formation of COP I vesicles, is appropriate to the needs of the cell. The resulting ARF-GTP complex then binds to ARF receptors on the Golgi membrane. Next, COP I coatomers bind to ARF and other proteins on the cytosolic face of the Golgi membrane, inducing budding of the transport vesicle. The final fission that creates the completed transport vesicle requires fatty acyl CoA, but how this molecule functions is unknown. When purified Golgi cisternae were incubated with coatomers, ARF, fatty acyl CoA, and a derivative of GTP that could not be hydrolyzed, researchers observed the formation and release of COP I–coated vesicles similar to those shown in Figure 17-56, providing support for this model of vesicle formation. Once the COP I vesicles are released from the donor membrane, the

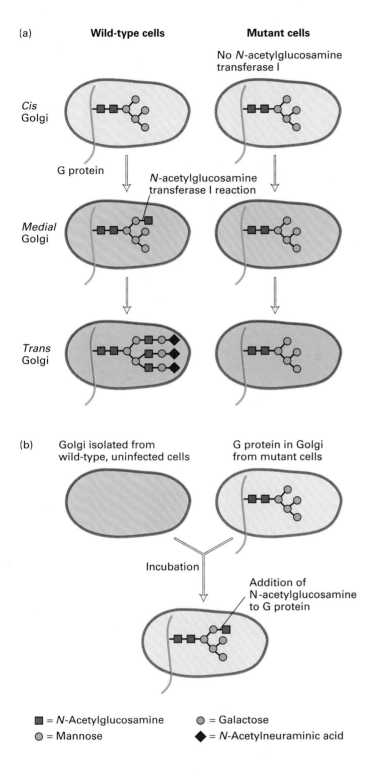

(a)

Wild-type cells   **Mutant cells**

No *N*-acetylglucosamine transferase I

Cis Golgi

G protein

N-acetylglucosamine transferase I reaction

Medial Golgi

Trans Golgi

(b)

Golgi isolated from wild-type, uninfected cells

G protein in Golgi from mutant cells

Incubation

Addition of N-acetylglucosamine to G protein

■ = *N*-Acetylglucosamine   ⦾ = Galactose

◉ = Mannose   ◆ = *N*-Acetylneuraminic acid

◀ **FIGURE 17-57 A cell-free system demonstrating movement of protein from one Golgi cisternae to another.** (a) A mutant line of cultured fibroblasts is essential in this type of assay. In this case, the cells lack the enzyme *N*-acetylglucosamine transferase I (step ② in Figure 17-38). In wild-type cells, this enzyme is localized to the *medial*-Golgi and modifies asparagine-linked oligosaccharides by the addition of one *N*-acetylglucosamine. In VSV-infected wild-type cells, the oligosaccharide on the viral G protein is modified to a typical complex oligosaccharide, as shown in the *trans*-Golgi panel. In infected mutant cells, however, the G protein reaches the cell surface with a simpler high-mannose oligosaccharide containing only two *N*-acetylglucosamine and five mannose residues. (b) When Golgi cisternae isolated from infected mutant cells were incubated with Golgi cisternae from normal, uninfected cells, the G protein produced contained the additional *N*-acetylglucosamine. Two interpretations of this finding are possible: (1) the G protein traveled from mutant *cis*-Golgi cisternae to wild-type *medial*-Golgi cisternae to receive the *N*-acetylglucosamine residue, or (2) the transferase enzyme moved in the retrograde direction from the wild-type *medial*- to the mutant *cis*-Golgi cisternae where it modified the VSV G protein. The latter explanation is now believed to be correct, with transport mediated by COP I vesicles. [See W. E. Balch et al., 1984, *Cell* **39**:405 and 525; W. A. Braell et al., 1984, *Cell* **39**:511; and J. E. Rothman and T. Söllner, 1997, *Science* **276**:1212.]

COP I proteins accumulate proteins in the rough ER at the restrictive temperature and thus are categorized as class B *sec* mutants (see Figure 17-14). Discovery of these mutants suggested that COP I vesicles mediate transport of proteins from the rough ER to the Golgi. Indeed, some COP I vesicles do form in cell-free extracts from purified rough ER vesicles, but other experiments showed that the major role of COP I vesicles is in the retrograde transport of proteins between Golgi cisternae and from the *cis*-Golgi back to the rough ER.

As described earlier, the KDEL receptor, which is located primarily in the *cis*-Golgi membrane, mediates the return of soluble luminal proteins containing the KDEL sequence from the Golgi to the ER (see Figure 17-29). The KDEL receptor and other membrane proteins that are transported back to the ER from the Golgi contain a Lys-Lys-X-X sequence at the very end of their C-terminal segment, which faces the cytosol (see Table 17-6). This sorting sequence, which binds to a complex of the COP α and β subunits, is both necessary and sufficient to direct retrograde transport of membrane proteins. Yeast mutants lacking COPα or COPβ not only are unable to bind this Lys-Lys-X-X motif but also are unable to retrieve proteins bearing this sequence back to the ER, indicating that COP I vesicles mediate Golgi-to-ER retrograde transport. Because COP I mutants cannot recycle membranes back to the rough ER, the ER gradually becomes depleted both of phospholipids and resident ER membrane proteins. Eventually vesicle formation from the rough ER grinds to a halt; secretory proteins continue to be synthesized but accumulate in the ER, the defining characteristic of class B *sec* mutants (see Figure 17-14).

COP I coat depolymerizes and dissociates. Hydrolysis of the GTP bound to ARF proteins in the vesicles triggers this step; as evidence of when COP I vesicles are formed in vitro in the presence of a nonhydrolyzable GTP analog, the COP I coats never dissociate.

Evidence for Retrograde Transport by COP I Vesicles Yeast cells containing temperature-sensitive mutations in

▶ **FIGURE 17-58 Model for formation of COP I–coated vesicles.** Budding is initiated when molecules of ARF protein exchange their bound GDP for GTP, a reaction catalyzed by an enzyme in the Golgi membrane. After ARF-GTP binds to ARF receptors on Golgi cisternae, coatomers bind to the cytosolic face of the Golgi cisterna and polymerize into a fibrous coat that induces vesicle budding. Because they can bind to coatomer, certain integral membrane proteins are incorporated into the vesicles. These include a V-SNARE, which functions in targeting vesicles to appropriate acceptor membranes. Soluble proteins in the lumen are selected for entry into these vesicles by binding to specific membrane receptor proteins. Fatty acyl CoA is essential for the final separation of the transport vesicle from the donor membrane, but how it functions is not known. Finally, hydrolysis of GTP bound to the ARF proteins causes depolymerization of the coat and release of coatomers and ARF-GDP. [Adapted from J. E. Rothman, 1994, *Nature* **372**:55, and J. E. Rothman and F. Wieland, 1996, *Science* **272**:227.]

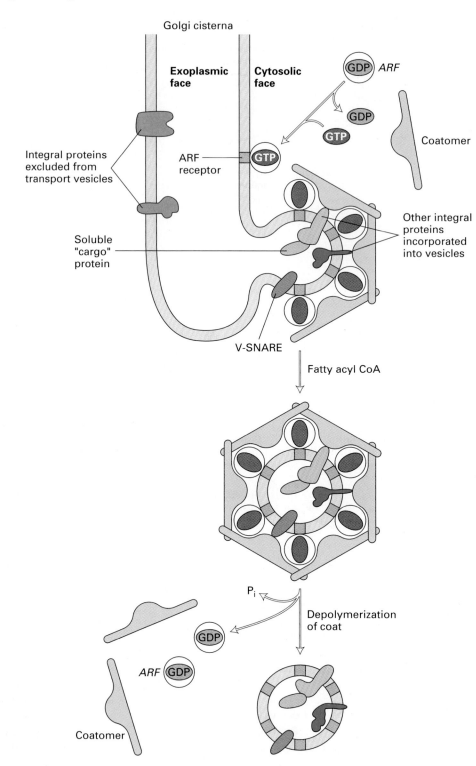

COP I vesicles also mediate retrograde transport within the Golgi cisternae. They selectively transport membrane and luminal proteins from the *trans-* to the *medial*-Golgi while leaving others behind in the *trans*-Golgi cisternae; similarly, they selectively move proteins from the *medial-* to the *cis*-Golgi (see Figure 17-50). As noted earlier, *cis*-Golgi vesicles are constantly moving forward through the Golgi stack by cisternal progression, eventually becoming *trans*-Golgi vesicles. Because COP I–coated vesicles selectively retrieve membrane proteins to earlier Golgi compartments, proteins such as the carbohydrate-modifying enzymes depicted in Figure 17-38 become localized in specific Golgi cisternae.

COP II Vesicles Mediate Transport from the ER to the Golgi

COP II vesicles were first recognized when cell-free extracts of yeast rough ER membranes were incubated with cytosol and ATP. The vesicles that formed from the ER membranes had a distinct coat, similar to COP I vesicles, but composed of different proteins, designated COP II proteins. Like COP I mutants, COP II mutants are class B *sec* mutants and accumulate proteins in the rough ER (see Figure 17-14). Analysis of such mutants has revealed several proteins required for formation of COP II vesicles.

Formation of COP II vesicles is thought to be generally similar to the process shown in Figure 17-58 for COP I vesicles. For instance, budding is triggered when Sec12 catalyzes the exchange of bound GDP with GTP on Sar1, a protein analogous to ARF (see Figure 17-51). This exchange induces binding of a complex of Sec23 and Sec24 proteins, followed by binding of a complex of Sec13 and Sec31 proteins. A large fibrous protein, called Sec16, which is bound to the cytosolic surface of the ER, interacts with the Sec13/31 and Sec23/24 complexes, and may act as a scaffold for assembly of the vesicle coat.

COP II vesicles also contain a family of 24-kDa membrane proteins that selectively bind soluble ER proteins destined for transport to the Golgi but not resident ER proteins such as the chaperone Hsc70. Certain integral membrane proteins also are specifically recruited from the ER into COP II vesicles for transport to the Golgi; the cytosolic segments of many of these proteins contain a di-acidic sorting signal (Asp-X-Glu) that binds to one or more COP II proteins and is essential for the selective export of these proteins from the ER (see Table 17-6). Additional biochemical and genetic studies are needed to elucidate the details of how COP II vesicles form, including the final separation step and subsequent loss of their coat proteins. How specific membrane proteins are recruited into these vesicles, while others are left behind, is also an active topic of current research.

To follow movement of COP II vesicles from the ER to the Golgi in living cultured mammalian cells, investigators constructed a recombinant gene encoding a chimeric protein consisting of green fluorescent protein (see Figure 5-7) fused to the cytosolic-facing C-terminus of the VSV G glycoprotein. Fluorescence microscopy revealed some cells in which small COP II–coated fluorescent vesicles, containing the chimeric protein, were seen to form from the ER, move less than 1 μm, and then fuse directly with the *cis*-Golgi. In other cells, in which the ER was located several micrometers from the Golgi complex, several COP II–coated vesicles were seen to fuse with each other shortly after their formation, forming what is termed the "ER-to-Golgi intermediate compartment" (see Figure 17-50). These larger structures then were transported along microtubules to the *cis*-Golgi, much in the way vesicles in nerve cells are transported from the cell body, where they are formed, down the long axon to the axon terminus (Chapter 19). Microtubules function much as "rail-road tracks" enabling these COP II–coated vesicles to move long distances to their *cis*-Golgi destination. At the time the ER-to-Golgi intermediate compartment is formed, some COP I–coated vesicles bud off from it, recycling some proteins back to the ER.

Specific Fusion of Intracellular Vesicles Involves a Conserved Set of Fusion Proteins

Despite the variety of coat proteins that mediate formation of transport vesicles, fusion of all vesicles with their target membranes exhibits several common features. In all cases, fusion occurs after the coats have depolymerized and seems to involve a conserved set of proteins that (a) mediates targeting of vesicles to the appropriate fusion partner and (b) triggers the fusion process itself.

Depolymerization of the vesicle coat proteins uncovers a unique type of integral membrane protein, called a **V-SNARE**, which is incorporated into a transport vesicle as it buds from the donor organelle (see Figure 17-58). The specific V-SNARE on each type of transport vesicle targets the vesicle to its correct membrane fusion partner. The membrane of each type of target membrane or organelle in a cell contains the ubiquitous fusion protein *SNAP25* and one or more integral membrane proteins, called **T-SNAREs**, which act cooperatively to specifically bind a particular type of V-SNARE. Yeast cells, like all eukaryotes, express several related V-SNARE and T-SNARE proteins, thereby permitting each type of transport vesicle to be targeted correctly.

Biochemical and Genetic Evidence for the Role of SNAREs and SNAPs Biochemical studies demonstrate that a V-SNARE, a T-SNARE, and SNAP25 are sufficient to mediate vesicle fusion. For instance, when liposomes containing a purified V-SNARE are incubated with other liposomes containing a preformed T-SNARE/SNAP25 complex, tertiary complexes form and the two classes of membranes fuse, albeit slowly. In cells, however, fusion occurs within seconds and several other cytosolic proteins also are required for vesicle fusion (Figure 17-59). One of these, NSF, is a tetramer of identical subunits that binds and hydrolyzes ATP. Other proteins, called α-, β-, and γ-SNAPs (soluble NSF attachment proteins) are required for NSF to bind to vesicle membranes. Precisely how these proteins participate in vesicle fusion is controversial; most likely they cause dissociation of the T-SNARE/V-SNARE/SNAP25 complexes that form as a result of fusion, allowing them to recycle and catalyze additional membrane fusion events. Similar proteins are involved in the fusion of synaptic vesicles with the plasma membrane of axon terminals, leading to release of neurotransmitters (Chapter 21.4).

An analysis of the various classes of temperature-sensitive yeast *sec* mutants described earlier has confirmed the involvement of several yeast Sec proteins in the fusion of transport vesicles. For instance, the mammalian NSF protein and the yeast Sec18 protein are similar in sequence, as are the mammalian α-SNAP and the yeast Sec17. Yeast with mutations

▶ **FIGURE 17-59 Model for targeting and fusion of transport vesicles with their acceptor membranes.** Step ①: The attachment of a transport vesicle to its target is initiated by binding of the cytosolic domain of the particular V-SNARE protein in the vesicle membrane to the cytosolic domains of a T-SNARE/SNAP25 protein complex in the acceptor membrane. As shown in the inset, four long α helices, two from SNAP25 and one each from the T- and V-SNARE proteins, interact to form a coiled coil that holds the vesicle close to the acceptor membrane. A Rab protein in the vesicle membrane may act as a timer for vesicle targeting and fusion. Step ②: Formation of a prefusion complex probably requires interaction of many T-SNARE and V-SNARE proteins and ATP hydrolysis. This step can be blocked selectively by *N*-ethylmaleimide (NEM), a chemical that reacts with an essential —SH group on NSF (hence the name, *NEM-sensitive factor*), suggesting that NSF is required. Step ③: Fusion of the two membranes immediately follows prefusion, but precisely how this occurs is not known. Step ④: Following fusion, the T-SNARE/V-SNARE/ SNAP25 complexes must dissociate so that these proteins can catalyze additional fusion events. The complex of the α-, β-, and γ-SNAP proteins, together with NSF, may catalyze this dissociation. Step ⑤: Vesicles containing V-SNARE proteins form and return to the original donor membrane. [See J. E. Rothman and T. Söllner, 1997, *Science* **276**:1212; A. Mayer and W. Wickner, 1997, *J. Cell Biol.* **136**:307; Y. Goda, 1997, *Proc. Nat'l. Acad. Sci.* **94**:769; and W. Weis and R. Scheller, 1998, *Nature* **395**:328. Inset adapted from R. Sutton et al., 1998, *Nature* **395**:347.]

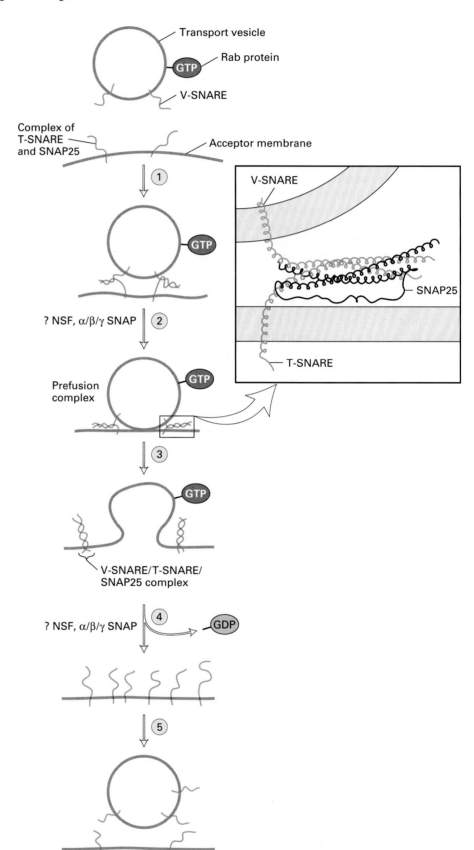

in the *sec17* or *sec18* genes accumulate ER-to-Golgi transport vesicles at the high, nonpermissive temperatures; when these class C mutant cells are placed at the lower, permissive temperature, the accumulated vesicles fuse with the *cis*-Golgi (see Figure 17-14). Thus, the yeast Sec17 and Sec18 proteins are required for vesicle fusion, in agreement with the biochemical demonstration that their mammalian counterparts, NSF and α-SNAP, are required for the fusion of transport vesicles with Golgi membranes.

GTP-Binding Rab Proteins as Regulators of Vesicular Traffic A family of GTP-binding proteins participates in the control of vesicular traffic in eukaryotic cells. All of these *Rab proteins* contain ≈ 200 amino acids and have an overall structure similar to Ras (see Figure 3-5). Like Ras, purified Rab proteins bind and hydrolyze GTP, and it is thought that the cycle of GTP binding and hydrolysis regulates the rate of vesicle fusion. Specifically, a cytosolic protein called GDI catalyzes the exchange of GDP, bound to cytosolic Rab, for GTP, inducing a conformational change in Rab that enables it to bind to a surface protein on a particular transport vesicle. After vesicle fusion occurs, the GTP bound to the Rab protein is hydrolyzed to GDP, triggering the release of the Rab protein, which then can undergo another cycle of GDP-GTP exchange, binding, and hydrolysis (see Figure 17-59). The rate of vesicle fusion is controlled by the absolute amount of Rab $\cdot$ GTP, which is modulated by unidentified protein regulators.

Several lines of evidence support the involvement of specific Rab proteins as timers of vesicle fusion events. For instance, Rab5 is localized to *early endosomes*—organelles that form from clathrin-coated vesicles just after they bud from the plasma membrane during receptor-mediated endocytosis (see Figure 17-46). The fusion of early endosomes with each other in cell-free systems requires the presence of Rab5; no other Rab protein can replace Rab5. Addition of Rab5 and GTP to cell-free extracts accelerates the rate at which these endosomes fuse with each other, suggesting that Rab5 $\cdot$ GTP acts as a timer for vesicle fusion. Similarly, Rab1 is essential for ER-to-Golgi transport reactions to occur in cell-free extracts, while other Rab proteins, such as Rab5 or Rab7, are not. The yeast *sec4* gene is a member of the Rab gene family, and yeast cells expressing mutant Sec4 proteins accumulate secretory vesicles that are unable to fuse with the plasma membrane (class E mutants in Figure 17-14). Thus, some individual Rab proteins are clearly essential for specific vesicle fusion reactions to occur, but we do not know whether Rab proteins interact with V-SNARE proteins to determine the specificity by which vesicles fuse with target membranes.

Conformational Changes in Influenza HA Protein Trigger Membrane Fusion

One of the best-understood examples of membrane fusion occurs during infection by enveloped animal viruses, particularly influenza virus. Such viruses have an outer phospholipid bilayer membrane, or envelope, surrounding the viral genetic material and protein coat. The virus is formed by budding from the plasma membrane of the host cell, and the viral membrane phospholipids are derived from the host's plasma membrane (see Figure 6-17). These viruses enter a host cell by receptor-mediated endocytosis following binding of one or more virus-specific surface glycoproteins with a host's cell-surface molecules. After a virus particle is endocytosed into an acidic endosome, the viral and endosomal membranes fuse, releasing the genetic material inside of the virus into the cytosol of the host cell and initiating replication of the virus. The molecular events of this fusion process have been elucidated in considerable detail in the case of influenza virus.

The predominant glycoprotein of the influenza virus is hemagglutinin (HA), which forms the larger spikes on the surface of the virus (see Figure 6-13). The HA spikes bind to sialic acid, a terminal sugar in the oligosaccharides of many cell-surface glycoproteins and glycolipids. There is considerable evidence that the low pH within the enclosing endosome triggers the fusion of its membrane with the viral membrane. For instance, viral infection is inhibited by the addition of lipid-soluble bases, such as ammonia or trimethylamine, which raise the pH of normally acidic endosomes. Also, a conformational change in the HA spike occurs over a very narrow change in pH (5.0–5.5) and is critical for infectivity.

As discussed earlier, HA is synthesized as a precursor protein, HA_0, that forms a trimer in the ER (see Figure 17-27). Each monomer in the trimeric precursor is cleaved in the Golgi, probably by the endoprotease furin, into an HA_1 and an HA_2 subunit. Thus each HA spike protein consists of three HA_1 and three HA_2 subunits. The N-terminus of HA_2, generated by the proteolytic cleavage, constitutes the *fusion peptide*—a strongly hydrophobic amino acid sequence: Glu-Leu-Phe-Gly-Ala-Ile-Ala-Gly-Phe-Ile-Glu.

At a pH of 7.0, the N-terminus of each HA_2 subunit is tucked into a crevice in the spike (Figure 17-60a). This is the normal HA conformation when a viral particle encounters the surface of a host cell. At the pH of 5.0 within an endocytic vesicle, HA undergoes several conformational changes that cause a major rearrangement of the subunits. As a result, the three HA_2 subunits twist together into a three-stranded coiled-coil rod that protrudes more than 13 nm outward from the viral membrane with the fusion peptides at the tip of the rod (Figure 17-60b). In this conformation, the highly hydrophobic fusion peptides are exposed and can insert into the lipid bilayer of the endocytic vesicle membrane, triggering fusion of the viral and endosomal membranes. Thus at pH 7 HA can be said to be trapped in a metastable, "spring-loaded" state, which is converted to the lower-energy fusogenic state by shifting the pH to 5. This conversion can be induced in the test tube at pH 7 by treatment with urea or other denaturing agents.

Multiple pH 5–activated HA spikes are essential for membrane fusion to occur. Figure 17-61 suggests one way by which the protein scaffold formed by many HA spikes, possibly together with other cellular proteins, could link

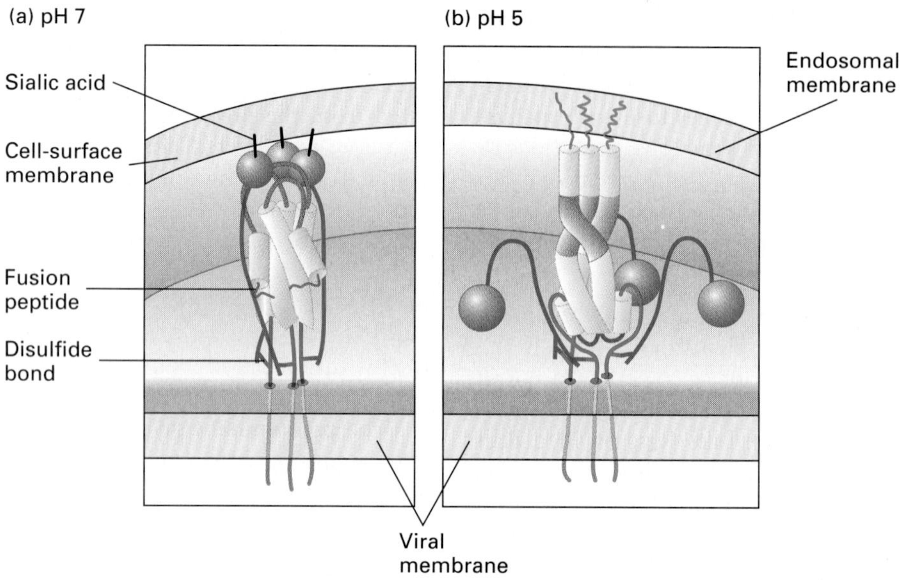

(a) pH 7

(b) pH 5

Sialic acid

Cell-surface membrane

Fusion peptide

Disulfide bond

Endosomal membrane

Viral membrane

▲ **FIGURE 17-60 Schematic models of the structure of influenza HA spikes at pH 7 and 5.** Three HA_1 and three HA_2 subunits compose a spike, which protrudes from the viral membrane. (a) At pH 7, part of each HA_1 subunit (green) forms a globular domain at the tip of the native spike. These domains bind to sialic acid residues on the host-cell plasma membrane, initiating viral entry. Each HA_1 subunit is linked to one HA_2 subunit by a disulfide bond at the base of the molecule near the viral membrane. Each HA_2 subunit contains a fusion peptide (red) at its N-terminus (only two are visible), followed by a short α helix (orange), a nonhelical loop (brown), and a longer α helix (light purple). The longer α helices from the three HA_2 subunits form a three-stranded coiled coil (see Figure 3-4). In this conformation, the fusion peptides are buried within the molecule. (b) At pH 5

within an endocytic vesicle, the binding of the fusion peptide to other segments of HA_2 is disrupted, inducing major structural rearrangements in the protein. First, the three HA_1 globular domains separate from each other but remain tethered to the HA_2 subunits by the disulfide bonds at the base of the molecule. Second, the loop segment of each HA_2 rearranges into an α helix (brown) and combines with the short and long α-helical segments to form a continuous 88-aa α helix. The three long α helices thus form a 13.5-nm-long three-stranded coiled coil that protrudes outward from the viral membrane. In this conformation, the fusion peptides are at the tip of the coiled coil. Their insertion into the endocytic vesicle membrane triggers fusion of the viral and endosomal membranes. [Adapted from C. M. Carr et al., 1997, *Proc. Nat'l. Acad. Sci.* **94**:14306; courtesy of Peter Kim.]

together the viral and endosomal membranes and induce their fusion. This figure also illustrates some likely but still hypothetical intermediates in the fusion process. Note that each HA molecule can undergo only one fusion event, whereas the cellular fusion proteins, such as V- and T-SNAREs, are recycled and catalyze multiple cycles of vesicle fusion.

SUMMARY Molecular Mechanisms of Vesicular Traffic

• Each of the three known types of coated vesicles mediates different transport routes (see Figure 17-50): clathrin vesicles, from the plasma membrane and *trans*-Golgi to endosomes; COP I vesicles, between Golgi cisternae and from the *cis*-Golgi back to the rough ER; and COP II vesicles, from the rough ER to the *cis*-Golgi.

• In addition to coat proteins, various adapter proteins and small GTP-binding proteins are required for formation of coated vesicles (see Figure 17-51).

• Formation of clathrin vesicles is assisted by assembly particles, which promote the polymerization of clathrin triskelions into cages and determine which proteins are specifically included in the budding transport vesicle. Dynamin, a cytosolic protein that binds and then hydrolyzes GTP, is essential for the final pinching off of a completed clathrin-coated vesicle (see Figure 17-54).

• Formation of COP I vesicles is initiated when a Golgi-attached enzyme catalyzes the exchange of GDP bound to cytosolic ARF with GTP. The ARF-GTP complex then binds to ARF receptors on the cytosolic face of the Golgi membrane, triggering polymerization of COP I protein complexes (coatomers), which induces vesicle budding (see Figure 17-58). COP II vesicles are formed by a similar process, but different proteins are involved.

• Once transport vesicles have budded off, the coat is depolymerized, releasing the coat proteins for reuse. Uncoating of clathrin vesicles is catalyzed by a cytosolic Hsc70 chaperone. Hydrolysis of the ARF-bound GTP depolymerizes COP I coats.

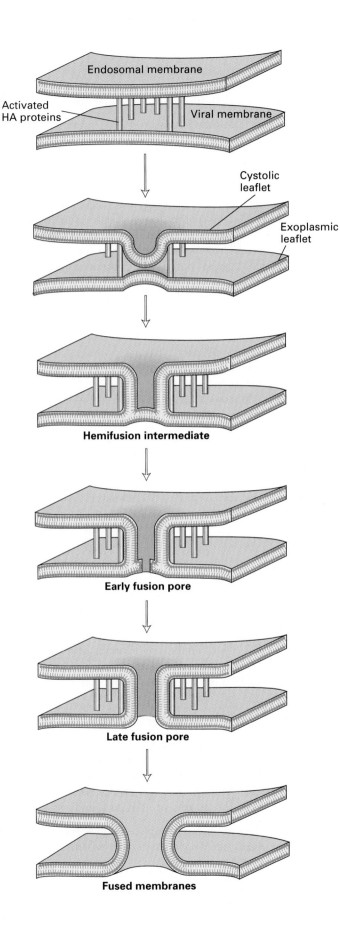

Endosomal membrane

Activated HA proteins **Viral membrane**

Cystolic leaflet

Exoplasmic leaflet

Hemifusion intermediate

Early fusion pore

Late fusion pore

Fused membranes

◀ **FIGURE 17-61 Model of membrane fusion directed by the HA protein.** A number of pH 5–activated HA spikes, possibly in concert with cellular membrane proteins, form a scaffold that connects a small region of the viral membrane and the endosomal membrane. By unknown mechanisms, the exoplasmic leaflets of the two membranes fuse and form a continuous phospholipid leaflet, shown as the hemifusion intermediate. Then the cytosolic leaflets of the two membranes fuse, forming an early fusion pore; the pore then widens until complete fusion of the two membranes is achieved. [Adapted from J. R. Monck and J. M. Fernandez, 1992, *J. Cell Biol.* **119**:1395.]

- Uncoating of transport vesicles exposes specific V-SNARE proteins on the surface of each type of vesicle. Each V-SNARE specifically binds to cognate T-SNARE proteins complexed with SNAP25 on the membrane of the target vesicle. NSF and α-, β-, and γ-SNAP proteins then bind to the T-SNARE/V-SNARE/SNAP25 complex, stabilizing the prefusion complex (see Figure 17-59) and/or catalyzing dissociation of the T-SNARE/V-SNARE/SNAP25 complex after vesicle fusion.

- GTP-binding Rab proteins serve as regulators of vesicle targeting and fusion.

- During infection by enveloped animal viruses, the membrane of an internalized viral particle fuses with the surrounding endosomal membrane. In the case of influenza virus, the acidic endosomal pH causes a conformational change in the HA spike protein that exposes the highly hydrophobic fusion peptides, which then insert into the endocytic vesicle membrane (see Figure 17-60). Insertion of multiple activated HA spikes triggers fusion of the viral and endosomal membranes in a process that probably involves several intermediates.

PERSPECTIVES for the Future

As this lengthy chapter has shown, we now possess, at least in outline, answers to many key questions about how proteins are sorted to their correct destinations and how proteins become secreted from cells. However, many fundamental issues remain unresolved. For instance, although the receptor and transport proteins that catalyze uptake of proteins into mitochondria, chloroplasts, and peroxisomes have been identified, we are only beginning to understand how organelle-targeted proteins actually cross one or more organelle membranes. Nor do we know how organelle receptor proteins manage to bind proteins with similar, but not identical, uptake-targeting sequences. Similarly, we have only a rudimentary understanding of how nascent secretory and membrane proteins actually cross the membrane of the rough endoplasmic reticulum. How, for instance, can the translocon channel accommodate hairpin structures, and how are long hydrophobic sequences of amino acids in the nascent chain actually recognized as membrane anchor and other topogenic sequences?

In a similar vein, we know the identity of many of the coat proteins and other types of proteins (e.g., those that bind and hydrolyze GTP) that are essential for formation of the various types of transport vesicles found in cells. We also know some of the proteins that target these transport vesicles to their correct destinations in the cell, and that catalyze fusion of the two membranes. Nonetheless, the actual process whereby a segment of a phospholipid bilayer is "pinched off" to form a pit and eventually a new vesicle is still not understood. How membrane and soluble proteins are selectively incorporated in or excluded from such a transport vesicle is a fundamental question in protein sorting, yet we have only a fragmentary understanding of these processes.

Moreover, the mechanism by which secretory and plasma membrane proteins move from the *cis-* to the *medial-* and then to the *trans-*Golgi is controversial. Most work now favors the cisternal progression model discussed in this chapter, but many workers still believe that secretory proteins move between Golgi compartments by vesicle budding and fusion: small transport vesicles with a cargo of secretory proteins would bud from the *cis-*Golgi and fuse with the *medial-*Golgi, for example, and others would bud from the *medial-*Golgi and fuse with the *trans-*Golgi. Perhaps most surprisingly, we do not know how fusion of two membranes actually occurs, and whether models such as that in Figure 17-61 are at all realistic.

Detailed understanding of these processes will require determination of the three-dimensional structures of many important proteins. This will not be easy since many of these proteins are normally made in small amounts and exist in very large, multi-protein complexes such as the translocon in the endoplasmic reticulum, the receptors and transport channels in the outer mitochondrial membrane, and the COP I coatomer. Also, many of the key proteins involved in protein sorting are integrally bound to a phospholipid membrane; such proteins are notoriously difficult to purify and crystallize. In parallel with this structural work, additional subcellular fractionation studies are needed to purify the many different types of transport vesicles and to identify all of their membrane and luminal proteins. Future genetic and biochemical studies, both in yeasts and in mammals, also will be essential for identifying the key proteins that catalyze each of the numerous steps in protein sorting.

PERSPECTIVES in the Literature

Several groups have developed cell-free systems to study the mechanism by which clathrin, COP I, and COP II vesicles form. These reactions contain protein-free liposomes, a small GTP-binding protein (either Sar1p or ARF1), and soluble coat proteins (either clathrin triskelia and adapter proteins,

COP I coatomer, or COP II protein subunits). Reactions also contain a nonhydrolyzable GTP analog to prevent vesicles that form from depolymerizing their coat. As you read the articles listed below think about these key questions:

1. How is vesicle budding measured experimentally?

2. How do the small GTP-binding proteins Sar1p and ARF1 bind to the donor vesicle?

3. Are integral proteins in the door membrane essential for vesicle budding? That is, do coats and their adapters first polymerize over areas of the donor vesicles and then bind to the cytosolic domains of integral proteins as they diffuse within the plane of the membrane? Or do the cytosolic domains of vesicle integral membrane proteins, together with GTP-binding proteins, recruit adapter proteins, thereby initiating coat assembly and linking this with cargo selection?

4. Does binding of coat proteins to specific membrane lipids promote coat polymerization and vesicle budding? Are interfaces between lipids and cytosolic proteins sufficient to deform lipid bilayers into buds and tubules, or are integral proteins required?

Matsuoka, K., et al. 1998. Cop II-coated vesicle formation reconstituted with purified coat proteins and chemically defined liposomes. *Cell* 93:263–275.

Spang, A., et al. 1998. Coatomer, Arf1p, and nucleotide are required to bud coat protein complex I-coated vesicles from large synthetic liposomes. *Proc. Nat'l. Acad. Sci USA* 95:11199–11204.

Takei, K., et al. 1998. Generation of coated intermediates of clathrin-mediated endocytosis on protein-free liposomes. *Cell* 94:131–141.

Zhu, Y., M. T. Drake, and S. Kornfeld. 1999. ADP-ribosylation factor 1 dependent clathrin-coat assembly on synthetic liposomes. *Proc. Nat'l Acad. Sci. USA* 96:5013–5018.

Testing Yourself on the Concepts

1. In the absence of a signal or targeting sequence, what is the expected location of any protein?

2. How does the role of chaperone proteins differ in the import of proteins into chloroplasts and mitochondria, the endoplasmic reticulum (ER), and the peroxisome?

3. What is the evidence that protein translocation into the endoplasmic reticulum is through a protein channel?

4. After much work, you isolate a small amount of a protein from a total homogenate of *C. elegans*. The protein contains disulfide bonds and no hydrophobic rich sequence longer than five amino acids. Based on these properties, where within *C. elegans* or its cells should this protein be found?

5. Protein glycosylation is a complicated process with many enzymes involved and a large array of structural variants.

What are some of the practical consequences of variations in protein glycosylation for a biotechnology protein product that is to be administered to humans by injection into blood vessels?

6. What is the role of transport vesicles in a cisternal maturation/cisternal progression model of Golgi apparatus function?

7. The fusion properties of some viral proteins are pH sensitive. For example, the HA fusion protein of influenza virus is activated at an acidic pH. Since the virus must first fuse with a membrane to enter a cell, what is the likely site of influenza virus entry into animal cells?

MCAT/GRE-Style Questions

Key Concept Please read the section titled "Analysis of Yeast Mutants Defined Major Steps in the Secretory Pathway" (p. 694) and refer to Figure 17-14; then answer the following questions:

1. Phenotypic classes of secretory (*sec*) mutants in yeast include all of the following *except*:

   a. An accumulation in rough ER phenotype.

   b. A loss of Golgi structure phenotype.

   c. An accumulation in Golgi phenotype.

   d. An accumulation in secretory vesicle phenotype.

2. When haploid yeast cells contain mutations in two different classes of *sec* mutants

   a. Proteins accumulate at the step blocked by the latest acting mutant.

   b. Proteins accumulate in the same step in the double mutant as in the single mutant.

   c. Proteins accumulate at the step blocked by the earliest acting mutant.

   d. Proteins accumulate in the cytosol because metabolic pathways always back up to their first committed step.

3. When haploid yeast cells contain two different mutations in the same class of *sec* mutants, sometimes a normal phenotype is observed. This can occur

   a. If the protein products of the two mutant genes interact in a complex in such a way that changes in each complement the other.

   b. Only in the very rare case in which spontaneous revertants in each mutation has happened during the cell plating.

   c. If only the mutations are in redundant genes for the same process.

   d. Only when parallel pathways exist within the same phenotypic class.

4. Over 60 different *sec* mutations have been isolated, yet only five phenotypic classes (A–E) have been observed. Based on the information presented in Chapter 17, what is the likely explanation for this?

   a. Multiple mutations have been isolated within individual *sec* genes and only five genes are required for the entire pathway.

   b. Morphologists have not applied techniques of a high enough resolution to describe the 60 different phenotypes that must result from 60 different mutations.

   c. The machinery required for the secretory pathway is organized into five megamachines acting at five different stages and therefore only five different phenotypes are possible.

   d. Multiple genes are required for the successful completion of each phenotypic class.

5. What is the expected phenotype of a *sec* gene required at multiple steps within the secretory pathway, e.g., Class C, D, and E?

   a. The mutation would produce accumulation at all three classes, C–E.

   b. The mutation would produce accumulation at the latest acting stage, class E.

   c. The mutation would produce accumulation at the intermediate stage, class D.

   d. The mutation would produce accumulation at the earliest acting stage, class C.

Key Experiment Please read the section titled "Two Proteins Initiate the Interaction of Signal Sequences with the ER Membrane" (p. 697) and refer to Figures 17-15, 17-16, and 17-17; then answer the following questions:

6. An extensively salt-washed canine microsomal preparation supplemented with SRP is unable to support the translocation of newly synthesized chicken secretory proteins. Why?

   a. The microsomal membranes are permanently damaged by the salt-washing step.

   b. Salt-washing activates a protease on the cytosolic face of the microsomes and the newly synthesized secretory proteins are degraded.

   c. Additional components must be added to the system, such as the α subunit of the SRP receptor.

   d. The fundamental machinery is species-specific, and mixing components from different species results in incompatibilities.

7. SRP selectively recognizes ER signal sequences on newly synthesized proteins. This is an outcome of

   a. Hydrogen bonding.

   b. Hydrophobic interactions.

   c. Ionic interactions.

   d. Formation of a covalent intermediate.

8. What is the effect of SRP receptor on the elongation of nascent polypeptides?

   a. SRP receptor inhibits elongation as the nascent chain becomes about 70 amino acids long.

   b. SRP receptor has no effect on nascent chain elongation. Rather, it is the P9 and P14 proteins of SRP that affect chain elongation.

c. The SRP receptor alone is incapable of affecting chain elongation one way or the other. It must first interact with an additional component to form a required binary complex.

d. SRP receptor relieves the inhibition by SRP of nascent chain synthesis.

9. Mild proteolytic treatment of microsomal membranes is likely to

a. Render the microsomes inactive in nascent chain translocation.

b. Render the microsomes more competent in nascent chain translocation by opening the translocon gate.

c. Have no effect on translocation or other microsomal properties.

d. Increase the efficiency of translocation by removing an inhibitor.

10. In a system containing SRP and a translation system for secretory proteins, translocation into the microsomal membranes will occur whether the microsomes are added at the time protein synthesis begins or later. What properties of the system make this possible?

a. Secretory proteins are all short proteins of about 70 amino acids and hence always translocate as complete proteins.

b. Translocation is not normally a cotranslational process.

c. In such a system, protein synthesis is arrested until the nascent chain/ribosome/SRP complex associates with the microsome membrane.

d. These are known but inaccessible as a trade secret.

Key Application Please read the section titled "Cytosolic Sequences in Some Cell-Surface Receptors Target Them for Endocytosis" (p. 728) and answer the following questions:

11. People diagnosed as having familial hypercholesterolemia have

a. Low levels of cholesterol in their bloodstream.

b. Defective transport of cholesterol across the intestinal epithelial cell layer.

c. Little, if any, deposition of cholesterol into atherosclerotic plaques.

d. High cholesterol levels in their bloodstream.

12. JD, a patient who presented to the University of Texas Southwestern Medical Center, was found to have an LDL receptor that bound cholesterol normally but did not collect in coated pits. The probable underlying molecular defect in JD's LDL receptor is in

a. The ability of the receptor to change conformation and hence transmit a signal to the cytosol.

b. The ability of the cytoplasmic domain of the receptor to interact with a clathrin-binding adaptor protein.

c. Not the LDL receptor itself but in a protein with which the LDL receptor interacts.

d. Its ability to interact with the COPII class of coat proteins.

13. Which of the following amino substitions in the ϕ position of the sequence Tyr-X-X-ϕ affects the internalization of LDL:

a. Asp.

b. Phe.

c. Leu.

d. Ile.

14. Only people homozygous for familial hypercholesterolemia display clinical symptoms. A heterozygote in LDL receptor should have all the following traits *except*:

a. Normal levels of blood plasma cholesterol.

b. No more than normal tendency to atherosclerosis.

c. Normal levels of functional LDL receptor.

d. Normal frequencies of heart attacks.

Key Terms

A, B, O antigens *715*

assembly particles *734*

chaperones *680*

cisternal progression *692*

clathrin-coated vesicles *734*

constitutive secretion *692*

COP I vesicles *733*

COP II vesicles *733*

cotranslational transport *697*

endocytic pathway *731*

GPI anchor *725*

late endosome *720*

mannose 6-phosphate (M6P) sorting pathway *720*

membrane-anchor sequence *703*

N- and O-linked oligosaccharides *712*

regulated secretion *694*

secretory pathway *677*

signal-recognition particle (SRP) *697*

stop-transfer sequences *685*

topogenic sequences *702*

transcytosis *725*

translocon *699*

T-SNAREs *741*

V-SNAREs *741*

References

Synthesis and Targeting of Mitochondrial and Chloroplast Proteins

Cline, K., and R. Henry. 1996. Import and routing of nucleus-encoded chloroplast proteins. *Ann. Rev. Cell Devel. Biol.* **12**:1–26.

Dalbey, R., and C. Robinson. 1999. Protein translocation into and across the bacterial plasma membrane and the plant thylakoid membrane. *Trends Biochem. Sci.* 24:17–22.

Heins, L., I. Collinson, and J. Soll. 1998. The protein translocation apparatus of chloroplast envelopes. *Trends Plant Sci.* 3:56–61.

Horst, M., A. Azem, G. Schatz, and B. S. Glick. 1997. What is the driving force for protein import into mitochondria? *Biochim. Biophys. Acta* 1318:71–78.

Martin, J., and F. U. Hartl. 1997. Chaperone-assisted protein folding. *Curr. Opin. Struct. Biol.* 7:41–52.

Neupert, W. 1997. Protein import into mitochondria. *Annu. Rev. Biochem.* 66:863–917.

Pfanner, N., and M. Meijer. 1997. The Tom and Tim machine. *Curr. Biol.* 7:R100–103.

Schatz, G. 1996. The protein import system of mitochondria. *J. Biol. Chem.* 271:31763–31766.

Schatz, G. 1998. Protein transport: the doors to organelles. *Nature* 395:439–440.

Schatz, G., and B. Dobberstein. 1996. Common principles of protein translocation across membranes. *Science* 271:1519–1526.

Schnell, D. 1998. Protein targeting to the thylakoid membrane. *Annu. Rev. Plant Physiol. Plant Mol. Biol.* 49:97–126.

Stuart, R. A., and W. Neupert. 1996. Topogenesis of inner membrane proteins of mitochondria. *Trends Biochem. Sci.* 21:261–267.

Synthesis and Targeting of Peroxisomal Proteins

Lazarow, P. B. 1995. Peroxisome structure, function, and biogenesis: human patients and yeast mutants show strikingly similar defects in peroxisome biogenesis. *J. Neuropathol. Exp. Neurol.* 54:720–725.

Moser, H. W. 1995. Adrenoleukodystrophy. *Curr. Opin. Neurol.* 8:221–226.

Rachubinski, R. A., and S. Subramani. 1995. How proteins penetrate peroxisomes. *Cell* 83:525–528.

Translocation of Secretory Proteins across the ER Membrane

Corsi, A. K., and R. Schekman. 1996. Mechanism of polypeptide translocation into the endoplasmic reticulum. *J. Biol. Chem.* 271:30299–30302.

Dalbey, R. E., and G. Von Heijne. 1992. Signal peptidases in prokaryotes and eukaryotes: a new protease family. *Trends Biochem. Sci.* 17:474–478.

Matlack, K., W. Mothes, and T. A. Rapoport 1998. Protein translocation: tunnel vision. *Cell* 92:381–390.

Rapoport, T. A., B. Jungnickel, and U. Kutay. 1996. Protein transport across the eukaryotic endoplasmic reticulum and bacterial inner membranes. *Annu. Rev. Biochem.* 65:271–303.

Insertion of Membrane Proteins into the ER Membrane

Englund, P. T. 1993. The structure and biosynthesis of glycosylphosphatidylinositol protein anchors. *Annu. Rev. Biochem.* 62:121–138.

Mothes, W., et al. 1997. Molecular mechanism of membrane protein integration into the endoplasmic reticulum. *Cell* 89:523–533.

Spiess, M. 1995. Heads or tails—what determines the orientation of proteins in the membrane. *FEBS Lett.* 369:76–79.

von Heijne, G. 1997. Getting greasy: how transmembrane polypeptide segments integrate into the lipid bilayer. *Mol. Microbiol.* 24:249–253.

Post-Translational Modifications and Quality Control in the Rough ER

Bergeron, J., M. Brenner, D. Thomas, and D. Williams. 1994. Calnexin: a new membrane-bound chaperone of the endoplasmic reticulum. *Trends Biochem. Sci.* 19:124–128.

Doms, R. W., et. al. 1993. Folding and assembly of viral membrane proteins. *Virology* 193:545–562.

Freedman, R. B. 1995. The formation of protein disulphide bonds. *Curr. Opin. Struct. Biol.* 5:85–91.

Gilbert, H. 1997 Protein disulfide isomerase and assisted protein folding. *J. Biol. Chem.* 272:29399–29402.

Hammond, C., and A. Helenius. 1995. Quality control in the secretory pathway. *Curr. Opin. Cell Biol.* 7:523–529.

Hebert, D. N., J. F. Simons, J. R. Peterson, and A. Helenius. 1995. Calnexin, calreticulin, and Bip/Kar2p in protein folding. *Cold Spring Harbor Symp. Quant. Biol.* 60:405–415.

Huppa, J. B., and H. Ploegh. 1998. The essence of –SH in the ER. *Cell* 92:145–148.

Jentsch, S., and S. Schlenker. 1995. Selective protein degradation: a journey's end within the proteasome. *Cell* 82:881–884.

Kopito, R.R. 1997. ER quality control: the cytoplasmic connection. *Cell* 88:427–430.

Sidrauski, C., R. Chapman, and P. Walker 1998. The unfolded protein response: an intracellular signaling pathway with many surprising features. *Trends Cell Biol.* 8:245–249.

Sommer, T., and D. Wolf. 1997. Endoplasmic reticulum degradation: reverse protein flow of no return. *FASEB J.* 11:1227–1233.

Suzuki, T., and W. Lennarz. 1998. Complex, two-way traffic across the membrane of the endoplasmic reticulum. *J. Biol. Chem.* 273:10083–10086.

Protein Glycosylation in the ER and Golgi Complex

Abeijon, C., E. C. Mandon, and C. B. Hirschberg. 1997. Transporters of nucleotide sugars, nucleotide sulfate and ATP in the Golgi apparatus. *Trends Biochem. Sci.* 22:203–207.

Driouich, A., L. Faye, and L. A. Staehelin. 1993. The plant golgi apparatus: a factory for complex polysaccharides and glycoproteins. *Trends Biochem. Sci.* 18:210–214.

Gabius, H.-J., and S. Gabius. 1997. *Glycosciences, Status and Perspectives.* Chapman and Hall.

Kornfeld, S. 1992. Structure and function of the mannose 6-phosphate/insulin-like growth factor II receptors. *Annu. Rev. Biochem.* 61:307–330.

Kornfeld, R., and S. Kornfeld. 1985. Assembly of asparagine-linked oligosaccharides. *Annu. Rev. Biochem.* 45:631–664.

Paulson, J. C. 1989. Glycoproteins: what are the sugar chains for? *Trends Biochem. Sci.* 14:272–276.

Silberstein, S., and R. Gilmore. 1996. Biochemistry, molecular biology, and genetics of the oligosaccharyltransferase. *FASEB J.* 10:849–858.

Golgi and Post-Golgi Protein Sorting and Proteolytic Processing

Berger, E. G., and J. Roth, eds. 1997. *The Golgi Apparatus.* Birkhäuser Verlag, Basel, Switzerland.

Douglass, J., O. Civelli, and E. Herbert. 1984. Polyprotein gene expression: generation of diversity of neuroendocrine peptides. *Annu. Rev. Biochem.* 53:665–715.

Farquahr, M., and G. Palade. 1998. The Golgi apparatus: 100 years of progress and controversy. *Trends Cell Biol.* 8:2–10.

Glick, B., and V. Malhotra. 1988. The curious status of the Golgi apparatus. *Cell* 95:883–889.

Golgi Centenary Issue. *Trends in Cell Biology* 8(1): entire issue.

Henkel, A. W., and W. Almers. 1996. Fast steps in exocytosis and endocytosis studied by capacitance measurements in endocrine cells. *Curr. Opin. Neurobiol.* 6:350–357.

Monck, J. R., and J. M. Fernandez. 1996. The fusion pore and mechanisms of biological membrane fusion. *Curr. Opin. Cell Biol.* 8:524–533.

Munro, S. 1998. Localization of proteins to the Golgi apparatus. *Trends Cell Biol.* 8:11–15.

Simons, K., and E. Ikonen. 1997. Functional rafts in cell membranes. *Nature* 387:569–572.

Sossin, W. S., and R. H. Scheller. 1991. Biosynthesis and sorting of neuropeptides. *Curr. Opin. Neurobiol.* 1:79–83.

Steiner, D. F., et al. 1996. The role of prohormone convertases in insulin biosynthesis: evidence for inherited defects in their action in man and experimental animals. *Diabetes Metab.* 22:94–104.

Thiele, C., H. H. Gerdes, and W. B. Huttner. 1997. Protein secretion: puzzling receptors. *Curr. Biol.* 7:R496–500.

Receptor-Mediated Endocytosis and the Sorting of Internalized Proteins

Brown, M. S., and J. L. Goldstein. 1986. Receptor-mediated pathway for cholesterol homeostasis. Nobel Prize Lecture. *Science* 232:34–47.

Brown, M. S., and J. L. Goldstein. 1997. The SREBP pathway: regulation of cholesterol metabolism by proteolysis of a membrane-bound transcription factor. *Cell* 89:331–340.

Hunziker, W., and I. Mellman. 1991. Relationships between sorting in the exocytic and endocytic pathways of MDCK cells. *Semin. Cell Biol.* 2:397–410.

Ihrke, G., and A. L. Hubbard. 1995. Control of vesicle traffic in hepatocytes. *Prog. Liver Dis.* 13:63–99.

Le Gall, A. H., C. Yeaman, A. Muesch, and E. Rodriguez-Boulan. 1995. Epithelial cell polarity: new perspectives. *Semin. Nephrol.* 15:272–284.

Mostov, K. E., et al. 1995. Regulation of protein traffic in polarized epithelial cells: the polymeric immunoglobulin receptor model. *Cold Spring Harbor Symp. Quant. Biol.* 60:775–781.

Riezman, H., P. Woodman, G. van Meer, and M. Marsh. 1997. Molecular mechanisms of endocytosis. *Cell* 91:731–738.

Robinson, M. S., C. Watts, and M. Zerial. 1996. Membrane dynamics in endocytosis. *Cell* 84:13–21.

Molecular Mechanisms of Vesicular Traffic

Bannykh, S. I., N. Nishimura, and W. E. Balch. 1998. Getting into the Golgi. *Trends Cell Biol.* 8:21–25.

Bean, A. J., and R. H. Scheller. 1997. Better late than never: a role for Rabs late in exocytosis. *Neuron* 19:751–754.

Bennett, M. K., and R. H. Scheller. 1993. The molecular machinery for secretion is conserved from yeast to neurons. *Proc. Nat'l. Acad. Sci. USA* 90:2559–2563.

Brodsky, F. M., et. al. 1991. Clathrin light chains: arrays of protein motifs that regulate coated-vesicle dynamics. *Trends Biochem. Sci.* 16:208–213.

Goodson, H. V., C. Valetti, and T. E. Kreis. 1997. Motors and membrane traffic. *Curr. Opin. Cell Biol.* 9:18–28.

Götte, M., and G. Fischer von Mollard. 1998. A new beat for the SNARE drum. *Trends Cell Biol.* 8:215–218.

Hay, J. C., and R. H. Scheller. 1997. SNAREs and NSF in targeted membrane fusion. *Curr. Opin. Cell Biol.* 9:505–512.

Hernandez, L. D., L. R. Hoffman, T. G. Wolfsberg, and J. M. White. 1996. Virus-cell and cell-cell fusion. *Annu. Rev. Cell Devel. Biol.* 12:627–661.

Herrmann, J., P. Malkus, and R. Schekman. 1999. Out of the ER: outfitters, escorts, and guides. *Trends Cell Biol.* 9:5–7.

Kirchhausen, T., J. S. Bonifacino, and H. Riezman. 1997. Linking cargo to vesicle formation: receptor tail interactions with coat proteins. *Curr. Opin. Cell Biol.* 9:488–495.

Kuehn, M. J., and R. Schekman. 1997. COPII and secretory cargo capture into transport vesicles. *Curr. Opin. Cell Biol.* 9:477–483.

Odorizzi, G., C. Cowles, and S. Emr. 1998. The AP-3 complex: a coat of many colours. *Trends Cell Biol.* 8:282–288.

Pelham, H. R. 1995. Sorting and retrieval between the endoplasmic reticulum and Golgi apparatus. *Curr. Opin. Cell Biol.* 7:530–535.

Pishvaee, B., and G. Payne. 1998. Clathrin coats—threads laid bare. *Cell* 95:443–446.

Rothman, J. E., and F. T. Wieland. 1996. Protein sorting by transport vesicles. *Science* 272:227–234.

Rothman, J. E., and T. H. Söllner. 1997. Throttles and dampers: controlling the engine of membrane fusion. *Science* 276:1212–1213.

Schekman, R., and I. Mellman. 1997. Does COPI go both ways? *Cell* 90:197–200.

Schekman, R., and L. Orci. 1996. Coat proteins and vesicle budding. *Science* 271:1526–1533.

Schimmöller, F., I. Simon, and S. Pfeffer. 1998. Rab GTPases, directors of vesicle docking. *J. Biol. Chem.* 273:22161–22164.

Schmid, S. 1997. Clathrin-coated vesicle formation and protein sorting: an integrated process. *Annu. Rev. Biochem.* 66:511–548.

Skehel, J. J., et al. 1995. Membrane fusion by influenza hemagglutinin. *Cold Spring Harbor Symp. Quant. Biol.* 60:573–580.

Cell Motility and Shape I: Microfilaments

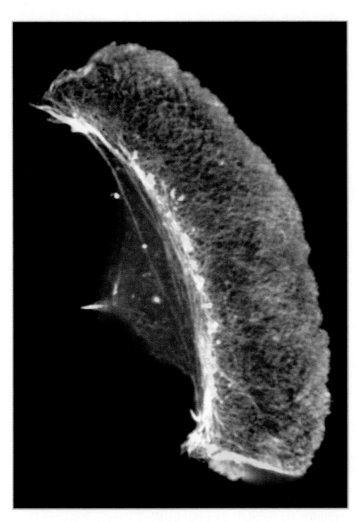

A fish epidermal keratinocyte.

ell motility is one of the crowning achievements of evolution. Primitive cells were probably immobile, carried by currents in the primordial milieu. With the evolution of multicellular organisms, primitive organs were formed by migrations of single cells and groups of cells from distant parts of the embryo. In adult organisms, movements of single cells in search of foreign organisms are integral to the host's defenses against infection; on the other hand, uncontrolled cell migration is an ominous sign of a cancerous cell.

Most cells in the body are stationary, but many of these exhibit dramatic changes in their morphology—the contraction of muscle cells, the elongation of nerve axons, the formation of cell-surface protrusions, the constriction of a dividing cell during mitosis. Perhaps the most subtle movements are those within cells—the active separation of chromosomes, the streaming of cytosol, the transport of membrane vesicles. These internal movements are essential elements in the growth and differentiation of cells, carefully controlled by the cell to take place at specified times and in particular locations.

All cell movements are a manifestation of mechanical work; they require a fuel (ATP) and proteins that convert the energy stored in ATP into motion. The **cytoskeleton**, a cytoplasmic system of fibers, is critical to cell motility. Like steel girders supporting the shell of a building, the cytoskeleton plays a structural role by supporting the cell membrane and by forming tracks along which organelles and other elements move in the cytosol. Unlike the passive framework of a building, though, the cytoskeleton undergoes constant rearrangement, which can produce movements.

MEDIA CONNECTIONS

Focus: Actin Polymerization

Technique: In Vitro Motility Assay

Focus: Myosin Crossbridge Cycle

Overview: Cell Motility

Classic Experiment 18.1: Looking at Muscle Contraction

In the electron microscope, the cytoskeleton appears as a dense and seemingly random array of fibers. However, we now recognize that this array consists of three types of cytosolic fibers: **microfilaments**, 7 to 9 nm in diameter; **intermediate filaments**, 10 nm in diameter; and **microtubules**, 24 nm in diameter. These cytoskeletal fibers are well-ordered polymers built from small protein subunits held together by noncovalent bonds. Instead of being a disordered array, the cytoskeleton is organized into discrete structures — primarily bundles, geodesic-dome-like networks, and gel-like lattices. Although the primary function of intermediate filaments is structural, to reinforce cells and to organize them into tissues, we discuss them in the next chapter because microtubules and intermediate filaments are often associated with one another.

Cells have evolved two basic mechanisms for generating movement. One mechanism involves a special class of enzymes called **motor proteins**. These proteins use energy from ATP to walk or slide along a microfilament or a microtubule. Some motor proteins carry membrane-bound organelles and vesicles along the cytoskeletal fiber tracks; other motor proteins cause the fibers to slide past each other. The other mechanism responsible for many of the changes in the shape of a cell entails assembly and disassembly of microfilaments and microtubules. A few movements involve both the action of motor proteins and cytoskeleton rearrangements. The structure and functions of microtubules are discussed in Chapter 19. In this chapter we focus on microfilaments and the **actin** subunits that compose them, which play a role in numerous types of movement from cell migration to cytosol transport.

We begin with a description of the actin cytoskeleton and its role in determining cell shape. Then we cover the biochemistry of actin assembly and cellular mechanisms for controlling this process. We next examine the structure of **myosin**, the actin motor protein, and how it transduces the energy of ATP into a sliding movement along actin filaments. With a basic understanding of the key components of the cytoskeleton and the interaction of actin and myosin, we can examine the mechanisms responsible for various types of cell movement, starting with the contraction of muscle and ending with the crawling movements of amebas and skin cells. The chapter ends with a discussion about the regulation of movement by cell-signaling pathways.

18.1 The Actin Cytoskeleton

Cell **locomotion** results from the coordination of motions generated by different parts of a cell. These motions are complex and difficult to describe, but we can pick out their major features using powerful fluorescent antibody-labeling methods combined with fluorescence microscopy. To provide a background for more detailed discussions of the mechanism of cell locomotion later, we first briefly describe movement by two types of crawling animal cells. One feature that characterizes all moving cells is polarity; that is, certain structures always form at the front of the cell, whereas others are found at the rear.

In the micrograph shown at the opening to this chapter, a skin cell (keratinocyte) displays motions typical of a fast-moving cell like an ameba. Initially, the keratinocyte forms a large, broad membrane protrusion, the *lamellipodium*, which moves forward from the cell. After contacting the substratum and forming special attachment structures, called *focal adhesions*, the lamellipodium quickly fills with cytosol and the rear of the cell retracts forward toward the body of the cell.

Locomotion of a slow-moving cell like a fibroblast involves several structures not seen in a fast-moving keratinocyte. As a fibroblast moves forward, it extends slender "fingers" of membrane, called *filopodia*, as well as lamellipodia (Figure 18-1a). Where the cell membrane contacts the substratum, focal adhesions assemble on the ventral surface of the cell. As the cell continues to travel, its tail is pulled forward, often leaving behind small patches of the cell still firmly attached to the substratum through the focal adhesions. Some areas of the cell membrane do not form stable adhesions at the leading edge of the cell; these areas project upward as a thin veil, or *ruffle*, that moves as an undulating ridge back along the dorsal cell surface toward the cell body.

The machinery that powers cell migration is built from the actin cytoskeleton, which is larger than any organelle. When a fibroblast is observed by fluorescence microscopy after the actin filaments are stained with a fluorescent dye, radially oriented actin filament bundles can be seen at the leading edge, and axial bundles, called *stress fibers*, are visible underlying the cell body (Figure 18-1b). In addition, a network of actin filaments fills the rest of the cell, but the individual filaments of this network are difficult to resolve in the light microscope.

Much of the discussion in this section focuses on the ability of huge actin filament structures to control the shape of a cell. Because the actin cytoskeleton is so big, it can easily change cell morphology just by assembling or disassembling itself. In previous chapters, we have seen examples of large protein complexes in which the number and positions of the subunits are fixed. For example, all ribosomes have the same number of protein and RNA components, and their three-dimensional geometry is invariant. However, the actin cytoskeleton is different—the lengths of filaments vary greatly, the filaments are cross-linked into imperfect bundles and networks, and the ratio of cytoskeletal proteins is not rigidly maintained. This organizational flexibility of the actin cytoskeleton permits a cell to assume many shapes and to vary them easily. In moving cells, the cytoskeleton must assemble rapidly and does not always have the chance to form well-organized, highly ordered structures. Keeping this in mind, we examine actin as a model for understanding how polymeric proteins form the structural framework of a cell and how the cell tailors this framework to carry out various tasks involving motion of the entire cell or subcellular parts.

(a)

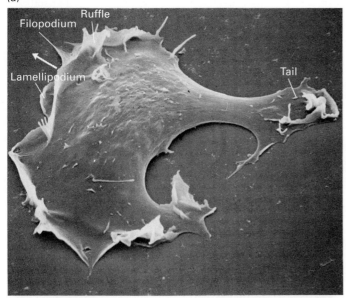

(b)

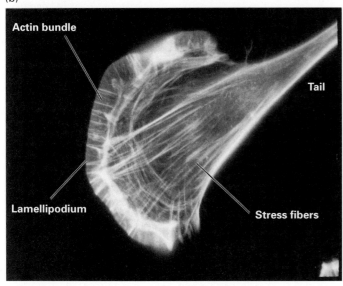

▲ **FIGURE 18-1 Actin structures in a fibroblast.** (a) Scanning electron micrograph of a cultured fibroblast. At the front of the cell, filopodia, lamellipodia, and ruffles project from the cell membrane. At the rear of the cell, the tail is firmly attached to the surface. The arrow indicates the direction of movement. (b) Fluorescence micrograph of a fan-shaped fibroblast, stained with rhodamine phalloidin. Visible are numerous actin bundles in the lamellipodia and stress fibers in the cell body. [Part (a) courtesy of J. Heath; part (b) courtesy of B. Hollifield.]

Eukaryotic Cells Contain Abundant Amounts of Highly Conserved Actin

Actin is the most abundant intracellular protein in a eukaryotic cell. In muscle cells, for example, actin comprises 10 percent by weight of the total cell protein; even in nonmuscle cells, actin makes up 1–5 percent of the cellular protein. A typical cytosolic concentration of actin in nonmuscle cells is 0.5 mM; in special structures like microvilli, however, the local actin concentration can be tenfold higher. To appreciate how much actin cells contain, consider a typical liver cell, which has 20,000 insulin receptor molecules but approximately half a billion (0.5×10^9) actin molecules. This predominance of actin compared with other cell proteins is a common feature of all cytoskeletal proteins. Because they form structures that must cover large spaces in a cell, these proteins are among the most abundant proteins in a cell.

A moderate-sized protein consisting of approximately 375 residues, actin is encoded by a large, highly conserved gene family. Some single-celled eukaryotes like yeasts and amebas have a single actin gene, whereas many multicellular organisms contain multiple actin genes. For instance, humans have six actin genes, which encode *isoforms* of the protein, and some plants have as many as 60. Sequencing of these actins has revealed that it is one of the most conserved proteins in a cell, comparable with histones, the structural proteins of chromatin (Chapter 9). Actin residues from amebas and from animals are identical at 80 percent of the positions. In vertebrates, the four α-actin isoforms present in various muscle cells and the β- and γ-actin isoforms present in nonmuscle cells differ at only four or five positions. Although these differences among isoforms seem minor, the isoforms have different functions: α-Actin is associated with contractile structures, and β-actin is at the front of the cell where actin filaments polymerize.

Recently, a family of *actin-related proteins* (Arps), exhibiting 50 percent homology with actin, has been identified in many eukaryotic organisms. As noted later, one group of Arps (Arp2/3) stimulates actin assembly; intriguingly, another group (Arp1) is associated with microtubules and a microtubule motor protein. They are discussed in the next chapter.

ATP Holds Together the Two Lobes of the Actin Monomer

Actin exists as a globular monomer called *G-actin* and as a filamentous polymer called *F-actin*, which is a linear chain of G-actin subunits. (The microfilaments visualized in a cell by electron microscopy are F-actin filaments plus any bound proteins.) Each actin molecule contains a Mg^{2+} ion complexed with either ATP or ADP. Thus there are four states of actin: ATP–G-actin, ADP–G-actin, ATP–F-actin, and ADP–F-actin. Two of these forms, ATP–G-actin and ADP–F-actin, predominate in a cell. We discuss later how the interconversion between the ATP and ADP forms of actin is important in the assembly of the cytoskeleton.

Although G-actin appears globular in the electron microscope, x-ray crystallographic analysis reveals that it is separated into two lobes by a deep cleft (Figure 18-2a). The lobes and the cleft compose the *ATPase fold*, the site where ATP and Mg^{2+} are bound. In actin, the floor of the cleft

(a)

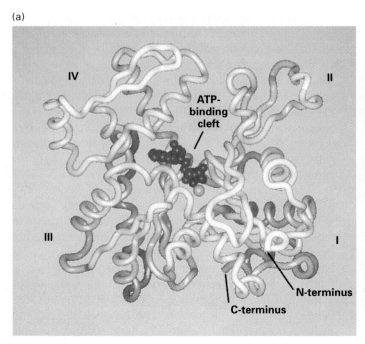

(b)

(c)

(−) end

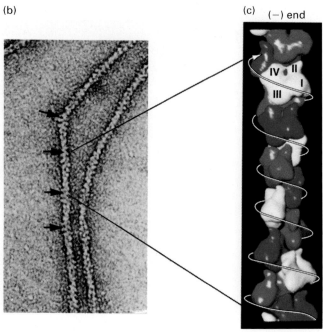

(+) end

▲ **FIGURE 18-2 Structures of monomeric G-actin and F-actin filament.** (a) Model of a β-actin monomer from a nonmuscle cell shows it to be a platelike molecule (measuring 5.5 × 5.5 × 3.5 nm) divided by a central cleft into two approximately equal-sized lobes and four subdomains, numbered I–IV. ATP (red) binds at the bottom of the cleft and contacts both lobes (the yellow ball represents Mg^{2+}). The N- and C-termini lie in subdomain I. (b) In the electron microscope, negatively stained actin filaments appear as long, flexible, and twisted strands of beaded subunits. Because of the twist, the filament appears alternately thinner (7 nm diameter) and thicker (9 nm diameter) (arrows). (c) In one model of the arrangement of subunits in an actin filament, the subunits lie in a tight helix along the filament, as indicated by the arrow. One repeating unit consists of 28 subunits (13 turns of the helix), covering a distance of 72 nm. Only 14 subunits are shown in the figure. The ATP-binding cleft is oriented in the same direction (*top*) in all actin subunits in the filament. As discussed later, this end of a filament is designated the (−) end; the opposite end is the (+) end. [Part (a) adapted from C. E. Schutt et al., 1993, *Nature* **365**:810, courtesy of M. Rozycki; part (b) courtesy of R. Craig; part (c) see M. F. Schmid et al., 1994, *J. Cell Biol.* **124**:341, courtesy of M. Schmid.]

acts as a hinge that allows the lobes to flex relative to each other. When ATP or ADP is bound to G-actin, the nucleotide affects the conformation of the molecule. In fact, without a bound nucleotide, G-actin denatures very quickly.

G-Actin Assembles into Long, Helical F-Actin Polymers

The addition of ions—Mg^{2+}, K^+, or Na^+—to a solution of G-actin will induce the **polymerization** of G-actin into F-actin filaments. The process is also reversible: F-actin depolymerizes into G-actin when the ionic strength of the solution is lowered. The F-actin filaments that form in vitro are indistinguishable from microfilaments isolated from cells. This indicates that other factors such as accessory proteins are not required for polymerization in vivo. The assembly of G-actin into F-actin is accompanied by the hydrolysis of ATP to ADP and P_i; however, as we discuss later, ATP hydrolysis affects the kinetics of polymerization but is not necessary for polymerization to occur.

When negatively stained by uranyl acetate for electron microscopy, F-actin appears as twisted strings of beads whose diameter varies between 7 and 9 nm (Figure 18-2b). From x-ray diffraction studies of actin filaments and the actin monomer structure shown in Figure 18-2a, scientists have produced a model of an actin filament in which the subunits are organized as a tightly wound helix (Figure 18-2c). In this arrangement, each subunit is surrounded by four other subunits, one above, one below, and two to one side. Each subunit corresponds to a bead seen in electron micrographs of actin filaments.

The ability of G-actin to polymerize into F-actin and of F-actin to depolymerize into G-actin is an important property of actin. In this chapter, we will see how the reversible assembly of actin lies at the core of many cell movements.

F-Actin Has Structural and Functional Polarity

All subunits in a filament point toward the same filament end (i.e., they have the same polarity). Consequently, at one

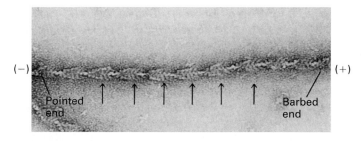

▲ **FIGURE 18-3 Experimental demonstration of polarity of an actin filament by binding of myosin S1 head domains.** When bound to all the actin subunits, S1 appears to spiral around the filament. This coating of myosin heads produces a series of arrowhead-like decorations, most easily seen at the wide views of the filament. The polarity in decoration defines a pointed (−) end and a barbed (+) end; the former corresponds to the top of the model in Figure 18-2c. [Courtesy of R. Craig.]

end of the filament, by convention designated the (−) end, the ATP-binding cleft of an actin subunit is exposed to the surrounding solution; at the opposite end, the cleft contacts the neighboring actin subunit (see Figure 18-2c).

Without the atomic resolution afforded by x-ray crystallography, the cleft in an actin subunit, and therefore the polarity of a filament, is not detectable. However, the polarity of actin filaments can be demonstrated by electron microscopy in so-called "decoration" experiments, which exploit the ability of myosin to bind specifically to actin filaments. In this type of experiment, an excess of myosin S1, the globular head domain of myosin, is mixed with actin filaments and binding is permitted to occur. Myosin attaches to the sides of a filament with a slight tilt. When all subunits are bound by myosin, the filament appears coated ("decorated") with arrowheads that all point toward one end of the filament (Figure 18-3). Because myosin binds to actin filaments and not to microtubules or intermediate filaments, arrowhead decoration is one criterion by which actin filaments are identified among the other cytoskeletal fibers in electron micrographs of thin-sectioned cells.

The Actin Cytoskeleton Is Organized into Bundles and Networks of Filaments

On first looking at an electron micrograph or immunofluorescence micrograph of a cell, one is struck by the dense, seemingly disorganized mat of filaments present in the cytosol (Figure 18-4). However, a keen eye will start to pick out areas, generally where the membrane protrudes from the cell surface, in which the filaments are concentrated into bundles. From these bundles the filaments continue into the cell interior, where they fan out and become part of a network of filaments. These two structures, *bundles* and *networks*, are the most common arrangements of actin filaments in a cell.

Functionally, bundles and networks have identical roles in a cell: both provide a framework that supports the plasma

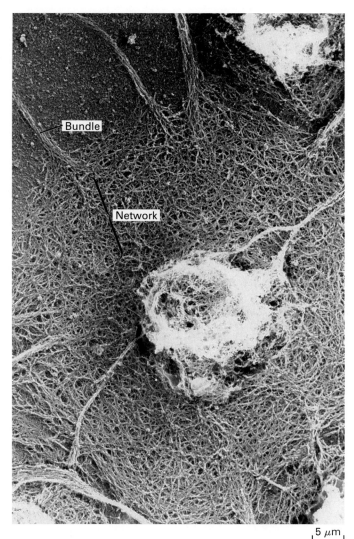

5 μm

▲ **FIGURE 18-4 Micrograph revealing bundles and networks of actin filaments in the cytosol of a spreading platelet treated with detergent to remove the plasma membrane.** Actin bundles project from the cell to form the spikelike filopodia. In the lamellar region of the cell, the actin filaments form a network that fills the cytosol. In contrast to the roughly parallel alignment of bundled filaments, the filaments in networks lie at various angles approaching 90°. [Courtesy of J. Hartwig.]

membrane and, therefore, determines a cell's shape. Structurally, bundles differ from networks mainly in the organization of actin filaments. In bundles the actin filaments are closely packed in parallel arrays, whereas in a network the actin filaments crisscross, often at right angles, and are loosely packed. Cells contain two types of actin networks. One type, associated with the plasma membrane, is planar or two-dimensional, like a net or a web; the other type, present within the cell, is three-dimensional, giving the cytosol gel-like properties.

In all bundles and networks, the filaments are held together by *actin cross-linking proteins*. To connect two

(a)

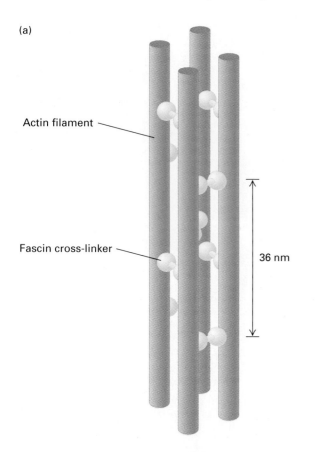

Actin filament

Fascin cross-linker

36 nm

(b)

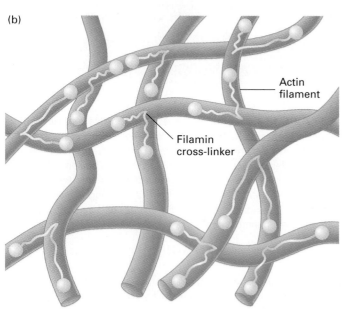

Actin filament

Filamin cross-linker

◄ **FIGURE 18-5 Actin cross-linking proteins bridging pairs of actin filaments.** (a) When cross-linked by fascin, a relatively short protein, actin filaments form a bundle. (b) Long cross-linking proteins such as filamin are flexible and thus can cross-link pairs of filaments lying at various angles.

able to adapt to any arrangement of actin filaments and tether orthogonally oriented actin filaments in networks (Figure 18-5b).

Many actin cross-linking proteins belong to the calponin homology–domain (CH-domain) superfamily (Table 18-1). Each of these proteins has a pair of actin-binding domains, whose sequence is homologous to calponin, a muscle protein. The actin-binding domains are separated by repeats of helical coiled-coil or β-sheet immunoglobulin motifs. Among the CH-domain proteins, the shortest (fimbrin and α-actinin) are found in actin bundles within cell extensions, and the longest (filamin, spectrin, and dystrophin) are found in actin networks in the cortical region adjacent to the plasma membrane. A smaller number of cross-linking proteins are classified into two other groups; these bind to different sites on actin than the CH-domain proteins.

Cortical Actin Networks Are Connected to the Membrane

The distinctive shape of a cell is dependent not only on the organization of actin filaments but also on proteins that connect the filaments to the membrane. These proteins, called *membrane-microfilament binding proteins,* act as spot welds that tack the membrane sheet to the underlying cytoskeleton framework. When attached to a bundle of filaments, the membrane acquires a fingerlike shape, as we discuss later. When attached to a planar network of filaments, the membrane is held flat. We focus on these network connections in this section. The simplest connections entail binding of integral membrane proteins directly to actin filaments. More common are complex linkages that connect actin filaments to integral membrane proteins through peripheral membrane proteins.

The richest area of actin filaments in a cell lies in the *cortex,* a narrow zone just beneath the plasma membrane. In this region, most actin filaments are arranged into a network that excludes most organelles from the cortical cytoplasm. Several ways to organize the cortical actin cytoskeleton are observed in different cell types. Perhaps the simplest cytoskeleton is the two-dimensional network of actin filaments adjacent to the erythrocyte plasma membrane. In more complicated cortical cytoskeletons, such as those in platelets, epithelial cells, and muscle, actin filaments are part of a three-dimensional network that fills the cytosol and anchors the cell to the substratum. We first describe the erythrocyte cytoskeleton and its linkage to the membrane and then examine the more complex cytoskeletons in platelets and muscle.

filaments, a cross-linking protein must have two actin-binding sites, one site for each filament. The length and flexibility of a cross-linking protein critically determine whether bundles or networks are formed. Short cross-linking proteins hold actin filaments close together, forcing the filaments into the parallel alignment characteristic of bundles (Figure 18-5a). In contrast, long, flexible cross-linking proteins are

TABLE 18-1 Actin Cross-Linking Proteins

Protein*	MW	Domain Organization†	Location
GROUP I			
30 kDa	33,000		Filopodia, lamellipodia, stress fibers
EF-1	50,000		Pseudopodia
Fascin	55,000		Filopodia, lamellipodia, stress fibers, microvilli, acrosomal process
Scruin	102,000		Acrosomal process
GROUP II			
Villin	92,000		Intestinal and kidney brush border microvilli
Dematin	48,000		Erythrocyte cortical network
GROUP III (CH-domain superfamily)			
Fimbrin	68,000		Microvilli, stereocilia, adhesion plaques, yeast actin cables
α-Actinin	102,000		Filopodia, lamellipodia, stress fibers, adhesion plaques
Spectrin	α: 280,000 β: 246,000–275,000		Cortical networks
Dystrophin	427,000		Muscle cortical networks
ABP 120	92,000		Pseudopodia
Filamin	280,000		Filopodia, pseudopodia, stress fibers

*Cross-linking proteins are classified into three groups. Each of the group I proteins has a unique actin-binding domain. Group II proteins have one or more common 7,000-MW actin-binding domains. CH-domain proteins have a pair of 24,000-MW actin-binding domains, whose sequence is homologous to calponin, a muscle protein.

†Actin-binding domains are shown in blue; calmodulin-like Ca^{2+}-binding domains, in red; α-helical repeats, in purple; β-sheet repeats, in green; and uncharacterized domains, in tan or yellow.

Erythrocyte Cytoskeleton A red blood cell must squeeze through narrow blood capillaries without rupturing its membrane. The strength and flexibility of the erythrocyte plasma membrane depends on a dense cytoskeletal network that underlies the entire membrane and is attached to it at many points. The primary component of the erythrocyte cytoskeleton is *spectrin*, a long fibrous protein. Two dimeric subunits of spectrin, each composed of an α and β polypeptide chain, associate to form head-to-head tetramers, which are 200 nm long. The entire cytoskeleton is arranged in a spoke-and-hub network (Figure 18-6). Each spectrin tetramer comprises a spoke, extending from and cross-linking a pair of hubs, called *junctional complexes*. As illustrated in Figure 18-7, each junctional complex is composed of a short (14-subunit) actin filament plus adducin, tropomyosin, and tropomodulin. The last two proteins strengthen the network by preventing the actin filament from depolymerizing. Because several spectrin molecules can bind the same actin filament, the erythrocyte cytoskeletal network has a spoke-and-hub organization. This polygonal arrangement acts as a lamination of the membrane.

To ensure that the erythrocyte retains its characteristic shape, the spectrin-actin cytoskeleton is firmly attached to the overlying erythrocyte membrane by two peripheral membrane proteins, each of which binds to a specific integral membrane protein (see Figure 18-7). *Ankyrin* connects the center of spectrin to band 3 protein, the anion-transporter protein in the membrane. *Band 4.1 protein*, a component of the junctional complex, binds to the integral membrane protein glycophorin. This dual binding ensures that the membrane is connected to both the spokes and the hubs of the spectrin-actin cytoskeleton.

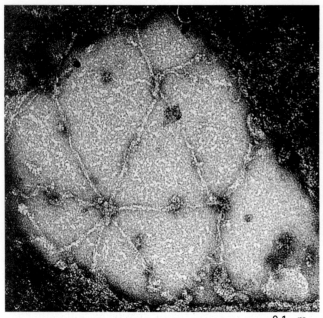

0.1 μm

◀ **FIGURE 18-6 Electron micrograph of human erythrocyte cytoskeleton.** The long "spokes" are composed mainly of spectrin and can be seen to intersect at the "hubs," or junctional complexes. The darker spots along the spokes are ankyrin molecules, which cross-link spectrin to integral membrane proteins.

Platelet Cytoskeleton Isoforms of spectrin and actin have been found in various nonerythroid cells, suggesting that these cell types have a cortical spectrin-actin cytoskeleton like that present in the erythrocyte. In nonerythroid cells, however, the actin cytoskeleton is more complicated than in erythrocytes. An example of this more complicated structure is seen in the platelet, a small, nonnucleated cell that is important in blood clotting and wound repair. The platelet cytoskeleton must undergo complicated rearrangements that are responsible for a repertoire of changes in cell shape during a blood clotting reaction (Figure 18-8).

These changes in platelet shape could not be generated by the simple cytoskeleton seen in the erythrocyte; thus additional components are needed in the platelet cytoskele-

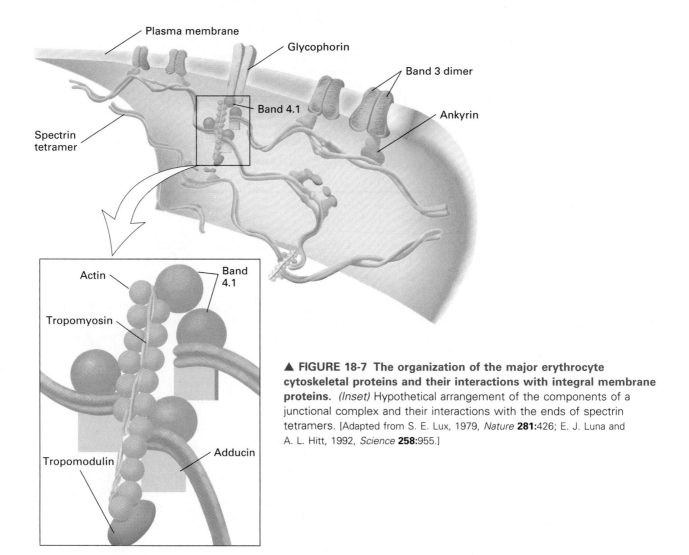

▲ **FIGURE 18-7 The organization of the major erythrocyte cytoskeletal proteins and their interactions with integral membrane proteins.** *(Inset)* Hypothetical arrangement of the components of a junctional complex and their interactions with the ends of spectrin tetramers. [Adapted from S. E. Lux, 1979, *Nature* **281**:426; E. J. Luna and A. L. Hitt, 1992, *Science* **258**:955.]

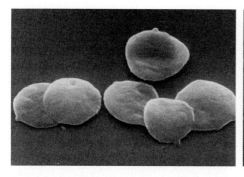 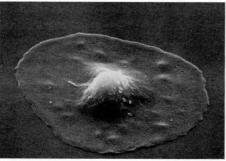 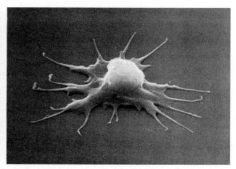

▲ **FIGURE 18-8 Changes in shape of platelets during blood clotting.** Resting cells have a discoid shape *(left).* When exposed to clotting agents, the cells settle and spread out on the substratum *(center).* During clot retraction, the cells assume a stellate shape and extend numerous filopodia *(right).* The changes in morphology result from complex rearrangements of the actin cytoskeleton, which is cross-linked to the plasma membrane. [Courtesy of J. White.]

ton. The cytoskeleton of an unactivated platelet consists of a rim of microtubules (the *marginal band*), a cortical actin network, and a cytosolic actin network. The cortical actin network in platelets consists of actin filaments cross-linked into a two-dimensional network by a nonerythroid isoform of spectrin and linked through ankyrin to an anion transporter, the Na/K ATPase, in the membrane. This network thus is somewhat similar to the erythrocyte cytoskeleton. A critical difference between the erythrocyte and platelet cytoskeletons is the presence in the platelet of the second network of actin filaments, which are organized by *filamin* cross-links into a three-dimensional gel (see Figure 18-5b). The gel fills the cytosol of a platelet and is anchored by filamin to the glycoprotein 1b-IX complex (Gp1b-IX) in the platelet membrane. Although both spectrin and filamin are CH-domain actin cross-linking proteins, filamin also can anchor the cytoskeleton to the plasma membrane.

In order to close a wound, a platelet must be able to transmit cytoskeletal changes within the cell to the blood clot outside the cell. This is effectively accomplished by linking the cytoskeleton to the same proteins in the membrane that also bind to the clot. For example, Gp1b-IX not only binds filamin but also is the membrane receptor for two blood-clotting proteins (Figure 18-9a). Contraction of the cell tightens the clot and closes the wound. The direct connection of the

▶ **FIGURE 18-9 Cross-linkage of actin filament networks to the plasma membrane in platelets, muscle cells, and epithelial cells.** (a) In platelets a three-dimensional network of actin filaments is attached to the integral membrane glycoprotein complex Gp1b-IX by filamin. Gp1b-IX also binds to proteins in a blood clot outside the platelet. Platelets also possess a two-dimensional cortical network of actin and spectrin similar to that underlying the erythrocyte membrane. (b) In muscle cells dystrophin attaches actin filaments to an integral membrane glycoprotein complex. This complex binds to laminin and agrin in the extracellular matrix (ECM). (c) In epithelial cells, the ERM protein, ezrin, and EBP50 crosslink an actin filament to the cystic fibrosis transmembrane conductance receptor. After activation, ezrin unfolds and oligomerizes to form head-to-tail dimers. The head domain binds EBP50, while the tail domain binds actin.

(a) Platelet
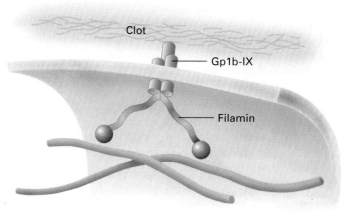

(b) Muscle
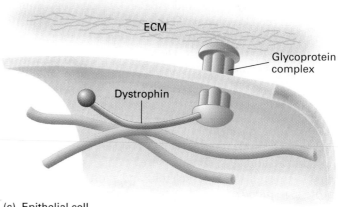

(c) Epithelial cell
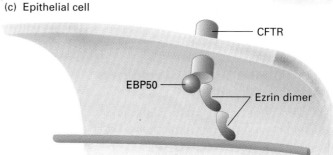

cytoskeleton to the extracellular matrix through shared membrane proteins also occurs in muscle and many other cells.

Membrane-Cytoskeleton Linkage in Muscle Cells The critical role of membrane-cytoskeleton linkages in muscle contraction was uncovered by studies on Duchenne muscular dystrophy (DMD), a fatal, degenerative sex-linked genetic disease of muscle that affects about 1 of every 3500 males born. Patients with this disease have a defect in the gene that encodes *dystrophin*, a very large protein that constitutes 5 percent of the membrane-associated cytoskeleton in muscle cells.

Like filamin in platelets, dystrophin cross-links actin filaments into a supportive cortical network and attaches this network to a glycoprotein complex in the muscle cell membrane (Figure 18-9b). This membrane complex also binds to proteins in the extracellular matrix. Thus the internal cytoskeleton is connected to the external matrix via the dystrophin–membrane glycoprotein linkage. In individuals with DMD, who lack functional dystrophin, the membrane of a muscle cell is not supported by a cortical cytoskeleton and presumably is easily damaged by the stress of repeated muscle contraction.

ERMs: Conformationally Regulated Membrane-Microfilament Proteins

In addition to the membrane linkages by CH domain proteins, several proteins in the apical membrane are cross-linked to microfilaments through the ERM (Ezrin, Radixin, and Moesin) family of proteins. Found in a variety of cell types, ERMs bind membrane proteins, including the cystic fibrosis transmembrane conductance regulator (CFTR) and β2-adrenergic receptor. This interaction with different membrane proteins requires an adapter protein, EBP50, and a Band 4.1 homology domain located at the N-terminus. The crosslink to the cytoskeleton is completed through an actin-binding site located in a C-terminal domain (Figure 18-9c).

Recent evidence suggests that a membrane-microfilament linkage represents an open conformation of the molecule that is regulated by several signaling pathways possibly through both serine/threonine and tyrosine kinases. In the inactive state, the membrane and microfilament binding sites are masked by an intermolecular association of the N- and C-terminal domains. In this closed conformation, ezrin is unable to bind the cytoskeleton, and most of the inactive protein is found in the cytosol. However, the membrane-binding function is correlated with phosphorylation of specific serine and threonine residues in the C-terminal domain.

Actin Bundles Support Projecting Fingers of Membrane

The surfaces of cells in multicellular organisms are studded with numerous membrane projections. Two types of fingerlike membrane projections, microvilli and filopodia, are supported by an internal actin bundle to which the phospholipid

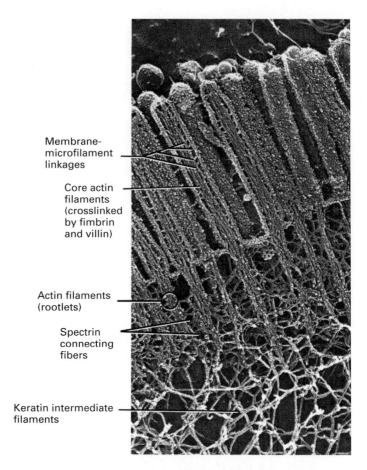

Membrane-microfilament linkages

Core actin filaments (crosslinked by fimbrin and villin)

Actin filaments (rootlets)

Spectrin connecting fibers

Keratin intermediate filaments

▲ **FIGURE 18-10 Micrograph of intestinal cell showing microvilli.** At the core of each 2-μm-long microvillus, a bundle of actin filaments, cross-linked by fimbrin and villin, stabilizes the fingerlike structure. The plasma membrane surrounding a microvillus is attached to the sides of the bundle by evenly spaced membrane-microfilament linkages consisting of myosin I. Each bundle continues into the cell as a 0.5-μm-long rootlet. The rootlets are cross-braced by connecting fibers composed of an intestinal isoform of spectrin, and the bases of the rootlets form attachment sites for keratin filaments. These numerous connections anchor the rootlets in a meshwork of filaments and thereby support the upright orientation of the microvilli. [Courtesy of N. Hirokawa.]

membrane is anchored. Because it is held together by protein cross-links, the actin bundle is stiff and provides a rigid structure that reinforces the fragile projecting membrane, enabling it to maintain its long, slender shape.

Microvilli range in length from 0.5 to 10 μm and are found where the cell membrane faces the fluid environment. As discussed in Chapter 15, a dense carpet of microvilli, the *brush border,* covers the surface of intestinal epithelial cells, greatly increasing the surface area of these cells, whose primary function is to transport nutrients. The cross-linking proteins that hold actin filaments in the core bundle differ in microvilli found on different cell types. Figure 18-10 shows the structure of intestinal microvilli and the various connections supporting them. The membrane is attached to the core bundle by membrane-microfilament linkages of myosin I.

Filopodia, which are much less common than microvilli, attach cells to a solid surface. These projections of the membrane typically are found at the edge of moving or spreading cells (see Figure 18-1). Filopodia are transient structures, present only during the time required to establish a stable contact with the underlying substratum.

SUMMARY The Actin Cytoskeleton

- A major component of the cytoskeleton, actin is highly conserved in all eukaryotes. F-Actin is a helical filamentous polymer of globular G-actin subunits (see Figure 18-2). An actin polymer, along with bound proteins, constitutes a microfilament, one of the three types of fibers that form the cytoskeleton.

- Actin filaments are organized into bundles and networks by a variety of bivalent cross-linking proteins, including spectrin in erythrocytes, filamin in platelets, dystrophin in muscle, and fimbrin and fascin in microvilli (see Table 18-1 and Figures 18-5, 18-7, and 18-9).

- Cortical actin networks are attached to the cell membrane by bivalent membrane-microfilament binding proteins. These link cortical actin filaments to integral membrane proteins, which in some cells also bind to the extracellular matrix (see Figures 18-7 and 18-9).

- Actin bundles form the core of microvilli and filopodia, fingerlike projections of the plasma membrane. A complex structure of cytoskeletal elements provides support for these membrane extensions.

18.2 The Dynamics of Actin Assembly

Thus far we have treated the actin cytoskeleton as if it were an unchanging structure consisting of bundles and networks of filaments. The microfilaments in a cell, however, are constantly shrinking or growing in length, and bundles and meshworks of microfilaments are continuously forming and dissolving. These changes in the organization of actin filaments cause equally large changes in the shape of a cell. In this section, we discuss the mechanism of actin polymerization and the regulation of this process, which is largely responsible for the dynamic nature of the cytoskeleton.

Actin Polymerization in Vitro Proceeds in Three Steps

As we mentioned earlier, addition of salts to a solution of G-actin induces polymerization, creating F-actin filaments.

The polymerization process can be monitored by viscometry, sedimentation, and fluorescence spectroscopy. When actin filaments become long enough to become entangled, the viscosity of the solution increases, which is measured as a decrease in its flow rate in a viscometer. The basis of the sedimentation assay is the ability of ultracentrifugation (100,000g for 30 minutes) to pellet F-actin but not G-actin. The third assay makes use of G-actin covalently labeled with a fluorescent dye; the fluorescence spectrum of the modified G-actin monomer changes when it is polymerized into F-actin. These assays are useful in kinetic studies of actin polymerization and during purification of actin-binding proteins, which cross-link or depolymerize actin filaments.

The polymerization of actin filaments proceeds in three sequential phases (Figure 18-11a). The first phase is marked by a *lag period* in which G-actin aggregates into short, unstable oligomers. Once the oligomer reaches a certain length (three or four subunits) it can act as a stable *seed*, or *nucleus*, which in the second phase rapidly *elongates* into a filament by the addition of actin monomers to both of its ends. As F-actin filaments grow, the concentration of G-actin monomers decreases until it is in equilibrium with the filament. This third phase is called **steady state** because G-actin monomers exchange with subunits at the filament ends but there is no net change in the total mass of filaments. The kinetic curves shown in Figure 18-11b show that the lag period can be eliminated by addition of a small number of F-actin nuclei to the solution of G-actin.

Once the steady-state phase is reached, the equilibrium concentration of the pool of unassembled subunits is called the **critical concentration** (C_c). This parameter is a measure of the ability of a solution of G-actin to polymerize. Under typical in vitro conditions, the C_c of G-actin is 0.1 μM. Above this value, a solution of G-actin will polymerize; below this value, a solution of F-actin will depolymerize (Figure 18-12).

After ATP–G-actin monomers are incorporated into a filament, the bound ATP is slowly hydrolyzed to ADP. As a result of this hydrolysis, most of the filament consists of ADP–F-actin, but ATP–F-actin is found at the ends. However, ATP hydrolysis is not essential for polymerization to occur, as evidenced by the ability of G-actin containing ADP or a nonhydrolyzable ATP analog to polymerize into filaments.

Actin Filaments Grow Faster at One End Than at the Other

We saw earlier that myosin decoration experiments reveal an inherent structural polarity of F-actin (see Figure 18-3). This polarity also is reflected in different rates of monomer addition to the two ends. One end of the filament, the (+) end, elongates five to ten times faster than does the opposite, or (−), end. The unequal growth rates can be demonstrated by a simple experiment using myosin-decorated actin filaments to nucleate polymerization of G-actin. Electron

(a)

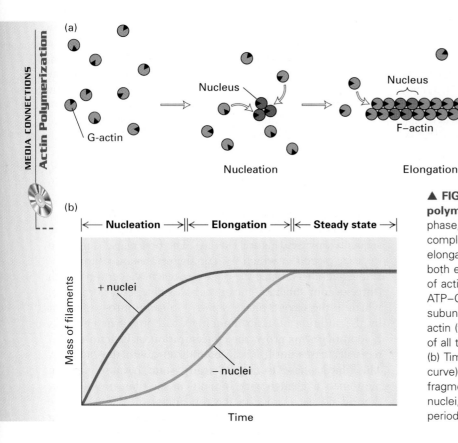

G-actin

Nucleation

Nucleus

F—actin

Elongation

Nucleus

F-actin

(−) end   (+) end

Steady state

(b)

|← Nucleation →||← Elongation →||← Steady state →|

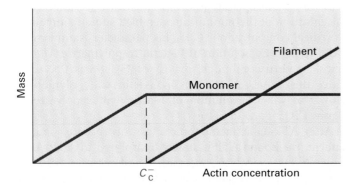

+ nuclei

− nuclei

Mass of filaments

Time

▲ **FIGURE 18-11 The three phases of G-actin polymerization in vitro.** (a) During the initial nucleation phase, ATP–G-actin monomers (pink) slowly form stable complexes of actin (purple). These nuclei are more rapidly elongated in the second phase by addition of subunits to both ends of the filament. In the third phase, the ends of actin filaments are in a steady state with monomeric ATP–G-actin. After their incorporation into a filament, subunits slowly hydrolyze ATP and become stable ADP–F-actin (white). Note that the ATP-binding clefts (black triangles) of all the subunits are oriented in the same direction in F-actin. (b) Time course of the in vitro polymerization reaction (pink curve) reveals the initial lag period. If some actin filament fragments are added at the start of the reaction to act as nuclei, elongation proceeds immediately without any lag period (purple curve).

Filament

Monomer

Mass

C_c^- Actin concentration

▲ **FIGURE 18-12 The critical concentration (C_c) is the concentration of G-actin monomers in equilibrium with actin filaments.** At monomer concentrations below the C_c, no polymerization occurs. At monomer concentrations above the C_c, filaments assemble until the monomer concentration reaches C_c.

microscopy of the elongated filaments reveals bare sections at both ends, corresponding to the added undecorated G-actin. The newly polymerized (undecorated) actin is five to ten times longer at the (+) end than at the (−) end of the filaments (Figure 18-13a).

The difference in elongation rates at the opposite ends of an actin filament is caused by a difference in C_c values at the two ends. This difference can be measured by blocking

one or the other end with proteins that "cap" the ends of actin filaments. If the (+) end of an actin filament is capped, it can elongate only from its (−) end; conversely, elongation occurs only at the (+) end when the (−) end of a filament is blocked (Figure 18-13b). Polymerization assays of such capped filaments have shown that the C_c is much lower for G-actin addition at the (+) end ($C_c^+ = 0.1$ μM) than for addition at the (−) end ($C_c^- = 0.8$ μM).

As a result of the difference in the C_c values for the (+) and (−) ends of a filament, we can predict the following: at G-actin concentrations below C_c^+, no filament growth occurs; at G-actin concentrations between C_c^+ and C_c^-, growth occurs only from the (+) end; and at G-actin concentrations above C_c^-, growth occurs at both ends, although it is faster at the (+) end than at the (−) end. Once the steady-state phase is reached at G-actin concentrations intermediate between the C_c values for the (+) and (−) ends, subunits continue to be added at the (+) end and lost from the (−) end (Figure 18-13c). The length of the filament remains constant, with the newly added subunits traveling through the filament, as if on a treadmill, until they reach the (−) end, where they dissociate. In the lamellipodia of cells, actin filaments probably turn over by a treadmilling type of mechanism. Subunits released from one end of the filament are rapidly recruited to assemble at the leading edge of the cell.

The C_c for assembly of actin filaments depends on whether the monomers are bound to ATP or ADP. When

(a)

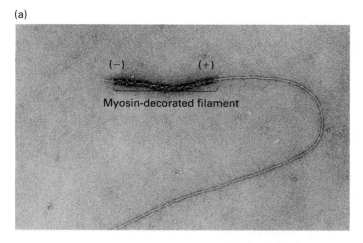

◀ **FIGURE 18-13 Experimental demonstration of unequal growth rates at the two ends of an actin filament.** (a) When short myosin-decorated filaments are the nuclei for actin polymerization, the resulting elongated filaments have a much longer undecorated (+) end than (−) end. This result indicates that ATP–G-actin monomers are added much faster at the (+) end than at the (−) end. (b) Blocking the (+) or (−) ends of a filament with actin-capping proteins permits growth only at the opposite end. In polymerization assays with capped filaments, the critical concentration (C_c) is determined by the unblocked growing end. Such assays show that the C_c at the (+) end is much lower than the C_c at the (−) end. (c) At G-actin concentrations intermediate between the C_c values for the (−) and (+) ends, actin subunits can flow through the filaments by attaching preferentially to the (+) end and dissociating preferentially from the (−) end of the filament, a phenomenon known as "treadmilling." The oldest subunits in a treadmilling filament lie at the (−) end. [Part (a) courtesy of T. Pollard.]

(b)

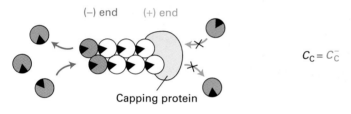

$$C_c = C_c^+$$

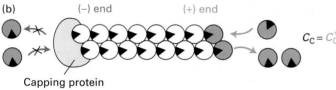

$$C_c = C_c^-$$

(c) $\quad C_c^- >$ G-actin concentration $> C_c^+$

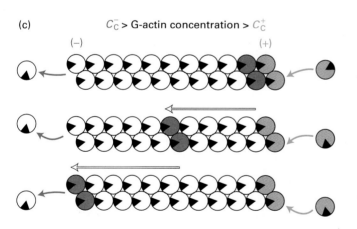

Cytochalasin D, a fungal alkaloid, depolymerizes actin filaments by binding to the (+) end of F-actin, where it blocks further addition of subunits. In contrast, latrunculin, a toxin secreted by sponges, binds G-actin and inhibits it from adding to a filament end. Exposure to either toxin shifts the monomer-polymer equilibrium in the direction of dissociation. When these toxins are added to live cells, the actin cytoskeleton disappears and cell movements like locomotion and cytokinesis are inhibited. These observations were among the first that implicated actin filaments in cell motility.

A third toxin, phalloidin, has the opposite effect on actin: It poisons a cell by preventing actin filaments from depolymerizing. Isolated from *Amanita phalloides* (the "angel of death" mushroom), phalloidin binds at the interface between subunits in F-actin and locks adjacent subunits together. Even when actin is diluted below its critical concentration, phalloidin-stabilized filaments will not depolymerize. Fluorescent-labeled phalloidin, which binds only to F-actin, is commonly used to stain actin filaments for light microscopy (see Figure 18-1b).

Actin Polymerization Is Regulated by Proteins That Bind G-Actin

In the perfect world of a test tube, experimenters can start the polymerization process by adding salts to G-actin or can depolymerize F-actin by simply diluting the filaments. Cells, however, must maintain a nearly constant cytosolic ionic concentration and thus employ a different mechanism for controlling actin polymerization. The cellular regulatory mechanism involves several actin-binding proteins that either promote or inhibit actin polymerization. Here we discuss two such proteins that have been isolated and characterized.

ADP-actin monomers are incorporated into actin filaments, the C_c at the (+) end becomes equal to that at the (−) end despite the inherent structural polarity of actin filaments.

Toxins Disrupt the Actin Monomer-Polymer Equilibrium

The equilibrium between actin monomers and filaments is easily perturbed by toxins. Two unrelated toxins, cytochalasin D and latrunculin, have two complementary effects.

Inhibition of Actin Assembly by Thymosin β_4 Calculations based on the C_c of G-actin (0.1 μM), a typical cytosolic total actin concentration (0.5 mM), and the ionic conditions of the cell indicate that nearly all cellular actin should exist as filaments; there should be very little G-actin. Actual measurements, however, show that as much as 40 percent of actin in an animal cell is unpolymerized. What keeps the cellular concentration of G-actin above its C_c? The most likely explanation is that cytosolic proteins sequester actin, holding it in a form that is unable to polymerize.

Because of its abundance in the cytosol and ability to bind ATP–G-actin (but not F-actin), *thymosin β_4* is considered to be the main actin-sequestering protein in cells. A small protein (5000 MW), thymosin binds ATP–G-actin in a 1:1 complex; in this complex, G-actin cannot polymerize. In platelets, the concentration of thymosin β_4 is 0.55 mM, approximately twice the concentration of unpolymerized actin (0.25 mM). At these concentrations, approximately 70 percent of the monomeric actin in a platelet should be sequestered by thymosin β_4.

Thymosin β_4 (Tβ_4) functions like a buffer for monomeric actin as represented in the following reaction:

$$\text{F-actin} \longleftrightarrow \text{G-actin} + \text{T}\beta_4 \longleftrightarrow \text{G-actin/T}\beta_4$$

In a simple equilibrium, an increase in the cytosolic concentration of thymosin β_4 would increase the concentration of sequestered actin subunits and correspondingly decrease F-actin, since actin filaments are in equilibrium with actin monomers. This effect of thymosin β_4 on the cellular F-actin level has been experimentally demonstrated in live cells.

Promotion of Actin Assembly by Profilin Another cytosolic protein, *profilin* (15,000 MW), also binds ATP-actin monomers in a stable 1:1 complex. At most, profilin can buffer 20 percent of the unpolymerized actin in cells, a level too low for it to act as an effective sequestering protein. Rather than sequestering actin monomers, the main function of profilin probably is to promote assembly of actin filaments in cells. It appears to do so by several mechanisms.

First, as a complex with G-actin, profilin is postulated to assist in addition of monomers to the (+) end of an actin filament. This hypothesis is consistent with the three-dimensional structure of the profilin-actin complex in which profilin is bound to the (+) end of an actin monomer, leaving the ATP-binding end, the (−) end, free to associate with the (+) end of a filament (Figure 18-14). After the complex binds transiently to the filament, the profilin dissociates from actin.

Second, profilin also interacts with membrane components involved in cell-cell signaling, suggesting that it may be particularly important in controlling actin assembly at the plasma membrane. Profilin binds to the membrane phospholipid phosphoinositol 4,5-bisphosphate (PIP$_2$); this interaction prevents binding of profilin to G-actin. (As we will see in Chapter 20, hydrolysis of PIP$_2$ in response to extra-

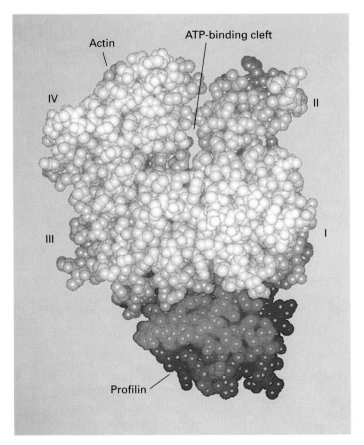

▲ **FIGURE 18-14 The three-dimensional structure of the profilin-actin complex.** Profilin (green) binds to the edge of subdomains I and III of G-actin (white) at the end opposite to the ATP-binding cleft. Profilin is unique among actin-binding proteins because it permits exchange of the nucleotide bound in the cleft. When actin is complexed with any other actin-binding protein, the nucleotide in the ATP-binding cleft of actin is nonexchangeable. [See C. E. Schutt et al., 1993, *Nature* **365**:810; courtesy of M. Rozycki and C. E. Schutt.]

cellular signals triggers intracellular signaling pathways.) In addition, profilin binds to proline-rich sequences that are commonly found in membrane-associated signaling proteins such as Vasp and Mena. This interaction, which does not inhibit profilin binding to G-actin, localizes profilin-actin complexes to the membrane. Figure 18-15 depicts how these properties of profilin could play a central role in stimulating actin polymerization in response to cell-cell signals.

Finally, profilin also promotes assembly of actin filaments by acting as a nucleotide-exchange factor. Profilin is the only actin-binding protein that allows the exchange of ATP for ADP. When G-actin is complexed with other proteins, ATP or ADP is trapped in the ATP-binding cleft of actin. However, because profilin binds to G-actin opposite to the ATP-binding cleft, it can recharge ADP-actin monomers released from a filament, thereby replenishing the pool of ATP-actin (see Figure 18-15d).

(a) Unactivated

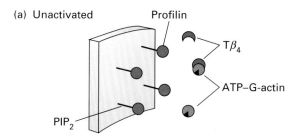

(b) Activated

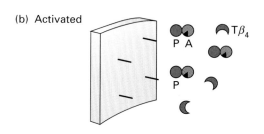

(c) Assembly

(d) Exchange

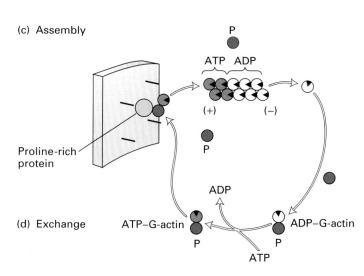

◀ **FIGURE 18-15 Model of the complementary roles of profilin and thymosin β4 in regulating polymerization of G-actin.** (a) At the cell membrane, profilin is bound to PIP_2, a membrane lipid, while most of the G-actin is complexed with thymosin β4 and thus cannot polymerize. (b) In response to an extracellular signal, such as chemotactic molecules that stimulate actin assembly, profilin is released from the membrane by hydrolysis of PIP_2. The released profilin displaces thymosin β4, forming profilin–G-actin complexes that can assemble into filaments. (c) The profilin-actin complexes interact with proline-rich proteins in the membrane, where profilin adds actin monomers to the (+) end of actin filaments. Eventually, the incorporation of monomers into filaments depletes the pools of profilin-actin and thymosin β4–actin complexes. (d) ADP–G-actin subunits that have dissociated from a filament are converted into ATP–G-actin by profilin, thus helping to replenish the cytoplasmic pool of ATP–G-actin.

severing protein changes the conformation of the subunit to which it binds, thereby causing strain on and subsequent breakage of the intersubunit bonds. In support of this hypothesis are electron micrographs showing that an actin filament with bound cofilin is severely twisted. After a severing protein breaks a filament at one site, it remains bound at the (+) end of one of the resulting fragments, where it prevents the addition or exchange of actin subunits, an activity called *capping*. The (−) ends of fragments remain uncapped and are rapidly shortened. Thus severing promotes turnover of actin filaments by creating new (−) ends and causes disintegration of an actin network, although many filaments remain cross-linked (Figure 18-16). The turnover of actin filaments promoted by severing proteins is necessary not only for cell locomotion but also for cytokinesis.

The capping and severing proteins are regulated by several signaling pathways. For example, both cofilin and gelsolin bind PIP_2 in a way that inhibits their binding to actin filaments and thus their severing activity. Hydrolysis of PIP_2 releases these proteins and induces rapid severing of filaments. The reversible phosphorylation and dephosphorylation of cofilin also regulates its activity, and the severing activity of gelsolin is activated by an increase in cytosolic Ca^{2+} to about 10^{-6} M. The counteracting influence of different signaling molecules, Ca^{2+}, and PIP_2 permits the reciprocal regulation of these proteins. At the end of this chapter, we discuss how extracellular signals coordinate the activities of different actin-binding proteins, including severing proteins, during cell migration.

Some Proteins Control the Lengths of Actin Filaments by Severing Them

A second group of proteins, which bind to actin filaments, control the length of actin filaments by breaking them into shorter fragments (Table 18-2). A valuable clue that led to the discovery of these severing proteins came from studies of amebas. Viscosity measurements and light-microscope observations demonstrated that during ameboid movement the cytosol flows forward in the center of the cell and then turns into a gel when it reaches the front end of the cell. As we discuss later, this "sol-to-gel" transformation depends on the assembly of new actin filaments in the front part of a moving ameba and the disassembly of old actin filaments in the rear part. Because the actin concentration in a cell favors the formation of filaments, the breakdown of existing actin filaments and filament networks requires the assistance of severing proteins such as *gelsolin* and *cofilin*.

Severing proteins can break the filaments in a network into shorter fragments. Currently, researchers believe that a

Actin Filaments Are Stabilized by Actin-Capping Proteins

Another group of proteins can cap the ends of actin filaments but, unlike severing proteins, cannot break filaments to create new ends. One such protein, *CapZ*, binds the (+) ends of actin filaments independently of Ca^{2+} and prevents the addition or loss of actin subunits from the (+) end.

TABLE 18-2 Some Cytosolic Proteins That Control Actin Polymerization

Protein	MW	Activity
Cofilin	15,000	Severing
gCAP39	40,000	Capping [(+) end]
Severin	40,000	Severing, capping
Gelsolin	87,000	Severing, capping [(+) end]
Villin	92,000	Cross-linking, severing, capping
CapZ	36,000 (α) 32,000 (β)	Capping [(+) end]
Tropomodulin	40,000	Capping [(−) end]

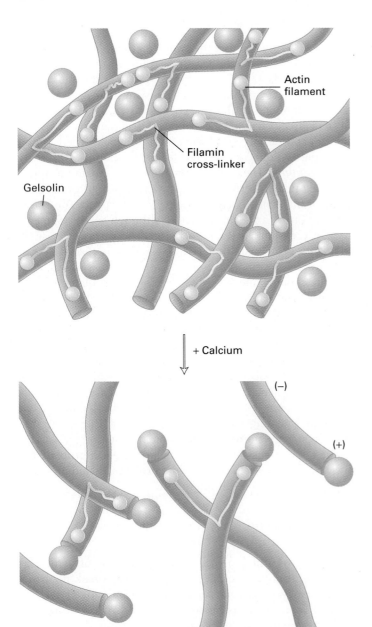

Capping by this protein is inhibited by PIP$_2$, suggesting that its activity is regulated by the same signaling pathways that control cofilin and profilin. *Tropomodulin*, which is unrelated to CapZ in sequence, caps the (−) ends of actin filaments. Its capping activity is enhanced in the presence of tropomyosin, which suggests that the two proteins function as a complex to stabilize a filament. An actin filament that is capped at both ends is effectively stabilized, undergoing neither addition nor loss of subunits. Such capped actin filaments are needed in places where the organization of the cytoskeleton is unchanging, as in a muscle sarcomere or at the erythrocyte membrane.

Many Movements Are Driven by Actin Polymerization

By manipulating actin polymerization and depolymerization, the cell can create forces that produce several types of movement. A classic example is the growth of a finger of membrane from the cell surface during the *acrosome reaction* of an echinoderm sperm cell (Figure 18-17). More recent examples suggest that actin polymerization stimulated by profilin is a common mechanism for generating the force for movement of intracellular pathogens and for cell locomotion.

◀ **FIGURE 18-16 Action of gelsolin in severing actin filaments.** At cytosolic Ca^{2+} levels below 10^{-6} M, gelsolin does not bind to actin filaments. At higher Ca^{2+} concentrations, binding of gelsolin to filaments causes a distortion that disrupts the noncovalent bonds between subunits. One consequence of severing is that actin networks are broken apart and new ends are created. Gelsolin remains bound to the (+) end of the fragments, preventing addition of monomers, while the uncapped (−) ends rapidly disassemble. Thus severing contributes to the turnover of actin filaments. Note that even though the filaments are shortened, they still remain cross-linked. The other major severing protein, cofilin, functions in a similar way, but it is regulated by reversible phosphorylation and dephosphorylation.

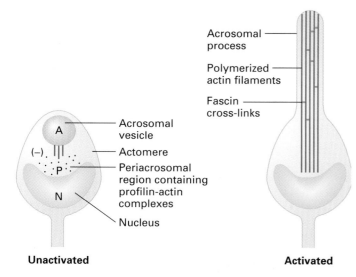

▲ FIGURE 18-17 The acrosome reaction in echinoderm sperm.
In an unactivated sperm, a membrane-bounded acrosomal vesicle
(A) lies within an indentation of the nucleus (N). Adjacent to
the vesicle is the periacrosomal region (P), which contains
unpolymerized profilin-actin complexes and a short actin bundle,
the actomere. When the sperm is activated by contact with the
egg jelly layer, the profilin-actin complexes dissociate and G-actin
adds to the (+) ends of the actomere filaments. In addition, the
filaments are cross-linked by fascin. The combined action of
polymerization and cross-linking causes outgrowth of the
acrosomal process, which pierces the jelly layer and provides
a pathway for passage of the sperm nucleus into the egg.

Intracellular Bacterial and Viral Movements Most infections are spread by bacteria or viruses that are liberated
when an infected cell lyses. However, some bacteria and
viruses escape from a cell on the end of a polymerizing actin
filament. Examples include *Listeria monocytogenes,* a bacterium that can be transmitted from a pregnant woman to
the fetus, and vaccinia, a virus related to the smallpox virus.
When such organisms infect mammalian cells, they move
through the cytosol at rates approaching 11 μm/min.
Fluorescent-microscopy experiments showed that a meshwork of short actin filaments follows a moving bacterium
or virus like the plume of a rocket exhaust (Figure 18-18).
These observations suggested that actin generates the force
necessary for movement. The first hints about how actin
mediates bacterial movement were provided by a microinjection experiment in which fluorescent-labeled G-actin was
injected into *Listeria*-infected cells. In the microscope, the
labeled monomers could be seen incorporating into the tail-
like meshwork at the end nearest the bacterium, with a
simultaneous loss of actin throughout the tail. This result
showed that actin polymerizes into filaments at the base of
the bacterium and suggested that as the tail-like meshwork
assembles, it pushes the bacterium ahead. Studies with mutant
bacteria indicate that interaction of cellular profilin with a
bacterial membrane protein promotes actin polymerization
at the end of the tail nearest the bacterium.

**Actin Polymerization at the Leading Edge of Moving
Cells** Understanding of the role of actin polymerization and
profilin in the acrosome reaction and intracellular bacterial
movement played a key role in elucidating how cells crawl
forward. Cell movement is led by changes in the position of
the plasma membrane at the front of the cell (*leading edge*);
video microscopy reveals that a major feature of this movement is the polymerization of actin at the membrane
(Figure 18-19). Profilin is thought to play a central role because it is located at the leading edge where polymerization
occurs. In addition, actin filaments at the leading edge are
rapidly cross-linked into bundles and networks in the projecting filopodia and lamellipodia. As we discuss in later
sections, these structures then form stable contacts with
the underlying surface and prevent the membrane from
retracting.

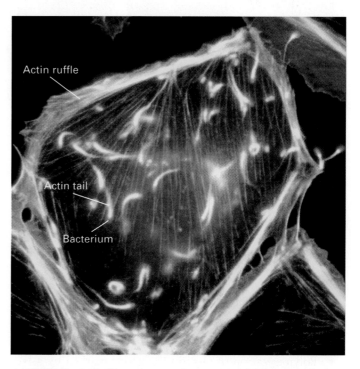

**▲ FIGURE 18-18 Fluorescent micrograph of *Listeria* in
infected fibroblasts.** Bacteria (red) are stained with an antibody
specific for a bacterial membrane protein that binds cellular
profilin and is essential for infectivity and motility. Behind each
bacterium is a "tail" of actin (green) stained with fluorescent
phalloidin. Numerous bacterial cells move independently within
the cytosol of an infected mammalian cell. Infection is transmitted to other cells when a spike of cell membrane, generated by
a bacterium, protrudes into a neighboring cell and is engulfed
by a phagocytotic event. [Courtesy of J. Theriot and T. Mitchison.]

(a)

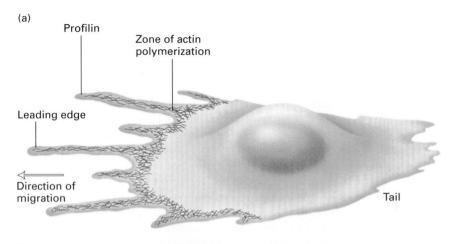

Profilin

Zone of actin polymerization

Leading edge

Direction of migration

Tail

(b)

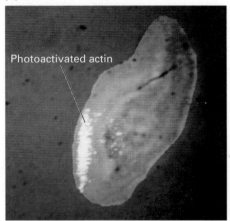

Photoactivated actin

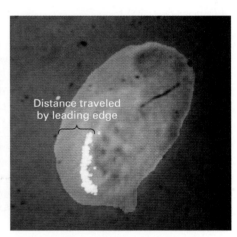

Distance traveled by leading edge

▲ **FIGURE 18-19 Assembly of actin filaments at the leading edge of migrating cells.** (a) As shown in this diagram of a fibroblast, profilin is located at the leading edge of the cell, the site where actin filaments are assembled. (b) In a microinjection experiment, the location of F-actin relative to the front of the cell is monitored by video microscopy. G-actin, modified with a caged fluorescent dye, is microinjected into a cell. (A caged dye is initially nonfluorescent, but UV irradiation induces a chemical change that renders the dye fluorescent.) When a narrow band of the cell at the leading edge is irradiated with UV light (first panel), the actin in that zone fluoresces (white band), while the remainder of the cell remains nonfluorescent. The G-actin has assembled into F-actin at the leading edge, and by 48 seconds (second panel) and 81 seconds (third panel) after UV irradiation, this zone moves away from the leading edge as the cell moves forward. The band becomes less intense, suggesting that F-actin dissociates as the zone moves back from the leading edge. [Part (b) see J. A. Theriot and T. J. Mitchison, 1991, *Nature* **352**:126; courtesy of J. Theriot.]

SUMMARY The Dynamics of Actin Assembly

• Within cells, the actin cytoskeleton is dynamic, with filaments able to grow and shrink rapidly.

• Polymerization of G-actin in vitro is marked by a lag period during which nucleation occurs. Eventually, a polymerization reaction reaches a steady state in which the rates of addition and loss of subunits are equal (see Figure 18-11).

• The concentration of actin monomers in equilibrium with actin filaments is the critical concentration (C_c). At a G-actin concentration above C_c, there is net growth of filaments; at concentrations below C_c, there is net depolymerization of filaments.

• Actin filaments grow considerably faster at their (+) end than at their (−) end, and the C_c for monomer addition to the (+) end is lower than for addition at the (−) end.

• The assembly, length, and stability of actin filaments are controlled by specialized actin-binding proteins. These proteins are in turn regulated by various mechanisms.

• The complementary actions of thymosin β_4 and profilin are critical to regulating the actin cytoskeleton near the cell membrane (see Figure 18-15).

• The regulated polymerization of actin can generate membrane projections and underlies the movement of certain bacteria and viruses within cells.

18.3 Myosin: The Actin Motor Protein

Although cells can harness polymerization of actin to generate some forms of movement, many cellular movements depend on interactions between actin filaments and **myosin,** an ATPase that moves along actin filaments by coupling the hydrolysis of ATP to conformational changes. This type of enzyme, which converts chemical energy into mechanical energy, is called a *mechanochemical enzyme* or, colloquially, a *motor protein.* Myosin is the motor, actin filaments are the tracks along which myosin moves, and ATP is the fuel that powers movement.

In this section we describe the structure and functions of the three major myosin classes and examine the mechanism of myosin-dependent movement. *Contraction,* a special form of movement resulting from actin and myosin interactions, is highly evolved in muscle cells, which we consider in the next section. Muscle contraction provides a basis for our subsequent discussion of contractile events involving less organized systems in nonmuscle cells. In the next chapter, we will discuss the microtubule motors kinesin and dynein, which exhibit many of the same properties as myosin.

All Myosins Have Head, Neck, and Tail Domains with Distinct Functions

Thirteen members of the myosin gene family have been identified by genomic analysis (Chapter 7). Myosin I and myosin II, the most abundant and thoroughly studied of the myosin proteins, are present in nearly all eukaryotic cells. A less-common isoform, myosin V, also has been isolated and characterized. Although the specific activities of these myosins differ, they all function as motor proteins. Myosin II powers muscle contraction and cytokinesis, whereas myosins I and V are involved in cytoskeleton-membrane interactions such as the transport of membrane vesicles. The activities of the remaining proteins encoded by the myosin gene family are now being discovered.

All myosins are composed of one or two heavy chains and several light chains. The heavy chains are organized into three structurally and functionally different domains (Figure 18-20a). The globular *head domain* contains actin- and ATP-binding sites and is responsible for generating force; this is the most conserved region among the various myosins. Adjacent to the head domain lies the α-helical *neck region,* which is associated with the light chains. The latter regulate the activity of the head domain. The *tail domain* contains the binding sites that determine the specific activities of a particular myosin, as discussed below.

Myosin II and myosin V are dimers in which α-helical sequences in the tail of each heavy chain associate to form a rodlike coiled-coil structure. Because the myosin I heavy chain lacks this α-helical sequence, the molecule is a monomer. The three myosins differ in the number and type

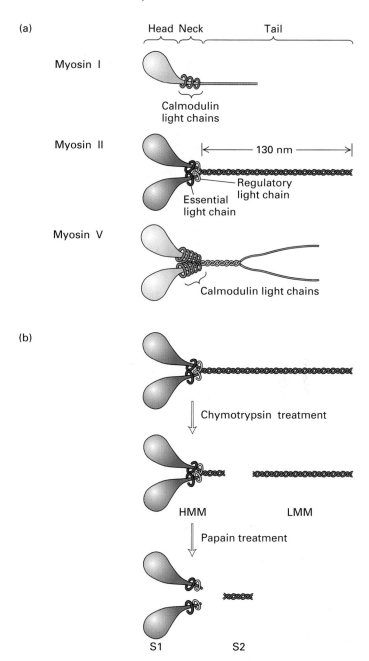

▲ **FIGURE 18-20 Structure of various myosin molecules.**
(a) The three major myosin proteins are organized into head, neck, and tail domains, which carry out different functions. The head domain binds actin and has ATPase activity. The light chains, bound to the neck domain, regulate the head domain. The tail domain dictates the specific role of each myosin in the cell. Note that myosin II, the form that functions in muscle contraction, is a dimer with a long rigid coiled-coil tail. (b) Proteolysis of myosin II reveals its domain structure. For example, most proteases cleave myosin II at the base of the neck domain to generate a paired-head and neck fragment, called *heavy meromyosin* (HMM), and a rodlike tail fragment, called *light meromyosin* (LMM). Further digestion of HMM with papain splits off the neck region (S2 fragment) and leads to separation of the two head domains into single myosin head fragments (S1 fragments).

(a)

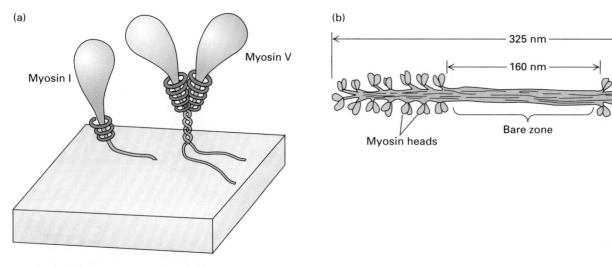

Myosin I

Myosin V

(b)

325 nm

160 nm

Myosin heads

Bare zone

Myosin tail

▲ **FIGURE 18-21 Functions of the myosin tail domain.**
(a) Myosin I and myosin V are localized to cellular membranes
by undetermined sites in their tail domains. As a result, these
myosins are associated with intracellular membrane vesicles or
the cytoplasmic face of the plasma membrane. (b) In contrast,

the coiled-coil tail domains of myosin II molecules pack side by
side, forming a thick filament from which the heads project. In
a skeletal muscle, the thick filament is bipolar. Heads are at the
ends of the thick filament and are separated by a bare zone,
which consists of the side-by-side tails.

of light chains bound in the neck region (see Figure 18-20a).
The light chains of myosin I and myosin V are **calmodulin,**
a Ca^{2+}-binding regulatory subunit in many intracellular
enzymes. Myosin II contains two different light chains
(called *essential* and *regulatory light chains*); both are
Ca^{2+}-binding proteins but differ from calmodulin in
their Ca^{2+}-binding properties. All myosins are regulated
in some way by Ca^{2+}; however, because of the differences
in their light chains, the different myosins exhibit different responses to Ca^{2+} signals in the cell.

The head, neck, and tail domains of the myosins are
clearly visible in electron micrographs. This domain organization has been confirmed by proteolytic studies of myosin
II, outlined in Figure 18-20b. The head domain corresponds
to the N-terminal half of the heavy chain, and the tail
domain to the C-terminal half.

Studies of the myosin fragments produced by proteolysis revealed the biochemical properties of the three domains.
In all myosins, the head domain is a specialized ATPase that
is able to couple the hydrolysis of ATP with motion. A critical feature of the myosin ATPase activity is that it is *actin-
activated*. In the absence of actin, solutions of myosin slowly
convert ATP into ADP and phosphate. However, when
myosin is complexed with actin, the rate of myosin ATPase
activity is four to five times faster than in the absence of
actin. The actin-activation step ensures that the myosin
ATPase operates at its maximal rate only when the myosin
head domain is bound to actin.

Although the head domain of all myosins exerts force
on actin, the role of a particular myosin in cells is related
to its tail domain. Because the tail domains of myosin I and

myosin V bind the plasma membrane or the membranes of
intracellular organelles, these molecules have membrane-
related activities (Figure 18-21a). For example, myosin I
serves as a linkage between the plasma membrane and the
microfilament bundles in brush-border microvilli and in filopodia. In contrast, the rodlike tail domains of multiple
myosin II dimers associate to form *thick filaments,* which
compose part of the contractile apparatus in muscle. A thick
filament from skeletal muscle has a bipolar organization—
the heads are located at both ends of the filament and are
separated by a central bare zone devoid of heads (Figure
18-21b). When packed tightly together in a thick filament,
many myosin head domains can interact simultaneously with
actin filaments.

Myosin Heads Walk along Actin Filaments

Studies of muscle contraction provided the first evidence that
myosin heads slide or walk along actin filaments. Unraveling the mechanism of muscle contraction was greatly aided
by development of in vitro motility assays that permit
movement of a single myosin molecule to be studied.

In one such assay, the *sliding-filament assay,* the movement of fluorescent-labeled actin filaments along a bed of
myosin molecules is observed in a fluorescence microscope.
Because the myosin molecules are tethered to a coverslip,
they cannot move; thus any force generated by interaction
of myosin heads with actin filaments forces the filaments to
move along the myosin (Figure 18-22a). If ATP is present,
added actin filaments can be seen to glide along the surface
of the coverslip; if ATP is absent, no filament movement is

► **FIGURE 18-22 The sliding-filament assay.** (a) Schematic diagram illustrates movement of actin filaments across myosin molecules attached to a coverslip. After myosin molecules are adsorbed onto the surface of a glass coverslip, excess myosin is removed; the coverslip then is placed myosin-side down on a glass slide to form a chamber through which solutions can flow. A solution of actin filaments, made visible by staining with rhodamine-labeled phalloidin, is allowed to flow into the chamber, and individual filaments are observed under a fluorescence light microscope. (The coverslip in the diagram is shown inverted from its orientation on the flow chamber to make it easier to see the positions of the molecules.) (b) Sliding movements of fluorescent actin filaments generated by the myosin head can be quantified by video microscopy. These photographs show the positions of three actin filaments (numbered 1, 2, 3) at 30-second intervals. In the presence of ATP, the actin filaments move at a velocity that can vary widely depending on the myosin tested and the assay conditions (ionic strength, temperature, ATP concentration, calcium concentration, etc.). [Part (b) courtesy of M. Footer and S. Kron.]

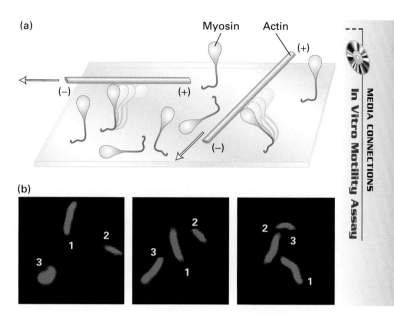

observed. Actin filaments always move with the (−) end in the lead. This movement is caused by a myosin head (bound to the coverslip) "walking" toward the (+) end of a filament.

From video camera recordings of sliding-filament assays, the velocities at which different myosins move an actin filament have been calculated (Figure 18-22b). These rates vary by more than a hundredfold, from 0.04 μm/s for myosin I in the brush border to 4.5 μm/s for myosin II in skeletal muscle. This variation reflects the specific functions of different myosins: Fast movement is associated with muscle contraction, whereas slow movement is adequate for transport in the cytosol.

Researchers have used the sliding-filament assay to identify the role of myosin domains, determine the effects of mutations on myosin, and study how myosin is regulated by other proteins. For example, the two-headed HMM fragment of myosin II, but not the LMM (tail) fragment or single-headed S1 fragment, was found to move actin filaments at velocities comparable with those achieved with intact myosin or myosin filaments. This and other observations pinpoint the myosin head as necessary and sufficient for movement; that is, it is the essential motor domain of the molecule.

Myosin Heads Move in Discrete Steps, Each Coupled to Hydrolysis of One ATP

Given that the myosin head domain is sufficient to cause movement, how much force is generated by a myosin head and how far does it travel when an ATP molecule is hydrolyzed? The forces generated by single myosin molecules can be measured with a device called an *optical trap*. In this device, the beam of an infrared laser focused by a light microscope on a polystyrene bead (or any other object that does not absorb infrared light) captures and holds the bead in the center of the beam. The strength of the force holding

the bead is adjusted by increasing or decreasing the intensity of the laser beam.

If a bead is attached to the end of an actin filament, then an optical trap can capture the filament, via the bead, and hold the filament to the surface of a myosin-coated coverslip (Figure 18-23a). The force exerted by a single myosin molecule on an actin filament is measured from the force needed to hold the bead in the optical trap. A computer-controlled electronic feedback system keeps the bead centered in the trap, and myosin-generated movement of the bead is counteracted by an opposing movement of the trap. The distance traveled by the actin filaments is measured from the displacement of the bead in the trap.

Such studies show that myosin II moves in discrete steps, approximately 5–10 nm long, and generates 1–5 piconewtons (pN) of force—the same force as that exerted by gravity on a single bacterium. This force is sufficient to cause myosin thick filaments to slide past actin thin filaments during muscle contraction or to transport a membrane-bounded vesicle through the cytoplasm. With a step size of 5 nm, myosin would bind to every actin subunit on one strand of the filament.

By using an optical trap with a novel imaging method, researchers have recently directly addressed the question of whether hydrolysis of one ATP molecule leads to a one-step movement of myosin (Figure 18-23b). However, the preliminary evidence that hydrolysis and movement are closely coupled is inconclusive; it is not clear whether myosin takes a discrete step for every ATP hydrolyzed.

Myosin and Kinesin Share the Ras Fold with Certain Signaling Proteins

X-ray crystallographic analysis of the myosin S1 fragment has revealed three key pieces of information about the myosin motor domain: its shape, the positions of the essential and

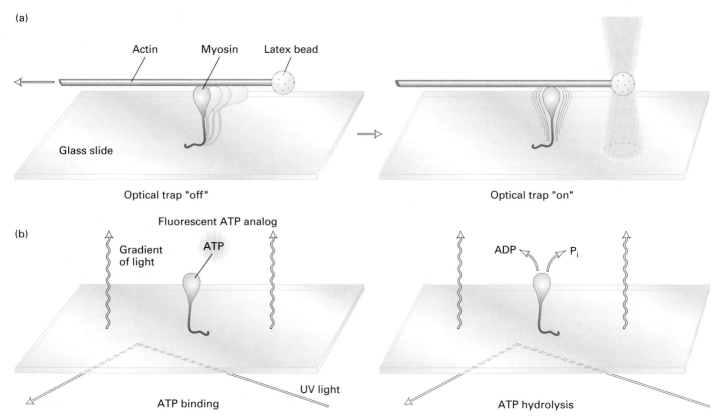

(a)

Actin   Myosin   Latex bead

Glass slide

Optical trap "off"

Optical trap "on"

(b)

Fluorescent ATP analog

Gradient of light

ATP

UV light

ATP binding

ADP   P$_i$

ATP hydrolysis

▲ **FIGURE 18-23 Use of the optical trap to determine myosin-generated force.** (a) The light from a single laser beam captures refractive objects—polystyrene beads, bacteria, or cell organelles. (Biological materials do not absorb infrared light and hence do not generate damaging heat.) In a motility assay, a polystyrene bead is fixed to the end of an actin filament. With the optical trap turned off, the actin filament and its attached bead move in response to the force generated by myosin adsorbed on the coverslip, as in Figure 18-22. When the optical trap is turned on and captures a bead, the trap will hold the bead immobile if the trapping force is great enough to resist the force generated by myosin. The minimum force needed to hold the bead in the trap is a measure of the force generated by myosin. (b) When the underside of a coverslip is illuminated with an extremely oblique angle of light, fluorescent molecules on the top surface of the coverslip radiate light, whereas the same molecules in solution do not. When a fluorescent ATP analog (Cy3) binds to a myosin molecule on the coverslip, a discrete fluorescent dot is seen at the position of the myosin. After ATP is hydrolyzed and the products dissociate, the myosin loses its fluorescent label. Experiments combining this imaging method with the optical trap technique depicted in part (a) are testing whether hydrolysis of one ATP molecule powers movement of myosin by one "step" along an actin filament.

regulatory light chains, and the locations of the ATP-and actin-binding sites. The elongated myosin head measures $16.5 \times 6.5 \times 4.5$ nm, and is attached at one end to the α-helical neck (Figure 18-24). Two light-chain molecules lie at the base of the head, wrapped around the neck like C-clamps. In this position, the light chains stiffen the neck region and therefore are able to regulate the activity of the head domain.

The surface of the myosin head is marked by a large cleft, extending from the actin-binding site on one side to the ATP-binding pocket on the opposite side. These two crucial binding sites are separated by 3.5 nm, a long distance in a protein. The presence of a surface cleft provides an obvious mechanism for generating large movements of the head domain. We can imagine how opening or closing of a cleft in the head domain, by binding or releasing actin or ATP, causes the head domain to pivot about the neck region. As discussed in detail in the next section, the ATP- and actin-binding sites are most likely coupled by large changes in the conformation of the head domain.

The core of the myosin motor domain is structurally similar to that of kinesin, a microtubule motor protein. Both motor domains contain the so-called *Ras fold*, in which a nucleotide molecule is bound to loops at one end of a β-sheet domain. The loops, called *P-loops*, are characteristic of ATPase active sites. This structural motif was initially identified in Ras protein, a GTP-binding protein that functions in certain intracellular signaling pathways (see Figure 3-5). The presence of the Ras fold in myosin and kinesin suggests that these motor proteins may have evolved from ancient nucleotide-binding proteins. The differences between the structures of myosin and kinesin reveal how they bind to actin and microtubules, respectively, and how binding is coupled to ATP hydrolysis.

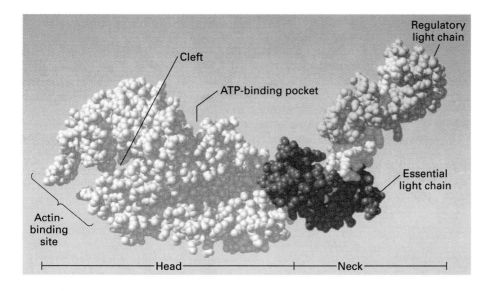

Cleft

ATP-binding pocket

Regulatory light chain

Essential light chain

Actin-binding site

Head — Neck

◄ **FIGURE 18-24 Three-dimensional structure of a myosin II S1 fragment (see Figure 18-20b).** X-ray crystallography reveals that the head domain has a curved, elongated shape (16.5 × 6.5 × 4.5 nm) and is bisected by a large cleft. The ATP-binding pocket lies on one side of the cleft about 3.5 nm from the actin-binding site on the other side near the tip of the head. A second prominent feature of myosin is the long α helix that extends from the head to form the neck. Wrapped around the shaft of the helix are a pair of calmodulinlike light chains, named *essential* and *regulatory light chains*. They act to stiffen the neck, which can then act as a lever arm for the head. Dimeric native myosin II has two head domains. [Adapted from I. Rayment et al., 1993, *Science* **261**:50; courtesy of I. Rayment.]

Conformational Changes in the Myosin Head Couple ATP Hydrolysis to Movement

Knowing the three-dimensional structure of the myosin head and the kinetics of the myosin ATPase, researchers could begin to understand how myosin harnesses the energy released by ATP hydrolysis to generate the force for movement. In the following discussion, we will consider the general case in which a myosin molecule walks along an actin filament. Because all myosins are thought to move using the same mechanism, we will ignore for the moment whether the myosin is bound to a vesicle or is part of a thick filament (as in muscle). One assumption in the model we describe is that the hydrolysis of a single ATP molecule is coupled to each step taken by a myosin molecule along an actin filament. As noted previously, some evidence indicates that this is true.

As shown in Figure 18-25, myosin undergoes a series of events during each step of movement. Repetition of this cycle causes myosin to slide relative to an actin filament. During one cycle, myosin must exist in at least three conformational states: a prehydrolysis ATP state unbound to actin, an ADP-P_i state bound to actin, and a state after the power stroke is completed. The major question to answer is how the nucleotide-binding pocket and the distant actin-binding site are mutually influenced and how changes at these sites are converted into force. Structural studies implicate the cleft as the physical link that communicates by domain movements. In structures of myosin bound to nucleotide analogs that mimic the prehydrolysis state and the transition state for hydrolysis, the presence of the γ-phosphate group of ATP and binding to actin control whether the cleft is open or closed. Binding of ATP to myosin opens the cleft, causing a disruption at the actin-binding site at the opposite end of the cleft. After ATP hydrolysis, the cleft closes partially. This conformational change traps the hydrolysis products and restores the actin-binding site. In addition, large rotations near the neck probably prepare myosin for the power stroke. An integral part of this model is that conformational changes in the head and neck are transmitted and amplified to other parts of the molecule through the light chains bound to the neck.

In the model depicted in Figure 18-25, myosin slides along an actin filament; however, the type of movement that occurs depends on how myosin or actin is anchored. For example, in a bipolar thick filament, the myosin II heads are firmly anchored to the thick filament backbone (see Figure 18-21b). Because the myosin heads in the two halves of a thick filament have opposite polarities, actin filaments slide toward the middle of the thick filament while the thick filament remains immobile. In contrast, a single myosin I molecule, bound to a membrane-bounded vesicle, moves along an actin filament because the actin filament is part of a massive structure, the cytoskeleton. Thus the frame of reference for movement changes depending on whether actin or myosin is immobile.

SUMMARY Myosin: The Actin Motor Protein

- Myosins are motor proteins that interact with actin filaments and couple hydrolysis of ATP to conformational changes that result in the movement of myosin and an actin filament relative to each other.

- Genomic analysis has revealed 13 different myosins. All consist of a highly conserved head (motor) domain, which is an actin-activated ATPase responsible for generating movement; a neck domain, which is associated with several regulatory light-chain subunits; and an effector tail domain, which is unique to each type of myosin and determines its specific functions in cells (see Figure 18-20).

- Myosin II, a dimeric molecule with a long rodlike tail domain, assembles into bipolar thick filaments that take part in muscle contraction.

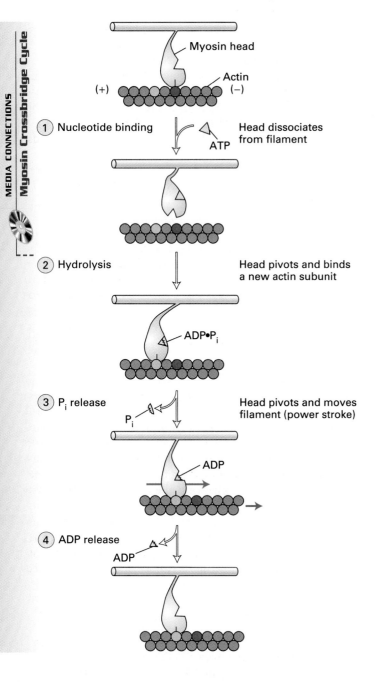

(+)   Myosin head

Actin

(−)

① Nucleotide binding   Head dissociates from filament

ATP

② Hydrolysis   Head pivots and binds a new actin subunit

ADP•P_i

③ P_i release   Head pivots and moves filament (power stroke)

P_i

ADP

④ ADP release

ADP

◀ **FIGURE 18-25 The coupling of ATP hydrolysis to movement of myosin along an actin filament.** In the absence of bound nucleotide, a myosin head binds actin tightly in a "rigor" state. When ATP binds (step ①), it opens the cleft in the head, disrupting the actin-binding site and weakening the interaction with actin. Freed of actin, the myosin head hydrolyzes ATP (step ②), causing a conformational change in the head that moves it to a new position, closer to the (+) end of the actin filament, where it rebinds to the filament. As phosphate (P_i) dissociates from the ATP-binding pocket (step ③), the myosin head undergoes a second conformational change—the power stroke—which restores myosin to its rigor conformation. Because myosin is bound to actin, this conformational change exerts a force that causes myosin to move the actin filament. The diagram shows the cycle for a myosin II head that is part of a thick filament, but other myosins attached to a membrane are thought to operate according to the same mechanism. [Adapted from I. Rayment and H. M. Holden, 1994, *Trends Biochem. Sci.* **19**:129.]

18.4 Muscle: A Specialized Contractile Machine

Muscle cells have evolved to carry out one highly specialized function—contraction. Muscle contractions must occur quickly and repetitively, and they must occur through long distances and with enough force to move large loads. In muscle, actin and myosin associate into a complex, called *actomyosin*, which is organized into a highly ordered structure having the ability to do work very efficiently.

Like any mechanical engine, muscles can be characterized by their power output, the rate at which they can work. Muscle power depends on several parameters: velocity of contraction, ability to contract repetitively, and force of contraction. Compared with that of other cellular systems, the power output of muscle is approximately equal to that of a flagellum but is 330,000-fold higher than the power output of the mitotic spindle, the microtubule machinery that separates chromosomes. Perhaps more instructive is to compare the power output of muscle with that of mechanical engines. When corrected for weight differences, a racing car engine and an aircraft engine are only tenfold to fortyfold more powerful than a muscle. A passenger car engine is only $1\frac{1}{2}$ times more powerful than muscle. In muscle, nature has obviously designed a very efficient and powerful engine.

In this section, we build on our discussion of myosin II as a motor protein by describing how myosin and actin are organized in muscle and how the motor properties of myosin are exploited to contract a muscle. Although we concentrate on the structure and function of skeletal muscle, the best-understood type of muscle, we also describe important details of smooth muscle, whose structure and activity are very similar to actin-myosin structures in nonmuscle cells. Thus when we later examine these structures in nonmuscle cells, we will understand how they generate movement but why they are less efficient than the muscle actomyosin system.

- In the presence of ATP, the dimeric head domain of myosin II alone can generate movement.

- The actin-binding site and ATP-binding pocket in the head domain of a myosin molecule are distant from each other but connected by a surface cleft. This cleft is thought to be critical in coupling ATP hydrolysis to myosin movement.

- Movement of myosin results from attachment of the myosin head to an actin filament, bending of the head, and its subsequent detachment in a cyclical ATP-dependent process (see Figure 18-25). During each cycle, myosin moves 5–25 nm and one ATP is hydrolyzed.

Some Muscles Contract, Others Generate Tension

Vertebrates and many invertebrates have two classes of muscle—*skeletal* and *smooth*—which differ in function. (A third class of muscle found in vertebrates, cardiac [heart] muscle, is not discussed here.) Skeletal muscles connect the bones in the arms, legs, and spine and are used in complex coordinated activities, such as walking or positioning of the head; they generate rapid movements by contracting quickly. In this type of contraction, termed *isotonic contraction,* the muscle shortens as force is generated. Another major function of skeletal muscle is to hold objects immobile. In the clenching of fists or tensing of muscles, for example, pairs of contracting muscles work to oppose each other and thus cancel out any movements. In such *isometric contraction,* the overall length of a muscle remains constant but its tension increases. Smooth muscles surround internal organs such as the large and small intestines, the uterus, and large blood vessels. The contraction and relaxation of smooth muscles controls the diameter of blood vessels and also propels food along the gastrointestinal tract. Compared with skeletal muscles, smooth muscle cells contract and relax slowly, and they can create and maintain tension for long periods of time.

Skeletal Muscles Contain a Regular Array of Actin and Myosin

A skeletal muscle comprises a bundle of muscle cells, or **myofibers** (Figure 18-26a). A typical muscle cell is cylindrical, large (1–40 mm in length and 10–50 μm in width), and multinucleated (containing as many as 100 nuclei). A myofiber is packed with **myofibrils,** bundles of filaments that extend the length of the cell. Myofibrils are further subdivided into alternating light and dark bands, which are aligned along the length of the muscle cell, giving the myofiber a striated appearance in the light microscope. Closer examination reveals that the dark bands, called *A bands,* are bisected by a dark region, the H zone, while the light bands, called *I bands,* are bisected by a different dark line, the Z *disk.* (The latter is also called the Z *line* because it appears as a line when seen in profile in electron micrographs.) The segment from one Z disk to the next, consisting of two halves of an I band and an A band, is termed a **sarcomere.**

A chain of sarcomeres, each about 2 μm long in resting muscle, constitutes a myofibril. The sarcomere is both the structural and the functional unit of skeletal muscle. During contraction, the sarcomeres are shortened to about 70 percent of their uncontracted, resting length. Electron microscopy and biochemical analysis have shown that each sarcomere contains two types of filaments: *thick filaments,* composed of myosin II, and *thin filaments,* containing actin (Figure 18-27). Near the center of the sarcomere, thin filaments overlap with the thick filaments in the *AI zone.*

Thick Filaments Salt extraction of skeletal muscle dissolves the thick filaments, causing loss of the A bands, but the thin filaments and I bands remain. Biochemical analysis shows that myosin but not actin is extracted, and dialysis of the extracted myosin into low-salt solutions reconstitutes thick filaments. Thus myosin is the primary structural component of the thick filaments. Because the myosin molecules

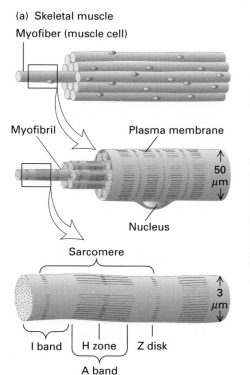

(a) Skeletal muscle

Myofiber (muscle cell)

Myofibril

Plasma membrane

50 μm

Nucleus

Sarcomere

3 μm

I band H zone Z disk

A band

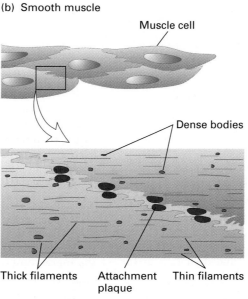

(b) Smooth muscle

Muscle cell

Dense bodies

Thick filaments   Attachment plaque   Thin filaments

◀ **FIGURE 18-26 General structure of skeletal and smooth muscle.** (a) Skeletal muscle tissue is composed of bundles of multinucleated muscle cells, or myofibers. Each muscle cell is packed with bundles of actin and myosin filaments, organized into myofibrils that extend the length of the cell. Packed end to end in a myofibril is a chain of sarcomeres, the functional units of contraction. The internal organization of the filaments gives skeletal muscle cells a striated appearance. (b) Smooth muscle is composed of loosely organized spindle-shaped cells that contain a single nucleus. Loose bundles of actin and myosin filaments pack the cytoplasm of smooth muscle cells. These bundles are connected to dense bodies in the cytosol and to the membrane at attachment plaques.

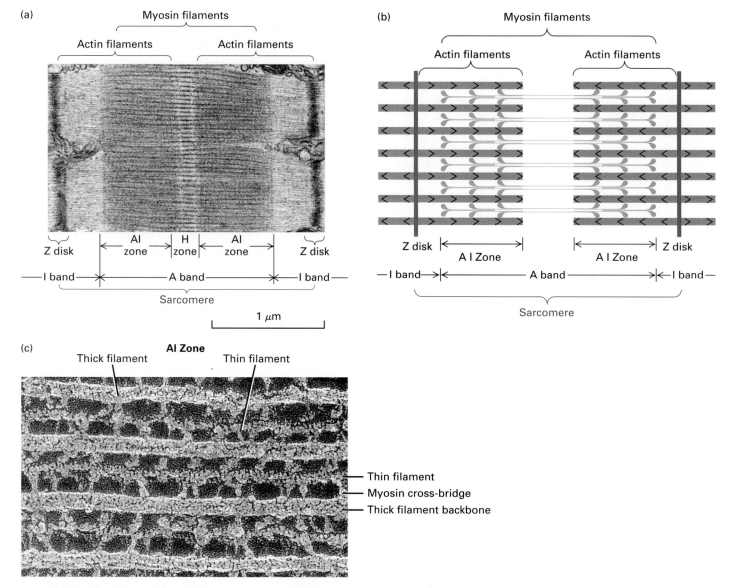

(a)

(b)

(c)

▲ FIGURE 18-27 Structure of the sarcomere. (a) Electron micrograph of mouse striated muscle in longitudinal section, showing one sarcomere. On either side of the Z disks are the lightly stained I bands, composed of actin filaments. These thin filaments extend from both sides of the Z disk to interdigitate with the dark-stained myosin thick filaments that make up the A band. The region containing both thick and thin filaments (the AI zone) is darker than the area containing only myosin thick filaments (the H zone). (b) Schematic diagram of a sarcomere.

The (+) ends of actin filaments are attached to the Z disks. (c) Electron micrograph showing actin-myosin cross-bridges in the AI zone of a striated flight muscle of an insect. This image shows a nearly crystalline array of thick myosin and thin actin filaments. The muscle was in the rigor state at preparation. Note that the myosin heads protruding from the thick filaments connect with the actin filaments at regular intervals. [Part (a) courtesy of S. P. Dadoune; part (c) courtesy of M. Reedy.]

composing a thick filament are oriented with their heads lying at the distal tips of the filament and their tails at the center, the filament is bipolar (see Figure 18-27b).

Thin Filaments The I band is a bundle of thin filaments. All the filaments in an I band are the same length, but can vary among different muscles. Biochemical studies show that a thin filament is basically an actin filament plus two additional proteins, tropomyosin and troponin, that are involved

in regulating actomyosin interactions. Evidence that actin is the main component of thin filaments is drawn from findings in three experiments. First, thin filaments appear identical with actin filaments reconstituted from purified G-actin monomers. Second, the thin filaments can be decorated with myosin S1 or HMM. Third, thin filaments but not thick filaments can be removed with the actin-severing protein gelsolin.

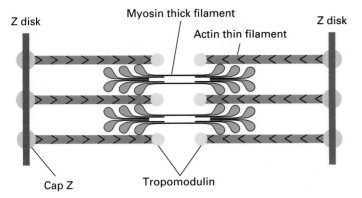

▲ FIGURE 18-28 Schematic diagram showing location of capping proteins that stabilize the ends of actin thin filaments. CapZ (green) caps the (+), or barbed, ends of filaments at the Z disk, and tropomodulin (yellow) caps the (−), or pointed, ends of thin filaments. The presence of these two proteins at opposite ends of a thin filament prevents actin subunits from dissociating during muscle contraction.

In electron micrographs, one end of a thin filament is associated with the Z disk, while the other end is near the center of the sarcomere (see Figure 18-27a, b). Myosin decoration experiments show that all thin filaments have the same polarity with respect to the Z disk: The barbed or (+) end of the filament is always closest to the Z disk. The heads of the myosin molecules protrude from the thick filaments, in the AI zone, forming cross-bridges with adjacent actin thin filaments (see Figure 18-27c).

The ends of a thin filament are associated with two actin-capping proteins, CapZ and tropomodulin (Figure 18-28). CapZ is present in the Z disk of skeletal muscle, where it helps prevent actin filaments from depolymerizing at their (+) end. CapZ probably also cross-links the (+) ends of actin filaments to other Z-disk proteins. At the opposite end of the thin filament lies tropomodulin, which protects the (−) end from depolymerization, as it does in the erythrocyte cytoskeleton. Capping at both ends causes thin filaments to be very stable.

The Z Disk The Z disk is a lattice of fibers whose major function is to anchor the (+) ends of actin filaments. How the filaments are attached is not certain, but scientists believe that the actin-capping protein CapZ and the actin cross-linking protein α-actinin play a role. A major component of isolated Z disks, α-actinin probably cross-links thin filaments in the I band, organizing them into a bundle of filaments.

Smooth Muscles Contain Loosely Organized Thick and Thin Filaments

A smooth muscle is composed of elongated spindle-shaped cells, each with a single nucleus. Although smooth muscle cells are packed with thick and thin filaments, these filaments are not organized into well-ordered sarcomeres and

myofibrils, as they are in skeletal muscle; for this reason, smooth muscle is not striated. Instead the filaments in smooth muscle are gathered into loose bundles, which are attached to dense bodies in the cytosol (see Figure 18-26b). Dense bodies apparently serve the same function as Z disks in skeletal muscle. The other end of the thin filaments in many smooth muscle cells is connected to *attachment plaques*, which are similar to dense bodies but are located at the plasma membrane of a muscle cell. Like a Z disk, an attachment plaque is rich in the actin-binding protein α-actinin; it also contains a second protein, *vinculin* (MW 130,000), not found in Z disks. Vinculin, which binds tightly to α-actinin in cell-free experiments, binds directly to an integral membrane protein in the plaque and to α-actinin, thereby attaching actin filaments to membrane adhesion sites.

Thick and Thin Filaments Slide Past One Another during Contraction

Previously, we examined the in vitro movement of myosin along actin filaments. These in vitro motility studies, combined with microscopy studies in the 1950s that showed thick and thin filaments did not change in length while the sarcomere shortened, led to a simple model of skeletal muscle contraction, called the *sliding-filament model* (Figure 18-29).

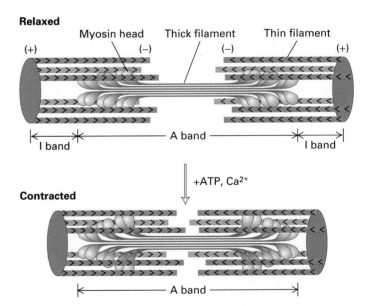

▲ FIGURE 18-29 The sliding-filament model of contraction in striated muscle. The arrangement of thick myosin and thin actin filaments in the relaxed state is shown in the top diagram. In the presence of ATP and Ca^{2+}, the myosin heads extending from the thick filaments pivot, pulling the actin thin filaments toward the center of the sarcomere. Because the thin filaments are anchored at the Z disks (purple), this movement shortens the sarcomere length in the contracted state *(bottom)*.

The central tenet of this model is that ATP-dependent interactions between thick filaments (myosin) and thin filaments (actin) generate a force that causes thin filaments to slide past thick filaments. The force is generated by the myosin heads of thick filaments, which form cross-bridges to actin thin filaments in the AI zone, where the two filament systems overlap. Subsequent conformational changes in these cross-bridges cause the myosin heads to walk along an actin filament, as discussed earlier. The sliding-filament model predicted that the force of contraction should be proportional to the overlap between the two filament systems.

To understand how a muscle contracts, consider the interactions between one myosin head (among the hundreds in a thick filament) and a thin filament as diagrammed in Figure 18-25. During these steps, also called the *cross-bridge cycle,* a myosin head has moved two subunits closer to the Z disk or the (+) end of the filament. Because the thick filament is bipolar, the action of the myosin heads at opposite ends of the thick filament draws the thin filaments toward the center of the thick filament and therefore toward the center of the sarcomere (see Figure 18-29). This movement shortens the sarcomere until the ends of the thick filaments abut the Z disk or the (−) ends of the thin filaments overlap at the center of the A band. Contraction of an intact muscle results from the activity of hundreds of myosin heads on a single thick filament, amplified by the hundreds of thick filaments in a sarcomere and thousands of sarcomeres in a muscle fiber.

The sliding-filament model, first proposed about 50 years ago, has been supported by more recent experimental results. In particular, the three-dimensional structure of the myosin head determined by x-ray crystallography and the force and step size measured during in vitro movement of single myosin molecules are compatible with the model (see Figures 18-22 and 18-23).

Titin and Nebulin Filaments Organize the Sarcomere

Muscle is elastic like a rubber band. Resting muscle can be stretched until the thick and thin filaments no longer overlap, developing a resisting force, or passive tension, which can be greater than the force normally developed by contracting muscle. If the stretching forces are removed, the muscle quickly resumes its normal resting length, and the regular arrangement of thick and thin filaments is restored. The source of this inherent elasticity was an enigma until scientists discovered a distinctive set of extremely long proteins, which organize the thick and thin filaments in their three-dimensional arrays and give muscle much of its elastic properties.

One component of this third filament system is the gigantic fibrous protein *titin* (also called *connectin*), which connects the ends of myosin thick filaments to Z disks and extends along the thick filament to the H zone (Figure 18-30). Titin appears to function like an elastic band, keeping the myosin filaments centered in the sarcomere when muscle

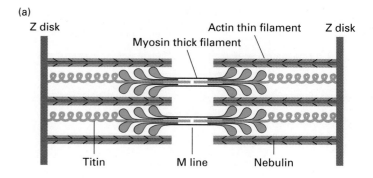

(a)

Z disk   Actin thin filament   Z disk

Myosin thick filament

Titin   M line   Nebulin

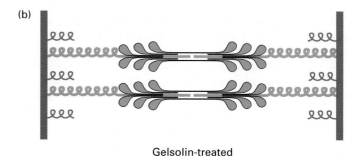

(b)

Gelsolin-treated

▲ **FIGURE 18-30 The titin-nebulin filament system stabilizes the alignment of thick and thin filaments in skeletal muscle.** (a) A titin filament attaches at one end to the Z disk and spans the distance to the middle of the thick filament. Thick filaments are thus connected at both ends to Z disks through titin. Nebulin is associated with a thin filament from its (+) end at the Z disk to its (−) end. The large titin and nebulin filaments remain connected to thick and thin filaments during muscle contraction and generate a passive tension when muscle is stretched. (b) To visualize the titin filaments in a sarcomere, muscle is treated with the actin-severing protein gelsolin, which removes the thin filaments. Without a supporting thin filament, nebulin condenses at the Z disk, leaving titin still attached to the Z disk and thick filament.

contracts or is stretched. As its name suggests, titin is huge ($\approx$2,700,000 MW; that is, over 25,000 amino acids); the protein is about 1 μm long, a length that spans half of a sarcomere. Most of its length is due to multiple repeats of immunoglobulin and fibronectin domains.

As noted earlier, salt treatment of muscle removes the thick filaments. Thin filaments, however, retain their regular organization in salt-treated muscle, suggesting that the lattice of thin filaments is maintained by a salt-resistant structure. Another large protein, called *nebulin* ($\approx$700,000 MW), is thought to perform this role. Nebulin forms long nonelastic filaments, consisting of a repeating actin-binding domain, that extend from each side of the Z disk and along the thin filaments (see Figure 18-30a). Each nebulin filament is as long as its adjacent actin filament; thus nebulin also may act as a molecular ruler by regulating the number of actin monomers that polymerize into each thin filament during the formation of mature muscle fibers. Treatment of muscle with

(a)

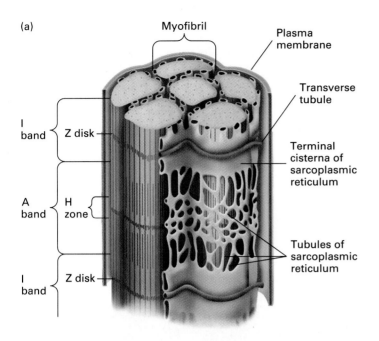

(b)

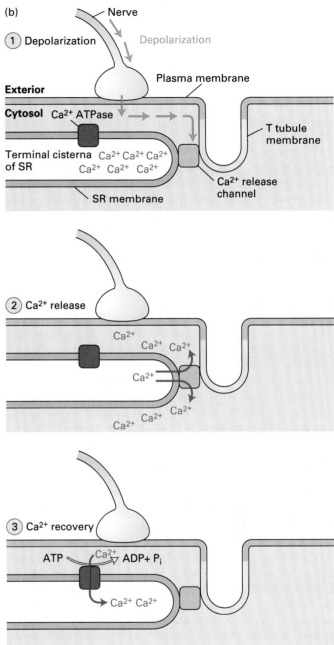

▲ **FIGURE 18-31 The sarcoplasmic reticulum (SR) regulates the cytosolic Ca^{2+} level in skeletal muscle.** (a) Three-dimensional drawing of a portion of a muscle cell (myofiber) composed of six myofibrils. The transverse (T) tubules, which are invaginations of the plasma membrane, enter myofibers at the Z disks, where they come in close contact with the terminal cisternae of the SR, forming triads. The terminal cisternae store Ca^{2+} ions and connect with the lacelike network of SR tubules that overlie the A band. (b) Release and recovery of Ca^{2+} ions by the SR. Depolarization of a muscle cell (step ①) induces the release of Ca^{2+} ions stored in the SR via Ca^{2+} release proteins in the SR membrane (step ②). Subsequently, Ca^{2+} ATPases in the SR membrane pump Ca^{2+} ions from the cytosol back into the SR, restoring the cytosolic Ca^{2+} concentration to its resting level within about 30 milliseconds (step ③).

the actin-severing protein gelsolin removes the thin filaments; without supporting thin filaments, nebulin filaments condense at the Z disk, leaving the titin filaments and myosin thick filaments (Figure 18-30b).

A Rise in Cytosolic Ca^{2+} Triggers Muscle Contraction

Like many cellular processes, muscle contraction is initiated by an increase in the cytosolic Ca^{2+} concentration. As described in Chapter 15, the Ca^{2+} concentration of the cytosol is normally kept low, below 0.1 μM. Nonmuscle cells maintain this low concentration by Ca^{2+} ATPases in the plasma membrane, which pump calcium out of the cell. In contrast, skeletal muscle cells maintain a low cytosolic Ca^{2+} level primarily by a unique Ca^{2+} ATPase that continually pumps Ca^{2+} ions from the cytosol into the sarcoplasmic reticulum (SR), a network of tubules in the muscle-cell cytosol. This activity establishes a reservoir of calcium in the SR.

As discussed in later chapters, when a nerve impulse reaches a skeletal muscle cell, it causes a change in the electric potential across the plasma membrane. Skeletal muscle cells can rapidly convert this electrical signal (called **depolarization**) into a rise in cytosolic Ca^{2+} (chemical signal), which then initiates contraction by a mechanism described later. The major anatomic features of this signaling pathway are invaginations of the plasma membrane, called *T (transverse) tubules,* that terminate next to the SR, forming structures called *triads* (Figure 18-31a). This system brings the membrane depolarization signal into the cytosol at a triad, where it stimulates the SR to release stored calcium into the cytosol through Ca^{2+} channels in the SR membrane (Figure 18-31b). Because of the close apposition of T tubules and SR membranes, depolarization of the plasma membrane induces an increase in cytosolic Ca^{2+} and contraction within milliseconds. Conversely, a muscle stops contracting when the channels close and Ca^{2+} is pumped back into the SR.

In smooth muscle cells, the SR membrane network is poorly developed and sparse, and much of the increase in cytosolic Ca^{2+} necessary for muscle contraction enters the cell via the plasma-membrane Ca^{2+} channel. As a result, changes in the cytosolic Ca^{2+} level occur much more slowly in smooth muscle than in skeletal muscle—on the order of seconds to minutes—thereby allowing the slow, steady response in contractile tension that is required by vertebrate smooth muscle.

Actin-Binding Proteins Regulate Contraction in Both Skeletal and Smooth Muscle

With the exception of cardiac muscle cells, which must contract rhythmically throughout the lifetime of the animal, all other muscles exhibit periods of activity followed by inactivity. In Chapters 20 and 21, we discuss in detail how cells recognize various external signals—electrical, hormonal, and chemical—that act to excite various intercellular responses, including the rise in cytosolic Ca^{2+} that initiates muscle contraction. So long as the Ca^{2+} concentration is sufficiently high and ATP is present in a muscle, the myosin-actin cross bridges will cycle continuously (see Figure 18-25), causing filament movement and contraction. Here we examine how actin-binding proteins mediate the Ca^{2+} signal to regulate skeletal and smooth muscle.

Role of Tropomyosin and Troponin in Skeletal Muscle Contraction In skeletal muscle, contraction is regulated by four accessory proteins on the actin thin filaments: tropomyosin and troponins C, I, and T. The cytosolic Ca^{2+} concentration influences the position of these proteins on the thin filaments, which in turn controls myosin-actin interactions.

Tropomyosin (TM) is a ropelike molecule, about 40 nm in length; TM molecules are strung together head to tail, forming a continuous chain along each actin thin filament (Figure 18-32a). Each TM molecule has seven actin-binding sites and binds to seven actin monomers in a thin filament. Associated with tropomyosin is *troponin* (TN), a complex of the three subunits, TN-T, TN-I, and TN-C. TN-C is the calcium-binding subunit of troponin. Similar in sequence to calmodulin and the myosin light chains, TN-C controls the position of TM on the surface of an actin filament through the TN-I and TN-T subunits.

Scientists currently think that, under the control of Ca^{2+} and TN, TM can occupy two positions on a thin filament—an "off" state and an "on" state. In the absence of Ca^{2+} (the off state), myosin can bind to a thin filament, but the TM-TN complex prevents myosin from sliding along the thin filament. Binding of Ca^{2+} ions to TN-C triggers a slight movement of TM toward the center of the actin filament

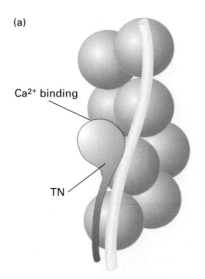

(a)

Ca²⁺ binding

TN

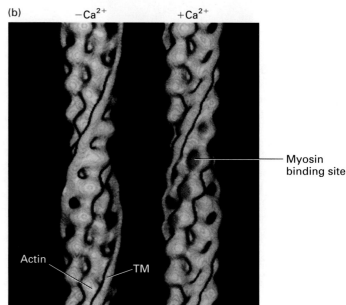

(b)   −Ca²⁺   +Ca²⁺

Myosin binding site

Actin   TM

▲ **FIGURE 18-32 Effect of Ca^{2+} ions on tropomyosin binding to actin filaments.** (a) Model of the tropomyosin-troponin (TM-TN) regulatory complex on a thin filament. TN, a clublike complex of TN-C, TN-I, and TN-T subunits, is bound to the long α-helical TM molecule. (b) Three-dimensional electron-microscopic reconstructions of the TM helix (yellow) on a thin filament from scallop muscle. TM in its "off" state (left) shifts to its new position (arrow) in the "on" state (right) when the Ca^{2+} concentration increases. This movement exposes myosin binding sites (red) on actin. (TN is not shown in this representation.) [Part (b) adapted from W. Lehman, R. Craig, and P. Vibert, 1993, *Nature* **123**:313; courtesy of P. Vibert.]

(the on state), which exposes the myosin-binding sites on actin (Figure 18-32b). Thus Ca^{2+} concentrations $> 10^{-6}$ M relieve the inhibition exerted by the TM-TN complex and contraction occurs. The Ca^{2+}-dependent cycling between on and off states in skeletal muscle is summarized in Figure 18-33a.

(a) Skeletal muscle

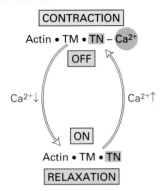

(b) Smooth muscle

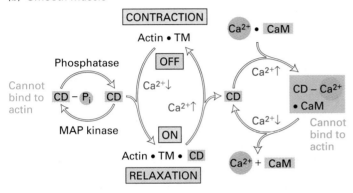

▲ **FIGURE 18-33 Ca²⁺-dependent mechanisms for regulating contraction in skeletal and smooth muscle.** (a) Regulation of skeletal muscle contraction by Ca²⁺ binding to TN. Note that the TM-TN complex remains bound to the thin filament whether muscle is relaxed or contracted. (b) Regulation of smooth muscle contraction by caldesmon. At low Ca²⁺ concentrations ($<10^{-6}$ M), caldesmon binds to TM and actin, reducing the binding of myosin to actin and keeping muscle in the relaxed state. At higher Ca²⁺ concentrations, a Ca²⁺-calmodulin complex binds to caldesmon, releasing it from actin; thus myosin can interact with actin and the muscle can contract. Phosphorylation by several kinases, including MAP kinase, and dephosphorylation by phosphatases also regulate caldesmon's actin-binding activity.

Role of Caldesmon in Smooth Muscle Contraction Several properties of smooth muscle account for its slow, steady contractile response. First, the actomyosin network is more disordered in vertebrate smooth muscle than in skeletal muscle. Second, as explained above, the cytosolic Ca²⁺ level rises and falls much more slowly in smooth muscle than in skeletal muscle. Finally, although smooth muscle contains tropomyosin (TM), it lacks troponin (TN), so the TM-TN system for rapidly turning skeletal muscle on and off cannot operate.

There are, in fact, several pathways that stimulate or inhibit smooth muscle contraction. One pathway, the smooth muscle equivalent of the TM-TN system, involves *caldesmon,* which binds to actin thin filaments at low Ca²⁺

concentrations (Figure 18-33b). Caldesmon (150,000 MW), an elongated protein about 75 nm in length, also interacts with the Ca²⁺-calmodulin complex at higher Ca²⁺ levels. Thus, when Ca²⁺ ions are in short supply, caldesmon forms a complex with TM and actin, thereby restricting the ability of myosin to bind to actin and preventing contraction. The binding of caldesmon to actin also is influenced by its phosphorylation by various kinases, including mitogen activated protein (MAP) kinase. The phosphorylated form of caldesmon does not bind well to thin filaments and is unable to inhibit myosin from binding to actin. During a prolonged contraction, MAP kinase activity is stimulated by pathways that signal through PK-C, Ras, and heterotrimeric G-proteins. As a result, MAP kinase directly stimulates smooth muscle contraction.

Myosin-Dependent Mechanisms Also Control Contraction in Some Muscles

So far we have examined control of actomyosin interactions by proteins that associate with actin filaments. However, smooth muscle and invertebrate skeletal muscle are also regulated by several mechanisms directed toward myosin rather than actin (Figure 18-34). In these muscles, Ca²⁺ activates myosin in two ways: by binding to the regulatory light chains of myosin or by stimulating calcium-dependent phosphorylation of those light chains. Various hormonal signals also activate or inhibit contraction of these muscles.

Calcium-Binding to the Regulatory Light Chain The simplest example of myosin-linked regulation is found in invertebrate muscle. Typically, in mollusks such as the scallop, the interaction between myosin heads and actin filaments is inhibited at low Ca²⁺ concentrations by the regulatory light chain (LC), one of the two pairs of LCs in the myosin neck region (see Figures 18-20 and 18-24). When the Ca²⁺ concentration rises, binding of Ca²⁺ ions to the regulatory LC in the neck region induces a conformational change in the myosin head that allows it to bind to actin; this, in turn, permits activation of the myosin ATPase and contraction of the muscle (see Figure 18-34a).

Activation of Myosin by Calcium-Dependent Phosphorylation Contraction of vertebrate smooth muscle is regulated primarily by a complex pathway involving phosphorylation and dephosphorylation of the myosin regulatory LC. As in mollusks, one of the two myosin LC pairs in smooth muscle inhibits actin stimulation of the myosin ATPase activity at low Ca²⁺ concentrations. This inhibition is relieved and the smooth muscle contracts when the regulatory LC is phosphorylated by the enzyme myosin LC kinase (see Figure 18-34b). This enzyme is activated by Ca²⁺; thus the Ca²⁺ level indirectly regulates the extent of LC phosphorylation and hence contraction. The Ca²⁺-dependent regulation of myosin LC kinase activity is mediated through calmodulin. Calcium first binds to calmodulin, and the Ca²⁺-calmodulin complex then binds to myosin LC

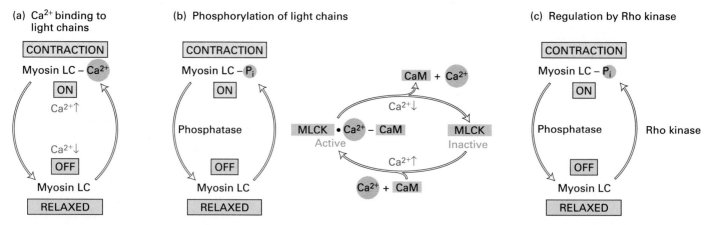

(a) Ca²⁺ binding to light chains   **(b)** Phosphorylation of light chains   **(c)** Regulation by Rho kinase

▲ **FIGURE 18-34 Three myosin-dependent mechanisms for regulating muscle contraction.** (a) In invertebrate muscle, binding of Ca²⁺ to the myosin regulatory light chain (LC) activates contraction. (b) In vertebrate smooth muscle, phosphorylation of the myosin regulatory light chains on site X by Ca²⁺-dependent myosin LC kinase activates contraction. At Ca²⁺ concentrations $< 10^{-6}$ M, the myosin LC kinase is inactive, and a myosin LC phosphatase, which is not dependent on Ca²⁺ for activity, dephosphorylates the myosin LC, causing muscle relaxation. (c) Activation of Rho kinase also leads to phosphorylation of the myosin regulatory LC at ser 19.

kinase and activates it. Because this mode of regulation relies on the diffusion of Ca²⁺ and the action of protein kinases, muscle contraction is much slower in smooth muscle than in skeletal muscle.

The role of activated myosin LC kinase can be demonstrated by microinjecting a kinase inhibitor into smooth muscle cells. Even though the inhibitor does not block the rise in the cytosolic Ca²⁺ level associated with membrane depolarization, injected cells cannot contract. The effect of the inhibitor can be overcome by microinjecting a proteolytic fragment of myosin LC kinase that is active even in the absence of Ca²⁺-calmodulin (this treatment also does not affect Ca²⁺ levels).

Activation of Myosin by Rho Kinase Unlike skeletal muscle, which is stimulated to contract solely by nerve impulses, smooth muscle is regulated by many types of molecules in addition to nervous stimuli. For example, factors such as norepinephrine, angiotensin, endothelin, and histamine as well as growth factors and hormones can modulate or induce contraction of smooth muscle by triggering various signal-transduction pathways that also affect nonmuscle cells (Chapter 20). These pathways regulate smooth muscle by modulating cytosolic Ca²⁺ levels and the activities of various enzymes, including myosin LC kinase and myosin phosphatase. As described above, for example, low levels of calcium stimulate myosin activity by activating myosin LC kinase. However, high levels of calcium inactivate the kinase through the action of calmodulin-dependent protein kinase II. More recent studies show how the Rho pathway stimulates myosin activity and leads to the formation of stress fibers. First, Rho kinase can phosphorylate myosin LC phosphatase and inhibit its activity. Because the phosphatase is inactivated, the level of myosin LC phosphorylation, and thus

myosin activity, increases. In addition, Rho kinase directly activates myosin by phosphorylating the regulatory light chain. Thus regulation of smooth muscle contraction is complex because it is responsive to many extracellular factors in addition to intracellular Ca²⁺ levels.

SUMMARY: Muscle: A Specialized Contractile Machine

• A skeletal muscle cell (myofiber) consists of multiple myofibrils. In each myofibril, actin thin filaments and myosin thick filaments are organized into a linear chain of highly ordered structures, called *sarcomeres* (see Figure 18-27a, b). One end of the thin filaments is attached to the Z disk, the demarcation between adjacent sarcomeres. Much of the length of the thin filaments overlaps the thick filaments.

• Two very large proteins, titin and nebulin, hold thin and thick filaments in the regular three-dimensional array of the sarcomere (see Figure 18-30).

• During contraction, each sarcomere shortens by as much as 30 percent as the thin actin filaments slide past the thick filaments (see Figure 18-29). ATP-dependent interactions between myosin heads and actin and the subsequent conformational change in the heads generates the sliding force that pulls the thin filaments toward the central A zone of each sarcomere.

• In a smooth muscle cell, actin and myosin filaments are packed in loose bundles rather than in highly ordered myofibrils. The filaments are attached to dense bodies in the cytosol and to the plasma membrane.

• A rise in the cytosolic Ca^{2+} concentration triggers muscle contraction. Stimulation of a skeletal muscle leads to the rapid release of stored Ca^{2+} from the SR into the cytosol (see Figure 18-31). Following stimulation of smooth muscle, the rise in cytosolic Ca^{2+} and hence contraction occurs much more slowly than in skeletal muscle.

• Both skeletal and smooth muscles contain proteins that interact with the actin filaments to regulate contraction (see Figure 18-33). In other muscles, contraction also is regulated by binding of Ca^{2+} to or phosphorylation of the myosin regulatory light chains, leading to activation or inhibition of the activity of the myosin thick filaments (see Figure 18-34).

18.5 Actin and Myosin in Nonmuscle Cells

Using our discussion of actin and myosin interactions in muscle cells as background, we now examine the function of actin-myosin structures in nonmuscle cells. At first, scientists thought that most cell movements were caused by a contractile mechanism similar to the sliding of actin and myosin filaments in muscle cells. This idea was based on several properties of at least some nonmuscle cells: the ability of cytosolic extracts to undergo contractile-like movements, the presence of actin and myosin II, and the existence of structures similar to muscle sarcomeres both in their organization and in their having ends anchored to the plasma membrane by proteins also found at the muscle Z disk.

Later biochemical studies led to the extraction of myosin I, which does not form thick filaments, from nonmuscle cells. As discussed earlier in the Chapter, 13 different myosins have been identified to date, but researchers have studied only 3—myosin I, myosin II, and myosin V—in most detail. In this section, we discuss some of the functions of these three myosins in various nonmuscle cells.

Actin and Myosin II Are Arranged in Contractile Bundles That Function in Cell Adhesion

Nonmuscle cells contain prominent bundles of actin and myosin II filaments in the cellular region that contacts the substratum or another cell. When isolated from cells, these bundles contract upon addition of ATP. The contractile bundles of nonmuscle cells differ in two ways from the noncontractile bundles of actin described earlier in this chapter (see Figure 18-5a). First, contractile bundles always are located adjacent to the plasma membrane like a sheet or belt, whereas noncontractile actin bundles form the core of membrane projections (microvilli and filopodia). Second, interspersed among the actin filaments of a contractile bundle is

myosin II, which is responsible for the contractility of the bundle. The noncontractile actin bundles in microvilli and filopodia sometimes are associated with myosin I, but myosin II usually is not a major component of these bundles. Despite their ability to contract, contractile bundles probably function primarily in cell adhesion rather than cell movement.

Contractile bundles in epithelial cells are most commonly found as a *circumferential belt,* which encircles the inner surface of the cell at the level of the adherens junction (Figure 18-35; see Figure 15-23). A circumferential belt resembles a primitive sarcomere in its organization and contains many proteins found in stress fibers and smooth muscle, including vinculin, tropomyosin, and α-actinin. As a complex with the adherens junction, the circumferential belt functions as a tension cable that can internally brace the cell and thereby control its shape. Contraction of the circumferential belt in epithelial cells surrounding a wound seals the gap in the sheet of cells and thus aids in wound healing.

As noted early in this chapter, long bundles of actin microfilaments, called *stress fibers,* lie along the ventral surfaces of cells cultured on artificial (glass or plastic) surfaces (see Figure 18-1b). Fluorescent-antibody techniques reveal that myosin and α-actinin are distributed in alternating patches in stress fibers, much like the pattern of alternating thick filaments and Z bands in muscle sarcomeres. Stress fibers also contain tropomyosin, caldesmon, and the regulatory protein

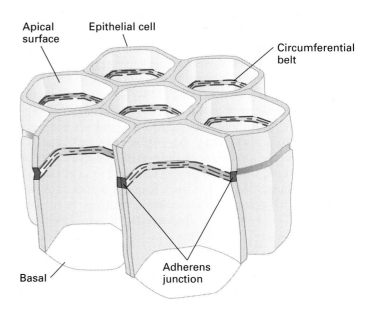

▲ FIGURE 18-35 The circumferential belt is located near the apical surface of epithelial cells. In epithelial tissue, a belt of actin and myosin filaments rings the inner surface of the cell adjacent to the adherens junctions, where cell-cell contacts are maintained. The circumferential belt is attached by linker proteins to cell-adhesion molecules in the plasma membrane (Chapter 22).

myosin LC kinase, which are found in smooth muscle. The ends of stress fibers terminate at focal adhesions, special structures that attach a cell to the underlying substratum. When stress fibers are separated from focal adhesions by cutting their ends with a laser beam, they contract on addition of ATP, thus demonstrating their contractility.

Although stress fibers are contractile and found in motile cells, several observations suggest that they function in cell adhesion rather than in movement. For one, migrating fibroblasts have few stress fibers during the time of rapid movement; however, once the cells stop migrating, the stress fibers increase in number. Also, when a cultured fibroblast is removed from its substratum, the cell becomes spherical and the stress fibers disappear. If the cell is returned to its substratum, stress fibers reappear within a few hours. In fact, stress fibers may be an artifact caused by culturing cells on glass or plastic surfaces, as they are rarely seen in cells in tissues. Apparently, the adhesion of cells to a substratum induces stresses on the cytoskeleton that cause the random assortment of actin and myosin filaments to align into stress fibers.

In Chapter 22, which covers the integration of cells into tissues, we describe the complex structure of adherens junctions and focal adhesions and how they are attached to contractile bundles.

Myosin II Stiffens Cortical Membranes

We saw earlier that cortical actin networks help support and stiffen the fluidlike plasma membrane. In addition to various actin cross-linking proteins, myosin II also is a component of the cortical cytoskeleton. Two observations support the hypothesis that myosin II molecules act as small tension rods that "tighten up" the cortical actin cytoskeleton. First, genetic and biophysical studies show that the membranes of mutant cells that lack myosin II are deformed more easily than the membranes of normal cells (Figure 18-36a). Second, capping experiments outlined in Fig 18-36b indicate that cortical myosin II is responsible for the movements of some cell-surface proteins. Although cell-surface proteins are normally immobile in the membrane, they will cluster, or "cap," at one region in the membrane in the presence of antibodies or lectins that bind them. Capping is inhibited in cells lacking myosin II, suggesting that myosin II in the cortex provides the force that aggregates the membrane proteins. In the last section of this chapter, we discuss how myosin-dependent tension in the cortex may contribute to cell locomotion.

Actin and Myosin II Have Essential Roles in Cytokinesis

Fluorescence microscopy shows that during mitosis actin and myosin II accumulate at the equator of a dividing cell, midway between the poles of the spindle (Figure 18-37a). There they align into a *contractile ring*, which is similar to a stress fiber or circumferential belt and encircles the cell. As division of the cytoplasm (cytokinesis) proceeds, the diameter of

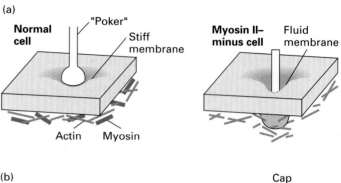

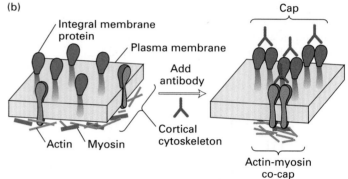

▲ **FIGURE 18-36 Myosin stiffening of the cortex.** (a) In cells containing a normal cytoskeleton, the membrane resists deformation when a bead-topped poker is pushed against it. In a mutant cell lacking myosin II, the membrane is easily deformed when the same force is applied with a poker. (b) Integral membrane proteins that are attached to the underlying cytoskeleton aggregate when bound by specific antibodies or lectins, forming a patch or "cap" on the cell surface. Actin and myosin filaments also collect beneath the membrane even though they are not bound by the antibodies or lecithin. In cells lacking functional myosin II, membrane proteins are unable to cap, suggesting that movement of membrane proteins depends on cortical myosin.

the contractile ring decreases, so that the cell is pinched into two parts by a deepening cleavage furrow. Dividing cells stained with antibodies against myosin I and myosin II show that myosin II is localized to the contractile ring, while myosin I is at the cell poles (Figure 18-37b). This localization indicates that myosin II but not myosin I is involved in cytokinesis.

Experiments in which active myosin II was eliminated from the cell demonstrated that cytokinesis is indeed dependent on myosin II (Figure 18-38). In one type of experiment, anti-myosin II antibodies were microinjected into one blastomere of a sea urchin embryo at the two-cell stage. In other experiments, expression of myosin II was inhibited in the *Dictyostelium* ameba by genetic deletion of the myosin gene or by antisense inhibition of myosin mRNA expression. The results were identical in all cases: cells lacking myosin II became multinucleated because cytokinesis but not karyokinesis (chromosome separation) was inhibited. Without myosin II, the cells failed to assemble a contractile ring, although other events in the cell cycle proceeded.

(a)

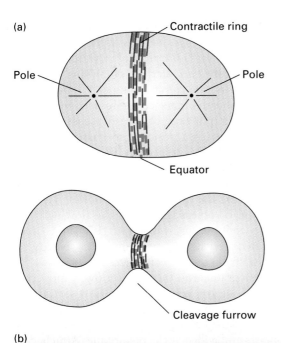

(b)

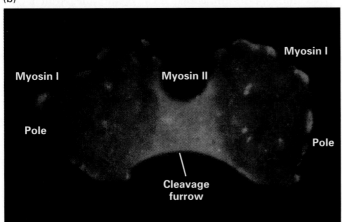

▲ FIGURE 18-37 Localization of myosin I and myosin II during cytokinesis. (a) As the spindle poles and aster microtubules separate during mitosis, myosin (blue) and actin (red) in the cortex of the cell assemble into an equatorial contractile ring around the cell. As the nuclei in the daughter cells start to re-form, the contractile ring constricts, causing the membrane to form a cleavage furrow. In the last step of cell division, cytokinesis, the cell is pinched into two parts. (b) During cell division of a *Dictyostelium* ameba, myosin II (red) is concentrated in the cleavage furrow, while myosin I (green) is localized at the poles of the cell. [Part (b) Courtesy of Y. Fukui.]

Membrane-Bound Myosins Power Movement of Some Vesicles

Among the many movements exhibited by cells, vesicle translocation has been one of the most fascinating to cell biologists. In early studies of the cytoplasm, researchers found that certain particles, now known to be membrane-bounded vesicles, moved in straight lines within the cytosol, sometimes stopping

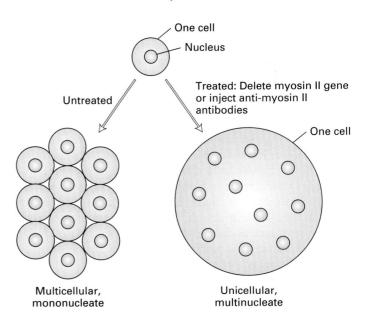

▲ FIGURE 18-38 Experimental demonstration that myosin II is required for cytokinesis. The activity of myosin II was inhibited either by deleting its gene or by microinjecting anti-myosin II antibodies into a cell. A cell that lacked myosin II was able to replicate its DNA and nucleus, but it failed to divide; this defect caused the cell to become large and multinucleate. In comparison, an untreated cell during the same period continued to divide and formed a multicellular ball of cells in which each cell contained a single nucleus.

and then resuming movement, at times after changing direction. This type of behavior could not be caused by diffusion because the movement was clearly not random. Therefore, researchers reasoned, there must be tracks, most likely actin filaments or microtubules, along which the particles travel, as well as some type of motor to power the movement.

Unlike cytokinesis, which involves myosin II, other motile processes, such as vesicle transport and membrane movements at the leading edge, possibly involve other myosins and, as discussed in the next chapter, some microtubule motor proteins also. Here we present evidence that some myosins, including myosins I and V, can move along an actin filament while carrying a membrane vesicle as cargo. In these processes, the interaction of actin and myosin differs from that in the sarcomere. We defer discussion of the role of myosin I in directing the movement of the leading edge of cells until the last section of this chapter.

Role of Myosins I and V in Moving Vesicles along Actin Filaments Studies with amebas provided the initial clues that myosin I participates in vesicle transport. Indeed the first myosin I molecule to be identified and characterized was from these organisms. The cDNA sequences of three myosin I genes have now been identified in *Acanthameba*, a common soil ameba. Using antibodies specific for each myosin I isoform, researchers found that they are localized to different membrane structures in the cell. For example, myosin IA is associated with small cytoplasmic vesicles. Myosin IC, by contrast, is found at the plasma membrane and at the contractile vacuole, a vesicle that regulates the osmolarity of the cytosol by fusing with the plasma membrane. The introduction of antibodies against myosin IC into a living ameba prevents transport of the vacuole to the membrane; as a result, the vacuole expands uncontrollably, eventually bursting the cell.

Myosin I is also implicated in vesicle transport in vertebrate cells. For example, in intestinal epithelial cells, myosin I co-purifies with vesicles derived from Golgi membranes. The presence of this motor on Golgi membranes suggests myosin I moves membrane vesicles between membrane compartments in the cytoplasm. In addition myosin I serves as a membrane-microfilament linkage in microvilli (see Figure 18-10), another example of a membrane-associated function.

Several types of evidence suggest that myosin V also participates in the intracellular transport of membrane-bounded vesicles. For example, mutations in the myosin V gene in yeast disrupt protein secretion and lead to an accumulation of vesicles in the cytoplasm. Vertebrate brain tissue is rich in myosin V, which is concentrated on Golgi stacks (Figure 18-39). This association with membranes is consistent with the effects of myosin V mutations in mice. Such mutations are implicated in defects in synaptic transmission and eventually cause death from seizures.

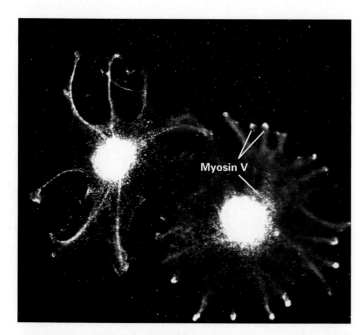

▲ **FIGURE 18-39 Immunofluorescence micrograph of astrocytes stained with a reagent specific for myosin V.** In these cells, which are found in nervous tissue, myosin V is concentrated in the Golgi stacks (bright central region) and at the tips of membrane processes that extend from the cell. This distribution indicates that myosin V is associated with membranes and suggests that the protein is involved in membrane transport from the Golgi to the cell periphery. [From E. M. Espreafico et al., 1992, *J. Cell Biol.* **119**:1541; courtesy of R. E. Cheney and M. Mooseker.]

Myosin-Generated Movements in Cytoplasmic Streaming Membrane-associated myosin also is critical in the phenomenon of *cytoplasmic streaming* in large, cylindrical green algae such as *Nitella* and *Chara*. In these organisms, the cytosol flows rapidly, at a rate approaching 4.5 mm/min, in an endless loop around the inner circumference of the cell (Figure 18-40a). The rapid flow of cytosol, sustained over millimeter-long distances, is a principal mechanism for distributing cellular metabolites, especially in large cells such as plant cells and amebas. This type of movement probably represents an exaggerated version of the smaller-scale movements exhibited during the transport of membrane vesicles.

Close inspection of objects caught in the flowing cytosol, such as the endoplasmic reticulum (ER) and other membrane-bounded vesicles, showed that the velocity of streaming increases from the cell center (zero velocity) to the cell periphery. This gradient in the rate of flow is most easily explained if the motor generating the flow lies at the membrane. In electron micrographs, bundles of actin filaments can be seen aligned along the length of the cell, lying across chloroplasts embedded at the membrane. Attached to the actin bundles are vesicles of the ER network (Figure 18-40b). The bulk cytosol is propelled by myosin attached to parts of the ER lying along the stationary actin filaments. Although the *Nitella* myosin has not been isolated, it must be one of the fastest known, because the flow rate of the cytosol in *Nitella* is at least 15 times faster than the rate produced by any other myosin.

SUMMARY Actin and Myosin in Nonmuscle Cells

- In nonmuscle cells, interactions of actin filaments with various myosins are important in such cellular functions as support, cytokinesis, and transport.

- Actin filaments and myosin II form contractile bundles that have a primitive sarcomerelike organization and function in cell adhesion. Common examples are the circumferential belt present in epithelial cells and stress fibers found along the ventral surface of cells cultured on plastic or glass surfaces. The latter rarely occur in cells in tissues and may be an artifact.

- Interaction of myosin II with cortical actin filaments helps stiffen the plasma membrane, reducing the likelihood of surface deformation.

- A transient actin–myosin II contractile bundle (the contractile ring) forms in dividing cells and pinches the cell into two halves during cytokinesis (see Figure 18-37).

- Myosins I and V power intracellular translocation of some membrane-limited vesicles along actin filaments. A similar process is responsible for cytoplasmic streaming although the identity of the myosin involved is unknown (see Figure 18-40).

(a)

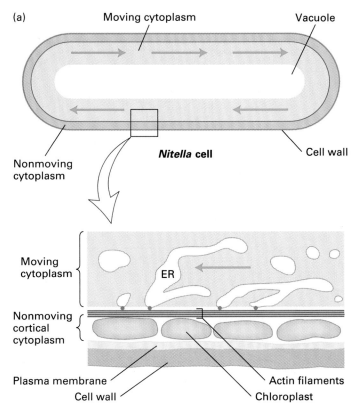

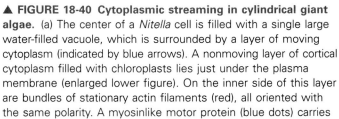

▲ **FIGURE 18-40 Cytoplasmic streaming in cylindrical giant algae.** (a) The center of a *Nitella* cell is filled with a single large water-filled vacuole, which is surrounded by a layer of moving cytoplasm (indicated by blue arrows). A nonmoving layer of cortical cytoplasm filled with chloroplasts lies just under the plasma membrane (enlarged lower figure). On the inner side of this layer are bundles of stationary actin filaments (red), all oriented with the same polarity. A myosinlike motor protein (blue dots) carries

(b)

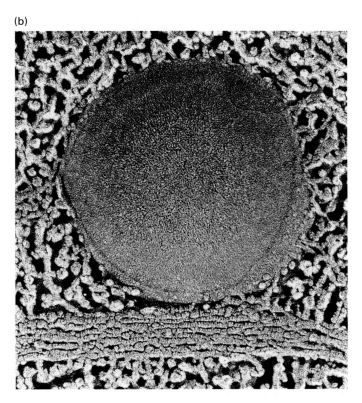

portions of the endoplasmic reticulum (ER) along the actin filaments. The movement of the ER network propels the entire viscous cytoplasm, including organelles that are enmeshed in the ER network. (b) An electron micrograph of the cortical cytoplasm shows a large vesicle connected to an underlying bundle of actin filaments. This vesicle, which is part of the endoplasmic reticulum (ER) network, contacts the stationary actin filaments and moves along them by a myosinlike motor. [Part (b) from B. Kachar.]

18.6 Cell Locomotion

We have now examined the different mechanisms used by cells to create movement—from the assembly of actin filaments and the formation of actin-filament bundles and networks to the contraction of bundles of actin and myosin and the sliding of single myosin molecules along an actin filament. These mechanisms are thought to represent the major processes whereby cells generate the forces needed to migrate. In this last section, we take a closer look at how cells employ the various force-generating processes to move across a surface. The major questions yet to be answered are how these actions of the cytoskeleton are coordinated and integrated.

Controlled Polymerization and Rearrangements of Actin Filaments Occur during Keratinocyte Movement

A moving keratinocyte and a moving fibroblast display the same sequence of changes in cell morphology—extension of

the cell membrane, attachment to the substratum, forward flow of cytosol, and retraction of the rear of the cell. These events are most easily observed in a fast-moving cell like a keratinocyte (Figure 18-41). The initial step, protrusion of a lamellipodium is accompanied by the controlled polymerization of actin filaments from sites at the leading edge and the subsequent cross-linking of those filaments into bundles and networks. But what propels the membrane forward? One hypothesis is that the membrane is extended forward by the pushing action of the polymerizing actin filaments, as in the extension of an acrosomal process or the movement of *Listeria* in the cytosol of an infected cell (Figure 18-42).

Whatever propels it, once the membrane is extended and the cytoskeleton is assembled, the membrane becomes firmly attached to the substratum. Time-lapse microscopy shows that actin bundles in the leading edge become anchored to the attachment site, which quickly develops into a focal adhesion. The attachment serves two purposes: it anchors the cell to the substratum and it prevents the leading lamella from retracting.

Direction of movement

Focal adhesion

① Extension

Lamellipodium

② Adhesion

New adhesion

Cell body movement

③ Translocation

④ De-adhesion

Old adhesion

▲ **FIGURE 18-41 Steps in keratinocyte movement.** In a fast-moving cell such as a fish epidermal cell, movement begins with extension of one or more lamellipodia from the leading edge of the cell (step ①); some lamellipodia adhere to the substratum via focal adhesions (step ②). Then the bulk of the cytoplasm in the cell body flows forward (step ③). The trailing edge of the cell remains attached to the substratum until the tail eventually detaches and retracts into the cell body (step ④). See the text for more discussion.

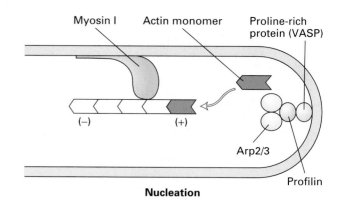

Myosin I   Actin monomer   Proline-rich protein (VASP)

(−)   (+)

Arp2/3

Profilin

Nucleation

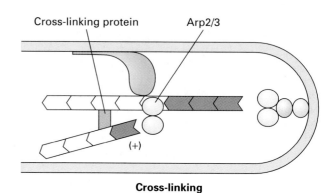

Cross-linking protein   Arp2/3

(+)

Cross-linking

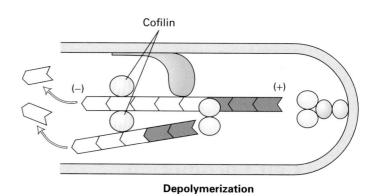

Cofilin

(−)   (+)

Depolymerization

▲ **FIGURE 18-42 A model of the molecular events at the leading edge of moving cells.** The polymerization of actin filaments at the (+) end, stimulated by profilin located at the leading-edge membrane, pushes the membrane outward. Other proteins like Vasp and Arp2/3 may participate in directing assembly. Simultaneously, cofilin induces the loss of subunits from the (−) ends of filaments. Arp2/3 and actin cross-linking proteins stabilize the actin filaments into networks and bundles. In addition, myosin I is thought to link actin filaments to the leading-edge plasma membrane.

After the forward attachments have been made, the bulk contents of the cell body are translocated forward. How this is accomplished is also unknown; one speculation is that the nucleus and the other organelles are embedded in the cytoskeleton and that myosin-dependent cortical contraction moves the cytoplasm forward. Finally, in the last step of movement (de-adhesion), the focal adhesions are broken and the rear of the cell, the tail, is brought forward. In the light microscope, the tail is seen to "snap" loose from its connections—perhaps by contraction of stress fibers in the tail or by elastic tension—but it leaves a little bit of its membrane behind, still firmly attached to the substratum.

Ameboid Movement Involves Reversible Gel-Sol Transitions of Actin Networks

Amebas are large, highly motile protozoans whose forward movement exhibits the same basic steps characterizing movement of keratinocytes (see Figure 18-41). Ameboid movement is initiated when the plasma membrane balloons forward to form a *pseudopodium,* or "false foot," which is similar to a lamellipodium in a vertebrate cell. As the pseudopodium attaches to the substratum, it fills with cytosol that is flowing forward through the cell. In the last step in movement, the rear of the ameba is pulled forward, breaking its attachments to the substratum.

Movement of an ameba is accompanied by changes in the viscosity of its cytosol, which cycles between sol and gel states. The central region of cytoplasm, the *endoplasm,* is a fluid-like sol, which flows rapidly toward the front of the cell, filling the pseudopodium. Here, the endoplasm is converted into the *ectoplasm,* a gel that forms the cortex, just beneath the plasma membrane. As the cell crawls forward, the ectoplasmic gel at the tail end of the cell is converted back into endoplasmic sol, only to be converted once again into ectoplasm when it again reaches the front of the cell. This cycling between sol and gel states continues only when the cell migrates.

The transformation between sol and gel states results from the disassembly and reassembly of actin microfilament networks in the cytosol. Several actin-binding proteins probably control this process and, hence, the viscosity of the cytosol. Profilin at the front of the cell promotes actin polymerization, and α-actinin and filamin form gel-like actin networks in the more viscous ectoplasm, as discussed earlier. Conversely, proteins like cofilin sever actin filaments to form the more fluid endoplasm.

Myosin I and Myosin II Have Important Roles in Cell Migration

Although the myosins provide the motive force for many cell movements, researchers are discovering the roles played by myosin I and myosin II in cell migration. Immunofluorescence micrographs of migrating amebas localize myosin I to the leading edge of the cell and myosin II to the rear (Figure 18-43a). This observation suggests that myosin I

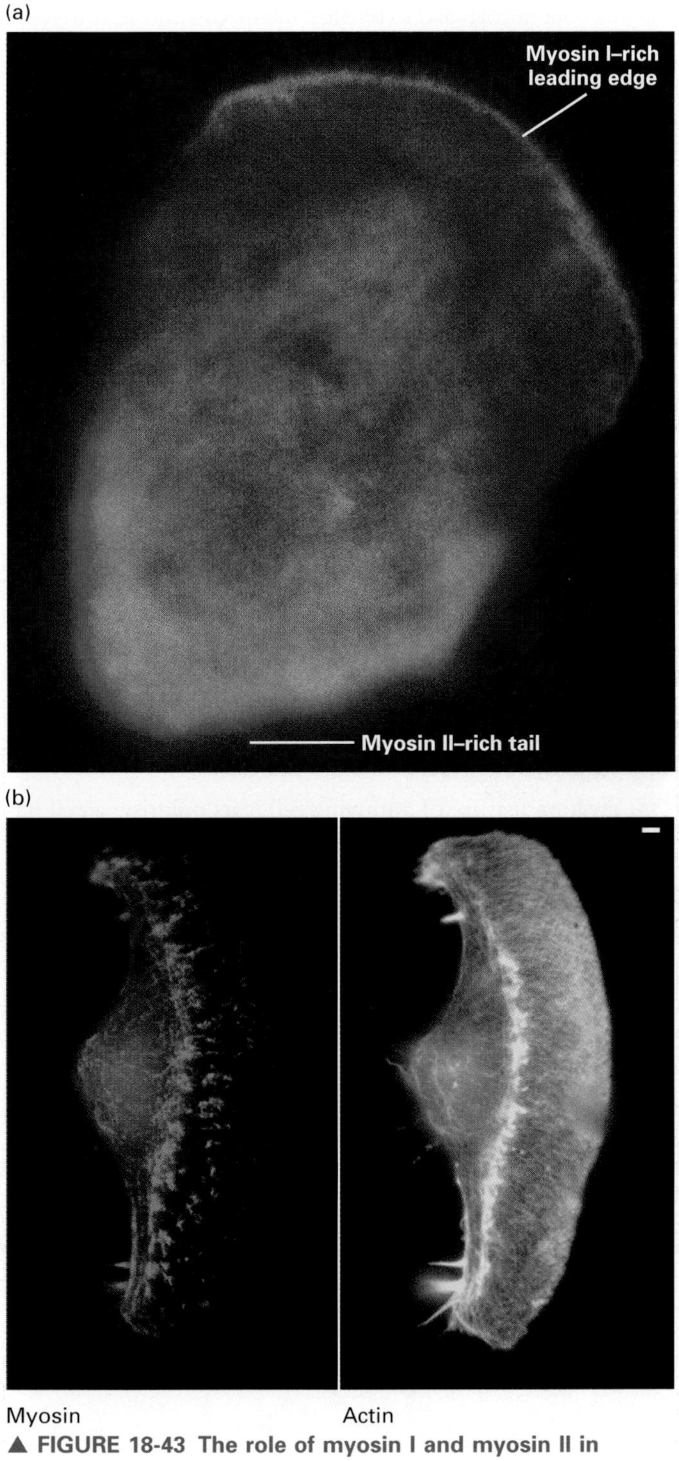

(a)

Myosin I–rich leading edge

Myosin II–rich tail

(b)

Myosin   Actin

▲ FIGURE 18-43 The role of myosin I and myosin II in locomoting cells. (a) In a moving ameba stained with antibodies to myosin I and myosin II, the myosin I (green) is located at the front, where active forward movement takes place. In the same cell, myosin II (red) is limited to the cortex and is enriched in the rear. (b) In a keratinocyte, myosin II is organized into a band located between the cell body and lamellipodia *(left)* and is largely absent from the actin-rich lamellipodia *(right)*. This finding suggests that myosin II does not play a major role in the extension phase of cell locomotion, but rather may participate in the translocation phase. [Part (a) courtesy of Y. Fukui; part (b) adapted from J. Verkhovsky.]

plays a role in forward extension of the cell and that myosin II functions in retraction of the cell body. Video-microscopy experiments with mutant amebas *(Dictyostelium)* provided evidence that myosin I and myosin II have some function in cell migration. Video micrographs show that a migrating cell normally sends out a single pseudopodium in the direction of movement. In contrast, mutant *Dictyostelium* cells lacking myosin I send out many pseudopodia, as if they were trying to move in all directions at once. This behavior makes it difficult to start sustained or persistent movement in one direction. By comparison, myosin II mutant cells crawl more slowly than wild-type cells. Adhesion experiments suggest that the mutant cell is less able to detract from the substrate.

The involvement of myosin-dependent cortical contraction in cell migration is supported by the localization of myosin II in a moving keratinocyte (Figure 18-43b). Immunofluorescence micrographs show that myosin II is organized into bands at the boundary between the lamellipodia and cell body. Contraction of the cortex is postulated to move the cell body forward during the retraction step.

Migration of Cells Is Coordinated by Various Second Messengers and Signal-Transduction Pathways

One striking feature of a moving cell is its polarity: a cell has a front and a back. When a cell makes a turn, a new leading lamella or pseudopodium forms in the new direction. If lamellae form in all directions, as in the myosin I mutants, then the cell is unable to pick a new direction of movement. To sustain movement in a particular direction, a cell requires signals to coordinate events at the front of the cell with events at the back and, indeed, signals to tell the cell where its front is. In this section, we present several examples of how external signals activate cell migration and control the direction of movement.

Activation of Filopodia, Membrane Ruffles, and Stress Fibers by Serum   Factors in a fresh wound stimulate a quiescent fibroblast to grow and divide by forming filopodia and lamellipodia at its leading edge and later to close the wound by assembling stress fibers and focal adhesions. These events require polymerization of actin filaments, activation of myosin molecules, and the assembly of actin bundles and networks. In Chapter 20 we describe in detail how binding of growth factors to cell-surface receptor tyrosine kinases triggers signal-transduction pathways. The cytoskeletal rearrangements that are a part of the wound healing response of fibroblasts involve pathways that are directed by three Ras-related proteins calls Rac, Rho, and Cdc42.

The roles of Ras-related proteins have been revealed by simple microinjection experiments. When Rac was microinjected into a fibroblast, the membrane immediately started to ruffle, and focal adhesions and stress fibers formed 5–10 minutes later. Injection of an inactive form of Rac inhibited all reorganization of the actin cytoskeleton when growth factors were added to the cell. When Rho, rather than Rac, was injected, it mimicked the mitogenic effects of lysophospholipid (also called *lysophosphatidic acid*, or *LPA*), a chemokine in serum and a potent stimulator of platelet aggregation. Both Rho and LPA induced the assembly of stress fibers and focal adhesions within 2 minutes but did not induce membrane ruffling.

These findings led to the hypothesis that extracellular factors trigger Ras-linked signal-transduction pathways that activate actin polymerization at the leading-edge membrane as an early event and formation of focal adhesions as a later event (Figure 18-44). If this model is correct, then the inhibition of stress-fiber assembly should not affect membrane

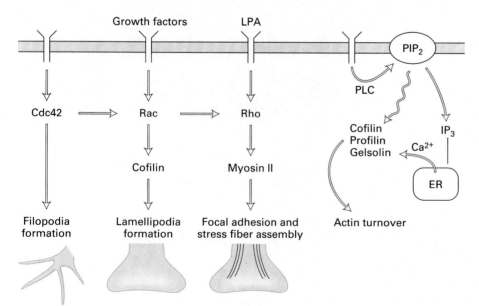

◀ **FIGURE 18-44 Role of signal-transduction pathways in cell locomotion and the organization of the cytoskeleton.** Extracellular signals are transmitted across the plasma membrane by receptors specific for different factors. One set of growth factors induces actin polymerization at the leading edge through a Rac-and Cdc42-dependent pathway *(left)*; another set of factors acts downstream through a Rho-dependent pathway to induce assembly of focal adhesions and cortical contraction *(center)*. Adhesion of a cell to the extracellular matrix triggers a parallel signaling pathway that induces activation of profilin, cofilin, and gelsolin *(right)*. Triggering of this pathway activates phospholipase C (PLC), which hydrolyzes PIP_2 in the membrane, and stimulates actin turnover.

ruffling. To test the model, Rac and ADP-ribosylase, an enzyme that inactivates Rho by covalently attaching ADP to it, were co-injected into a fibroblast. As predicted, membrane ruffles were formed, but the assembly of stress fibers was blocked. These observations suggest that Rho-dependent events like stress-fiber formation are "downstream" of control by Rac. Later experiments in which Cdc42 was microinjected into fibroblasts showed that this protein controlled an earlier step, the formation of filopodia. Thus the sequence of events in wound healing begins with the involvement of filopodia and lamellipodia during the locomotion of cells into the wound and the formation of focal adhesions and stress fibers to close the wound.

One important aspect of locomotion is how movement is coordinated in response to different stimuli. For example, the assembly of actin filaments at the membrane is enhanced by action of another signaling pathway involving PIP_2. As discussed previously, hydrolysis of PIP_2 by phospholipase C (PLC) releases profilin, cofilin, and gelsolin from the membrane. In addition, IP_3, a by-product of PIP_2 hydrolysis, stimulates release of Ca^{2+} ions from the endoplasmic reticulum into the cytosol; this increase in Ca^{2+} ions activates myosin II and the severing activity of gelsolin. This parallel pathway thus stimulates both actin severing and filament growth, thereby increasing actin turnover (Figure 18-44, *right*).

Steering of Cell Movements by Chemotactic Molecules

Under certain conditions, extracellular chemical cues guide the movement of a cell in a particular direction. In some cases, the movement is guided by insoluble molecules in the underlying substratum. In other cases, the cell senses soluble molecules and follows them, along a concentration gradient, to their source. This latter response is called *chemotaxis*. One of the best-studied examples of chemotaxis is the migration of *Dictyostelium* amebas along an increasing concentration of cAMP. Following cAMP to its source, the amebas aggregate into a slug and then differentiate into a fruiting body. Many other cells also display chemotactic movements. For example, leukocytes are guided by f-Met-Leu-Phe (*N*-formylmethionylleucylphenylalanine), a tripeptide secreted by bacterial cells.

Despite the variety of different chemotactic molecules— sugars, peptides, cell metabolites, cell-wall or membrane lipids—they all work through a common and familiar mechanism: binding to cell-surface receptors, activation of intracellular signaling pathways, and remodeling of the cytoskeleton through the activation or inhibition of various actin-binding proteins.

Coincident Gradients of Ca^{2+} and Chemotactic Molecules

When a cell's movement is directed by chemotaxis, how is the cell able to sense the difference between the concentrations of chemotactic molecules at the front and the back of the cell, a difference of only a few molecules? Optical microscopy using fluorescent dyes that act as internal Ca^{2+} sensors suggests that Ca^{2+} may play a role.

As a cell moves through a gradient of chemotactic molecules, a cytosolic gradient of Ca^{2+} is established, with the lowest concentration at the front of the cell and the highest concentration at the rear (Figure 18-45, *left*). This internal Ca^{2+} gradient can be disrupted by placing the cell in a new external chemotactic gradient. For example, if a pipette containing f-Met-Leu-Phe is placed to the side of a migrating leukocyte, diffusion of the peptide from the pipette will establish a new chemotactic gradient that immediately causes a general increase in the overall concentration of cytosolic Ca^{2+}. The Ca^{2+} gradient then reorients, with the lowest concentrations on the side of the cell closest to the pipette, causing the cell to turn toward the pipette (Figure 18-45, *right*). After the pipette is removed, the cell continues to move in the direction of its newly established Ca^{2+} gradient.

We have seen that many actin-binding proteins, including myosins I and II, gelsolin, α-actinin, and fimbrin, are

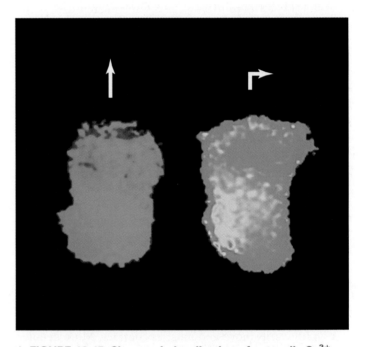

▲ **FIGURE 18-45 Changes in localization of cytosolic Ca^{2+} during cell location.** Fura-2, a fluorescent dye whose intensity changes at different Ca^{2+} levels, can be used to monitor the relative Ca^{2+} concentrations in moving cells. *(Left)* In a moving leukocyte, a Ca^{2+} gradient is established. The highest levels (green) are at the rear of the cell where cortical contractions take place, and the lowest levels (blue) are at the cell front where actin polymerization occurs. *(Right)* When the cell is induced to turn by placing a pipette filled with chemotactic molecules to the side of the cell, the Ca^{2+} concentration momentarily increases throughout the cytoplasm and a new gradient is established. The gradient is oriented such that the region of lowest Ca^{2+} (blue) lies in the direction the cell will turn, whereas a region of high Ca^{2+} (yellow) always forms at the site that will become the rear of the cell. [From R. A. Brundage et al., 1991, *Science* **254**:703; courtesy of F. Fay.]

regulated by Ca^{2+}. Hence the cytosolic Ca^{2+} gradient may regulate the sol-to-gel transitions that occur during cell movement. The low Ca^{2+} concentration at the front of the cell would favor the formation of actin networks by activating myosin I, inactivating actin-severing proteins, and reversing the inhibition of Ca^{2+}-regulated actin cross-linking proteins. The high Ca^{2+} concentration at the rear of the cell would cause actin networks to disassemble and a sol to form by activating gelsolin, or cause cortical actin networks to contract by activating myosin II. Thus an internal gradient of Ca^{2+} would contribute to the turnover of actin filaments in migrating cells.

SUMMARY Cell Locomotion

- Migrating cells undergo a series of characteristic changes in shape: extension of a lamellipodium or pseudopodium, adhesion of the extended leading edge to the substratum, forward flow (streaming of the cytosol), and retraction of the cell body (see Figure 18-41).

- Cell locomotion probably occurs through a common mechanism involving actin polymerization and myosin I–generated movement at the leading edge, assembly of adhesion structures, and cortical contraction mediated by myosin II.

- Assembly and organization of the cytoskeleton leading to cell locomotion are controlled by various external stimuli (e.g., growth factors and chemotactic molecules), which trigger intracellular signal-transduction pathways (see Figure 18-44), and by an internal Ca^{2+} gradient, marked by a low Ca^{2+} level at the leading edge and high Ca^{2+} level at the cell rear.

PERSPECTIVES for the Future

Future studies of the actin cytoskeleton will continue to focus on how cells move and muscles contract. More detailed x-ray crystallographic and electron-microscope analyses of the myosin head domain will soon reveal the key conformational states in generating force. However, the head domain contains only a part of the answer. The neck and tail domains, which differ substantially among the various myosins, hold the key to how the force is transmitted and regulated. In addition, the function of myosin is tied directly to its tail domain. How does a tail domain of myosin I or myosin V bind the membrane?

Additional studies of single myosin molecules as they hydrolyze ATP and slide along an actin filament will continue to stimulate exciting hypotheses. Some studies already hint that myosin and actin do not exhibit discrete structural states but a continuum of states. This range of states reflects the stochastic nature of molecular processes—that is, the structures of molecules fluctuate randomly. According to current paradigms, differences in structures should be exhibited by differences in function or activity. The quantal nature of the myosin step size is somewhat controversial. As more is learned about the behavior of single myosin molecules, today's models of force generation may evolve in unpredicted ways.

Studies of cell migration have been motivated in part by the desire to understand metastasis of transformed (cancer) cells. However, metastasis of cancer cells involves not only cell locomotion but also cell adhesion and cell signaling. Rapid progress is being made in identifying the components of the complicated protein circuits that control the assembly of the cytoskeleton and the regulation of cell movement. Just as we understand how nerve cells signal muscle contraction, we look forward to uncovering the molecular details of the pathways from extracellular signals to actomyosin contractions, actin polymerization, and the assembly of actin bundles and networks. Eventually, cell biologists will develop a much more complete description of how these complex mechanisms are integrated at the molecular level to produce the crawling movements of animal cells.

PERSPECTIVES in the Literature

Every new issue of a journal contains a report of a signaling pathway that controls the cytoskeleton. In this chapter, we discussed how Rac and Rho control the assembly of lamellipodia and stress fibers. Another GTPase, Cdc42, stimulates the assembly of filopodia. The paper below describes a protein, N-Wasp, that is implicated in interacting with the actin cytoskeleton. From your reading of this and related papers, describe a plausible pathway of how receptors at the cell membrane signal through N-WASP and other actin-binding proteins to stimulate the assembly of filopodia.

Miki, H., T. Sasaki, Y. Takai, and T. Takenawa. 1998. Induction of filopodium formation by a WASP-related actin-depolymerizing protein N-WASP. *Nature* 381:93–96.

Testing Yourself on the Concepts

1. In nonmuscle cells, actin filaments contribute to cell movement by two different mechanisms. What are these mechanisms, and what are examples of each?

2. In nonmuscle cells, filaments identical in structure and composed of identical types of actin subunits may be involved in different functions within the same cytoplasm. How is this possible?

3. There are at least 12 different types of myosin. What properties do all types share, and what makes them different?

4. Contraction of both skeletal and smooth muscle is triggered by an increase in cytosolic Ca^{2+}. Compare the mechanism by which each type of muscle converts the rise in Ca^{3+} into contraction.

MCAT/GRE-Style Questions

Key Concept Please read the section titled "Actin Polymerization Is Regulated by Proteins That Bind G-Actin" (p. 763) and answer the following questions:

1. Profilin normally binds to

a. The minus end of G-actin.

b. Thymosin β_4.

c. The ATP-binding cleft of G-actin.

d. The plus end of G-actin.

2. One of the main functions of profilin is to

a. Maintain actin as unassembled monomers.

b. Promote the conversion of ADP-actin to ATP-actin.

c. Inhibit the ATPase activity of G-actin.

d. Stabilize actin filaments.

3. Given the G-actin concentration in most cells, other cellular proteins must be required to ensure that

a. All of the cell's actin subunits are assembled into F-actin.

b. Some of the cell's actin subunits exist as unassembled G-actin.

c. All G-actin contains bound ADP.

d. All G-actin contains bound ATP.

4. PIP_2 is thought to inhibit cell movements by

a. Inhibiting the ATPase activity of F-actin.

b. Binding to profilin and gelsolin.

c. Promoting the ATPase activity of G-actin.

d. Preventing thymosin β_4 from binding to G-actin.

5. Injection of thymosin β_4 into living cells would be expected to

a. Increase the amount of F-actin.

b. Permanently stabilize existing F-actin.

c. Sever existing F-actin but not affect the total amount of F-actin.

d. Decrease the amount of F-actin.

Key Experiment Please read the sections titled "Myosin Heads Walk along Actin Filaments" (p. 770) and "Myosin Heads Move in Discrete Steps, Each Coupled to Hydrolysis of One ATP" (p. 771) and refer to figures 18-22 and 18-23; then answer the following questions:

6. The sliding-filament assay for myosin movement requires all of the following *except*:

a. That myosin is attached to the coverslip.

b. ADP.

c. Actin filaments.

d. A fluorescent microscope.

7. In the optical trap experiment described in Figure 18-23, the bead is used because

a. It is more refractive than myosin molecules or an actin filament.

b. It is less refractive than myosin molecules or an actin filament.

c. Unlike the proteins, it will not absorb heat from the laser light.

d. It can be attached easily to the coverslip.

8. During an optical trapping experiment like the one described in Figure 18-23, if the bead escapes from the trap one can infer that

a. The actin filament is generating greater force than the optical trap.

b. Myosin is generating greater force than the optical trap.

c. The optical trap is generating greater force than myosin.

d. The optical trap is generating greater force than the actin filament.

9. Analysis of different types of myosin molecules in the sliding-filament assay has demonstrated that

a. All move toward the minus end of an actin filament.

b. All move at the same rate.

c. Different myosins may move at different rates.

d. All bind different types of cargo.

10. If a myosin molecule hydrolyzes five molecules of ATP while moving along an actin filament, how far will the myosin travel?

a. The entire length of the filament.

b. Approximately 100 actin subunits.

c. 5 nm.

d. 25 nm.

Key Concept Please read the section titled "Skeletal Muscles Contain a Regular Array of Actin and Myosin" (p. 775) and answer the following questions:

11. The fundamental contractile unit of skeletal muscle is

a. The myofiber.

b. The sarcomere.

c. The myofibril.

d. The thick filament.

12. The sarcomere extends from

a. One end of the muscle cell to the other.

b. A band to Z disk.

c. A band to A band.

d. Z disk to Z disk.

13. Which of the following is closely associated with the Z disk?

a. The pointed or (−) end of each thin filament.

b. The bare zone of each thick filament.

c. The barbed or (+) end of each thin filament.

d. Tropomodulin.

14. Evidence that thick filaments are composed of myosin includes

a. Removal of thick filaments by gelsolin.

b. Decoration of thick filaments by CapZ.

c. Removal of thin filaments but not myosin by salt extraction.

d. Removal of thick filaments and myosin by salt extraction.

15. A defect in which of the following proteins might cause disruption of the I band without affecting the length of thin filaments:

a. Myosin.

b. CapZ.

c. α-Actinin.

d. Tropomodulin.

Key Terms

actin *752*	motor proteins *752*
critical concentration *761*	muscle *775*
cytoskeleton *751*	myofibers *775*
depolarization *779*	myofibrils *775*
intermediate filaments *752*	myosin *769*
locomotion *752*	polymerization *754*
microfilaments *752*	sarcomere *775*
microtubules *752*	steady state *761*
microvilli *760*	

References

General References

Bray, D. 1993. *Cell Movements.* Garland.

Web Sites

This site contains links to labs that study the cytoskeleton
http://expmed.bwh.harvard.edu/main/labs.html

A comprehensive list of cytoskeletal resources
http://expmed.bwh.harvard.edu/main/resources.html#cytoskeleton

The myosin home page
http://www.mrc-lmb.cam.ac.uk/myosin/myosin.html

The cytokinetic mafia home page—all things cytokinetic
http://www.unc.edu/depts/salmlab/mafia/mafia.html

Actin Cytoskeleton

Bretscher, A. 1999. Regulation of cortical structure by the ezrin-radixin-moesin protein family. *Curr. Opin. Cell Biol.* 11:109–116.

Furukawa, R., and M. Fechheimer. 1997. The structure, function, and assembly of actin filament bundles. *Int. Rev. Cytol.* 175:29–90

McGough, A. 1998. F-actin-binding proteins. *Curr. Opin. Struct. Biol.* 8:166–76

Sheterline, P., J. Clayton, and J. C. Sparrow. 1995. Protein profile. *Actin* 2:1–103.

Stradal, T., W. Kranewitter, S. J. Winder, and M. Gimona. 1998. CH domains revisited. *FEBS Lett.* 431:134–137.

Zigmond, S. H. 1998. Actin cytoskeleton: the Arp2/3 complex gets to the point. *Curr. Biol.* 8:R654–657.

Dynamics of Actin Assembly

Beckerle, M. C. 1998. Spatial control of actin filament assembly: lessons from *Listeria. Cell* 95:741–748

Kwiatkowski, D. J. 1999. Functions of gelsolin: motility, signaling, apoptosis, cancer. *Curr. Opin. Cell Biol.* 11:103–108.

Machesky, L. M., and M. Way. 1998. Actin branches out. *Nature* 394:125–126.

Theriot, J. A. 1997. Accelerating on a treadmill: ADF/cofilin promotes rapid actin filament turnover in the dynamic cytoskeleton. *J. Cell Biol.* 136:1165–1168.

Myosin: The Actin Motor Protein

Coluccio, L. 1997. Myosin I. *Am. J. Physiol.* 273:C347–C359.

Mermall, V., P. L. Post, and M. S. Mooseker. 1998. Unconventional myosin in cell movement, membrane traffic, and signal transduction. *Science* 279:527–533.

Rayment, I. 1996. The structural basis of the myosin ATPase activity. *J. Biol. Chem.* 271:15850–15853.

Simmon, R. 1996. Molecular motors: single-molecule mechanics. *Curr. Biol.* 6:392–394.

Muscle: A Specialized Contractile Machine

Bresnick, A. R. 1999. Molecular mechanisms of nonmuscle myosin-II regulation. *Curr. Opin. Cell Biol.* 11:26–33.

Gregorio, C. C., H. Granzier, H. Sorimachi, and S. Labeit. 1999. Muscle assembly: a titanic achievement? *Curr. Opin. Cell Biol.* 11:18–25.

Squire, J. M., and E. P. Morris. 1998. A new look at thin filament regulation in vertebrate skeletal muscle. *FASEB J.* 12:761–771

Actin and Myosin in Nonmuscle Cells

Field, C., R. Li, and K. Oegema. 1999. Cytokinesis in eukaryotes: a mechanistic comparison. *Curr. Opin. Cell Biol.* 11:68–80.

Rappaport, R. 1996. *Cytokinesis in Animal Cells.* Cambridge University Press.

Cell Locomotion

Hall, A. 1998. Rho GTPases and the actin cytoskeleton. *Science* 279:509–514.

Pettit, E. J., and F. S. Fay. 1998. Cytosolic free calcium and the cytoskeleton in the control of leukocyte chemotaxis. *Physiol. Rev.* 78:949–967.

Welch, M. D., A. Mallavarapu, J. Rosenblatt, and T. J. Mitchison. 1997. Actin dynamics in vivo. *Curr. Opin. Cell. Biol.* 9:54–61.

Cell Motility and Shape II: Microtubules and Intermediate Filaments

19

In Chapter 18, we looked at microfilaments, the smallest of the three types of cytoskeletal fibers, and their associated proteins. This chapter focuses on **microtubules** and **intermediate filaments** (IFs)—the other two cytoskeletal systems involved in cell motility and the determination of cell shape. Like microfilaments, both microtubules and intermediate filaments are long protein polymers. In the cell, microtubules and intermediate filaments fill the cytosol, spanning the distance between the nucleus and the cell membrane. In many places, the two types of cytoskeletal fibers overlap and follow each other through the cytosol; however, in spite of their similar distributions, they have different structures that carry out different functions in the cell.

Microtubules are responsible for various cell movements. Examples include the beating of cilia and flagella, the transport of membrane vesicles in the cytoplasm, and, in some protists, the capture of prey by spiny extensions of the surface membrane. These movements result from the polymerization and depolymerization of microtubules or the actions of microtubule motor proteins. Some other cell movements, such as the alignment and separation of chromosomes during meiosis and mitosis, involve both processes. Microtubules also direct the migration of nerve-cell axons by promoting the extension of the neuronal growth cone (Chapter 23).

In contrast to the motile functions of microtubules, the function of intermediate filaments is strictly structural. Intermediate filaments are usually

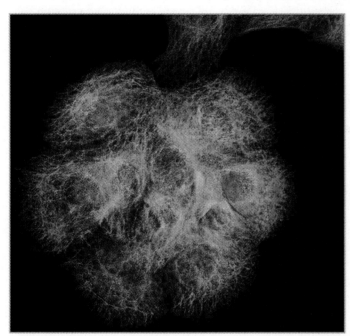

Tubulin and keratin in macrophages.

attached at one end to the plasma membrane through certain cell junctions or to integral membrane proteins through IF-binding proteins. Intermediate filaments integrate cells into tissues.

Thus there are many similarities and differences between the microfilament, microtubule, and IF cytoskeletal systems. In this chapter, we build on the general principles learned in the last chapter about the structure and function of the microfilament cytoskeleton and show how many of the same concepts also apply to microtubules and intermediate filaments. We begin the chapter by examining the structure and assembly of microtubules and then discuss how microtubule assembly and microtubule motor proteins can power cell movements. The discussion of microtubules concludes with a detailed examination of two important cell movements, the beating of flagella and cilia and the translocation of chromosomes during mitosis. In the last section, we consider the structure of intermediate filaments and their connection with other cell components. Intermediate filaments, which are found only in multicellular organisms, have an important role in organizing and strengthening tissues. We will thus revisit them in Chapter 22.

19.1   Microtubule Structures

A microtubule is a polymer of globular **tubulin** subunits, which are arranged in a cylindrical tube measuring 24 nm in diameter—more than twice the width of an intermediate filament and three times the width of a microfilament (Figure 19-1). Varying in length from a fraction of a micrometer to hundreds of micrometers, microtubules are much stiffer than either microfilaments or intermediate filaments because of their tubelike construction.

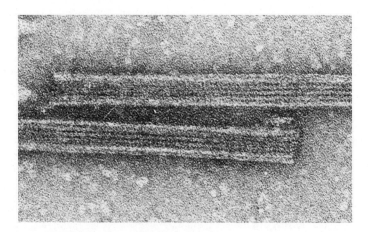

▲ **FIGURE 19-1 Electron micrograph of a negatively stained microtubule.** Globular tubulin subunits, each about 8 nm long, form the walls of this cylindrical structure. [Courtesy of E. M. Mandelkow.]

Heterodimeric Tubulin Subunits Compose the Wall of a Microtubule

The building block of a microtubule is the tubulin subunit, a heterodimer of α- and β-tubulin. Both of these 55,000-MW monomers are found in all eukaryotes, and their sequences are highly conserved. Although a third tubulin, γ-tubulin, is not part of the tubulin subunit, it probably nucleates the polymerization of subunits to form $\alpha\beta$-microtubules. Encoded by separate genes, the three tubulins exhibit an interesting, but not yet understood, homology with a 40,000-MW bacterial GTPase, called *FtsZ*. This bacterial protein has structural and functional similarities with tubulin, including the ability to polymerize and a role in cell division. Perhaps the protein carrying out these ancestral functions in bacteria was modified during evolution to fulfill the diverse roles of microtubules in eukaryotes.

The interactions holding α-tubulin and β-tubulin in a heterodimeric complex are strong enough that a tubulin subunit rarely dissociates under normal conditions. Each tubulin subunit binds two molecules of GTP. One GTP-binding site, located in α-tubulin, binds GTP irreversibly and does not hydrolyze it, whereas the second site, located on β-tubulin, binds GTP reversibly and hydrolyzes it to GDP. The second site is called the *exchangeable site* because GDP can be displaced by GTP. The recently solved atomic structure of the tubulin subunit reveals that the nonexchangeable GTP is trapped at the interface between the α- and β-tubulin monomers, while the exchangeable GTP lies at the surface of the subunit (Figure 19-2a). As we discuss later, the guanine bound to β-tubulin regulates the addition of tubulin subunits at the ends of a microtubule.

In a microtubule, lateral and longitudinal interactions between the tubulin subunits are responsible for maintaining the tubular form. Longitudinal contacts between the ends of adjacent subunits link the subunits head to tail into a linear *protofilament*. Within each protofilament, the dimeric subunits repeat every 8 nm. Through lateral interactions, protofilaments associate side by side into a sheet or cylinder—a microtubule. The exact arrangement of protofilaments in the wall of a microtubule is currently debated. In the model shown in Figure 19-2b, the heterodimers in adjacent protofilaments are staggered only slightly, forming spiraling rows of α- and β-tubulin monomers in the microtubule wall. In an alternative model, the α-tubulin and β-tubulin subunits are staggered enough to give the microtubule wall a checkerboard pattern.

Virtually every microtubule in a cell is a simple tube, a *singlet* microtubule, built from 13 protofilaments. In rare cases, singlet microtubules contain more or fewer protofilaments; for example, certain microtubules in the neurons of nematode worms contain 11 or 15 protofilaments. In addition to the simple singlet structure, doublet or triplet microtubules are found in specialized structures such as cilia and flagella (doublet microtubules) and centrioles and basal bodies (triplet microtubules). Each of these contains one

(a)

α-Tubulin   β-Tubulin

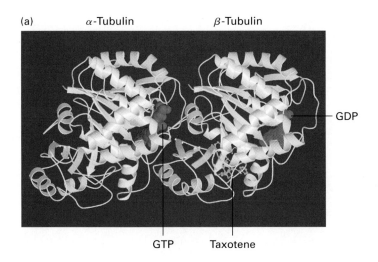

GTP   Taxotene

(b)

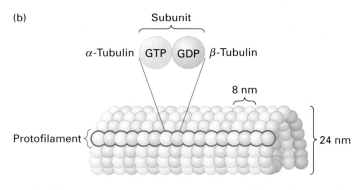

▲ **FIGURE 19-2 Microtubule structure.** (a) Ribbon diagram of the dimeric tubulin subunit, showing the α-tubulin and β-tubulin monomers and their bound nonexchangeable GTP (red) and exchangeable GDP (blue) nucleotides. An anticancer drug, taxotere (green), was used in the structural studies to stabilize the dimer structure. (b) The organization of tubulin subunits in a microtubule. The subunits are aligned end to end into a protofilament (magenta highlight). The side-by-side packing of protofilaments forms the wall of the microtubule. In this model, the protofilaments are slightly staggered so that α-tubulin in one protofilament is in contact with α-tubulin in the neighboring protofilaments. In an alternative model, the protofilaments are staggered by one-half subunit, forming a checkerboard pattern. In either structure, the microtubule displays a structural polarity in that addition of subunits occurs preferentially at one end, designated the (+) end. [Part (a) modified from E. Nogales, S. G. Wolf, and K. H. Downing, 1998, *Nature* **391**:199; courtesy of E. Nogales. Part (b) adapted from Y. H. Song and E. Mandelkow, 1993, *Proc. Nat'l. Acad. Sci. USA* **90**:1671.]

complete 13-protofilament microtubule (the A tubule) and one or two additional tubules (B and C) consisting of 10 protofilaments (Figure 19-3).

A microtubule is a polar structure, its polarity arising from the head-to-tail arrangement of the α- and β-tubulin dimers in a protofilament. Because all protofilaments in a microtubule have the same orientation, one end of a microtubule is ringed by α-tubulin, while the opposite end is ringed by β-tubulin. Microtubule-assembly experiments discussed

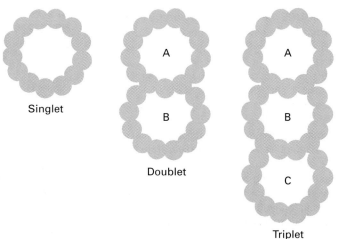

▲ **FIGURE 19-3 Arrangement of protofilaments in singlet, doublet, and triplet microtubules.** In cross section, a typical microtubule, a singlet, is a simple tube built from 13 protofilaments. In a doublet microtubule, an additional set of 10 protofilaments forms a second tubule (B) by fusing to the wall of a singlet (A) microtubule. Attachment of another 10 protofilaments to the B tubule of a doublet microtubule creates a C tubule and a triplet structure.

later show that microtubules, like actin microfilaments, have a (+) and a (−) end, which differ in their rates of assembly.

Microtubules Form a Diverse Array of Both Permanent and Transient Structures

As with microfilaments, there are two populations of microtubules: stable, long-lived microtubules and unstable, short-lived microtubules. Unstable microtubules are found when cell structures composed of microtubules need to assemble and disassemble quickly. For example, during mitosis, the cytosolic microtubule network characteristic of interphase cells disappears, and the tubulin from it is used to form the spindle-shaped apparatus that partitions chromosomes equally to the daughter cells (Figure 19-4a). When mitosis is complete, the spindle disassembles and the interphase microtubule network reforms.

In contrast to these short-lived, transient structures, some cells, usually nonreplicating cells, contain stable microtubule-based structures. These include the axoneme in the flagellum of sperm and the marginal band of microtubules in most red blood cells and platelets. Another example occurs in nerve cells (neurons), which are long-lived and seldom need to establish new connections in an adult. Neurons, however, must maintain long processes, called **axons,** and do so with the aid of an internal core of stable microtubules (Figure 19-4b). The disassembly of such stable structures would have catastrophic consequences—sperm would be unable to swim, a red blood cell would lose its springlike pliability, and axons would retract.

▶ **FIGURE 19-4 Diverse microtubule structures in cells.** (a) In cultured animal cells at interphase, microtubules are arranged in long fibers, which fill the entire cytosol. In a cell undergoing mitosis (inset), the network of micro-tubules disappears and is replaced with the spindle-shaped arrangement of microtubules in the mitotic apparatus. (b) Microtubules and intermediate filaments in a quick-frozen frog axon visualized by the deep-etching technique. There are several 24-nm-diameter microtubules running longitudinally. Thinner, 10-nm-diameter intermediate filaments also run longitudinally and form occasional connections with microtubules. [Part (a) courtesy of B. R. Brinkley and B. Scott, Baylor College of Medicine; part (b) from N. Hirokawa, 1982, *J. Cell Biol.* **94**:129; courtesy of N. Hirokawa.]

(a)

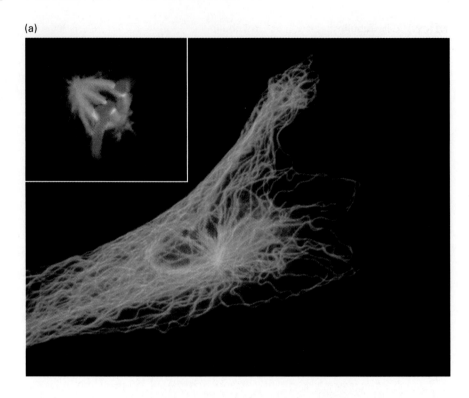

(b)

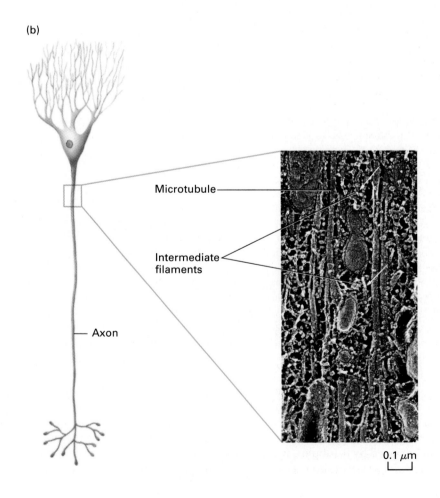

Microtubule

Intermediate filaments

Axon

0.1 μm

Microtubules Assemble from Organizing Centers

In an interphase fibroblast cell, a seemingly haphazard and random network of microtubules permeates the entire cytosol. However, upon closer analysis, we can see that the cytosolic microtubules are arranged in a hub-and-spoke array that lies at the center of a cell (Figure 19-5a). The microtubule spokes radiate from a central site occupied by the **centrosome**, which is the primary **microtubule-organizing center (MTOC)** in many interphase cells. We will use the term *MTOC* to refer to any of the structures used by cells to nucleate and organize microtubules. In animal cells, the MTOC is a centrosome, a lattice of microtubule-associated proteins that sometimes but not always contains a pair of centrioles (Figure 19-5b). The centrioles, each a pinwheel array of triplet microtubules, lie in the center of the MTOC but do not make direct contact with the ends of the cytosolic microtubules. Centrioles are not present in the MTOCs of plants and fungi; moreover, some epithelial cells and newly fertilized eggs from animals also lack centrioles. Thus, it is the associated proteins in an MTOC that have the capacity to organize cytosolic microtubules.

The mechanism whereby the MTOC organizes cytosolic microtubules was deduced from several microtubule-assembly studies. In cells treated with *colcemid*, a microtubule-depolymerizing drug, almost all the cytosolic microtubules, except those in the centrosome, are depolymerized. When colcemid is removed by washing the cells with a colcemid-free culture medium, tubulin repolymerizes to form new microtubules, which radiate from the MTOC (Figure 19-6). This result could arise by either of two mechanisms: the MTOC could nucleate polymerization of tubulin subunits or it could gather together the ends of microtubules that assembled independently in the cytosol. To identify the correct mechanism, centrosomes were purified and tested for their interaction with tubulin subunits or microtubules. The addition of purified centrosomes to a solution of tubulin dimers nucleated the assembly of microtubules whose (−)

(a)

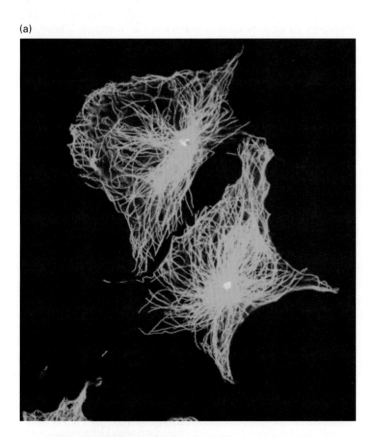

(b)

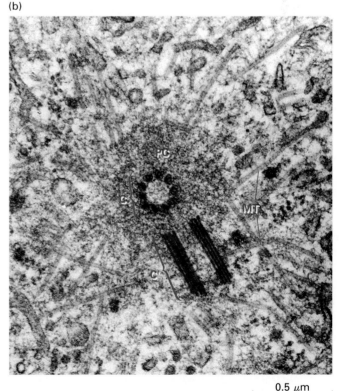

0.5 μm

▲ **FIGURE 19-5 Microtubule-organizing center.** (a) Fluorescence micrograph of a Chinese hamster ovary cell stained with antibodies specific for tubulin and a centrosomal protein. The microtubules (green) are seen to radiate from a central point, the microtubule-organizing center (MTOC), near the nucleus. The MTOC (yellow) is detected with an antibody to Cep135, a protein in the pericentriolar material. (b) Electron micrograph of the MTOC in an animal cell. The pair of centrioles (red), C and C′, in the center are oriented at right angles; thus one is seen in cross section, and one longitudinally. Surrounding the centrioles is a cloud of material, the pericentriolar (PC) matrix, which contains γ-tubulin and pericentrin. Embedded within the MTOC, but not contacting the centrioles, are the (−) ends of microtubules (MT; yellow). [Part (a) courtesy of R. Kuriyama; part (b) from B. R. Brinkley, 1987, in *Encyclopedia of Neuroscience*, vol. II, Birkhauser Press, p. 665; courtesy of B. R. Brinkley.]

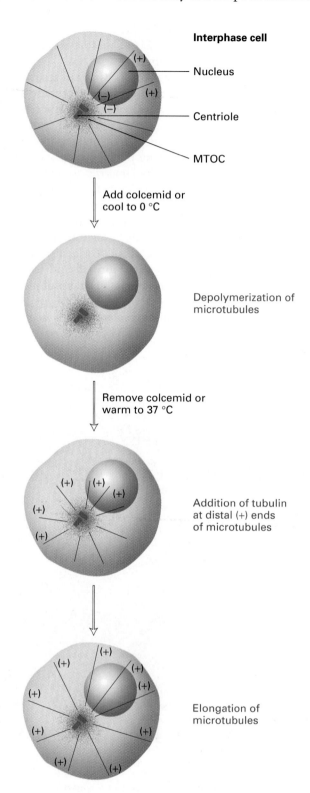

Interphase cell

Nucleus

Centriole

MTOC

Add colcemid or cool to 0 °C

Depolymerization of microtubules

Remove colcemid or warm to 37 °C

Addition of tubulin at distal (+) ends of microtubules

Elongation of microtubules

▲ **FIGURE 19-6 The disassembly and reassembly of microtubules in interphase cultured animal cells can be induced either by adding and subsequently removing colcemid or by cooling to 0 °C and subsequently rewarming to 37 °C.** Both the addition and the removal of tubulin occurs at the (+) ends of the microtubules.

ends remained associated with the centrosome. In the absence of centrosomes, the concentration of dimers was too low to permit spontaneous formation of microtubules. Thus the centrosome functions to nucleate the assembly of cytosolic microtubules.

Most Microtubules Have a Constant Orientation Relative to MTOCs

The MTOC, which probably is the major organizing structure in a cell, helps determine the organization of microtubule-associated structures and organelles (e.g., mitochondria, the Golgi complex, and the endoplasmic reticulum). In a nonpolarized animal cell such as a fibroblast, an MTOC is perinuclear and strikingly at the center of the cell. Although the centering mechanism is not well understood, it most likely involves microtubules that scout out the cell periphery. Because microtubules assemble from the MTOC, microtubule polarity becomes fixed in a characteristic orientation. In most animal cells, for instance, the (−) ends of microtubules are closest to the MTOC or basal body (Figure 19-7a). During mitosis, the centrosome duplicates and migrates to new positions flanking the nucleus. There the centrosome becomes the organizing center for microtubules forming the mitotic apparatus, which will separate the chromosomes into the daughter cells during mitosis (Figure 19-7b). The microtubules in the axon of a nerve cell, which help stabilize the long process, are all oriented in the same direction (Figure 19-7c).

In contrast to the single perinuclear MTOC present in most interphase animal cells, plant cells, polarized epithelial cells, and embryonic cells contain hundreds of MTOCs, which are distributed throughout the cell, often near the cell cortex. In plant cells and polarized epithelial cells, a cortical array of microtubules aligns with the cell axis (Figure 19-7d). In both cell types the polarity of the cell is linked to the orientation of the microtubules.

The γ-Tubulin Ring Complex Nucleates Polymerization of Tubulin Subunits

Despite its amorphous appearance, the pericentriolar material of an MTOC is an ordered lattice that contains many proteins that are necessary for initiating the assembly of microtubules. One of these, γ-tubulin, was first identified by genetic studies designed to discover proteins that interact with β-tubulin. Subsequent studies demonstrated that γ-tubulin and the lattice protein pericentrin are part of the pericentriolar material of centrosomes (Figure 19-8a); it has also been detected in MTOCs that lack a centriole. The finding that introduction of antibodies against γ-tubulin into cells blocks microtubule assembly implicates γ-tubulin as a necessary factor in nucleating polymerization of tubulin subunits.

Approximately 80 percent of the γ-tubulin in cells is part of a 25S complex, which has been isolated from extracts of frog oocytes and fly embryos. It is named the *γ-tubulin ring*

(a) Interphase animal cell

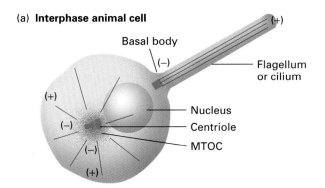

Basal body
(+)
(−)
Flagellum or cilium
(+)
Nucleus
(−)
Centriole
(−)
MTOC
(−)
(+)

(b) Mitotic animal cell

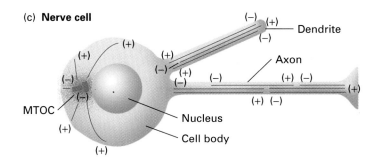

Chromosome
(+)
MTOC
(+)
(−)
(+)
(+)
(+)
(+)
(−)
(+)
(−)
(+)
(−)
(+)
(−)
(+)
(−)
(+)
(−)
(+)
(−)
(−)
(+)
(+)
(+)
Centriole
Spindle microtubule

(c) Nerve cell

(−) (+)
(−)
Dendrite
(+)
(+)
(+)
(−)
Axon
(−)
(+)
(+) (−)
(−)
(−)
(+)
MTOC
(−)
(+) (−)
(+)
Nucleus
Cell body
(+)
(+)

(d) Interphase plant cell

◄ **FIGURE 19-7 Orientation of cellular microtubules.** (a) In interphase animal cells, the (−) ends of most microtubules are proximal to the MTOC. Similarly, the microtubules in flagella and cilia have their (−) ends continuous with the basal body, which acts as the MTOC in these structures. (b) As cells enter mitosis, the microtubule network rearranges, forming a mitotic spindle. The (−) ends of all spindle microtubules point toward one of the two MTOCs, or poles, as they are called in mitotic cells. (c) In nerve cells, the (−) ends of axonal microtubules are oriented toward the base of the axon. However, dendritic microtubules have mixed polarities. (d) In plant cells, which contain numerous MTOCs, microtubules line the cell cortex. Webs of microtubules cap the growing ends of a plant cell.

(a)

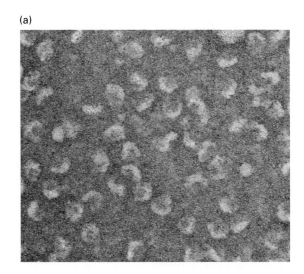

(b)

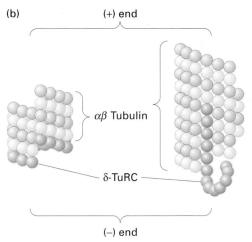

(+) end
$\alpha\beta$ Tubulin
δ-TuRC
(−) end

▲ **FIGURE 19-8 γ-Tubulin-mediated assembly of microtubules.** (a) Electron micrograph of γ-tubulin ring complexes (γ-TuRC). The complexes, isolated from *Xenopus* oocytes, consist of 8 polypeptides and measure 25nm in diameter. (b) Proposed models for γ-tubulin-mediated assembly of microtubules. In alternative models, γ-TuRC is thought to nucleate microtubule assembly by presenting a row of γ-tubulin subunits (left) or forming a protofilament (right), which can directly bind $\alpha\beta$-tubulin subunits. [Part (a) courtesy of Y. Zheng; part (b) modified from C. Wiese and Y. Zheng, 1999, *Curr. Opin. Struc. Biol.* **9**:250–259.]

complex (γ-TuRC) for its ringlike appearance in the electron microscope. In vitro experiments show that the γ-TuRC can directly nucleate microtubule assembly at subcritical tubulin concentrations, that is, at concentrations below which polymerization would not occur in the absence of the γ-TuRC (Figure 19-8b). These observations provide additional evidence that γ-tubulin plays a key role in directing microtubule assembly in vivo. Current biochemical, structural, and genetic studies are beginning to clarify the mechanism of nucleation.

SUMMARY Microtubule Structures

- Tubulins belong to an ancient family of GTPases that polymerize into hollow cylindrical structures, called *microtubules,* 24 nm in diameter.

- Dimeric αβ-tubulin subunits interact longitudinally to form protofilaments, which associate laterally into microtubules (see Figure 19-2b).

- Microtubules exhibit both structural and functional polarity.

- Microtubule-organizing centers (MTOCs), including centrosomes and basal bodies, nucleate the assembly of cytosolic microtubules.

- In most cases, the (−) end of a microtubule is adjacent to the MTOC from which it assembles and the (+) end is distal (see Figure 19-7).

- In some animal cells, the MTOC lies at the cell center, where it organizes cellular organelles.

- A γ-tubulin-containing complex is a major component of the pericentriolar material and is able to nucleate polymerization of tubulin subunits to form microtubules in vitro.

19.2 Microtubule Dynamics and Associated Proteins

Before proceeding further in our discussion of microtubule-containing structures and microtubule-based movements, we will take a close look at the assembly, disassembly, and polarity of microtubules. A microtubule can oscillate between growing and shortening phases. This complex dynamic behavior permits the cell to quickly assemble or disassemble microtubule structures. To appreciate fully the function of the microtubule cytoskeleton, we examine the details of microtubule assembly and disassembly, as well as the role of a group of proteins that are integrally associated with microtubules.

Microtubule Assembly and Disassembly Occur Preferentially at the (+) End

Microtubules assemble by polymerization of αβ-tubulin dimers. Once microtubules have assembled, their stability is temperature-dependent. For instance, if microtubules are cooled to 4 °C, they depolymerize into stable αβ-tubulin dimers (Figure 19-9a). When warmed to 37 °C in the presence of GTP, the tubulin dimers polymerize into microtubules.

(a)

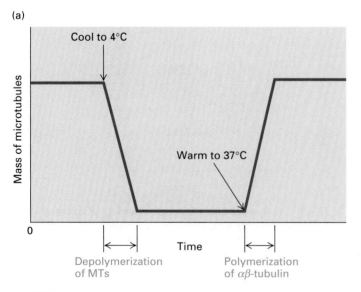

(b)

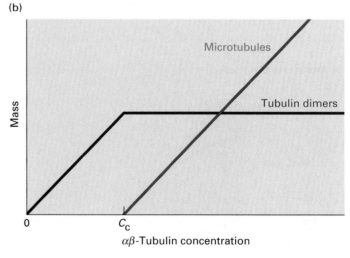

▲ **FIGURE 19-9 Effect of temperature and tubulin concentration on microtubule (MT) assembly and disassembly.** (a) At low temperatures, microtubules depolymerize, releasing αβ-tubulin, which repolymerize at higher temperatures in the presence of GTP. (b) The critical concentration (C_c) is the concentration of dimeric αβ-tubulin in equilibrium with microtubules. At dimer concentrations below the C_c, no polymerization occurs. At dimer concentrations above the C_c, tubulin polymerizes into microtubules.

The kinetics of tubulin polymerization and the structural intermediates observed during microtubule assembly or disassembly show that microtubule assembly is similar in many respects to microfilament assembly (see Figures 18-11 and 18-12). First, at $\alpha\beta$-tubulin concentrations above the critical concentration (C_c), the dimers polymerize into microtubules, while at concentrations below the C_c, microtubules depolymerize (Figure 19-9b). Second, the addition of fragments of flagellar or other microtubules to a solution of $\alpha\beta$-tubulin accelerates the initial polymerization rate by acting as nucleation sites. Third, at $\alpha\beta$-tubulin concentrations higher than the C_c for polymerization, dimers add to both ends of a growing microtubule, but the addition of tubulin subunits occurs preferentially at one end. This difference between the two ends of a growing microtubule is demonstrated by examining electron micrographs of microtubules that have assembled from the ends of nucleating flagellar fragments in vitro (Figure 19-10a). The electron micrographs show a tuft of microtubules sprouting from both ends of the fragment, but one tuft is much longer than the other. Using the same terminology as in actin assembly, the preferred assembly end is designated the $(+)$ end and the end that assembles more slowly is the $(-)$ end. When the tubulin

concentration is diluted below the C_c, the microtubules disassemble twice as rapidly at the $(+)$ end as at the $(-)$ end. Thus both assembly and disassembly occur preferentially at the $(+)$ end (Figure 19-10b). The dynamics of microfilament and microtubule assembly share these three properties.

One major difference between the assembly of microtubules and microfilaments is a consequence of the more complicated protofilament-based organization of a microtubule. Microtubule assembly involves three steps: protofilaments form from $\alpha\beta$-tubulin subunits; protofilaments associate to form the wall of the microtubule; and addition of more subunits to the ends of the protofilaments elongates the microtubule (Figure 19-11).

In the electron microscope, growing microtubules appear to have relatively smooth ends, although some protofilaments are longer than others, indicating that they elongate unevenly (Figure 19-12a). The appearance of microtubules undergoing shortening is quite different, suggesting that the mechanism of disassembly differs from that of assembly (Figure 19-12b). Under shortening conditions, the microtubule ends are splayed, as if the lateral interactions between protofilaments have been broken. Once frayed apart and freed from lateral stabilizing interactions, the protofilaments might depolymerize

(a)

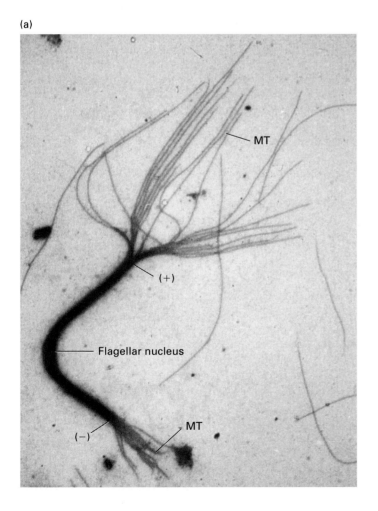

(b)

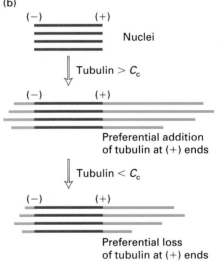

▲ **FIGURE 19-10 Polarity of tubulin polymerization.** (a) Fragments of flagellar microtubules act as nuclei for the in vitro addition of $\alpha\beta$-tubulin. The nucleating flagellar fragment can be distinguished in the electron microscope from the newly formed microtubules (MT), seen radiating from the ends of the flagellar fragment. Note that the microtubules at one end are longer than at the other end. (b) Addition and loss of tubulin subunits both occur primarily at one end of a microtubule, the $(+)$ end. [Part (a) courtesy of G. Borisy.]

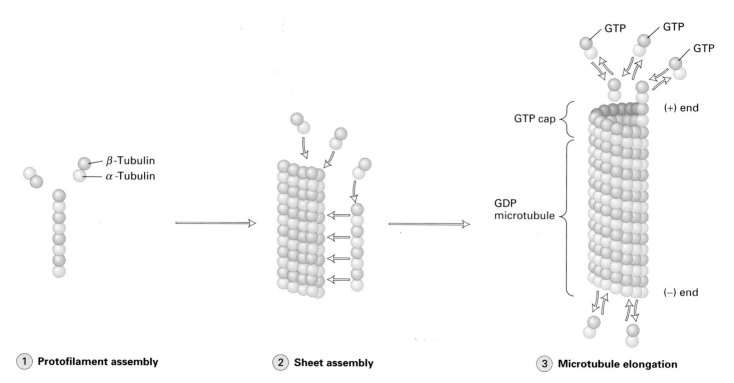

① **Protofilament assembly** ② **Sheet assembly** ③ **Microtubule elongation**

▲ **FIGURE 19-11 Assembly of microtubules.** Free $\alpha\beta$-tubulin dimers associate longitudinally to form short protofilaments (step ①). These probably are unstable and quickly associate laterally into more stable curved sheets (step ②). Eventually a sheet wraps around into a microtubule with 13 protofilaments. The microtubule then grows by the addition of subunits to the ends of protofilaments composing the microtubule wall (step ③). The free tubulin dimers have GTP bound to the exchangeable nucleotide-binding site on the β-tubulin monomer. After incorporation of a dimeric subunit into a microtubule, the GTP on the β-tubulin (but not on the α-tubulin) is hydrolyzed to GDP. If the rate of polymerization is faster than the rate of GTP hydrolysis, then a cap of GTP-bound subunits is generated at the (+) end, although the bulk of β-tubulin in a microtubule will contain GDP. The rate of polymerization is twice as fast at the (+) end as at the (−) end.

(a) Elongation

(b) Shrinkage

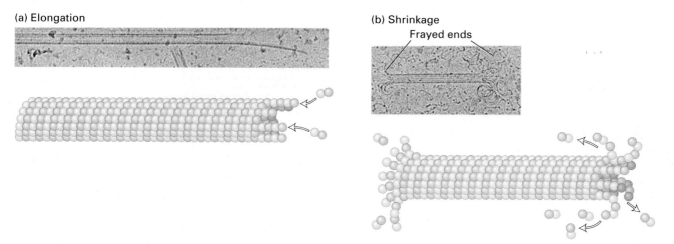

▲ **FIGURE 19-12 Appearance of microtubules undergoing assembly and disassembly.** Microtubules undergoing assembly or disassembly can be quickly frozen in liquid ethane and examined in the frozen state in a cryoelectron microscope. (a) In assembly conditions, microtubule ends are relatively smooth; occasionally a short protofilament is seen to extend from one end. (b) In disassembly conditions, the protofilaments splay at the microtubule ends, giving the ends a frayed appearance. [Micrographs courtesy of E. Mandelkow and E. M. Mandelkow.]

by endwise dissociation of tubulin subunits. The splayed appearance of a shortening microtubule provided clues about the potential instability of a microtubule.

Dynamic Instability Is an Intrinsic Property of Microtubules

So far in this discussion we have painted a simple picture of the dynamic behavior of microtubules; that is, above the C_c, tubulin subunits polymerize into microtubules, while below the C_c, microtubules depolymerize into tubulin subunits. However, this simple picture of the behavior of an individual microtubule is misleading. Under favorable in vitro conditions, microtubules exhibit the ability to *treadmill*, in which subunits add to one end and dissociate from the opposite end. Furthermore, a single microtubule can oscillate between growth and shortening phases (Figure 19-13). In all cases, the rate of microtubule growth is much slower than the rate of shortening. When first discovered, this more complex behavior of microtubules, termed *dynamic instability*, was surprising to researchers because they expected that under any condition all the microtubules in a solution or the same cytosol would behave identically.

Other studies have shown that individual cytosolic microtubules also display this opposing dynamic behavior in vivo. In these experiments, fluorescent αβ-tubulin subunits are microinjected into live cultured cells. The cells are chilled to depolymerize preexisting microtubules into tubulin dimers and are then incubated at 37 °C to allow repoly-

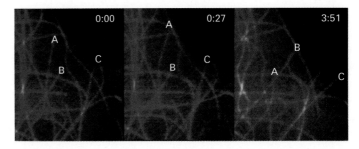

▲ **FIGURE 19-14 In vivo growth and shrinkage of individual microtubules.** Fluorescent-labeled tubulin was microinjected into cultured human fibroblasts. The cells were chilled to depolymerize preexisting microtubules into tubulin dimers and were then incubated at 37 °C to allow repolymerization, thus incorporating the fluorescent tubulin into all the cell's microtubules. A region of the cell periphery was viewed in the fluorescence microscope at 0 s, 27 s later, and 3 min 51 s later (left to right panels). During this period several microtubules elongate and shorten. The letters mark the position of ends of three microtubules. [From P. J. Sammak and G. Borisy, 1988, *Nature* **332**:724.]

merization, thus incorporating the fluorescent tubulin into all the cellular microtubules. In video records of a small region in labeled cells taken over a period of several minutes, some microtubules can be seen to lengthen, while others shorten (Figure 19-14). Also, within a few minutes, some microtubules appear alternately to grow and shrink. Since most microtubules in a cell associate by their (−) ends to MTOCs, we can conclude that their instability is largely limited to the (+) ends of microtubules.

Two conditions influence the stability of microtubules. First, the oscillations between growth and shrinkage in vitro occur at tubulin concentrations near the C_c. As we have discussed, at tubulin concentrations above the C_c, the entire population of microtubules grows, and at concentrations below the C_c, all microtubules shrink (see Figure 19-10b). At concentrations near the C_c, however, some microtubules grow, while others shrink. The second condition affecting microtubule stability is whether GTP or GDP occupies the exchangeable nucleotide-binding site on β-tubulin at the (+) end of a microtubule (Figure 19-15). A microtubule becomes unstable and depolymerizes rapidly if the (+) end becomes capped with subunits containing GDP–β-tubulin rather than GTP–β-tubulin. This situation can arise when a microtubule shrinks rapidly, exposing GDP–β-tubulin in the walls of the microtubule, or when a microtubule grows so slowly that hydrolysis of GTP bound to β-tubulin converts it to GDP before additional subunits can be added to the (+) end of the microtubule. Before a shortening microtubule vanishes, it can be "rescued" and start to grow if tubulin subunits with bound GTP add to the (+) end before the bound GTP hydrolyzes. Thus the one parameter that determines the stability of a microtubule is the rate at which GTP-tubulin subunits are added to the (+) end. Possible factors that switch a microtubule between growth and shrinkage have been

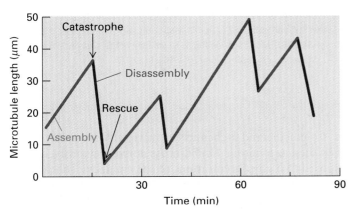

▲ **FIGURE 19-13 Dynamic instability of microtubules in vitro.** Individual microtubules can be observed in the light microscope, and their lengths can be plotted during stages of assembly and disassembly. Assembly and disassembly each proceed at uniform rates, but there is a large difference between the rate of assembly and that of disassembly, as seen in the different slopes of the lines. During periods of growth, the microtubule elongates at a rate of 1 μm/min. Notice the abrupt transitions to the shrinkage stage (catastrophe) and to the elongation stage (rescue). The microtubule shortens much more rapidly (7 μm/min) than it elongates. [Adapted from P. M. Bayley, K. K. Sharma, and S. R. Martin, 1994, in *Microtubules*, Wiley-Liss, p. 119.]

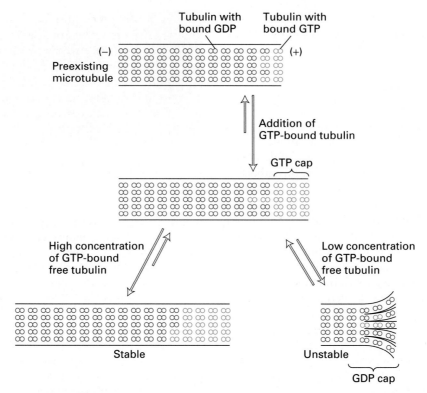

▲ **FIGURE 19-15 Dynamic instability model of microtubule growth and shrinkage.** Dimeric $\alpha\beta$-tubulin subunits with two bound GTP molecules (blue) add preferentially to the (+) end of a preexisting microtubule. After incorporation of a subunit, the GTP bound to the β-tubulin monomer is hydrolyzed to GDP, although the GTP bound to α-tubulin is not. This hydrolysis is apparently catalyzed by the microtubule itself but may be facilitated by cytosolic proteins. Only microtubules whose (+) end is associated with GTP-tubulin (those with a GTP cap) are stable and can serve as primers for polymerization of additional tubulin. Microtubules with GDP-tubulin (red) at the (+) end (those with a GDP cap) are rapidly depolymerized and may disappear within 1 min.

At high concentrations of unpolymerized GTP-tubulin, the rate of addition of tubulin is faster than the rate of hydrolysis of the GTP bound in the microtubule or the rate of dissociation of GTP-tubulin from the end; thus the microtubule grows. At low concentrations of unpolymerized GTP-tubulin, the rate of addition of tubulin is decreased; consequently, the rate of GTP hydrolysis exceeds the rate of addition of tubulin subunits and a GDP cap forms. Because the GDP cap is unstable, the microtubule end peels apart to release tubulin subunits. [See T. Mitchison and M. Kirschner, 1984, *Nature* **312**:237; M. Kirschner and T. Mitchison, 1986, *Cell* **45**:329; and R. A. Walker et al., 1988, *J. Cell Biol.* **107**:1437.]

identified. One is a microtubule-severing protein, katanin, which may generate nuclei at centrosomes. Another factor is Op 18, which increases the frequency of catastrophe, possibly by binding tubulin dimers.

Colchicine and Other Drugs Disrupt Microtubule Dynamics

Some of the earliest studies of microtubules employed several drugs that inhibit mitosis, a cell process that depends on microtubule assembly and disassembly. Three such drugs, colchicine, taxol, and vinblastine, all purified from plants, have proved to be very powerful tools for probing microtubule function, partly because they bind only to $\alpha\beta$-tubulin or microtubules and not to other proteins, and also because their concentrations in cells can be easily controlled (Figure 19-16).

Colchicine and a chemical relative, colcemid, have long been used as a mitotic inhibitor. As noted previously, in cells exposed to high concentrations of colcemid, cytosolic microtubules depolymerize, leaving an MTOC (see Figure 19-6). However, when plant or animal cells are exposed to low concentrations of colcemid, the microtubules remain and the cells become "blocked" at **metaphase,** the mitotic stage at which the duplicated chromosomes are fully condensed. When the treated cells are washed with a colcemid-free solution, colcemid diffuses from the cell and mitosis resumes normally. Thus experimenters commonly use colcemid to accumulate metaphase cells for cytogenetic studies; removal of the colcemid leaves a population of cells whose cell cycle is in synchrony. Such synchronous populations are advantageous for studies of the cell cycle (Chapter 13).

Each tubulin dimer has one high-affinity binding site for colchicine. In fact, colchicine binds tubulin irreversibly. This

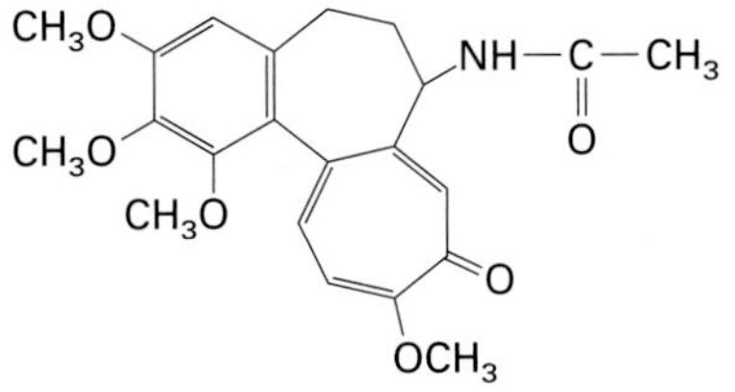

Colchicine

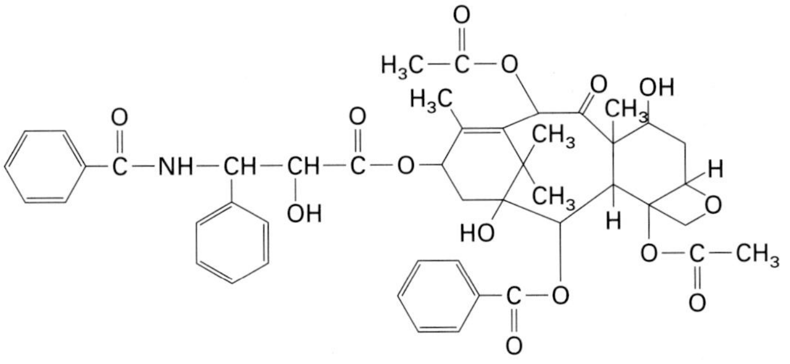

Taxol

▲ **FIGURE 19-16 Structures of colchicine and taxol.** These and other drugs that interfere with normal assembly and disassembly of microtubules have an antimitotic effect that is particularly devastating to rapidly dividing cells, such as cancer cells.

property was used as an early assay for tubulin during the purification of microtubules. Colchicine-bearing tubulin dimers, at concentrations much less than the concentration of free tubulin subunits, can add to the end of a growing microtubule. However, the presence of one or two colchicine-bearing tubulins at the end of a microtubule prevents the subsequent addition or loss of other tubulin subunits. Thus colchicine "poisons" the end of a microtubule and alters the steady-state balance between assembly and disassembly. As a result of this disruption of microtubule dynamics, the mitotic spindle does not form in cells treated with low concentrations of colchicine.

Other drugs bind to different sites on tubulin dimers or to microtubules and therefore affect microtubule stability through different mechanisms. For example, at low concentrations, taxol, its chemical derivative taxotere, and vinblastine bind to and stabilize microtubules by inhibiting microtubule dynamics—the lengthening and shortening of microtubules. High concentrations of vinblastine, however, promote depolymerization of microtubules and the assembly of tubulin dimers into nearly crystalline arrays called *vinblastine paracrystals*.

 Drugs that disturb the assembly and disassembly of microtubules have been widely used to treat various diseases. Indeed, more than 2500 years ago, the ancient Egyptians treated heart problems with colchicine. Nowadays, this drug is used primarily in the treatment of

gout and certain other diseases affecting the joints and skin. Other inhibitors of microtubule dynamics, including taxol and vinblastine, are effective anticancer agents, since blockage of spindle formation preferentially inhibits rapidly dividing cells like cancer cells. For instance, taxol treatment of ovarian cancer cells, which undergo rapid cell divisions, blocks mitosis but does not affect other functions carried out by microtubules.

Assembly MAPs Cross-Link Microtubules to One Another and Other Structures

Tubulin is typically isolated by pelleting microtubules from a cell lysate, depolymerizing the microtubules by cooling them to 4 °C, centrifuging the cooled solution to remove the insoluble material, and then polymerizing the tubulin-containing supernatant by warming to 37 °C. Highly enriched tubulin preparations obtained after several such assembly-and-disassembly cycles still contain small amounts of other proteins, which maintain their quantitative ratio to α- and β-tubulin through successive cycles. Co-purification of these proteins with tubulin suggested that they are not nonspecific contaminants but rather molecules that interact specifically with microtubules. Immunofluorescence micrographs of cultured cells have shown that the cellular localization of these co-purifying proteins, called **microtubule-associated proteins (MAPs)**, parallels that of microtubules (Figure 19-17). This finding is strong evidence that MAPs have microtubule-binding activity.

One major family of MAPs, called *assembly MAPs*, is responsible for cross-linking microtubules in the cytosol. These MAPs are organized into two domains: a basic microtubule-binding domain and an acidic projection domain (Table 19-1). In the electron microscope, the projection domain appears as a filamentous arm that extends from the wall of the microtubule. This arm can bind to membranes, intermediate filaments, or other microtubules, and its length controls how far apart microtubules are spaced (Figure 19-18).

Based on sequence analysis, assembly MAPs can be grouped into two types (see Table 19-1). Type I MAPs, MAP1A and MAP1B, contain several repeats of the amino acid sequence Lys-Lys-Glu-X, which is implicated as a binding site for negatively charged tubulin. This sequence is postulated to neutralize the charge repulsion between tubulin subunits within a microtubule, thereby stabilizing the polymer. MAP1A and MAP1B are large, filamentous molecules found in axons and dendrites of neurons and also in nonneuronal cells. Each of these proteins is derived from a single precursor polypeptide, which is proteolytically processed in a cell to generate one light chain and one heavy chain.

Type II MAPs include MAP2, MAP4, and Tau. These proteins are characterized by the presence of three or four repeats of an 18-residue sequence in the microtubule-binding domain. MAP2 is found only in dendrites, where it forms fibrous cross-bridges between microtubules and also

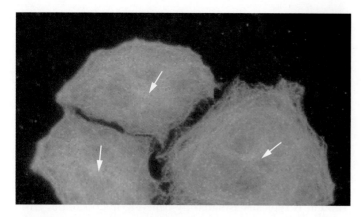

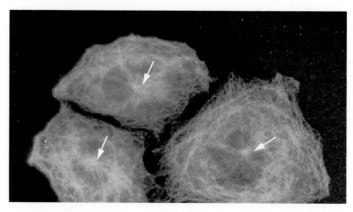

▲ **FIGURE 19-17 The organization of microtubules *(left)* and a microtubule-associated protein called MAP4 *(right)* in the same interphase HeLa cells.** Cells were stained with fluorescent-labeled antibodies against tubulin and MAP4 and then viewed in a fluorescence microscope. The colinear arrangement of MAP4 and microtubules, even at the MTOC (arrow), is suggestive of a binding interaction. [Courtesy of J. Chloe-Bulinski.]

▶ **FIGURE 19-18 Experimental demonstration that spacing of microtubules (MTs) depends on microtubule-associated proteins (MAPs).** Insect cells transfected with DNA expressing either long-armed MAP2 protein or short-armed Tau protein grow long axonlike processes. Shown here are electron micrographs of cross sections through the processes induced by expression of MAP2 *(left)* or Tau *(right)* in transfected cells. Note that the spacing between microtubules in MAP2-containing cells is larger than in Tau-containing cells. Both cell types contain approximately the same number of microtubules, but the effect of MAP2 is to enlarge the caliber of the axonlike process. [From J. Chen et al., 1992, *Nature* **360**:674.]

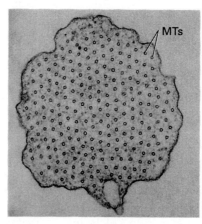

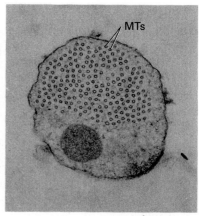

TABLE 19-1	**Major Microtubule-Associated Proteins**		
Protein	MW	Domain Organization*	Location
TYPE I			
MAP1A	300,000 heavy chain		Dendrites and axons
MAP1B	255,000		Dendrites and axons
TYPE II			
MAP2a	280,000		Dendrites
MAP2b	200,000		Dendrites
MAP2c	42,000		Embryonic dendrites
MAP4	210,000		Non-neuronal cells
Tau	55,000–62,000		Dendrites and axons

Yellow, microtubule-binding domain; pink, projection domain; green, 18 amino acid repeats.

links microtubules to intermediate filaments. MAP4, the most ubiquitous of all the MAPs, is found in neuronal and non-neuronal cells. As discussed later, MAP4 is thought to regulate microtubule stability during mitosis. Tau, which is much smaller than most other MAPs, is present in both axons and dendrites. This protein exists in four or five forms derived from alternative splicing of a *tau* mRNA. The ability of Tau to cross-link microtubules into thick bundles may contribute to the stability of axonal microtubules.

Transfection of cultured insect cells with either the *tau* gene or *MAP2* gene induces the growth of microtubule-filled processes. These observations indicate that both Tau and MAP2 accelerate the polymerization of tubulin subunits, as well as contribute to cross-linking of microtubules. The importance of Tau in promoting axon growth was further confirmed by the finding that cultured neurons microinjected with DNA encoding *tau* antisense RNA are unable to grow an axon. Expression of *tau* antisense RNA inhibits translation of *tau* mRNA and thus reduces the intracellular level of Tau (see Figure 11-46). Similar experiments with *MAP2* antisense RNA showed that MAP2 is critical to formation of dendrites.

Bound MAPs Alter Microtubule Dynamics

When MAPs coat the outer wall of a microtubule, tubulin subunits are unable to dissociate from the ends of a microtubule. Although the rate of microtubule disassembly is generally dampened by bound MAPs, the assembly of microtubules is affected to varying degrees: some MAPs, like Tau and MAP4, stabilize microtubules, whereas other MAPs do not.

Because of the effect of MAPs on microtubule dynamics, the length of microtubules can be controlled by modulating the binding of MAPs. In most cases, this is accomplished by the reversible phosphorylation of the MAP projection domain. Phosphorylated MAPs are unable to bind to microtubules; thus they promote microtubule disassembly. *MAP kinase*, a key enzyme for phosphorylating MAPs, is a participant in many signal-transduction pathways, indicating that MAPs are targets of many extracellular signals (Chapter 20). MAPs, especially MAP4, also are phosphorylated by a second kinase, *cdc2 kinase*, which plays a major role in controlling the activities of various proteins during the cell cycle (Chapter 13).

SUMMARY   Microtubule Dynamics and
Associated Proteins

- Assembly and disassembly of microtubules depends on the critical concentration, C_c, of $\alpha\beta$-tubulin subunits. Above the C_c, assembly occurs; below the C_c, disassembly occurs (see Figure 19-9). Addition and loss of subunits occur preferentially at one end, the (+) end.

- Microtubules exhibit two dynamic phenomena that are pronounced at C_c: treadmilling, the addition of subunits at one end and their loss at the other end, and dynamic instability, the oscillation between lengthening and shortening.

- At tubulin concentrations near the C_c, microtubules can alternately grow and shrink. The balance between growth and shrinkage depends on whether the exchangeable GTP bound to β-tubulin is exposed on the (+) end of a microtubule or whether it has been hydrolyzed to GDP (see Figure 19-15).

- Various drugs, including colchicine and taxol, disrupt microtubule dynamics and have an antimitotic effect. Some of these drugs are useful in the treatment of certain cancers.

- Microtubule-associated proteins (MAPs) co-purify with microtubules and have the same cellular localization.

- An important class of MAPs, the assembly MAPs, prevent cytosolic microtubules from depolymerizing, organize them into bundles, and cross-link them to membranes and intermediate filaments.

19.3   Kinesin, Dynein, and Intracellular Transport

Within cells, membrane-bounded vesicles and proteins are frequently transported many micrometers along well-defined routes in the cytosol and delivered to particular addresses. Diffusion alone cannot account for the rate, directionality, and destinations of such transport processes. Early video light microscopy studies showed that these long-distance movements follow straight paths in the cytosol, frequently along cytosolic fibers, suggesting that transport involves some kind of tracks. Subsequent experiments, using nerve cells and fish-scale pigment cells, first demonstrated that microtubules function as tracks in the intracellular transport of membrane-bounded vesicles and organelles, and that movement is propelled by microtubule motor proteins.

Fast Axonal Transport Occurs along Microtubules

As we discuss in Chapter 21, nerve impulses are transmitted from a neuron by release of neurotransmitters from the terminal of the axon, the very long process that extends from the cell body. The neuron must constantly supply new materials—proteins and membranes—to the terminal to replenish those lost by exocytosis at the junction (synapse) with another cell. Where do these new materials originate? Ribosomes are present only in the cell body and dendrites of nerve cells, so no protein synthesis can occur in the axons and synaptic terminals. Therefore, proteins and membranes must be synthesized in the cell body and then transported

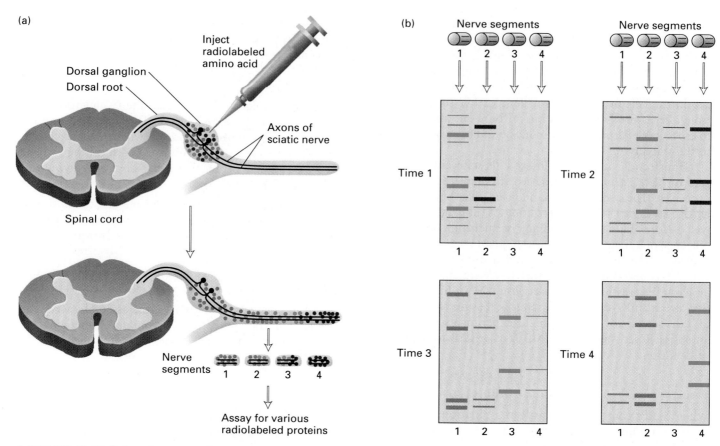

▲ **FIGURE 19-19 Pulse-chase experimental system for determining the in vivo rate of axonal transport and identifying the transported proteins.** (a) Radiolabeled amino acids are injected into a ganglion of an experimental animal *(top).* Animals are killed at different times after injection, and the sciatic nerve is dissected and cut into 5-mm segments. Initially, the ribosome-containing cell bodies in the ganglion incorporate the labeled amino acids into proteins, which then are transported down the axon. Soon after injection, labeled proteins are found only in segments closest to the cell body. With increasingly longer chase periods, labeled protein is detected more and more distal to the cell body *(bottom).* The red, blue, and purple dots represent groups of proteins that are transported down the axon at different rates, red most rapidly, purple least rapidly. (b) The amount of radiolabeled protein in each fragment is measured, and then the various proteins are resolved by gel electrophoresis. One set of polypeptides (red bands), characterized by molecular weight, are detected first in the segments proximal to the injection site and later in more distal segments. Lagging behind this fast-traveling component are other polypeptides whose rate of transport is slower (blue and purple bands). The distribution of labeled proteins in part (a) corresponds to the gel pattern at time 2. [Adapted from O. S. Ochs, 1981, in G. J. Siegel et al., eds., *Basic Neurochemistry,* 3d ed., Little, Brown, p. 425.]

down the axon, which can be up to several meters in length, to the synaptic regions. This process of *axonal transport,* first described in 1948, is now known to occur on microtubules. As noted earlier, the microtubules in axons are all oriented with their (+) ends toward the terminal, which is critical to axonal transport (see Figure 19-7c).

The rate at which proteins are transported along axons, and their identity, can be determined by a pulse-chase experiment, as outlined in Figure 19-19. Such experiments commonly are conducted on neurons in the mammalian sciatic nerve because their cell bodies are conveniently located in the dorsal root ganglion near the spinal cord and their nerve axons are very long. Studies like these have shown that axonal transport occurs in both directions. *Anterograde transport,* as depicted in Figure 19-19a, proceeds from the cell body to the synaptic junctions and is associated with axonal growth and the renewal of synaptic vesicles. In the opposite, *retrograde,* direction other substances move along the axon rapidly toward the cell body. These substances, which consist mainly of "old" membrane from the synaptic terminals, are destined to be degraded in lysosomes in the cell body.

Transported materials are divided into three groups according to the speed of movement. The fastest-moving material, consisting of membrane-bounded vesicles, has a velocity of about 250 mm/day, or about 3 μm/s. The slowest material, comprising mostly polymerized cytoskeletal proteins, moves only a fraction of a millimeter per day. Organelles such as mitochondria move down the axon at an intermediate rate.

(a)

(b)

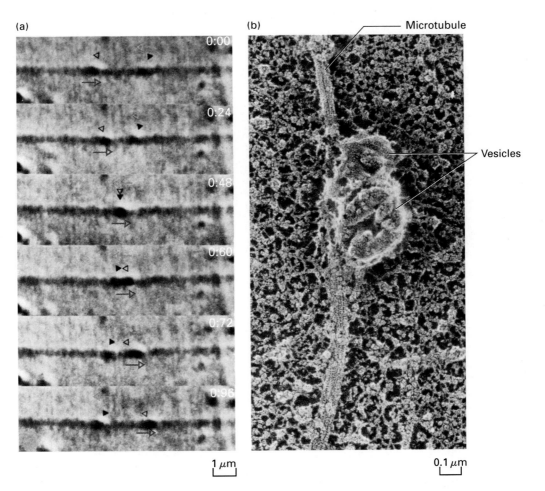

Microtubule

Vesicles

1 μm

0.1 μm

▲ **FIGURE 19-20 Video micrographs showing bidirectional movement of two vesicular organelles on a single transport microtubule filament.** (a) A piece of squid giant axon was dissected, the cytoplasm was extruded, and a buffer containing ATP was added. The preparation was then viewed in a differential interference contrast microscope, and the images were recorded on videotape. The two organelles (located at positions indicated by open and solid triangles) move in opposite directions (indicated by colored arrows) along the same filament, pass each other, and continue in their original directions. Elapsed time in seconds appears at the top-right corner of each video frame. (b) A region of cytoplasm similar to that shown in (a) was freeze-dried, rotary-shadowed with platinum, and viewed in the electron microscope. Two large structures attached to one microtubule are visible; these presumably are small vesicles that were moving along the microtubule when the preparation was frozen. [See B. J. Schnapp et al., 1985, *Cell* **40**:455; courtesy of B. J. Schnapp, R. D. Valle, M. P. Sheetz, and T. S. Reese.]

Axonal transport does not require an intact cell. In fact many studies on fast axonal transport are conducted with extruded axoplasm. In this type of preparation, the cytosol is squeezed from the axon with a roller onto a glass coverslip. When such an extract is provided with ATP, the movement of vesicles along microtubules can be observed by video microscopy. The rate of vesicle movement (1–2 μm/s) in this cell-free system is similar to that of fast axonal transport in intact cells. Movement may occur in both the anterograde and retrograde directions; in some cases, two organelles can be seen to move along the same fiber in opposite directions and to pass each other without colliding (Figure 19-20a). Electron microscopy of the same region of the axoplasm has revealed that the transport fibers are individual microtubules (Figure 19-20b). These in vitro experiments established that

fast axonal transport occurs along microtubules and that the movement requires ATP. As we will discuss shortly, these two observations led to the identification of microtubule motor proteins, which generate the movements.

Microtubules Provide Tracks for the Movement of Pigment Granules

Another system for observing rapid transport along microtubules is provided by the specialized pigment cells—called *melanophores*—that are found in the skin of many amphibians and on the scales of many fish. Nerves and hormones control the color of the skin by triggering the transport of membrane-enclosed pigment granules throughout the cell, to darken the color of the skin, or inward, toward the center

(a) (b) (c)

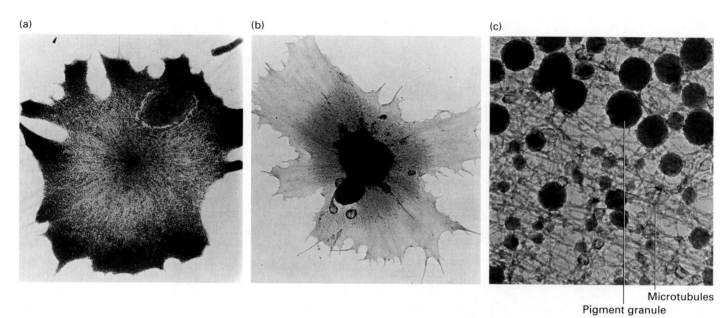

Microtubules
Pigment granule

▲ **FIGURE 19-21 High-voltage electron micrographs showing movement of pigment granules in a melanophore, or red-pigment cell, of the squirrelfish, *Holocentrus ascensionis.*** (a) The pigment granules are dispersed to the cell periphery.

(b) The granules are condensed around the nucleus. (c) A portion of a dispersing melanophore showing the pigment granules associated with tracks of microtubules. [Courtesy of K. Porter.]

of the cell, to lighten the color (Figure 19-21a, b). In this way, an animal can adjust its color.

The role of microtubules in color adjustment is studied by placing pigment cells in culture. If a melanophore-covered fish scale is placed in a culture medium, the pigment granules in the melanophores will be seen to move inward and outward spontaneously. In cases when individual granules could be followed, it has been observed that after movement to the cell periphery, each granule always returns to its prior location in the center of the cell. During this movement, microtubules serve as tracks along which the pigment granules can move in either direction (Figure 19-21c).

Intracellular Membrane Vesicles Travel along Microtubules

Having seen that membrane vesicles in specialized cells are transported between the cell body and the cell periphery, we can next ask whether membrane vesicles are transported along microtubules in every eukaryotic cell. The preliminary answer is a qualified yes: some types of vesicle transport are dependent on microtubules, though microfilaments may also be involved in some cases. The best-studied system is the intracellular movement of Golgi vesicles. In cultured fibroblasts, the Golgi complex is concentrated near the MTOC. During mitosis (or after the depolymerization of microtubules by colcemid), the Golgi complex breaks into small vesicles that are dispersed throughout the cytosol. When the cytosolic microtubules re-form during interphase (or after removal of the colcemid), the Golgi vesicles move along these

microtubule tracks toward the MTOC, where they reaggregate to form large membrane complexes.

In addition to the Golgi complex, microtubules are also associated with the endoplasmic reticulum (ER). Fluorescence microscopy, using anti-tubulin antibodies and $DiOC_6$, a fluorescent dye specific for the ER, reveals an anastomosing network of tubular membranes in the cytosol that colocalizes with microtubules (Figure 19-22). If microtubules are destroyed by drugs such as nocodazole or colcemid, then the ER loses its network-like organization. After the drug is washed from the cell, tubular fingers of ER grow as new microtubules assemble. In cell-free systems, the ER can be reconstituted with microtubules and an ER-rich cell extract. Even under this cell-free regime, ER membranes elongate along a microtubule. This close association between ER and intact microtubules suggests that certain proteins act to bind ER membranes to microtubules.

A key feature of the in vitro assembly of the ER is that the tubular membranes seem to elongate along microtubules. This outgrowth of the ER membrane is blocked by microtubule inhibitors like colchicine or nocodazole, suggesting that intact microtubules are required for membrane elongation. This movement along microtubules is one mechanism by which the MTOC organizes cell organelles such as the ER and Golgi.

Kinesin Is a (+) End—Directed Microtubule Motor Protein

To unravel the mechanism of axonal transport, cell biologists sought to identify the protein or proteins in neuronal

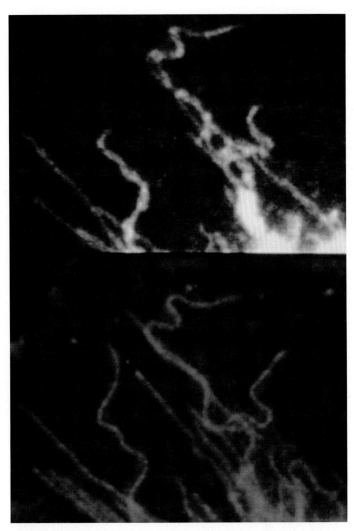

▲ FIGURE 19-22 In a frog fibroblast, the striking alignment of ER and microtubules is evident because the cell has sparse microtubules. In many but not all regions of the cytoplasm, the endoplasmic reticulum stained with DiOC$_6$ (green) colocalizes with cytoplasmic microtubules (red). [Courtesy of M. Terasaki.]

cytosolic extracts that can propel synaptic vesicles along microtubules assembled in vitro from purified tubulin subunits and stabilized by the drug taxol. When synaptic vesicles were added with ATP to these microtubules, the vesicles neither bound to the microtubules nor moved along them. However, the addition of squid nerve axoplasm (free of tubulin) caused the vesicles to bind to the microtubules and to move along them, indicating that a soluble protein in the nerve cytosol is required for translocation.

When researchers incubated vesicles, nerve cytosol (squid axoplasm), and microtubules in the presence of AMPPNP,

$$\text{Adenine} - \text{ribose} - \text{O} - \overset{\overset{\displaystyle O}{\|}}{\underset{\underset{\displaystyle O^-}{|}}{P}} - \text{O} - \overset{\overset{\displaystyle O}{\|}}{\underset{\underset{\displaystyle O^-}{|}}{P}} - \text{NH} - \overset{\overset{\displaystyle O}{\|}}{\underset{\underset{\displaystyle O^-}{|}}{P}} - \text{OH}$$

a nonhydrolyzable analog of ATP, the vesicles bound tightly to the microtubules but did not move. However, the vesicles did move when ATP was added. These results suggested that a motor protein in the cytosol binds to microtubules in the presence of ATP or AMPPNP, but movement requires hydrolysis of the terminal phosphoanhydride bond of ATP. To purify the soluble motor protein, scientists used AMPPNP to promote its tight binding to microtubules, which were used as an affinity matrix. A mixture of microtubules, brain extract, and AMPPNP was incubated; the microtubules with any bound proteins then were collected by centrifugation. Treatment of the microtubule-rich material in the pellet with ATP released one predominant protein back into solution; this protein was named **kinesin.**

Kinesin isolated from squid axoplasm is a dimer of two heavy chains, each complexed to a light chain, with a total molecular weight of 380,000. The molecule is organized into three domains, a pair of large globular head domains connected by a long central stalk to a pair of small globular tail domains, which contain the light chains (Figure 19-23a). Each domain carries out a particular function: the head domain, which binds microtubules and ATP, is responsible for the motor activity of kinesin, and the tail domain is responsible for binding to membrane vesicles. In light of the transport function of kinesin, a bound membrane vesicle is often referred to as kinesin's "cargo."

Recent x-ray crystallographic analysis of the kinesin head domain has revealed a surprising feature of its structure (Figure 19-23b). The three-dimensional structure of the kinesin head is similar to that of Ras, a guanine nucleotide–binding protein (see Figure 3-5) and to that of the myosin motor domains (see Figure 18-24). At the core of the head domain is a Ras-like fold, which binds ATP. Microtubule-binding sites lie on the surface of the domain in similar regions to the actin-binding loops of myosin. Thus, these motor proteins may have evolved from a common ancestor, becoming specialized to move along a microtubule or a microfilament.

Kinesin-dependent movement of vesicles can be tracked by an in vitro motility assay using microtubules nucleated from isolated centrosomes. In this assay, a vesicle or a plastic bead coated with kinesin is added to a glass slide along with a preparation of centrosome-nucleated microtubules. The kinesin-coated beads bind to the microtubules, and in the presence of ATP, kinesin carries the beads along a microtubule in one direction (Figure 19-24a). By determining the polarity of the microtubules, researchers found that the beads always moved from the (−) to the (+) end of a microtubule. Alternatively, kinesin alone, adsorbed onto a glass coverslip, causes microtubules to glide across the surface, with their (−) ends in the lead (Figure 19-24b); this direction of movement results from kinesin's walking toward the (+) end of the microtubule. Thus kinesin is a (+) end–directed microtubule motor protein. Because this direction corresponds to anterograde transport, kinesin is implicated by these studies as the motor protein responsible for anterograde and other (+) end–directed movements, such as the transport of secretory

(a)

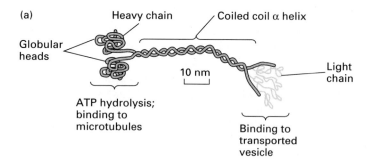

(b)

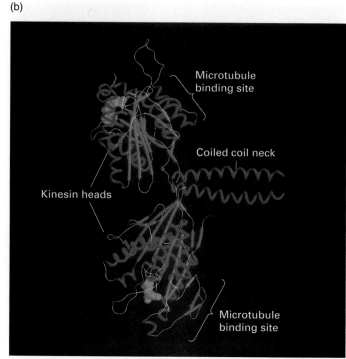

▲ **FIGURE 19-23 Structure of kinesin.** (a) Schematic model of kinesin showing the arrangement of the two heavy chains (each with a MW of 110,000–135,000) and two light chains (MW 60,000–70,000). (b) Ribbon trace of the kinesin dimer. Each head is attached to an α-helical neck region, which forms a coiled-coil dimer. Microtubules bind to the helix indicated, this interaction is regulated by the nucleotide (orange) bound at the opposite side of the domain. The distance between microtubule binding sites is 5.5 nm. [Part (b) courtesy of E. Mandelkow and E. M. Mandelkow, adapted from M. Thormahlen et al., 1998, *J. Struc. Biol.* **122**:30–41.]

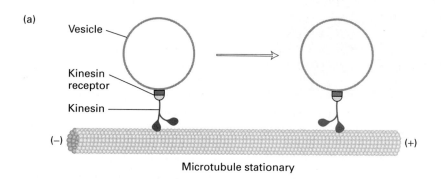

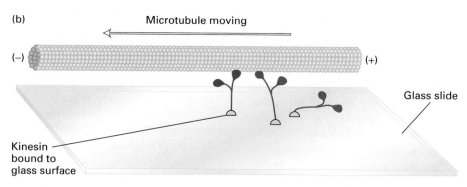

▲ **FIGURE 19-24 Model of kinesin-catalyzed anterograde transport.** (a) Kinesin-motored transport of vesicles along immobile microtubules. The kinesin molecules, attached to unidentified receptors on the vesicle surface, transport the vesicles from the (−) to the (+) end of a stationary microtubule. (b) Kinesin-catalyzed movement of microtubules. The kinesin molecules bound to the glass surface move toward the (+) end of the microtubule. Because the kinesin molecules are immobilized onto the coverslip, the sliding force is transmitted to the microtubule, which then moves in the direction of its (−) end. ATP is required for movement in both cases. [Adapted from R. D. Vale et al., 1985, *Cell* **40**:559; T. Schroer et al., 1988, *J. Cell Biol.* **107**:1785.]

vesicles to the plasma membrane and the radial movements of ER membranes and pigment granules.

In vitro motility experiments using an optical trap have determined two fundamental characteristics of the kinesin motor, its step size and force. In these experiments, similar to those performed on myosin (see Figure 18-23), a two-headed kinesin molecule was found to move in 8-nm steps and exert a force of 6 pN. The step size matches the distance between successive α- or β-tubulin monomers in a protofilament, suggesting that kinesin binds only to one or the other monomer. In other experiments, researchers have established that kinesin moves along a single protofilament, like walking a tightrope. However this poses a question. If the microtubule binding sites are separated by 5 nm in the kinesin dimer, then the kinesin structure must "give" during an 8-nm step. How this occurs is unknown.

Each Member of the Kinesin Family Transports a Specific Cargo

Like myosin, kinesin also belongs to a family of related motor proteins. (Table 19-2). To date, more than 12 different family members have been identified; all contain the kinesin motor domain, but they differ in their tail domains and several other properties. In most kinesins, the motor domain is at the N-terminus (N-type), but in some, the motor domain is central (M-type) or at the C-terminus (C-type). Both N- and M-type kinesins are (+) end–directed motors, whereas C-type kinesins are (−) end–directed motors. In addition, some kinesins are monomeric (i.e., have a single heavy chain); as noted earlier; however, most are dimeric. These two types of kinesins, differing in quaternary structure, may move along a microtubule by different mechanisms.

Kinesins can also be divided into two broad functional groups—*cytosolic* and *spindle* kinesins—based on the nature of the cargo they transport. The functional differences between kinesins may be related to their unique tail domains. Cytosolic kinesins are involved in vesicle and organelle transport; they include the classic axonal kinesin, implicated in transport of lysosomes and other membranous organelles. Some cytosolic kinesins, however, are responsible for transport of one specific cargo. For example, KIF1B transports mitochondria, and its relative KIF1A transports synaptic vesicles to the nerve terminal. Spindle kinesins, in contrast, participate in spindle assembly and chromosome segregation during cell division. (These are also known as *kinesin-related proteins*, or *KRP motors*.) This group comprises numerous proteins, including the kinetochore-associated protein CENP-E; the spindle pole protein BimC; and a (−) end–directed motor protein called *ncd*. We discuss the spindle kinesins in more detail in the later section on mitosis.

Dynein Is a (−) End—Directed Motor Protein

Some microtubule-directed movements, such as retrograde axonal transport or the transit of endocytotic vesicles of the plasma membrane to lysosomes, are in the direction opposite to most kinesin-dependent movements. A second group of motor proteins, the **dyneins**, were found to be responsible for movement toward the (−) end of microtubules. Dyneins are exceptionally large, multimeric proteins, with molecular weights exceeding 1,000,000. They are composed of two or three heavy chains (MW 470,000–540,000) complexed with a poorly determined number of intermediate and light chains. As summarized in Table 19-2, the dyneins are divided into two functional classes: *cytosolic* dynein, which is involved in the movement of vesicles and chromosomes, and *axonemal* dynein, which is responsible for the beating of cilia and flagella (discussed later).

TABLE 19-2 Functional Classes of Microtubule Motor Proteins

Class	Members	Cargo	Direction of Movement*
Cytosolic kinesins	Kinesin, Unc-104	Cytosolic vesicles	(+)
Spindle kinesins†	Ncd/KAR3, BimC/Eg5, CENP-E	Spindle and astral MTs, centrosomes, kinetochores	(+) or (−)
Cytosolic dyneins	Cytoplasmic dynein	Cytosolic vesicles, kinetochores during mitosis and meiosis	(−)
Axonemal dyneins	Outer-arm dyneins, inner-arm dyneins‡	A tubule of doublet microtubules in cilia and flagella	(−)

*Movement of motor protein toward the (+) end or (−) end of microtubules.
†Also known as kinesin related proteins (KRPs).
‡Outer-arm dyneins have three heavy chains, and inner-arm dyneins have two heavy chains.

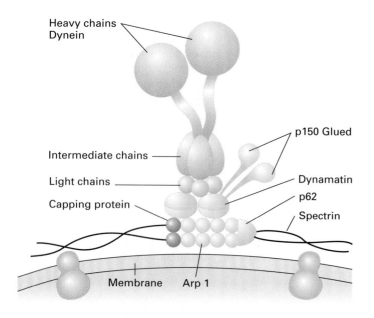

Heavy chains
Dynein

p150 Glued

Intermediate chains

Light chains

Capping protein

Dynamatin

p62

Spectrin

Membrane   Arp 1

◄ **FIGURE 19-25 Schematic diagram of cytosolic dynein and the dynactin heterocomplex.** Dynein (orange) is bound to the dynactin complex (green) through interactions between the dynein light chains and the dynamatin subunits of dynactin. The Arp1 subunits of dynactin are associated with spectrin underlying the cell membrane. The Glued subunits bind microtubules and vesicles. [Adapted from N. Hirokawa, 1998, *Science* **279**:519.]

Dynein-Associated MBPs Tether Cargo to Microtubules

Like kinesin, cytosolic dynein is a two-headed molecule, with two identical or nearly identical heavy chains forming the head domains. However, unlike kinesin, dynein cannot mediate transport by itself. Rather, dynein-related motility requires a large complex of *microtubule-binding proteins* that link vesicles and chromosomes to microtubules but by themselves do not exert force to cause movement. The best-characterized complex is *dynactin*, a heterocomplex of at least eight subunits, including a 150,000-MW protein called *Glued*, the actin-capping protein Arp1, and dynamatin (Figure 19-25). In vitro binding experiments show that dynactin enhances dynein-dependent motility, possibly through interaction with microtubules and vesicles. The microtubule binding site lies

in the N-terminal region of Glued. This region contains a 57-residue microtubule-binding motif that also is present in a vesicle-binding protein called *CLIP-170* and the yeast protein BIK1. Like dynactin, these proteins are implicated in the trafficking of endocytotic vesicles. The light chains of dynein interact with the dynamatin subunits of dynactin. One model proposes that dynactin tethers a vesicle to a microtubule, while dynein generates the force and polarity for movement.

As we discuss later, several lines of evidence suggest that the dynein-dynactin complex and another complex, the *nuclear/mitotic apparatus (NuMA) protein*, mediate association of microtubules with chromosomes during mitosis. In vitro studies show that truncated NuMA protein binds microtubules if the C-terminal region is retained. As in assembly MAPs, the C-terminal region of NuMA protein is highly acidic, and ionic interactions may mediate binding to microtubules.

Multiple Motor Proteins Are Associated with Membrane Vesicles

The identification of (+) and (−) end–directed microtubule motor proteins (kinesin and cytosolic dynein, respectively) explains not only how movement of vesicles is powered but also how the direction of movement is controlled (Figure 19-26). The direction of vesicle transport also depends on the

▶ **FIGURE 19-26 A general model for kinesin- and dynein-mediated transport in a typical cell.** The array of microtubules, with their (+) ends pointing toward the cell periphery, radiates from an MTOC in the Golgi region. Kinesin-dependent anterograde transport conveys mitochondria (carried by KIF1B), lysosomes, and an assortment of membrane vesicles to the endoplasmic reticulum (ER) or cell periphery. Cytosolic dynein–dependent retrograde transport conveys elements of the ER, late endosomes, and lysosomes to the cell center. [Adapted from N. Hirokawa, 1998, *Science* **279**:519.]

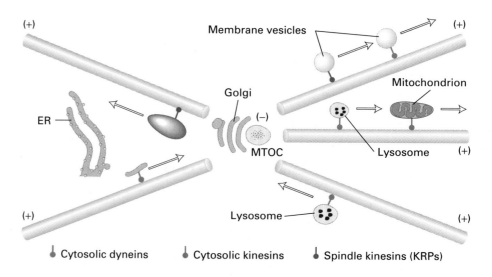

(+)   Membrane vesicles   (+)

Golgi

Mitochondrion

ER

(−)

MTOC

Lysosome   (+)

(+)

Lysosome

(+)

↓ Cytosolic dyneins   ↓ Cytosolic kinesins   ↓ Spindle kinesins (KRPs)

orientation of microtubules, which is fixed by the MTOC. Some cargoes, such as pigment granules, can alternate their direction of movement along a single microtubule. In this case, both anterograde and retrograde microtubule motor proteins must be associated with a microtubule. At any one time, however, only one motor protein is active or, alternatively, only one motor protein is bound to the vesicle.

Increasing evidence suggests that transport of some vesicles is more complicated than depicted in Figure 19-26. For example, during endocytosis vesicles from the plasma membrane are carried inward, while during secretion vesicles from the ER and Golgi are moved outward (Chapter 17). In both processes, a vesicle must traverse microtubule-poor but microfilament-rich regions in the cell. Several complementary experiments indicate that microtubule and microfilament motor proteins bind to the same membrane vesicles and cooperate in their transport. One piece of evidence was obtained from microscopy of vesicle movements in extruded squid axoplasm. As observed many times before, membrane vesicles traveled along microtubule tracks; however, at the periphery of the extruded axoplasm, movement continued even though no microtubules were present. This region contained microfilaments, and subsequent experiments demonstrated that a vesicle could move on a microtubule *or* a microfilament. Thus at least two motor proteins, myosin and either kinesin or dynein, must be bound to the same vesicle. The discovery that a given vesicle can travel along both cytoskeletal systems suggests that in a neuron, synaptic vesicles are transported at a fast rate by kinesin in the microtubule-rich axon and then travel through the actin-rich cortex at the nerve terminal on a myosin motor.

SUMMARY Kinesin, Dynein, and Intracellular Transport

- Two families of motor proteins, kinesin and dynein, transport membrane-bounded vesicles, proteins, and organelles along microtubules.

- Nearly all kinesins move cargo toward the (+) end of microtubules (anterograde transport), whereas dyneins transport cargo toward the (−) end (retrograde transport).

- The dimeric kinesin head domain binds microtubules and ATP, and the tail domain binds vesicles (see Figure 19-23). Although the structure of the kinesin head is similar to that of myosin, it lacks a rigid neck domain. Thus the model of myosin-dependent motility may not apply to kinesin.

- Each type of membrane vesicle is transported by its own kinesin motor protein. The specificity of binding may reside in the tail domain, which is unique to each kinesin.

- Cytosolic dyneins are linked to their cargoes (vesicles and chromosomes) by large complexes of microtubule-binding proteins (MBPs), such as dynactin (see Figure 19-25). Once tethered to a cargo, the dynein transports it to the final destination.

- Axonal transport is a well-studied model system for understanding anterograde transport by kinesin and retrograde transport by cytosolic dynein in a typical eukaryotic cell (see Figure 19-26).

- In microtubule-poor regions of the cell, vesicles probably are transported along microfilaments powered by a myosin motor.

19.4 Cilia and Flagella: Structure and Movement

Swimming is the major form of movement exhibited by sperm and by many protozoans. Some cells are propelled at velocities approaching 1 mm/s by the beating of **cilia** and **flagella,** flexible membrane extensions of the cell. Cilia and flagella range in length from a few microns to more than 2 mm in the case of some insect sperm flagella.

Although cilia and flagella are the same, they were given different names before their structures were studied. Typically, cells possess one or two long flagella, whereas ciliated cells have many short cilia. For example, the mammalian spermatozoon has a single flagellum, the unicellular green alga *Chlamydomonas* has two flagella, and the unicellular protozoan *Paramecium* is covered with a few thousand cilia, which are used both to move and to bring in food particles. In mammals, many epithelial cells are ciliated in order to sweep materials across the tissue surface. For instance, huge numbers of cilia (more than $10^7/mm^2$) cover the surfaces of mammalian respiratory passages (the nose, pharynx, and trachea), where they dislodge and expel particulate matter that collects in the mucus secretions of these tissues.

Ciliary and flagellar beating is characterized by a series of bends, originating at the base of the structure and propagated toward the tip. High-speed strobe microscopy allows the waveform of the beat to be seen (Figure 19-27). Beating can be planar or three-dimensional; like waves that you have studied in physics, it can be described by its amplitude, wavelength, and frequency. The bends push against the surrounding fluid, propelling the cell forward or moving the fluid across a fixed epithelium.

All Eukaryotic Cilia and Flagella Contain Bundles of Doublet Microtubules

Virtually all eukaryotic cilia and flagella are remarkably similar in their organization, possessing a central bundle of microtubules, called the **axoneme,** in which nine outer doublet microtubules surround a central pair of singlet microtubules (Figure 19-28). This characteristic "9 + 2" arrangement of microtubules is seen when the axoneme is viewed in cross

(a)

(b)

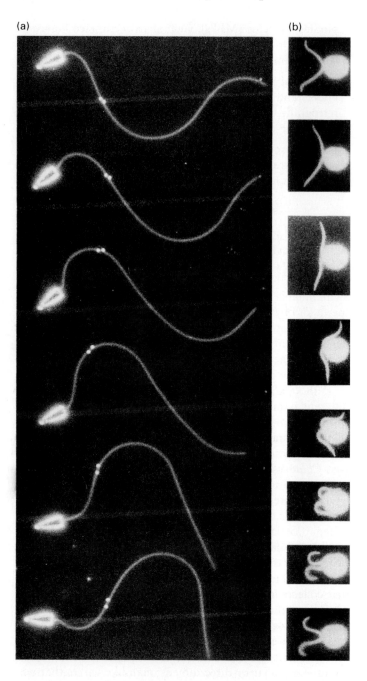

◀ **FIGURE 19-27 Flagellar motions in sperm and *Chlamydomonas*.** In both cases, the cells are moving to the left. (a) In the typical sperm flagellum, successive waves of bending originate at the base and are propagated out toward the tip; these waves push against the water and propel the cell forward. Captured in this multiple-exposure sequence, a bend at the base of the sperm in the first *(top)* frame has moved distally halfway along the flagellum by the last frame. A pair of gold beads on the flagellum are seen to slide apart as the bend moves through their region. (b) Beating of the two flagella on *Chlamydomonas* occurs in two stages, called the *effective stroke* and the *recovery stroke*. The flagella are stiff during the effective stroke (top three frames) and move in a way that pulls the organism through the water, similar to the breast stroke of a swimmer. During the recovery stroke (remaining frames), a different wave of bending moves outward from the bases of the flagella, pushing the flagella along the surface of the cell until they reach the position to initiate another effective stroke. Beating commonly occurs 5–10 times per second. [Part (a) from C. Brokaw, 1991, *J. Cell Biol.* **114**(6): cover photo; courtesy of C. Brokaw. Part (b) courtesy of S. Goldstein.]

Each triplet contains one complete 13-protofilament microtubule, the A tubule, fused to the incomplete B tubule, which in turn is fused to the incomplete C tubule (see Figure 19-3). The A and B tubules of basal bodies continue into the axonemal shaft, whereas the C tubule terminates within the transition zone between the basal body and the shaft. The two central tubules in a flagellum or a cilium also end in the transition zone, above the basal body. As we will see later, the basal body plays an important role in initiating the growth of the axoneme.

Within the axoneme, the two central singlet and nine outer doublet microtubules are continuous for the entire length of the structure. Doublet microtubules, which represent a specialized polymer of tubulin, are found only in the axoneme. Permanently attached to the A tubule of each doublet microtubule is an inner and an outer row of dynein arms (see Figure 19-28a). These dyneins reach out to the B tubule of the neighboring doublet. The junction between A and B tubules of one doublet is probably strengthened by the protein *tektin*, a highly α-helical protein that is similar in structure to intermediate-filament proteins. Each tektin filament, which is 2 nm in diameter and approximately 48 nm long, runs longitudinally along the wall of the outer doublet where the A tubule is joined to the B tubule.

The axoneme is held together by three sets of protein cross-links (see Figure 19-28a). The central pair of singlet microtubules are connected by periodic bridges, like rungs on a ladder, and are surrounded by a fibrous structure termed the *inner sheath*. A second set of linkers, composed of the protein *nexin*, joins adjacent outer doublet microtubules. Spaced every 86 nm along the axoneme, nexin is proposed to be part of a dynein regulatory complex. *Radial spokes*, which radiate from the central singlets to each A tubule of the outer doublets, form the third linkage system.

section with the electron microscope. As shown in Figure 19-3, each doublet microtubule consists of A and B tubules, or subfibers: the A tubule is a complete microtubule with 13 protofilaments, while the B tubule contains 10 protofilaments. The bundle of microtubules comprising the axoneme is surrounded by the plasma membrane. Regardless of the organism or cell type, the axoneme is about 0.25 μm in diameter, but it varies greatly in length, from a few microns to more than 2 mm.

At its point of attachment to the cell, the axoneme connects with the **basal body** (Figure 19-29). Like centrioles, basal bodies are cylindrical structures, about 0.4 μm long and 0.2 μm wide, which contain nine triplet microtubules.

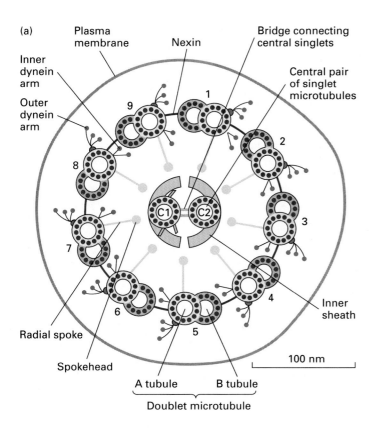

(a)

Plasma membrane · Nexin · Bridge connecting central singlets · Central pair of singlet microtubules · Inner dynein arm · Outer dynein arm · Inner sheath · Radial spoke · Spokehead · A tubule · B tubule · Doublet microtubule

100 nm

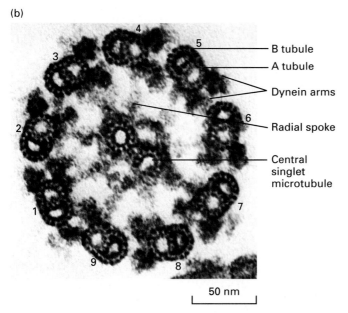

(b)

B tubule · A tubule · Dynein arms · Radial spoke · Central singlet microtubule

50 nm

▲ **FIGURE 19-28 Structure of ciliary and flagellar axonemes.**
(a) Cross-sectional diagram of a typical flagellum showing its major structures. The dynein arms and radial spokes with attached heads occur only at intervals along the longitudinal axis. The central microtubules, C1 and C2, are distinguished by fibers bound only to C1. (b) Micrograph of a transverse section through an isolated demembranated cilium. The two central singlet microtubules are surrounded by nine outer doublets, each composed of an A and a B subfiber. [Part (b) courtesy of L. Tilney; see U. W. Goodenough and J. E. Heuser, 1985, *J. Cell Biol.* **100**:2008.]

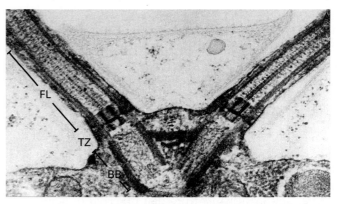

▲ **FIGURE 19-29 Electron micrograph of the basal regions of the two flagella in *Chlamydomonas reinhardtii*.** The bundles of microtubules and some fibers connecting them are visible in the flagella (FL). The two basal bodies (BB) form the point of a "V"; a transition zone (TZ) between the basal body and flagellum proper contains two dense-staining cylinders of unknown structure. [From B. Huang et al., 1982, *Cell* **29**:745.]

The biflagellated, unicellular alga *Chlamydomonas reinhardtii* has proved especially amenable to biochemical and genetic studies on the function, structure, and assembly of flagella. A population of cells, shorn of their flagella by mechanical or chemical methods, provide flagella in good purity and high yield, and the deflagellated cells quickly regenerate new flagella. Analysis of the sheared flagella by two-dimensional gel electrophoresis reveals approximately 250 discrete polypeptides, in addition to α- and β-tubulin. The functions of these polypeptides have been assessed by analysis of flagella from *Chlamydomonas* mutants that are nonmotile or otherwise defective in flagellar function. Some nonmotile mutants, for example, lack an entire substructure, such as the radial spokes or central-pair microtubules. Many mutants that are missing a particular flagellar substructure also have been found to lack certain specific proteins, thus permitting these proteins to be assigned to a specific substructure and associated with specific genes. Such studies have identified 17 polypeptides that are components of the radial spokes and spoke heads. The components of the inner and outer dynein arms, the central-pair microtubules, and other axonemal structures have been similarly identified.

Although the 9 + 2 pattern is the fundamental pattern of virtually all cilia and flagella, the axonemes of certain protozoans and some insect sperm show some interesting variations. The simplest such axoneme, containing three doublet microtubules and no central singlets (3 + 0) is found in *Daplius*, a parasitic protozoan. Its flagellum beats slowly (1.5 beats/s) in a helical pattern. Other axonemes consist of 6 + 0 or 9 + 0 arrangements of microtubules. These atypical cilia and flagella, which are all motile, show that the central pair of singlet microtubules is not necessary for axonemal beating and that fewer than nine outer doublets can sustain motility, but at a lower frequency.

Ciliary and Flagellar Beating Are Produced by Controlled Sliding of Outer Doublet Microtubules

Having examined the complex structure of cilia and flagella, we now discuss how the various components contribute to their characteristic motions. Cilia and flagella, from which the plasma membrane has been removed by nonionic detergents, can beat when ATP is added; this in vitro movement can be indistinguishable from that observed in living cells. Thus the forces that generate movement must reside within the axoneme and are not located in the plasma membrane or elsewhere in the cell body.

As in the movement of muscle during contraction, the basis for axonemal movement is the sliding of protein filaments relative to one another. In cilia and flagella, the filaments are the doublet microtubules, all of which are arranged with their (+) end at the outer tip of the axoneme. Axonemal bending is produced by forces that cause sliding between pairs of doublet microtubules. The active sliding occurs all along the axoneme, so that the resulting bends can be propagated without damping.

Sliding was seen in an activation-type experiment. Demembranated axonemes were briefly treated with proteolytic enzymes such as trypsin or elastase to digest the structural linkages and the radial spokes. Upon addition of ATP, the digested axonemes telescoped apart, but no bending was observed (Figure 19-30). The sliding was often nearly complete, so that the resulting structure was greater than five times longer than the original length of the axoneme. Clearly then, the ATP-dependent movement of outer doublets must be restricted by cross-linkage proteins in order for sliding to be converted into bending of an axoneme.

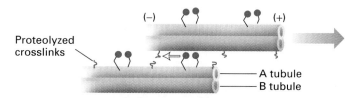

▲ **FIGURE 19-31 Model for dynein-mediated sliding of axonemal outer doublet microtubules.** The dynein arms attached to the A subfiber of one microtubule walk along the B subfiber of the adjacent doublet toward its (−) end (small arrow), moving this microtubule in the opposite direction (large arrow). When the nexin cross-links are broken, as shown here, sliding can continue unimpeded.

Dynein Arms Generate the Sliding Forces in Axonemes

Once it was clear that the doublet microtubules in axonemes slide past each other, researchers sought to identify the force-generating proteins responsible for this movement. The inner- and outer-arm dyneins, which bridge between the doublet microtubules, were the best candidates. The identity of dynein as the motor protein in axonemes is supported by various findings. For instance, cilia and flagella possess an active ATPase that is associated with the dynein arms. In addition, removal of outer-arm dyneins by treatment with high-salt solutions reduces the rate of ATP hydrolysis, microtubule sliding, and beat frequency of isolated axonemes by 50 percent. When the extracted outer-arm dyneins are added back to salt-stripped axonemes, both the ATPase activity and the beat frequency are restored, and electron microscopy reveals that the outer arms have reattached to the proper places.

Based on the polarity and direction of sliding of the doublet microtubules, we can propose a model in which the dynein arms on the A tubule of one doublet "walk" along the adjacent doublet's B tubule toward its base, the (−) end (Figure 19-31). The force producing active sliding requires ATP and is caused by successive formation and breakage of cross-bridges between the dynein arm and the B tubule. Successive binding and hydrolysis of ATP causes the dynein arms to successively release from and attach to the adjacent doublet. Although this general model most likely is correct, many important details such as the mechanism of force transduction by dynein are still unknown.

Axonemal Dyneins Are Multiheaded Motor Proteins

Axonemal dyneins are complex multimers of heavy chains, intermediate chains, and light chains. Isolated axonemal dyneins, when slightly denatured and spread out on an electron microscope grid, are seen as a bouquet of two or three "blossoms" (Figure 19-32a). Each blossom consists of a large globular domain attached to a small globular domain

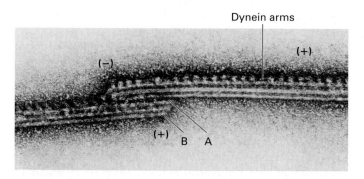

▲ **FIGURE 19-30 Electron micrograph of two doublet microtubules in a protease-treated axoneme incubated with ATP.** In the absence of cross-linking proteins, which are removed by protease, excessive sliding of doublet microtubules occurs. [Courtesy of P. Satir.]

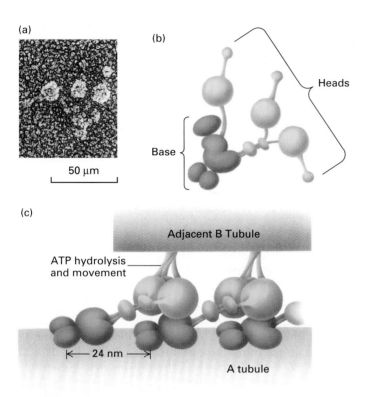

▲ FIGURE 19-32 Structure of axonemal dynein. (a) Electron micrograph of freeze-etched outer-arm dynein from *Tetrahymena* cilia reveal three globular "blossoms" connected by stems to a common base. (b) An artist's interpretation of the electron micrographs shows the arrangement of globular domains and short stalks. (c) Model showing the attachment of the outer dynein arm to the A tubule of one doublet and the cross-bridges to the B tubule of an adjacent doublet. The attachment to the A tubule is stable. In the presence of ATP, the successive formation and breakage of cross-bridges to the adjacent B tubule leads to movement of one doublet relative to the other. [Part (a) from U. W. Goodenough and J. E. Heuser, 1984, *J. Mol. Biol.* **18**:1083.]

(the "head") through a short stalk; another stalk connects one or more blossoms to a common base (Figure 19-32b). The base is thought to be the site where the dynein arm attaches to the A tubule, while the small globular heads bind to the adjacent B tubule (Figure 19-32c).

Each globular head and its stalk is formed from a single dynein heavy chain. The dynein heavy chain is enormous, approximately 4,500 amino acids in length with a molecular weight exceeding 540,000. Each heavy chain is capable of hydrolyzing ATP, and on the basis of sequences commonly found at the ATP-binding sites in other proteins, the ATP-binding domain of axonemal dynein is predicted to lie in the globular head portion of the heavy chain. The intermediate and light chains, thought to form the base of the dynein arm, help mediate attachment of the dynein arm to the A tubule and may also participate in regulating dynein activity. These base proteins thus are analogous to the MBP complexes associated with cytosolic dynein.

Axonemes contain at least eight or nine different dynein heavy chains. All inner dynein arms are two-headed structures, containing two heavy chains. The outer dynein arms have two heavy chains (e.g., in a sea urchin sperm flagellum) or three heavy chains (e.g., in *Chlamydomonas* flagella).

Conversion of Microtubule Sliding into Axonemal Bending Depends on Inner-Arm Dyneins

As we saw earlier, flagellar and ciliary beating is characterized by the propagation of bends that originate from the base of the axoneme (see Figure 19-27). On the other hand, the active sliding of microtubules relative to each other is a linear phenomenon (see Figure 19-31). How, then, is microtubule sliding converted to bending of a cilium or flagellum?

A bend is formed between a region of sliding and a region that resists sliding. Bending is regulated by controlling the regions where dynein is active along and around the axoneme. A close examination of the axoneme cross-section reveals that the nine outer doublets and their dynein arms are arranged in a circle so that, when viewed from the base of the axoneme, the arms all point clockwise. Since the dynein arms walk in only one direction, toward the (−) end, and each doublet slides down only one of its two neighboring doublets, active sliding in one half of the axoneme produces bending toward one side and active sliding in the other half produces bending toward the opposite side (see Figure 19-27a). By regulating the timing and location in which dynein arms are active, the axoneme can propagate bends in both directions from base to tip.

Genetic studies of mutant *Chlamydomonas* with abnormal motility reveal that the inner- and outer-arm dyneins contribute differently to the waveform and beat frequency of an axoneme. For example, the absence of one set of inner arms affects the waveform of flagellar beating. In contrast, mutant flagella lacking outer arms have normal waveform but slower beat frequencies. Thus the outer-arm dyneins accelerate active sliding of the outer doublets but do not contribute to bending. In contrast, the inner-arm dyneins are responsible for producing the sliding forces that are converted to bending; this suggests that inner-arm dyneins are essential for bending.

Proteins Associated with Radial Spokes May Control Flagellar Beat

Several lines of evidence indicate that the radial spokes and central-pair microtubules play a critical role in controlling the bending of a flagellum. First, mutant flagella lacking radial spokes are paralyzed. Further, a *dynein regulatory complex*, located at the junction between the radial spokes and inner dynein arms, has recently been identified by genetic suppressor studies. One hypothesis is that phosphorylation of the inner dynein arm inactivates it, while dephosphorylation activates it to cause sliding between outer doublet microtubules. A bend is propagated when inner-arm dynein is inactivated in one region and activated in a neighboring region.

Axonemal Microtubules Are Dynamic and Stable

For axonemes to participate in movement, they must be stable structures anchored by at least one end. As noted already, a cilium or flagellum is anchored at its cytosolic end to a basal body. In addition to its anchoring role, the basal body serves as a nucleus for the assembly of flagellar microtubules. Recall that the basal body has nine triplet microtubules. The nine A and B tubules of these triplets appear to initiate assembly of the nine outer doublet microtubules of the cilium or flagellum by growing outward from the basal body during elongation of the axonemal shaft.

Studies on the assembly of flagella and cilia provided the first evidence that a microtubule elongates by incorporating tubulin subunits at its tip. This model was originally proved by autoradiography of cells that were regenerating their flagella in the presence of radioactive tubulin subunits (and other axonemal components). More recent studies using a recombinant *Chlamydomonas* cell have confirmed that addition of tubulin subunits and incorporation of other axonemal components occur at the distal end of a flagellum (Figure 19-33). In these experiments, a *Chlamydomonas* cell expressing an epitope-tagged tubulin subunit was mated with a wild-type cell whose flagella had been amputated.

After mating, which involves fusion of the two cells and mixing of their cytoplasms, the diploid cell regenerated full-length flagella by incorporating the tagged tubulin subunits. Antibodies to the epitope tag showed that the recombinant tubulin was localized within the distal tips of the regenerated flagella. This pattern could arise only if elongation occurs at the tip and not the base.

SUMMARY Cilia and Flagella: Structure and Movement

- Flagellar beating propels cells forward, and ciliary beating sweeps materials across tissues.

- Despite their different names, flagella and cilia have the same axoneme structure, including nine doublet microtubules arranged in a circle around two central singlet microtubules (see Figure 19-28).

- Walking of dynein arms extending from one doublet toward the (−) end of a neighboring doublet generates a sliding force in the axoneme (see Figure 19-31). This force is converted into a bend by regions that resist sliding.

- Axonemal dynein is larger and more complex than cytosolic dynein. Instead of the vesicle cargo transported by cytosolic dynein, axonemal dynein moves one microtubule across the surface of its neighboring microtubule (see Figure 19-32c).

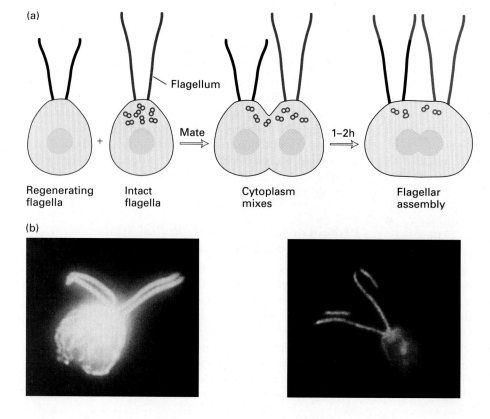

(a)

Regenerating flagella   Intact flagella   Flagellum   Mate   Cytoplasm mixes   1–2h   Flagellar assembly

(b)

◀ FIGURE 19-33 Assembly of flagellar microtubules. (a) Schematic diagram of an experiment in which two *Chlamydomonas* cells were mated. One cell has had its flagella amputated and is regenerating them, while the other cell contains a soluble pool of tubulin tagged with a small epitope. After mating, the cytoplasm mixes in the dikaryon, and the amputated flagella continue to regenerate, now using tubulin subunits from both parental cells. (b) When mated cells are treated with a tubulin-specific antibody, all four flagella are stained full-length *(left)*. However, when similar cells are stained with the antibody specific for the epitope tag on the tubulin, then the two flagella from the parental cell in which the epitope tag was expressed are labeled along their full length, but only the tips of the two regenerated flagella are labeled *(right)*. This staining pattern demonstrates that the epitope-tagged tubulin subunits have added to the tips, the (+) end, of the regenerating microtubules. [Part (b) courtesy of K. Johnson and J. Rosenbaum.]

- The (−) ends of the microtubules in a cilium or flagellum are anchored in the basal body and are extensions of microtubules located there. Elongation of cilia and flagella occurs by addition of αβ-tubulin subunits to the distal (+) end of axonemal microtubules.

19.5 Microtubule Dynamics and Motor Proteins during Mitosis

Mitosis is the process that partitions newly replicated chromosomes equally into separate parts of a cell. The last step in the cell cycle, mitosis takes about 1 hour in an actively dividing animal cell. During that period, the cell builds and then disassembles a specialized microtubule structure, the *mitotic apparatus.* Larger than the nucleus, the mitotic apparatus is designed to attach and capture chromosomes, align the chromosomes, and then separate them, so that the genetic material is evenly partitioned to each daughter cell. Fifteen hours later, the whole process is repeated by the two daughter cells. During **interphase,** the period between mitoses, chromosomes are decondensed and dispersed throughout the nucleus. Although the chromosomes are not visible by light microscopy during this period, DNA replication occurs during the S (synthesis) phase of interphase (see Figure 13-1).

Figure 19-34 depicts the characteristic series of events that can be observed by light microscopy during mitosis in a eukaryotic cell. Although the events unfold continuously, they are conventionally divided into four substages: **prophase, metaphase, anaphase,** and **telophase.** The beginning of mitosis is signaled by the appearance of visible condensed chromosomes, stainable as thin threads inside the nucleus. By late prophase, each chromosome is visible as two identical coiled filaments, the **chromatids** (often called *sister chromatids*), held together at a constricted region, the **centromere.** Each chromatid contains one of the two new daughter DNA molecules produced in the preceding S phase of the cell cycle; thus each cell that enters mitosis has four copies of each chromosomal DNA, designated 4*n.*

The mitotic cycle of chromosome separation is linked to two checkpoints in the cell cycle: breakdown of the nuclear envelope during late prophase and attachment of the microtubules to the kinetochores of sister chromatids at the metaphase-anaphase transition. We saw in Chapter 13 that the cell cycle, and hence cell replication, is controlled primarily by regulating the initiation of DNA synthesis, entry into mitosis, and chromatid separation. In this section, we focus on the mechanism by which microtubules separate chromosomes during mitosis in a "typical" animal cell. Mistakes in mitosis can lead to missing or extra chromosomes, causing abnormal patterns of development when they occur during embryogenesis and pathologies when they occur after

birth. To ensure that mitosis proceeds without errors during the billions of cell divisions that occur in the life span of an organism, a highly redundant mechanism has evolved in which each crucial step is carried out concurrently by microtubule motor proteins and microtubule assembly dynamics.

The Mitotic Apparatus Is a Microtubule Machine for Separating Chromosomes

The mitotic apparatus has no fixed structure: it is constantly changing during mitosis (Figure 19-35). For one brief moment at metaphase, however, when the chromosomes are aligned at the equator of the cell, the mitotic apparatus appears static. We will begin our discussion by examining the structure of the mitotic apparatus at metaphase and then discuss how it first organizes chromosomes during prophase, how it separates chromosomes during anaphase, and how it determines where cells are cleaved during telophase.

The mitotic apparatus at metaphase is organized into two parts (Figure 19-36): (1) a central **mitotic spindle**—a bilaterally symmetrical bundle of microtubules with the overall shape of a football, but divided into opposing halves at the equator of the cell by a plate of metaphase chromosomes—and (2) a pair of **asters**—a tuft of microtubules at each pole of the spindle.

In each half of the spindle, a single centrosome at the pole organizes three distinct sets of microtubules, whose (−) ends all point toward the centrosome (Figure 19-36b). One set, the *astral microtubules,* forms the aster; they radiate outward from the centrosome toward the cortex of the cell, where they help position the mitotic apparatus and later help to determine the cleavage plane during cytokinesis. The other two sets of microtubules compose the spindle. The *kinetochore microtubules* attach to chromosomes at specialized attachment sites on the chromosomes called **kinetochores.** The third set, *polar microtubules,* do not interact with chromosomes but instead interdigitate with polar microtubules from the opposite pole.

In an overlap zone at the equator, two types of interactions hold the spindle halves together to form the bilaterally symmetric mitotic apparatus: (1) lateral interactions between the (+) ends of the interdigitating polar microtubules from each pole and (2) end-on interactions between the kinetochore microtubules from each pole and the kinetochores of the sister chromatids.

The spindle pole–aster organization of the mitotic apparatus is basic to mitosis in all organisms, but the appearance of the mitotic apparatus can vary widely. The number of microtubules in a spindle, the overall size of the mitotic apparatus, and the timing and duration of mitotic movements all vary among different organisms. In addition, organisms differ in the length or number of their astral microtubules.

In the common baker's yeast, *S. cerevisiae,* for example, mitosis is carried out by a structurally simple mitotic apparatus that lacks centriole-based centrosomes and asters (Figure 19-37). Instead of a centrosome, the microtubules

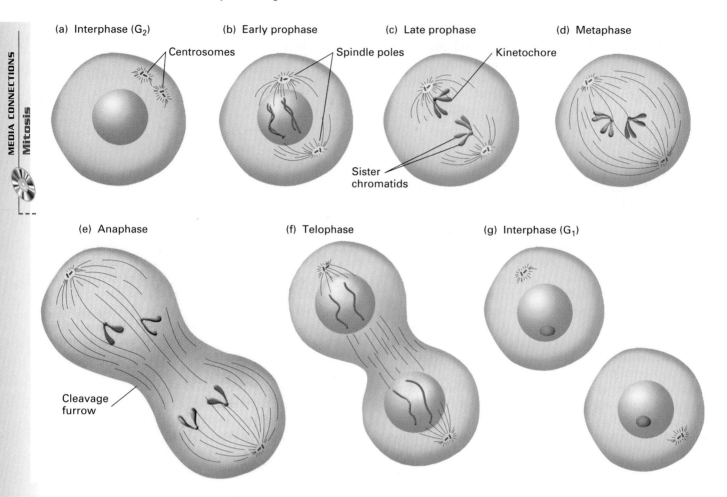

(a) Interphase (G$_2$) (b) Early prophase (c) Late prophase (d) Metaphase

Centrosomes Spindle poles Kinetochore

Sister chromatids

(e) Anaphase (f) Telophase (g) Interphase (G$_1$)

Cleavage furrow

▲ **FIGURE 19-34 The stages of mitosis and cytokinesis in an animal cell.** (Morphological types of chromosomes are distinguished by color.) (a) *Interphase:* The G$_2$ stage of interphase immediately precedes the beginning of mitosis and follows chromosomal DNA replication during the S phase. The chromosomes, each containing a sister chromatid, are still dispersed and not visible as distinct structures. During interphase, the centrioles also are replicated, forming small daughter centrioles. (b) *Early prophase:* The centrosomes, each with a daughter centriole, begin moving toward opposite poles of the cell. The chromosomes can be seen as long threads, and the nuclear membrane begins to disaggregate into small vesicles. (c) *Middle and late prophase:* Chromosome condensation is completed; each visible chromosome structure is composed of two chromatids held together at their centromeres. The microtubular spindle fibers begin to radiate from the regions just adjacent to the centrosomes, which are moving closer to their poles. Some spindle fibers reach from pole to pole; most go to chromatids and attach at kinetochores. (d) *Metaphase:* The chromosomes move toward the equator of the cell, where they become aligned in the equatorial plane. The sister chromatids have not yet separated. This is the phase in which morphological studies of chromosomes are usually carried out. (e) *Anaphase:* The two sister chromatids separate into independent chromosomes. Each contains a centromere that is linked by a spindle fiber to one pole, to which it moves. Simultaneously, the cell elongates, as do the pole-to-pole spindles. Cytokinesis begins as the cleavage furrow starts to form. (f) *Telophase:* New nuclear membranes form around the daughter nuclei; the chromosomes uncoil and become less distinct; and the nucleolus becomes visible again. Cytokinesis is nearly complete, and the spindle disappears as the microtubules and other fibers depolymerize. Throughout mitosis the "daughter" centriole at each pole grows, so that by telophase each of the emerging daughter cells has two full-length centrioles. Upon the completion of cytokinesis, each daughter cell enters the G$_1$ phase of the cell cycle and proceeds again around the cycle.

are organized around a *spindle pole body,* a trilaminated structure located in the nuclear membrane, which does not break down during mitosis. Furthermore, because a yeast cell is small, it does not require well-developed asters to assist in mitosis. Thus, the yeast mitotic apparatus comprises just a spindle, which itself is constructed from a minimal number of kinetochore and polar microtubules. In *S. cerevisiae,* which possesses 16 chromosomes, the spindle contains 32 kinetochore microtubules plus a few polar microtubules. (At metaphase, each kinetochore is attached to one microtubule; thus each chromosome is attached to two kinetochore microtubules.) Because the spindle pole body is functionally equivalent to, but structurally different from, an animal cell centrosome, the two structures must share centrosomal proteins

(a)

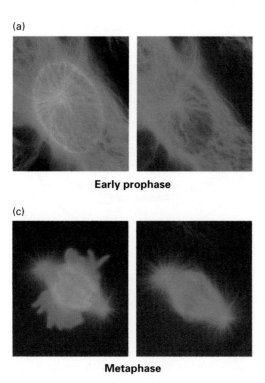

Early prophase

(b)

Late prophase

(c)

Metaphase

(d)

Anaphase

▲ **FIGURE 19-35 Fluorescence micrographs showing the organization of chromosomes and microtubules during four mitotic stages.** Cultured PtK2 fibroblasts were stained with a fluorescent anti-tubulin antibody (green) and the DNA-binding dye ethidium homodimer (purple). Both photographs in each panel show the same cell, stained with both reagents *(top)* or just anti-tubulin antibody *(bottom)*. Thus, in the top panels the areas of overlap are blue, and serve to highlight DNA. (a) During early prophase, the nucleus is surrounded by interphase microtubules. (b) By late prophase, the nuclear membrane has broken down and the chromosomes have condensed. The replicated centrosomes (centrioles) have migrated to the poles of the developing spindle. The microtubules radiate from the poles. (c) At metaphase, the chromosomes have aligned midway between the poles to form the metaphase plate. Dense bundles of microtubules connect the chromosomes to the poles. The purple stain of the chromosomes is visible outside the spindle, where there is no overlap with the green of the microtubules. (d) During late anaphase the chromosomes are pulled to the poles along the radiating microtubules. [From J. C. Waters, R. W. Cole, and C. L. Rieder, 1993, *J. Cell Biol.* **122:**361; courtesy of C. L. Rieder.]

such as γ-tubulin that act to organize the mitotic spindle.

Mitotic events in plant cells are generally similar to the events observed in animal cells. Although most higher plant cells do not contain visible centrioles, an analogous region of the plant cell acts as a microtubule-organizing center, from which the spindle microtubules radiate (Figure 19-38). Moreover, compared with an animal cell, the shape of a plant cell does not change greatly in mitosis because it is surrounded by a rigid cell wall. During telophase, the new cell membrane and cell wall are formed from membrane vesicles that fuse together in a plane perpendicular to a line separating the two nuclei.

The Kinetochore Is a Specialized Attachment Site at the Chromosome Centromere

The sister chromatids of a metaphase chromosome are transported to each pole along the kinetochore microtubules. In terms of cargoes delivered on microtubules, a chromosome is a much different piece of baggage than the membrane vesicles and organelles we have discussed in previous sections. Figure 19-39 depicts the attachment of kinetochore microtubules to the kinetochore, a platelike structure lying within the centromere, a small, highly specialized region of the chromosome. (Do not confuse *centromere* with *centrosome:* the centromere is a region of the chromosome, while the centrosome is a microtubule-organizing center.) The centromere is recognized as a constriction in the condensed chromosome where the sister chromatids are most closely associated. The location of the centromere and hence that of the kinetochore is directly controlled by a specific sequence of chromosomal DNA termed *centromeric DNA* (Chapter 9).

The kinetochore is first recognizable during late prophase, after the chromosomes have condensed but well before the mitotic apparatus has assembled. The complex structure of the kinetochore is revealed in ultrathin sections of chromosomes as a stack of disklike plates (see Figure 19-39a, inset). Some of the protein components of kinetochores have been

(a)

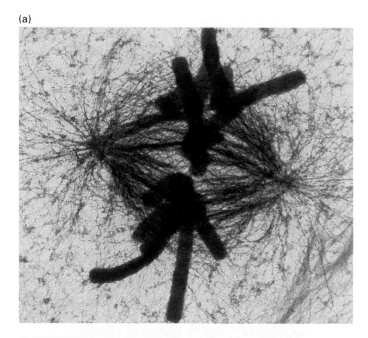

(b)

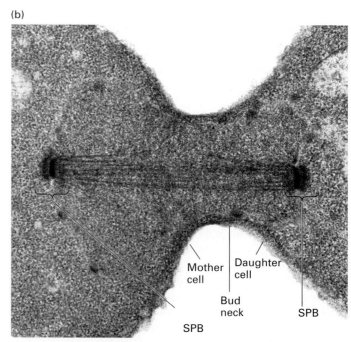

▲ **FIGURE 19-36 (a) High-voltage electron micrograph of the mitotic apparatus in a metaphase mammalian cell.** To visualize the spindle microtubules more clearly, biotin-tagged anti-tubulin antibodies were added to make microtubules more massive. The large cylindrical objects are chromosomes. (b) Diagram showing the three sets of microtubules (MTs) in the mitotic apparatus. Centered around the poles are astral microtubules, kinetochore microtubules, which are connected to chromosomes (blue), and polar microtubules. The (+) ends of these microtubules all point away from the centrosome at each pole. [Part (a) courtesy of J. R. McIntosh.]

(a)

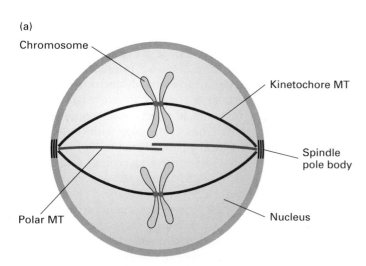

▲ **FIGURE 19-37 Mitotic apparatus in *S. cerevisiae.*** In yeast, the nucleus remains intact during mitosis; thus the chromosomes are isolated from direct interaction with the cytosol. (a) Spindle pole bodies, which are attached to the nuclear membrane, organize the spindle microtubules (MTs). Each of the 16 chromosomes is attached to two kinetochore microtubules. (b) A yeast cell divides by a process in which a bud emerges, enlarges, and eventually is pinched off, forming the daughter cell. This photomicrograph, taken late in mitosis, shows the two spindle pole bodies (SPB) and microtubules connecting them. Note the absence of astral microtubules.

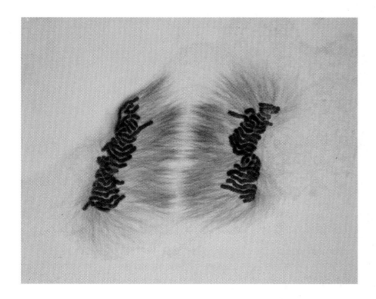

◀ FIGURE 19-38 Fluorescence micrograph showing arrangement of microtubules (red) and chromosomes (blue) during anaphase in a plant cell. At this stage, the sister chromatids have separated and are moving toward opposite poles of the spindle. [Courtesy of Andrew Bajer.]

identified by their reaction with antibodies that specifically recognize kinetochores. For unknown reasons, these antibodies are frequently produced by patients suffering from scleroderma (an autoimmune disease of unknown origin that causes fibrosis of connective tissue). Using these scleroderma autoantibodies, researchers have identified four proteins that are localized to the inner layer of the mammalian kinetochore. Studies with other antibodies have detected cytosolic dynein and a kinesin-related protein, CENP-E, in the fibrous corona on the surface of the kinetochore.

Our understanding of how kinetochores link centromeres to microtubules is most advanced in yeast. In Chapter 9, we described how centromeric (CEN) DNA sequences from yeast chromosomes can be identified by their ability to make self-replicating plasmids into artificial chromosomes that can be passed from mother to daughter at mitosis (see Figure 9-40). Sequence analysis of cloned CEN DNAs from yeast chromosomes reveals that they are generally organized into three regions, denoted CDEs, or *centromere DNA elements*, I, II, and III (see Figure 9-41). Of the three regions, mutational analysis implicates CDE III as the most critical for centromere function. This region appears to interact with microtubules via centromere-binding factor (CBF3), a multiprotein subunit complex, and other identified proteins (Figure 19-39b).

Centrosome Duplication Precedes and Is Required for Mitosis

Since each half of the metaphase mitotic apparatus emanates from a polar centrosome, its assembly depends on duplication of the centrosome and their movement to opposite halves of the cell. This process, known as the *centriole*

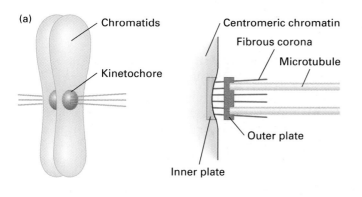

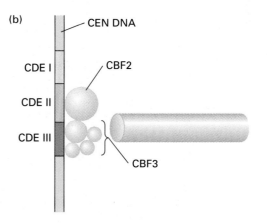

▲ FIGURE 19-39 Centromeric attachment of microtubules. (a) Schematic diagram of attachment of kinetochore microtubules to the sister chromatids of a metaphase chromosome. In animals and lower plants, the kinetochore is a three-layer, platelike structure lying within the centromere of each chromatid (inset). The (+) ends of microtubules insert into the outer layer of each kinetochore, and the microtubules extend toward one of the two poles of the cell. At anaphase, the sister chromatids separate, and the chromosomes are pulled to opposite poles of the cell by the kinetochore microtubules. (b) Association of a yeast centromere with components of the kinetochore. The centromeric (CEN) DNA is divided into three contiguous segments (CDE I–III). Two groups of centromere-binding factors, CBF2 and CBF3, are proteins associated with CDE II and CDE III, respectively. The CBFs mediate the attachment of a single microtubule to the centromere. Other microtubule-binding proteins form the rest of the kinetochore. [Part (a) adapted from A. F. Pluta et al., 1995, *Science* **270**:1591.]

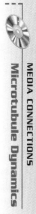

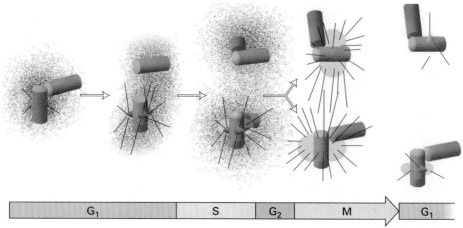

▲ FIGURE 19-40 Relation of centrosome duplication to the cell cycle. After the pair of parent centrioles (red) within the centrosome matrix separate slightly, a daughter centriole (blue) buds from each and elongates. By G_2 growth of the daughter centrioles is complete, but the two pairs remain within a single centrosomal complex. Early in mitosis, the centrosome splits, and each centriole pair migrates to opposite ends of the cell. That new centrioles arise de novo, as shown here, rather than by fission of a parent centriole and regrowth has been demonstrated in experiments with biotin-labeled tubulin.

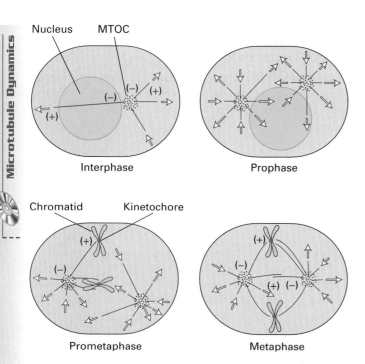

▲ FIGURE 19-41 Microtubule dynamics during mitosis. In both interphase and mitotic cells, most microtubules radiate from the microtubule-organizing centers (MTOCs), with the (−) ends of the microtubules closest to the MTOC and the (+) ends extending outward. A typical interphase cell has long microtubules; during mitotic prophase the microtubule-nucleating activity of the already replicated MTOCs (centrosomes in animal cells) increases, generating a larger number of shorter, more dynamic microtubules. In late prophase, some of the microtubules interact with kinetochores, causing the microtubules to be partially stabilized. In the metaphase mitotic apparatus, astral microtubules elongate and shorten.

cycle (or *centrosome cycle*) begins during G_1 when the centrioles and other centrosome components are duplicated (Figure 19-40). By G_2 the two "daughter" centrioles have reached full length, but the duplicated centrioles are still present within a single centrosome. Early in mitosis, the two pairs of centrioles separate and migrate to opposite sides of the nucleus, establishing the bipolarity of the dividing cell (see Figure 19-40). In some respects, then, mitosis can be understood as the migration of duplicated centrosomes, which along their journey pick up chromosomes, pause in metaphase, and during anaphase continue their movement to new locations in the daughter cells, where they release the chromosomes and organize the cytosolic microtubule.

Dynamic Instability of Microtubules Increases during Mitosis

The observation of high tubulin turnover in microtubules during mitosis is strong evidence that microtubule dynamics is critical to the mitotic process. During mitosis, the first indication of a change in microtubule stability occurs at prophase, when long interphase microtubules disappear and are replaced by a spindle and astral microtubules (Figure 19-41). Mitotic microtubules that are nucleated from the newly replicated centrosomes are more numerous, shorter, and less stable than interphase microtubules. The average lifetime of a microtubule decreases from 10 minutes in interphase cells to 30 seconds in the mitotic spindle. This increase in the turnover enables microtubules to assemble and disassemble more quickly during mitosis.

As we will see in the following sections, however, microtubule motor proteins, as well as dynamic microtubules, participate in the reorganization of the spindle microtubules, separation of centrosomes, capture and alignment of chromosomes, and subsequent movement of chromosomes poleward. We have already discussed how some microtubule-dependent movements in the cytosol are generated by microtubule dynamics or by microtubule motor proteins. The involvement of both mechanisms in mitosis perhaps ensures the fail-safe apportionment of chromosomes equally to each daughter cell.

Organization of the Spindle Poles Orients the Assembly of the Mitotic Apparatus

Genetic and cell biology studies, primarily in yeast and flies, have implicated several kinesin-related proteins (KRPs), or spindle kinesins, in the separation and migration of centrosomes, thereby orienting assembly of the spindle and spindle asters. For instance, antibodies against either a (+) or (−) end–directed KRP will inhibit the formation of a bipolar spindle when they are microinjected into a cell before but not after prophase. Such experiments suggest that a (−) end–directed motor protein in the nascent overlap region of the spindle aligns the oppositely oriented microtubules extending from each centrosome, and then a (+) end–directed motor protein pushes the centrosomes farther apart as the polar microtubules elongate. Cytosolic dynein, which is localized by antibodies to the centrosome and cortex of animal cells, also probably is involved in centrosome movement and spindle orientation. Genetic studies in yeast suggest that dynein at the cortex simultaneously helps tether the astral microtubules and orient one pole of the spindle. The combination of microtubule growth, polar microtubule cross-linking, and astral microtubule–cortex interactions provides a reasonable explanation of how the spindle poles separate and orient during prophase, as depicted in Figure 19-42.

Formation of Poles and Capture of Chromosomes Are Key Events in Spindle Assembly

Assembly of the metaphase spindle requires two types of events: attachment of spindle microtubules to the poles and capture of chromosomes by kinetochore microtubules. Experiments with *Xenopus* egg extracts suggest that cytosolic dynein may play a critical role in organizing spindle microtubules into a pole. In the presence of sperm nuclei, centrosomes, and microtubules, bipolar spindles form in egg extracts. However, addition of antibodies against cytosolic dynein releases and splays the spindle microtubules but leaves the centrosomal astral microtubules intact (Figure 19-43a). The same results are obtained when antibodies to dynein, dynactin, or NuMA are microinjected into a cell. According to a recently proposed model, cytosolic dynein and NuMA cross-link the (−) ends of the spindle microtubules forming a spindle pole, which is linked to the centrosome through other interactions perhaps involving KRPs (Figure 19-43b). These findings also suggest that formation of the spindle pole is not tightly linked with assembly of the aster. Consistent with this is the fact that cells lacking a cen-

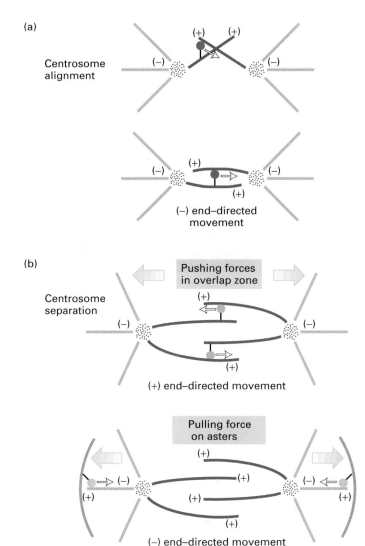

▲ FIGURE 19-42 Model for participation of microtubule motor proteins in centrosome movements during mitosis. (a) During late prophase centrosomes are aligned by (−) end–directed motors (dark green), which pull on polar microtubules and thus align them. (b) A family of (+) end–directed KRPs (pink), probably BimC kinesins, associated with the polar microtubules is involved in separation of the poles—beginning after centrosome alignment, continuing through the formation of the spindle poles, and ending, as we will see later, with spindle pole separation at anaphase. In addition, a (−) end–directed force exerted by cytosolic dynein (light green) located at the cortex may pull asters toward the poles in combination with the (+) end–directed motors in the spindle.

trosome (e.g., plant cells) form a spindle pole even though they lack astral microtubules.

At the opposite end of the spindle microtubules, rapid fluctuation in their length is used to capture chromosomes during prophase as the nuclear membrane begins to break down. By quickly lengthening and shortening at its (+) end, a dynamic microtubule acts like a poker that can probe into a chromosome-rich environment (Figure 19-44). Sometimes

(a)

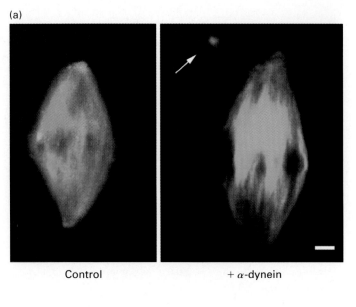

Control   + α-dynein

(b)

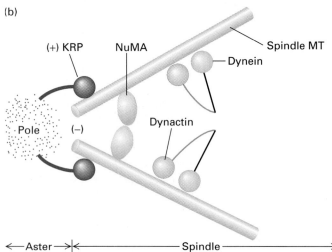

▲ **FIGURE 19-43 Spindle pole formation.** (a) Immuno-fluorescence micrographs showing in vitro reconstituted spindles stained with fluorescent-labeled antibodies to tubulin (green) and dynein (red). The control spindle *(left)* has cytosolic dynein at each tapered end, as well as at the equator. Addition of dynein antibody after the formation of the spindle *(right)* causes the poles to splay and disrupts protein localization. (b) Model of the interactions between spindle microtubules (MTs) and various proteins at the pole. [Part (a) from R. Heald et al., 1997, *J. Cell Biol.* **138**:615; courtesy of R. Heald. Part (b) adapted from T. Gaglio et al, 1995, *J. Cell Biol.* **135**:399.]

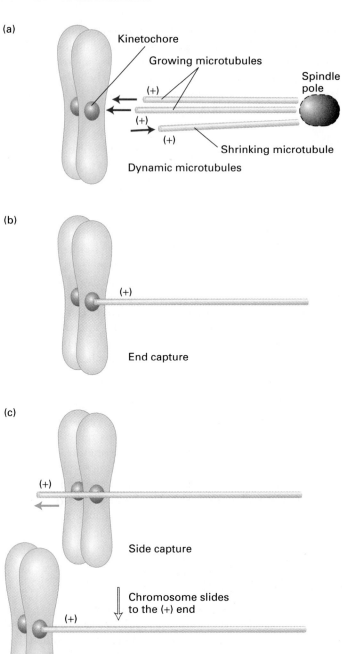

▲ **FIGURE 19-44 Dynamic instability and the capture of chromosomes.** (a) During mitotic prophase, some spindle microtubules are growing at their distal (+) end, while others are shrinking rapidly. (b) In late prophase, the ends of some microtubules interact with kinetochores (dark green), causing those microtubules to be stabilized. (c) In addition, some micro-tubules just miss the kinetochore, but the kinetochore binds to the side of the microtubule and then slides to the (+) end.

the end of a microtubule contacts a kinetochore, scoring a "bull's-eye." More commonly, a kinetochore contacts the side of a microtubule and then slides along the microtubule to the (+) end in a process that may involve kinesins on the kinetochore. Whether a chromosome attaches to the (+) end of a spindle microtubule by a direct hit or by the side capture/sliding process, the kinetochore "caps" the (+) end of the microtubule. Eventually, each sister chromatid in a chromosome is captured by microtubules arising from the nearest spindle poles. Each kinetochore also becomes attached to additional microtubules as mitosis progresses toward metaphase.

Kinetochores Generate the Force for Poleward Chromosome Movement

During late prophase (prometaphase), the newly condensed chromosomes attached to their kinetochore microtubules *congress,* or move, to the equator of the spindle (see Figure 19-34). Along the way, the chromosomes exhibit *saltatory behavior,* oscillating between movements toward and then away from the pole or equator. Until the kinetochore microtubules are attached to all kinetochores, the cell cycle is held in check. A single unattached kinetochore is sufficient to prevent entry into anaphase. A combination of microtubule motor proteins at the kinetochore and microtubule dynamics at the (+) end of kinetochore microtubules is thought to position the chromosomes equally between the two spindle poles. Figure 19-45 depicts several mechanisms that could position and hold chromosomes at the equatorial plate.

Experimental micromanipulation of chromosomes provides the best evidence of a force that pulls the two kinetochores on sister chromatids toward opposite poles. These studies suggest that the strength of the force is proportional to the distance from the chromosome to the pole. Thus, if a metaphase chromosome is displaced toward one pole by micromanipulation, then the force exerted from the opposite pole momentarily increases and quickly pulls the displaced chromosome back to the equator. An alternative explanation is that the pole closest to the chromosome exerts a pushing force that restores the chromosome to the

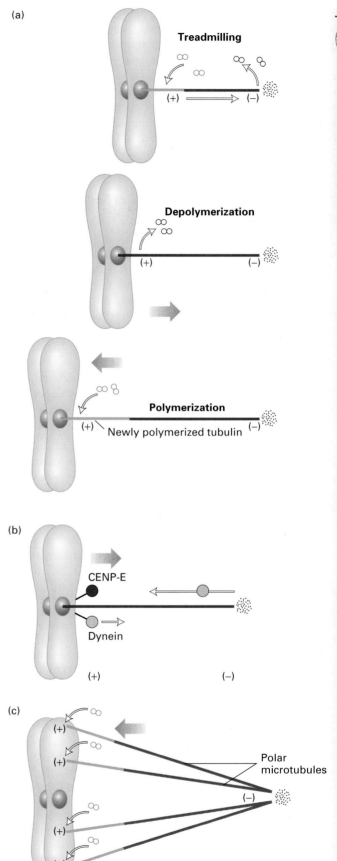

▶ **FIGURE 19-45 Proposed alternative mechanisms for chromosome congression.** A coupling of microtubule dynamics and microtubule motors may keep a chromosome positioned at the equator. (a) The flow of tubulin subunits through kinetochore microtubules can be used to pull or push the chromosome relative to the pole, especially if rapid polymerization or depolymerization occurs at the (+) end of a microtubule. (b) A (−) end–directed motor (dark green) at the kinetochore or a (+) end–directed motor (pink) at the spindle pole can pull a chromosome toward the pole, while CENP-E tethers the kinetochore to a shrinking microtubule. (c) Nonkinetochore microtubules can exert a pushing force on the chromatid arms of a chromosome by polymerization at their (+) ends.

equator. Whatever mechanism is in operation, these opposing forces are balanced when a chromosome is at the equator of the spindle, so the chromosome remains stationary there.

By metaphase, then, the chromosomes are aligned at the equator, their position fixed midway between each pole of the spindle. Although the lengths of kinetochore and polar microtubules have stabilized, there continues to be a flow, or treadmilling, of subunits through the microtubules toward the poles, but the loss of subunits at the (−) end is balanced by the addition of subunits at the (+) end.

During Anaphase Chromosomes Separate and the Spindle Elongates

The same forces that form the spindle during prophase and metaphase also direct the separation of chromosomes toward opposite poles at anaphase. Anaphase is divided into two distinct stages, *anaphase A* and *anaphase B* (or early and late anaphase). Anaphase A is characterized by the shortening of kinetochore microtubules, which pulls the chromosomes toward the poles. During anaphase B, the two poles move farther apart, bringing the chromosomes with them into what will become the two daughter cells.

In part because methods for studying anaphase A and anaphase B in certain cell-free extracts have been developed, we understand a good deal about their molecular mechanisms. Similar mechanisms may operate during other mitotic phases.

Microtubule Shortening during Anaphase A In vitro studies have indicated that the depolymerization of microtubules can generate sufficient force to move chromosomes. In one such study, purified microtubules were mixed with purified anaphase chromosomes, and as expected, the kinetochores bound preferentially to the (+) ends of the microtubules. To induce depolymerization of the microtubules, the reaction mixture was diluted, thus lowering the concentration

of free tubulin dimers. Video microscopy analysis then showed that the chromosome moved toward the (−) end, at a rate similar to that of chromosome movement in intact cells. Since no ATP (or any other energy source) was present in these experiments, chromosome movement toward the (−) end must be powered, in some way, by microtubule disassembly and must not be powered by microtubule motor proteins.

The in vivo fluorescence tagging experiment depicted in Figure 19-46 provides additional evidence that disassembly

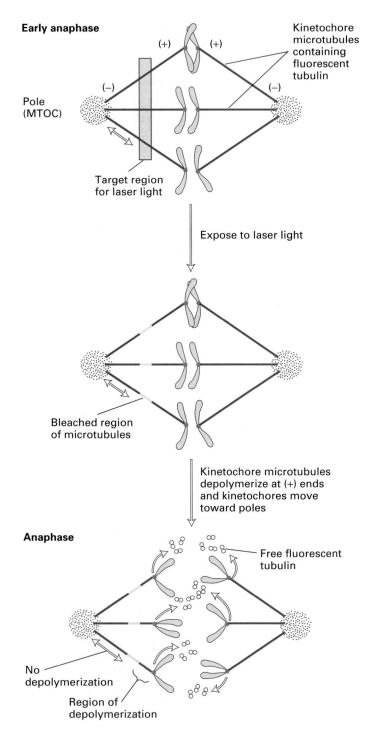

▶ **FIGURE 19-46 Experimental demonstration that during anaphase A chromosomes move poleward along stationary kinetochore microtubules, which coordinately disassemble from their kinetochore ends.** Fibroblasts are injected with fluorescent tubulin and then allowed to enter metaphase, so that all the microtubules are fluorescent. Only the kinetochore microtubules are shown. In early anaphase, a band of microtubules (yellow box) is subjected to a laser light, which bleaches the fluorescence but leaves the microtubules continuous and functional across the bleached region. The bleached segment of each microtubule thus provides a marker for the position of that part of the microtubule. During anaphase the distance of the bleached zone from the poles (measured in the diagrams by the black double-headed arrows) does not change, indicating that no depolymerization of the microtubules occurs at the poles. Rather, the kinetochore microtubules disassemble just behind the kinetochore, and the kinetochores move poleward along the microtubules. [Adapted from G. J. Gorbsky, P. J. Sammak, and G. Borisy, 1987, *J. Cell Biol.* **104**:9; and G. J. Gorbsky, P. J. Sammak, and G. Borisy, 1988, *J. Cell Biol.* **106**:1185.]

▶ **FIGURE 19-47 Model of spindle elongation and movement during anaphase B. Tubulin (light purple) adds to the
(+) ends of all polar microtubules, lengthening these fibers.**
Simultaneously, (+) end–directed KRPs (pink) bind to the polar
microtubules in the overlap region. Each KRP, bound to a microtubule in one half-spindle, "walks" along a microtubule in the
other half-spindle, toward its (+) end, utilizing the energy
from ATP hydrolysis. In cells that assemble an aster, a (−) end–
directed motor protein (light green) located in the cortex of the
plasma membrane pulls on the astral microtubules, which also
moves the poles farther apart. [Adapted from H. Masuda and
W. Z. Cande, 1987, *Cell* **49**:193.]

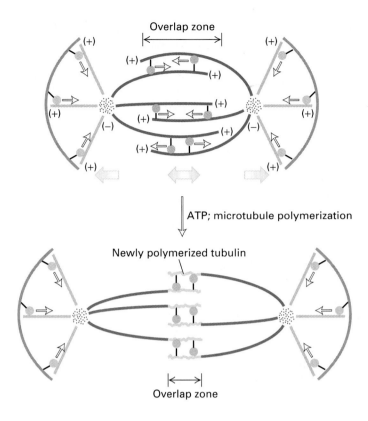

of kinetochore microtubules at their (+) ends coincides with
the poleward movement of chromosomes. The results of
these two experiments suggest that the proteins tethering the
kinetochore to the microtubule progressively interact with
increasingly distal portions of the microtubule as the (+)
end disassembles. In this way, the kinetochores move poleward by a passive process that does not require a motor
protein.

Spindle Elongation during Anaphase B In the second
stage of anaphase, polar microtubules slide past one another
and elongate, and pulling forces are exerted by the cellular
cortex on astral microtubules. Detergent treatment of mitotic cells, which causes ATP to leak out, does not affect
poleward chromosome movement (anaphase A), but prevents the separation of spindle poles that occurs in anaphase
B. Thus microtubule motor proteins clearly are involved in
separating the spindle poles, as they are in centrosome movement during prometaphase (see Figure 19-42).

Anaphase A and B movements can be reconstituted in
vitro by activating artificial spindles that assemble from frog
egg extracts. In the presence of calcium, the spindles elongate,
simulating anaphase B, and the zone of overlap between the
two half-spindles decreases in length by a distance similar
to the original length of the overlap. If we were to analyze the
direction of microtubule movement during anaphase B, we
would find that adjacent microtubules migrate in the direction of their pole-facing (−) ends. This polarity of movement
suggests that a (+) end–directed KRP is responsible for generating the force for spindle pole separation during anaphase
B. In one model, a KRP attached to a microtubule in the
overlap region walks toward the (+) end of a neighboring
but antiparallel microtubule, thus pushing the adjacent microtubule in the direction of its (−) end (Figure 19-47). This
model is supported by experiments in which antibodies
raised against a conserved region of the kinesin superfamily
inhibit ATP-induced elongation of diatom spindles in vitro.
The involvement of a kinesin-like protein and the requirement for ATP hydrolysis are two strong pieces of evidence
that a KRP could be responsible for anaphase B movements.

In addition to the sliding forces between polar microtubules, spindle elongation involves microtubule dynamics
and interactions at the cortex. In the presence of αβ-tubulin,

reactivated frog spindles and isolated diatom spindles add
tubulin subunits to the (+) end of polar microtubules (see
Figure 19-47). A third component of anaphase B is the interaction between astral microtubules and the cortex, which
generates a pulling force on the asters. This force can be
demonstrated by cutting the spindle in half with a mi-
croneedle during anaphase; the resulting half-spindles move
quickly to the poles, at a rate faster than usual during
anaphase. This observation suggests that a (−) end–directed
motor associated with the cortex, maybe cytosolic dynein
or a KRP, pulls the asters farther apart toward the poles of
the daughter cells (see Figure 19-47).

Astral Microtubules Determine Where Cytokinesis Takes Place

Once the chromosomes have been separated to the poles of
the cell during anaphase, the nuclear envelope re-forms around
each complete set of chromosomes during telophase and the
cytoplasm divides (**cytokinesis**) (see Figure 19-34). With completion of cytokinesis, the last event in mitosis, the life of a
daughter cell begins. In Chapter 18, we discussed how the
contractile ring of actin and myosin constricts the cell during
cytokinesis, but not the mechanism that determines the plane
of cleavage through the cell (see Figure 18-37). It is clear that
the contractile ring, and hence the cleavage furrow, always
develops where the chromosomes lined up during metaphase,
but it is not obvious which component of the mitotic apparatus dictates where the contractile ring will assemble.

Ingenious micromanipulation experiments with dividing sea urchin embryos have shown that the presence of two asters, not the spindle itself, is necessary to determine the cleavage plane. In one key experiment, illustrated in Figure 19-48, a hole is poked into a one-cell-stage sea urchin embryo before the mitotic apparatus has started to assemble, transforming it into a doughnut-shaped cell. During the first division, a spindle assembles on one side, and cytokinesis between the asters forms a single, binucleated, C-shaped cell. The two nuclei of this cell undergo another round of mitosis with assembly of a pair of mitotic apparatuses. If the spindle determines the cleavage plane, then we would expect the C-shaped cell to form two cleavage furrows and divide into three cells. In fact, four cells are generated in this experiment. The interpretation of this result is that an "extra" cell is generated by cleavage in the region between asters and lacking a spindle.

One hypothesis is that astral microtubules send a signal to the region of the cortex midway between asters. This signal activates the assembly of actin and myosin, resulting in the formation of the contractile ring followed by the development of the cleavage furrow. The signal is unidentified as yet, but an alluring candidate is cdc2 kinase. Not only is this protein kinase activated in the cell cycle (Chapter 13), but caldesmon and myosin are among its known substrates. As discussed in the previous chapter, phosphorylation of these two proteins will activate the assembly of myosin filaments and cause caldesmon to relieve the inhibition of actin-myosin interactions.

Plant Cells Reorganize Their Microtubules and Build a New Cell Wall during Mitosis

Mitotic events in plants are generally similar to those in animal cells, although a rigid cell wall must be constructed during cytokinesis (Figure 19-49). Interphase cells are girdled by a bundle of cortical microtubules that influence the pattern of deposition of cellulose in the cell wall. Despite the absence of a centrosome, plant cells reorganize their microtubules into a spindle during mitosis. At cytokinesis, in place of a contractile ring, a membrane structure, the *phragmoplast*, is assembled from vesicles in the interzone between the daughter nuclei. The vesicles originate from the Golgi complex and are first observed during metaphase, when they extend into the mitotic apparatus and in some cases even appear to contact the kinetochores.

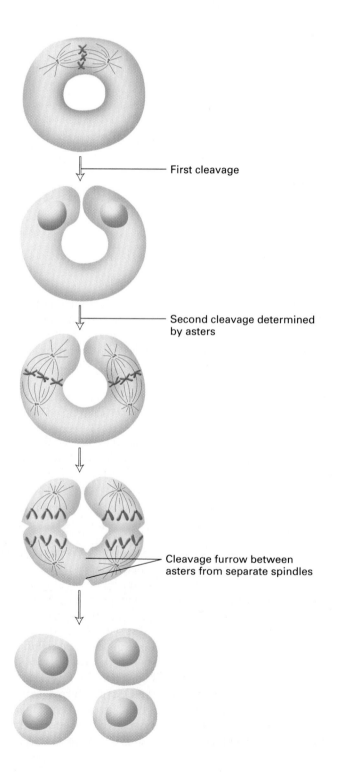

— First cleavage

— Second cleavage determined by asters

— Cleavage furrow between asters from separate spindles

◄ **FIGURE 19-48 Experimental demonstration that asters alone determine the cleavage plane during cytokinesis.** A small glass ball is pressed against a fertilized sea urchin egg until membranes from opposite sides of the cell touch and fuse. This changes the spherical egg into a doughnut shape. During the first cell division, a normal spindle develops and the doughnut-shaped cell divides between the two asters, producing a single C-shaped cell with two normal nuclei. At the next cell division, the two nuclei each produce a normal spindle. Two cleavage furrows bisect at the spindles, in the usual manner. But because the asters from the two separate spindles are close enough together, a third cleavage furrow also forms and divides the cell in a region not occupied by either spindle but lying between asters from the two separate spindles. Thus the cell divides in three planes, producing four daughter cells.

Interphase Prophase Metaphase Telophase

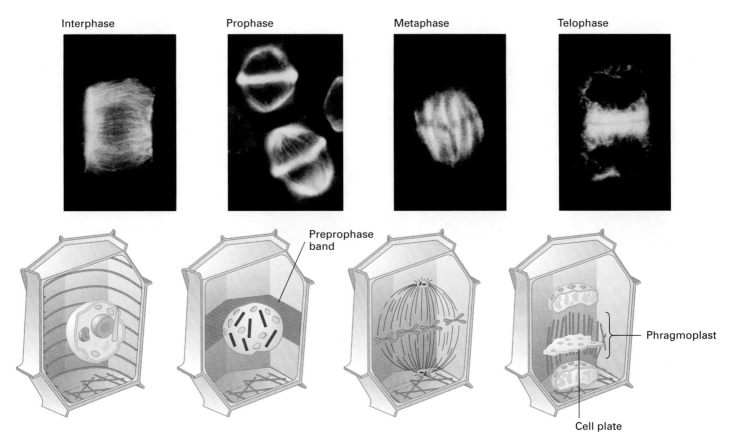

▲ **FIGURE 19-49 Mitosis in a higher plant cell.** Immunofluorescence micrographs *(top)* and corresponding diagrams *(bottom)* showing arrangement of microtubules in a plant cell. A cortical array of microtubules girdle the cell during interphase. As the cell enters mitosis, the microtubules are bundled around the nucleus and reorganized into a spindle. In late telophase, the nuclear membrane has re-formed, and the phragmoplast assembles at the equatorial plate. A set of small vesicles derived from the Golgi complex, which contain cellulose and other precursors of the cell wall, accumulate at the equatorial plate, and fuse to form the phragmoplast. Additional vesicles fuse with it to form the new cell plate. [Adapted from R. H. Goddard et al., 1994, *Plant Physiol.* **104:**1; micrographs courtesy of Susan M. Wick.]

Later in anaphase, by traveling along microtubules that radiate from each daughter nucleus, the vesicles line up near the center of the dividing cell, where they fuse to form the phragmoplast in telophase. The membranes of the vesicles become the plasma membranes of the daughter cells, and their contents form the cell plate. The vesicles also contain material for the future cell wall, such as polysaccharide precursors of cellulose and pectin.

**SUMMARY Microtubule Dynamics and
Motor Proteins during Mitosis**

• During mitosis, the replicated chromosomes are separated and evenly partitioned to two daughter chromosomes. This portion of the cell cycle is commonly divided into four substages: prophase, metaphase, anaphase, and telophase (see Figure 19-34).

• Early in mitosis, the chromosomes begin to condense, the nuclear membrane breaks down, and the mitotic spindle is assembled. These events culminate in the metaphase cell in which the fully condensed chromosomes, each composed of sister chromatids connected at the centromere, are aligned along the equatorial plate.

• The major components of the mitotic apparatus are the astral microtubules forming the asters; the polar and kinetochore microtubules forming the football-shaped spindle; the spindle poles, and chromosomes attached to the kinetochore microtubules (see Figure 19-36).

• A simple set of microtubule motor proteins—BimC, CENP-E, and cytosolic dynein—are conserved in all spindles. Additional motors are present in more complex organisms.

• The two centrioles in the centrosome, the MTOC in most animal cells, are replicated during interphase. As mitosis begins, the centriole pairs migrate to the two poles, where they organize the astral microtubules.

• Interactions between asters and the cell cortex and between opposing spindle microtubules separate and

(a) Intermediate filaments
(vimentin)

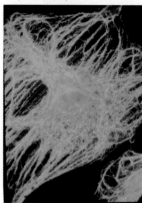

(b) Microtubules (tubulin)

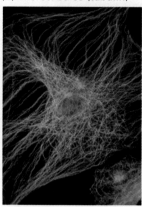

(c) Microfilaments (actin)

◀ **FIGURE 19-50 Immunofluorescence micrographs showing the distribution of the three networks of cytoskeletal fibers in the same cultured fibroblast.** Each type of fiber was detected with a specific fluorescent-labeled stain: intermediate filaments with anti-vimentin antibodies (green), microtubules with tubulin-specific antibodies (blue), and microfilaments with phalloidin (red). Many vimentin fibers are colinear with microtubules, suggesting a close association between the two filament networks. In contrast, the microfilaments in the cell's stress fibers exhibit a much different organization. [Courtesy of V. Small.]

align the centrosomes during prophase, establishing the bipolar orientation of the spindle (see Figure 19-42).

- Spindle microtubules capture the kinetochores of chromosomes and center the chromosomes at metaphase.

- The combined action of microtubule dynamics (polymerization and depolymerization) and microtubule motors positions and holds chromosomes at the metaphase plate and translocates chromosomes to the opposite poles at anaphase (see Figures 19-45 and 19-47).

- Cleavage during cytokinesis occurs between the asters and does not depend on the presence of a spindle or chromosomes. The signal for cleavage is unknown.

19.6 Intermediate Filaments

In the remainder of this chapter, we discuss intermediate filaments (IFs), the third set of cytoskeletal fibers in eukaryotic cells. Intermediate filaments are found in nearly all eukaryotic cells from multicellular organisms, but their presence in fungi and other lower unicellular eukaryotes is controversial. In epidermal cells and the axons of neurons, intermediate filaments are at least 10 times more abundant than microfilaments or microtubules. In immunofluorescence micrographs, we can see a network of intermediate filaments filling the entire cytosol of a cell in a pattern usually distinct from that of microfilaments but similar to that of microtubules (Figure 19-50).

Functions and Structure of Intermediate Filaments Distinguish Them from Other Cytoskeletal Fibers

The organization of intermediate filaments and their association with plasma membranes suggest that their principal function is structural—to reinforce cells and to organize cells into tissues. Indeed, the supportive role of intermediate filaments is carried to an extreme in claws and hair, the dead remnants of epidermal cells, which are composed largely of IF proteins. Perhaps the most important function of intermediate filaments is to provide mechanical support for the plasma membrane where it comes into contact with other cells or with the extracellular matrix. Unlike microfilaments and microtubules, intermediate filaments do not participate in cell motility. There are no known examples of IF-dependent cell movements or of motor proteins that move along intermediate filaments.

Several physical and biochemical properties distinguish intermediate filaments from microfilaments and microtubules. To begin with, intermediate filaments are extremely stable. Even after extraction with solutions containing detergents and high concentrations of salts, most intermediate filaments in a cell remain intact, whereas microfilaments and microtubules depolymerize into their soluble subunits. In fact, most IF purification methods employ these treatments to free intermediate filaments from other proteins. The intermediate filaments are then treated with urea, a denaturing solvent, and the depolymerized subunits purified by ion-exchange chromatography; the IF subunits repolymerize into filaments when the urea is removed by dialysis. Second, intermediate filaments differ in size from the other two cytoskeletal fibers. Indeed, their name derives from their 10-nm diameter—smaller than microtubules (24 nm) but larger than microfilaments (7 nm). Moreover, in contrast to the beaded filaments and the hollow tubules formed by the globular actin and tubulin subunits, respectively, IF subunits are α-helical rods that assemble into ropelike filaments. Finally, IF subunits do not bind nucleotides, and their assembly into intermediate filaments does not require nucleotide (GTP or ATP) hydrolysis, as does polymerization of G-actin and αβ-tubulin. However, many of the details concerning the assembly of intermediate filaments in cells remain speculative.

Much of the following discussion about intermediate filaments will seem familiar because their organization is similar to that of the microtubule and microfilament components of the cytoskeleton, and in some cases intermediate filaments perform functions analogous to those of actin filaments. We will point out the similarities and differences between the three cytoskeletal systems as we consider the role of intermediate filaments in the cell.

IF Proteins Are Classified into Six Types

In higher vertebrates, the subunits composing intermediate filaments constitute a superfamily of highly α-helical proteins that is divided into six major classes or types on the basis of similarities in sequence (Table 19-3). Unlike the actin and tubulin isoforms, IF protein classes are widely divergent in sequence and vary greatly in molecular weight. We introduce the five classes here and later consider their functions in various cells in more detail. The sixth class has little similarity to the other IF proteins and is not discussed. Members of this class are found in the lens of the eye and form beaded filaments. The expression of each IF protein is characteristic of a certain tissue or cell type.

Acidic and basic *keratins* (types I and II) are typically expressed in epithelial cells. As described later, they associ-ate in a 1:1 ratio to form heterodimers, which assemble into heteropolymeric keratin filaments; neither type alone can assemble into a keratin filament. The keratins are the most diverse classes of IF proteins, with a large number of keratin isoforms being expressed. These can be divided into two groups: About 10 keratins are specific for "hard" epithelial tissues, which give rise to nails, hair, and wool; about 20, called *cytokeratins*, are more generally found in the epithelia that line internal body cavities. Each type of epithelium always expresses a characteristic combination of type I and type II keratins.

Four proteins—vimentin, desmin, glial fibrillary acidic protein (GFAP), and peripherin—are classified as type III IF proteins. Unlike the keratins, the type III proteins can form both homo- and heteropolymeric IF filaments. The most widely distributed of all IF proteins is *vimentin*, which is typically expressed in leukocytes, blood vessel endothelial cells, some epithelial cells, and mesenchymal cells such as fibroblasts. Vimentin filaments help support cellular membranes. Vimentin

TABLE 19-3 Primary Intermediate-Filament Proteins in Mammals		
IF Protein	MW (10^{-3})*	Tissue Distribution
TYPE I[†]		
Acidic keratins	40–57	Epithelia
TYPE II[†]		
Basic keratins	53–67	Epithelia
TYPE III		
Vimentin	57	Mesenchyme
Desmin	53	Muscle
Glial fibrillary acidic protein	50	Glial cells and astrocytes
Peripherin	57	Peripheral and central neurons
TYPE IV[‡]		
NF-L	62	Mature neurons
NF-M	102	
NF-H	110	
Internexin	66	Developing central nervous system
NONSTANDARD TYPE IV		
Filensin	83	Lens fiber cells
Phakinin	45	
TYPE V		
Lamin A	70	Nucleus of all cells
Lamin B	67	
Lamin C	67	

*IFs show species-dependent variations in molecular weight (MW).
[†]More than 15 isoforms of both acidic and basic keratins are known.
[‡]NF=neurofilament; L, M, and H=low, medium, and high molecular weight.

networks also may help keep the nucleus and other organelles in a defined place within the cell. Vimentin is frequently associated with microtubules, and as noted earlier, the network of vimentin filaments parallels the microtubule network (see Figure 19-50). The other type III IF proteins have a much more limited distribution. *Desmin* filaments in muscle cells are responsible for stabilizing sarcomeres in contracting muscle. *Glial fibrillary acidic protein* forms filaments in the glial cells that surround neurons and in astrocytes. *Peripherin* is found in neurons of the peripheral nervous system, but little is known about it.

The core of neuronal axons is filled with *neurofilaments (NFs)*, each a heteropolymer composed of three type IV polypeptides—NF-L, NF-M, and NF-H—which differ greatly in molecular weight. In contrast to microtubules, which direct the elongation of an axon, neurofilaments are responsible for the radial growth of an axon and thus determine axonal diameter. As we see in Chapter 21, the diameter of an axon is directly related to the speed at which it conducts impulses. The influence of the number of neurofilaments on impulse conduction is highlighted by a mutation in quails named *quiver,* which blocks the assembly of neurofilaments. As a result, the velocity of nerve conduction is severely reduced.

In contrast with the cytosolic location of the other four classes of IF proteins, the type V proteins, called **lamins**, are found exclusively in the nucleus. Of the three nuclear lamins, two are alternatively spliced products encoded by a common gene, while the third is encoded by a separate gene. The nuclear lamins form a fibrous network that supports the nuclear membrane.

Intermediate Filaments Can Identify the Cellular Origin of Certain Tumors

 Familiarity with the various types of IF proteins not only is useful in comparing and contrasting their structural and functional properties, but also can help in the diagnosis and treatment of certain tumors. In a tumor, cells lose their normal appearance, and thus their origin cannot be identified by their morphology. However, tumor cells retain many of the differentiated properties of the cells from which they are derived, including expression of particular IF proteins. By use of fluorescent-tagged antibodies specific for those IF proteins, diagnosticians often can determine whether a tumor originated in epithelial, mesenchymal, or neuronal tissue.

For example, the most common malignant tumors of the breast and gastrointestinal tract contain keratins and lack vimentin; thus they are derived from epithelial cells (which contain keratins but not vimentin) rather than from the underlying stromal mesenchymal cells (which contain vimentin but not keratins). Because epithelial cancers and mesenchymal cancers are sensitive to different treatments, identifying the IF proteins in a tumor cell helps a physician select the most effective treatment for destroying the tumor.

All IF Proteins Have a Conserved Core Domain and Are Organized Similarly into Filaments

Besides sharing an ability to form filaments 10 nm in diameter, all IF subunit proteins have a common domain structure: a central α-helical core flanked by globular N- and C-terminal domains (Figure 19-51, *top*). The core helical domain, which is conserved among all IF proteins, consists of four long α-helices separated by three nonhelical regions. The positions of these "spacer" elements are also highly conserved among all IF proteins. The α-helical segments pair to form a coiled-coil dimer.

In electron micrographs, an IF-protein dimer appears as a rodlike molecule with globular domains at the ends. Labeling experiments with antibodies to the N- or C-terminal domains indicate that the polypeptide chains are parallel in a dimer. A pair of dimers associate laterally into a tetramer. Labeling experiments demonstrate that the dimers in a tetramer have an antiparallel orientation. Some evidence, discussed later, suggests that the tetramer and not the dimer is the subunit for assembly of intermediate filaments. Tetramers bind end to end, forming protofilaments 2–3 nm thick, which pair together into protofibrils. Finally, four protofibrils form a single intermediate filament that is 10 nm in diameter (Figure 19-51, *bottom*). The tails extending from the tetramers composing a protofilament often are visible as projections along the length of a fully assembled intermediate filament (Figure 19-52). Interestingly, because the tetramer is symmetric, an IF may not have a polarity like an actin filament or a microtubule. This idea is supported by experiments showing that vimentin subunits can incorporate along the length, as well as the ends, of a filament.

Although the α-helical core is common to all IF proteins, the N- and C-terminal domains of different types of IF proteins vary greatly in molecular weight and sequence. Because of this lack of sequence conservation and the observation that these domains project from the walls of intermediate filaments (see Figure 19-52), scientists initially speculated that the N- and C-terminal domains are not involved in IF assembly. Several subsequent experiments, however, proved this hypothesis to be partially incorrect. For instance, if the N-terminal domain of an IF protein is shortened, either by proteolysis or by deletion mutagenesis, the truncated protein cannot assemble into filaments. (Keratins are the exception; they form filaments even if both terminal domains are absent.) The prevailing view now is that the N-terminal domain plays an important role in assembly of most intermediate filaments. Although the C-terminal domain is dispensable for IF assembly, it seems to affect the stability of the filament. Thus these domains may control lateral interactions within an intermediate filament, as well as interactions between intermediate filaments and other cellular components.

Some IF proteins can form homopolymeric filaments, whereas others can form only heteropolymeric filaments. Vimentin, desmin, glial fibrillary acidic protein, and NF-L are able to form homodimers, which assemble into homopolymers;

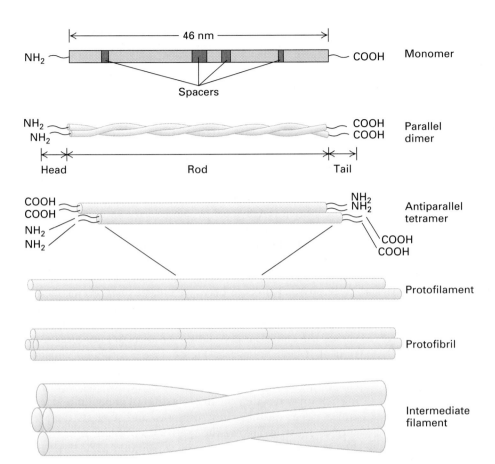

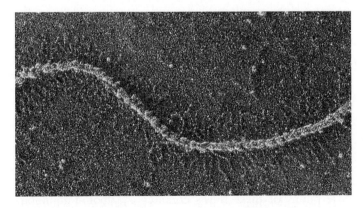

◀ FIGURE 19-51 Levels of organization and assembly of intermediate filaments. IF proteins form parallel homo- or heterodimers with a highly conserved coiled-coil core domain and nonhelical tails and heads, which are variable in length and sequence. The central core domain contains three nonhelical spacer elements. A tetramer is formed by antiparallel, staggered side-by-side aggregation of two identical dimers. Tetramers aggregate end to end, forming a protofilament; pairs of protofilaments then laterally associate into a protofibril. Lateral association of four protofibrils form a cylindrical filament 10 nm thick. [Adapted from D. A. D. Parry and P. M. Steinert, 1995, *Intermediate Filament Structure*, Springer-Verlag, pp. 27 and 101.]

▲ FIGURE 19-52 Electron micrograph of a neurofilament purified from spinal cord. After the filament is rotary-shadowed with metals, long whiskers are seen to project from its walls. These projections correspond to the N- and C-terminal domains of the neurofilament-protein tetramers. [Courtesy of U. Aebi.]

these proteins also can assemble into heteropolymers. In contrast, acidic and basic keratins, which form heterodimers with each other, form only obligate heteropolymeric keratin filaments. However, keratins do not form heteropolymers with other classes of IF proteins, whereas some IFs can. For

example, vimentin and NF proteins can assemble into a heteropolymer. NF-L self-associates to form a homopolymer, but NF-H and NF-M commonly co-assemble with the NF-L backbone, so that most neurofilaments contain all three proteins.

Whether an intermediate filament is a heteropolymer or a homopolymer probably is not dictated by simple hydrophobic interactions in the α-helical segments in its constituent IF dimers. Rather, spacer sequences in the coiled coil, sequences in the diverse N- or C-terminal domains, or both most likely are responsible for determining whether particular IF proteins assemble into heteropolymers or homopolymers. In fact, mutations in these regions generate mutated IF polypeptides that can form hetero-oligomers with normal IF proteins. These hybrid molecules often "poison" IF polymerization by blocking assembly at an intermediate stage. When a mutated IF protein is introduced into a cell by transfection or microinjection, the assembly of protofilament intermediates is inhibited. The ability of mutated IF proteins to block IF assembly has proved extremely useful in studies of the function of intermediate filaments in a cell. At the end of the chapter, we discuss how such mutations in keratins have revealed the role of keratin filaments in the epidermis.

Intermediate Filaments Are Dynamic Polymers in the Cell

Compared with our knowledge of microfilaments and microtubules, we understand very little about the assembly of IFs, either in vitro or in the cell. Scientists are still discussing, for example, whether the tetramer is the soluble subunit for assembly of an intermediate filament, analogous to the actin monomer and tubulin dimer. The main supporting evidence for involvement of the tetramer comes from cell fractionation experiments showing that although most vimentin in cultured fibroblasts is polymerized into filaments, 1–5 percent of the protein exists as a soluble pool of tetramers. The presence of a tetramer pool suggests that vimentin monomers are rapidly converted into dimers, which rapidly form tetramers.

Several experiments have demonstrated that IF proteins can exchange with the IF cytoskeleton. In one experiment, a biotin-labeled type I keratin is injected into an epithelial cell; within 2 hours after injection, the labeled protein has become incorporated into the already existing keratin cytoskeleton (Figure 19-53). In other experiments, the vimentin cytoskeleton is made fluorescent by incorporation of rhodamine-labeled vimentin, and then small patches of the fluorescent IF cytoskeleton are photobleached. Shortly after exposure to ultraviolet light, the bleached patches of filaments rapidly recover their fluorescence. (This FRAP technique is described in Figure 5-36). These experiments demonstrate that IF subunits in a soluble pool are able to add themselves to preexisting filaments and that subunits are also able to dissociate from intact filaments (since the bleached patches disappeared). Neither experiment, however, is able to reveal the oligomeric state of the subunit.

Although intermediate filaments exhibit dynamic properties within cells, they clearly are more stable than microtubules and microfilaments. The stability of intermediate filaments presents special problems in mitotic cells, which must reorganize all three cytoskeletal networks during the cell cycle. We have discussed previously how calcium and various kinases control the organization of the microfilament and microtubule cytoskeleton during the cell cycle. Similarly, under the control of cdc2 kinase, filaments of vimentin, desmin, and lamins disassemble prior to or early in mitosis and reassemble after cell division. The phosphorylation of serine residues in the N-terminal domain of lamin A and vimentin by cdc2 kinase induces the disassembly of intact filaments and prevents reassembly. The importance of this site for filament stability has been demonstrated in mutation studies in which the target serine residues in lamin A were converted to residues that cannot be phosphorylated. Without these sites, the lamin filaments remain intact and do not depolymerize early in mitosis as they normally do. In contrast, treatment of cells with drugs that inactivate protein phosphatases causes rapid collapse of the network-like lamin layer underlying the nuclear envelope. The opposing action of kinases and phosphatases thus provides a rapid mechanism for controlling the assembly state of intermediate filaments.

Various Proteins Cross-Link Intermediate Filaments and Connect Them to Other Cell Structures

Intermediate filament–associated proteins (IFAPs) cross-link intermediate filaments with one another, forming a bundle (also called a *tonofilament*) or a network, and with other cell structures, including the plasma membrane. Only a few IFAPs have been identified to date, but many more will undoubtedly be discovered as researchers focus attention on the proteins that control IF organization and assembly. Unlike actin-binding proteins or microtubule-associated proteins, none of the known IFAPs sever or cap intermediate filaments, sequester IF proteins in a soluble pool, or act as a motor protein. The organization of intermediate filaments and their supportive function in various cells types depends in large part on their linkage to other cell structures via IFAPs.

A physical linkage between intermediate filaments and microtubules can be detected with certain drugs. Recall that treatment of cells with high concentrations of colchicine causes the complete dissolution of microtubules after a period of several hours (see Figure 19-6). Although vimentin filaments in colchicine-treated cells remain intact, they clump into disorganized bundles near the nucleus. This finding demonstrates that the organization of vimentin filaments is dependent on intact microtubules and suggests the presence of proteins linking the two types of filaments.

One such protein is *plectin*, a 500,000-MW IFAP that has been shown to bind microtubules in vitro. Plectin also interacts with other cytoskeletal proteins, including spectrin, MAPs, and lamin B. Immunoelectron microscopy reveals gold-labeled antibodies to plectin decorating short, thin connections between microtubules and vimentin, indicating the presence of plectin in these cross-links (Figure 19-54). Similar types of cross-links between microtubules and neurofilaments are seen in micrographs of nerve cell axons (see Figure 19-4b). Although the identity of these connections in axons is unknown, they may represent IFAPs whose function is to cross-link neurofilaments and microtubules into a stable cytoskeleton. Alternatively, these connections to microtubules may be the long arms of the NF-H, which is known to bind microtubules. Interestingly, a family of proteins related to plectin is implicated in forming cross-links between microfilaments and intermediate filaments. The N-terminus of these IFAPs contains a calponin-homology (CH) domain similar to that in fimbrin and other actin cross-linking proteins. Thus IFAPs appear to play a role in integrating the IF cytoskeleton with both the microfilament and microtubule cytoskeletons.

IF Networks Support Cellular Membranes

A network of intermediate filaments is often found as a laminating layer adjacent to cellular membranes, where it provides mechanical support. The best example is the **nuclear lamina** along the inner surface of the nuclear membrane (see

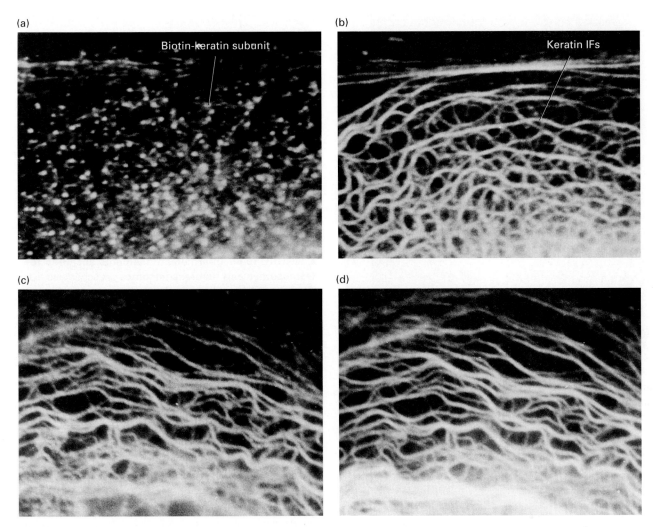

(a) Biotin-keratin subunit

(b) Keratin IFs

(c)

(d)

▲ **FIGURE 19-53 Assembly of biotin-labeled type I keratin in PtK2 fibroblasts.** Monomeric type I keratin was purified, chemically modified with biotin, and injected into living PtK2 cells. The cells were then fixed at different times after injection and stained with a fluorescent antibody to biotin and with antibodies to keratin. At 20 minutes after injection, the microinjected biotin-labeled keratin (a) is concentrated in small foci scattered through the cytoplasm, while the injected protein has not been integrated into the endogenous keratin cytoskeleton (b). By four hours, the biotin-labeled subunits (c) and the keratin filaments (d) display identical patterns, indicating that the microinjected protein has become incorporated into the existing cytoskeleton. [From R. K. Miller, K. Vistrom, and R. D. Goldman, 1991, *J. Cell Biol.* **113**:843; courtesy of R. D. Goldman.]

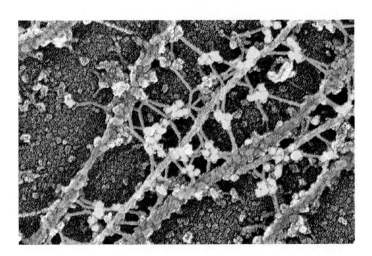

◄ **FIGURE 19-54 Immunoelectron micrograph of a fibroblast cell showing plectin cross-links between intermediate filaments and microtubules.** Microtubules are highlighted in red; intermediate filaments, in blue; and the short connecting fibers between them, in green. Staining with gold-labeled antibodies to plectin (yellow) reveals that these fibers contain plectin. See Figure 5-17 for a description of this microscopic technique. [From T. M. Svitkina, A. B. Verkhovsky, and G. G. Borisy, 1996, *J. Cell Biol.* **135**:991; courtesy of T. N. Svitkina.]

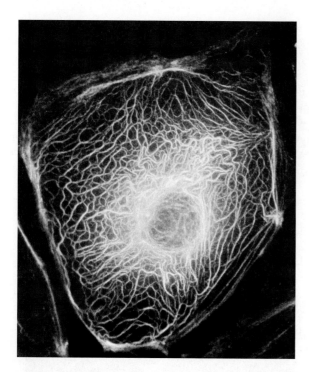

▲ **FIGURE 19-55 Fluorescent micrograph of a PtK2 fibroblast stained with anti-keratin antibodies.** Although the keratin filament network in these cells extends from the cell surface to the nucleus, it is densest at the nuclear periphery. [Courtesy of R. D. Goldman.]

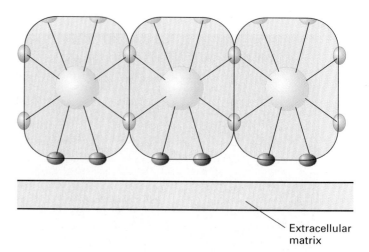

Extracellular matrix

▲ **FIGURE 19-56 Intermediate filaments anchored to desmosomes and hemidesmosomes.** A schematic diagram of cells in the basal layer of the epidermis shows the keratin filaments crisscrossing the cell and making connections to desmosomes (blue) between cells, hemidesmosomes (gray) at the basement membrane, and peripheral proteins (orange) at the apical membrane. [Adapted from E. Fuchs and D. Cleveland, 1998, *Science,* **279**:514.]

Figure 13-15b). This supporting network is composed of lamin A and lamin C filaments cross-linked into an orthogonal lattice, which is attached via lamin B to the inner nuclear membrane through interactions with a lamin B receptor, an IFAP, in the membrane.

In addition to forming the nuclear lamina, intermediate filaments are typically organized in the cytosol as an extended system that stretches from the nuclear envelope to the plasma membrane (Figure 19-55). Some intermediate filaments run parallel to the cell surface, while others traverse the cytosol; together they form an internal framework that helps support the shape and resilience of the cell. In vitro binding experiments suggest that at the plasma membrane, vimentin filaments bind two proteins: ankyrin, the actin-binding protein associated with the Na^+/K^+ ATPase in non-erythroid cells, and plectin. Plectin also binds lamin B in vitro. Through these two IFAPs, vimentin filaments are attached to the plasma membrane and the nuclear membrane, thereby providing these membranes a flexible structural support.

Intermediate Filaments Are Anchored in Cell Junctions

The epithelial cells of organs and skin are held together by specialized cell junctions called *desmosomes* and *hemidesmosomes* (see Figure 15-23). Desmosomes mediate cell-cell adhesion, and hemidesmosomes are responsible for attaching

cells to the underlying basement membrane. In the electron microscope, both junctions appear as darkly staining proteinaceous plaques that are bound to the cytosolic face of the plasma membrane and attached to bundles of intermediate filaments. The intermediate filaments in one cell thus are directly connected to the intermediate filaments in a neighboring cell via desmosomes or to the extracellular matrix via hemidesmosomes (Figure 19-56). As a result of the connections between intermediate filaments and these cell junctions, shearing forces are distributed from one region of a cell layer to the entire sheet of epithelial cells, providing strength and rigidity to the entire epithelium.

Without the supporting network of intermediate filaments, an epithelium remains intact, but the cells are easily damaged by abrasive forces. Like actin microfilaments, which also are attached to specialized regions of the plasma membrane, intermediate filaments form a flexible but resilient framework that gives structural support to an epithelium. Because of the important role of cell junctions in cell adhesion and the stability of tissues, we discuss their structure and relation to cytoskeletal filaments in detail in Chapter 22.

Desmin and Associated Proteins Stabilize Sarcomeres in Muscle

In muscle, a lattice composed of a band of desmin filaments surrounds the sarcomere (Figure 19-57). The desmin filaments encircle the Z disk and are cross-linked to the plasma membrane by several IFAPs, including paranemin and ankyrin.

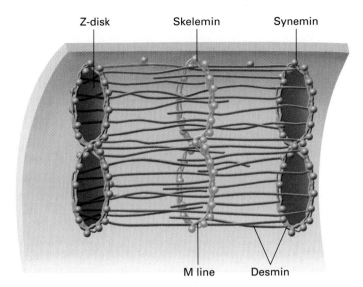

▲ **FIGURE 19-57 A diagram of desmin filaments in muscle.**
These type III intermediate filaments encircle the Z disk and make additional connections to neighboring Z disks. Longitudinally oriented desmin filaments bridge between Z disks in the same myofibril. The alignment of desmin filaments with the muscle sarcomere is held in place by numerous IFAPs, including skelemin at the M line and synemin at the Z disk.

Longitudinal desmin filaments cross to neighboring Z disks within the myofibril, and connections between desmin filaments around Z disks in adjacent myofibrils serve to cross-link myofibrils into bundles within a muscle cell. The lattice is also attached to the sarcomere through interactions with myosin thick filaments, possibly by skelemin in the H zone. Because the desmin filaments lie outside the sarcomere, they do not actively participate in generating contractile forces. Rather, desmin plays an essential structural role in maintaining muscle integrity. In transgenic mice lacking desmin, for example, this supporting architecture is disrupted and muscles are misaligned.

Disruption of Keratin Networks Causes Blistering

The epidermis is a tough outer layer of tissue, which acts as a water-tight barrier to prevent desiccation and serves as a protection against abrasion. In epidermal cells, bundles of keratin filaments are cross-linked by *filaggrin*, an IFAP, and are anchored at their ends to desmosomes. As epidermal cells differentiate, the cells condense and die, but the keratin filaments remain intact, forming the structural core of the dead, keratinized layer of skin. The structural integrity of keratin is essential in order for this layer to withstand abrasion.

In humans and mice, the K4 and K14 keratin polypeptides form heterodimers that assemble into protofilaments. A mutant K14 with deletions in either

the N- or C-terminal domains can form heterodimers in vitro but does not assemble into protofilaments. Expression of such mutant keratin proteins in cells causes IF networks to breakdown into aggregates. Transgenic mice who express a mutant K14 protein in the basal stem cells of the epidermis display gross skin abnormalities, primarily blistering of the epidermis, which resemble the human skin disease *epidermolysis bullosa simplex* (EBS). Histological examination of the blistered area reveals a high incidence of dead basal cells. Death of these cells appears to be caused by mechanical trauma from rubbing of the skin during movement of the limbs. Without their normal bundles of keratin filaments, the mutant basal cells become fragile and easily damaged, causing the overlying epidermal layers to delaminate and blister (Figure 19-58). Like the role of desmin filaments in

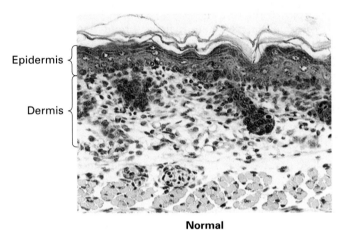

Normal

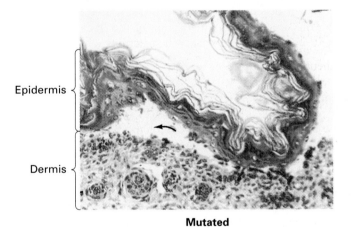

Mutated

▲ **FIGURE 19-58 Keratins and skin disease.** Histological sections through the skin of a normal and a transgenic mouse carrying a mutant keratin gene. In a normal mouse *(top)*, the epidermis consists of a hard outer epidermal layer covering the soft inner dermal layer. In contrast, the skin of mice expressing abnormal K14 keratin resembles the skin of humans suffering from epidermolysis bullosa simplex, a blistering disease. The mutation in keratin weakens the cells at the base of the epidermis, thus causing the two layers to separate (arrow). [From P. Coulombe et al., 1991, *Cell* **66**:1301; courtesy of E. Fuchs.]

supporting muscle tissue, keratin filaments appear to play a general role in maintaining the structural integrity of tissues by mechanically reinforcing the connections between cells.

SUMMARY Intermediate Filaments

- Intermediate filaments are present only in cells that display a multicellular organization. One essential role of intermediate filaments is to distribute tensile forces across cells in a tissue.

- Unlike microtubules and microfilaments, intermediate filaments are assembled from a large number of different IF proteins. These are divided into six types based on their sequences. The lamins are expressed in all cells, whereas the other types are expressed in specific tissues (see Table 19-3).

- Although intermediate filaments are dynamic polymers, which readily exchange subunits from a soluble pool, they are more stable than microfilaments and microtubules. Phosphorylation of the head domain depolymerizes intermediate filaments in vivo.

- Assembly of intermediate filaments probably proceeds through several intermediate structures, which associate by lateral and end-to-end interactions (see Figure 19-51).

- The organization of intermediate filaments into networks and bundles, mediated by various IFAPs, provides structural stability to cells. IFAPs also cross-link intermediate filaments to the plasma and nuclear membranes, microtubules, and microfilaments.

- Major degenerative diseases of skin, muscle, and neurons are caused by disruption of the IF cytoskeleton or its connections to other cell structures.

PERSPECTIVES for the Future

How a cell divides has been one of the fundamental questions in cell biology for the past 100 years. Because the pace of discovery in cell biology has accelerated, we are close to understanding several important aspects of this process. For example, 20 years ago biologists debated whether microtubule sliding or equilibrium was the causal force for chromosome movement. Since then, many motor proteins that contribute to chromosome movements have been identified, and the important role of microtubule dynamics has been confirmed. Within the next few years, we should understand how the spindle specifies the cleavage plane of a cell. In addition, complementary studies on the cell cycle should offer a glimpse of how mitotic checkpoints control the various mechanical steps of mitosis.

Determination of the atomic structure of kinesin and of tubulin has provided clues about the mechanism of motor-protein binding to microtubules and a possible mechanism of mechanochemical force transduction. However, the explanation of why some kinesins move toward the (+) end while others move toward the (−) end still eludes researchers. Understanding of this phenomenon will come from a combination of careful dynamic measurements of kinesin step size, single-molecule imaging of kinesin on microtubules, and kinetic experiments on the kinesins together with insightful mutagenesis experiments.

Dynein is the most complicated of the motor proteins and therefore the least understood. How does it select cargoes, and what is the role of MBPs in cargo selection? We do not know what sequences comprise the dynein motor domain. This fundamental clue to the mechanism of (−) end–directed movement will be learned from structural biology. Maybe dynein, like myosin and kinesin, has evolved from Ras. Whatever the speculation, determination of the dynein structure is eagerly awaited.

Relatively little is known about the assembly and disassembly of intermediate filaments. Although kinases and phosphatases influence the monomer-polymer equilibrium, researchers are looking for other proteins that regulate assembly by controlling the monomer pool. In addition, it is clear that the assembly and organization of intermediate filaments is linked to the assembly and organization of the microfilament and microtubule cytoskeletons. Studies in the next few years will begin to uncover the signaling pathways and common cross-linking proteins that mediate the complex interactions between the three cytoskeletal systems in a cell.

PERSPECTIVES in the Literature

A crucial step in mitosis is marked by the segregation of chromosomes at anaphase. Prior to this step, microtubules from the poles of a spindle must capture a kinetochore on the sister chromatids. At the checkpoint, mitosis is delayed until all kinetochores are captured by microtubules from the nearest pole. When this condition is met, the polar microtubules pull the sister chromatids to the opposite poles. Thus the checkpoint must employ a mechanism to detect and communicate that status of kinetochores on sister chromatids. Failure to satisfy this condition will cause chromosomes to missegregate and leave cells with incorrect numbers of chromosomes. In Chapter 20 we will discuss how the spindle assembly checkpoint involves the anaphase promoting complex. However, we do not fully understand how this complex is regulated by the machinery of the spindle.

The two papers referenced below investigate the intimate interactions between the kinetochore proteins that signal to the cell-cycle checkpoint. Using these papers as an entry into the literature, develop a model of the attachment-sensing mechanism from kinetochore to cell-cycle checkpoint. Include in the model the roles of microtubule dynamics and motor proteins in forming the attachment, as well as schemes that depict how regulatory proteins such as kinases and phosphatases might activate or inactivate spindle, kinetochore, and cell-cycle components. Because the experiments are conducted in various organisms, you might simplify your analysis by choosing one organism. Continue to read the literature on this topic—you will find that your model requires constant modification.

Rieder, C. L., et al. 1994. Anaphase onset in vertebrate somatic cells is controlled by a checkpoint that monitors sister kinetochore attachment to the spindle. *J. Cell Biol.* **127**:1301–10.

Waters, J. C., et al. 1998. Localization of Mad2 to kinetochores depends on microtubule attachment, not tension. *J. Cell Biol.* **141**:1181–91.

Testing Yourself on the Concepts

1. Microtubules are polar filaments; that is, one end is different from the other. What is the basis for this polarity, how is polarity related to microtubule organization within the cell, and how is polarity related to the intracellular movements powered by microtubule-dependent motors?

2. In cells, microtubule assembly depends on other proteins as well as tubulin concentration and temperature. What are some of the proteins that influence microtubule assembly in vivo, and how do they affect assembly?

3. Cell swimming depends on appendages containing microtubules. What is the underlying structure of these appendages, and how do these structures generate the force required to produce swimming?

4. The mitotic spindle is a highly organized machine designed to segregate chromosomes and ensure that each daughter cell receives equivalent genetic information. Spindle function relies heavily on microtubule motors. How do these proteins contribute to creation, maintenance, and function of the spindle?

5. Compared to microfilaments and microtubules, intermediate filaments are relatively stable. However, cells can induce intermediate filament disassembly when needed. How does this occur, and why is disassembly necessary?

MCAT/GRE-Style Questions

Key Concept Please read the section titled "Dynamic Instability Is an Intrinsic Property of Microtubules" (p. 805) and answer the following questions:

1. In cells, dynamic instability typically occurs

  a. At (+) ends only.

  b. At (−) ends only.

  c. At both (+) and (−) ends.

  d. At MTOCs.

2. The terms catastrophe and rescue describe

  a. Microtubule depolymerization.

  b. The process of microtubule severing.

  c. The type of microtubule assembly known as treadmilling.

  d. The transitions between growth and shortening.

3. For a given population of microtubules undergoing dynamic instability

  a. All of the microtubules will be growing at the same time.

  b. All of the microtubules will be shortening at the same time.

  c. All of the microtubules will be growing at one end and shortening at the other end.

  d. Some microtubules will be growing and some will be shortening at any given time.

4. The dependence of dynamic instability on tubulin concentration means that this type of microtubule assembly can take place only when

  a. The tubulin concentration is much greater than the C_c.

  b. The tubulin concentration is close to the C_c.

  c. The tubulin concentration is close to the microtubule concentration.

  d. The tubulin concentration is less than the microtubule concentration.

5. A growing microtubule has

  a. GTP bound to the β-tubulin subunits at the microtubule end.

  b. GTP bound to the β-tubulin subunits throughout the microtubule.

  c. GDP bound to the β-tubulin subunits at the microtubule end.

  d. GDP bound to the α-tubulin subunits throughout the microtubule.

Key Experiment Please read the section titled "Astral Microtubules Determine Where Cytokinesis Takes Place" (p. 833) and refer to Figure 19-48; then answer the following questions:

6. The experiment described in Figure 19-48 demonstrates that the formation of a contractile ring depends on

  a. A functional mitotic spindle.

  b. Two functional mitotic spindles.

  c. A single aster.

  d. Two asters.

7. Very infrequently, tripolar mitotic spindles mistakenly form instead of the normal bipolar spindle. Based on the experiment described in Figure 19-48, how many cleavage furrows will a cell with a tripolar spindle attempt to generate?

  a. One.

  b. Two.

  c. Three.

  d. Four.

8. Which of these spindle components determines where in the cell cortex the contractile ring will assemble:

    a. The (−) ends of polar microtubules.

    b. The (−) ends of astral microtubules.

    c. The (+) ends of polar microtubules.

    d. The (+) ends of astral microtubules.

9. Microtubule-depolymerizing drugs such as colchicine would be expected to

    a. Inhibit mitosis but allow cytokinesis.

    b. Inhibit both mitosis and cytokinesis.

    c. Induce formation of multiple contractile rings.

    d. Inhibit cytokinesis but allow mitosis.

10. Cells that divide asymmetrically to produce daughter cells of different sizes must first

    a. Position the mitotic spindle in the center of the cell.

    b. Form a contractile ring and then move the spindle to that position.

    c. Form mitotic spindles with different size asters.

    d. Position the mitotic spindle asymmetrically in the cell.

Key Application Please read the sections titled "Intermediate Filaments Can Identify Cellular Origin of Certain Tumors" (p. 838) and "Disruption of Keratin Networks Causes Blistering" (p. 843) and answer the following questions:

11. The outermost layer of the epidermis is composed of

    a. Undifferentiated epidermal cells.

    b. Basal stem cells.

    c. Cross-linked keratin filaments.

    d. Differentiated epidermal cells.

12. Detection of vimentin in cells of a malignant tumor indicates that the cancer probably originated in

    a. Neuronal tissue.

    b. Epithelial tissue.

    c. Mesenchymal tissue.

    d. Muscle tissue.

13. Blistering skin diseases such as epidermolysis bullosa simplex (EBS)

    a. Result from increased protofilament assembly.

    b. Demonstrate the importance of vimentin.

    c. Occur only in humans.

    d. Result from weakening of normal cell-cell interactions.

14. Diseases such as EBS point to importance of intermediate filaments in helping cells and tissues

    a. Resist mechanical stress.

    b. Resist chemical stress.

    c. Act as a barrier to water.

    d. Control differentiation.

Key Terms

anaphase *823*

asters *823*

axoneme *817*

basal body *818*

centromere *823*

centrosome *799*

cilia *817*

cytokinesis *833*

dyneins *815*

flagella *817*

intermediate filaments *795*

kinesin *813*

kinetochores *823*

lamins *838*

metaphase *806, 823*

microtubule-associated proteins (MAPs) *807*

microtubule-organizing center (MTOC) *799*

microtubules *795*

mitosis *823*

mitotic spindle *823*

prophase *823*

telophase *823*

tubulin *796*

References

Microtubule Structures

Baas, P. W. 1996. The neuronal centrosome as a generator of microtubules for the axon. *Curr. Topics Dev. Biol.* 33:281–298.

Erickson, H. P. 1998. Atomic structures of tubulin and FtsZ. *Trends Cell Biol.* 8:133–137.

Vaughn, K. C., and J. D. I. Harper. 1998. Microtubule-organizing centers and nucleating sites in land plants. *Int. Rev. Cytol.* 181:75–149.

Zimmerman, W., C. A. Sparks, and S. J. Doxsey. 1999. Amorphous no longer: the centrosome comes into focus. *Curr. Opin. Cell Biol.* 11:122–128.

Microtubule Dynamics and Associated Proteins

Desai, A., and T. J. Mitchison. 1997. Microtubule polymerization dynamics. *Annu. Rev. Cell Dev. Biol.* 13:83–117.

Maccioni, R. B.., and V. Cambiazo. 1995. Role of microtubule-associated proteins in the control of microtubule assembly. *Physiol. Rev.* 75:835–864.

Kinesin, Dynein, and Intracellular Transport

Block, S. M. 1998. Kinesin: what gives? *Cell* 93:5–8.

Goodson, H. V., C. Valetti, and T. E. Kreis. 1997. Motors and membrane traffic. *Curr. Opin. Cell Biol.* 9:18–28.

Hirokawa, N. 1998. Kinesin and dynein superfamily proteins and the mechanism of organelle transport. *Science* 279:519–526.

Holleran, E. A., S. Karki, and E. L. F. Holzbaur. 1998. The role of the dynactin complex in intracellular motility. *Int. Rev. Cytol.* 182:69–109.

Mandelkow, E., and K. A. Johnson. 1998. The structural and mechanochemical cycle of kinesin. *Trends Biochem. Sci.* 23:429–433.

Moore, J. D., and S. A. Endow. 1996. Kinesin proteins: a phylum of motors for microtubule-based motility. *Bioessays* 18:207–219.

Thaler, C. D., and L. T. Haimo. 1996. Microtubules and microtubule motors: mechanisms of regulation. *Int. Rev. Cytol.* **164**:269–327.

Vale, R. D., and R. J. Fletterick. 1997. The design plan of kinesin motors. *Ann. Rev. Cell Dev. Biol.* **8**:4–9.

Cilia and Flagella: Structure and Movement

Dutcher, S. K. 1995. Flagellar assembly in two hundred and fifty easy-to-follow steps. *Trends Genet.* **11**:398–404.

Microtubule Dynamics and Motor Proteins during Mitosis

Compton, D. A. 1998. Focusing on spindle poles. *J. Cell Sci.* **111**:1477–1481.

Inoue, S. 1996. Mitotic organization and force generation by assembly/disassembly of microtubules. *Cell Struc. Funct.* **21**:375–379.

Nicklas, R. B. 1997. How cells get the right chromosomes. *Science* **275**:632–637.

Rieder, C. L., and E. D. Salmon. 1998. The vertebrate cell kinetochore and its roles during mitosis. *Trends Cell Biol.* **8**:310–318.

Saunders, W. S. 1999. Action at the ends of microtubules. *Curr. Opin. Cell Biol.* **11**:129–133.

Intermediate Filaments

Chou, Y. H., O. Skalli, and R. D. Goldman. 1997. Intermediate filaments and cytoplasmic networking: new connections and more functions. *Curr. Opin. Cell Biol.* **9**:49–53.

Fuchs, E., and D. W. Cleveland. 1998. A structural scaffolding of intermediate filaments in health and disease. *Science* **279**:514–519.

Fuchs, E., and K. Weber. 1994. Intermediate filaments: structure, dynamics, function, and disease. *Ann. Rev. Biochem.* **63**:345–382.

Georgatos, S. D., et al. 1997. To bead or not to bead? Lens-specific intermediate filaments revisited. *J. Cell Sci.* **110**:2629–2634.

Herrmann, H. and U. Aebi. 1998. Intermediate filament assembly: fibrillogenesis is driven by decisive dimer-dimer interactions. *Curr. Opin. Struc. Biol.* **8**:177–183.

Houseweart, M. K., and D. W. Cleveland. 1998. Intermediate filaments and their associated proteins: multiple dynamic personalities. *Curr. Opin. Cell Biol.* **10**:93–101.

Inagaki, M., et al. 1996. Dynamic property of intermediate filaments: regulation by phosphorylation. *BioEssays* **18**:481–487.

Lee, M. K., and D. W. Cleveland. 1996. Neuronal intermediate filaments. *Ann. Rev. Neurosci.* **19**:187–217.

Parry, D. A. D., and P. M. Steinert. 1995. *Intermediate Filament Structure*. R. G. Landes.

20

Cell-to-Cell Signaling: Hormones and Receptors

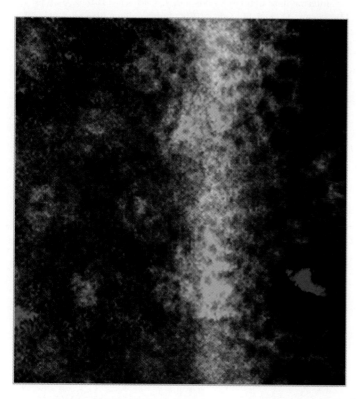

Activated MAP kinase (yellow) accumulates in clusters of assembling neurons in the developing eye of the fruit fly *Drosophila melanogaster.*

N o cell lives in isolation. In all multicellular organisms, survival depends on an elaborate intercellular communication network that coordinates the growth, differentiation, and metabolism of the multitude of cells in diverse tissues and organs. Cells within small groups often communicate by direct cell-cell contact. Specialized junctions in the plasma membranes of adjacent cells permit them to exchange small molecules and to coordinate metabolic responses; other junctions between adjacent cells determine the shape and rigidity of many tissues (Chapter 22). In addition, the establishment of specific cell-cell interactions between different types of cells is a necessary step in the development of many tissues (Chapter 23). In some cases a particular protein on one cell binds to a receptor protein on the surface of an adjacent target cell, triggering its differentiation.

In this chapter, we examine how cells communicate by means of extracellular **signaling molecules**. These substances are synthesized and released by *signaling cells* and produce a specific response only in *target cells* that have **receptors** for the signaling molecules. An enormous variety of chemicals, including small molecules (e.g., amino acid derivatives, acetylcholine), peptides, and proteins, are used in this type of cell-to-cell communication. The extracellular products synthesized by signaling cells can

MEDIA CONNECTIONS

Focus: Second Messengers in Signaling Pathways

Overview: Extracellular Signaling

Technique: Yeast Two-Hybrid System

Classic Experiment 20.1: The Infancy of Signal Transduction: GTP Stimulation of cAMP Synthesis

diffuse away or be transported in the blood, thus providing a means for cells to communicate over longer distances than is possible by chains of direct cell-cell contacts.

We begin with a general discussion of signaling molecules, cell-surface receptors, and the role of intracellular molecules in **signal transduction**, that is, the process of converting extracellular signals into cellular responses. We then describe techniques for detecting, purifying, and cloning cell-surface receptors. Next we examine in some detail several signal-transduction pathways that regulate many different aspects of cellular metabolism, function, and development. Specific signals promote changes in gene expression, cell morphology, and cell movements by modulating the activity of specific transcription factors (Chapter 10), by affecting the contacts between cells (Chapter 23) and between cells and the extracellular matrix (Chapter 22), and by remodeling the cytoskeleton (Chapters 18 and 19). Because there are far too many important signaling molecules and receptors to cover them all in one chapter, we present general schemes for classifying both and focus on those whose function in animal cells is reasonably well understood at the cellular and molecular level.

20.1 Overview of Extracellular Signaling

Communication by extracellular signals usually involves six steps: (1) synthesis and (2) release of the signaling molecule by the signaling cell; (3) transport of the signal to the target cell; (4) detection of the signal by a specific receptor protein; (5) a change in cellular metabolism, function, or development triggered by the receptor-signal complex; and (6) removal of the signal, which often terminates the cellular response.

In many eukaryotic microorganisms (e.g., yeast, slime molds, and protozoans), secreted molecules coordinate the aggregation of free-living cells for sexual mating or differentiation under certain environmental conditions. Chemicals released by one organism that can alter the behavior or gene expression of other organisms of the same species are called **pheromones.** Yeast mating-type factors discussed later in this chapter are a well-understood example of pheromone-mediated cell-to-cell signaling. Some algae and animals also release pheromones, usually dispersing them into the air or water, to attract members of the opposite sex. More important in plants and animals are extracellular signaling molecules that function *within* an organism to control metabolic processes within cells, the growth of tissues, the synthesis and secretion of proteins, and the composition of intracellular and extracellular fluids. This chapter focuses on such cell-to-cell signaling in single-celled eukaryotes and in a variety of higher eukaryotes, particularly mammals.

Signaling Molecules Operate over Various Distances in Animals

In animals, signaling by extracellular, secreted molecules can be classified into three types—endocrine, paracrine, or autocrine—based on the distance over which the signal acts. In addition, certain membrane-bound proteins on one cell can directly signal an adjacent cell (Figure 20-1).

(a) Endocrine signaling

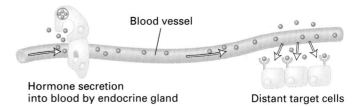

(b) Paracrine signaling

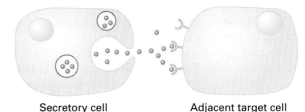

(c) Autocrine signaling

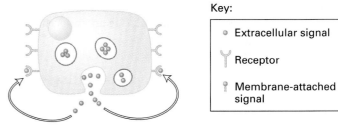

(d) Signaling by plasma membrane–attached proteins

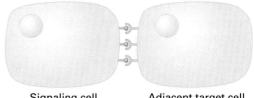

▲ **FIGURE 20-1 General schemes of intercellular signaling in animals.** (a–c) Cell-to-cell signaling by extracellular chemicals occurs over distances from a few micrometers in autocrine and paracrine signaling to several meters in endocrine signaling. (d) Proteins attached to the plasma membrane of one cell can interact directly with receptors on an adjacent cell.

In **endocrine signaling,** signaling molecules, called **hormones,** act on target cells distant from their site of synthesis by cells of endocrine organs. In animals, an endocrine hormone usually is carried by the blood from its site of release to its target.

In **paracrine signaling,** the signaling molecules released by a cell only affect target cells in close proximity to it. The conduction of an electric impulse from one nerve cell to another or from a nerve cell to a muscle cell (inducing or inhibiting muscle contraction) occurs via paracrine signaling. The role of this type of signaling, mediated by **neurotransmitters,** in transmitting nerve impulses is discussed in Chapter 21. Many signaling molecules regulating development in multicellular organisms also act at short range. Some of these molecules are discussed in Chapter 23.

In **autocrine signaling,** cells respond to substances that they themselves release. Many **growth factors** act in this fashion, and cultured cells often secrete growth factors that stimulate their own growth and proliferation. This type of signaling is particularly common in tumor cells, many of which overproduce and release growth factors that stimulate inappropriate, unregulated proliferation of themselves as well as adjacent nontumor cells; this process may lead to formation of tumor mass.

Some compounds can act in two or even three types of cell-to-cell signaling. Certain small amino acid derivatives, such as **epinephrine,** function both as neurotransmitters (paracrine signaling) and as systemic hormones (endocrine signaling). Some protein hormones, such as epidermal growth factor (EGF), are synthesized as the exoplasmic part of a plasma-membrane protein; membrane-bound EGF can bind to and signal an adjacent cell by direct contact. Cleavage by a protease releases secreted EGF, which acts as an endocrine signal on distant cells.

Receptor Proteins Exhibit Ligand-Binding and Effector Specificity

As noted earlier, the cellular response to a particular extracellular signaling molecule depends on its binding to a specific receptor protein located on the surface of a target cell or in its nucleus or cytosol. The signaling molecule (a hormone, pheromone, or neurotransmitter) acts as a **ligand,** which binds to, or "fits," a site on the receptor. Binding of a ligand to its receptor causes a conformational change in the receptor that initiates a sequence of reactions leading to a specific cellular response.

The response of a cell or tissue to specific hormones is dictated by the particular hormone receptors it possesses and by the intracellular reactions initiated by the binding of any one hormone to its receptor. Different cell types may have different sets of receptors for the same ligand, each of which induces a different response. Or the same receptor may occur on various cell types, and binding of the same ligand may trigger a different response in each type of cell. Clearly, different cells respond in a variety of ways to the same ligand.

For instance, acetylcholine receptors are found on the surface of striated muscle cells, heart muscle cells, and pancreatic acinar cells. Release of acetylcholine from a neuron adjacent to a striated muscle cell triggers contraction, whereas release adjacent to a heart muscle slows the rate of contraction. Release adjacent to a pancreatic acinar cell triggers exocytosis of secretory granules that contain digestive enzymes. On the other hand, different receptor-ligand complexes can induce the same cellular response in some cell types. In liver cells, for example, the binding of either glucagon to its receptors or of epinephrine to its receptors can induce degradation of glycogen and release of glucose into the blood.

These examples show that a receptor protein is characterized by *binding specificity* for a particular ligand, and the resulting hormone-ligand complex exhibits *effector specificity* (i.e., mediates a specific cellular response). For instance, activation of either epinephrine or glucagon receptors on liver cells by binding of their respective ligands induces synthesis of **cyclic AMP (cAMP),** one of several intracellular signaling molecules, termed **second messengers,** which regulate various metabolic functions; as a result, the effects of both receptors on liver-cell metabolism are the same. Thus, the binding specificity of epinephrine and glucagon receptors differ, but their effector specificity is identical.

In most receptor-ligand systems, the ligand appears to have no function except to bind to the receptor. The ligand is not metabolized to useful products, is not an intermediate in any cellular activity, and has no enzymatic properties. The only function of the ligand appears to be to change the properties of the receptor, which then signals to the cell that a specific product is present in the environment. Target cells often modify or degrade the ligand and, in so doing, can modify or terminate their response or the response of neighboring cells to the signal.

Hormones Can Be Classified Based on Their Solubility and Receptor Location

Most hormones fall into three broad categories: (1) small lipophilic molecules that diffuse across the plasma membrane and interact with *intracellular* receptors; and (2) hydrophilic or (3) lipophilic molecules that bind to *cell-surface* receptors (Figure 20-2). Recently, nitric oxide, a gas, has been shown to be a key regulator controlling many cellular responses. We discuss this important regulator later in this chapter. Here we briefly describe the three main types of hormones; later we discuss the mechanisms that regulate their synthesis, release, and degradation.

Lipophilic Hormones with Intracellular Receptors Many lipid-soluble hormones diffuse across the plasma membrane and interact with receptors in the cytosol or nucleus. The resulting hormone-receptor complexes bind to transcription-control regions in DNA thereby affecting expression of specific genes (see Figure 20-2a). Hormones of this type include the steroids (e.g., cortisol, progesterone, estradiol, and testosterone), thyroxine, and retinoic acid (see Figure 10-65).

(a) Intracellular receptors

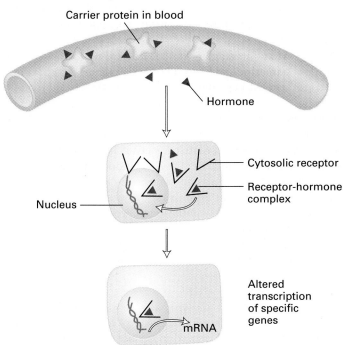

(b) Cell surface receptors

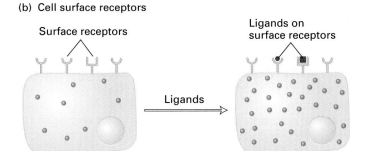

◀ **FIGURE 20-2 Some hormones bind to intracellular receptors; others, to cell-surface receptors.** (a) Steroid hormones, thyroxine, and retinoids, being lipophilic, are transported by carrier proteins in the blood. After dissociation from these carriers, such hormones diffuse across the cell membrane and bind to specific receptors in the cytosol or nucleus. The receptor-hormone complex then acts on nuclear DNA to alter transcription of specific genes. (b) Polypeptide hormones and catecholamines (e.g., epinephrine), which are water soluble, and prostaglandins, which are lipophilic, all bind to cell-surface receptors. This binding triggers an increase or decrease in the cytosolic concentration of second messengers (e.g., cAMP, Ca^{2+}), activation of a protein kinase, or a change in the membrane potential.

Progesterone

Thyroxine (tetraiodothyronine) and triiodothyronine—the principal iodinated compounds in the body—are formed in the thyroid by intracellular proteolysis of the iodinated protein thyroglobulin and immediately released into the blood.

Thyroxine

All steroids are synthesized from cholesterol and have similar chemical skeletons. After crossing the plasma membrane, steroid hormones interact with intracellular receptors, forming complexes that can increase or decrease transcription of specific genes (see Figure 10-68). These receptor-steroid complexes also may affect the stability of specific mRNAs. Steroids are effective for hours or days and often influence the growth and differentiation of specific tissues. For example, estrogen and progesterone, the female sex hormones, stimulate the production of egg-white hormones in chickens and cell proliferation in the hen oviduct. In mammals, estrogens stimulate growth of the uterine wall in preparation for embryo implantation. In insects and crustaceans, α-ecdysone (which is chemically related to steroids) triggers the differentiation and maturation of larvae; like estrogens, it induces the expression of specific gene products.

These two thyroid hormones stimulate increased expression of many cytosolic enzymes (e.g., liver hexokinase) that catalyze the catabolism of glucose, fats, and proteins and of mitochondrial enzymes that catalyze oxidative phosphorylation.

Retinoids are polyisoprenoid lipids derived from retinol (vitamin A). They perform multiple regulatory functions in diverse cellular processes. Retinoids regulate cellular proliferation, differentiation, and death, and they have numerous clinical applications. Their diverse effects reflect, at least in part, the multiplicity of retinoid derivatives, the existence of two different classes of receptors that form heterodimers, and differences in their cis-acting regulatory sites on DNA. During development retinoids act as local mediators of cell-cell interaction. For instance, during the formation of motor neurons in the chick, one class of motor neurons generates a retinoid signal which regulates the number and type of neighboring motoneurons.

Retinoic acid

Water-Soluble Hormones with Cell-Surface Receptors

Because water-soluble signaling molecules cannot diffuse across the plasma membrane, they all bind to cell-surface receptors. This large class of compounds is composed of two groups: (1) peptide hormones, such as **insulin,** growth factors, and **glucagon,** which range in size from a few amino acids to protein-size compounds, and (2) small charged molecules, such as epinephrine and histamine (see Figure 21-28), that are derived from amino acids and function as hormones and neurotransmitters.

Many water-soluble hormones induce a modification in the activity of one or more enzymes already present in the target cell. In this case, the effects of the surface-bound hormone usually are nearly immediate, but persist for a short period only. These signals also can give rise to changes in gene expression that may persist for hours or days. In yet other cases water-soluble signals may lead to irreversible changes, such as cellular differentiation.

Lipophilic Hormones with Cell-Surface Receptors

The primary lipid-soluble hormones that bind to cell-surface receptors are the **prostaglandins.** There are at least 16 different prostaglandins in nine different chemical classes, designated PGA–PGI. Prostaglandins are part of an even larger family of 20 carbon–containing hormones called eicosanoid hormones. In addition to prostaglandins, they include prostacyclins, thromboxanes, and leukotrienes. Eicosonoid hormones are synthesized from a common precursor, arachidonic acid. Arachidonic acid is generated from phospholipids and diacylglycerol.

Leukotrienes

↑

Arachidonate

↓ Aspirin

Prostaglandin H₂ (PGH₂)

Additional prostaglandins ← ↓ → **Thromboxanes** **Prostacyclin**

In both vertebrates and invertebrates, prostaglandins are synthesized and secreted continuously by many types of cells and rapidly broken down by enzymes in body fluids.

 Many prostaglandins act as local mediators during paracrine and autocrine signaling and are destroyed near the site of their synthesis. They modulate the responses of other hormones and can have profound effects on many cellular processes. Certain prostaglandins cause blood platelets to aggregate and adhere to the walls of blood vessels. Because platelets play a key role in clotting blood and plugging leaks in blood vessels, these prostaglandins can affect the course of vascular disease and wound healing; aspirin inhibits their synthesis by acetylating (and thereby irreversibly inhibiting) prostaglandin H₂ synthase. Other prostaglandins initiate the contraction of smooth muscle cells; they accumulate in the uterus at the time of childbirth and appear to be important in inducing uterine contraction.

Recent studies have shown that a family of plant steroids, called *brassinosteroids,* regulates many aspects of development. These lipophilic compounds, like prostaglandins, act through cell-surface receptors.

Cell-Surface Receptors Belong to Four Major Classes

The different types of cell-surface receptors that interact with water-soluble ligands are schematically represented in Figure 20-3. Binding of ligand to some of these receptors induces second-messenger formation, whereas ligand binding to others does not. For convenience, we can sort these receptors into four classes:

- **G protein–coupled receptors** (see Figure 20-3a): Ligand binding activates a **G protein,** which in turn activates or inhibits an enzyme that generates a specific second messenger or modulates an ion channel, causing a change in membrane potential. The receptors for epinephrine, serotonin, and glucagon are examples.
- *Ion-channel receptors* (see Figure 20-3b): Ligand binding changes the conformation of the receptor so that specific ions flow through it; the resultant ion movements alter the electric potential across the cell membrane. The acetylcholine receptor at the nerve-muscle junction is an example.
- *Tyrosine kinase–linked receptors* (see Figure 20-3c): These receptors lack intrinsic catalytic activity, but ligand binding stimulates formation of a dimeric receptor, which then interacts with and activates one or more cytosolic protein-tyrosine kinases. The receptors for many cytokines, the interferons, and human growth factor are of this type. These tyrosine kinase–linked receptors sometimes are referred to as the *cytokine-receptor superfamily.*

(a) G protein–coupled receptors (epinephrine, glucagon, serotonin)

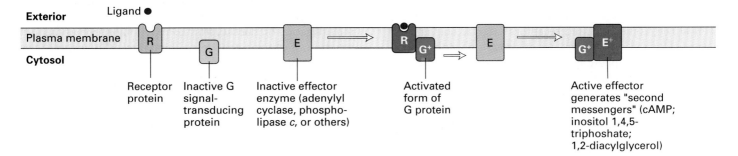

Exterior   Ligand ●

Plasma membrane

Cytosol

Receptor protein

Inactive G signal-transducing protein

Inactive effector enzyme (adenylyl cyclase, phospho-lipase *c*, or others)

Activated form of G protein

Active effector generates "second messengers" (cAMP; inositol 1,4,5-triphoshate; 1,2-diacylglycerol)

(b) Ion-channel receptors (acetylcholine)

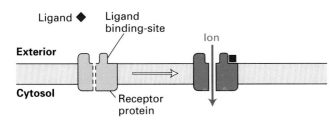

Ligand ◆   Ligand binding-site

Ion

Exterior

Cytosol

Receptor protein

(c) Tyrosine kinase–linked receptors (erythropoietin, interferons)

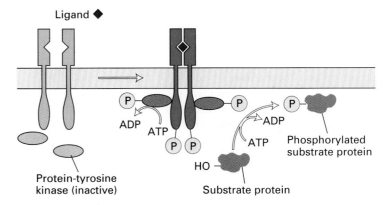

Ligand ◆

ADP   ATP   ADP   ATP

Phosphorylated substrate protein

HO

Protein-tyrosine kinase (inactive)

Substrate protein

(d) Receptors with intrinsic enzymatic activity

Ligand ●

Exterior

Cytosol

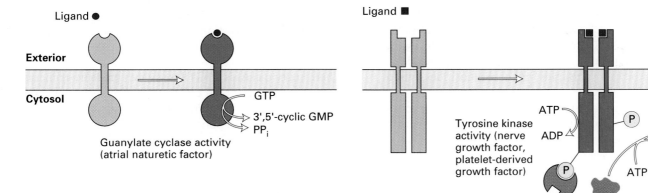

GTP

3',5'-cyclic GMP
PP_i

Guanylate cyclase activity (atrial naturetic factor)

Ligand ■

ATP

ADP

ATP

Tyrosine kinase activity (nerve growth factor, platelet-derived growth factor)

▲ **FIGURE 20-3 Four classes of ligand-triggered cell-surface receptors.** Common ligands for each receptor type are listed in parentheses. (a) G protein–linked receptors. Binding of ligand (maroon) triggers activation of a G protein, which then binds to and activates an enzyme that catalyzes synthesis of a specific second messenger. (b) Ion-channel receptors. A conformational change triggered by ligand binding opens the channel for ion flow. (c) Tyrosine kinase–linked receptors. Ligand binding causes formation of a homodimer or heterodimer, triggering the binding and activation of a cytosolic protein-tyrosine kinase. The activated kinase phosphorylates tyrosines in the receptor; substrate proteins then bind to these phosphotyrosine residues and are

phosphorylated. (d) Receptors with intrinsic ligand-triggered enzymatic activity in the cytosolic domain. Some activated receptors are monomers with guanine cyclase activity and can generate the second messenger cGMP *(left)*. The receptors for many growth factors have intrinsic protein-tyrosine kinase activity *(right)*. Ligand binding to most such receptor tyrosine kinase (RTKs) causes formation of an activated homodimer, which phosphorylates several residues in its own cytosolic domain as well as certain substrate proteins. [Part (c) see J. E. Darnell et al., 1994, *Science* **264**:1415. Part (d) see S. Schulz et al., 1989, *FASEB J.* **3**:2026; D. Garbers, 1989, *J. Biol. Chem.* **264**:9103; and W. J. Fantl et al., 1993, *Annu. Rev. Biochem.* **62**:453.]

- *Receptors with intrinsic enzymatic activity* (see Figure 20-3d): Several types of receptors have intrinsic catalytic activity, which is activated by binding of ligand. For instance, some activated receptors catalyze conversion of GTP to cGMP; others act as protein phosphatases, removing phosphate groups from phosphotyrosine residues in substrate proteins, thereby modifying their activity. The receptors for insulin and many growth factors are ligand-triggered protein kinases; in most cases, the ligand binds as a dimer, leading to dimerization of the receptor and activation of its kinase activity. These receptors—often referred to as receptor serine/threonine kinases or **receptor tyrosine kinases**—autophosphorylate residues in their own cytosolic domain and also can phosphorylate various substrate proteins.

The discussion in this chapter focuses primarily on signaling pathways initiated by G protein–coupled receptors (GPCRs) and receptor tyrosine kinases (RTKs). The general structure and mechanism of action of the intracellular receptors for steroid hormones are discussed in Chapter 10; ion channels are covered in detail in Chapters 15 and 21; and certain receptor serine/threonine kinases as well as other developmentally relevant cell-surface receptors are described in Chapter 23.

Effects of Many Hormones Are Mediated by Second Messengers

The binding of ligands to many cell-surface receptors leads to a short-lived increase (or decrease) in the concentration of the intracellular signaling molecules termed second messengers. These low-molecular-weight signaling molecules include 3′,5′-cyclic AMP (cAMP); 3′,5′-cyclic GMP (cGMP); 1,2-diacylglycerol (DAG); inositol 1,4,5-trisphosphate (IP_3); various inositol phospholipids (phosphoinositides); and Ca^{2+} (Figure 20-4).

The elevated intracellular concentration of one or more second messengers following hormone binding triggers a rapid alteration in the activity of one or more enzymes or nonenzymatic proteins. The metabolic functions controlled by hormone-induced second messengers include uptake and utilization of glucose, storage and mobilization of fat, and secretion of cellular products. These intracellular molecules also control proliferation, differentiation, and survival of cells, in part by regulating the transcription of specific genes. The mode of action of cAMP and other second messengers is discussed in a later section. Removal (or degradation) of a ligand or second messenger, or inactivation of the ligand-binding receptor, can terminate the cellular response to an extracellular signal.

Other Conserved Proteins Function in Signal Transduction

In addition to cell-surface receptors and second messengers, several types of conserved proteins function in signal-transduction pathways stimulated by extracellular signals. Here we introduce the three main classes of these intracellular signaling proteins; their structures and functions are described in detail in later sections.

GTPase Switch Proteins A large group of *GTP-binding* proteins act as molecular switches in signal-transduction pathways. These proteins are turned "on" when bound to GTP and turned "off" when bound to GDP (Figure 20-5a). In the absence of a signal, the protein is bound to GDP. Signals activate the release of GDP, and the subsequent binding to GTP over GDP is favored by the higher concentrations of GTP in the cell. The intrinsic GTPase activity of these GTP-binding proteins hydrolyzes the bound GTP to GDP and P_i, thus converting the active form back to the inactive form. The kinetics of hydrolysis regulates the length of time the switch is "on."

There are two classes of GTPase switch proteins: trimeric G proteins, which as noted already are directly coupled to certain receptors, and monomeric **Ras** and Ras-like proteins. Both classes contain regions that promote the activity of

MEDIA CONNECTIONS
Second Messengers in Signaling Pathways

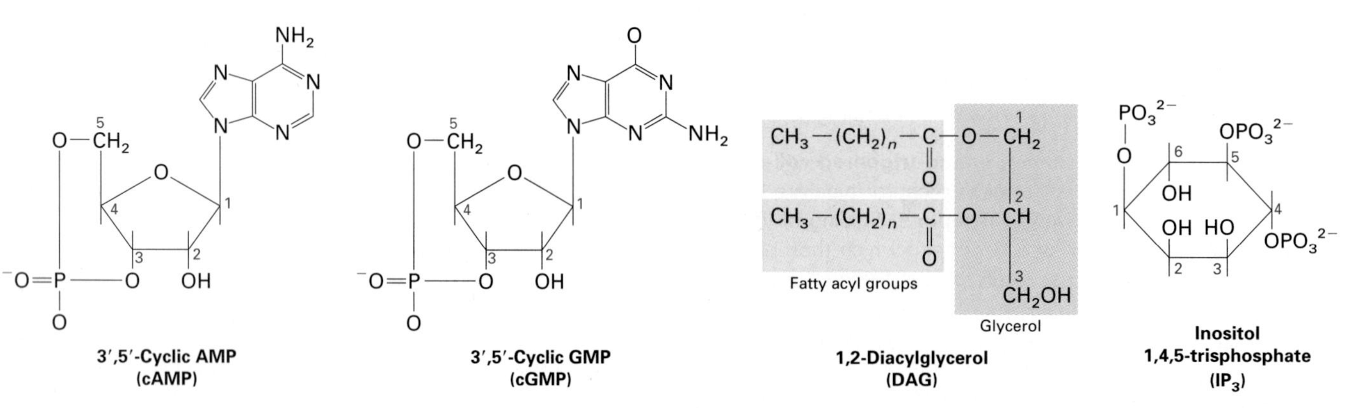

3′,5′-Cyclic AMP (cAMP)

3′,5′-Cyclic GMP (cGMP)

1,2-Diacylglycerol (DAG)

Inositol 1,4,5-trisphosphate (IP_3)

▲ **FIGURE 20-4 Structural formulas of four common intracellular second messengers.** Their abbreviations are indicated. The calcium ion (Ca^{2+}) and several membrane-bound inositol phospholipids (phosphoinositides) also act as second messengers but are not shown (see Figure 20-38).

(a) GTPase switch proteins

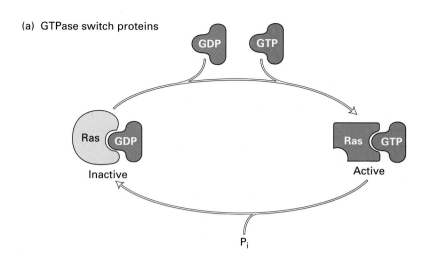

(b) Protein kinases

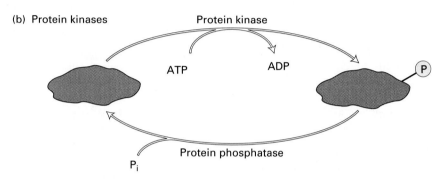

(c) Adapter proteins

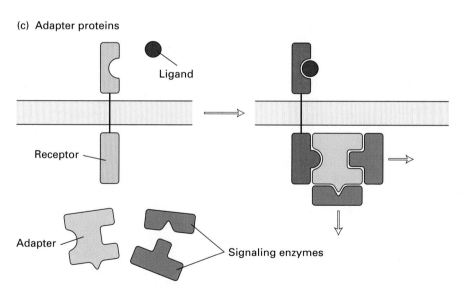

◀ **FIGURE 20-5 Common intracellular signaling proteins.** (a) GTP-binding proteins with GTPase activity function as molecular switches. When bound to GTP they are active; when bound to GDP, they are inactive. They fall into two categories, trimeric G proteins and Ras-like proteins. (b) Protein kinases modulate the activity or the binding properties of substrate proteins by phosphorylating serine, threonine, or tyrosine residues. The phosphorylated form of some proteins is active, whereas the dephosphorylated form of other proteins is active. The combined action of kinases and phosphatases, which dephosphorylate specific substrates, can cycle proteins between active and inactive states. (c) Adapter proteins contain various protein-binding motifs that promote the formation of multiprotein signaling complexes.

specific effector proteins by direct protein-protein interactions. These regions are in their active conformation only when the switch protein is bound to GTP. G proteins are coupled directly to activated receptors, whereas Ras is linked only indirectly via other proteins. The two classes of GTP-binding proteins also are regulated in very different ways.

Protein Kinases Activation of all cell-surface receptors leads to changes in protein phosphorylation through the activation of *protein kinases* (Figure 20-5b). In some cases kinases are part of the receptor itself, and in others they are found in the cytosol or associated with the plasma membrane. Animal cells contain two types of protein kinases: those directed toward tyrosine and those directed toward either serine or threonine. The structures of the catalytic core of both types are very similar. In general, protein kinases become active in response to the stimulation of signaling pathways.

The catalytic activities of kinases are modulated by phosphorylation, by direct binding to other proteins, and by changes in the levels of various second messengers. The activity of protein kinases is opposed by the activity of *protein phosphatases,* which remove phosphate groups from specific substrate proteins. The activities of kinases and phosphatases during cell cycle control are described in some detail in Chapter 13.

Adapter Proteins Many signal-transduction pathways contain large multiprotein signaling complexes, which often are held together by *adapter proteins* (Figure 20-5c). Adapter proteins do not have catalytic activity, nor do they directly activate effector proteins. Rather, they contain different combinations of domains, which function as docking sites for other proteins. For instance, different domains bind to phosphotyrosine residues (SH2 and PTB domains), proline-rich sequences (SH3 and WW domains), phosphoinositides (PH domains), and unique C-terminal sequences with a C-terminal hydrophobic residue (PDZ domains). In some cases adapter proteins contain arrays of a single binding domain or different combinations of domains. In addition, these binding domains can be found alone or in various combinations in proteins containing catalytic domains. These combinations provide enormous potential for complex interplay and crosstalk between different signaling pathways.

Common Signaling Pathways Are Initiated by Different Receptors in a Class

In general, different members of a particular class of receptors transduce signals by highly conserved pathways. Moreover, analogies are found in the signaling pathways associated with different receptor classes. Figure 20-6 illustrates the main components of the key signaling pathways downstream from G protein–coupled receptors (GPCRs) and receptor tyrosine kinases (RTKs), the two receptor classes that we consider in detail in this chapter. Although a GTPase switch protein occurs in both types of pathways, its position in the pathway relative to the receptor differs. Second messengers are critical components of most GPCR pathways and some RTK pathways. Adapter proteins function in all RTK pathways but not in the main GPCR pathways. Protein kinases, however, play a key role in all signaling pathways; ultimately an activated protein kinase phosphorylates one or more substrate proteins. The nature of the substrate proteins, which include enzymes, microtubules, histones, and transcription factors, plays an important role in determining the cellular response to a particular signal in a particular cell.

The Synthesis, Release, and Degradation of Hormones Are Regulated

Because of their potent effects, hormones and neurotransmitters must be carefully regulated. The release and degradation of some signaling compounds are regulated to produce rapid, short-term effects; others to produce slower-acting but longer-lasting effects (Table 20-1). In some cases, complex regulatory networks coordinate the levels of hormones whose effects are interconnected.

Peptide Hormones and Catecholamines Organisms must be able to respond instantly to many changes in their internal or external environment. Such rapid responses are mediated primarily by peptide hormones and the **catecholamines** epinephrine, norepinephrine, and dopamine (see Figure 21-28). The cells that produce these signaling molecules store them in secretory vesicles just under the plasma membrane (see Figure 21-30). The supply of stored, preformed signaling molecules is sufficient for 1 day in the case of peptide hormones and for several days in the case of catecholamines. All peptide hormones, including insulin and adrenocorticotropic hormone (ACTH), are synthesized as part of a longer propolypeptide, which is cleaved by specific proteases to generate the active molecule just after it is transported to a secretory vesicle (see Figure 17-61).

Stimulation of signaling cells causes immediate exocytosis of the stored peptide hormone or catecholamine into the surrounding medium or the blood. Secreting cells also are stimulated to synthesize the signaling molecule and replenish the cell's supply. Released peptide hormones persist in the blood for only seconds or minutes before being degraded by blood and tissue proteases. Released catecholamines are rapidly inactivated by different enzymes or taken up by specific cells (Section 21.4). The initial actions of these signaling molecules on target cells (the activation or inhibition of specific enzymes) also last only seconds or minutes. Thus the catecholamines and some peptide hormones can mediate short responses that are terminated by their own degradation.

Steroid Hormones, Thyroxine, and Retinoic Acid The pathways for synthesizing steroid hormones from cholesterol involve 10 or more enzymes. Steroid-producing cells, like those in the adrenal cortex, store a small supply of hormone precursor but none of the mature, active hormone. When stimulated, the cells convert the precursor to the active hormone, which then diffuses across the plasma membrane into the blood. Likewise, thyroglobulin, the iodinated precursor of thyroxine is stored in thyroid follicles. When cells lining these follicles are exposed to thyroid-stimulating hormone (TSH), they take up thyroglobulin; controlled proteolysis of this glycoprotein by lysosomal enzymes yields thyroxine, which is released into the blood.

Because the signaling cells that produce thyroxine and steroid hormones store little of the active hormone, release of these hormones takes from hours to days (see Table 20-1). These hormones, which are poorly soluble in aqueous solution, are transported in the blood by carrier proteins; the tightly bound active hormones are not rapidly degraded. Thus, cellular responses to thyroxine and steroid hormones take awhile to occur but persist from hours to days.

Retinol is stored in the liver and is found in high concentrations in blood in a complex with serum binding protein. Due to its lipophilic nature, retinol diffuses through the

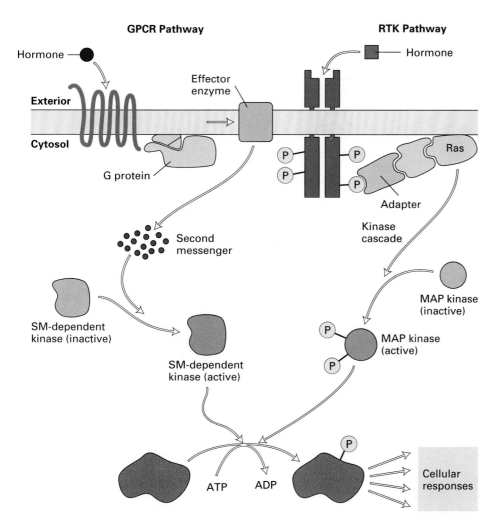

GPCR Pathway

Hormone

Exterior

Cytosol

G protein

Effector enzyme

Second messenger

SM-dependent kinase (inactive)

SM-dependent kinase (active)

ATP   ADP

RTK Pathway

Hormone

Adapter

Kinase cascade

Ras

MAP kinase (inactive)

MAP kinase (active)

Cellular responses

◀ **FIGURE 20-6 Schematic overview of common signaling pathways downstream from G protein–coupled receptors (GPCRs) and receptor tyrosine kinases (RTKs).** Hormone binding to the receptor initiates a series of events leading to phosphorylation of specific substrate proteins, which mediate the cellular responses such as changes in the activity of metabolic enzymes, gene expression, and cytoskeletal structures. The kinase cascade entails sequential activation of specific protein kinases induced by a signal from activated Ras protein. Second messengers (SM) play a role in some RTK signaling pathways, although not in the pathway depicted here. Likewise, some GPCR pathways do not involve second messengers; these lead to activation of MAP kinase. See text for discussion.

plasma membrane and forms a complex with a cytosolic retinol-binding protein called CRBP. Retinol is converted to retinal through the activity of retinol dehydrogenase, and retinal, in turn, is converted to retinoic acid by retinal dehydrogenase. Retinoic acid can act as a signal in the cell in which it is produced, or it can diffuse through the plasma membrane to influence the development of neighboring cells. Retinoic acid can also be further modified enzymatically to alter its signaling specificity.

Feedback Control of Hormone Levels The synthesis and/or release of many hormones are regulated by *positive* or *negative feedback*. This type of regulation is particularly important in coordinating the action of multiple hormones on various cell types during growth and differentiation. Often, the levels of several hormones are interconnected by feedback circuits, in which changes in the level of one hormone affect the levels of other hormones. One example is the regulation of estrogen and progesterone, steroid hormones that stimulate the growth and differentiation of cells in the endometrium, the tissue lining the interior of the uterus. Changes in the endometrium prepare the organ to receive and nourish an embryo. The levels of both hormones are regulated by complex feedback circuits involving several other hormones.

TABLE 20-1 Characteristic Properties of Principal Types of Mammalian Hormones

Property	Steroids	Thyroxine	Peptides and Proteins	Catecholamines
Feedback regulation of synthesis	Yes	Yes	Yes	Yes
Storage of preformed hormone	Very little	Several weeks	One day	Several days, in adrenal medulla
Mechanism of secretion	Diffusion through plasma membrane	Proteolysis of thyroglobulin	Exocytosis of storage vesicles	Exocytosis of storage vesicles
Binding to plasma proteins	Yes	Yes	Rarely	No
Lifetime in blood plasma	Hours	Days	Minutes	Seconds
Time course of action	Hours to days	Days	Minutes to hours	Seconds or less
Receptors	Cytosolic or nuclear	Nuclear	Plasma membrane	Plasma membrane
Mechanism of action	Receptor-hormone complex controls transcription and stability of mRNAs		Hormone binding triggers synthesis of cytosolic second messengers or protein kinase activity	Hormone binding causes change in membrane potential or triggers synthesis of cytosolic second messengers

SOURCE: Adapted from E. L. Smith et al., 1983, *Principles of Biochemistry: Mammalian Biochemistry,* 6th ed., McGraw-Hill, p. 358. Reproduced by permission of McGraw-Hill.

SUMMARY Overview of Extracellular Signaling

• Extracellular signaling molecules regulate interaction between unicellular organisms and are critical regulators of physiology and development in multicellular organisms.

• There are many different types of signals, including membrane-anchored and secreted proteins and peptides, small lipophilic molecules (e.g., steroid hormones, thyroxine), small hydrophilic molecules derived from amino acids (e.g., catecholamines), and gases. Signals can act at short range, long range, or both (see Figure 20-1).

• The multitude of cell-surface receptors fall into four main classes: G protein–coupled receptors, ion-channel receptors, receptors linked to cytosolic tyrosine kinases, and receptors with intrinsic catalytic activity (see Figure 20-3).

• Binding of extracellular signaling molecules to cell-surface receptors trigger intracellular pathways that ultimately modulate cellular metabolism, function, or development.

• The level of second messengers, such as Ca^{2+}, cAMP, and IP_3 are modulated in response to binding of ligand to cell-surface receptors. These intracellular signaling molecules, in turn, regulate the activities of enzymes and nonenzymatic proteins.

• Conserved proteins that act in many signal-transduction pathways include GTPase switch proteins (trimeric G proteins and monomeric Ras-like proteins), protein kinases, and adapter proteins (see Figure 20-6). The latter coordinate the formation of multicomponent signaling complexes.

• Extracellular signals are often integrated into complex regulatory networks in which the synthesis, release, and degradation of hormones are precisely regulated.

20.2 Identification and Purification of Cell-Surface Receptors

As noted earlier, hormone receptors bind ligands with great specificity and high affinity. Binding of a hormone to a receptor involves the same types of weak interactions—ionic and van der Waals bonds and hydrophobic interactions—that characterize the specific binding of a substrate to an enzyme (Section 2.2). The *specificity* of a receptor refers to its ability to distinguish closely related substances; the insulin receptor, for example, binds insulin and a related hormone called insulin-like growth factor 1, but not other peptide hormones.

Hormone binding usually can be viewed as a simple reversible reaction,

$$R + H \longrightarrow RH$$

which can be described by the equation

$$K_D = \frac{[R][H]}{[RH]} \tag{20-1}$$

where [R] and [H] are the concentrations of free receptor and hormone (ligand), respectively, and [RH] is the concentration of the receptor-hormone complex. K_D, the dissociation constant of the receptor-ligand complex, measures the *affinity* of the receptor for the ligand. This binding equation can be rewritten as

$$\frac{[RH]}{R_T} = \frac{1}{1 + K_D/[H]} \tag{20-2}$$

where R_T is the sum of free and bound receptors: [R] + [RH]. Equation 20-2 is similar in form to the Michaelis-Menten equation used to analyze enzymatic reactions (Section 3.3).

The lower the K_D value, the higher the affinity of a receptor for its ligand. The K_D value is equivalent to the concentration of ligand at which one-half of the receptors contain bound ligand. If [H] = K_D, then from Equation 20-2 we can see that [RH] = 0.5 R_T.

Hormone Receptors Are Detected by Binding Assays

Hormone receptors are difficult to identify and purify, mainly because they are present in such minute amounts. The surface of a typical cell bears 10,000–20,000 receptors for a particular hormone, but this quantity is only $\approx 10^{-6}$ of the total protein in the cell, or $\approx 10^{-4}$ of the plasma-membrane protein. Purification is also difficult because these integral membrane proteins first must be solubilized with a nonionic detergent (see Figure 3-38).

Usually, receptors are detected and measured by their ability to bind radioactive hormones to a cell or to cell fragments. When increasing amounts of a radiolabeled hormone (e.g., insulin) are added to a cell suspension, the amount that binds to the cells increases at first and then tapers off at higher concentrations (Figure 20-7, curve A). Much of the radiolabeled hormone is bound specifically to its receptor, but some is bound nonspecifically to the multitude of other

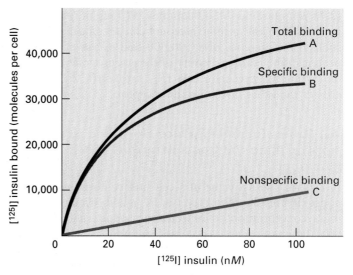

▲ **FIGURE 20-7 Identification of insulin-specific receptors on the surface of cells by their binding of radioactive insulin.** A suspension of cells is incubated for 1 hour at 4° C with increasing concentrations of ^{125}I-labeled insulin; the low temperature is used to prevent endocytosis of the cell-surface receptors. The total binding curve A represents insulin specifically bound to high-affinity receptors as well as insulin nonspecifically bound with low affinity to other molecules on the cell surface. The contribution of nonspecific binding to total binding is determined by repeating the binding assay in the presence of a 100-fold excess of unlabeled insulin, which saturates all the specific high-affinity sites. In this case, all the labeled insulin binds to nonspecific sites, yielding curve C. The specific binding curve B, which fits Equation 20-2, is calculated as the difference between curves A and C. For this insulin receptor, K_D is ≈ 20 nM (2×10^{-8} M), and the number of receptor molecules per cell, R_T, is $\approx 30,000$. [Adapted from A. Ciechanover et al., 1983, *Cell* **32**:267.]

proteins and phospholipids on the cell surface. Nonspecific binding of a labeled hormone can be measured by conducting the binding assay in the presence of a large excess of unlabeled hormone. Because the specific (high-affinity) binding sites are saturable, they all are filled by unlabeled hormone under these conditions and bind no labeled hormone. Nonspecific sites, however, do not saturate, so that binding of labeled hormone in the presence of excess unlabeled hormone represents nonspecific binding (Figure 20-7, curve C). Specific binding is calculated as the difference between total binding and nonspecific binding.

The number of hormone-binding sites per cell can be calculated from the saturation value of the specific binding curve (Figure 20-7, curve B). For example, a single hepatoma cell has $\approx 30,000$ insulin-binding sites. The K_D, the hormone concentration at which the receptor is half-saturated, also can be calculated from the specific binding curve, which is described by Equation 20-2. For the hepatoma-cell insulin receptor, K_D is 2×10^{-8} M, so the receptor will be half-saturated at an insulin concentration of 2×10^{-8} M

(0.12 μg/ml). Since blood contains $\approx$10 mg of total protein per milliliter, the insulin receptor can specifically bind insulin in the presence of a 100,000-fold excess of unrelated proteins. The binding affinity of the erythropoietin receptor on erythrocyte precursor cells ($K_D = \approx 1 \times 10^{-10}$ M) is even greater than that of the insulin receptor.

K_D Values for Cell-Surface Hormone Receptors Approximate the Concentrations of Circulating Hormones

In general, the K_D value of a cell-surface hormone receptor approximates the blood level of its ligand. Changes in hormone concentration are reflected in proportional changes in the fraction of receptors occupied. Suppose, for instance, that the normal (unstimulated) concentration of a hormone in the blood is 10^{-9} M and that the K_D for its receptor is 10^{-7} M; by substituting these values into Equation 20-2, we can calculate the fraction of receptors with bound hormone (RH):

$$\frac{[\text{RH}]}{R_T} = \frac{1}{1 + 10^{-7}/10^{-9}} = 0.0099$$

Thus about 1 percent of the total receptors will be filled with hormone. If the hormone concentration rises tenfold to 10^{-8} M, the concentration of receptor-hormone complex will rise proportionately, so that about 10 percent of the total receptors would have bound hormone. If the induced cellular response is proportional to the amount of RH, as is often the case, then the cellular responses also will increase tenfold.

For many hormone receptors, the ligand concentration needed to induce a maximal cellular response is lower than that needed to saturate all the receptor molecules on a cell. Likewise, the ligand concentration that induces a 50-percent maximal response is less than the K_D value for binding. In such cases, a plot of the percentage of maximal binding versus ligand concentration differs from a plot of the percentage of maximal cellular response versus ligand concentration (Figure 20-8).

Affinity Techniques Permit Purification of Receptor Proteins

Cell-surface hormone receptors often can be identified and followed through isolation procedures by *affinity labeling*. In this technique, cells are mixed with an excess of a radiolabeled hormone to saturate the hormone-binding sites on its specific receptor. After unbound hormone is washed away, the mixture is treated with a chemical agent that covalently cross-links the bound labeled hormone to the receptor. Most cross-linking agents contain two groups that react with free amino groups; by reacting with an amino group in the receptor and with one in the bound ligand, the cross-linking agent covalently joins the receptor and ligand. A radiolabeled ligand that is cross-linked to its receptor remains

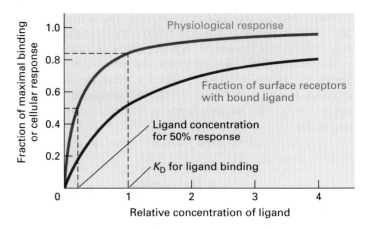

▲ **FIGURE 20-8 Comparison of binding curve and response curve for a cell-surface receptor and its ligand.** As illustrated here, the maximal physiological response to many hormones occurs when only a fraction of the cell receptors are occupied by ligand. In this example, 50 percent of the maximal response is induced at a ligand concentration at which only 18 percent of the receptors are occupied. Likewise, 80 percent of the maximal response is induced when the ligand concentration equals the K_D value, at which 50 percent of the receptors are occupied.

bound even in the presence of detergents and other denaturing agents that are used to solubilize receptor proteins from the cell membrane.

Another technique often used in purifying cell-surface receptors that retain their hormone-binding ability when solubilized is *affinity chromatography*. In this technique, a ligand of the receptor of interest is chemically linked to polystyrene beads. A crude, detergent-solubilized preparation of membrane proteins is passed through a column containing these beads. Only the receptor binds to the beads; the other proteins are washed through the column by excess fluid. When an excess of the ligand is passed through the column, the bound receptor is displaced from the beads and eluted from the column. This technique is similar in principle to antibody-affinity chromatography (see Figure 3-43c), except that a hormone ligand rather than an antibody is attached to the column beads. In some cases, a hormone receptor can be purified as much as 100,000-fold in a single affinity chromatographic step.

Many Receptors Can Be Cloned without Prior Purification

The number of receptors for many hormones on the surface of cells is too small to be purified by affinity chromatography. For example, each nucleated erythrocyte precursor cell possesses only about 1000 cell-surface receptors for erythropoietin, a hormone essential to the growth and differentiation of precursor cells into mature erythrocytes. Because the erythropoietin receptor constitutes only about 1 part per million (10^{-4} percent) of the total cellular protein, it is impossible to purify sufficient amounts of the receptor protein by conventional biochemical techniques in order to characterize or sequence it.

Key receptor proteins can now be obtained by DNA cloning and other recombinant DNA techniques, eliminating the need to isolate and purify them from cell extracts. The technique for identifying and cloning a cDNA encoding a desired receptor protein is summarized in Figure 20-9. A plasmid cDNA library prepared from cells that produce the receptor is screened by transfecting the cloned cDNAs into cells

that normally do not synthesize the receptor. Cells that take up a cDNA encoding the receptor are detected by their ability to bind radiolabeled or fluorescent-labeled ligand. Once a clone containing the receptor cDNA is identified, the sequence of the cDNA can be determined and that of the receptor protein deduced from the cDNA sequence. Special expression systems, discussed in Section 7.6, permit production of large amounts of the receptor from its cloned cDNA, providing enough protein to characterize its functional properties.

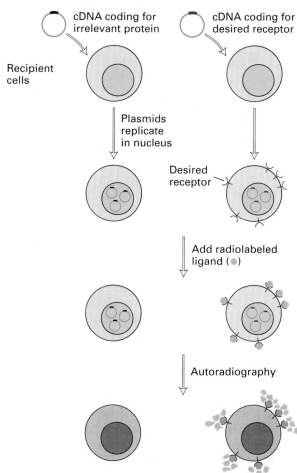

(a) **Plasmid expression vector**

(b) **Initial screening of cDNA pools**

▶ **FIGURE 20-9 Identification and isolation of a cDNA encoding a desired cell-surface receptor by plasmid expression cloning.** All mRNA is extracted from cells that normally express the receptor and reverse-transcribed into double-stranded cDNA. (a) The entire population of cDNAs is inserted into plasmid expression vectors in between a strong promoter and a terminator of transcription. The plasmids are transfected into bacterial cells that do not normally express the receptor of interest. The resulting cDNA library is divided into pools, each containing about 1000 different cDNAs. (b) Plasmids in each pool are transfected into a population of cultured cells (e.g., COS cells) that lack the receptor of interest. Only transfected cells that contain the cDNA encoding the desired receptor synthesize it; other transfected cells produce irrelevant proteins. To detect the few cells producing the desired receptor, a radiolabeled ligand specific for the receptor is added to the culture dishes containing the transfected cells; the cells are fixed and subjected to autoradiography. Positive cells synthesizing the specific receptor will be covered with many grains. Alternatively, transfected cells can be treated with a fluorescent-labeled ligand and passed through a fluorescence-activated cell sorter (see Figure 5-21). Cells expressing the receptor will bind the fluorescent label and be separated from those that do not. Plasmid cDNA pools giving rise to a positive signal are maintained in bacteria and subdivided into smaller pools, each of which is rescreened by transfection into cultured cells. After several cycles of screening and subdividing positive cDNA pools, a pure cDNA clone encoding the desired receptor is obtained. [See A. Aruffo and B. Seed, 1987, *Proc. Nat'l. Acad. Sci. USA* **84**:8573; A. D'Andrea, H. F. Lodish, and G. Wong, 1989, *Cell* **57**:277.]

SUMMARY Identification and Purification of Cell-Surface Receptors

- Receptors bind to one or a few related ligands. This specificity is determined by interactions between ligand determinants and specific amino acids in the receptor.

- The concentration of ligand at which half of the receptors are occupied, the K_D, can be determined experimentally and is a measure of the affinity of the receptor for the ligand.

- The amount of a particular receptor expressed on a cell is generally quite low, on the order of some 10,000 to 20,000 molecules. The maximal response of a cell to a ligand is generally achieved at concentrations at which most of its receptors are still not occupied (see Figure 20-8).

- Receptors can be purified directly using ligands as affinity reagents. Fractionation of receptors on columns to which ligands have been immobilized provides a powerful method for purification.

- In some cases, the genes encoding receptors for specific ligands can be isolated from cDNA libraries transfected into cultured cells (see Figure 20-9). Cells expressing the receptor are detected using labeled ligand as a probe.

20.3 G Protein–Coupled Receptors and Their Effectors

Many different mammalian cell-surface receptors are coupled to a trimeric signal-transducing G protein. As noted earlier, ligand binding to these receptors activates their associated G protein, which then activates an *effector enzyme* to generate an intracellular second messenger (see Figure 20-3a). All G protein–coupled receptors (GPCRs) contain seven membrane-spanning regions with their N-terminal segment on the exoplasmic face and their C-terminal segment on the cytosolic face of the plasma membrane (Figure 20-10). This large receptor family includes light-activated receptors (rhodopsins) in the eye and literally thousands of odorant receptors in the mammalian nose (Section 21.6), as well as numerous receptors for various hormones and neurotransmitters (Section 21.5). Although these receptors are activated by different ligands and may mediate different cellular responses, they all mediate a similar signaling pathway (see Figure 20-6).

To illustrate the operation of this important class of receptors, we discuss the structure and function of the catecholamine receptors that bind epinephrine and norepinephrine and of their associated signal-transducing G proteins. In this receptor system, we focus on the effector enzyme **adenylyl cyclase**, which synthesizes the second messenger

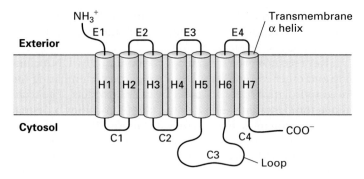

▲ **FIGURE 20-10 Schematic diagram of the general structure of G protein–linked receptors.** All receptors of this type contain seven transmembrane α-helical regions. The loop between α helices 5 and 6, and in some cases the loop between helices 3 and 4, which face the cytosol, are important for interactions with the coupled G protein. E1–E4 = extracellular loops; H1–H7 = transmembrane domains; C1–C4 = cytosolic loops.

cAMP. Later we describe how other receptors and other G proteins allow cells to integrate the actions of different types of receptors, to modify other enzymes, to control essential metabolic processes, and to alter gene expression.

Binding of Epinephrine to Adrenergic Receptors Induces Tissue-Specific Responses

Epinephrine and norepinephrine were originally recognized as products of the *medulla*, or core, of the adrenal gland and are also known as *adrenaline* and *noradrenaline*. Embryologically, nerve cells derive from the same tissue as adrenal medulla cells, and norepinephrine is also secreted by differentiated nerve cells. Both hormones are charged compounds that belong to the catecholamines, active amines containing a *catechol* moiety:

Epinephrine, which binds to two types of GPCRs, is particularly important in mediating the body's response to stress, such as fright or heavy exercise, when all tissues have an increased need for glucose and fatty acids. These principal metabolic fuels can be supplied to the blood in seconds by the rapid breakdown of glycogen in the liver (*glycogenolysis*) and of triacylglycerol in the adipose storage cells (*lipolysis*).

In mammals, the liberation of glucose and fatty acids can be triggered by binding of epinephrine (or norepinephrine) to β-*adrenergic receptors* on the surface of hepatic (liver) and adipose cells. Epinephrine bound to similar β-adrenergic receptors on heart muscle cells increases the contraction rate, which increases the blood supply to the tissues. Epinephrine bound to β-adrenergic receptors on smooth muscle cells of the intestine causes them to relax. Another type of epinephrine receptor, the *α₂-adrenergic receptor,* is found on smooth muscle cells lining the blood vessels in the intestinal tract, skin, and kidneys. Epinephrine bound to α_2 receptors causes the arteries to constrict, cutting off circulation to these peripheral organs. These diverse effects of epinephrine are directed to a common end: supplying energy for the rapid movement of major locomotor muscles in response to bodily stress. As discussed in more detail later, β- and α-adrenergic receptors are coupled to different G proteins. Both β_1- and β_2-adrenergic receptors are coupled to G proteins (G_s), which activate adenylyl cyclase. In contrast, α_1 and α_2 receptors are coupled to two other G proteins, G_q and G_i, respectively. G_i inhibits adenylyl cyclase, and G_q stimulates phospholipase C to generate IP3 and DAG as second messengers.

Stimulation of β-Adrenergic Receptors Leads to a Rise in cAMP

Many of the very different tissue-specific responses induced by binding of epinephrine to β-adrenergic receptors are mediated by a *rise* in the intracellular level of cAMP, resulting from activation of adenylyl cyclase. As a second messenger, cAMP acts to modify the rates of different enzyme-catalyzed reactions in specific tissues generating various metabolic responses. Binding of numerous other hormones to their receptors also leads to a rise in intracellular cAMP and characteristic tissue-specific metabolic responses (see later section).

Several types of experiments have been used to establish that binding of epinephrine to the β-adrenergic receptor induces a rise in cAMP. The data in Figure 20-11 show that the K_D for binding of epinephrine and other catecholamines to cell-surface β-adrenergic receptors is about the same as

the ligand concentration that induces a half-maximal activation of adenylyl cyclase. The experiment depicted in Figure 20-12 with purified receptors provided further evidence that the β-adrenergic receptor mediates induction of epinephrine-initiated cAMP synthesis. The same conclusion was reached from studies in which cloned cDNA encoding the β-adrenergic receptor was transfected into receptor-negative cells; the transfected cells acquired the ability to activate adenylyl cyclase in response to epinephrine.

Critical Features of Catecholamines and Their Receptors Have Been Identified

A variety of experimental approaches have provided information about which parts of catecholamine molecules and their receptors are essential for ligand binding and the subsequent activation of adenylyl cyclase. In many of these studies, chemically synthesized analogs of epinephrine have proved useful. These analogs fall into two classes: **agonists,** which mimic the function of a hormone by binding to its receptor and causing the normal response, and *antagonists,* which bind to the receptor but do not activate hormone-induced effects. An antagonist acts as an inhibitor of the natural hormone (or agonist) by competing for binding sites on the receptor, thereby blocking the physiological activity of the hormone.

Comparisons of the molecular structure and activity of various catecholamine analogs indicate that the side chain containing the NH group determines the affinity of the ligand for the receptor, while the catechol ring is required for the ligand-induced increase in cAMP level (Table 20-2). As is true for epinephrine, the K_D for binding of an agonist, such as *isoproterenol,* to β-adrenergic receptors generally is the same as the concentration required for half-maximal elevation of cAMP (see Figure 20-11). This relationship indicates that activation of adenylyl cyclase by isoproterenol is

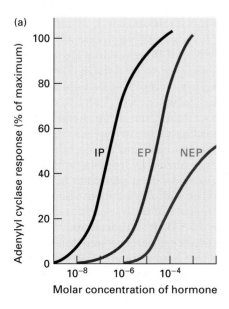

(a)

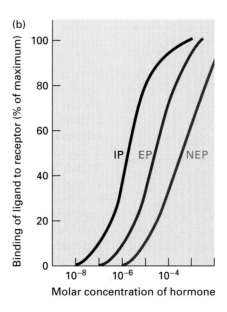

(b)

◀ FIGURE 20-11 Comparison of the abilities of three catecholamines to activate adenylyl cyclase, which catalyzes synthesis of cAMP, and to bind to cell-surface β-adrenergic receptors. The curves show that each ligand induces adenylyl cyclase activity (a) in proportion to its ability to bind to the receptor (b). Moreover, the concentration required for half-maximal binding of each ligand to the receptor is about the same as that required for activation of adenylyl cyclase. Note that the ligand concentration is plotted on a logarithmic scale ranging from 10^{-9} to 10^{-2} M. IP = isoproterenol; EP = epinephrine; NEP = norepinephrine.

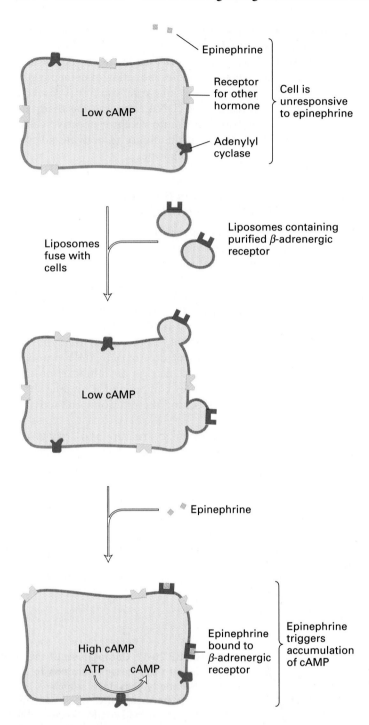

▲ **FIGURE 20-12 Experimental demonstration that β-adrenergic receptors mediate the induction of epinephrine-initiated cAMP synthesis.** Target cells lacking any receptors for epinephrine but expressing adenylyl cyclase and the appropriate signal-transducing G proteins were incubated with liposomes containing β-adrenergic receptors purified by affinity chromatography. Cells that fused with the liposomes became responsive to epinephrine, producing high levels of cAMP when the hormone was added to the medium. See Figure 15-4 for formation of liposomes containing membrane proteins. [See R. A. Cerione et al., 1983, *Nature* **306**:562.]

proportional to the number of β-adrenergic receptors filled with this agonist. Interestingly, the K_D for binding of isoproterenol to β-adrenergic receptors and subsequent induction of cAMP synthesis is about 10 times lower than the K_D for epinephrine; other agonists are even more potent having still lower K_D values. The affinity of various antagonists for the β-adrenergic receptor also varies over a wide range.

Humans possess two types of β-adrenergic receptors that are located on different cell types and differ in their relative affinities for various catecholamines. Cardiac muscle cells possess *β₁ receptors,* which promote increased heart rate and contractility by binding catecholamines with the rank order of affinities isoproterenol > norepinephrine > epinephrine. Drugs such as practolol, which are used to slow heart contractions in the treatment of cardiac arrhythmia and angina, are β₁-selective antagonists (see Table 20-2). These so-called **beta blockers** usually have little effect on β-adrenergic receptors on other cell types. The smooth muscle cells lining the bronchial passages possess *β₂ receptors,* which mediate relaxation by binding catecholamines with the rank order of affinities isoproterenol >> epinephrine = norepinephrine. Agonists selective for β₂ receptors, such as terbutaline, are used in the treatment of asthma because they specifically mediate opening of the bronchioles, the small airways in the lungs.

Although all GPCRs are thought to span the membrane seven times and hence to have similar three-dimensional structures, their amino acid sequences generally are quite dissimilar. For example, the sequences of the closely related β₁- and β₂-adrenergic receptors are only 50 percent identical; the sequences of the α- and β-adrenergic receptors exhibit even less homology. The specific amino acid sequence of each receptor determines which ligands it binds and which G proteins interact with it.

Studies with mutant forms of the β-adrenergic receptor generated by site-specific mutagenesis have identified four residues, located in transmembrane helices 3, 5, and 6, that participate in binding the agonist isoproterenol (Figure 20-13). Mutation of any one of these residues significantly reduces the ability of the receptor to bind the agonist. Binding of ligand to a β-adrenergic receptor is thought to cause several of the helices, particularly helices 5 and 6, to move relative to each other. As a result, the conformation of the long cytosolic loop connecting these two helices changes in a way that allows this loop to bind and activate the transducing G protein. The role of this hydrophilic loop in determining a receptor's specificity for a particular G protein was demonstrated in studies with recombinant chimeric receptor proteins containing part of an α₂ receptor and part of a β₂ receptor (Figure 20-14). It is thought that specific regions within the loop assume a unique three-dimensional structure in all receptors that bind the same G protein (e.g., G_s or G_i). Regions of the loop joining helices 3 and 4 in other GPCRs contribute to G protein binding. The chimeric

TABLE 20-2 Structure of Typical Agonists and Antagonists of the β-Adrenergic Receptor

Structure	Compound	K_D for Binding to the Receptor on Frog Erythrocytes
HO—, OH, HO—, CH—CH₂—NH—CH₃	Epinephrine	5×10^{-6} M
AGONIST — Isoproterenol structure	Isoproterenol	0.4×10^{-6} M
ANTAGONISTS — Alprenolol structure	Alprenolol	0.0034×10^{-6} M
Propranolol structure	Propranolol	0.0046×10^{-6} M
Practolol structure	Practolol	21×10^{-6} M

SOURCE: R. J. Lefkowitz et al., 1976, *Biochim. Biophys. Acta* **457**:1.

studies also showed that helix 7 functions in determining ligand specificity. This finding and the results with mutant β_2 receptors (see Figure 20-13) suggest that residues in at least four transmembrane helices in GPCRs participate in ligand binding.

Trimeric G$_s$ Protein Links β-Adrenergic Receptors and Adenylyl Cyclase

Having described which parts of GPCRs are necessary for interacting with ligand and their associated G protein, we now explain how G proteins function in signal transduction. Again, we focus our attention on β-adrenergic receptors,

which are coupled to G$_s$, or **stimulatory G protein**. As noted above, the initial response following binding of epinephrine to β-adrenergic receptors is an elevation in the intracellular level of cAMP. The increase in cAMP results from activation of adenylyl cyclase, which converts ATP to cAMP and pyrophosphate (PP$_i$). This membrane-bound enzyme has two catalytic domains on the cytosolic face of the plasma membrane that can bind ATP in the cytosol (Figure 20-15). The link between hormone binding to an exterior domain of the receptor and activation of adenylyl cyclase is provided by G$_s$, which functions as a signal transducer.

Cycling of G$_s$ between Active and Inactive Forms The G proteins that transduce signals from the β-adrenergic

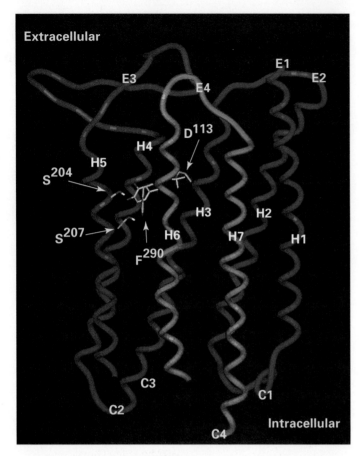

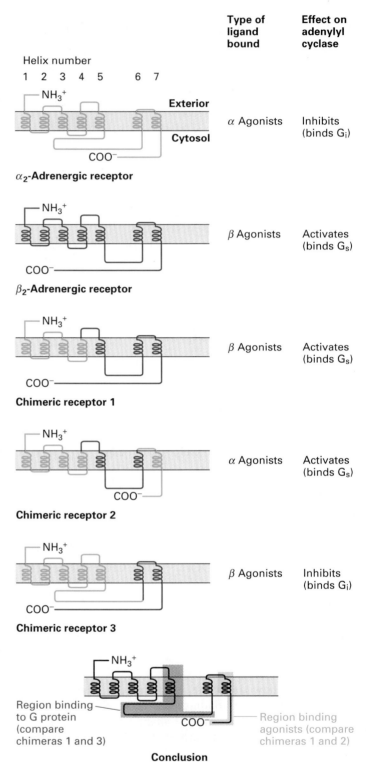

▲ **FIGURE 20-13 Model of complex formed between isoproterenol and the β₂-adrenergic receptor based on studies with mutant receptors expressed in cultured cells.** The polypeptide backbone of the receptor is shown in red and green. The transmembrane helices of the receptor are labeled H1–H7; the extracellular domains, E1–E4; and the intracellular domains, C1–C4. Most of the C3 domain (the long cytosolic loop between H5 and H6), C4 domain (the C-terminal tail), and N-terminal extracellular domain E1 are deleted. Isoproterenol (yellow) interacts with several residues in the receptor. Its amino group forms an ionic bond with the carboxylate side chain of aspartate 113 (D^{113}) in H3; the catechol ring engages in hydrophobic interactions with phenylalanine 290 (F^{290}) in H6; and two hydroxyl groups on the catechol ring hydrogen-bond to the hydroxyl groups in two serine residues (S^{204} and S^{207}) in H5. [See C. D. Strader et al., 1994, *Annu. Rev. Biochem.* **63**:101; courtesy of C. D. Strader and D. Underwood.]

receptor and other GPCRs contain three subunits designated α, β, and γ. As explained earlier, these GTPase switch proteins alternate between an "on" state with bound GTP and an "off" state with bound GDP (see Figure 20-5a). For example, when no ligand is bound to a β-adrenergic receptor, the α subunit of G$_s$ protein (G$_{s\alpha}$) is bound to GDP and complexed with the β and γ subunits (Figure 20-16). Binding of a hormone or agonist to the receptor changes its conformation, causing it to bind to the trimeric G$_s$ protein in such a way that GDP is displaced from G$_{s\alpha}$ and GTP is bound.

The G$_{s\alpha}$·GTP complex, which dissociates from the G$_{\beta\gamma}$ complex, then binds to and activates adenylyl cyclase. This activation is short-lived, however, because GTP bound to G$_{s\alpha}$ hydrolyzes to GDP in seconds, leading to the association of G$_{s\alpha}$ with G$_{\beta\gamma}$ and inactivation of adenylyl cyclase. The G$_{s\alpha}$ subunit thus relays the conformational change in

◀ **FIGURE 20-14 Demonstration of functional domains in G protein–coupled receptors by experiments with chimeric proteins containing portions of the β_2- and α_2-adrenergic receptors.** *Xenopus* oocytes microinjected with mRNA encoding the wild-type receptors or chimeric α-β receptors expressed the corresponding receptor protein on cell surfaces. Although *Xenopus* oocytes do not express adrenergic receptors, they do express G proteins, which can couple to the foreign receptors. Binding assays were conducted using agonists known to bind selectively to α or β receptors to determine the ligand-binding specificity of the chimeric receptors. The effects of the agonists on adenylyl cyclase activity were taken as a measure of whether the receptor protein bound to the stimulatory (G_s) or inhibitory (G_i) type of oocyte G protein. A comparison of chimeric receptor 1, which interacts with G_s, and chimeric receptor 3, which interacts with G_i, shows that the G protein specificity is determined primarily by the source of the cytosol-facing loop between α helices 5 and 6. A comparison of chimeras 1 and 2 indicates that α helix 7 plays a role in determining the ligand-binding specificity. [See B. Kobilka et al., 1988, *Science* **240**:1310; W. A. Catterall, 1989, *Science* **243**:236.]

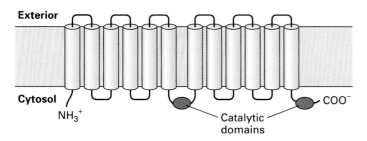

▲ **FIGURE 20-15 Schematic diagram of mammalian adenylyl cyclases.** The membrane-bound enzyme contains two similar catalytic domains on the cytosolic face of the membrane and two integral membrane domains, each of which is thought to contain six transmembrane α helices. The six adenylyl cyclase isoforms present in mammals are activated or inhibited by trans-ducing G proteins following hormone binding to an appropriate receptor. One isoform found mainly in the brain also is acti-vated by Ca^{2+} ions complexed to the protein calmodulin. [See W. -J. Tang and A. G. Gilman, 1992, *Cell* **70**:869.]

▶ **FIGURE 20-16 Activation of adenylyl cyclase following binding of an appropriate hormone (e.g., epinephrine, glucagon) to a G_s protein–coupled receptor.** Following ligand binding to the receptor, the G_s protein relays the hormone signal to the effector protein, in this case adenylyl cyclase. G_s cycles between an inactive form with bound GDP and an active form with bound GTP. Dissociation of the active form yields the $G_{s\alpha} \cdot$ GTP complex, which directly activates adenylyl cyclase. Activation is short-lived because GTP is rapidly hydrolyzed (step ⑤). This terminates the hormone signal and leads to reassembly of the inactive $G_s \cdot$ GDP form, returning the system to the resting state. Binding of another hormone molecule causes repetition of the cycle. Both the G_γ and $G_{s\alpha}$ subunits are linked to the membrane by covalent attachment to lipids. Binding of the activated receptor to $G_{s\alpha}$ promotes dissociation of GDP and its replacement with GTP.

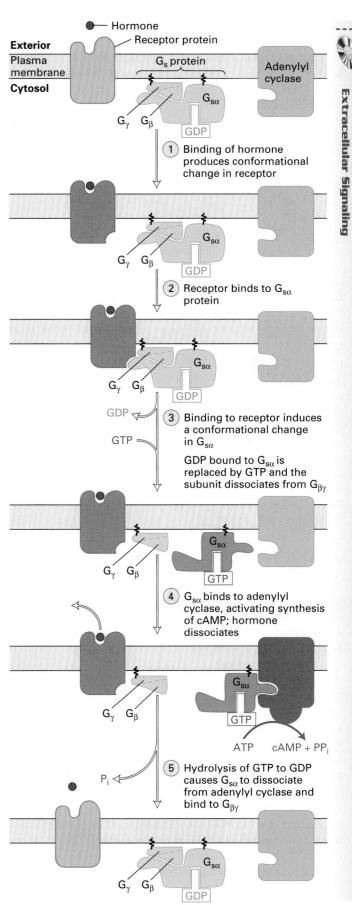

the receptor triggered by hormone binding to adenylyl cyclase.

Important evidence supporting this model has come from studies with a nonhydrolyzable analog of GTP called GMPPNP, in which a P–NH–P replaces the terminal phosphodiester bond in GTP:

$$\text{Guanine—ribose—O—}\overset{\overset{\displaystyle O}{\|}}{\underset{\underset{\displaystyle O^-}{|}}{P}}\text{—O—}\overset{\overset{\displaystyle O}{\|}}{\underset{\underset{\displaystyle O^-}{|}}{P'}}\text{—NH—}\overset{\overset{\displaystyle O}{\|}}{\underset{\underset{\displaystyle O^-}{|}}{P}}\text{—O}^-$$

Although this analog cannot be hydrolyzed, it binds to $G_{s\alpha}$ as well as GTP does. The addition of GMPPNP and an agonist to an erythrocyte membrane preparation results in a much larger and longer-lived activation of adenylyl cyclase than occurs with an agonist and GTP. Once the GDP bound to $G_{s\alpha}$ is displaced by GMPPNP, it remains permanently bound to $G_{s\alpha}$. Because the $G_{s\alpha} \cdot$ GMPPNP complex is as functional as the normal $G_{s\alpha} \cdot$ GTP complex in activating adenylyl cyclase, the enzyme is in a permanently active state.

Amplification of Hormone Signal The cellular responses triggered by cAMP may require tens of thousands or even millions of cAMP molecules per cell. Thus the hormone signal must be amplified in order to generate sufficient second messenger from the few thousand β-adrenergic receptors present on a cell. *Amplification* is possible because both receptors and G_s proteins can diffuse rapidly in the plasma membrane. A single receptor-hormone complex causes conversion of up to 100 inactive G_s molecules to the active form. Each active $G_{s\alpha} \cdot$ GTP, in turn, probably activates a single adenylyl cyclase molecule, which then catalyzes synthesis of many cAMP molecules during the time $G_s \cdot$ GTP is bound to it. Although the exact extent of this amplification is difficult to measure, binding of a single hormone molecule to one receptor molecule can result in the synthesis of at least several hundred cAMP molecules per receptor-hormone complex before the complex dissociates and activation of adenylyl cyclase ceases.

Termination of Cellular Response Successful cell-to-cell signaling also requires that the response of target cells to a hormone terminate rapidly once the concentration of circulating hormone decreases. *Termination* of the response to hormones recognized by β-adrenergic receptors is facilitated by a decrease in the affinity of the receptor that occurs when G_s is converted from the inactive to active form. When the GDP bound to $G_{s\alpha}$ is replaced with a GTP following hormone binding, the K_D of the receptor-hormone complex increases, shifting the equilibrium toward dissociation. The GTP bound to $G_{s\alpha}$ is quickly hydrolyzed, reversing the activation of adenylyl cyclase and terminating the cellular response unless the concentration of hormone remains high enough to form new receptor-hormone complexes. Thus, the continuous presence of hormone is required for continuous

activation of adenylyl cyclase. We discuss signal termination in more detail later in this chapter.

Some Bacterial Toxins Irreversibly Modify G Proteins

So far we have discussed the stimulatory G protein (G_s) that links β-adrenergic receptors and some other GPCRs to adenylyl cyclase. Cells, however, contain a number of other trimeric signal-transducing G proteins, including the inhibitory G_i coupled to α-adrenergic receptors. Studies with several bacterial toxins initially helped to unravel the functions of the various G proteins.

The cholera toxin is a hexameric protein (containing 1 α subunit and 5 β subunits) produced by the bacterium *Vibrio cholerae*, the bacterium that causes cholera. The classic symptom of cholera is massive diarrhea, caused by water flow from the blood through intestinal epithelial cells into the lumen of the intestine; death from dehydration is common. The α subunit of cholera toxin is an enzyme that penetrates the plasma membrane and enters the cytosol, where it catalyzes the covalent addition of the ADP-ribose moiety from intracellular NAD+ to $G_{s\alpha}$. ADP-ribosylated $G_{s\alpha} \cdot$ GTP can activate adenylyl cyclase normally but cannot hydrolyze the bound GTP to GDP; thus $G_{s\alpha}$ remains in the active on state, continuously activating adenylyl cyclase (Figure 20-17). As a result, the level of cAMP in the cytosol rises 100-fold or more. In intestinal epithelial cells, this rise apparently causes certain membrane proteins to permit a massive flow of water from the blood into the intestinal lumen. The studies with cholera toxin provided additional confirmation of the cycling of G_s described earlier.

Pertussis toxin is secreted by *Bordetella pertussis*, the bacterium causing whooping cough. The S_1 subunit of this toxin catalyzes addition of ADP-ribose to the α subunit of G_i. This irreversible modification prevents release of GDP, locking $G_{i\alpha}$ in the GDP-bound state. The identification and function of several other G proteins is discussed in later sections.

Adenylyl Cyclase Is Stimulated and Inhibited by Different Receptor-Ligand Complexes

The versatile trimeric G proteins enable different receptor-hormone complexes to modulate the activity of the same effector protein. In many types of cells, for example, binding of different hormones to their respective receptors induces activation of adenylyl cyclase. In the liver, glucagon and epinephrine bind to different GPCRs, but binding of both hormones activates adenylyl cyclase and thus triggers the same metabolic responses. Both types of receptors interact with and activate G_s, converting the inactive $G_s \cdot$ GDP to the active $G_{s\alpha} \cdot$ GTP form. Activation of adenylyl cyclase, and thus the cAMP level, is proportional to the total concentration of $G_{s\alpha} \cdot$ GTP resulting from binding of both hormones to their respective receptors.

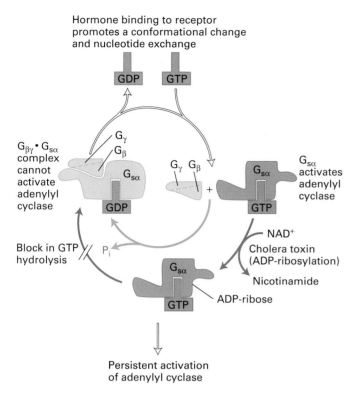

Hormone binding to receptor promotes a conformational change and nucleotide exchange

▲ FIGURE 20-17 Effect of cholera toxin on cycling of $G_{s\alpha}$ between the active and inactive forms. Normally, GTP in the active $G_{s\alpha} \cdot$ GTP is rapidly hydrolyzed (blue arrow), so that the activation of adenylyl cyclase and rise in cAMP persist only as long as hormone stimulation. Hydrolysis of GTP to GDP is catalyzed by $G_{s\alpha}$ itself. In the presence of cholera toxin, $G_{s\alpha}$ is irreversibly modified by addition of an ADP-ribosyl group; the modified $G_{s\alpha}$ can bind GTP but cannot hydrolyze it (red arrows). As a result, there is an excessive, nonregulated rise in the intracellular cAMP level.

In some cells, the cAMP level can be both up-regulated and down-regulated by the action of different hormones. In adipose cells, for example, epinephrine, glucagon, and ACTH all stimulate adenylyl cyclase, whereas prostaglandin PGE_1 and adenosine inhibit the enzyme. The receptors for PGE_1 and adenosine interact with inhibitory G_i, which contains the same β and γ subunits as stimulatory G_s but a different α subunit ($G_{i\alpha}$). In response to binding of an inhibitory ligand to its receptor, the associated G_i protein releases its bound GDP and binds GTP; the active $G_{i\alpha} \cdot$ GTP complex then dissociates from $G_{\beta\gamma}$ and inhibits (rather than stimulates) adenylyl cyclase (Figure 20-18). As discussed later, $G_{\beta\gamma}$ also can directly inhibit the activity of some isoforms of adenylyl cyclase.

GTP-Induced Changes in $G_{s\alpha}$ Favor Its Dissociation from $G_{\beta\gamma}$ and Association with Adenylyl Cyclase

Recent x-ray crystallographic studies have revealed how the G protein subunits interact with each other and with an activated receptor and adenylyl cyclase. These structural studies provide clues about how binding of GTP leads to dissociation of G_α from $G_{\beta\gamma}$, how G_α associates with adenylyl cyclase, and how $G_{s\alpha}$ and $G_{i\alpha}$ differ such that one activates adenylyl cyclase and the other inhibits it.

The three-dimensional structure of a trimeric G protein bound to GDP is shown in Figure 20-19. Two surfaces of the α subunit (G_α) interact with G_β: an N-terminal region near the membrane surface and two adjacent regions called switch I and switch II. Although G_β and G_γ also contact each other, G_γ does not contact G_α. Ligand binding to a GPCR causes the transmembrane helices in the receptor to slide relative to one another, revealing binding sites in the cytosolic extensions of these helices for the coupled trimeric G protein. Mutagenesis studies have shown that the N- and C-terminal domains of G_α interact with the receptor. X-ray

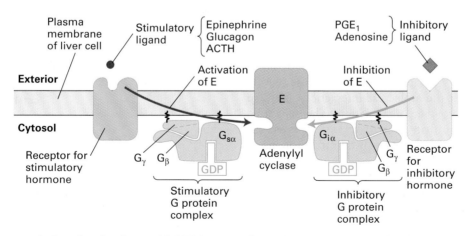

▲ FIGURE 20-18 Hormone-induced activation and inhibition of adenylyl cyclase is mediated by $G_{s\alpha}$ (blue) and $G_{i\alpha}$ (brown), respectively. Binding of $G_{s\alpha} \cdot$ GTP to adenylyl cyclase activates the enzyme (see Figure 20-16), whereas binding of $G_{i\alpha}$ inhibits adenylyl cyclase. The $G_{\beta\gamma}$ subunit in both stimulatory and inhibitory G proteins is identical; the G_α subunits and the receptors differ. Some isoforms of adenylyl cyclase are directly inhibited by binding to $G_{\beta\gamma}$. Others require coincident binding of dissociated $G_{s\alpha}$=GTP and $G_{\beta\gamma}$ subunits (see Section 20.7). [See A. G. Gilman, 1984 *Cell* **36**:577.]

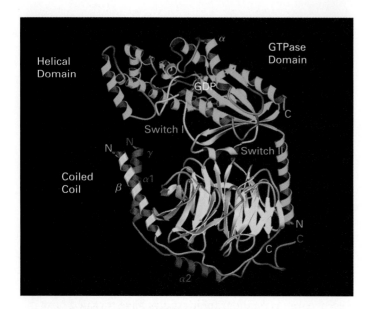

▲ **FIGURE 20-19 The structure of a trimeric G protein bound to GDP based on x-ray crystallographic analysis.** The N-terminus of the α subunit (green) and the C-terminus of the γ subunit (red) have hydrophobic lipid anchors, which tether the protein to the membrane. In the GDP-bound form, the α subunit and the β subunit (yellow) interact with each other, as do the β and γ subunits, but the γ subunit does not contact the α subunit. The C- and N-terminal domains of the α subunit are thought to interact with an activated receptor, causing a conformational change that promotes release of GDP and binding of GTP. Binding of GTP, in turn, induces large conformational changes in three switch regions of G_α, leading to dissociation of G_α from $G_{\beta\gamma}$. Only two switch regions (I and II) are visible in this orientation. See text for discussion. [From D. G. Lambright et al., 1996, *Nature* **379**:311.]

crystallographic studies suggest that these regions of G_α form a continuous surface with the lipid anchors at the C-terminus of G_γ and the N-terminus of G_α. The interactions promote the release of GDP from G_α and the subsequent binding of GTP. Upon binding GTP, G_α undergoes extensive conformational changes in three switch regions. These conformational changes, particularly those within switch I and II, disrupt the molecular interactions between G_α and $G_{\beta\gamma}$, leading to their dissociation. The separated $G_\alpha \cdot$ GTP and $G_{\beta\gamma}$ then drift apart from one another, anchored to the lipid bilayer, in search of effector proteins.

An understanding of how $G_{s\alpha} \cdot$ GTP interacts with adenylyl cyclase has come from x-ray crystallographic analysis of the complex between $G_{s\alpha} \cdot$ GTP and the catalytic domains of adenylyl cyclase. As noted earlier, adenylyl cyclase is a multipass transmembrane protein with two large cytosolic loops containing the catalytic domains (see Figure 20-15). Because such transmembrane proteins are notoriously difficult to crystallize, scientists prepared two protein fragments encompassing the catalytic domains of adenylyl cyclase and allowed them to associate in the presence of

$G_{s\alpha} \cdot$ GTP and forskolin, an agonist of adenylyl cyclase that stabilizes the catalytic fragments in their active conformations. The complex that formed was catalytically active and showed pharmacological and biochemical features similar to those of intact full-length adenylyl cyclase. In this complex, two regions of $G_{s\alpha} \cdot$ GTP contact the adenylyl cyclase fragments (Figure 20-20). These include the switch II helix and the α3-β5 loop. Hence, the GTP-induced conformation of $G_{s\alpha}$ that favors its dissociation from $G_{\beta\gamma}$ is precisely the conformation essential for binding of $G_{s\alpha}$ to adenylyl cyclase.

To understand how binding of $G_{s\alpha} \cdot$ GTP promotes adenylyl cyclase activity, scientists will first have to solve the structure of the adenylyl cyclase catalytic domains in their unactivated conformations (i.e., in the absence of bound $G_{s\alpha} \cdot$ GTP). One hypothesis is that binding of the switch II helix to a cleft in one catalytic domain of adenylyl cyclase leads to rotation of the other catalytic domain. This rotation is proposed to lead to a stabilization of the transition state, thereby stimulating catalytic activity.

Hydrolysis of GTP by the intrinsic GTPase activity of $G_{s\alpha}$ induces a conformational change that promotes its dissociation from adenylyl cyclase, leading to termination of the signal and reassociation of $G_{s\alpha}$ with $G_{\beta\gamma}$. Thus the GTPase activity of $G_{s\alpha}$ acts as a timer to control the length of time that it is associated with the effector.

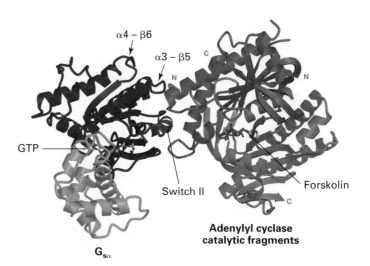

▲ **FIGURE 20-20 The structure of $G_{s\alpha} \cdot$ GTP complexed with two fragments encompassing the catalytic domain of adenylyl cyclase.** The α3-β5 loop and the helix in the switch II region (red) of $G_{s\alpha} \cdot$ GTP interact simultaneously with a specific region of adenylyl cyclase. The dark gray portion of $G_{s\alpha}$ is the Ras-like domain, which has GTPase activity, and the light gray portion is the helical domain (see Figure 20-19). The two adenylyl cyclase fragments are shown in purple and brown. Forskolin locks the cyclase fragments in their active conformations. Other studies indicate that $G_{i\alpha}$ binds to a different region of adenylyl cyclase, accounting for its different effect on the effector. [From J. J. G. Tesmer et al., 1997, *Science* **278**:1907; courtesy of Stephen Sprang.]

$G_{i\alpha}$ and $G_{s\alpha}$ Interact with Different Regions of Adenylyl Cyclase

Comparison of the structures of $G_{s\alpha} \cdot GTP$ and $G_{i\alpha} \cdot GTP$ suggests a molecular basis for the different effects of G_s and G_i on adenylyl cyclase. Although the overall conformation of the switch II and $\alpha 3$-$\beta 5$ loops are similar in $G_{s\alpha} \cdot GTP$ and $G_{i\alpha} \cdot GTP$, the relative position of these two regions differs in the two subunits. In $G_{s\alpha} \cdot GTP$, these regions are so positioned that they can bind simultaneously to a specific surface in adenylyl cyclase (see Figure 20-20). In $G_{i\alpha} \cdot GTP$, however, these two regions are displaced from each other in such a way that they cannot simultaneously bind to this surface of adenylyl cyclase.

The crystal structure of the complex between $G_{i\alpha} \cdot GTP$ and the adenylyl cyclase catalytic fragments has not been determined, but site-directed mutagenesis studies have identified sequences in both proteins that interact with each other. These studies indicate that the switch II region and probably the $\alpha 4$-$\alpha 6$ loop, but not the $\alpha 3$-$\alpha 5$ loop, in $G_{i\alpha}$ interact with adenylyl cyclase. Based on these and other studies, investigators have proposed that $G_{i\alpha}$ interacts with a site directly opposite the $G_{s\alpha}$-binding surface of adenylyl cyclase (see Figure 20-20); binding of $G_{i\alpha}$ to this site is thought to induce a conformational change in the active site that inhibits catalytic activity. As with $G_{s\alpha}$, the GTPase activity of $G_{i\alpha}$ rapidly hydrolyzes its bound GTP, thus terminating the inhibitory signal.

As discussed later, $G_{\beta\gamma}$ also regulates the activity of some adenylyl cyclase isoforms. Genetic studies suggest that the surface of adenylyl cyclase critical for binding of $G_{\beta\gamma}$ is distinct from the surfaces involved in binding of G_i and G_s.

Degradation of cAMP Also Is Regulated

The level of cAMP usually is controlled by the hormone-induced activation of adenylyl cyclase. Another point of regulation is the hydrolysis of cAMP to 5′-AMP by *cAMP phosphodiesterase*. This hydrolysis terminates the effect of hormone stimulation. As discussed later, the activity of many cAMP phosphodiesterases is stimulated by an increase in cytosolic Ca^{2+} (another intracellular second messenger), which often is induced by neuron or hormone stimulation. Some cells also modulate the level of cAMP by secreting it into the extracellular medium.

The synthesis and degradation of cAMP are both subject to complex regulation by multiple hormones, which allows the cell to integrate responses to many types of changes in its internal and external environments.

SUMMARY G Protein–Coupled Receptors and Their Effectors

- Many cell-surface receptors are linked to trimeric G proteins. Ligand binding to these receptors, which contain seven transmembrane domains, leads to activation of an associated signal-transducing G protein.

- All G proteins contain three subunits: α, β, and γ. Prior to activation, the α subunit is bound to GDP (see Figure 20-19). Binding of a trimeric G protein to an activated receptor leads to dissociation of GDP, binding of GTP to G_α, and dissociation of $G_\alpha \cdot GTP$ from $G_{\beta\gamma}$. $G_\alpha \cdot GTP$ and $G_{\beta\gamma}$ can specifically interact with effector proteins leading to changes in their activity (see Figure 20-16).

- The intrinsic GTPase activity of G_α inactivates $G_\alpha \cdot GTP$ by catalyzing GTP hydrolysis: P_i is released and the resulting $G_\alpha \cdot GDP$ then dissociates from its effector and reassociates with $G_{\beta\gamma}$.

- Binding of ligand to a G protein–coupled receptor causes a conformational change that permits the receptor to bind to a specific G protein. The long cytosolic loop between helices 5 and 6 in a receptor determines which G protein it binds (see Figure 20-14).

- Adenylyl cyclase, which catalyzes the formation of cAMP from ATP, is the best-characterized effector regulated by trimeric G proteins. All adenylyl cyclase isoforms are stimulated by $G_{s\alpha}$, but only specific isoforms are inhibited by $G_{i\alpha}$ and $G_{\beta\gamma}$. $G_{s\alpha}$, $G_{i\alpha}$, and $G_{\beta\gamma}$ interact with different regions of the catalytic domain of adenylyl cyclase.

20.4 Receptor Tyrosine Kinases and Ras

The receptor tyrosine kinases (RTKs) are the second major type of cell-surface receptors that we discuss in detail in this chapter (see Figure 20-3d, *right*). The ligands for RTKs are soluble or membrane-bound peptide/protein hormones including nerve growth factor (NGF), platelet-derived growth factor (PDGF), fibroblast growth factor (FGF), epidermal growth factor (EGF), and insulin. Binding of a ligand to this type of receptor stimulates the receptor's intrinsic protein-tyrosine kinase activity, which subsequently stimulates a signal-transduction cascade leading to changes in cellular physiology and/or patterns of gene expression (see Figure 20-6). RTK signaling pathways have a wide spectrum of functions including regulation of cell proliferation and differentiation, promotion of cell survival (Section 23.8), and modulation of cellular metabolism.

Some RTKs have been identified in studies on human cancers associated with mutant forms of growth-factor receptors, which send a proliferative signal to cells even in the absence of growth factor. One such mutant receptor, encoded at the *neu* locus, contributes to the uncontrolled proliferation of certain human breast cancers (Section 24.3). Other RTKs have been uncovered during analysis of developmental mutations that lead to blocks in differentiation of certain cell types in *C. elegans, Drosophila,* and the mouse.

In this section we discuss activation of RTKs and how they transmit a hormone signal to Ras, the **GTPase switch protein** that functions in transducing signals from many different RTKs. The second part of RTK-Ras signaling pathways, the transduction of signals downstream from Ras to a common cascade of serine/threonine kinases, is covered in the next section.

Ligand Binding Leads to Autophosphorylation of RTKs

All RTKs comprise an extracellular domain containing a ligand-binding site, a single hydrophobic transmembrane α helix, and a cytosolic domain that includes a region with protein-tyrosine kinase activity. Binding of ligand causes most RTKs to dimerize; the protein kinase of each receptor monomer then phosphorylates a distinct set of tyrosine residues in the cytosolic domain of its dimer partner, a process termed *autophosphorylation* (Figure 20-21). Autophosphorylation occurs in two stages. First, tyrosine residues in the *phosphorylation lip* near the catalytic site are phosphorylated. This leads to a conformational change that facilitates binding of ATP in some receptors (e.g., the insulin receptor) and binding of protein substrates in other receptors (e.g., FGF receptor). The receptor kinase activity then phosphorylates other sites in the cytosolic domain; the resulting phospho-

tyrosines serve as docking sites for other proteins involved in RTK-mediated signal transduction.

As described later, the subunits of some RTKs, including the insulin receptor, are covalently linked. Although these receptors exist as dimers or tetramers even in the absence of ligand, binding of ligand is required for autophosphorylation to occur. Presumably, ligand binding induces a conformational change that activates the kinase.

The phosphotyrosine residues in activated RTKs interact with adapter proteins containing SH2 or PTB domains. These proteins couple the activated receptors to other components of the signal-transduction pathway but have no intrinsic signaling properties. Before examining the structure and function of adapter proteins, we discuss the role of Ras, the other key signaling molecule in pathways triggered by activation of RTKs. As we will see later, several membrane-associated enzymes that function in signal transduction also bind to specific phosphotyrosines in activated RTKs.

Ras and G$_\alpha$ Subunits Belong to the GTPase Superfamily of Intracellular Switch Proteins

Ras is a GTP-binding switch protein that, like the G$_\alpha$ subunits in different G proteins, alternates between an active on state with a bound GTP and an inactive off state with a bound GDP. As discussed in the previous section, G$_\alpha$ is directly coupled to GPCRs and transduces signals to various effectors such as adenylyl cyclase. In contrast, Ras is not directly linked to RTKs.

Activation of both Ras and G$_\alpha$ is triggered by hormone binding to an appropriate cell-surface receptor. Ras activation is accelerated by a protein called *guanine nucleotide–exchange factor* (GEF), which binds to the Ras · GDP complex, causing dissociation of the bound GDP (Figure 20-22). Because GTP is present in cells at a higher concentration than GDP, GTP binds spontaneously to "empty" Ras molecules, with release of GEF. In contrast, formation of an active G$_\alpha$ · GTP complex does not require an exchange factor.

 Hydrolysis of the bound GTP deactivates both Ras and G$_\alpha$. The average lifetime of a GTP bound to Ras is about 1 minute, which is much longer than the lifetime of G$_\alpha$ · GTP. The reason for this difference is that the deactivation of Ras, unlike the deactivation of G$_\alpha$, requires the assistance of another protein: a *GTPase-activating protein* (GAP), which binds to Ras · GTP and accelerates its intrinsic GTPase activity by a hundredfold (see Figure 20-22). Mammalian Ras proteins have been studied in great detail because mutant Ras proteins are associated with many types of human cancer. These mutant proteins, which bind but cannot hydrolyze GTP, are permanently in the "on" state and cause neoplastic transformation (Chapter 24).

The differences in the cycling mechanisms of Ras and G$_\alpha$ are reflected in their structures. Ras ($\approx$170 amino acids) is smaller than G$_\alpha$ proteins ($\approx$300 amino acids), but its three-

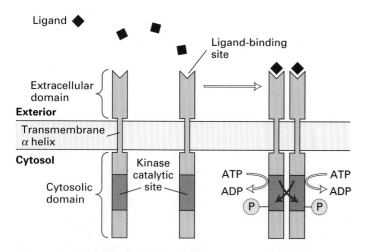

▲ **FIGURE 20-21 General structure and activation of receptor tyrosine kinases (RTKs).** The ligands for some RTKs, such as the receptor for EGF depicted here, are monomeric; ligand binding induces a conformational change in receptor monomers that promotes their dimerization. The ligands for other RTKs are dimeric; their binding brings two receptor monomers together directly (see Figure 20-4d). In either case, the kinase activity of each subunit of the dimeric receptor initially phosphorylates tyrosine residues near the catalytic site in the other subunit. Subsequently, tyrosine residues in other parts of the cytosolic domain are autophosphorylated. See text for discussion. [See G. Panayotou and W. D. Waterfield, 1993, *Bioessays* **15**:171; M. Mohammadi et al., 1996, *Cell* **86**:577.]

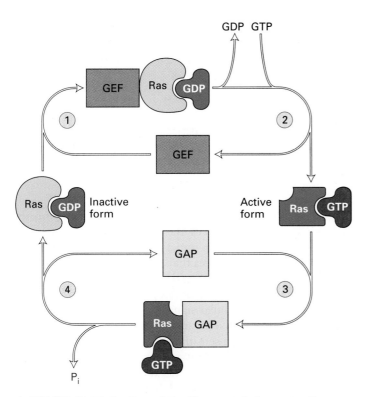

GDP GTP

▲ **FIGURE 20-22 Cycling of the Ras protein between the inactive form with bound GDP and the active form with bound GTP occurs in four steps.** By mechanisms discussed later, binding of certain growth factors to their receptors induces formation of the active Ras · GTP complex. Step ①: Guanine nucleotide–exchange factor (GEF) facilitates dissociation of GDP from Ras. Step ②: GTP then binds spontaneously, and GEF dissociates yielding the active Ras · GTP form. Steps ③ and ④: Hydrolysis of the bound GTP to regenerate the inactive Ras · GDP form is accelerated a hundredfold by GTPase-activating protein (GAP). Unlike G_α, cycling of Ras thus requires two proteins, GEF and GAP; otherwise, G_α and Ras exhibit many common features.

dimensional structure is similar to that of the GTPase domain of G_α. Recent structural and biochemical studies show that G_α also contains another domain, a helical domain that apparently functions like GAP to increase the rate of GTP hydrolysis by G_α (see Figure 20-19). In addition, the direct interaction between an activated receptor and G_α · GDP promotes release of GDP and binding of GTP, so that a separate exchange factor is not required.

Both $G_{s\alpha}$ and Ras are members of a family of intracellular GTP-binding switch proteins collectively referred to as the **GTPase superfamily.** This family also includes other G_α subunits (e.g., $G_{i\alpha}$), the Rab proteins which regulate fusion of vesicles within cells (Section 17.10), and the Rho family proteins which regulate the actin cytoskeleton (Section 18.2). The many similarities between the structure and function of Ras and $G_{s\alpha}$, and the identification of both proteins in all eukaryotic cells, indicate that a single type of signal-transducing GTPase originated very early in evolution. The gene encoding this protein subsequently duplicated and evolved to the extent that cells today contain a superfamily of such GTPases, comprising perhaps a hundred different intracellular switch proteins. These related proteins control many aspects of cellular growth and metabolism.

An Adapter Protein and GEF Link Most Activated RTKs to Ras

The first indication that Ras functioned downstream from RTKs in a common signaling pathway came from experiments in which cultured fibroblast cells were induced to proliferate by treatment with a mixture of platelet-derived growth factor (PDGF) and epidermal growth factor (EGF). Microinjection of anti-Ras antibodies into these cells blocked cell proliferation. Conversely, injection of a constitutively active mutant Ras protein (i.e., Ras[D]), which hydrolyzes GTP very inefficiently and thus persists in the active state, caused the cells to proliferate in the absence of the growth factors. These findings are consistent with studies showing that addition of fibroblast growth factor (FGF) to fibroblasts leads to a rapid increase in the proportion of Ras present in the GTP-bound active form.

But how does binding of a growth factor (e.g., EGF) to an RTK (e.g., the EGF receptor) lead to activation of Ras? Two cytosolic proteins—GRB2 and Sos—provide the key links (Figure 20-23). An *SH2 domain* in GRB2 binds to a specific phosphotyrosine residue in the activated receptor. GRB2 also contains two *SH3 domains*, which bind to and activate Sos. GRB2 thus functions as an adapter protein for the EGF receptor. Sos functions as a guanine nucleotide–exchange protein (GEF), which helps convert inactive GDP-bound Ras to the active GTP-bound form.

Genetic analysis of mutants blocked at particular stages of differentiation have provided considerable insight into RTK signaling pathways. Most of these genetic studies were done in the worm *C. elegans* and in the fly *Drosophila*. Mutants in these species in which development of specific cells is blocked were particularly useful in elucidating the pathway from an RTK to Ras shown in Figure 20-23. To illustrate the power of this experimental approach, we consider development of a particular type of cell in the compound eye of *Drosophila*.

The compound eye of the fly is composed of some 800 individual eyes called *ommatidia* (Figure 20-24a). Each ommatidium consists of 22 cells, eight of which are photosensitive neurons called retinula, or R cells, designated R1–R8 (Figure 20-24b). An RTK called Sevenless (Sev) specifically regulates development of the R7 cell. In flies with a mutant *sevenless (sev)* gene, the R7 cell in each ommatidium does not form (Figure 20-24c). Since the R7 photoreceptor is necessary for flies to see in ultraviolet light, mutants that lack functional R7 cells are easily isolated.

During development of each ommatidium, a protein called Boss (Bride of Sevenless) is expressed on the surface

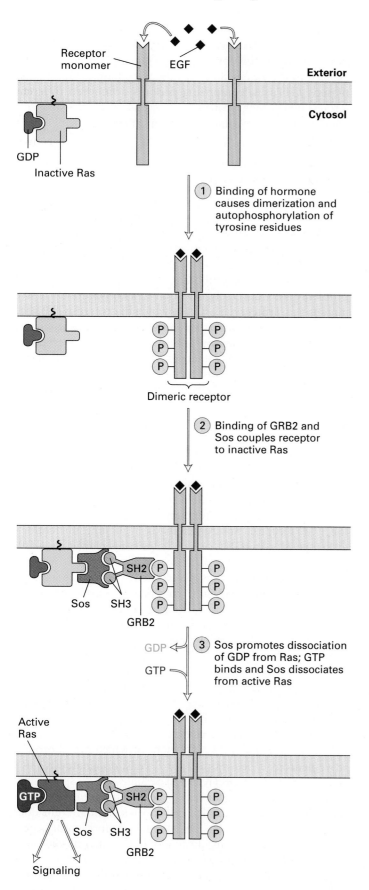

FIGURE 20-23 Activation of Ras following binding of a hormone (e.g., EGF) to an RTK. The adapter protein GRB2 binds to a specific phosphotyrosine on the activated RTK and to Sos, which in turn interacts with the inactive Ras · GDP. The guanine nucleotide–exchange factor (GEF) activity of Sos then promotes formation of the active Ras · GTP. Note that Ras is tethered to the membrane by a farnesyl anchor (see Figure 3-36b). [See L. Buday and J. Downward, 1993, *Cell* **73**:611; J. P. Olivier et al., 1993, *Cell* **73**:179; S. E. Egan et al., 1993, *Nature* **363**:45; E. J. Lowenstein et al., 1992, *Cell* **70**:431; M. A. Simon et al., 1993, *Cell* **73**:169.]

of the R8 cell. This membrane-bound protein binds to the Sev RTK on the surface of the neighboring R7 precursor cell, signaling it to develop into a photosensitive neuron (Figure 20-25a). In mutant flies that do not express a functional Sev RTK, interaction between the Boss and Sev proteins cannot occur, and no R7 cells develop (Figure 20-25b).

To identify intracellular signal-transducing proteins in the Sev RTK pathway, investigators produced mutant flies expressing a temperature-sensitive Sev protein. When these flies were maintained at a permissive temperature, all their ommatidia contained R7 cells; when they were maintained at a nonpermissive temperature, no R7 cells developed. At a particular intermediate temperature, however, just enough of the Sev RTK was functional to mediate normal R7 development. The investigators reasoned that at this intermediate temperature, the signaling pathway would become defective, and thus no R7 cells would develop, if the level of *another* protein involved in the pathway was reduced. A recessive mutation affecting such a protein would have this effect because, in diploid organisms like *Drosophila*, a heterozygote containing one wild-type and one mutant allele of a gene will produce half the normal amount of the gene product; hence even if such a recessive mutation is in an essential gene, the organism will be viable. However, a fly carrying a temperature-sensitive mutation in the *sev* gene and a second mutation affecting another protein in the signaling pathway would be expected to lack R7 cells at the intermediate temperature.

By use of this screen, researchers identified genes encoding three important proteins in the Sev pathway (see Figure 20-23):

- A Ras protein exhibiting 80 percent identity with its mammalian counterparts
- A GEF called Sos (Son of Sevenless) exhibiting 45 percent identity with its mouse counterpart
- An SH2-containing adapter protein exhibiting 64 percent identity to human GRB2

These three proteins have been found to function in other RTK signal-transduction pathways initiated by ligand binding to different receptors and used at different times and places in the developing fly. Recessive lethal mutations in these essential genes can be identified by the strategy described here much more easily than by the procedure described in Chapter 8 (see Figure 8-10).

(a)

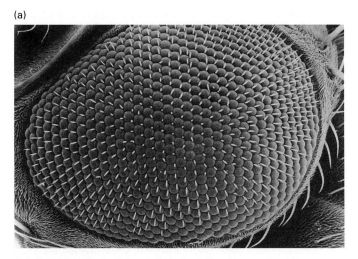

(b)

(c)

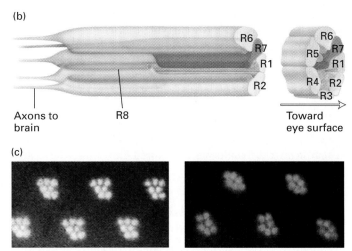

▲ FIGURE 20-24 The compound eye of *Drosophila melanogaster.* (a) Scanning electron micrograph showing individual ommatidia that compose the fruit fly eye. (b) Longitudinal and cut-away views of a single ommatidium. Each of these tubular structures contains eight photoreceptors, designated R1–R8, which are long, cylindrically shaped light-sensitive cells. R1–R6 (yellow) extend throughout the depth of the retina, whereas R7 (brown) is located toward the surface of the eye and R8 (blue) toward the backside, where the axons exit. (c) Comparison of

eyes from wild-type and *sevenless* mutant flies viewed by a special technique that can distinguish the photoreceptors in an ommatidium. Note that seven of the eight photoreceptors are easily seen in the wild-type ommatidia *(left)*, whereas only six are visible in the mutant ommatidia *(right)*. Flies with the *sevenless* mutation lack the R7 cell in their eyes. [Part (a) from E. Hafen and K. Basler, 1991, *Development* **1**(Suppl.):123. Part (b) adapted from R. Reinke and S. L. Zipursky, 1988, *Cell* **55**:321. Part (c) courtesy of U. Banerjee.]

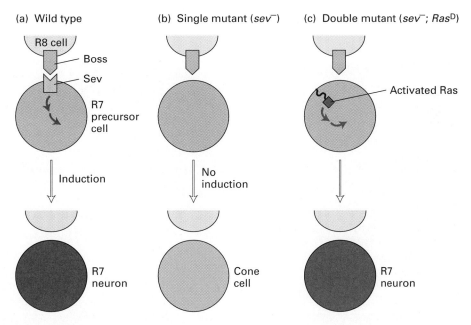

▲ FIGURE 20-25 Genetic analysis of induction of the R7 photoreceptor in the *Drosophila* eye. (a) During larval development of wild-type flies, the R8 cell (tan) in each developing ommatidium expresses a cell-surface protein, called Boss (orange), that binds to the Sev RTK (green) on the surface of its neighboring R7 precursor cell (light red). This interaction induces changes in gene expression that result in differentiation of the precursor cell into a functional R7 cell (dark red). (b) In fly embryos with a mutation in the *sevenless (sev)* gene, R7 precursor cells cannot

bind Boss and therefore do not differentiate normally into R7 cells. Rather the precursor cell enters an alternative developmental pathway and eventually becomes a cone cell (purple).(c) Double-mutant *(sev⁻; Ras^D)* larvae express a constitutively active Ras (Ras^D) in the R7 precursor cell, which induces differentiation of R7 precursor cells in the absence of the Boss-mediated signal. This finding shows that activated Ras mediates induction. [See M. A. Simon et al., 1991, *Cell* **67**:701; M. E. Fortini et al., 1992, *Nature* **355**:559.]

In subsequent studies, researchers introduced a mutant *ras*^D gene into fly embryos carrying the *sevenless* mutation. As noted earlier, the *ras*^D gene encodes a Ras protein that has reduced GTPase activity and hence is present in the active GTP-bound form even in the absence of a hormone signal. Although no functional Sev RTK was expressed in these double-mutants (*sev*^−; *ras*^D), R7 cells formed normally, indicating that activation of Ras is sufficient for induction of R7-cell development (Figure 20-25c). This finding is consistent with the results with cultured fibroblasts described earlier.

SH2 Domain in GRB2 Adapter Protein Binds to a Specific Phosphotyrosine in an Activated RTK

To identify proteins that associate with phosphotyrosine residues in the cytosolic domain of activated EGF receptors, scientists used an expression cloning strategy. cDNAs synthesized from mRNAs isolated from human brain-stem tissue were inserted into a λgt11 expression vector, which then was plated on a lawn of *E. coli* cells (see Figure 7-21). When the resulting cDNA library was screened using a fragment of phosphorylated human EGF receptor as the probe, two cDNA clones were identified. One encoded a subunit of PI-3 kinase that contains an SH2 domain and the other encoded GRB2, a homolog of the SH2-containing adapter protein identified in the *Drosophila* Sev pathway. Thus GRB2 and its *Drosophila* homolog are adapter proteins that function downstream from RTKs but upstream of Ras in both flies and mammalian cells.

GRB2 and similar adapter proteins bind to different phosphotyrosine residues on RTKs via the conserved SH2 domain. This domain derived its full name, the *Src homology 2 domain*, from its homology with a region in the prototypical cytosolic tyrosine kinase encoded by *src*. The three-dimensional structures of SH2 domains in different proteins are very similar. Each binds to a distinct sequence of amino acids surrounding a phosphotyrosine residue. The unique amino acid sequence of each SH2 domain determines the specific phosphotyrosine residues it binds. The SH2 domain of the Src tyrosine kinase, for example, binds strongly to any peptide containing the critical core sequence of phosphotyrosine–glutamic acid–glutamic acid–isoleucine (Figure 20-26a). These four amino acids make intimate contact with the peptide-binding site in the Src SH2 domain. Binding resembles the insertion of a two-pronged "plug"—the phosphotyrosine and isoleucine residues of the peptide—into a two-pronged "socket" in the SH2 domain. The two glutamic acids fit snugly onto the surface of the SH2 domain between the phosphotyrosine socket and the hydrophobic socket that accepts the isoleucine residue. Variations in the nature of the hydrophobic socket in different SH2 domains allow them to bind to phosphotyrosines adjacent to different sequences, accounting for differences in their binding specificity.

Activated RTKs also can recruit signaling molecules through a different domain called the phosphotyrosine-

(a)

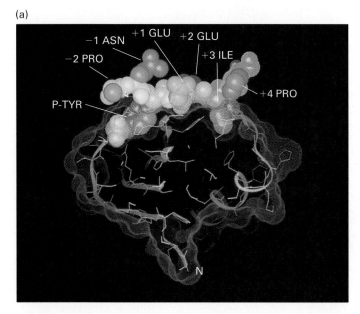

(b)

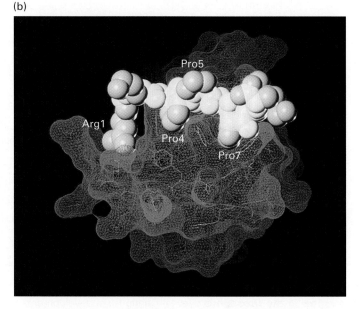

▲ **FIGURE 20-26 Models of SH2 and SH3 domains bound to short target peptides.** The peptides, positioned above the SH2 and SH3 domains, are shown as space-filling models with the backbone in yellow and the side chains in green. (a) SH2 domains interact with phosphotyrosine-containing sequences. In this target peptide, the phosphotyrosine (P-Tyr) and isoleucine (+3ILE) fit into a two-pronged socket on the surface of the SH2 domain. The phosphate group covalently attached to the tyrosine residue is light blue. (b) SH3 domains interact with proline-rich sequences. In this target peptide, two prolines (Pro4 and Pro7) fit into binding pockets on the surface of the SH3 domain. In both cases, interactions involving other residues in the target peptide determine the specificity of binding. [Part (a) from G. Waksman et al., 1993, *Cell* **72**:779. Part (b) from H. Yu et al., 1994, *Cell* **76**:933.]

binding (PTB) domain. While SH2-binding specificity is largely determined by residues C-terminal to the phosphotyrosine, PTB-binding specificity is determined by specific residues five to eight residues N-terminal to the phosphotyrosine residue.

Sos, a Guanine Nucleotide—Exchange Factor, Binds to the SH3 Domains in GRB2

In addition to one SH2 domain, which binds to phosphotyrosine residues in RTKs, GRB2 contains two SH3 domains, which bind to Sos, a guanine nucleotide–exchange factor. SH3 domains, which contain ≈55–70 residues, are present in a large number of proteins involved in intracellular signaling. Although the three-dimensional structures of various SH3 domains are similar, their specific amino acid sequences differ. SH3 domains selectively bind to proline-rich sequences in Sos and other proteins; different SH3 domains bind to different proline-rich sequences.

Proline residues play two roles in the interaction between an SH3 domain in an adapter protein (e.g., GRB2) and a proline-rich sequence in another protein (e.g., Sos). First, the proline-rich sequence assumes an extended conformation that permits extensive contacts with the SH3 domain, thereby facilitating interaction. Second, a subset of these prolines fit into binding "pockets" on the surface of the SH3 domain (Figure 20-26b). Several nonproline residues also interact with the SH3 domain and are responsible for determining the

binding specificity. Hence the binding of peptides to SH2 and SH3 domains follows a similar strategy: certain residues provide the overall structural motif necessary for binding, and neighboring residues confer specificity to the binding.

Following hormone-induced activation of an RTK (e.g., the EGF receptor), a complex containing the activated RTK, GRB2, and Sos is formed on the cytosolic face of the plasma membrane (see Figure 20-23). Formation of this complex depends on the dual binding ability of GRB2. Receptor activation thus leads to relocalization of Sos from the cytosol to the membrane, bringing Sos near to its substrate, membrane-bound Ras · GDP. Biochemical and genetic studies indicate that the C-terminus of Sos inhibits its nucleotide-exchange activity and that GRB2 binding relieves this inhibition. Binding of Sos to Ras · GDP leads to changes in the conformation of two regions of Ras, switch I and switch II, thereby opening the binding pocket for GDP so it can diffuse out (Figure 20-27). Because GTP is present in cells at a concentration some 10 times higher than GDP, GTP binding occurs preferentially, leading to activation of Ras. The activation of Ras and $G_{s\alpha}$ thus occur by similar mechanisms: a conformational change induced by binding of a protein— Sos and an activated GPCR, respectively—that opens the protein structure so bound GDP is released to be replaced by GTP. As we discuss in the next section, binding of GTP to Ras, in turn, induces a specific conformation of switch I and II that allow Ras · GTP to activate downstream effector molecules.

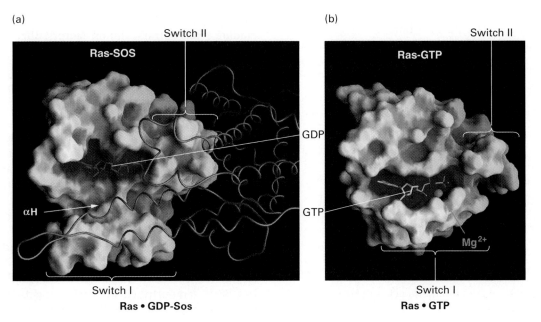

▲ **FIGURE 20-27 Structures of Ras · GDP-Sos complex and Ras · GTP determined by x-ray crystallography.** (a) Sos (shown as a trace diagram) binds to two switch regions of Ras · GDP, leading to a massive conformational change in Ras. In effect, Sos pries open Ras by displacing the switch I region, thereby allowing GDP to diffuse out. (b) GTP is thought to bind to Ras-Sos first through its base; subsequent binding of the GTP phosphates complete the interaction. The resulting conformational change in Ras displaces Sos and promotes interaction of Ras · GTP with its effectors (discussed later). Ras is in a slightly different orientation in parts (a) and (b). GDP and GTP are shown as small stick models in the center of Ras; the adjacent sphere is a Mg^{2+} ion. [From P. A. Boriack-Sjodin and J. Kuriyan, 1998, *Nature* **394**:341; courtesy of John Kuriyan.]

Several other proteins, including GAP, bind to specific phosphotyrosines in activated RTKs. This binding localizes GAP close to Ras · GTP, so it can promote the cycling of Ras (see Figure 20-22); exactly how GAP interacts with Ras and perhaps other components of the RTK-Ras pathway is unclear.

SUMMARY Receptor Tyrosine Kinases and Ras

- Receptor tyrosine kinases (RTKs), which bind to peptide/protein hormones, may exist as dimers or dimerize during binding to ligands.

- Ligand binding leads to activation of the kinase activity of the receptor and autophosphorylation of tyrosine residues in its cytosolic domain (see Figure 20-31). The activated receptor also can phosphorylate other protein substrates.

- Ras is an intracellular GTPase switch protein that acts downstream from most RTKs. Like $G_{s\alpha}$, Ras cycles between an inactive GDP-bound form and active GTP-bound form. Ras cycling requires the assistance of two proteins, GEF and GAP, (see Figure 20-22), whereas $G_{s\alpha}$ cycling does not.

- Unlike GPCRs, which interact directly with an associated G protein, RTKs are linked indirectly to Ras via two proteins, GRB2 and Sos (see Figure 20-23).

- The SH2 domain in GRB2, an adapter protein, binds to specific phosphotyrosines in activated RTKs. The two SH3 domains in GRB2 then bind Sos, a guanine-nucleotide exchange factor, thereby bringing Sos close to membrane-bound Ras · GDP and activating its exchange function.

- Binding of Sos to inactive Ras causes a large conformational change that permits release of GDP and binding of GTP.

- Normally, Ras activation and the subsequent cellular response is induced by ligand binding to an RTK. However, in cells that contain a constitutively active Ras, the cellular response occurs in the absence of ligand binding.

20.5 MAP Kinase Pathways

All Ras-linked RTKs in mammalian cells appear to utilize a highly conserved signal-transduction pathway in which the signal induced by ligand binding is carried via GRB2 and Sos to Ras, leading to its activation (see Figure 20-23). Activated Ras then induces a *kinase cascade* that culminates in activation of **MAP kinase**. This serine/threonine kinase, which can translocate into the nucleus, phosphorylates many different proteins including transcription factors that regulate expression of important cell-cycle and differentiation-

specific proteins. In this section, we first examine the components of the kinase cascade downstream from Ras in RTK-Ras signaling pathways in mammalian cells. Then we discuss the linkage of other signaling pathways to similar kinase cascades and recent studies indicating that both yeasts and cells of higher eukaryotes contain multiple MAP kinases.

Activation of MAP kinase in two different cells can lead to similar or different cellular responses, as can activation in the same cell by stimulation of different RTKs. The mechanisms controlling the response specificity of MAP kinases are poorly understood and are not considered in this chapter.

Signals Pass from Activated Ras to a Cascade of Protein Kinases

A remarkable convergence of biochemical and genetic studies in yeast, *C. elegans, Drosophila,* and mammals has revealed a highly conserved cascade of protein kinases that operate in sequential fashion downstream from activated Ras as follows (Figure 20-28):

1. Activated Ras binds to the N-terminal domain of Raf, a serine/threonine kinase.

2. Raf binds to and phosphorylates MEK, a dual-specificity protein kinase that phosphorylates both tyrosine and serine residues.

3. MEK phosphorylates and activates MAP kinase, another serine/threonine kinase.

4. MAP kinase phosphorylates many different proteins, including nuclear transcription factors, that mediate cellular responses.

Several types of experiments have demonstrated that Raf, MEK, and MAP kinase lie downstream of Ras and their sequential order in the pathway. For example, constitutively active mutant Raf proteins induce quiescent cultured cells to proliferate in the absence of hormone stimulation. These mutant Raf proteins, which initially were identified in tumor cells, are encoded by **oncogenes** and stimulate uncontrolled cell proliferation. Conversely, cultured mammalian cells that express a mutant, defective Raf protein cannot be stimulated to proliferate uncontrollably by a mutant, constitutively active Ras^D protein. This finding establishes a link between the Raf and Ras proteins. In vitro binding studies have shown that purified Ras · GTP protein binds directly to Raf. An interaction between the mammalian Ras and Raf proteins also has been demonstrated in the *yeast two-hybrid system,* a genetic system in yeast used to select cDNAs encoding proteins that bind to target, or "bait" proteins (Figure 20-29). The binding of Ras and Raf to each other does *not* induce the Raf kinase activity.

The location of MAP kinase downstream of Ras was evidenced by the finding that in quiescent cultured cells expressing a constitutively active Ras^D, activated MAP kinase is generated in the absence of hormone stimulation. More

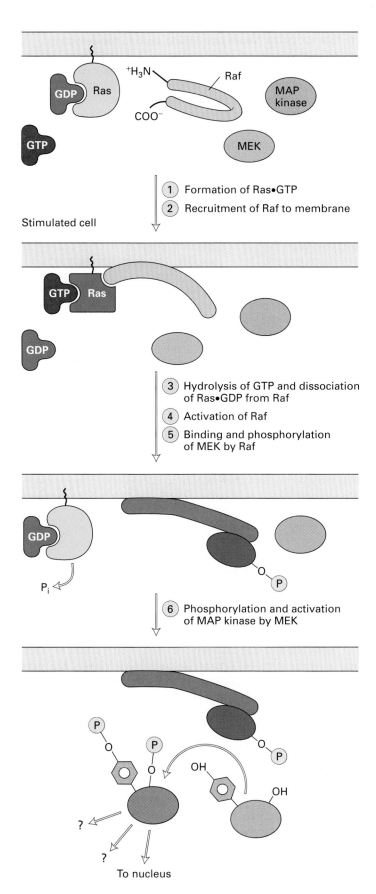

◀ **FIGURE 20-28 Kinase cascade that transmits signals downstream from activated Ras protein.** In unstimulated cells, most Ras is in the inactive form with bound GDP *(top)*; binding of a growth factor to its RTK leads to formation of the active Ras · GTP (see Figure 20-23). A signaling complex then is assembled downstream of Ras, leading to activation of MAP kinase by phosphorylation of threonine and tyrosine residues separated by a single amino acid. Phosphorylation at both sites is necessary for activation of MAP kinase. Several details not shown in the diagram are discussed in the text. [See A. B. Vojtek et al., 1993, *Cell* **74**:205; L. Van Aelst et al., 1993, *Proc. Nat'l. Acad. Sci. USA* **90**:6213; S. A. Moodie et al., 1993, *Science* **260**:1658; H. Koide et al., 1993, *Proc. Natl. Acad. Sci. USA* **90**:8683; P. W. Warne et al., 1993, *Nature* **364**:352.]

importantly, in *Drosophila* mutants that lack a functional Ras or Raf but express a constitutively active MAP kinase specifically in the developing eye, R7 photoreceptors were found to develop normally. This finding indicates that activation of MAP kinase is sufficient to transmit a proliferation or differentiation signal normally initiated by ligand binding to an RTK. Biochemical studies showed that Raf does not activate MAP kinase directly. The signaling pathway thus appears to be a linear one: activated RTK ⟶ Ras ⟶ Raf ⟶ (?) ⟶ MAP kinase.

Finally, fractionation of cultured cells that had been stimulated with growth factors led to identification of MEK, a kinase that specifically phosphorylates threonine and tyrosine residues on MAP kinase, thereby activating its catalytic activity. (The acronym MEK comes from MAP and ERK kinase, where ERK is another acronym for MAP.) Later studies showed that MEK binds to the C-terminal catalytic domain of Raf and is phosphorylated by the Raf serine/threonine kinase activity; this phosphorylation activates the catalytic activity of MEK. Hence, activation of Ras induces a kinase cascade that includes Raf, MEK, and MAP kinase.

Ksr May Function as a Scaffold for the MAP Kinase Cascade Linked to Ras

Two additional proteins not depicted in Figure 20-28 participate in the MAP kinase cascade downstream from Ras. Although the precise functions of these proteins, called 14-3-3 and Ksr, are not yet known, they both appear to play important roles in forming protein complexes necessary for signaling from Raf to MAP kinase.

In a resting cell prior to stimulation, Raf is present in the cytosol in an inactive conformation stabilized by a dimer of 14-3-3. Each 14-3-3 monomer binds to a phosphoserine residue in Raf, one to Ser-259 and the other to Ser-621. Ras · GTP, which is anchored to the membrane, recruits inactive Raf to the membrane and induces a conformational change in Raf that disrupts its association with 14-3-3. Ser-259 then is dephosphorylated, activating Raf's kinase activity. After Ras returns to the GDP form, it dissociates from

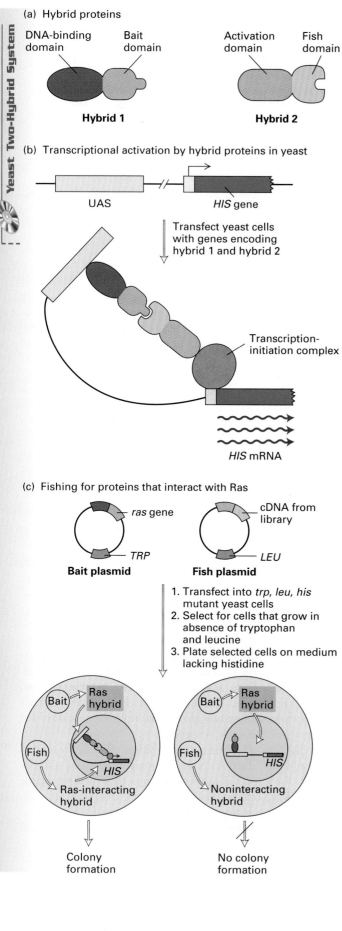

(a) Hybrid proteins

DNA-binding domain | Bait domain

Activation domain | Fish domain

Hybrid 1 | **Hybrid 2**

(b) Transcriptional activation by hybrid proteins in yeast

UAS | *HIS* gene

Transfect yeast cells with genes encoding hybrid 1 and hybrid 2

Transcription-initiation complex

HIS mRNA

(c) Fishing for proteins that interact with Ras

ras gene

cDNA from library

TRP

LEU

Bait plasmid | **Fish plasmid**

1. Transfect into *trp, leu, his* mutant yeast cells
2. Select for cells that grow in absence of tryptophan and leucine
3. Plate selected cells on medium lacking histidine

Bait → Ras hybrid

Fish | HIS

Ras-interacting hybrid

Colony formation

Bait → Ras hybrid

Fish | HIS

Noninteracting hybrid

No colony formation

▲ **FIGURE 20-29 Yeast two-hybrid system for detecting proteins that interact.** (a) Recombinant DNA techniques can be used to prepare genes that encode hybrid (chimeric) proteins consisting of the DNA-binding domain (purple) or activation domain (orange) of a transcription factor fused to one of two interacting proteins, referred to as the "bait" domain (pink) and "fish" domain (green). (b) If yeast cells are transfected with genes encoding both hybrids, the bait and fish portions of the chimeric proteins interact to produce a functional transcriptional activator. One end of this protein complex binds to the upstream activating sequence (UAS) of a test gene (in this example, the *HIS3* gene); the other end, consisting of the activation domain, stimulates assembly of the transcription-initiation complex (gray) at the promoter (yellow). (c) This strategy can be used to screen a cDNA library for clones expressing proteins that interact with a protein of interest, in this case Ras. This approach requires two types of plasmids: The bait plasmid includes a DNA sequence encoding the DNA-binding domain of a transcription factor (purple) connected to the coding sequence for Ras (pink). The fish plasmids contain individual cDNAs (green) from a library connected to the coding sequence for the activation domain (orange). Each type of plasmid also contains a wild-type selection gene (e.g., *TRP1* or *LEU2*). Both types of plasmids are transfected into yeast cells with mutations in genes required for tryptophan, leucine, and histidine biosynthesis (*trp1, leu2, his3* cells) and then grown in the absence of tryptophan and leucine. Only cells that contain the bait plasmid and at least one fish plasmid survive under these selection conditions. The cells that survive then are plated on medium lacking histidine; only cells that contain the bait plasmid and a fish plasmid encoding a protein that binds to Ras are able to grow, thus identifying cDNAs encoding Ras-binding proteins. [See A. B. Vojtek et al., 1993, Cell **74**:205; S. Fields and O. Song, 1989, Nature **340**:245.]

Raf and presumably can then be reactivated again, thereby recruiting additional Raf molecules to the membrane. In addition to its role in regulating Raf structure, the 14-3-3 dimer appears to have a more general function in linking together signaling components through phosphoserine residues.

Activation of Raf also requires Ksr, which contains binding sites for Raf, 14-3-3, MEK, and MAP kinase. Ksr may function as an adapter protein, providing a scaffold for formation of a large signaling complex that continues to operate after Raf dissociates from Ras · GDP. Although Ksr has a kinase domain, genetic studies in *Drosophila* have shown that this domain has a negative regulatory function, perhaps acting as part of a switch for regulating the turnover of components within the complex.

Phosphorylation of a Tyrosine and a Threonine Activates MAP Kinase

Biochemical and x-ray crystallographic studies have provided a detailed picture of how phosphorylation activates MAP kinase (Figure 20-30). In MAP kinase and other protein kinases, including the cytosolic domain of RTKs, the

(a)

(b)

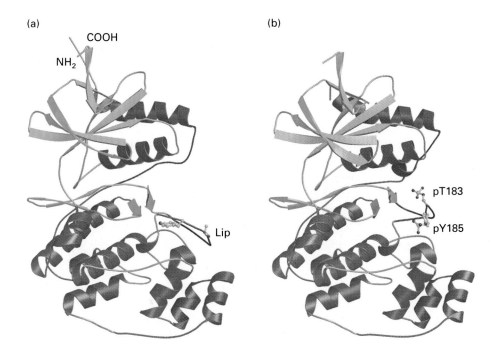

◀ **FIGURE 20-30 Structures of MAP kinase in its inactive, unphosphorylated form (a) and active, phosphorylated form (b).** Phosphorylation of MAP kinase by MEK at tyrosine 185 (pY185) and threonine 183 (pT183) leads to a marked conformational change in the phosphorylation lip (red). This change promotes dimerization of MAP kinase and binding of its substrates, ATP and certain proteins. [From B. J. Canagarajah et al., 1997, *Cell* **90**:859; courtesy of Bertram Canagarajah and Elizabeth Goldsmith.]

catalytic site in the inactive, unphosphorylated form is blocked by a segment of amino acids, the phosphorylation lip. Binding of MEK to MAP kinase destabilizes the lip structure, resulting in exposure of a critical tyrosine that is buried in the inactive conformation. Following phosphorylation of this tyrosine, MEK phosphorylates a neighboring threonine. Both the phosphorylated tyrosine and threonine residues in MAP kinase interact with additional amino acids, thereby conferring an altered conformation to the lip region, which in turn permits binding of ATP to the catalytic site. The phosphotyrosine residue also plays a key role in forming the binding site for specific substrate proteins on the surface of MAP kinase. Phosphorylation promotes not only the catalytic activity of MAP kinase but also its dimerization. The dimeric form of MAP kinase (but not the monomeric form) can be translocated to the nucleus where it regulates the activity of a number of nuclear localized transcription factors (see below).

Various Types of Receptors Transmit Signals to MAP Kinase

Although yeasts and other single-celled eukaryotes lack RTKs, they have been found to possess MAP kinase pathways. And in different cell types of higher eukaryotes, stimulation of receptors other than RTKs also can initiate signaling pathways leading to activation of MAP kinase. The mating pathway in *S. cerevisiae* is a well-studied example of a MAP kinase cascade that is linked to G protein–coupled receptors, in this case for two secreted peptide pheromones, the **a** and α factors. These pheromones control mating between

haploid yeast cells of the opposite mating type, **a** or α. An **a** haploid cell secretes the **a** mating factor and has cell-surface receptors for the α factor; and an α cell secretes the α factor and has cell-surface receptors for the **a** factor (see Figure 14-4). Thus each type of cell recognizes the mating factor produced by the other opposite type.

As with the GPCRs discussed earlier, ligand binding to the yeast pheromone receptors triggers the exchange of GTP for GDP on the α subunit and dissociation of $G_\alpha \cdot GTP$ from the $G_{\beta\gamma}$ complex. In contrast to most GPCR systems, however, the dissociated $G_{\beta\gamma}$ complex (not $G_\alpha \cdot GTP$) mediates all the physiological responses induced by activation of the yeast pheromone receptors. It does so by triggering a kinase pathway that is analogous to the one downstream from Ras. The components of this yeast kinase cascade were uncovered mainly through analyses of mutants that possess functional **a** and α receptors and G proteins but are defective in mating responses. The physical interactions between the components were assessed through immunoprecipitation experiments with extracts of yeast cells and other types of studies.

Based on these studies, scientists have proposed the kinase cascade depicted in Figure 20-31. $G_{\beta\gamma}$, which is tethered to the membrane via the γ subunit, binds to and activates Ste20, which in turn activates Ste11, a serine/threonine kinase analogous to Raf. Activated Ste11 then phosphorylates Ste7, a dual-specificity MEK, which in turn activates Fus3, a serine/threonine kinase equivalent to MAP kinase. After translocation to the nucleus, Fus3 promotes expression of target genes by activating nuclear transcription factors such as Ste12 by mechanisms that are not completely

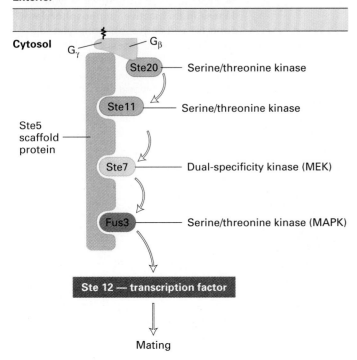

Exterior

Cytosol

Ste5 scaffold protein

Ste20 — Serine/threonine kinase

Ste11 — Serine/threonine kinase

Ste7 — Dual-specificity kinase (MEK)

Fus3 — Serine/threonine kinase (MAPK)

Ste 12 — transcription factor

Mating

▲ **FIGURE 20-31 Kinase cascade that transmits signals downstream from $G_{\beta\gamma}$ in the mating pathway in *S. cerevisiae*.** The receptors for yeast **a** or α mating factors are both coupled to the same trimeric G protein. Ligand binding leads to activation of the G protein and dissociation of $G_\alpha \cdot$ GTP from the $G_{\beta\gamma}$ complex as described previously (see Figure 20-16). In the yeast mating pathway, however, the physiological responses are induced by the dissociated $G_{\beta\gamma}$, which activates a protein kinase cascade similar to the cascade downstream of Ras in RTK pathways (see Figure 20-28). The final component, Fus3, is functionally equivalent to MAP kinase (MAPK) in higher eukaryotes. It phosphorylates transcription factors (e.g., Ste12) that control expression of proteins involved in mating-specific cellular responses. [See M. Whiteway et al., 1989, *Cell* **56**:467; G. F. Sprague and J. Thorner, 1992, in *The Molecular Biology of the Yeast Saccharomyces*, Cold Spring Harbor Press, p. 657; H. Liu et al., 1993, *Science* **262**:1741; M. Whiteway et al., 1995, *Science* **269**:1572.]

understood. The other component of the yeast mating cascade, Ste5, interacts with $G_{\beta\gamma}$ as well as Ste11, Ste7, and Fus3. Ste5 has no obvious catalytic function and acts as a scaffold for assembling other components in the cascade.

Multiple MAP Kinase Pathways Are Found in Eukaryotic Cells

In addition to the MAP kinases discussed above, both yeasts and higher eukaryotic cells contain other functionally equivalent proteins, including the Jun N-terminal kinases (JNKs) and p38 kinases in mammalian cells and six yeast proteins described below. Collectively referred to as MAP kinases, all these proteins are serine/threonine kinases that are activated in the cytosol in response to specific extracellular signals and can be translocated to the nucleus. Activation of all known MAP kinases requires *dual* phosphorylation of analogous residues in the phosphorylation lip of the protein (see Figure 20-30). Thus in all eukaryotic cells, binding of a wide variety of extracellular signaling molecules triggers highly conserved kinase cascades culminating in activation of a particular MAP kinase. The different MAP kinases mediate specific cellular responses, including morphogenesis, cell death, and stress responses.

Current genetic and biochemical studies in the mouse and *Drosophila* are aimed at determining which MAP kinases are required for mediating the response to which signals in higher eukaryotes. This has already been accomplished in large part for the simpler organism *S. cerevisiae*. Of the six MAP kinases encoded in the *S. cerevisiae* genome, five have been assigned by genetic analyses to specific signaling pathways triggered by various extracellular signals, such as pheromones, starvation, high osmolarity, hypotonic shock, and carbon/nitrogen deprivation. Each of these MAP kinases mediates very specific cellular responses (Figure 20-32).

In both yeasts and higher eukaryotic cells, different MAP kinase cascades share some common components. For instance, in yeast, Ste11 functions in the signaling pathways that regulate mating, filamentous growth, and osmoregulation. Nevertheless, each pathway activates only one MAP kinase: Fus3 in the mating pathway, Kss1 in the filamentation

▶ **FIGURE 20-32 Overview of five MAP kinase pathways in *S. cerevisiae*.** Each pathway is triggered by a specific extracellular signal and leads to activation of a single MAP kinase, which mediates characteristic cellular responses. [Adapted from H.D. Madhani and G.R. Fink, 1998, *Trends Genet.* **14**(4):152.]

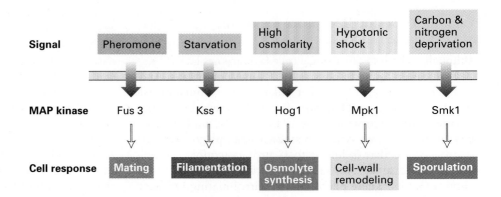

Signal	Pheromone	Starvation	High osmolarity	Hypotonic shock	Carbon & nitrogen deprivation
MAP kinase	Fus 3	Kss 1	Hog1	Mpk1	Smk1
Cell response	Mating	Filamentation	Osmolyte synthesis	Cell-wall remodeling	Sporulation

pathway, and Hog1 in the osmoregulation pathway (see Figure 20-32). Similarly, in mammalian cells, common upstream signal-transducing molecules participate in activating multiple JNK kinases. Given this sharing of components among different MAP kinase pathways, how is the specificity of the responses to particular signals achieved? Recent studies of the MAP kinase pathways in yeast suggest possible answers to this question.

Specificity of MAP Kinase Pathways Depends on Several Mechanisms

As we've just explained, Ste11 is a component of the yeast mating, filamentation, and osmoregulatory MAP kinase pathways. A complex set of regulatory interactions appears to limit intracellular signaling in response to a particular extracellular signal to the appropriate pathway. Here we describe recent findings about two mechanisms that can determine the specificity of MAP kinase pathways.

Pathway-Specific Signaling Complexes To illustrate one way multiple MAP kinases are segregated, we consider how a change in osmolarity activates Ste11 but does not lead to activation of downstream components in the mating pathway. There are two osmoregulatory MAP kinase pathways in *S. cerevisiae;* both lead to activation of the MAP kinase Hog1, but only one pathway requires Ste11. The dual-specificity MEK in this osmoregulatory pathway, called Pbs2, also functions as a scaffold for assembly of a large signaling complex. Pbs2 binds to Hog1, Ste11, and Sho1 (the osmolarity-sensitive receptor). Transmission of the signal from Sho1 to Hog1 occurs within the complex assembled by Pbs2. Recall that in the mating pathway, the scaffold protein Ste5 likewise stabilizes a large complex including Ste11. In both cases, the common component Ste11 is constrained within a large complex that forms in response to a specific extracellular signal, and signaling downstream from Ste11 is restricted to the complex containing it (Figure 20-33). As a result, exposure of yeast cells to mating factors induces activation of a single MAP kinase, Fus3, while exposure to a high osmolarity induces activation only of Hog1.

Kinase-Independent Functions of MAP Kinases Detailed genetic analysis of two different yeast MAP kinases, Fus3 (mating pathway) and Kss1 (filamentation pathway), have revealed another mechanism for restricting signaling to a single MAP kinase pathway. Mutant yeasts that express *no* Fus3 can still mate. In these "kinase-lacking" mutants, mating factors induce mating-specific genes, normally regulated by Fus3, by activating the catalytic activity of Kss1. Moreover, stimulation of these mutants by mating factors also induces filamentation-specific genes regulated by Kss1, which normally is activated in response to starvation. Kss1-mediated induction of mating-specific and filamentation-specific genes in response to pheromones requires Ste5, the scaffold protein in the mating pathway. Other yeast mutants, referred to as "kinase-dead" mutants, express a defective Fus3 without

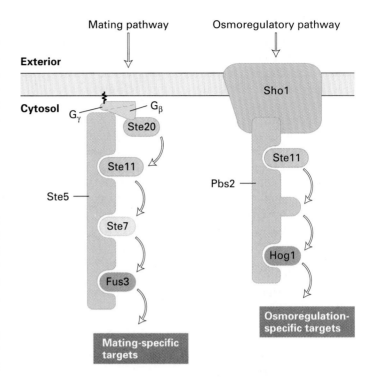

▲ **FIGURE 20-33 MAP kinase signaling complexes in the mating and osmoregulatory pathways in yeast.** Formation of such pathway-specific complexes prevents "cross-talk" between pathways that contain a common component, such as Ste11 in these two pathways. These large complexes are assembled on the molecular scaffolds Ste5 and Pbs2. Unlike Ste5, which has no catalytic function, Pbs2 has MEK activity (analogous to Ste7 in the mating pathway). Once phosphorylated by Ste11, activated Pbs2 phosphorylates Hog1. [See F. Posas and H. Saito, 1997, *Science* **276:**1703.]

catalytic activity. In kinase-dead mutants, mating factors induce neither mating-specific nor filamentation-specific genes. Apparently, Kss1 cannot be activated in the presence of an altered, catalytically inactive Fus3.

The results with these two types of yeast mutants suggest that Kss1 can bind to Ste5, but it does so less efficiently than wild-type or kinase-dead Fus3. Under normal conditions, Fus3 is recruited into a Ste5 signaling complex in response to mating-factor stimulation of cells (see Figure 20-31). If Fus3 is missing altogether, as in the kinase-lacking mutants, then Kss1 is recruited into the complex. In contrast, the inactive Fus3 in kinase-dead mutants is effectively recruited into the complex, but cannot be activated. Hence, kinase-dead Fus3 acts as a plug that inhibits signal flow by physically preventing recruitment of Kss1 into the complex. Thus, Fus3 has two separable functions, a *kinase-dependent function* necessary to induce its appropriate target genes and a *kinase-independent function* that prevents activation of the closely related MAP kinase, Kss1.

Genetic experiments reveal that Kss1 also has kinase-dependent and kinase-independent functions. In the absence of a filamentation-inducing signal, inactive Kss1 is bound in the cytosol to a transcription factor complex required for induction of filamentation-specific genes; in this form Kss1 inhibits transcription. Induction of filamentation-specific genes in wild-type yeast specifically requires activation of Kss1 kinase activity. In yeast mutants that lack Kss1, filamentation-specific genes are inappropriately induced by Fus3-mediated activation of the transcription factor complex in response to mating factors. However, kinase-dead Kss1, which can bind to the transcription factor complex, prevents this cross-talk presumably by physically preventing access of Fus3. It is not yet known, however, why activated Kss1 fails to activate Ste12 dimers promoting mating-specific gene expression in wild-type yeast.

Mammalian MAP kinases have been found to bind to specific proteins in a kinase-independent fashion, raising the possibility that MAP kinases in animals, like those in yeast, may restrict signal specificity through kinase-independent functions.

SUMMARY   MAP Kinase Pathways

- Activated Ras promotes formation of signaling complexes at the membrane containing three sequentially acting protein kinases and a scaffold protein Ksr. Raf is recruited to the membrane by binding to Ras · GTP and then activated. It then phosphorylates MEK, a dual specificity kinase that phosphorylates MAP kinase. Phosphorylated MAP kinase dimerizes and translocates to the nucleus where it regulates gene expression (see Figure 20-28).

- RTKs, GPCRs, and other receptor classes can activate MAP kinase pathways. Single-cell eukaryotes, such as yeast, and multicellular organisms contain multiple MAP kinase pathways that regulate diverse cellular processes (see Figure 20-32).

- Although different MAP kinase pathways share some upstream components, activation of one pathway by extracellular signals does not lead to activation of others containing shared components.

- In MAP kinase pathways containing common components, the activity of shared components is restricted to only a subset of MAP kinases by their assembly into large pathway-specific signaling complexes (see Figure 20-33).

- Some MAP kinases have kinase-independent functions that can restrict signals to only a subset of MAP kinases.

20.6   Second Messengers

In this section we consider in more detail the role of various second messengers in intracellular signaling (see Figure 20-4). As we saw earlier, hormone stimulation of G_s protein–coupled receptors leads to activation of adenylyl cyclase and synthesis of cAMP. The resulting rise in the cAMP level produces markedly different effects in various types of cells. So far as is known, cAMP does not function in signaling pathways initiated by RTKs. However, other second messengers function in signaling pathways initiated by both GPCRs and RTKs.

cAMP and Other Second Messengers Activate Specific Protein Kinases

The diverse effects of cAMP are mediated through the action of cAMP-dependent protein kinases (cAPKs). These enzymes, also referred to as protein kinases A (PKAs), modify the activities of target enzymes in various cell types by phosphorylating specific serine and threonine residues. Phosphorylation of many enzymes increases their catalytic activity, whereas phosphorylation of others decreases their activity.

The cAMP-dependent protein kinases are tetramers, consisting of two regulatory (R) subunits and two catalytic (C) subunits. In the tetrameric form, cAPK is enzymatically inactive. Binding of cAMP to the R subunits causes dissociation of the two C subunits, which then can phosphorylate specific acceptor proteins (see Figure 3-27a). Each R subunit has two distinct sites for binding cAMP, called A and B, each located in a different domain in the R subunit. Binding of cAMP to site B induces a conformational change that unmasks site A. Binding of cAMP to site A, in turn, leads to release of the C subunits. The binding of two cAMP molecules by a R subunit occurs in a cooperative fashion; that is, binding of the first cAMP molecule lowers the K_D for binding of the second. Thus, small changes in the level of cytosolic cAMP can cause proportionately large changes in the amount of dissociated C subunits and, hence, in the activity of a cAPK.

As discussed later, other second messengers also activate protein kinases via a similar mechanism: in the absence of the second messenger, the kinase shows low activity; binding of the second messenger increases the kinase activity. Each type of second messenger–dependent kinase is inhibited by the binding of a peptide sequence, called a *pseudosubstrate*, to the active site. The pseudosubstrate sequence can be part of a distinct regulatory subunit, as in cAPKs, or it can be part of a *regulatory domain* within the same polypeptide chain that contains the active site. Binding of a second messenger to the inactive kinase induces a conformational change that leads to release of the part of the regulatory domain bound to the kinase active site or to dissociation of the regulatory subunit; in both cases, the active sites of the kinase are unmasked and their enzymatic activities activated.

The catalytic subunits of some second messenger–dependent protein kinases are further activated by phosphorylation of specific residues within the phosphorylation lip. As in the case of MAP kinases, when the lip is unphosphorylated, it inhibits binding of ATP or protein substrates, or both. Phosphorylation induces a conformational change leading to effective binding of both ATP and protein substrates. X-ray crystallographic analyses have revealed that many different protein kinases share a common three-dimensional structure similar to that of MAP kinase shown in Figure 20-30.

cAPKs Activated by Epinephrine Stimulation Regulate Glycogen Metabolism

The cAMP-dependent protein kinases (cAPKs) induce many effects depending on the particular substrate proteins that they phosphorylate. The first cAMP-mediated cellular response to be discovered—the release of glucose from **glycogen**—occurs in muscle and liver cells stimulated by epinephrine or agonists of β-adrenergic receptors.

Glycogen, a large glucose polymer, is the major storage form of glucose in animals. Like most polymers, glycogen is synthesized by one set of enzymes and degraded by another. Three enzymes convert glucose into *uridine diphosphoglucose* (UDP-glucose), the primary intermediate in glycogen synthesis. The glucose residue of UDP-glucose is transferred

by *glycogen synthase* to the free hydroxyl group on carbon 4 of a glucose residue at the end of a growing glycogen chain (Figure 20-34a). Degradation of glycogen involves the stepwise removal of glucose residues from the same end by a phosphorolysis reaction, catalyzed by *glycogen phosphorylase*, yielding glucose 1-phospate (Figure 20-34b). In both muscle and liver cells, glucose 1-phosphate produced from glycogen is converted by phosphoglucomutase to glucose 6-phosphate. In muscle cells, this metabolite enters the glycolytic pathway (see Figure 16-3) and is metabolized to generate ATP for use in powering muscle contraction. Unlike muscle cells, liver cells contain a phosphatase that hydrolyzes glucose 6-phospate to glucose. Thus glycogen stores in the liver are primarily broken down to free glucose, which is immediately released into the blood and transported to other tissues, particularly the muscles and brain.

In liver and muscle cells, the epinephrine-stimulated elevation in the cAMP level enhances the conversion of glycogen to glucose 1-phosphate in two ways: by *inhibiting* glycogen synthesis and by *stimulating* glycogen degradation, as outlined in Figure 20-35a. The entire process is reversed when epinephrine is removed and the level of cAMP drops

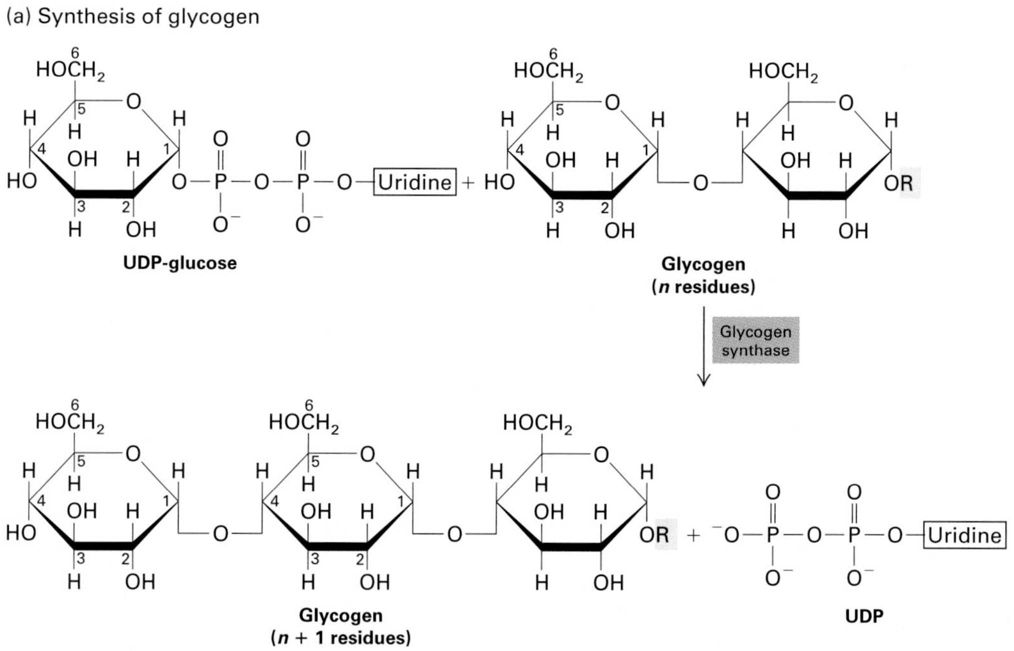

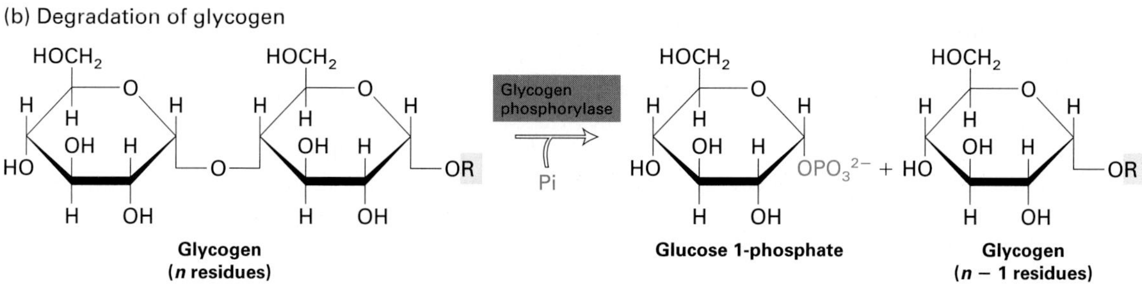

(a) Synthesis of glycogen

(b) Degradation of glycogen

◀ **FIGURE 20-34 Synthesis and degradation of glycogen.** (a) Incorporation of glucose from UDP-glucose into glycogen is catalyzed by glycogen synthase. (b) Removal of glucose units from glycogen is catalyzed by glycogen phosphorylase. Because two different enzymes catalyze the formation and degradation of glycogen, the two reactions can be independently regulated. R stands for the remainder of the glycogen molecule.

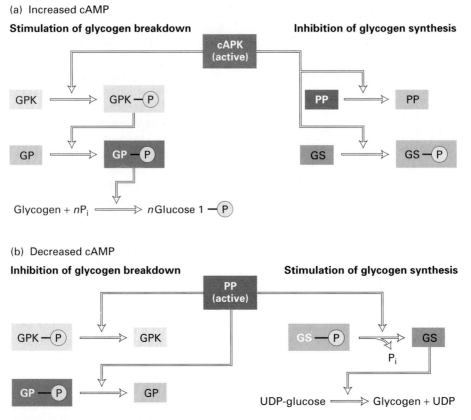

▲ **FIGURE 20-35 Regulation of glycogen breakdown and synthesis by cAMP in liver and muscle cells.** Active enzymes are highlighted in darker shades; inactive forms, in lighter shades. (a) An increase in cytosolic cAMP activates a cAMP-dependent protein kinase (cAPK) that triggers a protein kinase cascade involving glycogen phosphorylase kinase (GPK) and glycogen phosphorylase (GP), leading to breakdown of glycogen. The active cAPK also phosphorylates and thus inactivates glycogen synthase (GS), inhibiting glycogen synthesis. Phosphorylation of an inhibitor of phosphoprotein phosphatase (PP) by cAPK (see Figure 20-36) prevents PP from dephosphorylating the activated enzymes in the kinase cascade or inactive GS. (b) A decrease in cAMP inactivates the cAPK, leading to release of the active form of phosphoprotein phosphatase. This enzyme then removes phosphate residues from GPK and GP, thereby inhibiting glycogen degradation. The phosphatase also removes phosphate from inactive GS, thereby activating this enzyme and stimulating glycogen synthesis.

(Figure 20-35b). This reversal is mediated by *phosphoprotein phosphatase*, which removes the phosphate residues from the inactive form of glycogen synthase, thereby activating it, and from the active forms of glycogen phosphorylase kinase and glycogen phosphorylase, thereby inactivating them. The activity of phosphoprotein phosphatase also is regulated by cAMP, although indirectly. At high cAMP levels, cAPK phosphorylates an inhibitor of phosphoprotein phosphatase; the phosphorylated inhibitor then binds to phosphoprotein phosphatase, inhibiting its activity (Figure 20-36). At low cAMP levels, the inhibitor is not phosphorylated and phosphoprotein phosphatase is active. As a result, the synthesis of glycogen by glycogen synthase is enhanced and the degradation of glycogen by glycogen phosphorylase is inhibited.

The cAMP-dependent switch that regulates glycogen metabolism thus exhibits dual regulation: activation of the enzymes catalyzing glycogen degradation and inhibition of enzymes promoting glycogen synthesis. The coordinate regulation of stimulatory and inhibitory pathways provides an efficient mechanism for operating switches and is a common phenomenon in regulatory biology.

Kinase Cascades Permit Multienzyme Regulation and Amplify Hormone Signals

The cAMP-mediated stimulation of glycogen breakdown illustrates two important properties of a cascade, a series of reactions in which the enzyme catalyzing one step is activated (or inhibited) by the product of a previous step. Although such a cascade may seem overcomplicated, it has at least two advantages for the cell.

First, a cascade allows an entire group of enzyme-catalyzed reactions to be regulated by a single type of molecule. As we have seen, the three enzymes in the **glycogenolysis cascade**—cAMP-dependent protein kinase (cAPK), glycogen phosphorylase kinase (GPK), and glycogen phosphorylase (GP)—are

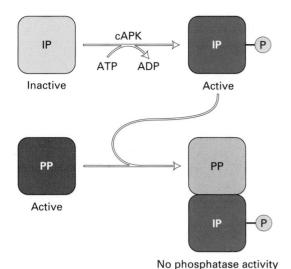

▲ FIGURE 20-36 Regulation of phosphoprotein phosphatase activity by cAMP is mediated by an inhibitor protein. At high levels of cAMP, a cAMP-dependent protein kinase (cAPK) phosphorylates an inhibitor protein (IP), which then binds to phosphoprotein phosphatase (PP), forming a complex that lacks phosphatase activity. When the cAMP level decreases, constitutive phosphatases dephosphorylate the inhibitor, releasing phosphoprotein phosphatase in its active form.

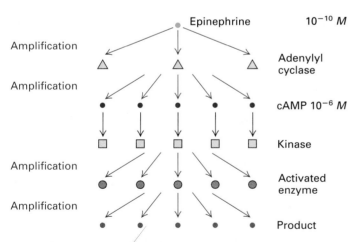

▲ FIGURE 20-37 Intracellular transduction of an extracellular signal via a cascade of sequential reactions produces a large amplification of the signal. In this example, binding of a single epinephrine molecule to one receptor molecule induces synthesis of a large number of cAMP molecules. These in turn activate multiple enzyme molecules, each of which produces multiple product molecules (e.g., active phosphorylated proteins). The more steps in such a cascade, the greater the signal amplification possible.

all regulated, directly or indirectly, by cAMP (see Figure 20-35a). Other metabolic pathways also are regulated by hormone-induced cascades, some mediated by cAMP and some by other second messengers.

Second, a cascade provides a huge amplification of an initially small signal (Figure 20-37). For example, blood levels of epinephrine as low as 10^{-10} M can stimulate liver glycogenolysis and release of glucose, resulting in an increase of blood glucose levels by as much as 50 percent. An epinephrine stimulus of this magnitude generates an intracellular cAMP concentration of 10^{-6} M, an amplification of 10^4. Because three more catalytic steps precede the release of glucose, another 10^4 amplification can occur. In striated muscle, the concentrations of the three successive enzymes in the glycogenolytic cascade (i.e., cAPK, GPK, and GP) are in a 1:10:240 ratio, which dramatically illustrates the amplification of the effects of epinephrine and cAMP.

Cellular Responses to cAMP Vary among Different Cell Types

The effects of cAMP on the synthesis and degradation of glycogen are confined mainly to liver and muscle cells, which store glycogen. However, cAMP also mediates the intracellular responses of many other cells to a variety of hormones that stimulate G_s protein–coupled receptors. In virtually all eukaryotic cells studied, the action of cAMP appears to be mediated by one or more cAPKs, but the nature of the metabolic response varies widely among different cells, as indicated in Table 20-3. The effects of cAMP on a given cell type depend, in part, on the specificity of the particular cAPK and on the cAPK substrates that it expresses.

For instance, in adipocytes, elevation of cAMP activates a cAPK that stimulates the production of fatty acids by controlling the activity of lipases. These fatty acids are released and taken up as an energy source by cells in other tissues such as the kidney, heart, and muscles. Likewise, the hormone-induced stimulation of cAPK in ovarian cells promotes the synthesis of estradiol and progesterone, two steroid hormones crucial to the development of female sex characteristics. cAMP acting through cAMP-dependent protein kinases also plays a critical role in mediating the communication between cells critical for the formation of specific tissues during development. The cAPK in nerve cells modulates the activity of specific ion channels important in short-term learning (Chapter 21) and can produce long-term changes in neurons (memory) through changes in the activity of specific transcription factors.

Anchoring Proteins Localize Effects of cAMP to Specific Subcellular Regions

In many cell types, a rise in the cAMP level may produce a response that is required in one part of the cell but is unwanted, perhaps deleterious, in another part. For instance, in migrating cells specific cAMP-dependent signals help reg-

TABLE 20-3 Metabolic Responses to Hormone-Induced Rise in cAMP in Various Tissues

Tissue	Hormone Inducing Rise in cAMP	Metabolic Response
Adipose	Epinephrine; ACTH; glucagon	Increase in hydrolysis of triglyceride; decrease in amino acid uptake
Liver	Epinephrine; norepinephrine; glucagon	Increase in conversion of glycogen to glucose; inhibition of synthesis of glycogen; increase in amino acid uptake; increase in gluconeogenesis (synthesis of glucose from amino acids)
Ovarian follicle	FSH; LH	Increase in synthesis of estrogen, progesterone
Adrenal cortex	ACTH	Increase in synthesis of aldosterone, cortisol
Cardiac muscle cells	Epinephrine	Increase in contraction rate
Thyroid	TSH	Secretion of thyroxine
Bone cells	Parathyroid hormone	Increase in resorption of calcium from bone
Skeletal muscle	Epinephrine	Conversion of glycogen to glucose
Intestine	Epinephrine	Fluid secretion
Kidney	Vasopressin	Resorption of water
Blood platelets	Prostaglandin I	Inhibition of aggregation and secretion

SOURCE: E. W. Sutherland, 1972, *Science* **177**:401.

ulate membrane cytoskeletal dynamics underlying motility at the leading edge of the cell, but similar cytoskeletal changes in other parts of the cell may be harmful. Recent biochemical and cell biological experiments have identified a family of anchoring proteins that localize inactive cAPKs to specific subcellular locations, thereby restricting cAMP-dependent responses to these locations.

This family of proteins, referred to as *cAMP kinase–associated proteins* (AKAPs), have a bipartite structure with one region conferring a specific subcellular location and another that binds to the regulatory subunit of cAPKs. One such protein, AKAP250, is localized to filopodia and presumably functions to integrate cAMP-dependent signals regulating the structure of the actin-based cytoskeleton (see Figure 18-1). Specific anchoring proteins may also function to localize other signaling proteins including other kinases and phosphatases, and thus may play an important role in integrating information from multiple signaling pathways to provide local control of specific cellular processes.

Modification of a Common Phospholipid Precursor Generates Several Second Messengers

In the rest of this section, we briefly discuss several other second messengers and the mechanisms by which they regulate various cellular activities. A number of these are derived from *phosphatidylinositol* (PI). The inositol group in this phospholipid, which extends into the cytosol adjacent to the membrane,

can be [reversibly] phosphorylated at various positions by the combined actions of various kinases and phosphatases, yielding several different **phosphoinositides,** which are membrane bound (Figure 20-38). The levels of PIs in cells are dynamically regulated by extracellular signals. For instance, in unstimulated cells the levels of PIs phosphorylated in the D3 position (PI-3Ps) are very low. In response to some signals (e.g., PDGF), there is an acute rise in PIs phosphorylated at this position through the activation of PI-3 kinase. Both GPCRs and RTKs stimulate the activity of PI-3 kinases. Some PIs [e.g., $PI(3,4,5)P3$] are rapidly regulated in response to signal, and others are not [e.g., $PI(3)P$]. As we discuss later in this chapter, PIs act as membrane docking sites for signaling molecules and also, in some cases, stimulate catalytic activity. Proteins bind to PIs through a PH domain. Different PH domains show different phosphoinositide binding specificities.

Phosphoinositides can be cleaved by the membrane-associated enzyme phospholipase C (PLC) to generate yet other second messengers. These cleavage reactions produce **1,2-diacylglycerol (DAG),** a lipophilic molecule that remains linked to the membrane, and free phosphorylated inositols, which can diffuse into the cytosol. The main pathway shown in Figure 20-38a generates DAG and inositol 1,4,5-trisphosphate (IP_3). Signaling pathways involving any of these second messengers sometimes are referred to as **inositol-lipid pathways.**

Activation of the β isoform of phospholipase C (PLCβ) is induced by binding of hormones to GPCRs containing

(a)

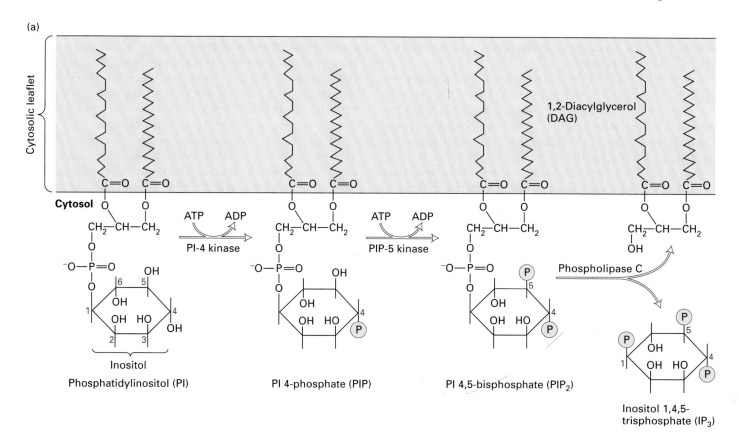

(b)

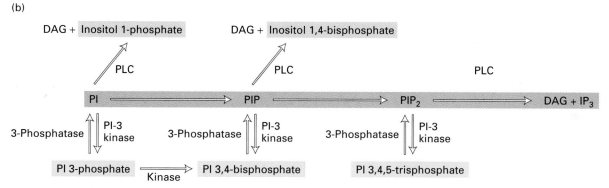

▲ **FIGURE 20-38 Several second messengers are derived from phosphatidylinositol (PI).** (a) Pathway for synthesis of DAG and IP$_3$, two important second messengers. Each membrane-bound PI kinase places a phosphate on a specific hydroxyl group on the inositol ring, producing the phosphoinositides PIP and PIP$_2$. Cleavage of PIP$_2$ by phospholipase C (PLC) yields DAG and IP$_3$.

(b) Formation of other phosphoinositides (yellow) and inositol phosphates (blue). These reactions are catalyzed by various kinases and PLC. The pathway shown in (a) is highlighted in red. See text for discussion. [See A. Toker and L. C. Cantley, 1997, *Nature* **387**:673-676 and C. L. Carpenter and L. C. Cantley, 1996, *Curr. Opin. Cell Biol.* **8**:153-158.]

either a G$_o$ or G$_q$ α subunit. Pertussis toxin locks G$_{o\alpha}$, but not G$_{q\alpha}$, in the inactive, GDP-bound form, preventing activation of PLC$_\beta$ even in the presence of hormone. The effect of pertussin toxin on G$_{o\alpha}$ thus is opposite to the effect of cholera toxin on G$_{s\alpha}$ discussed earlier (see Figure 20-17). Certain activated RTKs can increase the activity of the γ isoform of PLC. Thus hormone-induced stimulation of PLC activity and subsequent generation of DAG and phosphorylated inositols is mediated by both GPCRs and RTKs.

Hormone-Induced Release of Ca^{2+} from the ER Is Mediated by IP$_3$

Most intracellular Ca^{2+} ions are sequestered in the mitochondria and endoplasmic reticulum (ER) or other vesicles. Cells employ various mechanisms for regulating the concentration of Ca^{2+} ions free in the cytosol, which usually is kept below 0.2 μM. Ca^{2+} ATPases pump cytosolic Ca^{2+} ions across the plasma membrane to the cell exterior or into the lumens of the endoplasmic reticulum or other intracellular vesicles that store Ca^{2+} ions (see Figure 15-11). As we discuss below, a small rise in cytosolic Ca^{2+} induces a variety of cellular responses.

Binding of many hormones to their cell-surface receptors on liver, fat, and other cells induces an elevation in cytosolic Ca^{2+} even when Ca^{2+} ions are absent from the surrounding medium. In this situation, Ca^{2+} is released into the cytosol from the endoplasmic reticulum (ER) and other intracellular vesicles. The mechanism by which a hormone-receptor signal on the cell surface is transduced to the ER became clear in the early 1980s, when it was shown that a rise in the level of cytosolic Ca^{2+} often is preceded by an increase in IP3.

Since it is water soluble, IP3 can diffuse within the cytosol carrying a hormone signal from the cell surface to the ER surface. Here, IP3 binds to a Ca^{2+}-channel protein composed of four identical subunits, each containing an IP3-binding site in the large N-terminal cytosolic domain. IP3 binding induces opening of the channel allowing Ca^{2+} ions to exit from the ER into the cytosol (Figure 20-39). The resulting rise in the cytosolic Ca^{2+} level is only transient, however, because Ca^{2+} ATPases located in the plasma membrane and ER membrane actively pump Ca^{2+} from the cytosol to the cell exterior and ER lumen, respectively. Furthermore, within a second of its generation, IP3 is hydrolyzed to inositol 1,4-bisphosphate, which does not stimulate Ca^{2+} release from the ER.

Without some means for replenishing depleted stores of intracellular Ca^{2+}, a cell would soon be unable to increase the cytosolic Ca^{2+} level. Elegant patch-clamping studies have revealed that certain plasma-membrane Ca^{2+} channels, called **store-operated channels** (SOCs) open in response to depletion of intracellular Ca^{2+} stores (see Figure 20-39). Although the specific signal that promotes opening of SOCs has not yet been identified, opening of these channels is critical to cellular responses induced by elevated cytosolic Ca^{2+}.

When various phosphorylated inositols found in cells are added to preparations of ER vesicles, only IP3 causes release of Ca^{2+} ions from the vesicles. This simple experiment demonstrates the specificity of the IP3 effect. In addition, not all cells respond identically to IP3. This differential sensitivity may reflect expression of different isoforms of the IP3-sensitive Ca^{2+} channel in the ER membrane and/or variation in the Ca^{2+} content of the ER itself. Because of this variability, different types of cells may exhibit very different responses to the same extracellular signal.

Opening of the IP3-sensitive Ca^{2+} channels is potentiated by cytosolic Ca^{2+} ions, which increases the affinity of the receptor for IP3, resulting in greater release of stored Ca^{2+}. High concentrations of cytosolic Ca^{2+}, however, inhibit IP3-induced release of Ca^{2+} from intracellular stores by decreasing the affinity of the receptor for IP3. The complex regulation of the IP3 receptor in ER membranes can lead to rapid oscillations in the cytosolic Ca^{2+} level when the IP3 pathway in cells is stimulated. For example, stimulation of hormone-secreting cells in the pituitary by luteinizing hormone–releasing factor causes rapid, repeated spikes

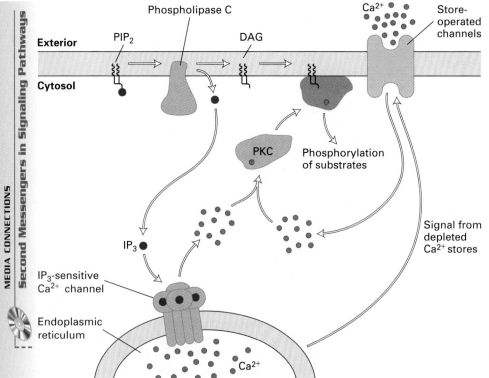

◀ **FIGURE 20-39 Elevation of cytosolic Ca^{2+} via the inositol-lipid signaling pathway.** This pathway can be triggered by ligand binding to RTKs or to GPCRs, as illustrated here. Binding of a hormone to its receptor leads to activation of the G protein (G_q), which in turn activates phospholipase C by a mechanism analogous to activation of adenylyl cyclase (see Figure 20-16). Phospholipase C then cleaves PIP2 to IP3 and DAG (see Figure 20-38a). The IP3 diffuses through the cytosol and interacts with IP3-sensitive Ca^{2+} channels in the membrane of the endoplasmic reticulum, causing release of stored Ca^{2+} ions, which mediate various cellular responses. Release of intracellular Ca^{2+} stores promotes influx of extracellular Ca^{2+} via store-operated channels. As discussed later, a rise in cytosolic Ca^{2+} recruits protein kinase C (PKC) from the cytosol to the membrane where it is activated by DAG. The activated kinase then phosphorylates several cellular enzymes and receptors, thereby altering their activity.

their activity so that the receptors either cannot bind ligand or form a receptor-ligand complex that does not induce the normal cellular response. In this section, we present several examples illustrating multiplex signaling and regulation of cell-surface receptors.

The Same RTK Can Be Linked to Different Signaling Pathways

We have seen that activated RTKs can initiate signaling via Ras and the downstream MAP kinase pathway (see Figure 20-6). These receptors also can trigger the inositol-lipid pathway by binding PI-3 kinase and PLC_γ, two of the enzymes needed to form the second messengers IP_3 and DAG (see Figure 20-38). The SH2 domains of PLC_γ, for example, bind to specific phosphotyrosines of certain RTKs, thus positioning the enzyme close to its membrane-bound substrate. In addition, the RTK phosphorylates tyrosine residues on the bound PLC_γ, enhancing its lipase activity. Thus activated RTKs promote PLC_γ activity in two ways: by localizing the enzyme to the membrane and by phosphorylating it.

The initiation of tissue-specific signaling pathways by stimulation of the *same* receptor in different cells is exemplified by EGF-stimulated signaling in *C. elegans*. The central importance of RTK-Ras-MAP kinase signaling, stimulated by EGF, in development of the vulva in *C. elegans* was demonstrated in studies analogous to those described earlier for development of R7 cells in *Drosophila* (see Figure 20-25). Other genetic studies, however, have shown that stimulation of the EGF receptor triggers a Ras-independent pathway in some tissues. For example, one of the many functions of EGF in *C. elegans* is to control contractility of smooth muscle, which in turn regulates the extrusion of oocytes from one compartment of the hermaphrodite gonad to another, where they are fertilized. Coupling of the EGF receptor to Ras is not required for the EGF-induced contractions of the gonad. Analysis of several different types of mutation led researchers to conclude that in *C. elegans* smooth muscle, the EGF receptor is linked to the inositol-lipid pathway. Ligand binding to the receptor leads to an increase in IP_3, which then promotes release of intracellular Ca^{2+} stores (see Figure 20-39). The increased cytosolic Ca^{2+} level then promotes muscle contraction.

Multiple G Proteins Transduce Signals to Different Effector Proteins

We noted previously that studies with bacterial toxins originally led to identification of several G proteins coupled to different effector proteins. More recently, molecular cloning has led to the isolation of a large number of proteins related to the α, β, and γ subunits of these previously characterized G proteins. In mammals, for instance, 16 distinct G_α subunits, 5 G_β subunits, and 12 G_γ subunits have been identified so far. Analysis of the *C. elegans* genome, whose entire sequence has recently been completed, revealed that this organism also encodes multiple G_α, G_β, G_γ subunits.

Clearly, eukaryotes generally produce multiple trimeric G proteins that link GPCRs to a variety of effector proteins, including ion channels, adenylyl cyclase, phospholipase C, and a cGMP-specific phosphodiesterase in photoreceptor cells (Table 20-5). Stimulation of the receptors coupled to these G proteins thus can modulate many cellular functions. The role of trimeric G proteins in these signaling pathways has been demonstrated by their sensitivity to nonhydrolyzable GTP analogs and, in some cases, by their sensitivity to cholera or pertussis toxin. Because the same cell may express diverse G proteins, it is often difficult to determine which specific G protein mediates the effect of a particular ligand.

The presence of multiple G_α subunits in a single cell raises the possibility that a single ligand could initiate signaling through more than one effector protein. Several examples of such multiplex signaling have been described, although the precise molecular details of which G proteins and which specific subunits mediate these effects are not yet known. In some cells, modulation of different effectors coupled to the same GPCR is observed at different ligand concentrations or when different concentrations of the receptor are expressed on the surface. The large number of possible combinations of different G protein subunits, coupled receptors, and effector proteins provide cells with the ability to respond in remarkably diverse ways to precisely control their development and function.

$G_{\beta\gamma}$ Acts Directly on Some Effectors in Mammalian Cells

In the signaling pathway stimulated by binding of mating factors to haploid yeast cells, the signal is transduced by the $G_{\beta\gamma}$ subunit complex, not by the $G_\alpha \cdot$ GTP complex (see Figure 20-30). This activation of a downstream signaling pathway by $G_{\beta\gamma}$ initially was thought to reflect an idiosyncrasy of yeast biology. More recent research has shown that in some mammalian cells $G_{\beta\gamma}$ can directly regulate certain effector proteins.

For example, one isoform of adenylyl cyclase (ACI) present in the brain is stimulated by $G_{s\alpha} \cdot$ GTP as described previously (see Figure 20-16); $G_{\beta\gamma}$ inhibits the activation of ACI by $G_{s\alpha}$. Another isoform (ACII), however, is stimulated by binding of $G_{\beta\gamma}$ but only if free $G_{s\alpha} \cdot$ GTP also is present. Yet other adenylyl cyclase isoforms (e.g., ACIII) are insensitive to $G_{\beta\gamma}$. Some brain cells contain two adenylyl cyclase isoforms that are regulated differently; in such cells adenylyl cyclase activity is subject to dual regulation. Similarly, some K^+ channel proteins in the heart are opened by binding of $G_{\beta\gamma}$ (see Figure 21-41). In general, considerably higher concentrations of $G_{\beta\gamma}$ than of $G_\alpha \cdot$ GTP are required to modulate the activity of an effector protein. Regulation of certain effector enzymes by $G_{\beta\gamma}$ and various $G_\alpha \cdot$ GTP complexes contributes to the integration of cellular metabolism (Figure 20-43).

TABLE 20-5   Properties of Mammalian G Proteins Linked to GPCRs

G_α Subclass*	Effect	Associated Effector Protein	2nd Messenger
G_s	↑	Adenylyl cyclase	cAMP
	↑	Ca^{2+} channel	Ca^{2+}
	↓	Na^+ channel	Change in membrane potential
G_i	↓	Adenylyl cyclase	cAMP
	↑	K^+ channel	Change in membrane potential
	↓	Ca^{2+} channel	Ca^{2+}
G_q	↑	Phospholipase C	IP_3, DAG
G_o	↑	Phospholipase C	IP_3, DAG
	↓	Ca^{2+} channel	Ca^{2+}
G_t	↑	cGMP phosphodiesterase	cGMP
$G_{\beta\gamma}$	↑	Phospholipase C	IP_3, DAG
	↓	Adenylyl cyclase	cAMP

*A given G_α may be associated with more than one effector protein. To date, only one major $G_{s\alpha}$ has been identified, but multiple $G_{q\alpha}$ and $G_{i\alpha}$ proteins have been described. In some cases (not indicated in this table) effector proteins are regulated by coincident binding to G_α and $G_{\beta\gamma}$.

KEY: ↑ = stimulation; ↓ = inhibition. IP_3 = inositol 1,4,5-trisphosphate; DAG = 1,2-diacylglycerol.

SOURCE: See A. C. Dolphin, 1987, *Trends Neurosci.* **10**:53; L. Birnbaumer, 1992, *Cell* **71**:1069.

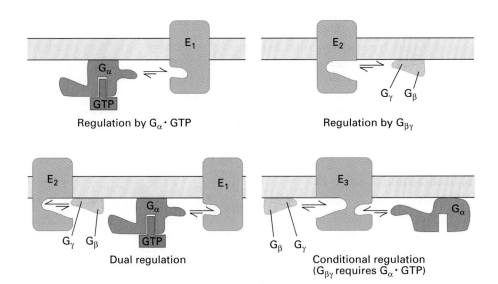

▲ **FIGURE 20-43 Multiple regulation of effector proteins mediated by G protein–coupled receptors.** Different isoforms of an effector protein (E), such as adenylyl cyclase or phospholipase C, have different binding affinities for the $G_\alpha \cdot$ GTP complex and $G_{\beta\gamma}$, leading to stimulation or inhibition by various G subunits. See text for details.

Glycogenolysis Is Promoted by Multiple Second Messengers

Earlier we saw that a rise in cAMP induced by epinephrine stimulation of β-adrenergic receptors promotes glycogen breakdown (see Figure 20-35). In both muscle and liver cells, other second messengers also produce the same cellular response.

Stimulation of muscle cells by nerve impulses causes release of Ca^{2+} ions from the sarcoplasmic reticulum and an increase in the cytosolic Ca^{2+} concentration. This rise not only triggers muscle contraction (Section 18.4) but also activates glycogen phosphorylase kinase (GPK), thereby stimulating degradation of glycogen to glucose 1-phosphate, which fuels prolonged contraction. Recall that phosphorylation by cAMP-dependent protein kinase also activates GPK. Thus this key regulatory enzyme in glycogenolysis is subject to dual regulation (Figure 20-44a).

In liver cells, hormone-induced activation of phospholipase C, which generates IP$_3$ and DAG, also regulates glycogen breakdown and synthesis by the two branches of the inositol-lipid signaling pathway (Figure 20-44b). DAG activates protein kinase C, which phosphorylates glycogen synthase, yielding the phosphorylated inactive form and thus inhibiting glycogen synthesis. IP$_3$ induces an increase in cytosolic Ca^{2+}, which activates GPK as in muscle cells, leading to glycogen degradation.

The dual regulation of GPK results from its multimeric subunit structure $(\alpha\beta\gamma\delta)_4$. The γ subunit is the catalytic protein; the regulatory α and β subunits, which are similar in structure, are phosphorylated by cAMP-dependent protein kinase; and the δ subunit is calmodulin. GPK is maximally active when Ca^{2+} ions are bound to the calmodulin subunit and at least the α subunit is phosphorylated; in fact, binding of Ca^{2+} to the calmodulin subunit may be essential to the enzymatic activity of GPK. Phosphorylation of the α and β subunits increases the affinity of the calmodulin subunit for Ca^{2+}, enabling Ca^{2+} ions to bind to the enzyme at the submicromolar Ca^{2+} concentrations found in cells not stimulated by nerves. Thus increases in the cytosolic concentration of Ca^{2+} or cAMP, or both, induce incremental increases in the activity of GPK. As a result of the elevated level of cytosolic Ca^{2+} after neuron stimulation of muscle cells, GPK will be active even if it is unphosphorylated; thus glycogen can be hydrolyzed to fuel continued muscle contraction even in the absence of hormone stimulation.

Insulin Stimulation Activates MAP Kinase and Protein Kinase B

Insulin is a prime example of a hormone that can initiate multiple signaling pathways, inducing both immediate and long-term cellular responses. The immediate effects of this hormone include an increase in the rate of glucose uptake from the blood into muscle cells and adipocytes and modulation of the activity of various enzymes involved in glucose metabolism. These effects occur within minutes, do not

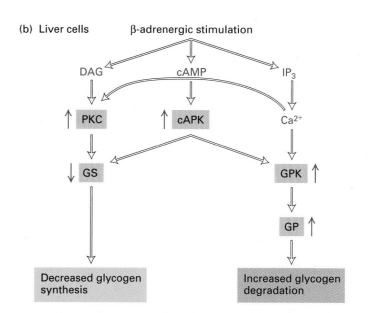

▲ **FIGURE 20-44 Multiplex regulation of glycogenolysis mediated by several second messengers (red type).** (a) Neuron stimulation of striated muscle cells or epinephrine binding to β-adrenergic receptors on their surface leads to increased cytosolic concentrations of Ca^{2+} and cAMP, respectively. The key regulatory enzyme, glycogen phosphorylase kinase (GPK), is activated by Ca^{2+} ions and by a cAMP-dependent protein kinase (cAPK). (b) In liver cells, β-adrenergic stimulation leads to increased cytosolic concentrations of cAMP, DAG, and IP$_3$. Enzymes are highlighted in green. Red arrows indicate activation (up arrow) or inhibition (down arrow) of enzyme activity. PKC = protein kinase C; GP = glycogen phosphorylase; GS = glycogen synthase.

require new protein synthesis, and occur at insulin concentrations of 10^{-9} to 10^{-10} M. Continued exposure to insulin produces longer-lasting effects including increased expression of liver enzymes that synthesize glycogen and of adipocyte enzymes that synthesize triacylglycerols; insulin also functions as a growth factor for many cells (e.g., fibroblasts). These effects are manifested in hours and require continuous exposure to $\approx 10^{-8}$ M insulin.

As noted earlier, the insulin receptor is an RTK, but unlike most RTKs it exists as a dimer in the absence of ligand. Binding of insulin can initiate two distinct signaling pathways: one that includes Ras and one that does not. Activation of Ras in the Ras-dependent insulin pathway, however, Ras differs somewhat from that for other RTKs depicted in Figure 20-23. Several lines of evidence suggest that insulin action via both the Ras-dependent and Ras-independent pathways depends on a 130-kDa polypeptide, called *insulin receptor substrate 1* (IRS1). For instance, injection of antibodies to IRS1 into cultured cells blocks the normal proliferative response induced by insulin.

Ras-Dependent Pathway IRS1 binds to the activated insulin receptor via its PTB domain and then is phosphorylated by the receptor's kinase activity. Phosphorylated IRS1, not the activated insulin receptor, binds to the **SH2 domain** of GRB2, which in turn binds to Sos protein. Insulin stimulation of target cells in liver, fat, and muscle leads to an increase in the proportion of active Ras · GTP and to activation of MAP kinase. These findings indicate that although activation of Ras induced by insulin binding requires an additional adapter protein, IRS1, which is not required by other RTKs, signal transduction downstream from Ras is similar in this insulin pathway as in other RTK-Ras pathways. Although IRS1 is not a substrate for most other RTKs, it is phosphorylated by the receptor for *insulin-like growth factor* (IGF-1), whose three-dimensional structure is similar to that of insulin.

Ras-Independent Pathway Phosphorylated IRS1 also binds PI-3 kinase, causing a tenfold stimulation in its kinase activity; this accounts for the rapid rise in phosphoinositides observed in insulin-stimulated cells. The increase in phosphoinositides leads to recruitment of **protein kinase B** (PKB) to the membrane (Figure 20-45). The N-terminal region of this kinase contains a PH domain, which binds to plasma-membrane phosphoinositides. Once localized to the membrane, PKB is phosphorylated and thereby activated by two membrane-associated kinases. After phosphorylated (active) PKB is released into the cytosol, it mediates many effects of insulin, including stimulation of glucose uptake and stimulation of glycogen synthesis. As we discuss in Section 23.8, protein kinase B, also called Akt, is a component of the signaling pathway that prevents cell death.

Another protein that binds to receptor-associated phosphorylated IRS1 is Syp, a tyrosine phosphatase. This binding causes a marked increase in the phosphatase activity of Syp. Activated Syp may dephosphorylate IRS1, thereby terminating insulin signaling, but the role of this phosphatase has not been conclusively demonstrated. In signaling pathways involving most other RTKs, GRB2, PI-3 kinase, and Syp bind directly to phosphotyrosine residues in the cytosolic domain of the receptor.

Insulin and Glucagon Work Together to Maintain a Stable Blood Glucose Level

During periods of stress, epinephrine plays a key role in inducing an increase in blood glucose. During normal daily living, however, the blood glucose level is under the dynamic control of insulin and glucagon. Both of these peptide hormones are produced by cells within the *islets of Langerhans*, cell clusters scattered throughout the pancreas. Insulin, which contains two polypeptide chains linked by disulfide bonds (see Figure 17-42b), is synthesized by the β cells in the islets; glucagon, a monomeric peptide containing 29 amino acids is produced by the α cells in the islets. Insulin acts to *reduce* the level of blood glucose, whereas glucagon acts to *increase* blood glucose. Each islet functions as an integrated unit, delivering the appropriate amounts of both hormones to the blood to meet the metabolic needs of the animal. Hormone secretion is regulated by a combination of neural and hormonal signals.

Insulin binding to receptors on muscle and adipocytes causes a rapid increase in the uptake of glucose and stimulation of glycogen synthesis via the Ras-independent pathway (see Figure 20-45). The glucagon receptor, found primarily on liver cells, is coupled to G_s protein, like the epinephrine receptor. Glucagon stimulation of liver cells activates adenylyl cyclase, leading to the cAMP-mediated cascade that leads to glycogenolysis and inhibition of glycogen synthesis (see Figure 20-35a).

The availability of glucose for cellular metabolism is regulated during periods of abundance (following a meal) or scarcity (following fasting) by the adjustment of insulin and glucagon concentrations in the blood (Figure 20-46). (Epinephrine is used only under stressful conditions.) After a meal, when blood glucose rises above its normal level of 80–90 mg/100 ml, the pancreatic β cells respond to the rise in glucose or amino acids by releasing insulin into the blood, which transports the hormone throughout the body. By binding to muscle and adipocyte cell-surface receptors, insulin causes glucose to be removed from the blood and stored in muscle cells as glycogen. Insulin also affects hepatocytes, primarily by inhibiting glucose synthesis from smaller molecules, such as lactate and acetate, and by enhancing glycogen synthesis from glucose. If the blood glucose level falls below ≈ 80 mg/100 ml, the pancreatic α cells start secreting glucagon. Glucagon binds to glucagon receptors on liver cells, triggering degradation of glycogen and the release of glucose into the blood.

Receptors for Many Peptide Hormones Are Down-Regulated by Endocytosis

The principal mechanism for down-regulating the receptors for many peptide hormones (e.g., insulin, glucagon, EGF, and PDGF) is ligand-dependent *receptor-mediated endocytosis*. In the absence of EGF ligand, for instance, the EGF receptor is internalized with bulk membrane flow. Binding to EGF induces a conformational change in the cytoplasmic tail of

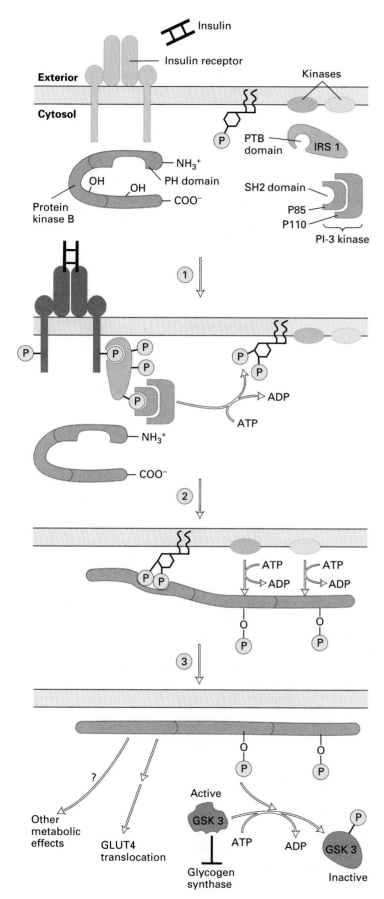

◄ **FIGURE 20-45 Activation of protein kinase B by the Ras-independent insulin signaling pathway.** The insulin receptor is a dimeric RTK. Step ①: Insulin binding to the receptor leads to a conformational change that induces autophosphorylation, similar to activation of other RTKs (see Figure 20-21). After IRS1 binds to a phosphotyrosine residue through a PTB domain, the activated kinase in the receptor's cytosolic domain phosphorylates IRS1. One subunit of PI-3 kinase binds to the receptor-bound IRS1 via its SH2 domain, and the other subunit then phosphorylates PI 4,5-bisphosphate and PI 4-phosphate to PI 3,4,5- trisphosphate and PI 3,4-biphosphate, respectively. Step ②: The phosphoinositides bind the PH domain of protein kinase B (PKB), thereby recruiting it to the membrane. Two membrane-bound kinases, in turn, phosphorylate membrane-associated PKB and activate it. Step ③: Activated PKB is released from the membrane and promotes glucose uptake by the GLUT4 transporter and glycogen synthesis. The former effect results from translocation of the GLUT4 glucose transporter from intracellular vesicles to the plasma membrane. The latter effect occurs by PKB-catalyzed phosphorylation of glycogen synthase kinase 3 (GSK3), converting it from its active to inactive form. As a result, GSK3-mediated inhibition of glycogen synthase is relieved, promoting glycogen synthesis. [See from J. Downward, 1998, *Curr. Opin. Cell Biol.* **10**:262.]

the receptor. This exposes a sorting motif that facilitates receptor recruitment into coated pits and subsequent internalization. After the receptor-hormone complex is internalized, the hormone is degraded in lysosomes—a fate similar to that of other endocytosed proteins, such as low-density lipoproteins (see Figure 17-64). Unlike the low-density lipoprotein (LDL) receptor, internalized receptors for many peptide hormones do not recycle efficiently to the cell surface.

In the presence of EGF, for instance, the average half-life of an EGF receptor on a fibroblast cell is about 30 minutes; during its lifetime, each receptor mediates the binding, internalization, and degradation of only two EGF molecules. Each time an EGF receptor is internalized with bound EGF, it has a high probability (about 50 percent) of being degraded in an endosome or lysosome. Exposure of a fibroblast cell to high levels of EGF for 1 hour induces several rounds of endocytosis, resulting in degradation of most receptor molecules. If the concentration of extracellular EGF is then reduced, the number of EGF receptors on the cell surface recovers, but only after 12–24 hours. Synthesis of new receptors is needed to replace those degraded by endocytosis, which is a slow process that may take more than a day.

The fewer hormone receptors present on the surface of a cell, the less sensitive the cell is to the hormone; as a consequence, a higher hormone concentration is necessary to induce the usual physiological response. A simple numerical example illustrates this important point. Suppose a cell has 10,000 insulin receptors on its surface with a K_D of 10^{-8} M. As noted earlier, in many cases only a fraction of the available receptors must bind ligand to induce the maximal

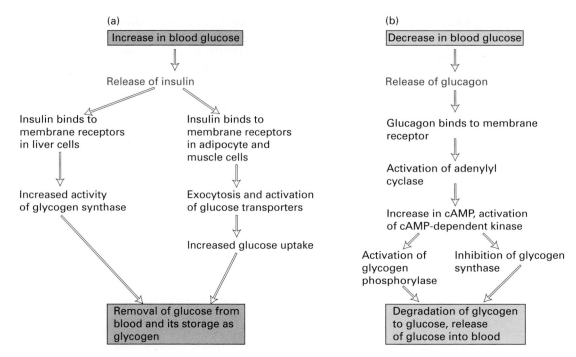

▲ **FIGURE 20-46 Regulation of blood glucose level by the opposing effects of insulin and glucagon.** (a) Insulin causes an increase in glucose uptake, mainly in muscle cells and adipocytes, and stimulates storage of glucose as glycogen, mainly in liver cells (see Figure 20-45). (b) Glucagon acts mainly on liver cells to stimulate glycogen degradation. This effect is mediated by the second messenger cAMP (see Figure 20-35).

physiological response (see Figure 20-7). If we assume only 1000 receptors must bind insulin to induce a physiological response (e.g., activation of glucose transport), we can calculate the insulin concentration [H] needed to induce this response from Equation (20-2) rewritten in the following form:

$$[H] = \frac{K_D}{\dfrac{R_T}{[RH]} - 1} \qquad (20\text{-}3)$$

where $R_T = 10,000$ (the total number of insulin receptors), $K_D = 10^{-8}$ M, and $[RH] = 1000$ (the number of insulin-occupied receptors). In this example, the necessary insulin concentrations is 1.1×10^{-9} M. If R_T is reduced to 2000/cell, then a ninefold higher insulin concentration (10^{-8} M) is required to occupy 1000 receptors and induce the physiological response. If R_T is further reduced to 1200/cell, an insulin concentration of 5×10^{-8} M, a 50-fold increase, is necessary to generate a response.

Experiments with mutant cell lines demonstrate that internalization of receptor tyrosine kinases plays an important role in regulating cellular responses to EGF and other growth factors. For instance, a mutation in the EGF receptor that makes it resistant to ligand-induced endocytosis or, in dynamin, that blocks formation of clathrin-coated endocytic vesicles substantially increases the sensitivity of cells to EGF as a mitogenic signal. Such mutant cells are prone to EGF-induced cell transformation. Interestingly, the mutant-dynamin

inhibition of internalization also causes a qualitatively different pattern of phosphorylation of substrate proteins by the activated EGF receptor, as well as quantitative changes in the phosphorylation of known components in the EGF signaling pathways. Interestingly, internalized receptors can continue to signal from intracellular compartments prior to their degradation. This raises the intriguing possibility that receptor activity can be spatially controlled. Hence, internalization may modulate both the nature of RTK-transmitted signals, their magnitude, and location.

Studies with RTKs that bind PDGF suggest that PI-3 kinase plays an important role in the endocytosis and down-regulation of this class of receptors. Mutations that abolish the ability of the PDGF receptor to bind PI-3 kinase but not other enzymes (e.g., PLCγ) cause a reduction in the rate of receptor degradation. Although the mutant receptor is internalized, its sorting to the lysosome for degradation is blocked by an unknown mechanism. The observation that yeast cells expressing a mutant PI-3 kinase exhibit defective sorting of proteins to the vacuole (the yeast lysosome) raises the intriguing possibility that this enzyme plays an important role in membrane trafficking both in yeasts and mammalian cells.

Phosphorylation of Cell-Surface Receptors Modulates Their Activity

The ability of many cell-surface receptors to transmit signals is either increased or decreased by phosphorylation, leading to sensitization or desensitization of cells to various hormones. For instance, when cultured cells are exposed to epinephrine for several hours, several serine and threonine residues in the cytosolic domain of the β-adrenergic receptor become phosphorylated. The phosphorylated receptor can bind epinephrine, but ligand binding does not lead to activation of adenylyl cyclase or a rise in the cAMP level; thus the receptor is desensitized.

Four residues in the cytosolic domain of the β-adrenergic receptor are phosphorylated by a cAMP-dependent protein kinase (cAPK). The activity of this kinase is enhanced by the high cAMP level induced by epinephrine, explaining the receptor desensitization observed after prolonged exposure to epinephrine. Because the activity of all G_s protein–coupled receptors, not just the β-adrenergic receptor, is reduced by cAPK-catalyzed phosphorylation, this process is called **heterologous desensitization**. Other residues in the cytosolic domain of the β-adrenergic receptor are phosphorylated by a receptor-specific enzyme, called β-adrenergic receptor kinase (BARK), when the receptor is occupied by ligand. Because BARK phosphorylates only the β-adrenergic receptor, this process is called **homologous desensitization**. Prolonged treatment of cells with epinephrine or other agonists results in BARK-catalyzed phosphorylation of the β-adrenergic receptor and inhibition of its ability to activate G_s and adenylyl cyclase. In this case, a protein called β-arrestin has been shown to bind to the phosphorylated receptor and to sterically block interaction of the receptor with G_s, thus preventing activation of adenylyl cyclase following hormone binding. A similar protein binding to phosphorylated rhodopsin has been identified in the visual system (Section 21.6).

Figure 20-47 illustrates the feedback loop for modulating the activity of the β-adrenergic and related G_s protein–coupled receptors. This loop permits a cell to adjust receptor sensitivity to the constant hormone level at which it is being stimulated, so as to maintain a normal physiological response. Because the phosphorylated receptors are constantly being dephosphorylated and resensitized by constitutive

phosphatases, the number of phosphates per receptor molecule reflects how much ligand has been bound in the recent past (1–10 min). If the hormone level is increased, the resulting rise in the intracellular level of cAMP leads to phosphorylation and desensitization of more receptors, so that production of cAMP and hence the response remain relatively constant. If the hormone is removed, the receptor is completely dephosphorylated and "reset" to a high sensitivity, in which case it can respond to very low levels of hormone.

A similar feedback loop regulates the activity of the EGF receptor, which is an RTK. Phosphorylation of this receptor by protein kinase C decreases its affinity for EGF, thereby moderating the growth-stimulating effect of EGF. As noted earlier, protein kinase C is activated by DAG, which is generated by hormone stimulation of the inositol-lipid signaling pathway. We can summarize the feedback loop modulating the activity of the EGF receptor as follows: binding of EGF to its receptor → activation of PLC$_\gamma$ → generation of DAG → activation of protein kinase C by DAG → phosphorylation of EGF receptor by protein kinase C → down-regulation of receptor activity.

Arrestins Have Two Roles in Regulating G Protein—Coupled Receptors

The observation that internalization of β-adrenergic receptors in response to agonists is stimulated by overexpression of BARK and β-arrestin suggested an additional function for β-arrestin in regulating cell-surface receptors. β-Arrestin binds not only to the β-adrenergic receptor but also to clathrin, which, in turn, promotes the formation of coated pits and vesicles. Immunofluorescence microscopy has shown that β-arrestin and clathrin are co-localized at the membrane in a punctate pattern with β-adrenergic receptors within 10 minutes of treating cultured cells with agonist. Endocytosis of the BARK-phosphorylated, inactive receptors is promoted by the co-localized β-arrestin, which acts as an adapter protein for clathrin polymerization and formation

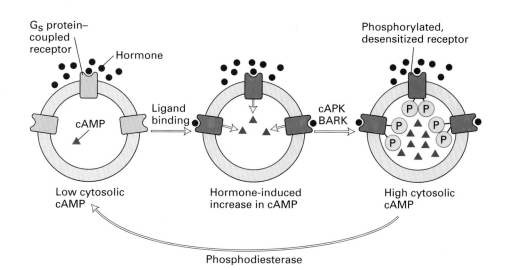

▶ **FIGURE 20-47 Schematic diagram of the regulatory feedback loop that controls the activity of G_s protein–coupled receptors by cyclical phosphorylation and dephosphorylation.** All receptors of this type are phosphorylated by cAMP-dependent protein kinase (cAPK). Additional residues are phosphorylated by receptor-specific kinases such as BARK, whose substrate is the β-adrenergic receptor.

of clathrin-coated vesicles (see Figure 17-54). Internalized receptors become dephosphorylated in endosomes, β-arrestin dissociates, and the resensitized receptors recycle to the cell surface, similar to recycling of the LDL receptor discussed in Chapter 17 (see Figure 17-46). Regulation of other GPCRs also is thought to involve dual-acting arrestins.

SUMMARY Interaction and Regulation of Signaling Pathways

- Many RTKs and GPCRs activate multiple signaling pathways, and different second messengers sometimes mediate the same cellular response.

- Some activated RTKs are coupled to the Ras-MAP kinase pathway or inositol-lipid pathway in a tissue-specific manner.

- Eukaryotes possess multiple G_α, G_β, and G_γ subunits. Different G_α subunits activate various effector proteins, leading to production of specific second messengers (see Table 20-5).

- The activity of some effector proteins, including certain adenylyl cyclase isoforms, is regulated by $G_{\beta\gamma}$.

- Glycogen breakdown and synthesis is regulated by multiple second messengers induced by neural or hormonal stimulation (see Figure 20-44).

- The insulin receptor, a dimeric RTK, can act through a Ras-dependent pathway, leading to activation of MAP kinase, or through a Ras-independent pathway involving phosphoinositides, leading to activation of protein kinase B (see Figure 20-45).

- Insulin stimulation of muscle cells and adipocytes leads to activation of protein kinase B, which promotes glucose uptake and glycogen synthesis, resulting in a decrease in blood glucose.

- Binding of glucagon to its GPCR promotes glycogenolysis and an increase in blood glucose via the cAMP-triggered kinase cascade.

- Ligand binding frequently induces phosphorylation of the cytosolic domain of a cell-surface receptor, thereby modulating its activity.

- At high ligand concentration, some cell-surface receptors are internalized by endocytosis, reducing the number of receptors on the surface and making cells less sensitive to ligand.

- Many internalized RTKs are degraded in lysosomes. In this case, resensitization depends on synthesis of additional receptor molecules.

- Internalization of phosphorylated (inactive) GPCRs leads to receptor dephosphorylation, β-arrestin dissociation, and recycling of active receptors to the cell surface.

20.8 From Plasma Membrane to Nucleus

As noted throughout this chapter, many cellular responses induced by hormones result from their effects on gene expression. The steroid hormones, which bind to intracellular receptors, provide the simplest example of the regulation of gene expression by extracellular hormones. The mechanism of action of steroid hormones was discussed in Section 10.7 and is not considered here. Many water-soluble hormones, growth factors, and neurotransmitters, which bind to cell-surface receptors, also induce long-term changes in cellular behavior that invariably involve changes in gene expression. The mechanisms by which binding of a ligand to a cell-surface receptor induce changes in gene expression have been thoroughly studied in several systems. In most cases, binding of a ligand to its receptor stimulates protein kinases that directly or indirectly phosphorylate serine, threonine, or tyrosine residues on specific transcription factors.

The most commonly used experimental approach in studying these signaling pathways is to identify a gene whose expression is induced (or repressed) by a hormone; then to define the cis-acting regulatory DNA sequences in that gene; and finally to identify and clone the cognate proteins that specifically bind to these DNA sequences (Chapter 10). In one cell surface → nucleus signaling pathway elucidated by this approach, binding of interferon γ to its cell-surface receptor induces membrane recruitment and activation of a cytosolic protein-tyrosine kinase called JAK (see Figure 10-68). The activated JAK then phosphorylates Stat1, a member of the Stat family of transcription factors. This promotes dimerization of Stat1 and its translocation to target genes in the nucleus. In a similar fashion, cytosolic Smad transcription factors are activated by receptor serine/threonine kinases that bind growth factors in the TGFβ superfamily (see Figure 23-3). In other cases, such as the cAMP and MAP kinase pathways, activated cytosolic kinases translocate to the nucleus where they directly modify transcription factors. We discuss these two mechanisms in this section. Activated cytosolic kinases have also been shown to regulate the stability and subcellular localization of transcription factors. One example of this mechanism is considered here.

CREB Links cAMP Signals to Transcription

In mammalian cells, an elevation in the cytosolic cAMP level stimulates the expression of many genes. For instance, increased cAMP induces production of somatostatin, a peptide that inhibits release of various hormones, in certain endocrine cells and of several enzymes involved in converting three-carbon compounds to glucose (gluconeogenesis) in liver cells. All genes regulated by cAMP contain a cis-acting DNA sequence, called *cAMP-response element* (CRE), which binds the phosphorylated form of a transcription factor called CRE-binding (CREB) protein. As discussed previously,

(a) G protein – cAMP pathway

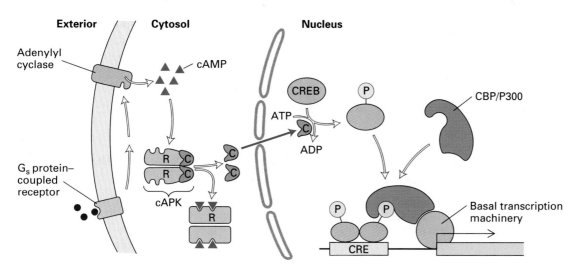

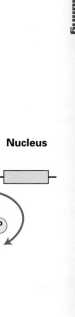

(b) RTK-Ras pathway

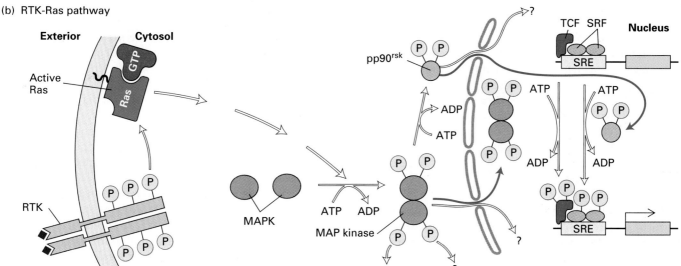

▲ **FIGURE 20-48 Signaling pathways leading to activation of transcription factors and modulation of gene expression following ligand binding to certain G$_s$ protein–linked receptors (a) and receptor tyrosine kinases (b).** In each case, receptor stimulation leads to activation of a protein kinase that translocates to the nucleus (red arrow) and there phosphorylates and activates transcription factors such as CREB, TCF, and SRF. As a result, transcription of genes controlled by DNA-regulatory elements (i.e., CRE and SRE) recognized by these transcription factors is stimulated. Phosphorylated CREB stimulates transcription through its association with the co-activator CBP/P300. cAPK = cAMP-dependent protein kinase. See text for details. [Part (a) see P. K. Brindle and M. R. Montminy, 1992, *Curr. Opin. Genet. Devel.* **2**:199; K. A. Lee and N. Masson, 1993, *Biochim. Biophys. Acta* **1174**:221; D. Parker et al., 1996, *Mol. Cell. Biol.* **16**(2):694. Part (b) see R. Marais et al., 1993, *Cell* **73**:381; V. M. Rivera et al., 1993, *Mol. Cell. Biol.* **13**:6260.]

binding of neurotransmitters and hormones to G$_s$ protein–coupled receptors activates adenylyl cyclase, leading to an increase in cAMP and subsequent activation of the catalytic subunit of cAMP-dependent protein kinase. The catalytic subunit then translocates to the nucleus where it phosphorylates serine-133 on CREB protein.

Phosphorylated CREB protein binds to CRE-containing target genes and also interacts with a *co-activator* termed CBP/300, which links CREB to the basal transcriptional machinery, thereby permitting CREB to stimulate transcription (Figure 20-48a). Earlier studies suggested that phosphorylation induced a conformational change in CREB protein, but more recent work indicates that CBP/P300 binds specifically to phosphoserine-133 in activated CREB. Interestingly, the intracellular receptors for steroid hormones, which function as transcription factors when bound to their ligands, also utilize CBP/P300. This finding suggests that CBP/P300 may play an important role in integrating signals from multiple signaling pathways and translating them into changes in transcription.

MAP Kinase Regulates the Activity of Many Transcription Factors

Addition of a growth factor (e.g., EGF or PDGF) to quiescent cultured mammalian cells in G_0 causes a rapid increase in the expression of as many as 100 different genes. These are called *early-response genes* because they are induced well before cells enter the S phase and replicate their DNA. One important early-response gene encodes the transcription factor c-Fos; together with other proteins, c-Fos induces expression of many genes encoding proteins necessary for cells to progress through the cell cycle (Section 13.6). Induction of the *c-fos* gene itself by various growth factors is mediated via several intracellular signaling pathways that lead to activation of different protein kinases: those involving cAMP (cAMP-dependent protein kinase), Ca^{2+}/DAG (protein kinase C), and Ras protein (MAP kinase).

The regulatory sequence of the *c-fos* gene contains a *serum-response element* (SRE), so named because it is activated by many growth factors in serum. Mutational analysis of the SRE in c-*fos* led to the surprising finding that different signaling pathways act through different sequences in this response element. For instance, cells carrying certain mutations in SRE cannot respond to extracellular signals that lead to activation of protein kinase C but can respond to signals that lead to activation of cAMP-dependent protein kinase or MAP kinase.

As discussed earlier in this chapter, hormonal stimulation of the RTK-Ras signaling pathway causes activation of MAP kinase. After dimerizing, activated MAP kinase translocates to the nucleus where it phosphorylates specific sites in the C-terminal domain of a protein called *ternary complex factor* (TCF). Phosphorylated TCF associates with two molecules of *serum response factor* (SRF), forming an active trimeric DNA-binding factor, which binds to SRE and activates transcription (Figure 20-48b). Mutational studies have shown that if TCF lacks the serine residues phosphorylated by MAP kinase, it does not activate gene expression in transfection experiments. Serine 103 in SRF also is rapidly phosphorylated in response to growth factors; this modification increases the rate and affinity of SRF binding to the SRE in vitro. A MAP kinase-activated kinase called pp90 has been shown to phosphorylate this serine in SRF in vitro.

Phosphorylation of transcription factors by MAP kinase can produce multiple effects on gene expression. For instance, Yan is a *Drosophila* transcription factor whose unphosphorylated form, present in the nucleus, inhibits development of R7 cells in the eye. Following signal-induced phosphorylation by MAP kinase, Yan accumulates in the cytosol and does not have access to the genes it controls, thereby relieving their repression. Mutant forms of Yan that cannot be phosphorylated by MAP kinase are constitutive repressors of R7 development. During development of the *Drosophila* eye, MAP kinase also activates a second transcription factor and indirectly promotes the degradation of yet a third, another transcriptional repressor. This example suggests that a complex interplay among multiple transcription factors, regulated by signal-activated kinases, is critical to cellular development.

Phosphorylation-Dependent Protein Degradation Regulates NF-κB

While signaling pathways frequently lead to phosphorylation of specific residues in transcription factors, thereby directly modulating their activity, studies in both mammalian cells and in Drosophila have led to the dissection of an evolutionarily conserved mechanism by which transcription factor activity is regulated by protein stability.

Stimulation of immune responses in a wide variety of mammalian cells leads to the activation of the NF-κB transcription factor (Figure 20-49). NF-κB was discovered in biochemical experiments on the basis of its requirement in B cells for transcription of the κ light chain gene of immunoglobulins (*n*uclear *f*actor κ chain transcription in *B* cells). Related proteins in Drosophila mediate the immune response and regulate expression of specific genes during early embryogenesis. Biochemical studies in mammalian cells and genetic studies in flies have provided important insights into how this transcription factor is activated by extracellular signals.

NF-κB is a heterodimer of two related proteins of 65 kD and 50 kD (p65 and p50). These proteins share a region of homology at their N-termini that is required for DNA binding and dimerization. A key insight into the mechanism regulating NF-κB came from subcellular localization studies. In resting cells, NF-κB is found in the cytoplasm. In response to an extracellular signal, NF-κB translocates to the nucleus, where it binds to specific sites in DNA and regulates transcription.

NF-κB is sequestered in an inactive state in the cytoplasm by direct binding to an inhibitor called I-κB. A single molecule of I-κB binds to the N-terminal domain of each subunit in the p50/p65 heterodimer, thereby masking the nuclear localization sequences. In response to signal, I-κB is phosphorylated at two N-terminal serine residues by an I-κB kinase complex. Phosphorylated I-κB is targeted for ubiquitination and degraded in the proteosome. A negative feedback loop regulates NF-κB activity as an early transcriptional target of NF-κB in the I-κB gene. As we learned in Chapter 13, phosphorylation-dependent degradation of the cyclin-kinase dependent inhibitor, Sic 1, plays a central role in regulation of G1-S progression in the yeast *S. cerevisiae*. It seems likely, then, that phosphorylation-dependent protein degradation may emerge as a common regulatory mechanism in many different cellular processes.

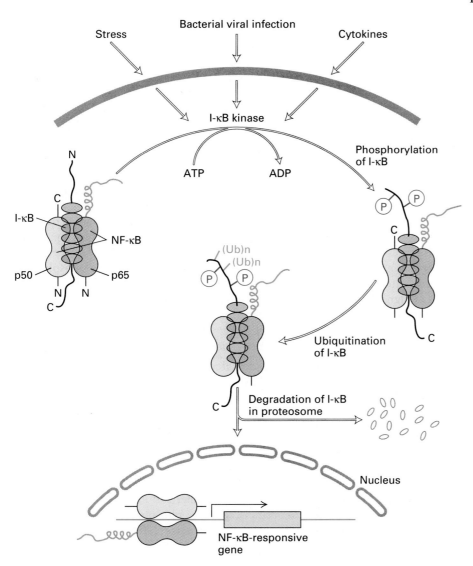

Stress

Bacterial viral infection

Cytokines

I-κB kinase

ATP   ADP

Phosphorylation
of I-κB

N

C

I-κB

NF-κB

p50   p65

N   N

C

(Ub)n
(Ub)n

P   P

C

P   P

C

Ubiquitination
of I-κB

C

Degradation of I-κB
in proteosome

Nucleus

NF-κB-responsive
gene

◀ **FIGURE 20-49 Activation of NF-κb.** Many different extracellular signals can induce activation of NF-κB. These signals activate an I-κB kinase complex. This complex phosphorylates two N-terminal serine residues in I-κB. Phosphorylated I-κB is ubiquitinated and subsequently degraded by the proteosome. Removal of I-κB unmasks the nuclear localization sites in both the p50 and p65 subunits of NF-κB. NF-κB enters the nucleus, binds to specific sequences in DNA and regulates transcription. [Adapted from S. Ghosh et al., 1998, *Annu. Rev. Immunol* **16**:225-260 and P. A. Baeuerle, 1998, *Cell* **95**:729-731.]

SUMMARY   From Plasma Membrane to Nucleus

• Protein phosphorylation plays a key role in regulating transcriptional activity in response to specific extracellular signals.

• The catalytic subunit of cAMP-dependent protein kinase translocates to the nucleus, where it phosphorylates CREB protein, which then interacts with the coactivator CBP/P300. The resulting trimeric complex binds to and activates transcription of target genes containing the CRE sequence.

• CBP/P300 also physically interacts with transcription factors whose activity is modulated by other signaling pathways.

• MAP kinase, activated via the RTK-Ras pathway, translocates to the nucleus, where it phosphorylates various transcriptional activators and repressors. MAP kinase phosphorylation promotes the activity of some transcription factors and inhibits the activity of others.

• NF-κB is localized to the cytoplasm in unstimulated cells bound to an inhibitor I-κB. In response to signal, I-κB is phosphorylated, ubiquitinated, and degraded in the proteosome. NF-κB translocates to the nucleus and regulates gene expression.

PERSPECTIVES   for the Future

The confluence of genetics, biochemistry, and structural biology has given us an increasingly detailed view of how signals are transmitted from the cell surface and transduced into changes in cellular behavior. The multitude of different extracellular signals, receptors for them, and intracellular signal-transduction pathways fall into a relatively small number of classes. Paradoxically, similar signaling pathways often regulate very different cellular processes. Indeed, the same signal and receptor in different cells can promote responses as diverse as proliferation, differentiation, and death.

Conversely, activation of the same signal-transduction component in the same cell through different receptors often elicits different cellular responses. How this specificity is determined remains an outstanding question in signal transduction. Exciting insights into this issue have come from recent studies on the multiple MAP kinase pathways in yeast. We can anticipate that similar genetic and molecular studies in flies, worms, and mice will lead to an understanding of the interplay between different pathway components and the underlying regulatory principles controlling specificity in multicellular organisms.

Researchers have determined the three-dimensional structures of various signaling proteins during the past several years. These recent advances permit more detailed analysis of signal-transduction pathways. The molecular structures of different kinases, for instance, exhibit striking similarities and important variations that impart to them novel regulatory features. By analyzing the structure of complexes of signaling proteins, such as Ras and Sos, the precise molecular interface between them can be uncovered.

Abnormalities in signal transduction underlie many different diseases, including the majority of cancers and many inflammatory conditions. Detailed knowledge of the signaling pathways involved and the structure of their constituent proteins will provide important molecular clues for the design of specific therapies. Despite the close structural relationship among different signaling molecules (e.g., kinases), recent studies suggest that inhibitors selective for specific subclasses can be designed. For instance, unregulated JAK-2 kinase activity promotes the formation of pre-B cell tumors in acute lymphoblastic leukemia. Proliferation of these cells can be blocked by a highly specific inhibitor of JAK-2 kinase. It may be possible to use such inhibitors to control abnormal activities of enzymes, like JAK-2, that are required in limited numbers of cell types: that is, inhibition of their normal activity may be tolerated. Conversely, the inhibition of proteins with more widespread function may be toxic, but the use of cell-type targeting systems to deliver inhibitors to affected cells only may circumvent this problem. For instance, in a recent study, a protein tyrosine kinase inhibitor was conjugated to an antibody that recognized a specific antigen expressed on the surface of tumor cells, so the inhibitor was selectively delivered to these cells. Detailed structural studies provide an exciting approach to the design of specific drugs that target inactivation of the catalytic function of specific signaling molecules (e.g., kinases), block normal subcellular localization (e.g., membrane localization of Ras by farnesyl anchors), or prevent interactions between signaling components (e.g., those between Ras and Sos).

PERSPECTIVES in the Literature

X-ray crystallographic studies have provided important insights into well-studied signaling mechanisms and clues to further biochemical and genetic studies of less well-characterized systems. A recent provocative example comes from a poorly understood family of receptors, the receptor tyrosine phosphatases. These receptors are single-pass transmembrane proteins with a large extracellular domain and two tandemly arranged tyrosine phosphatase catalytic domains in the C-terminal cytosolic tail. The relationship between the extracellular domain and its catalytic domains remain poorly understood, largely due to the lack of known ligands and assays of biologically relevant target molecules. The best-characterized receptor tyrosine phosphatase is CD45, which plays an important role in signaling in the immune system. Although the CD45 ligand is not known, biochemical and genetic studies indicate that the CD45 receptor activates a nonreceptor cytosolic tyrosine kinase by removing an inhibitory phosphate. A series of biochemical and structural studies on CD45 and other receptor tyrosine phosphatases have suggested a mechanism by which phosphatase activity is inhibited by ligand-induced dimerization. The articles referenced below describe the important experiments that have led to this model. How would you critically test whether ligand-induced dimerization is an important mechanism in vivo?

Bilwes A. M., J. den Hertog, T. Hunter, and J P. Noel. 1996. Structural basis for inhibition of receptor protein-tyrosine phosphatase-α by dimerization. *Nature* 382: 555–559.

Kokel M., et al. 1998. *clr-1* encodes a receptor tyrosine phosphatase that negatively regulates an FGF receptor signaling pathway in *Caenorhabditis elegans*. *Genes & Develop.* 12:1425–1437.

Majeti R., et al. 1998. Dimerization-induced inhibition of receptor protein tyrosine phosphatase function through an inhibitory wedge. *Science* 279:88–91.

Testing Yourself on the Concepts

1. Describe the cycle of events that occur after binding a ligand to a G protein–coupled receptor and activation of adenylyl cyclase.

2. Describe the cycle of events that occur after binding a ligand to receptor tyrosine kinases and activation of Ras.

3. Describe the cycle of events that lead to the stimulation of protein kinase C by phospholipase C activation.

MCAT/GRE-Style Questions

Key Application Please read the section titled "Critical Features of Catecholamines and Their Receptors Have Been Identified" (p. 863) and refer to Figure 20-14; then answer the following questions:

1. Which of the following statements is *not* true about agonists:

   a. An agonist mimics the action of a natural hormone.

b. An agonist binds with the same K_D as the natural hormone to the receptor.

c. An agonist competes with an antagonist for binding to the receptor.

d. An agonist competes with the natural hormone for binding to the receptor.

2. Which of the following statements is true about β-adrenergic receptors?

a. β_1-adrenergic receptors are present on smooth muscle cells lining the bronchial passages.

b. β_1- and β_2-adrenergic receptors show almost 100% amino acid identity.

c. β_1- and β_2-adrenergic receptors bind catecholamines with differing affinities.

d. β_2-adrenergic receptors are present on cardiac muscle cells.

3. Which of the following statements is *not* true about the action of β-adrenergic receptors?

a. Beta blockers are agonists of β_2-adrenergic receptors.

b. Antagonists of β_1-adrenergic receptors are used as a treatment for angina.

c. Stimulation of β_2-adrenergic receptors is used as a treatment for asthma.

d. Stimulation of β_1-adrenergic receptors increases heart rate and contractility.

4. An agonist of the β_1-adrenergic receptor would be expected to

a. Open bronchioles in the small airways in the lung.

b. Close bronchioles in the small airways in the lung.

c. Increase heart rate and contractility.

d. Decrease heart rate and contractility.

5. In Figure 20-14, the ligand-binding domain and G-protein specificity of the α- and β-adrenergic receptors were mapped using chimeric receptors. Based on this information, a chimeric receptor that contained helices 1–6 from the α-adrenergic receptor and helix 7 from the β-adrenergic receptor would be expected to

a. Bind α agonists and inhibit adenylyl cyclase.

b. Bind β agonists and inhibit adenylyl cyclase.

c. Bind α agonists and activate adenylyl cyclase.

d. Bind β agonists and activate adenylyl cyclase.

Key Concept Please read the section titled "Signals Pass from Activated Ras to a Cascade of Protein Kinases" (p. 878) and refer to Figure 20-28; then answer the following questions:

6. Which of the following events initiates the kinase cascade:

a. Binding of a growth factor to a receptor tyrosine kinase.

b. Internalization of a ligand-receptor complex.

c. Opening of a calcium channel.

d. Initiation of DNA synthesis.

7. Which of the following events in the kinase cascade would *not* occur in the presence of an MEK mutant lacking threonine and serine residues:

a. Dissociation of GDP from Ras.

b. Binding of GTP to Ras.

c. Activation of Raf.

d. Phosphorylation of MAP kinase.

8. Which of the following scenarios would still result in a proliferation signal via the kinase cascade in the absence of ligand binding to its receptor tyrosine kinase:

a. A constitutively active Ras and a defective MEK.

b. A constitutively active Raf and a defective MEK.

c. A constitutively active MEK and a defective Ras.

d. A constitutively active Ras and a defective Raf.

9. Regulation of the kinase cascade also requires the activity of an enzyme that

a. Removes phosphates.

b. Cleaves proteins into peptides.

c. Adds sugar groups to proteins.

d. Makes disulfide bonds.

Key Experiment Please read the section titled "cAPKs Activated by Epinephrine Stimulation Regulate Glycogen Metabolism" (p. 885) and refer to Figures 20-34, 20-35, and 20-36; then answer the following questions:

10. Which of the following statements about glycogen metabolism is *not* true:

a. Glycogen is the major storage form for glucose in animals.

b. The substrate for the synthesis of glycogen is UDP-glucose.

c. Glucose units are removed from glycogen by the action of glycogen phosphorylase.

d. Glycogen in the muscle is primarily broken down to free glucose.

11. Which of the following is *not* a step in the regulation of glycogen metabolism:

a. Epinephrine stimulates an increase in cAMP.

b. A cAMP-dependent kinase phosphorylates glycogen phosphorylase.

c. A cAMP-dependent kinase phosphorylates glycogen phosphorylase kinase.

d. A rise in cAMP activates a cAMP-dependent protein kinase.

12. Which of the following statements is true about the role of phosphorylation in regulating glycogen metabolism:

a. Phosphorylation of glycogen synthase activates the enzyme.

b. Phosphorylation of glycogen phosphorylase kinase inactivates the enzyme.

c. Phosphorylation of glycogen phosphorylase activates the enzyme.

d. Phosphorylation of an inhibitor protein leads to activation of phosphoprotein phosphatase.

13. Which of the following events occurs when cAMP levels drop after removal of epinephrine stimulation:

a. Glycogen synthase is inactivated.

b. Glycogen phosphorylase is activated.

c. Glycogen phosphorylase kinase is activated.

d. Phosphoprotein phosphatase is activated.

Key Terms

adenylyl cyclase *862*

agonist *863*

autocrine signaling *850*

beta blockers *864*

calmodulin *891*

cAMP *850*

cAMP-dependent protein kinases (cAPKs) *884*

endocrine signaling *850*

G protein–coupled receptors *852*

glycogenolysis cascade *886*

GTPase switch proteins *872*

heterologous desensitization *901*

homologous desensitization *901*

inositol-lipid pathways *888*

ligand *850*

MAP kinase *878*

paracrine signaling *850*

phosphoinositides *888*

prostaglandins *852*

protein kinase B *898*

protein kinase C *893*

receptor tyrosine kinases *854*

ryanodine receptors *891*

second messengers *850*

SH2 domain *898*

stimulatory G protein *865*

store-operated channels *890*

References

Overview of Extracellular Signaling

Baulieu, E. E., and P. A. Kelly. 1990. *Hormones: From Molecules to Disease.* Chapman and Hall (London).

Barritt, G. J. 1992. *Communication within Animal Cells.* Oxford Science Publications (Oxford, Eng.).

Wallis, M., S. L. Howell, and K. W. Taylor, eds. 1986. *The Biochemistry of Polypeptide Hormones.* Wiley.

Wilson, J. D., D. W. Foster, H. M. Kronenberg, and R. H. Williams. 1998. *Williams Textbook of Endocrinology,* 9th ed. W. B. Saunders.

Identification and Purification of Cell-Surface Receptors

Allen, J. M., and B. Seed. 1989. Isolation and expression of functional high-affinity Fc receptor complementary DNAs. *Science* **243**:378–381.

Flanagan, J. G., and P. Leder. 1990. The kit ligand: a cell-surface molecule altered in steel mutant fibroblasts. *Cell* **63**:185–194.

Simonsen, H., and H. F. Lodish. 1994. Cloning by function: expression cloning in mammalian cells. *Trends Pharmacol. Sci.* **15**:437–441.

Stadel, J. M., S. Wilson, and D. J. Bergsma. 1997. Orphan G protein–coupled receptors: a neglected opportunity for pioneer drug discovery. *Trends Pharmacol. Sci.* **18**:430–437.

G Protein—Coupled Receptors and Their Effectors

Birnbaumer, L. 1992. Receptor-to-effector signaling through G proteins: roles for βγ dimers as well as α subunits. *Cell* **71**:1069–1072.

Bourne, H. R. 1997. How receptors talk to trimeric G proteins. *Curr. Opin. Cell Biol.* **9**:134–142.

Gudermann, T., F. Kalkbrenner, and G. Schultz. 1996. Diversity and selectivity of receptor–G protein interaction. *Annu. Rev. Pharmacol. Toxicol.* **36**:429–459.

Hamm, H. E., and A. Gilchrist. 1996. Heterotrimeric G proteins. *Curr. Opin. Cell Biol.* **8**:189–196.

Lambright, D. G., et al. 1996. The 2.0 Å crystal structure of a heterotrimeric G protein. *Nature* **379**:311–319.

Sprang, S. R. 1997. G protein mechanisms: insights from structural analysis. *Annu. Rev. Biochem.* **66**:639–678.

Strader, C. D., et al. 1994. Structure and function of G protein–coupled receptors. *Annu. Rev. Biochem.* **63**:101–132.

Tesmer, J. J., R. K. Sunahara, A. G. Gilman, and S. R. Sprang. 1997. Crystal structure of the catalytic domains of adenylyl cyclase in a complex with $G_s\alpha.GTP\gamma S$. *Science* **278**:1907–1916.

Receptor Tyrosine Kinases and Ras

Boriack-Sjodin, P. A., S. M. Margarit, D. Bar-Sagi, and J. Kuriyan. 1998. The structural basis of the activation of Ras by Sos. *Nature* **394**:337–343.

Fantl, W. J., D. E. Johnson, and L. T. Williams. 1993. Signaling by receptor tyrosine kinases. *Annu. Rev. Biochem.* **62**:453–481.

Han, M., and P. W. Sternberg. 1990. *let-60,* a gene that specifies cell fates during C. elegans vulval induction. *Cell* **63**:921–931.

Hubbard, S. R., L. Wei, L. Ellis, and W. A. Hendrickson. 1994. Crystal structure of the tyrosine kinase domain of the human insulin receptor. *Nature* **372**:746–754.

Lowenstein, E. J., et al. 1992. The SH2 and SH3 domain-containing protein GRB2 links receptor tyrosine kinases to ras signaling. *Cell* **70**:431–442.

Mohammadi, M., J. Schlessinger, and S. R. Hubbard. 1996. Structure of the FGF receptor tyrosine kinase domain reveals a novel autoinhibitory mechanism. *Cell* **86**:577–587.

Rogge, R. D., C. A. Karlovich, and U. Banerjee. 1991. Genetic dissection of a neurodevelopmental pathway. *Son of sevenless* functions downstream of the *sevenless* and EGF receptor kinases. *Cell* **64**:39–48.

Simon, M. A., D. D. Bowtell, G. S. Dodson, T. R. Laverty, and G. M. Rubin. 1991. Ras 1 and a putative guanine nucleotide exchange factor perform crucial steps in signaling by the sevenless protein tyrosine kinase. *Cell* **67**:701–716.

Skolnik, E. Y., et al. 1991. Cloning of PI3 kinase-associated p85 utilizing a novel method for expression/cloning of target proteins for receptor tyrosine kinases. *Cell* **65**:83–90.

Waksman, G., S. E. Shoelson, N. Pant, D. Cowburn, and J. Kuriyan. 1993. Binding of high affinity phosphoryl peptide to the Src SH2 domain: Crystal structures of the complexed and peptide-free forms. *Cell* 72:779–790.

Yu, H., et al. 1994. Structural basis for the binding of proline-rich peptides to SH3 domains. *Cell* 76:933–945.

Zipursky, S. L., and G. M. Rubin. 1994. Determination of neuronal cell fate: lessons from the R7 neuron of *Drosophila*. *Annu. Rev. Neurosci.* 17:373–397.

MAP Kinase Pathways

Canagarajah, B. J., A. Khokhlatchev, M. H. Cobb, and E. J. Goldsmith. 1997. Activation mechanism of the MAP kinase ERK2 by dual phosphorylation. *Cell* 90:859–869.

Choi, K. Y., B. Satterberg, D. M. Lyons, and E. A. Elion. 1994. Ste5 tethers multiple protein kinases in the MAP kinase cascade required for mating in *S. cerevisiae*. *Cell* 78:499–512.

Karim, F. D., H. C. Chang, M. Therrien, D. A. Wassarman, T. Laverty, and G. M. Rubin. 1996. A screen for genes that function downstream of Ras1 during Drosophila eye development. *Genetics* 143:315–329.

Khokhlatchev, A. V., et al. 1998. Phosphorylation of the MAP kinase ERK2 promotes its homodimerization and nuclear translocation. *Cell* 93:605–615.

Madhani, H. D., C. A. Styles, and G. R. Fink. 1997. MAP kinases with distinct inhibitory functions impart signaling specificity during yeast differentiation. *Cell* 91:673–684.

Pawson, T., and J. D. Scott. 1997. Signaling through scaffold, anchoring, and adapter proteins. *Science* 278:2075–2080.

Posas, F., and H. Saito. 1997. Osmotic activation of the HOG MAPK pathway via Ste11p MAPKKK: scaffold role of Pbs2p MAPKK. *Science* 276:1702–1705.

Whitmarsh, A. J., and R. J. Davis. 1998. Structural organization of MAP-kinase signaling modules by scaffold proteins in yeast and mammals. *Trends Biochem. Sci.* 23:481–485.

Second Messengers

Berridge, M. J. 1997. Elementary and global aspects of calcium signaling. *J. Exp. Biol.* 200:315–319.

Hobbs, A. J. 1997. Soluble guanylate cyclase: the forgotten sibling. *Trends Pharmacol. Sci.* 18:484–491.

Hoth, M., and R. Penner. 1992. Depletion of intracellular calcium stores activates a calcium current in mast cells. *Nature* 355:353–356.

Jaken, S. 1996. Protein kinase C isozymes and substrates. *Curr. Opin. Cell Biol.* 8:168–173.

Rebecchi, M. J., and S. Scarlata. 1998. Pleckstrin homology domains: a common fold with diverse functions. *Annu. Rev. Biophys. Biomol. Struc.* 27:503–528.

Singer, W. D., H. A. Brown, and P. C. Sternweis. 1997. Regulation of eukaryotic phosphatidylinositol-specific phospholipase C and phospholipase D. *Annu. Rev. Biochem.* 66:475–509.

Toker, A., and L. C. Cantley. 1997. Signaling through the lipid products of phosphoinositide-3-OH kinase. *Nature* 387:673–676.

Wedel, B. J., and D. L. Garbers. 1997. New insights on the functions of the guanylyl cyclase receptors. *FEBS Letters* 410:29–33.

Interaction and Regulation of Signaling Pathway

Cadena, D. L., C.-L. Chan, and G. N. Gill. 1994. The intracellular tyrosine kinase domain of the epidermal growth factor receptor undergoes a conformational change upon autophosphorylation. *J. Biol. Chem.* 269:260–265.

Clandinin, T. R., J. A. DeModena, and P. W. Sternberg. 1998. Inositol trisphosphate mediates a RAS-independent response to LET-23 receptor tyrosine kinase activation in *C. elegans*. *Cell* 92:523–533.

Corvera, S., and M. P. Czech. 1998. Direct targets of phosphoinositide 3-kinase products in membrane traffic and signal transduction. *Trends Cell Biol.* 8:442–446.

Downward, J. 1998. Mechanisms and consequences of activation of protein kinase B/Akt. *Curr. Opin. Cell Biol.* 10:262–267.

Ferguson, S. S., and M. G. Caron. 1998. G protein–coupled receptor adaptation mechanisms. *Semin. Cell Develop. Biol.* 9:119–127.

Ferguson, S. S., et al. 1996. Role of β-arrestin in mediating agonist-promoted G protein–coupled receptor internalization. *Science* 271:363–366.

Krupnick, J. G., and J. L. Benovic. 1998. The role of receptor kinases and arrestins in G protein–coupled receptor regulation. *Annu. Rev. Pharmacol. Toxicol.* 38:289–319.

Lamaze, C., and S. L. Schmid. 1995. Recruitment of epidermal growth factor receptors into coated pits requires their activated tyrosine kinase. *J. Cell Bio.* 129:47–54.

Vieira, A. V., C. Lamaze, and S. L. Schmid. 1996. Control of EGF receptor signalling by clathrin-mediated endocytosis. *Science* 274:2086–2089.

From Plasma Membrane to Nucleus

Chen, X., et al. 1998. Crystal structure of a tyrosine phosphorylated STAT-1 dimer bound to DNA. *Cell* 93:827–839.

Ghosh, S., M. J. May, and E. B. Kopp. 1998. NF-κB and Rel proteins: Evolutionarily conserved mediators of immune responses. *Annu. Rev. Immunol.* 16:225–260.

Goldman, P. S., V. K. Tran, and R. H. Goodman. 1997. The multifunctional role of the co-activator CBP in transcriptional regulation. *Recent Prog. Hormone Res.* 52:103–119.

Parker, D., et al. 1996. Phosphorylation of CREB at Ser-133 induces complex formation with CREB-binding protein via a direct mechanism. *Mol. Cell. Biol.* 16:694–703.

Schindler, C., and J. E. Darnell, Jr. 1995. Transcriptional responses to polypeptide ligands: the JAK-STAT pathway. *Annu. Rev. Biochem.* 64:621–651.

Whitmarsh, A. J., P. Shore, A. D. Sharrocks, and R. J. Davis. 1995. Integration of MAP kinase signal transduction pathways at the serum response element. *Science* 269:403–407.

Nerve Cells

21

The nervous system regulates all aspects of bodily function and is staggering in its complexity. The human brain—the control center that stores, computes, integrates, and transmits information—contains about 10^{12} **neurons** (nerve cells), each forming as many as a thousand connections with other neurons. Millions of specialized neurons sense features of both the external and internal environments and transmit this information to the brain for processing and storage. Millions of other neurons regulate the contraction of muscles and the secretion of hormones. The nervous system also contains **glial (neuroglial) cells** that occupy the spaces between neurons and modulate their functions (see chapter opening figure).

The structure and function of individual nerve cells is understood in great detail, perhaps in more detail than for any other type of cell. The function of a neuron is to communicate information, which it does by two methods. *Electric signals* process and conduct information within a cell, while *chemical signals* transmit information between cells, utilizing processes similar to those employed by other types of cells to signal each other (Chapter 20). *Sensory neurons* have specialized receptors that convert diverse types of stimuli from the environment (e.g., light, touch, sound, odorants) into electric signals. These electric signals are then converted into chemical signals that are passed on to other cells called *interneurons*, which convert the information back into electric signals. Ultimately the information is transmitted to muscle-stimulating *motor neurons* or to other neurons that stimulate other types of cells, such as glands.

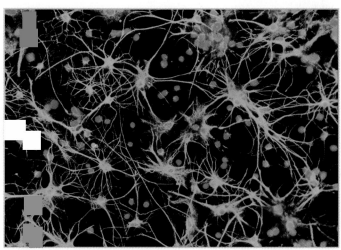

Rat cortical astrocytes stained for DNA (blue) and GFAP (green), an intermediate-filament protein. These glial cells, found in the brain, are marked by numeorus cytoplasmic processes.

The output of a nervous system is the result of its circuit properties, that is, the wiring, or interconnections, between neurons, and the strength of these interconnections. Complex aspects of the nervous system, such as vision and consciousness, cannot be understood at the single-cell level, but only at the level of networks of nerve cells that can be studied by techniques of systems analysis. The nervous system is constantly changing; alterations in the number and nature of the interconnections between individual neurons occur, for example, in the development of new memories.

In this chapter we focus on how individual neurons function and how small groups of cells function together. A great deal of information has been gleaned from simple nervous systems. Squids and sea slugs have large neurons that are relatively easy to identify and manipulate experimentally. Moreover, in these species, only a few identifiable neurons may be involved in a specific task; thus their function can be studied in some detail. Analyses of humans, mice, nematodes, and flies with mutations that affect specific functions of the nervous system have provided important insights, as have molecular cloning of key neuron proteins, such as ion channels and receptors. Genetic and molecular studies on the development of the nervous system, detailed in Chapter 23, have elucidated how neurons form and maintain specific connections with other neurons and other types of cells. Because the principles studied are basic, all of these findings are applicable to complex nervous systems, including that of humans.

21.1 Overview of Neuron Structure and Function

In this introductory section, we describe the structural features that are unique to neurons and the types of electric signals that they use to process and transmit information. We then introduce **synapses,** the specialized sites where neurons send and receive information from other cells, and some of the circuits that allow groups of neurons to coordinate complex processes. Each of these topics will be covered in more detail in later sections of the chapter.

Specialized Regions of Neurons Carry Out Different Functions

Although the morphology of various types of neurons differs in some respects, they all contain four distinct regions with differing functions: the cell body, the dendrites, the axon, and the axon terminals (Figure 21-1).

The *cell body* contains the nucleus and is the site of synthesis of virtually all neuronal proteins and membranes. Some proteins are synthesized in dendrites, but no proteins are made in axons and axon terminals, which do not contain ribosomes. Proteins and membranes that are required

for renewal of the axon and nerve termini are synthesized in the cell body and assembled there into membranous vesicles or multiprotein particles. By a process called *anterograde transport,* these materials are transported along microtubules down the length of the axon to the terminals, where they are inserted into the plasma membrane or other organelles (Chapter 19). Axonal microtubules also are the tracks along which damaged membranes and organelles move up the axon toward the cell body; this process is called *retrograde transport.* Lysosomes, where such material is degraded, are found only in the cell body.

Almost every neuron has a single **axon,** whose diameter varies from a micrometer in certain nerves of the human brain to a millimeter in the giant fiber of the squid. Axons are specialized for the conduction of a particular type of electric impulse, called an **action potential,** *outward,* away from the cell body toward the axon terminus. An action potential is a series of sudden changes in the voltage, or equivalently the electric potential, across the plasma membrane (Figure 21-2a). When a neuron is in the resting (nonstimulated) state, the electric potential across the axonal membrane is approximately -60 mV (the inside negative relative to the outside); the magnitude of this *resting potential* is similar to that of the **membrane potential** in most non-neuronal cells (Chapter 15). At the peak of an action potential, the membrane potential can be as much as $+50$ mV (inside positive), a net change of ≈ 110 mV. This **depolarization** of the membrane is followed by a rapid repolarization, returning the membrane potential to the resting value. These characteristics distinguish an action potential from other types of changes in electric potential across the plasma membrane and allow an action potential to move along an axon without diminution.

Action potentials move rapidly, at speeds up to 100 meters per second. In humans, axons may be more than a meter long, yet it takes only a few milliseconds for an action potential to move along their length. An action potential originates at the *axon hillock,* the junction of the axon and cell body, and is actively conducted down the axon into the axon terminals, small branches of the axon that form the synapses, or connections, with other cells (Figure 21-2b). A single axon in the central nervous system can synapse with many neurons and induce responses in all of them simultaneously.

Most neurons have multiple **dendrites,** which extend outward from the cell body and are specialized to receive chemical signals from the axon termini of other neurons. Dendrites convert these signals into small electric impulses and transmit them *inward,* in the direction of the cell body. Neuronal cell bodies can also form synapses and thus receive signals (Figure 21-3). Particularly in the central nervous system, neurons have extremely long dendrites with complex branches. This allows them to form synapses with and receive signals from a large number of other neurons, perhaps up to a thousand. Electric disturbances generated in the dendrites or cell body spread to the axon hillock. If the electric disturbance there is great enough, an action potential will originate and will be actively conducted down the axon.

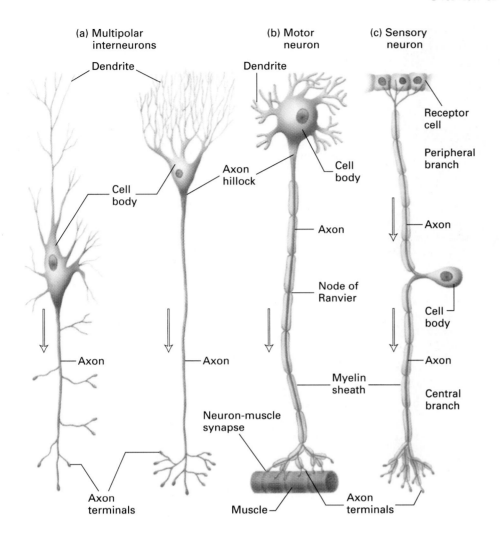

(a) Multipolar interneurons

Dendrite

Cell body

Axon

Axon terminals

(b) Motor neuron

Dendrite

Axon hillock

Cell body

Axon

Node of Ranvier

Axon

Myelin sheath

Neuron-muscle synapse

Muscle

Axon terminals

(c) Sensory neuron

Receptor cell

Peripheral branch

Axon

Cell body

Axon

Central branch

Axon terminals

◀ **FIGURE 21-1 Structure of typical mammalian neurons.** Arrows indicate the direction of conduction of action potentials in axons (red). (a) Multipolar interneurons. Each has profusely branched dendrites, which receive signals at synapses with several hundred other neurons, and a single long axon that branches laterally and at its terminus. (b) A motor neuron that innervates a muscle cell. Typically, motor neurons have a single long axon extending from the cell body to the effector cell. In mammalian motor neurons an insulating sheath of myelin usually covers all parts of the axon except at the nodes of Ranvier and the axon terminals. (c) A sensory neuron in which the axon branches just after it leaves the cell body. The peripheral branch carries the nerve impulse from the receptor cell to the cell body, which is located in the dorsal root ganglion near the spinal cord; the central branch carries the impulse from the cell body to the spinal cord or brain. Both branches are structurally and functionally axons, except at their terminal portions, even though the peripheral branch conducts impulses toward, rather than away from, the cell body.

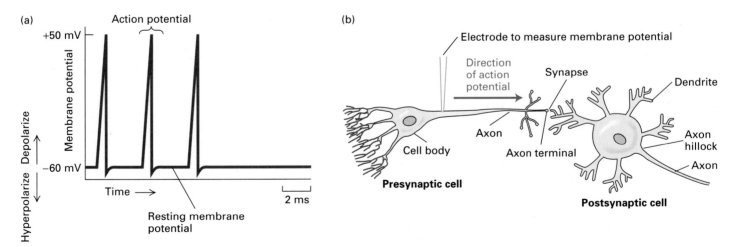

(a)

+50 mV

Action potential

Membrane potential

Depolarize / Hyperpolarize

−60 mV

Time →

2 ms

Resting membrane potential

(b)

Electrode to measure membrane potential

Direction of action potential

Synapse

Dendrite

Axon

Cell body

Axon terminal

Presynaptic cell

Axon hillock

Axon

Postsynaptic cell

▲ **FIGURE 21-2 (a) An action potential is a sudden, transient depolarization of the membrane followed by repolarization to the resting potential of about −60 mV.** This recording of the axonal membrane potential in a presynaptic neuron shows that it is generating one action potential about every 4 milliseconds. (b) The membrane potential across the plasma membrane of a presynaptic neuron is measured by a small electrode inserted into it. Action potentials move down the axon at speeds up to 100 meters per second. Their arrival at a synapse causes release of neurotransmitters that bind to receptors in the postsynaptic cell, generally depolarizing the membrane (making the potential less negative) and tending to induce an action potential in it.

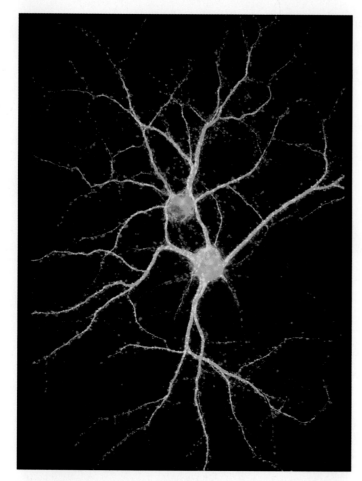

▲ **FIGURE 21-3 Typical interneurons from the hippocampal region of the brain makes about a thousand synapses.** The cells were stained with two fluorescent antibodies: one specific for the microtubule-associated protein MAP2 (green), which is found only in dendrites and cell bodies, and the other specific for synaptotagmin (orange-red) (see Figure 21-31), a protein found in presynaptic axon terminals. Thus the orange-red dots indicate presynaptic axon terminals from neurons that are not visible in this field. [Courtesy of O. Mundigl and P. deCamilli.]

Synapses Are Specialized Sites Where Neurons Communicate with Other Cells

Synapses generally transmit signals in only one direction: an axon terminal from the *presynaptic cell* sends signals that are picked up by the *postsynaptic cell* (see Figure 21-2b). There are two general types of synapse: the relatively rare *electric synapse*, discussed later, and, the *chemical synapse*, illustrated in Figure 21-4. In this type of synapse, the axon terminal of the presynaptic cell contains vesicles filled with a particular **neurotransmitter**. The postsynaptic cell can be a dendrite or cell body of another neuron, a muscle or gland cell, or, rarely, even another axon. When an action potential in the presynaptic cell reaches an axon terminal, it induces a localized rise in the level of Ca^{2+} in the cytosol. This, in turn, causes some of the vesicles to fuse with the plasma membrane, re-

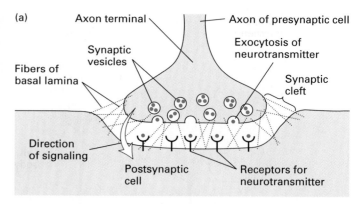

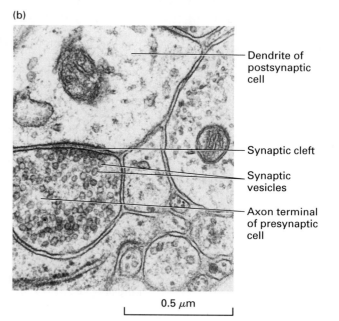

0.5 μm

▲ **FIGURE 21-4 A chemical synapse.** (a) A narrow region—the synaptic cleft—separates the plasma membranes of the presynaptic and postsynaptic cells. Transmission of electric impulses requires release of a neurotransmitter (red circles) by the presynaptic cell, its diffusion across the synaptic cleft, and its binding by specific receptors on the plasma membrane of the postsynaptic cell. (b) Electron micrograph showing a cross section of a dendrite synapsing with an axon terminal filled with synaptic vesicles. In the synaptic region, the plasma membrane of the presynaptic cell is specialized for vesicle exocytosis; synaptic vesicles, which contain a neurotransmitter, are clustered in these regions. The opposing membrane of the postsynaptic cell (in this case, a neuron) contains receptors for the neurotransmitter. [Part (b) from C. Raine et al., eds., 1981, *Basic Neurochemistry*, 3d ed., Little, Brown, p. 32.]

leasing their contents into the *synaptic cleft*, the narrow space between the cells. The neurotransmitters diffuse across the synaptic cleft; it takes about 0.5 millisecond (ms) for them to bind to receptors on postsynaptic cells.

Binding of the neurotransmitter triggers changes in the ion permeability of the postsynaptic plasma membrane, which, in turn, changes the membrane's electric potential at this point. If the postsynaptic cell is a neuron, this electric disturbance may be sufficient to induce an action potential. If the postsynaptic cell is a muscle, the change in membrane potential following binding of the neurotransmitter may induce contraction; if a gland cell, the neurotransmitter may induce hormone secretion. In some cases, enzymes attached to the fibrous network connecting the cells destroy the neurotransmitter after it has functioned; in other cases, the signal is terminated when the neurotransmitter diffuses away or is transported back into the presynaptic cell.

The postsynaptic neuron at certain synapses also sends signals to the presynaptic one. Such *retrograde* signals can be gases, such as nitric oxide and carbon monoxide, or peptide hormones. This type of signaling, which modifies the ability of the presynaptic cell to signal the postsynaptic one, is thought to be important in many types of learning.

Neurons Are Organized into Circuits

In complex multicellular animals, such as insects and mammals, various types of neurons form signaling circuits. In the simple type of circuit called a **reflex arc**, interneurons connect multiple sensory and motor neurons, allowing one sensory neuron to affect multiple motor neurons and one motor neuron to be affected by multiple sensory neurons; in this way interneurons integrate and enhance reflexes. For example, the knee-jerk reflex in humans involves a complex reflex arc in which one muscle is stimulated to contract while another is inhibited from contracting (Figure 21-5). Such circuits allow an organism to respond to a sensory input by the coordinated action of sets of muscles that together achieve a single purpose. However, such simple nerve systems do not directly explain higher-order brain functions such as reasoning and computation.

The sensory and motor neurons of circuits such as the knee-jerk reflex are contained within the peripheral nervous system (Figure 21-6). These circuits send information to and receive information from the **central nervous system (CNS)**, which comprises the brain and spinal cord and is composed mainly of interneurons. Highly specialized *sensory receptor cells,* which respond to specific environmental stimuli, send their outputs either directly to the brain (e.g., taste and odorant receptors) or to peripheral sensory neurons (e.g., pain and stretching receptors). The peripheral nervous system contains two broad classes of motor neurons. The *somatic motor neurons* stimulate voluntary muscles, such as those in the arms, legs, and neck; the cell bodies of these neurons are located inside the central nervous system, in either the brain or the spinal cord. The *autonomic motor neurons* innervate glands,

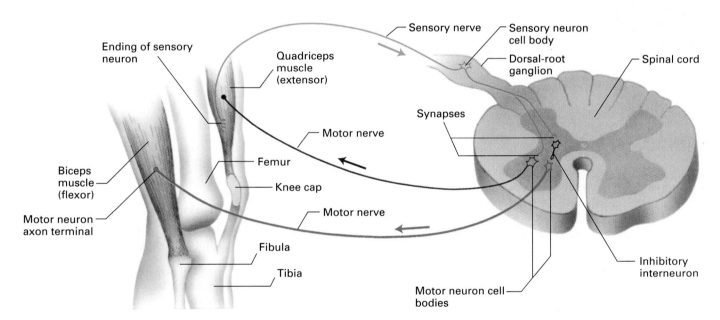

▲ **FIGURE 21-5 The knee-jerk reflex arc in the human.**
Positioning and movement of the knee joint are accomplished by two muscles that have opposite actions: Contraction of the quadriceps muscle straightens the leg, whereas contraction of the biceps muscle bends the leg. The knee-jerk response, a sudden extension of the leg, is stimulated by a blow just below the knee cap. The blow directly stimulates sensory neurons (blue) located in the tendon of the quadriceps muscle. The axon of each sensory neuron extends from the tendon to its cell body in a dorsal root ganglion. The sensory axon then continues to the

spinal cord, where it branches and synapses with two neurons: (1) a motor neuron (red) that innervates the quadriceps muscle and (2) an inhibitory interneuron (black) that synapses with a motor neuron (green) innervating the biceps muscle. Stimulation of the sensory neuron causes a contraction of the quadriceps and, via the inhibitory neuron, a simultaneous inhibition of contraction of the biceps muscle. The net result is an extension of the leg at the knee joint. Each cell illustrated here actually represents a nerve, that is, a population of neurons.

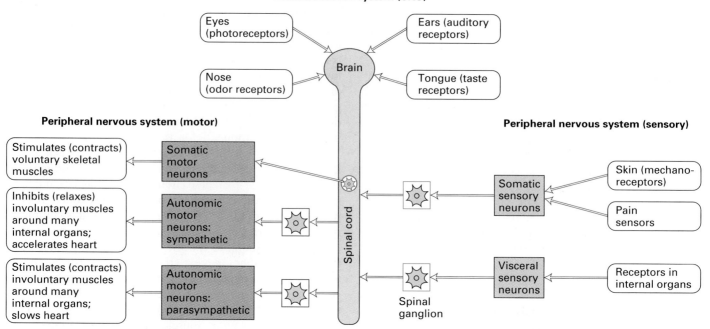

▲ FIGURE 21-6 A highly schematic diagram of the vertebrate nervous system. The central nervous system (CNS) comprises the brain and spinal cord. It receives direct sensory input from the eyes, nose, tongue, and ears. The peripheral nervous system (PNS) comprises three sets of neurons: (1) somatic and visceral sensory neurons, which relay information to the CNS from receptors in somatic and internal organs; (2) somatic motor neurons, which innervate voluntary skeletal muscles; and (3) autonomic motor neurons, which innervate the heart, the smooth involuntary muscles such as those surrounding the stomach and intestine, and glands such as the liver and pancreas. The sympathetic and parasympathetic autonomic motor neurons frequently cause opposite effects on internal organs. The cell bodies of somatic motor neurons are within the CNS; those of somatic sensory neurons and of autonomic motor neurons are in ganglia adjacent to the CNS (see Figure 21-5).

heart muscle, and smooth muscles not under conscious control, such as the muscles that surround the intestine and other organs of the gastrointestinal tract. The two classes of autonomic motor neurons, *sympathetic* and *parasympathetic*, generally have opposite effects: one class stimulates a muscle or gland, and the other inhibits it. Somatic sensory neurons, which convey information to the central nervous system, have their cell bodies clustered in **ganglia**, masses of nerve tissue that lie just outside the spinal cord. The cell bodies of the motor neurons of the autonomic nervous system also lie in ganglia. Each peripheral nerve is actually a bundle of axons; some are parts of motor neurons; others are parts of sensory neurons.

Having surveyed the general features of neuron structure, interactions, and simple circuits, let us turn to the mechanism by which a neuron generates and conducts electric impulses.

SUMMARY Overview of Neuron Structure and Function

• The cell body of a neuron contains the nucleus and lysosomes and is the site of synthesis and degradation of virtually all neuronal proteins and membranes.

• Axons are long processes specialized for the conduction of action potentials away from the neuronal cell body.

• Action potentials are sudden membrane depolarizations followed by a rapid repolarization. They originate at the axon hillock and move toward axon terminals, where the electric impulse is transmitted to other cells via an electric or chemical synapse (see Figure 21-2).

• Most neurons have multiple dendrites, which receive chemical signals from the axon termini of other neurons.

• When an action potential reaches a chemical synapse, a neurotransmitter is released into the synaptic cleft. Binding of the neurotransmitter to receptors on the postsynaptic cell changes the ion permeability and thus the electric potential of the postsynaptic plasma membrane (see Figure 21-4).

• Neurons are organized into circuits. In a reflex arc, such as the knee-jerk reflex, interneurons connect multiple sensory and motor neurons, allowing one sensory neuron to affect multiple motor neurons. One muscle can be stimulated to contract while another is inhibited from contracting (see Figure 21-5).

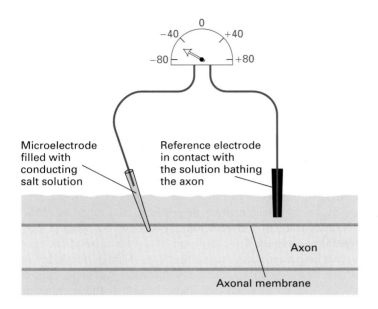

Microelectrode filled with conducting salt solution

Reference electrode in contact with the solution bathing the axon

Axon

Axonal membrane

◀ **FIGURE 21-7 Measurement of the electric potential across an axonal membrane.** A microelectrode, constructed by filling a glass tube of extremely small diameter with a conducting fluid such as KCl, is inserted into an axon in such a way that the surface membrane seals itself around the electrode. A reference electrode is placed in the bathing medium. A potentiometer connecting the two electrodes registers the potential. The potential difference maintained across the cell membrane in the absence of stimulation is called the resting potential, in this case, -60 mV. A potential difference is registered only when the microelectrode is inserted into the axon; no potential is registered if the microelectrode is in the bathing fluid. Recording of the changes in the membrane potential over time gives a trace such as that shown in Figure 21-2a.

21.2 The Action Potential and Conduction of Electric Impulses

We saw in Chapter 15 that a *voltage gradient*, also called an electric potential, exists across the plasma membrane of all cells. The potential across the plasma membrane of large cells can be measured with a microelectrode inserted inside the cell and a reference electrode placed in the extracellular fluid. The two are connected to a voltmeter capable of measuring small potential differences (Figure 21-7). In virtually all cases the inside of the cell membrane is negative relative to the outside; typical membrane potentials are between -30

and -70 mV. The potential across the surface membrane of most animal cells generally does not vary with time. In contrast, neurons and muscle cells—the principal types of electrically active cells—undergo controlled changes in their membrane potential (see Figure 21-2a).

The characteristic electrical activity of neurons—their ability to conduct, transmit, and receive electric signals—results from the opening and closing of specific *ion-channel proteins* in the neuron plasma membrane (Figure 21-8). Each open channel allows only a small number of ions to move from one side of the membrane to the other, yet these ion movements cause significant changes in the membrane potential. Here we explain the relationship between opening and closing of ion channels and the resultant changes in the voltage across the membrane that lead to propagation of action potentials. We examine the structure and operation of several types of ion channels critical to neuron functioning in more detail later in the chapter.

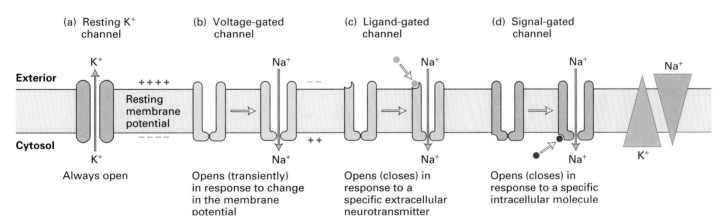

(a) Resting K$^+$ channel

(b) Voltage-gated channel

(c) Ligand-gated channel

(d) Signal-gated channel

Exterior

K$^+$ ++++   Na$^+$ --   Na$^+$ Na$^+$ Na$^+$

Resting membrane potential

Cytosol

K$^+$ ----   Na$^+$ ++   Na$^+$ Na$^+$ K$^+$

Always open

Opens (transiently) in response to change in the membrane potential

Opens (closes) in response to a specific extracellular neurotransmitter

Opens (closes) in response to a specific intracellular molecule

▲ **FIGURE 21-8 Ion channels in neuronal plasma membranes.** Each type of channel protein has a specific function in the electrical activity of neurons. (a) Resting K$^+$ channels are responsible for generating the resting potential across the membrane. (b) Voltage-gated channels are responsible for propagating action potentials along the axonal membrane. (c, d) Two types of ion channels in dendrites and cell bodies are responsible for generating electric signals in postsynaptic cells. One type (c) has a site for binding a specific extracellular neurotransmitter (blue circle). The other type (d) is coupled to a neurotransmitter receptor via a G protein; it responds to intracellular signals (red circle) induced by binding of neurotransmitter to a separate receptor protein (not shown). Signals activating different channels include Ca^{2+}, cyclic GMP, and the G$_{\beta\gamma}$ subunits of trimeric G proteins (Chapter 20).

The Resting Potential, Generated Mainly by Open "Resting" K⁺ Channels, Is Near E_K

The concentration of K^+ ions inside typical metazoan cells is about 10 times that in the extracellular fluid, whereas the concentrations of Na^+ and Cl^- ions are much higher outside the cell than inside; these concentration gradients are maintained by Na^+/K^+ ATPases with the expenditure of cellular energy (see Figure 15-13). As noted in Chapter 15, the plasma membrane contains abundant open "resting" K^+ channels that allow passage only of K^+. The resting potential—inside negative—is determined mainly by the movement of K^+ ions: Movement of a K^+ ion across the membrane down its concentration gradient leaves an excess negative charge on the cytosolic face and deposits a positive one on the exoplasmic face (Figure 21-9). Quantitatively, the usual resting potential of -60 mV is close to, but in magnitude less than, the value of E_K, the potassium equilibrium potential, calculated from the Nernst equation (see Equation 15-7) and the typical external and cytosolic K^+ concentrations ($[K_o]$ and $[K_i]$, respectively) given in Figure 21-9. If the concentration of K^+ surrounding a resting cell is changed, the measured membrane potential assumes a new value, again close to the calculated value of E_K; this is evidence that the resting potential is due mainly to movement of K^+ through open K^+ channels in the plasma membrane.

The situation in cells is complicated because there are some open Na^+ and Cl^- channels in the plasma membranes of the resting cell. Cells, of course, contain other ions, such as HPO_4^{2-}, SO_4^{2-}, and Mg^{2+}, but there are few channels that admit these ions. Furthermore, the membrane potential of electrically active cells such as neurons and muscle cells is affected mainly by opening and closing channels for K^+, Na^+, and Cl^-; thus these three ions are the only ones we need consider here. As we discuss later, Ca^{2+} channels are central to the release of neurotransmitters at synapses.

To calculate the membrane potential as a function of the concentrations of different ions, it is useful to define a **permeability constant** P for each ion. P is a measure of the ease with which an ion can cross a unit area (1 cm²) of membrane driven by a 1 M difference in concentration; it is proportional to the number of open ion channels and to the number of ions each channel can conduct per second (the channel *conductivity*). Thus P_K, P_{Na}, and P_{Cl} are measures of the "leakiness" of a unit area of membrane to these ions. Permeabilities are generally not measured directly; rather, the permeability of a membrane for a given ion is the product of the number of open ion channels and the conductivity of each channel; both parameters can be measured by techniques we describe later. Since the conductivities of the various ion channels are nearly the same, differences in permeability of a membrane for Na^+, K^+, and Cl^- largely reflect differences in the number of open channels specific for each ion.

What is important is not the absolute magnitude of the permeabilities for each ion, but the ratios of the permeabilities of Na^+ and Cl^- to that of K^+. The electric potential, E (in millivolts), across a cell-surface membrane is given by a more complex version of the Nernst equation in which the concentrations of the ions are weighted in proportion to the relative magnitudes of their permeability constants:

$$E = 59 \log_{10} \frac{[K_o] + [Na_o]\frac{P_{Na}}{P_K} + [Cl_i]\frac{P_{Cl}}{P_K}}{[K_i] + [Na_i]\frac{P_{Na}}{P_K} + [Cl_o]\frac{P_{Cl}}{P_K}} \quad (21\text{-}1)$$

where the "o" and "i" subscripts denote the ion concentrations outside and inside the cell. Because of their opposite charges (Z value in the Nernst equation), $[K_o]$ and $[Na_o]$

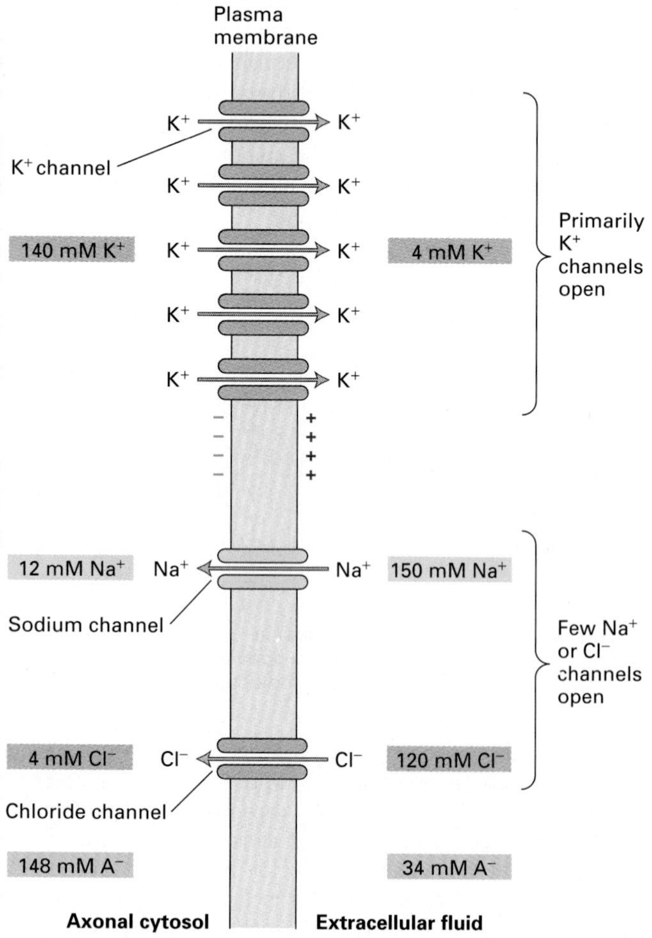

◀ **FIGURE 21-9 Origin of the resting potential in a typical vertebrate neuron.** The ionic compositions of the cytosol and of the surrounding extracellular fluid are different. A^- represents negatively charged proteins, which neutralize the excess positive charges contributed by Na^+ and K^+ ions. In the resting neuron there are about ten times more open K^+ channels than open Na^+ or Cl^- channels; as a consequence more positively charged K^+ ions exit the cell than Na^+ or Cl^- ions enter, and the outside of the plasma membrane acquires a net positive charge relative to the inside.

are placed in the numerator, but [Cl$_o$] is placed in the denominator; conversely, [K$_i$] and [Na$_i$] are in the denominator, but [Cl$_i$] is in the numerator. The membrane potential at any time and at any position in the neuron can be calculated with this equation if the relevant ion concentrations and permeabilities are known.

Note that if $P_{Na} = P_{Cl} = 0$, then the membrane is permeable only to K$^+$ ions and Equation 21-1 reduces to the Nernst equation for K$^+$ (see Equation 15-7). Similarly, if $P_K = P_{Cl} = 0$, then the membrane is permeable only to Na$^+$ ions and Equation 21-1 reduces to the Nernst equation for Na$^+$ (see Equation 15-6).

In resting neurons the ion concentrations are typically those shown in Figure 21-9, and the permeability of the membrane to Na$^+$ or Cl$^-$ ions is about one tenth that for K$^+$ (i.e., $P_{Na}/P_K = P_{Cl}/P_K = 0.1$). That is, there are about ten times more open K$^+$ channels than open channels for Na$^+$ or Cl$^-$. By substituting the typical ion concentrations and these permeability ratios into Equation 21-1, we can calculate the membrane potential as -52.9 mV, which is much closer to E_K (-91.1 mV) than to E_{Na} ($+64.7$ mV). Although E_{Cl} (-87.2 mV) is close to E_K, there are so few open Cl$^-$ channels that they contribute little to the resting potential. The resting potential is not equal to E_K because the membrane also contains some open Na$^+$ channels; influx of Na$^+$ ions down its concentration gradient adds positive charges to the inside of the cell membrane, making the membrane potential more positive (or less negative).

Opening and Closing of Ion Channels Cause Predictable Changes in the Membrane Potential

It is clear from Equation 21-1 that changes in the permeability of the membrane to various ions will cause the membrane potential to change. Figure 21-10 illustrates several quantitative changes. Here we summarize the direction of the predicted changes due to opening and closing of various channels:

1. *Opening of* Na$^+$ *channels (increasing P$_{Na}$) causes depolarization of the membrane;* the membrane potential becomes less negative, and if the increase in P_{Na} is large enough, the potential can become positive inside, approaching E_{Na}. Intuitively, Na$^+$ ions tend to flow inward from the extracellular medium, down their concentration gradient, leaving excess negative ions on the outer surface of the membrane and putting more positive ions on the cytosolic surface. Conversely, closing of Na$^+$ channels, decreasing P_{Na}, causes membrane hyperpolarization, a more negative potential.

2. *Opening of* K$^+$ *channels (increasing P$_K$) causes hyperpolarization of the membrane;* the membrane potential becomes more negative, approaching E_K. Intuitively, this occurs because more K$^+$ ions flow outward from the cytosol, down their concentration gradient, leaving excess negative ions on the cytosolic surface of the membrane and putting more positive ones on the outer surface. Conversely, closing of K$^+$ channels, decreasing P_K, causes depolarization of the membrane and a less negative potential.

3. *Opening of "nonspecific" cation channels that admit* Na$^+$ *and* K$^+$ *equally also causes membrane depolarization.* Such channels allow K$^+$ ions to flow outward from the cytosol and Na$^+$ ions to flow inward; the net effect is to drive the membrane potential toward zero.

4. *Opening of* Cl$^-$ *channels (increasing P$_{Cl}$) causes hyperpolarization of the membrane,* and the potential approaches E_{Cl}. Intuitively, Cl$^-$ ions tend to flow inward from the extracellular medium, down their concentration gradient, leaving excess positive ions on the outer surface of the membrane and putting more negative ions on the cytosolic surface. In

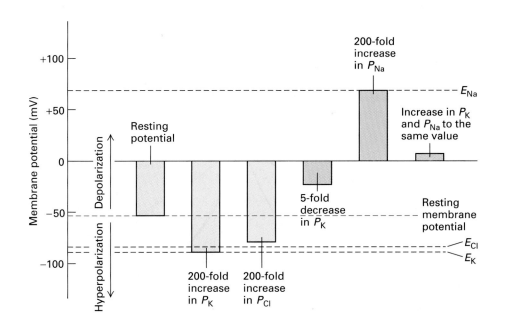

◀ **FIGURE 21-10 Effect of changes in ion permeability on membrane potential calculated with equation 21-1 using the permeability constants given in the text and the ion concentrations shown in Figure 21-9.** The resting membrane potential is -53 mV; E_{Na}, E_K, and E_{Cl} are the potentials calculated from the Nernst equation if the membrane contains only open channels for Na$^+$ or K$^+$ or Cl$-$, respectively.

muscle cells, resting Cl⁻ channels, not resting K⁺ channels, are the principal determinants of the inside negative resting potential. Conversely, closing of Cl⁻ channels, decreasing P_{Cl}, causes depolarization and a less negative potential.

At the resting potential, **voltage-gated ion channels** are closed; no ions move through them. However, when a region of the plasma membrane is depolarized slightly, voltage-gated Na⁺ channels open for a short period, allowing the influx of Na⁺ ions that causes the sudden and transient depolarization associated with an action potential. Following the opening (and closing) of voltage-gated Na⁺ channels during an action potential, the transient opening of voltage-gated K⁺ channels causes the membrane potential to return to the resting state and even become more negative (hyperpolarized) for a short time (see Figure 21-2a). The ability of axons to conduct action potentials over long distances without diminution thus depends on controlled opening and closing of voltage-gated Na⁺ and K⁺ channels (see Figure 21-8b). Binding of neurotransmitters to **ligand-gated ion channels** in postsynaptic cells triggers changes in the postsynaptic membrane potential during impulse transmission at synapses (see Figure 21-8c).

Membrane Depolarizations Spread Passively Only Short Distances

To understand how voltage-gated Na⁺ and K⁺ channels allow an action potential to be conducted down an axon in one direction, we first need to examine how a plasma membrane with only resting K⁺ channels would conduct an electric depolarization. In its electric properties, a nerve cell with only resting K⁺ channels resembles a long underwater telephone cable. It consists of an electrical insulator, the poorly conducting cell membrane, separating two media—the cell cytosol and the extracellular fluid—that have a high conductivity for ions.

Suppose that a single microelectrode is inserted into the axon and that the electrode is connected to a source of electric current (e.g., a battery) such that the electric potential at that point is suddenly depolarized and maintained at this new voltage. At this site the inside of the membrane will have a relative excess of positive charges, principally K⁺ ions. These ions will tend to move away from the initial depolarization site, thus depolarizing adjacent sections of the membrane. This is called the *passive spread of depolarization*. In contrast to an action potential, passive spread occurs equally in both directions. Also, the magnitude of the depolarization diminishes with distance from the site of initial depolarization, as some of the excess cations leak back across the membrane through resting cation channels (Figure 21-11). Only a small portion of the excess cations are carried longitudinally along the axon for long distances. The extent of this passive spread of depolarization is a function of two properties of the nerve cells: the permeability of the membrane to ions and the conductivity of the cytosol.

The passive spread of a depolarization is greater for large-diameter neurons than for small-diameter neurons, because the conductivity of the cytosol of a nerve cell depends on its cross-sectional area. The larger the area, the greater the number of ions there will be (per unit length of neuron) to conduct current. Thus K⁺ ions are able to move, on the

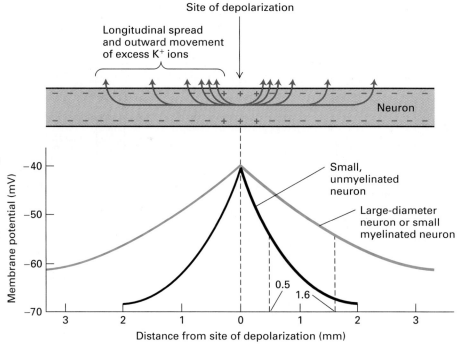

▶ **FIGURE 21-11 Passive spread of a depolarization of a neuronal plasma membrane with only resting K⁺ ion channels.** The neuronal membrane is depolarized from −70 to −40 mV at a single point and clamped at this value. The voltage is then measured at various distances from this site. Because of the outward movement of K⁺ ions through resting K⁺ channels, the extent of depolarization falls off with distance from the initial depolarization. Passive spread occurs equally in both directions from the site of depolarization. The *length* constant is the distance over which the magnitude of the depolarization falls to a value of 1/e (e = 2.718) of the initial depolarization. The length constant for a small neuron with a large number of resting K⁺ channels (black curve) can be as small as 0.1 mm; in this example it is about 0.5 mm. For a large axon, or a small one surrounded by a myelin sheath (blue curve), the length constant can be as large as 5 mm; in this example it is about 1.6 mm.

average, farther along a large axon than a small one before they "leak" back across the membrane. As a consequence, large-diameter neurons passively conduct a depolarization faster and farther than thin ones. Nonetheless, a membrane depolarization can spread passively for only a short distance, from 0.1 to about 5 mm. Depolarizations in dendrites and the cell body generally spread in this manner, though some dendrites conduct an action potential. Neurons with very short axons also conduct axonal depolarizations by passive spread. However, passive spread does not allow propagation of electric signals over long distances.

As we discuss below, some axons are surrounded by a **myelin sheath,** which impedes the leakage of excess cations associated with membrane depolarization. Thus small myelinated neurons and large unmyelinated ones have similar length constants for passive spread of membrane depolarizations.

Voltage-Gated Cation Channels Generate Action Potentials

The action potential is a cycle of membrane depolarization, hyperpolarization, and return to the resting value (Figure 21-12a). The cycle lasts 1–2 ms, and can occur hundreds of times a second. These cyclical changes in the membrane potential result from transient increases in the permeability of a region of the membrane, first to Na^+ ions, then to K^+ ions (Figure 21-12b). More specifically, these electric changes are due to voltage-gated Na^+ and K^+ channels that open and shut in response to changes in the membrane potential. The role of these channels in the generation and conduction of action potentials was elucidated in classic studies done on the giant axon of the squid, in which multiple microelectrodes can be inserted without causing damage to the integrity of the plasma membrane. However, the same basic mechanism is used by all neurons.

Voltage-Gated Na^+ Channels The sudden but short-lived depolarization of a region of the plasma membrane during an action potential is caused by a sudden massive, but transient, influx of Na^+ ions through opened voltage-gated Na^+ channels in that region. At the resting membrane potential these voltage-gated channels are closed. The depolarization of the membrane changes the conformation of the channel proteins, opening the Na^+-specific channels and allowing Na^+ influx through them.

During conduction of an action potential, the passive spread of depolarization to the adjacent distal region of membrane slightly depolarizes the new region, causing opening of a few voltage-gated Na^+ channels and an increase in Na^+ influx. A combination of two forces acting in the same direction drives Na^+ ions into the cell: the concentration gradient of Na^+ ions, and the resting membrane potential—inside negative—which tends to attract Na^+ ions into the cell. As more Na^+ ions enter the cell, the inside of the cell membrane becomes more positive and thus the membrane

(a) Depolarization (↑) and hyperpolarization (↓)

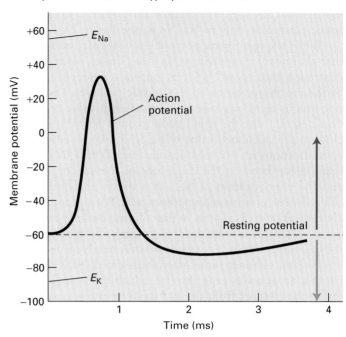

(b) Changes in ion permeabilities

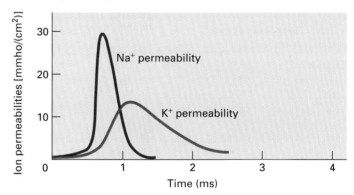

▲ **FIGURE 21-12 Kinetics of changes in membrane potential and ion permeabilities during an action potential in the giant axon of a squid.** (a) Following stimulation at time 0, the membrane potential rapidly becomes more positive, approaching the value of E_{Na}, and then becomes more negative. (b) A transient increase in Na^+ permeability, resulting from the transient opening of voltage-gated Na^+ channels, permits the Na^+ influx that causes the membrane to become depolarized. This precedes opening of voltage-gated K^+ channels and the resultant efflux of K^+ ions, which causes the membrane to become hyperpolarized for a brief period. [See A. L. Hodgkin and A. F. Huxley, 1952, *J. Physiol.* **117:**500.]

becomes depolarized further. This depolarization causes the opening of more voltage-gated Na^+ channels, setting into motion an explosive entry of Na^+ ions that is completed within a fraction of a millisecond. For a fraction of a millisecond, at the peak of the action potential, the permeability of this region of the membrane to Na^+ becomes vastly greater than that for K^+ or Cl^-, and the membrane potential approaches E_{Na}, the equilibrium potential for a membrane permeable only to Na^+ ions (see Figure 21-12).

When the membrane potential almost reaches E_{Na}, further net inward movement of Na^+ ions ceases, since the concentration gradient of Na^+ ions (outside>inside) is balanced by the membrane potential E_{Na} (inside positive). The action potential is at its peak. The measured peak value of the action potential for the squid giant axon is 35 mV, which is close to the calculated value of E_{Na} (55 mV) based on Na^+ concentrations of 440 mM outside and 50 mM inside. The relationship between the magnitude of the action potential and the concentration of Na^+ ions inside and outside the cell has been confirmed experimentally. For instance, if the concentration of Na^+ ions in the solution bathing the squid axon is reduced to one-third of normal, the magnitude of the depolarization is reduced by 40 mV, nearly as predicted.

Like all channel proteins, voltage-gated Na^+ channels contain an aqueous pore through which the ions flow. Entering the channel from the outside, one encounters first a wide vestibule, then a narrow pore that selects the type of ion allowed to pass. The pore leads into a large inner vestibule and, at the cytosolic end, a segment of the protein—the "gate"—that closes the pore in the resting state. As depicted in Figure 21-13, in the resting state voltage-gated Na^+ channels are closed but capable of being opened if the membrane is depolarized. The greater the depolarization, the greater the chance that any one channel will open. Opening is triggered by movement of *voltage-sensing α helices* in response to the membrane depolarization, causing a small conformational change in the gate that opens the channel and allows ion flow. Once opened, the channels stay open about 1 ms, during which time about 6000 Na^+ ions pass through them. Further Na^+ influx is prevented by movement of the *channel-inactivating segment* into the channel opening. As long as the membrane remains depolarized, the channel is inactivated and cannot be reopened. As we discuss later, this **refractory period** of the Na^+ channel is important in determining the unidirectionality of the action potential. A few milliseconds after the inside-negative resting potential is

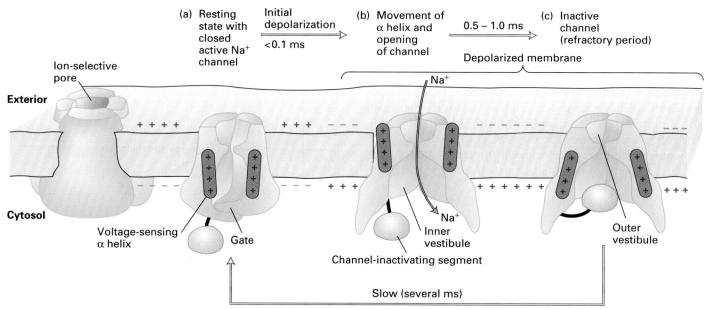

▲ FIGURE 21-13 Structure and function of the voltage-gated Na^+ channel. *(Left)* Like all voltage-gated channels, it contains four transmembrane domains, each of which contributes to the central pore through which ions move. The critical components that control movement of Na^+ ions are shown in the cutaway views. (a) In the closed, resting state, the gate obstructs the channel, inhibiting Na^+ movement, and the channel-inactivating segment is free in the cytosol. The channel protein contains four voltage-sensing α helices (maroon), which have positively charged side chains every third residue. The attraction of these charges for the negative interior of resting cells helps keep the channel closed. (b) When the membrane becomes depolarized (outside negative), the voltage-sensing helices move toward the outer plasma membrane surface, causing an immediate conformational change in the gate segment that opens the channel for influx of Na^+ ions. (c) Within a millisecond after opening, the voltage-sensing helices return to the resting position and the channel-inactivating segment (purple) moves into the open channel, preventing further ion movements. When the membrane potential is reversed so that the inside is again negative, the gate moves back into the blocking position (not shown). After 1–2 ms the channel-inactivating segment is displaced from the channel opening and the protein reverts to the closed, resting state (a) where it can be opened again by depolarization. [Adapted from C. Miller, 1991, *Science* **252**:1092, and C. Armstrong and B. Hille, 1998, *Neuron* **20**:371.]

reestablished, the channels return to the closed resting state, once again "primed" for being opened by depolarization.

Voltage-Gated K⁺ Channels During the time that the voltage-gated Na^+ channels are closing and fewer Na^+ ions are entering the cell, voltage-gated K^+ channel proteins open. This causes the observed increase in potassium ion permeability and an increased efflux of K^+ from the cytosol that repolarizes the plasma membrane to its resting potential. Actually, for a brief instant the membrane becomes hyperpolarized, with the potential approaching E_K, which is more negative than the resting potential (see Figure 21-12).

Opening of the voltage-gated K^+ channels is induced by the membrane depolarization of the action potential. Unlike the voltage-gated Na^+ channels, most types of voltage-gated K^+ channels remain open as long as the membrane is depolarized, and close only when the membrane potential has returned to an inside-negative value. Because the voltage-gated K^+ channels open a fraction of a millisecond or so after the initial depolarization, they are called *delayed K^+ channels*. Eventually all voltage-gated K^+ and Na^+ channels close. The only open channels are the non-voltage-gated K^+ channels that generate the inside-negative potential characteristic of the resting state; as a result, the membrane potential returns to its resting value.

Action Potentials Are Propagated Unidirectionally without Diminution

At the peak of an action potential, passive spread of the membrane depolarization is sufficient to depolarize a "downstream" segment of membrane. This causes a few Na^+ channels in this region to open, thereby increasing the extent of depolarization in this region, causing an explosive opening of more Na^+ channels. Thus, propagation of the action potential without diminution is ensured.

Because voltage-gated Na^+ channels remain inactive for several milliseconds after opening, those Na^+ channels immediately "behind" the action potential cannot reopen even though the potential in this segment is depolarized due to passive spread (Figure 21-14). The inability of Na^+ channels to reopen during the refractory period ensures that action potentials are propagated unidirectionally from the cell body to the axon terminus, and limits the number of action potentials per second that a neuron can conduct. Reopening of Na^+ channels "behind" the action potential is also prevented by the membrane hyperpolarization that results from opening of voltage-gated K^+ channels.

Movements of Only a Few Na⁺ and K⁺ Ions Generate the Action Potential

The changes in membrane potential characteristic of an action potential are caused by rearrangements in the balances of ions on either side of the membrane, *not* by changes in the concentrations of ions in the solutions on either side. The voltage changes are generated by the movements of Na^+ and

K^+ ions across the plasma membrane through voltage-gated channels, but the actual number of ions that move is very small relative to the total number in the neuronal cytosol. In fact, measurements of the amount of radioactive sodium entering and leaving single squid axons and other axons during a single action potential show that, depending on the size of the neuron, only about one K^+ ion per 3000–300,000 in the cytosol (0.0003–0.03 percent) is exchanged for extracellular Na^+ to generate the reversals of membrane polarity.

As discussed previously, the resting membrane potential in nerve cells results primarily from the gradient of K^+ ions that is generated and maintained by the Na^+/K^+ ATPase. This ATPase plays no direct role in impulse conduction. If dinitrophenol or another inhibitor of ATP production is added to cells, the membrane potential gradually falls to zero as all the ions equilibrate across the membrane. In large nerve cells such as in the squid this equilibration is extremely slow, requiring hours, but in smaller mammalian nerves this equilibration occurs in only 5 minutes. In either case, the membrane potential is essentially independent of the supply of ATP over the short time spans required for nerve cells to generate and conduct action potentials. Nerve cells normally can fire thousands of times in the absence of an energy supply because the ion movements during each discharge involve only a minute fraction of the cell's K^+ and Na^+ ions.

Myelination Increases the Velocity of Impulse Conduction

In man, the cell body of a motor neuron that innervates a leg muscle is in the spinal cord and the axon is about a meter in length (see Figure 21-5). Because the axon is coated with a myelin sheath (Figure 21-15), which increases the velocity of impulse conduction, it takes only about 0.01 second for an action potential to travel the length of the axon and stimulate muscle contraction. Various myelinated neurons conduct action potentials at velocities of 10 to 100 meters per second (m/s). Without myelin the velocity would be ≈1 m/s, and coordination of movements such as running would be impossible.

Myelin is a stack of specialized plasma membrane sheets produced by a glial cell that wraps itself around the axon. In the peripheral nervous system, these glial cells are called **Schwann cells;** in the central nervous system, they are called *oligodendrocytes*. Often several axons are surrounded by a single glial cell (Figure 21-16a). In both vertebrates and some invertebrates, axons are accompanied along their length by glial cells, but specialization of these glial cells to form myelin occurs predominantly in vertebrates. Vertebrate glial cells that will later form myelin have on their surface a *myelin-associated glycoprotein* and other proteins that bind to adjacent axons and trigger the formation of myelin.

A myelin membrane, like all membranes, contains phospholipid bilayers, but unlike many other membranes, it contains only a few types of proteins. The predominant myelin protein in the peripheral nervous system is P_o, which causes

▶ **FIGURE 21-14 Unidirectional conduction of an action potential due to transient inactivation of voltage-gated Na⁺ channels.** At time 0, an action potential (purple) is at the 2-mm position on the axon. The membrane depolarization spreads passively in both directions along the axon (Figure 21-11). Because the Na⁺ channels at the 1-mm position are still inactivated (green), they cannot yet be reopened by the small depolarization caused by passive spread. Each region of the membrane is refractory (inactive) for a few milliseconds after an action potential has passed. Thus, the depolarization at the 2-mm site at time 0 triggers action potentials downstream only; at 1 ms an action potential is passing the 3-mm position, and at 2 ms, an action potential is passing the 4-mm position.

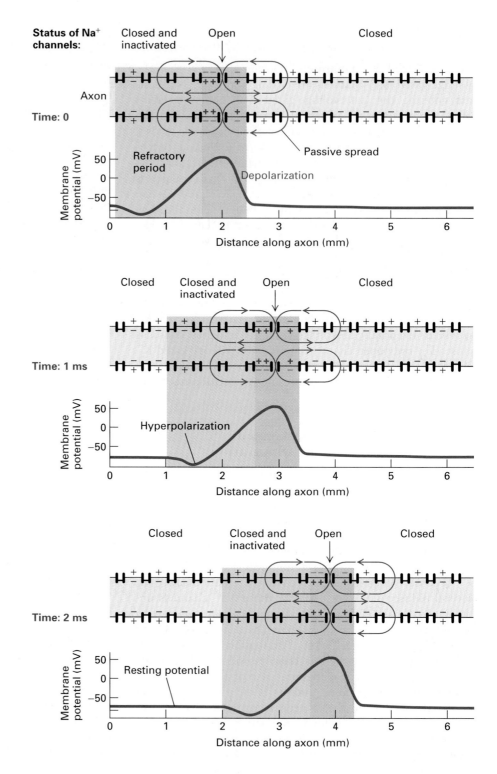

adjacent plasma membranes to stack tightly together (Figure 21-16b). Myelin in the central nervous contains a cytosolic and a membrane protein, termed *myelin basic protein* and *proteolipid*, respectively, that together function similarly to P_o.

The myelin sheath surrounding an axon is formed from many glial cells. Each region of myelin formed by an individual glial cell is separated from the next region by an unmyelinated area called the **node of Ranvier** (or simply, node); only at nodes is the axonal membrane in direct contact with the extracellular fluid (Figure 21-17). Because the myelin sheath prevents the transfer of ions between the axonal cytosol and the extracellular fluids, all electric activity in axons is confined to the nodes of Ranvier, where ions can flow across the axonal membrane. Glial cells secrete protein hormones that somehow trigger the clustering of Na⁺ chan-

(a)

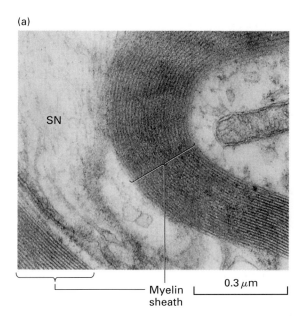

SN

Myelin sheath

0.3 μm

(b)

MS

Axon

SN

10 μm

▲ **FIGURE 21-15 Two views of the myelin sheath.** (a) Electron micrograph of a cross section of the axon of a myelinated peripheral neuron. It is surrounded by the Schwann cell (SN) that produced the myelin sheath, which can contain 50–100 membrane layers. (b) Freeze-fracture preparation of a cut section of the rat sciatic nerve viewed in a scanning electron microscope. The axon of each neuron in the nerve is surrounded by a myelin sheath (MS). The axonal cytoplasm contains abundant filaments—mostly microtubules and intermediate filaments—that run longitudinally and serve to make the axon rigid. [Part (a) from P. C. Cross and K. L. Mercer, 1993, *Cell and Tissue Ultrastructure, A Functional Perspective,* W. H. Freeman and Company, p. 137. Part (b) from R. G. Kessel and R. H. Kardon, 1979, *Tissues and Organs: A Text-Atlas of Scanning Electron Microscopy,* W. H. Freeman and Company, p. 80.]

nels at the nodes. As a result, the node regions contain a high density of voltage-gated Na^+ channels ($\approx 10,000$ per square micrometer of axonal plasma membrane), whereas the regions of axonal membrane between the nodes have few, if any, Na^+ channels. Na^+/K^+ ATPase, which maintains the ionic gradients in the axon, is also localized to the nodes. The fibrous cytoskeletal protein ankyrin binds to these proteins and keeps them in the nodal membrane.

Myelinated nerves have length constants of several millimeters for passive spread of depolarization because ions can move across the axonal membrane only at the myelin-free nodes (see Figure 21-11). Thus the excess cytosolic positive ions generated at a node during the membrane depolarization associated with an action potential spread passively through the axonal cytosol to the next node with very little loss or attenuation, causing a depolarization at one node to spread rapidly to the next node. This permits the action potential to "jump," in effect, from node to node (Figure 21-18). For this reason, the conduction velocity of myelinated nerves is about the same as that of much larger unmyelinated nerves. For instance, a 12-μm-diameter myelinated vertebrate axon and a 600-μm-diameter unmyelinated squid axon both conduct impulses at 12 m/s; the unmyelinated squid giant axon occupies several thousand times the space of this myelinated vertebrate axon and uses several thousand-fold more energy. Not surprisingly, myelinated nerves are used for signaling in circuits where speed is im-

portant. Evolution of the myelin sheath also allowed many more fast-conducting axons to occupy a smaller space, and clearly was essential for evolution of the vertebrate brain.

 One of the leading serious neurologic diseases among human adults is multiple sclerosis (MS), usually characterized by spasms and weakness in one or more limbs, bladder dysfunction, local sensory losses, and visual disturbances. This disorder—the prototype *demyelinating disease*—is caused by patchy loss of myelin in areas of the brain and spinal cord. In MS patients, conduction of action potentials by the demyelinated neurons is slowed and the Na^+ channels spread outward from the nodes. The cause of the disease is not known but appears to involve either the body's production of autoantibodies (antibodies that bind to normal body proteins) that react with myelin basic protein or the secretion of proteases that destroy myelin proteins.

SUMMARY The Action Potential and Conduction of Electric Impulses

• An electric potential exists across the plasma membrane of all eukaryotic cells because the ion compositions of the cytosol and extracellular fluid differ, as do the permeabilities of the plasma membrane to the principal cellular ions: Na^+, K^+, Cl^-, and Ca^{2+}.

(a)

(b)

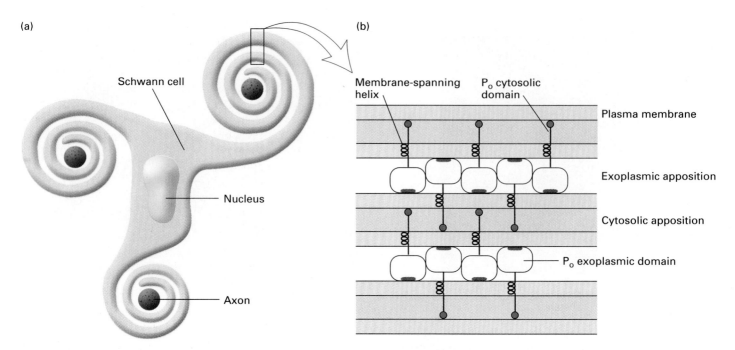

▲ **FIGURE 21-16 Formation and structure of a myelin sheath in the peripheral nervous system.** (a) By wrapping itself around several axons simultaneously, a single Schwann cell can form a myelin sheath around multiple axons. As the Schwann cell continues to wrap around the axon, all the spaces between its plasma membranes, both cytosolic and exoplasmic, are reduced. Eventually all cytosol is forced out and a structure of compact stacked plasma membranes is formed. (b) The compaction of these membranes is generated mainly by P_o protein, which is synthesized only in myelinating Schwann cells. The 124-amino acid exoplasmic domain of P_o protein, which is folded like an immunoglobulin domain, associates with similar domains emanating from the opposite membrane surface. These interactions "zipper" together the membrane surfaces, forming the close exoplasmic opposition. Membrane interactions are stabilized by a tryptophan residue on the tip of the exoplasmic domain, which binds to lipids in the opposite membrane. The close apposition of the cytosolic faces of the membrane may result from binding of the cytosolic tail of each P_o protein (green) to phospholipids in the opposite membrane. [Part (b) adapted from L. Shapiro et al., 1996, *Neuron* **17**:435, and G. Lemke, 1996, *Nature* **383**:395.]

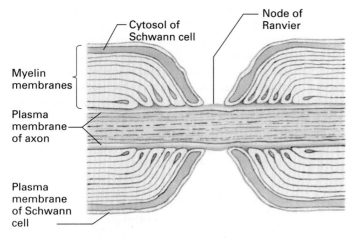

▲ **FIGURE 21-17 Structure of a peripheral myelinated axon near a node of Ranvier, the gap that separates the portions of the myelin sheath formed by two adjacent Schwann cells.** These nodes are the only regions along the axon where the axonal membrane is in direct contact with the extracellular fluid.

- In most nerve and muscle cells, the resting membrane potential is about 60 mV, negative on the inside; the potential is due mainly to the relatively large number of open K^+ channels in the membrane (see Figure 21-9).

- Without voltage-gated cation channels, membrane depolarizations would spread passively only short distances (0.1 to about 5 mm) before the membrane potential returns to its original value.

- An action potential results from the sequential opening and closing of voltage-gated cation channels. First, opening of Na^+ channels permits influx of Na^+ ions for about 1 ms, causing a sudden large depolarization of a segment of the membrane. The channel then closes and becomes unable to open (refractory) for several milliseconds, preventing further Na^+ flow (see Figure 21-13). Opening of K^+ channels as the action potential reaches its peak permits efflux of K^+ ions, which initially hyperpolarizes the membrane. As these channels close, the membrane returns to its resting potential.

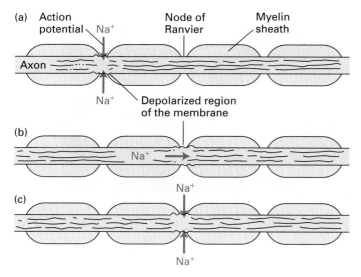

(a) Action potential, Na⁺ ... Node of Ranvier ... Myelin sheath

Axon

Na⁺ ... Depolarized region of the membrane

(b)

Na⁺

Na⁺

(c)

Na⁺

▲ **FIGURE 21-18 Regeneration of action potentials at the nodes of Ranvier.** (a) The influx of Na⁺ ions associated with an action potential at one node results in depolarization of that region of the axonal membrane. (b) Depolarization moves rapidly down the axon because the excess positive ions cannot move outward across the myelinated portion of the axonal membrane. The buildup of these cations causes depolarization at the next node. (c) This depolarization induces an action potential at that node. By this mechanism the action potential jumps from node to node along the axon.

- The depolarization associated with an action potential generated at one point along an axon spreads passively to the adjacent segment, where it triggers opening of voltage-gated Na⁺ channels and hence another action potential. Propagation of the action potential occurs in one direction only because of the short inactive period of the Na⁺ channels and the brief hyperpolarization resulting from K⁺ efflux (see Figure 21-14).

- Thick neurons conduct impulses faster than thin ones. Myelination increases the rate of impulse conduction up to a hundred-fold.

- In myelinated neurons, voltage-gated Na⁺ channels are concentrated at nodes of Ranvier. Depolarization at one node spreads rapidly with little attenuation to the next node, so that the action potential "jumps" from node to node (see Figure 21-18).

21.3 Molecular Properties of Voltage-Gated Ion Channels

The proteins that function as voltage-gated ion channels have three remarkable properties that enable nerve cells to conduct an electric impulse: (1) opening in response to changes in the membrane potential (voltage gating); (2) subsequent channel closing and inactivation; and (3) like all ion channels, exquisite specificity for those ions that will permeate and those that will not. In this section, we describe the molecular analysis of these voltage-gated ion-channel proteins, starting with an important technique for measuring their functional properties. One of the surprising results to emerge from molecular cloning is that all voltage-gated ion channels, be they Na⁺, K⁺, or Ca²⁺ channels, are related in structure and in function. We examine in some detail the voltage-gated K⁺ channel, the first to be cloned and the most widely distributed, and then briefly discuss other members of this family of ion channels, which also includes certain non-voltage-gated channels, such as the K⁺ channels that generate the resting membrane potential.

Patch Clamps Permit Measurement of Ion Movements through Single Channels

The technique of **patch clamping** enables workers to investigate the opening, closing, and ion conductance of a *single* ion channel. As illustrated in Figure 21-19, this technique measures the electric current caused by the movement of ions across a small patch of the plasma membrane. In general, the membrane is electrically depolarized or hyperpolarized and maintained (clamped) at that potential by an electronic feedback device. Thus the membrane potential cannot change, in contrast to the situation during an action potential. The patch-clamping technique can be used in various ways as shown in Figure 21-20.

The inward or outward movement of ions across a patch of membrane can be quantified from the amount of electric current needed to maintain the membrane potential at the designated "clamped" value. To preserve electroneutrality, the entry of each positive ion (e.g., a Na⁺ ion) into the cell across the plasma membrane is balanced by the entry of an electron into the cytosol from the electrode placed in it; the resulting current flow is measured by an electronic device. Conversely, the movement of each positive ion from the cell (e.g., a K⁺ ion) is balanced by the withdrawal of an electron from the cytosol.

The patch-clamp tracings in Figure 21-21 illustrate the use of this technique to study the properties of ion channels in the plasma membrane of muscle cells. In one study, two patches of muscle membrane, each containing one voltage-gated Na⁺ channel, were depolarized about 10 mV and clamped at that voltage. Under these circumstances, the transient pulses of current that cross the membrane result from the opening and closing of individual Na⁺ channels (see Figure 21-21a). Each channel is either open or closed; there are no graded permeability changes for individual channels. From such tracings, it is possible to determine the time that a channel is open (about 0.7 ms) and the ion flux through it (9900 Na⁺ ions/ms). Replacement of the NaCl within the patch pipette (corresponding to the outside of the cell) with KCl or choline chloride abolishes current through the channels, confirming that they conduct only Na⁺ ions.

(a)

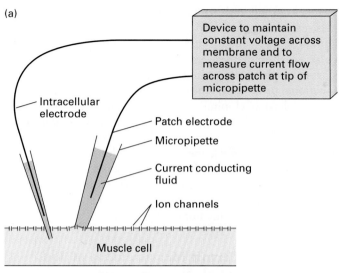

Device to maintain constant voltage across membrane and to measure current flow across patch at tip of micropipette

Intracellular electrode

Patch electrode

Micropipette

Current conducting fluid

Ion channels

Muscle cell

Intact cell

(b)

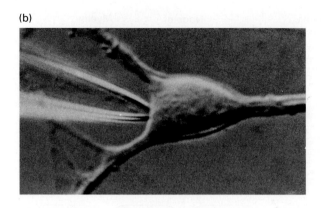

▲ **FIGURE 21-19 Outline of the patch-clamping technique.**
(a) Basic patch-clamping arrangement for measuring current flow through individual ion channels in the plasma membrane of a living cell. The patch electrode, filled with a current-conducting saline solution, is applied, with a slight suction, to the plasma membrane. The 0.5-μm-diameter tip covers a region that contains only one or a few ion channels. The second electrode is inserted into the cytosol. The recording device measures current flow only through the channels in the patch of plasma membrane. (b) Photomicrograph of the cell body of a cultured neuron and the tip of a patch pipette touching the cell membrane. [Part (b) from B. Sakmann, 1992, *Neuron* **8**:613 (Nobel lecture); also published in E. Neher and B. Sakmann, 1992, *Sci. Am.* **266(3)**:44.]

(a) On-cell measures effects of extracellular hormones on channels within the patch of plasma membrane still attached to a cell

(c) Whole-cell measures effects of extracellular hormones on all channels in the plasma membrane

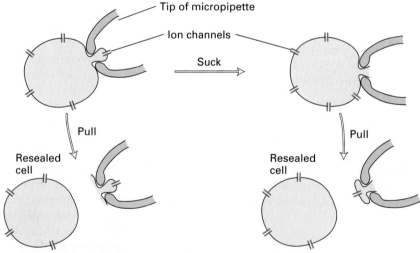

Tip of micropipette

Ion channels

Suck

Pull

Resealed cell

Pull

Resealed cell

(b) Inside-out measures effects of intracellular second messengers (cAMP, Ca^{2+}) on channels within the isolated patch

(d) Outside-out measures effects of extracellular hormones directly on channels within the isolated patch

▲ **FIGURE 21-20 Different patch-clamping configurations.**
The effects of different ion concentrations and other substances within and outside the cell on current flow through single channels can be measured on intact cells or isolated patches (a, b, d). In the whole-cell configuration (c), the piece of membrane in the patch is ruptured, allowing measurement of current flow through *all* of the ion channels. The effect of different solutes on channels is studied most easily with isolated, detached patches (b, d). [Adapted from B. Hille, 1992, *Ionic Channels of Excitable Membranes*, 2d ed., Sinauer Associates, p. 89.]

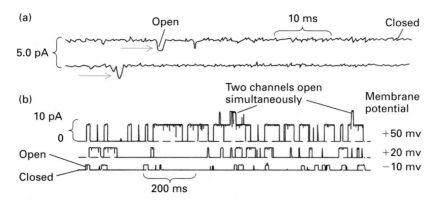

▲ **FIGURE 21-21 Current flux through individual voltage-gated channels determined by patch clamping of muscle cells.** (a) Two membrane patches, as shown in Figure 21-20(b), were depolarized by 10 mV and clamped at that value. The transient pulses of electric current in picoamperes (pA), recorded as large downward deviations (arrows), indicate the opening of a Na^+ channel and movement of Na^+ ions inward across the membrane. The smaller deviations in current represent background noise. The average current through an open channel is 1.6 pA, or 1.6×10^{-12} amperes. Since 1 ampere = 1 coulomb (C) of charge per second, this current is equivalent to the movement of about 9900 Na^+ ions per channel per millisecond:

$(1.6 \times 10^{-12} \text{ C/s})(10^{-3} \text{ s/ms})(6 \times 10^{23} \text{ molecules/mol}) \div 96,500 \text{ C/mol}$.

(b) Patches of membrane were clamped at three different potentials, +50, +20, and −10 mV. The upward deviations in the current indicate the opening of K^+ channels and movement of K^+ ions outward across the membrane. Increasing the extent of membrane depolarization from −10 mV to +50 mV increases the probability a channel will open, the time it stays open, and the amount of electric current (numbers of ions) that pass through it. [Part (a) see F. J. Sigworth and E. Neher, 1980, *Nature* **287**:447. Part (b) from B. Pallota et al., 1981, *Nature* **293**:471 as modified by B. Hille, 1992, *Ionic Channels of Excitable Membranes*, 2d ed., Sinauer Associates, p. 122.]

Several properties of voltage-gated K^+ channels can be deduced from the tracings in Figure 21-21b. At the depolarizing voltage of −10 mV, the channels in the membrane patch open infrequently and remain open for only a few milliseconds, as judged, respectively, by the number and width of the "upward blips" on the tracings. Further, the ion flux through them is rather small, as measured by the electric current passing through each open channel (the height of the blips). Depolarizing the membrane further to +20 mV causes these channels to open about twice as frequently. Also, more K^+ ions move through each open channel (the height of the blips is greater) because the force driving cytosolic K^+ ions outward is greater at a membrane potential of +20 mV than at −10 mV. Depolarizing the membrane further to +50 mV, such as at the peak of an action potential, causes the opening of more K^+ channels and also increases the flux of K^+ through them. Thus, by opening during the peak of the action potential, these K^+ channels cause the outward movement of K^+ ions and the repolarization of the membrane potential toward the resting value.

Voltage-Gated K^+ Channels Have Four Subunits Each Containing Six Transmembrane α Helices

The initial breakthrough in identification and cloning of voltage-gated ion channels came from analysis of fruit flies (*Drosophila melanogaster*) carrying the *shaker* mutation. These flies shake vigorously under ether anesthesia, reflecting a loss of motor control and a defect in excitable cells. The axons of giant nerves in *shaker* mutants have an abnormally prolonged action potential (Figure 21-22), suggesting that the mutation causes a defect in voltage-gated K^+ channels that prevents them from opening normally immediately upon depolarization. Thus the X-linked *shaker* mutation was thought to cause a defect in a K^+ channel protein.

The wild-type *shaker* gene was cloned by chromosome walking (see Figure 8-24). To show that the encoded protein indeed was a K^+ channel, the wild-type *shaker* cDNA was used as a template to produce *shaker* mRNA in a cell-free

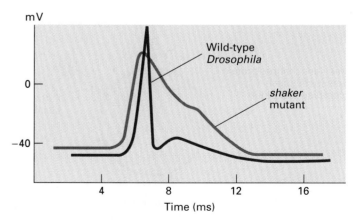

▲ **FIGURE 21-22 Action potentials in axons of wild-type *Drosophila* and *shaker* mutants.** The *shaker* mutants exhibit an abnormally prolonged action potential because of a defect in the voltage-gated K^+ channel that is required for normal repolarization. [See L. A. Salkoff and R. Wyman, 1983, *Trends Neurosci.* 6:128.]

system. Expression of this mRNA in frog oocytes and patch-clamp measurements on the newly synthesized channel protein showed that it had properties identical to those of the K^+ channel in the neuronal membrane, demonstrating conclusively that the *shaker* gene encodes that K^+ channel protein (Figure 21-23).

A functional Shaker K^+ channel is built of four identical subunits arranged in the membrane around a central pore. (These are equivalent to the four domains of the voltage-gated Na^+ channel shown in Figure 21-13.) The 656-amino-acid polypeptide encoded by the *shaker* gene, like most K^+-channel polypeptides, contains six membrane-spanning α helixes, designated S1–S6. Three regions in each subunit are critical to the functioning of the K^+ channel: a *P segment* between S5 and S6, which lines the pore of the channel; an amino-terminal segment, the "ball," which moves into the open pore,

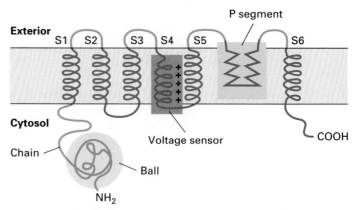

▲ **FIGURE 21-24 Proposed structural model of the subunits composing the Shaker voltage-gated K^+ channel from *Drosophila*.** Each of the four identical subunits in this channel protein is thought to contain six membrane-spanning α helixes, S1–S6. S4 contains several positively charged amino acids and is the voltage-sensing α helix. The pore-lining P segment lies between S5 and S6. The N-terminus of the polypeptide, located in the cytosol, contains a globular domain essential for inactivation of the open channel. [Adapted from C. Miller, 1992, *Curr. Biol.* **2**:573, and H. Larsson et al., 1996, *Neuron* **16**:387.]

inactivating the channel; and the S4 α helix, which acts as a voltage sensor (Figure 21-24). As we will see later, homologous regions are present in voltage-gated Na^+ and Ca^{2+} channels.

In *Drosophila*, at least five different *shaker* polypeptides (**isoforms**) are produced by alternative splicing of the primary transcript of the *shaker* gene. In the oocyte expression assay, these Shaker isoforms exhibit different voltage dependencies and K^+ conductivities. Thus differential expression of Shaker isoforms can affect the timing of repolarization during an action potential, accounting for differences in the electrical activity of different types of neurons. Also, if a single K^+ channel is constructed from two different isoforms, the properties of the resultant "hybrid" channel differ from those of both of the channels formed from four identical polypeptides.

Using the *shaker* gene as a hybridization probe, workers have isolated genes encoding more than 100 K^+-channel proteins from vertebrates. Although all the encoded channel proteins have a similar overall structure, they exhibit different voltage dependencies, conductivities, channel kinetics, and other properties. However, many open only at strongly depolarizing voltages, a property required for the channel to repolarize the membrane during an action potential. The *shaker* gene also has been used to isolate genes encoding a large number of K^+-channel proteins from humans, illustrating how research on invertebrates can advance understanding of the human nervous system.

P Segments Form the Ion-Selectivity Filter

The P segments of the **Shaker protein** were implicated as forming part of the pore in experiments with two inhibitors

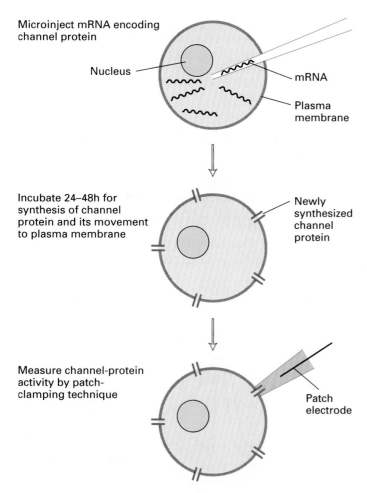

▲ **FIGURE 21-23 Oocyte expression assay.** The activity of normal and mutant channel proteins can be studied in frog oocytes, which do not normally express these proteins. A follicular oocyte is first treated with collagenase to remove the surrounding follicle cells, leaving a denuded oocyte, which is microinjected with mRNA encoding the channel protein under study. [Adapted from T. P. Smith, 1988, *Trends Neurosci.* **11**:250.]

that physically plug the channel mouth: a neurotoxin from scorpion venom (charybdotoxin) and tetraethylammonium ion. Site-specific mutation of any of several amino acids in the P segment rendered the channel, when expressed in oocytes, resistant to inhibition by these agents. In addition, K⁺ channels from different sources vary in their conductance (number of ions transported per millisecond) and sensitivity to tetraethylammonium. Chimeric proteins in which the P segment from one protein was replaced with that from another exhibited conductances and sensitivities to tetraethylammonium that correlated with the origin of the P segment, not with the origin of the rest of the channel protein. In the case of the K⁺ channel, four P segments, one from each subunit, are thought to form the ion-selective pore that constricts the center of the channel.

The recent determination of the three-dimensional structure of a bacterial K⁺ channel provides insight into how the remarkable ion specificity of channel proteins is achieved. Each subunit in this bacterial channel contains a P segment highly homologous in sequence to that in all known K⁺ channels. Eight transmembrane α helices, two from each subunit (analogous to the S5 and S6 helices in the Shaker channel), form an "inverted teepee," generating a water-filled cavity in the outer portion of the channel that leads to the actual selectivity filter (Figure 21-25a). The filter, which is lined by four extended loops that are part of the P segments, is wide enough to accommodate a K⁺ ion only if it is stripped of its water of hydration. The ability of this filter to select K⁺ over Na⁺ is due mainly to backbone carbonyl oxygens on glycine residues located in a Gly-Tyr-Gly sequence that is found in an analogous position in the P segment in every known K⁺ channel. Mutation of these residues disrupts the channel's ability to discriminate between K⁺ and Na⁺ ions.

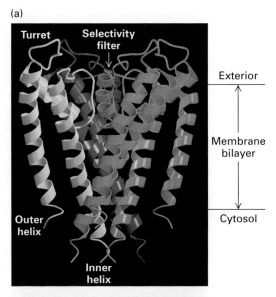

(a)

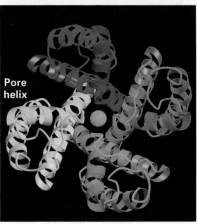

▶ FIGURE 21-25 Mechanism of ion selectivity in a resting K⁺ channel from the bacterium *Streptomyces lividans.*
(a) The three-dimensional structure of the pore in this channel, viewed from the side *(upper)* and from the top, or extracellular, side *(lower)*, reveals that the two membrane-spanning α helices in each of the four subunits are tilted relative to the vertical, forming an "inverted teepee." These helices, analogous to S5 and S6 in the Shaker channel, are connected by a P segment comprising a nonhelical "turret," which lines the outer part of the pore; a short α helix; and an extended loop, which forms the ion-selectivity filter. Selectivity of K⁺ over Na⁺ is due mainly to the carboxyl groups on two highly conserved glycine residues and one tyrosine on each of these loops, which protrude into the narrowest part of the channel. (b) Potassium ions, hydrated in solution, lose their bound water molecules as they pass through the selectivity filter and become coordinated instead to four backbone carbonyl oxygens, one from a glycine in the channel-lining loop of each P segment. Na⁺ ions, being smaller, cannot perfectly coordinate with these oxygens and therefore pass through the channel only rarely. [Part (a) from D. A. Doyle et al., 1998, Science **280**:69, courtesy of R. MacKinnon. Part (b) adapted from C. Armstrong, 1998, Science **280**:56.]

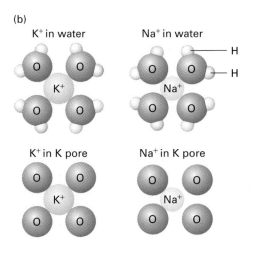

As a K^+ ion enters the narrow selectivity filter, it loses its water of hydration but becomes bound with a similar geometry to four backbone carbonyl oxygens, one from the loop in each P segment lining the channel (Figure 21-25 a, b); thus a relatively low activation energy is required for passage of K^+ ions. Because a Na^+ ion is smaller than a K^+ ion, after it loses its water of hydration it cannot become perfectly coordinated to the carbonyl oxygens that line the selectivity filter, so the activation energy for passage of Na^+ ions is relatively high. As a result, 1000 K^+ ions pass through the channel for every 1 Na^+ ion.

The S4 Transmembrane α Helix Acts as a Voltage Sensor

Since voltage-gated channels open when the membrane depolarizes, some segment of the protein must "sense" the change in potential. Sensitive electric measurements suggest that the opening of each voltage-gated K^+ (and Na^+) channel is accompanied by the movement of 10 to 12 protein-bound positive charges from the cytosolic to the exoplasmic surface of the membrane; alternatively, a larger number of charges may move a shorter distance across the membrane. The movement of these gating charges (or voltage sensors) under the force of the electric field triggers a conformational change in the protein that opens the channel. Several lines of evidence indicate that the voltage sensor is the S4 transmembrane α helix present in each Shaker polypeptide (see Figure 21-24); similar S4 helices are found in voltage-gated Na^+ and Ca^{2+} channel proteins. These voltage-sensing helices, often called *gating helices*, generally have a positively charged lysine or arginine every third or fourth residue. In the closed resting state, half of each S4 helix is exposed to the cytosol; when the membrane is depolarized, these amino acids move outward and become exposed to the exoplasmic surface of the channel.

The role of the S4 helix in voltage sensing has been demonstrated in studies with mutant Shaker K^+ channels. In one experiment, one or more arginine or lysine residues in the S4 helix were replaced with neutral or acidic residues. When these mutant proteins were expressed in frog oocytes, patch-clamping measurements showed that fewer positive charges than in wild-type channels moved across the membrane immediately in response to a membrane depolarization, indicating that the arginine and lysine residues in the S4 helix do indeed more across the membrane. In other studies, mutant proteins in which various S4 residues were converted to cysteine were tested for their reactivity with a membrane-impermeant cysteine-modifying chemical agent. The results indicated that in the resting state amino acids near the C-terminus of the S4 helix face the cytosol, but when the membrane is depolarized, some of these same amino acids become exposed to the exoplasmic surface of the channel. These experiments directly demonstrate movement of the S4 helix across the membrane, as depicted in Figure 21-13 for the analogous segment in Na^+ channels.

Movement of One N-Terminal Segment Inactivates Shaker K^+ Channels

An important characteristic of voltage-gated channels is inactivation: soon after opening they close spontaneously, forming an inactive channel that will not reopen until the membrane is repolarized. The N-termini of each of the four Shaker polypeptides forms a globular "ball," one of which swings into the open channel, inactivating it (Figure 21-26a). Two key experiments, explained in Figure 21-26b, have shown that inactivation depends on the ball domains, occurs after channel opening, and does not require the ball domains to be part of the protein. In addition, deletion of part of the ≈40-amino-acid segment (the "chain") connecting the ball to the first membrane-spanning helix increases the rate of inactivation. The shorter the chain, the more rapid the inactivation, as if a ball attached to a shorter chain can move into the open channel more readily. Conversely, addition of random amino acids to lengthen the normal chain slows channel inactivation.

All Pore-Forming Ion Channels Are Similar in Structure to the Shaker K^+ Channel

The pioneering research we've just described, starting with isolation of a mutant fruit fly that shakes under anesthesia, has led to the current detailed picture of the structure and operation of voltage-gated K^+ channels. Purification, molecular cloning, and analysis of other ion-channel proteins has revealed that many have a structure remarkably similar to that of voltage-gated K^+ channels (Figure 21-27). For instance, as we discuss later, both the visual and olfactory systems contain ion channels whose opening and closing are controlled by binding of cAMP or cGMP. Like the Shaker K^+ channel, these channels are tetramers, each subunit containing six transmembrane α helices and a pore-lining P segment. The resting K^+ channel, which generates the resting potential, also contains four identical subunits; these have only two membrane-spanning α helices, one on either side of a P segment, analogous to the two in the bacterial K^+ channel (see Figure 21-25a) and the S5 and S6 helices in the Shaker channel. However, none of the α helices in the resting K^+ channel or nucleotide-gated ion channels acts as a voltage sensor, since these channels are not voltage gated.

Although voltage-gated Na^+ and Ca^{2+} channels are monomeric proteins, both contain four homologous transmembrane domains, each similar in sequence and structure to the central part of the Shaker polypeptide (see Figure 21-27c,d). These domains are connected and flanked by nonhomologous cytosolic segments. Each of the four homologous domains contains a nonhelical P segment, between helices S5 and S6, and a voltage-sensing S4 α helix. Site-specific mutation studies have shown that the ion specificity of Na^+ and Ca^{2+} channels, like that of K^+ channels, is determined by amino acid side chains in the P segments. Unlike K^+ channels, however, Na^+ and Ca^{2+} channels have only a single channel-inactivation segment, located in the

(a)

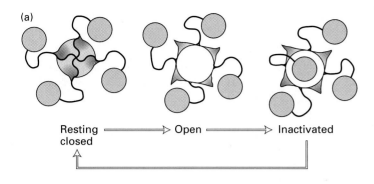

Resting ======⟶ Open ======⟶ Inactivated
closed

(b)

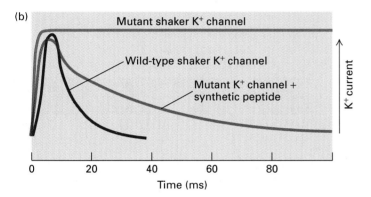

Mutant shaker K$^+$ channel

Wild-type shaker K$^+$ channel

Mutant K$^+$ channel + synthetic peptide

K$^+$ current

Time (ms)

◀ **FIGURE 21-26** (a) **The "ball and chain" model for inactivation of a voltage-gated K$^+$ channel.** In the resting state, the balls at the N-termini of the four subunits of the wild-type channel are free in the cytosol, and remain there during the first milliseconds after the channel is opened by depolarization. One ball then moves into the open channel, inactivating it. After a few milliseconds (depending on the exact type of channel), the ball is displaced from the channel and the protein reverts to the closed, resting state. The ball and chain in this model is equivalent to the channel-inactivating segment in Figure 21-13. (b) Experimental results supporting the ball and chain model. The wild-type Shaker K$^+$ channel and a mutant form lacking the amino acids composing the ball were expressed in *Xenopus* oocytes, and the activity of the channels was monitored by the patch-clamp technique (see Figure 21-23). When patches were depolarized from −0 to +30 mV, the wild-type channel opened for ≈5 milliseconds and then closed (red curve), whereas the mutant channel opened normally, but could not close (green curve). When a chemically synthesized ball peptide was added to the cytosolic face of the patch, the mutant channel opened normally and then closed (purple curve). This demonstrated that the added peptide inactivated the channel after it opened and that the ball does not have to be tethered to the protein in order to function. [Part (a) adapted from C. Armstrong and B. Hille, 1998, *Neuron* **20**:371. Part (b) from W. N. Zagotta et al., 1990, *Science* **250**:568.]

cytosol-facing loop between domains III and IV. The Na$^+$ and Ca^{2+} channels contain additional, essential regulatory subunits that affect the rate at which the channel opens and becomes inactivated, and the voltage dependence of channel inactivation. The subunit organization and number of transmembrane α helices varies in some other ion channels from that shown in Figure 21-27; nonetheless, they all contain four copies of the highly conserved P segment flanked by homologs of the S5 and S6 transmembrane α helices.

Although the density of voltage-gated K$^+$ and Na$^+$ channels is very low, both types of channel proteins have been isolated. As we saw earlier, isolation of a K$^+$ channel from *Drosophila* was greatly aided by discovery of *shaker* mutants. Likewise, certain neurotoxins facilitated the purification and study of Na$^+$ channels. For example, tetrodotoxin produced by the puffer fish and saxitoxin produced by certain red marine dinoflagellates specifically bind to and inhibit the voltage-gated Na$^+$ channels in neurons, preventing action potentials from forming. One molecule of either toxin binds to one Na$^+$ channel with exquisite affinity and selectivity. Measurements of the amount of radioactive tetrodotoxin or saxitoxin that binds to a typical unmyelinated invertebrate axon have shown that it contains 5–500 Na$^+$ channels per square micrometer of membrane, equivalent to about 1 part per million of total plasma-membrane protein molecules. This value agrees with the number of channels estimated from patch-clamp studies. The Na$^+$ channels in these unmyelinated membranes are thus spaced, on average, about 200 nm apart. However, as noted earlier, they are packed much closer together at the nodes of Ranvier.

Voltage-Gated Channel Proteins Probably Evolved from a Common Ancestral Gene

The structural and functional similarities among voltage-gated Na$^+$, Ca^{2+}, and K$^+$ channels suggest that all three proteins evolved from a common ancestral gene. The distribution of these channels among different organisms provides clues about a likely evolutionary pathway.

Voltage-gated K$^+$ channels have been found in all yeasts and protozoa studied, whereas only multicellular organisms have voltage-gated Na$^+$ channels. Intermediate in distribution are voltage-gated Ca^{2+} channels, which function in synaptic transmission (discussed below): These are present in multicellular organisms but in only a few of the more complex protozoa, such as *Paramecium*. Thus voltage-gated K$^+$ channel proteins probably arose first in evolution. The voltage-gated Ca^{2+} and Na$^+$ channel proteins are believed to have evolved by repeated duplication of an ancestral one-domain K$^+$-channel gene. Since all K$^+$ channels share a similar pore-lining P segment, it is probable that voltage-gated and non-voltage-gated K$^+$ channels also evolved from a common progenitor.

SUMMARY Molecular Properties of Voltage-Gated Ion Channels

- Patch-clamping techniques, which permit measurement of ion movement through single channels, are used to determine the potential at which a channel opens, its ion conductivity, and the rate of channel inactivation and closing (see Figure 21-21).

(a) Voltage-gated K$^+$ channel protein (tetramer)

(c) Voltage-gated Na$^+$ channel protein (monomer)

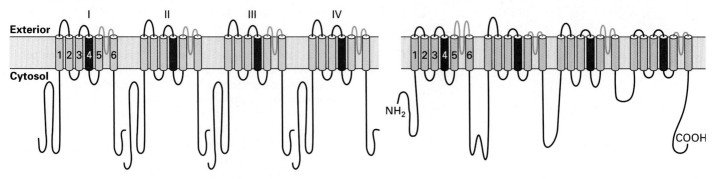

(b) Cyclic nucleotide-gated channel protein (tetramer)

(d) Voltage-gated Ca^{2+} channel protein (monomer)

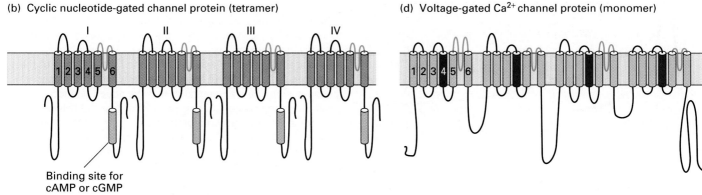

Binding site for
cAMP or cGMP

▲ **FIGURE 21-27 Proposed transmembrane structures of four types of gated ion-channel proteins.** (a) The Shaker voltage-gated K$^+$ channel isolated from *Drosophila* is a tetramer of four identical subunits each containing 656 amino acids and six transmembrane α helices (indicated by Arabic numerals). Helix 4 (maroon) acts as a voltage sensor, and the nonhelical P segments between helices 5 and 6 line the ion pore. (b) Cyclic AMP- or cyclic GMP-gated ion channels also contain four subunits. Since none of the transmembrane α helices act as a voltage sensor, these channels are not voltage-gated. Rather, binding of cAMP or cGMP to a cytosolic segment triggers opening of these channels, which are abundant in the sensing cells of the visual

and olfactory systems (see Figure 21-47). (c, d) Voltage-gated Na$^+$ and Ca^{2+} channels are monomeric proteins containing 1800–2000 amino acids organized into four homologous transmembrane domains (indicated by Roman numerals). About 64 percent of the residues are similar or identical in sequence in both channels. Each of the four homologous domains is thought to contain six transmembrane α helices similar in structure to those in the Shaker channel monomers in (a). These channels also contain essential regulatory subunits, which are not depicted here. [Adapted from T. M. Jessell and E. R. Kandel, 1993, in *Cell* vol. 72/*Neuron* vol. 10 (suppl.), pp. 1–3; R. MacKinnon, 1995, *Neuron* **14**:889; and F. Lesage et al., 1996, *EMBO J.* **15**:6400.]

- Recombinant DNA techniques allow the expression and study of cloned channel proteins in frog oocytes and other types of cells (see Figure 21-23).

- Voltage-gated K$^+$ channels, such as the Shaker protein from *Drosophila*, are assembled from four similar subunits, each of which has six membrane-spanning α helices and a nonhelical P segment that lines the ion pore. The S4 α helix in each subunit acts as a voltage sensor.

- Voltage-gated Na$^+$ and Ca^{2+} channels are monomeric proteins containing four homologous domains each similar to a K$^+$-channel subunit (see Figure 21-27).

- The ion specificity of channel proteins is due mainly to coordination of the selected ion with specific residues in the P segments, thus lowering the activation energy

for passage of the selected ion compared with other ions (see Figure 21-25).

- Voltage-sensing α helices have a positively charged lysine or arginine every third or fourth residue. Their movement outward across the membrane, in response to a membrane depolarization of sufficient magnitude, causes opening of the channel (see Figure 21-13).

- Voltage-gated K$^+$, Na$^+$, and Ca^{2+} channel proteins contain one or more cytosolic domains that move into the open channel thereby inactivating it (see Figure 21-26).

- Non-voltage-gated K$^+$ channels and nucleotide-gated channels lack a voltage-sensing α helix, but otherwise their structures are similar in many respects to voltage-gated K$^+$ channels.

- Most likely, voltage-gated channel proteins and possibly all K^+ channels evolved from a common ancestral gene.

21.4 Neurotransmitters, Synapses, and Impulse Transmission

As noted earlier, synapses are the junctions where neurons pass signals to other neurons, muscle cells, or gland cells. Most nerve-to-nerve signaling and all known nerve-to-muscle and nerve-to-gland signaling rely on chemical synapses at which the presynaptic neuron releases a chemical neurotransmitter that acts on the postsynaptic target cell (see Figure 21-4). In this section we discuss the types of molecules that function as transmitters at chemical synapses, their origin and fate, and their effects on postsynaptic cells. Because the ability of neurotransmitters to induce a response depends on their binding to specific receptors in the postsynaptic membrane, we introduce the major classes of receptors in this section; individual receptors are examined in more detail in the next section. We also briefly discuss electric synapses, which are much rarer, but simpler in function, than chemical synapses.

Many Small Molecules Transmit Impulses at Chemical Synapses

Numerous small molecules synthesized in the cytosol of axon terminals function as neurotransmitters at various chemical synapses. The "classic" neurotransmitters are stored in **synaptic vesicles**, uniformly sized organelles, 40–50 nm in diameter. With the exception of **acetylcholine,** the classic neurotransmitters depicted in Figure 21-28 are amino acids or derivatives of amino acids. Nucleotides such as ATP (see Figure 2-24) and the corresponding nucleosides, which lack phosphate groups, also function as neurotransmitters. Each neuron generally produces just one type of classic neurotransmitter. Following their exocytosis from synaptic vesicles into the synaptic cleft, neurotransmitters bind to specific receptors on the plasma membrane of a postsynaptic cell, causing a change in its permeability to ions.

Many neurons secrete **neuropeptides,** a varied group of signaling molecules that includes endorphins, vasopressin,

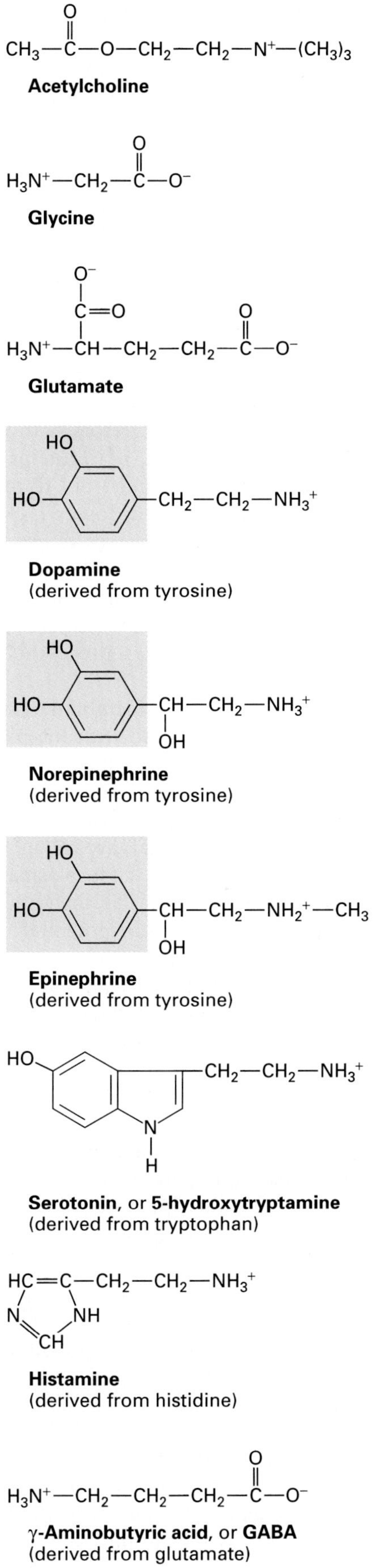

Acetylcholine

Glycine

Glutamate

Dopamine
(derived from tyrosine)

Norepinephrine
(derived from tyrosine)

Epinephrine
(derived from tyrosine)

Serotonin, or **5-hydroxytryptamine**
(derived from tryptophan)

Histamine
(derived from histidine)

γ-**Aminobutyric acid,** or **GABA**
(derived from glutamate)

▶ **FIGURE 21-28 Structures of several small molecules that function as neurotransmitters.** Except for acetylcholine, all of these are amino acids (glycine and glutamate) or derived from the indicated amino acids. The three transmitters synthesized from tyrosine, which contain the catechol moiety (blue highlight), are referred to as catecholamines.

oxytocin, and gastrin. Neuropeptides are stored in a different type of vesicle than classic neurotransmitters. Exocytosis of both types of transmitter is triggered by a localized rise in cytosolic Ca^{2+}, but neuropeptides are released outside the synaptic zone. The effects of neuropeptide transmitters are very diverse and often long-lived (hours to days). The following discussion deals mainly with the release and actions of the classic neurotransmitters such as those shown in Figure 21-28.

Influx of Ca^{2+} Triggers Release of Neurotransmitters

The exocytosis of neurotransmitters from synaptic vesicles involves vesicle-targeting and fusion events similar to those that occur at many points in the secretory pathway (Section 17.10). Indeed the same types of proteins—including T-SNARE and V-SNAREs, α, β, and γ SNAP proteins, and NSF—participate in both systems. However, exocytosis of neurotransmitters at chemical synapses differs from other secretory pathways in two critical ways: (a) Secretion is tightly coupled to arrival of the action potential at the axon terminus, and (b) synaptic vesicles are recycled locally after fusion with the plasma membrane, a process that takes less than one minute.

Depolarization of the plasma membrane cannot, by itself, cause synaptic vesicles to fuse with the plasma membrane. In order to trigger vesicle fusion, an action potential must be converted, or *transduced*, into a chemical signal—namely, a localized rise in the cytosolic Ca^{2+} concentration. The transducers of the electric signals are voltage-gated Ca^{2+} channels localized to the region of the plasma membrane adjacent to the synaptic vesicles. The membrane depolarization due to arrival of an action potential opens these channels, permitting an influx of Ca^{2+} ions into the cytosol from the extracellular medium. The amount of Ca^{2+} that enters an axon terminal through voltage-gated Ca^{2+} channels is sufficient to raise the level of Ca^{2+} in the region of the cytosol near the synaptic vesicles from <0.1 μM, characteristic of the resting state, to $1-100$ μM. As we describe below, the Ca^{2+} ions bind to proteins that connect the synaptic vesicle with the plasma membrane, inducing membrane fusion and thus exocytosis of the neurotransmitter. The extra Ca^{2+} ions are rapidly pumped out of the cell by Ca^{2+} ATPases, lowering the cytosolic Ca^{2+} level and preparing the terminal to respond again to an action potential.

The presence of voltage-gated Ca^{2+} channels in axon terminals has been demonstrated in neurons treated with drugs that block Na^+ channels and thus prevent conduction of action potentials. When the membrane of axon terminals in such treated cells is artificially depolarized, an influx of Ca^{2+} ions into the neurons occurs and exocytosis is triggered. Patch-clamping experiments show that voltage-gated Ca^{2+} channels, like voltage-gated Na^+ channels, open transiently upon depolarization of the membrane.

Synaptic Vesicles Can Be Filled, Exocytosed, and Recycled within a Minute

Two pools of neurotransmitter-filled synaptic vesicles are present in axon terminals: those "docked" at the plasma membrane, which can be readily exocytosed, and those in reserve in the *active zone* near the plasma membrane. Each rise in Ca^{2+} triggers exocytosis of about 10 percent of the docked vesicles. However, this is but one of the series of steps involved in forming synaptic vesicles, filling them with neurotransmitter, moving them to the active zone near the plasma membrane, docking them at the plasma membrane, and then, after vesicle fusion with the plasma membrane, recycling their membrane components by endocytosis (Figure 21-29).

Recycling of synaptic-vesicle membrane proteins is rapid, as indicated by the ability of many neurons to fire fifty times a second, and quite specific, in that several membrane proteins unique to the synaptic vesicles are specifically internalized by endocytosis. Endocytosis usually involves clathrin-coated vesicles, though non-clathrin-coated vesicles may also be used. After the endocytic vesicles lose their clathrin coat, however, they usually do not fuse with larger, low pH endosomes, as they do during endocytosis of plasma-membrane proteins in other cells (see Figure 17-46). Rather, the recycled vesicles are immediately refilled with neurotransmitter.

Multiple Proteins Participate in Docking and Fusion of Synaptic Vesicles

Because synaptic vesicles are small and homogeneous in size, they can be readily purified from the brain and their proteins isolated. The synaptic-vesicle membrane contains V-type ATPases, which generate a low intravesicular pH, and a proton-coupled neurotransmitter antiporter, which imports neurotransmitters from the cytosol (see Figure 21-29). Here we focus on synaptic-vesicle and plasma-membrane proteins that function in vesicle docking and fusion. Although the major proteins have been characterized in some detail, the exact functions of many and the order in which they interact still remain unclear.

Synapsin and Other Cytoskeletal Proteins The axon terminal exhibits a highly organized arrangement of cytoskeletal fibers that is essential for localizing synaptic vesicles to the plasma membrane at the synaptic cleft (Figure 21-30). The vesicles themselves are linked together by *synapsin*, a fibrous phosphoprotein structurally related to other cytoskeletal proteins that bind the fibrous proteins actin and spectrin (Section 18.1). Synapsin is localized to the cytosolic surface of all synaptic-vesicle membranes and constitutes 6 percent of vesicle proteins. Thicker filaments radiate from the plasma membrane and bind to vesicle-associated synapsin; probably these interactions keep the synaptic vesicles close to the part of the plasma membrane facing the synapse. Indeed, synapsin knockout mice, although viable, are prone to seizures; during repetitive stimulation of many neurons in such mice, the number of synaptic vesicles that fuse with

▶ **FIGURE 21-29 Release of neurotransmitters and the recycling of synaptic vesicles.** Vesicles import neurotransmitters (red circles) from the cytosol (step ①) using a H^+/neurotransmitter antiporter. The low intravesicular pH, generated by a V-type ATPase in the vesicle membrane, powers neurotransmitter import. The vesicles then move to the active zone near the plasma membrane (step ②) and "dock" at defined membrane sites by interacting with specific proteins (step ③). A rise in cytosolic Ca^{2+} triggers fusion of the docked vesicles and release of neurotransmitters into the synaptic cleft (step ④). Synaptic-vesicle membrane proteins are then specifically recovered by endocytosis, usually in clathrin-coated vesicles (step ⑤). The clathrin coat is depolymerized, yielding vesicles that are the same size as synaptic vesicles. These new synaptic vesicles then are filled with neurotransmitters (step ①), completing the cycle, which typically takes about 60 seconds. [Adapted from T. Südhof and R. Jahn, 1991, *Neuron* **6**:665; see also K. Takei et al., 1996, *J. Cell. Biol.* **133**:1237, and V. Murthy and C. Stevens, 1998, *Nature* **392**:497.]

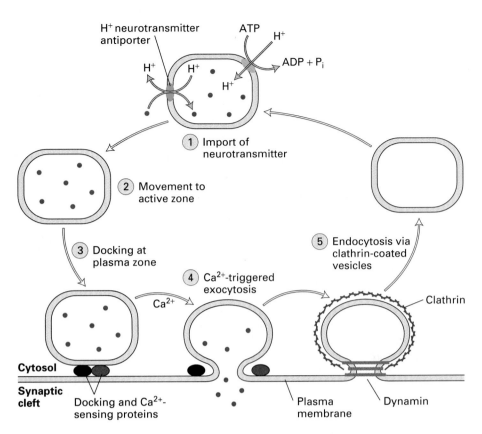

Axon terminal   Synapsin-containing fibers   Docked synaptic vesicle

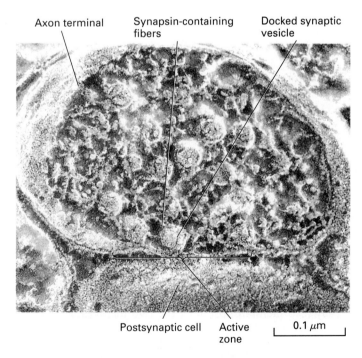

Postsynaptic cell   Active zone   0.1 μm

▲ **FIGURE 21-30 Micrograph of an axon terminal obtained by the rapid-freezing deep-etch technique.** Note the fibers, largely composed of synapsin, that interconnect the vesicles and also connect some to the active zone of the plasma membrane. Docked vesicles are ready to be exocytosed. Those toward the center of the terminal are in the process of being filled with neurotransmitter. [From D. M. D. Landis et al., 1988, *Neuron* **1**:201.]

the plasma membrane is greatly reduced. Thus synapsins are thought to recruit synaptic vesicles to the active zone. Because synapsins are substrates of cAMP-dependent and Ca^{2+}-calmodulin–dependent protein kinases, a rise in cytosolic Ca^{2+} triggers their phosphorylation. This apparently causes the release of synaptic vesicles from the cytoskeleton and increases the number of vesicles available for fusion with the plasma membrane.

Vesicle Targeting and Fusion Proteins The same types of proteins discussed in Chapter 17 mediate the targeting and fusion of neurotransmitter-filled vesicles at synapses (Figure 21-31). These include Rab3A, a neuron-specific GTP-binding protein similar in sequence and function to other Rab proteins that control vesicular traffic in the secretory pathway. Rab3A is located in the membrane of synaptic vesicles and appears to be essential for localization of vesicles to the active zone. Rab3A knockout mice, like synapsin knockout mice, are viable, but repetitive stimulation of certain neurons in such mice causes a reduction in the number of synaptic vesicles able to fuse with the plasma membrane.

The principal V-SNARE in synaptic vesicles is *VAMP* (*v*esicle-*a*ssociated *m*embrane *p*rotein), which also is called *synaptobrevin*. This V-SNARE tightly binds *syntaxin* and *SNAP25*, the principal T-SNAREs in the plasma membrane of axon terminals. As in the fusion of other intracellular vesicles with membranes, SNAP proteins and NSF assist in the disassociation of VAMP from

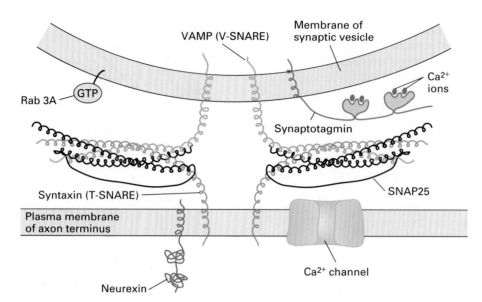

▲ **FIGURE 21-31 Synaptic-vesicle and plasma-membrane proteins important for vesicle docking and fusion.** Interaction of the T-SNAREs syntaxin and SNAP25 with the V-SNARE VAMP is aided by Rab3A. Neurexin, Ca^{2+} channels, and other plasma-membrane proteins localized to the synaptic region may also interact with synaptic-vesicle proteins. The vesicle protein synaptotagmin is the major Ca^{2+} sensor that triggers vesicle fusion. Not shown here is the synaptic-vesicle protein synaptophysin that may form part of the fusion pore between the synaptic vesicle and plasma membrane. The ubiquitous proteins NSF, α-SNAP, β-SNAP, and γ-SNAP assist in dissociation of the VAMP-syntaxin-SNAP25 complex after vesicle fusion. [See T. Südhof, 1995, *Nature* **375**:645; R. Scheller, 1995, *Neuron* **14**:893; and J. Littleton and H. Bellen, 1995, *Trends Neurosci.* **18**:177.]

T-SNAREs after vesicle fusion (see Figure 17-59). Strong evidence for the role of VAMP is provided by the mechanism of action of botulinum-B toxin, a bacterial protein that can cause the paralysis and death characteristic of *botulism,* a type of food poisoning. The toxin is composed of two polypeptides: One binds to motor neurons that release acetylcholine at synapses with muscle cells, facilitating entry of the other polypeptide, a protease, into the cytosol. The only protein this protease cleaves is VAMP; destruction of VAMP prevents acetylcholine release at the neuromuscular synapse, causing paralysis.

Ca^{2+}-Sensing Protein Another protein in the synaptic-vesicle membrane called *synaptotagmin* contains four Ca^{2+}-binding sites in its cytosolic domain (see Figure 21-31). Several types of evidence support the hypothesis that synaptotagmin is the key Ca^{2+}-sensing protein that triggers vesicle exocytosis. One piece of evidence is that in the presence of phospholipids the affinity of synaptotagmin for Ca^{2+} is 1 to 100 μM, consistent with concentrations of Ca^{2+} present in the active zone following opening of voltage-gated Ca^{2+} channels. Furthermore, injection of protein fragments derived from the cytosolic domain of synaptotagmin into a squid giant axon does not affect vesicle docking but does inhibit Ca^{2+}-stimulated vesicle exocytosis. Presumably these fragments compete with normal synaptotagmin protein for binding to critical target proteins. Additional evidence comes from synaptotagmin mutants of *Drosophila* and the nematode

C. elegans. Embryos that completely lack synaptotagmin fail to hatch and exhibit very reduced, uncoordinated muscle contractions. Larvae with partial loss-of-function mutations of synaptotagmin survive, but their neurons are defective in Ca^{2+}-stimulated vesicle exocytosis.

How synaptotagmin functions is beginning to be understood. At the low cytosolic Ca^{2+} levels found in resting cells, synaptotagmin apparently binds to a complex of the plasma-membrane proteins neurexin and syntaxin, perhaps facilitating vesicle docking with the membrane. The presence of synaptotagmin, however, blocks binding of other essential fusion proteins to the neurexin-syntaxin complex, thereby preventing vesicle fusion. When synaptotagmin binds Ca^{2+}, it is displaced from the complex, allowing other proteins to bind and thus initiating membrane docking or fusion. Thus synaptotagmin may operate as a "clamp" to prevent fusion from proceeding in the absence of a Ca^{2+} signal; by binding to proteins and phospholipids in the plasma membrane, it may also facilitate fusion of the two membranes.

Chemical Synapses Can Be Excitatory or Inhibitory

One way of classifying synapses is whether the action of the neurotransmitter tends to promote or inhibit the generation of an action potential in the postsynaptic cell. Binding of a neurotransmitter to an **excitatory receptor** opens a channel that admits Na^+ ions or both Na^+ and K^+ ions.

These non-voltage-gated ion channels can be part of the receptor protein or can be a separate protein that opens in response to a cytosolic signal generated by the activated receptor. Channel opening leads to depolarization of the postsynaptic plasma membrane, promoting generation of an action potential. In contrast, binding of a neurotransmitter to an **inhibitory receptor** on the postsynaptic cell causes opening of K^+ or Cl^- channels. The resulting membrane hyperpolarization inhibits generation of an action potential in the postsynaptic cell.

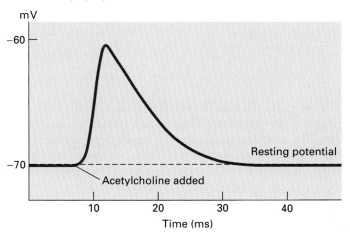

(a) Excitatory synapse

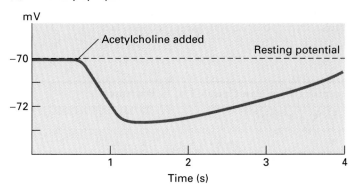

(b) Inhibitory synapse

▲ **FIGURE 21-32 Excitatory and inhibitory responses in postsynaptic cells stimulated by acetylcholine.** (a) Application of acetylcholine (or nicotine) to frog skeletal muscle produces a rapid postsynaptic depolarization of about 10 mV, which lasts 20 ms. The nicotinic acetylcholine receptors in these cells are ligand-gated cation channels; binding of acetylcholine opens the channel, admitting both Na^+ and K^+. (b) Application of acetylcholine (or muscarine) to frog heart muscle produces, after a lag period of about 40 ms (not visible in graph), a hyperpolarization of 2–3 mV, which lasts several seconds. These cells contain muscarinic acetylcholine receptors, which are coupled via a G protein to K^+ channels. Activation of the receptor leads to channel opening. Note the difference in time scales in the two graphs. [See H. C. Hartzell, 1981, *Nature* **291**:539.]

As illustrated in Figure 21-32, the same neurotransmitter (e.g., acetylcholine) can produce an excitatory response in some postsynaptic cells and an inhibitory response in others. Many nerve-nerve and most nerve-muscle chemical synapses are excitatory.

Two Classes of Neurotransmitter Receptors Operate at Vastly Different Speeds

Neurotransmitter receptors in the plasma membrane of postsynaptic cells fall into two broad classes: ligand-gated ion channels and G protein–coupled receptors. Synapses containing either type can be excitatory or inhibitory, but the two types vary greatly in the speed of their response.

Fast Synapses and Ligand-Gated Ion Channels The exoplasmic domain of a ligand-gated receptor possesses a neurotransmitter-binding site (see Figure 21-8c). Binding of the neurotransmitter causes an immediate conformational change that opens the channel portion of the protein, allowing ions to cross the membrane and causing the membrane potential to change within 0.1–2 milliseconds. Examples of excitatory and inhibitory ligand-gated receptors are listed in Table 21-1. Binding of ligand to excitatory receptors opens cation channels that allow passage of both Na^+ and K^+ ions, leading to rapid depolarization of the postsynaptic membrane; in contrast, binding of ligand to inhibitory receptors opens Cl^- channels, leading to hyperpolarization of the postsynaptic plasma membrane (see Figure 21-10).

Slow Synapses and Receptors Coupled to G Proteins Many functions of the nervous system operate with time courses of seconds or minutes; regulation of the heart rate, for instance, requires that action of neurotransmitters extend over several beating cycles measured in seconds. In general, the neurotransmitter receptors utilized in slow synapses are coupled to G proteins (Table 21-2). The sequence is similar to signaling in non-neuronal cells mediated by G protein–coupled receptors (Chapter 20). In a postsynaptic cell, binding of a neurotransmitter to this type of receptor activates a coupled G protein that, in most cases, directly binds to a separate ion-channel protein, causing an increase or decrease in its ion conductance. In other cases, the receptor-activated G protein activates adenylate cyclase or phospholipase C, triggering a rise in cytosolic cAMP or Ca^{2+}, respectively; these **second messengers** in turn affect the ion conductance of a linked ion channel. Examples of both types of G protein–coupled receptors are described in later sections. The postsynaptic responses induced by neurotransmitter binding to G protein–coupled receptors are intrinsically slower and longer lasting than those induced by ligand-gated channels. This is illustrated by the differing time scales of the responses to acetylcholine in skeletal muscle and heart muscle (see Figure 21-32).

As noted earlier, many neurons secrete neuropeptides whose effects tend to extend over a prolonged period. Not

TABLE 21-1 Neurotransmitter Receptors That Are Ligand-Gated Ion Channels

Functional Type	Ligand*	Ion Channel
Excitatory Receptors	Acetylcholine (nicotinic receptor)	Na^+/K^+
	Glutamate (NMDA class receptors)[†]	Na^+/K^+ and Ca^{2+}
	Glutamate (non-NMDA class receptors)[‡]	Na^+/K^+
	Serotonin (5HT$_3$ class receptors)	Na^+/K^+
Inhibitory Receptors	γ-Aminobutyric acid, GABA (A-class receptors)	Cl^-
	Glycine	Cl^-

*Most of these ligands also bind to receptors that are coupled to G proteins (see Table 21-2). Their receptors that are ion channels are indicated in parentheses.
[†]Opening of the cation channel in these receptors requires a slightly depolarized membrane and binding of glutamate (or N-methyl-D aspartate, NMDA).
[‡]Opening of the cation channel in these receptors only requires binding of glutamate.

SOURCE: H. Lester, 1988, *Science* **241**:1057; E. Barnard, 1988, *Nature* **335**:301; N. Unwin, 1993, *Cell* vol. 72/*Neuron* vol. 10 (suppl.), pp. 31–41; A. Maricq et al., 1991, *Science* **254**:432.

surprisingly, the receptors for this class of transmitters are coupled to G proteins (see Table 21-2). As we discuss in a later section, the receptor for light (rhodopsin) and the hundreds of different olfactory receptors that detect odorants also are linked to G proteins.

Acetylcholine and Other Transmitters Can Activate Multiple Receptors

The diversity of receptors for and responses to a single kind of neurotransmitter is illustrated by acetylcholine. Synapses in which acetylcholine is the neurotransmitter are termed *cholinergic* synapses. Acetylcholine receptors that cause excit-

atory responses lasting only milliseconds are called *nicotinic acetylcholine receptors*. They are so named because nicotine, like acetylcholine, causes a rapid depolarization (see Figure 21-32a). As noted already, these receptors are ligand-gated channels for Na^+ and K^+ ions. Other acetylcholine receptors are called *muscarinic acetylcholine receptors* because muscarine (a mushroom alkaloid) causes the same response as does acetylcholine. There are several subtypes of muscarinic acetylcholine receptors present in different cell types; all are coupled to G proteins, but they induce different responses. The M2 receptor present in heart muscle activates a G_i protein that causes the opening of a K^+ channel and thus a hyperpolarization lasting seconds (see Figure 21-32b).

TABLE 21-2 Some Neurotransmitter and Neuropeptide Receptors That Are Coupled to G Proteins

CLASSIC NEUROTRANSMITTERS

Acetylcholine* (muscarinic receptors)	GABA* (B-class receptors)
Adenosine	Glutamate*
ATP	Histamine
Dopamine	Serotonin* (5HT$_1$, 5HT$_2$, 5HT$_4$ receptors)
Epinephrine, norepinephrine	

NEUROPEPTIDES

Adrenocorticotropic hormone (ACTH)	Opiods (e.g., β-endorphin)
Bradykinin	Oxytocin
Cholecystokinin (CCK)	Tachykinins (e.g., substance P)
Endothelin	Thyrotropin-releasing hormone (TRH)
Gastrin	Vasoactive intestinal peptide (VIP)
Luteinizing hormone–releasing hormone (LHRH)	Vasopressin

*These neurotransmitters also bind to other receptors that are ligand-gated ion channels (see Table 21-1). Their G protein–coupled receptors are indicated in parentheses.

SOURCE: B. Hille, 1992, *Neuron* **9**:187.

The M1, M3, and M5 subtypes are coupled to other G proteins known as G_o or G_q and activate phospholipase C; the M4 subtype activates G_i and inhibits adenylate cyclase. Thus, a single neurotransmitter induces very different responses in different nerve and muscle cells, depending on the type of receptor found in the target cells.

Acetylcholine and the Neuromuscular Junction

Acetylcholine is released by motor neurons at synapses with muscle cells, often called *neuromuscular junctions*. Like other neurotransmitters, acetylcholine is synthesized in the cytosol of the presynaptic axon terminal and stored in synaptic vesicles. A single axon terminus of a frog motor neuron may contain a million or more synaptic vesicles, each containing 1000–10,000 molecules of acetylcholine; these vesicles often accumulate in rows in the active zone (Figure 21-33). Such a neuron can form synapses with a single skeletal muscle cell at several hundred points.

Acetylcholine is synthesized from acetyl coenzyme A (CoA) and choline in a reaction catalyzed by choline acetyltransferase:

$$CH_3-\overset{\overset{\displaystyle O}{\|}}{C}-S-CoA$$

Acetyl CoA

$$+ \ HO-CH_2-CH_2-N^+-(CH_3)_3 \ \xrightarrow{\text{choline acetyltransferase}}$$

Choline

$$CH_3-\overset{\overset{\displaystyle O}{\|}}{C}-O-CH_2-CH_2-N^+-(CH_3)_3 \ + \ CoA-SH$$

Acetylcholine

Synaptic vesicles take up and concentrate acetylcholine from the cytosol against a steep concentration gradient, using a H^+/acetylcholine antiporter in the vesicle membrane (see Figure 21-29). Curiously, the gene encoding this antiporter is contained entirely within the first intron of the gene encoding choline acetyltransferase, a mechanism conserved throughout evolution for ensuring coordinate expression of these two proteins.

Transmitter-Mediated Signaling Is Terminated by Several Mechanisms

Following release of a neurotransmitter or neuropeptide, it must be removed or destroyed to prevent continued stimulation of the postsynaptic cell. There are three main ways to end the signaling: The transmitter may (1) diffuse away from the synaptic cleft, (2) be taken up by the presynaptic neuron, or (3) be enzymatically degraded. Signaling by acetylcholine and neuropeptides is terminated by enzymatic degradation of the transmitter, but signaling by most of the classic neurotransmitters is terminated by uptake.

Hydrolysis of Acetylcholine After its release into the synaptic cleft, acetylcholine is hydrolyzed to acetate and choline by the enzyme *acetylcholinesterase*, which occurs

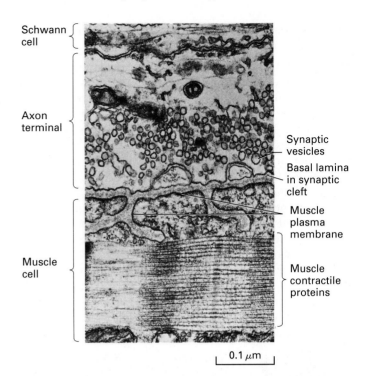

▲ **FIGURE 21-33 Longitudinal section through a frog nerve-muscle synapse (neuromuscular junction).** Synaptic vesicles in the axon terminal are clustered near the region where exocytosis occurs. The basal lamina lies in the synaptic cleft separating the neuron from the muscle membrane, which is extensively folded. Acetylcholine receptors are concentrated in the postsynaptic muscle membrane at the top and part way down the sides of the folds in the membrane. A Schwann cell surrounds the axon terminal. [From J. E. Heuser and T. Reese, 1977, in E. R. Kandel, ed., *The Nervous System*, vol. 1, *Handbook of Physiology*, Williams & Wilkins, p. 266.]

in several forms. The secreted form of this enzyme has a collagen-like subunit that anchors the enzyme to the basal lamina that fills the synaptic cleft between a neuron and muscle cell. The membrane-bound form is inserted into the plasma membrane by a glycosylphosphatidylinositol (GPI) anchor (see Figure 3-36a). Choline produced by hydrolysis of acetylcholine in the synaptic cleft is transported back into the nerve terminal by a Na^+/choline symporter, to be used in synthesis of more neurotransmitter. The operation of this transporter is similar to that of the Na^+/glucose symporters used to transport glucose into certain cells against a concentration gradient (see Figure 15-19).

During hydrolysis of acetylcholine by acetylcholinesterase, a serine at the active site reacts with the acetyl group forming an enzyme-bound intermediate. A large number of nerve gases and other neurotoxins inhibit the activity of acetylcholinesterase by reacting with the active-site serine. Physiologically, these toxins prolong the action of acetylcholine, thus extending the period of membrane depolarization. Such inhibitors can be lethal if they prevent relaxation of the muscles necessary for breathing.

Uptake of Neurotransmitters   With the exception of acetylcholine, all the neurotransmitters shown in Figure 21-28 are removed from the synaptic cleft by transport into the axon terminals that released them. Thus these transmitters are recycled intact. Transporters for GABA, norepinephrine, dopamine, and serotonin were the first to be cloned and studied. These four are encoded by a gene family, and are 60–70 percent identical in their amino acid sequences. Each transporter is thought to have 12 membrane-spanning α helices. All are Na^+/neurotransmitter symporters, and frequently Cl^- is transported along with the neurotransmitter. As with other Na^+ symporters, the movement of Na^+ into the cell down its electrochemical gradient provides the energy for uptake of the neurotransmitter (see Figure 15-9). Study and cloning of these transporters was facilitated by the observation that, following microinjection of mRNA from regions of the brain into frog oocytes, functional transporters for these neurotransmitters were expressed on the oocyte plasma membrane.

 The norepinephrine, serotonin, and dopamine transporters are all inhibited by cocaine. Binding of cocaine to the dopamine transporter inhibits dopamine uptake, thus prolonging signaling at key brain synapses; indeed, the dopamine transporter is the principal brain "cocaine receptor." Therapeutic agents such as the antidepressant drugs fluoxetine (Prozac) and imipramine block serotonin uptake,

and the tricyclic antidepressant desipramine blocks norepinephrine uptake. However, the precise role of transporters in the antidepressant action of these drugs is not yet clear.

Impulses Transmitted across Chemical Synapses Can be Amplified and Computed

Chemical synapses have two important advantages over electric ones in the transmission of impulses from a presynaptic cell. The first is *signal amplification,* which is common at nerve-muscle synapses. An action potential in a single presynaptic motor neuron can cause contraction of multiple muscle cells because release of relatively few signaling molecules at a synapse is all that is required to stimulate contraction.

The second advantage is *signal computation,* which is common at synapses involving interneurons, especially in the central nervous system. A single neuron can be affected simultaneously by signals received at multiple excitatory and inhibitory synapses (see Figure 21-3). The neuron continuously averages these signals and determines whether or not to generate an action potential. In this process, the various depolarizations and hyperpolarizations generated at synapses move by passive spread along the plasma membrane from the dendrites to the cell body and then to the axon hillock, where they are summed together. An action potential is generated whenever the membrane at the axon hillock becomes depolarized to a certain voltage called the **threshold potential** (Figure 21-34). Thus an action potential is generated in an

▶ **FIGURE 21-34 The threshold potential for generation of an action potential in a postsynaptic cell.** In this example, the presynaptic neuron is generating about one action potential every 4 milliseconds. Arrival of each action potential at the synapse causes a small change in the membrane potential at the axon hillock of the postsynaptic cell, in this example a depolarization of $\approx$5 mV. When multiple stimuli cause the membrane of this postsynaptic cell to become depolarized to the threshold potential, here approximately −40 mV, an action potential is induced in it.

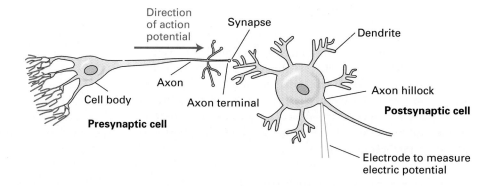

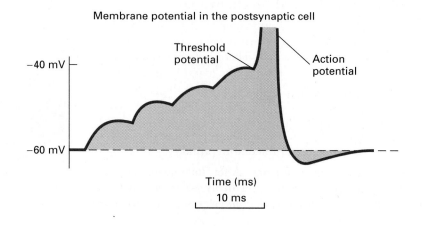

(a)

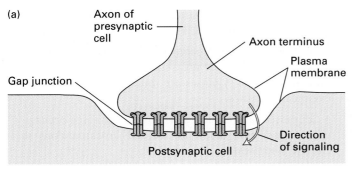

◀ FIGURE 21-35 An electric synapse. (a) The plasma membranes of the presynaptic and postsynaptic cells are linked by gap junctions. Flow of ions through these channels allows electric impulses to be transmitted directly from one cell to the other. (b) Negatively stained, electron microscopic image of the cytosolic face of a region of plasma membrane enriched in gap junctions; each "doughnut" forms a channel connecting two cells. [Part (b) courtesy of N. Gilula.]

all-or-nothing fashion: Depolarization to the threshold always leads to an action potential, whereas any depolarization that does not reach the threshold potential never induces it.

Whether a neuron generates an action potential in the axon hillock depends on the balance of the timing, amplitude, and localization of all the various inputs it receives; this signal computation differs for each type of interneuron. In a sense, each neuron is a tiny computer that averages all the receptor activations and electric disturbances on its membrane and makes a decision whether to trigger an action potential and conduct it down the axon. An action potential will always have the same *magnitude* in any particular neuron. The *frequency* with which action potentials are generated in a particular neuron is the important parameter in its ability to signal other cells.

Impulse Transmission across Electric Synapses Is Nearly Instantaneous

In an electric synapse, ions move directly from one neuron to another via gap junctions (Figure 21-35). The membrane depolarization associated with an action potential in the presynaptic cell passes through the gap junctions, leading to a depolarization, and thus an action potential, in the postsynaptic cell. Such cells are said to be electrically coupled. In the heart, for example, electric synapses allow groups of muscle cells to contract in synchrony. Gap junctions also connect nonexcitable cells, enabling small molecules such as cAMP and amino acids to pass from one cell to another (see Figure 22-8).

Electric synapses also have the advantage of speed and certainty; the direct transmission of impulses avoids the delay of about 0.5 ms that is characteristic of chemical synapses (Figure 21-36). In certain circumstances, this fraction of a millisecond advantage can mean the difference between life and death. Electric synapses in the goldfish brain, for example, mediate a reflex action that flaps the tail, which permits a fish to escape from predators. Examples also exist of electric coupling between groups of cell bodies and dendrites, ensuring simultaneous depolarization of an entire

group of coupled cells. The large number of electric synapses in many cold-blooded fishes suggests that they may be an adaptation to low temperatures, as the lowered rate of cellular metabolism in the cold reduces the rate of impulse transmission across chemical synapses.

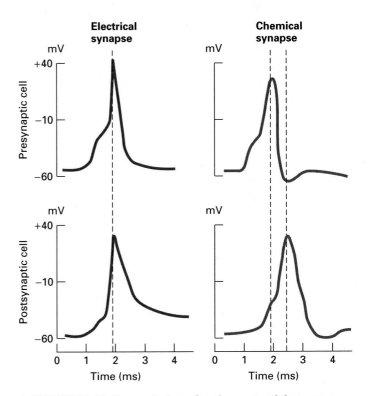

▲ FIGURE 21-36 Transmission of action potentials across electric and chemical synapses. In both cases, the presynaptic neuron was stimulated and the membrane potential was measured in both the presynaptic and postsynaptic cells (see Figure 21-7a). Signals are transmitted across an electric synapse within a few microseconds because ions flow directly from the presynaptic cell to the postsynaptic cell through gap junctions. In contrast, signal transmission across a chemical synapse is delayed about 0.5 ms—the time required for secretion and diffusion of neurotransmitter and the response of the postsynaptic cell to it.

SUMMARY Neurotransmitters, Synapses, and Impulse Transmission

- At chemical synapses, impulses are transmitted by the release of neurotransmitters from the axon terminal of the presynaptic cell into the synaptic cleft. Their subsequent binding to specific receptors on the postsynaptic cell causes a change in the ion permeability and thus the potential of the postsynaptic plasma membrane.

- Classic low-molecular-weight neurotransmitters are imported from the cytosol into synaptic vesicles by a proton-coupled antiporter. V-type ATPases maintain the low intravesicular pH that powers neurotransmitter import.

- Arrival of an action potential at a presynaptic axon terminal opens voltage-gated Ca^{2+} channels, inducing a localized rise in the cytosolic Ca^{2+} level that triggers exocytosis of synaptic vesicles. Following neurotransmitter release, vesicles are endocytosed and recycled (see Figure 21-29).

- Multiple cytosolic proteins including synapsin recruit synaptic vesicles to the active zone of the plasma membrane adjacent to the synaptic cleft.

- The principal V-SNARE in synaptic vesicles is VAMP (synaptobrevin), which tightly binds the principal plasma-membrane T-SNAREs—syntaxin and SNAP25—with the assistance of Rab3A and other docking and fusion proteins (see Figure 21-31). Synaptotagmin in the synaptic-vesicle membrane is thought to be the key Ca^{2+}-sensing protein that triggers exocytosis.

- Stimulation of excitatory receptors by neurotransmitter binding causes depolarization of the postsynaptic plasma membrane, promoting generation of an action potential. Conversely, stimulation of inhibitory receptors causes hyperpolarization of the postsynaptic membrane, repressing generation of an action potential.

- Neurotransmitter receptors that are ligand-gated channels induce rapid (millisecond) responses, whereas those that are coupled to G proteins induce responses that last seconds or more. Depending on the specific receptor, the same neurotransmitter can induce either an excitatory or inhibitory response.

- Removal of neurotransmitters from the synaptic cleft occurs by enzymatic degradation, re-uptake into the presynaptic cell, or diffusion.

- Chemical synapses allow a single postsynaptic cell to amplify, modify, and compute excitatory and inhibitory signals received from multiple presynaptic neurons. Such integration is common in the central nervous system.

- Postsynaptic cells generate action potentials in an all-or-nothing fashion when the plasma membrane at the axon hillock is depolarized to the threshold potential by the summation of small depolarizations and hyperpolar-

izations caused by activation of multiple neuronal receptors (see Figure 21-34).

- At electric synapses, ions pass directly from the presynaptic cell to the postsynaptic cell through gap junctions. These synapses are much less common than chemical synapses.

- Impulse transmission at chemical synapses occurs with a small time delay but is nearly instantaneous at electric synapses.

21.5 Neurotransmitter Receptors

The diversity of neurotransmitters is extensive, but as noted in the previous section, their receptors can be grouped into two broad classes: ligand-gated ion channels and G protein–coupled receptors (see Tables 21-1 and 21-2). In this section, we describe several important receptors in each class. By far the most-studied receptor is the muscle nicotinic acetylcholine receptor, the first ligand-gated ion channel to be purified, cloned, and characterized at the molecular level. The structure and mechanism of this receptor are understood in considerable detail, and it provides a paradigm for other neurotransmitter-gated ion channels. When activated, these receptors induce rapid changes, within a few milliseconds, in the permeability and potential of the postsynaptic membrane. In contrast, the postsynaptic responses triggered by activation of G protein–coupled receptors occur much more slowly, over seconds or minutes, because these receptors regulate opening and closing of ion channels indirectly. In many respects, neurotransmitter receptors in this class are structurally and functionally similar to the G protein–coupled receptors discussed in Chapter 20.

Opening of Acetylcholine-Gated Cation Channels Leads to Muscle Contraction

The nicotinic acetylcholine receptor, a ligand-gated cation channel, admits both K^+ and Na^+. Although found in some neurons, this receptor is best known for its role in synapses between motor neurons and skeletal muscle cells. Patch-clamping studies on isolated outside-out patches of muscle plasma membranes have shown that acetylcholine causes opening of a cation channel in the receptor capable of transmitting 15,000–30,000 Na^+ or K^+ ions a millisecond.

Since the resting potential of the muscle plasma membrane is near E_K, the potassium equilibrium potential, opening of acetylcholine receptor channels causes little increase in the efflux of K^+ ions; Na^+ ions, on the other hand, flow into the muscle cell. The simultaneous increase in permeability to Na^+ and K^+ ions produces a net depolarization to about -15 mV from the muscle resting potential of -85 to -90 mV. This depolarization of the muscle membrane generates an action potential, which—like an action potential in a neuron—is conducted along the membrane surface via voltage-gated Na$^+$ channels (see Figure 21-14). As detailed in Section 18.4, when the membrane depolarization reaches a specialized region, it triggers Ca^{2+} movement from its intracellular store, the sarcoplasmic reticulum, into the cytosol; the resultant rise in cytosolic Ca^{2+} induces muscle contraction (Figure 21-37).

Two factors greatly assisted in the characterization of the nicotinic acetylcholine receptor. First, this receptor can be rather easily purified from the electric organs of electric eels and electric rays; these organs are derived from stacks of muscle cells (minus the contractile proteins) and thus are richly endowed with this receptor. (In contrast, this receptor constitutes a minute fraction of the total membrane protein in most nerve and muscle tissues.) Second, α-bungarotoxin, a neurotoxin present in snake venom, binds specifically and irreversibly to nicotinic acetylcholine receptors. This toxin can be used in purifying the receptor by affinity chromatography and in localizing it. For instance, in autoradiographs of muscle-cell sections exposed to radioactive α-bungarotoxin, the toxin is localized in the plasma membrane of postsynaptic

striated muscle cells immediately adjacent to the terminals of presynaptic neurons.

Careful monitoring of the membrane potential of the muscle membrane at a synapse with a cholinergic motor neuron has demonstrated spontaneous, intermittent, and random ≈2-ms depolarizations of about 0.5–1.0 mV in the absence of stimulation of the motor neuron. Each of these depolarizations is caused by the spontaneous release of acetylcholine from a single synaptic vesicle. Indeed, demonstration of such spontaneous small depolarizations led to the notion of the *quantal release* of acetylcholine (later applied to other neurotransmitters) and thereby led to the hypothesis of vesicle exocytosis at synapses. The release of one acetylcholine-containing synaptic vesicle results in the opening of about 3000 ion channels in the postsynaptic membrane, far short of the number needed to reach the threshold depolarization that induces an action potential. Clearly, stimulation of muscle contraction by a motor neuron requires the nearly simultaneous release of acetylcholine from numerous synaptic vesicles.

All Five Subunits in the Nicotinic Acetylcholine Receptor Contribute to the Ion Channel

The acetylcholine receptor from skeletal muscle is a pentameric protein with a subunit composition of $\alpha_2\beta\gamma\delta$. Each molecule has a diameter of about 9 nm and protrudes about 6 nm into the extracellular space and about 2 nm into the cytosol (Figure 21-38a). The α, β, γ, and δ subunits have considerable sequence homology; on average, about 35–40 percent of the residues in any two subunits are similar. The complete receptor has a fivefold symmetry, and the actual cation channel is a tapered central pore, with a maximum diameter of 2.5 nm, formed by segments from each of the five subunits (Figure 21-38b).

The channel opens when the receptor cooperatively binds two acetylcholine molecules to sites located at the interfaces of the αδ and αγ subunits. Once acetylcholine is bound to a receptor, the channel is opened virtually instantaneously, probably within a few microseconds. Studies measuring the permeability of different small cations suggest that the open ion channel is, at its narrowest, about 0.65–0.80 nm in diameter, in agreement with estimates from electron micrographs. This would be sufficient to allow passage of both Na$^+$ and K$^+$ ions with their bound shell of water molecules (see Figure 2-14).

Although the structure of the central ion channel is not known in molecular detail, much evidence indicates that it is lined by five transmembrane M2 α helices, one from each of the five subunits. The M2 helices are composed largely of hydrophobic or uncharged polar amino acids, but negatively charged aspartate or glutamate residues are located at each end, near the membrane faces, and several serine or threonine residues are near the middle. If a single negatively charged glutamate or aspartate in one subunit is mutated to a positively charged lysine, and the mutant mRNA is injected

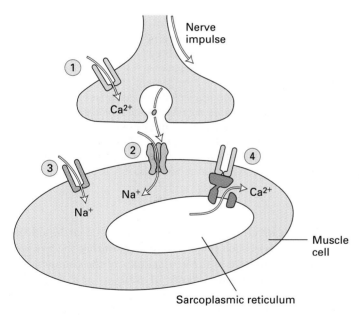

▲ **FIGURE 21-37 Sequential activation of gated ion channels at a neuromuscular junction.** Arrival of an action potential at the terminus of a presynaptic motor neuron induces opening of voltage-gated Ca^{2+} channels (step ①) and subsequent release of acetylcholine, which triggers opening of the ligand-gated nicotinic receptors in the muscle plasma membrane (step ②). The resulting influx of Na$^+$ produces a localized depolarization of the membrane, leading to opening of voltage-gated Na$^+$ channels and generation of an action potential (step ③). When the spreading depolarization reaches T tubules, it triggers opening of voltage-gated Ca^{2+}-release channels and release of Ca^{2+} from the sarcoplasmic reticulum into the cytosol (step ④). The rise in cytosolic Ca^{2+} causes muscle contraction by mechanisms discussed in Section 18.4

(a)

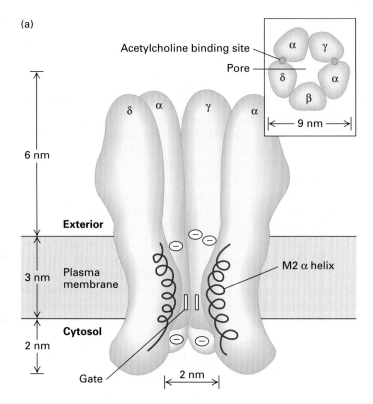

Acetylcholine binding site

Pore

α γ

δ α

β

9 nm

δ α γ α

6 nm

Exterior

3 nm — Plasma membrane

M2 α helix

Cytosol

2 nm

Gate

2 nm

(b)

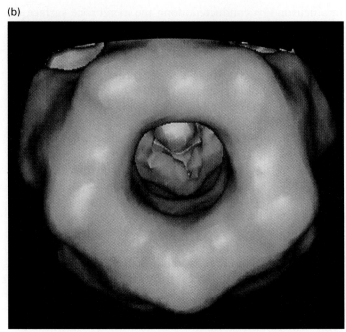

▲ **FIGURE 21-38 Three-dimensional structure of the nicotinic acetylcholine receptor based on amino acid sequence data, computer-generated averaging of high-resolution electron micrographs, and information from site-specific mutations.** (a) Schematic cutaway model of the pentameric receptor in the membrane; for clarity, the β subunit is not shown. Most of the mass of the protein protrudes from the outer (synaptic) surface of the plasma membrane. The M2 α helix (red) in each subunit is part of the lining of the ion channel. Aspartate and glutamate side chains at both ends of each M2 helix form two rings of negative charges that help exclude anions from and attract cations to the channel. The gate, which is opened by binding of acetylcholine, lies within the pore. *Inset:* Cross section of the exoplasmic face of the receptor showing the arrangement of subunits around the central pore. The two acetylcholine-binding sites are at the interfaces of the αδ and αγ subunits; they are located about 3 nm from the membrane surface. (b) A view from above, looking into the synaptic entrance of the channel. The tunnel-like entrance narrows abruptly after a length of about 6 nm. [Part (b) From N. Unwin, 1993, *Cell* vol. 72/*Neuron* vol. 10 (suppl.), p. 31.]

together with mRNAs for the other three wild-type subunits into frog oocytes, a functional channel is expressed, but its ion conductivity—the number of ions that can cross it during the open state—is reduced. The greater the number of glutamate or aspartate residues mutated (in one or multiple subunits), the greater the reduction in conductivity. These findings suggest that aspartate and glutamate residues —one residue from each of the five chains—form a ring of negative charges on the external surface of the pore that help to screen out anions and attract Na$^+$ or K$^+$ ions as they enter the channel (see Figure 21-38a). A similar ring of negative charges lining the cytosolic pore surface also helps select cations for passage.

The two acetylcholine-binding sites in the extracellular domain of the receptor lie ≈4 to 5 nm from the center of the pore. Binding of acetylcholine thus must trigger conformational changes in the receptor subunits that can cause channel opening at some distance from the binding sites. Receptors in isolated postsynaptic membranes can be trapped in the open or closed state by rapid freezing in liquid nitrogen. Images of such preparations suggest that the five M2 helices rotate relative to the vertical axis of the channel during opening and closing (Figure 21-39).

Two Types of Glutamate-Gated Cation Channels May Function in a Type of "Cellular Memory"

The hippocampus is the region of the mammalian brain associated with many types of short-term memory. Certain types of hippocampal neurons, here simply called postsynaptic cells, receive inputs from hundreds of presynaptic cells. In **long-term potentiation** a burst of stimulation of a postsynaptic neuron makes it more responsive to subsequent stimulation by presynaptic neurons. For example, stimulation of a hippocampal presynaptic nerve with 100 depolarizations acting over only 200 milliseconds causes an increased sensitivity of the postsynaptic neuron that lasts hours

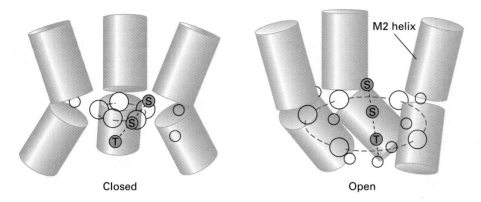

M2 helix

Closed   Open

◀ FIGURE 21-39 Schematic three-dimensional models of the pore-lining M2 helices in the closed and opened states. In the closed state, the kink in the center of each M2 helix points inward, constricting the passageway (red circle). In the open state, the kinks rotate to one side, so that the helices are farther apart. The green circles denote the hydroxyl groups of serine (S) and threonine (T) residues in the center of the M2 helices; in the open state these are parallel to the channel axis and allow ions to flow. [Adapted from N. Unwin, 1995, *Nature* **373**:37.]

to days. Changes in the responses of postsynaptic cells may underlie certain types of memory.

Two types of glutamate-gated cation channels in the postsynaptic neuron participate in long-term potentiation. Like other neurotransmitter-gated ion channels, both glutamate receptors have five subunits, each containing a pore-lining M2 helix; both are excitatory receptors, causing depolarization of the plasma membrane when activated. Because the two receptors were initially distinguished by their ability to be activated by the non-natural amino acid N-methyl-D-aspartate (NMDA), they are called *NMDA glutamate receptors* and *non-NMDA glutamate receptors*.

As illustrated in Figure 21-40, non-NMDA receptors are "conventional" in that binding of glutamate, released from the presynaptic cell, triggers their opening. NMDA glutamate receptors are different in two key respects. First, they allow influx of Ca^{2+} as well as Na^+. Second, and more important, *two* conditions must be fulfilled for the ion channel to open: glutamate must be bound and the membrane must be partly depolarized. In this way, the NMDA receptor functions as a **coincidence detector**; that is, it integrates activity of the postsynaptic cell—reflected in its depolarized plasma membrane—with release of neurotransmitter from the presynaptic cell, generating a cellular response greater than that caused by glutamate release alone. Once a postsynaptic cell becomes "sensitized," it takes fewer action potentials in the presynaptic neurons to induce a given depolarization in the postsynaptic neuron; in other words, the synapse "learns" to have an enhanced response to signals from the presynaptic cells.

Opening of NMDA receptors depends on membrane depolarization because of the voltage-sensitive blocking of the ion channel by a Mg^{2+} ion from the extracellular solution. A small depolarization of the membrane causes the Mg^{2+} ion to dissociate from the receptor, thereby making it possible for glutamate binding to open the channel. Mutagenesis of a single asparagine residue in the pore-lining M2 helix of the NMDA receptor abolishes the effect of Mg^{2+}, indicating that Mg^{2+} binds in the channel.

Since activation of a single synapse, even at high frequency, generally causes only a small depolarization of the membrane of the postsynaptic cell, long-term potentiation is induced only when many synapses simultaneously stimulate a single postsynaptic neuron. Thus the requirement for membrane depolarization explains why long-term potentiation depends on the simultaneous activation of a large number of synapses on the postsynaptic cell.

GABA- and Glycine-Gated Cl⁻ Channels Are Found at Many Inhibitory Synapses

Synaptic inhibition in the vertebrate central nervous system is mediated primarily by two amino acids, glycine and γ-aminobutyric acid (GABA); the latter is formed from glutamate by loss of a carboxyl group. The concentration of GABA in the human brain is 200—1000 times higher than that of other neurotransmitters such as dopamine, norepinephrine, and acetylcholine. Glycine is the major inhibitory neurotransmitter in the spinal cord and brain stem; GABA predominates elsewhere in the brain. Both glycine and GABA activate ligand-gated Cl⁻ channels.

The opening of Cl⁻ channels tends to drive the membrane potential toward the Cl⁻ equilibrium potential E_{Cl}, which in general is slightly more negative than the resting membrane potential (see Figure 21-10). In other words, the membrane becomes slightly hyperpolarized. If many Cl⁻ channels are opened, the membrane potential will be held near E_{Cl}, and a much larger than normal increase in the Na^+ permeability will then be required to depolarize the membrane. The effect of GABA or glycine on Cl⁻ permeability is induced rapidly (a fraction of a millisecond) but can last for a second or more, a long time compared with the millisecond required to generate an action potential. Thus GABA or glycine rapidly induces a fairly long-lasting inhibitory postsynaptic response.

GABA and glycine receptors have been purified, cloned, and sequenced. Although they are pentameric proteins, like the nicotinic acetylcholine receptor, GABA and glycine receptors

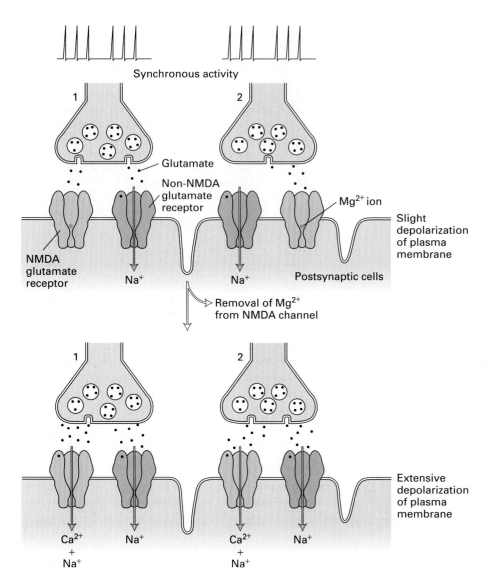

▶ **FIGURE 21-40 Different properties of two types of glutamate receptors found in the hippocampus region in the brain.** Because the ion channel in the NMDA receptor (green) normally is blocked by a Mg^{2+} ion, the glutamate released by firing of presynaptic neurons leads, at first, to opening of only the non-NMDA glutamate receptors (pink). The resultant influx of Na^+ partially depolarizes the membrane. If many presynaptic neurons (here two are shown) fire in synchrony, the membrane of the postsynaptic cell becomes sufficiently depolarized so that the Mg^{2+} ions blocking the NMDA receptors are removed; then both the NMDA and the non-NMDA glutamate receptors open in response to glutamate. Ca^{2+} ions as well as Na^+ ions enter through the open NMDA receptors, causing an enhanced response in the postsynaptic cells. [See T. M. Jessell and E. R. Kandel, 1993, *Cell* vol. 72/*Neuron* vol. 10 (suppl.), p. 1; and C. F. Stevens, 1993, *Cell* vol. 72/*Neuron* vol. 10 (suppl.), p. 55.]

are built of only one or two different types of subunits. Each subunit has a transmembrane M2 helix; these are thought to line the ion channel as in the nicotinic acetylcholine receptor. As mentioned previously, the negatively charged glutamate and aspartate side chains at the ends of the M2 helices in acetylcholine receptors may participate in selecting cations for passage (see Figure 21-38a). Strikingly, the M2 helices of the GABA and glycine receptor subunits have lysine or arginine residues at these positions; the positively charged side chains of these residues may attract Cl^- ions specifically and aid in repelling cations.

Cardiac Muscarinic Acetylcholine Receptors Activate a G Protein That Opens K^+ Channels

We saw earlier that binding of acetylcholine to muscarinic acetylcholine receptors in cardiac muscle generates a slow inhibitory response (see Figure 21-32b). Stimulation of the cholinergic nerves in heart muscle, which causes a long-lived (several seconds) hyperpolarization of the membrane, is one of the principal ways by which the rate of heart muscle contraction is slowed.

The muscarinic acetylcholine receptor is a G protein–coupled receptor whose activation leads to opening of K^+ channels and subsequent hyperpolarization of the plasma membrane. Like other G protein–coupled receptors, the muscarinic acetylcholine receptor has seven transmembrane α helices (see Figure 20-10). As depicted in Figure 21-41, binding of acetylcholine to the receptor activates a trimeric transducing G protein; the released $G_{\beta\gamma}$ subunit then directly binds to and opens a particular K^+ channel protein. That $G_{\beta\gamma}$ directly activates the K^+ channel has been shown by single-channel recording experiments in which purified $G_{\beta\gamma}$ was added to the cytosolic face of a patch of heart muscle plasma membrane (see Figure 20-21b). Potassium channels opened immediately on addition of $G_{\beta\gamma}$ in the absence of

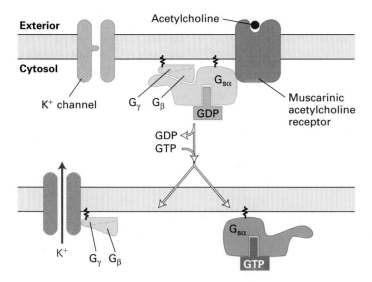

Exterior

Acetylcholine

Cytosol

K⁺ channel

G_γ G_β

$G_{s\alpha}$

GDP

Muscarinic
acetylcholine
receptor

GDP

GTP

K⁺

G_γ G_β

$G_{s\alpha}$

GTP

▲ **FIGURE 21-41 Acetylcholine-induced opening of K⁺ channels in the heart muscle plasma membrane.** Binding of acetylcholine by muscarinic acetylcholine receptors triggers activation of a transducing G protein by catalyzing exchange of GTP for GDP on the α subunit. The released $G_{\beta\gamma}$ subunit then binds to and opens a K⁺ channel. The increase in K⁺ permeability hyperpolarizes the membrane, which reduces the frequency of heart muscle contraction. Though not shown here, activation is terminated when the GTP bound to G_α is hydrolyzed to GDP and $G_\alpha \cdot$ GDP recombines with $G_{\beta\gamma}$. [See K. Dunlap et al., 1987, *Trends Neurosci.* **10**:241; E. Cerbai et al., 1988, *Science* **240**:1782; K. Ho et al., 1993, *Nature* **362**:31; and Y. Kubo et al., 1993, *Nature* **362**:127.]

acetylcholine or other neurotransmitters. The K⁺ channels coupled to muscarinic acetylcholine receptors are tetrameric proteins similar in structure to those that maintain the resting membrane potential.

The cardiac muscarinic receptor illustrates one way in which G protein–coupled receptors affect ion channels: the active $G_{\beta\gamma}$ subunit binds to a channel protein. Activation of other G protein–coupled neurotransmitter receptors affects the activity of enzymes that synthesize or degrade intracellular second messengers; these, in turn, can affect the activity of channel proteins. To illustrate this type of receptor, we examine the catecholamine receptors.

Catecholamine Receptors Induce Changes in Second-Messenger Levels That Affect Ion-Channel Activity

Epinephrine and norepinephrine function as both systemic hormones and neurotransmitters. Norepinephrine is the transmitter at synapses with smooth muscles that are innervated by sympathetic autonomic motor neurons (see Figure 21-6). Stimulation of these peripheral neurons increases the activity of the heart and internal organs in "fight or flight" reactions. Norepinephrine is also found at synapses in the central nervous system. Epinephrine is synthesized and released

into the blood by the adrenal medulla, an endocrine organ that has a common embryologic origin with neurons of the sympathetic system. Unlike neurons, the medulla cells do not develop axons or dendrites. Epinephrine, norepinephrine, and the related neurotransmitter dopamine are all synthesized from tyrosine and contain the catechol moiety; hence they are referred to as **catecholamines** (see Figure 21-28). Nerves that synthesize and use epinephrine or norepinephrine are termed *adrenergic*.

All known receptors for catecholamines are coupled to G proteins. Because different receptors are linked to different G proteins, their activation leads to changes in the levels of different intracellular second messengers. For instance, binding of norepinephrine to β-adrenergic receptors on nerve cells causes activation of G_s and an increase in cAMP synthesis, the same mechanism by which β-adrenergic receptors function in non-neuronal cells (see Figure 20-16). Other neuronal adrenergic receptors activate G_i, G_o, or other types of G proteins, resulting in a decrease in cAMP levels or increases in the levels of other intracellular second messengers, such as cGMP, inositol 1,4,5-trisphosphate (IP_3), diacylglycerol, and arachidonic acid (see Table 20-5). Some second messengers, such as cGMP and IP_3, act to directly open or close ion channels in neurons; IP_3, for example, opens Ca^{2+} channels in the membrane of the endoplasmic reticulum, causing an increase in cytosolic Ca^{2+}. Other second messengers have a more indirect effect on ion channels, as exemplified by the serotonin receptor, another G protein coupled receptor.

A Serotonin Receptor Indirectly Modulates K⁺ Channel Function by Activating Adenylate Cyclase

Often an axon terminal of one neuron synapses with the axon terminal of another neuron. Such a **modulatory synapse** may either inhibit or stimulate the ability of the second axon terminal to release its neurotransmitter and signal a third cell. The operation of one such modulatory synapse in the sea slug *Aplysia* demonstrates the effect of an increase in cAMP on ion-channel function. In this example, a particular type of interneuron, called a *facilitator neuron,* forms a synapse with the axon terminal of a sensory neuron that stimulates a motor neuron by releasing glutamate. Stimulation of the facilitator neuron increases the ability of the sensory neuron to stimulate the motor neuron.

As illustrated in Figure 21-42, when the facilitator neuron is stimulated, it secretes serotonin, which binds to serotonin receptors on the sensory neuron (steps 1 and 2). This binding activates adenylate cyclase, triggering the synthesis of cAMP in the sensory neuron (step 3). Subsequent activation of cAMP-dependent protein kinase leads to phosphorylation of a voltage-gated K⁺ channel protein or an associated protein, thereby preventing opening of the K⁺ channels during an action potential (steps 4 and 5). This inhibition decreases the outward flow of K⁺ ions that normally repolarizes

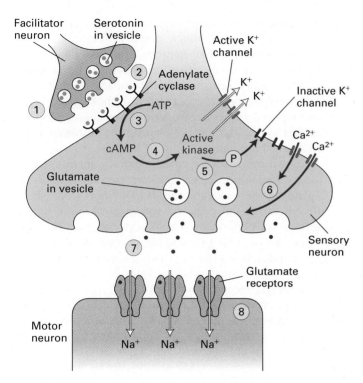

▲ **FIGURE 21-42 Action of a serotonin modulatory synapse in the sea slug *Aplysia punctata.*** Serotonin secreted by an activated facilitator neuron binds to the G protein-coupled serotonin receptors, leading to activation of adenylate cyclase and an increase in cAMP in the sensory neuron. Phosphorylation of the voltage-gated K+ channel protein or a channel-binding protein (circled P) prevents the K+ channels from opening, leading to prolonged depolarization during arrival of an action potential. This leads to enhanced secretion of the neurotransmitter glutamate, which stimulates the motor neuron. See text for discussion. [See E. R. Kandel and J. Schwartz, 1982, *Science* **218**:433; D. Glanzman, 1995, *Trends Neurosci.* **18**:30; and H. Lechner and J. Byrne, 1998, *Neuron* **20**:355.]

the membrane of the sensory neuron after an action potential reaches the axon terminal. The resulting prolonged membrane depolarization increases the influx of Ca^{2+} ions through voltage-gated Ca^{2+} channels (step 6). The increased Ca^{2+} level leads to greater exocytosis of glutamate-containing synaptic vesicles in the sensory neuron (step 7), and hence greater activation of the motor neuron (step 8) each time an action potential reaches the terminal.

As evidence for this model, direct administration of serotonin through a micropipette to the sensory neuron causes decreased efflux of K^+ ions and prolongs depolarization of the membrane induced during an action potential. Also, the *Aplysia* sensory neuron is large enough that the active catalytic subunit of the cAMP-dependent protein kinase can be injected into it. Such treatment mimics the effect of applying serotonin to the outside of the cell. Additional supporting evidence that serotonin acts by means of cAMP and a

protein kinase has come from patch-clamping studies on isolated inside-out pieces of sensory neuron plasma membrane (see Figure 21-20b). When both ATP and the purified active catalytic subunit of cAMP-dependent protein kinase are added to the cytosolic surface of the patches, the K^+ channels close. Thus the protein kinase indeed acts on the cytosolic surface of the membrane to phosphorylate the channel protein itself or a membrane protein that regulates channel activity. We shall return to this particular synapse at the last section of this chapter, as these modifications in synapse efficiency are part of a simple learning response.

Some Neuropeptides Function as Both Transmitters and Hormones

The receptors for many of the small neuropeptides found in nervous tissue have been cloned; all are coupled to G proteins and have the characteristic seven membrane-spanning α helices. Thus, the intracellular signaling pathways induced by neuropeptides are the same as those induced by the classical neurotransmitters that activate G protein–coupled receptors.

Many neuropeptides function as synaptic neurotransmitters; others act in a paracrine fashion as "diffusible" hormones that affect many nearby neurons (see Figure 20-1). Yet other neuropeptides act as regulators of nerve cell growth and division. Many of the neuropeptides listed in Table 21-2 are found both in the brain and in non-neural tissues. However, in contrast to capillaries in other parts of the body, capillaries in the brain are essentially impermeable to peptides. Thus, any peptide hormones traveling through the body in the blood will be excluded from the brain: this constitutes the blood-brain barrier. Hormones in the blood do not "confuse" the functioning of the central nervous system.

Neurons that secrete peptide hormones, called neurosecretory cells, were first discovered in the hypothalamus. Secretion of peptide hormones by the anterior cells of the pituitary gland is controlled by the hypothalamus, which in turn is regulated by other regions of the brain. The hypothalamus is connected to the anterior pituitary by a special closed system of blood vessels. Hypothalamic neurons secrete hypothalamic peptide hormones into these vessels, and the hormones then bind to receptors on the anterior pituitary cells. One such hypothalamic hormone, thyrotropin-releasing hormone (TRH), stimulates secretion by the anterior pituitary of prolactin and thyrotropin. Another hypothalamic hormone, luteinizing hormone–releasing hormone (LHRH), causes other cells in the anterior pituitary to secrete follicle-stimulating hormone (FSH) and luteinizing hormone (LH), which are important in regulating the growth and maturation of oocytes in the ovary.

In contrast to serotonin and catecholamines, but like acetylcholine, neurohormones and neurotransmitters are used only once and then are degraded by extracellular proteases; they are not recycled.

SUMMARY Neurotransmitter Receptors

- In general, neurotransmitter receptors that are ligand-gated ion channels mediate rapid postsynaptic responses, whereas G protein–coupled receptors mediate slow postsynaptic responses.

- At the synapse of a motor neuron and striated muscle cell, binding of acetylcholine to nicotinic acetylcholine receptors triggers a rapid increase in permeability of the membrane to both Na^+ and K^+ ions, leading to depolarization, an action potential, and then contraction (see Figure 21-37).

- The nicotinic acetylcholine receptor and other neurotransmitter receptors that are ligand-gated ion channels contain five subunits (see Figure 21-38). Each subunit contains a transmembrane α helix (M2) that lines the channel. Neurotransmitter binding to the receptor triggers a conformational change leading to channel opening.

- Glutamate, the principal excitatory neurotransmitter in the mammalian brain, binds to two types of ligand-gated cation channels: NMDA and non-NMDA glutamate receptors. Activation of NMDA receptors requires both partial membrane depolarization *and* glutamate binding; these receptors may function in a type of "cellular memory."

- GABA and glycine are the principal inhibitory neurotransmitters; their receptors are ligand-gated Cl^- channels.

- Binding of acetylcholine to muscarinic acetylcholine receptors in heart muscle causes dissociation of the coupled trimeric G protein; the released $G_{\beta,\gamma}$ subunit binds to and opens a K^+ channel protein (see Figure 21-41). The resulting influx of K^+ ions hyperpolarizes the cell membrane, slowing heart contraction.

- Stimulation of the G protein–coupled catecholamine receptors leads to an increase or decrease in cAMP or other intracellular second messengers. These and other G protein–coupled neurotransmitter receptors contain seven transmembrane α helices and induce signaling pathways similar to those in non-neuronal cells.

- Stimulation of G protein–coupled serotonin receptors in sensory neurons of the sea slug *Aplysia* increases the ability of the sensory neurons to activate postsynaptic motor neurons. This modulatory effect results from closing of K^+ channels induced by the rise in cytosolic cAMP following serotonin binding (see Figure 21-42).

- Many small peptides released by neurons function as paracrine hormones as well as neurotransmitters, affecting both nearby secretory cells and adjacent neurons. The receptors for these neuropeptides are coupled to G proteins.

21.6 Sensory Transduction

The nervous system receives input from a large number of sensory receptors (see Figure 21-6). Photoreceptors in the eye, taste receptors on the tongue, odorant receptors in the nose, and touch receptors on the skin monitor various aspects of the outside environment. Stretch receptors surround many muscles and fire when the muscle is stretched. Internal receptors monitor the levels of glucose, salt, and water in body fluids. The nervous system, the brain in particular, processes and integrates this vast barrage of information and coordinates the response of the organism.

Since the "language" of the nervous system is electric signals, each of the many types of receptor cells must convert, or *transduce,* its sensory input into an electric signal. A few sensory cells are themselves neurons that generate action potentials in response to stimulation. However, most are specialized epithelial cells that do not generate action potentials but synapse with and stimulate adjacent neurons that then generate action potentials (see Figure 21-1c). The key question that we will consider is how a sensory cell transduces its input into an electric signal.

Mechanoreceptors and Some Other Sensory Receptors Are Gated Cation Channels

Some sensory receptors are gated Na^+ or Na^+/Ca^{2+} channels that open in response to various stimuli; activation of such receptors causes an influx of ions, leading to membrane depolarization. Examples include the stretch and touch receptors that are activated by stretching of the cell membrane; these have been identified in a wide array of cells, ranging from vertebrate muscle and epithelial cells to yeast, plants, and even bacteria.

The cloning of genes encoding touch receptors began with the isolation of mutant strains of the nematode *Caenorhabditis elegans* that were insensitive to touch. Three of the genes in which mutations were isolated— *MEC4, MEC6,* and *MEC10*—encode three similar subunits of a Na^+ channel in the touch-receptor cells. These gated channels are necessary for touch sensitivity and may open directly in response to mechanical stimulation. Similar kinds of channels are found in prokaryotes and lower eukaryotes; by opening in response to membrane stretching, these channels may play a role in osmoregulation and the control of a constant cell volume. It is thought that such receptors were among the first sensory receptors to evolve, though no homologs have yet been identified in mammals.

Of the four taste stimuli used by vertebrates (salty, sweet, bitter, and sour), the "receptors" for salt are the best understood. These are simply ungated Na^+ channels in the apical membrane of taste-receptor cells; elevation of the extracellular Na^+ causes entry of Na^+ through these channels and thus depolarization of the membrane. A different sensory protein is the receptor for capsaicin*, the molecule that makes chili peppers seem hot. In this case, the receptor is a capsaicin-gated Na^+/Ca^{2+} channel that is found in many sensory pain neurons. When activated, the influx of Na^+ and Ca^{2+} ions depolarizes these cells, initiating a nerve impulse that goes to the brain. Interestingly, the capsaicin receptor is also activated by elevated temperatures ($\approx 48°C$) that produce pain, possibly explaining why we perceive foods that contain capsaicin as "hot."

More often, though, the connection between a sensory receptor protein and an ion channel is indirect; the sensory receptor activates a G protein that, in turn, directly or indirectly induces the opening or closing of ion channels. The light-sensing cells *(photoreceptors)* in the mammalian retina function in this manner, as do the chemical-sensing cells *(odorant receptors)* in the nose. We will discuss the light-sensing system in some detail, as it is one of the best-understood sensory systems and also illustrates how a sensory system can adapt to varying intensities of stimuli, in this case the level of ambient light.

Visual Signals Are Processed at Multiple Levels

The human retina contains two types of photoreceptors, *rods* and *cones,* that are the primary recipients of visual stimulation. Cones are involved in color vision, while rods are stimulated by weak light like moonlight over a range of wavelengths. In bright light, such as sunlight, the rods become inactive, for reasons that we will discuss.

The photoreceptors synapse on layer upon layer of interneurons that are innervated by different combinations of photoreceptor cells. Some, like bipolar and ganglion cells, are in the retina (Figure 21-43); others are in several places in the brain. Different interneurons have *receptive fields* of different shapes; that is, they are stimulated by only certain photoreceptors. This property allows distinct patterns of light to be recognized. Ganglion cells transmit distinct visual characteristics in multiple parallel pathways; such parallel processing allows the separation of different aspects of the visual stimulus, such as color, form, and motion. All of these signals are processed and interpreted by the part of the brain called the *visual cortex.*

Neurons in each subdivision of the cortex are responsive to different aspects of an object in the visual field, such as its orientation or its movement in a particular direction, its color, or its depth. At a variety of levels, from the retina up to the higher areas of the visual cortex, certain interneurons

are activated optimally in response to direct stimulation of their receptive field but are inhibited by stimulation coming from just outside their receptive field. This phenomenon, called *center-surround,* allows the appreciation of areas of contrast between adjacent areas of the visual field. Neurons in the visual system do not form a picture in the brain, but somehow we can interpret all of these multiple, parallel signals generated by "seeing" an object.

The Light-Triggered Closing of Na^+ Channels Hyperpolarizes Rod Cells

Membrane disks in the outer segments of rod cells contain **rhodopsin**, a light-sensitive protein also known as visual purple (Figure 21-44). In the dark, the membrane potential of a rod cell is about -30 mV, considerably less than the resting potential (-60 to -90 mV) typical of neurons and other electrically active cells. As a consequence of this depolarization, rod cells in the dark are constantly secreting neurotransmitters, and the bipolar neurons with which they synapse are continually being stimulated. Absorption of a pulse of light by rhodopsin in the outer segment of a rod cell causes the electric potential across the plasma membrane to become slightly *more* negative, to about -35 mV (Figure 21-45). This light-induced hyperpolarization in the outer segment spreads to the synaptic body and causes a decrease in neurotransmitter release.

The depolarized state of the plasma membrane of resting, dark-adapted rod cells is due to the presence of a large number of open *nonselective* ion channels that admit Na^+ and Ca^{2+} as well as K^+. The effect of light is to close these channels, causing the membrane potential to become more negative (see Figure 21-10). The more photons absorbed, the more channels are closed, the fewer Na^+ ions cross the membrane from the outside, the more negative the membrane potential becomes, and the less neurotransmitter is released.

Remarkably, a single photon absorbed by a resting rod cell produces a measurable response, a hyperpolarization of about 1 mV, which in amphibians lasts a second or two. Humans are able to detect a flash of as few as five photons; these dim flashes have a maximum effect within ≈ 150 ms and the response returns to baseline within ≈ 200 ms. A single photon blocks the inflow of about 10 million Na^+ ions due to the closure of hundreds of channels. Only about 30–50 photons need to be absorbed by a single rod cell in order to cause half-maximal hyperpolarization. The photoreceptors in rod cells, like many other types of receptors, exhibit the phenomenon of adaptation. That is, more photons are required to cause hyperpolarization if the rod cell is continuously exposed to light than if it is kept in the dark.

Let us now turn to three key questions: how is light absorbed; how is the signal transduced into the closing of ion channels; and how does the rod cell adapt to large variations in light intensity?

---

*Capsaicin is so potent that its effects are calibrated by the spice industry as Scoville heat units, a scale developed by Wilbur Scoville in 1912. He calibrated the potency of peppers by diluting alcoholic extracts until he could just detect the pungency after placing a drop on his tongue. According to this scale, Bell peppers have a potency of <1 unit; jalapeño peppers, 10^3; habenero peppers, 10^5; and pure capsaicin >10^7. [See D. Clapham, 1997, *Nature* 389:763.]

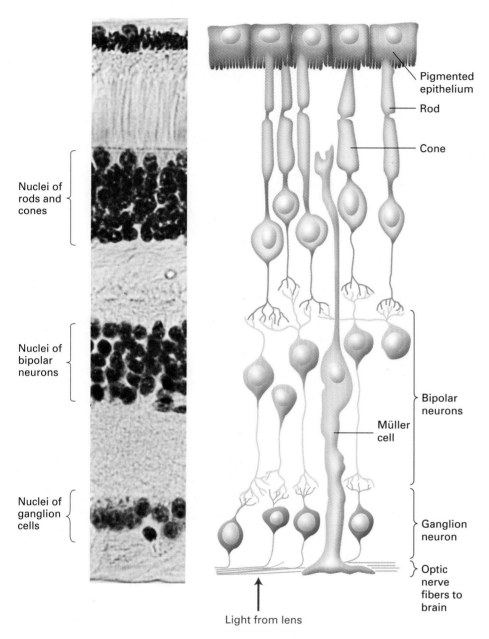

◀ **FIGURE 21-43 Some of the cells in the neural layer of the human retina.** The outermost layer of cells (in the rear of the eyeball) forms a pigmented epithelium in which the tips of the rod and cone cells are buried. Light focused from the lens passes through all of the cell layers of the retina and is absorbed by light-sensitive proteins in the rods and cones. The pigmented epithelium also absorbs light and prevents light reflection back to the cones and rods. The axons of each rod and cone synapse with many bipolar neurons, which integrate the responses of many photoreceptors. Some are involved in recognizing patterns of light that fall on the retina—for instance, a band of light that excites a set of rod cells in a straight line. Others respond to characteristics of visual images such as color, movement, and depth. Bipolar cells, in turn, synapse with cells in the ganglion layer that send axons—optic nerve fibers—through the optic nerve to the brain and that transmit distinct visual characteristics in multiple parallel pathways. Müller cells are supportive non-neural cells that fill much of the retinal spaces. Other types of cells are not depicted; all of the cells depicted here make many more synapses than are shown. [From R. G. Kessel and R. H. Kardon, 1979, *Tissues and Organs: A Text-Atlas of Scanning Electron Microscopy,* W. H. Freeman and Company, p. 87.]

Absorption of a Photon Triggers Isomerization of Retinal and Activation of Opsin

The photoreceptor in rod cells, rhodopsin, consists of the transmembrane protein *opsin* covalently bound to the light-absorbing pigment 11-*cis*-retinal (Figure 21-46). Opsin has seven membrane-spanning α helices, similar to other receptors that interact with transducing G proteins. Rhodopsin is localized to the thousand or so flattened membrane disks that make up the rod's outer segment; a human rod cell contains about 4×10^7 rhodopsin molecules.

The pigment 11-*cis*-retinal absorbs light in the visible range (400–600 nm). The primary photochemical event is isomerization of the 11-*cis*-retinal moiety in rhodopsin to

all-*trans*-retinal, which has a different conformation than the *cis* isomer; thus the energy of light is converted into atomic motion. The unstable intermediate in which opsin is covalently bound to all-*trans*-retinal is called *meta*-rhodopsin II, or *activated opsin*. The light-induced formation of activated opsin is both extremely efficient and rapid. At 500 nm, the wavelength of maximum rhodopsin absorption, ≈20 percent of the photons that strike the retina lead to a signal-transduction event, an efficiency comparable to that of the best photomultiplier tubes. Of the 57 kcal/mol of energy of photons of 500 nm, 27 kcal/mol (47 percent) is stored in the activated *meta*-rhodopsin II intermediate, making it an effective and reliable trigger of the

▶ **FIGURE 21-44 (a) Diagram of the structure of a human rod cell.** At the synaptic body, the rod cell forms a synapse with one or more bipolar neurons. Rhodopsin is a light-sensitive transmembrane protein located in the flattened membrane disks of the outer segment. (b) Electron micrograph of the region of the rod cell indicated by the bracket in (a); this region includes the junction of the inner and outer segments. [Part (b) from R. G. Kessel and R. H. Kardon, 1979, *Tissues and Organs: A Text-Atlas of Scanning Electron Microscopy*, W. H. Freeman and Company, p. 91.]

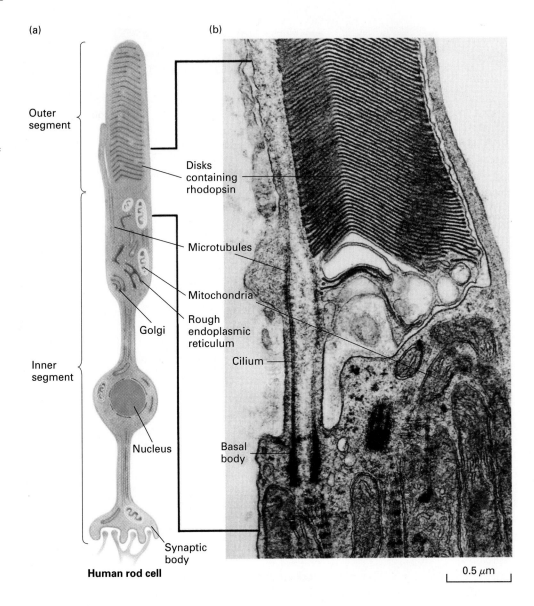

(a)   (b)

Outer segment

Disks containing rhodopsin

Microtubules

Mitochondria

Rough endoplasmic reticulum

Golgi

Cilium

Inner segment

Nucleus

Basal body

Synaptic body

Human rod cell

0.5 μm

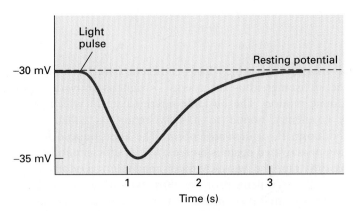

▲ **FIGURE 21-45 A brief pulse of light causes a transient hyperpolarization of the rod-cell membrane.** The membrane potential is measured by an intracellular microelectrode (see Figure 21-7).

next signaling step. An absorbed photon triggers opsin activation in less than 10 ms. In contrast, the spontaneous isomerization of 11-*cis*-retinal is extremely slow—about once per thousand years. Because there is very little spontaneous activation of opsin, the system has a very good ratio of signal to noise.

Activated opsin is unstable and spontaneously dissociates, releasing opsin and all-*trans*-retinal. In the dark, all-*trans*-retinal is converted back to 11-*cis*-retinal in several reactions catalyzed by enzymes in rod-cell membranes; the *cis* isomer can then rebind to opsin, re-forming rhodopsin.

Cyclic GMP Is a Key Transducing Molecule in Rod Cells

Rod outer segments contain an unusually high concentration of *3′,5′-cyclic GMP* (*cGMP*), about 0.07 mM. This

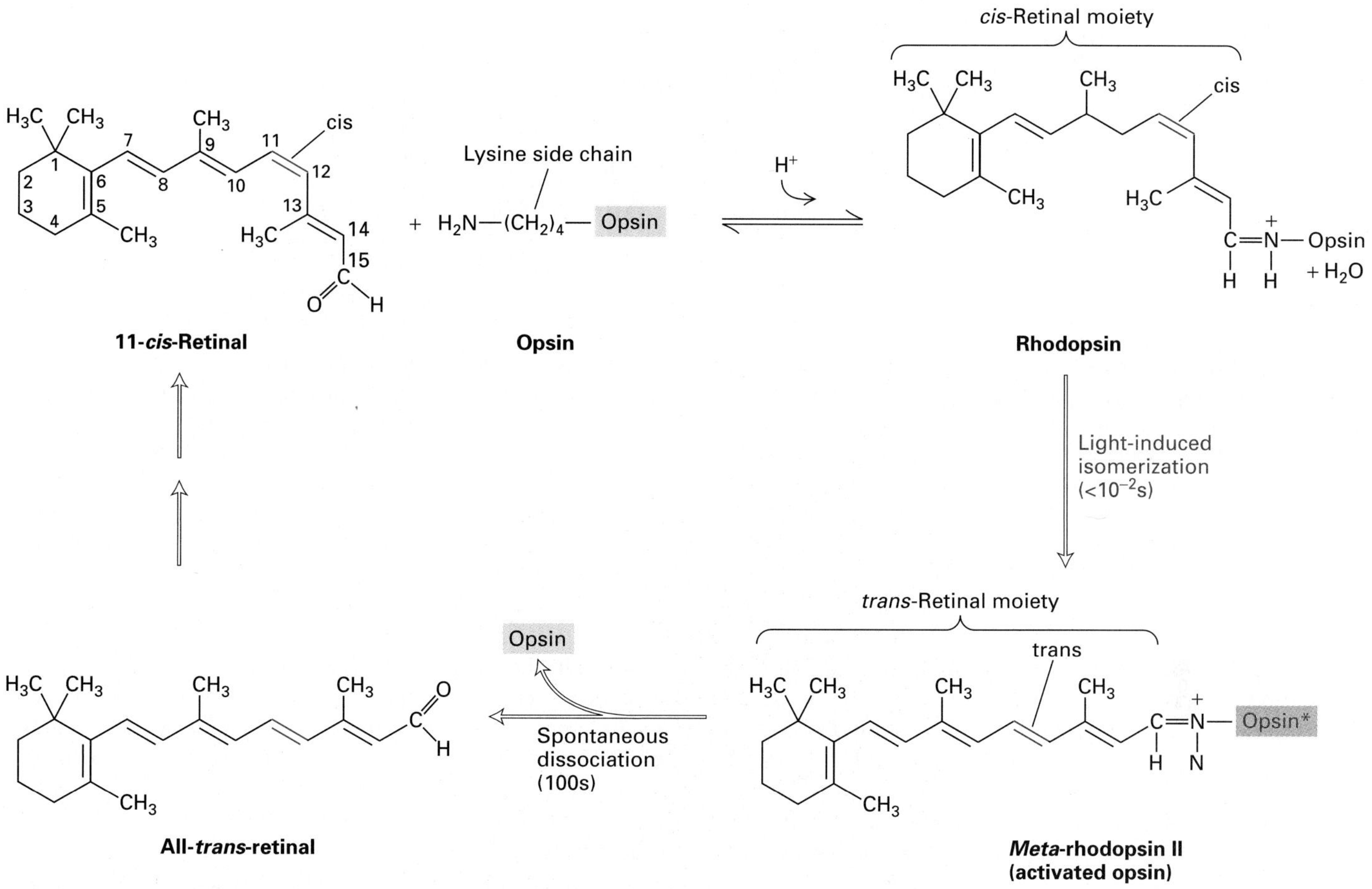

▲ **FIGURE 21-46 Rhodopsin, the photoreceptor in rod cells, is formed from 11-*cis*-retinal and opsin, a transmembrane protein.** Absorption of light causes rapid photoisomerization of the *cis*-retinal to the *trans* isomer, forming the unstable intermediate *meta*-rhodopsin II, or activated opsin. The latter dissociates spontaneously to give opsin and all-*trans*-retinal, which is converted back to the *cis* isomer by enzymes in rod cells. [See J. Nathans, 1992, *Biochemistry* **31**:4923.]

nucleotide is the key transducing molecule linking activated opsin to the closing of cation channels in the rod-cell plasma membrane. Formation of cGMP from GTP is catalyzed by guanylate cyclase, a reaction that appears to be unaffected by light. However, light triggers activation of a *cGMP-specific phosphodiesterase* that is present in rod outer segments:

$$3', 5' \text{ cGMP} + H_2O \xrightarrow[\text{phosphodiesterase}]{\text{cGMP}} 5' \text{ GMP}$$

As a result of this reaction, the cGMP concentration *decreases* upon illumination. Injection of cGMP into a rod cell depolarizes the cell membrane, and the effect is potentiated if an analog of cGMP that cannot be hydrolyzed is injected. Thus the high level of cGMP present in the dark acts to keep nucleotide-gated cation channels open; the light-induced decrease in cGMP leads to channel closing and membrane hyperpolarization.

Figure 21-47 depicts how light absorption by rhodopsin is coupled to activation of cGMP phosphodiesterase by transducin (G_T), a signal-transducing G protein that is found only in rods. Like other trimeric G proteins, transducin has three subunits—$G_{T\alpha}$, G_β, and G_γ—and cycles between active and inactive states (see Figure 20-19). In the resting (dark) state, the α subunit has a tightly bound GDP ($G_{T\alpha} \cdot$ GDP) and is incapable of affecting cGMP phosphodiesterase. Light-activated opsin catalyzes the exchange of free GTP for a GDP on the α subunit of transducin and the subsequent dissociation of $G_{T\alpha} \cdot$ GTP from the β and γ subunits. Free $G_{T\alpha} \cdot$ GTP then activates cGMP phosphodiesterase. (This pathway is similar to activation of adenylate cyclase by some G_s protein–coupled receptors; see Figure 20-16.) A single molecule of activated opsin in the disk membrane can activate 500 transducin molecules, which in turn activate cGMP phosphodiesterase; this is the primary stage of signal amplification in the visual system.

As with the α subunits of other G proteins, $G_{T\alpha}$ has an inherent GTPase activity, which slowly converts active $G_{T\alpha} \cdot$ GTP back to inactive $G_{T\alpha} \cdot$ GDP. Hydrolysis of GTP is accelerated when $G_{T\alpha} \cdot$ GTP is bound to the phosphodiesterase and enhanced further by the action of a GTPase-activating

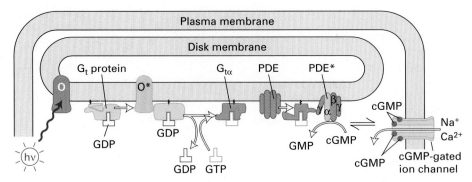

▲ **FIGURE 21-47 Coupling of light absorption by rhodopsin to activation of cGMP phosphodiesterase in rod cells.** In dark-adapted rod cells, a high level of cGMP acts to keep nucleotide-gated nonselective cation channels open and the membrane depolarized compared with the resting potential of other cell types. Light absorption leads to activation of opsin (O*) and conversion of inactive transducin (G_t) with bound GDP to the active state with bound GTP accompanied by dissociation of $G_{\beta\gamma}$ (not shown). The free $G_{\tau\alpha} \cdot$ GTP thus generated then activates cGMP phosphodiesterase (PDE) by binding to and dissociating its two inhibitory γ subunits; as a result, the released catalytic α and β subunits of activated PDE (PDE*) can convert cGMP to GMP. The resultant decrease in cGMP causes dissociation of cGMP from the nucleotide-gated channels in the plasma membrane; the channels then close and the membrane becomes transiently hyperpolarized. [Adapted from V. Arshavsky and E. Pugh, 1998, *Neuron* **20**:11.]

protein (GAP) specific for $G_{\tau\alpha} \cdot$ GTP (see Figure 20-22). Thus, in mammals $G_{\tau\alpha}$ remains in the active GTP-bound state only for a fraction of a second; cGMP phosphodiesterase rapidly becomes inactivated and the cGMP level gradually rises to its dark-adapted level. This allows rapid responses of the eye toward moving or changing objects. Once re-formed, $G_{\tau\alpha} \cdot$ GDP combines with G_β and G_γ, thus regenerating trimeric transducin.

Direct support for the role of cGMP in rod-cell activity has been obtained in patch-clamping studies using isolated patches of rod outer-segment plasma membrane, which contains abundant cGMP-gated cation channels. When cGMP is added to the cytosolic surface of these patches, there is a rapid increase in the number of open ion channels. The effect occurs in the absence of protein kinases or phosphatases, and cGMP acts directly on the channels to keep them open, indicating that these are nucleotide-gated channels. The channel protein contains four subunits each of which is able to bind a cGMP molecule (Figure 21-27b). Three or four cGMP molecules must bind per channel in order to open it; this allosteric interaction makes channel opening very sensitive to small changes in cGMP levels. Light closes the channels by activating cGMP phosphodiesterase, which acts to lower the level of cGMP.

Rod Cells Adapt to Varying Levels of Ambient Light

Cone cells are insensitive to low levels of illumination, and the activity of rod cells is inhibited at high light levels. Thus when we move from daylight into a dimly lighted room, we are initially blinded. As the rod cells slowly become sensitive to the dim light, we gradually are able to see and distinguish objects. This process of adaptation permits a rod cell to perceive contrast over a 100,000-fold range of ambient light levels;

as a result, differences in light levels, rather than the absolute amount of absorbed light, are used to form visual images.

One process contributing to **visual adaptation** affects the cGMP level and the affinity of gated cation channels for the nucleotide. Light, as we noted, causes a reduction in cGMP levels; this leads to a closing of cGMP-gated channels that admit both Na^+ and Ca^{2+} ions. Because Ca^{2+} is continuously extruded from the cells by Na^+/Ca^{2+} antiporters (see Figure 15-3b), the Ca^{2+} concentration in the cytosol falls at high ambient light levels. This drop in Ca^{2+} concentration "resets" the light-sensing system to a new, higher baseline level by two mechanisms:

- The enzyme that synthesizes cGMP is stimulated at low but not high concentrations of Ca^{2+} by a Ca^{2+}-sensing protein. Thus the light-induced drop in Ca^{2+} leads to an increase in cGMP concentration, causing the cGMP-gated Na^+/Ca^{2+} channels to remain open for longer periods.
- At high cytosolic Ca^{2+} concentrations, a calmodulin-like protein binds to the cGMP-gated Na^+/Ca^{2+} channels, reducing their affinity for cGMP. Conversely, a drop in Ca^{2+} causes the channels to bind cGMP more tightly and thus tend to remain open longer.

Both of these mechanisms favor opening of the cGMP-gated channels at high ambient light levels, so that a greater increase in light level is necessary to hydrolyze sufficient cGMP to close the same number of channels, and to generate the same visual signal than if the cells had not been exposed to light. In other words, at high ambient light, rod cells become less sensitive to small changes in levels of illumination.

A second process, affecting the protein opsin itself, participates in adaptation of rod cells to varying ambient light levels and also prevents overstimulation of the rod cell at very high ambient light (Figure 21-48). A rod-cell enzyme, *rhodopsin kinase*, phosphorylates light-activated opsin (O*)

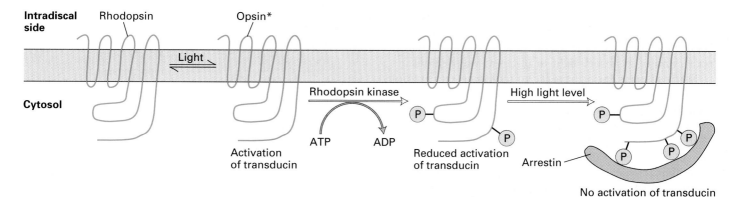

Intradiscal side   Rhodopsin   Opsin*

Light

Cytosol   Rhodopsin kinase   High light level

ATP   ADP

P

Activation of transducin   Reduced activation of transducin   Arrestin

No activation of transducin

▲ **FIGURE 21-48 Role of opsin phosphorylation in adaptation of rod cells to changes in ambient light levels.** Light-activated opsin (opsin*), but not dark-adapted rhodopsin, is a substrate for rhodopsin kinase. The extent of opsin* phosphorylation is directly proportional to the ambient light level, and the ability of an opsin* molecule to catalyze activation of transducin (see Figure 21-47) is inversely proportional to the number of sites phosphorylated. Thus the higher the ambient light level, the larger the increase in light level needed to activate the same number of transducin molecules. At very high light levels, arrestin binds to the completely phosphorylated opsin, forming a complex that cannot activate transducin at all. [See L. Lagnado and D. Baylor, 1992, *Neuron* **8**:995.]

but not dark-adapted rhodopsin (O). Each opsin molecule has seven phosphorylation sites; the more sites that are phosphorylated, the less able O* is to activate transducin and thus induce closing of cGMP-gated cation channels. Because the extent of O* phosphorylation is proportional to the amount of time each opsin molecule spends in the light-activated form, it is a measure of the background level of light. Under high light conditions, phosphorylated opsin is abundant and transducin activation is reduced; thus, a greater increase in light level will be necessary to generate a visual signal. When the level of ambient light is reduced, the opsins become dephosphorylated and transducin activation increases; in this case, fewer additional photons will be necessary to generate a visual signal.

At high ambient light (such as noontime outdoors), the level of opsin phosphorylation is such that the protein *arrestin* binds to opsin. Arrestin binds to the same site on opsin as does transducin, totally blocking activation of transducin and causing a shutdown of all rod-cell activity. The mechanism by which rod-cell activity is controlled by rhodopsin kinase is similar to adaptation (or desensitization) of the β-adrenergic receptor to high levels of hormone (see Figure 20-47). Indeed, rhodopsin kinase is very similar to β-adrenergic receptor kinase (BARK), the enzyme that phosphorylates and inactivates only the ligand-occupied β-adrenergic receptor, and each kinase can phosphorylate the other's substrate. Moreover, a homolog of arrestin binds to BARK-phosphorylated β-adrenergic receptors, blocking their interaction with G_s proteins.

Color Vision Utilizes Three Opsin Pigments

There are three classes of cone cells in the human retina. Each contains a different rhodopsin photopigment and absorbs light at a different wavelength (Figure 21-49). One absorbs mainly blue light, one green, and one red. As in rods,

the relative amount of light absorbed by each class of cones is translated into electrical signals that are transmitted to the brain. There the overall pattern of absorbed light of different wavelengths is converted into what we perceive as *color*. All cone opsins bind the same retinal as found in rods, and the three cone opsins are similar to rod opsin and to

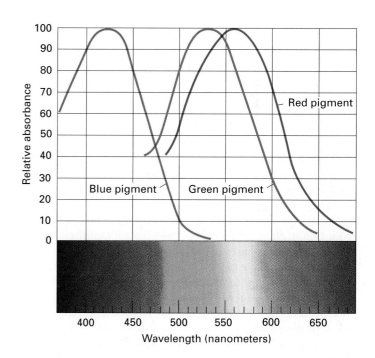

▲ **FIGURE 21-49 The absorption spectra of the three human opsins responsible for color vision.** Individual cone cells express one of the three cone opsins. The spectra were determined by measuring in a microspectrophotometer the light absorbed by individual cone cells obtained from cadavers. [From J. Nathans, 1989, *Sci. Am.* **260(2)**:44.]

each other. The unique absorption spectra of the three cone rhodopsins are due to different amino acid side chains that contact the retinal on the inside of rhodopsin and that affect its ability to absorb light of different wavelengths.

Study of cone opsins have led to molecular explanations of the different types of color blindness in humans. The "blue" opsin gene is located on human chromosome 7, while the "red" and "green" opsin genes are located next to each other, head-to-tail, on the X chromosome. The sequences of the red and green opsin genes are 98 percent identical, indicating that they arose from an evolutionary recent gene duplication. Furthermore, new world monkeys have only a single opsin gene on their X chromosome, whereas old world monkeys, which are more closely related to humans, have two. Two adjacent and almost identical genes can be expected to recombine unequally during gamete formation rather frequently, resulting in X chromosomes with only a green or only a red opsin gene. This results in red-green color blindness, a phenotype found in about 8 percent of males, who have a single X chromosome, but in only 0.64 percent of females, who have two X chromosomes.

Remarkably, because of polymorphisms in the red opsin genes, even individuals with "normal" color vision see colored objects differently. For instance, an alanine or serine may be located at position 180 of red opsin; this position is in the middle of the fourth membrane-spanning α helix, a region that contacts the bound retinal. The absorption maximum of the alanine form is ≈530 nm and of the serine form is ≈560 nm. Thus, individuals with serine at position 180 have a higher sensitivity to red light than others (see Figure 21-49); they "see" colors differently due to the change in a single nucleotide.

A Thousand Different G Protein–Coupled Receptors Detect Odors

The visual system functions efficiently with only four types of related photoreceptors, three cone opsins and one rod opsin. In contrast, the olfactory system utilizes a thousand homologous odorant receptors in responding to the millions of different chemicals we can smell.

Signal transduction in the olfactory system also differs from that in the visual system. The olfactory epithelium lining the air cavities in the nose contains thousands of sensory neurons, which have modified cilia extending from their apical (outward-facing) surface (Figure 21-50a). Each individual neuron expresses only one specific odorant receptor in the ciliary membrane and thus "senses" only one or a few odorants. All receptors are coupled to a single type of G protein, G_{olf}, unique to olfactory epithelia. Stimulation of G_{olf}, like G_s, activates adenylate cyclase, leading to an increase in the level of cAMP, which then binds to and opens a cAMP-gated cation channel unique to olfactory epithelia. (This channel is similar in structure to the cGMP-gated Na^+/Ca^{2+} channel in the visual system; see Figure 21-27b). Opening of the cAMP-gated channel induces depolarization of the

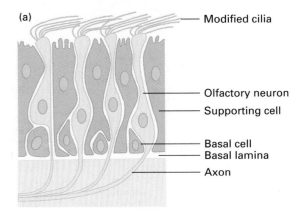

(a)
Modified cilia
Olfactory neuron
Supporting cell
Basal cell
Basal lamina
Axon

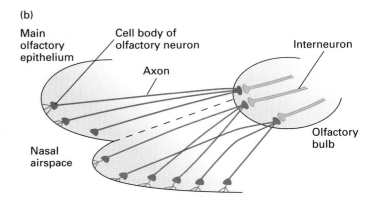

(b)
Main olfactory epithelium
Cell body of olfactory neuron
Axon
Interneuron
Nasal airspace
Olfactory bulb

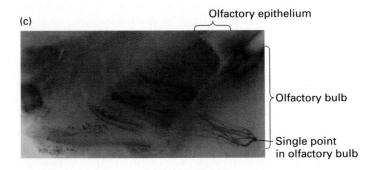

(c)
Olfactory epithelium
Olfactory bulb
Single point in olfactory bulb

▲ **FIGURE 21-50 General organization of the vertebrate olfactory system.** (a) The olfactory epithelium, lining the nasal air passages, contains olfactory neurons each of which expresses only one type of the ≈1000 different odorant receptors. These receptors are present on the cilia that extend from the apical surface of the neurons. (b) Axons from neurons expressing the same odorant receptor (indicated by color) project to the same point on the olfactory bulb in the brain. There they presumably synapse with one or a few interneurons that directly connect with higher levels in the brain, allowing the brain to detect each type of odorant separately. (c) Experimental demonstration that olfactory neurons expressing the same odorant receptor project to the same point. A transgenic mouse was generated in which the introduced DNA carried both the gene for a particular olfactory receptor and the gene for β-galactosidase. Each cell expressing the transgenic receptor also stains blue in a chromogenic assay for β-galactosidase. As revealed in this photomicrograph, all the dark-staining axons converge at one place in the olfactory bulb. [Part (b) adapted from C. Bargmann, 1997, *Cell* **90**:585. Part (c) from F. Ebrahimi et al., 1999, *Genes & Devel.* (in press); courtesy Dr. A. Chess.]

olfactory-cell membrane (rather than the hyperpolarization induced by activation of rhodopsin), initiating an action potential that is transmitted along the axon to the brain. Mice in which the gene encoding this cAMP-gated channel is disrupted are *anosmic* (cannot detect any odorant), attesting to the importance of the cAMP signaling pathway in olfaction.

To isolate genes encoding odorant receptors, researchers first identified sequences of amino acids that were conserved in many other known G protein–coupled receptors. Assuming that odorant receptors were also coupled to G proteins, the workers prepared primers for the polymerase chain reaction (PCR) that would allow amplification of cDNA sequences encoding novel G protein–coupled receptors. These sequences then were used to screen a cDNA library prepared from olfactory epithelia, leading to identification and cloning of several hundred receptor genes.

The diversity of odorant receptors is entirely encoded in the nuclear genome; there is no evidence that somatic recombination contributes to this diversity, as it does in the immune system. In situ hybridization has shown that each of the receptor genes is expressed in only a few of the millions of olfactory epithelial cells, as might be expected for a receptor that binds a specific kind of odorant. For many years, researchers could not determine which receptor bound which odorant molecule(s). Recent experiments in which one particular receptor is overexpressed in olfactory epithelial cells have provided an approach for matching up particular receptors and odorants. Cells expressing the recombinant receptor are exposed to many candidate odorants and the effects on membrane potential are monitored. In one such study, the receptor stimulated by *n*-octanal ($CH_3(CH_2)_6CHO$) was found to be unaffected by other molecules including the closely related octanoic acid ($CH_3(CH_2)_6COOH$) and octanol ($CH_3(CH_2)_6CH_2OH$). Thus, odorant receptors indeed can distinguish closely related smelly substances.

Neurons expressing a given odorant receptor are dispersed throughout the olfactory epithelium, and thus are able to sample all of the air in the nasal passages. Importantly, axons from neurons expressing each type of odorant receptor project to the same segment of the *olfactory bulb*, the part of the brain that collects signals from olfactory sensory neurons (Figure 21-50b,c). All the sensory neurons that respond to a given odorant are thought to synapse with one or a group of interneurons that sum these signals and transmit them to other parts of the brain. Apparently, the same receptors that bind odorants at one end of a sensory neuron in some way "target" the axons such that they synapse with only one set of interneurons. (Formation of such topographic maps, which also occurs in the visual system, is discussed in Chapter 23.) The brain determines which odorant receptors have been activated by examining the spatial pattern of electric activity in the olfactory bulb.

It is striking that so many different odorant receptors were selected during evolution. How each olfactory sensory neuron "chooses" to express only one of the thousand odorant-receptor genes is an interesting problem currently being studied.

SUMMARY Sensory Transduction

- Sensory transduction systems convert signals from the environment—light, taste, sound, touch, smell—into electric signals. These signals are collected, integrated, and processed by the central nervous system.

- The receptors that detect touch and stretch, heat, and capsaicin are gated cation channels that open in response to these stimuli. In contrast, the receptors that detect light or odor are coupled to G proteins.

- The retina contains rod cells, which respond to weak monochromatic light over a range of wavelengths, and three classes of cone cells, which respond to colors in bright light. Rhodopsin, the photoreceptor in rods and cones, occurs in four forms; each comprises one of four homologous opsin proteins linked to 11-*cis*-retinal.

- In rod cells, the light-induced isomerization of the 11-*cis*-retinal moiety in rhodopsin produces activated opsin, which then activates the signal-transducing G protein transducin (G_T) by catalyzing exchange of free GTP for bound GDP on the α subunit. $G_{T\alpha} \cdot$ GTP, in turn, activates cGMP phosphodiesterase, which acts to lower the cGMP level. This reduction leads to closing of cGMP-gated Na^+/Ca^{2+} channels, hyperpolarization of the membrane, and release of less neurotransmitter (see Figure 21-47).

- As the ambient light increases, the Ca^{2+} level in rod cells decreases, stimulating formation of cGMP and increasing the affinity of cGMP-gated Na^+/Ca^{2+} channels for ligand. Phosphorylation of light-activated opsin and subsequent binding of arrestin to phosphorylated opsin inhibits its ability to activate transducin (see Figure 21-48). As a result of these mechanisms, a greater increase in light level is necessary to generate a visual signal at high light levels than at low levels, permitting rod cells to function over a 100,000-fold range of illumination.

- Each sensory neuron in the olfactory epithelium expresses a single type of odorant receptor, which "senses" only one or a few odorants.

- Stimulation of odorant receptors, which are coupled to G_{olf}, leads to activation of adenylate cyclase. The resulting increase in cAMP opens cAMP-gated cation channels causing depolarization of the cell membrane and generation of an action potential.

- The axon of each odorant receptor is targeted to a particular point in the olfactory bulb, such that all cells expressing the same receptor synapse with one set of interneurons (see Figure 21-50). The spatial pattern of electric activity in the olfactory bulb reflects which receptors have been stimulated.

21.7 Learning and Memory

In its most general sense, learning is a process by which humans and other animals modify their behavior as a result of experience or as a result of acquisition of information about the environment. Memory is the process by which this information is stored and retrieved. Psychologists have defined two types of memory, depending on how long it persists: short term (minutes to hours) and long term (days to years). It is generally accepted that memory results from changes in the structure or function of particular synapses, but until recently learning and memory could not be studied with the tools of cell biology or genetics.

Most research indicates that *long-term memory* involves the formation or elimination of specific synapses in the brain and the synthesis of new mRNAs and proteins. Because *short-term memory* is too rapid to be attributed to such gross alterations, some have suggested that changes in the release and function of neurotransmitters at particular synapses are the basis of short-term memory. Indeed, several types of proteins that function as coincidence detectors have been shown to modify synaptic activity. These proteins, including the NMDA glutamate receptor discussed previously, integrate two (or more) coincident signals from different signaling pathways, generating an output that is different from that produced by either signal acting separately.

Study of the gill-withdrawal reflex in the sea slug *Aplysia* has provided insight into short-term learning processes (Figure 21-51). This simple behavior exhibits three of the most elementary forms of learning familiar in vertebrates: **habituation, sensitization**, and **classical conditioning**. Habituation is a *decrease* in behavioral response to a stimulus following repeated exposure to the stimulus with no adverse effect. For example, an animal that is startled by a loud noise may show decreasing responses on prolonged repetition of the

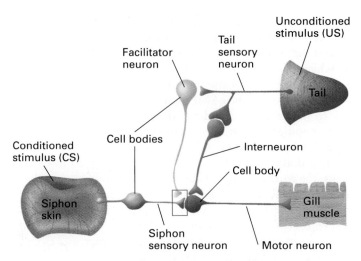

▲ **FIGURE 21-52 Neural circuits in the gill-withdrawal reflex of the sea slug *Aplysia*.** For simplicity, certain of the interneurons are omitted. This reflex exhibits habituation, sensitization, and classical conditioning depending on the extent and order in which the conditioned stimulus (gentle touching of the siphon) and unconditioned stimulus (sharp blow to the tail) are applied. A detailed depiction of the synapses indicated by the red box is shown in Figure 21-42; the synapse between the siphon sensory neuron and motor neuron contains the glutamate receptors depicted in Figure 21-40. [See T. W. Abrams and E. R. Kandel, 1988, *Trends Neurosci.* **11**:128; D. Glanzman, 1995, *Trends Neurosci.* **18**:30; and H. Lechner and J. Byrne, 1998, *Neuron* **20**:355.]

noise. Sensitization, in contrast, is an *increase* in behavioral response to a stimulus that does have an adverse effect. Classical conditioning is one of the simplest types of associative learning—the recognition of predictive events within an animal's environment. The animal learns that one event, termed the conditioned stimulus (CS), always precedes, by a critical and defined period, a second or reinforcing stimulus or event, the unconditioned stimulus (US).

When a sea slug is touched gently on its siphon, the gill muscles contract vigorously and the gill retracts into the mantle cavity. This behavior is mediated by a simple reflex arc in which sensory neurons in the siphon synapse with motor neurons that innervate the gill muscles (Figure 21-52). In this section, we describe the short-term changes in synaptic function that occur during habituation, sensitization, and classical conditioning of the gill-withdrawal reflex in *Aplysia*. This is one of the few cases in which the molecular events associated with learning responses are understood.

Repeated Conditioned Stimuli Cause Decrease in *Aplysia* Withdrawal Response

If the siphon of a sea slug is touched 10–15 times in rapid sequence, the gill-withdrawal response decreases to about one-third of its initial intensity. This habituative response is due to a progressive decrease in the amount of glutamate neurotransmitter released by the siphon sensory neurons at

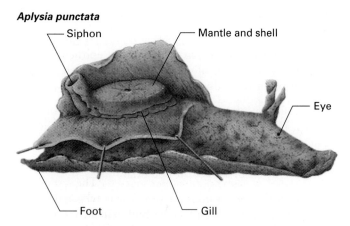

Aplysia punctata

— Siphon
— Mantle and shell
— Eye
— Foot
— Gill

▲ **FIGURE 21-51 The sea slug *Aplysia punctata*.** The gill, which lies under the protective mantle, can be seen if the over-lying tissue is pulled aside. When the siphon is touched gently, the gill is pulled in under the mantle. [Adapted from E. R. Kandel, 1976, *Cellular Basis of Behavior*, W. H. Freeman and Company, p. 76.]

their synapses with the motor neurons. In other words, repeated weak touching of the siphon (conditioned stimulus) leads to a *decrease* in the magnitude of the excitatory postsynaptic potential.

We have noted that release of neurotransmitters is triggered by a rise in the intracellular Ca^{2+} concentration following opening of voltage-gated Ca^{2+} channels. Measurements of Ca^{2+} movements in the *Aplysia* siphon sensory neuron have shown that habituation results from a decrease in the number of voltage-gated Ca^{2+} channels that open in response to the arrival of an action potential at the axon terminal, thus reducing the amount of glutamate neurotransmitter released. Habituation does not affect the generation of action potentials in the siphon sensory neuron or the response of the receptors in the postsynaptic cells.

Facilitator Neurons Mediate Sensitization of *Aplysia* Withdrawal Reflex

If a habituated sea slug is given a strong, noxious stimulus, such as a blow on the head or tail, it will respond to the next weak stimulus to the siphon by a rapid, enhanced withdrawal of the gill. The noxious stimulation (unconditioned stimulus) is said to sensitize the animal so that it exhibits an enhanced response to touching of the siphon (conditioned stimulus). *Aplysia* sensitization is mediated by interneurons called facilitator neurons, which are activated by shocks to the head or tail. Electron microscopy shows that the axon of a facilitator neuron synapses with the terminal of a siphon sensory neuron near the site where the siphon sensory neuron synapses with a motor neuron (see Figure 21-52). Stimulation of the facilitator neuron causes the siphon sensory neuron to release more neurotransmitter (glutamate) at its synapse with the motor neuron, thus increasing the magnitude of the gill-withdrawal response.

We already discussed the pathway by which stimulation of *Aplysia* facilitator neurons leads to enhanced activity of motor neurons and thus enhanced contraction of the gill muscle (see Figure 21-42). The effect of facilitator-neuron stimulation is mediated by cAMP and a cAMP-dependent protein kinase in the siphon sensory neuron terminal. Short-term sensitization persists as long as the concentration of cAMP is elevated and the kinase is activated, about 1 hour after each sensitizing stimulus.

Coincidence Detectors Participate in Classical Conditioning and Sensitization

A well-known example of classical conditioning is the Pavlovian response in dogs: if a bell—the conditioned stimulus (CS)—is rung a few seconds before food—the unconditioned stimulus (US)—is presented, the dogs soon learn to associate the two stimuli and to salivate in response to the bell alone. In such a learning process, it is essential that the conditioning stimulus always precede the unconditioned stimulus by a small and critical time interval.

Perhaps surprisingly, classical conditioning can be observed in the gill-withdrawal reflex of sea slugs. In the "training" process, a weak touch to the siphon (CS) is followed immediately by a sharp blow to the tail or head (US), which, of course, evokes a marked gill-withdrawal response (see Figure 21-52). After a series of such trials, the gill-withdrawal response to the CS alone is substantially enhanced, as if the animal "learns" that a weak siphon touch (CS) is followed by a noxious, sharp blow (US). As in conditioning in other animals, the CS must precede the US by a short and definite interval, in this case 1–2 seconds.

Both sensitization and classical conditioning of the gill-withdrawal reflex lead to an enhanced response. Sensitization occurs when the facilitator neuron is activated by the US (a blow to the head or tail) in the absence of the CS, whereas conditioning occurs when the sensory neuron is activated by the CS just before the US is applied.

Figure 21-53 outlines how an adenylate cyclase in the axon terminal of the siphon sensory neuron functions as a coincidence detector that integrates the signals triggered by the US and CS. This cyclase can be activated by $G_{s\alpha} \cdot GTP$, induced by stimulation of the facilitator neuron (US), or by Ca^{2+}-calmodulin, resulting from stimulation of the siphon sensory neuron (CS). However, activation of the cyclase, and hence the increase in cAMP, is greatest when the cyclase is first "primed" by an increase in Ca^{2+} concentration (generated by the CS) and then interacts, within 1–2 seconds, with $G_{s\alpha} \cdot GTP$ (generated by the US). In this way, the enhancement of adenylate cyclase activity triggered by the unconditioned stimulus makes the sensory neuron more sensitive to a conditioned stimulus: The animal "learns" to associate the CS with the US and to respond to a subsequent CS alone with an enhanced response that is greater than that produced if the animal had not been "trained" by paired application of a CS followed by a US. Indeed, in isolated membranes prepared from these *Aplysia* neurons, adenylate cyclase activity is greater when the membranes are exposed to elevated Ca^{2+} *before* exposure to $G_{s\alpha} \cdot GTP$ rather than vice versa. This biochemical asymmetry mirrors the key temporal requirement for conditioning in the intact animal: Conditioning is produced only when the CS precedes the US.

NMDA glutamate receptors in the motor neuron also act as coincidence detectors and contribute to enhancing the gill-withdrawal response. In addition to synapsing with facilitator neurons, the tail sensory neurons synapse with another set of interneurons that, in turn, form synapses with the motor neuron (see Figure 21-52). Activation of the tail sensory neurons—the US— leads to a partial depolarization of the motor neuron via these interneurons. The motor neurons contain NMDA glutamate receptors, which bind glutamate released by the siphon sensory neurons. Recall that activation of these glutamate-gated Na^+/Ca^{2+} channels requires *both* binding of glutamate and partial depolarization of the plasma membrane (see Figure 21-40). Thus the US stimulus, by partially depolarizing the motor-neuron membrane, causes a greater influx of Ca^{2+} and Na^+ ions than

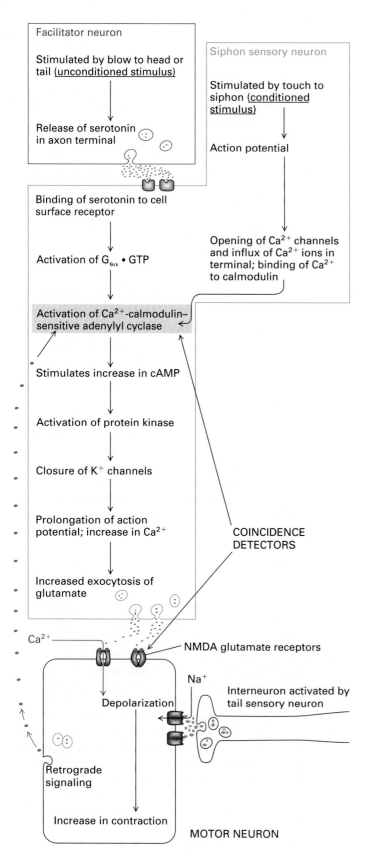

Facilitator neuron

Stimulated by blow to head or tail (<u>unconditioned stimulus</u>)

↓

Release of serotonin in axon terminal

Siphon sensory neuron

Stimulated by touch to siphon (<u>conditioned stimulus</u>)

↓

Action potential

↓

Binding of serotonin to cell surface receptor

↓

Activation of $G_{s\alpha} \cdot$ GTP

↓

Opening of Ca^{2+} channels and influx of Ca^{2+} ions in terminal; binding of Ca^{2+} to calmodulin

Activation of Ca^{2+}-calmodulin–sensitive adenylyl cyclase

↓

Stimulates increase in cAMP

↓

Activation of protein kinase

↓

Closure of K^+ channels

↓

Prolongation of action potential; increase in Ca^{2+}

↓

Increased exocytosis of glutamate

COINCIDENCE DETECTORS

Ca^{2+}

NMDA glutamate receptors

Na^+

Depolarization

Interneuron activated by tail sensory neuron

Retrograde signaling

Increase in contraction

MOTOR NEURON

◄ **FIGURE 21-53 Intracellular signaling pathways during sensitization and classical conditioning in the *Aplysia* gill-withdrawal reflex arc.** Sensitization occurs when the facilitator neuron is triggered by the unconditioned stimulus (US) in the absence of the conditioned stimulus (CS) to the siphon sensory neuron (see Figure 21-52). Classical conditioning occurs when the CS is applied 1–2 seconds before the US, and involves coincidence detectors in both the presynaptic siphon sensory neuron and the motor neuron. In the sensory neuron, the detector is an adenylate cyclase that is activated by both Ca^{2+}-calmodulin and by $G_{s\alpha} \cdot$ GTP (see Figure 21-42). In the motor neuron, the detectors are NMDA glutamate receptors (see Figure 21-40). Partial depolarization of the motor neuron induced by an unconditioned stimulus (via an unknown transmitter) from interneurons activated by the tail sensory neuron enhances the response to glutamate released by the siphon sensory neuron. See text for details. [See H. R. Bourne and R. Nicoll, 1993, *Cell* vol. 72/*Neuron* vol. 10 (suppl.), pp. 65–75; G. Murphy and D. Glanzman, 1997, *Science* **278**:467; and H. Lechner and J. Byrne, 1998, *Neuron* **20**:355.]

is induced by the CS acting alone (see Figure 21-53). Once "sensitized" by US-induced depolarization, the motor neuron can be activated by fewer action potentials generated in the presynaptic siphon sensory neurons; in other words, the synapse "learns" to have an enhanced response to signals from the presynaptic cells.

Increased Ca^{2+} in the motor neuron also results in secretion of an unknown retrograde signal, possibly CO or NO, that diffuses into the sensory neuron and acts to increase the sensitivity of the Ca^{2+}-calmodulin–sensitive adenylate cyclase. Thus, a paired combination of a conditioned stimulus and unconditioned stimulus results in enhanced responses both in the sensory neuron and motor neuron.

Long-Term Memory Requires Protein Synthesis

The short-term sensitization and conditioning responses in *Aplysia* can occur in the presence of inhibitors of protein synthesis, suggesting that no new proteins (or cells) are required for short-term learning responses (short-term memory). On the other hand, a series of closely spaced tail shocks (unconditioned stimulus) delivered over a few hours will produce a long-term sensitization (long-term memory), which can persist for days or even weeks. Both long-term and short-term sensitization affect the same synapses, and even the same K^+ channels. However, protein synthesis is essential for long-term sensitization, suggesting that certain new proteins must be made in these synapses in order for long-term memory to occur.

Long-term sensitization of the *Aplysia* gill-withdrawal reflex can be induced in cells in culture by the repeated application of serotonin, the neurotransmitter normally released by the facilitator neuron (see Figure 21-42). Repeated application of serotonin induces synthesis of several proteins in the sensory neuron and inhibits synthesis of others, including several N-CAMs, plasma-membrane proteins that function in cell-cell adhesion between neurons (Section 22.3).

Sensitization and classical conditioning of the gill-withdrawal reflex of *Aplysia* are among the few cases in which short-term changes in synaptic function are understood in molecular detail. These simple forms of learning are useful models for more complex forms of behavior, such as short-term and long-term memory in vertebrates. Increasingly, neuroscientists are identifying molecules that may function in memory in mammals.

SUMMARY Learning and Memory

- Modifications in the activity of certain synapses are associated with short-term memory, at least in some invertebrate and mammalian systems.

- In the sea slug *Aplysia*, the gill-withdrawal reflex exhibits habituation, sensitization, and classical conditioning—three forms of simple learning (see Figure 21-52).

- Habituation is linked to the closing of Ca^{2+} channels in the presynaptic axon terminals of siphon sensory neurons; this reduces the influx of Ca^{2+} in the axon terminals and reduces the amount of glutamate neurotransmitter released to the motor neurons.

- Sensitization and classical conditioning are mediated by facilitator neurons that synapse with the siphon sensory neurons. Stimulation of the facilitator neurons leads to prolonged depolarization and increased exocytosis of glutamate neurotransmitter in the sensory neuron.

- In classical conditioning, a conditioned stimulus, triggering the siphon sensory neurons and increasing Ca^{2+} levels, and an unconditioned stimulus, triggering the facilitator neurons, converge on an adenylate cyclase in the axon terminal of the siphon sensory neuron (see Figure 21-53). This enzyme, which is activated both by $G_{s\alpha} \cdot$ GTP and by Ca^{2+}-calmodulin, functions as a coincidence detector. NMDA glutamate receptors in the postsynaptic motor neurons also function as coincidence detectors, responding to both the unconditioned and conditioned stimulus.

- As a result of the action of both coincidence detectors, fewer action potentials in the presynaptic sensory neurons can induce the same response in the postsynaptic motor neuron; the synapse "learns" to have an enhanced response to the electric signals in the presynaptic cells.

PERSPECTIVES for the Future

A real understanding of the function of nerve cells requires knowledge of the three-dimensional structures of many types of membrane channel, receptor, and other proteins. Determining these structures will prove difficult because of many technical problems in purifying and crystallizing integral membrane proteins. The structure of a bacterial K^+ channel has been solved, providing insight into how such channels can distinguish between Na^+ and K^+ ions. Elucidation of the structure of the first voltage-gated K^+ channel should illuminate the mechanisms of channel gating, opening, and inactivation that may also apply to many other voltage-gated channels. Similarly, the nicotinic acetylcholine, glutamate, GABA, and glycine receptors are all ligand-gated ion channels, but it is disputed whether they all have the same overall structures in the membrane. Resolving this issue will also require knowledge of their three-dimensional structures, which also should tell us in detail how ligand binding leads to channel opening.

How does a neuron achieve its very long, branching structure? Why does one part of a neuron become a dendrite and another an axon? Why are certain key membrane proteins clustered at particular points—receptors in postsynaptic densities in dendrites, Ca^{2+} channels in axon termini, and Na^+ channels in myelinated neurons at the nodes of Ranvier? Such questions of cell shape and protein targeting also apply to other types of cells, but the vastly different structures of different types of neurons make these particularly complex questions to study. Development of pure cultures of specific types of neurons that maintain their normal properties would enable many of these problems to be studied by techniques of molecular cell biology. Perhaps the most difficult questions concern the formation of specific synapses within the nervous system; that is, how does a neuron "know" to synapse with one type of cell and not another. Studies on the olfactory system have already provided some clues, and ongoing research on the development of the nervous system, outlined in Chapter 23, is beginning to provide additional answers.

Few questions about the nervous system excite as much interest as those concerning human learning, intelligence, and memory. The types of studies on *Aplysia* described in the last section of this chapter are now being applied to organisms such as *Drosophila* and mice. Remarkable as it sounds, fruit flies can be trained to avoid certain noxious stimuli. Mutant flies defective in learning have been isolated; some of the mutations affect the same proteins (e.g., the adenylate cyclase activated by $G_{s\alpha} \cdot$ GTP and Ca^{2+}-calmodulin) that are implicated in *Aplysia* learning. Similarly, workers have generated mice defective in other coincidence detectors implicated in simple types of learning. Indeed these mice exhibit defects in very specific types of learning. As more such proteins are identified by molecular cloning, workers may identify other enzymes or receptors essential for other types of learning processes in mice.

The basis of long-term memory is particularly hard to understand. It involves changes in the pattern of gene expression by individual neurons that can persist for the lifetime of the animal. Furthermore, certain proteins become localized to certain dendrites, the ones that have "learned," but not to others on the same cell, and we have no idea how this happens. Determining whether specific types of human learning and intelligence relate to the amount or type of such proteins will be a difficult and controversial area of research.

PERSPECTIVES in the Literature

One way to study the role of a specific protein—such as the Fyn protein tyrosine kinase or an NMDA glutamate receptor—in learning and memory in mice is to generate animals in which the gene is knocked out and protein expression is blocked in just one region of the brain. Psychological studies on these mutant mice combined with electrophysiological studies on synapses in the affected region of the brain are used to draw conclusions about protein function. The papers below detail some of these very recent experiments. As you read them, consider the following questions:

How is learning measured in these animals?

Long-term potentiation (LTP) is a measure of a type of synaptic plasticity, or "learning." How is LTP measured? What alterations of LTP are seen in synapses from the mutant mice?

The experiments are interpreted in terms of the functions of the specific protein as a coincidence detector intimately involved in synaptic plasticity. However, another possibility is that mutations in these proteins affect the development of the cell—its shape or its formation of specific synaptic contacts with other neurons—rather than the function of the differentiated cell per se. Do these experiments adequately control for this possibility? What types of experiments might be performed to address this issue?

Kojima, N., et al. 1997. Rescuing impairment of long-term potentiation in Fyn-deficient mice by introducing *fyn* transgene. *Proc. Nat'l. Acad. Sci. USA* **94**:4761–4765.

Tsien, J. Z., et al. 1996. Subregion- and cell type-restricted gene knockout in mouse brain. *Cell* **87**:1317–1326.

Tsien, J. Z., P. T. Huerta, and S. Tonegawa. 1996. The essential role of hippocampal CA1 NMDA receptor-dependent synaptic plasticity in spatial memory. *Cell* **87**:1327–1338.

Wilson, M. A., and S. Tonegawa. 1997. Synaptic plasticity, place cells and spatial memory: study with second generation knockouts. *Trends Neurosci.* **20**:102–106. Review article.

Testing Yourself on the Concepts

1. Neurons must conduct signals along their length as well as transmit signals to other cells. How is each signaling process accomplished?

2. Vertebrate neurons are wrapped with myelin. What is myelin, and what advantage does myelination of the neuron provide? What happens if the myelin is lost?

3. In different chemical synapses, acetylcholine may function as an excitatory or an inhibitory neurotransmitter. How is this possible?

4. Detection of both light and odor requires the conversion of an external stimulus into an electrical signal. How are these sensory transduction processes similar, and how are they different?

MCAT/GRE-Style Questions

Key Concept Please read the section titled "Voltage-Gated Cation Channels Generate Action Potentials" (p. 921) and answer the following questions:

1. At rest, all of the following are closed except
    a. Voltage-gated Na^+ channels.
    b. Voltage-gated K^+ channels.
    c. Delayed K^+ channels.
    d. Non-voltage-gated K^+ channels.

2. During an action potential, which of the following is directly responsible for restoration of resting membrane potential:
    a. Influx of Na^+ ions.
    b. Influx of K^+ ions.
    c. Efflux of Na^+ ions.
    d. Efflux of K^+ ions.

3. The timed closing of voltage-gated Na^+ channels after about 1 ms is essential for
    a. The directional movement of the action potential.
    b. The efflux of K^+ ions.
    c. Preventing the efflux of Na^+ ions.
    d. Membrane depolarization.

4. Movement of Na^+ through open voltage-gated Na^+ channels is driven by
    a. The Na^+ concentration gradient across the membrane.
    b. The membrane potential at rest.
    c. The Na^+ concentration gradient across the membrane and membrane potential at rest.
    d. The Na^+ concentration gradient across the membrane and membrane potential at the peak of the action potential.

5. Open and closing of voltage-gated Na^+ channels
    a. Is triggered by an influx of K^+ ions.
    b. Involves small changes in protein conformation.
    c. Depends on whether other ions "plug" the channel.
    d. Is independent of membrane potential.

Key Experiment Please read the section titled "Patch Clamps Permit Measurement of Ion Movements Through

Single Channels" (p. 927) and refer to Figures 21-19, 21-20, and 21-21; then answer the following questions:

6. Experiments on a gated Na^+ channel in the membrane of a whole cell as described in Figures 21-19 and 21-20a require that Na^+ ions be present

   a. In the cytosol.
   b. In the solution surrounding the entire cell.
   c. In the solution inside the micropipette.
   d. In the intracellular electrode.

7. The term "clamp" is used to describe the technique described in the figures because

   a. One of the electrodes is clamped onto the surface of the plasma membrane.
   b. The membrane potential is clamped at a specific value.
   c. The cell must be clamped to hold it steady.
   d. The cell must be clamped inside an electrode.

8. In the tracing shown in Figure 21-21a, only two possible values for electric current (pA) are detected because

   a. The clamp is either on or off.
   b. The channel protein is either open or closed.
   c. Only two ion channels are present in the patch.
   d. Either Na^+ or K^+ is moving through the channels.

9. During the patch-clamp technique, ion movement is measured by the

   a. Amount of electric current needed to maintain a pre-set membrane potential.
   b. Amount of electric current needed to raise the membrane potential.
   c. Number of action potentials that occur during the experiment.
   d. Number of electrons that move through the ion channel.

10. Based on the figures, the patch-clamp technique can be applied to the study of

   a. Only voltage-gated ion channels.
   b. Only voltage-gated Na^+ channels.
   c. Only voltage-gated K^+ channels.
   d. Any gated ion channel.

Key Application Please read the section titled "Transmitter-Mediated Signaling Is Terminated by Several Mechanisms" (p. 941) and answer the following questions:

11. Uptake of many neurotransmitters

   a. Depends on a Na^+ gradient.
   b. Depends on ATP.
   c. Occurs by endocytosis.
   d. Is inhibited by nerve gas.

12. The antidepressant drug imipramine functions by inhibiting

   a. Dopamine uptake by the postsynaptic cell.
   b. Destruction of dopamine in the synapse.
   c. Serotonin uptake by the presynaptic cell.
   d. Serotonin release by the presynaptic cell.

13. Acetylcholinesterase

   a. Hydrolyzes acetylcholine inside the presynaptic cell.
   b. Is regulated by binding to cocaine.
   c. Must be secreted into the synaptic cleft.
   d. Normally functions to prolong stimulation of the postsynaptic cell.

14. Neurotransmitters must be removed from the synaptic cleft to

   a. Prevent prolonged stimulation of the presynaptic cell.
   b. Prevent prolonged stimulation of the postsynaptic cell.
   c. Promote prolonged stimulation of the presynaptic cell.
   d. Promote prolonged stimulation of the postsynaptic cell.

Key Terms

action potential 912
axon 912
catecholamines 949
classical conditioning 960
coincidence detector 947
dendrites 912
depolarization 912
excitatory receptor 938
habituation 960
inhibitory receptor 939
ligand-gated ion channels 920
long-term potentiation 946
modulatory synapse 949
myelin sheath 921
neurotransmitter 914
node of Ranvier 924
patch clamping 927
permeability constant 918
reflex arc 915
refractory period 922
resting K^+ channels 918
rhodopsin 952
Schwann cells 923
sensitization 960
Shaker protein 930
synaptic vesicles 935
synapse 912
threshold potential 942
transducin 955
visual adaptation 956
voltage-gated ion channels 920

References

Overview of Neuron Structure and Function

Bargmann, C. 1998. Neurobiology of the *Caenorhabditis elegans* genome. *Science* **282**:2028–2033.

Cell (vol. 72)/*Neuron* (vol. 10). 1993. *Signaling at the Synapse.* Supplement containing several excellent review articles.

Green, T., S. Heinemann, and J. Gusella. 1998 Molecular neurobiology and genetics: investigation of neural function and dysfunction. *Neuron* 20:427–444.

Hammond, C. 1996. *Cellular and Molecular Neurobiology.* Academic Press.

Hille, B. 1992. *Ionic Channels of Excitable Membranes,* 2d ed. Sinauer Associates.

Kandel, E. R., J. H. Schwartz, and T. M. Jessell. 1991. *Principles of Neural Science,* 3d ed. Elsevier.

Neuron. 1998. Review articles on six decades of neuroscience. Vol. 20, pp. 367–468.

Nicholls, J. G., A. R. Martin, and B. G. Wallace. 1992. *From Neuron to Brain.* Sinauer Associates.

Wallis, D., ed. 1993. *Electrophysiology. A Practical Approach.* IRL Press, Oxford, U.K.

The Action Potential and Conduction of Electric Impulses

Armstrong, C., and B. Hille. 1998. Voltage-gated ion channels and electrical excitability. *Neuron* 20:371–380.

Choe, S. 1996. Packing of myelin protein zero. *Neuron* 17:363–365.

Hodgkin, A. L. 1964. *The Conduction of the Nervous Impulse.* Liverpool University Press, Liverpool, U.K.

Keynes, R. D. 1979. Ion channels in the nerve-cell membrane. *Sci. Am.* 240(3):126–135.

Lemke, G. 1996. Unwrapping myelination [News]. *Nature* 383:395–396.

Salzer, J. L. 1997. Clustering sodium channels at the node of Ranvier: close encounters of the axon-glia kind. *Neuron* 18:843–846.

Molecular Properties of Voltage-Gated Ion Channels

Aldrich, R. W. 1990. Potassium channels: mixing and matching. *Nature* 345:475–476.

Catterall, W. A. 1992. Cellular and molecular biology of voltage-gated sodium channels. *Physiol. Rev.* 72:S15–48.

Clapham, D. 1999. Unlocking family secrets: K$^+$ channel transmembrane domains. *Cell* 97:547–550.

Cooper, E. C., and L. Y. Jan. 1999. Ion channel genes and human neurological disease: recent progress, prospects, and challenges. *Proc. Nat'l. Acad. Sci. USA* 96:4759–4766.

Doyle, D. A., et al. 1998. The structure of the potassium channel: molecular basis of K$^+$ conduction and selectivity. *Science* 280:69–77.

Finn, J., M. Grunwald, and K. W. Yau. 1996. Cyclic nucleotide-gated ion channels: an extended family with diverse functions. *Ann. Rev. Physiol.* 58:395–426.

Jan, L., and Y. Jan. 1997. Cloned potassium channels from eukaryotes and prokaryotes. *Ann. Rev. Neurosci.* 20:91–124.

Jones, S. W. 1998. Overview of voltage-dependent calcium channels. *J. Bioenerg. Biomembr.* 30:299–312.

Neher, E. 1992. Ion channels for communication between and within cells. Nobel Lecture reprinted in *Neuron* 8:605–612 and *Science* 256:498–502.

Neher, E., and B. Sakmann. 1992. The patch clamp technique. *Sci. Am.* 266(3):28–35.

Nichols, C., and A. Lopatin. 1997. Inward rectifier potassium channels. *Ann. Rev. Physiol.* 59:171–192.

Neurotransmitters, Synapses, and Impulse Transmission

Amara, S. G., and M. J. Kuhar. 1993. Neurotransmitter transporters: recent progress. *Ann. Rev. Neurosci.* 16:73–93.

Bajjalieh, S. M., and R. H. Scheller. 1995. The biochemistry of neurotransmitter secretion. *J. Biol. Chem.* 270:1971–1974.

Barondes, S. 1994. Thinking about Prozac. *Science* 263:1102–1103.

Bean, A. J., and R. H. Scheller. 1997. Better late than never: a role for rabs late in exocytosis. *Neuron* 19:751–754.

Bennett, M. K., and R. H. Scheller. 1993. The molecular machinery for secretion is conserved from yeast to neurons. *Proc. Nat'l. Acad. Sci. USA* 90:2559–2563.

Bennett, M. V., et al. 1991. Gap junctions: new tools, new answers, new questions. *Neuron* 6:305–320.

Betz, W., and J. Angleson. 1998. The synaptic vesicle cycle. *Ann. Rev. Physiol.* 60:347–364.

Brownstein, M. J. 1993. A brief history of opiates, opioid peptides, and opioid receptors. *Proc. Nat'l. Acad. Sci. USA* 90:5391–5393.

Fernandez, J. M. 1997. Cellular and molecular mechanics by atomic force microscopy: capturing the exocytotic fusion pore in vivo? [Comment]. *Proc. Nat'l. Acad. Sci. USA* 94:9–10.

Geppert, M., and T. Sudhof. 1998. Rab3 and synaptotagmin: the yin and yang of synaptic membrane fusion. *Ann. Rev. Neurosci.* 21:75–96.

Greengard, P., et al. 1993. Synaptic vesicle phosphoproteins and regulation of synaptic function. *Science* 259:780–785.

Kavanaugh, M. P. 1998. Neurotransmitter transport: models in flux [Comment]. *Proc. Nat'l. Acad. Sci. USA* 95:12737–12738.

Monck, J. R., and J. Fernandez. 1994. The exocytic fusion pore and neurotransmitter release. *Neuron* 12:707–716.

Neher, E. 1998. Vesicle pools and Ca^{2+} microdomains: new tools for understanding their roles in neurotransmitter release. *Neuron* 20:389–399.

Neimann, H., J. Blasi, and R. Jahn. 1994. Clostridial neurotoxins: new tools for dissecting exocytosis. *Trends Cell Biol.* 4:179–185.

O'Connor, V., G. J. Augustine, and H. Betz. 1994. Synaptic vesicle exocytosis: molecules and models. *Cell* 76:785–787.

Reith, M., ed. 1997. *Neurotransmitter Transporters: Structure, Function, and Regulation.* Humana Press.

Sakmann, B. 1992. Elementary steps in synaptic transmission revealed by currents through single ion channels. Nobel Lecture reprinted in *EMBO J.* 11:2002–2016 and *Science* 256:503–512.

Scheller, R. H. 1995. Membrane trafficking in the presynaptic nerve terminal. *Neuron* 14:893–897.

Schweizer, F. E., H. Betz, and G. J. Augustine. 1995. From vesicle docking to endocytosis: intermediate reactions of exocytosis. *Neuron* 14:689–696.

Stauffer, K. A., and N. Unwin. 1992. Structure of gap junction channels. *Semin. Cell Biol.* 3:17–20.

Sudhof, T. C. 1995. The synaptic vesicle cycle: a cascade of protein-protein interactions. *Nature* 375:645–653.

Taylor, P. 1991. The cholinesterases. *J. Biol. Chem.* 266:4025–4028.

Usdin, T. B., L. E. Eiden, T. I. Bonner, and J. D. Erickson. 1995. Molecular biology of the vesicular ACh transporter. *Trends Neurosci.* 18:218–224.

Neurotransmitter Receptors

Arshavsky, V., and E. Pugh. 1998. Lifetime regulation of G protein–effector complex: emerging importance of RGS proteins. *Neuron* 20:11–14.

Becker, C. M. 1990. Disorders of the inhibitory glycine receptor: the spastic mouse. *FASEB J.* 4:2767–2774.

Betz, H. 1991. Glycine receptors: heterogeneous and widespread in the mammalian brain. *Trends Neurosci.* 14:458–461.

Brown, A. M., and L. Birnbaumer. 1990. Ionic channels and their regulation by G protein subunits. *Ann. Rev. Physiol.* 52:197–213.

Changeux, J. P., et al. 1992. New mutants to explore nicotinic receptor functions. *Trends Pharmacol. Sci.* 13:299–301.

Hille, B. 1992. G protein–coupled mechanisms and nervous signaling. *Neuron* 9:187–195.

Karlin, A., and M. H. Akabas. 1995. Toward a structural basis for the function of nicotinic acetylcholine receptors and their cousins. *Neuron* 15:1231–1244.

Stahl, S. 1996. *Essential Psychopharmacology: Neuroscientific Basis and Practical Applications.* Cambridge University Press.

Unwin, N. 1993. Neurotransmitter action: opening of ligand-gated ion channels. *Cell* 72:31–41.

Sensory Transduction

Bargmann, C. 1997 Olfactory receptors, vomeronasal receptors, and the organization of olfactory information. *Cell* 90:585–587.

Baylor, D. 1996. How photons start vision. *Proc. Nat'l. Acad. Sci. USA* 93:560–565.

Buck, L. 1996. Information coding in the vertebrate olfactory system. *Ann. Rev. Neurosci.* 19:517–544.

Clapham, D. E. 1997. Some like it hot: spicing up ion channels [News; Comment]. *Nature* 389:783–784.

Corey, D. P., and J. Garcia-Anoveros. 1996. Mechanosensation and the DEG/ENaC ion channels [Comment]. *Science* 273:323–324.

Dowling, J. 1987. *The Retina: An Approachable Part of the Brain.* Harvard University Press.

Hamill, O. P., and D. W. McBride, Jr. 1994. The cloning of a mechano-gated membrane ion channel. *Trends Neurosci.* 17:439–443.

Hildebrand, J., and G. Shephard. 1997. Mechanisms of olfactory discrimination: converging evidence for common principles across phyla. *Ann. Rev. Neurosci.* 20:595–631.

Jentsch, T. J. 1994. Trinity of ion channels (a review of touch receptors). *Nature* 367:412–413.

Kernan, M., D. Cowan, and C. Zuker. 1994. Genetic dissection of mechanosensory transduction: mechanoreception-defective mutations of *Drosophila. Neuron* 12:1195–1206.

Khorana, H. G. 1992. Rhodopsin, photoreceptor of the rod cell. An emerging pattern for structure and function. *J. Biol. Chem.* 267:1–4.

Kinnamon, S. C. 1996. Taste transduction. A bitter-sweet beginning [News; Comment]. *Nature* 381:737–738.

Koutalos, Y., and K. W. Yau. 1996. Regulation of sensitivity in vertebrate rod photoreceptors by calcium. *Trends Neurosci.* 19:73–81.

Lamb, T. D. 1996. Gain and kinetics of activation in the G-protein cascade of phototransduction. *Proc. Nat'l. Acad. Sci. USA* 93:566–570.

Mollon, J. D. 1999. Color vision: opsins and options. *Proc. Nat'l. Acad. Sci. USA* 96:4743–4745.

Nathans, J. 1992. Rhodopsin: structure, function, and genetics. *Biochemistry* 31:4923–4931.

Nathans, J. 1994. In the eye of the beholder: visual pigments and inherited variation in human vision. *Cell* 78:357–360.

Polans, A., W. Baehr, and K. Palczewski. 1996. Turned on by Ca^{2+}: the physiology and pathology of Ca^{2+}-binding proteins in the retina. *Trends Neurosci.* 19:547–554.

Reed, R. 1998. Opening the window to odor space. *Science* 279:193.

Snider, W. D., and S. McMahon. 1998. Tackling pain at the source: new ideas about nociceptors. *Neuron* 20:629–632.

Stryer, L. 1996. Vision: from photon to perception. *Proc. Nat'l. Acad. Sci. USA* 93:557–559.

Tovee, M. 1996. *An Introduction to the Visual System.* Cambridge University Press.

Zuker, C. S. 1996. The biology of vision of *Drosophila. Proc. Nat'l. Acad. Sci. USA* 93:571–576.

Learning and Memory

Bailey, C. H., D. Bartsch, and E. R. Kandel. 1996. Toward a molecular definition of long-term memory storage. *Proc. Nat'l. Acad. Sci. USA* 93:13445–13452.

Carew, T. J. 1996. Molecular enhancement of memory formation. *Neuron* 16:5–8.

Chen, C., and S. Tonegawa. 1997. Molecular genetic analysis of synaptic plasticity, activity-dependent neural development, learning, and memory in the mammalian brain. *Ann. Rev. Neurosci.* 20:125–184.

Glanzman, D. L. 1995. The cellular basis of classical conditioning in *Aplysia californica*–it's less simple than you think. *Trends Neurosci.* 18:30–36.

Greenspan, R. J. 1995. Flies, genes, learning, and memory. *Neuron* 15:747–750.

Hawkins, R. D., E. R. Kandel, and S. A. Siegelbaum. 1993. Learning to modulate transmitter release: themes and variations in synaptic plasticity. *Ann. Rev. Neurosci.* 16:625–665.

Jessell, T. M., and E. R. Kandel. 1993. Synaptic transmission: a bidirectional and self-modifiable form of cell-cell communication. *Cell* 72:1–30.

Lechner, H., and J. Byrne. 1998. New perspectives on classical conditioning: a synthesis of Hebbian and non-Hebbian mechanisms. *Neuron* 20:355–358.

Milner, B., L. Squire, and E. Kandel. 1998. Cognitive neuroscience and the study of memory. *Neuron* 20:445–468.

Tully, T. 1996. Discovery of genes involved with learning and memory: an experimental synthesis of Hirschian and Benzerian perspectives. *Proc. Nat'l. Acad. Sci. USA* 93:13460–13467.

Tully, T. 1997. Regulation of gene expression and its role in long-term memory and synaptic plasticity. *Proc. Nat'l. Acad. Sci. USA* 94:4239–4241.

22

Integrating Cells into Tissues

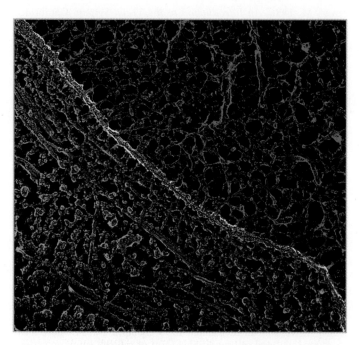

A cell-matrix interface.

The evolution of multicellular organisms permitted specialized cells and tissues to form; a flowering plant has at least 15 cell types, and a vertebrate hundreds. In both plants and animals, cells that are specialized to carry out a particular task are found together in the tissues in which the task is performed: a xylem or meristem; a liver, a muscle, or a nerve ganglion. Different types of cells in a tissue are often arranged in precise patterns of staggering complexity. For instance, the hundreds of different types of neurons in the human brain are interconnected to one another through a network of some 10^{15} synaptic connections! The coordinated functioning of many types of cells within tissues, and of multiple specialized tissues, permits the organism as a whole to move, metabolize, reproduce, and carry out other essential activities.

A key step in the evolution of multicellularity must have been the ability of cells to contact tightly and interact specifically with other cells. Various integral membrane proteins, collectively termed **cell-adhesion molecules** (**CAMs**), enable many animal cells to adhere tightly and specifically with cells of the same, or similar, type; these interactions allow populations of cells to segregate into distinct tissues (Figure 22-1). Following aggregation, cells elaborate specialized **cell junctions** that stabilize these interactions and promote local communication between adjacent cells. Animal cells also secrete a complex network of proteins and carbohydrates, the **extracellular matrix** (**ECM**), that creates a special environment in the spaces between cells. The matrix helps bind the cells in tissues together and is a reservoir for many hormones controlling cell growth and differentiation. The matrix also provides a lattice through which cells can move, particularly during the early stages of differentiation. Defects in these connections lead to cancer and developmental malformations.

MEDIA CONNECTIONS

Focus: Cell-Cell Adhesion in Leukocyte Extravasation

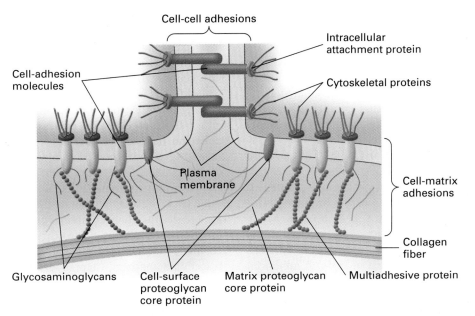

▲ FIGURE 22-1 Schematic overview of the types of molecules that bind cells to each other and to the extracellular matrix. Cell-adhesion molecules (CAMs) are integral membrane proteins. Some interact with similar molecules on other cells and, via intracellular attachment proteins, form anchors for cytoskeletal proteins. Other CAMs form connections with components of the extracellular matrix and also, via attachment proteins, with cytoskeletal proteins. Multiadhesive proteins bind to cell-surface receptor proteins and to other matrix components. Proteoglycans, consisting of a core protein, to which glycosamino glycan chains are attached, also participate in adhesion of cells to one another and to the protein components of the matrix. Together, these interactions allow cells to adhere to one another, interconnect the cytoskeletons of adjacent cells, and give tissues their strength and resistance to shear forces.

The extracellular matrix has three major protein components: highly viscous **proteoglycans,** which cushion cells; insoluble **collagen** fibers, which provide strength and resilience; and soluble **multiadhesive matrix proteins,** which bind these components to receptors on the cell surface. Different combinations of these components tailor the strength of the extracellular matrix for different purposes. For example, animals contain many types of extracellular matrices, each specialized for a particular function such as strength (in a tendon), cushioning (in cartilage), or adhesion. In the case of smooth muscle cells that surround an artery, the extracellular matrix must provide strong but flexible connections.

The extracellular matrix is not just an inert framework or cage that supports or surrounds cells. The matrix also communicates directly and indirectly with the intracellular signaling pathways that direct a cell to carry out specific functions. For example, the ability of hepatocytes—the principal cells in the liver—to express liver-specific proteins depends on their association with a matrix of appropriate composition. Specific ECM components can directly activate cytosolic signal-transduction pathways by binding to cell-adhesion protein receptors in the plasma membrane. Alternatively, by binding growth factors and other hormones, the matrix can either sequester these signals from cells or, conversely, present them to cells, thereby indirectly inducing or inhibiting signaling pathways. Morphogenesis—the

later stage of embryonic development during which form is achieved by cell movements and rearrangements—also is critically dependent on ECM components, which are constantly being remodeled, degraded, and resynthesized locally. Even in adults—in areas of wounding, for example —degradation and resynthesis of ECM components occurs. How the extracellular matrix regulates cell activities is considered in other chapters on signaling and development.

In this chapter, we focus on the structure of and interactions between ECM components, cell-adhesion molecules, and cell-adhesion junctions—the main structures that permit animal cells to form organized tissues. Plant cells are surrounded by a cell wall that is thicker and more rigid than the extracellular matrix. Although the plant cell wall and the extracellular matrix serve many of the same functions, they are structurally very different. Because of these differences, we discuss the plant cell wall and its interactions with plant cells in a separate section at the end of the chapter.

22.1 Cell-Cell Adhesion and Communication

Adhesion of like cells is a primary feature of the architecture of many tissues. A sheet of absorptive epithelial cells, for instance, forms the lining of the small intestine, and sheets of hepatocytes two cells thick make up much of the liver. A number of cell-surface proteins (the CAMs), mediate such homophilic (like-binds-like) adhesion between cells of a single type and heterophilic adhesion between cells of different types. Most CAMs are uniformly distributed along the regions of plasma membranes that contact other cells, and the cytosol-facing domains of these proteins are usually connected to elements of the cytoskeleton.

There are five principal classes of CAMs (Figure 22-2): cadherins, the immunoglobulin (Ig) superfamily, selectins, mucins, and integrins. Cell-cell adhesion involving cadherins and selectins depends on Ca^{2+} ions, whereas interactions involving integrin and Ig-superfamily CAMs do not. Many cells use several different CAMs to mediate cell-cell adhesion. The integrins mediate cell-matrix interactions whereas the other types of CAMs participate in cell-cell adhesion.

Adhesion of cells to one another generally is initiated by one or more of the cell-adhesion molecules described in Figure 22-2. In order for cells in tissues to function in an integrated manner, specialized junctions consisting of clustered cell-adhesion molecules are essential. There are four major classes of junctions: the tight junction, gap junction, cell-cell, and cell-matrix junctions (see Figure 15-23). In Chapter 15, we discussed the structure and function of **tight**

junctions, which connect epithelial cells (e.g., those lining the intestine) and prevent passage of fluids through the cell layer (see Figure 15-28). Here we consider the structure and specialized functions of cell-cell and gap junctions. Cell-cell and cell-matrix junctions perform a simple structural role, to hold cells into a tissue. They carry out this role by connecting the internal cytoskeleton directly to the cell exterior, either another cell or the extracellular matrix, via two cell-adhesion molecules—cadherins and integrins. Rather than inventing unique ways of connecting the actin and intermediate filament (IF) cytoskeletons to cadherin and integrin in the plasma membrane, the cell has instead evolved a common structure that adapts to different partners (see Figure 22-1). Despite their complexity, all cytoskeleton-associated junctions are organized into three parts: cell-adhesion molecules, which connect the cell to another cell or to the extracellular matrix; adapter proteins, which connect the CAMs to actin or keratin filaments; and lastly the bundle of cytoskeletal filaments itself.

In addition to their structural links, cells in tissues are in direct communication through gap junctions. Gap junctions are distributed along the lateral surfaces of adjacent cells and allow them to exchange small molecules. As we discussed in Chapter 21, gap junctions at electric synapses allow ions to pass from one nerve cell to the next, thereby

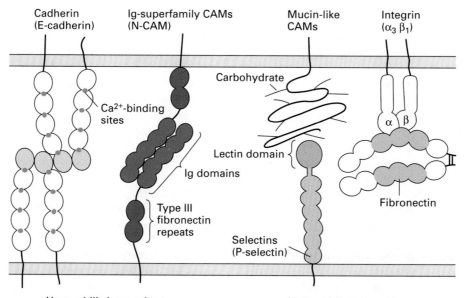

▲ **FIGURE 22-2 Major families of cell-adhesion molecules (CAMs).** Integral membrane proteins are built of multiple domains. Cadherin and the immunoglobulin (Ig) superfamily of CAMs mediate homophilic cell-cell adhesion. For cadherin, calcium binding to sites (orange) between the five domains in the extracellular segment is necessary for cell adhesion; the N-terminal domain (blue) causes cadherin to dimerize and to bind cadherin dimers from the opposite membrane. The Ig superfamily contains multiple domains (green) similar in structure to immunoglobulins and frequently contain type III fibronectin repeats (purple). In a heterophilic interaction, the lectin domain of selectins binds carbohydrate chains in mucin-like CAMs on adjacent cells in the presence of Ca^{2+}. The lectin domain is separated from the membrane by a series of repeated domains. The major cell-matrix adhesion molecule, integrin, is a heterodimer of α and β subunits. They bind to the cell-binding domain of fibronectin, laminin, or other matrix molecules.

allowing a presynaptic cell to induce an action potential in the postsynaptic cell without a lag period (see Figure 21-35). But gap junctions also are present in many non-neuronal tissues, where they help to integrate the metabolic activities of all the cells in a tissue by permitting exchange of ions and small molecules (e.g., cyclic AMP and precursors of DNA and RNA).

Cadherins Mediate Ca^{2+}-Dependent Homophilic Cell-Cell Adhesion

Cadherins, a family of Ca^{2+}-dependent CAMs, are the major molecules of cell-cell adhesion and play a critical role during tissue differentiation (Chapter 23). The most widely expressed, particularly during early differentiation, are the E-, P-, and N-cadherins. Over 40 different cadherins are known; some of the best understood cadherins are summarized in Table 22-1. The brain expresses the largest number of different cadherins, presumably due to the necessity of forming very specific cell-cell contacts.

Each cadherin is a type I integral membrane glycoprotein of 720–750 amino acids. The cadherin molecule consists of an N-terminal extracellular region, a single transmembrane spanning segment, and a C-terminal cytoplasmic tail. The extracellular domain contains repeated sequences that are sites necessary for Ca^{2+} binding and cell-cell adhesion. The cytoplasmic domain associates with the cytoskeleton. On average, 50–60 percent of the sequence is identical among different cadherins. Importantly, each cadherin has a characteristic tissue distribution. During differentiation and in some diseases, the amount or nature of the cell-surface cadherins changes, affecting many aspects of cell-cell adhesion and cell migration. For example, the metastasis of tumor cells is correlated with the loss of cadherin on their cell surface.

In adult vertebrates, E-cadherin holds most epithelial sheets together. Sheets of polarized epithelial cells, such as those that line the small intestine or kidney tubules, contain abundant E-cadherin at the sites of cell-cell contact along their lateral surfaces. When a monoclonal antibody to E-cadherin is added to a monolayer of cultured epithelial cells, the cells detach from one another, directly demonstrating the

requirement for E-cadherin in cell-cell adhesion. The removal of Ca^{2+} from the medium also disrupts cell-cell adhesion, showing that E-cadherin-mediated interactions require Ca^{2+}. If E-cadherin-mediated adhesion is blocked during cell aggregation, none of the specialized cell junctions between epithelial cells are generated. Later studies showed that calcium ions cause cadherin to dimerize and that cadherin dimers and not monomers are responsible for cell-cell adhesion.

E-cadherin, like the other cadherins, preferentially mediates homophilic interactions. This phenomenon was demonstrated in experiments with L cells, a line of cultured transformed mouse fibroblasts that express no cadherins and adhere poorly to themselves or to other cultured cells. Lines of transfected L cells that expressed either E-cadherin or P-cadherin were generated; such cells were found to adhere preferentially to cells expressing the same class of cadherin molecules. For instance, E-cadherin-expressing L cells adhere tightly to one another and to epithelial cells from embryonic lung that express E-cadherin; they do not attach to untransfected L cells or to L cells (or other cell types) expressing P-cadherin. L cells expressing P-cadherin adhere to one another, and to other types of cells that express this cadherin. Thus, cadherins directly cause homotypic interactions among cells.

The mechanism for cell-cell adhesion is explained by the atomic structures of the N-terminal domains from E- and N-cadherin. The current model proposes that cadherin molecules associate through their N-terminal domains into a parallel homodimer (see Figure 22-2). The Ca^{2+}-binding sites, located between the cadherin repeats, serve to rigidify the cadherin molecule and expose residues that form the dimer interface. Furthermore, cell-cell adhesion results from head-to-head contact between cadherin dimers in adjacent cell membranes. The two sets of interactions, head-to-head and side-to-side, are postulated to cause the clustering or "zipping" of cadherins in specialized adhesion junctions.

N-CAMs Mediate Ca^{2+}-Independent Homophilic Cell-Cell Adhesion

N-CAMs, a group of Ca^2-independent cell-cell adhesion proteins in vertebrates, belong to the Ig superfamily of CAMs (see Figure 22-2). Their full name—*nerve-cell adhesion molecule*—reflects their particular importance in nervous tissue. Like cadherins, N-CAMs primarily mediate homophilic interactions, binding together cells that express similar N-CAM molecules. Unlike cadherins, N-CAMs are encoded by a single gene; their diversity is generated by alternative mRNA splicing and by differences in glycosylation (Figure 22-3). Like N-cadherin, N-CAMs appear during morphogenesis, playing an important role in differentiation of muscle, glial, and nerve cells. Their role in cell adhesion has been directly demonstrated by use of specific antibodies. For instance, adhesion of cultured retinal neurons is inhibited by addition of antibodies to N-CAMs.

TABLE 22-1	Major Cadherin Molecules on Mammalian Cells
Molecule	Predominant Cellular Distribution
E-cadherin	Preimplantation embryos, non-neural epithelial tissue
P-cadherin	Trophoblast
N-cadherin	Nervous system, lens, cardiac and skeletal muscle

SOURCE: M. Takeichi, 1988, *Development* 102:639; M. Takeichi, 1991, *Science* 251:1451; H. Inuzuka et al., 1991, *Neuron* 7:69; and M. Donalies et al., 1991, *Proc. Nat'l. Acad. Sci. USA* 88:8024.

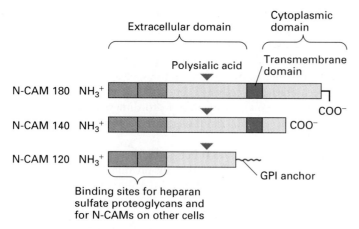

▲ **FIGURE 22-3 Three of the N-CAMs produced by alternative splicing of the primary transcript produced from the single N-CAM gene.** N-CAM 180 (180,000 MW) and N-CAM 140 are anchored in the membrane by a single hydrophobic α helix and differ in the length of their cytoplasmic domains. N-CAM 120 is attached to the membrane by a glycosylphosphatidylinositol (GPI) anchor (see Figure 3-36a). Each of these three N-CAMs can also vary in the length of the poly α (2→8) sialic acid chain, whose attachment site is indicated. [After G. M. Edelman, 1988, *Biochemistry* **27**:3533; and T. M. Jessell, 1988, *Neuron* **1**:3.]

The adhesive properties of N-CAMs are modulated by long chains of sialic acid, a negatively charged sugar (see Figure 17-31). N-CAMs that are heavily sialylated form weaker homophilic interactions than do less sialylated forms, possibly because of repulsion between the negatively charged sialic acid residues. In embryonic tissues such as brain, polysialic acid constitutes as much as 25 percent of the mass of N-CAMs; in contrast, N-CAMs from adult tissues contain only one-third as much sialic acid. The lower adhesive properties of embryonic N-CAMs enable cell-cell contacts to be made and then broken, a property necessary for specific cell contacts to form in the developing nervous system. The higher adhesive properties of the adult forms of N-CAM stabilize these contacts. Thus, the strength of cell-cell adhesions is modified during differentiation by differential glycosylation of the N-CAMs.

Selectins and Other CAMs Participate in Leukocyte Extravasation

Thus far we have considered cell interactions in solid tissues, such as epithelia and neuronal tissue. Once these interactions form during differentiation, they generally are stable for the life of the cells. In adult organisms, many types of cells that participate in defense against foreign invaders (e.g., bacteria and viruses) must move rapidly from the blood, where they circulate as unattached cells, into the underlying tissue at sites of infection or inflammation. Movement into tissue, termed *extravasation*, of four types of leukocytes

(white blood cells) is particularly important: monocytes, the precursors of macrophages, which can ingest foreign particles; neutrophils, which release several antibacterial proteins; and T and B lymphocytes, the antigen-specific cells of the immune system.

Extravasation requires the successive formation and breakage of cell-cell contacts between leukocytes in the blood and endothelial cells lining the vessels. These contacts are mediated by selectins, a class of cell-adhesion molecules that are specific for leukocyte–vascular cell interactions. A key protein in this process, *P-selectin*, is localized to the blood-facing surface of endothelial cells. Like other members of the selectin family of CAMs, P-selectin is a *lectin*, a protein that binds to carbohydrates. Each type of selectin binds to specific oligosaccharide sequences in glycoproteins or glycolipids. As with cadherins, binding of selectins to their ligands is Ca^{2+}-dependent. The sugar-binding lectin domain in selectins generally is at the end of the extracellular region of the molecule (see Figure 22-2). The ligand for P-selectin is a specific oligosaccharide sequence, called the *sialyl Lewis-x antigen*, that is part of longer oligosaccharides present in abundance on leukocyte glycoproteins and glycolipids.

As illustrated in Figure 22-4, in normal endothelial cells P-selectin is localized to intracellular vesicles and is not present on the plasma membrane. These cells are activated by various inflammatory signals released by surrounding cells in areas of infection or inflammation. Once endothelial cells are activated, the vesicles containing P-selectin undergo exocytosis within seconds, and P-selectin appears on the plasma membrane. As a consequence, passing leukocytes adhere weakly to the endothelium; because of the force of the blood flow, these "trapped" leukocytes are slowed but not stopped and seem to roll along the surface of the endothelium.

In order for tight adhesion to occur between activated endothelial cells and leukocytes, β_2-containing integrins on the surface of leukocytes also must be activated. For example, activation of the $\alpha_L\beta_2$ integrin, which is expressed by T lymphocytes, is induced by *platelet-activating factor* (PAF), a phospholipid released by activated endothelial cells at the same time that P-selectin is exocytosed. Binding of PAF to its receptor on T lymphocytes activates integrin $\alpha_L\beta_2$ through the Rho signaling pathway. The activated integrin then binds to *ICAM-1* and *ICAM-2*, which are Ig-superfamily CAMs expressed constitutively on the surface of endothelial cells. The tight adhesion mediated by the Ca^{2+}-independent interaction of $\alpha_L\beta_2$ and the ICAMs leads to spreading of T lymphocytes on the surface of the endothelium; soon the adhered T lymphocytes move between adjacent endothelial cells and into the underlying tissue.

Thus, the selective adhesion of T lymphocytes to the endothelium near sites of infection or inflammation depends on the sequential appearance and activation of several different CAMs on the surface of the interacting cells. Other leukocytes that express specific integrins containing the β_2 subunit move into tissues by a similar mechanism; $\alpha_M\beta_2$, for instance, is found primarily on macrophages. As might

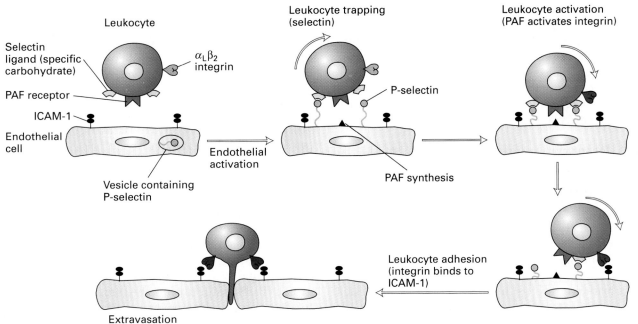

▲ FIGURE 22-4 Interactions between cell-adhesion molecules during the initial binding and tight binding of T cells, a kind of leukocyte, to activation endothelial cells. Once a T cell has firmly adhered to the endothelium, it can move (extravasate) into the underlying tissue. Activation of the endothelium requires signals, such as platelet-activating factor (PAF), that are released in areas of infection or inflammation; thus extravasation occurs only in such areas. See text for discussion. [Adapted from R. O. Hynes and A. Lander, 1992, *Cell* **68**:303.]

be expected, humans with a genetic defect in synthesis of the integrin β_2 subunit, termed *leukocyte-adhesion deficiency,* are susceptible to repeated bacterial infections.

Cadherin-Containing Junctions Connect Cells to One Another

Although hundreds of individual cell-cell adhesions are sufficient to cause cells to adhere, specialized junctions consisting of dense clusters of cell-adhesion molecules are required for the function of tissues. In electron micrographs, the plasma membranes of cell-cell junctions are parallel and only 15–20 nm apart. Concentrated in this region is E-cadherin, which links the plasma membranes of adjacent cells and via catenin adapter proteins attaches to actin filaments in adherens junctions or keratin filaments in desmosomes (Figure 22-5).

Adherens Junctions Epithelial cells contain a continuous band of cadherin molecules, usually located near the apical surface just below the tight junction, that connects the lateral membranes of epithelial cells (see Figure 15-23). Known as the *adherens junction*, this region contains α- and β-catenins that link E-cadherin in the plasma membrane to the circumferential belt of actin and myosin filaments (see Figure 18-35). Adherens junctions contain many of the same proteins found at focal adhesions, including vinculin, tropomyosin, and α-actinin. As a complex with the adherens junction,

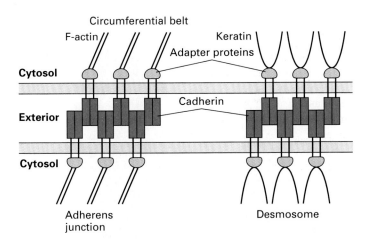

▲ FIGURE 22-5 Adhesion molecules in junctions involved in cell-cell adhesion. Adherens junctions and desmosomes are specialized cell-cell junctions that consist of clustered-cadherin dimers. Cadherin is connected to either the circumferential belt of actin filaments or bundles of keratin filaments in the cytoskeleton through the catenin adapter proteins.

the circumferential belt functions as a tension cable that can internally brace the cell and thereby control its shape.

Desmosomes A desmosome consists of proteinaceous adhesion plaques (15–20 nm thick) attached to the cytosolic face of the plasma membranes of adjacent cells and connected by transmembrane linker proteins (Figure 22-6). *Plakoglobin* is a major constituent of the plaques; it is very similar to β-catenin. The transmembrane linker proteins, called *desmoglein* and *desmocollin*, belong to the cadherin family of cell-adhesion molecules. They bind to plakoglobin and other proteins in the plaques and extend into the intercellular space, where they interact, forming an interlocking network that binds two cells together.

Desmoglein was first identified by an unusual but revealing skin disease called *pemphigus vulgaris*, an autoimmune disease. Patients with autoimmune disorders synthesize antibodies that bind to a normal body protein. In this case, the autoantibodies disrupt adhesion between epithelial cells, causing blisters of the skin and mucous membranes. The predominant autoantibody was shown to be specific for desmoglein, a major protein in the skin desmosomes; indeed, addition of such antibodies to normal skin induces formation of blisters and disruption of cell adhesion.

In epithelial cells, keratin intermediate filaments course near the cytoplasmic plaques of desmosomes and apparently are linked to them by the *desmoplakin proteins*. Some of these filaments run parallel to the cell surface, and others penetrate and traverse the cytoplasm. They are thought to be part of the internal structural framework of the cell, giving it shape and rigidity. If so, desmosomes also could transmit shearing forces from one region of a cell layer to the epithelium as a whole; they thus provide strength and rigidity to the entire epithelial cell layer.

Gap Junctions Allow Small Molecules to Pass between Adjacent Cells

Early electron micrographs showed that almost all animal cells that come in contact with each other have regions of junctional specialization characterized by an intercellular gap, which is filled by a well-defined set of cylindrical particles (Figure 22-7). Morphologists named these regions **gap junctions**, but in retrospect the gap is not their most important feature. The cylindrical particles, which are water-filled channels, are the key to the function of gap junctions. These channels directly link the cytosol of one cell with that of an adjacent cell, providing a passageway for movement of very small molecules and ions between the cells.

The size of these intercellular channels can be measured by injecting a cell with a fluorescent dye covalently linked to molecules of various sizes and using a fluorescence microscope

(a)

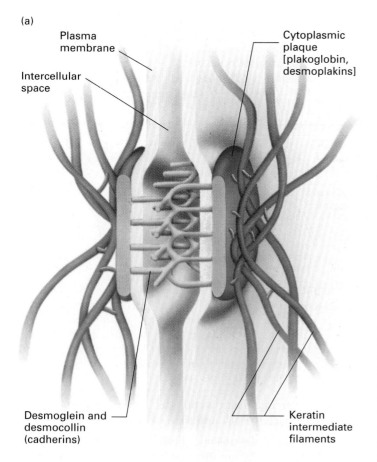

Plasma membrane

Intercellular space

Cytoplasmic plaque [plakoglobin, desmoplakins]

Desmoglein and desmocollin (cadherins)

Keratin intermediate filaments

(b)

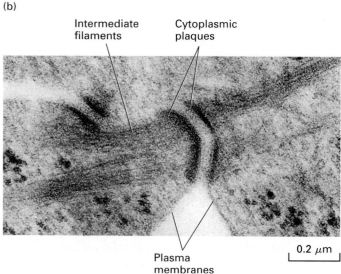

Intermediate filaments

Cytoplasmic plaques

Plasma membranes

0.2 μm

▲ **FIGURE 22-6 Desmosomes.** (a) Schematic model showing components of a desmosome between epithelial cells and attachments to the sides of keratin intermediate filaments, which crisscross the interior of cells. The transmembrane linker proteins, desmoglein and desmocollin, belong to the cadherin family. (b) Electron micrograph of a thin section of a desmosome connecting two cultured differentiated human keratinocytes. Bundles of intermediate filaments radiate from the two darkly staining cytoplasmic plaques that line the inner surface of the adjacent plasma membranes. See text for discussion. [Part (a) see B. M. Gumbiner, 1993, *Neuron* **11**:551; D. R. Garrod, 1993, *Curr. Opin. Cell Biol.* **5**:30; Part (b) courtesy of R. van Buskirk.]

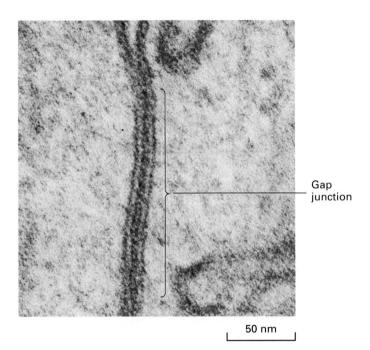

▲ **FIGURE 22-7 Electron micrograph of a thin section through a gap junction connecting two mouse liver cells.** The two plasma membranes are closely associated for a distance of several hundred nanometers, separated by a "gap" of 2–3 nm. [Courtesy of D. Goodenough.]

(labels in figure: Gap junction; 50 nm)

to observe whether they pass into neighboring cells. Gap junctions between mammalian cells permit the passage of molecules as large as 1.2 nm in diameter. In insects, these junctions are permeable to molecules as large as 2 nm in diameter. Generally speaking, molecules with a molecular weight less than 1200 pass freely, and those with a molecular greater than 2000 do not pass; the passage of intermediate-sized molecules is variable and limited. Thus ions and many low-molecular-weight building blocks of cellular macromolecules, such as amino acids and nucleoside phosphates, can pass from cell to cell.

A vivid example of this cell-cell transfer is the phenomenon of *metabolic coupling*, or *metabolic cooperation*, in which a cell transfers molecules to a neighboring cell that is unable to synthesize them. This phenomenon can be demonstrated with mutant cells unable to synthesize dATP, an immediate precursor of DNA, from hypoxanthine via a nucleotide-salvage pathway (see Figure 6-9). When cultured alone, these cells cannot incorporate radioactivity from hypoxanthine into their DNA. However, when the mutant cells are co-cultured with wild-type cells, radioactivity is frequently found in the nuclear DNA of the mutant cells. (The mutant and wild-type cells can be differentiated by their distinct morphologies or by feeding one of the cell lines carbon particles before mixing it with the other line.) The dATP derived from hypoxanthine is incorporated only into the DNA of mutant cells that are in direct or indirect contact (through an intermediate cell) with wild-type cells. It is thought that

labeled adenosine mono-, di-, or triphosphate is synthesized from the labeled hypoxanthine by wild-type cells and then passed through gap junctions to the mutant cells.

Another important compound transferred from cell to cell through gap junctions is cyclic AMP (cAMP), which acts as an intracellular second messenger. As discussed in Chapter 20, the amount of cellular cAMP increases in response to stimulation of cells by binding of many different hormones. The fact that cAMP can pass through gap junctions means that the hormonal stimulation of just one or a few cells in an epithelium initiates a metabolic reaction in many of them. For instance, binding of secretory hormones, such as secretin, to receptors on the basal plasma membranes of pancreatic acinar cells leads to increase in the intracellular concentration of either cAMP or Ca^{2+} ions, both of which trigger secretion of the contents of secretory vesicles. Because Ca^{2+} and cAMP can pass through the gap junctions, hormonal stimulation of one cell triggers secretion by many. As we saw in Chapter 18, an elevation in cytosolic Ca^{2+} in smooth muscle cells induces contraction. Gap junction–mediated transfer of Ca^{2+} ions between adjacent smooth muscle cells thus allows the coordinated contractile waves in the intestine during peristalsis and in the uterus during birth.

An important aspect of gap-junction physiology is that the channels close in the presence of very high concentrations of Ca^{2+} in the cytosol. Recall that the Ca^{2+} concentration in extracellular fluids in quite high (from 1×10^{-3} M to 2×10^{-3} M), whereas normally the concentration of Ca^{2+} free in the cytosol is lower than 10^{-6} M (see Table 15-1). If the membrane of one cell in an epithelium is ruptured, Ca^{2+} enters the cell, closing the channels that connect the cell with its neighbors and thus preventing leakage of the low-molecular-weight substances that are present in the cytoplasm of all epithelium cells. Even slight increases in the level of cytosolic Ca^{2+} ions or decreases in cytosolic pH can decrease the permeability of gap junctions. Thus cells may modulate the degree of coupling with their neighbors, but precisely why and how they accomplish this is a matter of debate.

Connexin, a Transmembrane Protein, Forms Cylindrical Channels in Gap Junctions

In the liver and many other tissues, large numbers of individual gap junction channels cluster together in an area about 0.3 mm in diameter. This property has enabled researchers to separate gap junctions from other components of the plasma membrane by shearing the purified plasma membrane into small fragments. Owing to their relatively high protein content, fragments containing gap junctions have a higher density than the bulk of the plasma membrane and can be purified on an equilibrium density gradient (see Figure 5-24). Electron micrographs of stained, isolated gap junctions reveal a lattice of hexagonal particles with hollow cores as intercellular channels (see Figure 21-35).

A current model of the structure of the gap junction is shown in Figure 22-8a. The transmembrane particles from

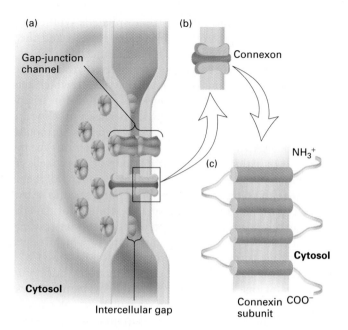

▲ **FIGURE 22-8 Structure of gap junctions.** (a) In this model, a gap junction is a cluster of channels between two plasma membranes that are separated by a gap of about 2–3 nm. (b) Both membranes contain connexon hemichannels, cylinders of six dumbbell-shaped connexin subunits. (c) Each connexin subunit has four transmembrane α helices. Two connexons join in the gap between the cells to form a gap-junction channel, 1.5–2.0 nm in diameter, that connects the cytoplasm of the two cells.

purified liver gap junctions are composed of *connexin* subunits, proteins with molecular weights between 25,000 and 50,000. Each hexagonal particle consists of 12 connexin molecules: 6 of the molecules are arranged in a connexon hemichannel, a hexagonal cylinder in one plasma membrane, and joined to a connexon hemichannel in the adjacent cell membrane.

The sequences of several connexin proteins, expressed in different tissues, have been determined from their cDNAs. All connexins have related amino acid sequences. Experiments suggest that each connexin polypeptide spans the plasma membrane four times (see Figure 22-8b) and that one conserved transmembrane α helix lines the aqueous channel. The connexins differ mainly in the length and sequence of their most C-terminal segment, which faces the cytosol. At least 12 different genes in the connexin family have been cloned; many are expressed in specific types of cells.

Some cells express a single connexin, consequently their gap junction channels are homotypic, consisting of identical connexons. However, most cells express at least two connexin genes. Consequently, different connexin polypeptides can assemble into hetero-oligomeric connexons, which in turn form heterotypic gap-junction channels. This diversity in channel composition leads to differences in permeability of the channels to different molecules.

SUMMARY   Cell-Cell Adhesion and Communication

• Cell-cell interactions involve multiple ligands and cell-adhesion molecules (CAMs), which are a diverse group of integral membrane proteins. CAMs fall into five main classes: the cadherins, Ig-superfamily CAMs, selectins, mucins and integrins (see Figure 22-2).

• Cadherins are responsible for Ca^{2+}-dependent homophilic interactions between cells.

• Ca^{2+}-independent homophilic interactions between cells are mediated by N-CAMs, which belong to the Ig superfamily.

• Selectins, which bind to carbohydrate groups on mucin-like CAMs via their distal lectin domain, mediate Ca^{2+}-dependent heterophilic cell-cell interactions. P-selectin on the surface of activated vascular endothelial cells plays an important role in the extravasation of leukocytes into tissues (see Figure 22-4).

• In all cell-adhesion junctions, clusters of transmembrane cell-adhesion molecules are linked via various adapter proteins in cytoplasmic plaques to the cytoskeleton.

• Adherens junctions and desmosomes are cadherin-containing junctions that bind the membranes of adjacent cells in a way that gives strength and rigidity to the entire tissue (see Figure 22-5).

• Gap junctions are constructed of 12 copies of a single protein, connexin, formed into a transmembrane channel that interconnects the cytoplasm of two adjacent cells (see Figure 22-8). Small molecules can pass through gap junctions, permitting metabolic coupling of adjacent cells.

22.2 Cell-Matrix Adhesion

The overall architecture of a tissue is determined by adhesion mechanisms that involve not only cell-cell interactions but also cell-matrix interactions. In animals, epithelia and most organized groups of cells like muscle are surrounded or underlain by an extracellular matrix (ECM) of collagen fibers, proteoglycans, and multiadhesive matrix proteins. This layer of matrix material serves several roles. It organizes cells into tissues and coordinates their cellular functions. In addition, the matrix provides a route for cell migrations, and molecules in the matrix activate classic signal-transduction pathways that induce cell growth, proliferation, and gene expression. These many effects of the matrix involve membrane-bound CAMs that bind directly to components of the ECM and the cytoskeleton. The principal class of CAMs that mediate cell-matrix adhesion is the integrins. However, other CAMs, including selectins and syndecan proteoglycans, also bind molecules in the matrix.

Integrins Mediate Weak Cell-Matrix and Cell-Cell Interactions

Integrins are heterodimers of α and β subunits, and the ligand-binding site is composed of parts of both chains. In mammals, at least 22 integrin heterodimers, composed of 17 types of α subunits and 8 types of β subunits, are known. A single β chain can interact with multiple α chains, forming integrins that bind different ligands. For example, the $\alpha_1\beta_1$ and $\alpha_2\beta_1$ integrins both bind a segment within the C-terminal domain of type IV collagen; these integrins, as well as the widely expressed $\alpha_6\beta_1$, also bind at least two different regions of laminin; and $\alpha_5\beta_1$ binds fibronectin. This diversity of integrins and their ligands in the matrix enables cells to migrate to their correct locations during tissue morphogenesis and helps sculpt the body plan of an embryo. We will discuss examples of cell-matrix adhesions during embryogenesis in Chapter 23.

Most integrins are expressed on a variety of cells, and most cells express several integrins, enabling them to bind to several matrix molecules. However, three integrins containing β_2 are expressed exclusively on leukocytes (white blood cells). One of these, $\alpha_L\beta_2$, mediates cell-cell interactions, rather than cell-matrix interactions. We saw how this integrin participates in the binding of leukocytes to specific ligands on endothelial cells at sites of infection or inflammation. A few other integrins mediate both cell-cell and cell-matrix interactions (Table 22-2).

Integrins typically exhibit relatively low affinities for their ligands (dissociation constants K_D between 10^{-6} and 10^{-8} mol/liter) compared with the high affinities (K_D values of 10^{-9} to 10^{-11} mol/liter) of typical cell-surface hormone receptors. However, the multiple weak interactions generated by binding of hundreds or thousands of integrin molecules to extracellular matrix proteins allow a cell to remain firmly anchored to the matrix. Alternatively, in situations where cells are migrating, it is essential that they be able to make and break specific contacts with the extracellular matrix; this is facilitated if individual contacts are weak. Cells that express several different integrins that bind the same ligand often can selectively regulate the activity of each type of integrin, thereby fine-tuning their interaction with the matrix.

Cell-Matrix Adhesion Is Modulated by Changes in the Activity and Number of Integrins

Platelets, the small cell fragments that circulate in blood and that are important for blood clotting, provide a good example of how cell-matrix interactions are modulated by controlling integrin activity. The $\alpha_{IIb}\beta_3$ integrin normally is present on the plasma membrane of platelets but is unable to bind the blood protein fibrinogen or the other protein ligands listed in Table 22-2, all of which participate in formation of a blood clot. Only after a platelet becomes "activated," by binding collagen or thrombin in a forming clot, can $\alpha_{IIb}\beta_3$ integrin bind fibrinogen; this interaction accelerates the formation of the clot. Platelet activation is accompanied by a conformational change in the $\alpha_{IIb}\beta_3$ integrin. The nature of this change is unknown, but as platelet activation also involves a major change in the platelet cytoskeleton (see Figure 18-8), this change probably involves binding of a cytoskeletal protein to the integrin cytosolic domain. Patients with genetic defects in the β_3 integrin subunit are prone to excessive bleeding, attesting to the role of this integrin in formation of blood clots.

Attachment of cells to matrix components also can be modulated by down-regulating the number of integrin molecules on the cell surface. The $\alpha_4\beta_1$ integrin, which is found on many hematopoietic cells (precursors of red and white blood cells) and binds fibronectin, offers an example of this mechanism. In order for these hematopoietic cells to proliferate and differentiate, they must be attached to extracellular-matrix fibronectin synthesized by supportive ("stromal") cells in the bone marrow. The $\alpha_4\beta_1$ integrin on hematopoietic cells binds to a Glu-Ile-Leu-Asp-Val (EILDV) sequence in the IIICS domain of stromal-cell fibronectin (see Figure 22-22), thereby anchoring the cells to the matrix. This integrin also binds to an EILDV sequence in VCAM-1, an integral membrane protein present on stromal cells of the bone marrow. Thus hematopoietic cells directly contact the stromal cells, as well as attach to the matrix. A decrease in the number of $\alpha_4\beta_1$ integrin molecules present on hematopoietic cells at a late stage in their differentiation is thought to allow mature blood cells to detach from the matrix and stromal cells in the bone marrow and subsequently enter the circulation.

TABLE 22-2		Some Vertebrate Integrins and Their Ligands*
Subunits		**Ligands**
β_1†	α_1	Collagens, laminin
	α_2	Collagens, laminin
	α_3†	Fibronectin, laminin
	α_4	Fibronectin; VCAM-1
	α_5	Fibronectin
	α_6†	Laminin
	α_7	Laminin
	α_V	Fibronectin, vitronectin
β_2	α_L	ICAM-1, ICAM-2
	α_M	C3b, fibrinogen, factor X; ICAM-1
	α_X	Fibrinogen, C3b
β_3†	α_{IIb}	Fibrinogen, fibronectin, von Willebrand factor, vitronectin, thrombospondin
	α_V	Same as $\beta_3\alpha_{IIb}$; also osteopontin, collagen

*The integrins are grouped in subfamilies sharing a common β subunit. Ligands shown in red are vascular ligands; all others are proteins in the extracellular matrix.
†These subunits can have multiply spliced isoforms with different cytosolic domains.

SOURCE: R. O. Hynes, 1992, *Cell* 69:11.

De-adhesion Factors Promote Cell Migration and Can Remodel the Cell Surface

Strong adhesion to the extracellular matrix (e.g., at basal laminae) prevents cells from migrating. In some cases, however, normally nonmotile cells must quickly become motile. For example, a wound to the skin is closed by the rapid migration of surrounding keratinocytes to the lesion area. This transition to a motile state requires *de-adhesion* of cells from the extracellular matrix by inhibition of cell-matrix interactions and by destruction of matrix components. One class of de-adhesion factors comprises small peptides called *disintegrins*, which contain the integrin-binding RGD sequence present in many ECM proteins. By binding to integrins on the surface of cells, disintegrins competitively inhibit binding of cells to matrix components. The disintegrins present in snake venoms, which prevent platelets from aggregating, are partly responsible for the anticoagulant property of venoms. In contrast to snake venom disintegrins, the second class of de-adhesion factors contains two types of proteases, fibrinogen and matrix-specific metalloproteinases (MMPs). Both proteases degrade matrix components, thereby permitting cell migration.

A family of membrane-anchored glycoproteins, termed *ADAM*, containing *a disintegrin and a metalloprotease* domain participate in a variety of events that depend on remodeling of the cell surface. Such events include determination of cell fates during embryogenesis, fusion of a sperm and egg during fertilization, fusion of myoblasts during myogenesis, and release of soluble tumor necrosis factor α (TNFα). For example, membrane-bound tumor necrosis factor α (TNFα) is released from the cell surface by a membrane-anchored converting enzyme containing a metalloproteinase domain. Soluble TNFα then activates the inflammatory response. In other cases, such as sperm-egg fusion, the protease is cleaved from ADAM, leaving the disintegrin on the sperm cell to mediate binding to an integrin on the egg cell. Such proteolytic processing of the extracellular domain of a membrane protein, termed "ectodomain shedding," permits the cell to inactivate membrane receptors or release soluble active proteins, such as cytokines, from the cell surface.

Integrin-Containing Junctions Connect Cells to the Substratum

Cells attach to the underlying extracellular matrix through two types of integrin-dependent junctions: *focal adhesions,* which attach the actin cytoskeleton to fibers of fibronectin, and *hemidesmosomes,* which connect intermediate filaments to basal laminae (Figure 22-9). Integrin-containing cell-matrix junctions are found in highly motile cells such as skin keratinocytes, which are weakly adherent, and in immobile, strongly adherent cells such as epithelia.

Focal Adhesions As discussed in Chapter 18, activation of fibroblasts in the Rho signaling pathway causes the formation of abundant actin-rich stress fibers, which help maintain the cells' shapes and adhere them to the substratum (see

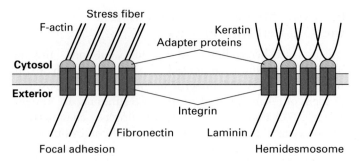

▲ **FIGURE 22-9 Adhesion molecules in junctions involved in cell-matrix adhesion.** Focal adhesions and hemidesmosomes are cell-matrix junctions that consist of clustered-integrin molecules. The extracellular domain of integrin binds matrix proteins, while the cytosolic domain interacts with adapter proteins. These proteins mediate attachment to stress fibers or to bundles of keratin intermediate filaments.

Figure 18-44). Actin filaments of the stress fibers are attached to the β subunit of integrins through adapter proteins; these peripheral membrane proteins, including vinculin, are generally located in the region of the cytosol just inside the plasma membrane (see Figure 22-9). Integrins cluster to assemble a focal adhesion. This process is mediated by tension exerted on actin filaments by myosin II. More than 20 proteins—actin-binding proteins, kinases, and membrane-binding proteins— are also localized in focal adhesions. They function to activate adhesion-dependent signals for cell growth and cell motility.

The role of integrin in mediating interactions with the cytoskeleton was first suggested by fluorescence microscopy of cultured fibroblasts showing the colocalization of stress fibers and $\alpha_5\beta_1$ integrin (Figure 22-10a). In some sections, exterior fibronectin fiber bundles appear continuous with bundles of actin fibers within the cell (Figure 22-10b). Thus integrin $\alpha_5\beta_1$, localized to focal adhesions, anchors the two types of fibers to the opposite sides of the plasma membrane.

Hemidesmosomes The second type of cell-matrix adhesion junction, the hemidesmosome, is found mainly on the basal surface of epithelial cells. These junctions firmly anchor epithelial cells to the underlying basal lamina. The cytosolic side of a hemidesmosome consists of a plaque composed of adapter proteins, which are attached to the ends of keratin filaments (see Figure 22-9). Integrin $\alpha_6\beta_4$ is localized to hemidesmosomes and is thought to bind to an adapter protein, plectin, within the plaques and to the extracellular-matrix protein laminin. By interconnecting the intermediate filaments of the cytoskeleton with the fibers of the basal lamina, these cell-matrix junctions increase the overall rigidity of epithelial tissues.

SUMMARY Cell-Matrix Adhesion

• Integrins primarily mediate cell-matrix interactions. These α, β heterodimers bind fibronectin, laminin, collagen, and other matrix proteins. Different isoforms of

(a)

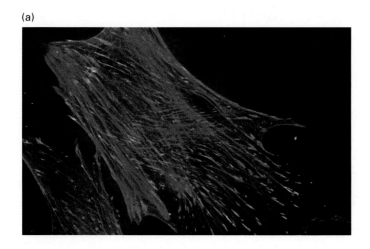

(b)

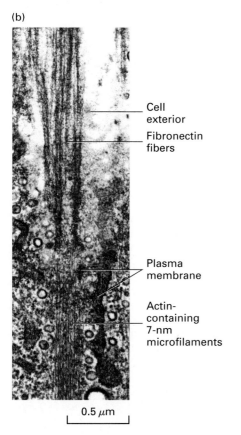

Cell
exterior

Fibronectin
fibers

Plasma
membrane

Actin-
containing
7-nm
microfilaments

0.5 μm

▲ **FIGURE 22-10 Immunofluorescent micrograph of a fixed,
stationary cultured fibroblast showing colocalization of the
$\alpha_5\beta_1$ integrin (green) and actin-containing stress fibers (red).**
At the ends of the stress fibers, where the cells contact the
substratum, there is a coincidence of actin and the fibronectin-
binding integrin (yellow). (b) Electron micrograph of the junction
of fibronectin and actin fibers in a cultured fibroblast. Individual
actin-containing 7-nm microfilaments, components of a stress
fiber, end at the obliquely sectioned cell membrane. The microfilaments appear in close proximity to the thicker, densely stained
fibronectin fibers on the outside of the cell. [Part (a) from J. Duband
et al., 1988, *J. Cell Biol.* **107**:1385. Part (b) from I. J. Singer, 1979, *Cell*
16:675; courtesy of I. J. Singer; copyright, 1979, M.I.T.]

the α and β chains determine the ligand-binding specificity of various integrins.

• Integrins that contain a β_2 chain bind CAMs of the
Ig superfamily, thereby mediating cell-cell adhesion.

• Soluble snake venom disintegrins and matrix-specific
proteinases interfere with cell adhesion, promoting cell
migration. Membrane-bound ADAMs with similar
catalytic activities are implicated in protein ectodomain
shedding and new cell adhesions.

• Focal adhesions and hemidesmosomes are integrin-
containing junctions that attach cells to fibronectin and
elements of the basal lamina in the extracellular matrix,
respectively.

22.3 Collagen: The Fibrous Proteins of the Matrix

Collagen is the major insoluble fibrous protein in the extracellular matrix and in connective tissue. In fact, it is the single most abundant protein in the animal kingdom. There are
at least 16 types of collagen, but 80–90 percent of the collagen in the body consists of types I, II, and III (Table 22-3).
These collagen molecules pack together to form long thin
fibrils of similar structure (see Figure 5-20). Type IV, in contrast, forms a two-dimensional reticulum; several other types
associate with fibril-type collagens, linking them to each
other or to other matrix components. At one time it was
thought that all collagens were secreted by fibroblasts in
connective tissue, but we now know that numerous epithelial cells make certain types of collagens. The various collagens and the structures they form all serve the same purpose, to help tissues withstand stretching.

The Basic Structural Unit of Collagen Is a Triple Helix

Because its abundance in tendon-rich tissue such as rat tail
makes the fibrous type I collagen easy to isolate, it was the
first to be characterized. Its fundamental structural unit is a
long (300-nm), thin (1.5-nm-diameter) protein that consists
of three coiled subunits: two $\alpha1(I)$ chains and one $\alpha2(I)$.*
Each chain contains precisely 1050 amino acids wound
around one another in a characteristic right-handed triple
helix (Figure 22-11a). All collagens were eventually shown
to contain three-stranded helical segments of similar structure; the unique properties of each type of collagen are due
mainly to segments that interrupt the triple helix and that
fold into other kinds of three-dimensional structures.

The triple-helical structure of collagen arises from an unusual abundance of three amino acids: glycine, proline, and

---

*In collagen nomenclature, the collagen type is in roman numerals
and is enclosed in parentheses.

(a)

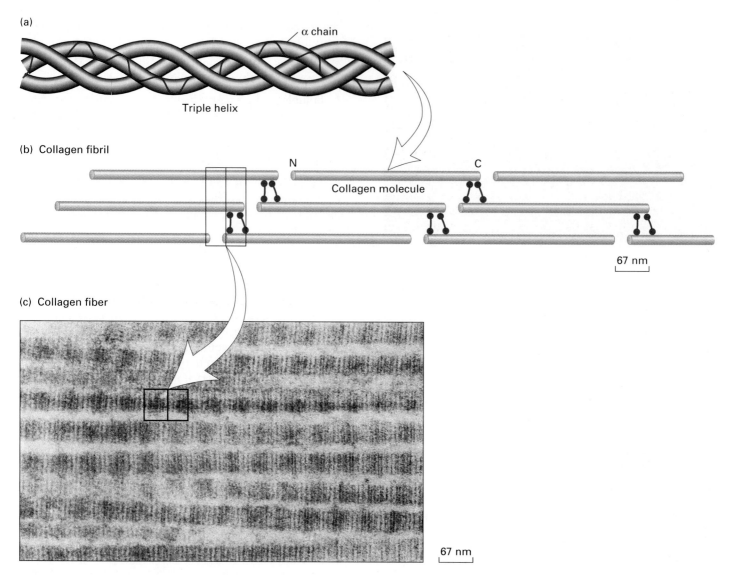

α chain

Triple helix

(b) Collagen fibril

N   C

Collagen molecule

67 nm

(c) Collagen fiber

67 nm

▲ **FIGURE 22-11 The structure of collagen.** (a) The basic structural unit is a triple-stranded helical molecule. Each triple-stranded collagen molecule is 300 nm long. (b) In fibrous collagen, collagen molecules pack together side by side. Adjacent molecules are displaced 67 nm, or slightly less than one-fourth the length of a single molecule. A small gap separates the "head" of one collagen from the "tail" of the next. The side-by-side interactions are stabilized by covalent bonds (red) between the N-terminus of one molecule and the C-terminus of an adjacent one. (c) An electron micrograph of calfskin collagen fibrils in an embryonic chick tendon. As indicated by the leaders, the striations created by the 67-nm periodic pattern of the packing are clearly visible. [Part (c) from D. E. Birk, F. H. Silver, and R. L. Trelstad, 1991, in E. D. Hay, ed., *Cell Biology of Extracellular Matrix,* Plenum, p. 222; courtesy of D. Birk.]

hydroxyproline. These amino acids make up the characteristic repeating motif Gly-Pro-X, where X can be any amino acid. Each amino acid has a precise function. The side chain of glycine, an H atom, is the only one that can fit into the crowded center of a three-stranded helix. Hydrogen bonds linking the peptide bond NH of a glycine residue with a peptide carbonyl (C=O) group in an adjacent polypeptide help hold the three chains together. The fixed angle of the C–N peptidyl-proline or peptidyl-hydroxyproline bond enables each polypeptide chain to fold into a helix with a geometry such that three polypeptide chains can twist together to form a three-stranded helix. Interestingly, although the rigid peptidyl-proline linkages disrupt the packing of amino acids in an α helix, they stabilize the rigid three-stranded collagen helix.

Collagen Fibrils Form by Lateral Interactions of Triple Helices

Many three-stranded type I collagen molecules pack together side-by-side, forming fibrils with a diameter of 50–200 nm. In fibrils, adjacent collagen molecules are displaced from one another by 67 nm, about one-quarter of their length (Figure

TABLE 22-3 Major Collagen Molecules

Type	Molecule Composition	Structural Features	Representative Tissues
Fibrillar Collagens			
I	$[\alpha1(I)]_2[\alpha2(I)]$	300-nm-long fibrils	Skin, tendon, bone, ligaments, dentin, interstitial tissues
II	$[\alpha1(II)]_3$	300-nm-long fibrils	Cartilage, vitreous humor
III	$[\alpha1(III)]_3$	300-nm-long fibrils; often with type I	Skin, muscle, blood vessels
V	$[\alpha1(V)]_3$	390-nm-long fibrils with globular N-terminal domain; often with type I	Similar to type I; also cell cultures, fetal tissues
Fibril-Associated Collagens			
VI	$[\alpha1(VI)][\alpha2(VI)]$	Lateral association with type I; periodic globular domains	Most interstitial tissues
IX	$[\alpha1(IX)][\alpha2(IX)][\alpha3(IX)]$	Lateral association with type II; N-terminal globular domain; bound glycosaminoglycan	Cartilage, vitreous humor;
Sheet-Forming Collagens			
IV	$[\alpha1(IV)]_2[\alpha2(IV)]$	Two-dimensional network	All basal laminaes

SOURCE: K. Kuhn, 1987, in R. Mayne and R. Burgeson, eds., *Structure and Function of Collagen Types,* Academic Press, p. 2; M. van der Rest and R. Garrone, 1991, *FASEB J.* 5:2814.

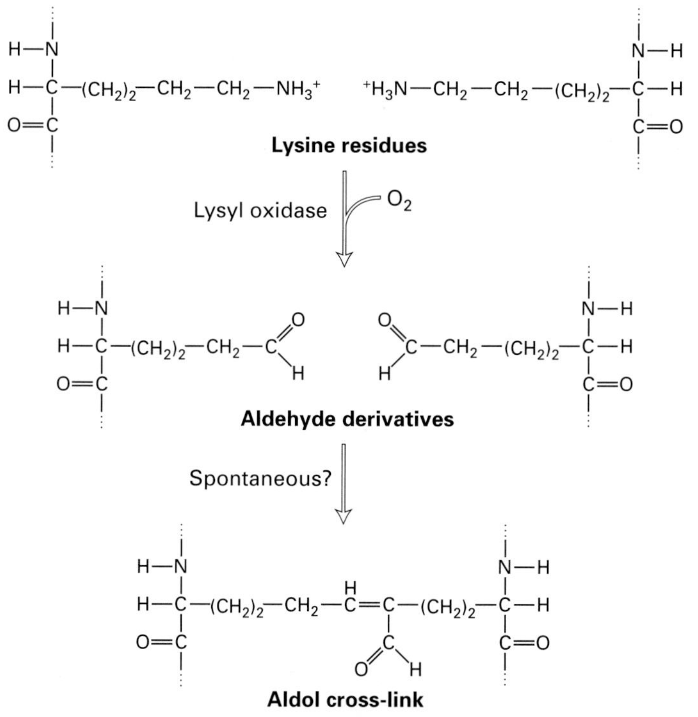

▲ **FIGURE 22-12 The side-by-side interactions of collagen helices are stabilized by an aldol cross-link between two lysine (or hydroxylysine) side chains.** The extracellular enzyme lysyl oxidase catalyzes formation of the aldehyde groups.

22-11b). This staggered array produces a striated effect that can be seen in electron micrographs of stained collagen fibrils; the characteristic pattern of bands is repeated about every 67 nm (Figure 22-11c). The unique properties of the fibrous collagens—types I, II, III, and V—are due to the ability of the rodlike triple helices to form such side-by-side interactions.

Short segments at either end of the collagen chains are of particular importance in the formation of collagen fibrils (see Figure 22-11). These segments do not assume the triple-helical conformation and contain the unusual amino acid *hydroxylysine* (see Figure 3-16). Covalent aldol cross-links form between two lysine or hydroxylysine residues at the C-terminus of one collagen molecule with two similar residues at the N-terminus of an adjacent molecule (Figure 22-12). These cross-links stabilize the side-by-side packing of collagen molecules and generate a strong fibril.

Type I collagen fibrils have enormous tensile strength; that is, such collagen can be stretched without being broken. These fibrils, roughly 50 nm in diameter and several micrometers long, are packed side-by-side in parallel bundles, called *collagen fibers,* in tendons, where they connect muscles with bones and must withstand enormous forces (Figure 22-13). Gram for gram, type I collagen is stronger than steel.

Assembly of Collagen Fibers Begins in the ER and Is Completed outside the Cell

Collagen biosynthesis and assembly follows the normal pathway for a secreted protein (see Figure 17-13). The collagen chains are synthesized as longer precursors called *procollagens;* the growing peptide chains are co-translationally

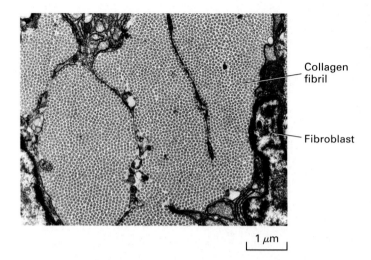

▲ **FIGURE 22-13 Electron micrograph of the dense connective tissue of a chick tendon.** Most of the tissue is occupied by parallel type I collagen fibrils, about 50 nm in diameter, seen here in cross section. The cellular content of the tissue is very low. [From D. A. D. Parry, 1988, *Biophys. Chem.* **29**:195.]

transported into the lumen of the rough endoplasmic reticulum (ER). In the ER, the procollagen chain undergoes a series of processing reactions (Figure 22-14). First, as with other secreted proteins, glycosylation of procollagen occurs in the rough ER and Golgi complex. Galactose and glucose residues are added to hydroxylysine residues, and long oligosaccharides are added to certain asparagine residues in the C-terminal *propeptide,* a segment at the C-terminus of a procollagen molecule that is absent from mature collagen. (The N-terminal end also has a propeptide.) In addition, specific proline and lysine residues in the middle of the chains are hydroxylated by membrane-bound hydroxylases. Lastly, intrachain disulfide bonds between the N- and C-terminal propeptide sequences align the three chains before the triple helix forms in the ER. The central portions of the chains zipper from C- to N-terminus to form the triple helix.

After processing and assembly of type I procollagen is completed, it is secreted into the extracellular space. During or following exocytosis, extracellular enzymes, the procollagen peptidases, remove the N-terminal and C-terminal propeptides. The resulting protein, often called *tropocollagen* (or simply *collagen*), consists almost entirely of a triple-stranded helix. Excision of both propeptides allows the collagen molecules to polymerize into normal fibrils in the extracellular space (see Figure 22-14). The potentially catastrophic assembly of fibrils within the cell does not occur both because the propeptides inhibit fibril formation and because lysyl oxidase, which catalyzes formation of reactive aldehydes, is an extracellular enzyme (see Figure 22-12). As noted above, these aldehydes spontaneously form specific covalent cross-links between two triple-helical molecules, which stabilizes the staggered array characteristic of collagen molecules and contributes to fibril strength.

Post-translational modification of procollagen is crucial for the formation of mature collagen molecules and their assembly into fibrils. Defects in this process have serious consequences, as ancient mariners frequently experienced. For example, the activity of both prolyl hydroxylases requires an essential cofactor, ascorbic acid (vitamin C). In cells deprived of ascorbate, as in the disease *scurvy,* the procollagen chains are not hydroxylated sufficiently to form stable triple helices at normal body temperature (Figure 22-15), nor can they form normal fibrils. Consequently, nonhydroxylated procollagen chains are degraded within the cell. Without the structural support of collagen, blood vessels, tendons, and skin become fragile. A supply of fresh fruit provides sufficient vitamin C to process procollagen properly.

Mutations in Collagen Reveal Aspects of Its Structure and Biosynthesis

Type I collagen fibrils are used as the reinforcing rods in construction of bone. Certain mutations in the α_1(I) or α_2(I) genes lead to *osteogenesis imperfecta,* or brittle-bone disease. The most severe type is an autosomal dominant, lethal disease resulting in death in utero or shortly after birth. Milder forms generate a severe crippling disease. As might be expected, many cases of osteogenesis imperfecta are due to deletions of all or part of the very long α_1(I) gene. However, a single amino acid change is sufficient to cause certain forms of this disease. As we have seen, a glycine must be at every third position for the collagen triple helix to form; mutations of glycine to almost any other amino acid are deleterious, producing poorly formed and unstable helices. Since the triple helix forms from the C- to the N-terminus, mutations of glycine near the C-terminus of the α_1(I) chain are usually more deleterious than those near the N-terminus; the latter permit substantial regions of triple helix to form. Mutant unfolded collagen chains do not leave the rough ER of fibroblasts (the cells that make most of type I collagen), or they leave it slowly. As the ER becomes dilated and expanded, the secretion of other proteins (e.g., type III collagen) by these cells also is slowed down.

Because each type I collagen molecule contains two α_1(I) and one α_2(I) chains, mutations in the α_2(I) chains are much less damaging. To understand this point, consider that in a heterozygote expressing one wild-type and one mutant α_2(I) protein, 50 percent of the collagen molecules will have the abnormal α_2(I) chain. In contrast, if the mutation is in the α_1(I) chain, 75 percent of the collagen molecules will have one or two mutant α_1(I) chains. In fact, even low expression of a mutant α_1(I) gene can be deleterious, because the mutant chains can disrupt the function of wild-type α_1(I) chains when combined with them. To study such mutations, experimenters constructed a mutant α_1(I) collagen gene with a glycine-to-cysteine substitution near the C-terminus. This mutant gene was used to create lines of transgenic mice with

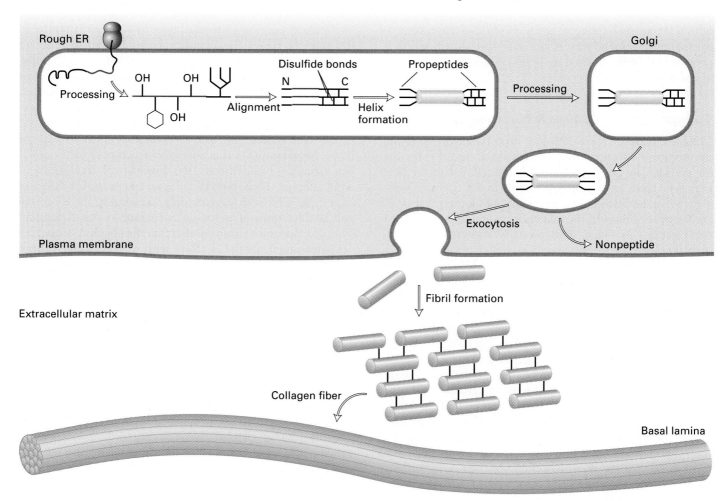

▲ FIGURE 22-14 Major events in the biosynthesis of fibrous collagens. Modifications of the procollagen polypeptide in the endoplasmic reticulum include hydroxylation, glycosylation, and disulfide-bond formation. Interchain disulfide bonds between the C-terminal propeptides of three procollagens align the chains in register and initiate formation of the triple helix. The process continues, zipperlike, toward the N-terminus. All modifications occur in a precise sequence in the rough ER, Golgi complex, and the extracellular space, and allow lateral alignment and formation of the covalent cross-linkers that enable helices to pack into 50-nm-diameter fibrils. The α-helical region is colored red. [After M. E. Nimni, 1993, in M. Zern and L. Reid, eds., *Extracellular Matrix*, Marcel Dekker, pp. 121–148.]

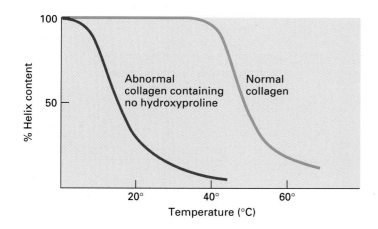

◀ FIGURE 22-15 Denaturation of collagen containing a normal content of hydroxyproline and of abnormal collagen containing no hydroxyproline. Without hydrogen bonds between hydroxyproline residues, the collagen helix is unstable and loses most of its helical content at temperatures above 20 °C. Such collagens are formed by experimental animals (or man) in the absence of ascorbic acid (vitamin C). Normal collagen is more stable and resists thermal denaturation until a temperature of 40 °C is reached.

otherwise normal collagen genes. High-level expression of the mutant transgene was lethal, and expression at a rate 10 percent that of the normal $\alpha_1(I)$ genes caused severe growth abnormalities.

Collagens Form Diverse Structures

Collagens differ in their ability to form fibers and to organize the fibers into networks. For example, type II is the major collagen in cartilage. Its fibrils are smaller in diameter than type I and are oriented randomly in the viscous proteoglycan matrix. Such rigid macromolecules impart a strength and compressibility to the matrix and allow it to resist large deformations in shape. This property allows joints to absorb shocks.

Type II fibrils are cross-linked to proteoglycans in the matrix by type IX, a collagen of a different structure (Figure 22-16a). Type IX collagen consists of two long triple helices connected by a flexible kink. The globular N-terminal domain extends from the composite fibrils, as does a heparan sulfate molecule, a type of large, highly charged polysaccharide (discussed later) that is linked to the $\alpha_2(IX)$ chain at the flexible kink. These protruding nonhelical domains are thought to anchor the fibril to proteoglycans and other components of the matrix. The interrupted triple-helical structure of type IX collagen prevents it from assembling into fibrils; instead, these three collagens associate with fibrils formed from other collagen types and thus are called *fibril-associated collagens* (see Table 22-3).

In many connective tissues, type VI collagen is bound to the sides of type I fibrils and may bind them together to form thicker collagen fibers (Figure 22-16b). Type VI collagen is unusual in that the molecule consists of relatively short triple-helical regions about 60 nm long separated by globular domains about 40 nm long. Fibrils of pure type VI collagen thus give the impression of beads on a string.

In some places, several ECM components are organized into a **basal lamina,** a thin sheetlike structure. Type IV collagen forms the basic fibrous two-dimensional network of all basal laminae. Three type IV collagen chains form a 400-nm-long triple helix with large globular domains at the C-termini and smaller ones of unknown structure at the N-termini. The helical segment is unusual in that the Gly-X-Y sequences are interrupted about 24 times with segments that cannot form a triple helix; these nonhelical regions introduce flexibility into the molecule (Figure 22-17a). Lateral association of the N-terminal regions of four type IV molecules yields a characteristic tetrameric unit that can be observed in the electron microscope (Figure 22-17b). Triple-helical regions from several molecules then associate laterally, in a manner similar to fibril formation among fibrous collagens, to form branching strands of variable but thin diameters. These interactions, together with those between the C-terminal globular domains and the triple helices in adjacent type IV molecules, generate an irregular two-dimensional fibrous network (Figure 22-17b). We will discuss the other components of the basal lamina and the functions of this specialized matrix structure in the next section.

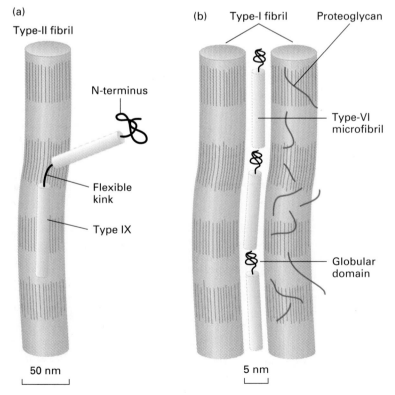

▶ **FIGURE 22-16 Interactions of fibrous and nonfibrous collagens.** (a) Association of types II and IX collagen in a cartilage matrix. Type II forms fibrils similar in structure to type I, with a similar 67-nm periodicity, though smaller in diameter. Type IX contains two long triple helices connected at a flexible kink. At this point a chondroitin sulfate chain (see Figure 22-24) is linked to the $\alpha_2(IX)$ chain. Type IX collagens are bound at regular intervals along type II fibrils, with an N-terminal nonhelical domain of type IX projecting outward. It is thought that these domains bind the collagen fibrils to the proteoglycan-rich matrix. (b) Organization of the major fibrous components in the extracellular matrix of tendons. Type I fibrils, with their characteristic 67-nm period, are all oriented longitudinally, that is, in the direction of the stress applied to the tendon. The fibrils are coated with an array of proteoglycans, as shown in blue on the right-hand fibril. Type VI fibrils bind to and link together the type I fibrils. Type VI collagen consists of thin triple helices, about 60 nm long, with globular domains at either end. The globular domains of several type VI molecules bind together, giving a "beads-on-a-string" appearance to the type VI fibril. [Part (a) after L. M. Shaw and B. Olson, 1991, *Trends Biochem. Sci.* **18**:191; part (b) after R. R. Bruns et al., 1986, *J. Cell Biol.* **103**:393.]

(a)

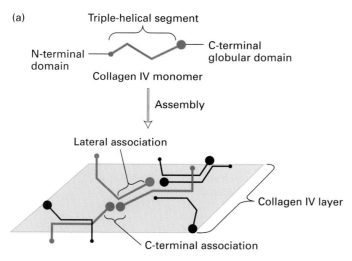

(b) Type IV network

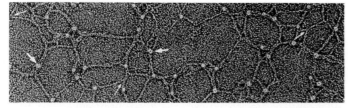

250 nm

▲ **FIGURE 22-17 Structure and assembly of type IV collagen.** (a) Schematic diagram of 400-nm-long triple-helical molecule of type IV collagen. This molecule has a noncollagen domain at the N-terminus and a large globular domain at the C-terminus; the triple helix is interrupted by several nonhelical segments that introduce flexible kinks in the molecule. Through lateral interactions of the triple-helical segments and head-to-head interactions between the C-terminal domains, collagen IV molecules are assembled into a sheetlike network. (b) An electron micrograph of an in vitro network. The lacy appearance of the network results from the flexibility of the molecule, the side-to-side binding between triple-helical segments (small arrows), and the interactions between C-terminal globular domains (large arrows). [Part (b) courtesy of P. Yurchenco; see P. Yurchenco and G. C. Ruben, 1987, *J. Cell Biol.* **105**:2559.]

SUMMARY Collagen: The Fibrous Proteins of the Matrix

- All 16 types of collagen contain a repeating Gly-Pro-X sequence and fold into a characteristic triple-helical structure.

- The various collagens are distinguished by the ability of their helical and nonhelical regions to associate into fibrils, to form sheets, or to cross-link different collagen types.

- Most collagen is fibrillar and composed of type I molecules. A two-dimensional network of type IV collagen is unique to the basal lamina.

- Fibrous type collagen molecules (e.g., types I, II, and III) assemble into fibrils that are stabilized by covalent aldol cross-links (see Figure 22-11).

- Procollagen chains are modified and assembled into a triple helix in the ER (see Figure 22-14). Helix formation is aided by disulfide bonds between N- and C-terminal propeptides, which align the polypeptide chains in register. Generally, the propeptides are removed after secretion, and then collagen fibrils form in the extracellular space.

- Fibrous collagen has specific structural requirements and is very susceptible to mutation, especially in glycine residues. Because mutant collagen chains can affect the function of wild-type ones, such mutations have a dominant phenotype.

22.4 Noncollagen Components of the Extracellular Matrix

In addition to the insoluble fibers of collagen, the extracellular matrix contains two major classes of soluble proteins: multiadhesive matrix proteins, which bind cell-surface adhesion receptors, and proteoglycans, a diverse group of macromolecules containing a core protein with multiple attached polysaccharide chains (see Figure 22-1). Another important component of the matrix, hyaluronan, is a large polysaccharide that forms a highly hydrated gel, making the matrix resilient to compression.

Multiadhesive matrix proteins are long flexible molecules that contain domains responsible for binding a variety of collagen types, other matrix proteins, polysaccharides, cell-surface proteins, and signaling molecules such as growth factors and hormones. Their major role is to attach cells to the extracellular matrix. The importance of these matrix proteins in initiating cellular responses through classic signal-transduction pathways has been recognized in recent years. Both roles are important for organizing the other components of the matrix and also for regulating cell attachment to the matrix, cell migration, and cell shape.

Proteoglycans are found in all connective tissues and extracellular matrices; they also are attached to the surface of many cells. Because of their high content of charged polysaccharides, proteoglycans are highly hydrated. The swelled, hydrated structure of proteoglycans is largely responsible for the volume of the extracellular matrix and also acts to permit diffusion of small molecules between cells and tissues.

We begin our discussion of these matrix components with the multiadhesive proteins because they bind collagen, and then we describe the unique structural features of proteoglycans and hyaluronan.

Laminin and Type IV Collagen Form the Two-Dimensional Reticulum of the Basal Lamina

As we've seen already, the basal lamina is a thin sheetlike network of ECM components, usually no more than 60–100 nm thick. Most epithelial and endothelial cells rest upon a basal lamina, which is linked to specific plasma-membrane receptor proteins and to fibrous collagens and other components of the underlying loose connective tissue (see Figure 15-23). Individual muscle cells and adipocytes also are surrounded by a basal lamina (Figure 22-18). After type IV collagen, **laminin**, a large multiadhesive matrix protein, is the most prevalent constituent of all basal laminae. The basal lamina is often called the type IV matrix after its collagen component. All the ECM components are synthesized by cells that rest on the basal lamina.

The laminins are a family of cross-shaped proteins that are as long as the basal lamina is thick. In adult animals, laminin is a heterotrimeric protein with a total molecular weight of 820,000 (Figure 22-19). Several laminin isoforms, containing slightly different A, B, or C chains, have been identified. Laminin has high-affinity binding sites for other components of the basal lamina, including collagen IV, and for certain cell-adhesion molecules on the surface of many cells. In vitro, laminin molecules assemble into a feltlike web, primarily via interactions between the ends of the arms.

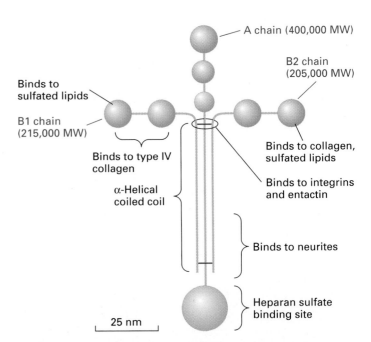

▲ **FIGURE 22-19 Structure of laminin, a large heterotrimeric multiadhesive matrix protein found in all basal laminae.** The cross-shaped molecule contains globular domains and a coiled-coil region in which the three chains are covalently linked via several disulfide bonds. Different regions of laminin bind to cell-surface receptors and various matrix components. [Adapted from G. R. Martin and R. Timpl, 1987, *Ann. Rev. Cell Biol.* **3**:57; and K. Yamada, 1991, *J. Biol. Chem.* **266**:12809.]

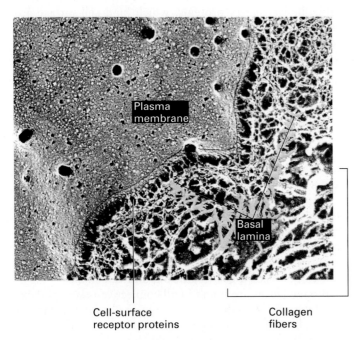

▲ **FIGURE 22-18 Electron micrograph showing the association of the plasma membrane of skeletal muscle with the basal lamina.** In this quick-freeze deep-etch preparation, the basal lamina is seen as a meshwork of filaments. Some of these filaments contact receptor proteins in the plasma membrane; others bind to the thick collagen fibers that form the connective tissue around the muscle. [From D. W. Fawcett, 1981, *The Cell,* 2d ed., Saunders/Photo Researchers; courtesy of John Heuser.]

Although laminin and type IV collagen form the basic reticulum of the basal lamina, most basal laminae in adult animals also contain *perlecan*, a heparan sulfate proteoglycan, and *entactin*, a small multiadhesive matrix protein that interacts with laminin and type IV collagen. The multiple interactions between these components connect the type IV and laminin networks and stabilize the overall structure of the basal lamina (Figure 22-20). Both type IV collagen and laminin also bind to specific **integrins**, an important class of cell-adhesion molecules present in the plasma membrane. These interactions attach a basal lamina to adjacent cells.

The basal lamina is structured differently in different tissues (Figure 22-21). For instance, the endothelial cells that line capillaries are polarized, with one surface facing the blood. The surface not facing the blood is surrounded by a basal lamina that forms a filter for regulating passage of proteins and other molecules from the blood into the tissues. The basal lamina underlying polarized epithelial cells lining the intestine likewise regulates passage of nutrients into the bloodstream. In smooth muscle, on the other hand, the basal lamina connects adjacent cells and maintains the integrity of the tissue. In the kidney glomerulus, a double-thickness basal lamina separates two cell sheets and acts as a filter in forming the urine.

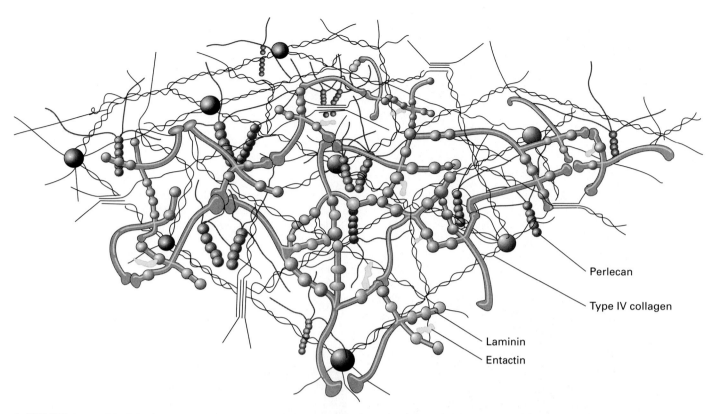

▲ FIGURE 22-20 Model of the basal lamina. [P. D. Yurchenco and J. C. Schittny, 1990, *FASEB J.* 4: 1577–1590.]

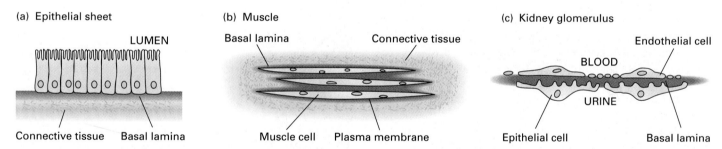

▲ FIGURE 22-21 Organization of the basal lamina in different tissues. (a, b) The basal laminae associated with endothelial cells and muscle cells separate these cells from the underlying or surrounding connective tissue. (c) In the kidney glomerulus, both the endothelium lining the capillaries and the epithelium lining the urinary space form a basal lamina. Thus the fused basal lamina between them is about twice as thick as that in other tissues. Because both cell sheets have gaps in them, this basal lamina is exposed to both the blood and the urine spaces and acts to filter capillary blood, forming in the urinary space a filtrate that ultimately becomes urine. [Adapted from B. Alberts et al., 1994, *Molecular Biology of the Cell*, 3d ed, Garland, p. 989.]

As discussed in Chapter 23, laminin and components of the basal lamina play important roles in embryonic development. For instance, the basal lamina helps four- and eight-celled embryos adhere together in a ball. During development of the nervous system, neurons migrate along extra- cellular-matrix pathways that contain laminin and other matrix components. Thus, the basal lamina not only is important for organizing cells into tissues but for guiding migrating cells during development.

Fibronectins Bind Many Cells to Fibrous Collagens and Other Matrix Components

Fibronectins are another important class of soluble multi-adhesive matrix proteins. Their primary role is attaching cells to all matrices that contain the fibrous collagens (types I, II, III, and V). The presence of fibronectin on the surface of nontransformed cultured cells, and its absence on transformed (or tumorigenic) cells, first led to the identification of fibronectin as an adhesive protein. By their attachments,

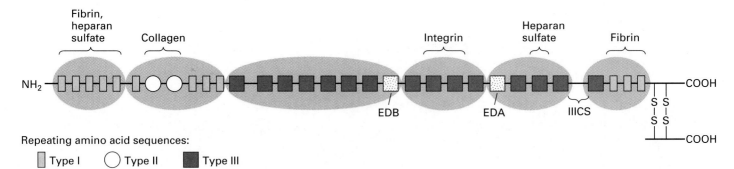

Repeating amino acid sequences:

☐ Type I   ◯ Type II   ■ Type III

▲ **FIGURE 22-22 Structure of fibronectin chains.** Only one of the two chains present in the dimeric fibronectin molecule is shown; both chains have very similar sequences. Each chain contains about 2446 amino acids and is composed of three types of repeating amino acid sequences. Circulating fibronectin lacks one or both of the type III repeats designated EDA and EDB owing to alternative mRNA splicing. At least five different sequences may occur in the IIICS region as the result of alternative splicing. Each chain contains six domains (tan ovals) containing specific binding sites for heparan sulfate, fibrin (a major constituent of blood clots), denatured forms of collagen, and cell-surface integrins. Binding to integrins is dependent on an Arg-Gly-Asp (RGD) sequence. Heparan sulfate and fibrin have binding sites in a shared domain, and each has another binding site in its own, unshared domain; these sites differ in their affinity for the ligand. [Adapted from G. Paolella, M. Barone, and F. Baralle, 1993, in M. Zern and L. Reid, eds., *Extracellular Matrix,* Marcel Dekker, pp. 3–24.]

fibronectins regulate the shape of cells and the organization of the cytoskeleton; they are essential for migration and cellular differentiation of many cell types during embryogenesis. Fibronectins also are important for wound healing because they facilitate migration of macrophages and other immune cells into the affected area.

Fibronectins are dimers of two similar polypeptides linked at their C-termini by two disulfide bonds; each chain is about 60–70 nm long and 2–3 nm thick. At least 20 different fibronectin chains have been identified, all of which are generated by alternative splicing of the RNA transcript of a single fibronectin gene (see Figure 11-24). Analysis of digests of fibronectin with low amounts of proteases shows that the polypeptides consist of six domains. Each domain, in turn, contains repeated sequences that can be classified into one of three types on the basis of similarities in amino acid sequence (Figure 22-22).

The multiadhesive property of fibronectin arises from the presence in different domains of high-affinity binding sites for collagen and other ECM components and for certain integrins on the surface of cells. Proteolytic digestion of the cell-binding domain in one of the type III repeats showed that a segment of about 100 amino acids could bind to integrins. Studies with synthetic peptides corresponding to parts of this segment identified the tripeptide sequence Arg-Gly-Asp (usually abbreviated RGD) as the minimal structure required for recognition by integrins in the plasma membrane. For example, when this tripeptide was covalently linked to an inert protein such as albumin and dried on a culture dish, it stimulated adhesion of fibroblasts to the surface of the dish similar to the effect of intact fibronectin. In the three-dimensional structure of this fibronectin type III repeat, the RGD sequence is at the apex of a loop that protrudes outward from the molecule, in a position to bind to an integrin or other protein (Figure 22-23).

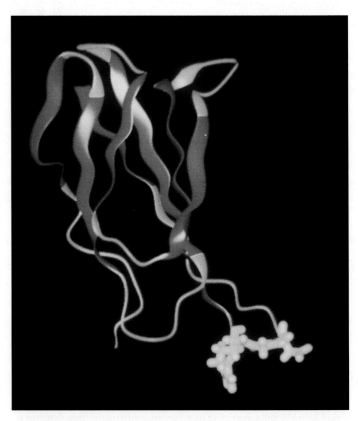

▲ **FIGURE 22-23 Three-dimensional structure of the type III repeat of fibronectin that contains the RGD integrin-binding sequence.** The core of seven β strands in this domain has a structure similar to that of the "immunoglobulin fold," a structural motif found in the light and heavy chains of antibody molecules. The side chains of the Arg^{78}-Gly^{79}-Asp^{80} sequence of the integrin-binding site are shown in yellow. [From A. L. Main et al., 1993, *Cell* **71**:671; courtesy of A. Main and I. D. Campbell.]

Although the RGD peptide is the minimal structure required for binding to several integrins, its affinity for integrins is substantially less than that of intact fibronectin or of the entire cell-binding domain. Thus, sequences surrounding the RGD sequence in fibronectin and other proteins apparently enhance binding to integrins. Further, the affinity of the RGD sequence for integrins is dependent on the absorbed state of fibronectin. Fibronectin in solution or circulating in blood binds to integrins on fibroblasts poorly. Absorption of fibronectin to a surface—in animals to a collagen matrix or the basal lamina surrounding an endothelial cell or, experimentally, to a tissue-culture dish—enhances its ability to bind to cells, probably because the segment of the protein containing the RGD sequence becomes more exposed.

Fibronectin circulating in the blood, which is secreted by the liver, lacks one or both of the type III repeats designated EDA and EDB (for *extra domain A and B*) present in fibronectin secreted by cultured fibroblasts (see Figure 22-22). Circulating fibronectin forms insoluble matrices somewhat less readily than fibronectin within tissues, but it does bind to fibrin, a constituent of blood clots. Following binding to fibrin, the immobilized fibronectin binds, via its exposed RGD-containing domain, to integrins expressed on passing, activated platelets. As a result, the platelets are localized to damaged regions of blood vessels and can participate in expansion of blood clots. As this example illustrates, both the polymerization of fibronectin into filaments and adhesion of cells to fibronectin are closely controlled to meet the organism's needs. Later we will return to this topic when we examine the role of fibronectins in cell adhesion and motility.

Proteoglycans Consist of Multiple Glycosaminoglycans Linked to a Core Protein

The proteoglycans have a much higher ratio of polysaccharide to protein than do collagen, fibronectin, and similar glycoproteins in the extracellular matrix. The polysaccharide chains in proteoglycans are long repeating linear polymers of specific disaccharides called **glycosaminoglycans** (GAGs). Usually one sugar is a uronic acid (either D-glucuronic acid or L-iduronic acid) and the other is either N-acetylglucosamine or N-acetylgalactosamine (Figure 22-24). One or both of the sugars contain one or two sulfate residues. Thus

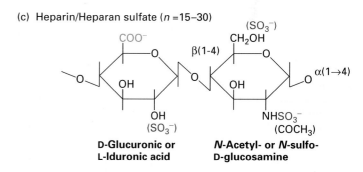

▶ **FIGURE 22-24 Structures of various glycosaminoglycans, the polysaccharide components of proteoglycans.** Each of the four classes of glycosaminoglycans is formed by polymerization of a specific disaccharide and subsequent modifications including addition of sulfate groups and inversion (epimerization) of the carboxyl group on carbon 5 of D-glucuronic acid to yield L-iduronic acid. Heparan sulfate, which is ubiquitous, and its derivative heparin, found mostly in mast cells, are actually complex mixtures resulting from the degree of sulfation. Hyaluron is unsulfated. The number *(n)* of disaccharides typically found in each glycosaminoglycan chain is given.

each GAG chain bears many negative charges. Frequently some of the residues in a GAG chain are modified after synthesis; dermatan sulfate is formed from chondroitin sulfate, for instance. Similarly, heparin (used medicinally as an anticlotting drug) is formed, only in mast cells, as a result of enzymatic addition of sulfate groups at specific sites in heparan sulfate. Proteoglycans commonly are named according to the structure of their principal repeating disaccharide in the attached GAGs.

In the synthesis of all proteoglycans, heparan sulfate or chondroitin sulfate chains are formed by the sequential addition of the repeating units to a three-sugar "linker" that is attached to serine residues in a core protein molecule (Figure 22-25). One of the "signal sequences" in a core protein that specifies addition of this linker sugar is Ser-Gly-X-Gly, where X is any amino acid. However, not all such sites in the core protein become substituted, and GAGs are attached to serines in other sequences. Thus the conformation of the core protein may be more important than localized primary sequences in determining where the GAG chains attach. In addition, the mechanisms determining the length of the chains are unknown.

Proteoglycans, which are remarkable for their diversity, are present both in the extracellular matrix and on the surface of many cells. A given extracellular matrix may contain proteoglycans with several different types of core proteins, and the number, length, and composition of the GAG chains attached to each core protein may vary. Thus, the molecular weight and charge density of a population of proteoglycans can be expressed only as an average; individual molecules can differ considerably. Nonetheless, a good deal is known of the structure and function of certain extracellular and cell-surface proteoglycans.

Extracellular Matrix Proteoglycans One of the most important extracellular proteoglycans is *aggrecan*, the predominant proteoglycan in cartilage. As its name implies, aggrecan forms very large aggregates, termed *proteoglycan aggregates*. A single aggregate, one of the largest macromolecules known, can be more than 4 mm long and have a volume larger than that of a bacterial cell. These aggregates give cartilage its unique gel-like properties and its resistance to deformation, essential for distributing the load in weight-bearing joints.

The central component of the cartilage proteoglycan aggregate is a long molecule of hyaluronan. Bound to it, tightly but noncovalently at 40-nm intervals, are aggrecan core proteins decorated with GAGs (Figure 22-26a). The aggrecan core protein ($\approx$250,000 MW) has one N-terminal globular domain that binds with high affinity to a decasaccharide sequence in hyaluronan; this binding is facilitated by a *link protein* that binds to the aggrecan core protein and hyaluronan (Figure 22-26b). Covalently attached to each aggrecan core protein, via the trisaccharide linker, are multiple chains of chondroitin sulfate and keratan sulfate. The molecular weight of an aggrecan monomer—that is, the core protein plus the bound glycosaminoglycans—averages 2×10^6. The entire proteoglycan aggregate, which may contain upward of 100 aggrecan monomers, has a molecular weight in excess of 2×10^8.

Not all extracellular proteoglycans form large aggregates like aggrecan. A class of proteoglycans present in the basal lamina, for example, consists of a core protein (20,000–40,000 MW) to which are attached several heparan sulfate chains. Such proteoglycans bind to type IV collagen and other structural proteins discussed later, thereby imparting structure to the basal lamina.

 The importance of the GAG chains that are part of various matrix proteoglycans is illustrated by the rare humans who have a genetic defect in one of the enzymes required for synthesis of dermatan sulfate. These individuals have many defects in their bones, joints, and muscles; do not grow to normal height; and have wrinkled skin, giving them a prematurely aged appearance.

Cell-Surface Proteoglycans Proteoglycans are attached to the surface of many types of cells, particularly epithelial cells. The most common proteoglycan on the plasma membrane is syndecan, whose general structure is illustrated in Figure 22-27. The core protein of cell-surface proteoglycans spans the plasma membrane and contains a short cytosolic domain, as well as a long external domain to which a small number of heparan sulfate chains are attached; some cell-surface proteoglycans also contain chondroitin sulfate. The GAG chains are attached to serine residues in the core protein via the same trisaccharide linker that is present in extracellular proteoglycans.

The extracellular and cytoplasmic domains of syndecan have separate functions. The heparan sulfate chains on the

SO$_4$

(GlcUA—GalNAc)$_n$ — GlcUA — Gal — Gal — Xyl — Ser

Core protein

Chondroitin sulfate (or keratan sulfate) repeats

Linking sugars

Gal = galactose
GalNAc = *N*-acetylgalactosamine

GlcUA = glucuronic acid
Xyl = xylose

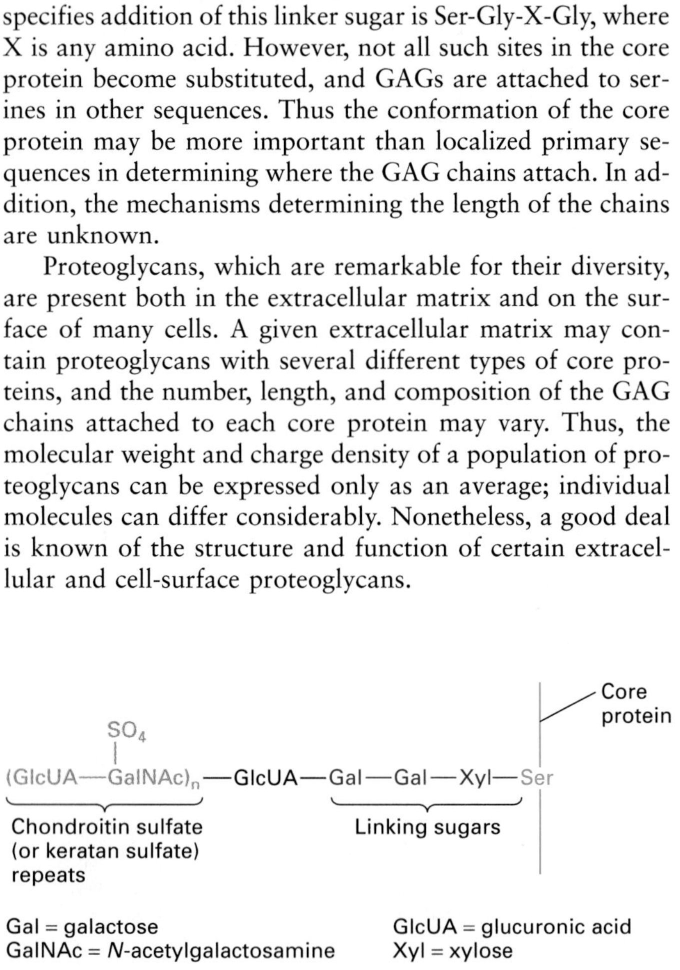

▲ **FIGURE 22-25 Glycosaminoglycan chains in proteoglycans.** Synthesis of a chondroitin sulfate chain is initiated by transfer of a xylose residue to a serine residue in the core protein, most likely in the Golgi complex, followed by sequential addition of two galactose residues. Glucuronic acid and *N*-acetylgalactosamine residues then are added sequentially, forming the chondroitin sulfate chain. In the Golgi, sulfate groups are added to sugars after their incorporation into the oligosaccharide. Heparan sulfate chains, which are connected to core proteins by the same three-sugar linker, are formed by the sequential addition of glucuronic acid and *N*-acetylglucosamine.

(a)

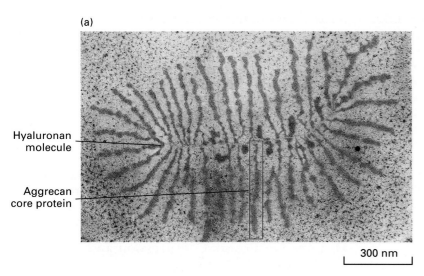

Hyaluronan molecule

Aggrecan core protein

300 nm

(b)

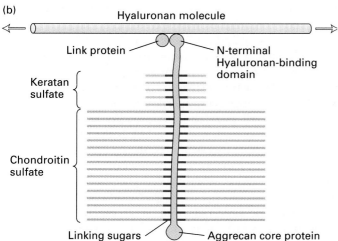

Hyaluronan molecule

Link protein

N-terminal Hyaluronan-binding domain

Keratan sulfate

Chondroitin sulfate

Linking sugars

Aggrecan core protein

◀ **FIGURE 22-26 Structure of cartilage proteoglycan aggregate.** (a) Electron micrograph of a proteoglycan aggregate from fetal bovine epiphyseal cartilage. Aggrecan core proteins are bound at ≈40-nm intervals to a molecule of hyaluronan. Numerous keratan sulfate and chondroitin sulfate chains are attached to the aggrecan core proteins. (b) Diagram of detailed structure of an aggrecan monomer. The N-terminal domain of the core protein binds to a hyaluronan (HA) molecule. Binding is facilitated by a link protein, which binds to both the hyaluronan disaccharide and the aggrecan core protein. Each aggrecan core protein has 127 Ser-Gly sequences at which the glycosaminoglycan chains are added; 30 short keratan sulfate chains and 97 longer chondroitin sulfate chains are added to each core protein molecule in aggrecan. [Part (a) from J. A. Buckwalter and L. Rosenberg, 1983, *Coll. Rel. Res.* **3**:489; courtesy of L. Rosenberg.]

▶ **FIGURE 22-27 Schematic diagram of the cell-surface proteoglycan syndecan-4.** The core protein in all syndecan proteoglycans (syndecan-1, -2, -3, and -4) spans the plasma membrane and dimerizes through the cytoplasmic domain. Syndecan core proteins range in size from 20,000 MW (syndecan-4) to 45,000 MW (syndecan-3) because of differences in their extracellular domains, but their membrane-spanning and cytosolic domains are similar. The syndecans contain three heparan sulfate chains and sometimes chondroitin sulfate.

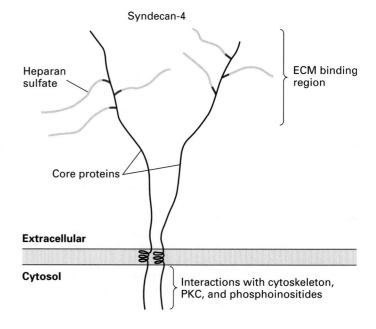

Syndecan-4

Heparan sulfate

ECM binding region

Core proteins

Extracellular

Cytosol

Interactions with cytoskeleton, PKC, and phosphoinositides

extracellular domain bind to fibrous collagens (types I, III, and IV) and to fibronectin, which are present in the interstitial matrix surrounding the basal lamina. In this way, cell-surface proteoglycans are thought to anchor cells to matrix fibers. Like many integral membrane proteins, the cytoplasmic domain of syndecan interacts with the actin cytoskeleton and in some cases phosphoinositides and protein kinase C. Thus, cell-surface proteoglycans function much like some of the cell-adhesion molecules in the plasma membrane.

Many Growth Factors Are Sequestered and Presented to Cells by Proteoglycans

Besides acting as structural components of the extracellular matrix and anchoring cells to the matrix, both extracellular and cell-surface proteoglycans, particularly those containing heparan sulfate, also bind many protein growth factors (Chapter 20). For instance, fibroblast growth factor (FGF) binds tightly to the heparan sulfate chains in extracellular proteoglycans. Since the bound growth factor is resistant to degradation by extracellular proteases, it serves as a reservoir of matrix-bound FGF (Figure 22-28). Active hormone is released by proteolysis of the proteoglycan core protein or by partial degradation of the heparan sulfate chains, processes that occur during tissue growth and remodeling or after infection. FGF also binds to cell-surface heparan sulfate

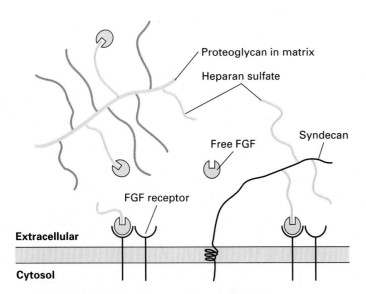

▲ **FIGURE 22-28 Modulation of activity of fibroblast growth factor (FGF) by heparan sulfate proteoglycans.** Free FGF cannot bind to FGF receptors in the plasma membrane. Binding of FGF to heparan sulfate chains such as those in syndecan, a cell-surface proteoglycan (see Figure 22-27), induces a conformational change that enables FGF to bind to its receptors. FGF bound to heparan sulfate chains released by proteolysis from matrix proteoglycans can also bind FGF receptors. Binding of FGF to extracellular heparan sulfate proteoglycan also protects the growth factor from degradation and forms a reservoir of active FGF. [Adapted from E. Ruoslahti and T. Yamaguchi, 1991, *Cell* **64**:867.]

proteoglycans such as syndecan, which then "present" the bound FGF to its receptor in the plasma membrane, inducing proliferation. Free FGF cannot interact with the FGF receptor, and cells that cannot synthesize heparan sulfate proteoglycans do not respond to FGF. Another example is transforming growth factor β (TGFβ), whose role in embryonic development is discussed in Chapter 23. The core protein of a cell-surface proteoglycan called beta-glycan binds TGFβ and then presents it to TGFβ receptors. These examples illustrate how proteoglycans commonly function as extracellular hormone reservoirs and facilitate binding of hormones to their cell-surface receptors, thus triggering intracellular signaling pathways.

Hyaluronan Resists Compression and Facilitates Cell Migration

Hyaluronan (HA), also called *hyaluronic acid* or *hyaluronate*, is a major component of the extracellular matrix that surrounds migrating and proliferating cells, particularly in embryonic tissues. It is also a major structural component of the complex proteoglycans that are found in many extracellular matrices, particularly cartilage (see Figure 22-26). Because of its remarkable physical properties, HA imparts stiffness and resilience as well as a lubricating quality to many types of connective tissue such as joints. HA is the only extracellular oligosaccharide that is not covalently linked to a protein.

Each hyaluronan molecule consists of as many as 50,000 repeats of the simple disaccharide glucuronic acid β(1→3) N-acetylglucosamine β(1→4) (see Figure 22-24). If stretched end-to-end, one molecule would be 20 mm long. Individual segments of an HA molecule fold into a stiff rodlike conformation because of the β linkages and extensive intrachain hydrogen bonding between adjacent sugar residues. Mutual repulsion between negatively charged carboxylate groups that protrude outward at regular intervals also contributes to these local rigid structures. Overall, however, HA is not a long, rigid rod as is collagen; rather, in solution it behaves as a random coil about 500 nm in diameter.

Because of the large number of anionic residues on its surface, HA binds a large amount of water and forms, even at low concentrations, a viscous hydrated gel. Given no constraints, an HA molecule will occupy a volume about 1000 times the space of the HA molecule itself. When placed in a confining space, such as in a matrix between two cells, the long HA molecules will tend to push outward. This creates a swelling, or *turgor pressure*, within the space; the HA molecules push against any fibers or cells that block their motion. Importantly, by binding cations, the COO⁻ groups on the surface increase the concentration of ions and thus the osmotic pressure in the HA gel. Large amounts of water are taken up into the matrix, contributing to the turgor pressure within the HA matrix. These swelling forces give connective tissues their ability to resist compression forces, in contrast to collagen fibers, which are able to resist stretching forces.

Hyaluronan is bound to the surface of many migrating cells by a 34-kDa receptor protein termed *CD44*, or by a homologous protein in the CD44 family. The domain in CD44 that binds HA is similar in sequence and structure to ones found in various extracellular proteoglycans that bind HA. This is one of many examples we shall encounter where a number of different matrix and cell-surface proteins contain domains, or "modules," of similar structure and function (see Figure 3-10). Almost certainly these arose during evolution from a single ancestral gene that encoded just this domain.

Because of its loose, hydrated, porous nature, the HA "coat" bound to cells appears to keep cells apart from one another, giving them the freedom to move about and proliferate. Cessation of cell movement and initiation of cell-cell attachments are frequently correlated with a decrease in HA, a decrease in the cell-surface molecules that bind HA, and an increase in the extracellular enzyme hyaluronidase, which degrades the matrix HA. These functions of HA are particularly important during the many cell migrations that facilitate differentiation.

SUMMARY Noncollagen Components of the Extracellular Matrix

• Laminin is a multiadhesive protein in the basal lamina that binds heparan sulfate, type IV collagen, and specific cell-surface receptor proteins.

• Fibronectins are multiadhesive proteins that link collagen and other matrix proteins to integrins in the plasma membrane, thereby attaching cells to the matrix.

• Glycosaminoglycans are linear chains of 20–100 sulfated disaccharides. The most common disaccharides are chondroitin sulfate, heparin and heparan sulfate, and dermatan sulfate (see Figure 22-24).

• Proteoglycans consist of multiple glycosaminoglycan chains that branch from a linear protein core. Extracellular proteoglycans are large, highly hydrated molecules that help cushion cells.

• In cartilage, a proteoglycan called *aggrecan* binds at regular intervals to a central hyaluronan molecule, forming a very large aggregate (see Figure 22-26).

• Smaller proteoglycans are attached to cell surfaces, where they facilitate cell-matrix interactions and help present certain hormones to their cell-surface receptors (see Figure 22-28).

• Hyaluronan is an extremely long, negatively charged polysaccharide. It forms viscous, hydrated gels that resist compression forces. When bound to specific receptors on certain cells, hyaluronan inhibits cell-cell adhesion and facilitates cell migration.

22.5 The Dynamic Plant Cell Wall

The cell wall surrounding plant cells serves many of the same functions as the extracellular matrix produced by animal cells, even though the two structures are composed of entirely different macromolecules and have a different organization. Like the extracellular matrix, the plant cell wall connects cells into tissues, signals a plant cell to grow and divide, and controls the shape of plant organs. In the past, the plant cell wall was viewed an inanimate rigid box, but it is now recognized as a dynamic structure that plays important roles in controlling the differentiation of plant cells during embryogenesis and growth.

Because the major function of a plant cell wall is to withstand the osmotic turgor pressure of the cell, the cell wall is built for lateral strength. Arranged into layers of **cellulose** microfibers embedded in a matrix of *pectin* and *hemicellulose*, the cell wall is 0.2 μm thick and completely coats the outside of the plant plasma membrane (Figure 22-29). The combination of pressure and strength contributes to the rigidity of a plant. Because the cell wall prevents a cell from expanding, some proteins in the matrix are responsible for loosening the structure of the wall when a cell grows. In addition, the porosity of the matrix permits soluble factors to diffuse across the cell wall and interact with receptors on the plant plasma membrane. However, the cell wall is a selective filter that is more impermeable than the matrices surrounding animal cells. Whereas water and ions diffuse freely in cell walls, diffusion of particles with a diameter greater than ≈4 nm, including proteins with a molecular weight less than 20,000, is reduced. This is one of the reasons that plant hormones are small, water-soluble molecules.

The cell wall undergoes its greatest changes at the *meristem* of a root or shoot tip. These are sites where cells divide and expand. Young cells are connected by thin primary cell walls (Figure 22-30), which can be loosened and stretched to allow subsequent cell elongation. After cell elongation ceases, the cell wall generally is thickened, either by secretion of additional macromolecules into the primary wall or, more usually, by formation of a secondary cell wall composed of several layers. In mature tissues such as the xylem—the tubes that conduct salts and water from the roots through the stems to the leaves (see Figure 16-53)—the cell body degenerates, leaving only the cell wall. The unique properties of wood and of plant fibers such as cotton are due to the molecular properties of the cell walls in the tissues of origin. We begin our discussion with the structure of the cell wall.

The Cell Wall Is a Laminate of Cellulose Fibrils in a Pectin and Hemicellulose Matrix

The strength of the cell wall is derived from layers of cellulose microfibrils that are extensively cross-linked by hemicellulose polysaccharide chains. Each microfibril consists of a bundle of linear polymers of glucose residues linked together

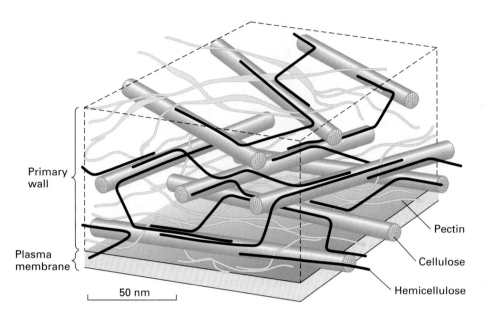

◀ FIGURE 22-29 Schematic representation of the cell wall of an onion. Cellulose and hemicellulose are arranged into at least three layers in a matrix of pectin polymers. The size of the polymers and their separations are drawn to scale. To simplify the diagram, most of the hemicellulose cross-links are not shown. [Adapted from M. McCann and K. R. Roberts, 1991, in C. Lloyd, ed., *The Cytoskeletal Basis of Plant Growth and Form*, p. 126.]

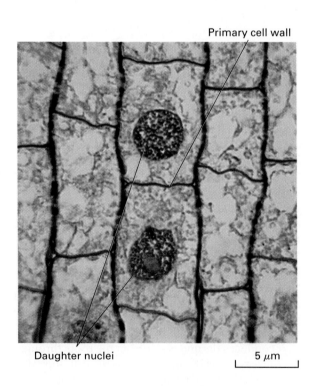

▲ **FIGURE 22-30 Light micrograph of young root tip cells of an onion.** A thin primary cell wall separates two recently separated cells. [Courtesy of Jim Solliday and Biological Photo Service.]

by β(1→4) glycosidic bonds into a straight glucan chain. In this bonding arrangement, each glucose residue is rotated by 180° around its (1→4) axis relative to an adjacent residue; thus a pair of residues, cellobiose, constitute a subunit (Figure 22-31). Microfibrils are 5–15 nm in diameter and can be many micrometers in length. Extensive hydrogen bonding within glucan chains and between adjacent chains makes the microfibril an almost crystalline aggregate. The layers of microfibrils prevents the cell wall from stretching laterally.

As a cell matures, it lays down an inner secondary wall (Figure 22-32). This inner wall may have several layers; within each layer the cellulose fibrils are parallel to one another, but the orientation differs in adjacent layers. Such a plywood-like construction adds considerable strength to the wall. Befitting its central structural role, cellulose makes up 20–30 percent of the wall's dry weight and is the most abundant molecule in the cell wall.

Two other polysaccharide molecules, hemicellulose and pectins, are major constituents of the cell wall. Cellulose microfibrils are cross-linked by hemicelluloses, highly branched polysaccharides with a backbone of about 50 β(1→4)-linked sugars of a single type. Hemicelluloses are linked by hydrogen bonds to the surface of cellulose microfibrils. The hemicellulose branches help bind the microfibrils to one another and to other matrix components, particularly the pectins. This interlinked network of pectin and hemicellulose helps bind adjacent cells to each other and cushion them. The gel-like property of the cell wall is derived in part from pectins. Like hyaluronan, pectin contains multiple negatively charged saccharides that bind cations such as Ca^{2+} and become highly hydrated. When purified, pectin binds water and forms a gel—hence the use of pectins in many processed foods. Pectins are particularly abundant in the middle lamella, the layer between the cell walls of adjacent cells. Treatment of tissues with pectinase or other enzymes that degrade pectin frequently causes cells with their walls to separate from one another.

(a)

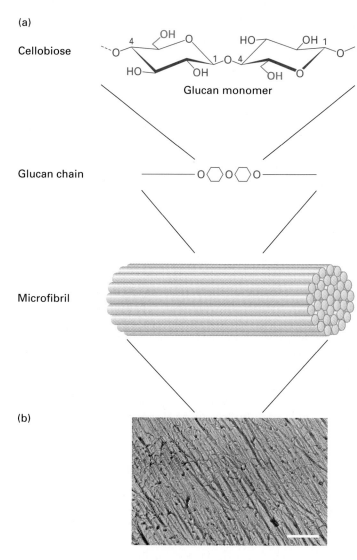

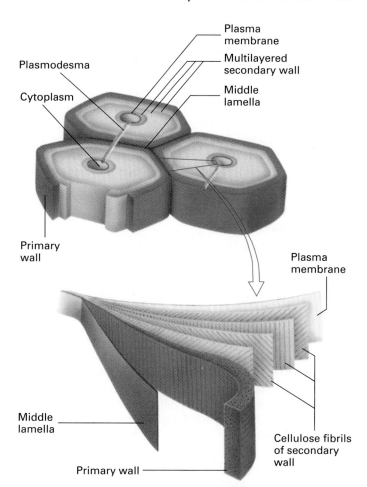

▲ **FIGURE 22-31 The structure of cellulose in the plant cell wall.** (a) Cellulose is a linear polymer consisting of 2000–20,000 glucose residues linked together by $\beta(1\rightarrow4)$ glycosidic bonds. Because the $\beta(1\rightarrow4)$ linkages cause alternating glucose residues to be rotated by 180°, a pair of residues constitute a repeating unit, the cellobiose monomer; these monomers polymerize into straight glucan chains. The chains pack together to form rodlike microfibrils, which are stabilized by hydrogen bonds between the chains. Each glucan chain is polar because its two ends are distinct, and all the chains in a microfibril have the same polarity. (b) A rotary shadowed platinum replica of a rapidly frozen, deep-etched onion cell wall shows the arrangement of cellulose fibers and thinner cross-links presumably composed of hemicellulose or pectin. The scale bar is 200 nm. [Part (b) from: M. C. McCann, B. Wells, and K. Roberts, 1990, *Journal of Cell Science* **96**:327; courtesy of Keith Roberts.]

Cell Walls Contain Lignin and an Extended Hydroxyproline-Rich Glycoprotein

As much as 15 percent of the primary cell wall may be composed of *extensin*, a glycoprotein made up of roughly 300 amino acids. Extensin, like collagen, contains abundant hy-

▲ **FIGURE 22-32 The structure of the secondary cell wall, built up of a series of layers of cellulose.** In each layer, the cellulose fibrils run more or less in the same direction, but the direction varies in different layers. As plant cells grow, they deposit new layers of cellulose adjacent to the plasma membrane. Thus the oldest layers are in the primary wall (the outer wall) and in the middle lamella (the pectin-rich part of the cell wall laid down between two daughter cells as they cleave during cell division). Younger regions of the wall—collectively the secondary cell wall —are laid down as successive layers, adjacent to the plasma membrane. The cytoplasms of adjacent cells are usually connected by plasmodesmata that run through the layers of the cell walls.

droxyproline (Hyp) and about half its length represents variations of the four-residue sequence Ser-Hyp-Hyp-Hyp. Most of the hydroxyprolines are glycosylated with chains of three or four arabinose residues, and the serines are linked to galactose. Thus extensin is about 65 percent carbohydrate, and its protein backbone forms an extended rodlike helix with carbohydrates protruding outward. Extensins, like other cell-wall proteins, are incorporated into the insoluble polysaccharide network and are believed to have a structural role, forming the scaffolding upon which the cell-wall architecture is formed.

Lignin—a complex, insoluble polymer of phenolic residues—associates with cellulose and is a strengthening material in all cell walls. It is particularly abundant in wood, where it accumulates in primary cell walls and in the secondary walls of the xylem. Like cartilage proteoglycans, lignin resists compression forces on the matrix. Particularly for soil-grown plants, lignin is essential for strengthening the xylem tubes to enable them to conduct water and salts over long distances. Lignin also protects the plant against invasion by pathogens and against predation by insects or other animals.

A Plant Hormone, Auxin, Signals Cell Expansion

Cell growth in higher plants frequently occurs without an increase in the volume of the cytosol. Because of the low ionic strength of the cell wall, water tends to leave it and enter the cytosol and vacuole, causing the cell to expand. A localized loosening of the primary cell wall, induced by *auxin*, allows the cell to expand in a particular direction; the size and shape of a plant are determined primarily by the amount and direction of this enlargement (Figure 22-33a). Individual plant cells can increase in size very rapidly by loosening the wall and pushing the cytosol and plasma membrane outward against it. The increase in cell volume is due only to the expansion of the intracellular vacuole by uptake of water. We can appreciate the magnitude of this phenomenon by considering that if all cells in a redwood tree were reduced to the size of a typical liver cell (≈ 20 mm in diameter), the tree would have a maximum height of only 1 meter.

The ability of auxin (indole-3-acetic acid) to rapidly induce cell elongation was first demonstrated in classical experiments on *coleoptiles* from grasses and oats. According to the *acid-growth hypothesis*, auxin stimulates proton secretion at the "growing" end of the cell by activating (directly or indirectly) a proton pump bound to the plasma membrane (Figure 22-33b). As a result, the pH of the cell wall near this region of the plasma membrane falls from the normal 7.0 to as low as 4.5. The low pH activates a class of wall proteins, termed *expansins*, that disrupt the hydrogen bonding between cellulose microfibrils, causing the laminate structure of the cell wall to loosen. With the rigidity of the wall reduced, the cell can elongate.

Expansins were discovered and purified using a novel biochemical assay on pure cellulose paper, since paper, like the plant cell walls from which it is made, derives its mechanical strength from hydrogen bonding between cellulose microfibrils. Extracts of plant cell walls were tested for their ability to mechanically weaken paper at pH values between 3.0 and 5.0, but not at pH 7 (Figure 22-34). The expansin-triggered loosening of the wall is reversed when the pH is raised back to 7.0, showing that expansin does not break covalent bonds in cellulose. Additional evidence for the acid-growth hypothesis stems from studies of the fungal compound *fusicoccin*. Like auxin, fusicoccin induces rapid cell elongation and triggers proton pumping out of sensitive cells, with accompanying localized wall loosening. The action of fusicoccin or auxin can be blocked by permeating the cell wall with buffers that prevent the extracellular pH from being lowered.

Cellulose Fibrils Are Synthesized and Oriented at the Plant Cortex

Cellulose microfibrils are synthesized on the exoplasmic face of the plasma membrane from UDP-glucose and ADP-glucose

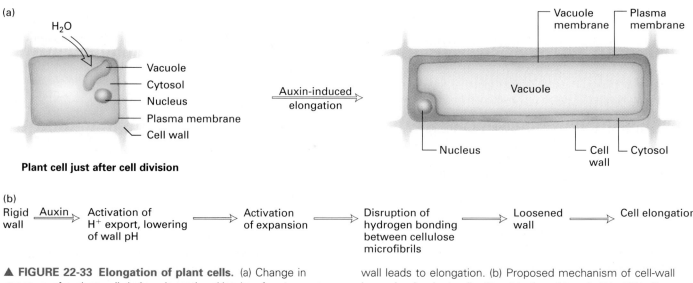

▲ FIGURE 22-33 Elongation of plant cells. (a) Change in structure of a plant cell during elongation. Uptake of water causes an internal pressure (turgor); in the presence of auxin, the cell wall is loosened, and the turgor pressure against the loosened wall leads to elongation. (b) Proposed mechanism of cell-wall loosening in plant cells. [Part (b) adapted from L. Taiz, 1994, *Proc. Nat'l. Acad. Sci. USA* **91**:7387.]

(a)

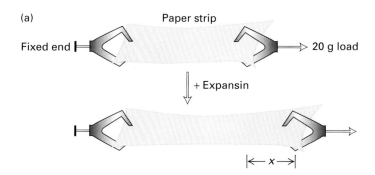

(b)

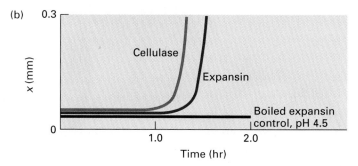

▲ **FIGURE 22-34 Experimental demonstration that expansin loosens hydrogen bonds.** (a) In an elastometer, a paper strip is clamped at both ends and immersed in a solution. One end is attached to a weight, while the other end is held fixed. Agents that break the covalent or hydrogen bonds between the cellulose fibers will cause the paper strip to elongate by x amount. The movement of the clamp is recorded. (b) Treatment of a paper strip with expansin at pH 4.5 (red) results in the reversible weakening of the cellulose molecule. In contrast, cellulase irreversibly weakens paper by breaking covalent bonds in the polymer. Control experiments show that the weakening is not caused by the pH 4.5 solution and is dependent on active protein.

(a)

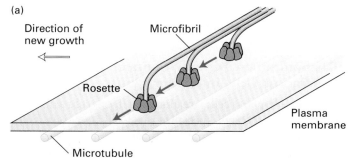

(b)

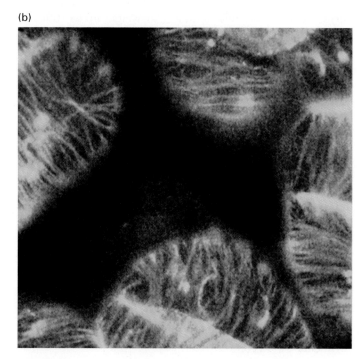

▲ **FIGURE 22-35 Microtubules and cellulose synthesis in an elongating root tip cell.** (a) Circumferential rings of microtubules lie just inside the plasma membrane, perpendicular to the direction of cell elongation. Glucan chains are synthesized by cellulose synthase, a large integral membrane protein arranged into a rosette of subunits, on the outer face of the plasma membrane and spontaneously assemble into cellulose microfibrils. As the long insoluble cellulose fibrils form, the synthase moves in the plasma membrane (red arrows) parallel to the underlying microtubule network. Thus in a growing cell, the new fibrils are arranged in circumferential rings perpendicular to the direction of elongation. (b) A fluorescence micrograph of GFP fused to the microtubule binding domain of MAP-4. The domain highlights the cortical belt of microtubules and shows their parallel orientation beneath the cell wall. [Part (b) adapted from R. Cyr, 1998, *Plant Cell* **10**:1927; courtesy of R. Cyr].

formed in the cytosol. The polymerizing enzyme, called *cellulose synthase,* is thought to be a large complex of many identical subunits, each of which "spins out" glucan chains that spontaneously form microfibrils (see Figure 22-31). The long microfibrils are insoluble, which probably explains why they are not formed within the cell. In contrast, soluble hemicellulose and pectin molecules are synthesized in the Golgi complex and secreted at the cell surface, where they cross-link the cellulose microfibrils into the matrix of the cell wall. In the primary cell wall of elongating cells, newly made cellulose microfibrils encircle the cell like a belt perpendicular to the axis.

Experiments with elongating root tip cells suggest that in the primary wall, at least, microtubules influence the direction of cellulose deposition. These cells have oriented bands or rings of microtubules located just under the plasma membrane; these microtubules are transverse to the direction of elongation but parallel to many of the cellulose microfibrils in the primary cell wall of the elongating cell (Figure 22-35). Moreover, disruption of the microtubular network by drugs eventually disrupts the pattern of cellulose disposition. Thus,

many investigators believe that cellulose synthase complexes move within the plane of the plasma membrane, as cellulose is formed, in directions determined by the underlying microtubule cytoskeleton. Any linkage, however, between the microtubules and cellulose synthase remains to be determined. Interestingly, in gliding bacteria, the synthase is immobile in the membrane. Consequently, a bacterial cell is thought to use the flow of cellulose molecules to affect motility.

Plasmodesmata Directly Connect the Cytosol of Adjacent Cells in Higher Plants

Even though plant cells are bounded by a cell wall, they communicate through specialized cell-cell junctions called **plasmodesmata**, which extend through the adjacent cell walls. Like gap junctions, plasmodesmata are open channels that connect the cytosol of adjacent cells and permit the diffusion of molecules with a molecular weight up to 1000, including a variety of metabolic and signaling compounds. However, during the trafficking of macromolecules, this limit increases to greater than 10,000 MW. The diameter of the cytosol-filled channel is about 60 nm, and plasmodesmata can traverse cell walls up to 90 nm thick. Depending on the plant type, the density of plasmodesmata varies from 1 to 10 per mm², and even the smallest meristematic cells (the growing cells at the tips of roots or stems) have more than 1000 interconnections with their neighbors. Plasmodesmata differ from gap junctions in two significant aspects. The plasma membranes of the adjacent cells extend continuously through each plasmodesma, whereas the membranes of cells at a gap junction are not continuous with each other. In addition, an extension of the endoplasmic reticulum called a *desmotubule* passes through the ring of cytosol, the *annulus,* connecting the cytosol of adjacent cells (Figure 22-36).

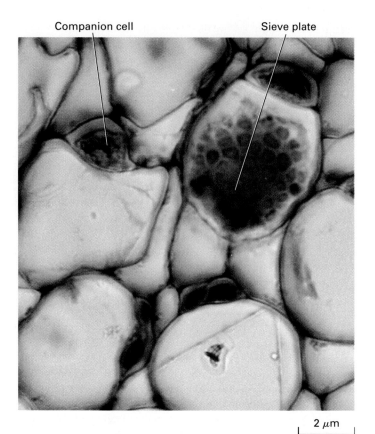

▲ **FIGURE 22-37 Cross section of the phloem from *Curcurbita*.** Note the large pores in the sieve plate and companion cells, which lie adjacent to phloem vessels. [Courtesy of J. R. Waaland and Biological Photo Service.]

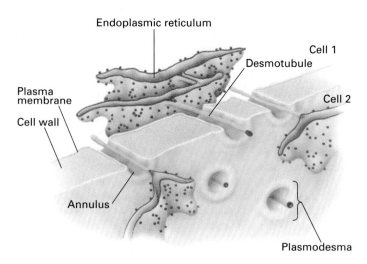

▲ **FIGURE 22-36 The structure of plasmodesmata.** A plasmodesma is a plasma membrane–lined channel through the cell wall. Note the desmotubule, an extension of the endoplasmic reticulum, and the annulus, a ring of cytosol that interconnects the cytosol of adjacent cells. Not shown is a gating complex that fills the channel and controls transport of materials through plasmodesmata.

Much evidence establishes that plasmodesmata are in fact used in cell-cell communication. For instance, fluorescent water-soluble chemicals microinjected into plant cells spread to the cytoplasm of adjacent cells but not into the cell wall. Many types of molecules spread from cell to cell through plasmodesmata, including proteins, nucleic acids, metabolic products, and plant viruses. Soluble molecules pass through the cytosolic annulus, but membrane-bound molecules may pass from cell to cell via the desmotubule. Transport of such substances is proportional to the number of plasmodesmata and does not occur between cells not connected by such junctions. The permeability of the plasmodesmata to these molecules is regulated in response to developmental, physiological, or environmental changes. As with gap junctions, transport through plasmodesmata is reversibly inhibited by an elevation in cytosolic Ca²⁺.

As discussed in Chapter 16, phloem vessels transport sucrose and other metabolites throughout a plant from their sites of synthesis in the leaves. In formation of the long, narrow *sieve-tube cells* composing a phloem vessel, the primary cell wall thickens and the nucleus, vacuole, and other internal organelles are lost, although the plasma membrane is retained. In each end wall, called the *sieve plate*, the plasmodesmata expand to form large pores that facilitate fluid movement (Figure 22-37; see also Figure 16-53). Numerous plasmodesmata connect sieve-tube cells with companion cells located along the length of a phloem vessel. Substances pass in and out of the sieve-tube cells through these plasmodesmata.

SUMMARY The Dynamic Plant Cell Wall

- Cellulose is a large, linear glucose polymer that assembles spontaneously into microfibrils stabilized by hydrogen bonding.

- The plant cell wall is a matrix of cellulose microfibrils cross-linked by hemicellulose, pectin, and extensin (see Figure 22-29).

- Plant cells grow by localized loosening of the cell walls and expansion of the vacuole resulting from osmotic influx of water.

- The acid-growth hypothesis states that cell-wall expansion is caused by an acid-dependent activation of the cell-wall protein expansin, which loosens hydrogen-bonded cellulose microfibrils.

- The plant hormone auxin induces acid secretion at the membrane, leading to activation of expansin.

- Cellulose synthase in the plasma membrane synthesizes new cellulose microfibrils on the exoplasmic face. Cortical microtubules underlying the membrane are thought to determine the orientation of the cellulose microfibrils (see Figure 22-35).

- Cell-cell communication takes place through plasmodesmata, which allow small molecules such as sucrose to pass between the cells.

PERSPECTIVES for the Future

Cell adhesion research will continue to advance in several areas. Only recently, we have learned that cell adhesion molecules signal to the cytoplasm and nucleus. Some feel that the future will show that communication through cell-cell and cell-matrix interactions is as important and frequent as through soluble molecules that bind the free cell surface. In the next few years, investigators will uncover not only more examples of signaling through cell adhesion molecules but also, more importantly, the mechanisms that distinguish this route from those that originate at the cell surface.

Just as signaling pathways are becoming more complex, the structure of an adhesion junction is not well defined. The large number of different proteins and their concentration at the cell membrane favor loose associations that may not withstand the rigors of gentle biochemical fractionation methods. One approach for identifying the core structure of

an adhesion junction is to look for clues from evolution. Here, we can look to the genome project for some clues about the origins of cell adhesions. The recent description of the nematode genome has revealed that of the 19,047 proteins, only 154 are elements of cell adhesions, including integrin, cadherin, and collagen. Although a nematode is a simple worm and is less complex than a vertebrate, it is nevertheless more complicated than the simplest multicellular organisms. The genomes of these organisms may reveal the first components of a cell adhesion.

The roles of the cell wall in directing plant development and in combating infection are two exciting areas of investigation. In contrast to the animal world, many of the extracellular signals are completely different, but they may nevertheless signal the initiation and growth of cell walls through familiar mechanisms. In addition, the mechanisms to combat plant disease are being uncovered. They involve imaginative schemes to surmount the plant defenses by mimicking a plant host or to defeat infection by unleashing a torrent of noxious chemicals.

PERSPECTIVES in the Literature

The key and distinguishing step in the assembly of a cell junction is the clustering of junction proteins. In the electron microscope, the aggregate of cadherins and integrin comprise the membrane-associated electron-dense plaque of hemidesmosomes, desmosomes, focal adhesions, and adherens junctions. Clusters of occludins and connexons in tight junctions and gap junctions are apparent in micrographs of freeze-fracture replicas of cell membranes. Because of the importance of cell adhesion and communication in development of an embryo, the onset of metastasis, and control of cell growth, the mechanism of junction assembly is an important and exciting area of research. Probably the assembly of a focal adhesion is the best model for understanding the general principles of the process. Extracellular matrix molecules and soluble factors activate a Rho–G-protein-dependent pathway that coordinates the organization of the actin cytoskeleton with the clustering of integrin receptors. A recent review discusses the current state of knowledge of this process.

S.M. Schenwaelder and K. Burridge. 1999. Bidirectional signaling between the cytoskeleton and integrins. *Curr. Opin. Cell Biol.* **11**:274–286.

From the cited literature, try to answer or provide a model for the following questions: What is the mechanism of integrin clustering? Does it require the action of the cytoskeleton or extracellular matrix components? How does signaling at the cell surface direct the formation of a focal adhesion? What effects would deletions of the extracellular or cytoplasmic domains of α or β integrins have on clustering?

Testing Yourself on the Concepts

1. Define the difference between homophilic and heterophilic interactions. Describe how cadherins mediate homophilic interactions.

2. Gap junctions are calcium regulated. What are the advantages of this calcium regulation to the organism?

3. Much of the extracellular matrix of animal cells consists of the collagen, while cellulose is a major component of the plant cell wall. Compare and contrast the biosynthesis of each.

4. The chemical composition of the extracellular matrix (ECM) is highly complex. Describe the general structure of aggrecan, a proteoglycan, and how it relates to function.

5. Binding of cells to ECM components is important to stabilizing tissues. Yet cell migration through the ECM is necessary for differentiation and when improperly regulated can contribute to cancer cell invasiveness. How can proteases function as de-adhesion factors?

MCAT/GRE-Style Questions

Key Concept Please read the section titled "Integrin-Containing Junctions Connect Cells to the Substratum" (p. 978) and refer to Figures 22-9 and 22-10; then answer the following questions:

1. One example of an integrin-containing cell-matrix junction is
  a. Gap junction.
  b. Desmosome.
  c. Hemidesmosome.
  d. Adherens junction.

2. Cytoskeletal proteins linked with integrin-containing cell-matrix junctions include all of the following *except*:
  a. Actin.
  b. Actin-binding proteins.
  c. Keratin.
  d. Tubulin.

3. Vinculin is an example of
  a. An extracellular adapter protein within the junctional complex.
  b. An intracellular adapter protein within the junctional complex.
  c. A protein with multiple binding domains that binds directly integrins and laminin.
  d. A protein with multiple binding domains that binds directly integrins and tubulin.

4. Functions of integrin-containing cell-matrix junctions include all of the following *except*:

  a. Communication between cells.
  b. Connecting cells to the substratum.
  c. Increasing the overall rigidity of tissues such as epithelial cell layers.
  d. Separating apical and domains of cells within tissues such as epithelial cell layers.

Key Application Please read the section titled "Fibronectins Bind Many Cells to Fibrous Collagens and Other Matrix Components" (p. 987) and refer to Figures 22-22 and 22-23; then answer the following questions:

5. Fibronectins are soluble multiadhesive matrix proteins because they
  a. Are able to bind to both ECM proteins and to cellular proteins.
  b. Bind to multiple cell types.
  c. Contain an RGD sequence, a characteristic motif of multiadhesive matrix proteins.
  d. Bind to both plastic substratum and to cells.

6. Which of the proteins or polysaccharides to which fibronectins are capable of binding is a normal plasma membrane component?
  a. Collagen.
  b. Fibrin.
  c. Heparan sulfate.
  d. Integrin.

7. You have synthesized an oligopeptide containing an RGD sequence as well as other amino acid sequences. What is the effect of this peptide when added to a fibroblast cell culture grown on a layer of fibronectin absorbed to the tissue culture dish?
  a. The peptide acts like a structural analogue of fibronectin and helps recruit collagen to the fibronectin layer.
  b. The peptide competes for binding to integrins and hence decreases the adhesion of the fibroblasts to the fibronectin layer.
  c. The peptide has no effect because the RGD sequence and the other sequences contained within the peptide are irrelevant to fibroblasts.
  d. The peptide displaces fibronectin from the surface of the tissue culture dish, producing almost instant cell death.

8. Which of the following is *not* a function of fibronectin:
  a. Regulation of the transmembrane conductance of cells.
  b. Regulation of cell shape and cytoskeleton organization.
  c. Regulation of cell migration and cellular differentiation.
  d. Facilitation of wound healing.

9. Blood clotting is a crucial function for mammalian survival. How do the multiadhesive properties of fibronectin lead to the recruitment of platelets to blood clots?

   a. Because fibronectin can bind to both fibrin in clots and to collagen forming a crosslinks, recruitment results.

   b. Because fibronectin can bind to both fibrin in clots and to heparan sulfate, recruitment results.

   c. Because fibronectin can bind to both fibrin in clots and integrins expressed on platelets, recruitment results.

   d. Because circulating fibronectin is free of either one or both extra domain A or B type III repeats, recruitment results, binding to nonfibrin blood clot components.

Key Experiment Please read the section titled "A Plant Hormone, Auxin, Signals Cell Expansion" (p. 996) and refer to Figures 22-33 and 22-34; then answer the following questions:

10. The expansion of plant cell walls signaled by auxin is

   a. Equal in all cell dimensions.

   b. An expansion of the cytosol relative to vacuole.

   c. Localized to regions of the plant cell wall.

   d. A delayed response due to the need to increase extracellular pH.

11. An auxin source placed on one side of a plant stem produces

   a. Bending of the stem away from the auxin source.

   b. Bending of the stem stem toward the auxin source.

   c. Upward elongation of the stem.

   d. Downward elongation of the stem toward the root.

12. Based on the cell extract experiment

   a. Hydrogen bonding between cellulose microfibrils is sensitive to acid pH.

   b. Expansin activity in disrupting hydrogen bonding between cellulose microfibrils is sensitive to acid pH.

   c. Expansin is a novel form of cellulase.

   d. Expansin produces an irreversible change in cell wall properties.

13. The laundry detergent industry wishes to include expansin in its detergent products as a way to rapidly wash out grass stains from blue jeans. Why might this not work?

   a. Because normal plant cell walls contain a greater variety of molecules than a piece of paper does, grass would be insensitive to expansin.

   b. Because expansin activity requires the expression of new genes and blue jeans as a manufactured product contain no active genes.

   c. Because market testing indicates that grass stains are prized by blue jeans wearers and hence effective stain removal would be counterindicated.

   d. Blue jeans are cotton products and hence cellulose based. Therefore, expansin, if active, would be expected to affect the blue jeans as well as the grass stain.

Key Terms

References

General References

Ayad, S., et al. 1994. *The Extracellular Matrix Facts Book.* Academic Press.

Chothia, C., and E. Y. Jones. 1997. The molecular structure of cell adhesion molecules. *Ann. Rev. Biochem.* **66**:823–862.

Gumbiner, B. M. 1996. Cell adhesion: the molecular basis of tissue architecture and morphogenesis. *Cell* **84**:345–357.

Hay, E. D. (ed.). 1991. *Cell Biology of Extracellular Matrix,* 2d ed. Plenum.

Pigott, R., and C. Power. 1993. *The Adhesion Molecule Facts Book.* Academic Press.

Cell-Cell Adhesion and Communication

Ben-Ze'ev, A., and B. Geiger. 1998. Differential molecular interactions of β-catenin and plakoglobin in adhesion, signaling and cancer. *Curr. Opin. Cell Biol.* **10**:629–639.

Christofori, G., and H. Semb. 1999. The role of the cell-adhesion molecule E-cadherin as a tumour-suppressor gene. *Trends. Biochem. Sci.* **24**:73–76.

Dunon, D., L. Piali, and B. A. Imhof. 1996. To stick or not to stick: the new leukocyte homing paradigm. *Curr. Opin. Cell Biol.* **8**:714–723.

Kowalczyk, A. P., et al. 1999. Desmosomes; intercellular adhesive junctions specialized for attachment of intermediate filaments. *Int'l. Rev. Cytol.* **185**:237–302.

Lo, C. W. 1999. Genes, gene knockouts, and mutations in the analysis of gap junctions. *Dev. Genet.* **24**:1–4.

Simon, A. M., and D. A. Goodenough. 1998. Diverse functions of vertebrate gap junctions. *Trends Cell Biol.* **8**:477–483.

Cell-Matrix Adhesion

Black, R. A., and J. M. White. 1998. ADAMs: focus on the protease domain. *Curr. Opin. Cell Biol.* **10**:654–659.

Burridge, K., and M. Chrzanowska-Wodnicka. 1996. Focal adhesions, contractility, and signaling. *Ann. Rev. Cell Dev. Biol.* **12**:463–519.

Hemler, M. E. 1998. Integrin-associated proteins. *Curr. Opin. Cell Biol.* **10**:578–585.

Jones, J. C. R., S. B. Hopkinson, and L. E. Goldfinger. 1998. The structure and assembly of hemidesmosomes. *BioEssays* **20**:488–494.

Werb, Z. 1997. ECM and cell surface proteolysis: regulating cellular ecology. *Cell* **91**:439–442.

Collagen

Adachi, E., I. Hopkinson, and T. Hayashi. 1997. Basement-membrane stromal relationships: interactions between collagen fibrils and the lamina densa. *Int'l. Rev. Cytol.* **173**:73–225.

Noncollagen Components of the Extracellular Matrix

Rapraeger, A. C., and V. L. Ott. 1998. Molecular interactions of the syndecan core proteins. *Curr. Opin. Cell Biol.* **10**:620–628.

Schwartz, N. B., et al. 1999. Domain organization, genomic structure, evolution, and regulation of expression of the aggrecan gene family. *Prog. Nucl. Acid Res. Mol. Biol.* **62**:177–225.

Timple, R. 1996. Macromolecular organization of basement membranes. *Curr. Opin. Cell Biol.* **8**:618–624.

Yurchenco, P. D., and J. C. Schittny. 1990. Molecular architecture of basement membranes. *FASEB J.* **4**:1577–1590.

The Dynamic Plant Cell Wall

Delmer, D. P., and Y. Amor. 1995. Cellulose biosynthesis. *Plant Cell* **7**:987–1000.

Fisher, D. D, and R. J. Cyr. 1998. Extending the microtubule/microfibril paradigm. *Plant Physiol.* **116**:10430–10451.

Kragler, R., W. J. Lucas, and J. Monzer. 1998. Plasmodesmata: dynamics, domains and patterning. *Ann. Bot.* **81**:1–10.

Kropf, D. L, S. R. Bisgrove, and W. E. Hable. 1998. Cytoskeletal control of polar growth in plant cells. *Curr. Opin. Cell Biol.* **10**:117–122.

McCann, M. C., B. Wells, and K. Roberts. 1990. Direct visualization of cross-links in the primary plant cell wall. *J. Cell Sci.* **96**:323–334.

McLean, B. G., F. D. Hempel, and P. C. Zambryski. 1997. Plant intercellular communication via plasmodesmata. *Plant Cell* **9**:1043–1054.

Pennell R. 1998. Cell walls: structures and signals. *Curr. Opin. Plant Biol.* **1**:504–510.

Reiter, W. D. 1998. The molecular analysis of cell wall components. **3**:27–32.

Wymer, C., and C. Lloyd. 1996. Dynamic microtubules: implications for cell wall patterns. *Trends Plant Sci.* **1**:222–228.

Cell Interactions in Development

23

In less than 9 months a single fertilized human egg gives rise to hundreds of different types of cells. Multiple cell-cell and cell-matrix interactions are required for formation of each type of differentiated cell and its specific organization into tissues and organs.

In Chapter 14, we learned that regionalization along the anteroposterior axis in the early *Drosophila* embryo is largely determined by gradients of transcription factors generated through translation of spatially restricted maternal mRNAs and subsequent diffusion of the encoded proteins through the common cytoplasm of the syncytial blastoderm. These transcription factors, in turn, control the patterned expression of specific target genes along the anteroposterior axis. In contrast, local interactions between cells, mediated by secreted or cell-surface signaling molecules, determine regionalization along the dorsoventral axis in *Drosophila* and along both major axes in early vertebrate embryos. Such local interactions also are the primary mechanism regulating the formation of internal organs such as the kidney, lung, and pancreas. Likewise, the vast number of highly specialized cells and their stereotyped arrangement in different tissues is a consequence of locally acting signals.

The importance of cellular interactions in **development** was demonstrated first in the early part of twentieth century through two complementary experiments. In one, destruction of an optic-vesicle primordium in developing frogs prevented formation of the lens from the overlying ectodermal cells. Conversely, transplantation of an optic-vesicle primordium to a region of ectoderm that normally does not give rise to a lens induced formation of a lens in an abnormal *(ectopic)* site (Figure 23-1). In modern biology we now use the term **induction** to refer to any mechanism whereby one cell population influences the development of neighboring cells.

MEDIA CONNECTIONS

Focus: TGF*β* Signaling Pathway

Focus: Apoptosis

Classic Experiment 23.1: Hunting Down Genes Involved in Cell Death

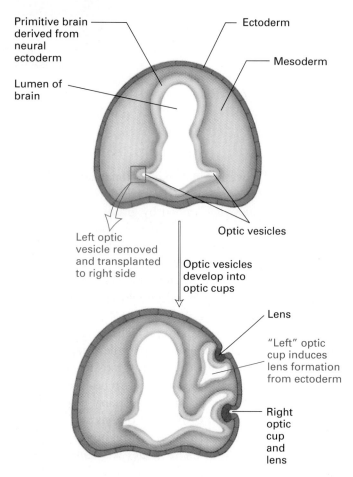

▲ FIGURE 23-1 Diagrams of a cross section through the head of a vertebrate embryo illustrating the interaction of the optic cup with the overlying ectodermal cells that become the optic lens. When the left optic vesicle is transplanted to another site in the head, it induces a lens in that inappropriate (ectopic) site. [After N. K. Wessels, 1977, *Tissue Interaction in Development*, Benjamin-Cummings, p. 46.]

In some cases, induction involves a binary choice. In the presence of a signal the cell is directed down one developmental pathway; in the absence of the signal, the cell assumes a different developmental fate or fails to develop at all. In other cases, signals can induce different responses in cells at different concentrations. For instance, a low concentration of an inductive signal causes a cell to assume fate A, but a higher concentration causes the cell to assume fate B. The concentration at which a signal induces a specific cellular response is called a *threshold*.

In many cases, an inductive signal induces an entire tissue containing multiple cell types. Two models have been proposed to account for these properties of extracellular signaling molecules. In the *gradient model*, a signaling molecule induces different fates at different threshold concentrations. A cell's fate, then, is determined by its distance from the signal source. In the alternative *relay model*, a signal induces a cascade of induction in

which cells close to the signal source are induced to assume specific fates; they, in turn, produce other inductive signals to pattern their neighbors.

Although inductive interactions often are unidirectional, they sometimes are reciprocal. Prominent examples of reciprocal induction include the formation of internal organs such as the kidney, pancreas, and lung. Many inductive interactions occur between *non-equivalent* cells; that is, the signaling and responding cells are already different. However, interactions between *equivalent* cells often are crucial in assuring that some cells in a developing tissue assume a specific fate and others do not. An evolutionarily conserved class of ligands and receptors regulates such interactions in *C. elegans*, *Drosophila*, and vertebrates.

Another feature that distinguishes various developmental pathways is the nature of the extracellular inductive signals. Many are freely diffusible and hence can act at a distance, whereas some are tethered to the cell surface and are available only to immediate neighboring cells. Still others are highly localized by their tight binding to the extracellular matrix. Early embryologists noted that cells differed in their ability to respond to inducing signals. Cells that can respond to such signals are referred to as *competent*. Competence may reflect the expression of receptors specific for a given signaling molecule, the ability of the receptors to activate specific intracellular signaling pathways, or the presence of the transcription factors necessary to stimulate expression of the genes required to implement the developmental program induced.

In this chapter, we first describe examples of various types of inductive signals and cellular interactions that regulate cell-type specification in several different developmental systems. Specific extracellular signals also control the migration of certain cells, which occurs during development of some tissues. As an example of this phenomenon, we discuss the role of extracellular signals in the assembly of connections between neurons. Another common feature of developmental programs is the highly regulated death of certain cells. In the final section of this chapter, we examine the conserved pathway leading to cell death and how it is controlled. The examples presented in this chapter were chosen to illustrate key concepts in this rapidly advancing field.

23.1 Dorsoventral Patterning by TGFβ-Superfamily Proteins

A number of related extracellular signaling molecules that play widespread roles in regulating development in both invertebrates and vertebrates constitute the *transforming growth factor β* (**TGFβ**) **superfamily**. The first members of the TGFβ superfamily were identified on the basis of their ability to induce a transformed phenotype of certain cells in culture, but these secreted growth factors are now known to have a remarkable spectrum of effects on normal growth and development. Some TGFβ-superfamily members have antiproliferative effects on certain tissues and proliferative effects on others. These growth factors also promote the production of cell-adhesion molecules, other growth factors, and extracellular-matrix molecules. We first describe the general structure of TGFβ proteins and the cell-surface receptors that bind them. Then we consider the signal-transduction pathway activated by binding of TGFβ proteins. Finally, we discuss the role of these molecules in dorsoventral patterning, focusing on the TGFβ homologs in *Drosophila* (Dpp protein) and the frog *Xenopus laevis* (BMPs).

TGFβ Proteins Bind to Receptors That Have Serine/Threonine Kinase Activity

Members of the TGFβ superfamily are derived from inactive secreted precursor proteins through proteolytic processing (Figure 23-2a). The precursors contain an N-terminal signal peptide, a central pro-domain containing 50–375 amino acids, and a C-terminal mature domain, which forms the active growth factor. The monomeric form of these growth factors contains 110–140 amino acids and has a compact structure with four antiparallel β strands and three intramolecular disulfide linkages forming a structure called a cystine knot (Figure 23-2b). The cystine-knot domain, which is relatively resistant to denaturation, may be particularly

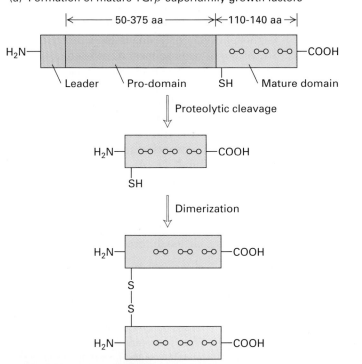

(a) Formation of mature TGFβ-superfamily growth factors

◀ **FIGURE 23-2 TGFβ-superfamily proteins function as inducers during vertebrate and invertebrate development.** (a) Schematic diagram of formation of mature dimeric TGFβ proteins from secreted monomeric TGFβ precursors. The mature domain contains six cysteine residues (yellow dots), which form three intrachain disulfide bonds. Another N-terminal cysteine in the mature domain forms an interchain disulfide bond that links monomers into the active homo- or heterodimeric proteins. (b) Ribbon diagram of mature TGFβ dimer based on x-ray crystallographic analysis. The two subunits are shown in tan and blue, with the hydrophobic residues at the interfaces indicated by balls. The white balls represent disulfide-linked cysteine residues. The three intrachain disulfide linkages in each monomer form a cystine-knot domain, which is resistant to degradation. [Part (b) from S. Daopin et al., 1992, *Science* **257**:369.]

(b) Structure of mature TGFβ

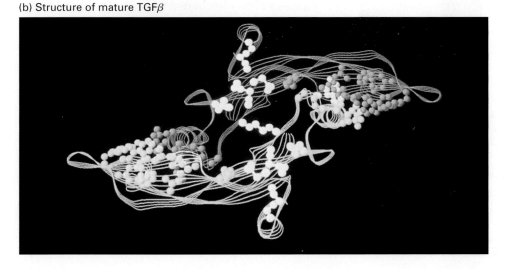

well suited for extracellular molecules. Much of the sequence variation among different TGFβ proteins is observed in the N-terminal regions, the loops joining the β strands, and the α helices. An additional N-terminal cysteine in each monomer links TGFβ monomers into functional homodimers and heterodimers. Different heterodimeric combinations may increase the functional diversity of these proteins beyond that generated by differences in the primary sequence of the monomer.

The primary sequence of TGFβ monomers exhibits less than 10 percent homology with nerve growth factor and platelet-derived growth factor. Nonetheless, the remarkable similarity in the three-dimensional structures of the monomers of these protein growth factors suggests a common ancestral origin with much sequence drift during evolution. The organization of the subunits in the dimeric proteins, however, varies among the three growth factors.

By cross-linking radio-iodinated TGFβ molecules bound to the surface of cells, investigators identified three different polypeptides with apparent molecular weights of 55, 85, and 280 kDa referred to as types I, II, and III TGFβ receptors, respectively. The type I and type II receptors are both transmembrane serine/threonine kinases. Binding of TGFβ induces the formation of multimeric receptors, most likely heterotetramers, containing both the type I and type II receptors. The type II subunit then phosphorylates serine and threonine residues in a highly conserved sequence motif in the juxtamembrane region of the type I subunit, thereby activating its kinase activity.

The type III TGFβ receptor is a cell-surface proteoglycan called β-glycan, which appears to regulate the accessibility of TGFβ to the signal-transducing heterotetramer of the type I and type II receptor. This phenomenon is similar to the binding of fibroblast growth factor (FGF) by a syndecan proteoglycan and presentation of the bound FGF to its receptors (see Figure 22-28).

Activated TGFβ Receptors Phosphorylate Smad Transcription Factors

Genetic studies similar to those used to dissect receptor tyrosine kinase pathways (Chapter 20) led to the identification of transcription factors downstream from the TGFβ family of receptors in *Drosophila*. The genes encoding these transcription factors in *Drosophila* and the related vertebrate genes are now called *Smads*. There are three types of **Smad proteins**: receptor-regulated Smads (R-Smads), co-Smads, and inhibitory or antagonistic Smads.

Residues near the C-terminus of R-Smads are phosphorylated by activated type I TGFβ receptors. As depicted in Figure 23-3, phosphorylated R-Smads then dimerize with co-Smads. The resulting heterodimers translocate to the nucleus and cooperate with other transcription factors to activate transcription of specific target genes. R-Smads can be divided into two domains, MH1 and MH2, separated by a flexible linker region. In the inactive state, the N-terminal MH1 domain suppresses the transcriptional activity of the C-terminal MH2 domain. In the active, phosphorylated state, the MH1 domain binds DNA, and the MH2 domain regulates interactions with co-Smads, promotes interaction with DNA-binding proteins, and provides a transcriptional activation function.

Binding of different TGFβ-superfamily growth factors to their specific receptors elicits different cellular responses.

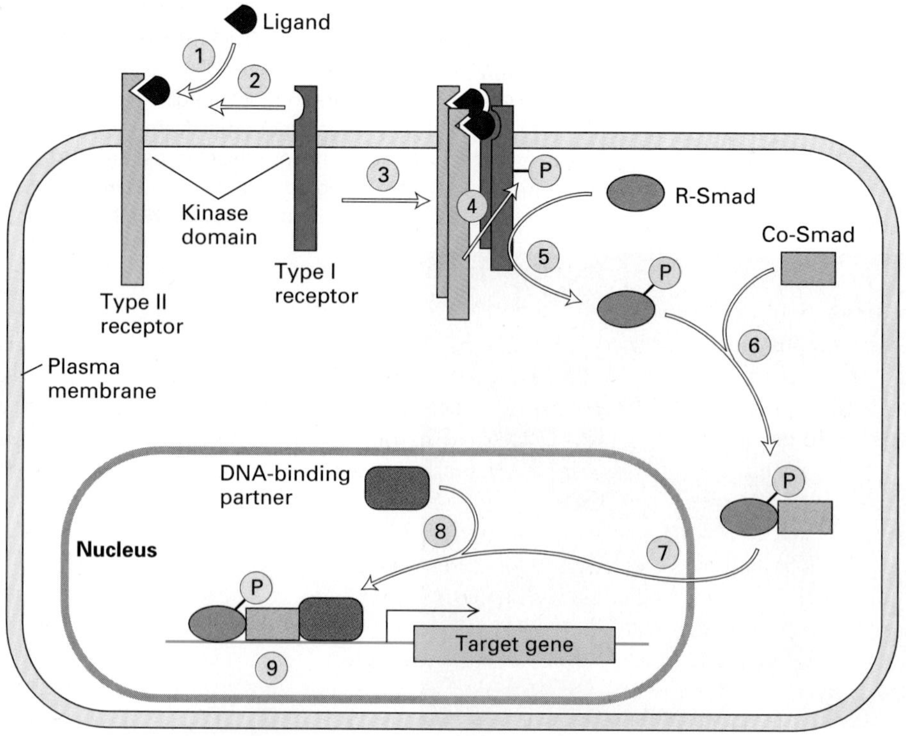

◀ **FIGURE 23-3 TGFβ signaling pathway.** Binding of ligand to the type I and type II receptors, which are serine/threonine kinases, induces formation of multimeric receptors. Type II receptors phosphorylate type I receptors in the juxtamembrane region. Activated type I receptors specifically phosphorylate receptor-regulated Smads (R-Smads), which then dimerize with Co-Smads in the cytosol. The R-Smad/Co-Smad complex translocates to the nucleus where it binds to regulatory sequences in combination with specific transcription factors, leading to transcription of specific target genes. [Adapted from J. Massagué, 1998, *Annu. Rev. Biochem.* **67**:753.]

The specificity exhibited by these related receptors is a common phenomenon in intercellular signaling. The TGFβ signaling pathway provides an excellent example of one strategy for achieving such response specificity. For example, binding of TGFβ to its receptor leads to phosphorylation of Smad2, its dimerization with Smad4, translocation of Smad2/Smad4 to the nucleus, and transcriptional activation of specific target genes. On the other hand, binding of BMP2, another member of the TGFβ superfamily, to its receptor results in phosphorylation of Smad1, its dimerization with Smad4, and activation of a specific transcriptional response distinct from that induced by Smad2/Smad4.

The response specificity of the TGFβ and BMP2 receptors is determined by three amino acids in the type I subunits of the receptors and complementary residues in R-Smads. The specificity of each receptor can be changed simply by swapping the amino acids at these three positions. Likewise, swapping the complementary sequences between Smad1 and Smad2 reverses the specificity of activation such that Smad1 is now activated by the TGFβ receptor and Smad2 is activated by the BMP2 receptor. Although these complementary sequences match specific receptors with specific R-Smads, another region within the C-terminal domain of R-Smads is critical for determining the specificity of target-gene induction, presumably through interactions with specific transcription factors.

Dpp Protein, a TGFβ Homolog, Controls Dorsoventral Patterning in *Drosophila* Embryos

In *Drosophila*, the *decapentaplegic (dpp)* gene encodes a TGFβ-superfamily member, the **Dpp protein**. The role of Dpp in embryonic induction is understood in some detail, and studies in vertebrate systems suggest that other members of the TGFβ superfamily play similar roles. In both *Drosophila* and vertebrates, TGFβ proteins have been shown to control cell fate along the dorsoventral axis during **embryogenesis.** Here, we discuss the mechanism of this patterning in flies.

Initial dorsoventral patterning in *Drosophila* is controlled by a set of extracellular signaling molecules encoded by maternal mRNAs, particularly the *Toll* and *dorsal* mRNAs. Generation of a signal in the most-ventral part of the embryo activates Toll protein, a cell-surface receptor, in the most-ventral region of the embryo. The resulting gradient of activated Toll leads to nuclear localization of Dorsal in the ventral region. In lateral regions, corresponding to lower levels of the Toll signal, Dorsal protein is evenly distributed between the cytoplasm and the nucleus. In the dorsal region of the embryo, Dorsal protein is found only in the cytoplasm. By mechanisms discussed in a later section, this gradient in the nuclear concentration of Dorsal, which both activates and represses gene transcription, leads to expression of the *dpp* gene only in the most-dorsal 40 percent of the embryo.

At this early stage of embryonic development, four distinct cell types are found in bands along the dorsoventral axis: the extraembryonic amnioserosa, dorsal ectoderm, ventral neurogenic region (VNR), and mesoderm (Figure 23-4a). These different cell identities can be assessed by examining the pattern of cuticular structures, as well as by the expression of specific marker genes. A combination of both genetic and molecular genetic evidence suggests that Dpp acts as a **morphogen** to induce establishment of different ectodermal cell types at different **threshold concentrations** at this stage in development. For instance, complete removal of Dpp function leads to the loss of all dorsal structures and their conversion into more-ventral ones (Figure 23-4b). Embryos carrying only one wild-type *dpp* allele show an increase in the number of cells assuming a ventral fate, whereas embryos with three copies of *dpp* form more dorsal cells. Microinjection experiments with mutant embryos genetically altered to give rise only to lateral structures (i.e., ventral neurogenic ectoderm destined to give rise to the central nervous system) have demonstrated that different levels of Dpp specify different cell fates along the dorsoventral axis (Figure 23-4c).

Several other *Drosophila* genes that affect dorsoventral patterning may do so by directly modulating Dpp function or by acting in combination with Dpp to specify different fates. For instance, dorsal-most structures are lost from embryos with mutations in the *screw* gene, which encodes another TGFβ homolog. Thus different cell fates along the dorsoventral axis may be specified by different homodimeric and heterodimeric combinations of Dpp and Screw and the receptors through which they act.

Genetic studies suggested that two other secreted proteins encoded by the *tolloid* and *short gastrulation (sog)* genes also play important roles in modulating Dpp activity. As the Dpp story unfolded in *Drosophila*, scientists discovered that the frog TGFβ homologs called **BMP2** and **BMP4** have inductive effects similar to Dpp protein; moreover, BMP activity is modulated by **chordin** (related to Sog) and by Xolloid (related to Tolloid). Before considering the crucial role of these proteins, we review the earliest stages in vertebrate embryogenesis.

Sequential Inductive Events Regulate Early *Xenopus* Development

The large size of the frog embryo and its ease of experimental manipulation make it a favored system for investigating the role of extracellular signaling molecules in regulating early pattern formation in vertebrate embryos. Most recent research, particularly at the molecular level, has concentrated on development of the frog *Xenopus laevis*.

As illustrated in Figure 23-5, the unfertilized *Xenopus* egg has an inherent asymmetry with a pigmented *animal* hemisphere (pole or cap) and an unpigmented, yolk-rich *vegetal* hemisphere. Entry of sperm into the animal pole induces the actin-rich cortical region to rotate about 30 degrees relative to the yolky cytoplasm of the vegetal pole. Cortical rotation results in a second asymmetry that leads to the future

(a) Wild-type embryo

(b) Dpp-deficient embryo

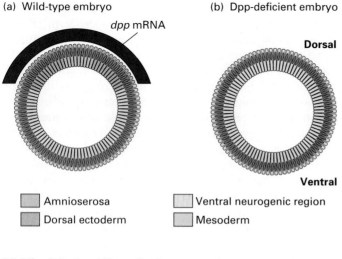

Amnioserosa

Dorsal ectoderm

Ventral neurogenic region

Mesoderm

(c) Microinjection of lateralized mutant embryos

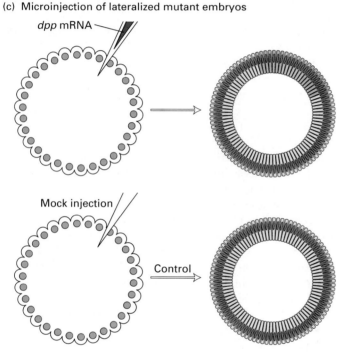

◀ **FIGURE 23-4 Effect of Dpp, a TGFβ-superfamily protein, on dorsoventral patterning in the *Drosophila* embryo.** Dorsal is toward the top; ventral, toward the bottom. (a) The four identifiable cell types present in wild-type fly embryos soon after cellularization are shown schematically in this cross-sectional diagram. The Dpp (maroon) concentration is highest in the most-dorsal region. (b) In mutant embryos lacking Dpp, cells in the dorsal and lateral region of the embryo all assume a ventral neurogenic fate. (c) Microinjection of *dpp* mRNA into mutant embryos in which the entire dorsoventral axis develops into ventral neurogenic region tissue demonstrates that Dpp acts as a morphogen. The area closest to the injection site, where the Dpp level is highest, develops into amnioserosa; as the distance from the injection site increases, more-ventral structures characteristic of dorsal ectoderm are induced. Even farther away, where the Dpp level falls below the threshold, cells differentiate along a ventral neurogenic pathway. In the control mock injection, no dorsal structures develop. [Adapted from J. M. W. Slack, 1993, *Mech. Devel.* **41**:91; see E. L. Ferguson and K. V. Anderson, 1992, *Cell* **71**:451.]

dorsoventral axis with the point of sperm entry defining the ventral side of the embryo. If cortical rotation is prevented— for instance, by UV irradiation—the establishment of dorsoventral polarity is blocked. Fertilization sets off a stereotyped pattern of rapid synchronous cell divisions. As these divisions occur, a cavity, called the *blastocoel*, forms in the animal hemisphere; the developing embryo at this stage is referred to as a **blastula.**

After the first 12 cell divisions, the cell-division rate slows and becomes asynchronous as zygotic transcription begins. The three germ layers become arranged in a simple layered pattern along the animal/vegetal axis by the midblastula stage. The animal pole will give rise to the **ectoderm** (e.g., skin and neural tissue); the vegetal pole, to the **endoderm** (e.g., gut);

and the marginal zone between the two poles, to the **mesoderm** (e.g., notochord, muscle, blood). In formation of the **gastrula,** the next stage, the mesodermal and endodermal cells move inside the embryo. The movement of cells into the embryo is initiated at a structure on the dorsal side of the embryo called the *dorsal lip of the blastopore.* At the completion of gastrulation, virtually the entire surface of the embryo is covered by ectoderm (see Figure 23-5). The endoderm is found towards the center of the embryo, and a layer of mesodermal cells separates the endoderm and ectoderm.

Early *Xenopus* development progresses as a series of inductive events. A popular model for mesodermal induction along the dorsoventral axis invokes three signals (Figure 23-6). In brief, during the blastula stage a signal from the ventral vegetal pole induces ventral mesodermal tissue. A second signal arising in the dorsal vegetal pole (a region called the Nieuwkoop center) induces a region of the overlying marginal zone to form dorsal mesoderm. This region of the dorsal mesoderm, called **Spemann's organizer** (or simply the organizer), specifies the organization of the mesoderm along the dorsoventral axis through the production of a third signal, the horizontal dorsalizing signal.

During gastrulation, the mesoderm induces the overlying ectoderm to form neural tissue and the endoderm to form various endodermal derivatives such as the pharynx and intestine (see Figure 23-5). In addition to specifying cell types, the mesoderm also regionalizes both the ectoderm and endoderm. Anterior regions of the mesoderm (regions of the blastula that fold into the gastrula first) induce anterior neural tissue in the overlying ectoderm, and more-posterior regions induce more posteriorly located neural structures. These inductive interactions have been demonstrated by transplanting explants from one developing embryo into an ectopic location in a recipient embryo. For instance, Spemann and Mangold showed some 75 years ago that transplantation of

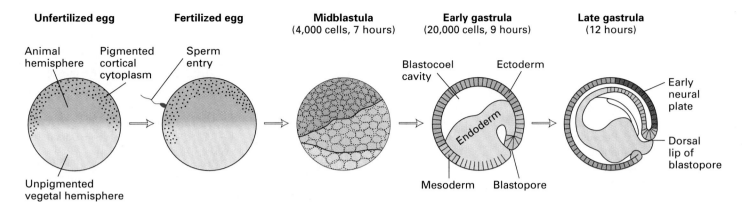

Unfertilized egg — **Fertilized egg** — **Midblastula** (4,000 cells, 7 hours) — **Early gastrula** (20,000 cells, 9 hours) — **Late gastrula** (12 hours)

▲ **FIGURE 23-5 Early embryogenesis of the frog *Xenopus laevis*.** The unfertilized egg is divided into two broad hemispheres, giving it an intrinsic asymmetry. The site of sperm entry defines the ventral side of the embryo and leads to rotation of the cortical cytoplasm. Fertilization leads to rapid cell divisions and formation of the early blastula (not shown), a hollow ball of 32 cells called blastomeres. Signals from the vegetal pole induce the formation of mesodermal cells in the marginal zone (blue) separating the vegetal and animal caps in the midblastula stage. During gastrulation, mesodermal cells fold into the embryo. Signals from the invaginating mesoderm induce the development of both the underlying endoderm and neural tissue from the overlying ectoderm. The anterioposterior axis is determined by the mesoderm; cells that invaginate first induce anterior structures. Subsequent interactions between different cell populations play an important role in organogenesis. [Adapted from E. M. De Robertis et al., *Sci. Am.* **263**(1):46.]

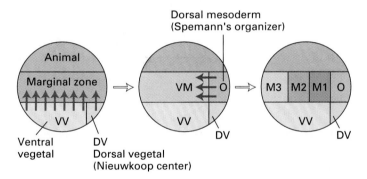

▲ **FIGURE 23-6 Model for induction at blastula stage in *Xenopus* embryogenesis.** Following cleavage, three signals (indicated by red arrows) are proposed to induce mesoderm formation in the marginal zone. Signals from the ventral vegetal (VV) pole induce ventral mesoderm (VM) tissue in the marginal zone. A signal from the dorsal vegetal (DV) pole, called the Nieuwkoop center, induces formation of the most-dorsal mesodermal tissue, called Spemann's organizer (O). A horizontal dorsalizing signal originating in the organizer further regionalizes the mesodermal tissue (M1, M2, M3) with more-dorsal cell fates specified in tissue closer to the organizer. [Adapted from S. C. Gilbert, 1991, *Developmental Biology*, Sinauer.]

the dorsal lip from one embryo into the ventral side of a recipient embryo leads to a twinned embryo (Figure 23-7). At the level of tissue induction, the dorsal lip (Spemann's organizer) has two main activities: it induces neural tissue from ectodermal tissues and it dorsalizes the mesoderm.

Inductive Effect of TGFβ Homologs Is Regulated Post-Translationally

To identify the molecules in the dorsal lip of vertebrate gastrulas that induce twinned embryos, scientists prepared cDNA libraries from isolated dorsal lips and screened them for genes selectively expressed in this region (Chapter 7). The inductive properties of these dorsal-lip specific cDNAs were tested by preparing the corresponding mRNAs in vitro and injecting them directly into cells on the ventral side of a recipient embryo prior to gastrulation. One of the mRNAs, encoding a protein called *chordin*, was found to induce formation of a second embryonic axis and eventually a twinned embryo that looked very similar to twinned embryos produced by transplanting the dorsal lip (Figure 23-8).

To determine whether cells injected with chordin mRNA gave rise to the twin themselves or whether the injected tissue recruited uninjected tissue from the host to form the twin, scientists co-injected chordin mRNA with a marker gene. The marker permits tracing the fate of the injected cells and their progeny. It was clear from these experiments that chordin recruited host tissue to form the twin. Thus, the activity of chordin mimicked the inductive properties of Spemann's organizer.

Inhibition of Signaling by Chordin and Sog The sequence of chordin revealed that it is a secreted protein closely related in structure to the *Drosophila* Sog protein. Genetic studies in flies had shown that *sog* mutant phenotypes are complementary to phenotypes associated with *dpp* mutations. For instance, *dpp* mutants show defects in formation of dorsal ectodermal structures, whereas in *sog* mutants dorsal ectodermal structures expand more ventrally. Experiments

▶ **FIGURE 23-7 Experimental demonstration that the dorsal blastopore lip is an inducer.**
(a) The blastopore lip from an early gastrula of a pigmented newt embryo (*Triturus taeniatus*) was transplanted to the ventral vegetal side of an unpigmented recipient embryo (*Triturus cristatus*). Because of the differences in pigmentation, the donor and recipient tissues could be distinguished visually. (b) The donor tissue induced the recipient tissue to form a secondary embryonic axis containing a notochord, neural tube, somites, and gut. (c) Eventually, a twinned embryo developed. [Adapted from S. C. Gilbert, 1994, *Developmental Biology*, 4th ed., Sinauer.]

(a)

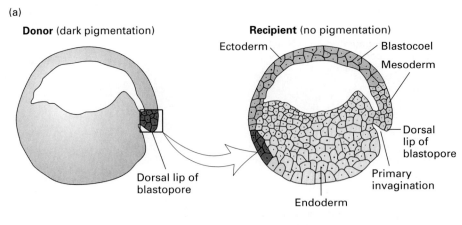

(b)

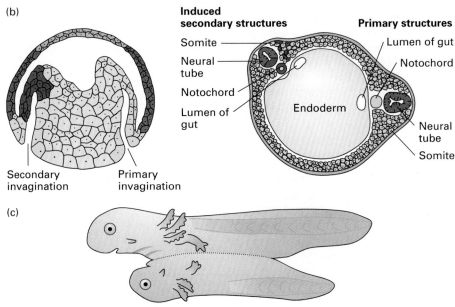

(c)

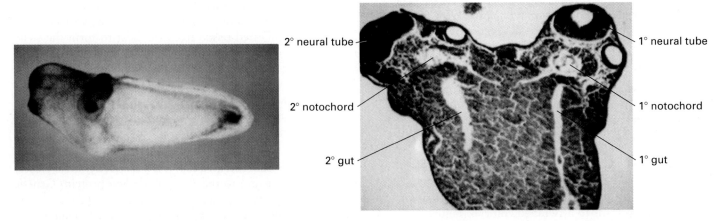

▲ **FIGURE 23-8 Inductive effect of chordin. Injection of chordin mRNA into the ventral vegetal pole of an early *Xenopus* gastrula leads to formation of a secondary (2°) neural tube, notochord, and gut *(right).* Eventually a twinned** embryo *(left)* develops, similar to that produced by transplanting the dorsal lip itself (see Figure 23-7). [See Y. Sasai et al., 1994, *Cell* **79**:779; courtesy of E. M. De Robertis.]

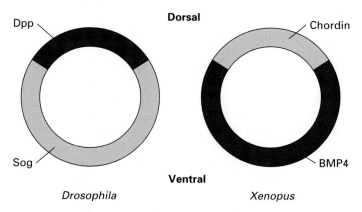

▲ **FIGURE 23-9 The proteins controlling the development of the dorsoventral axis are conserved between vertebrates and invertebrates, but their expression patterns along the dorsoventral axis are reversed.** Dpp and BMP4 (maroon) are TGFβ homologs expressed dorsally and ventrally, respectively, in *Drosophila* and *Xenopus*. Their antagonists (purple) are Sog, expressed ventrally in flies, and chordin, expressed dorsally in frogs.

in frog embryos demonstrated that high-level expression of BMPs (frog homologs of Dpp) prevents formation of dorsal structures, whereas chordin antagonizes this effect. These findings indicate that Sog and chordin inhibit signaling triggered by Dpp and BMPs in *Drosophila* and *Xenopus*, respectively.

In fly embryos Dpp is expressed dorsally and Sog ventrally; in frog embryos BMPs are expressed ventrally and

chordin dorsally (Figure 23-9). This inversion of the expression patterns of these homologous proteins along the dorsoventral axis between vertebrates and invertebrates parallels the inversion of tissue types that form along this axis during embryogenesis. For instance, in vertebrates the nervous system forms dorsally and in invertebrates it forms ventrally. This has led to the view that a common ancestor of the vertebrates and invertebrates used an ancestral Dpp/BMP system to pattern the dorsoventral axis. Some time shortly after the split between invertebrates and vertebrates, the signaling pathways became inverted along this axis in one of the two lineages.

To investigate the mechanism by which BMP signaling is regulated, scientists produced chordin protein and tested its ability to interfere with BMP activity. When explants of animal-cap tissue, which express endogenous BMP4, are incubated in culture, they do not form neural tissue. Addition of chordin protein to these explants induces neural tissue; this neural induction by chordin is reversed by addition of an excess of BMP4 protein. The earlier finding that *Drosophila* embryos lacking both Dpp and Sog were indistinguishable from embryos lacking Dpp suggested that the sole function of Sog is to inhibit Dpp activity. Together these data led to a simple model in which chordin (or Sog) prevents BMP (or Dpp) binding to its receptor. In principle inhibition could occur by direct binding of chordin to BMP receptors or to BMP molecules themselves. A series of biochemical studies demonstrated that chordin binds BMP2 and BMP4 homodimers or BMP4/BMP7 heterodimers with high affinity ($K_D = 3 \times 10^{-10}$ M) and prevents them from binding to their receptors (Figure 23-10a).

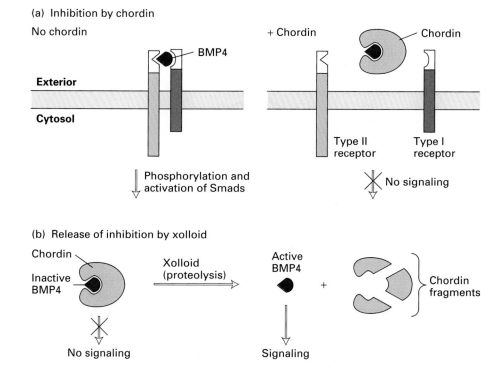

◀ **FIGURE 23-10 Modulation of BMP4 signaling in *Xenopus* by chordin and Xolloid.** (a) Chordin binds BMP4 and prevents it from binding to its receptor. (b) Xolloid specifically cleaves chordin in the chordin/BMP4 complex, releasing BMP4 in a form that can bind to its receptor and trigger signaling. Similar regulation of Dpp signaling in *Drosophila* involves Sog (related to chordin) and Tolloid (related to Xolloid). [See S. Piccolo et al., 1997, Cell **91**:407.]

Promotion of Signaling by Tolloid and Xolloid Specification of the most-dorsal tissues in *Drosophila* requires not only the *dpp* gene but also another gene called *tolloid*, which encodes a metalloprotease and is expressed only in the most-dorsal regions of the embryo. Genetic studies indicated that Tolloid promotes Dpp activity. In principle, Tolloid could act directly by cleaving Dpp, thereby generating a more active form. Alternatively, Tolloid could act indirectly by cleaving and thereby inactivating an inhibitor of Dpp such as Sog.

Biochemical experiments using both the fly enzyme Tolloid and the frog homolog Xolloid demonstrated that these metalloproteases specifically cleave and inactivate Sog and chordin, respectively. Hence, as a consequence of downregulation of an inhibitor, the activity of Dpp/BMPs is potentiated (Figure 23-10b).

A Highly Conserved Pathway Determines Dorsoventral Patterning in Invertebrates and Vertebrates

A large-scale screen for mutations affecting the development of the zebrafish uncovered many mutations affecting early embryogenesis, including loss-of-function mutations in fish homologs of chordin (called chordino), BMP2 (called swirl), and Xolloid (called minifin). By using tissue-specific markers, the effects of the swirl and chordino proteins on cellular patterning can be directly observed in the early gastrula prior to cellular differentiation. In one study three different markers were used: Eve1, normally expressed in the ventral mesoderm; Shh, in the dorsal mesoderm; and Fkd5, in the prospective neural plate (ectoderm). In mutants lacking chordino, expression of Eve1 expanded dorsally and expression of Shh contracted compared with their wild-type expression patterns (Figure 23-11a); in other words, dorsal mesoderm is replaced by ventral mesoderm. Similarly, expression of Fkd5 in the ectoderm contracts, presaging a re-

duction in the prospective neural plate. Complementary changes are seen in mutants lacking swirl, with dorsal mesoderm replacing ventral mesoderm. Thus the expression patterns of swirl and chordino in zebrafish, and the inhibitory effect of chordino, parallel BMP4 and chordin expression and function in the frog (Figure 23-11b). Loss-of-function mutations in *minifin* produce defects similar to loss of BMP2 function in *swirl* mutants. *Minifin* acts in late gastrulation when tissue that will give rise to the tail region, the ventral-most region of the marginal zone, is juxtaposed to the extreme dorsal marginal zone. In the absence of Minifin (Xolloid), the very high levels of Chordino (Sog) in dorsal marginal zone tissue inhibit Swirl (BMP2), signaling in the ventral region. In wild type, Minifin antagonizes Chordino at this late stage, thereby facilitating the ventralizing function of Swirl.

The combination of genetic and biochemical data in the fruit fly, frog, and zebrafish that we have considered indicate that a highly conserved pathway controls cell fates along the dorsoventral axis in vertebrate and invertebrate embryos. We can summarize this pathway as follows:

	Protease —	Inhibitor —	Signal
Fruit fly:	Tolloid	Sog	Dpp
Frog:	Xolloid	Chordin	BMP2/BMP4
Zebrafish:	Minifin	Chordino	Swirl

As noted earlier, the vertebrate TGFβ homologs (e.g., BMP2/BMP4) are expressed ventrally and chordin or its homologs are expressed dorsally, whereas the opposite expression pattern is found in invertebrates (see Figure 23-9). Likewise, the expression patterns of the homologous proteases that inactivate chordin and Sog are reversed, permitting maximal BMP signaling in the ventral-most regions of vertebrate embryos and maximal Dpp signaling in the dorsal-most regions of invertebrate embryos.

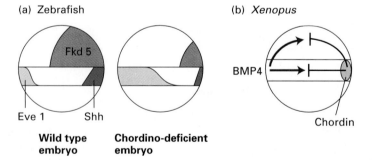

(a) Zebrafish

Fkd 5

Eve 1   Shh

Wild type embryo   **Chordino-deficient embryo**

(b) *Xenopus*

BMP4

Chordin

▲ **FIGURE 23-11 (a) Role of chordino in dorsoventral patterning in zebrafish embryos.** Shown here are schematic depictions of the expression patterns of three marker genes at the early gastrula stage in wild-type and chordino-deficient embryos. Ventral is toward the left; dorsal, toward the right. Eve1 is a marker for ventral mesoderm; Fkd5 and Shh are markers for the prospective neural plate and the dorsal mesoderm (including the blastopore lip), respectively. Ventralization occurs in mutants lacking chordino, which normally inhibits the activity of swirl in the dorsal region. (b) Expression and function of BMP4 and chordin in frogs. These proteins are functionally analogous to swirl and chordino in zebrafish. Ventrally expressed BMP4 diffuses dorsally where its activity is inhibited by chordin expressed in the dorsal mesoderm. [Adapted from M. Hammerschmidt et al., 1996, *Development* **123**:95.]

SUMMARY Dorsoventral Patterning by TGFβ-Superfamily Proteins

- Secreted growth factors in the TGFβ superfamily play widespread roles in development of both vertebrates and invertebrates.

- Dpp protein in *Drosophila* and BMP2/BMP4 in vertebrates are TGFβ homologs that function as morphogens in determining cell-type specification along the dorsoventral axis in early embryos.

- Binding of the active dimeric TGFβ proteins to their receptors activates the serine/threonine kinase activity in the receptor cytosolic domain. The activated receptor phosphorylates a receptor-regulated Smad, which associates with a co-Smad. This Smad complex translocates to the nucleus where it stimulates transcription of target genes (see Figure 23-3).

- Early in vertebrate embryogenesis, the three germ layers—ectoderm, mesoderm, and endoderm—are established (see Figure 23-5). During the subsequent gastrula stage, the mesoderm induces the overlying ectoderm to form neural tissue and the endoderm to form tissues such as the pharynx and intestine.

- The activity of TGFβ proteins is inhibited by Sog in *Drosophila* and chordin in vertebrates; these inhibitors directly bind to Dpp or BMP2/BMP4, preventing their interaction with their receptors (see Figure 23-10). Proteolytic cleavage of Sog (chordin) by a metalloprotease encoded by the *tolloid (xolloid)* locus downregulates the inhibitor activity and thus promotes TGFβ signaling.

- By regulating the activity of TGFβ proteins, signals originating in the dorsal lip of the blastopore (e.g., chordin) induce neural tissue in ectoderm and pattern mesodermal tissue along the dorsoventral axis.

23.2 Tissue Patterning by Hedgehog and Wingless

Genetic screens for mutations in *Drosophila* disrupting embryonic pattern formation led to the identification of a panel of genes involved in early development (Chapters 8 and 14). Although many of these genes, particularly those utilized in early embryogenesis prior to cellularization, encode transcription factors, several encode membrane and secreted proteins that act later in development during the process of establishing cellular identity within parasegments. The *hedgehog (hh)* gene, for instance, encodes a novel membrane-linked inductive ligand that plays an important role in the local patterning of many tissues in both vertebrates and invertebrates. We first discuss the mechanism by which the **Hedgehog** ligand is produced and the mechanism of signal transduction in responsive

cells. We then consider the role of Hedgehog in several developmental paradigms, as well as the Wingless pathway.

Modification of Secreted Hedgehog Precursor Yields a Cell-Tethered Inductive Signal

The primary Hedgehog (Hh) translation product includes a signal peptide, which is cleaved to yield a 45-kDa precursor protein. Cleavage of this secreted precursor produces a 20-kDa N-terminal fragment, which is associated with the plasma membrane and contains the inductive activity, and a 25-kDa C-terminal fragment. A series of elegant experiments demonstrated how the N-terminal Hh fragment, which does not contain any hydrophobic sequences, becomes tethered to the membrane. As shown in Figure 23-12, the thiol side

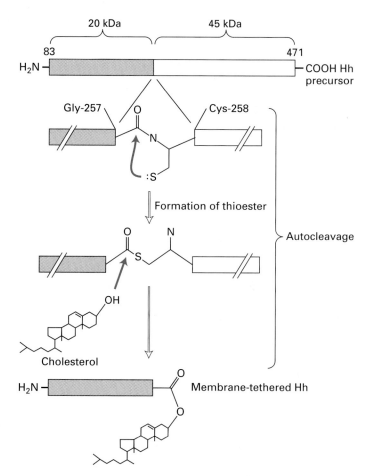

▲ **FIGURE 23-12 Processing of Hedgehog (Hh) protein.** Removal of the N-terminal signal peptide from the initial translation product yields the 45-kDa Hh precursor consisting of residues 83–471 in the original protein. Nucleophilic attack by the thiol side chain of cysteine 258 (Cys-258) on the carbonyl carbon of glycine 257 (Gly-257) forms a thioester intermediate. The C-terminal domain then catalyzes formation of an ester bond between the β-3 hydroxyl of cholesterol and glycine 257, cleaving the precursor into two fragments. The covalently attached cholesterol moiety tethers the N-terminal signaling fragment to the membrane. [Adapted from J. A. Porter et al., 1996, *Science* **274**:255.]

chain of the cysteine at position 258 attacks the carbonyl carbon of the glycine at position 257, forming an internal thioester linkage between Gly-257 and Cys-258. The C-terminal domain of the protein then facilitates the addition of cholesterol to Gly-257, splitting the molecule into two fragments. The N-terminal signaling fragment is then tethered to the membrane via its attached hydrophobic cholesterol moiety. Sequence analysis has identified a handful of other proteins with similar C-terminal domains that may function to promote membrane linkage of these proteins by the same autoproteolytic mechanism.

Is the cholesterol linkage necessary for the biological activity of Hh? This question has been addressed both in vivo and in vitro. An N-terminal Hh fragment lacking cholesterol, produced in E. coli using recombinant DNA techniques, is sufficient to induce specific neuronal cell fates in the vertebrate spinal cord. Studies in Drosophila using hh transgenes encoding the full-length molecule or an untethered N-terminal domain also indicate that the cholesterol linkage is not required for inductive activity. In wild-type flies, Hh is expressed in a single cell in each parasegment in the developing embryo and induces the expression of another patterning gene called wingless (wg) only in its immediate anterior neighboring cell. When normal membrane-tethered Hh was overexpressed within this cell using a cell-specific promoter, expression of Wg protein was still restricted to the immediate neighboring anterior cell, though it was expressed at increased levels. In contrast, expression of the untethered N-terminal Hh fragment in the same cell led to a high level of Wg expression in cells located more than a cell diameter away.

These studies show that although the cholesterol linkage is not necessary for the inductive activity of Hh, it functions to spatially restrict the Hh signal in the developing embryo. The spatial restriction of inductive signals plays a crucial role in patterning both vertebrate and invertebrate tissues.

Binding of Hedgehog to the Patch Receptor Relieves Inhibition of Smo

Genetic studies in Drosophila identified a gene, called smo, that acts in the Hh signaling pathway and encodes a transmembrane protein with seven membrane-spanning domains similar to G protein–coupled receptors (see Figure 20-10). The observation that the smo and hh mutant phenotypes in the fly embryo are very similar led to the hypothesis that Hh binds to Smo. Although biochemical studies have not provided evidence for a direct interaction between these two proteins, studies in tissue culture have shown that Hh can bind to cells overexpressing another membrane protein called Patched (Ptc). In in vitro cell culture systems, Hh does not bind to Smo directly, but it can form a complex with Smo and Ptc. These findings and genetic data suggest a model in which binding of Hh produced in one cell to Ptc on a neighboring cell relieves Ptc-mediated inhibition of Smo.

The pathway that transduces the Hh signal through the cytoplasm to transcription factors in a receiving cell is still poorly understood. Genetic studies, however, indicate that Hh signaling involves a complex set of proteins including Fused (Fu), a serine-threonine kinase; Costal-2 (Cos-2), a microtubule-associated protein; and Cubitis interruptis (Ci), a transcription factor. In the absence of Hh, when Ptc inhibits Smo, these three proteins form a complex that binds to microtubules in the cytoplasm, and Ci is cleaved, generating a fragment that translocates to the nucleus and inhibits target-gene expression (Figure 23-13a). In the presence of Hh, which relieves the inhibition of Smo, the complex of Fu, Cos-2, and Ci is not associated with microtubules, and an alternatively cleaved form of Ci is generated. After translocating to the nucleus, this Ci fragment binds to the transcriptional co-activator CBP (Chapters 10 and 20), promoting expression of target genes (Figure 23-13b).

One of the target genes up-regulated by Hh signaling is ptc. Thus not only is Ptc a component in the Hh signaling pathway, its expression also is controlled by Hh. As a consequence, the activity of Hh can be monitored in developing vertebrate and invertebrate embryos by following the pattern of Ptc expression. Ptc is expressed in cells close to a source of Hh, whereas it is not expressed in cells farther away.

Hedgehog Organizes Pattern in the Chick Limb and Drosophila Wing

Embryonic patterning in the vertebrate limb occurs along three axes: proximodistal (shoulders to finger tips), anteroposterior (thumb to little finger), and dorsoventral (upper surface of the hand to the palm). During early embryogenesis, mesodermal cells migrate from the lateral plate to specific regions beneath the epidermis causing a local bulge in the tissue called the limb bud. Mesodermal cells at the posterior edge of the limb bud form a region called the zone of polarizing activity (ZPA), which acts as an organizer, patterning cells along the anteroposterior axis (Figure 23-14a). This organizer activity was demonstrated by transplanting the ZPA from one limb bud to the anterior edge of a recipient limb bud, thus producing a bud with two ZPAs. Such transplantation resulted in formation of a limb with a mirror symmetric duplication of the digits (Figure 23-14b).

Until recently the molecular basis of patterning in the vertebrate limb was poorly understood. Reasoning that other secreted inductive proteins found in Drosophila (e.g., Dpp) are homologous to proteins in vertebrates with inductive capacities, investigators isolated vertebrate homologs of Drosophila Hedgehog. A chick homolog of hedgehog, called sonic hedgehog (shh), was found to be expressed in the ZPA. Insertion of chick fibroblasts infected with a retrovirus carrying a recombinant shh gene into the anterior region of a limb bud led to a mirror-image duplication of the limb (Figure 23-14c). Subsequent observations that shh

(a) No Hh

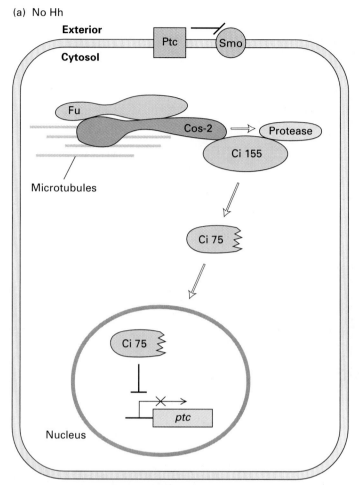

(b) Hh present

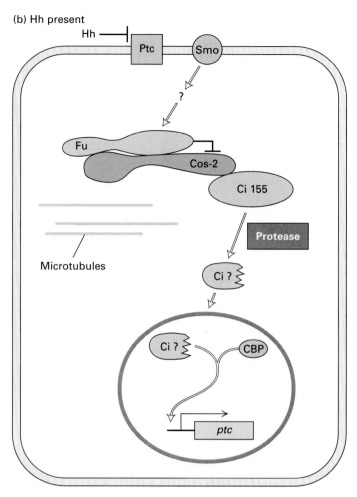

▲ **FIGURE 23-13 A model of the Hedgehog (Hh) signaling pathway.** (a) In the absence of Hh, the Ptc protein inhibits Smo at the cell surface. A complex containing the Fu, Cos-2, and Ci proteins binds to microtubules. An as-yet unidentified protease cleaves Ci generating Ci75, a transcriptional repressor. (b) In the presence of Hh, Ptc inhibition of Smo is relieved. The Fu, Cos-2, and Ci proteins are not associated with microtubules. Ci is cleaved by another protease, yielding a fragment that acts in conjunction with CBP to promote transcription. In the presence of Hh, transcription of *ptc* is markedly up-regulated. [Adapted from L. V. Goodrich and M. P. Scott, 1998, *Neuron* **21**:1243.]

knockout mice have striking defects in limb development, including defects in digit formation along the anteroposterior axis, confirmed the role of Sonic Hedgehog (Shh) in vertebrate limb patterning.

Recent studies have shown that limb growth along the proximodistal axis (e.g., from the tips of the fingers toward the body) is controlled by fibroblast growth factor (FGF) released from the apical epidermal ridge (see Figure 23-14a). This specialized signaling center within the epidermis of the limb bud is formed from mesodermal cells early in embryogenesis. Patterning along the dorsoventral axis is con-trolled by another secreted protein, Wnt7a, which is related to *Drosophila* Wingless (discussed below).

In *Drosophila*, Hedgehog not only functions in establishing segment polarity early in embryogenesis but also is critical for patterning adult appendages. The wings and legs of *Drosophila* are derived from epithelia called imaginal discs, which are transformed into appendages during metamorphosis. Since *hedgehog* is an essential gene for early embryogenesis, its role in later development has been studied using temperature-sensitive mutations and by ectopically expressing the gene during development. In wild-type fly

(a) Wild-type limb development

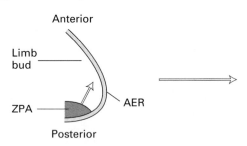

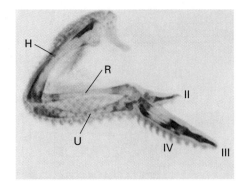

(b) ZPA transplant

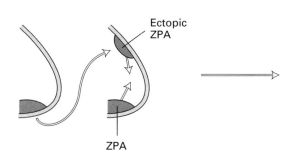

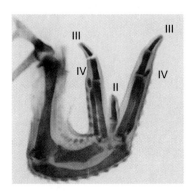

(c) *Sonic hedgehog* implant

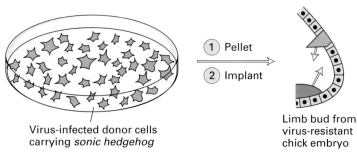

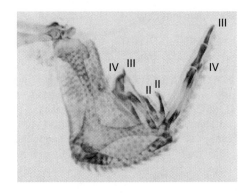

▲ **FIGURE 23-14 Patterning of limbs along the anteroposterior axis in the chick.** Diagrams of the embryonic limb bud are to the left; photographs of the adult structure are to the right. (a) Normal limb development is induced by the zone of polarizing activity (ZPA), a group of mesodermal cells at the posterior edge of the limb bud. The apical epidermal ridge (AER) is another signaling center that functions in proximodistal patterning. (b) Transplantation of the ZPA to the anterior region of a normal limb bud results in limb duplication. Considerable variation in the nature of the duplication is observed among limb-bud preparations. In this case, digit II was not duplicated. (c) When donor cells infected with a virus carrying the *sonic hedgehog* gene are implanted in a limb bud from a virus-resistant chick embryo, limb duplication occurs. These results indicate that Sonic Hedgehog, normally expressed in the ZPA, controls limb patterning in the chick. [Adapted from R. D. Riddle et al., 1993, *Cell* **75**:1401; photographs courtesy of C. Tabin.]

larvae, Hedgehog is expressed in the posterior compartment of the wing disc. Loss-of-function mutations in the *hedgehog* gene cause a dramatic decrease in the size of both the anterior and posterior portions of the wing and severe pattern abnormalities.

Grafting experiments similar to those in chick embryos cannot be performed in *Drosophila*. However, specific genes can be ectopically expressed by use of a site-specific yeast recombination system similar to the phage P1 loxP/Cre system described in other chapters (see Figures 8-35 and 12-37). The *hedgehog* interruption cassette construct, diagrammed in Figure 23-15a, is introduced into *Drosophila* along with another construct containing a heat shock–inducible promoter linked to the recombinase gene. When fly embryos

(a)

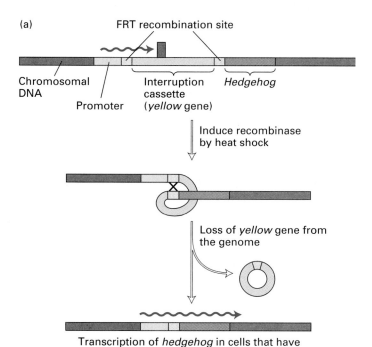

◀ **FIGURE 23-15 Effect of ectopic expression of Hedgehog on *Drosophila* wing development.** (a) A yeast interruption cassette construct can be used to express Hedgehog ectopically. This construct is introduced into *Drosophila* along with one containing a heat shock–inducible promoter linked to the recombinase gene. Expression of the *hedgehog* gene (orange) is disrupted by the intervening *yellow* gene (blue). Brief exposure to heat shock induces expression of the recombinase enzyme, which catalyzes recombination at the FRT sites, leading to excision of the *yellow* gene and fusion of the constitutive promoter to the *hedgehog* gene. (b) In the developing wing of a normal fly, Hedgehog is expressed in the posterior (P) compartment (orange). Expression of Hedgehog in a localized sector in the anterior (A) compartment during development leads to mirror image duplication of anterior structures. This effect is similar to that resulting from ectopic expression of Sonic Hedgehog in the anterior region of the chick limb bud (see Figure 23-14c). [See K. Basler and G. Struhl, 1994, *Nature* **368**:208.]

(b)

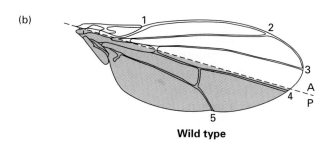

Wild type

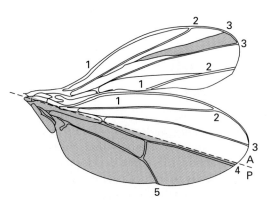

Ectopic Hedgehog

carrying these two constructs are exposed to heat shock early in development, recombination occurs randomly in a few cells, which then begin to express Hedgehog. If recombination occurs in cells in the anterior compartment of the wing disc, their proliferation leads to ectopic clones of Hedgehog-expressing cells. In the adult wing, loss of the *yellow* gene is easily detected by the lighter color of the tissue. These light patches are derived from cells that underwent recombination and thus expressed Hedgehog constitutively earlier in development. Ectopic expression of Hedgehog in the anterior compartment of the wing disc, where it normally is not expressed, leads to pattern duplication in *Drosophila* wings (Figure 23-15b). This effect is qualitatively similar to that caused by ectopic expression of Sonic Hedgehog in the chick limb.

Additional genetic experiments suggest that in some tissues Hedgehog does not directly pattern tissue but rather induces a secondary signal. In the fly wing, for instance, Hedgehog induces a narrow strip of Dpp expression at the boundary between the anterior and posterior compartments of the wing disc. Dpp, in turn, acts as a graded signal to pattern cells in both compartments.

Hedgehog Induces Wingless, Which Triggers a Highly Conserved Signaling Pathway

Another secreted patterning signal induced by Hedgehog is Wingless (Wg). As we described in Chapter 14, the interplay between transcription factors divides the early fly embryo into 14 parasegments, each composed of several bands of cells (see Figure 14-30). At the boundary between parasegments, cellular interactions establish and maintain a signaling center. Hedgehog, produced by the anterior-most cell band of each parasegment, specifically induces the expression of Wg in the cell band immediately anterior to it in the neighboring parasegment (Figure 23-16). Wg, in turn, feeds back on the Hh-expressing cells inducing Hh expression. Graded signals from this boundary play a critical role in patterning

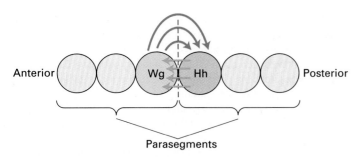

▲ **FIGURE 23-16 Hedgehog (Hh) acts with Wingless (Wg), a diffusible signaling molecule, to maintain a signal center at the parasegment boundary in the *Drosophila* embryo.** Hh is necessary to maintain Wg expression, and conversely Wg is required to maintain Hh. Signals from this center play a key role in patterning the epidermis. [See M. Hammerschmidt et al., 1997, *Trends Genetics* **13**:14.]

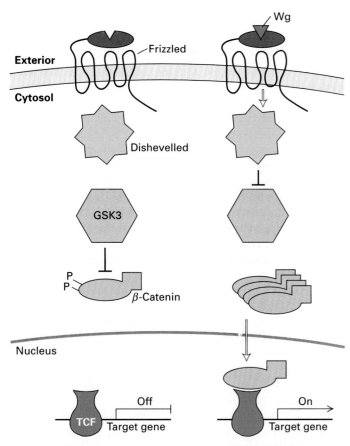

▲ **FIGURE 23-17 The Wingless (Wg) signaling pathway.** In the absence of Wg (or Wnt in vertebrates), the kinase GSK3 constitutively phosphorylates β-catenin. Phosphorylated β-catenin is degraded and, hence, does not accumulate in cells. Binding of Wg to its receptor Frizzled recruits Dishevelled to the membrane. Dishevelled, in turn, inhibits GSK3. As a result, unphosphorylated β-catenin accumulates in the cytosol, translocates to the nucleus, binds via its C-terminus to the transcription factor TCF, and activates transcription of target genes. [Adapted from T. C. Dale, 1998, *Biochem. J.* **329**:209.]

each parasegment. Like Hh and the TFGβ-superfamily proteins, Wg and its vertebrate counterparts, the *Wnts*, play crucial roles in regulating patterning in many developmental contexts. Defects in this pathway are associated with specific forms of human cancer.

Genetic studies in flies have led to the identification of several proteins that function in the Wg pathway (Figure 23-17). Wg binds to a receptor called Frizzled, which has seven membrane-spanning domains. Although the topology of Frizzled is similar to that of G protein–coupled receptors, it signals through a completely different pathway. Binding of Wg to Frizzled activates Dishevelled. Dishevelled, in turn, inhibits glutamine synthase kinase 3 (GSK3), which originally was named for its role in regulating intermediary metabolism. In unstimulated cells, GSK3 phosphorylates β-catenin; because phosphorylated β-catenin is rapidly degraded, it does not accumulate. In stimulated cells, however, the inhibition of GSK3 by Dishevelled permits accumulation of β-catenin in the cytosol. After translocating to the nucleus, β-catenin forms a complex with a DNA-binding protein called TCF, thereby activating transcription.

SUMMARY Tissue Patterning by Hedgehog and Wingless

- Hedgehog is an inductive signal that is tethered to the cell membrane by a cholesterol moiety (see Figure 23-12).

- Hedgehog on one cell appears to act through two membrane proteins—Ptc and Smo—on a receiving cell, triggering a novel signaling pathway that regulates the activity of Cubitis interruptis, a transcription factor (see Figure 23-13).

- In the early *Drosophila* embryo, Hedgehog and Wingless, a secreted protein, form a signaling center at the parasegment boundaries.

- Patterning of many different tissues is induced by graded expression of Hedgehog and Wingless in invertebrates and their homologs Sonic Hedgehog and Wnt proteins in vertebrates.

23.3 Molecular Mechanisms of Responses to Morphogens

In the previous sections, we described in some detail two key families of signaling molecules—TGFβ proteins (e.g., Dpp and BMPs) and Hedgehog—that pattern embryos through cell interactions. In some developmental contexts, these signals act locally to induce neighboring cells; in other contexts, they function as morphogens to induce different cell fates at different concentrations. In this section, we consider three aspects of such graded signals. Unfortunately, our examples must come from different systems, since as yet we do not have a complete picture of how morphogens function in any one system.

Hedgehog Gradient Elicits Different Cell Fates in the Vertebrate Neural Tube

Hedgehog (Hh) controls the development of four cell types in the chick ventral neural tube. These cells are found at different positions along the dorsoventral axis in the following order from ventral to dorsal: floor-plate cells, motoneurons, V2 interneurons, and V1 interneurons. During development Hh is initially expressed at high levels in the notochord, a mesodermal structure in direct contact with the ventral-most region of the neural tube (Figure 23-18). Upon induction, floor-plate cells also express Hh, forming a Hh-signaling center in the ventral-most region of the neural tube. Antibodies to Hh protein block formation of the different ventral neural-tube cells in the chick, and these cells fail to form in mice homozygous for mutations in the *sonic hedgehog* gene.

Are different ventral neural-tube cells induced by different concentrations of Hh or does Hh induce more-dorsal secondary signals that, in turn, induce different cell types? To address this question, scientists added different concentrations of Hh to chick neural-tube explants. In the absence

of Hh, no ventral cells formed. In the presence of very high concentrations of Hh, floor-plate cells formed, whereas at a slightly lower concentration, motoneurons formed. When the level of Hh was decreased another twofold, only V2 neurons formed. And finally, only V1 neurons developed when the Hh concentration was decreased another twofold. These data strongly suggest that in the developing neural tube different cell types are formed in response to a ventral → dorsal gradient of Hh.

Cells Can Detect the Number of Ligand-Occupied Receptors

Studies on mesodermal induction in *Xenopus* have provided insights into how cells determine the concentration of an inductive ligand. Different concentrations of activin, a member of the TGFβ superfamily, induce various mesodermal cell fates, as determined by the transcriptional activation of specific marker genes. At low concentration, activin induces expression of the *Xenopus brachyury* (*Xbra*) gene throughout the early mesoderm, whereas at higher concentrations, activin induces expression of the *Xenopus goosecoid* (*Xgsc*) gene in Spemann's organizer (see Figure 23-6). Although activin is a good candidate for being the natural inducer in early frog embryogenesis, this has not yet been proven.

Using ^{35}S-labeled activin, scientists demonstrated that *Xenopus* blastula cells each express some 5000 type II receptors. Additional experiments showed that receptor binding exhibits two different thresholds. Maximal Xbra expression was achieved when about 100 receptors were occupied. At a concentration at which 300 receptors were occupied, cells preferentially switched to express higher levels of Xgsc and lower levels of Xbra. These data indicate that cells respond to signals when only a small number of the receptors are occupied.

To assess whether cells respond to the absolute level of occupied receptors or the proportion of receptors occupied, scientists assessed induction in blastula cells expressing sevenfold higher levels of the type II receptor. As with the endogenous receptors, Xbra and Xgsc were induced when 100 and 300 receptors, respectively, were bound by ligand. Hence, cells measure the absolute number of receptors bound to ligand rather than the ratio of bound to unbound receptors. Although differences in activin-receptor occupancy can induce different cellular responses, it remains unclear how receptor occupancy is translated into transcription of different target genes. We turn to yet a different inductive signal to provide clues.

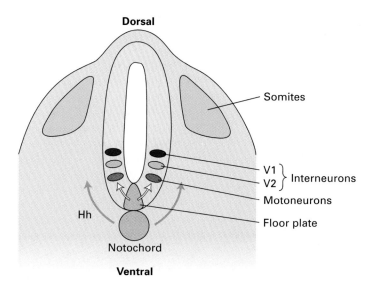

Dorsal

Somites

V1
V2 } Interneurons

Motoneurons

Floor plate

Hh

Notochord

Ventral

▲ **FIGURE 23-18 Different concentrations of Hh induce different neuronal cell types in the ventral spinal cord.** Hh expression in the notochord induces floor-plate development. The floor plate, in turn, expresses Hh (orange). Cells close to the ventral region experience high concentrations of Hh, while those farther away experience lower levels. [See J. Ericson et al., 1997, *CSHSQB* **LXII**:451.]

Target Genes That Respond Differentially to Morphogens Have Different Control Regions

Early in *Drosophila* embryogenesis, graded activation of the Toll receptor along the dorsoventral axis promotes the graded nuclear localization of Dorsal, a transcription factor. Activation of Toll by its ligand, Spaetzle, triggers a signaling pathway that leads to degradation of Cactus, a

▶ **FIGURE 23-19 Toll-mediated nuclear localization of Dorsal.** In the absence of Toll activation, Dorsal is retained in the cytosol due to binding of Cactus. Activation of Toll by binding of Spaetzle leads to degradation of Cactus, permitting translocation of free Dorsal to the nucleus where it promotes transcription of various target genes. A similar mechanism for controlling the activity of the NFκB transcription factor occurs in mammalian lymphocytes (see Figure 20-49). [See M. P. Belvin and K. V. Anderson, 1996, *Ann. Rev. Cell Dev. Biol.* **12**:393.]

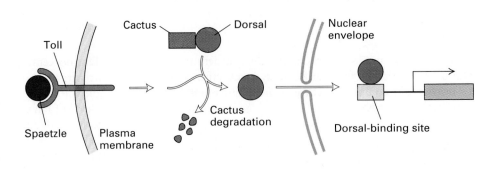

(a) Ventral cell: High Dorsal

Gene expression

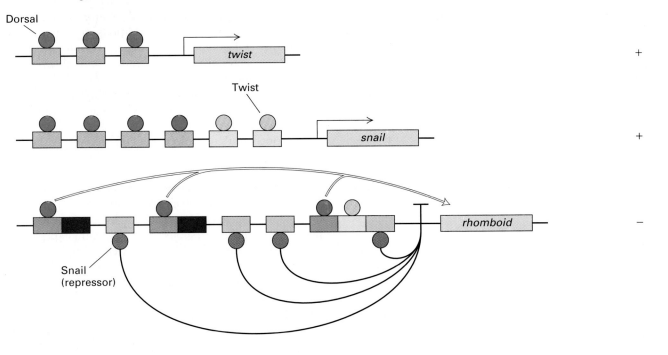

(b) Lateral cell: Low Dorsal

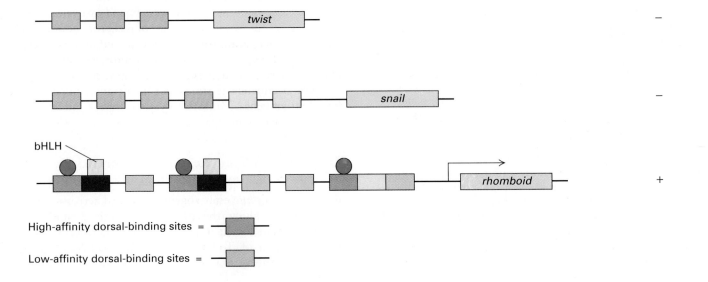

cytosolic inhibitor of Dorsal (Figure 23-19). The free Dorsal then can translocate to the nucleus. Dorsal, in turn, controls the expression pattern of specific target genes by acting through distinct high- and low-affinity binding sites and by interacting in a combinatorial fashion with other transcription factors. Dorsal represses transcription of three target genes—*tolloid*, *sog*, and *zerknult*—in the ventral half of the embryo. Conversely, it activates transcription of *twist*, *snail*, *single minded*, and *rhomboid* ventrally. Each of these genes contains a unique combination of cis-acting regulatory sequences to which both Dorsal and other transcription factors bind.

Figure 23-20 illustrates how the Dorsal gradient specifies different target-gene expression patterns. For instance, *twist*, which is expressed most ventrally, contains three low-affinity Dorsal-binding sites. As the concentration of Dorsal falls more dorsally, it falls beneath the threshold necessary to activate transcription of *twist*, whose encoded protein promotes expression of the *rhomboid* and *snail* genes. The *rhomboid* gene, which is expressed only in lateral regions, is controlled through a complex cis-acting regulatory region that contains three high-affinity Dorsal-binding sites. Two of these sites are adjacent to E-box sequences to which bHLH proteins expressed within this region bind (Section 14.2); the third is adjacent to a Twist-binding site. These proteins act cooperatively to induce expression of Rhomboid protein more dorsally. The *rhomboid* control region also contains four binding sites for Snail, a transcriptional repressor. Expression of Snail is induced only at high concentrations of Dorsal because the *snail* gene contains only low-affinity Dorsal-binding sites within its control region. Since Snail is localized ventrally, its repressor activity defines the sharp ventral boundary of *rhomboid* expression.

◀ **FIGURE 23-20 Activation of different target genes by Dorsal subsequent to Toll signaling.** The gradient of the active Toll ligand Spaetzle, high ventrally and low dorsally, leads to the graded nuclear localization of Dorsal (see Figure 23-19). The resulting graded concentration of nuclear-localized Dorsal, high ventrally and low dorsally, can lead to different patterns of gene expression. Shown here are three target genes that have either high-affinity or low-affinity Dorsal-binding sites. (a) In ventral regions where the concentration of Dorsal (dark green) is high, it can bind to low-affinity sites (light orange) in *twist* and *snail*, activating transcription of these genes. Twist (light green) also activates transcription of *snail*, which encodes a repressor (brown) that prevents transcription of *rhomboid* in this region. (b) In lateral regions, the Dorsal concentration is not high enough for binding of Dorsal to the low-affinity sites regulating *twist* and *snail*. Binding of Dorsal to *rhomboid* is facilitated by the presence of high-affinity sites and synergistic binding of bHLH heterodimeric activators to neighboring sites. [See A. M. Huang et al., 1997, *Genes & Devel.* **11**:1963.]

SUMMARY **Molecular Mechanisms of Responses to Morphogens**

- Morphogens are inductive signals that lead to qualitatively different developmental responses depending on their concentrations.

- Experiments with Hedgehog have directly demonstrated that it induces different ventral neuronal cell fates at different concentrations in the vertebrate spinal cord. These results confirm the importance of morphogens in development.

- The number of receptors occupied by activin, a graded TGFβ-like signal in *Xenopus*, determines which genes are preferentially induced in mesodermal cells in vitro. The cells appear to "count" the number of occupied receptors to determine the ligand concentration.

- As shown in the Toll/Dorsal system, a graded ligand can induce different patterns of gene expression in different cells as a consequence of differences in the number and type of binding sites in the promoter regions of different target genes and of the presence or absence of other transcription factors (see Figure 23-30).

23.4 Reciprocal and Lateral Inductive Interactions

So far we have discussed *unidirectional* inductive interactions; that is, the effect of signals produced by one type of cell or population of cells (the signaling cells) on the development of different neighboring cells (the responding cells). Often, however, communication between cells is a dialogue, a two-way conversation between cells. Many internal organs, for instance, develop through such **reciprocal inductive interactions** between different tissues. Cells in one tissue express a set of inductive signals to which cells in the other tissue respond. Conversely, the responding cells also produce a distinct set of signals to which the signaling cells respond. Hence, both populations send and receive signals. Examples of reciprocal inductive interactions have been known for many years, but only within the last few years have researchers begun to identify some of the signals and receptors controlling such interactions.

In contrast to unidirectional and reciprocal interactions, which occur between different cell populations, *lateral* interactions occur between developmentally equivalent cells. In this case, cells that are initially the same assume different fates as a consequence of the interactions between them. In this section, we first consider the role of reciprocal interactions in the development of the kidney and mammalian blood vessels and then discuss a highly conserved signaling pathway that mediates lateral interactions in both invertebrates and vertebrates.

Reciprocal Epithelial-Mesenchymal Interactions Regulate Kidney Development

An epithelium is a continuous sheet of polarized cells whose apical and basal regions are separated by tight junctions (see Figures 15-23 and 15-28). In contrast, **mesenchyme** comprises loosely associated nonpolarized cells. Epithelial cells in different organs are derived from one of the three germ layers: the ectoderm, mesoderm, or endoderm. Mesenchyme is derived from either the mesoderm or the ectoderm. Formation of internal organs such as the kidney, gut, pancreas, and lung is regulated by interactions between epithelial and mesenchymal cells. Such **epithelial-mesenchymal interactions** typically involve a series of reciprocal inductive events. As an example of such phenomena, we discuss development of the kidney.

The embryonic epithelium, called the *ureteric bud,* and mesenchyme that give rise to the kidney are both mesodermal derivatives (Figure 23-21). As embryonic development proceeds, the ureteric bud grows and branches, eventually differentiating into the collecting ducts of the kidney. The mesenchyme is transformed into a distinct epithelium that will form the proximal and distal tubules and the glomeruli. Several types of experiments have demonstrated reciprocal induction in the developing kidney. For example, if the developing kidney is removed from an 11- to 12-day-old mouse and placed in culture, the ureteric bud continues to grow and branch. Branching ceases, however, when the mesenchyme is separated from the epithelium by mild trypsin treatment; addition of kidney mesenchyme to the trypsin-treated culture restores branching and differentiation. Other in vitro and in vivo experiments involving disruption of kidney development have demonstrated that the epithelium is necessary for conversion of the mesenchyme into tubules. Interestingly, although kidney epithelium shows a specific requirement for kidney mesenchyme for induction of branching, kidney mesenchyme can be induced to form tubules by many different tissues. In other developing organs, however, the signal regulating epithelial development can be derived from other mesenchymal derivatives. The molecular basis for these differences is not known.

A rather large set of signaling molecules are necessary for normal kidney morphogenesis. These include the diffusible

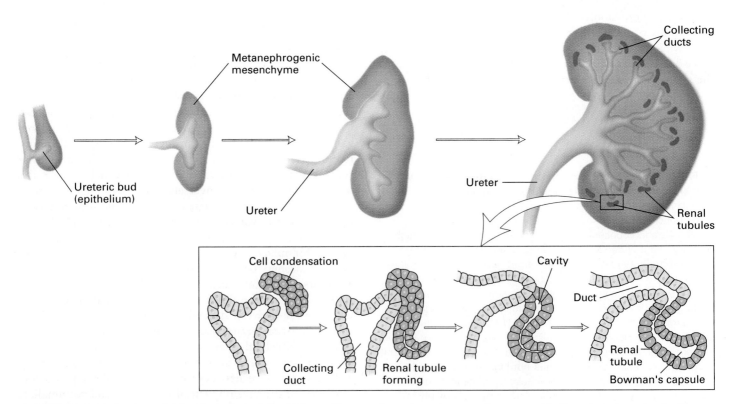

▲ **FIGURE 23-21 Embryonic development of the kidney.** The ureteric-bud epithelium is induced to branch by interactions with the metanephrogenic mesenchyme. A reciprocal induction drives the mesenchyme to form an epithelium that gives rise to the renal tubules.

BMPs and Wnts, discussed in previous sections, and **integrins**, which are attached to the cell surface (Section 22.3). Some signals are expressed in the epithelium and others are expressed in the mesenchyme. Recent genetic and biochemical studies have shown that signaling mediated by **Ret protein**, a receptor tyrosine kinase, which is expressed on the surface of the ureteric bud but not in the mesenchyme of the developing kidney, plays a critical role in kidney development.

Newborn mice carrying homozygous knockout mutations in the *ret* gene exhibit a variety of phenotypes, all involving highly abnormal rudimentary kidneys and in some cases blind-ended ureters (Figure 23-22). Developmental studies have shown that the ureteric bud often fails to form during development in *ret* knockout mice. This finding led some researchers to suggest that formation of the ureteric bud itself, not only its branching, may depend on a signal from the mesenchyme. Moreover, *ret* mRNA is specifically localized to the tip of the ureteric bud and to the tips of the branching epithelium. Taken together these observations

suggest that Ret functions as a receptor for a signaling molecule that is produced by the mesenchyme and induces budding of the kidney epithelium.

Activation of the Ret Receptor Promotes Growth and Branching of the Ureteric Bud

The ligand that binds to Ret was identified serendipitously in studies on a TGFβ-like growth factor called glial cell–derived neurotrophic factor (GDNF). Mice carrying homozygous knockout mutations in the *GDNF* gene showed defects in the nervous system and, remarkably, a kidney phenotype indistinguishable from that of *ret* knockouts. Although GDNF is expressed in the developing kidney at the same time as Ret, it is expressed in the mesenchyme not in the ureteric bud. The complementary patterns of expression of a signaling molecule and a receptor with indistinguishable knockout phenotypes suggested that GDNF is a ligand for Ret.

Using an expression cloning approach (see Figure 20-9), scientists identified a receptor protein that binds GDNF but clearly is not Ret. This protein, called GDNFR-α, is expressed on the surface of the ureteric bud, as is Ret. However, unlike Ret, GDNFR-α lacks an intracellular domain and is linked to the extracellular leaflet of the lipid bilayer by a *glycosylphosphatidylinositol (GPI) anchor* (see Figure 3-36a). The absence of an intracellular signaling domain in GDNFR-α suggested that it may act in combination with a *co-receptor* containing an intracellular signaling domain. Ret was an obvious candidate for this co-receptor since the *GDNF* and *ret* mutant phenotypes are virtually identical.

When GDNF is added to cultured mouse cells that express Ret, but not GDNFR-α, it neither binds to Ret nor stimulates its intrinsic tyrosine kinase activity. Moreover, Ret does not bind GDNFR-α. Surprisingly, addition of both GDNFR-α as a soluble protein unlinked to the plasma membrane and GDNF leads to formation of a ternary complex, Ret-GDNF-GDNFR-α, and stimulation of the Ret tyrosine kinase activity in cultured cells. Cells that express Ret but cannot respond to GDNF can be converted into GDNF-sensitive cells by transfecting them with the gene encoding GDNFR-α. Conversely, Ret-expressing cells that also respond to GDNF lose their responsiveness after treatment with a phospholipase C that cleaves the GPI anchor, releasing GDNFR-α from the membrane. Finally, the kidney phenotypes associated with GDNFR-α knockouts are indistinguishable from those of *ret* and *GDNF* knockouts.

Together, these data indicate that Ret and GDNFR-α, both expressed on the surface of ureteric bud, form a complex with GDNF released from closely opposed mesenchymal cells (Figure 23-23). This ternary interaction activates signaling by Ret, a receptor tyrosine kinase (see Figure 20-33), leading to growth and branching of the ureteric bud. The ureteric bud must, in turn, signal reciprocal events in the mesenchyme, although the signals and ligands involved have not been identified.

(a)

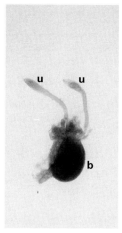

Wild type　　　　　　　　　**Ret knockout**

(b)

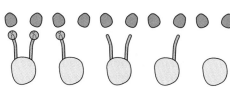

Wild type　　　**Homozygous mutant phenotypes**

▲ **FIGURE 23-22 Knockout mutations in *ret* produce severe defects in kidney morphogenesis in mice.** (a) Urogenital systems dissected from wild-type and mutant newborn mice. In this mutant, blind-ended ureters formed but no kidney development occurred. (b) Diagrammatic comparison of normal urogenital system and mutant phenotypes resulting from knockout of *ret*. a = adrenal; b = bladder; e = epididymis; k = kidney; u = ureter; v = vas deferens. [From A. Schuchardt et al., 1994, *Nature* **367**:382.]

▶ **FIGURE 23-23 Mechanism of mesenchymal inductive effect on the ureteric bud.** (a) Ret, a receptor tyrosine kinase, and GDNFRα, which binds GDNF, are expressed on the surface of the ureteric bud. GDNF, a diffusible TGFβ-like growth factor that is released from the mesenchyme, induces further branching and growth of the ureteric bud (see Figure 23-21). (b) Activation of Ret signaling requires formation of a ternary complex comprising GDNF, GDNFR-α, and Ret. [Part (b) adapted from J. Massagué, 1996, *Nature* **382**:29.]

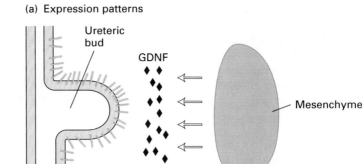

(a) Expression patterns

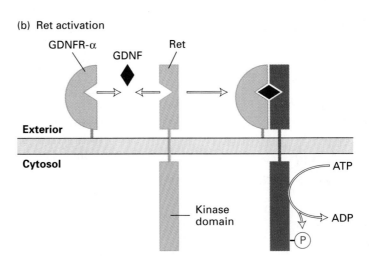

(b) Ret activation

The Basal Lamina Is Essential for Differentiation of Many Epithelial Cells

Epithelial and mesenchymal cell populations are separated by the **basal lamina**, which is composed of extracellular-matrix substances including laminin, type IV collagen, nidogen, and certain proteoglycans (Section 22.2). Several integrins on the surface of epithelial cells bind to components of the basal lamina, anchoring the cells to the extracellular matrix (see Table 22-2). The basal lamina is linked to the mesenchyme through a reticular network of extracellular components containing collagen (types I and III) and fibronectin.

Differentiation of many epithelial cells depends on interactions involving the basal lamina. For instance, in the developing salivary gland, groups of epithelial cells, derived from the endoderm, secrete proteins and proteoglycans that form a basal lamina, grow, and form clusters of cells that become the secreting cells of the mature gland. These epithelial-cell clusters together with the basal lamina can be separated from the underlying mesodermal cells by microdissection of embryos. The experiment depicted in Figure 23-24 demonstrates that two cell types—epithelial and mesodermal—*and* the basal lamina all interact and are necessary for the endodermal cells to differentiate into the mature secreting cells of the salivary gland. Although dramatic changes in the extracellular matrix have been observed during epithelial-mesenchymal interactions, the functions of specific matrix molecules in these developmental events have not yet been established.

Cell-Surface Ephrin Ligands and Receptors Mediate Reciprocal Induction during Angiogenesis

Perhaps the simplest type of reciprocal interaction is one in which two cells interact through two cell-surface proteins,

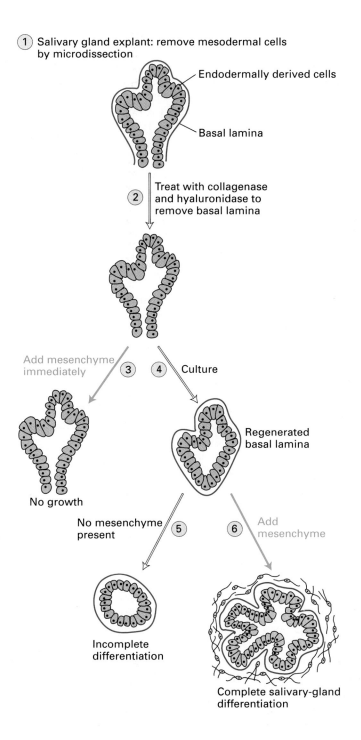

① Salivary gland explant: remove mesodermal cells by microdissection

Endodermally derived cells

Basal lamina

② Treat with collagenase and hyaluronidase to remove basal lamina

Add mesenchyme immediately ③   ④ Culture

No growth

Regenerated basal lamina

No mesenchyme present ⑤   ⑥ Add mesenchyme

Incomplete differentiation

Complete salivary-gland differentiation

▲ **FIGURE 23-24 Elements necessary for salivary-gland differentiation.** Salivary-gland explants can be dissected away from mesodermal cells (①) and treated with enzymes to remove the basal lamina (②). If the epithelial cells are then mixed immediately with mesenchymal cells, there is no growth or differentiation (③). If, on the other hand, the epithelial cells are cultured, they regenerate the basal lamina (④), but do not differentiate further (⑤) unless they are allowed to interact with mesenchymal cells (⑥). [Adapted from N. K. Wessels, 1977, *Tissue Interactions in Development*, Benjamin-Cummings, p. 225.]

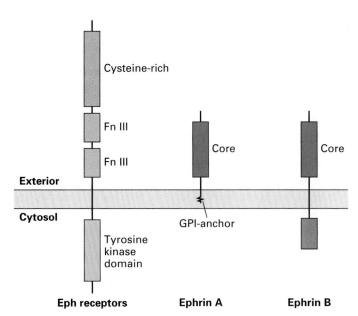

Cysteine-rich

Fn III

Fn III

Exterior

Cytosol

Core

Core

GPI-anchor

Tyrosine kinase domain

Eph receptors   **Ephrin A**   **Ephrin B**

▲ **FIGURE 23-25 General structure of Eph receptors and their ligands.** The cytosolic domain of Eph receptors has tyrosine kinase activity. These highly related receptors exhibit some 30–70 percent homology in their extracellular domains and 65–90 percent homology in their kinase domains. Their ligands, the ephrins, either are linked to the membrane through a hydrophobic GPI anchor (class A) or are single-pass transmembrane proteins (class B). The core domains of various ephrin ligands show 30–70 percent homology. [See J. G. Flanagan and P. Vanderhaegen, 1998, *Annu. Rev. Neurosci.* **21**:309.]

both of which can act as receptors and ligands. To illustrate this phenomenon, we discuss the role of a family of cell-surface ligands and receptors that are critical in the development of mammalian blood vessels.

The *Eph receptors* were identified first, detected by molecular screens, based on nucleic acid sequence similarity, for novel receptor tyrosine kinases. Using a clever variation on expression cloning described later in this chapter, scientists identified ligands for these receptors, the **ephrins**, which fall into two classes (Figure 23-25). Ephrin A ligands are tethered to the plasma membrane by a glycosylphosphatidylinositol (GPI) anchor. As we discuss later in this chapter, this class of ephrin ligands plays a crucial role in forming connections between neurons in the developing nervous system. Ephrin B ligands are single-pass transmembrane proteins. Biochemical experiments showed not only that ephrin B ligands induce tyrosine phosphorylation of EphB receptors, but also that these receptors stimulate tyrosine phosphorylation of the cytosolic tail of ephrin B ligands. These observations led to the intriguing notion that ephrin B/EphB ligand-receptor complexes promote bidirectional reciprocal interactions. Strong support for this hypothesis has come from study of the formation of blood vessels.

Blood vessels, arteries, and veins, form a complex network of branched structures in the adult. Blood vessels contain tubes of endothelial cells surrounded by support cells such as smooth muscle. The development of this network can be divided into two stages, an early stage called *vasculogenesis* and a late stage called *angiogenesis*. During vasculogenesis endothelial cells coalesce to form tubules leading to the construction of a vascular network with many small branches. This early network is remodeled during angiogenesis as larger branches assemble from smaller ones and vessels become surrounded by support cells.

Knockout mice lacking ephrin B2 exhibit a striking defect in the formation of blood vessels. Vasculogenesis appears normal in such mice, but angiogenesis is defective. This finding led scientists to explore the pattern of expression of ephrin B2 and its receptor, EphB4, in the developing embryo (Figure 23-26). In normal embryos, the devel-

oping arteries and veins are found in distinct portions of the yolk sac prior to angiogenesis. Ephrin B2 is expressed only on arteries; EphB4, only on veins. As angiogenesis proceeds, developing arterial capillaries fuse to form large branches, as do venous capillaries. Monitoring of the expression of ephrin B2 and EphB4 revealed that the arterial and venous capillaries form a densely intercalated structure. Moreover, although ephrin B2 is expressed only on arterial capillaries, venous capillaries also fail to undergo angiogenesis in *ephrin b2* knockouts. Together these data support a simple model in which ephrin B2 and EphB4 each function as both ligands and receptors to control the development of both veins and arteries.

The Conserved Notch Pathway Mediates Lateral Interactions

Now we shift our attention to **lateral interactions**, which cause developmentally equivalent cells to assume different fates. Genetic analyses in *Drosophila* and *C. elegans* have revealed that the highly conserved **Notch pathway** controls such lateral interactions. Here we describe the molecular components in this pathway; then we briefly consider two examples of its action during development in both vertebrate and invertebrate systems.

The *Drosophila* proteins *Notch* and *Delta* are the prototype receptor and ligand, respectively, in this signaling pathway (Figure 23-27). Both proteins are large transmembrane proteins whose extracellular domains contain multiple EGF-like repeats and binding sites for the other protein. The intracellular domain of Notch is essential for transmitting a signal to the nucleus initiated by Notch-Delta binding, establishing that Notch is the receptor and Delta, the ligand.

Upon ligand-induced activation, the Notch receptor undergoes a proteolytic cleavage that releases its intracellular domain, which then translocates to the nucleus. Although biochemical, genetic, and immunohistological studies are consistent with this cleavage/translocation model, the precise mechanism is not yet understood. Once in the nucleus, the intracellular domain of Notch forms a complex with a DNA-binding protein called Suppressor of Hairless, or Su(H), and stimulates transcription. As a consequence of this signaling, the stimulated cell up-regulates expression of the Notch receptor and down-regulates expression of the Delta ligand. Regulation of the receptor and ligand in this fashion is an essential feature of the interaction between initially equivalent cells, as illustrated by the cellular example discussed in the following section.

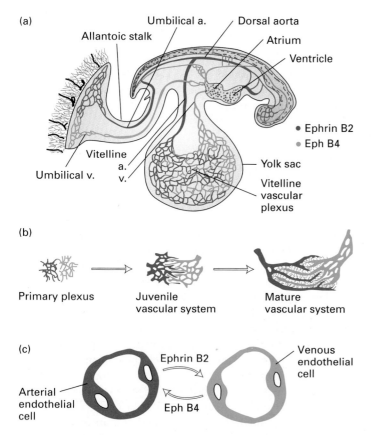

(a) Allantoic stalk — Umbilical a. — Dorsal aorta — Atrium — Ventricle — Ephrin B2 — Eph B4 — Vitelline a. — Umbilical v. — v. — Yolk sac — Vitelline vascular plexus

(b) Primary plexus — Juvenile vascular system — Mature vascular system

(c) Ephrin B2 — Venous endothelial cell — Arterial endothelial cell — Eph B4

▲ **FIGURE 23-26 Role of ephrin B2 and EphB4 in angiogenesis in the yolk sac.** (a) Ephrin B2 is expressed on arteries (red) and EphB4 (blue) on veins in the early mouse embryo. (b) Remodeling of the early vascular network occurs during angiogenesis. In *ephrin B2* knockouts angiogenesis is blocked at the primary plexus stage. (c) Intercalation between developing arterial and venous endothelial cells is mediated by ephrin B2 (arterial) and EphB4 (venous). [Adapted from H. U. Wang et al., 1998, *Cell* **93**:741.]

Interactions between Two Equivalent Cells Give Rise to AC and VU cells in *C. elegans*

The interactions between two equivalent precursor cells, designated Z1.ppp and Z4.aaa, have been studied in the soil nematode, *C. elegans*. Although each precursor cell can give

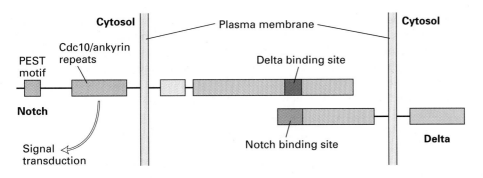

◀ **FIGURE 23-27 Notch and Delta are transmembrane proteins that mediate a wide spectrum of cellular interactions in *Drosophila*.** Both Notch and Delta contain numerous EGF-like repeats (green) in their extracellular domains. The PEST motif (pink) in the Notch cytosolic domain is found in many developmentally important molecules and targets a protein for rapid degradation. Binding of Delta on one cell to Notch on another cell triggers signal transduction via the Notch cytosolic domain. See text for discussion.

rise to AC and VU cells, interactions between them cause one to develop into an AC cell and the other into a VU cell; that is, each of the interacting precursor cells assumes a distinct fate. Laser ablation studies have shown that if either the Z1.ppp or Z4.aaa cell is removed, the remaining cell always becomes AC. In worms lacking functional LIN-12, the *C. elegans* homolog of Notch, both cells become AC. Conversely, constitutive activation of LIN-12 in the two precursor cells results in both cells becoming VU. Thus LIN-12 activity levels specify AC and VU cell fates.

Molecular studies have shown that both Z1.ppp and Z4.aaa express both the receptor, LIN-12, and its ligand, the Delta homolog LAG-2, at similar levels. As development proceeds, one cell begins to express more receptor through random fluctuations in protein levels or differences in the ambient level of signaling through the pathway. The cell receiving a slightly higher signal begins to up-regulate expression of the receptor and down-regulate expression of the ligand (cell Z1.ppp in Figure 23-28). The neighboring cell is now exposed to a reduced level of ligand; as a consequence, expression of the receptor falls and of the ligand increases in this cell (cell Z4.aaa in Figure 23-28). The regulatory circuit connecting expression of the receptor and ligand thus amplifies the initial asymmetry resulting from a random event, finally leading to the commitment of one cell as a pre-VU cell and its partner as a pre-AC cell.

Neuronal Development in *Drosophila* and Vertebrates Depends on Lateral Interactions

Loss-of-function mutations in the *Notch* or *Delta* genes produce a wide spectrum of phenotypes in *Drosophila*. The

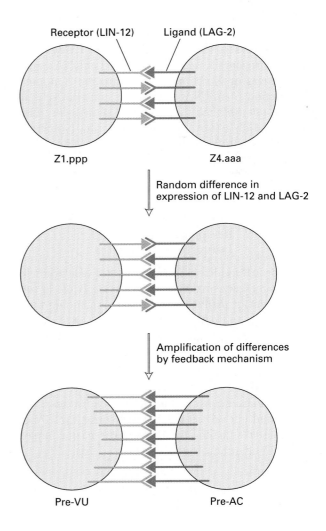

▶ **FIGURE 23-28 Lateral interactions leading to distinct cell fates in *C. elegans*.** LIN-12, a Notch homolog, and LAG-2, a Delta homolog, regulate interactions between equivalent precursor cells, designated Z1.ppp and Z4.aaa. Either cell can assume a VU or AC fate. See text for discussion. [Adapted from I. Greenwald, 1998, *Genes & Devel.* **12**:1751.]

hallmark of such mutations in either gene is neural hypertrophy, that is, an increase in the number of neuroblasts in the central nervous system or of sensory organ precursors (SOPs) in the peripheral nervous system of the developing fly. During neurogenesis of the *Drosophila* peripheral nervous system, a single cell within a group of cells, called a *proneural cluster,* develops into an SOP (see Figures 14-19 and 14-20). During normal development, cells not selected to become SOPs subsequently give rise to epidermis. Studies with temperature-sensitive loss-of-function mutations in either the *Notch* or *Delta* gene have shown that interactions between cells within a proneural cluster play an important role in this selection process. Disruption of these interactions by functional loss of either protein leads to development of additional SOPs (or neuroblasts in the CNS) from a proneural cluster. In contrast, in developing flies that express a constitutively active form of Notch (i.e., active in the absence of a normal ligand), all the cells in a proneural cluster develop into epidermal cells.

In the frog *Xenopus laevis,* neurons are generated from the neuroepithelium in waves. During the first wave, referred to as primary neurogenesis, neurons form in three longitudinal stripes on each side of the midline. These stripes are not continuous but rather contain scattered neurons. The cells that do not develop into neurons during primary neurogenesis will do so in subsequent waves or will give rise to glial cells. To assess the role of the Notch pathway during primary neurogenesis, scientists injected mRNA encoding different forms of Notch and Delta into the *Xenopus* embryo. Injection of mRNA encoding a constitutively active form of Notch inhibited the formation of neurons, whereas mRNA encoding a form of Delta disrupting Notch activation to the formation of too many neurons. These findings indicate that in vertebrates, as in *Drosophila,* Notch signaling acts to limit neuron formation.

SUMMARY   Reciprocal and Lateral Inductive Interactions

- Reciprocal inductive interactions between epithelial cells and mesenchymal cells play a crucial role in the development of many internal organs, including the kidney, lung, and pancreas.

- In development of the kidney, GDNF released from the mesenchyme induces proliferation and branching of the ureteric bud (epithelium). GDNFR-α, which directly binds GDNF, and the co-receptor Ret, which has a tyrosine kinase cytosolic domain, are expressed on the ureteric bud but not on the mesenchyme. GDNF triggers Ret-mediated signaling by forming a ternary complex with GDNFR-α and Ret (see Figure 23-23).

- The extracellular matrix, particularly components of the basal lamina, plays a crucial role in many reciprocal interactions between epithelial and mesenchymal cells.

- Ephrins are cell-surface signaling ligands whose receptors (Ephs) are receptor tyrosine kinases. Direct reciprocal interactions between ephrin B2 and its receptor EphB4 on adjacent arterial and venous endothelial cells control angiogenesis. Each transmembrane protein acts both as a ligand and as a receptor in this system.

- Notch and its cell-surface ligand Delta mediate lateral interactions between initially equivalent cells. Binding of Delta to the Notch receptor leads to cleavage of the Notch intracellular domain, which translocates to the nucleus where it associates with the DNA-binding protein Su(H) and stimulates transcription of target genes.

- An initial, random fluctuation in signaling between two equivalent cells triggers a regulatory circuit leading to one cell expressing the Delta ligand and the other, the Notch receptor (see Figure 23-28).

23.5   Overview of Neuronal Outgrowth

Proper functioning of the nervous system depends on the intricate array of connections, or **synapses,** that are formed during development. Within the vertebrate central nervous system, a highly specific network of synapses must be made among thousands of billions of neurons. Invertebrates, such as insects, have fewer neurons but face similar developmental tasks in constructing patterns of neuronal connectivity. During differentiation **dendrites** and **axons,** which may be very long, grow out from the cell body of each neuron. Axons grow outward toward those target cells with which they will form synapses, commonly on dendrites. To fathom how the nervous system is constructed, we need to answer several questions: How does an axon select the correct pathway along which to grow? How does it choose a specific target region within which to terminate? And how does it recognize certain cells with which to synapse?

The process of neuronal wiring can be divided into two different stages: an early stage that does not depend on neuronal activity (i.e., firing of an action potential) and a later stage that does, particularly in vertebrate systems. During the early stage of wiring, the growth cone at the leading edge of an axon navigates to the specific region where it will form synapses (Figure 23-29); subsequent fine tuning of the projection pattern on target cells then occurs. Because little is known about the molecular mechanisms by which neuronal activity regulates synaptic specificity, we will restrict our discussion to activity-independent processes. In this section, we discuss the general process of axon outgrowth and the evidence that it is targeted to specific regions. In the following two sections, we describe how the direction of outgrowth is guided and the role of graded signals in defining the spatial relationship between fields of neurons and their target neurons in the brain.

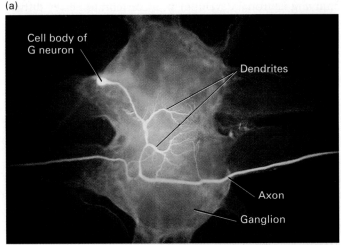

▲ FIGURE 23-29 The leading edge of the axon is called a growth cone. It is a sensorimotor structure that responds to signals in the developing embryo and leads the extending axon to its target. This micrograph shows a labeled photoreceptor axon and its growth cone in the developing *Drosophila* visual system.

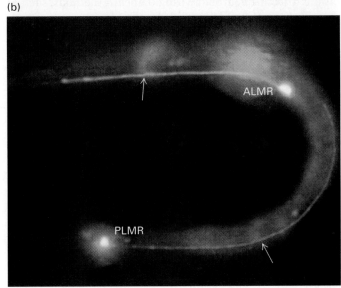

▲ FIGURE 23-30 Visualization of individual neurons. (a) The fluorescent dye Lucifer yellow was microinjected into the cell body of the G neuron, one of 2000 neurons in the second thoracic ganglion in the grasshopper embryo. The axon extends from the cell body, crosses the ganglion, then extends outward (to the right in this picture); a smaller axon branch extends in the opposite direction. (b) A jellyfish gene encoding a fluorescent protein was introduced into a living *C. elegans* worm. This gene was linked to a worm neuron-specific promoter that is highly active in the touch-sensitive neurons designated ALMR and PLMR. Expression of the fluorescent protein and observation by fluorescent microscopy reveals the cell bodies (light spots) and processes of these two neurons (arrows). In this photograph, the worm is bent in a U shape. [Part (a) from C. S. Goodman et al., 1984, *Sci. Am.* **251**(6):58. Part (b) from M. Chalfie et al., 1994, *Science* **263**:802.]

Individual Neurons Can Be Identified Reproducibly and Studied

A prerequisite for understanding how specific neural contacts form is identifying specific neurons in the developing embryo and observing them as their axons elongate and form contacts. Embryos of grasshoppers, *Drosophila*, and other insects in which the central nervous system is formed by a string of ganglia, one per body segment, provide excellent experimental systems. A single ganglion may contain a thousand or so nerve-cell bodies; some of these have a characteristic size and position, permitting precise identification of the same cell in different embryos. When a nerve cell is sufficiently large, it can be microinjected with a fluorescent dye (e.g., lucifer yellow) that spreads throughout its cytosol. The cell's projections then can be visualized in the fluorescence microscope (Figure 23-30a). Analogous techniques can be used to identify vertebrate motor neurons. For example, the enzyme peroxidase injected into a nerve near its terminus is transported back to the cell body by retrograde axonal transport (Chapter 19). The enzyme can then be detected in fixed tissue by histochemical staining. In this way, it is possible to localize in the spinal cord the cell body of a particular neuron that innervates a specific muscle. Lipophilic dyes (e.g., DiO and DiI) also have been particularly useful in studying neuronal development in vertebrates. They diffuse within the membrane highlighting the entire extent of the axon and neuritic processes. In contrast to peroxidase labeling, lipophilic dyes can be used to stain fixed specimens or to follow events in living tissue using confocal microscopy.

Important insights into neuronal wiring have come from study of the simple soil worm, *C. elegans*, which exists in two forms—a self-fertilizing hermaphrodite and a male that facilitates cross-fertilization. The adult hermaphrodite contains 959 somatic cells; 301 of these are neurons (see Figure 5-13c). Because of the small number of neurons in *C. elegans*, researchers have been able to reconstruct the complete neuronal wiring pattern from thin sections visualized by electron microscopy. Moreover, the short generation time of *C. elegans* (52 hours at 25 °C) and the ease of genetic analysis in this system have permitted identification of many different mutations affecting neuronal wiring. Labeling methods, such as that illustrated in Figure 23-30b, allow analysis of the effects of such mutations on outgrowth of individual neurons in living worms.

Growth Cones Guide the Migration and Elongation of Developing Axons

During differentiation of the nervous system, precursors of neurons, called *neuroblasts*, arise at particular locations at specific times. Such cells lack axonal or dendritic projections. Certain neuroblasts migrate to specific destinations where they form clusters called **ganglia**. Vertebrate cells migrate extensively in the central nervous system. As a newly born neuron begins to differentiate, it grows one or more axonal projections. At the leading edge of the elongating axon is the highly motile **growth cone**, which possesses cell-surface receptors for extracellular signals (see Figure 23-29).

As the growth cone moves outward, the cell body stays put and the axon elongates due, in part, to the polymerization of tubulin into microtubules, which give the axon its rigidity. Ligand binding to receptors on growth cones triggers intracellular signaling that regulates their motility, in part, by controlling actin polymerization in the filopodia and lamellipodia at the distal end. Recognition of signaling molecules by receptors located at different positions along the periphery of advancing growth cones is thought to steer them toward the appropriate target cells.

Different Neurons Navigate along Different Outgrowth Pathways

Although the nervous systems of vertebrates and invertebrates are different in their structure and complexity, similar principles of axon guidance are used: growth cones of elongating axons use a changing set of cell-surface receptors to move along specific extracellular-matrix fibers and also along specific cells. The notion that specific molecules guide growth cones to their targets was first proposed nearly a

century ago. Here we describe a few examples that argue persuasively for the importance of specific cues in directing growth cones to their targets.

In vertebrates, the cell bodies of motor neurons are located in the spinal cord, and axons extend out of the central nervous system by ventral roots. Axons that innervate a specific muscle are bound together to form a peripheral nerve. Obviously it is essential for each motor neuron to innervate only the appropriate muscle, and indeed this high degree of specificity is found throughout the nervous system.

In certain favorable experimental systems, scientists can observe individual motor neurons as their axons exit the spinal cord and elongate. During embryogenesis in the zebrafish, for example, the axons of three *pioneer neurons* are the first to emerge from the spinal cord in each segment of the embryo (Figure 23-31). Investigators identified the cell bodies of these three motor neurons and followed the trajectories of their axons by microinjection with a fluorescent

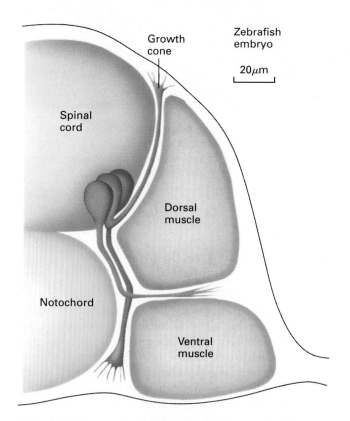

▲ **FIGURE 23-31 During development of the zebrafish embryo, growth cones of pioneer motor neurons follow different pathways.** A cross-sectional view of a trunk section of a 19-hour-old zebrafish embryo shows that axons of three adjacent motor neurons extend outward from the developing spinal cord at the same ventral root. They follow the same pathway out of the spinal cord, but then follow different pathways. One axon extends downward, innervating the ventral muscles; one upward, innervating the dorsal; and one laterally, innervating both. [Adapted from M. Westerfield and J. S. Eisen, 1988, *Trends Neurosci.* **11**:18.]

dye. These studies showed that the axons of all three pioneer neurons initially grow out of the spinal cord along a common pathway. They then diverge: one axon grows upward to innervate dorsal muscles; one downward to innervate ventral muscles; and one laterally to innervate both muscles. These pioneer axons are never seen to send branches or growth cones off in an inappropriate direction.

At a later stage of development, axons of other neurons grow out of the central nervous system to innervate the same muscles. Growth cones of these axons migrate along the surfaces of the three pioneer axons, which guide the *secondary neurons* to the correct muscles. However, experimental destruction of one of the pioneer neurons by an intense laser beam does not affect the ability of the secondary motor neurons to select the pathway that leads to the appropriate target muscle. This finding suggests that by the time a growth cone exits the spinal cord, the neuron is already programmed to follow a specific pathway. Experiments with the chick embryo—in which nerve transplantation experiments are easier—support this contention. Here, motor neurons from four adjacent segments of the spinal cord innervate various muscles of the hindlimb. Multiple axons destined for different muscles grow out of the spinal cord together and follow the same path to plexuses; from there the individual neurons follow different paths to the appropriate muscle. If, before nerve outgrowth, a piece of spinal cord containing motoneuron cell bodies is transplanted from one segment to an adjacent one, these motoneuron axons will exit the spinal cord at the "wrong" site. However, they will grow into the proper plexus and still innervate the correct muscle. A similar result is obtained if a segment is inverted, so that anterior neurons are moved to the posterior.

Clearly, the specificity in guidance of any axon does not depend on where it leaves the spinal cord, nor on its transplanted position within a segment. It seems to depend on its intrinsic properties, perhaps an inherent set of receptors on each growth cone that allow it to recognize the surrounding environment. For instance, in Figure 23-31 the neuron that grows along the ventral pathway does so because it specifically recognizes components of the extracellular matrix or glial or other cells located in this region. A growth cone most probably moves from one short-range target to another as it guides an axon to its destination.

Although we have chosen these examples from the vertebrate, similar principles probably apply to the outgrowth of the first axons in insect nervous systems. For instance, the growth cones of certain pioneer axons in the grasshopper central nervous system always make a turn at a specific glial cell. If the glial cell is destroyed by a laser beam, the growth cones do not turn but continue in the original direction. These growth cones use glial cells as landmarks or guides, as do those of other pioneer neurons in both vertebrate and invertebrate central nervous systems. Evidence suggests very intimate contact between pioneer neurons and guidepost glial cells; gap junctions, for instance, form transiently between the two contacting cells.

Various Extracellular-Matrix Components Support Neuronal Outgrowth

The extracellular matrix and some of its individual constituents (e.g., laminin) have been shown to regulate **neuronal outgrowth**. The experiment described in Figure 23-32 indicates that growth cones select specific matrix substrates over others for outgrowth. Some researchers concluded from this finding that the pathway taken by a growth cone to its target is determined by a series of choice points between substrates of different adhesiveness. Subsequent studies, however, showed that no simple relationship exists between the degree of adhesiveness and growth-cone motility on a variety of substrates. Indeed, under certain conditions, laminin has anti-adhesive properties. Hence, it is probably more accurate to view outgrowth as reflecting the ability of various extracellular-matrix components (or components expressed

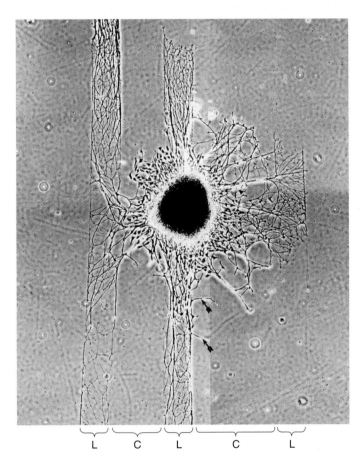

L   C   L   C   L

▲ **FIGURE 23-32 Stripe assay demonstrating that growth cones preferentially migrate on laminin rather than collagen IV.** Alternating stripes of laminin (L) and collagen IV (C), indicated by brackets, were affixed to a tissue-culture dish. When sensory neurites were cultured on the matrix materials, outgrowth was observed predominantly along the laminin stripes. The arrows indicate the infrequent outgrowth of small fascicles into a collagen stripe. [From R. W. Gundersen, 1987, *Devel. Biol.* **121**:423; courtesy of R. Gundersen.]

on the surfaces of cells along the pathway) to promote motility through intracellular signaling pathways regulating cytoskeletal dynamics in the growth cone (e.g., integrin-laminin interactions).

Although many common extracellular-matrix components support growth-cone movement, they do not appear capable of determining the *directionality* of outgrowth. For instance, neurons in culture do not show directed outgrowth even on a steep gradient of laminin. Before discussing how directionality is determined, we consider the outgrowth of axons along other axon tracts.

Growth Cones Navigate along Specific Axon Tracts

As pioneer axons extend using extracellular-matrix and cell-surface cues, more and more of the space in the central nervous system becomes occupied by axonal processes. As additional neurons differentiate, their growth cones navigate on the surfaces of other axons and their axons eventually bundle together forming *fascicles*. Different growth cones select different axonal surfaces in different fascicles on which to migrate. This can be seen vividly in the development of the grasshopper central nervous system.

Grasshopper and *Drosophila* embryos are divided into segments, most of which have a similar pattern of neurons. Each segment has two ganglia, one on each side of the embryo, which contain the neuron cell bodies. Bundles of axons *(longitudinal fascicles)* run along either side of the embryo and bundles of axons *(transverse fascicles)* cross the embryo, connecting the segmental ganglia in both directions. Of the hundreds of neurons in each segmental ganglion, many sprout axons whose growth cones migrate to the opposite side of the embryo by growing along a transverse fascicle. Six such identifiable neurons are depicted in Figure 23-33a. When these six growth cones reach the opposite side, each chooses a different longitudinal fascicle to follow or a different direction in which to migrate.

For instance, the G neuron growth cone initially makes contact with about 100 axons in 25 longitudinal fascicles, but it only follows the A-P fascicle, which is composed of the axons of four neurons called A1, A2, P1, and P2. More detailed studies have shown that the G growth cone actually moves along only two of the four axons in the fascicle, P1 and P2 (Figure 23-33b). If the A1 and A2 neurons are experimentally destroyed by a laser beam, the differentiation of the G neuron as well as most of the other neurons is unaffected; the G growth cone moves normally along the P1 and P2 axons. However, if the P1 and P2 neurons are destroyed, the G neuron grows abnormally, and its growth cone behaves as if it were undirected and does not bind to any other axon. Thus, migration of the G growth cone relies absolutely on binding to the P axons.

These observations suggest that the P1 and P2 axons bear a unique surface marker that is recognized by a receptor present on the G growth cone but not on growth cones

▶ FIGURE 23-33 **Different stereotyped pathways are taken by the axons of sister neurons.** (a) Each of the 17 segments of the grasshopper embryo have two segmental ganglia; these ganglia have a virtually identical pattern of nerve differentiation. One identifiable neuroblast in each half of each segment divides repeatedly to form about 50 ganglion mother cells, each of which divides to yield two sister neurons. The first six neurons formed, shown here, extend axons across the ganglion, forming part of a transverse fascicle. The growth cone of each neuron then recognizes a different longitudinal fascicle and moves along it, elongating the axon in a specific direction. (b) Details of the selective fasiculation of the growth cone of the identifiable G neurons. Each half segment has one G neuron, whose axon grows along a transverse fascicle to the opposite side of the embryo. There the G growth cones explore the surfaces of 25 fascicles with a total of 100 axons. The G growth cones adhere to and migrate along only the axons of the P1 and P2 neurons (blue) in the A-P fascicle (green and blue axons). [After C. S. Goodman and M. J. Bastiani, 1984, *Sci. Am.* **251**(6):58.]

of other neurons that do not follow the P axons. To identify these guide molecules, investigators prepared monoclonal antibodies to preparations of grasshopper neuronal membranes and screened them for their ability to stain specific subsets of fascicles. With these anti-fascicle antibodies, several cell-surface fascicle proteins were identified and purified; the genes encoding these proteins then were identified and cloned using techniques described in Chapter 7. Remarkably, when four of these proteins—*fasciclin I, II, III* and *neuroglian*—were individually expressed in transfected nonadherent tissue-culture cells, they each promoted aggregation of cells expressing the protein, suggesting that they function as cell-adhesion molecules, or CAMs (Section 22.3).

To further characterize these proteins and study their role in neuronal outgrowth, their *Drosophila* homologs were identified. All are transmembrane proteins with a large extracellular domain; fly neuroglian and fasciclin II contain immunoglobulin-type repeats and fibronectin type III repeats, whereas fasciclin III contains only immunoglobulin repeats. Researchers then introduced various mutations that disrupted the structure of these proteins. Surprisingly, these mutations had little or no effect on the formation of specific axon fascicles in the fly CNS. Similar results have been obtained on *N-CAM*, the vertebrate homolog of fasciclin II, which has long been viewed as a critical determinant of neuronal development, including formation of fascicles. For instance, knockout mice that produce no functional N-CAM are viable and fertile, and their neuronal organization exhibits few abnormalities. However, recent studies have demonstrated that L1 (a vertebrate neuroglian-like cell-adhesion molecule) is required for the formation of nerve tracts in mice and human. Mice lacking L1 show defects in some but not all axon tracts. In humans, mutations in L1 lead to neurological deficits and defects in some nerve tracts.

In view of the multitude of CAMs identified on fascicles and the often mild effects of mutations in any one protein,

(a)

Grasshopper embryo

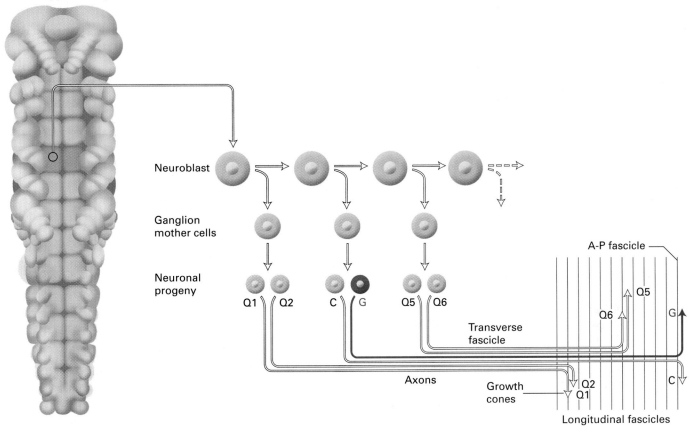

(b)

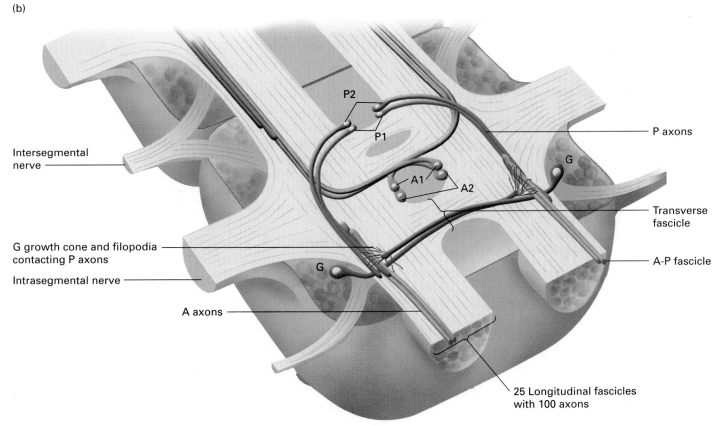

it is likely that considerable redundancy exists in the mechanisms guiding movement of growth cones along other axons. Indeed, such redundancy has been directly demonstrated in antibody-disruption experiments with cultured chick embryonic cells. In this system, the interactions between three different CAMs and their receptors on growth cones must be neutralized to prevent outgrowth of ciliary ganglion cells over Schwann-cell membranes. The emerging view is that selection of specific fascicles by a particular growth cone involves recognition of a unique combination of widely expressed cell-surface molecules rather than unique fascicle-specific molecules.

Soluble Graded Signals Can Attract and Repel Growth Cones

So far we have seen that components of the extracellular matrix and certain cell-adhesion molecules participate in neuronal outgrowth. Growth cones can also respond to soluble factors. Small molecules such as neurotransmitters and various soluble peptides have been shown to modulate growth cone motility in vitro. These studies, as well as the observation that axons often travel long distances toward their targets, support the long-held notion that specific secreted molecules can act as long-range attractants. In the next section, we discuss the recent identification of specific proteins that act as graded signals to regulate the directionality of axon outgrowth.

SUMMARY Overview of Neuronal Outgrowth

- During development, neurons elaborate remarkably specific and reproducible patterns of connections with other neurons (and with other target cells, e.g., muscle cells). Techniques for labeling individual neurons and their processes have been critical in investigating neuronal wiring.

- The leading edge of an elongating axon is a sensorimotor structure called a growth cone (see Figure 23-29).

- Elongation and migration of growth cones is promoted and guided by components of the extracellular matrix, cell-adhesion molecules (CAMs) on other neurons, and secreted, diffusible proteins.

- Each growth cone expresses a combination of cell-surface receptors on its surface that determines its pathway of choice. Ligand binding to these receptors is thought to trigger a signaling pathway that controls growth-cone motility and in some cases steers the growth cone toward its target.

- Graded diffusible signals that attract or repel growth cones are critical in determining the direction of neuronal outgrowth.

23.6 Directional Control of Neuronal Outgrowth

The convergence of genetic studies in *C. elegans* and elegant biochemical experiments in vertebrates has revealed a conserved class of proteins that provide directional information to growth cones. Here we consider the evidence that these attractants and repellents direct the migration of growth cones.

Three Genes Control Dorsoventral Outgrowth of Neurons in *C. elegans*

Numerous mutations affecting movement of *C. elegans* have been identified. Worms carrying these mutations, called *unc* (for *unc*oordinated body movement), are easily identified under a dissecting microscope. Many Unc mutants have been shown to have specific defects in neuronal organization and wiring. These defects fall into three classes: (1) those affecting basic processes of axonogenesis and outgrowth; (2) those affecting the ability of axons to form fascicles; and (3) those specifically affecting the directionality of neuronal outgrowth. Here we discuss the mutations producing the third class of defects.

The initial outgrowth of axons in the developing worm occurs between the epidermis and the basal lamina secreted by the epidermal cells. Some axons extend dorsally and others grow ventrally. Upon reaching the dorsal or ventral side of the body, these axons turn and extend along the anterioposterior axis. Mutational analysis has shown that three genes—*unc-5, unc-6,* and *unc-40*—are required for dorsoventral guidance. The *unc-5* gene is required for dorsal outgrowth; *unc-40,* for ventrally directed growth and in some neurons also for dorsal outgrowth, and *unc-6,* for both ventral and dorsal outgrowth (Figure 23-34). Mutations in these genes do not affect growth along the anterioposterior axis. Comprehensive phenotypic analysis of outgrowth and cell migration (these genes also regulate migration) show that these three genes are part of a common genetic pathway controlling these processes. We will discuss the molecular mechanisms by which *unc* genes function after we discuss the isolation through biochemical approaches of the vertebrate homolog of the UNC-6 protein.

Vertebrate Homologs of *C. elegans* UNC-6 Both Attract and Repel Growth Cones

During development of the vertebrate embryonic spinal cord, a subclass of neurons, called *commissural neurons,* grow from dorsal positions ventrally toward the floor plate found at the ventral midline (Figure 23-35a). By co-culturing floor-plate cells with explants of embryonic spinal cord in collagen gels, investigators showed that commissural neurons grew out of the explants and turned towards the floor-plate tissue (Figure 23-35b). This finding suggested that the oriented growth of commissural neurons ventrally

Outgrowth of PDE neuron / Frequency of outgrowth pattern

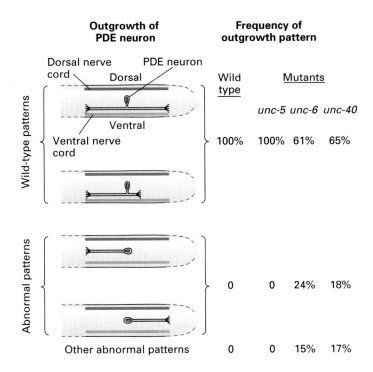

Wild type	Mutants		
	unc-5	unc-6	unc-40
100%	100%	61%	65%
0	0	24%	18%
Other abnormal patterns 0	0	15%	17%

◀ **FIGURE 23-34 Mutational analysis of directional outgrowth of the PDE neuron in *C. elegans*.** In wild-type worms, the PDE neuron grows ventrally, branches at the nerve cord and grows both anteriorly and posteriorly; in 5 percent of worms, the posterior branch is shorter. Mutations in *unc-6* and *unc-40*, but not in *unc-5*, cause errors in ventral outgrowth. In the case of *unc-6* and *unc-40* mutations, additional projection abnormalities differing from the two shown here were observed in some worms. Similar analyses with other identified neurons have shown that *unc-5* is required for dorsal outgrowth; *unc-6* for both dorsal and ventral outgrowth; and *unc-40* for ventral outgrowth and dorsal outgrowth for some neurons. [Adapted from E. M. Hedgecock et al., 1990, Neuron **2**:61.]

▶ **FIGURE 23-35 Experimental demonstration that chemoattractants direct neuronal outgrowth in vertebrates.** (a) In the embryonic spinal cord, cell bodies of commissural neurons (red) are located near the dorsal roof plate. During normal development, they grow ventrally toward the floor plate. (b) Explants of the dorsal spinal cord and floor plate were arranged in a collagen gel as diagrammed and co-cultured for 40 hours. Staining for a specific cell-surface marker on commissural neurons showed that outgrowth was directed toward the floor-plate tissue. (c) Growth cones of *Xenopus* retinal neurons growing in culture turned towards purified netrin released in pulses from a micropipette in the upper right-hand corner. Moving the position of the source of the netrin induced the growth cones to reorient toward the micropipette. [Part (c) from J. R. de la Torre et al., 1997, *Neuron* **19**:1211.]

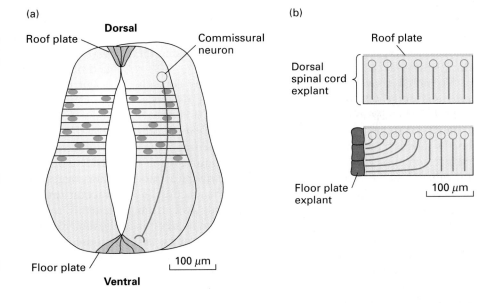

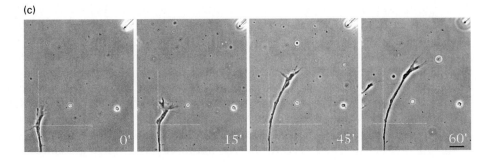

is due to secretion of a *chemoattractant* by the floor-plate cells. The floor-plate activity was also found in embryonic brain, which provided a more abundant source of material for purification.

Using outgrowth of commissural neurons as an assay, researchers purified two proteins called *netrin 1* and *netrin 2* (from Sanskrit *netr*, "one who guides"). The genes encoding these proteins then were cloned and sequenced. In knockout mice lacking netrin, commissural axons fail to extend to the floor plate, indicating that the netrins function as chemoattractants in vivo as well as in vitro. Moreover, expression of *netrin* cDNAs in tissue-culture cells promotes directed outgrowth of commissural axons toward these cells, supporting the idea that netrins are chemoattractants. More recent experiments have shown directly that growth cones can migrate towards a graded source of purified netrin protein (Figure 23-35c).

Netrin 1 is expressed in the floor plate and netrin 2 more weakly in the ventral and lateral regions of the embryonic spinal cord. Remarkably, the vertebrate netrins are highly related in structure to *C. elegans* UNC-6, which is expressed in the ventral region of the worm. Since UNC-6 and the netrins have similar structures, are expressed in similar regions, and act to promote directional outgrowth, they most likely are functional homologs. Both the netrins and UNC-6 also have structural homology to a region of the laminin B2 subunit, which binds to other components of the extracellular matrix (see Figure 22-19). This similarity suggests that these secreted chemoattractants may bind to the extracellular matrix, thereby stabilizing their gradients along the dorsoventral axis.

As noted above, *C. elegans* UNC-6 not only controls neuronal outgrowth in the ventral direction but also regulates dorsal migration, presumably by acting as a chemoattractant for some neurons and as a *chemorepellent* for others. By analogy, researchers hypothesized that **netrins**, which attract commissural neurons, may act as repulsive signals for other classes of vertebrate neurons. Since trochlear motoneurons located in the ventral region near the junction between the hindbrain and midbrain grow dorsally away from the floor plate, they were considered good candidates for neurons that would be repelled by netrins. By placing explants of tissue containing trochlear neurons in a collagen gel with either floor-plate tissue or COS cells expressing netrin, scientists demonstrated that trochlear neurons are, indeed, repelled by netrins. Hence, UNC-6 and its vertebrate homolog act as both a repellent and an attractant. How can the same signal be both an attractant and a repellent? Genetic and biochemical studies demonstrate that the specificity lies in the receptors, as we discuss next.

UNC-40 Mediates Chemoattraction in Response to Netrin in Vertebrates

The *unc-40* gene encodes a large transmembrane protein with four immunoglobulin repeats and six fibronectin III

repeats. Mutations in *unc-40* disrupt ventral outgrowth in *C. elegans* embryos. Similarly, in mouse knockouts lacking the vertebrate *unc-40* gene, commissural neurons fail to grow to the floor plate. The vertebrate UNC-40 protein, which is expressed on the axons and growth cones of commissural neurons, binds netrin with high affinity. Studies with *Xenopus* explants have shown that antibodies to UNC-40 prevent retinal neurons from turning towards a source of purified netrin. These studies demonstrate that UNC-40 is a growth-cone guidance receptor for the attractant response to netrin. Genetic studies in *C. elegans*, however, indicated that the role of UNC-40 in axon guidance is more complex.

UNC-5 and UNC-40 Together Mediate Chemorepulsion in Response to Netrin

We noted earlier that in *C. elegans* the *unc-5* gene is required specifically for dorsal neuronal outgrowth. For instance, the DA neuron projects dorsally in wild-type worms, but it fails to do so in an UNC-5 mutant. This finding led to the hypothesis that the UNC-5 protein is a receptor that specifically mediates the repellent response to UNC-6. Sequencing of the cloned *unc-5* gene showed that it encodes a transmembrane protein comprising a large intracellular domain and an extracellular domain containing two immunoglobulin repeats and two different repeats found in the protein thrombospondin. Biochemical studies have shown that like UNC-40, the vertebrate UNC-5 protein also binds netrin. Analysis of *C. elegans* loss-of-function mutants demonstrated that UNC-5 is required for dorsal outgrowth, and gain-of-function studies indicated that UNC-5 may be sufficient to specify dorsally directed growth. For instance, if the UNC-5 protein is misexpressed in a neuron that normally grows ventrally (e.g., the PVM neuron), then outgrowth is respecified to the dorsal direction.

Interestingly, respecification of PVM outgrowth is dependent on the proteins encoded by *unc-6* and *unc-40*. These data suggest that whereas UNC-40 mediates the chemoattractant response to UNC-6, UNC-5 and UNC-40 function together to mediate a chemorepellent response to the same ligand. Recent biochemical studies have shown that in response to netrin, vertebrate UNC-5 and UNC-40 form heterodimers (or heteromultimers). Netrin binding induces a conformational change that permits the intracellular domains of these proteins to interact. Furthermore, expression of UNC-5 protein in *Xenopus* retinal neurons, which normally express UNC-40 but not UNC-5, changed the attractant response to purified netrin in culture to a repellent response (Figure 23-36). Since antibodies to UNC-40 blocked both the repellent and attractant response of these neurons to netrin, both attraction and repulsion require UNC-40.

However, some *C. elegans* neurons whose dorsal outgrowth depends on UNC-5 are not dependent on UNC-40. This finding suggests that in these neurons UNC-5 either can act alone or collaborate with another, as-yet unidentified guidance receptor to promote dorsal outgrowth.

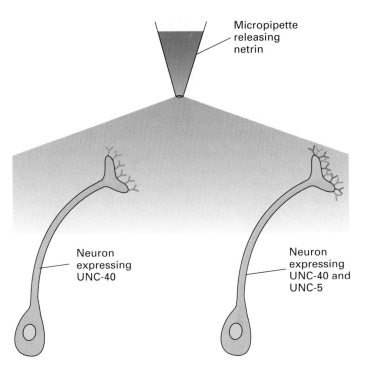

Micropipette releasing netrin

Neuron expressing UNC-40

Neuron expressing UNC-40 and UNC-5

▲ **FIGURE 23-36 Netrin can act both as a chemoattractant and as a chemorepellent.** *Xenopus* neurons expressing UNC-40 are attracted to netrin, whereas when the same neurons co-express UNC-5 and UNC-40, they are repelled by netrin. [See K. Hong et al., 1999, *Cell* **97**:927.]

Prior Experience Modulates Growth-Cone Response to Netrin

Studies in both vertebrates and invertebrates indicate that neurons often are guided to very distant targets in a stepwise fashion. A series of intermediate targets are thought to guide the growth cone through the developing embryo to its target. For instance, a cell may direct a growth cone toward it by producing a chemoattractant such as netrin. The growth cone grows up the netrin gradient to this intermediate target cell, but then must grow away from the cell as it continues its migration to the final target. A set of biochemical and tissue explant experiments suggest how a growth cone can grow away from the peak of a chemoattractant.

Desensitization of Commissural Neurons by the Floor Plate When explants from the metencephalic region of the rat embryo are grown in culture, commissural neurons in the explants grow across the floor plate at the midline and continue to grow laterally, as they do in vivo (Figure 23-37a). Since the concentration of netrin, which attracts these neurons, is highest at the midline floor plate, why do they not stop at the midline? One possible explanation is that upon reaching the midline, these neurons become desensitized to chemoattraction by netrin. Researchers tested this hypothesis by placing either an ectopic floor plate or netrin-secreting cultured cells contralateral to the commissural neurons in explants and assessing their ability to grow toward the

Explant

(a) No manipulation

Commissural neurons

Floor plate

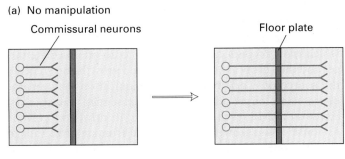

(b) Ectopic floor plate contralaterally

Ectopic floor plate

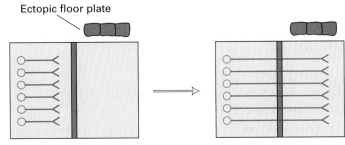

(c) Removal of floor plate and ectopic floor plate contralaterally

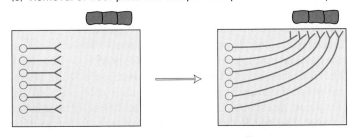

▲ **FIGURE 23-37 Experimental demonstration that the floor plate desensitizes commissural neurons to the floor-plate attractant signal.** (a) In explants from the metencephalic region of rat embryos, commissural neurons (red) continued growing beyond the floor plate (green). (b) Ectopic floor-plate tissue placed contralaterally did not attract neurons once they had crossed the floor plate. (c) When the floor plate was removed, neurons were attracted to the ectopically placed floor-plate tissue. Similar results were obtained using aggregates of cells expressing recombinant netrin in place of the ectopic floor plate. [See R. Shirasaki et al., 1998, *Science* **279**:105.]

netrin source. Remarkably, once commissural neurons crossed the midline, they were no longer attracted to netrin (Figure 23-37b,c).

Effect of cAMP on Growth-Cone Response As we described previously, *Xenopus* retinal ganglion cells exposed to a gradient of netrin turn towards it (see Figure 23-36). If these cells are incubated with an inhibitor of cAMP-dependent protein kinase (cAPK), they are no longer attracted to netrin but rather are repelled by it. Hence, high levels of cAMP promote growth-cone attraction and low levels promote repulsion.

While the in vivo significance of this regulatory mechanism is not yet known, it provides an intriguing model by which growth cones can be attracted to an intermediate target and then grow away from it. Once a growth cone reaches the source of a netrin signal, communication between the intermediate target and the growth cone may lead to a decrease in cAMP levels. This decrease, in turn, converts the response of the growth cone to netrin from an attractant to repellent one. As a result, the high concentration of the attractant, which initially drew the growth cone to it, is now interpreted as a high concentration of a repellent and drives the growth cone away from it. Although the cAPK inhibition is similar to the effect of misexpressing UNC-5 in these neurons, it is not known whether the molecular pathways underlying this switch are also similar.

Other Signaling Systems Can Both Attract and Repel Growth Cones

Studies in both vertebrates and invertebrates have identified other secreted factors that act both as attractants and repellents for neuronal outgrowth. For instance, in *Drosophila* a secreted protein called Slit is expressed at the midline. Binding of Slit to a receptor called Roundabout (Robo) prevents *Drosophila* axons that have crossed the midline once from crossing it again. In *robo* mutants, axons cross and recross the midline multiple times, forming swirls of axons that look like roundabouts on British roads. Hence, the Robo receptor mediates chemorepulsion in response to Slit. Slit also can act as an attractant that promotes branching of vertebrate axons. The Slit/Robo system controls aspects of midline guidance in worms, flies, and vertebrates; thus like the netrin system, it has been conserved during evolution. Another important class of guidance molecules, the *semaphorins,* are also found in both vertebrate and invertebrate; they too can act both as attractants and repellents.

SUMMARY Directional Control of Neuronal Outgrowth

- The best-characterized neuronal guidance system regulates growth-cone migration along the dorsoventral axis. In this system, a secreted signaling molecule called netrin in vertebrates and UNC-6 in *C. elegans* is expressed in a graded fashion, with the highest levels in ventral regions.

- Neurons that express UNC-40 on their growth cones are attracted to netrin. Those that express UNC-40 and UNC-5, which form a heteromultimer, are repelled by netrin. Thus the same guidance signal can act as a chemoattractant or chemorepellent depending on the receptors expressed on a growth cone (see Figure 23-36).

- A growth cone may change its responsiveness to a guidance signal, such as netrin, depending on its exposure to the signal or the activity of other signaling molecules in the growth cone. This phenomenon may explain the ability of neurons to migrate in a stepwise fashion in which they are attracted to an intermediate target cell and then grow past and away from this cell as they navigate to their final targets.

- Other signaling systems such as the semaphorins and Slit (ligand)/Robo (receptor) pathways also regulate neuronal outgrowth in both vertebrates and invertebrates. Like the netrin system, these systems can both attract and repel growth cones.

23.7 Formation of Topographic Maps and Synapses

Upon reaching a target region, growth cones must choose with which type of target cells to form connections. Typically many different types of neurons are present within a given target region, and growth cones from particular neurons will form synapses with only a subset of them. The ability of growth cones to select among different types of target neurons (or muscle cells) is referred to as *cell-type specificity.* In addition, in many regions of the nervous system, particularly in vertebrates, large populations of neurons of one cell type from one region of the nervous system connect to a population of another cell type in a target region. The ability of innervating neurons to maintain their spatial relationships in the field of target cells is referred to as *topographic specificity.* This phenomenon leads to a defined spatial relationship, or **topographic map**, between neurons and their targets. Although little is known about how specific neurons select their postsynaptic partners from a population of different cell types, considerable progress has been made in dissecting the molecular mechanisms regulating formation of topographic maps, particularly in the visual system of vertebrates.

Visual Stimuli Are Mapped onto the Tectum

Both electrophysiological and axon labeling experiments revealed the existence of topographic maps in the vertebrate visual system. For instance, stimulation of a specific region of the retinal field produces an electrical response in cells in defined regions of the *tectum,* the region of the brain to which the retinal ganglion cells project in lower vertebrates (e.g., frogs). By labeling cells in specific regions of the visual field with a dye that is transported along retinal ganglion cell axons, the map can be directly visualized in the tectum. For instance, in the chick retina, nasal axons (located near the nose) project to the posterior tectum, whereas temporal

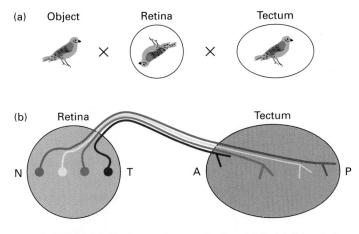

▲ FIGURE 23-38 Retinotopic map in the chick. (a) Stimulation of the visual space in the retina is directly mapped onto the tectum in the chick brain. (b) Retinal ganglion neurons in the retina elaborate a topographic map along both the anteroposterior axis and dorsoventral axis (not shown). Along the anteroposterior axis, nasal (N) axons project to posterior regions (P) of the tectum, and temporal (T) axons project to anterior (A) regions. [Adapted from J. F. Flanagan and P. Vanderhaegen, 1998, *Annu. Rev. Neurosci.* **21**:309.]

axons (located near the temple) project to the anterior tectum (Figure 23-38). Such experiments show that stimuli in the visual field are mapped in a smooth continuous manner in the tectum.

A classic series of surgical experiments in lower vertebrates produced evidence that retinal ganglion cells specifically recognize their targets in the tectum. In one set of experiments scientists removed the eye of a frog, cut the optic nerve that connects the eye to the tectum, rotated the eye 180 degrees and then placed it back into the eye capsule. When the frog was exposed to a visual stimulus, its behavioral response was directed about 180 degrees from the response elicited prior to the surgery or in control animals. The simplest interpretation of these studies, later validated using anatomical techniques, was that the retinal ganglion cells in the rotated eye connected to the position of the tectum reflecting their original position in the retinal field.

Temporal Retinal Axons Are Repelled by Posterior Tectal Membranes

The discovery of topographic maps in the visual system suggested that retinal ganglion cells in different regions of the visual field are directed to connect to different target regions in the tectum during development and that chemical differences between regions of the tectum are detected by the innervating retinal neurons. Because there are far too many neurons for each site in the tectum to express a different signal, it seemed likely that retinal axons from different regions of

the retina are able to detect quantitative differences in signals. That is, the graded distribution of a chemical signal in the target region specifies the retinotopic map.

Important progress toward identifying the signals responsible for establishing the retinotopic map came from studies of chick retinal ganglion cells growing in culture. To test whether retinal neurons can distinguish between molecular signals expressed by anterior and posterior tectal cells, scientists placed retinal tissue removed from chicks in a culture containing alternating lanes of anterior and posterior tectal membranes. Although nasal axons grew over the entire plate and did not discriminate between the lanes, the temporal axons extended only along the membranes prepared from the anterior tectum (Figure 23-39a).

By testing the response of individual growth cones to tectal membranes, scientists found that temporal growth cones collapse and fail to advance when incubated in the presence of posterior tectal membranes but are unaffected by anterior tectal membranes (Figure 23-39b). Nasal growth cones showed no response to either anterior or posterior tectal membranes. These studies led to the proposal that the graded distribution of a repellent affecting temporal growth cones plays an important role in establishing a retinotopic map.

Ephrin A Ligands Are Expressed as a Gradient along the Anteroposterior Tectal Axis

In an earlier section, we saw that ephrin B2 and its receptor, both expressed on the surface of cells, mediate reciprocal inductive interactions during angiogenesis. Another ephrin ligand/Eph receptor system has been shown to function in establishing the topographic map of retinal growth cones along the anteroposterior axis of the tectum. Although the cytosolic domain of the Eph receptors are homologous to other well-characterized receptor tyrosine kinases, there extracellular domains are novel (see Figure 23-25). One such receptor, designated EphA3, is widely expressed in the developing vertebrate brain. Since the ligands binding to the Ephs were unknown at first, these receptors were referred to as *orphan receptors*.

To identify the ligand(s) that binds to EphA3 (and other Ephs) scientists fused the region of the gene encoding its extracellular ligand-binding domain to a region of DNA encoding alkaline phosphatase and produced the encoded chimeric protein in cultured cells (Figure 23-40a). Researchers reasoned that the extracellular ligand-binding domain of the chimeric protein would bind tightly to ligands in tissue; the activity of the alkaline phosphatase domain then could be used as a histochemical stain to visualize the location of the bound ligands. This experimental approach detected a ligand for EphA3 that is expressed as a gradient in the tectum with high levels posteriorly and low levels anteriorly (Figure 23-40b). The receptor–alkaline phosphatase chimeric protein was also used as an affinity reagent to clone the ligand from a cDNA expression library (see Figure 20-9). This led to the isolation of the ligand, designated ephrin A2.

(a)

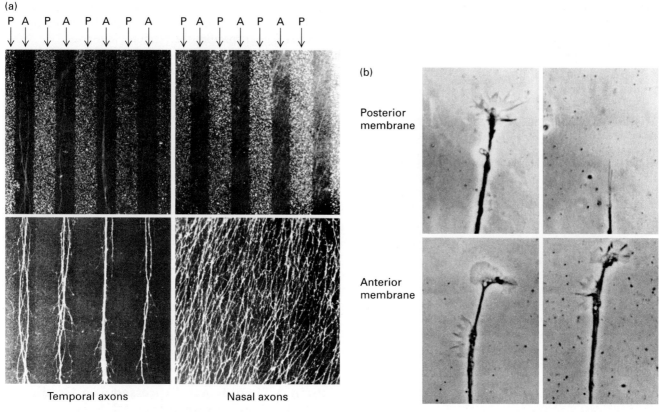

Temporal axons   Nasal axons

▲ **FIGURE 23-39 Effect of tectal membranes on outgrowth of chick retinal neurons.** (a) Temporal and nasal retinal tissues *(bottom)* were added to culture dishes to which stripes of anterior (A) and posterior (P) tectal membranes were affixed *(top)*. Temporal axons grew only along the anterior membranes, whereas nasal axons did not discriminate between anterior and posterior membranes. The membrane stripes were stained with a green-fluorescing dye (FITC); the axons, with a red-fluorescing dye (RITC). (b) Addition of posterior tectal membranes, but not anterior membranes, to individual cultured temporal neurons induced collapse of their growth cones within 25 minutes. Small numbers indicate minutes after addition of membranes. [Part (a) from: J. Walter et al., 1987 *Development* **101**:690. Part (b) from E. C. Cox et al., *Neuron* **4**:31.]

(b)

Posterior membrane

Anterior membrane

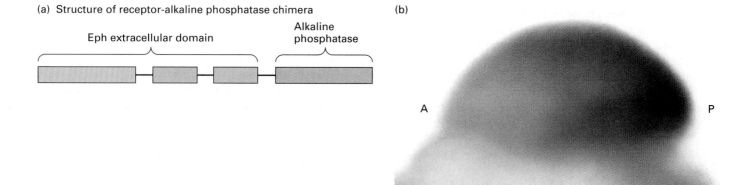

(a) Structure of receptor-alkaline phosphatase chimera

Eph extracellular domain   Alkaline phosphatase

(b)

A   P

▲ **FIGURE 23-40 Identification of a graded Eph-binding signal in the tectum.** (a) A DNA construct was engineered that encoded the extracellular domain of the Eph receptor (e.g., EphA3) fused to alkaline phosphatase. The chimeric protein, which binds any ligands for the receptor that are present in tissues, functions as an affinity reagent and can be used as a histological stain. (b) After the chimeric protein was incubated with the tectum, a soluble substrate of alkaline phosphatase was added. Hydrolysis of the substrate produces a purple precipitate over cells expressing the ligand. The gradation in the purple color in the stained tectum shown here indicates that the ligand for the Eph receptor is expressed as a gradient with high levels posteriorly (P) and low levels anteriorly (A). [Part (b) from H.-J. Cheng et al., 1995, *Cell* **82**:371.]

A biochemical approach was used to identify another ephrin ligand (A5) that is selectively expressed in the posterior tectum. Both ephrin A2 and A5 are expressed during the period in development when the repellent activity of posterior tectal membranes on temporal growth cones is detected. Both also are linked to the membrane by a GPI anchor (see Figure 23-25). Treatment of posterior tectal membranes with phospholipase C, which cleaves the GPI anchor, inactivates their repellent activity, further evidence that ephrin A2 and A5 act as chemorepellents.

Additional evidence that ephrin A2 acts as a repellent for temporal axons came from experiments in which membranes prepared from tissue culture cells overexpressing ephrin A2 were used in a stripe assay similar to that shown in Figure 23-39a. Temporal axons preferentially migrated along stripes lacking ephrin A2, but nasal axons showed no preference. In another study, anterior tectal membranes were infected with a retrovirus whose genome encoded ephrin A2, resulting in expression of ephrin A2 in patches of the chick retina. Temporal axons avoided these patches, whereas nasal axons grew over them. To assess why different retinal axons respond differently to ephrin A2, scientists examined the distribution of one of its receptors, EphA3.

The EphA3 Receptor Is Expressed in a Nasal-Temporal Gradient in the Retina

In situ hybridization experiments demonstrated that EphA3, a receptor for ephrin A2, is expressed as a gradient in the retina, with high levels in the temporal region and low levels in the nasal region. To determine the distribution of the EphA3 receptor on retinal growth cones in the tectum, scientists generated a ligand–alkaline phosphatase fusion protein similar to the receptor fusion protein depicted in Figure 23-40a. When researchers used the ephrin A2–phosphatase fusion protein as a histological reagent, the anterior tectum was heavily stained, reflecting the high-level expression of EphA3 on temporal retinal growth cones, and the posterior tectum was lightly stained, reflecting the low-level expression of EphA3 on nasal retinal growth cones.

These data suggest that the topographic map established along the anterioposterior tectal axis reflects the graded distribution of a repellent (ephrin A2) in the target tissue and the complementary pattern of expression of the receptor for it (EphA3) in the retina (Figure 23-41). Thus temporal retinal axons, which express high levels of EphA3, are repelled by posterior tectal cells, which express high levels of ephrin A2. Conversely, since nasal retinal axons express very low levels of the EphA3 receptor, they can innervate the posterior tectum. Although these studies provide insight into why temporal axons avoid the posterior tectum, they do not explain why nasal axons do not innervate the anterior tectum. One possible explanation is that a chemoattractant selective for nasal axons is expressed preferentially in the posterior tectum.

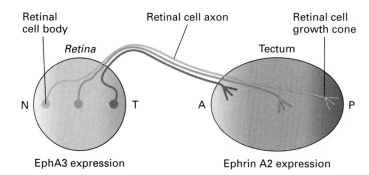

▲ **FIGURE 23-41 Graded expression of ephrin A2 and its receptor, EphA3, in the chick retina and tectum.** EphA3 is expressed at the highest levels on the surface of temporal retinal axons and their growth cones, whereas ephrin A2 is expressed at the highest levels on posterior tectal cells. Temporal axons with high levels of the EphA3 receptor avoid the high levels of ephrin A2 in the posterior tectum. As a result, they project to the anterior tectum. N = nasal; T = temporal; A = anterior; P = posterior. See text for discussion. [See H - J. Cheng et al., 1995, *Cell* **82**:371.]

Motor Neurons Induce Assembly of the Neuromuscular Junction

Once in the target region, the neuronal growth cone must select a specific target cell with which to form synapses. While little is known about the formation of nerve-to-nerve synapses, it is likely that the lessons learned from the detailed biochemical and genetic studies of nerve-muscle synapse, or neuromuscular junction (NMJ), formation will provide important insights into this issue. In this section, we discuss the role of an extracellular signal, agrin in NMJ formation.

The intricate cellular and molecular architecture of the NMJ is designed for rapid focal transmission of the nerve impulse to the muscle (see Figure 21-4). The highly specialized components of the synapse, including synaptic vesicles containing acetylcholine (ACh) in the presynaptic cell and acetylcholine receptors (AChR) in the postsynaptic muscle cell, must be concentrated and then assembled to lie in precise apposition from one another across the synaptic cleft. This process requires communication between the nerve and muscle cells.

Experiments on regeneration of damaged muscle and nerve, outlined in Figure 23-42, demonstrate that a specialized region of the basal lamina, which is present in the synaptic cleft, induces both nerve and muscle to form the specialized structures of the NMJ. Even if muscle regeneration is prevented, axons still return to precisely the original site on the basal lamina and form an axon terminus with synaptic vesicles. Conversely, if myofibers regenerate—but the axon does not—specialized regions of the plasma membrane enriched in AChRs form at the original synaptic site on the basal lamina.

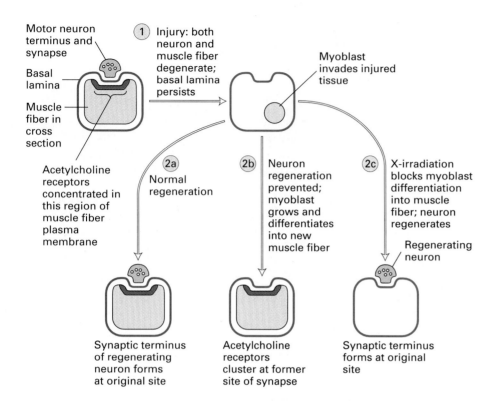

Motor neuron terminus and synapse

Basal lamina

Muscle fiber in cross section

Acetylcholine receptors concentrated in this region of muscle fiber plasma membrane

① Injury: both neuron and muscle fiber degenerate; basal lamina persists

Myoblast invades injured tissue

②a Normal regeneration

②b Neuron regeneration prevented; myoblast grows and differentiates into new muscle fiber

②c X-irradiation blocks myoblast differentiation into muscle fiber; neuron regenerates

Regenerating neuron

Synaptic terminus of regenerating neuron forms at original site

Acetylcholine receptors cluster at former site of synapse

Synaptic terminus forms at original site

◀ FIGURE 23-42 A specialized region of the basal lamina determines the site of the neuromuscular junction. When a striated frog limb muscle and innervating motor neurons are damaged, both the muscle and the nerve axon degenerate. All that remains are the muscle basal lamina and some surrounding Schwann cells. With time, myoblasts invade the space and differentiate into muscle fibers, and the motor neuron regenerates a new axon. The regenerated axons form synapses with the regenerating myofiber precisely at the sites of the original synapses. If muscle regeneration is prevented by x-irradiation, the axon regenerates synaptic termini at the original site. If nerve regeneration is prevented, the regenerating muscle fiber forms a synaptic specialization marked by a high concentration of acetylcholine receptors, again at the original site. [Adapted from S. J. Burden, 1987, in *The Vertebrate Neuromuscular Junction*, Liss, pp. 163–186.]

These findings led investigators to search for molecules in the basal lamina that direct where on the muscle cell a postsynaptic specialization will form. From the time of the initial contact between a neuron and a muscle cell to formation of a complete, functional synapse between them, more than 40 known molecules become concentrated specifically at the synapse. One of the earliest to become concentrated, and one of obvious physiological significance, is the AChR on the muscle-cell membrane. AChR subunits are synthesized as myoblasts fuse and differentiate into myotubes (see Chapter 14).

Prior to innervation, there are approximately 1000 AChR molecules/μm^2 on the muscle surface. In the mature NMJ, the density increases to 10,000 molecules/μm^2, while the density of AChR in the muscle membrane outside the NMJ falls to about 10 molecules/μm^2. At least three different signaling processes control this morphological differentiation: (1) Preexisting AChR on the surface of the uninnervated muscle are induced to aggregate by agrin, a signal released from the nerve. (2) The nerve releases a factor called ARIA (Acetylcholine Receptor Inducing Activity) that stimulates transcription of AChR subunits in nuclei that underlie the developing synapse. (3) AChR released from the nerve leads to voltage changes and Ca^{2+} entry, which represses transcription of AChR subunits in nuclei in the muscle syncitium that do not underlie synapses. The first step in synapse induction is the release of agrin from motor neurons as they approach muscle fibers. In the next two sections, we discuss agrin and its receptor.

Agrin Induces Acetylcholine Receptor Aggregation The localization of AChRs is easily monitored using fluorescent-labeled bungarotoxin, which directly binds to these receptors. In muscle cells cultured in the absence of motor neurons, AChRs are distributed randomly over the surface of the muscle cell. Contact with a co-cultured motor neuron results in rapid clustering of AChRs via lateral diffusion of receptor molecules in the muscle-cell plasma membrane to the site of nerve-muscle contact.

Cultured chick myotubes were used to assay the AChR–clustering activity of protein fractions of basal lamina prepared from a particularly rich source of cholinergic synapses, the electric ray *Torpedo californica*. From this source, investigators isolated and partially sequenced a protein they named *agrin*. Staining with antibodies specific for agrin showed that it is stably associated with the basal lamina at the NMJ. Furthermore, inclusion of anti-agrin antibodies in nerve-muscle co-cultures prevented the aggregation of AChRs (Figure 23-43). Agrin thus mediates aggregation of AChRs, as well as a dozen other components, at the NMJ.

Sequencing of agrin cDNA revealed that it encodes a 200-kDa protein related to several known proteins, including laminin, another component of the basal lamina. The N-terminal half of the agrin molecule contains matrix-binding sites; the C-terminal half contains membrane-binding sites and is required for the AChR-clustering activity of agrin. The protein has several repeats found in other extracellular-matrix proteins or signaling molecules. Alternative splicing of agrin transcripts results in a number of agrin isoforms,

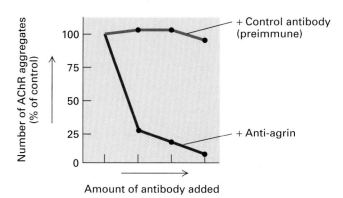

▲ FIGURE 23-43 Experimental demonstration that agrin, which is associated with the basal lamina, promotes aggregation of acetylcholine receptors (AChRs). Co-cultures of nerve and muscle cells were incubated in the presence or absence of agrin-specific antibody. Clustering of AChRs in the muscle-cell plasma membrane was determined by staining with fluorescent-labeled bungarotoxin, which binds specifically to acetylcholine receptors. Addition of anti-agrin antibodies prevented formation of AChR aggregates. [Adapted from N. E. Reist et al., 1992, *Neuron* **8**:865.]

some of which are inactive in AChR-clustering assays. Localization studies indicate that both muscle and motor neurons produce agrin, that motor neurons transport agrin to the nerve terminus, and that agrin from both cellular sources is localized at the neuromuscular junction. Scientists identified a neuron-specific form of agrin (formed by alternative splicing) and demonstrated that it is about 1000-fold more active in inducing AChR clustering than agrin isoforms expressed by muscle. AChRs do not aggregate and NMJs do not form in mice lacking agrin.

A Receptor Tyrosine Kinase Is Part of the Agrin Receptor Agrin causes AChR clustering at very low concentrations (0.1–1 ppm) and exhibits saturable, high-affinity binding to muscle-cell membranes. These data and the finding that one molecule of agrin mediates clustering of ≈200 AChRs suggest that agrin does not directly bind to the AChR. Rather, it most likely functions as a signaling molecule controlling the assembly of postsynaptic specializations at the site of motor neuron contact.

A key component of the agrin receptor was identified as a muscle-specific receptor tyrosine kinase. Scientists were interested in identifying receptors for novel factors regulating muscle differentiation or survival. They reasoned that genes encoding such receptors would be upregulated in response to denervation of skeletal muscle. A single muscle-specific cDNA clone encoding a novel receptor tyrosine kinase was identified from mRNA isolated from denervated muscle. This receptor, called MuSK (Muscle Specific Kinase), is expressed at low levels in myoblasts and is upregulated with the onset of myoblast fusion and differentiation. While overall levels of this receptor decrease in mature muscle, the receptors become concentrated in postsynaptic membranes at the NMJ.

Several lines of evidence indicate that MuSK is part of a receptor complex for agrin. Knockout mice lacking MuSK exhibit a phenotype similar to the agrin knockout. While agrin induces AChR aggregation in wild-type myotubes in vitro, it does not do so in myotubes from MuSK mutants. And finally, MuSK tyrosine kinase activity is stimulated specifically by the neuronal isoform of agrin. Nevertheless, agrin does not appear to bind directly to the extracellular domain of MuSK. These and other studies indicate that additional myotube components or myotube-specific modification of MuSK is necessary to generate an active agrin receptor.

How does MuSK promote AChR aggregation? Important clues came from the analysis of rapsyn, a 43-kD protein identified in biochemical studies as copurifying with the AChR. That rapsyn can aggregate AChR was shown in gain-of-function studies. While AChRs are expressed diffusely along the surface of nonmuscle cells transfected with AChR subunit genes, cotransfection with rapsyn induced receptor clustering. Conversely, rapsyn knock-out mice fail to aggregate receptors. The precise biochemical mechanisms by which activation of MuSK promotes rapsyn-dependent aggregation of AChR is as yet unclear.

These studies suggest the following pathway for establishing the NMJ: Agrin is released from motor neuron growth cone. MuSK activates rapsyn-dependent-AChR aggregation. Many other muscle components are induced to aggregate at the synapse including the receptors for ARIA. Myotube nuclei directly underlying the receptor are induced to up-regulate transcription of synapse-specific components by ARIA released from motor neurons. Transcription of AChR genes and other postsynaptic components are repressed in nuclei in regions away from the synapse. Repression requires depolarization of the muscle membrane by activation of AChR by presynaptic release of acetylcholine from motor nerve terminals.

SUMMARY Formation of Topographic Maps and Synapses

- A common feature of neuronal organization is that neuronal cell types of the same class arrayed in one field (e.g., retinal ganglion neurons in the retina) maintain their spatial relations to each other when they connect to their target cells in another field (e.g., the tectum). Such defined spatial arrangements, referred to as topographic maps, are found in the vertebrate visual system (see Figure 23-38).

- Topographic mapping in the visual system is regulated in part by ephrin A1, a cell-surface grow-cone repellent expressed in the tectum, and its receptor, EphA3, expressed on the growth cones of retinal neurons.

- Ephrin A1 and EphA3 are both expressed in a graded but complementary distribution (see Figure 23-41). Temporal retinal growth cones, which express high levels

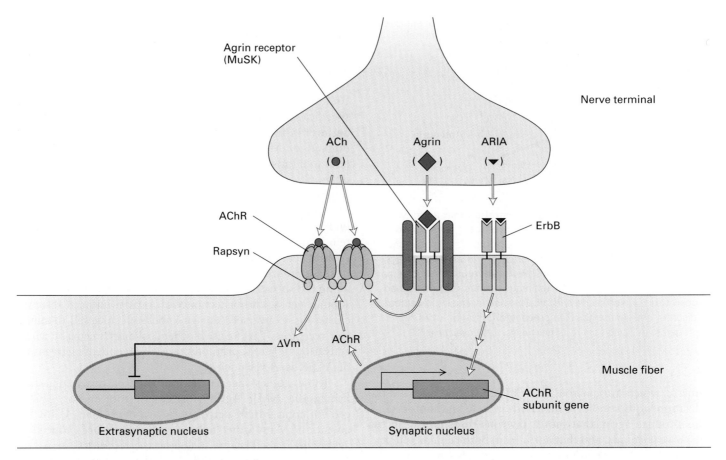

▲ **FIGURE 23-44 Signals from the motor neuron regulate the expression and localization of the acetylcholine receptor (AChR).** Agrin binds to the agrin receptor, which includes MuSK and additional components. Stimulation of the receptor promotes rapsyn-dependent aggregation of the AChR. The nerve also secretes ARIA. ARIA receptors (ErbB) also cluster beneath the nerve terminal. ErbB activation stimulates transcription of AChR subunits in synaptic nuclei. ACh release from the nerve propagates a voltage change in the muscle membrane that depresses the expression of the AChR in extrasynaptic nuclei. (Adapted from J. R. Sanes, 1997. *Cur. Opinion Neurobiol.* **7**:93–100).

of Eph3 receptor, are repelled by the high levels of ephrin A1 produced in the posterior tectum. Conversely, nasal retinal growth cones, which express low levels of Eph3 receptor, are less sensitive to the repelling activity of ephrin 1 in the posterior tectum, permitting them to form connections with target cells in this region.

• A specialized chemical synapse, the neuromuscular junction, is established between a motor neuron and its target muscle. Formation of this junction requires the concentration, assembly, and precise alignment of numerous components in both the innervating nerve and the muscle cell.

• Agrin, is released from the motor neuron and mediates clustering of acetylcholine receptors and other components in the neuromuscular junction. MuSK is a component of the agrin receptor on muscle membranes. Rapsyn is a cytosolic protein that binds to the AChR and mediates clustering.

23.8 Cell Death and Its Regulation

In the preceding sections, we have described various extracellular signaling molecules, and their intracellular signaling pathways, that play a role in regulating cell division, pattern formation, differentiation, morphogenesis, and motility. In this final section, we consider signaling pathways regulating cell survival. *Programmed cell death*, a central mechanism controlling multicellular development, leads to deletion of entire structures (e.g., the tail in developing human embryos), sculpts specific tissues by ablating fields of cells (e.g., tissue between developing digits), and regulates the number of neurons in the nervous system. In the mammalian nervous system, for instance, the majority of cells generated during development also die during development.

Cellular interactions regulate cell death in two fundamentally different ways. Most, if not all, cells in multicellular organisms require signals to stay alive. In the absence of such survival signals, frequently referred to as **trophic factors**, cells activate a "suicide" program. In some developmental contexts, including the immune system, specific signals induce a "murder" program that kills cells. Whether cells commit suicide for lack of survival signals or are murdered by killing signals from other cells, recent studies suggest that death is mediated by a common molecular pathway. In this final section, we first distinguish programmed cell death from death due to tissue injury, then consider the role of trophic factors in neuronal development, and finally describe the evolutionarily conserved effector pathway that leads to cell suicide or murder.

Programmed Cell Death Occurs through Apoptosis

The demise of cells by programmed cell death is marked by a well-defined sequence of morphological changes, collectively referred to as **apoptosis**, a Greek word that means "dropping off" or "falling off" as in leaves from a tree. Dying cells shrink and condense and then fragment, releasing small membrane-bound apoptotic bodies, which generally are phagocytosed by other cells (Figure 23-45). Importantly, the intracellular constituents are not released into the extracellular milieu where they might have deleterious effects on neighboring cells. The highly stereotyped changes accompanying apoptosis suggested to early workers that this type of cell death was under the control of a strict cellular program.

In contrast, cells that die in response to tissue damage exhibit very different morphological changes. Typically, cells that undergo this process, called **necrosis**, swell and burst, releasing their intracellular contents, which can damage surrounding cells and frequently cause inflammation.

Neutrophins Promote Survival of Neurons

The earliest studies demonstrating the importance of trophic factors in cellular development came from analyses of the

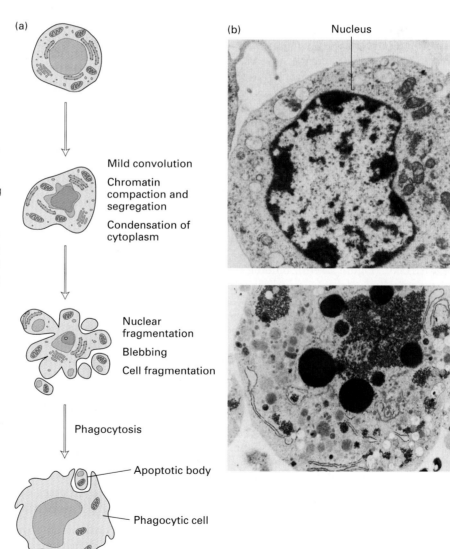

▶ **FIGURE 23-45 Ultrastructural features of cell death by apoptosis.** (a) Schematic drawings illustrating the progression of morphologic changes observed in apoptotic cells. Early in apoptosis, dense chromosome condensation occurs along the nuclear periphery. The cell body also shrinks although most organelles remain intact. Later both the nucleus and cytoplasm fragment, forming apoptotic bodies. These are phagocytosed by surrounding cells. (b) Photomicrographs comparing a normal cell *(top)* and apoptotic cell *(bottom)*. Clearly visible in the latter are dense spheres of compacted chromatin as the nucleus begins to fragment. [Part (a) adapted from J. Kuby, 1997, *Immunology*, 3d ed., W. H. Freeman & Co., p. 53. Part (b) from M. J. Arends and A. H. Wyllie, 1991, *Inter. Rev. Exp. Pathol.* **32**:223.]

(a)

Mild convolution

Chromatin compaction and segregation

Condensation of cytoplasm

Nuclear fragmentation

Blebbing

Cell fragmentation

Phagocytosis

Apoptotic body

Phagocytic cell

(b) Nucleus

MEDIA CONNECTIONS
Apoptosis

developing nervous system. In the early 1900s the number of neurons innervating the periphery was shown to depend upon the size of the target field. For instance, removal of limb buds from the developing chick embryo leads to a reduction in the number of sensory neurons and motoneurons innervating it, while grafting of ectopic limb tissue leads to an increase in the number of neurons in corresponding regions of the spinal cord and sensory ganglia. Indeed, incremental increases in target-field size is accompanied by commensurate incremental increases in the number of neurons innervating it. This relation was found to result from the selective survival of neurons rather than changes in their differentiation or proliferation. The observation that neurons die after reaching their target field in several regions of the nervous system suggested that neurons compete for survival factors produced by the target tissue.

Subsequent to the early observations that peripheral tissue promotes survival of sensory and motoneurons, scientists discovered that transplantation of a mouse sarcoma tumor into a chick led to a marked increase in cell number in sympathetic and spinal ganglion neurons. This finding implicated the tumor as a rich source of the presumed trophic factor. To isolate and purify this factor, known simply as *nerve growth factor* (NGF), scientists used an in vitro assay in which outgrowth of neurites from sensory ganglia was measured. The later discovery that the submaxillary gland in the mouse also produces large quantities of NGF enabled biochemists to purify it to homogeneity and to sequence it. A homodimer of two 118-residue polypeptides, NGF belongs to a family of structurally and functionally related trophic factors, collectively referred to as *neutrophins*. Brain-derived growth factor (BDNF) and neurotrophin-3 (NT-3) also are members of this protein family.

Neurotrophin Receptors   Neurotrophins bind to and activate a family of receptor tyrosine kinases called *Trks* (pronounced "tracks"). NGF binds to TrkA; BDNF, to TrkB; and NT-3, to TrkC (Figure 23-46). Binding of these factors to their receptors provides a survival signal for different classes of neurons. A second type of receptor called p75^{NTR} (NTR = neurotrophin receptor) also binds to neurotrophins, but with lower affinity. However, p75^{NTR} forms heteromultimeric complexes with the different Trk receptors; this association increases the affinity of Trks for their ligands. Recent studies indicate that the binding of NGF to p75^{NTR} in the absence of TrkA may promote cell death rather than prevent it.

Knockout of Neurotrophins and Their Receptors   To critically address the role of the neurotrophins in development, scientists produced mice with knockout mutations in each of the neurotrophins and their receptors. These studies revealed that different neurotrophins (and their corresponding receptors) were required for the survival of different classes of sensory neurons (Figure 23-47). For instance, pain-sensitive (nociceptive) neurons, which express TrkA, are selectively lost in the dorsal root ganglion of

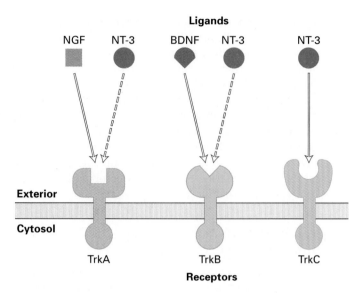

▲ **FIGURE 23-46 Neurotrophins bind to a family of receptor tyrosine kinases called Trks.** Each neurotrophin binds with high affinity to one receptor indicated by the solid arrow from the ligand to the receptor. NT-3 also can bind with lower affinity to both TrkA and TrkB as indicated by the dashed arrow. In addition, neurotrophins bind to a distinct receptor called p75^{NTR} either alone or in combination with Trks. [Adapted from W. D. Snider, 1994, *Cell* **77**:627.]

knockout mice lacking NGF or TrkA, whereas TrkB- and TrkC-expressing neurons are unaffected in such knockouts. In contrast, TrkC-expressing proprioceptive neurons, which detect the position of the limbs, are missing from the dorsal root ganglion in TrkC and NT-3 mutants. That different neurotrophins are required for the survival of different classes of sensory neurons is further supported by the requirement of BDNF and TrkB for the development of sensory neurons in the vestibular ganglia. These neurons innervate organs of the inner ear and are required for sensing motion.

Loss of these different classes of sensory neurons is associated with corresponding defects in behavior. In BDNF mutants, animals show defects in balance associated with the vestibular system. The loss of proprioceptors in NT-3 mutants correlates with postural and movement abnormalities. And finally, although neither NGF nor TrkA mutants survive long enough for behavioral studies, heterozygotes show a decreased sensitivity to pain consistent with a reduction in the number of nociceptive neurons in the dorsal root ganglion.

Three Classes of Proteins Function in the Apoptotic Pathway

Key insights into the molecular mechanisms regulating cell death came from genetic studies in *C. elegans*. Scientists have traced the lineage of all the somatic cells in *C. elegans* from

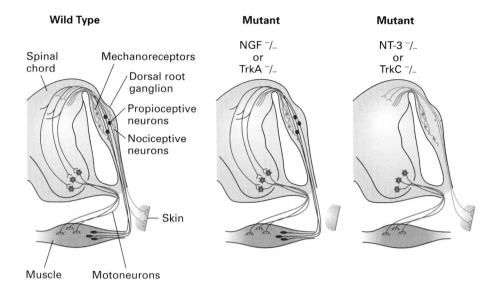

Wild Type

Spinal chord
Mechanoreceptors
Dorsal root ganglion
Propioceptive neurons
Nociceptive neurons
Skin
Muscle   Motoneurons

Mutant

NGF $^{-}/_{-}$
or
TrkA $^{-}/_{-}$

Mutant

NT-3 $^{-}/_{-}$
or
TrkC $^{-}/_{-}$

◀ **FIGURE 23-47 Experimental demonstration that different classes of sensory neurons in the dorsal root ganglion require different trophic factors for survival.** In animals lacking NGF or its receptor TrkA, small nociceptive neurons (blue) that innervate the skin are missing. These neurons express TrkA receptor and innervate NGF-producing targets. In animals lacking either NT-3 or its receptor TrkC, large propioceptive neurons (pink) innervating muscle spindles are missing. Muscle produces NT-3 and the propioceptive neurons express TrkC. Mechanoreceptors (brown), another class of sensory neurons in the dorsal root ganglion, are unaffected in these mutants. [Adapted from W. D. Snider, 1994, *Cell* **77**:627.]

the fertilized egg to the mature worm simply by following the development of live worms under Nomarski optics. Of the 1090 somatic cells generated during development, 131 cells undergo programmed cell death. Specific mutations have identified a variety of genes whose encoded proteins play an essential role in controlling this process. For instance, in worms carrying mutations in the *ced-3* or the *ced-4* genes, programmed cell death does not occur, and all 1090 cells survive (Figure 23-48). In contrast, in *ced-9* mutant animals, all 1090 cells die. These genetic studies indicate that the CED-3 and CED-4 proteins are required for cell death, that CED-9 suppresses apoptosis, and that the apoptotic pathway can be activated in all cells. Moreover, the finding that cell death does not occur in *ced-9/ced-3* double mutants suggests that CED-9 acts upstream of CED-3 to suppress the apoptotic pathway.

That apoptosis involved an evolutionarily conserved pathway was first suggested by the confluence of genetic studies in worms and studies on human cancer cells (Figure 23-49). The first apoptotic gene to be cloned, *bcl-2*, was isolated as a breakpoint rearrangement in human follicular lymphomas and was shown to act as an oncogene that promoted cell survival rather than cell proliferation. Not only are the Bcl-2 and CED-9 proteins homologous, but a *bcl-2* transgene can block the extensive cell death found in *ced-9* mutant worms. Thus both proteins act as regulators that suppress the apoptotic pathway. In addition, both proteins contain a single transmembrane domain and are localized to the outer mitochondrial, nuclear, and endoplasmic reticulum membranes.

(a)

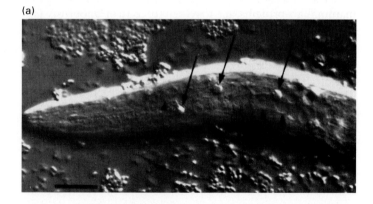

(b)

▶ **FIGURE 23-48 Mutations in the *ced-3* gene block programmed cell death in the nematode *C. elegans*.** During normal development in *C. elegans*, 131 cells die through apoptosis. The entire apoptotic process from the initiation of cell death to the disappearance of a particular cell takes about 1 hour. (a) Newly hatched larva carrying a mutation in the *ced-1* gene. Because mutations in this gene prevent engulfment of dead cells, highly refractile dead cells accumulate (arrows), facilitating their visualization. (b) Newly hatched larva with mutations in both the *ced-1* and *ced-3* genes. The absence of refractile dead cells in these double mutants indicates that no cell deaths occurred. Thus CED-3 protein is required for programmed cell death. This screen also has identified mutations in other genes required for programmed cell death. [From H. M. Ellis and H. R. Horvitz, 1986, *Cell* **91**:818; courtesy of Hilary Ellis]

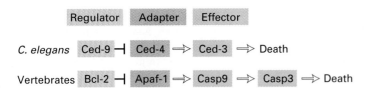

▲ FIGURE 23-49 Overview of the apoptotic pathway in
C. elegans and vertebrates. Three general types of proteins are
critical in this conserved pathway. Regulators either promote or
suppress apoptosis; the two regulators shown here, CED-9 and
Bcl-2, both function to suppress apoptosis in the presence of
trophic factors. Adapters interact with both regulators and effectors; in the absence of trophic factors, they promote activation of
effectors. A family of cysteine proteases serve as effector proteins; their activation leads to degradation of various intracellular
substrates and eventually cell death. [Adapted from D. L. Vaux and
S. J. Korsemeyer, 1999, *Cell* **96**:245.]

The effector molecules in the apoptotic pathway are a
family of enzymes called **caspases**, so named because they
are cysteine proteases that selectively cleave proteins at sites
just C-terminal to aspartate residues. These proteases have
specific intracellular targets such as proteins of the nuclear
lamina and cytoskeleton. Cleavage of these substrates leads
to the demise of a cell. Activation of caspases, discussed below, appears to be a common feature of most, if not all, cell-
death programs. The principal effector protease in *C. elegans* is CED-3. Mammalian cells contain multiple caspases.

Cell-culture studies have yielded important insights into
how the various CED proteins in *C. elegans* and the related
mammalian proteins act together to control apoptosis (see
Figure 23-49). Expression of *C. elegans* CED-4 in a human
kidney cell line induces rapid apoptosis. This can be blocked
by co-expression of CED-9 (or mammalian Bcl-2). CED-9
directly binds to CED-4 and relocalizes it from the cytosol
to intracellular membranes. Thus the pro-apoptotic function
of CED-4 is directly suppressed by the anti-apoptotic function of CED-9. CED-4 also binds directly to CED-3 (and related mammalian caspases) and promotes activation of its protease activity. Biochemical studies have shown that CED-4
can simultaneously bind both to CED-9 and CED-3.

Pro-Apoptotic Regulators Promote
Caspase Activation

Having introduced the major participants in the apoptotic
pathway, we now take a closer look at how the effector caspases are activated in mammalian cells. Although CED-9
and Bcl-2 suppress the cell-death pathway, other regulatory
proteins act to promote apoptosis. The first pro-apoptotic
regulator to be identified, named Bax, was found associated
with Bcl-2 in extracts of cells expressing high levels of Bcl-
2. Sequence analysis demonstrated that Bax is related in sequence to CED-9 and Bcl-2, but overexpression of Bax induces cell death rather than protecting cells from apoptosis,
as CED-9 and Bcl-2 do. Thus this family of homologous

regulatory proteins comprises both anti-apoptotic members
(e.g., CED-9, Bcl-2) and pro-apoptotic members (e. g., Bax).
All members of this family, which we refer to as the **Bcl-2
family**, are single-pass transmembrane proteins and can participate in oligomeric interactions. Thus the fate of a given
cell—survival or death—may reflect the particular spectrum
of Bcl-2 family members present in the cell and the intracellular signaling pathways regulating them.

Recent studies suggest that Bcl-2 family members can influence the subcellular distribution of cytochrome *c*; moreover, biochemical studies have implicated cytochrome *c* in
caspase activation. The current model of caspase activation
in mammalian cells, summarized in Figure 23-50a, accounts
for the involvement of cytochrome *c*. In normal healthy cells,
cytochrome *c* is localized between the inner and outer mito-
chrodrial membrane, but in cells undergoing apoptosis, cytochrome *c* is released into the cytosol. This release can be
blocked by overexpression of Bcl-2; conversely, overexpression
of Bax promotes release of cytochrome *c* into the cytosol and
apoptosis. In the cytosol, binding of cytochrome *c* to the adapter protein Apaf-1 (i.e., mammalian CED-4) promotes activation of a caspase cascade. Bax homodimers, but not Bcl-2
homodimers or Bcl-2/Bax heterodimers, permit influx of ions
through the mitochondrial membrane. It remains unclear how
this ion influx acts to trigger the release of cytochrome *c*.

Gene knockout experiments have dramatically confirmed
the importance of both pro-apoptotic and anti-apoptotic
Bcl-2 family members in neuronal development. Mice lacking the *bcl-xl* gene, which encodes another anti-apoptotic
protein, show massive defects in nervous system development with widespread cell death in the spinal cord, dorsal
root ganglion, and brain of developing embryos. In contrast,
bax knockouts exhibit a marked increase in neurons in some
regions of the nervous system.

Some Trophic Factors Prevent Apoptosis
by Inducing Inactivation of a
Pro-Apoptotic Regulator

We saw earlier that neutrophins such as nerve growth factor (NGF) protect neurons from cell death. The intracellular
signaling mechanisms linking such survival factors to inactivation of the cell-death machinery are not known in detail, but some intriguing clues are available. The finding that
trophic factors appear to work largely independent of protein synthesis suggested that the activity of one or more components of the cell-death pathway is altered post-translationally in response to intracellular signals activated by binding
of trophic factors to their receptors. Scientists demonstrated
that in the absence of trophic factors, the nonphosphorylated form of Bad is associated with Bcl-2/Bcl-xl at the mitochondrial membrane (see Figure 23-50a). Binding of Bad
inhibits the anti-apoptotic function of Bcl-2/Bcl-xl, thereby
promoting cell death. Phosphorylated Bad, however, cannot
bind to Bcl-2/Bcl-xl and is found in the cytosol complexed
to the phosphoserine-binding protein 14-3-3 (Chapter 20).

(a) Absence of trophic factor: Caspase activation

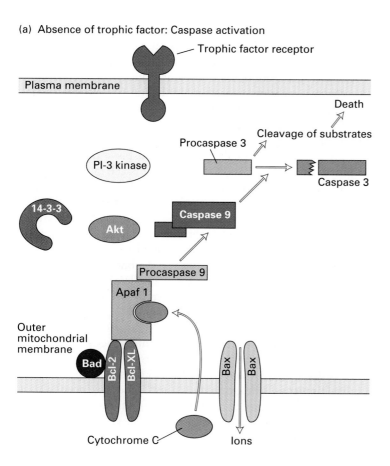

Trophic factor receptor

Plasma membrane

Death

Cleavage of substrates

Procaspase 3

PI-3 kinase

Caspase 3

14-3-3

Akt

Caspase 9

Procaspase 9

Apaf 1

Outer mitochondrial membrane

Bad Bcl-2 Bcl-XL

Bax Bax

Cytochrome C

Ions

(b) Presence of trophic factor: Inhibition of caspase activation

Trophic factor

Plasma membrane

PI-3 kinase

Procaspase 3

14-3-3 ATP

P ADP Akt

Bad

Procaspase 9

Apaf 1

Bad Bcl-2 Bcl-XL

Bax Bax

MEDIA CONNECTIONS
Apoptosis

◀ **FIGURE 23-50 Current models of the intracellular pathways leading to cell death by apoptosis or to trophic factor–mediated cell survival in mammalian cells.** The details of these pathways in any given cell type are not yet known. (a) In the absence of a trophic factor. Bad, a soluble pro-apoptotic protein, binds to the anti-apoptotic proteins Bcl-2 and Bcl-xl, which are inserted into the mitochondrial membrane. Bad binding prevents the anti-apoptotic proteins from interacting with Bax, a membrane-bound pro-apoptotic protein. As a consequence, Bax forms homo-oligomeric channels in the membrane that mediate ion flux. Through an as-yet unknown mechanism, this leads to the release of cytochrome *c* from the space between the inner and outer mitochondrial membrane. Cytochrome *c* then binds to the adapter protein Apaf-1, which in turn promotes a caspase cascade leading to cell death. (b) In the presence of a trophic factor such as NGF. In some cells, binding of trophic factors stimulates PI-3 kinase activity, leading to activation of the downstream kinase Akt, which phosphorylates Bad. Phosphorylated Bad then forms a complex with the 14-3-3 protein. With Bad sequestered in the cytosol, the anti-apoptotic Bcl-2/Bcl-xl proteins can inhibit the activity of Bax, thereby preventing the release of cytochrome *c* and activation of the caspase cascade. [Adapted from B. Pettman and C. E. Henderson, 1998, *Neuron* **20**:633.]

Hence, signaling pathways leading to Bad phosphorylation would be particularly attractive candidates for transmitting survival signals.

A number of trophic factors including NGF have been shown to activate PI-3 kinase, which in turn activates a downstream kinase called Akt (see Chapter 20). This kinase phosphorylates Bad at sites known to inhibit its pro-apoptotic activity. Moreover, a constitutively active form of Akt can rescue cultured neutrophin-deprived neurons, which otherwise would undergo apoptosis and die. These findings support the mechanism for the survival action of trophic factors depicted in Figure 23-50b. In other cell types, different trophic factors may promote cell survival through post-translational modification of other components of the cell-death machinery.

SUMMARY Cell Death and Its Regulation

- Cells die by murder or suicide through programmed cell death, often referred to as apoptosis.

- All cells require trophic factors to prevent apoptosis and thus survive.

- The best-characterized trophic factors are the neutrophins, including NGF, BDNF, and NT-3. During development, neurons compete for a limited supply of these trophic factors in their target fields. As a result, many cells undergo programmed cell death, so that the number of surviving neurons matches the target-field size.

- Genetic studies in *C. elegans* defined an evolutionarily conserved cell-death pathway with three major components (see Figure 23-49). The *C. elegans* anti-apoptotic protein, CED-9, is structurally and functionally homologous to Bcl-2 in vertebrates.

- The effectors of cell death are cysteine proteases, called caspases. Once activated, these proteases cleave specific intracellular substrates leading to the demise of a cell. Pro-apoptotic proteins promote caspase activation, and anti-apoptotic proteins suppress activation.

- Direct interactions between pro-apoptotic and anti-apoptotic proteins lead to cell death in the absence of trophic factors. Binding of extracellular trophic factors can trigger modulation of these interactions via phosphorylation or other post-translational modifications, resulting in cell survival (see Figure 23-50).

PERSPECTIVES for the Future

Enormous progress has been made during the past decade in identifying and characterizing specific developmental signals and their receptors as well as the intracellular signal-transduction pathways by which these receptors influence cellular development. Both genetic and biochemical studies have revealed a plethora of signals mediating the interactions between cells during multicellular development. Many,

indeed most, of these signals are extensively conserved in diverse organisms. Invertebrates often have one or only a few genes encoding a specific class of signaling molecules. Because of gene duplication and subsequent divergence in sequence, vertebrates express multiple isoforms of many signaling molecules; these isoforms are often expressed in different tissues and/or at different stages of development, act through different receptors, and commonly have different activities.

Extracellular signals regulate such basic developmental processes as proliferation, differentiation, cell migration, morphogenesis, and cell survival. Remarkably the same factor can influence different cells in different ways. That is, the response of cells to a specific signal is frequently context-dependent. In the intact developing organism, each cell must be listening to its neighbors and hence in all likelihood is detecting multiple signals, often simultaneously, and integrating them. We understand in some detail how cells detect individual signals, but the mechanisms by which cells integrate two or more signals remain an enigma.

 Abnormalities in cellular interactions can lead to severe health consequences in humans. For instance, loss of Hedgehog (Hh) signaling leads to a developmental abnormality called holoprosencephaly in which development of midline structures is disrupted. In mild cases of the disease, patients may have a single nostril; in severe cases, patients may have single central eye, a condition called cyclopia, and defects in brain development. Mutations in the human gene encoding Ptc, involved in controlling the response of cells to Hh, are associated with the development of many different tumors, including medullablastoma, an aggressive and deadly brain tumor in children. We can anticipate that further understanding of this signaling pathway and others implicated in cancer, such as the TGFβ superfamily, will lead to new effective therapeutic approaches for treating these diseases. As the details of more developmental pathways are uncovered in model organisms, they also will provide clues to the molecular basis of various human diseases.

A series of elegant biochemical and molecular genetic studies have revealed the signals that play important roles in establishing the formation of topographic maps, a central feature of the organization of neuronal projections in many different systems. Recent studies in genetically tractable systems such as *C. elegans* and *Drosophila* suggest that multiple signals act together to control the formation of specific patterns of neuronal wiring. Although enormous progress has been made in identifying signals and their receptors, we still know very little about the mechanisms by which these signals are translated into changes in growth-cone motility, let alone how the growth cone integrates multiple signals. A more complete understanding of the mechanisms underlying the formation of neuronal connections may lead to techniques for coaxing damaged neurons in the central nervous system to regenerate, thereby alleviating the devastating effects of spinal cord injuries.

PERSPECTIVES in the Literature

Scientists studying the organization of the olfactory system in mammals have shown that each olfactory neuron expresses only a single odorant receptor. The mouse genome encodes about 1000 different odorant receptors. These receptors are G protein–coupled receptors (Chapter 20). Each receptor is expressed by many different olfactory neurons that are scattered throughout the olfactory epithelium. Scientists discovered that neurons expressing the same receptor connect to the same group of postsynaptic cells clustered in a structure called a glomerulus. One hypothesis is that olfactory receptors have a dual function: (1) they bind and respond to different odorants in the olfactory epithelium and (2) they function in the developing growth cone to determine glomerular specificity. Read the articles referenced below. Then design an experiment to test the role of individual guidance receptors in targeting olfactory neurons to different glomeruli.

Mombaerts, P., et al. 1996. Visualizing an olfactory sensory map. *Cell* 87:675–686.

Wang, F., A. Nemes, M. Mendelsohn, and R. Axel. 1998. Odorant receptors govern the formation of a precise topographic map. *Cell* 93:47–60.

Testing Yourself on the Concepts

1. How can a gradient of an inductive signal produce different responses at various threshold concentrations? Consider the case of Hedgehog (Hh) as a morphogen.

2. Various inductive signals can be restricted to a local region by lipid modifications or binding to the extracellular matrix (ECM). As a specific example, describe how the diffusion of Hedgehog (Hh) is restricted.

3. Much of embryonic development is associated with cell multiplication, regulated cell migration, and local tissue formation. Programmed cell death, apoptosis, is also a normal part of development. Describe the gene products involved in apoptosis in *C. elegans*.

4. Axons form specific connections. As part of this, axons may migrate first to an intermediate target and then beyond it. Describe how this might occur for *Xenopus* retinal ganglion cells.

MCAT/GRE-Style Questions

Key Concept Please read the section titled "Interactions between Two Equivalent Cells Give Rise to AC and VU cells in *C. elegans*" (p. 1026) and refer to Figure 23-28; then answer the following questions:

1. Interactions can be restricted to adjacent cells if the ligand and receptor are

a. Subject to feedback inhibition.

b. Glycoproteins.

c. Processed properly by the Golgi complex.

d. Membrane proteins of the cells.

2. The default differentiation outcome for these cells is

a. AC.   b. Death.

c. VU.   d. Either AC or VU.

3. The cell expressing slightly more receptor:

a. Becomes the AC cell.

b. Fails to differentiate.

c. Becomes the VU cell.

d. Becomes neither AC nor VU.

4. Cell differentiation in this example is

a. A probabilistic process.

b. Regulated by diffusible factors.

c. Produced only by the interaction of cell surface proteins with no intracellular signaling events.

d. A determined process with only one possible outcome.

Key Application Please read the introductory paragraph in Section 23.1, "Dorsoventral Patterning by TGFβ-Superfamily Proteins," (p. 1004) and the section titled "TGFβ Proteins Bind to Receptors That Have Serine/Threonine Kinase Activity" (p. 1005); then answer the following questions:

5. Addition of purified TGFβ protein to the culture media of certain cells causes cell transformation. Cell transformation is often used as a cell culture model for cancer. Conversely, as described in Chapter 24, TGFβ inhibits the growth of many cells. Indeed, loss of growth inhibition by TGFβ contributes to tumor formation. These observations provide support for all of the following contentions *except*:

a. TGFβ can regulate the proliferative properties of cells.

b. TGFβ is a mutagen.

c. TGFβ likely interacts with one or more cell surface receptors.

d. TGFβ concentration must be highly regulated in the normal situation.

6. What is the likely effect of increased cellular phosphatase activity on the ability of members of the TGFβ superfamily to regulate cell patterning and cell proliferation?

a. No effect.

b. Promotion of TGFβ superfamily effects.

c. Inhibition of TGFβ superfamily effects.

d. Immediate cell death.

7. What would be the likely effect of overexpression of chordin-like molecules that bind to TGFβ on the growth of cells normally inhibited by TGFβ?

a. No effect.

b. Promotion of tumor formation.

c. Inhibition of cellular proliferation.

d. Cell death.

8. Mutations in type I or II TGFβ-superfamily receptors that interfere with the ability of the receptor molecules to oligomerize should have

   a. Negative effects on the ability of TGFβ superfamily members to regulate target cells.

   b. Little, if any, effect on the ability of TGFβ superfamily members to regulate target cells.

   c. Enhancing effects on the ability of TGFβ superfamily members to regulate target cells.

   d. Suppressive effects on mutations in BMPs.

Key Experiment Please read the section titled "The EphA3 Receptor Is Expressed in a Nasal-Temporal Gradient in the Retina" (p. 1041) and refer to Figure 23-41; then answer the following questions:

9. What evidence do these experiments provide on why retinal cells migrate toward the tectum?

   a. Much.   b. Some.

   c. Little.   d. None.

10. These experiments explain why EphA3-rich retinal cell axons

   a. Are more abundant in the anterior tectum than the posterior tectum.

   b. Are less abundant in the anterior tectum than the posterior tectum.

   c. Migrate out from the retina.

   c. Attach to the tectum rather than another tissue.

11. Ephrin A2 acts as a(n)

   a. Attractant for EphA3-rich axons.

   b. Protease that degrades EphA3-rich axons.

   c. Repellant for EphA3-rich axons.

   d. Extracellular matrix (ECM) component for EphA3-rich axons.

12. The chimeric protein containing alkaline phosphatase and ephrin A2 is used as a histological reagent to

   a. Localize the distribution of ephrin A2 protein.

   b. Localize the distribution of receptors binding to ephrin A2.

   c. Localize mRNA encoding ephrin A2.

   d. Localize other molecules related in structure to ephrin A2.

Key Terms

References

Dorsoventral Patterning by TGFβ-Superfamily Proteins

Chen, Y. G., et al. 1998. Determinants of specificity in TGF-β signal transduction. *Genes & Devel.* **12**:2144–2152.

Ferguson, E. L., and K. V. Anderson. 1992. Localized enhancement and repression of the activity of the TGF-β family member, *decapentaplegic*, is necessary for dorsal-ventral pattern formation in the *Drosophila* embryo. *Development* **114**:583–597.

Kishimoto, Y., et al. 1997. The molecular nature of zebrafish swirl: BMP2 function is essential during early dorsoventral patterning. *Development* **124**:4457–4466.

Marqués, G., et al. 1997. Production of a DPP activity gradient in the early *Drosophila* embryo through the opposing actions of the SOG and TLD proteins. *Cell* **91**:417–426.

Massagué, J. 1998. TGF-beta signal transduction. *Annu. Rev. Biochem.* **67**:753–791.

Piccolo, S., Y. Sasai, B. Lu, and E. M. De Robertis. 1996. Dorsoventral patterning in *Xenopus*: inhibition of ventral signals by direct binding of chordin to BMP-4. *Cell* **86**:589–598.

Tissue Patterning by Hedgehog and Wingless

Basler, K., and G. Struhl. 1994. Compartment boundaries and the control of *Drosophila* limb pattern by Hedgehog protein. *Nature* **368**:208–214.

Chiang, C., et al. 1996. Cyclopia and defective axial patterning in mice lacking *sonic hedgehog* gene function. *Nature* **383**:407–413.

Dale, T. C. 1998. Signal transduction by the Wnt family of ligands. *Biochem. J.* **329**:209–223.

Goodrich, L. V., and M. P. Scott. 1998. Hedgehog and Patched in neural development and disease. *Neuron* **21**:1243–1257.

Ng, J. K., K. Tamura, D. Büscher, and J. C. Izpisúa-Belmonte. 1999. Molecular and cellular basis of pattern formation during vertebrate limb development. *Curr. Topics Develop. Biol.* **41**:37–66.

Porter, J. A., K. E. Young, and P. A. Beachy. 1996. Cholesterol modification of Hedgehog signaling proteins in animal development. *Science* **274**:255–259. [Published erratum, 1996, *Science* **274**:5293.]

Riddle, R. D., et al. 1993. Sonic Hedgehog mediates the polarizing activity of the ZPA. *Cell* **75**:1401–1416.

Molecular Mechanisms of Responses to Morphogens

Dyson, S., and J. B. Gurdon. 1998. The interpretation of position in a morphogen gradient as revealed by occupancy of activin receptors. *Cell* **93**:557–568.

Huang, A. M., J. Rusch, and M. Levine. 1997. An anteroposterior Dorsal gradient in the *Drosophila* embryo. *Genes & Develop.* **11**:1963–1973.

Jiang, J., and M. Levine. 1993. Binding affinities and cooperative interactions with bHLH activators delimit threshold responses to the Dorsal gradient morphogen. *Cell* **72**:741–752.

Roelink, H., et al. 1995. Floor plate and motor neuron induction by different concentrations of the amino-terminal cleavage product of Sonic Hedgehog autoproteolysis. *Cell* **81**:445–455.

Reciprocal and Lateral Inductive Interactions

Cacalano, G., et al. 1998. GDNFR-α1 is an essential receptor component for GDNF in the developing nervous system and kidney. *Neuron* **21**:53–62.

Fehon, R. G., et al. 1990. Molecular interactions between the protein products of the neurogenic loci *Notch* and *Delta*, two EGF-homologous genes in *Drosophila*. *Cell* **61**:523–34.

Greenwald, I. 1998. LIN-12/Notch signaling: lessons from worms and flies. *Genes & Develop.* **12**:1751–1762.

Lipschutz, J. H. 1998. Molecular development of the kidney: a review of the results of gene disruption studies. *Am. J. Kidney Dis.* **31**:383–397.

Pan, D., and G. M. Rubin. 1997. Kuzbanian controls proteolytic processing of Notch and mediates lateral inhibition during *Drosophila* and vertebrate neurogenesis. *Cell* **90**:271–280.

Sánchez, M. P., et al. 1996. Renal agenesis and the absence of enteric neurons in mice lacking GDNF. *Nature* **382**:70–73.

Schroeter, E. H., J. A. Kisslinger, and R. Kopan. 1998. Notch-1 signalling requires ligand-induced proteolytic release of intracellular domain. *Nature* **393**:382–386.

Struhl, G., and A. Adachi. 1998. Nuclear access and action of Notch in vivo. *Cell* **93**:649–660.

Wettstein, D. A., D. L. Turner, and C. Kintner. 1997. The *Xenopus* homolog of *Drosophila* Suppressor of Hairless mediates Notch signaling during primary neurogenesis. *Development* **124**:693–702.

Overview of Neuronal Outgrowth

Garriga, G., C. Desai, and H. R. Horvitz. 1993. Cell interactions control the direction of outgrowth, branching and fasciculation of the HSN axons of *Caenorhabditis elegans*. *Development* **117**:1071–1087.

Neugebauer, K. M., K. J. Tomaselli, J. Lilien, and L. F. Reichardt. 1988. N-cadherin, NCAM, and integrins promote retinal neurite outgrowth on astrocytes in vitro. *J. Cell Biol.* **107**:1177–1187.

Raper, J. A., M. J. Bastiani, and C. S. Goodman. 1984. Pathfinding by neuronal growth cones in grasshopper embryos. IV. The effects of ablating the A and P axons upon the behavior of the G growth cone. *J. Neurosci.* **4**:2329–2345.

Tessier-Lavigne, M., and C. S. Goodman. 1996. The molecular biology of axon guidance. *Science* **274**:1123–1133.

Walsh, F. S., and P. Doherty. 1997. Neural cell adhesion molecules of the immunoglobulin superfamily: role in axon growth and guidance. *Annu. Rev. Cell Develop. Biol.* **13**:425–456.

Directional Control of Neuronal Outgrowth

Chan, S. S., et al. 1996. UNC-40, a *C. elegans* homolog of DCC (Deleted in Colorectal Cancer), is required in motile cells responding to UNC-6 netrin cues. *Cell* **87**:187–195.

de la Torre, J. R., et al. 1997. Turning of retinal growth cones in a netrin-1 gradient mediated by the netrin receptor DCC. *Neuron* **19**:1211–1224.

Hamelin, N., et al. 1993. Expression of the Unc-5 guidance receptor in the touch neurons of *C. elegans* steers their axons dorsally. *Nature* **364**:327–330.

Hedgecock, E. M., J. G. Culotti, and D. H. Hall. 1990. The *unc-5*, *unc-6*, and *unc-40* genes guide circumferential migrations of pioneer axons and mesodermal cells on the epidermis in *C. elegans*. *Neuron* **4**:61–85.

Hong, K., et al. 1999. A ligand-gated association between cytoplasmic domains of UNC-5 and DCC family receptors converts netrin-induced growth cone attraction to repulsion. *Cell* **97**:927–941.

Keino-Masu, K., et al. 1996. Deleted in Colorectal Cancer *(DCC)* encodes a netrin receptor. *Cell* **87**:175–185.

Serafini, T., et al. 1994. The netrins define a family of axon outgrowth-promoting proteins homologous to *C. elegans* UNC-6. *Cell* **78**:409–424.

Song, H., et al. 1998. Conversion of neuronal growth cone responses from repulsion to attraction by cyclic nucleotides. *Science* **281**:1515–1518.

Formation of Topographic Maps and Synapses

Cheng, H. J., M. Nakamoto, A. D. Bergemann, and J. G. Flanagan. 1995. Complementary gradients in expression and binding of ELF-1 and Mek4 in development of the topographic retinotectal projection map. *Cell* **82**:371–381.

Flanagan, J. G., and P. Vanderhaeghen. 1998. The ephrins and Eph receptors in neural development. *Annu. Rev. Neurosci.* **21**:309–345.

Frisén, J., et al. 1998. Ephrin-A5 (AL-1/RAGS) is essential for proper retinal axon guidance and topographic mapping in the mammalian visual system. *Neuron* **20**:235–243.

Ruegg, M. A., et al. 1992. The agrin gene codes for a family of basal lamina proteins that differ in function and distribution. *Neuron* **8**:691–699.

Sanes, J. R., and R. H. Scheller. 1997. Synapse formation: a molecular perspective. In W. M. Cowan, T. M. Jessell, and S. L. Zipursky, eds., *Molecular and Cellular Approaches to Neural Development*, Oxford, pp. 179–219.

Walter, J., B. Kern-Veits, J. Huf, B. Stolze, and F. Bonhoeffer. 1987. Recognition of position-specific properties of tectal cell membranes by retinal axons in vitro. *Development* **101**:685–696.

Cell Death and Its Regulation

Chinnaiyan, A. M., K. O'Rourke, B. R. Lane, and V. M. Dixit. 1997. Interaction of CED-4 with CED-3 and CED-9: a molecular framework for cell death. *Science* **275**:1122–1126.

Datta, S. R., et al. 1997. Akt phosphorylation of BAD couples survival signals to the cell-intrinsic death machinery. *Cell* **91**:231–241.

Ellis, H. M., and H. R. Horvitz. 1986. Genetic control of programmed cell death in the nematode *C. elegans*. *Cell* **44**:817–829.

Kerr, J. F. R., A. H. Wyllie, and A. R. Currie. 1972. Apoptosis: a basic biological phenomenon with wide-ranging implications in tissue kinetics. *Brit. J. Cancer* **26**:239–257.

Nagata, S. 1997. Apoptosis by death factor. *Cell* **88**:355–365.

Oltvai, Z. N., C. L. Milliman, and S. J. Korsmeyer. 1993. Bcl-2 heterodimerizes in vivo with a conserved homolog, Bax, that accelerates programmed cell death. *Cell* **74**:609–619.

Pettmann, B., and C. E. Henderson. 1998. Neuronal cell death. *Neuron* **20**:633–647.

Snider, W. D. 1994. Functions of the neurotrophins during nervous system development: what the knockouts are teaching us. *Cell* **77**:627–638.

Vaux, D. L., and S. J. Korsmeyer. 1999. Cell death in development. *Cell* **96**:245–254.

24

Cancer

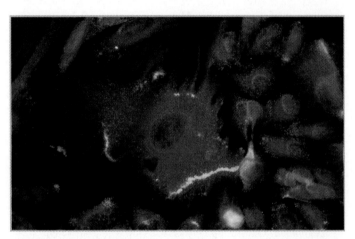

Human melanoma cells stained for a melanoma-specific cell-surface glycoprotein (green), and counterstained for myosin (red) and for DNA (blue).

The multiplication of cells is carefully regulated and responsive to specific needs of the body. In a young animal, cell multiplication exceeds cell death, so the animal increases in size; in an adult, the processes of cell birth and death are balanced to produce a steady state. For some adult cell types, renewal is rapid: intestinal cells have a half-life of a few days before they die and are replaced; certain white blood cells are replaced as rapidly. In contrast, human red blood cells have approximately a 100-day half-life, healthy liver cells rarely die, and, in adults, there is a slow loss of brain cells with little or no replacement.

Very occasionally, the exquisite controls that regulate cell multiplication break down. A cell in which this occurs begins to grow and divide in an unregulated fashion, without regard to the body's need for further cells of its type. When such a cell has descendants that inherit the propensity to proliferate without responding to regulation, the result is a clone of cells able to expand indefinitely. Ultimately, a mass called a **tumor** may be formed by this clone of unwanted cells. Some tumors do not have serious health consequences, but those composed of cells that spread throughout the body usually cause the disease. Cancer is caused by mutations, but there are two key differences between cancer and other genetic diseases. First, cancer is caused mainly by mutations in somatic cells, whereas other genetic diseases are caused solely by mutations in the germ line. Some individuals, however, have inherited genetic mutations that predispose them to develop specific types of cancer. Second, an individual cancer does not result from a single mutation, but rather from the accumulation of as few as 3 to perhaps as many as 20 mutations, depending on the type of cancer, in genes that normally regulate cell multiplication.

MEDIA CONNECTIONS

Overview: Cell Cycle Control

Focus: TGFβ Signaling Pathway

Classic Experiment 24.1: Studying the Transformation of Cells by DNA Tumor Viruses

Because years can be required for many mutations to accumulate, cancer is mainly a disease of the aged.

In this chapter we describe current knowledge about the genetic and physiological events that change a normally regulated cell into one that grows without responding to controls. We also discuss other important processes, such as the formation of new blood vessels in tumors and the escape of tumor cells from their site of origin, that also must occur for mutated cells to develop into cancer.

24.1 Tumor Cells and the Onset of Cancer

Although much research into the molecular basis of cancer utilizes cells growing in culture, we need first to consider tumors as they occur in experimental animals and in humans. In this way we can see the gross properties of the disease—the properties that ultimately must be explained by analysis of genes and cells.

Metastatic Tumor Cells Are Invasive and Can Spread

Tumors arise with great frequency, especially in older animals and humans, but most pose little risk to their host because they are localized and of small size. We call such tumors **benign**; an example is warts. It is usually apparent when a tumor is benign because it contains cells that closely resemble, and may function like, normal cells. The surface interaction molecules that hold tissues together keep benign tumor cells, like normal cells, localized to appropriate tissues. Benign liver tumors stay in the liver, and benign intestinal tumors stay in the intestine. A fibrous capsule usually delineates the extent of a benign tumor and makes it an easy target for a surgeon (Figure 24-1). Benign tumors become serious medical problems only if their sheer bulk interferes with normal functions or if they secrete excess amounts of biologically active substances like hormones.

In contrast, the cells composing a **malignant** tumor, or cancer, express some proteins characteristic of the cell type from which it arose, and a high fraction of the cells grow and divide more rapidly than normal. Some malignant tumors remain localized and encapsulated, at least for a time; an example is carcinoma in situ in the ovary or breast. Most, however, do not remain in their original site; instead, they invade surrounding tissues, get into the body's circulatory system, and set up areas of proliferation away from the site of their original appearance. The spread of tumor cells and establishment of secondary areas of growth is called **metastasis;** most malignant cells eventually acquire the ability to metastasize. Thus the major characteristics that differentiate metastatic (or malignant) tumors from benign ones are their invasiveness and spread.

Cancer cells can be distinguished from normal cells by microscopic examination. They are usually less well differentiated than normal cells or benign tumor cells. Liver cancers, for instance, express some of but not all the proteins characteristic of normal liver cells and may ultimately evolve to a state in which they lack most liver-specific functions. In a specific tissue, malignant cells usually exhibit the characteristics of rapidly growing cells, that is, a high nucleus-to-cytoplasm ratio, prominent nucleoli, many mitoses, and relatively little specialized structure. The presence of invading cells in an otherwise normal tissue section is the most diagnostic indication of a malignancy (Figure 24-2).

Malignant cells usually retain enough resemblance to the normal cell type from which they arose, as judged by morphology and by expression of cell-specific genes, that it is possible to classify them by their relationship to normal tissue. Normal animal cells are often classified according to their embryonic tissue of origin, and the naming of tumors

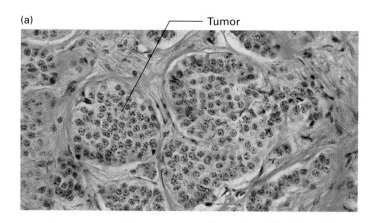

(a) — Tumor

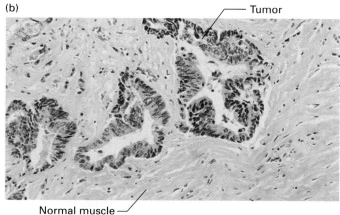

(b) — Tumor

Normal muscle —

▲ **FIGURE 24-1 Sections of two types of benign tumors.** (a) A tumor derived from cells that secrete neuroendocrine hormones. It is organized like a little gland in the midst of normal tissue. (b) A rectal epithelial tumor seen here as invaginations into the normal smooth muscle tissue of the rectum. [Photographs courtesy of Dr. J. Aster.]

(a)

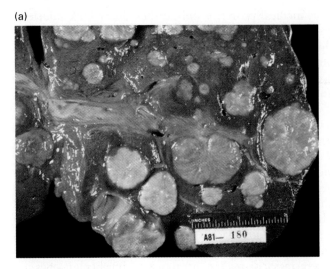

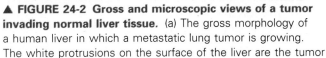

(b)

Tumor cells   Normal cells

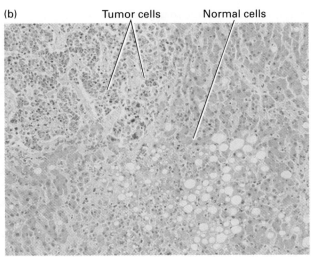

▲ **FIGURE 24-2 Gross and microscopic views of a tumor invading normal liver tissue.** (a) The gross morphology of a human liver in which a metastatic lung tumor is growing. The white protrusions on the surface of the liver are the tumor masses. (b) A light micrograph of a section of the tumor in (a) showing areas of small, dark-staining tumor cells invading a region of larger, light-staining, normal liver cells. [Courtesy of J. Braun.]

has followed suit. Cancers occur in most types of cells; compared with the 300 or so different types of cells in the human body, we can recognize 200 different types of human cancers. Malignant tumors are classified as **carcinomas** if they derive from endoderm or ectoderm and **sarcomas** if they derive from mesoderm. The **leukemias**, a class of sarcomas, grow as individual cells in the blood, whereas most other tumors are solid masses. (The name *leukemia* is derived from the Latin for "white blood": the massive proliferation of leukemic cells can cause a patient's blood to appear milky.)

Alterations in Cell-to-Cell Interactions Are Associated with Malignancy

The restriction of a normal cell type to a given organ or tissue is maintained by cell-to-cell recognition and by physical barriers. Primary among the physical barriers that keep tissues separated is the **basal lamina** (also called the *basement membrane*), which underlies layers of epithelial cells as well as surrounds the endothelial cells of blood vessels (see Figures 15-23 and 22-21). Basal laminae define the surfaces of external and internal epithelia and the structure of blood vessels.

Metastatic cells break their contacts with other cells in their tissue of origin and overcome the constraints on cell movement provided by basal laminae and other barriers. As a result, metastatic cells can enter the circulation and establish themselves in another site distant from their original location. In the process of metastasizing, they may invade adjoining tissue before spreading to distant sites through the circulation. Both these events require breach of a basal lamina.

Tumor cells often produce elevated levels of cell-surface receptors specific for the proteins and polysaccharides composing basal laminae (e.g., collagens, proteoglycans, and glycosaminoglycans) and secrete enzymes that digest these proteins. Many tumor cells also secrete a protease called *plasminogen activator,* which cleaves a peptide bond in the serum protein *plasminogen,* converting it to the active protease *plasmin.* Secretion of a small amount of plasminogen activator causes a large increase in protease concentration by catalytically activating the abundant plasminogen in normal serum. This increased protease activity promotes metastasis by helping tumor cells digest and penetrate the basal lamina. The normally invasive extraembryonic cells of the fetus secrete plasminogen activator when they are implanting in the uterine wall, a compelling analogy to invasion by tumor cells. As the basal lamina disintegrates, some tumor cells will enter the blood, but fewer than 1 in 10,000 cells that escape the primary tumor survive to colonize another tissue and form a secondary, metastatic tumor (see Figure 24-2). Such a cell first must adhere to an endothelial cell lining a capillary and migrate across or through it into the underlying tissue. To set up a metastasis, a tumor cell must be able to multiply without a mass of surrounding identical cells and to adhere to new types of cells. The wide range of altered behaviors that underlie malignancy may have their basis in new or variant surface proteins made by malignant cells.

Tumor Growth Requires Formation of New Blood Vessels

Tumors, whether primary or secondary, require recruitment of new blood vessels in order to grow to a large mass. In the absence of a blood supply, a tumor can grow into a mass of about 10^6 cells, roughly a sphere 2 mm in diameter. At this point, division of cells on the outside of the tumor mass is balanced by death of those in the center due to an inadequate supply of nutrients. Such tumors, unless they secrete hormones, cause few problems. However, most tumors induce the formation of new blood vessels that invade the tumor and nourish it, a process called *angiogenesis*. Although this complex process is not understood in detail, it can be described as several discrete steps: degradation of the basal lamina that surrounds a nearby capillary, migration of endothelial cells lining the capillary into the tumor, division of these endothelial cells, and formation of a new basement membrane around the newly elongated capillary.

Many tumors produce growth factors that stimulate angiogenesis; other tumors somehow induce surrounding normal cells to synthesize and secrete such factors. Basic fibroblast growth factor (bFGF), transforming growth factor α (TGFα), and vascular endothelial growth factor (VEGF), which are secreted by many tumors, all have angiogenic properties. New blood vessels nourish the growing tumor, allowing it to increase in size and thus increase the probability that additional harmful mutations will occur. The presence of an adjacent blood vessel also facilitates the process of metastasis.

One of the most mysterious aspects of angiogenesis is that a primary tumor will often secrete a substance that inhibits angiogenesis around secondary metastases. In this case, surgical removal of the primary tumor may stimulate growth of its metastatic secondary tumors. Several natural proteins that inhibit angiogenesis (e.g., angiogenin and endostatin) or antagonists of the VEGF receptor have excited much interest as therapeutic agents since they might be useful against many kinds of tumors. While new blood vessels are constantly forming during embryonic development, few form normally in adults; thus a specific inhibitor of angiogenesis might have few adverse side effects.

DNA from Tumor Cells Can Transform Normal Cultured Cells

The morphology and growth properties of tumor cells clearly differ from those of their normal counterparts. That mutations cause these differences was conclusively established by transfection experiments with a line of cultured mouse fibroblasts called 3T3 cells. These cells normally grow only when attached to the plastic surface of a culture dish and are maintained at a low cell density. Because 3T3 cells stop growing when they contact other cells, they eventually form a monolayer of well-ordered cells that have stopped proliferating and are in the G_0 phase of the cell cycle (Figure 24-3a). Although such *quiescent* cells in a saturated culture have stopped growing, they remain viable for a long time and can resume growth if they are released from contact inhibition and provided with growth factors present in serum. As is true for other cultured fibroblasts, the exact cell type that gives rise to 3T3 cells is uncertain, but they can differentiate into a range of mesodermally derived cell types, especially fat cells and endothelial cells (those that line blood vessels).

When DNA from a human bladder carcinoma, mouse sarcoma, or other tumor is added to a culture of 3T3 cells, about one cell in a million incorporates a particular segment of bladder carcinoma DNA that causes a distinctive phenotype. The progeny of the affected cell are more rounded and less adherent to one another and to the dish than are the normal surrounding cells, forming a three-dimensional cluster

(a)

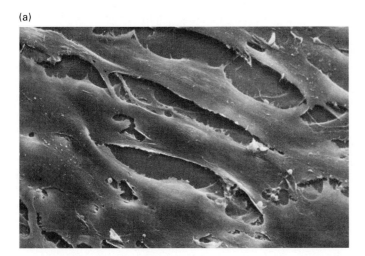

(b)

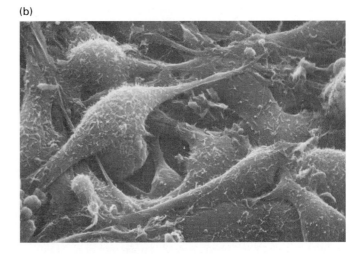

▲ FIGURE 24-3 Scanning electron micrographs of normal and transformed 3T3 cells. (a) Normal 3T3 cells are elongated and are aligned and closely packed in an orderly fashion. (b) 3T3 cells transformed by the *v-src* oncogene encoded by Rous sarcoma virus. The cells are much more rounded, and they are covered with small hairlike processes and bulbous projections. The cells grow one atop the other, and they have lost the side-by-side organization of the normal cells. These transformed cells have many of the same properties as malignant cells. [Courtesy of L.-B Chen.]

of cells (a focus) that can be recognized under the microscope (Figure 24-3b). Such cells, which continue to grow when the normal cells have become quiescent, have undergone **transformation** and are said to be *transformed*. Transformed cells have many properties similar to those of the cells composing malignant tumors, including changes in cell morphology, ability to grow unattached to a basal lamina or other extracellular matrix, reduced requirement for growth factors, secretion of plasminogen activator, and loss of actin microfilaments.

Figure 24-4 outlines the procedure for transforming 3T3 cells with DNA from a human bladder carcinoma and cloning the specific DNA segment that causes transformation. Subsequent studies showed that the cloned segment included a mutant version of the cellular *ras* gene, designated *ras*^{D}. Normal Ras protein, which participates in many intracellular signal transduction pathways activated by growth factors, cycles between an inactive, "off" state with bound GDP and an active, "on" state with bound GTP (see Figure 20-22). Because the mutated Ras^D protein hydrolyzes bound GTP very slowly, it accumulates in the active state, sending a growth-promoting signal to the nucleus even in the absence of the hormones normally required to activate the Ras–MAP kinase pathway (see Figure 20-28).

Expression of the Ras^D protein, however, is not sufficient to cause transformation of normal cells in a primary culture of human, rat, or mouse fibroblasts. Unlike cells in a primary culture, however, cultured 3T3 cells have undergone a loss-of-function mutation in the *p16* gene; as discussed later, the *p16* gene encodes a cyclin-kinase inhibitor that restricts progression through the cell cycle. Such cells can grow indefinitely if periodically diluted and supplied with nutrients. Transformation of these cells requires both loss of *p16* and expression of a constitutively active Ras protein; for this reason, transfection with the *ras*^{D} gene can transform 3T3 cells but not normal cultured primary fibroblast cells.

▶ **FIGURE 24-4 The identification and molecular cloning of the *ras*^{D} oncogene.** Addition of DNA from a human bladder carcinoma to a culture of mouse 3T3 cells causes about one cell in a million to divide abnormally and form a focus, or clone of transformed cells. To clone the oncogene responsible for transformation, advantage is taken of the fact that most human genes have nearby repetitive DNA sequences called *Alu* sequences. DNA from the initial focus of transformed mouse cells is isolated, and the oncogene is separated from adventitious human DNA by secondary transfer to mouse cells. The total DNA from a secondary transfected mouse cell is then cloned into bacteriophage λ; only the phage that receives human DNA hybridizes with an *Alu* probe. The hybridizing phage should contain part or all of the transforming oncogene. This expected result can be proved by showing either that the phage DNA can transform cells (if the oncogene has been completely cloned) or that the cloned piece of DNA is always present in cells transformed by DNA transfer from the original donor cell.

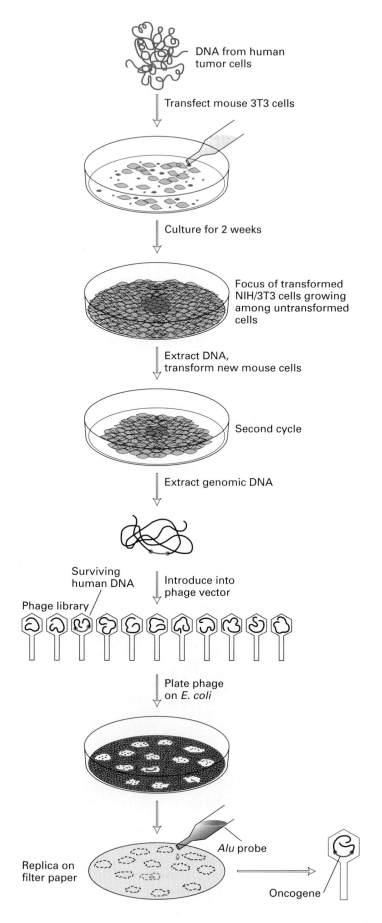

A mutant *ras* gene is found in most human colon, bladder, and other cancers, but not in normal human DNA; thus it must arise as the result of a somatic mutation in one of the tumor progenitor cells. Any gene, such as *ras*^{D} or v-*src*, that encodes a protein capable of transforming cells in culture or inducing cancer in animals is referred to as an **oncogene.** The normal cellular gene from which it arises is called a **proto-oncogene.**

Development of a Cancer Requires Several Mutations

Conversion of a normal body cell into a malignant one is now known to require multiple mutations. Three different types of experimental approaches all converged on this important conclusion: epidemiology of human cancers, analyses of DNA in cells at several stages in the development of cancers in humans and mice, and overexpression of oncogenes in cultured cells and transgenic animals.

Epidemiology Each individual cancer is a **clone** that arises from a single cell. Assuming that the rate of mutation is roughly constant during a lifetime, then the incidence of most types of cancer would be independent of age if only one mutation were required to convert a normal cell into a malignant one. In fact, however, the incidence of most types of human cancers increases markedly and exponentially with age (Figure 24-5). Although many explanations of this phenomenon have been considered, the incidence data are most consistent with the notion that multiple mutations are required for a cancer to form.

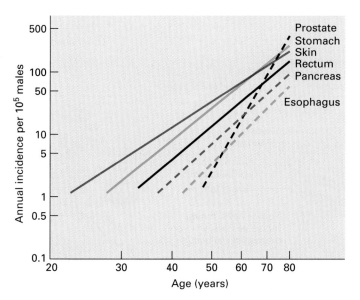

▲ **FIGURE 24-5 The incidence of several human cancers increases markedly with age.** Note that the logarithm of annual incidence is plotted versus the logarithm of age. [From B. Vogelstein and K. Kinzler, 1993, *Trends Genet.* **9**:101.]

According to this "multi-hit" model, cancers arise by a process of clonal selection not unlike the selection of individual animals in a large population. A mutation in one cell would give it a slight growth advantage. One of the progeny cells would then undergo a second mutation that would allow its descendants to grow more uncontrollably and form a small benign tumor; a third mutation in a cell within this tumor would allow it to outgrow the others, and its progeny would form a mass of cells, each of which would have these three mutations. An additional mutation in one of these cells would allow its progeny to escape into the blood and establish daughter colonies at other sites, the hallmark of metastatic cancer. Since decades are required for these multiple mutations to occur, the exponential increase in cancer incidence with age is predicted by the multi-hit model of cancer induction.

Somatic Mutations in Human Tumors Surgeons can produce fairly pure samples of many human cancers, but generally the cells that give rise to these tumors cannot be identified and analyzed. An exception is colon cancer, which evolves through distinct, well-characterized morphological stages (Figure 24-6). Because these intermediate stages—polyps, benign adenomas, and carcinomas—can be isolated by a surgeon, mutations that occur in each of the morphological stages can be identified. These studies have identified a series of mutations that commonly arise in a well-defined order, providing strong support for the multi-hit model. Invariably the first step in colon carcinogenesis involves loss of a functional *APC* gene; however, not every colon cancer acquires all the later mutations or acquires them in the same order.

Polyps are precancerous growths on the inside of the colon wall. Most of the cells in a polyp contain the same one or two mutations in the *APC* gene that result in its loss or inactivation; thus they are clones of cells in which the original mutations occurred. *APC* is one of many **tumor-suppressor genes,** most of which encode proteins that inhibit the progression of certain types of cells through the cell cycle. *APC* does so by inhibiting the ability of the Wnt protein to activate expression of the myc gene. The absence of functional APC protein thus leads to inappropriate activation of Myc, a transcription factor that induces expression of many genes required for the transition from the G_1 to the S phase of the cell cycle. Both alleles of the *APC* gene must carry an inactivating mutation for polyps to form, because cells with one wild-type *APC* gene express enough APC protein to function normally. Persons over 50 years of age are now advised to have a periodic colonoscopy, a procedure for scanning the wall of the colon. Any polyps that are present can be removed easily. Since polyps often evolve (or "progress") into a benign and then a metastatic tumor, the identification and removal of polyps often prevents development of colon cancer.

If one of the cells in a polyp undergoes another mutation, this time an activating mutation in the *ras* gene, its

▶ **FIGURE 24-6 The development and metastasis of human colorectal cancer and its genetic basis.** A mutation in the *APC* tumor-suppressor gene in a single epithelial cell causes the cell to divide, although surrounding cells do not, forming a mass of localized benign tumor cells called a *polyp*. Subsequent mutations leading to expression of a constitutively active Ras protein and loss of two tumor-suppressor genes, *DCC* and *p53*, generates a malignant cell carrying all four mutations; this cell continues to divide and the progeny invade the basal lamina that surrounds the tissue. Some tumor cells spread into blood vessels that will distribute them to other sites in the body. Additional mutations cause exit of the tumor cells from the blood vessels and growth at distant sites; a patient with such a tumor is said to have cancer. [Adapted from B. Vogelstein and K. Kinzler, 1993, *Trends Genet.* **9**:101.]

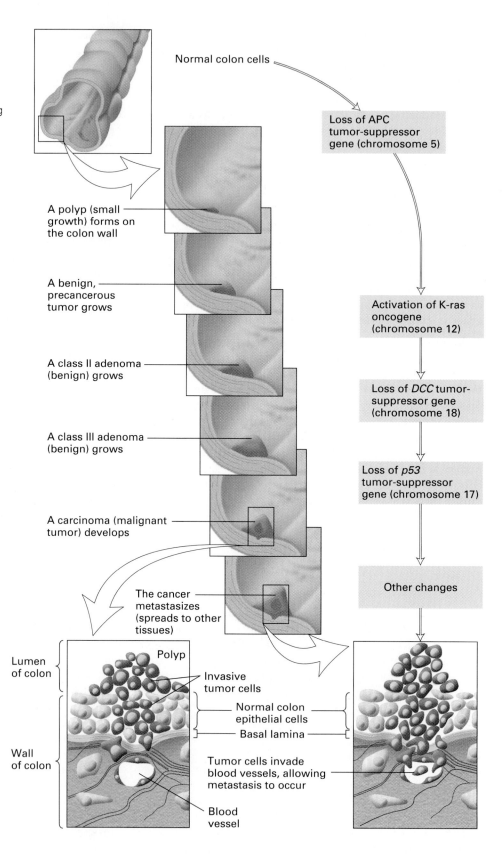

Normal colon cells

A polyp (small growth) forms on the colon wall

A benign, precancerous tumor grows

A class II adenoma (benign) grows

A class III adenoma (benign) grows

A carcinoma (malignant tumor) develops

The cancer metastasizes (spreads to other tissues)

Loss of APC tumor-suppressor gene (chromosome 5)

Activation of K-ras oncogene (chromosome 12)

Loss of *DCC* tumor-suppressor gene (chromosome 18)

Loss of *p53* tumor-suppressor gene (chromosome 17)

Other changes

Lumen of colon

Polyp

Invasive tumor cells

Normal colon epithelial cells

Basal lamina

Wall of colon

Tumor cells invade blood vessels, allowing metastasis to occur

Blood vessel

progeny divide in an even more uncontrolled fashion, forming a larger adenoma (see Figure 24-6). Mutational loss of another tumor-suppressor gene, designated *DCC*, followed by inactivation of the *p53* gene, results in a malignant carcinoma. DNA from different human colon carcinomas generally contains mutations in all these genes—*APC, p53,* K-*ras,* and *DCC*—establishing that multiple mutations in the same cell are needed for the cancer to form. Some of these mutations appear to confer growth advantages at an early stage of tumor development, whereas other mutations promote the later stages, including degradation of the basal lamina, which is required for the malignant phenotype.

Inherited Mutations That Increase Cancer Risk Most colon cancer patients have two normal *APC* alleles in their germ-line DNA, indicating that two somatic mutations, one in each *APC* allele, must have occurred in a single colon epithelial cell. However, in individuals who inherit a germ-line mutation in one *APC* allele, somatic loss or mutation of only the one remaining functional *APC* allele, rather than two, is required for a polyp to form. Thousands of precancerous polyps develop in these individuals; since there is a very high probability that one or more of these polyps will progress to malignancy, such individuals have a greatly increased risk for developing colon cancer before the age of 50.

As we detail below, individuals with inherited mutations in other tumor-suppressor genes have a *hereditary predisposition* for certain cancers. Such individuals inherit a germ-line mutation in one allele of the gene; somatic mutation of the second allele facilitates tumor progression. Although such cancers, which constitute about 10 percent of human cancers, are referred to as *inherited,* the inherited, germ-line mutation alone is *not* sufficient to cause tumor development. The inherited forms of many cancers are clinically similar to the noninherited form but occur earlier in life and often are marked by formation of multiple primary tumors, rather than a single one.

Overexpression of Oncogenes Overexpression of Myc protein is associated with many types of cancers, a not unexpected finding, since this transcription factor stimulates expression of many genes required for cell-cycle progression. But overexpression of Myc in transgenic mice is insufficient to induce tumor formation; other oncogenic mutations also must occur. This "cooperativity" of oncogenic mutations has been shown most dramatically in transgenic mice carrying both the *myc* gene and the mutant *ras*^{D} gene driven by a mammary cell–specific promoter/enhancer from a retrovirus (Figure 24-7). By itself, a *myc* transgene causes tumors only after 100 days, and then in only a few mice; clearly only a minute fraction of the mammary cells that overexpress the Myc protein become malignant. Expression of the mutant Ras^D protein alone causes tumors earlier but still slowly and with about 50 percent efficiency over 150 days. When the myc and ras^D transgenics are crossed, however, such that all mammary cells express both Myc and Ras^D, tumors arise much more rapidly and all animals succumb to cancer. Such

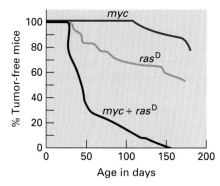

▲ **FIGURE 24-7 Kinetics of tumor appearance in female transgenic mice carrying transgenes driven by the mouse mammary tumor virus (MMTV) breast-specific promoter.** Shown are results for mice carrying either myc or *ras*^{D} transgenes as well as for the progeny of a cross of *myc* carriers with *ras*^{D} carriers that contain both transgenes. The percentage of tumor-free mice graphically depicts the time course of tumorigenesis. Females were studied because the hormonal stimulation of pregnancy activates expression of the MMTV-driven oncogenes. [See E. Sinn et al., 1987, *Cell* **49**:465.]

experiments emphasize the synergistic effects of multiple oncogenes. They also suggest that the long latency of tumor formation, even in the double-transgenic mice, is due to the need to acquire additional somatic mutations.

Similar cooperative effects can be seen in cultured cells. When normal cells are placed in medium with low amounts of growth factors such as platelet-derived growth factor (PDGF) or epidermal growth factor (EGF), they become blocked in the G$_0$ or G$_1$ stage of the cell cycle but remain viable. Recombinant cells that overexpress Myc also arrest their growth under these conditions, but soon undergo **apoptosis,** or programmed cell death (Chapter 23). Apparently the cell "senses" it is receiving an inappropriate "growth" signal from Myc in the absence of other "growth signals" from surface receptors and commits suicide. However, overexpression of Bcl-2, a protein that inhibits apoptosis, rescues these Myc-overexpressing cells from death. As a consequence, a cell that overexpresses both the Myc and Bcl-2 proteins can proliferate in the absence of normal growth factors. Consistent with these cell culture studies, overexpression of both Myc and Bcl-2 proteins is frequently found in human leukemias and lymphomas.

Cancers Originate in Proliferating Cells

In order for oncogenic mutations to induce cancer, they must occur in dividing cells so that the mutations are passed on to many progeny cells. When such mutations occur in nondividing cells (e.g., neurons and muscle cells), they generally do not induce cancer, which is why tumors of muscle and nerve cells are rare in adults. Nonetheless, cancer occurs in many tissues composed of nondividing differentiated cells such as erythrocytes and most white blood cells, absorptive cells that line the small intestine, and keratinized cells that form the skin. Although such differentiated cells cannot divide, they are continually replaced by differentiation of **stem cells.** This process is the key to understanding how cancers arise in these tissues.

Stem cells are capable of regenerating a particular tissue for the life of an organism. They are considered self-renewing in that some of their daughters become stem cells. *Unipotent* stem cells give rise to only a single differentiated cell type, whereas *pluripotent* stem cells can differentiate into several cell types that perform specialized functions (see Figure 14-7). The hematopoietic system provides a well-studied example of pluripotent stem cells. For example, hematopoietic stem cells can be purified from bone marrow with a fluorescent activated cell sorter (see Figure 5-21). This sorting depends on the presence of certain cell-surface proteins that distinguish hematopoietic stem and progenitor cells from other cell types and the absence on stem cells of other cell-surface proteins that are characteristic of differentiated hematopoietic cells. Each of the surface proteins binds a different monoclonal antibody, tagged with a different fluorescent dye, and thus FACS can separate those cells with particular combinations of cell surface proteins. When purified stem cells are injected into a mouse whose stem cells have been destroyed by whole-body gamma irradiation, the injected cells give rise to all the various types of blood cells. A similar approach is used to treat leukemia and breast cancer in humans: bone marrow or purified stem cells are injected into patients first subjected to lethal doses of radiation or cytotoxic drugs in order to kill the tumor cells as well as normal stem cells.

Detailed studies have elucidated the hematopoietic pathway shown in Figure 24-8. Under appropriate conditions, stem cells in the bone marrow divide to form two types of cells: another stem cell and a lineage-committed *progenitor cell.* The latter cells sometimes are referred to as *burst-forming units (BFUs)* and *colony-forming units (CFUs)* because after dividing several more times, each forms a clone (i.e., a colony) of differentiated cells. Numerous extracellular factors, called **cytokines,** are essential for formation of lineage-committed progenitors and the subsequent proliferation and differentiation of these cells. Many of these cytokines are secreted by stromal cells in the bone marrow or are on the surface of these cells. Some, like SCF, IL-3, or GM-CSF, support the proliferation and differentiation of progenitors for many blood cell types. Others, like Epo, exert their principal action on progenitors of a single lineage. In the absence of an essential cytokine, a progenitor cell undergoes apoptosis. Thus hematopoietic progenitor cells can undergo one of two fates: they can die, or they can give rise to a clone of a specific type of differentiated blood cell. Once formed in the bone marrow, differentiated blood cells enter the circulation.

Stem cells exist in other tissues, such as intestine, liver, skin, bone, and muscle; they give rise to all or many of the cell types in these tissues, replacing aged cells, by pathways

analogous to hematopoiesis in bone marrow. Because stem cells divide continuously over the life of an organism, generating additional stem cells, oncogenic mutations in their DNA can accumulate, eventually transforming them into cancer cells. Cells that have acquired these mutations have an abnormal proliferative capacity but generally cannot undergo normal processes of differentiation. Many oncogenic mutations, such as ones that prevent apoptosis or generate an inappropriate growth-promoting signal, also can occur in more differentiated but still replicating progenitor cells. Such mutations in hematopoietic progenitor cells can lead to various types of leukemia. Likewise, colon cancer arises from mutations in proliferating cells that continually are generated to replace worn-out epithelial cells lining the colon.

In humans, the kinetics of the appearance of tumors suggests that many different mutational events conspire together to cause cancer. Our longevity relative to that of mice, for instance, implies that the human species has evolved multiple ways to counter the tendency of cells to accumulate mutations, so that human cells have protections that rodent cells either lack or have in a less efficient form. Like the continual battle that we carry out with the infectious agents that surround us, we also are continually battling the tendency of cells to become transformed into cancer cells.

SUMMARY Tumor Cells and the Onset of Cancer

- Cancer is a fundamental aberration in cellular behavior, touching on many aspects of molecular cell biology. Most cell types of the body can give rise to malignant tumor (cancer) cells.

- Cancer cells can multiply in the absence of growth-promoting factors required for proliferation of normal cells and are resistant to signals that normally program cell death (apoptosis).

- Cancer cells also invade surrounding tissues, often breaking through the basal laminas that define the boundaries of tissues and spreading through the body to establish secondary areas of growth, a process called *metastasis.* Metastatic tumors often secrete proteases, which degrade the surrounding extracellular matrix.

- Both primary and secondary tumors require angiogenesis, the recruitment of new blood vessels, in order to grow to a large mass.

- Certain cultured cells transfected with tumor-cell DNA undergo transformation (see Figure 24-4). Such transformed cells share many properties with tumor cells.

- The requirement for multiple mutations in cancer induction is consistent with the observed increase in the incidence of human cancers with advancing age. Most such mutations are somatic and are not carried in the germ-line DNA.

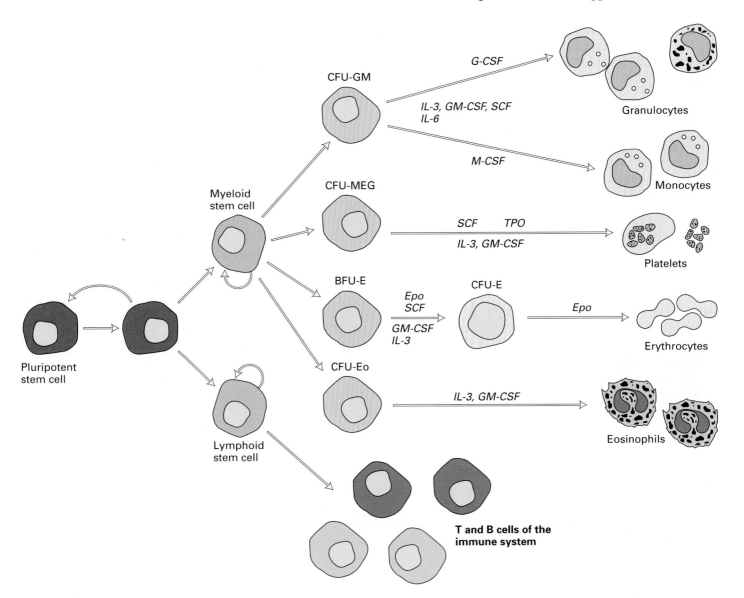

▲ **FIGURE 24-8 Formation of differentiated blood cells from hematopoietic stem cells in the bone marrow.** Pluripotent stem cells (dark red) may either self-renew or give rise to myeloid and lymphoid stem cells (light red). Although these stem cells retain the capacity for self-renewal, they are committed to one of the two major hematopoietic lineages. Depending on the types and amounts of cytokines present, the myeloid and lymphoid stem cells generate various progenitor cells, which are incapable of self-renewal (green). Further proliferation, commit-ment, and differentiation of the progenitor cells gives rise to the various types of blood cells. Among the cytokines that regulate hematopoiesis are granulocyte colony-stimulating factor (G-CSF); macrophage colony-stimulating factor (M-CSF); several inter-leukins (IL); stem cell factor (SCF); granulocyte-macrophage colony-stimulating factor (GM-CSF); erythropoietin (Epo); and thrombopoietin (Tpo). [Adapted from M. Socolovsky et. al., 1998, *Proc. Nat'l Acad. Sci. USA* **95**:6573.]

- Colon cancer develops through distinct morphological stages that commonly are associated with mutations in specific tumor-suppressor genes and oncogenes (see Figure 24-6).

- Cancer cells, which are closer in their properties to stem cells than to more mature differentiated cell types, usually arise from stem cells and other proliferating cells.

24.2 Proto-Oncogenes and Tumor-Suppressor Genes

As noted in the previous section, tumor cells differ from their normal counterparts in many respects: growth control, mor-phology, cell-to-cell interactions, membrane properties, cyto-skeletal structure, protein secretion, and gene expression. We also saw that two broad classes of genes—proto-oncogenes (e.g., *ras*) and tumor-suppressor genes (e.g., *APC*)—play a

▶ FIGURE 24-9 **The seven types of proteins that participate in controlling cell growth.** Cancer can result from expression of mutant forms of these proteins: growth factors (I), growth-factor receptors (II), signal-transduction proteins (III), transcription factors (IV), pro- or anti-apoptotic proteins (V), cell-cycle control proteins (VI), and DNA-repair proteins (VII). Mutations changing the structure or expression of proteins in classes I–IV generally give rise to dominantly active oncogenes. The class VI proteins mainly act as tumor suppressors; mutations in the genes encoding these proteins act recessively to release cells from control and surveillance, greatly increasing the probability that the mutant cells will become tumor cells. Class VII mutations greatly increase the probability of mutations in the other classes. Virus-encoded proteins that activate growth-factor receptors (Ia) also can induce cancer.

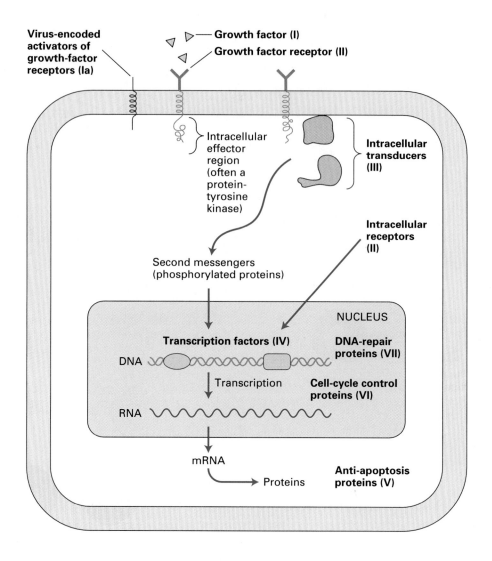

key role in cancer induction. These genes encode many kinds of proteins that help control cell growth and proliferation; mutations in these genes can contribute to the development of cancer (Figure 24-9). Most cancers have inactivating mutations in one or more proteins that normally function to restrict progression through the G₁ stage of the cell cycle (e.g., Rb and p16), although colon carcinomas usually do not. Virtually all human tumors have inactivating mutations in proteins such as p53 that normally function at crucial cell-cycle checkpoints, stopping the cycle if a previous step has occurred incorrectly or if DNA has been damaged. Likewise, a constitutively active Ras is found in several human tumors of different origin. Thus normal growth control and malignancy are two faces of the same coin.

In this section, we describe the types of mutations that are oncogenic and how oncogenes and tumor-suppressor genes were first discovered. In later sections, we discuss specific examples of precisely how oncogenic mutations cause the abnormalities characteristic of tumor cells. Certain viruses also cause cancer, and we indicate how certain virus-encoded proteins also can subvert normal cell control mechanisms.

Gain-of-Function Mutations Convert Proto-Oncogenes into Oncogenes

Recall that an oncogene is any gene that encodes a protein able to transform cells in culture or to induce cancer in animals. Of the many known oncogenes, all but a few are derived from normal cellular genes (i.e., proto-oncogenes) whose products participate in cellular growth-controlling pathways. For example, the *ras* gene discussed previously is a proto-oncogene that encodes an intracellular signal-transduction protein; the mutant *ras*^D gene derived from *ras* is an oncogene, whose encoded *oncoprotein* provides an excessive or uncontrolled growth-promoting signal. Because most proto-oncogenes are basic to animal life, they have been highly conserved over eons of evolutionary time.

Conversion, or activation, of a proto-oncogene into an oncogene generally involves a *gain-of-function* mutation. At least three mechanisms can produce oncogenes from the corresponding proto-oncogenes.

- Point mutations in a proto-oncogene that result in a constitutively acting protein product

- Localized reduplication (gene amplification) of a DNA segment that includes a proto-oncogene, leading to overexpression of the encoded protein
- Chromosomal translocation that brings a growth-regulatory gene under the control of a different promoter and that causes inappropriate expression of the gene

An oncogene formed by the first mechanism encodes an oncoprotein that differs slightly from the normal protein encoded by the corresponding proto-oncogene. In contrast, the latter two mechanisms generate oncogenes whose protein products are identical with the normal proteins; their oncogenic effect is due to their being expressed at higher-than-normal levels or in cells where they normally are not expressed. However they arise, the gain-of-function mutations that convert proto-oncogenes to oncogenes act dominantly; that is, mutation in only one of the two alleles is sufficient for induction of cancer.

Oncogenes Were First Identified in Cancer-Causing Retroviruses

Evidence that viruses could cause cancer first came from a series of studies by Peyton Rous beginning in 1911. He excised fibrosarcomas (connective tissue tumors) from chickens, ground them up, and removed cells and debris by centrifugation. After passing the supernatant through filters with very small pores, which retained even the smallest bacteria, Rous injected the filtrate into chicks. Most of the injected chicks developed sarcomas. The transforming agent in the filtrate eventually was shown to be a virus, called *Rous sarcoma virus (RSV)*. Some 50 years later, in 1966, Rous was awarded the Nobel prize for his pioneering work. The long delay in recognizing the importance of his discovery was due to the absence of any obvious molecular mechanism by which a virus could cause cancer, either in birds or in humans.

Later generations of molecular biologists showed that RSV is a **retrovirus** whose RNA genome is reverse-transcribed into DNA, which is incorporated into the host-cell genome (see Figure 6-22). Nontransforming retroviruses contain the genes *gag, pol,* and *env,* which encode the virus structural proteins and the reverse transcriptase. In addition to these "normal" retroviral genes, oncogenic transforming viruses like RSV contain the v-*src* gene. Subsequent studies with mutant forms of RSV demonstrated that only the v-*src* gene, not the *gag, pol,* or *env* genes, was required for cancer induction. One revealing mutation in the v-*src* gene was temperature-sensitive; transformed cells were generated at 30 °C, but these cells reverted to normal morphology at 39 °C. The v-*src* gene thus was identified as an oncogene.

The next breakthrough came in 1977 when Michael Bishop and Harold Varmus showed that normal cells from chickens and other species contain a gene that is closely related to the RSV v-*src* gene. This normal cellular gene, a proto-oncogene, commonly is distinguished from the viral gene by the prefix "c" (c-*src*). The landmark discovery of the close relationship between a viral oncogene and cellular proto-oncogene fundamentally reoriented thinking in cancer research because it showed that cancer may be induced by the action of normal, or nearly normal, genes. RSV and other oncogenic viruses are thought to have arisen by incorporating, or *transducing,* a normal cellular proto-oncogene into their genome. Subsequent mutation in the transduced gene then converted it into an oncogene.

As discussed below, v-Src protein is a constitutively active mutant form of c-Src protein, a protein-tyrosine kinase. In cells containing an integrated RSV genome, not only is v-*src* transcribed at inappropriately high rate levels, but the unregulated activity of v-Src protein causes continuous and inappropriate phosphorylation of target proteins. Because v-*src* can induce cell transformation in the presence of the normal c-*src* proto-oncogene, v-*src* is said to be a dominant gain-of-function mutant of c-*src,* analogous to the *ras*^{D} form of the *ras* proto-oncogene discussed previously. Many other oncogenes derived from cellular proto-oncogenes have been found in different retroviruses, implying that the normal vertebrate genome contains many potential cancer-causing genes.

Earlier we described the critical DNA transfection experiment establishing the existence of dominant gain-of-function oncogenes in human bladder tumors (see Figure 24-4), which led to the molecular cloning of a *ras* gene with a single point mutation. This oncogene, designated Ha-*ras,* also is present in Harvey sarcoma virus, a retrovirus. Similar experiments with DNA from many other tumors, both human and experimental, have led to the cloning of numerous oncogenes from tumor-cell DNA. Many of these cancer-causing genes are also found in various animal retroviruses.

Slow-Acting Carcinogenic Retroviruses Can Activate Cellular Proto-Oncogenes

Because its genome carries the v-*src* oncogene, Rous sarcoma virus induces tumors within days. Most oncogenic retroviruses, however, induce cancer only after a period of months or years. The genomes of the slow-acting retroviruses differ from those of transducing viruses such as RSV in one crucial respect: they lack an oncogene. Thus, slow-acting, or "long latency," retroviruses have no direct affect on growth of cells in culture.

The mechanism by which avian leukosis viruses cause cancer appears to operate in all slow-acting retroviruses. Like other retroviruses, avian leukosis virus DNA generally integrates into cellular chromosomes more or less at random. However, the finding that the site of integration in the cells from tumors caused by these viruses is near the c-*myc* gene suggested that these slow-acting viruses cause disease by activating expression of c-Myc. As noted earlier, c-Myc is required for transcription of many genes that encode cell-cycle proteins. These viruses act slowly both because integration near c-*myc* is a random, rare event and because additional mutations have to occur before a full-fledged tumor becomes evident.

(a) Promoter insertion

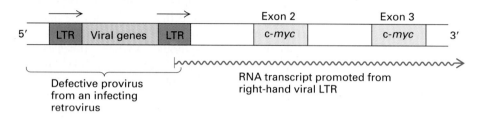

Defective provirus from an infecting retrovirus

RNA transcript promoted from right-hand viral LTR

(b) Enhancer insertions

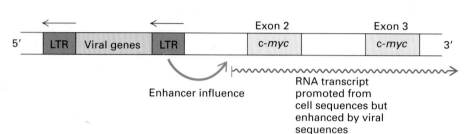

Enhancer influence

RNA transcript promoted from cell sequences but enhanced by viral sequences

◀ **FIGURE 24-10 Activation of the c-*myc* proto-oncogene by retroviral promoter and enhancer insertions.** (a) The promoter can be activated when the retrovirus inserts upstream (5′) of the c-*myc* exons. The right-hand LTR may then act as a promoter if the provirus has a defect preventing transcription through to the right-hand LTR. The c-*myc* gene is shown as containing two exons; there is a further upstream exon but it has no coding sequences. (b) The c-*myc* gene can also be activated when a retrovirus inserts upstream of the c-*myc* gene in the opposite transcriptional direction; a viral LTR acts as an enhancer, activating transcription from the c-*myc* promoter sequence. [Modified from actual cases of retroviral insertion described in G. G. Payne et al., 1982, *Nature* **295**:209.]

In some tumors, the avian leukosis proviral DNA is found at the 5′ end of the *myc* gene in the same transcriptional orientation. In such cases, the right-hand LTR of the integrated retrovirus—which usually serves as a terminator—is believed to act as a promoter, initiating synthesis of RNA transcripts from the c-*myc* gene (Figure 24-10a). In other tumors, the proviral DNA is found in the opposite transcriptional orientation; in this case, it is thought to exert an indirect enhancer activity (Figure 24-10b). Whether the inserted proviral DNA acts as a promoter or enhancer of c-*myc* transcription, the expressed c-Myc protein apparently is perfectly normal. The enhanced level of c-Myc resulting from the strong promoting or enhancing activity of the retroviral LTR partly explains the oncogenic effect of avian leukosis viruses. A second aspect is that c-*myc* expression is usually down-regulated when cells are induced to differentiate, but the LTR-driven expression of c-*myc* does not respond to such signals, and thus cells that normally would differentiate instead undergo DNA replication and cell division. These mechanisms of oncogene activation—called *promoter insertion* and *enhancer insertion*—operate in a variety of oncogenes and have been implicated in many animal tumors induced by slow-acting retroviruses.

In natural bird and mouse populations, slow-acting retroviruses are much more common than oncogene-containing retroviruses such as Rous sarcoma virus. Thus, insertional oncogene activation is probably the major mechanism whereby retroviruses cause cancer.

Many DNA Viruses Also Contain Oncogenes

Most animal cells infected by small DNA viruses such as SV40 are killed, but a very small proportion integrate the viral DNA into the host-cell genome. Although these cells survive infection, they become permanently transformed because the viral DNA contains one or more oncogenes. For example, many warts and other benign tumors of epithelial cells are caused by the DNA-containing papillomaviruses. Human genital warts are caused by one such virus that can induce stable transformation and transient mitogenic stimulation of a variety of cultured cells.

Unlike retroviral oncogenes, which are derived from normal cellular genes and have no function for the virus, the known oncogenes of DNA viruses are integral parts of the viral genome required for viral replication. As discussed later, the oncoproteins expressed from integrated viral DNA in infected cells act in various ways to stimulate cell growth and proliferation.

Loss-of-Function Mutations in Tumor-Suppressor Genes Are Oncogenic

Tumor-suppressor genes generally encode proteins that in one way or another inhibit cell proliferation. Loss of one or more of these "brakes" contributes to the development of many cancers. Five broad classes of proteins are generally recognized as being encoded by tumor-suppressor genes:

- Intracellular proteins, such as the p16 cyclin-kinase inhibitor, that regulate or inhibit progression through a specific stage of the cell cycle
- Receptors for secreted hormones (e.g., tumor-derived growth factor β) that function to inhibit cell proliferation
- Checkpoint-control proteins that arrest the cell cycle if DNA is damaged or chromosomes are abnormal

- Proteins that promote apoptosis
- Enzymes that participate in DNA repair

Although DNA-repair enzymes do not directly function to inhibit cell proliferation, cells that have lost the ability to repair errors, gaps, or broken ends in DNA accumulate mutations in many genes, including those that are critical in controlling cell growth and proliferation. Thus loss-of-function mutations in the genes encoding DNA-repair enzymes promote inactivation of other tumor-suppressor genes as well as activation of oncogenes.

Since generally one copy of a tumor-suppressor gene suffices to control cell proliferation, both alleles of a tumor-suppressor gene must be lost or inactivated in order to promote tumor development. Thus oncogenic loss-of-function mutations in tumor-suppressor genes act recessively. Tumor-suppressor genes in many cancers have deletions or point mutations that prevent production of any protein or lead to production of a nonfunctional protein.

The First Tumor-Suppressor Gene Was Identified in Patients with Inherited Retinoblastoma

We saw earlier that individuals who inherit a mutant allele of *APC,* a tumor-suppressor gene, have a high risk of developing colon cancer. Inheriting one mutant allele of another tumor-suppressor gene increases to almost 100 percent the probability that a person will develop a specific tumor. Indeed, genetic studies on cancer-prone families led to the initial identification of many tumor-suppressor genes. A classic case is *retinoblastoma,* which is caused by loss of function of *RB,* the first tumor-suppressor gene to be identified. As discussed later, the protein encoded by *RB* helps regulate progress through the cell cycle.

Children with *hereditary retinoblastoma* inherit a single defective copy of the *RB* gene, sometimes seen as a small deletion on chromosome 13. They develop retinal tumors early in life and generally in both eyes (Figure 24-11). Each tumor that develops is derived from a single transformed cell. The developing retina contains about 4×10^6 cells, but only about 1 in 10^6 cells actually gives rise to a tumor cell. This finding shows that the defective *RB* allele is acting recessively at the cellular level, and that other genetic events are needed to bring on the transformed state. One essential event is the deletion or mutation of the normal *RB* gene on the other chromosome, giving rise to a cell that produces no functional Rb protein (see Figure 8-7). Individuals with *sporadic retinoblastoma,* in contrast, inherit two normal *RB* alleles each of which has undergone a somatic loss-of-function mutation in a single retinal cell. Because this is an unlikely occurrence, sporadic retinoblastoma is rare, develops late in life, and usually affects only one eye.

If retinal tumors are removed before they become malignant, children with hereditary retinoblastoma often survive until adulthood and produce children. Molecular cloning of

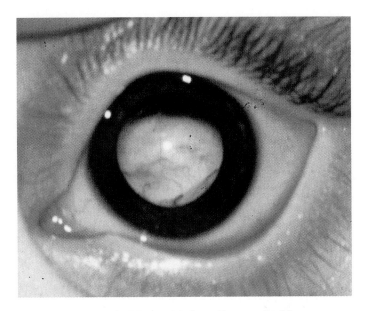

▲ **FIGURE 24-11 Children with hereditary retinoblastoma develop retinal tumors early in life and generally in both eyes.** They inherit one mutant allele of the *RB* gene. Somatic mutation of the other allele coupled with oncogenic mutations in other genes leads to tumor development. [Courtesy of T. Dryja.]

the *RB* gene established that these individuals inherited one normal and one mutant *RB* allele. On average, they will pass on the mutant allele to half their children and the normal allele to the other half. Children who inherit the normal allele are normal if their other parent has two normal *RB* alleles. However, those who inherit the mutant allele have the same enhanced predisposition to develop retinal tumors as their affected parent, even though they inherit a normal *RB* allele from their other, normal parent. Thus hereditary retinoblastoma is inherited as an autosomal dominant trait (Figure 24-12). As discussed below, many human tumors (not just retinal tumors) contain mutant *RB* alleles; most of these arise as the result of somatic mutations.

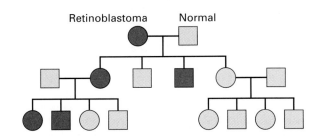

▲ **FIGURE 24-12 Hereditary retinoblastoma is inherited as an autosomal dominant trait, as illustrated in this pedigree.** Affected individuals (red), who inherit one mutant allele of *RB,* a tumor-suppressor gene, have a high probability of developing retinal tumors in childhood. Since the *RB* gene is located on chromosome 13, both males and females are affected equally.

A similar hereditary predisposition for breast cancer has been linked to *BRCA1*, another tumor-suppressor gene. Women who inherit one mutant *BRCA1* allele have a 60 percent probability of developing breast cancer by age 50, whereas those who inherit two normal *BRCA1* alleles have only a 2 percent probability of doing so. In women with hereditary breast cancer, loss of the second *BRCA1* allele, together with other mutations, is required for a normal breast duct cell to become malignant. However, *BRCA1* generally is not mutated in sporadic, noninherited breast cancer.

Loss of Heterozygosity of Tumor-Suppressor Genes Occurs by Mitotic Recombination or Chromosome Mis-segregation

In a somatic cell that contains one mutant and one normal allele of a tumor-suppressor gene, how is the normal allele lost or inactivated? That is, what mechanisms can result in *loss of heterozygosity (LOH)* of the normal allele? Point mutations are an unlikely cause because any such mutations that occur in the normal allele usually are repaired except in cells that are defective in certain DNA-repair systems (Chapter 12).

One common mechanism for LOH involves *mis-segregation* of the chromosomes bearing the heterozygous tumor-suppressor gene during mitosis. In this process, also referred to as *nondisjunction*, one daughter cell inherits only one normal chromosome (and probably dies), while the other inherits three, the other normal chromosome as well as two bearing the mutant allele. Such mis-segregation is caused by failure of a mitotic checkpoint, which would normally prevent a metaphase cell with an abnormal mitotic spindle from completing mitosis (see Figure 13-34). Subsequent loss of one chromosome often occurs, restoring the 2*n* complement; if the normal chromosome is lost, the resultant cell will contain two copies of the "mutant" chromosome (Figure 24-13a).

(a) Missegregation

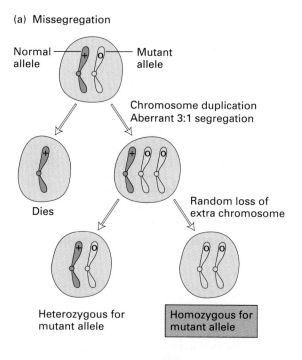

(b) Mitotic recombination

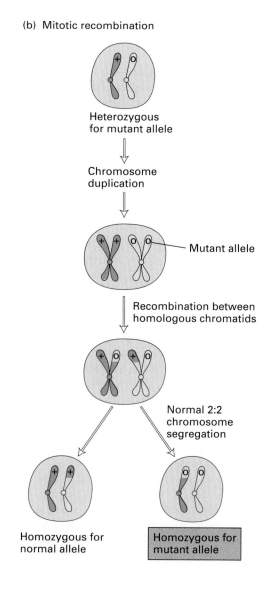

▲ **FIGURE 24-13 Loss of heterozygosity (LOH) of tumor-suppressor genes.** A cell containing one normal and one mutant allele of a tumor-suppressor gene is generally phenotypically normal. (a) If formation of the mitotic spindle is defective, then the duplicated chromosomes bearing the normal and mutant alleles may segregate in an aberrant 3:1 ratio. A daughter cell that receives three chromosomes of a type will generally lose one; sometimes the resultant cell will contain one normal and one mutant allele, but sometimes it will be homozygous for the mutant allele. (b) Mitotic recombination between a chromosome with a wild-type and a mutant allele, followed by chromosome segregation, will frequently result in a cell that contains two copies of the mutant allele and none of the wild-type.

Another likely mechanism for LOH is *mitotic recombination* between a chromatid bearing the wild-type allele and a homologous chromatid bearing a mutant allele. As illustrated in Figure 24-13b, the products of such a recombination are two normal chromatids and two mutant chromatids. Subsequent chromosome segregation can generate three types of daughter cells: one homozygous for the mutant tumor-suppressor allele; one homozygous for the normal allele; and one like the parental cell, heterozygous for the mutant allele.

SUMMARY Proto-Oncogenes and Tumor-Suppressor Genes

- Dominant gain-of-function mutations in proto-oncogenes and recessive loss-of-function mutations in tumor-suppressor genes are oncogenic.

- Among the proteins encoded by proto-oncogenes are positive-acting growth factors and their receptors, signal-transduction proteins, transcription factors, and cell-cycle control proteins (see Figure 24-9).

- An activating mutation of one of the two alleles of a proto-oncogene converts it to an oncogene, which can induce transformation in cultured cells or cancer in animals.

- Activation of a proto-oncogene into an oncogene can occur by point mutation, gene amplification, and gene translocation.

- The first recognized oncogene, v-*src*, was identified in Rous sarcoma virus, a cancer-causing retrovirus. Retroviral oncogenes arose by transduction of cellular proto-oncogenes into the viral genome and subsequent mutation.

- The first human oncogene to be identified encodes a constitutively active form of Ras, a signal-transduction protein. This oncogene was isolated from a human bladder carcinoma (see Figure 24-4).

- Slow-acting retroviruses can cause cancer by integrating near a proto-oncogene in such a way that gene transcription is activated continuously and inappropriately.

- Tumor-suppressor genes encode proteins that slow or inhibit progression through a specific stage of the cell cycle, checkpoint-control proteins that arrest the cell cycle if DNA is damaged or chromosomes are abnormal, receptors for secreted hormones that function to inhibit cell proliferation, proteins that promote apoptosis, and DNA repair enzymes.

- Inherited mutations causing retinoblastoma led to the identification of *RB*, the first tumor-suppressor gene to be recognized.

- Inheritance of a single mutant allele of many tumor-suppressor genes (e.g., *RB*, *APC*, and *BRCA1*) increases to almost 100 percent the probability that a specific kind of tumor will develop.

- Loss of heterozygosity of tumor-suppressor genes occurs by mitotic recombination or chromosome missegregation (see Figure 24-13).

24.3 Oncogenic Mutations Affecting Cell Proliferation

As summarized in Figure 24-9, genes encoding each class of cell regulatory protein have been identified as proto-oncogenes or tumor-suppressor genes. In the remainder of this chapter, we provide examples of each type of protein and examine in more detail how mutations in these proteins induce cancer. We begin with oncogenic mutations in genes that affect cell proliferation.

Misexpressed Growth-Factor Genes Can Autostimulate Cell Proliferation

Oncogenes rarely arise from genes encoding growth factors. In fact, only one naturally occurring growth-factor oncogene—*sis*—has been discovered. The *sis* oncogene, which encodes a type of PDGF, can aberrantly autostimulate proliferation of cells that normally express the PDGF receptor. Artificial class I oncogenes have been created. For example, if the gene encoding the granulocyte-macrophage colony-stimulating factor (GM-CSF) is inserted into a cell that has the GM-CSF receptor, the GM-CSF protein continually stimulates growth of the cell. Such autostimulation is called *autocrine induction* of cell growth (Figure 20-1).

Virus-Encoded Activators of Growth-Factor Receptors Act as Oncoproteins

Spleen focus-forming virus (SFFV) is a retrovirus that induces erythroleukemia (a tumor of erythroid progenitors) in adult mice. The proliferation, survival, and differentiation of erythroid progenitors into mature red cells absolutely requires Epo (made by cells in the kidney) and the corresponding Epo receptor (see Figure 24-8). SFFV encodes a mutant retrovirus envelope glycoprotein, termed gp55, that is missing much of the normal extracellular domain and all the cytosolic domain. Although gp55 cannot function as a normal retrovirus envelope protein in supporting virus budding and infection, it has acquired the remarkable ability to bind to and activate Epo receptors expressed in the same cell (Figure 24-14). By inappropriately and continuously stimulating the proliferation of erythroid progenitors, gp55 expression induces polycythemia, a condition reminiscent of

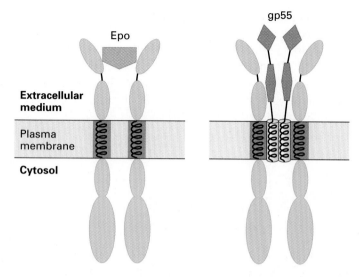

▲ **FIGURE 24-14 Activation of the erythropoietin (Epo) receptor.** Binding of the natural ligand, Epo, dimerizes the receptor and induces formation of erythrocytes from erythroid progenitor cells. Progenitor cells infected by the spleen focus-forming virus co-express viral gp55 and the Epo receptor. Binding of gp55 activates the Epo receptor in the absence of Epo, leading to inappropriate proliferation of progenitors and eventually to induction of erythroleukemia. Specific binding of dimeric gp55 to the Epo receptor is mediated by the membrane-spanning domains.[Courtesy of S. Constantinescu; see S. Constantinescu et al., 1999, *EMBO J.*, in press.]

the human disease *polycythemia vera* and characterized by a very high number of mature red cells. Malignant clones of erythroid progenitors emerge several weeks after SFFV infection as a result of further mutations in these aberrantly proliferating cells.

Another example of this phenomenon is provided by human papillomavirus (HPV), a DNA virus that causes genital warts. A papillomavirus protein designated E5, which contains only 44 amino acids, spans the plasma membrane and forms a dimer or trimer. Each E5 polypeptide can form a stable complex with one endogenous receptor for PDGF, thereby aggregating two or more PDGF receptors within the plane of the plasma membrane. This mimics hormone-mediated receptor dimerization and activation, causing sustained receptor activation and leading to cell transformation.

Activating Mutations or Overexpression of Growth-Factor Receptors Can Transform Cells

As discussed in Chapter 20, binding of a growth factor to its cell-surface receptor triggers an intracellular signal-transduction pathway leading to changes in gene expression. The receptors for many growth factors have intrinsic protein-tyrosine kinases in their cytosolic domains. Such receptor tyrosine kinases (RTKs) transmit the growth signal by phosphorylating tyrosine residues on themselves as well as

on one or more target proteins, thus initiating a cascade of events (see Figure 20-21).

Researchers have identified several types of mutations leading to production of constitutively active receptors, which transmit growth signals in the absence of the normal ligands. In some cases, a point mutation changes a normal RTK into one that dimerizes and is activated in the absence of ligand. For instance, a single point mutation converts the normal Her2 receptor into the Neu oncoprotein, which is found in certain mouse cancers (Figure 24-15, *left*). Similarly, human tumors called multiple endocrine neoplasia type 2 have been found to express a constitutively active dimeric GDNF receptor that results from a point mutation in the extracellular domain. In other cases, deletion of much of the extracellular ligand-binding domain produces a constitutively active oncoprotein receptor (Figure 24-15, *right*).

Another mechanism for generating an oncoprotein receptor is illustrated by the human *trk* oncogene. This oncogene, isolated from a colon carcinoma, resulted from a chromosomal translocation that replaced the sequences encoding most of the extracellular domain of the normal Trk receptor protein-tyrosine kinase with the sequences encoding the N-terminal amino acids of nonmuscle tropomyosin (Figure 24-16). Because tropomyosin forms a rope-like coiled coil of two long identical parallel α helices, this translocated segment can mediate dimerization of the resulting chimeric oncoprotein, leading to activation of the kinase domains. In contrast to the normal Trk protein, which is localized to the plasma membrane and is activated by a nerve growth factor, the constitutively active Trk oncoprotein is localized to the cytosol, since the N-terminal signal sequence directing it to the membrane has been deleted.

Mutations leading to overexpression of a normal receptor protein also can be oncogenic. For instance, many human breast cancers overexpress an otherwise normal Her2 receptor. As a result, the cells are stimulated to proliferate in the presence of very low concentrations of EGF and related hormones, concentrations too low to stimulate proliferation of normal cells. Based on this finding, a monoclonal antibody specific for Her2 is being evaluated as a new treatment for breast cancers that overexpress Her2. Injection of Her2 antibody into the blood has been found to selectively kill the cancer cells without any apparent effect on normal breast (and other) cells that express moderate Her2 levels.

Constitutively Active Signal-Transduction Proteins Are Encoded by Many Oncogenes

A large number of oncogenes are derived from proto-oncogenes whose encoded proteins act as intracellular transducers, proteins that aid in transmitting signals from a receptor to a cellular target. In this class are the *ras* oncogenes, discussed previously, which were the first nonviral oncogenes to be recognized. A common mutation by which a *ras* proto-oncogene becomes an oncogene causes only one change in the protein, substitution of any amino acid for

Proto-oncogene receptor proteins

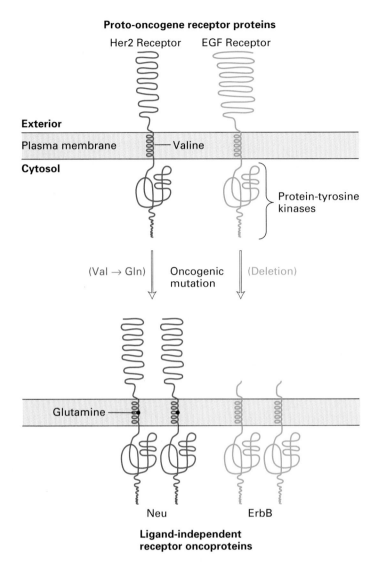

◄ FIGURE 24-15 Effects of oncogenic mutations in proto-oncogenes that encode cell-surface receptors. *(Left)* A mutation that alters a single amino acid (valine to glutamine) in the transmembrane region of the Her2 receptor causes dimerization of two receptor proteins in the absence of the normal EGF-related ligand, making the protein constitutively active as a kinase. *(Right)* A deletion that causes loss of the extracellular ligand-binding domain in the EGF receptor leads, for unknown reasons, to constitutive activation of the protein kinase.

glycine at position 12 of the sequence. This simple mutation reduces the protein's GTPase activity, thus linking GTP hydrolysis to the maintenance of normal, controlled Ras function. The oncogenic change causes only a slight alteration in the three-dimensional structure of Ras, but this is sufficient to change the normal protein into an oncoprotein. Constitutively active Ras oncoproteins are expressed by many types of human tumors including bladder, colon, mammary, skin, and lung carcinomas, neuroblastomas, and leukemias.

Ras proteins are anchored to the inner side of the cell's plasma membrane by a particular type of covalently attached fatty acid called a *farnesyl group* (see Figure 3-36b); this attachment is essential for either normal or oncogenic Ras proteins to function as signal transducers. Inhibitors of *farnesyl transferase*, the enzyme that adds the farnesyl group, prevent membrane localization of Ras. Such inhibitors, which have been shown to reduce the abnormal proliferation of cultured cells transformed by the *ras* oncogene, are now in clinical trials for treatment of several tumors, including colon cancer.

Ras is a key component in transducing signals from activated receptor tyrosine kinases (RTKs) to a cascade of protein kinases, eventually leading to alterations in cell growth and differentiation. In the first part of this pathway, a signal from an activated receptor is carried via two proteins (GRB2 and Sos) to Ras, converting it to the active GTP-bound form (see Figure 20-23). In the second part of the pathway, activated Ras transmits the signal, via two intermediate protein kinases, to MAP kinase (see Figure 20-28). The activated MAP kinase then phosphorylates a number of

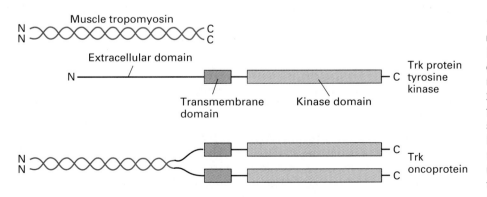

◄ FIGURE 24-16 Chimeric oncoprotein resulting from chromosomal translocation. In the Trk oncoprotein, most of the extracellular domain of the normal human Trk receptor tyrosine kinase is replaced with the 221 N-terminal amino acids of nonmuscle tropomyosin. Dimerized by the tropomyosin segment, the Trk oncoprotein protein-tyrosine kinase is constitutively active. Unlike the normal Trk, which is localized to the plasma membrane, the Trk oncoprotein is found in the cytosol. [See F. Coulier et al., 1989, *Mol. Cell. Biol.* **9**:15.]

transcription factors that induce synthesis of important cell-cycle and differentiation-specific proteins. Activating Ras mutations short-circuit the first part of this pathway, making upstream activation triggered by ligand binding to the receptor unnecessary.

Although a dominant gain-of-function mutation in the *ras* gene converts it into an oncogene, a recessive loss-of-function mutation in the *NF1* gene also leads to constitutive Ras activation. *NF1* encodes a GTPase-activating enzyme (GAP) that accelerates hydrolysis of GTP and the conversion of active GTP-bound Ras to inactive GDP-bound Ras (see Figure 20-22). The loss of GAP thus leads to sustained Ras activation of downstream signal transduction proteins. Individuals with neurofibromatosis have inherited a single mutant *NF1* allele; subsequent somatic mutation in the other allele leads to formation of neurofibromas, benign tumors of the sheath cells that surround nerves. Thus neurofibromatosis, like hereditary retinoblastoma, is inherited as an autosomal dominant trait (see Figure 24-12).

Oncogenes encoding other components of the RTK-Ras–MAP kinase pathway also have been identified. One example, found in certain transforming mouse retroviruses, encodes a constitutively activated Raf serine/threonine kinase, which is in the pathway between Ras and MAP kinase. Another is the *crk* (pronounced "crack") oncogene found in avian sarcoma virus, which causes certain tumors when overexpressed. The Crk protein, which has no known biochemical activity, contains one SH2 and two SH3 domains, and is similar to the GRB2 adapter protein that also functions between an RTK and Ras (see Figure 20-23). The SH2 and SH3 domains in GRB2 and other adapter proteins mediate formation of specific protein aggregates that normally serve as signaling units for cellular events. However, overexpression of Crk leads to formation of protein aggregates that inappropriately signal the growth and metastatic abilities characteristic of cancer cells.

Several oncogenes, some initially identified in human tumors, others in transforming retroviruses, encode nonreceptor protein-tyrosine kinases. One of these is generated by a chromosomal translocation that results in fusion of a portion of the *bcr* gene (whose function is unknown but whose N-terminal segment forms a coiled-coil domain that links several bcr polypeptides together) with part of the *c-abl* gene, which encodes a protein-tyrosine kinase whose normal substrates are not known. The chimeric polypeptides expressed from the resulting Bcr-Abl oncogene form a tetramer that exhibits constitutive Abl kinase activity. Although Abl is normally localized to the nucleus, addition of the Bcr segment causes the Bcr-Abl oncoprotein to be localized to the cytosol. Through the Abl SH2 and SH3 domains Bcr-Abl binds to many intracellular signal-transduction proteins and then phosporylates them, proteins that Abl would not normally activate. As a consequence, these signaling proteins, including JAK2 protein-tyrosine kinase, Stat5 transcription factor, and PI-3 kinase become activated in the absence of growth factors. A chromosome translocation that forms Bcr-Abl in

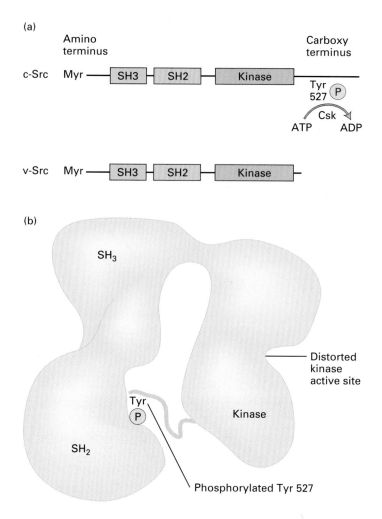

a hematopoietic stem cell forms the diagnostic "Philadelphia" chromosome (see Figure 9-38a) and results in the initial chronic phase of human chronic myelogenous leukemia (CML), characterized by an expansion in the number of well-differentiated granulocytes, a type of white blood cell (see Figure 24-8). A second mutation in one such cell (e.g., in *p53*) leads to acute leukemia, which often kills the patient.

Another oncogene that encodes a constitutively active cytosolic protein-tyrosine kinase is *src*. As we noted previously, v-*src* from Rous sarcoma retrovirus was the first oncogene to be discovered. The regulation of Src activity and the alterations by which the *c-src* proto-oncogene becomes an oncogene is known in great detail (Figure 24-17). The normal Src kinase (c-Src) contains an SH2, an SH3, and a protein-tyrosine kinase domain. The kinase activity of this 60-kDa protein is normally inactivated by phosphorylation of the tyrosine residue at position 527, which is six residues from the C-terminus. Hydrolysis of phosphotyrosine 527 by a specific phosphatase enzyme normally activates c-Src. Tyrosine 527 is often missing or altered in Src oncoproteins that have constitutive kinase activity; that is, they do not

◀ **FIGURE 24-17 Regulation of Src activity and its activation by an oncogenic mutation.** (a) Domain structure of c-Src and v-Src. Phosphorylation of tyrosine 527 by Csk, another cellular tyrosine kinase, inactivates the Src kinase activity. The transforming v-Src oncoprotein encoded by Rous sarcoma virus is missing the C-terminal 18 amino acids including tyrosine 527 and thus is constitutively active. (b) Effect of phosphorylation on c-Src conformation. Binding of phosphotyrosine 527 to the SH2 domain induces conformational strains in the SH3 and kinase domains, distorting the kinase active site so it is catalytically inactive. The kinase activity of c-Src is normally activated by removing the phosphate on tyrosine 527. [Adapted from T. Pawson, 1997, *Nature* **385**:582. See also W. Xu et al., 1997, *Nature* **385**:595; and F. Sichrei et al., 1997, *Nature* **385**:602.]

require activation by a phosphatase. In Rous sarcoma virus, for instance, the *src* gene has suffered a deletion that eliminates the C-terminal 18 amino acids of c-Src; as a consequence the v-Src kinase is constitutively active. Phosphorylation of target proteins by aberrant Src oncoproteins contributes to abnormal proliferation of many types of cells.

Deletion of the PTEN Phosphatase Is a Frequent Occurrence in Human Tumors

Since many proto-oncogenes are protein kinases, it has long been speculated that protein phosphatases might function as tumor suppressors; their deletion would overactivate signaling pathways induced by kinases. However, only in the past few years was the first phosphatase, PTEN, identified as a tumor suppressor; PTEN is deleted in multiple types of advanced human cancers. This phosphatase, which has an unusually broad specificity, can remove phosphate groups attached to serine, threonine, and tyrosine residues. However, its ability to dephosphorylate phosphatidylinositol 3,4,5-trisphosphate, the product of PI-3 kinase, is probably responsible for its tumor-suppressor effects. Cells lacking PTEN have elevated levels of phosphatidylinositol 3,4,5-trisphosphate and protein kinase B, which acts to prevent apoptosis. As discussed in Chapter 20, protein kinase B is itself activated by the lipid products of the PI-3 kinase signaling pathway (see Figure 20-44). Thus deletion of PTEN will reduce apoptosis and may also stimulate cell-cycle progression, two processes that are critical to the development of tumors.

Inappropriate Expression of Nuclear Transcription Factors Can Induce Transformation

By one mechanism or another, the proteins encoded by all proto-oncogenes and oncogenes eventually cause changes in gene expression. This is reflected in the differences in the proportions of different mRNAs in growing cells and quiescent cells, as well as similar differences between tumor cells and their normal counterparts. The most direct effect on gene expression is exerted by transcription factors, which bind to DNA and modify the rate of transcriptional initiation (Chapter 10). Consistent with this, many oncogenes encode transcription factors. Two examples are *jun* and *fos*, which initially were identified in transforming retroviruses and later found to be overexpressed in some humor tumors. The c-*jun* and c-*fos* proto-oncogenes encode proteins that sometimes associate to form a heterodimeric transcription factor, called AP1, that binds to a sequence found in promoters and enhancers of many genes (see Figure 10-60). Both Fos and Jun also can act independently as transcription factors. They function as oncoproteins by activating transcription of key genes that encode growth-promoting proteins or by inhibiting transcription of growth-repressing genes.

Many nuclear proto-oncogene proteins are induced when normal cells are stimulated to grow, indicating their direct role in growth control. For example, PDGF treatment of quiescent 3T3 cells induces an ≈50-fold increase in the production of c-Fos and c-Myc, the normal products of the *fos* and *myc* proto-oncogenes. Initially there is a transient rise of c-Fos and later a more prolonged rise of c-Myc (Figure 24-18). Although c-Fos rapidly disappears, c-Myc stays at a somewhat elevated level. In both cases, the proteins are maintained at high levels only briefly, probably because prolonged expression would be oncogenic. As discussed in Chapter 13, c-Fos and c-Myc stimulate transcription of genes encoding proteins that promote progression through the G_1 phase of the cell cycle and the G_1 to S transition. In tumors, the oncogenic forms of these or other transcription factors are frequently expressed at high and unregulated levels.

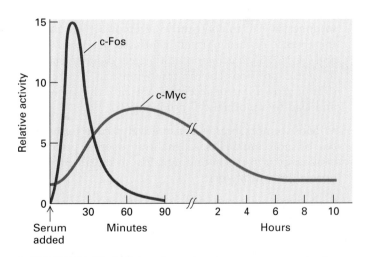

▲ **FIGURE 24-18 Activity of two proto-oncogene products, c-Fos and c-Myc, following serum stimulation of quiescent 3T3 cells.** Serum contains factors like platelet-derived growth factor (PDGF) that stimulate the growth of quiescent cells. One of the earliest effects of growth factors is to induce transcription of c-*fos* and c-*myc*. [See M. E. Greenberg and E. B. Ziff, 1984, *Nature* **311**:433.]

To assure the rapid loss of c-Fos and c-Myc after their induction in normal cells, both proteins and their corresponding mRNAs are intrinsically unstable. Some of the changes that turn c-*fos* from a normal gene to an oncogene involve loss of sequences in the gene that make the Fos mRNA and protein short-lived. Conversion of the c-*myc* proto-oncogene into an oncogene can occur by several different mechanisms. In cells of the human tumor known as Burkitt's lymphoma, the c-*myc* gene is translocated to a site near the heavy-chain antibody genes. An analogous translocation in the mouse genome is also involved in mouse myelomas. In both cases, the tumor cells arise from antibody-producing cells, which carry out DNA rearrangements during their maturation (see Figure 9-25). The c-*myc* translocation is a rare aberration of the normal rearrangement events, bringing it from its normal, distant chromosomal location into juxtaposition with the enhancer of the antibody genes. The translocated *myc* gene, now regulated by the antibody enhancer, is continually expressed, causing the cell to become cancerous. Localized reduplication of a segment of DNA containing the *myc* gene, which occurs in several human tumors, also causes inappropriately high expression of the otherwise normal Myc protein.

SUMMARY Oncogenic Mutations Affecting Cell Proliferation

- Certain virus-encoded proteins can bind to and activate host-cell receptors for growth factors, thereby stimulating cell proliferation in the absence of normal signals.

- Mutations or chromosomal translocations that permit growth factor receptor protein-tyrosine kinases to dimerize lead to constitutive receptor activation in the absence of their normal ligands (see Figures 24-15 and 24-16). Such activation ultimately induces changes in gene expression that can transform cells. Overexpression of growth factor receptors can have the same effect and lead to abnormal cell proliferation.

- Most tumors express constitutively active forms of one or more intracellular signal-transduction proteins, causing growth-promoting signaling in the absence of normal growth factors.

- A single point mutation in Ras, a key transducing protein in many signaling pathways, reduces its GTPase activity, thereby maintaining it in an activated state.

- The activity of Src, a cytosolic signal-transducing protein-tyrosine kinase, normally is regulated by reversible phosphorylation and dephosphorylation of a tyrosine residue near the C-terminus (see Figure 24-17). The unregulated activity of Src oncoproteins that lack this tyrosine promotes abnormal proliferation of many cells.

- Deletion of the PTEN phosphatase promotes the PI-3 kinase pathway and activation of protein kinase B, which inhibits apoptosis. Loss of this tumor suppressor, which is common in many human cancers, reduces apoptosis.

- Inappropriate expression of nuclear transcription factors, such as Fos, Jun, and Myc can induce transformation.

24.4 Mutations Causing Loss of Cell-Cycle Control

The entry of cells into the cell cycle from a quiescent state and the progression of cells around the cycle are precisely controlled events. This assures that cellular growth and the coordination of DNA synthesis with cell-size increase and cytokinesis are monitored and do not fall out of regulated synchrony. Once a cell progresses past a certain point in late G_1, called the *restriction point*, it becomes irreversibly committed to entering the S phase and replicating its DNA (see Figure 13-29). Cyclins, cyclin-dependent kinases (Cdks), and the Rb protein are all elements of the control system that regulate passage through the restriction point. The ability of these proteins to check cell-cycle progression, and hold cells in quiescence or even lead cells to commit suicide unless conditions are appropriate, means that they can prevent cells from becoming cancerous. Altered regulation of expression of at least one cyclin as well as mutation of several proteins that negatively regulate passage through the restriction point can be oncogenic (Figure 24-19).

Passage from G_1 to S Phase Is Controlled by Proto-Oncogenes and Tumor-Suppressor Genes

The expression of D-type cyclins in mammals is induced by many extracellular substances called growth factors or mitogens. These cyclins (D1, D2, and D3) assemble with their partners Cdk4 and Cdk6 to generate catalytically active kinases. Mitogen withdrawal prior to passage through the restriction point leads to accumulation of p27 or p16, which bind to and inhibit cyclin D–dependent kinase activity, thus causing G_1 arrest. Another key player in cell-cycle control is the Rb protein. In its unphosphorylated form, Rb binds transcription factors collectively called *E2F* and thereby prevents E2F-mediated transcriptional activation of many genes whose products are required for DNA synthesis, such as DNA polymerase. Also, the Rb-E2F complex acts as a transcriptional repressor for many of these same genes. Phosphorylation of Rb protein midway through G_1 causes it to dissociate from E2F, allowing E2F to induce synthesis of these DNA replication enzymes, which irreversibly commits the cell to DNA synthesis. Rb phosphorylation is initiated by an active Cdk4–cyclin D complex and is completed by other cyclin-dependent kinases (see Figure 13-31). Most tumors contain an oncogenic mutation of one of the genes in this pathway such that the cells are propelled into the S phase in the absence of the proper extracellular growth signals that regulate Cdk activity.

Overexpression of Cyclin D1 Gene amplification or a chromosome translocation that places cyclin D1 under control of an inappropriate promoter leads to overexpression of this cyclin in many human cancers, indicating that it can function as an oncoprotein. In certain tumors of the antibody-producing B cells, for instance, the cyclin D1 gene is translocated such that its transcription is under control of an antibody-gene enhancer, causing elevated cyclin D1 production throughout the cell cycle irrespective of extracellular signals. (This phenomenon is analogous to the c-*myc* translocation in Burkitt's lymphoma cells discussed earlier.) That cyclin D1 overexpression can directly cause cancer was shown by studies with a transgenic mouse in which the cyclin D gene was placed under control of an enhancer specific for mammary ductal cells. Initially the ductal cells underwent hyperproliferation, and eventually breast tumors developed in these transgenic mice. Amplification of the cyclin D1 gene and concomitant overexpression of the cyclin D1 protein is common in human breast cancer.

Loss of p16 Function The group of proteins that function as cyclin-kinase inhibitors (CKIs) play an important role in regulating the cell cycle (Chapter 13). Mutations, especially deletions of the *p16* gene, that inactivate the ability of p16 to inhibit cyclin D–dependent kinase activity are common in several human cancers. As Figure 24-19 makes clear, loss of p16 would mimic cyclin D1 overexpression, leading to Rb hyperphosphorylation and release of active E2F transcription factor. Thus p16 normally acts as a tumor suppressor.

Loss of Rb Function As already noted, loss of Rb function, whether by inherited or somatic mutation, leads to induction of many types of cancers, childhood retinoblastoma most notably. Tumors with inactivating mutations in Rb generally express normal levels of cyclin D1 and produce functional p16 protein. In contrast, tumor cells that overexpress cyclin D1 or have lost p16 function generally retain wild-type Rb. Thus, loss of only one component of this regulatory system for controlling passage through the restriction point is all that is necessary to subvert this aspect of normal growth control. We noted previously that the small DNA-containing human papillomavirus (HPV) is able to induce stable transformation and transient mitogenic stimulation of a variety of cultured cells. In addition to encoding a protein, E5, that activates the PDGF receptor, HPV encodes another protein, E7, that binds to and inhibits Rb function. The E7 protein is essential for HPV-mediated transformation of cells.

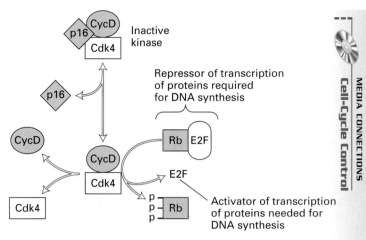

▲ **FIGURE 24-19 Restriction point control.** Overexpression or loss of the cell-cycle control proteins shown in pink commonly occurs in human cancers. See text and also Figures 13-37 and 13-29.

Loss of TGFβ Signaling Contributes to Abnormal Cell Proliferation and Malignancy

Until now we have considered growth factors that mainly stimulate proliferation of different types of cells. Tumor-derived growth factor β (TGFβ) is secreted by most body cells and has a diverse range of biological activities. Most relevant to the present discussion, however, is the ability of TGFβ to inhibit the growth of many types of cells, including most epithelial and immune system cells. Loss of TGFβ-mediated growth inhibition contributes to the development and progression of a variety of tumors. The BMP proteins discussed in Chapter 23 are homologs of TGFβ and control many important steps in development.

TGFβ signals through the sequential activation of two cell-surface receptors, termed *type I* and *type II*, both of which have intrinsic serine/threonine protein kinase activity (Figure 24-20). Binding of TGFβ induces formation of a complex of the type I and type II receptors and phosphorylation and activation of the type I receptor by the type II receptor kinase. *Smad proteins* are the key intracellular signal transducers in the pathway downstream from the TGFβ receptors. The ligand-activated type I receptor phosphorylates conserved serines at the C-terminus of either Smad2 or Smad3, which enables them to bind to one or more molecules of Smad4, a common partner for all phosphorylated Smads involved in signaling by both TGFβ and bone morphogenic proteins. These Smad complexes then enter the nucleus and activate transcription of a variety of genes. One important gene induced by TGFβ encodes p15. This G_1 cyclin-kinase inhibitor displaces p27 from the Cdk4-cyclin D complex, freeing p27 to bind to and inhibit the Cdk2-cyclin E complex, which is required for entry into the S phase (see Figure 13-29). Thus by inducing expression of p15, TGFβ causes the cell to arrest in G_1.

Many tumors contain inactivating mutations in either the TGFβ receptors or the Smad proteins, and thus are resistant

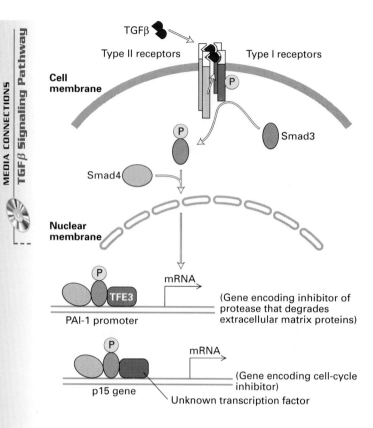

▲ **FIGURE 24-20 TGFβ signaling.** TGFβ signals through heteromeric complexes of types I and II serine/threonine kinase receptors, leading to phosphorylation of either Smad2 or Smad3. A complex of one of these phosphorylated Smad proteins and Smad4 then translocates to the nucleus, where it binds to other transcription factors to activate transcription of a variety of genes. Expression of p15 causes cells to arrest in G_1. See text for discussion. [See X. Hua et al., 1998, *Genes & Develop.* **12**:3084; and C. H. Heldin et al., 1997, *Nature* **390**:465.]

to growth inhibition by TGFβ. Most human pancreatic cancers contain a deletion in the gene encoding Smad4; this mutation-defined gene was originally called *DPC* (deleted in pancreatic cancer). Retinoblastoma, colon and gastric cancer, hepatoma, and some T- and B-cell malignancies are unresponsive to TGFβ growth inhibition. This loss of responsiveness correlates with loss of type I *or* type II TGFβ receptors; responsiveness to TGFβ can be restored by recombinant expression of the "missing" protein. Mutations in Smad2 also commonly occur in several types of human tumors.

The TGFβ signaling pathway also induces expression of the genes encoding many extracellular matrix proteins, such as collagens, and plasminogen activator inhibitor-1 (PAI-1), an inhibitor of a serum protease that degrades many matrix proteins (see Figure 24-20 and Section 24.1). An inability to synthesize these proteins may contribute to metastasis,

allowing tumor cells to "escape," since less matrix will be made by the cells and matrix that is present may be degraded by inappropriately activated proteases.

SUMMARY   Mutations Causing Loss of Cell-Cycle Control

• Overexpression of the proto-oncogene encoding cyclin D1 or loss of the tumor-suppressor genes encoding p16 and Rb can cause inappropriate, unregulated passage through the restriction point in late G_1, a key element in cell-cycle control. Such abnormalities are common in human tumors.

• TGFβ induces expression of p15, leading to arrest in G_1, and synthesis of extracellular matrix proteins such as collagens and plasminogen activator inhibitor-1.

• Loss of TGFβ receptors or Smad 4, a characteristic of many human tumors, abolishes TGFβ signaling. This promotes cell proliferation and development of malignancy.

24.5 Mutations Affecting Genome Stability

We have seen that the DNA of cancer cells contains insertions, deletions, and point mutations in multiple key growth and cell-cycle regulatory genes; in addition, several genes often are amplified and chromosome breaks and fusions (translocations) generate oncogenic chimeric proteins. Most cancer cells also lack one or more DNA-repair systems, which may explain the large number of mutations that they accumulate. Additionally, cancer cells are defective in one or more checkpoint controls, which normally prevent cells with damaged DNA or abnormal chromosomes from dividing or that cause such cells to undergo programmed death (see Figure 13-34).

Mutations in p53 Abolish G_1 Checkpoint Control

The p53 protein is essential for the checkpoint control that arrests human cells with damaged DNA in G_1, and mutations in the *p53* gene occur in more than 50 percent of human cancers (Figure 24-21a). As detailed in Chapter 13, cells with functional p53 arrest in G_1 when exposed to γ irradiation, whereas cells lacking functional p53 do not (see Figure 13-35). Unlike other cell-cycle proteins, p53 is expressed at very low levels in normal cells because it is extremely unstable and rapidly degraded. Mice lacking p53 are viable and healthy, except for a predisposition to develop multiple types of tumors. Rather, p53 is activated only in stressful situations, such as ultraviolet or γ irradiation, heat, and low

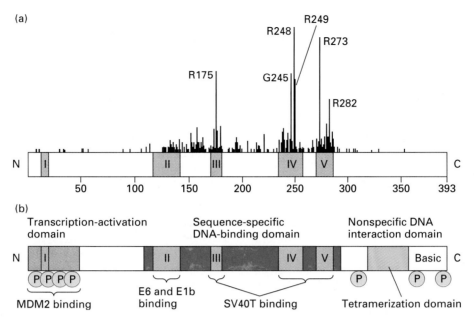

▲ FIGURE 24-21 The human p53 protein. (a) Mutations in human tumors that inactivate the function of p53 protein. Hatched boxes represent sequences highly conserved in evolution. Vertical lines represent the frequency at which mutations are found at each residue in various human tumors. These mutations are clustered in conserved regions II–V. (b) Structural organization of the p53 protein. Phosphorylation by various kinases at the sites indicated by Ⓟ stabilize p53. MDM2 protein binds at the indicated site and represses transcription activation by p53 as part of the normal control of p53 function. The activity of p53 also is inhibited by binding of viral proteins such as E6 from human papillomavirus and E1b from adenovirus. [Adapted from C. C. Harris, 1993, *Science* **262**:1980; and L. Ko and C. Prives, 1996, *Genes & Develop.* **10**:1054.]

oxygen. DNA damage by γ irradiation or by other stresses somehow leads to the activation of certain kinases, including ATM, which is encoded by the gene mutated in ataxia telangiectasia, and a DNA-dependent protein kinase. Phosphorylation of p53 by these and other kinases results in its stabilization and thus in a marked increase in its concentration (Figure 24-21b).

Although p53 has several functions, its ability to activate transcription of certain genes is most relevant to its tumor-suppressing function. Virtually all p53 mutations abolish its ability to bind to specific DNA sequences and activate gene expression. The most important protein relative to cell-cycle control whose transcription is induced by p53 is the cyclin-kinase inhibitor p21, which binds to and inhibits mammalian G_1 Cdk-cyclin complexes. As a result, cells with damaged DNA are arrested in G_1 until the damage is repaired and the levels of p53 and p21 fall; the cells then can progress into the S phase. In most cells, accumulation of p53 also leads to induction of proteins that promote apoptosis. While this may seem like a drastic response to DNA damage, it prevents proliferation of cells that are likely to accumulate multiple mutations. When the p53 checkpoint control does not operate properly, damaged DNA can replicate, producing mutations and DNA rearrangements that contribute to the development of a highly transformed, metastatic cell. p53 also is important for G_2-to-M checkpoint control, though the mechanism is only now being unraveled.

The active form of p53 is a tetramer of four identical subunits. A missense point mutation in one *p53* of the two alleles in a cell can abrogate almost all p53 activity because virtually all the oligomers will contain at least one defective subunit and such oligomers cannot function as a transcription factor. Oncogenic *p53* mutations thus act as "dominant negatives," in contrast to tumor-suppressor genes such as *RB*. Since the Rb protein functions as a monomer, mutation of a single *RB* allele has little functional consequence.

Proteins that interact with and regulate p53 are also altered in many human tumors. The gene encoding one such protein, MDM2, is amplified in many sarcomas and other human tumors that maintain functional p53. Under normal conditions MDM2 protein binds to a site in the N-terminus of p53, both repressing the ability of p53 to activate transcription of *p21* and other genes and mediating p53 degradation. Thus MDM2 normally inhibits the ability of p53 to restrain the cell cycle or kill the cell. Phosphorylation of p53 by ATM, as after γ irradiation, leads to displacement of bound MDM2 and thus stabilization of p53. Because the *Mdm2* gene is itself transcriptionally activated by p53, MDM2 functions in an autoregulatory feedback loop with p53, perhaps normally preventing excess p53 function. Enhanced MDM2 levels in tumor cells would cause a decrease in the concentration of functional p53 and abolish the ability of p53 to arrest a cell in response to irradiation.

The consequences of mutations in *p53* and *Mdm2* gene amplifications provide a dramatic example of the significance of cell-cycle checkpoints to the health of a multicellular organism. Germ-line *p53* mutations are also known; *p53* is mutated in the Li-Fraumeni syndrome of multiple inherited cancers.

Proteins Encoded by DNA Tumor Viruses Can Inhibit p53 Activity

We noted that human papillomavirus (HPV) is able to induce stable transformation and transient mitogenic stimulation of a variety of cultured cells. One HPV protein, E7, binds to and inhibits Rb, while another, E6, inhibits p53 (see Figure 24-21b). Acting together, E6 and E7 are sufficient to induce transformation in the absence of mutations in cell regulatory proteins. The HPV E5 protein, which causes sustained activation of the PDGF receptor, enhances proliferation of the transformed cells.

Similarly, SV40, a small transforming DNA monkey papovavirus, makes two early proteins called *T* (large T) and *t* (small t) formed by alternative splicing from the same reading frame. Large T, also called "T antigen," is a 90-kDa protein found in the nucleus of infected cells; different domains of large T bind to p53 and Rb, inhibiting their function. Because large T inhibits both proteins, expression of only the SV40 large T protein is sufficient to induce transformation of cultured cells. (As discussed in Chapter 12, another domain of large T binds to the origin of SV40 DNA and initiates replication of SV40 DNA.) SV40 was isolated as a contaminant of the first poliovirus vaccine. Despite being injected into millions of children, there is no evidence that SV40 can induce human tumors or that SV40 DNA is found in human tumor cells. Nonetheless, both HPV and SV40 provide illuminating examples of mechanisms by which small DNA viruses induce cell transformation.

Some Human Carcinogens Cause Inactivating Mutations in the *p53* Gene

As noted in Chapter 12, most carcinogenic agents cause DNA damage. Epidemiological studies established that cigarette smoking is the major cause of lung cancer, but precisely how this happens was not clear until the discovery that about 60 percent of human lung cancers contain inactivating mutations in the *p53* gene. The chemical benzo(*a*)pyrene, found in cigarette smoke, undergoes metabolic activation in the liver to a potent mutagen that causes mainly G to T transversion mutations. When applied to cultured bronchial epithelial cells, activated benzo(*a*)pyrene induces inactivating mutations at codons 175, 248, and 273 of the *p53* gene; these same positions are major mutational hot spots in human lung cancer (see Figure 24-21a). Thus, there is a direct link between a defined chemical carcinogen in cigarette smoke and human cancer; it is likely that cigarette smoke induces mutations in other genes as well.

Lung cancer is not the only major human cancer for which a clear-cut risk factor has been identified. Aflatoxin, a fungal metabolite found as a contaminant in moldy grains,

induces liver cancer, a disease whose incidence is high in China. Aflatoxin induces a G to T transversion at codon 249 of p53, leading to its inactivation. Exposure to other chemicals has been correlated with minor cancers. However, hard evidence concerning dietary and environmental risk factors that would help us avoid breast, colon, and prostate cancer, leukemias, and other cancers is generally lacking.

Defects in DNA-Repair Systems Perpetuate Mutations and Are Associated with Certain Cancers

A link between carcinogenesis and failure of DNA repair is suggested by the finding that humans with inherited genetic defects in certain repair systems have an enormously increased probability of developing certain cancers (Table 24-1). One such disease is *xeroderma pigmentosum,* an autosomal recessive disease. Individuals with this disease get the skin cancers called melanomas and squamous cell carcinomas very easily if their skin is exposed to the UV rays in sunlight. Cells of affected patients are unable to repair UV damage or to remove bulky chemical substituents on DNA bases. Such damage commonly is repaired by the excision-repair mechanism (see Figure 12-26). The complexity of mammalian excision-repair systems is shown by the fact that mutations in at least seven different genes lead to xeroderma pigmentosum lesions, all having the same phenotype and the same consequences. Genetic evidence suggests that in yeast several *RAD* genes encode proteins required for excision repair. The predicted protein product encoded by the yeast *RAD14* gene has considerable homology with the protein encoded by one of the genes that is mutated in some xeroderma pigmentosum patients, attesting to the conservation of the excision-repair system during evolution of eukaryotes.

Three Mut proteins participate in mismatch repair in bacteria (see Figure 12-24). Mutations in genes encoding the human homologs of the bacterial MutS or MutL repair proteins also have been associated with higher-than-normal cancer risks. Once cells lose the ability to repair mismatch errors in DNA, mutations in many other genes can accumulate, including those that are critical in controlling cell growth and proliferation. For instance, *hereditary nonpolyposis colorectal cancer,* one of the most common inherited predispositions to cancer, results from an inherited loss-of-function mutation in one allele of the gene encoding human MutSα, MutSβ, or MutL. Cells with one functional copy of any of these genes exhibit normal mismatch repair, but tumor cells frequently arise from those cells that have experienced a somatic mutation in the second allele and thus have lost the mismatch-repair system. Thus MutSα, MutSβ, and MutL are tumor-suppressor genes, and somatic inactivating mutations in these genes are also common in noninherited forms of colon cancer.

One gene frequently mutated in colon cancers because of the absence of mismatch repair encodes the type II receptor

TABLE 24-1 Human Hereditary Diseases Associated with DNA-Repair Defects

Disease*	Sensitivity	Cancer Susceptibility	Symptoms
Ataxia telangiectasia	γ irradiation	Lymphomas	Ataxia, dilation of blood vessels in skin and eyes, chromosome aberrations, immune dysfunction
Bloom's syndrome	Mild alkylating agents	Carcinomas, leukemias, lymphomas	Photosensitivity, facial telangiectases, chromosome alterations
Cockayne's syndrome	UV irradiation		Dwarfism, retinal atrophy, photosensitivity, progeria, deafness, trisomy 10
Fanconi's anemia	Cross-linking agents	Leukemias	Hypoplastic pancytopenia, congenital anomalies
Hereditary nonpolyposis colorectal cancer	UV irradition, chemical mutagens	Colon, ovary	Early development of tumors
Xeroderma pigmentosum	UV irradiation chemical mutagens	Skin carcinomas and melanomas	Skin and eye photosensitivity, keratoses

*Other human hereditary disorders that may be related to DNA-repair defects include dyskeratosis congenita (Zinsser-Cole-Engman syndrome), progeria (Hutchinson-Gilford syndrome), and trichothiodystrophy.

SOURCE: Modified from A. Kornberg and T. Baker, 1992, *DNA Replication*, 2d ed., W. H. Freeman and Company, p. 788.

for TGFβ. The gene encoding the type II receptor contains an A_{10} sequence that frequently undergoes mutation to A_9 or A_{11} because of "slippage" of DNA polymerase during replication. If unrepaired by the mismatch-repair system, these mutations cause a frame shift in the protein-coding sequence that abolishes production of the normal receptor protein. As noted earlier, such inactivating mutations make cells resistant to growth inhibition by TGFβ, thereby contributing to the unregulated growth characteristic of these tumors. This finding attests to the importance of mismatch repair in correcting genetic damage that might otherwise lead to uncontrolled cell proliferation. Additionally, recognition of certain DNA lesions by MutSα initiates a sequence of events that leads to cell death, and the absence of MutSα may allow the survival of cells with major DNA damage, cells that could progress to become tumors.

Chromosomal Abnormalities Are Common in Human Tumors

It has long been known that chromosomal abnormalities abound in tumor cells. Human cells ordinarily have 23 pairs of chromosomes, recognized by their well-defined substructure, but tumor cells are usually *aneuploid* (i.e., they have an abnormal number of chromosomes—generally too many), and they often contain *translocations* (fused elements from different chromosomes). Cells with abnormal numbers of chromosomes form when the S-phase or mitotic checkpoints are nonfunctional. The first condition normally prevents

entry into mitosis unless all chromosomes have completely replicated their DNA, while the latter causes arrest in anaphase unless all the replicated chromosomes are appropriately lined up to be segregated properly (see Figure 13-34). Defects in these checkpoint controls are common in tumor cells; the molecular basis for these defects is now being uncovered as the checkpoint control proteins themselves are being identified.

As a rule, these chromosomal abnormalities are not the same from tumor to tumor: each tumor has its own set of anomalies. Certain anomalies recur, however, and they point to the presence of oncogenes (Figure 24-22, see Figure 9-38a). The first to be discovered, the *Philadelphia chromosome*, is

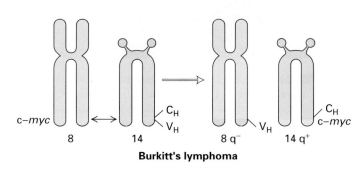

Burkitt's lymphoma

▲ FIGURE 24-22 Chromosomal translocation in Burkitt's lymphoma. This leads to overexpression of the Myc transcription factor.

found in hematopoietic cells of virtually all patients with the disease chronic myelogenous leukemia. This chromosome results from a translocation between chromosomes 9 and 22; which causes formation of the chimeric *bcr-abl* oncogene discussed earlier. We've also discussed the translocation of the c-*myc* gene seen in Burkitt's lymphoma, resulting in abnormal expression of the Myc transcription factor. In this case, c-*myc*, normally located on chromosome 8, is moved to a site on chromosome 14 near an antibody-gene enhancer. In other lymphomas, a different chromosomal translocation brings the anti-apoptotic gene *bcl-2* under the transcriptional control of an antibody enhancer. The resultant inappropriate overexpression of the Bcl-2 protein prevents normal apoptosis and allows survival of these tumor cells.

Another common chromosomal anomaly in tumor cells is the localized reduplication of DNA to produce as many as 100 copies of a given region (usually a region spanning hundreds of kilobases). This anomaly may take either of two forms: the duplicated DNA may be tandemly organized at a single site on a chromosome, or it may exist as small, independent chromosomelike structures. The former case leads to a *homogeneously staining region (HSR)* that is visible in the light microscope at the site of the duplication; the latter case causes double minute chromosomes to pepper a stained chromosomal preparation (Figure 24-23). Again, oncogenes have been found in the duplicated regions. Most strikingly, the *myc*-related gene called N-*myc* has been identified in both HSRs and double minute chromosomes of human nervous system tumors.

We noted that the *p16* tumor-suppressor gene is inactivated in many human cancers. In many cases the gene has been deleted, but in others the sequence of the entire *p16* gene is normal. To explain this, recall that gene expression is controlled at multiple levels, and that one of these involves the state of DNA methylation in long segments along a chromosome. Indeed, the *p16* promoter is inactivated by hypermethylation in many human tumors, including lung cancer. How this epigenetic change happens is not known, but it does lead to nonproduction of this important cell-cycle control protein.

(a)

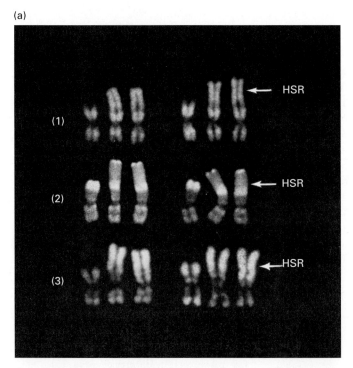

(b)

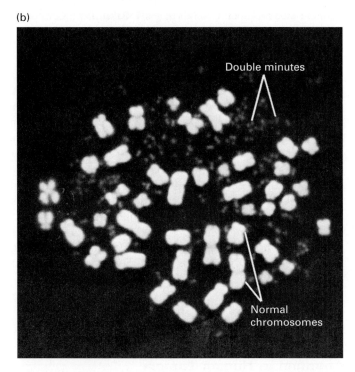

▲ **FIGURE 24-23 Visible DNA amplifications.** (a) Homogeneously staining regions (HSRs) in chromosomes from two neuroblastoma cells. In each set of three chromosomes, the left-most one is a normal chromosome 1 and the other two are HSR-containing chromosomes. The three preparations (1, 2, and 3) represent three different methods of staining the chromosomes. Method 1 is quinacrine staining, which highlights AT-rich regions; method 2 is staining with chromomycin A3 plus methyl green, which highlights GC-rich areas; and method 3 is 33258 Hoechst staining after a pulse of bromodeoxyuridine late during the S phase, which highlights the early replicating regions. In all three cases the HSRs stain homogeneously whereas the rest of the chromosomes are somewhat banded. (b) Quinacrine-stained double minute chromosomes from a human neuroblastoma cell. The normal chromosomes are the large white structures; the double minute chromosomes are the many small paired dots. Both the HSRs and the double minute chromosomes shown here contain the N-*myc* oncogene. [Part (a) see S. Latt et al., 1975, *Biopolymers* **24**:77; part (b) see N. Kohl et al., 1983, *Cell* **35**:359; photographs courtesy of Dr. S. Latt.]

Telomerase Expression May Contribute to Immortalization of Cancer Cells

Telomeres, the physical ends of linear chromosomes, consist of tandem arrays of a short DNA sequence, TTAGGG in vertebrates. Telomeres provide the solution to the end-replication problem—the inability of DNA polymerases to completely replicate the end of a double-stranded DNA molecule. In the germ line and in rapidly dividing somatic cells, such as stem cells, telomerase, a reverse transcriptase that contains an RNA template, adds TTAGGG repeats to chromosome ends (see Figure 12-13). Because most human somatic cells lack telomerase, telomeres shorten with each cell cycle. Complete loss of telomeres leads to end-to-end chromosome fusions and cell death.

Most tumor cells overcome this barrier by expressing telomerase. Many workers believe that telomerase expression is essential for a tumor cell to become immortal, and specific inhibitors of telomerase have been suggested as cancer therapeutic agents. Surprisingly, however, mice homozygous for a deletion of the RNA subunit of the telomerase gene are both viable and fertile. When treated with carcinogens, these telomerase-null mice develop tumors to the same extent as normal mice, establishing that in mice, activation of telomerase is not essential for tumorigenesis. Whether the same applies to humans is a matter of controversy, since mice have much longer telomeres than humans and mouse cells may not require continuous telomerase expression.

SUMMARY Mutations Affecting Genome Stability

- The p53 protein is essential for the checkpoint control that arrests human cells with damaged DNA in G_1. Replication of such cells would tend to perpetuate mutations. p53 functions as a transcription factor to induce expression of p21, an inhibitor of G_1 Cdk-cyclin complexes.

- Mutations in the p53 gene occur in more than 50 percent of human cancers.

- Because p53 is a tetramer, a point mutation in one p53 allele can be sufficient to inhibit all p53 activity.

- MDM2, a protein that normally inhibits the ability of p53 to restrain the cell cycle or kill the cell, is overexpressed in several cancers.

- Defects in cellular DNA-repair processes found in certain human diseases are associated with an increased susceptibility for certain cancers (see Table 24-1).

- Chromosome abnormalities, including aneuploidy and translocations, are common in human tumors, and often result in duplications of oncogenes such as myc.

- Cancer cells, like germ cells and stem cells but unlike most differentiated cells, express telomerase, which may contribute to their immortalization.

PERSPECTIVES for the Future

Studies of cancer, like those of many other human diseases, have provided profound insights into the behavior of normal cells and tissues. The cloning of the RB gene, based on its deletion from patients with inherited retinoblastoma, led to an understanding of how the product of this tumor-suppressor gene, the first to be discovered, functions to restrict passage through the cell cycle. Likewise, study of the mutant v-Src protein led to the recognition of the role of protein-tyrosine kinases as regulators of cell division and differentiation.

But the main motivation of studies on the molecular basis of cancer is to develop new therapies for tumors where there is no good treatment, or where existing therapies have significant side effects or problems. Current cancer therapies—mainly surgery, radiation therapy, and chemotherapy—all have drawbacks. Surgery will work for many primary tumors, but metastases are difficult to identify let alone remove at an early stage. Although γ irradiation kills many localized tumors, it also kills normal cells. Most drugs used in cancer therapy (e.g., taxol, 5-fluorouracil, and adriamycin) are designed to kill dividing cells. Not only do these drugs also kill intestine and skin stem cells, causing diarrhea and hair loss, they are ineffective against cells with mutations in the p53 gene, a common occurrence in cancer cells.

Transplantation of hematopoietic stem cells is being used as an experimental therapy for breast and other cancers: The patient is given sufficient drugs and radiation ideally to kill all the tumor cells, as well as normal hematopoietic stem cells, and then reinfused with stem cells to restore normal levels of blood cells. However, if the treatment fails, the patient is left with no white cells or immune cells. The introduction of hematopoietic cytokines such as erythropoietin and G-CSF, produced by recombinant DNA technology, into patients after irradiation greatly increases the rate of recovery of red and white blood cells and allows the patient to withstand these doses of radiation better.

Molecular insights into cancer biology have already resulted in some efficacious therapies, and many more can be anticipated in the coming years as pharmaceutical and biotechnology companies exploit these new discoveries. Cancers due mainly to dominant oncogenes are the most likely targets for therapy. We have mentioned that overexpression of the Her2 receptor in certain breast tumors can be attacked by an anti-Her2 monoclonal antibody, and inhibitors of farnesyl transferase show promise as therapies of tumors containing an activating Ras mutation. Specific inhibitors of telomerase activity or of angiogenesis are now being developed, but which, if any, tumors might be treated by such reagents is not yet clear.

Tumors caused by mutations in tumor-suppressor genes are harder to treat because these result from the loss of a normal protein. Among the approaches now being attempted is reintroduction of the gene encoding wild-type p53, PTEN, or a particular cyclin-kinase inhibitor into a tumor missing these proteins. One challenge of such gene therapy is targeting the introduced gene only to the tumor, not to normal cells, where it might inhibit their normal division. Another, one that particularly affects solid tumors, is getting the missing gene into all the tumor cells, as even a tiny number of uncorrected tumor cells can still kill the patient. Finally, the levels of proteins expressed from introduced genes must be near normal to be effective.

One immediately useful insight provided by molecular analyses is that many types of tumors that were formerly thought to be similar are, in fact, caused by different combinations of oncogenic mutations. Development of therapies specific for tumors of a specific genetic composition—the emerging discipline of pharmacogenetics—may allow even existing drugs to be used with greater success.

PERSPECTIVES in the Literature

As summarized in this chapter, cancer cells arise from normal cells following the mutation of cellular genes such as oncogenes and tumor suppressor genes. Inherited mutations in these genes can also predispose individuals to cancer. In an effort to model heritable cancer predisposition, as well as to develop tools to study the stages of tumor development and to evaluate potential anti-cancer therapies, investigators have created novel strains of mice using transgenic and gene targeting technologies. As you read the papers referenced below, related to such mouse models of tumor development, consider the following questions:

1. How can one explain species-specific differences in effects that germline mutations in tumor suppressor genes have on development and tumorigenesis?

2. How can in vivo models be used to examine cooperativity between cancer-associated mutations?

3. What is the effect of the genetic background of the mouse strains on tumor development?

4. In what settings are transgenic and gene knock-out strains useful for evaluating novel therapies?

Dietrich, W. F. 1993. Genetic identification of Mom-1, a major modifier locus affecting Min-induced intestinal neoplasia in the mouse. *Cell* 75:631–639.

Jacks, T., et al. 1992. Effects of an RB mutation in the mouse. *Nature* 359:295–300.

Kohl, N. E., et al. 1995. Inhibition of farnesyltransferase induces regression of mammary and salivary carcinomas in ras transgenic mice. *Nature Med.* 1:792–797.

Shibata, H., et al. 1997. Rapid colorectal adenoma formation initiated by conditional targeting of the APC gene. *Science* 27:120–123.

Williams, B. O., et al. 1994. Cooperative tumorigenic effects of germline mutations in Rb and p53. *Nature Genet.* 7:480–484.

Testing Yourself on the Concepts

1. Briefly describe the mechanisms by which changes in cell-surface receptors and signal-transduction pathways derange cell growth.

2. How do gain-of-function and loss-of-function mutations differ in how they cause cancer?

3. Explain this statement from page 1064 in the text: "[N]ormal growth control and malignancy are two faces of the same coin."

4. What hypothesis explains the observations that incidence of human cancers increases exponentially with age? Give an example of data that confirms the hypothesis.

MCAT/GRE-Style Questions

Key Concept Please read the section titled "Oncogenes Were First Identified in Cancer-Causing Retroviruses" (p. 1065) and answer the following questions:

1. Evidence supporting the idea that v-*src* is the oncogene of RSV includes all of the following *except*:
   a. A *ts* mutation in v-*src*.
   b. Homology with c-*src*.
   c. Its expression even in the presence of c-*src*.
   d. Mutation in v-*src* abrogates transformation.

2. Retroviruses are capable of causing cancer because they
   a. cause c-*onc*s to mutate to v-*onc*s.
   b. Contain point mutations in *pol*.
   c. Contain mutant versions of cellular genes that normally regulate cell growth.
   d. Pick up a mutated c-*onc* from the genome.

3. Advantages of using retroviruses to characterize viral oncogenes include all of the following *except*:
   a. A simple assay for presence of an oncogene in the viral genome.
   b. The virus prepackages the oncogene.
   c. Relative ease of sequencing viral genomes.
   d. They mimic the development of metastatic tumors.

4. The work of Bishop and Varmus
   a. Showed that many c-*onc*s could be converted to v-*onc*s.
   b. Characterized the transduction pathway of retroviruses.
   c. Showed the homology between c-*onc*s and v-*onc*s.
   d. Showed the difference between RSV and its slowly transforming variant.

Key Experiment Please read the section titled "DNA from Tumor Cells Can Transform Normal Cultured Cells"

(p. 1057) and refer to Figure 24-4; then answer the following questions:

5. The experiment outlined in Figure 24-4 is successful because of all of the following *except*:
    a. The oncogene is essentially isolated by the cell culture system.
    b. The added DNA is necessary and sufficient for transformation of cells from all mammals.
    c. Mouse and human *Alu* sequences can be distinguished from each other.
    d. Transformation is an in vitro model for oncogenesis.

6. Confirmation that a human gene isolated as shown in Figure 24-4 is an oncogene can be obtained by all of the following strategies *except*:
    a. Comparing the sequence with those of human genes known to regulate cell growth.
    b. Obtaining transformation when the gene is introduced into mouse fibroblasts.
    c. Comparing the sequence with that of human *p16*.
    d. Showing that the isolated gene hybridizes to DNA extracted from the originally transformed mouse cells or their progeny.

7. Characteristics of transformed fibroblasts include
    a. Contact inhibition.
    b. Entrance into G_1 upon exposure to growth factors.
    c. Ability to be detected by a focus-forming assay.
    d. A spindle-shaped morphology.

8. In theory, mouse fibroblasts could be transformed by
    a. Transfection with a λ phage containing an oncogene isolated as described in Figure 24-4.
    b. Mutation of *ras* to *ras*^{D}.
    c. Mutation to a loss-of-function state in *p16*.
    d. Mutation of *ras* to *ras*^{D} and mutation to a loss of function state in *p16*.

Key Concept Please read the section titled "Alterations in Cell-to-Cell Interactions Are Associated with Malignancy" and "Tumor Growth Requires Formation of New Blood Vessels" (p. 1056) and answer the following questions:

9. Establishment of a metastasis requires all of the following *except*:
    a. Breach of the basal lamina.
    b. Production of specific growth factors.
    c. Migration of many tumor cells in a mass.
    d. Movement through blood vessels.

10. Alternation in cell-surface proteins of tumor cells promotes metastasis by
    a. Secreting plasminogen activator.
    b. Producing TGFα.

c. Inhibiting interaction with protein such as collagen and proteoglycan, permitting cells to move within the organism.
    d. Allowing cells to multiply in a different tissue.

11. Breaching the basal lamina
    a. Requires production of plasmin by tumor cells.
    b. Is not required for angiogenesis.
    c. Utilizes degradative enzymes produced by normal cells in response to the presence of the tumor.
    d. Utilizes degradative enzymes produced by the tumor cells.

Key Terms

apoptosis *1061*	oncogene *1059*
basal lamina *1056*	proto-oncogene *1059*
benign *1055*	retrovirus *1065*
carcinomas *1056*	sarcomas *1056*
clone *1059*	stem cells *1062*
cytokines *1062*	transformation *1058*
leukemias *1056*	tumor *1054*
malignant *1055*	tumor-suppressor genes *1059*
metastasis *1055*	

References

Tumor Cells and the Onset of Cancer

Behrens, J., et al. 1992. Cell adhesion in invasion and metastasis. *Semin. Cell Biol.* 3:169–178.

Cavenee, W. K., and R. L. White. 1996. The genetic basis of cancer. *Sci. Am.* 272:72–79.

Coffin, J., S. Hughes, and H. Varmus, eds. 1998. *Retroviruses*. Cold Spring Harbor Laboratory Press.

Fields, B. N., and D. M. Knipe. 1990. *Virology*, 2d ed., 2 vols. Raven.

Folkman, J. 1996. Fighting cancer by attacking its blood supply. *Sci. Am.* 272:275–278.

Folkman, J., and D. Hanahan. 1991. Expression of the Angiogenic Phenotype during Development of Murine and Human Cancer. In *Origins of Human Cancer: A Comprehensive Review*. Cold Spring Harbor Laboratory Press.

Hartwell, L. H., et al. 1997. Integrating genetic approaches into the discovery of anticancer drugs. *Science* 278:1064–1068.

Kinzler, K. W., and B. Vogelstein. 1996. Lessons from hereditary colorectal cancer. *Cell* 87:159–170.

Klein, G. 1998. Foulds' dangerous idea revisited: the multistep development of tumors 40 years later. *Adv. Cancer Res.* 72:1–23.

Rouslahti, E. 1996. How cancer spreads. *Sci. Am.* 275:165–170.

Szabo, C. I., and M. C. King. 1995. Inherited breast and ovarian cancer. *Human Mol. Genet.* 1811–1817.

Varmus, H. E., and C. A. Lowell. 1994. Cancer genes and hematopoiesis. *Blood* 83:5–9.

Varmus, H., and R. A. Weinberg. 1993. *Genes and the Biology of Cancer*. Scientific American Library.

Vogelstein, B., and K. W. Kinzler. 1993. The multistep nature of cancer. *Trends Genet.* 9:138–141.

Weinberg, R. A. 1996. How cancer arises. *Sci. Am.* **275**:62–70.

Yancopoulos, G., M. Klagsburn, and J. Folkman. 1998. Vasculogenesis, angiogenesis, and growth factors: ephrins enter the fray at the border. *Cell* **93**:661–664.

Oncogenic Mutations Affecting Cell Proliferation

Barbacid, M. 1987. *ras* genes. *Ann. Rev. Biochem.* **56**:779–827.

Boguski, M. S., and F. McCormick. 1993. Proteins regulating Ras and its relatives. *Nature* **266**:643–654.

Cantley, L., and B. G. Neel. 1999. New insights into tumor suppression: PTEN suppresses tumor formation by restraining the phosphoinositide 3-kinase/AKT pathway. *Proc. Nat'l. Acad. Sci. USA* **96**:4240–4245.

Fieg, L. A. 1993. The many roads that lead to Ras. *Science* **260**:767–768.

Hesketh, R., ed. 1997. *The Oncogene and Tumor Suppressor Gene Facts Book,* 2d ed. Academic Press.

Hunter, T. 1997. Oncoprotein networks. *Cell* **88**:333–346.

Hunter, T. 1998. The Croonian Lecture 1997. The phosphorylation of proteins on tyrosine: its role in cell growth and disease. *Phil. Trans. Roy. Soc. Lond. B Biol. Sci.* **353**:583–605.

Martin, S., ed. 1998. *Apoptosis and Cancer.* Karger Landes Systems.

Wang, J. Y. J. 1993. Abl tyrosine kinase in signal transduction and cell-cycle regulation. *Curr. Opin. Genet. Dev.* **3**:35.

Mutations Causing Loss of Cell-Cycle Control

Dang. C. V. 1999. c-Myc target genes involved in cell growth, apoptosis, and metabolism. *Mol. Cell Biol.* **19**:1–11.

Elledge, S. 1996. Cell cycle checkpoints: preventing an identity crisis. *Science* **274**:1664–1672.

Hartwell, L. 1992. Defects in a cell cycle checkpoint may be responsible for the genomic instability of cancer cells. *Cell* **71**:543–546.

Hartwell, L. H., and M. B. Kastan. 1994. Cell cycle control and cancer. *Science* **266**:1821–1828.

Heldin, C. H., K. Miyazono, and P. ten Dijke. 1997. TGF-β signaling from cell membrane to nucleus through SMAD proteins. *Nature* **390**:465–471.

Hunter, T., and J. Pines. 1994. Cyclins and cancer. II: Cyclin D and CDK inhibitors come of age. *Cell* **79**:573–582.

Lees, J., and R. Weinberg. 1999. Tossing monkey wrenches into the clock: new ways of treating cancer. *Proc. Nat'l. Acad. Sci. USA* **96**:4221–4223.

McCormick, F., and P. Myers. 1994. From genetics to chemistry: tumor suppressor genes and drug discovery. *Chem. Biol.* **1**:7–9.

Murray, A. W. 1995. The genetics of cell cycle checkpoints. *Curr. Opin. Genet. Dev.* **5**:5–11.

Oliff, A., J. B. Gibbs, and F. McCormick. 1996. New molecular targets for cancer therapy. *Sci. Am.* **275**:144–149.

Paulovich, A. G., D. P. Toczyski, and L. H. Hartwell. 1997. When checkpoints fail. *Cell* **88**:315–321.

Planas-Silva, M. D., and R. A. Weinberg. 1997. The restriction point and control of cell proliferation. *Curr. Opin. Cell Biol.* **9**:768–772.

Sheer, C. J. 1996. Cancer cell cycles. *Science* **274**:1672–1677.

Weinberg, R. A. 1995. The retinoblastoma protein and cell cycle control. *Cell* **81**:323–330.

Weinberg, R. A. 1996. E2F and cell proliferation: a world turned upside down. *Cell* **85**:457–459.

Weinberg, R. A. 1997. The cat and mouse games that genes, viruses, and cells play. *Cell* **88**:573–575.

Westphal, C. H. 1997. Cell-cycle signaling: Atm displays its many talents. *Curr. Biol.* **7**:R789–R792.

White, R. 1998. Tumor suppressor pathways. *Cell* **92**:591–592.

Zhu, L., et al. 1994. Growth suppression by members of the retinoblastoma protein family. *Cold Spring Harbor Symp. Quant. Biol.* **59**:95–84.

Mutations Affecting Genome Stability

Brugarolas, J., and T. Jacks. 1997. Double indemnity: p53, BRCA and cancer. p53 mutation partially rescues developmental arrest in Brca1 and Brca2 null mice, suggesting a role for familial breast cancer genes in DNA damage repair [news]. *Nat. Med.* **3**:721–722.

Bogler, O., H. J. Huang, P. Kleihues, and W. K. Cavenee. 1995. The p53 gene and its role in human brain tumors. *Glia* **15**:308–327.

Fishel, R. 1998. Mismatch repair, molecular switches, and signal transduction. *Genes and Devel.* **12**:2096–2101.

Fishel, R., and T. Wilson. 1997. MutS homologs in mammalian cells. *Curr. Opin. Genet. Devel.* **7**:105–113.

Harris, C. C. 1993. p53: at the crossroads of molecular carcinogenesis and risk assessment. *Science* **262**:1980–1981.

Kipling, D. 1997. Mammalian telomerase: catalytic subunit and knockout mice. *Human Mol. Genet.* **6**:1999–2004.

Ko, L., and C. Prives. 1996. p53: puzzle and paradigm. *Genes and Devel.* **10**:1054–1072.

Kolodner, R. 1996. Biochemistry and genetics of eukaryotic mismatch repair. *Genes and Devel.* **10**:1433–1442.

Leaguer, C., K. Kinzler, and B. Vogelstein 1998. Genetic instabilities in human cancers. *Nature* **396**:643–649.

Levine, A. 1997. p53, the cellular gatekeeper for growth and division. *Cell* **88**:323–331.

Lindahl, T., ed. 1996. *Genetic Instability in Cancer.* Cold Spring Harbor Laboratory Press.

Loeb, L. A. 1998. Cancer cells exhibit a mutator phenotype. *Adv. Cancer Res.* **72**:25–56.

Modrich, P. 1997. Strand-specific mismatch repair in mammalian cells. *J. Biol. Chem.* **272**:24727–24730.

Prives, C. 1998. Signaling to p53: breaking the MDM2-p53 checkpoint. *Cell* **95**:5–8.

Wood, R. D. 1996. DNA repair in eukaryotes. *Ann. Rev. Biochem.* **65**:135–168.

Wood, R. D. 1997. Nucleotide excision repair in mammalian cells. *J. Biol. Chem.* **272**:23465–23468.

Zakian, V. 1997. Life and cancer without telomerase. *Cell* **91**:1–3

Glossary

Boldfaced terms within a definition are also defined in this glossary. Alternative names for main entries are in *italics*. Figures and tables that illustrate defined terms are noted in parentheses.

acetyl CoA Small, water-soluble metabolite comprising an acetyl group linked to coenzyme A (CoA); formed during oxidation of pyruvate, fatty acids, and amino acids. Its acetyl group is transferred to citrate in the **citric acid cycle**. (Figure 16-10)

acetylcholine (ACh) Neurotransmitter that functions at vertebrate neuromuscular junctions and at various neuron-neuron synapses in the brain and peripheral nervous system.

acid A compound that can donate a proton (H^+). The carboxyl and phosphate groups are the primary acidic groups in biological molecules.

actin Abundant structural protein in eukaryotic cells that interacts with many other proteins. The monomeric globular form (G actin) polymerizes to form actin filaments (F actin). In muscle cells, F actin interacts with **myosin** during contraction. See also **microfilaments**.

action potential Rapid, transient, all-or-none electrical activity that is propagated in the plasma membrane of excitable cells such as neurons and muscle cells. Action potentials, or nerve impulses, allow long-distance signaling in the nervous system. (Figure 21-14)

activation energy The input of energy required to (overcome the barrier to) initiate a chemical reaction. By reducing the activation energy, an enzyme increases the rate of a reaction. (Figure 2-27)

active site Region of an enzyme molecule where the substrate binds and undergoes a catalyzed reaction.

active transport Energy-requiring movement of an ion or small molecule across a membrane against its concentration gradient or electrochemical gradient. Energy is provided by the coupled hydrolysis of ATP or the **cotransport** of another molecule down its electrochemical gradient.

adenosine triphosphate See **ATP**.

adenylyl cyclase Membrane-bound enzyme that catalyzes formation of **cyclic AMP (cAMP)** from ATP; also called *adenylate cyclase*. Binding of certain ligands to their cell-surface receptors leads to activation of adenylyl cyclase and a rise in intracellular cAMP. (Figure 20-15)

aerobic Referring to a cell, organism, or metabolic process that utilizes O_2 or that can grow in the presence of O_2.

aerobic oxidation Oxygen-requiring metabolism of sugars and fatty acids to CO_2 and H_2O coupled to the synthesis of ATP.

allele One of two or more alternative forms of a gene located at the corresponding site (locus) on **homologous chromosomes**.

allosteric transition Change in the tertiary and/or quaternary structure of a protein induced by binding of a small molecule to a specific regulatory site, causing a change in the protein's activity. Allosteric regulation is particularly prevalent in multisubunit enzymes.

alpha (α) helix Common secondary structure of proteins in which the linear sequence of amino acids is folded into a right-handed spiral stabilized by hydrogen bonds between carboxyl and amide groups in the backbone. (Figure 3-6)

amino acid An organic compound containing at least one amino group and one carboxyl group. In the 20 different amino acids that compose proteins, an amino group and carboxyl group are linked to a central carbon atom, the α carbon, to which a variable side chain is bound. (Figure 3-2)

aminoacyl-tRNA Activated form of an amino acid, used in protein synthesis, consisting of an amino acid linked via a high-energy ester bond to the 3′-hydroxyl group of a tRNA molecule. (Figure 4-29)

amphipathic Referring to a molecule or structure that has both a **hydrophobic** and a **hydrophilic** part.

anabolism Cellular processes whereby energy is used to synthesize complex molecules from simpler ones. See also **catabolism**.

anaerobic Referring to a cell, organism, or metabolic process that functions in the absence of O_2.

anaphase Mitotic stage during which the sister chromatids (or paired homologs in meiosis I) separate and move apart (segregate) toward the spindle poles. (Figure 19-34)

antibody A protein that interacts with a particular site (**epitope**) on an antigen and facilitates clearance of that antigen by various mechanisms. See also **immunoglobulin**. (Figure 3-21)

anticodon Sequence of three nucleotides in a **tRNA** that is complementary to a **codon** in an mRNA. During protein synthesis, base pairing between a codon and anticodon aligns the tRNA carrying the corresponding amino acid for addition to the growing peptide chain.

antigen Any material (usually foreign) that elicits production of and is specifically bound by an **antibody**.

antiport A type of **cotransport** in which a membrane protein (antiporter) transports two different molecules or ions across a cell membrane in opposite directions. See also **symport**.

antisense RNA An RNA, with sequence complementary to a specific RNA transcript or mRNA, whose binding prevents processing of the transcript or translation of the mRNA. (Figure 11-46)

apoptosis Regulated process leading to cell death via a series of well-defined morphological changes; also called *programmed cell death*. (Figure 23-45)

archaea Class of **prokaryotes** that constitutes one of the three distinct evolutionary lineages of modern-day organisms; also called *archaebacteria* and *archaeans*. These prokaryotes are in some respects more similar to **eukaryotes** than to the so-called true bacteria (**eubacteria**). (Figure 1-5)

association constant (K_a) See **equilibrium constant**.

aster Star-shaped structure composed of microtubules (called *astral fibers*) that radiates outward from a **centrosome** during mitosis. (Figure 19-34)

asymmetric carbon atom A carbon atom bonded to four different atoms; also called *chiral carbon atom*. The bonds can be arranged in two different ways producing **stereoisomers** that are mirror images of each other. (Figure 2-6)

ATP (adenosine 5'-triphosphate) A nucleotide that is the most important molecule for capturing and transferring free energy in cells. Hydrolysis of each of the two high-energy phosphoanhydride bonds in ATP is accompanied by a large **free-energy change** (ΔG) of -7 kcal/mole. (Figure 2-24)

ATP synthase Multimeric protein complex bound to inner mitochondrial membranes, thylakoid membranes of chloroplasts, and the bacterial plasma membrane that catalyzes synthesis of ATP during oxidative phosphorylation and photosynthesis; also called F_0F_1 *complex*. (Figure 16-28)

ATPase One of a large group of enzymes that catalyze hydrolysis of **ATP** to yield ADP and inorganic phosphate with release of free energy.

autonomously replicating sequence (ARS) Sequence that permits a DNA molecule to replicate in yeast; a yeast DNA replication origin. (Figure 9-40)

autoradiography Technique for visualizing radioactive molecules in a sample (e.g., a tissue section or electrophoretic gel) by exposing a photographic film or emulsion to the sample. The exposed film is called an autoradiogram or autoradiograph. (Figure 3-45)

autosome Any chromosome other than a sex chromosome.

auxotroph A mutant cell or microorganism that grows only when the medium contains a specific nutrient or metabolite that is not required by the wild type.

axon Long process extending from the cell body of a neuron that is capable of conducting an electric impulse (**action potential**) generated at the junction with the cell body (called the axon hillock) toward its distal, branching end (called the axon terminal). (Figure 21-1)

axoneme Bundle of microtubules and associated proteins present in **cilia** and **flagella** and responsible for their movement. (Figure 19-28)

bacteriophage (phage) Any virus that infects bacterial cells. Some bacteriophage are widely used as **cloning vectors.**

basal body Structure at the base of **cilia** and **flagella** from which microtubules forming the axoneme radiate; structurally similar to a **centriole.**

basal lamina (pl. basal laminae) A thin sheetlike network of extracellular-matrix components that underlies most animal epithelial layers and other organized groups of cells (e.g., muscle), separating them from connective tissue.

base A compound, usually containing nitrogen, that can accept a proton (H^+). Commonly used to denote the purines and pyrimidines in DNA and RNA.

base pair Association of two complementary **nucleotides** in a DNA or RNA molecule stabilized by hydrogen bonding between their base components. Adenine pairs with thymine or uracil ($A \cdot T$, $A \cdot U$) and guanine pairs with cytosine ($G \cdot C$). (Figure 4-4b)

benign Referring to a tumor containing cells that closely resemble normal cells. Benign tumors stay in the tissue where they originate. See also **malignant.**

beta (β) sheet A planar secondary structure of proteins that is created by hydrogen bonding between the backbone atoms in two different polypeptide chains or segments of a single folded chain. (Figure 3-8)

bilayer See **phospholipid bilayer.**

biomembrane Permeability barrier, surrounding cells or organelles, that consists of a **phospholipid bilayer**, associated membrane proteins, and in some cases cholesterol and glycolipids.

blastula An early embryonic form produced by cleavage of a fertilized ovum and usually consisting of a single layer of cells surrounding a fluid-filled spherical cavity.

buffer A solution of the acid (HA) and base (A^-) form of a compound that undergoes little change in pH when small quantities of strong acid or base are added.

cadherin Protein belonging to a family of Ca^2-dependent **cell-adhesion molecules** that play roles in tissue differentiation and structure.

calmodulin A small cytosolic protein that binds four Ca^{2+} ions; the Ca^{2+}-calmodulin complex binds to and activates many enzymes. (Figure 20-40)

Calvin cycle The major metabolic pathway that fixes CO_2 into carbohydrates during photosynthesis; also called *carbon fixation*. It is indirectly dependent on light but can occur both in the dark and light. (Figure 16-49)

cAMP-dependent protein kinase (cAPK) Type of cytosolic enzyme that is activated by cAMP and functions to regulate the activity of numerous cellular proteins; also called *protein kinase A*. Generally is activated in response to a rise in cAMP level resulting from stimulation of **G protein–coupled receptors.** (Figures 3-24 and 20-47)

capsid The outer proteinaceous coat of a virus, formed by multiple copies of one or more protein subunits and enclosing the viral nucleic acid.

carbohydrate General term for certain polyhydroxyaldehydes, polyhydroxyketones, or compounds derived from these usually having the formula $(CH_2O)_n$. Primary type of compound used for storing and supplying energy in animal cells.

carbon fixation See **Calvin cycle.**

carcinogen Any chemical or physical agent that can cause cancer when cells or organisms are exposed to it.

carcinoma A malignant tumor derived from epithelial cells.

catabolism Cellular processes whereby complex molecules are degraded to simpler ones and energy is released. See also **anabolism.**

catalyst A substance that increases the rate of a chemical reaction without undergoing a permanent change in its structure. Enzymes are protein catalysts.

catecholamines Group of compounds derived from tyrosine that function as **neurotransmitters**; include epinephrine, norepinephrine, and dopamine. (Figure 21-28)

cDNA (complementary DNA) DNA molecule copied from an mRNA molecule by reverse transcriptase and therefore lacking the introns present in **genomic DNA**. Sequencing of a cDNA permits the amino acid sequence of the encoded protein to be deduced; expression of cDNAs in recombinant cells can be used to produce large quantities of their encoded proteins in vitro.

cell cycle Ordered sequence of events in which a cell duplicates its chromosomes and divides into two. Most eukaryotic cell cycles can be commonly divided into four phases: G_1 before DNA synthesis occurs; S when DNA replication occurs; G_2 after DNA synthesis; and M when **cell division** occurs, yielding two daughter cells. Under certain conditions, cells exit the cell cycle during G_1 and remain in the G_0 state as nongrowing, nondividing (quiescent) cells. Appropriate stimulation of such cells induces them to return to G_1 and resume growth and division. (Figure 13-1)

cell division Separation of a cell into two daughter cells. In higher eukaryotes, it involves division of the nucleus (mitosis) and of the cytoplasm (cytokinesis); mitosis often is used to refer to both nuclear and cytoplasmic division.

cell fusion Production of a hybrid cell containing two or more nuclei by various techniques that stimulate the fusion of the plasma membranes of two cells at the point of contact and intermingling of their cytoplasms. See also **hybridoma.**

cell junctions Specialized regions on the cell surface through which cells are joined to each other or to the extracellular matrix. (Figure 15-23)

cell line A population of cultured cells, of plant or animal origin, that has undergone a change allowing the cells to grow indefinitely, in contrast to a **cell strain**. Cell lines can result from chemical or viral transformation and are said to be immortal.

cell strain A population of cultured cells, of plant or animal origin, that has a finite life span, in contrast to a **cell line**. (Figure 6-5)

cell wall A specialized, rigid extracellular matrix that lies next to the plasma membrane, protecting a cell and maintaining its shape. It is prominent in most fungi, plants, and prokaryotes, but is not present in most multicellular animals. (Figures 22-29 and 22-32)

cell-adhesion molecules (CAMs) Integral membrane proteins that mediate cell-cell binding. The five major classes are the integrins, cadherins, selectins, immunoglobulin (Ig) superfamily, and mucins. (Figure 22-1)

cellulose A structural polysaccharide made of glucose units linked together by $\beta(1 \rightarrow 4)$ glycosidic bonds. It forms long microfibrils, which are the major component of plant cell walls. (Figure 22-31)

central nervous system (CNS) The part of the vertebrate nervous system comprising the brain and spinal cord; the main information-processing organ.

centriole Either of two cylindrical structures within the **centrosome** of animal cells and containing nine sets of triplet microtubules; structurally similar to a **basal body**. (Figure 19-5b)

centromere Constricted portion of a mitotic chromosome where sister **chromatids** are attached and from which **kinetochore** fibers extend toward a spindle pole; required for proper chromosome segregation during mitosis and meiosis.

centrosome (cell center) Organelle located near the nucleus of animal cells that is the primary microtubule-organizing center (MTOC) and contains a pair of **centrioles**. It divides during mitosis, forming the spindle poles.

chaperone Collective term for two types of proteins that prevent misfolding of a target protein (molecular chaperones) or actively facilitate its proper folding (chaperonins). (Figure 3-15)

checkpoint Any of several points in the eukaryotic **cell cycle** at which progression of a cell to the next stage can be halted until conditions are suitable. (Figure 13-34)

chemical equilibrium The state of a chemical reaction in which the concentration of all products and reactants is constant and the rates of the forward and reverse reactions are equal.

chemiosmosis Process whereby an electrochemical proton gradient (pH plus electric potential) across a membrane is used to drive an energy-requiring process such as ATP synthesis or transport of molecules across a membrane against their concentration gradient; also called *chemiosmotic coupling*. (Figure 16-1)

chimera An animal or tissue composed of elements derived from genetically distinct individuals; also a protein molecule containing segments derived from different proteins.

chlorophylls A group of light-absorbing porphyrin pigments that are critical in photosynthesis. (Figure 16-35)

chloroplast A specialized organelle in plant cells that is surrounded by a double membrane and contains internal chlorophyll-containing membranes (thylakoids) where the light-absorbing reactions of photosynthesis occur. (Figure 16-34)

cholesterol An **amphipathic** lipid containing the four-ring steroid structure with a hydroxyl group on one ring; a major component of many eukaryotic membranes and precursor of steroid hormones. (Figure 5-29)

chromatid One copy of a duplicated chromosome, formed during the S phase of the cell cycle, that is still joined at the **centromere** to the other copy; also called *sister chromatid*. During mitosis, the two chromatids separate, each becoming a chromosome of one of the two daughter cells.

chromatin Complex of DNA, histones, and nonhistone proteins from which eukaryotic **chromosomes** are formed. Condensation of chromatin during mitosis yields the visible metaphase chromosomes. (Figure 9-29)

chromatography, liquid Group of biochemical techniques for separating mixtures of molecules based on their mass (gel-filtration chromatography), charge (ion-exchange chromatography), or ability to bind specifically to other molecules (affinity chromatography). Commonly used technique for separating and purifying proteins. (Figure 3-43)

chromosome In eukaryotes, the structural unit of the genetic material consisting of a single, linear double-stranded DNA molecule and associated proteins. During mitosis, chromosomes condense into compact structures visible in the light microscope. In prokaryotes, a single, circular double-stranded DNA molecule constitutes the bulk of the genetic material. See also **karyotype.**

cilium (pl. cilia) Membrane-enclosed motile structure extending from the surface of eukaryotic cells. Cilia usually occur in groups and beat rhythmically to move a cell (e.g., single-celled organism) or to move small particles or fluid along the surface (e.g., trachea cells). See also **axoneme** and **flagellum**.

cis-acting Referring to a regulatory sequence in DNA (e.g., enhancer, promoter) that can control a gene only on the same chromosome. In bacteria, cis-acting elements are adjacent or proximal to the gene(s) they control, whereas in eukaryotes they may also be far away. See also **trans-acting**.

cisterna (pl. **cisternae**) Flattened membrane-bounded compartment, as found in the Golgi complex and endoplasmic reticulum.

cistron A genetic unit that encodes a single polypeptide.

citric acid cycle A set of nine coupled reactions occurring in the matrix of the **mitochondrion** in which acetyl groups derived from food molecules are oxidized, generating CO_2 and reduced intermediates used to produce ATP; also called *Krebs cycle* and *tricarboxylic acid cycle (TCA)*. (Figure 16-12)

clathrin A fibrous protein that with the aid of assembly proteins polymerizes into a lattice-like network at specific regions on the cytosolic side of a membrane, thereby forming a clathrin-coated pit, which buds off to form a vesicle. (Figures 17-53 and 17-54)

clone A population of identical cells or DNA molecules descended from a single progenitor. Also viruses or organisms that are genetically identical and descended from a single progenitor.

cloning vector An autonomously replicating genetic element used to carry a cDNA or fragment of genomic DNA into a host cell for the purpose of gene cloning. Commonly used vectors are bacterial plasmids and modified bacteriophage genomes. (Figures 7-3 and 7-12)

codon Sequence of three nucleotides in DNA or mRNA that specifies a particular amino acid during protein synthesis; also called *triplet*. Of the 64 possible codons, three are stop codons, which do not specify amino acids. (Table 4-2)

coenzyme Small organic molecule that associates with an enzyme and participates in the reaction catalyzed by the enzyme; also called *cofactor*. Some coenzymes form a transient covalent bond to the substrate; others function as carriers of electrons, acyl groups, or other activated groups. Generally, a coenzyme is bound less firmly to a protein than a **prosthetic group**.

coenzyme A (CoA) See **acetyl CoA**.

coiled-coil Stable rodlike quaternary protein structure formed by two or three α helices interacting with each other along their length; commonly found in fibrous proteins and certain transcription factors. (Figure 3-9a)

collagen A triple-helical protein that forms fibrils of great tensile strength; a major component of the extracellular matrix and connective tissues. The numerous collagen subtypes differ in their tissue distribution and the extracellular components and cell-surface proteins with which they associate.

complementary Referring to two nucleic acid sequences or strands that can form a perfect base-paired double helix with each other; also describing regions on two interacting molecules (e.g., an enzyme and its substrate) that fit together in a lock-and-key fashion.

complementary DNA (cDNA) See **cDNA**.

complementation In genetics, the restoration of a wild-type function (e.g., ability to grow on galactose) in diploid heterozygotes generated from **haploids**, each of which carries a mutation in a different gene whose encoded protein is required for the same biochemical pathway. If two mutants with the same mutant phenotype (e.g., inability to grow on galactose) can complement each other, then their mutations are in different genes. (Figure 8-11)

conformation The precise shape of a protein or other macromolecule in three dimensions resulting from the spatial location of the atoms in the molecule. A small change in the conformations of some proteins affects their activity considerably.

consensus sequence The nucleotides or amino acids most commonly found at each position in the sequences of related DNAs, RNAs, or proteins. See also **homology**.

constitutive Referring to cellular production of a molecule at a constant rate, which is not regulated by internal or external stimuli.

constitutive mutant (1) A mutant in which a protein is produced at a constant level, as if continuously induced; (2) a bacterial regulatory mutant in which an operon is transcribed in the absence of inducer; (3) a mutant in which a regulated enzyme is in a continuously active form.

cooperativity Property exhibited by some proteins with multiple ligand-binding sites whereby binding of one ligand molecule increases (positive cooperativity) or decreases (negative cooperativity) the binding affinity of successive ligand molecules.

cosmid A type of vector used to clone large DNA fragments. (Figure 7-16)

cotransport Protein-mediated transport of an ion or small molecule across a membrane against a concentration gradient driven by coupling to movement of a second molecule down its concentration gradient. See also **antiport** and **symport**.

covalent bond Stable chemical force that holds the atoms in molecules together by sharing of one or more pairs of electrons. Such a bond has a strength of 50–200 kcal/mol. (Table 2-1)

crossing over Exchange of genetic material between maternal and paternal chromatids during meiosis to produce recombined chromosomes. (Figure 8-18) See also **recombination**.

cyclic AMP (cAMP) A **second messenger**, produced in response to hormonal stimulation of certain G protein–coupled receptors, that activates cAMP-dependent protein kinases. (Figure 20-4)

cyclin Any of several related proteins whose concentrations rise and fall during the course of the eukaryotic cell cycle. Cyclins form complexes with cyclin-dependent kinases, thereby activating and determining the substrate specificity of these enzymes.

cyclin-dependent kinase (Cdk) A protein kinase that is catalytically active only when bound to a cyclin. Various Cdk-cyclin complexes trigger progression through different stages of the eukaryotic cell cycle by phosphorylating specific target proteins. (Figure 13-29)

cytochromes A group of colored, heme-containing proteins that transfer electrons during cellular respiration and photosynthesis. (Figure 16-21)

cytokine Any of numerous secreted, small proteins (e.g., interferons, interleukins) that bind to cell-surface receptors on certain cells to trigger their differentiation or proliferation.

cytokinesis The last stage of mitosis, where the two daughter cells separate, each with a nucleus and cytoplasmic organelles.

cytoplasm Viscous contents of a cell that are contained within the plasma membrane but, in eukaryotic cells, outside the nucleus.

cytoskeleton Network of fibrous elements, consisting primarily of **microtubules**, actin **microfilaments**, and **intermediate filaments**, found in the cytoplasm of eukaryotic cells. The cytoskeleton provides structural support for the cell and permits directed movement of organelles, chromosomes, and the cell itself.

cytosol Unstructured aqueous phase of the cytoplasm excluding organelles, membranes, and insoluble cytoskeletal components.

cytosolic face The face of a cell membrane directed toward the cytoplasm. (Figure 5-31)

dalton Unit of molecular mass approximately equal to the mass of a hydrogen atom (1.66×10^{-24} g).

degenerate In reference to the genetic code, having more than one **codon** specifying a particular amino acid.

denaturation Drastic alteration in the **conformation** of a protein or nucleic acid due to disruption of various noncovalent bonds caused by heating or exposure to certain chemicals; usually results in loss of biological function.

dendrite Process extending from the cell body of a neuron that is relatively short and typically branched and receives signals from **axons** of other neurons. (Figure 21-1)

deoxyribonucleic acid See **DNA**.

depolarization Change in the potential that normally exists across the plasma membrane of a cell at rest, resulting in a less negative membrane potential.

desmosomes Specialized regions of the plasma membrane, consisting of dense protein plaques connected to intermediate filaments, that mediate adhesion between adjacent cells (especially epithelial cells) and between cells and the extracellular matrix. (Figure 22-6)

determination In embryogenesis, a change in a cell that commits the cell to a particular developmental pathway.

development Overall process involving growth and **differentiation** by which a fertilized egg gives rise to an adult plant or animal, including the formation of individual cell types, tissues, and organs.

diacylglycerol (DAG) Intracellular signaling molecule produced by cleavage of **phosphoinositides** in response to stimulation of certain cell-surface receptors; functions as a membrane-bound **second messenger** in inositol-lipid signaling pathways. (Figures 20-4 and 20-37)

differentiation Process usually involving changes in gene expression by which a precursor cell becomes a distinct specialized cell type.

diploid Referring to an organism or cell having two full sets of **homologous chromosomes** and hence two copies (alleles) of each gene or genetic locus. Somatic cells contain the diploid number of chromosomes ($2n$) characteristic of a species. See also **haploid**.

disaccharide A small carbohydrate (sugar) composed of two monosaccharides covalently joined by a **glycosidic bond**. Common examples are lactose (milk sugar) and sucrose, a major photosynthetic product in higher plants.

dissociation constant (K_D) See **equilibrium constant**.

disulfide bond (–S–S–) A common covalent linkage between the **sulfhydryl groups** on two cysteine residues in different proteins or in different parts of the same protein; generally found only in extracellular proteins or protein domains.

DNA (deoxyribonucleic acid) Long linear polymer, composed of four kinds of deoxyribose **nucleotides**, that is the carrier of genetic information. In its native state, DNA is a double helix of two antiparallel strands held together by hydrogen bonds between complementary purine and pyramidine bases. (Figure 4-6)

DNA cloning Recombinant DNA technique in which specific **cDNAs** or fragments of **genomic DNA** are inserted into a cloning vector, which then is incorporated into cultured host cells (e.g., *E. coli* cells) and maintained during growth of the host cells; also called *gene cloning*. (Figures 7-3 and 7-15)

DNA library Collection of cloned DNA molecules consisting of fragments of the entire genome (genomic library) or of DNA copies of all the mRNAs produced by a cell type (cDNA library) inserted into a suitable **cloning vector**.

DNA polymerase An enzyme that copies one strand of DNA (the template strand) to make the complementary strand, forming a new double-stranded DNA molecule. All DNA polymerases add deoxyribonucleotides one at a time in the $5' \rightarrow 3'$ direction to a short pre-existing primer strand of DNA or RNA.

domain Region of a protein with a distinct tertiary structure (e.g., globular or rodlike) and characteristic activity; homologous domains may occur in different proteins.

dominant In genetics, referring to that allele of a gene expressed in the **phenotype** of a heterozygote; the nonexpressed allele is **recessive**. Also referring to the phenotype associated with a dominant allele. (See Figure 8-1)

dorsal Relating to the back of an animal or the upper surface of a structure (e.g., leaf, wing).

double helix, DNA The most common three-dimensional structure for cellular DNA in which the two polynucleotide strands are antiparallel and wound around each other with complementary bases hydrogen-bonded. (Figure 4-4a)

downstream For a gene, the direction RNA polymerase moves during transcription, which is toward the end of the template DNA strand with a 3' hydroxyl group. By convention, the +1 position of a gene is the first transcribed nucleotide; nucleotides downstream from the +1 position are designated +2, +3, etc. Also, events that occur later in a cascade of steps. See also **upstream**.

dynein Member of a family of ATP-powered motor proteins that move toward the (−) end of microtubules by sequentially breaking and forming new bonds with microtubule proteins. Dyneins can transport vesicles and are responsible for the movement of cilia and flagella. (Figure 19-32)

ectoderm Outermost of the three primary cell layers of the animal embryo; gives rise to epidermal tissues, the nervous system, and external sense organs. See also **endoderm** and **mesoderm**. (Figure 23-5)

electron carrier Any molecule or atom that accepts electrons from donor molecules and transfers them to acceptor molecules. Most are **prosthetic groups** (e.g., heme, copper, iron-sulfur clusters) associated with membrane-bound proteins.

electron transport Flow of electrons via a series of electron carriers from reduced electron donors (e.g., NADH) to O_2 in the inner mitochondrial membrane, or from H_2O to NADP in the thylakoid membrane of plant chloroplasts. (Figure 16-17)

electrophoresis Any of several techniques for separating macromolecules based on their migration in a gel or other medium subjected to a strong electric field. (Figure 3-41)

electrophoretogram An autoradiogram of a gel in which molecules have been separated by gel electrophoresis. (Figure 7-23b)

elongation factor One of a group of nonribosomal proteins required for continued **translation** of mRNA following initiation. (Figure 4-39)

embryogenesis Early development of an individual from a fertilized egg (zygote). Following cleavage of the zygote, the major axes are established during the blastula stage; in the subsequent gastrula stage, the early embryo invaginates and acquires three cell layers. (Figure 23-5)

endocytosis Uptake of extracellular materials by invagination of the plasma membrane to form a small membrane-bounded vesicle (early endosome). (Figure 17-46)

endoderm Innermost of the three primary cell layers of the animal embryo; gives rise to the gut and most of the respiratory tract. See **ectoderm** and **mesoderm**. (Figure 23-5)

endoplasmic reticulum (ER) Network of interconnected membranous structures within the cytoplasm of eukaryotic cells. The rough ER, which is associated with **ribosomes**, functions in the synthesis and processing of secretory and membrane proteins; the smooth ER, which lacks ribosomes, functions in lipid synthesis. (Figure 5-47)

endosome, late A sorting vesicle with an acidic internal pH in which bound ligands dissociate from their membrane-bound receptor proteins. Late endosomes participate in sorting of lysosomal enzymes and in recycling of receptors endocytosed from the plasma membrane.

endothelium Layer of highly flattened cells that forms the lining of all blood vessels and regulates exchange of materials between the bloodstream and surrounding tissues; it usually is underlain by a **basal lamina**.

endothermic Referring to a chemical reaction that absorbs heat (i.e., has a positive change in enthalpy).

enhancer A regulatory sequence in eukaryotic DNA (rarely in prokaryotic DNA) that may be located at a great distance from the gene it controls. Binding of specific proteins to an enhancer modulates the rate of transcription of the associated gene. (Figure 10-34)

enthalpy (H) Heat; in a chemical reaction, the enthalpy of the reactants or products is equal to their total bond energies.

entropy (S) A measure of the degree of disorder or randomness in a system; the higher the entropy, the greater the disorder.

enzyme A biological macromolecule that acts as a catalyst. Most enzymes are proteins, but certain RNAs, called ribozymes, also have catalytic activity.

epinephrine A catecholamine secreted by the adrenal gland and some neurons in response to stress; also called adrenaline. It functions as both a hormone and neurotransmitter, mediating "fight or flight" responses including increased blood glucose levels and heart rate. (Figure 21-28)

epithelium Coherent sheet comprising one or more layers of cells that covers an external body surface or lines an internal cavity. (Figure 6-4)

epitope The part of an antigen molecule that binds to an **antibody**; also called *antigenic determinant*.

equilibrium constant (K) Ratio of forward and reverse rate constants for a reaction. For a binding reaction, $A + B \rightleftharpoons AB$, it equals the association constant, K_a; the higher the K_a, the tighter the binding between A and B. The reciprocal of the K_a is the dissociation constant, K_D; the higher the K_D, the weaker the binding between A and B.

eubacteria Class of **prokaryotes** that constitutes one of the three distinct evolutionary lineages of modern-day organisms; also called the *true bacteria* or simply *bacteria*. Phylogenetically distinct from **archaea** and **eukaryotes**. (Figure 1-5)

euchromatin Less condensed portions of **chromatin**, including most transcribed regions, present in interphase chromosomes. See also **heterochromatin**.

eukaryotes Class of organisms, composed of one or more cells containing a membrane-enclosed nucleus and organelles, that constitutes one of the three distinct evolutionary lineages of modern-day organisms; also called *eukarya*. Includes all organisms except viruses and **prokaryotes**.

exocytosis Release of intracellular molecules (e.g., hormones, matrix proteins) contained within a membrane-bounded vesicle by fusion of the vesicle with the plasma membrane of a cell. This is the process whereby most molecules are secreted from eukaryotic cells.

exon Segments of a eukaryotic gene (or of its **primary transcript**) that reaches the cytoplasm as part of a mature mRNA, rRNA, or tRNA molecule. See also **intron**.

exoplasmic face The face of a cell membrane directed away from the cytoplasm. The exoplasmic face of the plasma membrane faces the cell exterior, whereas the exoplasmic face of organelles (e.g., mitochondria, chloroplasts, and the endoplasmic reticulum) face their lumen. (Figure 5-31)

exothermic Referring to a chemical reaction that releases heat (i.e., has a negative change in enthalpy).

expression See **gene expression**.

expression cloning Recombinant DNA techniques for isolating a cDNA or genomic DNA segment based on functional properties of the encoded protein and without prior purification of the protein. Also refers to techniques for producing high levels of a full-length protein once its cDNA or gene has been cloned.

expression vector A modified **plasmid** or virus that carries a gene or cDNA into a suitable host cell and there directs synthesis of the encoded protein. Some expression vectors are designed for screening DNA libraries for a gene of interest (Figures 7-21 and 20-9); others, for producing large amounts of a protein from its cloned gene (Figures 7-36 and 7-37).

extracellular matrix A usually insoluble network consisting of polysaccharides, fibrous proteins, and adhesive proteins that are secreted by animal cells. It provides structural support in tissues and can affect the development and biochemical functions of cells.

extrinsic protein See **peripheral membrane protein.**

F_0F_1 complex See **ATP synthase.**

facilitated transport Protein-aided transport of an ion or molecule across a cell membrane down its concentration gradient at a rate greater than that obtained by **passive diffusion;** also called *facilitated diffusion.* Such transport exhibits ligand specificity and saturation kinetics. The glucose transporter GLUT1 is a well-studied example of a protein that mediates facilitated diffusion. (Figure 15-7)

FAD (flavin adenine dinucleotide) A coenzyme that participates in oxidation reactions by accepting two electrons from a donor molecule and two H^+ from the solution. The reduced form, $FADH_2$, transfers electrons to carriers that function in oxidative phosphorylation. (Figure 16-8)

fatty acid Any hydrocarbon chain that has a carboxyl group at one end; a major source of energy during metabolism and precursors for synthesis of phospholipids. (Figure 2-18)

fertilization Fusion of a female and male gamete (both **haploid**) to form a **diploid** zygote, which develops into a new individual.

fibroblast A common type of connective-tissue cell that secretes collagen and other components of the extracellular matrix. It migrates and proliferates during wound healing and in tissue culture.

fibronectin An extracellular multiadhesive protein that binds to other matrix components, fibrin, and cell-surface receptors of the **integrin** family. It functions to attach cells to the extracellular matrix and is important in wound healing. (Figure 22-22)

flagellum (pl. flagella) Long locomotory structure, extending from the surface of a eukaryotic cell, whose whiplike bending propels the cell forward or backward. Usually there is only one flagellum per cell (as in sperm cells). Bacterial flagella are smaller and much simpler structures. See also **axoneme** and **cilium.**

fluorescein See **fluorescent staining.**

fluorescent staining General technique for visualizing cellular components by treating cells with a fluorescent-labeled agent that binds specifically to a component of interest and then observing the cells by fluorescence microscopy. For instance, an antibody specific for a protein of interest can be chemically linked to a fluorescent dye such as fluorescein, which emits green light, or rhodamine, which emits red light. Various fluorescent dyes that bind specifically to DNA are used to detect chromosomes or specific chromosomal regions.

footprinting Technique for identifying protein-binding regions of DNA or RNA. A radiolabeled nucleic acid sample is digested with a nuclease in the presence and absence of a specific binding protein. Because regions of DNA or RNA with bound protein are protected from digestion, the patterns of fragment bands separated by gel electrophoresis obtained from protected and unprotected samples differ, permitting identification of the protein-binding regions. (Figure 10-6)

free energy *(G)* A measure of the potential energy of a system, which is a function of the **enthalpy** *(H)* and **entropy** *(S).*

free-energy change *(ΔG)* The difference in the free energy of the product molecules and of the starting molecules (reactants) in a chemical reaction. A large negative value of ΔG indicates that a reaction has a strong tendency to occur; that is, at chemical equilibrium the concentration of products will be much greater than the concentration of reactants.

G protein Any of numerous heterotrimeric GTP-binding proteins that function in intracellular signaling pathways; usually activated by ligand binding to a coupled seven-spanning receptor on the cell surface. See also **GTPase superfamily.** (Table 20-5)

G protein–coupled receptor (GPCR) Member of an important class of cell-surface receptors that have seven transmembrane α helices and are directly coupled to a trimeric G protein. (Figure 20-16)

G_0, G_1, G_2 phase See **cell cycle.**

gamete Specialized haploid cell (in animals either a sperm or an egg) produced by **meiosis** of germ cells; in sexual reproduction, union of a sperm and an egg initiates the development of a new individual.

ganglion (pl. ganglia) Collection of neuron cell bodies located outside of the central nervous system.

ganglioside Any glycolipid containing one or more *N*-acetylneuraminic acid (sialic acid) residues in its structure. Gangliosides are found in the plasma membrane of eukaryotic cells and confer a net negative charge on most animal cells.

gap junction Protein-lined channel between adjacent cells that allows passage of ions and small molecules between the cells. (Figure 22-8)

gastrula An early embryonic form subsequent to the **blastula** characterized by invagination of the cells to form a rudimentary gut cavity and development of three cell layers.

gene Physical and functional unit of heredity, which carries information from one generation to the next. In molecular terms, it is the entire DNA sequence—including exons, introns, and noncoding transcription-control regions—necessary for production of a functional protein or RNA. See also **cistron** and **transcription unit.**

gene cloning See **DNA cloning.**

gene control All of the mechanisms involved in regulating **gene expression.** Most common is regulation of transcription, although mechanisms influencing the processing, stabilization, and translation of mRNAs help control expression of some genes.

gene conversion Phenomenon in which one allele of a gene is converted to another during meiotic **recombination.**

gene expression Overall process by which the information encoded in a gene is converted into an observable **phenotype** (most commonly production of a protein).

genetic code The set of rules whereby nucleotide triplets (**codons**) in DNA or RNA specify amino acids in proteins. (See Table 4-2)

genome Total genetic information carried by a cell or organism.

genomic DNA All the DNA sequences composing the **genome** of a cell or organism. See also **cDNA.**

genomics Comparative analysis of the complete genomic sequences from different organisms; used to assess evolutionary relations between species and to predict the number and general types of proteins produced by an organism.

genotype Entire genetic constitution of an individual cell or organism; also, the alleles at one or more specific loci.

germ cell Any precursor cell that can give rise to gametes. See also **somatic cell.**

germ line Lineage of germ cells, which give rise to **gametes** and thus participate in formation of the next generation of organisms; also the genetic material transmitted from one generation to the next through the gametes.

glial cells Nonexcitable supportive cells in the nervous system; also called *neuroglial cells*. Include astrocytes and oligodendrocytes in the vertebrate central nervous system and Schwann cells in the peripheral nervous system.

glucagon A peptide hormone produced in the α cells of the pancreas that triggers the conversion of glycogen to glucose by the liver; acts with **insulin** to control blood glucose levels. (Figure 20-45)

glucose Six-carbon monosaccharide (sugar) that is the primary metabolic fuel in most cells. The large glucose polymers, glycogen and starch, are used to store energy in animal cells and plant cells, respectively.

glycocalyx Carbohydrate-rich layer covering the outer surface of the plasma membrane of eukaryotic cells; composed of membrane glycolipids, the oligosaccharide side chains of integral membrane proteins, and absorbed peripheral membrane proteins.

glycogen A very long, branched polysaccharide, composed exclusively of glucose units, that is the primary storage carbohydrate in animal cells. It is found primarily in liver and muscle cells.

glycogenolysis Breakdown of glycogen to glucose 6-phosphate; stimulated by a rise in cAMP following epinephrine stimulation of cells and, in muscle, by a rise in Ca^{2+} following neuronal stimulation. (Figure 20-43)

glycolipid Any lipid to which a short carbohydrate chain is covalently linked; commonly found in the plasma membrane.

glycolysis Metabolic pathway whereby sugars are degraded anaerobically to lactate or pyruvate in the cytosol with the production of ATP; also called *Embden-Meyerhof pathway*. (Figure 16-3)

glycoprotein Any protein to which one or more oligosaccharide chains are covalently linked. Most secreted proteins and many membrane proteins are glycoproteins.

glycosaminoglycan (GAG) A long, linear, highly charged polymer of a repeating disaccharide in which one member of the pair usually is a sugar acid (uronic acid) and the other is an amino sugar and many residues are sulfated. Generally are covalently bound to core proteins forming **proteoglycans**, which are major components of the extracellular matrix. (Figure 22-24)

glycosidic bond The covalent linkage between two monosaccharide residues formed by a condensation reaction in which one carbon, usually carbon #1, of one sugar reacts with a hydroxyl group on a second sugar with the loss of a water molecule. (Figure 2-10)

glycosyl transferase An enzyme that forms a **glycosidic bond** between a sugar residue (monosaccharide) and an amino acid side chain of a protein or a residue in an existing carbohydrate chain.

Golgi complex Stacks of membranous structures in eukaryotic cells that function in processing and sorting of proteins and lipids destined for other cellular compartments or for secretion; also called *Golgi apparatus*. (Figure 5-49)

growing fork Site in double-stranded DNA at which the **template** strands are separated and addition of deoxyribonucleotides to each newly formed chain occurs; also called *replication fork*. (Figure 12-9)

growth factor An extracellular polypeptide molecule that binds to a cell-surface receptor triggering an intracellular signaling pathway leading to proliferation, differentiation, or other cellular response.

GTP (guanosine 5′-triphosphate) A nucleotide that is a precursor in RNA synthesis and also plays a special role in protein synthesis, signal-transduction pathways, and microtubule assembly.

GTPase superfamily Group of GTP-binding proteins that cycle between an inactive state with bound GDP and an active state with bound GTP. These proteins—including **G proteins, Ras proteins,** and certain polypeptide **elongation factors**—function as intracellular switch proteins. (Figure 20-22)

haploid Referring to an organism or cell having only one member of each pair of **homologous chromosomes** and hence only one copy (**allele**) of each gene or genetic locus. Gametes and bacterial cells are haploid. See also **diploid.**

HeLa cell Line of human epithelial cells, derived from a human cervical carcinoma, that grows readily in culture and is widely used in research.

helicase Any enzyme that moves along a DNA duplex using the energy released by ATP hydrolysis to separate (unwind) the two strands. Required for the replication and transcription of DNA.

helix-loop-helix A conserved structural motif found in many monomeric Ca^{2+}-binding proteins and dimeric eukaryotic transcription factors. (Figure 3-9b)

helix-turn-helix A DNA-binding motif found in most bacterial DNA-binding proteins.

heterochromatin Regions of **chromatin** that remain highly condensed and transcriptionally inactive during interphase.

heteroduplex A double-stranded DNA molecule containing one or more mispaired bases.

heterokaryon Cell with more than one functional nucleus produced by the fusion of two or more different cells.

heterozygous Referring to a diploid cell or organism having two different **alleles** of a particular gene.

hexose A six-carbon **monosaccharide.**

high-energy bond Covalent bond that releases a large amount of energy when hydrolyzed under the usual intracellular conditions. Examples include the phosphoanhydride bonds in ATP, thioester bond in acetyl CoA, and various phosphate ester bonds. (Table 2-7)

histones A family of small, highly conserved basic proteins, found in the chromatin of all eukaryotic cells, that associate with DNA in the **nucleosome.**

Holliday structure Intermediate structure in **recombination** between homologous chromosomes. (Figures 12-29 and 12-30)

homeobox Conserved DNA sequence that encodes a DNA-binding domain (homeodomain) in a class of transcription factors encoded by certain **homeotic genes.**

homeodomain A conserved DNA-binding motif found in many developmentally important transcription factors. See also **homeobox**.

homeosis Transformation of one body part into another arising from mutation in or misexpression of certain developmentally critical genes.

homeotic gene A gene in which mutations cause cells in one region of the body to act as though they were located in another, giving rise to conversions of one cell, tissue, or body region into another.

homologous chromosome One of the two copies of each morphologic type of chromosome present in a **diploid** cell; also called *homologue*. Each homologue is derived from a different parent.

homologue See **homologous chromosome**.

homology Similarity in the sequence of a protein or nucleic acid or in the structure of an organ that reflects a common evolutionary origin. Molecules or sequences that exhibit homology are referred to as homologs. In contrast, analogy is a similarity in structure or function that does not reflect a common evolutionary origin.

homozygous Referring to a diploid cell or organism having two identical **alleles** of a particular gene.

hormone General term for any extracellular substance that induces specific responses in target cells. Hormones coordinate the growth, differentiation, and metabolic activities of various cells, tissues, and organs in multicellular organisms.

Hox complex Clusters of homologous selector genes, which help determine the body plan in animals.

hyaluronan A large, highly hydrated polysaccharide that is a major component of the extracellular matrix; also called *hyaluronic acid* and *hyaluronate*. It imparts stiffness and resilience as well as a lubricating quality to many types of connective tissue.

hybridization Association of two **complementary** nucleic acid strands to form double-stranded molecules, which can contain two DNA strands, two RNA strands, or one DNA and one RNA strand. Used experimentally in various ways to detect specific DNA or RNA sequences.

hybridoma A clone of hybrid cells that are immortal and produce **monoclonal antibodies**; formed by fusion of normal antibody-producing B lymphocytes with myeloma cells. (Figure 6-10)

hydrogen bond A **noncovalent bond** between an electronegative atom (commonly oxygen or nitrogen) and a hydrogen atom covalently bonded to another electronegative atom. Particularly important in stabilizing the three-dimensional structure of proteins and formation of base pairs in nucleic acids.

hydrolysis Reaction in which a covalent bond is cleaved with addition of an H from water to one product of the cleavage and of an OH from water to the other.

hydrophilic Interacting effectively with water. See also **polar**.

hydrophobic Not interacting effectively with water; in general, poorly soluble or insoluble in water. See also **nonpolar**.

hydrophobic bond The force that drives nonpolar molecules or parts of molecules to associate with each other in aqueous solution. A type of **noncovalent bond** that is particularly important in stabilization of the phospholipid bilayer.

hypertonic Referring to an external solution whose solute concentration is high enough to cause water to move out of cells due to **osmosis**.

hypotonic Referring to an external solution whose solute concentration is low enough to cause water to move into cells due to **osmosis**.

immunoglobulin (Ig) Any protein that functions as an **antibody**. The five major classes of vertebrate immunoglobulins (IgA, IgD, IgE, IgG, and IgM) differ in their specific functions in the immune response.

in vitro Denoting a reaction or process taking place in an isolated cell-free extract; sometimes used to distinguish cells growing in culture from those in an organism.

in vivo In an intact cell or organism.

induction In embryogenesis, a change in the developmental fate of one cell or tissue caused by direct interaction with another cell or tissue or with an extracellular signaling molecule; in metabolism, an increase in the synthesis of an enzyme or series of enzymes mediated by a specific molecule (inducer).

initiation factor One of a group of proteins that promote the proper association of ribosomes and mRNA and are required for initiation of protein synthesis. (Figure 4-37)

initiator A eukaryotic promoter sequence for RNA polymerase II that specifies transcription initiation within the sequence.

insulin A protein hormone produced in the β cells of the pancreas that stimulates uptake of glucose into muscle and fat cells and with **glucagon** helps to regulate blood glucose levels (Figure 20-45). Insulin also functions as a growth factor for many cells.

integral membrane protein Any membrane-bound protein all or part of which interacts with the hydrophobic core of the **phospholipid bilayer** and can be removed from the membrane only by extraction with detergent; also called *intrinsic membrane protein*. (Figure 3-32)

integrins A large family of heterodimeric transmembrane proteins that promote adhesion of cells to the extracellular matrix or to the surface of other cells.

interferons (IFNs) Small group of cytokines that bind to cell-surface receptors on target cells inducing changes in gene expression leading to an antiviral state or other cellular responses important in the immune response.

intermediate filaments Cytoskeletal fibers (10 nm in diameter) formed by polymerization of several classes of cell-specific subunit proteins including keratins, lamins, and vimentin. They constitute the major structural proteins of skin and hair; form the scaffold that holds Z disks and myofibrils in place in muscle; and generally function as important structural components of many animal cells and tissues.

interphase Long period of the cell cycle, including the G_1, S, and G_2 phases, between one M (mitotic) phase and the next. (Figure 13-1)

intrinsic protein See **integral membrane protein**.

intron Part of a **primary transcript** (or the DNA encoding it) that is removed by splicing during RNA processing and is not included in the mature, functional mRNA, rRNA, or tRNA; also called *intervening sequence*.

ion channel Any transmembrane protein complex that forms a water-filled channel across the phospholipid bilayer allowing selective ion transport down its electrochemical gradient. See also **ion pump**.

ion pump Any transmembrane ATPase that couples hydrolysis of ATP to the transport of a specific ion across the phospholipid bilayer against its electrochemical gradient. (Table 15-2)

ionic bond A **noncovalent bond** between a positively charged ion (cation) and negatively charged ion (anion).

isoelectric focusing Technique for separating molecules by gel **electrophoresis** in a pH gradient subjected to an electric field. A protein migrates to the pH at which its overall net charge is zero.

isoelectric point (pI) The pH of a solution at which a dissolved protein or other potentially charged molecule has a net charge of zero and therefore does not move in an electric field.

isoform One of several forms of the same protein whose amino acid sequences differ slightly but whose general activity is identical.

isotonic Referring to a solution whose solute concentration is such that it causes no net movement of water in or out of cells.

karyotype Number, sizes, and shapes of the entire set of **metaphase** chromosomes of a eukaryotic cell. (Figure 9-33)

kinase An enzyme that transfers the terminal (γ) phosphate group from ATP to a substrate. Protein kinases, which phosphorylate specific serine, threonine, or tyrosine residues in target proteins, play a critical role in regulating the activity of many cellular proteins. See also **phosphatases**.

kinesin Member of a family of motor proteins that use energy released by ATP hydrolysis to move toward the (+) end of a microtubule, transporting vesicles or particles in the process. (Figure 19-24)

kinetochore A three-layer protein structure located at or near the **centromere** of each mitotic chromosome from which microtubules (kinetochore fibers) extend toward the spindle poles of the cell; plays an active role in movement of chromosomes toward the poles during anaphase.

K_m A parameter that describes the affinity of an enzyme for its substrate and equals the substrate concentration that yields the half-maximal reaction rate; also called the *Michaelis constant*. A similar parameter describes the affinity of a transport protein for the transported molecule or the affinity of a receptor for its ligand. (Figure 3-26)

knockin, gene Technique in which the coding sequences of one gene are replaced by those of another.

knockout, gene Technique for selectively inactivating a gene by replacing it with a mutant allele in an otherwise normal organism.

Krebs cycle See **citric acid cycle**.

label A fluorescent chemical group or radioactive atom incorporated into a molecule in order to spatially locate the molecule or follow it through a reaction or purification scheme. As a verb, to add such a group or atom to a cell or molecule.

lagging strand Newly synthesized DNA strand formed at the **growing fork** as short, discontinuous segments, called **Okazaki fragments**, which are later joined by DNA ligase. Although overall lagging-strand synthesis occurs in the $3' \rightarrow 5'$ direction, each Okazaki fragment is synthesized in the $5' \rightarrow 3'$ direction. See also **leading strand**. (Figure 12-9)

laminin A component of the **extracellular matrix** that is found in all basal laminae and has binding sites for cell-surface receptors, collagen, and heparan sulfate proteoglycans. (Figure 22-19)

lamins A group of intermediate filament proteins that form the fibrous network (**nuclear lamina**) on the inner surface of the nuclear envelope. (Figure 13-15)

leading strand Newly synthesized DNA strand formed by continuous synthesis in the $5' \rightarrow 3'$ direction at the **growing fork**. The direction of leading-strand synthesis is the same as movement of the growing fork. See also **lagging strand**. (Figure 12-9)

lectin Any protein that binds tightly to specific sugars. Lectins can be used in affinity chromatography to purify glycoproteins or as reagents to detect them in situ.

leucine zipper Common structural motif in some dimeric eukaryotic transcription factors characterized by a C-terminal coiled-coil dimerization domain and N-terminal DNA-binding domain. (Figure 10-43)

leukemia Cancer of white blood cells and their precursors.

library See **DNA library**.

ligand Any molecule, other than an enzyme **substrate**, that binds tightly and specifically to a macromolecule, usually a protein, forming a macromolecule-ligand complex.

ligase An enzyme that links together the 3' end of one nucleic acid strand with the 5' end of another, forming a continuous strand.

linkage In genetics, the tendency of two different loci on the same chromosome to be inherited together. The closer two loci are, the greater their linkage and the lower the frequency of **recombination** between them.

lipid Any organic molecule that is insoluble in water but is soluble in nonpolar organic solvents. Lipids contain covalently linked fatty acids and are found in fat droplets and, as phospholipids, in biomembranes.

lipophilic See **hydrophobic**.

liposome Spherical **phospholipid bilayer** structure with an aqueous interior that forms in vitro from phospholipids and may contain protein. (Figure 15-4)

locus In genetics, the specific site of a gene on a chromosome. All the **alleles** of a particular gene occupy the same locus.

lymphocytes Two classes of white blood cells that can recognize foreign molecules (**antigens**) and mediate immune responses. B lymphocytes are responsible for production of antibodies; T lymphocytes are responsible for destroying virus- and bacteria-infected cells, foreign cells, and cancer cells.

lysogenic cycle Series of events in which a bacterial virus (bacteriophage) enters a host cell and its DNA is incorporated into the host-cell genome in such a way that the virus (the prophage) lays dormant. The association of a prophage with the host-cell genome is called lysogeny. By various mechanisms, the prophage can be activated so that it enters the **lytic cycle**. (Figure 6-19.)

lysogeny See **lysogenic cycle**.

lysosome Small organelle having an internal pH of 4–5 and containing hydrolytic enzymes.

lytic cycle Series of events in which a virus enters and replicates within a host cell to produce new viral particles eventually causing lysis of the cell. See also **lysogenic cycle**. (Figure 6-17)

M (mitotic) phase See **cell cycle**.

macromolecule Any large, usually polymeric molecule (e.g., a protein, nucleic acid, polysaccharide) with a molecular mass greater than a few thousand daltons.

malignant Referring to a tumor or tumor cells that can invade surrounding normal tissue and/or undergo **metastasis**. See also **benign**.

MAP kinase Protein kinase that is activated in response to cell stimulation by many different growth factors and that mediates cellular responses by phosphorylating specific target proteins.

mapping Various techniques for determining the relative order of genes on a chromosome (genetic map), the absolute position of genes (physical map), or the relative position of restriction sites (restriction map).

meiosis In eukaryotes, a special type of cell division that occurs during maturation of germ cells; comprises two successive nuclear and cellular divisions with only one round of DNA replication resulting in production of four genetically nonequivalent haploid cells (**gametes**) from an initial diploid cell. (Figure 8-2)

membrane See **biomembrane**.

membrane potential Voltage difference across a membrane due to the slight excess of positive ions (cations) on one side and negative ions (anions) on the other.

mesenchyme Embryonic mesoderm tissue in animals from which are formed the connective tissues, blood vessels, and lymphatic vessels.

mesoderm The middle of the three primary cell layers of the animal embryo, lying between the **ectoderm** and **endoderm;** gives rise to the notochord, connective tissue, muscle, blood, and other tissues.

messenger RNA See **mRNA**.

metabolism The sum of the chemical processes that occur in living cells; includes **anabolism** and **catabolism**.

metaphase Mitotic stage at which chromosomes are fully condensed and attached to the mitotic spindle at its equator but have not yet started to segregate toward the opposite spindle poles. (Figure 19-34)

metastasis Spread of tumor cells from their site of origin and establishment of areas of secondary growth.

Michaelis constant See K_m.

microfilaments Cytoskeletal fibers ($\approx$7 nm in diameter) that are formed by polymerization of monomeric globular (G) actin; also called *actin filaments*. Microfilaments play an important role in muscle contraction, cytokinesis, cell movement, and other cellular functions and structures. (Figure 18-2)

microtubule-associated protein (MAP) Any protein, including **motor proteins,** that binds to microtubules in a constant ratio and determines the unique properties of different types of microtubules.

microtubules Cytoskeletal fibers (24 nm in diameter) that are formed by polymerization of α,β-tubulin monomers and exhibit structural and functional polarity. They are important components of cilia, flagella, the mitotic spindle, and other cellular structures. (Figure 19-2)

microvillus (pl. microvilli) Small, membrane-covered projection on the surface of an animal cell containing a core of actin filaments. Numerous microvilli are present on the absorptive surface of intestinal epithelial cells, increasing the surface area for transport of nutrients. (Figure 18-10)

mitochondrion (pl. mitochondria) Large organelle that is surrounded by two phospholipid bilayer membranes, contains DNA, and carries out **oxidative phosphorylation,** thereby producing most of the ATP in eukaryotic cells. (Figures 5-45 and 16-7)

mitogen Any extracellular substance, such as a growth factor, that promotes cell proliferation.

mitosis In eukaryotic cells, the process whereby the nucleus is divided to produce two genetically equivalent daughter nuclei with the diploid number of chromosomes. See also **cytokinesis** and **meiosis**. (Figure 19-34)

mitotic apparatus A specialized temporary structure, present in eukaryotic cells only during mitosis, that captures the chromosomes and then pushes and pulls them to opposite sides of the dividing cell. Consists of a central bilaterally symmetric bundle of microtubules with the overall shape of a football (the mitotic spindle) and two star-shaped tufts of microtubules (the asters), one at each pole of the spindle. (Figure 19-36)

mitotic spindle See **mitotic apparatus**.

mobile DNA element Any DNA sequence that is not present in the same chromosomal location in all individuals of a species.

monoclonal antibody Antibody produced by the progeny of a single B cell and thus a homogeneous protein exhibiting a single antigen specificity. Experimentally, it is produced by use of a **hybridoma**. (Figure 6-10)

monomer Any small molecule that can be linked with others of the same type to form a **polymer**. Examples include amino acids, nucleotides, and monosaccharides.

monomeric For proteins, consisting of a single polypeptide chain.

monosaccharide Any simple sugar with the formula $(CH_2O)_n$ where $n = 3$–7.

morphogen A molecule that specifies cell identity during development as a function of its concentration.

motif In proteins, a structural unit exhibiting a particular three-dimensional architecture that is found in a variety of proteins and usually is associated with a particular function. (Figure 3-9)

motor protein Any member of a special class of enzymes that use energy from ATP hydrolysis to walk or slide along a microfilament (**myosin**) or a microtubule (**dynein** and **kinesin**).

MPF (mitosis-promoting factor) A heterodimeric protein, composed of a **cyclin** and **cyclin-dependent kinase (Cdk)**, that triggers entrance of a cell into mitosis by inducing chromatin condensation and nuclear-envelope breakdown; originally called *maturation-promoting factor.*

mRNA (messenger RNA) Any RNA that specifies the order of amino acids in a protein. It is produced by **transcription** of DNA by RNA polymerase and, in RNA viruses, by transcription of viral RNA. In eukaryotes, the initial RNA product (primary transcript) undergoes processing to yield functional mRNA, which is transported to the cytoplasm. See also **translation**.

MTOC (microtubule-organizing center) General term for any structure (e.g., the centrosome) that organizes microtubules in non-mitotic (interphase) cells. (Figure 19-5)

multiadhesive matrix proteins Group of long flexible proteins that bind to other components of the extracellular matrix (collagen, polysaccharides) and to cell-surface receptors, thereby cross-linking the matrix to the cell membrane.

multimeric For proteins, containing several polypeptide chains (or subunits).

mutagen A chemical or physical agent that induces mutations.

mutation In genetics, a permanent, heritable change in the nucleotide sequence of a chromosome, usually in a single gene; commonly leads to a change in or loss of the normal function of the gene product.

myelin sheath Stacked specialized cell membrane that forms an insulating layer around vertebrate **axons** and increases the speed of impulse conduction. (Figures 21-15 and 21-16)

myofibril Long, highly organized bundle of actin and myosin filaments and other proteins that constitute the basic structural unit of muscle cells (myofibers) (Figure 22-26)

myosin One of a family of motor proteins with a globular head region and coiled-coil tail region that has actin-stimulated ATPase activity; drives movement along actin filaments during muscle contraction and cytokinesis (myosin II) and mediates vesicle translocation (myosins I and V). (Figure 18-20)

NAD$^+$ (nicotinic adenine dinucleotide) A widely used coenzyme that participates in oxidation reactions by accepting two electrons from a donor molecule and one H$^+$ from the solution. The reduced form, NADH, transfers electrons to carriers that function in **oxidative phosphorylation**. (Figure 16-4)

NADP$^+$ (nicotinic adenine dinucleotide phosphate) Phosphorylated form of NAD$^+$, which is used extensively as an electron carrier in biosynthetic pathways and during photosynthesis.

Nernst equation Mathematical expression that defines the electric potential E across a membrane as directly proportional to the logarithm of the ratio of the ion concentrations on either side of the membrane and inversely proportional to the valency of the ions.

neuron (nerve cell) Any of the impulse-conducting cells of the nervous system. A typical neuron contains a cell body; several short, branched processes (**dendrites**); and one long process (**axon**). (Figure 21-1)

neuropeptide A peptide secreted by neurons that functions as a signaling molecule either at a synapse or elsewhere. These molecules have diverse, often long-lived effects in contrast to neurotransmitters.

neurotransmitter Extracellular signaling molecule that is released by the presynaptic neuron at a chemical **synapse** and relays the signal to the postsynaptic cell. The response elicited by a neurotransmitter, either excitatory or inhibitory, is determined by its receptor on the postsynaptic cell. Examples include acetylcholine, dopamine, GABA (γ-aminobutyric acid), and serotonin. (Figure 21-28)

noncovalent bond Any relatively weak chemical bond that does not involve an intimate sharing of electrons. Multiple noncovalent bonds often stabilize the conformation of macromolecules and mediate highly specific interactions between molecules.

nonpolar Referring to a molecule or structure that lacks any net electric charge or asymmetric distribution of positive and negative charges. Nonpolar molecules generally are insoluble in water.

Northern blotting Technique for detecting specific RNAs separated by electrophoresis by hybridization to a labeled DNA **probe**. See also **Southern blotting**.

nuclear envelope Double-membrane structure surrounding the nucleus; the outer membrane is continuous with the endoplasmic reticulum and the two membranes are perforated by nuclear pores. (Figures 5-42 and 5-50)

nuclear lamina Fibrous network on the inner surface of the inner nuclear membrane composed of lamin filaments. (Figure 13-15)

nuclear pore complex (NPC) Large, multiprotein structure in the nuclear envelope through which ions and small molecules can diffuse and which mediates the active transport of ribonucleoproteins and large proteins between the nucleus and cytoplasm. (Figure 11-28)

nuclear receptor General term for intracellular receptors that bind lipid-soluble hormones (e.g., steroid hormones); also called *steroid receptor superfamily*. Following ligand binding, the hormone-receptor complex translocates to the nucleus and functions as a transcription factor. (Figure 10-67)

nucleic acid A polymer of **nucleotides** linked by phosphodiester bonds. DNA and RNA are the primary nucleic acids in cells.

nucleocapsid A viral **capsid** plus the enclosed nucleic acid.

nucleolus Large structure in the nucleus of eukaryotic cells where rRNA synthesis and processing occurs and ribosome subunits are assembled.

nucleoside A small molecule composed of a **purine** or **pyrimidine** base linked to a pentose (either ribose or deoxyribose). (Table 4-1)

nucleosome Small structural unit of **chromatin** consisting of a disk-shaped core of **histone** proteins around which a ≈146-bp segment of DNA is wrapped. (Figure 9-31)

nucleotide A **nucleoside** with one or more phosphate groups linked via an ester bond to the sugar moiety. DNA and RNA are polymers of nucleotides. (Figure 4-1a and Table 4-1)

nucleus Large membrane-bounded organelle in eukaryotic cells that contains DNA organized into chromosomes; synthesis and processing of RNA and ribosome assembly occur in the nucleus.

Okazaki fragments Short (<1000 bases), single-stranded DNA fragments that are formed during synthesis of the **lagging strand** in DNA replication and are rapidly joined by DNA ligase to form a continuous DNA strand. (Figure 12-9)

oncogene A gene whose product is involved either in transforming cells in culture or in inducing cancer in animals. Most oncogenes are mutant forms of normal genes (**proto-oncogenes**) involved in the control of cell growth or division.

oocyte Developing egg cell.

operator Short DNA sequence in a bacterial or viral genome that binds a repressor protein and controls transcription of an adjacent gene. (Figure 10-2)

operon In bacterial DNA, a cluster of contiguous genes transcribed from one promoter that gives rise to a **polycistronic** mRNA.

organelle Any membrane-limited structure found in the cytoplasm of eukaryotic cells.

osmosis Net movement of water across a semipermeable membrane from a solution of lesser to one of greater solute concentration. The membrane must be permeable to water but not to solute molecules.

osmotic pressure Hydrostatic pressure that must be applied to the more concentrated solution to stop the net flow of water across a semipermeable membrane separating solutions of different concentrations. (Figure 15-30)

oxidation Loss of electrons from an atom or molecule as occurs when hydrogen is removed from a molecule or oxygen is added. The opposite of **reduction.**

oxidation potential The voltage change when an atom or molecule loses an electron.

oxidative phosphorylation The phosphorylation of ADP to form ATP driven by the transfer of electrons to oxygen (O_2) in bacteria and mitochondria. This process involves generation of a **proton-motive force** during electron transport, and its subsequent use to power ATP synthesis. (Figure 16-9)

passive (simple) diffusion Net movement of a molecule across a membrane down its concentration gradient at a rate proportional to the gradient and the permeability of the membrane.

patch clamping Technique for determining ion flow through a single ion channel or across the membrane of an entire cell by use of a micropipette whose tip is applied to a small patch of the cell membrane. (Figures 21-19 and 21-20)

PCR (polymerase chain reaction) Technique for amplifying a specific DNA segment in a complex mixture by multiple cycles of DNA synthesis from short oligonucleotide primers followed by brief heat treatment to separate the complementary strands. (Figure 7-38)

pentose A five-carbon **monosaccharide.** The pentoses ribose and deoxyribose are present in RNA and DNA, respectively.

peptide A small polymer usually containing fewer than 30 amino acids connected by peptide bonds.

peptide bond Covalent bond that links adjacent amino acid residues in proteins; formed by a condensation reaction between the amino group of one amino acid and the carboxyl group of another with release of a water molecule. (Figure 3-3)

peripheral membrane protein Any protein that associates with the cytosolic or exoplasmic face of a membrane but does not enter the hydrophobic core of the phospholipid bilayer; also called *extrinsic protein.* See also **integral membrane protein.** (Figure 3-32)

peroxisome Small organelle in eukaryotic cells whose functions include degradation of fatty acids and amino acids by means of reactions that generate hydrogen peroxide, which is converted to water and oxygen by catalase.

pH A measure of the acidity or alkalinity of a solution defined as the negative logarithm of the hydrogen ion concentration in moles per liter: $pH = -\log [H^+]$. Neutrality is equivalent to a pH of 7; values below this are acidic and those above are alkaline. (Table 2-3)

phage See **bacteriophage.**

phagocytosis Process by which relatively large particles (e.g., bacterial cells) are internalized by certain eukaryotic cells.

phenotype The observable characteristics of a cell or organism as distinct from its **genotype.**

pheromone A signaling molecule released by an individual that can alter the behavior or gene expression of other individuals of the same species. The yeast **α** and **a** mating-type factors are well-studied examples.

phosphatase An enzyme that removes a phosphate group from a substrate by hydrolysis. Phosphoprotein phosphatases act with protein **kinases** to control the activity of many cellular proteins.

phosphoanhydride bond A type of **high-energy bond** formed between two phosphate groups, such as the γ and β phosphates and the β and α phosphates in ATP. (Figure 2-24)

phosphodiester bond A covalent bond in which two hydroxyl groups form ester linkages to the same phosphate group; joins adjacent nucleotides in DNA and RNA.

phosphoinositides A family of membrane-bound lipids containing phosphorylated inositol derivatives that are important in signal-transduction pathways in eukaryotic cells. (Figure 20-29)

phospholipid bilayer A symmetrical two-layer structure, found in all biomembranes, in which the **polar** head groups of phospholipids are exposed to the aqueous medium, while the **nonpolar** hydrocarbon chains of the fatty acids are in the center. (Figure 5-30)

phospholipids The major class of lipids present in biomembranes, usually composed of two fatty acid chains esterified to two of the carbons of glycerol phosphate, with the phosphate esterified to one of various polar groups. (Figure 5-27)

photosynthesis Complex series of reactions occurring in some bacteria and plant **chloroplasts** whereby light energy is used to generate carbohydrates from CO_2, usually with the consumption of H_2O and evolution of O_2.

pI See **isoelectric point.**

pinocytosis The nonspecific uptake of small droplets of extracellular fluid into endocytic vesicles.

plaque assay Technique for determining the number of infectious viral particles in a sample by culturing a diluted sample on a layer of susceptible host cells and then counting the clear areas of lysed cells (plaques) that develop. (Figure 6-14)

plasma membrane The membrane surrounding a cell that separates the cell from its external environment, consisting of a **phospholipid bilayer** and associated proteins. (Figure 3-32)

plasmid Small, circular extrachromosomal DNA molecule capable of autonomous replication in a cell. Commonly used as a **cloning vector.**

plasmodesmata (sing. plasmodesma) Tubelike cell junctions that interconnect the cytoplasm of adjacent plant cells and are functionally analogous to gap junctions in animal cells. (Figure 22-36)

point mutation Change of a single nucleotide in DNA, especially in a region coding for protein; can result in formation of a codon specifying a different amino acid or a stop codon, or a shift in the **reading frame.** (Figure 8-4)

polar Referring to a molecule or structure with a net electric charge or asymmetric distribution of positive and negative charges. Polar molecules are usually soluble in water.

polarity Presence of functional and/or structural differences in distinct regions of a cell or cellular component.

polymer Any large molecule composed of multiple identical or similar units (monomers) linked by covalent bonds.

polymerase chain reaction See **PCR.**

polypeptide Linear polymer of **amino acids** connected by peptide bonds. Proteins are large polypeptides, and the two terms commonly are used interchangeably.

polyribosome A complex containing several ribosomes all translating a single messenger RNA; also called *polysome.*

polysaccharide Linear or branched polymer of monosaccharides, linked by glycosidic bonds, usually containing more than 15 residues. Examples include glycogen, cellulose, and glycosaminoglycans.

positional cloning Isolation and cloning of the normal form of a mutation-defined gene (i.e., a gene identified by genetic analysis of mutants).

pre-mRNA Precursor messenger RNA; the **primary transcript** and intermediates in RNA processing.

pre-rRNA Large precursor ribosomal RNA that is synthesized in the nucleolus of eukaryotic cells and processed to yield three of the four RNAs present in ribosomes. (Figure 11-50)

primary structure In proteins, the linear arrangement (sequence) of amino acids and the location of covalent (mostly disulfide) bonds within a polypeptide chain.

primary transcript Initial RNA product, containing **introns** and **exons,** produced by transcription of DNA. Many primary transcripts must undergo RNA processing to form the physiologically active RNA species.

primase A specialized RNA polymerase that synthesizes short stretches of RNA used as primers for DNA synthesis.

primer A short nucleic acid sequence containing a free 3′ hydroxyl group that forms **base pairs** with a complementary **template** strand and functions as the starting point for addition of nucleotides to copy the template strand.

probe Defined RNA or DNA fragment, radioactively or chemically labeled, that is used to detect specific nucleic acid sequences by **hybridization.**

prokaryotes Class of organisms, including the **eubacteria** and **archaea,** that lack a true membrane-limited nucleus and other organelles. See also **eukaryotes.**

promoter DNA sequence that determines the site of **transcription** initiation for an RNA polymerase.

promoter-proximal element Any regulatory sequence in eukaryotic DNA that is located within ≈200 base pairs of the transcription start site. Transcription of many genes is controlled by multiple promoter-proximal elements. (Figure 10-34)

prophase Earliest stage in mitosis during which the chromosomes condense and the centrioles begin moving toward the spindle poles. (Figure 19-34)

prosthetic group A nonpeptide organic molecule or metal ion that binds tightly and specifically with a protein and is required for its activity, such as heme in hemoglobin. See also **coenzyme.**

proteasome Large multifunctional protease complex in the cytosol that degrades intracellular proteins marked for destruction by attachment of multiple **ubiquitin** molecules. (Figure 3-18)

protein A linear polymer of **amino acids** linked together in a specific sequence and usually containing more than 50 residues. Proteins form the key structural elements in cells and participate in nearly all cellular activities.

proteoglycans A group of glycoproteins that contain a core protein to which is attached one or more **glycosaminoglycans.** They are found in nearly all extracellular matrices, and some are attached to the plasma membrane. (Figures 22-26 and 22-27)

proton-motive force The energy equivalent of the proton (H^+) concentration gradient and electric potential gradient across a membrane; used to drive ATP synthesis by **ATP synthase,** transport of molecules against their concentration gradient, and movement of bacterial flagella.

proto-oncogene A normal cellular gene that encodes a protein usually involved in regulation of cell growth or proliferation and that can be mutated into a cancer-promoting oncogene, either by changing the protein-coding segment or by altering its expression.

pulse-chase A type of experiment in which a radioactive small molecule is added to a cell for a brief period (the pulse) and then is replaced with an excess of the unlabeled form of same small molecule (the chase). Used to detect changes in the cellular location of a molecule or its metabolic fate over time.

pump Any transmembrane protein that mediates the **active transport** of an ion or small molecule across a biomembrane. (Table 15-2)

purines A class of nitrogenous compounds containing two fused heterocyclic rings. Two purines, adenine and guanine, commonly are found in DNA and RNA. (Figure 4-2)

pyrimidines A class of nitrogenous compounds containing one heterocyclic ring. Two pyrimidines, cytosine and thymine, commonly are found in DNA; in RNA, uracil replaces thymine. (Figure 4-2)

quaternary structure The number and relative positions of the polypeptide chains in multisubunit proteins.

quiescent Referring to a cell that has exited the **cell cycle** and is in the G_0 state.

radioisotope Unstable form of an atom that emits radiation as it decays. Several radioisotopes are commonly used experimentally as **labels** in biological molecules.

Ras protein A monomeric GTP-binding protein that functions in intracellular signaling pathways and is activated by ligand binding to **receptor tyrosine kinases** and other cell-surface receptors. See also **GTPase superfamily.** (Figure 20-23)

reading frame The sequence of nucleotide triplets (**codons**) that runs from a specific **translation** start codon in a mRNA to a stop codon. Some mRNAs can be translated into different polypeptides by reading in two different reading frames. (Figure 4-21)

receptor Any protein that binds a specific extracellular signaling molecule (ligand) and then initiates a cellular response. Receptors for steroid hormones, which diffuse across the plasma membrane, are located within the cell; receptors for water-soluble hormones, peptide growth factors, and neurotransmitters are located in the plasma membrane with their ligand-binding domain exposed to the external medium.

receptor tyrosine kinase (RTK) Member of an important class of cell-surface receptors whose cytosolic domain has tyrosine-specific protein kinase activity. Ligand binding activates this kinase activity and initiates intracellular signaling pathways. (Figure 20-23)

recessive In genetics, referring to that allele of a gene that is not expressed in the **phenotype** when the **dominant** allele is present. Also refers to the phenotype of an individual (homozygote) carrying two recessive alleles. (Figure 8-1)

recombinant DNA Any DNA molecule formed by joining DNA fragments from different sources. Commonly produced by cutting DNA molecules with **restriction enzymes** and then joining the resulting fragments from different sources with DNA ligase.

recombination Any process in which chromosomes or DNA molecules are cleaved and the fragments are rejoined to give new combinations. Occurs naturally in cells as the result of the exchange (crossing over) of DNA sequences on maternal and paternal chromatids during meiosis; also is carried out in vitro with purified DNA and enzymes.

reduction Gain of electrons by an atom or molecule as occurs when hydrogen is added to a molecule or oxygen is removed. The opposite of oxidation.

reduction potential The voltage change when an atom or molecule gains an electron.

replication fork See **growing fork**.

replication origin Unique DNA segments present in an organism's genome at which DNA replication begins. Eukaryotic chromosomes contain multiple origins, whereas bacterial chromosomes and plasmids often contain just one.

replicon Region of DNA served by one replication origin.

resolution The minimum distance that can be distinguished by an optical apparatus; also called *resolving power*.

respiration General term for any cellular process involving the uptake of O_2 coupled to production of CO_2.

restriction enzyme (endonuclease) Any enzyme that recognizes and cleaves a specific short sequence, the restriction site, in double-stranded DNA molecules. These enzymes are widespread in bacteria and are used extensively in recombinant DNA technology. (Table 7-1 and Figure 7-5)

restriction fragment A defined DNA fragment resulting from cleavage with a particular restriction enzyme. These fragments are used in the production of recombinant DNA molecules and DNA cloning.

restriction point The point in late G_1 of the cell cycle at which mammalian cells become committed to entering the S phase and completing the cycle even in the absence of growth factors.

retrotransposon Type of eukaryotic mobile DNA element whose movement in the genome is mediated by an RNA intermediate and involves a reverse transcription step. See also **transposon**.

retrovirus A type of eukaryotic virus containing an RNA genome that replicates in cells by first making a DNA copy of the RNA. This proviral DNA is inserted into cellular chromosomal DNA, and gives rise to further genomic RNA as well as the mRNAs for viral proteins. (Figure 6-22)

reverse transcriptase Enzyme found in retroviruses that catalyzes synthesis of a double-stranded DNA from a single-stranded RNA template. (Figure 9-16)

ribosomal RNA See **rRNA**.

ribosome A large complex comprising several different **rRNA** molecules and more than 50 proteins, organized into a large subunit and small subunit; the site of protein synthesis. (Figures 4-32 and 4-34)

ribozyme An RNA molecule or segment with catalytic activity.

RNA (ribonucleic acid) Linear, single-stranded polymer, composed of ribose nucleotides, that is synthesized by transcription of DNA or by copying of RNA. The three types of cellular RNA—mRNA, rRNA, and tRNA—play different roles in protein synthesis.

RNA editing Unusual type of RNA processing in which the sequence of a pre-mRNA is altered.

RNA polymerase An enzyme that copies one strand of DNA or RNA (the template strand) to make the **complementary** RNA strand using as substrates ribonucleoside triphosphates.

RNA processing Various modifications that occur to many but not all **primary transcripts** to yield functional RNA molecules.

RNA splicing A process that results in removal of **introns** and joining of **exons** in RNAs. See also **spliceosome**. (Figure 11-16)

rRNA (ribosomal RNA) Any one of several large RNA molecules that are structural and functional components of **ribosomes**. Often designated by their sedimentation coefficient: 28S, 18S, 5.8S, and 5S rRNA in higher eukaryotes.

S (synthesis) phase See **cell cycle**.

sarcoma A malignant tumor derived from connective tissue.

sarcomere Repeating unit of a **myofibril** in striated muscle that extends from one **Z** disk to an adjacent one and shortens during contraction. (Figure 18-27)

sarcoplasmic reticulum Network of membranes that surrounds each **myofibril** in a muscle cell and sequesters Ca^2 ions. Stimulation of a muscle cell induces release of Ca^2 ions into the cytosol, triggering coordinated contraction along the length of the cell. (Figure 18-31)

Schwann cell Type of glial cell that forms the **myelin sheath** around axons in the peripheral nervous system.

second messenger An intracellular **signaling molecule** whose concentration increases (or decreases) in response to binding of an extracellular **ligand** to a cell-surface receptor. Examples include cAMP, Ca^2, diacylglycerol (DAG), and inositol 1,4,5-trisphosphate (IP_3). (Figure 20-4)

secondary structure In proteins, local folding of a polypeptide chain into regular structures including the α helix, β sheet, and U-shaped turns and loops.

secretory vesicle Small membrane-bound organelle containing molecules destined to be released from the cell.

segregation In genetics, the process that distributes an equal complement of chromosomes to daughter cells during mitosis and meiosis.

seven-spanning receptor See **G protein–coupled receptor (GPCR)**.

signal sequence A relatively short amino acid sequence that directs a protein to a specific location within the cell; also called *signal peptide* and *targeting sequence*.

signal transduction Conversion of a signal from one physical or chemical form into another. In cell biology commonly refers to the sequential process initiated by binding of an extracellular signal to a receptor and culminating in one or more specific cellular responses.

signaling molecule General term for any extracellular or intracellular molecule involved in mediating the response of a cell to its external environment or other cells.

silencer sequence A sequence in eukaryotic DNA that promotes formation of condensed chromatin structures in a localized region, thereby blocking access of proteins required for transcription of genes within several hundred base pairs of the silencer sequence.

simple-sequence DNA Short, tandemly repeated sequences that are found in **centromeres** and **telomeres** as well as at other chromosomal locations and are not transcribed.

somatic cell Any plant or animal cell other than a **germ cell** or germ-cell precursor.

Southern blotting Technique for detecting specific DNA sequences separated by electrophoresis by hybridization to a labeled nucleic acid **probe**. (Figure 7-32)

SPF (S phase–promoting factor) A heterodimeric protein, composed of a **cyclin** and **cyclin-dependent kinase (Cdk)**, that triggers entrance of a cell into the S phase of the cell cycle by inducing expression of proteins required for DNA replication and passage through the S phase.

spliceosome Large ribonucleoprotein complex that assembles on a pre-mRNA and carries out **RNA splicing**. (Figures 11-18 and 11-19)

starch A very long, branched polysaccharide, composed exclusively of glucose units, that is the primary storage carbohydrate in plant cells.

START A point in the G_1 stage of the yeast **cell cycle** that controls entry of cells into the S phase. Passage of a cell through START commits a cell to proceed through the remainder of the cell cycle.

stem cell A self-renewing cell that divides to give rise to a cell with an identical developmental potential and/or one with a more restricted developmental potential.

stereoisomers Two compounds with identical molecular formulas whose atoms are linked in the same order but in different spatial arrangements. In optical isomers, designated D and L, the atoms bonded to an **asymmetric carbon atom** are arranged in a mirror-image fashion. Geometric isomers include the *cis* and *trans* forms of molecules containing a double bond.

steroid A group of four-ring hydrocarbons including **cholesterol** and related compounds. Many important hormones (e.g., estrogen and progesterone) are steroids.

substrate Molecule that undergoes a change in a reaction catalyzed by an enzyme.

substrate-level phosphorylation Formation of ATP from ADP and P_i catalyzed by cytosolic enzymes in reactions that do not depend on a proton-motive force.

sulfhydryl group (–SH) A hydrogen atom covalently bonded to a sulfur atom; also called a *thiol group*. A substituent group present in the amino acid cysteine and other molecules.

supercoils, DNA Regions of DNA in which the double helix is twisted on itself.

suppressor mutation A mutation that reverses the phenotypic effect of a second mutation. Suppressor mutations are frequently used to identify genes encoding interacting proteins.

symport A type of **cotransport** in which a membrane protein (symporter) transports two different molecules or ions across a cell membrane in the same direction. See also **antiport**.

synapse Specialized region between an axon terminus of a neuron and an adjacent neuron or other excitable cell (e.g., muscle cell) across which impulses are transmitted. At a chemical synapse, the impulse is conducted by a neurotransmitter; at an electric synapse, impulse transmission occurs via gap junctions connecting the cytoplasms of the pre- and postsynaptic cells. (Figures 21-4 and 21-35)

syncytium A multinucleated mass of cytoplasm enclosed by a single plasma membrane.

TATA box A conserved sequence in the **promoter** of many eukaryotic protein-coding genes where the transcription-initiation complex assembles (Figure 10-50)

telomere End region of a eukaryotic chromosome containing characteristic telomeric (TEL) sequences that are replicated by a special process, thereby counteracting the tendency of a chromosome to be shortened during each round of replication. (Figure 12-13)

telophase Final mitotic stage during which the nuclear-envelope reforms around the two sets of separated chromosomes; the chromosomes decondense; and division of the cytoplasm (cytokinesis) is completed. (Figure 19-34)

temperature-sensitive (ts) mutant A cell or organism with a mutant gene encoding an altered protein that functions normally at one temperature (the permissive temperature) but is nonfunctional at another temperature (the nonpermissive temperature).

template A molecular "mold" that dictates the structure of another molecule; most commonly, one strand of DNA that directs synthesis of a **complementary** DNA strand during DNA replication or of an RNA during transcription.

termination factor One of several proteins that acts to terminate protein synthesis by recognizing a stop codon in mRNA and causing release of the ribosomal subunits. (Figure 4-40)

tertiary structure In proteins, overall three-dimensional form of a polypeptide chain, which is stabilized by multiple noncovalent interactions between side chains.

thylakoids Flattened membranous sacs in a chloroplast that are arranged in stacks forming the grana and contain the photosynthetic pigments. (Figure 16-34)

tight junction Ribbon-like bands connecting adjacent epithelial cells that prevent leakage of fluid across the cell layer. (Figure 15-28)

topoisomerase Class of enzymes that control the number and topology of **supercoils** in DNA. Type I enzymes cut one DNA strand, rotate it about the other, and reseal the ends. Type II enzymes cut and reseal both DNA strands.

trans-acting Referring to DNA sequences encoding diffusible proteins (e.g., transcription activators and repressors) that control genes on the same or different chromosomes. See also **cis-acting.**

transcript See **primary transcript.**

transcription Process whereby one strand of a DNA molecule is used as a template for synthesis of a **complementary** RNA by RNA polymerase. (Figure 4-15)

transcription factor (TF) General term for any protein, other than RNA polymerase, required to initiate or regulate transcription in eukaryotic cells. General factors, required for transcription of all genes, participate in formation of the transcription-initiation complex near the start site. Specific factors stimulate (or repress) transcription of particular genes by binding to their regulatory sequences.

transcription unit A region in DNA, bounded by an initiation (START) site and termination site, that is transcribed into a single **primary transcript.**

transcription-control region Collective term for all the cis-acting DNA regulatory sequences that regulate transcription of a particular gene.

transfection Experimental introduction of foreign DNA into cells in culture, usually followed by expression of genes in the introduced DNA.

transfer RNA See **tRNA.**

transformation Permanent, heritable alteration in a cell resulting from the uptake and incorporation of a foreign DNA. Also, conversion of a "normal" mammalian cell into a cell with cancer-like properties usually induced by treatment with a virus or other cancer-causing agent.

transgene A cloned gene that is introduced and stably incorporated into a plant or animal and is passed on to successive generations.

transgenic Referring to any plant or animal carrying a **transgene.**

translation The **ribosome**-mediated production of a polypeptide whose amino acid sequence is specified by the nucleotide sequence in an mRNA.

translocon Multiprotein complex in the membrane of the rough endoplasmic reticulum through which a nascent secretory protein enters the ER lumen as it is being synthesized. (Figure 17-16)

transport vesicles Small membrane-bounded organelles that carry secretory and membrane proteins in both directions between the rough endoplasmic reticulum (ER) and the Golgi complex, and from the Golgi to the cell surface or other destination. Form by budding off from the donor organelle and release their contents by fusion with the target membrane.

transposon A relatively long mobile DNA element, in prokaryotes and eukaryotes, that moves in the genome by a mechanism involving DNA synthesis and transposition. See also **retrotransposon.**

tricarboxylic acid cycle See **citric acid cycle.**

tRNA (transfer RNA) A group of small RNA molecules that function as amino acid donors during protein synthesis. Each tRNA becomes covalently linked to a particular amino acid, forming an **aminoacyl-tRNA.** (Figure 4-26)

tubulin A family of globular cytoskeletal proteins that polymerize to form microtubules.

tumor A mass of cells, generally derived from a single cell, that is not controlled by normal regulators of cell growth.

tumor-suppressor gene Any gene whose encoded protein directly or indirectly inhibits progression through the cell cycle and in which a loss-of-function mutation is oncogenic. Inheritance of a single mutant allele of many tumor-suppressor genes (e.g., *RB, APC,* and *BRCA1*) greatly increases the risk for developing certain types of cancer.

ubiquitin A small, highly conserved protein that becomes covalently linked to lysine residues in other intracellular proteins. Proteins to which a chain of ubiquitin molecules is added usually are degraded in a **proteasome.**

uncoupler An agent that dissipates the **proton-motive force** across the inner mitochondrial membrane and thylakoid membrane of chloroplasts, thereby inhibiting ATP synthesis.

upstream The direction on a DNA opposite to the direction RNA polymerase moves during transcription. By convention, the +1 position in a gene is the first transcribed base; nucleotides upstream from the +1 position are designated −1, −2, etc. See also **downstream.**

upstream activating sequence (UAS) Any protein-binding regulatory sequence in the DNA of yeast and other simple eukaryotes that is necessary for maximal gene expression; equivalent to an enhancer or promoter-proximal element in higher eukaryotes. (Figure 10-34)

van der Waals interaction A weak noncovalent attraction due to small, transient asymmetric electron distributions around atoms (dipoles).

vector In cell biology, an agent that can carry DNA into a cell or organism. See also **cloning vector** and **expression vector.**

ventral Relating to the front of an animal or lower surface of a structure (e.g., wing or leaf).

virion An individual viral particle.

virus A small parasite consisting of nucleic acid (RNA or DNA) enclosed in a protein coat that can replicate only in a susceptible host cell; widely used in cell biology research. (Table 6-3)

V_{max} Parameter that describes the maximal velocity of an enzyme-catalyzed reaction or other process such as protein-mediated transport of molecules across a membrane. (Figure 3-26)

Western blotting Technique for detecting specific proteins separated by electrophoresis by use of labeled antibodies. (Figure 3-44)

wild type Normal, nonmutant form of a macromolecule, cell, or organism.

x-ray crystallography Most commonly used technique for determining the three-dimensional structure of macromolecules (particularly proteins and nucleic acids) by passing x-rays through a crystal of the purified molecules and analyzing the diffraction pattern of discrete spots that results.

zinc finger Several types of conserved DNA-binding **motifs** composed of protein domains folded around a zinc ion; present in several types of eukaryotic **transcription factors.** (Figure 10-41)

zygote A fertilized egg; diploid cell resulting from fusion of a male and female **gamete.**

Index

Note: Page numbers in *italics* indicate illustrations; those in **boldface** indicate definitions; and those followed by t indicate tables.

MEDIA CONNECTIONS

The animations on the accompanying CD-ROM span topics in molecular cell biology in three different ways. Overview animations introduce key processes. Focus animations hone in on the details of those processes. Technique animations reveal the experimental techniques available to researchers today. Related materials are linked together on the CD-ROM. Below is a listing of the related Overview (blue) and Focus (red) Animations and their corresponding figure numbers:

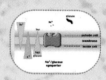

Biological Energy Interconversions Chapters 2, 15, 21 (Figures 2-25, 15-9, 15-13, 15-19, 21-9)

ATP Synthesis Chapter 16 (Figures 16-28,16-30)

Photosynthesis Chapter 16 (Figure 16-38)

Cell Cycle Control Chapters 13, 24 (Figures 13-2, 24-19)

Mitosis Chapter 19 (Figures 19-34, 19-35)

Bidirectional Replication of DNA Chapter 12 (Figures 12-2, 12-9)

Cell Motility Chapter 18 (Figure 18-41)

Actin Polymerization Chapter 18 (Figure 18-11)

Myosin Crossbridge Cycle Chapter 18 (Figure 18-25)

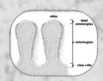

Life Cycle of a Cell Chapter 1 (Figures 1-9, 1-10)

Mitosis Chapter 19 (Figures 19-34, 19-35)

Apoptosis Chapter 23 (Figures 23-45, 23-50)

Life Cycle of an mRNA Chapters 4, 11 (Figures 4-19, 4-42, 11-7)

Basic Transcriptional Mechanism Chapter 4 (Figure 4-15)

Protein Synthesis Chapter 4 (Figure 4-39)

mRNA Splicing Chapter 11 (Figures 11-17, 11-19)

Life Cycle of a Protein Chapter 3 (Figure 3-18)

Chaperone-Mediated Folding Chapter 3 (Figure 3-15)

Protein Synthesis Chapter 4 (Figure 4-39)

Actin Polymerization Chapter 18 (Figure 18-11)

Life Cycle of a Retrovirus Chapter 6 (Figure 6-22)

Retroviral Reverse Transcription Chapter 9 (Figure 9-16)

Protein Secretion Chapters 5, 17 (Figures 5-48, 17-13)

Protein Synthesis Chapter 4 (Figure 4-39)

Synthesis of Secreted and Membrane-Bound Proteins Chapter 17 (Figures 17-16, 17-23)

Protein Sorting Chapter 17 (Figure 17-1)

Protein Synthesis Chapter 4 (Figure 4-39)